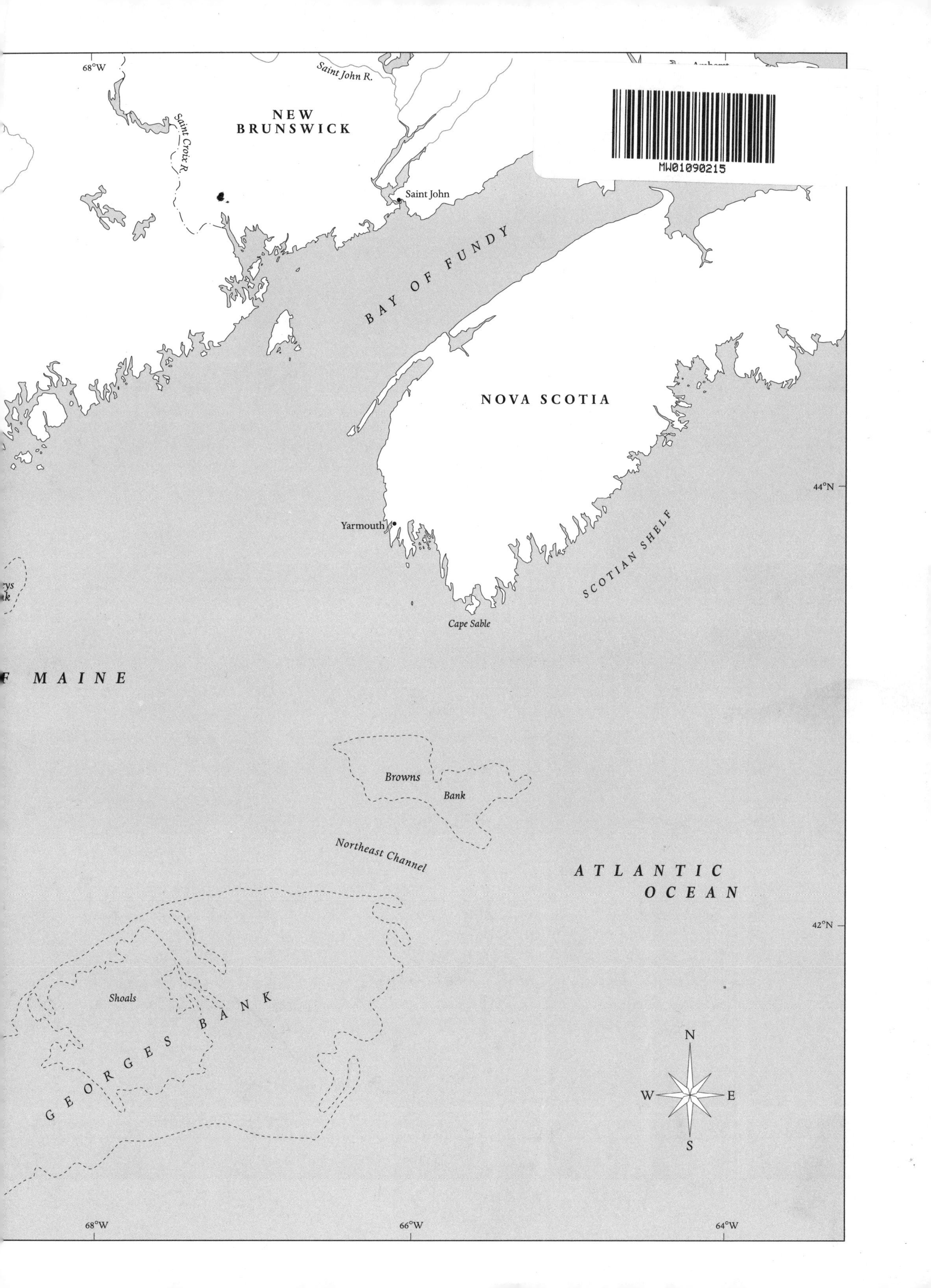
68°W
Saint John R.
NEW
BRUNSWICK
Saint Croix R.
Saint John
BAY OF FUNDY
NOVA SCOTIA
44°N
Yarmouth
SCOTIAN SHELF
Cape Sable
MAINE
Browns
Bank
Northeast Channel
ATLANTIC
OCEAN
42°N
Shoals
GEORGES BANK
N
W
E
S
68°W
66°W
64°W

FISHES OF THE GULF OF MAINE

BIGELOW AND SCHROEDER'S

FISHES

OF THE GULF OF MAINE

EDITED BY BRUCE B. COLLETTE AND GRACE KLEIN-MacPHEE

THIRD EDITION

SMITHSONIAN INSTITUTION PRESS • WASHINGTON AND LONDON

Publication of this book was supported by a grant from the Atherton Seidell Fund.

COPY EDITOR AND TYPESETTER: Princeton Editorial Associates, Inc.
PRODUCTION EDITOR: E. Anne Bolen
DESIGNER: Janice Wheeler

Library of Congress Cataloging-in-Publication Data

Bigelow and Schroeder's fishes of the Gulf of Maine / edited by Bruce B. Collette and Grace Klein-MacPhee.—3rd ed.
p. cm.
Rev. ed. of: Fishes of the Gulf of Maine / Henry Bryant Bigelow. 1953.
Includes bibliographical references (p.).
ISBN 1-56098-951-3 (alk. paper)
1. Fishes—Maine, Gulf of. I. Collette, Bruce B. II. Klein-MacPhee, Grace. III. Bigelow, Henry Bryant, b. 1879. Fishes of the Gulf of Maine.
QL621.5 .B54 2002
597′.09163′45—dc21 2001055005

British Library Cataloguing-in-Publication Data available

Manufactured in the United States of America
08 07 06 05 04 03 02 5 4 3 2

The paper used in this publication meets the minimum requirements of the American National Standard for Information Sciences—Permanence of Paper for Printed Library Materials ANSI Z39.48-1984.

We dedicate this book to the fishes of the Gulf of Maine with the hope that the information we have summarized will help those concerned with these fishes to make decisions that will restore the ecosystem and the fish populations to something resembling their original abundance.

CONTENTS

DISTRIBUTION MAPS **609**

FOREWORD

Early in my career as a fishery biologist, whether my feet were in the mud of a tidal flat or on the rolling deck of a fishing trawler, I always had handy a copy of the *Fishes of the Gulf of Maine* by Henry B. Bigelow and William C. Schroeder to help me identify species that were unfamiliar. In midcareer, when I had responsibility for managing and regulating fisheries, I frequently referred to "Bigelow and Schroeder" for historical information on the fisheries. I quickly learned that many fishermen also kept a copy in the wheelhouses of their vessels, and they often would cite "Bigelow and Schroeder" fishery information during fishing regulatory hearings!

I still have that original copy of the *Fishes of the Gulf of Maine*—sans its covers and with many pages stained by spilled food, coffee, and soda. There is also the distinct odor of fish and even a few scales interleaved. I am a bibliophile and have a nearly pristine copy of Fishery Bulletin 74, Volume 53, as well as a copy of the 1925 edition of *Fishes of the Gulf of Maine* by Henry B. Bigelow and William W. Welsh in my library. In 1964, Harvard's Museum of Comparative Zoology reprinted "Bigelow and Schroeder" after more than 3,000 copies of the 1953 edition had been distributed and it was no longer available. I have no idea of the total number of copies the museum distributed, but I'm sure there were many.

There is no question that "Bigelow and Schroeder" is a valued and useful reference for fishery scientists, biologists, and laypeople alike, but much has changed in regard to the fish and fisheries of the Gulf of Maine since 1953. There have been changes in the classification of some of the fishes, which are not only of academic interest. Witness the change of striped bass from *Roccus* to *Morone* and the resulting perplexity of fishermen who named their boats *Roccus*. Many species of fishes have been subjected to intensive fishing by foreign and domestic fisheries that did not even exist in 1953. The fishing management regime that was instituted by the United States in the late 1970s totally changed the nature of the fisheries in the Gulf of Maine. "Bigelow and Schroeder" became dated.

Sometime during the mid-1980s, during my tenure as science and research director of NOAA Fisheries at the Northeast Fisheries Science Center, Charles Sheldon (a fishery consultant and son of my graduate school mentor) and I were discussing these changes when Charlie mentioned the need to revise "Bigelow and Schroeder." Charlie was interested in taking on the project, but he did not have the resources to undertake it. I promised him that I would see what could be done to achieve our mutual objective.

I turned to Dr. Bruce Collette, then director of NOAA's National Systematics Laboratory, for his opinion on the need for a revision. He agreed, and most importantly expressed enthusiasm for taking on the job himself. We shook hands and I gave him the go-ahead. We both knew that this was a major undertaking and would take some time—maybe as long as five years to complete! In 1986 Dr. Grace Klein-MacPhee was brought in, on a part-time basis, to assist in the revision, which soon turned into a full-time effort. Subsequently the project grew to involve many other contributors.

The revision was completed and, with the cooperation of the Smithsonian Institution Press, is now available. Drs. Collette and Klein-MacPhee and the many contributors have done an outstanding job. We hope that this revised edition of Bigelow and Schroeder's *Fishes of the Gulf of Maine* will become as well known and widely used as the 1953 edition, and that perhaps someday it will become known as "Collette and Klein-MacPhee."

Charlie, I kept my word.

ALLEN E. PETERSON JR.

PREFACE

The Gulf of Maine is one of best known marine regions in the world owing in large part to early research by Henry Bryant Bigelow and his colleagues. Bigelow described the plankton of the Gulf of Maine in 1926 and its oceanography in 1927. Bigelow and Welsh published the first edition of *Fishes of the Gulf of Maine* in 1925, which was followed by the second edition, by Bigelow and Schroeder, in 1953. In his 1955 review of the second edition, the late ichthyologist Carl L. Hubbs noted that this work, "as now revised, is an extraordinarily full compilation of available information on the fishes that occur in the Gulf within the 150-fathom line." The popularity and importance of "Bigelow and Schroeder" is emphasized in Dobbs's recent book on the Gulf of Maine (Dobbs 2000). Summaries of the life and accomplishments of Henry Bigelow (Fig. 1) were published by Graham (1969) and Dick and Schroeder (1968), of Bill Schroeder (Fig. 2) by Merriman (1973).

Ichthyologists, fishery scientists, ecologists, marine biology students, and fishermen all need information about the composition of the Gulf of Maine's contemporary fish fauna and the natural history, life cycles, life histories, and ecology of the fish species. Changes in the abundance of many Gulf of Maine fishes as a result of commercial harvesting make an updated book even more valuable today than the 1953 edition was in its time. Collecting, particularly in the deeper waters of the Gulf, has added to its known fauna and adding these species brings the book up to date. We have undertaken the updating of Bigelow and Schroeder's accounts by incorporating new published and unpublished information. One measure of how the task has increased is to consider the number of species covered in each edition: Bigelow and Welsh (1925) included 83 families, 178 species; Bigelow and Schroeder (1953) dealt with 108 families, 219 species, an increase of 41 species; and our compilation has 118 families, 252 species, an increase of 33 species. A second measure is in the vast quantity of literature on many Gulf of Maine species, which made it necessary for us to rely on colleagues with expertise on many groups of fishes to assist by updating information on the groups with which they are familiar.

We have the same goal as Bigelow and Schroeder—to update their "handbook for the easy identification of the fishes that occur in the Gulf of Maine, with summaries of what is known of the distribution, relative abundance, and more significant facts in the life history of each."

We use the same geographical limits of the "Gulf of Maine" as did Bigelow and Schroeder—the oceanic bight from Nantucket Shoals and Cape Cod on the southwest to Cape Sable on the northeast. Thus we include the shorelines of northern Massachusetts, New Hampshire, Maine, and parts of New Brunswick and Nova Scotia. The eastern and western boundaries are 65° and 70° W longitude. The Gulf of Maine has a natural seaward rim formed by Nantucket Shoals, Georges Bank, and Browns Bank (Fig. 3A). We follow Bigelow and Schroeder in using the 150-fathom (300-m) contour as the arbitrary offshore boundary for including all the species likely to be caught within the Gulf of Maine and for excluding most of the so-called "deep-sea" species that are numerous in the Atlantic outside the Gulf.

The literature dealing with the fishes of the Gulf of Maine begins with the earliest descriptions of New England as pointed out by Bigelow and Schroeder. We have tried to build on Bigelow and Welsh, Bigelow and Schroeder, and Scott and Scott (1988) by adding references to the extensive recent literature on fishes in the area. We felt it necessary to utilize a large number of reports that may be considered as "gray literature" (Collette 1990), particularly NOAA Technical Reports and Woods Hole Laboratory Reference Series as well as unpublished M.S. and M.A. theses and Ph.D. dissertations.

The Magnuson-Stevens Fishery Conservation and Management Act, which was reauthorized and amended by the Sustainable Fisheries Act (1996), requires the eight regional fishery management councils to describe and identify essential fish habitat (EFH) in their respective regions. Drafts of relevant sections of the manuscript for this book were made available to Northeast Fisheries Science Center (NEFSC) to facilitate production of the required EFH documents. NEFSC compiled available information on the distribution, abundance, and habitat requirements for each of the species managed by the New England and Mid-Atlantic Fishery Management Councils (Reid et al. 1999a). A series of EFH reports was prepared, 25 of which concern species of fishes found in the Gulf of Maine and these are referred to in the relevant species accounts.

We have also drawn on many of the databases maintained by the NEFSC, some of which have only become available recently. Four examples of these databases that were of great value in preparing the habits and distribution, food, predator, and breeding habits sections of our revision are discussed below.

1. For our update, we have had the benefit of the more than 30 years of annual stratified random bottom trawl surveys conducted by the NEFSC (Grosslein 1969; Azarovitz 1988) (Map 1). This has produced perhaps the best marine bio-

Figure 1. Dr. Henry Bryant Bigelow at the helm of the *Grampus* in the Gulf of Maine, 1912. (Ernst Mayr Library, Museum of Comparative Zoology, © President and Fellows of Harvard College.)

Figure 2. William C. Schroeder on board the *Cap'n Bill II,* with chimaeras *Harriotta raleighana* in hand, 1953. (Photograph by Jan Hahn, © Sears Foundation for Marine Research, reprinted from Merriman 1973.)

diversity database available anywhere. It is superb for tracing changes in abundance through time (see the introductory chapter by Murawski et al.), particularly of large and conspicuous fishes and macroinvertebrates, many of which are of commercial importance. Data from these surveys have been used to prepare the distribution maps. Although the database may not reflect the true abundance of smaller, less commercially important organisms, NEFSC staff have saved many of the interesting oddities and most of these are vouchered at the Museum of Comparative Zoology (MCZ) at Harvard University and the National Museum of Natural History (USNM) in Washington.

2. Food of northwest Atlantic fishes has been studied and published on by Ray E. Bowman and colleagues, culminating in the recent paper (Bowman et al. 2000) summarizing stomach contents of 31,567 individuals of 180 species collected by the NEFSC bottom trawl and longline surveys made during 1977–1980. Their appendix tables provide a detailed listing of the overall stomach contents for each predator species and, for selected species, the stomach contents according to predator size or to both predator size and major geographic area of collection.
3. Food habits data for two periods, 1973–1980 and 1981–1990, separated because of differences in data collection, have been converted into a series of tables listing the species of fishes that prey upon the major northwest Atlantic fishes (Rountree 1999). These data are not yet available in a formal publication but can be seen on the Web.

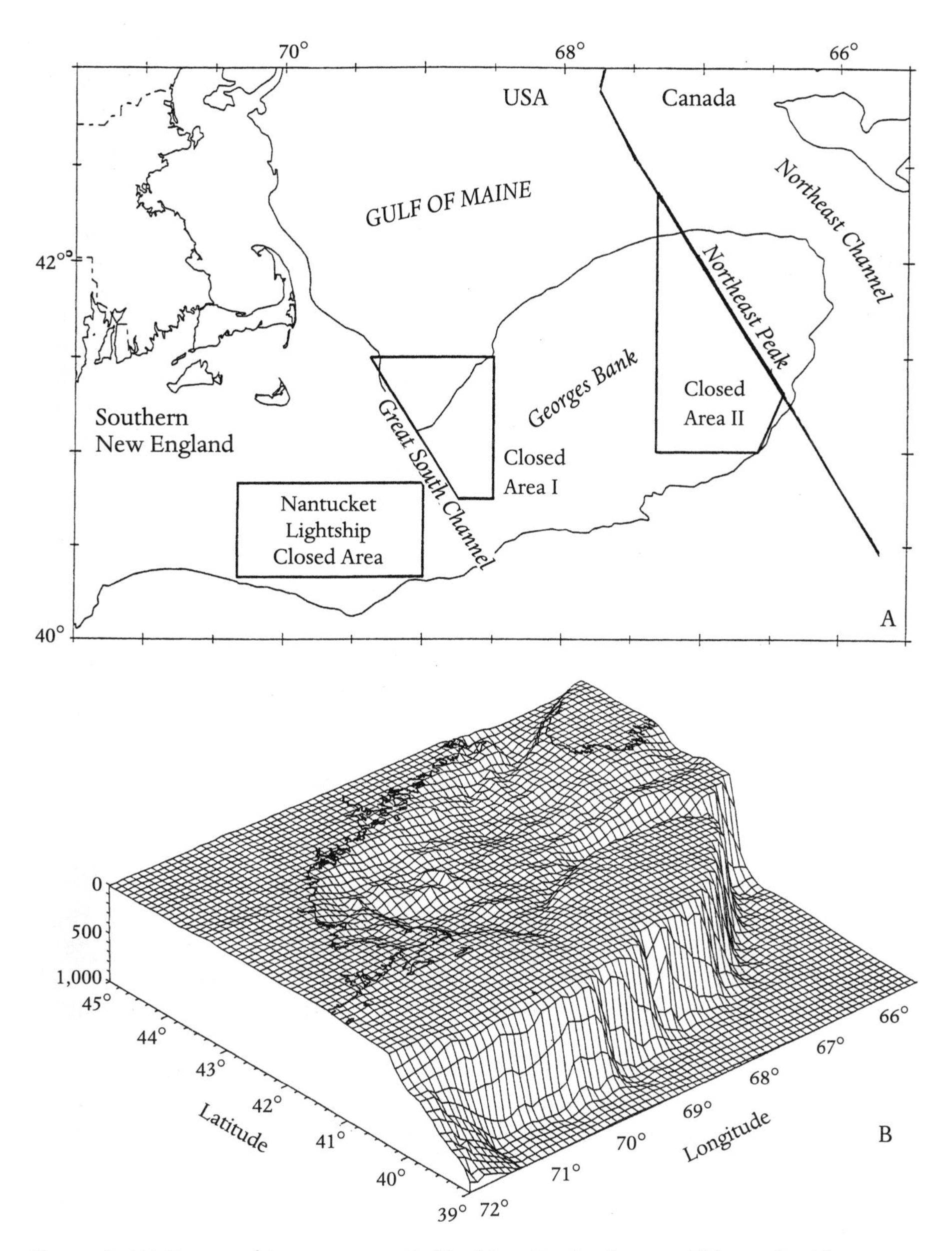

Figure 3. (A) Geographic areas occupied by New England groundfish stocks. Three outlined areas are currently closed to fishing with gears capable of catching groundfish. The line to the right of Closed Area II indicates the division between U.S. and Canadian waters. (B) Bathymetry of the Gulf of Maine area.

4. The NEFSC conducted a comprehensive fisheries ecosystem study in coastal waters off the northeastern United States from 1977 to 1987 known as MARMAP (Marine Resource Monitoring Assessment and Prediction; Sherman 1980). This study included standardized year-round collections of fish eggs and larvae that provide baseline information on the distribution of eggs to determine spawning times and locations. There were 81 MARMAP surveys conducted between February 1977 and December 1987. More than 10,000 samples from 73 surveys were examined for fish eggs. Information for 33 selected taxa is shown in graph and map form (Berrien and Sibunka 1999). These 33 taxa comprise 97.8% of all the eggs collected.

Families are arranged in the same order as in Nelson (1994). We employ the scientific names used by Eschmeyer (1998) unless there is a more recent revision. Common names generally follow the checklist of the American Fisheries Society (Robins et al. 1991).

Information in the species accounts falls into four general categories: description, biology, reproduction, and importance (to fisheries), which are divided into about 14 sections. Accounts

TABLE 1
COMPOSITION OF FISHES OF THE GULF OF MAINE

	Number	Percentage
Resident species		
Shallow-water species	55	21.8
Deeper-water species	23	9.1
Pelagic species	9	3.6
Subtotal	87	34.5
Visitors		
From south	92	36.5
From deep water	56	22.2
From north	12	4.8
Pelagic	5	2.0
Subtotal	165	65.5
Total	252	100.0

TABLE 2
NINE FISH SPECIES THAT OCCUR IN THE GULF OF MAINE REGION AND WERE FIRST DESCRIBED AFTER BIGELOW AND SCHROEDER (1953)

Family	Species and Author
Rajidae	*Bathyraja richardsoni* (Garrick 1961)
	Raja bigelowi Stehmann 1978
Sternoptychidae	*Maurolicus weitzmani* Parin and Kobyliansky 1993
	Polyipnus clarus Harold 1994
Hemiramphidae	*Hyporhamphus meeki* Banford and Collette 1993
Diretmidae	*Diretmichthys parini* (Post and Quero 1981)
Liparidae	*Liparis coheni* Able 1976
	Liparis inquilinus (Able 1973)
Trichiuridae	*Lepidopus altifrons* Parin and Collette 1993

begin with the generally accepted common name and current scientific name with author and date of original description. Following this is a reference to the pages where the species was treated in Bigelow and Schroeder's 1953 edition, and if the scientific name has changed since then, the name that Bigelow and Schroeder used for the species. A figure is included for almost all (244 species) fishes recorded from the Gulf of Maine. The first section of each species accounts is *Description,* a few paragraphs describing body shape, fins, head, and other morphological features of the species. This is followed by brief sections on *Color; Meristics,* the counts of fin rays, gill rakers, and vertebrae; and *Size,* which includes maximum length and weight and, for species important to recreational fisheries, the maximum weight recorded by the International Game Fish Association (IGFA 2001). The next section is *Distinctions,* where characters are given to differentiate the species in question from other Gulf of Maine fishes. Biological information, when available, is divided among sections such as *Habits,* including depth range, bottom type, and temperature and salinity preferences; *Food; Predators;* and *Parasites.* For some species that are rare in the Gulf of Maine, several of these sections may be combined under *Biology,* and for those few species with abundant information additional subcategories may be added to help the reader find the material being sought. *Reproduction* may be subdivided into *Breeding Habits, Fecundity, Early Life History, Eggs,* and *Larvae.* These sections are followed by *Age and Growth,* including methodology used in estimating age for a given species and, for 16 important commercial species, graphs of length-at-age curves. *General Range* gives the known geographic distribution of the species followed by *Occurrence in the Gulf of Maine,* which provides detailed information on local occurrences, supplemented by distribution maps in the appendix for 38 species based on the extensive bottom trawling data of the Northeast Fisheries Science Center. The last section is *Importance,* which may be subdivided into subsections such as *Commercial Fishery,* supplemented by graphs of commercial landings and an index of abundance for 30 of the most important commercial species based on the most recent *Status of the Stocks* document issued by the Northeast Fisheries Science Center. For many groups, an *Acknowledgments* section concludes the last species account in a family or suborder.

Cape Cod forms the southern border of the Gulf of Maine and is a major biogeographic boundary separating cold northern fishes from warm-water fishes to the south. Analysis of the list of fishes now known from the Gulf of Maine shows that only about a third of the species are resident in the Gulf (Table 1); another third are visitors from the south that travel around Cape Cod during the summer; and the final third are visitors from the north, deepwater or offshore. The largest proportion of the new records for the Gulf are deepwater visitors. Southern, northern, and deepwater strays are mentioned in the introductory accounts for each family and are sometimes included in the keys. Although the Gulf of Maine is relatively well known ichthyologically compared to most other regions of the world, it is interesting to note that nine fish species from the region have been described as new since 1953, mostly deepwater species (Table 2).

The Gulf of Maine fish fauna is characterized by relatively few species (252) in a relatively large number of families (118). The ten most speciose families account for nearly one-third of the Gulf species: Carangidae (jacks, 11 species), Gadidae (cods, 10), Scombridae (mackerels, 10), Rajidae (skates, 8), Clupeidae (herrings, 8), Pleuronectidae (righteye flounders, 7), Squalidae (dogfish sharks, 6), Myctophidae (lanternfishes, 6), Cottidae (sculpins, 6) and Carcharhinidae (requiem sharks, 5).

Based on temperature, depth, latitude, and ecology, the common Gulf of Maine fishes can be divided into four ecological groups (Murawski 1993): (1) shallow-water sedentary—23 species, such as little skate, winter skate, longhorn sculpin, and winter flounder; (2) deepwater sedentary—23 species, such as thorny skate, Arcadian redfish, and witch flounder; (3) warm-water migratory—92 species, mostly found in summer or autumn, such as butterfish and sea robin; and (4) pelagic—9 species. Some species, such as the spiny dogfish and the goosefish, do not fit neatly into any of the four categories.

Two generations of naturalists, ichthyologists, fishery biologists, and fishermen have grown up with and depended on the 1953 edition of Bigelow and Schroeder's *Fishes of the Gulf of Maine.* We have tried to achieve their goals of easy identification of the fishes of the Gulf with summaries of what is known about them. But much has happened to the fishes of the Gulf since the 1950s—many species are no longer abundant, others mature at smaller sizes and produce fewer eggs, and the fishery for many species has collapsed. More detailed information can be found in a volume edited by Boreman et al. (1997) and in *Proceedings of a Symposium: The History, Status, and Future of the New England Offshore Fishery,* held at Connecticut College in April 1999 (Dreyer 2000). There is too much information available today for most people to be able to find and utilize. We have tried to integrate all this information, both published and unpublished, into our revision. We hope that our new edition will be a worthy successor to "Bigelow and Schroeder" in providing information about the fishes that inhabit the Gulf. And we fervently hope that the information herein will help find solutions to the problems that confront the fishes and the people whose lives are involved with them in many different ways.

BRUCE B. COLLETTE AND GRACE KLEIN-MACPHEE

ACKNOWLEDGMENTS

We began this update of Bigelow and Schroeder at the request of Allen E. Peterson Jr., then Director of the Northeast Fisheries Science Center (NEFSC) of the National Marine Fisheries Service (NMFS) and have completed it with his support, and occasional prodding. This support (and prodding) have continued from Drs. Michael Sissenwine, current Center Director, and John Boreman, Assistant Center Director. It would be difficult to list all the other NEFSC staff, current and past, who have assisted us along the way but we must mention several that provided especially important help throughout the project: Tom Azarovitz, Ed Bowman, Don Flescher, John Galbraith, Steve Murawski, and Joan Palmer. Drafts of many sections, particularly those dealing with important commercial species, were reviewed and updated by specialists in the Population Dynamics group at the NEFSC Woods Hole Laboratory. Karsten Hartel of the Museum of Comparative Zoology, Harvard University, has been of constant help and encouragement from the very beginning of the project, alerting us to new material at the museum, answering numerous questions, reviewing accounts, and coauthoring most of the sections on "deep-sea" fishes that were found within the 300-m isobath.

Many of the illustrations will be familiar to users of Bigelow and Schroeder because they are based on the same original drawings done for the Bureau of Fisheries and the Smithsonian Institution many years ago. This treasure-trove of original illustrations is housed in the Division of Fishes of the National Museum of Natural History. We are indebted to Lisa Palmer, the current caretaker of these illustrations for facilitating access to them, and to Mollie Oremland, a summer intern of the Systematics Laboratory, who painstakingly searched the collection for suitable original illustrations and made POS 2 photographic copies of them to reproduce here. New illustrations of blackbelly rosefish, *Helicolenus dactylopterus,* and wolf eelpout, *Lycenchelys verrillii,* were expertly prepared by the NSL staff artist, Keiko Hiratsuka Moore. Distribution maps of species frequently collected by the NEFSC surveys are based on the survey data and were prepared at the Woods Hole Laboratory by Monica Holland under the supervision of Daniel Sheehan and Joan Palmer. Graphs of catch and estimates of abundance from the 1998 *Status of the Stocks* document (Clark 1998) were made available prior to their publication through the courtesy of Steven Murawski.

Literature cited sections were prepared by the authors of each section and then the 35–40 individual literature cited sections were integrated into an inclusive list by Joan Palmer. Assistance with checking literature cited sections against each section was provided by Nina Loewinger, Deborah Rothwell Schaner, and Everett Schaner. Copies of first and last pages of many references to check against the literature cited were compiled by Roosevelt McMillan.

Manuscript preparation, conversion of MacIntosh files to WordStar, and Wordstar to WordPerfect over the years of preparing this revision was assisted by Roosevelt McMillan, Alsane Shoumah, and Virginia Thomas. Assistance with various aspects of final manuscript preparation was provided by Ruth Gibbons, Sara Henley, and LaShaun Willis. The group of editors at the Smithsonian Institution Press assisted us and encouraged us to produce a better book. These include Anne Bolen, Vincent Burke, Nicole Sloan, and Peter Strupp and the staff of Princeton Editorial Associates. Producing such a large volume at a reasonable cost was made possible by a grant from the Atherton Seidell Fund of the Smithsonian Institution.

Whatever success we have had in completing this update of Bigelow and Schroeder was made possible by two factors: the first, the insight and dedication of many researchers and technicians over decades in amassing and making available incredible amounts of information on the biology of the fishes of the Gulf of Maine; and the second, the efforts of those of our colleagues that we persuaded to join us by writing or rewriting sections on the fish groups with which they were most familiar. These accounts were drafted over a 13-year period so any shortcomings are due more to our failure to properly update them than to the efforts of their authors. We thank all of our contributors for taking the time out of their busy schedules to write, rewrite, and then review revised accounts of their groups.

ACRONYMS AND ABBREVIATIONS

ARC—Atlantic Reference Collection, Huntsman Marine laboratory, St. Andrews, N.B.

ASMC—Atlantic States Marine Fishery Commission, Washington, D.C.

BIOME—An inshore fisheries ecological study of the coastal zone between Cape Cod and Cape Ann, extending from the tidal zone to the 100-m curve (see Lux and Kelly 1978).

CPUE—Catch per unit effort, a measure of abundance.

EEZ—Exclusive Economic Zone.

FL—Fork length, length of fish from tip of snout to fork of tail.

FMP—Fishery Management Plan.

GSI—Gonosomatic index.

ICNAF—International Commission for North Atlantic Fisheries.

IGFA—International Game Fish Association, Dania Beach, Fla.

MARMAP—Marine Resource Monitoring Assessment and Prediction, a comprehensive fisheries ecosystem study conducted in coastal waters off the northeastern United States from 1977 to 1987 by NEFSC, which included standardized year-round collections of fish eggs and larvae (see Berrien and Sibunka 1999).

MCZ—Museum of Comparative Zoology, Harvard University, Cambridge, Mass.

NEFSC—Northeast Fisheries Science Center, Woods Hole, Mass.

NL—Notochord length, length of larval fish from tip of snout to end of vertebral column, before flexion.

NMFS—National Marine Fisheries Service.

NOAA—National Oceanographic and Atmospheric Agency.

ROV—Remotely operated vehicle (underwater).

SL—Standard length, length of fish from tip of snout to base of caudal fin.

UMMZ—University of Michigan Museum of Zoology, Ann Arbor, Mich.

USNM—National Museum of Natural History, Washington, D.C.

VIMS—Virginia Institute of Marine Science, Gloucester Point, Va.

WHOI—Woods Hole Oceanographic Institution, Woods Hole, Mass.

YOY—Young-of-the-year.

YPM—Yale Peabody Museum, New Haven, Conn.

CONTRIBUTORS

Kenneth W. Able
Rutgers University Field Station
Tuckerton, New Jersey 08087-2004

Steven Branstetter
Fisheries Management Division
National Marine Fisheries Service
St. Petersburg, Florida 33702

Russell W. Brown
Woods Hole Laboratory
National Marine Fisheries Service
Woods Hole, Massachusetts 02543

George H. Burgess
Florida State Museum
University of Florida
Gainesville, Florida 32611

Steven X. Cadrin
Woods Hole Laboratory
National Marine Fisheries Service
Woods Hole, Massachusetts 02543

John H. Caruso
Department of Biological Sciences
University of New Orleans
New Orleans, Louisiana 70148-2960

Barry Chernoff
Division of Fishes
Field Museum of Natural History
Chicago, Illinois 60605

Bruce B. Collette
National Marine Fisheries Service
Systematics Laboratory
National Museum of Natural History
Washington, D.C. 20560-0153

Roy E. Crabtree
National Marine Fisheries Service
Southeast Regional Office
St. Petersburg, Florida 33702

James E. Craddock (retired)
P.O. Box 1347
West Falmouth, Massachusetts 02574
Formerly Woods Hole Oceanographic Institution
Woods Hole, Massachusetts 02543

Hugh H. DeWitt (deceased)
Formerly Department of Zoology
University of Maine
Orono, Maine 04469

Donald Flescher (retired)
1669 Georgetown Way
Salinas, California
Formerly Woods Hole Laboratory
National Marine Fisheries Service
Woods Hole, Massachusetts 02543

Kevin D. Friedland
National Oceanic and Atmospheric Administration/University of Massachusetts Cooperative Marine Education and Research Program
University of Massachusetts
Amherst, Massachusetts 01003

Paul Geoghegan
Normandeau Associates, Inc.
Bedford, New Hampshire 03110

Antony S. Harold
Grice Marine Biological Laboratory
College of Charleston
Charleston, South Carolina 29412

Karsten E. Hartel
Department of Ichthyology
Museum of Comparative Zoology
Harvard University
Cambridge, Massachusetts 02138

Grace Klein-MacPhee
Graduate School of Oceanography
University of Rhode Island
Narragansett, Rhode Island 02882

John F. Kocik
Northeast Fisheries Science Center
Maine Field Station
P.O. Box 190
Orono, Maine 04773

William H. Krueger
Department of Zoology
University of Rhode Island
Kingston, Rhode Island 02881

Frederic H. Martini
5071 Hana Highway
Haiku, Hawaii 96708
Formerly Shoals Marine Laboratory
Cornell University
Ithaca, New York 14853

Ralph K. Mayo
Woods Hole Laboratory
National Marine Fisheries Service
Woods Hole, Massachusetts 02543

Richard S. McBride
Florida Marine Research Institute
St. Petersburg, Florida 33701

John D. McEachran
Department of Wildlife and Fisheries
Texas A&M University
College Station, Texas 77843

Jon A. Moore
Honors College
Florida Atlantic University
Jupiter, Florida 33458

Thomas A. Munroe
National Marine Fisheries Service
Systematics Laboratory
National Museum of Natural History
Washington, D.C. 20560-0153

Steven A. Murawski
Woods Hole Laboratory
National Marine Fisheries Service
Woods Hole, Massachusetts 02543

John A. Musick
Virginia Institute of Marine Science
Gloucester Point, Virginia 23062

Martha S. Nizinski
National Marine Fisheries Service
Systematics Laboratory
National Museum of Natural History
Washington, D.C. 20560-0153

Loretta O'Brien
Woods Hole Laboratory
National Marine Fisheries Service
Woods Hole, Massachusetts 02543

William J. Overholtz
Woods Hole Laboratory
National Marine Fisheries Service
Woods Hole, Massachusetts 02543

Theodore W. Pietsch
School of Fisheries
University of Washington
Seattle, Washington 98195

Rodney A. Rountree
Department of Natural Resources
Conservation
University of Massachusetts
Amherst, Massachusetts 01003
Formerly Woods Hole Laboratory
National Marine Fisheries Service
Woods Hole, Massachusetts 02543

David G. Smith
Division of Fishes
National Museum of Natural History
Washington, D.C. 20560-0159

Joseph W. Smith
Beaufort Laboratory
National Marine Fisheries Service
101 Pivers Island Road
Beaufort, North Carolina 28516-9722

Katherine A. Sosebee
Woods Hole Laboratory
National Marine Fisheries Service
Woods Hole, Massachusetts 02543

Joseph N. Strube
Normandeau Associates, Inc.
Bedford, New Hampshire 03110

Bruce A. Thompson
Coastal Fisheries Institute
Louisiana State University
Baton Rouge, Louisiana 70803

Kenneth A. Tighe
Division of Reptiles
National Museum of Natural History
Washington, D.C. 20560-0162

FISHES OF THE GULF OF MAINE

AN INTRODUCTION TO THE HISTORY OF FISHES IN THE GULF OF MAINE

Steven A. Murawski, Russell W. Brown, Steven X. Cadrin, Ralph K. Mayo, Loretta O'Brien, William J. Overholtz, and Katherine A. Sosebee

Some groundfish resources off New England are now recovering from the record low stock sizes and landings observed in the early 1990s. However, other stocks continue to decline because of excessive fishing mortality and below-average recruitment. Species declines in some cases occurred steadily over time, whereas others happened more recently and abruptly. Resources more sensitive to overfishing declined and were supplanted by other target species in a sequential pattern of resource exploitation. The New England groundfish fishery is now supported by species most resilient to exploitation and others not heretofore considered marketable. Groundfish resources and their dependent fisheries are well documented by landings statistics dating back over a century and by standardized research vessel survey efforts that began over three decades ago. This chapter reviews the history of the New England groundfish fishery, its management, and prospects for long-term recovery and sustainability.

> There was no sound except the splash of the sinkers overside, the flapping of the cod, and the whack of muckles as the men stunned them. It was wonderful fishing.
>
> Rudyard Kipling, *Captains Courageous*

The fishing industry of New England has, for over 400 years, been identified economically and culturally with groundfishing. A mixture of bottom-dwelling fishes including cod, haddock, redfish, hakes, and flounders constitute the groundfish resource (Table 3). The complex history of the New England groundfish resource and its exploitation since the turn of the twentieth century can be divided into several periods. Some of the important technological developments and resource conditions associated with these various periods are described in the following.

CONVERSION FROM SAIL TO STEAM (1900–1920)

In the late nineteenth and early twentieth centuries large fleets of sailing vessels from Gloucester, Boston, and other New England ports ranged throughout the coastal areas and offshore banks from Cape Cod to the Grand Banks off Newfoundland (Fig. 3A). Catches, primarily of cod, supported over 800 dory schooners and a multitude of shoreside businesses, including salt mining, ice harvesting, and an active boat-building industry, the last necessitated by substantial losses of ships (and of men) to the vagaries of the North Atlantic. The catch from distant fishing grounds was generally salted (schooners fishing with these methods were called "salt bankers"), whereas catches from the Gulf of Maine and Georges Bank were generally stored on ice and sold fresh ("market" or "shack" boats).

At the turn of the twentieth century major technological innovations were introduced, which changed how fish were caught, handled, processed, distributed, and sold. The introduction of better handling (filleting and freezing) and distribution methods (train, refrigerated storage) meant that fresh and frozen fish could be sold in markets across the country, thereby reducing the dominance of salt cod as a preferred product.

Steam-powered trawl vessels were introduced to harvest flounders and haddock on smooth-bottom areas and rapidly replaced the traditional schooners. The first trawler, *Spray*, sailed out of Boston in 1906 (Fig. 4), and the trawler fleet quickly grew to over 300 vessels by 1930. Steam power was supplanted by diesel fuel after World War I, but the transition to otter trawling as the dominant fishing method was not without controversy. Objections to development of the otter trawl fishery centered on the potential for ecological damage to bottom-dwelling animals and plants and economic competition with existing fixed-gear fisheries. Management recommendations resulting from scientific investigations of the "trawler problem" included delimiting areas where trawls could and could not be used, but these recommendations were not implemented. Even before 1900, some species showed signs of decline owing to overfishing by hook-and-line fisheries, especially Atlantic halibut (Fig. 5). Halibut landings started to decline by the 1850s; by the 1890s almost all of the Atlantic halibut sold in Gloucester, Mass., were from Iceland; and by the turn of the century Pacific halibut were being shipped to Boston via train. Overfishing and resource decline accelerated with the increased intensity of fisheries and the expanding list of target species.

TABLE 3
SPECIES AND STOCKS COMPOSING THE NEW ENGLAND GROUNDFISH RESOURCE

Common Name	Scientific Name	Management Stocks	Included in NE FMP?
Atlantic cod	*Gadus morhua*	Georges Bank, southern Gulf of Maine	Yes
Haddock	*Melanogrammus aeglefinus*	Georges Bank, Gulf of Maine	Yes
Acadian redfish	*Sebastes fasciatus*	Gulf of Maine	Yes
Pollock	*Pollachius virens*	Gulf of Maine	Yes
White hake	*Urophycis tenuis*	Gulf of Maine	Yes
Red hake	*Urophycis chuss*	Gulf of Maine/northern Georges Bank, southern Georges Bank/ mid-Atlantic	Yes
Silver hake	*Merluccius bilinearis*	Gulf of Maine/northern Georges Bank, southern Georges Bank/ mid-Atlantic	Yes
Ocean pout	*Zoarces americanus*	Gulf of Maine/southern New England	Yes
Atlantic halibut	*Hippoglossus hippoglossus*	Gulf of Maine	Yes
Winter flounder	*Pseudopleuronectes americanus*	Georges Bank, Gulf of Maine, southern New England	Yes
Witch flounder	*Glyptocephalus cynoglossus*	Gulf of Maine	Yes
Yellowtail flounder	*Limanda ferruginea*	Georges Bank, southern New England, Cape Cod, mid-Atlantic	Yes
American plaice	*Hippoglossoides platessoides*	Gulf of Maine	Yes
Windowpane	*Scophthalmus aquosus*	Gulf of Maine/northern Georges Bank, southern Georges Bank/ southern New England	Yes
Cusk	*Brosme brosme*	Gulf of Maine	No
Atlantic wolffish	*Anarhichas lupus*	Gulf of Maine	No
Spiny dogfish	*Squalus acanthias*	Northeastern United States and Canada	No
Skates	Seven species	Gulf of Maine/mid-Atlantic	No
Goosefish	*Lophius americanus*	Gulf of Maine/northern Georges Bank, southern Georges Bank/ mid-Atlantic	No
Summer flounder	*Paralichthys dentatus*	Georges Bank/ mid-Atlantic	No

Figure 4. Otter trawl fishing vessels at Boston Fish Pier, ca. 1931. The vessel at the end of the pier is the *Spray.* Built in 1905, it was the first steam trawler in the U.S. groundfish fleet. (From the Oscar Sette Archives at the Northeast Fisheries Science Center, Woods Hole, Mass.)

Figure 5. Offloading of Atlantic halibut ca. 1930, at the Boston Fish Pier. Note traditional dories stacked aboard the vessel in the foreground. (From the Oscar Sette Archives at the Northeast Fisheries Science Center, Woods Hole, Mass.)

THE RISE OF THE TRAWL FISHERY (1920–1960)

The species composition of groundfish landings changed dramatically following the introduction of trawling, as the trawl fishery targeted haddock rather than cod (Fig. 6) and then other stocks. Prior to 1900, haddock landings were relatively low (20,000 mt·year^{-1}) since the species did not preserve well when salted. By the late 1920s, haddock landings increased to over 100,000 mt (Fig. 6). This level of catch, however, was not sustainable, and landings (primarily from Georges Bank) plummeted in the early 1930s.

By 1930 the groundfish fleet had grown too large relative to the capacity of the haddock stocks to produce increased yields. Growth overfishing was revealed by sampling of the catches by at-sea observers. In 1930, some 37 million haddock were landed at Boston, but, owing to the very small mesh used in the otter trawl nets, an estimated 70–90 million juvenile haddock were discarded dead at sea. Surprisingly, mesh regulations to protect haddock were not implemented until 1953. The crash in Georges Bank haddock landings prompted much new research to investigate the causes and resulted in management recommendations. Modern population dynamics research programs for New England groundfish date back to work initiated in the 1930s by Dr. William Herrington and his colleagues at Harvard University and the Bureau of Commercial Fisheries. Owing partly to a shift of the fishery to resources on Browns Bank off Nova Scotia, the Georges Bank haddock resource recovered in the mid-1930s and landings subsequently averaged about 50,000 mt·year^{-1} between 1935 and 1960 (Fig. 6). Haddock remained the mainstay of the Gulf of Maine groundfish fishery until the mid-1960s. Cod landings generally remained stable throughout the period 1915–1940 (Fig. 6), as haddock, redfish, and other species were of primary interest to consumers.

The years of World War II were prosperous for the industry as fish were canned for military use, and protein demands and rationing led to increased fish consumption at home. The fleet was reduced at this time, as many of the largest trawlers were requisitioned for war duty as mine sweepers. Development of new markets such as ocean perch (redfish), later marketed in the Midwest as a substitute for Great Lakes yellow perch, also sustained a portion of the offshore fleet during the war years, and many government subsidy programs were launched after the war, when demand for groundfish declined.

The redfish fishery, initiated in the 1930s, peaked in U.S. waters in the 1940s and expanded eastward to the Scotian Shelf, rising to about 120,000 mt·year^{-1} in the early 1950s. This long-lived, slow-growing resource was fished down to moderate levels in the Gulf of Maine during the 1930s and 1940s, and the stock collapsed following the return of the fleet from Canadian waters in the mid-1970s. Flatfish landings were dominated by catches of winter flounder, witch flounder, and American plaice until the 1940s. Thereafter, yellowtail flounder became the most important flatfish of New England, but declined greatly in abundance and landings through the 1940s and 1950s (Fig. 6). Reasons for the decline of yellowtail during this period are not known but recruitment declined steadily during the period. Other important groundfish stocks supporting the fishery prior to the 1960s included silver hake and pollock, with small numbers of red hake, white hake, and others. Because of the modest harvest rates on most stocks, recruitment overfishing either did not occur or was not persistent. When stocks declined, the fleet moved on to other targets or different stocks of the same species (e.g., off Canada).

DISTANT-WATER FLEETS (1960–1976)

Arrival of distant-water fleet effort off the northeast United States in the early 1960s included fleets of factory-based trawlers from eastern Europe, Asia, and elsewhere. Scouting vessels for Soviet fleets first ventured into New England waters

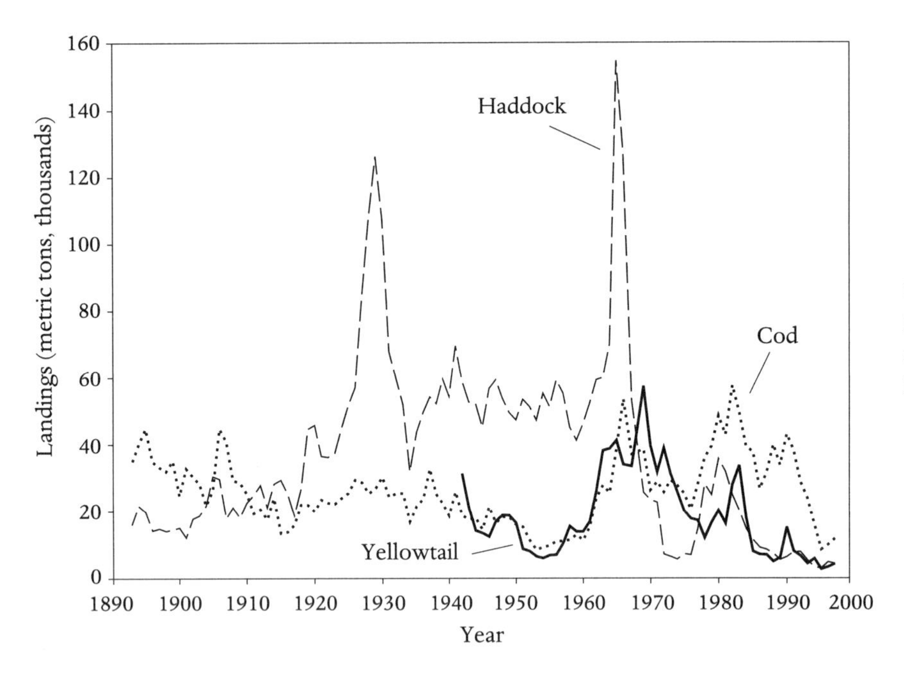

Figure 6. Total landings of Georges Bank cod and haddock and landings of yellowtail flounder from all New England waters, 1893–1997.

in 1961. Their initial target was Atlantic herring and the distant-water fleet caught about 63,000 mt that year. In subsequent years, catches of herring increased (peaking at 225,000 mt in 1963), and other species were targeted as well, including silver and red hake, haddock, and Atlantic mackerel. From 1960 to 1965 total groundfish landings increased from 200,000 to about 760,000 mt (Fig. 7). Landings of haddock reached a record high of 154,000 mt in 1965 and declined rapidly thereafter. Between 1964 and 1967 total groundfish landings were composed primarily of silver hake, haddock, red hake, flounders, and cod.

The intensified international fishery off the northeast United States prompted development of systematic, multispecies monitoring surveys, initiated in the autumn of 1963. Stratified-random bottom trawl surveys of the continental shelf waters from Nova Scotia to Hudson Canyon and later Cape Hatteras, N.C., have been conducted every autumn since 1963 and every spring since 1968. Abundance, as measured by these surveys, declined rapidly as various components of the demersal and pelagic systems were pulse-fished (Figs. 7 and 8). The parallel decline in groundfish abundance and landings was rapid and severe between 1966 and 1970 (Fig. 8).

Beginning in 1970, quota-based management was instituted for the offshore New England waters under the auspices of the International Commission for the Northwest Atlantic Fisheries (ICNAF). Quotas for each species were allocated by country, with the sum equal to the total recommended removals. Additionally, "second-tier" quotas, less than the sum of a country's species allocations, were intended to mitigate the effects of nontargeted by-catch, so that species quotas would not be exceeded. The quota system under ICNAF effectively ended directed distant-water fisheries on New England groundfish resources, as these resources were determined to have little capacity to support fisheries beyond the levels that

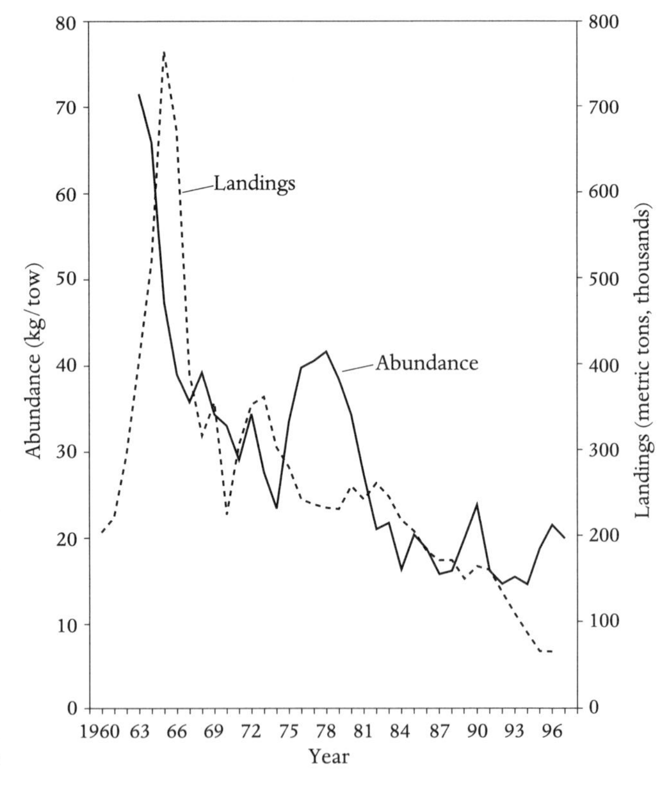

Figure 7. Landings (thousands of metric tons) and relative abundance (stratified mean catch per tow in kilograms from NEFSC bottom trawl surveys) for principal groundfish and flounder stocks off the northeast United States, 1960–1997.

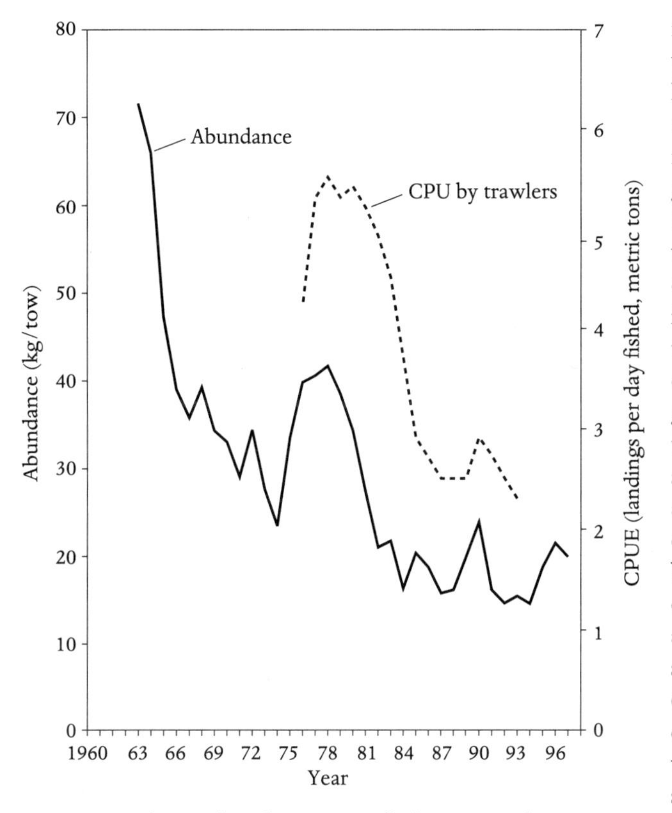

Figure 8. Relative abundance (stratified mean catch per tow in kilograms from NEFSC bottom trawl surveys) and commercial trawler CPUE (catch per unit of effort, in metric tons per day, standardized for vessel size) for principal groundfish and flounder stocks off the northeastern United States, 1963–1997.

would be taken by the United States and Canada. Quotas were progressively lowered on mackerel, herring, squids, and other species, as those resources declined as well.

In response to the declining abundance and landings of traditional New England offshore resources and elsewhere, the Magnuson Fishery Conservation and Management Act (MFCMA) was promulgated in 1976. This measure effectively ended distant-water fleet participation in New England fisheries, although some countries were allowed to harvest surpluses of squids, hakes, butterfish, and mackerel for a few years following enactment.

GROUNDFISH FISHERIES UNDER THE MAGNUSON ACT (1976–PRESENT)

No one knew exactly how many newcomers had arrived during the last four months of 1977, but according to one report, new boats entered the fishery at the astounding rate of about one every four days.

Margret Dewar, *Industry in Trouble*

With implementation of the MFCMA in 1977, the northeast groundfish fleet, once dominated by wooden side-trawlers, was replaced relatively quickly by steel stern-trawlers equipped with more modern technology for locating, catching, and handling fish. Relatively strong year-classes of cod, haddock, and some other groundfish stocks produced in 1975 and later resulted in improved resource conditions and increased groundfish abundance and effort in the late 1970s and early 1980s (Figs. 6 and 7). As a result of the elimination of distant-water fleet effort, U.S. and Canadian effort off New England expanded rapidly. Between 1976 and 1984 U.S. otter trawl fishing effort doubled. Fishery landings expanded quickly, with the Georges Bank component of the landings dominated by cod, haddock, and yellowtail flounder (Fig. 6). Trends in groundfish trawler catch per unit of effort (CPUE, in metric tons per day fished) paralleled the abundance indices from research vessel surveys (Fig. 8). Catch rates increased rapidly after 1976, but by the early 1980s had peaked and began to decline. By the mid-1980s, commercial catch rates had dropped by half, as had the overall abundance of the resource. Collapse of the Georges Bank haddock stock and then Georges Bank and southern New England yellowtail flounder resources resulted in an almost complete reliance by the fishery on cod (Fig. 9). Reduced landings of traditional groundfish stocks, combined with strong market demand for fish prompted development of fisheries for alternative species such as squids, spiny dogfish, skates, and goosefish (monkfish) (Fig. 9). Exploitation rates of most groundfish resources rose significantly, and spawning stock biomasses declined (Figs. 10–12).

The New England Fishery Management Council initially retained the quota-based fishery management system for groundfish that it inherited from the earlier management schemes adopted by ICNAF—eventually abandoning direct controls on fishing mortality in 1982 in favor of regulations based primarily on minimum mesh and fish sizes and other indirect fishery controls.

In addition to increases in domestic fishing effort, delimitation of the maritime boundary between the United States and Canada in 1985 ended fishing by New England fleets on the eastern portion of Georges Bank and on the Scotian Shelf off Canada and resulted in even greater pressure on stocks in U.S. waters.

Exploitation rates of groundfish reached their highest levels in the early 1990s, as stock biomasses fell, in many cases, to record lows (Figs. 10–12). Indirect controls had not resulted in sufficient conservation of the resources, and environmental groups sued the Department of Commerce over this failure. What emerged was a series of fishery management plan amendments, first implemented in 1994, that reduced days at sea by all fleet sectors to 50% of the pre-1994 levels. Additionally, these amendments closed over 5,000 nm^2 of prime groundfishing areas (see Preface, Fig. 3A), increased minimum net mesh sizes, implemented a moratorium on vessel entrants, and required mandatory vessel and dealer reporting of catches. The new regulations also implemented trip limits to reduce catches of depleted species and instituted "target" total allowable catches (TACS) to serve as a guide to measure the

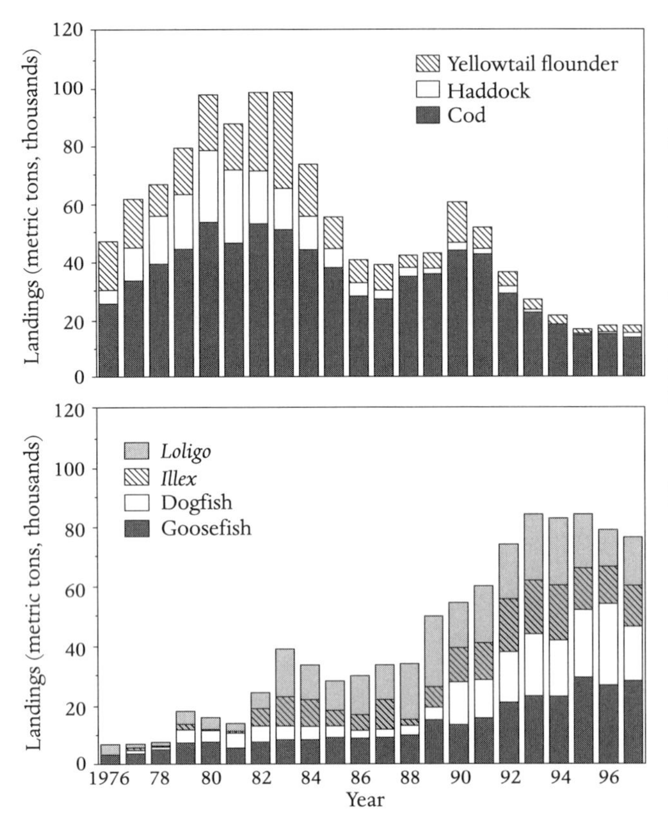

Figure 9. U.S. landings (in thousands of metric tons) of cod, haddock, and yellowtail flounder (top panel) and goosefish, spiny dogfish, and *Illex* and *Loligo* squid, 1976–1997 (bottom panel).

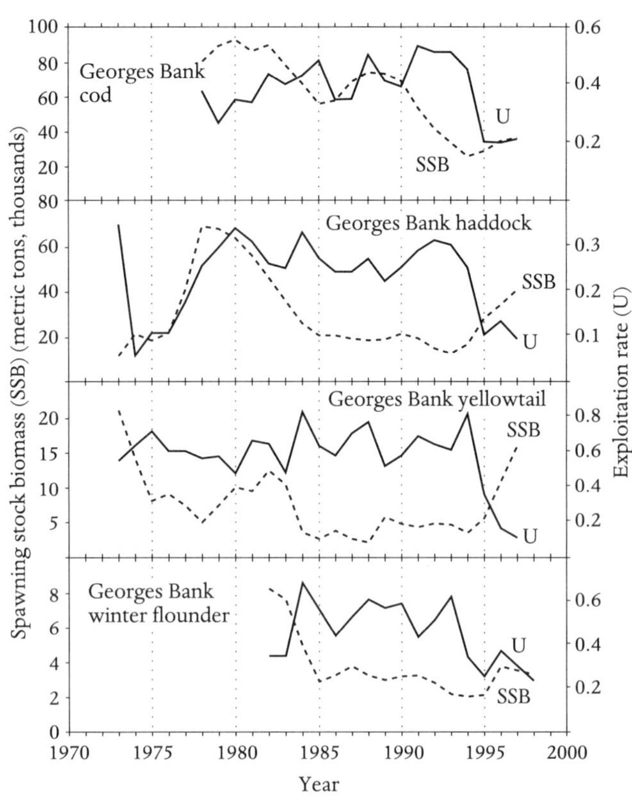

Figure 10. Spawning stock biomass (SSB, thousands of metric tons) and exploitation rate (U) for four Georges Bank groundfish stocks, 1973–1998.

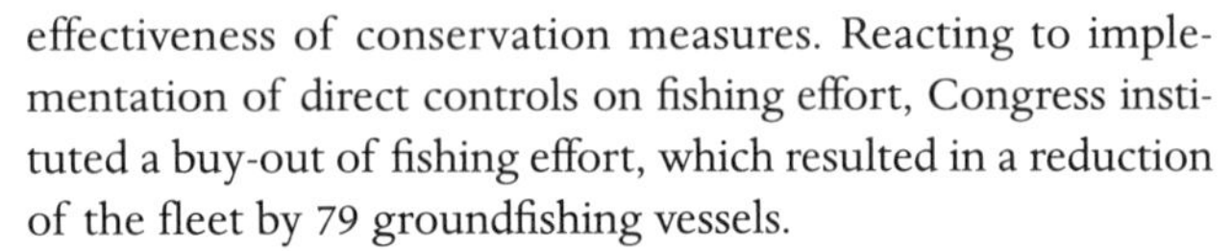

effectiveness of conservation measures. Reacting to implementation of direct controls on fishing effort, Congress instituted a buy-out of fishing effort, which resulted in a reduction of the fleet by 79 groundfishing vessels.

As a result of management measures enacted since 1994, exploitation rates, particularly on Georges Bank groundfish stocks, have declined to levels not seen in several decades (Fig. 10). Modest increases in spawning stock biomass for Georges Bank cod and haddock have occurred. Georges Bank yellowtail biomass has rebounded the fastest, to levels not seen since the early 1970s. Continued rebuilding of these stocks is contingent on improved recruitment, but year-classes spawned since 1994 have generally been well below average.

In the Gulf of Maine, exploitation rates have remained high, while spawning biomasses of cod, American plaice, and white hake have declined to near-record lows (Fig. 11). The lack of success in reducing exploitation for Gulf of Maine groundfish is due to several factors. Overall, groundfish effort has declined substantially; however, the large closed areas on Georges Bank (Fig. 3A), combined with days-at-sea regulations, have resulted in displacement of fishing effort to inshore areas and concentrations of trawl, gill net, and hook activity in the nearshore fishing grounds of the Gulf of Maine. As on Georges Bank, recruitment to most major groundfish stocks in the Gulf of Maine region has been below average in recent years.

Exploitation rates for southern New England flatfishes (winter and yellowtail flounder) have declined substantially since 1992 (Fig. 12). The spawning biomass for winter flounder has increased more than twofold over the time series low observed in 1994. Biomass of yellowtail flounder, although increasing, is well below levels necessary to sustain a significant fishery (Fig. 12).

Overall, the New England groundfish resource is beginning to increase in abundance (Fig. 8), and exploitation rates for many of the key stocks are at levels that should allow stock rebuilding. Recruitment has been generally poor in recent years, and exploitation of some stocks (e.g., in the Gulf of Maine) remains excessive.

PROSPECTS FOR THE RESOURCE AND FISHERY

Groundfish abundance and landings from the offshore New England region have varied considerably over the past 100 years, owing primarily to their exploitation history. Dramatic reductions in most offshore stocks were caused by the distant-water fleets, which pulse-fished the wide array of available

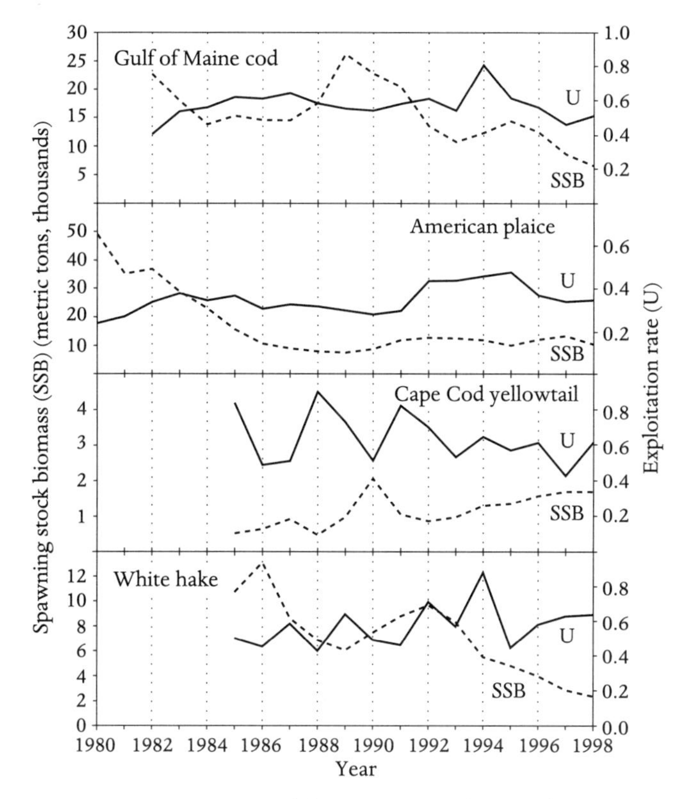

Figure 11. Spawning stock biomass (SSB, thousands of metric tons) and exploitation rate (U) for four Gulf of Maine groundfish stocks, 1980–1998.

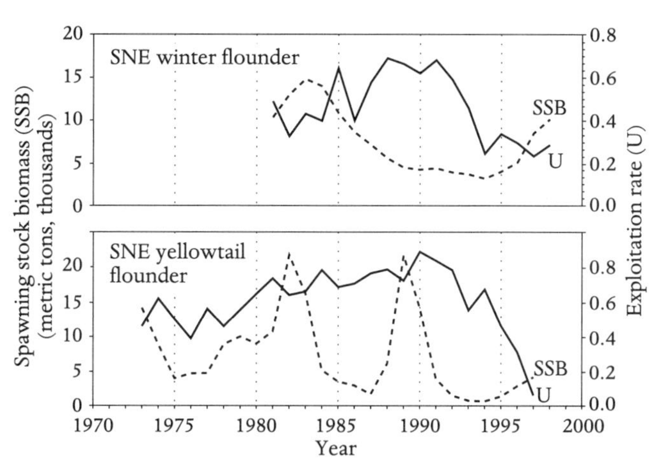

Figure 12. Spawning stock biomass (SSB, thousands of metric tons) and exploitation rate (U) for two southern New England groundfish stocks, 1973–1998.

species. After the foreign fleets were eliminated, some stocks rebounded to high levels, only to be overfished again by domestic fleets in the 1980s and early 1990s.

Projections of stock recovery included in recent groundfish fishery management plan amendments indicate that under exploitation rates such as those observed currently for Georges Bank stocks, recovery times of about a decade would be required for most species, with some stocks rebuilding sooner (yellowtail flounder) and some later (haddock). This process has clearly begun on Georges Bank and in southern New England, but additional conservation measures are required for many Gulf of Maine stocks. Managers are currently evaluating specific proposals for additional closed areas, effort reductions, and other measures to meet these goals.

The passage of the Sustainable Fisheries Act of 1996 has imposed stringent new requirements for conservation and management of all fishery resources, including New England groundfish. The new statutes require that overfished populations be rebuilt to levels that would produce maximum sustainable yields over the long term. Current rebuilding target exploitation rates for New England groundfish will, in many cases, also be the long-term management goals required under the new law.

Management measures enacted since 1994 have had significant positive benefits for many of the resources, reflected in reduced exploitation rates, increased spawning stock sizes, and more balanced population age and size structures. One factor implicated in the decline of many groundfish stocks was the increased reliance on first-time spawners, a consequence of high and increasing exploitation rates. Reduced effort on some stocks has resulted in greater proportions of fish spawning two or more times before capture. A more balanced age structure is an important element in rebuilding stocks (and fisheries) that can sustain normal year-to-year variations in recruitment, which may be extreme. Likewise, closed areas have been beneficial in promoting recovery of the western Georges Bank spawning components of cod and haddock, which were fished to very low levels prior to 1995.

The New England groundfish resource has shown remarkable resiliency to changes in fleet size, target-species shifting, and technological change over the past century. However, in some cases, stocks sensitive to overfishing (halibut, redfish) were "written off" in favor of more productive resources. New fishery management legislation requires that depleted resources be restored, and currently productive resources remain so. Fishing at sustainable exploitation rates will eventually result in much higher yields with less year-to-year variation in landings, more diverse catches (e.g., flounders, cod, haddock, redfish, pollock), and more stable catch rates in the fishery.

JAWLESS FISHES. CLASS AGNATHA

Modern hagfishes and lampreys are the most primitive living craniates. From the time they were originally described, these animals have been considered to be closely related, and they were combined as the Class Agnatha. The primary justification for this treatment was the fact that both groups lack features typical of gnathostomes. For example, neither hagfishes nor lampreys possess jaws, paired fins, or associated internal cartilages, ribs, or true bone, specialized characters found in other vertebrate groups. Characteristics shared by hagfishes and lampreys—such as an axial stiffening rod (the notochord), a cartilaginous structure surrounding the brain and associated sense organs (a cranium), pouched gills, and distinctive visceral organs such as heart, liver, and gallbladder—are primitive characteristics found in ancestral craniates. Much valuable information on hagfishes and lampreys is summarized in three references: *The Biology of Lampreys* (Hardisty and Potter 1972–1982), *The Biology of* Myxine (Brodal and Fänge 1960), and *The Biology of Hagfishes* (Jørgensen et al. 1998).

HAGFISHES. ORDER MYXINIFORMES

HAGFISHES. FAMILY MYXINIDAE

Frederic H. Martini and Donald Flescher

The hagfish skeleton is entirely cartilaginous and consists of a persistent notochord, a cranium with cartilage and fibrous connective tissue, cartilages associated with the barbels and face, cartilages associated with the finfold of the tail, the median cartilage plate of the caudal heart, an ancillary vascular pump located in the tail, and cartilages associated with the toothplates. The primary skeletal muscles of the body form approximately 110 simple, W-shaped segmental myomeres. More superficial muscle layers are found anteriorly, where a layer of branchial constrictors overlies the myomeres in the branchial region, and muscles associated with the barbels and mouth cover the segmental muscles of the prebranchial region. The muscles involved with the eversion and retraction of the toothplates form a separate series with distinctive structural and functional properties (Dawson 1960; Wright et al. 1984).

The skeletal muscles are separated from the skin by a large vascular sinus that extends on either side from the middorsal line to the ventrolateral surface just above the slime glands. When hagfish are active, these spaces contain significant amounts of blood plasma. Blood volume is roughly 18% of the total body weight, three times that in other fishes.

The digestive tract is a simple tube, without apparent regional specializations. It has a large diameter and the mucosa is thrown into longitudinal folds. There is a bilobed liver and a discrete gallbladder; pancreatic tissue is scattered within the submucosa of the gut. There is no spleen, although the connective tissues of the digestive tract contain extensive areas with apparent hematopoietic functions.

Many characteristics of modern hagfishes clearly represent primitive features shared with stem vertebrates (Martini 1998a): the cartilaginous internal skeleton, consisting of a flexible notochord, a cranium, and small axial cartilages associated with the longitudinal fins and tail; the tripartite brain; pouched gills; holonephric kidneys (a mesonephros with one nephron per body segment); and a variety of soft tissue features. Other hagfish characteristics probably represent specializations that have evolved within the lineage: reduction (or complete elimination) of the lateral line organs, and the presence of a single semicircular duct with its complex hair cell architecture. Whether characteristics such as the reduced or degenerative eyes and the lack of bone are primitive or derived remains controversial. The fossil record provides few clues to the evolutionary history of the hagfishes; a fossil hagfish (*Myxinikela siroka*) dated around 300 million years ago has the basic morphological characteristics of modern hagfishes (Bardack 1998). Hagfishes may in fact have diverged from the vertebrate lineage sometime in the Cambrian; they have teeth that are similar to but apparently not homologous with conodont teeth (Aldridge and Donoghue 1998).

Worldwide, 43 species in six genera were recognized by Nelson (1994), but Wisner and McMillan (1995) described nine new species from the New World and resurrected several others for a total of approximately 60 (Fernholm 1998). These species fall naturally into two subfamilies: the Myxininae, characterized by one pair of gill openings and the Eptatretinae, characterized by multiple gill openings (Fernholm 1998). Hagfishes in general are exclusively marine, although incursions into estuarine environments have been reported for some species. Limiting factors in their distribution appear to be, in order of importance, salinity, temperature, and substrate preference. Only one species, *Myxine glutinosa,* is known from the Gulf of Maine.

ATLANTIC HAGFISH / *Myxine glutinosa* Linnaeus 1758 / Bigelow and Schroeder 1953:10–12

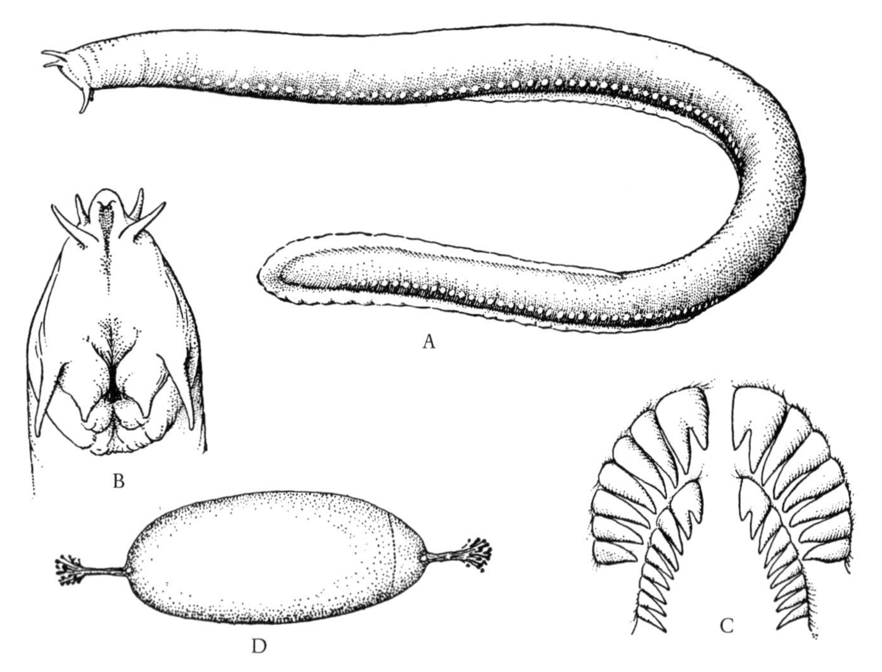

Figure 13. Hagfish *Myxine glutinosa.* (A) Adult, Gulf of Maine. Drawn by E. N. Fischer. (B) Lower view of head. (C) Tongue-teeth as seen from above, about 3× natural size. (D) Egg. (From Bigelow and Schroeder: Fig. 2.)

Description. Body slender and elongate (Fig. 13A). Nostril single, median, at tip of snout, framed by two pairs of barbels. Mouth funnel-shaped, below nostril, surrounded by four barbels (Fig. 13B). Feeding apparatus, an evertible *toothplate,* with two rows of horny *teeth* (Fig. 13C) on either side. A large, single curving tooth in roof of mouth. Eyes rudimentary, invisible externally. Two gill openings, one on each side located at about 25% of body length from the anterior end of the head. Left opening noticeably larger than the right. Ventrally, a single medial finfold begins immediately posterior to efferent branchial openings, extends caudally around the tail, rostrally to the level of the cloaca. Finfold not differentiated into dorsal, caudal, and anal fins. Skin smooth, scaleless, lubricated by mucus produced by superficial epithelial cells and by secretions of large, roughly segmental mucous glands. *Slime glands* on each side of the ventral midline visible as a series of small white dots; they begin in the anterior pharyngeal region and extend almost to the tip of the tail; each gland has its own duct, which empties onto the surface of the skin. Skin unusually loose and flexible, attached to underlying muscle of the trunk only along the middorsal line and along the ventral and ventrolateral surfaces.

Morphometric data on body proportions and the numbers of gills, tooth cusps, and slime gland pores are important in the identification of hagfishes (Table 1 in Martini et al. 1997a). The body is divided into prebranchial, branchial, trunk, and caudal regions. The prebranchial region extends from the tip of the snout to the first gill pouch; the branchial region extends to the posterior margin of the pharyngocutaneous duct; the trunk continues to the posterior margin of the cloaca; and the caudal region extends from there to the tip of the tail.

Color. Animals seen in the field are usually a bright pink or pink-gray color; animals struggling after capture rapidly change from pink to a deep burgundy. However, the color of an active hagfish depends primarily on the degree of oxygenation of the blood in the extensive network of dermal capillaries (Lametschwandtner et al. 1989) and in the capacious subcutaneous sinuses. The color of a resting or moribund animal reflects the basic skin coloration, and in preserved specimens the gray darkens considerably, becoming a deep slate color. Individuals may be marked by the presence of white patches, most often on the ventral surface or along portions of the finfold, or by scattered black spots ranging in size from a few millimeters to the size of a quarter. In preserved specimens, the skin is pale gray.

Meristics. Typically there are 6 gill pouches on each side for a total of 12 (75% of a Gulf of Maine sample of 94 specimens); less commonly there are 7 on each side (total 14, 16%), rarely a total of 10, 11, or 13.

Size. In American waters, hagfish have been recorded up to 950 mm TL (Kuenstner 1996), longer than in the eastern Atlantic where the maximum is only 420 mm (Bigelow and Schroeder 1948a) or elsewhere in the western Atlantic. Modal length for a sample of 311 specimens from the Gulf of Maine was 475 mm TL (Fig. 14A). A length-weight curve (Fig. 14B) shows a weight of about 300 g at 700 mm TL.

Distinctions. Hagfish are unlikely to be confused with any other fish from the Gulf of Maine. They are relatively slender, elongate fishes that resemble lampreys and true eels. How-

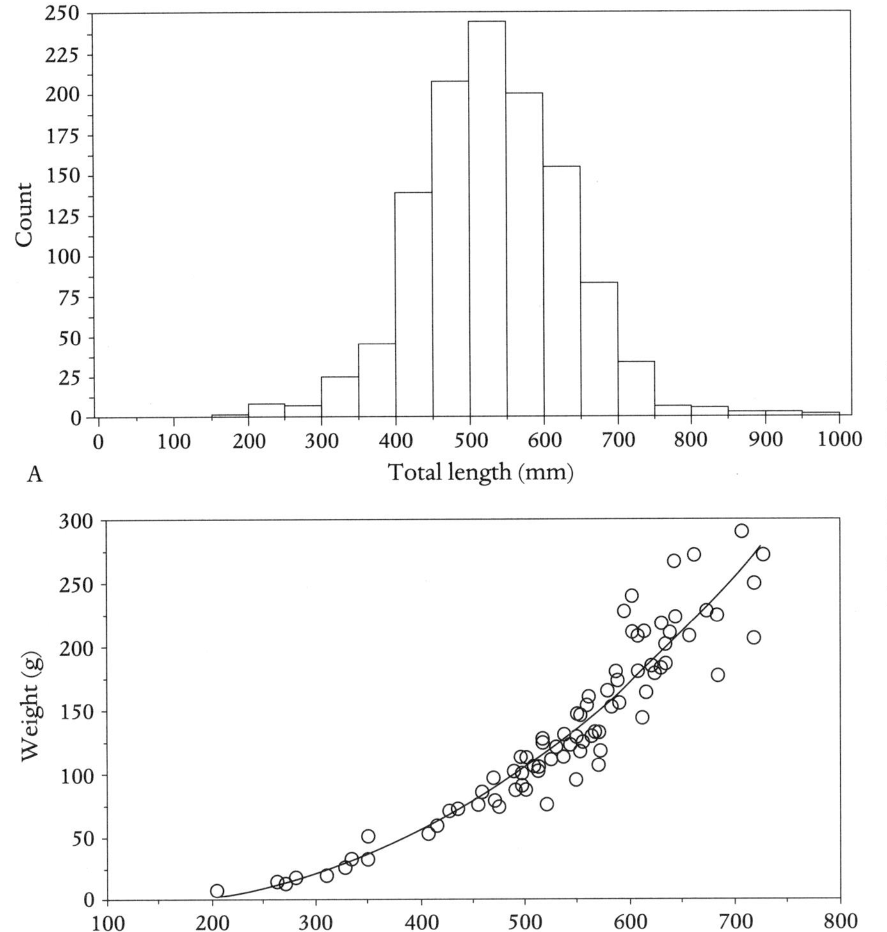

Figure 14. (A) Length-frequency for a sample of 311 hagfish, *Myxine glutinosa,* from the Gulf of Maine. (From Martini 1997a: Fig. 5.) (B) A length-weight curve for hagfish from the Gulf of Maine, regression equation $y = 13.699 - 0.222x + 0.001x^2$. (From Martini 1997a: Fig. 1.)

ever, they can easily be distinguished from lampreys by the lack of an oral hood and the presence of barbels at the tip of the snout, and from true eels by the absence of pectoral fins and bony jaws.

Taxonomic Remarks. The western Atlantic population was at one time assigned to a separate species, *Myxine limosa* (Girard), and Wisner and MacMillan (1995) suggested a return to this practice based on differences in size at maturity (eastern North Atlantic specimens are smaller) and color differences in preserved specimens. In the absence of other supporting data (there are no statistically significant differences in morphological characteristics), and because large specimens in the western North Atlantic are restricted to the Gulf of Maine, these features seem insufficient to justify dividing the eastern and western Atlantic populations into separate species (Martini et al. 1998).

Habits. Hagfish are found in areas of soft bottom sediments, clay, silt, sand, and gravel bottoms. However, a layer of soft, flocculent, muddy sediment of variable thickness often covers these different substrates, and it is this superficial layer that the hagfish appear to inhabit. Hagfish occupy shallow, transient burrows that parallel the surface. They are common in the deep basins of the Gulf of Maine and in the soft sediments surrounding rocky ledges, such as Jeffrey's Ledge or Old Scantum, closer inshore. The area of muddy substrate need not be large, as they have been reported in small patches of mud between rocks (Cole 1913). Hagfish are found in large numbers in benthic communities characterized by cerianthid anemones, tube worms, and the northern shrimp, *Pandalus borealis.* Based on substrate and depth profiles, suitable habitat covers roughly 60% of the bottom of the Gulf of Maine. Hagfish appear to require full-strength seawater. When suddenly exposed to lower salinities, an animal will release copious quantities of mucus and then become moribund. Sudden changes in temperature have a similar effect. There is general agreement (on both sides of the Atlantic) that hagfish can be held in captivity for extended periods at temperatures of 0°–4°C.

Water flows into the median nasal opening and reaches each pouch via a small afferent duct that originates from the pharynx. Each pouch is drained by an efferent duct that merges with its neighbors before reaching the exterior. As a result, there are only two efferent branchial ducts, one on each side. The left efferent opening is noticeably larger than the right, owing to the fusion of the common efferent duct on that side with the pharyngocutaneous duct, a direct connection between the posterior pharynx and the exterior. This passageway is thought to be used when clearing the pharynx of accumulated sediment or mucus.

Behavior. When preparing to burrow, the animal assumes an angle of 45–90° to the bottom and swims vigorously, driving its head into the substrate. Swimming movements continue for a variable period as the head moves into the substrate following a sinusoidal path that roughly parallels the surface. Complete disappearance of the animal into the substrate may take 5 min or more. Over time (as much as 1 h), a burrowed animal usually becomes positioned so that the tip of the snout projects from the substrate (Strahan 1963). This may make it easier to detect passing odors and to maintain a respiratory current over the gills. It is not known whether they can feed on passing invertebrates from this position. However, burrowed individuals may disappear for 10 min or more, traveling within the substrate to reach a new location. There is circumstantial evidence that feeding on polychaetes and other invertebrates may occur while hagfish are within the substrate (Adam and Strahan 1963).

Each hagfish fills the width of its burrow. Communal burrows containing several individuals are reported for Eptatretinae, but not for *Myxine*. The only known *Myxine* burrows are transitory. After its resident emerges, the burrow collapses, often leaving a sinusoidal depression in the seafloor. Hagfish are seldom seen motionless on the surface of the bottom. They are either residing in burrows or actively swimming within a few meters of the bottom.

Locomotion. *Myxine* swim at speeds of less than 2 knots using a form of anguilliform locomotion (Foss 1968). Their choice of substrates usually places them in regions where current velocities are relatively low (below 0.3 knots). The entire length of the body is involved in the undulatory movements, and the head makes far wider lateral oscillations than is the case, for example, in eels. Hagfish can swim backward virtually as rapidly as they can forward, and individuals will often do this if they encounter an obstacle or noxious stimuli.

While the animal is swimming, the nostril is flared and the four nasal barbels are directed forward. They are surprisingly active and agile animals, especially when aroused by the presence of attractive odors. Animals observed in submersible or remotely operated vehicle (ROV) work in the Gulf of Maine swam rapidly toward a bait station, following a linear or zigzag track (Martini, pers. obs.).

In one common hagfish behavior, called *knotting* (Adam 1960), the body forms a loop and the tail passes through it, forming a simple overhand knot. The knot tenses and through a wave of contraction moves toward the snout. Knotting may have three important functions: (1) Passage of a portion of the trunk through the knot cleans it of clinging mucus, slime, and associated debris, which is significant because a hagfish will suffocate if entrapped in its own slime for an extended period. (2) Combination of a mucous coating, knot formation, and movement can make the animal almost impossible to hold onto, and this may help it to avoid capture. (3) If the knot encounters an obstacle, it will stop its forward progression. As contractions continue, the more rostral portions of the body are forcibly retracted, which often plays a role in feeding. If the toothplates are grasping a piece of a large food item, such as the flank or visceral organ of a fish, when the knot contacts the prey surface the head is retracted, stretching and tearing the soft tissues held by the toothplates. Knotting can also assist hagfish in escaping from confinement, both in nature and in commercial traps. If a sufficient length of the tail can be extended through an opening, a knot will form, and as the knot braces against the container the rest of the animal will be pulled through the opening.

Although blind, hagfish are sensitive to light shining on the skin, especially to wavelengths below 600 nm, with peak sensitivity to blue-green light at 500–520 nm (Steven 1955), but the identity of the photoreceptor involved is not known. Data concerning the behavioral response to light exposure are inconsistent (Newth and Ross 1955; Strahan 1963; Foss 1963). Field studies in the Gulf of Maine have indicated that hagfish do not attempt to avoid high-intensity illumination by floodlights at depths of up to 150 m when swimming, burrowing, or feeding; and trapping, ROV surveys, and submersible observations have not shown day/night differences in hagfish distribution or activity.

Feeding. As far as is known, hagfish spend most of their time within burrows and emerge to feed. Release of bait in an area showing few if any hagfish on a superficial survey can result in the appearance of hundreds of animals in a matter of minutes. There is no indication of organized schooling, although the animals converge on their target en masse. It is likely that this kind of mass feeding occurs whenever a suitably large food source appears, often as a result of human activities. For example, trapping groundfish in gill nets, discard of by-catch, or disruption of benthic communities by trawling operations provide sudden, large-scale-feeding opportunities for hagfish. In the case of gill nets, this brings hagfish into direct conflict with humans. The extent of their impact on the gill net fishery in the Gulf of Maine is difficult to assess, but they are certainly a nuisance. For example, Bigelow and Schroeder reported that hagfish were damaging "a large proportion of the fish caught on longlines, unless the latter are tended frequently." They also noted that in the spring of 1913 hagfish gutted 3–5% of all the haddock caught in gill nets around Jeffrey's Ledge. It is very unlikely that groundfish or other vertebrate remains would form a substantial part of the hagfish diet without human assistance (Strahan 1963). Gut content analysis of hagfish indicates that their primary diet consists of invertebrates. Gustafson (1935) reported finding setae of the polychaetes *Eumenia crassa*, *Lumbrinereis fragilis*, and *Nereis* sp. Strahan (1963) examined fecal wastes and identified polychaete setae mixed with the remains of herring, hermit crabs, shrimp, and priapulids.

In the most detailed study of the dietary habits of hagfish (Shelton 1978), the gut contents of 129 trawled specimens from the North Sea were dominated by remains of the northern shrimp, *Pandalus borealis.* This is especially interesting because hagfish are abundant in the same benthic communities that harbor the commercially valuable *Pandalus* in the Gulf of Maine.

In addition to *Pandalus,* remains of small fishes and a scattering of benthic and burrowing invertebrates were found (Shelton 1978). Hagfish can be attracted to baits consisting of fish, shellfish, bird, or mammal remains, and they have been observed in the field feeding on bony and cartilaginous fishes trapped in gill nets, on whale carcasses, on shellfish at petroleum seeps, and on a variety of prey (a polychaete, herring remains, a hermit crab) presented by investigators (Strahan 1963).

When feeding on tough material, such as vertebrate tissues, the animal swims actively, forcing the head against the target while the toothplates are everted. When a superficial irregularity has been grasped, knotting occurs, and as the knot reaches the head a portion of tissue is torn away, as detailed above. This process is awkward and slow, and when feeding on dead or injured fishes hagfish often seek access to the relatively tender viscera by entering the mouth or anus. When feeding on soft material, such as a soft-bodied worm, knotting does not occur. First, the toothplates are everted and the object grasped and pulled into the mouth. The toothplates are then everted, and as this occurs the object is held in place by contact with the median fang located on the inferior surface of the palate. The everted toothplates grasp the object distally, and the cycle is repeated. This movement brings the teeth into opposition, making the complex effective for grasping or tearing at soft tissues or soft-bodied prey. In this way an elongate item can be ingested in a series of rapid grasp-retract-release-extend cycles. This behavior, which can easily be elicited in the laboratory, has also been observed in the field (Strahan 1963).

Energetics and Daily Ration. Preliminary work (Lesser et al. 1996) permits a rough estimate of the feeding requirements for *Myxine* at a temperature of 4°C. Resting oxygen consumption for an average (136 g) individual is approximately 5 μmol $O_2 \cdot day^{-1}$, which is the equivalent of 2.39 kJ (0.57 kC)$\cdot day^{-1}$. This is an extremely low metabolic demand, and it could explain why hagfish are able to survive in captivity for extended periods without feeding. The duration between feeding bouts in the wild is not known, but such a low basal metabolic rate may be of importance for a relatively sedentary opportunistic predator with a small home range. At 3.53 $kJ \cdot g^{-1}$ for shrimp flesh or 6.36 $kJ \cdot g^{-1}$ for fish flesh (cod), a single meal could provide enough calories to sustain a hagfish for a considerable period.

Predators. Hagfish have been identified in the stomach contents of cod, spiny dogfish, harbor seal (*Phoca vitulina*), and harbor porpoise (*Phocoena phocoena*) (Martini 1998b). Groundfish may also prey upon hagfish eggs; some of the first hagfish eggs described were from the stomach of a preserved cod. Intraspecific predation may also occur, as an unfertilized egg has been reported from the gut of a male hagfish (Holmgren 1946).

Hagfish are probably protected from predation by their burrowing habits and by the slime they secrete (Crystall 2000). A single individual hagfish can turn a 2-gal pail of water into a gelatinous mass of transparent slime within a few minutes. After roughly 15 min the slime begins to contract, ultimately forming white, viscous, stringy masses. The mechanism of contraction is not understood. Hagfish slime may serve multiple functions. Coating a food source with slime may deter other scavengers (Isaacs and Schwartzlose 1975). The slime may also interfere with the gill respiration of potential predators. A hagfish left entrapped in its own slime will soon expire; the removal of secreted slime appears to be one important function of knotting behavior.

Parasites. Perhaps surprising for a sedentary benthic animal of its size, there are no reports of external or internal parasites from hagfish.

Reproduction. In 1864 and 1865 the Royal Academy of Copenhagen offered a prize to anyone who could uncover the reproductive secrets of *M. glutinosa.* That prize was officially withdrawn in the late 1980s, without any indication that an answer was forthcoming. In the Gulf of Maine, *M. glutinosa* greater than 400 mm TL may have both ovarian and testicular tissues developed to some degree. The gonads develop within a mesenterial fold located to the right of the dorsal mesentery that supports the gut. The anterior two-thirds of the gonad may develop into ovarian tissue, and the posterior one-third may develop into testicular tissue. They are not simultaneous hermaphrodites; an individual may have an immature ovary and immature testis, a mature ovary and immature testis, or an immature ovary and a mature testis. The population contains a mixture of four morphotypes: (1) Individuals shorter than 400 mm are sexually immature. (2) Approximately 59% of the population is classified as female on the basis of egg development; testicular tissue is usually rudimentary in these animals. (3) Males represent a very small percentage of the population (less than 6%). (4) Roughly 25% of the adult population does not have macroscopically identifiable gonadal tissue; presence of large numbers of sterile individuals has also been reported for populations in the eastern North Atlantic (Holmgren 1946; Jespersen 1975; Schreiner 1955).

A mature female *Myxine* can produce up to 30 eggs, each measuring approximately 25 × 10 mm. As the eggs mature, they become suspended from the free edge of the mesentery. At ovulation each egg is surrounded by a horny shell that bears a series of short filaments ending in hooks at each end (Fig. 15A). There are no gonoducts, and the fully developed eggs are released into the coelomic cavity. The hooks, which serve to interlock the eggs within the body cavity, have been

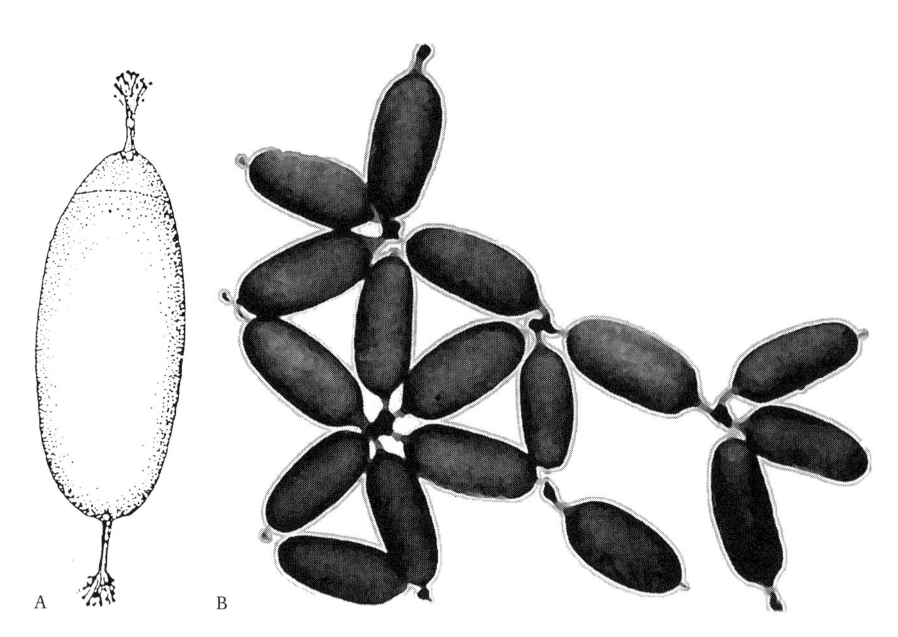

Figure 15. (A) A hagfish *Myxine glutinosa* egg that bears a series of short filaments ending in hooks at each end. (B) "Opened" cluster of eggs. (After Holmgren 1946: Fig. 1.)

compared to Velcro, since they make it difficult for the eggs to separate (Fig. 15B). This may be important in facilitating simultaneous ejection of the egg cluster.

The time required to produce a clutch of eggs is not known. Among hagfishes, only *Eptatretus burgeri* has been shown to have an annual breeding cycle. Unlike the situation with *E. burgeri, M. glutinosa* has no specific breeding season, and at any given time the population contains females at all stages of oogenesis.

The location of egg deposition remains a mystery. Over the last 150 years, fewer than 200 *M. glutinosa* eggs have been recovered. Only four of these eggs were fertilized, and none of the embryos was in an ideal state of preservation when examined. A trawled and damaged embryo was described by Dean (1900). Three other embryos, in somewhat better condition, served as the basis for later papers (Holmgren 1946; Fernholm 1969). Most of the *Myxine* eggs—fertilized or not—described in the literature were collected in the eastern North Atlantic, primarily from the nets of trawlers working soft bottom substrates. There has been only one report of hagfish eggs collected by trawl in the western Atlantic (Dean 1900); no egg clusters were seen during winter and summer ROV surveys or summer submersible dives in an area supporting a large hagfish population, nor was anything found on the adjacent ledges (Martini and Heiser 1989). One possible explanation for this is that *Myxine* embryos develop within burrows. Although the sediments are anoxic, calculations indicate that adult hagfish perfusing their gill pouches with oxygenated water at or near the burrow entrance could oxygenate the interior of the burrow through cutaneous exchange (Martini 2000).

Fertilization is assumed to be external, since: (1) males do not have copulatory organs, (2) the cloacal epithelium does not have folds or cilia that might assist in transporting sperm into the coelom of the female (Kosmath et al. 1981), and (3) self-fertilization can be ruled out as testicular tissue is rudimentary in females with fully developed eggs. [However, Jespersen (1975) felt that the sperm structure was consistent with internal fertilization.] Sperm are able to reach the surface of the egg only through a tiny micropyle at the base of the apical filaments (Fernholm 1975; Kosmath et al. 1981). The micropyle is barely the diameter of the head of a single sperm.

Because so few fertilized eggs have been recovered, we know very little about embryological development of *M. glutinosa*. There is general consensus that hagfishes, including *Myxine*, do not have a larval stage. The smallest specimen of *Eptatretus* reported, 6.5 cm in length, was assumed to be newly hatched (Worthington 1905) and was a miniature version of an adult.

Although our picture of reproduction in *M. glutinosa* in the Gulf of Maine remains incomplete, it is clear that the population reproduces very slowly. Of 121 animals surveyed in the summers of 1989–1990, only 5 were females with postovulatory follicles and only 1 had shelled eggs in the coelom. In animals with postovulatory follicles, the remaining ovarian tissue does not show evidence of advanced egg development. There must, therefore, be a significant time period between reproductive cycles for a given individual, probably longer than a year (Patzner and Adam 1981).

It appears likely that the reproductive potential of the population as a whole is relatively low, because: (1) many individuals are sexual neuters, (2) each mature female produces only 20–30 eggs, and (3) a relatively small proportion of females contain mature eggs at any given time. This has obvious implications for the development of a sustainable fishery for these animals.

General Distribution. Hagfish are widespread throughout the Atlantic Ocean at depths ranging from the surface to 1100 m or more. The Arctic limit appears to be the Kola Peninsula in Russia, and the furthest southern report is from the South Shetland Islands in the Scotia Sea, off the tip of the Antarctic Peninsula. Hagfish are found on both sides of the temperate North Atlantic. In the eastern Atlantic, *M. glutinosa* has been collected off the coasts of Norway, Great Britain, Ireland, France, and Portugal, as well as within the western Mediterranean Sea. In the western Atlantic, *M. glutinosa* is found in Davis Straits and off Greenland, offshore of the Labrador and Newfoundland coasts, within the Gulf of St. Lawrence, throughout the Gulf of Maine, and along the continental slope to the coast of Florida.

In the South Atlantic, *M. glutinosa* has been reported from the southern coasts of Argentina and Chile, including the Straits of Magellan, and off South Africa. It appears difficult to distinguish *M. glutinosa* from closely related species reported from southern waters, such as *M. capensis* or *M. australis,* but these southern forms are currently held to be valid species (Wisner and McMillan 1995).

Occurrence in the Gulf of Maine. Hagfish are found throughout the Gulf of Maine, where there are suitable combinations of depth/temperature, salinity, and substrate (Map 2). Bigelow and Schroeder (1948a, 1953c) reported hagfish from near Grand Manan, at the entrance to the Bay of Fundy, but apparently not within the Bay, to the inner and outer slopes of Georges Bank, the muddy bottom along the Maine coast, on the Boon Island–Isles of Shoals fishing grounds, around Jeffrey's Ledge, and in the deeper parts of Massachusetts, Passamaquoddy, and Penobscot bays.

There is evidence that the distribution and abundance of *M. glutinosa* has changed since 1953. For example, Bigelow and Schroeder noted that they were extremely common between Boon Island and the Isles of Shoals; however, no hagfish have been reported close to the Isles of Shoals in at least the last 25 years. It is not known whether this shift reflects alterations in benthic community structure that may be fisheries-related or if it occurred owing to other factors as yet unknown. There is no indication that any species of *Myxine,* including *M. glutinosa,* undertakes daily or seasonal migrations comparable to those reported for *Eptatretus burgeri* (Tsuneki et al. 1983). It seems likely that *Myxine* forms closed populations and that they may in some way be able to recognize locations. A tagged animal covered a distance of several kilometers to return to the place where it was originally captured (Walvig 1967).

The number of hagfish at other sites may have increased. Bigelow and Schroeder considered a catch of 11 hagfish in an hour or less as evidence that they were "plentiful" in the area sampled. They qualified that statement with, "But we question whether they ever occur in American waters in such numbers as seen in the fjords of western Sweden and southern Norway, where catches of 100 are usual in eel pots set overnight on suitable bottom." On research cruises from the Shoals Marine Laboratory, catches of 100 or more animals are common in sets of 30 min to 1 h, and a swarm of 300–400 hagfish began to assemble around the research submersible *Johnson Sea Link* within 5 min of its arrival on the bottom. Whether or not a population increase has occurred, human activities have certainly provided increased feeding opportunities for hagfish, through gill net and longline operations and the release of dead or dying fishes and other animals caught as by-catch in trawling operations.

Importance. Hagfish are important for several reasons. In terms of scientific knowledge, they are extremely interesting because they are the oldest extant clade among the craniates. A better understanding of their anatomical and physiological characteristics may reveal information about an early stage in vertebrate evolution. However, there are more immediate and compelling reasons why this species is important in the Gulf of Maine:

1. The substantial numbers present and their ongoing energy requirements suggest that they play a significant role in the benthic ecosystem throughout the Gulf of Maine, although they have thus far been ignored in energetics modeling. Preliminary calculations of minimum population densities based on trapping and ROV surveys indicated a peak density of 500,000 km^{-2} and average density of 59,700 km^{-2} (Martini 1998a). The average biomass was estimated to be 8119 $kg{\cdot}km^{-2}$ (Martini et al. 1997b).
2. Hagfish may have both direct and indirect effects on commercial fisheries in the Gulf of Maine. In areas of abundance their opportunistic feeding habits can reduce the value of the catches made by longline or fixed gill net fisheries. Bigelow and Schroeder reported that hagfish most often damaged haddock and hakes, relatively soft-bodied fishes often caught over soft bottom. Hagfish have also been known to feed on cod, herring, mackerel, spiny dogfish, marine mammals, and mackerel sharks caught in fisheries gear. Equally important, but even harder to quantify, feeding studies by Shelton (1978) suggest that hagfish could have a significant impact on *Pandalus* populations within the Gulf of Maine.
3. *Myxine* populations are now targeted by American and Canadian fishermen in the Gulf of Maine to meet the South Korean demand for "eelskin" used to manufacture expensive leather goods. In 1990, the sale of leather goods produced from hagfish skin, brought South Korea revenues of approximately $100 million (Gorbman et al. 1990). Details of the fishery are presented in the following section.

The Hagfish Fishery. Korea and Japan have long-established hagfish fisheries. The Pacific species *Paramyxine atami* and *Eptatretus burgeri* were hunted for food, using woven bamboo traps or baited longlines. Skins of hagfish collected in this way were sometimes treated and used to manufacture leather goods. In the early 1980s, "eelskin" products manufactured in South Korea gained popularity. By 1986–1987, the peak years of the fishery, a Korean hagfishing fleet of over 1,000 vessels provided skins to roughly 100 shoreside processing plants. Catches by each vessel initially averaged 1–5 tons per day (S. Kato, pers.

comm., 1989). To collect the animals, up to 4,500 cylindrical plastic traps baited with cut fish were set on the bottom using longline gear. In the years that followed there was a decrease in the number of processing plants, owing in part to the reduction in supply caused by overfishing, and the vessels began to work more distant waters. These vessels began to target additional species found around Japan, such as *Myxine garmani* and *Eptatretus okinoseana*.

Beginning in 1987, Korean buyers began purchasing Pacific hagfish, *Eptatretus stouti,* from fishermen between San Francisco and Monterey, California. By 1989, the fishery was active from Mexico to British Columbia and Alaska, and buyers were also accepting black hagfish, *Eptatretus deani*. In that year, there were over 100 vessels involved, with combined landings worth almost $5 million (Botta 1990). Thereafter, the California market began to decline owing to a combination of factors, including problems with skin quality and a reduction in the hagfish supply. In an attempt to regulate the fishery, California, Oregon, and Washington initiated permit systems. These state programs regulated the number of vessels, the number of traps, and the type and construction of traps (to prevent *ghost fishing*).

In 1989, Korean buyers approached fishermen in New England, with an interest in *M. glutinosa*. Within a year, the Canadian government offered permits for hagfishing by Canadian fishermen. The U.S. fishery typically involves setting a string of four to five baited traps on the bottom overnight. The traps consist of 45-gal plastic pickle barrels with funnels providing access to the interior. Data on the developing fishery have been collected by the National Marine Fisheries Service (NMFS) and by the New England Fisheries Development Association (NEFDA). The fishery has grown rapidly since its inception. In 1991, two boats made 32 trips over an 8-month period. During 1993, some 890,000 lb of hagfish were landed at Gloucester, Mass., Sandwich, Mass., Hampton, N.H., and Stonington, Maine. During 1994, five boats were involved in full-time hagfishing and made a total of 796 trips. This small fleet landed 2.7 million lb of hagfish, primarily at Gloucester, but with additional landings at Sandwich and Portland, Maine. Fishermen were paid an average price of $0.29 per lb, for a dockside value of roughly $780,000. The New England processing plants in turn were paid $0.43–0.45 per lb for frozen hagfish shipped to South Korea. Large hagfish were also processed for local New England consumption. Catches rose during 1997–1999, totaling 11.5 million lb and generating approximately $3.3 million (NMFS data).

Over the period 1991–1996, roughly 50 million hagfish were processed and shipped overseas (Martini 1998b); during 1997–1999, that number grew to roughly 212 million (NMFS data). The actual effect on the population is considerably greater, however, since hagfish shorter than 500 mm, the minimum length suitable for leather, are discarded into the surface waters, where they quickly become moribund. On some trips, over 50% of the catch (by weight) was discarded as unmarketable (Kuestner 1996). Under these conditions the number of individuals removed from the environment would be more than twice the number landed ashore. Because the fishing effort is not randomly distributed throughout the Gulf of Maine, the populations at sites targeted by this fishery can be expected to decline much more precipitously. (There is evidence that this is already occurring, according to Hall-Arber 1996). It is not known what effects such a decline will have on the benthic ecology in these regions. However, from a regulatory perspective it is obviously difficult to set defensible quotas or guidelines for a fishery when virtually nothing definitive is known about (1) the size of the population, (2) their reproductive potential, (3) their individual growth rates, or (4) their longevity. There is therefore an urgent need for increased research on the basic biology and ecology of this interesting species.

ACKNOWLEDGMENTS. The Shoals Marine Laboratory and the National Undersea Research Program provided financial and logistic support for this research. J. B. Heiser and Michael Lesser reviewed drafts of the manuscript. NEFDA (Boston) and NMFS (Woods Hole) provided catch records and statistics for the hagfish fishery in the Gulf of Maine.

LAMPREYS. ORDER PETROMYZONTIFORMES

LAMPREYS. FAMILY PETROMYZONTIDAE

Donald Flescher and Frederic H. Martini

Modern lampreys and hagfishes are the most primitive living vertebrates. Lampreys share a number of characteristics with hagfishes: the presence of cartilaginous internal skeletons but the absence of jaws, paired fins, ribs, and true bone. Yet lampreys share many other features with advanced fishes and tetrapods; for example, vertebral arches protect the spinal cord, placing them within the vertebrates. Their kidney structure, osmoregulatory strategies, sensory capabilities, and circulatory systems resemble those of bony fishes rather than hagfishes. Specializations unique to lampreys include: two pairs of semicircular canals in the inner ear (only one pair in hagfishes, three pairs in all other vertebrates); a branchial basket that surrounds the gill pouches; the oral hood and rasping tongue; and the focusing mechanism of the eye, which involves a muscle inserted on the cornea. The fossil record provides few clues to unravel the evolutionary history of lampreys. The earliest known fossil, *Mayomyzon* (approximately 220 million years old), closely resembles modern lampreys.

The paired kidneys project into the coelom from the dorsal wall of the body cavity. Histologically the kidneys of the adults resemble those of bony fishes, with multiple nephrons lacking any segmental organization. The larval form has a pronephros consisting of three or four large nephrons that open into the pericardial cavity. These nephrons are drained by a duct that extends caudally to the cloaca. The mesonephros, which develops when the larva reaches approximately 12 mm in length (Morris 1972), consists of multiple nephrons without any segmental organization. The entire larval kidney degenerates during transformation, when the adult (opisthonephric) kidneys develop. The body fluids at all ages have an osmolarity of approximately 361 mmol·liter^{-1} (Morris 1972). The ability to osmoregulate in seawater appears to develop at the time of larval transformation.

There are differences of opinion in lamprey classification, but Nelson (1994) recognizes four subfamilies, six genera, and some 41 species, including parasitic and nonparasitic forms. Of the parasitic forms (18 species, including the only species found in the Gulf of Maine, the sea lamprey, *Petromyzon marinus*), nine have what is assumed to be the ancestral lifestyle. The adults are marine animals who spawn in freshwater streams. There is an extended larval period of slow growth, supported by suspension feeding, followed by metamorphosis to a miniature version of the adult form. The metamorphosed larva heads downstream, returning to the sea. Growth is rapid, and the animals return to freshwater to spawn and die 2–4 years later. In the other 9 parasitic species, the adult phase is spent in freshwater lakes, rather than in the marine environment; this is presumably a result of the founder populations of larvae having been "cut off" from the ocean at some time in the past. The nonparasitic species, often called *brook lampreys*, typically complete sexual maturation prior to metamorphosis, and the adult form reproduces and dies without feeding.

SEA LAMPREY / *Petromyzon marinus* Linnaeus 1758 / Bigelow and Schroeder 1953:12–14

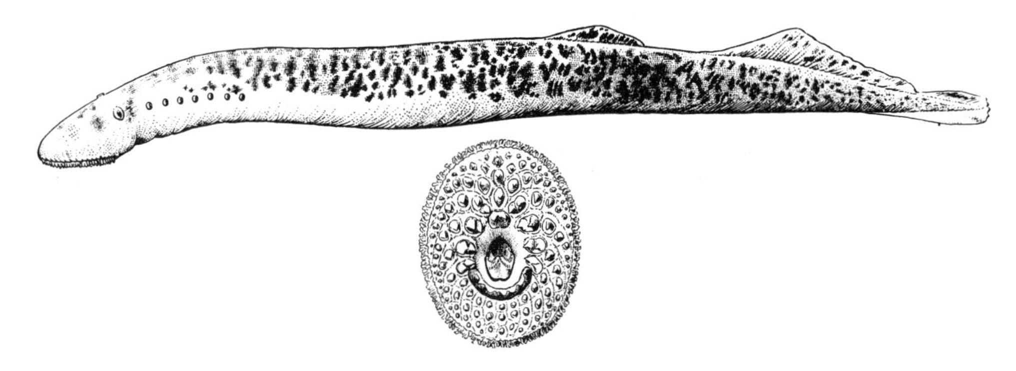

Figure 16. Sea lamprey *Petromyzon marinus.* Merrimac River. The arrangement of horny teeth is shown below. Drawn by E. N. Fischer. (From Bigelow and Schroeder: Fig. 3.)

Description. Body elongate, eel-like (Fig. 16). Jaws absent, mouth forming broad, elliptical oral hood armed with horny, hooked teeth arranged in 11 or 12 rows, innermost teeth largest. Two dorsal finfolds and a caudal fin present. Eyes small, dark, surrounded by branches of a lateral line system. A single median nasal passageway opens on surface of head between eyes and oral hood and leads to a small hypophyseal sac, not connected to pharynx. Seven gill pouches open on each side; internally they connect to a respiratory tube separate from the esophagus. Scales absent; skin lubricated by mucous secretions of epidermis and tightly bound to underlying muscle; outlines of individual myomeres visible from surface.

Color. Larvae are gray, dark brown, or olive green on the back, and paler ventrally. In recently metamorphosed animals, whether on their way downstream or in saltwater, the ventral surfaces are white and the sides silvery. The upper flanks and back may be mottled with darker shades of gray, brown, or olive; the color may become nearly black if the dark patches merge. In larger juveniles the lower surface may be white, gray, or of a pale shade of the same hue as the ground color of the back. During breeding season, adults become more colorful; for example, the background may take on bronze highlights, with lighter mottling. Landlocked forms may be brilliantly colored at spawning time, with bright yellow markings on a blue-black background.

Size. Mature individuals average about 720 mm TL and 880 g at the start of the spring migration (Scott and Scott 1988). The maximum recorded length for an adult is approximately 900 mm.

Distinctions. Lampreys are elongate, eel-like fishes that superficially resemble hagfishes and true eels. Lampreys can easily be distinguished from hagfishes by: the large oral hood, with prominent epidermal teeth; the presence of eyes; the absence of barbels or slime glands; the presence of lateral line pores on the head; and the heavily pigmented skin, which is tightly attached to the underlying muscle. Lampreys can easily be distinguished from true eels by: the oral hood; the absence of jaws and paired fins; the presence of two dorsal finfolds; and the presence of multiple gill openings.

Habits. Occasionally sea lamprey are found attached to driftwood, even to boats. When not clinging to anything, they are strong, vigorous swimmers. Sea lamprey usually keep their oral disc closed while swimming; as a result the head assumes the shape of a bullet. Applegate (1950) says they are capable of amazing bursts of speed when necessary, such as when surmounting a dam or when frightened. Such sudden spurts are certainly the maximum swimming velocity for lamprey and they probably cannot maintain such speeds for long. Migrating sea lamprey move upstream at ground speeds of 0.1–3.5 $km \cdot h^{-1}$. The mean rate, excluding rest periods, is 1.51 $km \cdot h^{-1}$ (Stier and Kynard 1986). During normal, nonmaximal swimming, sea lamprey move in pulses, followed by periods of rest. Rather than maintaining their position in the water column through constant swimming, as migrating bony fishes do, lampreys latch onto solid objects, such as logs or rocks. This behavior is seen even in still water.

Water can flow through the gill pouches in either of two ways. When the animal is free-swimming, water can enter the mouth, flow into the respiratory tube, and then through the gill pouches to the exterior. When the animal is attached to prey, water is pumped in and out of the pouches through the external gill openings. Contractions of pharyngeal muscles compress the pouches and expel water. The elastic recoil of the cartilages in the branchial basket reexpands the pouches and pulls water back in through the external gill openings.

Food. Sea lamprey prey on basking shark, sturgeons, eels, Atlantic herring, Atlantic menhaden, American shad, alewife, blueback herring, white sucker, brook trout, Atlantic salmon, Atlantic cod, haddock, pollock, hakes (*Urophycis*), bluefish, weakfish (Beamish 1980), swordfish, Atlantic mackerel, bluefin tuna, and sei whale (Halliday 1991). It is not known if or how often lamprey kill large pelagics; many seen with scars are still alive. Sometimes as many as three or four lamprey are attached at one time to a single shad, and they are said to be exceedingly aggressive in attacks on their prey.

Judging from their land-locked relatives and from the occasions on which they have been found fastened to marine fish, lamprey must be extremely destructive to the latter, which they attack by "sucking on" with their wonderfully effective mouths. Lamprey usually fasten onto the side of their victim, where they rasp away until they tear through the skin or scales and are able to suck the blood. For unknown reasons the attachment, at least in salmon, occurs significantly more often on the right side of the prey rather than on the left side (Beamish 1980). The victim is not invariably killed, and fishes are sometimes seen with scarring from two or three lamprey attacks. Lampreys are probably obligatory parasites specialized to feed on blood and body fluids. Nothing has been reported in the digestive tract except blood and occasionally fish eggs, which presumably were obtained by rasping into the peritoneal cavity of a gravid female.

Reproduction. Sexual dimorphism is marked in adults, and the difference is most apparent in mature individuals during the spawning run. As they ripen, males develop a strong ridge along the back. Females have a prominent ventral ridge, which may resemble a low finfold, beginning posterior to the cloaca and extending toward the caudal fin. This ridge may assist in the preparation of a nest at the spawning site.

It has been known from early times that sea lamprey breed in freshwater. However, they do not enter all the streams within their range indiscriminately. Lamprey require a gravelly bottom in rapid water for their spawning beds, near quiet water with patches of muddy or sandy bottom, where the larvae can develop. In many small streams, and in larger ones if these are blocked by dams or high falls, they may spawn only a short distance upstream; even within the influence of the tide, although invariably in freshwater. They are able to ascend falls, if these are not too steep and high, by throwing themselves upward and forward in short thrusts, and then attaching to the new location with their mouths, pausing briefly, and repeating the process. Spawning adults may run upstream for very long distances in large rivers, as they do in the Connecticut River. The longest spawning run reported for sea lamprey is 240 km (Beamish 1980); other species have spawning runs as long as 1,200 km (Hardisty and Potter 1972). It is not known if spawning adults return to their natal streams. The triggers for migration are not known, but Beamish (1980) reported a link to water temperature, with peak migratory activity occurring with a water temperature of 17°–19°C.

The gonads are suspended within a mesentery attached to the free edges of the kidneys. The sexes are separate. There are no gonoducts, and mature gametes are released into the coelom and discharged through genital pores that open into the cloaca. One female sea lamprey contained 305,000 ova, but the average is approximately 200,000 (Beamish 1980). In ripe, sexually mature females, the ovary is elongate and extends most of the length of the body cavity. The ovaries average 22.4% of the total weight of mature females (Applegate 1950).

Working in pairs, sometimes with a second female assisting, lampreys make nests in the stream bed (Applegate 1950). A typical nest consists of a shallow depression 60–90 cm in diameter and about 15 cm deep in an area of stony or pebbly bottom. The upstream border of the nest is created by latching onto stones with their suckerlike mouths and dragging them into a pile. They are able to move stones as large as a man's fist. The eggs are deposited within these depressions.

Lampreys do not feed after they start their spawning migration. Migration, nest-building, and spawning are supported by the metabolism of stored lipids. Over this period the lipid content of the body declines dramatically. The degree of lipid reduction is related to the duration of the spawning migration (Hardisty and Potter 1972). Sea lamprey entering Saint John River, N.B., in early April had an average total length of 83 cm, whereas spent animals averaged 66 cm (Beamish 1980). Overall, females shrank by an average of 24.3%, males by 18.6%. All lampreys die after spawning; they have already depleted their reserves of glycogen and lipids, and over the period during and immediately following spawning their eyes degenerate, their intestines atrophy, their immune systems weaken, and their kidneys become unable to osmoregulate in seawater.

Embryonic development to hatching takes 10–13 days (Piavis 1972). After another 4–7 days, a period that includes the appearance of gill clefts, the oral hood, and skin pigmentation, the larvae, called *ammocoetes,* leave the nest and travel downstream. On encountering suitable substrates, they form shallow burrows. Ammocoetes differ from juveniles and adults. There are no eyes but a light patch in the skin of the forehead marks the location of the photoreceptive median pineal eye. The mouth consists of a midventral slit surrounded by a small oral hood. Ciliated projections called oral cirri guard the mouth and prevent entrance of coarse sand and debris. A muscular velum pumps water into the mouth, along the pharynx, across the gill arches, and out of the seven paired gill slits. The inner surfaces of the gill arches are coated with mucus. Cilia move the mucus to the gill arches, where it traps particulate matter containing diatoms carried by the passing water. From the gill arches the mucus and diatoms are transported to the entrance to the digestive tube.

At the end of the larval period, the ammocoetes transform into juveniles, a process lasting 4–6 months (Beamish 1980), beginning in July and ending in October. Youson and Potter (1979) illustrated seven discernible stages in the metamorphosis. Virtually no system is unaffected, and the transformation is so extreme that ammocoetes were initially regarded as a separate genus. Eyes and eye muscles develop, the large oral hood and dentition appear, salivary glands develop, new kidneys form, and the characteristic coloration of the juvenile stage appears. Transformed lampreys, called macrophthalmia because of their relatively large eyes, are now capable of a parasitic lifestyle and are able to osmoregulate in seawater. When transformation is complete, the juvenile lamprey leaves the bottom and swims downstream toward the sea. This usually occurs in November or December, but some juveniles overwinter in the estuaries. Lampreys under 400 mm TL are collected in estuaries and in coastal waters to depths of approximately 200 m.

Age and Growth. Average length at transformation from the ammocoete stage has been reported as 100–200 mm (Bigelow and Schroeder), 143 mm (Halliday 1991), 160 mm (Beamish and Medland 1988), and 250 mm (Scott and Scott 1988). Mature individuals average about 720 mm TL and 880 g at the start of the spring migration (Scott and Scott 1988). A length-weight regression for 248 adult lampreys was reported as: log weight (g) = 5.9477 + 3.0518 × log length (mm).

Age studies using statoliths revealed that the larval period lasts 4 or (usually) 5 years (Beamish and Medland 1988). Transformation to the adult form usually lasts 4–6 months and the combined juvenile and adult stages comprise about 2.5 years; combining these stages, it is estimated that the total life span is 8 years with a few living to 7 or 9 (Beamish and Medland 1988).

General Range. Atlantic coasts of Europe and North America: on the eastern side of the Atlantic from northern Norway to the Mediterranean; on the western side of the Atlantic from the west coast of Greenland to Florida and the northern shore of the Gulf of Mexico (Vladykov and Kott 1980). Dempson and Porter (1993) provided capture records from the shelf edge of the Grand Banks and the first record of breeding in a Newfoundland river.

During the 1930s sea lamprey used the Welland Canal to penetrate the inner Great Lakes, where native fishes had no defense against the invader. By the 1940s sea lamprey were wreaking havoc, most notably on lake trout where commercial catches dropped precipitously (Smith 1972; Smith and Tibbles 1980).

Occurrence in the Gulf of Maine. Sea lamprey occur along the whole coastline of the Gulf of Maine; spawning-size individuals have been recorded in 31 rivers and tributaries of the Gulf (Beamish 1980). They have been taken during Canadian Department of Fisheries and Oceans surveys over the entire Scotian shelf including the seaward slopes between Sable Island and Brown's Bank (Beamish 1980). Lamprey regularly occur pelagically off the continental shelf in depths of 1000 m or more (Halliday 1991). They have also been recorded from the Great South Channel, Georges Bank, and over the continental slope off Nantucket and Martha's Vineyard (Halliday 1991).

Lamprey have long been known to run up New England rivers a little earlier in the spring than shad, perhaps beginning to work upstream as early as the beginning of April or even the end of March. In rivers tributary to the Gulf of Maine, the runs peak during May and early June, with few, if any, entering later than that. There are reports that spawning populations of lamprey are decreasing in number, perhaps linked to the decline in fish stocks within the Gulf of Maine.

Importance. Lampreys were esteemed as a great delicacy in Europe during the Middle Ages. (Historians tell us Henry I of England died by overindulging on them in the year 1135.) Considerable numbers were captured for food during Colonial times in the rivers of New England, particularly in the Connecticut and Merrimac rivers. At this time we know of no one in the United States who eats lampreys, although several popular cookbooks include lamprey recipes. A very small percentage of the lamprey population is currently used for medical and physiological research.

CARTILAGINOUS FISHES. CLASS CHONDRICHTHYES

Living cartilaginous fishes are considerably less diverse in shapes and in number of species than bony fishes. They are distinguished from bony fishes in that they have a single nostril on each side of the head that is only partially divided by one or two lobes of integument, a cartilaginous endoskeleton, and toothlike scales derived from mesoderm and ectoderm, and lack a swim bladder. Lack of bone cells in the endoskeleton is probably a primitive condition, but cartilaginous fishes have bone at the base of their spines, thorns, and teeth. The endoskeleton is also strengthened by calcification of cartilages and vertebral centra. The skull and jaws consist of single cartilaginous units, rather than numerous skeletal structures as in bony fishes. Gill arches number four (chimaeras) or five, six, or seven (sharks, skates, and rays) and open separately to the outside, but they are covered by a fleshy operculum in chimaeras. Teeth are embedded in the integument and are continually replaced in sharks, skates, and rays rather than being embedded in the jaws and very occasionally or irregularly replaced, as in bony fishes. In chimaeras the teeth are likewise embedded in the integument but are replaced only slowly or not all. The paired fins are supported by the pectoral and pelvic girdles, and unlike the case in bony fishes the pectoral girdle is free of the skull, except in skates and rays, where it is connected to the nasal capsules. The caudal fin is heterocercal (inclined dorsally) or diphycercal (straight), with the vertebral column extending to near the tip. Fin rays are hairlike and differ from those of bony fishes in that they are unpaired, lack segmentation, and greatly outnumber their underlying supports (radials). The alimentary canal is short, as in most predacious animals, but the intestine is coiled internally to form a spiral valve that increases the surface area for absorption. Metabolic wastes are partially converted to urea, which is stored in the tissues in high concentrations. The high urea cocentration combines with inorganic ions to make cartilaginous fishes slightly saltier than seawater thus aiding in osmoregulation. The high urea concentration may be the main reason that few cartilaginous fishes are able to live in freshwaters. Some that do occur frequently (bull shark) or permanently (South American freshwater stingrays) in freshwaters, temporarily or permanently lose the ability to conserve urea. Conversion of ammonia to urea is also advantageous for reproductive reasons. All cartilaginous fishes practice internal fertilization, encapsulate the fertilized eggs in membranes, and are either live bearers (viviparous) or deposit fertilized eggs in impervious horny egg capsules (oviparous) that are deposited on the substrate. If the nitrogenous wastes were not converted to urea, high concentrations of toxic ammonia (the typical nitrogenous waste product of fishes) would kill the embryos in the uteri or in the egg capsules. Males use rodlike extensions of their pelvic fins (claspers) to inseminate females.

Chondrichthyans are divided into two subclasses: (1) Holocephali (chimaeras), which have the upper jaw fused to the skull, teeth fused into grinding plates, and gills covered by an opercular flap, and (2) Elasmobranchii (sharks, skates, and rays), which have the jaws loosely connected to the skull or free of the skull, teeth of various shapes but not fused into grinding plates, and five to seven pairs of separate gill openings.

Chimaeras. Subclass Holocephali

CHIMAERAS. ORDER CHIMAERIFORMES

Chimaeras are distinguished from other cartilaginous fishes by a compressed trunk that tapers to a slender tail and often a thin filament. Eyes are located on the side of the head and spiracles are absent. Gill filaments are free at tips and are enclosed in a branchial chamber covered with a fleshy operculum. The upper jaw is fused to the cranium giving rise to the name Holocephali, or whole head. Teeth in the upper jaw consist of two pairs of dental plates and teeth of the lower jaw consist of a single pair of dental plates. The first dorsal fin is depressible, typically triangular, and preceded by a sharp, pointed spine. The second dorsal fin is low and extends from just behind the insertion of the first dorsal fin to the origin of the caudal fin. Pectoral and pelvic fins are well developed and triangular-shaped, with fleshy bases. Males have claspers on the pelvic fins. In addition, males have unique structures: a tentaculum or head clasper between the eyes and a pair of prepelvic claspers in pockets just ahead of the pelvic fins.

Chimaeras are benthic to benthopelagic fishes found from near shore to about 2,600 m. Food consists of a wide variety of invertebrates and fishes. Swimming is performed primarily by flapping of the pectoral fins and by undulations of the posterior part of the body. Fertilization is internal as in the other cartilaginous fishes and development is external. Eggs are deposited in brown, horny egg capsules, which are elliptical, oblong, or spindle-shaped.

Three families, six genera, and at least 34 species are recognized (Didier 1995). Only one species, the deepwater chimaera (family Chimaeridae), is really part of the Gulf of Maine ichthyofauna. However, two species of longsnout chimaeras (family Rhinochimaeridae), longnose chimaera, *Harriotta raleighana* Goode and Bean 1895 and knifenose chimaera, *Rhinochimaera atlantica* Holt and Byrne 1909 have been captured nearby, off the eastern slope of Georges Bank during a groundfish survey conducted by the Woods Hole Oceanographic Institution in 1952 and 1953 at depths of 670–970 m and are included in the key. Shortnose chimaeras (family Chimaeridae) have a rounded, conical snout, and a groove running from one side of the head to the other in front of the nostrils (nares). Longsnout chimaeras (family Rhinochimaeridae) have a long, pointed snout, and lack a groove running from one side to the other in front of the nostrils.

KEY TO GULF OF MAINE CHIMAERAS

1a. Snout short, front of snout rounded or conical; claspers of male bilobed or trilobed . . . . **Deepwater chimaera**
1b. Snout very long and pointed; claspers of male single-lobed 2
2a. Surface of dental tooth plates with ridges and rounded knobs; upper margin of caudal fin lacks knoblike tubercles. . . . . **Longnose chimaera**
2b. Surface of dental tooth plates smooth; upper margin of caudal fin of mature males, and to a lesser extent immature males and females, with a series of knoblike tubercles . . . . **Knifenose chimaera**

SHORTNOSE CHIMAERAS. FAMILY CHIMAERIDAE

JOHN D. MCEACHRAN

Shortnose chimaeras have short heads and greatly tapering trunks and tails. They are distinguished from other chimaeras in having a rounded snout and either bilobed or trilobed claspers in mature males. One species has been reported from the Gulf of Maine.

DEEPWATER CHIMAERA / *Hydrolagus affinis* (Capello 1868) / Bigelow and Schroeder 1953:79–80

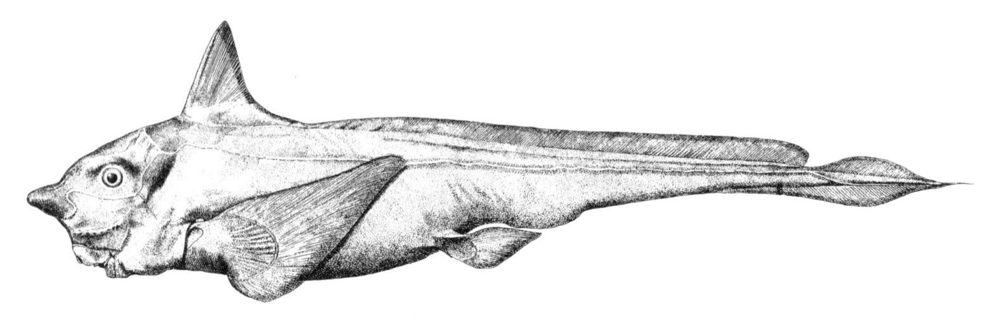

Figure 17. Deepwater chimaera *Hydrolagus affinis*. Southeast of La Have Bank. Drawn by H. L. Todd.

Description. Head short, trunk tapering from dorsal spine to tip of tail. Snout short, blunt; mouth small, with fleshy lips. Nostrils just anterior to upper jaw, connected to mouth by a covered groove, and preceded by a weak groove. Eyes oval, moderate sized. Upper jaw with two pairs of dental plates, anterior pair incisor-like, scalloped with seven or eight ridges anteriorly. Posterior pair flat on roof of mouth, about twice as long as anterior pair, about 1.6 times as long as broad, with a slightly irregular outer margin lacking definite ridges and an inner, anterior section with one low rounded prominence. Dorsal spine about as long as first dorsal fin, largely attached to anterior margin of dorsal fin. First dorsal fin triangular, higher than long. Second dorsal fin separated from first by a distance equal to one-half the length of first dorsal fin, margin free, more or less straight, about one-third the height of first dorsal fin. Caudal fin separated from second dorsal fin by a distinct notch, small, lanceolate-shaped, with a short filament. Anal fin continuous with ventral caudal fin lobe. Pectoral and pelvic fins with pointed tips. Pectoral fins extend nearly to origin of pelvic fins when depressed. Lateral line system distinct, a series of open canals on head and trunk. Dermal denticles absent except on claspers of mature males. Mature males possess trilobed claspers that originate from base of pelvic fins, a pair of prepelvic clasping organs anterior to pelvic fins, and a cartilaginous hook with recurved denticles on forehead.

Color. Entire body is lead-color, tan-brown, or dark sepia, but snout and branchial regions are grayish. Margin of first dorsal fin, posterior and inner margins of pelvic fins, and posterior margin of pectoral fins are dark (Stehmann and Bürkel 1984c).

Size. Maximum known size 125 cm TL.

Distinctions. Deepwater chimaera are distinguished from other species of chimaeras in the Atlantic by a combination of the following characters: snout is short and conical; dorsal spine does not exceed length of first dorsal fin; second dorsal fin is about one-third the length of first dorsal fin, with a straight free margin and is nearly continuous with dorsal lobe of caudal fin; pectoral fins extend nearly to origin of pelvic fins when depressed; anal fin is continuous with lanceolate caudal fin; caudal fin terminates as a short filament; males have trilobed claspers.

Biology. Deepwater chimaera are benthopelagic over continental slopes and abyssal plains at 300–2,400 m (Stehmann and Bürkel 1984c). Food consists of a variety of epibenthic macroinvertebrates such as hermit crabs and sea pens (J. Moore, pers. comm.). Parasites include cestodes and copepods (Scott and Scott 1988).

General Range. Occur on both sides of the North Atlantic. In the western Atlantic they are known from southeastern Grand Bank off Sable Island, offshore of Banquereau and La Have banks, and from southern Nova Scotia to Hudson Canyon. In the eastern Atlantic, they have been captured off Portugal, northern Bay of Biscay, and Rockall Trough (Stehmann and Bürkel 1984c).

Occurrence in the Gulf of Maine. Deepwater chimaera were reported as abundant on the off-shore slopes of the banks off the eastern part of the Gulf of Maine and off Nova Scotia several years subsequent to 1875 when fishermen longlining for halibut extended fishing operations to waters greater than 550 m. Few have been reported in recent years but several were captured near Block Canyon in 1995–1996 (J. Moore, pers. comm).

ACKNOWLEDGMENTS. Drafts of the chimaera section were reviewed by Dominique Didier and Jon A. Moore.

Sharks, Torpedoes, Skates, and Rays. Subclass Elasmobranchii

The most obvious external distinguishing characteristic of all sharks, skates, and rays is that they have five to seven gill openings instead of one. The tough collagenous skin is covered to varying degrees with placoid scales called denticles. Denticles differ from scales of bony fishes in that they originate from the dermis and epidermis instead of the dermis alone, and they are covered by both dentine and enamel homologous to teeth. In fact, teeth of sharks and rays represent modified placoid scales that are embedded in the gums alone, not in the jaw cartilages. They are produced continually, rotating forward to a functional position as older teeth wear down or break off.

An ornate pattern of pores, the ampullae of Lorenzini, which act as electroreceptive sensors, cover the head region and are especially dense on the underside of the snout anterior to the mouth. These pores detect the minute electrical charges emitted by the nervous system of potential prey buried in the bottom.

Chondrichthyans never developed a swim bladder to aid buoyancy. A large oil-filled liver aids buoyancy, although most species are still heavier than water and must swim continuously to avoid sinking. Some deep-sea species have huge livers, comprising as much as one-third of the total weight, with considerable amounts of squalene, a low-density oil that provides partial lift.

There are one or two median dorsal fins, an anal fin in many species, and two sets (pectoral and pelvic) of paired fins. All fins are supported by ceratotrichia, slender horny fibers. In sharks, the cartilages in the fins are highly prized for making shark-fin soup. The fins are covered by the same leathery skin that clothes the body. Among sharks, the tail is heterocercal, with the vertebral column extending out into its upper lobe, but it is whiplike in most of the skates and rays, with no definite caudal fin. The torpedo ray is an exception to this rule.

Modern representatives of the subclass may be separated into two groups: (1) sharks, and (2) torpedoes, skates, and rays. These are distinguished one from the other by the following external differences, as well as certain skeletal differences:

1a. Gill openings at least partly on sides of head, above or anterior to pectoral fins; edges of pectoral fins not attached to sides of head in front of gill openings; upper edges of orbits free from eyeballs, so that they form free eyelids . . . . **Sharks**

1b. Gill openings entirely on lower surface below pectoral fins and behind head; front edges of pectoral fins attached to side of head; upper edges of orbits attached to eyeballs so they do not form free eyelids . . . . **Torpedoes, skates, and rays**

Sharks. Superorder Selachimorpha

Sharks are cartilaginous fishes distinguished by having more or less rounded bodies, not flattened as in batoids; five to seven gill slits located on the side of the head; pectoral fins not attached to the side of the head; one or two dorsal fins with or without spines; an anal fin present or absent; and a distinct heterocercal caudal fin with elements of the vertebral column extending into the upper lobe.

Sharks are of interest to fishermen, mariners, and seaside visitors because of their purported ferocity, the large size of some species, and the damage they cause to fishermen's gear. Although numerous species have been sporadically recorded from the Gulf of Maine, the area is not rich in resident species. Only spiny dogfish are numerous enough for commercial importance. Otherwise, only two species—blue shark (*Prionace glauca*) and basking shark (*Cetorhinus maximus*)—occur in numbers sufficient for one to be fairly sure of seeing them during a summer's boating off the coast north of Cape Cod. With larger sharks generally so scarce, the danger of attacks on bathers is negligible in the Gulf of Maine, but as long as white shark (*Carcharodon carcharias*) seasonally visit the northern Gulf of Maine (Mollomo 1998), attacks are possible.

Sharks inhabiting the Gulf of Maine include representatives from several major groups. Squaloid sharks (spiny dogfishes, families Squalidae and Echinorhinidae) lack an anal fin, and most have spines associated with the dorsal fins. Lamnoid sharks (mackerel, thresher, basking, and sand tiger sharks) all have their gill slits anterior to the pectoral fin origin; mackerel sharks (Lamnidae) possess lateral keels along the caudal peduncle; thresher sharks (Alopiidae) are readily identifiable by the exaggerated upper lobe of the caudal fin; the monotypic basking shark (Cetorhinidae), while easily confused with the white shark at sea, is readily identifiable by its elongate gill slits; and sand tigers (Odontaspididae) have two large dorsal fins located posteriorly on the body and distinctive narrow, pointed teeth with cusplets. Carcharhinoid sharks (cat sharks, smooth dogfishes, requiem sharks, hammerheads) all possess a form of nictitating membrane (inner eyelid); cat sharks (Scyliorhinidae) are small species with both dorsal fins located posteriorly on the body, lacking a precaudal pit and having a low-angled caudal fin that has a poorly developed lower lobe; smooth dogfishes (Triakidae) have horizontally oval eyes and pavementlike teeth; requiem sharks (Carcharhinidae), probably the species most familiar to the average person, have well-developed fins, a dorsal and ventral precaudal pit, and distinct heterodont teeth; and hammerheads (Sphyrnidae) are easily distinguished by the laterally expanded head.

Sharks were once considered a nuisance because of the damage they did to nets and other gear, and historically there has been little market for sharks of the Gulf of Maine (attempts to introduce dogfish as a food fish had little success except as an export to Europe). Landings in Maine and Massachusetts were only about 110,000 kg in 1947 and about 140,000 kg in 1949, but they were in the hundreds of metric tons from 1960 to 1978 and increased to as much as 20,000 mt during the 1990s, peaking in 1993 with 20,360 mt (F. Serchuk, pers. comm., Nov. 1997).

Sharks of the Gulf of Maine can be identified by external morphological features such as the size and relative location of fins and tooth characters. The key to species includes those most likely to be encountered in the Gulf of Maine region but not all the species that might stray into the region. Should the user of this book have difficulty identifying a shark, reference should be made to one of two comprehensive works: Castro (1983) or Compagno (1984).

KEY TO GULF OF MAINE SHARKS

1a. Anal fin present **2**
1b. Anal fin absent **16**
2a. Head expanded sidewise, at level of eyes, in hammer- or shovel-form **3**
2b. Head of ordinary shape, with rounded or pointed snout **4**
3a. Outline of front of head with only a slight concavity near the nostrils, if at all; grooves (if any) from nostrils shorter than horizontal diameter of eyes; free tip of second dorsal fin not longer than forward margin of fin; rear margin of anal fin only weakly concave; teeth near outer corners of mouth rounded, without sharp cusps . . . . **Bonnethead**
3b. Outline of front of head deeply indented opposite each nostril; grooves from nostrils more than twice as long as horizontal diameter of eye; free tip of second dorsal fin considerably longer than front margin of fin; rear margin of anal fin deeply concave; teeth near corners of mouth like those near center of mouth, with sharp cusps . . . . **Smooth hammerhead**
4a. Caudal peduncle widely expanded sidewise as a lateral keel on either side; if low keel present, lower lobe of caudal fin not much shorter than upper lobe, suggesting the caudal fin of a mackerel or swordfish **5**
4b. Caudal peduncle not widely expanded sidewise as a lateral keel on either side; upper lobe of caudal fin much longer than lower lobe **8**
5a. Gill openings very large, first pair nearly meeting below throat; teeth tiny, many hundreds in number; gill arches with numerous horny gill rakers directed inward and rearward . . . . **Basking shark**
5b. Gill openings confined to sides of head; teeth large, few in number; gill arches without horny gill rakers **6**
6a. Upper teeth broadly triangular, with serrate edges; anal fin entirely behind second dorsal fin . . . . **White shark**
6b. Upper teeth with smooth-edged cusps; anal fin not entirely behind second dorsal fin **7**
7a. Origin of first dorsal fin over or in front of inner corner of pectoral fin; forward part of caudal fin has a small secondary lateral keel . . . . **Porbeagle**
7b. Origin of first dorsal fin behind inner corner of pectoral fin; no secondary keel on caudal fin . . . . **Shortfin mako**
8a. Upper lobe of caudal fin nearly as long as head and body combined . . . . **Thresher**
8b. Upper lobe of caudal less than one-half head and body combined . . . **9**
9a. Second dorsal fin nearly as high as first dorsal fin **10**
9b. Second dorsal fin less than one-half as high as first dorsal fin. **12**
10a. First dorsal fin at least partly anterior to pelvic fin origin **11**
10b. First dorsal fin wholly posterior to pelvic fin origin. . . . . **Chain dogfish**
11a. Teeth long, narrow, sharply pointed, not in mosaic arrangement; snout conical; fifth gill openings well in front of pectoral fins. . . . . **Sand tiger**

11b. Teeth small, low, rounded, in mosaic arrangement; snout flat, broadly rounded in front; fifth gill openings behind origins of pectoral fins . . . . **Smooth dogfish**

12a. Origin of first dorsal fin far behind inner corner of pectoral fin; upper surface brilliant blue in life, gray in death . . . . **Blue shark**

12b. Origin of first dorsal fin over or anterior to inner corner of pectoral fin; ground color of upper surface gray or brown, not bright blue. **13**

13a. Length of snout in front of mouth not more than one-half as great as breadth of mouth; large, coarsely serrate teeth similar in both upper and lower jaws; caudal peduncle with a low longitudinal keel on either side . . . . **Tiger shark**

13b. Length of snout in front of mouth more than two-thirds breadth of mouth; teeth, if similar in both jaws, not strongly serrate and recurved; caudal peduncle without longitudinal ridges. **14**

14a. Teeth alike in both jaws, oblique, margins smooth. . . . . **Atlantic sharpnose shark**

14b. Teeth only moderately oblique, their margins only finely serrate; lowers noticeably more slender than uppers (outer corners of mouth have no labial furrow on lower jaw and upper labial furrow is so short as to be hardly noticeable). **15**

15a. Origin of first dorsal fin over inner corner of pectoral fin, its vertical height less than distance from eye to first gill opening; scales imbricate, like shingles. . . . . **Dusky shark**

15b. Origin of first dorsal fin over axil of pectoral fin, its vertical height at least as great as distance from eye to third gill opening; scales nonimbricate, like cobble stones . . . . **Sandbar shark**

16a. Trunk much flattened dorsoventrally; eyes on top of head; front margins of pectorals overlap gill openings . . . . **Angel shark**

16b. Trunk subcylindrical; eyes on side of head; front margins of pectorals do not overlap gill openings **17**

17a. Each dorsal fin preceded by a stout and conspicuous spine **18**

17b. Dorsal fin-spines lacking or so nearly concealed in the skin that they can only be detected by touch **20**

18a. Upper teeth with five erect cusps; lower teeth bladelike, with only one cusp, directed outward, forming a nearly continuous horizontal cutting edge all along jaw. . . . . ***Etmopterus princeps***

18b. Upper and lower teeth alike in shape. **19**

19a. Upper teeth and lower quadrangular, with one cusp directed obliquely, forming a nearly continuous horizontal cutting edge along each jaw . . . . **Spiny dogfish**

19b. Upper and lower teeth each with three to five erect, triangular cusps . . . . **Black dogfish**

20a. First dorsal fin well in advance of pelvic fins; upper teeth noticeably different in shape from lower teeth **21**

20b. First dorsal fin over posterior part of pelvic fin base; upper teeth similar to lower teeth . . . . **Bramble shark**

21a. Lower teeth erect, triangular, and serrate . . . . ***Dalatias licha***

21b. Lower teeth quadrate, cusps directed outward, forming a nearly continuous horizontal cutting edge; outer margins deeply notched, edges smooth **22**

22a. Dermal denticles rounded, overlapping, and concealing skin; each dorsal fin preceded by a short spine, embedded nearly to its tip . . . . **Portuguese shark**

22b. Dermal denticles conical, only moderately close-set, skin visible between them; dorsal fins not preceded by spines . . . . **Greenland shark**

SAND TIGER SHARKS. FAMILY ODONTASPIDIDAE

STEVEN BRANSTETTER

The Odontaspididae consists of two genera and four species (Nelson 1994), only one of which occurs in the Gulf of Maine. A poorly known deepwater species may also occur in the New England Atlantic region. Species in this family possess two dorsal fins without spines. The second dorsal fin and anal fin are large, only slightly smaller than the first dorsal fin. The upper lobe of the caudal fin is much longer than the lower, occupying almost one-third of the total length of the fish. There are no lateral keels on the caudal peduncle; the gill slits are relatively long, all approximately equal in length; and the fifth gill openings are anterior to the pectoral fins. The teeth are awl-shaped, slender, and sharp-pointed, with cusplets present.

SAND TIGER / *Carcharias taurus* (Rafinesque 1810)

Figure 18. Sand tiger *Carcharias taurus.* Female, USNM 23278. Drawn by C. Phillips. (From Garrick and Schultz 1963: Fig. 30.)

Description. Large, stout shark with short, conical snout and small eyes (Fig. 18). Gill openings moderately large, nearly equal in length, all anterior to pectoral fins. Mouth large, ventral; teeth awl-shaped anteriorly, more bladelike posteriorly, all with single cusplet on each side. Upper jaw without symphyseal teeth; anteriorly three large teeth on each side separated from posterior teeth by a single row of small, intermediate teeth. Two dorsal fins nearly equal in size, first located closer to pelvic fins than pectoral fins. Pectoral fins short, triangular in shape, not much larger than first dorsal fin. Anal fin large, equaling second dorsal fin in size. Upper precaudal pit present; lateral keels on caudal peduncle absent. Caudal fin stout, upper lobe approximately one-third the precaudal length; lower lobe short but distinct.

Meristics. Precaudal vertebrae 80–87, caudal vertebrae 71–85, total vertebrae 156–170 (Springer and Garrick 1964).

Color. Light tan above, paling on sides to grayish white on lower surface; sides of trunk rearward from pectorals variously marked with round to oval brown to ocher yellow spots, of which there may be as many as 100. On some specimens the rear margins of the fins are edged with black.

Size. The greatest recorded length is 318 cm TL for a specimen from southwest Florida (Springer 1960). The all-tackle game fish record is a 158.8-kg fish caught in Charleston, S.C., on 29 April 1993 (IGFA 2001). Most sand tiger caught in the northern part of their American range, from Delaware Bay to Cape Cod, are immature, of perhaps 120–180 cm TL, but adults larger than 250 cm TL are reported there from time to time, especially in the vicinity of Nantucket.

Distinctions. The large awl-shaped teeth of sand tiger distinguish them from all other sharks in the Gulf of Maine except for the shortfin mako and porbeagle. The large size of the second dorsal and anal fins distinguish them from these two species.

Habits. This is a comparatively sluggish demersal species, more active at night. Primarily coastal, they are not likely to be found in waters exceeding 10–20 m depth, and can sometimes be seen moving slowly on the surface with fins showing above the water. They sometimes enter the mouths of rivers. This species can achieve neutral buoyancy by swallowing air into the stomach, which provides the necessary lift.

Food. Although appearing sluggish, sand tiger are known for their voracity as predators; groups may work together to feed on schools of fishes. Teleosts were the dominant prey of 42 western North Atlantic sand tiger on numerical, weight, and occurrence bases followed by elasmobranchs (Gelsleichter et al. 1999). The most important teleost were summer flounder, present in 14.3% of the stomachs. Other fishes found in sand tiger stomachs include skates and skate egg cases, goosefish, sea robin, scup, spot, bluefish, butterfish, tautog, and invertebrates such as lobsters, crabs, and squids (Gelsleichter et al. 1999; Bowman et al. 2000). A small, partially digested sand tiger (58 cm long) was present in the stomach of one adult (R. E. Bowman, pers. comm., Oct. 1993).

Reproduction. Sand tiger have an unusual reproductive strategy (Gilmore et al. 1983). Mating occurs in winter-spring, and females ovulate large numbers of eggs beginning in April. The large yolky ova are encapsulated, several (7–18) to a capsule, as they pass into the uterus. The first group of capsules ovulated is fertilized. Remaining encapsulated ova are generally not fertilized and provide nourishment for the young later in development. After the yolk for each developing embryo is depleted, the pups hatch from the capsule into the uterine cavity. The young then consume the other egg capsules (oophagy), and their stomachs become greatly distended. However, the embryos do not distinguish between unfertilized ova and their siblings. The largest embryo in each horn of the uterus, having the advantage of size, begins eating its smaller siblings, resulting in the final production of only one young in each horn. Pregnant sand tiger are most common from Florida and Louisiana. Those taken near Woods Hole have contained eggs only, making it likely that the small specimens common along southern New England have come from a more southerly birthplace.

It has been proposed that sand tiger have a 1-year reproductive cycle (Gilmore et al. 1983), but recent evidence (Branstetter and Musick 1994) suggests that the cycle is 2 years. This would be similar to many other carcharhinoid and lamnoid sharks, where there is a 1-year gestation period followed by a 1-year resting stage.

Age and Growth. Only one detailed field study on growth or longevity has been carried out (Branstetter and Musick 1994), but sand tiger have been commonly maintained in public aquariums for several years so additional age and growth information is available (Clark 1963; Gilmore et al. 1983; Schmid et al. 1991; Govender et al. 1991). In captivity, sand tiger have grown 30–35 cm·year^{-1} as neonates, 15–20 cm·year^{-1} as juveniles, and 6–9 cm·year^{-1} as adults. These captive growth rates have been corroborated for wild-reared fish (Branstetter and Musick 1994); individuals collected off Virginia grew 25–30 cm·year^{-1} for the first 2 years; growth then declined, becoming about 15 cm annually for adolescents and less than 10 cm annually for young adults. Males mature at 190–200 cm, corresponding to about 4 years of age and females at 220+ cm, at about 6 years of age (Gilmore et al. 1983; VIMS unpubl. data). The oldest individual on record, from a Durban, South Africa, aquarium, was 16 years old in 1991 (Govender et al. 1991). Von Bertalanffy growth coefficient estimates that best fit the growth of this species (Branstetter and Musick 1994) are: $L_\infty = 284$ cm, $K = 0.21$ for males; $L_\infty = 304$, $K = 0.18$ for females. These results corroborate previous work by Holden (1974), who used various life history parameters to estimate $K = 0.19$. In an update of Holden's work, Pratt and Casey (1990) estimated $K = 0.38$, a very fast growth rate for sharks.

Sand tiger are not particularly heavy sharks; a large adult of 250 cm TL weighs about 95 kg. A length-weight relationship for specimens collected off Virginia (Branstetter and Musick 1994) is: wt (kg) = 1.62×10^{-6} TL $(cm)^{3.24}$.

General Range. Cosmopolitan in coastal temperate and tropical waters but absent from the eastern Pacific.

Occurrence in the Gulf of Maine. Sand tiger are common in shoal waters at Woods Hole and Nantucket from June to November. Scattered records also exist along the outer beaches of Cape Cod, Massachusetts Bay, Cape Ann, Casco Bay, and near the mouth of the Bay of Fundy.

Importance. There were commercial fisheries for sand tiger around Nantucket during the first quarter of the twentieth century, but these were short-lived, reputedly because of exhaustion of the local stock. Western Atlantic populations may have declined by as much as 90% from overfishing from the early 1980s to mid-1990s so the population is considered *vulnerable* by the American Fisheries Society (Musick et al. 2000). From a recreational standpoint, sand tiger have been of interest to anglers fishing west of Cape Cod, who catch considerable numbers, both as objects of special pursuit or incidentally while surf-casting.

MACKEREL SHARKS. FAMILY LAMNIDAE

STEVEN BRANSTETTER

Currently, three genera and five species are included in the Lamnidae (Nelson 1994), of which three species occur in the Gulf of Maine. Sharks of this family are easily recognizable by the very firm, half-moon-shaped (lunate) caudal fin, with the lower lobe only slightly shorter than the upper. Their caudal fins resemble those of such fast-swimming bony fishes as mackerels, tunas, and swordfish. In addition, these sharks possess keels along the sides of the caudal peduncle. Basking shark have a similar caudal fin and caudal peduncle, but their teeth are minute and very numerous and their gill openings are so long that those of the two sides nearly meet on the lower surface of the throat.

Most lamnids are able to regulate their body temperatures above that of the surrounding seawater through a vascular system known as a *rete mirabile* (Carey et al. 1985). This highly vascularized tissue allows blood to be warmed by the exercising red muscle during swimming; this generated heat is then retained in the muscle tissue area by a close association of colder arterial and warmer venous blood. The arterial blood is warmed by the returning venous blood and is further heated as it passes through the muscle system. This results in a gradually increasing heat retention with increased activity, and warmer muscles are better utilized for swimming. In some species of lamnids, similar types of blood systems warm the brain and viscera, which helps these organs function at a more efficient rate.

WHITE SHARK / *Carcharodon carcharias* (Linnaeus 1758) / Bigelow and Schroeder 1953:25–28

Figure 19. White shark *Carcharodon carcharias.* Drawn by M. H. Wagner. (From Kato et al. 1967: Fig. 31.)

Description. Stout-bodied with moderately conical snout (Fig. 19). Teeth large and triangular, with heavily serrate cusps. Gill slits of moderate length, all anterior to pectoral fin origin. Triangular first dorsal fin of moderate size originates over axil of pectoral fins. Second dorsal and anal fins very small, second dorsal a little in advance of anal fin. Pectoral fins sickle-shaped, roughly twice as long as broad. Precaudal pits present above and below with strong lateral keels on caudal peduncle. Caudal fin lunate, lower lobe only slightly shorter than upper lobe, secondary keels absent.

Meristics. Precaudal vertebrae 103–108, caudal vertebrae 68–83, total vertebrae 172–187 (Springer and Garrick 1964).

Color. White shark are slaty brown or leaden gray above, sometimes almost black, shading more or less abruptly on the sides to cream white below. There is a black spot in the axil of each pectoral fin, and the lower surfaces of the pectoral fins are black toward their tips, usually with some black spots adjacent.

Size. This is one of the largest of sharks. Although reputed to reach lengths of 1,100 cm, investigations of the jaws of that particular specimen suggest that it was only 540 cm; there is no reliable evidence for specimens longer than 600 cm (Randall 1987). The all-tackle game fish record is a 1,208.38-kg fish caught off Ceduna, South Australia, in April 1959 (IGFA 2001), although heavier fish (1,247 and 1,263 kg) have been captured by other means. Age and growth information suggests a maximum attainable length of about 750 cm.

Distinctions. Unfortunately, there is no obvious field mark to distinguish a small white shark from a large porbeagle or shortfin mako from a distance. Once captured, however, there can be no confusion, for unlike the slim spikelike teeth of porbeagle and shortfin mako, the teeth of the white shark are broadly triangular.

Habits. Because of the white shark's semisolitary existence, little is known about their life history. Their distribution is temperature-related. Those that visit the Gulf of Maine may come very close inshore; some have been taken in seines, and others have been harpooned in less than 20–40 m of water. They are found at various depths, even oceanic situations. Usually occurring singly, they may congregate in numbers if a large food source, such a dead whale, is available (Pratt et al. 1982; Carey et al. 1982).

White shark can maintain their muscle temperature as much as 4°–5°C above the surrounding seawater and the presence of a large rete-type circulatory system around the viscera keeps the gut warm as well (Carey et al. 1985; Goldman 1997).

Food. White shark feed on numerous types of prey. Whereas juveniles feed primarily on demersal fishes and invertebrates, adults prefer marine mammals such as porpoises, pinnipeds, and whales (Casey and Pratt 1985). Fishes consumed by white shark in the western North Atlantic include smooth dogfish, blue shark, bluefish, and bluefin tuna (Bowman et al. 2000). A large individual recently captured off Beverly, Mass., was reported to contain a 14-kg porpoise.

Reproduction. Because of the difficulty in capturing such large fish, little is known about the breeding habits of white shark. Maturity is estimated to be at 450 cm TL (Casey and Pratt 1985). The number of pups is thought to be eight to ten (Branstetter 1990), which is unusual for lamnids in that most produce two or at most four young. Development is ovoviviparous, and the embryos are probably oophagous as are other lamnids. From estimates of growth rates and capture of small free-swimming juveniles, birth is estimated to be at about 125 cm TL (Casey and Pratt 1985; Cailliet et al. 1985).

Age and Growth. Given the limitations of studying this large shark, accurate age and growth information is limited to one Pacific study by Cailliet et al. (1985) of 21 sharks, none of which were exceptionally large individuals. They estimated rather consistent growth rates of 25–30 cm·year^{-1} at all ages for specimens taken off California. They also noted that growth estimates in the Atlantic were slightly lower, about 20 cm·year^{-1} (Pratt, pers. comm. in Cailliet et al. 1985). The largest Pacific specimens (490–510 cm TL) examined by Cailliet et al. were 14 and 15 years of age, although Pratt (pers. comm. in Cailliet et al. 1985) reported a 475-cm fish from the Atlantic that was aged at 20 years. The von Bertalanffy growth curve parameter estimates from the Pacific study were $L_\infty = 764$ cm TL, $K = 0.058$, $t_0 = -3.5$.

The weight-length relationship for 125 specimens (sexes combined) from the western North Atlantic (Kohler et al. 1996: Fig. 5) is: wt (kg) = 7.5763×10^{-6} FL (cm)$^{3.0848}$.

General Range. White shark are widespread in tropical and warm-temperate belts of all oceans. On the western side of the Atlantic, they have been recorded as far north as Newfoundland and as far south as Argentina, but are rare in equatorial latitudes.

Occurrence in the Gulf of Maine. White shark can be found in the Gulf of Maine from June through November, being most common in July and August (Scattergood 1962a; Casey and Pratt 1985; Mollomo 1998). Mollomo (1998) lists about 40 records apparently referring to white shark from Maine and Canadian Atlantic waters and concludes that large white shark are perennial, seasonal visitors to the northern Gulf of Maine. A relatively recent report from further south in the Gulf of Maine was of a 5.5-m, 909-kg specimen that became tangled in a gill net off the coast of Beverly, Massachusetts, in August 1996 (*Boston Globe,* 16 Aug. 1996).

Importance. Equipped with a terribly effective set of cutting teeth and with a reputation for attacking humans, white shark have borne an unsavory status from earliest times. More attacks are reported from white sharks than from any other shark species, although most shark attacks occur in tropical waters where white shark are rare, so some reports may be misidentifications. Increased use of their waters for human recreation has provided new evidence that presence of a white shark in the area does not always mean that an attack is imminent. Several cases have been reported in which white shark approached within just a few feet of divers and then departed. Many attacks, such as on people on surfboards and paddleboards, may be cases of mistaken identity on the part of the shark or simply exploratory attacks on potential food. In many cases, after the initial attack, which is not always a complete bite, the shark departs. Unfortunately, with its massive size and large teeth, the damage caused to the rather fragile human body may be significant. This is also perhaps the only shark for which "unprovoked" attacks on small boats are documented. Some of these records, however, represent attacks on a boat after the shark had been harpooned or otherwise agitated. This shark demands respect, and anyone who enters waters where it is common should be cognizant of the potential danger.

As a sport fishing target, white shark do not put up as much resistance when hooked as mako, tuna, or swordfish, neither running so fast nor jumping. Their size and power, however, will task a fisherman.

White shark are protected in the U.S. Atlantic, the state of California, and South Africa. The western Atlantic population is considered *conservation-dependent* by the American Fisheries Society (Musick et al. 2000).

SHORTFIN MAKO / *Isurus oxyrinchus* Rafinesque 1810 / Bigelow and Schroeder 1953:23–25

Figure 20. Shortfin mako *Isurus oxyrinchus.* Eastern Pacific, male, 180 cm TL. Drawn by M. H. Wagner. (From Kato et al. 1967: Fig. 32.)

Description. Body relatively slender, except in very large females (Fig. 20). Snout conical and sharp-pointed. Eyes of moderate size. Teeth awl-shaped, with single smooth cusps, alike in both jaws. First two teeth in both jaws much longer than others. Pectoral fins relatively short. Gill slits long, about equal in length, all located anterior to pectoral fin origin. First dorsal fin moderate-sized, originating over or behind inner corner of pectoral fins. Small second dorsal fin originates slightly anterior to equally small anal fin. Precaudal pits present dorsally and ventrally. Strong lateral keels on caudal peduncle. Caudal fin lunate with lower lobe only slightly shorter than upper lobe.

Meristics. Precaudal vertebrae 108–112, caudal vertebrae 79–86, total vertebrae 187–197 (Springer and Garrick 1964).

Color. Deep cobalt blue above, with distinct color change along the sides to snow-white below. The area of color change around the head varies from above the eye, through the eye, to below the eye.

Size. The maximum documented length is about 365 cm TL. The largest Atlantic specimens for which definite records exist are a 294-cm TL male from Great Britain and several females around 350 cm TL (Pratt and Casey 1983). Off the northeast U.S. coast, most specimens caught are 150–200 cm TL. The all-tackle game fish record is a 505.76-kg fish caught off Black River, Mauritius, in November 1988 (IGFA 2001).

Distinctions. In the Gulf of Maine, the species most likely to be confused with shortfin mako is porbeagle, but the former can be distinguished by the location of the origin of the first and second dorsal fins and the lack of secondary keels on the caudal fin. Shortfin mako differ from white shark in their awl-like teeth. Another deepwater tropical mako, longfin mako (*Isurus paucus*), which is unlikely to occur in the area, has very long pectoral fins and its dorsal coloration extends onto the ventral surface, especially in the region of the caudal peduncle.

Habits. Shortfin mako are among the most active and swift-swimming sharks. They are recognized as game fish and are famous for leaping clear of the water, not only when hooked but also under natural conditions.

Shortfin mako can elevate their body temperature above that of the surrounding seawater, but only about 4°C warmer; they may also be able to regulate their temperature according to their activity. This species is known to have countercurrent heat exchangers near the brain and viscera. They can keep their stomachs as much as 8°C warmer than the ambient temperature (Carey et al. 1985; Lowe and Goldman 2001).

Food. Shortfin mako prey chiefly on schools of smaller fishes: 97.9% by volume in 399 stomachs (Bowman et al. 2000). Bluefish occurred in 43% of the stomachs of mako taken in southern New England waters and the New York Bight and comprised over 75% of the food by volume; cephalopods made up a large part of the food in offshore waters (Stillwell and Kohler 1982; Stillwell 1990). Other fishes found in shortfin mako stomachs include blue shark, eel, menhaden, ocean pout, saury, redfish, butterfish, mackerel, skipjack tuna, and swordfish (Bowman et al. 2000).

Shortfin mako also attack larger fishes, such as swordfish, but may not always be the victor (Bozhkov 1975). A large (338 cm TL) female caught on a longline had been speared by a large swordfish, and the spear was intact in the mako's body. The rostrum of the swordfish had entered on the right side near the back of the body cavity and exited the left side near the first gill opening. Bozhkov suggested that the swordfish had aggressively attacked the shark, although it is just as probable that the shark had attacked the swordfish and had subsequently been speared.

Reproduction. Little is known about the reproductive biology of shortfin mako in the Atlantic. There are very few records of pregnant females and many are from photos supplied by commercial fishermen. Few have been examined by biologists. The best information on reproduction in this species is from specimens taken off Australia (Stevens 1983). The general life history characteristics of this species seem to be similar worldwide so these data are probably applicable to the population found in the northwest Atlantic. Embryos develop without the aid of a placenta. Young embryos have greatly dilated stomachs from feeding on unfertilized eggs that are passed into the uterus by the mother. This yolk is digested before birth, as near-term embryos do not exhibit the dilated condition. Young are born at about 70 cm TL, with litters ranging from 10–20 young. Parturition occurs in late spring–early summer (May–July) in the Northern Hemisphere and around November off Australia (Pratt and Casey 1983; Stevens 1983). Mollet et al. (2000) have suggested that shortfin mako have an 18-month gestation period followed by an 18-month resting stage.

Age and Growth. In the western North Atlantic, shortfin mako are estimated to grow about 50 cm·year^{-1} in the first year and 30 cm·year^{-1} for the second and third years. Males and females grow at similar rates, although females grow much larger. Males mature at about 3 years and females at 7 years. The oldest females that have been aged were 11.5 years (ca. 350 cm TL). The von Bertalanffy growth parameters (Pratt and Casey 1983) are: $L_\infty = 302$ cm TL, $K = 0.266$, $t_0 = -1$ year for males; $L_\infty = 345$ cm TL, $K = 0.203$, $t_0 = -1$ year for females.

Males mature at about 180 cm FL and females at about 250 cm FL (Kohler et al. 1996: Fig. 6). At birth shortfin mako weigh about 3 kg; at 180 cm the weight is about 60 kg and at 230 cm TL about 105 kg (Stevens 1983). The weight-length relationship for 2,081 specimens (sexes combined) from the western North Atlantic (Kohler et al. 1996: Fig. 6) is: wt (kg) = 5.2432×10^{-6} FL (cm)$^{3.1407}$.

General Range. Shortfin mako are cosmopolitan in tropical and warm-temperate belts of the world's oceans. In the Atlantic they are found from the United Kingdom to South Africa in the east and from Nova Scotia to southern Argentina in the west. Recent tagging studies indicate that mako may regularly cross the North Atlantic (Casey and Kohler 1990). Tag returns also point to regular movement between the New England coast and the southern Gulf of Mexico.

Occurrence in the Gulf of Maine. The center of abundance for shortfin mako lies in warmer seas south of the Gulf of Maine. Considerable numbers journey northward in summer along the mid-Atlantic and New England continental shelf. Thus stray individuals may be expected to visit the southern part of the Gulf in most summers. They have been reported as far north as southern Maine (Scattergood 1962b).

Importance. Shortfin mako are highly sought after as game fish because of their fast runs when hooked and their habit of leaping, but they are not plentiful enough in the Gulf of Maine to be targeted by fishermen there. Stimulated by increased landings in the swordfishery in the early 1980s, a strong commercial market has developed for the meat.

PORBEAGLE / *Lamna nasus* (Bonnaterre 1788) / Bigelow and Schroeder 1953:20–23

Description. Body stout, heavy-shouldered (Fig. 21). Teeth alike in upper and lower jaws; slender, pointed, smooth-edged, with a sharp cusplet near the base on each side (may be absent in young fish). First dorsal fin large, triangular, and about as high as long; originating a little rearward of pectoral fin axilla. Second dorsal and anal fins very small; pelvic fins a little larger.

Figure 21. Porbeagle *Lamna nasus.* Chile, male, 81 cm TL. Drawn by M. H. Wagner. (From Kato et al. 1967: Fig. 29.)

Second dorsal fin origin directly over anal fin origin. Pectoral fins large, one-half as broad as long. Strong keel present on caudal peduncle. Conspicuous precaudal pits present dorsally and ventrally. A small secondary keel on base of caudal fin on either side, below and behind primary keel of caudal peduncle. Caudal fin stout; lower lobe two-thirds to three-quarters as long as upper lobe.

Meristics. Precaudal vertebrae 84–91, caudal vertebrae 68–71, total vertebrae 150–162 (Springer and Garrick 1964).

Color. Dark bluish gray to bluish black above, including the upper surfaces of the pectoral fins. The color changes abruptly to white along the sides. The lower surfaces of the pectoral fins are dusky to black toward the apex and mottled toward their bases, with the anterior and posterior edges narrowly rimmed with black.

Size. Normally, porbeagle in the Gulf of Maine are 120–180 cm TL and less than 90 kg. Individuals longer than 200 cm TL are not common, but large individuals near 300 cm TL have been reported; this is near the maximum size for the species. Weight for specimens 100 cm long is about 9 kg, for 250 cm about 135 kg, and for 275 cm about 180 kg. The all-tackle game fish record is a 230.0-kg fish caught in Pentland Firth, Scotland, in March 1993 (IGFA 2001).

Distinctions. The only Gulf of Maine sharks that might be confused with porbeagle are white shark or shortfin mako. Porbeagle are easily told from the former by the spikelike smooth-edged teeth and by the position of the second dorsal fin directly over the anal fin. They can be distinguished from the latter by the presence of tooth cusplets and secondary caudal fin keels.

Habits. Porbeagle lead a pelagic life, rarely entering shallow, coastal waters. On calm days, they can be spotted on the surface when their triangular dorsal fins cut through the water. Although described as an active, strong-swimming shark, compared to other species they put up little resistance when hooked. They range from coastal surface waters to depths down to 300 m. Most are taken from August to November, but they do occur in the Gulf of Maine, as well as in north European waters throughout the winter. Apparently they descend into deeper water during the winter to escape low surface temperatures. They may reduce their feeding, as few have been taken in the winter longline fishery. Porbeagle are very efficient at warming their bodies, being able to keep their red muscle and stomach as much as 10°C warmer than the surrounding seawater (Carey et al. 1985; Lowe and Goldman 2001).

Food. In the Gulf of Maine, porbeagle feed chiefly on mackerel and herring, other sharks, other small fishes, and squids. Over 99% of the stomach contents (by volume) of six porbeagle examined from the northwest Atlantic was squid, mostly boreal squid, *Illex illecebrosus* (Bowman et al. 2000).

Reproduction. Porbeagle are ovoviviparous; the embryos develop within the uterus, but there is no placental connection between mother and young. The embryos are oophagous; after absorbing the yolk of their own egg, they receive nourishment by eating additional eggs that pass into the uterus. The stomach of each embryo becomes swollen by the masses of yolk that are eaten in this way. Embryos are not known to consume their siblings as sand tiger do. During development, the upper lobe of the caudal fin is much longer than the lower lobe; the lower lobe then increases in length as the embryo develops. The embryos are 60–70 cm TL at birth. Owing to their large size, litters consist of only one to four young.

In the Gulf of Maine, females contain developing embryos in the winter, but not in the summer. In the northwest Atlantic pupping may occur in late spring (Aasen 1963), which accounts for the absence of embryos in females during the summer fishing season. The collection of large embryos in European seas in summer suggests either a later pupping season in the eastern Atlantic or a single North Atlantic stock that migrates seasonally across the Atlantic. Based on the collection of ripening males in August, Aasen estimated that the gestation period is about 8 months, with females breeding again about 3 months later. Jensen et al. (in press) examined nearly 400 males and 400 females, and their results corroborate Aasen's findings that after a protracted fall mating period (September–November) females give birth to approximately four pups in the spring (May–June), with a 1-year reproductive cycle, with gestation lasting 8–9 months.

Age and Growth. Growth of about 20–30 cm·year^{-1} occurs for the first few years (Aasen 1963). At birth the pups are 60–70 cm TL, and total lengths at age are: 1 year = 98 cm; 2 years = 119 cm; 3 years = 137 cm; 4 years = 152 cm; and 5 years = 164 cm. The von Bertalanffy growth parameters are: $L_\infty = 280$ cm TL, $K = 0.1155$, $t_0 = 0.3$. Based on catches in the Norwegian fishery, which basically targeted a virgin stock, Aasen estimated that natural mortality for the porbeagle equaled about 0.18. In a recent study, Natanson et al. (2002) derived von Bertalanffy growth curve parameters (combined sexes) of $L_\infty = 289.4$ cm FL, $K = 0.07$, $t_0 = -6.06$. This is a slower growth rate than estimated by Aasen. Maximum age was estimated at 25 and 24 years for males and females, respectively. Longevity calculations, however, indicated a maximum age of 45–46 years in an unfished population.

The relationship between fork length and total weight (sexes combined) for 15 porbeagle from the western North Atlantic (Kohler et al. 1996: Fig. 7) is: wt (kg) = 1.4823×10^{-5} FL (cm)$^{2.9641}$. Jensen et al. (in press) found that males matured between 162 and 185 cm FL, with 50% mature at 174 cm FL. Females matured between 209 and 231 cm FL, with 50% mature at 218 cm FL.

General Range. Porbeagle prefer water colder than 18°C. They are distributed throughout the North Atlantic and the Southern Ocean. There is a congeneric species in the northwest Pacific.

Occurrence in the Gulf of Maine. Because porbeagle prefer boreal waters, they are more widely distributed in the area than other lamnoid sharks. Seemingly, the chief centers of the population in the western Atlantic are along outer Nova Scotia to the western Gulf of Maine. They apparently shun the cold waters of the Bay of Fundy, but are common in areas such as Platts Bank, Nantucket Shoals, the Isles of Shoals, Cape Ann, and Massachusetts Bay. The most recent record from the Gulf of Maine is of a 175-kg, 244-cm porbeagle taken about 5 miles southeast of the Isles of Shoals on the weekend of 1–2 August 1998 (*Boston Globe,* 4 Aug. 1998).

Importance. During the early nineteenth century, porbeagle liver oil was important. The meat became popular table fare in the 1940s, resulting in landings on the coast of Maine of about 20,900 kg in 1944 and 32,500 kg in 1945. This fishery soon died out, but another was initiated in the northwest Atlantic by the Norwegians in the early 1960s, which caught over 10,000 individuals (about 1,824 tons) from March through September 1961 (Aasen 1963; Anderson 1990). Expanding to over 8,000 tons by 1964, it declined to less than 300 tons by 1968; catch-per-unit-effort dropped from 9.1 sharks per 100 hooks fished in 1961 to 2.9 sharks per 100 hooks fished by 1964 (Anderson 1990). A similar effort by the Faroese during the mid-1960s also suffered quick reductions in catch/effort. The fact that this stock can be overfished so rapidly is often used as an example of how quickly sharks can be overexploited by fishing ventures (Anderson 1990).

BASKING SHARK. FAMILY CETORHINIDAE

STEVEN BRANSTETTER

The basking shark is the only representative of this family. Much of the updated information for this species account was derived from the comprehensive works by Castro (1983), Compagno (1984), and the study by Parker and Stott (1965).

BASKING SHARK / *Cetorhinus maximus* (Gunnerus 1765) / Bigelow and Schroeder 1953:28–32

Description. Body large and robust (Fig. 22). In adults, snout short and conical, relatively longer on small fish, projecting far beyond mouth, obliquely truncate in front, terminating above in a sharp point, head strongly compressed sideways in front of mouth. Large, subterminal mouth with numerous minute recurved teeth. Gill slits extremely large, nearly encircling body. Gill rakers very long, resembling whalebone of whales. Gills anterior to origin of pectoral fins. Pectoral fins moderate-sized, somewhat flexible. First dorsal fin located about midway along dorsal surface. Second dorsal fin originates

Figure 22. Basking shark *Cetorhinus maximus.* California, female, 820 cm TL. Drawn by M. H. Wagner. (From Kato et al. 1967: Fig. 22.)

anterior to anal fin, both fins similar in size. Precaudal pits present dorsally and ventrally. Strong lateral keels on caudal peduncle. Caudal fin lunate, lower lobe about two-thirds length of upper.

Meristics. Precaudal vertebrae 50, caudal vertebrae 60, total vertebrae 110 (Springer and Garrick 1964).

Color. Upper surface grayish brown, slate gray, or almost black; sometimes mottled with lighter patches. The lower surface is cream white to gray.

Size. Born at about 150–175 cm TL, basking shark reach an impressive size (Parker and Stott 1965), being the second largest fish, exceeded only by whale shark. Although there are numerous unsubstantiated reports of basking shark between 13–15 m, the largest documented fish have been about 980 cm (Bigelow and Schroeder 1948b), and most are shorter than 800 cm TL.

Distinctions. To an observer in the water, a large white shark could be easily mistaken for a basking shark; however the large gill slits and blotchy gray dorsal color pattern are diagnostic.

Habits. Basking shark are a sluggish, inoffensive species that spend much time sunning and swimming slowly at the surface, lying with their backs awash and dorsal fins high out of the water. They are also known to lie on their sides or even on their backs. They are often seen swimming slowly with their snouts out of the water, mouths open, feeding on plankton. They pay so little attention to boats that it is easy to approach them closely. They are reported to jump from the water. Those seen in the Gulf of Maine are usually traveling singly, but they are known to congregate in loose aggregations of as many as 60–100.

Food. Basking shark feed by filtering plankton with their greatly developed gill rakers. The mouth is very large and distensible at the corners; during feeding the mouth is opened wide and the gills are expanded, allowing plankton to be sieved by the numerous gill rakers. Excellent photographs have been taken of these sharks feeding. During the winter they reportedly shed their gill rakers and disappear from coastal areas, prompting the suggestion that they may move to deep water, where they do not feed and may hibernate. However, there have been winter captures of individuals on the surface, apparently feeding, but lacking gill rakers (Backus 1957b).

Reproduction. Although fisheries have existed for this species for a long time, almost nothing is known of their breeding. Males mature at about 500 cm; size at maturity for females is unknown, but may be 650–700 cm (Bigelow and Schroeder 1948b). Ovaries of nonpregnant females contain large numbers of small eggs, suggesting that development is ovoviviparous, with the young being nourished through oophagy, similar to other lamnoids. There is one modern record of the birth of basking sharks (Sund 1943); five living and one stillborn young were born to a harpooned female. These young were reported to swim actively and apparently began feeding; all were estimated to be 150–200 cm in length. Otherwise, the smallest free-swimming specimen collected was about 168 cm.

Age and Growth. Based on length-frequency analysis (Matthews 1950) and length/frequency by season in conjunction with vertebral ring analysis (Parker and Stott 1965), growth has been estimated to be about 150 cm·year^{-1} with individuals reaching maturity (500+ cm in males and 650+ cm in females) in 3–4 years. Matthews estimated that maximum size was attained (ca. 900 cm) in 5–6 years, but Parker and Stott estimated large individuals at 8–12 years of age. Based on these studies, a von Bertalanffy growth equation produces parameter estimates of: $L_{\infty} = 1{,}225$–$1{,}375$ cm, $K = 0.087$–0.092, $t_0 = 3.5$ years, and an estimated size at birth near 150 cm, which is about the size of the smallest known specimen (168 cm [Aasen 1966a]). Parker and Stott provided convincing, albeit circumstantial, evidence for the formation of two rings per year in the vertebral centra, which has since been proposed for other lamnoids (Pratt and Casey 1983; Branstetter 1986; Branstetter and Musick 1994). Pauly (1978) argued that there was no real evidence for two rings per year. Based on annual ring formation, he suggested that females mature at 18 years and males at 10–15 years. With these data, parameter estimates are: $L_\infty = 1{,}314$ cm, $K = 0.045$, $t_0 = 2.9$ years. He further estimated natural mortality M equal to 0.103 for juveniles and 0.048 for adults.

Exact weights for basking shark from the Atlantic are scarce, but 350–450 cm specimens from the Pacific weighed 450–800 kg, and large fish 850–900 cm weighed 3,000–3,900 kg.

General Range. The enormous basking shark is distributed in boreal to warm-temperate coastal pelagic waters throughout the world. They are common in the northern Atlantic, both east and west. The northern boundary appears to follow the line of transition from waters of predominantly Atlantic influence to those of Arctic origin; roughly from the outer coast of Nova Scotia and southern Newfoundland to western and southern Iceland, northward into Scandinavia. Off North America, they are common from Newfoundland to North Carolina (in winter), sometimes straying south to Florida in the winter.

Occurrence in the Gulf of Maine. At one time basking shark were common in the Gulf of Maine. During the 1800s, large numbers were taken in Massachusetts waters for their liver oil, which was used for lanterns. However, the local stock was depleted, and through the 1950s the population apparently had not recovered owing to continued low-level fishing. There have been few sightings or strandings in the area, and the population has fluctuated over the years. On 30 July 1998, a school of at least ten basking shark was seen at the surface over a gully north of Platts Bank at 43°16.3′ N, 69°35.8′ W (J. Moore, pers. comm., 21 Sept. 1998).

Importance. At one time, oil rendered from the liver of basking shark supported fisheries in numerous regions, including the Gulf of Maine area. However, in all cases, local stock depletion brought an abrupt end to the fishery. The western Atlantic population is considered *conservation-dependent* by the American Fisheries Society (Musick et al. 2000). Although sometimes used for human consumption, the flesh is primarily used for fishmeal. Fisheries were usually carried out using harpoons, but netting was also used. Basking shark are so sluggish and so unsuspicious of a boat that they are easy to approach for harpooning. Once struck, however, a large one is likely to put up an astonishingly active and enduring resistance. On several occasions remains of basking shark have been reported as sea serpents or modern-day marine dinosaurs. As the carcass rots, the gill arches and jaws, which are loosely attached to the skeleton, break away, leaving only the cranium, vertebral column, and pectoral and pelvic girdles. A photograph of a rotting basking shark (Koster 1977) did look much like a plesiosaur.

THRESHER SHARKS. FAMILY ALOPIIDAE

STEVEN BRANSTETTER

There is one genus with three species of thresher sharks (Nelson 1994). They are closely related to mackerel sharks and are well-known for the enormously long upper lobe of their caudal fin. Threshers use their long tails to stun and kill their prey.

THRESHER SHARK / *Alopias vulpinus* (Bonnaterre 1788) / Bigelow and Schroeder 1953:32–34

Description. Snout short and conical (Fig. 23). Eye moderate-sized. Head smoothly curved between eyes. Gills moderate in length; fourth and fifth gill slits over pectoral fins. First dorsal fin located midway along dorsal surface, its free tip ending anterior to origin of moderately large pelvic fins. Second dorsal fin originates anterior to origin of anal fin; both fins small. Pectoral fins long, falcate, and relatively pointed. Precaudal pits present dorsally and ventrally on relatively stout caudal peduncle. Upper lobe of caudal fin exaggeratedly long.

Meristics. Precaudal vertebrae 119–121, caudal vertebrae 240–298, total vertebrae 359–419, more than in any other species of shark (Springer and Garrick 1964).

Color. Dark brown, slate gray, or even nearly black above. Lower surface white, speckled with dark spots especially near the pelvic fins and around the caudal peduncle. The white ventral color extends forward above the base of the pectoral fins onto the head.

Size. Thresher shark may attain lengths of 600 cm TL; large females have been recorded at 575 cm TL and males at 420 cm TL. The all-tackle game fish record is a 348.0-kg fish caught in Bay of Islands, New Zealand, in February 1983 (IGFA 2001).

Distinctions. Threshers are easily distinguished from all other Gulf of Maine sharks by the long upper lobe of the caudal fin,

Figure 23. Thresher shark *Alopias vulpinus*. Eastern Pacific, male, 191 cm TL. Drawn by M. H. Wagner. (From Kato et al. 1967: Fig. 25.)

which is approximately the same length as the body. It could only be confused with the warmer water bigeye thresher shark, which has a large vertically oval eye and a V-shaped ridge on the head.

Habits. Thresher shark are an epipelagic species of both coastal and oceanic waters. Juveniles are more common in inshore waters and are often found in coastal bays. Adults are more common over the continental shelf, but have been documented from oceanic situations. They are not commonly taken by rod and reel, but are often taken on longlines and in nets. Because they use their long caudal fin to stun prey, they are often foul-hooked near the tail on longlines. When hooked, they are active swimmers and put up great resistance, often breaking free, especially when foul-hooked. Unlike the closely related mackerel sharks, thresher sharks do not seem to be able to regulate their body temperature (Carey et al. 1985).

Food. Thresher shark feed chiefly on small schooling fishes such as menhaden, herring, Atlantic saury, bluefish, sand lance, and mackerel, as well as on bonito and squids. Major food items of 19 thresher shark were bony fishes, 97% by volume, of which bluefish and butterfish were the most important (Bowman et al. 2000). Thresher shark often feed in groups, first herding a school of fish into a tight group, then stunning them with their whiplike tails.

Parasites. The copepod genus *Nemesis* contains nine species that parasitize sharks by attaching themselves to the gill filaments. *Nemesis robusta* causes tissue erosion of the gill filaments of thresher shark, potentially impairing proper respiratory functioning of the infected portions of the gills (Benz and Adamson 1990).

Reproduction. Development is ovoviviparous; the young develop without a placental attachment. The embryos are oophagous, feeding on eggs passed into the uterus. Depending on the population, two or four young develop during each pregnancy (Cailliet and Bedford 1983; Gubanov 1972, 1978). Size at birth varies considerably: 114–160 cm TL. Most young free-swimming specimens are longer than 150 cm. It is possible that young are born throughout the species range as young specimens are found in New England waters, but these may migrate into the area from a more southerly pupping ground. Young are common in the southeast U.S. waters. Reproduction appears to be an annual event; almost all mature females that are caught are pregnant.

Age and Growth. No aging studies have been carried out on the Atlantic population of this species; however they are common off California and studies of that population (Cailliet et al. 1983; Cailliet and Bedford 1983) are probably applicable to that of the Atlantic. Born at about 150 cm, they are estimated to grow at 50 cm a year as juveniles, and less than 10 cm a year as adults. The von Bertalanffy growth parameter estimates are: $L_\infty = 493$ cm TL, $K = 0.215$, $t_0 = -1.4$ for males; $L_\infty = 636$ cm TL, $K = 0.158$, $t_0 = -1.0$ for females.

Males mature at about 330 cm TL, 180 cm FL. There is some confusion as to size at maturity for females, which may differ by region, but it is estimated at 260–450 cm TL, 225 cm FL (Gubanov 1972; Cailliet et al. 1983; Kohler et al. 1996). A weight/length relationship for 88 specimens (sexes combined) from the western North Atlantic (Kohler et al. 1996: Fig. 4) is: wt (kg) = 1.8821×10^{-4} FL (cm)$^{2.5188}$.

General Range. Thresher shark are more common in temperate and subtropical seas. In the western Atlantic, they occur from Nova Scotia to northern Argentina, including the Gulf of Mexico. In the eastern Atlantic, they are known from southern Ireland, the North Sea, the Mediterranean, and along the west coast of Africa.

Occurrence in the Gulf of Maine. Thresher shark are common off the southern coast of New England in warmer months. They have often been reported along the coast of Maine and the outer coast of Nova Scotia.

Importance. Thresher shark do not play an important role in recreational or commercial efforts in the Gulf of Maine. Further south they are a component of the longline catch, both on swordfish/tuna and shark gear, and occasionally place as a winning fish in New York–New Jersey summer shark tournaments. The meat has good commercial value, as do the fins. The hides can be used for leather, although in the U.S. fisheries, fish are usually processed for their meat, not their skins. Thresher shark can also be a nuisance in the mackerel fisheries, often tearing or becoming tangled in the nets. A drift net fishery for thresher shark on the U.S. Pacific coast developed, especially in southern California, but apparently rapidly overfished the stock.

CATSHARKS. FAMILY SCYLIORHINIDAE

STEVEN BRANSTETTER

Catsharks comprise one of the most speciose families of sharks with 15 genera and nearly 100 recognized species (Nelson 1994). They are so named for their distinctively shaped eyes, which are elongate, oval or slitlike, with a length one and one-half times their height, with a rudimentary nictitating membrane. An anal fin is present and at least one-half the base of the first dorsal fin is posterior to the origin of the pelvic fins. The anterior margin of the nostrils does not bear a fleshy barbel. One species might venture into the Gulf of Maine.

CHAIN CATSHARK / *Scyliorhinus retifer* (Garman 1881) / Bigelow and Schroeder 1953:34

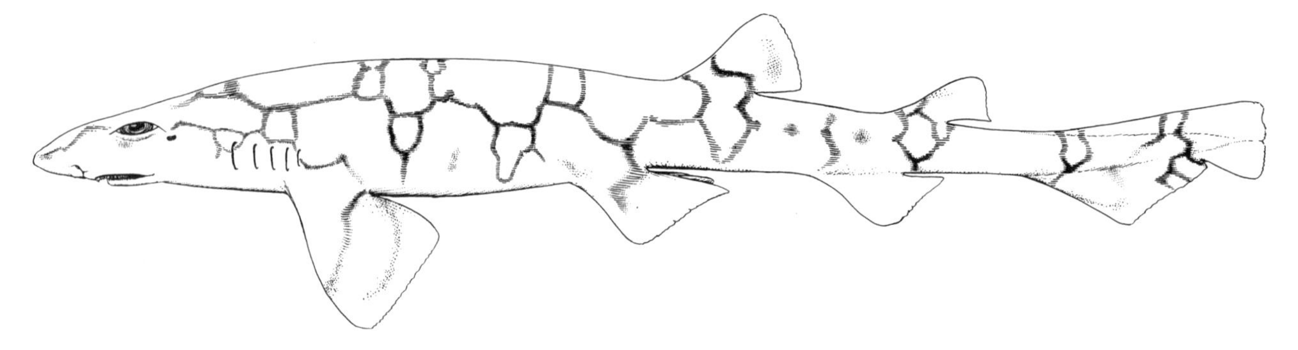

Figure 24. Chain catshark *Scyliorhinus retifer.* New Jersey, immature male, 428 mm TL, MCZ 33932. (From Bigelow and Schroeder 1948b: Fig. 33, © Sears Foundation for Marine Research.)

Description. First dorsal fin origin posterior to base of pelvic fins (Fig. 24). Second dorsal fin one-half as large in area as first. Caudal fin square-tipped, occupying only about one-fifth of length of fish. Teeth similar in both jaws; narrow-triangular with a small secondary cusp on either side.

Color. Chain catshark have a distinctive reddish brown, chainlike pattern along the back and sides. There are no light or dark spots on the body.

Size. Reach 580–600 mm TL.

Distinctions. Their chainlike pattern makes confusing them with any other shark in the Gulf of Maine unlikely.

Habits. This small demersal shark ranges from continental shelf waters to the outer continental slope at depths of 74–450 m (Springer 1979; Able and Flescher 1991). They prefer colder waters, 8.5°–14°C (Able and Flescher 1991). In northern regions, they occupy shallower depths than in the southern part of their range (Springer 1979; Able and Flescher 1991). However, specimens have been collected in waters as shallow as 40 m in the northern Gulf of Mexico.

Food. The stomach contents of 35 specimens contained, by weight, 49.5% bony fishes, 22.4% crustaceans, 20.7% cephalopods, and 6.3% polychaetes (Bowman et al. 2000). Smaller sharks (11–25 cm TL) consumed mostly crustaceans and polychaetes; larger ones (31–45 cm) consumed mostly fishes.

Springer (1979) noted that many specimens he examined had small pebbles in their stomachs and postulated that these may serve as ballast.

Reproduction. Chain catshark are oviparous, with the encapsulated eggs being attached to vertical relief structures on the bottom (Castro et al. 1988; Able and Flescher 1991; Sminkey and Tabit 1992). Eggs are deposited in pairs. Females may store sperm for as long as 2 years or more. Incubation is about 256 days. Pups hatch at about 100–110 mm TL. Size at maturity varies by latitude; in the northern part of their range, males and females are mature when they exceed 380 mm TL, but in more southerly latitudes, maturity may be delayed until about 500 mm TL.

Age and Growth. Born at 100 mm TL, individuals in the laboratory attain 170 mm by 1 year, and 250–300 mm TL in 2 years. They have been kept in captivity for as long as 9 years.

General Range and Occurrence in the Gulf of Maine. Chain catshark are distributed over outer continental shelf waters of the western North Atlantic from Cape Cod south through the Gulf of Mexico, and along the North American isthmus to Nicaragua (Springer 1979). The *Cap'n Bill II* trawled one in July 1952, south of Nantucket Lightship, 40°02′ N, 69°37′ W, at 150–180 m (Bigelow and Schroeder).

SMOOTH DOGFISHES. FAMILY TRIAKIDAE

Steven Branstetter

Smooth dogfishes are small- to medium-sized sharks, slender-bodied, with oval eyes. They have several functional rows of small teeth; some are flat and pavementlike, but others have three or four definite cusps. A nictitating membrane is variously developed. A spiracle is present. The pectoral fins are broadly triangular, the second dorsal fin is only slightly smaller than the first, and an anal fin is present. Precaudal pits are absent, and the caudal fin is strongly asymmetrical, its lower lobe poorly differentiated. Nine genera and 40 species are recognized (Nelson 1994; Heemstra 1997), but only one species is known from the Gulf of Maine.

SMOOTH DOGFISH / *Mustelus canis canis* (Mitchill 1815) /

Smooth Dog, Smooth Hound / Bigelow and Schroeder 1953:34–36

Figure 25. Smooth dogfish *Mustelus canis canis*. Long Island, N.Y., female, 68 cm TL. Drawn by E. Heemstra. (From Heemstra 1997: Fig. 4.)

Description. Body slender, snout softly rounded (Fig. 25). Teeth low, flat, pavementlike. Eyes large, horizontally oval. Nasal flaps large. Distinct spiracle present behind eye. Two large spineless dorsal fins; second dorsal fin slightly smaller than first. First dorsal fin originates at rear corner of broad pectoral fins; second originates well in advance of anal fin origin. Precaudal pits absent. Lower lobe of caudal fin poorly defined, rounded at tip.

Meristics. Precaudal vertebrae 90, caudal vertebrae 56, total vertebrae 146 (Springer and Garrick 1964).

Color. Upper surface gray to olive; lower surface yellowish or grayish white. Newborn specimens have the upper part of the first dorsal fin edged with dusky gray. All fins have a white trailing edge, and the apices of the second dorsal fin and upper caudal lobe are sooty. Smooth dogfish can change shade to suit their surroundings, paling to a translucent pearly tint above white sand, but darkening on a dark bottom.

Size. Maximum size about 150 cm TL. The all-tackle game fish record is a 12.15-kg fish caught off Galveston, Tex., in March 1998 (IGFA 2001).

Distinctions. The low pavementlike teeth and the large second dorsal fin are diagnostic of this small, gray-colored shark.

Habits. Smooth dogfish are common nearshore demersal sharks in the spring, entering bays and freshwater during the pupping period. Shortly after, females move back offshore to the outer part of the continental shelf, where they are commonly caught during fishing trips for such species as tilefish. They only come to the northern part of their range in midsummer (Rountree and Able 1996).

Food. Decapod crustaceans are the dominant food category of smooth dogfish (Moss 1972; Gelsleichter et al. 1999; Bowman et al. 2000). Rock and lady crabs are favorite prey, but jonah crabs, hermit crabs, and other decapods are also part of their diet. Other invertebrates of significance in the diet include squids (*Loligo* and *Lolliguncula*), lobster, bivalves especially razor clams, and gastropods such as moon snails (Rountree and Able 1996; Gelsleichter et al. 1999). Teleosts are the third most important category, including margined snake eel, round herring, anchovies, hakes, cuskeel, sea robins, scup, spot, rough scad, butterfish, and sand lance (Bowman et al. 2000).

Predators. A smooth dogfish was found in the stomach of a sandbar shark (Rountree 1999).

Parasites. The gastrointestinal tract parasite assemblage for smooth dogfish included four species of tapeworms (Cislo and Caira 1993).

Reproduction. Adults mature at about 85 cm (Castro 1983). Development is viviparous with a placental connection. Gestation takes 10–11 months with birth occurring in May. The number in a litter is 10–20 at 28–39 cm TL (Rountree and Able 1996). The inlets and tidal creeks from Chesapeake Bay to New York are a major nursery, especially the area along the coast of New Jersey.

Age and Growth. Growth rates of 15–20 cm·year^{-1} are common for the species in this genus (Francis 1981b; Yudin and Cailliet 1990; Cailliet et al. 1990; Rountree and Able 1996). Males and females grow at similar rates, maturing in 3–4 years at about 85 cm TL (Francis 1981b). Maturity may be reached in about 1 year and maximum size in 7 or 8 years (Moss 1972). Maximum life span may be only 10–15 years.

General Range. Bay of Fundy south to Florida, the Gulf of Mexico, and South America to Uruguay. Replaced in Bermuda, the Bahamas, and the Caribbean islands by *M. canis insularis* Heemstra 1997.

Occurrence in the Gulf of Maine. Although smooth dogfish are very common in the mid-Atlantic region from Cape Hatteras to New Jersey, Cape Cod and Nantucket Shoals appear to mark a boundary to their dispersal northward. On occasion, they stray into Massachusetts Bay, the Gulf of Maine, and the Bay of Fundy.

Importance. Smooth dogfish constitute part of the catch on commercial longlines, especially in more southern regions. They are taken in the directed shark fishery, and as by-catch in grouper/snapper and tilefish efforts. They are also common in coastal gill net and haul seine winter fisheries along North Carolina. The flesh has commercial quality, but is not often sold because of the small carcass size. Because of their size and adaptability, smooth dogfish are often used in laboratory studies.

REQUIEM SHARKS. FAMILY CARCHARHINIDAE

Steven Branstetter

This is a large family of 13 genera and some 58 species (Nelson 1994), which are distributed in warm-temperate to tropical waters and include a wide variety of small to large sharks. Five species in four genera might occur in the Gulf of Maine. Few characters are consistent among all members of this group. A nictitating membrane is present. The head is of normal shape, and the fourth and fifth gill slits are above the pectoral fins. Teeth vary from small single-cusped spikes to broadly triangular cusps. There are two spineless dorsal fins, the first usually much larger than the second and located well in advance of the pelvic fins. An anal fin is present, as are precaudal pits above and below the caudal peduncle.

TIGER SHARK / *Galeocerdo cuvier* (Peron and LeSueur 1822) / Bigelow and Schroeder 1953:37–38

Description. Snout short, broad, rounded, shorter than mouth width (Fig. 26). Long labial furrows present at corners of mouth; small spiracle present behind eye. Teeth similar throughout jaws, above and below; strongly serrate and

Figure 26. Tiger shark *Galeocerdo cuvier.* Gulf of California, female, 140 cm TL. Inset is ventral view of head. Drawn by M. H. Wagner. (From Kato et al. 1967: Fig. 55.)

oblique, distinct notch present on trailing edge of teeth. Pectoral fins rather small and triangular. First dorsal fin broadly triangular, originating over pectoral fin axil. Second dorsal fin small, originating anterior to origin of anal fin, which is strongly recurved. Low ridge along back between dorsal fins. Low longitudinal keel on caudal peduncle. Dorsal and ventral precaudal pits present. Upper caudal lobe rather long and flexible. Tiger shark become very heavy-shouldered with increased size.

Meristics. Precaudal vertebrae 105–108, caudal vertebrae 114–126, total vertebrae 222–231 (Springer and Garrick 1964).

Color. This is a handsomely colored shark, gray or greenish-gray above, contrasted with a stark white below. Young tiger sharks (less than 180 cm TL) have conspicuous dark spots along the back. These markings fade with increasing size, becoming dull bars. Tropical individuals may be very dark, almost black.

Size. Most tiger shark caught are less than 350 cm TL, but occasional individuals exceed 400 cm, and they may reach 550 cm TL (Bigelow and Schroeder 1948b). The all-tackle game fish record is an 807.40-kg individual taken off Cherry Grove, S.C. in June 1964 (IGFA 2001).

Distinctions. The short, rounded snout and heavily serrate broad-cusped teeth that are similar in both jaws readily distinguish this species from others in the region.

Habits. Tiger shark are found from nearshore environments to deep oceanic waters. They are often seen sunning at the surface and have been reported to depths of 1200 m.

Food. Numerous papers have been written recording the food of tiger shark from different regions (Randall 1992). They feed on a wider variety of prey than any other species of shark. Stomach contents contain other sharks, skates, bony fishes (including goosefish and bluefish), squids, horseshoe crabs, crabs, conchs, whelks, sea and land birds, marine mammals, and numerous indigestible items (Randall 1992; Bowman et al. 2000). Their powerful jaws and strong teeth are adapted for large prey; sea turtles and marine mammals are favorite foods.

Reproduction. Tiger shark are the only member of the family that is ovoviviparous but there is no placental connection between mother and young. Litters consist of up to 60–80 young, although 40 is more common. Gestation period is about 14–16 months (Clark and von Schmidt 1965), and pupping occurs in midsummer (Branstetter 1981). Young are born at about 70 cm TL and apparently do not occupy a defined nursery ground; they are taken throughout continental shelf waters.

Age and Growth. Tiger shark grow rapidly compared to other sharks (Branstetter et al. 1987). Juveniles nearly double in length in the first year of life. Growth of 25–30 cm·year^{-1} continues for about 4 years. Maturing at about 310 cm (8–10 years), growth slows to 5–10 cm·year^{-1} at lengths greater than 325 cm. The von Bertalanffy growth parameter estimates for this species are: $L_\infty = 390$–440 cm, $K = 0.107$–0.184, $t_0 = -1.1$ to -2.4.

Tiger shark are relatively small at birth compared to the mother, corresponding to the large litter size. They swim with an eel-like motion, which precludes great speed. Juvenile mortality may be high because they do not occupy a nursery ground (Branstetter 1990). Males mature at 310 cm TL and females at greater than 310 cm TL. A weight/length relationship (TL) for Atlantic specimens (Branstetter et al. 1987) is: wt (kg) = 1.41×10^{-6} TL (cm)$^{3.24}$. A weight/length relationship (FL) for 187 tiger shark (sexes combined) from the western North Atlantic (Kohler et al. 1996: Fig. 13) is: wt (kg) = 2.5281×10^{-6} FL(cm)$^{3.2603}$. Males reach maturity at about 180 cm FL and females at 225 cm FL.

General Range. Circumglobal in tropical and warm temperate seas (Randall 1992). In the Atlantic, tiger shark have been reported as far north as Iceland and Norway, and recent tagging data suggest that there might be a trans-Atlantic migration for this species. However, these eastern Atlantic movements may have been stray individuals that became trapped in Gulf Stream waters. Tiger shark prefer water temperatures higher than 18°C.

Occurrence in the Gulf of Maine. Tiger shark rarely come north of Cape Cod, although a stray individual may occasionally enter the Gulf of Maine.

Importance. Tiger shark have long been regarded as second only to white shark as the most dangerous in the sea and account for the second highest number of reported shark attacks (Randall 1992). Because of their large size, tiger shark are often targeted by recreational shark fishermen, and catches are commonly entered in tournaments. They occur regularly, in abundance, on longlines for sharks and other fishes. The meat is edible, but of low quality; the hide is of importance for leather. Catch rates for this species in the mid-Atlantic region have declined dramatically since the mid-1980s, indicating that fishing pressure adversely affected the size of the population (Musick et al. 1993b). Since then, an increasing number of juveniles have been taken on directed commercial shark longlines (Branstetter and Burgess 1997), which suggests that the population is recovering.

BLUE SHARK / *Prionace glauca* (Linnaeus 1758) / Bigelow and Schroeder 1953:38–40

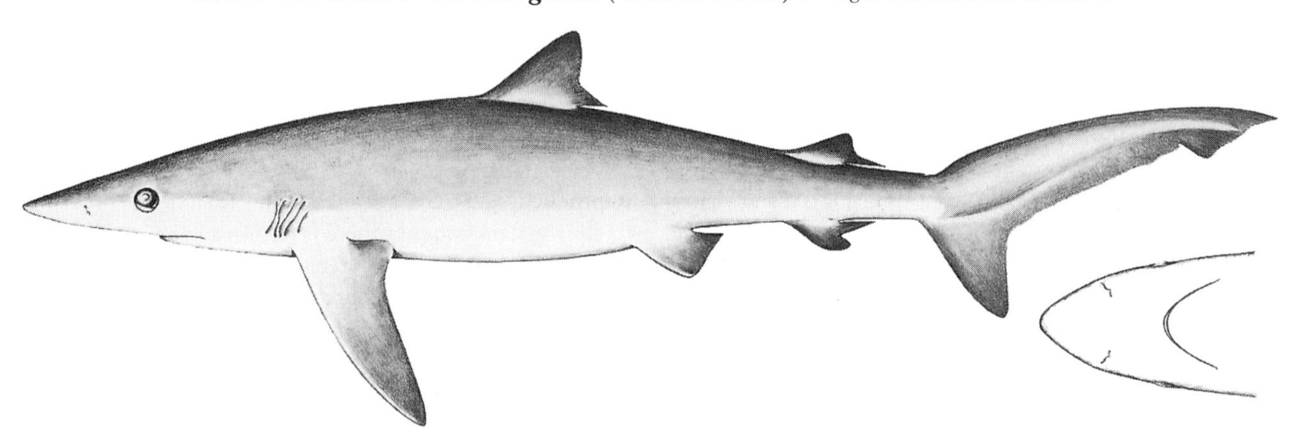

Figure 27. Blue shark *Prionace glauca.* Eastern Pacific, female, 80 cm TL. Inset is ventral view of head. Drawn by M. H. Wagner. (From Kato et al. 1967: Fig. 60.)

Description. Slender-bodied (Fig. 27). Snout long and conical. Teeth serrate, uppers slightly oblique, lowers erect, both with broad cusps. Gill slits short with papillose gillrakers on gill arches. First dorsal fin relatively small with a rounded tip, located about midway along back, slightly closer to pelvic fins than to pectoral fins. Interdorsal ridge absent. Pectoral fins long, greater than head length, slender and falcate.

Meristics. Precaudal vertebrae 142–149, caudal vertebrae 90–106, total vertebrae 237–252 (Springer and Garrick 1964).

Color. Living individuals are dark indigo blue along the back, shading to a clear bright blue along the sides; but this beautiful hue changes to a slate or sooty gray soon after death. The lower surface is snow-white, but with the tips of the pectoral fins dusky and the anal fin partly sooty.

Size. Maximum length 383 cm TL, with females 173–323 cm TL (Compagno 1984). The all-tackle game fish record is a 205.93-kg fish caught near Martha's Vineyard in July 1996 (IGFA 2001).

Distinctions. The long conical snout and slender body with the blue coloration readily identify this species.

Habits. Blue shark are most common along the outer edge of the continental shelf, but are also found in coastal waters. They are frequently seen sunning at the surface, swimming lazily with the first dorsal fin and the tip of the caudal out of water. They also are known to dive deeply on a regular basis (Carey and Scharold 1990). Tagging data suggest that over time blue shark may move through the entire North Atlantic gyre (Casey and Kohler 1990).

Food. About one-third of the stomach contents of 1,199 blue shark from the Mid-Atlantic area, by volume, were squids and octopods (Bowman et al. 2000). Squids also formed the major part of the diet of blue shark off Brazil (Vaske and Rincón-Filho 1998), in the northeast Atlantic southwest of Britain and Ireland (Henderson et al. 2001), and were important in the diet of blue shark off the Azores (Clarke et al. 1996). Half the stomach contents in the Middle Atlantic area were fishes such as skates, small blue shark, herring, lancetfish, cod, bluefish, scup, butterfish, mackerel, and yellowtail flounder (Bowman et al. 2000). Other prey include land and sea birds and marine mammals.

Parasites. The parasite assemblage in the spiral intestine of 24 blue shark caught off Montauk, N.Y., consisted of four species of tetraphyllidean tapeworms (Curran and Caira 1995): *Anthobothrium laciniatum, Paraorygmatobothrium prionacis, Platybothrium auriculatum,* and *Prosobothrium armigerum.*

Reproduction. Detailed information on reproduction of blue shark in the western North Atlantic was provided by Pratt (1979) and in the southwestern Atlantic by Hazin et al. (1994). Both sexes mature at 4–5 years, males at greater than 150 cm TL and females at greater than 180 cm TL. All females older than 5 years are mature. Blue shark are viviparous; embryos develop a placental connection to the uterine wall of the mother. As many as 80 young have been reported in a litter. Young are born at about 40 cm TL after a 9–12 month gestation period. Mating apparently occurs in May and June, and females can store sperm. Large eggs are present in postpartum females suggesting that there is no resting stage in the reproductive cycle. Gravid females are segregated from the rest of the population, possibly offshore.

Age and Growth. Blue shark grow relatively rapidly compared to many sharks. They nearly double in length the first year, growing about 30–40 cm·year^{-1} during the second and third years. Growth then gradually decreases. Large individuals may be more than 20 years old, although recent studies (Skomal 1990) indicate that the western Atlantic population may not reach this age. Skomal (1990) provided the following von Bertalanffy parameter estimates: $L_\infty = 286$ cm FL (ca. 355 cm TL), $K = 0.16$, $t_0 = -0.89$ for males; $L_\infty = 313$ cm FL (390 cm TL), $K = 0.15$, $t_0 = -0.87$ for females; maximum ages are found to be 13 years for males and 16 years for females. For the eastern Atlantic, von Bertalanffy growth parameter estimates are (Aasen 1966b, Stevens 1975): $L_\infty =$ ca. 400 cm TL, $K = 0.11–0.13$, $t_0 = -0.8$ to -1.04.

A weight-length relationship for 4,529 specimens (sexes combined) 60–290 cm FL from the western North Atlantic (Kohler et al. 1996: Fig. 14) is: wt (kg) = 3.1841×10^{-6} FL (cm)$^{3.1313}$.

General Range. Cosmopolitan on the high seas in the warmer parts of all oceans. Blue shark have the greatest geographic range of any elasmobranch and one of the largest of any marine vertebrate (Compagno 1984).

Occurrence in the Gulf of Maine. Blue shark are common along the New England coast in summer from New York to Nova Scotia but may be absent from the colder waters around the Bay of Fundy. Most are medium-sized or larger; there is little information on juvenile distributions in the western Atlantic (Pratt 1979).

Importance. Blue shark have little commercial value, although they are a major by-catch on longlines and drift gill nets. They are often caught by anglers and it is said that a large one will make long and powerful runs if hooked on rod and reel.

ATLANTIC SHARPNOSE SHARK / *Rhizoprionodon terraenovae* (Richardson 1836) /

Bigelow and Schroeder 1953:40–41 (as *Scoliodon terrae-novae*)

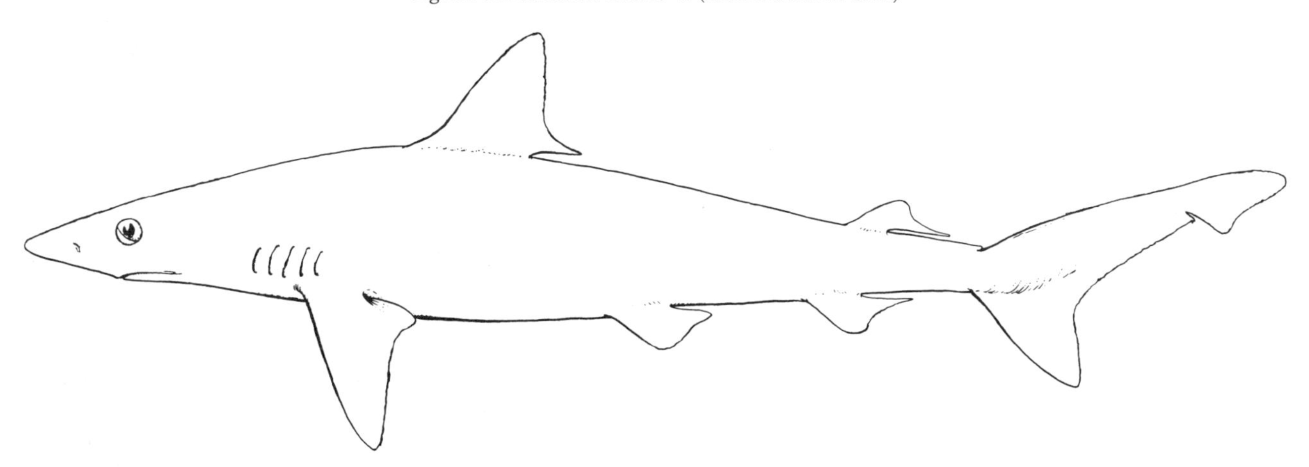

Figure 28. Atlantic sharpnose shark *Rhizoprionodon terraenovae.* Bahama Islands, female, 783 mm TL, MCZ 1144. Drawn by E. N. Fischer. (From Bigelow and Schroeder 1948b: Fig. 49A, © Sears Foundation for Marine Research.)

Description. A slender, little shark with a relatively long snout, variably rounded to pointed (Fig. 28). Long labial furrows present at corners of mouth. Eyes large. Teeth alike in both jaws, triangular, oblique, and smooth-edged. Origin of first dorsal fin just posterior to inner corners of pectoral fins. Small second dorsal fin originates at about midpoint of base of anal fin. Anal fin a little larger than second dorsal fin. Pectoral fins small and triangular.

Meristics. Precaudal vertebrae 58–66, caudal vertebrae 67–81, total vertebrae 126–144 (Springer and Garrick 1964).

Color. Gray to bronze above with irregular white spots scattered along the body; white below. The fins are white along their trailing edges but the second dorsal fin is dusky black.

Size. Reach a maximum size of 110 cm TL. The all-tackle game fish record is a 7.25-kg fish caught at Port Mansfield, Tex., in October 1994 (IGFA 2001).

Distinctions. The Atlantic sharpnose shark can be distinguished by the triangular oblique teeth, similar in both jaws, and by the location of the second dorsal fin, which originates posterior to the origin of the anal fin. The scattered white spots along the dorsal surface are also distinctive.

Habits. This is a common small shark of warm-temperate waters, often traveling in large, sexually segregated schools. Generally demersal, they can be caught at the surface.

Food. Stomach contents from two studies of Atlantic sharpnose shark—129 from Gelsleichter et al. (1999) and 85 from Bowman et al. (2000)—showed the primary food, about 80% by volume, to be small teleosts. The smallmouth flounder, *Etropus microstomus,* was the most important teleost in the first study, occurring in 16% of 129 stomachs. Other fish prey include round herring, striped anchovy, pipefish, sea robin, northern stargazer, Gulf Stream flounder, planehead filefish, and northern puffer. Crustaceans were the second most important prey group, including crabs (particularly lady crab, *Ovalipes ocellatus*), shrimps, and stomatopods. Other prey items included squids (particularly *Loligo*) and gastropods.

Reproduction. Males and females mature in about 3–4 years; males at about 80 cm TL and females at 85 cm TL. This species gives birth each year, unlike many other carcharhinids, which have a resting stage in the reproductive cycle. Mating occurs in late June, and the gestation period lasts about 11 months (Branstetter 1981; Parsons 1983). Pups develop with a placental connection; additionally filamentous villi present on the umbilical cord absorb a nutrient-rich *uterine milk* secreted from the uterine wall. Litters of four to six young are born at about 35 cm TL and 200 g in May. As they disappear from coastal waters at this time, females may seek tidal creeks and bays to give birth.

Age and Growth. This small shark grows rapidly, increasing about 22 cm in the first year, and 15 cm·year^{-1} thereafter. Maximum age may be about 10 years and maximum weight 6–7 kg (Parsons 1985; Branstetter 1987). The von Bertalanffy growth parameter estimates for this species are: $L_\infty = 110$ cm TL, $K = 0.40$–0.50, $t_0 = 0.95$.

General Range. Both sides of the tropical and subtropical Atlantic. Atlantic sharpnose are probably the most common coastal shark in the southeastern United States. They are replaced in the Caribbean by a similar species (*Rhizoprionodon porosus*) from which they can only be distinguished by vertebral counts (Springer 1964).

Occurrence in the Gulf of Maine. Not known to stray into the Gulf of Maine, although one was reported at the mouth of the Bay of Fundy in 1857.

Importance. This is a common shark taken by recreational fishermen in the southeastern United States and often eaten. Although taken in large numbers as by-catch of various commercial fisheries, they are not often sold because of their small size.

DUSKY SHARK / *Carcharhinus obscurus* (LeSueur 1818) / Bigelow and Schroeder 1953:41–43

Description. A large species, with a relatively short, rounded snout, approximately equal to mouth width (Fig. 29). Teeth serrate, uppers broadly triangular and slightly oblique, lowers with narrow cusps. Dorsal fin relatively low with a rounded tip, its origin behind or directly over trailing corner of pectoral fins; height of fin less than distance from eye to first gill slit. Prominent interdorsal ridge present. Second dorsal fin low, with a trailing edge about one to one and a half times fin height. Pectoral fins rather long and falcate.

Color. Gray to bronze above, white below. The trailing edge of the pectoral fin and the fin tips are dusky (hence the common name).

Size. Attain 350 cm TL. Large adults are usually 310–320 cm TL. The all-tackle game fish record is a 346.54-kg fish caught off Longboat Key, Fla., in May 1982 (IGFA 2001).

Distinctions. Dusky shark are a very ordinary-looking requiem shark. They can be distinguished from sandbar shark by location and the size of the first dorsal fin. In dusky shark, the first dorsal fin is located at the rear corner of the pectoral fin (over the pectoral axil in sandbar shark), and the height is less than the distance from the eye to the first gill slit (equal to eye-to-third gill distance in the sandbar shark). Another species, bignose shark, *C. altimus,* might occur as far north as Cape Cod. They can be distinguished by the forward first dorsal fin,

Figure 29. Dusky shark *Carcharhinus obscurus.* California, female, 89 cm TL. Inset is ventral view of head. Drawn by M. H. Wagner. (From Kato et al. 1967: Fig. 74.)

similar to sandbar shark, and the longer snout, with a length exceeding the width of the mouth. A third species that might be confused with dusky shark is silky shark, *C. falciformis,* but again, the first dorsal fin is distinctive. In silky shark, the fin is located posterior to the rear tip of the pectoral fin (over the tip in dusky shark), and silky teeth are oblique, notched on the trailing edge (triangular in dusky shark) with heavy serrations on the base of the cusp.

Habits. Dusky shark are a common large species of warm-temperate regions worldwide. They are taken in coastal areas and well offshore. Individuals are apparently highly migratory; tag returns in the western North Atlantic indicate that they may move from the New England coast into the Gulf of Mexico seasonally (Casey and Kohler 1990). They apparently avoid estuaries and areas of low salinity. The U.S. east coast from New Jersey to south of Cape Hatteras is a major nursery ground in the northwest Atlantic. Neonates and juveniles are common in the surf zone, but do not usually enter bays and inlets.

Food. Teleosts are the most important food category making up 85% of the diet of 59 dusky shark examined by Gelsleichter et al. (1999) and 58% by weight of 53 individuals reported by Bowman et al. (2000), followed by elasmobranchs, crustaceans, and squids. Teleosts include eels, menhaden, herring, anchovies, hakes, goosefish, black sea bass, scup, croaker, bluefish, sand lance, mackerel, tunas, and flatfishes (Gelsleichter et al. 1999; Bowman et al. 2000). Bluefish and summer flounder were the most important teleosts in the diet of the 59 dusky sharks; croaker, scup, and scombrids in the 53. Dusky shark also prey on small skates and invertebrates such as crabs (particularly lady crab, *Ovalipes ocellatus*), squids, lobster, and octopus.

Reproduction. The reproductive cycle of dusky shark is poorly understood. Males mature at 275 cm TL and females slightly larger. They are viviparous, producing about ten young per litter. Clark and von Schmidt (1965) suggested a 16-month gestation period with two distinct groups of females: one that gives birth in July and the other in November–December. However, interpreted in a different way, their data could be used to indicate a gestation period of 22–24 months (Musick et al. 1993b), which makes some sense in that the pups are extraordinarily large at birth, 90–100 cm TL (Branstetter 1981).

Age and Growth. Dusky shark are slow-growing (Lawler 1976; Schwartz 1983). The von Bertalanffy parameter estimates (Natanson 1990) are: $L_\infty = 351$ cm FL (420 cm TL), $K = 0.047$, $t_0 = -5.83$ for males; $L_\infty = 316$ cm FL (378 cm TL), $K = 0.061$, $t_0 = -4.83$. These numbers translate into a growth rate of 10 cm·year^{-1} for adolescents and less than 5 cm·year^{-1} for subadults and adults. Males and females mature at about 17 years. The oldest individuals aged have been about 35 years old.

Born at 80–100 cm TL, weights are 5–10 kg at birth (100 cm TL), 50 kg at 200 cm TL, and 250 kg at 300 cm TL (Lawler 1976).

A weight-length relationship for 247 specimens (sexes combined), 75–290 cm FL from the western North Atlantic (Kohler et al. 1996: Fig. 10) is: wt (kg) = 3.2415×10^{-5} FL (cm)$^{2.7862}$.

General Range. Cosmopolitan in warm-temperate and tropical waters.

Occurrence in the Gulf of Maine. Dusky shark do not normally come north of Cape Cod, although an occasional stray may enter the Gulf of Maine.

Importance. Owing to their large size, dusky shark have been a preferred target of recreational fishermen for years. They are also commonly taken on commercial longlines, both as a bycatch in the swordfish/tuna fishery and as part of the catch in the directed shark fishery. These fisheries, in combination, have apparently had a greater negative effect on the dusky shark population than on any other shark species in the western Atlantic. Catch rates of all sizes of dusky sharks are greatly reduced; large individuals are now rare in recreational catches and their occurrence on commercial gear has declined. Off Virginia, relative abundance of juvenile dusky sharks has declined from about 20% of the total shark catch on longlines to only 1–2% of the catch (Musick et al. 1993b). Western Atlantic populations may have declined by as much as 90% from overfishing during the 1980s so the population is considered *vulnerable* by the American Fisheries Society (Musick et al. 2000). Juveniles and adults are still common in the commercial fishery south of Cape Hatteras (Branstetter and Burgess 1997).

SANDBAR SHARK / *Carcharhinus plumbeus* (Nardo 1827) / Bigelow and Schroeder 1953:43–44 (as *Carcharhinus milberti*)

Figure 30. Sandbar shark *Carcharhinus plumbeus*. Atlantic, female, 150 cm TL. Drawn by M. H. Wagner. (From Kato et al. 1967: Fig. 73)

Description. Heavy-shouldered, with a relatively short, rounded snout shorter than mouth width (Fig. 30). Anterior edge of nostril expanded as a low but definite triangular lobe. Teeth broadly triangular and serrate in upper jaw, finely spiked in lower jaw. Pectoral fins and first dorsal fin large. First dorsal fin height equal to distance from eye to third gill slit. Second dorsal fin relatively large. Interdorsal ridge present. Origin of second dorsal fin roughly over origin of anal fin, its free rear corner only slightly longer than height of fin.

Color. Upper surface slate gray to brown; lower surface a paler tint of the same hue or white. Fins without any conspicuous black markings, although the pectoral fins are often dusky-colored at the tip and along the trailing edge.

Size. Maximum size in the northwest Atlantic is 225–250 cm TL. The all-tackle game fish record is a 117.93-kg fish caught off Gambia in the eastern Atlantic in January 1989 (IGFA 2001).

Distinctions. Sandbar shark differ from dusky in the more forward position and larger size of the first dorsal fin, in their broader pectoral fins, and in their stouter trunk. Sandbar shark can be told from bignose shark in that the snout length is less than mouth width, whereas in bignose the snout is longer than mouth width.

Habits. Sandbar shark are a common bottom-dwelling coastal shark of U.S. Atlantic and eastern Gulf of Mexico waters. Because of their abundance, they may be the best-studied shark of the area. They migrate seasonally along the eastern seaboard, moving north with warming water temperatures in the summer and southward again in the fall. Tagging studies show that sandbar shark commonly go as far south as the Caribbean and the Bay of Campeche in Mexico (Casey and Kohler 1990). Except during the mating season males and females remain in sexually segregated schools, with males usually occurring in deeper water. This shark prefers slightly cooler waters than many of its genus, so they are more common in the mid-Atlantic region than many other species, constituting over 50% of the longline catch off Virginia (Musick et al. 1993b). The mid-Atlantic region, especially Chesapeake Bay, is the primary nursery ground in the Atlantic; other nurseries occur in the northern Gulf of Mexico.

Food. Juveniles on the nursery ground prey heavily on blue crab in addition to numerous small fishes such as menhaden, sea bass, tonguefish, and other flatfishes (Medved and Marshall 1981; VIMS unpubl. data). Once they move to the continental shelf, food is primarily bony fishes (43% by frequency of occurrence), elasmobranchs (16%), cephalopods (3%), crabs, isopods, sand dollars, and trash (Stillwell and Kohler 1993). Fish prey include *Leucoraja erinacea* and other skates, *Lophius americanus*, cod, sculpins, *Pomatomus saltatrix, Scomber scombrus, Squalus acanthias,* and flatfishes such as *Limanda ferruginea.*

Reproduction. Sandbar shark are placentally viviparous producing six to ten young 50–60 cm TL after an 11- or 12-month gestation period. Pups are born at 50–60 cm TL. Females have at least a 1-year resting stage between pregnancies. Mating occurs in early June, and pups are born in late May–early June of the next year (Springer 1960; Clark and von Schmidt 1965; Dodrill 1977; Branstetter 1981). Pups occupy specific nursery grounds away from the adult population, which may reduce predatory mortality on the cohorts (Branstetter 1990).

Age and Growth. This is a very slow-growing species. Juveniles increase about 10 cm·year^{-1}; adolescents and subadults grow about 5 cm·year^{-1}, and adults may grow only 1–2 cm·year^{-1}. Juveniles remain in nursery areas until they are about 130 cm TL or 6–7 years of age. Males and females mature at 12–15 years of age; the oldest individuals aged have been 20–25 years old; however, some fish tagged as 2- to 5-year-old juveniles have been recaptured as much as 25 years later. Although the von Bertalanffy growth curve does not provide realistic estimates of all parameters, the Brody growth coefficient (K) generated in two studies was: $K = 0.05$ (Casey et al. 1985) for males; $K = 0.04$ (Casey et al. 1985), $K = 0.056$ (Lawler 1976) for females. In a more recent study (Sminkey and Musick 1995), these values were corroborated and a more realistic growth curve was generated.

Males reach maturity at about 170 cm TL and females at more than 180 cm TL. With increased fishing pressure on this species, most large individuals have been removed from the population; it is uncommon now to catch a sandbar shark longer than 200 cm TL. Weighing about 1–2 kg at birth, individuals at 100 cm TL are 7–10 kg, at 150 cm TL they are 25 kg, and at 200 cm TL they are 60–65 kg (Lawler 1976).

General Range. Cosmopolitan in warm-temperate coastal waters on both sides of the Atlantic, southern Indian Ocean, Indo–west Pacific from Australia to Japan, and also Hawaii. Apparently absent from the eastern Pacific.

Occurrence in the Gulf of Maine. The sandbar is the most numerous of the larger sharks along the coast of the mid-Atlantic as far north as Woods Hole. Some individuals occasionally stray into the limits of the Gulf.

Importance. Sandbar shark have long been important to recreational fisheries along the Atlantic coast and in the eastern Gulf of Mexico. They were the primary target of a Florida-based commercial shark fishery in the 1940s (Springer 1951), and are now a primary target of the directed commercial fishery along the east coast from Virginia to Florida and in the eastern Gulf of Mexico (Branstetter and Burgess 1997). The large fins and a recent high market value make sandbar shark a preferred target species. The meat is also highly marketable. Recent increased fishing pressure on this species has had a negative impact on its abundance. Catch rates in the mid-Atlantic region have declined about two-thirds since the mid-1970s (Musick et al. 1993b).

HAMMERHEAD SHARKS. FAMILY SPHYRNIDAE

STEVEN BRANSTETTER

The peculiar hammer-shaped head, with eyes far apart, sufficiently characterizes this family, which otherwise resemble requiem sharks and are included with them by some authors (Nelson 1994). Four species are known in the western Atlantic, all tropical and subtropical. Two of these have been reported from the Gulf of Maine, but only as strays.

BONNETHEAD / *Sphyrna tiburo* (Linnaeus 1758) / Shovelhead / Bigelow and Schroeder 1953:44–45

Description. Head only slightly expanded, shovel-shaped (Fig. 31). Anterior teeth with short, stout nonserrate cusps; posterior teeth low, forming a crushing plate. First dorsal fin relatively high, with origin over free margin of pectoral fins. Second dorsal fin moderately high, and anal fin large. Pectoral fins short and broadly triangular.

Meristics. Precaudal vertebrae 72–88, caudal vertebrae 70–87, total vertebrae 142–173 (Springer and Garrick 1964).

Color. Gray or grayish-brown above, and a paler shade of the same below; some are marked with a few small dark, roundish spots along the sides.

Size. Reach a maximum length of 120 cm TL, but most are less than 100 cm TL. The all-tackle game fish record is a 10.76-kg fish caught in Cumberland Sound, Ga. in August 1994 (IGFA 2001). Size at maturity is about 75 cm TL for males and 85 cm

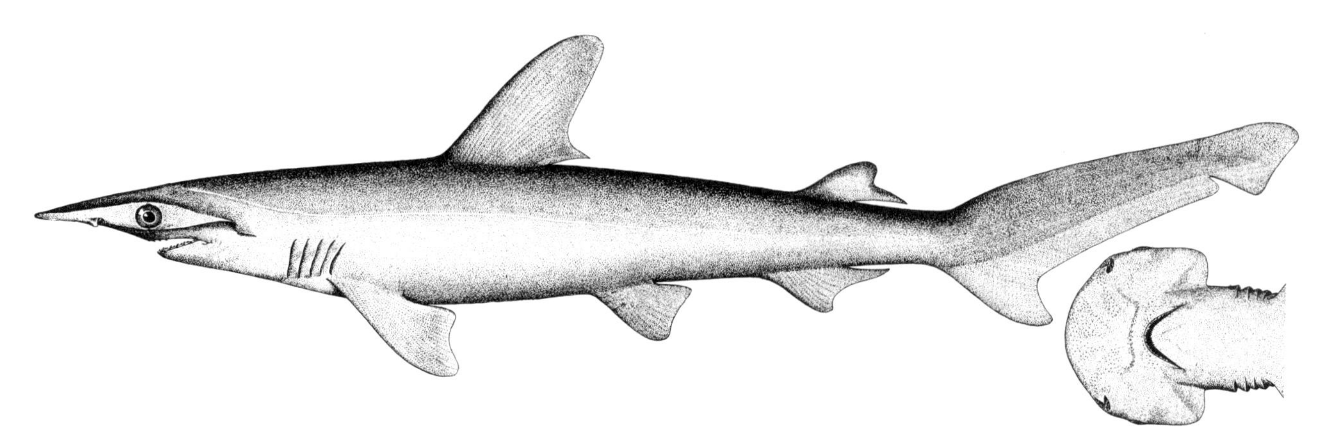

Figure 31. Bonnethead *Sphyrna tiburo.* Cedar Keys, Fla., 33 cm TL, USNM 26852. Inset is ventral view of head. Drawn by E. N. Fischer.

TL for females. Females attain slightly larger sizes than males. Adults weigh about 10–20 kg.

Distinctions. Among hammerheads, the shovel-shaped head readily identifies this species.

Habits. Bonnethead are common in southeast U.S. waters, often entering estuaries and bays; common out to 50 m. They usually occur in groups or schools of as many as 20 individuals. Because they do well in captivity, they have been used for behavioral, energetics, and reproductive studies (Myrberg and Gruber 1974; Parsons 1987). The schools exhibit several behavioral patterns such as *circling,* where several individuals follow head-to-tail in a circle pattern; other behaviors exhibited indicate that the group has a social hierarchy, with the largest individual, usually a female, being dominant. This species is heavily preyed upon by larger sharks.

Food. Bonnethead feed primarily on crustaceans such as shrimps and crabs, as well as on octopus and small fishes.

Reproduction. Size at maturity (2 years) is about 75 cm TL for males and 85 cm TL for females. The mating season is protracted, but generally occurs in the spring. Development is placentally viviparous, and the umbilical cord possesses numerous filamentous appendiculi that absorb nutrients secreted by the uterine wall. Females produce litters of six to nine pups 25–35 cm TL after a short gestation period of 4–5 months (Parsons 1987).

Age and Growth. Bonnethead grow rapidly, in excess of 20 cm·year^{-1}. Whereas a logistic growth curve based on weights describes growth more accurately (Parsons 1987), for comparison with other species, the von Bertalanffy parameter estimates are: L_{∞} = 81–89 cm TL, K = 0.53–0.58, t_0 = −0.64 to −0.77 for males; L_{∞} = 103–115 cm TL, K = 0.34–0.38, t_0 = −0.60 to −1.1 for females.

General Range. This is a New World species, found on both the east and west coasts of North and South America. In the western Atlantic, they occur from southern Brazil to North Carolina, straying to Massachusetts Bay. In the eastern Pacific, they occur from southern California to Ecuador.

Occurrence in the Gulf of Maine. Bonnethead have been reported to occasionally stray into Massachusetts Bay and Nantucket Sound.

Importance. Although not a targeted species, this small, common shark is often caught by recreational fishermen targeting other species and are by-catch in the directed commercial shark fishery. They are more commonly taken as by-catch in the shrimp trawl fishery and in trammel and gill net fisheries. Their meat is marketable and is sometimes sold.

SMOOTH HAMMERHEAD / *Sphyrna zygaena* (Linnaeus 1758) / Bigelow and Schroeder 1953:45–46

Description. Head greatly expanded, width greater than 25% TL (Fig. 32). Head convex, without a median notch (hence the common name). Teeth serrate and oblique in upper and lower jaw. First dorsal fin origin over pectoral fin axil; its height moderate. Second dorsal fin small, with origin over anal fin base. Trailing edge of anal fin strongly concave. Pectoral fins triangular, of moderate size.

Meristics. Precaudal vertebrae 99–102, caudal vertebrae 103–104, total vertebrae 202–206 (Springer and Garrick 1964).

Color. Leaden or brownish gray above; shading along sides to pure or grayish white below; tips and edges of the dorsal and caudal fins more or less dusky; a series of dark spots present on the lower part of the caudal peduncle.

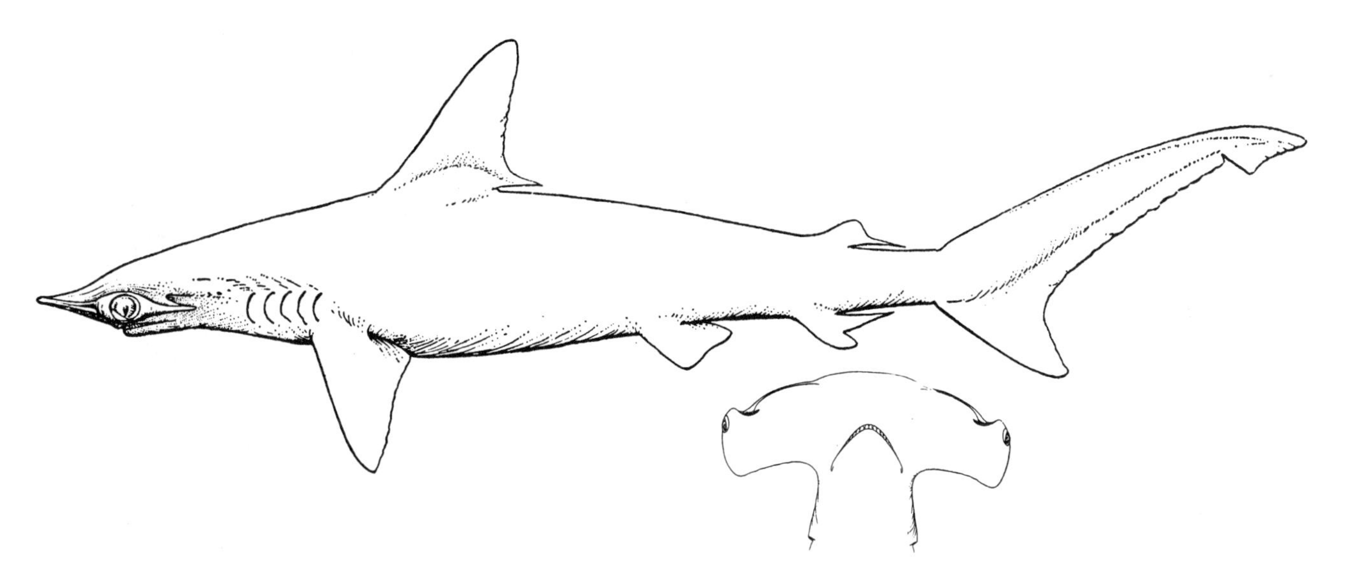

Figure 32. Smooth hammerhead *Sphyrna zygaena*. Nahant, Mass., female, 687 mm TL, MCZ 1159. Lower figure is ventral view of head. Drawn by E. N. Fischer.

Size. Reach a maximum size over 300 cm TL and may grow as large as 370 cm TL. The all-tackle game fish record is a 164.65-kg fish caught in the Azores in July 1999 (IGFA 2001).

Distinctions. Smooth hammerhead are readily distinguished from bonnethead simply by the head shape. However, another species, not included in this account, might stray as far north as Cape Cod. Scalloped hammerhead (*Sphyrna lewini*) are very similar morphologically to smooth hammerhead but are distinguished by a medial notch on the anterior edge of the head and the lack of dark spots on the caudal peduncle.

Biology. Very little is known about the biology of smooth hammerhead shark in the western Atlantic. They occur in outer continental shelf and offshore waters. They are the most tolerant of cool waters of any hammerhead species and usually do not occur in tropical areas. They are often found at the surface, but are also taken at depth. They feed on a variety of bony fishes such as herring, sea robins, dolphin, and butterfish (Bowman et al. 2000). Born at about 50–60 cm TL, they mature at about 250 cm TL. They are placentally viviparous, producing as many as 40 young per litter. Age and growth are unknown, but if growth is similar to that of scalloped hammerhead, then smooth hammerhead grow slowly: about 15 cm the first year, and 5–10 cm·year^{-1} through adulthood.

General Range. Cosmopolitan in the world oceans, usually in antitropical environments. They are documented from tropical regions, but are also easily confused with the more tropical scalloped hammerhead.

Occurrence in the Gulf of Maine. More common off Martha's Vineyard, Nantucket, and Woods Hole, smooth hammerhead have been taken occasionally within the Gulf of Maine from Massachusetts Bay to Nova Scotia.

BRAMBLE SHARKS. FAMILY ECHINORHINIDAE

George H. Burgess

Bramble sharks are large sharks that lack an anal fin. They are similar in appearance to spiny dogfishes of the family Squalidae (indeed some systematists refer them to that family), but differ in having the origin of the first dorsal fin located posterior to the pelvic origin. Both dorsal fins lack spines. Two species are known, with one, *Echinorhinus cookei,* confined to the Pacific (Nelson 1994). The cosmopolitan *E. brucus* is unique among sharks in possessing large, thornlike denticles called *bucklers* on its body and fins.

BRAMBLE SHARK / *Echinorhinus brucus* (Bonnaterre 1788) / Bigelow and Schroeder 1953:56–57

Description. Body relatively stout, soft and flabby; trunk cylindrical (Fig. 33). Snout short and depressed, gill openings (especially fifth) large. Underside of snout covered with large denticles. Mouth broadly arched, with short labial folds.

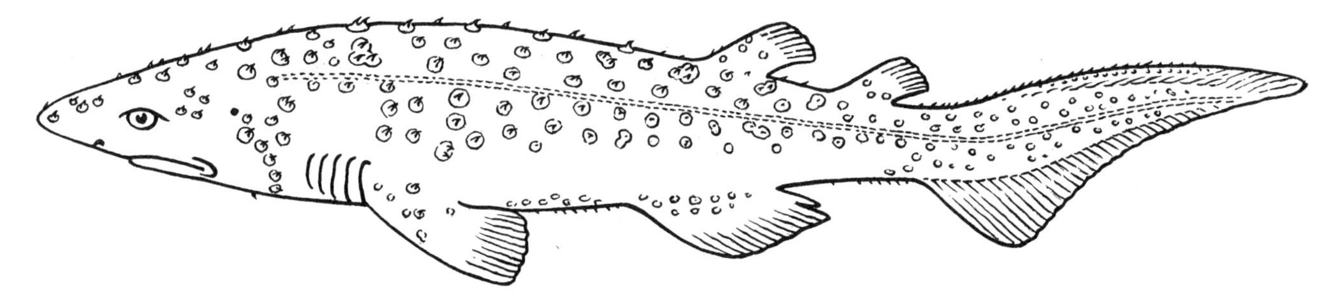

Figure 33. Bramble shark *Echinorhinus brucus.* Eastern Atlantic, 91 cm TL. Drawn by W. P. C. Tenison. (From Bigelow and Schroeder 1948b: Fig. 102A, © Sears Foundation for Marine Research.)

Nostrils widely separated with short anterior flaps; spiracles present behind eyes. Multicuspid teeth similar in both jaws, strongly compressed with a single cusp and up to three cusplets (cusplets lacking in juveniles) curving toward corners of jaws, forming a cutting blade. Two equal-sized, spineless dorsal fins located far back on body, just posterior to pelvic fin origin. Precaudal pits present; caudal fin lacks a subterminal notch. Skin thin and delicate. Body and fins irregularly covered with large to small bucklers. Single bucklers with round bases with radiating ridges, conically shaped; often a series (two or more) of bucklers coalesce to form large plates (15–25 mm) with multiple cusps.

Meristics. Upper jaw teeth 20–26; lower jaw teeth 22–26. Vertebral centra not calcified.

Color. Dark gray, olive, purple, black, or brown with metallic reflections dorsally, occasionally with darker blotches. Ventrally, pale brown or gray to white. The denticles have been described as luminescent, but there are no special luminous organs.

Size. By squaloid standards this is a relatively large species, reportedly reaching 3.1 m TL.

Distinctions. The presence of large, thornlike denticles should be enough to distinguish this species from all other Gulf of Maine sharks, but the combination of far-posterior placement of the two dorsal fins (both behind pelvic fin origins) and the lack of an anal fin are equally distinctive. Bramble shark might be confused with Greenland shark, but differ in having: (1) the first dorsal fin located just behind the origin of the pelvic fins (instead of about midway between the latter and the pectorals), (2) a very different tail shape, (3) large thornlike denticles, (4) larger gill openings, and (5) teeth that are similarly shaped in the two jaws (instead of unlike).

Biology. Bramble shark are primarily deepwater, bottom-dwelling sharks that are found on the deeper portions of the continental shelf and upper slope. The recorded depth range is 18–900 m, but are much more common in depths greater than 200 m. Bramble shark eat a variety of bony fishes, small sharks, and crabs. Females are ovoviviparous with 15–24 pups per litter. Males mature at 150–174 cm and females at 213–230 cm. Young are born between 29–90 cm. Total length-weights at selected sizes: 30 cm (embryo)/110 g, 150 cm/20 kg, 162 cm/29 kg, 170 cm/45 kg, 216 cm/78.2 kg, and 254 cm/136 kg.

General Range. Five western North Atlantic records from Cape Cod, off Virginia, and the northern Gulf of Mexico and in the eastern North Atlantic from the North Sea southward to Ivory Coast, including the Mediterranean Sea. South Atlantic from Argentina in the west and from Namibia to the Cape of Good Hope in the east. Elsewhere they have been caught in the Indian Ocean and in the western Pacific.

Occurrence in the Gulf of Maine. A single specimen of this little-known shark came ashore at Provincetown in December 1878.

SPINY DOGFISHES. FAMILY SQUALIDAE

GEORGE H. BURGESS

Spiny dogfishes are a large cosmopolitan family of predominantly small- to medium-sized sharks (but a few species reach great sizes if the deepwater Dalatiidae are included and not considered as a separate family). They are characterized by the absence of an anal fin, and in most species the two dorsal fins are preceded by a fixed spine, which may be long and conspicuous or so short that its presence can be detected only by touch. The first dorsal fin is located in front of the origin of the pelvic fins. The teeth are similarly shaped in the two jaws in some species, but markedly different in others. This family contains the smallest living sharks (*Etmopterus perryi* is mature at 160 mm TL (Springer and Burgess 1985), and other *Etmopterus, Squaliolus,* and *Euprotomicrus* species are nearly as small), as well as one of the largest (*Somniosus rostratus* exceeds

13 m). Bigelow and Schroeder reported capture of 50 specimens of *Etmopterus princeps* on the deeper slopes that front the Gulf so it is included in the key. Most species of spiny dogfishes inhabit moderately deep to very deep waters, with *Centrophorus* species occurring in depths of over 6,000 m. Only one species, the shallow-living spiny dogfish, *Squalus acanthias,* is of economic importance in the Gulf of Maine, but the species of *Centrophorus, Deania,* and *Centroscymnus* are sought elsewhere in the world for their liver oil. Recent information on spiny dogfish in the northwest Atlantic was summarized by McMillan and Morse (1999).

BLACK DOGFISH / *Centroscyllium fabricii* (Reinhardt 1825) / Bigelow and Schroeder 1953:51–52

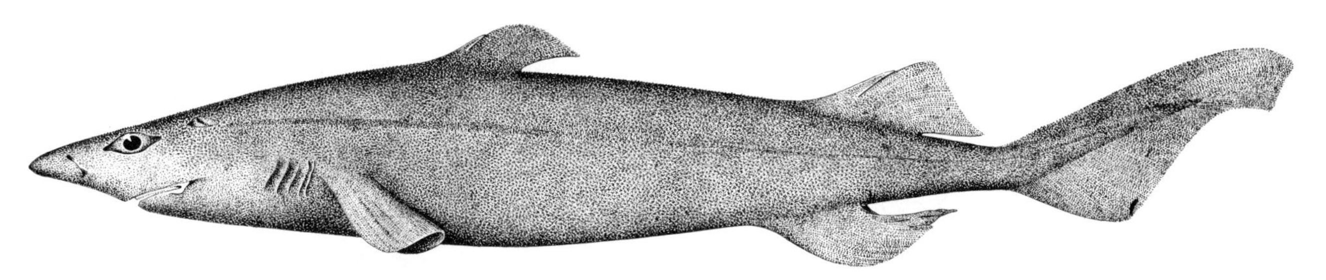

Figure 34. Black dogfish *Centroscyllium fabricii.* 44°23′N, 53°25′E, USNM 22879. Drawn by A. H. Baldwin.

Description. Body moderately stout and compressed, caudal peduncle short, and abdomen long (Fig. 34). Snout moderately long with an arched mouth, length about one-third of width. Upper and lower teeth numerous and similar in shape with a single large, erect cusp and two (occasionally four) smaller, erect cusplets. Second dorsal fin much larger than first, originating over midpelvic base. Both dorsal fins with prominent, grooved spines. Anal fin absent. Adpressed pectoral fins fall short of first dorsal origin. Close-set denticles short and conical with hooked cusps and ridged, stellate bases. Probably bioluminescent because skin bears minute, deeply pigmented dots similar to those seen on other luminescent squaloid species.

Meristics. Upper jaw teeth 68, lower jaw teeth 68; precaudal vertebrae 66 or 67, caudal vertebrae 29 or 30, total vertebrae 95–97 (Springer and Garrick 1964).

Color. Adults and subadults are uniformly blackish brown without any obvious black markings above the pelvic fins, ventrally on the caudal peduncle, or laterally on the upper caudal lobe. Juveniles are dark black ventrally and lighter black to brown dorsally with white-edged dorsal, pectoral, and pelvic fins. Spines are white at all sizes. Eyes are a reflective green when first caught.

Size. Reach a maximum length of 90 cm, but most common as adults at 60–75 cm.

Distinctions. Presence of numerous multicusped teeth in both jaws makes this species unique among Gulf of Maine sharks. They are most likely to be confused with great lanternshark, but differ in lower tooth shape (multicuspid vs. bladelike) and the origin of the second dorsal fin (at level of midpelvic base vs. near posterior pelvic base). They are distinguished from spiny dogfish in having deeply grooved dorsal spines (vs. rounded), in placement of the second dorsal fin (origin near rear pelvic base in *Squalus*), and in tooth shape (bladelike in both jaws in *Squalus*).

Habits. Although black dogfish are a deepwater, benthic species usually confined to the outer continental shelf and the upper continental slope, they have occasionally been taken near the surface in arctic waters and during periods of maximum cold and darkness. They have been collected from as deep as 1,600 m and temperatures as low as 1°C, but are most common at 550–1,000 m and temperatures of 3.5°–4.5°C. Some sexual and size segregation, schooling, and spring-winter inshore migration may occur.

Food. Black dogfish consume cephalopods, benthic and pelagic crustaceans, euphausiids, scyphozoans, and fishes, including small redfishes (*Sebastes*).

Parasites. Two parasites, a trematode (*Otodistomum cestoides*) and a copepod (*Lernaeopoda centroscyllii*), have been reported from this species.

Reproduction. The reproductive biology of 1,124 male (165–760 mm TL) and 1,476 female (175–898 mm TL) black dogfish from western Greenland was reported by Yano (1995). Size at maturity was about 550 mm TL in males and 650 mm TL in females. Litter size was 4–40. This is an ovoviviparous species with development of the embryos dependent solely on yolk reserves. The smallest free-swimming specimens were 165 mm for males and 175 mm TL for females.

General Range. In the western North Atlantic from Baffin Island and Greenland to the Gulf of Mexico and in the eastern North Atlantic from Iceland south to Senegal and Guinea to Namibia. Also known from southern Argentina in the western South Atlantic (Menni et al. 1993).

Occurrence in the Gulf of Maine. Black dogfish are often caught along the slopes of the offshore banks, from Grand to Browns and to the eastern part of Georges.

Importance. Although black dogfish are occasionally taken in large numbers in deepwater bottom trawls, they are not abundant enough to support a commercial fishery. Some may be returned to port for reduction into fish meal.

PORTUGUESE SHARK / *Centroscymnus coelolepis* Bocage and Capello 1864 /

Bigelow and Schroeder 1953:52–53

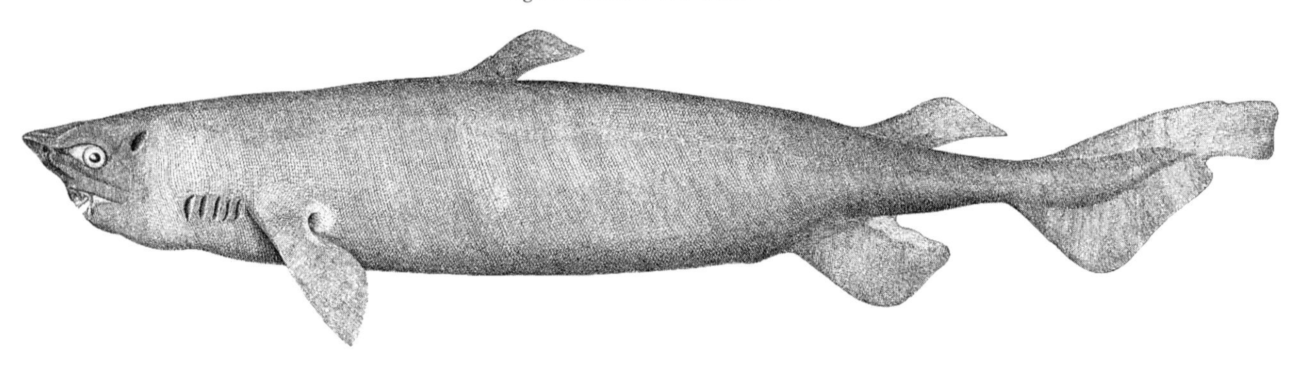

Figure 35. Portuguese shark *Centroscymnus coelolepis.*

Description. Body moderately stout (Fig. 35). Gill slits short. Snout short, blunt, and flattened. Mouth slightly arched with thick lips; upper and lower labial furrows present. Upper and lower teeth markedly different in shape; uppers awl-shaped, erect and thin, with one lanceolate cusp and no cusplets; lowers very short with oblique cusps and large, overlapping bases, forming a cutting edge. Very short, grooved spines barely protruding from origins of two equally small dorsal fins. First dorsal fin origin posterior to broadly round pectoral fins. Second dorsal and pelvic fins far posteriorly, second dorsal originating over midpelvic base. Anal fin absent. Upper caudal lobe much longer than lower, with deep subterminal notch. Lateral keels and precaudal pits absent. Trunk denticles of adults large, closely overlapping, concave and crown-shaped when viewed laterally; young with widely spaced, three-pointed, leaflike denticles.

Meristics. Upper jaw teeth 43–68, lower 29–41; precaudal vertebrae 75–79, caudal vertebrae 26–34, total vertebrae 105–109 (Springer and Garrick 1964).

Color. Adult Portuguese dogfish are uniformly blackish brown, without any obvious black markings above the pelvic fins, ventrally on the caudal peduncle, or laterally on the upper caudal lobe. Young tend to be bluish black and juveniles more black than brown.

Size. Average adult size is 70–100 cm; maximum size is about 120 cm. Males are smaller than females, reaching a maximum size of just over 90 cm.

Distinctions. Portuguese shark can be identified by the combination of the absence of an anal fin and the presence of small dorsal fins with very short spines, the broadly rounded pectoral fin, and the concave, crown-shaped denticles. They might be confused with a small Greenland shark or a kitefin shark, but both of the latter lack dorsal spines.

Habits. Portuguese shark are benthic and deepwater, captured most frequently at depths greater than 400 m. They have been taken from 270–3675 m at temperatures of 5°–13°C. Females tend to inhabit deeper waters than males.

Food. *Centroscymnus* feed primarily on benthic fishes and squids, as well as on octopods and gastropods, and are probably active scavengers since marine mammal remains have been found in stomachs. They feed mostly at night (Yano and Tanaka 1984).

Reproduction. Portuguese shark are ovoviviparous. Males mature at 700–750 mm TL; mature females have been reported at 900–1,000 mm TL (Yano and Tanaka 1988). Females ovulate mature ova 50–60 mm in diameter (Yano and Tanaka

1984). Embryos are nourished mainly by the yolk; there is no placental attachment. The largest observed embryo with an external yolk sac was 130 mm TL; the largest observed embryo with an internal yolk sac was 103 mm TL. Females bear 13–29 young, which are born at 270–300 mm.

Age and Growth. Weight/length equations, calculated by the least square method, for a Japanese population (Yano and Tanaka 1984) were $W = 6.06 \times 10^{-11}L^{3.71}$ for females; and $W = 2.31 \times 10^{-8}L^{2.81}$ for males. Length-weights: 1.64 kg at 633 mm TL, 5.51 kg at 924 mm TL, 11.75 kg at 1,085 mm TL.

General Range. In the western Atlantic from the Grand Banks, Newfoundland, to Virginia. Eastern Atlantic from Iceland southward to Sierra Leone, including the Mediterranean; also from Namibia to Quoin Point, South Africa. Also found in the west Pacific.

Occurrence in the Gulf of Maine. This species inhabits the troughs between the deep offshore banks and occasionally the deeper parts of the banks themselves, from Grand Bank southward.

Importance. Portuguese shark are occasionally captured on halibut lines and infrequently in deepwater trawls. They derive their name from their importance in the deepwater fishery off Portugal. They are not of commercial importance in the Gulf area, but are the object of fishing pressure in other parts of the world, primarily for the squalene in their liver oil, but also for human consumption and reduction into fish meal.

KITEFIN SHARK / *Dalatias licha* (Bonnaterre 1788) / Bigelow and Schroeder 1953:55–56

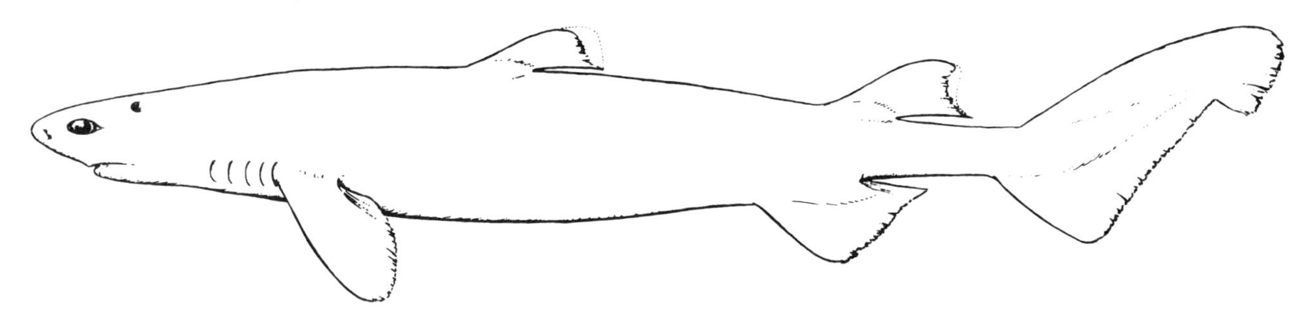

Figure 36. Kitefin shark *Dalatias licha*. Georges Bank, female, 147 cm TL, AMNH 14056. Drawn by E. N. Fischer.

Description. Moderate-sized, lacking an anal fin and dorsal fin spines (Fig. 36). Gill openings short. Snout short, conical, and blunt; mouth is mildly arched with thick, papillose lips. Teeth asymmetrical, small upper teeth narrow, smooth, and awl-shaped, curving somewhat outward toward corners of mouth; lower teeth large with erect, broadly triangular, serrate cusps and overlapping bases. First dorsal fin originates just behind rear tips of short, rounded pectoral fins; dorsal fin base closer to that of pectoral than pelvic bases. Second dorsal fin slightly larger than first, originating far posteriorly on body, over midpelvic bases. Large upper caudal lobe with very well-developed subterminal notch; lower lobe essentially nonexistent. No lateral keels or precaudal pits. Dermal denticles low and flattened with ridged, unicuspid crowns emerging from triangular bases.

Meristics. Upper jaw teeth 19, lower 19; precaudal vertebrae 50 or 51, caudal vertebrae 27 or 28, total vertebrae 78 or 79 (Springer and Garrick 1964).

Color. Uniformly dark chocolate brown, grayish black, cinnamon, or violet brown, often with poorly defined black spots on the dorsal surface; lacks obvious black markings above the pelvic fins, ventrally on the caudal peduncle or laterally on the upper caudal lobe. Fin margins are white or translucent, the tail tipped with black.

Size. Reach a maximum size of 159 cm (possibly to 182 cm). Most of those caught are between 100–140 cm.

Distinctions. Kitefin resembles Portuguese shark in the relative sizes and positions of their fins and in their denticles. However, the dorsal fins do not have any trace of spines, while the serrate margins of the lower teeth, in combination with their triangular shape, distinguish them from other sharks without an anal fin known from the Gulf area. Greenland shark, a much larger species, have a greatly expanded ventral lobe of the caudal fin and strongly oblique, smooth-edged lower teeth that form a cutting edge.

Habits. This is a solitary, epibenthic, tropical and warm-temperate species. They are most commonly found in the deepest shelf and upper slope waters, usually at depths of 200–600 m, but have been taken from depths of 37–1,800 m.

Food. *Dalatias* are formidable predators with large, serrate teeth and powerful jaws. They consume a variety of small deepwater fishes, skates, and sharks; squids and octopods; shrimps and lobsters; isopods and amphipods; siphonophores; and polychaetes. Fishes are their primary food.

Reproduction. Kitefin are ovoviviparous. A total of 10–16 young are born at about 30 cm. Adult females have been taken at 117–159 cm and adult males at 77–121 cm.

General Range. Western Atlantic from Georges Bank, the east coast of Florida, northern Gulf of Mexico, and off the southern Bahamas; eastern Atlantic, from the North Sea southward to Cameroon, including the western Mediterranean; central and western Pacific and the southwestern Indian Ocean.

Occurrence in the Gulf of Maine. The only record from the Gulf region is that of a female, about 1.5 m long, taken on the northern edge of Georges Bank on 19 August 1937.

Importance. Of no importance in the western Atlantic, but sought for its squalene-rich liver oil and flesh in Japan, for its oil in South Africa, and for reduction into fishmeal in the eastern Atlantic.

GREENLAND SHARK / *Somniosus microcephalus* (Bloch and Schneider 1801) /

Sleeper Shark, Ground Shark, Gray Shark, Gurry Shark / Bigelow and Schroeder 1953:53–55

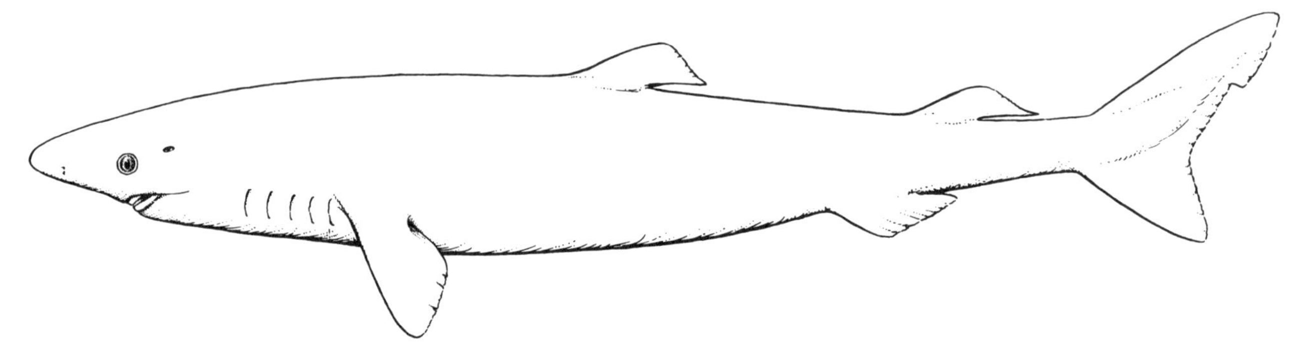

Figure 37. Greenland shark *Somniosus microcephalus.* Gulf of Maine, female, 174 cm TL. Drawn by E. N. Fischer.

Description. Large, heavy-set body (Fig. 37). Short, rounded snout and very small eyes, small pectoral fins, and two spineless, nearly equally small dorsal fins. Gill openings short, low down on sides of head. Lips thin. Upper and lower jaw teeth dissimilar: upper teeth thin and awl-like, smooth, without cusplets; interlocking lower teeth broad, squarish with short, smooth, outwardly directed cusps forming a cutting blade and very high roots. First dorsal fin inserted somewhat closer to pelvic bases than to pectoral bases; second dorsal over rear half of pelvic bases. Pectoral fins barely larger than pelvic fins, with rounded free rear tips and inner margins. Pelvic fins slightly larger than second dorsal fin. Ventral caudal lobe large, upper lobe small, with a subterminal notch, leading to a paddlelike appearance. Caudal peduncle short, without keels or precaudal pit. Denticles with narrow, recurved, cuspidate crowns and squarish bases.

Meristics. Upper teeth 48–52, lower teeth 50–52. Vertebral centra largely uncalcified.

Color. Uniformly blackish, coffee brown, or ashy, purplish, or slaty gray, changing to bluish gray if the epidermis is rubbed off, as is apt to happen when one is caught. The back and sides may be marked with many indistinct dark crossbars or white spots. Fins lack prominent light edges.

Size. Reach a maximum length of about 640 cm and may grow to 730 cm. Average size is 244–427 cm; females reach a larger size than males. The largest from the Gulf area include one of 5 m from the Grand Banks; one of 4.9 m from off Portland, Maine; and two of about 4.6 m from off Cape Ann. A 640-cm British specimen was said to weigh about 1,020 kg; two Gulf of Maine specimens, each about 335 cm long, weighed about 272 and 295 kg, respectively. The all-tackle game fish record is a 775.0-kg fish caught in Trondheimsfjord, Norway, in October 1987 (IGFA 2001).

Distinctions. Greenland shark resemble Portuguese shark in the sizes and relative positions of the dorsal and pelvic fins, in general form, and in their teeth, but they are easily distinguished by the lack of any trace of spines in their dorsal fins, by having thornlike and loosely spaced dermal denticles, and by having a more lunate tail. They also grow much larger than Portuguese shark. Bramble shark also are large and lack an anal fin, but have the first dorsal fin origin more posteriorly, over the midpelvic base, and have large dermal bucklers. Kitefin shark, a much smaller species, are similar in appearance, but lack the prominently expanded ventral caudal lobe and have diagnostic triangular, pointed lower teeth with serrate edges.

Habits. Greenland shark are very sluggish and offer little resistance when captured. In the northern part of its range, they are commonly found in nearshore waters, where they tend to approach the surface in the winter, often coming right up to the ice to chase prey. Most withdraw to depths of 180–550 m or deeper in the summer. In more southerly waters they are found epibenthically on the shelf and slope to depths of 1207 m, but may venture into shallower waters during the spring and summer. The few that visit the Gulf appear to frequent the bottoms of the deeper troughs, although a stray may occasionally come close to the shore. Water temperatures of Southern Hemisphere catches range from 0.6° to 12°C.

Food. Greenland shark are extremely rapacious. They devour any carrion, such as whales and seals, eagerly and congregate in large numbers around fishing, whaling, and sealing operations. Their diet includes a wide variety of large and small fishes, including small sharks, skates, eels, herring, capelin, Atlantic salmon, char, various gadoids, redfish, sculpins, lumpfish, wolffish, and flounders. Seals are a favorite food, and in view of their sluggishness, it is somewhat astonishing that they are able to capture prey as active as seals, porpoises, halibut, and salmon. A specimen from Cape Cod Bay contained half a dozen flounders and a large piece of seal (with hide and hair). They are also known to eat sea birds, sea urchins, brittle stars, crabs, amphipods, squids, large snails, and jellyfish. Objects as large as an entire reindeer (without horns), parts of horses, a whole seal, and a cod and a salmon about 1 m long have been found in Greenland shark stomachs. They will bite on any fish or meat bait, the more putrid and ill-smelling the better.

Parasites. A large (up to 60–70 mm long) parasitic lernaeopodid copepod, *Ommatokoita elongata,* is frequently found on the eyes of Greenland shark (and also the related eastern Pacific sleeper shark, *Somniosus pacificus,* Benz et al. 1998). All but 17 of 1,505 Greenland shark off Greenland were infested (Berland 1961). Ordinarily, only a single female attaches itself to the cornea of an eye, and usually only one eye is affected. Attachment of the copepod leads to corneal lesions, which suggests that severe impairment of vision is possible, although infested Greenland shark seemed otherwise in healthy condition (Borucinska et al. 1998). There is no scientific evidence to support the belief of some Norwegian fishermen (Berland 1961) that the copepod is bioluminescent and might work symbiotically with the shark in attracting prey close to its host (Borucinska et al. 1998).

Reproduction. This is an ovoviviparous species. Large numbers of soft eggs, without horny capsules, ranging in size up to that of a goose egg, have been found in female Greenland shark. A 5-m female had ten 38-cm long, full-term embryos in one uterus.

Age and Growth. As might be expected in such a cold-water species, growth is slow, averaging about 1 cm·year^{-1}. Tagged specimens grew from 271 to 272 cm in 2 years, from 285 to 300 cm in 14 years, and from 262 to 270 cm in 16 years.

General Range. Northern Atlantic and Arctic from Spitsbergen and the White Sea south to Iceland, the North Sea, and accidentally to Le Havre, France, and perhaps to Portugal in the east; from Greenland and Ellesmere Island south to Newfoundland and the northern part of the Gulf of St. Lawrence in the west, occasionally to Cape Cod and North Carolina. Also reported in the South Atlantic from Argentina and from Antarctic waters (Kerguelen Islands).

Occurrence in the Gulf of Maine. Greenland shark appear in the Gulf of Maine as irregular stragglers from the north. Records include two specimens taken in the neighborhood of St. Andrews in 1915 (one caught in a weir and the other on a longline); one reported off Eastport; six from off Cape Elizabeth between 1925 and 1948; one of about 4.6 m long-lined from Jeffreys Ledge on 16 February 1931; several from near Cape Ann, off Marblehead and Nahant, in Massachusetts Bay, off Barnstable in Cape Cod Bay, and at Provincetown; and one trawled in Cape Cod Bay off the entrance to the Cape Cod Canal in April 1924.

Recorded captures in the Gulf include small specimens, as well as large and have been from all four seasons of the year. Local records are distributed so widely that the species might be expected anywhere in the deeper parts of the Gulf. It has been suggested that they were more numerous in Gulf waters in early colonial times when Atlantic right whale were still being killed in numbers off the Massachusetts coast.

Importance. This shark is not plentiful enough in the Gulf to be of commercial value. They have long supported a hook-and-line, gaff, and longline fishery off northern Norway, around Iceland, and in West Greenland waters, chiefly for their liver oil. They are often free-gaffed by fishermen at the surface after being attracted to holes in the ice by meat or bread lowered on a gaff. They produce an intoxicating poison if eaten fresh, although they are wholesome if dried. In Greenland the flesh is also dried for sled-dog food and in Iceland to a small extent for human consumption. The skin has been used for boot leather, and the teeth have been utilized in making tools. Greenland shark inflict considerable damage on seal, whale, and fish nets in many areas.

SPINY DOGFISH / *Squalus acanthias* Linnaeus 1758 /

Dogfish, Grayfish, Piked Dogfish, Spurdog / Bigelow and Schroeder 1953:47–51

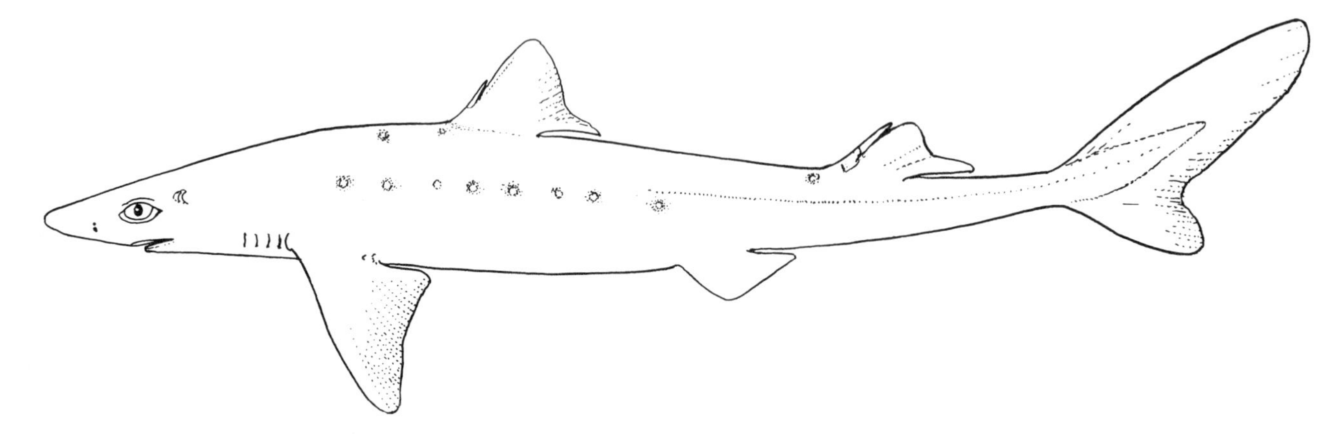

Figure 38. Spiny dogfish *Squalus acanthias.* Female, 69 cm TL. Drawn by E. N. Fischer.

Description. Body slender and elongate (Fig. 38). Head moderately flattened. Snout narrow, tapering to pointed tip. Eyes moderately large. Gill slits small. Narrow anterior nasal flaps and spiracles present. Mouth nearly straight with upper and lower labial furrows. Small upper and lower teeth similarly shaped, their sharp, oblique points bent toward outer corners of mouth forming a nearly continuous, smooth cutting edge along each jaw. First dorsal fin located about midway between pectoral and pelvic origins and behind rear tips of pectoral fins; second dorsal fin about two-thirds as large as first and posterior to pelvic fins. Sharp, ungrooved dorsal fin spines at anterior margins of dorsal fins, first about one-half as long and second nearly as long as anterior margins of their respective fins. Narrow, falcate pectoral fins form nearly equilateral triangles; rear tips rounded and posterior margins slightly concave. Pelvic fins closer to second dorsal than to first. Subterminal notch absent from long upper caudal lobe, lower lobe poorly developed. Well-developed upper precaudal pit and low lateral keels on caudal peduncle. Small, low denticles with tricuspid, leaf-shaped crowns in adults; central ridge robust; lateral extensions winglike.

Meristics. Upper teeth 28, lower teeth 22–24; precaudal vertebrae 79–85, caudal vertebrae 26–32, total vertebrae 108–117 (Springer and Garrick 1964).

Color. The upper surface is slate-colored, sometimes tinged brown, with a lateral row of small white spots on each side from the level of the pectoral fin to above the pelvic fins; a few additional white spots are found anterior and posterior to the first dorsal and anterior to the second dorsal fin. These spots are most conspicuous on small fish up to 30–36 cm in length and they fade with growth until they disappear altogether in some individuals. The margins of the first and second dorsals and of the caudal are more or less dusky at birth, but soon fade. The lower surface ranges from pale gray to pure white.

Size. Average size is about 70–100 cm. Most adult males are about 60–90 cm; adult females 76–107 cm. Males achieve a maximum length of 100 cm and females 124 cm. Mature females weigh 3.2–4.5 kg and reach a maximum of 9.8 kg. The all-tackle game fish record is a 7.14-kg fish caught in Kenmare Bay, County Kerry, Ireland, in May 1989 (IGFA 2001).

Distinctions. Any small gray or brownish shark with large, sharp dorsal spines caught within the Gulf or on the shallower portions of the offshore fishing banks is almost certain to be a spiny dogfish. They are most likely to be confused with two relatives, black dogfish (*Centroscyllium fabricii*) and great lanternshark (*Etmopterus princeps*), which are regular inhabitants of the deeper slopes of the offshore banks that front the Gulf. There is no danger of confusing abundant spiny dogfish with either of these rare species, for the latter are velvety black in color, their upper caudal lobes include a subterminal notch (not present in *Squalus*), and their upper teeth (and lower teeth in the *Centroscyllium*) are erect with three to five sharp points (oblique, with a single point in *Squalus*).

Habits. Spiny dogfish are the most abundant shark not only in the Gulf of Maine, but also in the entire western North Atlantic. They are usually found epibenthically, but move throughout the water column to surface waters. Dogfish are widely distributed in inshore and offshore shelf waters from nearshore shallows to depths of 900 m. Although they tolerate brackish water, they do not ascend estuaries into freshwaters and prefer full-strength seawater.

This is one of the most gregarious of Gulf fishes, swimming in large schools or packs. Dogfish of a size-class continue to associate together as they grow, the result being that any given school runs very even, consisting as a rule either of very large mature females, or of medium-sized fish (either mature males or immature females), or of small immature fish of both sexes in about equal numbers. Schools of juveniles are most common offshore, whereas schools of mature females are abundant inshore. Dogfish are not particularly swift swimmers, but they maintain a strong, slow pace.

Apart from their general seasonal migratory movements, dogfish are governed by the movements of the fishes on which they prey. They seldom stay in one place long, but there is seldom, if ever, a time during the summer when they are not common on some part of the Gulf of Maine coast. So erratic are their appearances and disappearances that where the fishing is good today, there may be only dogfish caught tomorrow, and nothing at all the day after, the other fishes having fled these predators and the latter departing in pursuit.

Dogfish use their dorsal spines for defense, curling around in a bow and striking, which makes them hard to handle on the hook. It is probable, too, that the spines are slightly poisonous, general reports to this effect being corroborated by the fact that the concave surfaces are lined with a glandular tissue resembling the poison glands of the venomous weever fish (*Trachinus draco*) of Europe.

Food. Voracious almost beyond belief, dogfish deserve their bad reputation among many fishermen. Not only do they harass and drive off mackerel, herring, and even fish as large as cod and haddock, but they consume vast numbers of them. Fishermen repeatedly have described packs of dogfish dashing among schools of mackerel and even attacking them within the seines, biting through the net, and releasing much of the catch. Fishes constituted the major food category in the stomachs of 2,662 spiny dogfish from the western Atlantic, 53.7% by weight, followed by mollusks, 26.6% (Bowman et al. 2000). Schooling, pelagic fishes such as herring, menhaden, river herrings, capelin, sand lance, and mackerel are heavily consumed, but such solitary, benthic fishes as wolffish and several species of flatfishes, routinely fall prey to this efficient predator (Bowman et al. 2000). In the Gulf of Maine, squid (24.7% by weight) and sand lance (12.2%) are of particular importance in the diet. The diet of spiny dogfish in the Irish Sea was similar (Ellis et al. 1996), with smaller individuals (<60 cm) consuming more crustaceans and fewer teleosts than larger size-groups. Dogfish on both coasts also take a variety of invertebrates, including ctenophores, jellyfish, polychaetes, sipunculids, amphipods, shrimps, crabs, snails, octopods, squids, and sea cucumbers. Dogfish are usually very thin when they reappear on the coast in spring, suggesting that they rarely feed during their winter stay in deep waters.

Predators. Spiny dogfish have been found in the stomachs of four species of fishes (Rountree 1999): spiny dogfish (five times), cod (three times), red hake (twice), and goosefish (six times). Larger species of sharks may be the primary natural predators of spiny dogfish. Seals and killer whales probably consume fewer numbers.

Reproduction. Fecundity increases with the size of females, but may drop owing to senescence at about 109 cm. Ovarian eggs develop during the year prior to the first fertilization, as well as during the last year of pregnancy, thereby allowing females the opportunity to mate again shortly after giving birth. Nonovulation of a certain proportion of ova occurs regularly, leading to the presence of very large, white ova in ovaries of mature females.

Copulation typically occurs offshore. Ova are fertilized internally and development is ovoviviparous. During early stages of development, ova in each horn of the uterus are encased in a membranous, horny capsule called a *candle*. Fecundity regressions for the number of candled embryos (C) on TL are calculated as: $\log_e C = 3.33 \log_e \mathrm{TL} - 13.22$, $r^2 = 0.36$.

Later, some 4–6 months into development, the membrane breaks down, leaving the 7–9 cm yolk-sac embryos free, without placental attachment to the uterine wall, for the remaining 17–19 months of development. Young are born headfirst. The number in a litter averages six or seven, but may range from one to fifteen. Litter size increases with increasing maternal length and age and with decreasing stock density. Fecundity regressions for number of postcandle embryos (E) on TL are calculated as: $\log_e E = 3.80 \log_e \mathrm{TL} - 15.49$, $r^2 = 0.40$

Young of both sexes are generally 26–27 cm long at birth, but range from 20 to 33 cm. Birth size of the young increases slightly with size (and age) of the female. The sex ratio of newborns is 1:1.

Females carry their young for 18–22 months. Accordingly, September-caught adult females from the Gulf contain either very early embryos, averaging only about 1.9 cm in length, or much larger, nearly full-term embryos, 18–28 cm long. Females bearing embryos 23–26 cm long have been taken in November on Cholera Bank near New York Harbor. Parturition occurs mostly on the offshore wintering grounds from November to January, but newborns are rarely taken along southern New England or in the Gulf in early summer, off Gloucester in July, or on Nantucket Shoals in August, showing that the season of production occasionally extends through the spring, or even into the summer. In Newfoundland waters, most of the young are born between January and May, perhaps because colder water temperatures slow embryonic growth rates.

Spiny dogfish have an extended adolescent period. Most females mature at 12 years and lengths of about 78 cm, but some may mature at 76 cm. Although a few males reach maturity at lengths as short as 58 cm, most mature at 6 years and about 60 cm.

Age and Growth. Spiny dogfish, like many sharks, are slow-growing and long-lived. Dorsal spine circuli are used to age dogfish: single rings are laid down annually. Males and females grow at similar rates until age 7, but thereafter females grow faster and larger than males. The calculated von Bertalanffy growth equations are: $L_t = 100.5(1 - e^{-0.1057(t + 2.90)})$ for females; $L_t = 82.49(1 - e^{-0.1481(t + 2.67)})$ for males.

Growth occurs at 1.5–3.5 cm·year^{-1}, according to tagging and spine-growth studies. Maximum ages reported for males and females are 35 and 40 years, respectively (Sosebee 1998a), but some have suggested longevity to 100 years. Both sexes show a well-defined length-weight inflection at 50 cm TL. Length-weight regressions are: $Y = 0.0119TL - 0.2676$, $r^2 = 0.91$, for TL < 50 cm females; $Y = 0.0293TL - 1.1714$, $r^2 = 0.96$, for TL > 50 cm females; $Y = 0.0114TL - 0.2532$, $r^2 = 0.92$, for TL < 50 cm males; $Y = 0.0235TL - 0.8535$, $r^2 = 0.91$, for TL > 50 cm males.

General Range. Spiny dogfish are a widely distributed, antitropical species that occurs in the temperate and boreal regions of the North and South Atlantic and in the North and South Pacific. In the western North Atlantic, they have been recorded from Greenland to northeastern Florida. A record of this species from Cuba probably refers to the related *Squalus cubensis*. In the eastern North Atlantic, they range from the Barents Sea and Iceland south to West Sahara and the Canary Islands, including the Mediterranean and Black seas. In the Southern Hemisphere they occur in South Africa, Uruguay, Argentina, Chile, New Zealand, and southeastern Australia. They are widely distributed throughout the North Pacific, from Baja California to the Bering Sea to Korea and northeastern China.

Occurrence in the Gulf of Maine. Spiny dogfish make up for the comparative rarity of other sharks in the Gulf of Maine by their overwhelming abundance. To mention all the localities from which they have been reported would be simply to list every seaside village and fishing ground from Cape Cod to Cape Sable (Map 3). They are also common on the offshore banks.

Migrations. Dogfish are one of the most highly migratory species of the Atlantic coast. They primarily occur north of Cape Cod in the summer, move southward to Long Island in the fall, and go as far south as North Carolina (rarely to northern Florida) in the winter. In the spring they migrate northward, reaching Georges Bank in March and April. In the inner parts of the Gulf of Maine the date of the first heavy run of dogfish varies widely from year to year and from place to place, but never before May. It is usually not until June that they arrive in numbers in the Massachusetts Bay region. They usually populate the northern Maine and western Nova Scotia coasts by the end of June, but few are seen in Passamaquoddy Bay until late in July. They have been recorded as early as 1 July near Raleigh, on the Newfoundland side of the Strait of Belle Isle, but they are not caught in any numbers in the inner parts of the Gulf of St. Lawrence until well into July, and they have not been reported from southeastern Labrador until early September. Schools of adult females are the first to enter inshore waters, followed later by adult males.

In the southern part of its range, from North Carolina to New York, spiny dogfish are merely spring and autumn migrants. West of Cape Cod, at Woods Hole, and along Long Island, they are mostly transients, passing north in spring and south in autumn; some do summer there, even in considerable numbers in certain years. It also seems that most of them withdraw from Massachusetts Bay during the warmest months, for few are taken there between June and September. But they are present all summer along outer Cape Cod and here and there throughout the northern and eastern parts of the Gulf, in varying abundance.

Most dogfish depart from the inner parts of the Gulf during October, few being caught on the coast north of Massachusetts Bay after November 1. But they sometimes stay later, as in 1903 and 1942, when they were abundant along the outer shore of Cape Cod as late as the first week of November. Ordinarily none are caught within the Gulf of Maine north of Georges Bank in winter. In 1882, schools were reported off Portsmouth, N.H., even as late as February, an exceptional event.

Dogfish appear earlier in spring and linger later into the winter on Georges Bank than in the inner parts of the Gulf. It is safe to say that there are few there in March, the earliest definite record being between the 20th and the 22nd, while some are trawled there all summer. A few were taken in November and December 1913, and 20–22 January 1914. Apparently dogfish reach Browns Bank later than Georges Bank, for none were taken there on 14 April 1913, although they are plentiful there in summer. It is also likely that they depart earlier, but a few lingered as late as 3–12 December 1913 on Western Bank off Halifax.

Spiny dogfish also exhibit inshore-offshore movements. They spend summers in inshore waters and overwinter in deeper water offshore. Considerable numbers have been trawled in the winter on the outer part of the continental shelf off Block Island, in 90–120 m, during the last week of January; off New York in November and January; and in February off the Middle Atlantic coast in 29–128 m, south as far as Cape Hatteras. On the other hand, the fact that numbers of them have been found washed ashore in January on the southwest coast of Newfoundland suggests that some of those that summer in that general region may survive the winter in the deep trough of the Gulf of St. Lawrence.

Seasonal inshore-offshore movements and coastal migrations are tied to bottom water temperatures. Spiny dogfish prefer temperatures of 6°–8°C and are seldom found in waters warmer than 15°C.

Tagging studies have shown that some dogfish cover long distances in their wanderings. Tagged individuals have traveled from Newfoundland to Virginia (2,092 km) and Iceland (1,421 km) and from Block Island to north Florida (1,802 km). Transoceanic journeys of 1,600–6,500 km have been documented in the North Atlantic and North Pacific.

Importance. During the years when the ground fishery was chiefly by hook and line, fishing often was actually prevented by dogfish in Massachusetts and Ipswich bays, unless moon snail (*Polynices*) were used for bait, for dogfish did not take these. Often, too, they bit groundfish from the hooks of long-

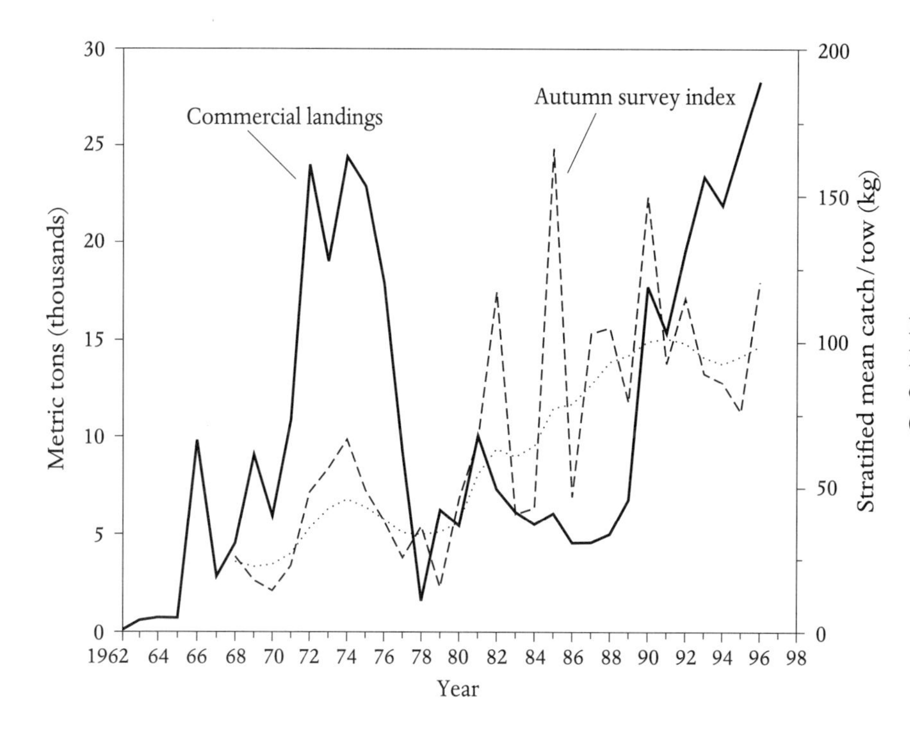

Figure 39. Commercial landings and NEFSC autumn survey index of spiny dogfish *Squalus acanthias,* 1962–1996. (From Sosebee 1998a.)

lines or took the baits and made it futile to fish with hook and line where they abounded. Replacement of hook-and-line fishing by the otter trawl has put an end to widespread complaints on this score. But when schools of dogfish get into a net or seine, they so snarl the twine that disentanglement and repair may be the work of days.

Total landings peaked at 24,700 mt in 1974, declined sharply to a fairly stable average of about 5,900 mt per year during 1977–1989, and then increased sharply to over 17,000 mt in 1990, and over 28,000 mt in 1996 (Fig. 39; Sosebee 1998a). Distant water fleets accounted for virtually all of the reported total from 1966–1977.

Historically, a longline, with 1,500 hooks, has been known to bring in a dogfish on nearly every hook and an average trawl catch of 6,000–8,000 per trip was made on Georges Bank in 1913 during their season of abundance. At the time of the 1904–1905 peak, it was estimated from recorded catches that at least 27,000,000 were being taken yearly off the coast of Massachusetts. The principal commercial fishing gears currently in use are otter trawls and sink gill nets (Sosebee 1998a).

Stocks. Spiny dogfish stocks must be managed very carefully because, like other sharks, they are slow to mature, have a prolonged gestation period, produce limited numbers of offspring, and grow slowly. They are typical "K" strategists, and as such they are extremely vulnerable to overfishing.

This is the only Gulf of Maine shark that rivals the important food fishes in numbers. The minimum swept-area total biomass estimates of spiny dogfish based on NEFSC spring bottom trawl surveys increased steadily from about 150,000 mt in 1968 to about 600,000 mt in 1990 and have since been stable (Sosebee 1998a). Declining abundance as shown by trends in commercial catch per unit effort and research vessel survey indices and declines in average length in commercial landings suggest that this stock is overexploited (Sosebee 1998a). There is evidence that recent fishing pressure on mature females has altered the size and sex structure of the population.

TORPEDOES, SKATES, AND RAYS. ORDER BATOIDEI

Batoid fishes are distinguished by their greatly flattened and expanded body form. The pectoral fins are attached to the sides of the head to form a disk, eyes and spiracles are on the top or the sides of the head, and the gill slits are located on the ventral surface behind the head. Pelvic fins are slightly to moderately expanded, the anal fin is absent, and dorsal fins are either present or absent. If present, dorsal fins are either one or two in number and are small to moderate in size. Four orders, 12 families, 70 genera, and about 456 species are known (Nelson 1994; McEachran and Dunn 1998). Three of the four major groups of batoids occur in the Gulf of Maine: torpedoes, skates, and stingrays. Torpedoes (family Torpedinidae) possess a nearly circular disk, a stout tail, and electric organs just posterior and lateral to the spiracles. Skates (family Rajidae) have a subcircular to trapezoid-shaped disk, a rather slender tail, and electric organs along the sides of the tail. Stingrays and manta rays (families Dasyatidae, Rhinopteridae, and Mobulidae) have a subcircular to trapezoidal disk, a whiplike tail, and generally a serrate spine near the base of the upper side of the tail.

Batoids are considered to be benthic fishes, except for one stingray and the manta rays, which are epipelagic. Food of the batoids consists of a large variety of benthic invertebrates and fishes, except for the manta rays, which consume small pelagic crustaceans and fishes. Swimming is accomplished by ripple-like undulations or flapping of the pectoral fins, except for the torpedoes, which swim by lateral undulations of their stout tails. Fertilization is internal as in sharks, and development ranges from oviparous to viviparous without a placenta. The inner edge of the pelvic fins in males is modified into a semitubular clasper used to inseminate the females. Batoids are generally associated with continental and insular shelves from the shoreline to depths of 3,000 m. Most species are associated with soft bottoms, although some stingrays occur around coral reefs.

KEY TO GULF OF MAINE TORPEDOES, SKATES, AND RAYS

1a. Front of head with a pair of separate, lobelike fins, extending forward . . . . **Manta**

1b. Front of head without a pair of separate lobelike fins extending forward **2**

2a. Tail rather stout, with two well-developed dorsal fins and a large triangular caudal fin. . . . . **Atlantic torpedo**

2b. Tail rather slender, with two relatively small or no dorsal fins and an indistinct or no caudal fin **3**

3a. No serrate spine(s) on tail (Rajidae). **4**

3b. One or more serrate spines on tail **16**

4a. Upper surface of disk marked with conspicuous black rosettes . . . . **Rosette skate**

4b. Markings on upper side of disk not in black rosettes **5**

5a. No conspicuous thorns along middorsal zone from level of spiracles to axil of pectoral fins **6**

5b. One or more conspicuous thorns along middorsal zone of disk rearward from the spiracles. **8**

6a. Rostral cartilage stout, extending to about one-half its length beyond most anterior pectoral rays; lower surface of disk marked with dark dots and dashes at openings of ampullar pores of electric receptors and lateral line pores . . . . **Barndoor skate**

6b. Rostral cartilage delicate, extending little beyond anterior most pectoral rays; lower surface of disk not marked with dark dots and dashes **7**

7a. Upper and lower surfaces of disk dark colored and bear dermal denticles; midline of tail with 12–22 thorns, no thorn between dorsal fins . . . . **Deepwater skate**

7b. Upper surface dark, bearing dermal denticles, lower surface light colored, lacks dermal denticles; midline of tail with 21–26 thorns, one or more thorns between dorsal fins . . . . **Spinytail skate**

8a. Upper surface of disk and all but lower midline of tail densely covered with fine dermal denticles; rear one-quarter to one-third of tail lacks thorns in medium to large-sized and some small specimens; small specimens with thorns on posterior section of tail have two pale yellow cross bars on tail **9**

8b. Upper surface of disk and upper and lateral sides of tail, at most, sparsely covered with dermal denticles; all specimens have one or more rows of thorns on posterior one-quarter to one-third of tail; small specimens lack pale yellow crossbars on tail **10**

9a. No thorns on disk posterior to shoulder girdle; small specimens lack yellow crossbars on tail . . . . **Soft skate**

9b. Thorns present along midline of disk and along anterior three-quarters to one-third of tail; small specimens have an entire row of midrow tail thorns and two pale yellow crossbars on tail . . . . **Smooth skate**

10a. Thorns on midrow of tail much larger than other tail thorns and have radiate bases **11**

10b. Thorns on midrow of tail not much larger or more conspicuous than other thorns of tail and lack radiate bases **12**

11a. Midrow thorns between nuchal region and origin of first dorsal fin 10–19 . . . . **Thorny skate**

11b. Midrow thorns between nuchal region and origin of first dorsal fin 24–31 . . . . **Shorttail skate**

12a. Only one row of large thorns along midzone of disk from nape to level of axil of pectoral fins; first and second dorsal fins separated by distinct interspace and one to several thorns; upper surface of disk marked with short, dark bars and roundish spots . . . . **Clearnose skate**

12b. At least three rows of thorns along midzone of disk from nape to level of pectoral fin axil; first and second dorsal fins not separated by distinct interspace and have one or more thorns; upper surface of disk not marked with dark bars but variously spotted or plain-colored **13**

13a. Upper and lower surfaces of disk and tail dark-colored . . . . **Chocolate skate**

13b. Upper surface of disk dark-colored, lower surface light-colored **14**

14a. Upper surface of disk plain, occasionally blotched or spotted; tail length 60% of total length; upper jaw tooth rows 30–38 . . . . **Round skate**

14b. Upper surface of disk covered with small round or oval spots; tail length less than 60% of total length; upper jaw tooth rows more than 35, except for specimens less than 160 mm TL . . . . **15** (note: there are 8 couplets based on size-classes)

15a. Specimens ranging from 90 to 160 mm TL. Snout blunt, extending little beyond anteriormost margin of pectoral fins; upper jaw tooth rows 44–55; generally 21 or more thorns along midline of tail from axil of pectoral fins to origin of first dorsal fin . . . . **Winter skate**

15b. Snout obtuse, extending clearly beyond anteriormost margin of pectoral fins; upper jaw tooth rows 30–48; generally less than 21 thorns along midline of tail from axil of pectoral fin to origin of first dorsal fin . . . . **Little skate**

15c. Specimens ranging from 161 to 214 mm TL. Snout blunt, extending little beyond anteriormost margin of pectoral fins; mouth slightly arched anteriorly; upper jaw tooth rows 50–60; generally 21 or more thorns along midline of tail from axil of pectoral fin to origin of first dorsal fin . . . . **Winter skate**

15d. Snout obtuse, extending clearly beyond anteriormost margin of pectoral fins; mouth strongly arched anteriorly; upper jaw tooth rows 36–53; generally fewer than 21 thorns along midline of tail from axil of pectoral fins to origin of first dorsal fin . . . . **Little skate**

15e. Specimens from 215 to 350 mm TL. Distance between orbits less than 12 times in tail length; mouth smoothly arched anteriorly; upper jaw tooth rows 58–70; generally 21 or more thorns along midline of tail from axil of pectoral fins to origin of first dorsal fin . . . . **Winter skate**

15f. Distance between orbits more than 12 times in tail length; mouth abruptly arched anteriorly; upper jaw tooth rows 43–52; generally fewer than 21 thorns along midline of tail from axil of pectoral fins to origin of first dorsal fin. . . . . **Little skate**

15g. Specimens 351 mm TL and longer. Distance between orbits less than 12 times in tail length; mouth smoothly arched anteriorly; upper jaw with 63 or more tooth rows; median row of tail thorns usually present; males lack alar spines on outer corners of disk; both sexes lack denticles on ventral side of pelvic fins; claspers of males do not extend posterior to posterior lobe of pelvic fins . . . . **Winter skate**

15h. Distance between orbits more than 12 times in tail length; mouth abruptly arched anteriorly; upper jaw with fewer than 64 and generally less than 54 tooth rows; median row of tail thorns behind scapular thorns partially or totally absent; males possess alar spines and claspers that extend beyond posterior lobe of pelvic fins; females possess denticles on ventral side of pelvic fins . . . . **Little skate**

16a. Small dorsal fin on upper side of tail, anterior to dorsal spine or spines; head elevated from disk, eyes and spiracles on sides of head; jaw teeth 7–9 series of flattened plates . . . . **Cownose ray**

16b. No dorsal fin on tail; head level with disk, eyes and spiracles on top of head; teeth in many series, in a mosaic arrangement 17

17a. Anterior margin of disk evenly convex; ventral surface of disk uniformly brown to black . . . . **Pelagic stingray**

17b. Anterior margin of disk subangular, tip of snout forming apex of angle; underside of disk white to whitish . . . . **Roughtail stingray**

TORPEDOES OR ELECTRIC RAYS. FAMILY TORPEDINIDAE

JOHN D. MCEACHRAN

Electric rays, or torpedoes, have a thick (fleshy) subcircular to oval-shaped disk and a stout tail bearing two relatively large dorsal fins and a large caudal fin. They are distinguished from the other batoids of the Gulf of Maine in possessing kidney-shaped electric organs along the anterior-lateral aspects of their disk. The electric organs give the overlying skin a honeycomb appearance, which is easily seen externally. The electric organs constitute about half of the lateral portion of the disk. There are two genera and some 14 species of electric rays (Nelson 1994), one of which occurs in the Gulf of Maine.

ATLANTIC TORPEDO / *Torpedo nobiliana* Bonaparte 1835 / Bigelow and Schroeder 1953:58–60

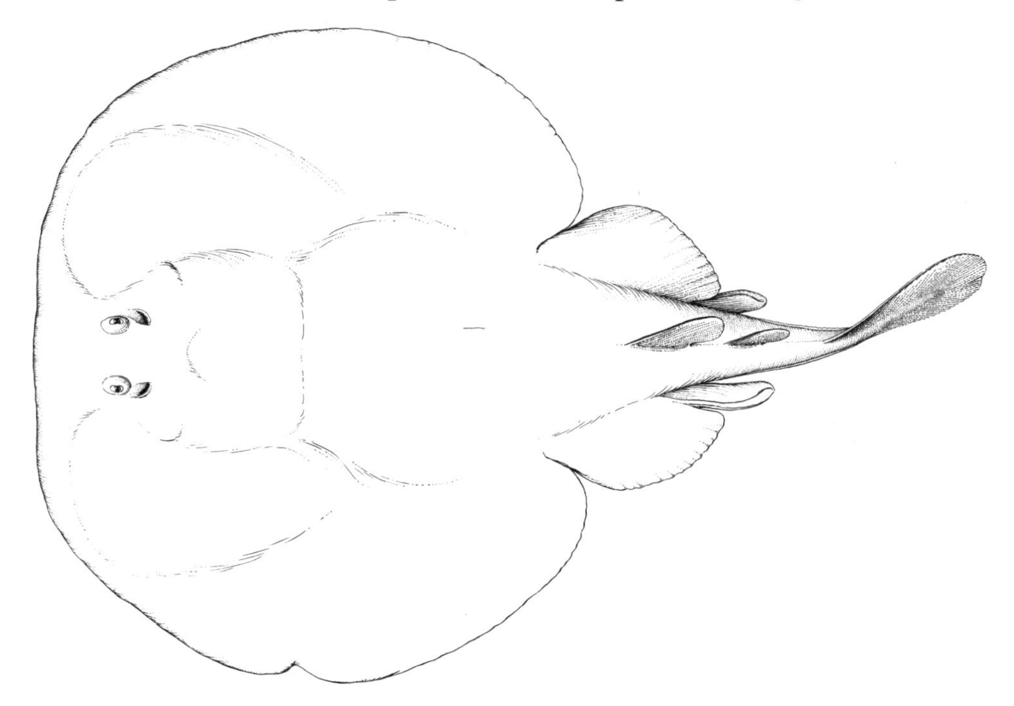

Figure 40. Atlantic torpedo *Torpedo nobiliana.* Off Plymouth, Mass., male, 835 mm, MCZ 36040. Drawn by E. N. Fischer. (From Bigelow and Schroeder 1953a: Fig. 22, © Sears Foundation for Marine Research.)

Description. Disk subcircular, snout short and blunt, tail relatively short and stout (Fig. 40). Eyes relatively small, located on top of head in front of well-developed spiracles. Nostrils immediately in front of relatively large mouth, partially covered by nasal curtain. Number of tooth rows in upper and lower jaws 38 and 66 in juveniles and adults, respectively. Teeth small, with broad oval bases and single sharp pointed cusp. Up to seven tooth rows exposed and functioning at any given time. Gill slits small. Electric organs kidney-shaped, occupying about one-half of lateral aspect of disk. Pelvic fins originate near axil of pectoral fins, slightly overlapped by distal part of pectoral fin. Dorsal fins relatively large, located on anterior one-half of tail. First dorsal fin about one-third to one-half larger than second dorsal fin, separated from second dorsal fin by a distance equal to about three-quarters of its base. Caudal fin relatively large and triangular-shaped with a vertical posterior margin. Body and fins lack denticles and thorns.

Color. The dorsal surface is dark to purplish brown, with or without obscure darker spots. The ventral surface is white with a dark brown to purplish brown margin and tail.

Size. Adult Atlantic torpedo range from 30 to 180 cm TL. Specimens captured off Woods Hole average 13.6 kg and most specimens captured in the western North Atlantic are less than 34 kg. The largest individuals known to Bigelow and Schroeder were a 120-cm TL, 45.4-kg individual captured off Nantucket and brought to Woods Hole; a 65.3-kg specimen captured off Nantucket; and a specimen estimated to be 77–90 kg captured near Provincetown.

Distinctions. The electric organs in the lateral aspects of the disk and the large tail distinguish Atlantic torpedo from other batoids in the Gulf of Maine.

Biology. Atlantic torpedo are benthic and generally restricted to the continental shelf, although they have been captured at depths to 580 m. There is little specific information about their behavior, but, like other electric rays, they are probably rather sluggish, spending considerable time partially buried in soft bottoms, where they are able to stun prey with their electric organs. The relatively large distensible mouth enables them to consume relatively large prey. Food apparently consists solely of fishes including silver hake (Bowman et al. 2000) and summer flounder. Individuals captured in British waters contained a 1-kg eel, a 0.5-kg flounder, a 1.9-kg salmon, a red mullet (*Mullus surmuletus*), a plaice (*Pleuronectes platessa*), and a small catshark (*Scyliorhinus*) (Day 1880–1884). The electric organs of Atlantic torpedo are capable of generating 50 volts of electricity in the air (Bennett 1971).

Reproduction. Development is viviparous without a placenta, and one female contained a single embryo. Based on slim evidence, the gestation period is about 1 year and birth takes place over the outer continental shelf. Size at birth is about 20–25 cm TL.

General Range. Atlantic torpedo occur in temperate to tropical Atlantic waters, from southern Nova Scotia (rare) to North Carolina, and rarely in the Florida Keys, northern Gulf of Mexico, Cuba, Trinidad, Panama, and Venezuela in the west; and from northern Scotland to the Mediterranean, Azores, Madeira, and tropical west Africa in the east (Bigelow and Schroeder 1962, 1965; Stehmann and Bürkel 1984a; Scott and Scott 1988; McEachran and Fechhelm 1998).

Occurrence in the Gulf of Maine. Atlantic torpedo are visitors to the Gulf of Maine, although they are more abundant along the Atlantic seaboard southwest of Cape Cod. They usually visit the Gulf of Maine during the summer. The most northern records of this species are from La Have Bank in 1890; St. Margaret Bay, N.S., around 1915; and from the lower Bay of Fundy off Eastport, Maine. They have also been recorded from off Williamsport, Maine; Seguin Island in 1880; from the mouth of Casco Bay; at Wood Island, near Cape Elizabeth in 1894; near Cape Ann; off Plymouth on the southern side of Massachusetts Bay; near Provincetown; and along the outer coast of Cape Cod.

Abundance of the Atlantic torpedo in the Gulf of Maine varies from year to year. Bigelow and Schroeder reported them as abundant off Provincetown from 1819 to 1823 or 1824 with 60–80 specimens captured per year. They were again abundant off Provincetown in 1845 when 12 specimens were taken. No further records were discovered until W.C. Kendall of the U.S. Fish Commission collected several specimens from the coast of Maine in 1896. Since then only one specimen is known to have been captured off Massachusetts, from Swampscott Harbor in 1981 (Collette and Hartel 1988).

Importance. Atlantic torpedo are of no commercial importance today but before the use of kerosene, the liver oil was considered as good as sperm whale oil for lighting. It was also used externally for muscle cramps and internally for stomach cramps.

SKATES. FAMILY RAJIDAE

JOHN D. MCEACHRAN

Skates have a relatively thin, subcircular to trapezoid-shaped disk and slender to very slender tail that is clearly marked off from the disk. The rostrum extends to the tip of the snout and the pectoral fins extend in front of the nostrils to near the tip of the snout. Pelvic fins are moderately laterally expanded, generally divided into anterior and posterior lobes and partially overlapped by the posterior extension of the pectoral fins. The tail is generally one-half or more of total length, usually slightly more slender distally, bearing two small dorsal fins and a poorly developed caudal fin near its tip. The caudal fin usually consists of a low dorsal lobe and a very short, low ventral lobe or a ventral ridge. Serrate dorsal spines are not present on the tail but lateral electric organs, which are not externally evident, are present. Most species bear small dermal denticles and larger thorns on various aspects of the upper, and occasionally lower, disk. Within species, females generally attain a larger size, and males generally have a more acute snout, undulated disk, more arched jaws, teeth with sharper cusps, and alar and malar thorns on the outer corners of the upper disk. The tips of the alar thorns are often embedded in the integument but are exposed when the tips of the pectoral fins are bent ventrally (McEachran and Konstantinou 1996).

The family Rajidae contains about 26 genera and more than 200 species (McEachran and Dunn 1998), nearly half the species in the order. Eight species of skates have been recorded from the Gulf of Maine but only three of these (winter, little, and thorny) are common (Fig. 41). Five more species have been captured in deep waters nearby and have been included in the key because they might occasionally wander into the Gulf. Two of these additional five species of skates have been reported from off southern Nova Scotia (Bigelow and Schroeder 1954; Schroeder 1955): spinytail skate, *Bathyraja spinicauda* (Jensen 1914) and chocolate skate, *Rajella bigelowi* (Stehmann 1978). Spinytail skate and three more species of skates have been reported from the eastern slope of Georges Bank (Bigelow and Schroeder 1954; Schroeder 1955; Templeman 1965a): round skate, *Rajella fyllae* (Lütken 1887); shorttail skate, *Amblyraja jenseni* (Bigelow and Schroeder 1950); and soft skate, *Malacoraja spinacidermis* (Barnard 1923). Until recently, all 13 species of skates known from the region (except for the deepwater and spinytail skates, *Bathyraja richardsoni* and *B. spinicauda*) were placed together in the genus *Raja*. However, in a comprehensive analysis of the family, McEachran and Dunn (1998) moved all but the clearnose skate into different genera, *Amblyraja* (shortail and thorny skates), *Dipturus* (barndoor skate), *Leucoraja* (little, rosette, and winter skates), *Malacoraja* (smooth and soft skates), and *Rajella* (chocolate and round skates).

Common skates are very similar in overall shape and are of little commercial importance, so fishermen rarely distinguish among them. For this reason little is known concerning specific life styles. All are closely associated with soft to gravelly bottoms and often are partially buried in the mud or sand. They swim by undulating the raylike supports of their pectoral fins and by pushing themselves over the bottom on their limblike anterior pelvic fin lobes. They maneuver by altering the rhythm of the right and left pectoral fins and by means of their long tails and dorsal and caudal fins. All species are thought to feed mostly on benthic organisms such as polychaetes, amphipods, isopods, shrimps, crabs, lobsters, bivalve mollusks, and fishes. Pelagic organisms such as euphausiids, squids, and fishes are also consumed, but presumably these are captured either near or on the bottom. Some species appear to have a greater preference for infauna while other species apparently prefer epifauna.

Skates, like other elasmobranchs, practice internal fertilization with the claspers, modified pelvic fins, but unlike most other elasmobranchs, skates are oviparous. Fertilized eggs are encapsulated in leathery egg capsules, *mermaid's purses,* which are subsequently deposited on the bottom. There is usually

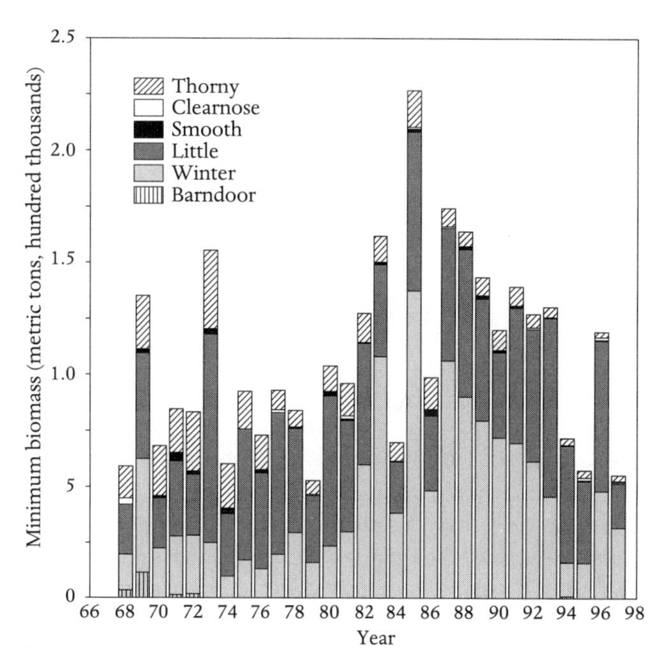

Figure 41. Species composition of skates, NEFSC spring surveys, 1968–1997. (From Sosebee 1998b.)

one egg per capsule but some species deposit several eggs in one capsule. The egg capsules are amber or light brown in color upon leaving the female but become greenish brown to blackish after prolonged exposure to seawater. They are oblong in shape with a hollow curved horn at each corner. The capsule walls are convex medially but flattened laterally. Adhesive threads are loosely attached to the surface of the capsules, and these, in addition to the horns, assist in fastening the capsules to the bottom. During incubation the embryos gradually develop a broader disk and shorter tail. External gills appear from the walls of the gill slits but these disappear before hatching (Pelster and Bemis 1992). Incubation lasts from several months to a year or more. It is likely that the local skates reproduce year-round but most intensely during the summer. Empty egg capsules are frequently seen along the shoreline.

Historically skates have been relatively abundant in the Gulf of Maine. Bigelow and Schroeder reported that several trawlers on 25 trips captured from 82 to 4,520 (an average of about 800) skates on 4- to 7-day trips to Georges Bank during January and February 1913. On a trip to the northeastern part of Georges Bank the trawler *Kingfisher* in September 1929 caught from 0 to 105 skates per haul (total 495) in 37 hauls; and the trawler *Eugene H* fishing from the Nantucket Lightship to the south-central part of Georges Bank caught an average of 146 skates per haul (total 6,130 skates) on 46 hauls. Records of the *Eugene H* suggest that there were 9 or 10 skates per acre on Georges Bank and they are probably equally abundant on Browns Bank. Bigelow and Schroeder also reported that skates were abundant inshore based on catches of 1 skate per 33 fishes on various kinds of inshore longline gear in the Gulf of Maine. In recent years, the abundance of skates has been reduced,

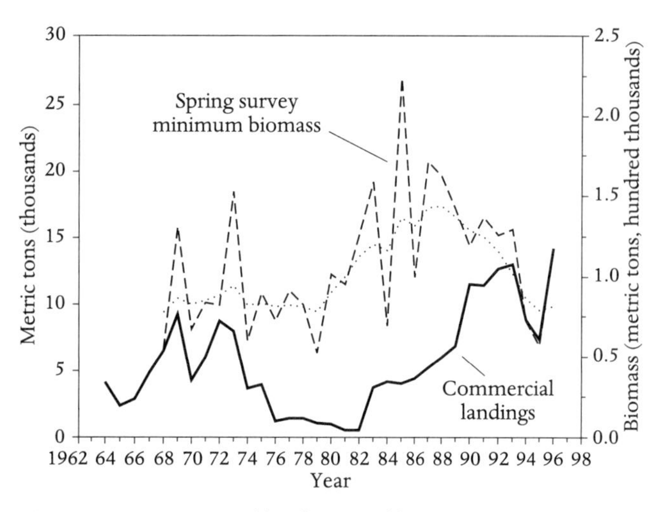

Figure 42. Commercial landings and biomass estimates of skates from the Gulf of Maine–Middle Atlantic region, 1963–1997. (From Sosebee 1998b.)

presumably by bottom-trawling operations, to the point where some researchers consider one species, the barndoor skate, to be "near extinction" (Casey and Myers 1998).

Skates used to be regarded as a nuisance in the Gulf of Maine and were seldom landed. Total landings of skates during 1947 and 1948 in Massachusetts and Maine were 12,800 kg and 26,900 kg, respectively. In northwestern Europe, skates are well regarded as a seafood with landings during the 1980s averaging 28,000–36,000 mt. In 1981, U.S. landings were 297 mt. With the development of European export markets for skate wings, U.S. landings increased to 4,000 mt in 1984 and peaked at 15,000 mt in 1996 (Fig. 42). The "wings" of skates are removed by fishermen at sea. Skates, like sharks, contain urea in their flesh and this can break down and release ammonia if the wings are not properly washed and iced. Skate wing fillets resemble an open fan with elongate, striated bands of muscle tissue. Cooked skate fillets have a texture similar to crab meat, with a very mild flavor like shellfish. The primary market for skate wings has been Europe but there appears to be a very slowly growing U.S. domestic market. Skates are generally very low in fat and cholesterol; 45–50% of the fat content is polyunsaturated.

THORNY SKATE / *Amblyraja radiata* (Donovan 1808) / Starry Skate /

Bigelow and Schroeder 1953:72–74 (as *Raja radiata*)

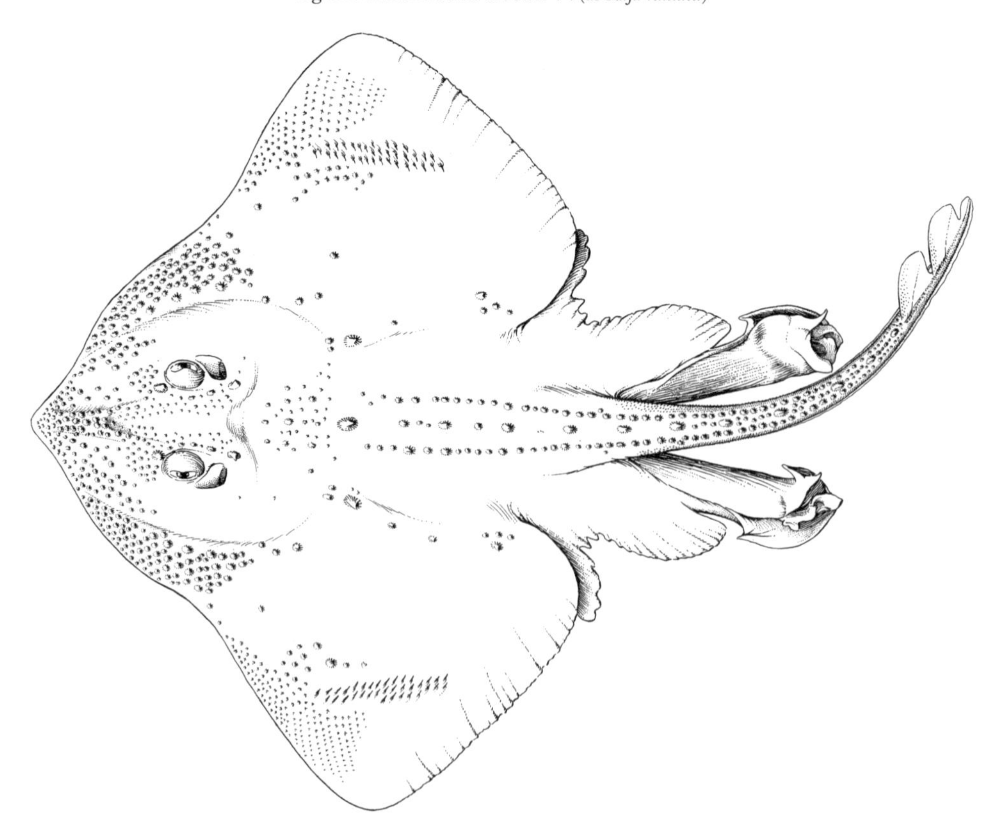

Figure 43. Thorny skate *Amblyraja radiata*. Massachusetts Bay, male, 585 mm TL, 420 mm disk width, MCZ 249. (From Bigelow and Schroeder 1953a: Fig. 56, © Sears Foundation for Marine Research.)

Description. Disk diamond-shaped, snout obtuse, tail moderately short (Fig. 43). Anterior margin of disk slightly concave from tip of snout to anterior extension of pectoral rays, convex from anterior extension of pectoral rays to level of anterior margin of orbits and concave from anterior margin of orbits to about level of first gill slits. Anterior margin of small specimens more or less straight. Outer corners of disk abruptly rounded. Posterior margins of disk nearly straight. Pelvic fins with a moderately broad and long anterior lobe and a moderately broad and short posterior lobe. Tail moderately broad, depressed at base, slightly longer or equal to distance from tip of snout to end of cloaca. Two small dorsal fins located near tip of tail, separated from each other by a short interspace. Small specimens nearly uniformly covered with dermal denticles. Large specimens lack denticles on central section, posterior section, and around axil of pectoral fins. Much of dorsal surface covered with medium- to large-sized thorns on radiate bases. Females more densely covered with thorns than males. Large thorns anterior to orbit, posterior to orbit, and medial to each spiracle. One or two thorns on nuchal region; one or two thorns over shoulder girdle. From 11 to 19 large thorns run in a continuous series from nuchal region to origin of first dorsal fin. Smaller thorns occur over all but center of each pectoral fin. Teeth of immature specimens in quincunx arrangement, with rounded cusps. Teeth of mature specimens arranged in rows, with sharp pointed cusps. Upper and lower jaws with 36–46 tooth rows.

Color. The dorsal surface is uniform brown, or brown-clouded, or spotted with darker or lighter pigment. The ventral surface is whitish with occasional dark blotches.

Size. Thorny skate reach 1,020 mm TL but maximum size varies over the range of the species. Individuals from Europe are smaller than those from Iceland and North America. Maximum recorded is 1,020 mm TL from Nova Scotia, 895 mm TL from Georges Bank, 800 mm TL from Massachusetts Bay, and 935 mm TL from off New Jersey (Scott and Scott 1988).

Distinctions. Thorny skate are distinguished from other skates of the Gulf of Maine by a combination of the following characters: Rostrum stout, extending distinctly anterior to the anteriormost pectoral rays. Thorns with radiate bases present in a single row along midline of disk and tail. Midrow thorns from nuchal region to origin of first dorsal fin range from 11 to 19.

Habits. Thorny skate are found over a wide variety of bottom types from sand, broken shell, gravel, pebbles to soft mud at depths of 18–1,200 m. The maximum depth recorded is 1,478–1,540 m in the northeastern Norwegian Sea (Stehmann and Parin 1994). They appear to be most common at 50–100 m (Stehmann and Bürkel 1984b) or 36–108 m (Scott 1982b). McEachran and Musick (1975) found them most abundant below 110 m. At the southern extreme of the range, they are limited to the continental slope. Off Virginia thorny skate occur at 300–1,200 m. Thorny skate have been captured at temperatures of −1.4° to 14°C (McEachran and Musick 1975). On the Nova Scotian shelf, they appear to prefer temperatures of 2°–5°C (Scott and Scott 1988). Based on tagging studies (Templeman 1984c), thorny skate live up to 20 years.

Food. Thorny skate feed on invertebrates and fishes that are associated with the bottom. Prey include hydrozoans, aschelminths, gastropods, bivalves, squids, polychaetes, pycnogonids, copepods, stomatopods (larvae), cumaceans, isopods, amphipods, mysids, euphausiids, shrimps, hermit crabs, crabs, holothuroideans, and fishes (McEachran et al. 1976; Pedersen 1995; Bowman et al. 2000). Individuals between 200 and 600 mm TL feed mostly on polychaetes (especially Aphroditidae), euphausiids, and decapods and those larger than 600 mm TL feed largely on fishes and squids (McEachran et al. 1976; Bowman et al. 2000). Fishes include herring, redfishes, sculpins, daubed shanny, wolffish, wrymouth, mackerel, sand lance, and flatfishes.

Predators. Thorny skate are eaten, at least as embryos in egg capsules, by halibut and Greenland shark (Jensen 1948) and by goosefish (Rountree 1999).

Parasites. Parasites include protozoans, myxosporid turbellarians, monogenean trematodes, cestodes, nematodes, and copepods, which occur in the skin, digestive tract, and body cavity (Scott and Scott 1988).

Reproduction. Thorny skate deposit a single fertilized egg in amber to brown egg capsules. The capsules are rectangular in outline. The dorsal surface is strongly convex and the ventral surface is nearly flat. The horns are stout and less than the length of the capsule excluding the horns. The anterior horns are curved inward and are shorter than the posterior horns. The surface of the capsule is covered with longitudinal rows of tubercles. Capsules range from 48 to 96 mm in length and 34 to 77 mm in width. Females with fully formed egg capsules are captured over the entire year, although the percentage of mature females with capsules is higher during the summer.

General Range. Thorny skate occur in the North Atlantic and eastern South Atlantic off South Africa. In the western North Atlantic, they range from western Greenland, Davis Straits, Hudson Straits, Hudson Bay, and Labrador to South Carolina. In the eastern North Atlantic, from Iceland, eastern Greenland, Barents Sea, and off the coast of Spitsbergen to the English Channel and the southwestern coasts of Ireland and England (Stehmann and Bürkel 1984b).

Occurrence in the Gulf of Maine. Thorny skate are the most common skate in the Gulf of Maine (McEachran and Musick 1975) but they avoid shallow waters (Map 4). They are frequently captured in the Bay of Fundy, along the entire coast of the Gulf of Maine, and on Georges Bank.

Migrations. This species does not appear to make seasonal migrations and is, in fact, apparently rather sedentary. Templeman (1984c) tagged over 700 individuals and noted that within 20 years most were recaptured no more than 97 km from the location at which they were tagged. A few were recaptured 161–87 km from the tagging site after 0.2–1 years. Similarly, in the North Sea 85% of tagged thorny skate were recaptured within 93 km of the release point and the longest distance traveled was 180 km (Walker et al. 1997).

Importance. Thorny skate are one of the five commercially important skates in the Gulf of Maine. Nutitional value: fat 0.7%, moisture 79.5%, protein 18.6%, ash 1.2%; cholesterol 55 mg% (Krzynowek et al. 1989).

DEEPWATER SKATE / *Bathyraja richardsoni* (Garrick 1961)

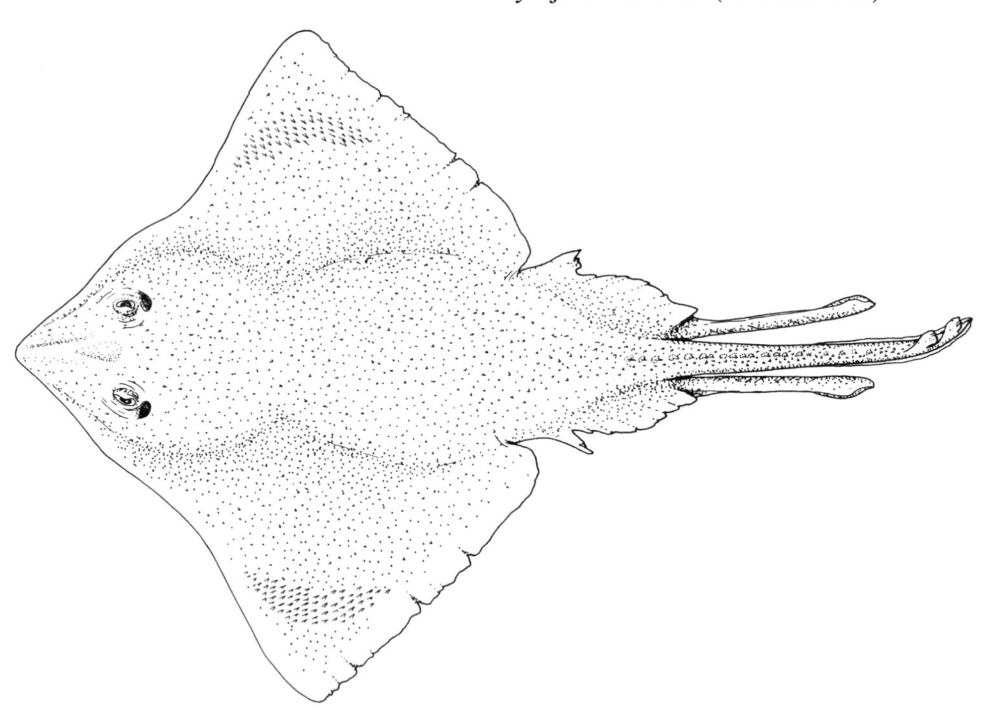

Figure 44. Deepwater skate *Bathyraja richardsoni.* Male. Drawn by J. D. Fechhelm.

Description. Disk diamond-shaped, snout slightly acute to slightly obtuse, tail short (Fig. 44). Snout tip blunt, anterior margins straight to slightly concave to level of anterior margin of orbits, convex from anterior margin of orbits to level of spiracles, and concave from level of spiracles to second or third gill slits. Anterior margins of small specimens more or less straight. Outer corners abruptly rounded and acute. Posterior margins straight to slightly concave; posterior corners abruptly rounded. Pelvic fins with narrow, short, and acutely tipped anterior lobe and a broad and relatively long posterior lobe. Tail considerably shorter than distance from tip of snout to end of cloaca, broad and depressed at base, narrowing distally, with two dorsal fins, confluent at bases. Dorsal and ventral surfaces of disk and tail covered with dermal denticles. Denticles present only on dorsal surface of small specimens. Moderate-sized thorns occur along midline of tail from base to origin of first dorsal fin. Teeth arranged in rows, with sharp clawlike cusps. Upper jaw with 21–32 tooth rows.

Color. The dorsal surface is pale brownish gray, light brown, or grayish brown (Templeman 1973a). The ventral surface is similarly colored except the area around mouth, nostrils, axil of pectoral and pelvic fins, cloaca, base of tail, and tip of claspers are usually white.

Size. This is one of the largest skates in the Gulf of Maine, reaching lengths of 1,740 mm TL. Mature males range from 1,201 to 1,422 mm TL and mature females from 1,590 to 1,740 mm TL.

Distinctions. Deepwater skate are distinguished from other skates of the Gulf of Maine by a combination of the following characters: Rostral cartilage delicate, extending little beyond the anteriormost pectoral rays. Thorns are absent from the disk but there is a complete row of thorns along the midline of the tail. Both dorsal and ventral surfaces of the disk and tail are covered with dermal denticles, and are dark-colored.

Habits. Deepwater skate are found on soft bottoms at depths of 1,370–2,400 m but are most abundant below 2,000 m. From 1965 to 1967 a total of 17 specimens were captured with bottom longlines aboard the *AT Cameron* from off southern Labrador to off Georges Bank (Templeman 1973a). Temperatures at depth of capture range from 3.26° to 3.85°C.

Food. Deepwater skate feed on alcyonarian corals, gastropods, shrimps, and fishes (Templeman 1973a). Shrimps include deep-sea red shrimp, *Pasiphaea tarda,* and fishes include redfish, *Sebastes mentella.*

Parasites. Parasites include copepods, trematodes, and cestodes (Templeman 1973a). Copepods, tentatively identified as *Lernaeopodina longimana,* were attached to the gill filaments. Trematodes thought to be *Otodistomum* and cestodes, thought to be *Onchobothrium pseudouncinatum,* were found in the stomach.

General Range. Deepwater skate have the broadest distribution of any species of skate. Originally described from a specimen caught in deep water off New Zealand, the next was reported from deep water in the Bay of Biscay in the eastern North Atlantic (Forster 1965). Recent reports are from the western North Atlantic from southern Labrador to Georges Bank (Templeman 1973a) and from Norfolk Canyon, off Virginia (Musick et al. 1975). Templeman thinks that this species may have a continuous distribution from the western North Atlantic through the areas off southern Greenland and south of Iceland–Faroes Ridge to the Bay of Biscay.

Occurrence in the Gulf of Maine. Deepwater skate have been reported within the Gulf of Maine only from the eastern edge of Browns Bank and Georges Bank. Templeman (1973a) reported three specimens at 41°45′15″ N, 65°06′45″ W and 41°11′45″ N, 65° 53′00″ W in 2,100 and 2,230 m.

BARNDOOR SKATE / *Dipturus laevis* (Mitchill 1818) / Bigelow and Schroeder 1953:61–63 (as *Raja laevis*)

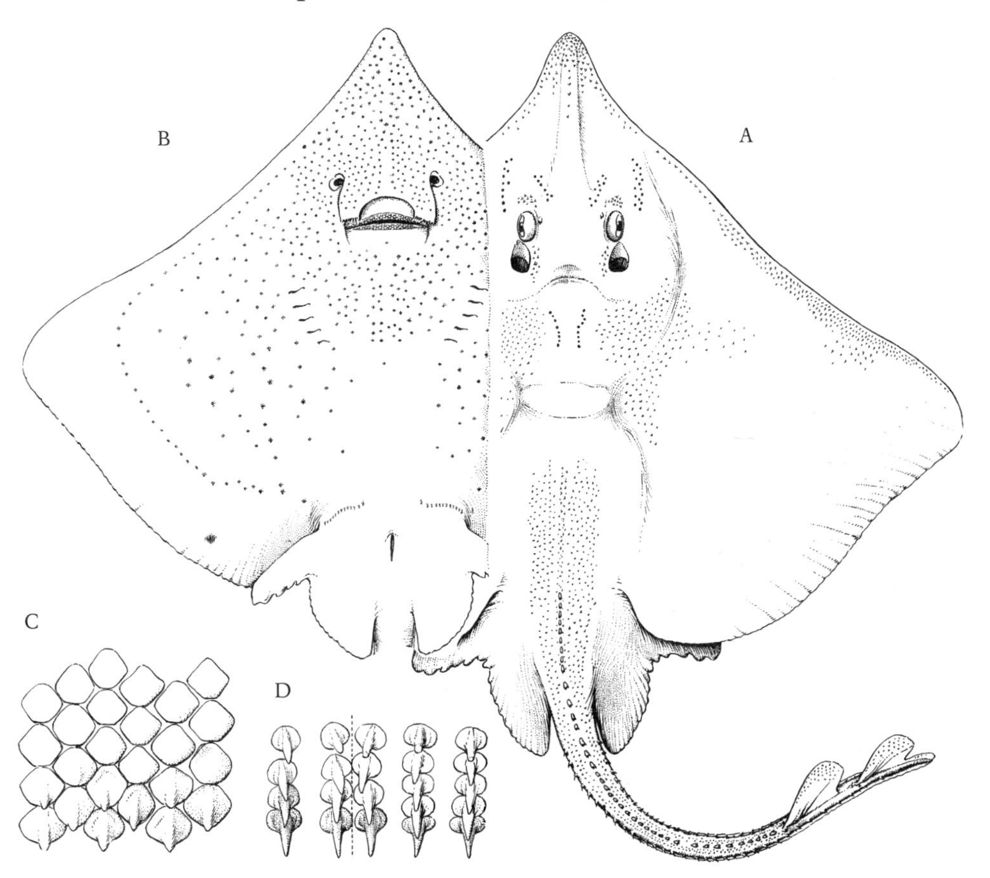

Figure 45. Barndoor skate *Dipturus laevis.* (A) Off Massachusetts, dorsal view of female, 119 cm TL. (B) Nantucket Shoals, ventral view of female to show pigmented mucous pores. (C) Upper jaw teeth from center of jaw of female, 127 cm TL. (D) Teeth from center of jaw of male, 132 cm TL. Drawn by E. N. Fischer. (From Bigelow and Schroeder 1953a: Fig. 47, © Sears Foundation for Marine Research.)

Description. Disk trapezoid-shaped, snout acute, tail moderately short (Fig. 45). Tip of snout blunt, anterior margins concave from tip of snout to about midlength of snout, convex from midsnout length to level of spiracles, moderately concave from level of spiracles to near outer corners of disk. Anterior margins of small specimens more or less straight. Outer corners of disk acute. Posterior margins of disk straight to slightly convex. Posterior corners broadly rounded. Pelvic fins with a narrow anterior lobe and a relatively broad and short posterior lobe. Tail shorter than distance from tip of snout to end of cloaca, narrowing distally, bearing two dorsal fins separated by a short interspace. Dorsal surface of disk free of dermal denticles except on snout, between orbits, and along anterior margin of dorsal fins. Denticles absent on small specimens. Relatively small thorns on anterior and posterior orbital rims, along midline and in a line along lateral aspect of tail and between dorsal fins. Thorns absent along midline of disk. Mature females, in addition, possess dermal denticles on head, along dorsal midbelt of disk and tail, and on shoulders. Teeth of females and immature males close-set, in quincunx arrangement, with rounded cusps. Teeth of mature males more widely spaced, arranged in rows, with sharp-pointed cusps. Upper jaw with 30–40 tooth rows; lower jaw 28–38 tooth rows.

Color. The upper surface is brown to reddish brown, usually with various-sized darker spots and blotches and lighter streaks and reticulations. Generally there is an oval spot or blotch on the center of each pectoral fin. The ampullar pores anterior to the eyes and on the nuchal region are darkly pigmented. The lower surface is white to gray, with gray blotches on the snout. Larger individuals have relatively more blotches and dark-pigmented ampullar pores on the snout and on most of the disk anterior to the pelvic girdle.

Size. Barndoor skate are among the largest skates found in the Gulf of Maine. Bigelow and Schroeder examined one specimen about 150 cm TL and stated that specimens have been reported, without confirmation, to reach 180 cm TL. Barndoor skate weigh about 2–3 kg when 71–76 cm long, about 4–5 kg at 91 cm, and 9–10 kg at 114–117 cm.

Distinctions. Barndoor skate are distinguished from other species of skates from the Gulf of Maine by a combination of the following characters: A straight line from the tip of the snout to the anterior margin of the outer corner of the disk does not intersect the disk. Snout angle acute. Thorns absent from dorsal midline of disk and from shoulder region to base of tail. Ventral surface is light except for ampullar pores, which are dark-pigmented.

Habits. Barndoor skate are found on soft muddy to sandy to gravelly bottoms. They range from the shoreline to about 750 m, although they are most abundant at depths less than 150 m (McEachran and Musick 1975; Scott and Scott 1988). The *Atlantis* found this species widespread but not abundant to 183 m in the Gulf of Maine, and the *Albatross IV* found it widespread from 38 to 351 m during trawl surveys conducted from 1967 through 1970 (McEachran and Musick 1975).

This species has a relatively broad temperature range, which may explain the wide depth distribution. In the Gulf of St. Lawrence, reports are from waters as cold as 1.2°C. Off the east coast of Nova Scotia they have been reported from 1.2° to 10.7°C and from Nova Scotia to Cape Hatteras they are reported from 3° to 20°C (McEachran and Musick 1975; Scott and Scott 1988). Salinity preference ranges from 31 to 35 ppt, although barndoor skate have been reported from the mouth of Chesapeake Bay, where salinities range from 21 to 24 ppt, and from the Delaware River near Philadelphia in brackish water.

Food. Food consists of invertebrates and fishes which are usually associated with the bottom. Prey include polychaetes, gastropods, bivalve mollusks, squids, crustaceans, and fishes (Scott and Scott 1988; Bowman et al. 2000). Smaller individuals apparently subsist mainly on benthic invertebrates, such as polychaetes, copepods, amphipods, isopods, crangon shrimp, and euphausiids, while larger specimens capture larger and more active prey, such as razor clams, large gastropods, squids, cancer crabs, spider crabs, lobsters, and fishes. Fish prey include spiny dogfish, alewife, sea herring, menhaden, hake, silver hake, sculpins, sea snails, cunner, tautog, sandlance, butterfish, and various flatfishes. Garman noted that the thorns on the snout of this species are usually worn smooth, as though the snout was used to dig in the mud or sand to obtain bivalve mollusks.

Predators. Nothing is known of predators of this species but it is probably eaten by sharks, and a similar species in the eastern North Pacific is preyed upon by sperm whale.

Parasites. Parasites include turbellarians, trematodes, cestodes, nematodes, and copepods, and are found in the gills, skin, and intestinal tract (Scott and Scott 1988).

Reproduction. Barndoor skate deposit single fertilized eggs in yellowish or greenish egg capsules. The capsules are rectangular, 124–132 mm long and 68–72 mm broad with a short horn at each corner (Vladykov 1936a). The capsules are smooth but possess fine filaments along their anterior and posterior margins. These egg capsules are considerably larger and have relatively shorter horns than those of other Gulf skates. Adult females with fully formed egg capsules in their uteri have been captured in December and January (Vladykov 1936a), although it is not known if egg capsule production and deposition is restricted to the winter. Reproduction apparently takes place over the entire range of the species. Young are thought to be 180–190 mm TL at hatching but small specimens are seldom captured.

General Range. Barndoor skate occur along the Atlantic coast of North America from the banks of Newfoundland, southern

Gulf of St. Lawrence, and the outer coast of Nova Scotia to North Carolina (Bigelow and Schroeder 1954; Schroeder 1955; McEachran and Musick 1975; Scott and Scott 1988). A similar species, *Dipturus teevani,* occurs along the coast south of Cape Hatteras (Bigelow and Schroeder 1962, 1968; Bullis and Thompson 1965; Struhsaker 1969a).

Occurrence in the Gulf of Maine. Barndoor skate are found throughout the Gulf of Maine. They were considered to be plentiful off the outer Nova Scotian coast; in St. Mary Bay; regularly found in the Bay of Fundy and Passamaquoddy Bay; occur from Eastport, Casco Bay, and the remainder of the coast of Maine; reported from Massachusetts Bay; and abundant on Georges Bank and Nantucket Shoals. During eight ground bottom trawl surveys conducted by the *Albatross IV* and the *Delaware II* between 1967 and 1970, barndoor skate were most abundant in the eastern Gulf of Maine and on the eastern section of Georges Bank; none were captured in the western Gulf of Maine (McEachran and Musick 1975). Casey and Myers (1998) consider the barndoor skate to be "close to extinction" on the continental shelf between southern New England and the Grand Banks of Newfoundland although they still survive on Browns and Georges banks in deep water (greater than 1,000 m) off Newfoundland.

Migrations. Part of the population of barndoor skate in the Gulf of Maine moves into shoal water during the summer. According to Huntsman they move into shallow waters of Passamaquoddy Bay from May to November. Bigelow and Schroeder caught a 150-cm specimen at Cohasset in Massachusetts Bay in less than 2 m of water; and they noted that individuals often strand on the beach. On the other hand, individuals have been caught in 37–110 m on Georges Bank and off Cape Cod throughout the year, and were captured as deep as 183 m by the *Atlantis* in summer. South of Cape Cod they occur in relatively shallower water during the spring and autumn.

Importance. Barndoor skate are one of the five species of skates with commercial value in the Gulf of Maine.

LITTLE SKATE / *Leucoraja erinacea* (Mitchill 1825) / Common Skate, Summer Skate /

Bigelow and Schroeder 1953:67–70 (as *Raja erinacea*)

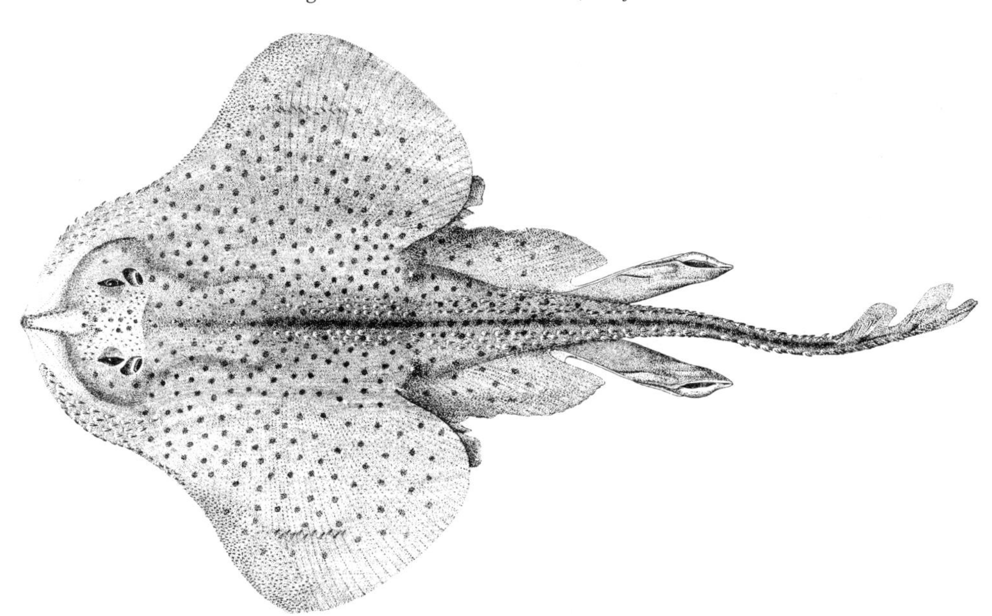

Figure 46. Little skate *Leucoraja erinacea.* Woods Hole, Mass., male.

Description. Disk heart-shaped, snout obtuse, tail moderately long (Fig. 46). Anterior margin slightly concave from tip of snout to anteriormost pectoral rays, convex from anteriormost pectoral rays to anterior margin of orbit, and concave from anterior margin of orbit to level of first gill slits. Anterior margin of small specimens slightly convex. Outer corners of disk broadly rounded, posterior margins slightly convex, posterior corners broadly rounded. Pelvic fins with a broad and relatively short anterior lobe and a broad and relatively long posterior lobe. Tail length greater than distance from tip of snout to end of cloaca, broad at base, narrowing distally, bearing two small dorsal fins that are confluent at bases. Dorsal surface of disk largely free of dermal denticles except along anterior margin and patches on dorsal surface of tail. Mature males have clawlike thorns (malar thorns), with diagonally oriented cusps lateral to orbits. Both sexes with thornlets on snout and on head and thorns on orbital rim, in a triangular patch on shoulder, and along midbelt of disk and tail to origin of first dorsal fin. Three or more irregular rows of thorns on midbelt posterior to shoulder region but median row lost at maturity. In mature females outer half of pectoral fins densely covered with thornlets. Ventral surface naked except for dermal denticles along anterior margin; mature females with a patch of dermal denticles on either side of cloaca. Teeth of females and juvenile males close-set, in a quincunx arrangement; teeth of mature males more widely spaced, arranged in rows, with pointed cusps. Upper jaw with 30–64 tooth rows, number increasing with growth.

Color. The dorsal surface is grayish to dark brown, usually with small round to oval darker spots scattered over the disk. Rarely there is a dark ocellar spot surrounded by white just posterior to the center of the pectoral fins (Templeman 1965b; McEachran and Musick 1973).

Size. Maximum known size 540 mm TL. Maturity is reached between 350–500 mm TL. Maximum size and size at maturity increase toward the northern end of the range (McEachran and Martin 1977).

Distinctions. Little skate are distinguished from all of the skates except for winter skate by a combination of the following characters: The snout is very obtuse. There are three to five rows of thorns along midbelt of disk and tail, and the midrow of thorns disappears with growth. The dorsal surface is patterned with small round to oval dark spots. Distinctions between little skate and winter skate are size-dependent (McEachran and Musick 1973) and are given in the key.

Habits. Little skate are generally found on sandy or gravely bottoms but also on mud. They range from the shoreline to 384 m, but are most abundant in water shallower than 111 m (McEachran and Musick 1975). Temperature range is 2°–21°C, although they are most frequently encountered between 2° and 15°C (McEachran and Musick 1975). Bottom trawl surveys conducted from 1967 through 1970 aboard the *Albatross IV* and *Delaware II* caught little skate in water between 3° and 12°C during the winter, 6°–14°C in the summer, and 7°–14°C in the autumn in the Gulf of Maine and on Georges Bank (McEachran and Musick 1975).

Food. Little skate feed on fishes and invertebrates that are generally associated with the bottom. Prey include hydrozoans, anthozoans, bryozoans, gastropods, bivalve mollusks, squids, polychaetes, copepods, cumaceans, isopods, amphipods, mysid shrimps, euphausiids, pandalid shrimps, crangon shrimps, hermit crabs, cancer crabs, portunid crabs, holothuroideans, and a wide variety of fishes (McEachran et al. 1976; Bowman et al. 2000). Fish prey include herring, alewife, tomcod, silver hake, sculpins, sea snail, silversides, wolffish, sand lance, cunner, and winter and yellowtail flounder. Decapod crustaceans (65.5% by weight), amphipods (15.7%), and polychaetes (12.9%) are the most important overall size-classes (Bowman et al. 2000). Small individuals (<41 cm TL) consume relatively more amphipods and fewer decapods, and large individuals (>41 cm TL) consume relatively more decapod crustaceans and fewer amphipods (McEachran et al. 1976; Bowman et al. 2000). Under laboratory conditions, evacuation rate varied among prey types: krill and clam tissue were digested faster than polychaetes and sand lance (Nelson and Ross 1995).

Predators. Little skate are preyed on by sharks (sandbar, smooth dogfish, and spiny dogfish); other skates, including winter skate; at least six species of bony fishes (cod, goosefish, sea raven, longhorn sculpin, bluefish, and summer flounder); and gray seal (McEachran et al. 1976, Scott and Scott 1988; Rountree 1999).

Parasites. Parasitized by protozoans, myxosporidians, nematodes, trematodes, and copepods on the skin, on the gills, and in the digestive tract (Scott and Scott 1988).

Reproduction. Little skate lay a single fertilized egg in a greenish brown egg capsule. The capsules are rectangular, 44–63 mm long and 30–45 mm broad, and have a moderately long horn at each corner (Vladykov 1936a; Fitz and Daiber 1963). The anterior horns are curved inward and are about half the length of the capsule. The posterior horns are more or less straight or slightly curved outwardly and are about as long as the capsule. Walls of the capsule are smooth but have longitudinal striations. Egg capsules are found partially to fully developed in mature females year-round but are most frequently encountered from late October though January and from June through July (Fitz and Daiber 1963; Richards et al. 1963a; Scott and Scott 1988). Young have external gill filaments from about 25–30 days to 90–95 days after spawning, when they are resorbed (Pelster and Bemis 1992). Gestation is 6 months or more. Young are 93–102 mm TL at hatching.

Age and Growth. Age and growth of little skate have been estimated from length frequency plots and by counting rings on vertebral centra (Richards et al. 1963a; Waring 1984). Specimens from Georges Bank to Delaware Bay average: 215 mm TL at age 1, 293 mm TL at age 2, 364 mm TL at age 3, 420 mm TL at age 4, 461 mm TL at age 5, 472 mm TL at age 6, 475 mm TL at age 7, and 481 mm TL at age 8 (Waring 1984).

General Range. Little skate occur from southeastern Newfoundland to Cape Hatteras, but are rare north of La Have Bank and probably do not occur in the Gulf of St. Lawrence (Schroeder 1955; McEachran and Musick 1975; McEachran and Martin 1977; Scott and Scott 1988).

Occurrence in the Gulf of Maine. Little skate are the most common skate inshore in the Gulf of Maine (McEachran and Musick 1975). They are very abundant along the entire coastline of the Gulf of Maine and on Georges Bank but are rare in the deepest troughs (Map 5). On eight bottom trawl surveys aboard the *Albatross IV* and *Delaware II* from 1967 through 1970, little skate were caught over the entire bank but were rare in the western Gulf of Maine (McEachran and Musick 1975). Bigelow and Schroeder noted that they were absent in the deeper basins of the Gulf of Maine, although they were caught at depths greater than 183 m on the *Albatross IV* and *Delaware II* surveys (McEachran and Musick 1975).

Migrations. Along the inshore edge of the range, little skate move onshore and offshore seasonally. They move into shoal water during April and May and into deeper water in December and January.

Importance. Little skate are growing in commercial importance as a source of skate wings and are one of the major species landed (Fig. 42). They are also used to bait lobster traps.

ROSETTE SKATE / *Leucoraja garmani virginica* (McEachran 1977) / Leopard Skate /

Bigelow and Schroeder 1953:66–67 (as *Raja garmani*)

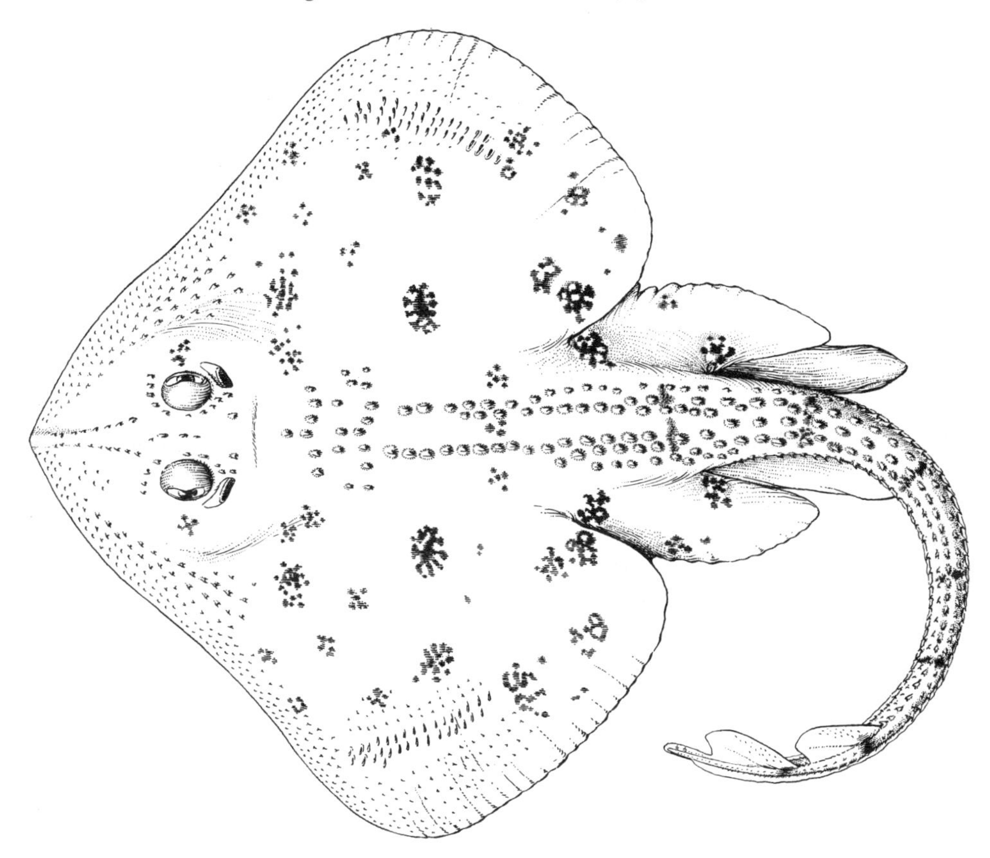

Figure 47. Rosette skate *Leucoraja garmani virginica.* Off North Carolina, male, 398 mm TL, MCZ 34762. (From Bigelow and Schroeder 1953a: Fig. 41, © Sears Foundation for Marine Research.)

Description. Disk heart-shaped, snout short, obtuse, tail relatively long (Fig. 47). Tip of snout projects as a short triangular process. Anterior margin convex from process to level of anterior margin of orbits, convex from level of anterior margin of orbits to level of first gill slits. Anterior margin of small specimens more or less straight. Outer corners broadly rounded. Posterior margins evenly convex; posterior corners broadly rounded. Pelvic fins with a relatively broad and short anterior lobe and a moderately long and broad posterior lobe. Tail length about 60% of distance from tip of snout to end of cloaca, narrowing distally, bearing two dorsal fins separated by a short interspace. Dorsal surface of disk more or less evenly covered with dermal denticles in juveniles; largely free of denticles in adults. Mature males have clawlike malar thorns lateral to orbits. Small thorns occur on rostrum and orbital rim. Large thorns in a triangular patch on shoulder and nuchal region and along midbelt in two to five rows from shoulder region to origin of first dorsal fin. In adults, midrow thorns on medial section of tail small to absent. Young specimens with a single row of relatively large thorns along midline from shoulder region to origin of first dorsal fin. Teeth of females and juvenile males close-set, in a quincunx arrangement; those of mature males more widely spaced, arranged in rows, and with sharp pointed cusps. Upper jaw with 44–55 tooth rows.

Color. The dorsal side is buff or brown and freckled with small light and dark spots forming dark rosettes surrounding a dark central spot. The ventral side is white to yellowish white.

Size. Maximum known size is 445 mm TL. Maximum size and size at maturity increase with latitude (McEachran 1977).

Distinctions. Rosette skate are distinguished from the other skates of the Gulf of Maine by the rosette color pattern.

Taxonomic Note. Populations north of Cape Hatteras and between Cape Hatteras and the Dry Tortugas have been considered as subspecies of the rosette skate but may be distinct species (McEachran 1977).

Habits. This skate is found along the outer continental shelf and upper slope on soft bottoms. They have been captured from 33 to 494 m but are most common between 74 and 274 m (Schroeder 1955; McEachran and Musick 1975). North of Cape Hatteras, temperatures at depth of capture range from 6 to 19°C but they are most abundant at temperatures between 9° and 13°C (McEachran and Musick 1975).

Food. Food consists of polychaetes, copepods, cumaceans, amphipods, crangon shrimp, cancer crabs, galatheids, squids, octopods, and small bony fishes (Stehmann and McEachran 1978). *Crangon septemspinosa* and *Cancer irroratus* are the most abundant items in the diet (Bowman et al. 2000).

Reproduction. Rosette skate lay single fertilized eggs in amber-colored egg capsules. The capsules are rectangular, 37–43 mm long and 26–30 mm broad, with a moderately short horn at each corner (McEachran 1970). Horns are shorter than the remainder of the capsule, and the posterior horns are longer than the anterior horns. Walls of the capsules are smooth but have longitudinal striations. North of Cape Hatteras egg capsules are found in mature females year-round but they are most frequent during the summer (McEachran 1970). The smallest known mature specimens are 335 mm TL, and all specimens greater than 370 mm TL are mature (McEachran 1977).

General Range. Rosette skate occur from Nantucket Shoals to the Dry Tortugas, Fla. (McEachran 1977; Stehmann and McEachran 1978). The northern subspecies occurs from Nantucket Shoals to Cape Hatteras (McEachran 1977). McEachran (1977) synonymized *L. lentiginosa* with the rosette skate but later (McEachran and Carvalho, in press) considered the two species distinct.

Occurrence in the Gulf of Maine. This species is rare in the Gulf of Maine. On 14 May 1950, one specimen was captured by the *Albatross III* in 95 m southeast of the Nantucket Lightship (40°05′ N, 69°22′ W and the *Cap'n Bill II* caught specimens from the eastern slope of Georges Bank to the offing of Nantucket in 1952 and 1953. However, no rosette skate were caught on the eastern slope of Georges Bank on eight groundfish surveys aboard the *Albatross IV* and *Delaware II* from 1967 to 1970 (McEachran and Musick 1975).

WINTER SKATE / *Leucoraja ocellata* (Mitchill 1815) / Big Skate, Eyed Skate /

Bigelow and Schroeder 1953:63–65 (as *Raja ocellata*)

Description. Disk heart-shaped, snout short, tail moderately short (Fig. 48). Tip of snout blunt; anterior margin of disk slightly concave from tip of snout to anterior extension of pectoral rays; convex from anterior extension of pectoral rays to level of anterior margin of orbits and concave from level of anterior margin of orbits to near outer corners of disk. Anterior margin of small individuals more rounded than those of large individuals. Outer corners of disk rather broadly rounded. Posterior margins of disk moderately convex. Pelvic fins with a moderately broad and blunt anterior lobe and a moderately broad and long posterior lobe. Tail relatively broad, varying in length from slightly longer to slightly shorter than distance from tip of snout to end of cloaca. Relative length of tail decreases with growth. Two small dorsal fins near end of tail, confluent at bases. Dermal denticles along anterior margin of disk. Mature males have clawlike malar thorns with diagonally oriented cusps lateral to eyes. Moderate-sized thorns on tip of snout, along lateral edges of rostral cartilage, along orbital rim, in a triangular patch over shoulder region, and in three or more irregular rows along midbelt from shoulder region to origin of first dorsal fin. Thorns along midline from shoulder region to origin of first dorsal fin entirely absent on specimens 69 cm TL or longer. As individuals approach maturity, they develop dermal denticles on ventral sides of pelvic fins. Teeth of females and immature males close-set, in a quincunx arrangement, with rounded cusps. Teeth of mature males more widely spaced, arranged in rows, with sharp pointed cusps. Upper and lower jaws with 44–110 tooth rows, number increasing with size.

Color. The upper surface is light brown with round dark spots. There is generally one large white eye spot with a black

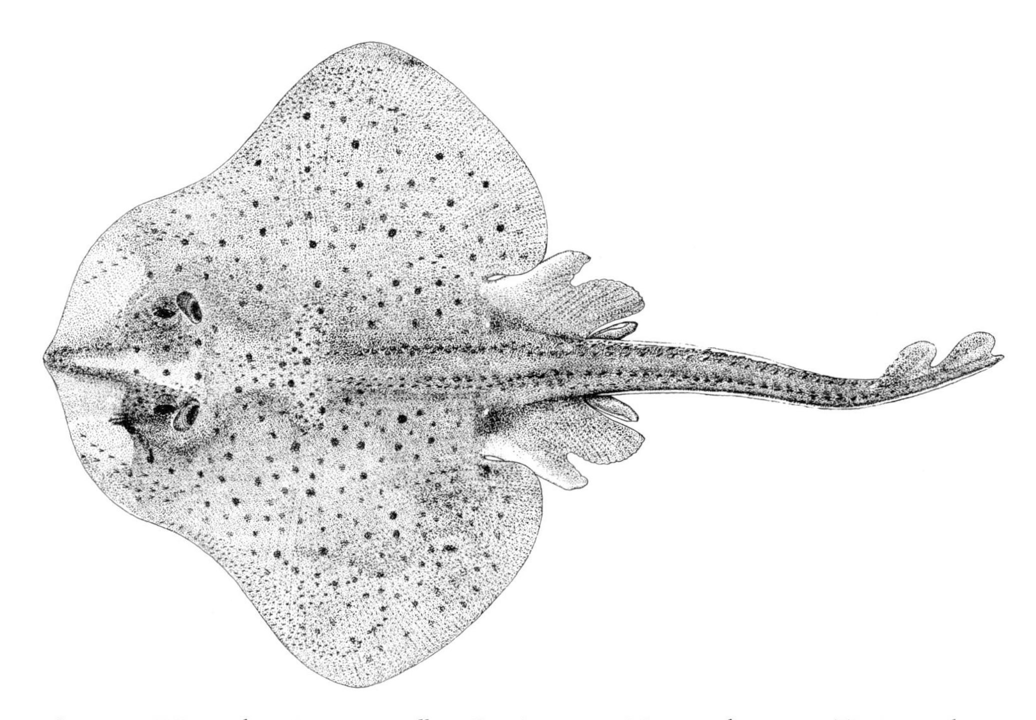

Figure 48. Winter skate *Leucoraja ocellata*. Provincetown, Mass., male, 64 mm TL. Drawn by H. L. Todd.

center near the rear corner of the pectoral fins and often one or more smaller ones near it. The region on the snout between the rostrum and the anterior extension of the pectoral rays is translucent. The lower surface is white. The eye spots have been used to distinguish winter skate from the very similar little skate, although, as Templeman (1965b) has demonstrated, both species may possess eye spots.

Size. One of the larger skates in the Gulf of Maine. Maximum known size is 150 cm TL with larger individuals more common at higher latitudes (McEachran and Martin 1977).

Distinctions. Winter skate are distinguished from all of the skates of the Gulf of Maine, with exception of little skate, by a combination of the following characters: Snout very obtuse. Midbelt of disk and tail bears three or more rows of thorns from shoulder region to origin of first dorsal fin, and mid row disappears with growth. Upper surface is patterned with round dark spots and generally with one or more eye spots near rear corner of pectoral fins. Characters given in the key distinguish it from little skate (McEachran and Musick 1973).

Habits. Winter skate are generally found on sandy to gravelly bottoms. They range from the shore line to 371 m, but are most abundant shallower than 111 m (McEachran and Musick 1975). In the Gulf of Maine, winter skate occasionally occur in less than 1 m of water and at depths of 285 m, but are rare at depths less than 2–4 m and are most abundant at 46–64 m. On the Scotian shelf winter skate are most frequently caught between 37–90 m (Scott 1982b).

Winter skate have been recorded over a temperature range of −1.2° to 19°C. Temperatures at depth of capture were −1.2° to 4.0°C in the Gulf of St. Lawrence, 1.1° to 12.7°C off northeastern Nova Scotia and 2° to 15°C from southern Nova Scotia to Cape Hatteras (McEachran and Musick 1975). On the Scotian Shelf, they were most frequently found at depths where temperatures range from 5° to 9°C (Scott and Scott 1988).

Food. Winter skate prey on fishes and invertebrates that are generally associated with the bottom. Prey include hydrozoans, gastropods, bivalves, squids, polychaetes, cumaceans, isopods, amphipods, mysids, euphausiids, pandalid shrimps, crangon shrimps, hermit crabs, cancer crabs, portunid crabs, echinoderms, and fishes (McEachran et al. 1976; Bowman et al. 2000). Amphipods and polychaetes are primary food items but fishes, decapod crustaceans, isopods, and bivalves are also important. Smaller individuals consume relatively more amphipods and cumaceans, and larger specimens consume relatively more decapods, polychaetes, and fishes (McEachran et al. 1976; Sedberry 1983). Fishes make up most of the diet of individuals larger than 70 cm TL (Bowman et al. 2000). Fish prey include skates, margined snake eel, herring, alewife, blueback herring, menhaden, round herring, silver hake, red hake, tomcod, cod, smelt, sculpins, redfish, sand lance, cunner, butterfish, chub mackerel, and summer and yellowtail flounders.

Predators. Winter skate are eaten by sharks, other skates, and gray seal (Scott and Scott 1988).

Parasites. Parasitized by protozoans, myxosporidians, trematodes, cestodes, and nematodes, which occur in the gills and intestinal tract (Scott and Scott 1988).

Reproduction. Size at maturity increases with latitude (McEachran and Martin 1977). On Georges Bank and in the Gulf of Maine, individuals mature between 70 and 109 cm TL. The Gulf of St. Lawrence population, however, matures at a smaller size and does not reach as large a size as other populations of the species (McEachran and Martin 1977). Winter skate deposit single fertilized eggs in amber to brown egg capsules. The capsules are rectangular in outline, upper and lower surfaces are about equally convex, and each corner of the capsule has a long slender horn (Vladykov 1936a; Scott and Scott 1988). Anterior horns are nearly as long as posterior horns and are the same length as the capsule. Capsules range from 55 to 196 mm in length and 35 to 53 mm in width. The capsules are smooth but are marked with fine longitudinal striations. Female winter skate with fully formed egg capsules are more abundant during the summer and fall but some reproduction may take place throughout the year (Vladykov 1936a; Scott and Scott 1988). Length of incubation is not known. Young hatch at 112–127 mm TL.

General Range. Winter skate occur along the Atlantic coast of North America from the south coast of Newfoundland and the southern Gulf of St. Lawrence to Cape Hatteras (Schroeder 1955; McEachran and Musick 1975; Scott and Scott 1988). The population in the southern Gulf of St. Lawrence is apparently isolated from those along the Atlantic seaboard (McEachran and Martin 1977).

Occurrence in the Gulf of Maine. Winter skate are common in the Gulf of Maine (Map 6) except for the deep troughs. They have frequently been reported from the Bay of Fundy and the coasts of Maine and Massachusetts, but because of the close resemblance to little skate, many of these records as well of those for little skate from the same localities are questionable. The are also frequently reported from Georges Bank. During eight bottom trawl surveys conducted by the *Albatross IV* and *Delaware II* from La Have Bank and the Gulf of Maine to Cape Hatteras during 1967 through 1970, winter skate were second in abundance to little skate on Georges Bank (McEachran and Musick 1975). Winter skate were most frequently captured shallower than 111 m but were occasionally taken at depths to 205 m in the southern Gulf.

Migrations. In the southern part of its range, winter skate apparently move seasonally (McEachran and Musick 1975). They are abundant south of Delaware Bay only during the winter and are more abundant in inshore waters near Woods Hole and in Massachusetts Bay during the winter than during the remainder of the year.

Importance. Winter skate are one of the five Gulf of Maine skates of commercial importance.

SMOOTH SKATE / *Malacoraja senta* (Garman 1885) / Bigelow and Schroeder 1953:70–72 (as *Raja senta*)

Description. Disk heart-shaped, snout clear and moderately obtuse, tail relatively long (Fig. 49). Tip of snout acute; anterior margin of disk concave from tip to anterior extension of pectoral rays, convex from anterior extension of pectoral rays to level of anterior margin of orbits, and concave from anterior margin of orbits to near outer corners of disk. Outer corners of disk broadly rounded. Posterior margins of disk slightly convex. Pelvic fins with a moderately narrow and long anterior lobe, and a moderately narrow and long posterior lobe. Tail relatively narrow, slightly depressed at base, distinctly longer than distance from tip of snout to end of cloaca. Two small dorsal fins near tip of tail, confluent at bases. Dermal denticles cover most of dorsal surface except for shoulder region in females and over center of pectoral fins in males. Ventral surface lacks denticles except for along snout in large specimens, on surface of tail in females, and immature males. Small specimens more or less evenly covered with denticles. Small thorns on orbital rim, on each side of midline on scapular process, and along midline from nuchal region to about midlength of tail. Mature males with a few thorns on rostrum and a triangular patch of malar thorns lateral and anterior to eyes. Small specimens with complete midrow of thorns from nuchal region to origin of first dorsal fin. Teeth of females and immature males close-set, in quincunx arrangement; cusps rounded. Teeth of mature males more widely spaced, arranged in rows, and with pointed cusps. Upper and lower jaws with 38–40 and 36–38 tooth rows, respectively.

Color. The dorsal surface is light brown with many obscure dark spots. Newly hatched specimens have two light yellow crossbars on the tail, and each is bordered anteriorly and posteriorly by darker crossbars or blotches. The ventral surface is whitish, either plain or with a few dark blotches. Occasionally the posterior part of the tail is uniformly dark.

Size. Smooth skate reach 577 mm TL.

Distinctions. Smooth skate are distinguished from other skates in the Gulf of Maine by a combination of the following characters: Rostrum relatively stout, extending anterior to anteriormost pectoral rays. Snout only moderately obtuse. Midline of disk and proximal one-half to two-thirds of tail bear an irregular row of small thorns. Distal half to third of tail of medium to large specimens lacks thorns. Dorsal surface dark-colored without a distinct color pattern. Ventral side of disk uniform white

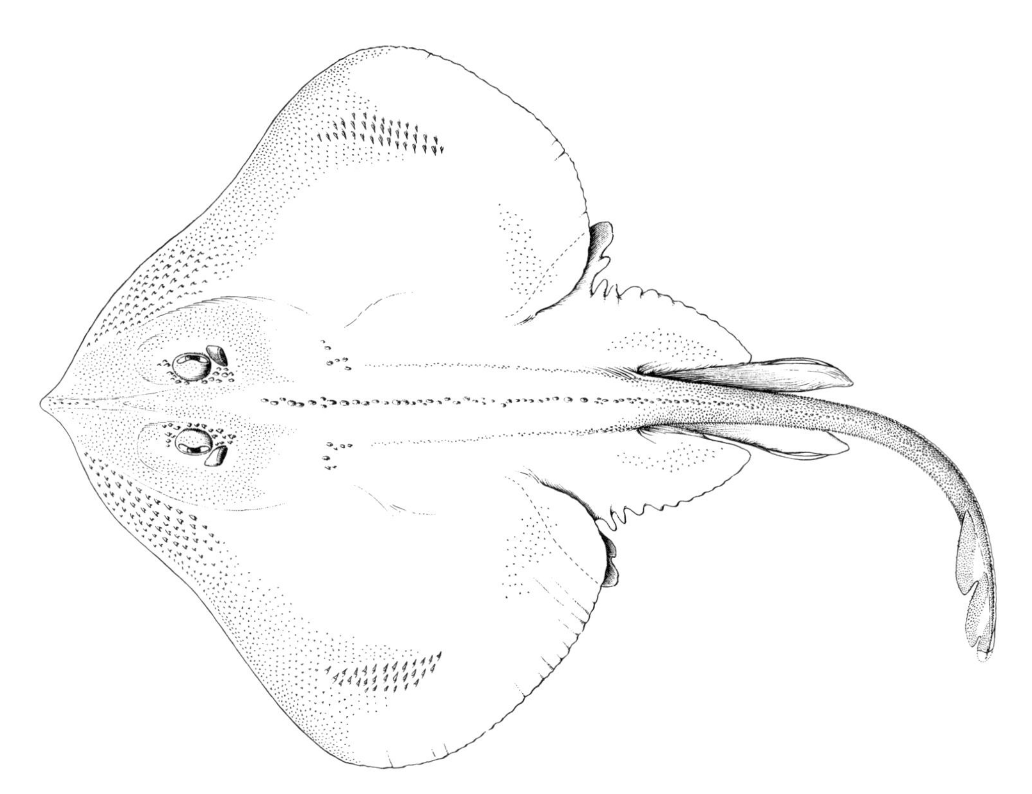

Figure 49. Smooth skate *Malacoraja senta.* Emerald Bank, Nova Scotia, male, 520 mm TL, MCZ 33919. Drawn by E. N. Fischer. (From Bigelow and Schroeder 1953a: Fig. 58A, © Sears Foundation for Marine Research.)

or mostly white with darker blotches. Small specimens have two light yellow bands on the dorsal side of the tail.

Habits. Smooth skate are found on both soft mud and clay bottoms of the deeper basins of the Gulf of Maine and on broken shell, sandy, or gravelly bottoms of the offshore fishing banks. They range from 31 to 874 m but are most abundant between 110 and 457 m (McEachran and Musick 1975). From southern Nova Scotia to Georges Bank, smooth skate have been captured at water temperatures of 2°–10°C. On the Scotian shelf they are usually caught at 73–163 m and 3°–8°C (Scott 1982b).

Food. Smooth skate feed on invertebrates and fishes that are generally associated with the bottom. Prey include cephalopods, polychaetes, copepods, larval stomatopods, isopods, amphipods, mysids, euphausiids, decapod shrimps and crabs, and fishes (McEachran et al. 1976; Bowman et al. 2000). Decapods (especially *Dichelopandalus leptocerus, Pandalus* spp., and *Crangon septemspinosa*) and euphausiids (*Meganyctiphanes norvegica*) are primary food items (Bowman et al. 2000). The diet shifts from amphipods, mysids, and euphausiids to crustaceans at about 30 cm TL. Compared to the other common skates in the Gulf of Maine (winter, little, and thorny skates), the diet of smooth skate is restricted in diversity of prey and habitat occupied by prey. Smooth skate appear to specialize on epifaunal crustaceans (McEachran et al. 1976).

Predators. Smooth skate are eaten by winter skate (Rountree 1999) and by other skates, at least as embryos in egg cases.

Parasites. Parasites include protozoans and cestodes in the blood and intestinal spiral valve.

Reproduction. Smooth skate deposit single fertilized eggs in amber- to brown-colored egg capsules (Vladykov 1936a). The capsules are rectangular in outline and have a strongly convex dorsal surface and a nearly flat ventral surface. The anterior horns are about half the length of the posterior horns, and the posterior horns are slightly shorter than the length of the capsule exclusive of the horns. The anterior horns are curved inward. The capsule is 50–61 mm in length and 35–46 mm in width. Capsules are striated and covered with fibrous tendrils. Females with fully formed egg capsules are found both in summer and winter.

General Range. Smooth skate occur along the Atlantic coast of North America from the St. Lawrence River estuary, Gulf of St. Lawrence, and southern Grand Bank to the Gulf of Maine and Georges Bank (Bigelow and Schroeder 1954; McEachran and Musick 1975).

Occurrence in the Gulf of Maine. Smooth skate are common throughout the Gulf of Maine, Bay of Fundy, and Georges Bank at depths greater than 31 m (Map 7). They are less abundant on Georges Bank than winter skate, little skate, and thorny skate. During eight bottom trawl surveys conducted aboard the *Albatross IV* and *Delaware II* from La Have Bank and the Gulf of Maine to Cape Hatteras from 1967 to 1970, smooth skate were fourth in abundance to little skate, winter skate, and thorny skate, and were caught over the lower part of the temperature range on Georges Bank (McEachran and Musick 1975).

CLEARNOSE SKATE / *Raja eglanteria* Bosc 1800 / Bigelow and Schroeder 1953:65–66

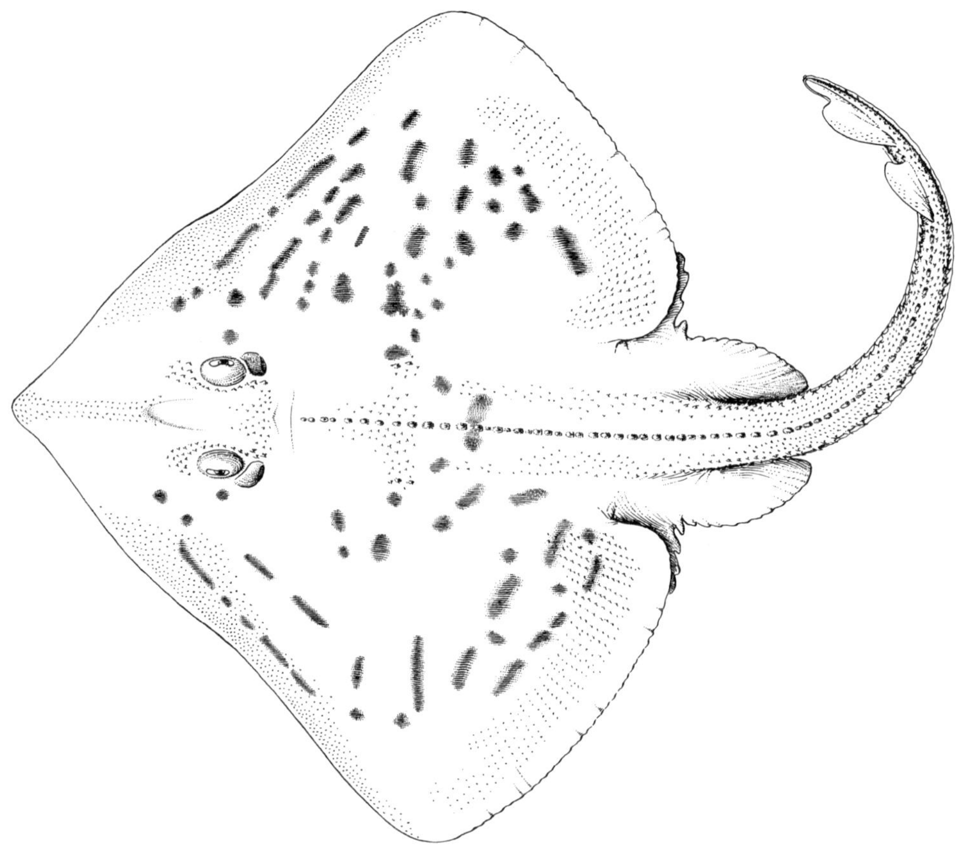

Figure 50. Clearnose skate *Raja eglanteria.* Woods Hole, Mass., female, 745 mm TL, MCZ 36234. Drawn by E. N. Fischer. (From Bigelow and Schroeder 1953a: Fig. 33A, © Sears Foundation for Marine Research.)

Description. Disk trapezoid-shaped, snout slightly acute, tail moderately long (Fig. 50). Anterior margin concave from tip of snout to anteriormost pectoral rays, convex from anteriormost pectoral rays to level of anterior margin of orbits, concave from anterior margin of orbits to level of first gill slits. Anterior margin of small specimens more or less straight. Outer corners of disk abruptly rounded. Posterior margin of disk slightly convex to straight. Posterior corners broadly rounded. Pelvic fins with a relatively broad and short anterior lobe and a moderately broad and long posterior lobe. Tail about equal to distance from tip of snout to end of cloaca, tapering slightly distally and bearing two small dorsal fins separated by a short interspace. Dorsal surface of disk more or less evenly covered with dermal denticles, except mature males have denticles only on head anterior to orbits and along anterior margin and clawlike malar thorns lateral to orbits. Medium-sized thorns on head anterior to orbits, on orbital rim, between spiracles; larger thorns on shoulder region and along midline from shoulder region to origin of first dorsal fin. Tail with irregular row of thorns on each side and several thorns between dorsal fin bases. Ventral side of disk naked except for denticles on snout and along anterior margin. Mature females also have a patch of dermal denticles on pelvic fins. Teeth of females and juvenile males close-set, in a quincunx arrangement. Teeth of males more widely spaced, arranged in rows, with sharp pointed cusps. Upper jaw with 46–54 tooth rows.

Color. Dorsal surface brown to gray with darker brown spots and transverse and diagonal bars. The area between the rostrum and the anterior pectoral rays is translucent, giving rise to the common name. Ventral surface whitish to yellowish.

Size. Maximum known size 949 mm TL (Schaefer 1967).

Distinctions. Distinguished from other species of skates in the Gulf of Maine by a combination of the following characters: Snout acute, three rows of equally sized thorns on dorsal and lateral surfaces of the tail, and dorsal surface of disk marked with a series of dark spots and bars.

Habits. Clearnose skate are found on soft bottoms along the continental shelf. They have been captured from the shore zone (in the northern part of its range) to 329 m (McEachran and Musick 1975), but are most abundant at depths less than 111 m. They occur over a temperature range of 9°–30°C, but are most abundant between 9° and 20°C in the northern part of the range (McEachran and Musick 1975), 19°–30°C in North Carolina (Schwartz 1996).

Food. Clearnose skate feed on polychaetes, amphipods, mysid shrimps, shrimps, crabs, hermit crabs, bivalve mollusks, squids, and small bony fishes (Fitz and Daiber 1963; Stehmann and McEachran 1978; Bowman et al. 2000). Crabs (particularly *Cancer* and *Ovalipes*) are very important up to 60 cm TL, fishes, such as weakfish and butterfish, become important about 50 cm TL (Bowman et al. 2000). In North Carolina, fish prey include striped anchovy, croaker, spot, and blackcheek tonguefish (Schwartz 1996).

Predators. Sharks, such as sand tiger, *Odontaspis taurus,* regularly prey on clearnose skate and one was found in the stomach of a greater amberjack, *Seriola dumerili* (Rountree 1999).

Reproduction. Clearnose skate deposit single fertilized eggs in amber to light brown egg capsules. The capsules are rectangular, 60–80 mm long and 40–51 mm broad, with a relatively short, medially curved horn at each corner (Fitz and Daiber 1963). The horns are shorter than the remainder of the capsule and the anterior horns are shorter than the posterior horns. The capsules are smooth but are marked with fine longitudinal striations. North of Cape Hatteras capsules are deposited in the spring and summer. Incubation time is about 3 months (Fitz and Daiber 1963).

Age and Growth. Age and growth of clearnose skate have been estimated by counting rings on vertebral centra and from length frequency distributions (Daiber 1960; Fitz and Daiber 1963) but vertebrae are difficult to read (Schwartz 1996). According to these techniques, clearnose skate are about 210 mm DW (disk width) at age 1, 280 mm DW at age 2, 340 mm DW at age 3, 400 mm DW at age 4, 420 mm DW (660 mm TL) at age 5, and 460 mm DW at age 6. Maximum age is 6 or 7 years. Size at maturity is 660 mm TL. Maximum size and size at maturity vary with latitude, the largest specimens occurring at the highest latitudes.

Linear regressions defining male and female total length/weight relationships in North Carolina (Schwartz 1996) are: log weight = −4.9320 + 2.8808 log TL for males; log weight = −5.7680 + 3.1869 log TL for females. Females grow longer and heavier than males.

General Range. Clearnose skate occur along the Atlantic coast of North America from Massachusetts to northeastern Florida, and in the northern Gulf of Mexico from northwestern Florida to Texas (McEachran and Musick 1975; Stehmann and McEachran 1978).

Occurrence in the Gulf of Maine. This species is rare off Massachusetts and in the Gulf of Maine. They have been reported from Gloucester and from Provincetown. Two specimens were captured on Nantucket Shoals near Round Shoals buoy by the trawler *Halcyon,* in July and September 1924. No specimens were captured off Massachusetts or in the Gulf of Maine during eight groundfish surveys aboard the *Albatross IV* and *Delaware II* from La Have Bank to Cape Hatteras in 1967–1970 (McEachran and Musick 1975).

Migrations. North of Cape Hatteras clearnose skate move inshore and northward along the continental shelf during the spring and early summer and offshore and southward during the autumn and early winter (McEachran and Musick 1975).

Importance. Clearnose skate are commercially important but major landings occur south of Cape Cod.

WHIP-TAIL STINGRAYS. FAMILY DASYATIDAE

JOHN D. McEACHRAN

Whip-tail stingrays have a relatively thin, oval to trapezoidal disk, and a very slender, whiplike tail that is clearly marked off from the disk. The pectoral fins extend to the tip of the snout. The anterior part of the head is not elevated from the disk, and the eyes and the spiracles are located on top of the head. Pelvic fins are laterally expanded, have a convex lateral margin, are not divided into an anterior and a posterior lobe, and are partially overlapped by the posterior section of the pectoral fins. The tail is usually considerably longer than the disk, and they generally possess one to several serrate spines and a longitudinal fold but lack dorsal fins and a caudal fin. The body is generally naked or has small to medium-sized thorns or large tubercles or bucklers on the dorsal surface of the disk and tail. Within species, females generally grow larger than males, and males generally have a more acute snout, undulated disk, more arched jaws, and teeth with sharper cusps than females. Alar and malar thorns are absent.

All but one species of whip-tail stingrays are benthic and most occur at to depths less than 110 m. Most species occur in brackish to marine waters but some species are apparently restricted to freshwater. Swimming is accomplished by rhythmic undulations of the raylike supports of the pectoral fin. Food consists of a wide variety of benthic invertebrates and fishes. Fertilization is internal and larval development is viviparous without a placenta.

There are six genera and about 50 species of whip-tail stingrays, subfamily Dasyatinae (Nelson 1994), of which five occur off the southeastern coast of the United States but only two reach the Gulf.

ROUGHTAIL STINGRAY / *Dasyatis centroura* (Mitchill 1815) / Stingaree, Clam Cracker /

Bigelow and Schroeder 1953:74–76

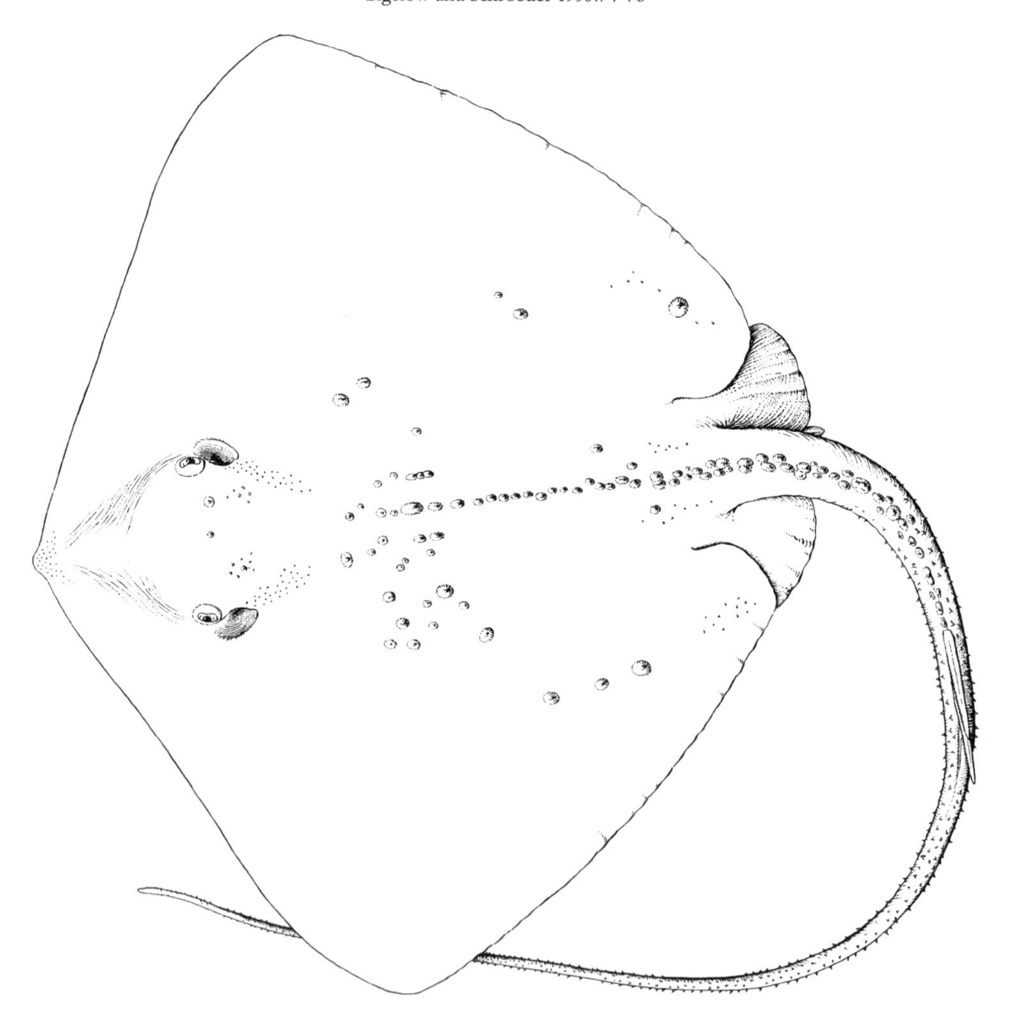

Figure 51. Roughtail stingray *Dasyatis centroura*. Woods Hole, Mass., immature male, 314 cm TL, MCZ 36513. Drawn by E. N. Fischer. (From Bigelow and Schroeder 1953a: Fig. 83, © Sears Foundation for Marine Research.)

Description. Disk trapezoid-shaped, snout moderately long and angular, tail long, slender, whiplike (Fig. 51). Tip of snout obtuse. Anterior margin of disk straight to slightly concave. Outer corners of disk abruptly rounded; posterior margins slightly convex. Anterior margin of pelvic fins nearly straight to slightly convex, outer corner abruptly rounded, posterior lateral margin straight to evenly convex. Tail about two and a half times distance from tip of snout to end of cloaca, without a dorsal fleshy keel posterior to serrate spine but with a narrow ventral fold, which at level of serrate spine is about half the height of tail. Body of specimens less than 50-cm disk width naked. Specimens greater than 50-cm disk width with a small patch of dermal denticles on snout tip, behind spiracles, on base of tail, and on posterior corners of disk. Larger specimens also have tubercles and bucklers on head, along midline from nuchal region to origin of serrate spine, and on central part of disk. Tail covered with thorns that decrease in size posteriorly. Ventral surface of disk naked.

Color. Freshly caught specimens are dark brown dorsally and white ventrally. The tail is black posterior to the serrate spine.

Size. This is the largest whip-tail stingray in the Atlantic; maximum known size 2,100 mm DW. Males reach maturity at about 1,400 mm DW and females at 1,500–1,600 mm DW.

Distinctions. Roughtail stingray are distinguished from all other whip-tail stingrays of the east coast of North America by a combination of the following characters: Anterior margin of disk subangular with snout forming the apex of the angle. Outer corners of disk are narrowly rounded. Snout does not project beyond remainder of anterior margin of disk, and anterior margin is straight to slightly concave. Longitudinal fold on ventral side of tail is about half the height of tail at level of serrate spine. Dorsal side of tail lacks a longitudinal ridge. Dorsal surface of specimens less than 50-cm DW is naked. Dorsal surface of specimens greater than 50-cm DW bears dermal denticles, tubercles, and bucklers. Lateral surface of tail posterior to serrate spines bears tubercles.

Biology. Roughtail stingray are found from near shore to 91 m but a specimen was reported from 274 m off the Bahamas. Temperatures at depth of capture are 15°–26°C (Struhsaker 1969b). They occur along the coast and in bays and estuaries but have not been reported from brackish or fresh water. Litter size ranges from two to six. Food consists of polychaetes, cephalopods, crabs such as *Ovalipes ocellatus,* and bony fishes such as scup and sand lance (Struhsaker 1969b; Bowman et al. 2000).

General Range. Roughtail stingray occur in temperate to tropical waters of the North Atlantic. In the western North Atlantic, they occur from Georges Bank to Florida and in the eastern and occasionally western Gulf of Mexico and also possibly off Uruguay and southern Brazil (McEachran and Capape 1984). In the eastern North Atlantic, they are known from the Mediterranean, Madeira, and the coastline from the Bay of Biscay to Zaire.

Occurrence in the Gulf of Maine. Roughtail stingray are summer stragglers in the Gulf of Maine. One was reported from the outer coast of Cape Cod at Chatham and specimens have been observed on shoal areas of Georges Bank. In 1953, a 914-mm TL specimen resembling this species was reported from the Bay of Fundy off New Brunswick (Scott and Scott 1988).

PELAGIC STINGRAY / *Dasyatis violacea* (Bonaparte 1832)

Description. Disk broad, wedge-shaped; snout moderately long, very obtuse; tail long, tapering (Fig. 52). Anterior margin of disk forms a broad arch. Outer corners abruptly rounded; posterolateral margins straight to slightly convex, sloping medially. Anterior margin of pelvic fins nearly straight, outer corners broadly rounded, posterolateral margin convex. Tail considerably longer than distance from tip of snout to end of cloaca, relatively broad at base but strongly tapered at level of serrate spine, lacks a fleshy dorsal ridge, and has a narrow ventral fold. Large specimens covered with dermal denticles dorsally, naked ventrally. A row of small thorns extends from nuchal region to origin of serrate spine.

Color. Dorsal surface dark purple to dark green; ventral surface slightly lighter grayish purple to greenish blue.

Size. Maximum known size is 800 mm DW. Males mature at about 480 mm DW and females at 400–500 mm DW (Wilson and Beckett 1970).

Distinctions. Pelagic stingray are distinguished from other whip-tail stingrays of the east coast of North America by a combination of the following characters: Anterior margin of disk forms a broad arch. Outer corners of disk are abruptly rounded. Longitudinal fold on ventral side of tail is narrow. Dorsal side of tail lacks a longitudinal ridge. Ventral surface of disk is uniformly dark in color.

Taxonomic Note. Some authors (e.g., McEachran and Fechhelm 1998) place this species in the monotypic genus *Pteroplatytrygon.*

Biology. Pelagic stingray occur within 90 m of the surface in the open ocean and appear to be most abundant near the edges of continental and insular shelves (Wilson and Beckett 1970). Temperatures at depth of capture range from 11.4° to 27.4°C. Food consists of squids, shrimps, and pelagic fishes. Pelagic stingray are apparently attacked by pelagic sharks as several rays caught on longlines in the Gulf Stream off Georges Bank (Wilson and Beckett 1970) had craterlike wounds resembling shark bites.

General Range. Pelagic stingray are probably circumtropical (McEachran and Fechhelm 1998). In the western Atlantic, they

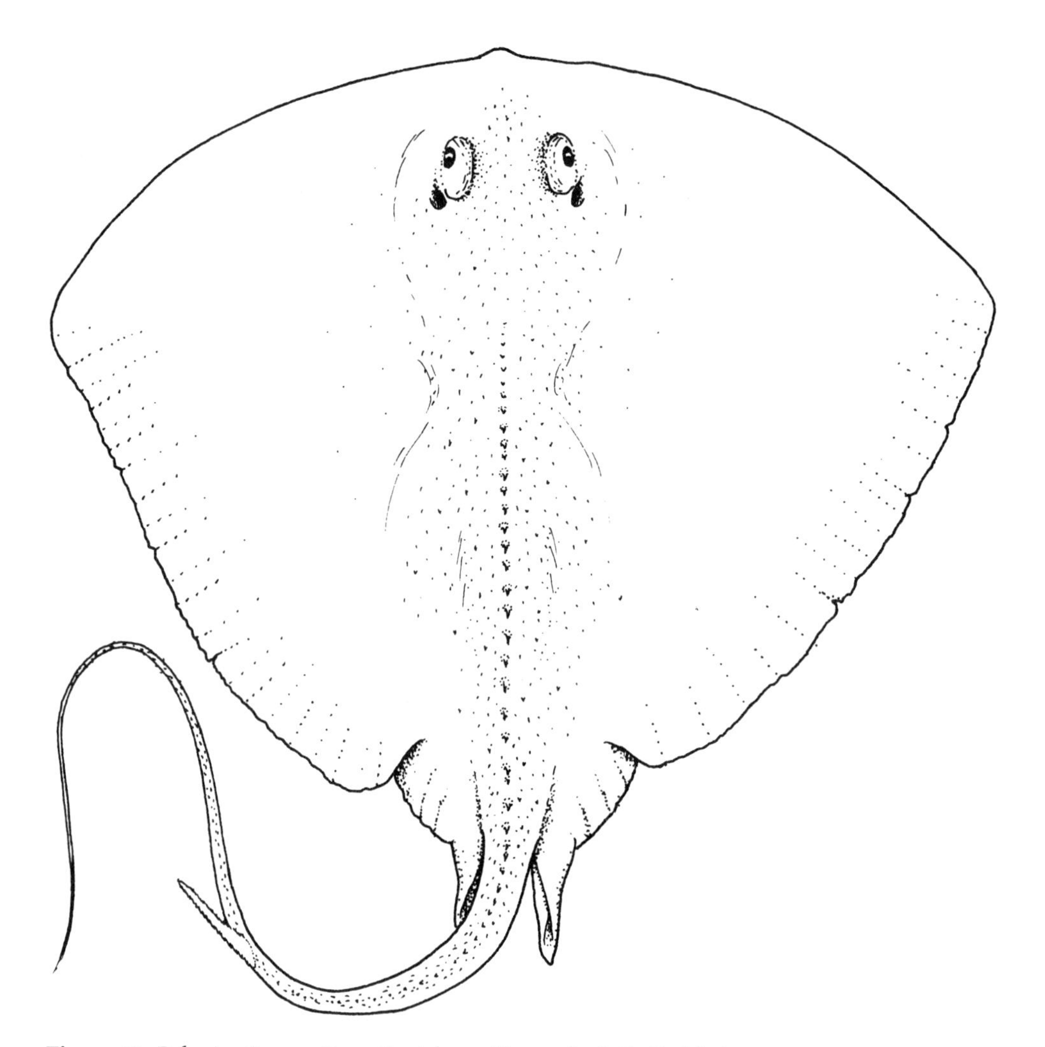

Figure 52. Pelagic stingray *Dasyatis violacea*. Drawn by J. D. Fechhelm.

have been reported from the Grand Banks, Flemish Cap, and Georges Bank to Cape Hatteras, the northern Gulf of Mexico, and the Lesser Antilles (Wilson and Beckett 1970; Branstetter and McEachran 1983). In the eastern Atlantic, they have been captured in the Mediterranean, Gulf of Guinea, and off South Africa (Wilson and Beckett 1970).

Occurrence in the Gulf of Maine. Pelagic stingray have been captured on the eastern slope of Georges Bank near the Gulf Stream from June through September (Wilson and Beckett 1970). From 1959 to 1968 eight specimens were reported from this area, which suggests they are regular summer visitors to the Gulf of Maine.

COWNOSE RAYS. FAMILY RHINOPTERIDAE

JOHN D. McEACHRAN

Cownose rays have a broad trapezoid-shaped disk, a deeply incised subrostrum below the snout and a very long, slender and whiplike tail that is sharply marked off from the disk. The head is elevated from the disk and the eyes and spiracles are located on the sides of the head. The pectoral fins are greatly laterally expanded, and are divided at the level of the head into an anterior section forming the subrostrum and a posterior section forming the disk. The anterior sections are deeply indented medially where they meet in front of the head to form two subrostral lobes. The pelvic fins are somewhat laterally expanded, have a convex posterior margin, are not divided into an anterior and a posterior lobe, and are largely covered by the posterior section of the pectoral fins. The whiplike tail is considerably longer than the disk width and there is a small dorsal fin near its base followed by one or two serrate spines. The body is naked in some species while other species are covered with stellate-based denticles. The teeth are in 6–10 series of flattened hexagonal plates. There is one genus with about five species (Nelson 1994), of which one visits the Gulf of Maine.

COWNOSE RAY / *Rhinoptera bonasus* (Mitchill 1815) / Bigelow and Schroeder 1953:76–77

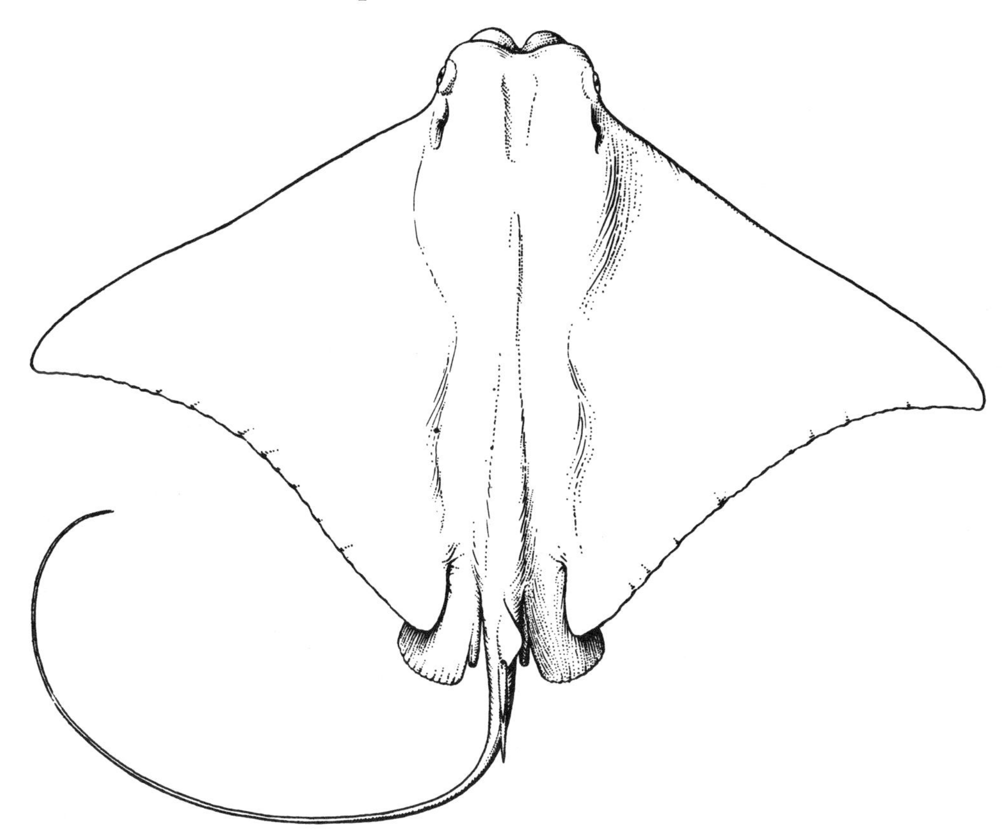

Figure 53. Cownose ray *Rhinoptera bonasus.* Newport, R.I., immature male, 567 mm disk width, MCZ 346. Drawn by E. N. Fischer. (From Bigelow and Schroeder 1953a: Fig. 107, © Sears Foundation for Marine Research.)

Description. Disk trapezoid-shaped, snout bilobed, tail very slender, long and whiplike (Fig. 53). Bilobed subrostrum short, projecting anterior to margin of disk. Anterior margin of cranium also indented. Disk considerably broader than long. Anterior margin straight to weakly concave. Outer corners of disk acute; posterior margins concave. Pelvic fins narrow, extending a moderate distance posterior to posterior corner of pectoral fins. Tail two to three times length of distance from subrostrum to end of cloaca. Dorsal fin originates near axil of pelvic fins. One or two serrate spines occur immediately posterior to insertion of dorsal fin. Body naked. Teeth of six to eight but generally seven series of flattened plates. Tooth plates of median series twice as wide as those of adjacent series, three to five times as wide as those of outer series.

Color. Dorsal surface brown; ventral surface white to yellowish. Some specimens are marked dorsally and ventrally with faint narrow dark lines radiating from the center of the disk.

Size. Cownose ray reach 91 cm DW and there are undocumented reports of over 200 cm DW.

Distinctions. Cownose ray are distinguished from other rays of the eastern seaboard of North America by a combination of the following characters: Disk considerably broader than long. Subrostrum divided into two lobes. Anterior margin of cranium indented. Teeth usually in seven series of flattened plates; plates of the median series twice as wide as those in adjacent series and three to five times as wide as those of outer series.

Biology. Cownose ray are most frequently encountered close to shore. They form large schools, up to 200,000 individuals, and are migratory (Schwartz 1990a). Food consists of bivalve and gastropod mollusks, mysids, decapod shrimps, hermit crabs, and fishes such as striped anchovy (Bowman et al. 2000). This ray is often reported from shallow areas in dense concentrations, where it decimates clam and oyster beds. Litters range from two to six young.

General Range. Cownose ray occur from southern New England to southern Brazil.

Occurrence in the Gulf of Maine. This is an occasional visitor to the southern elbow of Cape Cod during the summer. They have been reported from Nantucket and 145 specimens were captured in one day in fish traps near Woods Hole but they have not been reported from north or east of Cape Cod.

Importance. Cownose ray are presently of no commercial importance as a foodfish. However, there have been attempts to commercialize cownose ray "wings" in the Middle Atlantic region. The edible flesh is reddish in color and has a firm meaty texture. Taste-testing indicates that cownose ray fillets were acceptable in a number of typical seafood preparations and could even be substituted for chicken or beef in stews and soups (Licciardello and Ravesi 1988). Nutritional value: fat 0.8%, moisture 77.2%, protein 18.3%, ash 2.2%.

MANTAS AND DEVIL RAYS. FAMILY MOBULIDAE

JOHN D. MCEACHRAN

Mantas have a broad trapezoid-shaped disk, a broad head connected to cephalic lobes and a very slender, moderately long, whiplike tail. The head is a little elevated but is distinct from the disk. The snout is very broad, straight to slightly concave and very short. The anterior sections of the pectoral fins form narrow, vertically oriented lobes (*cephalic fins*) and are separated from the posterior sections that form the disk. Pelvic fins are a little laterally expanded, are not divided into anterior and posterior lobes, have a convex posterior margin and are largely overlapped by the posterior extensions of the pectoral fins. The whiplike tail is moderately long to very long, bears a small dorsal fin near its base and either has or lacks one to several serrate spines posterior to the dorsal fin. The skin is either naked or covered with small tubercles or dermal denticles. Minute teeth are present in one or both jaws.

Mantas are generally encountered at or near the surface over or near continental or insular shelves. Food consists of planktonic or small nektonic crustaceans and fishes that are guided into the mouth by the cephalic fins as the ray moves through the water. In the mouth the prey are strained from the water by means of gill plates located on the inner sides of the gill slits. Development is viviparous without a placenta.

There are two genera and 13 species (Nelson 1994), one of which is an occasional visitor to the Gulf of Maine.

MANTA / *Manta birostris* (Walbaum 1792) / Bigelow and Schroeder 1953:77–79

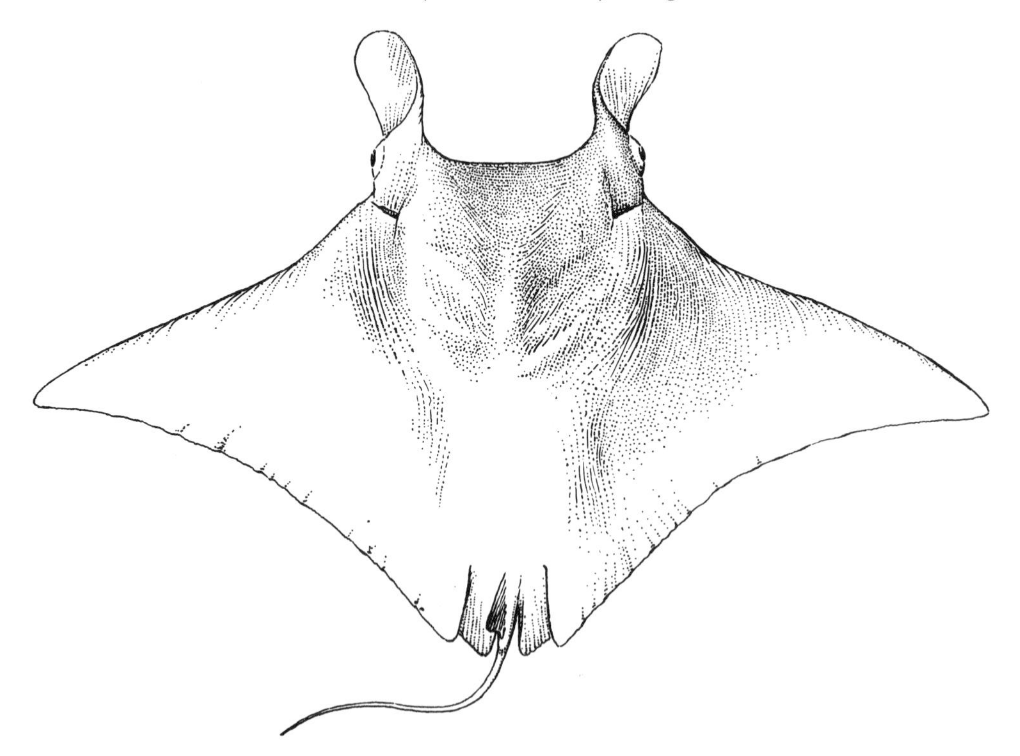

Figure 54. Manta *Manta birostris.* Bimini, Bahamas, juvenile male, 3.5 m disk width. Drawn by E. N. Fischer. (From Bigelow and Schroeder 1953a: Fig. 116A, © Sears Foundation for Marine Research.)

Description. Disk trapezoid-shaped; head broad bordered by large cephalic lobes; tail very slender, relatively short (Fig. 54). Disk slightly more than twice as wide as long. Cephalic lobes about half as broad at bases as long. Upper edge of lobes thick and fleshy, lower edge thin, tip rounded. Mouth terminal, lower jaw slightly projecting and wide, slightly wider than half the head width. Outer corners of disk tapering. Tail about as long as distance from front of head to end of cloaca. Dorsal fin originates anterior to axil of pectoral fins. One or more serrate spines may be present on tail. Body covered with small tubercles or dermal denticles. Teeth very small, present only in lower jaw.

Color. Dorsal surface varies from reddish or olive brown to bluish gray or black, either plain or with various white markings. Ventral surface white centrally and gray around the margin, with various dark blotches around gill slits and on the abdomen. The distal part of the tail is gray.

Size. Manta are reported to reach 6.7 m in width and 1,820 kg in weight. Individuals of 4.6–5.8 m DW are common. Maturity is reached at about 4.3–4.6 m DW.

Distinctions. Manta are distinguished from other devil rays of the Atlantic by a combination of the following characters: Head relatively broad, interorbital distance about one-quarter of maximum width of disk. Mouth terminal. Teeth confined to lower jaw. Cephalic lobes about one-half as broad at base as long.

Biology. Manta are generally encountered at or near the surface over or near continental or insular shelves. It is thought that individuals spend much of their time resting on the bottom because they are seldom seen. Food consists of planktonic and nektonic crustaceans and fishes. Litter size is unknown.

General Range. Manta occur worldwide in tropical to warm temperate waters. In the western Atlantic, they occur from southern New England and Georges Bank to North Carolina, and central Brazil, including Bermuda, the Gulf of Mexico, and the Bahamas.

Occurrence in the Gulf of Maine. This species is an occasional summer visitor to the Gulf of Maine. Two individuals judged to be 550–580 cm DW were sighted by Capt. H. W. Klimm on the southeast part of Georges Bank in late August 1949. A 580-cm DW specimen was harpooned by a swordfish fisherman off Block Island, southwest of Cape Cod in August 1921.

ACKNOWLEDGMENTS. An early draft of the skate section was reviewed by Gordon Waring, utilization comments were received from Robert Learson, and the food habits were reviewed by Ray Bowman.

RAY-FINNED FISHES. CLASS ACTINOPTERYGII

STURGEONS. ORDER ACIPENSERIFORMES

STURGEONS. FAMILY ACIPENSERIDAE

JOHN A. MUSICK

Sturgeons are a primitive family of anadromous and freshwater fishes confined to the Northern Hemisphere. The family contains four genera and 24 species (Nelson 1994) of which two, *Acipenser brevirostrum* and *A. oxyrinchus,* have been reported from the Gulf of Maine. Sturgeons, like sharks, have a heterocercal (uneven) tail with the vertebral column extending out into the upper lobe. Sturgeons have five rows of bony scutes running the length of the body, four barbels in front of the ventral protrusible mouth, no teeth in adults, a largely cartilaginous skeleton, and an intestine with a spiral valve.

Biology. The biology of *A. oxyrinchus* has been reviewed and summarized by Vladykov and Greeley (1963), Murawski and Pacheco (1977), Hoff (1980), and Van Den Avyle (1983). Hoff (1979) presented an annotated bibliography of the biology of *A. brevirostrum,* and Dadswell et al. (1984) provided a detailed synopsis of biological data. Proceedings of the International Conference on Sturgeon Biodiversity and Conservation held in New York in 1994 were published in *Environmental Biology of Fishes* (Birstein et al. 1997), and those from the Third International Symposium on Sturgeon, held in Piacenza, Italy, in 1997 were published in a special issue of the *Journal of Applied Ichthyology* (1999, vol. 15, no. 4–5).

KEY TO GULF OF MAINE STURGEONS

1a. Mouth large, width inside lips more than 62% (63–81%) interorbital width; no bony plates between anal fin and lateral scutes; intestine and peritoneum dark . . . . **Shortnose sturgeon**

1b. Mouth small, width inside lips less than 62% (43–66%) of interorbital width; 2–6 bony plates present between anal fin and lateral row of scutes; intestine and peritoneum pale . . . . **Atlantic sturgeon**

SHORTNOSE STURGEON / *Acipenser brevirostrum* LeSueur 1818 / Bigelow and Schroeder 1953:84–85

Description. A small sturgeon with a relatively short snout, posterior lower side lacking bony plates along and immediately above anal fin base, and with a dark intestine (Fig. 55). Body elongate, somewhat pentagonal in cross section; ventral surface flattened. Dorsal profile of head slightly convex, cephalic fontanelle covered by cartilage at ca. 160 mm TL; snout broad, flattened, and blunt or pointed, with four pendulate ventral barbels; mouth ventral. Mouth width (between inner edge of lips) usually greater than 62% of bony interorbital width; eye small. Dorsal fin origin slightly or moderately anterior to anal origin. Body armored with five major rows of large bony scutes; scutes generally sharp and closely spaced in juveniles, separated and lower in profile in adults. Caudal fin falcate. External secondary sexual differences lacking (Vladykov and Greeley 1963; Gorham and McAllister 1974; Dadswell et al. 1984). Skeletal variation in 13 specimens from the Connecticut River population was documented by Hilton and Bemis (1999).

Meristics. Dorsal fin rays 38–42; anal fin rays 18–24; mid-dorsal scutes 7–13; lateral scutes 21–35; ventrolateral scutes 6–11; gill rakers 22–32.

Color. Top of head and back dark; sides paler; ventral surface white (Vladykov and Greeley 1963). Dorsum blackish, tinged with olive, grading to reddish mixed with violet laterally; some young reported to be yellowish; head similar to body; iris golden. Fins similar to body; paired fins outlined in white. Young and juveniles have melanistic blotches until about 600 mm TL (LeSueur 1818; Scott and Crossman 1973; Dadswell et al. 1984).

Size. Shortnose sturgeon are the smallest species of *Acipenser.* A 1,430-mm TL Canadian fish is the size record for the species (Dadswell 1979; Dadswell et al. 1984). The all-tackle game fish record is a 5.04-kg fish caught in the Kennabacis River, N.B., on 31 July 1988 (IGFA 2001). Both sexes mature at 520–630 mm TL but females grow larger. Males and females attain maximum total lengths, respectively, of 694 mm (586 mm FL) and 995 mm (875 mm FL) in the Altamaha River, Ga., and 1,080 mm (970 mm FL) and 1,430 mm (1,220 mm FL) in the Saint John River, N.B.

Distinctions. Shortnose sturgeon may be distinguished from Atlantic sturgeon by the large mouth (more than 62% interorbital width), absence of bony scutes between the anal fin base and the lateral scute row, almost complete absence of dorsal scutes posterior to the dorsal fin, a single row of scutes anterior to the anal fin, and by the dark intestine.

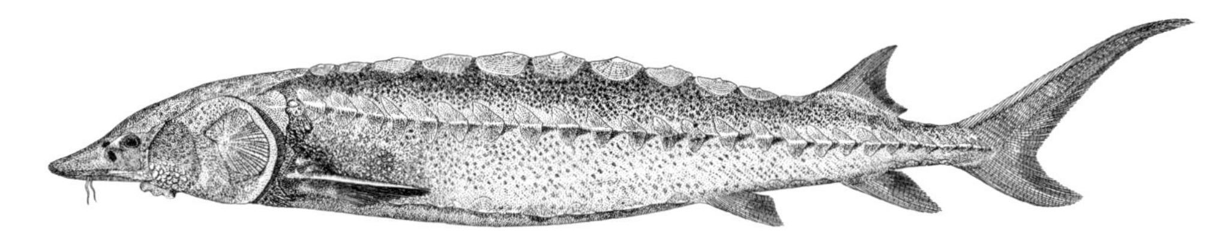

Figure 55. Shortnose sturgeon *Acipenser brevirostrum*. Woods Hole, Mass., 144 cm. Drawn by H. L. Todd.

Habits. Shortnose sturgeon are amphidromous, spending most of their lives in freshwater but periodically visiting salt water in estuaries or staying in freshwater throughout their lives. They typically occupy deep channels of Coastal Plain rivers, but foraging in shallow areas (2–10 m) during summer has been recorded. In some rivers, shortnose sturgeon winter in the lower estuary or, rarely, in nearby coastal waters. Adults have also been found to aggregate in deep upstream pools and channels near the spawning grounds (Dadswell et al. 1984; Hastings et al. 1987). Some populations, such as the one in the Connecticut River, spend their entire lives above the fall line in freshwater (Buckley and Kynard 1985a). Detailed studies of movements and habitat utilization of shortnose sturgeon in the Delaware River have been reported by Hastings et al. (1987) and O'Herron et al. (1993) and in the Savannah River by Hall et al. (1991). Seasonal distribution and abundance and size frequency of shortnose sturgeon in the Hudson River have been summarized by Dovel (1978) and Hoff et al. (1988). McCleave et al. (1977) used sonic telemetry to track several shortnose sturgeon (70–110 cm TL) in the Montsweag estuary (lower Kennebec system) in Maine. They found that daily swimming patterns varied among individuals and noted that fish rarely oriented to channels or to tidal currents. Most individuals made extensive use of shallow (<2 m) foraging grounds, and most remained in the lower salinity (0–24 ppt) portion of the estuary. Kieffer and Kynard (1993) and Kynard et al. (2000) used biotelemetry to study the movements of shortnose and Atlantic sturgeons in the Connecticut and Merrimack rivers, Mass. Sexually mature shortnose sturgeon moved upriver from freshwater wintering grounds to spawning areas in April when water temperatures reached 7°C. After spawning, the fish moved down river in late April or early May to a freshwater reach, where they remained the balance of the year. Some individuals moved down to brackish water for a short period (less than 6 weeks) after spawning, but then returned to the original freshwater foraging area.

Food. Shortnose sturgeon are opportunistic benthic foragers. Juveniles feed on small crustaceans and insects (Carlson and Simpson 1987; Curran and Ries 1937; Dadswell 1979; Dadswell et al. 1984). Adults in freshwater feed mostly on mollusks, benthic crustaceans, and insects, and in estuaries they eat mollusks, shrimps, and polychaete worms (Dadswell 1979; Dadswell et al. 1984). Information on diel and seasonal feeding periodicity is reviewed in Dadswell et al. (1984).

Parasites. Shortnose sturgeon parasite fauna is distinct from that of both mature and juvenile Atlantic sturgeon and reflects a greater resident time in freshwater (Appy and Dadswell 1978). Five species of helminths and arthropods were reported from shortnose sturgeon from the Saint John River estuary (Appy and Dadswell 1978): Platyhelminthes (*Diclybothrium armatum* and *Spirorchis* sp.); Nematoda (*Caballeronema pseudoargumentosus*); Acanthocephala (*Fessentis friedi*); Arthropoda (*Argulus alosa*). Three species of leeches, *Calliobdella vivida, Piscicola milneri,* and *P. punctata* were reported from shortnose sturgeon from the Connecticut River (Smith and Taubert 1980).

Reproduction. *Acipenser brevirostrum* exhibit sexual and latitudinal differences in maturation. Males and females usually mature, respectively, at ages 2–3 and age 6 in Georgia, ages 3–5 and 6–7 from South Carolina to New York, and ages 10–11 and age 13 in Saint John River (Dadswell et al. 1984). After maturity, females spawn every third year and males every year or every other year (Dadswell et al. 1984).

Shortnose sturgeon are amphidromous, ascending medium to large rivers to spawn as early as February in Georgia, in late April in Massachusetts, and as late as mid-May in Canada (Dadswell 1979; Dadswell et al. 1984; Kieffer and Kynard 1996). In the Delaware River, spawning occurs during the middle two weeks of April. Spawning temperatures range from 9° to 12°C. Spawning sites are deep, swift sections with rocky substrates; such areas are usually located beneath falls or rapids (Meehan 1910; Dadswell et al. 1984; Taubert 1980). Spawning grounds typically are located at or several kilometers above the fall line. In some South Carolina rivers where access to the fall line is barred by dams, spawning apparently occurs below the fall line adjacent to flooded hardwood swamps (D. E. Marchette, pers. comm.). Fecundity ranges from 10,000 to 16,000 eggs per kg of body weight, or about 27,000 to 208,000 eggs per fish (Dadswell et al. 1984).

Early Life History. Eggs are benthic, adhesive, and attached to bottom materials. They average about 3.0 mm in diameter, are bicolored brown and grayish white (Vladykov and Greeley 1963). Hatching occurs in 13 days at 10°C (Meehan 1910). Yolk-sac larvae are 8.0–12.1 mm TL, eye diameter 2.9–3.0% TL, snout length 3.0–4.0% TL. The eyes are unpigmented at less than12.0 mm TL, pigmented at greater than 12.5 mm TL. The entire body is covered with melanophores, particularly on the head, tail, and finfold. Pectoral fin buds are barely visible. Yolk-sac larvae 13–14.7 mm TL have well-pigmented eyes, a dark dorsum, and a light venter. The yolk sac is almost fully absorbed at 14.7 mm TL. Precursor of the dorsal fin is visible with 14 rays; the nares, mouth, barbels develop; the mouth width is 71% of head width, teeth are present in both jaws; eye diameter is 5.4–6.12% TL; preanal length 61% TL; snout length 5.16% TL. By 10–11 mm TL, larvae may be distinguished from those of Atlantic sturgeon by a larger mouth (mouth width greater than 65.5% head in *A. brevirostrum,* and less than 53.3% head width in *A. oxyrinchus*) (Bath et al. 1981; Taubert and Dadswell 1980). Shortnose sturgeon hatchlings (<1 day in age) were positively rheotactic, photonegative, and benthic and vigorously sought cover (Richmond and Kynard 1995). They retained this behavior until about 9 days of age when they became photopositive, left bottom cover, and swam in the water column. Thus, downstream migration away from the spawning site was not initiated by early embryos, but by late-stage larvae.

Age and Growth. The maximum age for shortnose sturgeon is 67 years from Saint John River (Dadswell 1984). Males seldom exceed 30 years. Sex ratios are even for small size-classes but most fish greater than 90 cm TL are females. The length-weight relationship in the Hudson was calculated as log w = 13.7 + 3.24 log TL (Hoff et al. 1988).

General Range. *Acipenser brevirostrum* range from the Saint John River, N.B., to the St. Johns River, Fla.

Occurrence in Gulf of Maine. Because of confusion with Atlantic sturgeon, it is difficult to determine the historic distribution of this species in the Gulf, but they probably inhabited most of the major rivers. They have been recorded from Provincetown (Bigelow and Schroeder) and there is a recent record from Ipswich Bay (Jerome et al. 1968). Viable populations of shortnose sturgeon are found currently in the Saint John River, Kennebec system (Sheepscot, Kennebec, and Androscoggin rivers and Montsweag Bay), Merrimac River, and perhaps in the Penobscot River (Fried and McCleave 1973; NMFS 1998a; Squiers and Smith 1979; K. E. Hartel, pers. comm.).

Management. Although classified as endangered by the National Marine Fisheries Service and as threatened in Canada (Vecsei and Peterson 2000b), *Acipenser brevirostrum* is relatively common in several major drainages, most notably the Saint John, Kennebec, Hudson, Delaware, Santee, and Altamaha rivers (Dadswell et al. 1984; Hastings et al. 1987). Total instantaneous mortality for the Saint John River population has been estimated at 0.12–0.15 (Dadswell 1984). Population estimates in Gulf of Maine rivers range from 18,000 adults in the Saint John (Dadswell, 1979) to 7,222 in the Kennebec system (Squiers et al. 1981) to 200 in the Merrimack (NMFS 1998a). In the Hudson River the population of subadults and adults has been estimated to be as high as 30,000 (Hoff et al. 1988).

Management options for recovery of shortnose sturgeon include continued protection from direct and incidental mortality from humans, habitat restoration, installation of fish lifts or ladders to allow fish to ascend dams and reach spawning grounds, and aquaculture and restocking (Dadswell et al. 1984; Smith et al. 1985; Friedland 1998a). If properly managed, shortnose sturgeon could support limited fisheries for gourmet items. Their flesh is excellent and they produce very good caviar (Dadswell 1984). Shortnose sturgeon management is guided by a recovery plan under the Endangered Species Act. The recovery plan is being revised to reflect the increased knowledge accumulated on shortnose sturgeon in recent years (Friedland 1998a). Some populations may be large enough to allow reclassification of their status.

ATLANTIC STURGEON / *Acipenser oxyrinchus oxyrinchus* Mitchill 1815 /

Bigelow and Schroeder 1953:81–84 (as *Acipenser sturio*)

Description. Large dermal scutes in five major rows, sharp and close together in juveniles, separated and lower in profile in adults (Fig. 56). Successive bucklers in dorsal row touch each other or even overlap; space between dorsal row of bucklers and uppermost of two lateral rows thickly set with coarse prickles. Head 26.5–27.6% FL, snout length 43–60% (mean 52.9%) of HL; interorbital width 22.4–28.4% (26%) of HL (Ryder 1890; Vladykov and Greeley 1963; Mansueti and Hardy 1967; Jones et al. 1978; and Hoff 1980).

Meristics. Dorsal fin rays 30–46; anal fin rays 23–27 (mean 24.9); dorsal scutes 7–13 (9.8), lateral scutes 24–35 (28.7), ventral scutes 8–11 (9.2); gill rakers 15–27.

Color. Canadian specimens have been described as bluish black dorsally, becoming progressively paler down the sides; ventrum white; dorsal and lateral shields very light (Vladykov and Greeley 1963). Specimens from the Chesapeake Bight and Hudson River tend to be brownish olive dorsally on the head and back, grading to pinkish tan on the sides and white ventrally. The dorsal and lateral scutes always are lighter than the surrounding skin (Musick et al. 1993a).

Size. A large species with adults usually ranging in size from 88–200 cm TL (Gruchy and Parker 1980). Although DeKay (1842) reported an Atlantic sturgeon of 18 ft (549 cm TL), the largest documented specimen was 427 cm TL caught off the mouth of the Saint John River (Vladykov and Greeley 1963).

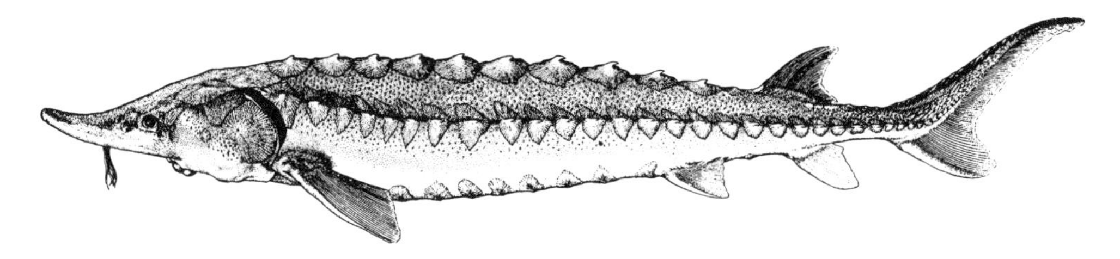

Figure 56. Atlantic sturgeon *Acipenser oxyrinchus oxyrinchus.*

Distinctions. Atlantic sturgeon may be distinguished from shortnose sturgeon by the small mouth (less than 62% interorbital width), presence of bony scutes between the anal fin base and the lateral scute row, a double row of dorsal scutes posterior to dorsal fin, and a double row of scutes anterior to anal fin, and by the pale intestine.

Taxonomic Note. Gilbert (1992) discovered that for over 100 years the name of the Atlantic sturgeon had been consistently been misspelled as *Acipenser oxyrhynchus* instead of *A. oxyrinchus* as given in the original description. This species is closely related to *A. sturio* Linnaeus of the Old World but, according to Magnin and Beaulieu (1963), can be separated by several characters. Two subspecies are recognized: *A. o. oxyrinchus* from the Atlantic coast of North America and *A. o. desotoi* Vladykov from the northern Gulf of Mexico (Vladykov 1955b; Vladykov and Greeley 1963). The taxonomic status of these populations was recently revaluated by Artyukhin and Vecsei (1999), who found more differentiation within European *A. sturio* than between *A. sturio* and *A. oxyrinchus* and recommended recognizing four subspecies within *A. sturio,* including the American subspecies as *A. s. oxyrinchus* and *A. s. desotoi.*

Habits. Atlantic sturgeon are anadromous. Juveniles may spend several years in freshwater in some rivers (Scott and Crossman 1973; Lazzari et al. 1986); however, in others they may move downstream to brackish waters when water temperatures drop in the fall (Dovel 1978; Hoff 1980). Although there are a few records of Atlantic sturgeon from offshore fishing banks, most captures at sea have been very near the coast (Vladykov and Greeley 1963). Most sturgeon taken by trawl in Virginia waters were from depths less than 20 m (Musick et al. 1993a).

Food. Atlantic sturgeon are opportunistic benthic feeders (Vladykov and Greeley 1963; Huff 1975; Scott and Crossman 1973; Johnson et al. 1997). In marine or estuarine waters, they have been reported to feed on polychaetes, isopods, decapod crustaceans (*Callinectes* sp., *Crago* sp.), amphipods, gastropods, bivalves, and fishes such as sand lance (*Ammodytes* sp.). Polychaetes were the major prey group eaten during all seasons off New Jersey (Johnson et al. 1997); eight species were identified, with *Lumbrineris fragilis* and *Pherusa affinis* being the most important. Isopods were the second most important prey group off New Jersey; *Politolana concharum* was the most important isopod consumed. In freshwater, food consists of aquatic insects (*Hexagenia* sp.), amphipods, and oligochaetes.

Parasites. Six species of helminth and arthropod parasites have been reported from Atlantic sturgeon in the Saint John River estuary (Appy and Dadswell 1978): Platyhelminthes (*Nitzschia sturionis, Derogenes varicus,* and *Dereopristis hispida*); Nematoda (*Truttaedacnitis sphaerocephala*); Acanthocephala (*Echinorhynchus "gadi"* complex); and Copepoda (*Dichelesthium oblongum*).

Reproduction. Fecundity and mature oocyte size of Atlantic sturgeon increase with age, body size, and iteroparity (Van Eenennaam et al. 1996). Fecundity ranges from 800,000 to 3.76 million eggs (Ryder 1890; Smith 1907; Vladykov and Greeley 1963). Size at maturity and spawning time varies with latitude. In the St. Lawrence River, males mature at 22–24 years and females at 27–28 years (Vladykov and Greeley 1963). In the Hudson River, they mature earlier: males at about 9 years (32 kg) and females at 10 years (68 kg). Atlantic sturgeon spawn from May to July in the St. Lawrence River (Scott and Crossman 1973). They spawn as early as February and March in northern Florida (Vladykov and Greeley 1963; Huff 1975). This species requires solid substrates upon which to spawn its adhesive eggs (Vladykov and Greeley 1963; Huff 1975; T. I. J. Smith 1985). Spawning temperatures range from 13.2° to 20.5°C (Borodin 1925; Huff 1975; T. I. J. Smith 1985). In the Hudson River, Atlantic sturgeon spawn in the oligohaline (slightly brackish) and tidal freshwater (Van Eenennaam et al. 1996) part of the tidal estuary (Greeley 1937; Dovel 1978). In the Kennebec River, Maine, they apparently spawned mostly above tidewater between Augusta and Waterville, as Atkins (1887) noted a great decline in the numbers of sturgeon after a dam was built in 1837 at Augusta. Male Atlantic sturgeon migrate to the spawning areas earlier in the year and spend a longer time in the Hudson River than females (Van Eenennaam et al. 1996). Males may spawn every year but females do not (Van Eenennaam et al. 1996) with interspawning intervals of about 3 years (Boreman 1997).

Early Life History. Early descriptions of Atlantic sturgeon eggs and development were summarized by Mansueti and Hardy (1967); Jones et al. (1978); and Snyder (1988). The eggs are benthic, often occur in strings, and are strongly adhesive, attaching to, for example, stones, shells, sticks, and weeds. Fertilized eggs average about 2.9 mm in diameter; are initially globular, becoming oval with development; and are gray to brown with stellate pigment at the animal pole (Ryder 1890; Dean 1893). Incubation takes 94 h at 20°C, 168 h at 17.8°C (Dean 1895; Vladykov and Greeley 1963).

The yolk-sac larval stage, 8.4–14.3 mm TL, lasts approximately 6 days (Vladykov and Greeley 1963). The yolk sac is yellowish, oval, vascular, about 30% TL. The mouth is small, mouth width 38.4–53.3% head width, and the head and tail are darkly pigmented (Ryder 1888, 1890; Bath et al. 1981). There is a continuous finfold from behind the head, dorsally, around the notochord, and ventrally to the posterior end of the yolk sac. The spiral valve and gills are visible at 9.6 mm TL. Barbel buds appear at 9.7 mm TL. The eye is fully formed at 11 mm; pectoral fin buds appear at 8.4 mm TL and are well developed by 14.0 mm TL. Pelvic fin buds and dorsal fin basal elements appear at 10.6 mm TL.

The yolk sac is absorbed at 14.2–14.3 mm TL when barbels, mouth, and head shape are easily identifiable as sturgeonlike. The fontanelle is visible on top of the head, the nasal capsules are visible anterior to the eyes, larger pectoral fin buds and small pelvic fin buds are apparent but not rayed. The finfold increases in size with TL, gradually differentiating into median fins and dorsal and anal scutes. Dorsal and anal basal supports are apparent at 16 mm TL, incipient dorsal scutes appear at 19.0 mm TL, and incipient lateral scutes appear at more than 29.0 mm TL. Teeth are present at 14.2–31.5 mm TL. Fin rays form gradually, first in the pectoral and dorsal fins at 19.0 mm TL and later in the pelvic and caudal fins. Metamorphosis to the juvenile phase is virtually complete at 31.5–37 mm TL (Bath et al. 1981).

Age and Growth. The maximum age recorded is apparently 60 years from a 267-cm TL specimen captured in the St. Lawrence River (Magnin 1964). Comparative age, length, and weight data for Atlantic sturgeon have been summarized by Hoff (1980) from compilations by Carlander (1969) and Murawski and Pacheco (1977). Information on age and growth of Atlantic sturgeon in the Gulf of Maine is very limited. A small sample (<18) of Atlantic sturgeon taken in the Kennebec River in 1980 yielded 15 males, 145–193 cm FL and 7–40 years of age, and 3 females, 170–208 cm FL and 25–40 years. Von Bertalanffy growth models for Atlantic sturgeon have been calculated (Magnin 1964; Dovel and Berggren 1983; Stevenson and Secor 1999). Growth coefficient (k) in all studies ranged from 0.16 to 0.03 for females (reflecting slow growth), but as high as 0.25 for males.

General Range. Occur along the northwest Atlantic coast from Hamilton River, Labrador, and Ungava Bay south to northern Florida (Gruchy and Parker 1980). Strays were recorded from Bermuda in the late nineteenth century (Smith-Vaniz et al. 1999).

Occurrence in Gulf of Maine. Historically, Atlantic sturgeon entered practically every stream of any size emptying into the Gulf of Maine. They were taken some distance upstream from the mouths of streams no larger than the Charles and Parker rivers in Essex County, Mass. Wood, writing of Massachusetts in 1634, described them as "all over the country, but best catching of them be upon the shoals of Cape Code and in the river of Merrimacke, where much is taken, pickled and brought for England, some of these be 12, 14 and 18 foot long."

Atlantic sturgeon may be expected anywhere off the coasts of the Gulf of Maine during their sojourn in saltwater. There are records of them on both sides of the Bay of Fundy; off Mt. Desert Island; in Penobscot Bay; in Casco Bay; at the mouth of the Piscataqua River; on the Boars Head Isles of Shoals fishing ground; at the mouths of the Essex and Ipswich rivers; Boston Harbor; at Provincetown; off Truro, Cape Cod; and at Nantucket, as well as west along the southern New England coast (Bigelow and Schroeder). The Kennebec, Androscoggin, and Sheepscot estuarine system in Maine may support the only present spawning population in New England (Squiers and Savoy 1998).

Movements and Migrations. Interestuarine migrations have been documented: Juvenile Atlantic sturgeon tagged in the Hudson River have been recaptured in lower Chesapeake Bay (Loesch et al. 1979). Kieffer and Kynard (1993) and Kynard et al. (2000) used biotelemetry to study the movements of both Atlantic and shortnose sturgeons in the Connecticut and Merrimack rivers, Mass. Although shortnose sturgeon were resident year-round in freshwater and low-salinity parts of the river, juvenile Atlantic sturgeon (70–156 cm TL) did not enter the river until mid-May, when water temperatures had warmed to 14.8°–19.0°C. They remained in the brackish part of the river until they emigrated in the fall when water temperatures were 13.0°–18.4°C. Juvenile sturgeon tagged along the coast between False Cape, Va., and Cape Lookout, N.C., migrated south from November to January and then north as far as Long Island in the late winter and early spring (Holland and Yelverton 1973). Juvenile sturgeon tagged off South Carolina migrate as far north as Pamlico Sound, N.C., and the Chesapeake Bay (T. I. J. Smith 1985).

Importance. The status of Atlantic sturgeon stocks was reviewed by Hoff (1980), T. I. J. Smith (1985), and Smith and Clugston (1997). Sturgeon were so reduced by overfishing, pollution, and dam construction that by 1994 directed fisheries in the United States were active only off New York and New Jersey (Waldman et al. 1996a), although North Carolina still had a sturgeon fishery up to 1991 (Schwartz 1997b). Total landings of Atlantic sturgeon have fluctuated from 30 to 120 mt from 1950 to 1996 (Fig. 57). The few taken in the Gulf of Maine are picked up accidentally in traps or weirs, in drift nets, or by trawlers. There are only two current fisheries: one very small, in the Saint John River (Smith and Clugston 1997) and the other, the only significant fishery, in the St. Lawrence estuary in Quebec (Caron and Tremblay 1999).

In former years, when streams were less obstructed and sturgeon more plentiful, the catch was of considerable value

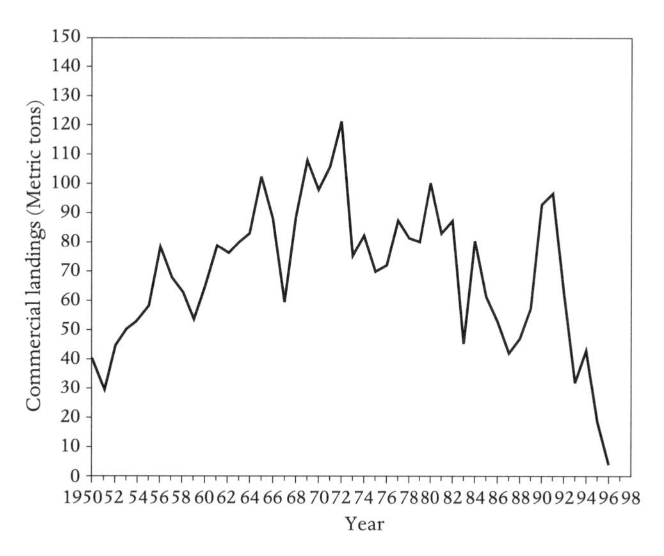

Figure 57. Commercial landings of sturgeons along the eastern coast of the United States, 1950–1996 (metric tons). (From Friedland 1998a.)

in some of the larger rivers, in excess of 3,000 mt·year^{-1} around the turn of the nineteenth century (Friedland 1998a). Bigelow and Schroeder summarized the early history of the fishery. For instance, sturgeon, doubtless from the Kennebec River and cured near what is now Brunswick, Maine, were shipped to Europe as early as 1628, and large quantities were also shipped from near Ipswich, Mass., in 1635. In the Kennebec, where an intermittent fishery was long maintained, the catch was about 250 fish in 1880, yielding 5,700 kg of meat, and not much less in 1898 (4,900 kg). The yearly landings were only about one-fourth as great there (1,300 kg) by 1919. Reported landings of sturgeon from the entire coastline of Maine (including what few were brought in from offshore) fell to only 136–182 kg in the years 1940–1947. Reported landings in Massachusetts of 2,400 kg in 1940 (all by otter trawlers) and of 3,000 kg (2,270 kg by otter trawlers, from offshore), corresponded to about 50–70 fish in 1947.

Stocks. Atlantic sturgeon exhibit strong genetic stock structure in both the Atlantic and Gulf of Mexico (Waldman et al. 1996b; Waldman and Wirgin 1998; Wirgin et al. 2000). The only population studied in the Gulf of Maine, Saint John River, was similar to the population from the St. Lawrence River, but very different from the Hudson River population and populations from Georgia. The two Canadian populations are monomorphic, whereas there is a pronounced latitudinal cline in the number of composite mtDNA haplotypes and in haplotypic diversity, which increased from north to south (Wirgin et al. 2000). This pattern seems due to the geographically recent colonization of Canadian Atlantic rivers by Atlantic sturgeon after the last glaciation (ca. 10,000 years ago).

Management. Young et al. (1988) used an age-structured population model to simulate predicted harvest and stock size for different fishing strategies for the Atlantic sturgeon population in the Hudson River. They noted that the slow growth, late maturity, and long life span of Atlantic sturgeon made them particularly vulnerable to overexploitation. This was well illustrated by Boreman (1997), who compared lifetime egg production in Atlantic and other sturgeons under differing levels of fishing mortality. Sturgeon populations in the Hudson and elsewhere are probably still in the process of recovering from overexploitation that occurred at the turn of the nineteenth century. Young et al. (1988) recommended that a minimum size limit of 72 in. be implemented for riverine waters, as suggested for coastal waters by Dovel (1978) and Hoff (1980), and that an annual catch quota be established for the Hudson River. Unfortunately ocean intercept fisheries decimated the Hudson River population in the early 1990s. More recent yield per recruit and eggs per recruit models have been used to estimate target fishing rates (*F*) and potential yield from this population (Kahnle et al. 1998). The Atlantic States Marine Fisheries Commission (ASMFC), which is charged with managing Atlantic sturgeon, finally closed all U.S. fisheries coastwide in 1996, recognizing that the stocks were severely depleted throughout the range of the species in the United States (Vecsei and Peterson 2000a). In addition, ASMFC has recommended that the moratorium remain in place for 20–40 years so that each spawning stock can be restored to a level where 20 protected year classes of adult females are present (ASMFC 1998). Populations of Atlantic sturgeon have declined so far that in some river systems, the species is even less abundant than the shortnose sturgeon, which is listed as endangered under the U.S. Endangered Species Act (ESA). The National Marine Fisheries Service and the U.S. Fish and Wildlife Service received a petition to list Atlantic sturgeon as endangered (Friedland 1998a) but refused (NMFS 1998b), citing the recent ASMFC moratorium. Restoration of Atlantic sturgeon in many river systems may depend on introduction of hatchery-reared juveniles (Secor et al. 2000)

ACKNOWLEDGMENTS. Drafts of the sturgeon section were reviewed by Don Flescher and Kevin D. Friedland.

TEN-POUNDERS AND TARPONS. ORDER ELOPIFORMES

The subdivision Elopomorpha is defined by the presence of a leptocephalus larva and contains two orders: Elopiformes and Anguilliformes. The Elopiformes contains two families, two genera, and about eight species (Nelson 1994). These fishes are characterized by having abdominal pelvic fins, a slender body, wide gill openings, a deeply forked caudal fin, seven hypural bones in the caudal skeleton, and a well-developed median gular plate. They were previously thought to be related to clupeoids, the herringlike fishes, but are now considered most closely related to the Anguilliformes because both groups have a leptocephalus larva. However, the leptocephalus larva of Elopiformes differs from that of eels in having a forked caudal fin.

KEY TO GULF OF MAINE ELOPIFORMES

1a. Scales small, lateral line scales 103–120; last dorsal fin ray not prolonged . . . . **Ladyfish**
1b. Scales very large, 40–48 in lateral series; last dorsal fin ray prolonged . . . . **Tarpon**

TEN-POUNDERS. FAMILY ELOPIDAE

ROY E. CRABTREE

Six species of *Elops* are recognized from tropical and subtropical seas worldwide (Whitehead 1962). Myomere counts of *Elops* larvae indicate the existence of two populations or species in the western Atlantic (Smith 1989c). Larvae from the Gulf of Mexico and the U.S. Atlantic coast have 79–86 myomeres and were referred to *E. saurus,* whereas those from the Caribbean and Bahamas had 74–78 myomeres and were referred to *Elops* sp. Any ladyfish from the Gulf of Maine is most likely to be the more northern form, *E. saurus.*

LADYFISH / *Elops saurus* Linnaeus 1766 / Bigelow and Schroeder 1953:86

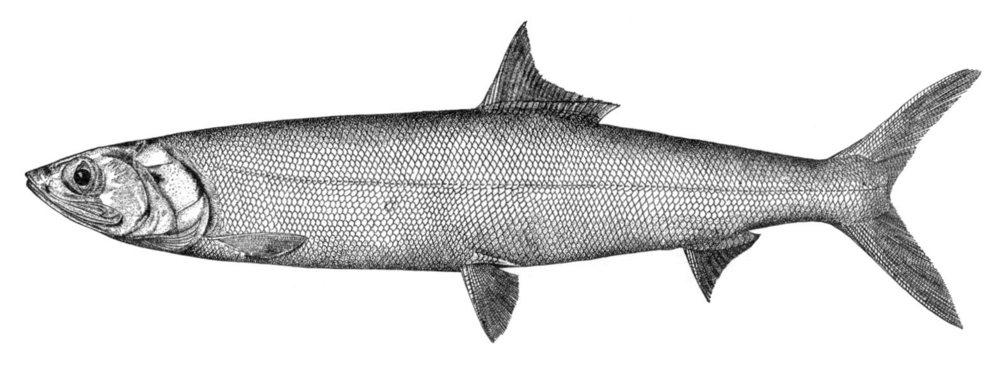

Figure 58. Ladyfish *Elops saurus.* Massachusetts. Drawn by H. L. Todd.

Description. Herringlike in fin arrangement, with a single, soft-rayed dorsal fin originating about midway along back, no adipose fin, pelvic fins located midway between tip of snout and fork of tail (Fig. 58). Scales small and thin, lateral line nearly straight, extending onto caudal base. Axillary scale of pectoral about one-half length of fin (Hildebrand 1963a).

Meristics. Dorsal fin rays 21–25; anal fin rays 14–17; pectoral fin rays 16–17; lateral line scales 103–120; gill rakers 6–8 + 10–15 (excluding rudiments); branchiostegal rays 27–35; vertebrae 73–80 (Hildebrand 1963a).

Color. Silvery all over, with the back bluish, the lower parts of the sides and the belly yellowish; the dorsal and caudal fins dusky yellowish and silvery; the pelvic and pectoral fins yellowish, speckled and dusky.

Size. Ladyfish are said to grow to 900 mm FL, but few reach more than about 500 mm FL (Hildebrand 1963a). There is a tie for the all-tackle world record game fish between two fish weighing 2.72 kg, one caught in the Loxahatchee River at Jupiter, Fla., in December 1997 and the second caught in Rio de Janeiro in January 1999 (IGFA 2001).

Distinctions. Among Gulf of Maine fishes, ladyfish are most similar to tarpon, but have much smaller scales and lack the prolonged last dorsal fin ray of tarpon.

Habits. Ladyfish are common in estuaries and coastal waters of tropical and subtropical latitudes. They can be extremely abundant and often occur in large schools. They are tolerant of a wide range of salinities but seldom occur in freshwater (Zale and Merrifield 1989). Ladyfish have been collected over a temperature range of 11°–35°C (Zale and Merrifield 1989). They are primarily nocturnal over shallow seagrass-covered banks in Florida Bay (Sogard et al. 1989).

Food. Ladyfish feed principally in midwater on pelagic prey, mainly fishes, but also decapod crustaceans (Sekavec 1974).

Reproduction. Ladyfish have a leptocephalus larva characterized by slender, fanglike teeth and an elongate, laterally compressed body consisting principally of an acellular mucinous material. Spawning appears to occur offshore (Hildebrand 1943; Gehringer 1959; Eldred and Lyons 1966; Smith 1989c). Larvae are common in the Gulf of Mexico and off the southern United States, where they have been reported as far north as Virginia (Smith 1989c). Fertilized eggs are undescribed (Jones et al. 1978; Zale and Merrifield 1989). Spawning may occur throughout the year but probably peaks during fall in Florida and the Gulf of Mexico (Smith 1989c; Zale and Merrifield 1989).

General Range. Ladyfish occur from Brazil northward to southern New England but are rare north of North Carolina.

Occurrence in the Gulf of Maine. One reason for including this southern fish is a specimen from Chatham, Mass., that could have been taken from the Gulf of Maine shore of Cape Cod (Bigelow and Schroeder 1940). A second specimen was taken just south of the Gulf from Pleasant Bay near Chatham in November 1954. Ladyfish are taken from time to time near Woods Hole and are most likely to occur in the area from July through September.

Importance. Ladyfish are often caught by recreational anglers but are seldom targeted. The species is fished commercially in Florida and sold both for food and as bait.

TARPONS. FAMILY MEGALOPIDAE

ROY E. CRABTREE

The family contains two species, *Megalops cyprinoides* of the Indo–West Pacific, and tarpon, *M. atlanticus* of the Atlantic. Some authors place tarpon in its own genus, *Tarpon,* but Nelson (1994) included both species in *Megalops.*

TARPON / *Megalops atlanticus* Valenciennes 1847 / Bigelow and Schroeder 1953:87 (as *Tarpon atlanticus*)

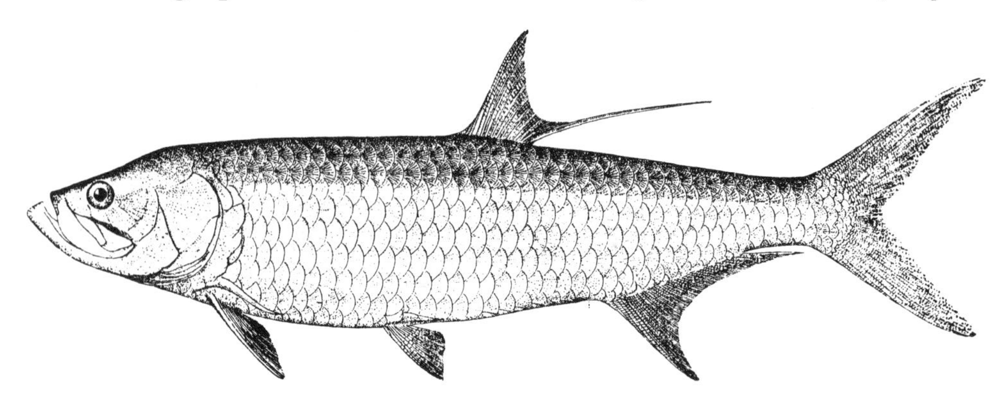

Figure 59. Tarpon *Megalops atlanticus.* New Jersey. Drawn by H. L. Todd.

Description. Body elongate, highly compressed (Fig. 59). Mouth oblique, lower jaw prominently projecting. Teeth small, bluntly villiform on jaws, vomer, palatines, pterygoids, tongue, and basibranchials (Hildebrand 1963a). Dorsal fin single, located behind pelvic fins but entirely before anal fin, with a greatly prolonged final ray. Terminal ray of anal fin somewhat elongate, but much less so than that of dorsal fin. Pectoral and pelvic fins with elongate axillary scales. Caudal fin deeply forked. Lateral line present, relatively straight. Scales large.

Meristics. Dorsal fin rays 13–15; anal fin rays 22–25; pectoral fin rays 13 or 14; lateral line scales 41–48; gill rakers 19–21 + 36–40; branchiostegal rays 23–27; vertebrae 53–57 (Hildebrand 1963a).

Color. Bright silvery all over, the back darker than the belly.

Size. Tarpon reach lengths of 1.8–2.2 m FL. The all-tackle world record game fish weighed 128.50 kg and was caught off Sierra Leone in April 1991 (IGFA 2001). Tarpon are sexually dimorphic; males rarely exceed 50 kg, but females reach larger sizes (Crabtree et al. 1995, 1997). In Florida, Central America, and the Caribbean, tarpon larger than 40 kg are common. Tarpon from Florida appear to grow larger than those from Costa Rica (Crabtree et al. 1997).

Distinctions. Tarpon are herringlike in general form and appearance but are easily recognized by the greatly prolonged last ray of the dorsal fin, which can be as long as or longer than the height of the dorsal fin, and by the presence of a bony gular plate between the branches of the lower jaw. They have much larger scales than ladyfish and an elongate terminal dorsal fin ray that ladyfish lack.

Habits. Tarpon are large, migratory fish that frequent coastal and inshore waters of the tropical and subtropical Atlantic Ocean. They occur in a wide variety of habitats ranging from freshwater lakes and rivers to offshore marine waters, but large tarpon targeted by recreational anglers are most abundant in estuarine and coastal waters. Young-of-the-year tarpon occur in small stagnant pools and sloughs of varying salinity and have been reported from North Carolina (Hildebrand 1934), Georgia (Richards 1968), Florida (Wade 1962, 1969), Texas (Simpson 1954; Marwitz 1986), Caribbean islands (Breder 1933), and Costa Rica (Chacon Chaverri and McLarney 1992). In tropical areas, juvenile tarpon typically occur in mangrove habitats, often in water with low dissolved oxygen levels. Tarpon occur in salinities ranging from freshwater to more than 45 ppt and are capable of surviving temperatures of at least 40°C, but they suffer mortalities at temperatures of 10°–12°C (Zale and Merrifield 1989).

Their habit of rising to the surface and breathing air is unusual among marine species (Breder 1942; Geiger et al. 2000). Anglers often detect the presence of schools of tarpon by observing individuals "rolling" at the surface. Air-breathing is accomplished by way of a highly vascularized swim bladder that functions as an air-breathing organ. This adaptation allows tarpon to survive in water with low dissolved oxygen concentrations such as is commonly encountered by juveniles. Experimental work suggests that tarpon are facultative air-breathers and in well-oxygenated waters are able to meet their oxygen requirements without breathing air (Geiger et al. 2000).

Food. Juvenile tarpon (16–75 mm SL) from Florida waters fed predominantly on cyclopoid copepods, fishes, caridean shrimp (*Palaemonetes*), and *Aedes* mosquitoes (Harrington and Harrington 1960). Juveniles from Brazil fed principally on insects (Hemiptera and Diptera larvae), fishes (poeciliids and larval fishes), and crustaceans, including cladocerans, copepods, and decapods (de Menezes 1968). No detailed studies have examined the feeding habits of large tarpon from Florida waters, but a wide variety of fishes are consumed, including mullet, pinfish (*Lagodon rhomboides*), ariid catfishes, needlefishes (*Strongylura* spp.), and herrings (*Harengula*), as well as crabs and shrimps (Zale and Merrifield 1989). Large tarpon from Brazil consume fishes (Carangidae, Clupeidae, and Scombridae) crustaceans (portunid crabs), as well as occasional mollusks and plants (de Menezes 1968).

Reproduction. Fecundity estimates range from 4.5 to 20.7 million oocytes, and fecundity has a significant positive relationship with fish size (Crabtree et al. 1997).

Tarpon have a leptocephalus larva characterized by slender, fanglike teeth and an elongate, laterally compressed body consisting principally of an acellular mucinous material. Based on distribution of larvae, tarpon are thought to spawn offshore (Smith 1980; Crabtree et al. 1992; Crabtree 1995). Spawning off Florida occurs from mid-May through mid-August with peak activity during June and July (Crabtree et al. 1992, 1997; Crabtree 1995). Spawning appears to be related to lunar phase; hatching date frequencies are consistently associated with new and full moons (Crabtree 1995). Fertilized eggs of tarpon are undescribed. Little is known about the processes that transport larvae from offshore spawning grounds to inshore juvenile habitats. The length of the larval period is estimated to be 2–3 months (Smith 1980; Cyr 1991; Crabtree et al. 1992). Metamorphic larvae are typically found inshore in mangrove-lined estuaries but also occur in temperate *Spartina* marshes (Harrington 1958; Erdman 1960; Wade 1962; Harrington 1966; Mercado and Ciardelli 1972; Tucker and Hodson 1976; Chacon Chaverri and McLarney 1992). Larvae have been reported as far north as North Carolina (Tucker and Hodson 1976) but are not known from the Gulf of Maine.

Age and Growth. Tarpon are relatively long-lived and can reach ages in excess of 50 years (Crabtree et al. 1995). By age 1, tarpon are about 400 mm FL and are common in rivers and the upper reaches of estuaries, where they remain until reaching sexual maturity. In Florida, sexual maturity is reached at 900–1,300 mm FL and about 10 years of age (Crabtree et al. 1997). After attaining sexual maturity, tarpon become more coastal in habit and are most numerous around inlets and off beaches. Large tarpon targeted by anglers in Florida are typically 15–35 years old.

General Range. Tarpon are restricted to the tropical and subtropical Atlantic. In the western Atlantic, they regularly occur from the eastern shore of Virginia to central Brazil and throughout the Caribbean Sea and Gulf of Mexico (Hildebrand 1934, 1963a; Wade 1962; de Menezes and Paiva 1966; Zale and Merrifield 1989). At least seven records exist from as far north as Nova Scotia (Hildebrand 1963a; Scott and Scott 1988), where tarpon have been reported from Isaac's Harbor (Halkett 1913), Harrigan Cove, Terence Bay, the Liscomb River, Sambro Light, Wine Harbor, and Porter's Lake in Halifax County (Scott and Scott 1988). Tarpon are also present in the eastern Atlantic from Senegal to Angola (Saldanha and Whitehead 1990).

Occurrence in the Gulf of Maine. A specimen 1,676 mm long, taken at Provincetown on 25 July 1915 (Radcliffe 1916), is the only record of the tarpon in the Gulf of Maine, which it reaches only as an accidental straggler from the south.

Importance. Tarpon are among the most highly esteemed recreational fish species in the world. Their large size, abundance in inshore waters, and spectacular leaps when hooked make them favorites of inshore anglers. Fisheries for tarpon are best developed in Florida and Costa Rica but occur throughout the Gulf of Mexico and Caribbean Sea. In south Florida and Costa Rica, tarpon support economically important charter boat fisheries. Tarpon are not utilized as food in the United States.

EELS. ORDER ANGUILLIFORMES

Eels have no pelvic fins or girdle; scales are either absent or small, embedded, and hardly visible; the fins are without spines; the gill openings are very small; the vertebrae extend in a straight line to the tip of the tail; and the dorsal, anal, and caudal fins are confluent over the back, around the tail, and forward on the ventral surface. All eels known from the Gulf of Maine have pectoral fins, but many others, including the well-known morays of warmer seas, lack pectorals. All eels have a leptocephalus larva, a transparent ribbonlike larva with dorsal and anal fins continuous around the tip of the tail, a larva that shrinks in metamorphosing into the juvenile form. Adults of several other fishes in the Gulf of Maine also have an eel-like form: hagfish and lampreys, rock gunnel (*Pholis*), snake blenny (*Lumpenus*), wrymouth (*Cryptacanthodes*), eelpout (*Zoarces*), and sand lance (*Ammodytes*). The jawless, suckerlike mouth immediately separates the first two from the true eels, whereas the others have either a well-marked separation between anal and caudal fins, pelvic fins (large or small), or a spiny dorsal fin.

There are 15 families, 141 genera, and about 738 species of eels (Nelson 1994), mostly living in tropical marine waters. Bigelow and Schroeder recorded only four species of eels from the Gulf of Maine: American eel, conger eel, margined snake eel, and slender snipe eel, all belonging to different families according to current usage. Northern cutthroat eel are a deepwater species to be expected in the deepest parts of the Gulf although they have not yet actually been recorded there. An additional three species were reported from stations in or near the Gulf of Maine (Musick 1973b): narrowneck eel, spoonbill eel, and stout sawpalate eel. However, Bigelow and Schroeder also included the deep-sea snubnose eel, *Simenchelys parasitica,* as possibly occurring in the Gulf of Maine. This species has been excluded here since no specimens have been recorded from the Gulf, and the shallowest capture reported by Robins and Robins (1989) is at 300 fathoms, considerably deeper than the deepest waters of the Gulf. Thus, representatives of seven families of eels have been taken in the Gulf of Maine or very close to it. In addition, leptocephalus larvae of at least 16 species of eels have been reported in continental shelf waters off Nova Scotia (Scott and Scott 1988) or collected by *Albatross IV* Cruise 94-02.

Common, conger, longneck, and northern cutthroat eels look much alike in general form, but differ from one another by the size of the mouth and by the relative lengths of the fins. Margined snake eel are very slender with a hard pointed tail. In snipe eel, both jaws are prolonged into a very long, slender beak, the tail is whiplike, the neck noticeably slimmer than the head, and the general form extremely slender and ribbonlike. Whereas stout sawpalate eel have their jaws prolonged into a long snout, they are not as prolonged as in slender snipe eel and are armed with numerous long, sharp teeth on the maxillary and mandible. In addition, stout sawpalate eel have two alternating series of large, laterally compressed vomerine teeth, resulting in the typical sawtooth dentition. Duckbill eel have the upper jaw prolonged into a spatulate snout.

KEY TO GULF OF MAINE EELS

1a. Both jaws elongate and slender 2
1b. Jaws not elongate or slender 4
2a. Jaws extremely elongate, diverging anteriorly . . . . **Slender snipe eel**
2b. Jaws slightly elongate, not diverging anteriorly. 3
3a. Upper jaw longer than lower and flattened into a spatulate tip; no median row of vomerine teeth. . . . . **Spoonbill eel**
3b. Upper and lower jaws pointed, about equal in length; median row of large vomerine teeth . . . . **Stout sawpalate eel**
4a. Anal fin origin near or in front of origin of dorsal fin . . . . **Northern cutthroat eel**
4b. Anal fin origin well behind origin of dorsal fin. 5
5a. Dorsal fin origin far behind (greater than head length) tips of pectorals . . . . **American eel**
5b. Dorsal fin origin close behind tips of pectorals. 6
6a. Mouth very small, its gape not reaching back as far as eye; body very soft . . . . **Snubnose eel**
6b. Mouth large, gaping back as least as far as middle of eye; body firm. . . 7
7a. Mouth gaping back considerably beyond eye; body slender, tip of tail hard and pointed . . . . **Margined snake eel**
7b. Mouth gaping back only about as far as middle or rear edge of eye, tip of tail soft, rounded 8
8a. Dorsal origin over tip of pectoral fin . . . . **Conger eel**
8b. Dorsal origin about pectoral fin length behind tip of pectoral fin . . . . **Longneck eel**

FRESHWATER EELS. FAMILY ANGUILLIDAE

David G. Smith and Kenneth A. Tighe

At least in terms of relationships to man, the Anguillidae is the most important family of eels. Freshwater eels are very important commercially, especially in Europe and Japan, where they are considered a delicacy. Anguillids are unusual among eels because adults live and grow in freshwater or estuaries but return to the sea to spawn. The leptocephalus larva returns to coastal areas, metamorphoses into an elver, and runs upstream to mature. There is one genus with 15 recognized species in the world (Smith 1989a).

AMERICAN EEL / *Anguilla rostrata* (LeSueur 1817) /

Common Eel, Silver Eel, Freshwater Eel, Elver (juvenile) / Bigelow and Schroeder 1953:151–154

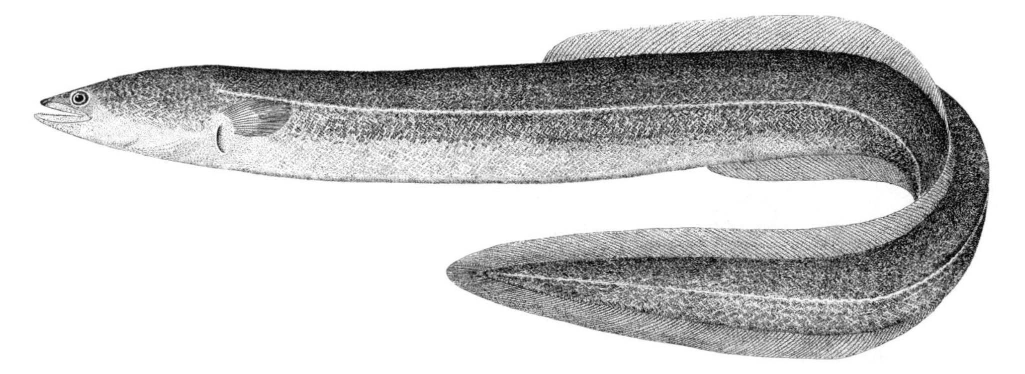

Figure 60. American eel *Anguilla rostrata.* Connecticut River, Mass. Drawn by H. L. Todd.

Description. Body moderately elongate, rounded in cross section anteriorly and somewhat compressed posteriorly (Fig. 60). Head rather long, about 15% of TL. Eyes small, round, well forward on head. Snout moderately short; lower jaw projecting beyond upper (or at least equal in length); mouth large, gaping back as far as middle of eye or past it. Gill slits set vertically on sides of neck, their upper corners abreast of center of pectoral fin base. Pectoral fins well developed, rounded. Dorsal fin origin far behind tip of pectorals. Scales small, elongate, embedded, arranged in a basket-weave pattern and easily overlooked.

Meristics. Pectoral fin rays 14–20; branchiostegal rays 9–13; precaudal vertebrae 41–45, total vertebrae 103–112 (Smith 1989a).

Color. Colors of American eel vary widely with the bottom over which they live. As a rule they are dark muddy brown or olive brown above, more or less tinged with yellow on the sides; the lower surface paler brown and yellower, with a dirty yellowish white belly. It is common knowledge that eels are dark if living on dark mud but much paler on pale sand. They can change from pale to dark in about 1.5 h and from dark to pale in a little more than 3 h, if moved from a white background to a black or vice versa, under a strong light (Parker 1945).

Size. Bigelow and Schroeder reported a maximum size of 120 cm TL and 7.5 kg. Females grow much larger than males and average about 60–100 cm. Any American eel more than 45 cm is probably a female, and one more than 60 cm would certainly be. The all-tackle game fish record is a 4.21-kg eel caught at Cape May, N.J., in November 1995 (IGFA 2001).

Distinctions. American eel are most similar to conger eel among Gulf eels but the dorsal fin origin is far behind the pectoral fins and the lower jaw projects beyond the upper or at least equals it in length.

Taxonomic Remarks. American eel are very closely related to the European eel (*Anguilla anguilla*) and have been considered conspecific with it by some authors (Meek 1883; Tucker 1959). American eel have fewer vertebrae (average about 107 compared with about 114 or 115 in European eel). There are also differences in proportional measurements, biochemical genetics, and chromosomal morphology that indicate that the species are separate. Smith (1989a) summarized the evidence for both views and followed the traditional course of considering them two species. The most recent molecular comparison using the randomly amplified polymorphic DNA (RAPD)-PCR technique (Lehmann et al. 2000) concluded that *A. rostrata* and *A. anguilla* were distinct but closely related species as indicated by their low genetic distance (0.384) and high degree of shared RAPD fragments (SF = 71.2%) compared with two Pacific species of *Anguilla.*

Habits. American eel are common inhabitants of streams, rivers, lakes, tidal marshes, and estuaries around the entire periphery of the Gulf of Maine. It is true in a general way that American eel seek muddy bottom and still water, but this is not always so whether in saltwater or in freshwater. Thus the rocky pool at the outer end of the outlet from Little Harbor, Cohasset, on the south shore of Massachusetts Bay, has been a good place to catch American eel, and large ones are common in swift-flowing, sandy trout streams on Cape Cod. The fact is, they can live and thrive wherever food is to be had in estuarine situations and in freshwater.

American eel are chiefly nocturnal, as every fisherman knows, usually lying buried in the mud by day to venture abroad at night. But American eel, large and small, are so often seen swimming about and so often bite by day that this cannot be a general rule. American eel can live out of water long enough to give rise to the story that they often travel overland. There is no positive evidence for this, but Sella (1929) proved, by experiments with marked European eel, that they can carry out journeys as long as 50 km along underground waterways. Doubtless it is this ability that explains the presence of American eel in certain ponds that have no visible outlet or inlet.

American eel tolerate a wide range of temperatures, but it is common knowledge that those inhabiting the salt marshes and estuaries of the Gulf and its tributary streams generally lie inactive in the mud during the winter.

Except, presumably, when they spawn, American eel are solitary in habit and exhibit little in the way of social behavior. They do not school in the true sense of the word, nor do they stake out and defend territories.

Food. Bigelow and Schroeder reported that American eel are both predators and scavengers and will eat almost anything they can swallow; that no animal food, living or dead, is refused; and that the diet of eel in any locality depends less on choice than on what is available. American eel in freshwater feed on insects, worms, crustaceans, frogs, and other fishes (Vladykov 1955c). In freshwater streams in New Jersey, smaller eel (less than 40 cm in length) feed primarily on insects while larger eel feed mostly on fishes and crustaceans (Ogden 1970). Polychaetes, crustaceans, and bivalves are important food items in the brackish waters of the lower Chesapeake Bay (Wenner and Musick 1975).

Predators. American eel serve as food for a variety of larger predators, although a large eel probably has few enemies. The only records of predation on American eel from the NEFSC food survey are spiny dogfish and haddock (Rountree 1999).

Reproduction and Early Life History. American eel are among the most prolific fishes: fecundity has variously been reported as 3–10 million eggs (Shulman and Paine 1974), 10–20 million (Vladykov 1955c), and 15–20 million (Eales 1968). Individuals migrating from Chesapeake Bay had lower fecundity estimates, the largest of 21 females having approximately 2.5 million eggs (Wenner and Musick 1974).

The life history of American eel remained a mystery until early in the twentieth century. It had long been common knowledge that young elvers run up into freshwater in spring, and that adults journey downstream in autumn. A host of myths grew up to explain the complete absence of ripe eel of either sex, either in freshwater or along the seacoast, but it was only through the persevering research of the Danish scientist Johannes Schmidt that the breeding places of European and American eels were discovered and the history of their larvae traced (Schmidt 1922). The life history of American eel is the antithesis of that of anadromous salmon, shad, and alewife, for eel breed far out at sea, but grow either in estuaries or in freshwater.

American eel may spend from 5 to 20 years in fresh or coastal waters feeding and growing. Sexual maturity takes place in the late summer or fall. At this point, eel stop feeding and the yellowish green color changes to a sharply bicolored pattern, gray to black above and pure white below. Often a silvery or bronze sheen appears on the flank, leading to the common name "silver eel" for this stage. The eyes and pectoral fins enlarge, and eel begin moving downstream, traveling mainly at night. Activity was greatest within the first 2 h of darkness and completed before midnight (Hain 1975); rainfall also seems to stimulate movement of silver eels. September to November is the peak period for this downstream migration.

Once American eel leave the shore, they drop wholly out of sight. What route they take and how they navigate remain mysteries, but by February they must arrive on the spawning grounds in the Sargasso Sea, south of Bermuda and east of the Bahamas. We know this because, although adults and eggs have never been taken, the young larvae appear in great numbers here at this time. The leptocephalus larva, like that of all eels, is very different in appearance from the adult, being ribbonlike and perfectly transparent, with a small pointed head; and it has very large teeth, although it is generally believed to take no food until metamorphosis. The leptocephali of American eel, living near the surface, have been found off the Gulf coasts as far north as the Grand Banks, but never east of longitude 50° W. Inasmuch as the breeding areas of American and European eels overlap, not the least interesting phase of the lives of the two is that the larvae of the American species work so consistently to the western side of the Atlantic and those of the European to the eastern side that no specimen of the former has ever been taken in Europe or of the latter in America.

Schmidt (1922) believed that American eel take only about a third as long as the European to pass through its larval stage (i.e., hardly 1 year, as against 2 or 3 years), and that this differential in growth rates resulted in the segregation of European and American species. However, leptocephali of both species grow at approximately the same rate (Boëtius and Harding 1985). Clearly, we do not yet understand the co-occurrence of these two closely related species in the Atlantic.

Leptocephali of American eel reach their full length of 60–65 mm by December or January, when metamorphosis takes place to the elver stage. The most obvious changes are a shrinkage in the depth and length of the body but an increase in thickness to cylindrical form, loss of larval teeth, and total alteration in the aspect of head and jaws, while the digestive tract becomes functional.

It is not until they approach the Gulf shores, however, that the elvers develop adult pigmentation or begin to feed. How such feeble swimmers as leptocephali find their way into the neighborhood of the land remains a mystery. It seems certain, however, that all young American eel bound for the Gulf of Maine complete the major part of their metamorphosis while they still are far offshore.

Young elvers, averaging 50–90 mm in length, appear along the Gulf shores in spring. They occur as early as March at Woods Hole; in both Narragansett Bay and in Passamaquoddy Bay at the mouth of the Bay of Fundy by mid- or late April; and may be expected in the mouths of most Gulf of Maine streams during May. They ascend streams in the Bay of Fundy region during the summer. A run may last for a month or more in one stream and only for a few days in another. There is a noticeable segregation even at this early stage, with some of the elvers remaining in tidal marshes, in harbors, in bays behind barrier beaches, and in other similar situations, with some even along the open coast, especially where there are beds of eelgrass (*Zostera*); still others go into freshwater, some ascending the larger rivers for tremendous distances.

Age and Growth. The smallest mature males are about 28–30 cm long and females about 45 cm. American eel grow slowly. Hildebrand and Schroeder (1928) concluded from a series of measurements taken at different seasons in lower Chesapeake Bay that those 6.4 cm long in April are about 13 cm long 1 year later, or about 2 years after their transformation. Winter rings on the scales have shown that full-grown adults of the European species are 5–20 years old, depending on food supply and other conditions. Ages of 9–17 years, based on otoliths, were found in females from Newfoundland (Gray and Andrews 1970).

General Range. Coasts and streams of west Greenland, eastern Newfoundland, Strait of Belle Isle, and the northern side of the Gulf of St. Lawrence south to the Gulf of Mexico, Bermuda, West Indies, Panama, and (rarely) to the north coast of South America; running into freshwater but returning to the sea to spawn.

Occurrence in the Gulf of Maine. Occurrence of American eel around the periphery of the Gulf can be described by the word *universal*. There is probably no harbor, stream mouth, muddy estuary, or tidal marsh from Cape Sable on the east to the elbow of Cape Cod on the west that does not support American eel in some numbers, and they run up every Gulf of Maine stream, large or small, from which they eventually find their way into the ponds at the headwaters unless barred by insurmountable barriers such as very high falls. A few eels (and some large ones) are sometimes seen along the open coast but always close in to the shoreline and in less than 1 m of water, where flounder fishermen catch them from time to time.

Movements. Although spawning migrations of American eel are now fairly well understood, less is known about their daily and seasonal movements. Marked and recaptured American eel in two Louisiana localities remained within some 60–140 linear meters of stream (Gunning and Shoop 1962). On the other hand, presence of American eel far up the St. Lawrence, Mississippi, and other rivers indicates that some of them travel great distances after entering freshwater. It has been suggested that the extent of inland migration depends on sex, with females occurring mainly in freshwater and males mainly in brackish, but many exceptions to this rule are found.

Examples of long journeys by American eel upstream in New England rivers are to the Connecticut Lakes, N.H., at the head of the Connecticut River; to the Rangeley Lakes, Maine, at the head of the Androscoggin River, and to Grand Matagamon Lake, Maine, at the head of the East Branch of the Penobscot River. American eel are even caught in certain ponds without outlets.

Importance. Schmidt suggested that American eel are not as plentiful in actual numbers as European, arguing from the facts that its larvae have not proven so common on the high seas and that the American catch of eels (about 2,000 tons yearly) was only a fraction as large as the European catch (about 10,000 tons annually). However, it is not safe to draw any conclusions from these statistics because the American catch was limited more by demand than by the available supply. Much of the commercial catch today is exported to Europe and Japan, where eel is considered a delicacy. The greater part of the catch is made in nets and eelpots although some are speared, mostly in late autumn and winter, often through the ice. Various aspects of the fishery and summaries of commercial landings for 1955–1973 and 1980–1985, were presented by Fahay (1978) and Facey and Van Den Avyle (1987), respectively, and aspects of the eel fisheries of eastern Canada were summarized by Eales (1968). In the 1970s, a fishery developed for elvers that were used for aquaculture (Scott and Scott 1988; Meister and Flagg 1997). The preponderance of data from resource and fisheries agencies along the east coast of the United States and Canada suggests that there has been a continent-wide decline in eel abundance over the last 10 years (Richkus and Whalen 2000). According to Meister and Flagg (1997), recent declines in landings of both adult eels and elvers may be due to overexploitation of the resource, especially elvers. Reasons for the decline, whether natural or anthropogenic, are unknown but possible contributing factors may include barriers to migration, habitat loss and alteration, hydroturbine mortality, oceanic conditions, overfishing, parasitism, and pollution (Haro et al. 2000; Richkus and Whalen 2000). Although American eel are not considered a sport fish, their ubiquity and readiness to take a bait lead them to be commonly caught by recreational fishermen. In addition, there is a small commercial market for American eel as bait for larger sport fish such as striped bass.

CUTTHROAT EELS. FAMILY SYNAPHOBRANCHIDAE

KENNETH A. TIGHE AND DAVID G. SMITH

Cutthroat eels are deep benthic species. The gill openings are low on the body, at or below the insertion of the pectoral fins. The leptocephalus larva differs from all other eels in having vertically or horizontally elongate (*tubular*) eyes (Robins and Robins 1989). Eleven genera and 29 species are recognized, placed in three subfamilies (Sulak and Shcherbachev 1997).

One species from the Synaphobranchinae, the northern cutthroat eel, occurs in deep waters close to the Gulf of Maine. Bigelow and Schroeder also included the deep-sea snubnose eel, *Simenchelys parasitica,* as possibly occurring in the Gulf of Maine. It is excluded because no specimens have ever been collected in the Gulf of Maine and the shallowest capture reported by Robins and Robins (1989), 300 fathoms, is considerably deeper than the deepest waters of the Gulf.

NORTHERN CUTTHROAT EEL / *Synaphobranchus kaupii* Johnson 1862 / Longnose Eel /

Bigelow and Schroeder 1953:158–159 (as *Synaphobranchus pinnatus*)

Figure 61. Northern cutthroat eel *Synaphobranchus kaupii*. La Have Bank. Drawn by H. L. Todd.

Description. Body elongate, moderately compressed anteriorly; greatly compressed posteriorly, tapering gradually to caudal fin (Fig. 61). Head relatively long (11–16% of TL); snout long and pointed; mouth extending well behind eye. Eyes relatively large, oval with greatest dimension horizontally. Dorsal fin originating at or behind anal origin. Pectoral fin well developed, about as long as snout, tapering to a point. Scales present, arranged in groups of three to six aligned at right angles in a basket-weave pattern; typical scale length-width ratio 3.1–4.1 (Sulak and Shcherbachev 1997: Fig. 9A).

Meristics. Pectoral fin rays 14–17; caudal fin rays 13; vertebrae 144–152 (Robins and Robins 1989).

Color. Grayish, darkest below, with the vertical fins darker behind but pale-edged in front, and with the inside of the mouth blue-black.

Size. Maximum size recorded is 813 mm TL (Grey 1956). Sexually mature individuals in waters off the Bahamas ranged from 430 to 640 mm TL (Robins and Robins 1989). Maturing males were 290–460 mm and sexually maturing females 370–660 mm TL in a population off northwest Africa (Merrett and Domanski 1985).

Distinctions. This deep-sea eel is readily identifiable by the fact that while the dorsal fin originates about as far back as in American eel, the point of origin is considerably behind the vent instead of in front of the latter, and the anal fin originates considerably in front of the dorsal fin instead of behind it as is the case in all other Gulf of Maine eels. Furthermore, the mouth is much wider, gaping far behind the eye, and the snout is pointed. The most interesting anatomical character is that the gill openings, opening longitudinally on the lower side of the throat, join together in front, apparently as a single V-shaped aperture, though actually they are separate.

Biology. *Synaphobranchus kaupii* is known from 493 captures (10,346 specimens) at depths of 274–2,869 m and bottom temperatures of −1.1° to 9.8°C (Sulak and Shcherbachev 1997). Stomachs of North Atlantic specimens often contain squids (Robins and Robins 1989). Circular superficial injuries, seemingly inflicted by squids, suggest that northern cutthroat eel in turn are preyed upon by larger squids. A broad diet of fishes, cephalopods, and crustaceans was reported in specimens from off the eastern Atlantic continental slope (Saldanha 1980; Merrett and Domanski 1985). Spiny dogfish are the only known predator (Rountree 1999).

Reproduction and Life History. Published information on life history is somewhat contradictory. Bigelow and Schroeder reported that specimens in spawning condition were taken in summer off New England. Robins and Robins (1989), working in the Tongue-of-the-Ocean, Bahamas, found ripe individuals all year long except in May, June, December, and January. Bruun (1937) found a pronounced seasonal pattern to the size distribution of leptocephali, with the smallest specimens (about 20 mm TL) present in the winter and increasing in average size as the year went on. In the spring and summer, two size-groups were found, indicating the presence of a second-year class. Thus the leptocephali seem to take about a year and a half to grow to full size and metamorphose. These data would suggest a late fall or early winter spawning season, contrary to the findings of Bigelow and Schroeder and Robins and Robins. Bruun also found that the leptocephali increased in size from west to east across the Atlantic, with those larger than 80 mm occurring only east of 30° W. Further studies are needed to reconcile these findings.

General Range. Widespread, mainly under cool to cold surface waters, but present in the Pacific Ocean only from the northwestern rim (Sulak and Shcherbachev 1997: Fig. 11). Their distribution may prove to be essentially antitropical, a pattern typical of several synaphobranchid species. In the western North Atlantic, they are found from 67° N south to the hump of Brazil.

Occurrence in the Gulf of Maine. This eel has not actually been reported within the limits of the Gulf, but they are to be expected in the Eastern Channel and possibly above 275 m along the slopes of Georges Bank. Many of them have been brought in by fishermen from deep water off the fishing banks east of 65° W, and many have been trawled along the continental slope westward from there so this eel must be common below 275–365 m, from the Grand Banks to New York.

SNAKE EELS. FAMILY OPHICHTHIDAE

DAVID G. SMITH AND KENNETH A. TIGHE

The posterior nostril is usually within or piercing the upper lip. Branchiostegal rays are numerous (15–49 pairs) and overlap along the midventral line. Ophichthids have stiff pointed tails and burrow into the bottom tail first. Snake eels and worm eels are the most diverse and speciose of the eels: 52 genera and more than 250 species (Nelson 1994), about one-third of the genera and species of eels. Their habitats range from intertidal to depths of 750 m and more, from coral reefs to sand and mud substrates (McCosker et al. 1989). One species of the snake eel subfamily Ophichthinae has been recorded from the Gulf of Maine.

MARGINED SNAKE EEL / *Ophichthus cruentifer* (Goode and Bean 1896) /

Bigelow and Schroeder 1953:159 (as *Omochelys cruentifer*)

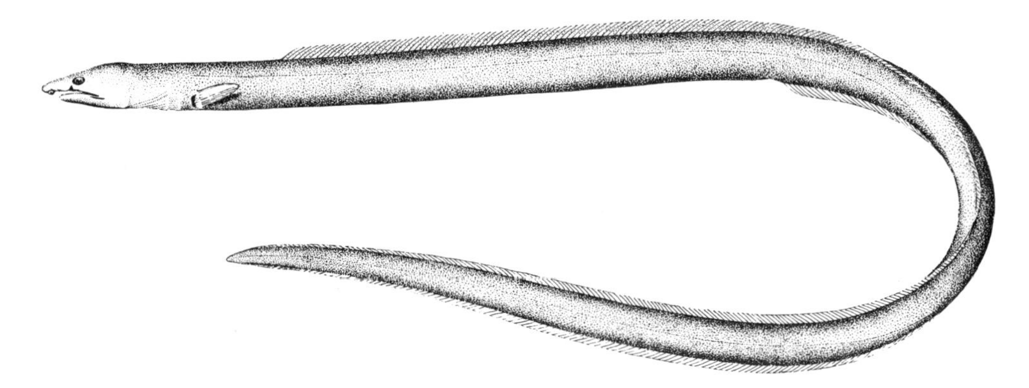

Figure 62. Margined snake eel *Ophichthus cruentifer.*

Description. Body extremely elongate, only about one thirty-seventh to one thirty-eighth as deep as it is long (Fig. 62). Head relatively short (7–9% of TL); snout bluntly pointed; mouth gapes rearward considerably beyond eyes; posterior nostrils open into upper lip. Gill openings short, crescent-shaped slits, close in front of bases of pectoral fins. Dorsal fin originates only a short distance behind pectoral fin tips; anal fin originates far behind dorsal fin; dorsal and anal fins end a little in front of tip of tail.

Meristics. Predorsal vertebrae 14–19; preanal vertebrae 56–61; total vertebrae 144–151 (McCosker et al. 1989).

Color. Uniform tan to grayish brown, with little or no countershading.

Size. Maximum size about 500 mm TL; the largest specimen examined by McCosker et al. (1989) was 467 mm TL.

Distinctions. Distinguished from all other Gulf of Maine eels by the hard and pointed tip of its tail.

Habits. Margined snake eel are bottom-dwellers, often using their hard, pointed tail to burrow into the sand or mud. They are most abundant between 250–350 m off Virginia (Wenner

1976). Specimens were observed from a submersible with only their heads exposed from the substrate; other individuals were seen resting on the bottom with their bodies in S-shaped curves (Wenner 1976). Specimens of this and other snake eel species are occasionally found mummified within the body cavities of larger fishes, a fact that has sometimes been interpreted as evidence of parasitism in the manner of snubnose eel and hagfish. Margined snake eel are not parasites; an eel, having been swallowed alive, bores through the thin walls of the alimentary tract and dies within the peritoneal cavity of its predator.

Food. Stomach contents of three margined snake eel showed the diet to be one-third polychaetes, mostly *Lumbrineris tenuis*, and two-thirds crustaceans, mostly two decapods, *Crangon septemspinosa* and *Munidaris* sp. (Bowman et al. 2000).

Predators. Four species of fishes have been recorded as one-time predators: sandbar shark, spiny dogfish, and northern and southern kingfishes (Rountree 1999).

Reproduction and Life History. Eggs of margined snake eel were caught during 11 months of the year, January to November, but were most abundant July to September, indicating predominantly summer spawning (Berrien and Sibunka 1999: Fig. 7). Eggs were first collected in scattered patches mostly in the southern part of the Middle Atlantic region followed by collections encompassing a broad area of outer shelf waters from Cape Hatteras to southern Georges Bank but not into the Gulf of Maine.

Eggs of margined snake eel were erroneously identified by Eigenmann (1902) as those of conger eel while the leptocephalus was described as a new species, *Leptocephalus mucronatus* by Eigenmann and Kennedy (1902). Richardson (1974) was the first to correctly identify the eggs and larvae (as *Pisodonophis cruentifer*) from material taken in the Chesapeake Bight, and referred *L. mucronatus* to this species. Fahay and Obenchain (1978) discussed the meristics, morphology, and distribution of larvae from the Mid-Atlantic Bight, and Naplin and Obenchain (1980) described embryonic and early larval development.

General Range. Margined snake eel are found from the Gulf of Maine to Florida in 36–1,350 m (McCosker et al. 1989).

Occurrence in the Gulf of Maine. Margined snake eel are apparently stragglers in the Gulf of Maine, being common only south of Cape Cod. A specimen from Jeffreys Bank (Goode and Bean 1896) is in the collection of the National Museum of Natural History (USNM 133872, 325+ mm TL) as is another from Georges Bank (USNM 116410, 330 mm TL).

LONGNECK EELS. FAMILY DERICHTHYIDAE

KENNETH A. TIGHE AND DAVID G. SMITH

The Derichthyidae is a small family of small mesopelagic to bathypelagic eels found in all major oceans. Scales are absent. Tail of adults equal to or shorter than remainder of total length. The nostrils are simple and porelike. There are three species in two genera (Robins and Robins 1989), two of which occasionally enter deep waters of the Gulf of Maine.

NARROWNECK EEL / *Derichthys serpentinus* Gill 1884 / Musick 1973:135

Description. Head deeper than neck; eyes large (Fig. 63). Dorsal fin originates about a head length behind base of pectoral fins; anal fin about midway along body; both dorsal and anal fins run back to tip of tail. Lateral line pores on head large, conspicuous, and associated with many straight, short sensory ridges arranged in groups.

Meristics. Pectoral rays 13; caudal fin rays 9 or 10; precaudal vertebrae 53–57; total vertebrae 128–134 (Robins and Robins 1989).

Color. Grayish to brown, head and dorsum darker and abdomen lighter-colored.

Size. Maximum known length 353 mm TL, a nearly ripe female (Robins and Robins 1989).

Distinctions. Narrowneck eel are easily recognizable by the attenuate "neck" region between the eye (and posterior margin of jaw opening) and the gill openings, and the posterior dorsal fin origin.

Biology. Narrowneck eel are a fish of the middepths of the ocean, not of the bottom. Adults seem to be most common between 200 and 700 m, whereas juveniles are often taken at depths as shallow as 75–100 m (Karmovskaya 1985). Little is known of their habits, but they seem to feed mainly, if not entirely, on pelagic shrimps (Beebe 1935). Castle (1970) identified the leptocephalus. Spawning probably occurs from April to August with a larval period of 2–3 months before onset of metamorphosis (Karmovskaya 1985).

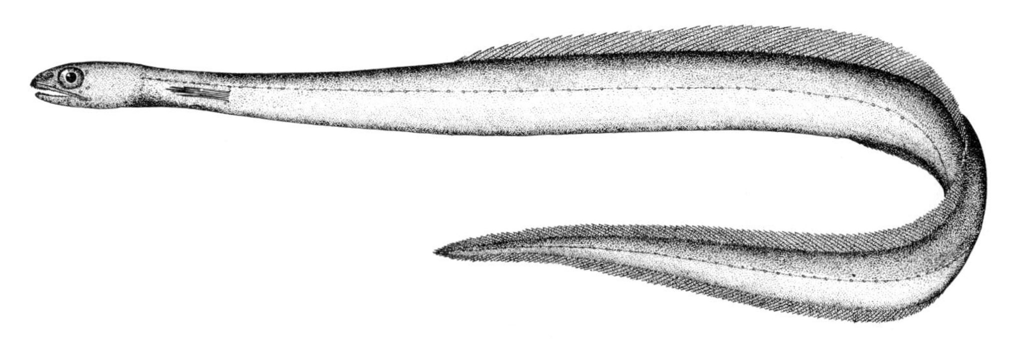

Figure 63. Narrowneck eel *Derichthys serpentinus.* Off North Carolina.

General Range. *Derichthys serpentinus* is widely distributed in the Atlantic, Indo-West Pacific and south central Pacific oceans (Robins and Robins 1989).

Occurrence in the Gulf of Maine. Narrowneck eel are occasional strays from deeper water. The first report from the Gulf of Maine was capture of a specimen in a midwater trawl off the mouth of Northeast Channel (Musick 1973b). Presence of this species in the Gulf was confirmed by capture of a specimen by *Albatross IV* at 42°32′ N, 67°17′ W at a depth of 210–270 m on 23 April 1994 (USNM 329919, 134 mm TL).

SPOONBILL EEL / *Nessorhamphus ingolfianus* (Schmidt 1912) / Musick 1973:135

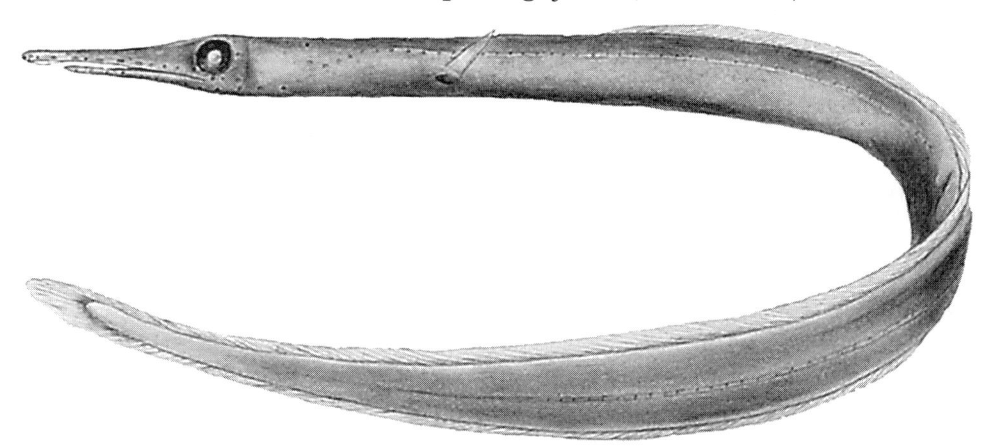

Figure 64. Spoonbill eel *Nessorhamphus ingolfianus.* Off North Carolina. Drawn by M. H. Fuges. (Reprinted from Robins and Robins 1989: Fig. 456. © Sears Foundation for Marine Research.)

Description. Snout long (about four times eye diameter) and narrow (less than one-half eye diameter) (Fig. 64). Upper jaw laterally flattened into a spatulate tip. Dorsal fin originates slightly behind pectoral fin tip; anal fin slightly behind midpoint of body; both dorsal and anal fins run back to tip of tail. Primary lateral line pores on head relatively large, conspicuous, and associated with an irregular number of smaller secondary pores and sensory ridges.

Meristics. Pectoral fin rays 12 or 13; caudal fin rays 10–12; branchiostegal rays 7; precaudal vertebrae 70–74, total vertebrae 147–159 (Robins and Robins 1989).

Color. Adults are grayish brown with a bluish tinge on the sides of fresh specimens (Beebe 1935); head, dorsum, and abdomen darker. Smaller specimens are normally paler.

Size. Reach about 600 mm TL (Robins and Robins 1989).

Distinctions. Spoonbill eel are distinguished from all other Gulf of Maine eels by the spatulate tip to the upper jaw.

Biology. Spoonbill eel are a fish of the middepths of the ocean, not of the bottom. Little is known of their habits, but they seem to feed mainly, if not entirely, on pelagic shrimps (Beebe 1935). Spawning occurs in the Sargasso Sea in the spring and early summer, metamorphosis takes place in August to October at a length of about 62–85 mm (Schmidt 1930). Larvae generally occur in the upper 100 m of the water column; juveniles were found at depths of 100–200 m, whereas adults were mainly found from 300 to 1,000 m (Karmovskaya 1985).

General Range. *Nessorhamphus ingolfianus* is widely distributed in the Atlantic, from about 60° N to 40° S. It also occurs widely in the Indian and Pacific oceans.

Occurrence in the Gulf of Maine. Spoonbill eel are not regular inhabitants of the Gulf of Maine. A specimen was reported from the deep basin north of Georges Bank and another off the mouth of Northeast Channel (Musick 1973b).

SNIPE EELS. FAMILY NEMICHTHYIDAE

DAVID G. SMITH AND KENNETH A. TIGHE

The Nemichthyidae is a small family of highly modified midwater eels. The jaws are extremely long (except in mature males), with the upper jaw longer than lower. The body is very elongate. Male snipe eels undergo a marked transformation at sexual maturity with the jaws undergoing a drastic shortening and loss of teeth. The family contains three genera and nine species of bathy- and mesopelagic fishes (Nelson 1994), one of which is a rare visitor to the Gulf of Maine.

SLENDER SNIPE EEL / *Nemichthys scolopaceus* Richardson 1848 / Bigelow and Schroeder 1953:159–160

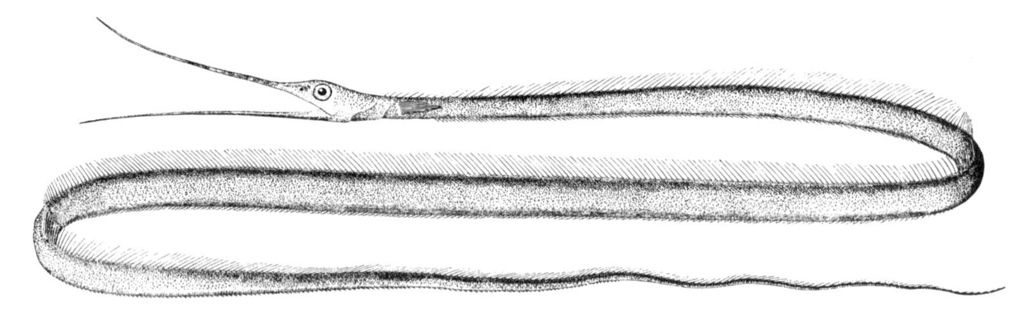

Figure 65. Slender snipe eel *Nemichthys scolopaceus.*

Description. Body extremely slender, about 75 times as long as deep (Fig. 65). Head much deeper than neck. Eyes large. Jaws elongate, slender, upper jaw curving upward, lower more nearly straight, with fine teeth pointing into throat. Dorsal fin originates in front of pectoral fins; anal about abreast of tip of pectoral fins; both dorsal and anal extend back to tip of tail. Anal fin higher than dorsal. Dorsal fin rays stiffer and somewhat more spinelike than those of anal. Lateral line with three longitudinal rows of small pores, arranged in a repeating pattern of a rectangle with one in the middle.

Meristics. Pectoral fin rays 10–14; branchiostegal rays 7–15; precaudal vertebrae 77–105, total vertebrae extremely numerous, 400–750 (Smith and Nielsen 1989).

Color. Variable, generally light brown above and darker below, with an intensification of pigment on the ventral midline below the stomach. Mature individuals of both sexes become uniformly dark brown or black.

Size. Reach about 1 m or more, but most of this length comprises the filamentous caudal region.

Distinctions. Slender snipe eel differ from all other eels in the Gulf of Maine by their extremely slender body (they may be 75 times as long as deep) and by their elongate slender jaws, the upper one curving upward, the lower more nearly straight.

Biology. Slender snipe eel are a fish of the middepths of the ocean, not of the bottom. They seem to be most common between 200 and 1,000 m. Little is known of their habits, but they seem to feed mainly on pelagic shrimps. Observers in submersibles have reported seeing snipe eel hanging vertically in the water; a video taken from the *Johnson Sea Link* in 1987 south of Long Island showed one following the craft as it descended. With steady undulations of its body, the eel kept perfect pace, maintaining its position for several minutes. Two predators are known: goosefish and pollock (Rountree 1999).

Reproduction and Life History. Male slender snipe eel undergo a remarkable transformation when they reach sexual maturity (Nielsen and Smith 1978). The characteristic beak disappears and the teeth are lost. The anterior nostrils enlarge into forwardly directed tubes, the pectoral fin is displaced posteriorly, the eyes enlarge, and the color darkens. So different in appearance are these males from the females and immatures that when first discovered they were thought to belong to a different genus and species, *Paravocettinops trilinearis.* The enlarged nostrils suggest that male slender snipe eel may detect pheromones released by the female and thus locate a mate. Loss of teeth and drastic shortening of the jaws suggest that this eel ceases feeding and dies after spawning. Females undergo similar changes, although not nearly as extreme; they retain the long jaws but lose the teeth and turn uniformly dark in color. The elongate leptocephalus larvae of slender snipe eel are found over deep, oceanic water. Unlike the adults, they stay relatively close to the surface, descending only as they begin metamorphosis.

General Range. *Nemichthys scolopaceus* is widely distributed in the Atlantic from about 55° N to 50° S, including the eastern Gulf of Mexico, the Caribbean, and the Mediterranean (Smith and Nielsen 1989). They also occur widely in the Indian and Pacific oceans, but the taxonomic status of these populations is uncertain.

Occurrence in the Gulf of Maine. Slender snipe eel are rare in the Gulf of Maine. Bigelow and Schroeder reported a specimen from the stomach of a cod caught on Georges Bank as the only Gulf of Maine record and indicated that it is more common off the seaward slope of Georges Bank. Since then, slender snipe eel have occurred as a regular, but rare catch in NMFS surveys. A juvenile specimen (USNM 329926, ca. 360 mm TL) was taken during *Albatross IV* 94-02 at Station 7 (approximately 42°15′ N, 68°10′ W). Several additional specimens from the Gulf of Maine are in the collection of the Museum of Comparative Zoology.

CONGER EELS. FAMILY CONGRIDAE

DAVID G. SMITH AND KENNETH A. TIGHE

The Congridae is a diverse family of marine eels found in tropical and temperate marine waters. Caudal and anal fins are present in most species. There are three subfamilies with 32 genera and roughly 150 species (Smith 1989b; Nelson 1994). One species of the subfamily Congrinae occurs in the Gulf of Maine.

CONGER EEL / *Conger oceanicus* (Mitchill 1818) / American Conger, Sea Eel /

Bigelow and Schroeder 1953:154–157

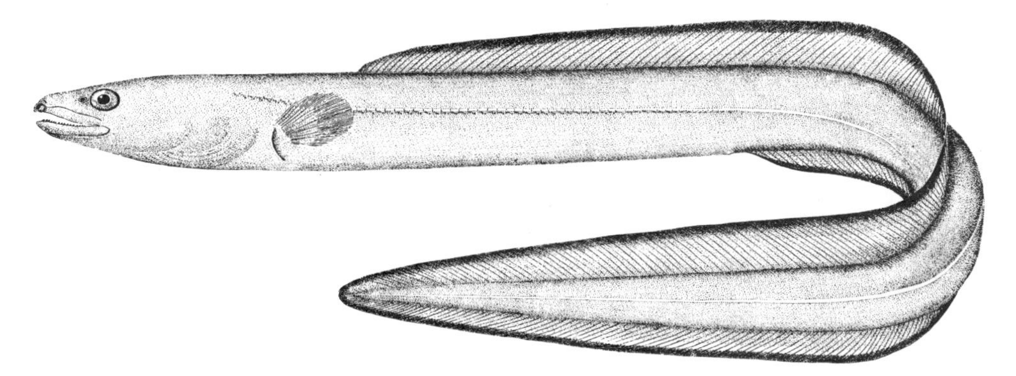

Figure 66. Conger eel *Conger oceanicus.* Connecticut. Drawn by H. L. Todd.

Description. Body thick and heavy (Fig. 66). Snout long, pointed; snout length one-quarter head length. Mouth large, opening back at least as far as middle of eye. Upper jaw usually projects beyond lower. Eyes large, oval. Distance from snout tip to dorsal fin about one-fifth TL. Origin of dorsal fin above or only very slightly behind tip of adpressed pectoral fin. Pectoral fin length a third to a quarter of the distance from dorsal fin to tip of snout. Scales completely absent.

Meristics. Pectoral fin rays 16–18; branchiostegal rays 5; precaudal vertebrae 50, total vertebrae 143–147 (Smith 1989b).

Color. Bluish gray or grayish brown above, sometimes of a reddish tinge, sometimes almost black; paler on the sides; dingy white below; vertical fins black-edged.

Size. This is a much larger fish than American eel. Bigelow and Schroeder reported a maximum size of 2 m and 10 kg. Large individuals taken off southern New England and New Jersey are said to measure 1.2–2.1 m TL. The general run of those caught weigh 1.8–5.4 kg. *Conger oceanicus* apparently never attains the enormous size reached by European *C. conger,* which has been reliably recorded up to 2.7 m TL and 73 kg (Jenkins 1925).

Distinctions. Conger eel differ from other eels in the Gulf in having the origin of the dorsal fin above or only very slightly behind the tip of the pectoral fin when the latter is laid back, the rather long, pointed snout, the large mouth opening back at least as far as the middle of the eye, and the scaleless skin. The upper jaw usually projects beyond the lower in conger, whereas in American eel the reverse is true, or at least the lower equals the upper. The eyes of conger eel are oval and larger than the round eyes of American eel.

Habits. Much less is known about conger eel than about American eel. They occur from the coastline (they are caught from the dock at Woods Hole) out to the edge of the continental shelf. Southern records tend to be deeper and farther offshore. The deepest reliable record is 577 m from the Gulf of Mexico (Smith 1989b); Bigelow and Schroeder trawled an individual at 260 m off southern New England. Although they occur around river mouths and estuaries, conger eel do not enter freshwater as American eel do. Conger eel live in tilefish (*Lopholatilus chamaeleonticeps*) burrows off the coast of New Jersey (Able et al. 1982). Evidently, conger eel do not tolerate either very cold or very warm water. They are taken near Woods Hole from July into the autumn and around Block Island from August until November, but they disappear in the winter, presumably moving offshore. In the southern part of their range, they avoid shallow coastal water, preferring to remain offshore in cooler water.

Food. Conger eel feed chiefly on fishes, 96.4% by weight gadids for seven individuals, 39–90 cm TL (Bowman et al. 2000). At Woods Hole, herring, American eel, and butterfish have been found in their stomachs. They also occasionally prey on shrimps and small mollusks.

Predators. Conger eel were found in the stomachs of six species of fishes: dusky shark, smooth dogfish, spiny dogfish, cod, sea raven, and cobia, of which only spiny dogfish are a significant predator (Rountree 1999).

Reproduction and Early Life History. Conger eel are extremely prolific; a European female may produce 3–6 million eggs. Like American eel, conger eel pass through a ribbonlike leptocephalus stage, very broad and thin and perfectly transparent, with a very small head. Also like American eel, they spawn at sea far from the coastal waters they inhabit as adults. It seems likely as well that they breed but once during their life and then perish, again like American eel. Ripe conger eel are never caught on hook and line, for they have ceased to feed, hence to bite, for some time previous. But males of the European species kept in aquariums have repeatedly been known to become fully ripe and females nearly so, and then invariably to die. Ripening of the sexual products is accompanied by changes in the shape of the head, loss of teeth, and jellification of the bones, while the eyes of the males become enormous and the females become much distended by the ovaries (Cunningham 1891–1892).

Although conger eel move offshore in the fall, it is not known whether this is part of a spawning migration or simply a seasonal migration into deeper and warmer water; there are probably elements of both involved. Absence of eggs and young larvae from waters off New England indicates that conger do not spawn in this area. The eggs described by Eigenmann (1902) and credited to this species were actually those of the margined snake eel, *Ophichthus cruentifer,* as shown by Naplin and Obenchain (1980). Schmidt (1931), however, found "quite tiny larvae" in the West Indian region and concluded that conger eel breed there.

The leptocephalus of conger eel is relatively more slender than that of American eel and grows larger (to a length of 150–160 mm). The vertebrae and muscle segments are far more numerous (140–149) than in American eel (about 107). They also have a series of tiny pigment spots along the underside of the body and, at least in larger individuals, spots along the midlateral body wall as well; leptocephali of American eel are without pigment except for the eye. Duration of the larval period of conger eel is unknown. Metamorphosis consists of thickening and narrowing of the body, enlargement of the head, formation of the swim bladder and permanent teeth, and development of pigment in the skin, a change that takes about 2 months.

General Range. Continental shelf of eastern America: adults are known north to the tip of Cape Cod, and larval stages to eastern Maine. The southern limit of its range is Florida and the eastern Gulf of Mexico; specimens have been recorded from just west of the Dry Tortugas and off the edge of Campeche Bank, north of the Yucatán Peninsula (Smith 1989b). There are no reliable records from the Caribbean or South America. It is replaced by the closely allied *Conger conger* in the eastern North Atlantic.

Occurrence in the Gulf of Maine. Conger eel are not regular residents of the Gulf. A few pass around Cape Cod, but adults are common only to the west and south of that point. Leptocephali commonly occur in coastal waters from Massachusetts to Maine, but are apparently unable to survive there as adults.

SAWTOOTH EELS. FAMILY SERRIVOMERIDAE

KENNETH A. TIGHE AND DAVID G. SMITH

The Serrivomeridae contains two genera and about ten species of marine midwater eels found worldwide in tropical and temperate waters (Nelson 1994). The jaws are very long and slender, the vomerine teeth are in two or more rows, and the gill openings are connected ventrally.

STOUT SAWPALATE EEL / *Serrivomer beanii* Gill and Ryder 1883 / Musick 1973b:135

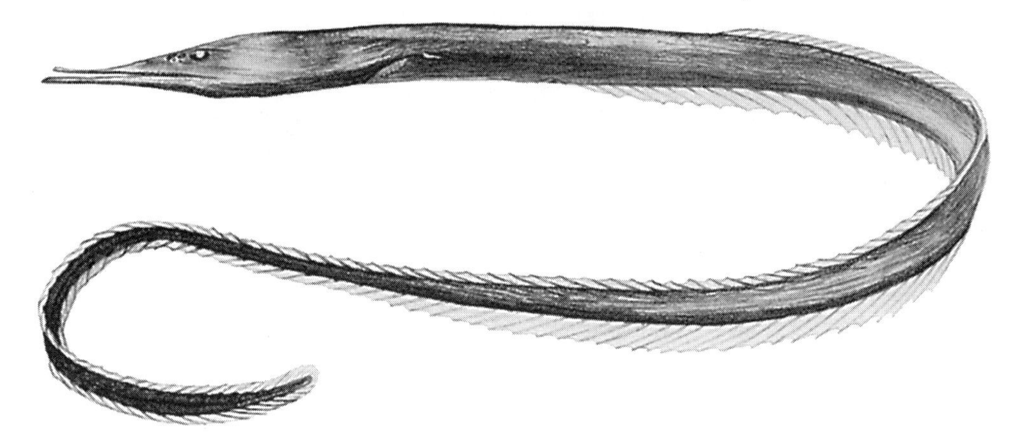

Figure 67. Stout sawpalate eel *Serrivomer beanii.* 495 mm TL, MCZ 64719. Drawn by M. H. Fuges. (Reprinted from Tighe 1989: Fig. 665. © Sears Foundation for Marine Research.)

Description. Body round (Fig. 67). Teeth in both jaws sharp. Double row of enlarged vomerine teeth give a saw-toothed effect. Gill openings relatively large, oblique, appearing united on ventral surface. Dorsal fin originates behind origin of anal fin, which originates at about one-third of body length; both dorsal and anal fins extend back to tip of tail. Pectoral fins extremely reduced.

Color. Dark brown to black with an iridescent silvery epidermal layer. The ventral surface is darker than the dorsal.

Size. Reach at least 745 mm TL (Tighe 1989).

Distinctions. The prolonged snout with its double row of enlarged vomerine teeth separate stout sawpalate eel from all other Gulf of Maine eels. The only other species of the family found in the western North Atlantic, *S. lanceolatoides,* the short-tooth sawpalate eel, is distinguished by fewer teeth, attachment of the branchiostegal rays to the hyoid arch, and more dorsal and anal rays (Tighe 1989). In addition, the body of *S. lanceolatoides* is more oval in cross section while that of *S. beani* is typically more round.

Habits. Stout sawpalate eel are a fish of the middepths of the ocean. They seem to feed primarily on pelagic shrimps and euphausiids, with occasional small fishes and other crustaceans included in the diet (Beebe and Crane 1936). Bathysphere observations indicate that they are fast swimmers, usually found singly, but occasionally in pairs or in groups of four or five.

Reproduction and Life History. Adult stout sawpalate eel undergo a prereproductive metamorphosis when they reach sexual maturity (Tighe 1989). The characteristic vomerine dentition becomes modified into a double sawtooth palate, and the maxillary and mandibular teeth are reduced to a single irregular row. The eye becomes enlarged and nearly circular in outline while the nostrils become slightly enlarged. These modifications may be adaptations for mate location in the mesopelagic environment, although the function of the modified teeth remains a mystery.

The leptocephalus larvae of stout sawpalate eel were identified by Bauchot (1959) and are one of the most common eel larvae over deep, oceanic water in the North Atlantic during the winter months. Unlike the adults, they stay relatively close to the surface, descending only as they begin metamorphosis. Detailed studies of their vertical distribution off Bermuda (Tighe 1975) showed that juveniles are common between 551 and 1,000 m, with subadult to adults occurring within the lower depth range of the juveniles. Breeding probably occurs in the late fall or early winter with metamorphosis the following spring (Tighe 1975).

General Range. *Serrivomer beanii* are widely distributed in the North Atlantic, from about 50° N to the equator, including the Caribbean, but not the Gulf of Mexico.

Occurrence in the Gulf of Maine. Stout sawpalate eel are not regular inhabitants of the Gulf of Maine. Two specimens were reported from the deep basin north of Georges Bank and two more from off the mouth of Northeast Channel (Musick 1973b). There are also several specimens from the Gulf of Maine and Georges Bank in the collection of the Museum of Comparative Zoology. All probably represent strays from deeper water.

HERRING AND HERRINGLIKE FISHES. ORDER CLUPEIFORMES

The order Clupeiformes comprises part of the Clupeomorpha, a group of fossil and Recent fishes commonly known as the herring and herringlike fishes (Grande 1985; Whitehead 1985a). As Grande (1985) pointed out, in earlier literature the Clupeomorpha (and Clupeiformes) were an artificial assemblage construed to contain any primitive teleost that did not fit into another, better-characterized teleost subgroup. This concept was evident in Bigelow and Schroeder, where Clupeiformes were allied with ten-pounders and tarpons (Elopidae). Greenwood et al. (1966) defined the Clupeomorpha rigorously, based on uniquely shared characters, and demonstrated that there was no common ancestry to any other groups and the Clupeomorpha. For example, ten-pounders and tarpons (*Elops, Megalops*) and ladyfishes (*Albula*) have a leptocephalus larva, which allies them with the eels in the Elopomorpha. Additional studies (Whitehead 1963; Patterson and Rosen 1977; Grande 1985, Whitehead 1985a,b) define the Clupeomorpha as those fishes sharing several uniquely derived features, including the presence of one or more abdominal scutes; an otophysic (inner ear to swim bladder) connection involving a diverticulum of the swim bladder that penetrates the exoccipital and then expands to form ossified bullae in the prootic and usually also the pterotic; and a supratemporal commissural sensory canal primitively passing through parietals and supraoccipital. Whitehead (1985a) pointed out that there is a membrane within each prootic bulla that separates gas (from the swim bladder) from perilymphatic liquid (surrounding the inner ear). This system, together with the head canal system and the *recessus lateralis,* probably functions in detecting and analyzing small vibrational pressures and displacements (Hoss and Blaxter 1982), thereby monitoring information necessary for schooling and other swimming activities and in the detection of predators and hazards.

Within the Clupeiformes, there are two suborders (Grande 1985), the Denticipitoidei (comprising two monotypic genera, one fossil and the extant African freshwater genus *Denticeps*) and the suborder Clupeoidei, containing the rest of the Clupeiformes. Of the clupeiform fishes, only members of the Clupeoidei are found in the Gulf of Maine.

The suborder Clupeoidei contains three superfamilies (Grande 1985), of which only two, the Clupeoidea and the Engrauloidea, occur in the Gulf of Maine. The Clupeoidei comprises a rather large group of fishes with roughly four families, 80 genera, and some 300 species. Most species in this group are marine, coastal, and schooling fishes, but some enter brackish or freshwaters and some live permanently in freshwaters (rivers or lakes).

The superfamily Clupeoidea includes all of the clupeid fishes commonly encountered in the western North Atlantic, such as herrings, sardines, sprats, shads, and menhadens, which are usually easily recognized externally by their keel of scutes along the belly, the small and often poorly toothed mouths, and their silvery appearance. Round herrings (subfamily Dussumieriinae) differ from other clupeids chiefly in their rounded belly, less deep body, and terminal position of the mouth. Anchovies (superfamily Engrauloidea) are usually distinctive because of their projecting, piglike snout, large mouth, and "underslung" lower jaw (which reflects externally the backward obliquely inclined suspensorium).

Clupeoid fishes are of major importance to fisheries (Blaxter and Hunter 1982). They represent the largest suborder, in terms of weight landings, of nondomesticated vertebrates harvested by man (Whitehead 1985a,b). Half the world catch of fishes comes from about 60 species of various groups, but a third of those prime species are clupeoids. More clupeid fishes are caught (by weight, but presumably also by number) than members of any other single group of fishes. The magnitude of clupeoid fisheries results from two main factors (Whitehead 1985a). First, most clupeoids feed close to the base of the food chain and thus benefit more directly from nutrient-rich areas, where there are strong seasonal or more continuous blooms of plankton. Second, clupeoids are almost always schooling pelagic fishes and thus very vulnerable to nets (especially purse seines), which can catch large volumes of fish in a short time. Since it is the cooler, high-latitude seas and the areas of upwelling that are richest in plankton, it is here that the major clupeoid fisheries exist. However, characteristic of those clupeoid species that dominate the fisheries of particular areas is a tendency to fluctuate rather drastically in their abundance. The production of good or bad year-classes (which may vary by a factor of ten or more) can be related to ecological factors affecting recruitment success for some species, while the role played by fisheries in regulating abundances of certain species can also be substantial.

Description and Diagnosis (from Whitehead 1985a). Moderate-sized, small, or very small fishes (2–100 cm SL) with no spines in the fins; dorsal fin single and short (11–23 fin rays), usually near midpoint of body; pelvic fins with 6–10 fin rays, slightly before, under, or slightly behind point equal with vertical through dorsal fin base; anal fin usually short or moderate (10–36 fin rays); caudal fin forked. Body usually fusiform, sometimes almost round in cross section (*Etrumeus*), but more often compressed, sometimes highly compressed. Typically, with pelvic scute with ascending arms just anterior to pelvic fins (W-shaped in Dussumieriinae and *Engraulis*); a series of similar scutes anterior and posterior to pelvic fins (absent in Dussumieriinae and all New World Engraulidae).

Mouth small, with lower jaw deep and triangular in Clupeidae, but slender and long in most Engraulidae. Premaxillae triangular (rectangular in Dussumieriinae); maxillae usually with an anterior (first) and posterior (second) supramaxilla along upper margin. Small conical teeth typically present in jaws and on vomer, palatines, and endo- and ectopterygoids (i.e., roof of mouth), but some or all may be absent. Gut short (carnivores) or long and coiled (phytoplankton feeders, filter feeders); some species with muscular stomach like a gizzard (Dorosomatinae; partially so in some Clupeinae); food is collected in a bolus by pharyngeal pouches in Dorosomatinae and some Clupeinae. Swim bladder present, sometimes double-chambered (some Engraulidae), with a pneumatic duct joined to esophagus or stomach.

Almost all species with a complete covering of cycloid scales on body, scales frequently deciduous; small scales occasionally cover bases of dorsal, anal, and/or caudal fins, and one or sometimes several axillary scales lie above bases of first pectoral and pelvic fin rays. No lateral line canal with pored scales along flanks (occasionally one or two behind gill opening). A branching, mainly cutaneous, sensory canal system covering top and sides of head; supraorbital, infraorbital, preopercular, and pterotic canals all meet in the *recessus lateralis,* a special chamber characteristic of clupeiform fishes, its inner wall being a membrane sealing the perilymphatic space that surrounds the inner ear.

KEY TO GULF OF MAINE CLUPEOID FAMILIES

1a. Articulation of lower jaw located well posterior to vertical line through posterior margin of eye, lower jaw narrow and elongate, usually very slender; snout piglike and projecting, lower jaw underslung .... **Engraulidae**

1b. Articulation of lower jaw under or only just slightly posterior to the vertical line through the posterior margin of eye; lower jaw deep .... **Clupeidae**

ANCHOVIES. FAMILY ENGRAULIDAE

THOMAS A. MUNROE

Anchovies are small to moderate-sized, elongate and somewhat moderately laterally compressed, silvery, herringlike fishes (Hildebrand 1963b; Scott and Scott 1988; Whitehead et al. 1988). They are closely related to the Clupeidae and have been described as essentially clupeids with a different head (Nelson 1984; Grande 1985; Grande and Nelson 1985; Whitehead et al. 1988).

Anchovies are characterized by a large, gaping, horizontal mouth situated low on the head, with a long, slender, underslung lower jaw whose point of articulation is located well beyond the posterior margin of the eye. Typically, anchovies have two supramaxillae. Jaw teeth are usually small or minute; with minute teeth also present on vomer, palatines, tongue, and pterygoids. Most species have a prominent piglike snout projecting beyond the lower jaw tip. The head is moderately long and the eye is prominent and located relatively far forward; sometimes adults have an adipose eyelid. Gill rakers are generally short and slender, 10–50 or more on the lower limb of the first arch (but long and up to 100 or more in *Anchovia*). Branchiostegals 7–19. All fins are soft-rayed; the dorsal fin is short (12–16 fin rays in American genera) and usually near the midpoint of the body; pectoral fins are situated low on the body; pelvic fins are ventral, with i, 6 fin rays, about midbody (before, under, or behind dorsal fin base); the anal fin is usually moderately long (in American genera), about 15–40 fin rays in American species; and the caudal fin is deeply forked. Scales are moderate in size, cycloid, about 30–60 in lateral series, often deciduous. Lateral line absent. A pelvic scute with lateral arms is always present; and New World anchovies lack pre- and postpelvic scutes. Vertebrae 40–44 (species occurring in the northwest Atlantic). Anchovies may reach sizes to 50 cm, but most species are usually less than 15 cm.

There are approximately 16 genera with about 139 species currently recognized in the family (Whitehead et al. 1988). Most species inhabit coastal waters in the tropical and temperate (from about 60° N to 50° S) Atlantic, Pacific, and Indian oceans. Some species enter brackish water or freshwater to feed or spawn and some live permanently there and are found even high up the Amazon (Whitehead et al. 1988). Most anchovies are strongly schooling species that feed on small planktonic animals (especially crustaceans), either by locating individual prey or by more indiscriminate filter-feeding. Larger anchovies may also consume small fishes and other invertebrates. Most, perhaps all, species are serial spawners that scatter relatively large numbers of eggs from which hatch planktonic larvae.

In areas of abundance, anchovies are fished commercially as food fishes and are sold fresh, cured, or canned, and a large proportion of the catch is also used for bait or processed into fish meal (Hildebrand 1963b; Scott and Scott 1988; Whitehead et al. 1988).

Only two species of anchovies, both in the genus *Anchoa,* are known from the Gulf of Maine. Most species of *Anchoa* occur in tropical or subtropical marine and estuarine habitats, and some species penetrate into freshwaters. They are found along Atlantic and Pacific coasts and lower parts of rivers of North, Central, and South America. There are approximately 33 species, 16 Atlantic, and 17 eastern Pacific. Bay anchovy, *Anchoa mitchilli,* and striped anchovy, *A. hepsetus,* are the anchovies known from the Gulf of Maine. Larvae of silver anchovy, *Engraulis eurystole* (Swain and Meek 1885) have been collected on the Scotian shelf (Markle et al. 1980), but thus far this species is unknown from the Gulf of Maine. Bay anchovy occur more frequently in the Gulf of Maine than other anchovies, and there is probably a spawning population in southern Cape Cod Bay. Striped and silver anchovies are more common and abundant in areas south of the Gulf of Maine and appearances of these species in the Gulf or on the Scotian shelf are only occasional, primarily during periods of higher water temperatures.

Gulf of Maine fishes with which one might possibly confuse an anchovy are juvenile herring, smelt, or silversides. Anchovies are easily distinguished from herring by their large, underslung mouths, which articulate well beyond the vertical line passing through the posterior margin of the eye (mouth smaller, terminal, and articulation point of jaws located at vertical line through anterior or midpoint of eye in herring); in lacking abdominal scutes (vs. scutes present in herring); in having the dorsal fin origin located entirely posterior to the vertical line passing through the origin of the pelvic fins (vs. dorsal fin origin located at vertical line equal with, or slightly anterior to, the pelvic fin origin in the herring); in having the tips of pectoral fin rays (when these are laid back against the body) nearly reaching the origin of the pelvic fins (vs. pectoral fins more widely separated from pelvic fin origin in herring); and the anal fins of anchovies are much longer than those of herrings. Anchovies lack a dorsal adipose fin, which is sufficient to distinguish them at a glance from smelt, whereas silversides (*Menidia* spp.) have two separate dorsal fins instead of the single dorsal fin characteristic of anchovies.

KEY TO GULF OF MAINE ANCHOVIES

1a. Anal fin origin under posterior rays of dorsal fin; silvery lateral band bright and well defined; anal fin rays 16–20 . . . . **Striped anchovy**
1b. Anal fin origin under anterior margin of dorsal fin; silvery lateral band diffuse; anal fin rays 24–30 . . . . **Bay anchovy**

STRIPED ANCHOVY / *Anchoa hepsetus* (Linnaeus 1758) / Bigelow and Schroeder 1953:119

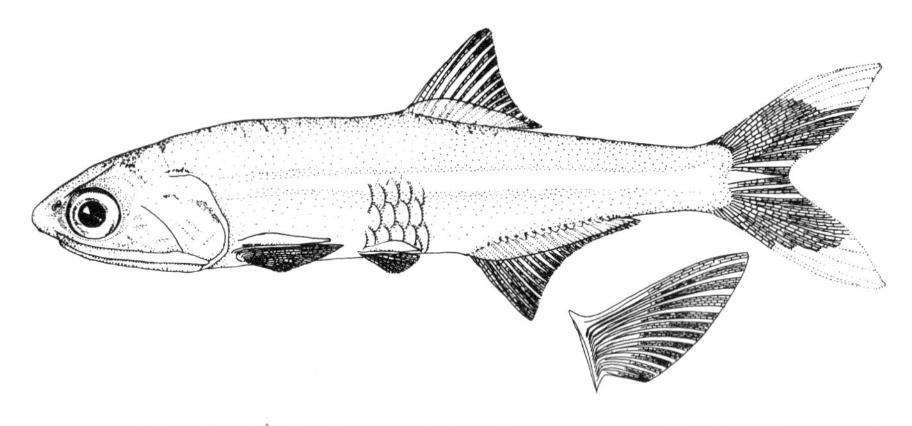

Figure 68. Striped anchovy *Anchoa hepsetus.* Colon, Panama, 70 mm TL, USNM 117664. Inset is enlarged view of pectoral fin.

Description. Body somewhat compressed, elongate, its depth about five times in SL (Fig. 68). Snout pointed, about three-quarters eye diameter; maxilla long, tip pointed, reaching beyond posterior border of preoperculum, almost to gill opening. Anal fin short, its origin below about midpoint of dorsal fin base. Anus nearer to anal fin origin than to pelvic fin tips (Hildebrand 1963b, Whitehead et al. 1988).

Meristics. Dorsal fin rays 13–16; anal fin rays iii, 16–20; pectoral fin rays 13–15; scales in lateral series about 37–43; lower gill rakers 15–25, usually 16–19; upper gill rakers 19–22; vertebrae 40–44.

Color. Fresh specimens pale gray and iridescent; upper surface of head with some green and yellow; dorsal regions with dusky dots. A prominent, broad, silver stripe along flank (a dark line above) of uniform width, except narrowed immediately behind gill opening, about three-quarters eye diameter. Dorsal and caudal fins more or less dusky on some specimens.

Size. Maximum about 150 mm SL; commonly 100–130 mm SL.

Distinctions. Striped anchovy closely resemble bay anchovy, but differ in having fewer anal fin rays (16–20 vs. 24–30 in bay anchovy) and in placement of the anal fin under the posterior rays of the dorsal fin.

Habits. Marine, pelagic, coastal, forming dense schools, often in shallow waters close to shore (but recorded down to 70 m); able to tolerate a wide range of salinities, from hypersaline to almost fresh water (Hildebrand 1963b; Whitehead et al. 1988).

Food. Striped anchovy feed on copepods (especially when young), mysids, gastropods, foraminifera, and, occasionally, ostracods and annelids (Stevenson 1958; Hildebrand 1963b). Mysids (mostly *Neomysis americana*) constituted 78.8% by weight of the food of 14 striped anchovy (Bowman et al. 2000). Summer and fall diets of striped anchovy in salt-marsh creeks of South Carolina consisted of a variety of small and large planktonic items (Allen et al. 1995), including copepods, crab zoeae and megalopae, shrimp larvae, and fish larvae. Virtually the only adult copepod consumed by striped anchovy was *Acartia tonsa*.

Predators. Striped anchovy are eaten by a wide variety of elasmobranchs and bony fishes. The NEFSC food habits survey recorded 18 species of fish predators, of which the two most important in order of numbers of occurrences are bluefish and weakfish (Rountree 1999).

Parasites. Principal parasites are nematodes, cestode larvae (*Rhyncobothrium* sp.), and digenetic trematodes (Hildebrand 1963b).

General Range. Western North Atlantic north as a stray to Maine (Kendall 1931) and to the outer coast of Nova Scotia, where a single capture of five specimens, 54–59 mm TL, was taken (Vladykov 1935a; Scott and Scott 1988) in Bedford Basin, south to Fort Pierce, Fla. (but not Florida Keys), at least the northern part of the Gulf of Mexico, and west central and South Atlantic (Gulf of Venezuela south to Uruguay). In coastal waters along the eastern United States, striped anchovy are more abundant from Chesapeake Bay to Florida; overall, this is a more southerly fish than bay anchovy (Hildebrand 1963b).

Occurrence in Gulf of Maine. Inclusion of this species is based on one record off the mouth of the Penobscot River, near Portland, Maine, 8 October 1930 (Kendall 1931). One specimen was saved and identified, and Bigelow and Schroeder reported that the herring fishermen who brought it in stated that there were "lots of them" on that date.

BAY ANCHOVY / *Anchoa mitchilli* (Valenciennes 1848) / Whitebait / Bigelow and Schroeder 1953:118–119

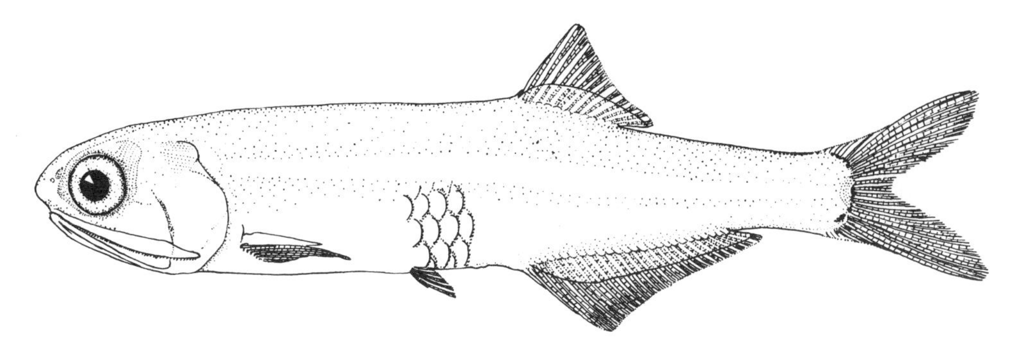

Figure 69. Bay anchovy *Anchoa mitchilli*. Woods Hole, Mass., 85 mm TL, USNM 125582.

Description. Body variable, rather slender, more slender in northern populations, its depth about four to five times in SL, moderately compressed, greatest thickness usually exceeding depth of caudal peduncle (Fig. 69). Body depth 16–27% SL. Head 22.0–26.5% SL. Snout short, fairly blunt, a little over one-half eye diameter; 3.7–7.2% SL, projecting not more than a quarter of its length beyond tip of mandible. Eye 5.8–8.2% SL. Postorbital length short, 11.0–13.5% SL. Maxilla long, tip pointed, reaching beyond posterior border of preopercle, extending nearly to margin of opercle, 17.5–21.0% SL. Mandible 16.5–18.5% SL. Cheek short and broad, about as long as eye, posterior angle approximately 60°. Dorsal fin rather low, with nearly straight margin, posteriormost fin ray scarcely longer than fin ray immediately before it; longest dorsal fin ray failing to reach tip of fin if depressed posteriorly; dorsal fin origin variable; usually located between a point slightly nearer to caudal fin base than to upper anterior angle of gill opening and the point equidistant between caudal fin base and posterior margin of eye. Anal fin moderate to long, its origin usually just somewhat posterior to vertical line through dorsal fin origin (below unbranched dorsal fin rays); anal fin base 25–30% SL. Anus nearer to pelvic fin tips than to anal fin origin. Pelvic fin very small, not quite reaching halfway point to origin of anal fin; inserted nearer to anal fin origin than to pectoral fin base by distance equal to diameter of eye. Pectoral fin length 12–20% SL, somewhat variable, failing to reach base of pelvic fins by distance equal to, or even greater than, diameter of eye in northern specimens; sometimes reaching base of pelvic fins in southern fish. Axillary scale of pectoral fin rather narrow. Caudal fin deeply forked. (Hildebrand 1963b; Whitehead et al. 1988).

Meristics. Dorsal fin rays 14–16, anal fin rays iii, 20–27 (rarely 28); pectoral fin rays 11 or 12; scales 38–44, relatively large, thin, easily detached; lower gill rakers 21–25 (higher part of range in northern populations), upper gill rakers 20–26; vertebrae 38–44.

Color. Whitish silvery to pale along sides of body, abdominal walls translucent; dorsalmost surface greenish with bluish reflections; sides with a narrow, ill-defined, silvery band scarcely wider than pupil of eye, extending from gill opening posteriorly to caudal fin base. Body and fins speckled with many small, rounded melanophores.

Size. Maximum length about 110 mm, but seldom more than 100 mm SL and usually only to about 75 mm SL (Hildebrand 1963b).

Distinctions. Bay anchovy are readily distinguished from striped anchovy in having the anal fin origin located at a vertical line passing through the anterior margin or the anteriormost (unbranched) dorsal fin rays (vs. anal fin origin located at vertical passing through middle or posterior dorsal fin rays in striped anchovy); in having a more diffuse silvery lateral band (vs. lateral band bright and well developed in striped anchovy); and in having more anal fin rays (24–30 vs. 16–20 in striped anchovy); in having fewer pectoral fin rays (i, 9–12, usually 10 or 11, in bay anchovy vs. 13–16, usually 14 or 15, in striped anchovy). In bay anchovy, the anus is also nearer to the pelvic fins than to the anal fin origin (vs. anus closer to anal fin origin in striped anchovy).

Biology. Much of the information presented is synthesized from species profiles by Robinette (1983); Morton (1989); and Houde and Zastrow (1991).

Habits. Throughout its range, bay anchovy are pelagic in nearshore coastal marine waters, but are most plentiful in shallow, estuarine tidal waters with muddy bottoms and brackish water (Hildebrand 1963b; Whitehead et al. 1988; Houde and Zastrow 1991). Bay anchovy also occur in the mouths of rivers, along exposed sandy beaches, and in shallow waters of intertidal creeks (Reis and Dean 1981). They seldom occur in waters deeper than 25 m (Byrne 1982), but have been collected to at least a depth of 36 m (Hildebrand 1963b). This species inhabits both clear and turbid waters and has been collected over all types of substrates, including muddy coves, grassy areas, surf zones, oyster bars, sandy beaches, and sand and silt bottoms (Vouglitois et al. 1987; Houde and Zastrow 1991).

Bay anchovy are pelagic in all life stages. In upper Chesapeake Bay, larval and juvenile bay anchovy were most abundant near the surface (upper 3 m) from May to October, but apparently moved to deeper waters as winter approached (Dovel 1971). In the York River and Chesapeake Bay, bay anchovy schools break up at night and individual fish are transported by currents, whereas in daylight the schools maintain a relatively fixed location (Luo 1993). Anchovies do swim at night but more slowly than in the daytime and not in any fixed direction.

Bay anchovy are euryhaline, tolerant of salinities from virtually freshwater to fully saline, or even hypersaline conditions (Hildebrand 1963b; Dovel 1971; Houde and Zastrow 1991). Further south, larvae and small juveniles are distributed throughout low-salinity subestuaries, remaining there until fall before dispersing to overwintering areas. In Virginia tributaries of Chesapeake Bay, juveniles have been collected as much as 64 km above brackish water (Massmann 1954).

In the northern parts of its range, bay anchovy occur inshore only during the summer (Hildebrand 1963b; Scherer 1984). There is no known substantial north-south seasonal migration. Rather the species undergoes seasonal inshore-offshore movements in the temperate parts of its range (Hildebrand 1963b; Vouglitois et al. 1987), moving out onto the inner continental shelf in fall and winter and returning to estuaries in spring. Further south, bay anchovy are a year-round resident in the larger estuaries. For example, in Chesapeake Bay, adults migrate to deeper waters in the Bay during winter (Hildebrand and Schroeder 1928; Setzler et al. 1981) and may migrate to the lower Bay regions during the coldest winters (Wang and Houde 1994), but are usually collected there throughout the entire year.

Food. Juvenile and adult bay anchovy feed opportunistically primarily on zooplankton, which are selected as individual particles (Cowan and Houde 1990). They are major consumers of plankton, with copepods comprising the dominant prey (Stevenson 1958; Hildebrand 1963b; Din and Gunter 1986; Vasquez 1989; Houde and Zastrow 1991). Larger fish also consume macrozooplankton such as mysids, larval fishes, crab larvae, isopods, other invertebrates including some benthic organisms (e.g., polychaetes and mollusks), and fish larvae (Allen et al. 1995). Small particulates (e.g., algae, detritus) may be found in stomachs of all anchovy length-classes (Houde and Zastrow 1991). Although copepods are the predominant food consumed, they may be replaced in the diet when other potential foods are abundant (Din and Gunter 1986). Allen et al. (1995) noted that the selective preference for larger crustaceans and fish larvae by adult anchovies may be an important factor in the recruitment success of some estuarine macrofauna, especially fiddler crabs. Crab megalopae were selected for even when their densities were low. Feeding may occur throughout the day, but during summer months in Chesapeake Bay, feeding is most intense from dawn to midmorning (Vasquez 1989).

Predators. Bay anchovy are a major prey item for many commercially and recreationally important predatory fishes (Hollis 1952; Merriner 1975; Chao and Musick 1977; Baird and Ulanowicz 1989; Safina and Burger 1989; Rountree 1999; Bowman et al. 2000), such as bluefish (Grant 1962; Friedland et al. 1988; Juanes et al. 1993), striped bass, summer flounder, and weakfish (Houde and Zastrow 1991). The NEFSC food habits survey recorded 22 species of fish predators, of which four are the most important: bluefish, weakfish, spiny dogfish, and smooth dogfish, in order of numbers of occurrences of bay anchovy in their diets (Rountree 1999). In Chesapeake Bay, bay anchovy are estimated to provide more than one-half of the total energy intake of predatory fishes, contributing 70, 90, and 60% to their diets in summer, fall, and spring, respectively (Baird and Ulanowicz 1989). In the Hudson River estuary, bay anchovy were the predominant food and were most abundant in stomachs of bluefish smaller than 150 mm. Small bluefish (<120 mm) fed predominantly on bay anchovy, and although prey size generally increased with bluefish size, even the largest bluefish examined consumed relatively small bay anchovy (Juanes et al. 1993). Bay anchovy are also eaten by seabirds, including the common tern (Safina and Burger 1985), and might also be an important food item for other waterfowl and other animals (Dovel 1971). In the laboratory, early life history stages of bay anchovy are preyed upon by the ctenophore, *Mnemiopsis leidyi* (Monteleone and Duguay 1988). This ctenophore predator co-occurs with spawning bay anchovy (Johnson 1987).

Reproductive Biology. Sexual maturity is attained at a relatively early age and small size. In Chesapeake Bay, some young-of-year anchovy may mature by late summer of their first year (Luo and Musick 1991), although most apparently overwinter before maturing the following year (Zastrow et al. 1991). In mid-Chesapeake Bay, male and female bay anchovy matured at 40–45 mm FL, corresponding to an average age at first maturity of approximately 10 months posthatch (Zastrow et al. 1991). Luo and Musick (1991) also found some female bay anchovy from the lower Chesapeake Bay that were mature at age 0+ and less than 40 mm FL, and Stevenson (1958) reported finding some sexually mature bay anchovy in the Delaware Bay region that were 35–40 mm SL. Age 1 females (50–55 mm in length) have been estimated to produce from 92% to more than 99% of the anchovy eggs spawned in Chesapeake Bay (Zastrow et al. 1991).

Bay anchovy, like other clupeiform fishes, are batch (i.e., serial) spawners (Zastrow et al. 1991; Luo and Musick 1991). Individual females in Chesapeake Bay spawned at least 50–55 times each season, producing from 442 to 2,026 (mean of 1,129) ova per batch (Zastrow et al. 1991; Luo and Musick 1991). On average, bay anchovy in lower Chesapeake Bay spawned every 4 days in June, every 1.9 days in the beginning of July, and every 1.3–1.4 days from early July to the end of August (Luo and Musick 1991).

Batch fecundity is a linear function of fork length and body weight (Luo and Musick 1991). Overall egg production for the season was estimated at 45,110 eggs per female (55 mm FL fish), equivalent to 346% of a female's body energy. Batch fecundity averaged 643–740 eggs per gram of female (Zastrow et al. 1991; Luo and Musick 1991). During the peak spawning period, daily spawning output was estimated to be 6.3% of body energy. Most spawning energy in bay anchovy was derived from daily feeding, not from fat reserves (Luo and Musick 1991). In Chesapeake Bay, the somatic weight component increased by 32–33% and total body weight by 26% during spawning season, indicating that feeding not only met energy requirements of daily spawning but also provided surplus energy for growth (Wang and Houde 1994).

Spawning takes place usually in the evening between the hours of 1800 and 2400 (Hildebrand and Cable 1930; Ferraro 1980; Zastrow et al. 1991; Luo and Musick 1991), although some spawning can occur somewhat later (at least until about 0100). Hydrated ova first appear in evening samples, beginning at about 1800 (Zastrow et al. 1991). Hydration occurred between the hours of 1700 and 1800, and most females had fully hydrated ova by 1800. Most spawning probably occurred between 2100 and 2400. No fish were found with hydrated ova after 0033, indicating that daily spawning activity was completed by that time. Daily spawning time varied from month to month (Luo and Musick 1991); spawning was delayed as the season progressed. It was estimated (Zastrow et al. 1991) that during the peak of the 1987 spawning season in mid-Chesapeake Bay, virtually all mature bay anchovy females spawned each night.

Bay anchovy have a protracted spawning season, from May to September in northern areas, and possibly extending throughout the year in the southern parts of the range (Houde and Lovdal 1984). In western Cape Cod Bay, eggs were collected from June to August and anchovy larvae (presumably this species) from July to October (Scherer 1984). In Barnegat Bay, N.J., and Great South Bay, N.Y., spawning may begin as early as April, peak in June and July, and be essentially complete in August (Vouglitois et al. 1987; Monteleone 1992). The spawning season in Great South Bay (late May to early August, with a shorter peak season during late June–late July) is among the shortest reported for this species (Castro and Cowen 1991) and comparable only to that reported for anchovy in Long Island Sound (Wheatland 1956). Some spawning begins in April in the lower Chesapeake Bay, with most spawning in this area extending from May to September (Dovel 1971; Olney 1983; Dalton 1987; Zastrow et al. 1991; Luo and Musick 1991; E. D. Houde, pers. comm.). The spawning season in mid-Chesapeake Bay extends from mid-May to mid-August (Zastrow et al. 1991). In this area, GSI values were low in March, increased in April and May, and peaked in July, before rapidly decreasing in August toward their lowest levels in fall and winter. At Beaufort, N.C., spawning in bay anchovy was reported (Kuntz 1914) to extend from late April to early September, while in Biscayne Bay, Fla., spawning may occur year round (Houde and Lovdal 1984).

Bay anchovy spawn where water depth is less than 20 m in salinities of 0–32 ppt (Dovel 1971; Olney 1983; Houde and Zastrow 1991). Peak spawning in Chesapeake Bay apparently occurs at 13–15 ppt (Dovel 1971) and at average surface water temperatures of 26.3°–27.8°C (Houde and Zastrow 1991). In the Delaware River estuary, Wang and Kernehan (1979) reported peak spawning at 22°–27°C.

Early Life History. An account of the embryology and larval development can be found in Kuntz (1913). Other general references on early life history and development are Lippson and Moran (1974), Mansueti and Hardy (1967), and Fahay (1983). Tucker (1988) reported on energy utilization in eggs and larvae. Eggs are pelagic and transparent, barely elliptical with a long axis of 0.84–1.11 mm. The shell is smooth and transparent; the yolk is segmented. The perivitelline space is narrow and there are no oil globules. Hatching occurs at 1.8–2.7 mm. The body is long and slender. The yolk sac is greatly elongate and tapers posteriorly. Larvae are transparent and show no pigmentation. Yolk-sac absorption is completed 15–18 h posthatching. The mouth is apparently functional 36 h after hatching. The mouth is large and terminal and extends to the middle of the eye, becoming subterminal as development progresses. Flexion occurs at 7–8 mm, and transformation occurs at about 20 mm. Larvae 7–8 mm in length have developing dorsal and anal fins and some pigmentation in the thoracic region and at the base of the anal fin. Dorsal, caudal, and anal fins develop at the same time; the pectoral fin forms as a bud but is not complete until later. The pelvic fins form late in development. At 12 mm length, the dorsal and anal fins have fin rays formed.

The projecting snout is not developed until the fish reaches 20–25 mm in length (Hildebrand 1963b). Juveniles differ from adults in having a terminal mouth and a short rounded maxilla that does not reach the opercular margin. Juveniles also lack the silvery band of adults. Larval and juvenile stages may be completed in as little as 2.5 months (Hildebrand 1963b; Zastrow et al. 1991; Luo and Musick 1991). Growth rates of larvae based on field estimates ranged from 0.25 mm·day^{-1} (Fives et al. 1986), to 0.43–0.56 mm·day^{-1} (Leak and Houde 1987), to 0.60–0.75 mm·day^{-1} (Rilling 1996). In the laboratory, reported mean growth rates of 0.48–0.54 mm·day^{-1} were estimated (Saksena and Houde 1972) in direct relationship to prey concentrations, while in another study (Houde 1978) growth rates of 0.32–0.63 mm·day^{-1} were reported. In mesocosm studies (Cowan and Houde 1990), growth rates of 0.39–0.63 mm·day^{-1} have been recorded. Instantaneous mortality rates of larvae reared in these mesocosms were high (0.08–0.23 per day), but were lower than those recorded for bay anchovy larvae from estuaries with gelatinous zooplankton and fish predators present. Larval growth rates tended to be higher in Great South Bay, N.Y. (Castro and Cowen 1991), than for those reported for anchovies occurring further south.

Larval anchovies (bay and striped) constituted a major portion of the ichthyoplankton collected throughout Narragansett Bay, R.I. (Bourne and Govoni 1988). During a 2-year period in Great South Bay, N.Y. (Monteleone 1992), bay anchovy were the most abundant ichthyoplankter, accounting for more than 96% of the eggs and more than 69% of the larvae collected. In several other East Coast localities including Long Island Sound (Wheatland 1956), Mystic River estuary (Pearcy and Richards 1962), Barnegat Bay (Vouglitois et al. 1987), lower Hudson River (Dovel 1971), and Chesapeake Bay (Dovel 1971; Olney 1983), bay anchovy have been reported as the numerically dominant member of the ichthyoplankton.

The first food of larval bay anchovy is microzooplankton, including copepod nauplii, rotifers, and tintinnids (Detwyler and Houde 1970; Houde and Lovdal 1984). Older larvae feed upon larger copepodites and adult copepods. In Great South Bay, N.Y. (Castro and Cowen 1991), and in Chesapeake Bay (Rilling 1996), positive correlations between anchovy egg or larval densities and microzooplankton abundances were found, suggesting that larval food availability determines intensity and timing of the peak spawning period in this system.

Age and Growth. Adult bay anchovy may live to be slightly more than 3 years old, although it appears that few otolith-aged individuals survive to that age (Newberger and Houde 1995). Mean lengths-at-age of adults in mid-Chesapeake Bay (Newberger and Houde 1995) were 55.0 mm FL at age 1, 70.7 mm FL at age 2, and 83.1 mm FL at age 3. For anchovy in lower Chesapeake Bay, size at 50% maturity has been estimated to be 36.9 mm FL with age at 50% maturity occurring at about 80 days (Luo and Musick 1991). Other studies (E. D. Houde, pers. comm.) from this area, however, report that few anchovy smaller than 40 mm are mature and very few of these fish mature before they have overwintered, i.e., age at maturity is generally 10 months or older. Average annual mortality rates for bay anchovy are high, ranging from 89 to 95% per year (Newberger 1989).

General Range. Western North and central Atlantic coastal regions from Maine and Cape Cod Bay, Mass., south to the Florida Keys and westward around the Gulf of Mexico south to Yucatán, but not in West Indies (Hildebrand 1963b; Scherer 1984; Lawton et al. 1984). In coastal waters around Woods Hole and to the westward and southward of this area, this species occurs in much greater abundance (Hildebrand 1963b; Vouglitois et al. 1987).

Occurrence in Gulf of Maine. Bay anchovy have been taken in Casco Bay, Maine, at Provincetown, Mass. (Bigelow and Schroeder), and western Cape Cod Bay (Lawton et al. 1984; Scherer 1984). They do not appear to be abundant in the Gulf of Maine and seldom stray north of Cape Cod Bay. Generally, bay anchovy occur only in low abundance in the southern Gulf of Maine in western Cape Cod Bay (Lawton et al. 1984) and are rare in the middle and northern Gulf (Bigelow and Schroeder; Hildebrand 1963b). Eggs and larvae have also been collected in ichthyoplankton samples taken during June through October in this same region (Scherer 1984). Larvae of *Anchoa* spp. (presumably *A. mitchilli*) were not very abundant in the samples, ranking only 29th of the 35 most abundant species in the study.

Importance. Bay anchovy are small, unexploited anchovies. They have limited commercial use primarily as baitfish and have been used to a limited extent in the preparation of anchovy paste (Hildebrand 1963b).

In areas where abundant, they are extremely important in estuarine and coastal food webs (Baird and Ulanowicz 1989; Houde and Zastrow 1991). In fact, bay anchovy may have the largest numbers of any estuarine fish found along the South Atlantic and Gulf coasts of the United States (Baird and Ulanowicz 1989). Its production links secondary plankton production to fisheries output and is very important to the energetic processes of the ecosystem (Luo and Musick 1991).

ACKNOWLEDGMENT. A draft of the anchovy section was reviewed by Ed Houde.

HERRINGS. FAMILY CLUPEIDAE

THOMAS A. MUNROE

Herrings are among the most familiar of northern sea fishes and are certainly among the most abundant in terms of numbers of individuals. This is a rather large family of mostly marine or anadromous fishes comprising nearly 200 species placed in 56 genera (Whitehead 1985a,b). Body typically deep, oval in cross section, often deep-bodied and laterally compressed; moderately large (about 40–50 in lateral series) deciduous scales on the body (usually absent from the head region); a complete series of scutes usually present along the belly (pelvic scute always present); all fins soft-rayed, entirely lacking spines. Dorsal fin short and near midpoint of body; no adipose fin; pelvic fins abdominal and located far behind pectoral fins, near a vertical through dorsal fin base; anal fin short, its origin well behind last dorsal fin ray; caudal fin deeply forked. Mouth usually terminal, with two supramaxillae, with small or minute jaw teeth. Eye moderate, usually with distinct adipose eyelids. Gill rakers long and numerous.

Clupeids display a surprising diversity in biological and ecological life-history traits (Whitehead 1985a,b). Some species enter freshwater to feed, some are anadromous (river herrings), and some live permanently in freshwater or marine ecosystems. Many are partial or full-time filter feeders (gizzard shads), some prey on fishes, and some produce only small numbers of eggs or attach their eggs to the substrate. Typically, clupeids are coastal marine schooling fishes found in all seas from 70° N to about 60°S, feeding on small planktonic animals (mainly crustaceans), forming large schools at or near the surface, and producing large numbers of pelagic eggs and planktonic larvae. Adults of most species reach 10–20 cm SL.

Nine species of clupeids now occur in the Gulf of Maine area: Atlantic herring, alewife, blueback herring, and American shad are regular components of the Gulf of Maine fauna; Atlantic menhaden appear periodically in the Gulf, but may be abundant; hickory shad, round herring, and Atlantic thread herring occur much less often and generally in low abundance, and gizzard shad have invaded the Connecticut and Merrimack rivers and might be found in estuaries. A detailed summary of biological and ecological information for eight species of Gulf of Maine herrings is presented in Munroe (2000).

KEY TO GULF OF MAINE HERRINGS

1a. Last dorsal fin ray prolonged. 2
1b. Last dorsal fin ray not prolonged. 3
2a. Mouth terminal . . . . **Atlantic thread herring**
2b. Mouth subterminal . . . . **Gizzard shad**
3a. Belly rounded, without scutes; mouth relatively small, upper jaw extending to a point anterior to, or under, anterior margin of eye; pelvic fins posterior to vertical line through end of dorsal fin base. . . . . **Atlantic round herring**
3b. Belly sharp-edged, with prominent scutes; mouth relatively large, upper jaw extending to a point equal with or beyond a vertical through midpoint of eye; pelvic fins at point between verticals through anterior or middle of dorsal fin 4
4a. Predorsal scales forming a ridge on either side of dorsal midline; head large, equal to one-third standard length; pelvic fin rays i,6; body scales serrate or pectinate . . . . **Atlantic menhaden**
4b. Predorsal scales not forming ridge on either side of dorsal midline; head relatively small, equal to about one-quarter of standard length; pelvic fin rays i, 8; body scales rounded 5
5a. Ventral edge of belly hardly saw-toothed, though sharp; body comparatively shallow; dorsal fin origin about midbody; a cluster of small teeth present on roof of mouth . . . . **Atlantic herring**
5b. Ventral edge of belly more or less strongly saw-toothed, especially between pelvic and anal fins; body comparatively deep; dorsal fin origin ahead of midbody; no teeth on roof of mouth 6
6a. Tip of lower jaw extending noticeably beyond the upper when mouth is closed. . . . . **Hickory shad**
5b. Tip of lower jaw not extending appreciably beyond the upper when mouth is closed 7
7a. Upper outline of anterior part of lower jaw (visible if mouth is opened) nearly straight, without pronounced angle; upper jaw extending to vertical through rear margin of eye; cheek wider than deep . . . . . **American shad**
7b. Upper outline of anterior part of lower jaw concave, with pronounced angle; upper jaw reaching only to vertical through center of eye; cheek as wide as deep. 8
8a. Eye diameter relatively large, greater than snout length; body distinctly gray-green; peritoneum pale. . . . . **Alewife**
8b. Eye diameter relatively small, about equal to snout length; body distinctly blue-green; peritoneum sooty or black . . . . **Blueback herring**

River herrings, *Alosa,* several species of which were formerly placed in the genus *Pomolobus,* are anadromous fishes that have supported one of the oldest documented fisheries in North America (Kocik 1998a). The principal fishing gears used to catch river herrings are fish weirs, pound nets, and gill nets. This was exclusively a U.S. inshore fishery until the late 1960s, when distant-water fleets began fishing for river herring off the Mid-Atlantic coast (Neves 1981; Kocik 1998a). The U.S. nominal catch averaged 24,800 mt annually between 1963 and 1969, declined to 4,000–5,000 mt in the mid-1980s to record lows of 423 mt in 1994 and 464 mt in 1996 (Kocik 1998a) (Fig. 70). In response to the observed decline in catch, the Atlantic States Marine Fisheries Commission prepared a comprehensive coastwise management plan for shad and river herring to facilitate cooperative management and restoration efforts between the states (Kocik 1998a). The emphasis here is on the life history of river herrings while they are at sea. Detailed biological and ecological information on the freshwater part of their life is presented in Munroe (2000).

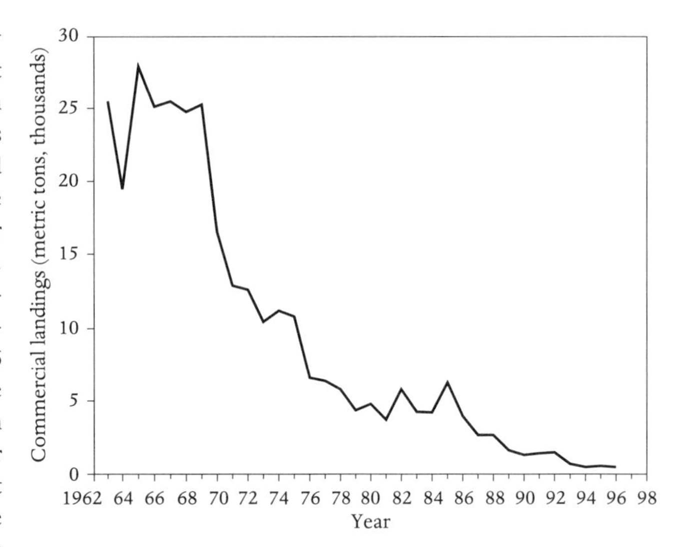

Figure 70. Commercial landings of river herrings (blueback, hickory shad, and alewife) from the Gulf of Maine–Middle Atlantic region, 1962–1996. (From Kocik 1998a.)

BLUEBACK HERRING / *Alosa aestivalis* (Mitchill 1814) /

Glut Herring, Summer Herring, Blackbelly, River Herring / Bigelow and Schroeder 1953:106–107 (as *Pomolobus aestivalis*)

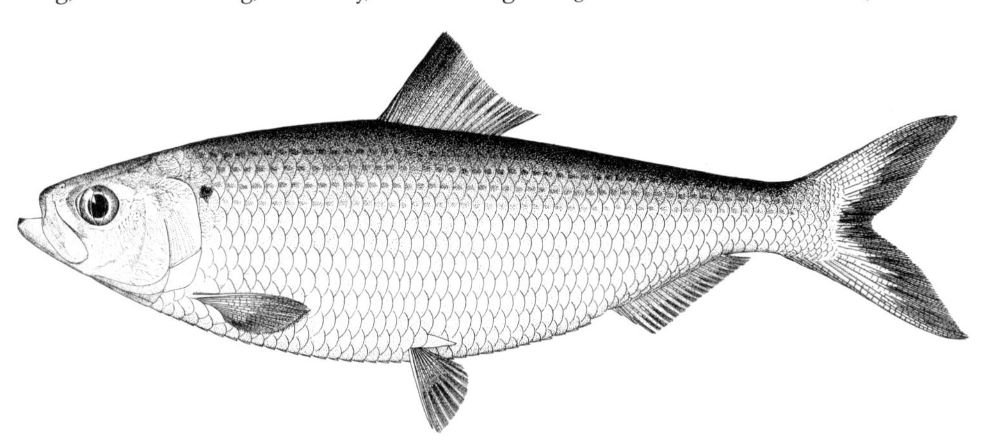

Figure 71. Blueback herring *Alosa aestivalis.* Washington, D.C., 265 mm TL. Drawn by H. L. Todd.

Description. Body deep, moderately laterally compressed, margin of abdomen with distinct keel of sharp saw-toothed scutes; caudal peduncle slender (Fig. 71). Head relatively small, pointed; cheek as wide as deep. Mouth terminal, lower jaw projecting slightly, rather thick at end, extending beyond upper jaw when mouth closed and not fitting into a groove in upper jaw. Upper outline of anterior part of lower jaw concave, with a pronounced angle. Maxilla extending to below middle of eye. Teeth small, weak, few in number on upper and lower jaws, disappearing with age; no teeth on vomer. Eye relatively large, equal to snout length. Dorsal fin small, distal margin concave; anal fin slightly longer than dorsal; pectoral fin moderate; pelvic fins small, abdominal; caudal fin forked. Scales cycloid, large, deciduous; ventral scutes strong.

Meristics. Dorsal fin rays 15–19; anal fin rays 15–21; pelvic fin rays 8–11; pectoral fin rays 12–18; scale rows on side about 41–54; ventral scutes anterior to pelvic fins 18–22, posterior ventral scutes 12–17, total scutes 31–37; gill rakers on lower limb of first gill arch 41–52 in adults (fewer in fishes under 10 cm SL); branchiostegal rays 7; pyloric caeca numerous; vertebrae 49–53 (Hildebrand 1963c; Whitehead 1985a; Scott and Scott 1988).

Color. Freshly caught fish are dark blue, bluish green, or sometimes bluish gray above, sides and abdomen silvery and iridescent; olive black longitudinal lines above midline of sides sometimes evident on adults and a dark or black spot posterior to gill cover at eye level. The fins are pale yellow to green. The peritoneum is usually sooty, brown, or black.

Size. Hildebrand (1963c) reported a maximum length of 38 cm SL but few individuals exceed 30 cm or about 1 kg in weight (Ross 1991). Females are longer than males (Scherer 1972).

Distinctions. Blueback herring resemble alewife and are difficult to distinguish, especially as juveniles, but even as adults. Adult blueback herring can be distinguished from alewife by differences in eye diameter, body depth, and peritoneum color (Bigelow and Schroeder; Loesch 1987). Eye diameter in blueback is generally about equal to snout length, whereas in alewife eye diameter is larger than snout length. The peritoneum of blueback is generally uniformly dark brown or blackish or sooty, sometimes with darker spots (melanophores), while that of alewife is pale (pearly gray to pinkish white). In fresh specimens, the dorsum of blueback is generally dark blue vs. dark green in alewife but the coloration fades soon after capture, and there is variation with changes in ambient light (MacLellan et al. 1981). Dorsal color is associated with differences in vertical distribution between the two species (Neves 1981). Blueback have more gill rakers on the lower arch than alewife, more vertebrae, and fewer dorsal and anal fin rays (Messieh 1977). The meristic differences between these two species are small and overlap (Hildebrand 1963c; Messieh 1977). There are additional differences in scale imbrication (O'Neill 1980; MacLellan et al. 1981); otolith shape (Scott and Crossman 1973; Price 1978); and electrophoretic patterns of muscle myogen (McKenzie 1973).

Blueback herring are distinguishable from young American shad by the dorsal outline of the lower jaw margin, which is deeply concave in blueback but nearly straight in shad. They have fewer (41–52) gill rakers on the lower arch than American shad (59–76) and also differ in mtDNA composition (Chapman et al. 1994). They are distinguished from Atlantic herring by the position of the dorsal fin, which is ahead of midbody rather than at it as in Atlantic herring. Blueback also differ from Atlantic herring in lacking teeth on the roof of the mouth. The anterior part of the body in blueback is more heavily built than that of Atlantic herring, and serrations on the abdominal midline scutes are considerably stronger and sharper.

Habits. Blueback herring, like alewife, are anadromous, euryhaline, schooling, coastal pelagic fish that spend most of their adult lives in the sea, approaching the shore and returning to freshwater only to spawn late in spring. Spent fish return to the sea shortly after spawning. Recent syntheses of the biology of blueback are contained in Loesch (1987), Scott and Scott (1988), Klauda et al. (1991), DesFosse et al. (1994), and Munroe (2000).

At sea, blueback herring exhibit seasonal migrations and movements in conjunction with changes in temperature and photoperiod, but direct evidence as to the relative importance of the factors affecting migration is unavailable. New Jersey inshore waters, to at least 8 km offshore, appear to be an important overwintering area for juveniles (<120 mm SL) originating from rivers in that region (Milstein 1981).

In summer and fall, blueback are confined to shelf areas north of 40°N, such as Nantucket Shoals, Georges Bank, and the perimeter of the Gulf of Maine, usually in water less than 13°C (Neves 1981). In the fall, catches occur along the northwestern edge of the Gulf of Maine. Winter catches were between 40 and 43° N, and spring catches were distributed over most of the continental shelf between Cape Hatteras and Nova Scotia as fish began their migration toward their spawning rivers. On the continental shelf between Cape Hatteras and Nova Scotia, blueback occupied water shallower than that inhabited by alewife during the same periods. Blueback were more frequently captured at depths of 27–55 m and alewives at 56–110 m (Neves 1981).

Blueback herring, like alewife, are vertical migrators at sea and follow the upward (night) and downward (day) movements of their planktonic food supply (Neves 1981). Juvenile blueback in tidal freshwaters of Virginia also undertake diel migrations, and a high proportion (90%) of night surface trawls took blueback, while relatively few (1%) had alewife (Loesch et al. 1982).

Several investigators have noted the effects of temperature on behavior of blueback herring (Bigelow and Welsh 1925; Collins 1952; Loesch and Lund 1977). Collins (1952) reported that adults respond to temperature differentials of about 0.5°C. Marcy (1976a) collected blueback in the Connecticut River over a temperature range of 6.7°–32.5°C. Mass mortalities of adults in the Connecticut River were correlated with a lethal combination of low dissolved oxygen content and high water temperatures (Moss et al. 1976).

Food. Blueback herring are plankton feeders, subsisting chiefly on ctenophores, calanoid copepods, amphipods, mysids and other pelagic shrimps, and small fishes while at sea (Bigelow and Schroeder; Brooks and Dodson 1965; Neves 1981; Stone 1986; Stone and Daborn 1987; Scott and Scott 1988; Bowman et al. 2000). Feeding efficiency of blueback is strongly inhibited even by the presence of small amounts of weed (Janssen 1982). Adults on their spawning migration continue to feed on a variety of planktonic organisms as well as terrestrial insects (Creed 1985).

Juvenile blueback in nursery areas in the Holyoke Dam region of the Connecticut River feed predominantly below the surface on copepods and cladocerans (Domermuth and Reed 1980) with selection for daphnids and bosminid cladocerans. Only small amounts of benthic prey are taken, indicating that feeding is mostly in the water column. Diet of blueback is somewhat restricted compared with that of co-occurring American shad. Juvenile blueback in the James River, Va., also consume a low diversity of prey, mostly copepods (Burbidge 1974).

Blueback search for food as they swim, sighting most prey above the horizontal course, and swimming upward to take individual prey (Janssen 1982). They apparently do not distinguish between motile and nonmotile prey.

Feeding activity of juvenile blueback was found to vary directly with, but lag behind (6–8 h), illumination levels (Jessop 1990a). Feeding begins after dawn, increases during the day to a maximum near dusk, then declines or ceases overnight, at which time stomachs empty (Burbidge 1974; Weaver 1975; Jessop 1990a).

Predators. Little is known of predation on blueback in the sea (Scott and Scott 1988), although they are important forage for a variety of predators, including spiny dogfish, eel, cod, silver hake, white hake, Atlantic halibut, and many larger species of schooling predators such as bluefish, weakfish, and striped bass (Dadswell 1985; Ross 1991; Rountree 1999; Bowman et al. 2000). They are also preyed upon by seals, gulls, and terns. During spawning runs, blueback are undoubtedly eaten by predacious fishes and birds, but data on actual predation rates are unavailable. Young-of-the-year fall prey to a variety of predators such as eel, yellow perch, and white perch. A large variety of predators are listed for anadromous and landlocked river herring by Loesch (1987), including turtles, snakes, birds, and mink. Included were 18 species of marine fishes such as silver hake, striped bass, bluefish, and salmon. Juvenile blueback in estuaries, such as the Hudson River (Juanes et al. 1993), are common items in diets of young bluefish.

Parasites. Little is known about parasites or diseases of blueback herring (Scott and Scott 1988). In the Woods Hole region, the acanthocephalan *Echinorhynchus acus* was listed (Sumner et al. 1913), and the parasitic copepod *Clavellisa cordata* also infects the gills of this species (Rubec and Hogans 1987). Landry et al. (1992) recovered 13 species of parasites from blueback in the Miramichi River, N.B. Among these were one species of monogenetic trematode, four species of digenetic trematode, one species each of Cestoda, Acanthocephala, Annelida, Copepoda, and Mollusca (glochidia), and three species of nematodes.

Breeding Habits. First spawning occurs between age 3 and age 6, but the composition of virgin spawners is strongly dominated by age-4 fish (Messieh 1977; Loesch 1987). Males tend to dominate age-classes 3–5; females live longer and dominate older age-classes. Recruitment to the spawning population is usually complete by age 5.

Spawning Location. In portions of the geographic range where blueback herring and alewife co-occur, the two species are, to a large degree, spatially isolated with respect to their spawning grounds (Loesch 1987). Blueback in the sympatric range prefer to spawn over hard substrates, where the flow is relatively swift, and actively avoid lentic sites (Bigelow and Welsh 1925; Marcy 1976b; Loesch and Lund 1977; Johnston and Cheverie 1988). The two species may occur together if further upstream migration is prohibited (Loesch 1987). At such sites, blueback concentrate and spawn in the main stream flow, while alewife favor shorebank eddies or deep pools for spawning (Loesch and Lund 1977). Although northern stocks do not usually spawn in ponds, they may (Loesch 1987), as evidenced by blueback successfully spawning after being released into the head pond of the Mactaquac Dam in the Saint John River system. Selection of lotic spawning sites by blueback in the north but lentic sites in the south suggests a spawning pattern that reduces competition with alewives where the two species are sympatric (Loesch 1987).

Blueback herring spawn in freshwater or brackish habitats above the head of the tide and can undergo extensive migrations to reach upstream spawning habitats (Nichols and Breder 1927; Hildebrand 1963c). In a coastal stream in the Gulf of St. Lawrence, blueback spawned in fast-flowing waters where eggs were spread over the bottom and where they adhered to sticks, stones, gravel and aquatic vegetation (Johnston and Cheverie 1988). The larvae are highly tolerant of salinity early in life, allowing the species to utilize both freshwater and marine nurseries (Chittenden 1972b). Fish as small as 34–47 mm reportedly tolerate water of 28 ppt salinity.

Earlier reports (Hildebrand and Schroeder 1928; Bigelow and Schroeder; Hildebrand 1963c) that on their spawning migration blueback do not ascend rivers as far as alewife are not entirely accurate. Studies in the Connecticut River (Crecco 1982) indicate that blueback herring, not alewife, migrate farther upriver. Both species occur at the Mactaquac Dam, 148 km from the mouth of the Saint John River (Messieh 1977; Jessop et al. 1982), and some fish passed above the dam proceed another 100 km upstream. Distributions of young-of-the-year fish further substantiate that blueback migrate far upstream (references in Loesch 1987). The upstream distribution of gravid blueback may only be a function of habitat suitability and hydrological conditions permitting access to such sites (Loesch and Lund 1977). The premise of a shorter blueback spawning migration developed because early studies were primarily in northern areas where alewife entered the head ponds (Loesch 1987).

Blueback herring, like alewife, presumably return to spawn in natal streams (Messieh 1977; Loesch 1987), but some individuals may stray to adjacent streams. Olfaction appears to be the major sensory mechanism used by alewife, and perhaps blueback, to find and migrate into natal watersheds (Thunberg 1971). Meristic differences among fish from different river systems supports the theory that blueback home to natal streams (Messieh 1977). Other evidence comes from the establishment or reestablishment of spawning runs after gravid fish are placed in ancestral or new systems lacking runs (Bigelow and Welsh 1925). Blueback herring will also occupy new systems or increase in abundance within systems when changes in physical or hydrological conditions permit or enhance entry (Loesch 1987), as evidenced by the huge increase in the numbers that passed above the Holyoke Dam on the

Connecticut River after improvements to the lift facilities (Moffitt et al. 1982).

Populations of blueback herring spawning in the northern parts of the range show increasing length at age (Richkus and DiNardo 1984), a smaller gonad weight relative to body size, a lower fecundity, and higher egg weight, which are components of a life history strategy that serves to maximize individual reproductive potential in highly variable environments (Glebe and Leggett 1981a,b; Jessop 1993).

Spawning Seasonality. Onset of spawning is related to water temperature (Loesch 1987) and may vary annually by 3–4 weeks in a given locality. Blueback spawning generally begins at 10°–15°C (Loesch 1987). Optimal spawning temperatures are 21°–25°C (Cianci 1969; Marcy 1976b; Klauda et al. 1991). The minimum temperature in which spawning has been reported is 14°C, and spawning ceases when temperatures exceed 27°C (Loesch 1968). Coincident with this observation, Edsall (1970) and Marcy (1971, 1973) have recorded minimal survival of river herring larvae held at temperatures above 28°C. In rivers of Nova Scotia, blueback spawning migrations occur primarily in June at 13°–21°C, but spawning does not occur until the water warms to 20° or 22°C (Crawford et al. 1986). In the southwest Margaree River, N.S., the blueback run began when the water temperature was 13.3°C–21 days later and at a temperature some 4.4°C higher than the earlier alewife run. This difference in seasonal timing (about a month or so) in peak spawning activity between alewife and blueback seems to happen wherever these species occur sympatrically. Several authors have noted that although blueback spawn about a month or so later than alewife, their spawning peaks differ only by 2–3 weeks (Hildebrand and Schroeder 1928; Loesch 1987).

In northern sections of its distribution, blueback reproduce from April to as late as August (Bigelow and Schroeder; Marcy 1976b). In tributaries to the Saint John River, N.B., blueback herring are present as early as May (Messieh 1977; Jessop et al. 1982), but do not spawn until June when water temperatures have increased (Scott and Scott 1988). Blueback herring were collected in mid-April in the lower Connecticut River in water temperatures as low as 4.7°C, but spawning did not commence until about mid-May (Loesch and Lund 1977) and may have continued until August (Marcy 1976b). It occurs much earlier in the year in populations reproducing in the southern portions of the range.

Spawning Behavior. Males generally arrive in spawning streams before females but the proportion of females increases in later runs (Loesch and Lund 1977; Loesch 1987). Spawning time for a wave of migrants is about 4–5 days (Klauda et al. 1991). Estimates of sex ratios vary because they are affected by spatiotemporal differences in the occurrence of the sexes and by sampling location and effort (Loesch and Lund 1977). The proportion of male blueback on the spawning grounds after the day of arrival can change because males tend to remain longer than females and, after exiting, some males may actually return with the succeeding wave of upstream migrants (Loesch 1969).

Blueback herring are reported to spawn in the late afternoon in the Connecticut River (Loesch and Lund 1977) or between dusk and 0100 on Prince Edward Island (Johnston and Cheverie 1988). During spawning, a female and two or more males swim circularly about 1 m from the surface. Swimming speed gradually increases and the group dives to the bottom and releases gametes (Loesch and Lund 1977). Eggs and sperm are broadcast over the substrate. Spent adults then migrate rapidly downstream.

Repeat spawning occurs in blueback at an average rate of about 30–40% (Richkus and DiNardo 1984). Some 75% of blueback spawning in Nova Scotia were repeaters (O'Neill 1980), as were 44–65% in Chesapeake Bay tributaries (Joseph and Davis 1965), including 21% that had spawned twice, 7% three times, and 1% four times (Krauthamer and Richkus 1987).

Fecundity. Fecundity is relatively high but variable in blueback herring and is related to age and size of the female (Loesch 1987). Length may be the best predictor of fecundity for this species (Loesch 1981; Jessop 1993). Total fecundity estimates for anadromous blueback range from 30,000 to 400,000 eggs (Loesch 1981; Jessop 1993). However, total fecundity exceeds fertility because postspawning females often retain unripe eggs, which may account for 23–44% of the total fecundity (Loesch and Lund 1977; Jessop 1993).

Fecundity is positively correlated with age in some populations. Maximum fecundity occurs at about age 6 (Loesch and Lund 1977) but age was found to be a statistically nonsignificant predictor of fecundity for blueback in Canadian Maritime rivers (Jessop 1993). Fecundity may decline chronologically or physiologically in older fish but no evidence of fecundal senility was found in blueback taken in Nova Scotian waters (Jessop 1993).

Early Life History. Eggs are yellowish, semitransparent, 0.87–1.11 mm in diameter, pelagic or semidemersal, and are adhesive during the water-hardening stage, becoming less so afterward (Johnston and Cheverie 1988). Both unfertilized and fertilized eggs of blueback differ from those of alewife (Kuntz and Radcliffe 1917; Norden 1967).

Incubation requires only about 3–4 days at 20°–21°C (Kuntz and Radcliffe 1917; Jones et al. 1978) and 58–55 h at 22.2°–23.7°C (Cianci 1969; Klauda et al. 1991). Young blueback are about 3.1–5.0 mm TL at hatching (Jones et al. 1978). Eggs and larvae can tolerate salinities as high as 18–22 ppt (Johnston and Cheverie 1988), and small juveniles (34–47 mm) reportedly tolerate water of 28 ppt salinity. Yolk absorption takes 72 h at 23.6°C (Cianci 1969) and occurs by about 5.0–9.8 mm (Marcy 1976b). Feeding starts 3–4 days posthatch (Klauda et al. 1991). Transformation to the juvenile stage is usually complete by about 20 mm TL (Klauda et al. 1991). Growth is fairly rapid and the young are 30–50 mm long within a month

(Bigelow and Schroeder). Essig and Cole (1986) used daily growth rings on otoliths to estimate lengths-at-age for larvae.

Larvae of blueback and alewife are difficult to distinguish but do differ in the number of myomeres between the insertion of the dorsal fin and the anus: 11–13 for blueback vs. 7–9 myomeres for alewife (Chambers et al. 1976).

Yolk-sac larvae have limited swimming ability and are carried passively by currents and swept downstream to slower-moving water, where they grow and develop into juveniles (Johnston and Cheverie 1988). Marcy (1976b) suggested that larvae become more pelagic with downstream drift. Larval blueback are photosensitive, and the density of larval blueback at the surface gradually increases from day through dusk and night, with maximum density occurring in surface waters at dawn (Meador 1982).

First-feeding (5–12 mm) larvae in the Connecticut River (Crecco and Blake 1983) consumed mostly rotifers (67% of the diet), while larger larvae fed extensively (27% of the diet) on cladocerans (*Bosmina* spp.). Blueback herring larvae have a smaller mouth gape at length compared with that of American shad, and tended to select smaller prey sizes at length compared with those consumed by shad larvae (Crecco and Blake 1983). Prey widths were also consistently smaller for blueback herring than for shad. Temporal changes in prey selection among blueback larvae generally followed changes in river zooplankton. There was a linear relationship between mouth gape and body length among blueback larvae. Niche-breadth values for blueback larvae varied considerably among collection periods and appeared to be partially governed by prey-switching. The ability of blueback larvae to utilize the abundant rotifer community may be instrumental in their numerical dominance over American shad in the Connecticut River (Marcy 1976c), as well as in other east coast rivers (Loesch and Kriete 1980).

Age and Growth. In general, female blueback are larger and heavier, and grow somewhat faster, than males of the same age (Loesch 1987). Blueback are generally smaller and shorter than alewife of the same age. Males and females reach a maximum age of about 11 years (Jessop et al. 1983). Mean lengths of blueback caught off Georges Bank (otoliths used for aging) are: 24.0 cm at age 3, 26.9 cm at age 4, 28.1 cm at age 5, 29.2 cm at age 6, 30.2 cm at age 7, and 31.3 cm at age 8 (Netzel and Stanek 1966). Lengths of male and female Connecticut River blueback are similar.

General Range. Blueback herring occur in the northwest Atlantic from Cape Breton, N.S., to the St. John's River, Fla. (Bigelow and Schroeder; Scott and Scott 1988). They are not as abundant in the northern parts of their range as in southern New England. Blueback have a more southerly distribution than alewife.

Occurrence in the Gulf of Maine. Although fishermen have recognized the existence of two distinct species since at least 1816, owing to morphological similarities, blueback and alewife have not always been correctly identified.

Bigelow and Schroeder speculated that schools of blueback could be expected anywhere between Cape Sable and Cape Cod. Scott and Scott (1988) reported this species as occurring in many rivers along the coast of Nova Scotia, in the Saint John and Kennebecasis rivers, and probably in other rivers of the Bay of Fundy drainage of New Brunswick. Bigelow and Schroeder gave many specific Gulf of Maine localities for blueback based on specimens or reliable citations.

Importance. No commercial distinction is made between blueback herring and the more abundant alewife; the species are equally useful for bait and for food and are harvested and marketed in similar fashion. Ross (1991) and Kocik (1998a) provide a discussion of management strategies being used to restore spawning populations of blueback and alewife (see alewife account), and Ross (1991) presents an excellent discussion of preparing blueback herring as table fare.

HICKORY SHAD / *Alosa mediocris* (Mitchill 1814) / Fall Herring, Shad Herring /

Bigelow and Schroeder 1953:100–101 (as *Pomolobus mediocris*)

Description. Body moderately slender, moderately compressed, its greatest thickness notably less than one-half its depth (Fig. 72). Body depth (26.7–32.8% SL), greater in large than in small specimens, usually exceeding head length. Belly with a distinct keel of scutes. Snout length 5.9–7.5% of SL. Maxilla length 10.4–13.4% SL. Lower jaw very prominent, projecting beyond upper when mouth is closed; not rising steeply within mouth. Teeth present in jaws, reduced or absent in upper jaw in larger fishes (over 23 cm SL); teeth absent on vomer; those on tongue minute, in a small elongate patch. Scales only moderately adherent, with definitely crenulate margins, preceded by longitudinal striae.

Meristics. Dorsal fin rays 15–20, most frequently 17 or 18; anal fin rays 19–23, most frequently 20 or 21; pectoral fin rays 15 or 16; pelvic fin rays 9; scales about 45–50 in a lateral series; about 16 longitudinal rows of scales on body between base of pelvic fin and anterior dorsal rays; ventral scutes moderately developed, 19–23 (usually 20–23) prepelvic ventral scutes, 12–17 (usually 13–16) postpelvic ventral scutes; lower gill rakers 18–23 (usually 20 or 21), apparently not increasing in number with age; branchiostegal rays 7 or 8; pyloric caeca numerous; vertebrae 54 or 55 (Bigelow and Schroeder; Hildebrand 1963c; Whitehead 1985a).

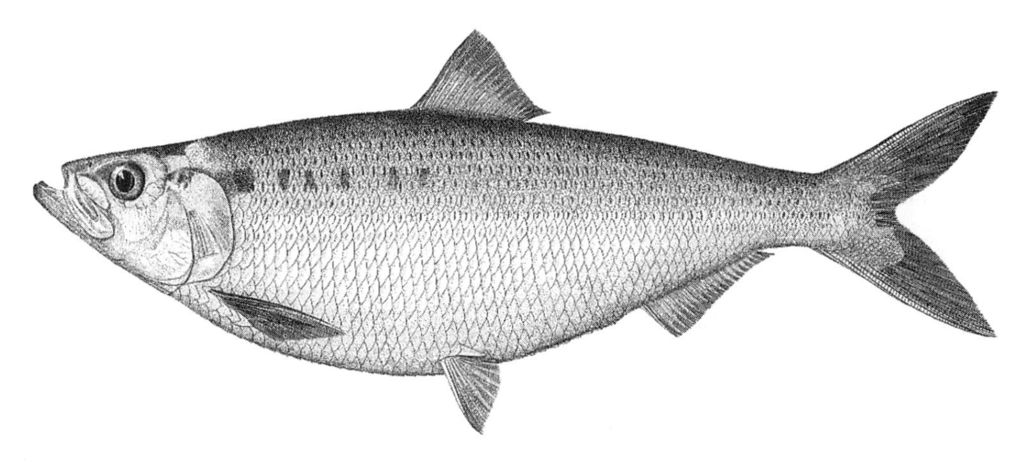

Figure 72. Hickory shad *Alosa mediocris.* Washington, D.C. Drawn by H. L. Todd.

Color. Hickory shad are grayish green above, shading somewhat gradually into the iridescent silver of the sides. The nape is green; the sides of the head brassy. The tip of the lower jaw and snout are dusky. The tongue is dusky to blackish. The dorsal, anal, and caudal fins are dusky. The outer edge of the pelvic fins is dusky or black, the inner rays are translucent. The pectoral fins are dusky to darkly pigmented. There are narrow dark lines along the rows of scales on the upper part of the sides, most distinct in large specimens that have lost their scales. There is a dark spot on the shoulder and several obscure dark spots along the flank (sometimes missing). The peritoneum is pale with scattered dusky punctulations (Hildebrand 1963c).

Size. Hickory shad are relatively large anadromous herrings reaching 457 mm TL and 0.9 kg (Hildebrand and Schroeder 1928). An earlier report of about 610 mm cannot be verified (DesFosse et al. 1994). Burgess (1980) gave sizes of 287–414 mm TL for males and 320–452 mm TL for females. The average size in Chesapeake Bay populations was 381 mm TL and 0.45 kg.

Distinctions. Hickory shad differ from Atlantic herring in that the dorsal fin origin is anterior to the body midpoint (vs. at about body midpoint in Atlantic herring), in lacking the cluster of teeth on the vomer (present in Atlantic herring), and in the presence of a strongly saw-toothed keel along the margin of the abdomen (weakly saw-toothed keel in Atlantic herring).

Hickory shad are similar to Atlantic shad, alewife, and blueback herring in general body shape, dorsal fin position, deep body, strongly saw-toothed belly, and lack of vomerine teeth. They are readily distinguished from all three species by the fact that when the mouth is closed, the lower jaw projects strongly beyond the upper (vs. lower jaw not projecting beyond upper when mouth is closed in the other species). They have fewer gill rakers (18–23 on the lower limb of the first gill arch) than any of the other three American river herrings (38–76). The faint dusky longitudinal stripes on the sides and the dusky snout tip (vs. sides without dusky longitudinal stripes and snout not dusky) are also diagnostic in fresh specimens.

Habits. Hickory shad are euryhaline and anadromous and spend most of their adult lives in the sea, entering brackish and fresh waters only to spawn. Little is known concerning the habits of hickory shad while in the sea or during their spawning migration.

Food. Hickory shad are the most piscivorous river herring in the Gulf of Maine. Small fishes, including sand lance, anchovies, cunner, herring, scup, and silversides, together with squids, fish eggs, small crabs, and a variety of pelagic crustaceans, constituted significant parts of the diets of fish examined at Woods Hole (Bigelow and Schroeder; Hildebrand 1963c).

Predators. Little is known about predators of this species (Hildebrand 1963c).

Parasites. Linton (1901b) listed a variety of parasites infecting hickory shad, including nematodes, cestodes, and digenetic trematodes. The parasitic copepod *Clavellisa cordata* has been found on the gills (Wilson 1915).

Breeding Habits. Hickory shad are anadromous (Mansueti 1962a) and ascend coastal streams during spring spawning runs. They do not reproduce in the New England region. In Chesapeake Bay, they spawn in tidal freshwaters during spring, with peaks in early May, and spawning continuing through early June (Mansueti 1962a). In Virginia and more southern parts of its range, hickory shad have been found in rivers as early as February and as late as May (Smith 1898c; Davis et al. 1970). There is evidence of spawning in main channels, flooded swamps, and sloughs (Mansueti 1962a; Davis et al. 1970; Pate 1972; DesFosse et al. 1994). The spawning period may be relatively long, as inferred from the large variation in the size of young fish captured on the same day at one locality

(Mansueti 1962a; DesFosse et al. 1994). Fecundity is 43,000–348,000 eggs per female (Pate 1972). Most hickory shad mature at 3–5 years but a small percentage of each sex matures at age 2 (Mansueti 1958b; Pate 1972). In spawning runs on the Patuxent River, males were 287–414 mm TL and females 320–452 mm TL (Mansueti 1962a).

Early Life History. Mansueti (1962a) described eggs, larvae, and early stages of specimens taken in freshwater systems of Chesapeake Bay. The eggs are slightly adhesive and semi-demersal. Hickory shad larvae were reviewed by Mansueti and Hardy (1967), Lippson and Moran (1974), Jones et al. (1978), and Wang and Kernehan (1979).

Age and Growth. Little is known concerning growth rates, age structure, or population dynamics of this species. Bigelow and Schroeder reported that a hickory shad of about 38 cm weighed slightly less than 0.5 kg, whereas one of 46 cm weighed about 0.91 kg. Growth of juveniles may be more rapid than in other east coast *Alosa,* with total lengths of 140–190 mm attained by age-1 fish.

General Range. Hickory shad inhabit coastal waters and rivers of North America from the Gulf of Maine (perhaps to the mouth of the Bay of Fundy) southward to St. John's River, Fla. (Hildebrand 1963c). This is the least abundant and least common river herring on the east coast (DesFosse et al. 1994). It is most abundant in the Chesapeake Bay and North Carolina.

Occurrence in the Gulf of Maine. The Gulf of Maine is the extreme northern limit of the range of hickory shad, and they are not commonly found there. Most captures are during the autumn. Bigelow and Schroeder did not directly observe any specimens from the Gulf, although they noted that in 1932, anglers trolling off the Merrimack River caught hickory shad while fishing for striped bass and mackerel. Historical, documented localities compiled by Bigelow and Schroeder include North Truro, Provincetown, Brewster, Boston Harbor, and Casco Bay off Portland. They also noted that hickory shad might occur as far north as the mouth of the Bay of Fundy. Hildebrand (1963c) listed it from Campobello Island, N.B.

Importance. Hickory shad are of minor importance to commercial and recreational fisheries (Whitehead 1985a). The recorded commercial catch for 1983 was 34 tons (Whitehead 1985a). They are taken in seine nets, pound nets, and in lesser quantities in gill and fyke nets. They have become a popular sport fish in recent years (Burgess 1980), will strike a small spinner or other artificial lure, and are reported to give a good fight when hooked (Bigelow and Schroeder). In some regions, such as the Chesapeake Bay, their roe is prized above that of other river herrings.

ALEWIFE / *Alosa pseudoharengus* (Wilson 1811) / Gaspereau, Sawbelly, Branch Herring, Freshwater Herring, Grayback, River Herring / Bigelow and Schroeder 1953:101–106 (as *Pomolobus pseudoharengus*)

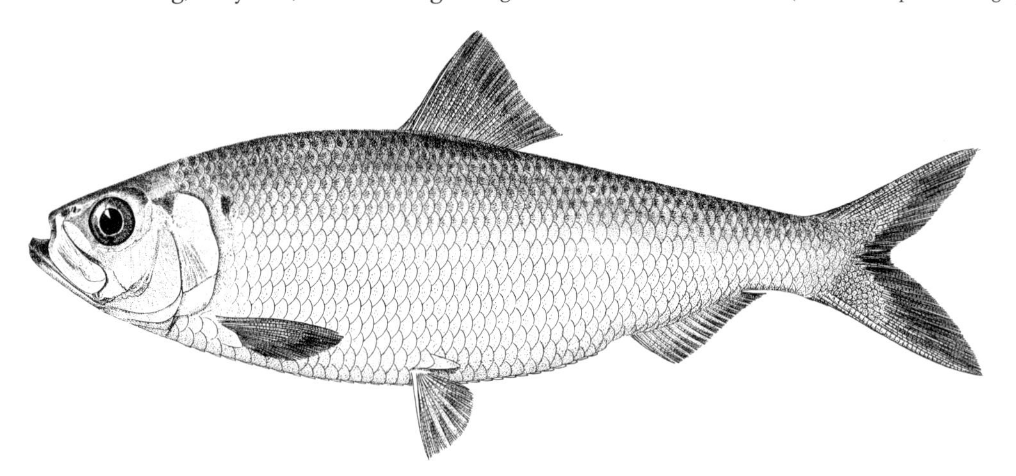

Figure 73. Alewife *Alosa pseudoharengus.* Chesapeake Bay. Drawn by H. L. Todd.

Description. Body relatively deep, moderately laterally compressed (Fig. 73). Ventral margin of abdomen with a distinct keel of saw-toothed scutes; caudal peduncle slender. Head relatively small, pointed; mouth terminal, lower jaw rising steeply, rather thick at end, projecting slightly beyond upper jaw when mouth closed and not fitting into a groove in upper jaw; maxilla extending to below middle of eye; a few minute teeth present on mandible and premaxilla (disappearing with age); no teeth on vomer. Eye large, adipose eyelid well developed. Pelvic fins small, abdominal; pectorals low on sides; caudal fin forked. Scales cycloid, large, deciduous.

Meristics. Dorsal fin rays 12–18 (usually 13–16); anal fin rays 15–20 (usually 16–18); pelvic fin rays 10; pectoral fin rays 14–

16; scale rows along side about 42–50; ventral scutes strong, 17–21 anterior to pelvic fins, 13–16 posteriorly; gill rakers on lower limb of gill arch 38–44, increasing in number with age; branchiostegal rays 7, rarely 6; pyloric caeca numerous; vertebrae 46–50 (Hildebrand 1963c; Whitehead 1985a; Scott and Scott 1988).

Color. Alewife are grayish green above, darkest on the back, paler and silvery on the sides and abdomen. Adults often have a dusky spot on either side of the body just behind the margin of the gill cover at eye level; longitudinal lines are sometimes present on the sides above the midline in larger fish. The sides are iridescent in life, with shades of green and violet. Colors change to some extent in shade from darker to paler, or vice versa, to match the bottom as the fish migrate upstream into shallow water. A golden or brassy cast is evident on sea-run fish. The peritoneum is pale to dusky.

Size. The maximum length is 36–38 cm TL (Hildebrand and Schroeder 1928; Kocik 1998a), but few individuals exceed 30 cm or about 1 kg in weight (Ross 1991).

Distinctions. Alewife are distinguishable from Atlantic herring by the deeper body, which is three and a third times as long as deep and by the dorsal fin origin, which is considerably nearer to the tip of the snout than to the caudal base. Usually there is a dusky spot on either side of the body just behind the margin of the gill cover, and the upper side may be faintly striped with dark longitudinal lines in large fish (both lacking in Atlantic herring). Furthermore, alewife are much more heavily built forward than herring, and the serrations on the midline of the belly are much stronger and sharper (hence the local name *sawbelly*), so much so that a practiced hand can separate Atlantic herring from alewives in the dark. Alewife also differ from Atlantic herring in lacking teeth on the roof of the mouth.

Alewife are distinguishable from young American shad by their smaller mouths with shorter upper jaws, by the fact that the lower jaw projects slightly beyond the upper when the mouth is closed, and by the outline of the edge of the lower jaw, the forward part of which is deeply concave in the alewife, but nearly straight in the shad. Other differences between alewife and shad are that shad have a gently rising lower jaw and more (59–73) lower arch gill rakers. Alewife and American shad differ in mtDNA composition (Chapman et al. 1994).

Adult alewife can be distinguished from blueback herring by differences in eye diameter, body depth, and peritoneum color (Loesch 1987). The diameter of the alewife eye is generally greater than the snout length, whereas it is about equal in blueback. The peritoneum of alewife is pale (pearl gray to pinkish white), sometimes with small dark spots (melanophores), while that of blueback is generally uniformly dark brown or blackish or sooty gray with darker spots. In fresh specimens, the dorsum is generally dark green in alewife (dark blue in blueback), although coloration fades soon after capture and there is substantial variation in dorsal coloration with ambient light changes (MacLellan et al. 1981). The two species also differ in scale patterns (O'Neill 1980; MacLellan et al. 1981), meristics (fewer gill rakers on the lower arch, fewer vertebrae, and more dorsal and anal fin rays in alewife [Messieh 1977]), otolith shapes (Scott and Crossman 1973; Price 1978), and electrophoretic patterns of muscle myogen (McKenzie 1973).

Habits. Alewife is an anadromous, highly migratory, euryhaline, pelagic, schooling species that spends most of its life at sea and enters freshwaters to spawn. Some alewife populations are landlocked in freshwater systems, including the Great Lakes and some of the Finger Lakes of New York (Scott and Crossman 1973) and man-made reservoirs in other areas. Along the U.S. Atlantic seaboard, alewife were taken most frequently in coastal waters ranging between 56 and 110 m at temperatures of 3°–17°C (Neves 1981). Adult and immature alewife are also taken in abundance in weirs set in shallow coastal waters. Most catches of alewife in offshore locations are usually made at depths less than 100 m. Offshore catches in the Gulf (Vladykov 1936b) include one that recorded up to 1,800 kg per haul made by otter trawlers some 125 km off Emerald Bank, N.S. (ca. 43°15′ N, 63° W), between 111 and 148 m in March 1936. Off the Atlantic coast of Nova Scotia, Bay of Fundy, and Gulf of Maine (Stone and Jessop 1992), most catches of river herring (principally alewife) occurred at bottom temperatures of 7°–11°C offshore at middepths (101–183 m) in spring, in shallower nearshore waters in summer (46–82 m), and in deeper offshore waters in the fall (119–192 m).

Alewife undertake seasonal oceanic migrations, possibly in conjunction with changing patterns of water temperature (Neves 1981). Spring movement is generally inshore and northward from the overwintering areas (Stone and Jessop 1992). In the fall, river herrings generally move offshore and southward (Stone and Jessop 1992). Winter catches in the northwest Atlantic are made between 40 and 43°. Spring catches are widespread but occur most frequently over the continental shelf area between Nova Scotia and Cape Hatteras, N.C. During summer and fall, catches are confined to three general areas in the region north of 40° latitude: Nantucket Shoals, Georges Bank, and the perimeter of the Gulf of Maine. Catches of alewife in specific areas in the Gulf of Maine and elsewhere may also be related to zooplankton abundance (Neves 1981; Stone and Jessop 1992). Alewife congregate in schools of thousands of individuals of similar size and sometimes form mixed schools with other herring species. Apparently a given school holds together during most of its sojourn in saltwater.

Alewife are light-sensitive and tend to be found deeper in the water column rather than at the surface during daylight hours (Loesch et al. 1982). Diel migratory activities are also evident in the young-of-the-year (Loesch et al. 1982; Jessop and Anderson 1989). Adult alewife in the sea may undertake vertical migrations corresponding to diel movements of zooplankton in the water column (Neves 1981). In landlocked populations, they concentrate in bottom waters during the day and migrate upward at night, possibly coinciding with the vertical migrations of their major food item, *Mysis relicta* (Lindenberg 1976).

For nearly their entire lives, alewife are in the sea, where most of their growth takes place, but upon reaching sexual maturity they enter freshwater to spawn. Adults migrate up rivers and even small tributary streams, spawning in lakes and quiet stretches of rivers during late April or May in Maine and Canada and somewhat earlier in more southern regions (March in the Chesapeake Bay area). Landlocked populations also ascend affluent rivers and streams on spawning migrations (Scott and Crossman 1973).

In summarizing spatial and temporal patterns of distribution off Nova Scotia, Stone and Jessop (1992) noted that these patterns were greatly influenced by oceanographic features. In all seasons, river herring occurred in the Bay of Fundy and off southwestern Nova Scotia, regions characterized by strong tidal mixing and upwelling, but were rarely present on the Scotian Shelf. In spring, alewife were most abundant in the warmer, deeper waters of the Scotian Gulf, particularly along the edges of Emerald and Western banks and within the channel separating them, and in regions of warm slope water intrusion along the Scotian Slope, the western and southern edges of Georges Bank, and the eastern Gulf of Maine. River herring were not present in colder regions on the eastern and western Scotian Shelf.

Alewife may prefer and be better adapted to cooler water than blueback herring (Loesch 1987; Klauda et al. 1991). Northern populations may exhibit more tolerance to cold temperatures, and flexibility to thermal selection than might be expected of a typical migratory anadromous fish (Stone and Jessop 1992). Antifreeze activity was detected in blood serum of an alewife collected off Nova Scotia (Duman and DeVries 1974a), and alewife from Nova Scotia have a much lower serum freezing point than those from Virginia (no antifreeze detected in their serum).

Food. Alewife is chiefly a particulate-feeding planktivore that consumes a wide variety of zooplankton. They feed on zooplankton either selectively on individual prey or nonselectively by filter-feeding with their gill rakers (Janssen 1976). Choice of feeding mode depends mostly on prey density, prey size, and visibility (turbidity), as well as on predator size (Janssen 1976, 1978a,b). In the sea they also take small fishes, such as Atlantic herring, eel, sand lance, cunner, and their own species, as well as a variety of fish eggs and larvae. During the spawning season, adult alewife may also eat their own eggs (Edsall 1964; Carlander 1969), although most are reportedly eaten by immature alewife.

During development, changes in gill raker morphology result in variable prey capture efficiency (MacNeill and Brandt 1990). At lengths smaller than 95 mm FL, alewife switch from feeding primarily on microzooplanton to ingesting large quantities of macrozooplankton (Stone and Jessop 1993). Unlike Atlantic herring, adult alewife often include diatoms in their diets.

Alewife generally feed most heavily during the day (Jessop 1990a), while nighttime predation may be restricted to larger macrozooplankton that produce detectable silhouettes (Janssen 1978b; Janssen and Brandt 1980; Stone and Jessop 1993). During night feeding, alewife can also utilize the lateral line sense (Janssen et al. 1995). Differences in prey selected in Minas Basin, N.S., a turbid macrotidal estuary, suggest that alewife utilize a particulate feeding strategy, whereas blueback herring are predominantly filter feeders (Stone and Daborn 1987).

At sea, alewife consume a variety of zooplankton, including euphausiids, calanoid copepods, mysids, hyperiid amphipods, chaetognaths, pteropods, decapod larvae, and salps (Edwards and Bowman 1979; Neves 1981; Vinogradov 1984; Stone and Daborn 1987; Bowman et al. 2000). Euphausiids, particularly *Meganyctiphanes norvegica,* were the most important prey item and represented more than 82% by volume of total stomach contents seasonally and geographically off Nova Scotia (Stone and Jessop 1993) and 70.9% by weight in the Gulf of Maine (Bowman et al. 2000). Proportionally, other prey groups were small and varied temporally and spatially. The high proportion of euphausiids in diets of alewife from the Scotian Shelf are higher than the contributions (37–56% by weight) reported for alewife collected off the U.S. Atlantic seaboard (Edwards and Bowman 1979; Vinogradov 1984; Stone and Jessop 1993).

The proportion of euphausiids in alewife stomachs from the Scotian shelf (winter) and Bay of Fundy (summer) tended to increase with depth and coincided with an increased relative abundance of euphausiids with increasing depth (Stone and Jessop 1993). The greater proportion and number of other prey in diets of alewife taken at depths less than 101 m on the Scotian Shelf and Bay of Fundy probably resulted from decreased euphausiid abundance rather than an absolute increase in the abundance of other zooplankters. Diet among alewife between 95 and 305 mm was relatively homogeneous, although alewife greater than 200 mm FL generally consumed the largest-sized *M. norvegica* available (Stone and Jessop 1993).

Differences in the feeding activity of alewife at sea, as evidenced by stomach fullness, varies on both a diel and seasonal basis (Stone and Jessop 1993). Consistent with observations on other visual particulate feeders, summer feeding activity peaks at midday and occurs during midafternoon during the winter. Feeding activity in all areas was reduced at night and was more apparent for fish taken during winter than summer. Daily ration was estimated at 1.2% of body weight during the winter and 1.9% during the summer (Stone and Jessop 1993). Alewife do not feed when they are migrating upstream to spawn, but when spent fish reach brackish water on their return downriver they feed ravenously on small mysids.

Much is known about the food of alewife in freshwater. Cladocerans (mainly *Cyclops* and *Limnocalanus*) constitute 75%

or more of the organisms eaten by larval alewife (Norden 1968; Johnson 1983). Juveniles tend to eat zooplankton until about 12 cm TL, but larger fish eat increasing amounts of the more benthic amphipod *Pontoporeia* (Morsell and Norden 1968). Juvenile anadromous alewife are opportunistic feeders rather than determined planktivores (Gregory et al. 1983). Cladocerans and copepods were principal food items of young alewife until zooplankton densities decreased in August, and then alewife ate more insects.

Predators. Alewife are eaten by a variety of predators, particularly schooling species such as bluefish, weakfish, and striped bass (Bigelow and Schroeder; Ross 1991), and other marine fishes, such as dusky shark, spiny dogfish, salmon, goosefish, cod, pollock, and silver hake (Rountree 1999; Bowman et al. 2000). At sea, spiny dogfish were the predators most often found with alewife in their stomachs (Rountree 1999). Juvenile bluefish (81–197 mm FL), from the Marsh River estuary, Maine, actively fed on alewife and other clupeids (Creaser and Perkins 1994). Young alewife in freshwater fall prey to a variety of predators such as eel, yellow perch, and white perch (Loesch 1987). Predators on anadromous and landlocked river herrings are diverse, including some 18 species of fishes, as well as turtles, snakes, birds, and mink (Loesch 1987).

Parasites. Parasites from alewife taken near Woods Hole (Sumner et al. 1913), include an acanthocephalan (*Echinorhynchus acus*), cestode (*Rhynchobothrium imparispine*), digenetic trematodes (*Distomum appendiculatum, D. bothryophoron, D. vitellosum, Monostomum* sp.), argulid *Argulus alosae,* and copepods (*Caligus rapax* and *Lepeophtheirus edwardsi*). Nematodes (*Anisakis simplex* and *Thynnascaris adunca*) were reported from alewife taken in western North Atlantic localities (Gaevskaya and Umnova 1977). The parasitic copepod *Clavellisa cordata* has also been reported to parasitize the gills of this host species (Wilson 1915; Rubec and Hogans 1987). Landry et al. (1992) recovered 12 species of parasites from alewife in the Miramichi River. Among these were one species of monogenetic trematode, four species of digenetic trematodes, one cestode larva, three nematode species, and one species each of annelid, copepod, and glochidia of a freshwater mussel. Alewife are host species for glochidia of the rare freshwater mussel, *Anodonta implicata,* which has disappeared where river herrings have disappeared (Davenport and Warmuth 1965). Piscine erythrocytic necrosis (PEN), a blood disease of fishes, was reported from anadromous alewife from Maine coastal waters (Sherburne 1977).

Reproduction. Alewife usually spawn in quiet waters of ponds and coves, including those behind barrier beaches (if there are openings to the sea, natural or artificial), and in sluggish stretches of streams above the head of the tide (Smith 1907; Belding 1921; Bigelow and Schroeder; Marcy 1976b). Where further upstream migration is barred by dams, alewife will spawn in shore-bank eddies or deep pools (Loesch and Lund 1977). Landlocked freshwater populations also spawn in streams or in shallow-water areas near shore on sand or gravel bottoms (Galligan 1962) and often in areas with some vegetation. Alewife do not usually spawn in swift-running water, but many eggs were found in rapids on the Miramichi River (McKenzie 1959).

In the Gulf of Maine region, alewife run indifferently up rivers as large as the Saint John or Merrimack or into small tributary streams only a few centimeters deep (Bigelow and Schroeder). In some large rivers they run far upstream to reach spawning grounds, as happens at the Mactaquac Dam, 148 km from the mouth of the Saint John River, where some fish pass upstream of the dam to proceed another 100 km upriver (Messieh 1977; Jessop et al. 1982). In other environments, their journey may be only a few meters long, as it is in the artificial cuts that are kept open through barrier beaches to allow fish access to freshwater ponds immediately behind the beach. The shortest alewife stream known to Bigelow and Schroeder was at Boothbay Harbor, where a considerable number of alewife migrated upstream annually to spawn in Campbell's Pond, a small body of water that is dammed off from the harbor and reached by a fishway only 5 m long.

During their spawning migration, alewife are much more successful than American shad in navigating fishways of suitable design. They do not generally jump over obstructions although they easily negotiate white water in rapids and fishways. Negotiating swift water apparently does not stress them because increases in blood lactic acid levels were not very great when tested during spawning runs in a fishway in Gaspereau River, N.S. (Dominy 1973).

Most alewife are believed to return to spawn in their stream of origin (Bigelow and Schroeder; Loesch 1987). This theory is supported by meristic data (Messieh 1977), by establishment or reestablishment of spawning runs by stocking gravid adults (Belding 1920, 1921; Bigelow and Schroeder; Havey 1961), and by olfaction experiments (Thunberg 1971).

Little energy is required for gonadal maturation during the freshwater phase of migration since gonads of prespawning alewife are near full maturation when the fish first enter river systems (Crawford et al. 1986). In prespawning migration, lipid depletion, and not protein utilization, apparently serves as the sole source of energy, and adult alewife may lose a substantial portion of their body weight during migration. After spawning, they are noticeably thin, but apparently recover body weight rapidly upon reaching saltwater.

Spawning Seasonality. Onset of spawning runs in alewife is related to water temperature; thus it varies with latitude and may vary annually by 3–4 weeks in a given locality (Loesch 1987). Alewife generally initiate spawning runs when water temperatures reach about 5°–10°C (Loesch 1987). Temperatures below 8°C and above 18°C (24-h average) generally result in little adult movement into spawning streams (Richkus 1974).

Spawning runs start in late March or early April south of Cape Cod (Hildebrand and Schroeder 1928; Cooper 1961;

Marcy 1969), but usually not until early to mid-April in Massachusetts (Belding 1921), and late April to mid-May in Maine (Rounsefell and Stringer 1943; Bigelow and Schroeder). Alewife spawning runs in tributaries in the Bay of Fundy begin in late April or early May and may last for 2 months, while those in the Gulf of St. Lawrence occur about a month later (Scott and Scott 1988). In the Miramichi River system, spawning usually occurs in June (McKenzie 1959).

Alewife spawning generally precedes that of blueback herring in the same watershed by about 3–4 weeks but there is considerable overlap in the spawning seasons of the two species (Loesch 1987). Their spawning peaks may differ by only 2–3 weeks (Hildebrand and Schroeder 1928).

Spawning has been reported to occur at temperatures ranging from 10° to 22°C (Rounsefell and Stringer 1943; Carlander 1969). The bulk of alewife spawning in the Gulf of Maine region reportedly takes place when water temperatures are about 12°–15°C (Bigelow and Schroeder). In the lower Connecticut River, the majority of alewife probably spawn between 7.0° and 10.9°C (Marcy 1976b). Spawning in the Connecticut River ceases when waters warm to 27.2°C (Kissil 1974), and Cooper (1961) reported that upstream migration of alewife in a Rhode Island spawning stream stopped at 21°C.

Spawning Behavior. Males arrive in spawning streams before females and usually outnumber females early in the season, but the ratio decreases as the season progresses. Male predominance is attributed to males maturing a year earlier than females (Havey 1961; Kissil 1974) and ripening earlier in the season (Cooper 1961). Upstream movements are influenced by light intensity (most movement occurring during daylight hours), water flow (more movement during higher flows), and temperature (Collins 1952; Richkus 1974). Alewife spawn during the day or night, but apparently more at night (Graham 1956).

Characteristically, larger and older fish spawn first, while smaller and younger fish spawn progressively later (Cooper 1961; Kissil 1974; Rideout 1974; and Libby 1981). In the Saint John River system, early spawners are older, with some 9- and 10-year-old fish that may have spawned as many as five times or more (Jessop et al. 1982). However, alewife age 3 to age 5 appeared to be the dominant age-groups on the spawning grounds.

Spawning lasts only a few days for each wave of arriving fish, after which the spent fish move rapidly downstream and pass later migrants on their way upstream to the spawning grounds (Bigelow and Schroeder; Cooper 1961; Kissil 1974). Fish on their return journey to freshwater are familiar sights in every alewife stream (Bigelow and Schroeder).

During the spawning act, a solitary female alewife usually swims close to shore accompanied by as many as 25 males (Belding 1921; McKenzie 1959; Cooper 1961). Groups of spawning fish are often seen swimming rapidly in a circle 1–2 m wide just below the surface. In a matter of seconds this so-called nuptial dance or swim results in the simultaneous extrusion of eggs and sperm that are randomly broadcast into the water column and over the substrate by the mating fishes. Spawning ends abruptly with the fish creating a large splash and discontinuing the circling swimming behavior.

There is considerable variation in the amount of repeat spawning that occurs in different populations of anadromous alewife. In some populations, fish spawn only once during their lifetimes, while in other populations individuals spawn over several years [up to seven or eight (Jessop et al. 1982)]. The amount of repeat spawning may indicate a clinal trend, increasing from south to north (Klauda et al. 1991). As many as 60% of repeat spawners are found in populations occurring in Nova Scotian waters (O'Neill 1980), Virginia (Joseph and Davis 1965), and Maryland (Weinrich et al. 1987; Howell et al. 1990), whereas it was estimated that fewer than 10% of spawning fish in North Carolina repeat (Tyus 1974). Others (Richkus and DiNardo 1984) disagree with the clinal hypothesis and estimate that typical values may be 30–40% repeat spawners for alewife populations throughout their range.

Fecundity. Alewife are prolific and may produce 60,000–467,000 eggs per female annually (Loesch 1987): in Maine 60,000–100,000 (Havey 1950) and in Nova Scotia 68,000–457,000 (Smith 1907; Hildebrand and Schroeder 1928; Breder and Nigrelli 1936). Total fecundity generally exceeds fertility by the extent of postspawning unripe egg retention (Jessop 1993). It was estimated that spent alewife from the Parker River, Mass., retain less than 1% of the total of ripe eggs (Huber 1978), while unripe eggs constituted 30% of the total fecundity. Jessop (1993) estimated that 38–52% of the total fecundity of alewife in Nova Scotia is attributable to unripe eggs.

Fecundity is related to age and size of females, but is highly variable (Mayo 1974; Loesch and Lund 1977; Huber 1978; Loesch 1987; Jessop 1993). There is also a corresponding decline in fecundity for similar-sized fish with increasing latitude (Jessop 1993). However, for alewife in Nova Scotia (Jessop 1993), age was found to be a nonsignificant predictor of fecundity, and no fecundal senility was evident.

Early Life History. Fertilized eggs are 0.80–1.27 mm in diameter, pink, semidemersal to pelagic, and slightly adhesive, but adhesive properties are lost after several hours (Mansueti 1956; Jones et al. 1978). Alewife eggs are slightly larger than those of blueback herring and do not contain oil globules (Kuntz and Radcliffe 1917; Norden 1967; Wang and Kernehan 1979). When spawned in flowing streams, alewife eggs (after loss of adhesion) are transported downstream (Wang and Kernehan 1979). Marcy (1976b) noted that in the lower Connecticut River alewife eggs were generally more abundant nearer the bottom than at the surface.

Incubation takes 2–3 days at 22°C, 3–5 days at 20.0°C, and about 6 days at 15.6°C (Rounsefell and Stringer 1943; Mansueti 1956; Jones et al. 1978) with optimal temperatures between 17 and 21°C. Maximum hatching success occurs at 20.8°C, but declines significantly at higher temperatures and ceases entirely at 29.7°C (Edsall 1970; Marcy 1971; 1973; Kellogg 1982). A high proportion (69%) of deformed larvae are produced from eggs incubated below 11°C (Edsall 1970; Kellogg 1982).

At hatching, yolk-sac larvae are about 2.5–5.6 mm TL (Mansueti 1956; Jones et al. 1978), and begin exogenous feeding at 3–5 days posthatch (Jones et al. 1978). Yolk-sac absorption (Cianci 1969) occurs in 72 h at 11.7°C at about 10 mm (Marcy 1976b). Post-yolk-sac larvae are positively phototropic and alternate active vertical movements toward the surface with passive, vertical descents (Odell 1934; Cianci 1969). Larvae form schools at about 2 weeks posthatching (Cooper 1961) and in the Gulf of Maine area grow to about 15 mm by the time they are a month old (Bigelow and Schroeder). Larvae transform gradually to juveniles at about 20 mm TL, and are usually fully covered with scales by about 45 mm TL (Norden 1967).

At some stages, alewife larvae can be distinguished from those of blueback herring by the number of myomeres between the insertion of the dorsal fin and the anus (7–9 myomeres for alewives vs. 11–13 for blueback herring) (Chambers et al. 1976). Alewife larvae can also be separated from those of American shad using myomere numbers (Marcy 1976b).

Age and Growth. Growth rates, age at sexual maturity, and longevity vary greatly with geography. Different methods have been used to back-calculate lengths-at-age derived from analyses of scale annuli for alewife, creating problems associated with subsequent comparisons between studies (Loesch 1987). Females may grow slightly faster and live longer than males (Rounsefell and Stringer 1943; Havey 1961), and growth in both length and weight continues for most alewife populations after sexual maturation, but at a rate that decreases with age.

No significant differences in length-at-age were reported for alewife from 21 coastal watersheds of Maine (Walton and Smith 1974). Mean lengths and weights at ages 4–10 for alewife caught during the spring spawning run on the Saint John River at the Mactaquac Dam in 1981 were provided by Jessop et al. (1982). Lengths-at-age of Connecticut River specimens 3–8 years old are 26.4 cm at age 4, 27.7 cm at age 5, 29.0 cm at age 6, and 30.2 cm at age 7.

Age at sexual maturity is generally higher in populations spawning in the northern portions of the species range. Maturity for alewife is reached in 3 years in populations spawning in Massachusetts tributaries and ranges from 3 to 5 years for fish spawning in Maine rivers, with the majority reaching maturity in about 4 years (Rounsefell and Stringer 1943; State of Maine 1982). In Long Pond, Maine, some fish matured as early as age 3, others not until age 4, and some females spawned in as many as three different years (Havey 1961). Experiments in which adult alewife were stocked in ponds in which there were none before led to the conclusion that alewife generally mature sexually at age 3 or age 4, for none of their progeny returned until 3 or 4 years after the original stocking (Belding 1921). Sexual maturity of alewife populations in the Saint John River is usually reached at ages 3, 4, or 5, with males tending to mature earlier than females (Jessop et al. 1983). Only about 5% of males in this region mature at age 2. Loesch (1987) summarized information on age at spawning, noting that first spawning generally occurs between ages 3 and 6, but the composition of virgin spawners is strongly dominated by age-4 fish. Modal age for spawning alewife across the geographic range of anadromous populations is generally 4 or 5, but the modality is readily affected by the presence of a strong year-class or by recruitment failure. Annual mean lengths of fish entering Damariscotta Lake to spawn were between 300 and 309 mm TL (Walton 1987). Alewife caught during spawning runs in Atlantic Canada average 25.4–30.5 cm FL and occasionally to 35.6 cm FL (Scott and Scott 1988).

Little data are available on age composition of sea-caught alewife (Scott and Scott 1988). Netzel and Stanek (1966) gave average lengths and weights of alewife comprising a subsample from some 13 tons of alewife and blueback caught by otter trawl in October off Georges Bank.

Stock Recruitment. The abundance of alewife returning to freshwater areas to spawn is generally positively correlated with surface area of the spawning grounds and nursery habitats (Walton 1987). Havey (1973) reported a mean relationship value of 0.7 female spawners and 407 juvenile emigrants per hectare of freshwater habitat, whereas Walton (1987) calculated a much higher value of 1.3 female spawners and 8,157 emigrants per hectare of lake surface.

For anadromous alewife populations in Maine lakes, no significant relationships have been found between brood stock size and numbers of progeny produced (Havey 1973; Walton 1987). During one study of Damariscotta Lake alewife (Walton 1987), annual harvests decreased by an order of magnitude, yet reproductive success was not affected by this apparent stock decline. Although a positive and significant relation between the numbers of females entering the lake and the estimated egg deposition was recorded, no significant relationship between estimated egg deposition and number of juvenile emigrants was found. Despite annual variability in estimated egg deposition, the number of juveniles emigrating from the lake remained relatively constant and independent of spawning escapements of adult fish (11–38 fish·ha^{-1}) throughout the study period (1977–1984). These data supported the hypothesis that the number of spawning females and the number of juvenile emigrants are asymptotically related over the observed range of adult escapement from the lake.

Growth of juvenile alewife may be influenced both by intraspecific competition and abiotic factors as in shad (Crecco et al. 1984; Crecco and Savoy 1985; Jessop 1990c, 1994). Year-class abundance of anadromous populations of alewife that spawn in Maine lakes appears to be established prior to emigration of juveniles from freshwater nursery areas (Havey 1973; Walton 1983, 1987). Intraspecific competition for zooplankton during the freshwater growth phase may be a major factor affecting growth, survival, and eventual reproductive success in established populations of anadromous alewife in Maine, such as those of Damariscotta Lake (Walton 1983). Juvenile alewife abundance from Mactaquac Lake, a head-pond environment of the Saint John River system, was significantly and negatively correlated with spring discharge (May–June) from the lake (Jessop 1990c, 1994). However, as Jessop pointed out, the relative importance of biotic and abiotic factors in controlling growth and mortality rates of juvenile alewife is uncertain. A high collinearity among environmental factors such as water temperature and discharge may obscure the mechanisms by which hydrographic fluctuations influence larval and juvenile growth and mortality rates (Crecco and Savoy 1984, 1985).

General Range. Northwest Atlantic and tributary waters along the coast from Labrador and northeastern Newfoundland (Winters et al. 1973), southward in the Gulf of St. Lawrence, south to South Carolina (Loesch 1987). Alewife have become landlocked in many parts of eastern North America, including the Great Lakes, lakes Seneca and Cayuga in the Finger Lakes of New York, and other freshwater lakes and reservoirs (Scott and Crossman 1973).

Occurrence in the Gulf of Maine. Prior to European settlement, there was probably no stream from Cape Sable to Cape Cod that did not have an annual alewife run unless it was barred by impassable falls. Occurrence in this region (Map 8) is from March to December, peaking during the summer. Although alewife are still familiar fish throughout the Gulf of Maine, stocks have declined substantially during the past two centuries and the range of spawning sites utilized by this species has been severely reduced owing to overfishing, pollution, and the erection of dams that prevent fish from reaching suitable spawning grounds. Presently, alewife still enter small and large rivers throughout the Gulf of Maine, including those of the Bay of Fundy, coastal Maine, New Hampshire, and Massachusetts, provided that these allow passage to spawning areas.

Movements. In spring, Gulf of Maine alewife are found predominantly in three areas: the Scotian Gulf, the southern Gulf of Maine, and off southwestern Nova Scotia from the Northeast Channel north to the central Bay of Fundy (Stone and Jessop 1992). Some catches also occur along the southern edge of Georges Bank and in the canyon between Banquereau and Sable Island banks. Relative abundance was highest in the Scotian Gulf between Emerald and Western banks and on the southern slope of Georges Bank. Summer distributions were less extensive than those in spring and were concentrated mainly in the eastern Gulf of Maine (off southwestern Nova Scotia) and the Bay of Fundy. Fall distributions were also more extensive than those in summer.

Importance. Historically, alewife have been harvested for food, bait, and fertilizer. Their scales commanded a high price for use in the manufacture of artificial pearls for a brief period during World War I and for a few years afterward. Currently alewife have little commercial value except as bait for cod, haddock, pollock, striped bass, and lobster and snow crab (Scott and Scott 1988). The major use of this resource (State of Maine 1982) is by commercial fishermen, who harvest and sell them for lobster bait. The fishery takes place in the spring from late April through late June as the fish ascend rivers to spawn. In recent years, mean annual landings of 3 million lb have provided about 6% of the annual bait needs for Maine's 10,000 licensed lobster fishermen. However, as a seasonal bait source for the spring lobster fishery, the alewife resource provides 30–50% of the total bait needs for the coast-wide lobster fishery. In some regions alewife are also used in the production of fish meal and oil (Scott and Scott 1988).

Alewife are caught commercially in weirs, traps, gill nets, and dip nets set in harbors, river mouths, or lakes upriver (Scott and Scott 1988; Jessop 1990b; Kocik 1998a). They are considered one of the easiest fish to catch. Today, recreational fishing accounts for modest harvests, with the greatest effort occurring in the Mid-Atlantic states. Much of this harvest is used as bait for predatory sport fishes. Ross (1991) presented a detailed discussion of recreational angling for alewife.

Alewife are an excellent food fish and they used to be marketed both fresh and salted (Bigelow and Schroeder). They have been marketed for human consumption, fresh, frozen, smoked, and salted or pickled, and canned for pet food (Scott and Scott 1988). The flesh is white and sweet, but bony. Only a very small part of the total catch is smoked or pickled and sold for human consumption in Maine (State of Maine 1982). Ross (1991) described preparation of alewife as table fare.

Historically, many more coastal streams in the Gulf of Maine yielded an abundant catch of alewife than do so today (Belding 1921; Bigelow and Schroeder). Recorded landings of alewife in Maine peaked in 1956 when 4,587,925 lb valued at $41,800 was reported (State of Maine 1982). Alewife landings for commercial and recreational fisheries in the State of Maine for recent years are summarized by Squiers and Stahlnecker (1994). In 1994–1996, an average of only 500 mt of river herrings were landed (Kocik 1998a).

Management. Owing in part to the prolonged depletion of river herring and shad stocks in the Middle and southeastern Atlantic states, the Atlantic States Marine Fisheries Commission established a coastal management plan for river herrings and American shad (Kocik 1998a). Objectives of this plan include regulating harvests, improving habitat quality and accessibility, and initiating stocking programs to restore populations in rivers where they historically but do not presently occur. Since most of the alewife harvest of New England traditionally occurred in or adjacent to estuaries and river mouths, management of these species has focused on small geographic areas (Ross 1991). The State of Maine has a management plan that sets regulations on a county-by-county basis (State of Maine 1982). There are presently 34 coastal municipalities in

Maine that operate fisheries on 35 rivers and streams annually and use the income to defray costs of municipal government. These fisheries are managed under joint cooperative management plans approved by the Department of Marine Resources. All other alewife fisheries in the state are under the general jurisdiction of the Department of Marine Resources and Inland Fisheries and Wildlife, which coordinates management of these resources on a watershed basis.

Current management methods in Maine waters include control of commercial fishing effort, maintenance and improvement of water quality in spawning and nursery areas, construction of fish passage facilities, and dam removal to allow fish access to additional spawning areas (State of Maine 1982). Alewife readily utilize most types of fishways, such as the vertical slot, pool and weir (overflow and chute type), and denil. Recent efforts to restore, enhance, and monitor river herring populations in Maine systems were summarized by Squiers and Stahlnecker (1994). Alewife restoration is being attempted at several locations within Maine and New Hampshire (Grout and Smith 1994).

AMERICAN SHAD / *Alosa sapidissima* (Wilson 1811) / Bigelow and Schroeder 1953:108–112

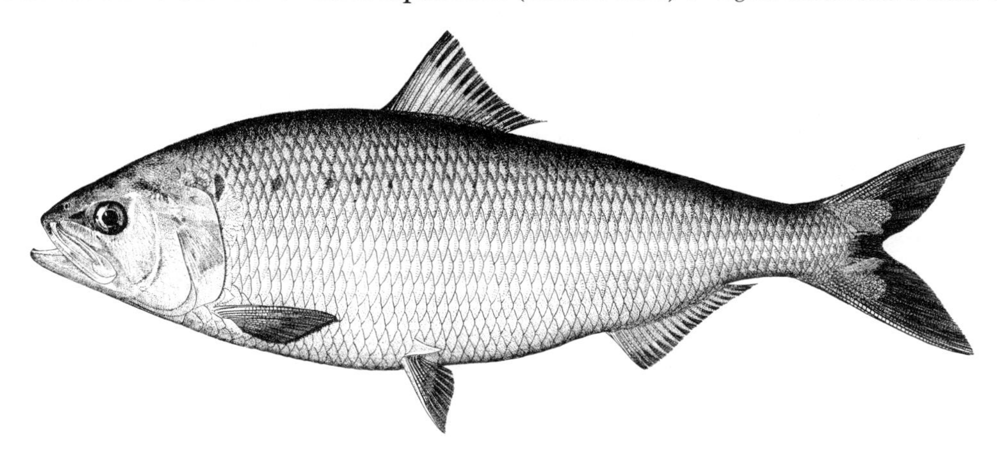

Figure 74. American shad *Alosa sapidissima*. Norfolk, Va., 280 mm TL. Drawn by H. L. Todd.

Description. Body relatively deep, but depth variable and increasing with age (30.2–36.8% SL); moderately compressed, its greatest thickness generally about one-half of its depth (Fig. 74). Margin of abdomen with sharp saw-toothed scutes forming a distinct keel; caudal peduncle slender. Head relatively small (23–28% SL), broadly triangular; snout length 4.8–7.1% SL; cheek as wide as deep. Mouth terminal and relatively large; lower jaw not excessively thickened and not rising steeply within mouth, fitting into a deep notch in the upper jaw; jaws about equal when mouth closed; maxilla reaches a vertical through posterior margin of eye. Teeth small (missing in adults, minute in juveniles to 15 cm SL), weak, few in number on premaxilla and mandible, and median line of tongue; no teeth on vomer. Eye relatively small (4.0–5.9 in HL), adipose eyelid well developed. Dorsal and anal fins moderately sized; dorsal fin somewhat elevated anteriorly, its margin slightly concave, situated above pelvic fins and anterior to body midpoint. Pelvic fins abdominal, with axillary scale equal to or slightly larger than one-half length of fin; pectoral fins low on sides. Caudal fin deeply forked; lobes of nearly equal width, somewhat shorter than head. Scales cycloid, large, deciduous, crenulate on posterior margin (Bigelow and Schroeder; Hildebrand 1963c; Whitehead 1985a; Scott and Scott 1988; DesFosse et al. 1994).

Meristics. Dorsal fin rays 15–20, usually 17 or 18; anal fin rays 18–25, usually 20–22; pelvic fin rays 8–10, usually 9; pectoral fin rays 14–20 (usually 16); lateral scale rows about 50–64; prepelvic ventral scutes strong, 18–24 (usually 20–22); postpelvic scutes 12–19, usually 15–17; total scutes 34–39, usually 35–38; lower gill rakers numerous, 59–76 in adults (fewer in young, 26–43 in specimens smaller than 125 mm); branchiostegal rays 7, rarely 6; pyloric caeca numerous, usually clustered on right side; vertebrae 53–59. Additional meristic data for hatchery-reared larvae (Johnson and Loesch 1983) and comparisons of fin ray counts of several populations of juvenile American shad are also available (Nichols 1966; Carscadden and Leggett 1975a; DesFosse and Loesch 1985).

Color. The back is dark blue or has a blue-green metallic luster, gradually shading to white and silvery on the lower sides and belly. There is a large black spot on the shoulder immediately posterior to the gill cover, followed by several (4–27) smaller dark or indistinctly dusky spots in an irregular longitudinal row, with a second row of spots (1–16) occasionally below the first, and rarely a third row of spots (2–9) ventral to the second row. No dark lines are present along the rows of scales. The fins are pale to greenish, the dorsal and caudal somewhat dusky in large specimens, and the tips of the caudal lobes are dark in some specimens. The peritoneum is pale to silvery.

Size. American shad are the largest clupeid species in the Gulf of Maine. Adults are sexually dimorphic in size (Walburg and Nichols 1967), with females being larger than males at all ages. Presently, an average individual reaches about 50 cm SL, typically 0.9–1.4 kg, and females 1.4–1.8 kg (DesFosse et al. 1994). However, at the height of their abundance (DesFosse et al. 1994 and references therein), American shad reached 5.4–6.3 kg. The all-tackle game fish record is a 5.10-kg fish taken in the Connecticut River, at South Hadley, Mass., in May 1986 (IGFA 2001). Carlander (1969) cited a 658-mm FL specimen.

Distinctions. American shad are similar to hickory shad, alewife, and blueback herring in having a deep body and sharply saw-edged belly. They differ from hickory shad by having a long upper jaw that reaches a vertical through the posterior margin of the eye, and when the mouth is closed the tip of the lower jaw is entirely enclosed within the tip of the upper jaw (in hickory shad the lower jaw only reaches a vertical through the mideye and the lower jaw extends beyond the upper jaw when the mouth is closed). American shad also have many more lower arch gill rakers (59–73) than hickory shad (18–23).

The upper outline of the lower jaw is very slightly concave, without a sharp angle, which separates it from alewife, blueback herring, and hickory shad, all of which have the outline of the lower jaw deeply concave with a pronounced angle. Shad have more lower arch gill rakers (59–73) than alewife (38–43) or blueback herring (41–51). The mouth is larger in American shad than in either of these other species (upper jaw reaching only to a vertical through midpoint of eye in alewife and blueback herring). American shad further differ from alewife in that the lower jaw does not extend slightly beyond the upper when the mouth is closed and the eye is smaller (about equal to snout length in alewife). The peritoneal lining in American shad is very pale vs. dusky or dark black in blueback herring. The three species also differ in mtDNA composition (Chapman et al. 1994).

Habits. American shad are an anadromous, highly migratory, coastal pelagic, schooling species that spends most of its life in the sea. The length of time spent at sea varies depending on sex and latitude of the home river (Leggett 1976). Adult fish apparently do not reenter freshwater until they return to spawn for the first time, although they may sometimes be found in lower reaches of estuaries at other times of the year. In the ocean, shad are found from the surface to about 220 m at bottom temperatures of 3°–13°C (Walburg and Nichols 1967; Neves and Depres 1979). American shad are commonly taken in shallow coastal waters in spring, summer, and early fall, but in late fall and winter they move deeper and further offshore. Records indicate captures 80–95 km offshore of eastern Nova Scotia (Vladykov 1936a), 65–80 km off the coast of Maine, 40–145 km off southern New England, and even the southern part of Georges Bank (40°52′ N, 67°40′ W), 175 km from the nearest land (Bigelow and Schroeder; Dadswell et al. 1987).

While at sea, American shad form relatively large schools, sometimes numbering in the thousands. Shad are vertical migrators that follow the daily upward (night) and downward (day) movement of large zooplankters on which they feed (Neves and Depres 1979). During spawning runs, they ascend rivers and migrate considerable distances upstream. After spawning, adult fish return to oceanic waters, followed later in the year by juveniles emigrating seaward from spawning grounds and nursery areas. Postspawning fish up to 4.5 kg and averaging about 2.2 kg have been reported in the Gulf of Maine and Bay of Fundy (Hildebrand 1963c), and schools have been seined occasionally in summer and autumn, even into December, at various places along the Maine coast, where they have been the object of a frozen-fish industry in some years. Immature fish (0.2–1.2 kg) sometimes congregate along the Maine coast and are more or less common in the Massachusetts Bay region during late summer and fall. Probably, young-of-the-year American shad overwinter near the mouths of their parent streams.

Young shad are sensitive to temperature changes and behaviorally avoid temperatures that could prove lethal (Moss 1970). Lower lethal temperatures for American shad are estimated to be about 4°–6°C (Tagatz 1961; Chittenden 1972a), and few if any shad enter estuaries during spawning migrations when water temperatures are below 4°C (Leggett 1973). Temperatures above 22°C approach the upper limit of tolerance for adult shad in the Connecticut River (Marcy 1976a). Minimal survival of river herring larvae (including American shad) was noted at temperatures above 28°C (Edsall 1970; Marcy 1971, 1973). Occurrence of juvenile shad in rivers is also strongly influenced by temperature. Marcy (1976c) caught juvenile shad in the Connecticut River between 10° and 31°C, but only one fish occurred in water warmer than 30°C.

Food. American shad are opportunistic predators that tend to eat whatever is most readily available (Willey 1923; Leim 1924; Bigelow and Schroeder; Maxfield 1953; Massmann 1963; Levesque and Reed 1972; Marcy 1976c). In the ocean, they feed primarily on zooplankton, various copepods (*Calanus, Arcatia, Temora,* and other smaller species), mysid shrimps, ostracods, amphipods, isopods, euphausiids, larval barnacles, and fish eggs. Although copepods and mysids are eaten by shad of all sizes, diets of fish over 40 cm usually consist of larger zooplankters, predominantly mysids. Occasionally, shad eat small fishes, such as smelt, sand lance (Leidy 1868), and silver hake (Bowman et al. 2000), but generally fishes are only a minor dietary component. Adults seldom feed in freshwater during their spring migration, although they will often strike an artificial fly or a live minnow. Juvenile fish have a broad diet in rivers, where they consume copepods, cladocerans, aquatic insect larvae, especially chironomid larvae, and adult aquatic and terrestrial insects (Leim 1924; Maxfield 1953; Massmann 1963; Levesque and Reed 1972; Marcy 1976c). Generally, shad select individual food items during prey capture. However, in turbid waters, such as those of the inner Bay of Fundy, they will switch to filter-feeding (Dadswell et al. 1983).

First-feeding shad larvae (10–13 mm) are 4–9 days old (Crecco et al. 1983), and larvae of all sizes eat mostly crustacean zooplankton (cyclopoid copepodites) and immature insects, such as chironomid larvae (Leim 1924; Mitchell et al. 1925; Maxfield 1953; Crecco and Blake 1983). Shad larvae exhibit some feeding selectivity, as most in this area feed predominantly on less abundant crustaceans (such as the cladoceran *Bosmina* sp.) and immature insects (Crecco and Blake 1983). Size of prey items consumed generally increases with larvae size, and the range of prey widths also increases steadily with body length.

Predators. A variety of predators include American shad in their diets, for example, spiny dogfish and angel shark (Rountree 1999). Seals feed on shad (Melvin et al. 1985; Scott and Scott 1988), and goosefish caught in weirs also eat adult shad. It is likely that young shad fall prey to predators in freshwaters, but Leim (1924) found no evidence of predation by eels or striped bass in the Shubenacadie River. In the lower reaches of the Hudson River (Juanes et al. 1993), juvenile shad are common in the diet of young bluefish. American shad are also attacked by sea lamprey (Warner and Katkansky 1970; Melvin et al. 1985).

Parasites. Early information on parasites infecting shad included the acanthocephalan, *Echinorhynchus acus*, the nematodes *Ascaris adunca*, *Ascaris* sp., and the copepod *Caligus radiatus* (Sumner et al. 1913). Of 110 American shad examined, infestation rates of 35%, 51%, and 7% for nematodes, trematodes, and acanthocephalans, respectively, were noted (Leim 1924). Twenty-six species of parasites were identified in a recent study of 695 fish examined from seven Atlantic localities ranging from Florida to the Bay of Fundy (Hogans et al. 1993). Only three of these parasites had previously been reported from American shad. Species identified were predominantly boreal marine forms that have low host specificities and widespread distributions. The greatest parasite diversity was found among fish from the Cumberland Basin and the Shubenacadie River. The seven most abundant parasites included the monogenetic trematode *Mazocreoides georgei*, digenetic trematodes *Lecithaster confusus*, and *Genitocotyle atlantica*, the cestodes *Diplostomum spathaceum*, *Scolex pleuronectis*, and an unidentified pseudophyllidean plerocercoid, as well as the nematode *Hysterothylacium aduncum*. High levels of typically southern parasite species, such as *Genitocotyle atlantica*, in shad from the Cumberland Basin and most or all river populations indicate that shad become infested with these species near the southern extent of their reported range, near Cape Hatteras (Hogans et al. 1993). The copepod *Clavellisa cordata* also parasitizes gills of shad (Rubec and Hogans 1987). The leech *Calliobdella vivida* has also been reported from this host (Appy and Dadswell 1981). Additional information on parasites of American shad was provided in Hoffman (1967).

Spawning Location. In the spring, American shad move from the sea into coastal rivers to spawn, when water temperatures range from 16.5° to 19.0°C (Leim 1924; Massmann 1952; Walburg 1960; Leggett and Whitney 1972; Marcy 1976c; Williams and Daborn 1984). They spawn in river areas dominated by broad flats with relatively shallow water (1–6 m) with moderate (0.3–1.0 $m \cdot s^{-1}$) current (Smith 1907; Bigelow and Welsh 1925; Massmann 1952; Marcy 1972; Leggett 1976; State of Maine 1982). In the Connecticut River, shad eggs were found from 0.6 to 7.3 m deep (Marcy 1976c). Walburg and Nichols (1967) reported that about 40% of spawned eggs occurred in water less than 3 m. In Gulf of Maine tributaries, viable eggs may be found on river bottom types ranging from fine sand to coarse rubble to ledge, but never on silty or muddy bottom areas (State of Maine 1982). Highest survival rates of shad eggs reportedly occur over gravel and rubble substrates (Layzer 1974).

There does not appear to be any required distance above brackish water for spawning shad (Massmann 1952). In larger rivers, shad may migrate far upstream to reach favorable spawning grounds. In the Saint John River, they ascend about 320 km to Grand Falls. Currently, they can only run up about 55 km in the Penobscot River, where they formerly ascended some 145 km, or 70 km (to Augusta) in the Kennebec River, which they could formerly ascend 174 km (to Carratuk Falls). The dams at Lawrence, only 31 km upstream, now stop any stray American shad that may still enter the Merrimack, which they formerly ascended for 200 km to Lake Winnipesaukee (Stevenson 1899). In the Connecticut River, shad spawn in areas up to 174 km from the river mouth (Marcy 1976c).

Males arrive at spawning grounds first, followed by females (Leim 1924). Spawning takes place in the evening after sundown and may continue until midnight or later (Leim 1924; Massmann 1952; Walburg and Nichols 1967; Chittenden 1969; Williams and Daborn 1984). Spawning may also occur on dark afternoons, as was observed for shad in the Connecticut River (Marcy 1976c). During spawning, the female is accompanied by one or several males; spawning fish swim close to the surface, sometimes with their backs exposed leaving a visible wake and splashing (Medcof 1957; Marcy 1976c). Eggs are released in open water, where they are fertilized.

During upriver migrations individual American shad apparently spawn repeatedly as they progress upriver thereby undergoing a gradual reduction both in weight and energy content of testes and ovaries (Glebe and Leggett 1981b). This spawning mode may be necessary to facilitate the high fecundity of American shad. An estimated two- to fourfold increase in body cavity volume would be required in American shad to accommodate an equivalent number of eggs if all of them reached full volume and were spawned at one time (Shoubridge and

Leggett 1978). Late-run migrants may spawn sooner in the upriver migration (Chittenden 1969; Glebe and Leggett 1981b), because higher water temperatures may force spawning.

Spent and very emaciated fish begin their return journey to the sea immediately after spawning. In the Kennebec River, they are first seen on their way downstream in late June and constantly thereafter throughout July, whereas in the Saint John River, they migrate downstream in July and August (Bigelow and Schroeder). Fish migrating downstream may begin feeding before reaching saltwater and recover a good deal of fat before moving out to sea (Atkins 1887). The young descend downriver in the autumn of their first year of life.

Spawning Seasonality. Water temperature controls timing of migration into natal rivers, spawning activity, and survival of eggs and larvae (Massmann and Pacheco 1957; Walburg 1957; Leggett and Whitney 1972; Leggett 1976). Throughout their geographic range, adult shad enter streams in their home river in spring or early summer, when water temperatures have reached about 10°–12°C. Consequently, American shad spawning runs occur at correspondingly later times in the year passing from south to north along the coast. For Gulf of Maine localities (Atkins 1887), such as the Kennebec River, the first American shad appeared late in April, with the main run occurring in May and June. The first ripe females were caught during the last week in May, and spawning began in early June, with most fish doing so during that month. A few shad spawn as late as July, and in some years there may be an occasional shad spawning as late as August. These dates probably applied equally well to timing of spawning in the Merrimack River in days when American shad were plentiful there (Bigelow and Schroeder), but the season was somewhat later in the Saint John River (Leim 1924) and in the Shubenacadie (i.e., from mid-May until the end of June). In Canadian rivers, spawning normally occurs in May, June, or even July (Scott and Scott 1988). Spawning fish enter rivers as early as November in Florida, with peak migration occurring in January (Leggett 1976); peak migration occurs in January in Georgia; in March in waters tributary to Pamlico and Albemarle sounds, N.C.; in April in the Potomac River; and generally in May and June in streams from Delaware to Canada (Walburg 1957; Leggett and Whitney 1972).

American shad usually enter rivers at temperatures around 13°–16°C and cease to migrate when water temperatures reach about 20°C. Initiation of spawning depends on water temperatures, with peak runs usually occurring at about 18.5°C. Mansueti and Kolb (1953) reported that shad ovaries developed more slowly at 12.8°C than at 20°–25°C. Temperatures below about 16°C prolong the developmental period and reduce egg survival (Leim 1924; Leach 1925; Mansueti and Kolb 1953; Bradford et al. 1968). In the Annapolis River, eggs were collected at water temperatures of 10.5°–22.7°C, but most were taken at 13°–18°C (Williams and Daborn 1984). Peak spawning occurred at about 18.3°C. Spawning apparently ceased when water temperature dropped below 13°C. In the Shubenacadie River, N.S., American shad spawn mostly in temperatures higher than about 12°C, and spawning is temporarily interrupted if water chills below this temperature (Leim 1924). In the Connecticut River, Marcy (1976c) collected eggs at temperatures 7.5°–24°C, with peak spawning occurring at 22°C in 1968 and at 14.8°C in 1969. Onset of spawning in the Connecticut River begins at temperatures of 13°–18°C (Leggett and Whitney 1972). Based on records at fish lifts, peak movement in east coast rivers occurs in temperatures at 16.5°–21.5°C (Leggett 1976). The temperature range for maximum hatch and survival of eggs is 15.5°–26.5°C (Leim 1924; Massmann 1952; Walburg 1960; Bradford et al. 1968; Marcy 1972).

Levels of dissolved oxygen (DO) also strongly influence the location and success of spawning by shad (Ellis et al. 1947; Sykes and Lehman 1957; Bradford et al. 1968; Chittenden 1969, 1973; Thurston-Rogers and Baren 1978; Miller et al. 1982; Maurice et al. 1987). DO levels of 4–5 mg·liter^{-1} are suitable for juveniles (Burdick 1954) and result in the successful hatch of healthy larvae (Bradford et al. 1968). Adult shad undergo rapid respiratory movements as DO decreases below 4 mg·liter^{-1} (Tagatz 1961); DO levels of about. 3.5 mg·liter^{-1} cause sublethal effects (Chittenden 1973); levels at 2–3 mg·liter^{-1} result in about a 33% mortality (Dorfman 1970); and high mortalities occur at DO levels less than 2 mg·liter^{-1} (Tagatz 1961; Chittenden 1969).

Homing Behavior. American shad returning to spawn are guided by olfactory and rheotactic cues imprinted during their initial migration to the ocean (Hollis 1948; Mansueti and Kolb 1953; Dodson and Leggett 1973, 1974; Carscadden and Leggett 1975a, b; Williams and Daborn 1984; Melvin et al. 1986). Mark-recapture studies (Hollis 1948; Mansueti and Kolb 1953; Williams and Daborn 1984; Melvin et al. 1986) have shown that American shad have low straying rates (about 3%). Meristic studies also suggest that American shad not only home to their natal rivers, but even to specific tributaries within some of the major northern spawning rivers (Carscadden and Leggett 1975a,b).

Repeat Spawning. South of Cape Hatteras, N.C., American shad are semelparous (they die after spawning), whereas in rivers to the north of this region, they become increasingly iteroparous, with individuals surviving and returning to the ocean, living to spawn in subsequent years (Leggett and Carscadden 1978; Glebe and Leggett 1981a,b). In the northern part of their range, shad may return to spawn as many as five times (Carscadden and Leggett 1975b). The percentage of adults that live to be repeat spawners increases northward along the Atlantic coast. In the York River, Va., 24% of shad are repeat spawners, whereas in the Connecticut River, shad have a 63% incidence of repeat spawning (Leggett and Carscadden 1978), and 73% and 64% of American shad in the Saint John and Miramichi rivers are estimated to be repeat spawners.

Fecundity. Fecundity is generally highest and often severalfold higher in southern populations and decreases in those spawning in rivers at the northern end of the geographic range. Since northern shad spawn more than once, reciprocal latitudinal trends in annual fecundity and repeat spawning cause average lifetime fecundity to be roughly constant over the total Atlantic coast range (Bentzen et al. 1989).

American shad are prolific spawners, producing up to 600,000 eggs (Cheek 1968), and fecundity generally increases with the size of the female. In the St. Lawrence River, the number of eggs ranged from 58,534 to 390,633 per female, with an average of 125,166 eggs for 48 females (Roy 1969). In the Saint John River population, the number of eggs ranged from 118,929 to 165,776 (Carscadden and Leggett 1975a). Most Canadian shad (Leim and Scott 1966) produce 20,000–150,000 eggs per female, which is probably representative of fecundity of shad spawning in Maine (State of Maine 1982). Estimated fecundity is 256,000–384,000 eggs per female for shad spawning in the Connecticut River (Leggett 1969).

Early Life History. Fertilized eggs are 2.5–3.5 mm in diameter, transparent, pale pink or amber, semibuoyant, and not adhesive. After fertilization, the eggs sink to the bottom, where water-hardening of eggs and their resultant increase in diameter may lodge them in bottom rubble (Jones et al. 1978), or fertilized eggs may drift with the current for several kilometers downstream from the spawning site. Whitworth and Bennett (1970) followed movement of shad eggs from the time they were broadcast until they sank or lodged on the bottom and found that this occurred about 5–35 m downstream. Most shad eggs travel only 1.6–6.4 km downstream from where spawned (Carlson 1968; Chittenden 1969; Marcy 1976c).

Hatching occurs in 12–15 days at 12°C and in 6–8 days at 17°C. Development of shad eggs in the Connecticut River is prolonged and mortality increases if water temperatures drop below 16°C (Marcy 1972). Larvae are found in the Connecticut River from May to August (Marcy 1976c) but June is when most larvae emerge, with peaks occurring at water temperatures of 20°–22°C (Marcy 1976c; Crecco et al. 1983; Crecco and Savoy 1987). No shad eggs occurred in water with a DO content of less than 5 ppm (Marcy 1976c), and Carlson (1968) reported that the LC_{50} for DO is between 2.0–2.5 ppm. In the Connecticut River, larvae reportedly grow well at pH values between 6.0–7.0 (Crecco and Savoy 1987). Larval development reportedly is more successful in brackish than in pure fresh water, with salinities of about 7.5 ppt being about the most favorable for development (Leim 1924).

Development and ontogenetic changes associated with eggs and early yolk-sac larvae have been described by a series of authors: Hildebrand (1963c), Watson (1968), Chittenden (1969), and Marcy (1976b). Johnson and Loesch (1983, 1986) described morphological development from yolk-sac absorption to postflexion stage. A complete description of morphological development from hatching through the adult stage has been presented by Mansueti and Hardy (1967), Lippson and Moran (1974), Jones et al. (1978), Johnson and Loesch (1983), and Howey (1985).

Larvae are about 5.7–10.0 mm TL at hatching (Marcy 1976c) and are transparent and extremely slender. Some larvae drift into brackish water shortly after hatching, while others remain in freshwater throughout the summer months (State of Maine 1982). Yolk-sac absorption occurs in 3–5 days at 17°C and in about 7 days at 12°–14°C, with fish length ranging between 9–15.5 mm TL at the end of this stage (Hildebrand 1963c; Watson 1968; Marcy 1976c). During development, the vent moves anteriorly as in other larval Clupeidae. Median fin formation is complete by 17–21 mm SL (Bigelow and Welsh 1925; Johnson and Loesch 1983), and paired fin development is complete at 23–28 mm SL, or in about 21–28 days, when metamorphosis is nearly complete. Illustrations and descriptions of larvae and juveniles are provided in Jones et al. (1978) and Johnson and Loesch (1983).

American shad eggs and larvae are readily distinguished from other species of *Alosa* (Mansueti and Hardy 1967). Larvae are extremely slender, with the anus situated almost as far back as the caudal fin base (Leim 1924). Ventral pigmentation is one of the most important characteristics for identification of larval American shad from yolk absorption to about 13 mm SL (Leim 1924; Hildebrand 1963c; Jones et al. 1978; Johnson and Loesch 1983). Specimens from freshwater are usually more pigmented than those taken in brackish water (Leim 1924; Jones et al. 1978; Johnson and Loesch 1983). Marcy (1976a) noted that American shad in the Connecticut River had 41 or more myomeres compared with 37–40 myomeres in alewife and blueback herring. A table comparing pigmentation characteristics, and morphometric and meristic information useful in identifying American shad, alewife, and blueback herring is presented in Johnson and Loesch (1983).

Age and Growth. American shad can be aged using scales and otoliths (Leim 1924; Cating 1953; Lapointe 1957; Judy 1961; Melvin et al. 1985). American shad spawning in New England waters generally do not live beyond 5–7 years of age, although age 10 and age 11 individuals are caught infrequently. Based on scale studies and length frequencies, shad in the upper Bay of Fundy range to about 4.1–4.5 cm in 9–10 weeks, 7.5–15.5 cm by the end of the first growing season, 12.8–15.0 cm long when 1 year old; 23.0–25.4 cm long at 2 years; 33–36 cm at 3 years; 38–41 cm at 4 years; and 47–49 cm at 5 years (Leim 1924). Individuals 62 and 63 cm long appeared to be 7 and 6 years old, respectively. In the Bay of Fundy (Leim 1924), American shad weigh about 91 g at 20 cm; about 273 g at 31 cm; about 605 g at 38–42 cm; about 1.14 kg at about 52 cm; and 2.04 kg at 59–61 cm, but weights varied with condition. Shad may grow somewhat faster in waters of the open Gulf of Maine (Bigelow and Schroeder).

Females may reach a greater age than males. Among older fish (7–11 years old) examined by Borodin (1924), only 7 were males as compared to 86 females. In the Shubenacadie, and presumably in other Gulf of Maine rivers, the oldest shad are estimated to be 8–9 years old, but shad runs in the northeastern United States and Canadian Maritimes are dominated by 4- and 5-year-old fish (State of Maine 1982). Annapolis River populations are slow-growing, longer-lived, and larger (Melvin et al. 1985) than those of surrounding areas. In this system, the oldest male was estimated to be 12 years old and the oldest female 13. The largest shad caught was a female 61.7 cm FL, weighing 1.92 kg. The usual size was about 50 cm FL, and none were over 62 cm FL. The large individual size and older age of the population is attributable to a very low fishing mortality: about 5% per year.

In the Bay of Fundy, a few shad spawn at age 4 and most are 5 years old and 45.7–48.3 cm long at first spawning, but in New England waters, spawning males reach sexual maturity between age 3 and age 5 and females between age 4 and age 6. Males enter the Connecticut River from the sea in their fourth year when they are 300–350 mm long. The smallest females returning to the Connecticut River to spawn were between 400 and 430 mm and were in their fifth, sixth, and seventh years (Marcy 1972).

General Range. American shad occur in Atlantic coastal waters and in streams and rivers of North America from northern Labrador (Dempson et al. 1983), southward to the St. John's River, Fla. (Bigelow and Schroeder). In the Canadian Atlantic region (Scott and Scott 1988), shad occur from Labrador near Nain and at Sand Hill River in southern Labrador (Hare and Murphy 1974), and from Newfoundland (a few collections), but apparently there are no spawning populations north of the St. Lawrence River (Leggett 1976). Historically, shad were believed to be most abundant in central portions of the Atlantic coast range (Leim 1924). American shad were introduced into the Sacramento and Columbia rivers of the Pacific coastal system in 1871 and are now successfully established from Kamchatka Peninsula (Asia) and Cook Inlet, Alaska, to Baja California.

Occurrence in the Gulf of Maine. When the first settlers arrived in New England, they found seemingly inexhaustible multitudes of American shad migrating annually in all of the larger rivers and many of the smaller streams, with the tributaries of the Gulf of Maine hardly less productive than those of the Hudson or Delaware rivers (Bigelow and Schroeder; State of Maine 1982). With increases in human population and industrial development, one stream after another was rendered impassable by construction of dams, degradation of spawning areas, or diminished water quality through pollution. Access to spawning rivers in the Gulf of Maine region was almost completely denied through construction of dams during the early part of the twentieth century (Walburg and Nichols 1967), resulting in a drastic decline in spawning populations (Bigelow and Welsh 1925; Taylor 1951; Mansueti and Kolb 1953). Consequently, American shad stocks in the Gulf of Maine are merely remnants in comparison with those of colonial days (Taylor 1951; Bigelow and Schroeder).

Migrations. One of the most significant features of the life cycle of American shad is a pronounced seasonal migration (McDonald 1884; Talbot and Sykes 1958; Leggett 1976; Dadswell et al. 1987). Although they spawn in Atlantic coastal rivers from Florida to Quebec, juveniles and adults originating throughout this range form seasonally migratory aggregations in the sea (Talbot and Sykes 1958; Leggett 1977a,b; Dadswell et al. 1987; Melvin et al. 1992). For example, mark-recapture studies of fish tagged during the summer in the Bay of Fundy have provided recaptures from all major shad spawning rivers along the Atlantic coast of North America including those from Labrador to the St. John's River, Fla., about 3,000 km from the tagging location (Dadswell et al. 1987). During a mean life span of about 5 years at sea, an American shad could migrate up to 20,000 km (Dadswell et al. 1987). Each spring juveniles and adults leave offshore wintering grounds and take part in extensive seasonal migrations along the Atlantic coast, returning generally southward and eastward in the fall to overwintering grounds (Talbot and Sykes 1958; Dadswell et al. 1983, 1987). The large spring migratory schools consist not only of spawning adults but also of nonspawning younger fish. Seasonal migrations are thought to occur mainly in surface waters, but in the sea American shad have been caught at depths to 220 m (Walburg and Nichols 1967).

The northward, inshore coastal migration each spring of American shad was postulated over a century ago (McDonald 1884), but not demonstrated until the 1950s (Talbot and Sykes 1958). Historically, the timing of American shad migrations was believed to be regulated by water temperature, both at sea and in the rivers (Leggett and Whitney 1972). While at sea, fish from all Atlantic coast populations were previously thought to move together as a single group (Hollis 1948; Vladykov 1936b, 1956; Talbot and Sykes 1958) along a coastal corridor in areas of preferred temperature range (13°–18°C) (Leggett and Whitney 1972). Temperature was thought to provide both the migration cue (McDonald 1884; Leggett and Whitney 1972) and also to regulate ocean swimming speeds (Leggett 1977a). Neves and Depres (1979) demonstrated that not all shad moved in a single aggregation. While some fish were migrating north within a few kilometers of the coast, another group was offshore along the continental shelf over depths of 50–200 m in temperatures of 3°–15°C. This finding required revision of the hypothesized temperature controlled—migration model (Leggett and Whitney 1972; Leggett 1976; Neves and Depres 1979). The expanded hypothesis postulated that seasonal movements of American shad were broadly controlled by climate and that fish maintained themselves within migration corridors or oceanic paths of "preferred" isotherms.

In contrast to a temperature controlled–migration model, evidence from the analysis of 50 years of tagging studies (Dadswell et al. 1987) suggests that American shad cross thermal barriers, remain for extended periods in temperatures outside their "preferred" range, and migrate rapidly between regions regardless of currents and temperatures (Melvin et al.

1986; Dadswell et al. 1987). These results also indicate that shad are not all concentrated into a relatively small geographic area at any one time, nor do they migrate together at the same rate. Ocean aggregations represent a heterogeneous mixture of American shad from many rivers. Discrete aggregations exist seasonally at widely separated marine locations. Summary by season of tag returns, occurrence records, and trawl survey information illustrates a pattern consistent with three winter sites and three summer terminus points for American shad during their annual migration along the Atlantic coast (Dadswell et al. 1987). During January–February, American shad are off Florida, the Mid-Atlantic Bight, and Nova Scotia, and are entering rivers to spawn from Florida to South Carolina. In March–April, movement is onshore and northward, both in the Mid-Atlantic Bight and off Nova Scotia, and spawning runs are underway from North Carolina to the Bay of Fundy. By late June, American shad are concentrated in the inner Bay of Fundy, inner Gulf of St. Lawrence, and off Newfoundland, but spawning fish are still upstream in coastal rivers from the Delaware River north to the St. Lawrence River. During autumn, American shad leaving the St. Lawrence estuary are captured across the southern Gulf of St. Lawrence, but at the same time, those departing the Bay of Fundy are found from Maine to Long Island and some have already arrived off Florida and Georgia.

Migration rates of shad in inshore areas calculated from tag return information indicate that they cross "thermal barriers" (Dadswell et al. 1987). Prespawning American shad moved north along the coast rapidly, some traversing 2,500 km from Delaware Bay to the Gulf of St. Lawrence in 60 days. Prespawning migrants traveled at speeds estimated at about 30.2 km·day^{-1}, which, as noted by Dadswell et al. (1987), were similar to the theoretical optimal migration rate of 29 km·day^{-1} (Leggett and Trump 1978). In contrast, postspawning adults had a mean migration rate of only about 8.8 km·day^{-1}. South of Cape Cod, prespawning shad migrate close inshore (Leggett and Whitney 1972; Dadswell et al. 1987), but north of there, tag returns are fewer and the migration corridor is less clear. Some fish perhaps migrate along the coast of Maine, whereas others may migrate offshore around the edges of Georges Bank (Dadswell et al. 1987). Prespawning fish arrive at rivers south of Cape Cod at water temperatures of about 13°–18°C (Leggett and Whitney 1972), but shad move to northern rivers almost as quickly before ocean temperatures are above 10°C (Melvin et al. 1986). Dadswell et al. (1987) noted that nonreproductive shad migrating from wintering sites in the Mid-Atlantic Bight must cross the Gulf of Maine during May–June where a constant subsurface temperature of 6°C prevails, to arrive annually in the Bay of Fundy during June–July. A counterclockwise migration pattern was evident in shad in the Bay of Fundy. Fish entered the Bay during April–May on the Nova Scotian side and later departed on the New Brunswick side from August to October. In the inner Bay, the shad run appeared to divide by chance, with portions going to both Minas and Cumberland basins. Once fish had committed to either route, however, their migration pathway was rigid. American shad that migrated first to Cumberland Basin moved along the northern shore (New Brunswick) and left the Bay during August–September on either side of Grand Manan Island. Fish that moved first to Minas Basin, however, migrated through Cumberland Basin before leaving the Bay of Fundy by the same route as those migrating into Cumberland Basin only. Migration rates of all recaptured Bay of Fundy shad during summer were similar, about 3.6 km · day $^{-1}$, and the direction of movement was the same as the direction of residual current flow in the Bay of Fundy (Dadswell et al. 1987).

If not temperature, then what cues or clues are shad using to regulate where or when they move in the sea? Shad definitely move north or south seasonally, but tag information does not suggest that certain temperatures provide the major cue. For example, shad migrating off Virginia experience a totally different suite of temperature stimuli from those wintering on the Scotian shelf, yet fish from both sites arrive at the Annapolis River, N.S., to spawn during the same time period (Melvin et al. 1986). Likewise, shad migrating from Delaware Bay in April to spawn in the Miramichi River in June will encounter a variety of ocean temperatures along their migration routes, ranging from 10° to 12°C in some areas to 4°C in others. Similarly, fish leaving the Bay of Fundy in September and entering the St. John's River, Fla., to spawn in December would leave 18°C water, migrate through waters of 10°–12°C in the Gulf of Maine, and traverse water as warm as 20°C before arriving in Florida. In each case, the same behavioral response occurs under different and diverse temperature stimuli. Temperature change or some aspect of seasonality strongly influences migratory direction. During increasing temperatures and day length, fish move north; when these decrease, shad move south. However, when shad are within large semi-enclosed coastal regions, they follow the direction of residual currents and the coastline. Temperature change with an additional stimulus such as photoperiod may initiate migratory behavior, but timing of such behavior by different fish appears related to their origin or life history stage. Tagging data and information from population discrimination studies suggest that origin, life-history, and chance also play a role in the seasonal migrations of American shad. American shad alternate between extrinsic and intrinsic cues to direct migration, depending on their physiological state. This species may use a physiological optimizing strategy (Leggett 1977a) while it is nonreproductive, but uses a bicoordinate navigation system with map (geographic contours), compass (magnetic capabilities), and clock (timing of tides) (Quinn 1982) when it must reach a specific goal at a certain time (Dadswell et al. 1987). Extrinsic factors related to ocean climate, seasonality, and currents may provide cues and clues for portions of non-goal-oriented migration. Intrinsic cues and bicoordinate navigation appear to be important during goal-oriented stages of migration.

Importance. Extensive commercial and recreational fisheries developed for American shad along the entire east coast in the

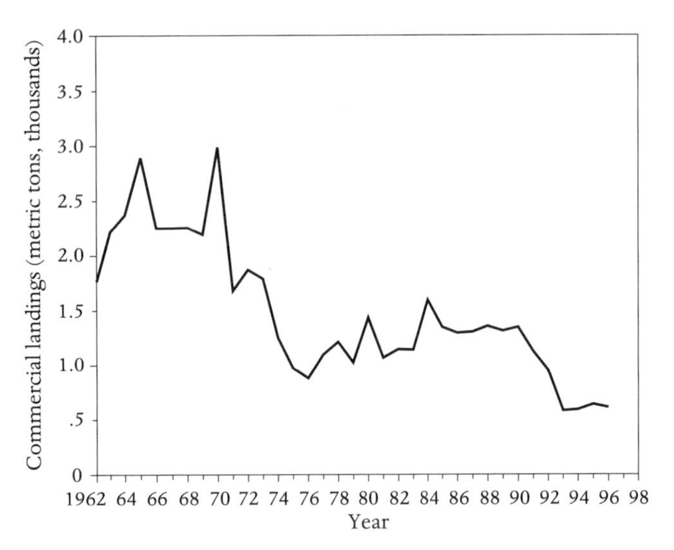

Figure 75. Commercial landings of American shad *Alosa sapidissima* in the Gulf of Maine–Middle Atlantic region, 1962–1996. (From Kocik 1998b.)

nineteenth century. Currently, American shad are only of moderate commercial importance, with landings occurring chiefly in the lower rivers of the Mid-Atlantic region and in the Hudson River (Whitehead 1985a). American shad have little current commercial importance in the Gulf of Maine. Commercial fishing methods for taking shad include pound nets, gill nets, and seines. Small quantities are also taken with fyke nets, otter trawls, purse seines, traps, and dip nets.

From the beginning of recorded landings in 1887, the fishery remained relatively stable at about 462 mt annually until 1911, rose to a peak of 1,452 mt in 1912, dropped to an average of about 51 mt annually from 1928 to1933, essentially were commercially extinct from 1934 to 1940, and then increased to a high of 499 mt in 1946 (State of Maine 1982). Commercial landings of 2,000–3,000 mt in the 1960s decreased to about 500 mt from 1992 to 1996 (Fig. 75).

American shad are a highly regarded game fish in many parts of the coastal United States (Ross 1991), but less so in Canada (Scott and Scott 1988). Presently, there are no sport fisheries for American shad in U.S. Gulf of Maine tributaries, and the Annapolis River has one of the few sport fisheries for American shad in Canada (Scott and Scott 1988). Historically, substantial quantities of shad were taken in recreational fisheries, which were especially targeted where migrating fish were enroute to spawning grounds. Many American shad were caught with flies and lures, and by snagging, by crowds of anglers seeking this species on its migratory journey. Ross (1991) provided a detailed account of sport fishing and preparation techniques for enjoying American shad as table fare. American shad are marketed fresh and salted; the flesh is white and flaky and of fine flavor, but it is bony. Good-quality caviar can be made from the roe. American shad liver oil contains 500–800 USP of vitamin A and 50–100 IU of vitamin D per gram, but is not as rich in vitamins as the oil from cod or haddock livers. Nutritional values for shad are: fat 3–17% (varies seasonally); moisture 65–73%; protein 16–20%; ash 1–2% (Sidwell 1981).

Stocks. Despite intermixing of shad populations in the marine habitat, evidence that fish home to their natal rivers to spawn (Melvin et al. 1986) suggests that adults ascending particular rivers might form genetically discrete populations. In trying to identify distinct river populations, previous investigators of Atlantic shad have utilized a wide variety of approaches: meristics and morphometrics (Warfel and Olson 1947; Fischler 1959; Hill 1959; Nichols 1966; Carscadden and Leggett 1975b); multivariate analysis of meristic characters (Carscadden and Leggett 1975b); latitudinal variation in phenotypic reproductive traits (Leggett 1969; Carscadden and Leggett 1975b; Leggett and Carscadden 1978; Shoubridge 1978; Melvin et al. 1992); otoliths (Williams 1985); protein electrophoresis (Sismour and Birdsong 1986); a linear discriminant function of 10 meristic and 16 morphometric characters (Melvin et al. 1992); and mtDNA variation (Bentzen et al. 1988, 1989; Nolan et al. 1991; Chapman 1993).

Data from these studies support three different theories about stocks of American shad: (1) Each major shad-producing river along the Atlantic seaboard has its own distinctive spawning stock (Carscadden and Leggett 1975a,b). (2) There are three broad geographic groups: Canadian Atlantic, mid-U.S.–Atlantic, and South U.S.–Atlantic rivers or Miramichi, Hudson-Delaware, and St. John's rivers (Williams 1985; Dadswell et al. 1987; Bentzen et al. 1989; Nolan et al. 1991; Melvin et al. 1992). · (3) No definition of stocks can currently be made that is of value for management purposes although there are genetic differences between populations in different rivers. Analysis of mtDNA variation in American shad from different river drainages (Bentzen et al. 1988, 1989; Nolan et al. 1991; Chapman 1993) suggest that overall levels of mtDNA sequence divergence among shad are low.

Management. Intensified fishing efforts combined with effects of pollution and loss of access to spawning and nursery habitat have markedly reduced New England shad stocks (Ross 1991). American shad restoration projects, including shad culture, aimed at reversing the declines in tributaries along the east coast, have been underway for some time (Leach 1925; Chittenden 1971a; St. Pierre 1977; Howey 1985; Meinz 1978; Wiggins et al. 1985, 1986; Barry and Kynard 1986; Ross 1991). In addition to habitat restoration, the number of adult American shad lifted over dams to gain access to additional spawning areas has improved reproductive success and subsequent recruitment (Moffitt et al. 1982). The Atlantic States Marine Fisheries Commission has implemented a coastwide management plan for American shad and river herring to facilitate cooperative management and restoration plans among states (Kocik 1998b). Despite improved returns in some major river systems such as the Susquehanna, Delaware, and Connecticut rivers, abundance of American shad is well below historic levels. Overall, this resource is considered to be fully exploited (Kocik 1998b).

ATLANTIC MENHADEN / *Brevoortia tyrannus* (Latrobe 1802) / Pogy, Mossbunker, Bunker, Fat Back /

Bigelow and Schroeder 1953:113–118

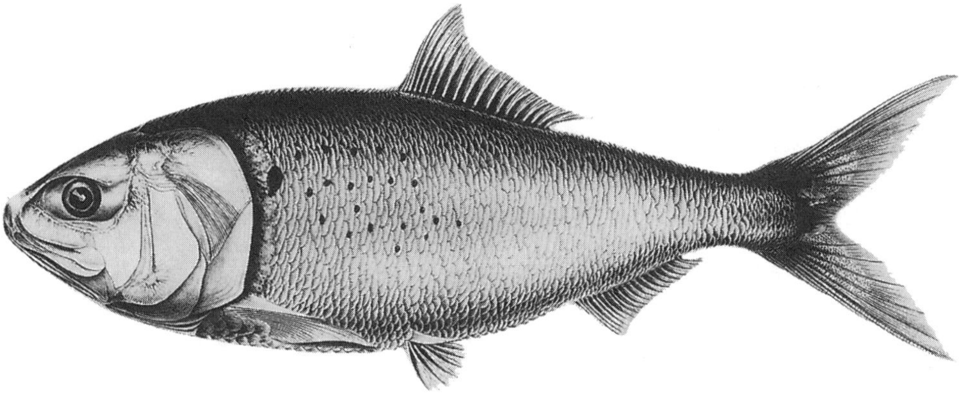

Figure 76. Atlantic menhaden *Brevoortia tyrannus.*

Description. Body strongly laterally compressed and deep, about three times as deep as long (Fig. 76). Deepest at dorsal fin origin or the region slightly anterior to this point; ventral margin of abdomen with sharp-edged scutes; caudal peduncle slender. Head large, cheeks deeper than long; mouth large, oblique; maxilla extending to point equal with verticals through posterior margin of eye or posterior to this point, lower jaw projecting slightly and fitting into notch in upper; teeth absent. Snout blunt. Eye small. Gill rakers very long and numerous, close-set, forming effective basket-like sieve, gill raker numbers increasing with size of fish. Dorsal and anal fins moderately long; dorsal fin origin about midway between snout and base of caudal fin. Anal fin origin at or slightly posterior to vertical through base of posteriormost dorsal fin ray. Pelvic fins relatively small, abdominal, axillary process present. Pelvic fin origin at vertical equal with or slightly posterior to dorsal fin origin. Pectoral fin slightly falcate, axillary process present. Caudal fin deeply forked. Scales adherent, cycloid, exposed field deeper than long, posterior margin serrate or pectinate. Scales on dorsum, above base of anal fin, and at base of tail much smaller and irregularly placed.

Meristics. Dorsal fin rays 19 or 20; anal fin rays 20–24; pelvic fin rays i, 6; pectoral fin rays 16 or 17; predorsal scales in two rows (ca. 33–39 specialized scales) on either side of body midline; prepelvic scutes 19–21; postpelvic scutes 11–13; scales in lateral series 40–58 (usually about 45–52); gill rakers 100–150 on lower limb of first arch; vertebrae 45–50, usually 47–49 (Bigelow and Schroeder; Whitehead 1985a; Scott and Scott 1988).

Color. Dark blue, green, blue-gray, or blue-brown above, with silvery sides and abdomen, and with a strong yellow or brassy luster. Conspicuous dusky or black spot on each side of body immediately posterior to gill cover, followed along flank by varying number of irregularly arranged smaller dark spots forming up to six approximate lines. Fins usually with yellowish cast; caudal sometimes dusky at base and on free margin of fin. Peritoneum black.

Size. Maximum sizes have been reported to 50–51 cm TL (Whitehead 1985a; Scott and Scott 1988), but the largest menhaden on record are two 8-year-old fish, one from Rhode Island measuring 470 mm TL and weighing 1.67 kg (Cooper 1965a) and the other measuring 418 mm FL and weighing over 1.36 kg (Reintjes 1982). Adult menhaden average 20–30 cm FL and 0.25–0.60 kg in weight.

Distinctions. The very large scaleless head, nearly one-third of the total body length, gives menhaden an appearance so distinctive that they cannot be mistaken for any other Gulf of Maine fish. They are easily distinguished from other clupeids in the Gulf by the two rows of modified predorsal scales, forming a ridge on either side of the body midline (lacking in other species). They are further distinguished by having the posterior margins of the scales nearly vertical (not rounded), and pectinate (vs. smooth margins in other clupeids). The upper jaw has a distinct median notch and lacks teeth.

Taxonomic Remarks. *Brevoortia patronus,* Gulf menhaden, from the northern Gulf of Mexico is closely related to *B. tyrannus* and has been accorded specific status based on differences in meristic features and life-history characteristics (Hildebrand 1948; Dahlberg 1970). However, genetic distinctness of Gulf menhaden is questionable (Avise et al. 1989), as analysis of mtDNA genotype frequencies indicated that recent gene flow has occurred between populations of Atlantic and Gulf menhadens (Bowen and Avise 1990). Atlantic and Gulf menhaden populations are apparently disjunct, with a gap on the Atlantic coast of south Florida and the Florida Keys (Dahlberg 1970). However, at the southern ends of their respective ranges, both species hybridize with yellowfin menhaden, *B. smithi* (Reintjes 1960; Hettler 1968; Dahlberg 1970), providing a possible avenue of gene flow between Atlantic and Gulf forms (Bowen and Avise 1990).

Habits. Recent overviews of biology, ecology, fishery, and utilization of menhadens appear in Ahrenholz (1991), Smith

(1991), Waters (1994), and Munroe and Smith (in Munroe 2000).

Menhaden inhabit pelagic, euryhaline waters of estuaries and bays, as well as polyhaline coastal waters on the inner continental shelf. When found offshore, they are seldom far from land and rarely leave continental shelf waters (Hildebrand 1963c; Scott and Scott 1988). Along eastern North America, Atlantic menhaden stratify by age and size, with older and larger fish ranging farther north (Nicholson 1978).

Atlantic menhaden are pelagic, marine, schooling fish, forming large and compact schools, both as juveniles and as adults. Schools are composed of fish of similar size. In calm weather, schools are often detected by a distinctive ripple action on the surface or by frequent whips as the dorsum and the caudal fin break the surface of the water, and the fish are occasionally seen with their snouts out of water. These behaviors distinguish menhaden from schools of Atlantic herring and mackerel. Menhaden schools can also be visually identified by the brassy hue of the sides of the fishes, which is often readily visible to the careful observer who approaches the school cautiously.

Atlantic menhaden inhabit waters varying substantially in salinity from almost freshwater (3.5 ppt) to full-strength ocean salinity (Engel et al. 1987; Reintjes 1982). Adult Atlantic menhaden are generally found in meso- and polyhaline estuarine and neritic waters, but juveniles in particular occur throughout most of the larger estuaries along the coast, including low-salinity waters of estuarine tributaries. Juvenile menhaden are estuarine-dependent and can tolerate sudden salinity shifts (Engel et al. 1987). Larval and juvenile menhaden are frequently found in great abundance in shallow, low-salinity, estuarine waters. Juveniles remain within these protected estuarine waters throughout the summer, moving down-estuary into coastal rivers and open bays and into nearshore oceanic waters as they grow.

Atlantic menhaden are usually found in waters warmer than about 10°C, and Bean (1903) reported that they cannot survive in water temperatures much below that. However, juveniles have been collected in Narragansett Bay, R.I., at temperatures as low as 1.2°C (Herman 1963), although the condition of juveniles raised in mesocosms in Rhode Island declined noticeably in January when temperatures dropped below the range at which menhaden normally live (Keller et al. 1990). Mortalities during winter cold periods have occurred in Canadian waters, particularly in the Kennebecasis River, where menhaden kills triggered by low winter temperatures reportedly occur nearly annually between December and February (Scott and Scott 1988). Studies on the effects of thermal effluents on juvenile Atlantic menhaden concluded that rapid mortality occurs in effluents with temperatures exceeding 33°C (Lewis and Hettler 1968; Young and Gibson 1973). Relationships between survival time and various temperatures are presented in Lewis (1965) and Hoss et al. (1974).

The ecological role of menhaden cannot be overstated. They convert energy derived from phyto- and zooplankters, and possibly also vascular plant detritus, into hundreds of thousands of tons of fish flesh. In turn, they are of paramount importance as a prey species for numerous piscivorous fishes, mammals, and seabirds. Larval and juvenile menhaden are seasonally very important components of estuarine fish assemblages (Tagatz and Dudley 1961; Pacheco and Grant 1965; Bozeman and Dean 1980). Given the tremendous numbers of individual menhaden, individual growth rates, filtering and feeding capacity, and seasonal movements, they consume and redistribute significant amounts of energy and biomass on an annual basis, both within and between estuarine and continental shelf waters. In Narragansett Bay, for example, menhaden were shown to have local impacts on plankton communities (Oviatt et al. 1971), where notable reductions in chlorophyll-*a* levels within menhaden schools were due to their consumption of phytoplankton. Also, within the immediate vicinity of schools, dissolved oxygen tended to be lower than ambient levels, and ammonia levels were higher than in areas without schools of menhaden.

Food. Atlantic menhaden are a pelagic, filter-feeding fish. They feed at the primary production level, chiefly consuming large quantities of phytoplankters, particularly diatoms (Peck 1894). As they increase in size, there is a gradation, or trend, in feeding repertoire that changes from a predominantly herbivorous to a more omnivorous diet, with wide overlap between size-classes (Friedland et al. 1984). Menhaden are adaptable, capable of grazing on planktonic organisms, including several species of benthic diatoms (Edgar and Hoff 1976), calanoid copepods, mysid shrimps, euphausiids, and annelid worms, as well as organic detritus (Bigelow and Schroeder; June and Carlson 1971; Jeffries 1975; Bowman et al. 2000). Few other adult fishes have the ability to feed directly on phytoplankton. Recent evidence suggests that juvenile Atlantic menhaden readily digest cellulose and other vascular plant material, and that detritus of vascular plant origin forms an important component of their diets (Lewis and Peters 1984). Their digestive system is characteristic of herbivores, with a strong, muscular gizzardlike stomach. However, distribution of juvenile menhaden appears to respond to gradients of phytoplankton cells of sufficient size to be filtered by the fish, suggesting a chemosensory preference for plant rather than detrital particles and a foraging strategy patterned by the efficiency of their gill raker feeding structure (Friedland et al. 1989).

Atlantic menhaden feed by filtering prey items from the water with a straining apparatus in the shape of successive layers of comblike gill rakers. They feed by swimming with mouth agape and gill openings spread, filtering out the entrapped organisms with their uniquely efficient, feathery gill rakers. No other Gulf of Maine fish has a comparable filtering apparatus, nor has this species any rival in the Gulf in its utilization of the phytoplanktonic pasture. Bigelow and Schroeder watched small fish in Chesapeake Bay swimming downward as they fed, then turning upward to break the surface with their snouts, still with open mouths. The mouth and pharyngeal sieve act exactly as a tow net, retaining whatever is large enough to enmesh, with no voluntary selection of particular plankton units. Menhaden parallel whalebone whales, basking shark, and giant manta rays in their mode of feeding, except that their diet consists of smaller organisms because the filter-feeding apparatus is closer-meshed.

Atlantic menhaden capture food particles by mechanically sieving food items through their branchial basket as opposed to aerosol entrapment mechanisms, that is, structures that are sticky or mucus-covered (Friedland 1985). Food particles are retained at various positions on the gill rakers and then transported to the base of the raker blade before being passed farther back in the bucchal cavity for ingestion. Particles larger than the spaces between raker blades tumble along the leading edges of the rakers to the base, whereas those smaller than the spaces between blades, but nonetheless filterable, may be transported by a different mechanism. Sites of small-particle capture are the branchiospinules, which lack mucous cells, suggesting that food is captured primarily by a mechanical sieving at these sites. Gill arch sections, where most filtration takes place, have a cutaneous fold over the mesial raker elements. Epithelium on the lateral surface of this cutaneous fold is rich in mucous cells. Positioning of this surface relative to the base of the filtering rakers suggests that particles coming off the raker blades are held in dynamic balance and are transported back into the bucchal cavity over, or complexed with, a mucous layer and then ingested.

Taste buds are located on the tongue and in the fifth branchial arch. Taste seems to be a two-stage process (Friedland 1985). Both food and nonfood particles incite gulping and in turn initiate feeding. It is possible that taste buds on the tongue are mechanoreceptors responsible for initial detection of potential food items. If particles ingested are usable food, then feeding continues; conversely, if particles are not usable, feeding stops. Taste buds located on the gill arches are chemosensory in nature, probably enabling the fish to determine if ingested particles are food and whether feeding should proceed.

An adult menhaden can filter 15–28 liters of water per minute (Peck 1894; McHugh 1967). Fish do not feed continuously, but assuming a steady filtering rate over a 6-month period, an individual fish could filter plankton from more than 3.9 million liters of water.

Historically Atlantic menhaden were thought to feed primarily on phytoplankton (Peck 1894; Darnell 1958; June and Carlson 1971), but questions have been raised recently as to whether zooplankton, bacteria, and detritus were also important foods (Friedland et al. 1984). The widely held concept was that menhaden were unable to filter small phytoplankton and depended mostly on large phytoplankton and zooplankton (Durbin and Durbin 1975, 1981; Durbin et al. 1981; Blaxter and Hunter 1982). This view was supported by analyses of stomach contents in which zooplankton was the primary constituent (Richards 1963b; Jeffries 1975). Bacteria had been suggested as a potential food resource based on anatomical considerations (Reintjes and Pacheco 1966) and analyses that found anomalously high nitrogen levels in stomach contents (Peters and Kjelson 1975). Detritus was also suggested as an important energy source for menhadens in allochthonous systems (Peters and Schaaf 1981; Lewis and Peters 1984; Deegan et al. 1990).

In laboratory studies, minimum sizes of particles that adult menhaden (257 mm FL) could filter were 13–16 μm, and filtering efficiencies increased with the size of the food particles (Durbin and Durbin 1975). In the absence of detritus, minimum particle sizes filtered by menhaden (138 mm FL) were 7–9 μm (Friedland et al. 1984). Filtration efficiencies also increased with increasing size of prey particles. For example, the dinoflagellate, *Prorocentrum minimum,* was filtered at an average of 16.5% for the 12-μm morph, whereas an average of 41.5% was filtered for the 17-μm morph.

Detritus in combination with phytoplankton resulted in capture of prey particles below the minimum-size threshold and enhanced filtration efficiencies of prey particles above the threshold. However, effects of detritus on filtering efficiency were not consistent among various algae tested. For some species, filtering efficiency with detritus present increased, but for others it was slightly lower or unchanged (Friedland et al. 1984).

The difference in minimum-size threshold and a more rapid increase in filtration efficiency with increased prey size for small menhaden (138 mm FL) suggest differences in functional morphology of the menhaden feeding apparatus as the fish increase in size. Perhaps more important than the number and size of gill rakers is the spacing of the branchiospinules, as it is probably the spacing that determines filtering efficiency (Magnuson and Heitz 1971). The rate of increase in filtering efficiency as prey size increased was greater for smaller menhaden than for larger ones, suggesting that raker gaps in juveniles are in a narrower size range. As prey size increases, proportionally more rakers of a smaller fish would be capable of filtering prey. Maximum filtration efficiencies should correspond to prey sizes exceeding most of the raker gap dimensions. The maximum occurs at prey sizes of about 100 μm for 138-mm FL menhaden, whereas for large menhaden this maximum occurs at a prey size of about 200 μm (Durbin and Durbin 1975).

Predators. Oil-laden menhaden, swimming in schools of closely ranked individuals, are prey for nearly every piscivorous fish, marine mammal, and seabird in the Gulf of Maine. Whales and porpoises devour them in large numbers, and sharks often follow schools of Atlantic menhaden. Pollock, cod, silver hake, striped bass, and swordfish also feed on menhaden in the Gulf of Maine, as do weakfish south of Cape Cod. Bluefin tuna also feed on menhaden (Crane 1936). Bluefish prey heavily on menhaden (Grant 1962; Friedland et al. 1988), even in the Gulf of Maine when both bluefish and menhaden are abundant there. Other predators include smooth dogfish, spiny dogfish, angel shark, and goosefish (Rountree

1999). In Chincoteague Bay, Va., menhaden were found in 13% of stomachs, and were the second most abundant prey of sandbar shark (*Carcharhinus plumbeus*), ranging in size from 40 to 80 cm FL (Medved et al. 1985). During winter in Chesapeake Bay, newly metamorphosed sea lamprey attack Atlantic menhaden (Mansueti 1962b). Menhaden are also important in the diets of seabirds, herons, egrets, ospreys, and eagles (Hildebrand 1963c).

Yolk-sac and first-feeding larvae of Atlantic menhaden are prey to a variety of predators including adults of the comparatively large-sized copepod *Anomalocera ornata* (Turner et al. 1985). Yolk-sac larvae were also consumed by the smaller copepod (*Centropages typicus*), but this crustacean was unable to feed upon larger, and more active, first-feeding larvae.

Parasites. Menhaden are host to a variety of parasites, including protozoans (Hardcastle 1944), monogenetic (McMahon 1963) and digenetic (Linton 1905) trematodes, nematodes, cestodes, copepods (Wilson 1932), and isopods (Kroger and Guthrie 1972), and play an important role in their life cycles. Westman and Nigrelli (1955) provided a list of 13 species of parasites and Reintjes (1969) listed 5 more. A barnacle (*Balanus venustus*) was reported on a juvenile menhaden taken off South Carolina (Guthrie and Schwartz 1990).

Mass Mortalities. Mass mortalities of menhaden have been reported in many parts of the range along the coastal United States, usually associated with oxygen depletion in semi-enclosed, shallow-water habitats (small coves and heads of tidal creeks) during late summer. Menhaden often strand in enormous numbers in shoal water, either in their attempt to escape their enemies or for other reasons. The teeming numbers of fish milling about in the warm shallow waters may exhaust the dissolved oxygen, resulting in deaths of hundreds to thousands of fish (Reintjes and Pacheco 1966). Algal blooms and bacterial respiration associated with active or decaying plankton probably contribute to dissolved oxygen depletion and mass mortalities of menhaden. Dissolved oxygen tolerance studies indicated that significant mortalities occurred at a concentration of 1.1 $mg{\cdot}liter^{-1}$ dissolved oxygen when menhaden were acclimated at 28°C (Burton et al. 1980).

In southern regions, annual menhaden kills or die-offs are well documented (Westman and Nigrelli 1955; Stephens et al. 1980; Ahrenholz et al. 1987a). During the spring, mass mortalities of menhaden occurred from *spinning disease,* which is caused by a virus and named for the erratic swimming behavior and disorientation of infected fish (Stephens et al. 1980).

Noga and Dykstra (1986) noted that in the spring of 1984, menhaden in the Pamlico Sound area of North Carolina began to develop deep, penetrating ulcers on their flanks and near the anus. An increasing number of fish became affected, until a massive die-off occurred in November. The necrotic ulcers were infected with oomycetes of the fungal genera *Saprolegnia* and *Aphanomyces;* however, these fungi may not be the primary cause of the disease, but only secondary to some other stressor. Ulcerative mycosis is a serious regional problem, observed on fish from Delaware to Florida (Hargis 1985; Levine et al. 1990). At times 100% of the fish collected in trawls are infected with this pathogen (Dykstra et al. 1989). The genera of fungi involved are normally associated with freshwater environments and it is uncertain what predisposes menhaden to infection by these fungi. High precipitation levels and subsequent increased runoffs, with the associated low salinities, suspended organic and inorganic materials, and sediment and detritus loads, were suggested as possible causative agents altering responses of menhaden to the fungal infection (Dykstra et al. 1989). The suspected pathogen (*Saprolegnia* sp.) recovered from menhaden was more salt-tolerant than previously believed, suggesting that the fungus could be pathogenic in menhaden, at least under mesohaline conditions (Shafer et al. 1990).

Recently, an insidious dinoflagellate was identified (Burkholder et al. 1992) as the agent responsible for major kills of Atlantic menhaden (also affecting striped bass, flounders, sciaenids, and eels) in the Pamlico Sound system, N.C. The so-called "phantom" alga, *Pfiesteria piscida,* requires live finfish excreta for excystment from the bottom sediments and release of its potent exotoxin. Even low algal cell densities cause neurotoxic signs in fish and eventual death. Within hours of the fish's death, the vegetative form of the alga encysts and returns to the sediments. This toxic phytoplankter may be an undetected source of fish mortalities in nutrient-enriched estuaries.

Another source of mass mortalities of Atlantic menhaden occurs in the vicinity of effluent water from power plants generating electricity, where water supersaturated with dissolved gases adversely affects finfishes, causing *gas bubble disease.* Mass mortalities, involving an estimated 5,000–43,000 menhaden were recorded near the Pilgrim Nuclear Power Station in western Cape Cod Bay (Bridges and Anderson 1984).

Breeding Habits. Atlantic menhaden begin to mature at age 2 at a length of about 180 mm FL, and all fish are mature by age 3 (Lewis et al. 1987). Individuals of a given age-class tend to be larger toward the northern half of their range, although they mature at smaller sizes in more southerly areas (Reintjes et al. 1979). The minimum FL of potential spawners was 230 mm in the North and Middle Atlantic areas and 210 mm in the Chesapeake Bay and the South Atlantic areas. Reproductive life may extend for 6 or 7 years (Higham and Nicholson 1964).

Fecundity is related to the size of the female (Higham and Nicholson 1964; Dietrich 1979; Lewis et al. 1987), although there is high intra- and interyear variability in the relationship between length and potential number of ova produced. Fecundity estimates range from 48,000 ova for a 180-mm FL female to over 500,000 ova for females of 360 mm FL. A maximum of about 6.3×10^5 ova per female was calculated.

Menhaden are probably determinate, multiple spawners that spawn over a broad geographical and temporal range (Lewis et al. 1987). They probably spawn during every month of the year, although not in all areas at the same time (Nicholson 1972; Lewis et al. 1987). Spawning occurs during the winter on the continental shelf south of Cape Hatteras (Judy and Lewis 1983; Checkley et al. 1988) and is thought to be maximal during storms in water upwelled near the western edge of the Gulf Stream (Govoni 1993). During November and December, most menhaden of spawning age occur off Virginia and North Carolina (Nicholson 1971). Apparently, little contribution to the overall stock is made from spawning that takes place in areas north of the northern New Jersey–Long Island, N.Y., region. The majority of new recruits are probably produced in the estuaries of the Carolinas, Virginia, and north to New Jersey. North of Long Island sexually active fish have been collected from May to October, except for July (Lewis et al. 1987).

Spawning occurs over open-shelf ocean waters as well as near major sounds and bays from Long Island northward (Reintjes and Pacheco 1966; Ferraro 1981a,b; Judy and Lewis 1983; Ahrenholz et al. 1987b; Ahrenholz 1991). In Peconic Bays, N.Y., spawning took place at water temperatures of 12.1°–25.0°C and salinities of 20.0–30.0 ppt, but spawning was most intense at water temperatures of 15°–18°C (Ferraro 1981a,b). In Narragansett Bay, menhaden may spawn as early as April (Oviatt 1977); in southeastern Cape Cod Bay, from about May to early October (Scherer 1984); and in the Gulf of Maine, during July and August. Larvae are present in southern Cape Cod Bay from May to December (Scherer 1984), and have been reported from several other areas north of Martha's Vineyard (Marak and Colton 1961). There is a paucity of information on spawning in Canadian waters (Scott and Scott 1988), but ripe males and females were taken in the Saint John River as late as 24 August, so some spawning may occur in this vicinity.

Eggs, which are spawned in the open ocean, and larvae depend on Ekman transport and ocean and tidal currents for transport into the estuaries (see below) (Nicholson 1972; Nelson et al. 1977). Larvae metamorphose into juveniles within the estuaries, where in southern estuaries the juveniles spend up to one full year, after which they tend to join the coastal migratory population of the adults.

Early Life History. Eggs of Atlantic menhaden are buoyant, spherical, and highly transparent. They usually occur in the upper water column to depths of 10 m (Reintjes 1969; Judy and Lewis 1983). They are easily distinguished from eggs of other Gulf of Maine fishes by their large size (1.3–1.9 mm in diameter), broad perivitelline space, small oil globule (0.15–0.17 mm), and very long embryo. Menhaden eggs tend to be larger at higher latitudes (Powell 1993), but there is considerable variation in size at any given latitude, and egg size is positively correlated with the size of the female. Incubation is rapid, with hatching usually occurring in less than 48 h at incubation temperatures ranging between 15° and 20°C (Jones et al. 1978) and at 66 h at 15°C (Hettler 1981).

Embryology of Atlantic menhaden was described by Ferraro (1980), who noted temperature and salinity effects on the rate of embryonic development of artificially fertilized eggs from one female. Hettler (1981) successfully spawned and reared larvae to the juvenile stage. Larvae are 2.4–4.5 mm SL on hatching and grow to 5.7 mm in 4 days (Bigelow and Schroeder; Fahay 1983; Powell 1993). Estimates of yolk utilization, yolk and oil volumes, and growth rates during development were provided by Powell (1993) for Atlantic menhaden reared under laboratory conditions.

Dorsal and caudal fins first become visible at a length of 9 mm; at 23 mm all fins are well developed; scales are present at 33 mm; metamorphosis commences at about 35 mm; and at 41 mm fry have most characters of the adult, except that their eyes are proportionately much larger.

The development of sensory systems in menhaden larvae was described by Hoss and Blaxter (1982). At hatching, larvae have unpigmented eyes. In early stages, the larva possesses a row of 8–11 prominent neuromast organs on either side of the body, which become less obvious at 12 mm. The anlage of the swim bladder is present at 10 mm and prootic bullae first appear at 12.5 mm. Pterotic bullae develop at about 30 mm. The swim bladder first contains bubbles of gas at a body length of 13 mm. The lateral line first appears in the region of the lateral recess at a body length of about 17 mm. The lateral recess membrane is functional at 18 mm. Three neuromast organs can be identified within the lateral line on each side by 26 mm; by 32 mm the secondary canal system starts to develop.

The youngest menhaden larvae resemble those of Atlantic herring, but in menhaden the fins are formed, the tail becomes forked, and the body deepens at a much smaller size than in Atlantic herring. A menhaden of 20 mm is as far advanced in development as is a larval herring of 35 mm, which makes it easy to distinguish older larvae of these two clupeids.

Larval Atlantic menhaden are pelagic and those spawned offshore may spend up to several months in continental shelf waters before being transported to estuaries at lengths of 10–22 mm (Massmann et al. 1962; Nelson et al. 1977). In the ocean, larvae appear to be most concentrated in the upper water column (Kendall and Reintjes 1975; Nelson et al. 1977; Judy and Lewis 1983). With movement inshore, larval menhaden are found closer to the bottom (Kjelson et al. 1976). They are transported and move into lower-salinity waters in estuarine tributaries, where they are found in great abundance, and metamorphose into juveniles, usually at a length of about 34 mm (McHugh 1967; Wilkens and Lewis 1971). Metamorphosis apparently occurs only in the estuary, as no metamorphic larvae or prejuveniles have been collected at sea (Kendall and Reintjes 1975).

From offshore areas off the southeast United States, eggs and larvae of Atlantic menhaden drift shoreward in the warm surface stratum of a density-driven circulation maintained by a large sea-air heat flux, bringing the larvae to within about 20 km of shore on average and at times much closer (Checkley et al. 1988). Transport of larval Atlantic menhaden across the southeastern continental shelf to bays and estuaries is dependent on many physical and biological conditions (De Vries et al. 1995a,b). Selective pressures to obtain sufficient food, escape predation, and remain in water warm enough for survival combine with those for enhanced shoreward transport to affect that portion of larval depth distribution under behavioral control. Menhaden larvae may avoid offshore-flowing waters; are associated with specific water masses off North Carolina (Govoni and Pietrafesa 1994); and have sustainable swimming speeds (1–2 body lengths·s^{-1}, which roughly equals ca. 0.5–3.0 cm·s^{-1}) too insignificant to conduct active transport to nurseries. Subsequent movement of menhaden larvae to and through the inlets is believed to result from their vertical movement combined with nearshore and estuarine circulations (Wilkens and Lewis 1971).

Salinity and temperature gradients may be cues for eliciting depth-selective behavior (De Vries et al. 1995a,b). In the laboratory, larval menhaden display an ascent response when exposed to salinity increases (De Vries et al. 1995a,b). In the field, they would typically experience an increase in salinity upon descent in the water column, which would then cause them to swim upward. Utilizing salinity gradients could maximize larval residence in onshore-flowing water and minimize transit time across the continental shelf. In another study, laboratory-reared larval menhaden of two different ages were exposed to varying relative rates of temperature increases and decreases from both above and below. Temperature decreases from below caused an ascent response in both age-groups, but neither responded to this cue from above. The minimum absolute decrease for a response to occur was 0.1° and 0.05°C, respectively, for each size-group. Young larvae did not respond to a temperature increase, while older larvae ascended regardless of whether the increase was presented from above or below. On the continental shelf, detectable temperature gradients appear common, both for temperature decreases that would occur upon descending and also for temperature increases that would occur upon ascending.

Atlantic menhaden larvae larger than 12 mm SL apparently undertake vertical migrations to the surface at sunset to replenish gas in the swim bladder (Hoss et al. 1989; Forward et al. 1993), and this behavior may also be important in relation to the transport mechanisms that move menhaden from offshore spawning grounds toward estuarine nurseries. Larval menhaden inflate and deflate the swim bladder on a diurnal cycle that appears to be a light-dark function and not endogenous, because it is disrupted in constant light and can be initiated during daylight by darkening. Inflation occurs rapidly, beginning within 5 min of the onset of darkness. This behavior is believed to increase buoyancy of the larvae at night, allowing them to expend less energy staying in the middle to upper water column (Hoss et al. 1989; Forward et al. 1993). At sunrise, air is expelled from the swim bladder, giving larvae more swimming agility and likely maintenance of a deeper position in the water column. Deflation is less well studied, but appears to be cued by an increase in light intensity (Forward et al. 1994; Govoni and Pietrafesa 1994). Since larvae are relatively transparent, the difference in refractive index between air and water increases the contrast between an inflated swim bladder and the surrounding water. An increase in visibility could lead to increased detection by predators, so deflation of the swim bladder at sunrise may be a predator avoidance response (Forward et al. 1994).

Atlantic menhaden larvae begin feeding on zooplankton about 2–4 days after hatching (Reintjes 1982; Powell 1993), depending on water temperatures during development. Size at first feeding is about 4.8 mm SL (Powell 1993). They are size-selective plankton feeders and, although there is no direct evidence of the food they ingest before entering the estuary, it is quite possible that they feed on pteropods and bivalve larval stages, as well as crustacean nauplii, which are food sources for other members of the herring family.

As postlarval menhaden metamorphose into prejuveniles (ca. 30 mm) they develop a functional branchial filtering apparatus, which enhances their ability to graze on phytoplankton and suspended detritus. Late-stage juveniles and adults are primarily herbivores, but they also retain the ability to feed on zooplankton (Edgar and Hoff 1976).

Age and Growth. Atlantic menhaden, which can normally live to 8 years, are relatively fast-growing until at least age 4, at which time growth slows (Henry 1971). During the 1950s and 1960s, Reintjes (1969) noted that of over 100,000 fish examined, fish 8–10 years old were uncommon and only one 12-year-old was noted. June and Roithmayr (1960) presented the following age-size relationships for menhaden (age in years; size in mm FL): 1 = 135 mm; 2 = 215 mm; 3 = 250 mm; 5 = 270 mm; 5–7 = 300–350 mm. Contemporary (1993) average fork lengths and weights for coast-wide port samples are 178 mm FL and 104 g for age-1 fish, 235 mm FL and 237 g for age-2 fish, 258 mm FL and 325 g for age-3 fish, and 279 mm FL and 393 g for age-4 fish (unpublished data, NMFS, Beaufort, N.C.).

Growth rates of Atlantic menhaden vary in different fishing areas as well as from year to year (Reish et al. 1985). Fish caught in the North Atlantic are not only older but also larger for a given age than fish caught further south (Nicholson 1971). During the summer, menhaden stratify by age along the coast, with younger, smaller fish occurring in the southern part of the range and older, larger fish predominating in the north. Age-1 menhaden were most abundant from Chesapeake Bay to New Jersey; age 2 from New Jersey to the south shore of Long Island; age 3 from Long Island Sound to Nantucket Sound; and age 4+ from Nantucket Sound to Maine (Nicholson 1971). Typical age compositions of commercial port samples in recent years are: age-1 and age-2 Atlantic menhaden in Chesapeake Bay catches, age-2 and age-3 fish in catches from the Middle Atlantic, and age-3 to age-5 fish in the Gulf of Maine (unpublished data, NMFS, Beaufort, N.C.). Fish taken by purse seiners in Narragansett Bay (Durbin et al. 1983), were predominantly age 2 and age 3, and in all age groups represented, fish from Narragansett Bay were significantly smaller than those caught from Long Island Sound to the Gulf of Maine during 1955–1971. Average size of individuals within each age group also increased with latitude, especially with age-1 and age-2 fish. This size stratification was much less pronounced for age-3 and older menhaden. There is no information on growth of menhaden in the Canadian Atlantic area (Scott and Scott 1988).

The age structure of Atlantic menhaden on the east coast of the United States, as indicated by changes in age composition of the fishes taken by the commercial fishery, has been dramatically altered by fishing activities and recruitment patterns. Prior to 1966, age-4 and older menhaden contributed significantly in numbers and biomass to the North Atlantic catch (Nicholson 1975). During the 1960s, the stock structure became truncated and fish older than age 3 were uncommon; and older fish virtually disappeared from the Gulf of Maine. The stock rebuilt during the 1970s and commercial quantities of the older age groups were again available in the Gulf by the late 1970s to mid-1980s.

Instantaneous growth rates of 0.043 mm·day^{-1} have been recorded for larvae at 20°C for 21 days with abundant food supply (Powell and Phonlor 1986). In experimental situations (Powell 1993), growth rates at 16°C (0.27 mm·day^{-1}) are lower than for those at 20°C (0.047 mm·day^{-1}) and 24°C (0.049 mm·day^{-1}).

Growth of juvenile menhaden is relatively rapid, but variable depending on a variety of factors, including time of hatching and length of growing season (function of latitude). Growth of juveniles in lower-salinity waters (5–10 ppt) was faster than that noted (Hettler 1976) for juveniles maintained at higher salinities (28–34 ppt). Estimated rates of daily growth of juveniles raised in mesocosms ranged from 0.85 to 1.10 mm·day^{-1} (Keller et al. 1990). Reintjes (1969) also reported increases in mean length of juveniles of about 1 mm·day^{-1} over intervals of 2–3 days. Ahrenholz et al. (1995) used known-age fish to provide continuous validation from first feeding through metamorphosis to juveniles up to 9 months old. On average, larval and juvenile Atlantic menhaden form one growth increment per day on their sagittal otoliths (Ahrenholz et al. 1995). Juvenile menhaden up to 200 days old were reliably aged within a confidence interval of 7 days and up to 250 days old within a confidence interval of about 16 days. One experimental group displayed growth rates (0.67–0.95 mm·day^{-1}) similar to the higher rates observed for juveniles captured from estuarine nursery areas. Menhaden hatched in summer are 6–8 cm long by their first winter and average about 16 cm by their second winter; fall-hatched fish are 3 cm and about 13 cm long in their first and second winters, respectively, with every gradation between the two depending on the precise season in which they were spawned (Ahrenholz 1991).

General Range. Atlantic menhaden occur in coastal waters along the Atlantic coast of North America from the Gulf of St. Lawrence and Nova Scotia to Indian River, Fla. Occurrence of menhaden in Canadian waters is sporadic and unpredictable; sometimes they occur in relatively enclosed or restricted areas, such as St. Margaret Bay, N.S., and in inland tidal waters such as the Kennebecasis River, a tributary of the lower Saint John River, N.B., and even as far as 24 km upstream from the mouth of the Saint John River (Scott and Scott 1988). Fair numbers have been taken in Passamaquoddy Bay, N.B., the Gulf of St. Lawrence, off Nova Scotia, and in the Bay of Fundy (Leim and Scott 1966; Scott and Scott 1988).

Occurrence in the Gulf of Maine. Atlantic menhaden are a summer seasonal species in the Gulf of Maine, and the Gulf is the northerly limit for commercial quantities. Since menhaden tend to stratify along the east coast of the United States by size and age, it is generally accepted that fish taken on the New England coast are larger and fatter than those caught farther south. In years of peak abundance, menhaden occur throughout nearshore waters of the Gulf from Cape Cod to Penobscot Bay and the Mount Desert Isle region. Their chief centers of abundance lie in Massachusetts Bay within a mile or so of land, particularly off Barnstable and in the mouths of Boston and Salem harbors, in Casco Bay, and among the islands, and northward to Penobscot Bay.

Migrations and Movements. Atlantic menhaden are temperate, summer and fall seasonal species in the Gulf of Maine. Adults undergo extensive north-south seasonal migrations along the coast and stratify latitudinally by size and age from late spring to early autumn (Nicholson 1971). Locally, adults undergo frequent movements in and out of bays and inlets, depending on tides, season, and weather. Generally speaking, throughout their range, Atlantic menhaden move north and inshore in summer and at least some of the population moves south and into deeper water during winter. Juvenile (Kroger et al. 1971) and adult menhaden leave northern New England waters by autumn, migrating southward and wintering off Virginia and the Carolinas. The following spring adults migrate northward into the Chesapeake Bay area and New England waters (Henry 1971). Seasonal movements of Atlantic menhaden were described by June and Reintjes (1960), June and Nicholson (1964), Nicholson (1971), and Dryfoos et al. (1973).

Seasonal appearance and disappearance of menhaden into and out of the Gulf of Maine in spring and fall, respectively, is a result of migration around Cape Cod and is a well-documented annual event. In years when menhaden reach the Gulf of Maine, they usually appear in Massachusetts Bay about mid-May, when coastal waters have warmed to 10°C or more; off the Maine coast during the last half of May or first part of June (June and Reintjes 1976); and usually reach peak abundance between July and early September. In response to falling temperatures, most depart coastal areas of Maine by late September (Nicholson 1971) and the Massachusetts Bay region by early November. They may persist into fall in the southern Gulf, as Lawton et al. (1984) reported catching menhaden (85–349 mm FL) in western Cape Cod Bay from March to December.

Importance. Atlantic menhaden are one of the most commercially important fishes occurring along the Atlantic coast of the United States, especially from Massachusetts to the Carolinas They are rarely eaten by humans because of the oily and bony flesh, although the roe is highly prized in some areas of coastal North Carolina. The majority of the catch is processed into fish oil, fish solubles, and fish meal for poultry, livestock, and aquaculture feeds. Most menhaden oil is exported to refine into edible cooking oils. Domestic uses include paints, soaps, pharmaceuticals, and lubricants. Menhaden are an excellent source of polyunsaturated fatty acids and contain a high percentage of omega-3 fatty acids (Krzynowek and Murphy 1987). Nutritional values for menhaden are: fat 8–16%; moisture ca. 71%; protein ca. 16.5%; and ash ca. 4%. Considerable amounts are used locally for bait in lobster and crab fisheries, and minor quantities are ground up into chum, which is used by sport fishermen to attract game fish. Practically the entire commercial catch of menhaden is taken by purse seines, pound nets, and gill nets.

Perhaps the most interesting aspect of menhaden occurrence in the Gulf of Maine is the tremendous interannual fluctuations in abundance. In their account of historical cycles of menhaden abundance in the Gulf of Maine, Bigelow and Schroeder noted that there were periods of great abundance followed by times when this species was scarce or entirely absent. For example, after being absent from the Gulf of Maine for several years, menhaden were plentiful there in 1889 (when more than 10 million lb were caught). In 1890, they were so numerous that four fertilizer factories were established, and nearly 90 million fish were taken during that season. This period of plenitude was short-lived as less than half as many fish were caught in Maine waters (ca. 41 million) in 1891, and in 1892 few menhaden were taken or seen north of Cape Cod. Since the 1950s, the Atlantic menhaden stock has undergone several similar periods of expansion and contraction. Coast-wide landings peaked during the 1950s (712,000 tons in 1956), and up to 23 plants operated from Florida to Maine. Through the 1960s, stock size decreased owing to exploitation and poor recruitment (Ahrenholz et al. 1987b), the age structure became truncated, and fish became scarce in the northern half of their range (Vaughan 1990). Through the 1970s and 1980s recruitment levels increased and the age structure again broadened. The Atlantic menhaden population supported a large commercial fishery and also managed to steadily increase its size despite dramatic changes in characteristics of its spawning stock (Ahrenholz et al. 1987b). Population regrowth in the 1970s occurred during a period when an estimated 80–90% of potential spawners were harvested by the fishery and average size at age was declining. Landings improved to 418,600 tons in 1983, and several factories in New England reopened (Smith 1991). By the 1980s, the character of the fishery in the Gulf of Maine changed. Despite an abundance of fish in coastal waters, factories in South Portland (in 1983), Gloucester (in 1984), and Rockland (in 1988) closed, owing primarily to recurring odor problems associated with fish reduction adjacent to populated urban areas (Smith 1991). In 1987, a fish factory at Blacks Harbour, N.B., began processing menhaden caught in southern Maine and transported by U.S. steamer to Canada. In 1988, a firm in Portland entered into an agreement (an Internal Waters Processing, or IWP, venture) with Russia, which sited a foreign factory ship in Maine territorial waters (less than 3 miles from shore), for the purpose of processing menhaden caught by American owned and operated vessels. Through 1993 up to three factory ships (under two IWPs) operated along the Maine coast, and in 1992 a second Canadian factory in Saulnierville, N.S., processed menhaden caught in Maine. Between 1990 and 1993 combined menhaden landings for the IWPs and in Canada averaged about 37,000 tons annually, although landings for 1993 declined to about 10,000 tons (unpublished data, NMFS, Beaufort, N.C.). During years of peak abundance, and as recently as 1990 and 1991, numerous menhaden fish kills occurred in isolated coves of coastal Maine. It is hypothesized that menhaden schools are chased into blind coves by schools of predators such as bluefish and striped bass (Conniff 1992) or that schools of menhaden are feeding on a local bloom of plankters. The large numbers of menhaden in a small area deplete the available dissolved oxygen, and the fish school dies *en masse*. Lobsters held in pounds for future markets may also succumb in affected coves.

Stocks. Debates have arisen over whether this species consists of a single panmictic population or two or more subpopulations (Epperly 1989). Certain meristics and morphometrics, growth rates, movements, and the spatiotemporal nature of spawning have suggested the existence of two or more subpopulations (June 1958, 1965; Sutherland 1963; Higham and Nicholson 1964; June and Nicholson 1964; Dahlberg 1970). However, after reviewing available data, Nicholson (1971, 1978) rejected the multiple subpopulation hypothesis and attributed differences in meristics of fish from different regions to differences in water temperatures during larval development. Juveniles collected north of 40° N have lower numbers of total and trunk vertebrae, ventral scutes, and interhemal spines as well as shallower heads, smaller eyes, and shorter predorsal lengths (Epperly 1989). In considering the possibility of at least two subpopulations of Atlantic menhaden, Epperly (1989) noted that meristic and morphometric characteristics of juveniles of a geographic area are homogeneous with respect to their parents and are temporally consistent. There is also an accumulation of significant differences in the frequencies of transferrin alleles between juveniles caught north and south of 40° N, which may be consistent with a division of the gene pool. Despite increasing population sizes and numbers of older fish, which migrate farthest north (Smith et al. 1987), the numbers of juvenile Atlantic menhaden in estuaries of the northern North Atlantic area declined over 98% in the 1970s (Ahrenholz et al. 1989); concomitant declines in yield per recruit and size at age (more than can be explained by density-dependent growth) indicate biological changes in characteristics of the stock or, alternatively, in the relative proportions of contributing subpopulations (Ahrenholz et al. 1987b). Although evidence is still inconclusive, support still exists for at least two subpopulations: one that spawns in summer and is responsible for primary recruitment in the northern areas, and one that spawns in autumn through spring and contributes the majority of recruitment in the Middle and Southern Atlantic areas.

NOTE. The menhaden account was written by Thomas A. Munroe and Joseph W. Smith.

ATLANTIC HERRING / *Clupea harengus* Linnaeus 1758 /

Herring, Sea Herring, Labrador Herring, Sardine / Bigelow and Schroeder 1953:88–100

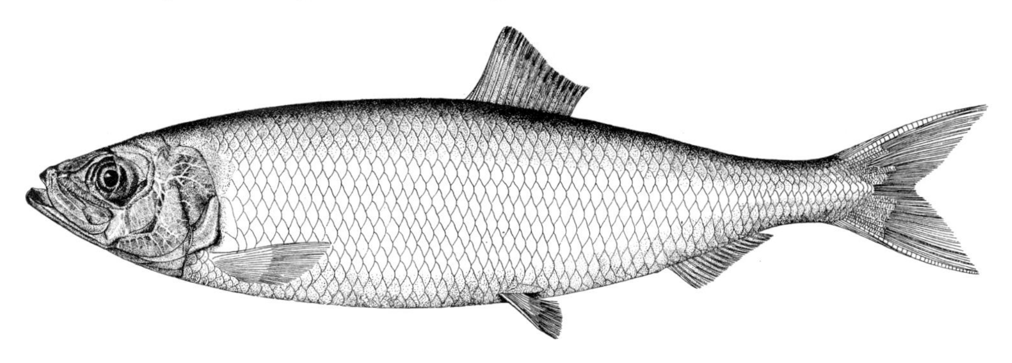

Figure 77. Atlantic herring *Clupea harengus.* Drawn by H. L. Todd.

Description. Body elongate, strongly laterally compressed; ventral margin of abdomen slightly rounded, scutes without prominent keel; caudal peduncle slender (Fig. 77). Head relatively small; snout pointed; mouth terminal, relatively large, without median notch in upper jaw; lower jaw projecting slightly beyond upper when mouth closed; maxilla extending posteriorly almost to vertical through middle of eye. Teeth absent on upper jaw, but with minute teeth on maxilla, small teeth on lower jaw, elongate patch of teeth on middle of tongue and elongate patch of a few stronger teeth on vomer. Eye moderate. Dorsal fin origin approximately at body midlength, directly dorsal to origin of much smaller pelvic fins; height of anal fin relatively low compared with that of dorsal fin; pelvic fins relatively small, with distinct axillary process; caudal fin deeply forked. Scales large, cycloid, deciduous, with rounded posterior margin.

Meristics. Dorsal fin rays 17–22; anal fin rays 12–21; pectoral fin rays 14–22; pelvic fin rays i, 8; scales about 57 in lateral series; abdominal scutes 26–33 anterior to pelvic fins, 11–17 posterior to pelvic fins; gill rakers long, 40–49, fewer on fish less than 100 mm SL; branchiostegal rays 8 or 9; vertebrae 51–60, usually 55–57 (Hildebrand 1963c; Whitehead 1985a; Scott and Scott 1988; see Scott 1975 for additional information on meristics).

Color. Color greenish blue to steel blue on dorsum; sides with green reflections; abdomen and lower sides silvery; the change from dark back to pale sides often marked by a greenish band; gill covers sometimes golden or brassy; freshly caught fish generally iridescent with shades of blue, green, or violet, these colors fading soon after capture, leaving only dark coloration on back and silvery pigment on sides. Pelvic and anal fins translucent; pectoral fins dark at bases and along upper edges; caudal and dorsal fins dark grayish or shading into green or blue. No distinctive dark spots on body or fins. Peritoneum dusky.

Size. Maximum size about 43 cm TL (usually 20–25 cm) and 0.68 kg. The all-tackle game fish record is 0.48 kg for a herring caught at Long Beach, N.Y., in April 1995 (IGFA 2001).

Distinctions. Primary characters separating herring from shad and other river herrings (*Alosa* spp.) are an oval patch of small teeth on the vomer (absent in *Alosa*) and the absence of a median notch in the upper jaw (present in *Alosa*). Conspicuous field characters separating herring from shad, hickory shad, blueback herring, and alewife are that the dorsal fin origin in herring is about at the body midlength (vs. considerably farther forward in the others); the shallower body depth (deeper in the other species); and the weakly saw-toothed scutes compared with the usually strongly saw-toothed (especially those on the abdomen between pelvic and anal fins) ventral scutes in *Alosa*.

Taxonomic Remark. Two morphologically similar but genetically distinct species with different life history patterns and different otolith shapes (Bird et al. 1986) are currently recognized in the genus (Grant 1986), Atlantic herring, *Clupea harengus,* and Pacific herring, *C. pallasi.*

Biology. Whitehead (1985a,b) estimated that perhaps as many as a hundred or more papers appear annually on various aspects of herring biology and fisheries and that perhaps more information has been published on *C. harengus* than on any other fish species. Important review articles on Atlantic herring biology and fisheries are found in Sindermann (1979), Blaxter and Hunter (1982), Whitehead (1985a,b), and Blaxter (1985, 1990). Literature of particular interest on herring in western Atlantic regions includes that of Bigelow and Schroeder, Sindermann (1979), Anthony and Fogarty (1985), Scott and Scott (1988), Townsend (1992), Reid et al. (1999b), and Munroe (2000). A bibliography of earlier studies on herring in the northwest Atlantic was compiled by Messieh (1980).

Habits. Atlantic herring are a marine, coastal pelagic species often occurring in shallow inshore waters or taken offshore from the surface to depths of about 200 m (Whitehead 1985a,b). They are a strongly schooling species that undertakes complex feeding and spawning migrations, whose times and extent correlate with the various more or less distinct spawning populations (Whitehead 1985a). Prior to spawning, juveniles and mature fish may form huge schools. Although found primarily in coastal and neritic locations, some populations may enter, or even be confined to, brackish water in bays or saline lakes.

Atlantic herring are fish of open waters, traveling as a rule in schools of hundreds or thousands; single fish or even small schools are seldom seen. The magnitude of some of these schools is evidenced in perceptible small-scale ecological changes in the water column through which a school has passed. It is sometimes possible to plot the passage of herring schools through an area because the water downstream from the school can be depleted of oxygen and food items, is heavy with feces and mucus, and has even been reported to have a characteristic cucumber odor (Blaxter 1990).

Schooling is one of the most important behavioral attributes of herrings, and this behavior develops very early. Schooling begins at about the onset of metamorphosis, usually when fish are between 35 and 40 mm, is well established by the end of metamorphosis, and persists throughout the lifetime of the individual (Sindermann 1979; Gallego and Heath 1994a). Herring are obligate schoolers (Breder 1967) that are always found in polarized schools (Blaxter and Parrish 1965; Shaw 1970), and individuals become extremely agitated if isolated from conspecifics (Parr 1927b). Schools are usually composed of similar-sized fish (Pitcher et al. 1985; Pitcher and Parrish 1993), which to a large extent are fishes of the same year-class, although in any school there may be some representation of other sizes and year-classes (Sindermann 1979). Schools may subdivide or coalesce, and it is unknown how long any given school may preserve its identity as such.

Next to schooling, the most important behavioral response of herring is probably vertical movement in response to changing light intensity (Sindermann 1979). Atlantic herring generally undertake diurnal vertical migrations, rising toward the surface at dusk and sinking toward the seabed at dawn (Blaxter 1990). Herring schools are sometimes visible at the water's surface during the daytime, especially on calm days, when the school can often be detected by a fine rippling of the water (Bigelow and Schroeder). Atlantic herring reportedly do not "fin" or lift their noses above the surface as menhaden often do, and they do not jump unless frightened; however, smaller-sized fish are often seen jumping when pursued from below by larger predatory fishes, such as silver hake or striped bass.

Nighttime behavior of herring and other clupeids is less well known (Blaxter and Holliday 1963; Blaxter and Hunter 1982). They are often active at the surface at night, where their presence may be betrayed by luminescent wakes. Experimental work with juveniles clearly demonstrated that herring activity has a diurnal pattern, with maxima just after sunrise and just before sunset, and that vertical diurnal movements occur at all seasons, except that they go down to greater depths in winter (Brawn 1960a,b; Tibbo 1964; Stickney 1972). Juveniles generally move up in the water column at twilight and remain near the surface if light intensity is low enough. Field observations and catch information also attest to the importance of light intensity, especially with respect to the behavior of juveniles. Moonlight and the phase of the moon are important determinants of the success of the juvenile fishery, so much so that Anthony (1971) was able to demonstrate successive monthly peaks in the Maine sardine fishery coinciding with dark cycles of the moon.

Activity of herring is also controlled in great part by water temperatures. Early research showed that adult herring could distinguish between temperatures varying by as little as 0.2°–0.6°C (Shelford and Powers 1915). Field studies (Zinkevich 1967) of seasonal distribution of adult herring on Georges Bank suggested a preferred temperature range of 5°–9°C. Catches of adult herring in the NEFSC bottom trawl survey were greatest at 5°C and 30–50 m in spring, 6°C and 20–130 m in summer, 5°–6°C and 60–170 m in fall, and 7°–8°C and 70–100 m in winter (Reid et al. 1999b). Juvenile herring were found to prefer temperatures of 8°–12°C (Stickney 1969), and Brawn (1960a) demonstrated that physiological stress occurred below 4°C and above 16°C, whereas temperatures below about –1.1°C and above 20°C were generally lethal for this species. When given a choice, herring larvae demonstrated a preference for warmer temperatures than those to which they were acclimated, to a maximum of about 13°C (Batty 1994). High environmental temperatures are detrimental to developing larvae.

Atlantic herring have been observed to survive winter temperatures as low as −1.1°C (Brawn 1960a). One adaptation allowing these fish to withstand such cold is the production of antifreeze proteins (AFPs) in the blood. Serum of herring taken off Nova Scotia contained antifreeze proteins that were similar to those found in smelt and sea raven (Ewart and Fletcher 1990). The presence of AFPs in herring indicates they have the capacity to survive in icy seawater. Herring off Newfoundland, particularly those caught in the under-ice fishery off northeast Newfoundland, also appear to be quite freeze-tolerant, judging by the low water temperatures in which they are found and their substantial plasma hysteresis. Fish in these populations may also generate antifreeze proteins as adults.

Plasma-freezing points are significantly lower and antifreeze activity significantly higher in juveniles than in adults (Chadwick et al. 1990). Mean thermal hysteresis of juveniles was found to be nearly three times higher than that of adults, and juveniles were also more prepared than adults to overwinter under freezing conditions at an earlier date. The significantly higher levels of antifreeze in juveniles as compared with that in adults suggest that juveniles are better adapted to overwinter in colder water, which is more typically found near the coast, in habitats occupied by juveniles. By migrating offshore during the winter, adults avoid the severely low winter temperatures found closer inshore. Capacity for antifreeze production, or the level of activity of AFPs, may also be related to the population from which the herring originates (Chadwick et al. 1990). Accordingly, herring from Brown's Bank, where water is relatively warmer in winter (ca. 7°C), showed no evidence of AFP production, which directly contrasts with the situation found in herrings from colder areas within the region.

Salinity is probably a less critical factor than temperature in influencing overall movements and distribution of herring. With increasing age, there is an increasing preference for higher salinities. Larval and 0-age-group herring can frequently be found in inshore coves and estuaries, where salinities are markedly reduced (Townsend 1992). However, older juveniles reportedly avoid these brackish estuarine conditions (Recksieck and McCleave 1973). Some experimental evidence suggests that juveniles respond to salinity changes and that they can tolerate such changes (Brawn 1960c,d). Although 28–32 ppt salinity was preferred by juveniles, salinities as low as 5 ppt could be tolerated for brief periods. In another study (Stickney 1969), juveniles were found to exhibit a salinity preference in excess of 29 ppt at temperatures below 10°C, but no preference was demonstrated when environmental temperatures exceeded 10°C. Although adult herring regularly enter bays and estuaries in the Gulf of Maine, they are not usually reported from water that is appreciably brackish. Bigelow and Schroeder suggested that salinities of about 28 ppt were probably the lower limit for adult herring.

Food. Atlantic herring are facultative, zooplanktivorous, filter feeders (Blaxter 1990). Food items in the diet vary markedly with fish and prey size, season, and geography (in terms of what food organisms may be present in abundance in a specific location). Larvae, juveniles, and adult herrings are selective opportunistic feeders, taking advantage of concentrations of whatever prey of appropriate size is available in their immediate environment (Sherman and Honey 1971; Sherman and Perkins 1971). Thus, in early spring, dense swarms of barnacle larvae or cladocerans constitute the principal prey, whereas later in the season copepods and euphausiids may dominate. Herring are primarily visual particulate feeders that consume a variety of planktivorous organisms, especially crustaceans, during daylight and twilight hours (Battle 1934; Blaxter 1966), although there may be little feeding during the day when the fish are in tight schools at depth or near the seabed. As herring grow, their eyes become increasingly adapted to twilight (Blaxter 1964, 1966, 1968). James (1988) noted that herring are primarily twilight foragers whose peak feeding activity occurs at dusk and dawn in the upper water layers, with feeding intensity decreasing during the darkest hours.

Herring display a diversity of feeding behaviors, with fish size, light intensity, biting, filtering, prey size, and prey density having interacting roles (Batty et al. 1990). Until recently, it was thought that herring were obliged to feed by biting, and biting was observed to cease at low light intensities, with a visual threshold of 0.007 lux when fish were feeding singly and 0.036 lux when they were feeding as a group (Blaxter 1964). Herring will also switch to filter-feeding in the light if food particles of a suitable size and concentration exceed a critical level (Gibson and Ezzi 1985, 1990). In the light, juvenile herring have been observed to use two modes of feeding, particle biting and filtering, but in the dark only filtering was observed (Batty et al. 1986). When offered natural zooplankton, juvenile herring consumed the larger organisms first, by biting, but only when light intensities were above a threshold of 0.001 lux. Biting was possible at lower light intensities, but only when prey were large.

In darkness, herring apparently are not able to feed by biting, but if conditions are appropriate they can filter feed at night. Filtering commenced in the dark at high prey densities when prey were small but at low prey densities when prey were large. In the light, fish continued to school while feeding in both modes; however, in darkness, juvenile herring stopped schooling, they swam more slowly in tight circular paths, and fed only by filtering. It is possible that this circling behavior was a response to a chemical stimulus. Herring larvae respond to chemical stimuli, including extracts of food organisms, such as barnacle nauplii and *Artemia* spp., by increasing their activity and swimming up a concentration gradient (Dempsey 1978). If this behavior persists throughout the life of the fish, it might enable herring to remain within food patches and exploit them in the dark (Batty et al. 1986). Filtering fish swam faster ($0.11\ m \cdot s^{-1}$) in the dark than nonfiltering fish ($0.07\ m \cdot s^{-1}$). In the light, no difference in speed was measured between filtering and nonfiltering fish ($0.34\ m \cdot s^{-1}$). Owing to the lower filtering speeds in the dark, the removal rate of nauplii from the water was much lower than in the light, except at the highest prey concentrations. This suggests that if nighttime filter-feeding takes place in the sea, it will be of importance only when exploiting dense patches of food.

When filter-feeding, herring utilize their gill rakers as menhaden do (Moore 1898), although the former species does not include smaller microscopic plants, either diatoms or peridinians, beyond fish sizes more than 15–20 mm, probably because the microstructure of the gill rakers is not fine enough to retain these small plankters. Gill rakers first appear on gill arches of herring at a total length of about 16 mm (Gibson 1988). Their number then increases rapidly until the fish are about 50 mm TL, when the rate of addition becomes much slower. Rakers on the first gill arch account for almost 60% of the entire filtering area. Particle retention capabilities of herring when filter-feeding were lower than those expected on the basis of the estimated spaces between the rakers (Gibson 1988). The length of individual rakers and the spacing between them continue to increase throughout the life of the fish. Change in filtration area with increasing fish length is thus due mainly to increasing length of the rakers, with increase in filtration area accompanied by an increase in mesh size of the filter.

Ontogenetic Changes in Diet. Larvae begin exogenous feeding before the yolk sac disappears. Copepod eggs, nauplii, copepodids, mollusk larvae, peridinians, diatoms, and other algae comprise much of the diet of early life-history stages (Sindermann 1979). Seasonal differences occur in diet composition of larvae. For example, during winter, small copepods such as *Pseudocalanus minutus* and copepodid stages are primary food items for larvae, whereas in spring, somewhat larger copepods (especially small *Pseudocalanus elongatus*) and copepodid stages, cirriped larvae, crustacean eggs, and tintinnids are principal food organisms (Bigelow and Schroeder; Sherman and Honey 1971; Checkley 1982; Cohen and Lough 1983; Munk and Kiørboe 1985; Kiørboe and Munk 1986). As larvae grow, copepods make up a large proportion of the items eaten.

As juveniles grow they feed mainly on larger copepods (especially *Calanus finmarchicus* and *Temora longicornis*), but will also feed opportunistically on hyperiid amphipods, euphausiids, mysids, cladocerans, barnacle larvae, bivalve larvae, small fishes, arrowworms, ctenophores, pteropods, and decapod crustacean larvae (Battle et al. 1936; Bigelow and Schroeder; Legaré and MacClellan 1960; Sherman and Perkins 1971; Maurer 1976). In Maine coastal waters, diets of juvenile herring are varied, but copepods are often the dominant prey, especially in summer (Sherman and Perkins 1971). Legaré and MacClellan (1960) found copepods of the genera *Calanus, Pseudocalanus, Eurytemora, Acartia,* and *Tortanus* to be important dietary items for herring from the Quoddy region of New Brunswick.

Between age 2 and age 3, herring change from a predominantly copepod diet to a predominantly euphausiid diet (Sherman and Perkins 1971; Maurer 1976). Adult herring feed principally on euphausiids (*Meganyctiphanes norvegica*), chaetognaths (Maurer and Bowman 1975; Maurer 1976; Bowman et al. 2000), and to a lesser degree copepods, particularly *Calanus finmarchicus.* Stomachs of adult herring off the Maine coast (Moore 1898) contained copepods and pelagic euphausiid shrimps (*M. norvegica*). Fish less than 10 cm long depended on the former alone, whereas larger herring consumed both types of prey organisms. However, a few larger fish, as well as the smaller ones, fed almost entirely on copepods, even when both shrimp and copepods abounded. Shrimp are an important prey item of larger herring, for even in winter, when shrimp are rarely seen at the surface, they can be an important component of the herring diet (Moore 1898). Preference for this prey organism was such as to suggest to Bigelow and Schroeder that local appearances and disappearances of schools of large herring in the open Gulf were probably related to presence or absence of euphausiid shrimps.

In the absence of shrimp, copepods are the chief prey item for herring of all sizes. Relatively large-sized amphipods of the genus *Euthemisto* are an important food for herring in European seas, but are not found in stomachs of herring from the western Atlantic (Moore 1898; Bigelow and Schroeder), undoubtedly owing to the comparative scarcity of this large active crustacean in the coastal waters of the Gulf of Maine.

In addition to consuming a variety of Crustacea, herring also feed on molluscan larvae, fish eggs, annelids, and even on such microscopic prey as small as tintinnids and *Halosphaera.* Although herring are not usually piscivorous, larvae of sand lance, silverside, and young herring have been found in herring stomachs, and Templeman (1948) also reported that off Newfoundland herring consumed quantities of small capelin during the winter. Principal prey organisms of over 5,700 herring from the North Sea (Last 1989) were copepods (*Calanus, Temora*), but Euphausiacea and postlarval stages of *Ammodytes* spp. and clupeoids accounted for a large percentage of the weight. Fish eggs, chiefly those of plaice (*Pleuronectes platessa*), were also consumed, but not in large numbers.

Spring and summer are the most intensive feeding times for juveniles and adults. Adults cease feeding when spawning begins (Pankratov and Sigajev 1973). Off New Brunswick, the most active feeding period for herring is from September to November (Legaré and MacClellan 1960). Off Newfoundland, herring eat very little during the winter (December to April), apparently living on their accumulated fat during this season (Hodder 1972).

Larval, juvenile, and adult herring have been shown to select larger prey organisms from those available in the plankton (Sandström 1980; Checkley 1982; Batty et al. 1986). Herring ordinarily target individual food objects (Battle 1934) during feeding in daylight. When feeding on euphausiids, they have been seen pursuing individual shrimps, which frequently leap clear of the water in an effort to escape (Bigelow and Schroeder).

Predators. Herring are preyed upon by nearly all pelagic predators and are extremely important forage species because of their numerical abundance and schooling behavior. They are a basic food for many fishes, marine birds (Gaskin and Smith 1979; Braune and Gaskin 1982; Hislop and MacDonald 1989), short-finned squid (Lidster et al. 1994), and seals, porpoises, and whales (Watkins and Schevill 1979; Hain et al. 1982; Wallace and Lavigne 1992; Lawson et al. 1994). Herring of all sizes are preyed upon by predacious fishes, including spiny dogfish, porbeagle shark, thorny and winter skates, salmon, goosefish, cod, pollock, haddock, hakes (red, white, and silver), sculpins, sea raven, striped bass, sea bass, bluefish, mackerel, swordfish, tuna, billfish, and winter flounder (Reid et al. 1999b; Rountree 1999; Bowman et al. 2000). Three predators account for most of the prey records: spiny dogfish, silver hake, and cod (Rountree 1999).

Predation by fishes can be a significant mortality factor for adult herring, with particular pressure at spawning time. Diver observations (Cooper et al. 1975) disclosed that bluefish, cod, and pollock were voracious predators on spawning concentrations in the southwestern Gulf of Maine, with maximum predation activity occurring at night. Silver hake, in particular, are reported to drive schools of herring up onto beaches in the Gulf, where pursued and pursuers alike strand in the shallow waters. One such feeding frenzy was described, which occurred on an October morning at Cohasset in Massachusetts Bay many years ago, in which hake and herring were so intermingled in shallow water that at the height of the carnage Bigelow and Schroeder filled a dory with both species of fish using only their bare hands to capture them.

Finback whales also devour herring in great quantities. In the West Quoddy Head area, humpback were observed occasionally feeding on herring close inshore and in coves using a bubble cloud and lunge feeding method (Hain et al. 1982). In this feeding behavior, whales dive underwater and swim in a circle beneath the schooling fishes. While swimming in this manner approximately 15 bubble bursts are released, which rise to the surface as columns and appear to form an effective corral. As the bubble corral nears completion, the whale pivots on the axis of its flippers and then banks to the inside and turns sharply into and through the center of the corral below the surface of the water. With mouth agape and lower jaw region distended, it then feeds on the fish that have been concentrated into a tight school within the bubble curtain.

Short-finned squid (*Illex*) consume multitudes of young herring. On one occasion near Provincetown, in June 1925, Bigelow and Schroeder reported that packs of perhaps 10–50 squid circled around a school of juvenile herring, bunching them into a compact mass. Individual squid then darted into the school, seized one or two herring, ate a small part, and then darted back for more. A silvery streak of fragments of dead herring remaining along the beach bore witness to the extent of that carnage.

Herring eggs and larvae are also cannibalized by adult herring. Spent herring on spawning grounds have been observed to have eggs in their stomachs, and adult herring have had larvae in their stomachs (Sindermann 1979). Herring eggs and spawn are also subject to predation by a variety of bottom predators. Winter flounder are a major predator (Tibbo et al. 1963; Pottle et al. 1981; Messieh et al. 1985), but other species also take their toll, including cod, haddock, and red hake (Caddy and Iles 1973), sand lance (Fuiman and Gamble 1988; Rankine and Morrison 1989), sculpins, skates, and smelt (Messieh et al. 1985; Scott and Scott 1988), tomcod, pollock, cunner, mackerel, and even herring themselves (Messieh 1988; Fuiman and Gamble 1988). In one area, fish predation rates varied, with estimates of between 45 and 69% of the initial herring spawn being consumed (Messieh 1988). These estimates were conservative and represented only a portion of the total predation on the eggs, because predation mortalities owing to invertebrate predators such as lobsters and starfish were not included. On Georges Bank, about 8% of the herring spawn was estimated to have been removed by predation within 1–2 days after spawning (Caddy and Iles 1973). Sand lance are clearly a most effective predator on early life stages of herring, with perhaps to as many as 400 eggs and larval herring being consumed per individual sand lance (Rankine and Morrison 1989). Since sand lance are associated with the coarse sublittoral sediments actively selected as herring spawning sites, they could easily be present in the same areas as herring egg masses (Rankine and Morrison 1989). In one study, sand lance actively hunted herring larvae in preference to more numerous copepods, which, from stomach analyses, provide their usual staple food (Christensen 1983). Additionally, a scyphozoan medusa *Aurelia aurita* has also been shown to be an important predator on larval herring (Bailey 1984; Bailey and Batty 1984; Möller 1984).

Mass Mortalities. Herring are susceptible to wholesale destruction by stranding on beaches during storms and in attempts to evade predators, by ingestion of toxic dinoflagellates, and by anthropogenic sources such as water pollution and impacts associated with power generating stations. Instances of mass mortalities, especially of young herring, have been reported in several Gulf of Maine harbors. Bigelow and Schroeder, for example, reported a mass mortality of herring that took place at Cohasset, on the south shore of Massachusetts Bay, in October 1920. On 5 October, a large school of juveniles, 10–13 cm in length, ran up the harbor (which is nearly landlocked), having been driven into the harbor, according to local fishermen, by silver hake. Once in the harbor, the herring were trapped by the falling tide and stranded on the mudflats. The herring were so numerous that the flats were entirely covered with them, and an estimated 20,000 barrels of fish perished. During the next few days, the fish (alternately covered and uncovered by the tide) decayed and, despite tidal circulation, so fouled the water that lobsters impounded in floating cars nearby also died. On 10 October there was a second, smaller run of herring, and on 15 October there was a third run as numerous as the first. The newcomers died soon after they entered the harbor. Altogether, approximately 50,000 barrels of fish perished, of which more than 90% were sperling (young herring), 5–10% were large adults, and a few were small mackerel and silver hake; there were also a large number of smelt among the dead fish. So many herring died in this area that the flats at low tide were silvery with herring scales even to the last half of October, and residents in the harbor area found the stench from the decaying fish almost unbearable. During the ensuing winter, the fish decomposed and the water purified itself.

Mass mortalities of herring sometimes also occur when they feed on herbivorous pteropods and other zooplankton (cladocerans) that have been grazing on toxic dinoflagellates (*Gonyaulax excavata*), which produce paralytic shellfish (PSP) toxins (White 1977, 1980, 1981). Larval herring are also susceptible to PSP toxins when they feed on PSP-producing dinoflagellates. Toxic blooms of *Gonyaulax excavata* (*tamarensis*) have a long history in the Bay of Fundy (White 1980), and since 1972 this organism has spread southward along the New England coast, causing annual red tides as far south as Cape Cod (Anderson and Wall 1978).

Scott and Scott (1988) reported that a series of extensive herring mortalities occurred in the Placentia Bay region of Newfoundland during the winter of 1969. The dead herring were red in color (fins and body) and attracted much publicity. Mortality was first thought to be caused by a disease, but was later shown to be the result of industrial pollution, mainly phosphorus, from a nearby industrial plant (Jangaard 1970).

Herring are also affected by operations associated with coastal power plants. At the Pilgrim Nuclear Power Station in western Cape Cod Bay (Lawton et al. 1984), herring constituted 49% of the fishes that impinged on intake screens. Induced stress and debilitation in the narrow confines of the station's intake forebay were also implicated as potential causative agents for a mass mortality (tubular necrosis of the kidneys) of herring in this area.

Parasites and Diseases. Atlantic herring are host to a diverse parasite fauna of over 81 species of parasitic organisms (Appy and Dadswell 1981; Arthur and Arai 1984; MacKenzie 1987; Bray and MacKenzie 1990). Of these, only about five species, all with direct life cycles (Protozoa and Monogenea), are wholly dependent on herring for their survival (McKenzie 1987). Herring serve as host (MacKenzie 1987) for the following groups of parasitic organisms (number of species in parentheses): Protozoa (11); Monogenea (17); Digenea (18); Cestoda (9); Nematoda (12); Acanthocephala (10); Hirudinea (1); Branchiura (2); and Copepoda (7). Not only do adult herring serve as hosts, but larval herring are also utilized as intermediate hosts in life cycles of cestodes and digenetic trematodes (Courtois and Dodson 1986; Heath and Nicoll 1991).

Parasites utilizing herring as an intermediate host include larval nematodes, *Anisakis simplex* recovered from herring taken off Newfoundland (Pippy and Van Banning 1975; Threlfall 1982) and *Contracaecum osculatum,* commonly reported from herring in the eastern Atlantic (Smith and Wootten 1978; Valtonen et al. 1988). Although *A. simplex* is potentially pathogenic (Van Thiel et al. 1960), infection rates in herring from the western Atlantic are relatively low, especially among coastal migratory herring, and raw, salted, smoked, or pickled fillets of most Canadian Atlantic herring are probably safe for human consumption (McGladdery 1986).

Parasites have been used as biological tags more for herring than for any other marine fish (MacKenzie 1987). Because they tend to have longer life spans, larval and preadult stages of helminths, for which the herring serves as a second intermediate host, have proved to be more useful for this purpose than adult helminths. Differences in prevalence of a protozoan parasite and an anisakid nematode larvae between consecutive age-groups of young herring in the Gulf of Maine were interpreted in terms of age-dependent migrations of young herring within the Gulf (Sindermann 1957, 1961). Trypanorhynch metacestodes have also been used in attempts to separate stocks of adult herring from a wider area of the northwest Atlantic. Variations in prevalence and intensity of *Anisakis* spp. larvae were employed to separate stocks of herring in Canadian and U.S. Atlantic waters (Parsons and Hodder 1971; Lubieniecki 1973; Beverley-Burton and Pippy 1977; Chenoweth et al. 1986). The potential usefulness of seven parasite species as biological indicators of different aspects of herring biology were also demonstrated (McGladdery and Burt 1985). Among these species, four were found to be potentially useful to reflect changes in seasonal migrations of hosts between different parts of a study area comprising the Bay of Fundy, the Nova Scotian shelf, and the Gulf of St. Lawrence. Biological information from these parasites suggested that seasonal variation in prevalences could be better explained by changes in herring stock composition than by changes within parasite populations in the same stock of herring.

The life cycle of a sporozoan, *Eimeria sardinae,* is closely linked to the reproductive dynamics of male herring (McGladdery 1987). The infective oocyst stage is released with sperm during spawning, and subsequent infection occurs directly via ingestion of these oocysts by other herring on and around the spawning grounds (Lom 1970). Infections with *E. sardinae* may prove useful in separating groups of herring that spawn at different times of the year (McGladdery and Burt 1985; McGladdery 1987), but infection levels are not useful for distinguishing between first- and repeat-spawning herring (McGladdery 1987).

Some parasites cause mass mortalities in herring (Hodder and Parsons 1971; Morrison and Hawkins 1984). Outbreaks of the systemic fungus pathogen *Ichthyosporidium hoferi* caused large-scale mortalities in the Gulf of Maine in 1932 and 1947 and in the Gulf of St. Lawrence in 1898, 1916, 1940, and 1955 (Sindermann 1958, 1963, 1965, 1970, 1979). From disease prevalences in 1955 and 1956, Sindermann estimated that at least one-half of the herring stock of the Gulf of St. Lawrence was killed, an estimate supported by reductions in herring catches in the years immediately following these mortalities. Later examination of herring in the Gulf of St. Lawrence (Tibbo and Graham 1963) indicated that spring-spawning stocks were more severely affected than autumn-spawning stocks. Herring may also be infected by the piscine erythrocytic necrosis virus (PEN) (Reno et al. 1978).

Parasitic disease is undoubtedly an important contributing factor in the natural mortality in herring and, as MacKenzie (1987) noted, ignorance of the nature of this contribution represents a major gap in our knowledge of herring biology. Heath and Nicoll (1991), for example, noted that despite a number of studies on feeding biology of herring larvae, few of these noted the occurrences of parasites. They found that the incidence of feeding for larval herring infected with cestode larvae was reduced by as much as 50% when compared with levels of prey consumption for noninfected larvae. They also discussed the implications of reduced feeding by larvae and host and parasite distributions relative to recruitment success.

Reproduction. Atlantic herring are unusual among clupeoid fishes in that they are synchronous spawners, producing a single batch of eggs (Blaxter and Hunter 1982). However, on a populational level, they may be considered asynchronous or serial spawners, similar to other Clupeiformes (McQuinn 1997). Herring are also quite unusual in the wide range of seasonality exhibited in time of spawning for different populations. Some populations spawn in spring, others in summer or autumn, and, according to locality, populations of both spring- and autumn-spawning herring may occur in the same area.

The evolutionary origin and ecological significance of multiplicity of herring spawning locations and spawning times and the relationship between variations in reproductive biology and larval production have been the foci of much interesting discussion (Cushing 1975; Blaxter and Hunter 1982 for eastern Atlantic herring; Graham 1982; Lambert 1984; Lambert and Ware 1984; Sinclair 1988; Townsend 1992; McQuinn, 1996, 1997 for western Atlantic herring). McQuinn (1996) found that his data did not support the concept of discrete sympatric seasonal-spawning populations in Atlantic herring. He suggested that progeny of a given seasonal-spawning population may recruit to a local population that has a different reproductive season. Crossover of a large number of individuals from one seasonal spawning population to another results in year-class twinning. McQuinn further suggests that the spawning season of an individual is established at the time of first maturation and then maintained for the remainder of adult life.

Spawning Location. Throughout its range, herring spawn in a range of water depths from 0–5 m off Greenland to 200–300 m for some populations living in the North Sea. In general, spring spawning takes place in inshore shallows, whereas summer and fall spawning occurs in deeper, offshore waters (Messieh 1980, 1988). In the Gulf of Maine (including the Bay of Fundy), spawning takes place in water from about 4–6 m down to about 90 m. It does not usually occur in the littoral zone, nor has herring spawn ever been reported as cast up by the surf onto beaches of New England, a fate that often overtakes it in the Gulf of St. Lawrence. Spawning by herring in deep waters (>300 m) has not been observed in western Atlantic localities. In the Gulf of Maine, Bigelow and Schroeder suggested that deepwater spawning would probably be possible in the eastern basin, where the seafloor is hard, but that it was not likely to take place in basins on the western side of the Gulf, where soft, muddy substrates predominate.

Herring spawning has occurred at many places in coastal and near-coastal waters around the periphery of the Gulf of Maine to various shoals and ledges that lie 8–40 km off the coast, although there is considerable interannual variation in precise location and intensity (Sindermann 1979). Spawning has been inferred at several locations from capture of fully ripe females (Boyar et al. 1973a), from egg-bed and larval surveys (Boyar et al. 1973a; Cooper et al. 1975; Graham 1982; Iles and Sinclair 1982; Townsend et al. 1986), and from the presence of eggs on lobster traps (Stevenson, 1984). From these observations, it is evident that spawning occurs from the Canada–United States boundary to about Jonesport (44°32′ N), in Canadian waters south of Grand Manan Island, on various shoals and ledges off central Maine, on Jeffreys Ledge, in the coastal waters of western Maine, New Hampshire, and Massachusetts, in the west side of Passamaquoddy Bay, and in a number of locations on the Nova Scotia coast from Yarmouth to Halifax (Ridgway 1975). Long-term trends indicate a reduction in spawning sites along the immediate New England coast. Historical locations of spawning grounds in the Gulf and nearby environs were discussed by Bigelow and Schroeder. Present major spawning sites (Graham 1982; Stevenson 1989) are Lurcher Shoal and Trinity Ledges, Jeffrey's Ledge (Boyar et al. 1973a) and Stellwagen Bank, and Georges Bank and Nantucket Shoals. Minor sites such as Grand Manan, Matinicus, and Pumpkin ledges, have been reported, but their relative contribution is probably small.

In western Cape Cod Bay, Scherer (1984) reported catching larval herring from October to May, but he thought that larvae collected in this region might have originated primarily from outside Cape Cod Bay. He argued that their long larval period (i.e., 100 days or more) and slow growth rates (i.e., 1.5–2.1 mm·week^{-1} (Townsend and Graham 1981) could allow for dispersion over relatively long distances (Boyar et al. 1973b). Rocky, pebbly, or gravelly bottoms, preferred spawning substrates for herring, are found only in small isolated patches in Cape Cod Bay, in particular off Duxbury Beach, Billingsgate, at Provincetown, and at the southern tip of Stellwagen Bank, which was identified as a major spawning area for herring (Graham et al. 1972).

On Georges Bank, spawning centers shifted from historic sites on the northeastern side, where 86% of the total spawning occurred in 1974, to Nantucket Shoals, where 97% of the spawning took place in 1976 (Cohen and Lough 1983). Several reasons that might account for this apparent shift in distribution were discussed by Lough et al. (1980) and Anthony and Waring (1980).

Spawning Substrate. After eggs are released by the female they sink to the bottom, where they are fertilized and remain until hatching. Eggs are spawned on rocky, pebbly, or gravelly bottoms, shell substrates, and on clay to some extent, but probably never on soft mud. No eggs were found on sandy bottoms in the spawning area studied by Messieh (1988). Eggs are adhesive and stick in layers or clumps to sand or clay, seaweeds, stones, or to any other object on which they chance to settle. They are often found massed on net warps, anchors, and anchor ropes.

Direct observations of spawning sites reveal that herring choose a shell, gravel, or bedrock substrate, which ensures stability of deposited eggs (Messieh 1988). Using a submersible, Caddy and Iles (1973) examined sites used by autumn-spawning herring on Georges Bank in water averaging about 40 m in depth (Drapeau 1973). Spawn was observed on gravel patches devoid of sand, and tidal currents were relatively intense (1.2 $m \cdot s^{-1}$). The high-energy environment of spawning beds with strong currents prevents silt accumulation that could smother eggs and provides better circulation to supply oxygen and remove accumulated metabolites (Hempel 1971). Three of six grounds used by autumn-spawning herring on Jeffrey's Ledge were characterized by very rough boulder-rock substrate, with slope gradients ranging from 0° to 40° (Cooper et al. 1975). Bottom water currents throughout a tidal cycle at the ledge ranged from 0 to 2 $km \cdot h^{-1}$, with an average of about 0.3–0.5 $km \cdot h^{-1}$.

Egg deposition sites of spring-spawning herring in Canadian waters examined by divers evidenced a direct relationship between intensity of egg deposition and degree of algal cover (Tibbo et al. 1963; Pottle et al. 1980, 1981; Messieh et al. 1985). Messieh (1988) reported that correlations between egg densities and macrophyte abundances were not statistically significant for three of four cases he examined. Most eggs were found attached to bottom vegetation at depths of 0.9–4.3 m, with the greatest egg concentrations occurring at 1.4–4.0 m. In some areas, larger concentrations of eggs, proportionally, were found on *Phyllophorus* sp. and *Fucus* sp., than on Irish moss *Chondrus crispus* found in the same area, suggesting a preference for these algae. In other areas, a high proportion of eggs were attached to Irish moss, which may be due to the widespread distribution of the plant rather than to behavioral preference (Messieh 1988).

Spawning Seasonality. Throughout its range, at least one population of herring is spawning during any one month of the year, with each population having a different spawning time and place. Off the Atlantic coast of the United States, the major herring spawning event occurs from late August through November (Boyar et al. 1973b; Colton et al. 1979). Neither Scattergood (1952) nor Bigelow and Schroeder found evidence of spring-spawners in the Gulf of Maine. Subsequent studies (Watson 1964; Anthony and Waring 1980; Kornfield et al. 1982) supported their conclusion, although others (Tibbo et al. 1958; Boyar 1968; Boyar et al. 1973b) suggested that whereas there was some indication of minor spring spawning occurring in the Gulf of Maine, its contribution was probably negligible compared with the vast volumes of eggs and larvae produced during late summer and autumn spawning in this region. Throughout the Canadian Atlantic area, herring are probably spawning somewhere every month from April to November, with each stock having its own characteristic spawning time (Scott and Scott 1988). Both spring-spawning and summer-fall spawning schools of herring were formerly reported in the Bay of Fundy (Scott and Scott 1988). Spring spawners were present in the south (Nova Scotia) side of the bay from Bier Island at the mouth in as far as Digby Gut, and also in the Parrsboro region on the New Brunswick shore near the head of the bay, with spawning occurring during April and May. Spring spawners were never very numerous, and it is not known whether any now spawn in the bay before summer (Scott and Scott 1988). Spring-spawning as well as autumn-spawning herring have also been reported by fishermen along the west coast of Nova Scotia, although springtime spawning activity in this region has not been verified (Scott and Scott 1988).

Kelly and Stevenson (1985) noted that a major distinguishing feature of herring populations that spawn at the extreme ends of the U.S. Gulf of Maine coast is the difference in their spawning times. There is a generally southwest progression of the onset of spawning time, from late August off Nova Scotia to September–October on Georges Bank. Spawning begins in eastern Maine waters during middle to late August and apparently continues into October (Moore 1898; Graham and Sherman 1984; Stevenson 1984), with the heaviest runs of summer–autumn spawners usually arriving in July, August, and September. Spawning begins on Jeffreys Ledge in middle to late September (Cooper et al. 1975; McCarthy et al. 1979) and, judging from the presence of fully mature fish in October, continues at least into November. Further south in Ipswich and Massachusetts bays spawning occurs chiefly during October (Allen 1916); while in the vicinity of Woods Hole spawning takes place in late October and early November.

How different spawning times for herring populations have evolved has been the source of much speculation and study (Iles 1964; Cushing 1967, 1969, 1973, 1975; Iles and Sinclair 1982; Sinclair and Tremblay 1984; Sinclair 1988; McQuinn 1997). Initially, timing was thought to be linked to primary production cycles (Iles 1964; Cushing 1967) but as the time of spawning tended to be less variable than the timing of primary production blooms, Cushing (1975) postulated that larval survival and the subsequent extent of recruitment might depend on the closeness of the match between the two events (match-mismatch theory). Cushing's attractive match-mismatch theory was challenged by Sinclair and Tremblay (1984), who, developing further the larval retention hypothesis of Iles and Sinclair (1982), suggested that timing of spawning of herring is determined by two other constraints: herring larvae of a particular population develop within a discrete retention area at a rate attuned to the particular oceanographic situation of that area, and metamorphosis only takes place within a specific period of the year. Timing of spawning and egg size are thus geared to achieve metamorphosis at the appropriate time under certain specific and reasonably predictable environmental circumstances. McQuinn (1996, 1997) proposed a different hypothesis: that spawning season is established at the time of first maturation and is maintained for the rest of adult life.

Spawning Behavior. Direct observations of herring behavior during spawning in the North Atlantic are limited (Blaxter and Hunter 1982). Haegele and Schweigert (1985) reviewed the subject for both Atlantic and Pacific herring. Hay (1985), using a beach seine, took small samples on site as fish were spawning. Almost all males were totally spent, but most females were partially spent, suggesting that males probably initiated the spawning act, after which females deposited their eggs on the substrate. Direct observations of spawning *in situ* provided evidence that female herring do not release their eggs until males have released their milt (Messieh 1988). This sequence in the spawning act ensures proper fertilization of the eggs before they adhere to the substrate. Using underwater video to study spawning behavior, Messieh (1988) observed milt on the spawning bed prior to deposition of eggs, thereby providing evidence that females do not release their eggs until the milt is released. A few hours after the milt was discovered, the school of spawning fish was seen moving in a highly organized manner about 30 cm above the bottom. The school was watched swimming in the milt for about 4 h while the vessel was still in the center of the milt patch. The vessel was allowed to drift at about 4 $km \cdot h^{-1}$ so that the substrate could be searched, but no spawn was detected during this time. Deposition of eggs was observed the next day, however, and spawning was completed. For the first time during spawning bed surveys, spawning by herring was observed during the daytime.

Spring-spawning herring often arrive at the spawning grounds in runs or waves. Discrete batches of eggs deposited by these waves give rise to a succession of larval cohorts (Hourston 1958; Lambert et al. 1982; Dempsey and Bamber 1983). No pattern of rhythmicity has been discerned in autumn-spawning fish (Lambert 1984), and, in contrast to spring-spawning fish, it has been suggested (Bigelow and Schroeder; Jean 1956) that fall spawners leave the spawning grounds and move to deeper water as soon as they have spawned. Lambert (1984) suggested that segregation into runs or spawning waves was probably more commonplace in herring spawning groups (i.e., spring, summer, and autumn spawners) than has been observed.

Spawning Temperatures and Salinities. In the Gulf of Maine region, autumn-spawning herring usually reproduce at relatively high temperatures (10°–15°C) and at high salinities. Around Grand Manan and in the northern part of the Gulf, practically all spawning is carried out in waters of about 8°–12°C (Bigelow and Schroeder). Spawning in southern Massachusetts Bay and along Cape Cod, where autumnal cooling of surface waters is not as rapid as it is farther north, may take place in slightly warmer waters to 12°–14°C. A temperature of 17.5°C, at which autumn spawning occurred in the Gulf of St. Lawrence (Messieh 1988), appears to be near the optimal temperature known for herring (Blaxter and Hunter 1982), whereas a spring temperature of 3.7°C in the same area is near the lower end of the temperature range for spawning (Jean 1956).

In the Gulf of Maine, herring spawn at salinities ranging from about 31.9 to 33.0 ppt. They do not spawn in brackish water within the limits of the Gulf, although other populations are known to do so at the mouths of certain European rivers in water that is nearly fresh.

Fecundity. Fecundity of individual female herring varies according to age and size, as well as to the stock to which the female belongs. At the population level, there is a generally positive correlation between adult stock biomass and egg production (Hempel 1971). Values for individual females range from about 12,000 to upward of 260,000 eggs per spawning. Just prior to spawning, gonads in a sexually mature fish may account for up to one-fifth of its total weight.

In general, up to a maximum age, larger females produce more eggs, after which egg numbers begin to decline with age (Scott and Scott 1988). Fecundities among Georges Bank, southwestern Gulf of Maine, and Nova Scotia autumn-spawning populations are approximately similar (Perkins and Anthony 1969). Eggs per female ranged from 17,000 to 141,000 for 25- to 33-cm fish, respectively, and no significant differences in fecundity or egg size were found among the three stock complexes studied. A length-specific fecundity estimate (Kelly and Stevenson 1985) for herring taken in three spawning areas of the Gulf of Maine indicated that, in general, mean fecundities increased with fish size (24–35 cm), ranging from about 35,000 eggs in 24-cm females to nearly 191,000 in fish measuring 35 cm. For spring, summer, and fall spawners in the Gulf of St. Lawrence region, egg numbers per female ranged from 23,000 to 261,000 (Messieh 1976), whereas in the Newfoundland area they are estimated to be from 12,750 for females 27.8 cm TL to 241,630 for females 37.0 cm TL (Hodder 1972).

Variations in fecundity among spawning populations of herring have been attributed to both genetic and environmental factors (Blaxter and Holliday 1963; Parrish and Saville 1965; Hodder 1972; Messieh 1976). Comparative fecundity data, by lengths, for females from different spawning populations in the Gulf of St. Lawrence and Nova Scotia waters indicate that, in general, spring spawners produce fewer eggs (up to 50% fewer) than do autumn-spawning females of comparable size (Hodder 1972; Messieh 1976). Fecundity also varies inversely with egg size, which is related to egg weight and number (Parrish and Saville 1965; Hempel and Blaxter 1967; Messieh 1976; Blaxter and Hunter 1982). Lower fecundity of spring spawners is related to an increase in egg weight owing to an increase in yolk size, which, in turn, is possibly an adaptation to colder water and the need by the young for more energy in order to survive the lower water temperatures of spring (Messieh 1976).

Size and Age at Maturity. A median size at maturity of 25.4 and 25.3 cm TL, respectively, was recorded for female and male herring taken off the northeastern coast of the United States during recent autumn trawl surveys (O'Brien et al. 1993). Median age at maturity was 3.0 years for females and 2.9 years for males. Maturity at age is similar to that obtained for Gulf of Maine herring from monthly commercial sampling during 1985–1989 (0.39 of age-3 fish, 0.98 of age-4 fish, 1.00 of age-5+ fish). Extensive observations of adult herring from Georges Bank, the Gulf of Maine, and southern Nova Scotia during the 1960s led Boyar (1968) to the conclusion that herring in those areas spawned at age 4 and at an average total length of 27.5 cm. Fewer herring in that study spawned at age 3 and at sizes around 26.0 cm. Sinclair et al. (1982) noted a positive correlation between juvenile growth rates and L_{50}, and Winters (1976) reported decreases in A_{50} with decreasing adult biomass for herring in the Gulf of St. Lawrence, suggesting that density-dependent processes may influence maturation rates.

The percent contribution of age-3 fish to spawning stocks is variable from year to year (probably related to abundance of the particular year-class being recruited). In 1960, for example, age-3 fish accounted for 62% of the Georges Bank stock. Livingstone and Hamer (1978) reviewed data on age at maturity and found that from 1960 to 1965 about 29% of age-3 herring on Georges Bank and about 9% of Gulf of Maine age-3 herring were mature. For the period 1966–1970, 34% of age-3 herring were mature. Early maturation (at age 3) of the 1970 year-class was also reported by Dornheim (1975). Samples available to Livingstone and Hamer (1978) during the period 1973–1977 were too small to calculate meaningful percentages, and they were unable to detect any change in age and length at which 50% of herring were mature (M_{50}) from the limited data. However, they did find a decrease in relative numbers of age-3 herring in samples as well as an increase in the mean age and length of fishes in the spawning population.

Early Life History. Eggs are demersal, adhesive, and 1.0–1.4 mm in diameter, depending on the size of the parent fish and also, perhaps, on the population involved. Eggs have a segmented yolk, a wide perivitelline space, and no oil globules (Fahay 1983).

Development of herring embryos can occur over a wide range of salinities (Holliday and Blaxter 1960; McQuinn et al. 1983). Developmental rate is governed predominantly by, and is inversely related to, temperature (Messieh 1988). Time to hatching may require as much as 40 days at 4°–5°C, 15 days at 6°–8°C, 11 days at 10°–12°C, and 6–8 days at 14.4°–16.0°C, with 10–15 days being an average incubation period for autumn-spawned herring in the Gulf of Maine (Bigelow and Schroeder; Messieh 1988). Hatching success is also temperature-dependent. In experimental situations, all eggs maintained at 15°C hatched, none hatched at 0°–5°C, and all eggs held at 20°C died (MacFarland 1931).

In the Gulf of St. Lawrence, spring-spawned eggs required about 30 days at 5°C to hatch. Incubation times measured *in situ* indicated that spring-spawned eggs at bottom temperatures of 6.5°–9.3°C in the southwestern Gulf of St. Lawrence required 14–16 days (Messieh et al. 1985) to hatch and 17–20 days to hatching at temperatures of 6.0°–7.5°C (Messieh 1988). Fall-spawned eggs develop more quickly in the warmer water temperatures prevailing during this spawning season, with estimates of time to hatching of about 10 days at 15°C (Scott and Scott 1988) and 11 days at 10°C in waters off Nova Scotia (Jean 1956).

The pelagic larval phase is relatively long in herring, varying from 3 to 11 (commonly 6) months for different populations (Sinclair and Tremblay 1984). Herring are about 5–10 mm total length at hatching and have a small yolk sac and pigmented eyes. Yolk-sac absorption occurs at about 10 mm. Autumn-spawned herring larvae in the southern Gulf of St. Lawrence (Messieh et al. 1987) require an estimated 15–17 days from yolk-sac absorption to the exogenous feeding stage. Larvae are elongate and have a long straight gut, with the vent always situated posterior to the vertical through the dorsal fin. The preanal length is some 80% of the total length. There are 47 preanal myomeres until larvae exceed 20 mm TL, at which time preanal myomere numbers are reduced to 41–46. Flexion occurs at 16–17 mm, and transformation takes place at about 30 mm. A swim bladder forms at 10–15 mm, but is not noticeable until larvae are about 30 mm. Sequence of fin formation in developing fishes is: pectoral fins form first as buds, but are not completely formed until transformation; the dorsal fin forms at about 10 mm, the anal fin at about 16 mm, with both being complete by transformation; principal caudal rays are complete at about 20 mm; pelvic fins form between 20–30 mm and migrate posteriorly at transformation.

Metamorphosis is a gradual transition to adult characteristics, which is generally accomplished by the time fish are 45–55 mm, although some studies report metamorphosis occurring at lengths as small as 30–35 mm (Blaxter and Staines 1971; Boyar et al. 1973b; Ehrlich et al. 1976; Doyle 1977; Saila and Lough 1981). At metamorphosis, the developing fish resembles a juvenile herring; the body deepens, scales and pigment appear in the skin, and hemoglobin is present in the blood. While spawning of various herring populations in the North Atlantic can occur throughout the year, metamorphosis from the larval to the juvenile stage is restricted to a period from April to October (Sinclair and Tremblay 1984), and, as a consequence, relative durations of larval developmental periods are dependent on spawning time. In the northwest Atlantic, herring larvae from large winter-spring spawned eggs metamorphose within 3–6 months, whereas those from smaller summer-winter eggs overwinter as larvae and metamorphosis does not occur until 7–8 months later.

Pronounced behavioral and physiological changes accompany metamorphosis. The most obvious change in behavior is the development of schooling responses (Sindermann 1979; Gallego and Heath 1994a). In herring, metamorphosis appears to be a physiologically demanding event accompanied by an increase in hemoglobin synthesis as a result of increased activity (De Silva 1974). Fat levels also increase prior to this life-history transformation (Marshall et al. 1937). These and other related growth processes might explain the seasonal restriction of metamorphosis to the productive period of the year (Sinclair and Tremblay 1984).

Atlantic herring larvae are unique among clupeid larvae in the Gulf of Maine in that they have a high myomere count and the anal fin forms relatively late in development (Fahay 1983). The larvae are very slender and can easily be distinguished from larvae of all other Gulf of Maine fishes of similar form (e.g., sand lance, smelt, or rock eel) by the location of the vent, which is situated more posteriorly and closer to the caudal fin base.

Age and Growth. Growth has been well studied in Atlantic herring. Growth rates among and within populations are variable depending on a number of factors, including temperature (Day 1957a; Saville 1978; Moores and Winters 1982), food availability, and population size. Historically, scales were used to age this species (Huntsman 1919; Lea 1919), but since the 1960s otoliths have been used (Hunt et al. 1973). Using otoliths to age young herring from the Gulf of Maine was validated by Watson (1964), but difficulties with aging older fish persisted through the mid-1970s (Dery and Chenoweth 1979; Dery 1988a), with poor levels of agreement between estimates of fish aged by scales and by otoliths (Messieh and Tibbo 1970) and between otolith age readings by different readers (Parsons and Winters 1972). Otoliths of postlarvae as small as 0.31 mm are discoid but adopt the adult shape when they reach about 1.0 mm (Harkonen 1986). A description of otolith shape and an otolith length–fish length relationship for herring taken off Newfoundland was described by Lidster et al. (1994). Otolith nuclei have also been found useful in discriminating between individuals belonging to spring- and autumn-spawning populations (Einarsson 1951).

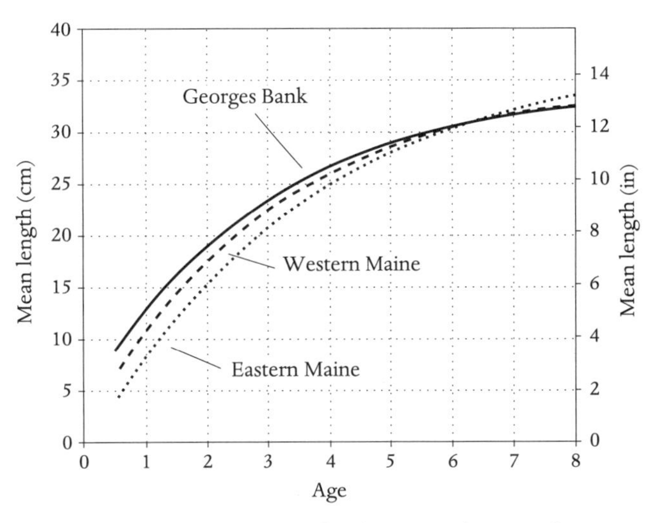

Figure 78. Age-length curves for three populations of Atlantic herring *Clupea harengus:* Georges Bank, western Maine, and eastern Maine. (From Penttila et al. 1989:9.)

In U.S. waters, herring attain an overall maximum length of about 39 cm TL and an age of about 15–18 years (Anthony 1972), although some reports list herring as old as 20 years, and older. Male and female herring grow at about the same rate and become sexually mature beginning at age 3, with most maturing by age 4 or age 5. Beverton (1963) and Anthony and Waring (1980) tabulated von Bertalanffy growth parameters for many Atlantic herring stocks. All showed rapid growth in the first years of life, with a marked slowing at the onset of maturity, normally at age 4 (Fig. 78). Estimated K values for the Georges Bank population were 0.35–0.36, while those for Canadian Atlantic herring populations were between 0.17 and 0.30 (Anthony and Waring 1980). In all cases, 50% maturity occurred at lengths near 80% of the asymptotic length L_{∞}. During periods of low population levels, herring may mature at smaller sizes. In the western Atlantic, growth rates in several studies were found to increase progressively in a gradient from Nova Scotia to Georges Bank, with western Maine herring intermediate between Nova Scotia and Georges Bank, but with only slight differences (summarized in Sindermann 1979). Georges Bank herring grew more rapidly, but the maximum size attained was only 35 cm at 14 years; Nova Scotia and eastern Maine herring grew more slowly, but reached a greater maximum size (39 cm in 16–18 years).

Juveniles grow to about 90–125 mm by the end of their first year of life in the Bay of Fundy and in the coastal waters of Maine. Juveniles, 50–65 mm in length, predominate among small herring at Provincetown at the end of June and are 54–100 mm on Nantucket Shoals in mid-July. At Woods Hole, herring spawned in October and early November are 76–125 mm long by the following autumn. Growth of juveniles is very rapid at age 1 and age 2 (Anthony 1972). In western Maine, age-1 herring grew as much as 10 cm from early spring to November, and age-2 herring as much as 8 cm. When herring reach about age 2 and lengths of 190–200 mm they accumulate large amounts of fat among the body tissues and viscera during warmer months of the year when growth is rapid. They lose this fat in winter and also at the approach of sexual maturity. This "fat" stage is characteristic of herring in American as well as in European waters, where "fat" herring are the targets of extensive fisheries.

There was great variation in growth between year-classes, with size of age-2 herring varying in one area by as much as 6 cm between one year-class and another (Anthony 1971). Herring in western Maine grew faster than those in eastern Maine and averaged nearly 3 cm longer at the end of their second year. Although herring in eastern Maine grew very slowly at age 1, they increased their rate of growth at age 3 and were the same lengths as western Maine herring at age 5 (Anthony 1972). Because growth varies significantly between year-classes, composition of age-groups entering the juvenile fishery is not the same each year. Since the fishery is selective to 17-cm fish, fast-growing age-1 fish will enter the fishery in late summer, while slow-growing age-3 fish will enter in early summer.

Density-dependent growth in herring has been observed by a number of authors (Iles 1967; Anthony 1971; Saville and Jackson 1974; Lett and Kohler 1976; Burd 1984; Molloy 1984). Although there is some evidence of density-dependent growth of young herring, the literature regarding density-dependent growth of mature herring is contradictory (Sinclair et al. 1982). Anthony (1971) concluded that growth of juvenile herring along the Maine coast during the first two years of life was density-dependent but also mediated by regional differences in plankton abundance and water temperature. Growth appeared to be related to both abundance of age-2 fish and water temperature. When fish abundance was great, its effect overcame positive effects of temperature (or other factors influenced by temperature). Anthony and Waring (1978) reported that while density-dependent growth was exhibited by Gulf of Maine juveniles during some years, in other years this density-dependent relationship completely disappeared. Lett and Kohler (1976) also suggested that growth in herring was density-dependent during the first year and that this influence determined subsequent growth rates. Any such density dependence must be weak, however, since, as Doubleday (1985) pointed out, even a doubling of growth rates would, in most cases, imply less than 20% annual increase in weight for older fishes. An apparent increase in growth rate in juvenile and adult fish from Georges Bank and the Gulf of Maine was observed beginning with the 1968 year-class (Anthony and Waring 1978). It was thought that an increase in growth rate for all year-classes was possibly a response to declines in overall biomass in that population. There was no change in maximum size attained.

General Range. Atlantic herring are confined to cold-temperate and boreal waters of the Northern Hemisphere on both sides of the Atlantic (Whitehead 1985a). In the western North Atlantic, they range from southwestern Greenland and northern Labrador, regularly and commonly as far south as Cape Cod and Block Island, and in winter are occasionally seen in small numbers as far south as Cape Hatteras and South Carolina. In the eastern Atlantic, herring are found from the northern Bay of Biscay northward to Iceland and southern Greenland, eastward to Spitsbergen and Novaya Zemlya, and south to the Straits of Gibraltar. They are replaced in the North Pacific by a morphologically similar, closely related species, *C. pallasi* (Whitehead 1985a).

Occurrence in the Gulf of Maine. Atlantic herring appear at one season or another along nearly the entire coastline of the Gulf, as well as on the offshore fishing banks (Map 9). All stages in the herring's life history occur in Gulf of Maine waters, and this species is one of the most common and abundant elements of the Gulf's fish fauna.

Seasonal distribution of juveniles in the Gulf of Maine has been well studied. Seasonal variations in abundance of juvenile herring in this area are simply a matter of their local availability. During springtime (March-May), juveniles 30–50 mm long become widely distributed in the lower Bay of Fundy, around the entire periphery of the open Gulf (east as well as west), out over the basin, and on the northern and eastern parts of Georges Bank (Bigelow and Schroeder). Sardine-size herring, 12–15 cm long and generally 1 and 2 years old, are usually abundant all summer east of Penobscot Bay, particularly in the Passamaquoddy Bay region, where they support the famous sardine fishery. Summer concentrations of juveniles along coastal southwestern New Brunswick and eastern Maine are a mixture of Nova Scotia and Gulf of Maine stocks (Sindermann 1979). Age-1 herring tagged in western and central Maine waters during the third and fourth calendar quarters (Creaser and Libby 1986) contributed to commercial catches of age-2 fish during the second and third quarters of the following year east of the area where they were tagged. Age-1 herring tagged in easternmost Maine and western New Brunswick waters during the third and fourth quarters remained in the same area, where they contributed to the commercial fishery for age-2 fish during the second and third quarters of the following year.

Bigelow and Schroeder and others (Creaser and Libby 1986) have noted that juveniles apparently overwinter along the entire coasts of Maine, New Hampshire, and Massachusetts. The relative quantity of fish overwintering in these areas is unknown because tag recovery information is available only from regions where winter fisheries presently exist. Herring tagged as overwintering juveniles in eastern and western Maine remained in close proximity to the area where they were tagged throughout the following summer. Some herring tagged as summer-feeding juveniles in southwestern Maine overwintered in the region of Massachusetts Bay and off New Hampshire, whereas juveniles tagged at the same time in eastern Maine had a greater tendency to overwinter in eastern Maine.

Movements. Herring are capable of making extensive journeys. Tagging studies have demonstrated the existence of annual migratory patterns, such as movements to spawning grounds, to overwintering areas, and to feeding areas. Some populations have a more persistent migratory pattern than others, and some populations also intermingle more than others. Adults have generally been recovered from a wider range of locations than juveniles, indicating that they are apparently capable of covering greater distances in their seasonal movements than younger stages (Creaser et al. 1984; Creaser and Libby 1988). Some adults tagged in the southwestern Gulf of Maine have been found at least as far east as Mount Desert Island on the Maine coast, while others from this spawning population may possibly overwinter in the Mid-Atlantic Bight intermingled with fish from the Georges Bank population (Ridgway 1975). Adults from the Gulf of Maine that overwinter south of Cape Cod may move through Great South Channel and Cape Cod Canal to summer feeding grounds along the Maine coast. Seasonal movements of herring within the Gulf of Maine are discussed in greater detail below.

Circumstantial evidence and data based on tagging studies indicate that progeny of a spawning component of herring have a homing tendency and return to the parental spawning ground (Messieh and Tibbo 1971; Wheeler and Winters 1984b). Analysis of tagging data from Newfoundland waters supported the hypothesis that the majority of herring return to that same area to spawn in successive years. Average annual homing rates, defined as the number of fish returning to the same area to spawn in successive years, was estimated at about 73%. Herring captured and tagged on Jeffreys Ledge and in the Great South Channel showed that adults moved from tagging sites to coastal waters of Maine and the Bay of Fundy during summer and early autumn and returned to Massachusetts and Rhode Island coastal waters in late autumn and winter

(Creaser and Libby 1988). Extremes of movements for adult fish tagged in the Gulf of Maine are Point Judith, R.I., and Sydney Bight, N.S. Little is known of causal mechanisms or environmental cues responsible for herring migration and homing. Herring may not directly recognize a particular spawning ground, but innately recognize it as a place suitable for spawning (Harden-Jones 1968). However, some cue must exist to aid the fish in returning to this general area. Current direction may serve as a general cue allowing herring populations to return to an area of olfactory sensitivity, which then permits return to the parental spawning ground. In the Newfoundland area, herring indicate a general denatant (away from spawning ground) migration to overwintering areas, followed by a contranatant (toward spawning ground) migration during the prespawning period (Pinhorn 1976). However, McQuinn (1996, 1997) argues that sufficient data exist that make it difficult to accept a discrete population concept for Atlantic herring. Spring-type otoliths are found in autumn-spawning herring and vice versa, and the lack of genetic divergence between seasonal spawning populations does not support the discrete population hypothesis.

Seasonal movements of adult herring on Georges Bank (Sindermann 1979) comprise three apparent phases: (1) a late summer–early autumn spawning migration of ripening fish; (2) a rapid postspawning migration to warmer waters to the south for overwintering; and (3) a spring–early summer northward feeding migration. Postspawning adult herring from Georges Bank move southwest to off Chesapeake Bay in November and overwinter there, with the largest and oldest fish moving furthest south. A feeding migration back to Georges Bank begins in May or early June and continues to shallower spawning sites on the northern edge of the Bank in September. The waters off Cape Cod seem to constitute a mixing area, with different groups passing at different times of the year.

Stock intermixture is a seasonal phenomenon. Anthony (1977) summarized general information about movements of adult stocks in the Gulf of Maine. Tagging studies showed that for fish off southeast Newfoundland there is substantial intermingling of local populations from different bays (Moores and Winters 1984; Wheeler and Winters 1984a,b). This intermingling is mainly due to northward feeding migrations in summer and southward migrations in autumn to overwintering areas. Movement to spawning grounds occurs in the spring, at which time populations of these various bays tend to be very discrete. Relationships of populations outside the spawning season are dynamic, with the degree of intermingling being partially dependent on the size of the population. Off the coast of Nova Scotia, herring begin spawning in August and continue to October. Postspawners migrate offshore, and then move northward or southward (even to Cape Cod). Nova Scotia spawners that move south undoubtedly form part of the mixed stocks taken in the U.S. winter–early spring adult fishery in southern New England. A return migration begins in the spring, and some adults reach the Bay of Fundy by June.

Importance. Atlantic herring are a significant resource along the eastern coast of the United States and Canada (Kornfield et al. 1982), where both spring- and autumn-spawning herring populations support major fisheries (Messieh 1988). Herring usually occur in commercial quantities along the coast of southern Labrador, around the coast of Newfoundland and the offshore banks, in the Gulf of St. Lawrence, along the coast of Nova Scotia and the offshore banks, the Bay of Fundy, including Passamaquoddy Bay, Gulf of Maine, Georges Bank, and during the winter to the coastal waters of Rhode Island, and south to Virginia. The waters of southwest Nova Scotia support a herring population estimated to be on the order of 500,000 tons, which is the basis of the largest herring fishery in the western Atlantic (TAC, total available catch, in excess of 100,000 tons) (Stephenson et al. 1987). The larval patch that results from spawning of this population is the largest and best defined of those in the Bay of Fundy and Gulf of Maine (Stephenson and Power 1989). Annual surveys of this larval area are used as an index of abundance in stock assessment (Iles et al. 1985; Stephenson et al. 1987).

Herring have been continuously exploited since pre-Colonial times when prodigious populations were commonly observed by coastal communities. At one time, herring populations were considered inexhaustible, but in recent years stocks in both Atlantic and Pacific waters have declined owing to overexploitation (Whitehead 1985a,b; Scott and Scott 1988). Although stocks have been badly overfished and depleted in recent years, *Clupea harengus* still ranks as the third most heavily exploited clupeid fish in the world (Whitehead 1985a).

Historical trends and locations of commercial catches within the Gulf of Maine were discussed by Bigelow and Schroeder. A detailed account of fishing activity and fisheries exploitation of *C. harengus* in the western Atlantic was provided by Sindermann (1979) and is summarized below.

The earliest organized fishery for herring in the western North Atlantic was probably conducted by Indians using brush weirs (Earll 1887). Captain John Smith's account of herring in the Gulf of Maine provides some indication of the former abundance of this species: "The savages compare the store in the sea with the hair of their heads, and surely there are an incredible abundance upon this coast." With the appearance of Europeans on American shores and in American waters, herring fisheries developed in a series of phases: (1) European vessels fishing for cod visited waters of the western North Atlantic beginning about 1500. The Europeans discovered very early that herring for bait could be taken near the cod grounds at night by gill nets. (2) Beginning with the earliest permanent settlement, herring were used as food by the colonists, as well as for cod bait. Following the Indian example, the Plymouth colony built and operated a herring weir as early as 1641. (3) During the nineteenth century, salt herring, either as food or bait, was much in demand, and substantial fisheries developed off Newfoundland, in the Gulf of St. Lawrence, and off New England. (4) With the introduction of trawling for cod and other groundfish, demand for herring as bait declined in the later part of the nineteenth century, but at about that time (the 1870s) canning juveniles as sardines began on the Maine and New Brunswick coasts and was profitable. Additionally, the lobster fishery expanded after 1860, and herring was a major source of bait. The sardine fishery for juveniles was the principal herring fishery during the first half of the twentieth century. (5) Beginning in 1961, exploitation of offshore adult herring stocks by distant-water fleets began, and it increased annually until 1969, when the combined catch by all nations fishing in waters of the western North Atlantic was almost a million tons. (6) During the 1970s, concern about overexploitation led to the imposition of international catch limitations of increasing severity as herring stocks continued to decline. Almost total failure of the autumn 1977 adult herring fishery on Georges Bank was the most disturbing event during this period of intensive fishing and resultant decline in stock size.

In terms of relative impact on herring stocks, the entire history of the fishery can be divided into two phases: pre- and post-1961. Before 1961, exploitation by man was minimal, with some stocks (such as Georges Bank) untouched, and others (such as those off southwest Nova Scotia and in the Gulf of St. Lawrence) harvested minimally and inefficiently by gill nets and other fixed gear. The only herring fishery that could be described as intensive prior to 1961 was the sardine fishery for juveniles on the Maine and New Brunswick coasts, and even this fishery was conducted principally with fixed gear. After 1961, fishing pressure on all stocks increased enormously, with mobile gear (otter trawls, paired trawls, midwater trawls, and especially purse seines) accounting for dramatic increases in annual landings from all known herring stocks of the northwest Atlantic region.

Landings from 1920 to 1940 were relatively stable at about 80,000–100,000 mt, then increased gradually through the 1940s to a peak of 242,000 mt in 1948. During the 1950s catches stabilized at about 160,000–200,000 mt. The herring fishery on Georges Bank was initiated by distant-water fleets. Landings peaked in 1968 at 773,600 mt and subsequently declined to 43,500 mt in 1976 as the fishery collapsed (Fig. 79) (Friedland 1998b). There has been no directed fishery for herring on Georges Bank since then.

Herring, especially smaller fishes fresh from the water, are among the most delicious of fishes. Their only drawback is that they do not keep well, being rich-meated and oily, and larger-sized fish have many hairlike bones that are troublesome. Herring are especially rich in oil, with seasonal values of oil content of raw herring averaging 5–9% for spring-caught fish and 10–15% for those caught in summer (Leim 1957). Nutritional values for herring are: fat 5–12% (varies seasonally); moisture 67–73%; protein 20–22%; and ash 1–2% (Krzynowek et al. 1989). Braune (1987) found significant positive correlations between mercury concentrations in herring and the age, weight, and length of the fish.

Herring represent the raw material for the Maine sardine industry, which is centered in the Bay of Fundy, particularly in

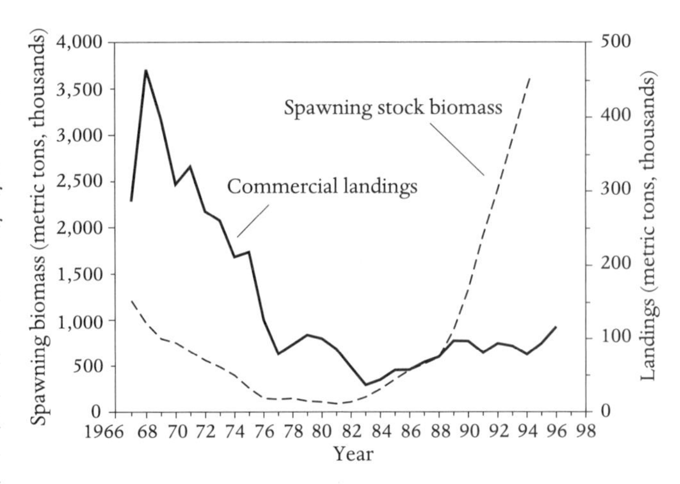

Figure 79. Commercial landings and estimate of spawning stock biomass of the coastal stock complex of Atlantic herring *Clupea harengus,* 1966–1998. (From Friedland 1998b.)

the Passamaquoddy Bay region, where the largest sardine cannery in the world is located. Fish 4–8 in. long are packed as Maine sardines. Larger fish are prepared as canned steaks, with a variety of flavoring sauces. The Maine sardine industry peaked in the late 1940s with total production exceeding 3 million standard cases. In 1980, production fell to less than 1 million cases, and in 1990 the total was only 820,000 cases.

Mature herring are sold fresh, frozen, smoked, salted, pickled, and canned. Small herring are packed as sardines, whereas larger herring are canned as kipper snacks and fillets. Probably the most important products are frozen fillets, sardines, and pickled and cured dressed herring and fillets. Smoked herring are sold as kippers or bloaters. Herring roe has recently found a ready market as a delicacy in Japan, where the final product is called *kazunoko* (Scott and Scott 1988). The market for herring roe arose because of the decline of Japanese herring stocks. Large fish are usually exported to Europe for production of smoked and marinated products.

Herring are sometimes canned for pet food, and there is a market for adult herring as bait and feed for zoo and aquarium animals. During the 1950s and 1960s, use of raw herring for reduction to oil and fish meal grew steadily, until by 1968–1969 most landings of Canadian herring were being processed for this market (Scott and Scott 1988). With the decline in world landings of herring, their use for fish meal and oil production ceased and catches increased markedly in value. Herring scales are processed for the production of *pearl essence,* which is used in paints and cosmetics to provide glitter. During the 1940s there was a big demand for pearl essence derived from herring scales for making high-quality paints for aircraft.

Stocks. No other teleost species has such a complex population structure or exploits such a wide range of reproductive environments in space and time as does the Atlantic herring (Blaxter 1990). Different populations have colonized environments ranging from high-salinity–long-migration–oceanic niches in the Atlantic to low-salinity–local-migration–inshore niches, such as those in the Baltic Sea. Each population seems to have preferred spawning grounds and feeding and wintering areas. Fluctuations in biomass in many populations, resulting from natural conditions as well as from increased fishing efforts, have created a need for accurate information on the population biology of this species (Kornfield et al. 1982).

The most interesting biological differences among the various populations of Atlantic herring relate to reproduction (Blaxter 1990). In general, herring, at the time of spawning, can be separated into populations that have a characteristic spawning time and place. Historically, throughout the western Atlantic and other regions, a number of specific and subspecific categories have been described to recognize the spawning populations of herring, including sibling species, subspecies, populations, stocks, substocks, or races, with little distinction being drawn among these categories (Parrish and Saville 1965; Haegele and Schweigert 1985; Smith and Jamieson 1986). Biologists have attempted to use a variety of techniques to discriminate herring populations and stocks, including differences in morphology, morphometrics, and meristics (Svetovidov 1952; Parsons 1972, 1973; Parsons and Hodder 1974; Lough 1976; Anthony 1981b; Bird et al. 1986), tagging (Anthony 1981a; Waring 1981; Stobo 1983) and biological tags (McGladdery and Burt 1985), electrophoretic studies (Ridgway et al. 1970; Truveller 1971; Odense 1980; Odense and Annand 1980; Kornfield et al. 1982; Grant 1984), mtDNA studies (Kornfield and Bogdanowicz 1987), and life-history patterns (Côté et al. 1980; Iles and Sinclair 1982; King 1985).

Attempts to define populations and stocks of herring have a long and controversial history (Kornfield et al. 1982; McQuinn 1997). Since large numbers of mature adults spawn at relatively discrete geographic locations (Boyar et al. 1973a), individual spawning aggregations were historically considered by fisheries managers to be distinct populations or stocks. Parrish and Saville (1965) divided Atlantic herring into five groups: three in the northeastern Atlantic and two in the northwestern Atlantic. Western Atlantic groups were distinguishable by times and locations of spawning that geographically overlapped in the Gulf of St. Lawrence, but were not distinguishable by differences in body size and number of vertebrae (Parsons and Hodder 1981; Messieh and Tibbo 1971; but see Côté et al. 1980). A southern group that spawned in autumn on offshore banks extended from Virginia to the southern portion of the Gulf of St. Lawrence (Anthony and Waring 1980). A northern group of herring spawned in spring, but did not undertake extensive migrations (Tibbo 1956; Day 1957a,b).

In the northwest Atlantic, a mixture of spawning seasons occurs in the populations, with a general clinal trend of spring spawners predominating in the north and autumn spawners predominating in the south (Haegele and Schweigert 1985). These two seasonal spawning groups also differ in a number of other respects. Winter- and spring-spawning populations have large eggs and low fecundity, whereas summer and autumn spawners have small eggs and high fecundity. A simplistic explanation for these differences cites a limited food supply and low predator pressure in winter and spring vs. a good food supply and large predator population in summer and autumn (Blaxter 1990). In addition to differences in spawning time, reported differences between members of these two broad seasonal groups were also found in age-frequency distributions, growth rates, otolith morphologies, and mean numbers of pectoral and dorsal fin rays.

Intermixing of seasonal spawners from different spawning areas has been the focus of repeated observations (Messieh 1972; Vernberg 1977), and tagging studies in the western North Atlantic have shown extensive migration and mixing of stocks during nonreproductive periods (Creaser et al. 1984). Fish tagged in the Gulf of Maine, Georges Bank, and Nova Scotia mix during nonspawning migratory periods (Speirs 1977; Stobo 1976). Earlier investigations reported that members of different herring populations in the western Atlantic could sometimes be separated outside the spawning season by anatomical characteristics such as counts of vertebrae, fin rays, gill rakers, and scales; or by length or age at first maturity; or by asymptotic (theoretical maximum) length. Herring spawning on Jeffreys Ledge and along the U.S. coast of the Gulf of Maine were recognized as a single stock (Kelly and Stevenson 1985) distinguished from the Georges Bank and southwest Nova Scotia stocks on the basis of differences in phenotypic characters such as growth rates and numbers of fin rays and vertebrae, as well as differences in sizes of spawning populations, larval drift patterns, seasonal adult migrations, and parasite occurrences (see Sindermann 1979, for summary). Anthony and Boyar (1968), noting significant differences in pectoral fin ray and vertebral numbers among samples of herring from the northwest Atlantic, concluded that two general complexes of herring existed within the Gulf of Maine. They interpreted the meristic differences observed in the 2 years sampled as indicating a change in the distribution of herring. Stevenson et al. (1989) contended that there were three stocks of herring that spawn in the summer and fall in the Gulf of Maine off southwest Nova Scotia, on Georges Bank, and at various locations along the U.S.-Canadian coast between Cape Cod and Grand Manan Island.

Earlier researchers had considered spring- and fall-spawning herring to represent a single population (Jean 1956; Tibbo and Graham 1963), whereas others strongly supported the idea of separate populations (Messieh and Tibbo 1971; Messieh 1975; Côté et al. 1980) for herring spawning in different seasons. Clinal trends indicate an environmental trigger to spawning, and there is firmer evidence that the spawning period is not genetically fixed. Smith and Jamieson (1986) viewed Atlantic herring as comprised of dynamic and relatively unstructured assemblages. For example, in the northwest Atlantic, autumn- and spring-born fish, identified by otolith characteristics, have been found spawning in their opposite seasons (Messieh 1972). These results suggest that the spawning period is not highly heritable and that fish born in one season do not necessarily spawn in the same season (see also McQuinn 1996, 1997). A group of herring within an area may also change its spawning pattern in response to changes in environmental conditions. Thus, apparent differences in spawning times among populations are not necessarily evidence of discrete stocks. Although adults annually concentrate at specific locations to spawn, fidelity of particular stocks to specific spawning areas is not absolute. Tagging studies in the northwest Atlantic indicate a homing rate of 66–93% (Wheeler and Winters 1984b), with a movement to neighboring stocks of 7–34%. McQuinn (1997) summarized literature that points out that since early in the development of herring population structure theory, many authors have cited evidence that is difficult to reconcile with the notion of discrete spawning populations.

Stephenson and Kornfield (1990) concluded that the reappearance of spawning herring on Georges Bank was due to a resurgence of the population found there and not recolonization from other areas. They argued that persistence of this population, in spite of considerable potential for recolonization by herring from other areas, supported the discrete population concept in herring. McQuinn (1997), however, pointed out that the spawning grounds were recolonized by a single strong year-class encountered throughout the Gulf of Maine, and that this pattern of recolonization by a successful cohort is consistent with predictions of the metapopulation hypothesis of population structure.

Evidence for genetic differentiation among North Atlantic herring populations is weak. For over 20 years, biologists have studied biochemical variation in Atlantic herring in attempts to define population structure (Kornfield et al. 1982). Although allozyme differences between Atlantic and Pacific herring were found to be great enough to recognize them as distinct species (Altukhov and Salmenkova 1981; Grant and Utter 1984; Grant 1986), genetic differences between herring populations from eastern and western regions of the Atlantic are minimal (Odense et al. 1973; Odense and Annand 1980; Andersson et al. 1981; Grant 1984). Much more of the gene diversity of Atlantic herring is found contained within populations (98.9%) and much less is due to population differences (1.2%) (Grant 1984).

Kornfield and Bogdanowicz (1987) using mtDNA did not find a single stock identifier among the three spawning groups they examined. Spawning groups were not fully distinguishable by composite mtDNA digestion patterns; no absolute stock markers were present. They concluded that despite the availability of a larger number of polymorphic markers and adequate sample sizes, significant genetic heterogeneity among Atlantic herring stocks has not been demonstrated (Andersson et al. 1981; Kornfield et al. 1981; Grant 1984; Riviere et al. 1985). The magnitude of differentiation between groups is small, which implies that the groups are probably of relatively recent common origin. Statistically significant differences in genetic markers between seasonal spawners do not provide suitable biochemical markers for accurate discrimination of individual fish. Safford and Booke (1992) also found little genetic variation using starch-gel electrophoresis and morphometric analysis of herring in the western North Atlantic and concluded that Atlantic herring there form a single panmictic population.

In spite of the wealth of data, the taxonomic and evolutionary status of herring populations remains problematical and continues to intrigue biologists and perplex fisheries managers (Smith and Jamieson 1986; Kornfield and Bogdanowicz 1987; McQuinn 1997). Regularity of spawning (both geographic and temporal) (Sinclair and Tremblay 1984), tag evidence for homing (Harden-Jones 1968; Wheeler and Winters 1984b), differential population dynamics of neighboring groups (Sinclair and Iles 1985), and discrete larval distributions (Iles and Sinclair 1982) all suggested that herring spawning units are distinct populations, that is, self-sustaining, and geographically and genetically discrete. On the other hand, much traditional evidence for discrete stocks, based on morphometrics, spawning times, and tag returns, is shown to be weak and based on a typological concept (Smith and Jamieson 1986). The lack of demonstrable differences in traditional stock identification methods and genetic (particularly isozyme) characteristics has been interpreted as indicating significant gene flow among neighboring spawning aggregations. Results of allozyme surveys have shown little or no genetic divergence between herring stocks (Kornfield et al. 1982; Grant 1984; Grant and Utter 1984; Ryman et al. 1984; Safford and Booke 1992). Because of the weight of established opinion on herring stocks, allozyme markers have been interpreted as being insensitive to stock events (Kornfield et al. 1982; Grant 1984; Grant and Utter 1984; Ryman et al. 1984; Cushing 1985).

McQuinn (1997) concluded that neither the discrete population concept nor the dynamic balance concept adequately explain all the data associated with herring population structure dynamics, including meristic and morphometric characters, life-history traits, homing, year-class twinning, and biochemical analyses. Instead, he feels that available information suggests that Atlantic herring population structure and dynamics are well described within a metapopulation concept. Within this concept, population integrity is maintained though behavioral isolation, that is, repeat rather than natal homing to spawning areas, whereas local population persistence is ensured through social transmission of migration patterns and spawning areas from adults to recruiting individuals.

Kornfield and Bogdanowicz (1987) concluded that consistent, significant genetic differences among spawning groups of Atlantic herring are sufficient, but not necessary, conditions to regard populations as discrete stocks. Although their results did not support the hypothesis that discrete Atlantic herring stocks exist throughout the Gulf of Maine, the absence of such differences did not allow them to rigorously conclude that there was gene flow among the populations in question. They recommended that in order to preserve variability, Atlantic herring should be managed under the assumption that every spawning group is a semidiscrete genetic entity.

ATLANTIC ROUND HERRING / *Etrumeus teres* (DeKay 1842) /

Bigelow and Schroeder 1953:87–88 (as *Etrumeus sadina*)

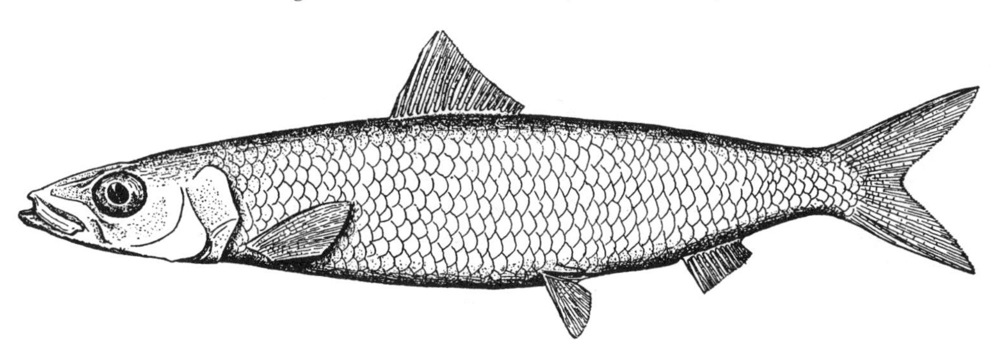

Figure 80. Atlantic round herring *Etrumeus teres.* (From Bigelow and Schroeder: Fig. 40.)

Description. Body slender, only about one-sixth as deep as long, body depth 16.0–18.5% SL (Fig. 80). Belly rounded, without pre- and postpelvic scutes. Head triangular; head length 24.0–29.0% SL. Eye diameter 6.5–8.0% SL. Snout length 6.0–9.5% SL. Mouth terminal and relatively small, posterior extent of jaws not extending posteriorly beyond anterior margin of eye. Maxilla length 7.0–10.5% SL. Premaxillae rectangular, teeth minute. Isthmus with distinct notch ("shoulders") anteriorly on each side. Dorsal fin located at about body midpoint. Pectoral fins moderately developed, reaching notably less than halfway to base of pelvic fin, with long membranous axillary process present. Pelvic fins situated posterior to vertical through posterior dorsal fin base. A peculiar, unkeeled, W-shaped pelvic scute is located immediately anterior to the pelvic fins with the arms extending laterally around the base of the fin rays. Caudal fin forked. Scales cycloid, relatively large and deciduous.

Meristics. Dorsal fin rays 15–22, usually 18 or 19; anal fin rays 10–13, most frequently 11; pectoral fin rays 14–16; pelvic fin rays i, 8; lateral scales about 48–55 (usually missing); gill rakers 30–38 on lower limb, about 14 on upper limb; branchiostegal rays 11–15; pyloric caeca numerous; vertebrae 49 (Hildebrand 1963c; Fahay 1983; Whitehead 1985a).

Color. Olive green above with silvery sides and belly.

Size. Adults range from 20 to 25 cm. Age estimates for the eastern Atlantic *E. whiteheadi* show that fish smaller than 120 mm are less than 1 year old and that analysis of growth increments is unreliable for larger fish (Waldron et al. 1991).

Distinctions. The W-shaped pelvic scute, numerous branchiostegal rays (11–15), and rectangular premaxillae distinguish *Etrumeus* from all other genera of clupeids occurring in the Gulf of Maine (Whitehead 1985a). The most distinctive feature of Atlantic round herring is that the belly is rounded, without a keel of sharp-edged scutes. The dorsal fin is situated entirely anterior to the vertical line through the origin of the pelvic fins (vs. being directly above the latter in all other Gulf of Maine clupeids), and there are fewer anal fin rays (only about 13) compared with herring (about 17), alewife (about 19), and American shad (about 21).

Habits. Round herring are marine, pelagic, fish, which are rarely taken inshore. It is a schooling species found mainly in deep waters along the continental shelf and slope or deeper, with schools usually concentrated between 37–92 m (Shaw and Drullinger 1990, and literature cited therein). On the southeast U.S. continental shelf, adults have been taken between 10–366 m, and a seasonal shift in depth distribution may occur, with fish moving offshore (56–183 m) during summer-fall, and inshore (10–27 m) during winter-spring (Barans and Burrell 1976). In southern areas, they do not occur in nearshore or estuarine habitats, but only offshore in water at least several meters in depth (Hildebrand 1963c; Shaw and Drullinger 1990). They have not been observed to make long-distance migrations along the South Atlantic coast of the United States (Reintjes 1979).

Round herring apparently undertake diurnal vertical migrations (Bullis et al. 1971), with fish occurring at the surface at night and 9–37 m off the bottom during the day. They form dense schools, often with other pelagic schooling species, including Spanish sardine, rough scad, and chub mackerel (Bullis et al. 1971; Crawford 1981). Bullis et al. (1971) reported a mixed school of round herring and Spanish sardine west of Tampa, Fla., that was estimated to be 80 km long, 16 km wide, and about 3.7 m thick.

Food. Round herring feed mainly on zooplankton (62.6% by weight) and larval fishes (32.8%, Bowman et al. 2000). Zooplankton prey include copepods (27.7%), stomatopods, amphipods, mysids (particularly *Neomysis americana*, 7.8%), and decapod larvae (20.6%).

Predators. Round herring are consumed by a wide variety of pelagic predators, including fishes, seabirds, and marine mammals. The NEFSC food habits survey recorded 15 species of fishes preying on round herring, of which 4 had multiple predation records, in order of frequency: bluefish, smooth dogfish, summer flounder, and Spanish mackerel (Rountree 1999).

Parasites. Little information is available about parasites of round herring. Takao (1990) reported that larval nematodes utilize round herring as an intermediate host.

Reproduction. Reproductive biology of round herring and details relevant to distribution and population dynamics of this species in the western Atlantic are presently sketchy (Shaw and Drullinger 1990; Chen et al. 1992). Larvae of round herring have been collected off the South Atlantic Bight during January and February (Fahay 1975). Spawning occurs from late January to early June in the Gulf of Mexico (Fahay 1983; Shaw and Drullinger 1990). In the eastern Gulf of Mexico, round herring spawn from mid-October to the end of May offshore between the 30- and 200-m isobaths (Houde 1977). One major spawning area in the Gulf is located about 150 km west-southwest of Tampa Bay, Fla., with a minor area just north of the Dry Tortugas (Houde 1977). Off Texas and Louisiana, round herring spawn 50–200 km offshore and may also spawn at the edge of the continental shelf (Fore 1971). High concentrations of round herring larvae were found on the outer shelf (depths of 40–182 m) and near the shelf break during sampling off the Mississippi River delta region (Shaw and Drullinger 1990).

Round herring scatter pelagic eggs at night, with peak spawning estimated to occur at 2200 (Houde 1977). Eggs have been collected at the surface at temperatures of 18.4°–26.9°C and salinities of 34.5–36.5 ppt (Houde 1977). Shaw and Drullinger (1990) collected recently hatched larvae at surface temperatures and salinities ranging from 16.7° to 23.8°C and 33.8 to 36.5 ppt, respectively. For all sizes of larvae collected, temperatures at time of collection were 15.0°–30.0°C, and salinities were 28.7–37.5 ppt.

Early Life History. Eggs are pelagic, spherical, about 1.17–1.37 mm in diameter (Houde and Fore 1973). The yolk is segmented and there is a small perivitelline space, but the egg lacks oil globules.

Hatching occurs in about 2.1 days at 21°–22°C (Houde 1977). At hatching planktonic larvae are 3.8–4.8 mm TL and have unpigmented eyes. Larvae are elongate with a long straight gut with the vent always posterior to the dorsal fin (Houde and Fore 1973). Flexion occurs at 8–10 mm TL; transformation begins at about 18 mm SL and is completed at 28–33 mm TL (Hildebrand 1963c).

In the eastern Gulf of Mexico, greater numbers of larvae are caught at night (daytime net avoidance); larger larvae (13.0–18.0 mm) apparently are also able to avoid nets at night (Houde 1977; Shaw and Drullinger 1990).

Larval diet consists primarily of copepod nauplii, copepodites, and adult copepods, with pteropods, tintinnids, invertebrate eggs, and *Eucalanus* nauplii contributing lesser percentages (Chen et al. 1992).

Age estimates, based on daily otolith rings, averaged 0.71 mm·day^{-1} for December (Chen et al. 1992). The fastest growth rate (ca. 0.85 mm·day^{-1}) for round herring larvae occurred at about 15 days.

General Range. Round herring occur in the northwest Atlantic from the mouth of the Bay of Fundy southward to Florida, the Gulf of Mexico, Venezuela, and the Guianas (Whitehead 1985a). Occasionally, this species is common as far north as Woods Hole, Mass., but usually occurs only as strays north of Cape Cod.

Occurrence in the Gulf of Maine. Adults are infrequently recorded from the Gulf of Maine. Specimens have been taken at Provincetown, Mass. (two specimens at MCZ); one round herring taken in the Yarmouth River, which empties into Casco Bay, and one in Casco Bay itself on 15 September 1924 (Bigelow and Schroeder). They were also reported from Jonesport and Eastport, Maine, in 1908 (Bigelow and Schroeder). In 1937 a number of specimens were taken at Campobello Island, at the mouth of Passamaquoddy Bay, in September, and in weirs at Campobello and Grand Manan islands in October (Leim 1937; McKenzie 1939). Along the coast of eastern Maine, in August-September 1953, something like 91,000 kg of round herring were landed (Scattergood 1953). Adults are reported infrequently from the Canadian Atlantic region, and these are almost always from the lower Bay of Fundy (including Passamaquoddy Bay) and southern Nova Scotia (Scott and Scott 1988). Some captures appear to have corresponded with periods of unusual warming.

Importance. Round herrings account for important fisheries in other regions of the world (primarily Japan, southern Africa, and the Red Sea). An evaluation of the fishery potential of round herring resources in the eastern Gulf of Mexico estimated stocks from 3.3×10^4 to 4.2×10^5 mt (Houde 1977). For the entire Gulf of Mexico, estimates (Reintjes 1980) range from 1.1×10^5 to 1.1×10^6 mt. The total catch worldwide for all species of round herrings in 1983 was 110,084 mt (Whitehead 1985a).

ATLANTIC THREAD HERRING / *Opisthonema oglinum* (LeSueur 1818) /

Bigelow and Schroeder 1953:112–113

Description. A rather thin fish, with body depth about 32–37% SL; belly sharp and saw-edged (Fig. 81). Small or moderate-size herringlike fish with a pelvic scute without ascending arms; abdominal scutes present before and behind the pelvic fins. Mouth terminal; lower jaw sometimes projecting slightly; teeth small, conical. Upper jaw rounded and not notched in

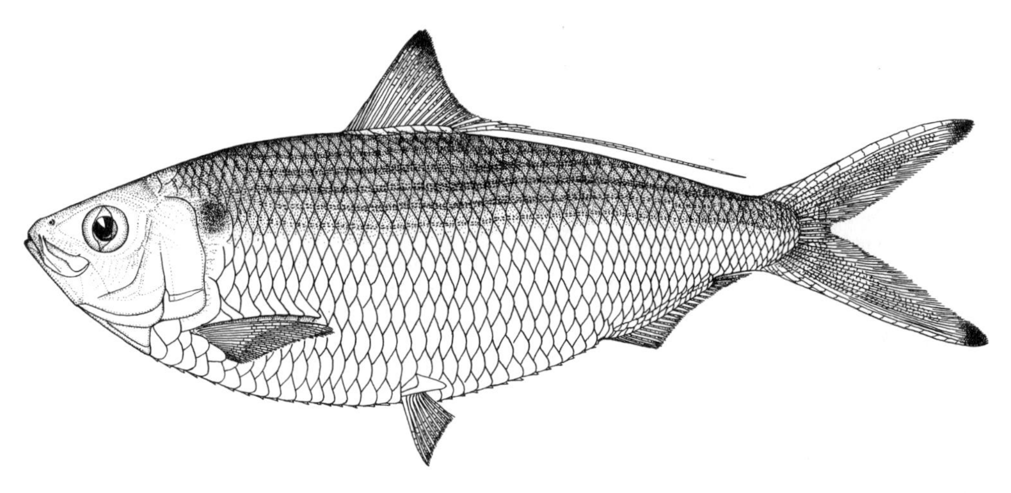

Figure 81. Atlantic thread herring *Opisthonema oglinum*. Beaufort, N.C., 220 mm TL. Drawn by A. S. Green.

anterior view; two supramaxilla present, the anterior usually elongate and the posterior paddle-shaped. Dorsal fin at about midpoint of body, short; posteriormost dorsal fin ray prolonged into a long filament. Anal fin short, its origin usually well posterior to vertical through base of posteriormost dorsal fin ray. Pelvic fins on abdomen under midpoint of dorsal fin. Bony capsule present in pterotic bone. Opercle smooth; gill rakers usually present on posterior face of third epibranchial; upper margin of ceratohyal smooth. Posterior border of gill opening with two distinct fleshy outgrowths. No hypomaxillary bone. Caudal fin deeply forked.

Meristics. Dorsal fin rays 19–22; anal fin rays 22–25; pelvic fin rays i, 7; scales in longitudinal series 32–35; lower gill rakers increasing with size, but stable at 28–46 (usually 30–37) after 8 cm SL (Whitehead 1985a).

Color. Bluish above, silvery on sides and belly. Scales along the dorsum with dark centers, forming longitudinal streaks, and with a faint dark spot just behind the dorsal margin of the gill cover. Dorsal and caudal fins black-tipped.

Size. Maximum length about 31 cm.

Distinctions. Thread herring are distinguishable at a glance from all other herrings inhabiting the Gulf of Maine by the prolonged last dorsal fin ray (usually about as long or longer than the body is deep). They resemble gizzard shad (*Dorosoma cepedianum*) of fresh and brackish waters in this respect, but the two differ rather conspicuously. In Atlantic thread herring, the mouth is terminal and the upper edge of the caudal fin is about one and a half times as long as the head (vs. mouth inferior and caudal fin only about as long as the head in the gizzard shad); the dorsal fin origin is slightly anterior to a vertical through the pelvic fin origin (vs. slightly posterior to this vertical in the gizzard shad); distance from pelvic fin origin to anal fin origin is at least one and a half times as long as the anal fin base (vs. only about 0.75–0.80 in gizzard shad); the anal fin is very low, with its first few rays slightly shorter than the diameter of the eye (vs. about one and a half times as long as the eye diameter in gizzard shad).

Habits. A coastal, pelagic, schooling species, which forms dense, surface schools (but solitary individuals reported) and probably does not enter low-salinity waters. Studies in the Gulf of Mexico report that Atlantic thread herring prefer "bluer" water, higher salinities, and higher water temperatures than menhaden (Finucane and Vaught 1986). Schools of Atlantic thread herring are exceptionally fast and agile (Butler 1961) and are more difficult to catch in purse seines than menhaden.

Food. Thread herring feed by filter-feeding plankton (copepods), but small fishes, crabs, and shrimps are also included in their diet.

Predators. Atlantic thread herring are important food for several coastal pelagic fishes including weakfish, bluefish, crevalle jack, and king and Spanish mackerels (Smith 1994; Rountree 1999). In ocean inlet areas of North Carolina, juveniles leaving estuarine waters in fall are fed upon voraciously by bluefish and Spanish mackerel (Smith 1994).

Reproduction. Spawning occurs in May-June off North Carolina (Hildebrand 1963c; Smith 1994), while spawning in the eastern Gulf of Mexico is protracted from April through September (Houde 1977). Richards et al. (1974) described egg and larval development, and Houde (1977) described aspects of early life history for this species in the eastern Gulf of Mexico.

Age and Growth. Age estimates using sagittal otoliths indicated that off North Carolina this species reaches ages to 8 years, with most fish being 4 years old or younger (Smith 1994). Mean fork lengths (in mm) at age were: 0+ = 76; age 1 = 155; age 2 = 172; age 3 = 178; age 4 = 180; age 5 = 183; age 6 = 180; age 7 = 186; and age 8 = 175.

General Range. Atlantic thread herring occur in the western Atlantic primarily in tropical and subtropical latitudes, including Bermuda, the Gulf of Mexico, the Caribbean and the West Indies, straying northward to Chesapeake Bay and occasionally to southern Massachusetts, and only rarely to the Gulf of Maine. They are not numerous north of Cape Hatteras (Hildebrand 1963c). They also occur in the South Atlantic to about Santa Catarina, Brazil.

Occurrence in the Gulf of Maine. Atlantic thread herring are occasionally caught off southern New England, and they were even reported as rather common in Buzzards Bay and Vineyard Sound during the summer of 1885 (Bigelow and Schroeder). The only record within the Gulf of Maine is that of a specimen about 17 cm long, taken off Monomoy Point, Cape Cod, in August 1931 (MacCoy 1931).

Movements. During the summer along the southeastern Atlantic coast, thread herring are ubiquitous in coastal waters from North Carolina to northern Florida to depths of about 9 m, whereas they concentrate off Georgia and north Florida during spring, fall, and winter (Wenner and Sedberry 1989). Tag recoveries indicate that schools migrate south along the southeastern Atlantic coast of the United States in the fall, traveling up to 11 km·day^{-1}.

Importance. Along the northern Gulf of Mexico and southeastern Atlantic coasts of the United States, Atlantic thread herring are harvested mainly with purse seines (Smith 1994). Most of the catch of coastal herrings, including this species, is sold as bait or processed into pet food (Finucane and Vaught 1986). Small numbers of thread herring are harvested off Florida as bait and, at present, commercial landings of this species for reduction to fish meal and fish oil are restricted to coastal waters between Cape Hatteras and Cape Fear, N.C., during late summer and fall (Smith 1994). Landings between 1965–1994 fluctuated widely, with an average of 1.9 million kg harvested annually. The Atlantic thread herring resource, especially the stock in the eastern Gulf of Mexico (Houde 1976), has been extolled as a latent fishery resource with estimates of population size off the southeastern United States ranging between 22,000 and 92,000 mt.

ACKNOWLEDGMENTS. Joseph W. Smith provided valuable information for the threadfin herring account. Drafts of the alewife and blueback accounts were reviewed by Gary Shepherd; the American shad and hickory shad accounts, by John F. Kocik; and the herring account, by William J. Overholtz.

SMELTS. ORDER OSMERIFORMES

Maxilla included in gape, adipose fin present or absent, radii absent from scales, basisphenoid and orbitosphenoid bones lost. Includes two suborders, 13 families, 74 genera, and some 236 species (Nelson 1994). Osmeriforms, except the Argentinoidei and *Osmerus eperlanus,* spawn in freshwater. One species from the suborder Argentinoidei and two species from the suborder Osmeroidei have been reported from the Gulf of Maine.

KEY TO GULF OF MAINE SMELTS

1a. Mouth small, not gaping back as far as eye; dorsal fin completely in front of pelvic fins; pelvic fin rays 12 or 13; scales large, 64–69 along lateral line (Argentinidae) . . . . **Argentine**
1b. Mouth large, gaping back as far as eye; dorsal fin above pelvic fin origin; pelvic fin rays 8; scales smaller, 62–205 along lateral line (Osmeridae) . . 2
2a. Teeth on jaws and tongue weak; dorsal fin rays 12–14; anal fin rays 19–21; scales in lateral series 175–205 . . . . **Capelin**
2b. Well-developed fanglike teeth on jaws and tongue; dorsal fin rays 8–11; anal fin rays 12–16; scales in lateral series fewer than 62–72 . . . . **Rainbow smelt**

ARGENTINES. FAMILY ARGENTINIDAE

Grace Klein-MacPhee

A small family of marine fishes with two genera and some 19 species (Nelson 1994). Eyes not tubular as in some other argentinoids; adipose fin over anal fin base. Among Gulf of Maine fishes, argentines are most closely related to smelts (Osmeridae). One species present in the Gulf of Maine.

ATLANTIC ARGENTINE / *Argentina silus* (Ascanius 1775) / Bigelow and Schroeder 1953:139–140

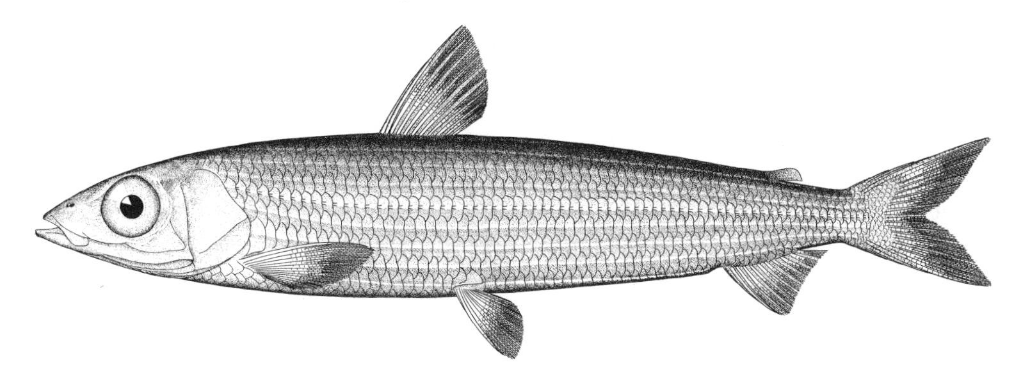

Figure 82. Atlantic argentine *Argentina silus.* Biddeford Pool, Maine. Drawn by H. L. Todd.

Description. Body slender, compressed (about one-fifth as deep as long), tapering toward both head and tail (Fig. 82). Sides flat, back and belly broad, making body nearly rectangular in cross section instead of oval. Head and snout pointed. Mouth small, terminal, not reaching to very large eye. Jaws toothless, but palate and tongue armed with small teeth. Dorsal fin short but high, inserted midway between snout and adipose fin; anal fin inserted below dorsal fin; pelvic fins abdominal, ventral, inserted behind dorsal; pectoral fins inserted just behind gill opening; adipose fin very small. Tail deeply forked. Scales large, usually with spines, a median row along back and belly. Lateral line straight.

Meristics. Dorsal fin rays 11–13; anal fin rays 11–17; pelvic fin rays 12 or 13; pectoral fin rays 15–18; gill rakers 13; branchiostegals 6; lateral line scales 64–69; pyloric caeca 21–28; vertebrae 65–67 (Scott and Scott 1988).

Color. The back is brownish or olivaceous, the sides silvery or with iridescent golden or brassy luster, and the belly white. The adipose fin is yellowish; swim bladder silvery.

Size. Maximum length of western Atlantic argentine is considered to be 45.7 cm, but Cohen (1964) reported that the largest specimen he examined was only 32.1 cm SL. Norwegian specimens have been reported to reach 49 cm.

Distinctions. Argentine resemble smelt and capelin but have much larger eyes; a much smaller mouth, not gaping back as far as the eye; and a dorsal fin that is completely in front of the pelvic fins, instead of above them. The scales are larger than those of smelt, only 64–69 rows along the lateral line.

Habits. Atlantic argentine are deepwater fish concentrated mainly along the continental shelf and in deepwater basins. On the Scotian Shelf, they were most often taken in depths of 183–256 m, temperature 7°–10°C, and mean salinity 34 ppt (Scott 1982b). The young frequent shallower waters, although rarely in depths less than 128 m. There is no information on seasonal movements. They form aggregations and are rarely captured singly (Cohen 1964).

Food. Food of Atlantic argentine on Georges Bank consisted primarily of euphausiids (particularly *Meganyctiphanes norvegica*), amphipods, arrowworms, and *Thermisto* (Emery and McCracken (1966), which agrees with dietary information from other authors (Cohen 1958; Borodulina 1964; Wood and Raitt 1968; Bowman et al. 2000). Squids, ctenophores, and fish remains were also found but these items occurred infrequently. Fish prey include a bristlemouth, *Cyclothone braueri* (Cohen 1964) and a pearlside, *Maurolicus weitzmanni* (Bowman et al. 2000). Most mature Atlantic argentine do not feed at spawning time (Emery and McCracken 1966). Feeding began following spawning and was more intensive in the fall than in the winter (Borodulina 1964).

Predators. An Atlantic argentine was reported from the stomach of a redfish, *Sebastes* sp. (Cohen 1964), one from a hake (Goode and Bean 1896), and two from white hake (Rountree 1999). Spiny dogfish, cod, silver hake, and goosefish have been found with prey identified only as Argentinidae and most of these are probably Atlantic argentine.

Parasites. Four species of trematodes, *Lecithophyllum botryophorum, Derogenes varicus, Hemiurus levinseni,* and *Lampritrema miescheri,* were found in Atlantic argentine from waters off Nova Scotia and Newfoundland (Scott 1969a,b,c). The intensity of parasite infestation is related to the vertical distribution and feeding of mature argentine on food organisms that are intermediate hosts of the parasites.

Reproduction. In western Atlantic waters, Atlantic argentine are sexually mature at age 6 or 7 (onset at 6, fully mature at 10) with a few maturing at 4 and 5 years (Emery and McCracken 1966). In Europe, maturation was estimated to be at 8–15 years, a third of the population first maturing at age 10. Median length at maturity for argentine on Georges Bank was 33 cm for males and 25 cm for females, with both sexes 100% mature by about 37 cm. Variability in length and age at which 50% of the individuals mature has been shown for argentine caught in different areas on the Scotian Shelf. From 1965 to 1969, median age at maturity was 5.3 years for males and 4.6 years for females (Beacham 1983a). Median length at maturity generally declined with time in recent years, possibly owing to exploitation, and was more pronounced in females than males. Males tended to mature at older ages, although maturity of both sexes was attained at similar lengths (Beacham 1983a).

Gonad condition suggests that spawning takes place on the Scotian Shelf in March and April (Emery and McCracken 1966). They spawn to some extent in the Gulf, for on 17 April 1920, a tow-net haul on *Albatross I* from 61 m in the southeastern part of the Gulf basin yielded 43 argentine eggs. Spawning in European waters occurs from April to July (Borodulina 1964). In the Skagerrak, eggs and larvae were found year-round with a peak in spring at depths of 150–550 m (Bergstad and Gordon 1994).

Fecundity increased from 9,300 eggs in 33-cm females to 28,400 eggs in 51-cm females (Keysler 1968). The relationship between fecundity and weight was 2,700 eggs per 100 g in weight, which is relatively low. In a sample of 50 specimens, a 37-mm female contained 10,000 eggs, while one 44.5 cm long had 38,599 eggs (Borodulina 1968). Diameter of ovarian eggs ranged from 1.6 to 2.8 mm. Atlantic argentine eggs are found chiefly in deeper water layers, seldom rising to the surface, and are among the largest of buoyant fish eggs (3–3.5 mm in diameter) containing a flattened oil globule (0.95–1.16 mm) and vacuolated yolk (Bigelow and Schroeder).

When hatched, the larvae, which are colorless, are 6–9 mm in length and have a large yolk sac, which is absorbed when they have grown to 12 mm. At this length, a line of spots appears along the belly. The fin rays are formed by the time the fish has reached 45 mm, the anus has moved forward, and the forked outline of the tail is apparent, but the pelvic fins do not appear until the larva is about 50 mm long. A few larvae have been taken at localities as widely separated as the offing of Mt. Desert Rock and the northwestern edge of Browns Bank. Cohen (1958) gave a detailed account of larval development.

Age and Growth. Growth of argentine from the Scotian Shelf was rapid during the first year, decreased markedly after the second year of life, and continued at a slower rate until maturity (Emery and McCracken 1966). Differing habitat conditions affect rate of growth (Borodulina 1968). Females grew faster and reached a larger size on the southwestern slope of the Grand Bank than on the southwestern region of the Scotian Shelf. Variability in size and age attained on the Scotian Shelf was also noted by Beacham (1983a).

The von Bertalanffy growth equations calculated from otolith data in the Scotian Shelf area (Zukowski 1972) are: $L_t = 37.9(1 - e^{-0.145(t+0.75)})$ for males; $L_t = 41.4(1 - e^{-0.129(t+0.75)})$ for females. Argentine grow rapidly up to 7–8 years of age and reach a maximum age of 29. Natural mortality was low and the coefficient of total mortality was −0.28; therefore an increase in fishing effort and excessive fishing intensity could rapidly reduce stock size (Zukowski 1972).

General Range. Widely dispersed on both sides of the North Atlantic Ocean in deep waters along the continental slopes, usually between 44–167 m. They are found from off southern Labrador east of Hamilton Inlet (44°38′ N, 43°38′ W), and the Nova Scotia Banks southward to the edge of Georges Bank (Cohen 1958). On the European coast, they range from northern Norway to the northern North Sea, with populations occurring off the Faroes and Iceland. Populations in Denmark Strait near the west coast of Greenland may be strays, not permanent populations (Cohen 1958).

Occurrence in the Gulf of Maine. Atlantic argentine were once considered rare in the Gulf of Maine but they are fairly common around the edges of Georges Bank and in the deeper waters of the Gulf (Map 10).

Importance. There is no U.S. or Canadian commercial fishery for Atlantic argentine and those captured are taken as a by-catch when fishing for demersal species. Surveys indicate that stock biomass remains stable and foreign fleets, particularly those of Japan and Russia, have taken considerable quantities. Decreased catches by foreign fleets are due to a large extent to the restrictions placed on the fishery by the Canadian and U.S. governments (Scott and Scott 1988).

Atlantic argentine are taken in trawls on offshore grounds of northwest Europe and the British Isles (Wheeler 1969). They are sometimes marketed but are not considered of prime commercial importance. There has been a small directed fishery for argentine in Iceland since 1991, which peaked in 1993 (Magnusson 1996). The experimental fishery showed that a commercial fishery will always be problematic because of the high percentage of other species, especially redfish, in the catch.

SMELTS. FAMILY OSMERIDAE

GRACE KLEIN-MACPHEE

A small family of small marine, freshwater, and anadromous fishes inhabiting coastal and inland waters. Distributed around the Northern Hemisphere in waters of the Atlantic, Arctic, and Pacific oceans, often occurring in large numbers. Seven genera with 13 species, including the Plecoglossidae (Nelson 1994). Body compressed, slender. Head pointed, mouth large, with well-developed (*Osmerus*) or weakly developed teeth (*Mallotus*). Branchiostegal rays 5–10; dorsal fin rays 7–14, adipose fin present; anal fin rays 14–23; pelvic fins abdominal; pelvic axillary process absent; pectoral fins low on the sides. Scales thin, cycloid. Lateral line present.

CAPELIN / *Mallotus villosus* (Müller 1776) / Bigelow and Schroeder 1953:134–135

Figure 83. Capelin *Mallotus villosus.* Grosswater Bay. Drawn by H. L. Todd.

Description. Body slender, elongate, and laterally compressed, of nearly uniform depth from gill cover to anal fin (Fig. 83). Head pointed; eye large; snout elongate. Mouth large, upper jaw gaping back to center of eye, terminal, oblique; tip of lower jaw protruding noticeably beyond upper. Minute teeth on jaws, tongue, vomer, palatine, maxilla, and premaxilla. Gill rakers long and slender. Dorsal fin located more than halfway back on body; anal fin low, with a longer base than dorsal fin; pectoral fins inserted just behind gill opening; pelvic fins abdominal with no axillary process; low long-based adipose fin over middle of anal fin; caudal fin deeply forked. Scales cycloid, tiny, deciduous. Lateral line complete and straight.

Meristics. Dorsal fin rays 12–14; anal fin rays 19–21; pectoral fin rays 18–20; pelvic fin rays 8; gill rakers 8–10 + 24–27; lateral line scales 175–205; pyloric caeca 5–8; vertebrae 62–69 (Scott and Scott 1988).

Color. Capelin are transparent to bottle green above and the sides are uniformly silvery below the lateral line. The scales are dotted at the margins with minute dusky specks; the belly is white. The back and head darken at spawning time.

Size. The largest capelin recorded was a 5-year-old female, 25.2 cm TL, weighing 59 g, captured in Trinity Bay, Nfld. (Winters 1970). Capelin are generally 13–20 cm long.

Distinctions. Capelin most closely resemble smelt but they are more slender, the body being only about one-sixth to one-seventh as deep and about one-twelfth as thick as it is long, and of nearly uniform depth from gill cover to anal fin, whereas smelt are usually deepest about midlength, which gives the two species different aspects. The scales are minute, much smaller than those of smelt and more numerous (about 200 per row on the sides), and the tongue fangs so characteristic of smelt are lacking. The outline of the adipose fin is low in capelin and about half as long as the anal, but short and high in smelt. The pectoral of capelin is broader, usually with 15 or more rays.

Capelin superficially resemble silversides but can be distinguished by the presence of an adipose fin instead of two dorsal fins; the position of the pectorals, which are abdominal in capelin and higher on the body in silversides; and the much larger mouth.

Taxonomic Note. Capelin occurring off the Pacific coast have been considered a separate species, *Mallotus catervarius,* or subspecies, *M. v. catervarius,* by authors such as Bigelow and Schroeder (1963) but Atlantic and Pacific forms are now considered conspecific (McAllister 1963).

Habits. Capelin are marine fish of cold, deep waters, found in the Atlantic Ocean on the offshore banks and in coastal areas, occasionally spending the winter and early spring months in deep bays off the east coast of Newfoundland (Scott and Scott 1988). Capelin are most in evidence during the spawning season, when they come inshore in multitudes along arctic-subarctic coasts. They are schooling fish, especially during the spawning season. Schooling behavior was noted in the laboratory 2 days posthatch, and time in schools increased with age (Morgan et al. 1995). Disturbance seemed to have little effect on schooling development. As larvae grew older, those in high concentrations spent more time in schools than those in low concentrations.

Food. Food consists of planktonic organisms. Euphausiids dominate by weight, although copepods, especially *Calanus,* occur in greater numbers and are more frequently found in stomachs. Amphipods and other planktonic invertebrates are also taken (Templeman 1948; Jangaard 1974; Vesin et al. 1981). Capelin also devour their own eggs (Bigelow and Schroeder). Feeding is seasonal, intensifying in later winter and early spring in the prespawning period, declining as the spawning season approaches, and virtually ceasing during spawning. Several weeks after spawning, surviving capelin commence feeding and continue until cessation in early winter (Winters 1969).

Predators. The role of capelin as food for other fishes, birds, and marine mammals is most significant and has been estimated to be on the order of 5 million tons annually at virgin population levels (Winters 1975). Ten fish species, five marine mammals, and nine marine bird species are known to forage extensively on capelin (Bailey et al. 1977).

Atlantic cod are probably their chief predator. They are followed inshore during the spawning migration by cod and haddock, which eat capelin and their eggs (Templeman 1965d, 1968). From June to early August, 98% of the food of Atlantic cod was capelin. Offshore, cod also feed heavily on capelin. They are also consumed in large quantities by haddock on the spawning grounds off Norway (Björke et al. 1972) and by Greenland halibut off southern Labrador and Norway (Bowering and Lilly 1992); they are also the chief food of the Atlantic salmon in the northwest Atlantic (Lear 1972). Capelin form an important part of the diet of shortfin squid, *Illex illecebrosus* (Dawe et al. 1997).

Juveniles and larvae often form part of the food of smaller fishes such as Atlantic herring. Many other fishes, such as dogfish, sculpins, eelpout, and flounders prey on them. Winter flounder feed heavily and systematically on capelin eggs, supporting up to 25% of their annual somatic growth from this food source (Frank and Leggett 1984). The amphipod *Calliopus laevisculus* may consume from 5 to 30% capelin eggs on the spawning beaches (Deblois and Leggett 1993).

Marine mammals, notably minke whale (*Balaenoptera acutorostrata*) (Sergeant 1963), fin whale (*B. physalus*), harp seal (*Phoca groenlandica*) (Sergeant 1973; Beck et al. 1993; Lawson and Stenson 1997), ringed seal (Lowry and Frost 1978), and harbor porpoises (*Phocoena phocoena*) (Fontaine et al. 1994) eat capelin as do seabirds such as the thick-billed murre (Gaston et al. 1985), common murre, Atlantic puffin (Piatt 1987), black-legged kittiwake (Regehr and Montevecchi 1997), and double-crested cormorant (Rail and Chapdelaine 1998). Capelin were found in 59.1% of harp seal stomachs and constituted more than 40% of the prey biomass (Lawson and Stenson 1997).

Parasites and Disease. Capelin appear to be lightly parasitized compared to other marine fishes (Palsson and Beverley-Burton 1983). A more recent study on capelin on the spawning grounds in Canada (Arthur et al. 1995) listed additional parasites: Protista, *Microsporidium* sp., *Trichodina* sp.; Monogenea, *Gyrodactyloides crenatus, G. petruschewskii;* Digenea, *Brachyphallus crenatus, Derogenes varicus, Hemiurus levinseni, Lecithaster gibbosus;* cestodes, *Bothrimonus sturionis,* Pseudophyllidea, *Scolex pleuronectes;* nematodes, *Ascarophis* sp., *Hysterothylacium aduncum, Pseudoterranova decipiens,* Spirurida, Nematoda; and Acanthocephala, *Echinothynchus laurentianus.*

Capelin larvae have been shown to be very sensitive to the toxic dinoflagellate *Protogonyaulax tamarensis* in the laboratory. At cell concentrations that compare to local bloom densities, mortality reached 92% in capelin larvae and postlarvae and was correlated with ingestion of the dinoflagellate cells (Gosselin et al. 1989).

Reproduction. The number of eggs increases with the size of the female, with large females producing up to 50,000 eggs (Scott and Scott 1988). Spawning begins in the third year of life, mass spawning taking place when fish are 3–4 years old.

Capelin exhibit pronounced sexual dimorphism. The male has much longer pectoral fins; the base of his anal fin is elevated on a pronounced hump, whereas it follows the general outline of the belly in the female. In males, the scales in one of the longitudinal rows immediately above the lateral line and in another row along each side of the belly are pointed, distinctly larger than the other scales, and become still longer at spawning time when each pushes up the skin in a fingerlike process, making males appear angular during breeding season.

Mating occurs on beaches and is most intensive during periods of intermediate tide, when gravid females and ripe males begin the spawning act. Spawning takes place at night or during times of heavy overcast and ceases during hours of sunlight. Rolling and swimming inshore near the crest of the waves bring the spawning fish onto the beaches, where the eggs are deposited and fertilized on the substrate. A male presses against the side of the female, and on many occasions a second male takes a position on the opposite side of the female. Spent fish, sometimes stranded, appeared immobile for a short period before reentering the water with the following waves. Mass mortalities often occur, leading to the mistaken belief that capelin may spawn only once, but repeated spawning of females has been noted (Templeman 1948). There is some evidence that males may be semelparous because all males collected postspawning and held in the laboratory died within 6 weeks, whereas over half the females survived up to 20 weeks and showed evidence of prior spawning and progressive development of oocytes for the next year's spawning. Spent males collected in the wild showed almost empty testes with no evidence of developing spermatocytes (Burton and Flynn 1998; Huse 1998).

Capelin spawn from May to August on the bottom, close below the tide line, many of them in the wash of waves on the beach; however, eggs have also been reported offshore in 46–49 m (Bigelow and Schroeder) and as deep as 80 m (Pitt 1958). Beach-spawning capelin spawn on coarse sand or fine gravel, where the eggs are buried by wave action and are presumably safe from flushing by tidal exchange and predation while they develop. Substrate characteristics of capelin spawning beaches are quite specific; for example, at Holyrood Beach, Nfld., preferred pebble diameter was 5–15 mm (Templeman 1948), whereas at Bryant's Cove, Nfld,, the preferred pebble size ranged from 1 to 4 mm (Frank and Leggett 1981a). Spawning takes place at water temperatures from 2.5° to 10.8°C (Frank and Leggett 1981b). During the spawning season, capelin are influenced by water temperatures and tidal conditions. Most capelin are found in the upper 20–30 m of water from the surface to the 5°C isotherm. Capelin abundance off eastern Newfoundland peaked 1–2 weeks after nearshore temperatures increased from 0° to above 6°C and at or near maximum tidal oscillation. Thus, low temperatures of bottom waters can reduce the volume of water column suitable for occupancy by capelin. Annual population variations corresponded to the volume of cold water (<0°C) and sea-ice transport by the Labrador Current (Methven and Piatt 1991).

The largest concentration of spawning areas in Canadian waters is from Newfoundland and Labrador south to the Gulf of St. Lawrence (Scott and Scott 1988). Presence of a spawning population on the Southeast Shoal of the Grand Bank was noted by Pitt (1958). This population may remain in deep water in this region throughout life, moving close to the bank to spawn. There are no reported spawning areas in the Gulf of Maine (Bigelow and Schroeder).

Early Life History. Capelin eggs are reddish, spherical, 0.97 mm (0.9–1.16 mm) in diameter with many tiny oil globules (Templeman 1948; Bigelow and Schroeder). Eggs are demersal, and adhesive, becoming attached to beach gravel or to bottom substrate, where they develop and hatch. They can be buried 15 cm or more beneath the surface of the beach (Frank and Leggett 1981b).

The specificity of beach spawning sites used by capelin results in dense concentrations of eggs (>800 cm^{-1}) in the beach gravel (Frank and Leggett 1981a, 1984). Egg mortality studies showed that mortalities vary annually with changes in biological, meteorological, and hydrological conditions. Lack of required oxygen resulting from egg density, water and air temperature effects, substrate characteristics, amount of rainfall, and accumulation of excretory products are hypothesized to affect egg development and egg mortality (Frank and Leggett 1981a).

Hatching takes place in beach sediments averaging 9–15 days in a high-tide zone and 22–24 days in a low-tide zone, with hatching inversely related to incubation temperature (Frank and Leggett 1981b).

A detailed description of larval development is given by Templeman (1948). Larvae are 5–7 mm at hatching, long, and very slender, with a preanal length 75% TL and 48–51 preanal myomeres present at hatching. The yolk sac may be absorbed prior to hatch. Pigment consists of a double row of ventral spots anterior to the yolk sac but posterior to the pectoral fin base, a very large spot over the anus, and a single row of spots from yolk sac to caudal fin base. A row of spots is added to each lateral surface dorsal to the gut as the larva grows. Flexion occurs at 11–16 mm TL; by 16 mm an adipose fin is present; by 40 mm TL all fin rays are present.

The larvae resemble those of herring, except that they are smaller at hatching, 6–7 mm vs. 7–10 mm, have a shorter gut (preanal length 75% TL vs. 83% TL in herring); have no median streak of pigment on the isthmus; and have an adipose fin in later development stages. They differ from argentine in having a larger mouth and more pigment. They differ from sand lance and rock gunnel larvae in having longer dorsal and anal fin bases and from anchovies by having dorsal and anal fins that do not overlap (Fahay 1983). They closely resemble smelt, but are found in more saline waters. Capelin are encountered so seldom in the Gulf of Maine that their larvae are not likely to be seen there.

Age and Growth. Rate of growth for Newfoundland capelin was calculated from otolith and scale readings by Templeman (1948). These figures correspond closely to those reported for Newfoundland and Grand Banks capelin (Sleggs 1933; Pitt 1958). Estimates of capelin age, and hence growth, may be biased as a result of mistaken interpretation of a metamorphic check as a true annulus (Bailey et al. 1977). Larval growth rates average 25 mm·day^{-1} (Jacquaz et al. 1977; C. Taggart, unpubl. data in Scott and Scott 1988).

Capelin exhibit sexual dimorphism in size. Males grow faster than females until they reach maturity, after which time their rates of growth are approximately the same. There is a general north-south cline. Those from the Grand Banks and southern Newfoundland areas grow more quickly than those from the Labrador area, until similar maximum size is attained. Labrador capelin tend to mature 1 year later than those from the Grand Bank (Winters and Carscadden 1978).

The majority of capelin do not live longer than 5 years, but in Greenland, where the growth rate is slower, 7-year-old fish are known (Hansen 1943).

General Range. Capelin are circumpolar in distribution. They occur on the east coast of North America from Hudson Bay to Nova Scotia and the Grand Banks; greater quantities occur off the shores of Newfoundland and Labrador, but they extend infrequently as far as Cape Cod. The northernmost record in the western Atlantic was in James Bay at 51°53′ N, 80°45′ W (Zalewski and Weir 1981). They frequent eastern Atlantic waters from northern Russia and Finnmark south to Trondheim Fjord off western Norway, and are found sporadically in the White and Kara seas, throughout the Barents Sea from Novaya Zemlya to Spitsbergen and Bear Island. In the western Pacific, they extend south from Cape Barrow, Alaska, around the Bering Sea, and along the coast to Juan de Fuca Strait. Along the Asian coast they occur from the Sea of Okhotsk south to Hokkaido, Japan, and Tumen River, Korea (Winters 1969).

Occurrence in the Gulf of Maine. Capelin are sub-Arctic fish that visit the Gulf of Maine occasionally. Apparently a period of this sort occurred about the middle of the nineteenth century, for Perley, writing in 1852, reported them from a number of points in the neighborhood of Saint John, N.B. They then seem to have disappeared from the Gulf of Maine until 1903, when they were common in the Bay of Fundy in May. A few were again taken off Passamaquoddy Bay in May 1915, which was the prelude to a period of local abundance from 1916 to 1917 and another in 1919 off Passamaquoddy Bay and the Penobscot River (Bigelow and Schroeder). They were reportedly caught in great abundance in herring weirs in the Passamaquoddy Bay region of the Bay of Fundy from 23 March to 12 May 1965 (Tibbo and Humphreys 1966).

Migration. An intensive migration inshore by coastal populations takes place prior to spawning activities on beaches (Templeman and Fleming 1962). The migration occurs from June to August depending on latitude, tides, winds, and water temperature. After spawning the adults move out to sea again (Templeman 1948; Frank and Leggett 1981a,b).

Importance. Capelin seldom occur in the Gulf of Maine and have no commercial or recreational value there, but they are an important commercial species in Canada, used in the early part of the nineteenth century as a fertilizer, as bait for the cod line fishery, and as dog food. Although most of the world's supply is now being processed into fish meal and oil, capelin and capelin products are considered high in nutritional value. They are relished by the Japanese when dried and smoked and by the Greenlanders when dried and used during the winter months. Their use is also growing in significance as food for cage-reared fishes such as trout and salmon. Capelin stocks in the Canadian Atlantic area have received increased attention since about 1980 because they are a valuable resource not only in terms of landed value, but because the decline in stocks of capelin is of great concern to the cod fishery since capelin is such an important prey of the Atlantic cod (Scott and Scott 1988).

RAINBOW SMELT / *Osmerus mordax* (Mitchill 1814) / Saltwater Smelt, Smelt, Atlantic Smelt /

Bigelow and Schroeder 1953:135–139

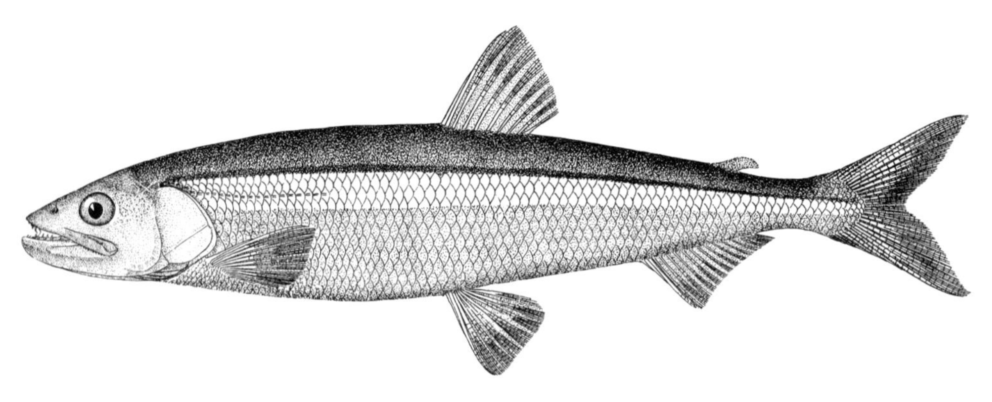

Figure 84. Rainbow smelt *Osmerus mordax*. Woods Hole. Drawn by H. L. Todd.

Description. Body slender, elongate, laterally compressed; depth about five times in SL; back broadly rounded, laterally flattened enough to be egg-shaped in cross section (Fig. 84). Head moderately long; snout elongate, pointed. Mouth terminal, large, gaping behind eye; lower jaw protruding, many teeth on vomer, palatine, pterygoid, basibranchial, dentary, maxilla, premaxilla, and tongue, especially large on tongue and front of vomer. Eye moderately large. Dorsal fin about midbody, above pelvic fins; pectoral fins low on sides; pelvic fins abdominal and fanlike; well-developed adipose fin present; caudal fin deeply forked. Scales cycloid, thin, deciduous. Lateral line straight and complete.

Meristics. Dorsal fin rays 8–11; anal fin rays 12–16; pelvic fin rays 8; pectoral fin rays 11–14; gill rakers 8–11 + 18–24; branchiostegals usually 7; lateral line scales 62–72; vertebrae 58–70; pyloric caeca 4–8 (Scott and Scott 1988).

Color. Transparent olive to bottle green above; the sides are a paler cast of the same hue but with a broad longitudinal silvery band with pink, blue, and purple iridescent reflections on sides of freshly caught fish (hence, rainbow smelt). The belly is silvery, whereas the fins and body are more or less flecked with tiny dusky dots (Scott and Scott 1988). Scott and Crossman (1973) compare color of anadromous and landlocked populations.

Size. Maximum recorded length is 35.6 cm (Scott and Scott 1988) but most smelt average 18–23 cm.

Distinctions. Rainbow smelt are distinguishable from all other fishes in the Gulf of Maine by their slender form combined with a long pointed head, large mouth with many teeth, a small but evident adipose fin above the rear part of the anal fin, and a deeply forked tail. The dorsal fin is positioned above the pelvic fins instead of in front of them, and the much larger mouth and small eye separate them from argentine. The large, fanglike teeth on the tongue, larger scales, narrower-based adipose fin, lower jaw projecting only slightly beyond the upper, and deciduous scales separate them from capelin. The color pattern is similar to silversides, but they have two distinct dorsal fins.

Taxonomic Note. Rainbow smelt of the northwest Atlantic are considered to be distinct from European smelt, *Osmerus eperlanus* (Linnaeus), rather than a subspecies of the latter (Bigelow and Schroeder 1963). Some authors have recognized smelt of the Arctic and North Pacific as a separate species or subspecies, *Osmerus mordax dentex,* but we follow McAllister (1963) in recognizing only a single species.

Habits. Rainbow smelt are pelagic and anadromous, usually found in coastal waters. Life history information on rainbow smelt was summarized by Buckley (1989). Many smelt spend the whole year in estuarine situations. Landlocked populations of smelt, as many anadromous species, can live successfully in freshwater. Their summer habitat varies in different parts of the Gulf of Maine, depending on summer temperature of the water and perhaps on the food supply. Most of them desert the harbors and estuaries of the Massachusetts Bay region and the southern coast of Maine during the warmest season, but they probably move out only far enough to find cooler water at a slightly greater depth, and a few may be found in harbors throughout the summer. East of Penobscot Bay, where the surface temperature does not rise as high as off Massachusetts, smelt are found in the harbors, bays, and river mouths all summer. Smelt overwinter in nearshore locations, where they encounter subzero seawater temperatures (McKenzie 1964b), and antifreeze activity has been detected in their blood (Duman and De Vries 1974a). Antifreeze proteins have been characterized as cystine-rich AFP type II, similar to those found in sea raven (Ewart and Fletcher 1990). They also produce glycerol in their livers as antifreeze (Driedzic et al. 1998), and synthesize TMO (trimethylamine oxide) and urea, which produces a small but significant depression of the freezing point of body fluids of winter-acclimatized smelt (Driedzic et al. 1998).

Smelt school and are often seen in large shoals in the estuaries and rivers during spawning migrations. Young smelt certainly, and old ones probably, travel in schools that are mostly composed of fish of a similar size. Smelt exhibit diurnal vertical migration with a change in aggregation patterns (Appenzeller and Leggett 1995). Smelt in a Canadian lake were typically concentrated in the upper 10–15 m during daytime and in deeper waters (30–40 m) at night. They dispersed at night but formed dense schools during the day. The pattern of vertical migration was modified by water temperature. In colder waters (<18°C), the upper boundary of migration approached the surface, whereas at temperatures above 18°C, smelt aggregated just below the thermocline.

Food. Rainbow smelt are carnivorous and voracious. Amphipods, euphausiids, mysids, shrimps, and marine worms, and, as they grow, any available small fishes (silverside, mummichog, and herring) make up the bulk of their food (Scott and Scott 1988). Smelt taken in the Sheepscot River, Maine, in May were packed full of young herring, while cunner, anchovy, sand lance, sticklebacks, silversides, and alewife have been identified from smelt stomachs at Woods Hole. The diet at Woods Hole also includes shellfish, squid, annelid worms (*Nereis*), and crabs (Bigelow and Schroeder). Stomach analyses indicate that although smelt are suspected of consuming large quantities of small fishes, these usually do not constitute a great proportion of their food (Scott and Crossman 1973). Smaller individuals are selective feeders. Indications are that smelt do not feed heavily immediately before spawning. In freshwater habitats, especially in the Great Lakes, the primary food is opossum shrimp (*Mysis relicta*).

Predators. Rainbow smelt are a source of food for many larger species of fishes and for aquatic birds such as mergansers, cormorants, gulls, terns, and, as postspawning mortality, crows. They are also preyed upon by their own species. In the sea, smelt are eaten by Atlantic cod and Atlantic salmon, and in coastal regions are taken by large brook trout, striped bass, bluefish, and seals (Clayton et al. 1978). Smelt eggs may be eaten by sticklebacks and trout (Baird 1967). In freshwater, smelt are a prime food supply of lake trout and are also eaten by burbot, walleye, and perch (Scott and Scott 1988). Landlocked salmon in Maine lakes (Rupp 1968; Sayers et al. 1989) and Canadian coastal waters consume large quantities of smelt.

Parasites and Disease. In marine and freshwater areas, smelt are host to many parasites. Those reported from the Atlantic include trematodes *Brachyphallus crenatus, Derogenes varicus,* and *Hemiurus appendiculatus;* cestodes *Proteocephalus* sp. (Margolis and Arthur 1979) and *Diphyllobothrium sebago* (Frèchet et al. 1983); nematodes *Phocanema decipiens* and *Pseudoterranova decipiens* (Appleton and Burt 1991); acanthocephalans *Leptorhynchoides thecatus* in the musculature, viscera, body cavity, and mesenteries (Margolis and Arthur 1979) and *Echinorhynchus salmonis* (Frèchet et al. 1983); and a copepod, *Ergasilus funduli,* which attacks the gills. The microsporidian *Glugea hertwigi* was reported from rainbow smelt in coastal waters from Massachusetts to the Canadian Maritimes (Sherburne and Bean (1979) and in freshwater. An acute epidemic of this parasite occurred in Loon Pond, N.H. (Haley 1952): 23% of smelt examined from the Great Bay region, N.H., were infected (Haley 1954), and there was an infestation in the Bras d'Or Lakes, N.S., in 1971 (Scott and Scott 1988). In 1970 a heavy mortality of parasitized rainbow smelt in Loch Lomond, N.B., was caused by a parasite of the genus *Dermocystidium,* the first record of this parasite in North America. The parasite heavily infected tissues of fins, operculum, head, and cornea of the smelt (Scott and Scott 1988).

Rainbow smelt populations from the Canadian Provinces to Massachusetts are also infected by piscine erythrocytic necrosis (PEN), a viral disease. This infection has a high incidence of occurrence within populations, but at low levels in individuals (Sherburne and Bean 1979; Jimenez et al. 1982).

Epidermal tumors, probably of viral origin, have been described in rainbow smelt in freshwater during the spring spawning run in Nova Scotia. These tumors apparently progress from small preneoplastic lesions to papillomas, and to carcinomas in some males (Morrison and MacDonald 1995). Epithelial skin tumors thought to be caused by herpeslike virus particles occurred in landlocked smelt populations in several New Hampshire lakes during spawning runs. Lesions occurred in both males and females. Prevalence varied annually, reaching as high as 30% (Herman et al. 1997).

A freshwater population of rainbow smelt from a pond in Maine showed incidences of enzootic squamous cell carcinoma, which occurred in the jaws of males and was destructively invasive (Harman 1988).

Reproduction. A female smelt can produce 7,000–75,570 eggs (Chen 1970; Clayton 1976). The larger the female the greater the number of eggs produced. Estimated fecundity for size ranged from 33,373 eggs at 160 mm FL to 75,570 eggs at 229 mm FL for anadromous fish collected in Grande Rivière, Gaspè, Quebec. The relationship between fecundity and fork length (mm) was: $\log_{10} F = -2.0043 + 2.9090 \log_{10} L$ (Chen 1970).

Rainbow smelt usually begin spawning by age 2, but this shows clinal variation along the east coast, increasing with latitude. Age-1 males and females comprised 26% of the spawning run in the Parker River, Mass. (Murawski and Cole 1978), and 15% in the Jones River, Mass. (Lawton et al. 1990), whereas only a few precocious age-1 males were found occasionally on the spawning grounds in Canada (McKenzie 1964b) and New Hampshire (Warfel et al. 1943). In the Miramichi River, N.B., all spawners were age 2 and older (McKenzie 1964b). An introduced population in Lake Superior was not fully recruited until age 3 (Bailey 1964). Early maturation in the southern part of the range may be related to growth as maturity appears related to size.

Spawning takes place in spring, and timing may be related to increasing photoperiod, water temperature, and time of ice breakup (Rupp 1959). Spawning can begin in late February in southern Massachusetts coastal streams (Crestin 1973) and mid-March farther north when water temperature rises to about 4°C (Lawton et al. 1990). Smelt spawn at water temperatures of 4°–9°C with the exception of the Miramichi estuary, N.B., where spawning may occur up to 15°C (McKenzie 1964b), and some freshwater populations, where it may occur up to 18°C (Jilek et al. 1979). Maximum numbers of fish occur in the run about 1–2 weeks after it begins, and the peak may last for about 1 week (Murawski 1976; Clayton 1976). East of Portland, smelt seldom begin to run before April and continue through May. In the colder streams on the southern shores of the Gulf of St. Lawrence they do not spawn until June (Bigelow and Schroeder).

There is sexual dimorphism during the breeding season. Small epidermal breeding tubercles, like sandpaper to the touch, develop extensively over head, body, and fins on males, seldom on females. Breeding tubercles probably help maintain contact during spawning.

Males appear first on the spawning grounds, followed by females, generally at night, when spawning occurs (McKenzie 1964b). Fish move up from estuarine areas to the spawning grounds, which are characterized by rock, rubble, sand, and gravel substrates and fast-flowing freshwater (Clayton 1976; Hulbert 1974). As a rule smelt do not journey far upstream; many, indeed, go only a few hundred yards above tidewater, whether the stream is small or large. Smelt cannot ascend falls or other obstructions higher than 0.5 m. Some spawn in slightly brackish water in certain ponds behind barrier beaches, but flooding with salt water, which sometimes happens, kills the eggs (Bigelow and Schroeder). Usually several males attend one female during spawning (Clayton 1976). Sex ratios are highly skewed to females on the spawning grounds (Murawski et al. 1980; Lawton et al. 1990). Males may utilize spawning areas for up to 14 days and females for 1–3 days (Murawski 1976). Individual fish may spawn in several streams in an estuary during the spawning period, and this appears to be related to the distance between streams (Rupp 1968; Murawski et al. 1980; Frèchet et al. 1983). Adult smelt return to saltwater after spawning to spend the summer either in the estuary into which the stream they spawned in empties or in the sea close by. On the Massachusetts coast north of Cape Cod all spent fish leave freshwater by the middle of May, earlier in some years. On the Maine coast, too, a good proportion of the spent fish are in saltwater by the first weeks in May (Bigelow and Schroeder).

Early Life History. Smelt eggs average about 1 mm (0.9–1.2 mm) in diameter; they sink to the bottom, where they stick in clusters to pebbles, to each other, or to any stick, root, grass, or water weed they chance to touch (Bigelow and Schroeder). They become pedicellate after water-hardening (Clayton 1976).

Times of incubation at various temperatures are as follows (McKenzie 1964b): 51–63 days, 3°–9°C; 29 days, 6°–7°C; 25 days, 7°–8°C; 19 days, 9°–10°C; 16–21 days, 11.7°C; 11 days, 12.2°C; 8–10 days, 20°C.

Water velocity, substrate type, and egg density appear to be important factors in survival (Sutter 1980). Obstructions in spawning streams may prevent upstream spawning migration. Below these obstructions, heavy egg mortality often occurs when the bottoms of streams become carpeted with large patches of eggs and fungal infections occur (McKenzie 1964b). If the obstructions are sufficiently far downstream, saltwater intrusion may kill the eggs since salinities greater than 12% are fatal to smelt eggs (Baird 1967).

Smelt larvae are about 5 mm in length when hatched. They are negatively phototactic (Rupp 1965). Newly hatched larvae are long and slender with a single oil globule located near the anterior end of the yolk sac, pectoral buds are present, and the mouth is well formed (Dorr et al. 1976; Cooper 1978). The body is transparent with two rows of pigment on the ventral side extending from the pectoral buds to the anterior end of the yolk sac, a single row of pigment on the midventral line between the posterior end of the yolk and the anus, and several widely spaced, elongate chromatophores on the ventral side between the anus and the base of the caudal fin. Yolk-sac absorption occurs by about 7 mm. The paired rows of melanophores become a single row between the pelvic buds and the anus. By 22–33 mm, anal and caudal fin rays are complete; the swim bladder is distinct, forcing the gut downward anterior to the pelvic buds; the adipose fin is developed; and teeth are present on both jaws and tongue (Cooper 1978). Pigment consists of numerous melanophores on the dorsal part of the swim bladder and melanophores on the caudal fin arranged in lines following individual rays. The full complement of fin rays is formed between 35 and 45 mm (Cooper 1978). After hatching, larvae are carried downstream to brackish water, where they are found early in May. Retention of larvae within an estuary may result from passive mechanisms utilizing the landward moving salt wedge (Krochmal 1949) or by active tidal vertical migrations in well-mixed estuaries (Laprise and Dodson 1989b). Later, larvae become more concentrated in bottom waters (McKenzie 1964b; Crestin 1973; Clayton 1976). Juvenile smelt also occurred in eelgrass beds in the Weweantic River (Crestin 1973). As water temperatures drop in the fall, young-of-the-year fish move into the upper estuarine areas, concentrating in channels, and intermix with schools of adult smelt (McKenzie 1964b; Clayton 1976). They return to the sea sometime during their first year of life.

Larvae eat rotifers (McCullough and Stanley 1981), copepods and other planktonic forms; in the Great Lakes they eat dipteran larvae and crustaceans (Gordon 1961; Burbidge 1969).

Smelt were artificially propagated and large numbers of larvae were produced yearly in Massachusetts and Long Island, N.Y., in the 1950s. It was claimed to be possible to reestablish smelt by introducing the eggs or larvae into streams from which they had been extirpated (Bigelow and Schroeder).

Age and Growth. Larvae that hatch early grow to a length of 6.3 cm by November. The von Bertalanffy growth equation for the first year for rainbow smelt in the Parker River, Mass., was $TL = 102.14(1 - e^{-2.7769(t - 0.0673)})$, where TL is the total length in mm and t is the time in years (Clayton 1976). Subsequent growth is rapid in saltwater and mature 2- and 3-year-old smelt, 12.7–20.3 cm in length, make up the major part of the commercial catch, but individuals 4 and 5 years old are also caught. Growth in freshwater is also fairly rapid. Young may be 20–40 mm long in a few months and 51 mm long by August (Scott and Crossman 1973). Age can be determined by examination of scales. Size attained varies from location to location. Saltwater forms reach greater lengths than most inland forms and McKenzie (1964b) reported lengths of 18.0 cm for 5-year-old males and 20.6 cm for females of the same age. Occasional specimens measuring 35.6 cm have been taken in Maritime coastal waters and in Lake Ontario. Females grow faster, live longer, and grow larger than males.

General Range. Rainbow smelt occur on both the Atlantic and Pacific coasts. Along the North American Atlantic coast, they are found in suitable bays and estuaries from the northern limit in the Hamilton Inlet–Lake Melville estuary of Labrador southward to New Jersey (Scott and Crossman 1973). Their former range extended to the head of Delaware Bay (Robins and Ray 1986) and Virginia (Bigelow and Schroeder). The centers of abundance of anadromous smelt are the southern Maritime Provinces and Maine (Clayton et al. 1978). Throughout their range along the Atlantic coast, smelt are normally anadromous, although landlocked populations occur naturally in many freshwater lakes and ponds in New Hampshire and in Maine, in Lake Champlain, and in various Canadian lakes. This dispersal appears to be closely associated with areas inundated by postglacial marine waters from post-Wisconsin glacial lakes and their outlets (Dadswell 1975). The species was introduced into Crystal Lake, Mich., in the 1920s and has subsequently spread throughout the Great Lakes (Scott and Crossman 1973) and into the Mississippi-Missouri drainage (Burr and Mayden 1980). Captures of rainbow smelt from the lower Nelson River represent a range extension in the Hudson Bay drainage (Remnant et al. 1997). A summary of range extensions and possible effects of smelt in the Hudson Bay drainage is given by Franzin et al. (1994).

Occurrence in the Gulf of Maine. Smelt are familiar fish around the entire coast of the Gulf of Maine, but vary greatly in abundance from place to place according to the accessibility of streams suitable for spawning, from which they seldom wander far from shore. Spawning populations are found in Massachusetts waters from the Palmer River to the Merrimac (Clayton et al. 1978); northward and eastward all along the coast of Maine, in the region of Passamaquoddy Bay, and more so along the western shore of Nova Scotia; but they are less plentiful passing inward along the Nova Scotia shore of the Bay of Fundy (Bigelow and Schroeder).

Migrations. Although there is evidence of migrations in the sea, little is known of this period of rainbow smelt life history. One smelt, tagged at Portage Island in Miramichi Bay, N.B., was caught 161 km away in upper Chaleur Bay, but such information is rare. Rainbow smelt enter estuaries in late fall and winter months, avoiding cold waters (McKenzie 1964b). Smelt, like alewife, shad, and salmon, grow in saltwater, but run up into freshwater to spawn. Adult smelt gather in harbors and brackish estuaries in early autumn. The schools then tend to move into the smaller harbors on the flood tide and out again on the ebb, especially if the tidal current is strong. Some smelt remain over the ebb in the deeper basins, and some have run as far as the head of the tide by the time the first ice forms in December. Most winter between harbor mouths and brackish water farther up; maturing fish commence their spawning migration into freshwater as early in the spring as the ice goes out of the streams (Bigelow and Schroeder). In the Gulf of St. Lawrence, smelt arrive in all areas at the same time and the runs are relatively short. Timing of the runs did not correlate with environmental data, suggesting that cues for migration might be photoperiod rather than temperature or water flow (Chadwick and Claytor 1989).

Importance. Rainbow smelt are caught in both commercial and sport fisheries. Along the coast and during spawning migrations in rivers and estuaries, they are eagerly sought and are an excellent food fish. There are three types of recreational fishing in Massachusetts: fall hook-and-line sport fishery in most coastal areas; a winter ice fishery in many coastal estuaries of the Gulf of Maine, particularly north of Boston; and a dipnetting fishery restricted to the Weweantic River (Clayton et al. 1978).

Commercially, smelt have supported a successful fishery for over 100 years in Canada. In the eastern Canadian provinces the major fishery is located in the Gulf of St. Lawrence region, N.B., especially in the Miramichi River estuary, where they are caught through the ice in box nets (McKenzie 1964b). There is significant by-catch of white hake, striped bass, and winter flounder in the autumn "open water" fishery (Bradford et al. 1997). At times, smelt have supported a substantial fishery in Newfoundland. Fish are taken in bag nets, gill nets, seines, and, when abundant, by angling. Inland, a successful smelt fishery, with otter trawls, developed in the late 1950s and early 1960s in the Canadian waters of the Great Lakes, particularly in Lake Erie.

Smelt are not as numerous as they were even 50 years ago around the Massachusetts shoreline of the Gulf of Maine, where various streams either have been closed to them or have been rendered uninhabitable by pollution, and there is no commercial fishery for them now (Clayton et al. 1978). Acid rain is likely to be detrimental to egg survival, and this may explain the disappearance of runs throughout the Gulf of Maine (S.A. Murawski, pers. comm., 30 Dec. 1994).

Enough still remain to provide sport for thousands of anglers, and smelt have great recreational value, smelt fishing being a favorite pastime for home consumption. This applies equally to many localities along the coast of Maine and in Great Bay, N.H. Smelt provide much pleasure to those seining at night when a run occurs. They are caught at other times by hook and line from docks and wharves throughout the area. This highly palatable fish is usually cooked whole. The characteristic odor, likened to that of freshly cut cucumber, is undoubtedly the feature responsible for the common name—smelt (Bigelow and Schroeder).

ACKNOWLEDGMENTS. Early drafts of the smelt account were reviewed by Steven A. Murawski and Karsten E. Hartel, and the argentine account, by Daniel Cohen.

SALMONS AND TROUTS. ORDER SALMONIFORMES

Recent views on the relationships of this order, now restricted to the single family Salmonidae, vary in placing it near the Osmeriformes (smelts and allies), Esociformes (pickerels), Ostariophysi, or neoteleosts (Nelson 1994).

SALMONS AND TROUTS. FAMILY SALMONIDAE

JOHN F. KOCIK AND KEVIN D. FRIEDLAND

Salmonids are soft-rayed fishes that have no spines in any of the fins, pelvic fins situated on the abdomen far behind the pectoral fins, and a fleshy rayless adipose fin posterior to a rayed dorsal fin. Other characteristics are dorsal rays fewer than 16; scales small with 115–200 along the lateral line; teeth present on the jaws and vomer (Nelson 1994). Presence of an adipose fin and its location separate salmonids from other Gulf of Maine fishes except for Osmeriformes (rainbow smelt, capelin, and Atlantic argentine) and some deep-sea fishes, lanternfishes (Myctophidae), viperfish (Stomiidae), and lancetfish (Alepisauridae). Absence of luminescent organs distinguishes them from Stomiiformes. Blunt snouts, stout bodies, and nearly square tails of salmonids distinguish them at a glance from the sharper-nosed, slender, forked-tailed Argentinidae and Osmeridae.

The Salmonidae includes three groups, herein considered as subfamilies: Salmoninae (salmon, trout, and char), Coregoninae (whitefishes), and Thymallinae (grayling) with approximately 66 species in 11 genera (Nelson 1994). Salmonids are indigenous to the Arctic and north temperate zones of the Northern Hemisphere. Many species are commercially and recreationally important, which has led to their introduction and establishment in waters throughout the world. About a third of the species are anadromous, spawning in freshwater but moving to the sea to grow and mature. Five salmonids (all Salmoninae) can be considered common or occasional inhabitants of the Gulf of Maine. Atlantic salmon and brook trout are native to tributaries to the Gulf of Maine and utilize the Gulf during feeding migrations. Coho salmon, pink salmon, and rainbow trout have been introduced from Pacific drainages and are encountered only rarely in this region. In addition, Chinook salmon (*Oncorhynchus tshawytscha*) and brown trout (*Salmo trutta*) have also been stocked in rivers draining into the Gulf of Maine but are extremely rare in the open waters of the Gulf, and so are not included in this section.

KEY TO GULF OF MAINE SALMONS AND TROUTS

1a. Scales so small that they are hardly visible; back with vermiculate markings; teeth on midline of roof of mouth confined to anterior of boat-shaped vomer (not shaft). . . . . **Brook trout**

1b. Scales large enough to be easily visible; back without vermiculate markings; a row of teeth runs back along the anterior and shaft of flat vomer **2**

2a. Anal fin with 8–11 rays **3**

2b. Anal fin with 13–19 rays. **4**

3a. Caudal fin distinctly marked with radiating rows of black spots; body never with red spots; adipose fin often with a black margin; scale rows 120–180 . . . . **Rainbow trout**

3b. Caudal fin usually unspotted, never with regular rows of black spots; reddish spots sometimes on body; scale rows usually 110–130 . . . . **Atlantic salmon**

4a. Back and lower half of caudal fin, as well as its upper half, conspicuously marked with large black spots, largest as large as eye; scales small, 166–229 in row above lateral line; total gill rakers 24–35. . . . . **Pink salmon**

4b. Back with very small black spots, largest as large as pupil of eye, or none at all; no black spots on lower half of caudal fin; scales moderate, fewer than 154 in row above lateral line; total gill rakers 18–25. . **Coho salmon**

PINK SALMON / *Oncorhynchus gorbuscha* (Walbaum 1792) / Humpback, Humpbacked Salmon /

Bigelow and Schroeder 1953:131–133

Figure 85. Pink salmon *Oncorhynchus gorbuscha*.

Description. Pink salmon have the familiar salmon shape while at sea, but are stouter-bodied than Atlantic salmon (Fig. 85). The head is naked but the body is covered with scales large enough to be seen easily. The dorsal fin is located about midway along the body, above the pelvic fins, and the adipose fin is over the end of the anal fin. Male pink salmon (like all Pacific salmons, and Atlantic salmon to a lesser degree) undergo a remarkable change in form prior to spawning, with the body deepening and the development of a prominent hump in front of the dorsal fin. In addition, the jaws elongate, and become hooked at the tip, and the teeth increase in size.

Meristics. Dorsal fin rays 14; anal fin rays 13–17; pelvic fin rays 11; pectoral fin rays 16; scales small, 166–229 in row above lateral line; total gill rakers 24–35; vertebrae 68 (Scott and Scott 1988).

Color. The back and tail of pink salmon are bottle green with faint black spots while it is in the sea. These spots are particularly conspicuous on the tail, where they are oval in outline and as much as 8 mm in diameter. The spots are among the distinctive marks by which pink salmon can be distinguished from other salmonids. The sides and belly are silvery, with a faint pinkish tinge. Pink salmon fry are unique among salmonids in having practically adult coloration.

Size. Pink salmon are the smallest of the Pacific salmons, much smaller than Atlantic salmon, adults averaging only about 1.4 kg and 50–60 cm FL. Males are typically larger and can reach a weight exceeding 5 kg. The all-tackle game fish record is a 5.94-kg pink salmon caught in the St. Mary's River, Ontario, in September 1992 (IGFA 2001).

Distinctions. Pink salmon resemble Atlantic salmon closely but have 13 or more anal fin rays, whereas Atlantic salmon have only 8–11 rays.

Habits. Pink salmon are semelparous; that is, they spawn only once. They ascend rivers to spawn from August through October and die a week or two after spawning. Pink salmon have the least-variable and marine-dependent life history among the salmonids. Eggs hatch in late winter and fry emerge from their redds and out-migrate, without an extended freshwater residence, during May and June. Pink salmon typically reside in the ocean for 15–16 months. This life-history pattern results in genetically isolated even- and odd-year runs.

General Range. The native range is the Pacific coast of North America from central California to the Mackenzie River, Canada, which flows into the Beaufort Sea, and in Asia from North Korea to the Lena River, Russia, which flows into the Arctic Ocean.

Occurrence in the Gulf of Maine. The history of early attempts to introduce this west coast salmon into New England was summarized by Neave (1965) and Ricker (1972). Substantial adult returns were noted in three Maine rivers in the late 1910s. Natural reproduction sustained these populations for at least four generations. Bigelow and Schroeder reported that few had been seen since 1927. Later attempts (1959–1966) to introduce pink salmon into the Maritime Provinces resulted in some returns and subsequent natural reproduction. It was thought that these populations were extirpated by 1977 (Scott and Scott 1988; Heard 1991), but a ripe male of unknown origin was caught in the Miramichi River, N.B., in September 1983. It is important to note that unsuccessful attempts to introduce pink salmon in Hudson Bay during the 1950s from hatcheries based on Lake Superior resulted in the accidental establishment of populations in the Great Lakes. Pink salmon were introduced into the Great Lakes watershed by a single release of only 21,000 fry into Lake Superior in 1956. Following this single introduction of an odd-year stock, pink salmon increased in abundance and by 1979 had colonized all five Great Lakes (Kocik et al. 1991). Even-year runs of pink salmon became evident in Lake Superior in 1976, presumably the result of these typically biennial fish taking an additional year to mature. Great Lakes populations have been extremely variable, reaching levels of abundance in excess of 10,000 fish in several rivers and falling to low abundance in others (Kocik et al. 1991). Given the speed of colonization, high straying rates, and documented Pacific salmon catches in the St. Lawrence at Montreal, it is possible that pink salmon from the Great Lakes are the source of individuals seen in the Gulf of Maine in recent years. However, since pink salmon are not abundant in Lake Ontario (Kocik and Jones 1997) that lake may serve as a buffer, reducing the threat of their colonization.

Importance. On the Pacific coast, pink salmon are the most abundant salmon comprising 40% by weight and 60% by number of the commercial catch (Heard 1991).

COHO SALMON / *Oncorhynchus kisutch* (Walbaum 1792) / Coho, Silver Salmon /

Bigelow and Schroeder 1953:133

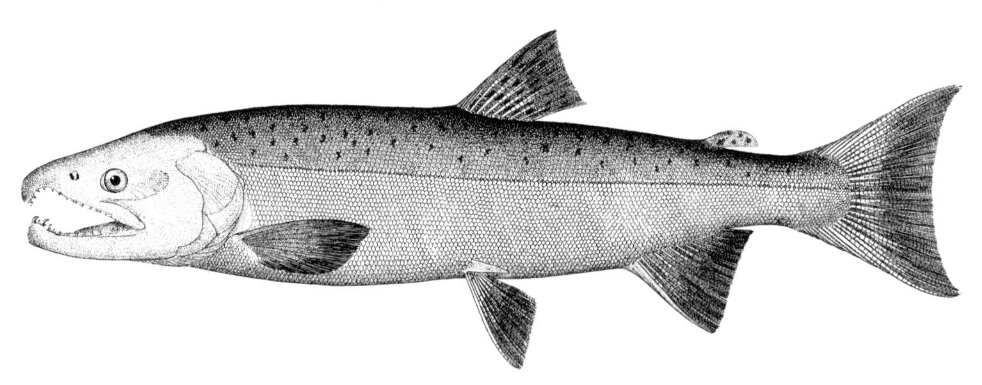

Figure 86. Coho salmon *Oncorhynchus kisutch*. Unalaska.

Description. Coho salmon resemble stout Atlantic salmon in general shape, relative size, and position and shape of the fins (Fig. 86).

Meristics. Dorsal fin rays 9–12; anal fin rays 12–17; pectoral fin rays 13–16; pelvic fin rays 9–11; scales moderate, fewer than 150 in row above lateral line; total gill rakers 18–25.

Color. Coho salmon are silvery, like Atlantic salmon, but more closely sprinkled with small black spots along the back and on the upper part of the caudal fin. These spots are roundish or oval on coho salmon, never in the form of crosses as in Atlantic salmon. Black spots are restricted to lower half of the tail in coho, whereas the entire tail is spotted in pink salmon.

Size. In their native range, coho commonly average 55–69 cm; individuals weighing 6 kg are common but those over 9 kg are rare (Sandercock 1991). The all-tackle game fish record is a 15.08-kg coho caught in the Salmon River, Pulaski, N.Y., in September 1989 (IGFA 2001).

Distinctions. Coho salmon have at least 12 anal fin rays, whereas most Atlantic salmon have only 8 or 9 anal rays, and never more than 10.

Habits. Coho are semelparous, ascending rivers to spawn from late summer through late autumn, spawning shortly after arrival on spawning ground, and dying a week or two after spawning (Sandercock 1991). They have variable life-history patterns but in general are similar to Atlantic salmon. Eggs hatch in late winter and fry emerge from their redds and reside in streams for 1 or 2 years prior to smoltification. Seaward migration typically occurs in May and June. In the sea, they grow rapidly and typically return after 18 months. Like Atlantic salmon, some males, called jacks, return sooner, after spending only 4–6 months in the ocean. Rarely, females will return after a short time at sea, and these are termed jennies.

General Range. The native range extends from the Sacramento River in California to Point Hope, Alaska. In Asia, they occur from North Korea to the Anadyr River in Siberia (Sandercock 1991).

Occurrence in the Gulf of Maine. Coho salmon have a long history of attempted introductions into the Gulf of Maine region. Early efforts between 1901 and 1930 met with limited success (Sandercock 1991). Bigelow and Schroeder included coho salmon based on stocking activities in the Ducktrap River, a tributary on the western side of Penobscot Bay, which resulted in a modest return during the early 1950s (Ricker 1954). Apparently, this population slowly died out. Attempts to provide put-and-take fisheries for coho salmon in tributaries to the Gulf of Maine, primarily in New Hampshire and Massachusetts, have been ongoing since that time. Based on unpublished reports and conversations with biologists in the region, the greatest effort was expended in New Hampshire, where between 80,000 and 355,000 coho salmon smolts were stocked each year from 1967 to 1988. Return rates ranged from 0.2 to 2.4% with 60–1,250 fish being caught by anglers. In addition, some coho were used experimentally for mariculture programs in Maine in the early 1980s. Straying from these programs was documented by adults collected in Maine (K. Beland, Maine Atlantic Salmon Authority, Bangor, Maine, pers. comm.) and recovery of wild coho salmon parr in New Brunswick (Symons and Martin 1978). The latter was the most intriguing as it represented documented evidence of establishment and reproduction in nontarget systems. Stocking of coho in New Hampshire and Massachusetts was replaced with an experimental chinook salmon program that ran from 1989 to 1993; this was discontinued because of poor returns. Stocking programs have ceased but there are still occasional reports of coho in Gulf of Maine tributary streams.

Importance. Coho salmon account for less than 10% of the commercial catch of Pacific salmon, but this represented a peak of 10.6 million fish in 1986. In addition, they are locally important components of recreational catch.

RAINBOW TROUT / *Oncorhynchus mykiss* (Walbaum 1792) / Steelhead

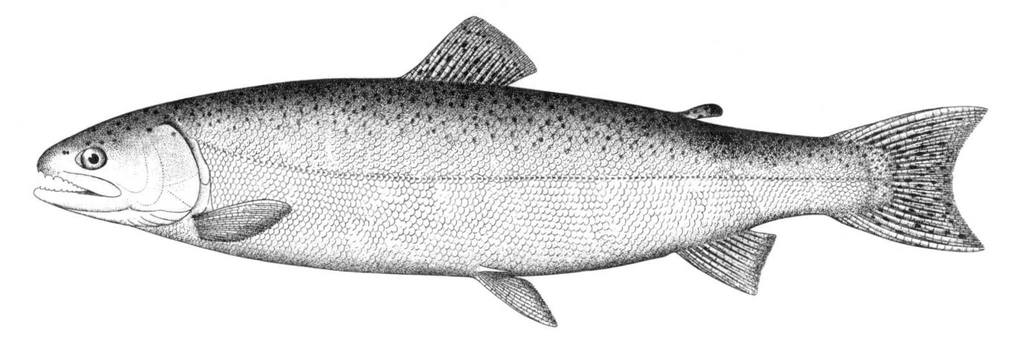

Figure 87. Rainbow trout *Oncorhynchus mykiss.*

Description. Rainbow trout are highly variable in body proportions throughout their range but are characteristically salmonlike in shape (Fig. 87). The general arrangement of the fins, including the adipose, parallels that of Atlantic salmon, but the pelvic fin origin is closer to the anterior end of the dorsal, further forward than in Atlantic salmon. The caudal fin of rainbow trout is less forked than that of young Atlantic salmon of equal size.

Meristics. Dorsal fin rays 10–12; anal fin rays 8–12; pectoral fin rays 11–17; pelvic fin rays 9 or 10; gill rakers 6–8 on upper limb, 6–9 on lower limb; scales small but variable between stocks, about 100–150 rows at lateral line; branchiostegals 9–13; pyloric caeca 27–80; vertebrae 60–66 (Scott and Crossman 1973).

Color. In freshwater, rainbow trout are readily identified by the prevalent fine black spotting all over the body but particularly above the lateral line and the characteristic red band along the lateral line. Background colors are blue to green on the dorsal area and white below. Rainbow trout living in salt water become predominantly silver but the spotting patterns and countershading are still rather obvious. The back and the top of the head become steel blue or blue-green—hence the name steelhead.

Size. Growth depends upon both genetics and habitat. Stream-resident fish are typically smaller than forms found in lake or marine environments. The anadromous form can reach sizes of 122 cm and 16 kg but typical sizes are more in the range of 4 kg. The all-tackle game fish record is a 19.10-kg rainbow trout caught in Bell Island, Alaska, in June 1970 (IGFA 2001).

Distinctions. Rainbow trout are distinguished from brook trout by larger scales, a back without vermiculate markings, and the arrangement of teeth on and the shape of the vomer. The number of anal fin rays, usually 8–11 and always fewer then 12, distinguish them from other Pacific salmons. Rainbow trout are easily distinguished from Atlantic salmon by the distinctly marked radiating rows of spots on the caudal fin in all environments. In addition, rainbow trout never have red spots on its body and the adipose fin often has a black margin.

Taxonomic Note. For many years, rainbow trout had the scientific name of *Salmo gairdneri* but because it is more similar to other Pacific salmons of the genus *Oncorhynchus* it was transferred to that genus, and as it is the same species as the Asian *O. mykiss,* which was described first, the species name of rainbow trout was also changed (Smith and Stearley 1989).

Habits. The life history of rainbow trout is exceedingly variable, ranging from nonmigratory lotic forms to highly migratory anadromous forms with various patterns of freshwater residency (MacCrimmon 1971). However, anadromous forms typically share a similar life-history pattern with Atlantic salmon: residing in freshwater for two growing seasons and spending two growing seasons at sea. The primary difference is that rainbow trout are typically spring spawners and begin migrating into tributaries in the preceeding September and continue through May. Most steelhead spawn in March, April, or May, although winter spawning has been noted (Biette et al. 1981). Fry emerge from their redds in June or July and remain in nursery streams until out-migration from May through June (MacCrimmon 1971; Seelbach 1993).

Food. In their native marine range, the diet of rainbow trout is predominantly fishes and squids with amphipods, copepods, and euphausiids encountered in smaller quantities (LeBrasseur 1966; Manzer 1968). In freshwater streams, their diet is primarily aquatic insects (Johnson and Ringler 1980; Nakano and Kaeiryama 1995).

General Range. Rainbow trout were native to Asia and western North America, north to the Kuskokwim River, Alaska, and southward to extreme northern Baja California, west of the continental divide (MacCrimmon 1971; Scott and Crossman 1973). They are so popular for fish culture and angling that they have been stocked almost worldwide and naturalized populations are common. Their current range in the United States is in practically "all suitable locations" (Scott and Crossman 1973).

Occurrence in the Gulf of Maine. Rainbow trout are included here because of their utilization in mariculture. The first successful year-round sea-cage culture of Atlantic salmon and rainbow trout began in 1978 in the southwestern Bay of Fundy, N.B., and Maine (Saunders 1991). While this industry is dominated by production of Atlantic salmon, some rainbow trout are still raised. As recently as 1995, a large number of rainbow trout that escaped from a netpen in Maine resulted in recreational catches in several Gulf of Maine rivers. Given their performance in other ecosystems, that is, North American Great Lakes, the ability of rainbow trout to develop self-sustaining populations should not be discounted. However, little evidence of long-term survival or reproduction of these escapees has been documented to date. In addition to mariculture escapees, there are nonmigratory stocks in several Gulf of Maine rivers.

Importance. Rainbow trout are the most important of the Pacific salmonids that have been introduced into the Gulf of Maine region, primarily as sport fish.

ATLANTIC SALMON / *Salmo salar* Linnaeus 1758 / Sea Salmon, Silver Salmon, Black Salmon /

Bigelow and Schroeder 1953:121–131

Description. Sea-run Atlantic salmon are sleek fish, about one-quarter as deep as long, deepest below dorsal fin, where they taper toward both head and tail; and oval in cross section (Fig. 88). Head small, 20–26% SL; snout blunt, 7–11% SL; eye rather small; mouth gaping back to below eye. Dorsal fin about midway between tip of snout and base of tail; pelvic fin origin under posterior end of dorsal fin; anal fin similar in form to dorsal but with only about nine rays. Tail very slightly emarginate in adults, and almost square in large fish but more forked in juveniles.

Meristics. Dorsal fin rays 10–12; anal fin rays 8–11; pectoral fin rays 14 or 15; pelvic fin rays 9 or 10; total gill rakers 15–20; scales moderate, 109–124 in lateral line; branchiostegals usually 11 or

Figure 88. Atlantic salmon *Salmo salar.* Susquehanna River, Md. Drawn by H. L. Todd.

12; pyloric caeca 40–74, mean 55.4; vertebrae 58–61, mean 58.5 (Dymond 1963; Scott and Scott 1988).

Color. Atlantic salmon are silvery all over while at sea, with a brownish back and numerous small black crosses and spots on head and body (chiefly above the lateral line). These marks become more pronounced and the silver coloration dissipates with extended freshwater residence. Prior to seaward migration, parr are conspicuously marked with 10 or 11 dark crossbars while in fresh water, alternating with bright red and black spots.

Size. The all-tackle game fish record Atlantic salmon weighed 35.9 kg and was caught in the Tana River, Norway, in 1928 (IGFA 2001). None approaching this size is recorded from the northwest Atlantic. A 25.1-kg salmon taken in the Grand Cascapedia River, Quebec, in 1939 may be the Canadian record (Scott and Scott 1988). Fish of 18 kg were not uncommon in Bigelow and Schroeder's time in some of the larger rivers emptying into the Gulf of St. Lawrence. In recent years, anadromous Atlantic salmon have averaged about 57 cm (1.3 kg) for fish that spent 1 year at sea, 75 cm (4 kg) for fish that spent 2 years at sea, and 88 cm (7 kg) for fish that spent 3 years at sea (Baum 1997).

Distinctions. Teeth and scales distinguish small Atlantic salmon from brook trout. The scales of Atlantic salmon are large enough to be seen easily, whereas those of brook trout are much finer. The caudal fin is heavily spotted in rainbow trout, whereas Atlantic salmon have little or no spotting. Atlantic salmon can be easily distinguished from other Pacific salmon in the Gulf of Maine by the low number of anal fin rays; a count exceeding 12 indicates that the specimen is a pink or coho salmon.

Habits. Atlantic salmon have been the subject of much scientific study and a wealth of literature has grown up about them. Bigelow and Schroeder drew freely from the work of Huntsman (1931) in their original manuscript. Much more has been learned about their ecology since that time and this a[illegible]ount u[illegible]ilize[illegible] the reviews of Mills (19[illegible]) and Colligan et al. (1999) for details of typical Atlantic salmon life history. Atlantic salmon have a relatively complex life history, which extends from spawning and juvenile rearing in freshwater streams to extensive feeding migration in the high seas. As a result, Atlantic salmon go through several distinct phases in their life history, and these are identified by specific behavioral and physiological changes.

Adult Atlantic salmon typically return to their natal river beginning in April and continuing into October, often producing both a spring and a fall run. Spawners stray to neighboring rivers at a rate less than 5% annually. Spawning occurs from late October through November over riffles with coarse gravel. Fish that return after 24 months at sea are called two sea-winter fish. Those that return after only 12 months at sea are typically males and are called grilse. Fish that spend 2 years at sea account for more than 90% of the spawning runs in New England rivers, whereas fish that spend only 1 year at sea are more prevalent in Canada. In addition, a few large fish that spend 3 years at sea (virgin spawners) and repeat spawners are among returning adults. Atlantic salmon typically cease feeding during their upriver migration to spawning grounds, and darken in color. After spawning, Atlantic salmon either return to the sea immediately or overwinter in the river. A spawned-out salmon in freshwater is termed a kelt or black salmon. Unlike Pacific salmon, Atlantic salmon are iteroparous, that is, they may spawn more than once, but mortality is substantial for postspawn fish. Each spawning run may include several age-groups, ensuring some level of genetic exchange between generations.

Atlantic salmon deposit their eggs in gravel depressions (redds). Redds are dug by females. Males fertilize the eggs as they are deposited, after which the female covers the eggs by further digging just upstream of the redd. Females produce 1,500–1,800 eggs per kg of body weight. Eggs hatch in late March or April into a stage called an alevin. Alevins remain in the gravel for about 6 weeks and are nourished by their yolk sac. When they emerge and begin active feeding about mid-May, they are designated fry. When vertical bars (parr marks) are visible on their sides, the fish are termed parr. Atlantic salmon may reside in freshwater streams from 1 to 8 years, but 1[illegible]4 ye[illegible] is [illegible] Gulf of Maine region. Parr reach

13–19 cm in length in the autumn, and begin to smoltify in winter—a process of physiological changes that prepares parr for migration to the ocean and life in saltwater.

After smoltification the parr loses its parr markings and becomes a smolt with a streamlined, silvery body and a pronounced fork in the tail. Survival from fry to smolt ranges from about 1 to 12% (Bley and Moring 1988). As smolts out-migrate to the sea, they must adapt to changes in water temperature, pH, dissolved oxygen, pollution levels, and predation. A source of out-migration mortality in many rivers is the impact of impoundments and obstructions. Most smolts in Gulf of Maine rivers enter the sea during May and June to begin their feeding migration. Smolt movement in tidal freshwater is relatively passive, progressing seaward on ebb tides and neutral or upstream on flood tides (Fried et al. 1978; Thorpe et al. 1981). As smolts enter more saline portions of an estuary, their movements are more directed and less affected by tides, and they move rapidly seaward at speeds averaging two body lengths per second (La Bar et al. 1978).

Upon entering the sea, most Atlantic salmon are highly migratory, moving from the mouth of natal rivers to the northwest Atlantic Ocean, where they are distributed seasonally over much of the region (Reddin 1985) (Fig. 89). At this point, the juveniles are termed postsmolts. Migrations place numerous stocks from North America in the Gulf of Maine at various times. Salmon of European origin may also be present, but their numbers would be expected to be low. Within the Gulf of Maine watershed, only stocks in the Bay of Fundy are not highly migratory with significant components that do not venture out of the Bay (Ritter 1989). Postsmolts grow rapidly and move in small schools and loose congregations close to the surface (Dutil and Coutu 1988), but the postsmolt stage is probably the least understood life-history period. Most of the U.S.-origin postsmolt tag recoveries come from incidental catch in herring and mackerel weirs in the Bay of Fundy and the south shore of Nova Scotia during July (Meister 1984). Tag recoveries in seabird colonies indicate that U.S. postsmolts are also present off eastern Newfoundland by August (Montevecchi 1988; Reddin and Short 1991). Upon entry into the nearshore waters of Canada, U.S. postsmolts mix with stocks from various North American streams. Postsmolts in the northern Gulf of St. Lawrence stay near shore for much of the first summer. Decreasing nearshore temperatures in autumn appear to trigger offshore movements (Dutil and Coutu 1988). Postsmolts also occur off the Grand Bank and further north in the Labrador Sea during the late summer and autumn (Reddin 1985; Reddin and Short 1991; Reddin and Friedland 1993), where North American stocks intermix with fish from Europe and Iceland.

Food. In the marine environment, postsmolts appear to feed opportunistically and primarily near the surface. Their diet includes invertebrates, terrestrial insects, amphipods, euphausiids, gammarids, and fishes (Hislop and Youngson 1984; Jutila and Toivonen 1985; Fraser 1987; Hislop and Shelton 1993). As salmon grow, fishes become more important in their diet. Among the fish species noted are Atlantic herring, alewife, rainbow smelt, capelin, mummichog, haddock, small sculpins, small Atlantic mackerel, sand lance, and even flatfishes (Comeau 1909; Kendall 1935; Hislop and Youngson 1984; Jutila and Toivonen 1985; Fraser 1987; Hislop and Shelton 1993).

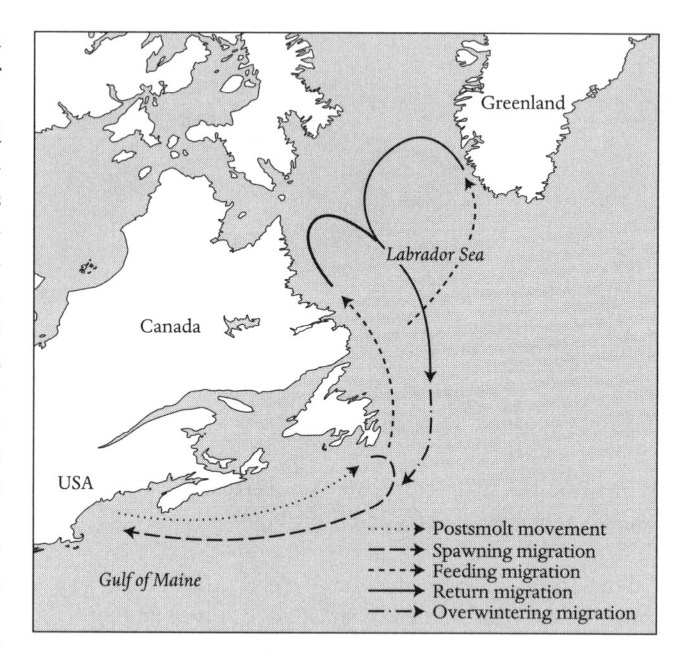

Figure 89. Migratory patterns of Atlantic salmon *Salmo salar* in the western North Atlantic.

Predators. Atlantic salmon postsmolts are preyed upon by gadids, whiting, cormorants, ducks, terns, gulls, and many other opportunistic predators (Hvidsten and Mokkelgjerd 1987; Hvidsten and Lund 1988; Montevecchi et al. 1988; Hislop and Shelton 1993). Predation rates are difficult to estimate because of the wide spatial and temporal distribution of Atlantic salmon and the large number and variety of potential predators. As salmon grow, only fishes as large as tuna, swordfish, or larger sharks can menace them. At this point, most predation may come from harbor and gray seals that prey on Atlantic salmon near shore and in the lower rivers (Dymond 1963).

Parasites. Atlantic salmon are susceptible to a number of parasites. In freshwater, external parasites of Atlantic salmon are gill maggots, freshwater lice, and leeches. Internal parasites include trematodes, cestodes, acanthocephalans, and nematodes (Bakke and Harris 1998). Once in the sea, Atlantic salmon lose many of their freshwater parasites but acquire others. The variety of parasites may increase in the sea owing to the number of intermediate hosts found in the ocean; the area of migration, which increases exposure; the tendency of fishes to school; and/or the increase in size of the host body (Colligan et al. 1999). In marine ecosystems, two species of copepods are common and widespread on Atlantic salmon: sea lice (*Lepeophtheirus salmonis*) and gill maggots (*Salminicola salmoneus*). In addition, reports of internal parasites include 25 species of trematodes, 38 cestodes, 25 nematodes, 22 acanthocephalans, and 8 protozoans (Dymond 1963). Sea lamprey also attack Atlantic salmon.

Age and Growth. Freshwater growth rates of parr depend on the productivity and climate of the river system, as is evidenced by the variation in age required to meet smolt size thresholds throughout their range. For the Gulf of Maine, freshwater residence averages 2 years and average smolt size is about 17 cm. Size and age at maturity also varies by stock in the Gulf of Maine, reflecting the combined influences of genetics and environment. These influences can be seen in historical accounts of variation in size and age of maturity in the region. The average weight of salmon caught in the Penobscot (Radcliffe 1922) was about 5.2 kg in 1905 (6,378 fish) and 4.0–4.5 kg in 1919 and 1920 (3,920 fish). The heaviest Penobscot fish of those times weighed a little more than 15.8 kg (Kendall 1935). The fish in the rivers flowing into the head of the Bay of Fundy run much smaller and most of them spawn first as grilse. This, Huntsman (1931) pointed out, contrasts with the prevalent 6-year-old fish in the Miramichi discharges into the southern side of the Gulf of St. Lawrence, and with the 7- or 8-year-old fish in the Grand Cascapedia, tributary to the Bay of Chaleur. Various explanations have been advanced to account for these differences from river to river. For stocks transiting the Gulf of Maine as juveniles, postsmolt growth appears to be an important determinant of maturation rate (Friedland et al. 1996). However, variation in age of first spawning for an individual stock may mediated by overwintering conditions when the fish are in the Labrador Sea (Friedland et al. 1998).

General Range. Coastal waters of both sides of the North Atlantic, entering rivers to spawn. On the European side, the range extends from the boundary between Spain and northern Portugal northward to well within the Arctic Circle. Atlantic salmon occur in a few rivers in western Greenland (Jensen 1948). On the North American side, they were indigenous to all suitable rivers from northeastern Labrador to the Housatonic River, which flows into Long Island Sound. The range of Atlantic salmon on both continents has been drastically reduced by human activities.

Occurrence in the Gulf of Maine. Atlantic salmon stocks of the Gulf of Maine have been among the most affected by human activities. By the early 1800s, Atlantic salmon runs in New England had been severely depleted. The earliest impacts were from fishing, and these were quickly followed by water-quality degradation and barriers to migration caused by the Industrial Revolution. Restoration efforts were initiated in the late 1800s but had little success owing to the presence of dams and the inefficiency of early fishways and fish-culture practices. By 1865 Atlantic salmon had been eliminated from southern New England rivers (Colligan et al. 1999). Loss of these stocks shifted the southern extent of the species range to Maine. Remnant stocks are limited to several small river systems or parts of larger river systems north of the Kennebec River, Maine (Colligan et al. 1999). Currently, efforts to rehabilitate remnant stocks and restore Atlantic salmon to select New England rivers are coordinated among federal and state agencies and the private sector. Strategic plans for Atlantic salmon restoration have been developed for the Connecticut, Pawcatuck, Merrimack, Saco, Penobscot, and coastal Maine rivers with remnant stocks.

Stocks in Canadian tributaries to the Gulf of Maine have also been affected and unique genetic resources have been lost. However, the magnitude of these declines is less than in U.S. waters. Some 20 Nova Scotia and 10 New Brunswick streams are still presumed to have Atlantic salmon stocks, the largest of which are the Stewiacke and the Big Salmon rivers. Collectively these stocks are known as the "Inner" Bay of Fundy stocks and generally produce runs of grilse and repeat spawners. These stocks are not believed to undergo extensive ocean migrations and thus probably utilize the Gulf of Maine most months of the year. Impacts to these rivers have for the most part been related to habitat degradation, as, for example, what has occurred on the Petitcodiac River involving changes in flow management. Outer Fundy stocks of western Nova Scotia, such as the Tusket, Barrington, and Clyde, may also occur in the Gulf of Maine, but probably as transients during migration. These rivers, unlike the "Inner" group, have been significantly affected by acid rain, which has depressed the abundance of salmon in recent decades. The declines exhibited in both U.S. and Canadian Atlantic salmon stocks of the Gulf of Maine mirror a continental collapse of Atlantic salmon throughout their range (Anderson et al. 2000). This collapse poses a threat to all populations but is particularly critical for those of the Gulf of Maine, which were already at low levels when the continental collapse began.

Movements. Atlantic salmon in the Gulf of Maine fall into two general categories: migratory stocks (long ocean migrations) and resident stocks (limited ocean migrations). All New England stocks and those on the south shore of Nova Scotia are migratory. Those in the Bay of Fundy are primarily resident stocks (Huntsman 1948). Bigelow and Schroeder recognized this to some extent but did not have the information that we have now to better understand the habits of migratory stocks (see Habits section above). This is evident in their description of Atlantic salmon caught hundreds of kilometers from their natal rivers: "It is not likely that these wandering salmon return at all to their home rivers; probably they are lost permanently from the breeding population." They did, however, give interesting accounts of Atlantic salmon catch in the Gulf of Maine, which are not recorded elsewhere. Following is an edited version of these reports, contrasted with recent observations.

1. Three reports of salmon caught on Western Bank appeared in the daily press between 1925 and 1951, and Kendall (1935) reported one caught on La Have Bank 100 miles from Halifax and another 60 miles off Cape Sable.
2. Otter trawlers pick up odd salmon from time to time in the South Channel, and even on Georges Bank up to 160 miles or more at sea from Cape Cod (Kendall 1935).
3. The great majority of the salmon that are caught in the Gulf are taken within 20 miles of the land.
4. The average catch for the coast of Maine east of Casco Bay was about 5,400 kg (some 1,200 fish) from 1933 to 1946, whereas the corresponding 10-year average for the whole western side of the Gulf from Cape Elizabeth to the elbow of Cape Cod was only 270–315 kg (some 60–80 fish) at its highest levels.
5. A few Atlantic salmon as far as Cape Cod Bay; catches of one to five or six fish (4.5–24.8 kg) in 14 out of 16 years by eight traps, at North Truro, Cape Cod, during the period 1935–1950, in the months of May, June, July, September, and November.
6. Sometimes, a considerable number have been taken off the coast of Massachusetts. In 1937, floating traps along the North Shore of Massachusetts Bay picked up nearly 2,000 kg of salmon. All of these were taken close inshore. But the 1,600 or so salmon (7,200 kg) that were reported for Massachusetts in 1928 seemingly were farther out at sea, for all of them either hooked on long lines (4,560 kg), or were taken in otter trawls. These must have come from as far as the Penobscot, if not from the Bay of Fundy, which is equally true of the salmon that are caught around Martha's Vineyard from time to time.
7. In the spring of 1915 about 75 (including fish up to 16 kg) were taken at Gay Head and in the neighborhood of Woods Hole. One, however, of about 4.5 kg, reported in the North River, Marshfield, in the summer of 1938 and a few seen jumping in the Parker River (also in Massachusetts) in the summer of 1951 may have been the product of attempts to stock these streams.
8. Occasional salmon that have been taken along the New Jersey coast and off Delaware (Smith 1895) may have been the product of attempts to stock the Hudson.

Results of a tagging study with the Penobscot stock gives a more comprehensive picture of the movements of migrating fish through the Gulf of Maine. These experiments were carried out during 1962–1992 using externally tagged smolts. Most tags were recovered in distant water fisheries in Canada and Greenland and in the Penobscot itself when the fish returned to spawn. However, some recaptures were made in the Gulf of Maine area, including recaptures of postsmolts and maturing salmon. The postsmolts were recaptured from June through July in the inshore areas of the Bay of Fundy and along the eastern coast of Nova Scotia as the fish migrated to feeding areas north of Newfoundland. It is reasonable to assume that many postsmolts also utilized the more open waters of the Gulf of Maine during this migration. Adults on return spawning migrations were recaptured by Nova Scotia and Bay of Fundy fisheries, but other tags were also returned from the Gulf of Maine during the months prior to the spawning run. The fish appear to stage in the Gulf of Maine before completing their spawning migrations, often distributing to the south of their natal rivers, as evidenced by the recovery of stray tags in neighboring river systems and the anecdotal accounts of Bigelow and Schroeder summarized above. For long-distance migrating stocks like the Penobscot, the Gulf of Maine is clearly an important transit area to and from the feeding grounds.

Importance. Atlantic salmon have long been highly esteemed as a table fish; their bones have been found in caves occupied by prehistoric man in western Europe. They are prized as a game fish as well as a commercial species. Accounts of their decline have been given by Bigelow and Schroeder, Dymond (1963), and Netboy (1974).

Extirpation of Atlantic salmon from their southern North American range resulted in a great decrease in the abundance of Atlantic salmon in the open Gulf, clearly reflected in the catches. Data are not available for the years when all the rivers had viable populations. Bigelow and Schroeder reported on sporadic catches of Atlantic salmon in Massachusetts and their subsequent absence from that fishery. Baum (1997) recorded the specifics of the rise and fall of the Penobscot River and Bay Atlantic salmon fishery. This fishery built up through the late 1800s from a reported 183 weirs and nets capturing 7,320 salmon in 1867 to a total of 230 weirs and 36 gill nets capturing over 10,016 in 1880. Catch for this fishery peaked in 1889 at over 17,000 salmon. With the exception of brief peaks in 1912–1915 and 1930–1933, the catch steadily declined until only 40 fish were caught in 1947. The commercial fishery was finally closed in 1948.

Since that time, Atlantic salmon in the United States have been treated primarily as a recreational species. Unfortunately, their demise continued through the 1960s until restoration efforts were reinvigorated and modern hatchery practices became more effective at enhancing the runs (Baum 1997; Colligan et al. 1999). Despite these actions, the number of naturally produced Atlantic salmon in the remnant populations in Maine has remained low through the present day (Colligan et al. 1999) (Fig. 90).

Catches of Atlantic salmon in Canadian waters of the Gulf remained at high levels thanks to conservation measures such as limiting netting at the mouths of the rivers and keeping streams more accessible by fishways at the dams. The average yearly catches from 1870 to 1946 for the west coast of Nova Scotia and for the Bay of Fundy combined ranged from 295,000 kg in 1870–1879 down to 124,000 kg in 1940–1946. The Canadian catch in the open Gulf and in the Bay of Fundy ran about 180,000–270,000 kg in the early 1950s, or 40,000–60,000 fish,

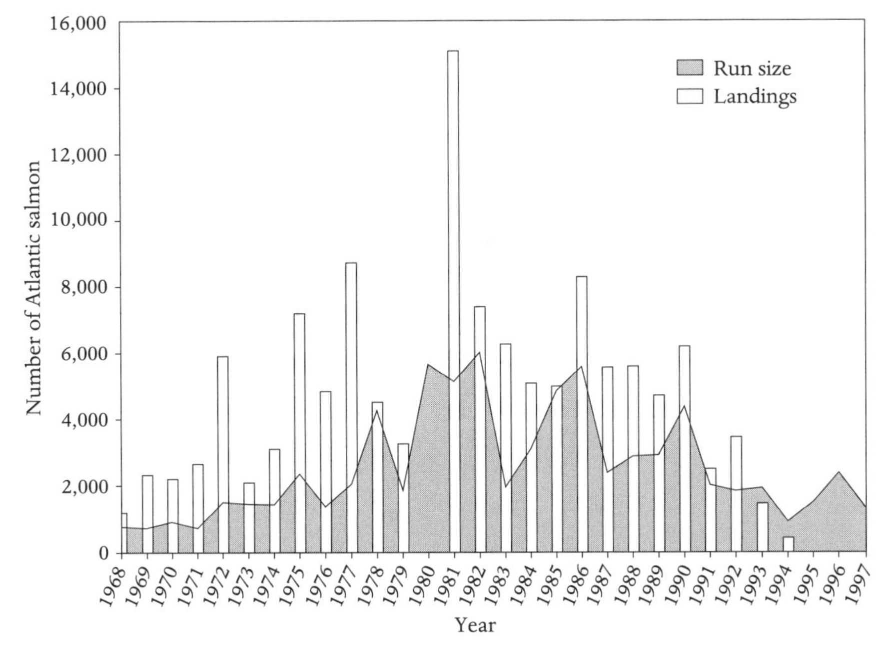

Figure 90. Spawning run size of Atlantic salmon *Salmo salar* returning to Maine rivers and total catch of these stocks by U.S. anglers from Maine rivers and at-sea foreign catches (landings since 1994 are not estimated because of fishery closures).

which is perhaps 100 times as great as that for the entire coastline of Maine and of Massachusetts. The distribution of the catches shows that the Saint John River contributed four-fifths of this, or a yearly average of about 50,000 fish. As abundance of Atlantic salmon in North America declined during the 1980s and 1990s, commercial harvests were progressively phased out in many Canadian provinces including New Brunswick and Nova Scotia. Today there is no commercial fishing for Atlantic salmon in the Gulf of Maine area. Recreational fisheries for Gulf of Maine rivers have continued under strict management.

Despite the great decline of wild Atlantic salmon populations in the Gulf of Maine, the region remains an important Atlantic salmon fishery resource as a result of the cage-culture of farmed Atlantic salmon. The production of farmed Atlantic salmon in the entire North Atlantic was 326,785 tons in 1994. This represents over 85 times the catch of wild Atlantic salmon in the entire North Atlantic area. In North America, combined U.S. and Canadian production during 1994 was 18,262 tons. This amount far exceeds the biotic potential of wild populations and the growth rate of this industry has been exponential from its infancy in 1978 to the present. It is ironic that farmed Atlantic salmon provide both a buffer to overexploitation and a threat to native populations owing to disease and genetics interactions (Jonsson 1997; Youngson and Verspoor 1998; Colligan et al. 1999). It will be a challenge for fisheries managers to balance the needs of wild and domestic stocks of Atlantic salmon, sustaining wild populations while enhancing the economic benefits of mariculture operations.

BROOK TROUT / *Salvelinus fontinalis* (Mitchill 1814) / Salter, Sea Trout, Charr /

Bigelow and Schroeder 1953:120–121

Description. Although brook trout from different streams vary somewhat in form, they are generally fusiform (Fig. 91). When taken in saltwater, they are characteristically salmon-like in shape, about one-quarter as deep as long, tapering to a small head. Head 23–25% SL, snout bluntly rounded, 24–30% of HL. The general arrangement of the fins, including the adipose, parallels that of Atlantic salmon but the pelvic fin origin is under the middle of the dorsal, further forward than in its larger relative. All the fins are relatively larger, particularly the pelvics. The caudal fin of the brook trout is less forked than that of a young Atlantic salmon of equal size.

Meristics. Dorsal fin rays 10–14; anal fin rays 9–13; pectoral fin rays 11–14; pelvic fin rays 8–10; gill rakers 4–7 on upper limb, 10–15 on lower limb; scales small, about 230 rows at lateral line, 110–130 pores above lateral line; branchiostegals 9–13; pyloric caeca 23–55, average 38.4; vertebrae 58–62 (Dymond 1963; Scott and Scott 1988).

Color. Brook trout living in saltwater almost wholly lack the yellow and red tints and dorsal vermiculations so conspicuous in freshwater. The back is steel blue or bottle green, cheeks and sides are silvery, like an Atlantic salmon, and the belly is white. The sides above the lateral line are more or less dotted with pale yellow spots. Wavy crossbars are present on the dorsal fin and on the corners of the caudal fin. The sides and flanks below the level of the lateral line usually are marked with small pale vermilion dots. The pelvic fins are often plain

Figure 91. Brook trout *Salvelinus fontinalis.*

white; at most, the margins that are so conspicuous in freshwater brook trout are faint on fish in saltwater.

Size. Brook trout usually enter saltwater after attaining a fork length greater than 15 cm, typically at age 1 or 2 in southern New England or age 3 or greater in the northern portion of their range (Ritzi 1959; Naiman et al. 1987; Ryther 1997). Sea-run individuals exhibit growth rates more than twice as great as those of fish that remain in streams, attaining a weight of as much as 5 kg (Dymond 1963; Dutil and Power 1980; Naiman et al. 1987). The all-tackle game fish record is a 6.57-kg fish caught in the Nipigon River, Ontario, in July 1916 (IFGA 2001).

Distinctions. Scales and teeth provide the best means of distinguishing brook trout from young Atlantic salmon and other Gulf of Maine salmonids. Teeth on the roof of the mouth of brook trout are confined to a cluster on the head of the vomer, and their scales are so tiny as to be hardly visible, whereas those of the Atlantic salmon are large and easily seen.

Habits. Sea-run brook trout typically do not venture very far from their natal stream. On their seaward migration, they typically exit the river system in April and May and reside in the upper estuary for a time period as brief as a few days up to 4 months before returning to the river system (Mullan 1958; Smith and Saunders 1958; Bergin 1984; Ryther 1997). Given the range in marine residence period, the timing of their upstream migrations is quite variable and is affected by complex interactions among water temperatures, discharge, and other environmental factors that are not fully understood (Ryther 1997). In subsequent migrations, older fish generally travel in schools in the lower estuary or along the coast close to their natal river but have been encountered up to 45 km from their natal streams in open ocean and other estuarine systems (White 1942; Naiman et al. 1987). Ocean survival appears to be quite high (for anadromous fish) with reported return rates of 42% for age 2+ fish and 61% for age 3+ (Whoriskey et al. 1981). Surviving brook trout will migrate to the ocean in subsequent years following a similar pattern. In general, survival in marine ecosystems is lower than in freshwater but growth rates are greatly accelerated (Ryther 1997). Chemosensory factors are thought to be used in homing to their natal stream but the exact mechanisms are not known (Keefe and Winn 1991).

Food. During their marine sojourn, brook trout are opportunistic feeders with fishes and crustaceans comprising more than 90% of their diet (Whoriskey et al. 1981). Recorded brook trout prey includes at least 10 fish species (alewife, small eels, smelt, young hake, silversides, killifishes, sticklebacks, young sea raven, rock gunnel, and sand lance) and several crustaceans (shrimps, amphipods, and isopods) (White 1942; Dymond 1963; Whoriskey et al. 1981). In freshwater, they exhibit a seasonally diversified diet that includes insects, annelids, mollusks, and fishes. A Quebec study found that insect larvae were the most important prey group in spring, whereas in September and October brook trout fed mainly on juvenile alewives emigrating to the sea (Verreault and Courtois 1989).

General Range. Native to eastern North America, north to the outer coast of Labrador, west to Minnesota, and southward to Georgia along the Appalachian chain but widely introduced into other parts of the world. Sea-run populations were indigenous from the Atlantic Provinces of Canada southward to Long Island, N.Y. (Scott and Crossman 1973).

Occurrence in the Gulf of Maine. Sea-run brook trout are most accurately classified as amphidromous, that is, making regular feeding migrations to the sea while still undergoing significant freshwater growth. Ryther (1997) considered sea-run brook trout rare in tributaries to the Gulf of Maine. However, migratory populations have not been thoroughly assessed.

These migratory individuals require linkages between both riverine and coastal ecosystems to complete their life history and constitute a unique resource. The earliest historical accounts of anadromous brook trout in North America come from 1770 (Bergin 1984). These accounts suggest that relatively productive populations occurred in smaller river systems near the southern extent of their range. These and other historical accounts seem to indicate that amphidromous brook trout were common in the coastal portion of their geographic range prior to the 1700s (Ryther 1997). While their historical abundance and demise has not been well-recorded, given the history of other anadromous fishes, it appears that most of the coastal brook trout populations were lost in the past 200 years. Bigelow and Schroeder noted that only a couple of small streams on the Massachusetts Bay side of Cape Cod and one or two small brooks tributary to Ipswich Bay still supported amphidromous brook trout in 1953. They noted only one population present between Cape Ann and Cape Elizabeth. North of this point, they noted populations in the tidal portions of many of the brooks that empty into Casco Bay and in the Belfast River, a tributary to upper Penobscot Bay (Evermann 1905; Towne 1940). In New Brunswick and Nova Scotia, local inquiry determined that a few populations were present but that many had been "fished out long since." Since their writings, coastal brook trout populations have been further affected by overharvest, habitat degradation, and stocking of hatchery-reared brook trout and other nonnative salmonids (Ryther 1997). Despite these factors, inquiries among regional fish biologists suggest that brook trout populations with amphidromous components still exist within pockets of the general areas noted above.

Bigelow and Schroeder believed that sea-run brook trout are simply individuals that have the habit of running down to saltwater and that most of a population never leaves fresh water, even in streams offering free access to the sea. Alternately, this behavior could have a hereditary basis. Population level data seem to suggest that a variable portion, up to 35%, of a river population moves to sea (Ryther 1997). In addition, the current body of meristic, morphometric, physiological, and genetic evidence suggests that amphidromous behavior is an intrinsic element of some brook trout populations (Wilder 1952; McCormick et al. 1985; Ryther 1997). Wilder (1952) concluded that they arise more commonly in river systems that lack coldwater lakes or deep water refugia. The lack of a genetic race of sea-run brook trout should not be taken as evidence of the homogeneity of all brook trout stocks; whereas variability is low in brook trout compared to other salmonids, genetically discrete stocks do occur throughout their range and the populations with a sea-run element may be distinct from inland populations. Future research may help clarify these relationships for remnant coastal brook trout populations.

ACKNOWLEDGMENT. A draft of the salmonid section was reviewed by Karsten E. Hartel.

DRAGONFISHES. ORDER STOMIIFORMES

The order Stomiiformes is a very large and morphologically diverse group of deep-sea fishes, which has been classified in several ways. They have luminescent organs (photophores, see Figure 92 for terminology) whose structure has been used, among other characters, to diagnose the order (Harold and Weitzman 1996). Some species have chin barbels, often with luminescent tissue. The premaxilla and maxilla both bear teeth and enter into the gape of the mouth. Nelson (1994) recognized four families with 51 genera and about 320 species. The Gonostomatidae and Sternoptychidae form the suborder Gonostomatoidei, and the Phosichthyidae (= "Photichthyidae") and Stomiidae the suborder Phosichthyoidei (= Photichthyoidei). We follow Harold and Weitzman (1996) at the family level, but for convenience we follow Nelson (1994) in using subfamily groups within the Stomiidae, recognizing that some of these may not be monophyletic (Fink 1985; Harold and Weitzman 1996).

KEY TO NORTH ATLANTIC STOMIIFORM FAMILIES

1a. Normal gill rakers present; teeth small, variable in size, but never fanglike; no postorbital photophore **2**
1b. Normal gill rakers present in larvae only, adult gill arches are naked or have teeth or spines; jaw teeth moderate to large, usually with one or more large fangs present in each jaw; postorbital photophore present, often large . . . . **Stomiidae**
2a. AC photophores (those from anal fin origin to caudal fin base) joined in clusters . . . . **Sternoptychidae**
2b. AC photophores never in clusters **3**
3a. Photophores on isthmus, except in the "gonostomatids" *Diplophos*, *Manducus*, and *Triplophos*, which have more than 65 IP (isthmus photophores) . . . . **Phosichthyidae**
3b. No photophores on isthmus . . . . **Gonostomatidae**

BRISTLEMOUTHS. FAMILY GONOSTOMATIDAE

JAMES E. CRADDOCK AND KARSTEN E. HARTEL

The small family Gonostomatidae contains some of the world's most abundant fishes. All species in the family are oceanic, mostly mesopelagic and bathypelagic, but at least one species is benthopelagic over deep slopes. All except one are bioluminescent with at least one row of photophores on each side running the entire length of the body. Some species

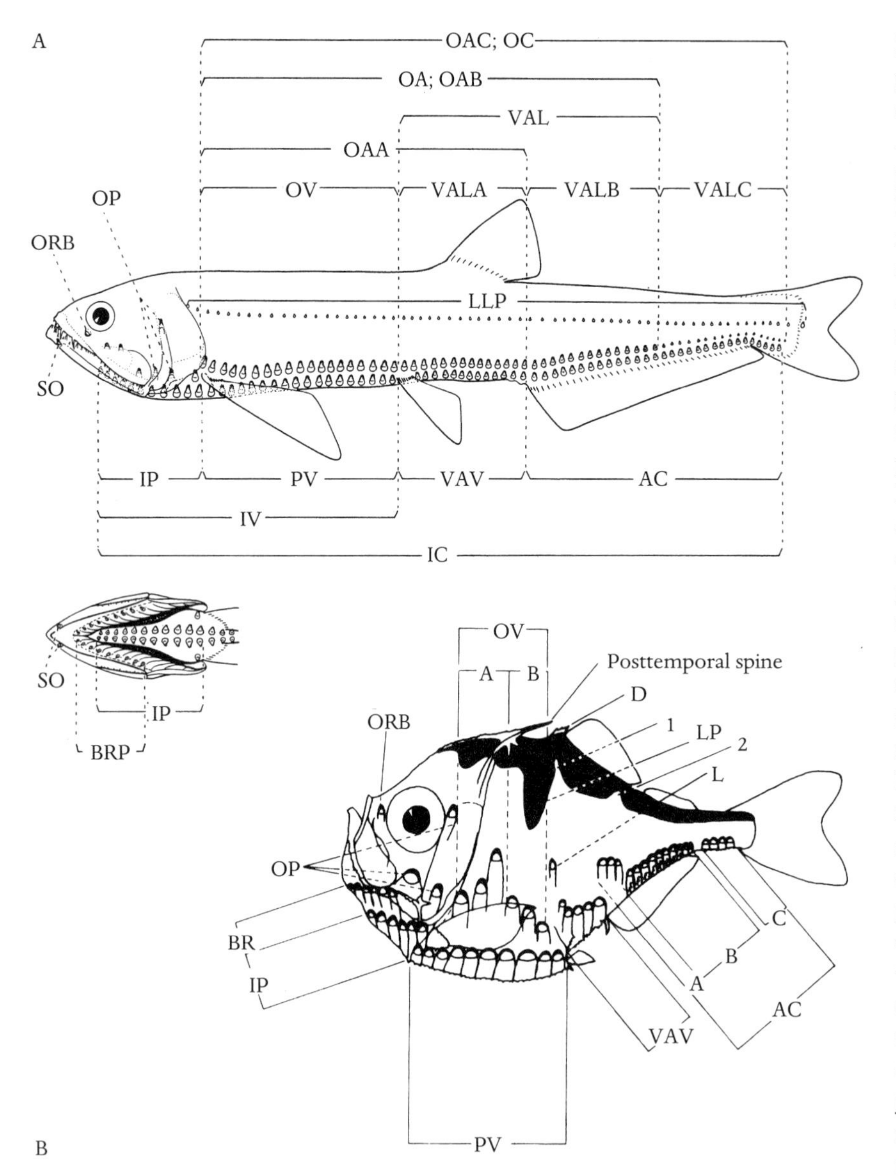

Figure 92. Diagram of stomiiform photophore terms. (A) Generalized stomiiform fish. *Photophore terminology:* AC, posterior part of IC series; BRP, photophores on branchiostegal membranes; IC, entire ventral photophore row extending from anterior end of isthmus to posterior termination on caudal peduncle, includes IP, PV, VAV, and AC; IP, anterior part of IC, from anterior end of isthmus to origin of pectoral fin; IV, part of IC from anterior end of isthmus to a vertical with pelvic fin origin, includes IP and PV; LLP, small lateral photophores from opercular margin to base of middle caudal fin rays; OA or OAB, large photophores of the OAC series, includes OV, VALA, and VALB; OAA, part of OAC from opercular margin to a vertical with origin of anal fin, includes OV and VALA; OAC or OC, entire lateral photophore series on body from opercular margin to caudal fin base, includes OV, VALA, VALB, and VALB; OP, three photophores associated with gill cover; ORB, photophores associated with eye; OV, part of OAC from opercular margin to a vertical with pelvic fin origin; PV, part of IC series between origins of pelvic and pectoral fins; SO, photophore near anterior end of symphysis of lower jaw; VAL, part of OAC between pelvic fin origin and end of large photophores, includes VALA and VALB; VALA, part of OAC from anal fin origin to end of large photophores; VALC, small photophores of OAC series; VAV, part of IC between origins of pelvic and anal fins. (From Weitzman 1986:227.) (B) Photophores of a hatchetfish, *Polyipnus* sp. (From Harold 1994: Fig. 1.)

mature at less than 20 mm, whereas others reach about 300 mm. Harold and Weitzman (1996) restrict the family to four genera: *Margrethia, Bonapartia, Gonostoma,* and *Cyclothone,* with 21 species. Nelson (1994) recognizes this group as the subfamily Gonostomatinae and includes the three problematical genera *Diplophos, Manducus,* and *Triplophos* in the subfamily Diplophinae. We follow Nelson in including these genera in the key to gonostomatids because there is presently no other group in which to place them. The genus *Gonostoma* has recently been divided by Miya and Nishida (2000) into *Gonostoma* (including the western Atlantic *G. atlanticum*) and *Sigmops* (including the Atlantic *S. elongatum* and *S. bathyphilum.* Many gonostomatid species are geographically widespread, some occurring in all three major oceans. Fifteen species of these fishes occur in the western North Atlantic and, of them, 13 (in six genera) have been collected in the Slope Water south of New England. Any (or all) of these could stray into the Gulf of Maine but only two of the seven species of *Cyclothone* known from the western Atlantic have been verified as occurring in the Gulf of Maine.

KEY TO GONOSTOMATIDAE OCCURRING OFF NEW ENGLAND

1a. More than two rows of photophores on body; IV photophores 20 or more, IC photophores 42 or more; photophores present on isthmus . . . 2
1b. Only one or two rows of photophores on body; IV photophores 17 or less, IC 44 or less; photophores absent on isthmus 3
2a. Dorsal fin origin behind middle of body; IV photophores 30–32, IC photophores 70–75 . . . . **_Manducus maderensis_**
2b. Dorsal fin origin slightly before middle of body; IV photophores more than 40, IC photophores 111–113 . . . . **_Diplophos taenia_**
3a. Body photophores in one row 4
3b. Body photophores in two rows 5

4a. Dorsal fin origin in advance of anal fin origin; adipose fin present . . . . *Margrethia obtusirostra*
4b. Dorsal fin origin opposite or behind anal fin origin; adipose fin absent . . . . *Bonapartia pedaliota*
5a. Eye moderate (usually 6–10 in HL); lateral photophores (OA) 11–21; adipose fin usually present 6
5b. Eye small (about 10–15 in HL); lateral photophores (OA) 6–10; adipose fin absent *Cyclothone* 7
6a. Body dark (black to blue-black); total gill rakers more than 18. . . . . *Sigmops*
6b. Body light (silvery in life, tan when preserved); total gill rakers 18 or fewer . . . . *Gonostoma*
7a. Body white, almost transparent in life, with a few dorsolateral melanophores; scales absent . . . . **Brauer's bristlemouth**
7b. Body dark all over; scales present . . . . **Small-toothed bristlemouth**

BRAUER'S BRISTLEMOUTH / *Cyclothone braueri* Jespersen and Täning 1926

Figure 93. (A) Brauer's bristlemouth *Cyclothone braueri,* 38°58′ N, 68°18′ W, 23 mm SL, MCZ 143624. (B) Small-toothed bristlemouth *Cyclothone microdon,* 26°01′ N, 77°30′ W, 44 mm SL, MCZ 91005. Images by K. E. Hartel.

Description. Body elongate (Fig. 93). Head small with tiny eye (14 in HL) placed far forward. Mouth large, extending more than 10 eye diameters behind eye. Opposing dorsal and anal fin origins just posterior to midbody; anal fin slightly longer than dorsal fin. Scales absent.

Meristics. Dorsal fin rays 13–15; anal fin rays 18–20; pectoral fin rays 9 or 10; pelvic fin rays 6 or 7; gill rakers 15–18; branchiostegal rays 12–14; vertebrae 30–32; photophores: BRP 8–10, IV 12–13, VAV 4, AC 13–15, IC 29–31, OA 6 or 7.

Size. In intensive studies of warm-core Gulf Stream rings, the maximum size of *C. braueri* is typically 28 mm SL but one 31-mm SL specimen was found (Craddock et al. 1992). Males seldom exceed 20 mm SL (Badcock and Merrett 1976). In the eastern Atlantic this species reaches 38 mm SL (Badcock 1984a).

Distinctions. Distinguished from *Cyclothone microdon* by various pigment characteristics, most obvious of which is the fact that *C. braueri* is essentially white (almost transparent in life) with a few dorsolateral, stellate melanophores while *C. microdon* is dark all over (Fig. 93). In *C. braueri,* branchiostegal pigment is limited to streaks along the rays and along the free and attached edges while *C. microdon* has dark pigment distributed evenly over the whole branchiostegal membrane. Scales are absent in *C. braueri* and present in *C. microdon.*

Habits. *Cyclothone braueri* is a very abundant species, often making up 30% of the biomass and numerically over 50% of all fishes in the upper 1,000 m of the subtropical gyre in the North Atlantic. They normally live between 400–600 m and do not undergo daily vertical migration.

Food. Small crustaceans, copepods, and ostracods are typical prey items (Roe and Badcock 1984).

Reproduction. Mature *C. braueri* are highly sexually dimorphic with the adult male developing a very large nasal rosette, which is presumably used to locate females. In the eastern Atlantic, this species spawns mainly between April and October with 100–900 eggs per female, mean diameter 0.5 mm (Badcock 1984a). Larvae metamorphose between 11–14 mm SL.

General Range. This species occurs throughout the Atlantic from near Iceland to about 40° S, but reaches its greatest abundance in the subtropical gyres (unpublished WHOI data). In the Indian and South Pacific oceans it is found between 20 and 50° S (Mukhacheva 1974; Gon 1990a; Miya and Nishida 1996).

Occurrence in the Gulf of Maine. Two specimens (18 and 19 mm SL, MCZ 151174, 151401) were taken in opening/closing midwater trawls fished between 300–200 m over Georges Basin in April and May 1997 by L. P. Madin. *Cyclothone braueri* is the second most abundant species in the Slope Water off New England (Craddock et al. 1992). Bigelow and Schroeder recorded two small (23 mm SL) *Cyclothone* as *C. signata* (a Pacific species) from the Fundy Deep and off Browns Bank. These specimens (which have been lost) most probably were *C. braueri*, which is superficially similar to *C. signata*.

SMALL-TOOTHED BRISTLEMOUTH / *Cyclothone microdon* Günther 1878

Description. Body elongate. Head small with tiny eye (14 in HL) placed far forward. Mouth large, extending more than 10 eye diameters behind eye. Opposing dorsal and anal fin origins just posterior to midbody; anal fin slightly longer than dorsal fin. Scales present.

Meristics. Dorsal fin rays 13–15; anal fin rays 17–20; pectoral fin rays 9 or 10; pelvic fin rays 5 or 6; gill rakers 19–23; branchiostegal rays 12–15; vertebrae 31–33; photophores: BRP 9 or 10, IV 12 or 13, VAV 5, AC 13–15, IC 29–31, OA 6 or 7.

Size. In the Slope Water south of New England, *C. microdon* commonly reaches 50 mm SL (largest seen by us 53 mm SL). In the eastern Atlantic, it reaches 66 mm SL (Badcock 1984a), and Gon (1990) noted a 72-mm SL specimen from the Southern Ocean.

Distinctions. Distinguished from *C. braueri* by being darkly pigmented all over (Fig. 93) and by other pigment characteristics as noted under *C. braueri*. Scales are present in *C. microdon*.

Habits. *Cyclothone microdon* is the most abundant species in the Slope Water off New England, where it was found to be concentrated between 800 and 1,000 m (Craddock et al. 1992). Elsewhere it has been recorded down to 2,700 m (Badcock 1984a). This species does not undergo daily vertical migration; larger individuals live in deeper water.

Food. *Cyclothone microdon* feeds mainly on copepods (Badcock 1984a) and euphausiids.

Reproduction. *Cyclothone microdon* is a protandrous hermaphrodite, changing sex from male to female, with growth between 22 and 42 mm SL (Badcock and Merrett 1976; Badcock 1984a). This species may reproduce twice in a lifetime with a fecundity of between 2,000 and 10,000 eggs, depending on latitude and depth. The eggs are 0.5 mm in diameter. Larvae metamorphose at 11–14 mm SL (Badcock 1984a).

General Range. This is an exceptionally wide-ranging species; in the Atlantic it is found from north of Iceland to the edge of the Antarctic continent, but is most abundant in the subtropical and temperate areas (unpublished WHOI records). In the Indian and Pacific oceans it is found from 20° S to beyond 65° S (Mukhacheva 1974; Gon 1990a).

Occurrence in the Gulf of Maine. Confirmation of this species in the Gulf is based on two specimens (15 and 27 mm SL, MCZ 151175 and 151400) from Georges Basin and a 32-mm SL specimen (MCZ 151176) from the Northeast Channel. These were collected with an opening/closing midwater trawl fished at 100–300 m by L. P. Madin (pers. comm., WHOI) in April and May 1997.

HATCHETFISHES AND RELATIVES. FAMILY STERNOPTYCHIDAE

Antony S. Harold, Karsten E. Hartel, James E. Craddock, and Jon A. Moore

Marine hatchetfishes are bioluminescent pelagic fishes, rarely exceeding 100 mm SL, that are predominantly found in tropical to temperate seas. The family includes deep-bodied hatchetfishes, *Argyropelecus, Polyipnus,* and *Sternoptyx,* and the elongate "maurolicine" genera, such as *Maurolicus* and *Valenciennellus* (Weitzman 1974). Of the 57 or more species (in ten genera) in the family, only three are known from the Gulf of Maine, but nine others occur in the Slope Water off New England. Most species are mesopelagic, although some, like the common genera in the Gulf of Maine (*Maurolicus* and *Polyipnus*), are benthopelagic. Several peripheral species are included in the key but species accounts are not given.

In addition to certain osteological characters, this family has a unique type of photophore development (united clusters formed by budding). Photophores are generally located ventrally and ventrolaterally in these clusters. Terminology for the photophores is shown in Fig. 92B (also see Ahlstrom et al. 1984; Weitzman 1986; Harold 1994). In the species accounts, photophore counts in parentheses indicate that individual photophores are joined in a composite organ or cluster.

KEY TO STERNOPTYCHIDAE FOUND OFF NEW ENGLAND

1a. Body laterally compressed and deep, depth 1–2 in SL. **2**
1b. Body somewhat elongate, depth more than 3.5 in SL. **4**
2a. Dorsal blade reduced, not obvious; some photophores from three groups (OV, L, ACA) elevated almost to midbody; L photophore present; 5 VAV photophores; a dark dorsal triangular pigment bar present (*Polyipnus*). **5**
2b. Dorsal blade obvious; at most 1 ACA photophore elevated; L photophore absent; 3 or 4 VAV photophores; no dark dorsal triangular pigment bar **3**
3a. Eyes tubular, directed upward; 12 PV photophores; 4 VAV photophores; ACA photophores absent; dorsal blade composed of several elements (*Argyropelecus*). **7**
3b. Eyes lateral, not directed upward; 10 PV photophores; 3 VAV photophores; 1 elevated ACA photophore; dorsal blade a single element (*Sternoptyx*). **11**
4a. First AC photophore isolated and elevated, followed by two groups of level AC photophores (14–17 and 8–9); SO photophores present . . . . ***Maurolicus weitzmani***
4b. AC photophores in four or five clusters, each with 2–4 photophores, none elevated; SO photophores absent. . . . . . . ***Valenciennellus tripunctulatus***
5a. ACB photophores (in specimens larger than 25 mm SL) 11–13; ventral preopercular spine prominent and free of margin of preopercle; first ACA photophore slightly ventral in relation to second and third, and nearly in contact; posttemporal spine relatively long (5.7–8.6% SL) . . . . ***Polyipnus laternatus***
5b. ACB photophores 7–10; ventral preopercular spine reduced, embedded within margin of preopercle; first ACA photophore located well ventral in relation to second and third, not in contact; posttemporal spine shorter **6**
6a. Lateral pigment bar narrow, long, and tapered, extending to lateral midline; anal fin pterygiophore spines short and basally expanded; gill rakers 20–24. . . . . ***Polyipnus asteroides***
6b. Lateral pigment bar wide, short and triangular, extending less than half way to lateral midline; anal fin pterygiophore spines long and needlelike; gill rakers 19–21 . . . . ***Polyipnus clarus***
7a. OV, VAV, and AC photophores in a nearly continuous straight line; ACC photophores isolated from each other, none joined in a common cluster **8**
7b. OV, VAV, and AC photophores not in a straight line, but discrete clusters at various levels; ACC photophores united in a cluster, without gaps separating the reflectors **9**
8a. Dorsal blade not well developed, its height less than one-third of its length; profile of dorsal body margin not markedly arched along base of dorsal fin rays; sphenotic spine reduced or absent. . . . . . . ***Argyropelecus affinis***
8b. Dorsal blade well developed, its height greater than one-third of its length; profile of dorsal body margin markedly arched along base of dorsal fin rays; sphenotic spine prominent. . . . . ***Argyropelecus gigas***
9a. A single, serrate iliac spine, directed posteriorly; caudal peduncle long and not deep, length of ACB photophore cluster less than the gap between ACB and ACC photophores; dorsal body profile posterior to dorsal fin horizontal and elongate; two or three patches of large chromatophores along lateral midline; dorsal fin rays 8. . . . . ***Argyropelecus hemigymnus***
9b. Two separate smooth iliac spines; caudal peduncle moderate, length of ACB photophore cluster greater than the gap between ACB and ACC photophores; dorsal body profile posterior to dorsal fin shorter and oblique; dorsal fin rays 9 . . . . **10**
10a. Lower jaw with prominent canine teeth; posterior iliac spine much larger than anterior spine and directed posteroventrally; anterior and posterior walls of last 3 PV photophores sloping anteroventrally; body deep, greater than 0.75 SL; ACC photophore scales with long, spinelike denticles . . . . ***Argyropelecus aculeatus***
10b. Lower jaw without prominent canine teeth; iliac spines of similar size, directed anteroventrally and ventrally; anterior and posterior walls of last 3 PV photophores approximately vertical; body moderately deep, less than 0.70 SL; ACC photophore scales without denticles . . . . ***Argyropelecus sladeni***
11a. First AC photophore (ACA) slightly above level of ACB cluster of photophores, less than halfway to trunk midline . . . . ***Sternoptyx diaphana***
11b. First AC photophore (ACA) well above level of ACB cluster of photophores, near or at trunk midline. . . . . ***Sternoptyx pseudobscura***

SILVER HATCHETFISH / *Argyropelecus aculeatus* Valenciennes in Cuvier and Valenciennes 1849 /

Bigelow and Schroeder 1953:149–150

Description. Body compressed and deep, depth three-quarters of SL (Fig. 94). Body profile anterior to dorsal fin a posteriorly tapered, vertically elongate rectangle. Head very deep, compressed with a moderately large, vertical mouth. Eyes dorsal, large (3 in HL), vertically directed and telescopic; separated by narrow interorbital. Premaxillary and dentary teeth quite variable in length, usually with a pair of large "canine" teeth in lower jaw. Dorsal blade (the "hard triangular plate" of Bigelow and Schroeder, actually an extension of supraneurals), prominent, composed of several elements. Bases of dorsal and anal fins short; anal fin divided into two distinct sections with gap below ACB 2–4. Long-based adipose fin present. Spines present at posterior angle of fused posttemporal plus supracleithrum, at posteroventral angle of preopercle, as a complex of two spines associated with iliac process, in anal fin hiatus, and on platelike scales that cover VAV, ACB, and ACC photophores. Short spinelike structure present on cleithrum at ventral body margin. Caudal fin broad with narrow lobes, not deeply forked. Lateral line restricted to few canals on head.

Meristics. Dorsal fin rays 9; anal fin rays 12; pectoral fin rays 9 or 10; pelvic fin rays 6; vertebrae 34–36; gill rakers 15–17; photophores: ACB (6), ACC (4), BRP (6), IP (6), OV (2) + 6, OP 3, ORB 1, PV (12), VAV (4)

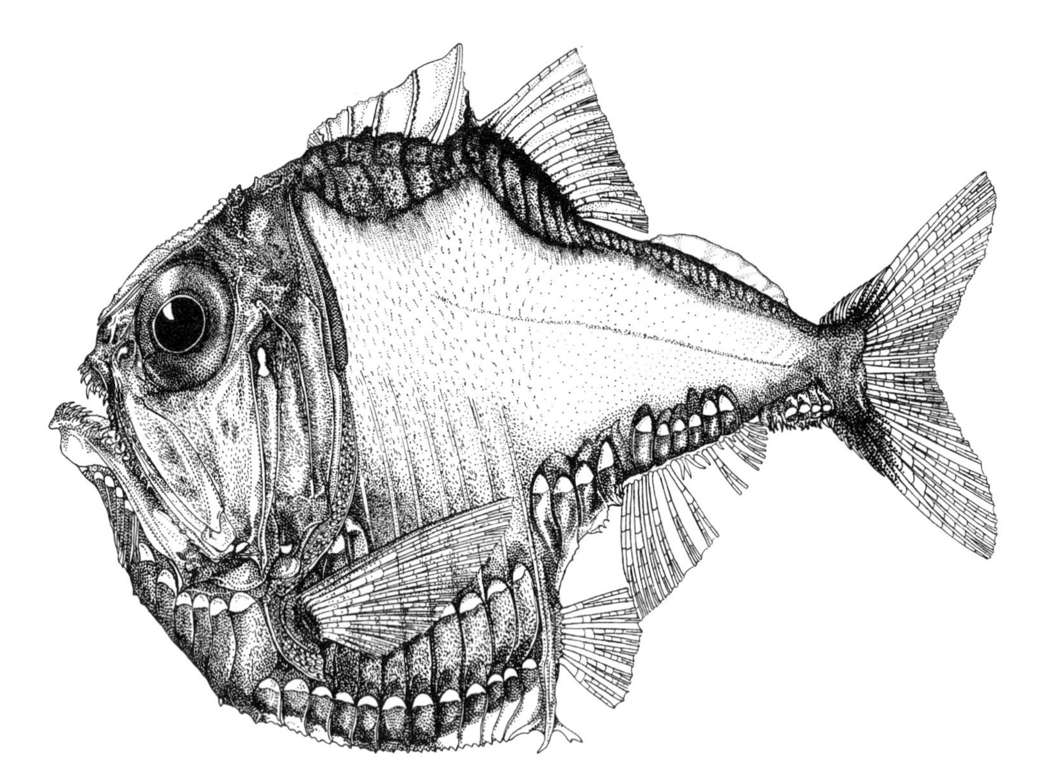

Figure 94. Silver hatchetfish *Argyropelecus aculeatus*. 62 mm SL, USNM 196700. Drawn by M. J. Johnson. (From Weitzman 1974: Fig. 12.)

Color. Most of sides of the body and head are very silvery. The entire dorsum is blue-black in life (dark brown in alcohol) with pigment that extends ventrally on the lateral surface of the head and the dorsal portions of the flank. On the latter area the pigment terminates sharply along an undulating margin. The ventral and ventrolateral surfaces are very darkly pigmented in association with the photophores.

Size. One of the largest hatchetfishes, reaching at least 83 mm SL. Individuals of 40–60 mm are quite common.

Distinctions. Distinguished from other Gulf of Maine hatchetfishes by tubular eyes, four VAV photophores, presence of six ACB (anal) photophores above the anal fin, minute spines on the ACC (subcaudal) photophores, and a maximum body depth of about 75% SL. The lower jaw canine teeth and the external spines originating in the anal fin gap distinguish it from all congeners.

Habits. Mesopelagic and found at 200–550 m during the day and 80–200 m at night, with a daily vertical migration of about 100–200 m. The greatest daytime concentrations are apparently between 350 and 450 m (Baird 1971; Howell and Krueger 1987). Preferred water temperature ranges from 10° to 21°C (Baird 1971) owing to its marked vertical migration. Captures are usually of small numbers of specimens, indicating little schooling.

Food. Diet consists of zooplankton, dominated by crustaceans such as copepods and mysids. Large individuals will occasionally eat other fishes, including other hatchetfishes.

Reproduction. Silver hatchetfish live 2 or 3 years, with most spawning taking place in the summer or fall of the second year (Howell and Krueger 1987), but some individuals may spawn in the first year. Most individuals die after spawning, although some survive and spawn the next season. Presence of larvae and ripe adults throughout the year indicates that there is always some spawning activity, at least in warmer waters. Larvae have not been found in the Gulf of Maine and it is doubtful that any spawning takes place there.

Early Life History. The eggs of *A. aculeatus* are unknown, but they are probably similar to those of *A. hemigymnus* (see Ahlstrom et al. 1984), with a segmented yolk. Larvae are like adults, but with a moderately elongate body trunk. By about 7 mm SL, PV, and AC photophores reach the full adult complement. Other photophore groups develop in the sequence VAV and OV (counts of four and six, respectively, by 12–13 mm SL), followed by ACB (a count of six by 16–17 mm SL). Paired iliac spines and a dorsal blade develop in postlarvae.

General Range. *Argyropelecus aculeatus* are widely distributed, mesopelagically throughout the Atlantic, Indian, and Pacific oceans between 47° N and 40° S.

Occurrence in the Gulf of Maine. This species is rare and generally peripheral to the Gulf of Maine (e.g., Grand Banks, Scotian Shelf, and off Georges Bank). Bigelow and Schroeder cited a single capture by *Albatross* (sta. 2063, 42°22′ N 66° 23′ W) between Georges and Browns banks at 266 m in August 1883.

WEITZMAN'S PEARLSIDE / *Maurolicus weitzmani* Parin and Kobyliansky 1993 /

Bigelow and Schroeder 1953:144–145 (as *Maurolicus pennanti*)

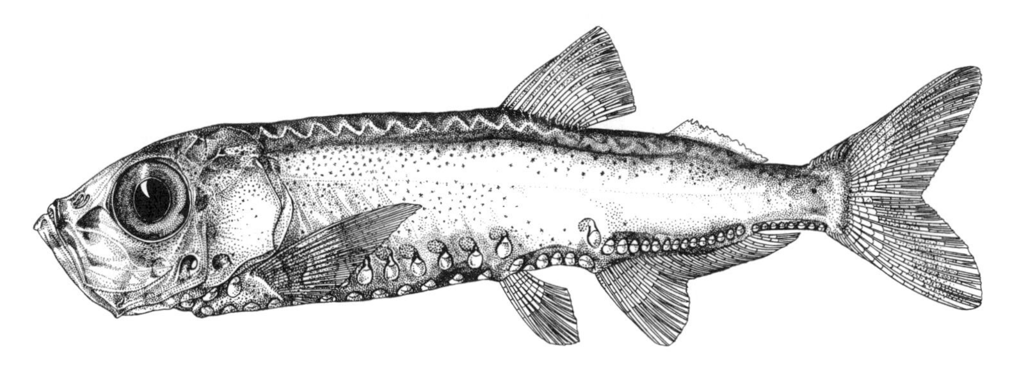

Figure 95. Weitzman's pearlside *Maurolicus weitzmani*. Off Florida, 26°24′ N, 79°50′ W, 45.5 mm SL, USNM 201138. Drawn by M. J. Johnson. (From Weitzman 1974: Fig. 5.)

Description. Body laterally compressed, not deep (Fig. 95). Head moderately deep, compressed, with large, vertical mouth. Jaw teeth minute, uniserial, and numerous on premaxilla and along posterior half of maxilla. Anterior half of lower jaw with a row of 12–14 small, recurved teeth. Usually two short parallel rows of two to four minute teeth near symphysis of lower jaw (Grey 1964; Weitzman 1974). Eyes lateral and large, one-third of HL. Base of dorsal fin short; that of anal fin quite long and lacking a hiatus; anal fin with an anterior lobe formed by elongation of first six or seven rays; a long-based adipose fin present; pectoral fin quite long, extending posteriorly about three-quarters of distance to pelvic fin base; caudal fin deeply forked, lobes rounded. OV photophores widely separated from one another, except first two joined.

Meristics. Dorsal fin rays 10 or 11; anal fin rays 19–24; pectoral fin rays 17–19; pelvic fin rays 7; vertebrae 32 or 33; gill rakers 17–19 + 4–7 = 22–25; photophores: AC 1 + (15 or 16) + (7–9) = 23–26, BRP (6), IP (6), OV (2) + 7 = 9, OP 3, ORB 1, PV (12), SO 1, VAV (4).

Color. Color in life is greenish, greenish blue, or brown dorsally and silvery laterally and ventrally (Grey 1964). Dorsal pigment is confined to a sharply defined area along the length of the body. Color in alcohol is similar but the dark dorsal area is greenish brown or black. Melanophores are scattered over the opercle and lateral surfaces of body. Dark pigment overlying guanine covers the photophore bulbs and interconnecting tissue. The peritoneum is dark brown to black. The photophores luminesce yellowish green.

Size. Maximum size 52 mm SL; individuals of 40–50 mm are most common.

Distinctions. Pearlsides differ from all other stomiiforms by having an isolated, elevated AC photophore present immediately anterior to the two long posterior groups in specimens larger than 19 mm SL. *Maurolicus* species are also distinguished by paired SO (symphyseal) photophores, absent in all other North Atlantic sternoptychids.

Taxonomic Note. *Maurolicus* has been considered monotypic, but Parin and Kobyliansky (1993, 1996) recognized 15 species and described the western North Atlantic species as new. In the past *M. weitzmani* has been treated under the name *M. muelleri* or *M. pennanti* (Bigelow and Schroeder; Scott and Scott 1988).

Habits. Several authors considered the species rare (Bigelow and Schroeder; Steele 1967; Musick 1973b; Scott and Scott 1988) but based on NMFS bottom trawl data, it is far more abundant than previously thought. Overall abundance is probably greatly underestimated owing to the large mesh used in most sampling. On a 1992 NMFS survey in the Gulf of Maine, the largest number of pearlside was taken pinched between the large wing-mesh of the net, which probably snapped shut as it came off the bottom (K. E. Hartel, pers. obs.). Data from NMFS specimens at MCZ show depth ranges of about 50–400 m with peak abundance at depths of 120–240 m. Grey (1964) reported a maximum depth of capture in the western Atlantic of 549 m, while Scott and Scott (1988) give a record of

1,000 m, but this may have been from an open net. Pearlside are most common over the shelf and upper continental slope (MCZ specimens). There is considerable evidence of association with the bottom, as supported by the numbers taken in bottom trawls off southern New England. However, the species may undergo diel vertical migration, as does *M. walvisensis* off the west coast of South Africa (Armstrong and Prosch 1991). Formation of dense scattering layers by various species of *Maurolicus* has been documented in the Atlantic, Indian, and Pacific oceans, indicating that schooling behavior may be common in the genus. Such "schools" have not been documented in the Gulf of Maine.

Food. The diet is made up mainly of various planktonic organisms, especially copepods and euphausiids (Badcock 1984b; Scott and Scott 1988; Bowman et al. 2000).

Predators. Found in the stomachs of *Argentina silus*, cod, pollock, cod, and silver hake (Rountree 1999). Cod and Atlantic herring are known to prey on pearlside in the eastern Atlantic. Grey (1964) listed a variety of midwater predators. In the southern Gulf of Maine, harbor porpoise (*Phocoena phocoena*) feed extensively on *Maurolicus;* of 85 stomachs examined, 3,600 otoliths were found in one and over 1,000 in five others (Gannon et al. 1998).

Reproduction. Presence of eggs, larvae, and ripe adults throughout the year indicates that there is always some spawning activity, at least in warmer waters. Length at maturity is about 25 mm SL, and fecundity in the eastern South Atlantic *M. walvisensis* is 200–500 eggs per female (Armstrong and Prosch 1991). Spawning usually takes place in the summer or fall of the second year but some may spawn in the first year. Most die after spawning, but some survive and spawn the next season.

Early Life History. Pearlside eggs are 1.32–1.63 mm in diameter with a single 0.25–0.28 mm oil globule (Ahlstrom et al. 1984). The yolk is segmented and the exterior has a hexagonal pattern (Sanzo 1931a; Mito 1961). Adult body form and fin placement are attained by about 11 mm SL. The first photophores to appear at 5.5 mm SL are BRP, followed by PV. Other light organs develop in the following sequence: OP, ORB, IP and ACB, VAV and ACC, OA (Ahlstrom 1974; Ahlstrom et al. 1984).

Eggs were collected by the MARMAP surveys throughout the year (Berrien and Sibunka 1999), with a maximum from September to November. In September, egg distribution was almost continuous along the shelf edge from off Cape Hatteras to southeastern Georges Bank, and in scattered locations in the Gulf of Maine (Berrien and Sibunka 1999: Fig. 19). Eggs were found in the Gulf of Maine from September through December.

General Range. Caribbean Sea and Gulf of Mexico north to southern Newfoundland. Populations range across the equatorial Atlantic from the Gulf of Guinea to off Brazil (Parin and Kobyliansky 1996).

Occurrence in the Gulf of Maine. Weitzman's pearlside is common throughout the Gulf of Maine and around Georges Bank in depths usually greater than 100 m. It is usually not found near shore but specimens are often washed ashore, especially on beaches in the Provincetown area at the tip of Cape Cod (Collette and Hartel 1988). In areas adjacent to the Gulf of Maine, *M. weitzmani* is very common to abundant. Up to 300 individuals are not uncommon in some 30-min NMFS groundfish trawls fished deeper than 100 m south of Cape Cod. Two larger size-classes are usually present based on MCZ specimens. The largest adults (probably ca. 3+ years) seem to drop out of the population in early spring. An additional smaller size-class (<20 mm SL) is also taken but these were undersampled owing to mesh size.

SLOPE HATCHETFISH / *Polyipnus clarus* Harold 1994

Description. Body laterally compressed, moderately deep (1.5–1.7 in SL) (Fig. 96). Body profile anterior to dorsal fin broadly elliptical and posteriorly tapered at 45–60°. Head large (2.9–3.4 in SL). Eyes lateral and large (15% of body length). Mouth, when closed, appears small and vertical but can be greatly protruded (Weitzman 1974). Jaw teeth generally recurved, quite variable in length. Single row of slightly recurved teeth present along maxillary margin. Frontal and parietal bones ornamented with serrate ridges or keels. A small posterodorsally directed spine present at posterior angle of fused posttemporal and supracleithrum. Two spines present at posteroventral angle of preopercle; ventral spine highly reduced, embedded within distal flange or lamella of bone; smaller spine located at base of ventral spine, directed laterally. Lateral line restricted to the head. Rays of all fins short and usually damaged upon capture. Bases of dorsal and anal fins moderate in length, continuous. A modified pterygiophore with an external bispinous process at origin of dorsal fin. Long-based adipose fin present. Caudal fin moderately broad, quite deeply forked. Cleithrum exits ventral body margin in a short spinelike structure. Two subequal iliac spines exit ventral body margin immediately lateral to pelvic fin. Ventral photophores (BRP and IC groups) arranged in horizontal rows or clusters. OV groups dorsal to IC, arranged as follows: OVA (2) + 1, increasing in elevation from anterior to posterior; OVB 3, in an "L" or "V" configuration. Other elevated photophores: L, anterior VAV, and the three ACA.

Meristics. Dorsal fin rays 15 or 16; anal fin rays 16 or 17; pectoral fin rays 13–15; pelvic fin rays 7; vertebrae 32 or 33; gill

Figure 96. Slope Hatchetfish *Polyipnus clarus.* 33°59′ N, 76°05′ W, 43.6 mm SL, USNM 273283. (From Harold 1994: Fig. 31.)

rakers 19–21; photophores: ACA 3, ACB (8–10), ACC (4), BRP (6), IP (6), L 1, OP 3, ORB 1, OVA (2) + 1, OVB 3, PV (12), VAV 1 + (4).

Color. The dorsum is pale brown in alcohol, with pigment terminating sharply in a straight margin except for a triangular bar extending ventrally immediately anterior to the dorsal fin. Dark pigment is also present on the dorsal parts of the photophores and along myosepta.

Size. Maximum body size about 60 mm SL; specimens over 50 mm are rare.

Distinctions. Distinguished from all other Gulf of Maine fishes by its deep body, presence of a bispinous predorsal process, OV photophores in a nonlinear arrangement, an L photophore, and AC photophores in the formula 1 + (2) + (8–10) + (4). No other species of *Polyipnus* is known from the Gulf of Maine but *P. laternatus* has been recorded as far north as Cape Hatteras. It is distinguished by more numerous ACB photophores (11–13 compared with 8–10), a longer posttemporal spine, and its free (not embedded) ventral preopercular spine (see Harold 1994).

Taxonomic Remarks. This species was described (Harold 1994) as distinct from the Caribbean *P. asteroides* Schultz 1938. Previous authors (e.g., Baird 1971; Scott and Scott 1988) treated *P. clarus* under the name *P. asteroides.*

Habits. Species of *Polyipnus* appear to be benthopelagic, associated with the continental or island slope fauna and are usually captured by bottom trawl rather than by midwater tow. *Polyipnus clarus* is most often captured near the continental slope at depths of 300–400 m, although it has been caught shallower than 100 m in the Gulf of Maine and in the Gulf of Mexico. It is not known whether *P. clarus* undergoes diel vertical migration, but two Indo-Pacific species appear to migrate between 50 and 100 m at night (Harold, unpubl. data).

Food. *Polyipnus* are planktivorous, feeding on copepods, euphausiids and mysids of the upper mesopelagic and epipelagic zones near the slope.

Predators. *Polyipnus* otoliths have been found in the stomachs of common dolphin (*Delphinus delphis*) from just over the shelf break (Craddock, unpubl. data).

Reproduction. Spawning may occur throughout the year, but peak spawning is likely in winter, based on the few captures of larvae (6–11 mm SL) collected between February and May. Larvae are captured at depths similar to those of juveniles and adults.

Early Life History. Larvae are deep and laterally compressed like the adults (see illustration of *P. polli* larva in Ahlstrom et al. 1984). The body trunk is not elongate as in *Argyropelecus* species. The smallest specimens (6.0–9.5 mm SL) have well-developed bispinous posttemporals, one spine being reduced or lost in adults. Long anterior anal fin pterygiophore spines and a lateral pigment bar are also present at this size and retained in adults. By 14 mm SL, most photophores reach adult complements. Development of the ACB photophores is protracted and the full number of nine or ten is not attained until 24 or 25 mm SL.

General Range. This species occurs from off Venezuela, through the Gulf of Mexico and Antilles, and northward off the east coast of the United States to the Scotian Shelf (Harold 1994).

Occurrence in the Gulf of Maine. Three of the four known records from the Gulf were taken in the Wilkinson Basin in

190–253 m (MCZ 94480, 97211, 146456). The fourth record is from deep in the Gulf at 43°10′ N, 68°54′ W from 100 m (MCZ 97210, Aug. 1991). The species is probably much more common in the Gulf than these data indicate, given that the near-bottom habitat has been poorly sampled. Captures in the Gulf of Maine and on the Scotian Shelf (Musick 1973b; Scott and Scott 1988) are limited to one or two specimens per tow. Immediately south of Cape Cod, *P. clarus* is more abundant, with up to eight specimens taken in a single tow (MCZ 91623).

LIGHTFISHES. FAMILY PHOSICHTHYIDAE

JAMES E. CRADDOCK AND KARSTEN E. HARTEL

Phosichthyidae is a small family of bioluminescent, pelagic fishes containing 22 species in seven genera. Before Weitzman (1974), species in the Phosichthyidae and Gonostomatidae were included in the family Gonostomatidae. The seven genera appear to share no uniquely derived characters (Harold and Weitzman 1996), but the family Phosichthyidae is used here for convenience. Eight species are benthopelagic over continental and island slopes; the other 14 species are mesopelagic. They have photophores on the isthmus, two ventrolateral rows of photophores running the length of the body, and a short dorsal fin located about midbody; they lack large fangs or barbels. Seven species of phosichthyids are found in the deeper water adjacent to the Gulf of Maine and off Georges Bank. *Polymetme thaeocoryla* is the only phosichthyid known from the Gulf of Maine. *Yarrella blackfordi* has recently been found at 360–800 m on the south slope of Georges Bank (MCZ 124870, 126580) but has not been found in the Gulf of Maine.

KEY TO PHOSICHTHYIDAE OCCURRING OFF NEW ENGLAND

1a. Two orbital (ORB) photophores; premaxillary teeth uniserial 2
1b. One orbital photophore; premaxillary teeth biserial 4
2a. Anal fin origin well behind end of dorsal fin base; branchiostegal (BRP) photophores 11–18. . . . . *Ichthyococcus ovatus*
2b. Anal fin origin under or just behind end of dorsal fin base; branchiostegal photophores 8 or 9. 3
3a. Length of anal fin base equal to length of dorsal fin base; anal rays 12–16; AC photophores 12–16, 6 or 7 over anal fin base . . . . *Vinciguerria*
3b. Length of anal fin base twice the length of dorsal fin base; anal rays 22–30; AC photophores 19–21, 13–15 over anal fin base . . . . *Pollichthys mauli*
4a. Two rows of photophores on body; photophores large; adipose fin present; VAV photophores 7 or 8; OA photophores ending before anal fin origin . . . . *Polymetme thaeocoryla*
4b. More than two rows of photophores on body; photophores small; adipose fin absent; VAV photophores 9–12; OA photophores extend to caudal fin base . . . . *Yarella blackfordi*

Polymetme thaeocoryla Parin and Borodulina 1990

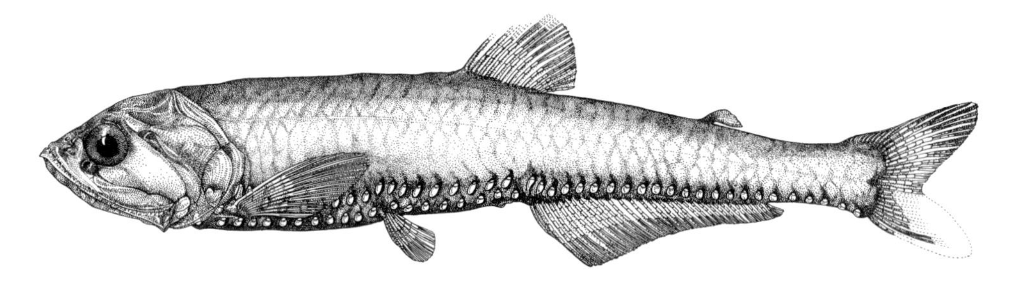

Figure 97. *Polymetme thaeocoryla.* Caribbean Sea, 162 mm SL, USNM 203281. (From Weitzman 1974: Fig. 15.)

Description. Body slightly elongate (Fig. 97). Dorsal fin base short, origin at about midbody; longer anal fin originating under posterior edge of dorsal fin; adipose fin opposite posterior quarter of anal fin. Two rows of large, highly visible photophores on body.

Meristics. Dorsal fin rays 12–14; anal fin rays 30–34; pectoral fin rays 9–11; pelvic fin rays 7; photophores: BRP 9, IV 21, VAV 8, AC 24–25, IC 53–54, OA 17.

Color. In life, the body is bright silver and highly reflective but in alcohol specimens fade to tan.

Size. Maximum reported length is 216 mm SL, from the eastern Atlantic (Badcock 1984c); the largest from the western Atlantic is 207 mm SL (Grey 1964).

Distinctions. This species is distinguished from *Pollichthys, Vinciguerria,* and *Ichthyococcus* by lacking a second orbital photophore and by having biserial (instead of uniserial) premaxillary teeth. It differs from *Yarrella* and *Triplophos* by having an adipose fin and lacking more than two longitudinal rows of photophores.

Taxonomic Remark. Until Parin and Borodulina (1990) recognized and described *P. thaeocoryla,* it had been included with the Indo-Pacific *P. corythaeola* (Grey 1964; Scott and Scott 1988).

Biology. Almost nothing is known about the biology of *P. thaeocoryla*. It is benthopelagic on the upper continental slope. Extreme range of depth is 230–880 m off southern New England (MCZ specimens). We have identified a 9.5-mm larva (MCZ 93241), which looks very much like larvae of *Yarrella* collected by Michael Fahay (NMFS) from a Bongo net fished from 0 to 200 m on 22 October 1985 off Georges Bank. Two juveniles (72 and 82 mm; MCZ 91007-008) were observed and collected just above the bottom in the Bahamas by the *Johnson Sea Link I* at depths of 578 and 598 m (P. Herring, pers. comm., 1989); both fish were in a horizontal attitude when seen.

General Range. In the western Atlantic, *P. thaeocoryla* occurs along the continental slope from as far north as the Scotian Shelf (Scott and Scott 1988) almost to the equator. It occurs irregularly north of Cape Hatteras; we know of only 15 records between Cape Hatteras and Georges Bank (MCZ and USNM specimens). In the eastern Atlantic it occurs from Ireland to the Gulf of Guinea; two specimens have been reported from the western South Atlantic (Parin and Borodulina 1990).

Occurrence in the Gulf of Maine. We know of only a single record from the Gulf of Maine, a 162-mm specimen (MCZ 144434) taken 10 October 1995 by *Albatross IV* at 285 m in the Northeast Channel.

DRAGONFISHES AND VIPERFISHES. FAMILY STOMIIDAE

KARSTEN E. HARTEL AND JAMES E. CRADDOCK

Stomiids form a large group of almost 200 species of deep-ocean fishes that were recently united by joining several related families (Fink 1985; Harold and Weitzman 1996). They are generally elongate, have large teeth, and a single barbel on the chin (absent in *Photostomias* and adult *Chauliodus*). Photophores and other light organs are present on the body, head, barbel, and sometimes on the fins. Most species are mesopelagic, living at 100–1,000 m. Almost all are predators that swallow their prey whole. Stomiids constitute a significant portion of the mesopelagic ecosystem. A total of 58 species is known from the Slope Water off New England (MCZ specimens) and 83 species have been reported from the Gulf of Mexico (Sutton and Hopkins 1996). As other mesopelagic fishes, species in this family are only casual wanderers into the Gulf of Maine. Occasionally they may be found in the stomachs of deep-feeding fishes or mammals.

KEY TO STOMIID FAMILIES AND SELECTED GENERA FROM OFF NEW ENGLAND

1a. Body silvery, iridescent, and covered with a hexagonal scalelike pattern . 2
1b. Body color variable, black to brown and sometimes metallic, but never with hexagonal scalelike patterns 3
2a. Dorsal fin far anterior, near or over tip of pectoral fins; teeth extremely elongate, lower reaching above eye level with jaws closed; barbel absent in adults (Chauliodontinae) . . . . ***Chauliodus***
2b. Dorsal fin far back on body, opposite anal fin; teeth not very elongate, largest about equal to eye diameter; barbel present at tip of lower jaw (Stomiinae) . . . . ***Stomias***
3a. Dorsal fin origin at about midbody, well in advance of anal fin 4
3b. Dorsal and anal fins opposite, and far posterior 5
4a. Dorsal fin short with fewer than 25 rays . . . . **Astronesthinae** (not known from the Gulf of Maine)
4b. Dorsal fin long, more than 50 rays; body elongate and eel-like . . . . **Idiacanthinae** (not known from the Gulf of Maine)
5a. Membranous floor of mouth absent . . . . **Malacosteinae** (not known from the Gulf of Maine)
5b. Membranous floor of mouth present (Melanostomiinae). . ***Trigonolampa***

SLOAN'S VIPERFISH / *Chauliodus sloani* Bloch and Schneider 1801 / Bigelow and Schroeder 1953:145–146

Description. Body slender and compressed, covered with a hexagonal scalelike pattern and coated in a jacket of clear, gelatinous material in fresh specimens (Fig. 98). Head deep. Eyes large. Both jaws armed with large widely spaced teeth, those at lower jaw tip extending outside of closed mouth to a point well above level of eye. Dorsal fin placed far forward, over fifth to eighth OA (lateral) photophore. First dorsal fin ray elonga and filamentous. Dorsal adipose and anal fins far posterior. Two rows of large photophores along lower side.

Meristics. Dorsal fin rays 6; anal fin rays 10 or 11; pectoral fin rays 12 or 13; pelvic fin rays 7; photophores: BRP 13–16, IP 9–11, PV 18–21, VAV 24–28, AC 9–13, IC 62–72, OA 42–49; vertebrae: 54–[illegible] (Mor[illegible] 1964).

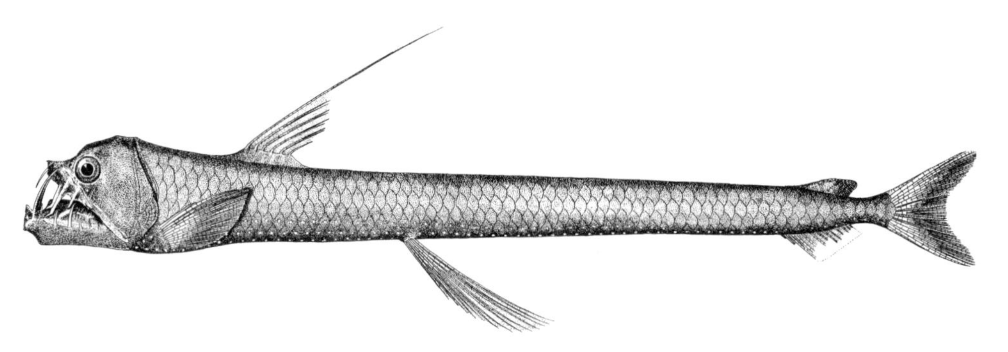

Figure 98. Sloan's viperfish *Chauliodus sloani.* 42°08′ N, 65°35′ W, USNM 23420. Drawn by H. L. Todd. (From Goode and Bean 1896: Fig. 115.)

Size. To about 300 mm TL (Gibbs 1984).

Distinctions. A combination of hexagonal scalelike pattern, forward placed dorsal fin, huge teeth, and photophores distinguishes *Chauliodus* from all other Gulf of Maine fishes. It differs from the only other western Atlantic species in the genus, *C. danae,* by having its dorsal fin farther forward, over the fifth to eighth OA photophore instead of farther back, over the ninth to eleventh photophore.

Biology. Sloan's viperfish is fairly common off the continental slope below 300 m. Observations from submersibles in recent years show that it orients in midwater in a horizontal position, with its large pectoral and pelvic fins spread wide and the long second ray of the dorsal fin thrown far forward with its tip dangling in front of the head (R. Harbison, pers. comm., 1994). Although it has never been seen to strike at prey, Tchernavin (1953) showed that the specialized anatomy of the head allows the viperfish to lift its head far back and depress its lower jaw in order to open its mouth wide enough to allow food to pass by its enormous teeth. Regardless of its large teeth, it seems to feed mostly on small midwater fishes and crustaceans, which are swallowed whole. Viperfish migrate vertically. In the Gulf of Mexico, they have an asynchronous vertical migration pattern, residing deep in the mesopelagic zone (450–700 m) during the day with about 70% of the individuals vertically migrating into the upper 200 m at night and the others remaining at depth (Sutton and Hopkins 1996).

Reproduction. Viperfishes have spherical pelagic eggs. Larvae are about 7 mm at hatching, grow to about 44 mm, and then shrink to 27 mm before growth resumes (Fahay 1983). Larvae have a chin barbel, which is lost during transformation.

General Range. Sloan's viperfish is found from 55° N to 50° S in the Atlantic and is also widespread in the Indian and Pacific oceans (Gibbs 1984).

Occurrence in the Gulf of Maine. This species is a stray to the Gulf of Maine. Bigelow and Schroeder knew of only two local records: one from a cod stomach taken on Georges Bank in 1871 and the other from a swordfish harpooned between Browns and Georges banks in 1931. Musick (1973b) reported a juvenile (89 mm SL) taken in a midwater trawl fished between 20 and 160 m inside Northeast Channel (42°20′ N, 60°45′ W). We know of five additional specimens from the edges of Georges Bank between 150 and 380 m collected by NMFS surveys.

BOA DRAGONFISH / *Stomias boa ferox* Reinhardt 1843 / Bigelow and Schroeder 1953:147 (as *Stomias ferox*)

Description. Dorsal and anal fins opposite each other, set far back on elongate body, depth 10–15 in SL (Fig. 99). Head small, 10 in SL. A short chin barbel present, with two to four distal filaments. Large scalelike hexagons set in six rows along body.

Meristics. Dorsal fin rays 17–21; anal fin rays 19–23; pectoral fin rays 6; pelvic fin rays 5; photophores: BRP 16 or 17, IP 10–13, PV 46–51, VAV 10–13, AC 15–19, IC 85–91, OA 57–63; vertebrae 77–83 (Morrow 1964b).

Size. Maximum size about 300 mm (Gibbs 1984).

Distinctions. *Stomias* can be distinguished from other Gulf stomiids by the combination of opposite dorsal and anal fins set far back on an elongate body, a short chin barbel with two to four distal filaments, and large scalelike hexagons set in five to six rows along the body. *Stomias boa ferox* is the only Atlantic species commonly found north of 40° N but *S. brevibarbatus* has been recorded as far as 42° N off the Scotian Shelf (Coad 1986). *Stomias brevibarbatus* has five rows of hexagons above the lateral photophores (six in *S. boa ferox*) and ten or more premaxillary teeth (fewer than eight in *S. boa ferox*). *Stomias affinis,* which is also found in Slope Water, has six rows of hexagons like *S. b. ferox* but it has five to eight VAV photophores vs. ten to thirteen in *S. b. ferox.*

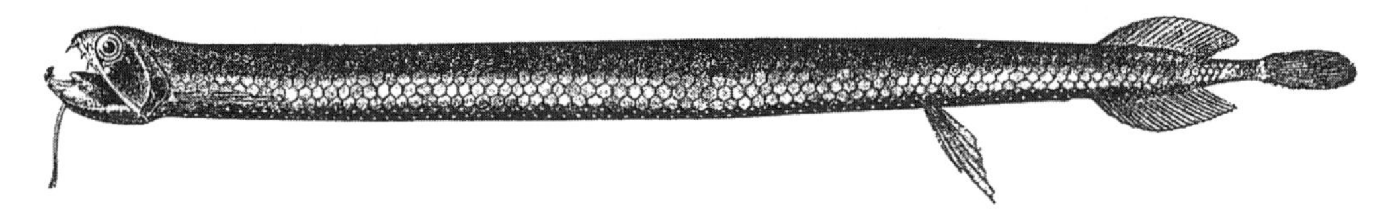

Figure 99. Boa dragonfish *Stomias boa ferox*. (Reprinted from Morrow 1964: Fig. 75. © Sears Foundation for Marine Research.)

Habits. This dragonfish is a mesopelagic species that lives deeper than 500 m during daylight hours and migrates each night to the upper 200 m. Recent observations from submersibles off New England show that they position themselves horizontally in the water column with pelvic and pectoral fins spread wide and the barbel pointing forward (R. Harbison, pers. comm., 1994). They are presumed to be lie-and-wait predators capable of quick bursts of speed.

Reproduction. Eggs are unknown. Larvae are translucent, grow to about 44 mm, shrink to about 23 mm, then resume growth, develop photophores, and transform into the adult form (Fahay 1983).

General Range. *Stomias boa ferox* is found across the North Atlantic north of 30° N, where it is quite common below 500 m. This subspecies is replaced off northern Africa and in the Mediterranean by *S. boa boa* (Gibbs 1969).

Occurrence in the Gulf of Maine. A rare stray in the Gulf of Maine but resident in the Slope Water. Bigelow and Schroeder reported a "12-in." specimen from about 180 m just north of Georges Bank (42°10′ N, 67°05′ W). Ten juveniles (70–166 mm SL) were collected inside Northeast Channel (42°20′ N, 66°45′ W) with a midwater trawl fished at 20–160 m (Musick 1973b). We have seen many specimens from depths of 220–800 m from the south face of Georges Bank collected by NMFS surveys. Bigelow and Schroeder's record of *Stomioides nicholsi* Parr from a swordfish harpooned on the southeast edge of Browns Bank in 1932 is this species.

THREELIGHT DRAGONFISH / *Trigonolampa miriceps* Regan and Trewavas 1930 /

Bigelow and Schroeder 1953:148

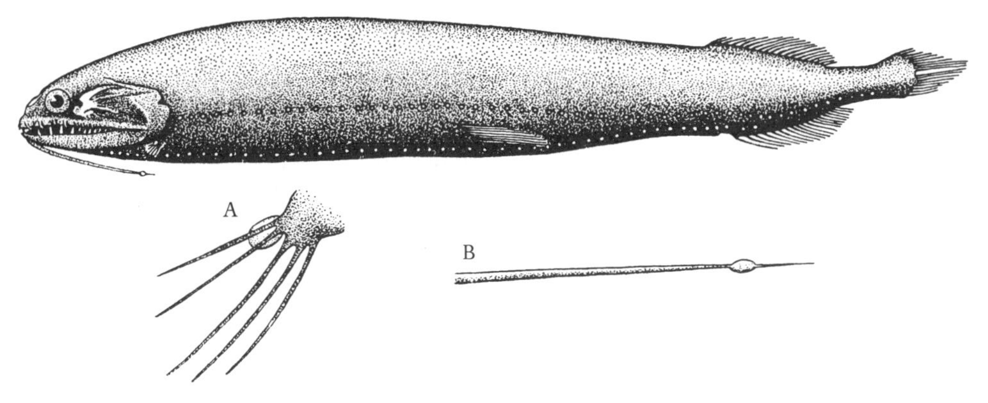

Figure 100. Threelight dragonfish *Trigonolampa miriceps*. MCZ 37161. (A) Base of left pectoral fin to show arrangement of rays and lump of luminous tissue. (B) Tip of barbel. Drawn by S. G. Hartman. (Reprinted from Morrow and Gibbs 1964: Fig. 96. © Sears Foundation for Marine Research.)

Description. Body elongate, depth six to ten in SL (Fig. 100). Head moderate, five to six in SL; chin barbel shorter than head; terminal luminescent bulb with a distal filament (Gon 1990b). Large photophore near eye. Two rows of photophores along lower body.

[illegible]teris[illegible] Do[illegible] fin r[illegible] 9 o[illegible] 20; an[illegible] fi[illegible] ays 18 [illegible] 19; [illegible] ral [illegible] ys [illegible] lvic [illegible] ays [illegible] hot[illegible] s: BR [illegible] 1–16 [illegible] 10–12, PV 21–26, VAV 20–24, AC 10–14, IC 65–70, OA 47–50; vertebrae: 61 or 62 (Morrow and Gibbs 1964; MCZ specimens).

Color. Body black to dark brown with one or two luminous lines on flanks that may extend as far as the dorsal and anal fin bases (Gibbs 1984).

[illegible]ze. This dra[illegible]o[illegible] fish [illegible] ea[illegible] es [illegible] o[illegible] t 320 mm (Gibbs 1984).

Distinctions. *Trigonolampa* is distinguished from other North Atlantic stomiids most readily by the large area of luminescent tissue behind the eye, which often has white lines radiating back toward the opercle. It is made up of two parts: a large photophore lower and smaller than the eye and a large patch of luminescent tissue. *Trigonolampa* most closely resembles *Photonectes margarita* and *Melanostomias bartonbeanii,* both of which occur commonly in the Slope Water. *Photonectes* has a lower jaw that is longer than the upper and curves distinctly upward. *Melanostomias* has many tightly packed teeth, some of which are depressible, whereas *Trigonolampa* teeth are widely spaced and nondepressible (Morrow and Gibbs 1964).

Biology. Essentially nothing is known about the life history of *Trigonolampa.* It probably lives at depths of 300–1,000 m (MCZ data; Gon 1990b). A juvenile (43 mm SL) was taken as shallow as 65 m just north of the Azores (MCZ 114691). Eggs and larvae are unknown.

General Range. *Trigonolampa* is one of the rarest of the scaleless dragonfishes and the species is known from only a handful of specimens. It is found in temperate to polar waters in the North Atlantic from north of about 40° N and in the South Atlantic south of 30° S. Two specimens from just outside the Gulf of Maine mentioned below, three adults (MCZ 124654–55, 137986) taken by the *Contender* in 1995 at 430–1,830 m on the south slope of Georges Bank, and one specimen (YPM 11438) from off Delaware are the only six specimens known from the western North Atlantic (Scott and Scott 1988; MCZ records).

Occurrence in the Gulf of Maine. The species is included here based on two MCZ specimens from swordfish stomachs taken near Georges Bank in 1913 and 1922 (Bigelow and Schroeder).

ACKNOWLEDGMENTS. Drafts of stomiiform sections were reviewed by William H. Krueger, Jon A. Moore, Kenneth A. Tighe, and Stanley Weitzman.

AULOPIFORM FISHES. ORDER AULOPIFORMES

Aulopiformes are defined by several unique specializations of the gill arches (Rosen 1973; Johnson 1992) as well as features of the intermuscular bones, internal soft anatomy, larval pigmentation, and morphology of the pelvic girdle (Baldwin and Johnson 1996). Many aulopiforms are synchronous hermaphrodites. The order contains 13 families of pelagic, bathypelagic, or benthic fishes with about 219 species (Nelson 1994), of which three species from three families are found in the Gulf of Maine. Scott and Scott (1988) report 20 species in nine families (including the three Gulf of Maine species) from Atlantic waters of Canada. Several of these (such as lizardfishes, family Synodontidae) could occur as strays in the Gulf of Maine, but have not yet been reported from there.

KEY TO GULF OF MAINE AULOPIFORM FISHES

1a. Dorsal fin extends nearly entire length of body . . . . **Longnose lancetfish**
1b. Dorsal fin short, ending about midway along body 2
2a. Dorsal fin origin anterior, over pectoral fin; body fusiform; teeth small and in bands on jaws and palatines. . . . . **Shortnose greeneye**
2b. Dorsal fin origin well behind pectoral fin; body elongate and compressed; lower jaw and palatines with large canine teeth . . . . **White barracudina**

GREENEYES. FAMILY CHLOROPHTHALMIDAE

KENNETH A. TIGHE

Single elongate supramaxilla. Eyes large and normal. Pseudobranch and pyloric caeca present. Worldwide in moderately deep marine waters. Two genera with 20 species (Nelson 1994), of which one has been reported from the Gulf of Maine.

SHORTNOSE GREENEYE / *Chlorophthalmus agassizi* Bonaparte 1840

Description. Body elongate, compressed posteriorly, but round in cross section anteriorly (Fig. 101). Head relatively large, approximately 30% of SL. Eye large, approximately 10% of SL; pupil horizontally elongate and keyhole-shaped. Snout depressed. Dorsal fin originates just anterior to pelvic fin origin. Anal fin originates below and slightly anterior to well-developed adipose fin. Anus black; slightly posterior to pelvic origin.

Meristics. Dorsal fin rays 10 or 11; anal fin rays 7–9; pectoral fin rays 15–17; pelvic fin rays 8 or 9; gill rakers on lower limb of first arch 19–22; vertebrae 46–48 (Mead 1966).

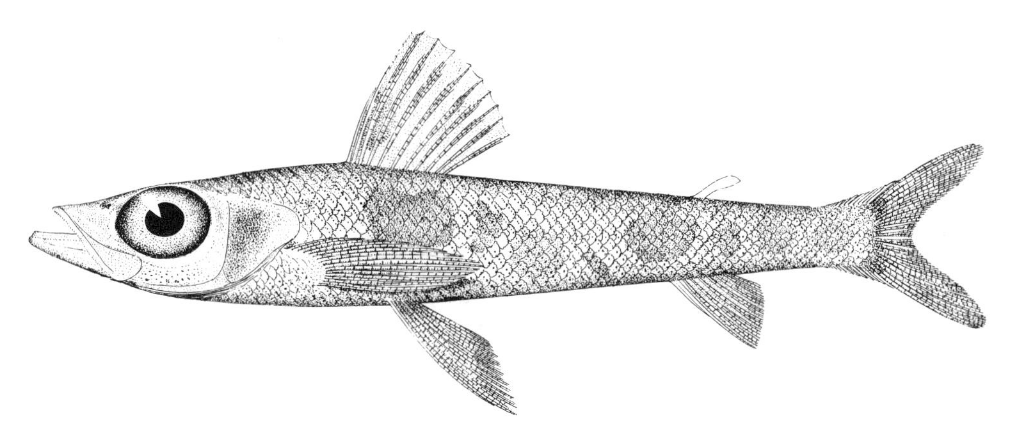

Figure 101. Shortnose greeneye *Chlorophthalmus agassizi*. USNM 41600. Drawn by A. H. Baldwin.

Color. In life, shortnose greeneye are silvery with dark dorsal blotches. When alive (or recently dead), the tapetum of the eye reflects a light green, hence their common name.

Size. Reach more than 160 mm, but the average size of adults caught with bottom trawls is about 115 mm (Mead 1966).

Distinctions. The large head, large eyes with keyhole shaped pupil, and presence of an adipose fin distinguish shortnose greeneye from all other Gulf of Maine fishes.

Biology. Shortnose greeneye are small benthonic predators that occur in continental slope and in deeper coastal waters. Mead (1966) suggested that they are usually found on mud or clay bottoms, or rarely on fine sand. They apparently feed primarily on small benthic invertebrates. Larvae and juveniles have been described from the Mediterranean Sea (Sanzo 1915; Tåning 1918). Shortnose greeneye are occasionally eaten by cod, goosefish, and fourspot flounder (Rountree 1999).

General Range. In the western North Atlantic, shortnose greeneye are known from continental slope waters ranging from off Nova Scotia to off northern South America, including the Gulf of Mexico and some Caribbean islands (Mead 1966; Scott and Scott 1988). They are also known from the eastern North Atlantic and the Mediterranean Sea, where they are commercially utilized.

Occurrence in the Gulf of Maine. Although they have not previously been recorded from the Gulf of Maine, shortnose greeneye are occasional inhabitants of deeper waters of the Gulf. Although they are regularly taken in the NMFS groundfish surveys on the continental slope just outside of the Gulf (Map 11), they are rare components of the catch within deeper waters between Georges Bank and Browns Bank. A total of seven specimens was collected at three stations in Northeast Channel at depths of 193–265 m on *Albatross IV* cruise 94-02 in April 1994 (USNM 329909, 329913, 329915).

BARRACUDINAS. FAMILY PARALEPIDIDAE

KENNETH A. TIGHE

Barracudinas superficially resemble barracudas. The dorsal fin origin is in the middle of the body and the anal fin base is long. Maximum length about 1 m. Worldwide in all oceans from the Arctic to the Antarctic. There are 12 genera and about 56 species (Nelson 1994), of which one extends into deep waters of the Gulf of Maine.

WHITE BARRACUDINA / *Arctozenus rissoi* (Bonaparte 1840) / Musick 1973b:135 (as *Notolepis rissoi kroyeri*)

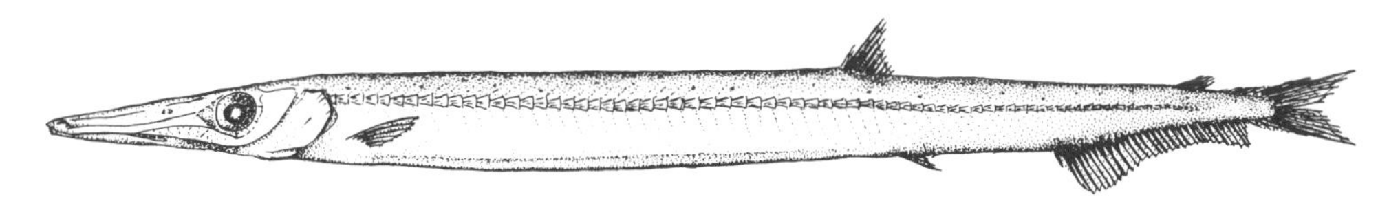

Description. Body elongate, somewhat compressed (Fig. 102). Head long, with a sharply pointed snout and a large, long mouth. Jaws equal or lower jaw projecting slightly. Teeth on both jaws large depressible canines accompanied by shorter fixed canines. Eye large (3–4% of SL), round, with well-developed adipose eyelid. Dorsal fin located just posterior to midpoint of body. Pelvic fin origin behind vertical from posterior of dorsal fin base. Anal fin located far back on body. Adipose fin well developed, located over posterior anal rays. Anus slightly posterior to pelvic fin origin, well anterior to anal origin. Lateral line well developed, within a series of enlarged, embedded scales.

Color. In life, white barracudina are a brilliant iridescent silver with a dark dorsal. The deciduous scales are usually lost upon capture and the silver color is lost, especially when preserved. In preservative, the head is dark brown dorsally and over the snout, and the body is mottled dark brown.

Meristics. Dorsal fin rays 8–10; anal fin rays 30–33; pectoral fin rays 11 or 12; pelvic fin rays 9; vertebrae 80–85 (Rofen 1966).

Size. Reach about 300 mm in length (Rofen 1966).

Distinctions. White barracudina might be confused with argentine, capelin, or smelt. They can be distinguished from argentine by the smaller mouth and toothless jaws of the latter; from capelin by the dentition (large depressible canines vs. small teeth); and from smelt by the enlarged lateral line scales (vs. relatively small) and the broad lateral silvery band of smelt. White barracudina have more anal rays (30–33) than argentine (11–17), capelin (18–20), or smelt (12–16). White barracudina is the only paralepidid known from within the Gulf of Maine, but Scott and Scott (1988) reported three additional species within Canadian waters just outside the Gulf. White barracudina can be distinguished from other barracudinas that might occur in the Gulf by the posterior pelvic origin (posterior to dorsal fin base vs. before or below dorsal fin base) and the greater number of anal fin rays (30–32 vs. fewer than 26).

Biology. White barracudina are fish of the open oceans, mainly at middepths. They are carnivorous, feeding primarily on pelagic shrimps and larval fishes. Rofen (1966) suggested that spawning in the North Atlantic occurred southwest of Ireland and in the Bay of Biscay. Rofen (1966) also described development of a series of postlarval specimens.

Predators. White barracudina are eaten by many fishes, including cod, pollock, and redfish (Scott and Scott 1988) and are a major component of the diet of swordfish (Scott and Tibbo 1968, 1974). They are also eaten by tunas and lancetfishes (Matthews et al. 1977) and occasionally by harp seal (Lawson and Stenson 1997).

General Range. Wide-ranging in the Atlantic, Pacific, and Indian oceans as well as the Mediterranean Sea.

Occurrence in the Gulf of Maine. Musick (1973b) was the first to report *Arctozenus rissoi* (as *Notolepis rissoi kroyeri*) from the Gulf of Maine. White barracudina are regular (but rare) inhabitants of deeper waters of the Gulf. They occur as rare components of the catch in NMFS surveys, and there are numerous specimens in the collections of the Museum of Comparative Zoology from the Gulf of Maine.

LANCETFISHES. FAMILY ALEPISAURIDAE

KENNETH A. TIGHE

Lancetfishes have a huge dorsal fin that extends above most of the body. Scales and photophores are absent. The mouth is large with very large teeth. Marine fishes of the Atlantic and Pacific oceans. One genus with two species.

LONGNOSE LANCETFISH / *Alepisaurus ferox* Lowe 1833 / Bigelow and Schroeder 1953:161

Description. Body slender, somewhat compressed, deepest at gill covers, tapering back to slender caudal peduncle (Fig. 103). Snout long, one-third to one-half of HL, pointed. Mouth wide, gape extends posterior to eye. Palatines and lower jaws with two or three large fangs besides smaller teeth. Dorsal fin origin over posterior edge of opercle; fin occupying most of dorsum, highest near anterior portion with the distal tips of longest rays free from fin membrane, rays slightly lower in mid portion and tapering rapidly posteriorly; about twice as high as fish is deep; depressible in a groove along back. Adipose fin well developed. Anal fin originates under posterior portion of dorsal fin, deeply concave in outline. Pelvic fins about as long as body is deep, situated about halfway between anal origin and tip of snout. Pectoral fins considerably longer than body depth, situated very low on sides below dorsal fin origin. Fins exceedingly fragile. Caudal fin very deeply forked, upper lobe prolonged as a long filament (lost in most specimens).

Meristics. Dorsal fin rays 36–45; anal fin rays 15–17; pectoral fin rays 14 or 15; pelvic fin rays 8–10; gill rakers 23–28.

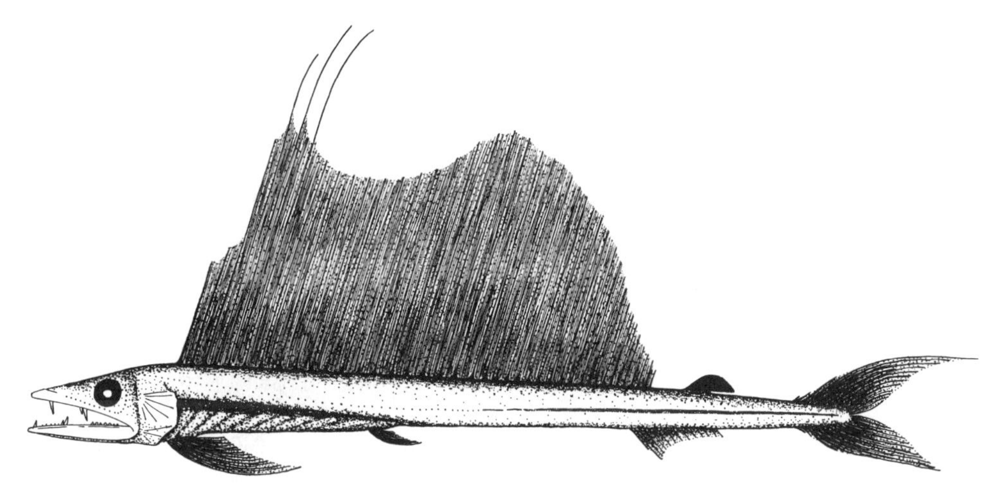

Figure 103. Longnose lancetfish *Alepisaurus ferox*. Drawn by G. G. Pasley. (From Gibbs 1960: Fig. 1, lower.)

Color. Iridescent silvery and generally rather pale, darkest dorsally. All fins are dark brown to black.

Size. Reach a length of at least 1,515 mm (Gibbs and Wilimovsky 1966). A specimen of *Alepisaurus* (not identified to species) of 2,080 mm washed ashore at Monterey Bay, Cal., on 4 July 1949.

Distinctions. Combination of a long and very high soft-rayed dorsal fin with the presence of an adipose fin and huge fang-like teeth distinguish longnose lancetfish from all other Gulf of Maine fishes. The only possible source of confusion could be with shortnose lancetfish, *A. brevirostris,* which have not been reported from the Gulf of Maine but do occur off Georges Bank (where the holotype and some paratypes were collected). The two species were confused by workers prior to the description of shortnose lancetfish by Gibbs (1960), so early records mentioned by Bigelow and Schroeder and discussed below could be of either species. Scott and Scott (1988) reported a shortnose lancetfish from off the Newfoundland Shelf, northwest of Flemish Cap, so both species could occur in the Gulf of Maine.

Biology. Longnose lancetfish were considered by Bigelow and Schroeder to be an oceanic species of the middepths, appearing only as strays in depths of less than 400 m. However, Gibbs (1960) reported longline captures of both species at depths of 40 m or less, and Gibbs and Wilimovsky (1966) reported longline captures in the Gulf of Mexico at depths of 20 m. In addition, Sæmundsson (1949) reported a shoal of 30–40 individuals in surface waters off Iceland in the spring.

Food. Longnose lancetfish are an opportunistic and voracious predator. Stomachs from specimens from the western North Atlantic contained a large variety of deepwater fishes, as well as mollusks and crustaceans (Haedrich 1964; Matthews et al. 1977). Studies of the diet of this species in the southeastern Pacific (Haedrich and Nielsen 1966), the Indian Ocean (Parin et al. 1969), and other parts of the range (Kubota and Uyeno 1970; Grandperrin and Legand 1970) showed similar food habits.

Predators. Utilized as prey by several commercially important species including tunas of the genus *Thunnus* and swordfish (Scott and Tibbo 1968). Lancetfishes are cannibalistic and are known to feed on large numbers of their own species (Scott and Scott 1988).

Reproduction and Early Life History. There is little published information on the reproduction or early life history of lancetfishes. Rofen (1966) described and illustrated a series of early larval specimens from off Bermuda, and Maul (1946) and Haedrich (1964) described small juveniles of both species.

General Range. Worldwide (Gibbs 1960; Francis 1981a). Distributed over deep waters of the western North Atlantic from Greenland to the Caribbean Sea and the Gulf of Mexico. Also known from the eastern Atlantic off Madeira and the Canary Islands and the Mediterranean Sea.

Occurrence in the Gulf of Maine. A specimen brought in by a fishermen from Georges Bank about 1878 or 1879 was the first evidence of occurrence in the Gulf of Maine. Scott and Scott (1988) reported three specimens from Le Have Bank.

Importance. Although there is no commercial fishery for them, the flesh is sweet and of excellent flavor, although somewhat soft (Gibbs and Wilimovsky 1966).

LANTERNFISHES. ORDER MYCTOPHIFORMES

The superorder Scopelomorpha now contains only the order Myctophiformes, which differs from the Aulopiformes in having the upper pharyngobranchials and retractor muscles like those of generalized paracanthopterygians. Adipose fin present. The order includes two families, 36 genera, and about 245 species of deep-sea pelagic and benthopelagic fishes (Craddock and Hartel, in press).

LANTERNFISHES. FAMILY MYCTOPHIDAE

JAMES E. CRADDOCK, KARSTEN E. HARTEL, AND DONALD FLESCHER

Lanternfishes are small to medium-sized pelagic fishes, from 30 to 300 mm as adults, which live primarily in the open ocean. Most species inhabit the upper 1,000 m of the water column, the mesopelagic zone, and many migrate into the upper 100 m at night. A few species are bathypelagic and some are epibenthic on continental and island slopes.

Lanternfishes are conservative in body form and superficially resemble young salmonids, with a large head, a short dorsal fin near midbody, abdominal pelvic fins, and a dorsal adipose fin. Typically, they have large eyes and large mouths with a toothless maxilla excluded from the gape by a toothed premaxilla. Myctophids have several types of luminous organs arranged in characteristic patterns (Fig. 104), which are used for identification down to the species level. The light-producing organs range from small, spherical organs (called photophores) on the head and body, to luminous scales, to large organs above and/or below the caudal peduncle. As far as is known, all light produced by myctophids is the result of the relatively simple oxidation of luciferin in the presence of the enzyme luciferase (Hastings and Morin 1991).

Worldwide, there are about 240 species of lanternfishes in two subfamilies and 33 genera (Craddock and Hartel, in press), with 82 species occurring in the North Atlantic (Nafpaktitis et al. 1977; Backus et al. 1977). At least 59 species (Scott and Scott 1988; WHOI/MCZ unpublished data) occur just outside the Gulf. Only a few species of lanternfishes occur in the Gulf of Maine proper, and none reproduce there. All individuals in the Gulf are the result of intermittent passive transport from offshore waters.

Six species of lanternfishes have now been recorded from the Gulf of Maine, but accounts of *Diaphus effulgens* and *Myctophum affine,* the only lanternfishes mentioned in the 1953 edition of Bigelow and Schroeder, are not presented here because their inclusion was based on a small number of records from outside the Gulf of Maine proper. Neither has ever been taken in the Gulf.

KEY TO GULF OF MAINE LANTERNFISHES

[Note: Lanternfishes from just outside the Gulf of Maine may not be correctly identified using this key; readers should consult keys in Nafpaktitis et al. (1977).]

1a. Two Prc photophores **2**
1b. More than two Prc photophores **3**
2a. Prc photophores close together, separated by about one photophore diameter; Prc2 close to ventral profile; SAO photophores in an almost vertical straight line (60° from horizontal). . . . . **Spotted lanternfish**
2b. Prc photophores well separated, at least two photophore diameters apart; Prc2 just below lateral line; SAO photophores in an almost horizontal straight line (30° from horizontal). . . . . **Glacier lanternfish**
3a. VO1 through VO3 photophores in oblique ascending line; no luminous tissue on procurrent caudal rays **4**
3b. VO1 through VO3 photophores not in oblique straight line; luminous tissue on procurrent caudal rays **5**
4a. No VN photophore; luminous tissue on either ventral or dorsal caudal peduncle in adults greater than about 20 mm SL; SAO3, Pol, and Prc4 far below lateral line. . . . . **Doflein's false headlightfish**
4b. VN photophore present but small and round at anterior ventral margin of eye; no luminous tissue on either ventral or dorsal caudal peduncle; SAO3, Pol, and Prc4 all near lateral line . . . . . . **Dumeril's headlightfish**
5a. Forward-directed supraorbital spine present; five VO photophores; no cheek photophores; luminous scales present on dorsal midline, below

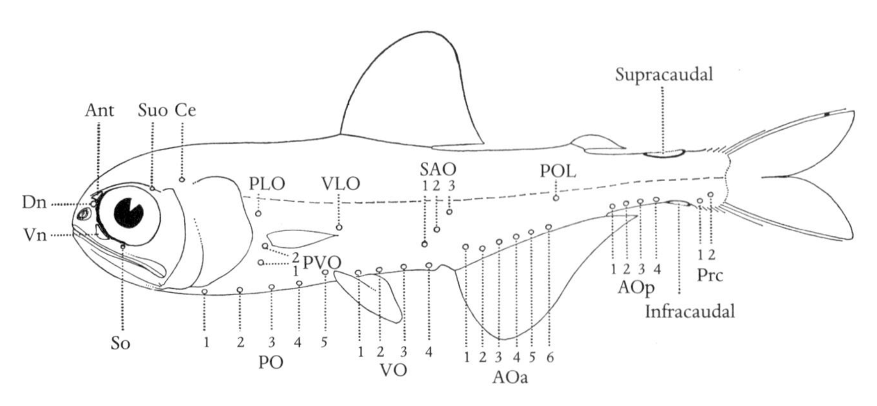

Figure 104. General distribution of photophores and other luminous tissue and their abbreviations in lanternfishes. Ant, antorbital; AOa, anterior anal; AOp, posterior anal; Ce, cervical; Dn, dorsonasal; PLO, suprapectoral; PO, thoracic; POL, posterolateral; Prc, precaudal; PVO, subpectoral; SAO, supra-anal; So, suborbital; Suo, supraorbital; VLO, supraventral; Vn, ventronasal; VO, ventral. (Reprinted from Nafpaktitis et al. 1977: Fig. 3. © Sears Foundation for Marine Research.)

pectoral base, on each side of anus, and along anal fin base, as well as luminous tissue on the caudal peduncle and procurrent caudal rays. . . . . **Horned lanternfish**

5b. No supraorbital spine; four VO photophores; three photophores on cheeks; luminous tissue confined to caudal peduncle, procurrent caudal rays, and base of adipose fin . . . . **Crocodile lanternfish**

GLACIER LANTERNFISH / *Benthosema glaciale* (Reinhardt 1837)

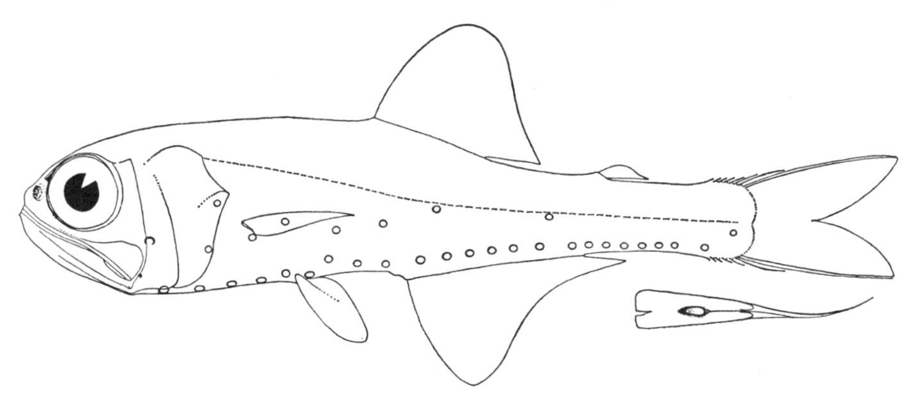

Figure 105. Glacier lanternfish *Benthosema glaciale,* young female, 54 mm. Inset shows supracaudal luminous gland of male, 64 mm. (Reprinted from Nafpaktitis et al. 1977: Fig. 18. © Sears Foundation for Marine Research.)

Description. Head moderately large (3.5 times in SL), eye large (2.4–2.7 in HL); jaw long (1.5 times in HL) and oblique, posterior maxilla widely expanded, reaching well beyond eye (Fig. 105). Dorsal fin about midbody, its origin behind pelvic fins; anal fin origin before midpoint of dorsal; dorsal adipose fin just behind end of anal base; procurrent caudal fin rays not spiny. Males with small supracaudal organ covering less than one-half of dorsal caudal peduncle; females usually have two very small patches of luminous infracaudal tissue.

Meristics. Dorsal fin rays 12–14; anal fin rays 17–19; gill rakers 15–18; photophores: PO 5, level; VO 4, second slightly elevated; AO 12–14; Prc 2, second just below lateral line; Pol 1.

Size. The largest known specimen is a 103-mm SL individual from off Norway (Gjösaeter 1973a).

Distinctions. Of the North Atlantic myctophid species with two widely spaced Prc photophores and with Prc2 at or near the lateral line, only *Benthosema* has one Pol. *Benthosema glaciale* differs from the smaller *B. suborbitale* by having the SAO and VO in a slightly arching, almost horizontal line and by lacking the suborbital photophore.

Habits. This is the northernmost occurring lanternfish in the Atlantic Ocean and is found at water temperatures ranging from 4° to 18°C (Halliday 1970). It is a vertical migrator, found between 275 and 850 m during the day and in the upper 200 m at night, sometimes coming to the surface. The entire popula-

Food. *Benthosema glaciale* feed principally on copepods off Nova Scotia (Sameoto 1988) and in the eastern Atlantic (Gjösaeter 1973b; Roe and Badcock 1984). Off Nova Scotia, they feed mostly at night in the upper 200 m (Sameoto 1988).

Predators. In the Slope Water (Backus et al. 1977), *B. glaciale* is preyed upon by the common dolphin *Delphinus delphis* (Craddock, unpubl. data).

Reproduction. Off New England and the Scotian Shelf, spawning occurs along the continental slope and further offshore during late winter and early spring (Halliday 1970; Morse et al. 1987). Although reproduction has not been documented in the Gulf of Maine proper, a small number of larvae have been taken in the Northeast Channel (Morse et al. 1987). Larvae reach flexion stage at about 6–7 mm and transformation (full photophore development) at 14–15 mm (Fahay 1983; Moser et al. 1984).

Age and Growth. Adult *B. glaciale* in the western Atlantic typically reach 50–70 mm SL, which is larger than those in the Mediterranean but smaller than individuals in the northern populations off Greenland and Norway (Gjösaeter 1973a; Halliday 1970). Maturity is reached at age 2 and this species lives for 4–5 years (Halliday 1970).

General Range. This is a subpolar-temperate species (Nafpaktitis et al. 1977), which seldom ranges south of the Gulf Stream in the western Atlantic. It is found from the Slope Water across the Atlantic north of about 40° and into the

extremely abundant throughout most of its range and has made up more than half of all the myctophids collected north of the subtropical gyre (Backus et al. 1977).

Occurrence in the Gulf of Maine. *Benthosema glaciale* is the most frequently encountered lanternfish in the Gulf of Maine. In addition to Musick's report (1973b), MCZ has records of 89 specimens collected in bottom trawls at 18 localities by NMFS research vessels from depths of 200–300 m. An opening/closing midwater trawl fished in Georges Basin and the Northeast Channel in April 1997 by L. P. Madin (pers. comm.) caught 33 specimens at 100–300 m.

HORNED LANTERNFISH / *Ceratoscopelus maderensis* (Lowe 1839)

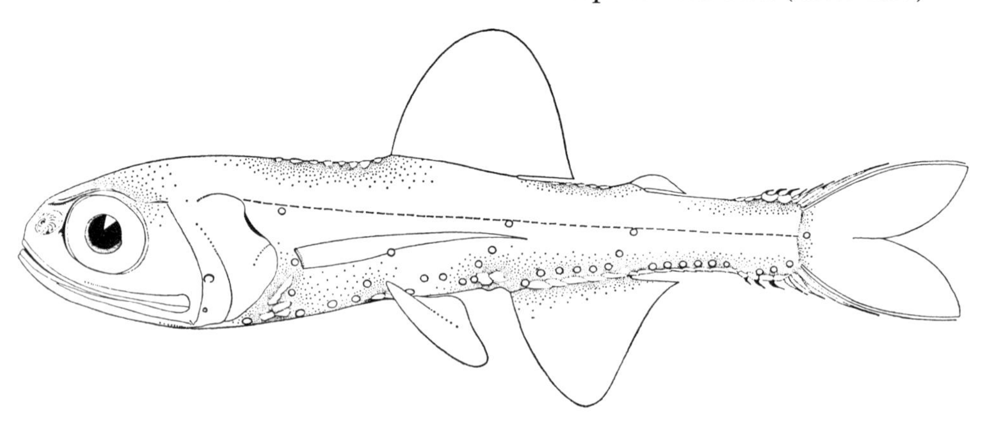

Figure 106. Horned lanternfish *Ceratoscopelus maderensis,* male, 62 mm. (Reprinted from Nafpaktitis et al. 1977: Fig. 169. © Sears Foundation for Marine Research.)

Description. Head large (3 times in SL), forming deepest part of body, body tapering gradually behind occiput; eye large (3 in HL); jaw long (1.25 in HL) extending almost one eye diameter behind rear margin of eye (Fig. 106). Dorsal origin just behind pelvics; anal origin just in advance of end of dorsal base; dorsal adipose fin over end of anal base; pectoral fin long, reaching past anal fin origin; procurrent caudal rays produced as strong spines. A sharp, forward-directed supraorbital spine present at forward corner of orbit. Luminous tissue along dorsal midline, below pectoral base, at base of procurrent rays, along base of anal, and along ventral caudal peduncle.

Meristics. Dorsal fin rays 13 or 14; anal fin rays 13–15; gill rakers 18–22; photophores: PO 5, fifth slightly raised; VO 5, second and third slightly raised; AO 14–18; Prc 4, fourth at caudal base; Pol 2.

Size. Adult *Ceratoscopelus maderensis* range from 50 to 70 mm SL off the New England shelf (Backus et al. 1968). The largest recorded specimen is 81 mm SL (Hulley 1981).

Distinctions. In the western North Atlantic, only three genera of lanternfishes (*Lampanyctus, Lepidophanes,* and *Ceratoscopelus*) have at least several luminous scales on the body, spinelike procurrent caudal fin rays, and four Prc photophores. Of these *Ceratoscopelus* is the only genus in which the PO4 photophore is not highly elevated. *Ceratoscopelus maderensis* can be distinguished from its only other Atlantic congener, *C. warmingii,* by the presence of a prominent forward-directed spine at the anterodorsal margin of the eye and by the absence of a midventral row of luminous scales between the pelvic and anal fins. *Ceratoscopelus warmingii* also lacks the luminous tissue that is found on the lower principal caudal rays in *C. maderensis.*

Biology. Like many lanternfishes, *C. maderensis* is a vertical migrator. Its maximum daytime abundance is between 325 and 500 m; at night, it is found from the surface to 175 m (Nafpaktitis et al. 1977). Massive midwater aggregations have been observed in Slope Water south of New England at daytime depths of 330–600 m from the submersible *Alvin* (Backus et al. 1968).

Food. *Ceratoscopelus maderensis* feeds on small planktonic crustaceans (Hulley 1984).

Predators. Off Cape Breton Island in February, pollock, cod, and hake were found to feed almost exclusively on *C. maderensis* (Scott and Scott 1988). Along the continental slope south of New England, *C. maderensis* is eaten in very large numbers by swordfish, bigeye tuna, opah, common dolphin (*Delphinus delphis*), and harbor porpoise (*Phocoena phocoena*) (Craddock, unpubl. data).

Reproduction. On the continental slope south of New England, *C. maderensis* larvae are found most often from June through October (Morse et al. 1987; Linkowski et al. 1993).

Larvae reach flexion stage at 6 mm and transform at 16 mm (Fahay 1983). No larvae have been taken in the Gulf of Maine.

Age and Growth. Based on otolith studies in the eastern North Atlantic, Linkowski et al. (1993) found: (1) growth is very rapid during the first 10 months of life for early spawned individuals but is slower for those spawned late in the season; (2) some individuals reach 50 mm SL and become sexually mature by 10 months; a size of 70 mm SL is reached in less than 2 years; and (3) most of the population dies during the second year of life, which may be correlated with postspawning mass mortality.

General Range. *Ceratoscopelus maderensis* lives across the temperate Atlantic from Cape Hatteras on the west to the eastern Mediterranean; it also lives in the northern half of the subtropical gyre (north of 30° N) and to about 50° N (Nafpaktitis et al. 1977).

Occurrence in the Gulf of Maine. One specimen was reported from Georges Basin and 20 from the Northeast Channel (Musick 1973b). Since that report, eight additional specimens have been collected by NMFS in the Great South and Northeast channels, at depths of 145–300 m. *Ceratoscopelus maderensis* otoliths were found in a harbor porpoise taken just south of Jeffreys Ledge (J. E. Craddock, unpubl. data). In areas just south of Georges Bank, *C. maderensis* is one of the most abundant lanternfishes between the continental margin and the Gulf Stream.

DUMERIL'S HEADLIGHTFISH / *Diaphus dumerilii* (Bleeker 1856)

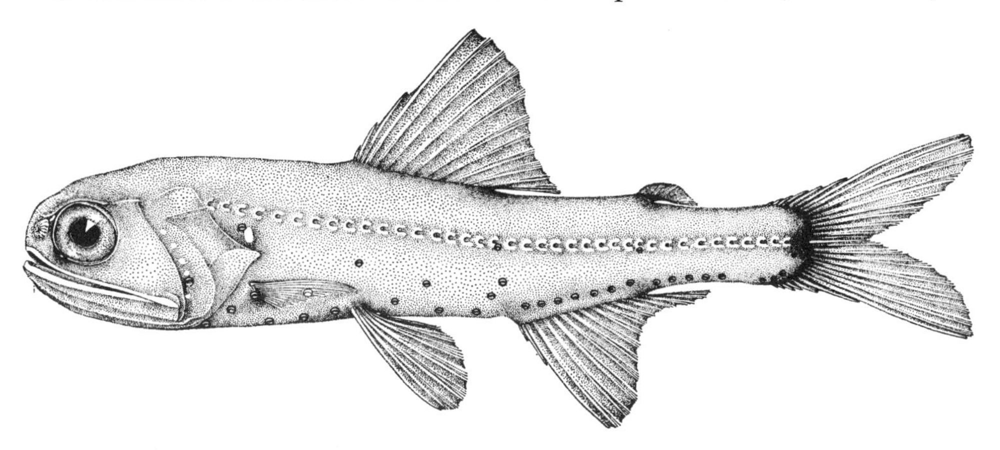

Figure 107. Dumeril's headlightfish *Diaphus dumerilii,* female, 61.5 mm. (Reprinted from Nafpaktitis et al. 1977: Fig. 56. © Sears Foundation for Marine Research.)

Description. Head large (3.0–3.3 times in SL); jaw long, extending about one eye diameter behind eye; eye large (3.2–3.7 in HL) (Fig. 107). Dorsal origin over pelvic fins; anal origin behind end of dorsal fin; adipose over end of anal base. Supra- and infracaudal glands absent. A luminous scale present above tip of opercle adjacent to PLO photophore. Dorsal nasal (Dn) photophore moderately well developed in males, small in females.

Meristics. Dorsal fin rays 14 or 15; anal fin rays 14 or 15; gill rakers 23–26; photophores: PO 5, fourth highly elevated; VO 5, first through third in a straight oblique line; AO 10–14; Prc 4; Pol 1.

Size. *Diaphus dumerilii* matures at about 50 mm SL and can grow to 87 mm (Hulley 1981).

Distinctions. The genus *Diaphus* is unique among lanternfishes that have four Prc photophores and lack spiny procurrent caudal rays... tophores, and VO1 through VO3 in a straight ascending line. *Diaphus dumerilii* has the least well-developed nasal photophores of the species in the genus. It can be distinguished from other western North Atlantic *Diaphus* by its small Vn photophore situated under the anterior margin of the eye, pointing upward and inward toward the eye. It is also unique in having SAO3 anterior to SAO2.

Biology. *Diaphus dumerilii* is a tropical species, and specimens from off New England and Canada were once thought to be waifs from the Gulf Stream (Nafpaktitis et al. 1977). Based on NMFS bottom trawl surveys, however, we know now that there is a viable reproducing population at 200–300 m along the upper slope south of New England. We have also seen a number of specimens from bottom trawls fished as shallow as 81 m. These trawl data and observations off the Bahamas suggest that *D. dumerilii* may be common over shallow bottoms. Schooling has been observed from submersibles off Cape

range, *D. dumerilii* is most commonly found between 400 and 500 m during the day. At night it moves to a depth of about 50 m and sometimes to the surface.

Predators. Fairly large numbers of *D. dumerilii* otoliths have been found in stomachs of common dolphin (*Delphinus delphis*) taken over the continental slope south of New England (J. E. Craddock, unpubl. data).

General Range. Present across the tropical Atlantic with temperate populations occurring to 50° N off Newfoundland. An analogous temperate population occurs off Uruguay and Argentina (Nafpaktitis et al. 1977).

Occurrence in the Gulf of Maine. Rare in the Gulf of Maine proper but relatively abundant between 150–300 m along the south slope of Georges Bank. NMFS collections at MCZ contain 14 specimens taken in Georges Basin in April 1994, two from the Northeast Channel and one from 188 m off Cashes Ledge.

CROCODILE LANTERNFISH / *Lampanyctus crocodilus* (Risso 1810)

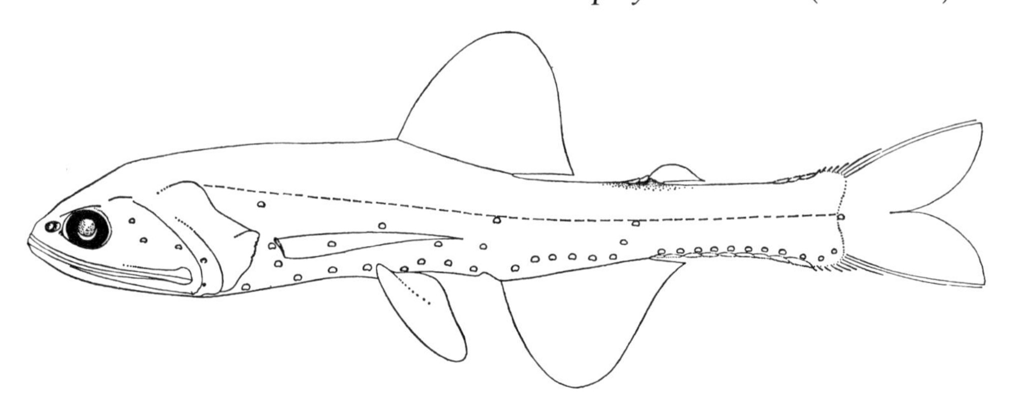

Figure 108. Crocodile lanternfish *Lampanyctus crocodilus,* young, 34 mm. (Reprinted from Nafpaktitis et al. 1977: Fig. 142. © Sears Foundation for Marine Research.)

Description. Head large (3.5 times in SL); jaw very long (1.5 in HL) and nearly horizontal, almost reaching two eye diameters behind rear margin of eye; eye relatively small (4.5 in HL) (Fig. 108). Dorsal origin slightly behind pelvics; anal origin under end of dorsal base; adipose over end of anal base; pectoral fins long, reaching midpoint of dorsal. Luminous scalelike tissue at base of adipose, before upper procurrent rays, and in a long band almost covering the entire ventral caudal peduncle midline.

Meristics. Dorsal fin rays 14 or 15; anal fin rays 17 or 18; gill rakers 16 or 17; photophores: PO 5, fourth highly elevated; VO 4; AO 13–16; Prc 4; Pol 2; three cheek photophores.

Size. *Lampanyctus crocodilus is* the largest species of lanternfish, adults reaching more than 300 mm SL.

Distinctions. *Lampanyctus crocodilus* is more elongate and has less solid muscle development than any of the other genera that might be expected in the Gulf of Maine. The combination of an elevated PO4 photophore, cheek photophores, and luminous scales above and below the caudal peduncle and at the base of the adipose fin but nowhere else on the body should identify the genus. Eleven species of *Lampanyctus* occur off the Gulf of Maine; as this genus is difficult to identify, especially if the specimen is damaged, specimens should be reviewed using Nafpaktitis et al. (1977) or be brought to a museum specialist.

Taxonomic Remarks. Recently, Stefanescu et al. (1994) restricted *L. crocodilus* (Risso 1810) to populations in the Mediterranean Sea and resurrected *L. gemmifer* (Goode and Bean 1879) for populations in the Atlantic Ocean proper. Their conclusion was based on analysis of a few meristic characters on only 147 specimens, so we retain use of *L. crocodilus* for both Atlantic and Mediterranean populations, as have all recent revisions (Nafpaktitis et al. 1977; Hulley 1981; Bekker 1983).

Biology. Adult *L. crocodilus* are probably benthopelagic on the slope to a depth of at least 1,200 m (Stefanescu and Cartes 1992). This species does not mature until it is larger than 100 mm SL. It feeds on a wide variety of pelagic and benthopelagic invertebrates and occasionally small fishes (Stefanescu and Cartes 1992). *Lampanyctus* larvae are deep-bodied and almost football-shaped until about 15 mm SL; transformation occurs at 19 mm SL (Fahay 1983).

Predators. Otoliths of *L. crocodilus* have been taken from stomachs of Cuvier's beaked whale (*Ziphius cavirostris*) and Sowerby's beaked whale (*Mesoplodon bidens*), both of which had been feeding near the bottom on the continental slope south of New England (Craddock, unpubl. data).

General Range. *Lampanyctus crocodilus* is found only in the North Atlantic. It is abundant throughout the temperate North Atlantic (Nafpaktitis et al. 1977), with scattered records to Ungava Bay (Dunbar and Hildebrand 1952) and Greenland.

Occurrence in the Gulf of Maine. Presence of this species in the Gulf of Maine is based on a 43-mm SL juvenile collected with a midwater trawl at 160 m by *Albatross IV* in 1965 over Georges Basin (Musick 1973b).

DOFLEIN'S FALSE HEADLIGHTFISH / *Lobianchia dofleini* (Zugmayer 1911)

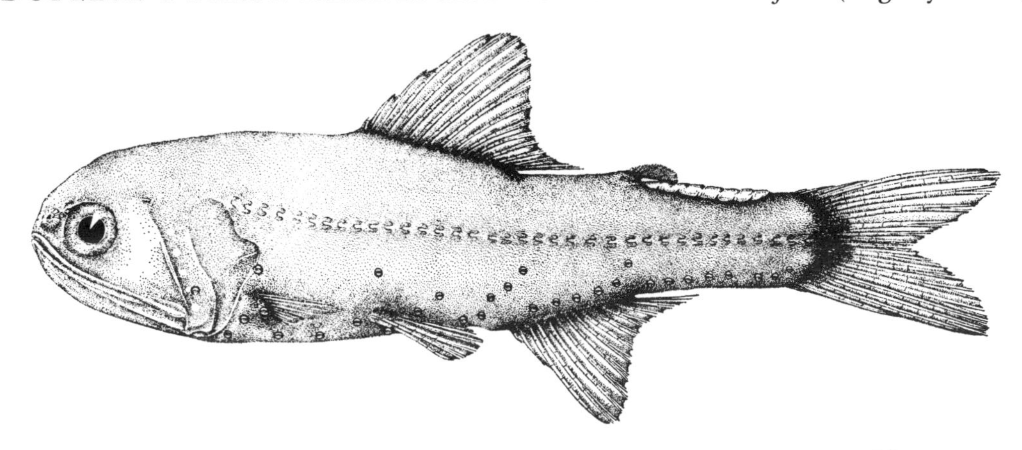

Figure 109. Doflein's false headlightfish *Lobianchia dofleini*, male, 27 mm. (Reprinted from Nafpaktitis et al. 1977: Fig. 50. © Sears Foundation for Marine Research.)

Description. Head large (3–3.5 times in SL), jaw long (1.3 in HL), eye moderately small (3.5–4.7 in HL) (Fig. 109). Dorsal fin slightly in advance of pelvic fin; anal origin under or slightly in advance of end of dorsal base; adipose over end of anal base. Adult males have a large supracaudal organ, consisting of seven or eight segments, covering most of dorsal caudal peduncle; females have a smaller infracaudal organ with three or four segments.

Meristics. Dorsal fin rays 15–17; anal fin rays 13–15; gill rakers 18–21; photophores: PO 5, fourth highly elevated; VO 5, first through third in a straight oblique line; AO 9–12; Prc 4; Pol 1.

Size. *Lobianchia dofleini* is the smallest lanternfish known from the Gulf of Maine. The majority of the spawning population in Slope Water is 30–35 mm SL; maximum size 50 mm SL (Nafpaktitis et al. 1977).

Distinctions. *Lobianchia,* like *Diaphus,* has four Prc photophores, lacks spiny procurrent caudal rays, and has the VO1 through VO3 in a straight ascending line. *Lobianchia* differs from *Diaphus* by lacking a Vn photophore and by adults, larger than 23 mm SL, having a luminous supracaudal organ in males and an infracaudal organ in females. *Lobianchia dofleini* can be distinguished from *L. gemellarii* by the large space between Prc3

Biology. *Lobianchia dofleini* is a vertical migrator but depths vary with the seasons. Maximum abundance at night is in the upper 200 m and between 400 and 600 m during the day (Karnella and Gibbs 1977). Spawning occurs from January to June near Bermuda (Karnella 1987), and larvae transform to adults at 10–11 mm SL (Fahay 1983). It has a 1-year life cycle (Karnella 1987).

Predators. *Lobianchia dofleini* otoliths have been taken from stomachs of common dolphin (*Delphinus delphis*) taken over the continental slope south of New England (Craddock, unpubl. data).

General Range. *Lobianchia dofleini* is found from about 50° N to about 50° S in the Atlantic Ocean, but does not occur in the western tropics (Nafpaktitis et al. 1977). It is also found in a narrow band in the temperate southern Indian and Pacific oceans (Bekker 1983).

Occurrence in the Gulf of Maine. Occurrence of this species in the Gulf of Maine is based on two adults, 32 and 34 mm SL, taken in a midwater trawl by *Albatross IV* in October 1965 over Georges Basin (Musick 1973b). The species is also known from the southern slope of Georges Bank, in Slope Water (MCZ records), and from the edge of Scotian Shelf (Scott and Scott

SPOTTED LANTERNFISH / *Myctophum punctatum* Rafinesque 1810

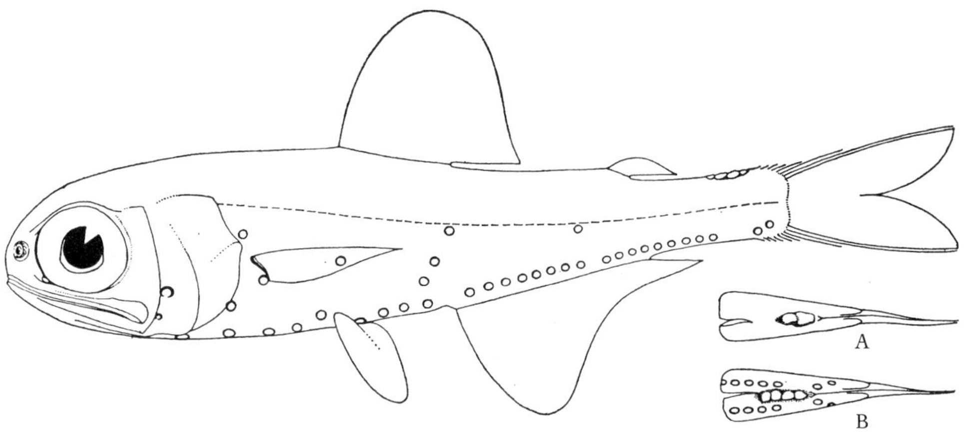

Figure 110. Spotted lanternfish *Myctophum punctatum,* male, 51 mm. (A) Supracaudal luminous gland of male in dorsal view. (B) Infracaudal luminous gland of female in ventral view. (Reprinted from Nafpaktitis et al. 1977: Fig. 24. © Sears Foundation for Marine Research.)

Description. Head large (3.3–3.6 times in SL), jaw somewhat oblique (1.5 in HL), eye very large (2.5 in HL) (Fig. 110). Scales cycloid. Dorsal origin over pelvic fin; anal origin under end of dorsal fin base; adipose origin in advance of end of anal fin base. Males have a small supracaudal luminous organ, females an infracaudal organ.

Meristics. Dorsal fin rays 13 or 14; anal fin rays 20–22; gill rakers 24–27; photophores: PO 5, level; VO 4; AO 16 or 17; Prc 2, close together; Pol 1.

Size. This species matures by 50 mm SL and commonly grows to just larger than 100 mm SL.

Distinctions. Of the western North Atlantic lanternfishes, only *Myctophum* and *Symbolophorus* have two Prc photophores, which are close together; one Pol; and a PLO situated well above the pectoral base. *Myctophum* has its SAO photophores in a straight line, whereas in *Symbolophorus* the SAO photophores are almost at a right angle. *Myctophum punctatum* can be distinguished from all other *Myctophum* by having three or four AOp photophores over the posterior anal base, seven or more total AOp, and having the Pol anterior to the base of the adipose fin. *Myctophum affine,* which is known from Browns Bank (Bigelow and Schroeder), has ctenoid scales.

Biology. *Myctophum punctatum* is the only lanternfish known from the Gulf of Maine that regularly comes to the surface at night, with part of the population spread throughout the upper 125 m. During the day it is caught between 225 and 750 m. This species does not spawn in the Canadian Atlantic region, and individuals found there are expatriates from spawning populations to the east (Zurbrigg and Scott 1972). WHOI data show that Slope Water populations are also expatriates. In the eastern Atlantic, *M. punctatum* spawns in late winter and early spring and in the Mediterranean throughout spring.

Food. *Myctophum punctatum* feeds on small planktonic organisms, including fish larvae (Hulley 1984).

Predators. Like most lanternfishes, *M. punctatum* are important prey for fishes and mammals; they are eaten by swordfish (Scott and Tibbo 1974) and cod (Popova 1962). R. H. Backus (pers. comm. 1995; MCZ specimens) saw "thousands" of *M. punctatum* from the stomach of a 18-m male fin whale (*Balaenoptera physalus*) landed at Hawke Harbour, Labrador, in August 1951. Although not a relatively abundant species when compared with some other lanternfishes, *M. punctatum* are easy prey when concentrated at the surface (upper 0.25 m) at night.

General Range. *Myctophum punctatum* is a temperate-subpolar species from Cape Hatteras east into the Mediterranean (Nafpaktitis et al. 1977).

Occurrence in the Gulf of Maine. One 47-mm SL specimen was collected in a midwater trawl over Georges Basin in 1966 and six specimens (66–83 mm SL) were taken from the Northeast Channel in 1965 (Musick 1973b). NMFS trawls have only added one additional record since that time, from near the mouth of the Northeast Channel in 1983. We have seen numerous specimens from the outer slope off Georges Bank.

ACKNOWLEDGMENTS. Drafts of the myctophid section were reviewed by William H. Krueger, John Paxton, and Kenneth A. Tighe.

LAMPRIDIFORM FISHES. ORDER LAMPRIDIFORMES

The Lampridiformes contains seven families of marine fishes with 12 genera and about 19 species (Nelson 1994). They lack true spines in their fins and have a unique type of protrusible upper jaw, the maxilla sliding forward with the premaxilla during jaw protrusion. They belong to the acanthomorph clade (spiny-rayed fishes) and are primitive with respect to percomorphs, but their precise placement among basal acanthomorphs is not clear (Olney et al. 1993).

OPAHS. FAMILY LAMPRIDAE

GRACE KLEIN-MACPHEE

Opahs belong to one of the two families of deep-bodied lampridiforms. There are two species of opahs, one restricted to cold and temperate waters of the Southern Hemisphere, the other worldwide, occasionally wandering into the Gulf of Maine.

OPAH / *Lampris guttatus* (Brünnich 1788) / Moonfish / Bigelow and Schroeder 1953:247–248 (as *Lampris regius*)

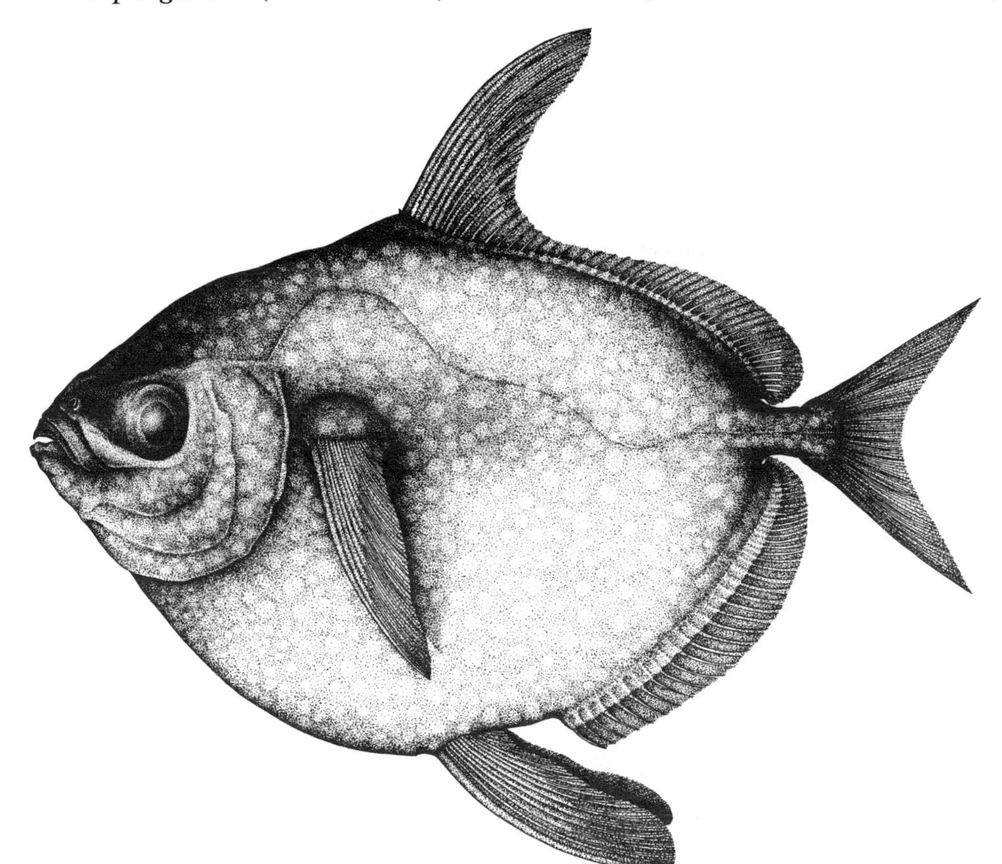

Figure 111. Opah *Lampris guttatus.* Drawn by N. Williams.

Description. Body short, compressed and very deep, 2 times in TL (Fig. 111). Caudal peduncle moderately slender; head 3.75 times in TL, compressed; mouth terminal, small, not extending beyond front of eye, toothless. Eye large, four in HL. Anterior part of single dorsal fin high, falcate. Anal fin shorter than dorsal, about equal in height to low part of dorsal throughout its length. Both anal and dorsal fins extend posteriorly close to caudal fin base, each depressible into groove,

Pectoral fins conspicuously pointed, their bases horizontal instead of vertical. Pelvic fins very long, falcate, inserted ventrally under anterior part of dorsal fin. Caudal fin emarginate. Vent midway between base of pelvics and anal. Scales minute. Lateral line strongly arched upward above pectoral fin.

Meristics. Dorsal fin rays 48–56; anal fin rays 33–42; pelvic fin rays 12–17; gill rakers on the first arch 13 or 14; vertebrae 43–46 (Hart 1973; Nelson 1994).

Color. Living opah are spectacular fish. The body is brilliant pink with iridescent silvery spots that fade to a dull pink after death (Bane 1965). The region between the head and middorsal fin is bright blue, and the fins are deep scarlet, edged with gold.

Size. Opah reach 91–183 cm but most are 91–122 cm. The all-tackle game fish record weighed 73.93 kg and was caught at Port San Luis Obispo, Cal., in October 1998 (IGFA 2001). There are reports of 228- to 273-kg specimens from the Pacific.

Distinctions. The thin, deep body form with moderately slender caudal peduncle without longitudinal keels and the rather pointed snout suggest an enormous butterfish, except for the very long falcate pelvic fins, which butterfish lack. They are also distinguished by the scarlet to vermilion jaws and fins, and silvery spots on the body.

Taxonomic Remarks. The name *Lampris guttatus* (Brünnich 1788) replaces *L. regius* (Bonnaterre 1788) based on the actions of the first reviser (see Palmer and Oelschläger 1976).

Habits. Opah are large predatory fish of open oceans occurring from near the surface to 500 m (Gudger 1930). Average depth of capture in the hook-and-line fishery off Madeira is 90–180 m but catch increases with depth to 280 m in the Japanese longline fishery in the central and eastern Pacific (Nakano et al. 1997).

Opah swimming behavior is unusual in that they cruise by pectoral swimming (Rosenblatt and Johnson 1976). The shoulder girdle is massive and provides attachment for large red adductor and abductor muscles, which account for 16% of the total weight. There is also an unusual layer of thick adipose tissue present only over the pectoral musculature. The pectoral fins are inserted horizontally and can be completely adpressed. The hypothesis is that low swimming speed is a function of the caudal fin; acceleration occurs by lateral white muscles and the caudal fin, with the pectoral fins maintaining cruising speed by flapping.

Morphology of the olfactory organ and its functional correlation with the lachrymal bone and jaw apparatus during respiration provide a very efficient exchange of water flow.

Food. An opah caught off Nova Scotia contained a small octopus and the remains of 25 fishes, probably argentines (McKenzie and Tibbo 1963). A 963-mm specimen from Puerto Rico (Bane 1965) contained 82% squids (by volume) and 19% fishes (dolphin, berycoids, and elongate fishes, perhaps gempylids or paralepidids).

Parasites. A Puerto Rican specimen contained nematodes, digenetic trematodes, and cestodes in the gut (Bane 1965).

Breeding Habits. Spawning probably takes place in spring in the Pacific Ocean (Hart 1973).

Early Life History. Fecundity estimates from a 963-mm female from Puerto Rico ranged from 7.2 to 9.7 million eggs (Bane 1965). Ovarian eggs have a thick chorion with an amber tint (Olney 1984).

Larvae of lampriform fishes in general possess well-developed protrusible jaws; differentiated guts with an open lumen and little or no yolk material; elongate anterior dorsal elements that insert between the posterior eye margin and the shoulder, and well-developed pelvic elements. They are slender but rapidly increase in body depth and by 10.6 mm SL assume characteristics of the adult deep-body form (Olney 1984).

Two juveniles (46.5 and 51.5 mm SL) collected from the southwest Atlantic were pelagic, had simple coloring, and were half transparent (Oelschläger 1974). They are very deep-bodied, laterally compressed, and nearly disklike. The scales and lateral line system are not developed. The eye is large and the protrusible jaw mechanism is already well developed. The pectoral fins are also well developed, with the insertion horizontal and the pectoral girdle powerful. There are teeth in the lower jaw. Dorsal and pelvic fins elongate anteriorly into a long filament, probably as flotation devices. As the juvenile develops, the deep body of the postlarvae decreases in height to the subadult, then increases again in adulthood; the truncate throat contours of the young fish becomes rounded; and the dorsal and pelvic filament is reduced.

General Range. Widespread in the open waters of the world oceans. In the Atlantic, recorded from 79° N in the Barents Sea south to 54° S off Shagg Rocks in the Scotian Sea; from Iceland, Newfoundland, Scandinavia, the British Isles, Maine, Cape Cod, Cuba, Puerto Rico, and Madeira in the North Atlantic; also in the Gulf of Mexico off the west coast of Florida. Common off Brazil and southwest Africa (Parin and Kukuev 1983: Fig. 5). In the Pacific they are found from Cape San Lucas, Baja California, to Icy Bay in the Gulf of Alaska and in Japanese and Western Australian waters (Hart 1973).

Occurrence in the Gulf of Maine. Several specimens of this oceanic wanderer have been reported within the limits of the Gulf, one caught on a longline on Browns Bank in the spring of 1932 and another taken in an otter trawl on the northeastern part of Georges Bank, in August 1947. One was captured 10 September 1962 by the *Sen Sen,* which was long-lining for halibut south of Halifax, N.S. (McKenzie and Tibbo 1963); another was caught near Hyannis, Cape Cod, in 1928.

Importance. Opah are excellent food fish and are caught on hook-and-line off Madeira, but because they are mostly solitary, they are not caught in sufficient quantities to be of commercial importance. Opah are also taken as by-catch in the Japanese longline fishery in the central and eastern tropical Pacific (Nakana et al. 1997).

ACKNOWLEDGMENT. A draft of the lampridiform section was reviewed by Carole Baldwin.

BEARDFISHES. ORDER POLYMIXIIFORMES

This order is composed of one family, which in the past was placed in either the Beryciformes or the Paracanthopterygii. More recent evidence (Johnson and Patterson 1993) suggests that this family might be the sister-group to all other acanthomorphs (spiny-rayed fishes).

BEARDFISHES. FAMILY POLYMIXIIDAE

Jon A. Moore

Beardfishes are a small family (ten species) of tropical to temperate fishes found worldwide (Kotlyar 1984, 1992). They derive their name from a pair of long, conspicuous hyoid barbels that extend from just behind the lower jaw symphysis. This is the only acanthomorph group retaining two sets of intermuscular bones: epineurals and epipleurals (Johnson and Patterson 1993). Beardfishes generally are groundfishes on the shelf and upper slope, caught to depths of 700 m. There are two species in the Atlantic, *Polymixia nobilis* and *Polymixia lowei*, the latter occurring in the Gulf of Maine.

BEARDFISH / *Polymixia lowei* Günther 1859

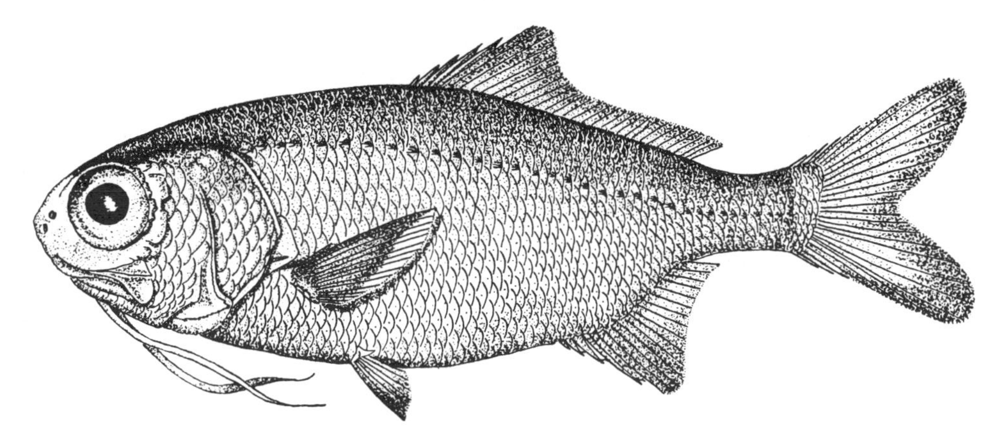

Figure 112. Beardfish *Polymixia lowei* Günther 1859. Western Caribbean, FMNH 66216. (Reprinted from Woods and Sonoda 1973: Fig. 4. © Sears Foundation for Marine Research.)

Description. Body moderately elongate, compressed (Fig. 112). Snout rounded, short. Eye relatively large. Jaws with two supramaxillae. Jaw teeth minute, in bands; teeth also on tongue, vomer, palatines, and endopterygoid. Dorsal fin long; anal fin short; caudal fin forked. Lateral line complete. Scales cover body and much of head, including lower jaw, cheek, op-

Meristics. Dorsal fin rays V or VI, 26–32; anal fin rays III or IV, 13–17; pectoral fin rays 15–17; pelvic fin rays 7; lateral line scales 31–36; gill rakers 4–8 + 1 + 9–13 = 14–22; vertebrae 29 or 30.

Color. Body is dark bluish gray; abdomen is lighter. Cheek and lower jaw are white and snout is brownish gray. The bar-

black area at distal ends of longest dorsal rays. Sexual dimorphism in adults larger than 12 cm has been documented, especially in the fin coloration. Males have intense black on the distal ends of the longest anal rays and the posterior edge of the caudal fin, whereas females are pale or slightly dusky in those areas (Lachner 1955).

Size. The largest recorded is 19.8 cm (Lachner 1955).

Distinctions. Distinguished from the other Atlantic species, *P. nobilis,* by fewer soft rays in the dorsal fin and more total gill rakers on the first gill arch: 26–32 soft rays and 14–22 total rakers vs. 34–38 rays and 10–13 total rakers in *P. nobilis.*

Biology. Benthopelagic fishes living at a wide range of depths down to 700 m. The few stomach samples studied contained crustacean, worm, and fish remains (Woods and Sonoda 1973). Very little information is available on the life history.

General Range. Found in the western North Atlantic, from southern Brazil throughout the Caribbean and Gulf of Mexico north to Bermuda (Smith-Vaniz et al. 1999) and the Gulf of Maine.

Occurrence in the Gulf of Maine. One 8-cm specimen was collected on the edge of Browns Bank in the Northeast (Fundian) Channel (Scott and Scott 1988).

Commercial Fishery. No commercial fishery targets this species, although they are reported to be edible.

CUSK-EELS. ORDER OPHIDIIFORMES

The family Ophidiidae is currently placed in the order Ophidiiformes, but there has been much disagreement about the phylogenetic relationships of the group, with some ichthyologists placing it within the Gadiformes and others in the Perciformes (Nelson 1994). The order contains two suborders, four families, 93 genera, and some 367 species (Nielsen et al. 1999). If present, the pelvic fins are always on the throat at a point under the preopercle or farther anterior, jugular or mental, at least as adults.

CUSK-EELS. FAMILY OPHIDIIDAE

Bruce B. Collette and Grace Klein-MacPhee

The family Ophidiidae includes oviparous ophidiiform fishes lacking a vexillifer larva and possessing a supramaxillary bone (Nielsen et al. 1999). Cusk-eels are distinguished from true eels by the presence of pelvic fins and the large conventional opercular opening. They resemble eelpouts in general form, but can be distinguished by the forked barbel-like shape of the pelvic fins. There are four subfamilies, 48 genera, and about 218 species (Nielsen et al. 1999), of which only two species have been recorded from the Gulf of Maine.

KEY TO GULF OF MAINE CUSK-EELS

1a. Snout armed with short, sharp spine concealed in skin; upper part of body marked with 14–23 roundish, pale spots; head extensively covered with imbricate rows of cycloid scales; gill rakers on lower arch 5–7; dorsal fin rays 129–142; anal fin rays 109–121 . . . . **Fawn cusk-eel**

1b. No spine on snout; body with dark horizontal stripes; head entirely naked; gill rakers on lower arch 4 or 5; dorsal fin rays 138–162; anal fin rays 116–131 . . . . **Striped cusk-eel**

FAWN CUSK-EEL / *Lepophidium profundorum* (Gill 1863) /

Bigelow and Schroeder 1953:517–518 (as *Lepophidium cervinum*)

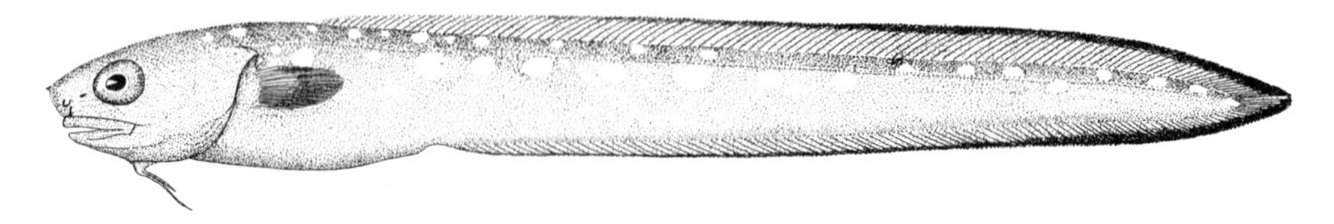

Figure 113. Fawn cusk-eel *Lepophidium profundorum.*

Description. Body long, slender, eel-shaped, about 12–13 times as long as deep (Fig. 113). Head slender, somewhat compressed; snout sharp, conical, armed with a short sharp spine pointing forward and downward and nearly concealed in skin. Teeth villiform, in narrow bands on jaws, vomer, and palatines. Gill rakers short; gill chamber and gill bars black (Robins 1986). Fins soft and continuous, no separation between dorsal, caudal, and anal. Pelvic fins short, located on throat far anterior to pectoral fins, modified to forked, barbel-like structures with two simple rays. Scales small, in regular rows, covering body except snout, undersurface of head, and fins.

Meristics. Dorsal fin rays 129–142; anal fin rays 109–121; pectoral fin rays 22–24; caudal fin rays 4 + 5; gill rakers 2 + 5–7 = 7–9 on the first arch; vertebrae 15–17 precaudal + 56–61 caudal = 73–78 total (Miller and Jorgenson 1973; Gordon et al. 1984; Robins 1986; Fahay 1992).

Color. Fawn cusk-eel are brownish yellow, darker above than below, the upper part of each side marked with a row of 14–23 roundish white or pale brown spots. The dorsal and anal fins have narrow black or dusky margins. Adult color patterns probably develop later because neither the spots nor the edging of the fins were visible in a young specimen 6.6 cm long.

Size. Maximum length about 26.2 cm.

Distinctions. Fawn cusk-eel differ from striped cusk-eel in having a sharp spine on the snout, pale spots on the body instead of horizontal stripes, more gill rakers (7–9 on lower arch vs. 4 or 5), fewer fin rays (dorsal fin rays 129–142 vs. 138–162; anal fin rays 109–121 vs. 116–131), and more vertebrae (73–78 vs. 68–70).

Habits. Fawn cusk-eel occur mostly in shelf waters at depths of 55–365 m (Nielsen et al. 1999). They are found on the outer shelf primarily in fall from Cape Cod to Cape Hatteras (Colvocoresses and Musick 1984). Cusk-eels collected near Georges Bank were in water 5.6°–9.1°C (McKenzie 1967).

Fawn cusk-eel were never observed during daytime submersible dives in the Mid-Atlantic Bight (Levy et al. 1988). However, night dives northwest of Veatch Canyon and in the vicinity of Hudson Canyon revealed large numbers of fawn cusk-eel emerging from the substrate or swimming over the bottom.

Food. For 165 fawn cusk-eel (Bowman et al. 2000), major food items included polychaete worms (26% by weight); amphipods (23%), such as *Unciola irrorata;* decapods (24%), such as *Crangon septemspinosa;* echinoderms (9%), such as ophiuroids; and fishes (10%), such as unidentified flatfishes and *Brosme brosme.* Fawn cusk-eel are the top-ranked predator of both flatfishes (6%) and *B. brosme* (2%) based on percent diet composition (Rountree 1999). Echinoderms predominate at the smallest size-classes (75% at 11–15 cm) but decline to 4% of the diet by the largest size-class (26–30 cm), whereas fish consumption increases with size until it accounts for 22% of the diet of the largest individuals (Bowman et al. 2000).

Predators. Fawn cusk-eel are an important component of the diet of conger eels in offshore stations in the Mid-Atlantic Bight (Levy et al. 1988). Other predators include skates (little, winter, and thorny), spiny dogfish, angel shark, cod, pollock, hakes (red, white, spotted, and silver), goosefish, longhorn sculpin, sea raven, and summer and fourspot flounders; spiny dogfish and sea raven are the most frequent predators (Rountree 1999).

Reproduction. Little is known of the reproductive habits. Eggs are unknown. Larvae collected by NEFSC ichthyoplankton surveys from 1984 to 1987 ranged from 1.9 to 32.0 mm NL (Fahay 1992). The largest pelagic larva examined was 42.0 mm SL. Flexion begins at about 8 mm and is complete by 13 mm. Larvae are primarily distributed over the outer half of the continental shelf, as far north as the southern flank of Georges Bank during the summer (Fahay 1992: Fig. 11). A series of prominent melanophores along the ventral edge of the body, with a single dorsal edge melanophore over the most posterior one in the ventral series (Fahay 1992: Fig. 6), and early formation of pectoral fin rays distinguish larval *Lepophidium* from species of *Ophidion.*

General Range. Along the outer part of the continental shelf from Georges Bank to northern Florida and the Gulf of Mexico (Nielsen et al. 1999).

Occurrence in the Gulf of Maine. Common on the outer slope of Georges Bank and wandering into the Gulf of Maine on occasion (Map 12). There are only three historical reports from the Gulf of Maine region: one from 137 m off Nantucket Shoals in the late 1800s (Goode and Bean 1896), two from the stomach of a white hake taken on the southwestern part of Georges Bank in 1951 (Bigelow and Schroeder), and five collected at five stations, 120–180 m deep, from Lydonia Canyon, Georges Bank, northeastward to Sable Island (McKenzie 1967).

STRIPED CUSK-EEL / *Ophidion marginatum* (DeKay 1842)

Description. Body elongate, compressed, of about uniform depth from nape to midanal fin base (Fig. 114). Head compressed, snout moderately pointed, 3.0–4.05 times in head posterior margin of eye. Teeth in jaws pointed, in bands; vomer and palatine bones with bands of blunt teeth. Scales small, elongate, alternating in direction to form basket-weave

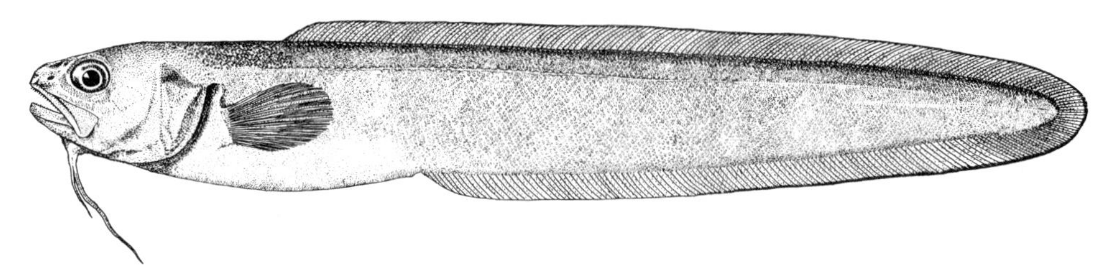

Figure 114. Striped cusk-eel *Ophidion marginatum.*

Meristics. Dorsal fin rays 138–162; anal fin rays 116–131; pectoral fin rays 22; caudal fin rays 4 + 5; gill rakers 4 or 5 on lower limb of first arch; vertebrae 15 precaudal + 53–55 caudal = 68–70 total (Hildebrand and Schroeder 1928; Miller and Jorgenson 1973; Fahay 1992; Schwartz 1997d).

Color. In life, grayish green, sides golden, belly snow white (Hildebrand and Schroeder 1928). Two or three dark stripes, which may be indistinct, along sides. Dorsal and anal fins dark-edged.

Size. Reaches 250 mm (Robins and Ray 1986), the largest among many from the Cape Fear River, N.C., was a 245-mm TL female (Schwartz 1997d).

Distinctions. Differ from fawn cusk-eel in having two or three horizontal stripes, no spine in the snout, higher fin ray counts (dorsal fin rays 138–162 vs. 129–142; anal fin rays 116–121 vs. 109–121), and fewer vertebrae (68–70 vs. 73–77).

Habits. A coastal species occurring in bays, estuaries, and the nearshore oceanic zone. They are burrowing bottom-dwellers that are most active at night (Levy et al. 1988; Able and Fahay 1998; Rountree and Bowers-Altman 2002), but can be taken by trawling during the day (Schwartz 1997d).

Sexual dimorphism occurs in the sound-producing mechanisms of many cusk-eels. Adult male striped cusk-eel (and some females [Schwartz 1997d]) develop a hump on the nape produced by enlargement of the supraoccipital bone and attachment of a large pair of dorsal sound-producing muscles (Courtenay 1971). The swim bladder of males is partly ossified, differing in morphology from that of females. Males have two pairs of sonic muscles and females have three pairs. For illustrations and a complete account of the complex sound-producing mechanisms of the striped cusk-eel, see Courtenay (1971). Sound production is associated with reproduction (Mann et al. 1997; Rountree and Bowers-Altman 2002).

Food. Food was found in 132 of 549 males and 204 of 854 females examined from the Cape Fear River, N.C. (Schwartz 1997d). Crustaceans were the dominant food category by volume, number, and frequency of occurrence (Schwartz 1997d: Table 6). Fishes were also important by volume, particularly in females, and included anchovies, gobies, and tonguefish. Crustaceans and a goby were found in three specimens from Chesapeake Bay (Hildebrand and Schroeder 1928). Hamer (1966) found striped cusk-eel in the body cavity of striped bass on two separate occasions. He suggested that the striped bass died after consuming the cusk-eel which broke into the body cavity causing damage to internal organs of the striped bass.

Predators. Found in stomachs of winter and thorny skates, spiny dogfish, goosefish, cod, windowpane, and summer flounder; spiny dogfish is the most frequent predator (Rountree 1999).

Reproduction. Spawning occurs from June through November in the Mid-Atlantic Bight based on occurrence of larvae (Fahay 1992; Able and Fahay 1998). However, Schwartz (1997d) collected the smallest mature females (160 mm TL) in December in the Cape Fear River, N.C., and some females were full of eggs in February suggesting an extended spawning season in the southern part of the range. Spawning of specimens trawled at night in Great Bay, N. J., occurred in the laboratory within a few hours after sunset (Fahay 1992; Mann et al. 1997; Rountree and Bowers-Altman 2002). Spawning occurred nightly in the laboratory and was initiated by the onset of calling just before sunset, often while the fish were still buried. Sounds consisted of 1–27 pulses between 500 and 1,800 Hz (Mann et al. 1997). Typically, a male would approach a buried female and nudge her with his mouth and barbel-like pelvic fins until she emerged from the sand. At other times a male would approach a female resting on the bottom. If the female was receptive, courtship would continue with rapid swimming around the tank in tandem. Pair swimming terminated in a rapid rush to the surface with the female arching over to expose her vent. The male would twist around the female to accomplish a brief ventral mount from above lasting a fraction of a second. After a few seconds, the female would turn over and swim slowly off with the egg mass still attached. It is uncertain whether the egg mass simply falls off or is rubbed off (Rountree and Bowers-Altman 2002).

Early Life History. Eggs are held together in a buoyant, viscous, gelatinous, oval-rectangular mass 26 mm in diameter and 60 mm long (Fahay 1992). After 12 h, the egg mass increased in size to 52 × 104 mm. Incubation at ambient temperatures (24°–26°C) took 36 h. Eggs are slightly off-round, lack an oil globule, and have a narrow perivitelline space. Larvae were collected during the NEFSC MARMAP surveys from July to December, with peak concentrations off the coast of New Jersey from August through September (Able and Fahay 1998).

Larvae ranged in length from 2.0 to 21.7 mm NL. Flexion begins at about 6.5 mm and is complete at about 16.0 mm. Larvae are more elongate than those of *Lepophidium profundorum*. Later larvae differ from other cusk-eel larvae in the Middle Atlantic Bight by the presence of a midlateral streak of pigment on the posterior part of the tail (Fahay 1992: Fig. 5). Settlement probably occurs on the shelf between 20–40 m at a size of about 22 mm TL. At this time the pelvic fins move from the typical chest position to a location near the tip of the lower jaw. Movement of the pelvic fins may determine time of settlement (Able and Fahay 1998). Juveniles have been collected in the vicinity of estuarine inlets in the Mid-Atlantic Bight from March to April, but are not known from other habitats (Able and Fahay 1998).

Age and Growth. Back-calculating lengths at each apparent annulus on vertebrae of 518 specimens from the Cape Fear River (Schwartz 1977d: Table 5) indicated that the largest males, at 196 mm TL, were 3 years old and the largest females, at 204 mm TL, were 4 years old. Juveniles grow 1 mm·day^{-1} during the summer, reaching 100–200 mm TL by the end of age 1, but apparently do not grow during the winter period based on length frequency data (Able and Fahay 1998).

General Range. Known from the lower Hudson River, N.Y. (C. L. Smith 1985) to northeastern Florida (Nielsen et al. 1999). Larvae occur as far north as Block Island Sound, R.I. (Fahay 1992: Fig. 9). Recently suggested to be common in Narragansett Bay, R.I., based on sound recordings made from 1965 through 2000 (Perkins 2001). An ongoing passive acoustic study, validated with the collection of voucher specimens, has similarly found striped cusk-eel to be abundant on Cape Cod (R. Rountree, pers. comm.).

Occurrence in the Gulf of Maine. Added to the fish fauna of the Gulf of Maine based on one 176-mm TL specimen (MCZ 152986) entrapped in the cooling water system of Seabrook Station, N.H. (42°54′17″ N, 70°47′12″ W) in April 1997 (Geoghegan et al. 1998). Striped cusk-eel have been heard calling in Barnstable Harbor on the north shore of Cape Cod so the single record from New Hampshire may not be as unusual as first thought (R. Rountree, pers. comm.).

ACKNOWLEDGMENTS. Drafts of the ophidiiform section were reviewed by Michael Fahay, Thomas A. Munroe, C. Richard Robins, Rodney A. Rountree, and Frank Schwartz.

CODLIKE FISHES. ORDER GADIFORMES

In this book the order Gadiformes is restricted to the cods, hakes, grenadiers, and their immediate relatives (Cohen et al. 1990; Nelson 1994). An expanded Gadiformes including the ophidioids (cusk-eels) and zoarcoids (eelpouts) has been proposed by some ichthyologists, but no one has yet found external or internal characters that uniquely define the order. Gadiform taxonomy suffers from additional weaknesses. There is no general agreement on the arrangement of genera into subfamilies, families, and suborders (Cohen 1989; Cohen et al. 1990). Of the eight families and approximately 500 species recognized by Cohen et al. (1990), 18 species from four families may occur in the Gulf of Maine. They are mostly soft-finned fishes, although in two Gulf species, silver and offshore hakes, the basal parts of the dorsal and anal fin rays are so stiff that they feel like spines to the touch, but they are distinguishable from all other soft-rayed Gulf of Maine fishes by the fact that their large pelvic fins are situated under or in front of the pectoral fins them and not behind them, as they are in herrings and salmons. They differ from most typical spiny-rayed fishes by the structure of the tail. Characters of gadiform larvae are treated by Fahay (1983) and Fahay and Markle (1984).

KEY TO GULF OF MAINE HAKES, CODS, AND GRENADIERS

2a. Three separate dorsal fins and two anal fins 3
2b. Two separate and well-developed dorsal fins 6
2c. Only one well-developed dorsal fin 13
3a. Lateral line black; black blotch on each shoulder . . . . **Haddock**
3b. Lateral line pale; no shoulder blotch 4
4a. Lower jaw projects beyond upper; chin barbel very small, if present . . . . **Pollock**
4b. Upper jaw projects beyond lower; chin barbel long 5
5a. Pelvic fin tip narrow, prolonged as filamentous feeler that is as long as the rest of the fin; eyes small . . . . **Tomcod**
5b. Pelvic fins broad, their filamentous tips less than one-third as long as remainder of fin; eyes large . . . . **Cod**
6a. Anal fin originates considerably in front of the origin of second dorsal fin; caudal peduncle very narrow . . . . **Hakeling**
6b. Anal fin originates under origin of second dorsal fin or behind it; caudal peduncle not very narrow 7
7a. Pelvic fins short and of ordinary form (Silver hakes) 8
7b. Pelvic fins very long and feelerlike 9
8a. Gill rakers on first arch 16–20; lateral line scales 101–110; cheek pigment absent; inner mouth and gill chamber may be dark, but not jet black . . . . **Silver hake**
8b. Gill rakers on first arch 8–11; lateral line scales 104–119; cheek pigment present; inner mouth and gill chamber jet black . . . . **Offshore hake**
9a. First dorsal fin hardly higher than second dorsal; no dorsal fin rays prolonged or filamentous . . . . **Spotted hake**
9b. First dorsal fin much higher than second dorsal, with one or two long filamentous rays 10
10a. Pelvic fins reach nearly as far back as rear end of anal fin. **Longfin hake**
10b. Pelvic fins do not extend to middle of anal fin 11

11b. Anal fin not notched, about equal in height from end to end 12
12a. Two gill rakers on upper limb of first gill arch; 119–148 lateral line scales; upper jaw usually reaches back to below rear edge of eye . . . . **White hake**
12b. Three gill rakers on upper limb of first gill arch; 95–117 lateral line scales; upper jaw usually reaches back only as far as rear edge of pupil . . . . **Squirrel hake**
13a. No isolated rays in front of dorsal fin; no barbels on top of snout . . **Cusk**
13b. Dorsal fin preceded by a fringe of short rays and one long ray; top of snout and chin bear barbels . . . . **Fourbeard rockling**
14a. Leading edge of elongate spinous ray in first dorsal fin perfectly smooth . . . . **Longnose grenadier**
14b. Elongate spinous ray in first dorsal fin serrate, with teeth that can be felt if not seen 15
15a. Vent considerably in front of origin of anal fin; skin surrounding vent naked and black; elongate dorsal fin spinous ray strongly serrate . . . . **Marlin-spike**
15b. Vent close to origin of anal fin; skin around vent scaly and pale-colored; serrations on elongate dorsal fin spinous ray so fine that they are hardly visible, though they can be felt . . . . **Roughhead grenadier**

GRENADIERS. FAMILY MACROURIDAE

GRACE KLEIN-MACPHEE

Grenadiers or rattails are characterized externally by having large heads, projecting snouts, and slender bodies that taper to whiplike tails, with no definitely demarked caudal fin. They have two dorsal fins, the first high and the second very low but occupying the greater part of the back. The anal fin is approximately the length of the second dorsal fin. A chin barbel is usually present. A light organ is present on the ventral midline of the trunk in some species. Most species have a well-developed swim bladder. General features of macrourid eggs and larvae are given in Fahay and Markle (1984) and Merrett (1986, 1989).

Grenadiers are allied to cods but differ in having one stout spinous ray in the first dorsal fin and in lacking a caudal fin. Most species live close to the bottom in the deep sea. Over 300 species in 34 genera are known (Cohen et al. 1990), but only three of them have ever been taken within the Gulf of Maine.

Besides the species described below, three species of *Coryphaenoides, C. rupestris, C. carapinus,* and *C. armatus,* have been taken on the continental slope abreast of the Gulf and off southern New England often enough to show that they are common there below 700 m. They are typical inhabitants of the deep sea, never likely to rise shallow enough to come within the limits of the Gulf of Maine. If any rattail collected in the Gulf proves difficult to identify, one should refer to Marshall and Iwamoto (1973) or Cohen et al. (1990).

LONGNOSE GRENADIER / *Caelorinchus caelorhincus carminatus* (Goode 1880) /

Bigelow and Schroeder 1953:246–247 (as *Coelorhynchus carminatus*)

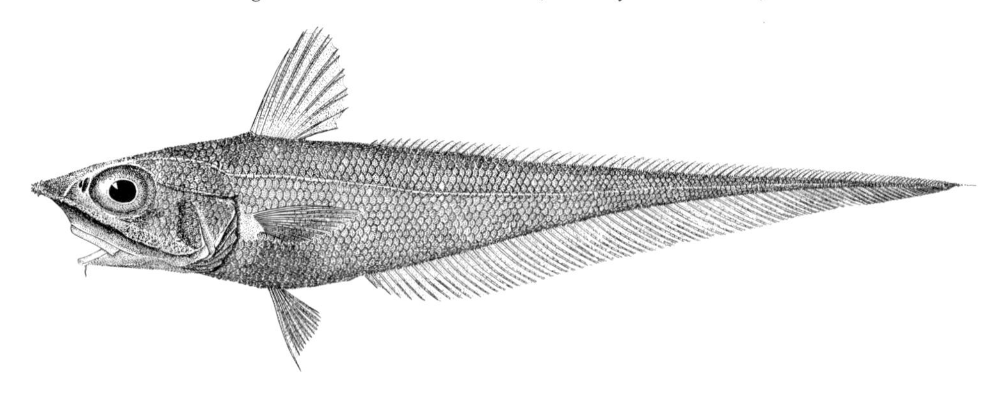

Figure 115. Longnose grenadier *Caelorinchus caelorhincus carminatus.* Continental slope off Martha's Vineyard. Drawn by H. L. Todd.

Description. Body elongate, stout anteriorly tapering to long pointed tail (Fig. 115). Jaws small and inferior, bearing two to several rows of small, pointed teeth on the premaxilla and dentary. Chin barbel present. Snout triangular, moderately pointed, with scutelike scales. Eyes large; largest diameter usually much longer than upper jaw length; orbits oval in outline. Elongate second spinous ray of rounded first dorsal fin with a smooth leading edge. Scales on trunk and tail with prominent spinules. Large, naked, black area in shallow fossa on chest extending posterior to pelvic fin bases covers an internal light organ (Marshall and Iwamoto 1973).

Meristics. First dorsal fin 2 spinous and 8 or 9 soft rays; pectoral fin rays 17–20; pelvic fin rays 7; scale rows below second dorsal fin origin 4–6; gill rakers on first arch 7–12 (Cohen et al. 1990).

Color. Silvery gray to pale tan seemingly dependent on the bottom over which they are taken. Three dark, saddle-shaped marks have been noted on some specimens, one each under the origins of the first and second dorsal fin and a third equally spaced from the other two on the tail. The oral cavity is pale to dark. The first dorsal fin is dusky to blackish; the pectoral, second dorsal, and anal fins are dusky. The inner rays of the pelvic fins are black; the outer ray is whitish.

Size. Reaches at least 38 cm TL (Cohen et al. 1990).

Distinctions. Resemble marlin-spike in general appearance and size but differ in that the snout not only overhangs the mouth slightly farther but is thinner-tipped, its dorsal spine is perfectly smooth, and its dorsal fin is rounded, not triangular.

Habits. Benthopelagic in 200–500 m, but they have been captured as shallow as 90 m and as deep as 850 m (Cohen et al. 1990).

Food. Feed on a variety of benthic organisms such as polychaetes, gastropods, cephalopods, crustaceans, and fishes (Cohen et al. 1990). Polychaetes and crustaceans are their principal food in the Gulf area (Langton and Bowman 1980; Bowman et al. 2000). Polychaetes include *Nephtys, Ophelina, Glycera, Lumbrineris,* and *Ninoe;* crustaceans are mostly shrimps (*Crangon*), amphipods (*Ampelisca* and *Unciola*), and copepods (*Calanus*).

General Range. In the western Atlantic from about 45 to 7° N and in the eastern Atlantic from about 60 to 18° N (Cohen et al. 1990: Fig. 367). Four subspecies are recognized, *C. caelorhincus carminatus* in the western Atlantic and three other subspecies in the eastern Atlantic (Cohen et al. 1990).

Occurrence in the Gulf of Maine. Included here because it was recorded once off Nantucket in 81 m.

Importance. A common by-catch in trawls operating deeper than 200–400 m. Often taken in moderate quantities, mostly for reduction to fishmeal and oil (Cohen et al. 1990).

ROUGHHEAD GRENADIER / *Macrourus berglax* Lacepède 1810 / Bigelow and Schroeder 1953:245–246

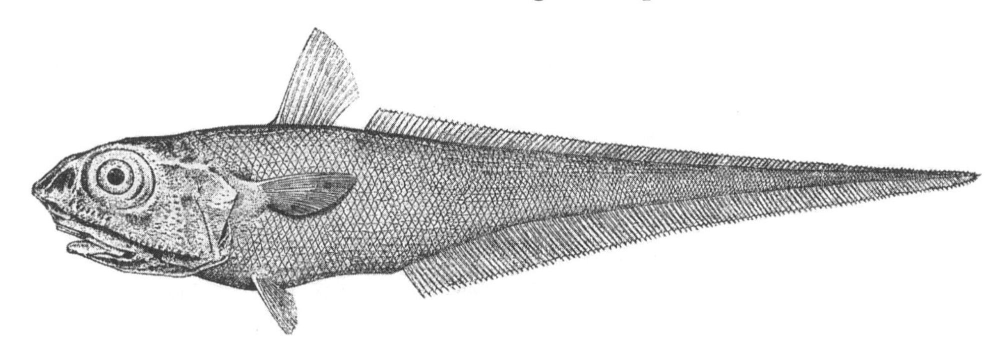

Figure 116. Roughhead grenadier *Macrourus berglax.* Banquereau Bank. Drawn by H. L. Todd.

Description. Body elongate, stout anteriorly, tapering to a point posteriorly, compressed behind vent; body depth 60–80% of HL (Fig. 116). Head 4.5 times in TL, compressed, hexagonal in cross section; upper profile convex; four to six ridges on top of head, snout projecting and overhanging mouth; lower jaw shorter than upper; angle of mouth under posterior edge of pupil of eye; prominent suborbital ridge runs from tip of snout to lower corner of operculum. Small teeth in an irregular double row in lower jaw and in a band in upper jaw; a small barbel on lower jaw about 10% HL. Upper parts of head covered with scales that do not overlap, larger on crests and ridges, smaller between; underside almost or entirely naked. Head scales armed with one or more rows of spines. Eyes large and oval. Two dorsal fins, first higher than long; second spinous ray armed with very fine teeth; second fin to posterior tip of body. Anal fin extending from vent to posterior tip of body, only slightly shorter than dorsal fin, and with slightly longer rays. Pectoral fins moderate, inserted on middle of sides behind gill opening, reaching almost to below beginning of second dorsal. Pelvic fins small and located on ventral part of body directly below pectoral fins. Body covered with small, firmly attached ctenoid scales, each with a median keel of several spines; scales on head not overlapping. (Scott and Scott 1988; Cohen et al. 1990).

Meristics. First dorsal fin segmented rays 9 or 10; second dorsal fin rays about 124; anal fin rays 148; pectoral fin rays 16–19; pelvic fin rays 8, rarely 7 or 9; gill rakers on first arch and inner series of second arch 8–10; scale rows below second dorsal fin origin 5.5–6.5; pyloric caeca about 19 or 20 (Cohen et

Color. Freshly caught specimens ash gray above and below, with chest a little darker. Rear edges of scales on rear part of the body still darker. Anal fin narrowly dark-edged; first dorsal, pectoral, and pelvic fins sooty, except that the outermost rays of the pelvics are white after preservation in alcohol.

Size. A large grenadier reaching 1 m TL and 1.8–2.3 kg. The largest seen in the Gulf of Maine was 73.6 cm TL (Bigelow and Schroeder).

Distinctions. Closely resemble marlin-spike, the most obvious differences are: the snout is shorter and blunter, with a more highly arched dorsal profile; there are four to six distinct ridges on the top of the head; the head is relatively larger (about one-fourth to one-seventh in marlin-spike); the trunk is relatively stouter (about six times as long as it is deep); the vent is close to the point of origin of the anal fin with the skin scaly around it, and no darker-colored than on the back; and the serrations on the large spine in the first dorsal fin are so fine that they are hardly visible. There are fewer rays in the second dorsal fin, but more rays in the anal fin than in marlin-spike; the first dorsal fin has a different outline and is relatively higher, with its membrane more developed, whereas the filamentous prolongation of the outer ray of the pelvic fins is shorter. The structure of the scales is diagnostic: those on the head and shoulders are armed with either one longitudinal row of spines (10–12 rows on each scale) or with up to three or four radiating ridges of spines, whereas those farther back each have a single row of spines, which together form conspicuous lateral ridges along each side of the rear part of the body.

Habits. Benthopelagic in 100–1,200 m, with greatest concentrations in 300–500 m. Temperature preferences range from about 1° to 4°C, although bottom temperatures below 0°C have been recorded (Cohen et al. 1990). They have a swim bladder, and males have well-developed drumming muscles, which may be used to produce sounds for courtship (Marshall 1965).

Food. Stomach contents of 144 fish collected from 235 to 490 m in the Barents Sea and the northwest Atlantic show that this grenadier feeds primarily on benthos (Geistdorfer 1976), mainly ophiurids (23.3%), gammaridean amphipods (20.4%), and polychaetes (16.7%). Decapod crustaceans, in particular *Pandalis borealis,* were also found in the stomachs, especially those of larger fish. Roughhead grenadier from the Carson Canyon region of the upper continental shelf of the Grand Banks (Houston and Haedrich 1986) had eaten mainly, by number, cumaceans (58.7%), amphipods (18%), polychaetes (7.9%), and euphausiids (5.7%). Bivalves, isopods, and ctenophores are important at times (Cohen et al. 1990).

Parasites. Twenty-one species of parasites were found in roughhead grenadier from the Flemish Cap and northern Grand Banks, many of which were intermediate hosts of benthic animals and fishes (Zubchenko 1981). The most common groups were nematodes, trematodes, acanthocephalans, and crustaceans. The parasite fauna of fish caught off south Labrador was much more diverse, and more fish were infested than were those caught off the deeper-water Flemish Cap, indicating that the two groups are separate populations. Myxosporidian parasites of fish from Newfoundland and Iceland were studied by Lom et al. (1975), and blood protozoans, very prevalent in fish from Davis Strait, by Khan et al. (1991).

Breeding Habits. Nearly ripe females have been collected off Norway in May and July (Marshall 1965). A ripe 705-mm female collected at 17°14′ N, 15°45′ E in 470–680 m in January contained 25,000 eggs (Yanulov 1962a). There were three size-groups of eggs in the ovary (0.5–1.10 mm, mean 0.96; 2.3–2.75 mm, mean 2.52, and 3.4–3.85 mm, mean 3.62), so it is believed to be a serial spawner. Off the Norwegian coast, roughhead grenadier form spawning populations with a predominance of males at depths of 700–800 m from winter to spring. The relative proportions of females increases from December to January, and spawning probably continues from December to May (Savvatimsky 1989a).

Age and Growth. Age determination based on scale annuli shows a life span of at least 25 years (Savvatimsky 1971). Growth appears to be linear and although females are larger than males, length-weight relationships of the sexes are similar.

General Range. Arctic to temperate waters of the North Atlantic, from Norfolk Canyon off the coast of Virginia and Georges Bank north to Labrador, Davis Strait, eastern and western Greenland, and Iceland, and from the Atlantic Irish slope north to the Faeroe Islands, Norwegian coast, Spitsbergen, and into the Barents Sea to 82° N (Cohen et al. 1990: Fig. 535).

Occurrence in the Gulf of Maine. Three-quarters of a century ago, when halibut were plentiful in the Gulf of Maine and vessels long-lining from Gloucester still regularly fished the deep channel between Georges Bank and Browns Bank, as well as the deep gullies that interrupt the Nova Scotian Banks, large grenadiers were often hooked. Fishermen described them as common enough to be a nuisance because they stole bait meant for other fish and were considered of no commercial value. It was on the strength of such reports that Goode characterized them as "exceedingly abundant on all of our offshore banks." Reexamination of three specimens collected locally in the late 1800s has proved that earlier identification was correct (Bigelow and Schroeder). They have been listed from deep waters of Massachusetts Bay and one was taken off Cape Ann, Mass. (Collette and Hartel 1988).

Importance. This is one of two grenadiers actively targeted by fishermen in the North Atlantic but catch statistics are not separated from those for *Coryphaenoides rupestris* (Cohen et al. 1990). Landings of grenadiers in the North Atlantic have shown a steady decline since the high in 1972 of more than 83,000 mt down to 3,567 mt in 1983.

MARLIN-SPIKE / *Nezumia bairdii* (Goode and Bean 1877) / Common Grenadier, Rattail /

Bigelow and Schroeder 1953:243–245 (as *Macrourus bairdii*)

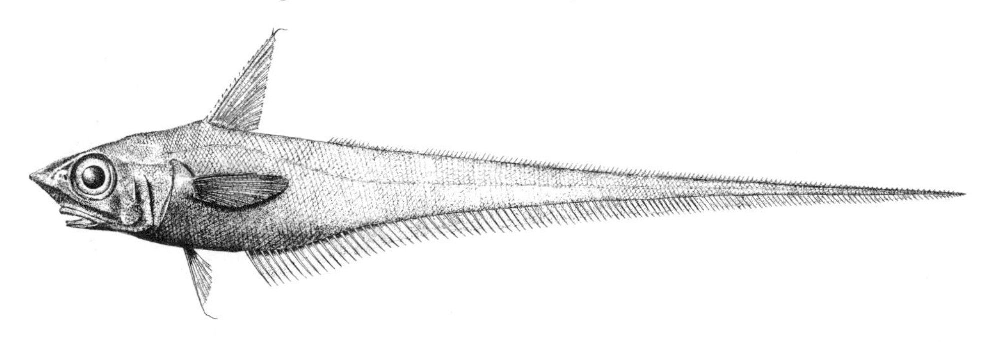

Figure 117. Marlin-spike *Nezumia bairdii*. Off Cape Ann. Drawn by H. L. Todd.

Description. Body slender (compressed behind vent and tapering to a whiplike tail with no caudal fin), pointed snout that overhangs mouth; very large eyes; a high first dorsal fin with one large spine and a very low second dorsal fin (Fig. 117). Broad bands of small teeth in both jaws; outer series in upper jaw enlarged. Small chin barbel present. First dorsal fin triangular, about twice as high as long, originating over pectoral fins, close behind gill openings. Second element of first dorsal fin a spinous ray, serrate along its front edge with about 15 sharp and very noticeable teeth pointing upward. Space between two dorsal fins about as long as height of first dorsal fin. Second dorsal fin extends back to tip of tail, so low that its membrane is hardly visible, and tapers to practically nothing at tip. Anal fin considerably longer than second dorsal fin and more than twice as high as the second dorsal. Pectoral fins rounded at tip. Pelvic fins inserted under or a little behind pectorals, triangular with first ray prolonged as a threadlike filament. Exposed parts of scales on body, including head and shoulders, rough with minute, sharp spines closely crowded together. Vent situated a considerable distance in front of origin of anal fin, and skin immediately surrounding it scaleless and black. A presumed light organ lies in the black naked area between pelvic fin bases (Marshall and Iwamoto 1973).

Meristics. First dorsal fin with 2 spinous rays, the first very short, and 9–11 shorter rays; second dorsal fin rays about 137; anal fin rays about 120; pectoral fin rays 17–20; pelvic fin rays 7; scales below second dorsal fin 7–9; inner gill rakers on first arch 8–10 (Cohen et al. 1990).

Color. The many specimens that Bigelow and Schroeder saw were uniform gray above, silvery below, with dark bluish or blackish belly. The lower surface of the snout is pink, the throat is deep violet, the first dorsal is pink with blackish spines, and the eyes are dark blue.

Size. The largest seen by Bigelow and Schroeder was 40.6 cm TL; the average is 16.7–31.6 cm.

Distinctions. ... by the vent, which is well in front of the origin of the anal fin, not close to it. Skin surrounding the vent is naked and black, not scaly and pale-colored like the roughhead grenadier.

Habits. Benthopelagic, most commonly in 90–700 m, but taken in much shallower depths in areas with cold surface waters. The shallowest record is 16.5 m in Vineyard Sound, trawled by the *Fish Hawk* many years ago. Records deeper than 1,000 m should be viewed skeptically (Cohen et al. 1990).

Preferred depth on the Scotian shelf was 270–360+ m and temperature and salinity preferences were 5.8°C and 34 ppt, but they were also found at temperatures of 4°–13°C and salinities of 31–34 ppt (Scott 1982b). Catches were confined to deep water, over 180 m, and they were usually distributed in comparatively warm saline water on the shelf slope, not in the deep basins of the shelf.

Catches off Newfoundland and Labrador increased with depth, and peak catches were made in daytime (Savvatimsky 1989b). Average length of fish in northern Newfoundland was greater than in southern Newfoundland. Females were larger and more abundant than males. Length-weight relationships of males and females of equal length were similar.

Marlin-spike have a size-related depth distribution in the Hudson Canyon area (Haedrich et al. 1980). Smaller fish (<17 cm) predominate in the shallowest zone (300–600 m), intermediate fish (17–30 cm) at moderate depths (700–1,000 m), and the largest fish (>25 cm) in the deepest zone (1,100–1,900 m). The area of highest fish abundance is in moderate depths.

Food. Feed primarily on euphausiids, amphipods, shrimps, bivalves, ophiuroids, and polychaetes (Savvatimsky 1989b; Cohen et al. 1990). The diet of 23 individuals collected in the northwest Atlantic (Langton and Bowman 1980) consisted mainly of crustaceans (47.3%) and polychaetes (27.5%). Crustaceans included hippolytid shrimp, a mysid *Neomysis americanus*, an isopod *Cirolina*, and a euphausiid *Meganyctiphanes*. Bowman and Grosslein (1988) confirmed the importance of these organisms as prey.

of the Gulf of Maine on 18 August were nearly ripe, and Bigelow and Schroeder reported a fully ripe male from South Channel in the last week of September. The eggs of this fish have not been seen, but it is probable that they resemble other macrourid eggs in being buoyant at least for the first part of the incubation period, with a large oil globule, wide perivitelline space, and raised hexagonal surface sculpturing.

Parasites. Marlin-spike from the Flemish Cap and northern Grand Banks had 33 species of parasites (Zubchenko 1981). Trematodes and nematodes were the most widely represented groups followed by cestodes, Myxosporidia and acanthocephalans. The parasite faunas of fish from the two areas were quite different, indicating that the fish were from separate populations. Many of the parasites had planktonic intermediate hosts, indicating that marlin-spike feed on plankton at some stage, but there was a species that had a benthic intermediary host. Eleven species of parasites were found off the New York Bight, and 301 out of 337 fish examined were infested (Campbell et al. 1980). Nematodes were the most common group (65%) followed by digenetic trematodes (12%), monogenetic trematodes (10%), cestodes (5%), and acanthocephalans (1%). An aegid isopod, *Syscenus infelix,* was commonly seen attached to the dorsal midline immediately behind the first dorsal fin in marlin-spike observed from submersibles off Virginia and North Carolina (Ross et al. 2001).

General Range. Western North Atlantic from Newfoundland to the northern end of the Straits of Florida (Cohen et al. 1990: Fig. 622). Known from the Laurentian Channel, the Scotian Banks, the Gulf of Maine, Vineyard Sound, and Hudson Canyon off New York Bight (Wilk et al. 1978). Records from the Gulf of Mexico and West Indies probably refer to *Nezumia suilla,* and a record from the Azores is also questionable (Cohen et al. 1990).

Occurrence in the Gulf of Maine. Marlin-spike were formerly regarded as rare strays in the inner parts of the Gulf of Maine for few have been recorded there. Bigelow and Schroeder reported one found floating near the surface at Eastport, Maine, one taken in a weir in Lubec, Maine, one from the slope of Jeffreys Ledge at 50 m, one from the western basin in 160 m, and another from off Gloucester, the last two taken many years ago. They must be very common on muddy bottoms of deeper parts of the Gulf in 84–124 m, as Bigelow and Schroeder took more than 100 of them at various localities on trawling trips. Marlin-spike, along with longfin hake, were the most abundant fish on the continental slope abreast of the Gulf below 100 m (Goode and Bean 1896).

Importance. While there is no directed fishery for this species, tens of thousands have been taken in the Soviet bottom trawl fishery for cod, redfish, and flounders in the areas off Labrador and Newfoundland (Savvatimsky 1989b).

DEEP-SEA CODS. FAMILY MORIDAE

GRACE KLEIN-MACPHEE

The genera and species now included in the Moridae were placed in the Gadidae for a long time. Morids are relatively elongate fishes, many with a narrow caudal peduncle and a distinct caudal fin. Anterior paired projections of the swim bladder attach to a membranous area at the back of the skull. Several hypural bones attach to the last vertebra (Cohen et al. 1990). There are about 100 species but there is no agreement as to how many genera should be recognized within the family (Cohen 1989; Cohen et al. 1990). One species, hakeling, was recorded from off Nantucket in 142 m and another, blue hake, *Antimora rostrata* (Günther 1878) might possibly occur in the Gulf although records at the latitude of the Gulf of Maine are deeper than 500 m.

HAKELING / *Physiculus fulvus* Bean 1884 / Bigelow and Schroeder 1953:233–234

Description. Hakelike in appearance and in arrangement of fins (Fig. 118). Two dorsal fins, first triangular and much shorter than second, of nearly uniform height from end to end. Single long anal fin similar to second dorsal in shape. Pelvic fins situated in front of pectoral fins. Pelvic fins short; longest ray (second filamentous at tip) hardly reaches back as far as middle of pectoral fins. Snout blunt, barbel on lower jaw. A single band of brushlike teeth on upper and lower jaws; teeth absent from vomer and palatines. Caudal fin very small, at end of elongate caudal peduncle. No dorsal fin rays prolonged. Lateral line indistinct, located high on body. Dark, naked area on belly between pelvic fin bases covers an internal light organ.

Meristics. First dorsal fin rays 10 or 11; second dorsal fin rays 50–58; anal fin rays 60–64; pelvic fin rays 5; lateral line with about 62 medium-sized scales; vertebrae 48 or 49 (Miller and Jorgenson 1973).

Color. Light yellowish brown with the lower surface of the head, the abdomen, and the margins of the dorsal and anal

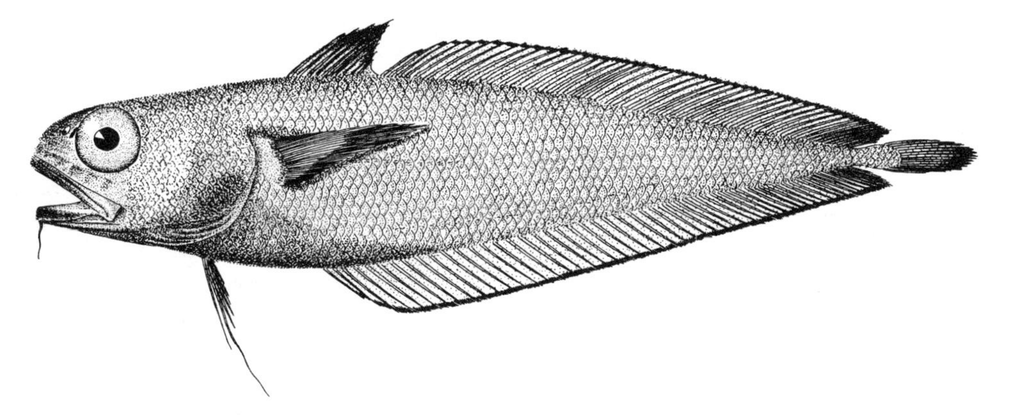

Figure 118. Hakeling *Physiculus fulvus.* Outer edge of continental shelf. Drawn by H. L. Todd.

fins very dark brown, and with a dark brown blotch on each cheek (on the subopercular bone).

Size. Reaches at least 12.2 cm (Miller and Jorgenson 1973).

Distinctions. Separable from white, red, and longfin hakes in that: the anal fin originates in front of the origin of the second dorsal fin instead of considerably behind the latter and the pelvic fins have five rays each instead of two and are much shorter than those of hakes, with the longest ray (the second, which is filamentous at the tip) hardly reaching back as far as the middle of the pectoral fins. Furthermore, the snout is blunter than that of any true hake; the caudal fin is much smaller; the body tapers more abruptly to a very narrow caudal peduncle; and no rays of the first dorsal fin are prolonged.

Habits. Nothing is known of the habits of hakeling except they are deep-water fish, having been taken at 142–1,719 m, and they live on or near the bottom.

General Range and Occurrence in the Gulf of Maine. Hakeling have been taken at several localities in the Gulf of Mexico and on the continental slope off the eastern United States. One was collected off Nantucket (40°01′ N, 69°56′ W) in 142 m (Bigelow and Schroeder), and it is on this record that hakeling are included here. The most northerly record is of a juvenile taken at the surface at 43°16′ N off Canada (Markle et al. 1980).

SILVER HAKES. FAMILY MERLUCCIIDAE

GRACE KLEIN-MACPHEE

Silver hakes and cods have been grouped together as subfamilies of a single family by some but are considered separate here following most recent authors (Cohen et al. 1990; Nelson 1994). Four genera are recognized, of which two of the 13 species of *Merluccius* may be found in the Gulf of Maine. Two dorsal fins and one anal fin. Chin barbels absent. Strong, pointed teeth in both jaws. Well-developed caudal fin in Gulf of Maine species. Differ from other gadiform fishes in having a pair of more or less well-developed ridges forming a V-shape on top of the skull.

OFFSHORE HAKE / *Merluccius albidus* (Mitchill 1818) / Ginsburg 1954a:192–194

Description. Very similar to silver hake. Cheek, preopercle, and interopercle fully scaly; mouth large with numerous teeth. Base of tongue with teeth. Lower jaw extends beyond upper jaw. Peritoneum solid dark brown or black in larger specimens; in smaller fish, dark brown or almost black, stippled with darker dots.

Meristics. First dorsal fin with 1 spine and 10–12 rays; second 40; first arch gill rakers 8–11; lateral line scales 104–119; total vertebrae 51–55 (Markle 1982; Cohen et al. 1990).

Color. Offshore hake are dusky blackish blue above or brown with a brassy hue. The belly is white or silvery. The iris of the eye is yellow and the pupil black.

Size. The largest recorded male and female measured 40 and 70 cm, respectively (Cohen et al. 1990); common to 30 cm

Distinctions. Offshore hake closely resemble silver hake but differ in having fewer gill rakers on the first arch, 8–11 vs. 16–20, and more lateral line scales, 104–119 vs. 101–110. In addition, cheek pigment is absent in silver hake and the snout is broader in offshore hake (Markle et al. 1980).

Habits. Offshore hake inhabit the outer part of the continental slope and the upper parts of the slope between 80 and 1,170 m and are most abundant between 160 and 640 m (Chang et al. 1999a). Adults probably live at or near the bottom, and there is evidence of diel vertical migration. On the continental slope of the Gulf of Mexico they segregate by size and sex. Juveniles, young adult females, and adult males occupy the upper slope, down to a depth of 550 m, and larger mature females are concentrated on the lower slope. Gulf of Mexico fish appear to feed on or near the bottom, and at the same rate throughout day and night (Rohr and Gutherz 1977).

Food. Offshore hake prey most heavily on fishes (particularly clupeids, anchovies, and lanternfishes) and, to a lesser extent, on crustaceans and squids (Cohen et al. 1990). Euphausiid prey include *Meganyctiphanes* and *Thysanoessa.* Pandalid shrimps such as *Pandalus* and *Dichelopandalus,* the pelagic shrimp, *Pasiphaea,* and squids have also been identified as food (Maurer and Bowman 1975; Langton and Bowman 1980; Bowman et al. 2000).

Parasites. Helminths (Scott 1987, most intensive infestations marked *), four Digenea: **Derogenes varicus,* **Hemiurus levinseni, Lecithaster gibbosus,* and *Lepidapedon elongatum;* two cestodes: **Clestobothrium crassiceps* and *Scolex* sp.; and three nematodes: *Anisakis* sp., *Copellaria gracilis,* and **Hysterothylacium aduncum.*

Predators. Offshore hake are seldom reported as prey, probably because of confusion with silver hake, but small juveniles are prey for goosefish (Chang et al. 1999a). Cannibalism has been noted at low levels, that is, it was in evidence in 2% of the specimens examined (Rohr and Gutherz 1977).

Breeding Habits. Spawning occurs near the bottom at depths of 330–550 m from April to July in New England (Cohen et al. 1990), but eggs were collected every month of the year by the MARMAP surveys (Berrien and Sibunka 1999). Fecundity is estimated at 340,000 eggs per female.

Early Life History. The eggs are pelagic, spherical, and transparent, 1.04–1.24 mm in diameter, perivitelline space 0.04–0.16 mm, oil globule 0.28–0.54 mm (Markle and Frost 1985: Table 3). Offshore hake eggs are distinctly larger than those of silver hake (0.82–1.00 mm). Small stellate melanophores appear on the body, yolk, and oil globule.

Larvae hatch after 6–8 days at 8.9°–10.6°C and are 3.0–3.8 mm long (Marak 1967). The pigmented oil globule lies in the posterior part of the yolk; the anus opens laterally on the finfold. There is a group of melanophores on the jaw region and posterior part of the head. Four concentrations of pigment are found on the body: dorsal to the yolk sac, over the vent, at the midpoint of the trunk, and two-thirds of the distance to the tail. Pelvic buds appear at 5 mm, and the fin rays become moderately long. Pectoral and pelvic fins are usually heavily pigmented. Transformation occurs at 20 mm (Fahay 1983). Offshore hake larvae are distinguished from those of silver hake by the large size; lack of pigment over the posterior part of the gut; yolk pigment; two well-developed bands of pigment dorsally between the head and the anus; and the fourth pigment band on the tail, which extends onto the dorsal and ventral finfold (Hardy 1978a).

Age and Growth. There is no information for the Gulf of Maine, but in the Gulf of Mexico fish grew rapidly until age 3, at which time the growth rate slowed; females grew faster than males and appeared to live longer (Rohr and Gutherz 1977).

General Range. From the southern edge of the Grand Banks as far north as the Laurentian Trough (Chang et al. 1999a) and southeastern slope of Georges Bank (Markle et al. 1980) south to Suriname and French Guiana (Cohen et al. 1990: Fig. 724). Offshore hake are a very important component of the Florida slope community (Rohr and Gutherz 1977) but are comparatively rare in the North Atlantic.

Occurrence in the Gulf Maine. Regularly caught near the outer edge of the Scotian Shelf, but easily confused with silver hake and so have often been misidentified (Markle et al. 1980). Bigelow and Schroeder (1955) reported that they did not catch any north of Georges Bank in 1953. Larvae occur occasionally over the Scotian Shelf. Eggs were reported in June from Salem Harbor (Elliott and Jimenez 1981). Both adults and juveniles were caught on the slopes of several deep basins within the Gulf of Maine by NMFS trawl surveys (Chang et al. 1999a).

Importance. Offshore hake are the object of minor local fishing and negligible catch statistics have been reported by Cuba and the United States (Cohen et al. 1990). Offshore hake are taken as by-catch by otter trawls in the silver hake fishery.

Stocks. There appear to be several stocks. Caribbean and Gulf of Mexico populations cannot be distinguished meristically but there are local population differences (Karnella 1973). The northern Gulf population is divergent from the northern Atlantic stock, but it is also divergent from the southern Gulf population, and this intergrades with the Atlantic stock. The southern populations were separated by Ginsburg (1954a) into two species, *Merluccius albidus* and *Merluccius magnoculus,* but Karnella (1973) showed that they are all *M. albidus.*

SILVER HAKE / *Merluccius bilinearis* (Mitchill 1814) / Whiting; New England Hake /

Bigelow and Schroeder 1953:173–182

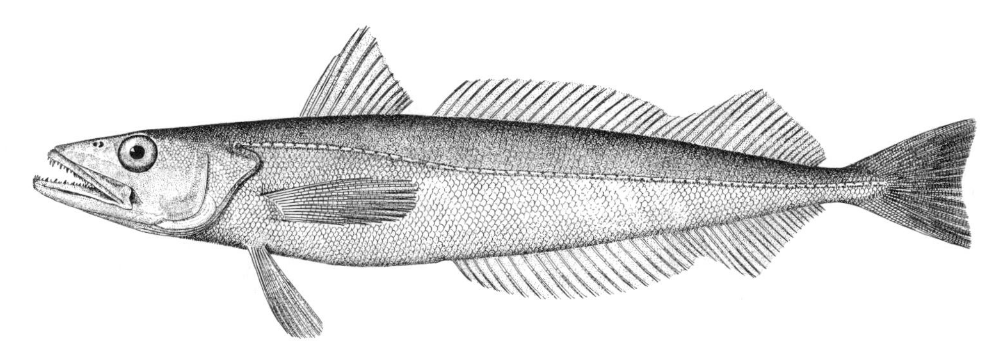

Figure 119. Silver hake *Merluccius bilinearis.* Nova Scotia. Drawn by H. L. Todd.

Description. Slender, five to six times as long as deep; body rounded in front of vent but flattened sidewise behind it, with large, flat-topped head occupying about 25% TL; large eyes (Fig. 119). Lower jaw projecting beyond upper; wide mouth armed with two or more rows of sharp recurved teeth, which also occur on vomer. First dorsal fin originates close behind gill openings, roughly an equilateral triangle in shape and separated by a short space from second dorsal fin. Second dorsal fin about four times as long as first dorsal, but hardly more than half as high, and deeply emarginate two-thirds of the way back, with rear section higher. Anal fin corresponds in height and shape to second dorsal, under which it is located. Caudal fin square-tipped when widespread; otherwise its rear margin weakly concave. Pectoral fins rather narrow, their tips slightly rounded, and extending far enough to slightly overlap second dorsal. Pelvic fins situated slightly in front of pectorals and perceptibly shorter than the latter. Lateral line prominent, appearing double (Scott and Scott 1988).

Meristics. First dorsal fin with 1 spine and 10–12 rays; second dorsal fin rays 37–42; anal fin rays 37–42; total caudal fin rays 34–37; lateral line scales 101–110; gill rakers on first arch 16–20; precaudal vertebrae 26–29 + 27–29 caudal = 53–57 total (Markle 1982; Cohen et al. 1990)

Color. Silver hake are dark gray above with five to seven irregular, darkened vertical bars on the sides. They have a silvery iridescent cast, as the name implies, or golden reflections. The lower part of the sides and belly are silvery. The inside of the mouth is dusky blue; the lining of the belly is black or brown speckled with black. They are brightly iridescent when taken from the water, but fade soon after death.

Size. Maximum size about 760 mm TL and about 2.3 kg. A 760-mm TL individual caught on *Delaware II* cruise 78-06, station 421 at 43°28′ N, 66°26′ W was the largest taken by the NEFSC surveys, 1963–1986. The largest individuals from 54 bottom trawl surveys ranged from 490 to 760 mm TL. The all-tackle game fish record is 2.04-kg fish caught in Perkins Cove, Ogunquit, Maine, in August 1995 (IGFA 2001).

Distinctions. Presence of two separate and well-developed dorsal fins, both of them soft-rayed, the second much longer than the first, combined with location of the pelvic fins anterior to the pectoral fins distinguish silver hake from all other Gulf of Maine fishes except true hakes and offshore hake. They differ from the former by the lack of a chin barbel and the ordinary shape of the pelvic fins, which in true hakes are attenuated into long feelers. They are distinguished from the latter by having more gill rakers (16–20 vs. 8–11) and fewer lateral line scales (101–110 vs. 104–119).

Habits. The lowest temperatures in which silver hake have been taken by NEFSC trawling surveys 1963–1986 are 3°–6°C. Most winter and spring records for silver hake have been from areas where the bottom temperature was warmer than 6°C. Maximum temperatures ranged from 7° to 17°C. Similarly, they were not reported in any numbers where the water was warmer than 18°C (Bigelow and Schroeder). More detailed habitat requirements are given in Morse et al. (1999).

Silver hake are strong, swift swimmers, well-armed with teeth and extremely voracious. Although they do not form definite schools, multitudes of them often swim together. Silver hake are said to rest on the bottom by day, on sandy or pebbly bottom or on mud but seldom over rocks. Adults observed from the deep-sea submersible *Alvin* were seen 2 m or so off the bottom or resting in shallow depressions at depths of 37–46 m during the day (Edwards and Emery 1968). Young-of-the-year silver hake, 1.5–5.0 cm TL, occurred at higher densities on bottoms with greater amphipod tube cover at a 55-m site in the Mid-Atlantic Bight (Auster et al. 1997). Observations at other sites showed that young-of-year silver hake occurred only on silt-sand bottom with amphipod tubes at bottom temperatures of 8.7°–11.4°C. Adults hunt by night, and it is usually at night that they run into the shallows and enter traps. Their movement off bottom at night also makes them more available to bottom trawls (Bowman and Bowman 1980). Silver hake are wanderers, independent of depth within wide limits. Sometimes they swim close to the bottom, sometimes in the upper levels of the water, their vertical movements being governed chiefly by their pursuit of prey. Their upper limit is the tide line; at the other extreme they were caught down to 400 m in 54 NEFSC bottom trawl surveys. There are reports as deep as 900 m (Scott and Scott 1988), but it seems likely that such deep records were based on misidentified specimens of *M. albidus.*

Food. Silver hake are crustacean/fish predators with a size-dependent shift in predation from crustaceans to fishes at about 20–25 cm (Langton 1982). Silver hake less than 1 year old feed primarily on copepods, amphipods, euphausiids, and small decapod shrimps (Bowman 1981; Bowman et al. 2000). Euphausiids (mostly *Meganyctiphanes*) are an especially important food (about half of the diet) at about 15 cm in FL. Shrimps such as *Pandalus, Dichelopandalus, Crangon,* and *Pasiphaea* are regular prey in the Gulf.

When silver hake reach 20 cm, they begin to depend more on fishes and squids for food, which together make up about 80% of their diet (Langton and Bowman 1980; Bowman 1984; Bowman and Grosslein 1988). Fish species preyed upon include anchovies, herrings, silver hake, lanternfishes, Atlantic mackerel, sand lance, butterfish, and any other small fishes abundant in a particular area. Bigelow and Schroeder reported that Vinal Edwards observed cunner, sand lance, scup, silversides, and smelt in the diet of fish collected near Woods Hole. In the offshore waters of the Gulf of Maine, herrings, mackerel, sand lance, and silver hake are the principal fish prey. Cannibalism is common, especially in the Mid-Atlantic area.

Composition and quantity of food consumed differs between males and females of the same size. Females tend to eat more food and take primarily fishes as prey. Males have much larger proportions of crustaceans in their stomachs than females (Bowman 1984). As growth is directly related to food intake, it is not surprising that females grow faster than males (Penttila and Dery 1988).

Silver hake feed mainly at night from just after dusk until predawn. Intensive feeding occurs between dusk and midnight, and during the spring they may also feed around noon. In spring, just prior to summer spawning, they eat larger quantities of food. During spawning food intake is curtailed; in autumn it increases again, although not to the extent of spring. Over the winter months feeding is reduced compared to the spring and autumn (Bowman 1984). The average daily ration has been estimated at 2.9–3.2% of the body weight for hake smaller than 20 cm and 0.8–2.2% for fish larger than 20 cm (Durbin et al. 1983). From the point of view of total food consumption, silver hake represent one of the most important predators in the Mid-Atlantic Bight (Grosslein et al. 1980). Lists of prey species and further details of silver hake feeding are given by Langton and Bowman 1980; Bowman and Michaels 1984; and Bowman et al. 2000.

Diet overlap calculated as the percentage similarity between the diets of silver hake and 16 other northwest Atlantic fishes showed that the greatest overlap in diet was among species of gadoid fishes rather than between gadoids and fishes of other orders (Langton 1982). The greatest overall potential for interaction exists between smaller stages of silver hake (11–30 cm) and slightly larger (16–65 cm, depending on species) stages of gadids such as white hake, red hake, spotted hake, and pollock, largely owing to predation on the euphausiid *Meganyctiphanes norvegica* and the shrimps *Dichelopandalus leptocerus* and *Crangon septemspinosa* (Langton 1982).

Predators. Silver hake are preyed upon by numerous other fishes, and older silver hake commonly eat juvenile silver hake. A partial list of fishes that eat silver hake includes spiny dogfish, little skate, goosefish, pollock, cod, haddock, spotted, red, and white hakes, Acadian redfish, sea raven, bluefish, mackerel, swordfish, and flounders (Maurer and Bowman 1975; Bowman and Michaels 1984; Bowman et al. 2000). Silver hake are also common prey of harbor porpoise in the Gulf of Maine (Gannon et al. 1998).

Parasites. Helminths (Scott 1987, most severe infestations marked *), four Digenea: **Derogenes varicus, Hemiurus levinseni, Lecithaster gibbosus,* and *Podocotyle atomon;* one Monogenea: *Anthocotyle merlucii;* two cestodes: **Clestobothrum crassiceps* and *Grillotia* sp. (larvae); three nematodes: *Anisakis* sp., *Capillaria gracilis,* and **Hysterothylacium aduncum;* and three copepods: *Caligus curtus, C. elongatus,* and *Chondrocanthus merluccii.*

Breeding Habits. Median length at maturity for female and male silver hake from the Gulf of Maine–northern Georges Bank stock was 23.1 and 22.3 cm, respectively (O'Brien et al. 1993). Stock abundance was the most important predictor of sexual maturation of 2- and 3-year-olds in the northwest Atlantic between 1973 and 1990 (Helser and Almeida 1997). Median age at maturity for both sexes was 1.7 years. Average fecundity is 343,000 eggs in females 25–30 cm and 391,700 eggs in females 30–35 cm (Sauskan and Sererbryakov 1968). Silver hake are serial spawners; three successive generations of oocytes develop in the ovary and all are spawned in a single season (Sauskan and Serebryakov 1968).

Concentrations of spawning stock of silver hake in the Gulf of Maine were located in the coastal area between Cape Cod and Grant Manan Island (Almeida 1987). Major spawning grounds on the continental shelf are along the southeastern and southern slopes of Georges Bank (Sauskan and Serebryakov 1968), around Nantucket Shoals, and south of Martha's Vineyard (Fahay 1974) as far as Cape Hatteras (Anderson 1982: Map 32). Significant spawning also occurs in inshore waters of the Gulf of Maine during the summer (Berrien and Sibunka 1999).

Egg and larval collections show that spawning extends throughout the year but eggs are most abundant from May to

November (Berrien and Sibunka 1999). The earliest egg record north of Cape Cod was 11 June; egg production is at its height in July and August; and 22 October appeared to be the latest date. Spawning occurs near the bottom, but has been recorded from surface to bottom both inshore and offshore (Sauskan 1964) at salinities of 31.5–32.5 ppt (Svetovidov 1962).

Early Life History. The eggs are buoyant, transparent, 0.82–1.00 mm in diameter, perivitelline space 0.05–0.13 mm, with a single yellowish or brownish oil globule 0.19–0.34 mm (Markle and Frost 1985: Table 3). Incubation is rapid, 48 h at Woods Hole (Kuntz and Radcliffe 1917) and 39 h at 22.1°C in the Mid-Atlantic Bight (Kendall and Naplin 1981). Larvae are 2.64–3.52 mm long at hatching. The vent is located on one side near the base of the larval finfold, and the trunk behind the vent is marked with two black crossbars. Dorsal, anal, and caudal fins assume their definite outlines by the time the fish is 10–11 mm long, and larvae of 20–25 mm begin to resemble their parents (Bigelow and Schroeder).

Silver hake can be distinguished from offshore hake, with which they occur in some places, by the scarcity or lack of pigment on the paired fins and by the two major melanophores found laterally on the trunk. In the latter the paired fins are usually heavily pigmented and the posterior lateral flank melanophores dominate the anterior. In larger offshore hake larvae, a line of dense large spots covers the flank (Fahay 1983). Silver hake have more total caudal fin rays than offshore hake (34–37 vs. 40, Markle 1982: Table 6).

Larvae drift at the surface or in undercurrents and are present down to 40 m. In the Mid-Atlantic Bight, most larvae occurred at 30 m during the day but at 15 m during the night, implying that they avoided the net during the day or migrated upward at night (Kendall and Naplin 1981). They are inshore in Maine during August and September (Graham and Boyar 1965). Juveniles are reported to associate with jellyfish. They become benthic at 17–20 mm according to Fahay (1974), but Fritz (1965) and Bigelow and Schroeder reported they descend to the bottom at 25–75 mm in autumn. They overwinter in deep depressions (Bigelow and Schroeder). Vertical distribution of young-of-the-year silver hake on the Scotian Shelf (Koeller et al. 1986) showed that they were present throughout the water column with relatively high concentrations above the thermocline; they were not caught during the day because they migrated to the bottom before dawn. This behavior begins when the fish are 20–30 mm long and continues throughout life. Cumulative distribution plots of juvenile silver hake catches during the years 1973–1976 by *Albatross IV* are given in Bowman et al. (1987).

Calanoid copepods are the primary food of larval silver hake and metamorphosed juveniles (19–40 mm). At lengths greater than 40 mm, more noncopepod items such as amphipods, mysids, and euphausiids are included in the diet although calanoid copepods are still important. At sizes over 46 mm, cannibalism became evident and increased in importance. Juveniles of other fish species were rare at this time. An abrupt change in feeding pattern occurs as larvae metamorphose into juveniles, and a strong diel vertical migration pattern begins in which juveniles are near the bottom during the day and throughout the water column at night (Koeller et al. 1989).

Age and Growth. Silver hake produce daily growth rings on their otoliths, and both fast-growth zones and slow ones (spawning zones) are apparent (Pannella 1971). Based on otolith analysis (Hunt 1980), there is very rapid growth during the juvenile stage and a divergence between males and females greater than 25 cm, with females growing faster. Observed maximum growth for males and females was 37 and 65 cm, respectively (Ross and Almeida 1986), and the maximum age appears to be 14 years (Penttila et al. 1989). The first annulus is the most difficult to determine. Growth is faster in the Gulf of Maine than further south (Fig. 120). Growth rates in the northern Georges Bank–Gulf of Maine stock appear to be density-dependent after the fish become piscivorous (Ross and Almeida 1986). Weight was inversely correlated with the estimate of stock density for all ages, although it was lower for fish of age-group 2 than for older ones.

General Range. Atlantic coast of Canada and the United States. from Belle Isle Channel (52° N) to the Bahamas; most common from southern Newfoundland to South Carolina (Cohen et al. 1990: Fig. 730).

Occurrence in the Gulf of Maine. Silver hake are familiar all around the Gulf of Maine (Map 13) from Cape Cod to the Bay of Fundy and to the west coast of Nova Scotia, but their center of abundance is in the southwestern part of the Gulf. Distribution in the Mid-Atlantic Bight is shown in Anderson (1982: Map 32), for the Gulf of Maine in Halliday et al. (1986).

Migrations. Silver hake migrate seasonally, inhabiting waters shallower than 90 m in summer and autumn and deeper offshore waters in winter and spring (Anderson 1982). They also move north and south (Fritz 1962). Water temperature appears to be an important factor governing timing of migration and fish distribution. Silver hake prefer progressively warmer waters as they increase in size in the first year of life (Edwards 1965).

Importance. Silver hake are a delicious fish if eaten fresh; however they soften so fast that there was no regular market demand for them and those caught incidentally were thrown overboard or used as fertilizer (Bigelow and Schroeder). A commercial market first developed in the 1920s as fillets for fried fish shops. They were also used for canned pet food and fish meal (Anderson et al. 1980). They have an enzyme system that causes rapid breakdown of flesh after capture, which renders them unsuitable for fillets after frozen storage. Commercial processing trials have shown that they make excellent surimi when processed without belly flaps, and this should

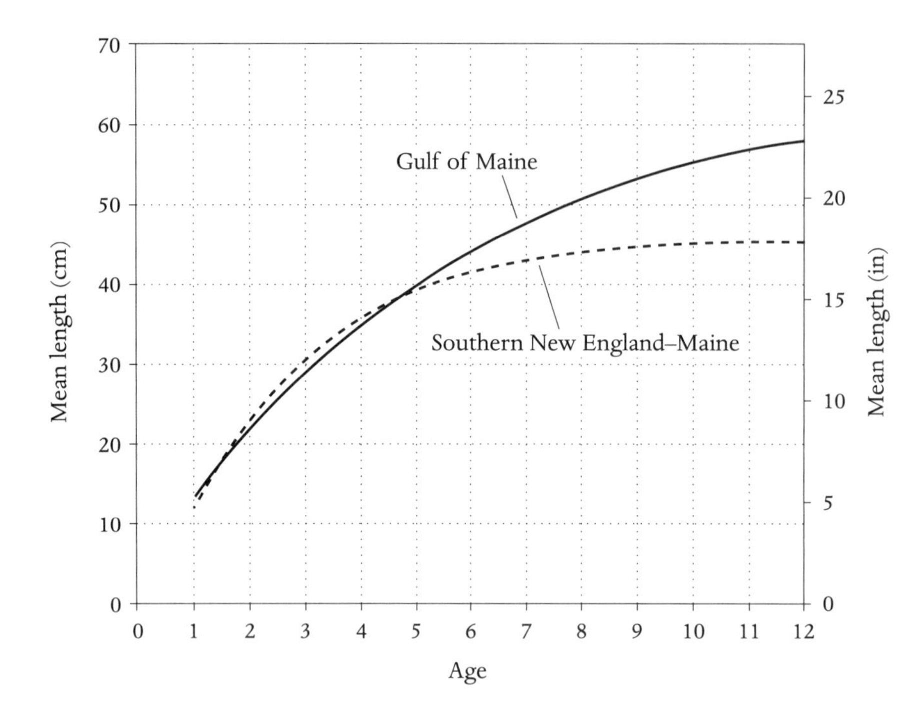

Figure 120. Length-age relationship in silver hake *Merluccius bilinearis.* (From Penttila et al. 1989:9.)

Following the arrival of distant-water fleets in 1962, total landings increased rapidly to a peak of 94,500 mt in 1964, dropped sharply in 1965, and declined for 13 years, reaching their lowest level (3,400 mt) in 1979 (Mayo 1998a). Landings increased slightly during the early 1980s and have varied between 4,400 and 6,800 mt since 1987 (Fig. 121). Neither the Gulf of Maine–Northern Georges Bank nor the Southern Georges Bank–Mid-Atlantic stocks can support increased fishing, and both stocks must be considered fully exploited.

The Canadian-directed silver hake fishing activity developed from 1987 to 1996 (Showell and Cooper 1997) and became a commercial fishery in 1995–1996 in Emerald and La Have basins.

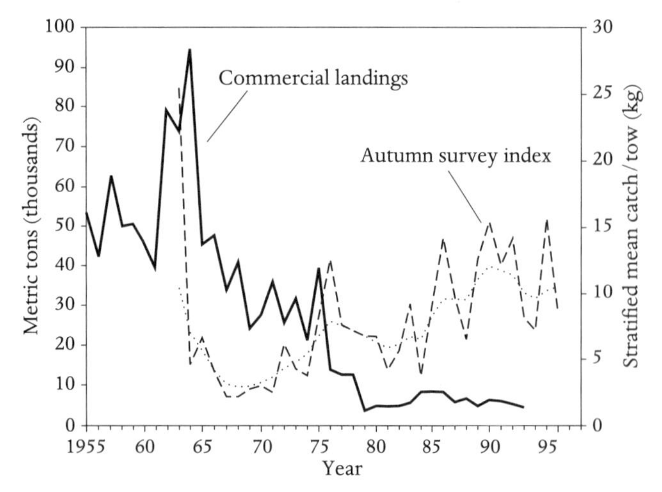

Figure 121. Catch and index of abundance of silver hake *Merluccius bilinearis,* Gulf of Maine–northern Georges Bank stock. (From Mayo 1998a.)

Trawler discards of silver hake in the northern shrimp fishery in the 8- to 31-cm size range indicate that 93% per tow was discarded, and this could have a negative effect on the species in the Gulf of Maine (Howell and Langan 1992). Regulations have been implemented to reduce finfish by-catch in the shrimp fishery.

There is a recreational hook-and-line fishery for silver hake from southern Massachusetts to New Jersey. They rank low as a sport fish, because even though they bite greedily, they put up only feeble resistance when hooked (Bigelow and Schroeder). They are caught on rod and reel in the fall at night off docks and piers (Fritz 1962).

Stocks. Two distinct stocks of silver hake are recognized (Mayo 1998a; Bolles and Begg 2000): Gulf of Maine–northern Georges Bank and southern Georges Bank to the Mid-Atlantic area. The Gulf of Maine stock differs from populations south of Cape Cod morphometrically (Conover et al. 1961); head length and pelvic fin length are significantly different between the two groups. Growth patterns of otoliths of immature silver hake (ages 0 and 1) from the Gulf of Maine and southern New England waters differ in growth increments (Nichy 1969). Biochemical analysis indicates significant differences in the enzyme systems of spawning silver hake (Schenk 1981). Use of a statistical technique (discriminant analysis) based on 13 morphometric characters, bottom trawl survey data, and commercial fishery catch statistics support earlier conclusions that there are two distinct stocks, one on northern Georges Bank and the Gulf of Maine and another from southern Georges Bank south to Cape Hatteras (Almeida 1987). The greatest group separation occurs during spawning season. Some mixing occurs on Georges Bank and Nantucket Shoals, but the shoal part of Georges Bank appears to separate the two stocks.

CODS. FAMILY GADIDAE

Grace Klein-MacPhee

Cods are small to very large fishes, ranging in size from 15 cm to 2 m. They are found in circumpolar to temperate waters, mainly in the Northern Hemisphere. Cods are primarily marine fishes but a few inhabit estuaries and one is restricted to freshwater. The family is hard to diagnose because it has no unique characters. It is characterized by an externally symmetrical caudal fin, the lack of a V-shaped ridge on top of the head, rounded cycloid scales, and a swim bladder that is not connected to the skull (Cohen et al. 1990). The family is a diverse assemblage of about 50 species in three subfamilies (Gadinae, Phycinae, and Lotinae) and 21 genera. Several authors recognize these subfamilies as either two families (Gadidae and Phycidae [Nelson 1994]) or three (Lotidae removed from Gadidae [Eschmeyer 1998]), but this account follows the more conservative approach utilized by Cohen et al. (1990) in the FAO Species Catalogue. The global catch of gadid fishes in 1992 was about 10.5 million mt. Essential fish habitat source documents summarizing information on western North Atlantic populations of five species of Gadidae have been published: Atlantic cod (Fahay et al. 1999a), haddock (Cargnelli et al. 1999b), pollock (Cargnelli et al. 1999c), red hake (Steimle et al. 1999c), and white hake (Chang et al. 1999b).

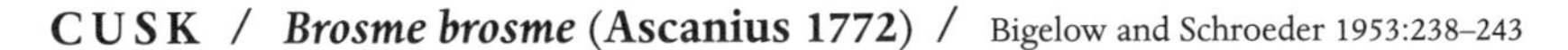

CUSK / *Brosme brosme* (Ascanius 1772) / Bigelow and Schroeder 1953:238–243

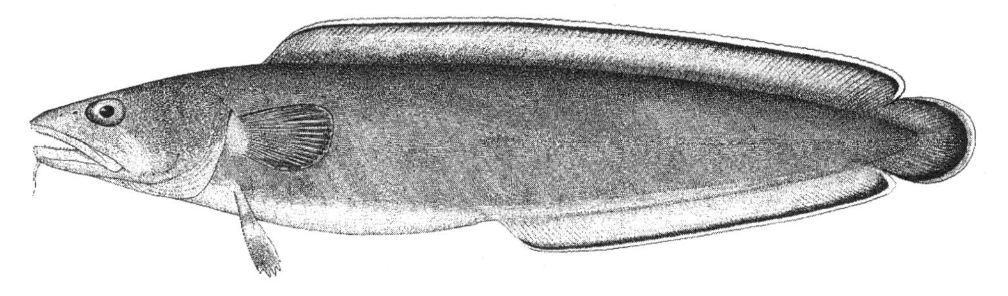

Figure 122. Cusk *Brosme brosme.* Boston market. Drawn by H. L. Todd.

Description. Slender, about one-fifth to one-sixth as deep as long, round-bodied in front of vent but compressed behind vent, and tapering evenly backward to base of caudal fin (Fig. 122). Mouth large, gaping back to opposite rear third of eye, set slightly oblique, and armed with a wide patch of small, sharp, curved teeth. A row of fairly large teeth on V-shaped vomer. Snout blunt. Upper jaw encloses lower with mouth closed. Eye of moderate size. Chin with one barbel. Dorsal fin runs length of back from nape of neck, of uniform and moderate height from end to end. Anal fin similar in outline, but only about two-thirds as long. Pectoral fins rounded, about half as long as head. Pelvic fins about as long as pectoral fins, with their five rays free at tips, situated below or a little in front of pectorals. All fins so thick and fleshy at bases that it is only near their margins that rays can be seen. Entire head and trunk clad with small cycloid scales.

Meristics. Dorsal fin rays 85–108; anal fin rays 62–77; total caudal fin rays 45–48; gill rakers about 7 low bumps on lower arch; pyloric caeca about 15; vertebrae 63–66 (Markle 1982; Fahay and Markle 1984).

Color. Cusk vary in color, probably conforming to the bottom: dull reddish brown to pale yellow, paling to gray on the lower part of the sides and to dirty white on the belly. Old fish are plain-colored. The sides of small ones are often cross-barred with about half a dozen yellowish bands. The pectoral and pelvic fins are the same color as the sides, and the pelvic fins are sooty at their tips. The most characteristic color mark is the black margin (narrowly edged with white) of the dorsal, caudal, and anal fins, which are of the general body tint at their bases.

Size. Maximum size 110 cm (Cohen et al. 1990). The largest two cusk caught by NEFSC bottom trawl surveys were 109 and 108 cm taken on *Albatross IV* cruises 70-6, station 188 at 42°10′N, 69°37′ W and 69-11, station 243 at 43°14′ N, 69°15′ W. Those caught in the Gulf of Maine average only 45.7–76.2 cm long and 2.3–4.5 kg in weight. An individual 112 cm long was collected in the Barents Sea (Lukmanov et al. 1985). In European waters the average size is 30–70 cm, with few being greater than 90 cm. The all-tackle game fish record is a 16.30-kg fish from Langesund, Norway, caught in April 1998 (IGFA 2001).

Distinctions. Cusk are separable from all other Gulf of

The relationship of the anal and dorsal fins to the caudal and the outline of the latter are distinctive, for both dorsal and anal are continuous with the caudal at the base but are separated from it by notches so deep that the rounded caudal is obviously distinct.

Habits. Cusk are so completely groundfish that they are not known to swim up to the upper waters, as cod and hake so often do. They are sluggish and weak swimmers, but have powerful bodies; when cusk are hooked, they are likely to twine themselves around the line in a bothersome way.

More or less solitary, cusk are not as abundant anywhere as cod, haddock, or hake. They also appear not to move much from bank to bank. Thus the "Massachusetts fishermen tell me," wrote Goode (1884), "that these fish are usually found in considerable abundance on newly discovered ledges, and that great numbers may be taken for a year or two, but that they are soon all caught. Sometimes, after a lapse of years, they may be found again abundant on a recently deserted ground." There is no definite evidence that cusk perform inshore or offshore migrations with the seasons, at least not in the Gulf of Maine.

Cusk are fish of at least moderately deep water. They are rarely taken in less than 18 m of water within the Gulf of Maine, and there are few cusk living below 180 m or so in the deep basins of the Gulf. They range down to 460–550 m on the continental slope off southern New England (Goode and Bean 1896), and they have been caught down to 970 m in the Faroe Channel.

Cusk are found chiefly on hard ground (Svetovidov 1986), especially where the seafloor is rough with rocks or boulders; on gravelly or pebbly grounds; occasionally on mud with hakes; but seldom on smooth clean sand. In Norwegian waters they often lurk among gorgonian corals and may have this same habit on the parts of offshore banks where these corals are plentiful.

Cusk are cool-water fish; minimum temperatures from NEFSC bottom trawl surveys were 3°–7°C, maximum 6°–14°. Their temperature range on the Scotian Shelf was 2°–12°C with most occurring between 6° and 10°C; and a salinity range of 32–34 ppt (Scott 1982b). They were almost absent from the northeastern area of the shelf except in deeper, warmer water, which indicates that their distribution may be determined by water temperature.

Food. Examination of stomachs of 49 cusk (Bowman et al. 2000) showed the three main components of the diet to be crustaceans (51.4% by weight), mostly decapods; fishes (15.5%); and echinoderms (15.0%). There are regional differences in diet (Langton and Bowman 1980; Scott and Scott 1988): in western Nova Scotia, 98.2% was fishes; in three other areas, the major prey were either crustaceans or echinoderms; and in the Gulf of Maine, crustaceans made up 90%, mainly toad crabs and pandalid shrimps. Cusk also eat squids, polychaetes, euphausiids, shrimps (mainly *Dichelopandalus* and *Crangon*), hermit and toad crabs, and brittle stars (*Ophiura*). Stomachs of fish less than 51 cm in length contained large quantities of polychaetes, *Crangon,* and hermit crabs (Bowman et al. 2000). Larger fish eat mostly euphausiids, pandalid shrimps, *Ophiura,* and fishes such as sculpins (*Artediellus uncinatus*).

Predators. Cusk are preyed on by spiny dogfish, winter skate, cod, white hake, goosefish, fawn cusk-eel, sea raven, and summer and windowpane flounders, with spiny dogfish being the most frequent predator (Rountree 1999). Other predators include hooded seal (Jensen 1948) and gray seal (Bowen et al. 1993).

Parasites. Two parasites have been recorded, a trematode, *Prosorhynchus squamatus* (Margolis and Arthur 1979) and a nematode, *Pseudoterranova decipiens* (Jensen et al. 1994).

Breeding Habits. Male cusk begin to mature at age 5 and females at age 6, and all of both sexes are mature by age 10 (Oldham 1972). Cusk are very prolific. Fecundity studies showed a range from 100,000 eggs for a 56-cm female to 3,927,000 eggs for a 90-cm fish (Oldham 1972). Fecundity depends on body length and, to a lesser extent, on age.

Eggs were collected by the MARMAP surveys primarily in the Gulf of Maine and Georges Bank from March to November and were most abundant in late spring and summer (Berrien and Sibunka 1999: Fig. 22). During May and June spawning increased, with eggs occurring extensively throughout the Gulf of Maine and over the southern portion of Georges Bank. In the Barents Sea, spawning occurred at depths greater than 100–200 m at temperatures of 2.5°–7.5°C and salinities of 34–35 ppt (Lukmanov et al. 1985). Spawning aggregations do not appear to form.

Early Life History. Cusk eggs are buoyant and spherical, 1.09–1.44 mm in diameter, perivitelline space 0.08–0.21 mm, oil globule 0.20–0.46 mm (Markle and Frost 1985: Table 3). They may be recognized by the brownish or pinkish color of the oil globule, and the entire surface of the egg has greatly modified and enlarged chorion pits (Markle and Frost 1985: Fig. 2B–C).

Larval development was summarized by Bigelow and Schroeder based largely on European references (Schmidt 1905; Ehrenbaum 1909). Larvae are about 4 mm long at hatching. The vent is situated at the base of the ventral finfold as in other gadoids. They are distinguishable from all other gadoid larvae that occur in the Gulf of Maine by the pinkish oil globule at the posterior end of the yolk. The yolk is absorbed about a week after hatching, when the larvae are about 5 mm long. The pelvic fins of the larva elongate as it grows and become heavily pigmented with black. Cusk larvae are distinguishable from those of hakes and rockling because their three pelvic fin rays are separate from each other, as well as by the presence of three patches of black pigment (one on the top of the head, a second over the gut, and a third at the tip of the tail) and two vertical black bands that divide the trunk behind the head into three nearly equal sections. Cusk have more total caudal fin rays (45–50) than *Merluccius, Urophycis,* or *Phycis* (29–40, Markle 1982: Table 6).

The first traces of vertical fin rays are visible at about 12.5 mm. Dorsal and anal fins are differentiated at about 28 mm, and it is at this stage that the pelvics are relatively longest. Larvae of 40 mm and upward show most of the characters of the adult, and the presence of only one dorsal and anal fin is enough to identify them. Young pelagic cusk are associated with the jellyfish *Cyanea* on Georges Bank (Colton and Temple 1961). Distributional maps of cusk eggs and larvae are given in Colton and St. Onge (1974) and more recent maps of egg distributions in Berrien and Sibunka (1999).

Young cusk drift near the surface until they are about 50 mm and then move to the bottom, where they become sedentary and rather solitary in habit (O'Brien 1998a).

Age and Growth. Cusk appear to grow rapidly from June to November and more slowly in winter and spring. Rates of growth of male and female cusk do not differ significantly. Cusk from the Scotian Shelf were aged by otoliths (Oldham 1972), and the oldest was 14 years old. Growth rate was linear but this may have been biased by gear selectivity, since only fish larger than 35 cm were captured. Cusk collected from the La Have area had a higher percentage of males among larger fish, indicating that males might have better survival rates.

General Range. Cusk occur on both sides of the North Atlantic, chiefly in moderately deep water and on hard bottoms. On the American coast from New Jersey to the Strait of Belle Isle and the Grand Banks of Newfoundland (Cohen et al. 1990: Fig. 65). In the eastern Atlantic, they are found off Iceland, in the northern part of the North Sea, and along the coasts of Scandinavia to the Murmansk coast. They reach the southern tip of Greenland only as rare strays from the south. The northernmost record in the eastern Atlantic is off Spitsbergen at 78°10′ and 9°20′ E (Lukmanov et al. 1985); the southernmost, 4 miles off Northumberland, Great Britain (Davis 1990).

Occurrence in the Gulf of Maine. Surveys show cusk to be distributed fairly evenly in waters deeper than 18 m (Map 14), except in the inner parts of the Gulf of Maine, where high concentrations may be found (Halliday et al. 1986). Cusk are rarely taken in Cape Cod Bay or in the deeper holes in Massachusetts Bay, and none has been taken on the soft mud of the deep bowl west of Jeffreys Ledge. Considerable numbers were once caught on the ledges off Chatham, Cape Cod, on Stellwagen Bank, on the broken grounds between the latter and

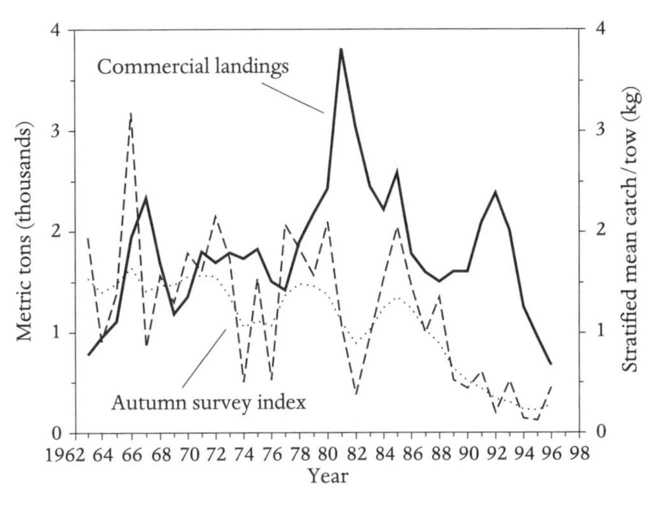

Figure 123. Catch and index of abundance of cusk *Brosme brosme,* Gulf of Maine–Georges Bank stock. (From O'Brien 1998a.)

Cape Ann, and on Jeffreys Ledge, the last being one of the most productive cusk grounds in the Gulf of Maine, along with the rocky slopes off Cashes Ledge. Cusk are also caught on Fippenies and Platts banks and are said to be plentiful on the rather indefinite ground off Penobscot Bay that is known as Jeffreys Bank. Some cusk are caught at the mouth of the Bay of Fundy, but not in the inner parts. Small rocky patches along the west Nova Scotian shore and off Seal Island also yield some cusk, and they are regularly taken on Grand Manan Bank. Large catches are taken on Browns Bank and fair numbers on the rougher spots on Georges Bank.

Importance. Cusk are good food fish and there is a ready market for all that are caught. The principal fishing gears used are line trawl, otter trawl, gill net, and longline (O'Brien 1998a). Bigelow and Schroeder summarized early catch statistics for the Gulf of Maine. During the late 1960s and early 1970s, annual landings were relatively stable at about 1,700 mt·year^{-1} but increased in the late 1970s and early 1980s to peak at 3,800 mt in 1981 (O'Brien 1998a). Historically, most of the U.S. catch has been from the Gulf of Maine, but since 1993 landings from the Gulf of Maine and Georges Bank have been nearly equal. Almost all Canadian landings are taken on Georges Bank.

U.S. commercial landings have declined from an average of 1,700 mt in 1977–1986 to a low of 500 mt in 1996 (O'Brien 1998a) (Fig. 123). Although the NEFSC autumn bottom trawl survey biomass index has fluctuated considerably (Fig. 123), a declining trend has been evident since the late 1960s (O'Brien 1998a). Mean length of cusk caught on NEFSC surveys has declined from a long-term average of 61 cm during 1964–1993 to 38 cm during 1994–1996. The stock appears to be overexploited and at a low biomass level. The fishery is not currently under management.

FOURBEARD ROCKLING / *Enchelyopus cimbrius* (Linnaeus 1766) /

Bigelow and Schroeder 1953:234–238

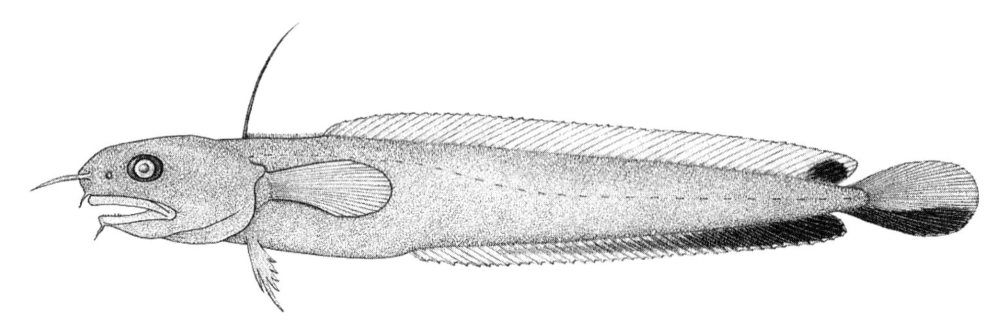

Figure 124. Fourbeard rockling *Enchelyopus cimbrius*. Off Cape Cod. Drawn by K. H. Moore. (From Cohen and Russo 1979: Fig.1A.)

Description. Body slender, tapering back from shoulders; rounded in front of vent but compressed behind it (Fig. 124). Upper jaw longer than lower; teeth small, first six to eight teeth on upper jaw slightly elongate. Snout short and blunt; eyes small; dorsal profile of head somewhat rounded. Four barbels, a pair of long ones at anterior nostril, a shorter one on chin and one at tip of snout. Pectoral fins rounded, narrow pointed pelvic fins situated well in front of pectorals. First section of dorsal fin consists of one ray, nearly as long as head, located over upper corners of gill openings, followed by a series of about 50 very short, separate, hairlike rays without a connecting membrane, which can be laid down in a groove on back. Second dorsal fin originates over midlength of pectoral fins, runs back nearly to base of caudal fin, and is equally high from end to end, with a rounded rear corner. Anal fin similar to second dorsal in shape, but shorter. Caudal fin oval when spread. Lateral line slightly curved and interrupted along its length.

Meristics. First dorsal fin a single long ray followed by a series of about 50 very short, hairlike rays; second dorsal fin rays 45–55; anal fin rays 37–48; pectoral fin rays 15–17; total caudal fin rays 31–35; lateral line scales 51–57; gill rakers 5–12; precaudal vertebrae 16 + caudal 38 or 39 = total vertebrae 54 or 55 (Hardy 1978a; Cohen and Russo 1979; Markle 1982). Populations from the South Atlantic off Florida and the Gulf of Mexico vary in meristic characters (Cohen and Russo 1979).

Color. The back is dark yellowish olive or dusky brown, the sides are paler, and the belly is white dotted with brown. On some individuals the sides behind the vent are more or less clouded with a darker shade of the general body hue. The first dorsal ray, the posterior edges of the second dorsal fin and the anal fin, the lower half of the caudal fin, and the pectoral fins are sooty or bluish black. Otherwise the vertical fins are gray or bluish brown. Pelvic fins are pale, and the lining of the mouth is dark purple or blue. There is geographic variation in color with those found in the South Atlantic and Gulf of Mexico having more dark blotches on dorsal and anal fins (Cohen and Russo 1979).

Size. Fourbeard rockling have been described as growing to a length of 42 cm in Scandinavian waters, but 32.8 cm TL was the longest of 727 specimens from the Gulf of Maine (Deree 1999).

Distinctions. Rockling differ from hakes in having short pelvic fins, with five to seven rays, and the first dorsal fin consists of a single ray followed by the series of short, hairlike rays.

Taxonomic Note. The generic name *Rhinonemus* has been used by many European ichthyologists for this species although Cohen and Russo (1979) clearly showed that *Rhinonemus* Gill 1863 is a junior synonym of *Enchelyopus* Bloch and Schneider 1801.

Habits. The name rockling is a misnomer for these fish because they are found most often on soft mud or sand bottoms. They live in burrows during the daytime and forage at night (Keats and Steele 1990). In the Gulf of Maine, they occur in 79–254 m (Deree 1999). Occasionally they have been found in very shallow water, on Nahant Beach in Massachusetts Bay, for example; in water less than 1 m deep at Woods Hole; in 11–13 m, both in St. Mary's Bay, N.S., and in Buzzards Bay, Mass. They appear to be more plentiful in depths of 45–55 m or more. There are rockling in the deep gully off Halifax and in the deep trough of the Gulf of St. Lawrence. They have been taken on the continental slope off southern New England to 146 m (Wheatland 1956). The depth record is 650 m (Cohen and Russo 1979).

Food. The diet of 36 fourbeard rockling from the Gulf of Maine consisted primarily of bivalves, copepods, and decapods (Deree 1999). Prey composition changed with age. Numerical indices showed that copepods composed 71.2% of the diet of 1-year-old fish, bivalves were second in importance (19.2%), followed by cumaceans, amphipods, and decapods.

The importance of copepods decreased to 28.7% and they were replaced by bivalves (56.3%) in 2- to 7-year-old fish. The bivalve was *Yoldia,* family Nuculanidae.

Other food studies showed similar results. The most important components of rockling diet in Passamaquoddy Bay were a euphausiid *Meganyctiphanes,* two amphipods, *Maera* and *Unicola,* and a polychaete *Nephtys* (Tyler 1972). The principal food of 48 specimens, about half of them from Georges Bank, was crustaceans (57.4% by weight), primarily *Crangon,* but also pandalid shrimp, euphausiids, and amphipods (Langton and Bowman 1980). Polychaetes accounted for 12.5% of the stomach contents. In Newfoundland, feeding was observed only at night and consisted mostly of polychaetes, especially a scale worm *Harmothoe imbricata.* The importance of polychaetes increased with increasing size of the fish (Keats and Steele 1990). In a Norwegian fjord, food was mainly crustaceans, polychaetes, fishes, and bivalves (Mattson 1981). The crustaceans were primarily decapods, mysids, amphipods, and cumaceans.

Predators. Predators include spiny dogfish, little skate, goosefish, cod, pollock, red and white hakes, and sea raven, of which sea raven and white hake are the most frequent predators (Rountree 1999).

Parasites. The parasite fauna of fourbeard rockling includes two protozoans, a trichodinid (Lom and Laird 1969) and a coccidian, *Eimeria gadi* (Odense and Logan 1976); a larval nematode of the subfamily Raphidascarinae; a nematode, *Pseudoterranova decipiens* (Des Clers and Andersen 1995); a trypanorhynch cestode, *Grillotia erinaceus;* and three digenean trematodes, *Genolinea laticauda, Gonocerca phycidis,* and *Bucephaloides gracilescens* (Karlsbakk 1995; Deree 1999). The definitive hosts of adult trypanorhynchs are skates, suggesting that skates prey upon fourbeard rocklings.

Breeding Habits. Eggs were collected throughout the year by the MARMAP surveys, April through November in the Gulf of Maine (Berrien and Sibunka 1999: Fig. 25). Eggs were collected in deep water in winter months and in shallower waters during the summer. Peak spawning occurred in June, with most eggs occurring in the western Gulf of Maine.

Early Life History. Eggs are pelagic, spherical, and transparent. The overall size is 0.73–0.86 mm, perivitelline space 0.04–0.14 mm, and the oil globule 0.16–0.27 mm (Markle and Frost 1985: Table 6). There are variable numbers of oil droplets, most commonly a single one of 0.13–0.25 mm. The oil globule is usually pigmented, but this varies geographically from clear to green, yellowish, pink, blackish, or cream (Hardy 1978a). In Salem Harbor, eggs were untinted (Elliott and Jimenez 1981). The eggs differ from those of hakes in lacking melanophores on the yolk sac and from those of butterfish in having ventral pigmentation (Markle and Frost 1985).

Descriptions of developmental stages at 15° and 13°C are given in Hardy (1978a) from Battle (1929). At 15°C, hatching occurred at 108 hr and at 13°C at 129.6 h. Larvae average 2.03 mm TL at hatch and range from 1.6 to 2.4 mm (Colton and Marak 1969). The body is short and stocky with rounded head and snout. The vent opens laterally on the finfold, not at the margin. The yolk is absorbed at about 3.6 mm and later larval stages up to about 10 mm are characterized by very large black pelvic fins; by the presence of only one postanal band of black pigment; and by the short, stocky body form. Young hake are more slender and have scattered pigment; young cusk have two postanal bands; and all other Gulf of Maine gadoids have short pelvic fins. Barbels develop on the lower jaw at 10 mm. After the rockling is 17–20 mm long, the structure of the first dorsal fin serves to identify it. Transformation occurs at about 20 mm. These larger larvae are silvery, awaiting their descent to the bottom before assuming the dull colors of the adult.

Ichthyoplankton collections from Georges Bank, Nantucket, and the Gulf of Maine showed that the center of larval distribution was the northwestern Gulf of Maine not far offshore and that another concentration occurred within 100 km of Cape Cod. The larvae drift in the upper layers until they reach about 35 mm and then begin their descent. Planktonic forms were collected up to 48 mm TL. They were collected at the same time as hake (*Urophycis* spp.) larvae, but their distribution only overlapped at the temperature range 11°–13°C; they were usually caught at the surface but, unlike hakes, showed no evidence of vertical migration (Hermes 1985). Larval and egg distribution maps were given by Colton and St. Onge (1974) and more recent maps of egg distribution by Berrien and Sibunka (1999). Rockling larvae were among the dominant species in the lower estuaries and outer parts of the Boothbay Harbor area in Maine from June to October (Chenoweth 1973).

Age and Growth. Gulf of Maine fourbeard rockling were aged from sectioned sagittal otoliths (Deree 1999). Nine age-classes were found. The regression of observed age-class data against total length (mm) was $y = 25.194x + 81.331$. Linear growth was significantly different in males and females. Males weighed more than females (means 26.5 vs. 24.2 g, respectively).

General Range. Fourbeard rockling occur on both sides of the North Atlantic. The American range is from the northern part of the Gulf of Mexico to Newfoundland and western Greenland. In the eastern Atlantic, Iceland and the coasts of the British Isles and Europe from the Barents Sea to the northern Bay of Biscay plus an isolated record from Cape Blanc, Mauritania (Cohen et al. 1990: Fig. 78).

Occurrence in the Gulf of Maine. Fourbeard rockling are common bottom fish in the deeper parts of Massachusetts Bay and much of the inner Gulf of Maine (Map 15).

ATLANTIC COD / *Gadus morhua* Linnaeus 1758 / Cod, Rock Cod /

Bigelow and Schroeder 1953:182–196 (as *Gadus callarias*)

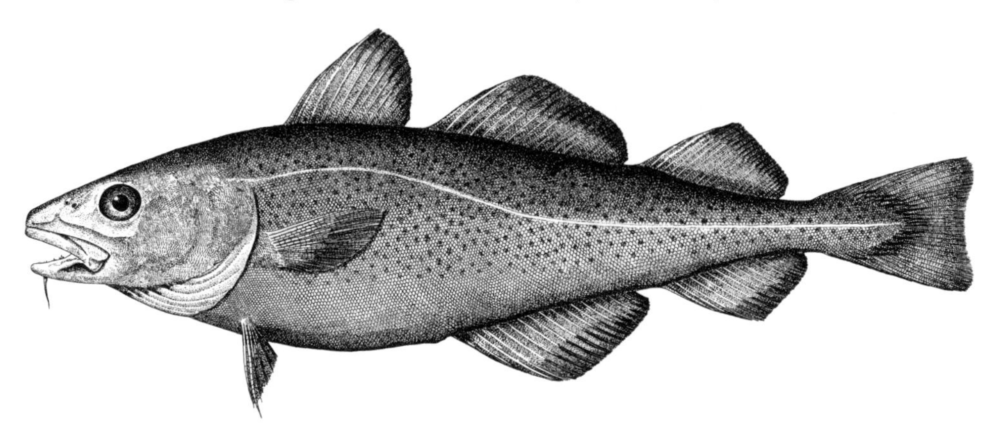

Figure 125. Atlantic cod *Gadus morhua*. Eastport, Maine. Drawn by H. L. Todd.

Description. Heavy-bodied, only slightly compressed, body deepest under first dorsal fin, tapering to a moderately slender caudal peduncle (Fig. 125). Head large, about 25% in TL but relatively narrow, interorbital space 15–22% of HL. Snout conical and blunt at tip; upper jaw protrudes beyond lower; mouth wide with angle of jaw reaching back to anterior part of eye; many very small teeth in both jaws. Teeth also present on V-shaped vomer. Three dorsal fins. First dorsal fin usually originating well in front of midlength of pectoral fins; highest of three dorsal fins, triangular, with a rounded apex and convex margin. Second dorsal fin nearly twice as long as first dorsal and about twice as long as high, decreasing in height from front to rear with slight convex margin. Third dorsal fin a little longer than first dorsal and similar to second dorsal in shape. Two anal fins originate below second and third dorsals, to which they correspond in height, length, and shape. Pectoral fins, set high up on sides, reach back as far as rear of first dorsal fin. Pelvic fins anterior to pectoral fins; nearly as long as pectoral fins in young cod but relatively shorter in large fish, with second ray extending beyond general outline as a filament for a distance almost one-fourth as long as entire fin. Caudal fin nearly square, broom-shaped, and about as broad as third dorsal fin is long. Head and body clothed with small scales. Lateral line pale and slightly arched.

Meristics. First dorsal fin rays 13–16; second dorsal fin rays 19–24; third dorsal fin rays 18–21; first anal fin rays 20–24; second anal fin rays 17–22; total caudal fin rays 50–56; lateral line scales 150–170; total gill rakers on first arch 18–28; precaudal vertebrae 18–20 + caudal 32–35 = total vertebrae 49–55; pyloric caeca about 280, many of which are branched (Hildebrand and Schroeder 1928; Hardy 1978a, Markle 1982; Scott and Scott 1988).

Color. Cod vary so widely in color that many of the color phases have been named, but all of them fall into two main groups: gray-green and red. The back and upper sides of the former range from almost black through dark sooty or brownish gray, olive gray, olive brown, sepia brown, mouse gray, ashy gray, clay-colored, and greenish to pale pearl (darker on the back than on the sides). The fins are of the general body tint and the belly is whitish, usually tinged with the general ground color. Red or "rock" cod vary from dull reddish brown to orange or brick red, with a white belly tinged with red and red, olive, or gray fins. In most cod the upper surface of the body, sides of the head, and fins and tail (but not the snout or belly) are thickly speckled with small, round, vague-edged spots. On "gray" fish these are brown or yellowish, darker than the general body color, whereas they are usually reddish brown or sometimes yellowish on "red" fish. The lateral line is paler than the general body tint, pearly gray or reddish, according to the hue of the particular fish, standing out against the darker sides. Individual fish are able to change color, shading, and pattern readily. Occasionally a spotless cod is seen.

Size. Cod sometimes grow to a tremendous size. A huge one of 96 kg and more than 183 cm long was caught on a longline off the Massachusetts coast in May 1895; Svetovidov (1962) recorded a female of 162 cm and 40 kg; and Goode (1884) noted several others of 45–73 kg caught off Massachusetts. Cod of 45 kg are exceptional. The all-tackle game fish record is a 44.79-kg fish taken at the Isle of Shoals, N.H. in June 1969 (IGFA 2001). Even a 34-kg fish is now a rarity, but 23- to 27-kg fish are not unusual. The largest cod taken on 54 NEFSC bottom trawl surveys was 150 cm TL. The average commercially caught cod are 2.5–4.0 kg.

Distinctions. Cod are easily distinguished from large tomcod by their relatively broad pelvic fins with slender filaments, location of the first dorsal fin, and larger eyes. The pale lateral line readily distinguishes cod from haddock; and the square-tipped tail, projecting upper jaw, and spotted color pattern distinguish them from pollock.

Habits. Cod range from the surface down to depths greater than 450 m, but few cod living in the Gulf of Maine occur much deeper than 180 m and the 10–135 m contour probably includes most cod during both summer and winter. As a rule, larger cod are found in deeper water.

Cod occur on rocky and pebbly grounds, on gravel, on sand, and on a particularly gritty type of clay with broken shells. They also frequent deeper slopes of ledges along shore, where they forage among Irish moss (*Chondrus crispus*), and among seaweeds of other kinds. Young red ones are especially common in these situations. Good catches are sometimes taken on mud, as off Mount Desert, where large and medium-sized cod are regularly caught on soft ground in winter.

Cod are typical groundfish, usually found within 2 m or so of the bottom. As a rule large ones keep closer to the bottom than smaller ones, but even large ones sometimes follow herring up to the surface. They come to the surface more commonly on the Grand Banks and along the eastern coast of Labrador when they are following capelin.

Adult cod are at home in any temperature from 0° to 13°C, in all but the most superficial layers of the Gulf of Maine, at all seasons. Experience at the Woods Hole hatchery in the late 1890s and early 1900s proved that freezing may be fatal owing to the formation of anchor ice. On the other hand, whereas large cod tend to avoid water temperatures warmer than 10°C, they are at times abundant in temperatures as high as 15°C on Nantucket Shoals. Small cod are less sensitive to heat than large ones, a fact reflected in the presence of greater numbers of them in shoal water in summer than larger fish. The temperature range is reported as -2° to 20°C, and 20°C may be lethal (Hardy 1978a). Newfoundland cod contain antifreeze glycoproteins in their blood, which enables them to withstand water temperatures below freezing when they are inshore in winter (Hew et al. 1981). These proteins vary seasonally, reaching maximum concentrations in the winter and disappearing in the summer. Production of these proteins appears to be initiated by exposure to water temperatures of 1°C or lower. Photoperiod has no effect on appearance of the proteins. It is not known what determines when biosynthesis of antifreeze proteins ceases (Fletcher et al. 1987). No evidence of antifreeze proteins was found in cod from Nova Scotia in the winter (Duman and DeVries 1974a).

Cod in the northwestern Atlantic have definite vertical movements. From mid-July to fall, they move into midwater at night, and from May to July there are as many in midwater during the day as at night (Brunel 1965; Beamish 1966a). Activity rhythms of cod were monitored by an underwater television camera deployed from an oil platform in the North Sea using baited hooks to stimulate longline fishing efforts (Lokkeborg et al. 1989). Cod showed two peaks of activity in December, one in the morning and one in the afternoon. Periods of low activity occurred at night and at noon. In December, there was only one peak of activity and that was at noon.

Cod have directional hearing and can orient toward a sound

the swim bladder–inner ear complex. They produce sounds with muscles that originate on the pleural ribs and insert on the swim bladder causing the latter to vibrate and produce sounds in the form of growls, thumps, and low-frequency grunts (Fish and Mowbray 1970). The swimming muscles are sexually dimorphic, with males having bigger muscles than females (Engen and Folstad 1999). The muscles are used during threat displays by both sexes and during spawning preliminaries by males to stimulate females and intimidate intruders.

Cod are visual feeders for the most part, but they are able to detect odors from many live algae, invertebrates, and fishes (Brawn 1969). They can also detect food on the bottom by taste buds on the trailing barbel and pelvic fin rays. Food can be detected below gravel and stones by smell and is uncovered by removing gravel in the mouth or rolling stones aside with the head. Food covered by sand was not found. Cod groups fed more effectively than individuals and feeding behavior by one fish attracted others. The combined digging and rooting of several cod served to uncover more food. Schooling behavior may be adaptive by enabling more food to be obtained from a given area.

Food. Algae appear to be the first food of cod larvae and they seem to concentrate algae by filter-feeding. In 7-day-old larvae, 39.2% of the food was algae with a diameter greater than 8 μm. These appear to be green 10-μm spheres, a naked dinoflagellate, and short chains of *Skeltonema costatum*. Twelve-day old larvae contained 12.6% algae in their stomachs (van der Meeren 1991). After yolk-sac absorption, small zooplankters are preferred food. Cod larvae eat all life stages of several species of copepods along with lamellibranch larvae and phytoplankton. Feeding intensity reaches a peak shortly before sunset, declines at night, and increases again during the day. Cod larvae are visual feeders and require a minimum light intensity for active feeding (Ellertsen et al. 1980).

Larval cod and haddock occur together throughout much of their range. Field and laboratory studies indicate that the two species compete for food resources. Dietary niche breadth and overlap indicated severe competition between and among similar-sized individuals of cod and haddock larvae from Georges Bank (Kane 1984; Auditore et al. 1994). However, neither species appeared to be starving or in weakened condition, and laboratory results appeared to be caused by sieving zooplankton through a fine mesh net so that the smaller items that the haddock fed upon were excluded. Cod larvae (<11 mm SL) consume larger prey items than haddock larvae, but haddock compensate by consuming more smaller prey (Auditore et al. 1994). A comparison of prey selectivity of cod and haddock from yolk sac through demersal juveniles is given by Auditore et al. (1994). Laurence et al. (1981) showed that when larvae of the two species were raised together in aquariums, cod mortality was lower and they grew much faster than haddock, although growth, respiration, and delayed feeding

Age-0 cod feed predominantly on copepods and amphipods (mostly pelagic prey). Age-1 cod eat a more diverse diet (Bowman et al. 1987; Neilson et al. 1987) including isopods, invertebrate eggs, shrimps, decapods, polychaetes, and fishes (mostly benthic prey). There is a rapid transition from pelagic to benthic prey related primarily to an increase to a body size of 60–100 mm SL. During this time, the mouth size increases, allowing capture of larger prey (Lomond et al. 1998). There is also a rapid proliferation of cutaneous taste buds preceding adoption of a benthic habitat. A high density of taste buds on the barbel and pelvic fins appears to be correlated with feeding behavior (Harvey and Batty 1998). Transition to benthic prey also has a diel component. Nearshore regions occupied by age-0 cod also harbor larger conspecifics. Cod exhibited a size-related shift (age 0 to age 1) from feeding predominantly on zooplankton by day to benthos by night. Intercohort cannibalism occurred when age-1 cod were approximately three times larger than their prey. Concentrations of age-0 cod foraged in the water column during the day but ceased feeding at night and appeared to disperse to the bottom. Although seasonally decreasing day length and prey size contributed to a decline in daily ration, age-0 cod still maintained a strict diurnal foraging cycle. The nocturnal increase in feeding coincided with an increase in the catch of age-2 and age-3 conspecifics and increased foraging activity of age-1 cod, suggesting that avoidance of older conspecifics is an important factor influencing the diel foraging and activity cycles of age-0 cod (Grant and Brown 1998).

Examination of 718 stomachs (Bowman et al. 2000) showed the primary food categories to be fishes (56.8% by weight); crustaceans (21.1%), mostly decapods; and mollusks (14.9%), mostly squids. Fish prey include herring, silver hake and other gadoids, redfish, sand lance, mackerel, and flounders. Cod will pursue and gorge on squids anytime they are available. Any shellfish a cod encounters that can be swallowed whole is likely to be consumed. Feeding on scallop viscera discarded by fishermen is not uncommon. Cod also eat fish eggs, sea clams (empty shells may be found neatly nested in their stomachs), cockles (*Polynices*), sea mussels (*Modiolus*), rock crabs (*Cancer*), hermit crabs (*Pagurus*), lobsters, pandalid and other types of shrimps, brittle stars, sea urchins, sea cucumbers, worms, and tunicates. Surprisingly, cod even feed on ctenophores (*Pleurobrachia*), and rocks are commonly seen in their stomachs (they probably swallow the latter for the anemones and other organisms growing on them). Objects as indigestible as pieces of wood, rope, boots, jewelry, and clothing have also been found in their stomachs. Large cod have even been known to eat a wild duck on occasion. Despite being so rapacious, cod generally fast while spawning.

Based on examination of the stomach contents of 4,102 fish from 1969 to 1980, the most important cod prey were Atlantic herring, sand lance, Atlantic mackerel, *Loligo* and *Illex* squids, and rock crabs (*Cancer*) (Langton and Bowman 1980; Bowman and Michaels 1984; Bowman et al. 2000). In the Gulf of Maine, in addition to the prey listed above, silver hake and redfish were also a principal food of cod.

There is no apparent diel feeding periodicity in adult cod, and they may feed at any time of night or day (Tyler 1971a). They do, however, exhibit a diel activity rhythm, with higher swimming speeds and a larger range during the day than at night (Lokkeborg and Ferno 1999). During the period of high activity, more fish localized food olfactorily (baited hooks) and the time to localization was 50% shorter, indicating that high swimming activity increased the probability of encountering the odor plume and the odor source. No diel variations in the response threshold to olfactory stimuli were found. The probability of cod detecting prey by taste receptors, encountering the odor plume of prey, or localizing a stationary food source after olfactory stimulation should be relatively independent of light, which explains why cod are active throughout the 24-h cycle. On the other hand, visual detection distance and the ability to capture active prey are influenced by light level, resulting in higher activity during the day.

Estimated upper and lower daily rations for cod larger than 30 cm were 1.5 and 0.9% body weight per day (Durbin et al. 1983).

Predators. Predators of cod include 14 species of fishes, of which the four most frequent are cod, fourspot flounder, spiny dogfish, and sea raven (Rountree 1999). Skates (winter and thorny) and hakes (red, white, spotted, and silver) are occasional predators. Cod are also eaten by short-finned squid (Bowman et al. 2000) and harp seal (Beck et al. 1993). Adult cod have few natural enemies other than large sharks and seals. They account for 18.5% of the diet by occurrence of gray seal in eastern Canada (Benoit and Bowen 1990b). The size range of cod eaten averaged 24.1 cm, 28 g for onshore populations of seal vs. 33.7 cm, 61 g for offshore seal populations (Bowen et al. 1993).

Parasites. Cod have been reported to harbor 43 different kinds of parasites, many of which are specific to gadoids (Appy 1979). Parasites useful as indicators of cod migrations and mixing of stocks include *Lernaeocera branchialis, Cryptocotyle lingua, Phocanema decipens, Corynosoma* sp., *Grillatia erinaceus,* and mesenteric nematodes. The increasing degree of infestation with the seal worm acanthocephalan *Corynosoma wegeneri* and the nematode *Pseudoterranova decipiens* is probably due to the dramatic growth of the seal population, especially on Sable Island near the Scotian Shelf (Marcogliese and McClelland 1992).

Cod larvae had a greater prevalence and number of copepod parasites (*Caligus* sp.) than haddock in two locations sampled on Georges Bank (Neilson et al. 1987). Preferred sites of attachment also differed, with cod parasites most often attached toward the caudal region. There was no direct evidence of reduced fish condition or that parasites were a direct source of mortality for cod. The hematophagous copepod *Lernaeocera branchialis* appears to affect the growth rate of cod by interfering with food conversion efficiency (Khan and Lee 1989). Adults infested with young parasites consume more food and gain more weight, but in the long term, when the parasites mature, the cod show low food conversion efficiency and weight gain is transitory. Young cod infested by the parasite consumed less food and exhibited less growth than uninfested fish.

Breeding Habits. Median length at maturity for female and male Atlantic cod from the Gulf of Maine was 32.1 cm and 36.0 cm, respectively (O'Brien et al. 1993). Median age at maturity was 2.1 and 2.3 years for females and males, respectively. Current median maturity values are noticeably lower for both sexes compared with previous studies (O'Brien et al. 1993). Median length at maturity of cod collected on Georges Bank in 1972 was 51.5 cm (females) and 44.0 cm (males) and median age at maturity was 2.9 years for females and 2.6 years for males (Livingstone and Dery 1976). Cod sampled over the entire region during 1977 (Morse 1979) had a median size at maturity of 49.6 cm (females) and 53.7 cm (males). A more comprehensive analysis of cod from Georges Bank and the Gulf of Maine showed that from 1970 to 1997, there were significant declines in median size and age at maturation, indicating compensatory changes in maturation over time in response to declining stock abundance and temperature (O'Brien 1998c). To investigate potential causes for this decline, stepwise logistic regression was used to estimate the effect of stock density and temperature (O'Brien 1998c). Stock density accounted for much of the variation in maturation for both sexes from Georges Bank and the Gulf of Maine, and temperature was responsible to a significant degree from Georges Bank but to a lesser extent from the Gulf of Maine. Median age and size at maturity for slower-growing Atlantic cod from the Scotian Shelf also declined considerably from the 1960s to the 1970s. In 1963, the average length at maturity for females was 52 cm and for males 51 cm, and by 1978 the values were 35 and 38 cm, respectively. Over the same time period, median age at maturity declined from 3.7 and 4.8 years in females and males, respectively, to 2.9 and 2.8 years (Beacham 1983c).

Cod are among the more prolific fishes. A 50-cm female may produce 250,000–500,000 eggs and a 100-cm female may deposit 4–8 million eggs (May 1967). Maximum recorded fecundity was 12,000,000 eggs for a 140-cm female (Powles 1958). Experiments monitoring egg and larval production of captive northwest Atlantic cod indicated that first-time spawners perform poorly compared to second-timers (Trippel 1998). The former breed for a shorter period, produce fewer egg batches, exhibit lower fecundity, and produce smaller eggs with lower fertilization and hatching rates; moreover, their larvae are less likely to hatch in environmental conditions favorable for survival. Seasonal composite hatching rates of all eggs spawned by first- and second-time spawners were 13 and 62%, respectively. Larval production per maternal gram was an order of magnitude greater for second-time than for first-time spawners (200 vs. 20 larvae·g^{-1}). These results have a direct bearing on the development of mathematical models of spawner-recruitment relationships, and they suggest that conventional approaches may overestimate the reproductive

Eggs were collected throughout the year by the MARMAP surveys (Berrien and Sibunka 1999: Fig. 28). They were found over a wide geographic area from Nova Scotia to Cape Hatteras, primarily at locations where the water depth was less than 100 m. Egg densities were minimal during August and September and expanded during autumn and winter throughout the principal spawning areas, which are the Gulf of Maine, Georges Bank, and southern New England. Although cod eggs are among the most common collected on the U.S. northeast continental shelf, a downward trend in egg abundance was observed over 9 years of sampling from 1977 to 1987 (Berrien and Sibunka 1999).

Spawning takes place mostly at night and may be crepuscular (dawn and dusk). Temperature ranges from –1° to 12°C with an optimum of 5° to 7°C (Hardy 1978a; Grosslein and Azarovitz 1982). The salinity range is from 10 to 35 ppt, but the lower range is typical of cod from Baltic waters (Hardy 1978a), and the average in the Gulf of Maine is 32 ppt. Under experimental conditions complete mortality of the eggs occurred at 9.9 ppt, and sperm became immotile at 7.5 ppt (Hardy 1978a).

Three weeks before spawning occurred in tanks, aggressive behavior was displayed by males, who established territories and defended them against intruders by swimming rushes, threat displays, and grunts (Brawn 1961). One male was dominant over all the others. Behavior culminating in the spawning act was initiated by a female swimming into a dominant male's territory. Males recognized females and made courtship displays by raising their dorsal fins and making exaggerated lateral bends of their bodies. This was accompanied by a loud grunting sound. The female swam normally and did not produce any displays. She followed the male while he was courting and eventually swam up to the surface. The male swam on top of the female, then underneath her, belly to belly, and spawning occurred high up in the water column. After spawning the female swam away and rested. Brawn noted color changes at the beginning of the spawning period: most males became dark brown and females turned pale gray.

Reproductive behavior of cod at a field-reported spawning density was observed under experimental conditions at ambient photoperiod and temperature in a large tank (Hutchins et al. 1999). Agonistic interactions appeared to maintain a size-based dominance hierarchy among males. Multiple paternity per spawning bout suggested a link between dominance and fertilization success. Interactions between sexes were dominated by circling of females by males. After descending to the bottom, a motionless female would be circled up to 17 times, often by one male per spawning bout but by several males throughout the spawning period. Although circling frequency increased with male dominance and male body size, initiation and termination of this behavior appeared to be under the control of the female. Circling provides opportunities for males to gain individual access to reproductive females and, by male-male competition and display, for females to assess the

Eggs. Cod eggs are pelagic, buoyant, transparent, without oil globules, and measure 1.20–1.69 mm in diameter with a narrow perivitelline space, 0.01–0.17 mm (Fahay 1983; Markle and Frost 1985: Table 3). The chorion is smooth and the yolk is homogeneous. The eggs may vary widely in color and size over their geographic range: from clear to cream-colored, pale green, and yellowish red (Hardy 1978a). Studies on laboratory-raised female cod showed that they spawned repeatedly for 3 months, egg size was positively correlated with fish length and decreased over time for each female, and the numbers of eggs spawned followed a dome-shaped distribution (Kjesbu 1989).

The period of incubation depends on temperature, and varies from 8 to 60 days. A summary of the literature on hatch times at various temperature/salinity combinations is given in Hardy (1978a).

Early cod eggs are indistinguishable from those of haddock, which also lack an oil globule and are about the same size, but just before hatching, pigment of cod gathers in four or five distinct patches (one over the region of the pectoral fins, one above the vent, and the others equally spaced behind the latter); whereas in haddock the pigment cells are arranged in a row along the ventral side of the trunk. Newly spawned cod eggs resemble those of witch flounder; but the black pigment of cod eggs identifies them as soon as it appears because the embryonic pigment of witch is yellow.

Cod eggs have a broad range of temperature tolerance. The highest percentage of viable hatches with parent stock from Narragansett Bay, R.I., occurred at 2°–10°C and 28–36 ppt (Laurence and Rogers 1976). Time to 50% hatch was inversely related to temperature and salinity. Highest mortality occurred just prior to hatching.

Larvae. The average hatching length of cod larvae is 4.38 mm, with a range of 3.3–5.71 mm. Larval length is positively correlated with egg size (Knutsen and Tilseth 1985). Larval cod can be identified as a gadoid by the vent that opens laterally at the base of the finfold. Larval cod resemble haddock and pollock larvae but are distinguishable by the lower number of total caudal fin rays (50–56 compared to 57–60 in haddock and 66–70 in pollock [Markle 1982: Table 6]). Also, the pigment is in two dorsal and three (rarely two) ventral bars, with the dorsal bars shorter than the ventral bars opposite them, whereas the dorsal bars are longer than the opposing ventral bars in pollock larvae up to 10 mm long. Haddock larvae are not barred but have a continuous row of pigment cells along the ventral margin of the trunk behind the vent and other patches on the nape and in the lining of the abdomen.

Young cod float when first hatched, yolk uppermost, but they assume the normal position in about 2 days. The yolk is absorbed and the mouth forms in 6–12 days, depending on the temperature, when the larvae are about 4.5 mm long. As they grow, the pigment bars gradually fuse and at 8–10 mm a median streak forms. Cod 10–20 mm long may be distinguished from pollock by the fact that pigment extends to the tail, whereas in pollock it ends abruptly some distance in front of the tail. The most reliable character is the number of primary caudal fin rays articulating with the superior hypural: four in cod and five in pollock (Markle 1982). Haddock of this size have much less pigment. Cod larvae 15–30 mm are recognizable by the location of the vent under the second dorsal fin, combined with dense pigmentation. At 20 mm the dorsal and anal fin rays attain their final number and the separate fins are outlined, and at 30 mm the larvae begin to show the spotted color pattern characteristic of adult cod. A summary of characteristics of larval gadoid fishes and how to distinguish them is presented by Fahay (1983) and Dunn and Matarese (1984). The most obvious characters distinguishing Georges Bank cod and haddock larvae and juveniles (Auditore et al. 1994) are lack of dorsal pigmentation in newly hatched cod, presence of midline pigment in cod by 7 mm, development of a chin barbel in cod by 35 mm, and the appearance of the shoulder blotch in haddock by 40 mm.

The yolk is usually absorbed by the sixth day after hatch. Morphological studies of the jaws and digestive tract showed that cod larvae are able to absorb digested food well before the yolk sac is exhausted. The foregut and midgut were active in lipid absorption and the hindgut showed pinocytotic activity. In starved larvae, degeneration of gut tissue was pronounced by day 9 posthatch and larvae starved longer would probably not survive (Kjorsvik et al. 1991).

Cod eggs and larvae spawned in the Gulf of Maine tend to drift south and eastward toward Georges Bank because of the counterclockwise Gulf of Maine gyre. The extent of exchange between the Gulf of Maine and Georges Bank is not known (Serchuk et al. 1994). Cod spawned on Georges Bank tend to drift in a gyre and depending on their depth distribution and swimming ability drift along the southern flank extending onto western Georges Bank and the Great South Channel (Werner et al. 1993). Larvae in the top 25 m tend to be advected off the bank to the deep ocean or to the southern New England Shelf. Those below 25 m were generally advected along the southern flank more slowly and were more likely to be retained via recirculation in the Great South Channel. Wind-driven advection caused by winds with a significant along-shelf component toward the northeast when larvae are on the Northeast Peaks or southern flank is likely to cause greater losses (Lough et al. 1989).

Cod descend to the bottom at a minimum size of 24 mm, with an average of about 45 mm, but this varies geographically (Hardy 1978a). It also depends on whether the larvae are near land or far out at sea and partly on whether they are floating over deep water or over shoal. Pelagic life is not likely to last for more than 2 months for fish that hatch on the inshore spawning grounds of the Gulf of Maine, where the bottom is within easy reach. On Georges Bank, cod assume a demersal lifestyle by 60 mm but the transition may begin as early as 30–40 mm. Larvae as small as 6–8 mm begin to make vertical migrations and sampling depth differences are evident at 9–13 mm (Lough and Potter 1993). Recently settled 0-group cod inhabited primarily a pebble-gravel deposit on the northeastern edge of Georges Bank at 70–100 m. Their color resembles the pebble bottom and appears to make them less visible and, therefore, less vulnerable to predation. They do not occur in any numbers on the surrounding light-colored sandy parts of the bank. They are present from late July to September. During the day, juveniles are within a few centimeters of the bottom but at night they rise off the bottom to feed on invertebrates. While near the bottom, they orient and swim into the strong tidal currents that sweep across the banks (Lough et al. 1989).

During the first year after the young cod take to the bottom, many of them live in very shoal water, even along the littoral zone; many young have been taken at Gloucester and elsewhere along the shores of New England, while many small cod are caught about the rocks only a few meters deep even in summer. Many juveniles also take to the bottom on the offshore banks, for young have been trawled at many localities between Nantucket Shoals and Browns Bank.

Juvenile cod were more abundant in inshore waters off eastern Newfoundland with a fleshy macroalgal canopy than at barren sites (Keats et al. 1987a). This canopy was primarily *Desmarestia* spp., but other areas contained various kelp species and some ephemeral algae. Cod appeared to be using the algae as cover, not primarily as a source of food; the smallest cod were eating zooplankton and they retreated into the algae when chased.

Age and Growth. Age and growth of cod on Georges Bank and in the Gulf of Maine was studied by otolith analysis (Penttila and Gifford 1976). Use of otoliths has been validated for the Gulf of Maine (Jensen 1970) and the western Gulf of St. Lawrence (Kohler 1964). Length-age curves for cod from Georges Bank and the Gulf of Maine (Penttila et al. 1989) show that cod grew larger on Georges Bank, but the von Bertalanffy curve was very similar to that of the Gulf of Maine fish (Fig. 126). Cod from Browns Bank and the Scotian Shelf exhibit the slowest growth rates of any areas (Penttila et al. 1989).

Young cod grow very rapidly. Growth rates for Gulf of Maine age-0 cod increase from 0.13 mm·day^{-1} after hatching to about 1.0 mm·day^{-1} at 100 days of age (Lough et al. 1989). In later life, cod grow at varying rates in different seas, as shown by the structure of their scales. Studies on cod caught at the mouth of the Bay of Fundy and Nantucket Shoals suggest that they grow more rapidly in the Gulf of Maine than in European waters.

Cod reach a maximum age of 26 and maybe 29 years (Scott and Scott 1988). On NEFSC survey cruises from 1970 to 1988, the maximum age recorded was 18 years (Penttila et al. 1989).

General Range. Atlantic cod occur from Cape Hatteras to Ungava Bay along the North American coast, east and west coasts of Greenland, around Iceland, and the coasts of Europe from the Bay of Biscay to the Barents Sea (Cohen et al. 1990: Fig. 88).

Occurrence in the Gulf of Maine. Cod used to rank as one of the most plentiful of the important food fishes in the Gulf of Maine and were the mainstay of commercial fisheries from earliest colonial times. There was no patch of hard bottom, rock, gravel, or sand with broken shells, from Cape Sable in the east to Cape Cod on the west, that did not support cod at one time or another. Cod were even caught on soft mud bottoms, though they were not common there. Although cod are essentially fish of the open sea, they appeared regularly in various river mouths in Maine and Massachusetts during the late autumn and winter.

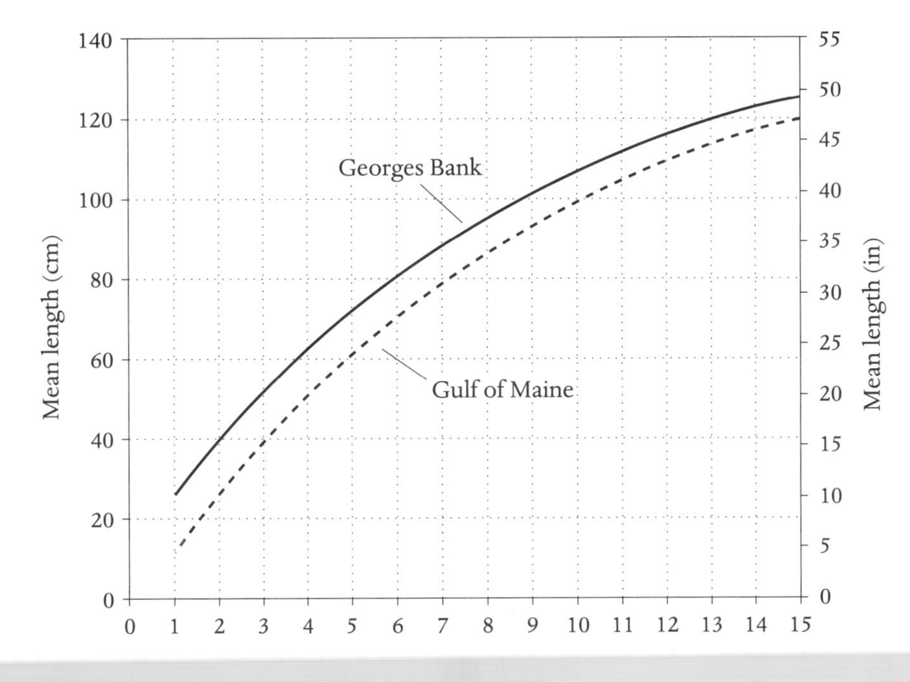

Figure 126. Length-age relationship in Atlantic cod *Gadus morhua*, Gulf of Maine–Georges Bank stock. (From Penttila et al. 1989:7.)

The eastern half of Georges Bank has always been a most productive cod ground (Map 16) and one of the most famous south of the Grand Banks of Newfoundland. Significant concentrations of cod also occur in the South Channel–Nantucket Shoals region in the southwestern part of the Gulf, and from Browns Bank and Stellwagen in the eastern part, the latter being particularly productive in winter. The broken bottom off Seal Island, N.S., the ground near Lurcher Shoal, and Grand Manan Bank were all famous cod grounds. Other well-known inshore grounds were certain hard patches off Chatham, between Provincetown and Plymouth and off the latter port, Jeffreys Ledge, Ipswich Bay, Cashes Ledge, Platt Bank, and Fippenies. They were also common from the Isles of Shoals to Casco Bay, Matinicus Island, Mt. Desert Island, and on ridges into the Bay of Fundy.

Migrations. It has long been known that cod carry out extensive migrations in some regions, but are more nearly stationary in others (Lear and Green 1984). Those that move have travels falling into three categories: (1) vertical movements by older cod in search of food; (2) spawning migrations; and (3) regular seasonal migrations between different regions that are suitable for cod during different times of the year.

Most cod that live off coasts from Cape Cod to northern Nova Scotia, in the southeastern part of the Gulf of St. Lawrence, and on the southern part of the Grand Banks can be termed "nonmigratory" in a broad sense. However, juveniles are concentrated at depths less than 100 m between Jeffreys Ledge and Cape Cod and in the Nantucket Shoal region and widely distributed over Georges Bank in the spring. In the autumn, the distribution is more restricted, occurring inshore of the 100 m isobath in Massachusetts Bay, southeast of Cape Cod, and the Northeast Peak, Georges Bank (Wigley and Gabriel 1991).

In the extreme northern and southern fringes of their geographic range cod are regularly migratory. Thus it is only in summer and early autumn that they visit the waters of the polar current along the eastern coast of Labrador, from which they withdraw again later in the autumn, to pass the winter and spring either to the south or in deeper water. Cod along the north shore of the Gulf of St. Lawrence showed movements correlated with sea temperatures and currents. Peak catches coincided with decreases in sea temperature (0.5°–8.5°C) coupled with favorable wind-driven longshore currents. A southwest wind tended to force warmer water offshore and form an upwelling of colder water that the cod followed inshore. A northeast wind brought warm surface water onshore and forced colder water out deeper and cod were also caught deeper (Rose and Leggett 1988).

Cod in the Middle Atlantic region, favoring temperatures between 0° and 10°C, migrate north and east in summer and autumn reaching Nantucket Shoals when the shallower waters of the New York Bight exceed 20°C. They range southward again to coastal New York, New Jersey, and Chesapeake Bay during winter and spring (Heyerdahl and Livingstone 1982).

Tagging studies in the Gulf of Maine area in 1984–1997 showed little exchange between the area east of Browns Bank and Georges Bank and the inner Gulf of Maine (Hunt et al. 1999). There is exchange between areas within the Gulf of Maine and cod tagged on Georges and Browns banks during the winter spawning season showed widespread postspawning dispersal within their respective divisions and to adjacent divisions (based on NAFO divisions 4 and 5). There is evidence of immigration and emigration and an apparent net loss from the Georges Bank area to Browns Bank. The area is not a closed system and there is exchange among the Bay of Fundy, the inshore area of southern Nova Scotia, and Browns and Georges banks. The results are consistent with earlier tagging studies (which are reviewed). These results demonstrate substantial interaction of cod from different management areas and may have implications for stock assessment models and management practices.

The record for long-distance travel is held by a European fish. A cod tagged in June 1957 in the North Sea was caught on Grand Banks on January 1962, having made a transatlantic journey of about 3,200 km (Gulland and Williamson 1962).

Small cod found along the coast of Maine do not shift ground much from season to season. Further offshore, cod are always on the move over the bottoms of their chosen banks. Although cod cannot be described as schooling in the same sense as herring or mackerel, these traveling cod often hold closely together. When cod travel, they often rise to middepths.

Importance. Cod are one of the most important commercial species and were intimately involved with the early history of New England (Jensen 1972; Kurlansky 1997). At least 100 years before the landing of the *Mayflower* in Massachusetts, European fishermen were catching cod on hook and line along the coast of North America. The importance of cod for the early settlers is attested to by the model of the cod that has hung in the Massachusetts statehouse since 1784. During the early days of the fishery, the entire Gulf of Maine catch of cod was made on hook and line; on handlines at first, but with long or trawl lines coming into general use about the middle of the nineteenth century.

The earliest important addition to fishing methods came during the winter of 1880–1881, when gill nets based on the Norwegian system were introduced. Since about 1908, when otter trawls came into general use, an increasing proportion of the catch has been taken by this method. Today most of the Gulf of Maine catch is made by otter trawls and gill nets. An historical review of the Georges Bank cod fishery from 1890 to 1990, including assessment and management, was presented by Serchuk and Wigley (1992). Total commercial landings in the Gulf of Maine (exclusively United States) in 1996 were 7,200 mt (Fig. 127), a 60% decrease from the record-high 1991 total of 17,800 mt and well below the 1977–1986 average of 12,100 mt (Mayo and O'Brien 1998).

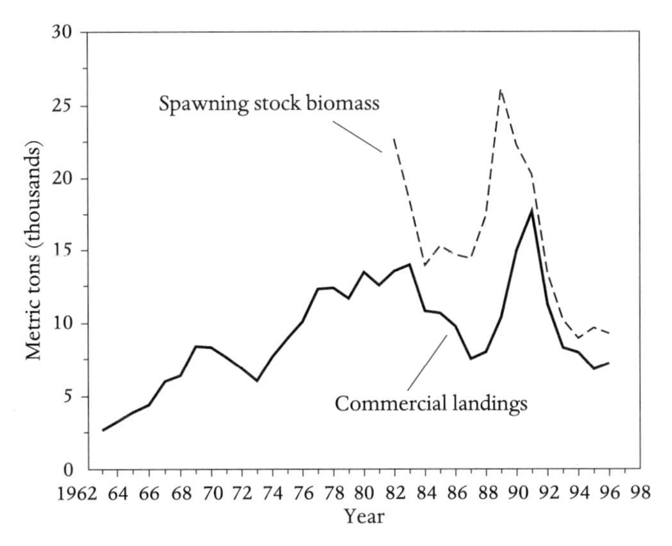

Figure 127. Catch and index of abundance of Atlantic cod *Gadus morhua,* Gulf of Maine stock. (From Mayo and O'Brien 1998.)

Recreational Fishing. Recreational fishing is also very important and occurs year-round, although peak activity is seen during the summer in the lower Gulf of Maine and in fall and winter in the inshore waters from Massachusetts southward. Party and charter boats, shore fishing, and private boat angling make up the recreational fishery. Recreational catches declined from the 1977–1986 average of 3,100 mt to 2,100 mt in 1996 (Mayo and O'Brien 1998).

Stocks. In U.S. waters, cod are assessed as two stocks: Gulf of Maine, and Georges Bank and southward (Mayo and O'Brien 1998). Studies on genetic variation in cod stocks throughout their geographical range (North America to the Baltic Sea) by Mork et al. (1985) indicate little genetic differentiation among the stocks and these findings are supported by more recent research on stock structure in juvenile cod from the Newfoundland Shelf (Pepin and Carr 1993). Atlantic cod appear to be a species with sufficient gene flow to prevent significant genetic divergence.

Distribution of the cod copepod parasite *Lernaeocera branchialis* in the New England area has been used to separate stocks (Sherman and Wise 1961). There were heavy infestations (20%) in the northern coastal region of the Gulf of Maine and moderate ones (10%) in the central Gulf. On Georges Bank and the South Channel, infestations were low (2%). Migrating cod taken off Rhode Island had no parasites. These results, together with data from tagging studies, growth rate analyses, and recruitment patterns, suggest that cod in the Gulf of Maine are distinct from the more southerly groups of the species.

Tagging data collected since 1887, tagging studies from 1955 to 1959, and evidence from other sources suggested that there are four major groups of cod in the New England area (Wise 1962): (1) offshore banks (Georges and Browns) closely related to fish of the southeastern Nova Scotian coast; (2) Gulf of Maine, probably divided into many subgroups and receiving considerable recruitment from the south; (3) southern New England, south and west of Nantucket Shoals; and (4) Mid-Atlantic coast (New York, New Jersey, Chesapeake Bay), which spend part of the year mixed with the southern New England fish. Based on evidence from growth analyses, research survey catch composition, relative abundance indices, and commercial catch composition data, Serchuk and Wood (1979) documented strong affinities between cod populations in the southern New England–Mid-Atlantic region and Georges Bank cod and suggested that the cod from Georges Bank and south composed a single stock. Since 1977, these cod have been assessed as a unit stock (Serchuk and Wigley 1986) and Gulf of Maine cod as another unit stock.

Management. Currently, U.S. commercial and recreational fisheries for cod are managed under the New England Fishery Management Council's Multispecies Fishery Management Plan. Both Gulf of Maine and Georges Bank and South stocks are overexploited and remain at a low biomass level (Mayo and O'Brien 1998). Canadian populations are considered *vulnerable* by the American Fisheries Society, and other populations are categorized as *not at risk but overfished* (Musick et al. 2000).

During the 1850s, cod eggs and larvae were collected in hatcheries and released into the sea in attempts to increase the adult yield both in the United States and Europe (history summarized by Solemdal in Dahl et al. 1984). There have been efforts to rear cod to market size in artificial enclosures (see the volumes resulting from an international symposium on artificial propagation of cod held in 1983 at Arendal, Norway [Dahl et al. 1984]). These volumes also provide an excellent summary of what is currently known of the biology, ecology, and early life history of the cod and are a testimony to the economic importance of this species. Recently, efforts to rear cod for stock enhancement have been renewed (Folkvord et al. 1994). The first cod-farming experiments were conducted in coastal Labrador to demonstrate that grow-out of cod is technically feasible, and cod farming is now being carried out along the northeast coast of Newfoundland (Wroblewski et al. 1999).

HADDOCK / *Melanogrammus aeglefinus* (Linnaeus 1758) / Bigelow and Schroeder 1953:199–213

Description. Body heavy and laterally compressed (Fig. 128). Upper jaw projecting over lower, which has a small barbel. Snout wedge-shaped in side view and rather pointed. Three dorsal fins, first pointed and higher than second and third.

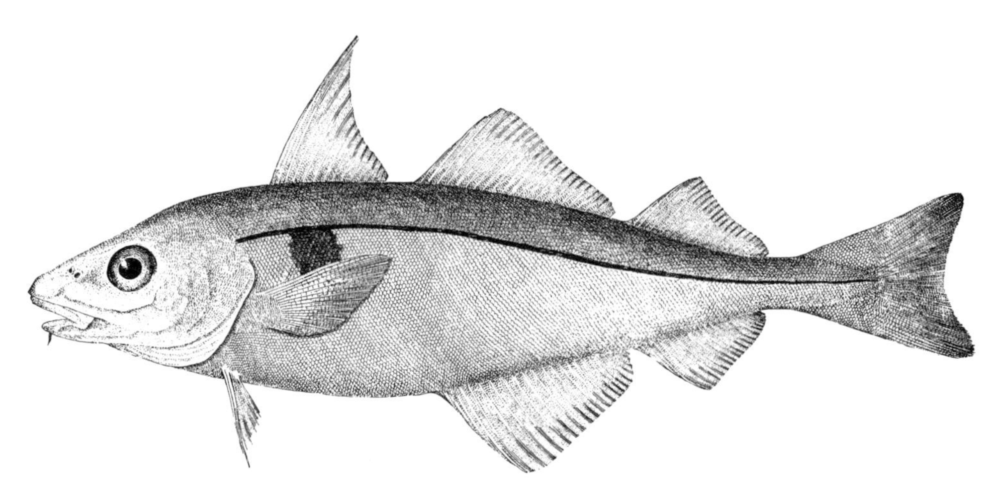

Figure 128. Haddock *Melanogrammus aeglefinus*. Eastport, Maine. Drawn by H. L. Todd.

slightly behind it; second anal fin shorter. Pelvic fins located in front of pectoral fins, second pelvic ray slightly elongate. Caudal fin with a concave margin.

Meristics. First dorsal fin rays 14–17; second dorsal fin rays 20–24; third dorsal fin rays 19–22; first anal fin rays 21–25; second anal fin rays 20–25; total caudal fin rays 57–60; gill rakers 24–27; lateral line scales about 160; pyloric caeca about 186, some of them branched; vertebrae 53 or 54 (Svetovidov 1962; Markle 1982; Scott and Scott 1988).

Color. When a live haddock is first taken from the water, the top of its head, back, and sides down to the lateral line are dark purplish gray, paling below the lateral line to a beautiful silvery gray with pinkish reflections and with the black lateral line and sooty shoulder patch standing out vividly. This patch, the "devil's mark," is indefinitely outlined and varies in size and in distinctness, but only rarely does a haddock fail to show it. The belly and lower sides of the head are white. The dorsal, pectoral, and caudal fins are dark gray; the anal fins pale like the lower part of the sides and black speckled at the base. The pelvics are white, more or less dotted with black. Haddock are usually quite uniform in color, but occasionally one shows from one to four dark transverse bars or splotches in addition to the black shoulder blotch. Several of these serially striped fish have been taken in Passamaquoddy Bay and one near Mount Desert. Occasionally a haddock may be golden on the back and sides, with the lateral line golden, and such fish may not have dark blotches.

Size. The largest haddock on record, from Icelandic waters, was 111.8 cm long and weighed 16.8 kg (Thompson 1929). The heaviest recorded in American waters was 13.6 kg. Welsh collected a fish near Gloucester in 1913 that measured 90 cm and weighed 7.5 kg. The largest haddock taken on 54 NEFSC bottom trawl cruises 1963–1986 was a 98-cm TL fish taken on *Albatross IV* cruise 63-7, station 17 at 42°51′ N, 69°14′ W in November 1963. The largest haddock taken on the other 53 cruises were 78–92 cm TL. Commercially caught fish average 35.5–58.5 cm long and 0.5–2 kg. The all-tackle game fish record is a 6.80-kg fish caught in Saltraumen, Germany, in August 1997 (IGFA 2001).

Distinctions. Haddock can be distinguished from cod, pollock, and tomcod by their black lateral line, the sooty black patch on each side over the middle of the pectoral fin, and the pointed first dorsal fin, which is rounded in the other three species.

Habits. Haddock generally live in deeper waters than cod; few are caught in less than 20 m of water and they are most common at depths of 45–135 m (Brown 1998). The maximum depth record for NEFSC bottom trawl surveys is 380 m. Haddock seldom come into shoal water; the difference in habitat among haddock, cod, and pollock holds from the time the young become benthic.

Haddock are most common at temperatures of 2°–10°C (Brown 1998). Minimum temperatures from NEFSC bottom trawl surveys were 2°–6°C, maximum 7°–16°C. Few are caught where the bottom water is colder than about 1.7°–2.2°C, although good catches are sometimes made in temperatures as low as 1.1°C.

It is evident that at the depths they frequent, the water temperatures in the entire Gulf of Maine, are suitable for haddock, but the uppermost stratum may be too warm from late summer through early autumn and too cold from late winter through early spring. High mortalities resulted when temperatures fell below 1°C in Newfoundland (Templeman 1965c). Salinities at the localities and depths at which haddock live in the Gulf range from about 31.5 ppt inshore to a maximum of about 34.5 ppt on the offshore edge of Georges Bank, with most of the catch made in water more saline than about 32 ppt. Haddock enter bays and reaches between the islands along the coast of Maine in some numbers, but they never run up estuaries into brackish water.

Haddock are more exclusively a groundfish than cod, and

although adults sometimes pursue herring and other small fishes, they have not been reported coming to the surface when so engaged. Haddock are also more selective than cod in the type of bottom they frequent, being rarely caught over ledges, rocks, or kelp or on the soft oozy mud that hake prefer. They are chiefly taken on broken ground, gravel, pebbles, clay, smooth hard sand, sticky sand of gritty consistency; and where there are broken shells. They are particularly partial to smooth areas between rocky patches. Haddock on the Scotian Shelf showed a strong association with coarser sandy sediments and tended to avoid silt and clay (Scott 1982a).

Both male and female haddock produce sounds (Hawkins and Chapman 1966; Hawkins and Amorim 2000). The sounds are short sequences of repeated "knocks" or pulses repeated singly or in short volleys that resemble grunts. Sounds are produced by contractions of striated muscles attached to the anterior end of the swim bladder. These muscles are more highly developed in sexually mature male fish. Sounds are produced within groups of haddock concomitant with aggressive displays, usually by mature or maturing males, and in groups of immature haddock when they are hungry or engaged in competitive feeding. Haddock chased by a net in tanks or in the presence of large cod also produced sounds. This suggests that sound is involved in aggression and self-defense. Sound is also a part of courtship behavior, with the male producing sounds that vary in their characteristics as courtship proceeds (Hawkins and Amorim 2000) (also see the section on breeding habits).

Food. Larval haddock on Georges Bank ate phytoplankton, eggs, nauplii and copepodite stages of copepods, and lamellibranch larvae (Kane 1984). Smaller larvae eat smaller food organisms and are more euryphagous than larger ones. Copepod eggs were the most abundant food item. Larval haddock are sight feeders. Feeding intensity reaches a peak shortly before sunset, declines steadily until after sunrise, and then increases during the day. There is severe competition between cod and haddock, and cod is the more aggressive predator. Cod out-compete haddock larvae under experimental conditions (Laurence et al. 1981), as discussed under cod. In the field, there appears to be enough food to sustain both species under normal conditions.

Juvenile haddock eat primarily small crustaceans. Amphipods are their single most important food (over 25% of their diet by weight). Euphausiids, small decapod shrimp (*Crangon*), mysids, isopods, cumaceans, and copepods make up the rest of the crustacean prey. Polychaetes (particularly *Eunice*) and small fishes such as silver hake and herring also contribute to their diet (about 15 and 5.0% by weight, respectively). Haddock smaller than 8 cm FL feed predominantly on copepods and euphausiids. When larger than 8 cm they feed intensively on benthos (small decapods and echinoderms) (Bowman et al. 1987). Echinoderms assume increasing importance in the diet of haddock larger than 30 cm (Bowman et al. 2000).

imals devoured by the haddock would doubtless include nearly all the species belonging to the fauna" of the ground on which the haddock were living. They begin an adult diet when they are still quite small. Bivalve and gastropod mollusks, small worms, amphipods, larger crustaceans such as hermit, spider and common crabs, shrimps, and amphipods; starfish, sea urchins, sand dollars, brittle stars; and sea cucumbers all enter regularly into the diet of the haddock according to locality. Food in the Gulf of Maine and western Nova Scotia is quite different from that on Georges Bank, southern New England, and the Mid-Atlantic Bight (Langton and Bowman 1980). In the Gulf of Maine–Nova Scotia area, about half of the diet, by weight, consists of echinoderms, mainly brittle stars; crustaceans and polychaetes account for 10–15% of the diet, mainly shrimps, amphipods, and various polychaete worms; mollusks (mostly bivalves) make up 1–3%. In the Georges Bank area, fishes account for 28.4% of the diet, largely a result of the consumption of herring eggs. Polychaetes make up 23.5%, much more than in other areas. The third major prey group is Crustacea (16%) of which 7.1% are amphipods. Echinoderms, primarily brittle stars, represent only 7–8% of the diet. Lastly mollusks, especially pelecypods, are up to 2% of the diet. Results of recent studies are similar (Bowman et al. 2000).

In general, haddock are more exclusively benthic feeders than cod or hake (Kohler and Fitzgerald 1969; Mattson 1992). The importance of herring-egg predation has been noted for Georges Bank; they have also been reported as feeding on capelin eggs in Newfoundland (Templeman 1965d). They are known to eat argentine, silversides, sand lance (Scott and Scott 1988), herring, small mackerel, young hake, young eel (Bigelow and Schroeder), and are cannibalistic on their own young (Wigley 1956). Lists of food organisms are given by Baird (1889), Homans and Needler (1944), Wigley (1956), and Bowman et al. (2000).

In the laboratory, adult haddock feed mainly during the day, with a peak in the late afternoon to dusk and a smaller peak at dawn (S. J. Hall 1987). At these times larger meals are most likely to be eaten, although foraging continues during the day with consumption of smaller meals. Comparison of their maximum daily rations with estimates of consumption in the field indicate that wild haddock feed at 15–22% of their maximum capacity.

Predators. Juvenile haddock are occasionally eaten by spiny dogfish, skates (thorny, winter, barndoor, and little), cod, haddock, pollock, cusk, hakes (red, white, and silver), goosefish, sea raven, bluefish, and halibut (Bowman 1980; Rountree 1999). Haddock are also prey of gray seal (Benoit and Bowen 1990b).

Parasites. A list of internal parasites (Blacker 1971) includes seven species of protozoans, nine Platyhelminthes, five nematodes, and one cestode. Twelve species of copepods were

Georges Bank are externally parasitized by a copepod, *Caligus* sp. The preferred site of attachment is under the eye. Although there was no direct evidence of reduced fish condition owing to parasitism, there was circumstantial evidence that it may be a source of mortality of young haddock because of damage done to the host's skin or occlusion of its vision (Neilson et al. 1987). A gill copepod, *Lernaeocera obtusa,* reduced the weight of the host by 10–30% (Kabata 1958).

Breeding Habits. In addition to changes in growth since the early 1960s, maturation has shifted by about a year (Brown 1998). Formerly, females age 4 and older were fully mature and about 75% of age 3 females were mature (Clark 1959). Currently, nearly all age 3 and 35% of age 2 females are mature. Median length at maturity for females and males from Georges Bank was 29.7 and 26.8 cm TL, respectively (O'Brien et al. 1993). In the Gulf of Maine region, 50% of females were mature at 34.5 cm and 50% of males at 35.0 cm. Median age at maturity was 1.5 years for females and 1.3 years for males from Georges Bank, 1.8 and 2.1 years, respectively, for the Gulf of Maine.

Fecundity ranges from 12,000 to 3,000,000 eggs (Hodder 1963, 1965). Averages vary from 31,000 in 2-year-olds to 2,158,000 in older fish. Fecundity increases with age and size, but may vary considerably in the same size fish from year to year; and this may be correlated with water temperature (Hodder 1965).

Prior to spawning in an aquarium, males engaged in aggressive displays toward other males (Hawkins et al. 1967), which consisted of an extension of the vertical fins, intense sound, and rapid movement about the tank. One male was dominant, and he courted a single female with further fin extensions and sounds. The female then approached the male with her fins held against her body and proceeded to follow him around the tank. He increased the repetition frequency of the sounds to give a humming noise (Hawkins and Amorim 2000) and changed his coloring so that two accessory pigment spots developed along each flank behind the permanent patch. The male then swam alongside the female and mounted her dorsally or laterally. He slid around her to a ventral position and both swam vertically vent to vent while they released eggs and milt. Sound production stopped during these activities. After spawning the female went to a corner of the tank and the male began making noises again. The interval between successive spawnings was determined by the female. Other males who were sexually mature showed mating behavior but did not mate with the female.

Haddock spawn over rocks, gravel, smooth sand, or mud (Hardy 1978a). Ripe fish were found chiefly over broken ground wherever sand, gravel, mud, and rocks alternated in the area between Cape Ann and Cape Elizabeth (Welsh in Bigelow and Schroeder). At times haddock in the Gulf may deposit their eggs at depths a few meters from the surface, but this is unusual; 27–36 m is usually the shallowest bottom water depth for spawning.

Major spawning concentrations occur on eastern Georges Bank, the Scotian Shelf, the coast of Maine, Emerald-Western Banks, Grand Banks, and St. Pierre Bank (Begg 1998). In most years, there has been a definite spawning center on the northeastern part of the bank, just east of Georges Shoals, and there may be a second spawning center in the South Channel. Browns Bank is also a productive spawning center (Cargnelli et al. 1999b).

Eggs were collected by the MARMAP surveys from January to August (Berrien and Sibunka 1999: Fig. 31). Eggs were found on Nantucket Shoals and adjacent waters, Georges Bank, the Scotian Shelf, and the Gulf of Maine. Spawning began in January on the Scotian Shelf and Georges Bank and as the season advanced, increased in intensity and areal extent, south to Nantucket Shoals and north into the Gulf of Maine. Peak activity is controlled by temperature, preferred temperatures ranging between 2.5° and 6.6°C (Colton and Temple 1961). Salinity ranges are 32–32.5 ppt (Svetovidov 1962). On the Grand Banks, successful year-classes are correlated with low numbers of icebergs, which probably affects salinity (Templeman 1965c).

Early Life History. Haddock eggs are buoyant, without an oil globule, 1.32–1.60 mm in diameter with a perivitelline space of 0.05–0.19 mm (Markle and Frost 1985: Table 3). They cannot be distinguished from those of cod by traditional methods in the early stages of development. When they are newly spawned, there is even a danger of confusing them with witch flounder, except that black pigment soon forms in gadoid eggs and the embryonic pigment of witch is yellow. Haddock eggs can be distinguished from cod eggs biochemically (Mork et al. 1983; Knutsen et al. 1985).

Incubation time varies with temperature from 6 to 42 days at extreme temperature ranges. Average hatch times are about 17–21 days from controlled temperature studies summarized in Hardy (1978a). Marak and Livingstone (1970) described six major stages of development in fish from Georges Bank at 3.3°C: (1) Fertilization to early blastodermal cap formation: 0–72 h. (2) Complete blastodermal cap to development of segmentation cavity: 4–7 days. (3) Appearance of early embryonic axis to approach of germinal ring to equator: 7–9 days. (4) Equatorial position of germinal ring to just before closure of blastopore: 10–13 days. (5) Closure of blastopore to early scattered pigment: 14–17 days. (6) Formation of characteristic pattern to hatching: 18–21 days.

The highest percentage of viable hatches of haddock in the laboratory occurred at 4°–10°C and 30–36.5 ppt (Laurence and Rogers 1976). Haddock hatching at intermediate temperatures (6°–10°C) were largest, and there appeared to be no association with salinity. Time to 50% hatch was inversely related to temperature. Highest mortality occurred during the gastrula stage.

Larvae are 2.0–4.99 mm at hatching with some geographic variation. The mean for Georges Bank–Gulf of Maine fish is 4.08 mm. Length at hatching decreases over the spawning period (Colton and Marak 1969). The newly hatched larva is very plump and blunt with the head deflected down over the round yolk sac. The mouth is not developed and in most cases the eye is unpigmented. The vent opens laterally at the base of the finfold, typical of gadoids. There are scattered chromatophores at the back of the head, and heavy pigmentation over the gut. Postanal pigmentation consists of a ventral row of fine melanophores extending from the vent to the tip of the tail. Haddock larvae closely resemble cod larvae; differences are discussed under cod. The yolk sac is absorbed in about 10 days when the fish is about 5.5 mm long. Dorsal and anal fins are fully formed at 16–20 mm, and young haddock begin to take on the general aspect of the adult by about 30–40 mm. Arrangement of larval pigment serves to differentiate the little haddock until about 12 mm. Larger larvae are distinguishable from both cod and pollock by their pale pigmentation and the greater height of their first dorsal fin. Young haddock spend 5–6 months in the pelagic stage and begin to go benthic at about 100 mm in length (Colton and Temple 1961).

In the Gulf of Maine, haddock eggs are concentrated in the surface layers with a decrease in abundance with increasing depth (Colton 1965). Larvae and pelagic juveniles are concentrated within a limited depth stratum defined by the thermocline, which is usually between 10 and 40 m. On Browns Bank, haddock eggs are concentrated at the surface in early stages of development but become more evenly distributed throughout the water column in later stages (Page et al. 1989).

The swim bladder appears on day 11 of embryogenesis as a dorsal outgrowth of the gut just posterior to the liver (Schwartz 1971). Its true position becomes visible just anterior to the liver in later stages. Inflation occurs soon after yolk-sac absorption, when the larvae have begun to feed. Initial gas volume may be derived from glycogenesis. The swim bladder in larvae less than 8 mm decreases larval density and carries them up toward the level of the developing thermocline. By the time the thermocline has become established, the larvae have a much improved swimming ability and can maintain themselves within the thermocline. This has important implications for feeding.

During haddock spawning, surface circulation patterns in New England waters are dominated by variable gyres flowing counterclockwise in the Gulf of Maine and clockwise over Georges Bank (Lough and Bolz 1989). A large anticyclonic gyre and rotary tidal current hold larvae over the bank. This carries larvae westward along the southern half of the bank. Some are advected as far west as Nantucket Shoals, but the majority are retained east of Great South Channel. Factors affecting distribution and retention of larvae on the bank are discussed under cod. Larvae do not mix with haddock spawned in the Gulf of Maine or the western part of Nova Scotia and, hence, are geographically isolated (Begg 1998).

Haddock eggs and larvae, like those of many other fishes, may drift for considerable distances from where they were spawned; these involuntary drifts may be greatly exaggerated bells of larger jellyfishes. Welsh found many small haddock (60–70 mm) in company with the common red jellyfish, *Cyanea,* on Georges Bank and off Nantucket Island. Wiley and Huntsman (in Schroeder 1942) found young haddock about 6 cm long under *Cyanea* in the Bay of Fundy. Most (90%) positive tows of young haddock about 10 cm long were made in areas of Georges Bank where *Cyanea* were present (Colton and Temple 1961). Even though the young fish can move independently of currents, their net movement is controlled in part by drift of the jellyfish.

There were no significant diurnal differences in depth distribution of pelagic larvae and juveniles (Colton 1965). Young-of-the-year and 1-year-old benthic haddock were caught more frequently at night while age-2+ haddock were caught more frequently during the day. Juvenile haddock are more available to bottom trawls by day (Bowman 1980).

Circulation patterns over Georges Bank help create a thermocline in the spring in the deeper waters surrounding the bank, whereas shallower water over the bank is well-mixed. Haddock larvae and prey organisms concentrate above the thermocline, and this provides a rich feeding ground. Growth rates, condition of haddock larvae, and prey densities were compared from three sites across the bank, two stratified and one well-mixed (Buckley and Lough 1987). Using RNA-DNA ratios as a measure of larval condition, larvae from the stratified sites were found to be in good condition; 50% of the larvae from the well-mixed site had RNA-DNA ratios similar to larvae that had been starved in the laboratory; and cod were in better condition than haddock in well-mixed sites. This supports the hypothesis that stratified conditions in the spring favor good growth and survival of haddock and that cod were better adapted to survive in well-mixed waters with less available food, which means they can out-compete haddock when food is limited.

Laurence (1974, 1978) studied growth and survival of haddock larvae in relation to food concentration in the laboratory. Haddock fed plankters in concentrations of 0.5, 1, and 3 ml^{-1} at 7°C grew at similar rates, but the greater the food concentration the larger the larvae. Larvae fed less than 0.1 ml^{-1} died after 3 weeks. Growth was positively correlated with temperature at 7° and 9°C but suppressed below 4°C. Larvae could survive without food and still initiate feeding until 8 days at 7°C and 10 days at 9°C, which was the point of no return.

Juvenile haddock live and feed in the epipelagic zone until they are ready to assume a demersal existence (Mahon and Neilson 1987). At this time, they make forays to the bottom to feed. When they locate a suitable bottom, they become demersal and feed almost exclusively on benthic prey, including polychaetes, amphipods, mollusks, decapod crustaceans, and echinoderms (Mahon and Neilson 1987). Transition from pelagic to benthic habitat takes about a month for a cohort (Koeller et al. 1986) but is probably very abrupt for individuals (Mahon and Neilson 1987).

Age and Growth. Aging has been done by counting scale an-

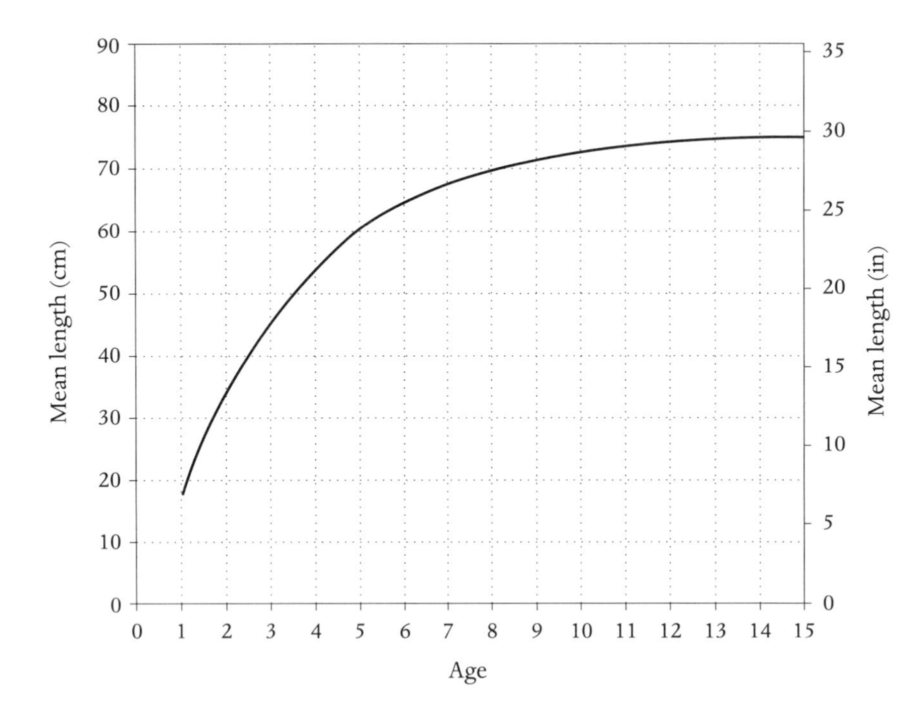

Figure 129. Length-age relationship in haddock *Melanogrammus aeglefinus,* Gulf of Maine–Georges Bank stock. (From Penttila et al. 1989:17.)

and recapture studies and length-frequency curves. One annulus is laid down each year during August–April (Jensen and Wise 1962). No significant differences were found between otolith and scale aging techniques up to the age of 7; then scale readings were lower (Kohler and Clark 1958). Length-age data for Gulf of Maine–Georges Bank is presented in Fig. 129. Maximum age documented from NEFSC survey cruises in 1970–1988 is 14 years (Penttila et al. 1989), although ages in excess of 9 years are uncommon. A maximum age of 22 years was reported in the Barents Sea (Blacker 1971). Haddock from different areas grow at different rates. Fish from the Bay of Fundy grew faster than those from Nova Scotia (Hennemuth et al. 1964). Blacker (1971) gave a composite age-length graph for haddock from Europe and the northwest Atlantic. They appear to grow largest in Icelandic waters and the Barents Sea.

Growth rate of haddock has changed substantially over the past 30–40 years, possibly in response to changes in abundance (Brown 1998). Prior to 1960, when haddock were considerably more abundant than at present, the average size of an age-4 fish was about 48–50 cm. Presently, haddock reach this size at age 3.

General Range. Haddock are found on both sides of the North Atlantic. American haddock are found from Cape May, N.J., to the Strait of Belle Isle (Cohen et al. 1990: Fig. 102). In the eastern Atlantic, the range is from the Bay of Biscay to Spitsbergen, in the Barents Sea to Novaya Zemlya, around Iceland, and rare at the southern tip of Greenland.

Occurrence in the Gulf of Maine. Haddock were once very plentiful all around the open Gulf (Map 17), as well as on all the offshore banks, especially on Georges Bank. Good haddock grounds are less extensive close inshore than good cod grounds, haddock being confined more to depths greater than 18 m and being more selective of the bottoms they frequent. Considerable numbers of haddock have been caught on German Bank, and on the broken grounds off Lurcher Shoal. Whereas haddock are less plentiful than other groundfish on Grand Manan Bank at the mouth of the Bay of Fundy, there was a rich center of population at the mouth of the Bay of Fundy on the Nova Scotian side. Haddock diminish in numbers inward into the Bay, but were plentiful on the New Brunswick side of the Bay near its mouth and within Passamaquoddy Bay. The most productive small grounds in the western side of the Gulf of Maine were Cashes Ledge, Jeffreys Ledge north of Cape Ann, Stellwagen Bank, and several areas off Chatham, Cape Cod (Chang 1990).

Migrations. Tagging studies in the Gulf of Maine by the U.S. Bureau of Fisheries (Schroeder 1942) and in Nova Scotia by the Biological Board of Canada (Needler 1930) have shown that most of the movements of Gulf of Maine haddock are of short duration. Few returns were reported from distant grounds. Haddock of the coasts of Massachusetts and of western Maine with the off-lying banks may be less stationary, for only two fish that were tagged on Stellwagen Bank and between Boone Island and Boothbay were recaptured locally. Tagging experiments do not suggest that the Gulf of Maine haddock that do wander follow any regular migratory routes, but rather that they fan out in all directions. Although they do not migrate on Georges Bank, they do seek shallower water during the spawning season (Colton 1955).

Generally haddock spend the winter in deep water and move shoreward in summer into warmer, shallower coastal water. When the water temperature rises above 10°–11°C in

Maine and Massachusetts in summer, they tend to withdraw from the shallower grounds; but those in deeper channels may remain there year-round. Except for shifts in depth, apparently associated with temperature, haddock are year-round residents as far east as the offing of southeastern Nova Scotia; many of them as far east as Halifax and Sable Island Bank. There is a southern migration from Passamaquoddy Bay in March and April; individuals from New England may move as far south as New York, New Jersey, or Cape Hatteras in winter (Carson 1943). Immature haddock are most concentrated at depths near or less than 100 m south of Jeffreys Ledge to Cape Cod and along the southern edge of Georges Bank in the spring. In autumn, these groups are augmented by concentrations on Georges Bank extending from Northern Peak and along the northern edge to the Great South Channel region (Wigley and Gabriel 1991).

More extensive migrations occur in Canadian waters. In Nova Scotia there is a spawning migration in early spring to the offshore banks and back again in late spring (Scott and Scott 1988). Distribution appears to be determined by season, temperature, depth, and size of the fish. Haddock in the southern Grand Banks region form a distinct population, being separated from those of Nova Scotia by the deep Laurentian Channel. They make a summer inshore migration to the southwest coast of Newfoundland, avoiding regions where the bottom water is colder than about 1.1°C (Thompson 1929).

There is some evidence that haddock undergo vertical migrations. Midwater catches of haddock in European waters and the Barents Sea increase during the day in November, presumably because the fish leave the bottom (Woodhead 1965). The greatest diurnal differences in catch changes occur in smaller fish in the 20–50 cm range. These differences are apparently not associated with feeding. Responses to baited hooks showed no clear diel activity patterns (Lokkeborg et al. 1989).

Importance. Haddock were once much less in favor than cod. Expansion of the fresh-fish trade brought an increasing acceptance of haddock on the market because of their good keeping qualities and convenient size for the table. In 1919 the Gulf of Maine, inshore and offshore, yielded 85 million lb. Development of filleting and packaging of fresh and frozen haddock brought an increase to 206 million lb in 1929. This was the high point and the catch varied in the 1930s, 1940s, and 1950s. Early history of the haddock fishery was summarized by Bigelow and Schroeder.

There was a huge year-class of haddock on Georges Bank in 1963, which began to enter the fishery in the late 1960s (Brown and Patil 1986). This year-class was six times the size seen since 1930. Large increases in fishing effort by foreign fleets, the United States and Canada reduced the stock drastically by 1969. Catches increased between 1977 and 1980, reaching about 28,000 mt, but then declined steadily to 4,400 mt in 1989. Commercial landings in the Gulf of Maine decreased

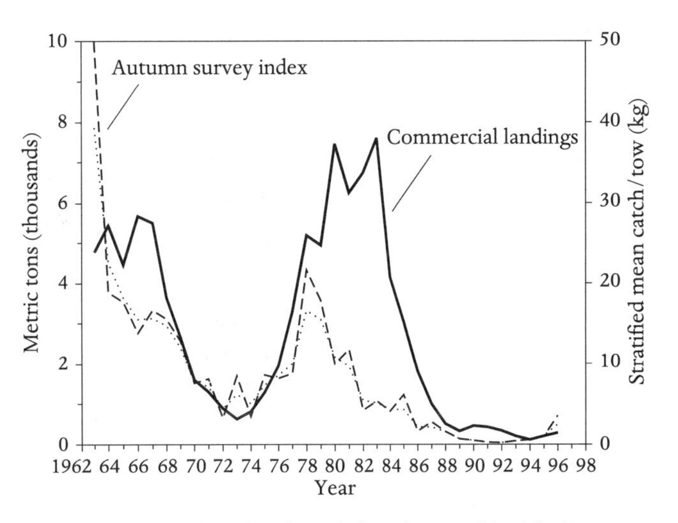

Figure 130. Catch and index of abundance of haddock *Melanogrammus aeglefinus,* Gulf of Maine stock. (From Brown 1998.)

1973 (Brown 1998). Catch then increased to an annual average of 7,000 mt from 1980 to 1983 but subsequently plummeted to historic lows (Fig. 130).

Previous to general adoption of the otter trawl in American waters, haddock were caught mostly on handlines; some in gill nets, especially at spawning time inshore between Cape Ann and southern Maine. Today all but a very small part of the catch is made by otter trawls: only about 3% on longlines and a fraction of 1% on handlines and gill nets.

The New Brunswick Department of Fisheries and the Canadian Department of Fisheries and Oceans initiated a research and development program for the development of haddock aquaculture beginning in 1995, with emphasis on development of early-rearing techniques (Henry 1997).

While haddock are of primary interest from a commercial standpoint, they also deserve a word from the angler's viewpoint, for they bite on almost any bait, and a haddock of fair size is likely to prove astonishing to anybody who is lucky enough to hook one while fishing with a light sinker. Recreational catches are insignificant (Brown 1998).

Stocks. A review of the stock structure of North American haddock populations (Begg 1998) showed that there are major aggregations on Georges Bank and the Scotian Shelf. This review examined all available information on haddock stock structure and investigated areas requiring further research. Combined information from tag-recapture, demographic, meristic, parasitic, and genetics studies provides evidence for identification of haddock stocks, with major population divisions occurring between New England, Nova Scotia, and Newfoundland waters. Within each of these major divisions, a number of discrete stocks appear to exist, although uncertainty remains as to the amount of separation found within each region. The Georges Bank and Gulf of Maine stocks have little mixing (Smith and Morse 1985). Recent DNA studies revealed significant heterogeneity in frequencies of four mitochondrial DNA control region haplotypes between 1975 and 1985 cohorts, indicating that haddock from other geographic regions episodically contribute to the Georges Bank gene pool. Thus, the population of haddock spawning on Georges Bank may not be genetically discrete and, with respect to Atlantic haddock, Georges Bank may not be a closed system (Purcell et al. 1996).

New England and Nova Scotian populations are separated by the so-called "Eastern Channel" more than 180 m deep, to the mouth of the Bay of Fundy. Depth may be an actual barrier in this case, there being no evidence that haddock normally cross channels that are deeper than 180 m once they have taken to the bottom. Only within the Bay of Fundy, where there is no intervening water as deep as 180 m, have tagging experiments provided any evidence of a mixture between the two adult populations. There was evidence of a seasonal northward movement in the western Gulf of Maine in spring followed by a reverse migration in early winter (McCracken 1965). The greater depths of the Laurentian Channel probably make it an even more effective barrier between Nova Scotian and Newfoundland populations. The Gulf of Maine and Nantucket Shoals group also appears to be separated from the Georges Bank group as interchange between these areas does not appear to be extensive (Clark et al. 1982).

Management. Fishing is managed under the New England Fishery Management Council's Multispecies Fishery Management Plan. The sharp decline in landings observed since 1983 (from 7,600 to 300 mt) and the corresponding decline in the autumn survey index (Fig. 130) reflect the severely depleted state of the Gulf of Maine stock (Brown 1998). Spawning stock biomass is below maintenance level and is likely to remain so for the foreseeable future. Fishing mortality must be reduced significantly in order to allow resource recovery.

ATLANTIC TOMCOD / *Microgadus tomcod* (Walbaum 1792) / Tomcod, Frostfish /

Bigelow and Schroeder 1953:196–199

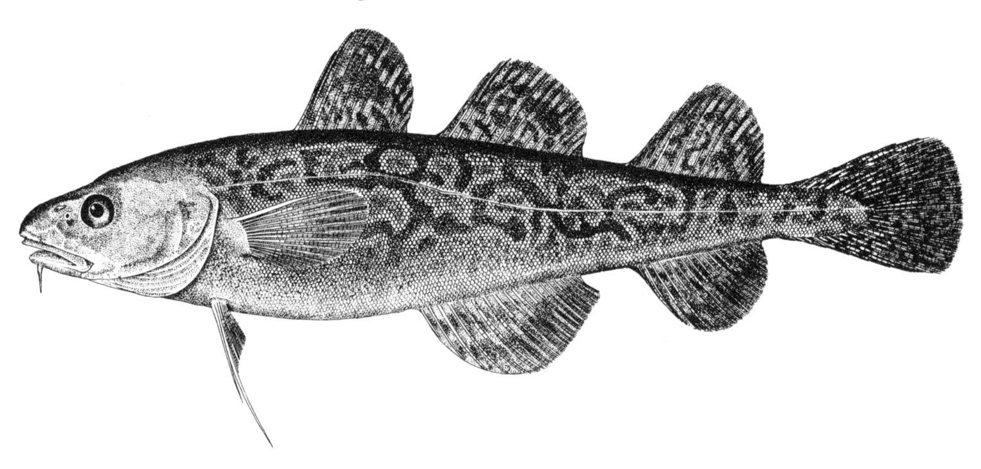

Figure 131. Atlantic tomcod *Microgadus tomcod.* Woods Hole. Drawn by H. L. Todd.

Description. Body moderately elongate, only slightly compressed (Fig. 131). Body depth greatest under first dorsal fin. Head 4.6 times in TL, upper jaw extends beyond lower, snout rounded, barbel on lower jaw. Eye small, about two times in snout. Three moderately high and rounded dorsal fins, first dorsal inserts over middle of pectoral fin. Two anal fins. Pelvic fins narrow and tapering, anterior ray extended as a filament as long as rest of fin. Vent under space between first and second dorsal fins. Margin of caudal fin rounded. Scales small, cycloid, and somewhat embedded. Lateral line pale, moderately arched over pectoral fins.

Meristics. First dorsal fin rays 11–15; second dorsal fin rays 15–20; third dorsal fin rays 16–23; first anal fin rays 17–21; second anal fin rays 16–20; pectoral fin rays 16–19; total caudal fin rays 46–50; gill rakers 16–21; vertebrae 53–57 (Hardy 1978a; Markle 1982; Scott and Scott 1988).

Color. Tomcod are olive or muddy green above, with a yellowish tinge, darkest on the back, paling on the sides, and mottled with indefinite dark spots or blotches. Lower parts of the sides usually show a decided yellowish cast in large fish; the belly is grayish or yellowish white; the dorsal and caudal fins are the same color as the back; the anal fins are pale at the base but olive at the margin; and the fins are more or less darkly mottled.

Size. Maximum length reported by Vladykov (1955a) as 44.7 cm and by Bigelow and Schroeder as 38 cm, 0.57 kg. Average size is 22.8–30 cm.

Distinctions. Tomcod closely resemble small cod but have a rounded tail, filamentous pelvic fins, and small eyes. Small cod have a square or concave tail, the pelvic ray is not as elongate, and the eye is larger (Scott and Scott 1988).

Habits. Tomcod are inshore fish; few probably ever descend more than 6 m or stray as much as 1.5 km outside the outer headlands. In the Gulf of Maine they frequent mouths of streams and estuaries, as well as shoal, muddy harbors such as Duxbury Bay. As often as not they are in brackish water to spawn, and run up into fresh water in winter. They are caught 20 km above the head of the tide in the Petitcodiac River, N.B. Tomcod are less plentiful in harbors where there is no stream drainage, but occasionally are caught off open shores such as Nahant, and such fish are usually large. Sometimes they are found over eelgrass beds (Howe 1971) and in salt marshes (Dutil et al. 1982). There are landlocked populations in Nova Scotia, Quebec, and Newfoundland and maybe in other freshwater lakes in Canada (Scott and Crossman 1973).

Tomcod can tolerate salinities of 0–31.4 ppt (Fiske et al. 1967) and temperatures of −1.2° to 25.5°C (Gordon et al. 1962). Tomcod produce glycopeptide antifreeze proteins, which enable them to tolerate water temperatures below freezing (O'Grady et al. 1982). Nova Scotian tomcod lost their antifreeze when they were acclimated for 5 weeks to 14°C and a long photoperiod (Duman and DeVries 1974a). Antifreeze production in a Long Island population started in November at about 10°C, peaked in January and February at minimum temperatures, and decreased during March to reach a minimum volume in June (Reisman et al. 1984, 1987). The antifreeze protein is similar to that found in some Arctic and Antarctic fishes.

Because of their year-round residency in estuaries, tomcod are subject to many stresses from pollutants. More than 45% of 1-year-old fish and over 90% of 2-year-old fish from the Hudson River were found to have hepatocellular carcinoma (liver cancers) (Klauda et al. 1981; Dey et al. 1984; Cormier 1986). Tissues of Hudson River adult tomcod, especially livers, contained elevated levels of PCBs, which are known carcinogens and are present at high levels in the river sediments. The liver tissue also contained several pesticides (DDT and metabolites, chlordane, and dieldrin), and several heavy metals. Juvenile tomcod from the Hudson 4–5 months prior to winter spawning did not have hepatic lesions, suggesting that the lesions formed rapidly coincident with gonadal maturation. The age structure of the Hudson River population consists almost entirely of age-1 and age-2 fish with age-3 fish rare, and this may be related to pesticide levels. In the Pawcatuck River, Conn., lesion prevalence is an order of magnitude lower, and age-3 fish constitute almost 5% of the population, whereas in the Hudson River fewer than 0.1% of the fish are age 3 (Dey et al. 1993). There may possibly be a causal relationship since tomcod are largely benthic and feed in the sediments. According to Scott and Crossman (1973), in Maine and Canada, tomcod cause eggs and larvae succumb to chemical pollution and low oxygen levels. They hypothesize that whereas other fishes may be able to return to an area after pollution abates, tomcod disappear entirely from affected regions.

Food. Tomcod larvae and young-of-the-year eat mainly copepods. They change to a diet of amphipods, mysids, and isopods at about 90 mm in the Hudson River from July to December (Grabe 1978). Growth paralleled feeding intensity, which was elevated in June, October, and November, depressed in July–September, and further decreased prior to spawning in December. Feeding and growth were inhibited at temperatures higher than 24°C and oxygen levels less than 7 mg·liter^{-1}. Age-1 and age-2 tomcod from the Hudson River are opportunistic feeders with the primary prey being amphipods (*Gammarus* sp.) and a decapod *Crangon septemspinosa* (Grabe 1980). They feed on smaller organisms such as copepods in the winter. Feeding intensity is greatest during fall and spring, prior to and subsequent to spawning. Cannibalism on eggs, larvae, and juveniles occurs during winter and spring.

Adult tomcod also feed on small crustaceans, especially shrimps and amphipods; worms; small mollusks; squids; and larvae of fishes such as alewife, menhaden, herring, anchovies, smelt, mummichog, silverside, sculpins, sticklebacks, sand lance, cunner, winter flounder, and tomcod. Although infaunal polychaetes and amphipods are of secondary importance in the diet of Hudson River tomcod, polychaetes are of primary importance in Montsweag Bay, Maine (Alexander 1971). Tomcod are not as keen-sighted as pollock nor as active as hake (Herrick 1904) but they are able to recognize concealed bait by their sense of smell. They search the bottom by dragging the chin barbel and the sensitive tips of the pelvic fins as they swim to locate prey.

Peak feeding occurs in early morning hours in Montsweag Bay, Maine (Alexander 1971). Stomach fullness decreases by about 50% during each of two 6-h intervals from morning to evening. There are no significant tidal effects on the feeding rhythm.

Predators. Tomcod are preyed upon by yearling striped bass in the Hudson River (Dew and Hecht 1994b) and by bluefish (Wilk 1977); they accounted for 59% of the diet by weight of young-of-the-year bluefish between 150–278 mm (Juanes et al. 1993).

Parasites. Tomcod are parasitized by a protozoan, a myxosporidian, two trematodes, a cestode, three nematodes, two acanthocephalans, a brachiuran, and three copepods (Margolis and Arthur 1979). Tomcod have a low infestation rate by sealworm *Phocanema decipiens* in the Newfoundland area (Templeman et al. 1957).

Breeding Habits. Tomcod in the Hudson River mature at 11–12 months (Hardy 1978a), but not until the end of the third

The smallest mature female in a sample of 100 fish measured 17 cm and the largest was 35 cm. The mean number of ova produced off Hull, Mass., was 20,000; the range was 6,000–30,000 (Schaner and Sherman 1960). A 35.5-cm female had 67,780 eggs (Vladykov 1955a).The gonadal biomass of young-of-the-year fish from the Hudson River just prior to spawning was 15% for males and 30% for females (Grabe 1978). Feeding rate is high in October, but there is a decline in body mass that begins at the onset of the reproductive cycle, implying that food was not sufficient to meet energy demands for growth and maturation of gonads (Salinas and McLaren 1983). Energy is transferred from muscle tissue during winter. After spawning, feeding increases and food is used more efficiently for growth.

Tomcod migrate up rivers to spawn. In Muddy Creek, Harwich, Mass., fish spawned close to shore in shallow areas of the estuary on the windward side of the creek in emergent *Spartina* beds and under mats of floating debris (Howe 1971). They rolled together in groups of three to seven. Typically, they spawn over sand and gravel bottoms (Scott and Crossman 1973), usually in association with ice cover (Booth 1967). Spawning lasts from November to February with the peak in January. Temperatures at this time are 0°–5°C; salinity is 0–16 ppt. Sperm are inactivated at salinities higher than 19–20 ppt (Booth 1967).

Channel morphology and flow velocity at low water indicate that formation of ice cover in winter combined, with presence of sand bars near the confluence of the Ste. Anne River, Quebec, with the Saint-Laurent River, causes a decrease in cross-sectional area of the channel and an increase in flow velocity at the mouth of the river, where velocities higher than 30 $cm \cdot s^{-1}$ were measured at low water (Bergeron et al. 1998). Underwater video observations of tomcod movements in the Ste. Anne River indicate that such flow velocities limit access of upstream migrating fish to the spawning site. Upstream migrants avoid the downstream flow velocities of the falling tide and favor the short period of flow reversal associated with high rising tides in order to move upstream.

Early Life History. Tomcod eggs are about 1.5 mm in diameter (1.39–1.7 mm). They sink to the bottom, where they stick together in masses, or to seaweed, stones, or any available support. Incubation takes 24–30 days at temperatures of 4°–6°C. There are some reports of no oil globules and others of 3–12 small ones (Hardy 1978a). Developmental stages of the embryo are described and illustrated in Hardy (1978a). The eggs are not likely to be confused with other demersal eggs deposited at the same time of year. Winter flounder have smaller eggs (0.9 mm), and sculpins have more heavily pigmented embryos. Sculpins spawn in lower reaches of an estuary or in nearshore areas (Elliott and Jimenez 1981; Pearcy and Richards 1962).

Larvae were collected in the Mystic River, Conn., in January–April (Pearcy and Richards 1962); appeared in the Weweantic River, Mass., in February to mid-April with the peak in March (Howe 1971); and appeared in the Woods Hole region from December to April (Fish 1925).

Length at hatching is 4.12–6.45 mm, with an average of 5 mm. The swim bladder is present, the eyes pigmented, and the mouth unformed (Hardy 1978a) or formed (Howe 1971). The yolk sac is large and elongate in the newly hatched and round in older larvae. The vent opens behind the yolk sac and to the right of the finfold between the margin of the finfold and the ventral edge of the myomeres. There are large stellate melanophores in a row above the gut, and two or three rows ventrally between the vent and tip of the tail. There are sometimes a few punctate melanophores on the head and a few mediolaterally (Hardy 1978a).

Tomcod larvae can be distinguished from winter flounder by their large size and heavy pigmentation. They differ from grubby sculpin by the absence of spines on the head, the position of the vent laterally on the finfold, and by having less pigmentation on the gut.

Newly hatched larvae swim up to the surface within 24 h of hatching to gulp surface air before the pneumatic duct closes (Booth 1967; Peterson et al. 1980), then sink head downward. Larvae are usually collected close to the bottom and in the upper parts of the estuary (Pearcy and Richards 1962). They occur at 0.5–20 ppt salinity (average 12 ppt) and temperatures of 1.1°– 11.7°C (average 5.8°C) and are reported to be able to withstand increases of 14.4°C above ambient of 1.1°C for 30 min (Hardy 1978a). In the Hudson River, larvae hatch on or near the bottom and are passively dispersed throughout the water column by density-induced circulation drift downriver and are returned to the bottom at the salt front by downwelling, which is probably enhanced by active behavioral orientation of the larvae toward darker, more saline waters (Dew and Hecht 1994a). Tomcod larvae in the St. Lawrence estuary use net upstream currents of deeper waters to move passively upstream. Older tomcod occurred nearer the surface and were transported downstream as the season progressed (Laprise and Dodson 1989a). Post-yolk-sac larvae exhibit endogenous activity and a tidal component for 3 weeks (Massicotte and Dodson 1991). They change to a benthic habitat between 12 and 20 mm in April or May in Massachusetts (Howe 1971) and during May and June in Maine (Lazzari et al. 1999).

Juveniles are more than 23 mm long. The barbel is well developed, pigment is in parallel narrow blotches sometimes forming a chainlike pattern. They inhabit shoal areas in coves near mouths of rivers and subtidal flats over eelgrass, sand, and silt bottoms (Howe 1971). In the Sheepscot estuary, Maine, the community in the rocky habitat contained a larger proportion of tomcod than the vegetated and muddy habitats (Tort 1995). Young-of-the-year may remain in brackish river water for the first spring and summer of their lives. They remain in the last pond at Kennebunk Point, Maine, through October or November (Lazzari et al. 1999). Juveniles larger than 76 mm probably undergo fall and winter migrations with the adults.

Age and Growth. Age-1 tomcod average 13.1 cm, age-2, 18.3 cm (Grabe 1980). They averaged 16–18 cm TL in coastal Maine (Lazzari et al. 1999). They probably do not exceed age 4.

General Range. North American coastal waters from southern Labrador to Virginia (Cohen et al. 1990: Fig. 109), running up into freshwater.

Occurrence in the Gulf of Maine. Tomcod are locally common around the entire coastline of the Gulf. They have been reported from Pubnico and in St. Mary's Bay, on the west coast of Nova Scotia; at various locations on both shores of the Bay of Fundy; Eastport; from almost every river mouth along the Maine coast; in the vicinity of Boothbay Harbor; at stations in Casco Bay; and in Portland Harbor, Maine. They are found in practically every estuary around the Massachusetts Bay region.

Migrations. Tomcod make short seasonal migrations into streams and rivers in October and November (Pearcy and Richards 1962). These movements are associated with spawning. In the St. Lawrence River they move upriver from September to December and downriver from February to May (Vladykov 1955a). In the Weweantic estuary, Mass., they move into deeper water at the mouth of the estuary in June, apparently in response to rising temperatures (Howe 1971).

Importance. Tomcod are delicious little fish, but they were more highly valued a century ago, when 2,300–4,500 kg were caught in the Charles River, Boston. Today it is unusual to see them for sale in any market. Tomcod are not plentiful enough anywhere around the Gulf to support a regular fishery. Bigelow and Schroeder gave an historical summary of the fishery in the Gulf of Maine and the Bay of Fundy. The fishery in New Brunswick is coincident with the smelt fishery (McKenzie 1964b). In Maine and Canada they are taken in bag or pocket nets set in the rivers and a few are taken in weirs. Landings reported to FAO were 304 mt in 1978, 255 mt in 1986, and only 10 mt in 1987, all from Canada (Cohen et al. 1990).

Many tomcod are caught in autumn on hook and line by smelt fishermen and anglers along the shores of New England and New York for home consumption. There is a winter and summer sport fishery in Connecticut (Sampson 1981). Tomcod take bait greedily. Clams, shrimp, sea worms, or cut fish will serve, and they afford amusement to many anglers in harbors and stream mouths.

LONGFIN HAKE / *Phycis chesteri* Goode and Bean 1878 / Bigelow and Schroeder 1953:232–233 (as *Urophycis chesteri*)

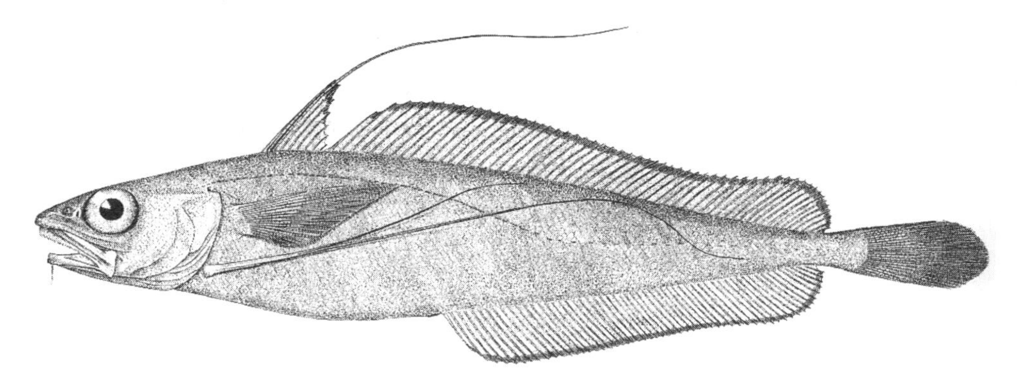

Figure 132. Longfin hake *Phycis chesteri*. Off Cape Ann. Drawn by H. L. Todd.

Description. Body elongate, greatest depth at origin of second dorsal fin 6 times in TL, tapering to a small caudal peduncle (Fig. 132). Head pointed, 5.5 times in TL; mouth large, with fine teeth, upper jaw projecting, small barbel on lower jaw, angle of mouth under pupil of large eye. Two dorsal fins; first dorsal triangular, third ray greatly elongate, five times length of next longest ray. Second dorsal fin a little higher at either end than in the middle. These two fins extend from above pectoral fin base to caudal peduncle. Single anal fin almost uniform in height, extending from vent to caudal peduncle. Pectoral fins moderate, pointed, situated on midside behind gill opening reaching to below seventh ray of second dorsal fin. Pelvic fins very long and filamentous, longest ray almost reaching posterior part of anal fin; next longest reaching thirteenth anal fin ray. Caudal fin rounded. Lateral line broadly arched in first half, broken in posterior half.

Meristics. First dorsal fin rays 8–12; second dorsal fin rays 50–63; anal fin rays 43–56; pectoral fin rays 16–18; total caudal fin rays 28–37; lateral line with 90 rows of scales; gill rakers 4 or 5 + 13–17 = 17–21; pyloric caeca 11–14; vertebrae 45–52 (Comyns and Grant 1993; B. B. Collette, pers. obs.).

Color. Freshly preserved specimens are olive above and on the sides, with a silvery white belly. The fins are olive with dusky markings on the dorsal filament, the outer edge of the dorsal fins, the caudal fin, and the pelvics.

Size. Longfin hake reach nearly 40 cm TL (Cohen et al. 1990). The average size is 18.8–23.9 cm (Methven and McKelvie 1986).

Distinctions. The most distinctive character of longfin hake is the very long pelvic fins. The filamentous dorsal ray is also longer than in other Gulf of Maine hakes. There are only about 90 lateral line scales that are relatively larger than in white or red hake. The eye is much larger; the outline of the anal fin is slightly concave instead of straight; the pectoral fins are more slender and pointed; and the caudal fin is narrower with more strongly convex margins.

Habits. Longfin hake are bottom dwellers living on the continental slope over soft, silty bottoms (Scott 1982a). Longfin hake were seen from the submersible *Alvin* lying on the bottom curled in depressions next to rubble at depths of 500–630 m (Sedberry and Musick 1978). Individuals actively swimming over the bottom held their pelvic fins spread forward, tips dragging over the bottom, possibly indicating that they use these fins to detect prey, as has been observed in species of *Urophycis.*

They are more abundant at shallower depths in the northern part of their range than in corresponding depths in the southern part. The largest catches off the Virginia coast at depths of 160–1290 m were between 500–699 m (Wenner 1983a). On the continental slope south of New England at 200–900 m, the largest catches were at 600 m (Haedrich et al. 1975). Most were caught between 300–450 m from the southern edge of the Grand Bank to the Mid-Labrador Shelf at sampling depths of 195–748 m (Methven and McKelvie 1986).

They have been collected at temperatures of 1.6°–10°C, and are most abundant at stations with water temperatures of 5°–9°C (Methven and McKelvie 1986). They appear to avoid the 10°C isotherm (Wenner 1983a). Since this occurs at greater depths in southern latitudes, it might account for the deeper distribution there than in higher latitudes where the 10° isotherm is shallower. Larger fish were collected at greater depths, and there appeared to be a bathymetric segregation between small demersal juveniles and adults, typical of gadids (Methven and McKelvie 1986). This appears to be a life-history change in depth preference, as adult males are smaller than females but occur at the same depths. Longfin hake were collected at all hours, but most were taken at night (1800–0800).

Food. Crustaceans make up over 95% of longfin hake diet (Langton and Bowman 1980; Bowman et al. 2000). The most important prey organisms from western Nova Scotia to the Mid-Atlantic Bight were a euphausiid, *Meganyctiphanes norvegica,* and a pandalid shrimp, *Pandalus.* On the continental slope and rise off the Middle Atlantic states, longfin hake eat polychaetes, shelled mollusks and cephalopods, crustaceans, and fishes (Sedberry and Musick 1978). Decapod crustaceans and fishes occurred regularly in the diet, but euphausiids were numerically the most important food item; fishes were the most important volumetrically. Fishes eaten were mostly lanternfishes. Both benthic and pelagic organisms were eaten, but pelagic prey were most important. Longfin hake probably feed as active predators on and just above the bottom.

Predators. Spiny dogfish, white hake, and goosefish prey on longfin hake (Maurer and Bowman 1975; Rountree 1999).

Parasites. On the Scotian Shelf, longfin hake are parasitized by 14 gut helminths (Scott 1987): 9 digenes, 2 cestodes, 2 nematodes, and an acanthocephalan. Three of the digenes are characteristic of deepwater host species. One digene, *Hysterothylacium aduncum,* was common in longfin hake guts and was also the only helminth common to all five hake species examined. A digene, *Lepidapedon merretti,* was described later from longfin hake (Campbell and Bray 1993). Longfin hake also have a gill monogene, *Diclidophoroides macallumi* (Rubec 1991).

Breeding Habits. Males mature at a smaller average size than females (Wenner 1983a). The mean standard length of mature males was 22.4 cm (14.5–33.8 cm) and of females 29.2 cm (20.5–39 cm). Since it was impossible to age them, it is not known at what age they mature. Longfin hake have the highest relative fecundity of any gadid for which data are available. The average number of eggs per gram body weight is 2,039, an order of magnitude greater than any gadid except white hake. The maximum number of eggs recorded was 1,305,700 for a 36-cm fish. Wenner believed that this high fecundity is selected for because of the restricted area of suitable habitat (it is confined to the continental slope). On the continental slope off the Virginia Coast, longfin hake spawn from late autumn to early winter (Wenner 1983a). Spawning began in late September and was completed by April, with the peak in December and January. Ripe fish were collected in depths of 300–1000 m. A 340-mm TL female collected in February on *Delaware II* was nearly ripe (B. B. Collette, pers. obs.).

Early Life History. Eggs are pelagic and spherical, with a sculptured surface, unsegmented yolk, and a single oil globule (Wenner 1983a). They have a mean diameter of 0.76 mm (0.73–0.79 mm), a perivitelline space of 0.06 mm (0.02–0.10 mm), and an oil globule 0.24 mm (0.17–0.29 mm) (Markle and Frost 1985: Table 3). No figures of the eggs are available.

Longfin hake closely resemble red and white hakes (Methven 1985), and the way to distinguish these three is discussed under white hake early life history. Methven believed that lack of temporal spines and three (instead of four) pelvic fin rays may mean longfin hake belong in *Urophycis* rather than *Phycis.* Because of the difficulty in differentiating the three species, there is little information on the larvae and juveniles of longfin hake. Wenner (1983a) first collected what he believed to be longfin hake juveniles, 3–7 cm SL, in a trawl in June. Another size-class of 8–15 cm SL was present at this time, and he thought that these were 18-month-old fish. Bigelow and Schroeder reported that they took pelagic young 8–35 mm long off Martha's Vineyard in late August, 57–71 mm in April, and 74–110 mm in July, but these may actually have been white hake. Pelagic juveniles were collected from the Grand Banks and Labrador Shelf from February to June (Methven and McKelvie 1986). The smallest collected was 4.5 mm and the largest 62.9 mm. Growth was estimated at 6.5 mm·month^{-1}. Pelagic specimens were most abundant along the edge of the continental shelf, but a few were collected 500 km from the shelf in the Gulf Stream. The smallest demersal juvenile collected was 136 mm, indicating that they begin descending from the water column by the time they reach this size.

Age and Growth. Aging by otolith ring counts was attempted by Wenner (1983a) but proved to be impossible as the rings were variable and often unclear, and counts of the same section could not be duplicated.

General Range. Longfin hake are deep-water fish, occurring in great abundance on the outer continental shelf and slope off North America from Carson Canyon on the Grand Banks to the Labrador Shelf (Methven and McKelvie 1986) and south to the Straits of Florida (Cohen et al. 1990: Fig. 126).

Occurrence in the Gulf of Maine. Longfin hake are plentiful all along the seaward slopes of Browns Bank, of Georges Bank, and of Nantucket Shoals at depths greater than 180 m; on the slope between the south-central part of Georges Bank; and off the eastern end of Long Island, N.Y. Up to 1931, the only definite records from the inner parts of the Gulf Maine were three specimens taken off Cape Ann in 200–256 m in 1878 and a few others that were trawled on the northern edge of Georges Bank in 155–180 m. Captures of a number of longfin hake to the west along Georges Bank and the central basin of the Gulf in 1931 at depths of 128–256 m showed they are more numerous in the deeper parts of the Gulf than had been previously suspected.

Importance. An abundant but soft-bodied fish that has not been fished commercially but might support a fishery (Cohen et al. 1990).

POLLOCK / *Pollachius virens* (Linnaeus 1758) / American Pollock, Coalfish, or Saith (in Great Britain) /

Bigelow and Schroeder 1953:213–221

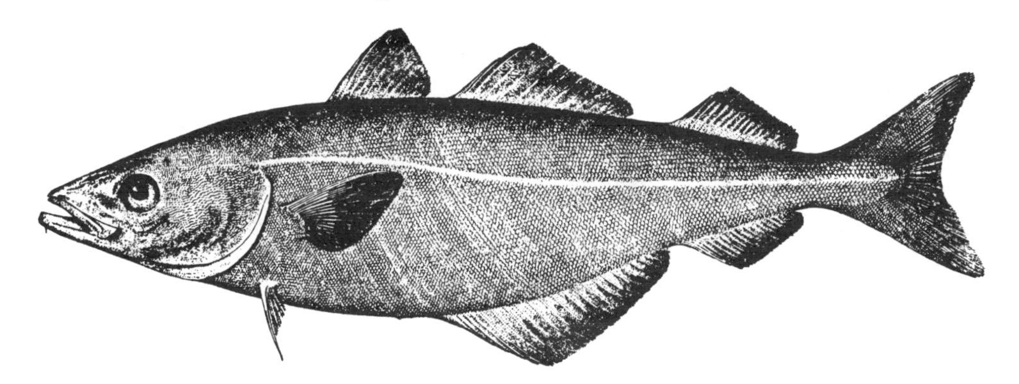

Figure 133. Pollock *Pollachius virens.* Eastport, Maine. Drawn by H. L. Todd.

Description. Body deep and plump (about 4.5 times as long as deep), tapering to a pointed snout and to a slender caudal peduncle (Fig. 133). Mouth moderate, teeth present on jaws and vomer, small, equal, pointed, and cardiform. Lower jaw projecting, with a small chin barbel, which may be absent in large fish. First dorsal fin originates slightly behind pectoral fins, triangular, and highest of three dorsals. Second dorsal fin also triangular, longest of dorsal fins, and separated by a considerable space from third dorsal, which is more rhomboid in outline. Second anal fin corresponds in shape and size to third dorsal, under which it originates, but first anal fin considerably longer than second dorsal though similar in shape. Pelvic fins small, located a little in front of or directly under pectoral on sides, longer than first dorsal fin, but shorter than second dorsal; lower corners rounded and tips bluntly pointed. Caudal fin noticeably forked, with angular corners. Lateral line almost straight.

Meristics. First dorsal fin rays 13 or 14; second dorsal fin rays 21 or 22; third dorsal fin rays 19 or 20; first anal fin rays 24–28; second anal fin rays 20 or 21; pectoral fin rays 19–22; superior procurrent caudal fin rays 30–32 + inferior 31–33 = 66–70 total caudal fin rays; lateral line scales 154–156; gill rakers on the first gill arch 35–40; precaudal vertebrae 23–25 + caudal vertebrae 29–32 = total vertebrae 53–57 (Hildebrand and Schroeder 1928; Jensen 1948; Svetovidov 1962; Hardy 1978a; Markle

Color. Pollock are a greenish hue, usually deep rich olive green above, paling to yellowish or smoky gray on the belly. The lateral line is white or very pale gray, contrasting strongly with the dark sides. The dorsal, caudal, pectoral, and anal fins are olive, the latter pale at the base. The pelvic fins are white with a reddish tinge. Young fish are darker than older ones, and many are more tinged with yellow on their sides.

Size. Pollock reach nearly 130 cm TL and are common from 30 to 110 cm (Cohen et al. 1990). The largest pollock taken on 54 NEFSC bottom trawl surveys from 1963 to 1986 was recorded as 160 cm TL, taken on *Albatross IV* cruise 77-07, station 279 at 42°09′ N, 67°27′ W. The next largest was only 120 cm, taken on cruise 73-08, station 212 at 42°41′ N, 67° 50′ W, so confirmation of the 160-cm fish is needed. Maximum size at other stations was 95–110 cm. The all-tackle game fish record is a 22.7-kg fish from Salstraumen, Norway, caught in November 1995 (IGFA 2001) displacing a 21.14-kg fish caught in Perkins Cove, Ogunquit, Maine, in October 1990.

Distinctions. Pollock are distinguished from cod, tomcod, and haddock by their forked tail, projecting lower jaw, and greenish color without spots. Pollock have more caudal fin rays (total 66–70) than other western Atlantic gadids (27–60) (Markle 1982: Table 6).

Habits. Pollock are active fish, living at any level between bottom and surface depending on the food supply and the season, often schooling and sometimes gathering in bodies so large that Bigelow and Schroeder reported that a purse seiner once took 60,000 fish from one school. In the Gulf of Maine, their depth range is from the surface down to 280 m at temperatures of 5°–8°C. They may descend somewhat deeper in the deepest troughs in the Bay of Fundy and occur from 35 to 380 m on the Scotian Shelf at temperatures of 7.2°–8.6°C (Scott 1982a). Pollock are found at the edges of shoals and banks and in tidal rips in the Bay of Fundy (Steele 1963). In the Gulf of Maine they may occur close inshore and on offshore banks. They are relatively unselective of bottom type and on the Scotian Shelf are associated with sediments ranging from gravel to clay (Scott 1982a).

Pollock are cool-water fish; minimum temperatures recorded on NEFSC survey cruises were 3°–7°C, maximum temperatures 6°–14°C. Large pollock are not caught at the surface when the temperature is higher than about 11.1°C, although there may be plenty of them deeper down where the water is cooler. Even the little "harbor pollock" of 2 cm or so do not appear in any great numbers at times or places where the water is warmer than 15.5°C. At the other extreme, pollock of all sizes must experience temperatures as low as 0°C on the fishing grounds in the southern side of the Gulf of St. Lawrence and on the more easterly of the Nova Scotian banks during the late winter or early spring, unless they descend to considerable depths. Collections on the Scotian Shelf and Gulf of St. Lawrence were successful only on the slopes and channels where temperatures were above 1.1°C (Steele 1963).

Another factor influencing distribution is size. Offshore pollock have a regular size gradient across the Bay of Fundy (Steele 1963). Large fish (65–85 cm) are caught on the New Brunswick side, medium-sized fish (60–75 cm) around Grand Manan, and smaller fish (45–60 cm) off western Nova Scotia. Segregation by size into schools is an important feature of pollock behavior.

Pollock produce sounds (Fish and Mowbray 1970) by means of a multilobed swim bladder vibrated by contractions of associated muscles, as well as general body contractions. Pollock produce thumps during handling and escape. Their hearing range is similar to that of cod.

Food. The first food of larval pollock at Woods Hole is phytoplankton and copepod nauplii (Marak 1960). Older larvae eat copepods in various life-history stages, amphipods, cumaceans, isopods, and larval fishes. The most numerous organisms in the diet are the copepods *Centropages* and *Tortanus.* Larger larvae eat larger organisms; from 4 to 13 mm larvae eat mostly larval copepods, from 13 to 23 mm, small copepods, and over 25 mm, amphipods and euphausiids.

Juvenile pollock consume large quantities of euphausiids (primarily *Meganyctiphanes norvegica*), which are a staple food. Small quantities of amphipods, decapod larvae, isopods, copepods, polychaetes, and fishes have also been found in their stomachs (Bowman 1981; Bowman et al. 1987). Juvenile pollock collected in the intertidal zone near Pemaquid Point, Maine, fed on algae; crustaceans, especially amphipods; fishes, especially herring; and insects (Ojeda and Dearborn 1991).

Adult pollock feed chiefly on pelagic crustaceans, especially large euphausiids, small fishes, and squids. Examination of stomachs of 82 pollock (Bowman et al. 2000) showed three main categories of food: fishes (54.1% by weight), cephalopods (23.5%), and crustaceans (20.9%), mostly euphausiids. Below 31 cm TL, chaetognaths are the major food category, followed by euphausiids and amphipods. Between 31 and 60 cm, euphausiids are the most important. At 61 cm and larger, fishes and cephalopods become much more important. Pollock eat great quantities of small herring, cod, haddock, hake, silver hake, sand lance, and other small fishes in the Gulf of Maine. Pollock chasing schools of herring are a familiar sight; pollock up to 2 kg commonly run up estuaries in pursuit of smelt in autumn. Sometimes they consume considerable quantities of ctenophores. They have been reported as gorging themselves on herring spawn. As far as is known they never take shelled mollusks. Experiments on fish in captivity at Woods Hole (Herrick 1904) have shown that pollock capture food more by sight than by scent.

The major prey of pollock on Georges Bank and in western Nova Scotia is crustaceans, principally the euphausiid, *Meganyctiphanes norvegica,* which contributed 46% by weight to the diet (Langton and Bowman 1980). Other euphausiids, pandalid shrimps, and miscellaneous decapods are also eaten. Fishes account for the bulk of the remaining food (22.8% Georges Bank, 37.1% Nova Scotia). In the Gulf of Maine, fishes eaten include silver hake, wolffish, and rock gunnel; on Georges Bank, snake eel, lanternfish, pearlside, silver hake, pollock, and sand lance; in Nova Scotia, blueback herring, haddock, silver hake, redfish, and sand lance. Squids such as *Loligo* and *Illex* are taken as prey when available (Bowman and Michaels 1984).

In the Gulf of Maine, the principal component of the diet is fishes (64.9%). Atlantic herring make up the bulk of the fish prey, along with smaller quantities of pollock, redfish, and silver hake. Crustaceans are the second most important food (33.8%). Decapod shrimp (*Pasiphaea*) and euphausiids account for most of the crustacean prey. Diets of Atlantic cod, pollock, silver hake, offshore hake, cusk, red hake, and spotted hake were very similar. On Georges Bank there was a high degree of similarity between the diets of pollock and longfin hake.

Predators. Predators include spiny dogfish and goosefish (Rountree 1999), lobster (Ojeda and Dearborn 1991), minke whale in northern Norway (Nordøy and Blix 1992), gray seal in Canada (Benoit and Bowen 1990a), and harbor seal (Bowen and Harrison 1996). Pollock are also cannibalistic (Rountree 1999).

Parasites. Pollock are host to a protozoan, *Haemogregarina aeglefini;* 2 trematodes, *Derogenes varicus* and *Diclidophora denticulata;* a nematode, *Anisakis* sp.; an acanthocephalan, *Echinorynchus gadi* (Margolis and Arthur 1979); 15 helminths (Scott 1985), and 4 copepods, *Caligus curtus, C. elongatus, Clavella aduna,* and *Clavella* sp. (Margolis and Kabata 1988). Many larval pollock are parasitized by *Caligus* sp., which are mostly found on the opercle and dorsal fin. The parasite does not appear to hamper the larva's ability to feed (Neilson et al. 1987).

Breeding Habits. Median length at maturity for female and male pollock from the Gulf of Maine were 39.1 and 41.8 cm, respectively (O'Brien et al. 1993). Median age at maturity was 2.0 years for females and 2.3 years for males. These values are lower than median values presented by earlier workers: 47–48 cm for females and 50.5 cm for males, 3.2 and 3.5 years for U.S. and Canadian samples (Mayo et al. 1989) and 62.5 and 58 cm, 5–6 and 4–5 years for the Bay of Fundy (Steele 1963). Average fecundity of a female pollock is 225,000 eggs, but more than 4,000,000 eggs were reported in a 10.7-kg fish by Bigelow and Schroeder.

Pollock eggs were collected by MARMAP surveys from October to June and were most abundant during winter (Berrien and Sibunka 1999: Fig. 40). Spawning began in October at scattered locations on Georges Bank and in western and eastern portions of the Gulf of Maine. During November spawning increased and became more widespread. At peak spawning during December and January, egg densities were concentrated in western Gulf of Maine and Georges Bank. Eggs were collected in the Gulf of Maine each month from October to June. Egg distribution continued to expand toward the south-

A major spawning area exists in the western Gulf of Maine and several others have been identified on the Scotian Shelf (Bolz and Lough 1983; O'Boyle et al. 1984; Mayo 1998b). Pollock spawn in great numbers at the mouth of Massachusetts Bay, especially on broken bottom southeast of Gloucester and along the seaward (eastern) slope of Stellwagen Bank. On the Massachusetts Bay grounds, breeding commences when the whole water column has cooled to about 8.3°C and is at its climax (late in December) at temperatures of 4.4°–6.1°C, while major production of eggs takes place long before the water has cooled to its winter minimum of 1.7°–2.2°C at the level where the fish lie. Thus pollock spawn on a falling temperature, with most of the eggs produced within a comparatively narrow range. Jeffreys Ledge is another major spawning area within the Gulf. Important spawning areas occur on the Northwest Peak of Georges Bank and in the Great South Channel (Mayo et al. 1989).

Pollock larvae were collected for the first time in the Bay of Fundy in 1979 (Scott 1980). The numerical distribution and length composition of catches indicated that the larvae originated outside the Bay and dispersed from the south toward the inner part of the Bay. The population in the Bay of Fundy is believed to derive from spawning grounds in the southern Gulf of Maine and the Scotian Shelf.

Early Life History. The egg is buoyant, spherical, and transparent, no oil globule, a narrow perivitelline space, 0.05–0.17 mm, and a homogeneous yolk. Egg size of 1.04–1.20 mm in diameter is smaller than the eggs of cod and haddock (1.20–1.69 and 1.32–1.60) (Markle and Frost 1985: Table 3). Incubation takes 9 days at a temperature of 6.1°C, 6 days at 9.4°C. Times to hatch at different temperatures are summarized in Hardy (1978a). Incubation proceeds normally, and the resultant larvae are strong and active over a temperature range of 3.3°–8.9°C. Massachusetts Bay spawning takes place in salinities ranging from 32–32.8 ppt depending on precise locality, depth, and season.

The larvae are 3.4–3.8 mm long at hatch, slender, with a large yolk sac, and with the vent situated on one side of the body at the base of the ventral finfold. They are sprinkled with black pigment cells. After about 5 days, the yolk sac is absorbed and the mouth forms; in the meantime pigment of the postanal section of the trunk becomes grouped in longitudinal bars, two dorsal and two ventral, the former longer than the latter. Pollock have more caudal fin rays (total 66–70) than other western Atlantic gadids (27–60) (Markle 1982: Table 6). Caudal fin rays appear at about 9 mm; all dorsal and anal rays and pelvic fin rays at about 15 mm. The dorsal and anal fins are separate from one another at 20 mm. Larvae of 25–30 mm show most of the characters of adults.

Juveniles larger than 50 mm develop small rudimentary chin barbels. Smaller fish are darker than larger ones, sometimes with a yellow tinge on the sides. Juvenile pollock (age 1 or 1+) have been taken in tow nets near the surface at Woods

and weighing less than 1 kg swarm inshore after early April, when thousands were taken from traps at Gloucester and Magnolia. In the southern part of Massachusetts Bay these "harbor pollock," as they are called locally, move out in June, probably to avoid rising temperatures, to return again in autumn. Immature pollock are distributed in the northern Gulf of Maine and between Cape Cod and Jeffreys Ledge at depths less than 90 m in spring and autumn. Juveniles also appear along Northeast Peak of Georges Bank in spring (Wigley and Gabriel 1991). They continue to be abundant all summer and autumn in the harbors and bays and among the islands along the coast northward from Cape Ann and eastward to Nova Scotia. Most of them seek slightly deeper water in winter, probably to avoid the cold.

Patterns of distribution and foraging by young-of-the-year pollock in the rocky intertidal zone were investigated (Rangeley and Kramer 1995). Pollock were sampled by beach seine in fucoid macroalgae and in open habitats at all stages of the tide, day and night, throughout the summer. Their presence in shallow water at high tidal stages indicated that at least part of the pollock population migrated across the full width of the intertidal zone (150 m) each tide. Densities in shallow water were much higher at low than at high tidal stages suggesting that a large influx of pollock moved in from the subtidal zone at low tidal stages and them dispersed into intertidal habitats at high tidal stages. There were few differences in pollock densities between algal and open habitats but abundances probably increased in the algal habitat at higher tidal stages when changes in habitat availability are taken into account. Densities were higher at night and there was an order of magnitude decline in pollock densities from early to late summer. Tidal migrations of juvenile pollock observed in this study and their use of macroalgae as forage and possibly a refuge habitat strongly suggests that the rocky intertidal zone may be an important fish nursery area. Other habitat preferences are summarized in Cargnelli et al. (1999c).

Further investigations on the algal habitat as a refuge from predators showed, as previously suggested, that juvenile pollock school in open water at low tide and disperse in beds of intertidal algae at high tide (Rangeley and Kramer 1998). In large manipulated arenas, pollock preferred the algal habitat, and the proportion of fish in the algae increased as the amount of algal habitat increased but the proportion decreased (i.e., they dispersed within the algae). Following exposure to a predator model, a stuffed cormorant, the proportion of fish in the algae increased further but fish continued to be dispersed. Fish in the open were aggregated most of the time, but there was a trend toward increased aggregation of those fish remaining in the open following exposure to the predator. Thus, the use of two alternative antipredator tactics can produce very dynamic spatial distributions. Because benefits of aggregations are positively density-dependent whereas those of refuges are likely negatively density-dependent, species that use both tactics are likely to show dramatic shifts in habitat distribution with changes in population size and refuge availability.

Young harbor pollock were rare or absent in the Gulf of St. Lawrence and on the west, north, and east coast of Newfoundland. Based on their size and the current patterns young pollock found at the mouth of the Bay of Fundy appear to have come from the Scotian Shelf or Massachusetts Bay.

Fish of age-class 0 move inshore in June near New Brunswick, remain for the summer, and then winter offshore in large pelagic schools. Not much is known about 2-year-olds. Based on catch records, some appear to be inshore and some offshore in 130–150 m (Steele 1963).

Age and Growth. Age and growth measurements for pollock in the Gulf of Maine were done by Bigelow and Schroeder using scale annuli. American fish appear to grow a bit larger than European fish of the corresponding age, so size was extrapolated for older fish, which Bigelow and Schroeder were unable to collect. Growth of pollock in the Gulf of Maine (chiefly Georges Bank) using scale increments and fork lengths (Hoberman and Jensen 1962) was very similar to that noted by Bigelow and Schroeder except that the latter found that pollock grew faster from ages 3–5. Age-growth from Georges Bank–Gulf of Maine is presented in Fig. 134. This growth is similar to European fish but Rojo's (1955) data on fish from the Grand Banks indicated they grew faster than other pollock. Growth of pollock from the Bay of Fundy (Steele 1963) is similar to other areas for immature and young mature fish but slower for older fish, perhaps owing to biased sampling of older fish. Results from U.S. and Canadian data sets from 1970 to 1984 were very similar. No differences were apparent between growth of males and females in those years (Mayo et al. 1989).

The maximum recorded age collected during NEFSC bottom trawl surveys from 1970 to 1988 was 19 years (Penttila et al. 1989), which is also the maximum recorded for European fish (Damas 1909).

Growth of small pollock in the Bay of Fundy based on otolith analysis was compared (Steele 1963) with growth in the Barents Sea (Mironova 1957) and western Norway (Lie 1961). Barents Sea fish grew more slowly, but this may be due to differences in spawning time.

Daily growth increments on otoliths were used to estimate the age of larval and juvenile pollock collected on Emerald and Sable Island banks, eastern Canada, between March 1991 and May 1993 (Quiñonez-Velázquez 1999). Daily periodicity of increments was validated from observations of reared larvae. The first increment was deposited the day after hatching and thereafter daily increments were added. A Laird-Gompertz growth curve fitted to length-age data showed that growth rates varied significantly in different years. The 1993 cohort had the highest growth rate. Average growth rate was 0.18 mm·day^{-1} for the first month and 0.23 mm·day^{-1} for the second month. Growth continued exponentially after transition from a primarily pelagic life to a predominantly demersal one, which occurred at an age of about 40–50 days. No indication of a cessation in growth was observed. Analysis of

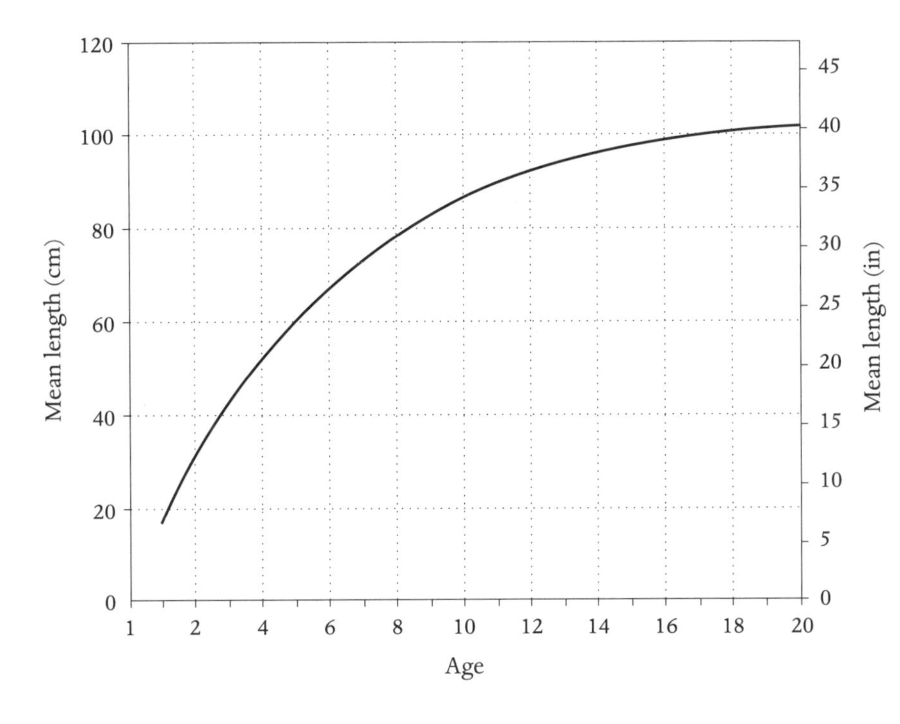

Figure 134. Length-age relationship in pollock *Pollachius virens,* Gulf of Maine–Georges Bank stock. (From Penttila et al. 1989:19.)

length-age data indicated that accelerated growth of juveniles after 50 days of age could have reflected exploitation of a more abundant food resource after settlement. Thus, pelagic and early demersal growth appear to represent distinct stanzas in the growth history of pollock. The hatch-date–frequency distributions (HFD) of pollock larvae sampled in 1992 and 1993 (Fortier and Quiñonez-Velásquez 1998) were reconstructed for different age intervals (0–20, 21–40, 41–80 days). The ratio of the HFD at a given age to the HFD at an earlier age was used as an index of the relative survival of larvae grouped into 5-day hatch-date cohorts. Seasonal variations in relative survival of the cohorts over the age intervals 21–40 and 41–80 days were correlated with strong variations in growth, with slow growth resulting in low survival but fast growth resulting in either low or high survival, indicating that fast growth is a necessary but not sufficient condition for survival.

General Range. Pollock live in continental waters on both sides of the North Atlantic in cool-temperate and boreal latitudes. They are found from Hudson and Davis straits to North Carolina, although rare at the extremes of the range (Cohen et al. 1990: Fig. 134). In the eastern Atlantic, they occur from the Bay of Biscay to the Barents Sea, around Iceland, and southwest Greenland.

Occurrence in the Gulf of Maine. In the western Atlantic, pollock have their center of abundance on the southwest Scotian Shelf and in the Gulf of Maine (Map 18), where they are caught in large numbers both on offshore banks, especially Georges Bank, and all around the coastline, from Nantucket Shoals and Cape Cod to Cape Sable. The only regional exception is the New Brunswick shore, where so few pollock are taken that they do not appear in the landings. They are plentiful at the mouth of the Bay along the outer Nova Scotian coast and banks and the Atlantic coast of Cape Breton Island east of the Gut of Canso.

Migrations. Tagging studies suggest considerable movement of pollock between the Scotian Shelf and Georges Bank but less between the Scotian Shelf and the Gulf of Maine (Mayo et al. 1989, 1998b); there are also onshore-offshore migrations in response to temperature changes and north-south movements for spawning purposes. From December through March, they concentrate in areas less than 100 m in the Gulf of Maine from the Great South Channel to Jeffreys Ledge. Adults move to deeper (100–200 m) offshore areas of the Gulf of Maine during spring and summer. Juvenile harbor pollock reside in shallow coastal waters during their first summer and move offshore during the following winter (Steele 1963). On the Scotian Shelf, the highest concentrations of adults have traditionally occurred between Browns and Emerald banks and off western Nova Scotia south of the Bay of Fundy. Since 1986, significant numbers of relatively large pollock have been observed on the Scotian Shelf north of the Gully (Mayo et al. 1989). Many that were tagged by the U.S. Bureau of Fisheries were recaptured within short distances of the localities where they were marked after long periods of time. A few of the marked fish are known to have made considerable journeys eastward: one, for example, from Jeffreys Ledge to Sable Island (Bigelow and Schroeder). Commercial catches in the Gulf of Maine suggest that fish caught in the northern Gulf in the summer migrate south to spawn in the southern Gulf in winter. Tagging studies conducted on Bay of Fundy pollock in July 1960 (Steele 1963) showed that they remain separate from pol-

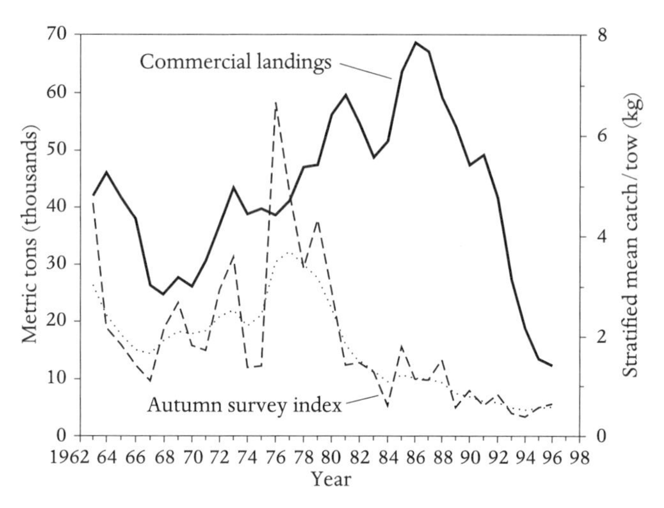

Figure 135. Catch and index of abundance of pollock *Pollachius virens* from the Scotian Shelf, Gulf of Maine, and Georges Bank. (From Mayo 1998b.)

summer but migrate south to spawn in the winter and probably mix with the other pollock.

Importance. Traditionally, pollock were taken as by-catch in the demersal otter trawl fishery, but directed otter trawl effort increased steadily during the 1980s, peaking in 1986 and 1987 (Mayo 1998b). Directed effort by Canadian and U.S. trawlers has since declined substantially and similar trends have occurred in the winter gill net fishery. Nominal commercial catches from the entire Scotian Shelf, Gulf of Maine, and Georges Bank region increased from an annual average of 38,200 mt during 1972–1976 to 68,800 mt in 1986 (Mayo 1998b) (Fig. 135). U.S. catches peaked at 24,500 mt in 1986 and declined to 3.0 mt in 1996. High landings during the 1980s and later years (in excess of 63,000 mt·year^{-1} between 1985 and 1987) resulted in relatively high fishing mortality rates, ranging from 0.62 (42% exploitation rate) to 0.85 (53% exploitation rate) during the late 1980 and early 1990s (Mayo 1998b). Subsequent projections indicate a substantial reduction in fishing mortality owing to the combined effect of reduced catch and effort in the Canadian sector and continued recruitment of the large 1988 and 1989 year classes. Over the full range of the stock, spawning biomass is increasing but the stock is considered fully exploited. Within the Gulf of Maine, however, stock abundance and biomass remain low.

There is significant by-catch discard mortality of pollock in the Gulf of Maine fishery for northern shrimp (Ross and Hokenson 1997). The highest rate of seabird predation occurs when pollock are discarded (65%), and there is also very high handling-induced mortality.

Pollock will take an artificial lure and put up strong resistance. Small ones will take a bright artificial fly freely. A pollock rises so fiercely to the fly and makes so long and strong a run when it is hooked that it gives fully as good sport as a trout. Pollock of all sizes will also bite on clams, minnows, or cut bait fish. Pollock are caught by anglers from party boats out of various ports along the New England coast. Estimated U.S. recreational catches have declined from an average of 800 mt for 1979–1986 to less than 100 mt (Mayo 1998b).

Stocks. Although there are some meristic and morphometric differences between populations on the Scotian Shelf and the western Gulf of Maine, electrophoretic analyses showed no significant differences, and tagging studies suggest considerable mixing of pollock between the Scotian Shelf and Georges Bank, and, to a lesser extent, the Gulf of Maine (Mayo et al. 1989). Therefore, pollock from Cape Breton and south are assessed as a unit stock (Mayo 1998b).

Management. The domestic portion of the fishery is managed under the New England Fishery Management Council's Multispecies Fishery Management Plan (Mayo 1998b). Under this FMP, pollock are included in a complex of ten groundfish species that have been managed by time/area closures, gear restrictions, minimum size limits, and, since 1994, direct effort controls including a moratorium on permits and days-at-sea restrictions under Amendments 5 and 7. The Canadian fishery is managed under fleet-specific quotas.

RED HAKE / *Urophycis chuss* (Walbaum 1792) / Squirrel Hake, Ling / Bigelow and Schroeder 1953:223–230

Description. Body rounded in front of vent, somewhat compressed behind, and tapering uniformly to a slender caudal peduncle (Fig. 136). Head pointed; upper jaw projecting. Teeth on vomer and jaws, upper jaw teeth in two indefinite rows, lower jaw teeth very irregular (Hildebrand and Schroeder 1928). A small barbel on lower jaw. Angle of mouth extends posterior to rear margin of large eye. First dorsal fin triangular, third ray at least twice as long as others. Second dorsal fin long. Anal fin low, uniform in height. Pectoral fins large, reaching third ray of second dorsal. Pelvic fins inserted below and in front of pectorals, reduced to two elongate feelerlike rays, longer ray reaching to or slightly beyond vent. Caudal fin rounded.

Meristics. First dorsal fin rays 9–12; second dorsal fin rays 52–64; anal fin rays 45–57; pectoral fin rays 14–17; total caudal fin rays 29–34; lateral line scales 95–117; gill rakers on upper limb of first arch 3; precaudal vertebrae 14–17 + caudal 33 = total vertebrae 45–50 (Musick 1973a; Markle 1982; Comyns and Grant 1993).

Color. Red hake are reddish, muddy, or olive brown on sides and back, darkest above; sometimes almost black, sometimes more or less mottled, and sometimes plain, with a pale lateral line. The lower parts of the sides are usually washed with yellow and sometimes marked with dusky dots. Belly and lower

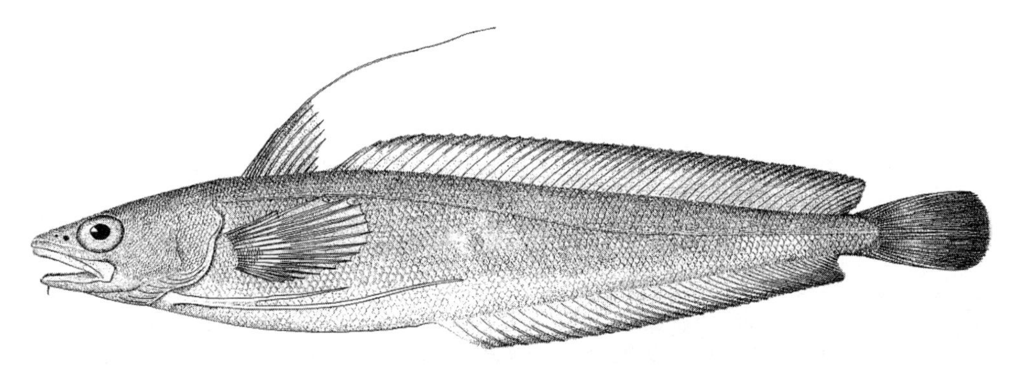

Figure 136. Red hake *Urophycis chuss.* Off Martha's Vineyard. Drawn by H. L. Todd.

parts of the sides of the head are pure white, gray, or yellowish; the dorsal, caudal, and anal fins are the same color as the back except the anal is pale at the base. The pelvic fins are very pale pink or yellow.

Size. Red hake do not grow as large as white hake, normally reaching a maximum of 50 cm and a weight of 2 kg (Musick 1967). Records of larger fish are probably white hake. Females are both longer and heavier than males of the same age. The all-tackle game fish record is a 3.6-kg fish taken in Mud Hole, N.J., in March 1994 (IGFA 2001).

Distinctions. Red hake closely resemble white hake and differ primarily in having fewer lateral line scales (95–117 vs. 119–148) and three gill rakers on the upper gill arch instead of only two (Musick 1973a).

Habits. As adults, red hake are a coastal species found in relatively deep water. The recorded depth range is 35 m (Svetovidov 1962) to 980 m (Bigelow and Schroeder). Average depths at which the greatest concentrations occur are 110–130 m in summer, and 183–457 m in winter off New England; and 36–358 m on the Scotian Shelf, with a preferred depth of 72–124 m (Grosslein and Azarovitz 1982). Red hake haunt soft bottoms, sand and mud, few being caught on gravelly, shelly, or rocky grounds in the Gulf of Maine, but they are reported to be abundant on hard bottoms in temperate reef areas of Maryland and northern Virginia (Eklund and Targett 1990). Adult fish prefer a water temperature range of 5°–12°C (Grosslein and Azarovitz 1982) and a salinity range of 33–34 ppt (Scott 1982a).

Behavioral responses of red hake to a decrease in oxygen concentration were tested because anoxia has become a periodic problem in such areas as Long Island Sound and Chesapeake Bay (Bejda et al. 1987). Of the three age-groups tested, 0, 1, and 2–3+, young-of-the-year (0) were the most sensitive. At oxygen concentrations below 4.2 mg·liter^{-1} they leave the substrate and their scallop shelters and swim in the water column. Age-1 fish showed variable responses indicating a transition stage; and fish age 2–3+ are the least responsive, only increased movement could expose the youngest fish to increased predation as they leave their shelters (see the section on early life history).

Red hake produce sounds (Fish and Mowbray 1970); those produced in an aquarium under electronic stimulation consisted of a single or double series of low thumps and knocks. The mechanism for sound production is the large multilobed swim bladder, which is vibrated by contraction of associated, extrinsic musculature as in other gadids.

The modified pelvic fins, chin barbel, and free filament of the dorsal fin contain chemo- and mechanosensory organs (Bardach and Case 1965) that function in food gathering (Herrick 1904). Hake swim in midwater with the pelvic fins drawn close to the body, but lower them upon approaching the bottom (Bardach and Case 1965) and swim along the bottom with the pelvic fins spread apart to an angle of 45°. The entire fin is swept from the snout back toward the flank. Commonly, when the barbel touches a morsel of food, the hake backs up and ingests it. When a hake nears a piece of clam, it further extends its pelvic fins and lowers its head so its barbel is in contact with the substrate (Pearson et al. 1980). The head is often moved from side to side. There are three basic feeding behaviors: emerging from shelter, swimming, and extending the pelvic fins to contact the bottom. Variation in the strength of the chemical food cue determined the onset and duration of food searching and feeding behavior. Detection is accomplished through olfactory and tactile cues, and the final process of seizing the food is visual (Herrick 1904).

Food. Juvenile red hake from the outer continental shelf in the Mid-Atlantic Bight feed primarily on amphipods, especially in the fall when they account for almost all of their food (Sedberry 1983). Decapods and polychaetes are also dominant prey. Copepods are important in fall and winter; arrowworms are eaten in spring and summer. There are two distinct size ranges: 49–147 mm in fall and summer and 238–251 mm in winter and spring. These are undoubtedly juveniles and young adults. Fish over 300 mm change their feeding strategy to larger and fewer prey items. The smallest red hake (about 5 cm) eat large quantities of chaetognaths, along with smaller

Bowman et al. 1987). Hake 6–10 cm ingest mostly decapod shrimps such as *Dichelopandalus* and *Crangon,* but amphipods, polychaetes, and chaetognaths are also preferred food. When the fish are 11–20 cm long, they feed on the same basic food types as 6–10 cm fish, but they include euphausiids (mostly *Meganyctiphanes*) as a staple food.

In the laboratory, juvenile hake are discontinuous feeders that consume as much food as possible in a short period of time (Luczovich and Olla 1983). Laboratory-held hake ate 7.4% of their body weight per day; and the specific growth rate was 0.94% TL per day. In the field, it was 0.93% TL per day. Growth rate is inhibited in the laboratory at low feeding levels.

Examination of stomach contents of 1,482 red hake (Bowman et al. 2000) shows crustaceans to be the major component of the diet (63.3% by weight) followed by fishes (21.4%). Earlier studies also found crustaceans to be their major prey (Langton and Bowman 1980; Bowman and Michaels 1984; Bowman and Grosslein 1988). The most important crustaceans are euphausiids, mostly *Meganyctiphanes norvegica,* and decapods, such as *Dichelopandalus leptocerus, Crangon septemspinosa,* and rock crab, *Cancer irroratus.* Other invertebrate prey includes bivalves, squids, polychaetes, amphipods, galatheid crabs, and munid crabs. Fish prey include haddock, silver hake, sea robins, sand lance, and mackerel (Bowman et al. 2000). Red hake are cannibalistic on their young. Mollusks make up about 6% of the diet, but most were not identifiable except for squids and *Buccinum.* Polychaetes are a minor dietary component (2.9%). In all five geographic regions sampled (Middle Atlantic, southern New England, Georges Bank, Gulf of Maine, and western Nova Scotia), crustaceans make up at least 50% of the diet. More crustaceans (88.5%), particularly pandalid shrimps, are found in the diet of fish from western Nova Scotia than from other areas. Gastropods are proportionately most important in the diet of fish from Georges Bank (17.7%). The most important prey of 681 red hake trawled in the Gulf of Maine were euphausiids (*Meganyctiphanes*), pandalid shrimps (*Dichelopandalus* and *Pandalus*), haddock, and silver hake (Bowman et al. 2000).

Predators. Red hake were found in the stomachs of 15 species of fishes in the NEFSC food habits survey, of which four species had more than ten occurrences: spiny dogfish, cod, goosefish, and silver hake, in order of frequency of occurrence (Rountree 1999). Other fish predators included skates (little, winter, and thorny), hakes (red and white), sea raven, longhorn sculpin, and bluefish.

Parasites. Red hake are parasitized by a protozoan, *Haemogregarina aeglefina;* a monogene, *Diclidophoia maccallumi;* a digene, *Podocotyle simplex,* and 2 nematodes, *Anisakis* sp. and *Hysterothylacium aduncum* (Margolis and Arthur 1979). Scotian Shelf red hake contained 19 species of helminths in their guts (Scott 1987): 12 digenes, 2 cestodes, 3 nematodes, and 2 acanthocephalans. Red hake 20–49 cm long were heavily parasitized by a digene, *Derogenes varicus,* and 2 nematodes, *Hysterothylacium aduncum* and *Spinitectus cristalus.*

Breeding Habits. Median length at maturity for female and male red hake from the Gulf of Maine–northern Georges Bank region was 26.9 and 22.2 cm, respectively (O'Brien et al. 1993). Median age at maturity was 1.8 years for females and 1.4 years for males. There is no information on fecundity.

Spawning is prolonged, from May through November, with major spawning areas located on the southwest part of Georges Bank and in the southern New England area south of Montauk Point, Long Island (Colton and Temple 1961; Colton and Marak 1969; Sosebee 1998c). The southernmost record for a ripe female is Chesapeake Bay (Hildebrand and Schroeder 1928). Hake larvae occur as far south as Cape Hatteras (Morse et al. 1987) but species were not distinguished. Red hake spawn at depths shallower than 47–108 m at temperatures of 5°–10°C (Hardy 1978a). Larvae have been reported from the coast of Maine in June (Graham and Boyar 1965) but some of their records may have been misidentifications of white hake (M. Fahay, pers. comm.).

Early Life History. The eggs are pelagic, buoyant, spherical, and transparent (Miller and Marak 1959). The diameter is 0.64–0.78 mm, perivitelline space 0.02–0.10 mm, oil globule 0.13–0.27 (Markle and Frost 1985: Table 3). There is one large oil globule with several smaller ones. Incubation takes 30 h at 21°C.

Pigment patterns of the egg are distinctive; one of the most characteristic features is development of black chromatophores on the embryo, yolk, and oil globule. In late stages of incubation, this feature, combined with the small size of the egg and multiple oil globules, distinguishes eggs of red hake from other species in the Gulf of Maine except for rockling eggs, which may have a pigmented oil globule.

Eggs are most numerous in May and June in the New York Bight and off Georges Bank (Colton and St. Onge 1974). Larvae are most numerous in September and October on Georges Bank and in the New York Bight, but are present from May to December. Because few larvae have been collected in the Gulf of Maine, it has been suggested that spawning further south supplies most of the recruits to the Gulf of Maine (Steimle et al. 1999c).

First cleavage occurred in 1.5 h at 15.6°C; by 50 h eyes were evident, the embryo was halfway around the yolk and there were melanophores on the body and yolk; by 74 h the oil globule was pigmented and the larvae began to move; hatching occurred between 96 and 98 h (Hildebrand and Cable 1938). Length at hatching is 1.76–2.2 mm. The larva has a large yolk mass extending far forward under the head; the oil globule is in the posterior part of the yolk sac. The mouth is not yet developed, and fins are present only as finfolds and a small pectoral fin bud. The anus is located laterally and at the base of the finfold. Lateral line sensory organs are present as delicate, transparent membranous extensions from the body. Elongate pelvic fins are noticeable at 5 mm and are black at the tips. The chin barbel becomes evident at 15 mm and scales form at 25 mm.

Methven (1985) distinguished red, white, and longfin hake larvae. Red hake larvae differ from spotted hake by the closely crowded pigment blotches on the head. Spotted hake have two widely separated blotches on the crown and the snout. Older spotted hake larvae lack pigment on the pelvic fins.

Red hake spend their first months drifting at or near the surface, and larvae 12.7–100 mm are often taken in summer under floating eelgrass or rockweed. On calm days they have been seen darting to and fro on the surface. Larvae become demersal at a length of 25–30 mm (Methven 1985). In New Jersey waters, this occurs from August through December. Most descend after the autumn breakdown of the thermocline (Steiner and Olla 1985).

Following descent from the plankton, red hake juveniles are commonly found within mantle cavities of sea scallops, *Placopecten magellanicus*. They appear to maintain this association until they reach a maximum size of 136 mm (Musick 1969). In the laboratory and in the field, red hake live in symbiotic association within or underneath sea scallops, sheltering from predators (Steiner et al. 1982). It is not clear what benefit, if any, the scallop receives in return. Hake inhabiting scallops ranged in size from 23 to 116 mm. Small scallops (<100 mm) contained mostly small hake (25–65 mm), but large scallops contained a wide size range (26–116 mm). Hake were primarily nocturnal, but there was some variation in this. In laboratory preference tests, hake selected nonliving shelters over living scallops.

Recruitment of hake from the plankton to the benthos occurred from September to December, with the highest rates from October to November. Most left the scallop beds by February, but a few late recruits stayed until May. Hake are believed to leave the scallops in response to decreasing temperatures and their increasing size, and they then move to deeper, warmer waters offshore.

Hake appear to associate with scallops whenever they are available, but are also associated with other objects, and structure appears to be critical to their survival (Able and Fahay 1998). A small hake was found curled inside a moon snail egg collar, and other hake were curled around or sheltered under surf clams (Ogren et al. 1968). Larger hake have been seen hiding under shells, sponges, and rocks at depths of 40–50 m (Edwards and Emery 1968).

Juveniles have been recorded at salinities of 31–32.8 ppt and temperatures of 4.2°–7.5°C (Richards and Castagna 1970). Steiner and Olla (1985) exposed prejuveniles to a thermocline in the laboratory as they were just reaching the size at which they would be expected to descend from the plankton. Water temperatures in the experimental thermocline ranged from 20°C at the top to 10°C at the bottom. The fish did not descend through the thermocline immediately, but remained in the water column above 15°C. Those in a tank without the thermocline (20°C) went to the bottom immediately. Those in the thermocline began to descend and then went back up as that the fish must undergo an acclimation period while descending through thermoclines to the bottom.

Juvenile hake that have migrated offshore return with adults in April and become mature in summer (J. A. Musick 1974). Juveniles of several year-classes, 50–225 mm long, in Chesapeake Bay in late fall and spring move to offshore waters by the end of June (Hildebrand and Cable 1938).

Red and white hakes have quite different life-history strategies (Markle et al. 1982). Red hake avoid predation by associating with scallops, concentrate growth in the juvenile stage, and mature early. White hake achieve large sizes quickly by delaying maturity and concentrating efforts on body growth.

Age and Growth. Maximum age of red hake taken by NEFSC survey cruises from 1970 to 1985 was 14 years (Penttila et al. 1989), but few fish survive beyond 8 years (Sosebee 1998c). A combined age-length curve for red hake from the Gulf of Maine and the Mid-Atlantic is shown in Fig. 137. Age determination using otoliths is easy for fish from southwest Georges Bank to Cape Hatteras but more difficult for fish from northern Georges Bank because they exhibit anomalous patterns (Penttila and Dery 1988).

General Range. Red hake are exclusively North American, occurring in continental waters from North Carolina to Nova Scotia, straying to the Gulf of St. Lawrence (Cohen et al. 1990: Fig. 162).

Occurrence in the Gulf of Maine. Red hake are present throughout the Gulf of Maine, but are particularly abundant in the Great South Channel and on Georges Bank (Map 19).

Migrations. Red hake make seasonal inshore-offshore migrations, which are apparently governed by temperature since they avoid temperatures below 5°C. In New England these movements are generally inshore in April and May and again in October and offshore to the edge of the continental shelf in winter (Musick 1973a).

Importance. Hakes have soft meat and poor keeping qualities, but larger ones find a market for human consumption and smaller ones for animal food. Commercial processing trials have shown that red hake make excellent surimi when prepared without the belly flaps (Lanier 1984), and this should enhance their commercial desirability. Bigelow and Schroeder summarized historical fisheries data and noted that red and white hake landings were not reported separately until after 1944.

Trends in landings of the Gulf of Maine–northern Georges Bank stock have shown three distinct periods (Sosebee 1998c). The first period, from the early 1960s through 1971, was characterized by relatively low landings of 1,000–5,600 mt. The second period, 1972–1976, showed a sharp increase, with landings ranging from 6,300 to 15,300 mt. During this time about 93% of the total annual landings were taken by distant-water

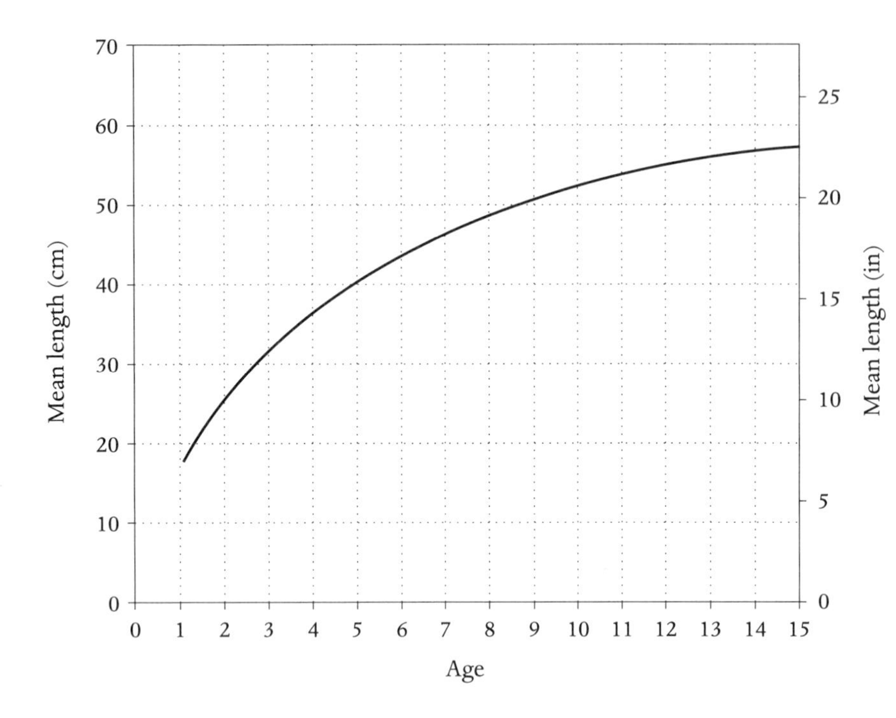

Figure 137. Length-age relationship in red hake *Urophycis chuss,* all areas combined (From Penttila et al. 1989:23.)

of the Magnuson Fisheries Conservation and Management Act in 1977, both total landings and the proportion of landings by the distant-water fleet dropped sharply. From 1977 to the present, annual landings of this stock have averaged 1,100 mt or less (Fig. 138). Otter trawls are the principal commercial fishing gear. The fishery is managed under the New England Fishery Management Council's Multispecies Management Plan under the "nonregulated multispecies" category. Both stocks are underexploited and could support substantially higher catches (Sosebee 1998c).

Hakes are such inactive fish that they are of no special interest to anglers, and recreational catches are insignificant. However, a good many fair-sized ones are caught handlining from party boats for they bite readily, and small hake are caught from small boats in harbors and bays along the coast of Maine.

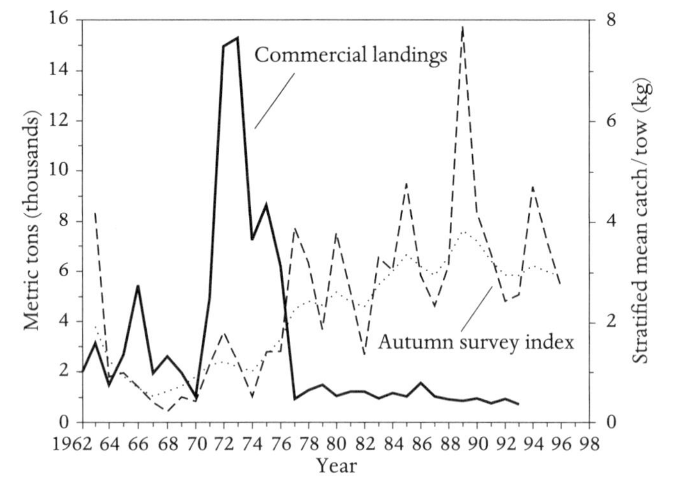

Figure 138. Catch and index of abundance of red hake *Urophycis chuss,* Gulf of Maine–northern Georges Bank stock. (From Sosebee 1998c.)

Stocks. Two stocks of red hake have been assumed, divided north and south in the central Georges Bank region (Sosebee 1998c): Gulf of Maine–northern Georges Bank and southern Georges Bank–Mid-Atlantic.

SPOTTED HAKE / *Urophycis regia* (Walbaum 1792) / Bigelow and Schroeder 1953:230–232 (as *Urophycis regius*)

Description. Body elongate and compressed (Fig. 139). Head scarcely depressed with depth about equal to width, snout blunt, mouth large, its angle extended to behind eye; upper jaw projecting slightly, small barbel on lower jaw. Small teeth on jaws and vomer. First dorsal fin without prolonged rays, about same height or shorter than second dorsal. Second dorsal fin extends to caudal peduncle. Pectoral and pelvic fins extend as far as anal fin origin. Pelvic fins filamentous with two rays.

Meristics. First dorsal fin rays 8–10; second dorsal fin rays 43–52; anal fin rays 41–50; pectoral fin rays 16; total caudal fin rays 30–32; lateral line scales 83–97; gill rakers on upper arm of first arch 3, rarely 2, total 14 or 15; pyloric caeca 18–21, usually in two groups of 9 or 10 on each side of intestine; precaudal vertebrae 13 or 14 + caudal 30–34 = total vertebrae 44–48 (Cohen et al. 1990; Comyns and Grant 1993; B. B. Collette, pers. obs.).

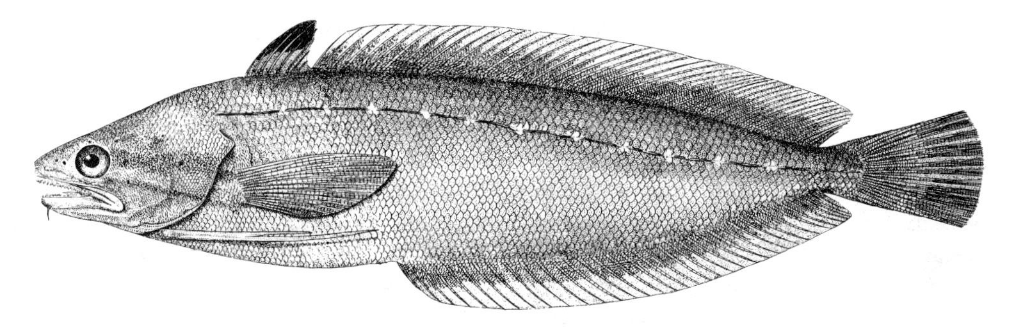

Figure 139. Spotted hake *Urophycis regia.* Drawn by H. L. Todd.

Color. Spotted hake are pale brown above, tinged with yellow; the lateral line is dark brown or black interrupted by conspicuous pale spots; pores of mucous canals on the sides of the head are marked with prominent dark spots. The first dorsal fin is light at the base, black in the middle portion, with white between the light and black areas and on the distal tip; the second dorsal is olive colored with irregular dark spots; the pelvics and lower edge of the pectorals are whitish (Hardy 1978a).

Size. Maximum recorded length is 41.7 cm (Nichols and Breder 1927). Welsh measured one 40.6 cm and 0.5 kg collected off New Jersey (Bigelow and Schroeder). Average length is usually less than 30 cm.

Distinctions. Spotted hake can be distinguished from other hakes by the fact that there is no prolonged ray in the first dorsal fin; the outer half of the first dorsal fin is black with a whitish margin; pectoral fins reach back to the origin of the anal fin, whereas they are considerably shorter in both white and red hakes; and the lateral line is darker brown than the general body color and is interrupted by a series of distinct whitish spots. They also have fewer lateral line scales (83–97) than red (95–117) and white (119–148) hakes.

Habits. Spotted hake are coastal bottom fish found over the continental shelf, but also enter rivers and harbors (Hardy 1978a). They are associated with objects on the bottom, but less so than red hake (Edwards and Emery 1968). The depth range is inshore to 426 m (Goode and Bean 1883). There are no tagging studies, but spotted hake travel north-south and inshore-offshore. They are inshore in Massachusetts and Sandy Hook Bay, N.J., in summer and primarily south of Chesapeake Bay in winter. In spring they begin moving north, and have their major concentration north of the Bay in summer and fall (Hardy 1978a).They produce sounds via a large multilobed swim bladder vibrated by associated musculature and body contraction (Fish and Mowbray 1970), and weak thumps are heard upon electrical stimulation.

Food. Small spotted hake (1–5 cm TL) eat mostly copepods

phipods, mysids, euphausiids, shrimps (mostly *Crangon* and *Dichelopandalus*), and small fishes (Hildebrand and Schroeder 1928; Bowman et al. 1987). Young-of-the-year (25–225 mm SL) feed largely on crustaceans (mysids and decapods) and fish larvae (clupeid and sciaenid) in the Cape Fear estuary, N.C. (Burr and Schwartz 1986).

Stomachs of 40 spotted hake 11–25+ cm TL (Bowman et al. 2000) contained fishes (39.1% by weight), mollusks (30.9%, mostly cephalopods), and crustaceans (28.3%). Predominant crustaceans are an amphipod, *Ampelisca verrillii,* and crabs of the genus *Cancer.* The most important fishes were cusk-eels and sand lance. Spotted hake also eat bivalves, polychaetes, isopods, and other species of fishes such as lanternfishes, silver hake, red hake, Atlantic mackerel, and Gulf Stream and yellowtail flounders (Langton and Bowman 1980; Bowman and Michaels 1984; Bowman et al. 2000).

On the outer continental shelf off the Mid-Atlantic Bight, decapod crustaceans were numerically the most important food, followed by fishes and amphipods (Sedberry 1983). The number of fishes eaten decreased with an increase in predator size, but larger fishes were taken as prey by larger spotted hake. In terms of numbers of prey, decapods increased in importance and amphipods decreased in importance with an increase in predator size.

Predators. Spotted hake were identified in stomachs of 16 species of fishes by the NEFSC food habits survey, of which spiny dogfish and goosefish are the most frequent predators (Rountree 1999). Other fishes with multiple occurrences of spotted hake include smooth dogfish, silver hake, windowpane, and summer flounder.

Breeding Habits. Little is known about breeding habits of spotted hake. Size at maturity is 31 cm for females and 21 for males (Barans and Barans 1972). Fecundity is not known. They spawn in offshore waters from the Gulf of Maine to Cape Hatteras from August to April with the peak in October (Colton et al. 1979); in the Chesapeake Bay region from February to March with the peak in October (Barans and Barans 1972); off the Carolinas from November to February (Hildebrand and Cable 1938). Serebryakov (1978) collected ripe males

of 125 m. In the New York Bight, they appear to have a bimodal spawning pattern with peaks in the gonosomatic index occurring in March and September. Ripe females were collected from February to April and again from August to November. However, the apparently divided spawning season may be attributable to dissimilar distribution patterns of specific age-groups since females collected in February–April were 283–396 mm long whereas females collected in August–November were 191–366 mm (Wilk et al. 1990). Ripe ovaries were observed in females from the Mid-Atlantic Bight from August to October, with a peak in September (Eklund and Targett 1990).

Early Life History. The eggs are buoyant, transparent, spherical, 0.67–0.81 mm in diameter, and contain 6–30 oil droplets that coalesce into a single oil globule after blastodisc formation. Egg and larval development have been described by Barans and Barans (1972) but the most detailed description is by Serebryakov (1978). Cleavage begins 2 h after fertilization at a mean temperature of 18.1°C; the blastodisc forms at 9 h; gastrulation is at 19 h; the embryo can be seen at 26 h; pigment cells appear on the body and yolk at 36 h; and hatching occurs at 60 h. Events proceeded similarly at incubation temperatures of 22°–23°C with hatching at 54–60 h (Barans and Barans 1972).

Newly hatched larvae range in size from 1.5 to 2.05 mm. The yolk sac is large and pigmented, and the oil globule is located at the posterior part of the yolk and is also pigmented. The anal opening is lateral and at the base of the ventral finfold. There is a prominent melanophore on the anterior tip of the head and a postorbital melanophore located slightly above the eye. There are five or six pairs of melanophores on the side from nape to vent, one or two on the dorsal surface behind the vent, and a single large melanophore on the ventral surface posterior to these. There are six or seven melanophores on the yolk sac. The yolk sac is absorbed at about 7 days when the larva is 3.3 mm long. The chin barbel appears at 15 mm, fin development begins at 7 mm and is completed by 15 mm. The pelvic fin has three filamentous rays at this time. Scales are present at 25 mm (Barans and Barans 1972).

Eggs and larvae are very similar to red hake, with which it overlaps in range. Identification is possible by comparing size of the oil droplet to the egg diameter: the droplet exceeds one-quarter the diameter of the egg in spotted hake but not in red hake. There is also a pigment spot near where the tip of the snout will be in spotted hake and none in red. This persists through hatching and can be used to distinguish the larvae. In later stages the pigment spots on the caudal portion are one above the other in red hake, whereas the upper pigment spot is in advance of the lower in spotted hake (Serebryakov 1978). Once the pelvic fins have developed, spotted hake are identifiable by the usual lack of pigment on these fins (Fahay and Markle 1984).

Young-of-the-year spotted hake were present in the York River and lower Chesapeake Bay from March to June at temperatures of 6.5°–25°C (Barans 1972). They were not found in salinities less than 10 ppt and were caught in progressively higher salinities as water temperature increased. Growth in the estuary was variable, but one estimate was that they grew from a mean of 117 mm in March to 191 mm in June, with the fastest growth occurring between April and May. Average monthly growth increments in the Cape Fear estuary varied from 12 to 26 mm SL (Burr and Schwartz 1986). In Little Egg Harbor, N.J., spotted hake juveniles appear to be associated with a yellow sponge-peat habitat (Szedlmayer and Able 1996). Young-of-the-year are found in subtidal marsh areas in New Jersey, primarily in the spring (Rountree and Able 1992a).

General Range. Spotted hake live on the coast of the United States from southern New England and New York to Cape Hatteras and range south as far as the offing of northern Florida in deep water. They are also found in the northeastern Gulf of Mexico (Cohen et al. 1990: Fig. 170).

Occurrence in the Gulf of Maine. Spotted hake seldom stray past Cape Cod. They reach the coast of Maine as rare strays; they were reported over a century ago off Halifax, N.S.; and a single pelagic specimen was taken near Sable Island in 1931. Specimens were taken off Seguin Island many years ago, and some are occasionally trawled on the southwestern part of Georges Bank. They may be overlooked among the many young hakes of the two common species (red and white) that are caught in the southwestern part of the Gulf.

Importance. Spotted hake occur in too small quantities to be of commercial interest but are probably included with other hakes in the catch.

WHITE HAKE / *Urophycis tenuis* (Mitchill 1814) / Bigelow and Schroeder 1953:221–223

Description. Body rounded in front of vent, compressed behind vent, four and a half to five and a half times as long as deep (Fig. 140). Mouth large, gaping back to below eyes, upper jaw usually projecting beyond lower, chin with small barbel. First dorsal fin originates close behind pectoral fins, shorter than pectorals, triangular, with third ray prolonged as a filament longer than fin is high. Second dorsal fin runs entire length of trunk from close behind first dorsal to caudal peduncle, of about equal height from end to end, with rounded corners, only about one-half to three-quarters as high as first dorsal. Anal fin similar in outline to second dorsal but shorter. Pectoral fins rounded when spread. Pelvic fins situated considerably in front of pectorals, each reduced to two very prolonged rays, the lower (longer) ray falling slightly short of vent. Scales small and numerous.

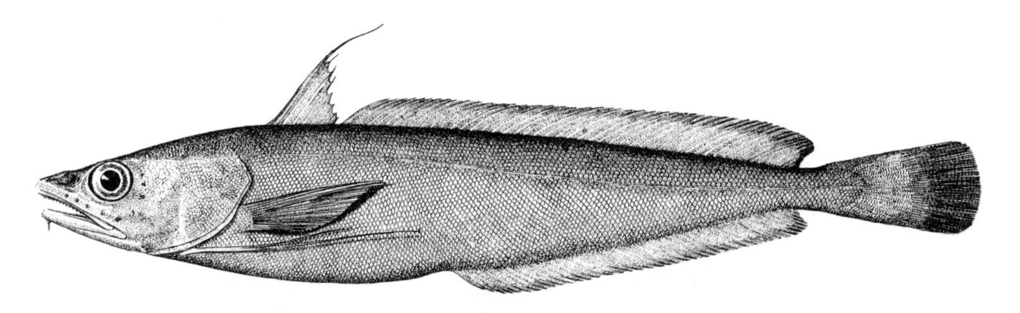

Figure 140. White hake *Urophycis tenuis.* Halifax, Nova Scotia. Drawn by H. L. Todd.

Meristics. First dorsal fin rays 9–12; second dorsal fin rays 50–62; anal fin rays 41–53; pectoral fin rays 16; total caudal fin rays 33–40; lateral line scales 119–148; 2 gill rakers on the epibranchial of the first gill arch, occasionally 3 on one side + 9–12 = 11–14; pyloric caeca 44–62; precaudal vertebrae 15 or 16 + caudal 34 or 35 = total vertebrae 47–51 (Musick 1973a; Markle 1982; Cohen et al. 1990, Comyns and Grant 1993; B. B. Collette, pers. obs.).

Color. White hake vary in color. Usually they are muddy or purple-brown above, sometimes almost slaty, the sides sometimes bronze or golden, and the belly dirty or yellowish white peppered with tiny black dots. The dorsal fins are the same color as the back; the anal fin is the same as the belly. Dorsal, anal, and caudal fins are edged with black. The pelvic fins are pale like the belly, but usually pinkish.

Size. Maximum length is 135 cm and maximum weight 22.3 kg. The former was collected in the Bay of Fundy (Markle et al. 1982) and the latter off Beaver Harbor, N.B. (Scattergood 1953). Five large white hake taken on NEFSC bottom survey cruises ranged from 130 to 134 cm. A 130-cm individual weighing 21.4 kg was taken on *Albatross IV* cruise 76-9 at station 352, 43°26′ N, 66°45′ W. The average size is 71 cm and 3.6 kg. The all-tackle game fish record is a 20.97-kg fish caught in Perkins Cove, Ogunquit, Maine, in October 1986 (IGFA 2001).

Distinctions. Hakes can be distinguished from cod, haddock, and pollock by the fact that they have only two dorsal fins and one anal fin and by elongation of the pelvic fin into thin, narrow feelers. White hake very closely resemble red hake, and the two were frequently confused in literature until the paper by Musick (1973a). Two characteristics previously given to distinguish them are not reliable. The pelvic fins do not always reach beyond the vent in red hake. The upper jaw of white hake is supposed to reach the posterior orbit of the eye, whereas in red hake it is supposed to reach the posterior edge of the pupil, but this is not always true. The most reliable characters are the number of scales in the lateral line (119–148 in white hake, 95–117 in red hake) and the number of gill rakers on the epibranchial of the first arch (white hake 2, red hake 3) of caudal fin rays, 34–38 in white hake and 29–34 in red (Markle 1982: Table 6).

Habits. White hake are demersal, and their habits are very similar to those of red hake, with which they co-occur. Most trawl catches are taken at depths of 110 m or more, although white hake are taken in waters as shallow as 27 m during summer gillnetting (Sosebee 1998d). The greatest amount of co-occurrence of white and red hake off the Canadian Maritime Provinces was between late-stage juveniles and adults (Markle et al. 1982). White hake are more eurythermal and eurybathic, but their habitats overlap to a great extent. White hake had no statistically identifiable preferred depth along the Scotian Shelf and Bay of Fundy, but were caught consistently in depths of 144–358 m in water of 5°–9°C and 33–34 ppt salinity. Depth ranges were about the same along the shore except for a concentration in shallow water (36–106 m) in the Bay of Fundy (Scott 1982a). They were caught in temperatures ranging from 0° to 13°C and salinities 31–34 ppt overall. Juveniles occur at temperatures from 2° to 15°C; preferred temperatures appear to be 4°–10°C (Hardy 1978a). Their preferred and mean temperatures increased from north to southeast along the shelf, whereas their preferred salinities decreased over the same area. White hake prefer bottoms with fine-grained sediments (Scott 1982a). On the Scotian Shelf they were absent from gravel and sand deposits, but catch rates increased over fine deposits and peaked on the clays of the La Have facies, which are found at the bottoms of basins on the shelf.

Depth distribution varies by age and season; juveniles typically occupy shallower areas than adults, but individuals of all ages tend to move inshore or shoalward in summer, dispersing to deeper areas in winter (Sosebee 1998d).

Food. Stomachs of 238 white hake from smaller than 11 to 80+ cm TL examined by the NEFSC food habits survey showed 57.1% by weight bony fishes, 21.4% cephalopods, and 20.6% crustaceans (Bowman et al. 2000). Crustaceans were dominant to 40 cm TL, and fishes and squids thereafter.

White hake less than 1 year old feed mainly on shrimp (*Crangon*), a mysid (*Neomysis*), amphipods, other small crustaceans, and polychaetes (Bowman 1981; Bowman et al. 1987). Postjuvenile fish (20–40 cm TL) eat large quantities of deca-

euphausiid, *Meganyctiphanes,* along with some fishes (e.g., silver hake and other gadoids).

Adults prey mostly on bony fishes, euphausiids, and squids. Fish prey includes Atlantic herring, argentine, hakes (red, white, longfin, and silver), Atlantic cod, haddock, redfish, Atlantic mackerel, northern sand lance, and winter flounder. In the Gulf of Maine, their food includes squid (*Illex*), euphausiids, and fishes such as argentine, red hake, silver hake, offshore sand lance, and winter flounder (Langton and Bowman 1980; Bowman and Michaels 1984; Bowman et al. 2000).

Predators. White hake are eaten by seven species of fishes among which sandbar shark, larger white hake, and cod are the most frequent predators (Bowman et al. 2000). They are also prey of Atlantic puffin and arctic tern (Kress in Fahay and Able 1989).

Parasites. White hake are infested by 2 protozoans (*Haemogregarina aeglefini* and *Haemohormidium terraenovae*); 3 myxosporidians (*Ceratomyxa acadiensis, C. urophysis,* and *Myxidium bergense*); a trematode (*Podocotyle reflexa*); a nematode larva (*Phocanema decipiens*); and a copepod (*Caligus elongatus*) (Margolis and Arthur 1979). On the Scotian Shelf, white hake are host to 25 species of helminths (Scott 1987): 13 digenes, 4 cestodes, 6 nematodes, and 2 acanthocephalans. There were heavy infestations by a digene, *Derogenes varicus,* in hake 20–50 cm long and a nematode, *Hysterothylacium aduncum;* moderate infestations with a digene, *Genolinea laticauda,* a nematode, *Capillaria gracilis,* and acanthocephalans; and light infestations with a cestode, *Clestobothrium crassiceps,* in fish over 50 cm long and a nematode, *Spinitectus cristatus,* in fish 30–50 cm and longer.

Breeding Habits. Mean length at sexual maturity for female and male white hake was 35.1 and 32.7 cm, respectively (O'Brien et al. 1993). Median age at maturity was 1.4 years for both sexes. Maturation of white hake in the Gulf of Maine–Georges Bank region occurs at a smaller size and younger age than for slower-growing white hake in Canadian waters, 40.1–53.9 cm for females and 36.6–44 cm for males (Beacham and Nepszy 1980). Based on growth curves, this would be about 3 years for males and 4 for females.

Fecundity of white hake is extremely high, 100,000–3,000,000 eggs over a size range of 60–110 cm (Beacham and Nepszy 1980); or a mean of 1,772 eggs per g body weight (Wenner 1983a). Three sizes of eggs were present in the ovaries prior to spawning; thus white hake may be serial spawners (Battle 1951). Canadian Atlantic white hake seem to spawn sporadically, but there is a tendency for mass seasonal spawning in the southwestern Gulf of St. Lawrence in the summer (Markle et al. 1982). White hake may not spawn in the Gulf of Maine–Georges Bank region. Eggs and larvae are absent from the Gulf of Maine so the local population may be sustained primarily by recruits from a spring-spawning continental slope population and secondarily by recruits from the summer-spawning and fall-spawning Scotian Shelf and southern Gulf of St. Lawrence populations (Fahay and Able 1989).

Early Life History. White hake eggs are 0.70–0.79 mm in diameter, the perivitelline space is 0.04–0.12 mm, and the oil globule is 0.19–0.28 mm (Markle and Frost 1985: Table 3). Low chorionic ridges are visible on SEM photographs of the eggs (Markle and Frost 1985: Fig. 2D–F). White hake are among ten species of gadoids included in the key to planktonic eggs of the Scotian Shelf (Markle and Frost 1985).

Methven (1985) described the larvae of three species of co-occurring hakes: white, red and longfin. The smallest white hake collected were 4 mm SL and were past yolk-sac absorption. They can be identified by differences in caudal fin ray and epibranchial gill raker counts, body depth, and pigmentation. In specimens 7–10 mm SL, white hake have 34–38 total caudal fin rays as opposed to 29–34 in red and longfin hakes (Markle 1982: Table 6). The adult complement of epibranchial gill rakers separates these larvae at lengths greater than 16 mm. White hake have two gill rakers, red have three and longfin have four or more. Longfin hake are deeper-bodied than white, which are deeper-bodied than red. A key to hake larvae is given in Methven (1985).

Prior to settlement, white hake were caught more frequently at night in neuston samples, and red during the day (Markle et al. 1982). The hypothesis is that pelagic juveniles occurring regularly in Georges Bank–Gulf of Maine waters during May and June and recruiting to nearshore areas come from spawning populations along the slope of Georges Bank, southern New England, and the Mid-Atlantic Bight (Fahay and Able 1989). Juvenile white hake become demersal at 50–60 mm TL and are found in very shallow inshore water and estuaries. White hake of this size range are separated from juveniles over 150 mm by depth, the larger ones occurring at depths greater than 50 m. Juvenile white hake are not associated with scallops as are red hake juveniles, but young have been reported under floating or attached vegetation and are sometimes associated with jellyfish. Early juveniles (25 mm and up) are sometimes seen with pollock at the surface under weeds and eelgrass. Otter trawl collections found white hake demersal juveniles primarily in eelgrass beds or from a station characterized by seasonal occurrence of drift algae in Nauset Marsh, Cape Cod (Heck et al. 1989). Habitat parameters were summarized by Chang et al. (1999b).

Age and Growth. Growth of white hake in the southern Gulf of St. Lawrence (Beacham and Nepszy 1980) appears to be linear between age 3 and 10. Growth was faster in one section of the Gulf of St. Lawrence than in the other, but this may have been due to sampling or aging bias or to temperature differences. Growth rate of demersal juveniles was calculated at a mean of 1.02 mm·day^{-1} in the first summer postsettlement (Fahay and Able 1989). Ages of more than 20 years have been documented (Sosebee 1998d).

General Range. White hake occur from Labrador and the Grand Banks of Newfoundland south to the coast of North Carolina (Cohen et al. 1990: Fig. 172), straying to Florida in the south and Iceland in the east.

Occurrence in the Gulf of Maine. White hake occur throughout the Gulf of Maine, but avoid shallow waters (Map 20).

Migrations. Based on tag returns in the southern Gulf of St. Lawrence, the white hake population appears to be resident (Kohler 1971), although no collections were made in the winter. Only three of the more than 600 recaptures tagged near the eastern end of Prince Edward Island occurred outside the Gulf on the Scotian Shelf. There is some seasonal movement away from Prince Edward Island in the winter as a few recaptures were made on the west coast of Cape Breton. White hake of all ages migrate inshore in warmer months and disperse to deeper waters in colder months (Chang et al. 1999b).

Importance. White hake are taken in the western Gulf of Maine both incidentally to directed operations for other demersal species and as an intended component in mixed-species fisheries (Sosebee 1998d). Since 1968, U.S. vessels have accounted for about 90% of the Gulf of Maine–Georges Bank catch. Canadian landings averaged 600 mt from 1977 to 1991, increased to 1,700 mt in 1993, and then declined to former levels.

Total landings increased from about 1,000 mt during the late 1960s to 8,300 mt in 1985 (Sosebee 1998d). Landings then declined to 5,100 mt in 1989, rose sharply to an historic high of 9,600 mt in 1992, and have steadily declined since (Fig. 141). Small white hake are difficult to distinguish from red, resulting in an unknown degree of bias in reported landings. Recreational catches are insignificant (<0.1 mt·year^{-1} [Sosebee 1998d]). Fishing is managed under the New England Fishery Management Council's Multispecies Fishery Management Plan. The Gulf of Maine–Georges Bank white hake stock is considered at a low biomass level and is overexploited (Sosebee 1998d).

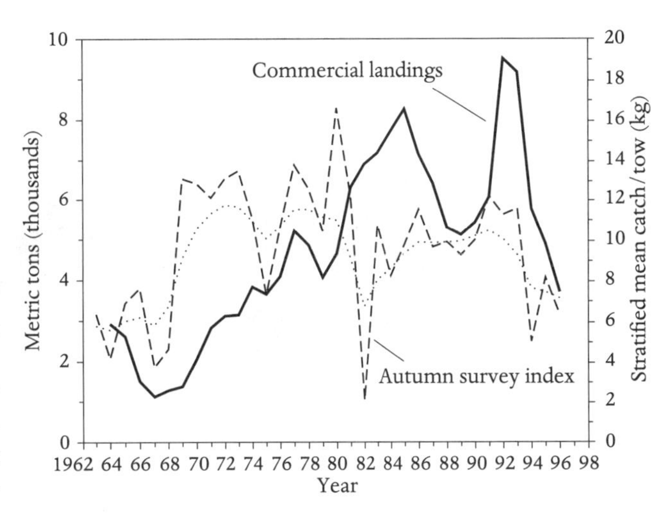

Figure 141. Catch and index of abundance of white hake *Urophycis tenuis,* Gulf of Maine–Georges Bank stock. (From Sosebee 1998d.)

ACKNOWLEDGMENTS. Drafts of the entire gadiform section were reviewed by Daniel Cohen and Michael Fahay; the food habits sections, by Ed Bowman; and selected species accounts, by Frank Almeida (silver and red hake), Karsten E. Hartel (tomcod), Ralph K. Mayo (pollock), Nigel Merrett (Macrouridae), William J. Overholtz (haddock), and Fred Serchuk (cod).

TOADFISHES. ORDER BATRACHOIDIFORMES

The order contains only the family Batrachoididae and seems most closely related to the Lophiiformes (Nelson 1994). The classical name for the order, Haplodoci, is derived from the Greek, in reference to the undivided posttemporal bone. Toadfishes superficially resemble sculpins, but their pelvic fins are situated on the throat well in front of the pectoral fins ("jugular") and they have only three gills and three gill arches.

TOADFISHES. FAMILY BATRACHOIDIDAE

BRUCE B. COLLETTE

Both the soft and spiny portions of the dorsal fin are well developed as separate fins, the former much longer than the latter. Upper hypural bones have a peculiar intervertebral basal articulation with the posterior caudal vertebrae. There are family (D. W. Greenfield, pers. comm.). Most species are benthic inhabitants of warm seas with only one species of the subfamily Batrachoidinae occasionally reaching north to the Gulf of Maine.

OYSTER TOADFISH / *Opsanus tau* (Linnaeus 1766) / Bigelow and Schroeder 1953:518–520

Figure 142. Oyster toadfish *Opsanus tau*. Woods Hole, Mass., 27 cm. Drawn by C. H. Hudson.

Description. Large flat head, round snout, tremendous mouth, tapering body with plump belly, and fanlike pectoral fins (Fig. 142). Fleshy flaps with irregular outlines on tip of upper jaw, along edge of lower jaw, on cheeks, and over each eye. Teeth large, blunt, in a single row on vomer, palatines, and jaws. Two short spines at upper angle of gill cover, hidden in thick skin. All fins fleshy. Two dorsal fins extend length of trunk from nape nearly to caudal fin base. Second dorsal fin five to six times as long as first dorsal, separated from it by a deep notch. Anal fin somewhat shorter than second dorsal, originating under about eighth dorsal fin ray; similar to dorsal fin in outline except rays more or less free at outer ends, especially in anterior half. Pelvic fins covered by thick fleshy skin, jagged in outline, first ray stouter than others. Axillary pore present on body behind pectoral fins. Glands (6–16) between upper pectoral fin rays on posterior surface of fin. Caudal fin rounded. Scales absent, skin covered with thick layer of slimy mucus.

Meristics. Dorsal fin rays III, 25 or 26; anal fin rays 20–22; pectoral fin rays 19–21; papillae along upper lateral line 26–30, along lower lateral line 21–25; teeth on upper jaw 20–28, lower jaw 13–21; precaudal vertebrae 11 + 23–25 caudal = 34–36 total (Collette 2001).

Color. The general ground tint ranges from dark muddy olive green to brown or yellow, darker on back and sides, paler below, and variously and irregularly marked with darker bars and marblings, which may be restricted to head and fins or may extend over the whole fish, belly as well as back. Toadfishes, like many other bottom fishes, can change color to match the bottom on which they lie.

Size. Maximum reported size 381 mm. The all-tackle game fish record weighed 2.23 kg and was caught at Oracoke, N.C., in June 1994 (IGFA 2001).

Distinctions. Toadfishes differ from sculpins, and all other spiny-finned fishes of the Gulf of Maine except blennylike fishes, in the location of the pelvic fins, which are under the throat well in front of the pectorals ("jugular") instead of below the latter or behind them. Confusion with blennylike fishes is unlikely because the dorsal fin of toadfish is mostly soft-rayed and that of blennylike fishes is spiny throughout.

Habits. Toadfish are negatively buoyant and live on the bottom in shallow water. They are resident year-round wherever found, but migrate into deeper water and probably become torpid in winter in the northern part of the range. They are most common on sandy or muddy substrates, hiding among eelgrass or under stones and cans, where they hollow out dens in which to lie in wait for prey. Toadfish are more active at night than during daytime (Phillips and Swears 1981), although they may call, mate, and bite on hook and line during the day. Schwartz (1974) reported that tagged fish were mostly recaptured at the original tagging site within a few months of release and that if displaced they returned to the original site. Two fish were caught by winter oyster dredgers 8.9 km from the release site, suggesting migration into deeper waters during the winter months, so toadfish may not be as sedentary as many writers have indicated (Gudger 1910).

Toadfish erect their spiny dorsal fin, produce open-mouth threats, and may snap when caught, and they often fight among themselves by jaw-locking. Their bite is painful. Despite their clumsy appearance, they can dart out of their hiding places and back again with considerable speed. They can live out of water for an astonishingly long time—up to 45 h if kept moist (Schwartz and Robinson 1963).

Toadfish produce a short-duration agonistic grunt and a longer, higher-pitched tonal courtship boat-whistle call (Gill 1907; Gray and Winn 1961; Fine 1978). Sounds are produced by synchronous contraction of paired intrinsic muscles on the heart-shaped swim bladder (Skoglund 1961). The swim bladder is two-chambered and contains an unusually high concentration of oxygen for a shallow water fish (Fänge and Wittenberg 1958). Swim bladders and sonic muscles grow larger in males than in females, but the bladder-weight and bladder-width regressions of males and females are not significantly different, pointing to a morphological similarity of the bladders between sexes (Fine 1975; Fine et al. 1990). Buoyancy is similar for both sexes, suggesting that a larger bladder in males provides an acoustic advantage (Fine et al. 1995). Both sexes grunt, but only males produce the boat-whistle call (Gray and Winn 1961), which attracts females to the nest and increases the calling rate of nearby males (Winn 1972). Grunts are most often produced by a male in his nest at the approach of other males, spent females, or blue crab (Gray and Winn 1961).

There is a large axillary gland in the pectoral fin axil (Wallace 1893; Vernick and Chapman 1968) containing cells and structures suggesting secretory and ion regulatory roles (Maina et al. 1998). Several hypotheses of its function, such as being antibacterial (Hamlett and Schwartz 1979) or toxic to other fishes, seemed to have been ruled out, leaving its exact function unknown (Maina et al. 1998).

Food. Toadfish are carnivores, and Vinal Edwards's diet list for this species at Woods Hole includes sea worms (*Nereis*), amphipods, shrimps, crabs, hermit crabs, a variety of mollusks, both univalve and bivalve, ascidians, squids, and fish fry such as alewife, menhaden, smelt, silversides, mummichog, sculpins, scup, cunner, winter flounder, and puffers. In Delaware and Chesapeake bays and at Hilton Head Island, S.C., crustaceans are the main category of food, particularly xanthid crabs such as *Panopeus herbstii* and *Eurypanopeus depressus,* common inhabitants of oyster reefs (McDermott 1965; Schwartz and Dutcher 1963; Wilson et al. 1982). Squids, stone crab, anchovies, and juvenile blue crab were incidental toadfish foods in these studies.

Parasites. There are records of a nematode, an argulid, a flagellate, and a digenetic trematode from this toadfish. The intestinal nematode *Thynnascaris habena* (Linton) utilizes *O. tau* as its definitive host and can reach high incidences (Norris and Overstreet 1975). It has been reported (as *Contracaecum habena*) in several studies: Linton (1905, Woods Hole), Gudger (1910), Schwartz and Dutcher (1963, 64% of Chesapeake Bay specimens), McDermott (1965, 80% of Delaware Bay specimens), Wilson et al. (1982, Hilton Head Island, S.C.). Dutcher and Schwartz (1962) published an excellent study of the host-parasite relationship of *O. tau* and the branchiuran crustacean parasite *Argulus laticauda* in Chesapeake Bay (later described as a new species *A. chesapeakensis* by Cressey 1971). The highest incidence of parasitism was found in August with one fish having 284 parasites. The mouth was the most heavily parasitized area of the toadfish. There are also records of an argulid identified as *Argulus laticauda* from toadfish from Woods Hole (Meehean 1940). Pearse (1949) reported a digenetic trematode, *Paracryptogonimus americanus,* from the intestine of a toadfish from Beaufort, N.C. Hematozoan kinetoplastids of the genus *Trypanoplasma* have been found in the blood of toadfish from Long Island Sound and Chesapeake Bay transmitted by leeches *Piscicola funduli* and *Calliobdella vivida* (Nigrelli et al. 1975; Burreson and Zwerner 1990).

Reproduction. There are several early descriptions of the life history of the oyster toadfish (Ryder 1887a; Gill 1907; Gudger 1910). Toadfish spawn in June and early July in the northern part of the range. Eggs were laid in Chesapeake Bay from 8 May to 15 July at water temperatures of 17.5° to 27°C (Gray and Winn 1961). The very large eggs (about 5 mm in diameter) are laid in holes under stones, under large shells, in old tin cans, among sunken logs, or among eelgrass, where they adhere to whatever serves as a nest, which the male guards during incubation. Incubation takes 5–12 days in Chesapeake Bay (Gray and Winn 1961). After hatching, the tadpole-shaped larvae remain attached to the nest by the yolk sac (cling young) for an extended period until the latter is absorbed at a length of 15–16 mm, when they break free. Males protect and fan the eggs, attached young, and free young for 23–46 days (Gray and Winn 1961). A series of ten larvae, from a 7.4-mm TL newly hatched to a 17.1-mm TL specimen 20 days after hatching was illustrated by Dovel (1960). This series was expanded to include Ryder (1887a) and other references by Martin and Drewry (1978: Figs. 179–185).

Age and Growth. Length and weight of eight large specimens collected in Chesapeake Bay were 308–344 mm and 681–795 g (Schwartz and Dutcher 1963). A larger Chesapeake Bay sample contained 80 males, 112–367 mm TL, 23–1,098 g, and 55 females, 90–347 mm TL, 13–740 g (Swartz and Van Engel 1968), clearly showing that males reach larger sizes than females. Their length-weight relationship was: $\log W = -5.223 + 3.223 \log TL$. Wilson et al. (1982: Fig. 3) gave the length-weight relationship for 278 toadfish from South Carolina as: $W_L = 4.9e^{0.017L}$. Radtke et al. (1985: Fig. 1) presented a length-weight diagram of 73 Virginia toadfish with $W = 6.098L^{3.18}$. South Carolina fish grow faster and die younger than Chesapeake Bay fish. Based on annular vertebral rings, Schwartz and Dutcher (1963) found that females lived only 7 years and reached 272 mm TL; males lived 12 years and reached 344 mm. Based on otolith increments, Radtke et al. (1985) found that York River, Va., females reached 9 years and 260 mm and males 11 years and 340 mm. They found the von Bertalanffy growth curve to be: $TL_t = 407.46 \text{ mm} \times (1 - e^{-0.15(t+0.33)})$ for 41 males; $TL_t = 271.53 \text{ mm} \times (1 - e^{-0.39(t+0.41)})$ for 20 females. Sex ratio was 2:1 in favor of males in both Delaware and Chesapeake bays (Schwartz and Dutcher 1963; McDermott 1965), although this was likely an artifact of sampling, males being more likely to enter traps.

General Range. Shallow water, typically in estuaries, along the east coast of North America from Florida to Cape Cod, straying northward to Maine. There appear to be northern and southern genetic populations with overlap at Cape Hatteras, N.C. (Avise et al. 1987).

Occurrence in the Gulf of Maine. Toadfish are common about Woods Hole and southward. They so seldom venture around Cape Cod that none of the fishermen in Massachu-

or heard of them. There are only four Gulf of Maine records: "Maine" (Storer 1846); "Massachusetts Bay" (Goode and Bean 1879); Cohasset (Bigelow and Schroeder); and Plymouth (Lawton et al. 1984).

Importance. Not generally used for food, although the trunk and tails are sometimes eaten (Gudger 1910; F. J. Schwartz, pers. comm.) and live toadfish have been for sale in Chinese restaurants in Washington, D.C. (B. B. Collette, pers. obs.). There is an annual recreational toadfish tournament at Hilton Head Island, S.C. (Wilson et al. 1982), and toadfish are among the nongame species taken by the recreational fishery in the New York Bight (Buchanan et al. 1988). *Opsanus tau* is an important experimental subject for studies on behavior, sound production, physiology, endocrine analyses, insulin and diabetes investigations, and much other research, as can be seen from the bibliographies of Robinson and Schwartz (1965: 261 papers) and Schwartz and Bright (1982: 380 papers). More than 290 additional papers on oyster toadfish were published between 1982 and 1994 (F. J. Schwartz, pers. comm.).

ACKNOWLEDGMENTS. Drafts of the toadfish section were reviewed by Michael Fine, Thomas A. Munroe, and Frank Schwartz.

ANGLERFISHES. ORDER LOPHIIFORMES

The order Lophiiformes comprises an assemblage of 18 families, 64 genera, and 297 species of bizarre and fascinating fishes known as anglerfishes (Bertelson and Pietsch 1998). The name is derived from the fact that the first dorsal fin spine (the illicium), which is located far forward on the head just behind the mouth, is disconnected from the other spines posterior to it (if they are present at all) and functions as an angling apparatus to lure potential prey within reach of the mouth. In most anglerfishes, the illicium bears a fleshy appendage called the esca at its tip, which serves as a lure. The Greek root of the names Lophiiformes and Lophiidae (the goosefish family), *lophos* (*lophox*), means crest and refers to the crest on the top of the head formed by the first few spines of the highly modified spinous dorsal fin. The older name for the anglerfish order, Pediculati, refers to the highly modified structure of the base of the pectoral fin that resembles an arm (*pediculus* is Latin for little foot) and is formed by elongation of pectoral radial bones, which are so short in all other bony fishes that they are not noticeable externally. Apart from this peculiar structure of the pectorals, the gill openings are reduced to small apertures in or near the axils ("armpits") of these fins. One species of each of four families has been recorded from the Gulf of Maine.

GOOSEFISHES OR MONKFISHES. FAMILY LOPHIIDAE

JOHN H. CARUSO

Lophiids are distinguished from their other lophiiform relatives by a very large and very much flattened head; by an enormous mouth with well-developed teeth; and by the fact that they have only two pectoral radials in each "arm." The mobile illicium has a fleshy esca at its tip, which acts as a lure to attract prey within range of the large mouth. There are four genera and 25 species (Caruso 1985; Nelson 1994) of which only one occurs in the Gulf of Maine. A summary of recent information on goosefish was published by Steimle et al. (1999a).

GOOSEFISH / *Lophius americanus* Valenciennes 1837 /

Monkfish, Angler, Allmouth, Molligut / Bigelow and Schroeder 1953:532–541

Description. Body greatly flattened dorsoventrally, much like a skate or a ray (Fig. 143). Head rounded from above, about as broad as long, enormous compared to body, which is narrow and tapering back of pectoral fins, giving the fish a tadpolelike appearance when viewed from above. Mouth enormous, directed upward, lower jaw projecting so far beyond upper jaw that most lower teeth are freely exposed even when mouth is closed. Both jaws armed with long, slender, curved teeth, alike in form but of various sizes, very sharp, mostly depressible, all pointing inward toward gullet. Some teeth as long as a few centimeters in a large fish. Lower jaw teeth mostly large, in one to three rows. Upper jaw teeth few in middle (a toothless space in the midline) largest, with a single row of smaller ones flanking them. Several rows of thornlike teeth on roof of mouth. Gill openings below, behind, and just slightly in front of pectoral fins. Eyes on top of head, directed upward. Pectoral fins distinctive, their bases like thick fleshy arms bearing fins proper at outer edge. Finlike parts fanlike when spread, so

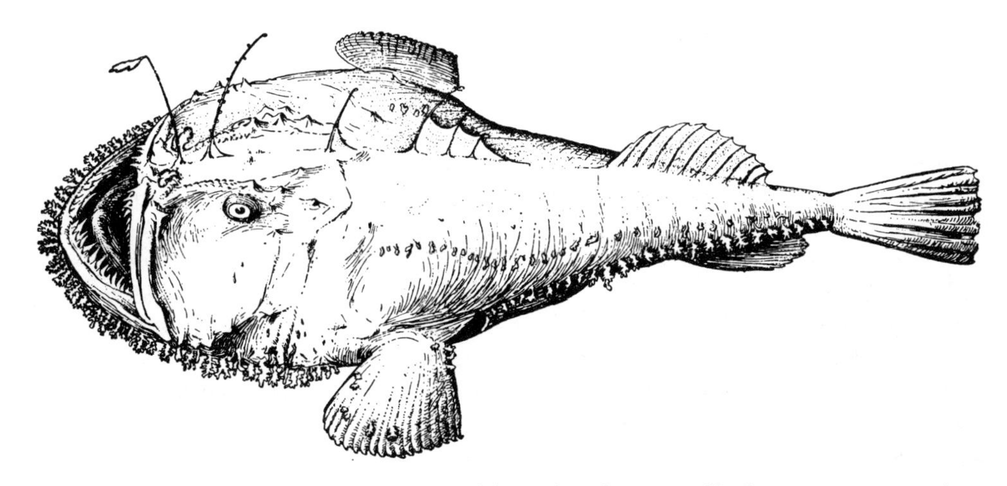

Figure 143. Goosefish *Lophius americanus,* oblique-dorsal view. Gulf of Maine. (From Bigelow and Welsh 1925.)

thick-skinned that rays hardly visible except in scalloping of margins. Top of head with three stiff, slender spines (representing anterior part of spinous dorsal fin) hardly thicker than bristles; first (illicium) close behind tip of snout; second a little in front of eyes; third on nape of neck. First and second spines movable from recumbent to erect; third sloping backward with its basal half or more embedded in skin. Relative lengths of spines vary; first two about equal in length on most fish, or second a little longer, third shortest. Illicium bears an irregular, leaflike flap of skin (esca) at tip, playing important role as lure for prey; second and third spines with small triangular membranes at bases, and one or both may be fringed with short lobes of skin. Behind these spines are two additional, well-developed dorsal fins: the postcephalic spinous dorsal fin of three spines situated at level of pectoral fins, and second, or soft dorsal fin, on rear part of trunk. Anal fin below soft dorsal fin; pelvic fins on lower surface of head, well in front of pectorals. Caudal fin small and broom-shaped. Dorsal fins with thin delicate membranes; caudal, anal, and pelvic fins thick and fleshy, like the pectoral fins. Skin naked, very smooth, and slippery; a row of fleshy flaps of irregular shape runs around margin of head and around edge of lower jaw, besides smaller tags that fringe sides of trunk back to base of caudal fin. Upper side of head bears numerous low conical tubercles or spines, which vary in prominence from fish to fish (for detailed discussions of cranial spine morphology, see Caruso 1981, 1983).

Meristics. Dorsal fin rays VI, 9–12; anal fin rays 8–10; pectoral fin rays 25–28; pelvic fin rays I, 5; branchiostegals 6; vertebrae 26–31 (Caruso 1983; Fahay 1983; Scott and Scott 1988).

Color. The numerous goosefish observed by Bigelow and Schroeder were chocolate brown above, variously and finely mottled with pale and dark. The dorsal fins, the upper sides of the pectoral fins, and the caudal fin are of a darker shade of whole lower surface of the fish is white or dirty white, except the distal third of the pectoral fins gradually darkens to the color of the dorsal surface, and the pelvic fins become dusky distally. The esca is greenish in life and has a dark spot at its base. Very small goosefish are described as mottled and speckled with green and brown. In an aquarium, the European *L. piscatorius* is able to match both its color and its color pattern closely to the sand and gravel on which it lies (Wilson 1937).

Size. Adults run from 610 to 1,220 mm, weighing up to at least 27 kg (Scott and Scott 1988). One 965 mm long, caught at Woods Hole in July 1923, weighed 14.5 kg alive. The all-tackle game fish record is a 22.56-kg fish caught in Perkins Cove, Ogunquit, Maine, in July 1991 (IGFA 2001).

Distinctions. The depressed body and huge mouth distinguish goosefish from all other fishes in the Gulf of Maine.

Taxonomic Remarks. Goosefish of the western North Atlantic were once thought to be identical with the widespread eastern Atlantic angler (*L. piscatorius*). In 1923 Täning observed that the late larval stages of *L. americanus* do not resemble those of *L. piscatorius* as closely as they do those of the other European angler, *L. budegassa*. For this reason Berrill (1929) recognized goosefish in the western Atlantic as a distinct species and resurrected the name *L. americanus*. There are distinct morphological differences between the European anglers *L. budegassa* and *L. piscatorius,* and the American *L. americanus* (Wheeler et al. 1974; Caruso 1983).

Habits. The depth range extends from just below the tide line (Bigelow and Schroeder) to depths of at least 840 m (Markle and Musick 1974), but few large individuals are taken below 400 m (Wenner 1978). Adults are found on hard sand, pebbly bottoms, the gravel and broken shells of good fishing grounds, and soft mud, where they have been trawled in the deep basins

European goosefish kept in the aquarium at Plymouth, England, spent most of the time resting quietly (Wilson 1937). When they swam they did so slowly, and they used their paired fins for walking on the bottom. Wilson described one digging a small hollow in the bottom when it settled down, using its pelvic fins to shovel the sand and pebbles forward and outward and using its pectorals, almost like webbed hands, to push the sand away to either side until its back was almost flush with the surrounding bottom.

American goosefish are at home through a very wide range of temperatures, 0°–24°C (Wood 1982). It appears unlikely that they can survive temperatures much colder than 0°C, since many were seen floating dead in Narragansett Bay and on the shore during the winter of 1904–1905, apparently killed by the unusually severe cold (Tracy 1906). They have been found to be most abundant in Canadian waters at 3°–9°C (Jean 1965), in the Mid-Atlantic Bight at around 9°C (Edwards 1965), and on the continental slope off Virginia at 7°–11°C (Wenner 1978).

At the other temperature extreme, goosefish caught by commercial fishermen in shoal water near Cape Lookout, N.C., are exposed to temperatures higher than 21°C for part of the season, perhaps as high as 24°C. Reports that the inshore contingent of goosefish population of Rhode Island waters moves offshore (i.e., deeper) in July and inshore again in October suggest that they tend to avoid extreme summer heat if they can do so by moving into deeper water (Tracy 1906). In his study of the seasonal distribution of *L. americanus* in the Gulf of St. Lawrence, Jean (1965) found the greatest winter concentrations at depths of 180–225 m (3°–6°C), and the greatest summer concentrations at 25–220 m, with the greatest abundance at 25–92 m (5°–9°C).

Although they may appear tolerant to a relatively wide range of salinities, occurring from estuaries out to the upper part of the continental slope, Bigelow and Schroeder noted that "we have never heard of one in brackish water." Since their kidneys are aglomerular, it seems unlikely that they could tolerate low-salinity conditions for very long.

Food. Goosefish feed mostly on fishes after they take to the bottom: 65.1% by weight bony fishes, 24.9% cephalopods, and 8.8% elasmobranchs (Bowman et al. 2000). Over two dozen fish species and two squid species are listed from 872 goosefish captured from the Scotian Shelf to the Middle Atlantic states between 1977 and 1980. Goosefish caught in the Gulf of Maine consumed 15 fish species and one squid species, the most important of which were squid, unidentified clupeids, silver hake, squirrel hake, and American plaice. Off southern New England, the most important of the 16 species of fishes in the diet of goosefish were little skate, red hake, goosefish, and sand lance (Armstrong et al. 1996).

Goosefish capture seabirds, as their vernacular name implies: cormorant, herring gulls, widgeon, scoter, loon, guillemot, and razor-billed auk are in its recorded diet. Bigelow and Schroeder also found grebe and other diving fowl, such as scaup duck and merganser, in goosefish from Pamlico Sound, N.C. It is questionable, however, whether even the largest of them would be able to master a live goose, as rumor has it, nor do the local fishermen believe they ever do so in Pamlico Sound, although the abundance of wild geese there in winter would afford them every opportunity. Goode (1884), however, tells of one that a fisherman saw struggling with a loon, and one was even found with a sea turtle (Schroeder 1947).

Goosefish have often been cited for their remarkable appetites. Bigelow and Schroeder reported one that had made a meal of 21 flounders and a dogfish, all of marketable size; of half a pailful of cunner, tomcod, and sea bass in another; of 75 herring in a third; and of one that had taken seven wild ducks at one meal. They stated that "in fact it is nothing unusual for one to contain at one time a mass of food half as heavy as the fish itself." With its enormous mouth (about 1 m long gaping about 230 mm horizontally and 203 mm vertically), it is able to swallow fish of almost its own size. Fulton (1903b), for instance, found a codling 585 mm long in a British goosefish of only 660 mm, and Field (1907) took a winter flounder almost as big as its captor from an American specimen. One that Bigelow and Schroeder gaffed at the surface on Nantucket Shoals contained a haddock 787 mm long, weighing 5.5 kg; they cited Capt. Atwood as having long ago seen one attempting to swallow another as large as itself. Wilson's (1937) observations, however, indicate that they are no more gluttonous than any other rapacious fish, for those that he watched in the aquarium usually refused food for 2 or 3 days after a meal. His observation that they evidently preferred small fishes is in line with their normal habits, for they feed mostly on small fishes, not on large, and even the largest of them take very small fry on occasion.

Goosefish, especially at smaller sizes, are also known to eat invertebrates such as lobsters, crabs of several species, hermit crabs, squids, annelid worms, shellfish, starfish, sand dollars, and even eelgrass. The most important invertebrates in small (up to 200 mm TL) goosefish off southern New England (Sedberry 1983; Armstrong et al. 1996) were red shrimp (*Dichelopandalus leptocerus*), sand shrimp (*Crangon septemspinosus*), and long-finned squid (*Loligo pealeii*). Invertebrates became less important at 200–400 mm TL, and the diet of goosefish over 400 mm TL was dominated by teleosts (Armstrong et al. 1996; Bowman et al. 2000).

Part of the angler's reputation for gluttony is due to anglers swallowing food in the cod-end of trawls after capture. Goosefish frequently open their mouths widely when disturbed, a response that could account for some of the more unlikely stomach contents that have been recorded (Caruso 1977; Armstrong et al. 1996). The projecting lower jaw, dorsally directed mouth, and sharp slender teeth of lophiids are ill-suited for capturing benthic organisms, and most of these organisms are not likely to strike at a rapidly moving lure.

The most interesting habit of goosefish is the use of the highly modified first dorsal fin spine as an angling apparatus to lure small fishes within seizing distance, much as Aristotle described. W. F. Clapp (the first observer to watch American goosefish feeding, according to Bigelow and Schroeder) described individuals in Duxbury Harbor as lying motionless among the eelgrass, with the "bait" or esca at the tip of the first dorsal fin spine (illicium) swaying to and fro over the mouth. When a tomcod (the only fish he saw them take) chances to approach, it usually swims close up to the esca, but never (in his observation) actually touches it, for the goosefish opens its vast mouth as soon as the victim comes within a few inches and closes it again, instantly engulfing its prey .

Further details added by observations on European anglers in aquariums at Port Erin, Isle of Man (Chadwick 1929), and Plymouth, England (Wilson 1937) are that the illicium, with its terminal esca, is held down along the top of the head, to be raised at the approach of a prospective victim; the esca may be jerked to and fro quite actively in front of its owner's head; the victim is usually taken in headfirst; a fish swimming close enough may be snapped up without the bait being brought into play; and some anglers use the bait often, others seldom. Wilson also made the interesting observation that touching the esca does not cause a reflex snapping of the jaws, showing that angler feed by sight. Gudger (1945) gave an interesting and readable survey of observations on use of the bait.

Predators. Adult goosefish cannot have many enemies, but small goosefish are no doubt eaten by various predacious fishes, including swordfish (Scott and Tibbo 1968) and larger goosefish (Armstrong et al. 1996). Other predators include sharks (dusky, sandbar, spiny dogfish, and smooth dogfish), of which spiny dogfish are the most significant, and cod (Rountree 1999). Goosefish larvae in aquarium jars at Plymouth, England, were devoured by larvae of spiny lobster (*Palinurus*), large copepods, ctenophores, and hydroids when they came close enough to the walls of the jar to be seized (Lebour 1925).

Parasites. Goosefish parasites include a protozoan, six trematodes, and larvae of a nematode *Phocanema* sp. (Margolis and Arthur 1979). A protozoan, *Haemogregarina* infects immature goosefish in the Gulf of Maine (Bridges et al. 1975; Khan and Newman 1982). The microsporidian *Glugea americanus* (often referred to as *Spraguea lophii*, a similar species that occurs in *L. piscatorius*) infects spinal and cranial ganglia of American goosefish (Takvorian and Cali 1986)

Reproduction. Both sexes begin to mature at about 30 cm TL at ages between 3 and 4; males generally attain 100% maturity by about 50 cm, females by about 60 cm (Almeida et al. 1995). Mean lengths at which 50% of the males matured averaged about 40 cm and females about 44 cm.

Goosefish spawn in spring, summer, and early autumn, according to the latitude, and through a long season. Eggs and larvae have been taken near Cape Lookout, N.C., in March and April (Hildebrand, pers. comm. to Bigelow and Schroeder); in May off Cape Hatteras (Täning 1923); and as early as May summer in the Gulf of Maine, for 24 June is the earliest date eggs have been seen north of Cape Cod (Connolly 1921); 18 September (off Seguin Island, Maine) is the latest recorded date for American waters.

The locality of spawning of *L. americanus* has been the subject of discussion, whether inshore in shoal water or offshore in deeper water. Eggs reported from the Bay of Fundy (Connolly 1921), from Passamaquoddy Bay (Berrill 1929), and from Frenchman Bay near Mount Desert (Procter 1928) were in such early stages of incubation that they must have been spawned close at hand. This also applies to some isolated eggs that were collected at about the 20-fathom (36.6 m) contour line off northern North Carolina by the *Dana* (Täning 1923). Neither is there any reason to suppose that eggs farther advanced in incubation that have been taken in the inner parts of the Gulf of Maine, at Woods Hole, and at Newport, had come from any great distance. Furthermore, large adult fish are present in abundance inshore throughout the spawning season, which would not be the case if they moved offshore or into deep water to spawn. Recently spawned eggs have been reported near the 2,000 m contour line of the continental slope south of the Newfoundland Banks (Murray and Hjort 1912), and over similar depths off North Carolina (36°16′ N, 74°33′ W; Täning 1923). However, the latter could also have been produced by *L. gastrophysus.*

Presence of lophiid eggs off North Carolina, near Newport (Agassiz 1882), near Woods Hole, in the Gulf of Maine, and over the continental slope south of the Newfoundland Bank, and the capture of a very small (10-cm) specimen on the Grand Banks show that American goosefish breed throughout the range. However, egg veils found off North Carolina could also have been produced by *L. gastrophysus.*

As in nearly all anglerfishes for which the mode of egg production is known, eggs are shed in remarkable, buoyant, ribbonlike, nonadhesive, mucoid veils, which in goosefish may be 6–12 m long, and 0.15–1.5 m wide (Martin and Drewry 1978). Within each veil eggs are arranged in a single layer, lying one to three or even four in separate hexagonal compartments, with an oil globule uppermost. Each compartment has an opening that provides water circulation (Fulton 1898; Gill 1905; Rasquin 1958; Ray 1961; and Armstrong et al. 1992). In an egg veil found near St. Andrews, N.B., between 10 and 11 m long, about 200 mm wide and 3 mm thick, and about 26.5 liters in volume, about 5% of the eggs were single, about 80% were in pairs, and about 5% were in threes, per compartment. This veil was estimated to contain about 1,320,000 eggs (Berrill 1929), and Fulton (1898) estimated roughly the same numbers (1,345,848 and 1,317,587) in the ovaries of two European goosefish from Scottish waters.

The veils are light violet-gray or purplish brown, made more or less blackish by embryonic pigment of the eggs, according to their stage of development. They are so conspicuous when floating at the surface that fishermen have long been familiar with them, although it was not until about 1871

true parentage. The eggs occasionally become isolated, perhaps when a storm shreds the mucous veil to pieces, and when this happens they float like any ordinary buoyant fish eggs. Bigelow and Schroeder did not actually find them in this condition in the Gulf of Maine, but Agassiz and Whitman (1885) saw isolated eggs at Newport, and Täning reported others from North Carolina waters.

Egg veils have been reported within the Gulf of Maine from Campobello Island at the entrance to the Bay of Fundy; from Passamaquoddy Bay (Connolly 1921; Berrill 1929); in Frenchman Bay, Maine (Proctor 1928); about 24 km off Seguin Island, Maine, 18 September 1925 (with eggs nearly ready to hatch), found by Capt. Greenleaf of the U.S. Bureau of Fisheries; and at Provincetown, Mass., where Bigelow and Schroeder found a veil within 1 m of the shore, on 26 June 1925. There have been fewer captures of pelagic larvae within the Gulf; , three were taken near Brazil Rock off southwestern Nova Scotia and two very small ones (5 and 6.5 mm) were collected by Bigelow and Schroeder on the *Grampus* in Massachusetts Bay, one in July 1912, the other in September 1915.

Early Life History (Fig. 144). The eggs are 1.61–1.84 mm in longest diameter (Fahay 1983) as they lie in their mucous compartments (Fig. 144A). The yolk is straw-colored, and they have either one copper-colored or pinkish oil globule of 0.4–0.56 mm or several smaller ones. Incubation takes 7–22 days at temperatures as low as 5°C to as high as 17.5°C, and probably at higher temperatures. The larvae, which float with the yolk uppermost at first, are 2.5–4.5 mm long at hatching.

The first of the dorsal fin spines (which will become the second cephalic dorsal fin spine of the adult) appears within 4 days or so after hatching, as a lobe at the margin of the embryonic finfold on the nape of the neck. The pectoral fins form at about 7 days, when the larva is 5.5 mm long; the pelvic fins have now appeared as two long conical processes below and behind the pectorals; and pigment has become congregated in three or four masses behind the vent (Fig. 144D), the last being a very conspicuous feature that the larvae of the European species do not share. The yolk is absorbed at a length of 6–8 mm, a second dorsal spine forms behind the first, and the pelvic fins become two-rayed. The third and fourth dorsal spines appear while there are still only two pelvic rays in the American goosefish, but not until the third rays have developed in the pelvic fins in the European species.

A fifth dorsal spine next appears behind those that have developed already, and a sixth in front of these, all of them being interconnected with membrane at their bases but free at their tips. The pectoral fins assume a great breadth and fanlike outline; the second dorsal, anal, and caudal fins take definite form; the pelvic rays become filamentous at their tips, streaming far out behind the tail; and a complete row of teeth appears in the lower jaw, with a few in the upper. The goosefish pictured at this stage by Agassiz (Fig. 144F) was 30 mm long, and one much like it taken off Brazil Rock, described by Connolly, was 27 mm long, but, according to Stiasny (1911, 1913), larvae of Mediterranean goosefish attain this stage when they are only 13–18 mm long.

Older postlarval stages of American goosefish develop in a fashion similar to *L.budegassa;* that is, the foremost dorsal fin spine becomes bristlelike with the esca appearing at its tip; the last three of the free spines on the nape of the neck join together to form the postcephalic spinous dorsal fin; lappets of skin appear around the margin of the lower jaw and along the cheeks; and the head broadens and flattens while the young fish are still pelagic, with enormous pectoral fins and with threadlike pelvic fins (Fig. 144F).

The largest free-swimming Mediterranean larva seen by Stiasny (1911) was 50 mm long. Probably the young take to the ground shortly after this stage, for Bowman (1919) described European goosefish fry of about 65 mm that were trawled on the bottom off Scotland as of adult form, except that their pectorals were proportionately larger. To attain this state entails growth of the head out of proportion to the rest of the body; enlargement of the mouth; "shrinkage" of all the fins (of the pelvic fins most of all); alteration of the second and third free dorsal rays into spines (they are soft previously); and a general flattening of the whole fish. Young of 76 mm taken at Halifax, one of 114 mm from Campobello (both pictured by Connolly), and others as small as 100–114 mm that were trawled by Bigelow and Schroeder, were at about this stage in their development.

Age and Growth. Capture of a 64-mm specimen in October and of another of 76 mm (date not recorded [Connolly 1921]), both in Halifax Harbor, suggests that goosefish may reach about that length by the onset of their first winter in Gulf waters. One 114 mm long from Halifax studied by Connolly seemed from the thickness of its otoliths to have been in its second summer or autumn; that is, one full year old, which probably applies to three others of 100–114 mm trawled in August and seen by Bigelow and Schroeder.

Fry of European goosefish may be 127–140 mm long by November off Scotland (Fulton 1903b), where spawning commences in March or earlier; this is as large as fry of the American species in their second summer in Gulf waters, where the first growing season is at least 3 or 4 months shorter. Fulton's measurements also point to more rapid growth by the larger Scottish fish than by American goosefish in the Bay of Fundy, namely to 229–406 mm at 1.5 years; to 368–470 mm at 2.5 years; and to about 533 mm at 3 years.

One of the larger fish studied by Connolly showed four concentric rings in its vertebrae; one 787 mm long seemed to have nine rings; one 940 mm seemed to have ten rings; and one 1,016 mm seemed to have twelve rings. It is not certain if these vertebral rings are laid down regularly, one per year. Both European species of *Lophius* have been aged from sections of the illicium (Duarte et al. 1997). In their samples, *L. piscatorius* reached a length of about 70 cm at age 8, *L. budegassa* grew more slowly, reaching a length of a little over 60 cm at age 16.

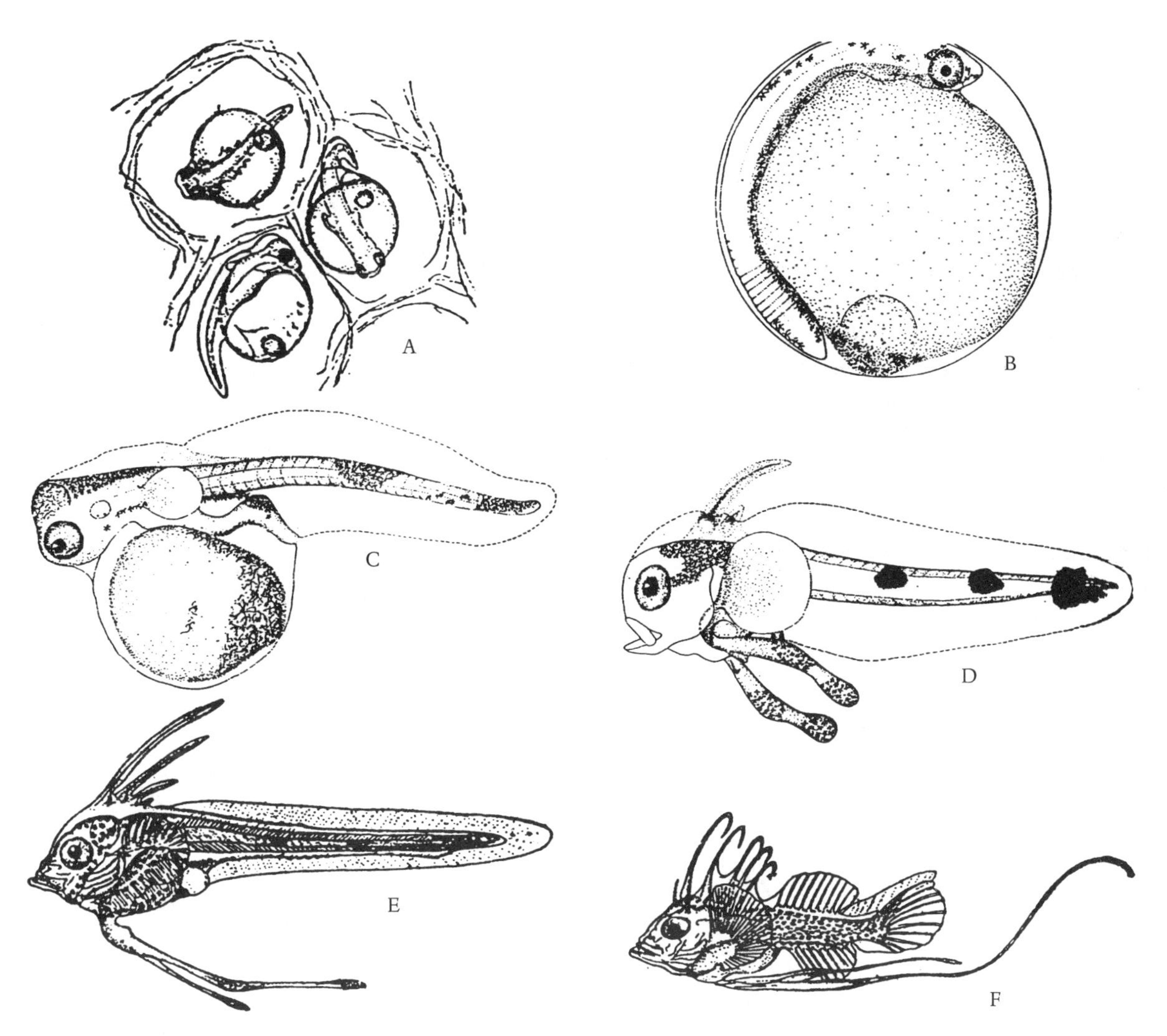

Figure 144. Development of goosefish *Lophius americanus.* (A) Eggs in veil. (B) Egg with advanced embryo. (C) Larva, about 5 days old. (D) Larva, about 12 days old. (E) Older larva. (F) Larva, 30 mm. (From Bigelow and Schroeder: Fig. 285.)

The relationship between total length and total weight (in grams) calculated from 1,216 individuals (Almeida et al. 1995) was $W = 0.0000410L^{2.849}$, $r^2 = 0.983$.

General Range. Coast of eastern North America from the southern and eastern parts of the Grand Banks off Newfoundland and the northern side of the Gulf of St. Lawrence south to the coast of Florida (approximately 29° N [Caruso 1983]). Widespread throughout the Mid-Atlantic Bight out almost to the 1,000 m line (Wood 1982: Map 30). Records from the Gulf of Mexico, Caribbean Sea, and Atlantic Ocean south of Florida (Goode and Bean 1896; Bigelow and Schroeder; Longley and Hildebrand 1941) represent misidentifications (Caruso 1983); specimens collected by Longley and reported as *L. piscatorius* (Longley and Hildebrand 1941) are *Lophiodes*

Occurrence in the Gulf of Maine. This is a familiar fish throughout the Gulf of Maine both along the shore and on the outer fishing banks (Map 21).

Importance. In 1953, Bigelow and Schroeder stated: "No regular commercial use has been made of the goosefish in America up to the present time. But it is an excellent food fish, white-meated, free of bones, and of pleasant flavor, as Dr. Connolly assures us from personal experience. In 1948 (most recent year for which the international fisheries statistics are readily available), English and Scottish vessels landed about 7 million lb of the European species, as 'monk' which fetched nearly as high a price as haddock in English markets, though it brings only about one-half as high a price as haddock in Scotch ports." However, for many years, some New England restau-

meat. In the early 1970s goosefish began appearing on the market "legitimately" as "mock-lobster," and shortly thereafter under the British epithet as "monkfish." By the middle to late 1970s, its popularity as a food fish in its own right, and not as a lobster substitute, had spread across the northeastern and midwestern United States.

Total landings remained at low levels until the mid-1970s, increasing from a few hundred metric tons to around 6,000 mt in 1978 (Idoine 1998a). Landings remained stable at between 8,000 and 10,000 mt until the late 1980s and then increased to a peak level of 26,800 mt in 1996 (Fig. 145). Usually only the tails are landed. Reported landings of tails have increased dramatically since about 1972 (Almeida et al. 1995); landings rose to about 2,300 mt in 1980, and to more than 5,300 mt by 1991–1992.

An export market for goosefish livers developed, increasing steadily from 10 mt in 1982 to 600 mt in 1996. Ex-vessel prices for livers rose from an average of $0.97 per lb to more than $5.00 per lb, with seasonal variations as high as $19.00 per lb (Idoine 1998a), but this export market has dropped off recently.

The NEFSC autumn bottom trawl survey biomass index has declined sharply over the last 15 years (Idoine 1998a) (Fig. 145).

Average size of goosefish has decreased in almost all areas, and size at maturity has also been reduced significantly, leading to the conclusion that the goosefish stock is overexploited and at low levels of abundance (Almeida et al. 1995; Idoine 1998a).

Goosefish are not under management in federal waters but a management plan is under development by the New England and Mid-Atlantic Fishery Management councils.

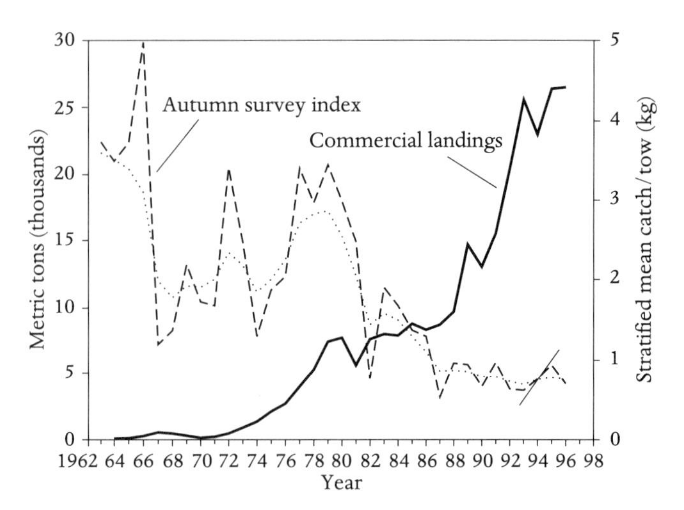

Figure 145. Commercial landings and autumn survey index for goosefish, *Lophius americanus,* from the Gulf of Maine–Mid-Atlantic, 1962–1996. (From Idoine 1998a.)

FROGFISHES. FAMILY ANTENNARIIDAE

THEODORE W. PIETSCH

Frogfishes resemble their relatives, the goosefishes (Lophiidae), in having peculiar armlike pectoral fins, large fleshy pelvic fins (length greater than 25% SL), and the anterior part of the dorsal fin represented by a series of separate spines. They differ strikingly from goosefishes in general appearance, the head and body being laterally compressed instead of dorsoventrally flattened; the soft dorsal fin being much longer than the anal fin; the second and third dorsal fin spines enveloped by fleshy skin so thick that it obscures their true nature; and their mouth being much smaller than that of goosefishes. Of the 12 genera and 41 species of frogfishes (Pietsch and Grobecker 1987), all are benthic except for the one species that reaches the Gulf of Maine, the sargassumfish.

SARGASSUMFISH / *Histrio histrio* (Linnaeus 1758) / Sea Mouse, Mousefish /

Bigelow and Schroeder 1953:541–542 (as *Histrio pictus*)

Description. Head and body appear as one, each gill opening being a small pore on lower margin of pectoral fin near its base, so small that it is likely to be overlooked (Fig. 146). Three detached dorsal fin spines: the first, on snout beyond anterior margin of eye, a tiny slender tentacle, the illicium, bearing a terminal swelling, the esca or bait. The whole structure functions to serve as a lure to attract prey. Second and third spines (inserted close behind first and second, respectively) considerably longer and thicker than first, each enveloped by thick skin and bearing numerous cutaneous appendages and filaments. Other fins also fleshy: soft dorsal fin with approximately twice the number of rays as anal fin, detached tips of both dorsal and anal fin rays often bear short cutaneous appendages. Posterior margin of caudal fin almost straight to conspicuously rounded. Skin nearly always with numerous, variously shaped cutaneous appendages, but otherwise naked, smooth to the touch (some specimens with tiny embedded dermal spinules, visible microscopically in cleared and stained specimens).

Meristics. Dorsal fin rays 11–13; anal fin rays 6–8; pectoral fin rays 9–11; pelvic fin rays I, 5; caudal fin rays 9 (the outermost

Figure 146. Sargassumfish *Histrio histrio.* Drawn by L. E. Cable. (From Bigelow and Schroeder: Fig. 287.)

rays simple, the 7 innermost rays bifurcate); vertebrae 18 or 19, of which 13 or 14 are caudal vertebrae.

Color and color pattern. Highly variable, capable of rapid change from a gray-white, bleached-out appearance to a pattern of streaks and mottling of browns, olive, and yellow, with numerous small brown to black spots and sometimes narrow, irregular white lines; pectoral and pelvic fins sometimes edged with orange; cutaneous filaments white; basidorsal spot rarely present; illicium without banding; dark bars or streaks radiating from eye usually continuous with mottling of head and body (Pietsch and Grobecker 1987:199).

Size. The largest known specimen is a 141-mm SL specimen captured in the Sea of Japan near Obama (UMMZ 204179).

Distinctions. The peculiar armlike pectoral fins, the long fleshy pelvic fins situated on the throat slightly anterior to the pectoral fins, the soft flabby skin, and a laterally compressed body distinguish this fish from any other that is known from the Gulf of Maine. Goosefish, the only close relative inhabiting the Gulf, are of very different appearance, with extremely flattened head and body and enormous mouth.

Habits. Under natural conditions, and except during periods of spawning, *H. histrio* appears to be a nongregarious, solitary predator. But in the laboratory, individuals will not tolerate the near approach of another. Thus to avoid casualties and through cannibalism, individuals must be kept separately or their aquariums must be large and the animals kept well fed.

A number of observers have commented on the aggressive behavior of this species. Gudger (1905) described daily combats between a pair of *H. histrio,* the smaller of the two suffering considerably, its filamentous appendages and the ends of its fins continually bitten off until the animal was finally killed by its tank mate. Gill (1909) referred to *H. histrio* as "in fact, a quarrelsome fish." Mosher (1954) witnessed numerous ferocious attacks of males against females following courtship and spawning behavior, which often resulted in death of the female. Finally, Gordon (1955) wrote that captive specimens of *H. histrio* "bite and tear each other, their fleshy head and body ornaments are ripped to shreds. Their hand-like and foot-like fins become frayed and their delicate fin-rays protrude like broken bones."

Numerous authors have commented on the ability of frogfishes to expand their stomachs enormously by swallowing large quantities of air or water (see references in Pietsch and Grobecker 1987:342–343). In the most significant of these many reports, Gordon (1938) wrote that *H. histrio* sometimes "uses the quick gulping technique for self-defense. If it is attacked by a larger fish, . . . [it] throws open its jaws, swallows water as it is on the point of being devoured, and instantly pumps itself up to an unexpected size. Thus the swallower is forced to cough up the swallowee." Günther (1861) went further by suggesting that body inflation provides a mechanism for dispersal: frogfishes are "enabled, by filling the spacious

water. . . . They are therefore found in the open sea as well as near the coasts, and being bad swimmers, are driven with the currents into which they happen to fall." Although *H. histrio* may drift on the surface in an inflated state for short distances, it seems very unlikely that geographic distributions have been altered substantially in this way.

Food. Frogfishes are among the most voracious of fishes, and anyone who has maintained frogfishes in aquariums knows that they will eat nearly anything and everything that moves, from prey items much smaller than themselves to fishes considerably greater than their own lengths (Pietsch and Grobecker 1987). In describing the feeding habits, Gill (1909) wrote that if "careless or unlucky animals approach too near . . . , the quiescent but hungry fish is stirred instantaneously into vigorous action. It leaps upon its prey as quickly as a tiger would upon its own." Straughan (1954) observed individuals of this species "foolishly try to swallow a fish twice their size and then finally give up, spitting out the lifeless victim and swimming about the aquarium in a rage." *Histrio* are major predators of the sargassum complex, second only to jacks, filefishes, and triggerfishes (Dooley 1972). Frogfish are such indiscriminate feeders that they will consume large numbers of their own kind; reports of cannibalism are numerous (Pietsch and Grobecker 1987). It is not at all unusual to find a dozen or more small individuals of *H. histrio* in the stomach of a larger one.

Reproduction and Early Life History. Like those of all other lophiiform fishes, the eggs of *H. histrio* are spawned encapsulated within a nonadhesive mucoid mass or, more typically, a continuous, ribbonlike sheath of gelatinous mucous, often referred to as an "egg raft" or "veil." These rafts reflect the form and structure of the ovaries (Gill 1909). The egg raft was described as "a soft jelly-like mass, quivering to the touch, but withal rather tenacious and [after full expansion in the water] 90–120 cm long by 5–10 cm or thereabouts in breadth. . . . The entire mass is thickly permeated with eggs, which appear to be in several irregular layers, or at least more than one." For a detailed description of ovaries and egg masses, one should refer to Rasquin (1958), and for courtship and spawning behavior, with figures of egg rafts and courtship, to Mosher (1954). Pietsch and Grobecker (1987) provide excellent summaries of these papers.

Age and Growth. Little is known about growth beyond the juvenile stage or about age and size at maturity. By 15 mm SL, growth becomes relatively isometric (Adams 1960). The length-frequency histogram for *H. histrio* collected under sargassum in the Florida Current during 1966–1967 suggests that growth from 5 to 45 mm SL took 4–5 months (Dooley 1972; Martin and Drewry 1978).

General Range. Sargassumfish have the broadest longitudinal and latitudinal range of any frogfish. Their distribution is largely dependent on the dispersal capabilities of floating sargassum weed with which they are apparently an obligate associate (Dooley 1972). They are found across the Atlantic, Indian, and western Pacific oceans as far east as the Hawaiian Islands (Pietsch et al. 1992). In the western Atlantic, they range from the Gulf of Maine to the mouth of the Rio de la Plata, Uruguay. On the eastern side of the Atlantic, they are apparently quite rare; specimens are known from the Azores and off west Africa. A record from Vardø, northern Norway (Düben and Koren 1846), is no doubt based on a straggler taken northward by the North Atlantic and Norwegian currents.

Occurrence in the Gulf of Maine. A 12-cm specimen taken in a purse seine near the surface over the west central part of Georges Bank, by the schooner *Old Glory* on 15 September 1930 (Firth 1931), and a second of 5 cm, taken off the southeastern slope of Georges Bank, by the swordfisherman *Leonora C*, on 15 June 1937, are the only known records from the Gulf of Maine. They represent the northernmost occurrences of *H. histrio* in the western Atlantic. They have been picked up from time to time near Woods Hole (Smith 1898a; Sumner et al. 1913).

GAPERS, SEA TOADS, OR COFFINFISHES. FAMILY CHAUNACIDAE

JOHN H. CARUSO

The family Chaunacidae is a poorly known, rarely encountered group of bottom-dwelling anglerfishes that occurs in all but polar seas at depths of 90–2,680 m. Like other anglerfishes, they have the first dorsal fin spine modified as an angling apparatus (the illicium) but in these fishes it is very short and has a terminal bait or esca comprised of a dense cluster of short cirri, giving the apparatus the appearance of a short-handled mop. The illicium is retractable and can be retracted into an ovoid, scaleless, patch or depression immediately behind it. Two additional cephalic dorsal fin spines are present as embedded vestiges, and postcephalic dorsal fin spines are absent. Chaunacids have a large, oblique to nearly vertical mouth with relatively small, sharp, slender teeth, a large globose head with a conspicuous network of open sensory canals. A single open lateral-line canal extends posteriorly along a slightly compressed trunk and tail. They are moderate-sized (usually <30 cm total length), generally pink or reddish in background color, and have very loose, flaccid skin densely covered with minute, spinelike scales that are somewhat similar both in shape and feel to the placoid scales of some sharks.

Chaunacids may be distinguished from other anglerfishes by their short illicium and moplike esca, their pink or reddish skin with small, spinelike scales, their complete lack of external dorsal fin spines behind the illicium (a character they share with batfishes, Ogcocephalidae), and their large, globose head with its conspicuous network of open sensory canals and its large, oblique, nearly vertical mouth that bears numerous, small, sharp teeth. Of the two genera and 16 species in the family (Caruso 1989), only a single species occurs in the Gulf of Maine. A second species, *Bathychaunax roseus,* was recently taken in deep water (904 m) near the Gulf of Maine off Nova Scotia (43° 40.4′ N, 61° 24.4′ W [ARC 9713145]).

REDEYE GAPER / *Chaunax stigmaeus* Fowler 1946

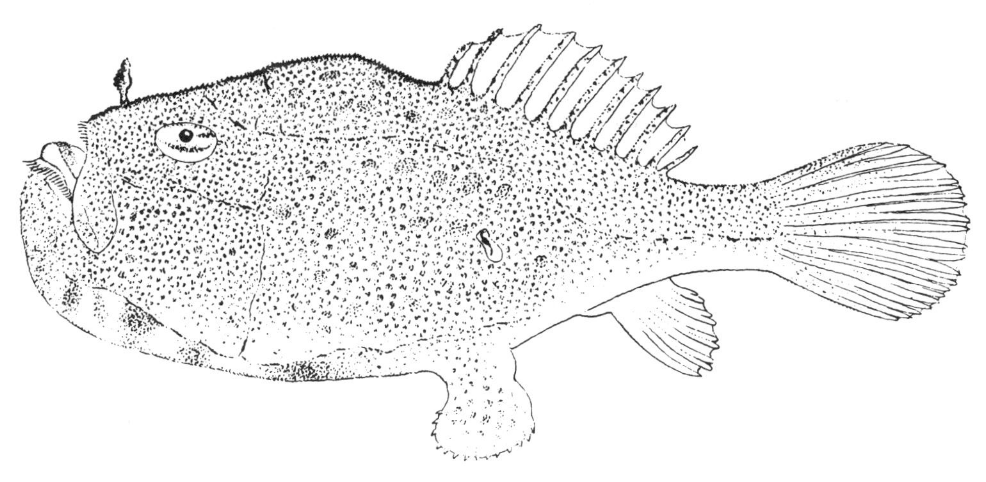

Figure 147. Redeye gaper *Chaunax stigmaeus.* (From Fowler 1946: Fig. 2.)

Description. Body short, very slightly compressed, tapering rapidly (Fig. 147). Head large, globose, roughly rectangular in cross section. Prominent network of open sensory canals on head. Mouth large, oblique to nearly vertical, upwardly directed, with numerous, small, sharp, slender teeth. Eyes covered by oval, scaleless, transparent patch of skin. Spinous dorsal fin represented only by a short moplike angling apparatus. Illicium, when retracted, lies in a concave, darkly pigmented cavity; other external dorsal fin spines absent. Soft dorsal and anal fins rather far back on trunk. Pectoral fins narrow, paddle-like, at about midbody. Structure of pectoral radials similar to that of other anglerfishes, but pectoral fins not as armlike as those of other anglerfishes because loose, flaccid skin covers all but the fins themselves. Like most other anglerfishes (some lophiids excepted), gill openings are small circular holes, just behind and well above pectoral fins. Pelvic fins narrow, just protruding from loose skin on ventral surface, well anterior to pectoral fins, just behind level of eyes. Body covered with minute, densely packed, spinelike scales. Lateral line system distinctive, consisting of a series of open channels that are especially prominent on head; a single, open channel on body. Within each channel is a linear series of exposed neuromasts, each partially covered by a modified scale forming an open arch over neuromast.

Meristics. Dorsal fin rays 11 or 12; anal fin rays 6 or 7; pectoral

unbranched (Caruso 1989). There are 34–39 neuromasts in the lateral line proper, 2 or 3 (usually 2) in the upper preopercular canal, 3 or 4 (usually 3) in the lower preopercular canal, 3 in the hyomandibular canal, and 10–15 (usually 11) in the pectoral canal (for discussion of the chaunacid lateral line system and its taxonomic significance, see Caruso 1989).

Color. In preservative, redeye gaper are uniform pale tan above with darker tan spots or blotches of varying sizes. The large spots are approximately half the eye diameter, and the small spots are one-quarter to one-third the diameter of large spots (the name *stigmaeus* is derived from the Greek *stigma* for spotted). The dorsal pattern continues onto the proximal portions of the caudal and pectoral fins. Ventral surfaces of the body, pectoral, pelvic, and anal fins are pale tan and devoid of spots. The interradial membrane of the soft dorsal fin is transparent. The illicial cavity is very dark brown or black; the illicium is pale tan with a dark ring or band just below the esca; and the esca is black anterodorsally and white or translucent posteroventrally. The peritoneum is dusky.

In his original description of *Chaunax stigmaeus,* Fowler (1946) described the coloration as follows: "Color when fresh in alcohol with entire back, sides and upper surfaces dull olivaceous brown, marked or specked with innumerable and variable small dark olive markings. These markings also extend on the dorsal, caudal and pectoral fin rays, though their

the back and sides also have variable suffusions of shades of red. Iris deep rose red. Naked skin around maxillary and lips all more or less red or orange. Under surface of head and abdomen with brighter red and orange to rose tints, along lower side of head 4 vivid rose red blotches with adjacent vermilion and orange tints. Upper basal portion of pectoral with bright light olive blotches, the separating intervals more or less vermilion and on underside of each fin a transverse indistinct rose red bar. Anal red."

Size. The largest of 17 known specimens (MCZ 41908) is a 235-mm SL specimen from Georges Bank (Caruso 1989).

Distinctions. No other Gulf of Maine anglerfish have the spinous dorsal fin represented only by a short moplike angling apparatus, and no other Gulf of Maine fish have such loose, flaccid, skin densely covered with fine, spinelike scales.

Habits. Little is known of the biology other than depth range for seven lots, 90–699 m (mean 309 m, Caruso 1989).

General Range. *Chaunax stigmaeus* occurs off the Atlantic coast of the United States from Georges Bank (ca. 43° N) to the Norfolk Canyon (ca. 36° N).

Occurrence in the Gulf of Maine. Redeye gaper are rare throughout their range, and there are only three records from the Gulf of Maine (Caruso 1989): one from Georges Bank (MCZ 41908) at the seaward limit of the Gulf of Maine (the 275-m contour [Bigelow and Schroeder]), and two off Cape Sable near 43° N, 63° W at depths of 189 m (ARC uncat.) and 366 m (the latter [ARC 42694]), is technically beyond the seaward limit of the Gulf of Maine).

SEADEVILS OR DEEP-SEA ANGLERFISHES. FAMILY CERATIIDAE

THEODORE W. PIETSCH

Ceratiids, together with members of several other closely related families of deep-sea anglerfishes (suborder Ceratioidei), differ from other anglerfishes in lacking pelvic fins and the armlike structure of the pectoral fins; but, more significantly, these deep-sea forms differ strikingly in having a bioluminescent bacterial bait and an extreme sexual dimorphism in which males are dwarfed (reaching only a small fraction of the size of the females) and become permanently and parasitically attached to relatively gigantic females. For a full synopsis of the 11 families, 34 genera, and approximately 135 species of ceratioid fishes, the reader is referred to Bertelsen (1951). The head and body of the female is laterally compressed (not dorsoventrally flattened as in goosefishes); the soft dorsal and anal fins are very short, the dorsal containing three to five rays (nearly always four), the anal with four rays; the caudal fin contains nine rays (the ninth or ventralmost ray reduced to a small remnant), the four innermost rays bifurcate. The mouth is oblique or nearly vertical when closed. Associated with a meso- or bathypelagic habitat, the body is noticeably soft and flabby. Eyes of adult females are very small, and those of large specimens (greater than approximately 30 cm SL) become covered with pigmented skin so that the animal appears to be blind.

Ceratiids are oceanic as a group, living at middepths, primarily between approximately 400–2,000 m. They are typically dark brown to deep black in color as are so many other pelagic fishes of that same depth zone. The genus *Ceratias* contains three species, one of which is known from the Gulf of Maine.

NORTHERN SEADEVIL / *Ceratias holboelli* Krøyer 1845 / Bigelow and Schroeder 1953:543–545

Description. Body of female strongly compressed; eyes very small (sometimes covered with darkly pigmented skin), set high on head; mouth nearly vertical when closed (Fig. 148). Fishing apparatus very long and extremely slender. Emerging anteriorly from top of head, this structure consists of an elongate basal pterygiophore, which lies in a deep groove along dorsal surface of head and back, and a thin, highly flexible dorsal fin spine (illicium, length 14.5–37.8% SL), bearing a distal bioluminescent bait (esca) (see Bertelsen 1951; Pietsch 1974; Munk and Bertelsen 1980). Esca with a single slender escal appendage, usually simple but sometimes with as many as three short filaments on each side (Pietsch 1986: Fig. 2). The fishing apparatus of *C. holboelli* corresponds to the whiplike head spine of the goosefish, but in the former it emerges farther back, approximately above the eyes. The basal pterygiophore is provided with retractor muscles by which it can be withdrawn posteriorly within a tunnel-like sheath along the head and back, allowing the "bait" to approach the mouth. In this retracted position, the posterior end of the pterygiophore extends backward, enveloped by elastic skin, emerging from the rear end of the fish, just before the insertion of the soft dorsal fin (Bertelsen 1943). When the basal pterygiophore is pulled forward by protractor muscles, its protruding posterior end is withdrawn into the sheath, either partially, when the posterior extension appears as a short fingerlike process, or wholly, leaving an indentation or pore on the midline of the back, as in the Gulf of Maine specimen pictured in Figure 148. Close behind the posterior extension of the pterygiophore (or behind the

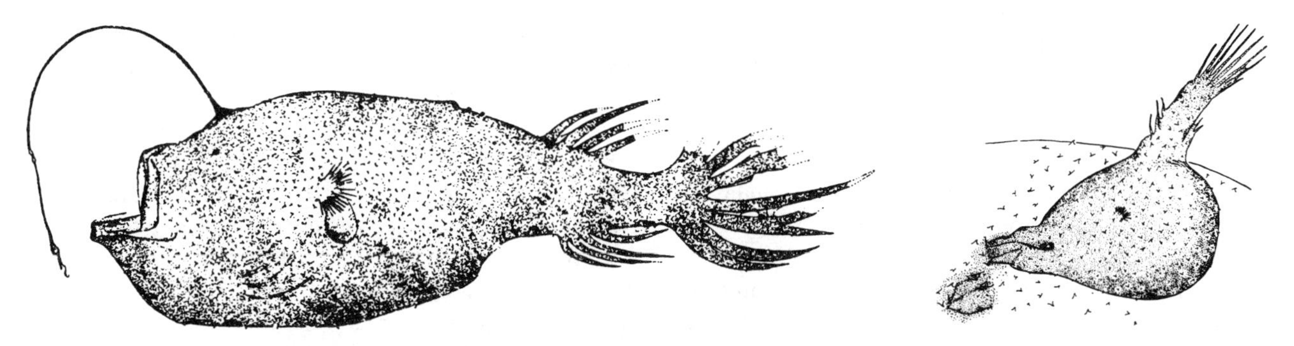

Figure 148. Northern seadevil *Ceratias holboelli.* Off Mt. Desert Rock, adult female (left) and parasitic male (right) that was attached to her. (From Bigelow and Schroeder: Fig. 288.)

pore representing the latter) are a pair of bulbous, bioluminescent appendages called "caruncles," scarcely noticeable on large specimens but more conspicuous on small ones (Bertelsen 1951:16, 239). These glandular structures, each bearing a small terminal opening, differ from the escal light organ in that the wall of the gland and the surrounding skin are entirely covered with pigment, so luminescence is possible only by the emission of a secretion through the terminal opening. Luminescence was first observed from caruncles of a living ceratiid by Bertelsen (1951:239) while at sea aboard the Danish research vessel *Dana* in 1947. The skin of females is covered everywhere with small, close-set dermal prickles in very small individuals, but with broad-based, conical thorns in larger specimens. Eyes of females are minute, apparently functional in small individuals, but covered over by darkly pigmented skin in large specimens. The gill opening is very small and oval in shape, situated below and slightly behind the base of the pectoral fin. Jaw teeth of females are slender, recurved (directed into the mouth), and depressible, those in lower jaw considerably larger and slightly more numerous than those in upper. Vomerine teeth (one to three on each side) are nearly always present in females smaller than about 80 mm SL, but rarely present in larger specimens (Pietsch 1986: Table 1). The upper pharyngobranchials are heavily toothed.

Free-living adolescent males of *C. holboelli* are equipped with a pair of large denticular teeth on the snout, fused at the base and articulating with the pterygiophore of the illicium; two similar pairs of teeth are situated on the tip of the lower jaw (Bertelsen 1951:129). Jaw teeth are absent and the premaxillae are degenerate. The eyes are large, bowl-shaped, directed laterally, the axis short, the pupil much larger than the lens. Olfactory organs are minute; caruncles, characteristic of females, are absent; and the skin is naked. Adult males become permanently attached to females as parasites; their skin becomes spinulose and their eyes and gut degenerate. For a full description of the males of *Ceratias,* see Bertelsen (1951:137–138).

Color. Females in preservative are dark red-brown to black over the entire surface of the body (except for the distal portion of the escal bulb) and oral cavity. The skin is unpigmented in adolescent males, but darkly pigmented in the parasitic stages.

Size. Females of *C. holboelli* represent the largest known ceratioids, attaining at least 770 mm SL. All known free-living males measure less than 12 mm SL, whereas parasitically attached males range from 9.8 to 118 mm SL (Pietsch 1976, 1986).

Distinctions. This deep-sea anglerfish is so bizarre in appearance that there is no danger of confusing it with any other fish in the Gulf of Maine, except for other members of the suborder.

Habits. Adolescents and adults of *C. holboelli* may be captured anywhere between 150 and 3,400 m, but most known specimens were taken between 400 and 2,000 m. A number of large adults have been found in relatively shallow water in high latitudes of the North Atlantic and in the Bering Sea, some of them from unknown depths and others in as little as 120–680 m. The 770-mm SL holotype of *Reganichthys giganteus* (MCZ 36042), a junior synonym of *Ceratias holboelli,* was captured by otter trawl between the surface and 230 m. The average maximum depth for known captures was 1,143 m (Pietsch 1986).

Reproduction. *Ceratias holboelli* are among a number of deep-sea anglerfishes that display a bizarre and unique mode of reproduction in which males are dwarfed and become permanently and parasitically attached to much larger females. Males of most species are equipped with large nostrils, apparently for homing in on a female-emitted, species-specific pheromone; normal jaw teeth are lost during metamorphosis, but are replaced by a set of pincerlike denticles at the anterior tips of the jaws for grasping and holding fast to a prospective mate. In some forms, attachment appears to be followed by fusion of

for blood-transported nutriment, while the female becomes a kind of self-fertilizing hermaphroditic host. Since the time of its discovery 70 years ago, the story of sexual parasitism in ceratioid anglerfishes has become a part of common scientific knowledge. However, the known facts concerning this remarkable reproductive strategy have not been thoroughly and satisfactorily analyzed, despite the elegant research of Bertelsen (1951:251) and Munk and Bertelsen (1983).

In 1922, the Icelandic ichthyologist Bjarni Sæmundsson published a description of two small fish attached by their snouts to the belly of a large female deep-sea anglerfish identified as *C. holboelli*. Not recognizing them as dwarf males, Sæmundsson described them as the young of the same species: "I can form no idea of how, or when, the larvae, or young, become attached to the mother; I cannot believe that the male fastens the egg to the female. This remains a puzzle for some future researcher to solve" (Sæmundsson 1922, translation in Regan 1925). Three years later, Regan dissected a small fish attached to a newly discovered female *C. holboelli* and concluded that the small fish must be a male, parasitic on the female. The male fish is "merely an appendage of the female, and entirely dependent on her for nutrition. . . . So perfect and complete is the union of husband and wife that one may almost be sure that their genital glands ripen simultaneously, and it is perhaps not too fanciful to think that the female may possibly be able to control the seminal discharge of the male and to ensure that it takes place at the right time for fertilization of her eggs" (Regan 1925; Bertelsen 1951:244–250).

All evidence indicates that the sexual parasitic mode of reproduction is obligatory in *Ceratias*. Examination of numerous specimens shows that free-living males and nonparasitized females never have well-developed gonads. Males thus do not mature unless they are in parasitic association with a female, and females probably never become gravid until stimulated by the permanent attachment of a male. The jaw apparatus of free-living males seems to be unsuited to capturing prey and the alimentary canal is undeveloped, indicating that males do not feed after metamorphosis and thus depend on a parasitic association with a female for long-term survival. These two lines of evidence suggest that spawning and fertilization in *Ceratias* occur only during a permanent parasitic association of male and female (for a more detailed review, see Pietsch 1976).

Early Life History. Little is known about early life history of *C. holboelli*. No spawned, fertilized eggs of this species (or of any ceratioid) have been described, but it is assumed that eggs of *Ceratias*, like those of other lophiiform fishes, are expelled in free-floating mucoid egg rafts or veils. Larvae and metamorphosed stages are known and have been described in some detail: larvae have a peculiar "humpbacked" appearance (Bertelsen 1951: Fig. 90, 1984: Fig. 168); the head and body are enveloped by inflated, transparent skin; the mouth is subvertical; the pectoral fins are unusually small, not reaching beyond the dorsal and anal fins; pelvic fins are absent; female larvae have two caruncles on dorsal surface of trunk. Subdermal pigmentation is present in the larvae of many ceratioids, but *Ceratias* is characterized by a complete lack of pigment (Bertelsen 1951, 1984).

General Range. *Ceratias holboelli* has a broad distribution in all three major oceans of the world, but is replaced in the Southern Ocean by *C. tentaculatus*. It is known from throughout the North Atlantic, ranging as far north as the Strait of Denmark and extending south to approximately 8° S in the central Atlantic.

Occurrence in the Gulf of Maine. A female of about 770 mm SL, with an attached male, showing a scar from the attachment of another (Fig. 148), was taken about 19.3 km south of Mount Desert Rock, off Portland, Maine, somewhere between the surface and 230 m, on 1 October 1943, by the schooner *Dorothy and Ethel II*, Capt. Harold Paulsen. This specimen is the holotype of *Reganichthys giganteus* (MCZ 36042), described by Bigelow and Barbour (1944a) and replaced by *Reganula giganteus* by Bigelow and Barbour (1944b). A second female, about 480 mm SL, trawled on the southeast part of Georges Bank, between 275 and 366 m, on 9 February 1927, appears to belong to this same species, but was so badly damaged that positive identification was impossible. This specimen (MCZ 35771) was described by Barbour (1942) as a new species, *Typhloceratias firthi*. The head and back of the specimen were so badly damaged that the fishing apparatus was torn away; thus having lost the primary characters of the esca that allow for proper identification, Barbour's name has been designated a *nomen dubium* (see Pietsch 1986:487). A third probable Gulf of Maine record of *C. holboelli*, a female approximately 90 cm TL and weighing about 9 kg, was taken by the trawler *Ebb* in about 468 m on Georges Bank in June 1936. Photographs appeared in the *Boston Globe* and the *Boston Post* for June 29 of that year but the specimen itself was lost.

ACKNOWLEDGMENTS. Drafts of the entire lophiiform section were reviewed by the authors of each part, John H. Caruso and Theodore W. Pietsch.

MULLETS. ORDER MUGILIFORMES

There is considerable disagreement concerning the phylogenetic relationships of the one family placed in this order, with Stiassny (1993), Parenti (1993), and Johnson and Patterson (1993) presenting conflicting views. Johnson and Patterson (1993) place the mullets, along with four other groups in the Smegmamorpha close to the Atherinomorpha, similar to the arrangement proposed by Stiassny (1993). Nelson (1994) places the mugiloids as the first order in the superorder Acanthopterygii, the spiny-rayed fishes, right before the atherinoids and did not recognize Smegmamorpha.

MULLETS. FAMILY MUGILIDAE

BRUCE A. THOMPSON

The Mugilidae contains about 17 genera and between 66 and 80 species (Harrison and Howes 1991; Nelson 1994). They inhabit coastal marine and brackish waters in all tropical and temperate seas. Some species are freshwater. Most species feed on vegetable, mud, and detritus. They have long gill rakers; their teeth are small or absent; and their alimentary tract consists of a gizzard, a thick-walled stomach, and a long, coiled intestine reflecting their herbivorous life style. The coelomic lining is black. Mullet possess two dorsal fins, the first of weak spines and the second of all soft rays, although the first element is an unbranched ray that has frequently been considered to be a spine. Only two species of *Mugil* occur along the Atlantic coast of the United States north of Georgia (Collins and Stender 1989), and both stray into the Gulf of Maine.

KEY TO GULF OF MAINE MULLETS

1a. Anal fin with 11 total elements, 3 spines and 8 rays in juveniles and adults, 2 spines and 9 rays in young less than 40 mm; spot on each scale forms stripes along body; no distinct colors on head or body except for diffuse yellow-gold wash; second dorsal and anal fin nearly scaleless. . . . . **Striped mullet**

1b. Anal fin with 12 total elements, 3 spines and 9 rays in juveniles and adults, 2 spines and 10 rays in young less than 40 mm; no distinct stripes along body; fresh specimens with distinct yellow-gold spot on upper opercle; second dorsal and anal fin nearly covered with rows of fine scales . . . . . **White mullet**

STRIPED MULLET / *Mugil cephalus* Linnaeus 1758 / Black Mullet, Gray Mullet /

Bigelow and Schroeder 1953:305–306

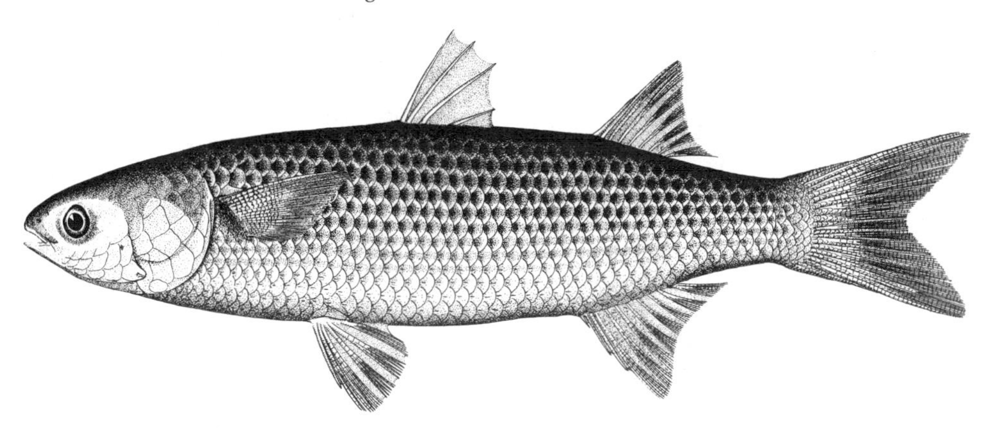

Figure 149. Striped mullet *Mugil cephalus.* Woods Hole, Mass., 182 mm. Drawn by H. L. Todd. (From Bigelow and Schroeder: Fig. 164.)

Description. Body fusiform, robust, somewhat compressed and oval in cross section (Fig. 149). Snout blunt, pointed, shorter than orbit diameter. Head large, broader than deep. Eye large, with a prominent adipose eyelid covering much of eye. Body and much of head covered with large ctenoid scales (Roberts 1993). Reports of cycloid scales (Hildebrand and Schroeder 1928) are related to size. First dorsal fin naked; second dorsal fin with some scales anteriorly and basally. Spinous dorsal inserts slightly behind insertion of pelvic fins; second size as second dorsal. Pelvic fins anterior on abdomen. Pectoral fins pointed, high on side of body near upper margin of opercle. Enlarged pectoral and pelvic fin axillary scales present. Lateral line poorly developed. Mouth moderate-sized, resembling a curved inverted v when closed. Upper lip with an outer row of small unicuspid teeth and usually two inner rows of bicuspid teeth. Lower lip with an outer row of unicuspid teeth, sometimes with an inner row of bicuspid teeth.

scales 37–43; predorsal scales 23–26; gill rakers numerous, 24–36 on upper arch, 50–76 on lower arch, number increasing with size; vertebrae 11 +13 or 12 +12, total 24; pyloric caeca 2 (Pillay 1962; Thomson 1963, 1966, 1981; Martin and Drewry 1978; Ditty et al. 2000).

Color. Coloration often varies with habitat and salinity, but usually silver-white on sides and venter with dorsum darker gray or blue-black. Spot on each scale forms stripes along body (from which common name is derived). Fins usually clear or slightly dusky. Pectoral fin with large black blotch at base. Little chromatic color present except for a yellow wash on pelvic fins and yellow in eye. Small juveniles can be bright silver.

Size. Maximum size of females 622 mm and males 599 mm (Thomson 1963). Gopalakrishnan (1971) listed a 914-mm TL striped mullet from India. Large females seldom exceed 450–550 mm FL; males average slightly smaller. The all-tackle game fish record is 3.14 kg for a striped mullet caught near Chiba City, Japan, in September 1995 (IGFA 2001).

Distinctions. Combination of an adipose eyelid, two dorsal fins with the first with only four spines, pectoral fins high on the body, pelvic fins anterior on the abdomen, the strongly forked tail, and black peritoneum distinguish mullets from other fishes in the Gulf of Maine. Atlantic silversides resemble mullets, but the second dorsal fin is much shorter than the anal fin. Silversides have more anal rays than do mullets (15–24 vs. 8 or 9). Striped mullet have one fewer anal element than white mullet for a total of 11 rather than 12.

Habits. Striped mullet are euryhaline, ranging in coastal waters from open water, inland in estuaries and lagoons, readily penetrating far up into riverine freshwaters. Movement offshore is related to spawning. They have a wide range of temperature and salinity tolerance: temperature 5°–37°C, salinity 0–81 ppt. Immense schools can form at times feeding either at the surface or more often gulping mouthfuls of sediment.

Food. In the Gulf of Mexico, striped mullet are strongly associated with bottom feeding (Leard et al. 1995). They feed on detritus, microscopic algae, and small particulate matter high in organic materials (Odum 1970).

Predators. Predators include birds and a large variety of sharks and other fishes (Leard et al. 1995).

Parasites. Striped mullet are susceptible to a wide range of parasites and diseases (Paperna and Overstreet 1981; Leard et al. 1995). Parasitic isopods are common in the pharyngobranchial cavity of mullets (I. Harrison, pers. comm., March 1998).

Breeding Habits. Striped mullet spawn at night, at the surface, about 60–80 km offshore in the Gulf of Mexico (Arnold and Thompson 1958). Some reports suggest estuarine spawning but Thomson (1963) and Render et al. (1995) argue against this. They are winter spawners (Render et al. 1995). Spawning occurs from November through April off the coast of Georgia and Florida, peaking in January and February (Collins and Stender 1989).

Early Life History. Striped mullet do not spawn in the Gulf of Maine. Extensive descriptive reviews of eggs, larvae, and prejuveniles are presented by Anderson (1958), Martin and Drewry (1978), Fahay (1983), and Ditty et al. (2000).

General Range. Striped mullet are worldwide, with permanent populations confined between 51° N and 42° S (Thomson 1963, 1966). They are found on both sides of the Atlantic, in the western Atlantic from Nova Scotia to Brazil, including the Gulf of Mexico (Leard et al. 1995). Northern records consist primarily of juveniles. Striped mullet are quite common on the south side of Cape Cod, where they sometimes enter coastal rivers (Clayton et al. 1978).

Occurrence in the Gulf of Maine. There are very few records of striped mullet in the Gulf of Maine proper with only a 28-mm SL juvenile caught in the general area on the outer edge of Georges Bank at 41°45′ N, 66°02′ W (Scott and Scott 1988). All earlier Canadian records (Vladykov 1935a; Leim and Scott 1966; Gilhen 1969) are thought to be *Mugil curema* (Gilhen 1972; Scott and Scott 1988) and the reports of Bigelow and Schroeder, Jackson (1953), and Scattergood and Coffin (1957) for the U.S. part of the Gulf are also probably *M. curema,* but voucher specimens are not available so identifications cannot be verified. However, Fairbanks and Lawton (1977) reported a remarkable occurrence of several hundred large striped mullet in the discharge canal of the Pilgrim Nuclear Generating Station at Plymouth in late November and early December 1975. During this period, mullet were also reported from elsewhere in Massachusetts Bay. A sample of 14 specimens (MCZ 52713) measured 338–572 mm TL and scales indicated that they were from age-classes 1–4 (K. Hartel, pers. comm.).

Importance. Striped mullet are of no commercial importance in the area but are important food fishes in southern parts of their range and are of interest to mariculture (Oren 1981).

WHITE MULLET / *Mugil curema* Valenciennes 1836 / Silver Mullet

Description. Smaller, more slender than striped mullet (Fig. 150). Much of the description given for striped mullet applies also to white mullet. Upper lip with an outer row of small unicuspid teeth, an inner row of smaller unicuspid teeth sometimes present. Lower lip with a single row of small unicuspid teeth.

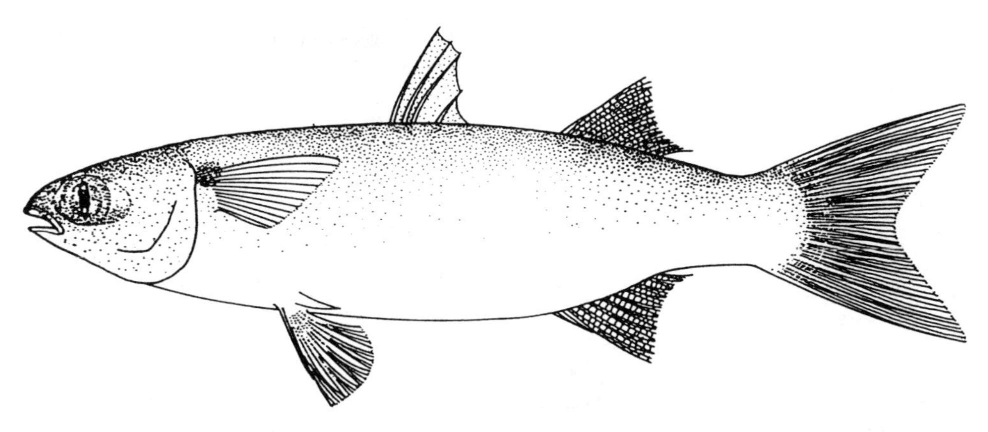

Figure 150. White mullet *Mugil curema*. (© Food and Agricultural Organization of the United Nations.)

Meristics. Dorsal fin rays IV, 9; anal fin rays III, 9, changing from II, 10 with growth; pectoral fin rays i, 14–17, usually i, 15 or 16; lateral line scales 33–39; gill rakers 46–84, 46 in a 34-mm specimen, number increasing with size; vertebrae 11 +13 or 12 +12, total 24; pyloric caeca 2 (Hoese and Moore 1977; Martin and Drewry 1978; Thomson 1981; Ditty et al. 2001).

Color. Coloration varies, but usually silver-white on sides and venter and bluish, dark gray or blue-black on back. This species usually lacks the streaks present on the sides of the body in striped mullet. Small dark spot at base of pectoral. Opercle with distinct yellow-gold spot. Yellow and gold pigment in iris of eye confined in narrow vertical area above and below pupil. Small juveniles often bright silver.

Size. Smaller than striped mullet. Böhlke and Chaplin (1968) suggest a maximum size of 910 mm, but most literature on Atlantic specimens gives adult sizes of 250–350 mm. The all-tackle game fish record is 0.68 kg for a white mullet caught in Rio de Janeiro in April 2001 (Anon. 2001).

Distinctions. White mullet are smaller and more slender than striped mullet and have one more element in the anal fin, 12 fin rays instead of 11. Perez-Garcia and Ibanez-Aguirre (1992) compared white mullet and striped mullet morphologically.

Habits. White mullet tend to be found in warmer waters and in higher salinity than striped mullet, preferring salinities of 20–35 ppt and being more closely associated with passes and open nearshore waters (Moore 1974). Unlike striped mullet they do not penetrate freshwaters. Movement offshore is related to spawning.

Food. Food habits (Moore 1974) are similar to those of striped mullet. Inorganic particles and detritus make up most of the diet in Cuba (Gonzalez and Alvarez-Lajonchere 1978).

Parasites. Ciliates, copepods, digenetic trematodes, cestodes, nematodes, spiny-headed worms, and leeches have been listed as parasites of *M. curema* (Paperna and Overstreet 1981).

Breeding Habits. White mullet are open-water spawners, with small young found 80–90 km offshore (Anderson 1957; Caldwell and Anderson 1959). The spawning season extends from late March to September with a peak from April to June (Anderson 1957; Collins and Stender 1989). No spawning occurs in the Gulf of Maine.

Early Life History. Extensive descriptions of eggs, larvae, and prejuveniles are presented by Anderson (1957), Martin and Drewry (1978), and Ditty et al. (2000).

General Range. White mullet have a more restricted distribution than striped mullet. In the western Atlantic, from Brazil to Massachusetts (Martin and Drewry 1978), including Bermuda, but the range actually extends north to Nova Scotia (Gilhen 1972). In the eastern Atlantic, they occur off the west coast of Africa south to Namibia. They are also found in the eastern Pacific from southern California to Chile.

Occurrence in the Gulf of Maine. Although white mullet are quite common in bays and estuaries on the south side of Cape Cod, they are rare north of the Cape. Gilhen (1972) reported over 300 white mullet from Nova Scotia and questioned many of the earlier reports of striped mullet (Bigelow and Schroeder; Jackson 1953; Scattergood and Coffin 1957). These reports were probably based on white mullet but voucher specimens were not saved for identification. New records include three juveniles (25–39 mm SL) taken in the mouth of the Webhannet River at Wells, Maine, in mid-August 1988 by S. Murphy and deposited in the University of Massachusetts collection at Amherst. As is the case with striped mullet, presence of young-of-the-year white mullet is probably the result of waifs

Importance. White mullet are caught throughout their range. They are one of the most abundant species of gray mullets in some areas, especially along the Caribbean coast and some islands of the West Indies, where they are regularly harvested. White mullet are of no importance in the Gulf of Maine.

ACKNOWLEDGMENTS. Drafts of the mugilid section were reviewed by Ian Harrison and Karsten E. Hartel. This is contribution number LSU-CFI-96-07.

ATHERINIFORM FISHES. ORDER ATHERINIFORMES

The order Atheriniformes contains about 285 species (Nelson 1994) divided into two suborders, six families, and 49 genera (Dyer and Chernoff 1996). Usually two, well-separated dorsal fins, the first, when present, with flexible spines. Weak or flexible spine present as first or outer element of second dorsal, anal, and pelvic fins. Lateral line absent or very poorly developed. Generally slender fishes with body depth less than 20% SL. Scales usually cycloid. Ventral face of vomer concave. Lacrimal tendon of the *adductor mandibulae 1* muscle long, attaching to subnasal shelf of lacrimal. Pelvic girdle to rib attachment well developed. Median plate not reaching anterior tip of pelvic bone. Larvae with a single mid-dorsal row of melanophores. Preanal length less than 40% of body length at hatching-flexion stage. Two species of the suborder Atherinopsoidei, family Atherinopsidae reach the Gulf of Maine.

NEW WORLD SILVERSIDES. FAMILY ATHERINOPSIDAE

BARRY CHERNOFF

New World silversides range in size from less than 20 mm to greater than 1 m SL and inhabit subarctic to tropical marine, estuarine, and freshwater environments of the New World. The majority of species north of the equator, including those in the Gulf of Maine, are relatively small silvery fishes, less than 150 mm SL, whose appearances are often likened to those of smelts, mullets, or herrings. Silversides of the Gulf of Maine may be distinguished from nonatherinopsid fishes by the following combination of external characters: two well-separated dorsal fins; spinous dorsal relatively weak and second dorsal fin with a single, flexible spine at its origin; anal fin with a single, weak spine; pelvic fins I, 5 and intermediate in position between ventral portion of cleithrum (pectoral girdle) and anal fin origin; lacking a continuous lateral line; cycloid scales; small teeth in both jaws; and forked caudal fin. Two species of *Menidia* are known from the Gulf of Maine. Because both species exhibit extreme geographic variation, counts are presented for only northern populations: New York to Maine for *Menidia beryllina* and New York to Prince Edward Island for *Menidia menidia*. Methods for counting and measuring these species are given in Chernoff et al. (1981).

KEY TO GULF OF MAINE NEW WORLD SILVERSIDES

1a. Spinous dorsal fin anterior to vertical through anus; segmented anal fin rays 13–19, modally 16 (only 2 of 300 with 19); predorsal scales 12–17; lateral scales 36–42 . . . . **Inland silverside**

1b. Spinous dorsal fin posterior to vertical through anus; segmented anal fin rays 19–29, modally 23–25 (only 2 of 855 specimens with 19); predorsal scales 18 or more; lateral scales 43–55 . . . . **Atlantic silverside**

INLAND SILVERSIDE / *Menidia beryllina* (Cope 1867) / Bigelow and Schroeder 1953:304–305

Description. Body slender, slightly compressed laterally to terete (Fig. 151). Dorsal and ventral profiles only slightly convex, tapering to a thin caudal peduncle. Snout bullet-shaped. Mouth terminal; posterior margin of premaxilla not extending beyond anterior margin of orbit. Head small, approximately 25% SL. Diameter of eye greater than snout length. Spinous dorsal fin anterior to anus. Second dorsal fin with convex posterior margin; originating over first third to middle of anal fin. Anal fin with convex distal margin, anterior rays forming a small lobe. Pectoral fin pointed but not falcate, distal margins of rays straight. Scales cycloid. Trunk lateralis system with canals or open pits; almost continuous line from origin of second dorsal to base of caudal peduncle. Ranges of selected morphometric characters as thousandths of SL: snout to spinous dorsal fin origin 452–555; snout to anal fin origin 545–640; snout to pectoral fin insertion 229–265; pectoral fin length 155–208; pelvic fin length 119–143; head length 234–274; snout length 60–77; orbit diameter 63–80; maximum body depth 14–21 (Robbins 1969; Martin and Drewry 1978).

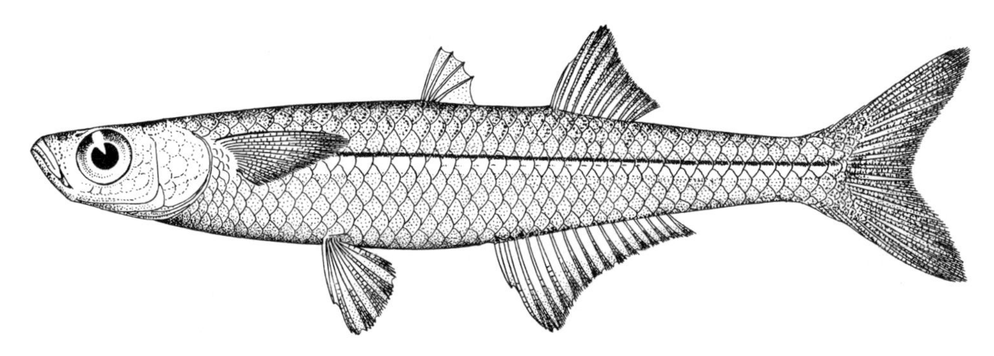

Figure 151. Inland silverside *Menidia beryllina*. Potomac River, 80 mm. Drawn by A. H. Baldwin.

Meristics. Dorsal fin rays IV–VII, I, 8–11; anal fin rays I, 13–19; pectoral fin rays 11–14; lateral scales 36–42; predorsal scales 12–17; total vertebrae 37–40.

Color. Pale greenish on the back with top of head bright yellow-greenish in pond or clear water populations. Lightly colored to silvery below the well-defined silver stripe, which is bounded above by a dark line. Scales on the back and sides with numerous brown dots, occasionally producing a cross-hatched pattern above the lateral stripe. Dorsum stippled. Anal, second dorsal, caudal, and pelvic fins may be pale orange to light red in populations from clearer, semi-isolated ponds; in estuarine populations, fins clear or slightly dusky without color.

Size. Maximum length just over 100 mm SL. However, specimens from Massachusetts and the Gulf of Maine are substantially smaller: maximum 67 mm SL (Hartel et al. in press).

Distinctions. The dorsal fin originates anterior to a vertical through the anus in *M. beryllina,* posterior in *M. menidia*. There are fewer anal rays, predorsal scales, and lateral scales in *M. beryllina* than in *M. menidia* (13–19, 12–17, and 36–42 vs. 19–29, more than 18, and 43–55, respectively).

Habits. Where common, *M. beryllina* aggregate in large schools and can often be among the most abundant species present in the upper portion of estuaries (e.g., at Waquoit Bay, Eel Pond estuary [Curley et al. 1970]). Inland silverside have a wide salinity tolerance, but prefer less saline conditions (Robbins 1969; Martin and Drewry 1978; Bengtson 1984), and can be collected over a wide variety of bottom types. There is little direct data to support an offshore winter migration of northern *M. beryllina* to deeper waters, despite the suggestion, for example, by Schwartz (1964a). Curley et al. (1970) captured inland silverside from March through December in Waquoit Bay. Bengtson (1984) collected throughout the year in Rhode Island and noted that inland silverside moved only into the lower estuaries in October, after the major predators had departed. This is consistent with studies by Hoff et al. (1968) and Hoff (1972) in the Slocum River, Buzzards Bay, which showed that inland silverside resided year-round and have a very limited

Food. *Menidia beryllina* are omnivorous, feeding on copepods, mysids, isopods, amphipods, barnacle nauplii, ostracods, mollusks, worms, larval fishes, fish eggs, insects, other components of the zooplankton, some algae, and small amounts of detritus (Darnell 1958; Bengtson 1984; Hartel et al. in press).

Breeding Habits. Spawning and reproductive ecology is much less studied than that of *M. menidia*. In the northern portion of its range, *M. beryllina* spawn during the summer months of June and July (Bengtson 1984). Although there is evidence for twice annual spawning for inland silverside south and west of Cape Hatteras, there is no evidence that this occurs in New England. Inland silverside spawn in shallow waters of the intertidal zone of the upper estuary at high tide; eggs are attached to vegetation, detritus, roots, or leaves (Martin and Drewry 1978; Bengtson 1984; Hartel et al. in press). In the laboratory, *M. beryllina* spawned throughout the day whether or not tidal or light signals were presented (Middaugh et al. 1986).

Early Life History. The eggs are 0.9–1.0 mm in diameter with a tuft of four to nine long adhesive filaments (Martin and Drewry 1978); one filament is thickened. Hatching is strongly determined by temperature and requires 10 days to 2 weeks above 20°C. Hatching was successful with 73–78% at 5, 15, or 30 ppt salinity, but optimum growth and survival of larvae occurred at 15 ppt salinity (Middaugh et al. 1986). Hatching takes 8 days at 17°–25°C (Wang and Kernehan 1979). Hatching size is 3.5–4.0 mm SL and all fins are complete by 24 mm (Martin and Drewry 1978). A 10.3-mm SL juvenile was illustrated in Able and Fahay (1998: Fig. 34.1)

Juvenile *M. beryllina* grew at an average of only 0.19 mm·day^{-1} for the first 74 days in Rhode Island (Bengtson 1984). Postlarval and juvenile inland silverside required only 5–60% of their body weight as a minimum ration for survival (Bengtson 1985), and maximum growth occurred with rations of 80–160% of body weight (cf. Atlantic silverside). A series of laboratory and field experiments on larvae and juveniles demonstrated that food was not limiting for growth or survival and, therefore, that predation is likely the primary source of mortality of young-of-the-year *M. beryllina* (Gleason and

General Range. *Menidia beryllina* range from the Gulf of Maine to the Laguna de Tamiahua, Mexico. Although inhabiting the upper portion of estuaries, inland silverside ascend tidal creeks and rivers throughout the range. They are found 72 km above tidal interface in Virginia (Martin and Drewry 1978) and in inland lakes of Florida (e.g., Lake Eustis). In the Mississippi River Basin, inland silverside can be found as far north as Illinois, and west in the Red River system into Lake Texoma (Chernoff et al. 1981). Presence in the Pecos River and Rio Conchos appear to be due to introductions.

Occurrence in the Gulf of Maine. *Menidia beryllina* have been reported from the southern portion of the Gulf of Maine north to Wells Harbor, Maine. Bigelow and Schroeder reported only Kendall's (1902) Cohasset record for north of the Cape. Curley et al. (1972) recorded inland silverside from Wellfleet Harbor, but that is also just inside Cape Cod. They are abundant in bays, estuaries, and tidal rivers around Nahant, Mass. (Collette and Hartel 1988; Hartel et al. in press). Recent surveys in the Charles River (1981–1989) failed to locate this species where they were formerly abundant (Hartel et al. in press). Three specimens were reported from the Wells Harbor estuary, Maine, 43°22′ N, 70°30′ W (Ayvazian et al. 1992) but identification has not been verified.

Importance. Inland silverside are an important food resource for adults and juveniles of many commercial or sport fishes that frequent the upper portions of estuaries, such as striped bass and bluefish (Hildebrand and Schroeder 1928; Hartel et al. in press). Kendall (1902) and Bayliff (1950) reported that inland silverside were marketed commercially as whitebait for human consumption. *Menidia beryllina* is a useful indicator organism used in environmental toxicology (e.g., Fournie et al. 1990; Hemmer et al. 1990).

ATLANTIC SILVERSIDE / *Menidia menidia* (Linnaeus 1766) / Green Smelt, Sand Smelt, Whitebait /

Bigelow and Schroeder 1953:302–304

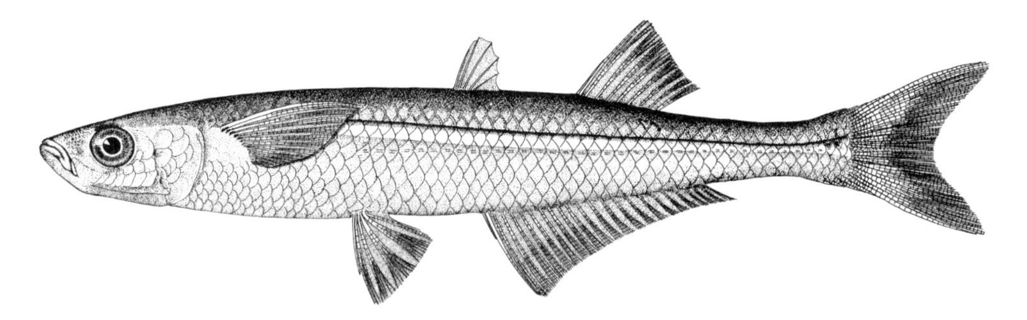

Figure 152. Atlantic silverside *Menidia menidia*. Woods Hole, Mass. Drawn by H. L. Todd.

Description. Body slender, laterally compressed (Fig. 152). Dorsal and ventral profiles slightly convex, tapering to relatively stout caudal peduncle. Snout blunt to slightly angular. Mouth terminal; posterior margin of premaxilla not extending beyond anterior margin of orbit. Head relatively small, less than 25% SL. Eye small, its diameter less than length of snout. Spinous dorsal fin posterior to anus or just over anal fin origin. Second dorsal fin with straight to slightly convex posterior margin; originating over anterior third to middle of anal fin. Anal fin with convex distal margin and lacking anterior lobe. Pectoral fin pointed but not usually falcate. Scales cycloid. Trunk lateralis system present as a continuous line posterior to insertion of pelvic fin with well-formed canals. Ranges of selected morphometric characters as thousandths of SL: snout to spinous dorsal fin origin 474–531; snout to anal fin origin 516–553; snout to pectoral fin insertion 229–265; pectoral fin length 160–195; pelvic fin length 100–122; orbit diameter 53–62; snout length 64–76; maximum body depth 144–232 (Robbins 1969; Martin and Drewry 1978).

Meristics. Dorsal fin rays IV–VII, I, 7–11; anal fin rays I, 19–29; pectoral fin rays 12–16; lateral scales 43–55; predorsal scales 16–31, for populations in Massachusetts and north more than 18; total vertebrae 41–47, more vertebrae in northern populations (Kendall 1902; Robbins 1969; Billerbeck et al. 1997).

Color. Translucent sea green above with top of head slightly yellow-green. Top of head, snout, and chin dusky. Scales are speckled posteriorly with dark brown melanophores forming cross-hatched pattern above lateral silver stripe. On dorsum melanophores almost uniformly distributed, appearing dusky or dark. The silver stripe, or stole, extends from the insertion of the pectoral fin to the base of the caudal fin; the stole is outlined above by a narrow black streak. The belly is light in color, almost white. Fins are clear to dusky but have no color.

Size. In the Gulf of Maine at Essex Bay, Mass., Atlantic silverside can exceed 125 mm SL (Conover and Ross 1982), but are usually less than 110 mm SL. Females are larger than males;

the mean difference in SL between the sexes can be as large as 10 mm (Conover and Ross 1982).

Distinctions. The dorsal fin originates posterior to a vertical through the anus in *M. menidia,* anterior in *M. beryllina.* There are more anal rays, predorsal scales, and lateral scales in *M. menidia* than in *M. beryllina* (19–29, more than 18, and 43–55 vs. 13–19, 12–17, and 36–42, respectively).

Habits. During spring, summer, and fall, Atlantic silverside are common inhabitants of intertidal creeks, marshes, and shore zones of estuarine embayments. They often congregate in large schools over sand, gravel, mud, or peat bottoms. During high tides Atlantic silverside can be found among stands of sedge grass, *Spartina alternaflora* (Bigelow and Schroeder; Bayliff 1950; Conover and Ross 1982) and eelgrass, *Zostera marina* (Mattila et al. 1999).

Food. *Menidia menidia* is an omnivorous visual feeder, preying on copepods, barnacle nauplii, horseshoe crab larvae, mysids, shrimps, small decapods, amphipods, Cladocera, fish eggs (including their own), young squid, annelid worms, molluscan larvae, insects, algae, and diatoms (Bigelow and Schroeder; Bengtson 1984; Crawford and Daborn 1986; Scott and Scott 1988; Spraker and Austin 1997). Most active feeding takes place during daylight as indicated by an index of stomach fullness, but feeding continues all night (Spraker and Austin 1997). In Rhode Island, hatchlings through juveniles less than 50 mm SL prey almost exclusively on zooplankton (Bengtson 1984); they require 160–640% of their body weight per day in order to achieve maximum growth rates and 20–135% as a minimum survival ration (Bengtson 1985). Bengtson (1984) concluded that Atlantic silverside are primarily responsible for the June decline of zooplankton in the upper portions of two Rhode Island estuaries.

Predators. Atlantic silverside are important as forage fishes for many species of fishes (e.g., bluefish, striped bass, and striped sea robin), as well as other organisms (e.g., dolphin, aquatic birds, and blue crab [Bigelow and Schroeder; Richards et al. 1979; Middaugh 1981; Hartel et al. in press]).

Breeding Habits. Spawning is dependent upon temperature and varies with latitude. In the southern portion of the Gulf of Maine as in New England, *M. menidia* reproduce from May through July, corresponding to water temperatures of 9°–12°C (Bigelow and Schroeder; Kendall 1902; Middaugh and Lempesis 1976; Bengtson 1984; Hartel et al. in press). In the northern portion of the Gulf of Maine, Nova Scotia, and Prince Edward Island, the spawning season extends from June through July (Scott and Scott 1988).

Reproductive ecology and behavior have been studied extensively in South Carolina (Middaugh 1981; Middaugh et al. 1981; Middaugh and Takita 1983) and in the Gulf of Maine those noted for New England and Gulf of Maine populations (Bigelow and Schroeder; Bengtson 1984; Scott and Scott 1988; Hartel et al. in press). During the spawning season, large schools of *M. menidia* move from the lower estuary into the intertidal zone of upper estuaries during daytime high tides in synchrony with the new and full moons. Maximum spawning activity occurs every 13–16 days when the tidal velocity drops after peak high tide; no spawning activity takes place in darkness (Middaugh and Takita 1983). Eggs are laid in vast quantities 1.5–1.8 m above mean low water on stems or roots of *Spartina* or on detrital mats. In the Gulf of Maine, however, eggs are deposited on filamentous intertidal algae such as *Enteromorpha* and *Pilayella.* Eggs remain exposed for about 10 h between tidal cycles. Upon hatching the larvae develop in the upper portions of estuaries until they attain about 50 mm SL and then move into the lower estuarine environment (Bengtson 1984).

Early Life History. The eggs are 1.0–1.2 mm in diameter and each has a tuft of adhesive filaments that adhere readily to the substrate (Bigelow and Schroeder; Martin and Drewry 1978). Incubation and hatching periods are strongly dependent upon temperature and can range from 4 days at 30°C to 27 days at 15°C (Austin et al. 1975; Middaugh and Lempesis 1976; Martin and Drewry 1978). Optimum salinity for hatching was 30 ppt; hatching was delayed 10 h at 20 ppt and 42 h at 10 ppt (Middaugh and Lempesis 1976). The yolk is absorbed before hatching, at which time the larvae are about 3.85–5 mm long (Bigelow and Schroeder; Middaugh and Lempesis 1976). Dorsal, anal, and caudal fins form in larvae of 12–15 mm. A 17.4-mm SL juvenile is illustrated in Able and Fahay (1998: Fig. 35.1). The young may reach 20 mm within the first month (Conover and Ross 1982; Bengtson 1984).

Age and Growth. The life cycle of this species is essentially annual: less than 1% of breeding adults are 2+ years old (Conover and Ross 1982). Populations in the Gulf of Maine grow rapidly, 20 mm per month (Conover and Ross 1982), which is at least 5 mm per month greater than average rates found for populations in Rhode Island or on the north shore of Long Island (Mulkana 1966; Austin et al. 1973). However in Rhode Island, the earliest spawned individuals exhibited growth rates up to 0.84 $mm \cdot day^{-1}$ (Bengtson 1984), but this rate declines for individuals hatching in June or July. The accelerated growth rate and more rapid accumulation of energy reserves in Gulf of Maine and northern populations compensate for a shorter growing season, enhance survival, and have a genetic basis (Conover and Present 1990; Schultz and Conover 1997).

General Range. Atlantic silverside are common from the southern Gulf of St. Lawrence and the outer Nova Scotian coast to northern Florida (Johnson 1975; Scott and Scott 1988).

Occurrence in the Gulf of Maine. Except for the winter

throughout the Gulf of Maine (Clayton et al. 1976; Collette and Hartel 1988; Scott and Scott 1988; Ayvazian et al. 1992; Hartel et al. in press). A series of fishery surveys conducted by the State of Massachusetts between 1963 and 1974 demonstrated that this species was among the most dominant numerically of all fishes collected by beach seines (e.g., at Gloucester Harbor and Anisquam River [Jerome et al. 1969]). Needler (1940) also reported that Atlantic silverside were abundant in Malpeque Bay, Prince Edward Island. Atlantic silverside are seldom seen or collected along stretches of rocky coast exposed to the open sea, which make up a large part of the northern shoreline of the Gulf of Maine.

Movements. In winter, populations of this species north of Cape Hatteras migrate offshore to continental shelf waters beginning in November in the Gulf of Maine (Conover and Ross 1982). Atlantic silverside were captured in bottom trawls up to 170 km from the shore and at depths to 126 m (Conover and Murawski 1982). Most offshore captures, however, were within 50 km of the shoreline and 10–50 m deep at bottom temperatures of 2°–6°C. When offshore (Conover and Murawski 1982), silverside undergo diel vertical migrations like other planktivorous fishes (e.g., Atlantic herring). The pattern of seasonal appearance and disappearance by Atlantic silverside from Gulf of Maine estuaries is readily apparent from a series of fisheries surveys conducted by the State of Massachusetts. Based upon data presented in Jerome et al. (1965, 1966, 1967, 1968, 1969), Fiske et al. (1966), Chesmore et al. (1971, 1972, 1973), Curley et al. (1972), and Iwanowicz et al. (1973, 1974) including more than 50,000 specimens of *M. menidia,* more than 99% of all individuals were present in 12 estuaries and embayments north of Cape Cod between April and November when the water temperature ranged between 2.2°–25.6°C. Fewer than 1% of the individuals were present in March and December and only a total of 10 individuals were captured in January or February when recorded water temperatures were –2.8° to 1.1°C. South of Cape Cod, however, Atlantic silverside were more commonly captured in March, December, and even January, but water temperatures were also noticeably warmer (e.g., 1.6°C in the Westport River [Fiske et al. 1968] and 2.8°C in Waquoit Bay [Curley et al. 1970], both in January). Thus, although there is a pronounced migration from estuaries in the late fall, some individuals may remain. Needler (1940) reported capturing *M. menidia* under the ice in Malpeque Bay, Prince Edward Island, and Hildebrand and Schroeder (1928) collected them at depths of 10–49 m in Chesapeake Bay. Movement to deeper water may serve to escape critically low temperatures. Hoff and Westman (1966) reported lower lethal temperatures of 1°–2°C for this species. Populations south of Cape Hatteras apparently remain within coastal estuaries during the winter (Conover and Ross 1982 and included references).

Importance. In addition to serving as prey for commercial and sport fishes such as mackerel, bluefish, and striped bass, Atlantic silverside have been of occasional commercial importance. At the turn of the nineteenth century, there was a fishery for them (Kendall 1902); silverside, called whitebait, were fried and eaten in New England and Long Island. Atlantic silverside were canned for human consumption during World War II (Scott and Scott 1988). The Division of Marine Fishery Monographs (cited above) refer to many of the estuaries and embayments supporting large fisheries operations for baitfish, particularly in the 1700s and 1800s. These reports do not mention the composition of the "bait" other than alewife. However, given the abundance, if not dominance, of *M. menidia,* along the shores of the same estuaries, its hard to imagine that they were not included in the bait. Bigelow and Schroeder reported that they were used in Rhode Island to bait eel pots. A commercial trap net fishery was established in Canada to export Atlantic silverside to Japan as a food fish. This fishery lasted from 1973 to 1982, peaking in 1979, landing 319 tons with a value of $91,000 Canadian (Scott and Scott 1988).

Atlantic silverside have also been used as an indicator organism for the biological effects of a number of estuarine contaminants, such as soaps (Eisler and Deuel 1965) and a variety of insecticides (Eisler 1970; Weis and Weis 1976).

ACKNOWLEDGMENT. A draft of the atheriniform section was reviewed by David Conover.

FLYINGFISHES AND ALLIES. ORDER BELONIFORMES

Beloniformes (or Synentognathi) is an order of atherinomorph fishes containing two suborders, five families, 37 genera, and about 200 species. Species of Adrianichthyoidei inhabit Asian fresh and/or brackish waters. Features common to the suborder Exocoetoidei include dorsal and anal fins on the rear half of the body, abdominal pelvic fins with six soft rays, no fin spines, a lateral line running along the ventral edge of the body, an open nasal pit, and lower pharyngeal bones fused into a triangular plate (leading to the name Synentognathi). Most species of the four families are tropical epipelagic marine fishes, but several genera of Belonidae and Hemiramphidae are restricted to freshwater and a few other genera include estuarine and freshwater, as well as marine species. A few representatives of all four families of Exocoetoidei stray into the Gulf of Maine.

KEY TO GULF OF MAINE BELONIFORMES

1a. Both upper and lower jaws elongate; scales small **2**
1b. Neither jaw or only lower jaw elongate; scales large **3**
2a. Jaws slightly elongate with small, weak teeth; several finlets present behind dorsal and anal fins . . . . **Sauries**
2b. Jaws very elongate with larger, sharp teeth; no finlets present . . . . **Needlefishes**
3a. Lower jaw elongate; pectoral and pelvic fins not enlarged . . . . **Halfbeaks**
3b. Jaws not elongate; pectoral and pelvic fins enlarged . . . . **Flyingfishes**

SAURIES. FAMILY SCOMBERESOCIDAE

BRUCE B. COLLETTE

Both jaws (of adults) are elongate and form a slender beak in sauries as in needlefishes, and the anal, dorsal, and pelvic fins are set far back. Presence of four to seven finlets between the dorsal and anal fins and the caudal in sauries (absent in needlefishes) is a ready field mark for their identification. The teeth are small and weak, and their bodies only moderately slender. Two genera and four species are known (Collette et al. 1984a); one species ranges into the Gulf of Maine.

NORTH ATLANTIC SAURY / *Scomberesox saurus saurus* (Walbaum 1792) / Billfish, Skipper /

Bigelow and Schroeder 1953:170–171

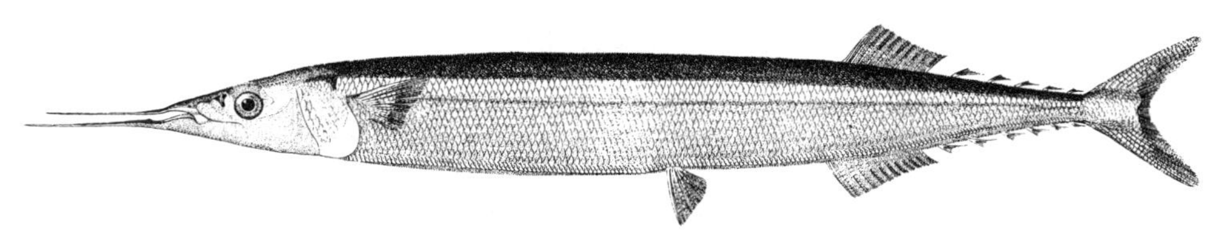

Figure 153. North Atlantic saury *Scomberesox saurus saurus.* Woods Hole, Mass., 277 mm. Drawn by H. L. Todd.

Description. Upper and lower jaws prolonged (Fig. 153). Five or six small separate finlets, dorsally between dorsal fin and caudal, ventrally between anal fin and caudal. Body about nine times as long as deep, laterally flattened, tapering toward head and tail, with slender caudal peduncle. Lower jaw projecting a little beyond upper; teeth pointed but small. All fins small. Dorsal fin originates slightly behind origin of anal; these fins similar in outline, originating posteriorly. Pelvic fins situated about midway along body. Caudal fin deeply forked and symmetrical, similar to a mackerel tail. Trunk covered with small scales. Patch of scales on each gill cover.

Meristics. Dorsal fin rays 9–12 + 5 or 6 finlets; anal fin rays 12 or 13 + 5–7 finlets; pectoral fin rays 12–15; predorsal scales 73–81; gill rakers on first arch 34–45; precaudal vertebrae 39–43, caudal vertebrae 24–28, total 64–70 (Hubbs and Wisner 1980)

Color. Olive green above with a silver band on each side at the level of the eye and about as broad. There is a dark green spot above the base of each pectoral fin; the dorsal fin is greenish; the lower parts are silvery with golden gloss. Young fry, which live in the surface waters of the open Atlantic, have dark-blue backs and silvery sides.

Size. Maximum size 762 mm.

Distinctions. Saury resemble needlefish in the slender form and in the fact that both upper and lower jaws are prolonged, but differ in having a series of five or six small separate finlets, both dorsally, between the dorsal fin and the caudal, and ventrally, between the anal fin and the caudal.

Habits. Saury spend most of their lives in warm homogeneous surface layers of the open sea, far from shallow continental shelf waters (Zilanov 1970, 1977). They live close to the surface; so close that in English waters, where they are plentiful in summer, few are caught in nets as deep as 2 m.

Food. Atlantic saury are among the most abundant epipelagic planktivores inhabiting the open part of the Atlantic Ocean and feed mainly on siphonophores, copepods, euphausiids, and amphipods (Dudnik et al. 1981; Bowman et al. 2000). Larvae of polychaetes, decapods, isopods, ostracods, cirripeds, and siphonophores, fish eggs and larvae, protozoans, and algae were also present in the diet, but in smaller quantities (Nesterov and Shiganova 1976). Qualitative composition of food items vary with seasonal distribution of saury and their prey. In spring and summer, copepods (mainly *Calanus finmarchicus*), siphonophores, and fish larvae constitute the main component of the diet on Georges Bank and the Scotian Shelf; secondary items are euphausiids and decapod larvae. In autumn, copepods (*Centropages* sp.), various calanoids, and euphausiids are the major prey organisms in the same areas. In winter, most saury caught in the southern part of the area (34° N) feed on larvae of decapods, hyperiid amphipods, mollusks, and foraminiferans.

Predators. Atlantic saury serve as food for many inhabitants of the sea, including squids, swordfish, marlins, sharks, tunas, hakes, cod, pollock, dolphin, whales, and birds. The great abundance of saury and their wide distribution make them an important link in the epipelagic food chain of the ocean as they enable the transfer of energy from lower to higher trophic levels. When saury strand on beaches, as often happens, it is probably while they are fleeing from their enemies. At sea they attempt to escape by leaping, schools of them breaking the surface together as has often been described, and as Bigelow and Schroeder witnessed in Massachusetts Bay.

Breeding Habits. From distribution data of Atlantic saury eggs, larvae, and juveniles, spawning occurs south of the frontal zone of the Gulf Stream (Nesterov and Shiganova 1976). Data on saury maturity stages by month indicate that spawning occurs mainly during the winter-spring period. Spawning in both males and females occurs when they reach a minimum length of 26 cm, and most spawning fish belong to age-groups 2 and 3 (Dudnik et al. 1981). It is not likely that they ever spawn in the cool Gulf of Maine waters, for Bigelow and Schroeder never collected their larvae in tow nets, although larval saury are among the most numerous of young fish in the open Atlantic between 11° or 12° N and 40° W.

Early Life History. Atlantic saury eggs are spherical, nearly transparent and 2.5–3.2 mm in diameter. The eggs lack oil globules and are covered with short bristles that apparently represent remnants of the typical beloniform chorionic filaments (Boehlert 1984: Fig. 20B). Incubation requires more than 2 weeks (Sanzo 1940).

Development was described and illustrated by Sanzo (1940) and Nesterov and Shiganova (1976) and summarized by Hardy (1978a). Hatching occurs at 6.0–8.5 mm. Jaws begin to elongate at 15–17 mm and by 20.5–21.6 mm the lower jaw is considerably longer than the upper (Nesterov and Shiganova 1976). Larvae of 100–150 mm look more like halfbeaks (halfbeak stage) than their own parents. Fin rays may begin to form as early as 6.4 mm (Nesterov and Shiganova 1976). Finlets first become evident at 15–18 mm (Hardy 1978a). Other morphological and pigment changes during ontogeny were summarized by Hardy (1978a).

Age and Growth. Four age-classes were distinguished in the northwest Atlantic based on scales (Dudnik et al. 1981), five age-classes were found in the Mediterranean based on otoliths (Potoschi 1996). In the North Atlantic, saury mature at 2 years (Sauskan and Semenov 1969). Some males mature at 230 mm; all are mature by 320 mm; some females mature at 230 mm; all are mature by 340 mm (Zilanov and Bogdanov 1969).

General Range. Antitropical in temperate parts of the Atlantic, Pacific, and Indian oceans (Hubbs and Wisner 1980: Figs. 14, 15). Two subspecies are recognized with the nominal subspecies broadly distributed in the North Atlantic Ocean (mostly north of 30° N) and throughout the Mediterranean Sea. In the northwest Atlantic, they are found from Cape Hatteras to Newfoundland. In summer and autumn, they are encountered on the Scotian Shelf, on Georges Bank, in the Gulf of Maine, and southward around Cape Cod. The area to the west of the Gulf Stream core is the main habitat of Atlantic saury in the open sea of the northwest Atlantic, although they have been taken east of the Gulf Stream. Generally, the area of distribution extends from coastal waters eastward to 40° W and from 32° N northward to 50° N.

Occurrence in the Gulf of Maine. While saury are stragglers to the Gulf from warmer waters offshore or farther south, they have been taken along the northern coasts off New England more often than have any of their relatives—specifically along Cape Cod; at Provincetown; at several locations in Massachusetts Bay, where Bigelow and Schroeder saw schools of them; at Annisquam a few miles north of Cape Ann; at Old Orchard; in Casco Bay; at Monhegan Island; in the central part of the Gulf; among the islands at the northern entrance to the Bay of Fundy; and on the northern part of Georges Bank, where one was gaffed from the *Albatross II* on 20 September 1928. Bigelow and Schroeder found no records along the Nova Scotia shore of the Gulf of Maine. The inner curve of Cape Cod from Provincetown to Wellfleet seems to be a regular center of abundance, as Storer remarked long ago, for schools of saury were picked up in traps along that stretch of beach almost every year, the catch occasionally amounting to hundreds of barrels, and hosts of them have been known to strand there. Hundreds of saury stranded in a marsh area in Rock Harbor, Eastham, along the inner elbow shoreline of Cape Cod Bay in November 1998 (Leaning 1998).

In the Gulf, they are likely to be taken any time from mid-June to October or November, the largest catches usually being made late in summer. Bigelow and Schroeder reported several schools skipping, as is their common habit, off the Scituate shore on the southern side of Massachusetts Bay. Saury are so much less common farther within Massachusetts Bay that some fishermen have never heard of them there. They appear only as strays north of Cape Ann. A large school could be encountered anywhere within the Gulf, as documented by their occasional abundance off northern Nova Scotia. Cornish (1907) reported that large schools are often seen at Canso, N.S., skipping over the water as they flee from pollock. When saury do invade the waters of the Gulf, they may be expected in multitudes, for they usually travel in vast schools.

Importance. Saury are valuable food-fishes in some parts of the world; Atlantic saury are important in the Mediterranean (Potoschi 1996). At present, there is no fishery for saury in the northwest Atlantic, but an experimental fishery was conducted by Russian vessels during 1969–1974. Saury were caught in nets suspended from booms along the side of the vessels, the fish being attracted by bright lights (Zilanov 1977). A review of published statistics indicates that the bulk of the catches during 1970–1974 was taken by USSR vessels (Dudnik et al. 1981). The size of Atlantic saury in catches on Georges Bank during 1970–1973 ranged from 18 to 39 cm and from 19 to 200 g in weight (Dudnik et al. 1981). Two size-groups, 23–30 cm and 31–39 cm, were distinguishable in the length frequencies that are characteristic of the sizes of saury caught during the autumn–winter period. The corresponding average weights for these two groups were 55 and 110 g.

NEEDLEFISHES. FAMILY BELONIDAE

BRUCE B. COLLETTE

The most noticeable feature of adult needlefishes is that both jaws are prolonged to form a long slim beak well-armed with teeth. Juveniles of most species go through a *halfbeak stage,* where only the lower jaw is elongate (Boughton et al. 1991) and they feed on planktonic organisms. There are no finlets between the dorsal and anal fins and the caudal, their absence being the readiest field mark to distinguish needlefishes from sauries (*Scomberesox*). Needlefishes are swift-swimming predacious fishes represented by 32 species, seven in the western Atlantic (Collette 1978a). Only two species have been recorded in the Gulf of Maine, but agujon (*Tylosurus acus* Lacepède 1803) have been taken at Nantucket. They may appear as strays from the south and can be easily distinguished from Atlantic needlefish by the deeply forked tail, the presence of a black keel on the caudal peduncle, and by the fact that the dorsal and anal fins are much longer, the former with 23–26 rays and the latter with 20–24.

KEY TO NEEDLEFISHES IN THE GULF OF MAINE

1a. Body as thick as it is deep; dorsal, anal, and caudal fins only moderately concave; anal fin rays 16–20; no prominent enlarged black lobe in posterior part of dorsal fin . . . . **Atlantic needlefish**
1b. Body less than one-half as thick as it is deep; dorsal, anal, and caudal fins deeply concave; anal fin rays 24–28; prominent enlarged black lobe in posterior part of dorsal fin . . . . **Flat needlefish**

FLAT NEEDLEFISH / *Ablennes hians* (Valenciennes 1846) / Bigelow and Schroeder 1953:168–169

Figure 154. Flat needlefish *Ablennes hians.* Gulf of Guinea, USNM 202853, 418 mm BL. Drawn by M. H. Carrington.

Description. Body strongly compressed, less than one-half as thick as deep (Fig. 154). Dorsal fin arises farther back relative to anal fin than in Atlantic needlefish. Dorsal and anal fins concave; pectoral fins falcate; caudal fin deeply forked. Females lack a right gonad; males either lack it or have it greatly reduced.

Meristics. Dorsal fin rays 23–26; anal fin rays 24–28; pectoral fin rays 13–15; gill rakers present in juveniles, but lost in adults;

Color. Back bluish green, lower part of sides bright silvery, dorsal fin mostly greenish, but with the rays black-tipped; caudal fin greenish; tip of lower jaw red. Sides usually marked with 12–14 prominent dark vertical bars. Juveniles and adults have an elongate black lobe in the posterior part of the dorsal fin.

Size. Reaches at least 63 cm body length (without head and beak) and 82.5 cm SL, common to 70 cm SL. The all-tackle game fish record is a 4.80-kg fish caught at Zavora Island,

Distinctions. The body is so strongly compressed as to be less than one-half as thick as it is deep, instead of about as thick as deep, or thicker, as it is in the Atlantic needlefish; the 12–14 vertical bars are diagnostic.

Habits. This is an offshore epipelagic species.

Parasites. Copepods, isopods, and cestodes have been found on western Atlantic *Ablennes* (B. B. Collette, unpubl. data). Two of five species of copepods found on western Atlantic specimens (Cressey and Collette 1970) were found as far north as Woods Hole: *Parabomolochus bellones* and *Lernanthropus tylosuri.* Two species of the parasitic isopod genus *Mothocya* were reported by Bruce (1986) but only *M. longicopa* was found in the Atlantic. The single cestode record is of a postlarval *Ptychobothrium belones* in the intestine of a specimen from Bermuda.

Predators. Several species of fishes feed on *Ablennes* in the western Atlantic (B. B. Collette, unpubl. data): barracuda, *Sphyraena barracuda;* cero, *Scomberomorus regalis;* sailfish, *Istiophorus americanus;* and blue marlin, *Makaira nigricans.*

Reproduction. *Ablennes* probably spawn offshore (Berry and Rivas 1962). A 278-mm female had 660 eggs (3.0–3.15 mm in diameter) in its right ovary (the only one that develops). The eggs are covered with uniformly spaced tufts of filaments longer than the diameter of the egg (Collette et al. 1984a).

General Range. Worldwide offshore in tropical seas, mostly within the 23.9° isothere (Cressey and Collette 1970: Fig. 175), Brazil to Chesapeake Bay in the western Atlantic, and northward as strays to Cape Cod.

Occurrence in the Gulf of Maine. A specimen, 405 mm body length, was taken in a fish trap on the shore of Cape Cod Bay at North Truro, Mass., on 15 August 1949 (MCZ 37040).

ATLANTIC NEEDLEFISH / *Strongylura marina* (Walbaum 1792) / Silver Gar /

Bigelow and Schroeder 1953:167–168 (as *Tylosurus marinus*)

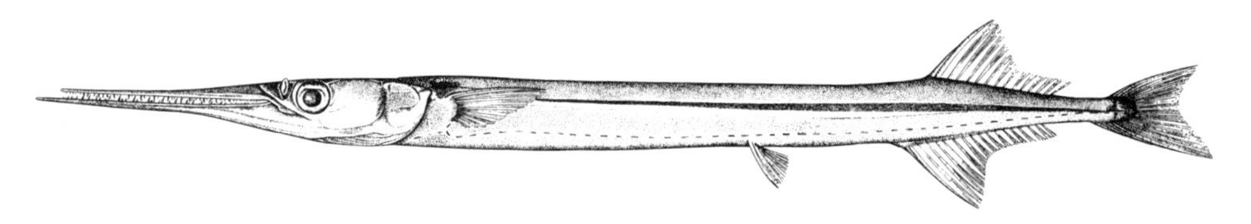

Figure 155. Atlantic needlefish *Strongylura marina.* Lake Pontchartrain, La., TU 6950, 272 mm BL. Drawn by M. H. Carrington.

Description. Head of adult nearly one-third of TL (Fig. 155). Upper jaw, from eye forward, twice as long as rest of head; both jaws armed with sharp teeth; eyes large. Body long, slender, about one-twentieth as deep as long, rounded (not laterally flattened) in cross section, and thicker than deep. Dorsal and anal fins similar in outline, anterior rays of both much longer than those toward rear. Both fins situated far back, dorsal arising a little behind forward end of anal. Pelvic fins originate about halfway between a point below eye and caudal fin base. Caudal fin margin moderately concave. Both body and sides of head scaly.

Meristics. Dorsal fin rays 14–17; anal fin rays 16–20; pectoral fin rays 10–12, usually 11; predorsal scales numerous and tiny, 213–304; gill rakers absent; total number of vertebrae 69–77 (Collette 1978a).

Color. Bluish green above, silvery on the sides, white below; a bluish silvery stripe along each side becoming broader and less distinct toward the tail; snout dark green; there is a blackish blotch deeper than long on the upper part of the cheek. The fins are generally without markings; the dorsal fin may be somewhat dusky, and the caudal bluish at its base.

Size. Reaches 42 cm body length (without head); 64 cm SL. The all-tackle game fish record is a 1.47-kg fish caught at Brigantine, N.J., in July 1990 (IGFA 2001).

Distinctions. The long bill and slender body give Atlantic needlefish so peculiar an aspect that they are not likely to be confused with any other Gulf of Maine fish other than a halfbeak (*Hyporhamphus*), saury (*Scomberesox*), or its own close relative *Ablennes.* They are easily distinguishable from the first of these because both jaws are prolonged instead of only the lower, and from the second by the lack of detached finlets behind the dorsal and anal fins. The most conspicuous differences between Atlantic needlefish and *Ablennes* are that the body of the former is thicker than it is deep and its fins are only moderately concave, whereas the latter is so strongly flattened laterally that it is less than one-half as thick as it is deep, with deeply concave fins. The margin of the caudal fin is emarginate, this being the readiest field mark to separate this needlefish from the forked-tail *Tylosurus acus,* which has not been taken in the Gulf of Maine. There is a distinct longitudinal ridge or low keel along either side of the caudal peduncle in species of *Tylosurus.*

Habits. A surface-dwelling species found in coastal areas and estuaries. It enters freshwater rivers, especially in the southern United States.

Food. Needlefish are voracious carnivores, feeding largely on small surface fishes, which they catch sideways in their beaks and then manipulate to swallow headfirst. Fishes consumed include *Fundulus* and *Menidia.* Juveniles in the halfbeak stage (35–50 mm SL) feed on small crustaceans such as shrimps, mysids, and copepods, but after the transition to both jaws being elongate, the diet is almost completely piscivorous (Carr and Adams 1973).

Parasites. A wide variety of parasites has been found on *Strongylura marina:* copepods, branchiurans, isopods, acanthocephalans, monogenes, digenes, leeches, cestodes, nematodes, and myxosporidia (B. B Collette, unpubl. data). Of the six species of copepods found on *S. marina* (Cressey and Collette 1970), only *Parabomolochus bellones* has been reported from Massachusetts waters. One acanthocephalan, *Neoechinorhynchus agile,* has been found in the intestine of a Woods Hole fish (Linton 1901b). Two monogenes were reported from *S. marina* taken at Woods Hole, *Ancyrocephalus parvus* and *Nudaciraxine gracilis* (Linton 1940), as well as three monogenes, *Podocotyle olssoni, Rhipidocotyle transversale,* and *R. lintoni.* Both species of *Rhipidocotyle* employ the same intermediate and final hosts: metecercariae in silverside, *Menidia menidia,* and adults in *S. marina* (Stunkard 1976).

Breeding Habits. Spawning takes place in bays, estuaries, and river mouths. Some spawning occurs well up some rivers in strictly freshwater. The spawning season is probably May and June in Rhode Island and New York (Hardy 1978a). The right gonad is absent in both males and females. Fecundity is unknown, but a 294-mm body length specimen had 1,000 large ovarian eggs with an average diameter of 2.75 mm in its single (right) ovary (B. B. Collette, pers. obs.).

Early Life History. The eggs are demersal, attached to aquatic vegetation and other objects by numerous elongate filaments. Eggs are 3.5–3.6 mm in diameter and lack oil globules. Development and pigmentation in a series from yolk-sac larvae (9.2 mm) to the halfbeak stage (30 mm) were illustrated in Hardy (1978a), based partly on Ryder's pioneering study (1882). Incubation time is unknown. Length at hatching is 9.2–14.4 mm. The lower jaw grows more rapidly than the upper one so that juveniles of 23-mm body length are in a halfbeak stage. Relative development of the upper and lower jaws and the dorsal and anal fins was illustrated by Breder (1932: Pls. 1 and 3).

General Range. Maine to Brazil (Collette 1968: Fig. 1); abundant along the south Atlantic and Gulf coasts of the United States, often running up rivers such as the Susquehanna, Potomac, and Pamunkey above tide water (Hardy 1978a).

Occurrence in the Gulf of Maine. Atlantic needlefish are common along the southern shores of New England, for example, in Rhode Island waters and at Woods Hole, where quite a few are found from June to October. Like many other southern fishes, however, they seldom journey eastward past Cape Cod; the only definite records within the Gulf of Maine are of several collected by William C. Kendall at Monomoy Island, forming the southern elbow of Cape Cod; at Wolfs Neck, Freeport, and Casco Bay, Maine; and of one reported by Crane (1936) from the stomach of a tuna that she examined in Portland, Maine, in July 1936. Crane's record, however, is more likely to have been based on the superficially similar saury, which are much more abundant in the region. Bigelow and Schroeder did not find the Atlantic needlefish in the Gulf nor did they hear even a rumor of its presence there from fishermen, good evidence that it is a rare straggler. No specimens from the Gulf were located in museum collections during the course of a systematic revision of the family.

HALFBEAKS. FAMILY HEMIRAMPHIDAE

BRUCE B. COLLETTE

Halfbeaks are close allies of flying fishes (Exocoetidae), sharing union of the third pair of pharyngeal bones into a plate and loss of the fourth upper pharyngeal bone, but in halfbeaks the lower jaw is elongate and the upper jaw short, whereas in flyingfishes both jaws are short. Marine halfbeaks are largely herbivorous, feeding mainly on green algae and seagrasses. There are 13 genera and more than 100 species, mostly in warm seas, but several genera live in southeast but only one is known to reach the Gulf of Maine. Several other species might reach the Gulf as strays either via the Gulf Stream route or from offshore (Collette 1978b). *Euleptorhamphus velox* has been taken at Nantucket and Newport (Kendall 1908). Its lower jaw is even longer and more slender than that of Meek's halfbeak, its body is more compressed, its pectoral fins are much longer, and there are many more rays in the dorsal (21–25 vs. 12–16) and anal (19–24 vs. 14–18)

MEEK'S HALFBEAK / *Hyporhamphus meeki* Banford and Collette 1993 /

Bigelow and Schroeder 1953:169 (as *Hyporhamphus unifasciatus*)

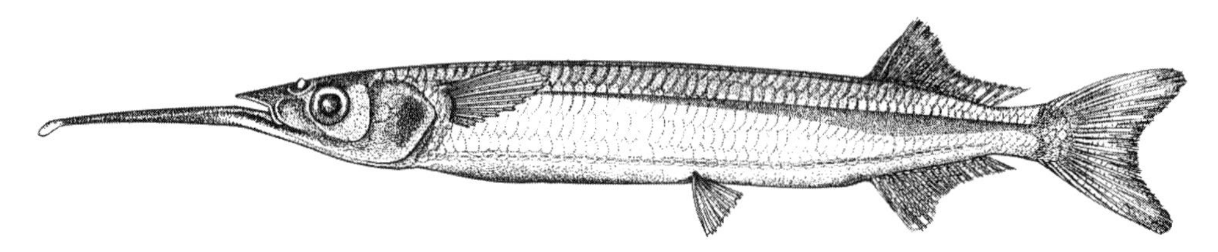

Figure 156. Meek's halfbeak *Hyporhamphus meeki*. Morehead City, N.C., 160 mm SL, USNM 294369. Drawn by M. H. Carrington.

Description. Body slender, only one-sixth to one-tenth as deep as long (younger fish are still more slender), tapering slightly toward head and tail (Fig. 156). Lower jaw very long; upper jaw short. Dorsal and anal fins originate far back and opposite each other, about equal in length and alike in outline. No separate finlets. Pelvic fins originate about midway between a point below the eye and caudal fin base. Teeth small, tricuspid. Scales large, largest on upper surface of back.

Meristics. Dorsal fin rays 12–17, usually 14 or 15; anal fin rays 14–18, usually 16; pectoral fin rays 10–13, usually 11 or 12; gill rakers on first arch 31–40, usually 33–36; on second arch 20–30, usually 25–27; precaudal vertebrae 32–35, caudal vertebrae 16–19, total 49–53 (Banford and Collette 1993).

Color. Sides and back translucent green with a silvery tinge, each side with a narrow but well-defined silvery band running from the pectoral fin origin to the caudal fin base, the sides darkest above and paler below this band. Three narrow dark streaks run along the middle of the back. The fleshy tip of the lower jaw is crimson in life. The forward parts of the dorsal and anal fins and the tips of the caudal fins are dusky. The lining of the belly is black.

Size. Adults attain a maximum size of 179 mm SL.

Distinctions. The most striking feature of Meek's halfbeak, which sets it off from every other fish known from the Gulf of Maine, is the fact that while the lower jaw is very long, the upper jaw is short.

Taxonomic Remark. Until recently (Banford and Collette 1993), *H. meeki* had been confused with its southern relative, *H. unifasciatus*. The Gulf of Maine *Hyporhamphus* differs from *H. unifasciatus* in having more gill rakers on both the first and second gill arches and a larger ratio of preorbital length to orbit diameter, usually greater than 0.70 rather than less.

Habits. Meek's halfbeak is an inshore surface schooling species found over sandy bottoms in bays, harbors, and estuaries. They are frequently found in proximity of submerged aquatic vegetation such as eelgrass. Salinity ranges from 7.5 to 42.9 ppt. Halfbeaks move north into shallow coastal waters from July to November in Long Island Sound, N.Y. (Latham 1917), and New England (Sumner et al. 1913).

Food. Meek's halfbeak is largely a herbivore as an adult, feeding on algae and seagrasses plus a few small invertebrates (Linton 1905; Hildebrand and Schroeder 1928). Carr and Adams (1973) found 51% epiphytic algae and detritus and 49% seagrasses in 77 specimens (130–199 mm SL) from Crystal River, Fla.

Predators. Preyed on by a wide variety of game fishes such as bluefish and king mackerel (Yarrow 1877; Schwartz 1964b; DeVane 1978).

Parasites. Nematodes, cestodes, trematodes, and an isopod occur on *H. meeki* (B. B. Collette, unpubl. data). Bruce (1986) reported an isopod parasite, *Mothocya nana*, from its gill arches.

Early Life History. Eggs are attached to floating eelgrass blades in Chesapeake Bay (Olney and Boehlert 1988). They are about 2.0 mm in diameter, almost transparent, with several long attachment filaments (Hardy 1978a). No information is available on egg development, incubation time, or size at hatching. The smallest known specimen was 3.0 mm. No beak is present at 7.0–11.6 mm, but it develops by 15 mm SL (Hardy and Johnson 1974). Morphological and pigment changes with development were described and illustrated by Hardy (1978a).

General Range. Found along the Atlantic coast from Chamcook, N.B., south to Miami, the Gulf of Mexico from Everglades Park, Fla., to Galveston, Tex., and at Yucatán (Banford and Collette 1993: Fig. 1). Abundant off the south Atlantic coast of the United States, not uncommon northward to Cape Cod, and straying to the coast of Maine and New Brunswick (Leim and Day 1959).

Occurrence in the Gulf of Maine. This halfbeak is a rare stray from the south, recorded only twice from the Gulf of Maine in early literature, from Machias and Casco Bay, Maine (Kendall 1914). Bigelow and Schroeder added one taken in Quincy Bay, Boston Harbor, on 10 July 1951, another off Revere (also in

Boston Harbor) on the 19th of the month; a third, taken in a trap at Sandwich on 24 September of the same year (MCZ 37326, 37327, 37342), and several dozen taken in a pound net at Small Point, Maine, 14–15 July, reported by Leslie Scattergood. The northernmost records are a 114-mm TL specimen collected southeast of Halifax, N.S., in August 1978 (NMC 82-0187; Scott and Scott 1988) and a second Canadian specimen, 128 mm SL, caught in the Annapolis Tidal Generating Station, N.S., by Jamie Gibson on 23 September 1999, in an unusually warm year (J. Gibson, pers. comm., 20 March 2000).

FLYINGFISHES. FAMILY EXOCOETIDAE

BRUCE B. COLLETTE

Flyingfishes are easily identified by their exceptionally large, winglike pectoral fins, which are used for gliding. The pelvic fins of some species are also enlarged. Their caudal fins are very deeply forked with the outer tips rounded and the lower lobe longer. They have small mouths and large, rounded scales. The family contains seven genera and about 60 species (Collette et al. 1984a). Their most distinctive feature is that they can plane through the air on their pectoral fins, a meter or more above the water, which they do in attempts to escape their enemies. This so-called "flight" is really not flight, for a flyingfish does not flap its wings; it glides on them after a thrust by the caudal fin has launched it into the air. (For a more detailed study of their "flying," based on first-hand observations, see Hubbs 1933). Flyingfishes are probably better known to voyagers in tropical seas than any other fishes. They are also often seen in the warm ultramarine blue waters of the Gulf Stream abreast of the northeastern coast. None are to be expected in the boreal waters of the Gulf except as the rarest of strays.

ATLANTIC FLYINGFISH / *Cheilopogon melanurus* (Valenciennes 1847) /

Bigelow and Schroeder 1953:172 (as *Cypselurus heterurus*)

Figure 157. Atlantic flyingfish *Cheilopogon melanurus.* 35°50′ N, 72°35′ W, USNM 198389.

Description. Snout shorter than eye, blunt (Fig. 157). No teeth on palatine bone in roof of mouth. Dorsal and anal fins set far back on body, their bases short. Anal fin origin three rays or more behind dorsal fin origin. Pectoral and pelvic fins very long, overlapping anal fin when laid back. Only first ray of pectoral fins unbranched.

Meristics. Dorsal fin rays 12 or 13; anal fin rays 8–10; pectoral [illegible]

rakers 5–8 + 15–18; vertebrae 45–48, usually 45 or 46 (Gibbs 1978).

Color. Dark blue-gray on the back and upper part of the sides, silvery lower down on the sides, and below; the dorsal fin is plain gray, without a black spot; pectoral fins lightly pigmented with a paler basal triangle; upper and lower caudal fin

Size. A comparatively small species reaching a maximum known size of 260 mm SL (Gibbs and Staiger 1970).

Biology. Western Atlantic females caught in June and August contain ovarian eggs 1.8–1.9 mm in diameter, the same size as living, just-fertilized eggs (Gibbs and Staiger 1970). The eggs are covered with filaments. Juveniles have a pair of slender barbels that persist until about 100 mm SL.

General Range. Found on both sides of the Atlantic (Gibbs and Staiger 1970: Fig. 8). The western Atlantic population occurs close inshore throughout the Antilles, inshore and offshore in the Caribbean Sea and Gulf of Mexico, and in the Gulf Stream and adjacent waters to about 40° N, 65° W.

Occurrence in the Gulf of Maine. A flyingfish, 213 mm SL (MCZ 37990), seemingly of this species but not in good enough condition for certain identification, was taken in a trap of the Pond Village Cold Storage Co. at North Truro, on the Massachusetts Bay shore of Cape Cod on 4 August 1952. This is the only record of a flyingfish from the Gulf. The only record of a flyingfish from Nova Scotian coastal waters is by Jones (1879), from Sable Island, in 1859. Flyingfishes are taken now and then at Woods Hole, *C. melanurus* perhaps more often than any other, according to published reports, but several species resembling one another very closely might be expected in the Gulf Stream off the coasts of the Gulf of Maine (see Gibbs 1978).

ACKNOWLEDGMENTS. Drafts of the beloniform section were reviewed by Thomas A. Munroe and Rodney A. Rountree.

KILLIFISHES. ORDER CYPRINODONTIFORMES

Monophyly of this order is based on several derived characters such as symmetrical caudal fin supported internally by one epural bone and first pleural rib on the third or fourth vertebra as occurs in other atherinomorphs (Parenti 1981; Nelson 1994). The lateral line is present on the body as pitted scales; nasal openings paired; and protrusible upper jaw bordered only by premaxilla. A number of species of this order are popular aquarium and experimental fishes. Eight families with 88 genera and about 807 species (Nelson 1994).

KILLIFISHES. FAMILY FUNDULIDAE

KENNETH W. ABLE

Killifishes are small, stout fishes of shallow water habitats. They are recognizable by a small mouth, single dorsal fin that originates about halfway along the body, a relatively deep caudal peduncle, and abdominally placed pelvic fins. The family Fundulidae (following Parenti 1981) contains five genera and about 48 species occurring in fresh, brackish, and coastal marine areas from southern Canada to Yucatán, Bermuda, and Cuba (Nelson 1994). Of the approximately 32 species of *Fundulus*, two occur in the Gulf of Maine. A related species, *Cyprinodon variegatus*, was included in Bigelow and Schroeder based on an account by Storer. To the best of our knowledge it has not been collected in the area subsequently, but it is common on the southern shores of Massachusetts.

KEY TO KILLIFISHES OF THE GULF OF MAINE

1a. Preorbital scales present; short, narrow, irregular bars on sides of body . . . . **Mummichog**

1b. Preorbital scales absent; 14–20 long and dark vertical bars on sides of males; adult females with horizontal stripes . . . . **Striped killifish**

MUMMICHOG / *Fundulus heteroclitus macrolepidotus* (Walbaum 1792) / Bigelow and Schroeder 1953:162–164

Description. Stout-bodied, with a flattened head, deep caudal peduncle and rounded caudal fin (Fig. 158). Mouth small, oriented somewhat dorsally. Dorsal and anal fins originate just behind midpoint of body. Head and body covered with large, round scales except for preorbital region, which is naked. Sexual dimorphism pronounced (Newman 1907, 1909a; Relyea 1983). Females have a sheath or ovipositor on the anterior edge of the anal fin, which is especially prominent during the spawning season. Males have numerous dermal contact organs on the posterior rim of selected scales and on the rays of the dorsal, anal, and pelvic fins during the breeding season.

Meristics. Dorsal fin rays 10–15; anal fin rays 9–12; pectoral fin rays 16–21; caudal fin rays 14–21; branchiostegal rays 5. Details of meristic and morphometric variation can be found in Relyea (1983).

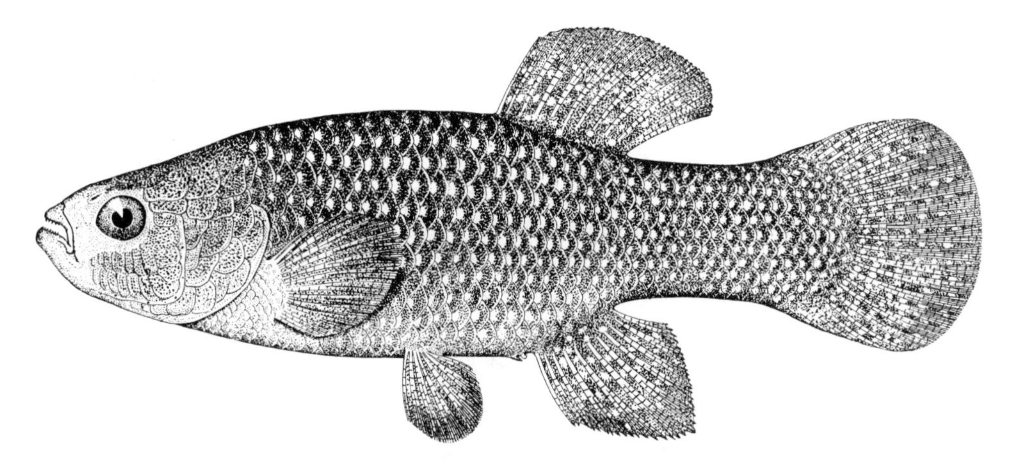

Figure 158. Mummichog *Fundulus heteroclitus macrolepidotus.* Drawn by A. H. Baldwin.

Color. Color is especially variable and can differ between the sexes, in and out of the reproductive season, and dependent upon the substrate over which they are found (Parker 1925). Females are olive green, darker on the dorsal surface. They sometimes show subtle vertical bars on the posterior half of the body. Males become much brighter and more colorful during the spawning season. Prior to the spawning season they are dark green with whitish or silvery spots and dark irregular vertical bars on the lateral surfaces and dark mottling on the dorsal surface. The fins are dark and the dorsal fin may have an ocellus on the posterior portion. The belly is much lighter.

Size. Maximum size attained in Canadian waters is 10–13 cm (Needler 1940; Fritz and Garside 1975) but they are seldom more than 10 cm in the Gulf of Maine or in the New Jersey and North Carolina marshes (Kneib and Stiven 1978; Able 1990b).

Distinctions. *Fundulus heteroclitus* can be distinguished from *F. majalis,* the only other *Fundulus* in the Gulf of Maine, because *F. heteroclitus* lack preorbital scales while *F. majalis* have them (Relyea 1983). Typically, the vertical bars on the body of *F. majalis* are longer and darker than those of *F. heteroclitus* and only adult female *F. majalis* have horizontal stripes. In addition, the anal sheath of female *F. heteroclitus* is much longer than in *F. majalis,* reaching to almost half the length of the anal fin ray.

Taxonomic Remarks. Variation in adult and egg morphology, reproductive behavior, and genetic variation (Able and Felley 1986; Marteinsdottir and Able 1988; Gonzalez-Villasenor and Powers 1989; Powers et al. 1993) has resulted in resurrection of two subspecies as recognized by Jordan and Evermann (1896). The northern subspecies, *F. h. macrolepidotus,* is present in the Gulf of Maine. Characteristics of the more frequently studied southern subspecies should not be assumed for the northern subspecies.

Habits. This shallow water species is euryhaline, with some Good 1977; Denoncourt et al. 1978; Samaritan and Schmidt 1982). Under laboratory conditions, adults and larvae can survive salinities higher than 110 ppt (Griffith 1974; Joseph and Sakesena 1966). They occupy an array of marsh habitats from salt marshes, where they are especially abundant, to eelgrass beds, open shores, and many altered and impacted habitats where few other fish can survive (Bigelow and Schroeder; Nixon and Oviatt 1973; Able and Fahay 1998). They are especially tolerant of low dissolved oxygen (Packard 1905; Smith 1995; Layman et al. 2000).

Food. Larval, juvenile, and adult *F. heteroclitus* feed, apparently opportunistically, on small crustaceans, annelids, and gastropods, and although detritus is ingested it does not appear to have any nutritional value (Clymer 1978; White et al. 1986; Kneib 1987; Petrillo 1987). Food habits change during early life history with diets of the smallest individuals (<20 mm SL) composed of harpacticoid copepods and annelid worms whereas the diets of larger individuals (up to 30 mm SL) comprised primarily *Spartina* detritus and aggregates of other food particles (Smith et al. 2000). All life-history stages depend to a large degree on marsh surface food sources (Weisberg and Lotrich 1982, 1986). The high annual production and mortality of *F. heteroclitus* (Valiela et al. 1977; Meredith and Lotrich 1979) implies that they are important in trophic dynamics of salt marshes, yet their roles in salt marsh ecosystems are hard to quantify (Clymer 1978; Kneib 1987, 1997).

Parasites. Mummichog are parasitized by every major parasite group including dinoflagellates, sporozoans, monogenetic trematodes, digenetic trematodes, cestodes, nematodes, and copepods (Dillon 1966; Marcogliese 1995).

Predators. They are probably preyed upon by a variety of birds and fishes but data are often lacking (Daiber 1982; Kneib 1982; Tupper and Able 2000), particularly for the Gulf of Maine. There are reports of predation by herons, terns, king-

1957; Clymer 1978). Other predators include a variety of fishes, crabs, and shrimps, as well as other *F. heteroclitus* (Chidester 1916; Clymer 1978; Kneib 1982, 1987; Able and Hata 1984).

Reproduction. Reproductive characteristics vary between the two subspecies (Marteinsdottir and Able 1988, 1992; Able and Fahay 1998). The following account is based on Gulf of Maine populations or the northern subspecies whenever possible. Maturation, at least for some populations, occurs as individuals enter their second year at sizes as small as 40–50 mm TL (Able 1990b). Maturation appears to occur under conditions of increasing photoperiod and temperature (Taylor 1986). Fertilization appears to be affected by salinity (Able and Palmer 1988) but can be altered by acclimation (Palmer and Able 1987). Spawning occurs in early spring through summer, but timing and duration of spawning may vary between subspecies (Marteinsdottir and Able 1992). Reproduction occurs in Massachusetts from May to July (Wallace and Selman 1981). All populations appear to spawn intertidally (Taylor 1986). Populations of *F. h. macrolepidotus* may spawn daily during the reproductive season in Massachusetts (Wallace and Selman 1981; Taylor 1986) and on Long Island (Conover and Kynard 1984). They do have semilunar peaks in reproduction (Taylor 1986) although perhaps not as pronounced as in more southern populations.

Field observations of spawning of a *F. h. macrolepidotus* population from Massachusetts (Able and Hata 1984) differ in several respects from earlier accounts (Newman 1907; Foster 1967). In a salt marsh creek at low tide, males and females concentrated in isolated pools (Able and Hata 1984). As the tide began to flood a small tributary of the creek, adults moved into the area for spawning. By this time males assumed typical spawning coloration with the eyes becoming very dark and the posterior portions of the dorsal, anal and caudal fins becoming yellowish green. The dominant male in a given area was always the brightest and usually defended a territory around a mat of dead eelgrass (*Zostera marina*) and decaying algae. Courtship consisted of a male following a female and contacting her lightly on the top of her head with his lower jaw. At this time the female often nipped the spawning site and then the male attempted to wrap his dorsal and anal fins around the female while she inserted her anal fin ovipositor into the substrate. While clasping, the male and female vibrated for 1 or 2 s as eggs and sperm were released. Females, regardless of subspecies, deposit eggs in small interstices, a behavior that is facilitated by the female's anal sheath or ovipositor (Able 1984b). Populations of *F. h. macrolepidotus* deposit their eggs in the substrate or in mats of vegetation in intertidal or shallow subtidal areas (see Able 1984b for summary).

Early Life History. The demersal eggs are smaller (*F. h. macrolepidotus* [1.6–1.9 mm diameter] than those of *F. majalis* [2.0–3.0 mm]). In addition, the eggs have longer chorionic filaments than those of *F. majalis,* which are very short and difficult to observe except with scanning electron microscopy (Able 1984a). Embryonic development of *F. heteroclitus* from Woods Hole has been described in detail (Armstrong and Child 1965). Daily increment formation in otoliths of embryos, larvae and juveniles has been validated (Radtke and Dean 1982). Hatching normally occurs in 9–18 days but may be a function of temperature and elevation at the intertidal spawning site, in part, because hatching is cued to immersion on high tides (Di Michele and Taylor 1980; Di Michele and Westerman 1997; Williamson and Di Michele 1997). If hatching is delayed (Taylor et al. 1977; Di Michele and Powers 1982) the larvae are more advanced developmentally (fin ray formation, pigmentation [K. W. Able, pers. obs.]). After hatching, larvae resemble those illustrated by Armstrong and Child (1965). They differ from recently hatched *F. majalis* in the smaller size (*F. h. macrolepidotus* [4.8–5.1 SL] [Marteinsdottir and Able 1992] vs. *F. majalis* [7.0 mm] [Hardy 1978a]), as well as the number of branchiostegals, with *F. heteroclitus* typically having five and *F. majalis* possessing six (Richards and McBean 1966). At larger sizes (>10 mm) they differ in pigment patterns, head scale pattern, and a number of body proportions (Richards and McBean 1966). Recently hatched individuals are distributed on the marsh surface in shallow pools, ponds, ditches, and depressions (Kneib and Stiven 1978; Wang and Kernehan 1979; Taylor et al. 1979; Talbot and Able 1984; Smith 1995; Able et al. 1996; Able and Hagan 2000). On several occasions in Nauset Marsh on Cape Cod, schools of small individuals were observed at the edge of larger marsh creeks in eddies behind peat reefs (K. W. Able, pers. obs.). Many of the young-of-the-year remain in marsh pools during the first summer (Talbot and Able 1984; Able and Hagen 2000). Recently hatched individuals have a wide salinity tolerance (<1 to >100 ppt) but growth was retarded at the extremes (Joseph and Sakesena 1966).

Age and Growth. Age of larvae and adults has been determined from otoliths (days) and scales (years), respectively, as well as modal increases in length (Valiela et al. 1977; Kneib and Stiven 1978; Radtke and Dean 1982). In New Jersey salt marshes, young-of-the-year grew approximately 1.8–3 $mm \cdot week^{-1}$ (Clymer 1978; Rountree 1992) and attained 24–58 mm TL by the fall and 35–63 mm TL by the end of the first year (Able 1990b). Most studies commonly report 2- to 3-year-classes and occasionally age 4 (Fritz and Garside 1975; Valiela et al. 1977; Kneib and Stiven 1978; Able 1990b). This species probably spawns for the first time at about 12 months (Kneib and Stiven 1978; Able 1990b). Young-of-the year are estimated to suffer an annual mortality rate of 99.5% (Meredith and Lotrich 1979), whereas mortality estimates for older year-classes are considerably lower (Meredith 1975; Valiela et al. 1977). Abundance of this species makes it a dominant component of many shallow water estuarine systems especially salt marshes (Nixon and Oviatt 1973). Annual production from Delaware marsh creeks was estimated as 8.14 gdw (grams dry weight)$\cdot m^{-2} \cdot year^{-1}$ (Meredith and Lotrich 1979) to 7.37 $gdw \cdot m^{-2} \cdot year^{-1}$ (Teo and

Able, unpublished data). Estimates from a Massachusetts marsh, after recalculation, were 1.6 gdw·m^{-2} year^{-1} (Valiela et al. 1977; Teo and Able, unpublished data).

General Range. Mummichog occur from southwestern Newfoundland and Prince Edward Island in the Gulf of St. Lawrence (Scott and Scott 1988) to northern Florida (Relyea 1983). The northern subspecies occurs throughout the Gulf of Maine as far south as Long Island and northern New Jersey, where it intergrades with the southern form, *F. h. heteroclitus* (Able and Felley 1986; Marteinsdottir and Able 1988, 1992).

Occurrence in the Gulf of Maine. Found in shallow waters throughout the Gulf. See Collette and Hartel (1988) for details of distribution in Massachusetts Bay.

Movements. Although mummichog are among the most sedentary of fishes, especially during the summer (Horton 1965; Lotrich 1975; Clymer 1978; Murphy 1991; Halpin 1997), they appear to move greater distances during the winter (Fritz et al. 1975). More recent observations indicate that the summer home range can be as much as 15 hectares (Teo and Able, unpublished data). A spring migration from high-salinity waters into brackish and freshwaters has also been reported (Chidester 1920). More recent observations in southern New Jersey indicate that they leave marsh creeks to overwinter in pools on the marsh surface (Smith and Able 1994), where they are reported to bury 15–20 cm in the mud (Chidester 1916). During the summer, homing is evident in marsh pools and creeks (Horton 1965; Lotrich 1975; Murphy 1991). Movements within marshes have a strong tidal component, with juveniles and adults moving up creeks and onto the marsh surface on high tides (Clymer 1978; Werme 1981) and retreating on low tides usually to the same channel or creek (Butner and Brattstrom 1960; Kneib 1987). Access to the marsh surface is usually by intertidal creeks (Murphy 1991).

Importance. Mummichog are of some economic importance as baitfish (Hildebrand and Schroeder 1928; Livingstone 1953), but also have great value as an experimental laboratory animal in a variety of disciplines (Di Michele et al. 1986; Atz 1986) and for pollution studies (Eisler 1986).

STRIPED KILLIFISH / *Fundulus majalis* (Walbaum 1792) / Bigelow and Schroeder 1953:164–165

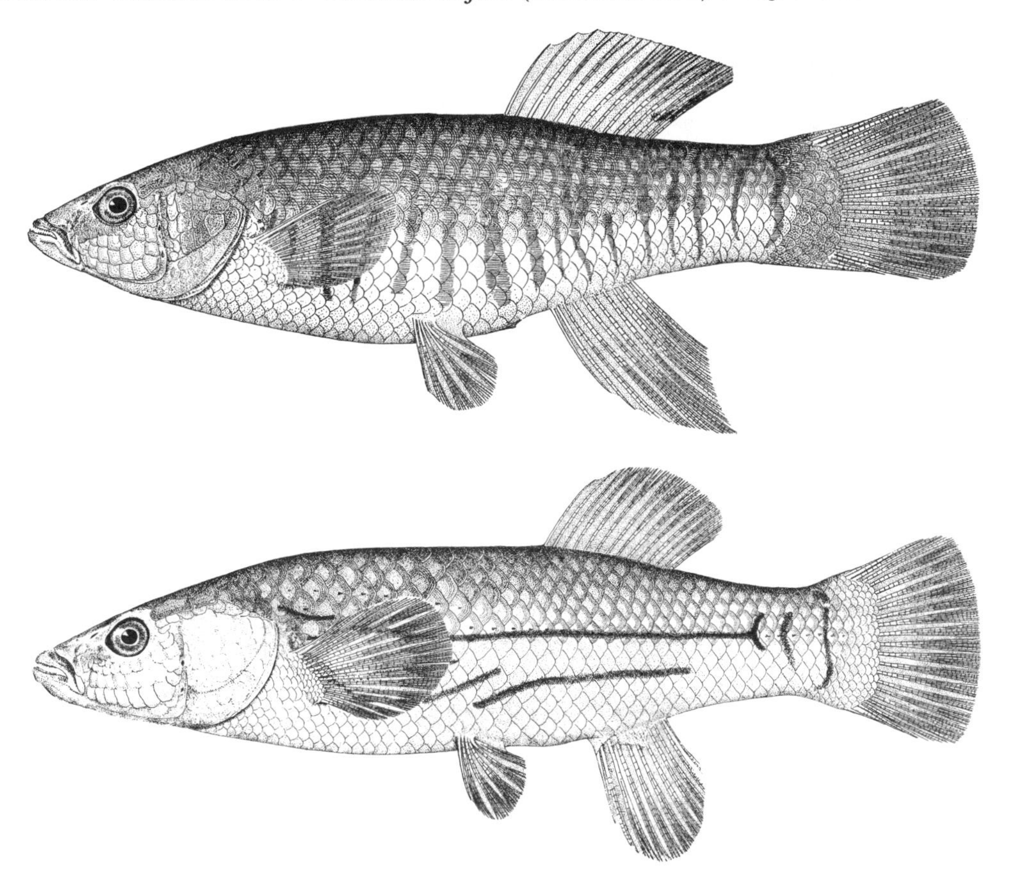

Figure 159. Striped killifish *Fundulus majalis.* Woods Hole, Mass. Male, 130 mm (above);

Description. This fish resembles *F. heteroclitus* in relatively small body size, flattened head, small mouth, and deep caudal peduncle (Fig. 159). Dorsal and anal fins originate at same distance from snout, just behind midpoint of body. Both head and body covered with large, round scales. Adult males differ markedly from females (Newman 1907, 1909a; Relyea 1983) in color pattern and in absence of a short ovipositor on anterior edge of anal fin.

Meristics. Dorsal fin rays 11–16; anal fin rays 9–13; pectoral fin rays 16–21; caudal fin rays 14–20; branchiostegal rays 6. Details of meristic and morphometric variation can be found in Relyea (1983).

Color. Juveniles have vertical bars on the lateral surfaces. Males have 7–12 bars when young and 14–20 as adults. Transition from juvenile to adult pattern occurs with growth (Newman 1907). The adult color pattern is sexually dimorphic (Newman 1907). The vertical bars typical of juveniles are retained by the male, but at approximately 50 mm an ocellus appears on the posterior portion of the dorsal fin (Smith 1907). In females, the black vertical bars transform into horizontal stripes that first appear at approximately 32 mm. The belly is lighter than the dorsal surface.

Size. Adults reach 152–178 mm in the Gulf of Maine and a similar size in Chesapeake Bay (Hildebrand and Schroeder 1928; Young 1950; Clemmer and Schwartz 1964) and Long Island (Briggs and O'Connor 1971).

Distinctions. *Fundulus majalis* differs from the only other killifish in the Gulf of Maine (*F. heteroclitus*) in having preorbital scales (Relyea 1983). The vertical bars on the body are large and darker than in *F. heteroclitus,* and only females have the horizontal stripes. The ovipositor on the anal fin is much shorter in this species.

Habits. Found in a variety of shallow habitats from open beaches to coves and bays. Seldom collected on the marsh surface (Weisberg 1986) but can be abundant in marsh creeks and along subtidal shorelines especially at higher salinities (Rountree and Able 1992; Able et al. 1996). They may prefer sandy sediments (Briggs and O'Connor 1971). They can bury completely in sandy substrate and frequently do so when surrounded by a seine. They can be abundant in small protected bodies of water, where they travel in schools of a few individuals to several hundred or more. On an ebbing tide they may be found on shallow flats in a few centimeters of water, but on a flood tide they are often abundant at the water's edge (Foster 1967; Weinstein 1979; Weisberg 1986). Lethal temperature–salinity experiments on juveniles and nonripe and ripe adults indicated that survival time and salinity tolerance varied with temperature, with the greatest survival at 20°C (Schmelz 1970). Ripe adults and larvae were least resistant to temperature and salinity extremes. Adults have survived salinities up to 106 ppt (Griffith 1974).

Earlier experiments suggested an internal tidal rhythm and some ability to home (Mast 1915). A seasonal change in distribution from shallow habitats in the summer to deeper ones (1–3 m) in the winter has been reported (Able and Fahay 1998). Juveniles (as small as 24 mm TL) and adults have been observed to be cleaned by *Cyprinodon variegatus* while posing in an exaggerated head-down position (Able 1976).

Food. Includes a wide variety of prey including mollusks, crustaceans, fishes, and insects (Hildebrand and Schroeder 1928; Clymer 1978). Generally more carnivorous than *F. heteroclitus* (Jeffries 1972). Juveniles in a Connecticut marsh fed on, in order of abundance, cumaceans, copepods, bivalves, and gastropods, whereas adult stomachs contained polychaetes, gastropods, and bivalves (Petrillo 1987). In Delaware, they consume eggs of horseshoe crab, *Limulus polyphemus* (de Sylva et al. 1962; Clymer 1978). *Fundulus majalis* are capable of feeding on buried prey, especially in sand (Hals 1981), much more so than the frequently co-occurring *F. heteroclitus* (Werme 1981).

Parasites. A broad spectrum has been reported including dinoflagellates, sporozoans, monogenetic and digenetic trematodes, cestodes, nematodes, and copepods (Dillon 1966). Individuals parasitized with monogenetic trematodes (*Swingleus* sp.) and unidentified leeches posed for cleaning by other fishes (Able 1976).

Predators. Predators in a New Jersey estuary included *Anguilla rostrata, Pomatomus saltatrix, Synodus foetens,* and *Prionotus carolinus* (Clymer 1978).

Reproduction. Adults probably mature in the second year (Tracy 1910); at 76 and 63 mm for females and males, respectively (Hildebrand and Schroeder 1928) or as small as 55 mm (Able and Fahay 1998). During spawning, the lateral surfaces of the head of a male are black and the ventral surface and anal fin are yellow (Newman 1907; K.W. Able pers. obs.). Contact organs on the posterior rim of the scales are especially prominent at this time (Newman 1909a). Spawning occurs from April to September in Chesapeake Bay (Hildebrand and Schroeder 1928) and from May through July in New Jersey (Able and Fahay 1998), and has been reported from nearshore shallow water (Nichols and Breder 1927) and in intertidal pools (Able and Fahay 1998). Eggs are buried in the substrate during spawning (Newman 1909b) sometimes as much as 7–10 cm deep (Sumner et al. 1913).

Early Life History. The eggs are 2.0–3.0 mm in diameter, much larger than those of *F. heteroclitus,* with approximately 50 oil droplets, and are often amber in color (Hardy 1978a). Chorionic filaments have been variously reported as absent or present, but many of the former accounts are incorrect (Able 1984a). The filaments are notable in that they possess microfilaments that give individual filaments a bottlebrush appearance (Able 1984a). Embryos tolerate a wide range of temperature, salinity, and dissolved oxygen (Schmelz 1970). Hatching takes 22 or 23 days (Newman 1908, 1914). Descriptions of larval development are summarized by Hardy (1978a). The larvae differ from recently hatched *F. heteroclitus* in the larger size (7.0 mm, see Hardy 1978a), as well as in the number of branchiostegals with *F. majalis* typically having six and *F. heteroclitus* having five (Richards and McBean 1966). Larger individuals of these two species differ in pigment pattern and a number of body proportions (Richards and McBean 1966). A series of experiments has examined the influences of temperature and crowding on meristic characters of embryos and larvae (Fahy 1978, 1979, 1980, 1982). Presence of multiple-size cohorts among recently hatched individuals in a New Jersey population suggests two or three spawning peaks that may be associated with spring tides (Able and Fahay 1998).

Age and Growth. Scales have been used for a Chesapeake Bay population (Clemmer and Schwartz 1964) to determine that males reached approximately 65 mm the first year and subsequently grew 4.4–21.0 mm·year^{-1}. Females reached approximately 61 mm the first year and then grew at a rate of 8–28 mm·year^{-1}. Another estimate indicated similar sizes (Able and Fahay 1998). Males older than age 1 were consistently smaller than females. Males and females lived 6 and 7 years, respectively. These estimates have been questioned (Foster 1967) based on comparisons with the lengths of presumed young-of-the-year collected by Warfel and Merriman (1944a).

General Range. Striped killifish (including the nominal *F. similis*) occur from New Hampshire to the Gulf of Mexico (Relyea 1983).

Occurrence in the Gulf of Maine. Striped killifish may have extended their range into the Gulf of Maine. Little evidence of their presence was found in the Gulf of Maine prior to 1937. Subsequently, they have become more abundant in Cape Cod Bay, perhaps as a result of completion of the Cape Cod Canal (Schroeder 1937). Their later occurrence as far north as New Hampshire has been verified (Jackson 1953). Recent collections have indicated that they are abundant in Waquoit Bay on the south shore of Massachusetts (Ayvazian et al. 1992) and in Nauset Marsh on the outer Cape Cod as well (K. W. Able, unpubl. data). See Collette and Hartel (1988) for details of their distribution in Massachusetts Bay.

ACKNOWLEDGMENT. A draft of the cyprinodontiform section was reviewed by Lynne Parenti.

BERYCOID FISHES. ORDER BERYCIFORMES

Beryciforms have traditionally been viewed as a morphologically diverse, but primitive, lineage of acanthomorph or spiny-finned fishes. Most beryciforms are deepwater species, as is the case for all taxa found in the vicinity of the Gulf of Maine. Large lateral line canals extend across the head, usually divided by narrow bony septae that often feature serrations or spines. Seven species from four beryciform families have been recorded from the Gulf of Maine. It would not be surprising if more beryciforms, such as other species of Diretmidae, Trachichthyidae, Anoplogastridae, or Melamphaeidae, are found someday in deeper portions of the Northeast Channel within the Gulf of Maine.

KEY TO BERYCIFORM FAMILIES FROM THE GULF OF MAINE

1a. Two supramaxillae; pelvic fin I, 7–13. 2
1b. One supramaxilla; pelvic fin I, 6 3
2a. Pelvic fins I, 7; dorsal spines XI or XII . . . . **Holocentridae**
2b. Pelvic fins I, 7–13; dorsal spines IV . . . . **Berycidae**
3a. No spines in dorsal or anal fins; no lateral line. . . . . **Diretmidae**
3b. Dorsal fin with 3–8 spines, anal fin with 2 or 3 spines; lateral line present . . . . **Trachichthyidae**

SQUIRRELFISHES AND SOLDIERFISHES. FAMILY HOLOCENTRIDAE

JON A. MOORE, PAUL GEOGHEGAN, AND JOSEPH N. STRUBE

Squirrelfishes and soldierfishes are small to medium-sized (10–30 cm SL) benthopelagic fishes associated with rocks and coral reefs (0–500 m depth). They are recognizable by their large eyes; hard scales with spinous posterior edges; a single vertical row of scales along anterior edge of the operculum; radiating bony ridges on the skull above the eyes; and laterally species. They are active nocturnal predators on benthic invertebrates or zooplankton. They typically hide in crevices, caves, or under ledges during the day. Larvae to young juveniles (<40 mm SL) have prominent spines on the head projecting forward from the snout (nasal bone), backward from the crown (supraoccipital bone), and backward from the lower

are pelagic and lack the reddish colors of adults; instead, they usually have bluish backs and silvery sides. Kotlyar (1998) most recently reviewed the family. He recognized eight genera with about 75 species worldwide, of which seven genera with 11 species are found in the Atlantic (Woods and Sonoda 1973).

BIGEYE SOLDIERFISH / *Ostichthys trachypoma* (Günther 1859)

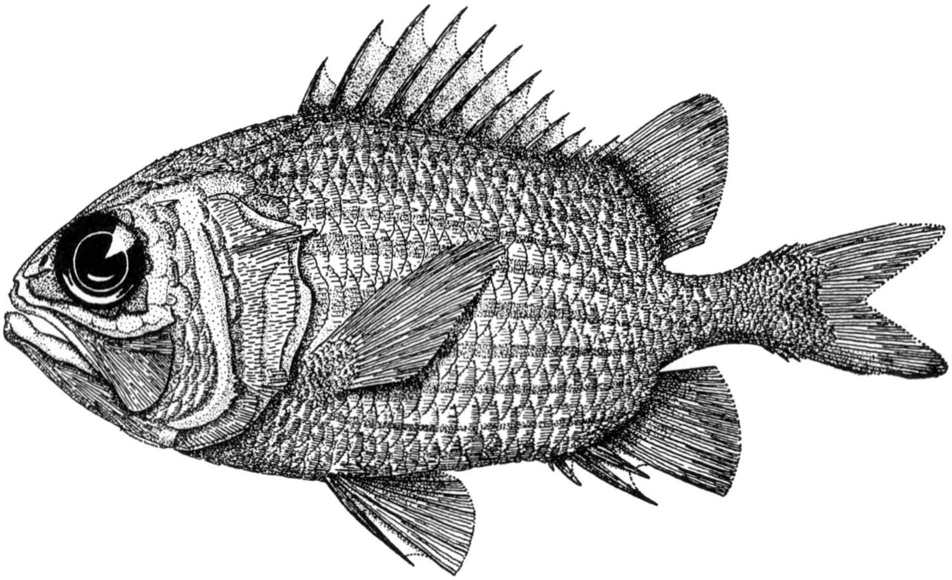

Figure 160. Bigeye soldierfish *Ostichthys trachypoma*. Off Guianas-Brazil, 141 mm SL, FMNH 64677. Drawn by M. A. Holloway. (Reprinted from Woods and Sonoda 1973: Fig. 53. © Sears Foundation for Marine Research.)

Description. Body deep and compressed, greatest body depth 45–50% of SL (Fig. 160). Eye very large, 13.7–17% SL. Prominent spine on operculum and small spinules along edges of most skull bones. Mouth large, slightly oblique. Jaws with small villiform teeth in bands. Dorsal fin deeply notched between spinous and soft portions. Anal fin with four spines, the third spine much larger than the others. Caudal fin forked. Scales with numerous flat spines along posterior edge.

Meristics. Dorsal fin rays XI, I, 13 or 14; anal fin rays IV, 10–12; pectoral fin rays 14–16; pelvic fin rays I, 7; lateral line scales 28–30; gill rakers 8–11 + 14–17 = 23–25; vertebrae 11 + 15 = 26.

Color. Head, iris, upper body, and all fins are bright red; fins lack any distinctive markings; sides of the body have an alternating series of reddish and whitish stripes (Woods and Sonoda 1973).

Size. Reach 19 cm SL (Woods and Sonoda 1973).

Distinctions. Distinguished from all other Atlantic holocentrids by the combination of 12 dorsal spines, 14–16 pectoral rays, 28–30 lateral line scales, and absence of large spines on the suborbital bones in adults.

Biology. Bigeye soldierfish are typically found in deeper waters between 100 and 500 m. They are benthopelagic, taking refuge in caves or crevices during the day and coming out at night to feed on benthic invertebrates and small fishes. Very little biological information is available for this species.

General Range. Adults are known from deep-reef habitats from northern Brazil, throughout the Caribbean, in the Gulf of Mexico, and as far north as Cape Hatteras. Larvae and postlarvae are carried further north by the Gulf Stream. Two postlarvae (39°44.5′ N 70°56′ W [USNM 194246]) were dipnetted from the surface directly south of Martha's Vineyard ("100 miles SE of Montauk Pt., N.Y.," Anderson and Gutherz 1964).

Occurrence in the Gulf of Maine. In September 1999, a 37-mm SL juvenile (MCZ 156942) was found in an impingement sample at the Seabrook Nuclear Power Station. The fish entered the cooling water system through the intakes located in 12 m of water approximately 2 km off Hampton Beach, N.H.

(42°54′17″ N, 70°47′27″ W). This specimen represents a northward range extension and first record for this species and family in the Gulf of Maine.

Commercial Fishery. No commercial fishery exists for this species but they are occasionally caught and eaten in the Caribbean.

ALFONSINOS. FAMILY BERYCIDAE

Jon A. Moore

Alfonsinos are medium-sized (25–50 cm SL) deepwater pelagic fishes found near the bottom along the edges of deep banks and the upper continental slope (100–1,200 m). They may prefer hard bottom. They are recognizable by their laterally compressed ovate bodies, red to silvery pinkish color, bright red fins, large eyes, a large oblique mouth with bands of small teeth in the jaws, two supramaxillae, a conspicuous laterally directed spine anterior to the eye, cheek and operculum largely scaly, a deeply forked caudal fin, and 7–13 soft rays in the pelvic fin. In young individuals (<75 mm SL), the anterior dorsal and pelvic soft rays may be elongate and filamentous. These species are well regarded as food fishes in Europe and Asia. The family was revised by Busakhin (1982). Two genera with about nine species are recognized (Nelson 1994), of which two species of *Beryx* have been caught in the Gulf of Maine.

KEY TO THE FAMILY BERYCIDAE IN THE GULF OF MAINE

1a. Second dorsal fin rays 16–20; total gill rakers 22–24, greatest body depth 44–50% of SL; lateral line scales 61–73 (counting scales on caudal fin) . . . . ***Beryx decadactylus***

1b. Second dorsal fin rays 13–15; total gill rakers 25–27; greatest body depth 33–40% of SL; lateral line scales 69–82 . . . . ***Beryx splendens***

ALFONSINO / *Beryx decadactylus* Cuvier 1829 / Red Bream

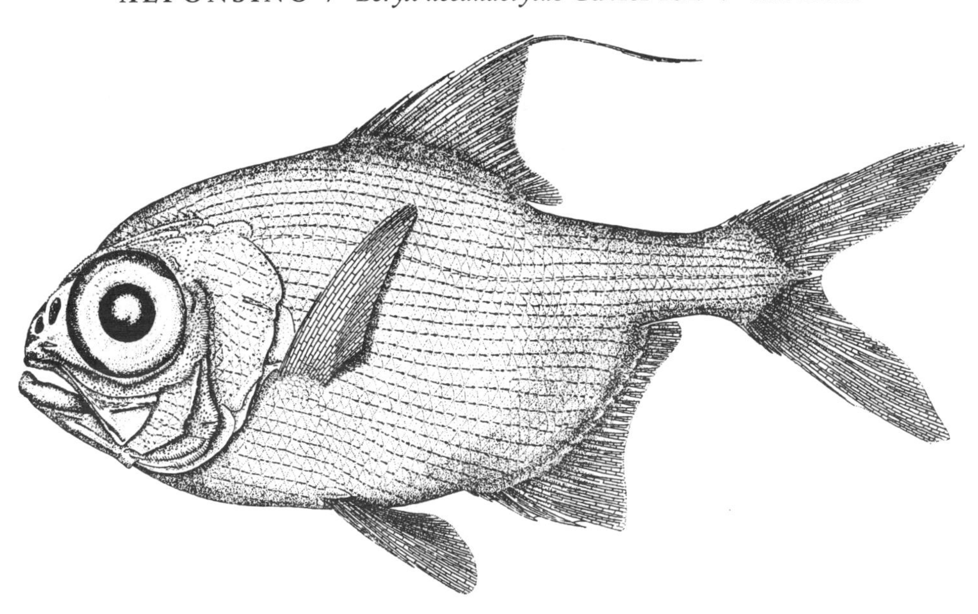

Figure 161. Alfonsino *Beryx decadactylus*. Madeira, 260 mm SL, FMNH 47969. Drawn by M. A. Holloway. (Reprinted from Woods and Sonoda 1973: Fig. 7. © Sears Foundation for Marine Research.)

Description. Body deep and compressed, greatest body depth 44–50% of SL (Fig. 161). Eye very large, 14–17% SL. Mouth large, oblique. Jaws with small villiform teeth in bands. Dorsal fin single, short. Anal fin origin below middle of dorsal fin. Caudal fin deeply forked. Scales spinose.

Meristics. Dorsal fin rays III or IV (usually IV), 16–20; anal fin rays IV, 25–29; pectoral fin rays 15–17; pelvic fin rays I, 9 or 10; lateral line scales 61–73; gill rakers 15–17 + 1 + 6 or 7 = 22–24; pyloric caeca 74–100; vertebrae 10 + 14 = 24.

Color. The upper part of the head, iris, inner edges of jaws, base of preopercle, upper body, and all fins are bright red; side of head is a shiny white; side of the body is silvery with thin reddish stripes; breast is yellowish (Woods and Sonoda 1973). The peritoneum is black.

Size. Reach 60 cm (Heemstra 1986a).

Distinctions. Distinguished from slender alfonsino (*B. splendens*) by the deeper body, relatively larger eye (14–17% SL), more soft dorsal fin rays (16–20 vs. 13–15), fewer total number of gill rakers (22–24 vs. 25–27), generally fewer lateral line scales, and more pyloric caeca (74–100).

Biology. Alfonsino are benthopelagic living near the bottom during the day. At night, they move up from the bottom to feed on crustaceans and small fishes. Very little biological information is available for this species. Larvae have large spines on the first infraorbital (Mundy 1990).

General Range. Known from deep waters throughout the temperate and tropical Atlantic, western and central Pacific, and the southwest Indian oceans.

Occurrence in the Gulf of Maine. Occasionally taken in trawls at depths of 100–200 m in various areas, including Jones Ground (Woods and Sonoda 1973), east of Jeffreys Ledge (MCZ 65741), and in Jordan Basin (MCZ 61978, 65711).

Commercial Fishery. No commercial fishery exists for this species in the Gulf of Maine but they are commercially fished in the Azores and Japan by longline and trawled in New Zealand.

SLENDER ALFONSINO / *Beryx splendens* Lowe 1833 / Golden Eye Perch

Figure 162. Slender alfonsino *Beryx splendens. Albatross* sta. 2415, 30°44′ N, 79°26′ W. Drawn by M. M. Smith.

Description. Body ovate and compressed, greatest body depth 33–40% SL (Fig. 162). Eye large, 13.9–14.4% SL. Mouth large, oblique. Jaws with small villiform teeth in bands. Dorsal fin single, short. Anal fin origin below or behind posterior end of dorsal fin base. Caudal fin deeply forked. Scales spinose.

Meristics. Dorsal spines IV, dorsal fin rays 13–15; anal spines IV, anal fin rays 25–30; pectoral fin rays 16–18; pelvic fin rays I, 10–12; lateral line scales 69–82; gill rakers 18–20 + 1 + 6–7 = 25–27; pyloric caeca 23–30; vertebrae 10 + 14 = 24.

Color. Color is as described for *B. decadactylus.*

Size. Reach 59 cm FL (Vinnichenko 1996).

Distinctions. Slender alfonsino can be distinguished from *B. decadactylus* by the less deep body, relatively smaller eye (13.9–14.4% SL), fewer soft dorsal fin rays (13–15 vs. 16–20), more gill rakers (25–27 vs. 22–24), generally more lateral-line scales, and fewer pyloric caeca (23–30 vs. 74–100).

Biology. Slender alfonsino are benthopelagic living near the bottom at a wide range of depths (10–1,200 m), although reported to be most abundant at 200–680 m (Busakhin 1982; Maul 1986a; Horn and Massey 1989). At night, they move up from the bottom to feed on squids, shrimps, euphausiids, mysids, tunicates, amphipods, and lanternfishes. Off New Caledonia, females grow faster than males, and maturity begins at age 6 for females and at age 7–8 for males (Lehodey and Grandperrin

1996). On the Mid-Atlantic Ridge, the size at maturity is 18 cm (ICES 1995). Larvae have large spines on the first infraorbital (Mundy 1990). Maximum age is 17 years in New Zealand (Horn and Massey 1989) and about 20 years off New Caledonia (Lehodey and Grandperrin 1996). Predators include swordfish.

General Range. *Beryx splendens* have a similar distribution to *B. decadactylus* and are known from deep waters throughout much of the temperate and tropical Atlantic, Pacific (including the eastern Pacific off Chile [Nakamura 1986]), and Indian oceans.

Occurrence in the Gulf of Maine. Collected at least once at the mouth of the Northeast Channel (Scattergood 1958; MCZ 39550) and most likely range into the Gulf over deeper hard bottoms.

Commercial Fishery. Dense concentrations sometimes occur over seamounts and other topographic features. Although no commercial fishery exists in the Gulf of Maine, they are commercially fished by longline or trawl at the Corner Rise Seamounts, Azores, New Caledonia, Japan, and New Zealand.

SPINYFINS. FAMILY DIRETMIDAE

Jon A. Moore

Spinyfins, also known as discfishes, are moderate-sized, meso- and bathypelagic fishes, caught very infrequently in deepwater trawls. They are externally recognizable by their deep, disc-shaped or oval, and compressed body, which lacks a lateral line. The eye is large. The mouth is large, oblique, with minute teeth on the jaws.

Only one supramaxilla is present on the upper jaw. Color is dark brown or black. No spines are present in either the dorsal and anal fins. Scales are spinoid, deciduous, and cover the body and cheek. Clumps of epidermal glands are found on the median wall of the gill chamber.

Three species of discfishes in three genera occur in the western North Atlantic, and at least two are known from within the Gulf or just outside along Georges Bank. The family was recently revised by Post and Quéro (1981), and Kotlyar (1987, 1990).

KEY TO THE FAMILY DIRETMIDAE IN THE GULF OF MAINE AND VICINITY

1a. Body with roundish outline; anus immediately before anal fin origin; ventral keel of scutes in midline anterior to pelvic fins; 7–20 bony lamina radiating across upper opercle . . . . ***Diretmus argenteus***

1b. Body outline ovate; anus roughly midway between pelvic and anal fins; no ventral keel anterior to pelvic fins; 3–6 bony lamina radiating across upper opercle **2**

2a. Dorsal fin with 24–27 (usually 25) fin rays; pelvic fin short, does not reach to origin of anal fin; 12–16 gill rakers on first arch . . . . ***Diretmoides pauciradiatus***

2b. Dorsal fin with 26–29 (usually 27 or 28) fin rays; pelvic fin reaches to or beyond anal fin origin; 16–20 (usually 18 or 19) gill rakers on first arch. . . . . ***Diretmichthys parini***

BLACK SPINYFIN / *Diretmichthys parini* (Post and Quéro 1981) / Black Discfish

Description. Body ovate, body depth 40–60% of SL (Fig. 163). Eyes very large, 14–19% SL. Spinous rays absent in long low dorsal and anal fins. Characteristic small windows present in fin membranes between bases of dorsal and anal fin rays; a series of small spinules point laterally from dorsal and anal fin bases. Pelvic fin spine present. Caudal fin forked. Body scales small, except for enlarged row of ventral scutes in midline between pelvic and anal fins.

Meristics. Dorsal fin rays 26–29; anal fin rays 20–23; pectoral fin rays 17–20; pelvic fin rays I, 6; gill rakers 5–7 + 11–14 = 16–20; vertebrae 29–31.

Color. Adults all black or very dark brown; adolescents silvery with black horseshoe-shaped dark area on side.

Distinctions. The compressed dark, ovate body, lack of dorsal and anal spines, ventral keel posterior to the pelvic fins, and placement of the anus midway between the pelvic and anal fins distinguish this fish from all other fishes in the Gulf of Maine.

Biology. Black spinyfin are primarily mesopelagic, but are occasionally caught in the upper bathypelagic zone. Juveniles are usually taken from the surface down to 200 m. Adults are mostly found below those depths down to 1,300 m. Post (1986b) reported that they reproduce year-round in tropical to subtropical waters. Larvae and juveniles are easily distinguishable by the presence of two pairs of prominent head spines. One pair projects posterodorsally from the parietals and another pair projects anteroventrally from the preoperculum. They metamorphose by 20 mm length and lose their head

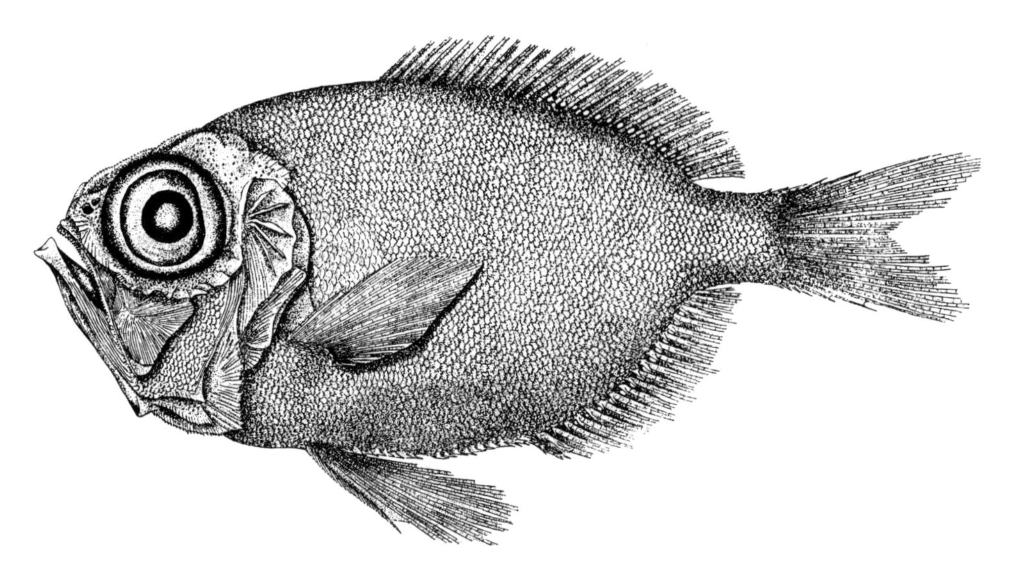

Figure 163. Black spinyfin *Diretmichthys parini.* Gulf of Mexico, 245 mm SL, FMNH 66792. Drawn by M. A. Holloway. (Reprinted from Woods and Sonoda, 1973: Fig. 12, as *Diretmus argenteus.* © Sears Foundation for Marine Research.)

General Range. Known from throughout much of the tropical and temperate Atlantic, Indian, and the central and western Pacific oceans (Post and Quéro 1981; Kotlyar 1987).

Occurrence in Gulf of Maine. One 24-cm specimen (MCZ 145258) was collected in 1991 from Georges Basin in the Gulf.

ROUGHIES. FAMILY TRACHICHTHYIDAE

JON A. MOORE

Roughies, also known as slimeheads, are medium-sized fishes (up to 60 cm TL) and benthopelagic at depths from 10 to 1,900 m, although most species are abundant from 250 to 1,000 m. The body is ovate and compressed; the head has a very large canal system covered by a membrane of skin. Preopercular and posttemporal bones usually feature a prominent posteriorly directed spine. The mouth is large and oblique, with very small teeth arranged in bands on the jaws. One supramaxilla is present. Three to eight dorsal fin spines are present and the pelvic fin has one spine and six rays. An abdominal series of enlarged scales or scutes is present between the pelvic and anal fins, often forming a distinct median keel.

The family consists of eight genera and about 35 species (Kotlyar 1996), with three genera and five species reported from the western Atlantic Ocean (Woods and Sonoda 1973).

The much sought after orange roughy (*Hoplostethus atlanticus*) is the basis for major commercial fisheries in New Zealand, Australia, and Namibia. Smaller fisheries for this species also exist off Iceland, the Mid-Atlantic Ridge, and west of Britain. Woods and Sonoda (1973) listed 14 specimens of orange roughy from the Gulf of Maine, and subsequent authors have included it in the Gulf (Scott and Scott 1988; Pohle et al. 1992). However, examination of some of these specimens at the MCZ shows that the Gulf of Maine is not mentioned on the labels nor in the catalogue and that the longitude and latitude given on the labels corresponds to the Laurentian Channel between Nova Scotia and Newfoundland. It, therefore, appears that orange roughy are not validly recorded from the Gulf of Maine, although they are occasionally taken off the Scotian Shelf (Pohle et al. 1992), and it would not be surprising to find them near canyons or seamounts off Georges Bank. For this reason they are included in the key.

KEY TO THE FAMILY TRACHICHTHYIDAE IN THE GULF OF MAINE AND VICINITY

1a. Dorsal fin with 8 spines (very rarely 7); lateral line scales roughly same size as other body scales; prominent spines projecting forward on nasal bones . . . . *Gephyroberyx darwini*
1b. Dorsal fin with 4–7 spines; lateral line scales very enlarged and diamond-shaped, usually twice as high (dorsoventrally) as wide (anteroposteriorly); nasal spines very small or absent 2
2a. Dorsal fin with 15–18 soft rays; anal fin with 10–12 soft rays; pectoral fin with 17–20 rays . . . . *Hoplostethus atlanticus*
2b. Dorsal fin with 12–14 soft rays; anal fin with 9 or 10 soft rays; pectoral fin with 14–16 rays 3
3a. Anal fin with 9 or 10 rays (almost always 10); ventral scutes 8–12; gill rakers 7 or 8 + 1 + 15–17 = 23–26 . . . . *Hoplostethus mediterraneus*
3b. Anal fin with 8–10 (very rarely 10); ventral scutes 10–17; gill rakers 5 or 6 + 1 + 12–14 = 18–22 . . . . *Hoplostethus occidentalis*

DARWIN'S ROUGHY / *Gephyroberyx darwini* (Johnson 1866) / Big Roughy

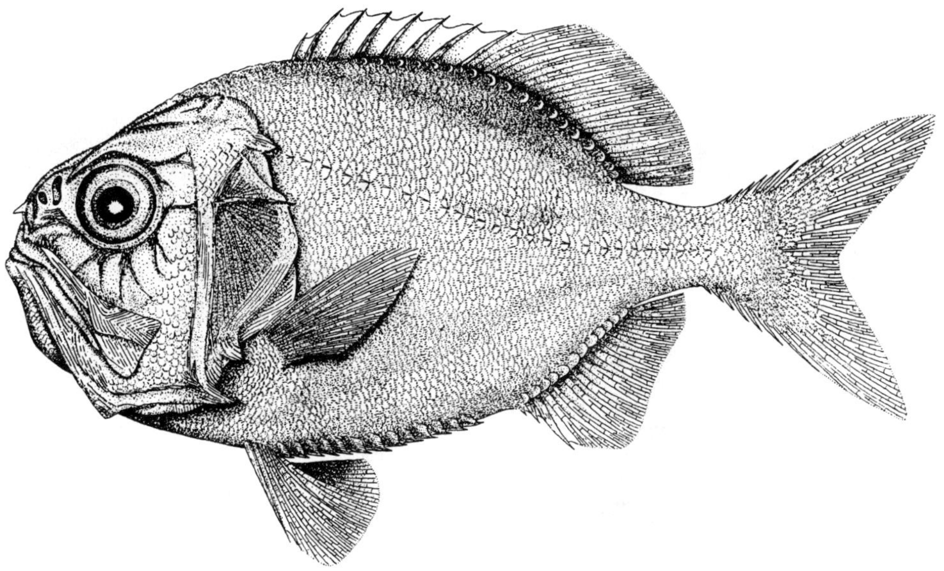

Figure 164. Darwin's roughy *Gephyroberyx darwini*. Caribbean Sea, 174 mm SL, FMNH 69534. Drawn by M. A. Holloway. (Reprinted from Woods and Sonoda 1973: Fig. 16. © Sears Foundation for Marine Research.)

Description. Body, deep ovate (Fig. 164). Head large with low ridges separating shallow sensory canals. Head canals covered by thick membrane. Snout short, with anteriorly pointing spine on each nasal bone. Mouth large, oblique. Posttemporal, opercle, and preopercle with spines, the last large and prominent. Abdomen with ventral keel of 10–12 thick, enlarged scutes. Lateral line scales only slightly larger than other body scales, large centrally placed spine on each lateral line scale.

Meristics. Dorsal fin rays VII or VIII + 13 or 14; anal fin rays III, 11 or 12; pectoral fin rays 13–15; pelvic fin rays I, 6; lateral line scales 26–31; scutes 9–13; gill rakers 15–20; vertebrae 26.

Color. Head and fins red, upper body brownish red, and lower flanks pinkish silvery.

Size. Reach 58 cm SL (Quéro 1982).

Distinctions. Distinguished from other Atlantic roughies by presence of 8 dorsal fin spines (rarely 7), only slightly enlarged lateral line scales that feature a centrally placed spine, 11 or 12 anal fin rays, and generally fewer gill rakers.

Biology. A benthopelagic fish found from 10 to 1,210 m (Kotl- hard bottoms and canyons. Feeds on mysids, shrimps, and small fishes (Woods and Sonoda 1973; Van Guelpen 1993). A 15.3-cm juvenile collected off Browns Bank indicates that it may reproduce in the vicinity of the Gulf of Maine (Van Guelpen 1993).

General Range. Widely distributed in the western Atlantic from Brazil to the Gulf of Maine, in the eastern Atlantic from Iceland to the southern tip of Africa, in the Indian Ocean along the African coast, near Sri Lanka, in the Bay of Bengal, and around the southern half of Australia.

Occurrence in the Gulf of Maine. Darwin's roughy have been collected along Browns Bank in the Northeast Channel and most likely range into deeper parts of the Gulf over hard bottoms. Sightings of this species in Oceanographer Canyon (Van Guelpen 1993), on the southern slope of Georges Bank, indicate that they may be more common in this area than previously recognized.

Importance. No fishery presently targets this species, but they are sometimes caught incidentally in deep-trawling operations and are edible. Seventy pounds of Darwin's roughy were caught in a gill net set at 1,000 m in a canyon on Georges Bank and sold at the Portland Fish Exchange in May 2000. This has prompted consideration of exploratory fishing

SILVER ROUGHY / *Hoplostethus mediterraneus* Cuvier 1829 / Mediterranean Roughy

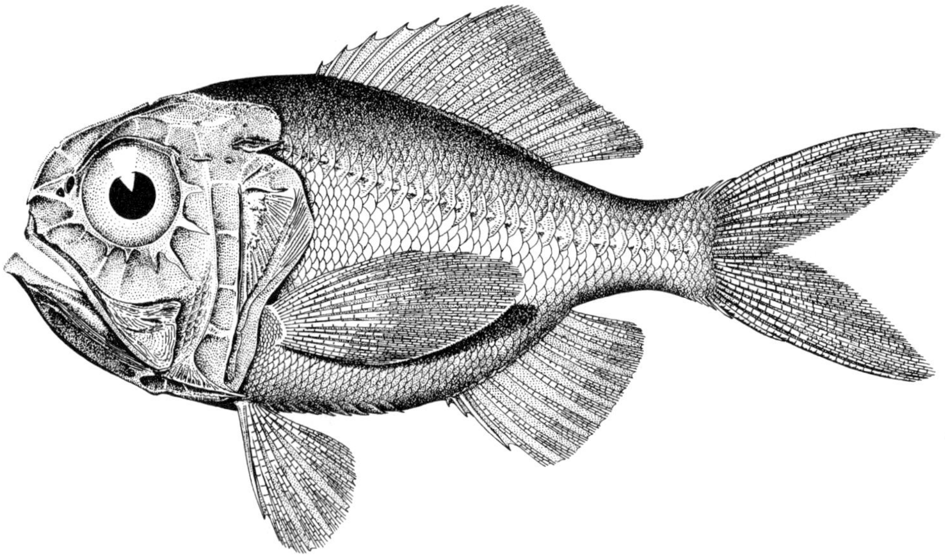

Figure 165. Silver roughy *Hoplostethus mediterraneus*. *Albatross* sta. 2659. Drawn by A. H. Baldwin.

Description. Small to moderate-sized (generally 15–25 cm SL), with a deep body (44–49% SL) (Fig. 165). Head large, 37–43% SL, with extensive and deep sensory canals covered by thin membrane of skin. Eye relatively large, 13–16% SL. Lateral line scales large, diamond-shaped. Body scales smaller, relatively smooth and deciduous, except abdominal keel with 8–12 enlarged stout scutes.

Meristics. Dorsal fin rays VI or VII + 12–14; anal fin rays III, 9 or 10; pectoral fin rays 14–16; pelvic fin rays I, 6; lateral line scales 28–29; scutes 8–12; gill rakers 7 or 8 + 1 + 15–17 = 23–26; vertebrae 26.

Color. Pale pinkish with dark silvery lower flanks, rosy brown above the lateral line, cheek and head pearly silver with colorless to bright red ridges on head, fins orangy pink, dark distal tips to caudal fin, inside of mouth reddish black. Peritoneum and stomach jet black.

Size. Reach 32 cm SL (Vinnichenko 1996).

Distinctions. Distinguished from other Atlantic roughies by the combination of 12–14 dorsal fin rays, more gill rakers (23–26), fewer ventral scutes (8–12), and distinctive silvery lower flanks and opercular bones.

Biology. A benthopelagic fish found at 150–1,200 m, most abundant at 600–800 m. Silver roughy from off Britain and West Africa feed on mysids, sergestid shrimps, amphipods, and small fishes (Merrett and Marshall 1981; Gordon and Duncan 1987). Maturity is reached by 10 cm SL and eggs are shed in one batch. Spawning in the Mediterranean is in September (Cau and Deiana 1982) and west of Britain it probably occurs in autumn (Gordon and Duncan 1987). A 34-mm juvenile taken off Georges Bank (MCZ 38241) indicates that they may reproduce in the vicinity of the Gulf.

General Range. Widely distributed in the Atlantic, along the coast of North America, from Iceland to the Azores, around the southern tip of Africa into the Indian Ocean, and the southwestern Pacific (Kotlyar 1986; Paxton and Hanley 1989).

Occurrence in the Gulf of Maine. Taken several times on the southern slope of Georges Bank (MCZ 38241; YPM 11068) and quite likely extend into the Gulf.

Importance. No U.S. or Canadian fishery targets this species, but they are sometimes caught incidentally in deep-trawling operations over seamounts (e.g., Corner Rise and New Zealand) and are edible.

WESTERN ROUGHY / *Hoplostethus occidentalis* Woods 1973

Description. Small to moderate-sized (up to 17.3 cm SL), with a deep body (47–54% SL) (Fig. 166). Head large, 40–43% SL, with extensive and deep sensory canals covered by thin membrane of skin. Eye relatively large, 13–16.3% SL. Lateral line scales large, diamond-shaped. Body scales relatively smooth and deciduous, except abdominal keel of 10–17 enlarged stout scutes.

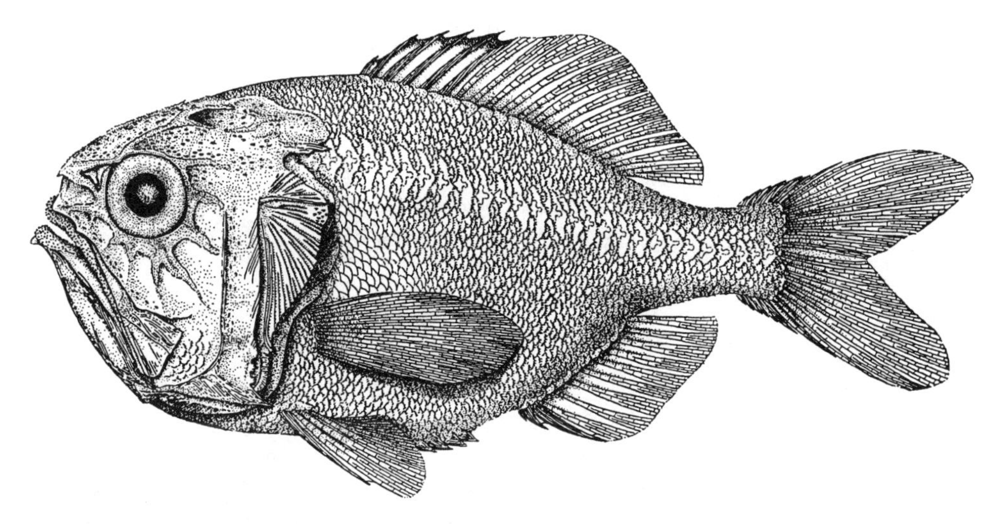

Figure 166. Western roughy *Hoplostethus occidentalis.* Gulf of Mexico, 173 mm SL, FMNH 66790. Drawn by M. A. Holloway. (Reprinted from Woods and Sonoda 1973: Fig. 24. © Sears Foundation for Marine Research.)

Meristics. Dorsal fin rays IV–VII + 12–14; anal fin rays II or III, 8–10; pectoral fin rays 14–16; pelvic fin rays I, 6; lateral line scales 27–31; scutes 10–17; gill rakers 5 or 6 + 1 + 12–14 = 18–22; vertebrae 26.

Color. All fins and ridges on the head are pale pink. The body above the lateral line is brownish pink; the flank below the lateral line is silvery (but less so than in *H. mediterraneus*). The head is silvery underneath the infraorbital and opercular bones. The mouth, gill cavity, and underside of the gill covers are charcoal to black. There is a dark gular patch between the lower jaw bones. The peritoneum and stomach are both jet black.

Size. Reach 17.3 cm (Woods and Sonoda 1973).

Distinctions. Western roughy can be distinguished from other Atlantic roughies by the combination of 12–14 dorsal fin rays, few gill rakers (18–22), and 10–17 ventral scutes.

Biology. Western roughy are benthopelagic fish at 250–550 m, most abundant at 366–460 m. New England individuals feed on mysid shrimps and other small decapods (J. Moore, unpubl. data).

General Range. Endemic to the western Atlantic, where they are widely distributed from northern South America, the Caribbean, Gulf of Mexico, and the eastern coast of North America to Browns Bank.

Occurrence in the Gulf of Maine. They have been collected from Browns Bank at the mouth of the Northeast Channel (ROM 26228) and most likely range into deeper parts of the Gulf. This specimen was misidentified by Scott and Scott (1988:324) as *H. mediterraneus.*

Importance. No fishery targets them, but they are sometimes caught incidentally in deep-trawling operations and are edible.

ACKNOWLEDGMENT. A draft of the beryciform section was reviewed by Karsten E. Hartel.

DORIES. ORDER ZEIFORMES

Order Zeiformes consists of marine fishes, most of which are deep-bodied and very compressed. Zeiformes are usually considered primitive acanthomorphs placed near Beryciformes. The pelvic fin is with or without a spine and with five to ten soft rays. Zeiformes inhabit tropical to temperate waters at various depths. There are six families with some 20 genera and 39 species (Nelson 1994). Three families are included here as Gulf of Maine zeiforms, although there is still some question as to whether the Caproidae properly be-

KEY TO ZEIFORMES IN THE GULF OF MAINE

1a. Scales thin and vertically elongate; anal fin consists of 2 spines widely separated from remainder of anal fin (Grammicolepididae) 4
1b. Scales may be present, absent, or replaced by bony bucklers, but never as described above; anal fin spines more than 2, separated from the rest of the anal fin by a small gap 2
2a. Scales small, spinoid; mouth very small and terminal; pelvics I, 5; branchiostegal rays 5–7 (Caproidae) 5
2b. Scales minute, absent, or replaced by bony bucklers; mouth large and

3a. Spinous dorsal fin with long filaments; dorsal fin spines 8 or 9; large bony bucklers along bases of dorsal and anal fins; 3 anal fin spines; pelvic fin I, 5 or 6; body silver-gray, with diffuse black spots in juveniles .... ***Zenopsis conchifera***
3a. Spinous dorsal fin without long filaments; dorsal fin spines 7 or 8; no large bony bucklers along bases of dorsal and anal fins; anal fin spines 1 or 2; pelvic fin with no spine and 9 or 10 rays; body red or pink, with black pelvic fins .... ***Cyttopsis rosea***
4a. Total number of dorsal spines plus soft rays 39–41; several widely spaced spiny scutes on sides of body........... ***Grammicolepis brachiusculus***
4b. Total number of dorsal spines plus soft rays 32–35; no spiny scutes on sides of body .... ***Xenolepidichthys dalgleishi***
5a. Body depth equal to or greater than body length; second dorsal fin rays 31–37; anal fin rays 29–34 .... ***Antigonia capros***
5b. Body depth less than body length; second dorsal fin rays 26–30; anal fin rays 23–28 .... ***Antigonia combatia***

DORIES. FAMILY ZEIDAE

Jon A. Moore and Hugh H. DeWitt

Dories are mesopelagic or benthopelagic fishes found on the continental shelf and upper slope to about 600 m. The body is typically oval-shaped and compressed. The jaws are large and highly protrusible. Seven genera and 13 species are known (Nelson 1994). The red dory, *Cyttopsis rosea,* has been collected off Georges Bank and may eventually be found in the Gulf so it is included in the key.

BUCKLER DORY / *Zenopsis conchifera* (Lowe 1852) / American John Dory /

Bigelow and Schroeder 1953:297–299 (as *Zenopsis ocellata*)

Figure 167. Buckler dory *Zenopsis conchifera*. Provincetown, Mass. Drawn by H. L. Todd.

Description. Body truncate, oval, strongly compressed, dorsal profile of head noticeably concave (Fig. 167). Mouth oblique, with small jaw teeth in one or two rows. Anterior dorsal fin spines very long, filamentous at tips. Soft dorsal fin low. Anal fin preceded by three spines, soft rays low and symmetrical with soft dorsal fin. Caudal peduncle slender; caudal fin small compared to rest of body. Pelvic fins large and long, with free tips to fin rays. Lateral line lacks scales, greatly arched over midbody. Enlarged bony scales (bucklers) along base of dorsal and anal fins.

Meristics. Dorsal fin rays VIII or IX, 24–26; anal fin rays III, 22–25; pectoral fin rays 12; pelvic fin rays I, 5 or 6; gill rakers 3 + 7–10 = 10–13; vertebrae 12 + 22 or 23 = 34 or 35.

Color. Adults are silvery all over. Juveniles and subadults have a number of dark spots irregularly arranged over the body. These spots fade in adults leaving one indistinct blotch in the center of each side.

Size. Reach 75 cm SL (Quéro 1986a).

Distinctions. Buckler dory are easily recognized by the very compressed body, long dorsal fin spines with trailing filaments, large bony bucklers along the ventral midline and the margins of the dorsal and anal fins, tiny tail, and large head and jaws. They are most similar to red dory, which are distinguished by the relatively short dorsal fin spines without filaments, serrate bony ridge over the eye, and pink or red body.

Taxonomic Remarks. Buckler dory have been referred to as the American John dory because of their resemblance to John dory (*Zeus faber*) found in the eastern Atlantic.

Biology. They are found both over the bottom and in midwater mainly at depths of 100–400 m. They probably school (Berry 1978). Diet consists mostly of other fishes (e.g., scup and offshore hake), and also some crustaceans and squids (Bowman et al. 2000, J. A. Moore, unpubl. data). Known predators include *Mustelus canis* and *Amblyraja radiata* (Rountree 1999). Little is known about reproduction in the Gulf. Specimens collected off Martha's Vineyard in late January or early February had well-developed ovaries with eggs 1.2–1.4 mm in diameter (Bigelow and Schroeder). Larvae are described by Weiss et al. (1987).

General Range. Known from southern Brazil to Nova Scotia in the western Atlantic, in the eastern Atlantic from Ireland to South Africa, and from the Indian Ocean off southern India (Heemstra 1980).

Occurrence in the Gulf of Maine. Occasionally found in the Gulf of Maine. One specimen was collected at Provincetown at the tip of Cape Cod and another was found at Campobello Island at the mouth of the Bay of Fundy (Bigelow and Schroeder). Individuals were taken in Massachusetts Bay 40 km east of Cape Ann (MCZ 36658), in Cape Cod Bay (MCZ 37541), off Mt. Desert Rock (MCZ 101171), and in Ipswich Bay (Collette and Hartel 1988). They are also frequently taken off the southern slope of Georges Bank.

Importance. No directed fishery exists for this species, although large numbers are occasionally taken and larger individuals are sometimes landed at local ports. Closely related species are highly regarded as food in Europe and a fishery for *Zenopsis nebulosus* is starting in Australia.

DIAMOND DORIES AND TINSELFISHES. FAMILY GRAMMICOLEPIDIDAE

Jon A. Moore and Hugh H. DeWitt

The Grammicolepididae are small to medium-sized marine fishes usually found at moderate depths (100–800 m) in tropical and temperate waters (Karrer and Heemstra 1986). The body is deep and very compressed. The head is very small with a relatively large eye, especially in juveniles. The mouth is small and oblique, bearing tiny, weak teeth on the jaws. The scales are unusually narrow, vertically elongate strips. There is a row of bony knobs, each bearing a spine, along the base of the dorsal and anal fins. The caudal fin has 13 branched rays vs. 11 in other zeiforms (Heemstra 1980). Juveniles have elongate anterior dorsal and anal spines. Three genera and four species are known (Nelson 1994).

THORNY TINSELFISH / *Grammicolepis brachiusculus* Poey 1873 /

Bigelow and Schroeder 1953:299–300 (as *Xenolepidichthys americanus*)

Description. Body deep and strongly compressed in young (body depth 62–71% SL), becoming less so in adults (43– [illegible]

tened spiny scutes protrude from body. Anterior fin spines in dorsal and anal fins elongate in young. Metamorphosis to

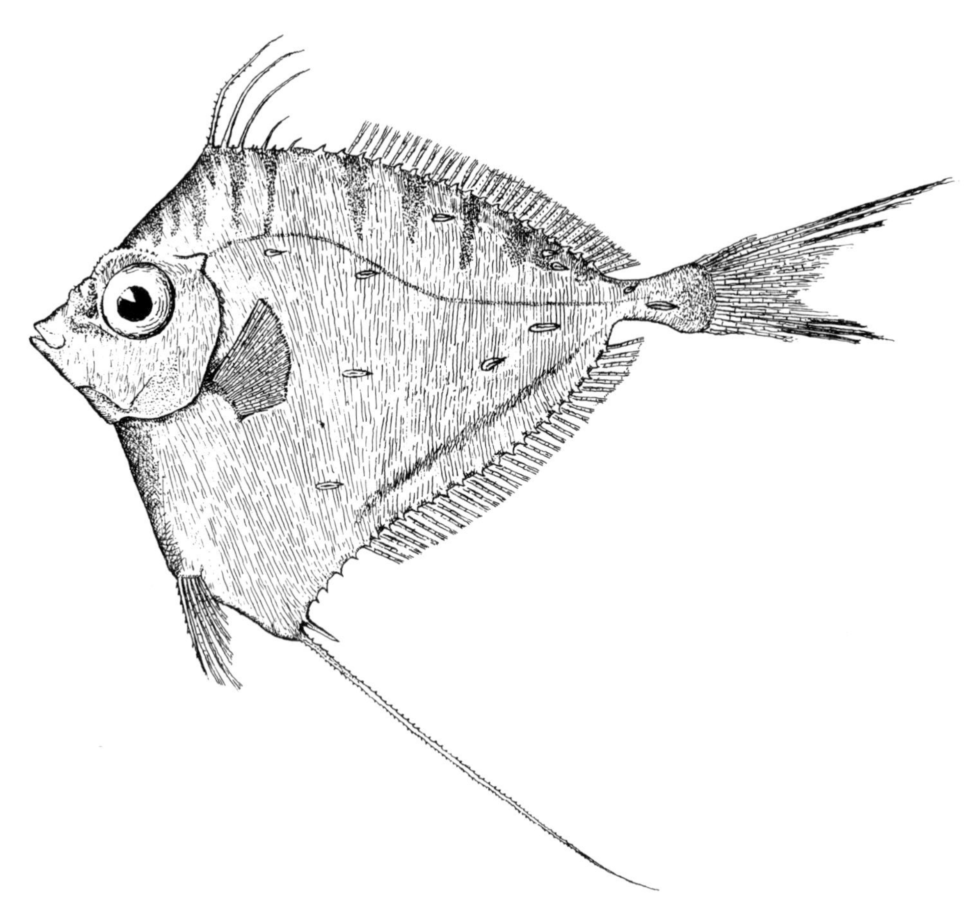

Figure 168. Thorny tinselfish *Grammicolepis brachiusculus.* Georges Bank. (From Bigelow and Schroeder: Fig. 157.)

becomes more elongate, spiny scutes are lost and prolonged fin spines shorten (Karrer and Heemstra 1986). Eyes large in relation to head size. Teeth minute, one or two rows on jaws.

Meristics. Dorsal fin rays VI or VII, 32–34; anal fin rays II, 33–35; pectoral fin rays 14 or 15; pelvic fin rays I, 6; gill rakers rudimentary (1 or 2) + 12; vertebrae 43–46.

Color. The body is silvery in adults. Young have numerous irregular blotches and bars on body and anal and caudal fins.

Size. Maximum size 64 cm (Karrer and Heemstra 1986).

Distinctions. Spiny scutes protruding from the body and dark bars below the dorsal fin or on the anal fin make the young very distinctive. Adults are distinguished by the large total number of dorsal fin elements and large number of vertebrae.

Taxonomic Remarks. Thorny tinselfish have previously been reported off North America as *Xenolepidichthys americanus* (by Nichols and Firth 1939 [Bigelow and Schroeder]) or *Daramattus americanus* (by Robins and Ray 1986).

Biology. Typically found in midwater and over bottom at depths of 300–900 m (Quéro 1986b). Very little is known about their biology. Changes in body proportions during growth are discussed by Quéro (1979).

General Range. Found on both sides of the Atlantic and in the Indian and Pacific oceans (Karrer and Heemstra 1986; Quéro 1986b).

Occurrence in the Gulf of Maine. Bigelow and Schroeder described a specimen collected at the surface on Georges Bank (AMNH 14107), the holotype of *Xenolepidichthys americanus.*

SPOTTED TINSELFISH / *Xenolepidichthys dalgleishi* Gilchrist 1922

Description. Body very deep, strongly compressed (body depth 70–125% SL with lower values in larger specimens) (Fig. 169). No flattened spiny scutes protrude from body. Anterior fin spines in dorsal and anal fins elongate in young. Eyes large in relation to head size. Teeth minute, in one or two rows on jaws.

Figure 169. Spotted tinselfish *Xenolepidichthys dalgleishi.*

Meristics. Dorsal fin rays V, 27–30; anal fin rays II, 27–29; pectoral fin rays 14 or 15; pelvic fin rays I, 6; gill rakers rudimentary; vertebrae 36 or 37.

Color. The body is silvery with numerous round dark spots on the sides and a dark band along the posterior edge of the caudal fin.

Size. Maximum size 15 cm (Karrer and Heemstra 1986).

Distinctions. The lack of spiny scutes protruding from body, deeper body, fewer total dorsal fin elements, fewer vertebrae, and round spots on the sides distinguish spotted from thorny tinselfish.

Biology. Spotted tinselfish are found near the bottom at 90–900 m, most commonly at 100–300 m. [illegible]

about their biology. Fourmanoir (1976) illustrated a 2.5-cm SL juvenile from New Caledonia.

General Range. Known from widely separated localities in the western Atlantic and Caribbean, off South Africa, Japan, Philippines, Australia, and New Caledonia (DeWitt et al. 1981; Karrer and Heemstra 1986).

Occurrence in the Gulf of Maine. Single individuals were reported from 10 km south of Small Point and 19 km south of South Bristol, Maine (DeWitt et al. 1981). Another specimen is known from the Northeast Channel (USNM 329914). After reexamination, a specimen from off northwestern Browns Bank (ARC 8600554) considered to be *Daramattus americanus* by Scott and Scott (1988) actually proved to be *Xenolepidichthys* [illegible]

BOARFISHES. FAMILY CAPROIDAE

Jon A. Moore and Grace Klein-MacPhee

The Caproidae are deepwater fishes found at 40–900 m (Quéro 1986c). The body is deep and very compressed. The head is relatively small with a pronounced occipital crest; the eye is relatively large. The mouth is small and protrusible. The pelvic fin has one spine and five rays. Small, rough spinoid scales cover the body. Coloration is reddish or pink overall. There are two genera with eight species (Nelson 1994), seven of which belong to the genus *Antigonia* (Zehren 1987). Deepbody boarfish were included in Bigelow and Schroeder even though no specimens were known from the Gulf at that time. Subsequent trawling, primarily by NMFS vessels, showed frequent captures on Georges Bank. Another species, shortspine boarfish (*Antigonia combatia*) is also included in the key because it is known from the area south of Nantucket Shoals at 160–165 m (MCZ 37604).

DEEPBODY BOARFISH / *Antigonia capros* Lowe 1843 / Bigelow and Schroeder 1953:438–439

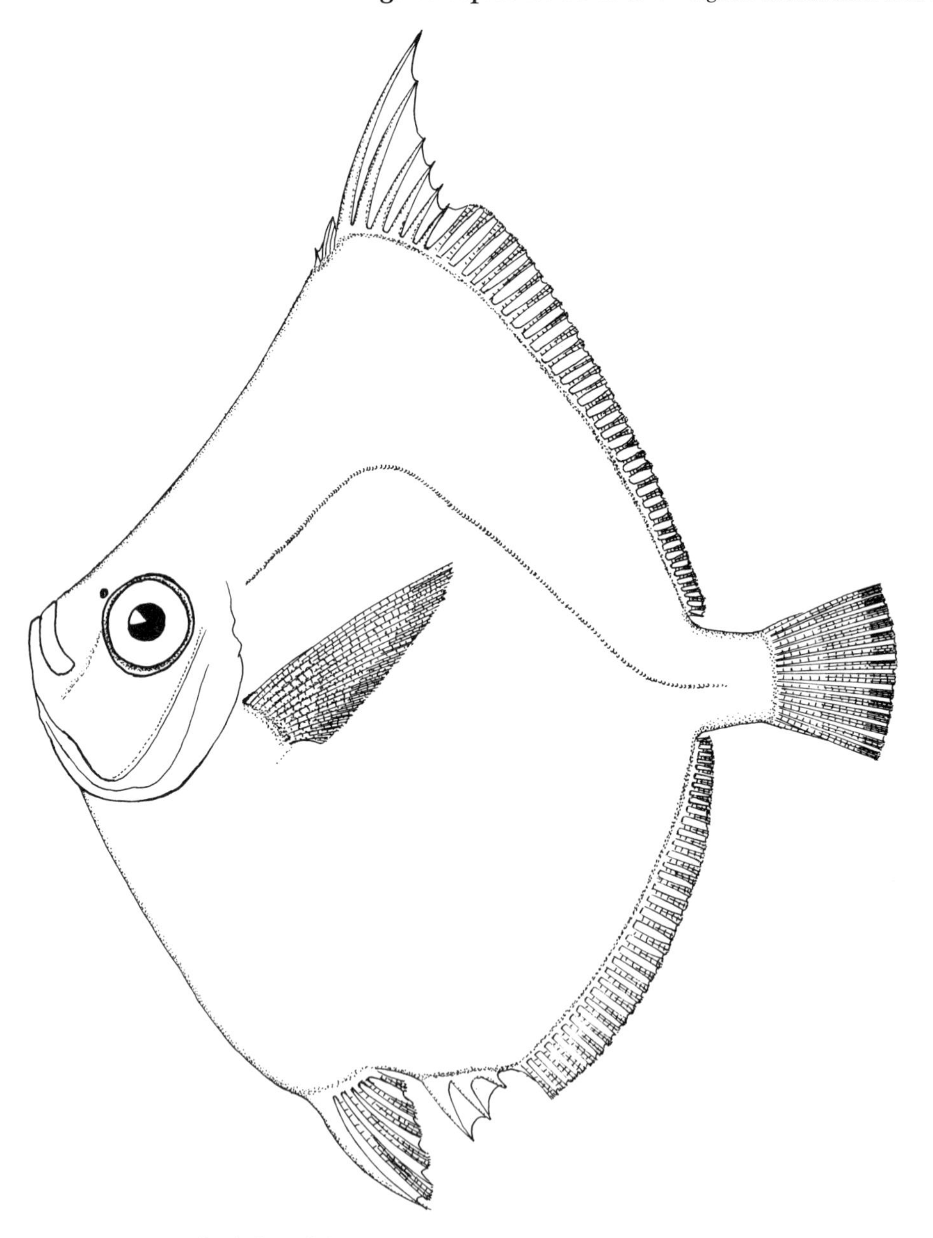

Figure 170. Deepbody boarfish *Antigonia capros*. Caribbean Sea, USNM 159600. Drawn by K. H. Moore.

Description. Body very compressed, rhomboidal in shape and deeper than long (Fig. 170). Eye large; mouth very small. Spines in dorsal fin long and strong. Soft-rayed portion of dorsal and anal fins long and low.

Meristics. Dorsal fin rays VII–IX, 31–37; anal fin rays III, 29–34; pectoral fin rays 12–14; pelvic fin rays I, 5; lateral line scales 56–57; gill rakers 5 or 6 + 13–16 (Berry 1959a).

Color. The body and fins are red or pink.

Size. Reach 30 cm SL (Quéro 1986c).

Distinctions. Deepbody boarfish are distinguished from all other Gulf of Maine fishes by body color and the rhomboid-shaped body that is deeper than it is long. Body shape is similar to that of young spotted tinselfish, but that fish has elongate striplike scales, a silvery body with dark spots, and elongate pelvic spines.

Biology. Deepbody boarfish live at or near the bottom on the deeper continental shelf and upper slope, especially near ledges and rocky outcroppings at depths of 40 to over 600 m, but mainly at 100–300 m. Boarfish sometimes form small schools. Diet consists of small crustaceans (e.g., copepods, cumaceans, and amphipods) and mollusks (Bowman et al. 2000). Several boarfish were observed swimming in proximity to an isocrinid (*Cenocrinus asterius*) and one may have been feeding off invertebrates caught in the fan (Baumiller et al. 1991). Divers off Discovery Bay, Jamaica, also observed boarfish that appeared to be sheltering near crinoids (Colin 1974). In Hawaii, deepbody boarfish were seen behaving as if they were cleaning other fishes, the latter swimming over to the boarfish and assuming a stationary posture (Chave and Mundy 1994).

Early Life History. Larvae have been collected in plankton in the vicinity of Rio de Janeiro at depths of 133–229 m in April and June (Weiss et al. 1987). At 2.7 mm, boarfish larvae are robust with a short gut. Several serrate ridges extend from the top of the head to the occipital region with one strong spine at each end. There are also serrate ridges on the upper and lower jaws. A strong, slightly serrate spine and three smaller spines extend backward from the angle of the preopercle.

At 4.5 mm, body depth further increases to 60% of SL and larvae are armed with a strong occipital spine and preopercular spines that reach the anus. Spines and serrate ridges develop on the supraorbital, frontal ridge, and preopercle. Notochord flexion is underway; and dorsal, anal, and caudal fin rays are incipient. These are distinctive larvae because of their heavy spination, and are unlikely to be confused with other genera.

General Range. Widespread in the western Atlantic from Brazil to the Gulf of Maine and in the eastern Atlantic from the Bay of Biscay to the Congo estuary (Quéro 1986c). They are also found in the western and central Pacific and Indian oceans.

Occurrence in the Gulf of Maine. Bigelow and Schroeder cited eight specimens that were trawled in 30–45 m south of Nantucket Lightship in 1950 (MCZ 37155). Numerous specimens are now known from Georges Bank (MCZ collections).

ACKNOWLEDGMENT. A draft of the zeiform section was reviewed by Karsten E. Hartel.

GASTEROSTEIFORM FISHES. ORDER GASTEROSTEIFORMES

Some authors have recognized Gasterosteiformes and Syngnathiformes as separate orders but Nelson (1994) accepted Johnson and Patterson's (1993) view that the two groups are probably each other's closest relatives and so they are considered suborders here. Body often covered with an armor of bony plates; mouth usually small. The two suborders contain 11 families, 71 genera, and 257 species (Nelson 1994). Most species are marine but about 20 are restricted to freshwater and another 40 species are found in brackish waters. Four sticklebacks, family Gasterosteidae, represent Gasterosteoidei and several species from Syngnathidae, Fistulariidae, and Macroramphosidae represent Syngnathoidei in the Gulf of Maine.

KEY TO FAMILIES OF GULF OF MAINE GASTEROSTEIFORMES

1a. Snout not tubular; pelvic fins with a very strong spine; first dorsal fin consisting of 3–10 separate spines . . . . **Gasterosteidae**
1b. Snout tubular; pelvic fins small or absent, without a strong spine; first dorsal fin not consisting of separate spines **2**
2a. Two dorsal fins; first dorsal fin with long, strong, and saw-edged second spine . . . . **Macroramphosidae**
2b. A single dorsal fin, without spines **3**
3a. Snout more than six times as long as dorsal fin; anal fin about the same size as dorsal fin; pelvic fins present, but small; caudal fin forked with an elongate filament between lobes of the fin . . . . **Fistulariidae**
3b. Snout not longer than dorsal fin; anal fin very small; pelvic fins absent; caudal fin rounded or absent . . . . **Syngnathidae**

STICKLEBACKS. FAMILY GASTEROSTEIDAE

William H. Krueger

Sticklebacks are closely related to tube-snouts, trumpetfishes, cornetfishes, snipefishes, pipefishes, and seahorses. They differ in being smaller and less elongate, with well-developed dorsal and pelvic fin spines. Sticklebacks lack scales, but three of the four Gulf species have bony lateral plates. Sticklebacks are small fishes, adults usually 30–70 mm SL; body oblong to moderately elongate, compressed; scales absent; bony lateral plates present or absent; dorsal and pelvic fin spines usually stout and sharp; two or more free, depressible spines in front of the dorsal fin; pelvic fins (rarely absent) with a large spine and one to three rays; three branchiostegal rays (rarely four); circumorbital ring incomplete posteriorly; epipleurals present; 27–42 vertebrae. Five genera and seven species are usually recognized (Wootton 1976; Nelson 1994), but Keivany and Nelson (2000) recognize three valid species of *Pungitius,* and the *Gasterosteus aculeatus* complex includes several unnamed forms that may or may not represent distinct biological species; the white stickleback of Nova Scotia (Blouw and Hagen 1990; Haglund et al. 1990) is a Gulf example. Although mostly marine in the Gulf of Maine area, all four Gulf species are at least occasionally found in freshwater as well. A fifth North American species (*Culaea inconstans,* brook stickleback) is restricted to freshwater and is not considered here.

In addition to their taxonomic complexity (see, e.g., Hubbs 1929; Bell 1976; Wootton 1976), sticklebacks are best known for their ritualized breeding behavior, which can readily be observed in an aquarium. Males defend territories, build nests of plant materials glued together with mucous secretions from their kidneys, court females, care for the eggs, and usually guard the young until they are free-swimming. A volume on the evolutionary biology of sticklebacks (Bell and Foster 1994) focuses on the most studied species, the threespine stickleback *G. aculeatus.* There have been three international conferences on stickleback behavior and evolution, the most recent at the University of British Columbia in June 1999 (Taylor 2000).

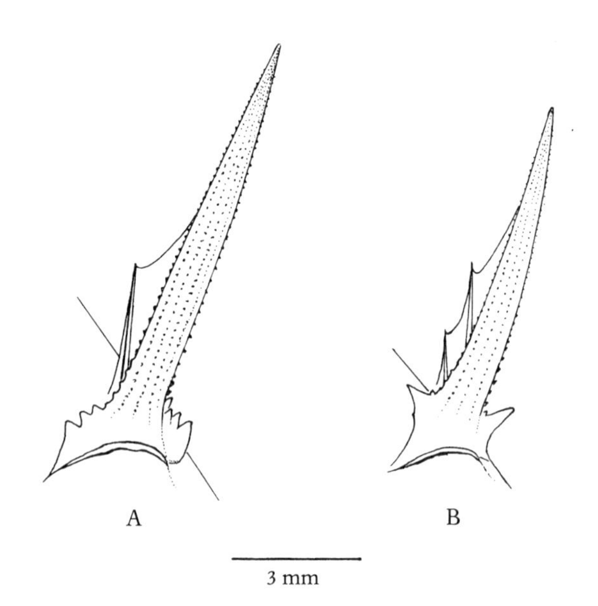

Figure 171. Cusps at base of pelvic spine in (A) *Gasterosteus aculeatus* and (B) *G. wheatlandi.* Drawn by K. H. Moore.

KEY TO GULF OF MAINE GASTEROSTEIDAE

1a. Dorsal spines usually 8 or more. . **North American ninespine stickleback**
1b. Dorsal spines usually 6 or fewer **2**
2a. Dorsal spines usually 4 or 5; no bony lateral plates on sides; pelvic bones form a ridge on each side of abdomen; trunk triangular in cross section . . . . **Fourspine stickleback**
2b. Dorsal spines usually 3; sides armored with few to many bony plates along lateral midline; no ridges on sides of abdomen; trunk oval in cross section **3**
3a. A cusp on dorsal surface of pelvic spine base only (Fig. 171A); lateral plates usually 30–35, forming a continuous row from pectoral girdle to end of caudal peduncle; posterior plates usually expanded to form a conspicuous lateral keel . . . . **Threespine stickleback**
3b. Cusps on both dorsal and ventral surfaces of pelvic spine base (Fig. 171B); lateral plates usually 5–13, anterior only, with only occasional much smaller plates on caudal peduncle; no caudal peduncle keel. . . . . **Blackspotted stickleback**

FOURSPINE STICKLEBACK / *Apeltes quadracus* (Mitchill 1815) / Bloody Stickleback /

Bigelow and Schroeder 1953:311–312

Description. Body somewhat triangular in cross section, abdominal region fairly flat; deepest near midpoint, depth about 25% of SL, tapering to a slender caudal peduncle that lacks lateral keels (Fig. 172). First three dorsal fin spines inclined alternately left and right, first two longest; last spine separated from first three by a distinct gap. Anal fin spine attached to soft-rayed portion; length equal to last dorsal spine. Each pelvic fin with two rays preceded by a single spine half again as long as the longest dorsal spine, stout, serrate anteriorly, and exceptionally sharp. All spines can be locked in the forward position. Dorsal and anal fins taper only slightly posteriorly; anal origin somewhat behind dorsal origin. Caudal fin truncate or slightly rounded.

Meristics. Dorsal fin rays I–VII, usually IV or V, 9–14; anal fin rays I, 7–11; pectoral fin rays 11 or 12; pelvic fin rays I, 2; gill rakers 4–9; vertebrae 29–33 (Blouw and Hagen 1981).

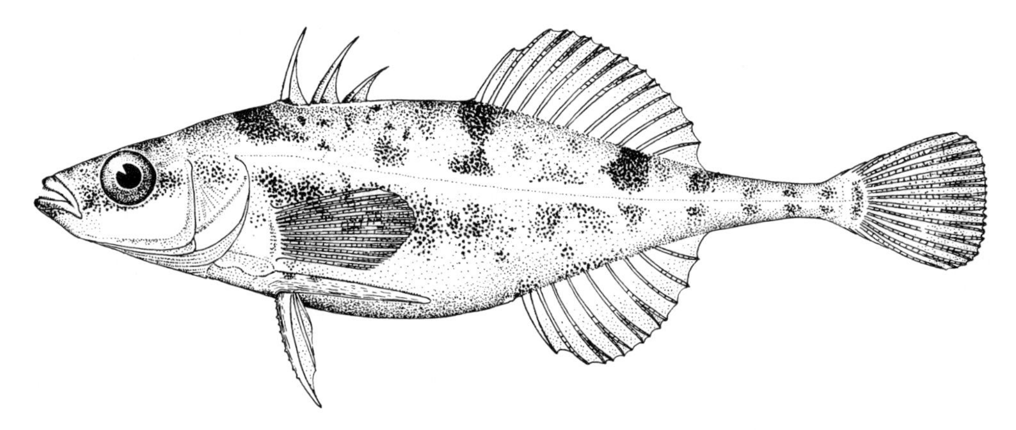

Figure 172. Fourspine stickleback *Apeltes quadracus.* Woods Hole. Drawn by A. H. Baldwin.

Color. Variable and generally cryptic in brackish water populations, with back and sides ranging from almost black to light olive green or brown with irregular dark brown blotches, abdomen dull silver or light brass; males have red pelvic membranes, and in some populations pelvic and dorsal fins are red in both sexes (Blouw and Hagen 1981). In a freshwater population, pelvic fins were crimson in males and orange in females, and both sexes showed dark pigment on the isthmus and the area ventral to and between the lower jaws (Coad and Power 1973c).

Size. Maximum size about 53 mm SL. In large samples from coastal sites in New Brunswick and Nova Scotia, mean body lengths of adults ranged from 22.9 to 33.6 mm for males and 23.0 to 44.1 mm for females, which averaged larger than males in 44 of 45 samples (Blouw and Hagen 1984b).

Distinctions. The triangular body cross section, four or five dorsal spines, lack of bony lateral plates, and slender, unkeeled caudal peduncle distinguish this stickleback from all other Gulf of Maine fishes.

Habits. Delbeek and Williams (1987a) found *Apeltes* in a shallow New Brunswick inlet with a mud and silt substrate, salinity range 0–24 ppt and temperature 5°–22°C. At Isle Verte, Quebec, they were found in brackish portions of a tidal creek, where they were the most abundant stickleback in July and August; however they were absent from nearby tidal marsh pools, from which they may have been excluded by *Pungitius* (Worgan and FitzGerald 1981a). Preference for 7 ppt in a salinity gradient of 0–35 ppt may also explain their absence from the pools, which are highly saline (Audet et al. 1985). In contrast, among individuals from freshwater populations, fourspine had a higher salinity tolerance than ninespine or brook sticklebacks (Nelson 1968a).

Fourspine do not move long distances, in keeping with the size of their pectoral fins, which are the smallest among all the Gulf sticklebacks (Delbeek and Williams 1987b). Sustained swimming by sticklebacks is accomplished by labriform motion, a rapid sculling

Food. Adults from brackish sites in New Brunswick contained ostracods, copepods, *Gammarus* spp., dipterans, and fish eggs (Delbeek and Williams 1987a). The proportions of prey types eaten by both adults and juveniles reflected prey species abundance; adults were more efficient than *Gasterosteus* spp. at capturing benthic prey (Delbeek and Williams 1988).

Predators. They have been found in stomachs of American eel, brook trout, chain pickerel, Atlantic tomcod, grubby sculpin, white perch, striped bass, windowpane flounder, and great blue heron (Blouw and Hagen 1981, 1984b).

Parasites. Margolis and Arthur (1979) reported one digenetic and three monogenetic trematodes, three cestodes, one nematode, and one mollusk from *Apeltes.*

Breeding Habits. In a tidal creek population at Isle Verte, females 31–52 mm SL (mean 41.1 mm) contained 16–61 eggs (mean 36.1), 1.27–1.47 mm in diameter (Craig and FitzGerald 1982).

Courtship and nesting behavior (Rowland 1974) differs from that of other sticklebacks in several ways. Males may build nests on the substrate or at plant stem junctions 10–15 cm above the bottom. The nest has a distinctive cup shape, and a male may build and tend more than one at a time. Many males in New Brunswick cared simultaneously for two to five nests containing zero to four clutches of eggs (Courtenay 1985). Nests are sometimes stacked on top of one another.

When a female enters his territory the male initiates courtship by prodding her on the body several times in rapid succession. He then darts away and swims around her in a rapid spiral. During the leading sequence, the female often abandons her position near the male, interrupting the courtship. The male then usually hovers, waiting for the female to catch up. After she enters the nest, the male holds her gently by the caudal fin or peduncle and repeatedly quivers in 1-s bursts, triggering egg-laying. The male "fans" the eggs by a unique sucking action, using contractions of the orobranchial

eggs have developed completely. Unlike other sticklebacks, the fourspine male pays no attention to the young.

Age and Growth. Otoliths and pelvic spines were unreliable for aging (Blouw and Hagen 1981). Newly hatched larvae are about 4.5 mm TL (Kuntz and Radcliffe 1917). Length-frequency distributions of adults indicate that most live 1 year, with a small proportion of each sex reaching the age of 2.

General Range. Newfoundland and the Gulf of St. Lawrence to Virginia (Scott and Crossman 1973).

Occurrence in the Gulf of Maine. Fourspine are found all along the Gulf coast in calm, well-vegetated waters, including brackish estuaries and lagoons, shallow, protected coves and tidal creeks, often penetrating into adjacent freshwaters.

THREESPINE STICKLEBACK / *Gasterosteus aculeatus* Linnaeus 1758 / Bigelow and Schroeder 153:308–310

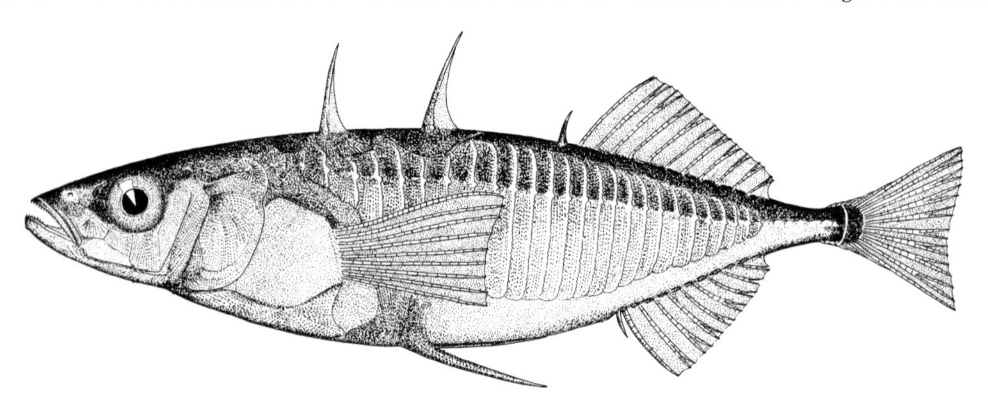

Figure 173. Threespine stickleback *Gasterosteus aculeatus.* Alaska, 82 mm. Drawn by H. L. Todd.

Description. Largest and most heavily armored among Gulf sticklebacks (Fig. 173). Body about a quarter deep as long, oval in cross section, with a moderately slender caudal peduncle usually bearing a pronounced lateral keel. Almost always three dorsal fin spines arranged in a straight line, third attached to soft dorsal fin; first two spines two to four times longer than third, with small, triangular posterior membranes. Pelvic spines stouter than dorsal spines, at least as long. Anal fin preceded by a single small spine about the size of last dorsal spine. All spines depressible and lockable in forward position. Soft dorsal and anal fins about equal in height, tapering posteriorly, dorsal fin originating somewhat ahead of anal fin. Caudal fin weakly forked or truncate.

This species is noted for variation in the lateral bony plates. Most populations comprise one or more of three distinct plate morphs: complete, partial, or low (Hagen and Gilbertson 1972). Complete morphs have an uninterrupted row of plates from the pectoral girdle to the end of the caudal peduncle; partial morphs have anterior and posterior plates separated by a plateless gap; and lows have anterior plates only. In completes, several posterior plates are expanded to form a strong lateral keel on the caudal peduncle, whereas the keel is usually reduced in partials and always absent in low morphs. The complete morph gave rise to partial and low morphs via paedomorphosis (Bell 1981). European authors have long used the terms *trachurus, semiarmatus,* and *leiurus* for these three morphs, respectively, with *semiarmatus* usually considered to represent hybrids between *trachurus* and *leiurus* (Wootton 1976). In North America, partials are rarely the result of hybridization between completes and lows, and many populations in the northeast are monomorphic for partials or dimorphic with partials and completes (Hagen and Moodie 1982). Although some North American authors have used the European terminology, complete, partial, and low are used here following Bell (1984) and Wootton (1984).

Marine and anadromous populations in the Gulf of Maine seem to be monomorphic for the complete morph, which has 30–36 (usually 32–34) plates, the last one often a tiny sliver of bone. These sticklebacks generally resemble the streamlined marine form shown in Fig. 1 of Hubbs (1929). The "less extreme marine type" (Hubbs's Fig. 2) is a more chunky fish with a shorter, deeper caudal peduncle. Perlmutter (1963) took both forms in the same seine hauls on eastern Long Island, N.Y. The significance of these marine types has never been determined.

Not all marine populations are fully plated. Along the coast of Norway, only 15% of samples were monomorphic for the complete morph, and polymorphic populations occurred even in outer fjord habitats (Klepaker 1996).

In the northeast, the partial morph is largely restricted to freshwater in northern New England and Atlantic Canada and to marine sites in Hudson, James, and Ungava bays; the low morph is almost entirely restricted to inland freshwaters of Arctic Canada (Hagen and Moodie 1982; Coad 1983). However, an adult partial morph was taken 645 km from the nearest land in the eastern North Atlantic (Jones and John

1978), and a trimorphic freshwater population lives within the city limits of Boston (Bell and Baumgartner 1984).

Meristics. Dorsal fin rays III or IV, 10–14; anal fin rays I, 8–10; pectoral fin rays 10; pelvic fin rays I, 1; lateral plates 30–36; gill rakers 14–25; vertebrae 29–35.

Color. When adults first enter spawning areas in spring they are bright polished silver with blue-black backs and orange pelvic membranes (pers. obs.). Males soon develop their courtship coloration, which includes blue irises and red pigment that usually covers the throat region and may tint almost the entire ventral third of the fish. Some males develop bright red pigmentation inside the mouth, with little red elsewhere. The orange of the pelvic membrane fades in both sexes as the breeding coloration develops. When courtship ends and the parental phase begins, the male's coloration turns drab and cryptic. Spawning females retain their silvery flanks but have a pale olive-tan dorsum; as they become distended with eggs their sides develop a transient dark-barred or mottled pattern, which signals sexual receptivity to the male (Rowland et al. 1991).

Young 14–28 mm SL taken at the surface off Long Island were dark blue dorsally with silvery sides (Cowen et al. 1991).

Size. Adults in the Gulf area are usually 40–70 mm SL, but may exceed 90 mm in northern Quebec. Females typically average longer than males, but there is considerable overlap (see below).

Distinctions. The combination of three dorsal spines and a strong lateral keel on the caudal peduncle distinguishes this species from almost all other Gulf of Maine fishes. Presence of the keel and absence of a ventral cusp on the base of the pelvic spine separates it from *G. wheatlandi*.

Habits. Gulf coastal populations spend most of their existence in the open sea, entering inshore marine habitats to spawn and die in spring of the second or (usually) third year of life (age 2+). Adults are found in tidal marsh pools and ditches and shallow coves (ca. 1 m deep at low tide), usually close to vegetation, particularly eelgrass and filamentous green algae. Anadromous populations (which spawn in full freshwater) have received remarkably little attention in the northeast. One such population spawns below an impassable dam in Riviere des Vases, Quebec, 2.5 km from the St. Lawrence estuary (Worgan and FitzGerald 1981a; Kedney et al. 1987). Water depth was 0.5–2.5 m, and temperature in May and June was 0°–25°C, averaging 15.5°C (Worgan and FitzGerald 1981a) and 7.1°–15.7°C (Kedney et al. 1987) in different years. Shallow (<0.5 m) tidal marsh pools near the mouth of the river had salinities and mean temperatures of 19.1–26.7 ppt and 14.0°–22.3°C (Worgan and FitzGerald 1981a) and 21.6–26.2 ppt and 9.7°–20.7°C (Kedney et al. 1987) in different (Coad and Power 1973a). They are widespread along the coast of Nova Scotia, spawning at salinities of 3–32 ppt (Blouw and Hagen 1990). In New Brunswick, they spawned in two coastal sites in May and June at 5°–28°C and 5–28 ppt (Delbeek and Williams 1987a).

Upper lethal temperatures in the laboratory ranged from 21.6° to 28.8°C, with a mean of 26.2°C (Jordan and Garside 1972). Adults acclimated to 21 ppt preferred 7 ppt and 14 ppt in a salinity gradient and avoided 0 and 35 ppt (Audet et al. 1985). Young, 5 weeks old, were euryhaline and showed no preference in a salinity gradient of 0–35 ppt (Campeau et al. 1984).

Food. Stomach contents of adults from brackish waters in eastern Canada included copepods, gammarids, oligochaetes, hemipterans, chironomids, and stickleback eggs (Delbeek and Williams 1987a, 1988; Worgan and FitzGerald 1981b; Walsh and FitzGerald 1984). In Amory Cove, Quebec, adult stomachs contained mostly small crustaceans and stickleback eggs, but also chironomids, mosquito larvae and pupae, Collembola, and Araneae (Coad and Power 1973a). Egg cannibalism is common, especially among females (Whoriskey and FitzGerald 1985b).

Young in Quebec tidal marsh pools ate copepods, ostracods, branchiurans, rotifers, and dipterans (Poulin and FitzGerald 1989). Young taken at the surface in the Bay of Fundy contained mostly calanoid copepods, copepod nauplii, cladocerans, and planktonic eggs (Williams and Delbeek 1989).

Predators. Threespine stickleback have been found in the stomachs of shorthorn sculpin in Amory Cove, Quebec (Coad and Power 1973a), in salmonid stomachs in Moosehead Lake, Maine (Cooper and Fuller 1945), and in Atlantic cod stomachs at depths greater than 200 m in the open Atlantic (Brown and Cheng 1946), suggesting that any number of piscivorous fishes may prey on them. Several birds are known to prey on threespine, including herons, mergansers, gulls, and loons. In tidal marsh pools at Isle Verte, Quebec, where predacious fishes were absent, bird predation removed 30% of the sticklebacks; three species (black-crowned night heron, bronzed grackle, and ring-billed gull) accounted for 80% of the captures (Whoriskey and FitzGerald 1985a).

Parasites. About 50 species of parasites have been reported, including protozoans, acanthocephalans, myxosporidians, monogeneans, trematodes, cestodes, nematodes, mollusks, branchiurans, and copepods (Wootton 1976; Margolis and Arthur 1979). Poulin and FitzGerald (1987) studied effects of a parasitic copepod, *Thersitina gasterostei*, and a fish louse, *Argulus canadensis*, at Isle Verte, Quebec. Although some sticklebacks were heavily infested with the copepod, their condition factors did not differ from those of unparasitized fish. Heavy infestations of the fish louse decreased stickleback survival in the laboratory.

Breeding Habits. This is the most fecund of the Gulf stickle-

116–838 eggs (mean 366), with diameters of 1.3–1.5 mm (mean 1.39 mm); egg number increased with body weight, but egg diameter did not increase with SL (Craig and FitzGerald 1982). In nearby tidal marsh pools, mean clutch size of females 51–77 mm SL (mean 68.8 mm) ranged from 189 to 313 and decreased with the advancing season (Bolduc and FitzGerald 1989). In Amory Cove, Quebec, females averaging 61 mm SL had 120–515 eggs (mean 265) 1.01–1.60 mm in diameter (mean 1.25 mm) (Coad and Power 1973a).

The following synopsis of spawning behavior is based largely on Wootton (1976). After establishing a territory, the male builds a nest on the substrate, gluing together bits of vegetation and sand grains with mucus and kidney secretions. Nest construction ends with the male creeping through to form a tunnel, which marks the beginning of the courtship phase. When a gravid female enters his territory, the male approaches her and performs his zigzag dance, which consists of a series of jumps away from and toward the female, usually with his mouth open and spines erect. A receptive female adopts a head-up posture and swims toward the male, who then returns to the nest, exhibits gluing behavior, and fans the nest with his pectoral fins. He then zigzags back to the female, turns away, and swims in a straight line back to the nest. If the female follows, the male turns on his side and shows her the nest by pointing his snout toward the nest opening while moving rapidly back and forth. The female enters the nest with her caudal peduncle protruding. The male then quivers against the female's flank. She deposits her eggs and leaves the nest. The male immediately moves through the nest, fertilizing the eggs enroute, and drives the female away. He pushes the eggs deeper into the nest, flattening the egg mass, repairs the nest, and begins fanning the eggs. The male may then induce additional females to spawn in the same nest. Mean nest contents (eggs or fry) in Quebec tidal marsh pools indicated that most males spawned with two or three females (FitzGerald 1983). Incubation time (days) was 17.1 ± 2.8 in 1985 and 11.4 ± 5.1 in 1986 (Poulin and FitzGerald 1989). When the eggs hatch, the male tears the nest apart, guards the young, and retrieves any that stray from the nest area. When his progeny become free-swimming, he may build a new nest and start another cycle of courtship.

Age and Growth. Using otolith annuli, three age-classes were found in a marine threespine population in Quebec (Coad and Power 1973a): 0+ and 1+ fish were 10–21 and 28–33 mm SL, respectively; 2+ females (39–71 mm SL, mean 60.2 mm) averaged significantly larger than 2+ males (42–67 mm SL, mean 54.0 mm). A freshwater population in British Columbia includes individuals larger than 80 mm SL and 5–8 years old (Reimchen 1992).

The mean TL of juveniles in tidal marsh pools at Isle Verte, Quebec, increased from 9.10 ± 1.31 mm on 18 June to 28.85 ± 3.37 mm on 25 September (Poulin and FitzGerald 1987). Juveniles taken in surface nets as far as 110 km off the Long Island coast were 14–28 mm SL (mean 20 mm) and had 29–56 daily otolith increments (Cowen et al. 1991).

General Range. Threespine stickleback are found on nearly all coasts and in near-coastal fresh waters in temperate portions of the Northern Hemisphere. In North America from southern California north to Alaska, east across Hudson and James bays to Baffin Island, then south to North Carolina (Wootton 1976; Ross et al. 1981). They are also found in large inland lakes including Ontario and Champlain. They are often taken at the surface of the open ocean, even hundreds of kilometers from shore (see, e.g., Quinn and Light 1989 and Cowen et al. 1991)

Occurrence in the Gulf of Maine. They occur in tidal marsh pools and ditches and shallow protected coves all along the coast from Sable Island, N.S. (Marcogliese 1992) south, and also ascend tidal creeks and rivers, often into full freshwater. Both adults and young have been taken at the surface offshore, in the Bay of Fundy (Williams and Delbeek 1989), the St. Lawrence estuary (Picard et al. 1990), and along the coast of Maine (Kendall 1896).

Migrations. The streamlined anadromous form of the threespine stickleback is well adapted for sustained swimming, which it accomplishes by rapidly "rowing" with its large pectoral fins (labriform motion) at speeds of up to 5 lengths·s^{-1}. It ascends rivers as far as 100 km inland (Taylor and McPhail 1986), but total migration distances may be much greater, for this form has been found 645 km from the nearest land in the North Atlantic (Jones and John 1978) and up to 1,074 km from land in the North Pacific (Quinn and Light 1989). The latter authors noted that a 70-mm stickleback could swim 800 km in under 2 months at a speed of only 3 lengths·s^{-1}. A 20-mm stickleback swimming at 3 lengths·s^{-1} could travel 50 km in about 10 days, which explains the occurrence of juveniles only 29–56 days old well off the Long Island coast (Cowen et al. 1991).

BLACKSPOTTED STICKLEBACK / *Gasterosteus wheatlandi* Putnam 1867 /

Twospine Stickleback / Bigelow and Schroeder 1953:310–311

Description. Somewhat stouter than threespine stickleback, body depth about 25–30% of SL (Fig. 174). Caudal peduncle shorter than that of threespine, deeper than wide, and lateral keel always absent. Three dorsal spines in a straight line, second the longest, last one about half length of first; all can be folded into a groove in dorsal midline. Spines I and II bear a small, triangular membrane posteriorly. Pelvic fin spine longer and stouter than dorsal spines. Anal fin preceded by single

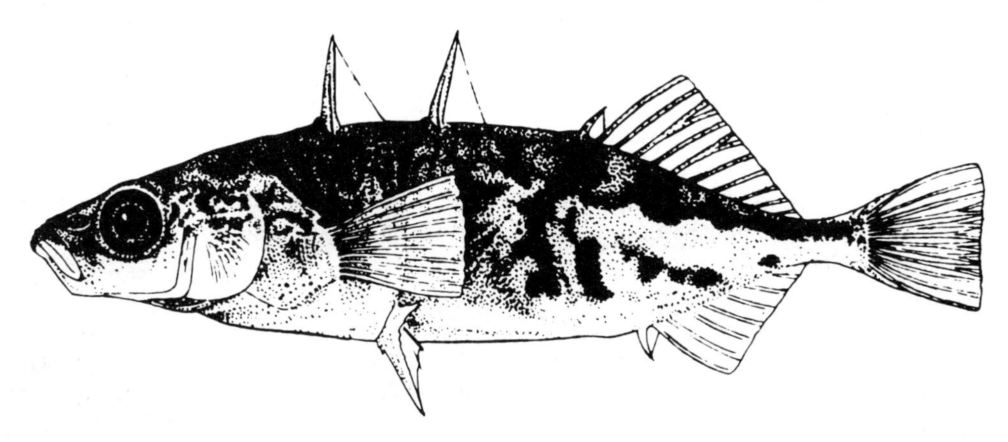

Figure 174. Blackspotted stickleback *Gasterosteus wheatlandi*. Port au Port, Nfld., 45 mm. Drawn by A. Odum. (From Scott and Scott 1988:341.)

spine equal in length to last dorsal spine, originating somewhat behind soft dorsal. All spines lockable in forward position. Caudal fin truncate or emarginate.

Lateral plate phenotypes are similar to those of threespine, but there is essentially continuous variation between low and complete phenotypes, with no distinct partial phenotype. Thus Sargent et al. (1984) recognized only low (anterior plates only) and partial-complete phenotypes, the latter encompassing all specimens with one or more plates on the caudal peduncle. Frequency of the low phenotype equals or approaches 100% in populations north of Cape Cod, with only occasional specimens having a few extremely small posterior plates. Frequency of the partial-complete phenotype increases to the south, as does the mean number of plates in both phenotypes. Rate of plate development in a low phenotype population is only 23% of that in a partial-complete population indicating that neoteny is responsible for the low plate counts (Krueger 1994).

Meristics. Dorsal fin rays III or IV, 8–10; anal fin rays I, 6–8; pectoral fin rays 9 or 10; pelvic fin rays I, 1–3; lateral plates 5–13 (to 31 in southern populations); gill rakers 11–20; vertebrae 26–31.

Color. When adults first enter shallow coastal waters in spring, their coloration is very similar to that of threespine stickleback, with blue-black backs, polished silver sides and orange pelvic membranes; at this time black spots are absent or inconspicuous. The sides of spawning females remain silvery, while the dorsal coloration turns pale olive; they may develop a transient barred or mottled pattern that signals sexual receptivity to the male (Rowland et al. 1991). Males develop nuptial coloration usually described as greenish gold (McInerny 1969) or yellowish green (Rowland 1983) with black spotting or mottling, particularly on the posterior third of the trunk. Rowland noted that the males nest in or very close to vegetation that matches their coloration, and thus are inconspicuous. Juveniles, which are abundant in green vegetation

specimens dipnetted under floating rockweed off the Maine coast were described as grass green dorsally and golden with dark blotches laterally (Kendall 1896). Thus cryptic coloration may be important at several stages in the life of the blackspottted stickleback.

Size. Males are usually 29–34 mm SL, females 33–42 mm SL, but adults as small as 21 mm have been reported from sites north of Cape Cod (Coad and Power 1973a; Craig and FitzGerald 1982; Sargent et al. 1984); they may represent fish that hatched very late in the spawning season. Rowland (1983) found little difference between mean SL of adults from Long Island and New Brunswick. New York males were 32 ± 2.45 mm, New Brunswick males 33 ± 1.15 mm; New York females were 35 ± 2.55 mm, New Brunswick females 37 ± 2.09 mm. The maximum length (76 mm) and mean length (51 mm) reported by Scott and Scott (1988) for Atlantic Canada are much greater than those reported by other authors.

Distinctions. Very similar to threespine stickleback, but distinguished by absence of a caudal peduncle keel and presence of a ventral cusp on the base of the pelvic spine (Fig. 171B).

Habits. The blackspotted stickleback is the most marine of all the Gulf stickleback species. They spend most of their approximately 12- to 14-month life span offshore, entering inshore marine habitats in spring to spawn and die in the second year of life (age 1+). Although individuals have been taken in freshwater, spawning has been reported only in brackish water (see, e.g., McInerny 1969 and Worgan and FitzGerald 1981a). Spawning sites include tidal marsh pools and shallow (~1 m) protected coves, which they share with other sticklebacks, particularly threespine. Because they may spawn slightly later than their congener (Perlmutter 1963; Rowland 1983), temperatures and salinities may average somewhat higher than those reported above for *G. aculeatus*. In a laboratory salinity gradient, adults preferred 21 ppt (Audet et al. 1985), whereas young pre-

Food. Adults from two brackish water sites in New Brunswick ate mostly oligochaete worms and harpacticoid and calanoid copepods; there was little overlap with adults of the other three sticklebacks (Delbeek and Williams 1987a). Adults in Amory Cove, Quebec, contained mostly small crustaceans and stickleback eggs (Coad and Power 1973a). Young in Quebec tidal marsh pools ate copepods, amphipods, ostracods, Branchiura, rotifers, and Hemiptera; there was considerable dietary overlap with young threespine sticklebacks (Poulin and FitzGerald 1989). Young taken at the surface in the Bay of Fundy had fuller guts than those of *G. aculeatus* and contained mostly calanoid copepods (Williams and Delbeek 1989).

Predators. Remarkably, there is no record of predation on this species by fishes. FitzGerald and Dutil (1981) concluded that birds preyed selectively on threespine and not on blackspotted stickleback in Quebec tidal pools, but Whoriskey and FitzGerald (1985a) found that several birds, particularly black-crowned night heron, bronzed grackle, and ring-billed gull preyed on both species, taking significantly more males than females.

Parasites. There are reports of four monogenean species, one trematode, three cestodes, and one copepod from adults (Margolis and Arthur 1979). Young in tidal marsh pools had fewer parasitic copepods (*Thersitina gasterostei*) than other sticklebacks; heavy infestations of a crustacean *Argulus canadensis* decreased survival of young in aquariums (Poulin and FitzGerald 1987).

Breeding Habits. Females 28–40 mm SL (mean 32.9 mm) from Amory Cove, Quebec, had 75–168 eggs (mean 126) 0.86–1.21 mm in diameter (mean 1.05 mm); the relationship between fecundity (Y) and fish size (X) in mm SL was described by $Y = 21.2 + 0.10X$, $r = 0.66$ (Coad and Power 1973a). Females 27–41 mm SL (mean 33 mm) from a tidal creek adjacent to the Gulf of St. Lawrence contained 45–142 eggs (mean 80) 1.10–1.43 mm in diameter (mean 1.25 mm); the relationship between fecundity (F) and somatic weight (SW) in grams was described by $F = 67.0SW - 47.5$, $r = 0.54$ (Craig and FitzGerald 1982). Large females from Newfoundland (mean 38 mm SL) contained 140–276 eggs (mean 186) averaging 1.30 mm (Scott and Crossman 1964).

The mean number of eggs per nest and incubation times in tidal marsh pools at Isle Verte, Quebec, were 341.5 ± 244.6 eggs and 13.7 ± 5.2 days in 1985, and 300.4 ± 120.0 eggs and 8.9 ± 1.4 days in 1986 (Poulin and FitzGerald 1989). Each nest probably contained the eggs of two or three females (FitzGerald 1983).

Reproductive behavior is similar to that described above for *G. aculeatus,* but differs in some important details (McInerny 1969). The male begins the leading sequence by adopting a head down (30°) posture on or near the substrate, with the body bowed so that the ventral surface is concave and quivers rapidly. The female then assumes a head-up posture and nudges the male between his pelvic spines. He then moves forward, quivering rapidly and continuously, and the pair takes a circuitous route (1–2 min or longer) to the nest while in almost constant contact with each other and with the substrate. This leading sequence is very different from the simple, brief, direct-route movement exhibited by threespine stickleback. McInerny found that although males of both species courted females of either species with equal vigor, females rarely responded to the courtship display of heterospecific males and never followed them to the nest. This combination of markedly different male breeding colors and leading sequences may ensure reproductive isolation, for no *aculeatus* × *wheatlandi* hybrid has ever been reported.

As the end of incubation nears, the male piles considerable amounts of plant material on top of the nest, providing a loose meshwork within which the newly hatched fry can move (McInerny 1969).

This remarkable stickleback coexists with its larger, more aggressive congenor (Rowland 1983), often spawning in the same tide pools (see, e.g., FitzGerald 1983; FitzGerald and Whoriskey 1985). Cleveland (1994) suggested that it succeeds by behaving as a fugitive species (Hutchinson 1953), using less preferred resources when competition with *G. aculeatus* is severe.

Age and Growth. All adults are age 1+ based on otolith rings (Coad and Power 1973a). In tidal marsh pools at Isle Verte, Quebec, the mean TL of juveniles increased from 17.13 ± 2.30 mm on 3 July to 28.86 ± 2.30 mm on 25 September (Poulin and FitzGerald 1987). Young taken in the Bay of Fundy averaged 20.3 ± 1.1 mm SL at the beginning of August and 29.0 ± 1.2 mm SL in mid-November (Williams and Delbeek 1989). In contrast, young taken as far as 55 km off the Long Island coast in June were only 11–19 mm SL (mean 15 mm) and 22–40 days old; their lateral plate development was complete (Cowen et al. 1991), suggesting that southern young develop rapidly and leave their natal areas at an earlier age than northern young.

General Range. Blackspotted stickleback are known only from Newfoundland to southern New York (Wootton 1976), one of the smallest ranges of any Gulf of Maine fish. Their offshore movements may be less extensive than those of their congener, for they were found within 55 km of the Long Island coast, with most captures within 10 km, whereas *G. aculeatus* was taken up to 110 km from land (Cowen et al. 1991).

Occurrence in the Gulf of Maine. They are found in tidal marsh pools and shallow protected coves all along the coast in spring and early summer, and have been taken at the surface in offshore waters, including the Bay of Fundy (Williams and Delbeek 1989), the St. Lawrence estuary (Picard et al. 1990), and near the Maine coast (Kendall 1896).

NORTH AMERICAN NINESPINE STICKLEBACK / *Pungitius pungitius occidentalis* (Cuvier 1829) / Bigelow and Schroeder 1953:307–308 (as *Pungitius pungitius*)

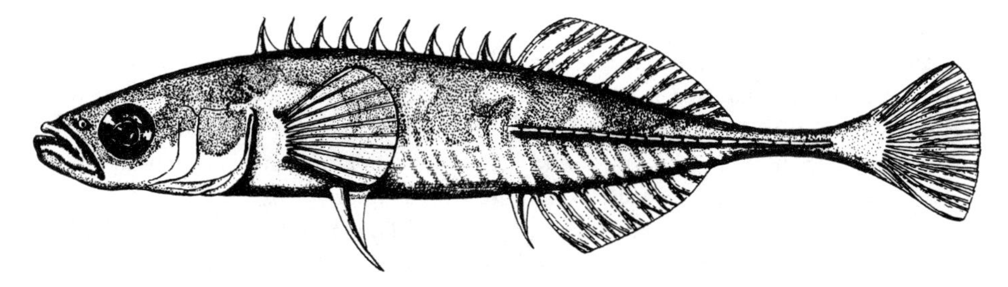

Figure 175. North American ninespine stickleback *Pungitius pungitius occidentalis.* (From Bigelow and Schroeder: Fig. 166.)

Description. The most slender Gulf stickleback, SL about five to six times its depth, with a very slim caudal peduncle (Fig. 175). Dorsal spines set in a slightly zigzag line, leaning alternately to each side. Spines weakly curved rearward, about half as long as anterior dorsal rays, except last one, attached to soft dorsal fin, longer and stouter than others; each spine with a small triangular membrane at its base. Spines depressible into a shallow groove along back. Pelvic fin spine (absent in some inland populations [Nelson 1971]) stout, curved, thicker and longer than last dorsal spine. Second dorsal and anal fins originate directly opposite one another, and both taper toward rear; anal fin spine equal in size to last dorsal spine. Caudal fin weakly forked. Caudal peduncle wider than deep, with a well-developed lateral keel formed by 11–22 rectangular bony plates.

Most Gulf of Maine specimens ("partial" phenotypes) with 2–11 inconspicuous, rounded lateral plates behind pectoral girdle, separated from caudal plates by a plateless gap, as in the partial morph of *G. aculeatus;* some specimens (complete phenotypes) with uninterrupted row of 28–35 plates from pectoral girdle to end of caudal peduncle, as in the complete morph of *G. aculeatus* (see Hagen and Gilbertson 1972). The caudal phenotype (Ayvazian and Krueger 1992), with plates only on the caudal peduncle, predominates in inland populations but is rare in coastal waters; this phenotype has no counterpart in *Gasterosteus.*

Meristics. Dorsal fin rays VI–XIII, usually VII–XI, 9–12; anal fin rays I, 7–11; pectoral fin rays 10; pelvic fin rays I, 0–2; anterior lateral plates 2–15 (rarely absent); caudal plates 11–22; total plates (anterior + caudal) 14–32 in partial phenotypes, 28–35 in completes; gill rakers 9–15; vertebrae 31–35 (Blouw and Hagen 1984a).

Color. Usually dull olive-brown or gray above, the upper part of the sides faintly barred or blotched darker, the abdomen white or silvery. Breeding males are black ventrally or entirely black, with white or pale blue pelvic membranes. In Nova Scotia, the combination of a jet black body and glossy blue-white pelvics, found only in sites where predation risk was low, appeared to make the males conspicuous (Blouw and Hagen 1984a). The black body coloration can be turned on or off very rapidly (W. J. Rowland, pers. comm.). Gravid females are black or brown dorsally and silvery ventrally (Foster 1977; Blouw and Hagen 1984a; pers. obs.). Some breeding adults may have reddish tints (Scott and Scott 1988).

Size. Adults in the Gulf area are usually 35–55 mm SL, with a maximum of 76 mm (Scott and Scott 1988; Ayvazian and Krueger 1992). They average about 10–20 mm longer in inland lakes (Nelson 1968c; McKenzie and Keenleyside 1970; Griswold and Smith 1973).

Distinctions. Combination of a slender body profile, slim, keeled caudal peduncle, and numerous but short dorsal fin spines distinguish them from other Gulf of Maine fishes.

Taxonomic Remarks. Ninespine stickleback are part of a widespread species complex that has long been known as *P. pungitius* (Wootton 1976; Nelson 1994); all of the pre-1992 works cited below are under that name. Based on allozyme variation, Haglund et al. (1993) divided this complex into three species on three continents: *P. pungitius* (Europe), *P. sinensis* (Asia), and *P. occidentalis* (North America). Their *P. occidentalis* includes a coastal (Bering) and an inland (Mississippian) form; coastal populations have more dorsal spines and anterior lateral plates but fewer gill rakers (McPhail 1963). Lateral plate phenotypes and plate ontogeny also differ (Ayvazian and Krueger 1992), and Foster's (1977) experiments provided clear evidence for reproductive isolation between these forms. Although McPhail (1963) attributed their morphological differences to isolation in and subsequent dispersal from two widely separated glacial refugia, the differences are most likely results of natural selection (Bell 1984; Blouw and Hagen 1984a). Thus coastal and inland forms may also represent distinct species. Keivany and Nelson (2000) reviewed the genus, recognized three species, and divided *P. pungitius* into five geographic subspecies, with *P. pungitius occidentalis* being the North American subspecies.

Habits. North American ninespine occur in shallow estuaries and tidal marsh pools, usually in dense vegetation (Worgan and FitzGerald 1981a; FitzGerald 1983; Blouw and Hagen 1984a). They are also widespread in freshwater, both near the coast and well inland (McPhail 1963). They are readily taken in seine hauls (depth ca.1 m) in both tidal waters (see, e.g., Worgan and FitzGerald 1981a; Ayvazian and Krueger 1992) and coastal fresh waters (Coad and Power 1973b), but are found as deep as 92 m in Lake Superior (Griswold and Smith 1973). In laboratory choice tests males from both Lake Huron and coastal New Brunswick chose nesting depths of 90 or 135 cm vs. 45 cm (Foster 1977).

Adults from an Indiana lake showed moderate salinity tolerance in the laboratory (Nelson 1968a). Adults taken from a Quebec river mouth, with salinities of 3–20 ppt, then acclimated to 20 ppt in the laboratory, preferred 14 ppt in a salinity gradient (Audet et al. 1985). In tidal marsh pools at Isle Verte, Quebec, with salinities of 22–38 ppt and mean daily temperatures 13.6°–19.5°C in May–August (Poulin and FitzGerald 1989), *Pungitius* generally spawned later and at higher temperatures than *Gasterosteus* spp., but showed bimodal temperature preferences of 9°–10°C and 15°–16°C in the laboratory (LaChance et al. 1987). Also, some *Pungitius* continued spawning during a cold snap in May when water temperatures were 2°–9°C and *Gasterosteus* spp. ceased spawning (Whoriskey et al. 1986). Spawning adults and developing young were taken at 13°–24°C and 24–34 ppt in Rhode Island (Ayvazian and Krueger 1992).

Food. In brackish New Brunswick pools, ninespine stickleback fed mainly in algal mats, eating copepods, gammarids and oligochaetes; adult females fed in groups of three or four, whereas juveniles formed schools of ten or more (Delbeek and Williams 1987a). In the laboratory, adults fed more efficiently on gammarids than did *Gasterosteus* spp., and swallowed *Gammarus* tail first (Delbeek and Williams 1988). In tidal marsh pools in Quebec, they fed mainly on copepods, ephydrid larvae, adult corixids, gammarids, and chironomids (Worgan and FitzGerald 1981b; Walsh and FitzGerald 1984). Males fed sporadically throughout the day and consumed much less food than females. Differences in diel feeding activity of females between these two studies were attributed to differences in water temperature and weather conditions.

Predators. Bronze grackle feed on this species; however, as the ninespine quickly hide when disturbed and are slow to reemerge, they are less susceptible to predation than threespine or blackspotted sticklebacks (Whoriskey and FitzGerald 1985a).

Parasites. There are reports of 33 species of parasites from ninespine stickleback in Atlantic Canada, including protozoans, myxosporidians, monogeneans, trematodes, cestodes, acanthocephalans, mollusks, a branchiuran, and copepods (Margolis and Arthur 1979).

Breeding Habits. This is one of the less fecund of the Gulf sticklebacks. Females 36–53 mm SL from a tidal creek in Quebec contained 37–136 eggs (mean 76) averaging 1.14 mm in diameter (Craig and FitzGerald 1982), and females 30–54 mm SL from the freshwater Matamek River had only 10–71 eggs (mean 31) averaging 0.70 mm (Coad and Power 1973b). Spawning is in spring and early summer, usually in protected coves and tidal marsh pools densely vegetated with filamentous algae and/or eelgrass.

The following synopsis of reproductive behavior is based on Wootton (1976) and Foster (1977). After establishing territories, males build their nests in vegetation, usually 2–14 cm above the substrate. Most nests are tubular with two entrances that are formed by the male creeping through, the behavior that signals the beginning of the sexual phase. When a gravid female enters a male's territory he begins his zigzag dance, swimming head down toward the female in a series of complex jumps, each of which has forward, upward, and sideways components. A receptive female responds by moving toward the male with her head up, thus displaying her distended abdomen, and then places her snout close to the male's erect pelvic spines. The male then dances toward the nest, and the female follows close behind and below the male with her tail down. The male shows the nest to the female by placing the tip of his snout just above the nest entrance while in a head-down position and fanning with his pectoral fins. The female enters the nest and stops with her head protruding from the exit and her tail from the entrance. The male then quivers against her caudal peduncle, triggering release of the eggs. When the female leaves the nest, the male swims straight through, fertilizing the eggs enroute, and immediately drives the female away from his territory.

The male pushes the eggs deeper into the nest, fans them, and rebuilds the nest to prepare it to receive another clutch of eggs from another female. In the ensuing parental cycle, he spends much time fanning the eggs, removes any dead ones, and retrieves eggs that have fallen out of the nest. After the eggs hatch, he may construct a loose meshwork of vegetation above the nest to serve as a nursery. Young reared in the laboratory from Lake Superior stock hatched in 5 days at 12°C, averaging 5.65 mm TL (Griswold and Smith 1972). Several days after hatching the young fill their swim bladders at the surface and soon become free-swimming.

Age and Growth. Age, based on otolith annuli, has been determined for inland and coastal freshwater populations but not for any tidal population. In Quebec they attain age 1+ in the Matamek River and 2+ in Matamek Lake (Coad and Power 1973b); mean SL was 22.2–25.6 mm and 35.2–38.1 mm for four samples of riverine fish aged 0+ and 1+ and 28.4, 36.7, and 46.9 mm for lacustrine fish aged 0+, 1+, and 2+, respectively. Lake Superior ninespine stickleback are older as well as larger (Griswold and Smith 1973). In tidal marsh pools at Isle Verte, Quebec, juveniles increased from 25.98 ± 3.59 mm SL on 13 August to 36.45 ± 8.32 mm SL on 7 November (Poulin and FitzGerald 1987).

General Range. The coastal form ranges from southern Alaska east to Hudson's Bay, Baffin Island, and southern Greenland, and south along the east coast to New Jersey; the inland form ranges from Alberta, Manitoba, and Saskatchewan eastward to the Great Lakes region and the upper St. Lawrence Valley (McPhail 1963), with the southernmost population in the Mississippi drainage of Indiana (Nelson 1968b).

Occurrence in the Gulf of Maine. The ninespine stickleback may be expected in shallow, protected marine or brackish water habitats all around the shores of the Gulf of Maine from Nova Scotia and the Bay of Fundy to Massachusetts. They are also widespread in freshwater, and marine populations may be restricted to habitats adjacent to freshwater. Thus although they spawn in tidal marsh pools adjacent to freshwater in Quebec (Worgan and FitzGerald 1981a; Craig and FitzGerald 1982), they are rare or absent in isolated rocky tidal pools in Maine (Moring 1990) and Massachusetts (Collette 1986), and I have not taken any in several Cape Cod embayments where other sticklebacks abound, but where adjacent freshwater habitats are lacking. They are primarily freshwater in insular Newfoundland (Scott and Crossman 1964).

Migrations. McDowall (1988) labeled this species marginally anadromous, but gave no documentation. In contrast, there is indirect evidence that the coastal form is catadromous. Like *Gasterosteus* spp., they invade inshore marine habitats such as tidal marsh pools each spring to spawn, remaining there until summer or fall (see, e.g., Poulin and FitzGerald 1989). Yet they are never taken offshore, where *Gasterosteus* spp. abound (Quinn and Light 1989; Williams and Delbeek 1989; Cowen et al. 1991), suggesting that the fall migration is a return to freshwater. Indeed, in Quebec *Pungitius* are taken in freshwater rivers adjacent to their tidal marsh spawning pools until freeze-up in November, when sampling is terminated (G. J. FitzGerald, pers. comm.). They were taken only at freshwater and adjacent brackish stations from October through January in a Massachusetts estuary (Hoff and Ibara 1977; J. G. Hoff, pers. comm.). Thus as the spring spawning migration appears to be from freshwater to marine habitats, the Gulf ninespine stickleback may share with *Anguilla rostrata* the distinction of being the only catadromous species in the western North Atlantic.

PIPEFISHES AND SEAHORSES. FAMILY SYNGNATHIDAE

GRACE KLEIN-MACPHEE

The body is very slender and elongate, with or without a prehensile tail; armored with rings of bony plates, with or without an external fleshy integument. The neurocranium and suspensorium are extremely elongate with a small mouth (pipette) at the tip of the tubular snout. Prey is captured during a sudden upswing of the head that brings the mouth close to the prey, which is then drawn into the buccal cavity by suction caused by hyoid depression (Bergert and Wainwright 1997). The jaws are edentate but with odontoid processes present in some genera (Dawson and Fritzsche 1975). The gills, which form tufts of small, rounded lobes instead of the familiar filaments, have an opening reduced to a pore at the top of the opercular membrane, and the gill membranes are otherwise fused to the isthmus or body. There are no pelvic fins and only one dorsal fin (soft-rayed); other fins variously present or absent. There is no pyloric sphincter or distinct stomach. Males care for the eggs, which are attached to them by the female in a special area under the trunk or tail, which may be developed into a pouch.

There are two subfamilies, pipefishes (Syngnathinae) and seahorses (Hippocampinae), with 52 genera and about 215 species (Nelson 1994), many in shallow warm seas, but only the lined seahorse and the northern pipefish regularly inhabit the Gulf of Maine. Four other pipefishes that have been recorded from nearby as strays from the south are included in the key: seahorse pipefish, *Acentronoura dendriticum* (Barbour 1905)—two juveniles (31.5 and 39.3 mm), Canadian Atlantic, August 1977, associated with floating brown algae (Markle et al. 1980); pugnose pipefish, *Bryx dunckeri* (Metzelaar 1919)—48-mm juvenile, neuston net, east of Georges Bank, 40°35′ N, 66°00′ W, 18 June 1978 (Scott and Scott 1988); Sargassum pipefish, *Syngnathus pelagicus* Linnaeus 1758—89 mm, Georges Bank, 42° 09′ N, 66°41′ W, 20 September 1927, *Albatross II;* bull pipefish, *Syngnathus springeri* Herald 1942—105-mm juvenile, neuston net, east of Georges Bank, 41°06′ N, 66°31′ W, 18 June 1978 and a second, Isaacs-Kidd midwater trawl, 42°11.75′ N, 65°29.05′ W, 23 June 1980 (Scott and Scott 1988). Western Atlantic pipefishes and seahorses were comprehensively reviewed by Dawson 1982) and Vari (1982), respectively.

KEY TO GULF OF MAINE PIPEFISHES AND SEAHORSES

1a. Caudal fin absent; tail coiled ventrad, prehensile; longitudinal axis of head angled at about 70–90° from axis of body; brood pouch under tail, saclike and opening only through an anteromesial pore . . . . . . **Lined seahorse**

1b. Caudal fin present; tail straight and not prehensile; longitudinal axes of head and trunk essentially parallel; brood pouch under trunk or tail, but not saclike or opening only through an anteromesial pore (Pipefishes) **2**

2a. Tail curled . . . . **Seahorse pipefish**
2b. Tail straight 3
3a. Anal fin absent . . . . **Pugnose pipefish**
3b. Anal fin present 4
4a. Trunk rings 14–19, modally 18 or fewer . . . . **Sargassum pipefish**
4b. Trunk rings 18–24, modally 19 or more 5
5a. Trunk rings 22–24 . . . . **Bull pipefish**
5b. Trunk rings 18–21 . . . . **Northern pipefish**

LINED SEAHORSE / *Hippocampus erectus* Perry 1810 / Seahorse /

Bigelow and Schroeder 1953:315–316 (as *Hippocampus hudsonius*)

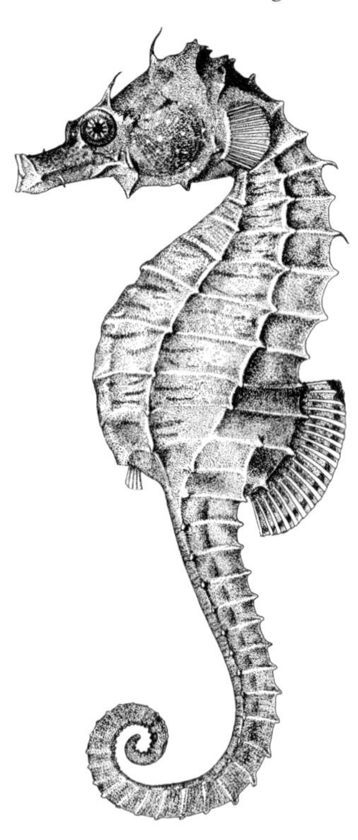

Figure 176. Lined seahorse *Hippocampus erectus*. Virginia. Drawn by H. L. Todd.

Description. Resemble the knight in a set of chessmen in the compressed body, deep convex abdomen, curved neck and curious horselike head carried at right angles to general axis of body (Fig. 176). Head surmounted by a pentagonal star-shaped coronet, snout tubular with small oblique toothless mouth at tip. A blunt horn between nostrils; a sharp spine on each side above eye and one behind it, a third over gill cover, and a fourth on side of throat, which sometimes terminates in short fleshy filaments. Neck, body, and tail covered with rings of bony plates, each body ring armed with four blunt spines or tubercles. Branched dermal flaps sometimes present, most developed on tubercles of superior trunk and tail ridges. Body tapers suddenly behind anal fin to a long tail, four-cornered in cross section, curled inward, and strongly prehensile. Lower surface of fore part of tail of males with brood pouch, opening by a slit in front. Dorsal fin originates about midbody, opposite vent, running backward over three and a half rings to within half a ring of commencement of tail sector of trunk. Anal fin very small, opposite rear part of dorsal fin. Pectoral fins moderate-sized, broad-based, and round-tipped. Pelvic and caudal fins absent. Sexually dimorphic. HL 0.20–0.27 of TL; snout 0.33–0.46 of TL; orbit diameter 0.12–0.18 HL; depth between deepest trunk rings 0.11–0.22 of TL in females, 0.16–0.24 of TL in males; trunk length 0.36–0.40 of TL in females, 0.33–0.43 of TL in males; tail length 0.54–0.64 of TL in females, 0.57–0.67 in males; brood pouch in egg-carrying males extending along five to eight tail rings (Vari 1982).

Meristics. Dorsal fin rays 16–20; pectoral fin rays 14–17; anal fin rays 3 or 4; 10–12, usually 11, rings on the trunk, 32–38 on the tail; vertebrae 13 + 36–38 (Miller and Jorgenson 1973; Vari 1982).

Color. Body light brown or dusky to ashen gray or yellow, sometimes almost black, and rarely bright red; variously mottled and blotched paler and darker, sometimes spangled with silver dots, sometimes plain-colored (Bigelow and Schroeder). Larger specimens often have irregular transverse bars or transverse bars and longitudinal stripes on the tail and a series of diagonal stripes on the opercle (Vari 1982). European seahorses change color according to their surroundings, tints of red, yellow, brown, and white all being within their capability, and it is likely that the American species is equally adaptable.

Size. Adults are usually 7.6–15.2 cm long; the largest examined by Vari (1982) was 17.3 cm TL; one of 20.3 cm (Beebe and Tee-Van 1933) is the largest on record.

Distinctions. The horselike head, bony rings, and prehensile tail distinguish seahorses. The prehensile tail, head carried at a pronounced angle to the body, and lack of a caudal fin separate them from pipefishes.

Taxonomic Note. This species shows a great deal of variation in degree of tuberculation and spination on the head and body and in the presence and extent of development of dermal flaps. This variability, along with allometric and sex-related differences, has resulted in descriptions of several species and subspecies; however, examination of a large series of *H. erectus* did not reveal any characters supporting other nominal forms (Vari 1982).

Habits. Lined seahorse dwell in both shallow and deeper waters where vegetation is present, chiefly among eelgrass, seaweed, and *Sargassum* weed, where they cling with their prehensile tails, monkeylike, to some stalk. They have been captured as deep as 73 m (Vari 1982). They inhabit bays (Schwartz 1961), beaches (Tracy 1910), salt marshes (Tracy 1910), oyster beds (Linton and Soloff 1964), weed-covered banks (Beebe and Tee-Van 1933), and offshore waters out to the 183-m contour (Fahay 1975). They tolerate a wide range in temperature and salinity, 5.0°–29.9°C (Christmas and Waller 1973) and 10–36.6 ppt (Christensen 1965; Christmas and Waller 1973).

Seahorse usually swim in a vertical position by undulations of the dorsal and pectoral fins (Consi et al. 2001), not with the tail, the trunk being too stiff for much sidewise motion. They produce sounds for communication.

Food. Seahorse feed on minute crustaceans, copepods, amphipods, and various larvae. They suck in their prey after a sudden upswing of the head and depression of the hyoid, as described for northern pipefish (Bergert and Wainwright 1997).

Predators. Lined seahorse are eaten by a variety of predators including Atlantic cod (Langton and Bowman 1980), bluefish (Buckel et al. 1999b), remoras (Longley and Hildebrand 1941), spiny and smooth dogfishes (Rountree 1999), and offshore by lancetfish and bluefin and yellowfin tunas (Matthews et al. 1977).

Parasites. In Florida Bay, seahorse were infected by a microsporidan, *Glugea heraldi*, a myxosporidan, *Sphaeromyxa* sp., and an unidentified nematode (Vincent and Clifton-Hadley 1989).

Breeding Habits. Usually 250–400 eggs are deposited in the males' brood pouch. The number of eggs in female ovaries and the number of developing young in the male brood pouch increase with adult size. The smallest male with a developed brood pouch was 7.8 cm and the smallest with brood pouch eggs and embryos was 8.9 cm (Vari 1982). Females have paired ovaries; unusual in that they consist of a cylindrical tube with two dorsally located germinal ridges (Selman et al. 1991).

Courtship consists of the sexes following each other around, males presenting their pouches toward the females' genitals (Breder and Rosen 1966). The fish rise in the water free from attachments to objects at the time of egg transfer, and there are considerable changes in coloration during sexual excitement (Sharrock 1934). Seahorse breed in summer and breeding habits resemble those of pipefish: the female deposits her eggs in the male's brood pouch a few at a time in repeated pairings, and the eggs develop in the brood pouch (Ryder 1882a). Developing embryos absorb calcium and probably oxygen from brood pouch fluids (Linton and Soloff 1964).

Spawning may be year-round in Florida (Reid 1954). Gravid males have been found in late August in New Jersey and North Carolina (Gudger 1906) and young 6–33 mm in Chesapeake Bay from June to September (Pearson 1941).

The eggs are pear-shaped, 3.1–3.9 mm, light orange (Hudson and Hardy 1975) or orange-yellow (Ryder 1882a) in color. The yolk contains one or more large oil droplets in early eggs and numerous small deep orange ones in advanced eggs (Hudson and Hardy 1975).

Incubation in the brood pouch takes at least 12–14 days Ryder 1882a). There is no true larval period (Hardy 1978a) and tiny miniature seahorse (about 6 mm) are extruded from the pouch. At extrusion, the anal fin has four hyaline rays; ventral dermal plates are not developed; and the swim bladder is evident at the level of the pectoral fin (Ryder 1882a). Cirri develop on the body at about 7 mm; spines are longer in young than in adults and better developed in females than in males (Ginsburg 1937). The tail becomes prehensile at 1 day (Lockwood 1867). Pigmentation is dark-colored with light-colored blotches around the base of the dermal spines more or less coalescent (Ginsburg 1937). Age at maturity is unknown but is more than 3 months (Herald 1951). The largest immature male was 95 mm long (Ginsburg 1937).

Newborn lined seahorse swim in a cluster near the surface (Smith 1907), and have been recorded from rivers entering the Potomac (Bigelow and Schroeder; Lippson and Moran 1974), with young 6.0–33 mm found in masses of floating seaweed in lower Chesapeake Bay (Pearson 1941). Specimens up to 95 mm are pelagic in comparatively deep offshore waters (de Sylva et al. 1962).

Age and Growth. The age-length relationship of fish from the Gulf of Mexico was given as $Y = 21.4 + 0.107X$, where Y = TL in mm and X = time in days (Matlock 1992). The range in size was 20–160 mm TL. Rate of growth was about 0.11 mm·day^{-1}.

General Range. Atlantic coast of North America, occurring regularly from South Carolina to Cape Cod, and to Nova Scotia as strays. They occur south along the Atlantic coast of South America to Uruguay (Vari 1982).

Occurrence in the Gulf of Maine. Lined seahorse are not common north of New York. A few are found each year around Woods Hole, chiefly in July, August, and September. They so rarely stray past the elbow of Cape Cod that there are only a few definite records, one from Nahant on a lobster trapline at 20 m (Collette and Hartel 1988); two from the Maine coast in 1953 (Taylor et al. 1957), one from Provincetown, and one from Georges Bank (Bigelow and Schroeder).

Importance. Lined seahorse have no commercial importance except in the aquarium trade but the demand for seahorses is so great in many parts of the world that several species are becoming rare and perhaps endangered from overharvesting

NORTHERN PIPEFISH / *Syngnathus fuscus* Storer 1839 / Pipefish, Common Pipefish /

Bigelow and Schroeder 1953:312–314

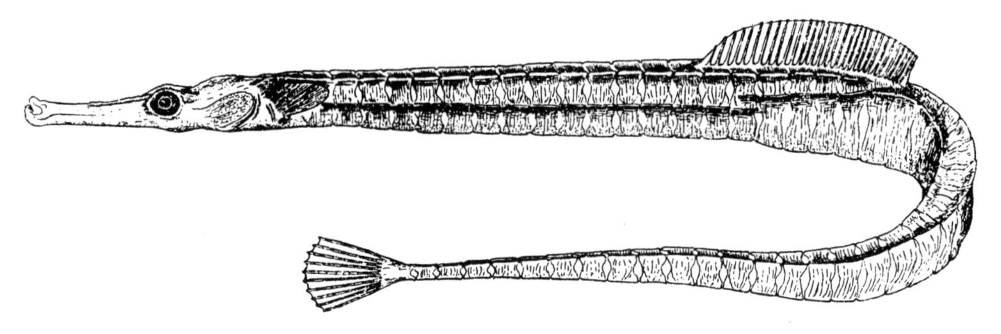

Figure 177. Northern pipefish *Syngnathus fuscus.* (From Bigelow and Welsh 1925.)

Description. Body very slender, particularly behind vent, distinctly ridged (Fig. 177). Males about 35 times as long as deep, females about 30 times. Snout tubelike, blunt-ended, with a small toothless obliquely oriented mouth at its tip. Gill openings very small. Body hexagonal in cross section in front of vent, four-sided behind dorsal fin, and enclosed in an armor of bony plates connected in rings. Abdomen of male wider just back of vent than elsewhere, with two lateral flaps that meet along midline to form the so-called "marsupial" or brood pouch; females lack these. The brood pouch occupies 13–20 rings (Lippson and Moran 1974; Dawson 1982). Fish from Nova Scotia and Massachusetts have 17–20 pouch rings; those from Virginia to Florida have 15–18 (Dawson 1982). Median ventral trunk ridge often prominent in subadults and adults, sometimes with a fleshy keel in large females; mature females often with a deep trunk, particularly in northern populations (Dawson 1982). Dorsal fin originates a little behind first third of body and extends along four or five bony rings in front of vent and as many behind it. Anal fin very small, close behind vent. Pectoral fins moderate-sized. Pelvic fins absent. Caudal fin rounded, middle rays longest. Head occupies one-eighth to one-ninth of TL (7.1–9.5 times in SL). Snout depth 2.3–7.9 in snout length. Dorsal fin five to six times as long as high. Body depth at anal ring 3.1–10.1 in HL, trunk depth 1.7–5.4 in HL (Dawson 1982).

Meristics. Dorsal fin rays 33–49; pectoral fin rays 12–15; trunk rings 18–21 and tail rings 34–42; vertebrae 19–21 + 36–39 (Miller and Jorgenson 1973; Dawson 1982). There is a south-north cline in the frequency of dorsal fin rays and subdorsal trunk rings; counts are higher northward (Dawson 1982).

Color. The body is greenish, brownish, or olive above, rarely brick red, cross-barred and mottled with darker markings. Lower parts of the gill covers are silvery. The lower parts of the sides are sprinkled with many tiny white dots, and the longitudinal angles separating sides from abdomen are marked by 12 or 13 longitudinal brown bars. The lower body surface is colorless to golden yellow, with the marsupial flaps flesh-colored. The dorsal and pectoral fins are pale, sometimes with diagonal bands, but the caudal is brown. Pipefishes can change color according to the color of their surroundings.

Size. Maximum length is 30.5 cm (Nichols and Breder 1927), but they rarely exceed 20.8 cm (Dawson 1982).

Distinctions. Northern pipefish have some resemblance to cornetfish but have a much shorter snout, lack pelvic fins, and have a rounded rather than a forked caudal fin. They can be distinguished from pelagic pipefish by the greater number of trunk rings (18–29 vs. 16 or 17).

Habits. This pipefish is found among eelgrass or seaweeds, in shallow bays, salt marshes, harbors, creeks, and river mouths, where they often go up into brackish water. They are reported over bottoms of mud, sand, and gravel (Smith 1971). Inshore, pipefish are found in greater densities in grass beds than in adjacent unvegetated substrate and appear to actively select seagrass habitats (Sogard and Able 1991; Roelke and Sogard 1993). They sometimes stray out to sea near the surface and have often been found under floating rockweed along the Maine coast. Depth range is from intertidal (Bleakney and McAllister 1973) to 49.4 m (Hildebrand and Schroeder 1928). They are euryhaline and tolerate a salinity range of 0–38.8 ppt (Christensen 1965; Dahlberg 1972), but are most commonly found at 13–20 ppt (Mercer 1973). The range of temperature tolerance is 3°–35°C (de Sylva et al. 1962). In the vicinity of Long Island, at a temperature of 10.6°C, they were found in a torpid state lying motionless on the bottom or partially buried in sand or curled around sand dollars (Wicklund et al. 1968). Whether this is a typical winter behavior pattern in northern populations or just a response to low temperature is unknown.

Northern pipefish usually propel themselves with the dorsal fin, but they can travel swiftly when alarmed, with eel-like strokes of the tail from side to side. They are able to roll their eyeballs separately.

Food. Northern pipefish are diurnal feeders (Ryer and Boehlert 1983). They feed chiefly on minute copepods, amphipods, fish eggs, very small fish larvae, and no doubt indiscriminately on any small marine animals (Hildebrand and Schroeder 1928). Studies of pipefish in a seagrass community of lower Chesapeake Bay showed that gammarid and caprellid amphipods and calanoid copepods were dominant food items. Other organisms eaten were polychaetes and mysid shrimp. An amphipod (*Gammarus mucronatus*), calanoid copepods, and a caprellid (*Erichsonella attenuata*) were dominant food items in spring, summer, and fall, respectively. Diet appears to be controlled by seasonal abundance patterns of different prey; predation on certain components such as *G. mucronatus* may, in turn, partly control seasonal patterns of that species abundance. Pipefish preyed upon the smallest size-classes of each species, and there were ontogenetic shifts in diet.

Pipefish dart toward the prey, snap their heads upward, and then suck the prey into their mouths by depressing a peglike hyoid (Bergert and Wainwright 1997). In foraging experiments (Ryer and Orth 1987), the attack probability of large fish (180–200 mm TL) on amphipods was negatively correlated to amphipod size in an artificial narrow-leaf seagrass habitat, but positively correlated in wide-leaf habitat. The attack probability of small fish (110–130 mm) was negatively correlated to amphipod size in both habitats. Foraging success was negatively correlated to prey size and greater for large than for small fish. Mean prey size was positively related to fish size (Ryer and Orth 1987). Pipefish appeared to select amphipods that were exposed or moving about in the vegetation (Ryer 1988).

Daily rations were 4.0 and 4.4% body weight per day. Evacuation rate was temperature-dependent and positively correlated with gut content. Time to completely evacuate a meal was 30.2 h at 15°C, 14.1 h at 23°C, and 10.3 h at 27°C (Ryer and Boehlert 1983).

Predators. Known fish predators include smooth dogfish, cod, sea raven, black sea bass, and weakfish (Rountree 1999); oyster toadfish (Roelke and Sogard 1993); and bluefish (Buckel et al. 1999b). There is also a report of entanglement in the tentacles of *Physalia pelagica* (Gordon 1956).

Parasites. Northern pipefish are known to be infected by a cestode *Rhynchobothrium heterospine* (Linton 1901b) and digenetic trematodes *Cymbephallus vitellosus* and *Opecoeloides manteri* (Hunninen and Cable 1940; Linton 1940; Doss and Farr 1969).

Breeding Habits. Reported maximum fecundity is 860 eggs; the brood pouch capacity is 104–570 (Hildebrand and Schroeder 1928). Sexual maturity is reached at about 1 year (Bigelow and Welsh 1925). The smallest brooding male was 83 mm (Herald 1943).

Male northern pipefish incubate the eggs in a brood pouch, the flaps of which ordinarily lie flat but are swollen with their edges cemented together during the breeding season. The protruding oviduct of the female is inserted into the opening of the male's pouch to transfer a dozen or more eggs. This occurs several times in succession, with [illegible]

the pouch is filled. Fertilization is thought to take place during the transfer of eggs from one parent to the other. The eggs become embedded in the lining of the brood pouch (Bigelow and Schroeder). Eggs are deposited in a single row on each side of the pouch in males 82–100 mm long. Larger males have two to four rows and two or three layers on each side (Hardy 1978a). In New Jersey estuaries, there was a skewed sex ratio of 9:1 females to males, males apparently a limiting reproductive resource for females because most males collected were gravid (Roelke and Sogard 1993).

Northern pipefish breed from January to October, and those from New England to the Chesapeake Bay region from March to October (Lippson and Moran 1974; Hardy 1978a). In Georgia, brooding males were collected from January to March and in Florida during April and June (Dawson 1982). While males are gravid they move around less than females, possibly to reduce risk of predation and conserve energy (Roelke and Sogard 1993).

Eggs are 0.75–1.0 mm in diameter (Ryder 1887b; Herald 1943); the yolk is lemon or orange-yellow with deeper-colored oil globules (Ryder 1882a, 1887b). Details of development are given in Hardy (1978a) from Ryder (1887).

Incubation takes about 10 days (Bigelow and Schroeder), and the young are retained in the brood pouch until they are 8–12 mm long, by which time the yolk sac has been absorbed (Lippson and Moran 1974). The young pipefish are then ready for independent existence and once they leave the pouch they never return to it. Young pipefish about 8.5 mm have five dark bands of pigment behind the dorsal fin; by 13.2 mm there is a series of chromatophores along the dorsal and ventral edges of the gut, pigment over the developing swim bladder, and large chromatophores on the snout in front of the eye (Scotton et al. 1973; Lippson and Moran 1974).

Young pipefish are common in plankton, ranking eleventh in abundance in Beverly-Salem Harbor (Elliott et al. 1979). They appeared to show differential distribution and were more abundant at night in flood tides than during the day in ebb tides (Williams 1960).

Growth. Pipefish larvae kept in aquariums grow from about 10–70 mm in length within about 2 months after hatching (Tracy 1910), and length-frequency data indicated average length increments of 60–120 mm from June to October in the Delaware River estuary (de Sylva et al. 1962).

General Range. Coast of eastern North America, in salt and brackish water, from the southern side of the Gulf of St. Lawrence, the Bay of Fundy, and outer Nova Scotia at Halifax (Scott and Scott 1988), to Jupiter Inlet and lower reaches of the Loxahatchee River, Fla. (Christensen 1965). Gulf of Mexico records refer to *S. affinis* (Dawson 1982).

Occurrence in the Gulf of Maine. Northern pipefish can be expected anywhere under suitable conditions. They are com-

eelgrass (Collette and Hartel 1988) and ranked sixth of 40 species in abundance in Salem Harbor (Anderson et al. 1975). They are common in mixing and tidal zones of 17 estuaries along the northwestern Atlantic coast from Passamaquoddy to Cape Cod bays, being ranked as rare only in Englishman-Machias and Narragaugus Bay, Maine (Jury et al. 1994). They are common in the Bay of Fundy, have been reported from outer Nova Scotian waters, and are common locally on the southern side of the Gulf of St. Lawrence.

Migrations. In the Chesapeake Bay area, northern pipefish move into deeper water in autumn and return to the shallows in spring (Hildebrand and Schroeder 1928), but they were reported to be resident in the eelgrass at Woods Hole throughout the year (Bigelow and Schroeder). Data from several areas in the northern Mid-Atlantic Bight indicate that northern pipefish undergo seasonal inshore-offshore migrations. They move from estuaries where they are resident in spring-fall into nearshore continental shelf waters off Cape Cod in late September–October and off Long Island and New Jersey in November and return in March and April. Most collections occurred at water temperatures of 10°–15°C in the fall (September–November), and 3°–6°C in spring (March–May). Most collections were made within 20 km of the coast at depths between 10 and 20 m. Both young-of-the-year and older fish migrate (Lazzari and Able 1990).

ACKNOWLEDGMENT. A draft of the syngnathid section was reviewed by Richard Vari.

CORNETFISHES. FAMILY FISTULARIIDAE

GRACE KLEIN-MACPHEE

Body and snout long and slender with an elongate caudal filament between the upper and lower lobes of the caudal fin. The long tubular snout is adapted for preying on fishes among reefs. Lack the chin barbel found in trumpetfishes (Aulostomidae). Worldwide in tropical marine waters. One genus with four species (Fritzsche 1976), two of which occur in the western Atlantic, one straying into the Gulf of Maine.

BLUESPOTTED CORNETFISH / *Fistularia tabacaria* Linnaeus 1758 / Trumpetfish, Cornetfish /

Bigelow and Schroeder 1953:316–317

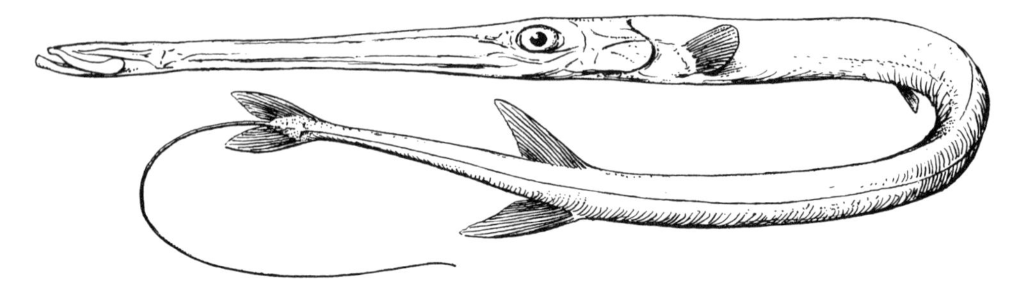

Figure 178. Bluespotted cornetfish *Fistularia tabacaria*. (From Bigelow and Schroeder: Fig. 174.)

Description. Body 30–35 times as long as deep (not including the caudal filament) and only about two-thirds deep as wide (Fig. 178). Head almost one-third and snout one-quarter of body length. Mouth small, oblique; lower jaw projects a little beyond upper. Teeth minute, crowded together in rows on jaws. Snout ridges smooth, interorbital narrow with a smooth depression (Fritzsche 1976). Eye oval, orbits with an angular process or spine. Dorsal and anal fins triangular, higher than long, located over one another rather far back on body, about three-quarters of distance from eye to base of caudal fin. Pelvic fins very small, located closer to head, about one-third the way from base of eye. Caudal fin forked; middle rays prolonged in a filament about as long as snout, although likely to be broken off during capture. Body of adults scaleless, but small juveniles with rows of spinules. A row of embedded bony plates or shields along either side of midline, more conspicuous near dorsal and anal fin.

Meristics. Dorsal fin rays 13–18; anal fin rays 13–17; pectoral fin rays 15 or 16; vertebrae 4 + 49 + 34 = 76–87 (the first four elongate and fused [Fritzsche 1976]).

Color. Greenish brown above, the back and sides with many large, oblong, pale blue spots and about ten dark crossbars. A 200-mm specimen was greenish crossed with a number of light lines (Jungerson 1910). The lower surface is pale and silvery, the caudal filament deep blue. Cornetfish can change their color pattern almost instantly to match their background (Böhlke and Chaplin 1968).

Size. Reported to reach 1.8 m, not including the caudal filament (Robins and Ray 1986).

Distinctions. The very long snout and protrusible mouth and long caudal filament distinguish this species from any other caught in the Gulf of Maine.

Habits. Adults live inshore, often associated with reefs in the tropics. They are usually taken over sandy or stony bottom among shells and seafans to depths of 200 m (Robins and Ray 1986). In the Bahamas, they are found drifting or darting above seagrass beds or rubble bottom (Böhlke and Chaplin 1968). This species sometimes enters estuaries in the northern part of its range (Gordon 1960), and was once collected at river mile 26 in the Hudson River at a salinity of 10 ppt (Young et al. 1982).

Food. Food consists of small fishes and shrimps (Scott and Scott 1988). In aquariums, juvenile cornetfish fed on mummichog, and small mullet (Burgess 1976), swallowing them whole, headfirst.

Predators. Found in stomachs of dusky shark, *Carcharhinus obscurus,* off North Carolina (Schwartz and Jensen 1996) and black grouper, *Mycteroperca bonaci,* in the West Indies (Randall 1967).

Reproduction. Nothing is known of its reproductive behavior, but other species of *Fistularia* spawn pelagic eggs in open water (Leis and Rennis 1983). The smallest juvenile collected (16 mm) had a blunt head, caudal filament, and developing pectoral fins (Zhudova 1971). Larger juveniles (43 mm) have minute serrae on the lateral ridges of the snout, and the body is covered with hooked spinules (Jungerson 1910: Pl. 7, Fig. 1). Snout serrae become obsolete with growth (Fritzsche 1976).

General Range. Both sides of the tropical Atlantic. In the western Atlantic south through the Gulf of Mexico, Caribbean and West Indies to Brazil (Fritzsche 1976). Young are occasionally found as far north as Nova Scotia (Scott and Scott 1988).

Occurrence in the Gulf of Maine. There are two records of this species from Massachusetts Bay (Collette and Hartel 1988), one from Rockport, Mass., in 1805; and one found on the northern edge of Georges Bank in 1947. Juveniles are taken at Woods Hole almost every year in late summer.

SNIPEFISHES. FAMILY MACRORAMPHOSIDAE

BRUCE B. COLLETTE

Snipefishes have a long, tubular snout with a small toothless mouth at its tip and a very long, stout second spine in the dorsal fin. The Macroramphosidae contains three genera and about 12 species of tropical to subtropical marine fishes (Nelson 1994). This family is sometimes considered to be a subfamily of the shrimpfishes, Centriscidae (Eschmeyer 1998). The correct original spelling of the type genus is *Macroramphosus,* not the commonly used *Macrorhamphosus* (Wheeler 1973).

LONGSPINE SNIPEFISH / *Macroramphosus scolopax* (Linnaeus 1758) /

Bigelow and Schroeder 1953:301 (as *Macrorhamphosus scolopax*)

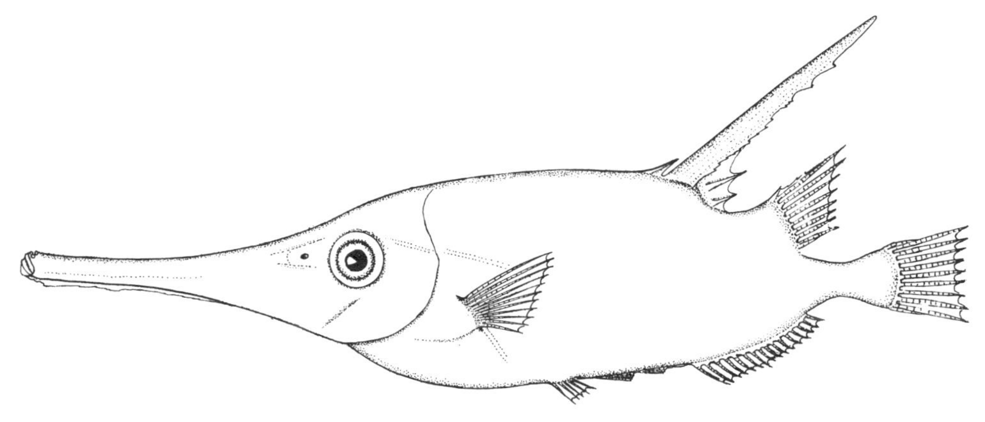

Figure 170. [illegible]

Description. Body compressed, deep, 16.7–36% SL (Fig. 179). Snout tubular, very long, 28.6–40% SL. Eye large, diameter about four to five times in snout length. Two dorsal fins originate far behind middle of trunk. First dorsal fin short; second dorsal spine much longer than others, 8.6–38.9% SL; serrate on rear margin. Second dorsal fin longer, pointed; dorsal fins separated by an interspace nearly as long as base of first dorsal fin. Anal fin much longer than second dorsal fin, but lower. Pelvic fins very small, located behind pectoral fins. Caudal fin emarginate, moderate-sized. Snout, head, and sides clothed with small, rough scales. Body stiffened with bony plates, two longitudinal rows of four each, high up on each side behind gill opening; also three longitudinal series of six each along lower breast and belly in front of pelvic fins, followed by three pairs behind pelvic fins and by a single plate close in front of anal fin, these last forming a sharp keel.

Meristics. Dorsal spines 4–8; second dorsal fin rays 10–14; anal fin rays 19–21; pectoral fin rays 18–21; pelvic fin rays I, 4; vertebrae 23 (Ehrich 1976, 1986).

Color. Pinkish or reddish on sides above, fading to silvery white below. Sometimes described as golden above. In a large aquarium, individuals that were actively courting changed color (de Oliveira et al. 1993). The brownish ventral area and the base of the dorsal fin became darker. The posterior part of the body became brick red, masking the silvery hue of the flanks.

Size. Maximum size 228 mm TL, a 6-year-old female taken off Aveiro, Portugal (Borges 2001).

Distinctions. The long tubular snout with a small toothless mouth at the tip combined with the very long, stout dorsal fin spine distinguish longspine snipefish from all other Gulf of Maine fishes.

Taxonomic Note. There is a question as to whether there are one or two species, at least in the eastern Atlantic. Ehrich (1976) concluded that *M. gracilis* (Lowe 1839) was a form of *M. scolopax* but Brêthes (1979), Clarke (1984), and Assis (1993) concluded that the ecological, biological, and morphological differences between the two forms justified maintaining them as separate species.

Habits. Juveniles to about 10 cm live pelagically in oceanic waters; adults live close to the bottom at depths of 50–150 m (Ehrich 1986). In a large aquarium, snipefish were scattered around the tank with no obvious schooling behavior (de Oliveira et al. 1993). Most of the time, the fish were stationary in the water column in a head-down position.

Food. Juveniles feed on pelagic invertebrates, particularly copepods, bivalve and gastropod larvae, foraminifera, and polychaete eggs (Ehrich 1976). Intermediate-sized snipefish feed on a wide variety of crustaceans, mollusk larvae, polychaetes, and foraminifera. Adults eat both pelagic and bottom invertebrates, particularly mysids and foraminifera.

Predators. Stomach contents of 195 blue shark from the Azores were examined (Clarke et al. 1996). Of 23 stomachs with food, six contained 160 individuals of *M. scolopax,* the second most abundant fish taxon in these stomachs.

Reproduction. In the eastern Atlantic, snipefish spawn on the shelf, over seamounts, or near islands between October and March (Ehrich and John 1973; Ehrich 1986). Courtship and spawning were described for a group of 200 fish observed in a large aquarium (de Oliveira et al. 1993). Oogenesis, development, and fecundity were studied off Portugal (Arruda 1988). Maximum growth of gonads occurred in February and March. Fecundity $= 0.094 \times TL^3 + 21.107$. The eggs are pelagic. Larvae and juveniles up to about 50 mm live in the surface layers at night, moving to deeper water in the daytime. At about 50 mm and an age of about 3 months, they leave midwater and move to the bottom.

Age and Growth. Based on specimens collected in Portuguese continental waters (Borges 2000), the von Bertalanffy growth curves provided parameters of $L_\infty = 15.7$ cm, $K = 0.654$, and $t_0 = -1.066$. Maximum age, based on otoliths, was estimated at 6 years.

General Range. Worldwide, mainly in temperate latitudes between 20 and 40° N (Ehrich 1976, 1986). In the western Atlantic known from off the Greater Antilles and the east coast of the United States north to Nantucket and Massachusetts Bay (Wheeler 1977). In the eastern Atlantic, abundant off Portugal and Morocco, present in the Mediterranean, northward to southern England and south to 21° N (Ehrich et al. 1987).

Occurrence in the Gulf of Maine. Oddly enough, the first records from the western Atlantic were all from the Gulf of Maine region: namely, one reported from Massachusetts Bay (Goode and Bean 1896); a second trawled south of Nantucket, at the 238-m contour line, both many years ago; and eight specimens trawled in that same general vicinity (39°59′ N, 69°47′ W) at 146 m, by the *Albatross III* on 14 May 1950. Evidently they reach the inner parts of the Gulf only as strays, and at long intervals, although they have been taken from time to time by otter trawlers along the southwestern edge of Georges Bank in 137–155 m.

Importance. Of no commercial importance in the western Atlantic, but considered to be potentially important off Portugal and Morocco (Belveze and Bravo de Laguna 1980; Morais 1981).

ACKNOWLEDGMENT. A draft of the entire gasterosteiform section was reviewed by Karsten E. Hartel.

MAIL-CHEEKED FISHES. ORDER SCORPAENIFORMES

Order Scorpaeniformes, the so-called "mail-cheeked fishes," is distinguished by the bony suborbital stay, a posterior extension of the third infraorbital bone that extends across the cheek to the preopercle and is usually firmly attached to that bone. However, it is not certain that this character defines a monophyletic unit (Johnson and Patterson 1993). Head and body tend to be spiny or have bony plates; pectoral fin is rounded; membranes between ventral rays are often incised; and caudal fin is usually rounded. As recognized by Nelson (1994), the order contains seven suborders, 25 families, 266 genera, and about 1,271 species. The 21 species found in the Gulf of Maine are placed in three suborders, Dactylopteroidei (Dactylopteridae—one species); Scorpaenoidei (Scorpaenidae and Triglidae—six species); and Cottoidei (Cottidae, Psychrolutidae, Hemitripteridae, Agonidae, Cyclopteridae, and Liparidae—15 species).

KEY TO SUPERFAMILIES OR FAMILIES OF GULF OF MAINE SCORPAENIFORMES

1a. First (lower) few rays of pectoral fin separate from rest of fin 2
1b. First (lower) few rays of pectoral fin not separate from rest of fin. 3
2a. First (lower) 2 or 3 pectoral fin rays in the form of separate feelers; first few dorsal fin spines not separate from rest of fin . . . . **Triglidae**
2b. First (lower) 5 or 6 pectoral fin rays not separate in the form of feelers; first 2 dorsal fin spines free, separated from second dorsal fin by deep notch. . . . . **Dactylopteridae**
3a. Pelvic fins united to form ventral sucking disk . . . . **Superfamily Cyclopteroidea**
3b. Pelvic fins separate, not forming a sucking disk 4
4a. Anal fin with 3 well-developed spines; body covered with scales . . . . **Scorpaenidae**
4b. Anal fin without spines, only soft rays present; body without scales, naked or with prickles or bony plates . . . . **Superfamily Cottoidea**

FLYING GURNARDS. FAMILY DACTYLOPTERIDAE

GRACE KLEIN-MACPHEE

The Dactylopteridae is included here as a scorpaeniform family in its own suborder Dactylopteroidei following Nelson (1994), although Johnson and Patterson (1993) place it in its own order next to the Scorpaeniformes. More recently, Imamura (2000) showed that the posterior extension of the infraorbital in Dactylopteridae is not homologous with the suborbital stay of Scorpaeniformes so he combined the Dactylopteridae with the percoid family Malacanthidae under the former name. The old Dactylopteridae contains two genera and about seven species (Nelson 1994), only one of which occurs in the Atlantic Ocean.

FLYING GURNARD / *Dactylopterus volitans* (Linnaeus 1758) / Bigelow and Schroeder 1953:472–473

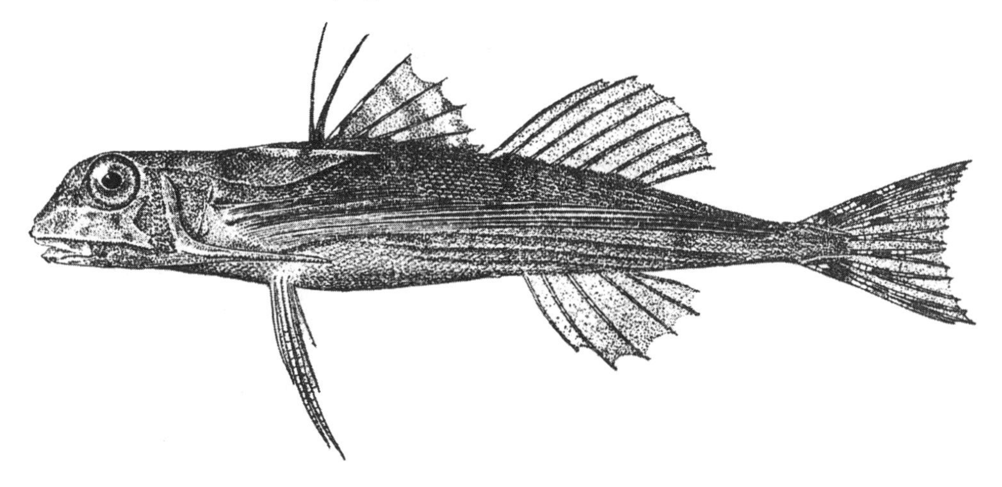

Figure 180. Flying gurnard *Dactylopterus volitans.* Key West, Fla. Drawn by H. L. Todd.

Description. Body moderately elongate, covered with scute-like scales (Fig. 180). Large head with top and sides encased in and jaws with a band of small granular teeth. Long spine extends from nape posteriorly below midline of dorsal fin base.

to slightly beyond pectoral fin base. Both spines more conspicuous in young fish. First dorsal fin with first two spines free, separated from second dorsal fin by a deep notch. Anal fin slightly posterior to origin of second dorsal. Pectoral fins divided into two sections, anterior part short, posterior long. Pectoral fins enormous when half grown or larger, reaching nearly to base of caudal fin when laid back; when spread, they appear as enormous rounded fanlike wings. Pelvic fins long, abdominal. Caudal fin emarginate. No lateral line.

Meristics. Dorsal fin rays VI or VII, the first few free, 8; pelvic fin rays I, 4; pectoral fin rays, anterior part 6 rays, posterior 26–30 rays; branchiostegal rays 6; vertebrae 22 (Miller and Jorgenson 1973; Scott and Scott 1988).

Color. A brilliant fish, varying widely in color; most are some shade of brownish to greenish olive above, with the lower side paler, but marked irregularly with reddish salmon or salmon yellow. The winglike pectoral fins are marked with bright blue streaks near their bases with blue spots and bars toward their tips. The caudal fin usually has about three brownish red crossbars.

Size. Attain at least 403 mm SL and 495 mm TL. The all-tackle game fish record is a 1.81-kg fish caught at Panama City, Fla., in June 1986 (IGFA 2001).

Distinctions. Flying gurnard superficially resemble sea robins but differ in that the first five or six rays of each pectoral fin are not separate in the form of feelers, but are continuous with the remainder of the fin; the first two spines of the dorsal fin are separate; the bony armor covering the front and top of the head reaches rearward considerably beyond the origin of the dorsal fin on either side to end in a stout spine; each gill cover is extended rearward as a stout spine about as far as the axil of the pectoral fin, and the scales are much larger, each armed with a short stout spine.

Habits. Adults are associated with sandy or muddy bottoms in 5–294 m (Schwartz and Lindquist 2000). They frequently occur in clear coastal waters, often around coral reefs (Robins and Ray 1986).

They "walk" on the bottom by projecting the pelvic fins straight down like stilts and moving them alternately (Böhlke and Chaplin 1968). They are also reported to leap from the water and "glide" with their large pectoral fins (Fish and Mowbray 1970), but this has been disputed as unproven (Böhlke and Chaplin 1968). However, this author was given a specimen by a commercial fisherman from Point Judith, R.I., who said the fish leaped onto the deck of his boat while he was at sea. Flying gurnard produce sounds when under stress or when being handled. The sonic mechanism is a two-lobed swim bladder with lateral musculature (Fish and Mowbray 1970). They have also been reported to emit sounds by stridulation using the hypomandibular bone (Smith-Vaniz 1978).

Predators. There is little information on predators; however a flying gurnard was found in the stomach of a striped sea robin (Marshall 1946).

Breeding Habits. Ripe females were collected in June and July in the Mediterranean (Sanzo 1933).

Early Life History. The eggs are pelagic, rosy coral or bright coralline when unfertilized, slightly oval, transparent, small (about 0.8 mm), smooth, yolk vesiculate, single reddish coral oil droplet 0.14 mm diameter, and no obvious perivitelline space (Sanzo 1933). Egg development is rapid and hatching occurs in about 1 day (Sanzo 1933).

Yolk-sac larvae are 1.8–1.86 mm. They are short, stout, and unpigmented with 22 myomeres, posterior coralline-colored oil globule, and rounded membranous pectoral fins. The mouth opens 3 days after hatching when the yolk is absorbed. A spine appears at the inferior angle of the preopercle at 1.86 mm.

Pigmentation begins to develop 2 days posthatch on the head and posterior on the body; eyes are pigmented; there is intense black pigment on the abdomen, a spot at the base of the pectoral fin and on the opercle, and a yellowish tinge on the nape at 1.86 mm.

Larvae of 2.4–9.12 mm have a deep body, short snout, narrow mouth, large eye, elevated orbital crest, and three large spines on the head. The second dorsal fin, anal fin, and hypurals form by 3.48 mm. Pigment is present on the interorbital, postorbital, and sides of the body; chromatophores on the dorsum evenly spaced; some pigment ventrally on caudal peduncle and behind the anus; pigment absent posteriorly. By 7.48 mm, the head has large, evenly spaced chromatophores, numerous tiny spots, opercle with pigment; trunk is golden with ventral surface spotted.

Juveniles (16.5–100 mm SL) have a deep head, short snout, small mouth, big eyes and very big head spines. The pectoral fins are long and reach to the caudal base by 100 mm. Pigmentation intensifies by 32 mm. The dorsolateral pigment is brown; by 46 mm pectoral and anal fin origins are black-pigmented, bases whitish; pelvics unpigmented; dorsal rays spotted (Sanzo 1933, 1939).

General Range. Tropical to warm temperate latitudes of both coasts of the Atlantic; south to Argentina (Smith-Vaniz 1978) including the Gulf of Mexico; and north rather commonly to North Carolina on the American coast; a few to New York and the southern coast of Massachusetts.

Occurrence in the Gulf of Maine. The northernmost records are from Nova Scotia: one found dead on a beach on Sable Island, N.S., one captured in surface waters in a neuston net (Markle et al. 1980); a juvenile taken during scuba diving at Shelburne Harbor, N.S.; one picked up by hand in the shallows at Martinique Beach, Halifax Co.; Annapolis Tidal Generating Station (ARC 200016249); and several ARC lots from between 42°46′ and 46° N, 52°02′ and 64°40′ W (L. van Guelpin, pers. comm., April 2000).

SCORPIONFISHES. FAMILY SCORPAENIDAE

Grace Klein-MacPhee and Bruce B. Collette

Scorpionfishes are perchlike or basslike in general appearance but are related to sculpins (Cottidae) and sea robins (Triglidae) in having a bony suborbital stay. Their cheeks are spiny and, in most species, the top of the head has ridges that end in spines. Both spiny and soft portions of the dorsal fin are well developed, either as one continuous fin or subdivided into two fins by a deep notch. The pelvic fins are thoracic. Scorpionfishes derive their name from venomous glandular tissue that is associated with fin spines of some species. Wounds from these spines in some Indo-Pacific species can be very painful but are no more than a mild bee sting in Atlantic species. There are 12 subfamilies, at least 56 genera, and about 390 species (Nelson 1994), mostly in the Indo–West Pacific. The subfamily Sebastinae, with four genera and about 128 species, contains all scorpaenids found in the Gulf of Maine. Other subfamilies are oviparous but the Sebastinae contains species in which the eggs are fertilized internally and retained within the body of the female until after they hatch (Wourms 1991). The livebearing genus *Sebastes* is the largest in the family with about 110 species, mostly occurring in the North Pacific. Two species, blackbelly rosefish, *Helicolenus dactylopterus*, and Acadian redfish, *Sebastes fasciatus*, are common in the Gulf of Maine, and two more species of *Sebastes* occur in deeper water nearby and may venture into the Gulf. Names of Atlantic redfishes follow Robins et al. (1986).

The three species of redfishes found in or near the Gulf of Maine are very difficult to distinguish in the field, although fin ray counts, extrinsic swim bladder muscle passage, and vertebral counts have been used (Hallacher 1974; Litvinenko 1980; Ni 1981a, 1981b, 1982, 1984; Barsukov et al. 1991). They can be separated reliably by biochemical and genetic means such as electrophoresis (Payne and Ni 1982), polyacrylamide gel isoelectric focusing (Trottier et al. 1988) and malate dehydrogenase mobility patterns (Rubec et al. 1991). For practical purposes, the commercial catch is called redfish, and no attempt is made to separate the species. A summary of recent information about redfishes was published by Pikanowski et al. (1999). Acadian redfish are the most abundant species in the Gulf of Maine and are probably the only species of the genus found there. Table 4 gives general characteristics of the three species.

TABLE 4
CHARACTERS TO DISTINGUISH THREE SPECIES OF *SEBASTES*

Character	*S. fasciatus*	*S. mentella*	*S. norvegicus*
Symphyseal tubercle (beak)	Well-developed, long, sharp	Well-developed, long, sharp	Poorly-developed, blunt
Dorsal rays	13–14	14–15	15 (13–17)
Anal rays	7 (6–8)	8 (8–10)	8 (7–10)
Vertebrae	29–30	30–31	30–31
Extrinsic gas-bladder muscle			
Number of muscle heads	1 or 2	1 or 2	3 or 4
Overall passage of tendons	Between ribs 3 and 4 or 4 and 5	Between ribs 2 and 3	Between ribs 2 and 3 or between ribs 2 and 3, and 3 and 4
Appearance of muscle	Narrow	Narrow	Wide
Tendon branches	3 branches, attached to vertebrae 8, 9, and 10	Not branched, attached to vertebra 7	Multibranched (usually 6)
Color	Orange-red, green-black blotches on body, iridescent green flecks	Bright red, silvery between origin of pectorals and operculum, cheeks, in pelvic region	Orange-yellow or golden yellow

KEY TO GULF OF MAINE SCORPIONFISHES

1a. Lower 7–9 pectoral fin rays free for the outer half of their length; first dorsal spines 11–13; anal fin rays 3–5 . . . . ***Helicolenus dactylopterus dactylopterus***
1b. Lower pectoral rays connected by membranes nearly to their tips; first dorsal spines 14–16; anal fin rays 6–11 (*Sebastes*) **2**
2a. Dorsal fin rays 12–15, usually 13 or 14; anal fin rays 6–8, usually 7; most common dorsal–anal fin ray combinations 13–7 or 14–7 (Table 5); body color in life orange-red, frequently with dark blotches. . . ***Sebastes fasciatus***
2b. Dorsal fin rays 14 or 15; anal fin rays 7–10, usually 8 or more; body usually lacking dark blotches. **3**
3a. Symphyseal tubercle (beak) well developed, sharp . . . . ***S. mentella***
3b. Symphyseal tubercle (beak) poorly developed, blunt . . . . . . ***S. norvegicus***

TABLE 5
SECOND DORSAL AND ANAL FIN RAY COUNTS IN 235 GULF OF MAINE SPECIMENS OF *SEBASTES FASCIATUS*[a]

		Second dorsal fin rays				
		12	13	14	15	Total
Anal fin rays	6	1	1	4	—	6
	7	2	81	105	6	194
	8	1	13	21	—	35
Total		4	95	130	6	235

BLACKBELLY ROSEFISH / *Helicolenus dactylopterus dactylopterus* (Delaroche 1809) /

Bigelow and Schroeder 1953:437–438

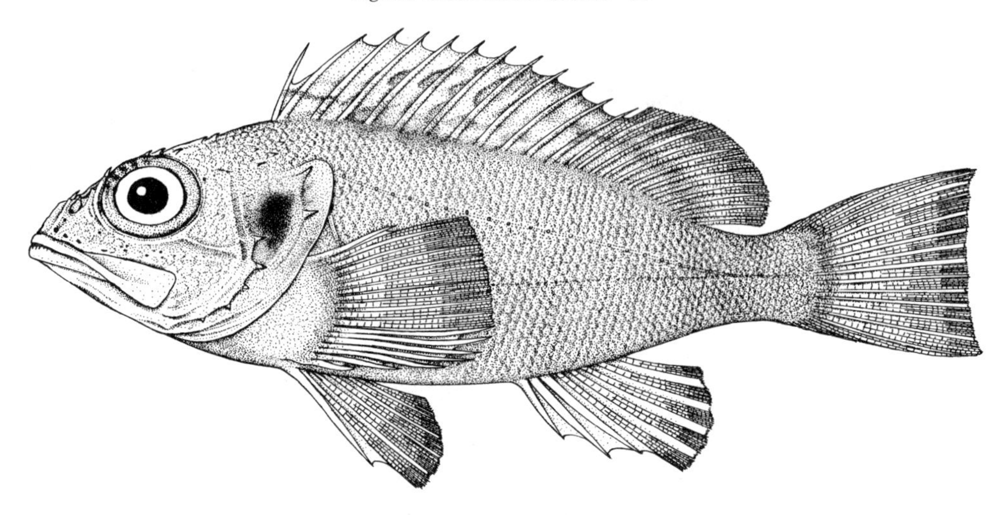

Figure 181. Blackbelly rosefish *Helicolenus dactylopterus dactylopterus.* Gulf of Maine. Drawn by K. H. Moore.

Description. Body deep, compressed (Fig. 181). Head large with conspicuous spines associated with many head bones; five preopercular spines most obvious, second from top largest. Upper profile low and curved with equal jaws; mouth terminal, slightly oblique, containing small villiform teeth on jaws, palatines, and vomer. Eyes large and oval. Dorsal fin with spiny and soft portions, no gap between them. Pectoral fins large, ventral seven to nine rays free for outer half of their length. Caudal fin small, emarginate. Ctenoid scales cover body and most of head. Lateral line complete, with tubed scales (Scott and Scott 1988).

Meristics. Dorsal fin rays XI–XIII (usually XII), 9–12 (usually 12); anal fin rays III, 3–5; pectoral fin rays 17–20; gill rakers 6–10 lower + 15–24 upper; pored lateral line scales usually 28 or 29; vertebrae 23–25, excluding urostyle, usually 24 (Eschmeyer 1969; Barsukov 1979; Kenchington 1986; Scott and Scott 1988).

Color. More or less vivid reddish or pale pinkish, usually with some brown and green along the back and irregular crossbands of darker or brighter scarlet on some specimens; the upper part of the sides marked with a sparse pattern of narrow, dusky vermiculations, roughly following the edges of the scales, and each gill cover generally has a leaden or dusky patch caused by the black inner surface of the gill chamber showing through the bone. The lower surface of the belly is without dark markings. All fins are pinkish, the spiny part of the dorsal mottled with white, and the soft portion of the dorsal, the pelvics, and the anal edged with white. The anterior roof of the mouth is silvery, the posterior jet black. The lining of the belly cavity is black, hence one of its common names.

Size. Maximum length 412 mm TL (White et al. 1998). The all-tackle game fish record is 1.55 kg for a fish caught in Langesund, Norway, in September 1998 (IGFA 2001).

Distinctions. This species closely resembles the redfish species in general form and in the outline and arrangement of the fins, but the lower seven to nine pectoral fin rays are free from the fin membrane along the outer half to third of their length, and the upper margin of the pectoral fins is nearly straight. Its caudal fin is relatively larger than that of redfishes; the space between the eyes is grooved whereas it is flat in redfishes; the eyes are closer together, the distance between them being less than half as great as the diameter of the eye (about two-thirds to three-quarters in redfishes); the maximum body depth is somewhat less than the distance from tip of the upper jaw to the upper corner of the gill cover; and the scales are larger. Blackbelly rosefish have only 23–25 vertebrae, redfishes 29–31, and there is no swim bladder. The back of the mouth, pharynx, gill chamber, and lining of body cavity are mostly black. Blackbelly rosefish can be distinguished electrophoretically from *Sebastes* species, 12 enzymes being diagnostic (McGlade et al. 1983; Johansen et al. 1993).

Taxonomic Note. Morphological differences between allopatric populations of blackbelly rosefish have been treated in several ways. Eschmeyer (1969) recognized four populations within *H. dactylopterus dactylopterus:* Atlantic coast of the United States, Gulf of Mexico, and Caribbean; northeastern Atlantic and Mediterranean; Gulf of Guinea; and South Africa, and considered the population in the southwestern Atlantic off Argentina and Uruguay to be another subspecies, *H. dactylopterus lahillei* Norman 1937. Barsukov (1979) recognized a total of six subspecies of *H. dactylopterus,* considering the western North Atlantic population to be a distinct subspecies, *H. dactylopterus maderensis* Goode and Bean 1895.

Habits. Catch records show that blackbelly rosefish sometimes occur in the middepths and sometimes on the bottom or close to it. They occur at depths of 219–225 m in the Fundian Channel (Scott and Scott 1988), and at depths of 124–682 m from the Gulf of Maine to Florida, deeper farther south (Bigelow and Schroeder). The largest catches from the Rockall Trough region in the northeastern Atlantic were at 650–950 m (Kelly et al. 1999). They appear to school in the Gulf of Maine.

Food. Blackbelly rosefish feed on a wide variety of benthic organisms including Foraminifera, polychaetes, amphipods, isopods, decapod crustaceans, ostracods, Tanaidacea, mollusks (bivalves and cephalopods), ophiuroids, tunicates (*Pyrosoma*), and fishes (Wertz 1977; MacPherson 1979, 1985; Merrett and Marshall 1981; Bowman et al. 2000). In the western North Atlantic, decapod crustaceans, particularly *Dichelopandalus leptocerus,* and salps were the most important components (by percent weight) of the diet of 80 specimens (Bowman et al. 2000). Off Namibia, the diet of 647 specimens was composed primarily of benthic crustaceans such as *Calocaris barnardi, Bathynectes piperitus,* and *Perosquilla armata capensis* (MacPherson 1985). Decapod crustaceans and fishes constituted the most important part of the diet of 808 *Helicolenus* from the western Mediterranean (MacPherson 1979). In the Ligurian Sea (Wertz 1977), 32 kinds of organisms were identified, the principal ones being amphipods (73.3%), an ophiuroid *Ophiacantha abyssicola* (68.1%), isopods (59.5%), and decapod crustaceans (53.4%). Off northwestern Africa, *O. abyssicola* was present in 92% of specimens examined (Merrett and Marshall 1981).

Blackbelly rosefish are daytime predators (9 A.M. to 3 P.M.), feeding for a short period and then remaining inactive. Daily ration was 0.16–0.68% of body weight. Levels of daily consumption were minimal in June because mean weight of ingested prey was lower than in other months (MacPherson 1985).

Predators. Predators of blackbelly rosefish include thorny skate, goosefish, and red, white, and silver hakes (Rountree 1999); conger eel in the Mid-Atlantic Bight (Levy et al. 1988); a deep-sea angler, *Cryptopsaras couesi,* in the northeast Atlantic (Minchin 1988); and wreckfish off South Carolina (G. Sedberry, pers. comm., 29 April 1998).

Parasites. The only parasite reported from blackbelly rosefish was a monogenetic trematode, *Microcotyle* sp., recovered from the gills of a specimen taken in the Rockall Trough, eastern Atlantic (Pascoe 1987).

Breeding Habits. Off the Carolinas, males mature by about 256–275 mm TL at 14 or 15 years; females by 215–289 mm at 50% of males were mature at 260 mm TL at age 15 or 16; 50% of females at 230 mm, age 13 (Kelly et al. 1999).

In the genus *Helicolenus,* fertilization is internal (Krefft 1961; Wourms 1991; White et al. 1998). The mode of reproduction in blackbelly rosefish is a zygaparous form of oviparity (Wourms 1991), intermediate between oviparous and viviparous. Free spermatozoa were noted from July through early December in resting ovaries of specimens caught of North and South Carolina, with a delay of 1–3 months before fertilization (White et al. 1998). The spermatozoa remain inside richly vascularized crypts during the long storage period (Muñoz et al. 2000). Early celled embryos were released January through April (White et al. 1998). Eggs were collected in the plankton on the outer shelf and slope off Uruguay and Buenos Aires. Identification of the eggs was confirmed by examination of adult *H. dactylopterus laheillei* caught by trawling at 104 m depth off Uruguay 2 months later. These females contained egg masses embedded in a clear, elastic, gelatinous matrix. The fertilized eggs were cultured until identifiable embryos developed. Planktonic eggs and females bearing fertilized egg masses were collected in September and November, respectively, off Uruguay, indicating a reproductive period of at least 2 months (Sánchez and Acha 1988).

Early Life History. Ripe ovarian eggs are nonspherical with long and short axis measuring 0.91 mm (range 0.84–1.04) and 0.78 mm (range 0.18–0.22 mm) and a narrow but distinct perivitelline space. Planktonic eggs are nonspherical (long axis 0.96, short axis 0.78) with segmented yolk and an oil globule 0.20 mm. Planktonic eggs collected off Uruguay contained embryos at the tail-bud formation stage. Eggs from trawl-captured adults incubated at 14°C showed blastodisc formation after 6 h; gastrulation after 12 h; blastopore closure, head development, and Kuppfer's vesicle formation after 42 h; and tail separation, optic vesicle, brain differentiation, and 16 somites by 60 h. At this point, the embryos died. Eggs at the tail-bud stage had been found floating free in the plankton so it is possible that eggs developed to this stage in the female, and were then released (Sánchez and Acha 1988).

Larvae hatch at about 2.8 mm; flexion occurs at 6–7.9 mm. The body is moderately slender with a short gut. Preanal length increases from 49% SL in preflexion to 58% SL in postflexion. The third anal and first dorsal spines start as rays and change into spines in the early juvenile stage. There is a mass of spongy tissue appearing at 4.0 mm in the region of the spiny dorsal fin and persisting at least through 18 mm. There are 24 or 25 myomeres and vertebrae. Pigmentation consists of spots on top of the head and lower jaw, dorsolateral pigment on the gut, a few spots on the trunk above the pectoral base, and a solidly pigmented medial surface with a few spots inside the distal edge of fin and on the ray bases and a few spots on the ventral midline anterior to the caudal fin. Blackbelly rosefish larvae can be distinguished from other scorpaenid larvae by the lack of dense pigmentation on the pectoral fins and by

Benthic juveniles of 60–85 mm SL have been collected off Uruguay, but size at first settlement is unknown. Juveniles are red with dorsal and lateral body surfaces mottled with brown; pectoral fin base and dorsal fin membranes have dark blotches and black peritoneum. They can be distinguished from juvenile *Sebastes* by presence of free pectoral fin rays, numbers of dorsal spines and anal rays, shorter snout to anus distance, and larger dorsal fin spines and rays. The longest preopercular spine in *Helicolenus* is the second, whereas the third is the longest in *Sebastes* (Sánchez and Acha 1988).

Age and Growth. Based on otoliths, ages ranged from 7 to 30 years off North and South Carolina (White et al. 1998) and up to 43 years for males and 37 years for females in the northeastern Atlantic (Kelly et al. 1999). Mean lengths increase with increasing age for the first 25 years and then level off. White et al. (1998) were unable to fit the von Bertalanffy growth equation to their mean length at age data but Kelly et al. (1999) did so for their northeastern Atlantic data. Males grow significantly faster than females, reaching greater lengths and weights, L_∞ = 372 mm, W_∞ = 747 g vs. L_∞ = 310 mm, W_∞ = 700 g.

General Range. The subspecies *H. dactylopterus dactylopterus* is found on both sides of the North Atlantic. In the western Atlantic from the Scotian Shelf south to Florida, Venezuela, and Argentina (Eschmeyer 1969); from Norway to the southern tip of Africa in the eastern Atlantic; and in the Mediterranean Sea.

Occurrence in the Gulf of Maine. This fish is probably generally distributed over the outer part of the continental shelf and along the upper part of the continental slope as far east as the offing of Nantucket, for they have been reported from 27 stations between longitude 72′ and a few miles east of longitude 70′, including one catch of more than 100. One was trawled on the eastern edge of Georges Bank, at 320 m, 6 October 1929. Subsequent records from the Gulf of Maine are of 24 fish trawled at five stations south of Nantucket in 124–439 m; one brought in by a trawl from 220 m on the northern slope of Georges Bank; and a catch of about 136 kg made in the southeastern part of the basin of the Gulf, at 216–252 m. This last catch indicates that blackbelly rosefish may occasionally enter the Gulf via the deep channel between Georges and Browns banks. Six collections of blackbelly rosefish made by NMFS vessels north of 42°20′ N were reported by Musick (1966).

Importance. Although blackbelly rosefish are captured along with species of *Sebastes* in Canada, they are not large enough to have a directed fishery (McGlade et al. 1983). They are too scarce in the Gulf of Maine to be of commercial importance but there is a relatively new, small fishery for them off the southeast coast of United States, predominantly by the bottom longline and vertical line fisheries (White et al. 1998). They are also commercially important in the eastern Atlantic from the Rockall Trough region (Kelly et al. 1999), the Azores (Esteves et al. 1997), and off Namibia.

ACADIAN REDFISH / *Sebastes fasciatus* Storer 1854 /

Ocean Perch, Redfish, Labrador Redfish, Beaked Redfish / Bigelow and Schroeder 1953:430–437 (as *Sebastes marinus*)

Description. Perchlike in general appearance, moderately compressed, about a third as deep as long, with a large bony head; body tapering back from shoulders to a moderately slender caudal peduncle (Fig. 182). Dorsal profile of head slightly concave, mouth large, very oblique, gaping to below eyes, lower jaw projecting beyond upper. A bony knob at tip of lower jaw fitting into a corresponding notch in upper jaw. Knob may be very well developed and sharp, sometimes referred to as the schnable or beak. Both jaws armed with many small teeth. Eyes very large, set high. Sides of head armed with spines, most prominent of which are two near rear angle of each gill cover, and a series of five confluent ones on each cheek. Third preopercular spine directed backward and upward or backward and downward (Litvinenko 1979). These spines, with a ridge behind and above each eye socket, give the head a characteristic bony appearance. Gill openings very wide, with pointed gill covers. One continuous dorsal fin running from nape of neck to caudal peduncle; spiny part considerably longer than soft part, but the latter higher than the former. Anal fin, consisting of three graduated spines and seven or eight longer rays, shorter than soft portion of dorsal, under which it stands. Pectoral fins very large; smaller pelvic fins originate below them. Caudal fin noticeably small, its rear edge moderately concave, with angular corners. Both head and body clad with scales of moderate size. About 60–70 oblique rows of scales from gill opening to origin of caudal fin, just below lateral line. A detailed description of 30 specimens from Eastport, Maine, was presented by Litvenenko (1979) in the description of a new subspecies, *Sebastes fasciatus kellyi*.

Internal morphology of the extrinsic swim bladder muscle (EGM) passage patterns between ventral ribs show that in Acadian redfish, the main EGM passes between ventral ribs three and four in 94% of the fish and between ribs four and five in the remaining 6% (Ni 1981a).

Meristics. First dorsal fin rays XIV–XVI, 12–16 (the last two rays counted as one), most often 13 or 14; anal fin rays III, 6–8, usually 7, the last ray double; pectoral fin rays 18–20, usually 19; gill rakers on first arch 34–38, usually 36 or more; vertebrae 30 or 31, usually 30 (including the urostyle) (Litvinenko 1979; Ni 1984).

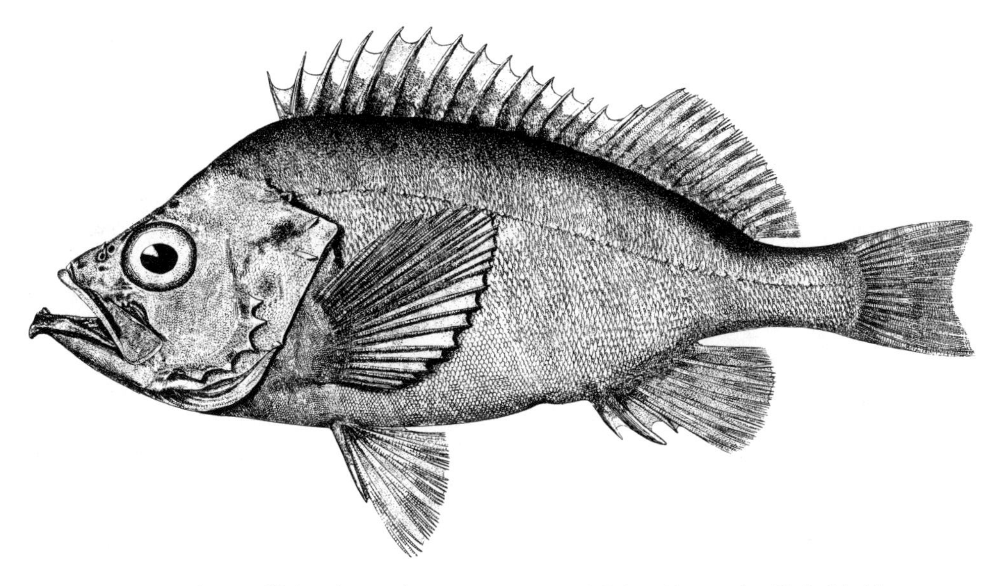

Figure 182. Acadian redfish *Sebastes fasciatus*. Eastport, Maine. Drawn by H. L. Todd.

Color. Acadian redfish are reddish orange to flame red, occasionally grayish red or brownish red, with the belly a paler red that fades to white after death. The black eyes contrast vividly with the brightly colored body. Medium-sized fish usually have a dusky blotch on each gill cover and several irregularly broken greenish black patches along the back and below the dorsal fin. All fins are red, but the pelvics and anal are more intense (Scott and Scott 1988).

Size. Maximum size of Gulf of Maine Acadian redfish is about 45.7 cm and 1.36 kg (Sandeman 1969). Inshore individuals average 20–30.5 cm, but specimens 45–50 cm have been taken in the Gulf of Maine (Mayo et al. 1983). Larger ones occur off Newfoundland, in the eastern North Atlantic, and in Arctic seas. Maximum weight reported is 11.2 kg for a specimen collected off Canada (skeleton in Royal Ontario Museum [Scott and Scott 1988]).

Distinctions. Acadian redfish resemble cunner, tautog, and sea bass in union of spiny and soft portions of the dorsal into a single long fin and in the general perchlike conformation. They differ from the first two by color, much larger mouth, spiny head, large eyes, more slender caudal peduncle, and larger pectorals; and from sea bass by the large spiny head, the shape and small size of the caudal fin, and the fact that the anal fin and the soft portion of the dorsal are relatively much lower. The brilliant red color is a conspicuous field mark.

Acadian redfish can be distinguished from blackbelly rosefish, *Helicolenus dactylopterus,* by the pale-colored mouth, pharynx, gill chamber, and body cavity, presence of a swim bladder, by having 14 or more dorsal spines and by having the pectoral fin rays united to the margin of the fin.

Habits. A summary of habitat characteristics of redfishes in

Pikanowski et al. (1999). Acadian redfish are found chiefly on rocky or hard grounds or on mud, but seldom on sand. On the Scotian Shelf and in the Bay of Fundy they are strongly associated with the fine-grained, clay-silt bottom (Scott 1982a). Their depth range on the bottom is from within 1 m of the tide line down to 366 m, but they have been collected as deep as 592 m (Powers in Scott and Scott 1988). They are most numerous in the Gulf of Maine from 128 to 366 m. The proportion of larger fish increases with depth (Perlmutter 1953; Atkinson 1984).

Acadian redfish in the Gulf of Maine prefer temperatures of 2.8°–8.3°C (Kelly et al. 1972), in the Gulf of St. Lawrence, 7°C (Atkinson 1984). On the Scotian shelf, preferred temperatures and salinities were 5°–9°C and 33–34 ppt. Temperature and salinity ranges at which they were found were 0°–13°C and 31–34 ppt (Scott 1982b). These preferences were given for *Sebastes* sp., and probably refer to Acadian redfish and deepwater redfish (*S. mentella*).

Food. The most extensive information on the food of redfishes (466 stomachs examined [Bowman and Michaels 1984]) is based on a mix of species of redfishes although most of the information probably refers to *S. fasciatus.* Most of the diet (90% by weight) is composed of free-swimming shrimps, mostly a euphausiid *Meganyctiphanes norvegica* (55%) and decapods such as *Pasiphaea multidentata, Dichelopandalus leptocercus,* and *Pandalus borealis.* Acadian redfish are believed to rise off the bottom to feed at night (Scott and Scott 1988). *Meganyctiphanes* is the primary food for all sizes of adult redfish throughout all seasons. Incidental food includes hyperiid amphipods, mysids, copepods, and fishes. Fishes identified from redfish stomachs include pearlside, *Maurolicus,* silver hake, longhorn sculpin, wrymouth, and plaice (Bowman and Michaels 1984; Bowman et al. 2000).

Stomach content analysis of larval redfish ranging from 9

copepods, euphausiids, fish eggs, and invertebrate eggs. Over 50% of the diet of larvae in the size-group 9–13 and 14–18 mm in July consisted of the larval copepods *Oithona similis* and *Pseudocalanus minutus* and invertebrate eggs. Larvae up to 44 mm ate *Centropages typicus* and *Calanus finmarchicus*. Larvae 44–48 mm ate the euphausiid *Thysanoessa inermis*. Redfish larvae appear to be opportunistic feeders on what is available, but they avoid crab zoea. Larvae do not appear to feed only in the daytime because many taken in both day and night samples had full stomachs. There were no discernible vertical migration patterns in larval collections; but stomachs of fish from 60 to 110 m contained mainly surface-dwelling zooplankton and many were empty, so the larvae appeared not to be feeding at this depth. The only identifiable stomach contents of two juvenile Acadian redfish from the Gulf of Maine (11–15 cm TL) were copepods (Bowman et al. 1987).

Predators. Acadian redfish are eaten by larger predacious fishes such as little skate, cod, pollock, silver hake, white hake, goosefish, Acadian redfish, bluefish, and Atlantic wolffish, of which cod and white hake are the most frequent predators (Maurer and Bowman 1975; Bowman and Michaels 1984; Rountree 1999).

Parasites. Fifty Acadian redfish from the Gulf of St. Lawrence were host to 22 species of parasites (Moran et al. 1996): one Coccidia, two Myxosporea, nine Digenea, two Cestoda, five Nematoda, and three Crustacea. Eight of these were considered common: *Ceratomyxa macrospora, Myxidium incurvatum, Derogenes varicus, Lecithophyllum botryophorum, Podocotyle reflexa, Anisakis simplex,* a larval contracaecine nematode, and *Hysterothylacium aduncum*. Generally the parasites of Acadian redfish were similar to those of 50 *S. mentella* from the same study. There were significant differences for eight of 25 parasite taxa; in seven of these cases, parasite prevalence or abundance was higher in Acadian redfish than in *S. mentella*. Only the copepod *Sphyrion lumpi* was significantly more abundant in *S. mentella*. These differences were attributed to ecological differences between the host species, in particular depth differences. Acadian redfish were taken on average 100 m shallower than *S. mentella* (mean depths of 227 vs. 331 m). Earlier studies from other areas in the northwest Atlantic Ocean (Bourgeois and Ni 1984; Scott 1988) found similar parasite faunas. Some 87% of Acadian redfish collected by Bourgeois and Ni (1984) were parasitized. The nematode *Contracaecum* sp. was the most prevalent parasite (40% occurrence in all infected fish). The copepod parasite, *Sphyrion lumpi,* has also been found on redfish in the Gulf of Maine and off the coast of Canada (Nigrelli and Firth 1939; Templeman and Squires 1960). Larvae collected in September had cestode larvae in their guts, but none from July contained cestodes (Marak 1973).

Breeding Habits. Redfishes are ovoviviparous; eggs develop and hatch within the oviduct of the mother, and the number produced by large females is 25,000–50,000 yearly, but only about 15,000–20,000 are extruded as larvae (Kelly et al. 1972).

Age at sexual maturity varied in Acadian redfish from the Gulf of Maine–Georges Bank region. Some males were mature at age 2 and females at age 3, but others did not mature until age 10. The median age at maturity was 5.5 for both males and females (O'Brien et al. 1993). The proportion of mature individuals within a given age-group increased with size. Median length at maturity for female and male Acadian redfish from the Gulf of Maine was 22.3 and 20.9 cm, respectively. Estimates of both median age and size at maturity were considerably lower for redfish from shallow inshore areas than for those from offshore. In offshore males and females and inshore females, size seemed to be the main determinant of maturity, whereas for inshore males, age was the driving factor (Mayo et al. 1990). Morse (1979) reported length at 50% maturity for pooled redfish from the Gulf of Maine and Scotian Shelf as 21.5 cm for males and 23.6 cm for females. Perlmutter and Clarke (1949) indicated median sizes at maturity of 23 cm for males and 25 cm for females in pooled redfish from the Gulf of Maine. In a sample taken in the Gulf of Maine on *Albatross IV* on 26 April 1994, 76 males ranged from 16.5 to 29.5 cm TL, mean 24.8, while 24 females that extruded young while we were measuring them ranged from 26.0 to 32.5 cm TL, mean 29.0. Criteria for determining the state of sexual maturation by visual examination of gonads were given by Mayo et al. (1990) for Gulf of Maine redfish.

There is a geographic cline in size at maturity (Ni and Sandeman 1984) with a decreasing trend from Grand Banks to the Nova Scotian Shelf. The size at which 50% of beaked redfishes (including both *S. fasciatus* and *S. mentella*) reached sexual maturity was 18.5 cm FL for males and 29.5 cm FL for females.

Nothing is known about breeding behavior, but fertilization is internal. Copulation most likely occurs from October to January, but fertilization is delayed until February to April for northeast Atlantic redfish (Ni and Templeman 1985).

Larval extrusion has two peak periods in the Gulf of Maine, an early period centered at mid to late March and a later one in early June. Redfish larvae collected before June occurred in eastern regions of the Gulf of Maine close to the Fundian Channel and on the southeast flank of Georges Bank and may have originated on the Scotian Shelf. Larvae taken after June were confined to central and western parts of the Gulf of Maine and were produced by the adult population resident in the Gulf. This suggested a peak time of larval extrusion between late May and early June (Mayo et al. 1990). As we were measuring redfish collected on board *Albatross IV* on 26 April 1994, half of the females (26.0–32.5 cm TL) extruded swimming larvae. Females with well-developed eggs and males with well-developed milt are taken commonly by mid-May, both within the Gulf and on Georges Bank. The earliest that larvae were taken in any numbers in tow nets was 8 July. Production of young continues through July and August, for *Albatross II* trawled many gravid females in the central basin of the Gulf in July, one containing about 20,000 young, 6–7 mm long, practically ready for birth. Newly extruded larvae (6.5–7 mm) have been collected in one part of the Gulf or another in July and August, but it is not likely that many young are produced after the first week in September.

MARMAP surveys showed larval redfish first appearing in the Gulf of Maine in April–May, peak abundance in July–August, and few occurring after September (Morse et al. 1987: Fig. 18). In southern Newfoundland, spawning (extrusion) occurs from March to July, but the peak is in May and June (Ni and Templeman 1985).

Early Life History. Eggs are incubated and hatch in the females' body with the yolk mostly absorbed, the mouth already formed, and first traces of the caudal rays already visible. Larvae are extruded at 5.8 mm SL. The body is slender, with a short gut. Flexion occurs at 8.5–10 mm. At 12 mm, dorsal and anal fin rays appear, pelvic fins are visible, and head spines are prominent. All but the very youngest larvae are recognizable as redfish by their large spiny heads, large eyes, short tapering bodies, very short digestive tract, and the presence of both a dorsal and a ventral row of postanal pigment cells. They resemble mackerel larvae but lack teeth at sizes smaller than 9 mm, are more slender, and have a shorter preanal length and opercular spines (Fahay 1983).

It is very difficult to distinguish Acadian from deepwater redfish larvae. Descriptions of larval Acadian redfish (Moser et al. 1977; Fahay 1983) were based on specimens taken in the Gulf of Maine that were presumed to be Acadian redfish because of their collection site. Preextrusion larvae collected from adult females of known parentage from southern Newfoundland were described by Penney (1985). Acadian redfish had dorsal and ventral body pigmentation that tended to begin and end more anteriorly and an overall longer length of pigmentation pattern compared to that of deepwater redfish. Acadian redfish had more head pigment but it was diffuse and amorphous compared to deepwater redfish, whose head pigment consisted of distinct and expanded melanophores. Morphometric differences between the two species include a more slender body, a longer snout, and a greater pectoral fin base depth for Acadian redfish. Acadian redfish were also smaller than golden redfish (7.34 mm TL vs. 7.89 mm TL). Nevertheless, there was a great deal of overlap among characteristics; and larvae of the two species could only be classified correctly 95% of the time by using discriminant analysis on a suite of morphometric, meristic, and pigmentation variables (Penney 1985).

Redfish larvae are ready to go to the bottom when they are about 25–30 mm long because larvae larger than 27 mm have not been taken in tow nets, whereas larvae of 38 mm and upward are plentiful on the bottom, both in the Bay of Fundy and in deep water off southern New England. Failure to take any young redfish in tow nets off Massachusetts Bay in November or anywhere in the Gulf in winter is evidence that their descent to the bottom takes place early in their first autumn.

Age and Growth. Acadian redfish are a slow-growing, long-lived species. Maximum age recorded is 58 years (Penttila et al. 1989). Both scales and otoliths have been used to determine age. Whereas Europeans prefer scales, North Americans prefer otoliths. Annual growth marks on the otolith include a white opaque zone representing fast summer growth and a darker hyaline zone representing slow winter growth (Gifford and Crawford 1988). These growth marks have been validated as annual events (Mayo et al. 1981).

Larval redfishes (all three species of redfishes occur on the Flemish Cap) were aged by otolith increment analysis (Penney and Evans 1985). A few incomplete marks form after hatching within the female's body, but these are irregular and do not represent daily marks. Only after extrusion do daily increments form on the otoliths. Larvae grow slowly for 10–15 days after extrusion and then rapidly for 60–70 days before slowing as they enter the pelagic juvenile stage. Mean daily growth rate varied between years being 0.16 mm·day^{-1} in 1980 and 0.11 mm·day^{-1} in 1981.

Growth curves of adult Acadian redfish from the Gulf of Maine show that males and females grow at about the same rate for the first 6 years, after which females grow faster than males (Fig. 183).

General Range. Acadian redfish commonly occur from Iceland to New Jersey (Atkinson 1987; Scott and Scott 1988). The southernmost record is from Virginia, 37°38′ N, 74°15′ W at 192 m (reported as *Sebastes marinus* by Davis and Joseph 1964). They are most common from the Newfoundland Banks to Georges Bank and the Gulf of Maine, including the Nova Scotian Banks, Flemish Cap, and the south and northeast slope of the Grand Banks (Atkinson 1987).

Occurrence in the Gulf of Maine. Acadian redfish are most common in the relatively deep waters of the Gulf of Maine, on the northern and southeastern slopes of Georges Bank to depths of 400 m, and on Browns Bank (Map 22).

Seasonal Migrations and Movements. Redfish undergo extensive seasonal migrations in the Gulf of St. Lawrence, as indicated by their distribution patterns (Atkinson 1984). They occurred throughout the deepwater areas of the Gulf during summer and early autumn, then migrated southward and eastward to become concentrated in the southeastern region of the Gulf during winter. Juvenile fish preferred shallower depths than adults (181–260 m vs. 221–300 m and greater). Golden redfish were excluded from this survey; Acadian and deepwater redfish were not separated to species, but were combined as "beaked redfish."

In the Gulf of Maine, vertical range is correlated with temperature. Thus, redfish are found deeper than 30 m during the warm half of the year in the southwestern Gulf, but have been known to run up into Gloucester Harbor in numbers in winter.

Importance. The fishery for redfishes peaked at 60,000 mt in 1942 and there has been a downward trend in landings since then, reaching 322 mt in 1996, the lowest recorded level since the directed fishery began in the early 1930s (Mayo 1998c). The Gulf of Maine population declined by about 99% from

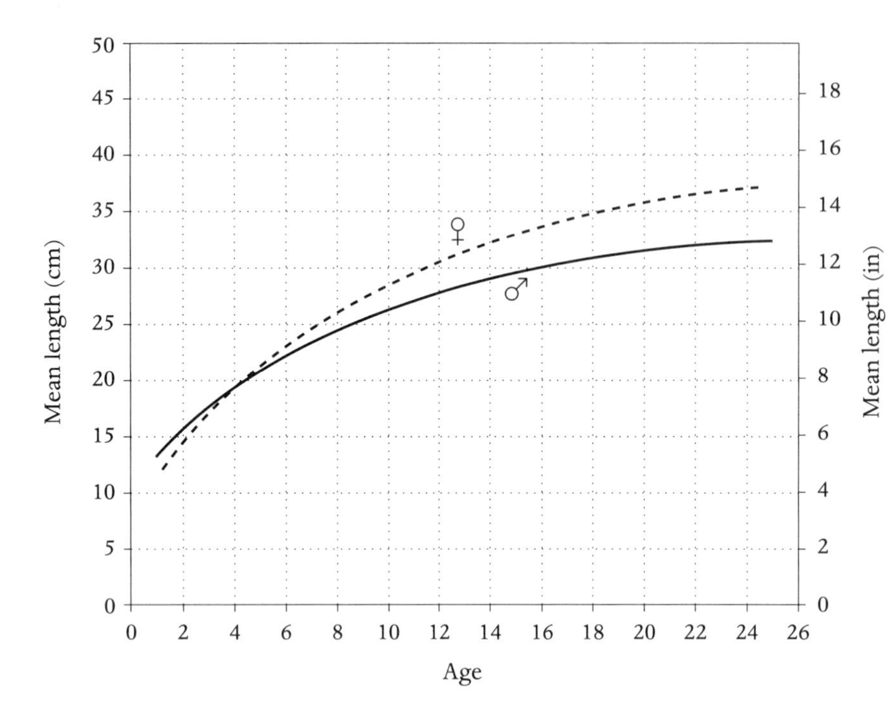

Figure 183. Length-age relationship of Gulf of Maine Acadian redfish *Sebastes fasciatus.* (From Penttila et al. 1989:21.)

The fishery has been supported by a few strong year-classes, two of which occurred in 1971 and 1978 and a few others that have recruited more recently. In general, commercial catches have declined since the 1950s with a small increase in 1971 and 1979. The NEFSC autumn bottom trawl survey biomass index declined from 40.4 kg per tow in 1968 to an average of 3.8 kg per tow in 1982–1984 (Mayo 1998c). This index increased to 10.0 kg per tow during 1990–1993, decreased again in 1994 and 1995 but increased substantially to 30.6 kg per tow in 1996 (Mayo 1998c; Fig. 184).

Stock biomass increased steadily through the mid-1980s, substantially so based on 1996 data (Mayo 1998c). However, most redfish supporting the recent increase are small, immature fish produced in the early 1990s, and have yet to realize their full growth and reproductive potential. If fishing mortality increases significantly in the near-term, stock biomass may decline to the low levels observed during the 1980s. For recovery to continue, catches must remain low. The stock is fully exploited at present (Mayo 1998c). The recreational catch is insignificant and there is no directed sport fishery.

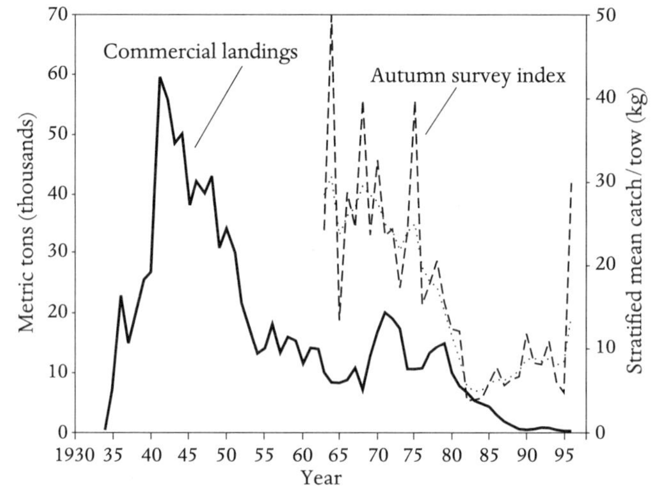

Figure 184. Commercial landings and autumn survey index, Acadian redfish *Sebastes fasciatus,* Gulf of Maine–Georges Bank, 1932–1995. (From Mayo 1998c.)

ACKNOWLEDGMENTS. A draft of the scorpaenid section was reviewed by Thomas A. Munroe, and drafts of the *Helicolenus* account were reviewed by George Sedberry and David Wyanski.

SEA ROBINS. FAMILY TRIGLIDAE

GRACE KLEIN-MACPHEE AND RICHARD S. MCBRIDE

Sea robins and their European relatives, gurnards, suggest sculpins in their broad heads, slender bodies, large fanlike pectoral fins, separate dorsal fins (spiny and soft-rayed), and location of their pelvic fins under the pectorals. Sea robins differ from sculpins by encasement of their entire head in bony plates; their smaller mouth; flat, depressed dorsal snout profile; large pelvic fins, and three modified pectoral rays. Sea robins also resemble flying gurnard except that gurnard do not have three separate pectoral rays; the first two dorsal fin spines are separate; the bony armor covering the head reaches back beyond the origin of the dorsal fin to end in a stout spine; each gill cover is extended backward as a stout spine to the axil of the pectoral fin; and the scales are much larger, many armed with a stout spine.

Two subfamilies are currently recognized although some researchers consider each worthy of family status. The subfamily Triglinae includes ten genera and about 70 species (Nelson 1994), of which two species of *Prionotus* reach the Gulf of Maine. Keys to western Atlantic species of Triglinae are provided by Russell et al. (1992). Armored sea robins, subfamily Peristediinae, include four genera and about 30 species (Nelson 1994). They are close relatives of sea robins but differ from the latter in four very noticeable ways: (1) entire body enclosed in an armor of bony plates, each plate with a spine (except abdominal plates); (2) only two instead of three lower pectoral fin rays modified into separate feelers; (3) each side of the front of the skull projects forward as a long, flat process, so that the snout appears to be double; and (4) two long barbels on the chin. Armored sea robins live in deeper waters than species of Triglinae. One species of *Peristedion* occurs in the Gulf of Maine.

KEY TO GULF OF MAINE SEA ROBINS

1a. Front of snout so deeply concave that it seems to be double when seen from above; two long barbels on chin . . . . **Armored sea robin**

1b. Front of snout only slightly concave as seen from above; no barbels on chin **2**

2a. Pectoral fin short, ending between origin and center of anal fin base, with two broad dusky blotches; no prominent longitudinal stripe on side of body. . . . . **Northern sea robin**

2b. Pectoral fin long, ending near or beyond distal end of anal fin base, with only one broad dusky blotch; a prominent longitudinal dark brown stripe on each side of body. . . . . **Striped sea robin**

ARMORED SEA ROBIN / *Peristedion miniatum* Goode 1880 / Bigelow and Schroeder 1953:471–472

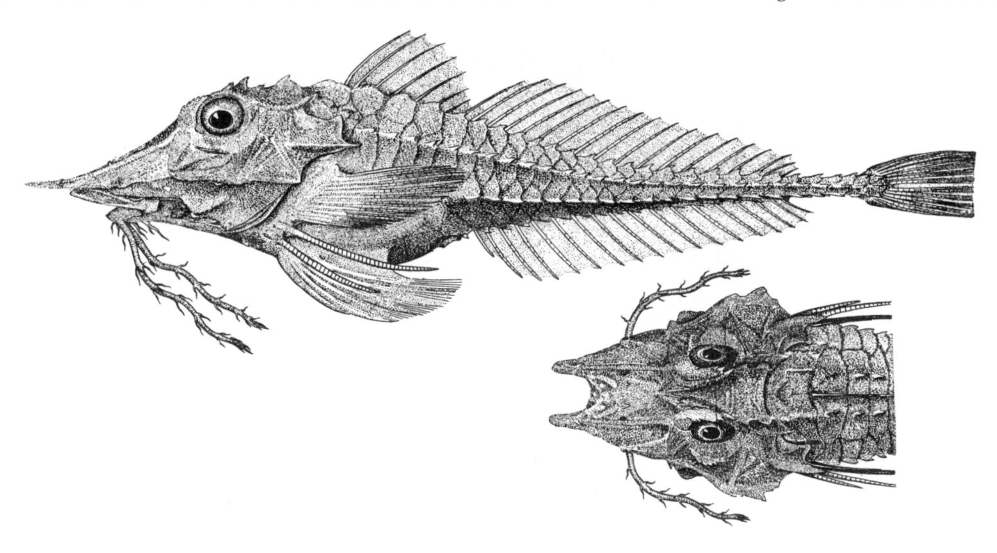

Figure 185. Armored sea robin *Peristedion miniatum*. Side view (above) and top view of head (below). Slope off Martha's Vineyard. Drawn by H. L. Todd.

Description. Body moderately low and broad, head short and depressed, snout long and narrow, mouth small, eyes small, a number of ridges and spines on the head, opercular spine very long (Fig. 185). Lower jaw with eight to ten barbels; two elongate barbels with short side branches at each side of mouth, extending to pectoral fin base. Front of head with two projections, rostral spines short and narrow, usually shorter than eye diameter. Pectoral fins relatively short, two lower rays forming feelers. Body tapers to very narrow caudal peduncle. Caudal fin rather small. Body covered with large, bony plates, four rows on each side from vent rearward, and an equal number of rows of thornlike spines, close-set and directed rearward.

Meristics. Dorsal fin rays VII or VIII, 17 or 18; anal fin rays 17–19; pectoral fin rays 10 + 2 free rays; gill rakers on the lower limb of the first arch 16–18; vertebrae 33 (Teague 1961; Miller and Jorgenson 1973).

Color. Bright crimson above and below with a reddish-black border on the dorsal fins.

Size. Average size is 30 cm (Robins and Ray 1986), maximum 33–35.5 cm.

Distinctions. Armored sea robin can be distinguished from

covering the body, presence of two rostral exsertions, and two long barbels on each side of their mouth.

Habits. Armored sea robin are bottom fish recorded from 64 to 210 m, maximum depth 421 m. They are associated with the edge of the continental shelf at temperatures of 6.7°–7.2°C.

Food. They feed on crustaceans (Bowman et al. 2000), decapods (38.8% by weight), especially *Munida* sp., amphipods (14.9%), and stomatopods (10.1%), largely *Heterosquilla armata*.

Predators. Found in the stomach of a spiny dogfish (Rountree 1999).

General Range. Known from the Canadian Atlantic to the northern Gulf of Mexico (Briggs 1958) and Brazil. They are taken on the outer part of the continental shelf and upper part of the continental slope from Georges Bank to Charleston, S.C., 32°24′ N, 78°44′ W. The northernmost record is off Sable Island on the Scotian Shelf (Scott and Scott 1988).

Occurrence in the Gulf of Maine. They are sometimes trawled on the southwestern part of Georges Bank and are fairly common south of Nantucket at depths of 37–103 m.

NORTHERN SEA ROBIN / *Prionotus carolinus* (Linnaeus 1771) /

Common Sea Robin, Sea Robin, Robin / Bigelow and Schroeder 1953:467–470

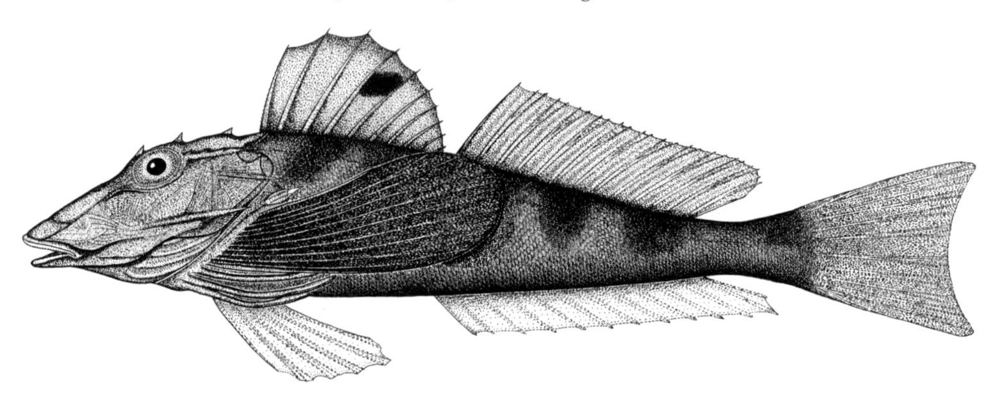

Figure 186. Northern sea robin *Prionotus carolinus.* Cape Charles, Va., 150 mm. Drawn by I. L. Towers.

Description. Head broad, enclosed in bony plates, body slender and tapering, triangular in cross section (Fig. 186). Anterior margin of upper jaw slightly concave when viewed from above; teeth small in villiform bands in each jaw (Hildebrand and Schroeder 1928). Lachrymal plate moderately projecting, moderately serrate; head plates rough; one sharp spine on each eye pointing backward; a spine on either side of neck and one on each shoulder above pectoral fin base. Head spines become reduced with growth (Ginsburg 1950). Spiny and soft-rayed portions of dorsal fin separate but close together at bases, almost in contact. Spiny dorsal rounded in outline and higher than soft dorsal; soft dorsal much longer. Anal fin similar in outline and size to soft dorsal, located under it. Pectoral fins large, fanlike, partly overlapping anal and second dorsal fins when laid back. Lower three pectoral rays separated and modified into fingerlike feelers with slightly dilated tips. Caudal fin of moderate size, its margin slightly concave. Chest and body covered with scales. Lateral line high on body.

Meristics. Dorsal fin rays X, rarely IX, 13 or 14; anal fin rays 11–13 (usually 12); pectoral fin rays 13–15 (usually 14), plus 3 separate rays modified as feelers; lateral line scales about 61; gill rakers 10–14 on lower limb of first arch; vertebrae 26 (Ginsburg 1950; Teague 1951; Miller and Jorgenson 1973).

Color. The body is usually grayish or reddish brown above with five, dark saddlelike blotches along the back, and dirty white or pale yellow below. Dorsal fins are grayish marked with pale spots and stripes with a distinctive black spot between the fourth and fifth spines. The caudal fin is uniform grayish or brown, the anal fin plain brown, the pelvic fins plain yellow to brown, the pectoral fins yellow or orange, strikingly marked with two broad, dusky blotches, one of them crossing its outer third. The pectoral filaments are orange. The color pattern is variable and changes with growth (Ginsburg 1950).

Size. Maximum length is 45 cm TL based on NEFSC groundfish surveys, average less than 20 cm (Wong 1968; McBride 1994).

Distinctions. Northern sea robin resemble striped sea robin, but can be distinguished by the absence of the two dark stripes on the sides of the body; by having a rounded snout; a shorter pectoral fin reaching back to the fifth or sixth dorsal fin ray instead of the ninth or tenth as in striped sea robin; usually having one more anal fin ray (12 vs. 11) and one more pectoral fin ray (14 vs. 13); pectoral feelers with broader tips, and a concave caudal fin rather than a square one. They are distinguished from armored sea robins by having no armored plates on the body, lacking fleshy barbels on the lower jaw, having three instead of two free pectoral fin rays, and by not having two sharp anterior projections on the snout.

Habits. Northern sea robin are a coastal species closely associated with the bottom and occurring from the tide line to the edge of the continental shelf (McBride and Able 1994; McBride et al. 1998). In the Chesapeake Bay area they are associated with deeper flats and channel edges (Musick 1972). They occur at temperatures of 4°–28°C in continental shelf habitats (McBride et al. 1998) and have been reported from Long Island Sound at temperatures as low as 1.4°C (Richards et al. 1979). They have been found in salinities from 5 ppt (Musick 1972) to 32.3 ppt (Marshall 1946; Richards and Castagna 1970) but generally above 20 ppt (McBride and Able 1994). They have been collected as deep as 413 m but their average depth in NEFSC surveys was 78 m in the spring and 31 m in fall (McBride 1994)

Although closely associated with the bottom, sea robins swim actively, usually with their pectoral fins held close to the body. When on the bottom they often rest with their fanlike pectoral fins spread. They can bury themselves with only the top of the head and eyes protruding (Bardach and Case 1965). Northern sea robin show no evidence of schooling, although occasionally large catches are made in trawl surveys (Roberts 1978).

Sea robins produce sound by vibrating the large swim bladder with lateral intrinsic muscles (Fish and Mowbray 1970; Evans 1973). Occipital nerve roots carry sonic motor axons to the drumming muscles from the brain (Bass et al. 1987). Details of muscle and swim bladder functions are given by Fish (1954) and Evans (1973); and of the muscle innervation and brain neuromotor connection by Bass et al. (1987) and Bass (1990).

Sounds produced are vibrant squawks, barks, growls, and a series of rapid clucks. The volume of the outburst is directly proportional to the degree of stimulation (Fish 1954). In the field, sound level may be correlated with the abundance of sea robins (Moulton 1958). Loud barks were heard in exchanges with sounds of other species (toadfish), probably in competition for food or territory. Sounds are used for defense, warning, and courtship. At the height of the spawning season, sounds were more frequent at dusk (Fish and Mowbray 1970).

Food. Stomach contents of 41 northern sea robin (Bowman et al. 2000) indicate that the three major components of the diet are decapod crustaceans (51.5% by weight), polychaetes (15.2%), and fishes (14.7%). The most important decapods are polychaetes, the family Spionidae. Other prey include bivalves such as *Tellina agilis,* cumaceans such as *Oxyurostylis smithi,* and isopods such as *Unciola irrorata.* Other studies have described the diet of adult northern sea robin to consist principally of invertebrates, particularly mysids, amphipods, decapods, and bivalve mollusks (Marshall 1946; Richards 1963b; Richards et al. 1979). Other diet items tabulated in these studies include annelids, cumaceans, gastropods, pelecypods, fishes, and, for juvenile sea robins, copepods.

Sea robins often rest on the seafloor on the tips of their free fin rays, moving them like fingers and digging into the bottom to dislodge invertebrate prey (Morrill 1895). These modified pectoral rays contain free nerve endings that have a chemosensory function. Experiments on blinded anosomic fish showed that they respond to extracts of rotten clam, squid, worms, milk, and the amino acids phenylalanine, tryptophane, and aspartic acid. They also have taste buds on the lips and tongue (Bardach and Case 1965). Innervation of the fin rays is by spinal nerves that terminate in accessory spinal lobes (Finger 1982; Finger and Kakil 1985). The fin rays are sensitive to mechanical as well as olfactory stimulation (Silver and Finger 1984).

Predators. Northern sea robin are eaten by dusky shark (Gelsleichter et al. 1999), sandbar shark (Casey 1964), goosefish, weakfish (Maurer and Bowman 1975), and bluefish (Lux and Mahoney 1972). Other fish predators include spiny dogfish, smooth dogfish, angel shark, white hake, and sea raven (Rountree 1999). Juveniles are sometimes eaten by adult northern and striped sea robins (Marshall 1946; Richards et al. 1979).

Parasites. Parasites of northern sea robin from Woods Hole (Linton 1901b) included nematodes, immature ascarids encapsulated in the peritoneum; cestodes, *Rhynchobothrium* encysted on viscera and *Tetrarhynchus bisulcatus* encysted in stomach and intestine; and trematodes, *Distomum appendiculatum, Distomum* sp., and *Diplostomum* sp. from the intestine.

Breeding Habits. Northern sea robin reach sexual maturity at 2–3 years or at about 20 cm in the Chesapeake Bight (Wong 1968) and maybe as early as age 1 and 14 cm in New England, but the ovaries are small and there are only a few eggs (Marshall 1946). There is no information on fecundity. Little is known of their courtship, but they emit a staccato call during breeding season.

Ripe females were collected in the New York Bight from May through September, and the highest mean gonadosomatic index was in July (Wilk et al. 1990). Mean bottom water temperature was 15.1°C (range 13.8°–16.5°C) and mean depth 20 m (range 15–27 m) (Wilk et al. 1990). Spawning occurs from Block Island, R.I., through Cape Hatteras, N.C., from May to November (Colton et al. 1979; McBride 1994; Able and Fahay 1998). In southern New England, spawning takes place from June to August (Kuntz and Radcliffe 1917; Richards et al. 1979). Eggs were collected in Salem Harbor, Mass., from July

of 12.8°–24.4°C were reported in southern New England (Perlmutter 1939; Marshall 1946). In southern New Jersey, *Prionotus* spp. eggs were present from May to October, principally in inlets and nearshore habitats (McBride and Able 1994). Spawning takes place at night; estimated time of spawning was 19.2 h after sunrise (Ferraro 1980).

Early Life History. Northern sea robin eggs are pelagic and transparent with faint yellow coloration, and the chorion is unpigmented and lightly sculptured. There are numerous oil globules of varying sizes. Descriptions in the literature vary widely and are often conflicting. Egg sizes have been reported to range from 0.86 to 1.25 mm and oil globule numbers from 10 to 37 (Kuntz and Radcliffe 1917; Perlmutter 1939; Marshall 1946; Wheatland 1956; Yuschak and Lund 1984). The most recent description (Yuschak and Lund 1984) was based on laboratory-reared eggs stripped from adults collected in Long Island Sound. Fertilized eggs ranged from 0.86 to 0.97 mm (mean 0.92 mm) with 11–37 oil droplets. Kuntz and Radcliffe (1917) described laboratory-reared eggs as 1.0–1.5 mm with 10–20 oil droplets.

Yuschak and Lund (1984) found that the oil globules were scattered within one hemisphere of the egg. At 15°C, cell division began 2 h after fertilization and oil globules migrated to the area of newly developed cells, then dispersed throughout the yolk until gastrulation after 27 h, when they aggregated in a band near the egg equator. The early embryo was visible after 37 h and the oil droplets dispersed throughout the yolk periphery and remained scattered. Hatching occurred 112–155 h after fertilization. Kuntz and Radcliffe (1917) reported faster development at 22°C.

The eggs resemble those of striped sea robin, and occur in coastal waters at about the same time. Comparisons between laboratory-incubated eggs of the two species showed that northern sea robin eggs are smaller than those of striped sea robin, 0.86–0.97 mm vs. 1.05–1.25 mm (Yuschak 1985). Previously, patterns of oil globules were used to distinguish the two species (Perlmutter 1939; Wheatland 1956), but this was found to be unreliable. Keirans et al. (1986) used oil globule characters as described in Perlmutter (1939) to separate *Prionotus* spp. eggs and found that it was only 78% reliable compared to immunochemical identifications. Sea robin eggs also resemble hogchoker (*Trinectes maculatus*) eggs, which have multiple oil globules and occur from May to September. Hogchoker eggs are smaller in full-strength seawater (0.66–0.84 mm), but in lower-salinity water they overlap in size (0.86–1.2 mm) with sea robin eggs (Dovel et al. 1969). Hogchoker eggs sometimes have a greenish tinge (Martin and Drewry 1978), whereas sea robin eggs are yellowish (Yuschak and Lund 1984).

Northern sea robin larvae were found over the entire width of the continental shelf from Martha's Vineyard to Cape Hatteras, N.C., with highest concentrations from New Jersey to Chesapeake Bay (Morse et al. 1987; Able and Fahay 1998). They are typically more abundant than striped sea robin at egg, larval, and juvenile stages, but the two species have overlapping distributions (McBride 1994). In general, young stages of northern sea robin are present in estuaries from Cape Cod to South Carolina but are more abundant on the continental shelf of this region (Roberts 1978; McBride 1994).

Newly hatched larvae range in size from 2.6 to 3.1 mm NL (Kuntz and Radcliff 1917; Yuschak and Lund 1984). At hatch they have 23 myomeres, 12–19 oil globules located in the posterior half of the yolk, unpigmented eyes, and an undeveloped mouth. Pigmentation consists of stellate melanophores and xanthophores scattered sparsely on the head and trunk; a transverse band of pigment extending onto the finfold halfway between the vent and notochord tip; a row of melanophores and xanthophores along the ventral margin of the vent and just beyond the transverse band; a few large melanophores on the finfold above the pectoral bud, along the vent, and on the ventral margin near the notochord tip; and contracted melanophores and xanthophores on the surface of the yolk sac. By 3 mm NL, the pectoral fins are prominent, the head is large, the yolk sac is absorbed, the eyes are pigmented, and the gut is looped. Xanthophores disappear and pigmentation increases around the gut and over the brain. A row of melanophores is present along the lower jaw. By 4 mm a row of teeth is present in the upper jaw and head spines begin to appear. Pigmentation increases in general, but the transverse band midway between the vent and notochord disappears. By 5.4 mm NL the head begins to flatten and the body assumes its triangular cross-sectional shape, teeth are present on the upper and lower jaws, the pectoral fin is densely pigmented, and melanophores intensify on the head and trunk. Flexion of the notochord is complete at 6–7 mm SL, about 20 days posthatch (McBride 1994). By 7.0 mm SL young sea robin exhibit the adult complement of rays and spines; pigmentation increases on the head, lower jaw, pectoral and pelvic fins; head spines increase; and the three lower pectoral rays separate from the rest of the fin. Settlement occurs at 8–9 mm SL at about 25 days posthatch (McBride 1994).

Age and Growth. Length-weight equations for northern sea robin from trawl surveys in the New York Bight are: $\log_{10} W = -4.9951 + 2.9935(\log_{10} L)$ for males; $\log_{10} W = -5.0242 + 3.0135(\log_{10} L)$, females, where W = weight (g) and L = FL (mm). Mature females are heavier than mature males of the same length (Wong 1968; Wilk et al. 1978).

Northern sea robin from the Chesapeake Bight reached 11 years based on otolith analysis (Wong 1968). Fish from Connecticut and Long Island Sound, aged by scale annuli, reached 6 years. The asymptotic length determined for the von Bertalanffy growth equation was 300 mm SL (approximately 390 mm TL). They were smaller than sympatric striped sea robin and did not appear to live as long (Richards et al. 1979).

General Range. Northern sea robin occur in the western North Atlantic from the Bay of Fundy (Scott and Scott 1988) south to Florida but are most common from Cape Ann to South Carolina. They are found from estuaries to the edge of the continental shelf.

Occurrence in the Gulf of Maine. Northern sea robin are plentiful off southern New England, uncommon north of Cape Cod (McBride et al. 1998), but do occur in Massachusetts Bay (Collette and Hartel 1988) and on Georges Bank (Map 23).

Seasonal Migrations and Movements. Seasonal movements away from estuaries are closely associated with the temperature cycle (Marshall 1946; Richards et al. 1979; McBride and Able 1994; McBride et al. 1998). North of Cape Hatteras, sea robin make seasonal migrations offshore in October when temperatures fall below 15.5°C and reappear in late April when temperatures rise above 4.4°C (Marshall 1946). There is a slight north-south component to their seasonal movements; fish move well south of Block Island in winter and return to Cape Cod Bay in summer (McBride et al. 1998). They occur year-round in coastal habitats less than 10 m depth south of Cape Fear, but frequency of occurrence is much lower in fall and winter (Wenner and Sedberry 1989). In the New York Bight, the general pattern is onshore to offshore, where fish are found in less than 28 m from June to September, widely scattered over the shelf in May and October–November and deeper than 28 m from January to April (McBride and Able 1994).

Importance. Sea robins are generally considered trash fish and are taken incidentally in both commercial trawl and recreational hook-and-line fisheries (Roberts 1978; McBride et al. 1998). They have long been recognized as tasty (Goode 1888) but are not numerous or large enough in the Gulf of Maine to be of any economic importance.

STRIPED SEA ROBIN / *Prionotus evolans* (Linnaeus 1766) / Bigelow and Schroeder 1953:470–471

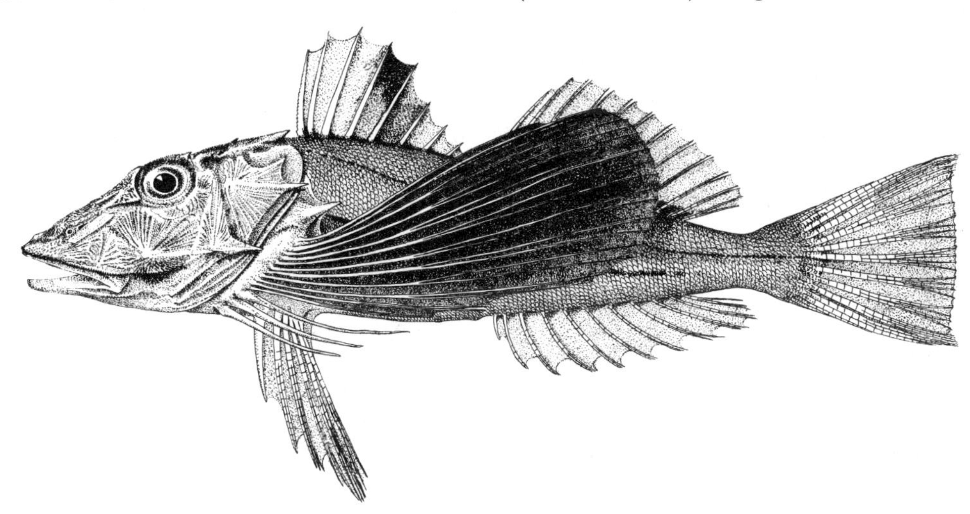

Figure 187. Striped sea robin *Prionotus evolans.* Woods Hole, Mass., 199 mm. Drawn by H. L. Todd.

Description. Striped sea robin resemble northern sea robin in general appearance except that they have two prominent lateral stripes, a larger mouth, flatter head, longer pectoral fin, and stouter body (Fig. 187).

Meristics. Dorsal fin rays IX–XI (most commonly X), 11–13; anal fin rays 10–12 (usually 11); pectoral rays 12–14 (usually 13), plus 3 separate feeler rays; lateral line scales about 55; gill rakers on the lower limb of the first arch 13–20 (Ginsburg 1950; Teague 1951; Yuschak 1985).

Color. The body is dark yellowish brown above, lighter below. There are two distinct dark stripes, one along the lateral line and the other below it extending from the humeral spine to the caudal peduncle. There is a large, dark spot on the dorsal

The pectoral fin is orange to brown, usually with very narrow transverse wavy brown lines close together and a single broad dusky blotch. The body often has three to five brown crossbars extending ventrally (Ginsburg 1950). The free pectoral rays are pale brown or orange, sometimes barred with darker color (Teague 1951).

Size. Striped sea robin range from 12 to 41.5 cm but most are between 26 and 33 cm (McEachran and Davis 1970). A maximum size of 48.5 cm TL reported by Briggs (1977a) equals the largest size (49 cm) found in the NEFSC groundfish surveys. Striped sea robin are larger and heavier than northern sea robin (Richards et al. 1979; McBride et al. 1998). Mature females are heavier than mature males (Wilk et al. 1978). The all-tackle game fish record is a 1.55-kg fish caught on Long Island, N.Y.,

Distinctions. Refer to northern sea robin.

Habits. Striped sea robin are bottom dwellers, occurring in coastal waters between the tide line and 100 m (McEachran and Davis 1970; McBride and Able 1994; McBride et al. 1998). They occur in the same general area at the same times as northern sea robin, but because they are more frequently caught in pound nets vs. trawls in the Woods Hole region it is believed that they inhabit waters closer to shore (Marshall 1946). A multivariate data set (depth, temperature, salinity, dissolved oxygen, and turbidity) showed that striped sea robin were captured more frequently in seasonably warmer, less oxygenated, and more turbid habitats than northern sea robin (McBride and Able 1994). Striped sea robin appear in New England waters when the temperature reaches 4°C and leave when it drops below 15°C (Marshall 1946). They have been reported from salinities of 7–32 ppt (Musick 1972) and temperatures of 1.4°–28.2°C (Richards and Castagna 1970; Richards et al. 1979) but they are generally captured in warmer and more saline waters within this range (McBride and Able 1994; McBride et al. 1998). They have been trawled as deep as 259 m but average depth of capture is 60 m in the spring and 20 m in the fall (McBride 1994).

Striped sea robin produce sounds similar to those made by northern sea robin, such as grunts, growls, barks, and bursts of clucks; however, their sounds are less staccato (Fish and Mowbray 1970). The swim bladder is vibrated by lateral intrinsic muscle masses; its anatomy and function is identical to that of northern sea robin except that the lobes end more bluntly (Fish 1954; Evans 1973).

Food. Both striped and northern sea robins are bottom or near-bottom feeders (Morrill 1895). Predominant food items are small crustaceans, particularly amphipods, mysids, and crangon; also annelids, cumaceans, crabs, megalops larvae, mollusks, fishes, and eggs (Marshall 1946). They appear to consume larger prey and more fishes than northern sea robin (Marshall 1946; Richards et al. 1979). Fishes identified in striped sea robin stomachs include bay anchovy, pipefish, Atlantic silverside, northern and striped sea robins, flying gurnard, scup, and winter and windowpane flounders. Striped sea robin have a larger mouth than northern sea robin, which might account for their more piscivorous habits (Marshall 1946; Richards et al. 1979). As found for juvenile northern sea robin, copepods were a principal prey item of juvenile striped sea robin (Marshall 1946; Richards et al. 1979)

Predators. Striped sea robin may be cannibalistic on their own juveniles (Richards et al. 1979). Several species of sharks feed on northern sea robin (Gelsleichter et al. 1999) and probably eat striped sea robin as well; spiny dogfish do (Rountree 1999).

Parasites. There is no specific information on parasites.

Breeding Habits. Fecundity estimates ranged from 90,000 to 218,000 eggs and showed a general increase with age (Yuschak 1985). The youngest sexually mature females and males were 2 years of age (McEachran and Davis 1970). Courtship behavior is unknown, but striped sea robin produce sounds during spawning season. The gonadosomatic index (GSI) peaks in June and July, and ripe females are present in the New York Bight from May to August (Wilk et al. 1990); GSI in Long Island Sound peaks in late May (Richards et al. 1979). Spawning takes place at night or in the evening over a sandy bottom (Ferraro 1980). Temperatures recorded when ripe fish were collected (Wilk et al. 1990) are 8.4°–14.1°C (mean 10.9°C), depths 9–21 m (mean 16 m).

Early Life History. Eggs are pelagic, transparent pale yellow with a lightly sculptured chorion. Fertilized eggs range from 1.05 to 1.25 mm (mean 1.15 mm) and contain 16–37 oil globules (mean 24). At incubation temperatures of 19°–20°C eggs develop rapidly with gastrulation occurring at 15 h; embryonic shield formation at 20 h; embryo one-third of the way around the yolk with Kupffer's vesicle and two somites at 31 h; embryo halfway around the yolk, optic cups, brain, notochord, melanophores, and xanthophores scattered over the body and yolk at 39 h; prior to hatching tail slightly overlaps head, pectoral fin buds are elevated; hatching at 80–90 h (Yuschak 1985).

Newly hatched larvae are about 2.8 mm NL with unpigmented eyes, undeveloped mouth, and a large yolk sac with oil globules distributed throughout the yolk. Larvae have a large head; short gut; noticeable pectoral fin buds, and pigmentation consisting of moderate numbers of melanophores and xanthophores on the head, trunk, and yolk-sac surface and a horizontal band of pigment about halfway between the vent and notochord tip. As larvae develop, yellow pigmentation is reduced, the transverse band disappears, and pigmentation increases around the dorsal, lateral, and anterior surface of the gut. Pigment appears on the head as a single expanded melanophore over the brain and a row of melanophores under the brain. There is also a row of melanophores along the margin of the upper jaw and along the midline of the lower jaw. Pigment intensifies around the gut, trunk, and pectoral fins; head spines appear by 6.2 mm (Yuschak 1985). Flexion of the notochord is complete at about 6–7 mm SL and 15 days posthatch (McBride 1994). Fin formation is complete by 7.8 mm. Settlement occurs at 8–9 mm SL, about 21 days posthatch (McBride 1994). The very prominent pectoral fin increases in size, and the three free rays develop by 11 mm (Yuschak 1985).

The prominent pectoral fins and large mouth distinguish striped sea robin from all other larvae except northern sea robin. The most noticeable differences between striped and northern sea robins are head spine formation and pigmentation. The nuchal spine is moderately developed in the latter but develops into an elongate ridge with numerous spines by 8.3 mm SL in the former. There is also a distinctive absence of pigment on the caudal peduncle of striped sea robin (Yuschak 1985; McBride 1994; Able and Fahay 1998).

Distribution of larval striped sea robin is similar to that of northern sea robin but the former are uncommon across the continental shelf (Able and Fahay 1998). Behavior and food of the two species are similar. Diet of young-of-the-year striped sea robin 41–56 mm from Woods Hole consisted of 89% copepods, 8% decapods (crangon and megalops larvae), 1% amphipods, and less than 1% annelids and fishes (Marshall 1946).

Age and Growth. Striped sea robin from the Chesapeake Bight live to 8 years, based on otoliths (McEachran and Davis 1970). Striped sea robin from Long Island Sound aged by scale annuli reached 9 years (Richards et al. 1979). Asymptotic length determined from the von Bertalanffy growth equation is 384 mm (approximately 474 mm TL). Weight-length relationships of fish collected in the Chesapeake Bay and Bight (McEachran and Davis 1970) are: $\log_{10}W = -5 + 3.19 \log_{10}L$ during spawning, and $\log_{10}W = -4.175 + 2.72 \log_{10}L$ after spawning. Weight-length relationships of fish collected by trawl in the New York Bight (Wilk et al. 1978) are: $\log_{10}W = -4.3727 + 2.7943 \log_{10}L$ for males; $\log_{10}W = -4.8289 + 2.9928 \log_{10}L$ for females, where W = weight (g) and L = FL (mm). Mature females are heavier than mature males of the same length.

General Range. Striped sea robin occur from the Bay of Fundy (Leim and Day 1959; Scott and Scott 1988) to the northeast coast of Florida (Bullis and Thompson 1965; Gilmore 1977), but are most common from Cape Cod to South Carolina.

Occurrence in the Gulf of Maine. Although common in the Woods Hole region during warmer months, they occur in the Gulf of Maine only as strays (McBride et al. 1998). Striped sea robin have occasionally been found at Monomoy, North Truro, Salem, Gloucester, and Monhegan Island (Collette and Hartel 1988).

Seasonal Migrations and Movements. In Long Island Sound, striped sea robin appear in April and May and leave in December. This seems to be in response to water temperature, as they appear when it rises above 6°C and leave when it falls below 8°C (Richards et al. 1979). In the New York Bight, they have been collected at less than 28 m from May to September, are widely scattered across the shelf in October and November, and migrate south of New Jersey from January to April (McBride and Able 1994). They are found year-round in the Chesapeake Bight (McEachran and Davis 1970; McBride et al. 1998).

Importance. Although generally larger and heavier than northern sea robin, striped sea robin are not found in enough numbers to be commercially important in the Gulf of Maine (Roberts-Goodwin 1981; McBride et al. 1998).

SCULPINS. SUPERFAMILY COTTOIDEA

The superfamily Cottoidea is regarded as monophyletic based on a combination of 23 synapomorphies (Yabe 1985). Sculpins and relatives form a homogeneous group characterized by large spiny heads, very wide gill openings, very broad mouths; slender bodies, separate spiny and soft-rayed dorsal fins (united in some rare species), large fanlike pectoral fins but small caudal fins, and pelvic fins that are reduced to three long rays. All have a habit of spreading the gill covers and flattening the head when taken in the hand. They produce grunting sounds, and some can inflate themselves with air or water when they are molested. The only other Gulf of Maine fishes resembling them in general form are sea robins, toadfish, and goosefish. The entire head of sea robins is armed with bony plates, different from the soft-skinned head of sculpins; in toadfish the soft portion of the dorsal fin is many times as long as the spiny part (at most twice as long as the spiny part in sculpins); and not only are the fins of goosefish small and weak as compared with sculpins, but their lower jaw projects far beyond the upper and their mouth is full of very large, pointed teeth, whereas in sculpins the teeth are small and the upper and lower jaws are of approximately equal length. Sculpins lay eggs. Males of Arctic sculpins, including *Artediellus, Cottunculus, Gymnocanthus,* and *Icelus,* have a long urogenital supposition is that this papilla serves as a copulating organ, with fertilization taking place within the female and the fertilized eggs being laid soon after (Jensen and Volsøe 1949).

Six families of the superfamily Cottoidea are recognized by Yabe (1985), four of which occur in the Gulf of Maine: Cottidae (six species), Psychrolutidae (one), Hemitripteridae (one), and Agonidae (one).

KEY TO GULF OF MAINE SCULPINS

1a. Body elongate, covered with smooth plates arranged in longitudinal rows . . . . . **Alligatorfish**
1b. Body not particularly elongate, without plates, naked or with prickles or small plates present only along lateral line 2
2a. Only one dorsal fin, spiny and soft parts continuous; fin low and inconspicuous 3
2b. Two separate dorsal fins 4
3a. Skin rough; eye small, diameter less than interorbital width; head spotted; body usually with at least three broad bands of blotching . . . . . **Polar sculpin**
3b. Skin almost naked and smooth except for a few spines; eye large, diameter greater than interorbital width; body color gray-brown, without blotches . . . . . **Pallid sculpin**
4a. First dorsal fin deeply notched between the spines; lower jaw and top of head adorned with fleshy tags . . . . . **Sea raven**
4b. First dorsal not deeply notched between the spines; no fleshy tags about

5a. Longest spine on each cheek branched at tip... **Arctic staghorn sculpin**
5b. Longest spine on each cheek simple, not branched at tip............ **6**
6a. Anal fin long (25 rays); a series of bony plates along each side of body.. .... **Moustache sculpin**
6b. Anal fin short (14 rays or fewer); no bony plates along sides of body .. **7**
7a. Long spine on cheek hooked upward........ **Atlantic hookear sculpin**
7b. Long spine on cheek straight, not hooked **8**
8a. Longest (uppermost) cheek spine four times as long as one below it, reaching back to margin of gill cover; all head spines very sharp....... .... **Longhorn sculpin**
8b. Uppermost cheek spine not more than twice as long as one below it, not reaching more than about halfway to margin of gill cover; head spines blunter.. **9**
9a. Total length more than 23 cm.................. **Shorthorn sculpin**
9b. Total length less than 20 cm **10**
10a. Anal fin with 13 or 14 rays; soft skin of each side of throat pierced by a minute pore close behind lower part of last gill arch .... **Shorthorn sculpin (young specimens)**
10b. Anal fin with only 10 or 11 rays; sides of throat behind last gill arch without a pore .... **Grubby**

SCULPINS. FAMILY COTTIDAE

GRACE KLEIN-MACPHEE

The Cottidae is regarded as monophyletic based on one autapomorphy, presence of a lateral process on the hyomandibular bone, and a combination of nine synapomorphies (Yabe 1985). It is the largest family of the superfamily Cottoidea with about 70 genera and 300 species (Nelson 1994). Most are marine fishes, but there are freshwater sculpins in North America and Europe. Six species of Cottidae have been recorded from the Gulf of Maine.

ATLANTIC HOOKEAR SCULPIN / *Artediellus atlanticus* Jordan and Evermann 1898 /

Hookear, Arctic Sculpin / Bigelow and Schroeder 1953:440–441 (as *Artediellus uncinatus*)

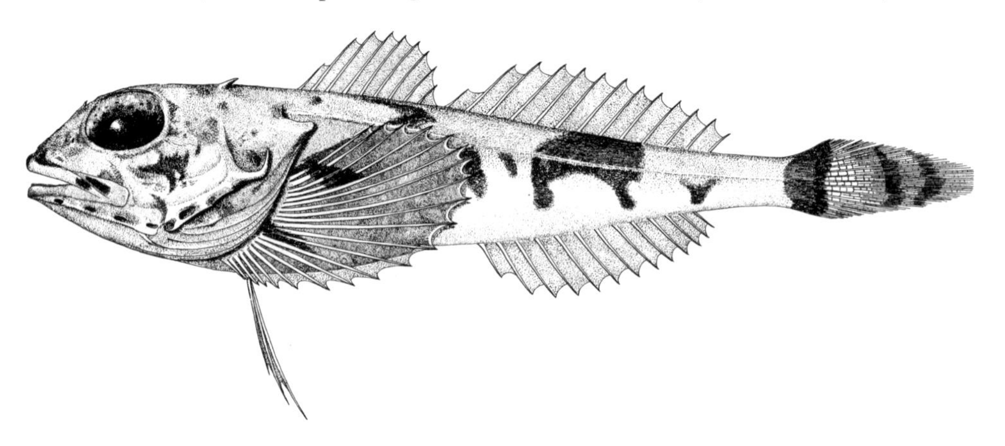

Figure 188. Atlantic hookear sculpin *Artediellus atlanticus*. Massachusetts Bay, 41 mm. Drawn by H. L. Todd.

Description. Large head, mouth and tapering body of usual sculpin form (Fig. 188). Head cirri often present. Eyes moderately large, high on head. Gill rakers poorly developed, branchiostegal membranes broadly joined across isthmus. Jaws, vomer, and palatines armed with series of small, bristlelike teeth. Long hooklike spine on each cheek pointing backward and upward, a short backward-pointing spine covered by flap of skin at upper corner of each gill cover; two short spines between the two pairs of nostrils; and a pair of blunt knobs above eyes. Spiny dorsal fin short and rounded; soft dorsal fin about twice as long as spiny dorsal fin; dorsal fins higher in males than in females. Anal fin slightly shorter than soft dorsal under which it originates. Pelvic fin rays long, reaching almost to vent. Pectoral fins wide at base, rounded in outline, and reach beyond origin of soft dorsal fin when laid back. Caudal fin narrow. Skin smooth and naked.

Meristics. Dorsal fin rays VIII, 12–16, usually 14; anal fin rays 9–13, usually 12; pectoral fin rays 19–24, usually 21; pelvic fin rays I, 3; lateral line pores 18–28; vertebrae 29–32 (Scott and Scott 1988).

Color. Preserved specimens mottled with dark and pale brown, sometimes with a reddish tinge. Most have a dark band on the caudal peduncle. All fins are grayish or blackish, with oblique or vertical pale crossbands especially obvious in mature males.

Size. Atlantic hookear are one of the smallest sculpins. Maximum size is 10.9 cm SL for males and 10.6 cm SL for females; size varies with geography (Van Guelpen 1986). The largest specimens were collected in Greenland waters (Jensen 1952), and smaller fish from Labrador (maximum 10.1 cm SL) and Newfoundland (maximum 9.4 cm SL). The mean size of fish collected in the Gulf of Maine was 5.3 cm SL (Van Guelpen 1986).

Distinctions. The backward-pointing hooked spine on the cheek distinguishes Atlantic hookear from all other Gulf of Maine sculpins. The caudal fin is also narrower than in most of the other sculpins, except for moustache sculpin.

Taxonomic Remarks. Bigelow and Schroeder confused this sculpin with *Artediellus uncinatus,* snowflake sculpin. For a detailed discussion of the taxonomic problem see Van Guelpen (1986).

Habits. Atlantic hookear sculpin live in cool subarctic waters usually over a soft bottom at depths of 0–384 m; maximum depth 795 m (Jensen 1952). Large fish tend to occur in deeper, cold water except in Saguenay Fjord, Gulf of St. Lawrence, where larger fish were found in relatively shallow water (Van Guelpen 1986). The minimum temperature at which this species was collected was –1.7°C off west Greenland (Jensen 1952); no upper temperature limits have been given.

Food. Hookear sculpin feed mostly on polychaetes (54.2% by weight, especially Nephtyidae), small crustaceans (25.3%) such as amphipods (17.6%) and cumaceans, and bivalves (12.1%) (Bowman et al. 2000).

Predators. Occasionally found in stomachs of gadids such as cusk, cod, pollock, and white hake (Rountree 1999).

Breeding Habits. Males grow larger (to 10.9 cm SL) than females (to 10.6 cm SL) and have higher and darker dorsal fins (Jensen 1952; Van Guelpen 1986). The smallest female with visible eggs was 2.3 cm; the smallest male with diagnostic fin coloration was 4.3 cm (Van Guelpen 1986). In the eastern Atlantic, numbers of eggs ranged from 48 to 216 for females 35–123 mm SL; mean number of eggs for three populations 77–156 (von Dorrien 1996). In the western North Atlantic, spawning appears to occur from May to November (Van Guelpen 1986).

Early Life History. In aquariums, males have always been observed close to egg clusters, which may be interpreted as brood protection (von Dorrien 1996). Hatching required up to 200 days at 0°C. At hatching (mean 11.3 mm TL), larvae were well developed with all fins formed and most fin rays present. Larvae grew at a rate of 0.02 mm SL·day^{-1} for the first 100 days reaching 11.3 mm SL. The smallest fish collected in the western Atlantic (2.1–2.4 cm SL) were similar to adults morphologically and in coloration (Van Guelpen 1986). The only difference appeared to be in the prominence of the parietal spines, which were large and sharp in the smallest individuals and became partially covered with skin by 30 cm SL.

General Range. Known from Labrador and the west coast of Greenland to Cape Cod in the western Atlantic; also in littoral waters of Arctic Europe, Siberia, and Greenland eastward to the Barents and Kara seas (Jensen 1952).

Occurrence in the Gulf of Maine. Hookear were previously thought to be rare in the Gulf of Maine, but are now known to be generally distributed in depths greater than 11 m. They were dredged in numbers in deeper parts of Massachusetts Bay many years ago and have been taken regularly near Mt. Desert, off Cape Elizabeth, in the trough between Jeffreys Ledge and the coast, around Cashes Ledge, along the northern slopes of Georges Bank, in the southeastern part of the basin of the Gulf, and at the entrance to the deep gully between Georges and Browns banks in depths down to 83 m. Eastward and northward they have been taken off Cape Sable, the outer coast of Nova Scotia, the Newfoundland Banks, the Gulf of St. Lawrence, the St. Lawrence estuary near Trois Pistoles, Hamilton Inlet on the outer coast of Labrador, and from Baffin Island (Van Guelpen 1986).

ARCTIC STAGHORN SCULPIN / *Gymnocanthus tricuspis* (Reinhardt 1830) /

Bigelow and Schroeder 1953:452–453

Description. Body elongate, tapering to a slender caudal peduncle (Fig. 189). Head broad, upper surface prickly, very small spines on ridges on top of head, although these may be absent in some individuals. Three blunt spines on preopercle, upper broad and flat with three (variable) short, sharp branches at tip. Upper corner of gill cover rounded; gill rakers poorly developed; mouth terminal; lower jaw projecting slightly; teeth small, on jaws only; eyes large and located high on head. Dorsal fins separated by a notch, anal fin originates shaped. Scales restricted to body below pectoral fins, otherwise skin naked and smooth. Lateral line high on body. Males with well-developed urogenital papilla.

Meristics. Dorsal fin rays X–XII, 14–17; anal fin rays 15–19; pectoral fin rays 17–21; pelvic fin rays I, 3; vertebrae 36–40 (Backus 1957b; Scott and Scott 1988).

Color. Dark brownish or gray above, the sides marked with

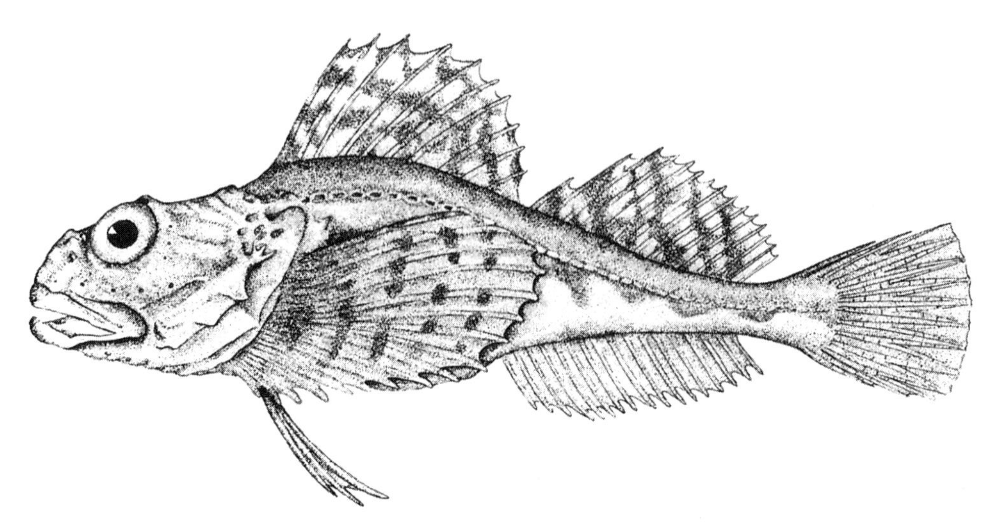

Figure 189. Arctic staghorn sculpin *Gymnocanthus tricuspis*. Drawn by D. R. Harriott. (From Scott and Scott 1988:495.)

spots; the lower surface white or yellowish with an irregular line of demarcation between dark sides and pale belly. Dorsal and pectoral fins are pale, the former with three and the latter with four or five irregular dark brown or black crossbands. Pelvic and anal fins are yellow-rayed, with membranes of the same color as the belly. Males are more brightly colored than females (Scott and Scott 1988).

Size. Up to 25.6 cm in Greenland (Jensen 1952).

Distinctions. Arctic staghorn are easily distinguishable from other sculpins by shape of the uppermost of the three cheek spines, which is broad and flat, with three short, sharp branches at its tip, instead of cylindrical and single-pointed; the anal fin (16–18 rays) originates well in front of the second dorsal fin instead of behind it; and the two dorsal fins are separated by a distinct space instead of being practically continuous at the bottom of the notch that separates them. The spines characteristic of the top of the head and shoulders of other sculpins are either lacking on staghorn or are very short, and the corners of the gill covers are rounded instead of sharp. Distinctive also, if less obvious, is the fact that the top of the head is more or less prickly or warty.The three-rayed pelvic fins reach only to about the vent in young of 4.5–5 cm, but they become relatively longer with growth until in mature fish they reach considerably past the anal fin origin; they extend further in males of breeding age than in females.

Habits. Arctic staghorn sculpin are a benthic species confined to cold waters. They have only been taken at water temperatures of −1.8° to 5°C. The lowest salinity at which they have been collected is 21 ppt (Backus 1957b). They are generally found over rocky and/or sandy bottoms, usually below 18.3 m, but a few have been taken at depths of 1.8–3.7 m. Maximum depth of capture was 174 m (Backus 1957b).

Food. Invertebrates and small fishes, including sand lance, have been found in stomachs (Jensen 1952).

Predators. Eaten by black guillemot in Hudson Bay (Gaston et al. 1985) and harp seal off west Greenland (Jensen 1952).

Parasites. The copepod *Diocis gobinus* is found on the fins (Margolis and Arthur 1979; Margolis and Kabata 1988). In Greenland waters, they are heavily parasitized by tapeworms (Jensen 1952).

Breeding Habits. This sculpin probably reproduces in the fall as this is when gravid females have been collected off Newfoundland (Ennis 1968).

Early Life History. A clump of developing eggs was collected in the Barents Sea from between 90–120 m (Norvillo and Zhuravleva 1989). The eggs were hatched in the laboratory, and the resulting larvae monitored for the 18 days they survived, at which time they were identified. Eggs are benthic; spherical; 1.79–1.88 mm in diameter; with a dense, smooth, semitransparent membrane and a single straw-colored oil globule 0.41–0.52 mm in diameter. The perivitelline space is narrow (about 0.02 mm) and the yolk is homogeneous. Embryos have heavily pigmented eyes but weakly pigmented bodies. Melanophores occur on the peritoneum and yolk sac. Blood vessels are visible on the yolk sac.

Hatching occurred 75–80 days after the egg mass was collected (Norvillo and Zhuravleva 1989). Day-old larvae were 7.2–7.3 mm long with a round head and spindle-shaped body. There were eight precaudal and 23–29 caudal myomeres. The yolk sac was small with the oil globule located posteriorly. Pectoral fins were developed and jaws formed. Pigmentation consisted of intense melanophores on the peritoneum; strongly branched, long melanophores on the lower edge of the yolk; one large branched melanophore beneath the base of the pectoral fins; one to seven star-shaped melanophores on top of the head; a subcaudal row began at the fourth myomere, club-shaped up to the twenty-sixth myomere and then branched. The end of the tail was free of pigment. Pigment appeared and increased on the lower edge of the stomach and auditory capsule. Pelvic melanophores formed a characteristic V-shape and spines appeared after 5 days. By day 14 when larvae were 8–8.7 mm long, the yolk was fully absorbed and preopercular spines developed.

General Range. Arctic and North Atlantic oceans. On the American coast, south along the outer coast of Labrador (Backus 1957b) to the Gulf of St. Lawrence, where they are generally distributed along the north shore and are characteristic of icy water on the banks of the southern side. On the European coast south to northern Norway.

Occurrence in the Gulf of Maine. One was reported from Eastport, Maine (Kendall 1909), the most southerly record for this Arctic sculpin and the only one for the Gulf of Maine.

GRUBBY / *Myoxocephalus aenaeus* (Mitchill 1814) / Little Sculpin /

Bigelow and Schroeder 1953:443–445 (as *Myoxocephalus aeneus*)

Figure 190. Grubby *Myoxocephalus aenaeus.* Staten Island, N.Y. Drawn by H. L. Todd.

Description. Typical sculpin form, proportionately stouter than either shorthorn or longhorn sculpins, about one-quarter as deep as long (Fig. 190). Head broad, covered with smooth skin but showing head ridges and spines typical of the genus. Most noticeable are a ridge with two spines running along top of head over each eye, a pair of spines between nostrils, and six short spines on each side of head between snout and gill opening. All cheek spines short. No slit or pore behind last gill arch. Spiny dorsal fin originates slightly in front of upper corner of gill opening, shorter than second dorsal fin; its longest spines, from base to tip, about same length as longest rays of second dorsal; the two fins so close together there is no free space between them. Anal fin slightly shorter than second dorsal fin, under which it originates. Pectoral fins large and fanlike. Lateral line prominent and located high on anterior part of body.

Meristics. Dorsal fin rays VIII–XI, 13 or 14; anal fin rays 10 or 11; pectoral fin rays 14–17; pelvic fin rays I, 3; vertebrae 30–34 (Fahay 1983; Scott and Scott 1988).

Color. Grubby vary in color according to the bottoms on ish gray above, with darker shadings or irregular barrings that are most evident on the sides and fins. The sides of the head are usually mottled light and dark; the belly is pale gray or white. There is frequently an uninterrupted pale band of considerable length along the lower sides of the caudal peduncle.

Size. This is the smallest of the common sculpins, usually reaching a maximum length of 12.7–15.2 cm. The largest grubby on record was 19.4 cm (Ennis 1969).

Distinctions. The most distinctive features of grubby are the short, simple head spines combined with the small size at maturity. They differ from shorthorn sculpin in the number of anal rays (10 or 11 vs. 9–16) and in the absence of a pore behind the last gill arch. Determination of this character requires close examination, and grubby resemble young shorthorn so closely in all other respects that it is hard to tell one from the other.

Habits. On the southern shores of New England, grubby are found from tide mark down to 27 m and have been taken as

sorts of bottoms, most abundantly among eelgrass (*Zostera*). Off Newfoundland, they have been found on sand, mud, or gravel bottoms in shallow waters in protected locations and on rocky bottoms at depths greater than 4 m off exposed shores. It is the only sculpin that summers in very shoal water along southern New England and near New York Harbor. Grubby are found in estuaries in the Gulf of St. Lawrence, on the Nova Scotian coast, and in the southern part of the range, but seem more restricted to open coasts in the Gulf of Maine. They are tolerant of a broad range of salinities and of water temperatures from 0° to 21.1°C. Grubby produce antifreeze proteins (polypeptides) (Chakrabartty et al. 1988). The levels of these proteins in the plasma of grubby in Long Island Sound start to increase in November (peak levels occur in January), begin to disappear when water temperatures begin to rise, and become insignificant by late spring (Reisman et al. 1987).

Grubby produce weak, low-pitched growls by rapidly vibrating the pectoral muscles (Barber and Mowbray 1956).

Food. Grubby are omnivorous like their relatives, feeding on all sorts of small animals that they find on the bottom, such as annelid worms, shrimps, crabs, copepods, snails, nudibranch mollusks, ascidians, and small fishes including alewife, cunner, eel, mummichog, sand lance, silverside, sticklebacks, and tomcod. They also scavenge animal refuse. Food in Nauset Marsh consisted primarily of crustaceans, particularly *Crangon septimspinosus*. Fishes were a minor component in the diet (Lazzari et al. 1989).

In January and February, major foods of grubby larvae (8.5–9.5 mm SL) in the Damariscotta River estuary, Maine (Laroche 1982), were a harpacticoid copepod *Microstella norvegica* and a diatom *Coscinodiscus;* in March the overwhelmingly dominant prey of all three species of *Myoxocephalus* were nauplii of the barnacle *Balanus balanoides*. Next in importance was *Microstella norvegica*. Other food items included turbellarians, other harpacticoid copepods, and unidentified invertebrate eggs.

Predators. Grubby are eaten by spiny dogfish, little skate, cod, longhorn sculpin, and sea raven (Rountree 1999). Larvae are eaten by sea raven larvae (Laroche 1982).

Parasites. Grubby from Newfoundland are parasitized by trichodinids, protozoan parasites found on the gills (Lom and Laird 1969).

Breeding Habits. Grubby probably breed throughout their range, certainly as far north as the southern part of the Gulf of St. Lawrence. The spawning season lasts all winter throughout most of the range, but between late fall and early winter around Newfoundland (Ennis 1969). Ripe fish were found in the Mystic River, Conn., from December to March. By May all were spent (Lund and Marcy 1975). In the field, clumps of eggs have been found attached to vegetation, shells, stones, bryozoans, and wooden traps (Lund and Marcy 1975).

Early Life History. The eggs are spherical, transparent, and adhesive (Lund and Marcy 1975). Color is variable but consistent within a single female; light green, red, white, and sometimes yellow. Average diameter of the eggs is 1.58 mm (range 1.5–1.7 mm). The chorion is thin and transparent, with a smooth surface. The perivitelline space is narrow, about one-sixth of the egg radius. There are two large oil globules with an average diameter of 0.2 mm and a few smaller globules scattered throughout the yolk. Cleavage begins about 8.5 h after fertilization, blastula stage at 98–196 h, gastrula about 174 h, midembryo (halfway around the egg, somites evident, eye lens thickening) 13.5 days, late embryo (retinal pigment seen, finfold evident, heartbeat) 17.3 days, hatching 40–44 days.

Newly hatched larvae range between 4.7–6.3 mm TL, with yolk sac about 0.9 mm. There is a large oil globule containing a small bubble. The eyes are pigmented; stellate chromatophores are present scattered in two bands over the ventral yolk sac with 20–24 spots along the lower ventral line of the tail. The yolk sac and oil globule are absorbed within 5 days after hatching. There are large, dark chromatophores over the gut area, with 30–34 spots along the lower ventral line of the tail, and a few chromatophores extend onto the finfold in the caudal region. There are a few chromatophores at the base of the pectoral fin and a few scattered behind the auditory vesicles. Two head spines develop on the dorsal bulge of the auditory capsule and four on each side of the head along the operculum. Transformation to the juvenile stage occurs approximately 55 days after hatch (Lund and Marcy 1975).

Grubby larvae are found in estuaries from Maine (Chenoweth 1973; Laroche 1982) to Connecticut (Pearcy and Richards 1962). In the Damariscotta River estuary, Maine, diet overlap was considerable among grubby, longhorn, and shorthorn larvae (Laroche 1982). Vertical distribution differences may reduce competition for food between grubby and longhorn sculpin. More grubby were located 1.5 m above the bottom than longhorn, the latter occurring more frequently in the upper 1.5 m (Laroche 1982).

General Range. North American coastal waters from New Jersey to northern Nova Scotia and to the Gulf of St. Lawrence and the west coast of Newfoundland. The northern limit is Raleigh, Strait of Belle Isle (Jeffries 1932).

Occurrence in the Gulf of Maine. Grubby are probably found in suitable localities all around the shores of the Gulf of Maine, for they are reported as common along both shores of the Bay of Fundy and at various places in the Massachusetts Bay region, such as Cape Ann, Gloucester, Salem, Cohasset, Nauset Marsh, Provincetown, and off the outer coast of Cape Cod abreast of Chatham. They seem to be decidedly local in distribution, for the only places where they have been definitely reported along the coast between Cape Ann and the Bay of Fundy are Casco Bay and Passamaquoddy Bay.

LONGHORN SCULPIN / *Myoxocephalus octodecemspinosus* (Mitchill 1814) /

Gray Sculpin, Hacklehead / Bigelow and Schroeder 1953:449–452 (*Myoxocephalus octodecimspinosus*)

Figure 191. Longhorn sculpin *Myoxocephalus octodecemspinosus*. New Jersey. Drawn by H. L. Todd.

Description. Body elongate, slender, tapering posteriorly to a slender caudal peduncle (Fig. 191). Head blunt, heavy, flattened, and adorned with a number of spines. One pair of short nasal spines; one pair of spines above and behind eyes; one pair well behind those on top of head; three on preopercle, upper one four times as long as one below it, sharp and naked at tip, middle spine short, third turned downward; two sharp spines on upper corner of gill cover and one short, sharp spine behind upper part of gill opening. Mouth long, terminal, low on head. Bands of small teeth on jaws and vomer. Eye large, high on head. Gill rakers poorly developed, represented by short bumps. Two dorsal fins somewhat rounded, no appreciable space between them. Anal fin somewhat rounded, located under second dorsal. Pectoral fins large, fanlike. Pelvic fins under middle of pectoral fin base. Caudal fin moderate-sized, truncate to rounded. Scales absent or nearly so; skin naked except for lateral line that is marked by a series of poorly calcified plates (Scott and Scott 1988).

Meristics. Dorsal fin rays VII–X, 15–17; anal fin rays 12–15; pectoral fin rays 16–19; pelvic fin rays I, 3 or 4; branchiostegals 6; vertebrae 34–44 (Scott and Scott 1988).

Color. Longhorn, like other sculpins, vary in color with their surroundings. The ground tint of the back and sides ranges from dark olive to pale greenish yellow, greenish brown, or pale mouse color. As a rule they are marked with four irregular, obscure, dark crossbars, but these are often broken up into blotches and may be indistinct. The coarseness of pattern often corresponds to that of the bottom, as does the degree of contrast between pale and dark. On mud and sand bottoms they may be nearly plain-colored, but when they are lying on pebbles with white corallines, their back is often invisible. The first dorsal fin is pale sooty with pale and dark mottlings or spots; the second dorsal is paler olive with three irregular oblique dark crossbands; the caudal is pale gray, the pectoral fins yellowish. Caudal and pectorals are marked with three to six rather narrow but distinct dark crossbands. The anal is pale yellowish with dark mottlings; and there is often an obscure yellowish band along the lower part of the sides, marking the transition from the dark upper parts to the pure white belly.

Size. Maximum reported size 45.7 cm; average 25.4–35.6 cm.

Distinctions. Longhorn sculpin resemble shorthorn sculpin. The major differences are the great length of the uppermost cheek spine, which is usually about four times as long as the spine just below and reaches at least as far back as the edge of the gill cover. This also serves to distinguish young longhorn from grubby, which are short-horned. All the head spines of longhorn are very sharp and naked at the tip. The first dorsal fin is higher than the second, whereas in shorthorn sculpin these two fins are of about equal height. The lateral line of longhorn sculpin is marked by a series of smooth plates instead of by prickly scales as it is in shorthorn, and the body is more slender (about five and a half times as long as it is deep).

Habits. Longhorn sculpin are benthic, rather slow-moving fish that inhabit coastal waters from the shoreline to the offshore banks. They are abundant in many shoal harbors and bays and run up into estuaries, salt creeks, and river mouths, but never into fresh water. At the other extreme, they have been caught in considerable numbers down to 90 m and have been reported as deep as 192 m. The preferred depth range on the Scotian shelf was 53–90 m (Scott 1982b). They can toler-

Longhorn sculpin may be bothersome for the angler to unhook because they spread their needle-sharp spines and erect their spiny dorsal fin. They grunt when pulled out of the water. They also produce sounds under water spontaneously and when under duress. The sonic mechanism is movement of the pectoral girdle actuated by contraction of the lateral muscles (Fish and Mowbray 1970).

Food. Bigelow and Schroeder described longhorn sculpin as voracious scavengers feeding on shrimps, crabs, amphipods, hydroids, annelid worms, mussels, and sundry other mollusks, squids, and ascidians, as well as a considerable number of fish larvae.

The most important food category of 149 longhorn was crustaceans (82.4% by weight) followed by fishes (10.5%) (Bowman et al. 2000). Smaller longhorn (6–15 cm TL) ate more amphipods such as *Leptocheirus pinguis, Unciola irrorata,* and Podoceridae. Larger longhorn (16–40 cm TL) switched to decapods such as rock crab (*Cancer irroratus*), *Dichelopandalus leptocerus, Crangon septemspinosa, Pagurus acadianus,* and fishes. Fishes eaten include skates, Atlantic herring, eels, sculpins, sand lance, radiated shanny, eelpout, wrymouth, and yellowtail flounder (Maurer and Bowman 1975; Bowman and Michaels 1984; Langton and Watling 1990; Bowman et al. 2000). *Cancer irroratus* and *Leptocheirus pinguis* were the dominant food of a large series of longhorn sculpins from Block Island Sound (Morrow 1951). Nearly all of them had eaten small amounts of shrimps (*Crangon*); a few contained small lobsters; and spider crabs (*Libinia*) were a regular diet item in winter, but not in summer. There was dietary overlap between longhorn sculpin and cod in particular and to a lesser degree among sculpin, red hake, little skate, yellowtail flounder, and winter flounder as a result of the dependence of these species on decapod and amphipod prey (Hacunda 1981). In Sheepscott Bay, crustaceans were also the most important food category with *Mysis stenolepis* being the largest contributor by weight (13.6%), although an amphipod, *Unicola inermis,* was most frequent in occurrence (29.7%).

In January and February, the major food of longhorn sculpin larvae (7.5–12.4 mm SL) in the Damariscotta River estuary, Maine (Laroche 1982), was a harpacticoid copepod *Microstella norvegica;* in March the overwhelmingly dominant prey of all three species of *Myoxocephalus* were nauplii of the barnacle *Balanus balanoides.* Other food items included a calanoid copepod *Temora longicornis,* a diatom *Coscinodiscus,* and unidentified invertebrate eggs. Stomachs of 199 juvenile longhorn sculpins up to 10 cm TL from Georges Bank were filled almost entirely with crustaceans (Bowman et al. 1987): decapods (50% by weight, mostly Crangonidae and Pandalidae), amphipods (26–30%), and isopods (3–5%).

Predators. Longhorn sculpin have been found in the stomachs of 14 species of fishes of which the eight most frequent predators are, in order of frequency: cod, spiny dogfish, winter skate, sea raven, little skate, goosefish, white hake, and other longhorn sculpin (Rountree 1999). In the Bay of Fundy, they are eaten by cormorants (Scott and Scott 1988).

Parasites. Longhorn sculpin are parasitized by five species of protozoans; three myxosporidians; two nematodes; four hirudineans (Margolis and Arthur 1979); a monogenic trematode, *Gyrodactylus nainum* (Cone and Wiles 1983); a digene, *Neophasis burti* (Bray and Gibson 1991), and a fungus located in the kidney (Hendricks 1972). They were lightly infected with blood parasites (Khan et al. 1980).

Breeding Habits. An average female produces about 8,000 chocolate brown eggs. Sexual maturity is attained in the third year (Morrow 1951). Off the southern New England coast longhorn deposit their eggs from late November through January, and perhaps into February, with the chief production from late December to mid-January. Presumably the spawning season is the same in the Gulf of Maine. Spawning occurs inshore in estuaries and shallow enclosed areas where the bottom is rocky (Scott and Scott 1988). Egg masses have been found free on the bottom, in empty clamshells or other cavities, or among branches of the finger sponge, *Haliclona,* and they are sometimes washed up on the beach.

Presence of longhorn sculpin of all sizes, from larvae to adults, supports breeding all along the coasts of Massachusetts and of Maine, probably along western Nova Scotia as well, but breeding in the Bay of Fundy seems to be restricted to the Scotian side.

Early Life History. Ripe eggs are about 0.85 mm in diameter before being laid, but they swell when they come into contact with the water. They vary in color from coppery green to reddish brown, orange, or purple. Unfertilized, water-hardened eggs are spherical with a mean diameter of 2.2 mm (2.1–2.3 mm) and a mean yolk diameter of 2.1 (2.0–2.2 mm). Oil droplets vary in number and size from several small droplets to two or more large ones. The perivitelline space is less than 10% of the egg capsule radius, and the chorion is colorless, translucent, and leathery. The yolk surface is slightly irregular.

In laboratory-incubated eggs at 5°C (Walsh and Lund 1983), the first division occurred 8 h after fertilization, blastulation at 132 h, early embryogenesis at 7 days, the tail-free stage at 17.7 days, movement within the eggs at 22.7 days, and circulation at 33.3 days. Hatching occurred between 36 and 65 days after fertilization; newly hatched larvae were 6.2–7.8 mm TL. There was a single oil droplet 0.4 mm (0.3–0.5) located at the anterior end of the yolk. Pigment consisted of a few stellate chromatophores above and behind the auditory vesicles, many dense chromatophores covering the dorsal surface of the yolk, and a series of 18–28 spots along the ventral line of the tail. The anus was just anterior to the ventral origin of the finfold. Larvae were relatively well developed with pigmented eyes, functional mouths, large pectoral fins, and 37 or 38 myomeres. Yolk was absorbed at 10 days posthatch, when larvae were about 8.7 mm TL. Pigmentation consisted of large stellate melanophores scattered over the top of the head and on the midline above and between the nostrils with several stellate chromatophores on the isthmus; additional stellate melanophores developed almost to the anus; several contracted chromatophores posterior and parallel to the cleithrum and ventral spots on the tail (16–24). Four preopercular spines were present on each cheek and one pair of spines on the crown above the auditory vesicles. Flexion occurs at 9–11 mm TL. Metamorphosis occurs at 51–58 days at about 12 mm TL. Each cheek had five spines and the uppermost began to elongate more than the others, the two pairs of head spines fusing to a single large flesh-covered structure. Pigmentation increased and darkened on the head. At 15.2 mm TL, juveniles had the characteristic four crossbar marks of adults (Walsh and Lund 1983).

Larvae resemble those of grubby, but are larger at hatch, flexion, and metamorphosis. Early pigmentation in longhorn has fewer spots along the midline. In grubby larvae, spots extend to the anus, and there are no spots on the head or nape. They are distinguished from moustache sculpin by the smaller number of myomeres, 33–44 vs. 43–46 (Fahay 1983).

Juveniles have been taken in February and March off southern New England, in April on the eastern part of Georges Bank and in the channel between the latter and Browns Bank. These young stages have longer cheek spines than corresponding stages of shorthorn sculpin, they are more slender, and they differ in the outline of the dorsal fin, which is continuous from end to end. Only the largest of them shows a shallow notch between spiny and soft portions, whereas in shorthorn the two sections are separate from the time the fin first takes definite form.

Age and Growth. A study based on otoliths (Morrow 1951) suggested that longhorn sculpin off southern New England were 5.5 cm at age 1, 18 cm at age 2, 21 at 3, 25 at 4, 27 at 5, and 30 at 6 years.

General Range. Coastal waters of eastern North America from the Strait of Belle Isle off eastern Newfoundland and the north shore of the Gulf of St. Lawrence, south regularly to New Jersey, and reported from the Atlantic coast of Virginia.

Occurrence in the Gulf of Maine. Longhorn are the commonest sculpin, caught anywhere and everywhere along the entire coast line of the Gulf of Maine. There is not a bay, harbor, estuary, or a fishing station from Cape Sable to Cape Cod where they are not found. Not only are they more plentiful in most places than their short-horned relative, but they occupy a wider depth zone. They also occur plentifully on Georges Bank. They are very common along the Nova Scotian coast and banks eastward from Cape Sable, in suitable depths, and are widely but irregularly distributed around the southern shores and islands of the Gulf of St. Lawrence, the Bay of Fundy, and St. Mary's Bay (Scott and Scott 1988).

Seasonal Migrations and Movements. The only periodic movements of longhorn sculpin are off and onshore, and of short extent, combined with movements to and from particular grounds. Off New York, they are commonest near shore from September to May, and are seen only occasionally in summer. In Long Island Sound, they appear to carry out east-west movements about which little is known; in Block Island Sound they are plentiful on productive fishing grounds from November through April, but most withdraw during May either offshore or onto more rocky grounds nearby, not to return en masse until the following October.

All that is known of their movements in the Gulf of Maine is that in partially enclosed and very shallow situations where the water on the flats heats to more than 20°C in the warmest part of the season but ice forms in the winter (Duxbury Bay, e.g.), longhorn seek slightly deeper water for the summer. They generally continue right up to the low tide line all summer in localities where the surface does not become so warm in summer or so cold in winter. This is the general rule northward and eastward around the coast of Maine, including the Passamaquoddy region.

Importance. The only commercial value this sculpin had was as bait for lobster pots, for which they were speared in some localities and caught on hook and line in others. Very few of them are now used in this way.

SHORTHORN SCULPIN / *Myoxocephalus scorpius* (Linnaeus 1758) / Bigelow and Schroeder 1953:445–449

Description. Large flat head, vast mouth, weak tapering body, and batlike pectorals, typical of sculpins (Fig. 192). A longitudinal ridge with three knobs or spines running along each side of its crown, one before eyes, one behind them; also five to seven short, triangular blunt spines on each side of cheek between snout and gill opening, uppermost less than twice as long as the one below it, and reaching not much more than halfway to edge of gill cover. A short, sharp spine at upper corner of each gill cover, pointing rearward and lying on a flap of skin, two thornlike spines on each shoulder [illegible]

per corner of gill cover. A pore, or a small slit, piercing soft skin low down on each side of throat close behind last gill arch, easily seen on large specimens and detectable on close examination even on small ones. Gill rakers poorly developed. Mouth moderately large with small teeth on jaws and vomer. Eyes very large, at least as wide as space between them, set high up on sides of head with upper edges close to dorsal profile, and directed a little upward, as well as outward. Two parts of dorsal fin entirely separated by a deep notch; no gap [illegible]

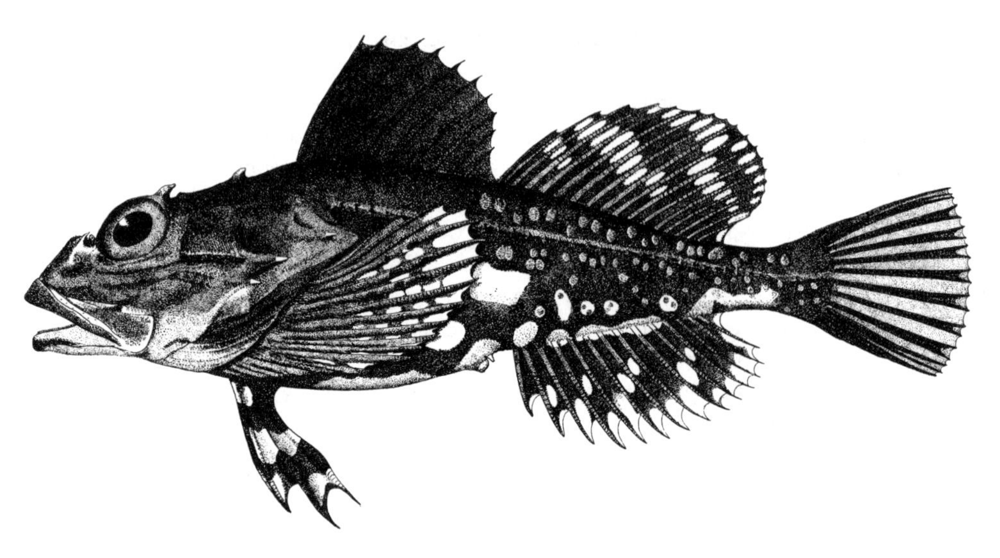

Figure 192. Shorthorn sculpin *Myoxocephalus scorpius.* Eastport, Maine, 372 mm. Drawn by H. L. Todd.

Anal fin similar to second dorsal fin in shape, but a little smaller; originates under fourth or fifth soft dorsal ray. Fanlike pectoral fins reach back about as far as vent. On large specimens, dorsal, anal, and pectoral fins noticeably thick and fleshy. Caudal fin small, rear margin weakly rounded. Two series of prickly platelike scales along each side of body, one above lateral line, the other below it. Lateral line pronounced and complete.

Males and females differ in appearance, the former being more slender and brightly colored with higher fins. Each of the scales also differs along the lateral line with three or more prickles in males, but with only one or two at most in females, whereas some of the latter have no scales. Inner edges of rays of pectoral and pelvic fins armed with teeth or prickles on males, but not on females.

Meristics. Dorsal fin rays VII–XII, 12–20; anal fin rays 9–16, but there is considerable regional variation (Dunbar and Hildebrand 1952; Backus 1957b); pectoral fin rays 14–19; pelvic fin rays I, 3; vertebrae 32–41 (Fahay 1983; Scott and Scott 1988).

Color. The basic hue of the upper parts is usually some shade of brown, from a warm reddish tint to almost black, with the top and sides of the head marked with pale blotches and the back and sides of the body with broad dark bars on individuals on which the ground tint is pale. Lower parts of the sides are more or less spotted with yellow. Belly whitish or yellowish in females, usually reddish orange with large, round white spots in males, a good field mark for distinguishing the sexes. Dorsal fins are mottled dark and pale, the second dorsal often marked with three or four definite crossbars, and the caudal fin with various dark mottlings. Rays of the pectoral and anal fins are yellow with two or three irregular dark crossbars on many specimens, but they are uniformly dark on some. Males are more brightly colored than females in the breeding season, when their red and yellow tints become very brilliant and intensification of the red or coppery ground color of the belly brings out the white spots more clearly than in other seasons.

Size. Shorthorn are the largest Gulf of Maine sculpin. The average adult taken in the Gulf is about 20.3–35.6 cm. Average size increases moving from south to north, with Greenland fish averaging much larger than those taken off New England or off the Maritime Provinces. The maximum observed length in Greenland for females was 50.6 cm, and for males 42.2 cm (Ennis 1970).

Taxonomic Note. Based on morphological features in a detailed study, Cowan (1971) suggested that there are three genetically different forms: European, Arctic, and Gulf of Alaska.

Distinctions. Large shorthorn could hardly be mistaken for any other Gulf of Maine fish except for longhorn sculpin, and a cursory look is enough to separate one from the other: in shorthorn the upper cheek spine is less than twice as long as the one below it and does not reach more than halfway to the edge of the gill cover, whereas in longhorn it is about four times as long as the one below itand reaches back at least as far as the edge of the gill cover.

Young shorthorn, up to 15.2–17.8 cm, closely resemble grubby. Shorthorn have a pore piercing each side of the throat close behind the last gill arch, whereas grubby have no such pore.

Habits. Bays and the vicinity of ledges that rise from comparatively smooth bottom in shoal water are the chief haunts of shorthorn sculpin, and they are found there on mud, sand, pebbles, and bare bottom, or among weeds. Many are also caught off piers and along rocky shores by cunner fishermen.

Offshore, most live shoaler than 18 m. Few are caught on longlines set deeper than 27–36 m. The deepest record in American waters is 80–110 m in Maine (Backus 1957b). They do not move far up in estuaries, and never enter brackish water.

This is a cold-water fish. Even in summer they are most plentiful at localities and depths where temperatures are lower than 13°–16°C. In winter they endure temperatures close to the freezing point of saltwater. Antifreeze peptides isolated from the skin of shorthorn sculpin are likely the first line defense against ice formation during contact with ice crystals in the water (Schneppenheim and Theede 1982; Low et al. 1998).

Shorthorn are sluggish, are often seen lying motionless, and as a rule they hug the bottom so closely that dangling bait hardly tempts them to rise as much as a meter. Shorthorn sculpin usually swim slowly with undulating motion, spreading their great pectoral fins like bat wings. They move only a little way when disturbed, but on occasion they dart ahead with folded "wings." They grunt or gurgle when drawn out of the water, particularly when handled; they are also known to grunt in the water. Sounds are probably produced by vibrations of the pectoral muscles (Fish and Mowbray 1970).

Food. Shorthorn sculpin are voracious, feeding chiefly on crustaceans, particularly crabs and shrimps; sea urchins; worms; mollusks, *Littorina littorea, Margaretes umbilicus,* and *Modiolaria discors;* and planktonic and benthic amphipods (Moore and Moore 1974; Lydersen et al. 1989; Bowman et al. 2000). They are eager scavengers after any kind of refuse, congregating around fish wharves and lobster cars to feed on the debris.

In January and February, major foods of shorthorn sculpin larvae (8.5–13.4 mm SL) in the Damariscotta River estuary, Maine (Laroche 1982), were a diatom, *Coscinodiscus,* and a harpacticoid copepod, *Microstella norvegica;* in March the overwhelmingly dominant prey of all three species of *Myoxocephalus* were nauplii of the barnacle *Balanus balanoides.* Next in importance was a calanoid copepod, *Temora longicornis.* Other food items included other harpacticoid copepods and unidentified invertebrate eggs.

Predators. Shorthorn sculpin are eaten by smooth skate, clearnose skate, cod, haddock, sea raven, and black sea bass (Maurer and Bowman 1975; Bowman and Michaels 1984; Rountree 1999). Black guillemot is another predator (Gaston et al. 1985).

Parasites. Shorthorn sculpin are heavily parasitized. Margolis and Arthur (1979) listed two protozoans; two myxosporidians; six trematodes, including *Derogenes varicus* and *Hemiurus appendicullatus;* one cestode; three nematodes; and four hirudinoideans. Margolis and Kabata (1988) reported an isopod, *Aega psora,* and an amphipod, *Lafystius morhuanus.* The monogenean *Gyrodactylus groenlandicus* inhabits the gills, fins, and body surface (Cone and Wiles 1983). Parasitological investigations were also conducted by Khan et al. (1980) and Margolis

Breeding Habits. Males in Newfoundland begin to mature at age 2 and all are mature by age 6; females begin to mature at age 3 and are all mature by age 8. Fecundity varies from 4,205 to 60,976 eggs in females 20.1–50.5 cm long (Ennis 1970).

The spawning season is from November to February, with the chief egg production in December, both about Woods Hole and in northern European waters. The season is late November to early December in waters around Newfoundland. At this season, adults have been described as gathering in schools on sandy or weedy bottom, with females greatly outnumbering males.

Discussion has centered on the manner in which eggs are fertilized, it being generally agreed that this takes place externally, but in some parts of the Baltic Sea the eggs may be fertilized within the body of the female. In either case, the eggs sink and stick together in irregular spongy masses through which the water circulates and these masses retain a considerable amount of moisture even if they are left bare by the ebbing tide, as often happens. These egg masses are deposited on sandy bottoms, in pools on the rocks, among seaweeds, or in any crevice or hollow (in a tin can, e.g., or in an old shoe). Sometimes males make a nest of seaweed and pebbles; they have been described as sometimes clasping the egg mass with their pectoral and pelvic fins and have been photographed that way. The males guard the eggs until they hatch while the females move to deeper waters (Ennis 1970).

The eggs are of varying shades of pinkish red or yellow, 2.0–2.5 mm in diameter (Fahay 1983). Incubation is so slow (occupying 4–12 weeks, according to temperature) that egg masses with advanced embryos have often been found as late in the spring as April or even May. Newly hatched larvae are about 7.9–8.6 mm long (Fahay 1983). In 1 month they are about 10 mm long and the yolk sac has been absorbed. The young larvae soon rise to the surface, where quantities of them have been taken in tow nets in British waters in March, April, and May. By May and June some have grown to a length of 22–25 mm. They abandon their drifting life at about this size, or soon after, for the bottom, and they may be 38 mm long by July, showing all the distinctive characters of adults. This timetable, compiled from European sources, probably applies to the Gulf of Maine as well, for larvae are found as early as February in the Bay of Fundy and thereafter throughout the spring.

Age and Growth. In Newfoundland, shorthorn sculpin live to age 15 and attain a size just over 50 cm (Ennis 1970). Females grow faster than males after age 4. The von Bertalanffy growth equations are: $L_t = 42.75(1 - e^{-0.1845(t + 0.897)})$ for males, and $L_t = 52.61(1 - e^{-0.1693\,(t - 0.0437)})$ for females.

General Range. This wide-ranging fish is known from Great Britain northward along the coasts of Europe; in Arctic seas generally, including Spitsbergen, Novaya Zemlya, north Siberia, Alaska, Hudson Bay, Baffin Island, west Greenland, and northern Labrador; southward along the American coast

Occurrence in the Gulf of Maine. Shorthorn sculpin are one of the most familiar shore fishes, common around the entire coastline of the Gulf of Maine. They are not as abundant as longhorn sculpin, but there is hardly a suitable situation from Cape Cod to Cape Sable where some shorthorn are not to be found, except perhaps at the head of the Bay of Fundy.

Seasonal Migrations and Movements. Shorthorn sculpin are one of the most stationary of Gulf of Maine fishes and have been characterized as the only fish that remains near shore during the coldest part of the year. They have been described as more plentiful along the shores of Massachusetts Bay in winter than in summer, as they are south of Cape Cod. Certain shallow bays such as Duxbury Harbor, where broad expanses of flats are exposed at low tide to heating by the sun in summer and the formation of ice in winter are an exception. Here shorthorn sculpin tend to keep to deeper channels through the coldest part of the winter, as well as during the heat of midsummer, but there is no evidence that they undertake any more extensive seasonal migrations.

Importance. Although edible, they have no commercial value at present. They were once of some value as bait for lobster pots, for which they were speared in some localities and caught on hook and line in others. They are useful as research animals in cell structure and pollution studies (Fletcher et al. 1982; Johnston 1983).

MOUSTACHE SCULPIN / *Triglops murrayi* Günther 1888 / Mailed Sculpin /

Bigelow and Schroeder 1953:441–443 (as *Triglops ommatistius*)

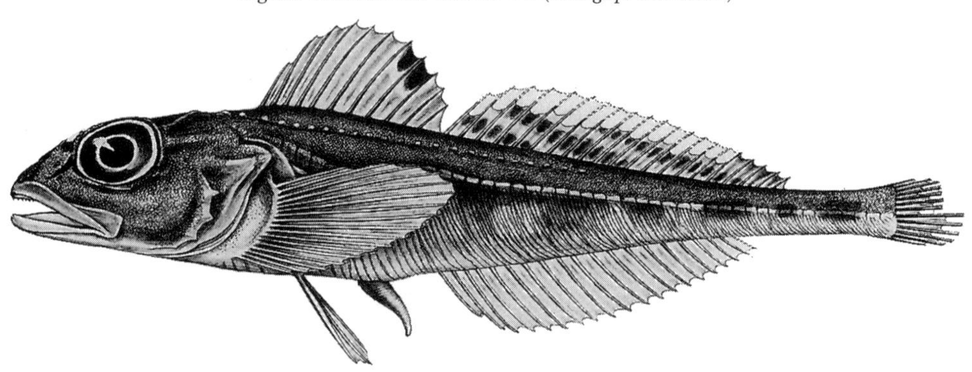

Figure 193. Moustache sculpin *Triglops murrayi*. Gloucester, Mass., male. Drawn by K. Ito.

Description. Body elongate, tapering posteriorly to a slender caudal peduncle (Fig. 193). Head large with a blunt snout and more or less equal jaws, maxilla extending posteriorly to posterior margin of eye, which is large with a diameter equal to or greater than postorbital head length. Interorbital width 6.5–10.0% HL (Pietsch 1993). Gill rakers poorly developed, membranes broadly joined across isthmus (Scott and Scott 1988). First dorsal fin originates over base of broad pectoral fins, higher than second dorsal but only half as long. Anal fin similar to second dorsal in form but a little shorter and originating below it. Pelvic fins reach back to rear end of first dorsal. Pectoral fins fan-shaped. Caudal fin slightly rounded and truncate. Males have a very large, prominent anal papilla. Lateral line straight, with a row of broad, platelike scales; smaller spiny scales on body below dorsal fins and on oblique folds of skin below lateral line.

Meristics. Dorsal fin rays X–XII, 19–24; anal fin rays 18–23; pectoral fin rays 16–20; pelvic fin rays I, 3, the inner third longest in females, second longest in males; lateral line scaleplates 45–48; vertebrae 42–47 (Pietsch 1993).

Color. Moustache sculpin are olive above; white, yellowish, or orange below. There is a pronounced black bar, the moustache, on the snout just above the jaws and under the anterior half of the eye. There are four dusky blotches above the lateral line on each side, one on the caudal peduncle, one passing through the second dorsal fin. The fins are variously marked with yellow and gray-black. The first dorsal of males has a dusky blotch between the first and second spines; the second dorsal is marked with three horizontal olivaceous bars. Prespawning males have immaculate white patches on the breast and along the sides. Females lack blotches on the first dorsal fin, and their second dorsal is marked with narrow lines of dots.

Size. This is a small species, usually 9–10 cm. The largest recorded for the Gulf of Maine was 15.2 cm.

Distinctions. The most distinctive feature of the moustache sculpin apart from the very long anal fin, is the row of about 45 broad platelike scales along the lateral line on each side, with smaller spiny scales below the dorsal fins and the oblique folds of skin below the lateral line. The body is more tapering than those of other sculpins, the caudal peduncle more slender, and the caudal fin smaller. The head is smaller and smoother than in any of the sculpins common in the Gulf of Maine, and they have short spines and many prickles.

Habits. This is a cool-water species living at depths of 18–110 m and caught as deep as 320 m (Scott and Scott 1988). Like other sculpins, they are bottom fish.

Food. Crustaceans constitute the major portion of the diet (83.5% by weight) followed by polychaetes (16.1%). Polychaetes, amphipods, the decapod *Dichelopandalus leptocerus,* and cumaceans are relatively more important in smaller moustache sculpin, 6–10 cm TL, and decapods, particularly pandalid shrimps, in larger fish, 11–15 cm TL (Bowman et al. 2000). Feeding may continue during spawning.

In January and February, the major food of moustache sculpin larvae (8.5–14.4 mm SL) in the Damariscotta River estuary, Maine (Laroche 1982), were adults of a calanoid copepod *Pseudocalanus minutus;* in March, the dominant prey items were *P. minutus* and nauplii of the barnacle *Balanus balanoides.* Other food items included other copepods and unidentified invertebrate eggs.

Predators. Predators include nine species of fishes, four of which were multiple occurrences: cod, white hake, sea raven, and longhorn sculpin (Rountree 1999); a bird, thick-billed murre (Gaston et al. 1985); and harp seal (Lawson and Stenson 1997).

Parasites. Moustache sculpin in Newfoundland waters were infested with a protozoan *Trichodina domerguei saintjohnsi.* Those in the Bay of Fundy did not have this parasite (Lom and Laird 1969). They were also host to larvae of sealworm, *Phocanema decipiens* (Scott 1954).

Reproduction. Ripe females were caught in the Bay of Fundy–Gulf of Maine in October (Musick and Able 1969) and in July in Canadian waters (Scott and Scott 1988). Fahay (1983) suggested that spawning occurs in autumn–winter. A ripe female 98 mm SL contained 1,965 eggs and a 106-mm female 2,739 eggs (Musick and Able 1969).

Eggs have not been described in detail but Cox (1921) described eggs from a ripe female as pinkish, 2 mm in diameter with many oil globules. Ovarian eggs were described as 2.0–2.2 mm in diameter, transparent, and amber-colored with 3–15 oil globules (Musick and Able 1969). Newly hatched larvae were 7–8 mm long; flexion occurred at 12 mm (Fahay 1983).

General Range. Restricted to the North Atlantic and the Atlantic sector of the Arctic Ocean (Pietsch 1993), from Hudson Bay and the southern end of Baffin Island south to Cape Cod along the American coast and to the White Sea on the European side of the Atlantic in rather deep water.

Occurrence in the Gulf of Maine. From the scarcity of records, it would seem that this cold-water fish is uncommon in the Gulf of Maine. Specimens have been recorded from the neighborhood of St. Andrews in the Bay of Fundy, Mt. Desert, a few from Massachusetts Bay and from off Race Point, Cape Cod, Gloucester, and Georges Bank. The most southerly record for it was about 10 miles east of Chatham, Mass.

The fact that Gilbert (1913) found some morphometric differences between Gulf of Maine and Newfoundland specimens with others from Nova Scotia and Georges Bank intermediate between them suggests that moustache sculpin are permanent residents of inner parts of the Gulf rather than there only as strays.

FATHEAD SCULPINS. FAMILY PSYCHROLUTIDAE

GRACE KLEIN-MACPHEE

The family Psychrolutidae was defined as monophyletic based on two autapomorphies, the infraorbital sensory canal connected with the operculomandibular canal and a highly specialized supratemporal commissure, and a combination of four synapomorphies (Yabe 1985). The family contains 7 genera and about 29 species (Nelson 1994). One species has been reported from the Gulf of Maine, the polar sculpin, *Cottunculus microps,* but a second species, known as the pallid sculpin, *C. thomsoni,* lives nearby, off Nova Scotia and Georges Bank, south to Rhode Island (Scott and Scott 1988) and is included in the key.

POLAR SCULPIN / *Cottunculus microps* Collett 1875 / Arctic Sculpin / Bigelow and Schroeder 1953:453–454

Description. Body tadpole-shaped, somewhat elongate tapering to a slender caudal peduncle (Fig. 194). Head very large with four bony knobs on top and several on sides; interorbital oblique; teeth on jaws and on vomer small, conical, absent on palatines; gill rakers poorly developed (Scott and Scott 1988). Two parts of dorsal fin united into one continuous fin, spiny

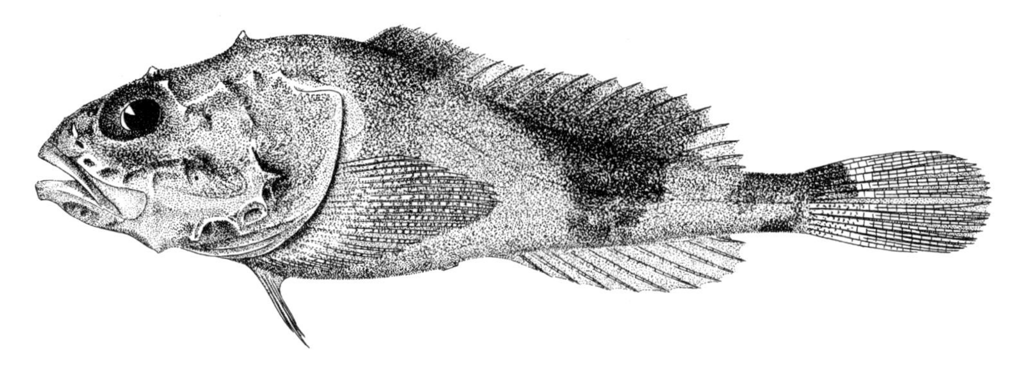

Figure 194. Polar sculpin *Cottunculus microps.* Continental slope off southern New England. Drawn by H. L. Todd.

of dorsal fin; pectoral fins large, fan-shaped; pelvic fins small; caudal fin small and rounded. Scales absent, skin loose, rough and warty. Lateral line straight and high on body.

Meristics. Dorsal fin rays VI–VIII, 13–15; anal fin rays 10; pectoral fin rays 17–19; pelvic fin rays I, 3; vertebrae 28 or 29 (Wheeler 1969; Scott and Scott 1988).

Color. Pale with dusky crossbars, one on the head, two on the body and fins, and one at the caudal fin base. The head is also spotted. Scandinavian specimens have been reported as having another band across the tip of the caudal and having the anal and pectoral fins darkly mottled.

Size. Up to about 20.3 cm; off west Greenland to 30 cm (Jensen 1952).

Distinctions. The head spines so characteristic of most sculpins are reduced in this species to four bony knobs on the top of the head and several on its sides. The two portions of the dorsal fin are united into one continuous fin, and the spiny part is shorter and lower than the soft part, distinguishing it from all other sculpins in the Gulf of Maine. The pallid sculpin, *C. thomsoni,* has almost naked skin, a larger eye, diameter greater than interorbital width, and it lacks body blotches.

Habits. An Arctic deepwater species found at depths of 220–877 m.

Food. On the continental slope of Norway, stomach contents of 14 polar sculpin (8–21 cm TL) were dominated by benthic prey such as pycnogonids (*Nymphon*) and polychaetes, which comprised 67% and 14% of weight, respectively (Bjelland et al. 2000). In addition, 14 amphipods, mostly hyperbenthic species, were present.

Parasites. Margolis and Kabata (1988) listed the copepod *Chondracanthus cottunculi* from the gills of polar sculpin.

General Range. This Arctic species is known off east Greenland and about Spitsbergen in the Arctic Ocean and from both sides of the North Atlantic. Off the American coast, they have been taken at numerous localities on the continental shelf and slope south to New Jersey. On the European side, they have been reported from northern Iceland, from Norwegian waters southward to the Channel, and doubtfully from the Skagerrak.

Occurrence in the Gulf of Maine. Polar sculpin have been collected sporadically in the Gulf; one in the extreme southeast corner of the basin (42°23′ N, 65°23′ W) in 254 m; one in the eastern channel between Browns and Georges banks (42°15′ N, 65°48′ W) in 220 m (Bigelow and Schroeder); and one on the northern slope of Georges Bank, in 216 m of water, which shows that they might be expected anywhere in deep basins of the Gulf at depths greater than 180 m.

SEA RAVENS. FAMILY HEMITRIPTERIDAE

GRACE KLEIN-MACPHEE

The Hemitripteridae is considered as the sister group of the Agonidae and defined as a distinct family based on the autapomorphy of having minute spines all over the body and a combination of seven synapomorphies (Yabe 1985). It includes one species that is found in the Gulf of Maine.

SEA RAVEN / *Hemitripterus americanus* (Gmelin 1789) / Red Sculpin, Sea Sculpin, Raven /

Bigelow and Schroeder 1953:454–457

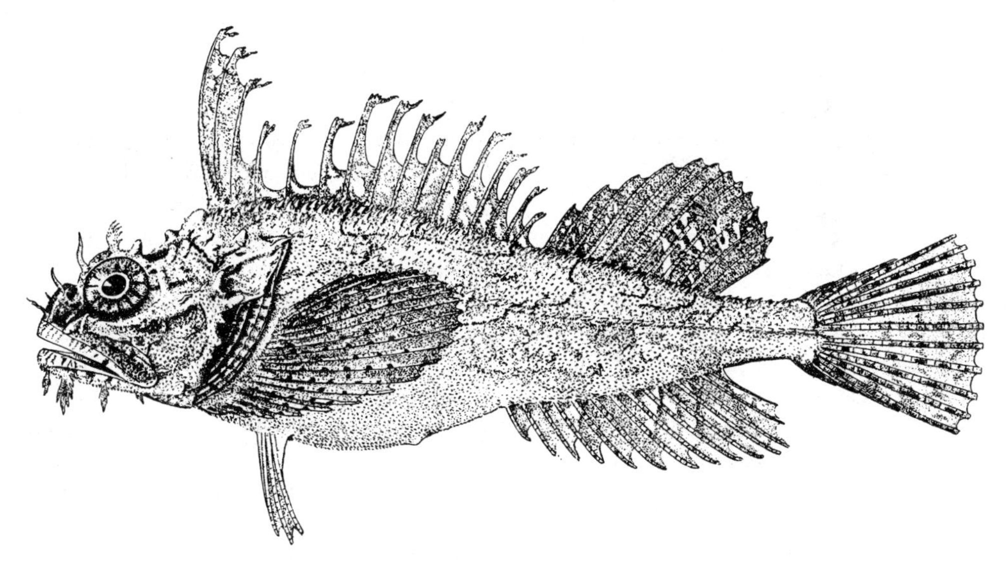

Figure 195. Sea raven *Hemitripterus americanus*. Halifax, N.S. Drawn by H. L. Todd.

Description. Body elongate, heavy in front, tapering to a moderate caudal peduncle (Fig. 195). Head large, with numerous bony humps and ridges on upper surface including a short, high keel on top of snout with a deep hollow behind it, another high ridge above each eye, a lower one below eye, and two short spines on preopercle. A series of seven to eleven simple or branched fleshy tabs along lower jaw and on head. Mouth large, terminal, and oblique with stout, sharp teeth in several rows on jaws, vomer, and palatines. Eye large, high on head. First dorsal fin irregular and ragged, first two or three spines longest, fourth and fifth spines shorter than those further back. Dorsal fin membrane deeply emarginate between every two spines from third spine backward, but expanded at tip of each spine as an irregular flap of skin. Anal fin margin similar to that of dorsal fin, but less deeply scalloped between rays. Second dorsal fin rounded and separated from first by a short space. Pectoral fins fanlike; caudal fin brush-shaped. Skin prickly, with prickles largest on back and along lateral line; smallest prickles, still obvious to the touch, on lower parts of sides and on belly.

Meristics. Dorsal fin rays XV–XVII, I, 12; anal fin rays 13 or 14; pectoral fin rays 18 or 19; pelvic fin rays I, 3; vertebrae 38 or 39.

Color. Sea raven vary in color from blood red to reddish purple, chocolate, or yellowish brown, but are invariably paler below than above, and usually have a yellow belly. Many are uniformly plain, others are mottled with a paler or darker cast of the general ground tint, or even with white. The fins are variously barred with light and dark, and the pectorals and anal are often yellow-rayed.

Size. One of the largest on record was 64 cm and 3.2 kg. Many sea raven are 46–51 cm long with weights of 1.4–1.8 kg. The all-tackle game fish record is a 1.47-kg fish caught in Manasquan Inlet, N.J., in January 1996 (IGFA 2001).

Distinctions. Sea raven are stouter-bodied than other sculpins, about three and three-quarter times as long as deep with a very large head. The most distinctive features are the fleshy tabs, simple and branched, on the head, the prickly texture of the skin and the curiously ragged outline of the first dorsal fin. Both jaws of the wide mouth are armed with several rows of sharp teeth that are noticeably longer and stouter than the teeth of either longhorn or shorthorn sculpins.

Habits. Sea raven inhabit rocky ground (their chief haunt from Massachusetts Bay northward), pebbles, hard sand, or clay, which they frequent off Cape Cod and on offshore banks. There is no definite upper limit to their vertical wanderings other than the surface, but most live deeper than 2 m, and their usual range extends down to about 91 m. The deepest one was taken was at 192 m. Preferred depths on the Scotian Shelf were 37–108 m (Scott 1982b). They are seldom caught within smaller estuaries, perhaps never on the tidal flats at any time of year.

They prefer cooler temperatures (6°–9°C on the Scotian Shelf), but in the Gulf of Maine they can tolerate a range of temperatures from near-freezing seawater to about 15.6°C. Sea raven possess an antifreeze protein that is a high-molecular-weight peptide (Kao et al. 1986). Unlike many other fishes that produce antifreeze proteins, sea raven do not appear to produce them seasonally, but rather maintain them year-round, although some individuals elevate production during the winter (Fletcher et al. 1984).

Sea raven, alone among Gulf of Maine sculpins, can inflate its stomach with water, like a bladder, when captured. If released in this condition it drifts helplessly, feebly waving its tail to and fro. It probably produces some type of sound, because the body vibrates similarly to that of longhorn during capture (Fish and Mowbray 1970).

Food. Sea raven are voracious consumers of benthic and near-benthic invertebrates, such as mollusks (both bivalves and gastropods), various crustaceans, sea urchins, worms, and fishes.

Over all size groups, fishes (73.8% by weight) and crustaceans (24%) are the most important components of the diet (Bowman et al. 2000). Smaller sea raven (6–15 cm TL) feed mostly on crustaceans, especially euphausiids and *Crangon septemspinosa*. Larger sea raven (16–45+ cm TL) eat mostly fishes. Fish prey include skates, silver hake, haddock, *Liparis* spp., cusk-eel, shorthorn and longhorn sculpins, rock gunnel, *Lumpenus lampraetiformis*, ocean pout, and wolffish (Maurer and Bowman 1975; Bowman et al. 2000).

In January and February, the major food of sea raven larvae (12.0–16.4 mm SL) in the Damariscotta River estuary, Maine (Laroche 1982), was fish larvae. This diet was unlike that of larvae of other sculpins (*Myoxocephalus* and *Triglops*) which fed on diatoms and crustaceans. Prey included larvae of three of the other four species of sculpins found with sea raven larvae, *Myoxocephalus aenaeus* (most important prey), *M. octodecemspinosus*, and *Triglops murrayi*, as well as larval *Pholis gunnellus* and *Lumpenus lampraetiformis*.

Predators. Occasionally eaten by little skate, spiny dogfish, goosefish, cod, longhorn sculpin, and halibut (Rountree 1999).

Parasites. Sea raven host several endo- and ectoparasites, including two protozoans; two myxosporidians; at least nine trematodes; a cestode, *Bothriocephalus scorpii;* and two nematodes (Margolis and Arthur 1979). Sea raven from the Labrador-Newfoundland area also had blood parasites (Khan et al. 1980).

Breeding Habits. Adult females contain 15,000 maturing eggs on the average and occasionally as many as 40,000. Presumably sea raven breed throughout their geographic range. Off southern New England, eggs are deposited from early October until late December, and probably in autumn and early winter in the more northern part of its range as well. Spawning takes place 0.6–1.8 km from shore at 18–27 m depth. Spawning temperatures were recorded at 2.9°–14°C (Warfel and Merriman 1944b). The eggs are deposited chiefly at bases of the branches of the finger sponge (*Haliclona*) and less often on the smaller sponge *Halichondria panicea,* where they stick together in clusters and to the sponge. Since the eggs average only about 242 per cluster (minimum 141, maximum 478, among many clusters counted), it appears that a female does not lay all her eggs at one time, but deposits many clusters during each spawning season. Larvae have been taken from the Bay of Fundy to Cape Cod so they presumably breed in the Gulf wherever they occur.

Early Life History. The eggs are large, averaging 3.9 mm in diameter (3.75–4.09 mm), with a tough egg membrane 0.1 mm thick, yellow when first spawned, but soon changing to an amber or light orange hue. There is one unpigmented oil globule of about 0.8 mm diameter. Eggs brought into the laboratory by Warfel and Merriman (1944b) hatched a few at a time, and some of those in a cluster collected on 23 January and left thereafter in a bottle fastened to a buoy at ambient temperatures in Long Island Sound did not hatch until 12 March.

Larvae are 10–14 mm at hatching. The eyes are pigmented and the mouth is well developed. The body is moderately deep, snout moderately pointed, gut long and bulging out even more than is usual for sculpins. Larval sea raven can be distinguished from those of other sculpins by the degree of lateral pigment. The body is almost completely pigmented except for the peduncle area and a narrow area posterior to the gut. Spines begin forming 3 months after hatching (Fuiman 1976): four preopercular spines, two parietal, two supracleithral, and two opercular (develop in juveniles). The caudal fin develops soon after hatching and all fin rays are usually complete at 18–20 mm TL. Total myomeres number 38 or 39. The first dorsal fin (16 or 17 spines) is much longer than that in other sculpins (Fahay 1983).

Trophic studies of larvae in the Damariscotta estuary, Maine, showed that sea raven were present from January to April, but were the rarest of the five species of sculpins collected (grubby, longhorn, shorthorn, and moustache). Sea raven larvae were larger than larvae of the other sculpins and selected the largest prey, mainly fish larvae and decapod zoea, mostly larger than 800 μm (Laroche 1982).

The few young sea raven that have been taken in the Gulf of Maine suggest that they reach lengths of 61–102 mm by the middle of their first summer, when 6–8 months old, and 152 mm by the following April, at an age of about 1.5 years. Subsequent growth rate is not known.

General Range. Atlantic coast of North America, from Hamilton Inlet, Labrador (Backus 1957b), south to Chesapeake Bay. To the eastward and northward, sea raven are described as common all along outer Nova Scotia to Canso; they have been reported on Sable Island Bank and on Banquereau Bank in depths of about 36–54 m, have been taken on the Gulf of St. Lawrence coast of Cape Breton, and reported from Anticosti and in the Belle Isle Strait. There is only one report from the Grand Banks and one, Trinity Bay, for the Atlantic coast of Newfoundland. Sea raven are rather common south to New York and New Jersey.

Occurrence in the Gulf of Maine. Sea raven are found all around the coast of the Gulf (Map 24), but are not known to occur on the soft mud bottoms of its deep troughs and basins. Although generally distributed in the Gulf, sea raven are not as numerous as shorthorn and longhorn sculpins, which is true also in the Bay of Fundy, Massachusetts Bay, and Georges Bank.

Seasonal Migrations and Movements. Off the southern shores of New England, sea raven work inshore in autumn and out again into slightly deeper water in spring, but no seasonal movement of this sort has been reported in the cooler waters of the Gulf of Maine.

Importance. Although sea raven are said to be a good table fish, there is no more market for them than for other sculpins in New England or Canada, although they are generally considered excellent bait for lobster pots.

ALLIGATORFISHES. FAMILY AGONIDAE

Grace Klein-MacPhee

Alligatorfishes are curious little fishes anatomically related to sculpins, although their general appearance gives no hint of it. They are defined as monophyletic based on three autapomorphies, including the body being covered with several rows of overlapping plates and a combination of seven synapomorphies (Yabe 1985). Some agonids have a spiny dorsal fin that others lack, whereas the pelvic fins are situated far forward (only a little rearward of the pectorals) in all of them. There are 20 genera and 44 species in the family (Kanayama 1991; Nelson 1994), most of which occur in the North Pacific; only two species are known from the eastern coast of North America.

One of the Atlantic species (*Leptagonus decagonus* Bloch and Schneider 1801), with two dorsal fins, is Arctic, ranging south only to northern Nova Scotia; the other, with only one dorsal fin (*Aspidophoroides monopterygius*) is a regular member of the Gulf of Maine fish fauna.

ALLIGATORFISH / Aspidophoroides monopterygius (Bloch 1786) / Sea Poacher /

Bigelow and Schroeder 1953:457–459

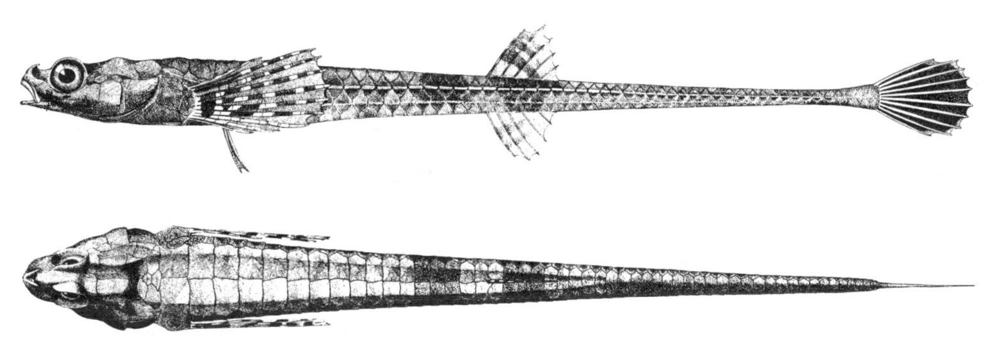

Figure 196. Alligatorfish *Aspidophoroides monopterygius.* Halifax, N.S. Lateral view (above) and dorsal view (below). Drawn by H. L. Todd.

Description. Body elongate, very slender (12–13 times as long as deep), broader than deep, tapering posteriorly from head to very slender caudal peduncle (Fig. 196). Trunk octagonal in front of unpaired fins, hexagonal behind them. Head small. Eyes very large, with prominent ridges above them. Two sharp recurved spines on top of nostrils. Mouth small with minute teeth on jaws, vomer, and palatines. Dorsal and anal fins fan-shaped, one over the other, situated about midbody. Pectoral fins larger and fanlike, pelvic fins reduced. Caudal fin small and rounded. Two large plates and several small ones in front of each pectoral fin. Body covered with smooth plates arranged in longitudinal rows, eight dorsal and ventral rows

Meristics. Dorsal fin rays 5 or 6; anal fin rays 4–6; pelvic fin rays I, 2, longer in adult males than females; pectoral fin rays 10 or 11; scales occur in the form of plates in 46–49 dorsal rows; lateral line straight.

Color. Light to dark brown above, lighter brown to white below, with two darker crossbands between the pectoral fins and the dorsal fin; one crossband under the dorsal, and two or three crossbands between the dorsal and caudal fins. Dorsal and pectoral fins are more or less barred, sometimes with milky white pigment down distal end (Scott and Scott 1988); the caudal is dusky with a white border.

Distinctions. Alligatorfish somewhat suggest pipefish in their armor and slender form, but there is no danger of confusing one with the other, as the mouth of the former is of the ordinary form and pelvic fins are present.

Habits. Nothing is known of the life of alligatorfish except that they are bottom fish. They have been trawled on pebbly bottom, on sand and broken shells, and on soft mud in the Gulf of Maine and over sand and mud off Labrador. So far as is known, adults never stray into water shoaler than 18–27 m, and the deepest record is 332 m off west Greenland (Jensen 1942). They were collected in depths of 40–88 m in Massachusetts Bay (Collette and Hartel 1988). Their upper temperature limit is about 9°–11°C, and their lower limit is close to the freezing point of saltwater.

Food. Only crustaceans were found in 24 stomachs (Bowman et al. 2000), mostly caprellid (49.3% by weight) and aorid (36.0%) amphipods, with some decapods (3.4%).

Predators. Alligatorfish have been found in stomachs of spiny and smooth dogfishes, hakes (red, white, and silver), cod, and halibut, of which cod is the only significant predator (Rountree 1999).

Breeding Habits. Breeding habits are unknown. The pelvic fins are longer in males than in females. A medium-sized female contained 600 large eggs (Jensen 1942). Presence of larvae in Passamaquoddy Bay, off Boothbay, and near Seal Island, N.S., from April to June points to late autumn and early winter as the spawning season in the Gulf of Maine. Larvae are present in Salem Harbor from March through April (Elliott and Jimenez 1981). They occur in the plankton up to 29 mm long.

General Range. Found from west Greenland and the east coast of Labrador south to Cape Cod, and to northern New Jersey as a stray. They are apparently widespread over the eastern half of the Grand Banks, along eastern Newfoundland, and off southeastern Labrador. They are numerous enough in the southern part of the Gulf of St. Lawrence to be described as "characteristic" of the ice-cold banks water there. They have been reported in the St. Lawrence River estuary near Trois Pistoles and at several localities along the west coast of Newfoundland. The only records to the west of Cape Cod are of the head of one that was dredged off Watch Hill, N.J., in 1874; and one that was taken off Sandy Hook, N.J., in 1864.

Occurrence in the Gulf of Maine. Alligatorfish have been taken throughout the Gulf of Maine from the Bay of Fundy to east of Cape Cod (Map 25). They may be expected anywhere in the Gulf in depths of 18–180 m. They ranked 27th in abundance of 43 species in the BIOME station collections in Massachusetts Bay (Collette and Hartel 1988) and are perhaps more plentiful along the Nova Scotian shelf eastward and northward from Cape Sable.

ACKNOWLEDGMENTS. The cottoid section was reviewed by Thomas A. Munroe, and the food habits were checked by Ray Bowman.

Lumpfishes and Snailfishes. Superfamily Cyclopteroidea

Lumpfishes and snailfishes form a monophyletic unit and are sometimes combined into one family. Two families, Cyclopteridae and Liparidae, both having the pelvic fins (if present) modified into a thoracic sucking disk, are recognized (Nelson 1994). A large group of benthic and pelagic marine fishes found mostly in cold waters of the Arctic and Antarctic, these families comprise 26 genera and more than 230 species (Nelson 1994). Larval characters and comparative information on lumpfishes and snailfishes was presented by Able et al. (1984).

KEY TO GULF OF MAINE CYCLOPTERIDS AND LIPARIDS

1a. Two dorsal fins, first dorsal fin embedded in the skin and not visible in adults; skin usually with prominent conical tubercles; caudal fin distinct from dorsal and anal fins (Cyclopteridae) 2
1b. Single, long dorsal fin; skin naked (with small dermal prickles in males during breeding season); caudal fin usually confluent with dorsal and anal fins (Liparidae) 3
2a. Small irregularly spaced tubercles on body; first dorsal fin distinguishable only in juveniles; body triangular in cross section . . . . **Lumpfish**
2b. Large tubercles on body; first dorsal fin distinguishable in juveniles and adults; body nearly round in cross section . . **Atlantic spiny lumpsucker**
3a. Pelvic disk absent 7
3b. Pelvic disk present 4
4a. Disk length many times larger than eye diameter; nostrils with double opening; anal fin rays 25–35 (snailfishes, *Liparis*) 5
4b. Disk length approximately equal to eye diameter; each nostril with a single opening; anal fin rays 42–49 . . . . **Flatdisk snailfish**
5a. Dorsal fin rays anterior to prominent notch (rays 4–7) elongate (not as prominent in female *L. atlanticus*); each nostril with a double opening; anal fin rays 25–31 6
5b. Anterior dorsal fin rays not elongate, notch in dorsal fin absent; anal fin rays 30–35 . . . . **Gulf snailfish**
6a. Dorsal and anal fins reaching only to base of caudal fin; pyloric caeca 23–45 . . . . **Atlantic sea snail**
6b. Dorsal and anal fins overlapping one-quarter to one-half of caudal fin; pyloric caeca 14–21 . . . . **Inquiline snailfish**
7a. Teeth in single row . . . . ***Paraliparis copei***
7b. Teeth in multiple rows . . . . ***Paraliparis calidus***

LUMPFISHES. FAMILY CYCLOPTERIDAE

GRACE KLEIN-MACPHEE

Lumpfishes inhabit cool marine regions of the Northern Hemisphere. The body is globose and usually covered with tubercles. Two dorsal fins are present but the first dorsal in adults may be covered by skin. Of the seven genera and 28 species (Nelson 1994), two occur in the Gulf of Maine.

LUMPFISH / *Cyclopterus lumpus* Linnaeus 1758 / Lump, Lump Sucker / Bigelow and Schroeder 1953:459–463

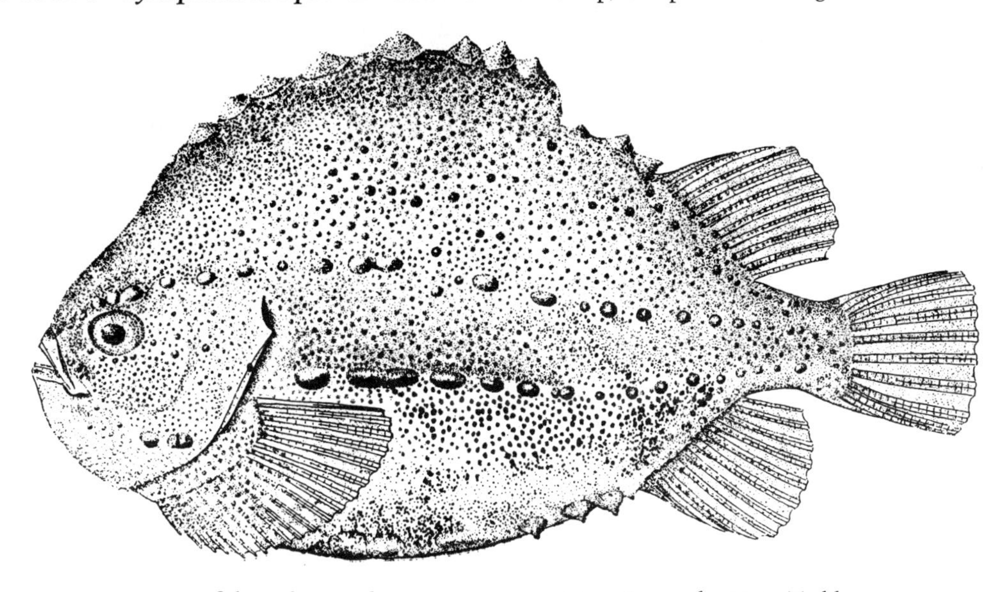

Figure 197. Lumpfish *Cyclopterus lumpus.* Eastport, Maine. Drawn by H. L. Todd.

Description. Body twice as long (including caudal fin) as deep, with short head, dorsal profile of trunk much more arched then ventral (Fig. 197). Seven longitudinal ridges on body. One ridge runs along back as a partly cartilaginous, partly gelatinous hump that encloses first dorsal fin in adults and continues rearward as two ridges from first dorsal fin to second. Three ridges along each side, one over eye, one close above level of pectoral fin, and one marking transition between side and belly. Trunk roughly triangular with flat belly and sharp back. Body scaleless, but each ridge marked by line of large, pointed tubercles; skin between ridges thickly studded with small knobs. Head profile characteristic: concave above, convex below, mouth at tip of snout. Jaws with small teeth in a single row. Eyes moderate-sized. Gill openings large. First dorsal fin visible only on very small specimens. Second dorsal fin and anal fin similar in outline. Caudal fin broad and square-tipped or slightly convex. Pectoral fins large, rounded, and broad-based, nearly meeting on throat; larger on males than females. Pelvic fins not visible as such, altered into six pairs of fleshy knobs, surrounded by a roughly circular flap of skin modified into an adhesive "sucking" disk. Entire disk about as wide as head width, situated close behind throat.

Meristics. First dorsal fin spines, when visible, VI–VIII; second dorsal and anal fin rays 9–11; pelvic fin rays 6 fleshy knobs; pectoral fin rays 20 or 21; vertebrae 28 or 29 (Scott and Scott 1988).

Color. Descriptions of this fish credit it with a great variety of tints. On adults the ground tint may be bluish gray, olive, brownish or yellow green, chocolate or kelp brown, or slaty blue, the belly usually being of a paler or more yellowish cast of the same hue, but sometimes whitish. On some specimens the back and sides are marked with dark blotches and more or less dotted with black. Others, however, are plain-colored or nearly so, except that the tubercles are usually dark-tipped. Young lumpfish often match their surroundings very closely in color, usually being bright green, mottled olive green, and ochre yellow with silvery dots and stripes. Color changes depend on light intensity (Davenport and Bradshaw 1995), and color matching depends largely on changes in color value (lightness/darkness). The default color is light green. Males, when mature, are more vividly colored than females, and their bellies turn red (brightest near the sucking disk) during the breeding season (Davenport and Thorsteinsson 1989). Coloration of adult lumpsucker off Iceland was described as predominantly green or yellow-green when on offshore feeding grounds; as they move into shallow water breeding grounds both sexes become darker dorsally and females lose their green color. Many males become purple, some remain green, and all acquire red breeding colors.

Size. The longest lumpfish so far recorded from the American coast measured 58.4 cm and weighed 6 kg; the heaviest weighed 9.1 kg but measured only 54.6 cm (both from Orient Point, N.Y.). Canadian and European records are up to 70 cm (Cox and Anderson 1922) and 9.64 kg (Collins 1976). The average size of lumpfish from commercial catches around Labrador were 35.5 cm for males (range 31–39 cm) and 44.8 cm for females (range 36–53 cm) (Stevenson and Baird 1988). Few are longer than 35.5–40.6 cm or heavier than 1.4–2.7 kg. Females average larger than males.

Distinctions. Lumpfish can be distinguished from other Gulf of Maine fishes by their short, thick, high-arched body; tubercle-covered skin; and pelvic fins reduced to knobs surrounded by the bony sucking disk. They are distinguishable from spiny lumpfish by the concealment of the first dorsal fin by skin in the adult; the smaller, more irregularly spaced tubercles; and the more triangular shape of the body compared to the rounder spiny lumpfish.

Habits. Adult lumpfish are primarily fish of cold and temperate waters. They were thought to be bottom fish dwelling in rocky nearshore areas but several studies describe them as semipelagic during early life and as hiding in floating masses of rockweed or swimming in the pelagic zone as adults outside of breeding season in the eastern North Atlantic (Blacker 1983) and in the St. Lawrence River estuary and Gulf of St. Lawrence (Able and Irion 1985). In the Gulf of Maine they range from one to several meters, but have been netted in winter by commercial trawlers off Newfoundland in depths from 182 to 329 m (Collins 1976). In the Gulf of St. Lawrence they ranged from 78 to 383 m, but most were found between 141 and 240 m (Able and Irion 1985). In European seas they range from the tide mark down to 83–111 m. Sometimes they strand on the beach (Bigelow and Schroeder) and have been reported stranded in Minas Basin, N.S., during extreme low tide (Bleakney and McAllister 1973).

Although lumpfish appear ungainly, they can swim more rapidly for a short distance by vigorous tail strokes than their shape might suggest, and their pelagic larvae are very active. Males and females swim differently: females with aid of their pectoral fins and males with a stronger tail stroke of greater velocity and amplitude (Davenport and Kjorsvik 1986). They are adapted for a pelagic lifestyle both anatomically and physiologically, particularly females, by having an almost completely cartilaginous skull, vertebral column, and branchial skeleton; an extensive thick, stiff, subcutaneous gelatinous tissue; reduced-density muscles (high fat, watery tissue); and low-density ovarian fluid (Davenport and Kjorsvik 1986).

Large lumpfish are often found hiding among rockweed or holding fast by the sucker to stones or other objects. About Massachusetts Bay, lobster pots are favorite resorts for them when set on stony bottom. Occasionally one is found clinging to one of the poles of a trap or weir, although this is a much less common event in the Gulf of Maine than in Scottish waters, where they are frequently caught in salmon nets set along the shore. They have (rarely) been found clinging to floating logs or inside a floating box or barrel, and there is at least one record of a lumpfish clinging to a mackerel.

Food. Cox and Anderson (1922) reported pelagic prey such as euphausiid shrimps (*Meganyctiphanes*), fragments of jellyfish (*Aurelia*), amphipod crustaceans (*Hyperia*), and caprellid crustaceans, with the remains of small fishes in stomachs of lumpfish from Passamaquoddy Bay. Large numbers of young clupeids have occasionally been found in their stomachs. This is one of the few fish that regularly feeds on ctenophores and medusae; 25 specimens examined at Woods Hole by Vinal Edwards contained nothing but ctenophores. In British waters food has been found to consist chiefly of isopods, amphipods, euphausiids, and other small crustaceans, with various other invertebrates, including worms, ctenophores, and soft-bodied mollusks, as well as fish eggs (Garrod and Harding in Blacker 1983). Lumpfish cease feeding during the spawning season and males do not feed while brooding egg masses (Davenport and Thorsteinsson 1989).

Food of larval and juvenile lumpfish in tide pools consisted of amphipods, copepods, isopods, cumaceans, acarinas, polychaetes, and mysids. Amphipods and mysids were the most important (Moring 1989). Food of pelagic juveniles consisted of harpacticoid and calanoid copepods, amphipods, and crab megalops. There was a change in diet with size: smaller lumpfish fed extensively on harpacticoid copepods and an amphipod *Calliopius* and larger juveniles (>25 mm) fed on amphipods, crab megalops, isopods, polychaetes, and insects. The harpacticoid copepod *Harpacticus chelifer* was particularly important to small lumpfish (Daborn and Gregory 1983). This switch in diet from harpacticoid copepods to crab megalops with increasing size was also noted in pelagic juveniles in Galway Bay, Ireland (Tully and O'Ceidigh 1989).

Predators. Lumpfish are said to be a favorite food of seals and also have been found in stomachs of sperm whale off Iceland (Roe 1969). Ocean pout and cunner prey on the eggs in their nests as does a sea urchin, *Strongylocentrotus droebachiensis* (Goulet et al. 1986). Lumpfish are occasionally found in stomachs of spiny dogfish and cod (Rountree 1999).

Parasites. Lumpfish parasites include protozoans *Cryptobia dahli* in the stomach and urinary bladder, *Trichodina domerguei saintjohnsi,* and *T. galyae;* and copepods *Caligus elongatus* and *Lernaeocera branchialis* (Margolis and Arthur 1979).

Breeding Habits. Fecundity is 15,000–200,000 eggs (Zhitenev 1970). A 45.7-cm female produced 136,000 eggs. In Europe they mature at about 5 years old, although this is size- rather than age-dependent (Davenport and Thorsteinsson 1989). They were said to mature at 12.7 cm in the Bay of Fundy (Cox 1920). In Scotland, size at maturity was 22.9 cm, 0.3 kg for males and 35.6 cm, 0.3–0.36 kg for females (Fulton 1906).

Courtship, spawning behavior, and parental care of lumpfish have been described from field observations in Newfoundland (Goulet et al. 1986). Males arrived inshore in early spring before females and established nest sites. Females arrived later asynchronously so males had the opportunity of courting and mating with more than one female. Males displayed nuptial coloration while courting. The entire body darkened to a deep grayish black; a small metallic silver patch became visible along the flanks posterior to the pectoral fin margin and below the second row of body tubercles; the ventral surface, pectoral fins, sucking disk, and anal and caudal fins became bright orange-red. This coloration was noted several days before spawning but faded about 20 min after spawning.

When a female approached a male's nest, he became visibly active and attentive. The male initiated courtship, which could last several hours. Courtship typically consisted of nest cleaning, fin brushing of the females body, and quivering. Females remained relatively passive at the nest throughout courtship, but sometimes participated in nest cleaning.

During spawning the female approached the nest site, turned on her side, released a batch of bright pink eggs, and then turned upright and swam away. The male immediately swam to the nest and released milt over the eggs. After fertilization, the male pushed the eggs into the nest crevice, molded them with his snout to form a characteristic mass with a funnel-like depression, and then stationed himself beside the egg mass to guard it.

Male lumpfish remain with the eggs during the entire incubation period. Parental care consists of aeration and antipredator behavior. Aeration is accomplished by pectoral, dorsal, and caudal fanning, which creates water currents over the egg mass, and puffing, which involves expelling water from the mouth onto the eggs, presumably forcing aerated water deeper into the mass. Fanning may also disperse the ammonia that is released by the eggs in considerable quantities during the first 2–3 days of development, probably as a consequence of the hardening process (Davenport 1983; Davenport et al. 1983). Oxygen uptake of the eggs is low until they are close to hatching and then increases dramatically (Davenport 1983). Antipredator behavior consists of removing invertebrates such as sea urchins and periwinkles from the eggs and chasing fishes such as cunner and ocean pout away from the nest. Cunner often attack the nest in groups, and a lumpfish male may be unable to repel them. When he abandons a nest, the eggs are eaten. Although parental care is essential for hatching success, neither the size of the guarding male nor the degree of his aeration efforts appears to influence success. When the eggs hatched, the male's fanning and puffing behavior swept emerging larvae from the nest, and water currents carried them away.

There was no evidence of female mate selection involving size of males or the characteristics of the nest sites in the courtship process (Goulet and Green 1988).

In Scottish waters, spawning takes place from February until near the end of May; and evidence from ichthyoplankton collections suggests an equally protracted spawning season in the Gulf of Maine. Larvae were collected in Salem Harbor from April to September (Elliott and Jimenez 1981). The season is April to June in Nova Scotia (Cox and Anderson 1922) and May to July in Newfoundland (Goulet et al. 1986; Goulet and Green 1988). Late May is probably the height of the breeding season.

Spawning occurs nearshore in shallow water in Newfoundland and Nova Scotia and probably along the coast of the Gulf of Maine. Capture of recently hatched larvae over Georges Bank is evidence that it serves as a spawning ground in 27–46 m. The lower depth limit to spawning has yet to be determined. In Europe there is evidence that breeding lumpfish return to the same spawning locality each year (Davenport and Thorsteinsson 1989). Spawning may be intermittent with two or three spawning events taking place based on collections of unspawned and partially spawned individuals (Gregory and Daborn 1982). Egg masses are deposited in nests consisting of crevices or depressions in the substrate, frequently on rocks in seaweed beds. The crevices include rock interstices and cracks in boulders or bedrock. Nests have never been found on a flat smooth surface of a boulder or bedrock (Goulet and Green 1988).

Spawning off Newfoundland may be temperature-dependent, occurring when the water warms to 4°C (Collins 1976); however, egg masses have been collected at 3°C (Benfey and Methven 1986) and spawning was observed at −1°C (Goulet et al. 1986).

Early Life History. The eggs are 2.2–2.6 mm in diameter, demersal, adhesive, pink when first laid with multiple oil droplets, which fuse into a single drop as development proceeds (Cox and Anderson 1922). The chorion is very thick and hard (Lonning et al. 1988). The color of the egg mass is variable and usually changes after deposition to pale green or yellow, deepening in tint as development progresses. Unfertilized egg masses have also been reported as reddish, salmon, lilac, pale violet, pale brown (Fritzsche 1978), and reddish purple (Davenport and Thorsteinsson 1989). Cleavage takes place 12 h after fertilization (Lonning et al. 1988), and development is slow. Gastrulation may not take place until 9–10 days after fertilization (Davenport 1983). During egg development the vascular system of the yolk forms at the time of embryo segmentation, with the heart starting to function shortly thereafter; erythrocytes are found at the beginning of eye pigmentation; and embryonic paired fins develop (Zhitenev 1970).

Incubation takes 10 (Mochek 1973) to 70 (Cox 1920) days. Goulet et al. (1986) reported 25–53 days in Newfoundland. Newly hatched lumpfish average 5.6 mm SL and 2.4 mg, were well developed with the adult complement of 20–21 pectoral fin [illegible] were about 4–7.4 mm long at hatching, shaped like tadpoles with a large head and slender tail, swimming actively, and soon able to cling to any bit of weed (Bigelow and Schroeder). Pigmentation was present on the head and some larvae had diffuse and scattered pigment on the trunk (Benfey and Methven 1986). Pigment was yellowish green with brassy luster. Head, pectoral fins, and caudal fin base are dotted with dark spots. As fins grow coloring increases, and there is one unpigmented band between the eyes and another between the snout and gill opening (Cox 1920; Cox and Anderson 1922). By day 3 (5.6 mm SL) adult complements of anal and second dorsal fin rays formed; the notochord was not flexed but hypural elements were beginning to form. By day 9 (6.1 mm SL) the adult complement of first dorsal and caudal fin rays was formed (Benfey and Methven 1986). The yolk sac was absorbed by 15 days posthatch (Brown 1986). Tubercles appeared behind and over the eye at 18 mm; the two side rows reach midbody and the ventral rows of flattened papillae appear at 22 mm; the upper lateral series is complete and numerous small tubercles are scattered thickly over the body at 25 mm (Cox 1920; Cox and Anderson 1922).

Larvae 10–23 mm have a light brown head with a darker band extending from the nostrils above the eye to the base of the first dorsal; a light blue band extends from the rear of the orbit to the top of the operculum and in front of the eyes to the nostrils; there is a blue spot at posterior base of dorsal fin. The remainder of the body is straw-colored but at a later stage it is usually bright olive green, darkest toward the dorsal side with the same blue band extending from behind the orbit to the operculum, with one or two round blue spots above the level of the pectoral fin along the lateral line. Some specimens are a bluish neutral slate tint uniformly spotted with darker pigment cells with the same blue band on the head (Fritzsche 1978).

Juveniles 23–34 mm have the adult complement of dorsal, anal, and pectoral fin rays. At 34 mm tubercles appeared to be well developed and they show most of the characters of the adult, except for the large first dorsal fin and slender form. Pigmentation is green, olive, brown, or yellow, and they often take on the color of their surroundings (Fritzsche 1978).

Larval lumpfish have rarely been reported in ichthyoplankton collections in the western North Atlantic, but early juveniles occur in near-surface waters of the Bay of Fundy from July to September (Daborn and Gregory 1983). Juveniles have also been taken at the surface in the Gulf of Maine in tow nets wherever there were floating masses of rockweed (a refuge in which all but the smallest regularly hide or cling to the fronds). Lumpfish taken in tow nets or dipped up were usually less than 4 cm long. Most young lumpfish left the surface by winter; indeed very few were taken at any depth in the Gulf of Maine during the cold months (Bigelow and Schroeder). Juveniles have also been found associated strongly with floating drift weed in Galway Bay off Ireland (Tully and O'Ceidigh 1989) but not in the Bay of Fundy (Daborn and Gregory 1983).

Larvae and juveniles in Maine use tide pools as a nursery area (Moring 1990). Lumpfish young ranging in size from 5 to 80 mm begin entering tide pools in June and occupy them through December. They were among the most common species found in tide pools along with threespine stickleback, rock gunnel, and silverside. Marking experiments showed that many lumpfish remain in individual tide pools for several days and even weeks (Moring and Moring 1991). Fish transplanted to a different tide pool left the site the next day and some returned to the original pool, perhaps indicating some homing. Disappearance from tide pools in winter appeared to be the result of low water temperatures and reduced algal cover. Lumpfish were not collected in pools until water temperatures reached 12.7°C, and they remained until temperatures dropped below 9.3°C (Moring 1990). Lumpfish young have also been collected in tide pools in Nahant, Mass. (Collette 1986) and have been described as abundant in rock pools along the littoral zone in the British Isles (Russell 1976).

Larvae, young-of-the year, and age-1 lumpfish in Maine tide pools were associated with algae and seagrass (*Zostera marina*), most frequently with a brown alga *Laminaria* when this species was present, but in its absence with *Zostera*. They also attached to brown algae *Fucus vesiculosis* and *Ascophyllum nodosus,* red algae *Rhodymenia palmata* and *Agarum cribosum,* and to a lesser extent to 12 other species of algae and to blue mussel. In tide pools containing both *Zostera* and *Ascophyllum,* the relationship changed with lumpfish size. Individuals smaller than 20 mm were more often found attached to *Zostera* and larger ones to *Ascophyllum.* These associations with algae and seagrass appear to be for protection from predators and direct wave action, as well as for proximity to invertebrates living on the vegetation (Moring 1989).

Ontogeny of feeding behavior (Brown 1986) is not typical larval feeding behavior, but is influenced by presence of the ventral adhesive disk. The larvae cling to surfaces during much of the first 2 weeks of free-swimming life. They do not take the S-shaped feeding posture that most larval fish use to initiate prey capture, but undertake prey search and capture from a clinging position. During weeks 1–5 larvae spent more time clinging than swimming; swimming increased afterward, but the feeding repertoire remained relatively unchanged throughout the first year of life.

Age and Growth. Lumpfish have been aged using vertebrae, but this was not considered successful because of faintness of marks (Cox and Anderson 1922) and irregularity in patterns (Blackwood 1983 in Stevenson and Baird 1988). Thorsteinsson (1981) used otoliths from lumpfish in Icelandic waters successfully. Most individuals from the roe fishery were 5–9 years old; the oldest was 12. Size ranges of female lumpfish collected for the roe fishery in Newfoundland were similar to those from Iceland so the age structure may be similar. Age and growth was plotted from average lengths of fish from the Bay of Fundy (Cox and Anderson 1922; Daborn and Gregory 1983) and from Iceland (Thorsteinsson 1981).

The length-weight relationships for juvenile lumpfish in the Bay of Fundy was $W = 8.7L^{3.36} \times 10^{6}$ (Daborn and Gregory

1983). The length-weight relationship for females taken in the Labrador roe fishery was $W = 0.1993L^{-4.5476}$ for a size range of 37–50 cm (Kean 1979 in Stevenson and Baird 1988).

General Range. Both sides of the North Atlantic. In the western North Atlantic, they are found northward from western Greenland, Davis Strait, Hudson and James bays, along the coast of Labrador, the Gulf of St. Lawrence, Newfoundland, Nova Scotia southward to New Jersey, and to Chesapeake Bay as strays. In the eastern North Atlantic from east Greenland, Iceland, Faroes, White Sea, northern Norway, the Norwegian Sea, northern and western parts of the Barents Sea and off the northwest coast of Spitsbergen, over most of the North Sea to north of Shetland and the Faroes, and off the British coast to the Bay of Biscay and occasionally to Portugal in the east (including the Baltic).

Occurrence in the Gulf of Maine. Lumpfish are common along the outer coast of Nova Scotia and are found all around the shores of the Gulf of Maine. They have been reported at Yarmouth and in St. Mary's Bay on the Nova Scotian side and are abundant in all stages at various localities in the Bay of Fundy. There are many records from the Maine coast, including Eastport, Penobscot Bay, vicinity of Boothbay, the offing of Seguin Island, and Casco Bay, and from Massachusetts waters, where they have been reported regularly at Nahant, Swampscott, Cohasset, Plymouth, Truro, along Cape Cod, and at Monomoy. They were once picked up in the deep bowl between Jeffreys Ledge and the coast. They even enter river mouths, but are never found where water is appreciably brackish.

Migrations and Movements. So far as is known the only regular migration carried out by lumpfish in the Gulf of Maine is a general movement of adults into shoal water at spawning time followed by an offshore movement afterward. In European waters they come inshore to spawn from February to August and then migrate offshore and are captured at sea from August to February in pelagic trawls (Blacker 1983). Tagging of lumpfish in Newfoundland showed that most fish were collected within 19 miles of the tagging site and suggested that they return to the same spawning grounds each year (Blackwood 1983 in Stevenson and Baird 1988).

Importance. Lumpfish are not eaten in the United States, but in Canada they were formerly fare in several communities in Newfoundland and were used as food for animals and as lobster bait. There is at present a growing inshore fishery for roe in Newfoundland waters conducted by small vessels (35–65 ft long) (Stevenson and Baird 1988). The season begins in April and ends in July, coincident with the lumpfish breeding season. It is primarily a gill net fishery with incidental catches in cod traps and salmon nets. Since the flesh is not marketable, the ovaries are extracted from females at sea and the carcasses are discarded. Males are released alive. There is an optimum stage of development for lumpfish eggs to be used commercially (Dewar et al. 1971). Mature eggs that are purple or red in color are at the desired stage; gray-white eggs are immature and reddish orange eggs are overripe. The average size of lumpfish taken in the fishery varies from 30.3 to 49.4 cm (Blackwood 1982 and Keene 1979 in Stevenson and Baird 1988).

Newfoundland and Iceland are major exporters of lumpfish roe, although many countries do not specify what type of roe they market (Stevenson and Baird 1988).

Utilization. Roe is used in the production of caviar. Studies to determine if carcasses were suitable for reduction into glue or fish meal showed that the water content was too high and the protein and oil content too low (Paradis et al. 1975), but others to determine if the flesh could be used in surimi seemed more promising (Mouland and Voigt 1990). Lumpfish is also being considered as a likely candidate for aquaculture for roe production (Brown et al. 1989, 1992).

Other Information. A species synopsis for lumpfish (Davenport 1985) gives much information pertinent to European populations of this species.

ATLANTIC SPINY LUMPSUCKER / *Eumicrotremus spinosus* (Müller 1776) /

Pimpled Lumpsucker / Bigelow and Schroeder 1953:463

Description. Body short, stout, deepest before dorsal fin, tapering to narrow caudal peduncle (Fig. 198). Head large, snout bluntly rounded, mouth small, lips thick, gill openings small, eye large. Small teeth on jaws. Pelvic fins modified to form an adhesive disk. Pectoral fins with a long base. Body scaleless but covered with large, rough conical tubercles.

Meristics. Dorsal fin rays VI or VII, 10–12; anal fin rays 10–12; vertebrae 26 or 27 (Scott and Scott 1988).

Color. Olivaceous to brownish with some striping or banding

Size. Up to 11.5 cm (Jensen 1944c); average 3.5–6.1 cm (Backus 1957b).

Distinctions. Spiny lumpsucker are easily distinguished from lumpfish by the fact that their skin tubercles are relatively much larger and are irregularly and closely scattered over the body and head. Furthermore, the gill openings are much shorter, while the body is not so high-arched and is nearly round in cross section rather than triangular. The first dorsal fin (though fleshy in some) retains a finlike appearance through life instead of becoming entirely concealed by the

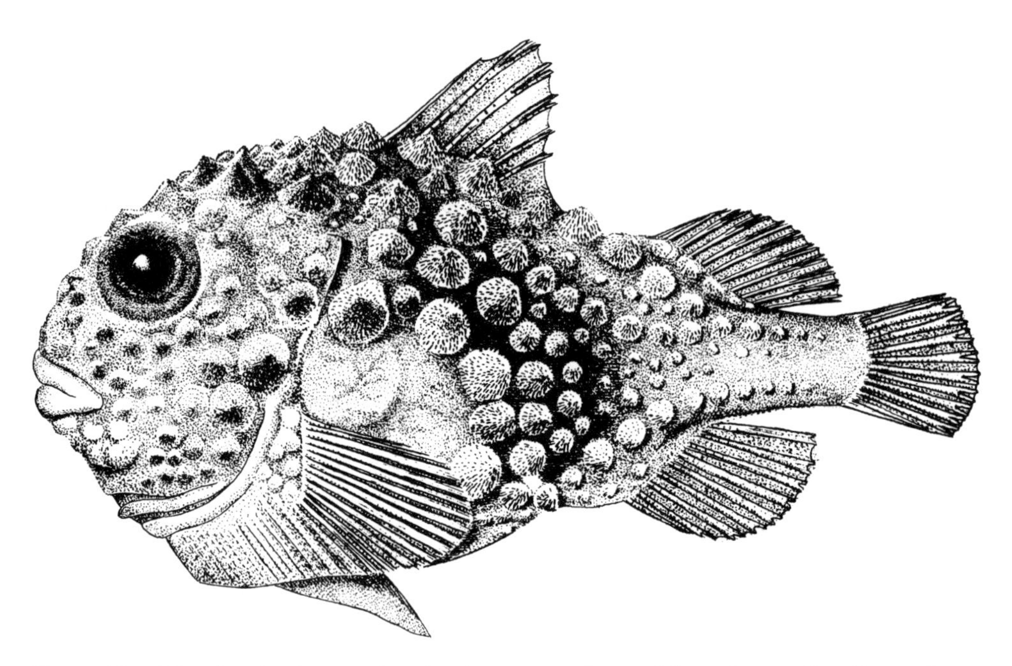

Figure 198. Atlantic spiny lumpsucker *Eumicrotremus spinosus*. Massachusetts Bay. Drawn by H. L. Todd.

Biology. A benthic species occurring in cool northern seas, living on mud, gravelly, or rocky bottoms at depths of 5–82 m (Dunbar and Hildebrand 1952; Backus 1957b). Temperature range is −1.4° to 3°C (Scott and Scott 1988). Spiny lumpsucker from European waters eat amphipods (Scott and Scott 1988) and are eaten by Atlantic cod (Jensen 1944c) and thick-billed murre (Gaston et al. 1985). One specimen was infected with a blood parasite (Khan et al. 1980). Little is known about reproduction, but eggs measuring 2.3–2.5 mm in diameter of a golden yellow color were found in a Greenland specimen (Scott and Scott 1988). Andriashev (1954) reported eggs 3.2–4.5 mm in diameter deposited on a rocky bottom in August or September.

General Range. Spiny lumpsucker inhabit the Canadian Arctic and Greenland east to Nova Zemlya in the Barents Sea and northern parts of the Atlantic Ocean, from Hudson Bay south to the Gulf of Maine as strays (Scott and Scott 1988).

Occurrence in the Gulf of Maine. Stray specimens of this northern fish have been reported from Eastport, Maine; from off Cape Ann; from Salem, and on the north side of Massachusetts Bay; three small specimens were collected about 15 miles southeast of Cape Ann by the U.S. Fish Commission in 1878.

ACKNOWLEDGMENT. A draft of the cyclopterid section was reviewed by Kenneth W. Able.

SEA SNAILS. FAMILY LIPARIDAE

KENNETH W. ABLE

Sea snails are tadpole-shaped, soft-bodied, naked, generally small fishes; like lumpfish, most of them have a sucking disk on the ventral surface. But, unlike lumpfishes, snailfishes have a single dorsal fin (vs. two) with numerous dorsal rays (28–82), more anal rays (24–76), and more vertebrae (38–86). The more than 150 species of snailfishes (Liparididae or Cyclopteridae of some authors, see Able et al. 1984) are known in all oceans from the Arctic to the Antarctic and from the intertidal zone down to greater than 7 km (Andriashev 1954, 1986). The Gulf of Maine harbors four species.

There has been considerable confusion regarding the taxonomy of these forms. In addition, realization that there are more species (three *Liparis* and one *Careproctus*) in the Gulf of Maine than were recognized by Bigelow and Schroeder necessitates a complete reexamination of their distribution, biology, and life history. Most notably, it is clear that *Liparis liparis* does not occur in the western North Atlantic (Able 1990a). Prior accounts of this form are probably due to confusion with two previously undescribed species, *L. inquilinus* and *L. coheni* (Able 1973, 1976b). Two other deepwater snailfishes that lack disks (*Paraliparis calidus, P. copei*) are reported from north and south of the Gulf of Maine (Able et al. 1986) and might occur here.

FLATDISK SNAILFISH / *Careproctus ranula* (Goode and Bean 1879)

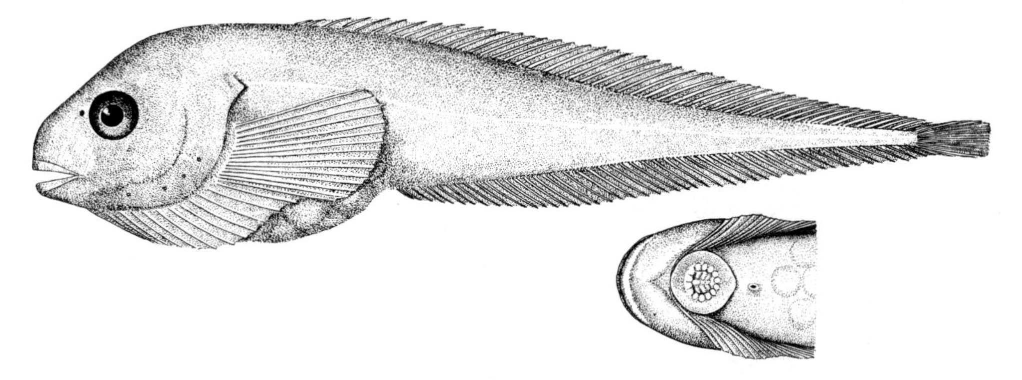

Figure 199. Flatdisk snailfish *Careproctus ranula*. Off the mouth of Halifax Harbor, female, 52 mm SL, USNM 22310. Inset is ventral view of sucking disk. Drawn by H. L. Todd. (From Goode and Bean 1896: Fig. 251.)

Description. Anterior part of body deepest and broadest, tapering gradually to small caudal fin (Fig. 199). Pectoral fin bilobed, with upper, larger lobe reaching back to about anal fin origin. Pectoral fin notch shallow. Sucking disk flattened; disk diameter generally greater than eye diameter but not exclusively so, contrary to previous reports (Burke 1930). Skin naked, soft, flaccid, and unpigmented.

Meristics. Dorsal fin rays 46–55, anal fin rays 42–49, pectoral fin rays 24–30, caudal fin rays 9–12; total vertebrae 53–62; pyloric caeca 8–13.

Color. Unpigmented except for the black eye.

Size. All specimens examined, including adults, are less than 75 mm TL.

Distinctions. Distinguished from closely related species (i.e., those with a sucking disk) by the presence of a single pair of nostrils (as opposed to two pairs in *Liparis* and *Cyclopterus*), more median fin rays, and the lack of a notch in the dorsal fin as found in *L. atlanticus* and *L. inquilinus*.

Reproduction. The smallest female with ripe ovaries was 54 mm TL (Able and Irion 1985). Eggs that appeared mature were 3.0 mm in diameter, and females with eggs of this size typically had ten to twenty eggs in the ovary. A 44-mm specimen from the Gulf of Maine (MCZ 9948) taken in April had seven eggs about 2.0 mm in one ovary and ten eggs in the other. Mature eggs have been observed in several seasons, suggesting a protracted spawning period.

General Range. This species has been collected from 100 to 253 m from the Gulf of St. Lawrence, off Nova Scotia, and in the Gulf of Maine (Able and Irion 1985). Data from unpublished collections indicate that it occurs further south, at similar depths, from the edge of Georges Bank to off New Jersey and off Chesapeake Bay (K. W. Able, pers. obs.), but these individuals have somewhat different meristic features and should be investigated further.

Occurrence in the Gulf of Maine. Most collections are from deep water in the mouth of the Bay of Fundy, but individual specimens have been collected in deep water off the central coast of Maine, in the central basin, and the Northeast Channel (K. W. Able, pers. obs.). *Albatross IV* collected one (MCZ 99446) at 42°53′ N, 70°05′ W at 101–106 m depth.

ATLANTIC SEA SNAIL / *Liparis atlanticus* (Jordan and Evermann 1898) /

Bigelow and Schroeder 1953:464–466 (as *Neoliparis atlanticus*)

Description. Body broad anteriorly, laterally compressed behind anal fin origin (Fig. 200). Head profile relatively flat with gradual slope to snout. Anterior and posterior nostrils in tubes, anterior nostril much longer. Eyes and mouth small. Posterior jaw teeth strongly trilobed, with lobes of similar size. Dorsal fin originates behind suprabranchial pores. Notch in dorsal fin especially prominent in males during breeding season when anterior rays become elongate and separate at tips. Posterior dorsal and anal fin rays barely reach caudal fin base. Pectoral fin bilobed, longest ray in upper lobe not quite reaching anal fin origin; rays in lower lobe fleshy at tips. Disk relatively large.

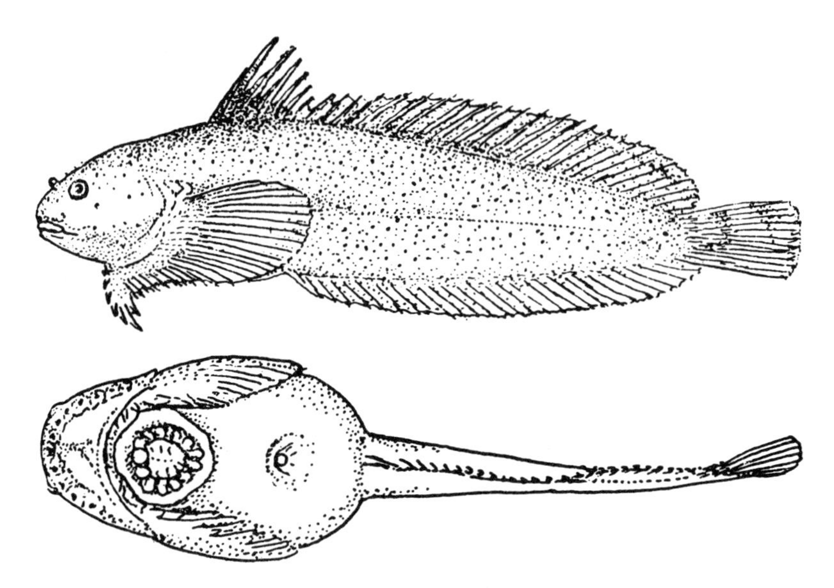

Figure 200. Atlantic sea snail *Liparis atlanticus.* Side view (above) and ventral view (below). (From Bigelow and Schroeder: Fig. 243.)

Meristics. Dorsal fin rays 32–35, anal fin rays 25–29, pectoral fin rays 27–31, 0–2 pectoral rays over gill slit; total vertebrae 38–41, precaudal 10 or 11, caudal 28–30; pyloric caeca 23–45, usually 24–35. For details of meristic and morphometric characters, see Able and McAllister (1980).

Color. In preserved specimens, dorsal surface uniformly brown, somewhat lighter ventrally, lightest on disk. Caudal fin often with one to three vertical bars. Live specimens were black, gray, or dark brown on the body with fins barred with white, blue, or pink (Detwyler 1963).

Size. Reach 144 mm TL (128 mm SL) and 28.4 g. The largest specimen from extensive collections in the Gulf of Maine was 97 mm TL (Detwyler 1963).

Distinctions. Differ from other snailfishes, except *L. inquilinus,* in the presence of a notched dorsal fin, and from that species in that the dorsal and anal fins only reach the base of the caudal fin (vs. overlapping) and in having fewer pyloric caeca (23–45 for *L. atlanticus* vs. 14–21 for *L. inquilinus*).

Habits. This benthic species is intertidal to subtidal along the coast, although some are apparently found on Georges Bank as well (Able et al. 1986). Juveniles and adults are often found under stones or attached to kelp and other seaweeds (Bigelow and Schroeder; Detwyler 1963), as well as on hard, sandy-mud bottom (Gordon and Backus 1957). *Liparis atlanticus* have been found stranded in the extensive intertidal zone of Minas Basin, N.S. (Bleakney and McAllister 1973). Prior records from sea scallops appear to be misidentifications (Able 1973; Able and Musick 1976). *Liparis atlanticus* have been reported from tide pools (6–75 cm deep) in Newfoundland at 6.0°–12.5°C and 29.5–30.0 ppt (Van Vliet 1970). At one extensively collected location in the Gulf of Maine they were rare at temperatures above 12°C (Detwyler 1963).

Food. Stomachs of fish (65–88 mm TL) from the New Hampshire coast contained primarily crustaceans with fewer polychaetes. Feeding may occur in early morning and evening (Detwyler 1963).

Predators. Found in the stomachs of one winter skate and three sea raven (Rountree 1999).

Parasites. A tapeworm, *Spathebothrium simplex,* was common in fish from New Hampshire (Munson 1970) but other parasites including trematodes, acanthocephalans, nematodes, and protozoans were found (Detwyler 1963). A protozoan, *Haemogregarina* sp., was recorded from Quebec specimens (Margolis and Arthur 1979) and haemotozoa from specimens collected in New Brunswick (Laird and Bullock 1969).

Reproduction and Early Life History. In New Hampshire, this species probably reaches sexual maturity at 60–70 mm in its second year (Detwyler 1963). Mature adults begin migrating into the intertidal zone in October prior to spawning in March. Although these had as many as 1,400–3,000 eggs, only 475–700 reached maturity at any one time. In aquariums, females typically induced the male to select and prepare a spawning site

(Detwyler 1963). Prior to spawning the male drove conspecifics away from the spawning area. The male cleaned the area, a stone with attached Irish moss, *Chondrus crispus,* by rubbing it with his snout. Eventually he took a position near the site, expanded his fins, and began to quiver. The female, now with a prominent bulge in the area of the vent, swam over the site prepared by the male and deposited a small mass of eggs. The male fertilized the eggs and then nuzzled them against the algae, butting them into place with his snout. The female deposited four more egg masses and these were added to the initial mass. The female eventually left the site and became inactive while the male took a position next to the egg mass and fanned it with his fins. He guarded the eggs for several days and attacked any intruders. In another instance, Detwyler (1963) observed that a single egg mass could result from a single pair of adults or one female and several males. Adults left the intertidal spawning area in June (see Detwyler 1963 for additional details).

Fertilized eggs are 0.8–1.4 mm in diameter. Embryonic development up to yolk-sac absorption has been described (Detwyler 1963). The larvae are distinct from all other western North Atlantic *Liparis,* except *L. inquilinus,* in the small size at comparable stages of development, presence of a notched dorsal fin, and fin ray counts (see Able et al. 1986). *Liparis atlanticus* larvae differ from *L. inquilinus* in having lateral melanophores on the tail in preflexion larvae (<8 mm) and a smaller eye diameter–disk length ratio.

The average length-weight relationship can be expressed as $\log W = -13.1546 + 3.4181 \log L$ for females; $\log W = -11.5067 + 3.0004 \log L$, $r = 0.92$ for males (Detwyler 1963).

General Range. Along the coast from Ungava Bay to as far south as Rhode Island, Connecticut, and the south shore of Long Island (Able and McAllister 1980), including Georges Bank (Able et al. 1986).

Occurrence in the Gulf of Maine. Extensively collected from the coast of New Hampshire and Massachusetts Bay (Detwyler 1963; Collette 1986; Collette and Hartel 1988). Larvae have been collected frequently from southern Nova Scotia, near the mouth of the Bay of Fundy (Able et al. 1986), and the coast of Maine (Hauser 1973), suggesting that adults also occur in these areas.

GULF SNAILFISH / *Liparis coheni* Able 1976 / Bigelow and Schroeder 1953:466–467 (as *Liparis liparis*)

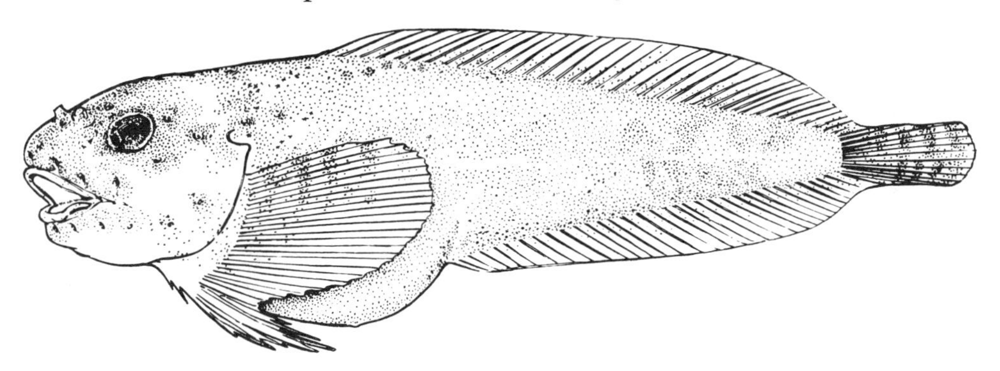

Figure 201. Gulf snailfish *Liparis coheni.* 44°37.5′ N, 66°30.5′ W, 70.7 mm TL, USNM 212261. Drawn by R. Bradley. (From Able 1976: Fig. 1.)

Description. Anterior part of body broad, then compressed behind anal fin origin (Fig. 201). Head broader than high, sloping gradually to snout. Anterior nostril in tube; posterior a raised slit. Mouth extending to below front of small eyes. Teeth trilobed; anterior teeth with weakly defined lobes. Dorsal fin origin behind vertical through suprabranchial pores. Anterior dorsal rays somewhat shorter, longer posteriorly. Notch in dorsal fin absent. Dorsal fin overlaps one-quarter to one-half of caudal fin. Pectoral fin bilobed, rays in upper lobe extending to near origin of anal fin, rays in lower lobe separate and fleshy at tips. Disk slightly longer than wide.

Meristics. Dorsal fin rays 36–41, anal fin rays 30–35, pectoral

42–46, precaudal 10 or 11, caudal 32–36; pyloric caeca 14–29 (Able 1976b).

Color. Preserved specimens uniformly light brown, darker brown on posterior of dorsal and anal fins. Caudal fin vertically barred (Able 1976b). Some specimens striped as in *L. inquilinus* (Able 1973).

Size. Maximum recorded size is 89.0 mm TL.

Distinctions. Distinguished from *Careproctus ranula* by the larger disk, which is several times greater than eye diameter. *Liparis coheni* differs from other *Liparis* in the Gulf of Maine in lacking a notch in the anterior dorsal fin. This species was

Habits. Juveniles and adults are benthic at depths from 4 to 210 m in estuaries and near shore (Able 1976b; Able and Irion 1985). Little else is known because it is seldom collected and in the past has been confused with other snailfishes.

Reproduction and Early Life History. Spawning occurs over a prolonged period during the winter and early spring, based on collection of mature males and females, as well as larvae (Able 1976b). Larvae can be distinguished from other *Liparis* in the Gulf of Maine by larger sizes at comparable stages of development, lack of a dorsal fin notch, and higher fin ray counts (Able et al. 1986). Most larvae have been collected in estuaries or nearshore waters of the central Gulf of Maine and off southern Nova Scotia.

General Range. *Liparis coheni* have been collected from the Gulf of Maine, northern Georges Bank, on Roseway Bank off Nova Scotia, and in the Gulf of St. Lawrence, especially in the St. Lawrence estuary and Chaleur Bay (Able 1976b; Able and Irion 1985).

Occurrence in the Gulf of Maine. Known to occur in Massachusetts Bay (Collette and Hartel 1988), Boothbay Harbor, the mouth of the Damariscotta River, the mouth of the Bay of Fundy, Passamaquoddy Bay, and northern Georges Bank (Able 1976b; Able and Irion 1985).

INQUILINE SNAILFISH / *Liparis inquilinus* Able 1973

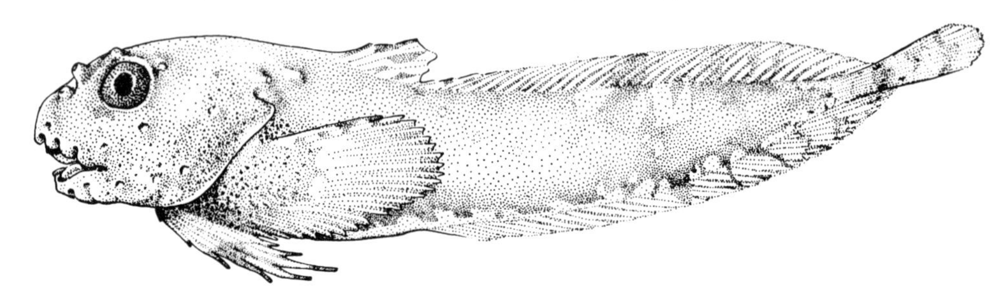

Figure 202. Inquiline snailfish, *Liparis inquilinus.* 39°30′ N, 73°30′ W, 42.5 mm TL, USNM 208466. Drawn by J. Davis. (From Able 1973: Fig. 1.)

Description. Tadpole-shaped, body compressed behind anal fin origin (Fig. 202). Head broader than high, sloping to snout, then rounded. Anterior nostrils in tubes; posterior nostrils appear as raised pores. Mouth extends to front of small eye. Lips with scalloped appearance. Teeth trilobed, anteriorly with weakly defined lobes. Dorsal fin origin behind suprabranchial pores, notched at fifth to seventh ray; rays in notch about one-half of longest ray before notch. Dorsal fin overlaps one-tenth to one-quarter and anal fin overlaps one-quarter to one-half of caudal. Pectoral fin bilobed; rays in upper lobe extend to anal fin origin; rays in lower lobe separate and fleshy at tips. Disk slightly longer than wide.

Meristics. Dorsal fin rays 33–38 rays, anal fin rays 28–31, pectoral fin rays 30–35, 0–4 pectoral rays over gill slit; total vertebrae 38–42, pyloric caeca 14–21 (Able 1973).

Color. Extremely variable; spotted, striped, mottled, or plain (Able 1973). These patterns do not seem to be correlated with sex, season, or geographical location. The eye in live specimens can be black, brown, or silvery. The caudal fin is barred.

Size. The maximum size of adults is 71 mm TL.

Distinctions. *Liparis inquilinus* have a larger disk than *C. ranula* (greater than eye diameter vs. less than eye diameter). They differ from *L. coheni* in the notched dorsal fin and from *L. atlanticus* in having more dorsal (33–38 vs. 31–34), anal (28–31 vs. 24–27), and pectoral fin rays (30–35 vs. 26–30).

Habits. *Liparis inquilinus* have been collected from 3 to 97 m, where they are often commensal with sea scallop, *Placopecten magellanicus* (Able 1973; Able and Musick 1976). This association appears to be distinctly seasonal (Able and Musick 1976; Luczkovich et al. 1991), with many individuals (12–55 mm) in scallops through the summer but fewer in fall and winter, apparently because they move inshore to spawn. When *L. inquilinus* were introduced into aquariums with scallops they swam over and around the scallops but concentrated most of their activity along the scallop's mantle. Eventually they placed their heads against the mantle and forced their way into the mantle cavity with exaggerated swimming motions. Inside the scallop, fish attached to the mantle by their disks, usually in an inverted position. As many as 32 juveniles have been found inside a single scallop (Able and Musick 1976).

Use of sea scallops may follow a diel pattern (Able and Musick 1976). In a single 24-h period 3,595 fish, averaging 21.0 mm TL, were collected from 616 of 841 scallops examined in a single area in the New York Bight. Most occurred during the day. No noticeable damage to scallop tissues has been observed as a result of their presence. *Liparis inquilinus* and red hake, *Urophycis chuss,* may co-occur in sea scallops (Bigelow and Schroeder; Musick 1969; Able and Musick 1976). In aquariums, without scallops, *L. inquilinus* preferred an inverted resting position with the disk attached to any smooth substrate such as bivalve shells, rocks, or the side of the aquarium. Once attached, a fish usually flexed the tail alongside the head.

Food. *Liparis inquilinus* is apparently a nocturnal feeder (Able and Musick 1976). In aquariums, fish appeared to depend on reception of tactile and/or gustatory stimuli received by the head or pectoral fins and not visual cues. When swimming, the fleshy fin rays in the lower lobe of the pectoral fin were extended toward the bottom. If these fin rays or the head touched live amphipods, the amphipods were eaten. Histological examination demonstrated the presence of taste buds on the surface of the lower pectoral fin rays. Observations over a 24-h period indicated that fish probably left the scallops at night to feed: the fullest stomachs, with the least-digested prey items, occurred after the fish returned to the scallops. Food was primarily benthic crustaceans, as indicated by presence of sand grains in the stomach along with the prey, which were usually amphipods (Able and Musick 1976).

Predators. Sea scallops may offer *L. inquilinus* a refuge from most predation. The only known predators of large sea scallops, that is, those large enough to be occupied by *L. inquilinus* (Able and Musick 1976; Luczkovich et al. 1991), are Atlantic wolffish and Atlantic cod (Bourne 1964), and these eat scallops only occasionally. *Liparis inquilinus* may maximize protection by living in scallops most of their demersal life, entering as soon as they leave the plankton and remaining in association until they begin to move inshore to spawn. There is no evidence of predation by the co-occupant, *U. chuss* (Luczkovich et al. 1991). *Liparis inquilinus* have been found in stomachs of cod and sea raven (Rountree 1999).

Reproduction and Early Life History. In the laboratory, individuals deposited eggs in unguarded small clumps of 20–80 on the bottom of the aquarium. Once, eggs were collected in the field attached to hydroids (Able and Musick 1976). Females may spawn more than once, based on laboratory observations and field-collected fish with multiple modes of egg diameters in the ovary. The number of maturing eggs varied from 105 to 1,135 for fish in the laboratory and from 231 to 563 for fish collected off New Jersey. Ripe eggs are 1.0–1.3 mm in diameter with oil globules.

In the Mid-Atlantic Bight spawning probably occurs at lengths of 41–72 mm TL from February through April (Able and Musick 1976) and over a similar period in the southern Gulf of St. Lawrence (Able and Irion 1985). Larvae are planktonic from April through June and have been collected during May in the Gulf of Maine. The larvae can be distinguished from other members of the genus in the western North Atlantic by the small sizes at comparable stages of hatching, disk formation, notochord flexion, presence of a notch in the anterior dorsal fin during late flexion and beyond, and generally lower fin ray counts (Able et al. 1986). Larvae larger than 13 mm TL were not usually found in the plankton.

General Range. Known from Newfoundland to Cape Hatteras including the Gulf of St. Lawrence and Georges Bank (Able 1973; Able et al. 1986).

Occurrence in the Gulf of Maine. Collections of larvae and juveniles from sea scallops have been reported from Georges Bank and shallow inner portions of the Gulf of Maine, the Bay of Fundy, and Browns Bank off Nova Scotia (Able and Musick 1976; Able et al. 1986).

PERCIFORM FISHES. ORDER PERCIFORMES

The Perciformes are the most diversified of all fish orders (Nelson 1994). Perciforms are the dominant vertebrate group in the oceans and the dominant fish group in many tropical and subtropical freshwaters. Although Johnson and Patterson (1993) presented evidence that the order as presently constituted is not monophyletic, the classical arrangement utilized by Nelson (1994) will be followed here with a few exceptions. According to Nelson (1994), the Perciformes contain 18 suborders, 148 families, about 1,496 genera, and about 9,293 species. There are representatives of 6 of the 18 suborders in the Gulf of Maine: Percoidei, Labroidei, Zoarcoidei, Trachinoidei, Scombroidei, and Stromateoidei.

Perchlike Fishes. Suborder Percoidei

Percoidei is the largest suborder of the Perciformes containing 71 families, 528 genera, and about 2,860 species (Nelson 1994). formes, were derived but definition of the suborder is based on primitive characters and so is probably not monophyletic as

fins are thoracic in position; and the premaxilla is included in the gape of the upper jaw. Representatives of 13 percoid families are known from the Gulf of Maine: temperate basses (Moronidae), wreckfishes (Polyprionidae), sea basses (Serranidae), bigeyes (Priacanthidae), tilefishes (Malacanthidae), bluefish (Pomatomidae), remoras (Echeneidae), dolphins (Coryphaenidae), jacks (Carangidae), pomfrets (Bramidae), porgies (Sparidae), croakers (Sciaenidae), and butterflyfishes (Chaetodontidae). Many of these are southern species that just visit the Gulf of Maine in the summer.

KEY TO FAMILIES OF PERCOIDEI IN THE GULF OF MAINE

1a. Oval adhesive disk with transverse ridges (lamellae) on top of head .... **Echeneidae**
1b. No adhesive disk on top of head **2**
2a. Single continuous dorsal fin, or if notched, notch extends only about half way to fin base, and first dorsal fin spines not short and isolated in adults **3**
2b. Dorsal fins separate, or deeply divided by notch that extends to or almost to fin base, or at least half of dorsal fin spines short and isolated (at least in adults) **10**
3a. Dorsal fin spines, if present, weak and flexible, not clearly differentiated from soft rays (except anterior 1–3 rays stout and unsegmented in Bramidae) **4**
3b. Dorsal fin spines rigid and sharp, clearly differentiated from soft rays **5**
4a. Maxilla scaly in adults; scales large, keeled in adults; lateral line absent or obsolete; dorsal fin elevated anteriorly, originates posterior to head, usually falcate in adults .... **Bramidae**
4b. Maxilla naked; scales small to moderate; lateral line well developed; dorsal fin origin on top of head, in advance of upper end of gill opening .... **Coryphaenidae**
5a. Dorsal and anal fins naked, except a scaly basal sheath may be present; jaws with posterior molariform teeth .... **Sparidae**
5b. Dorsal and anal fins covered with scales; jaws without posterior molariform teeth **6**
6a. Bands of bristlelike teeth in each jaw; body highly compressed, disk-shaped .... **Chaetodontidae**
6b. Dentition not as above; body fusiform to elongate or only slightly compressed, not disklike **7**
7a. Dorsal margin of upper jaw fully exposed, not slipping under preorbital bone when mouth is closed; opercle with 2 or 3 flattened spines **8**
7b. Dorsal margin of upper jaw slipping under preorbital bone for most of its length when mouth is closed; opercle with a single posterior spine **9**
8a. Opercle with 3 flattened spines posteriorly, but only the middle one obvious; no rough bony ridge across upper part of opercle; dorsal fin spines 9–11 .... **Serranidae**
8b. Opercle with 2 spines posteriorly; rough bony ridge across upper part of opercle; dorsal fin spines 11 or 12 .... **Polyprionidae**
9a. Pelvic fins broadly joined to body by a membrane; no adipose flap on nape; lower jaw scaly; in life, body and fins red .... **Priacanthidae**
9b. Pelvic fins not broadly joined to body by a membrane; adipose flap on nape, ahead of dorsal fins; lower jaw naked; coloration variable .... **Malacanthidae**
10a. Lateral line extending to end of caudal fin .... **Sciaenidae**
10b. Lateral line not extending to end of caudal fin **11**
11a. Anal fin preceded by 2 small anal fin spines; lateral line may have scutes along posterior portion .... **Carangidae**
11b. Anal fin not preceded by anal fin spines; scutes never present along posterior portion of lateral line **12**
12a. First and second dorsal fins about equal in height; first dorsal with 8–10 long spines; teeth of moderate size .... **Moronidae**
12b. First dorsal fin much lower than second dorsal; first dorsal with 7 or 8 short spines; teeth large and very sharp .... **Pomatomidae**

TEMPERATE BASSES. FAMILY MORONIDAE

GRACE KLEIN-MACPHEE

Temperate basses have two well-developed dorsal fins, the first with eight to ten spines and the second with one spine and 10–13 soft rays. The anal fin has three spines and 9–12 soft rays. The opercle has two spines. The lateral line extends almost to the tip of the tail and has auxiliary rows above and below the main row. Temperate basses were included with the sea basses in the Serranidae by Bigelow and Schroeder, transferred to the Percichthyidae, which is now considered to include only South American and Australian freshwater species, and then placed in their own family (Johnson 1984; Nelson 1994). Sea basses have three opercular spines and the lateral line does not extend out onto the caudal fin. There are two genera of temperate basses, *Morone* with four species in North America (two confined to freshwater) and *Dicentrarchus* with two species from off Europe and North Africa (Nelson 1994).

KEY TO GULF OF MAINE TEMPERATE BASSES

1a. Dorsal fins separated by a distinct space; sides distinctly striped .... **Striped bass**
1b. Dorsal fins joined at their bases; sides not distinctly striped .... **White perch**

WHITE PERCH / *Morone americana* (Gmelin 1789) / Bigelow and Schroeder 1953:405–407

Description. Body short and deep, three and a half times in total length, compressed, with a thick caudal peduncle (Fig. 203). Head pointed; mouth terminal and short. Lower jaw projects slightly. Small unequal teeth on jaws and sides of tongue. Eye large, contained about five times in HL. Edges of gill covers slightly serrate. First dorsal fin rounded, third and fourth spines

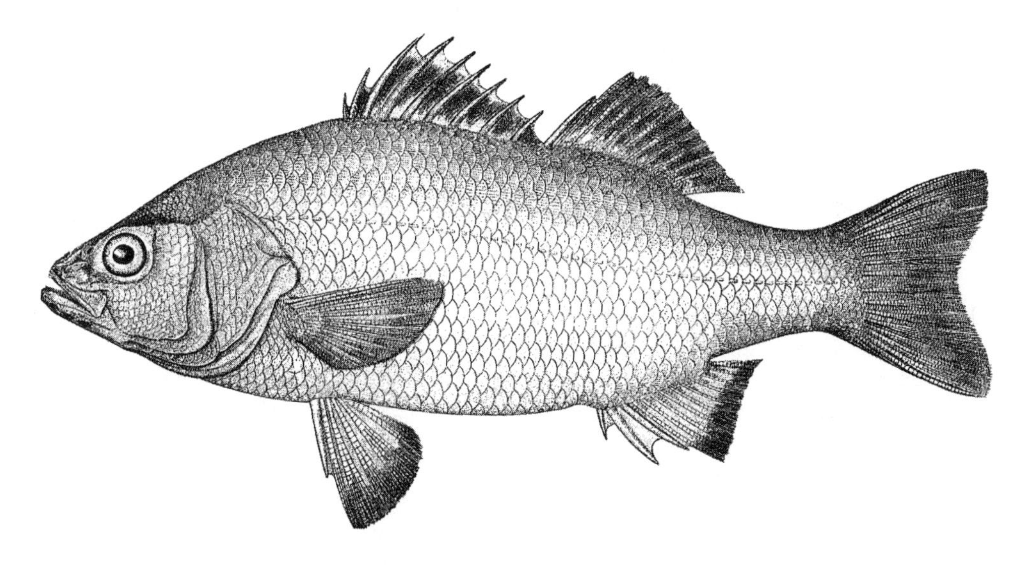

Figure 203. White perch *Morone americana*. Drawn by H. L. Todd.

longest. Two dorsal fins separated by deep notch. Second dorsal fin rhomboid in outline, leaving rather long caudal peduncle. Pelvic fins originate a little behind pectoral fins. Both fins large. Anal fin originates under middle of second dorsal, of same shape as that fin. Third anal spine largest. Lateral line distinct and slightly arched. Scales large, covering body, sides of head, and space between eyes and nostrils.

Meristics. Dorsal fin rays VII–IX, I, 11–13; anal, III, 8–10; lateral line scales 44–55; vertebrae 11 + 14 = 25.

Color. Upper surface is variously olive, dark grayish green, or dark silvery gray, shading to paler olive or silvery green on the sides and to silvery white on the belly; large fish often show a bluish luster on the head. Fins often more or less dusky. Pelvic and anal fins sometimes rose-colored at the base. Sides of young specimens marked with pale longitudinal stripes, which usually fade with growth.

Size. Reach 48 cm (Leim and Scott 1966) and 0.9 kg or more, but the average is 20.3–25.4 cm and 0.5 kg, or less. The all-tackle game fish record is a 2.15-kg fish taken in Messalonskee Lake, Maine, in June 1949 (IGFA 2001).

Distinctions. White perch resemble their larger relative, striped bass, in number, outline, and arrangement of the fins and in the deep caudal peduncle without longitudinal keels. White perch are deeper-bodied (only about 2.5–3 times as long as deep, not counting the caudal fin); and are more laterally compressed. The dorsal profile of the body is more convex than that of striped bass, the head is noticeably concave, and the mouth is smaller. There is no free space between the two dorsal fins of white perch, whereas they are separated by a short interspace in striped bass. White perch have fewer lateral line scales than striped bass (about 48 in white perch, 60 with the second and third about equal in length (graduated in striped bass). They usually have only one spine (sometimes two) at the margin of the gill cover. Both pectoral and pelvic fins of white perch are larger in comparison with the size of the fish. Finally, there is the difference in color.

Habits. White perch are much more restricted in their seaward range than striped bass, for while they are taken in undiluted seawater along southern New England and at various other localities westward and southward, they are much more plentiful in ponds connected with the sea, in brackish water bays behind barrier beaches, in estuaries, and in river mouths. They have been recorded as far offshore as 16.1 km at Block Island, R.I. (Hildebrand and Schroeder 1928). White perch are also landlocked in freshwater ponds in many places and have entered the Great Lakes (see Boileau 1985).

White perch are schooling fish, ordinarily found in shallow water, usually not deeper than 4 m, but sometimes in 18–38 m in Chesapeake Bay, the maximum depth being 42.1 m. They are relatively sedentary, probably adapted for swimming efficiently at low speed (Neumann et al. 1981). They are not bottom fish (except in winter), but wander from place to place in small schools. Apart from this, they are resident throughout the year wherever found. In winter they congregate in deeper parts of bays and creeks, where they pass the cold season in a sluggish condition. They prefer level, firm bottom of silt, mud, clay, or sand with little cover. They are eurythermal and euryhaline, having been reported from temperatures of 2.0°–32.5°C and from freshwater to 30 ppt (Smith 1971).

Food. When living in salt or brackish water, white perch feed on small fishes of all kinds, shrimps, crabs, and various other invertebrates, as well as on eggs of other fishes (Schaeffer and Margraf 1987). Swarms of young white perch have been seen following alewife around the shores of ponds on Mar-

oligohaline region of the Hudson River estuary (salinity 0–10 ppt), the five most frequently taken food items of adult white perch were gammarid amphipods, isopods, insects, annelid worms, and the mud shrimp *Crangon*. Immature white perch eat planktonic organisms—cladocera, fish eggs, and fish larvae (Bath and O'Connor 1985)—feeding mainly during the early evening.

Predators. In the Hudson River, juvenile white perch are prey for yearling and older striped bass, adult white perch, and bluefish (Bath and O'Connor 1985). White perch larvae from the Chesapeake and Delaware Canal were preyed upon by the copepod *Cyclops bicuspidatus thomasi*. Many larvae exhibited damage to the fins and yolk sac severe enough to cause death (Smith and Kernehan 1981).

Parasites. The most abundant gill parasites in the Hudson River were a monogenetic trematode, *Protocleidus nactus;* freshwater mussel glochidia; a copepod, *Ergasilus labricus;* and an isopod, *Lironeca ovalis* (Liquori and Insler 1985). Fish collected while they were migrating downstream after spawning were heavily infested, but the numbers of parasites decreased during residence in more saline waters. Mussel glochidia were particularly sensitive to higher salinity (10 ppt). Copepod parasites disappeared in response to higher water temperatures in August (25°C) and reappeared in October when the temperature was 16°C. A review of parasites in white perch from freshwater is given in Thoits (1958).

Breeding Habits. In the Hudson River, white perch males begin to mature at age 2 and all are mature by age 3. Females mature a year later, at age 3, and are mostly mature by age 4 (Holsapple and Foster 1975). In the Bay of Quinte, Lake Ontario, findings were similar (Sheri and Power 1968); but in the Patuxent estuary, Maryland, males mature by age 2 and females begin to mature at age 2 and are mostly mature at age 3 (Mansueti 1961a). There is a great deal of variability in fecundity estimates, which range from 5,210 to 321,000 eggs (Goode 1888b; Taub 1969). Fecundity studies of white perch in the Hudson River estuary ranged from 15,726 to 161,441 eggs per female with a mean of 50,000 (Bath and O'Connor 1982). Spawning takes place in June and July in the Hudson River when the water temperature is 18°–20°C and salinity is below 1 ppt (Bath and O'Connor 1982). In Lake Ontario, peak spawning occurs at 10°C (Sheri and Power 1968). Along southern New England white perch spawn in April, May, and June. Presumably the season commences a few weeks later around the Gulf of Maine, but definite data are lacking. Mansueti (1961a) observed white perch spawning in the Patuxent River, Maryland. Several larger individuals (presumably females) were pursued by more than a dozen smaller fish. The fish milled around constantly with the larger ones swimming to the surface and churning up the water. Release of eggs occurred and milt emission followed.

Early Life History. The eggs are amber-colored, 0.73–1.04 mm in diameter, with a large oil globule. They sink and stick together in masses or to any object on which they chance to rest. Fertilized eggs have an attachment disc and are 0.65–1.09 mm in diameter. They may be pelagic in fast-flowing streams. The chorion is thick, tough, and rough-surfaced. The perivitelline space is narrow, being about 24% of the egg diameter (Mansueti 1964). Developmental events at 18.3°C (Mansueti 1964) are summarized as follows: (1) 1 h: perivitelline space fully formed, 4–16 cell stage; (2) 6 h: morula stage, blastoderm granular; (3) 14 h: blastocoel formed, periblast thickened; (4) 18 h: embryo developed, neural ridge visible; (5) 24 h: embryo pigmented, somites barely evident, tail movements begin; (6) 30 h: oil globule and yolk sac pigmented, eye well formed, tail free; (7) 44–50 h: hatching.

Newly hatched larvae are about 2.3 mm long (1.7–3.0 mm) with the vent some distance behind the yolk sac and have very little pigment. The head is flexed over the yolk sac, which has a large oil globule located at the anterior end. Between 5 and 6 days after hatching, the head begins to project forward; the yolk sac is partly absorbed; and branched pigment cells appear on the oil globule, the anterior part of the yolk sac, head, and ventral edges of the hind gut and trunk. Orange and brown chromatophores are also present except on the oil globule. Orange chromatophores are concentrated one-third the distance from the caudal fin (Mansueti 1964). Fin rays are complete by 30 mm TL and scales begin to form at 16–35 mm (Marcy and Richards 1974).

Larvae resemble striped bass and are differentiated in that species account. They also resemble yellow perch, but perch have more myomeres (30 vs. 25) (Lippson and Moran 1974).

Temperature, but not salinity, influenced hatching (Morgan and Raisin 1982). Optimum temperature for hatching was 14°C at 10 ppt salinity. Eggs incubated at 20°C or higher developed quickly, but most died at the gastrula stage. The largest larvae hatched from eggs incubated between 16° and 18°C.

Kellogg and Gift (1983) studied growth rates and temperature preferences of white perch 6–8 weeks old. The difference between optimum temperature and final preferendum was less than 2°C. Optimum temperature for growth was 28.5°C, the preferred temperature was 30°C, and the upper lethal temperature was between 34° and 35°C.

Respiration rates of white perch swimming at two speeds were tested (Neumann et al. 1981). The average oxygen consumption rate of a 50-g fish swimming at 8.6 $cm \cdot s^{-1}$ was 17.6 mg $O_2 \cdot h^{-1}$, whereas the rate for a fish swimming at 31.7 $cm \cdot s^{-1}$ was 24.5 mg $O_2 \cdot h^{-1}$; 150-g fish consumed 23.5 mg $O_2 \cdot h^{-1}$ at the lower speed and 39 mg $O_2 \cdot h^{-1}$ at the higher speed. White perch were more oxygen-efficient than striped bass.

Age and Growth. Annulus formation in the Hudson River occurred each year starting the first week in May and finishing the end of July (Bath and O'Connor 1982). The most rapid growth occurred in the first 3 years of life and accounted for

78% of the total growth. Females were slightly larger in total length than males of the same age. Published reports of growth rates of white perch from several geographic areas were compared with data from the Hudson River (Bath and O'Connor 1982). Rates were similar except for fish from the Connecticut River (Marcy and Richards 1974). These fish grew faster and larger, resembling white perch from freshwater impoundments (Thoits 1958). White perch live on the average 6–7 years. The maximum age recorded was 17 years (Thoits 1958).

General Range. Atlantic coast of North America from the Gulf of St. Lawrence and Nova Scotia to South Carolina, with peak abundance in the Hudson River and Chesapeake Bay; breeding in fresh or brackish water and permanently landlocked in many freshwater ponds and streams. They have been introduced into many lakes and ponds in New York and the New England states and have invaded the Great Lakes and the Mississippi drainage system (Scott and Christie 1963; Schaeffer and Margraf 1986b). They were accidentally introduced into a flood control reservoir in Nebraska and spread from there into the Missouri and the Platt River drainage systems (Hergenrader 1980).

Occurrence in the Gulf of Maine. White perch inhabit salt, brackish, and fresh water along the shores of southern New England. Although they are familiar fish in many ponds throughout northern New England, New Brunswick, and Nova Scotia, they are found regularly in only a few estuarine situations north of Cape Cod and are rarely found in the open Gulf of Maine. They have been reported infrequently in Duxbury Bay and in the North and South rivers in Marshfield. In the summer of 1950, white perch running upstream to a pond were reported in salt creeks around Cohasset, Mass. Storer long ago described white perch brought to Boston market from the mouths of neighboring rivers and from ponds to which the sea had access. They were found in Floating Bridge Pond in Salem and Flax Pond in Lynn (Goode and Bean 1879), Fresh Pond in Cambridge (Kendall 1908), and in the lower Mystic River (Haedrich and Haedrich 1974). White perch run in salt and brackish reaches of the Parker River in northern Massachusetts, providing fishing for many small-boat anglers in spring and summer.

White perch were generally so scarce along the open coast from Cape Cod northward that they did not figure in the statistics of the shore fisheries of any part of Massachusetts Bay from 1907 to 1928. They are not common along the coast of Maine: none was reported from shore fisheries of Maine in 1905 or 1919; only 400 lb in 1902; and none at all of late years. However, on rare occasions they appeared locally in unusual numbers. Large catches were reported for shore fisheries of the short coastline of New Hampshire in 1912; Casco Bay saw a run of them in the summer of 1901 when local fishermen, not knowing the fish, dubbed them "sea bass"; and they were reported at Eastport, Maine. Apparently they do not occur around the shores of the Bay of Fundy, in either salt or brackish water.

Migrations. Marine or estuarine white perch move shoreward and upstream in spring to spawn. After spawning, some move into deeper water and back into estuaries (Bath and O'Connor 1982). Fish tagged in the Patuxent estuary, Md., at spawning time moved farther than fish tagged in summer, fall, or winter (Mansueti 1961a). The mean distance traveled for fish tagged in spring was 25 km, but distances up to 72.5 km or were more recorded. In all other seasons, fish traveled less than 16 km from the tagging site. In Nova Scotia, there is an extensive seawater migration in summer.

Importance. White perch are of considerable commercial importance wherever they are abundant in tidal waters. A review of commercial fishery landings in New York and New Jersey from 1880 to 1975 is given in McHugh and Williams (1976). Several million artificially hatched larvae are released in the Chesapeake Bay yearly. They have been hybridized with striped bass for commercial and recreational purposes. They afford good sport to many anglers wherever they are plentiful, both in brackish water and fresh, but they are not important in the open Gulf of Maine in either of these respects.

STRIPED BASS / *Morone saxatilis* (Walbaum 1792) / Striper, Rockfish /

Bigelow and Schroeder 1953:389–404 (as *Roccus saxatilis*)

Description. Body 3.25–4 times as long, to base of caudal fin, as deep, thick throughout, back slightly arched (Fig. 204). Caudal peduncle moderately stout. Head almost as long as body depth. Two spines on margin of gill cover. Mouth oblique, gaping back to eye; snout moderately pointed; lower jaw projecting. Teeth small, in bands on jaws, vomer, palatines, and in two parallel patches on tongue. Gill rakers long and slender. Young fish more slender than older fish. Two dorsal fins of about equal length; first triangular, originating over middle of to rear and separated from first by distinct short space. Anal about same size and form as second dorsal, originating below middle of latter. Caudal fin moderately wide, slightly forked. Pectoral and pelvic fins moderate-sized, pelvics inserting somewhat behind pectorals.

Meristics. Dorsal fin rays VIII–X, 9–14; anal III, 7–13; pectoral rays 13–19; lateral line scales 50–72; gill rakers on first arch

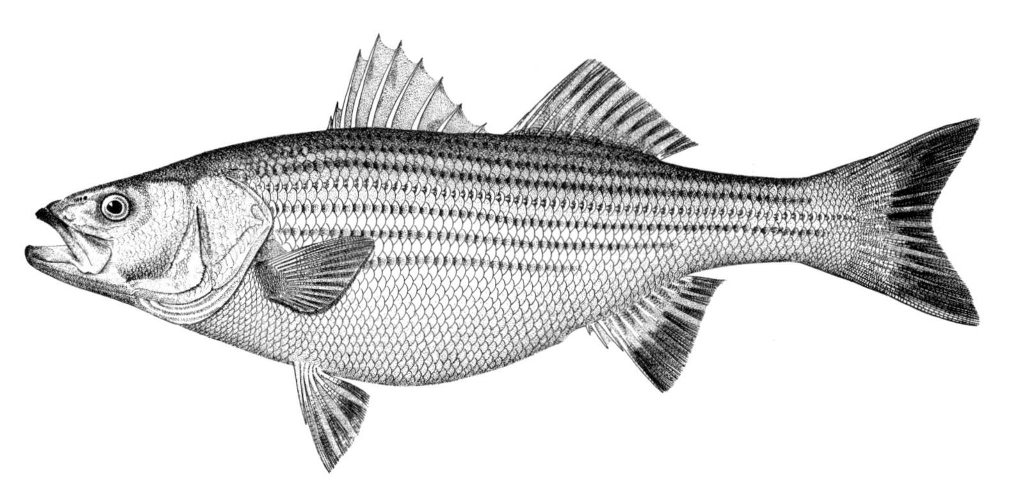

Figure 204. Striped bass *Morone saxatilis.* Washington, D.C., market. Drawn by H. L. Todd.

Color. Dark olive green varying to bluish above, paling on the sides, and silvery on the belly, sometimes with brassy reflections. Sides barred with seven or eight narrow, sooty, longitudinal stripes, one of which always follows the lateral line. Others also may follow rows of scales and may be variously interrupted. The highest stripe is the most distinct, and all of them except the lowest are above the level of the pectoral fins. Dorsal, caudal, and anal fins are somewhat dusky.

Size. Striped bass grow to a great size, the heaviest on record being several of about 56.8 kg taken at Edenton, N.C., in April 1891. One of 50.9 kg, which must have been at least 1.8 m long, was caught at Orleans, Mass., many years ago. One of 45.7 kg is said to have been taken in Casco Bay, Maine, and fish of 22.7–27.3 kg are not exceptional. Bass usually weigh from 1.5 to 16–18 kg. The all-tackle game fish record is a 35.6-kg fish taken off Atlantic City, N.J., in September 1982 (IGFA 2001).

Distinctions. The rather deep and keel-less caudal peduncle, stout body, presence of two well-developed dorsal fins (spiny and soft-rayed, of about equal length), lack of dorsal or anal finlets, and a moderately forked tail distinguish striped bass from mackerels, bluefish, and jacks. The anal fin having three spines and being almost as long as the second dorsal and the maxilla not being sheathed by the preorbital bone distinguish them from croakers. The two dorsal fins are entirely separate, whereas in sea bass, cunner, and tautog, spiny and soft-rayed parts are continuous as a single fin. White perch come closest to striped bass in general appearance but the two dorsal fins of white perch have no free space between them and the fin spines are stiffer. Hybrids and backcrosses have been produced between striped bass and other closely related species. These have been stocked in reservoirs and rivers and may have interbred with wild populations of the parent stock.

Habits. Striped bass are inshore fish usually occurring no more than 6-8 km from the nearest point of land, although migrating schools doubtless pass much farther out in crossing the mouths of larger coastal indentations, such as Delaware Bay and Long Island Sound. Few fish have been caught more than 16 km offshore, maximum distance for strays being 97–113 km (Raney 1954). Most of the total population of striped bass frequent the coastline, except at breeding season and perhaps during the winter. Among these, smaller sizes, up to 7 kg or so, are found within enclosed bays, in small marsh estuaries, in the mouths of rivers, and off the open coast. Most large bass, 14 kg or more, hold to the open coast, except at spawning time and perhaps in winter, but this is not invariable. Many run up into estuaries and into river mouths. In some rivers, good numbers (large as well as small) are caught far upstream and may remain in the rivers year-round. Striped bass off the open coast are most likely to be found along sandy beaches, in shallow bays, along rocky stretches, over and among submerged or partially submerged rocks and boulders, and at the mouths of estuaries, the precise situations that they occupy being governed by availability of food and time of day. Off outer beaches they may be anywhere right up to the breakers. When they are close in they frequent troughs that are hollowed out by the surf behind off-lying bars and gullies through which water rushes in and out across the bars as rollers break, for it is in such situations that baitfish are most easily caught and that crabs, worms, and clams are most likely to be tossed about in the wash of the breakers.

When the tide is high, bass often lie on a bar or even in white water along a beach if there is a good surf running. When the tide falls they drop down into troughs or move farther out, according to the precise topography. They also lie under rafts of floating rockweed at times, probably to prey on small animals they find among the weeds. The best spots along rocky shores are in the surf generally, in the wash of breaking waves behind off-lying boulders and among them, or where a tidal current flows most swiftly past some jutting point. In mouths of estuaries they are apt to hold to the side where the current is the strongest and in breakers out along the bar on that side. In shallow bays, they often pursue small fish among the submerged sedge grass when the tide is high, dropping back into deeper channels on the ebb. They frequent mussel beds, both in enclosed waters and on shoal grounds outside, perhaps because these are likely to harbor an abundance of sea worms (*Nereis*).

Striped bass of various life-history stages can tolerate from 0.1°C to 35°C (Davies 1970). Adult bass are active over a temperature range of 6°–7.8° to 21°C. Present indications are that if the temperature falls lower they either withdraw to somewhat warmer water if they are off the outer coast or lie on the bottom in a more or less sluggish state if they are in some estuary. On the other hand, it is not likely that they can long survive temperatures higher than about 25°–27°C, for many were found dead in shallow estuaries in Connecticut and Massachusetts during the abnormally hot August of 1937. In general, the temperature range appears to be 13°–22°C for larvae, 5°–30°C for juveniles, and 6°–25°C for adults (Setzler et al. 1980; Coutant 1986). They are equally at home in fresh or slightly brackish water and in full oceanic salinities (35 ppt). Dissolved oxygen requirements for juveniles and adults are 3 mg·liter^{-1} or higher at temperatures of 16°–19°C (Chittenden 1971b). Coutant (1985, 1986) hypothesized that, in certain situations, striped bass population size can be limited by an unsuitable summer habitat for adults with temperatures above 18°–25°C and dissolved oxygen concentrations below 2–3 mg·liter^{-1}. Juveniles have a higher temperature preference of 24°–26°C, which decreases as they age and grow.

Striped bass juveniles will orient to and swim against an oncoming current, but given a choice will select the path with the least current (Bibko et al. 1974). Larger stripers 25–75 mm long are able to swim against a current velocity up to 61 cm·s^{-1}. The maximum velocity of juveniles 8.9–11.4 cm at a temperature of 22°C is 67 cm·s^{-1} or 5.9 body lengths (BL)·s^{-1} (endurance to 50% exhaustion). Fish 2–14 cm can swim for 10 min at 22°C at a maximum velocity of 35–87 cm·s^{-1} or 7.6–12.6 BL·s^{-1} (Kerr 1953).

Striped bass swim steadily using both symmetric and asymmetric thrusts. Thrusts are symmetric at slower velocities up to 1 BL·s^{-1}, become slightly asymmetric up to 1.5–2 BL·s^{-1}, and are very asymmetric above these speeds. Maximum sustainable velocities for adults are 2.9–3.3 BL·s^{-1} (76–86 cm·s^{-1}) at 45 and 30 min, respectively (Freadman 1979). Bass use cyclic ventilatory movements for respiratory gas exchange at rest and at slow swimming speeds and shift to ram gill ventilation at intermediate and high velocities. The result is a substantial metabolic savings within their cruising range. Metabolic cost of active ventilation just prior to ram ventilation is estimated at 8.1% of total metabolism, so the changeover results in important hydrodynamic advantages (Freadman 1981).

During the first 2 years they live mostly in small groups. Later they are likely to congregate in larger schools; this applies especially to those up to 4.5 kg or so, which are often spoken of as school fish. Larger fish may school, but the very biggest, 13.6–18.1 kg and upward, are more often found singly or a few together. They are most likely to be in schools while migrating, but more scattered while feeding in one general locality. creeks and estuaries from North Carolina to the Canadian Maritimes. Small fish (2 and 3 years old), in particular, tend to school densely and they travel considerable distances without scattering, but it is not likely that a given school holds together for any long period. Fish of various sizes (i.e., ages) up to the very large ones often school together, which shows that different ages intermingle.

Food. Striped bass are voracious, feeding on whatever smaller fishes may be available and on a wide variety of invertebrates. Lists of stomach contents include sea lamprey, eel, shad, alewife, herring, menhaden, threadfin shad, anchovy, smelt, silver hake, tomcod, silverside, mosquito fish, mummichog, mullet, sculpins, white perch, yellow perch, croakers, weakfish, channel bass, flounders, northern puffer, rock gunnel (*Pholis gunnellus*), sand lance, tautog, lobsters, crabs of various kinds, shrimps, isopods, gammarid amphipods, various worms, squid, soft clams (*Mya*), and small mussels. In the Gulf of Maine, larger bass prey chiefly on eel, herring, smelt, silver hake, sand lance, squid, large and small crabs, lobsters, and sea worms (*Nereis*), whereas small ones are said to feed to a considerable extent on gammarid amphipods and shrimps. Diets were composed primarily of fishes, the species depending on what was commonly available at the time, reflecting seasonal changes in the fish population. In Chesapeake Bay (Hollis 1952), it consisted mainly of fishes (95.5% by weight of the total diet). In summer, bay anchovy and menhaden were the principal species eaten; in winter, larval and juvenile spot and Atlantic croaker were predominant; in early spring white perch were the most important prey; and in late spring and early summer alewife and blueback herring predominated. Clupeids in general are very important food (Dovel 1968).

An exception to the primarily piscivorous diet was noted in striped bass from surf waters of Long Island Sound (Schaefer 1970): 85% by volume of food of bass smaller than 399 mm FL consisted of invertebrates, primarily *Gammarus* spp., haustorid amphipods (45%), and mysid shrimp, *Neomysis americanus* (33%). Bass 400–599 mm FL ate 46% fishes (bay anchovy and Atlantic silverside) and 53% invertebrates, mostly amphipods. The largest bass, 600–940 mm FL, ate more fishes (65%), including white hake, striped mullet, tautog, and puffer but still ate large numbers (35%) of invertebrates, including amphipods, mysids, and lady crabs. Schaefer (1970) attributed the importance of invertebrates in the diet to turbidity of the surf environment, which made it more difficult for striped bass to pursue and eat fast-swimming vertebrates. Food habit studies of migrating young-of-the-year and adult striped bass in the upper Bay of Fundy (Rulifson and McKenna 1987) also showed high percentages of invertebrates. Young-of-the-year striped bass (69–94 mm FL) fed exclusively on sand shrimp (*Crangon septimspinosa*), whereas individuals 142–240 mm FL ate mainly sand shrimp and fish (hake). Larger bass (271–360 mm FL) ate mainly fishes and sand shrimp, but also polychaete worms. The largest bass (381–520 mm FL) ate mostly fishes (hake,

Subadult striped bass that gathered below a hydroelectric dam on the Connecticut River were found to scavenge fishes that were injured or killed while attempting to pass through or around the dam, as well as fish discarded by sports fishermen. These age-2 and age-3 striped bass contained body parts of adult American shad, blueback herring, and sea lamprey. Bass caught later in the season had fewer body parts of large fish, probably because the runs of those fish had decreased by then (Warner and Kynard 1986). Striped bass schools of different ages feed at the same time (Raney 1952). When bass are feeding on any one particular prey, they are likely to ignore food of other sorts, and they appear to follow and feed on schools of fishes (Scofield 1928). On the whole, they seem to be more active, and to feed more actively, in the evening just after dark and just before sunrise than while the sun is high. Striped bass feed more in summer and fall and less during winter and spring (Hollis 1952). Temperature may be a determining factor. Feeding frequency declined sharply below 5°C. Adults ceased feeding below this temperature and juveniles ate smaller amounts less frequently. Feeding of subadults and adults also declines at temperatures over 26°C in seawater (Rogers and Westin 1978). Fasting occurs for a brief period during the spawning season beginning slightly before and ending slightly after spawning occurs (Trent and Hassler 1966).

Several studies have been conducted on food conversion ratios of young striped bass fed commercial fish foods (Westin and Rogers 1978; Setzler et al. 1980). Striped bass held in cages in Rhode Island coastal waters and fed a diet of ground hake had a gross feeding efficiency of 17–21% when fed 6–12% body weight per day (Rogers and Westin 1978).

Respiratory metabolism for juvenile striped bass was studied at five temperatures (8°, 12°, 16°, 20°, and 24°C) and three water velocities (0, 5, and 10 $cm \cdot s^{-1}$) (Kruger and Brocksen 1978). Differences between standard and active metabolism over a limited range of water velocities were used to calculate scope for activity. This was very similar at 8°, 12°, and 16°C, coming to 55, 44, and 45 mg $O_2 \cdot kg^{-1} \cdot h^{-1}$ for an average 40-g bass. At 20° and 24°C, scope for activity increased to 99 and 143 mg $O_2 \cdot kg^{-1} \cdot h^{-1}$. The temperature optimum for juveniles appeared to be about 16°C.

Striped bass are less oxygen-efficient than white perch or spot (Neumann et al. 1981). For two sizes of bass at three different water velocities at a temperature of 15°C, oxygen consumption increased with both fish weight and water velocity. Average respiration rates for a 50-g fish at 8.6, 31.7, and 49 $cm \cdot s^{-1}$ water velocity were 19, 24.3, and 33.7 mg $O_2 \cdot liter^{-1}$; for a 150-g fish at the same water velocities respiration rates were 31.4, 41.3 and 63.7 mg $O_2 \cdot liter^{-1}$. Oxygen consumption rates for different sizes of striped bass at various temperatures and salinities have been summarized by Westin and Rogers (1978). Metabolic rate dependence on temperature was investigated by Klyashtorin and Yarzhombek (1975). For striped bass weighing between 1 and 3 g acclimated to 22°C, Q_{10} values varied from 2.1 to 1.8 between 15° and 30°C. Oxygen consumption ranged from 250 to 1,000 $mg \cdot kg^{-1} \cdot h^{-1}$.

Predators. Although direct information is lacking, larger bluefish and weakfish probably prey on small striped bass, but man is undoubtedly the most important predator. Cannibalism by large stripers on smaller ones has been reported during the spawning season in the Hudson River (Dew 1981). Silver hake and cod may also prey upon juvenile striped bass (Scott and Scott 1988).

Laboratory predation studies were conducted using potential fish and invertebrate predators that occur on striped bass spawning grounds in the Pamunkey River, Va. A cyclopoid copepod, *Acanthocyclops vernalis,* attacked and killed striped bass larvae. Fish species that ate striped bass larvae were juvenile and adult satinfin shiner, spottail shiner, tesselated darter, white perch, striped bass, bluegill, pumpkinseed, channel catfish, and white catfish. However, neither eggs nor larvae of striped bass were positively identified in the guts of these fishes collected in the field (McGovern and Olney 1988). A free-living copepod, *Cyclops bicuspidatus thomasi,* was found attached to a number of striped bass yolk-sac larvae 3.2–6.4 mm TL in the Chesapeake and Delaware Canal. Damage ranged from constriction of tissue around the point of attachment to missing parts of the finfold and ruptured yolk sacs. Most of the larvae had damage severe enough to cause death (Smith and Kernehan 1981).

Parasites. Striped bass are host to a large variety of parasites. Field studies conducted on Chesapeake Bay stripers age 0+ to 3+ (Paperna and Zwerner 1976) showed the most common parasites to be protozoans *Colpanema* sp. (at age 0+); *Myxosoma morone* (age 0+); *Trichodina davisi* (all ages); *Glossatella* sp. (age 0+); Platyhelminthes, Cestoda *Scolex pleuronectes,* protocephalid larvae (type A and B); acanthocephalans *Pomphorhynchus rocci* (larvae, age 0+, and adults, ages 1–3+); Aschelminthes, Nematoda *Philometra rubra* (all ages); and crustaceans, Copepoda *Ergasilus labracis* (age 1–3+).

Although an earlier report of striped bass parasites (Merriman 1941) did not find direct evidence that pathogenicity was associated with parasitic infection, Paperna and Zwerner (1976) found distinct pathological processes associated with the most common and abundant parasites. These included gill hyperplasia caused by *Ergasilis labracis;* visceral granuloma and adhesions associated with *Philometra rubra;* perforations of the gut and resulting histological changes in the gut wall by *Pomphorhynchus rocci;* extreme fibrosis of visceral organs, especially the liver and spleen, caused by proliferation by larval helminths; skin lesions from *Argulus bicolor;* and lymphocystis. These conditions probably do not cause mass mortality, but do increase susceptibility of infected fish to predation and stresses caused by water quality deterioration.

Extensive lists of parasites found on striped bass have been compiled by Setzler et al. (1980) and Westin and Rogers (1978). Bonn et al. (1976) listed parasites of cultured striped bass, which are often quite different from those found in the field.

Diseases. Cultured striped bass (Hawke 1976) and wild populations suffer from a number of diseases. Fin rot, which is believed to be caused by a combination of environmental stress and bacterial infection, was reported by Mahoney et al. (1973) in stripers from the New York Bight. It paralleled seasonal temperatures, with the rate of infection being lowest in winter, increasing in spring, peaking in summer, and decreasing in fall. One of the common infectious agents in fin rot is *Vibrio.*

Pasteurellosis, an infection by the bacteria *Pasteurella,* was noted in Chesapeake Bay bass in 1963 (Snieszko et al. 1964). Paperna and Zwerner (1976) reported mass mortalities of age-1+ and age-2+ fish from pasteurellosis. They described the pathological conditions as appearance of necrotic foci visible as white spots in the spleen. Moribund fish showed progressive liquefaction necrosis in liver, kidney, and finally the intestine.

Columnaris disease is caused by the bacteria *Flexibacter columnaris* and manifests itself as a moldlike growth on the skin and fins. Kelley (1969) reported that it depressed the hematocrit value of infected striped bass by 12–15%.

Lymphocystis disease has been reported in 2-year-old striped bass from the Chesapeake (Paperna and Zwerner 1976). This is a viral disease manifested by wartlike nodules that may reach a size of 2 mm and often cover the entire fish, but despite its alarming appearance it is rarely fatal.

Kudo cerebralis, a myxosporidian disease of connective tissue that is associated with nervous tissue, was reported by Paperna and Zwerner (1976). This disease appears as cysts embedded on the ventral and lateral surfaces of the brain and inside distal cranial nerve branches.

Epitheliocystis lesions were found on gill filaments and arches of striped bass in the York River (Paperna and Zwerner 1976). Wolke et al. (1970) found epitheliocystis disease in striped bass from Connecticut waters.

Tubercular lesions caused by an acid fast bacillus (*Mycobacterium* sp.) have been found in striped bass from four areas in California and in Coos Bay, Ore. (Sakanari et al. 1983). Prevalence of infection was 25–68% in California and 46% in Oregon. These lesions have been reported from striped bass in aquariums on the east coast (Nigrelli and Vogel 1963) and from wild populations in the Chesapeake Bay (D. E. Zwerner, VIMS, pers. comm.). Although the fish collected did not appear to be seriously affected, this species of *Mycobacterium* is pathogenic and can cause skin lesions in humans who handle infected fish (Snieszko 1978). Caution should be used in handling any fish with lesions.

Intestinal contents of striped bass collected in the Hudson River and Long Island Sound were examined for presence of bacterial flora that could be opportunistic pathogens (produce disease in fish when they are stressed). Several of these bacteria were common in the gut flora: *Aeromonas, Vibrio, Pseudomonas, Enterobacter,* and *Alcaligenes.* The total number of gut bacteria was consistently higher (100–1000 times) in striped bass from the Hudson River, perhaps reflecting the higher organic content of Hudson River water, than those from Long Island Sound. The enteric bacteria were reflective of migratnance of *Aeromonas hydrophila* were similar to those of most marine fishes (MacFarlane et al. 1986).

Breeding Habits. Striped bass are very fecund, with the number of eggs produced highly correlated with age, weight, and length. Fecundity estimates range from 15,000 eggs in a 46-cm fish (Mansueti and Hollis 1963) to 4,010,325 in an age 13, 14.5-kg fish (Jackson and Tiller 1952). Westin and Rogers (1978) and Hardy (1978b) summarized fecundity data for striped bass of various ages collected in a number of areas. Fecundity per kilogram of body weight has been estimated at 173,000 eggs for Hudson River fish (Texas Instruments 1973), 176,000 for Roanoke River fish (Lewis and Bonner 1966), and 318,000 for offshore North Carolina fish (Holland and Yelverton 1973). The last authors estimated fecundity for striped bass 77–110 cm FL and 7–13 years of age and found the following linear relationships between fecundity and length, weight and age: $F = 9.33 \times 10^4\text{FL} - 6.24 \times 10^6$, $r = 0.85$; $F = 2.18 \times 10^5 W - 1.17 \times 10^4$, $r = 0.86$; and $F = 4.33 \times 10^5 A - 1.78 \times 10^6$, $r = 0.66$, where F = fecundity × 10^6, FL = fork length in cm, W = weight in kg, and A = age in years.

Two or three distinct sizes of oocytes occur in striped bass ovaries so either there is batch spawning or more than 1 year's oocytes are present in the ovaries (DeArmon 1948; Jackson and Tiller 1952; Zolotnitskij and Romanenko 1981). This pattern of oocyte development is group synchrony, the most common teleost pattern, where ovaries contain a heterogeneous population of smaller oocytes and a group of more synchronized developing oocytes (Specker et al. 1987).

Hermaphroditism has been noted (Schultz 1931). Westin (1978) reported a 52-cm, 16.3-kg immature hermaphrodite striper from Rhode Island. Morgan and Gerlach (1950) found that 3% of striped bass sampled in Coos River, Ore., were hermaphrodites. Moser et al. (1983) found there were increasing signs of pathology associated with egg retention on older (7–10 years) hermaphrodites.

Among Connecticut fish, Merriman (1941) found that, "approximately 25 percent of the female striped bass first spawn just as they are becoming 4 years old, that about 75 percent are mature as they reach 5 years of age, and that 95 percent have attained maturity by the time they are 6 years old." A large percentage of the males had matured at 2 years, and probably nearly all by the time they were 3. It is probable that this applies equally to Maine bass. This remains the best information on age at maturity gathered from ocean migrating striped bass; however, its validity has been questioned. Hudson River striped bass were found to mature at 4 years of age (68%) with 28% of the males age-class 4 to 6 still immature. Females began to mature at age 4 (22%), but only 54% were mature by age 6 and 80% by age 7 (Dew 1988). Both sexes were larger when they attained maturity than fish analyzed by Merriman (1941).

Different stocks mature at different ages. Specker et al. (1987) described oocyte development in striped bass and related this to factors influencing estimates of age at maturity.

and a population impounded in a saltwater cove in Massachusetts (salinity 35 ppt). Ovaries of migrating fish collected in summer had a heterogeneous group of small oocytes some of which appeared to be forming yolk vesicles. Fish collected in the fall were of two types. One had ovaries with oocytes similar to those of fish collected in summer and the other had ovaries with a batch of small heterogeneous oocytes and a batch of larger oocytes that were shown histochemically to have recruited into the gonadotropin-dependent growth phase (maturing oocytes). Striped bass from Cat Cove had oocytes that underwent the same developmental process as the migrant stock. Migrating fish collected in November with small oocytes were age-classes 3 and 4, those with small and large oocytes were from age-classes 5–7. However, it is not known if final maturation takes 1 year or more. Since Merriman (1941) established maturity from oocyte size and sampled primarily from the spring to November, when it is most difficult to judge maturity, he might have miscalculated the ages. A complete description of the stages of oocyte development in mature females throughout the year is needed so that one can assess sexual maturity of fish collected at any time of the year (Specker et al. 1987). Two studies are currently in progress to determine age at maturity, one in Maryland on the spawning grounds and one in Rhode Island on migrating stocks.

Striped bass are assumed to return to their natal rivers to spawn, and this is supported by tagging data (Mansueti 1961b; Nichols and Miller 1967; Moore and Burton 1975). Males are the first to arrive on the spawning grounds, and females move in later. Bass spawn either in brackish water at the heads of estuaries or in freshwater rivers, not off the open coast in saltwater. Those that enter freshwater rivers may deposit their eggs only a short distance above the head of the tide, as they do in the Potomac, or they may run much farther upstream. Most spawning occurs within the first 40 km of freshwater in the river (Tresselt 1952). Within an individual river, spawning grounds may change from year to year (Farley 1966).

The major spawning area for striped bass on the east coast is now the Chesapeake Bay and its tributaries (Merriman 1941; Raney 1957; Able and Fahay 1998). Other important areas are the Hudson, Roanoke, and Delaware rivers (Richards and Deuel 1987). In the twentieth century the Delaware River ceased to produce striped bass because pollution from the Philadelphia area caused oxygen depletion, which prevented stripers from reaching their freshwater spawning areas (Chittenden 1971b). Recent improvements in water quality have resulted in increased reproduction and survival of juvenile striped bass in the Delaware River (Weisburg and Burton 1993). The Roanoke River stock appears to have a limited migratory range (Boreman and Lewis 1987). In the seventeenth and eighteenth centuries, striped bass spawned in almost every river on the coast of New England, supporting large fisheries, until these populations were extirpated (Little 1995). Striped bass still spawn in the Saint John River, and it is probable that they also spawn in small streams tributary to Minas Basin and Cobequid Bay at the head of the Bay of Fundy; in Grand Lake at the head of the Shubenacadie River, and in the Annapolis River. Great numbers of young-of-the-year fish (4.9–7.6 cm) were caught, for example, in winter in the 1880s in the Kennebec, where ripe fish also have been reported from the end of June into July. The only Maine or Massachusetts streams where there has been evidence of spawning bass in the past 50 years are the Mousam, Maine, where fishermen reported taking females with ripe eggs on several occasions; and the Parker, Mass., where Merriman took three young-of-the year 7.1–8.5 cm long on 4 August 1937. Davis (1966) found no evidence of spawning in Maine and no evidence of discrete populations in the estuaries he sampled: the Piscataqua, Saco, Cousins, Kennebec, Sheepscott, Penobscot, and St. Croix rivers. It is probable that spawning in Maine waters today is relatively rare.

The spawning season is from late April to early May in North Carolina, chiefly in May in the Chesapeake Bay region, and mid-May to June in the Hudson River (Raney 1952). Any bass that may spawn in the rivers of Massachusetts or Maine, or in the Bay of Fundy probably do so in June, and those of the southern shore of the Gulf of St. Lawrence and of the lower St. Lawrence River in June and July. A summary of striped bass spawning seasons was given by Hardy (1978b) and Westin and Rogers (1978).

Bass are broadcast spawners and have no special courtship sequence. During spawning a large female may be surrounded by many small males, and the latter are described as fighting fiercely with one another. The female frequently swims close to or breaks the surface of the water (Merriman 1941). Diel spawning patterns are very variable. Peaks occur most often at dawn or dusk, but spawning has been recorded at all hours during the day, as well as at night (Setzler et al. 1980).

Spawning is triggered by spring increases in temperature and varies according to the weather for that year. The number of spawning peaks also varies depending on the weather. Westin and Rogers (1978) summarized the time of spawning, temperature, and date and temperature of peak spawning for major areas from the Hudson River, N.Y., to the Savannah River, Ga., and two California rivers. In general, temperatures ranged between 10° and 26°C, with 14°–15°C usually initiating the spawning. Salinity on the spawning grounds ranged from 0 to 4 ppt (Tresselt 1952). Higher values have been recorded but it was probable that eggs drifted to those areas from farther upstream. An important requirement for successful spawning is a current turbulent enough to prevent the eggs from settling on the bottom, where they would be in danger of being silted over and smothered.

Early Life History. Eggs are spherical, nonadhesive, semibuoyant, and transparent to greenish or golden green. The eggs have a clear tough chorion, a wide perivitelline space, a single large oil globule, and a lightly granulated yolk (Raney 1952; Mansueti 1958a). Eggs average 1.25–1.80 mm in diameter when they are deposited (Pearson 1938; Raney 1952; Mansueti and Mansueti 1955). The perivitelline membrane swells during the first hours after fertilization to 1.3–4.6 mm or an average diameter of about 3.6 mm (Murawski 1969; Albrecht 1964). The yolk diameter is 0.90–1.50 (mean 1.18 mm) and accounts for 35% of the egg diameter. The oil globule diameter is 0.40–0.85 (mean 0.61), and it may fragment into several smaller ones (Mansueti 1958a). The only other pelagic eggs of that size found in estuaries in spring are those of shad and, rarely, menhaden. Shad eggs have no oil globule and menhaden have a very small one, so their eggs are easily distinguished from those of striped bass (Lippson and Moran 1974). The eggs sink in quiet water but are swept up from the bottom by the slightest disturbance, so that they tend to drift downstream with the current. Consequently, eggs that are produced far upstream may not hatch until they have reached tidewater.

Egg development at 16.7°–17.2°C as described by Mansueti (1958a) is as follows: (1) Fertilization–5 min: perivitelline space begins to form; (2) 20–40 min: first cell divisions; (3) 2 h: 4- to 32-cell stage, perivitelline space fully formed; (4) 8 h: 16-cell morula stage, blastoderm appears granular; (5) 12 h: gastrulation begins, blastoderm covers half of yolk; (6) 16 h: 64 cells, gastrula, blastocoel forms, germ ring thickens; (7) 20 h: embryo developed, neural ridges and eyes visible, pigmentation present on embryo and oil globule; (8) 24 h: embryo about halfway around the yolk, pigment intensified on dorsolateral region of body and adjacent blastoderm; (9) 36 h: embryo about 1.6–2 mm long, posterior part of body free from yolk, eye well differentiated but not pigmented, embryo floats free and fully extended in egg; and (10) 48 h: hatching.

Hatching time in relation to temperature varies from 109 h at 12°C (Rogers et al. 1977) to 25 h at 26.7°C (Shannon and Smith 1967), with an average of 48 h at 17°–19°C (Westin and Rogers 1978).

Optimum conditions for striped bass embryonic growth and survival have been reported as temperatures of 14°–24°C (Albrecht 1964; Bayless 1972; Morgan and Raisin 1973) with a narrower range of 15°–18°C observed by Rogers et al. (1977). Critical oxygen levels are 3–5 mg·liter^{-1} at 18°C (Rogers and Westin 1978; Turner and Farley 1971). Egg survival is highest at low salinities; no significant effect on survival to hatch was observed at salinities of 0–8 ppt (Morgan and Raisin 1973). Booker et al. (1969) found pH ranges from 6.6 to 9 to be satisfactory for hatching. Acidification of surface waters (low pH, high aluminum concentrations, and low water hardness) have been shown to adversely affect survival of striped bass larvae by *in situ* experiments conducted in the Nanticoke River, a tributary of Chesapeake Bay (Hall et al. 1985). Laboratory experiments confirmed that acidification levels found in the field, pH of 6.5, aluminum concentration of 0.12 mg·liter^{-1} and water hardness of 30 mg·liter^{-1}, were 90–99% toxic to striped bass larvae (Mehrle et al. 1984). These conditions can occur in several tributaries of Chesapeake Bay, but other tributaries have salinity and alkalinity conditions that buffer the effects of acidification (Hall 1987).

Newly hatched larvae are 2.0–3.7 mm TL at hatch with a mean of 3.1 mm (Mansueti 1958a). They are transparent with is at the anterior end of the yolk sac, which projects beyond the head or at least anterior to the eye. The mouth is unformed and the eyes unpigmented.

The rate of yolk-sac absorption is variable, ranging from 3 days at 24°C (Albrecht 1964) to 9 days at 12°C (Rogers et al. 1977). The average is about 8 days at 18°C (Rogers et al. 1977). A summary of yolk-sac absorption times at different temperatures was provided by Setzler et al. (1980). At this time the larvae are referred to as finfold stage and when fin rays form they are called postfinfold. These stages are sometimes collectively called larval stages by some authors and postlarvae by others. Duration of the larval stage varies from 23 days at 24°C to 68 days at 15°C, with an average of 33 days at 18°C (Rogers et al. 1977).

Hardy (1978b) gives a detailed description of developmental events up to the adult stage, and Westin and Rogers (1978) provide a table of important events from hatch to transformation of striped bass larvae reared in their laboratory at about 17°C. The following is a list of some important events condensed from Mansueti (1958), Hardy (1978b), and Westin and Rogers (1978): (1) Day 1: Eye pigmentation develops. (2) Day 2–5 (4.5–5.2 mm TL): Differentiation of jaws and gut begins, pectoral buds form, pelagic swimming begins, chromatophores are present and intensifying into the three areas characteristic of the species—a series of stellate chromatophores along the posterior two-thirds of the trunk and tail, concentrations along the dorsal peritoneal wall and on the dorsolateral and ventrolateral wall of the yolk and along the gut and a heavy concentration on the oil globule. (3) Day 6–8 (5.5–7.5 mm): Yolk absorption, differentiation of stomach, active pelagic swimming and feeding begins, swim bladder visible and filled in most cases; a large chromatophore present on the upper surface of swim bladder, an almost continuous line of pigment on ventral part of body from opercle to midway between anus and tip of tail. (4) Day 10–15 (10.5–12.5 mm): Teeth visible, oil globule absorbed, swim bladder filled, finfold divided into three parts. (5) Day 18–30 (12–16 mm): Differentiation of rays in caudal, anal, and dorsal fins; myotomes correlated with vertebral number, three preopercular spines present, number of teeth increases. (6) Day 30–50 (16–35 mm): Initial formation of lateral line scales, soft dorsal, anal, and caudal fins well differentiated; spinous and pelvic fins partially differentiated; body shaped like adult; pigmentation, small black dots all over body. (7) Day 50–80 (35–50 mm): Early juvenile coloration of vertical dark bars on dorsal part of body from opercle to caudal peduncle, pigmentation stronger, covered with scales, three anal spines and full complement of meristic characters. (8) Day 80–100 (50–80 mm): Development of horizontal stripes, fully developed fin rays.

Larvae are similar to white perch. Yolk-sac larvae may be distinguished by total length in relation to developmental stage, striped bass being about 1 mm longer at each stage, and the relative position of anus and gut extension beyond the yolk. In striped bass the vent is closer to the yolk with only

the vent while in white perch there are five or six. In striped bass the gut angles directly downward to the vent whereas in white perch it extends along the tail before angling down. When the two species begin developing their fin rays (at 5–6.5 mm), it becomes difficult to separate them as differences are not well defined. In general, white perch have shorter, stouter bodies with a thicker caudal peduncle; preopercular spines are well defined and appear earlier than in striped bass. Between 6 and 12 mm, striped bass have teeth, whereas white perch have none (Lippson and Moran 1974). Bass also show a distinct dark pelvic spot on the liver between the pelvic fins as they develop (Drewry 1981).

More accurate identification can be obtained by clearing and staining specimens. Unique characteristics in position and shape of median ethmoid and predorsal bones, dorsal and anal pterygiophores (fin skeletal supports), vertebral column, and caudal skeleton were found in laboratory-reared striped bass and white perch (Fritzsche and Johnson 1980). Olney et al. (1983) further refined this for field-collected larvae using the same staining technique to show pterygiophore interdigitation patterns.

Behavior. Yolk-sac larvae are positively phototactic and alternate between swimming to the surface and sinking between swimming efforts (Pearson 1938; Raney 1952; Doroshev 1970). They begin to swim horizontally between 4–5 days of age (McGill 1967). Post-yolk-sac larvae can resist water currents, and exhibit nocturnal migration patterns strongly oriented toward the bottom. This intensifies as the juvenile stage approaches. Juveniles are collected in water more than 6 m deep in the early part of the season and later migrate to shoal waters and toward the shore zone. These movements are triggered by increasing water temperatures. Falling water temperatures bring downstream movements into deeper water, so that by December juveniles leave the shore zone and migrate into deeper water or move out of estuaries entirely. Day-night beach seine comparisons show onshore movements at night, offshore movements by day, and dispersion closer to the bottom by day and throughout the water column at night. Movements are also affected by tidal stage, temperature, and salinity (Westin and Rogers 1978). Distribution of striped bass eggs, larvae, and juveniles in the Potomac estuary (Setzler-Hamilton et al. 1981) showed peak abundances of larval and juvenile bass either in areas of peak spawning or upstream in spite of the downstream river flow. Three phenomena were suggested to account for this distribution: continued upstream migration of the spawning stock, differential mortality of eggs and larvae throughout the spawning season, and the ability of older larvae and juveniles to maintain longitudinal position within the estuary. Densities of striped bass from egg to juvenile stage reported for different areas from the Hudson River to Albemarle Sound were summarized by Westin and Rogers (1978).

Because of declines in some striped bass populations, which may be linked to increasing pollution in nursery areas, numerous studies have been done on exposure of larvae and juveniles to various toxicants. A summary of the results of tests using pesticides, heavy metals, dyes, pharmaceuticals, organic substances, and fish anesthetics is given in Westin and Rogers (1978) and Bonn et al. (1976). Hall (1988) described *in situ* prolarval and yearling striped bass survival studies conducted for several years at spawning sites in the Nanticoke River, Chesapeake and Delaware Canal, and Potomac River. He described water quality conditions, test results, and procedures for relating these to young-of-the-year abundance indices. He also explained why these procedures may often give conflicting results, for example, poor prolarval survival occurred in *in situ* tests in the Potomac in 1986, but a fair juvenile index was reported. The consensus among most biologists is that overfishing was the primary cause of striped bass decline, probably exacerbated by poor spawning conditions. Other factors that have been investigated are power plant effects such as thermal pollution, impingement, and entrainment. Results of these experiments were summarized by Setzler et al. (1980).

Comparison of survivorship of laboratory-reared larvae and juvenile striped bass in various temperature-salinity combinations (Otwell and Merriner 1975) showed significant effects owing to all three factors (temperature, salinity, age). The greatest mortality rates occurred at lowest temperature and highest salinity combinations, and younger larvae had lower rates than older ones. Temperature was more limiting to growth and survival than salinity; the lowest temperature (12°C) produced 50% mortality when fish were introduced vs. 3.5% and 7.2% at 24° and 18°C, respectively. Growth experiments at three temperatures (17°, 21°, 28°C) of larvae from six native anadromous stocks spanning most of the geographic range of the species supported a countergradient variation in growth (Conover et al. 1997). Northern fish (New York, Maryland, and Nova Scotia) had higher growth rates than southern fish (North Carolina, South Carolina, and the Gulf of Mexico).

Experiments on digestive enzyme activities in striped bass from first feeding through larval development (Baragi and Lovell 1986) showed that larvae begin feeding 4 days posthatch when all digestive enzymes except pepsin were present. Enzyme activities were 25–60% of those at day 32, increased until day 12, then decreased until day 16, at which point all enzyme activity increased again. The stomach is not formed until day 16 (Gabaudan 1984) when pepsin activity was detected, probably reflecting further development of gastric glands, which secrete this enzyme.

Food and feeding of larvae were studied in the laboratory as they related to mortality, point of no return, development, and energetics (Eldridge et al. 1981). Survival was directly related to the density of *Artemia salina* nauplii fed. The importance of the oil globule as an endogenous food source was demonstrated, as highest mortality was seen after absorption of the oil globule, which occurred much later than yolk-sac absorption. The rate of oil utilization was inversely related to food density. Starved larvae survived an average of 31 days after fertilization and did not exhibit a point of no return. Larvae starved for varying periods lost weight but recovered when provided with food, thus showing no clear point of no return (Rogers and Westin 1981).

The first food of larval striped bass is copepod and cladoceran nauplii and adults. Larval striped bass from the Potomac estuary (Beaven and Mihursky 1980) ate the largest prey items they could capture, and the food most frequently occurring in larval stomachs were adults and copepodites of *Eurytemora affinus* and cyclopoid copepod species. Other food positively selected were cladocerans *Bosmina longirostris* and *Daphnia* spp. Rotifers were selected against, probably because of size.

Some field studies (Kernehan et al. 1981; Setzler-Hamilton et al. 1981) showed circumstantial evidence that food densities influence survival of striped bass larvae. Martin et al. (1985) attempted to assess starvation in wild larval striped bass in the Potomac River estuary and to correlate this with prey densities. Nutritional state was determined by morphometrics, histology, RNA:DNA ratios, and fatty acid composition. All four techniques showed evidence of poor nutritional state early in the season but not in the latter part. Significant correlations among nutritional indices and copepod and cladoceran densities were found, indicating that the switch in predominance from copepod nauplii, copepodites, and rotifers to cladocerans (especially *Bosmina*) later in the season caused changes in the index values to a better nutritional status.

Juveniles 25–100 mm in the Potomac were nonselective feeders (Boynton et al. 1981). They ate mostly insect larvae, polychaete worms, larval fishes, mysids, and amphipods. The diet reflected changes in the estuarine community composition associated with salinity changes; insect larvae being predominant where salinity was less than 5 ppt. The most important food items were crustaceans (*Neomysis americanus, Crangon septimspinosus,* and *Palaemonetes pugio*) and fishes (naked goby, *Gobiosoma bosci,* bay anchovy, Atlantic silverside, weakfish, and menhaden). Juveniles over 100 mm ate mostly fishes (Markle and Grant 1970; Bason 1971).

Age and Growth. Striped bass are long-lived; one kept in the New York Aquarium lived to be 23 years old.

Growth of striped bass up to 70 cm length is calculated from scale annuli with the formula $l = [(L - 1)l'/L'] + 1$, where L = TL of fish; l' = ratio of radius to annulus in question; L' = scale radius; and l = unknown TL (Scofield 1931; Merriman 1941). Annuli form on scales of bass caught in Virginia between April and June (Grant 1974) and in North Carolina from October to January (Trent and Hassler 1966). Growth from Maine to North Carolina was studied by Davis 1966 (Maine), Frisbie 1967 (Massachusetts), Westin and Rogers 1978 (Rhode Island), Merriman 1941 (Connecticut), Texas Instruments 1973 (Hudson River, N.Y.), Bason 1971 (Delaware), Jones et al. 1977 (Potomac River), Mansueti 1961b (Chesapeake Bay), Marshall 1976 (Albemarle and Pamlico sounds, N.C.), Holland and Yelverton 1973 (offshore North Carolina).

Growth rates are similar for fish from different geographical areas. Females grow larger than males; most bass of 13.5 kg and centimeters and weight in kilograms is as follows: 30–33 cm/0.3 kg; 45–50 cm/1.3–1.4 kg; 61 cm/2.3 kg; 76–81 cm/4.5–6.8 kg; 83–91 cm/8.2–9 kg; 100 cm/13.6 kg; 109 cm/18 kg.

Growth of young-of-the-year striped bass in the Hudson River followed an S-shaped curve with peak growth rates (0.8–0.9 mm·day^{-1}) occurring from mid-June to mid-August (Dey 1981). Growth was positively correlated with water temperature in the early stages of development but showed no relationship in the later juvenile stages. Neither freshwater flow nor juvenile abundance showed significant correlations with growth rate. Growth of striped bass larvae in the Hudson River was less than that reported from the Patuxent River (Mansueti 1958a) and Chesapeake Bay (Vladykov and Wallace 1952) probably because spawning occurs earlier and the growing season is longer in these more southerly locations. Jones and Brothers (1987) examined the otolith increment aging technique for larval striped bass in the laboratory under optimal and suboptimal feeding conditions. They found that daily rings were deposited and were discernible with light microscopy from day 4 posthatch (corresponding to initiation of feeding) through the first 2 months in well-fed larvae. Larvae reared under restricted feeding regimes yielded ring counts that underestimated their age by several days. Counts made using scanning electron microscopy came closer to reflecting the true age. Therefore, one cannot rely on otolith aging techniques using the light microscope in field-collected larvae without making random checks using scanning electron microscopy preparations for verification of the results.

General Range. Atlantic coast of eastern North America, from the lower St. Lawrence River and the southern side of the Gulf of St. Lawrence to northern Florida; along the northern shore of the Gulf of Mexico to Alabama and Louisiana; running up into brackish or fresh water to breed. In the last quarter of the nineteenth century they were introduced on the Pacific coast, where their range now extends from British Colombia to Ensenada, Mexico (Forrester et al. 1972). They have also been introduced into Russia, France, and Portugal. Striped bass have been stocked in many rivers and reservoirs throughout the United States. Information on transplanted stocks can be obtained from Setzler et al. (1980) and Westin and Rogers (1978).

Occurrence in the Gulf of Maine. The range includes the coastline of the Gulf from Cape Cod to western Nova Scotia. Distribution is determined by the very evident preference for surf-swept beaches and for particular stretches of rocky or bouldery shoreline, as well as for shallow bays, inlets, and estuaries. The geographic status of bass in the Gulf also depends on whether it is a good bass year or a poor one. When bass are reasonably plentiful, the outer shore of Cape Cod provides the most productive surf casting, with Monomoy Island, the general vicinity of Nauset Inlet, and the tip of the Cape northward from Highland Light being perhaps the warmest stretches.

in Pleasant Bay, within Nauset Marsh, and in Town Cove, Orleans. The most productive trolling grounds are along the eastern and southern sides of Cape Cod Bay in most summers, especially off the Eastham shore a few miles southward from Wellfleet, and off the mouth of Scorton Creek, Barnstable, and the Sandwich shore. The shores of Cape Cod and Cape Cod Bay have, in fact, been the chief centers of abundance for bass within the Gulf from as far back as the records go. Few bass are reported along the rocky stretch from the Cape Cod Canal to the entrance to Plymouth Harbor, but many are caught in Plymouth Harbor, especially off Eel Creek, and up Duxbury Bay to the salt marsh creeks that open into its head.

Surf casters account for some along Duxbury Beach on the outside and a few in the boulder-strewn area at the western end of Humarock Beach. The North and South rivers in Marshfield yield considerable numbers in good years. Anglers casting from the shore take a few on boulder-strewn stretches along the Scituate shore, and Glades Point is famous for large bass. The Cohasset shoreline yields a few yearly, occasionally a very large one. In seasons when there is a good run of the smaller sizes, considerable numbers are taken at various places within the limits of Boston Harbor; Hull Gut, Weir River in Hingham, and Wollaston Beach are well-known localities. In years in which there is a run of little fish, many are caught from the docks and bridges to the head of Boston Harbor.

The north shore of Massachusetts Bay seems not to be as attractive for bass as its succession of inlets, beaches, and rocky headlands might suggest, for catches reported are small and scattered in most summers. But the beaches and enclosed waters from a few miles north of Cape Ann to the mouth of the Merrimac River are productive enough to rank second to the Cape Cod–Cape Cod Bay region. Bass are taken in the surf from Ipswich Beach, Cranes Beach, and along the entire length of Plum Island Beach. Many are caught by boat fishermen over the flats within the mouth of the Merrimac, as well as around the jetties at its entrance. Schools are often reported in Plum Island Sound, and the Parker River, emptying into it, is well known for bass.

Some are caught in Hampton Harbor, N.H. The next important bass waters (moving northward) are the lower reaches of the Piscataqua River system, marking the boundary between Maine and New Hampshire. In good years bass are to be caught in several streams that drain the southern part of the Maine coast, especially in the York, the Mousam, and the Saco, which is the most productive. Schools are sighted and a few fish are caught along intervening beaches and some in the shallows of Biddeford Pool.

In the Bay of Fundy region, striped bass are confined to large warm estuaries and neighboring fresh water, that is, to those of the Saint John, Minas Basin–Cobequid Bay, and Shubenacadie River systems and of the Annapolis. Information suggests that bass were always more plentiful in Saint John River waters than anywhere along the eastern part of the coast of Maine. Bass are well known in the Minas Basin–Cobequid region. The status of bass is especially interesting in the Shubenacadie River, for they are not only caught in freshwater there and in Shubenacadie Lake, where they are known to spawn, but some large fish remain throughout the year in the lake; that is, they behave like a landlocked population. A thousand or so are caught yearly by anglers in the lake and in the river; and it is said that fish as large as 23 kg have been taken, although most of them run small. Bass are also to be caught in various bays and river mouths along the western shore of Nova Scotia; but there is no definite information as to how plentiful they are or how large.

Localities along the outer coast of Nova Scotia where stripers have been reported are the head of Mahone Bay, Chedabucto Bay, Mira Bay, and other harbors of Cape Breton. The shoal estuaries of the Richibucto Bay region and the estuary of the Miramichi River harbor isolated populations of bass plentiful enough to have yielded commercial catches. There is also a population (or populations) below Quebec in the lower St. Lawrence River that winters in that same general region, as proved by marking experiments carried out by Vladykov as reported by Bigelow and Schroeder. There are enough bass around Isle d'Orleans for bass fishing to be a favorite sport there, but commercial catches are so small as to suggest that the stock is not very large.

Migrations. No phase of the life history of striped bass generates as much discussion among fishermen as their migrations, and the picture remains puzzling. It seems certain that striped bass do not ordinarily travel far until they are 2 years old. Migratory patterns vary from local seasonal movements within a river system or estuary to extensive coastal migrations. In general, striped bass in Canadian waters remain within the Gulf of St. Lawrence and its river systems. Some interchange may take place among populations found in various bays and rivers around the outer coast of Nova Scotia, but it is doubtful whether these have any regular migratory association, either with Gulf of St. Lawrence fish or with those of more southern waters, except in occasional years. South of Cape Hatteras and in the Gulf of Mexico striped bass remain in the home river for their entire life cycle and migrate up and down river instead of going to sea (Raney 1957).

From Cape Hatteras north to New England some fish remain within a river system, whereas others migrate regularly along the coast following the shoreline northward and eastward as far as New England in spring, to return westward and southward in autumn. This was verified for bass 2 and 3 years old by returns from tagging experiments by Merriman (1941) at the eastern end of Long Island and in Connecticut from 1936 to 1938. Recaptures of fish that had been tagged there in May came mostly from farther east along southern New England, one from Cape Cod Bay and another from Cohasset on the southern shore of the inner part of Massachusetts Bay. Recaptures from fish tagged in summer were mostly from nearby (evidence of a stationary population), whereas those for autumn-tagged fish were scattered along the coast from the eastern end of Long Island to Chesapeake Bay, with one from Croatan Sound, one from Albemarle Sound, and one from Pamlico Sound in North Carolina. Chesapeake Bay harbors both migratory bass, as proved by tagging experiments and other evidence, and nonmigratory, as proved by the fact that fish of all sizes are taken there both in summer and winter, though not as many of them as in spring and fall. Similarly, some bass winter in northern waters though most of the fish appear to be migrants there, and a considerable percentage do so in the lower reaches of the Hudson River estuary.

The coastal migrating stock is predominantly female (only about 10% of the bass of northern waters are males), but males are nearly as numerous as females southward from Delaware Bay (Merriman 1941; Oviatt 1977). This is probably because larger fish migrate farther and females grow larger than males.

In the salt estuaries and open waters of the Gulf bass are taken only from late spring through the summer and until late in the fall. In years when they are plentiful enough to attract attention, they are likely to be reported about equally early in the season all along from Cape Cod to the Merrimac River. It has long been known, too, that pound nets on Long Island and along southern New England ordinarily make large catches only in spring (peak in May) and again from early October into November and that large spring catches are made progressively later in the season, proceeding from south to north, the reverse being true in autumn. Bass are generally distributed along the Massachusetts coast of the Gulf in May or by the first days of June. The first bass were reported in and off Hampton Harbor and in the Piscataqua River about the beginning of the second week in June (1950) and in Casco Bay about the middle of the month. They are said to appear as early as the end of May in Bangor Pool at the head of the estuary of the Penobscot in some years. In 1950 they were scattered all along Penobscot Bay before the end of June. It is probable that the seasonal schedule is about the same for the bass at the head of the Bay of Fundy, but information is scant.

Once bass have appeared, they continue in evidence until well into the autumn. During this part of the year, bass off the coasts of Massachusetts and most of those in Maine are in salt and brackish waters, except for those that enter freshwater to spawn. But they are caught all summer in freshwater far above the head of the tide in the Shubenacadie and the Annapolis in Nova Scotia. Part of the stock may have a similar habit in various rivers of Maine, as the Kennebec, where they ran up as far as Waterville until they were prevented from doing so by the construction of the dam at Augusta.

In rivers where bass winter, they may be taken in any month from late autumn into spring. As autumn approaches they vanish from the open coast. Most of them have disappeared along the outer coasts of Maine by mid-October or the end of that month in most years, but they may be in evidence in Maine rivers until later in autumn. Farther southward in the Gulf, they may linger equally late off open beaches. In 1950, a late season, Cape Cod Bay eastward from the Cape Cod Canal until the third week in October, schools of small fish were reported on 9 November, a half a dozen were landed from the surf on 18 November, and one was caught on 3 December.

Striped bass in saltwater may be in evidence until equally late in the season in the Minas–Cobequid Bay region at the head of the Bay of Fundy, for fishermen report taking them there through October and into November. Knight (in Bigelow and Schroeder) noted that as the weather becomes colder, bass of the southern side of the Gulf of St. Lawrence penetrate into bays and arms of the sea and ascend rivers at some distance, where they spend the winter resting on the mud in a half-torpid state. According to Atkins (in Bigelow and Schroeder), bass in Maine "pass the winter in quiet bays and coves of freshwater in the rivers." It has been known for many years that some bass winter in the Parker River in northern Massachusetts. Local fishermen also say that a few bass winter in deeper parts of the North and South rivers in Marshfield, on the southern side of Massachusetts Bay, apparently in saltwater.

Capture in 1949 of a 46-cm striped bass some 97 km south of Martha's Vineyard in 128 m of water in February seems to support the view that at least some of the bass of the Cape Cod region may only move offshore to winter on bottom well out on the continental shelf in localities where otter trawlers do not ordinarily operate, as has been found to be true of summer flounder. If true, this would mean that, as Merriman suggested, some Chesapeake-hatched bass that spread northward to Massachusetts and Maine when 2 or 3 years old may never return to their home waters.

Importance. Striped bass were a familiar fish all along the coast from Cape Cod to the Bay of Fundy when New England and the Maritime Provinces were first colonized (Little 1995). Plentiful and easy to capture because of their large size and their habit of coming into the mouths of streams and creeks, they were an important food supply for early settlers. Bass were caught, dried, and eaten by New England Indians before colonists arrived (Fearing 1903). Nothing regarding bass is of greater interest to commercial fishermen and to anglers than the great fluctuations in their numbers in the Gulf within historic times. Population fluctuations from colonial times to the 1950s were summarized by Bigelow and Schroeder. Houseman and Kernehan (1976) produced an indexed bibliography of striped bass from 1670 to 1976.

Commercial Fishery. Striped bass have not been plentiful enough in the Gulf of Maine to support a commercial fishery of any great magnitude at any time during the past 100 years. It is illegal to take striped bass commercially in New Hampshire, Maine, or New Jersey as they are considered a sport fish. Commercial fishing is allowed in Massachusetts with hook and line. The gear used varies geographically, depending on local preferences and restrictions. The most frequently used gear are gill nets, haul seines, floating traps, pound nets, and

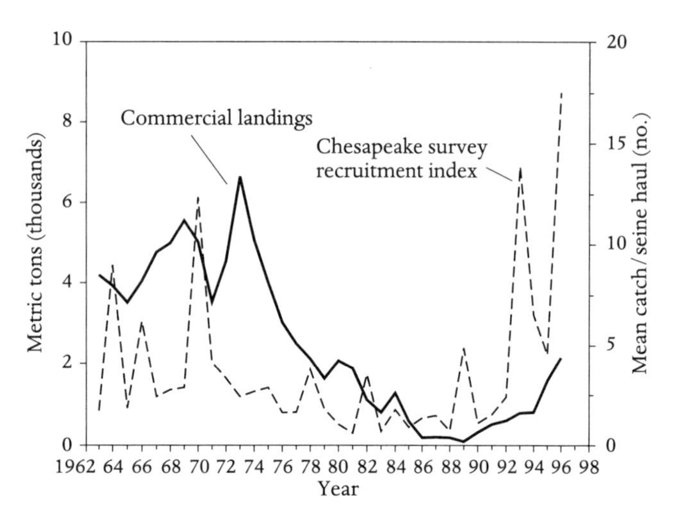

Figure 205. Commercial landings (in thousands of metric tons) and Chesapeake survey recruitment index (mean catch/seine haul) of striped bass *Morone saxatilis* from the Gulf of Maine–Mid-Atlantic, 1962–1996. (From Shepherd 1998a.)

from Maine to North Carolina were summarized by Setzler et al. (1980). Commercial landings from 1962 to 1996 are given in Fig. 205.

Recreational Fishery. During periods of plenty, striped bass are the leading game fish in the Gulf of Maine all along the coast, from the outer shore of Cape Cod to New Hampshire waters. The number of anglers who cast for them in the surf along the beaches of Cape Cod, northward from Cape Ann to the mouth of the Merrimac, and at scattered spots elsewhere is certainly in the thousands. Many charter boats troll daily for bass in Cape Cod Bay; some also troll along the Plum Island shore and at the mouth of the Merrimac; and many fish are caught by trolling, live-line fishing, and even by still fishing in the various inlets. So far as is known, the Shubenacadie River and Lake and the Annapolis River are the only waters on the Canadian shores of the Gulf where stripers attract attention as game fish.

Striped bass are one of the most important sport fishes along the entire U.S. Atlantic coast, prized by both surf fishermen and boat anglers for their large size and fighting qualities (Lyman and Woolner 1954). In the 1800s anglers established large clubs for striped bass fishermen from New Jersey to Massachusetts (Alperin 1987).

Stripers are voracious feeders and strike almost any type of bait or lure. Baits most frequently used are menhaden, squid, eel, crab, clams, bloodworm, plugs, spoons, flies, and casting lures. They are caught by casting, trolling, or bait fishing. Most angling occurs in inland waters but there is considerable variation in the percentages caught in the ocean and inland. In the ocean, catch by boat nearly always exceeds catch from shore. Most boat fishing is from private or rental boats rather than charters (Richards and Deuel 1987). Much has been written about the techniques of surf casting, trolling, choice of lines, and baits; but it is interesting, in comparison, to read, in Wood's *New England's Prospect,* published in 1634 (p. 37), that "the way to catch them is with hook and line, the fisherman taking a great cod line to which he fasteneth a peece of lobster and threwes it into the sea. The fish biting at it, he pulls her to him and knockes her on the head with a sticke."

Culture. The desirability of striped bass for commercial and recreational purposes has led to many stocking and culture programs. Striped bass were artificially propagated in the 1800s (Worth 1882). The first transplant of striped bass stock from the Navesink River, N.J., to the west coast (lower Sacramento River) took place in 1879 (Mason 1882). When it was discovered that striped bass could complete their life cycle in freshwater (Scruggs and Fuller 1955), inland states became interested in stocking them. Striped bass culture was rapidly expanded after the development of techniques for artificial induction of ovulation by hormone injections (Stevens 1966). Today bass are produced in state and federal hatcheries for stocking lakes, reservoirs, and impoundments for sport fishing, as well as for the control of threadfin and gizzard shad populations. Hatcheries are located in 16 different states (Westin and Rogers 1978).

Management. Striped bass management has a long history. In 1693 the General Court of Massachusetts Bay Colony ordered that they not be used for fertilizer. In 1776 New York and Massachusetts passed laws prohibiting the sale of stripers in the winter (Setzler et al. 1981). Regulations restricted netting in the Hudson River in 1892 (Richards and Deuel 1987). In the 1800s there were large numbers of stripers but they nearly disappeared from New England and the Mid-Atlantic coasts by the early 1900s.

By 1938 striped bass reappeared in force, products of the huge 1934 year-class and the need was recognized for interstate cooperation for utilization of the resource. In 1942 Congress created the Atlantic States Marine Fisheries Commission (ASMFC) to develop a joint program for the promotion and protection of marine and anadromous fisheries of the Atlantic seaboard. One of their first recommendations was adoption of a 16-in. FL (40.6-cm) minimum size for striped bass (a 3-year-old fish). All the New England states, New York, New Jersey, and Pennsylvania adopted this measure, but from Delaware south the states continued to allow the taking of smaller fish.

Coastwise commercial landings fell from 6.8 million kg to 1.6 million kg by 1983 and juvenile production in Chesapeake tributaries remained poor throughout most of the 1970s and 1980s (Field 1997). The decline in the east coast striped bass stocks has been attributed to overfishing, nutrient enrichment of the habitat with the resulting temperature oxygen squeeze on subadults, deterioration of the nearshore habitat for juvenile bass from loss of submerged vegetation, decreased survival of larvae owing to environmental pollution, poor nutrition, fluctuations in the physical environment, and predation (reviewed by Setzler-Hamilton et al. 1988 for Chesapeake Bay).

In 1979 congressional action established the Emergency Striped Bass Research Study (ESBS) directing the U.S. Fish and Wildlife Service and the National Marine Fisheries Service to monitor the status of striped bass stocks and determine the causes of the decline and its economic impact (Richards and Deuel 1987). The role of the various agencies and the management perspectives were discussed in the November–December issue of *Fisheries* (Vol. 12, 1987). The ASMFC was asked to administer an interjurisdictional management plan known as the Interstate Fisheries Management Plan (ISFMP) for striped bass of the Atlantic coast from Maine to North Carolina. This was adopted in 1981 and called for a minimum size of 14 in. TL (35.6 cm) in nursery rivers and bays and 24 in. TL (61 cm) along the coast, and banned fishing in spawning rivers during the spawning period. This plan proved inadequate and has been amended several times; by 1984 all the states within the migratory range of striped bass were ordered to adopt the new plan (Alperin 1987). In 1985 an additional amendment was adopted to protect the 1982 and subsequent year-class females until 95% have spawned at least once. This resulted in a 33-in. (83.8-cm) minimum size limit in all areas by 1987.

In 1995 spawning stock biomass in the Chesapeake Bay reached healthy levels so striped bass were formally declared to be a restored stock and commercial and recreational management restrictions were somewhat relaxed (Field 1997; Shepherd 1998a). Male spawning stocks increased from three to ten spawning age-classes between 1985 and 1995. Coastwise recreational catch climbed from an estimated 514,000 fish in 1985 to a record high of 8.5 million. The Hudson River population had not suffered the same decline in abundance as the Chesapeake stock, and the Delaware and Roanoke populations continued to produce strong year-classes. States adopted Amendment 5 to the original management plan in 1995 to allow expanded state fisheries on the recovered populations.

Stocks. Striped bass stocks commingle along the northeastern U.S. and Canadian coasts supporting mixed-stock fisheries in which stock compositions fluctuate widely (Wirgin et al. 1997). Attempts have been made to discriminate stocks for over 50 years. The methods, results, strengths, and weaknesses of the techniques and suggestions for future research and application of these results were reviewed by Waldman et al. (1988). In general, techniques applied included those based on behavior and ecology (tagging, catch data, parasites); phenotype (meristics, morphometrics, trace element uptake, scale morphology, and isoelectric focusing of eye lens proteins); and genotype (cytogenetics, protein electrophoresis, restriction endonuclease analysis of mitochondrial DNA, and immunogenetics). Most techniques can classify stocks correctly at the 70–80% level but none is 100% accurate. All have advantages and disadvantages, and results have been further complicated by widespread stocking of nonnative stocks to supplement wild populations (e.g., stocking the Kennebec River, Maine, with Hudson River, N.Y., striped bass). Waldman et al. (1988) recommended establishment of a library of genetic information on striped bass stocks so that continual adjustments do not have to be made for environmental and ontogenetic changes that affect phenotypic stock discrimination, a reanalysis of the stability and usefulness of meristic characters for determining the contributions of various stocks to the coastal migrating population, and an assessment of hatchery contributions to wild populations.

Four major spawning stocks of striped bass make up the Atlantic coast migratory stock: the Roanoke stock, which spawns in the Roanoke River, N.C.; the Chesapeake stock, which spawns in Maryland and Virginia tributaries of Chesapeake Bay; the Delaware River stock; and the Hudson River stock. The Chesapeake Bay and Hudson River stocks are the most important in a mixed-stock model (Wirgin et al. 1997). Chesapeake Bay contributed the major portion of the stock (over 90%) up to 1975, as shown by discriminant analysis of five morphological characters (Berggren and Lieberman 1978). Decline in the Chesapeake stock size after the dominant 1970 year-class and the relative stability of the Hudson River stock meant that until recently the Hudson River contributed a higher percentage of the migrating stock than it did previously. The Delaware River, once an important spawning area, had low production for most of the twentieth century, but the stock has increased tremendously in recent years. Striped bass collected in Rhode Island and separated into Hudson and Chesapeake Bay stock on the basis of isoelectric focusing of eye lens proteins (Fabrizio 1987) showed that about 54% were from Chesapeake Bay and 46% from Hudson River stock.

ACKNOWLEDGMENTS. Drafts of the white perch account were reviewed by John Galbraith and John Waldman, and drafts of the striped bass account, by John Boreman, Gary Shepherd, John Waldman, and John Walter.

WRECKFISHES. FAMILY POLYPRIONIDAE

GRACE KLEIN-MACPHEE

Wreckfishes superficially resemble large groupers but differ in having only two opercular spines. Some authors have assigned the wreckfishes to the Percichthyidae or to the Acropomatidae but they are placed here in the Polyprionidae following Heemstra (1986b) and Roberts (1993). The family contains two gen- Important commercially, wreckfishes are found in depths of 50–800 m on continental shelves and seamounts, where they often occur in large concentrations, which makes them especially vulnerable to overfishing. Juveniles are pelagic until they are quite large (60 cm TL) and are often found associated with

WRECKFISH / *Polyprion americanus* (Bloch and Schneider 1801) / Bigelow and Schroeder 1953:409–410

Figure 206. Wreckfish *Polyprion americanus.* Grand Bank, 110 mm. Drawn by H. L. Todd.

Description. Body deep, laterally compressed, two and a half to three times as long as deep. Mouth large, lower jaw projecting. Teeth slender, in bands on jaws, vomer, palatines, and tongue. Tongue teeth in a single oval patch or in two or three patches arranged in a triangle. Gill cover with prominent longitudinal ridge and strong spines. Bony protuberance over eye. Scales small, rough, extending over bases of soft-rayed fins. Spiny part of dorsal fin continuous with soft-rayed part. Soft-rayed part of dorsal fin similar to anal fin in outline. Caudal fin gently rounded in juveniles, truncate in adults (Roberts 1977). Pectoral fins about half as long as head, inserted above pelvic fins. Lateral line gently curved.

Meristics. Dorsal fin rays XI or XII, 11 or 12; anal fin rays III, 9 or 10; pectoral fin rays 17 or 18; gill rakers on lower limb of first arch 8–15, including rudiments; lateral line scales 70–90; vertebrae 13 precaudal + 14 caudal = 27 (Heemstra 1986b).

Color. Grayish, blackish brown, or reddish brown; caudal fin edged with white in juveniles. The pelvic fins have white rays, bluish black membranes and a whitish border. Young fish are mottled with gray and cream on head and body.

Size. Reach 2 m and over 100 kg (Roberts 1989). The all-tackle world record wreckfish weighed 71.0 kg and was caught at White Island, Whakatane, New Zealand, in November 1990 (IGFA 2001).

Distinctions. The combination of a sea-bass-like body, a very rough head with a prominent ridge and strong spines on each gill cover, and a bony protuberance over the eye and on the nape gives wreckfish an aspect so different from that of any other Gulf of Maine fish that they should be easily recognized. They differ from striped bass and white perch in having the dorsal fins united and from redfishes by a rounded tail.

Habits. Small wreckfish are pelagic until they reach a length of about 50 cm and are most likely to be found under floating logs or wreckage (Roberts 1977), as the common name implies. When larger, they take to the bottom in water 100–1,000 m deep along rocky continental slopes, over seamounts, and off oceanic islands. Several adults 69–91 cm (mean 80 cm) viewed and photographed by an unbaited camera on a peak of the Crestal Mountains of the Mid-Atlantic Ridge (Ryall and Hargrave 1984) were on the bottom at depths of 445–740 m. Schooling and aggressive behavior was suggested by the numbers of fish and the fact that they attacked the camera.

Food. Stomachs of wreckfish collected off the Carolinas contained nemichthyid and ophichthid eels, *Beryx splendens,* blackbelly rosefish, snake mackerels, and mesopelagic organisms such as *Anoplogaster cornuta, Gonostoma* sp., caridean shrimps, and squid beaks (G. Sedberry, pers. comm., May 1998). In Argentina, wreckfish fed mainly on hake and some shrimps (Menni and López 1979).

Reproduction. Male wreckfish on the Blake Plateau in the western North Atlantic are in spawning condition from December to May, females from December to April, and most spawning occurs in February and March (D. Wyanski, pers. comm., May 1998). Eggs and larvae are pelagic. In New Zealand, the sex ratio is close to unity and females reach a slightly larger size than males (Roberts 1989). Suggestions that wreckfish are hermaphroditic and undergo sex change were disproved by Roberts (1989).

General Range. Both sides of the Atlantic; the Mid-Atlantic Ridge and Atlantic islands (Bermuda, Azores, Madeira, Canaries, Tristan da Cunha); Atlantic seamounts; Mediterranean Sea; southern Indian Ocean (St. Paul and Amsterdam islands); and western South Pacific (Roberts 1977; Heemstra 1986b; Sedberry et al. 1996). In the western Atlantic, from the Grand Banks of Newfoundland and Nova Scotia (Gilhen 1986) to the La Plata River, Argentina, but absent from tropical waters from the Caribbean to northern Brazil. Populations from eastern and western North Atlantic and the Mediterranean are genetically similar and differ from wreckfish from Brazil and the South Pacific (Ball et al. 2000).

Occurrence in the Gulf of Maine. The only report from the Gulf of Maine is of a fish 65 cm long, weighing 4.2 kg (dressed), taken on the northern edge of Georges Bank on 13 August 1951 by the trawler *Winthrop*. Another, 15.4 cm long, was caught on the surface off No Man's Land Island, near Martha's Vineyard on 21 August 1925; and two have been brought in from the Grand Banks, one of them many years ago and the second in 1929 (Schroeder 1930).

Importance. Owing to its rarity in the Gulf of Maine, wreckfish is of no importance there. However, large concentrations are found in depths of 200–600 m on continental shelves and seamounts, where they are very vulnerable to overfishing. There are fisheries for wreckfish on the Blake Plateau off the southeastern U.S. coast, Brazil, the Azores, Madeira, and the Mediterranean Sea (Sedberry et al. 1996; Sedberry et al. 1999).

ACKNOWLEGMENTS. Drafts of the wreckfish account were reviewed by Clive Roberts and George Sedberry.

SEA BASSES. FAMILY SERRANIDAE

GRACE KLEIN-MACPHEE

Sea basses are a generalized group of fishes difficult to define but all have three spines on the opercle, a main spine with another above it and one below it. Scales are usually ctenoid. Dorsal fins are generally continuous but there may be a notch between spiny and soft portions of the fin. They have three anal fin spines. Sea basses are hermaphroditic, with some species beginning reproductive life as females and then changing to males and others having both testes and ovaries at the same time. Composition of this family has varied widely over the years but is now thought to comprise a monophyletic group of three subfamilies, about 62 genera, and 449 species (Nelson 1994). The subfamily Anthiinae has 20 genera and about 170 species, including yellowfin bass, a species rarely reported from the Gulf of Maine, and the subfamily Serraninae contains about 13 genera and 75 species, also with one species, black sea bass, in the Gulf of Maine. Recent information about the northern population of black sea bass has been summarized by Steimle et al. (1999d).

KEY TO GULF OF MAINE SERRANIDAE

1a. Caudal fin rounded, not forked; pelvic fins longer than pectoral fins; dorsal soft rays 14 or 15; lateral line scales 22 + 7, interrupted on middle of caudal peduncle . . . . **Yellowfin bass**

1b. Caudal fin forked; pectoral fins longer than pelvic fins; dorsal soft rays 11; lateral line scales 46–49, lateral line continuous . . . . **Black sea bass**

YELLOWFIN BASS / *Anthias nicholsi* Firth 1933

Description. Body moderately elongate, deepest at base of pelvic fins, depth 2.3–2.6 times in SL, caudal peduncle deep (Fig. 207). Profile slightly concave, humped at nape. Mouth oblique, upper jaw reaching just past middle of eye. Teeth small, in narrow bands in jaws, outer series larger, canines at front corners of jaws, those of lower jaw larger. Teeth also in bands on vomer and palatines. Caudal fin forked. Pelvic fins long, pointed, reaching almost to anterior anal rays (Firth 1933). Scales large, ciliated, covering body, sides, and top of head. Lateral line follows curve of back to an interruption just before caudal peduncle, continuing on middle side of caudal peduncle to caudal base (Gilhen and McAllister 1981).

8–12 + 27–29 = 35–41; vertebrae 26 (Firth 1933; Gilhen and McAllister 1981).

Color. Rose red or pink with three or four lengthwise yellow stripes on the body and a deep blue blotch in the middle of the back at the base of the first dorsal. Belly silvery white with a median lemon yellow band from the symphysis to the pelvics; head with radiating stripes of yellow and red. Iris crimson with a lemon yellow ring around the pupil. Dorsal fin brilliant red basally, yellow distally; caudal chrome yellow, outer half of each lobe pink; pectoral fins salmon; pelvic fins yellow with pink or red anterior edge (Firth 1933; Fowler 1937; Hardy 1978b).

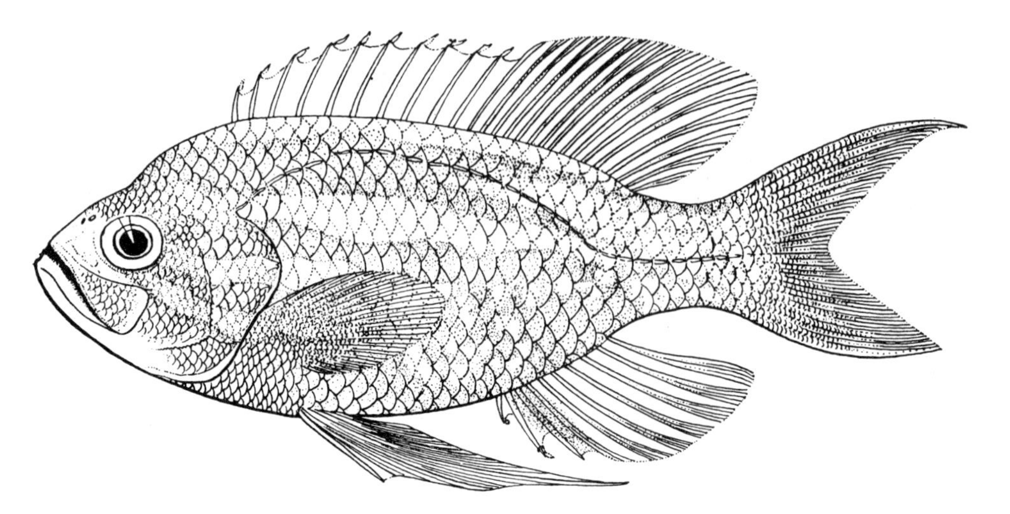

Figure 207. Yellowfin bass *Anthias nicholsi.* Drawn by H. W. Fowler. (From Fowler 1937: Fig. 4.)

Distinctions. Distinguished from other serranids by large scales, moderately elongate body, large eyes, and very long pelvic fins. Distinguished from fishes in other families by the opercular spines, long pelvic fins, and high interrupted lateral line. Distinguished from short bigeye by coloration, fin ray counts (14 or 15 dorsal rays, not 11, and 7 anal rays, not 10), and smaller eyes.

Habits. Adults are recorded from depths of 73–190 m. Examination of gonads of 20 specimens showed 12 females (52.0–125 mm SL), one individual (73.0 mm) transforming from female to male, and seven secondary males (99.4–134 mm), strongly indicating that yellowfin bass are protogynous hermaphrodites (Anderson and Baldwin 2000).

General Range. Nova Scotia, New Jersey, and Virginia.

Occurrence in the Gulf of Maine. The first confirmed Canadian record was a 125-mm SL specimen caught on the Scotian Shelf between Baccaro and La Have banks off southeastern Nova Scotia at 42°38′ N, 64°41′ W in 190 m (Gilhen and McAllister 1981). A 21.5-mm larva, originally identified as *Anthias* sp., was collected at 42°07′ N, 65°34′ W in 435 m in August 1976 (Markle et al. 1980; Scott and Scott 1988).

BLACK SEA BASS / *Centropristis striata* (Linnaeus 1758) / Bigelow and Schroeder 1953:407–409 (as *Centropristes striatus*)

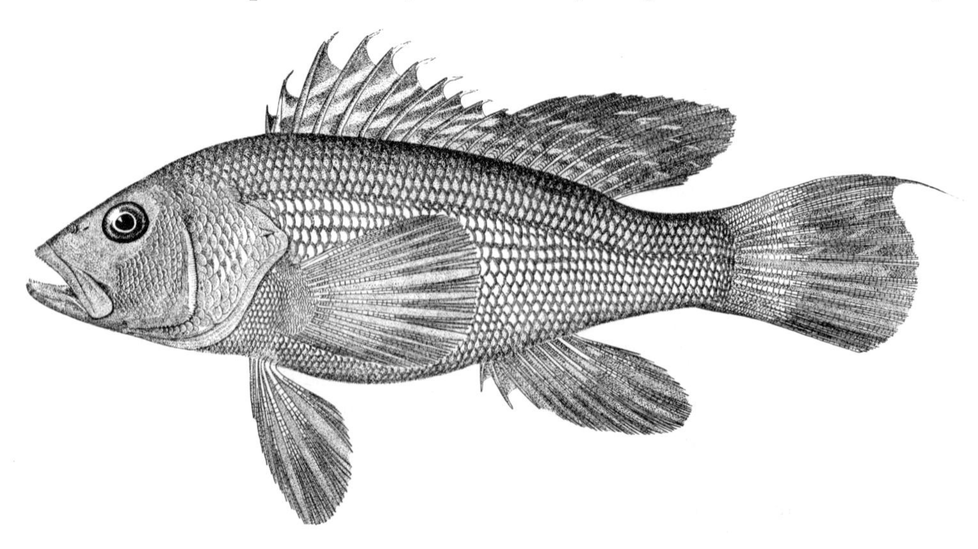

Figure 208. Black sea bass *Centropristis striata.* Connecticut. Drawn by H. L. Todd.

Description. Body moderately stout, depth 25–33% SL, with a rather high back (Fig. 208). Head flat-topped, snout moderately pointed. Mouth large, oblique. Eyes set high on head. Jaw teeth in wide bands, outer and inner teeth slightly enlarged; vomerine teeth in crescent-shaped patch; palatine teeth in long, narrow patch. Spiny and soft portions of dorsal fin (which originates slightly in front of the rear corner of the gill covers) separately rounded, the latter higher than long. Caudal fin rounded posteriorly, distinctly three-lobed in large fish. Anal fin originates under or very slightly behind origin of soft portion of dorsal fin, which it resembles in rounded outline. Both anal fin and soft part of dorsal fin noticeably soft and flexible. Pectoral fins so long that they reach back almost to anal fin, broad and round tipped. Pelvic fins large, originating in front of pectorals. Scales rather large; top of head naked. Adult males develop a hump in front of dorsal fin.

Meristics. Dorsal fin rays X, 11; anal fin rays III, 7; pectoral fin rays usually 18–20; total gill rakers 23–29; vertebrae 10 precaudal + 14 caudal (Miller 1959; Kendall 1977).

Color. Black sea bass, like most fishes that live on rocky bottoms, vary widely in color, the general ground tint ranging from smoky gray to dusky brown or blue black, usually more or less mottled. Belly only slightly paler than sides. Bases of exposed parts of the scales are paler than their margins, giving the fish the appearance of being barred with longitudinal series of dots of a lighter tint of brown than the general hue on dark fish, but pearl gray on pale ones. Dorsal fin marked with several series of whitish spots and bands; other fins mottled with dusky. Young fish 5–7 cm long greenish or brownish with dark side stripe passing from eye to caudal fin, with dark crossbars on sides.

Size. Reach 60 cm (Robins and Ray 1986). The all-tackle game fish record is a 4.65-kg fish caught at Virginia Beach, Va., in January 2000 (IGFA 2001).

Distinctions. Black sea bass are easily distinguished from striped bass and white perch by the fact that spiny and soft-rayed portions of the dorsal fin are continuous, so that there is only one long fin instead of two short separate fins. The general form, rounded caudal and pectoral fins, and the short but high anal fin are sufficient to separate them from scup; their color precludes confusing them with redfishes, and they should not be taken for tautog or cunner because the mouth and the pectoral fins are much larger, the caudal fin has a different outline, and the soft portion of the dorsal is as long as the spiny portion. They differ from wreckfish in many respects, especially in their much larger scales, smoothness of head and gill covers, and shape of the tail.

Taxonomic Note. Bigelow and Schroeder referred to black sea bass as *Centropristes striatus* but Briggs (1960b) showed which is feminine so the species name has to be emended to *striata*.

Habits. Black sea bass contrast with striped bass in being strictly confined to saltwater although juveniles are common in higher-salinity portions of estuaries, usually greater than 14 ppt. Life history information was summarized by Mercer (1989). The inshore-offshore range extends from close in to the coastline in depths of only 1 m out to about 165 m (Musick and Mercer 1977), according to the season of the year. During the part of the year that sea bass are inshore they are most plentiful on hard bottom, in water less than 36 m or so, often around submerged wrecks and pilings of wharves. Offshore they prefer ledges and banks, generally over hard bottoms of rock or coral (Hildebrand and Schroeder 1928). In the South Atlantic Bight, they prefer the inshore sponge-coral habitat (Wenner et al. 1986).

They can tolerate temperatures from 6° to 29.8°C (Musick and Mercer 1977; Hardy 1978b) but are rarely found below 11°C (Miller 1959). Minimum salinity tolerated is 7.7 ppt (Hardy 1978b), but in estuarine areas they are not common in salinities lower than 12 ppt (Musick and Mercer 1977).

Black sea bass travel in small schools when migrating inshore and offshore. They produce sounds that include single weak grunts, small thumps, and possible scrapes; the thumps appear to be associated with escape and competitive feeding. They have a large, elliptical single-chambered swim bladder, which can be vibrated by general body contraction and striking of opercula against the body. Paired patches of pharyngeal teeth can generate stridulatory sounds (Fish and Mowbray 1970).

Food. Black sea bass feed on a wide variety of crustaceans, fishes, mollusks, and worms (Miller 1959; Maurer and Bowman 1975; Steimle and Ogren 1982; Sedberry 1988; Bowman et al. 2000). Crustaceans, particularly decapods, are important in the diet across all size categories: *Crangon septemspinosa* up through 10 cm TL and *Cancer irroratus* from 6 cm on (Bowman et al. 2000). The euphausiid *Meganyctiphanes norvegica* is important at intermediate sizes, 11–25 cm. Fishes, including anchovies and herring, first become a major component of the diet at 21 cm and account for 69% of the food consumed at sizes of 40 cm and larger. Other fishes eaten include margined snake eel, seahorse, pipefish, cusk-eel, scup, sand lance, and windowpane.

Predators. Among fishes sampled for food content by the NEFSC bottom trawl surveys, predators of black sea bass include little skate, spiny dogfish, goosefish, spotted hake, windowpane, and summer flounder (Rountree 1999).

Breeding Habits. Black sea bass are protogynous hermaphrodites, changing sex from female to male (Lavenda 1949). Median length at maturity for both males and females is about

ages 2 and 3 but males may mature as early as age 1 (Mercer 1978). Sexual succession occurs at ages 1–8 (Mercer 1978; Wenner et al. 1986). Increased ovarian activity and degeneration coincide with proliferation of testicular tissue during the change. There has been no instance of active ovarian development concurrent with testicular degeneration. Sexual transition begins in the posterior region of the ovary and proceeds anteriorly. Wenner et al. (1986) determined that 3% of the fish showed simultaneously developed hermaphroditic gonads at all maturity stages, but they could not tell whether the fish could self-fertilize. Sex ratios were significantly skewed in favor of females up to an intermediate age and then favored males. No females were found among fish over age 7. Although some fish may always be males, most males are sexually reversed females.

Black sea bass are sexually dimorphic. Males larger than 25 cm have an adipose hump; they are bright blue at spawning time especially around the hump and eyes. Females are darker and duller. After spawning they become brownish and some even become almost white (Lavenda 1949).

Spawning progresses seasonally from south to north starting as early as April off North Carolina and Virginia and in June off New Jersey and Long Island (Able et al. 1995; Berrien and Sibunka 1999: Fig. 49). Peak spawning is in August; eggs are found from south of Chesapeake Bay to off Long Island, but not in the Gulf of Maine. Spawning occurred predominantly in the inner half of shelf waters in depths of 18–45 m (Kendall 1972; Berrien and Sibunka 1999).

Early Life History. The eggs, which are buoyant, transparent, and clear, have a diameter of 0.9–1.0 mm, a single oil globule with a diameter of 0.13–0.19 mm, and a narrow perivitelline space. The egg membrane is smooth and unsculptured (Fahay 1983).

Embryonic development was described by Wilson (1891) at an unspecified temperature, which was probably close to 16°C. Incubation takes 5 days at 15°C (Kendall 1972) and 3 days at 23°C (Hoff 1970). Length at hatching is 1.5–2.0 mm SL. Larvae are deep-bodied with large heads (30–40% of SL). The vent is approximately at midbody; the pigmented oil globule is located in the anterior part of the yolk sac, which is absorbed after 3 days. There are ventral pigment spots from the throat to the caudal fin base, a medial spot posterior to the lower jaw, one on each angular, one at the junction of the cleithra, and one at the vent. Most key larval characteristics are present by 5 mm SL.

Black sea bass larvae resemble those of bluefish, but the latter have a pigment pattern of three rows of melanophores (dorsal, midlateral, and ventral). They also resemble spot (*Leiostomus*) and croaker (*Micropogonias*) except that sea bass have much larger pectoral fins and the anus is closer to the middle of the body. The young are easily identifiable as sea bass by the time they have grown to a length of 60 mm.

Larvae are distributed across much of the Mid-Atlantic Bight continental shelf north to New Jersey and Long Island; greatest abundance occurred between Cape Hatteras and Delaware Bay in August (Able et al. 1995: Fig. 3; Steimle et al. 1999d: Fig. 11). A single 6.4-mm larva was collected on Georges Bank in November 1982 (Able et al. 1995), and a few have been reported from Cape Cod Bay (Scherer 1984). Larvae have been found in inlets (Kendall 1972), bays (Herman 1963), and offshore waters (Perlmutter 1939; Able et al. 1995), at a distance of 4–82 km from the shore at maximum depths of 18–33 m in waters 15–51 m deep. Large larvae are found deeper than small ones. Based on NEFSC MARMAP surveys, larvae were most abundant between 13°–21°C (Steimle et al. 1999d: Fig. 5). Sea bass larvae migrate into estuaries at an early stage and become benthic at 13–24 mm (Kendall 1972). They enter high-salinity estuarine areas in August and September in southern Massachusetts, in August in New York and on Long Island, and in July in Delaware Bay (Nichols and Breder 1927; de Sylva et al. 1962; Lux and Nichy 1971). They remain in these nursery areas until temperatures go below 14°C, then migrate into deeper waters. Young-of-the-year in the Chesapeake Bay area move southwestward at depths less than 55 m. They winter at depths of 55–110 m and migrate north with spring warming (Musick and Mercer 1977).

Age and Growth. Lavenda (1949) gave age-length curves for male and female sea bass from New Jersey based on scale analysis. Sea bass from the South Atlantic Bight were aged by otolith increments (Wenner et al. 1986). Annulus formation occurred in April and May and might have been associated with spawning. Growth is almost linear to age 6 and then slows. Sea bass from the Mid-Atlantic Bight are larger at a given age than those from the Southern Atlantic Bight (Mercer 1978). Northern individuals are seldom heavier than 2.3 kg and average about 0.7 kg; a 30-cm fish weighs about 0.5 kg; and one of 46–51 cm weighs about 1.4 kg. Growth of juveniles (34–111 mm TL) in a southern New Jersey estuary, based on recaptured fish, was fastest during summer (0.74 $mm \cdot d^{-1}$), but averaged 0.45 $mm \cdot d^{-1}$ from spring through fall (Able and Hales 1997).

General Range. Atlantic coastal waters of the United States, from Cape Canaveral, Fla., to Cape Cod, occasionally to Nova Scotia and the Bay of Fundy (Scott and Scott 1988).

Occurrence in the Gulf of Maine. Black sea bass enter the Gulf only as rare strays from the south, Pemaquid Point and Martinicus Island off Penobscot Bay, Maine, being the northernmost known outposts. They occur occasionally in Passamaquoddy Bay and the Bay of Fundy, N.B. (Scott and Scott 1988). They have been taken in Casco Bay; near Gloucester; off Nahant, Salem, Beverly, and Cohasset in Massachusetts Bay; at North Truro and Monomoy on Cape Cod; and 8 km east of Pollock Rip Lightship (Bigelow and Schroeder; Read 1975; Collette and Hartel 1988).

Migrations and Movements. Trawl surveys made in the Mid-Atlantic Bight (Musick and Mercer 1977) showed that black sea bass migrate south and offshore to the Chesapeake Bight, where the population spends the winter. Larger and older fish move offshore earlier than young-of-the-year, and winter in deeper water (73–165 m). In the spring, sea bass migrate inshore and to the north. Adults move to their coastal spawning grounds and juveniles to estuarine nursery areas, most likely in response to water temperature. Juveniles leave estuaries when temperatures fall below 14°C (Richards and Castagna 1970; Lux and Nichy 1971). Off southern New England, Long Island, and New Jersey, they appear inshore during the first or second week in May, withdrawing again late in October or early in November. Most adults are captured in waters 9°C or warmer. In South Carolina, where temperatures rarely go below 10°C except in shallow estuaries, adult sea bass in given areas are year-round residents and do not migrate.

Stocks. Two stocks are recognized, divided at Cape Hatteras (Shepherd 1998b). Meristic and morphometric studies indicate that some variation is present within the northern stock (Shepherd 1991), which migrates seasonally in response to temperature changes and spawns June to October. The southern stock does not appear to migrate and spawns in April and May (Mercer 1978).

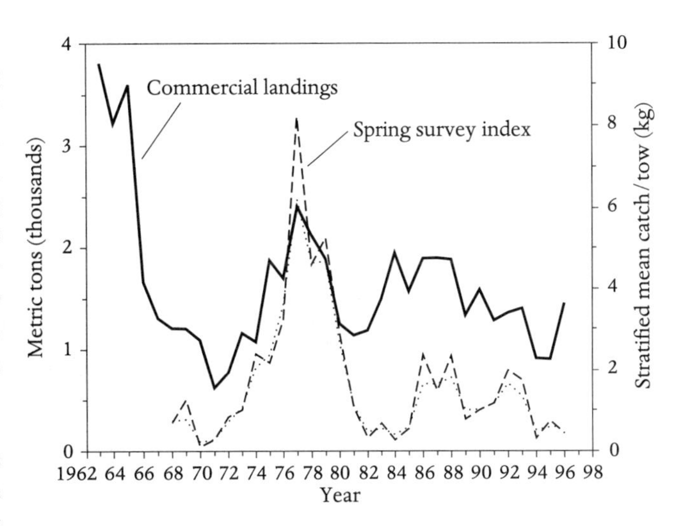

Figure 209. Commercial landings and spring survey index for black sea bass *Centropristis striata*, Gulf of Maine–Mid-Atlantic, 1962–1996. (From Shepherd 1998b.)

Importance. Black sea bass are too scarce to be of any importance in the Gulf, but they are very valuable food and game fish in southern New England and the Mid-Atlantic Bight. The commercial otter trawl fishery is primarily a winter fishery and produces 60% of the total catch of black sea bass, scup, and summer flounder (Shepherd and Terceiro 1994). The inshore trap fishery, which occurs between late spring and late autumn, consists mostly of black sea bass. Reported commercial landings north of Cape Hatteras fluctuated around 2,600 mt from 1887 until 1948, when landings increased to 6,900 mt, peaking at 9,900 mt in 1952 (Shepherd 1998b). Since 1962, commercial landings have fluctuated between 4,000 and 1,000 mt·year^{-1} (Fig. 209).

Estimated recreational landings, primarily in the Middle Atlantic states, are comparable to the commercial catch. The recreational fishery takes place between late spring and autumn (Shepherd and Terceiro 1994). U.S. recreational landings have ranged from 700 to 6,300 mt since 1979, averaging 2,100 mt, accounting for 31–87% of the total landings (Shepherd 1998b).

Black sea bass are managed under Amendment 9 to the Summer Flounder, Scup, and Black Sea Bass Fishery Management Plan developed in 1996. Management measures include gear restrictions, minimum fish sizes, a coastwise commercial quota, and a recreational harvest limit (Shepherd 1998b).

Hormone-induced ovulation of black sea bass and rearing of larvae were described by Tucker (1984) and Hoff (1970). This fish may have mariculture potential as other species of sea basses have shown promise. Hettler and Powell (1981) described a culture system for rearing sea bass.

ACKNOWLEDGMENT. A draft of the black sea bass account was reviewed by Linda Mercer.

BIGEYES. FAMILY PRIACANTHIDAE

GRACE KLEIN-MACPHEE

Bigeyes comprise a relatively small circumtropical family of four genera and about 18 species of marine percoids (Starnes 1988; Nelson 1994). They have extremely large eyes, deep bodies, rough scales, and are primarily bright red in color. Bigeyes are carnivorous and nocturnal. They are especially noted for their "eyeshine," emanating from the brilliant reflective layer of their large eyes. Juveniles of one species of bigeye have been reported from the Gulf of Maine.

SHORT BIGEYE / *Pristigenys alta* (Gill 1862) / Bigelow and Schroeder 1953:410–411 (as *Pseudopriacanthus altus*)

Description. Body ovate, thin as juveniles, relatively deep- pointed, in narrow bands on jaws, vomer, and palatines; outer

Figure 210. Short bigeye *Pristigenys alta*. Key West, Fla., 131 mm, USNM 37772. Drawn by H. L. Todd.

angle. Spiny and soft portions of dorsal fin continuous, extending back from nape nearly to caudal fin base. Anal fin originates under eighth or ninth dorsal spine, soft portion about same form as soft portion of dorsal, except outer angle somewhat more rounded. Pelvic fins originate a little in front of pectorals, much larger than pectorals, round tipped. Caudal fin slightly convex.

Meristics. Dorsal fin rays X, 10–12; anal fin rays III, 9–11; pored lateral line scales 31–39; gill rakers on upper arm of first arch 6–9, total 23–30; vertebrae 10 + 13 (Caldwell 1962a).

Color. Bright red or orange-red in life, below and above, with four vertical light bars often evident, faint or absent in large adults; dorsal fin red, the spinous part edged with yellow, a few blackish dots on the soft rays; sometimes two rows of orange spots, especially in juveniles; caudal fin pale, with blackish blotches in juveniles, only a black marginal band in adults; anal fin red, edged with black; pelvics red at base, dusky to black on outer part; pectorals plain red. Iris red or gold with white spots.

Size. Reach 334 mm TL.

Distinctions. The most striking characters are their very large eyes and their brilliant red color. Short bigeye are distinguishable from sea basses by the fact that the whole head, as well as the body, is clothed in rough scales and the anal fin is longer than the soft-rayed portion of the dorsal fin. The compressed body, unusually stout dorsal fin spines, very large pelvic fins, and small pectoral fins separate them from redfishes, the only common Gulf of Maine fishes of similar appearance that rival them in color.

Habits. Adults are bottom-dwelling, secretive, and usually hide solitarily in rock niches and crevasses (Parker and Ross 1986), but may form aggregations. They are found over bottoms of rock, coral, gravel, shell, oyster beds, and gray mud (Caldwell 1962a). Transformed juveniles and adults occur at depths of 5–125 m (Caldwell 1962a).

Reproduction and Early Life History. Nothing is known of age at maturity, fecundity, breeding habits, or the appearance of the eggs. However, larvae ranging in size from 2.2 to 8.0 mm TL have been described (Caldwell 1962b; Starnes 1988; Powell 2000). The smallest had a single medial cranial crest with eight serrations, a supraocular crest with three serrations, and a strong preopercular spine extending nearly to the anal opening. The crest gradually tilts backward and the serrations first increase then are lost. Supraocular spine serrations also increase, and the preopercular spine becomes shorter. The full complement of fin rays is acquired by 4.8 mm. Scales appear at 5.3 mm above and in front of the anus. Pigmentation consists of a few internal chromatophores in the gut region, a dark region under the cranial spine, a small patch of pigment between the eyes across the surface of the forebrain. Prejuveniles have been described from 8.2 to 65 mm. The eye is enormous; crest serrations are weak and eventually absorbed. Fin ray and spine development begins by 8 mm and is complete by 20–34 mm. Pigmentation may be bluish or silvery but a 22-mm specimen was reddish with three clear-cut dark bars and a 28-mm individual was red-orange with 12 black blotches above the lateral line. A 19-mm specimen was bright orange-red with black spots along the lateral line and no pigment in the caudal fin (Schwartz and Purifoy 1997: color Fig. 1). Vertical bars are evident around 34 mm. In juveniles 63.2 mm and up the caudal fin acquires a black border and the soft rays of the dorsal fin are black-tipped; by 75 mm the barred pattern is no longer evident (see color photograph in Starnes 1988:135). Pelvic fins are very large in prejuveniles and juveniles (Caldwell 1962a).

General Range. Southern Gulf of Maine south along the Atlantic coast and into the Gulf of Mexico to Yucatán; Bermuda, Virgin Islands, and Caribbean Sea south to Venezuela in rather deep water.

Occurrence in the Gulf of Maine. Postlarvae are transported northward in Gulf Stream waters and juveniles have been recorded inshore from the Gulf of Maine from August to November (Nichols and Breder 1927). There are three records for Massachusetts Bay (Collette and Hartel 1988): one specimen found alive on Marblehead Beach on 3 September 1859; a second at Scituate, Mass., in 1932; and a third, about 38 mm long, picked up in a tide pool at Cohasset, Mass., in September 1937. The northernmost record is a 28-mm specimen collected in a lobster trap 18 m deep at Southport, Maine (Scattergood and Coffin 1957).

ACKNOWLEDGMENT. A draft of the bigeye account was reviewed by Wayne Starnes.

TILEFISHES. FAMILY MALACANTHIDAE

Kenneth W. Able

Following Nelson (1994), sand tilefishes (Malacanthinae) and tilefishes (Branchiosteginae or Latilinae) are included in one family. Recently, Imamura (2000) combined these two subfamilies with Dactylopteridae and Hoplolatilidae to form a restructured Dactylopteridae with four subfamilies, but for this book, Dactylopteridae is maintained as a separate scorpaeniform family. Members of Malacanthinae and Branchiosteginae have a long continuous dorsal fin and a relatively long anal fin. The subfamily Branchiosteginae contains three genera and 28 species (Nelson 1994), all of which have a predorsal ridge, lack an enlarged spine at the angle of the opercle, and have a greater body depth than species of sand tilefishes. One species of tilefish is abundant off the southern edge of Georges Bank. Recent information on tilefish has been summarized by Steimle et al. (1999f).

TILEFISH / *Lopholatilus chamaeleonticeps* Goode and Bean 1879 / Bigelow and Schroeder 1953:426–430

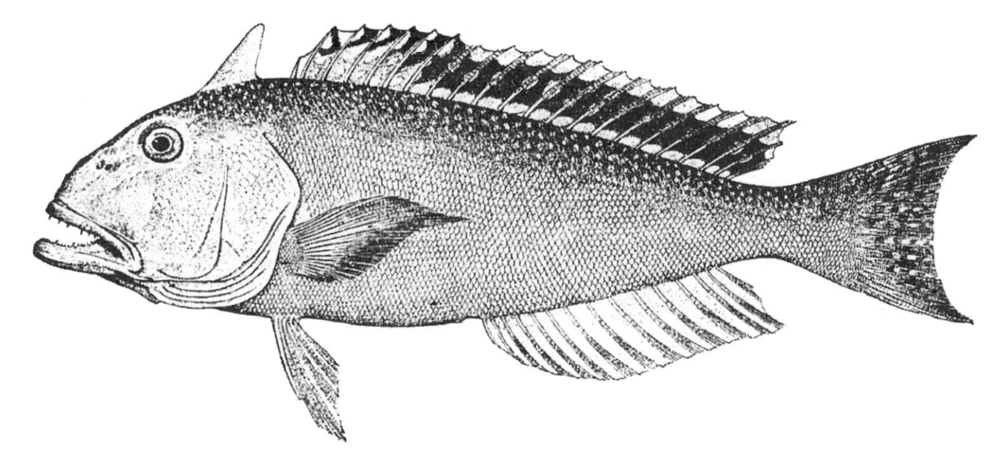

Figure 211. Tilefish *Lopholatilus chamaeleonticeps.* Off Martha's Vineyard. Drawn by H. L. Todd.

Description. Body compressed, deep (Fig. 211). Head large, mouth wide, jaws with large conical teeth and inner rows of just anterior to dorsal fin. Flap absent in small juveniles (<80 mm SL), increasing in size with age, most prominent in

margin of lower jaw. Spinous- and soft-ray portions of dorsal fin continuous. Both dorsal and anal fins with longest rays in posterior portion of fin. Pectoral fins elongate, inserted high on sides of body. Pelvic fins originate directly below pectorals. Caudal fin shallowly lunate and broad. Prominent scales present on lateral surfaces of head and operculum. General body proportions are given by Dooley (1978).

Meristics. Dorsal fin rays VII or VIII, 14 or 15; anal fin rays I, 13 or 14; pectoral fin rays 16–18; total gill rakers 22–26; pored lateral line scales 66–75; vertebrae 10 precaudal + 14 caudal.

Color. One of the most colorful fishes along the east coast of the United States. Upper body steely blue fading to milky white below midline and covered with yellow spots. Smaller yellow spots on head to interorbital and on cheek to operculum. Lower jaw and branchiostegal membrane milky white. Distal and proximal margin of dorsal and anal fin light near body, darker in the center of the fin. Yellow prominent on dorsal fin membranes. Caudal fin with yellow spots; light blue on margin. Adipose flap yellow with dark edge, the latter more prominent in juveniles.

Size. Maximum size of females 95 cm FL; males 112 cm FL (Turner et al. 1983). The all-tackle game fish record is a 9.18-kg fish caught at Murrell's Inlet, S.C., in April 2000 (IGFA 2001).

Distinctions. Tilefish can be distinguished from all other Gulf of Maine fishes by the prominent adipose flap anterior to the dorsal fin, the barbels at the posterior margin of the jaw, and the unique color pattern, characters that also differentiate this species from all other tilefishes (Dooley 1978).

Habits. In the northern part of its range, tilefish occur as shallow as 80 m, but extend to 305 m, whereas they are generally found deeper than 200 m in the South Atlantic Bight (Grossman et al. 1985; Able et al. 1993; Barans and Stender 1993) and in depths of 162–450 m in the Gulf of Mexico (Nelson and Carpenter 1968; Jones et al. 1989). Depth preference appears to be largely a function of temperature, with most populations occupying a rather narrow range of 9°–14°C along the east coast (Grimes et al. 1986; Barans and Stender 1993).

The population dynamics of tilefish appear to be heavily influenced by environmental factors, especially temperature. As Bigelow and Schroeder reported, it is astonishing that the very existence of so large a fish so close to the Gulf of Maine should have remained unsuspected until May 1879, when a Capt. Kirby, cod fishing in 274 m of water south of Nantucket Shoals Lightship, caught the first specimens (Goode and Bean 1880b). Others were caught at 159 m nearby during the following July. Trips by the U.S. Fish Commission during the next two summers proved that tilefish were plentiful enough to support an important new fishery. These early and subsequent investigations (Twitchell et al. 1985; Grimes et al. 1986) proved that tilefish occupy a very definite environment, along the upper part of the continental slope and on the outer edge of the shelf, where a narrow band of seafloor is bathed with a belt of warm water (about 9°–14°C), varying by only a couple of degrees from season to season. Within 3 years (1882) of its discovery, a massive mortality occurred (Collins 1884; Marsh et al. 1999). Numerous vessels reported multitudes of dead tilefish floating on the surface from the Great South Channel to north of Delaware Bay. It has been estimated that at least 1.5 billion dead tilefish were sighted (Collins 1884).

It is generally believed that this destruction was caused by a temporary flooding of the bottom along the warm zone by abnormally cold water. Consonant with this explanation is the fact that other species of fishes suffered as well, and dredgings carried out during the following autumn proved that the peculiar invertebrate fauna that had been found in abundance along this warm zone in previous summers had likewise been exterminated (Verrill 1882).

This mortality of tilefish was so devastating that fishing trials carried out off southern New England by the Fish Commission later in 1882, in 1883, 1884 (when a particularly careful search was made), 1885, 1886, and 1887 did not yield a single fish (Bigelow and Schroeder). But the species was not quite extinct, as eight of them were taken off Martha's Vineyard in 1892 and 53 were caught in 1893; they were more numerous in 1898 (Bumpus 1899).

The northern stock is known to occupy a variety of habitats, including scour basins around rocks and boulders and "pueblo habitats" (Warme et al. 1977; Valentine et al. 1980; Able et al. 1982; Grimes et al. 1986). The dominant habitat type, at least in U.S. waters, is a vertical burrow in the substrate (Able et al. 1982, Grimes et al. 1986). Based on numerous observations from submersibles, size and shape of burrows vary considerably: the smallest juveniles observed (10–20 cm TL) occupy simple vertical shafts in the substrate, and larger fish occupy much larger burrows (up to 4–5 m in diameter and 2–3 m deep) that are funnel-shaped in cross section. The upper margins usually contain numerous small secondary burrows of associated fishes and crabs. The increase in burrow size with fish size suggests a regular sequence of burrow construction by tilefish and their burrow associates. Several associates, such as galatheid crabs (*Munida*), lobsters (*Homarus americanus*), and *Chacellus filiformis,* may help in burrow construction by removing sediments from the upper margin of the burrow. Others such as conger eels appear to be consistent residents of these burrows (Hood et al. 1988; Levy et al. 1988). The burrows, based on submersible and sidescan sonar observations (Able et al. 1987), appear to occur in aggregations or clusters (Twitchell et al. 1985; Grimes et al. 1986) at least near Hudson Submarine Canyon, where they are abundant (average density in 1982 was 2,500 km^{-2}).

Substrate type apparently plays a role in tilefish distribution as well. In most areas where the northern stock has been studied, particularly Hudson Submarine Canyon, burrows are found in semi-lithified, silty clay but not in sand substrates. When a large fish with a correspondingly large burrow is removed through natural or fishing mortality from the longline fishery (Grimes et al. 1980, 1982), the burrow fills with sediment over several months and is abandoned by the associates (Grimes et al. 1986; Able et al. 1993). These abandoned burrows became much more common in tilefish grounds off New York, New Jersey, and Florida after intensive fishing in the 1980s.

The behavior of tilefish is intimately tied to their burrows (Able et al. 1982; Grimes et al. 1986). Burrow construction probably occurs through a combination of oral excavation of chunks of sediment, secondary bioerosion by associated species, and tilefish swimming motions to flush finer sediments from the burrow. Observations from submersibles strongly suggest that there is one fish per burrow although a single 24-h time lapse series indicated that pairing may occur. The fish appear to be away from their burrows foraging during the day and around or in them at night. Tilefish always enter a burrow head first and exit tail first, usually with slow deliberate movements of the caudal fin. The burrow may be a means of predator avoidance.

Food. The food of tilefish is varied and includes numerous invertebrates such as shrimps, crabs, mollusks, polychaetes, sea cucumbers, brittlestars, urchins, anemones, and tunicates and occasional fishes (Bigelow and Schroeder; Collins 1884; Linton 1901a,b; Bowman et al. 2000). Of these, galatheids (*Munida* spp.), spider crabs (*Euprognatha*), and ophiuroids appear to be the most important. The diet appears to change with age, with mollusks and echinoderms found mostly in the stomachs of smaller tilefish (Freeman and Turner 1977b).

Predators. Small tilefish are sometimes preyed upon by spiny dogfish and conger eel and have been found in stomachs of larger tilefish (Freeman and Turner 1977b). There is one record from the stomach of a goosefish over 90 cm TL (Rountree 1999; Bowman et al. 2000). Large sharks may prey upon free-swimming tilefish as sharks often attack tilefish caught on longlines (Bigelow and Schroeder; Freeman and Turner 1977b; Grimes et al. 1982).

Parasites. Tilefish parasites include cestodes, trematodes, nematodes, and acanthocephalans (Linton 1901a,b). Tilefish are also attacked by sea lamprey (Freeman and Turner 1977b).

Reproduction and Early Life History. Size and age at maturation vary with sex (Harris and Grossman 1985; Erickson and Grossman 1986; Grimes et al. 1988). Maturation of females in the northern part of the range may begin at approximately 50 cm FL and 5 years of age; by 60–65 cm FL and 8–9 years all fish are sexually mature. Comparisons of visual and histological staging techniques for males suggest that males produce sperm at 65–70 cm TL and 7–8 years. They do not develop a large testicular mass until 80–85 cm FL and an age of 10–11 years. There is some evidence that the rapidly expanding fishery during 1978–1982 reduced population density by one-half system by causing males to spawn at smaller sizes and younger ages in 1982 relative to 1978 (Grimes et al. 1988).

Sex ratios are skewed in favor of males at larger sizes; however, both sexes are equally abundant at most ages (Grimes et al. 1988). The predorsal adipose flap is a striking sexually dimorphic character, so much so that it can easily be used to determine sex in individuals larger than about 70 cm FL. The flap is typically larger in males than in females but its size decreases for both sexes in more southern populations (Katz et al. 1983).

Tilefish in the Mid-Atlantic–southern New England area appear to spawn from March to November, with a peak in activity between May and September. Eggs have been taken by the NEFSC MARMAP ichthyoplankton surveys in each of these 9 months along the shelf break from Georges Bank to Cape Hatteras (Berrien and Sibunka 1999), but not from the Gulf of Maine. Females release small batches of eggs; that is, they are fractional spawners. Estimates of fecundity, made early in the season before spawning occurred, ranged from approximately 195,000 for a 53-cm FL female to 10 million for a 91-cm FL individual, with a mean of 2.28 million (Grimes et al. 1988).

Eggs that were artificially fertilized and raised to hatching (Fahay and Berrien 1981) were 1.16–1.25 mm in diameter with a thin, colorless chorion with reticulations. The yolks were amber with a single oil globule 0.18–0.20 mm in diameter. Incubation at 22.0°–24.6°C was 40 h. Size at hatching was 2.6 mm NL. Larvae are characterized by spinous scales covering most of the body, with 13 series of cranial processes, and by well-developed preopercle spination (Fahay and Berrien 1981). The largest pelagic specimen captured was 8.7 mm SL, and the smallest demersal juvenile was 15.5 mm SL. Tilefish larvae in the Mid-Atlantic Bight occur in the plankton from July to September, with the center of abundance between Hudson and Baltimore canyons (Fahay and Berrien 1981).

Age and Growth. During intensive sampling at a relatively unexploited phase of the fishery, maximum sizes were 95 cm FL for females and 112 cm FL for males, and maximum ages were 35 and 26 years, respectively (Turner et al. 1983). Based on validated otoliths, both sexes grew about 10 cm FL $year^{-1}$ for the first 4 years. At age 4, males and females averaged 43 and 41 cm FL, respectively. By the 9th year males averaged 74 cm FL and females averaged 64 cm FL. Females had a much smaller L_∞ (90 cm FL) and a larger K (0.153) than males (L_∞ = 111 cm FL and K = 0.130). Growth models (von Bertalanffy) for males [$L_\infty = 111.3(1 - e^{-0.130(t - 0.216)})$] and females [$L_\infty = 90.2(1 - e^{-0.153(t - 0.026)})$] were significantly different with faster growth for males.

General Range. Tilefish are a demersal, shelter-seeking species that is distributed from Nova Scotia to Suriname, excluding the Caribbean Sea (Dooley 1978; Markle et al. 1980; Freeman and Turner 1982; Barans and Stender 1993). They appear most abundant over the outer continental shelf and upper slope from southern Georges Bank and south along the

Occurrence in the Gulf of Maine. Stray tilefish have been frequently encountered off Nova Scotia including off La Have, Emerald, Roseway, and Yankee banks (Scott and Scott 1988). They are abundant off the southern edge of Georges Bank, particularly between Veatch and Lydonia submarine canyons and there are accounts of tilefish from fishermen as far east as Corsair Canyon on the eastern edge of Georges Bank.

Migrations. Limited information suggests that tilefish do not undergo extensive migrations (Grimes et al. 1983, 1986). In one set of observations, 12 fish that were tagged with breakaway, labeled hooks were at liberty from 115 to 577 days. All recaptures were made less than 1 nautical mile from the release location. Anecdotal accounts from fishermen also suggest relatively little movement (Freeman and Turner 1977b). However, there is evidence that there may be some seasonal movements in response to cooler winter temperatures in shallower and more northern portions of the Mid-Atlantic Bight (Grimes et al. 1986).

Stocks. Two stocks are recognized based on meristic, morphometric, and electrophoretic characters (Katz et al. 1983). The northern stock is distributed south to Cape Hatteras and the southern stock occurs from south of that area to at least the Yucatán Peninsula. Size of the dorsal flap also varies clinally with southern populations having smaller flaps (Katz et al. 1983).

Importance. In 1915, the Bureau of Commercial Fisheries undertook to popularize tilefish in the market, believing it numerous enough to support an important fishery and knowing it to be an excellent food fish. It proved so plentiful and so easily caught that a record 4,500 mt were landed in 1916, but only 5 mt were reported for 1920 (Shepherd 1998c). Beginning in the early 1970s, a directed commercial longline fishery expanded rapidly in the Mid-Atlantic and longlines have been the predominant gear used since then (Shepherd 1998c). Landings varied widely from 30 to 3,800 mt between 1962 and 1996, with a peak of 3,800 mt in 1979 (Fig. 212). Status of the populations, especially relative to the recent longline fisheries, has been summarized for the northern Mid-Atlantic Bight (Turner 1986; Shepherd 1998c) and off South Carolina and Georgia (Low et al. 1983; Hightower and Grossman 1988; Barans and Stender 1993). Landings and catch-per-unit-effort data indicate that tilefish were overexploited during the height of the fishery (1977–1982) and remain overexploited (Shepherd 1998c) owing in part to their complex life history and unusual habitat (Grimes and Turner 1999).

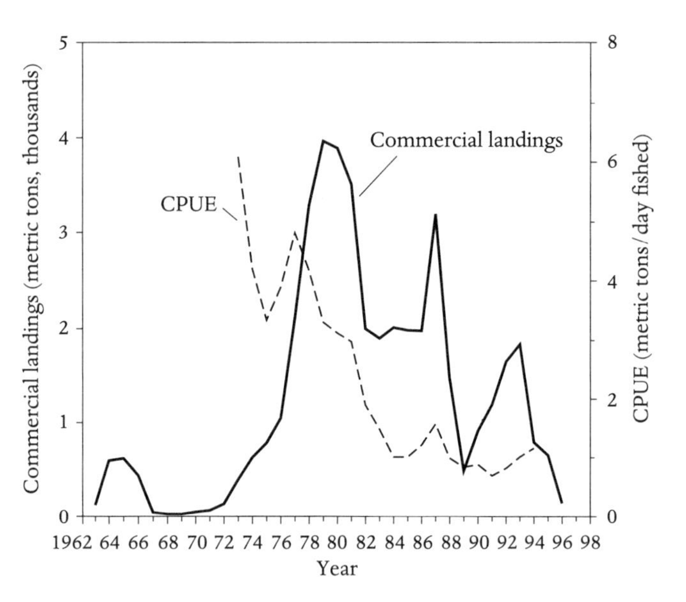

Figure 212. Commercial landings and catch-per-unit-effort (CPUE) for tilefish *Lopholatilus chamaeleonticeps*, Georges Bank–Mid-Atlantic, 1962–1996. (From Shepherd 1998c.)

BLUEFISH. FAMILY POMATOMIDAE

GRACE KLEIN-MACPHEE

The first dorsal fin of bluefish contains seven to nine spines that are much shorter than the rays of the second dorsal fin. Bluefish are well known for their large sharp teeth and ferocious feeding habits on schooling fishes. Pomatomidae is usually considered to be a monotypic family (Johnson 1984), although some authors (e.g., Nelson 1994) include *Scombrops* within it. Recent information on bluefish in the western North Atlantic has been summarized by Fahay et al. (1999b).

BLUEFISH / *Pomatomus saltatrix* (Linnaeus 1766) / Bigelow and Schroeder 1953:383–389

Description. Moderately stout-bodied (large individuals about one-fourth as deep as long) (Fig. 213). Belly flat-sided but blunt-edged below; caudal peduncle moderately stout; head deep; snout moderately pointed; mouth large and oblique, with projecting lower jaw and prominent teeth. "Snappers" (young-of-the-year bluefish) relatively deeper. First dorsal fin originates over middle of pectoral fins, low, rounded, and depressible in a groove. First dorsal separated by a very short interval from second dorsal, which is more than twice as long as first and about twice as high, tapering backward with slightly

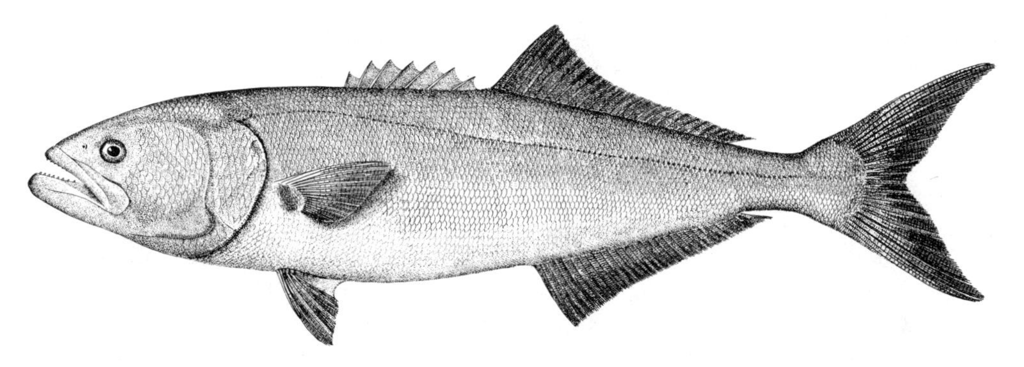

Figure 213. Bluefish *Pomatomus saltatrix.* 707 mm. Drawn by H. L. Todd.

concave margin. Anal fin similar to second dorsal but with a somewhat less concave outer margin, originating somewhat farther back, preceded by two very short detached spines often hidden in skin. Pelvic and pectoral fins moderate-sized. Caudal fin broad and moderately or deeply forked. Body, most of head, and second dorsal and anal fins clothed with medium-sized scales. Lateral line complete, slightly arched over pectoral fin.

Meristics. Dorsal fin rays VII–IX, I, 23–26; anal fin rays II or III, 25–28; pectoral fin rays 18; vertebrae, 11 precaudal + 15 caudal = 26.

Color. Sea green above, silvery below. Second dorsal, caudal, and pectoral fins are of the general body tint, with black blotch at base of pectoral fin.

Size. Maximum length about 115 cm. The all-tackle game fish record was a 14.40-kg fish caught at Cape Hatteras, N.C., in January 1972 (IGFA 2001). A 114.3-cm fish weighing 12.3 kg was caught off Nantucket in 1903 and one of 9 kg was taken off Montauk, N.Y., in August 1951. It is said that fish of 13.6 or even 22.7 kg were not unheard of during the last half of the eighteenth century, but these huge fish may not have been weighed. The general size of the largest fish that are currently caught off the American coast is approximately 4.5–6.8 kg.

Distinctions. Bluefish are separable from jacks (Carangidae) by their jaws: in bluefish, the upper and lower jaws are armed all around with a single series of stout, conical, canine teeth (3–6 mm long in a 5-kg fish), whereas crevalle jack have canines, and only two of them. Furthermore, the caudal peduncle of bluefish is stouter than that of any jack. They are sharply differentiated from mackerels (Scombridae) by the absence of dorsal and anal finlets. They superficially resemble sea trouts (Sciaenidae) in general body form and in arrangement of the fins, but they are readily separable from them by the fact that the anal fin is nearly as long as the second dorsal. They differ from sea bass (Serranidae) in that the first dorsal fin is much lower than the second.

Habits. Bluefish are oceanic, found both inshore and offshore, as well as in many parts of the ocean, usually in continental shelf waters. They are warm water fish and are never found in any numbers in temperatures lower than about 14°–16°C (at least in summer), appearing along the U.S. coast as warm-season migrants. They can tolerate temperatures of 11.8°–30.4°C, but exhibit signs of stress at both extremes (Olla and Studholme 1971). They can survive temporarily in waters of 7.5°C but juveniles cannot survive below 10°C (Lund and Maltezos 1970). Juveniles lose equilibrium at temperatures of 34.5°–35.6°C (Olla et al. 1985). Bluefish enter Long Island Sound waters when temperatures reach 12°–15°C and leave when they drop below 15°C in the fall (Lund and Maltezos 1970).

Bluefish swim continuously in a definite rhythm, moving faster by day than by night (Olla and Studholme 1972). This activity rhythm is characterized by a sharp increase in speed from the last hour before sunrise to the first after sunrise, and a more gradual increase during the next 6–7 h. During the afternoon, swimming speed decreases rapidly and then tapers off within 1–2 h after dark, and swimming is slower throughout the night. This rhythm appears to have an endogenous component as it persisted under constant low light for several days.

Temperature also affects bluefish activity. Response to both high- and low-temperature stress was an increase in swimming speed that was 3.5 and 1.5 times higher than acclimation speeds at high and low temperatures, respectively.

Bluefish are powerful swimmers. Maximum velocity was calculated as 3.8 m·s^{-1} when the fish were startled. Average speeds varied from 0.18 to 1.6 m·s^{-1}. They use two different swimming patterns. At speeds between 0.5 and 1 m·s^{-1}, the two sides of the body being used symmetrically for thrust; at higher speeds they swim using a more asymmetric thrust (Dubois et al. 1976). A bluefish tagged in New Jersey was recaptured 379 km to the south 20 days later for an average distance of 18 km·day^{-1} (Deuel 1964a).

Bluefish are schooling fish that travel in similar-sized groups, often in schools of many thousands; in 1901, for example, a school 8–9 km long was reported in Narragansett Bay. There is a high degree of group interaction during the day, favoring large schools, but these tend to break down into smaller schools

Bluefish produce weak clicks, thumps, and knocks when disturbed. These sounds are probably made by bands of villiform teeth on the vomer, palatines, and tongue and on the series of strong teeth on the jaws (Fish and Mowbray 1970).

Food. Bluefish are considered to be among the most ferocious and bloodthirsty fish in the sea, leaving in their wake a trail of dead and mangled menhaden, herring, alewife, mackerel, and other fishes on which they prey. Goode wrote long ago, that bluefish, "not content with what they eat, which is itself of enormous quantity, rush ravenously through the closely crowded schools, cutting and tearing the living fish as they go, and leaving in their wake the mangled fragments." In contrast to most piscivores, bluefish are capable of severing prey into pieces (Juanes et al. 1994). The effects of bluefish predation are as significant as Goode noted. The prey biomass consumed by bluefish annually along the U.S. Atlantic coast is equal to eight times the biomass of the bluefish population (Buckel et al. 1999c). Bluefish consume a much higher biomass of squid and butterfish than is currently harvested by commercial fisheries for these species.

On the continental shelf and Georges Bank, bluefish feed on a wide variety of prey, 28 species of bony fishes and more than 10 species of invertebrates (Buckel et al. 1999b: Table 1). Both spring- and summer-spawned young-of-the-year bluefish diets on the shelf are dominated by bay anchovy (Buckel et al. 1999b). The adult diet is dominated by schooling species such as squids, clupeids, and butterfish (Buckel et al. 1999b; Bowman et al. 2000). On Georges Bank in 1994, percent frequency and percent weight figures for squids were 29.6% *F*, 23.7% *W* for *Loligo pealei* and 11.1% *F*, 8.4% *W* for *Illex illecebrosus;* for butterfish 27.8% *F*, 18.6% *W;* and for Atlantic herring 7.4% *F*, 11.3% *W* (Buckel et al. 1999b). From Cape Hatteras to Montauk, bay anchovy were the single most important item at 25.9% *F*, 21.4% *W*. Fish prey include eel, menhaden, round herring, alewife, anchovies, killifishes, silversides, silver hake, mullet, longhorn sculpin, sea robins, spot, Atlantic croaker, scup, weakfish, butterfish, cunner, sand lance, mackerel, gobies, and flatfishes (Wilk 1977; Naughton and Saloman 1984; Bowman et al. 2000). Bluefish also eat squids, amphipods, decapods, and polychaetes. In the Indian River estuary, Del. (Grant 1962), mummichog made up 40.5% of the diet followed in importance by menhaden (15.8%), silversides (13.9%), and anchovy (8%). Some of the menhaden consumed were half the size of the bluefish predator. Bluefish in the New York Bight ate primarily anchovy, menhaden, round herring, silversides, sand lance, mackerel, and butterfish. Invertebrate prey included shrimps, squids, crabs, mysids, and annelid worms (Wilk 1982). Deuel (1964b) found a sea lamprey in a bluefish stomach.

Feeding behavior studies have shown that bluefish rely primarily on vision to locate and capture prey, although they are responsive to olfactory stimuli (Olla et al. 1970). Responses to live mummichog were classified into six categories: (1) initial perception (schooling ceased and swimming speed increased when prey were sensed); (2) visual fixation; (3) pursuit-chasing (at swimming speeds as high as 0.8–1 $m \cdot s^{-1}$); (4) capture and ingestion (accomplished by lowering the mandible, arching the head, extending the opercles, and making a sharp 90–180° turn in the direction of the attack while ingesting the prey); (5) feeding intention and searching (swimming toward the point where prey had previously been sighted); and (6) satiation (characterized by more selectivity and capture of only larger prey).

Early juveniles (18–74 mm FL) from continental shelf waters of the Mid-Atlantic Bight feed largely on copepods until about 60–100 mm TL, when they switch to larval teleosts (Marks and Conover 1993; Creaser and Perkins 1994). Food of juveniles inshore in Sandy Hook Bay, N.J., consisted of crustaceans, teleosts, and polychaetes (Friedland et al. 1988). Prey that dominated the diets in percent frequency of occurrence, numerical abundance, or weight included three shrimps, *Neomysis americana, Crangon septemspinosa,* and *Palaeomonetes vulgaris,* and three fishes, *Anchoa mitchelli, Fundulus majalis,* and *Menidia menidia.* During their first summer in the Hudson River estuary, juvenile bluefish consume a variety of prey fish species such as striped bass, American shad, and bay anchovy (Juanes et al. 1993, 1994; Scharf et al. 1997). Bluefish predation may account for 50–100% of the total estimated loss of young-of-the-year striped bass in the Hudson River estuary (Buckel et al. 1999a). On a percent weight basis, the dominant prey species of juvenile bluefish (81–200 mm FL) in Maine included mud shrimp (*Crangon septemfasciata*), alewife, silverside (*Menidia menidia*), mummichog, and unidentified fish remains (Creaser and Perkins 1994). Consumption and growth rates of juvenile bluefish increased with increasing temperature and decreased with increasing fish size in short-term experiments (Buckel et al. 1995). Daily ration estimates for spring-spawned bluefish are 3.7–9.0 $g \cdot [(g \times day) \times 100]^{-1}$ (Buckel et al. 1999b).

Predators. Several species of sharks prey on bluefish, including bigeye thresher, white, shortfin mako, longfin mako, tiger, blue, sandbar, smooth dogfish, spiny dogfish, and angel shark (Medved et al. 1985; Kohler 1988; Rountree 1999; Bowman et al. 2000). Bluefish are the main component of the diet of shortfin mako shark, constituting 77.5% of the diet by volume (Stillwell and Kohler 1982). They are more important in the diet of sharks caught inshore than offshore. Mako may consume 4.3–14.5% of the available bluefish resource in the area between Cape Hatteras and Georges Bank, and is the most important predator on bluefish except for man.

Cod, bluefish, and summer flounder are also predators (Rountree 1999). Bluefish ranked fourth in number and occurrence and third in volume in swordfish diets; however, most swordfish containing bluefish remains were collected off the Carolinas, whereas overall samples were taken from Cape Hatteras to the Grand Banks (Stillwell and Kohler 1985).

In southwestern Maine young-of-the-year bluefish are caught by the birds Atlantic puffin (*Fratercula a. arctica*), Arctic tern (*Sterna paradioaea*), and roseate tern (*Sterna d. dougalli*) to feed their young (Creaser and Perkins 1994). Common tern compete for food with bluefish off Long Island. Both feed on sand lance and bay anchovy (Safina and Burger 1985). The tern season is from early May to November; bluefish enter the area in late May and leave in September.

Parasites. A total of 37 parasites of bluefish are known, 16 of which occur frequently (Anderson 1970). These include an isopod *Lironeca ovalis;* copepods *Lernanthropus pomatomi* and *Lernaenicas longiventris;* acanthocephalan *Serrasentis socialis;* platyhelminthes *Scolex pleuronectes, Nybelinia bisulcata, Callitetrarhynchus gracilis, Otobothrium crenacolle,* and *Pterobothrium filicolle;* trematodes *Microcotyle pomatomi, Bucephaloides arcuatus, Distoma finestrum,* and *Trypanorhynchas* sp., and unidentified nematodes.

Breeding Habits. A bluefish 58 cm long contained about 1.1 million eggs, and one 53 cm long about 900,000 (Lassiter 1962). The average from European specimens is 112,000–195,000 (Hardy 1978b). Bluefish become sexually mature at age 2 (Wilk 1982). Fecundity data on fish collected off New Jersey indicated a linear relationship between fork length and number of eggs per female in the size range 56–80 cm FL (Boreman 1983). Females with large ova approaching ripeness are taken off North Carolina in spring and off various parts of the coast farther north in summer. Ripe males have been taken inside Chesapeake Bay in June and July. There were thought to be two major areas and seasons of spawning along the east coast: offshore near the edge of the Gulf Stream from southern Florida to North Carolina mainly in April and May and the Mid-Atlantic Bight in summer from June through August (Wilk 1977, 1982; Kendall and Walford 1979). However, more recent data indicate that bluefish spawn continuously from about March to at least September as they migrate northward along the coast (Hare and Cowen 1993). Offspring spawned in the middle of the spawning season may have a lower probability of recruitment, thereby creating a bimodal pattern of survival.

Biologists have not observed spawning by bluefish; however, sports fishermen off No Man's Land in Rhode Island Sound, Coney Island, N.Y., and in the eastern parts of Boston Harbor have made observations. Females swam slowly, at depths of less than 30 m, escorted by several males. They rolled on their sides and extruded eggs, and the males did the same at a faster rate while extruding milt. Males and females were not seen to make physical contact with one another (Lyman 1987). Judging from egg collections, most spawning probably takes place offshore, deep in the water column. Off Chesapeake Bight, spawning occurs mainly over the mid-continental shelf and seaward at temperatures between 18°–25°C; no spawning took place below 18°C, and maximum spawning occurred at 25°C. Distribution was also related to salinity, which ranged from 26.6 to 34.9 ppt, but eggs were most numerous in waters that were 30 ppt or higher. Optimum temperature and salinity for spawning are 25.6°C and 31 ppt. The minimums were 18°C at 31.7 ppt and 20.5°C at 26.6 ppt (Norcross et al. 1974). Peak spawning occurs near

Early Life History. Bluefish eggs are pelagic and spherical, 0.9–1.2 mm in diameter with an oil globule ranging from 0.22 to 0.30 mm. The egg is transparent and colorless with a thin tough membrane and a narrow perivitelline space (Fahay 1983). It has a pale amber yolk and a deep amber oil globule. Bluefish eggs have been collected from Cape Hatteras to Long Island from June to August (Norcross et al. 1974; Fahay et al. 1999b: Fig. 3) at water temperatures of 18°–22°C and salinities higher than 31.0 ppt. Although eggs of scup, weakfish, and butterfish are similar in appearance to those of bluefish, the latter can be distinguished from scup eggs by the larger oil globule, and their larger size distinguishes them from weakfish (0.75–0.87 mm) and butterfish eggs (0.68–0.82).

Development was observed on laboratory-spawned eggs at about 20°C. The egg develops to the blastula stage within 12 h and gastrulation quickly follows. The embryo appears at about 17 h and develops to the tail-free stage between 20 and 30 h. Black pigment appears on the oil globule in two rows along each side of the notochord and encircling the eyes. By 31–40 h, pigment increases in stellate patches and a row of melanophores appears along the ventral aspect of the tail. The heart develops and begins to beat at 34 h, and the embryo begins to twitch at 37 h. The tail never completely encircles the yolk.

Hatching occurs after 46–48 h at 18.0°–22.2°C (Deuel et al. 1966). Newly hatched larvae are 2.0–2.4 mm. The oil globule is posterior in the yolk sac, which is more than 50% of the larval length. Larvae float head down at a 45° angle near the surface of the water. Occasionally they swim by rapid tail movements for about 40 s. They remain near the surface for about 1 h and then sink to midwater. Pigmentation at hatching consists of two rows of stellate melanophores widely separated on the head, but converging posteriorly. The eyes are unpigmented. The first day after hatching, yellow and yellow-green pigment spots appear on the dorsal and ventral finfold margin. Later, black pigment appears on the yolk sac and oil globule; melanophores enlarge on the head, nape, and dorsal and ventral edges of the body. There is a distinctive melanophore at the curve of the gut above the anus. By 4.0 mm there is a stripe of black pigment on the midline, as well as dorsal and ventral lines of melanophores. Morphologically, bluefish larvae have a relatively large head and slender body, which deepens as development proceeds. The swim bladder is visible in larvae of 3–10 mm and then is obscured. The finfold begins to break up into fins at 5 mm, when the caudal fin begins to form and flexion occurs. Dorsal and anal fins maintain their relative positions after formation. The space between the end of the first and beginning of the second dorsal fin is small. The origin of the second dorsal fin is vertical to that of the anal fin and both are equal in length. Small preopercular spines develop in 3-mm-long larvae (Norcross et al. 1974).

Several other similar species of larvae co-occur with bluefish but they can be separated by myomere numbers (bluefish 24, scup 30, red hake 40) and by body pigmentation. Bluefish

a ventral line and red hake have dorsal and lateral lines. Atlantic mackerel larvae, which spawn earlier and are larger at hatching, resemble bluefish but can be distinguished by a myomere count of 30 and by the presence of finlets after dorsal and anal fins develop. Some carangids (*Trachurus*) resemble bluefish, but they have more pronounced preopercular spines.

Larvae attain their full complement of adult fin rays at a length of 14 mm (Norcross et al. 1974). Scales form above and below the lateral line anterior to the caudal peduncle at 12.5–14 mm. By 36 mm, specimens developed the adult scale pattern, and scales begin to develop ctenii around 66 mm (Silverman 1975).

Bluefish larvae are found offshore between Cape Cod and Palm Beach, Fla., during every season of the year, and are closely associated with the surface (Kendall and Walford 1979). They have been taken in lower Chesapeake Bay (Pearson 1941) and Narragansett Bay (Herman 1963) but they are much more common offshore. There are two major concentrations of larvae, with distinct temperature and salinity regimes. One is south of Chesapeake Bay near the Gulf Stream in spring at 20°–26°C and 35–38 ppt and the other is north of Cape Hatteras over the middle of the continental shelf in summer at 18°–26°C and 30–32 ppt. Movement is inshore as the season progresses. South of Cape Hatteras juveniles spend most of the summer in estuaries, and then migrate south along the shore in fall. In the Mid-Atlantic Bight, juveniles appear in coastal waters and estuaries in late summer and early fall and move southward out of the bight in late fall. No one knows, as yet, where they go in fall and winter (Kendall and Walford 1979). They are found along ocean beaches, tidal inlets, estuaries, creeks, and rivers in Florida in early summer (Padgett 1967).

When they do come inshore, multitudes of little ones or snappers, run up into harbors and estuaries along the coast, from Delaware Bay to Cape Cod. Larger bluefish, arriving somewhat later, also often come close enough to beaches west and south of Cape Cod for many to be caught by anglers casting in the surf.

Young bluefish of 1.9–7.6 cm, which have often been taken along shore in summer not only south of Cape Cod but even in the Gulf of Maine in some years, are presumably the product of that season's spawning. They grow to a length of 10.6–22.8 cm by autumn, fish of that size being common in October, and achieve a length of 20.3–30.5 cm by the following spring.

Age and Growth. Maximum age is about 12 years (Terceiro 1998a). Richards (1976a) determined age by scale analysis and back-calculated a growth curve on fish from central Long Island Sound. Boreman (1983) presented growth curves of fish from the Gulf of Mexico and North Carolina. Mean length at age from Penttila et al. (1989) is presented in Figure 214. Growth of juveniles was studied by Nyman and Conover (1988), Chiarella and Conover (1990), McBride and Conover (1991), and Hare and Cowen (1994).

General Range. Found in eight major isolated populations in coastal temperate and subtropical waters of all ocean basins except the eastern Pacific (Goodbred and Graves 1996). The known range includes the eastern coast of the Americas, northward regularly to Cape Cod, occasionally to outer Nova Scotia, south to Brazil and Argentina; Bermuda; eastern Atlantic, Azores, off Spain, and northwestern Africa; the Mediterranean and Black seas; both coasts of southern Africa and Madagascar; eastern Indian Ocean and Malay Peninsula; southwestern and southeastern Australia.

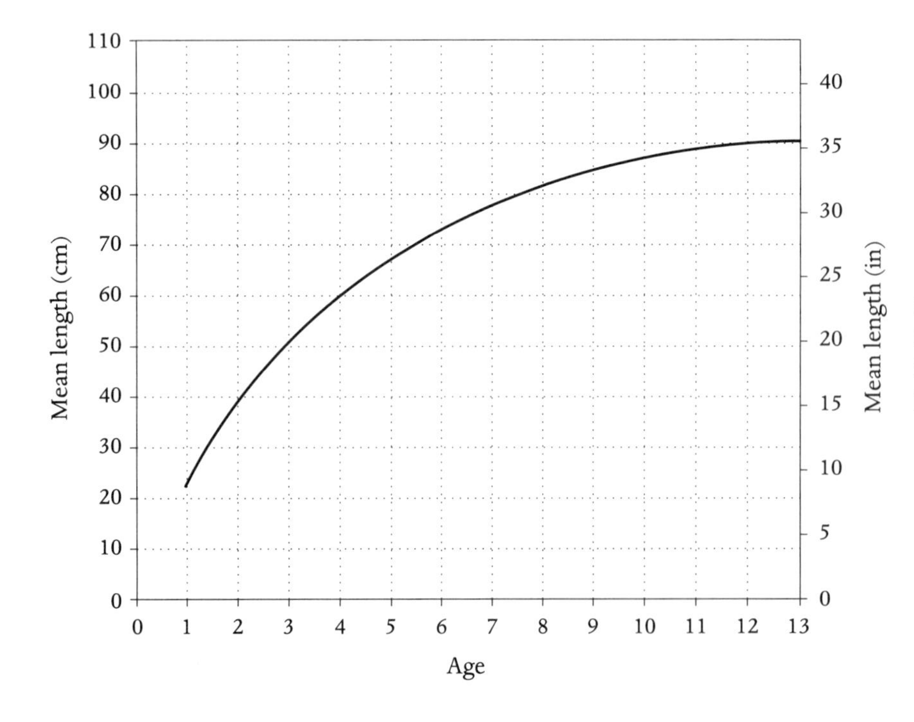

Figure 214. Length at age for bluefish from the Atlantic coast of the United States, all areas combined. (From Penttila et al. 1989:13.)

Occurrence in the Gulf of Maine. Bluefish have been found at one time or another all around the western side of the Gulf. They have seldom been seen east of Penobscot Bay (reported at Mt. Desert in 1889); Bigelow and Schroeder heard of only one taken in the Bay of Fundy, a fish caught in Minas Basin in July 1951, and there are no records of bluefish off the Nova Scotian coast or the open Gulf of Maine. One was caught off Halifax in 1925, another was taken near Liverpool on the outer coast of Nova Scotia, and they were reported "common" near Port Medway, N.S., in the summer of 1951. In recent years, there have been more reports from Maine east into Canadian waters. Juveniles (39–218 mm TL) have been reported from southwestern Maine from August to September (Creaser and Perkins 1994).

In the Gulf, too, they seem to be confined to the vicinity of the coast, but since the mid-1970s many have been caught by research vessels on Georges Bank and in offshore Mid-Atlantic waters. Snappers run up into brackish water, as in the Parker River, Mass., but larger bluefish (1 kg or more) keep to outside waters.

The geographic distribution of the localities where they have been recorded would suggest at first glance that bluefish are practically universal in the western side of the Gulf. But this is true only for brief terms of years and at long intervals, for whereas they have been known to swarm there for several summers in succession, they may then be so rare over periods of many years that the capture of a single fish is notable. It is only in the northern part of its range that the bluefish falls periodically to a very low level, which is to be expected as that is near the boreal boundary. Bigelow and Schroeder give a detailed chronology of these population fluctuations from the 1600s to the early 1950s.

In the years when bluefish pass Cape Cod in any numbers, they usually appear in Cape Cod and Massachusetts bays about the middle of June, sometimes as early as the first of that month, and they are seen off and on all summer. Most of them depart late in September, but an occasional fish lingers into late autumn. Bluefish have been caught around Provincetown even as late as December.

Migrations and Movements. In general, bluefish travel north in spring and summer and south in autumn and winter (Wilk 1982). These movements are probably environmentally induced. Two important factors are temperature and photoperiod (Olla and Studholme 1971, 1972). The destination of these migrations during the summer centers in the New York Bight and southern New England and the northern section of North Carolina. During the winter, the center is in the southeastern part of Florida. "Bluefish," wrote Lyman (1987), "appear off the southern coast of Florida in midwinter," and by "late March anglers take them off the Florida coast in good quantities. Large schools pass the Carolinas during March and April, appear off Delaware during April, and are first taken off New Jersey and Long Island, N.Y., during April and May," by commercial catches are reported off southern Massachusetts in late May. But it is not until about a month later that they work inshore in any numbers. Except for an occasional belated individual, bluefish disappear wholly from the entire coast northward from Maryland by early November. The winter home of this northern contingent has long been the subject of speculation. But the fact that one was trawled in 105 m off Martha's Vineyard in mid-January in 1950 and that several hauls of 80–636 kg per trip were brought in from the region of the Hudson Gorge by otter trawlers early that same February makes it likely that most of the northern contingent merely move offshore on bottom to pass the winter in the warm zone along the outer edge of the continent. It is certain, however, that some migrate far southward (as has often been suggested for the stock as a whole), for one that was tagged off New York in August 1936 was recaptured off Matanzas, Cuba, in January 1939.

Tagging returns for adults from near Long Island Sound show that they return to the same general area year after year, but this pattern was not found among juveniles. In fact juvenile migration patterns differ from those of adults in that the former have a coastal southward migration in the fall whereas adults have an inshore-offshore migration (Lund and Maltezos 1970).

Importance. Bluefish are an excellent table fish, but have never been plentiful enough to support a fishery of any magnitude in the Gulf of Maine. Nevertheless, its presence or absence there may be a matter of direct importance to fishing interests, for it may drive away mackerel, if not herring and menhaden as well.

A favorite game fish, many anglers troll for bluefish in Cape Cod Bay in seasons when there are enough of them to be worth following, and many are caught in the surf in good years by anglers casting from the beach as far northward along the coast as the outer shore of Cape Cod. Bluefish and scup dominated the recreational fishery in the North Atlantic from 1979 to 1986, accounting for over 57% of the total catch in numbers. The most important method of fishing is from private/rental boats followed by beach/bank (Boreman 1983). A very interesting book (Lyman 1987) on bluefishing also contains an accurate summary of bluefish biology and ecology in addition to everything an angler might want to know about how to fish for them.

In 1985, a Federal survey found that polychlorinated biphenyls (PCBs) in larger bluefish (over 500 mm FL) exceeded the U.S. Food and Drug Administration tolerance level of 2 ppm (Eldridge and Meaburn 1992). Health risks can be minimized by avoiding consumption of too many large bluefish.

Bluefish are managed under a fishery management plan developed by the Mid-Atlantic Fisheries Management Council and the Atlantic States Marine Fisheries Commission (Terceiro 1998a). Total landings of bluefish along the Atlantic coast peaked in 1981 at an estimated 51,400 mt (Terceiro 1998a). Landings have since declined substantially: the 1994–1996

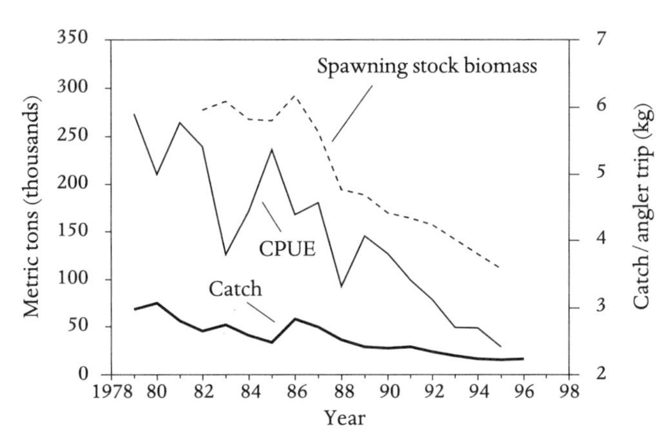

Figure 215. Catch and spawning stock biomass for bluefish from the Atlantic coast of the United States, 1979–1997. (From Terceiro 1998a.)

of 41,600 mt. Bluefish spawning stock biomass declined to 22,700 mt in 1997 (Fig. 215). Atlantic coast bluefish have been overexploited since 1979 and the stock is currently well below levels needed to produce maximum sustainable yield (Terceiro 1998a).

Stocks. No mitochondrial DNA haplotypes were shared among samples from six widely separated populations of bluefish from different parts of the world (Goodbred and Graves 1996), but no significant genetic differences were detected among spring-spawned bluefish and yearling bluefish from different geographic locations along the Mid-Atlantic coast (Graves et al. 1992b).

Lund (1961) identified six stocks of bluefish along the Atlantic Coast based on the number of gill rakers on the first branchial arch. In warmer months these six stocks were found as follows: Massachusetts to New York; New Jersey; Delaware; Chesapeake Bay to Cape Lookout, North Carolina; Cape Lookout to Georgia; and Florida. Wilk (1977) used morphometric characters and scale peculiarities and concluded that there were two spawning stocks off the Atlantic coast: one in the Mid-Atlantic Bight, from Cape Hatteras to Cape Cod, from June to July (summer spawning season), and the other in the South Atlantic Bight, from southern Florida to Cape Hatteras, from March to May (spring spawning season) off North Carolina. Some early life-history studies (Kendall and Walford 1979; Collins and Stender 1987; Nyman and Conover 1988; McBride and Conover 1991) supported the two-stock conclusion but it is now believed that there is a single, migratory spawning stock along the east coast of the United States (Graves et al. 1992; Hare and Cowen 1993, 1996; Smith et al. 1994; Able and Fahay 1998; Terceiro 1998a).

ACKNOWLEGMENTS. Drafts of the bluefish account were reviewed by Thomas A. Munroe, Mark Terceiro, and Stuart Wilk.

REMORAS AND SHARKSUCKERS. FAMILY ECHENEIDAE

Bruce B. Collette

Remoras are easily distinguished from all other fishes by the fact that the spiny part of the dorsal fin is modified into a flat oval sucking plate, composed of a double series of cartilaginous crossplates with serrate-free edges, and is situated on the top of the head. Remoras are fusiform and elongate, with the lower jaw projecting well beyond the upper. Their jaws and vomer are armed with many small pointed teeth. Their long soft dorsal and anal fins are about the same in form and size, one above the other; and their pectoral fins are set high up on the sides. The lower surface of the head is convex, the upper flat (a very conspicuous feature) with the lower surface of the body nearly as deeply colored as the upper so that the back is often mistaken for the belly. Members of this family attach themselves to other fishes, to sea turtles, or to boats by their sucking disk. They usually cling to the sides of their hosts, but also occur within the mouth or gill cavities of billfishes, large sharks, and giant rays. They are carried about in this way and were reputed to feed on scraps from the meals of their transporters. It is now evident that parasitic copepods, removed from the host fish, also constitute an important component of the diet of remoras (Cressey and Lachner 1970). Debelius (1997:156) includes a remarkable photograph of eight remoras peeking out of the cloaca of a whale shark.

The Echeneidae is divisible into two subfamilies, four genera, and eight species, all of which occur worldwide, except *Echeneis neucratoides,* which is restricted to the western Atlantic (Collette in press). Remoras are tropical; they appear only as strays in boreal seas, usually attached to sharks or to spearfishes. Three species are known from the Gulf of Maine.

KEY TO GULF OF MAINE REMORAS

1a. Pectoral fins pointed; disk laminae 18–28; pelvic fins attached to the belly for less than one-third of their length; anal fin rays 29–41. . . . **Sharksucker**

1b. Pectoral fins rounded; disk laminae 14–16; pelvic fins attached to the belly for more than one-half of their length; anal fin base short; anal fin rays 18–28 . . . . **2**

2a. Dorsal fin rays 27–34; anal fin rays 22–28; disk laminae 15–18 . . . . **Spearfish remora**

2b. Dorsal fin rays 21–27; anal fin rays 20–24; disk laminae 16–20 . . . . **Remora**

SHARKSUCKER / Echeneis naucrates Linnaeus 1758 / Bigelow and Schroeder 1953:485–486

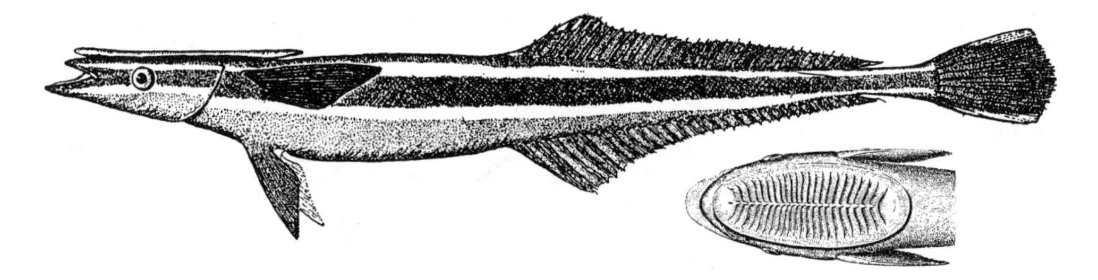

Figure 216. Sharksucker *Echeneis naucrates.* Tortugas, 283 mm. Inset is dorsal view of head, Boca Grande Pass, Fla., 462 mm. Drawn by H. L. Todd.

Description. A very slim fish, body depth contained 8–14 times in SL, nearly round in cross section, and tapering to a very slender caudal peduncle (Fig. 216). Sucking disk extends from close behind snout tip over nape of neck to middle of pectoral fin, about as broad as head, flat, oval, and with very conspicuous transverse plates. Soft dorsal and anal fins both originate about midbody and extend nearly to caudal fin base. Both fins taper from front to rear, anal fin more concave than dorsal. Pelvic fins pointed, originating below pectoral fins; pelvic inner rays attached to abdomen for a short distance. Pectoral fins broad-based, pointed, located high on sides, upper margins close below overlapping edges of sucking disk. Caudal fin almost truncate in adult fish but lanceolate in young ones with middle rays elongate and filamentous.

Meristics. Disk laminae 18–28; second dorsal fin rays 32–41; anal fin rays 29–41; vertebrae 30 (Lachner 1986).

Color. The general ground tint is slaty or dark brownish gray, with the belly nearly as dark as the back. Each side is marked by a broad darker brown or sooty stripe with white edges that runs from the angle of the jaw to the base of the caudal fin, interrupted by both the eye and the pectoral fin. The caudal fin is velvety black with white corners. Dorsal and anal fins are dark slate color or black, more or less margined with white. Pectoral and pelvic fins are black, either plain or more or less pale-edged. The upper and lower fin margins of juveniles are white.

Size. Reach about 1 m SL. The all-tackle game fish record sharksucker weighed 2.30 kg and was caught in Papua New Guinea in March 1994 (IGFA 2001)

Distinctions. The pectoral fin is pointed in sharksucker and rounded in the other two remoras that might be found in the Gulf. The anal fin is longer, 29–41 rays compared with 18–28 rays in the other two species.

General Range. Cosmopolitan in tropical and temperate seas, north as a stray to Halifax, N.S.

Occurrence in the Gulf of Maine. Sharksucker are the rarest of strays, clinging to some ship or shark, taking it north of Cape Cod. The only positive records from the Gulf are for one taken from the bottom of a fishing boat in Boston Bay before 1839 (Storer 1863), a second from Salem Harbor (Wheatland 1852), a third reported by Goode and Bean (1879) from the mouth of the Merrimac River in June 1870, and the most recent, an individual attached to a 2.7-m blue shark (*Prionace glauca*) caught off the Lynnway Marina in Massachusetts Bay in August 1972 (Collette and Hartel 1988). Scott and Scott (1988) summarized the few Atlantic Canadian records from Nova Scotia and Halifax.

SPEARFISH REMORA / *Remora brachyptera* (Lowe 1839) / Bigelow and Schroeder 1953:486–487

Description. Stouter than sharksucker, depth contained only 5–8 times in SL and about as thick through the shoulders as deep, with a thicker caudal peduncle (Fig. 217}. Pectoral fins relatively shorter than those of sharksucker, softer, and rounded instead of pointed; upper margins of these fins not so close to the edge of the sucking disk. Distal two-thirds of pectoral fin rays flexible. Pelvic fins attached to skin of abdomen along inner margins for at least half their length.

Meristics. Disk laminae 15–18; second dorsal fin rays 27–34; anal fin rays 22–28; pectoral fin rays 23–27 (Lachner 1986).

Color. Light reddish brown above and darker below, with paler dorsal and anal fins.

Size. Reach 260 mm SL.

Distinctions. The long dorsal fin (27–34 rays) of spearfish remora separate it from remora (21–27 rays). Spearfish remora lack the lateral stripe and white fin edgings characteristic of sharksucker.

Habits. Billfishes are preferred hosts but other hosts are known and they are rarely observed free-swimming. Spearfish

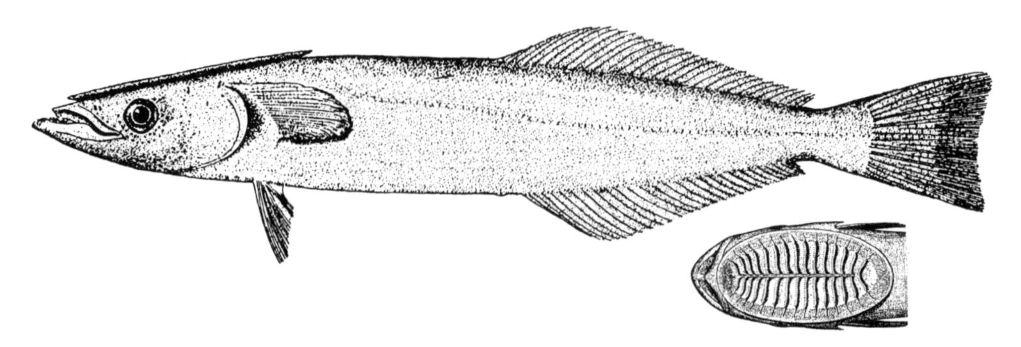

Figure 217. Spearfish remora *Remora brachyptera*. Halifax, N.S. Inset is dorsal view of head. Drawn by H. L. Todd.

remora have been taken from sailfish, several species of marlins, swordfish, sharks, and sharptail mola, *Mola lanceolata* (Cressey and Lachner 1970). Several species of parasitic copepods were found in stomachs of spearfish remora (Cressey and Lachner 1970).

General Range. Warm and warm-temperate oceanic seas, probably paralleling the distribution of spearfishes.

Occurrence in the Gulf of Maine. Goode and Bean's (1879) description of this remora as not infrequently accompanying swordfish into Massachusetts Bay probably applies to the whole Gulf, except the Bay of Fundy, for specimens have been brought in from near Matinicus Rock and near the Isles of Shoals; fishermen occasionally speak of seeing "suckers" clinging to swordfish harpooned on the offshore banks, sometimes several fastened to a single swordfish. But they also report far more swordfish lacking these uninvited guests than carrying them, and this was the case with the few fish harpooned by the *Grampus* during Bigelow and Schroeder's cruises in the Gulf. Spearfish remora are described by eyewitnesses as usually holding fast to the shoulder of the swordfish and also occurring within the gill cavity of the latter.

COMMON REMORA / *Remora remora* (Linnaeus 1758) / Bigelow and Schroeder 1953:487

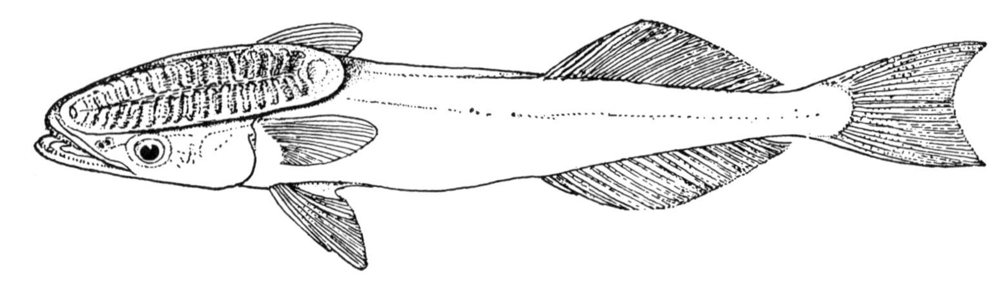

Figure 218. Common remora *Remora remora*. (From Bigelow and Schroeder: Fig. 253.)

Description. Similar to spearfish remora, stouter than sharksucker, pelvic fins similarly attached to the skin of the abdomen along their inner edges (Fig. 218).

Meristics. Disk laminae 16–20; second dorsal fins rays 21–27; anal fin rays 20–24; pectoral fin rays 25–32; gill rakers, including rudiments 28–37; vertebrae 27 (Lachner 1986).

Color. Uniform brownish, blackish, or sooty, both above and below.

Size. Reach 618 mm SL.

Distinctions. Common remora have more laminae in the sucking disk than spearfish remora, usually 18 (16–20) vs. 14–16 and there are usually fewer rays in the dorsal fin, 22–25, whereas spearfish remora usually have 29–32.

Habits. Very little is known of the life history of remoras. The young of this and of other species of *Remora* have been taken in the open Atlantic, usually in June or July, suggesting a limited spawning period. A remora may join a shark, or other host, when only about 3 or 4 cm long. It is not yet known how long or how constantly one may accompany a single shark or how often it may transfer from one host to another. Most remora hosts are sharks with records from 12 species (Cressey and Lachner 1970), the most frequented shark hosts being the white-tip (*Carcharhinus floridanus*) and the blue (*Prionace glauca*). Stomach contents of 147 stomachs showed that parasitic copepods removed from host sharks constitute a significant portion (70%) of the diet (Cressey and Lachner 1970).

General Range. Tropical seas generally; very common in the West Indies, occasionally north to New York and to Woods Hole, and only a stray north of Cape Cod. It is usually found attached to large sharks or to sea turtles.

Occurrence in the Gulf of Maine. The only Gulf of Maine records for remora up to 1925 were of one taken many years ago in Salem Harbor (Goode and Bean 1879) and one in the Museum of Comparative Zoology that was taken at Provincetown (Bigelow and Schroeder). Other records are of one found clinging to the bottom of a lobster trap in Portland Harbor in 1931, probably brought in by some West Indian schooner, several of which had been in the harbor; of a second found attached to the gills of a blue shark (*Prionace glauca*) caught on the northeast edge of Georges Bank, 1 August of that same year; and of a third fastened to a shark of the genus *Carcharhinus* that was caught at the surface over the southeast slope of Georges Bank in July 1939. The most recent nearby record is of a 110-mm SL specimen (MCZ 151395) from a *Manta birostris,* 246-cm disk width, caught at the east end of Georges Bank near Nygren Canyon in July 1991.

Records from farther east and north along the American coast are of one taken from a blue shark, 16 km off Cape Sable, 1 June 1933; of another (also from a blue shark) west of Sable Island, 9 September 1934; and of two taken from sharks on St. Pierre Bank, south of Newfoundland, one of them on 13 August 1936 and the other on 7 October 1937. Scott and Scott (1988) summarized recent records from Nova Scotia.

ACKNOWLEDGMENT. A draft of the echeneid section was reviewed by Thomas A. Munroe.

DOLPHINFISHES. FAMILY CORYPHAENIDAE

BRUCE B. COLLETTE

The family Coryphaenidae contains two species, one of which is a rare visitor to the Gulf of Maine, found worldwide in offshore tropical waters (Gibbs and Collette 1959). Biological information on both species has been summarized by Palko et al. (1982), and a summary of recent information on *Coryphaena hippurus* in the western Atlantic is found in Oxenford (1999). Dolphins have a very long dorsal fin that originates on the head. Based on osteology and larval development, Johnson (1984) suggested that Coryphaenidae form a monophyletic group with Echeneidae and Rachycentridae.

DOLPHINFISH / *Coryphaena hippurus* Linnaeus 1758 / Dolphin, Mahi Mahi / Bigelow and Schroeder 1953:360–361

Figure 219. Dolphinfish *Coryphaena hippurus.* Fulton Fish Market, New York. Drawn by H. L. Todd.

Description. Body long and tapering, most massive and deepest close behind head; greatest body depth less than 25% SL (Fig. 219). Head profile becomes vertical in larger males (30–200 cm). Bands of teeth on jaws, vomer, and palatines; tooth patch on tongue small and oval. Dorsal fin originates on nape and extends nearly to base of deeply forked caudal fin.

Anal fin originates about midbody, about half as long and half as high as dorsal fin. Dorsal and anal fins lack sharp spines. Pelvic fins moderately long, located under pectoral fins. Scales small and cycloid. Lateral line curved upward above pectoral fin.

Meristics. Dorsal fin rays 58–66; anal fin rays 25–31; pectoral fin rays 19 or 20; total gill rakers 8–10; lateral line scales 200–320; vertebrae (13 or 14) + (17 or 18) = 30 or 31, usually 13 + 18 = 31 (Gibbs and Collette 1959; Collette et al. 1969; Potthoff 1980).

Color. Dolphin are famous for their brilliant hues and for the vivid waves of color that flash when they are first taken from the water. Alive, in the sea, their sides are largely vivid blue, variously mottled and washed with gold; their tail largely golden yellow. After death, these colors fade rapidly and become uniformly silvery.

Size. Maximum length about 2 m, but common to 1 m. The all-tackle game fish record is a 39.91-kg individual from Exuma, Bahamas, taken on 5 May 1998 (IGFA 2001).

Distinctions. Dolphin differ from related fishes in that the long tapering body is most massive and deepest close behind the head, and the dorsal fin, originating over the gill cover, extends back nearly to the base of the deeply forked tail. These characters, with the compressed form, notably steep forehead, deeply forked tail, and large pelvic fins, separate dolphin at a glance from the few other Gulf of Maine fishes that have long dorsal fins with bodies that are deepest forward.

Habits. Dolphin, despite the blunt snout, are among the swiftest of fishes. Voyagers on tropic seas often see them leaping in pursuit of small fishes or when they are pursued by larger fishes. In sailing-ship days, they were often caught by trolling from the stern.

Food. Dolphin feed on fishes, crustaceans, and squids. Stomachs of 46 adults taken in the Gulf Stream contained at least 32 species of fishes from 19 families plus the Sargassum crab, a shrimp, and four cephalopods, including *Illex* and octopods (Gibbs and Collette 1959). Jacks, triggerfishes, and filefishes, all associated with Sargassum weed, represented the three families found most frequently. A dolphinfish taken off Ocean City, Md., contained 31 sand lance. The individual mentioned by Bigelow and Schroeder from Sandwich on the southern shore of Cape Cod Bay had silversides in its stomach. Flying fishes are a well-known component of the diet of dolphins in some areas, and constituted 26% of the total prey weight of 396 dolphins taken off Cape Hatteras, N.C. (Rose and Hassler 1974). Other important components of the diet in that study were Scombridae (22%), Carangidae (12%), Balistidae (9%), and Coryphaenidae (5%). An index of relative importance of the diet of 2,632 dolphin, 250–1,530 mm FL, from the southeastern and Gulf coast of the United States (Manooch et al. 1984) showed the most important foods to be Balistidae, crustaceans, Carangidae, Exocoetidae, teuthid squids, Syngnathidae, *Coryphaena*, stomatopods, and Diodontidae. A comparison of diet studies from different parts of the western Atlantic was given by Oxenford (1999).

Predators. All life stages of dolphin, particularly juveniles, serve as prey for oceanic pelagic fishes such as dolphin, tunas, billfishes, and sharks (Palko et al. 1982; Oxenford 1999).

Parasites. Monogenetic and digenetic trematodes, nematodes, isopods, cestodes, acanthocephalans, and copepods have been reported from *Coryphaena* (Palko et al. 1982: Table 6). One species of digenetic trematode and eight species of copepods were found infesting the gills and buccal cavity of common dolphin in the Straits of Florida area (Burnett-Herkes 1974). Two kinds of worms were found in intestinal tracts of a large series of dolphin from the southeastern and Gulf coasts of the United States, the nematode *Hysterothylacium pelagicum* and a digenetic trematode, probably *Hirudinella* sp. (Manooch et al. 1984).

Reproduction. Both sexes reach sexual maturity in the first year of life. In the Straits of Florida, females begin to mature at 350 mm FL (6–7 months old) and all are mature by 550 mm FL (Beardsley 1967). The smallest mature male was 427 mm FL. In Barbados (Oxenford 1999), females also mature at smaller sizes (667 mm FL) than males (805 mm FL). Dolphin spawn year-round in the Gulf of Mexico, at least where water temperatures remain above 24°C (Ditty et al. 1994).

Western Atlantic dolphin typically have two or three size-class of eggs in their ovaries, indicative of multiple spawnings (Beardsley 1967; Oxenford 1999). Fecundity increases sharply with size and if they spawn three times a year, total annual egg production per female would range from 240,000 to 3,000,000 eggs (Beardsley 1967).

Early Life History. The ovarian egg is buoyant, colorless, and spherical, 1.2–1.6 mm in diameter, with a single light yellow oil globule, 0.3–0.4 mm in diameter (Mito 1960). During development, many melanophores and xanthophores appear on the embryonic body, yolk, and oil globule. Hatching took 2 days at 24°–26°C in Japanese waters (Mito 1960); 48–50 h at 24°–25°C in Hawaiian waters (Uchiyama et al. 1986); and 38 h at 25°C in the Gulf of Mexico (Ditty et al. 1994). At hatching, the larva was 3.9–4.0 mm TL. Four days after hatching, it reached 5.70 mm and the yolk and oil globule were consumed. The body became relatively deeper during flexion, about 7.5–9.0 mm (Ditty et al. 1994). In the Gulf of Mexico, larvae were collected primarily at water temperatures equal to or greater than 24°C and salinities equal to or greater than 33 ppt.

Larval morphology in the two species of *Coryphaena* has been compared and illustrated by several authors (Gibbs and Collette 1959; Aoki and Ueyanagi 1989; Ditty et al. 1994).

Development and structure of fins and fin supports were compared by Potthoff (1980). Juveniles of the two species of dolphin are so distinctly patterned that they are easier to identify than adults. Juvenile common dolphin tend to have the entire body marked with a pattern of alternating dark and light bars (Gibbs and Collette 1959; Ditty et al. 1994). This is particularly evident at 10–20 mm SL, when specimens with erect dorsal and anal fins resemble miniature feathers.

Age and Growth. Growth is rapid, reaching 1,000–1,200 mm FL in a year (Uchiyama et al. 1986). Life span is very short; maximum age appears to be 4 years (Beardsley 1967; Rose and Hassler 1968; Palko et al. 1982), although most (96%) die before 2 years (Rose and Hassler 1968; Oxenford 1999). Based on scales of 738 dolphin from off North Carolina (Rose and Hassler 1968), 1-year-olds ranged from 650 to 1,100 mm FL (mean 868 mm), 2-year-olds from 900 to 1,300 mm (mean 1,108), and 3-year-olds from 1,100 to 1,430 mm (mean 1,269). Males are heavier than females throughout their growth (Rose and Hassler 1968: Fig. 1; Uchiyama et al. 1986): $W = 0.50 \times 10^{-7} L^{2.75}$ for males; $W = 1.27 \times 10^{-7} L^{2.59}$ for females, where W = weight in kg and L = length in mm FL. There appears to be little difference in length-weight relationships among different localities in the western Atlantic (Oxenford and Hunte 1983).

General Range. Cosmopolitan in warm seas (Gibbs and Collette 1959). In the western Atlantic, south to Rio de Janeiro (Shcherbachev 1973) and northward along the Atlantic coast to southern New England, where it is rare inshore, occasionally straying as far as the outer coast of Nova Scotia (Vladykov and McKenzie 1935).

Occurrence in the Gulf of Maine. A dolphin about 1 m long (from the collection of the Boston Society of Natural History) and weighing 10.5 kg taken 60 miles SW of Cape Sable by the trawler *Natalie Hammond* on 15 August 1930 was the first record for the Gulf of Maine (MacCoy 1931); a second was taken in a trap at North Truro on Cape Cod Bay in August 1949 (Schuck 1951b), a season when many were taken off Martha's Vineyard; a third was caught at Sandwich, on the southern shore of Cape Cod Bay in mid-July 1951 (specimen seen by Bigelow and Schroeder); and a fourth was taken at Cape Elizabeth, Maine, in 1952 (Scattergood 1953). Eight more were captured on handlines on the eastern edge of Georges Bank at about 41°01′ N, 65°54.5′ W (Tibbo 1962).

Importance. Dolphin are important and highly appreciated commercial and recreational fish wherever they occur. They are particularly important to the recreational charter boat fisheries of North Carolina and Florida. The 1978 catch was 52,480 fish (6.2 fish per trip) weighing 124,847 kg (Manooch et al. 1981). Key West, Fla., charter boat anglers spent 39–43% of their efforts fishing for dolphin (Browder et al. 1981). Dolphin appears to be an excellent candidate species for mariculture, and significant advances are being made in holding and rearing them (Hassler and Hogarth 1977; Palko et al. 1982). Dolphin are usually marketed under their Hawaiian name of mahi mahi.

ACKNOWLEDGMENT. A draft of the dolphinfish section was reviewed by Thomas A. Munroe.

JACKS. FAMILY CARANGIDAE

GRACE KLEIN-MACPHEE

Jacks are similar to mackerels in body shape: both have deeply forked tails, very slender caudal peduncles, and pelvic fins situated below the pectorals. Jacks have two dorsal fins, the first with four to eight sharp spines and the second with one spine and 17–44 soft rays. They are readily separable from mackerels by the fact that their spiny first dorsal fin is much shorter than the soft-rayed second dorsal and may be reduced to a series of very short spines or even completely skin-covered and apparently absent in large individuals of some jacks. They either lack the dorsal and anal finlets characteristic of mackerels, except for the leatherjack, or have only one of each. They further differ from mackerels in the number of vertebrae (only 24–27 vs. 31 or more), in having their upper jaws (fixed in mackerels) protractile (except in adult leatherjack), and in the fact that their anal fin is preceded by two free spines. Some scales along the lateral line are modified into spiny scutes in many species. Warm seas support about 32 genera and 140

more than an accidental stray to the Gulf of Maine. There are records of 11 species of jacks from the Gulf of Maine, and Atlantic bumper, *Chloroscombrus chrysurus,* may be taken there and so is included in the key. In addition to the banded rudderfish, two more species of amberjacks (*Seriola*) occur south of the Cape (Mather 1952) and might wander into the Gulf.

KEY TO GULF OF MAINE JACKS

1a. Body scales needlelike and partially embedded; rear parts of soft dorsal fin, from seventh ray backward, and of anal fin from sixth ray backward so deeply indented between every two rays as to form a series of 11–15 nearly separate low finlets . . . . **Leatherjack**

1b. Body scales normal, not needlelike; rear parts of soft dorsal and anal fins even-edged, not forming finlets **2**

2a. Body very much compressed; nearly or quite half as deep as it is long to base of caudal fin **3**

2b. Body moderately stout, less than two-fifths as deep as it is long to base of

3a. Imaginary horizontal line from tip of upper jaw passes through middle of eye; black saddle spot present on upper part of caudal peduncle; anal fin rays 25–28 . . . . **Atlantic bumper**
3b. Imaginary horizontal line from tip of upper jaw passes well below middle of eye; no black saddle spot on upper part of caudal peduncle; anal fin rays 15–20 4
4a. Second dorsal and anal fins conspicuously falcate in shape, very high in front, tapering abruptly toward the rear . . . . **Lookdown**
4b. Second dorsal and anal fins only moderately high in front, tapering rearward gradually 5
5a. Upper anterior profile of head concave; pelvic fins very small; anterior rays of soft dorsal and anal not elongate . . . . **Atlantic moonfish**
5b. Upper anterior profile of head convex; pelvic fins as long as head or longer; anterior rays of soft dorsal and anal fins elongate, threadlike. . . . . **African pompano**
6a. One well-developed dorsal fin (the soft-rayed), first (spiny) dorsal reduced to a few short spines, without separate fin membranes . . . . **Pilotfish**
6b. Two well-developed dorsal fins, although first (spiny) is smaller than second 7
7a. A detached finlet behind dorsal fin and one behind anal fin . . . . **Mackerel scad**
7b. No finlets behind dorsal and anal fins 8
8a. No enlarged bony plates (scutes) along straight part of lateral line; anal fin only about one-half as long as soft dorsal . . . . . . **Banded rudderfish**
8b. Enlarged bony plates (scutes) present at least along straight part of lateral line; anal fin nearly or as long as soft dorsal. 9
9a. Anterior part of lateral line scarcely arched; vertical limb of pectoral girdle with a large papilla and a furrow or groove near its lower edge . . . . **Bigeye scad**
9b. Anterior part of lateral line strongly arched; vertical limb of pectoral girdle smooth 10
10a. Breast naked, except for a small patch of scales in front of pelvic fins . . . . **Crevalle jack**
10b. Breast completely covered with scales. 11
11a. Body (to base of tail) not more than three times as long as deep; soft dorsal fin rays 22–25 . . . . **Blue runner**
11b. Body to base of tail more than three times as long as it is deep; soft dorsal fin rays 28–34. . . . . **Rough scad**

AFRICAN POMPANO / *Alectis ciliaris* (Bloch 1787) / Threadfin /

Bigelow and Schroeder 1953:381–382 (as *Alectis crinitus*)

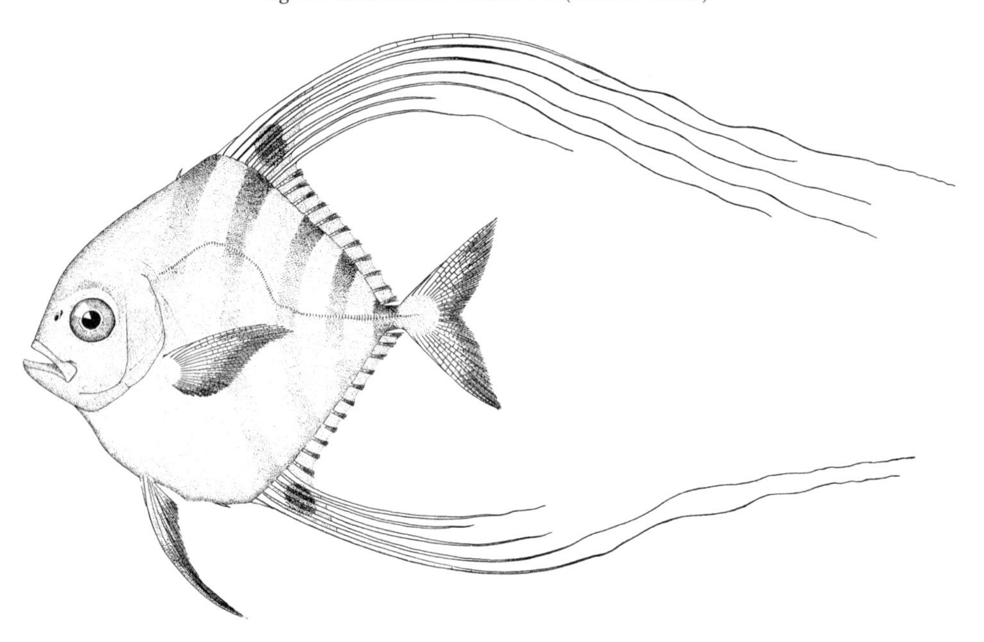

Figure 220. African pompano *Alectis ciliaris.* Woods Hole, Mass., 115 mm.

Description. Only juvenile African pompano are seen in northern waters, so the juvenile description here precedes that of the adult. Trunk nearly as high as long (to caudal peduncle), dorsal profile of head strongly convex and laterally compressed (Fig. 220). Dorsal fin preceded by six short, separate, inconspicuous spines; anal fin preceded by two very short spines, likely to be overlooked. First few rays of soft dorsal and anal fins extremely long and threadlike, giving rise to the name threadfin by which it is known in the northern part of its range. Lateral line strongly arched over pectoral fin; rear part armed with a series of bony platelike scutes that form a well-developed keel. Body scales minute, embedded; body and cheeks partly scaly. Caudal fin deeply forked. Pelvic fins larger than in most jacks. Pectoral fins longer than head. Adults more fusiform; dorsal profile from origin of first dorsal to over eye straight and nearly horizontal. Profile of head blunt and nearly vertical, mouth medium, inclined, lower jaw moderately projecting. Teeth small, in a band at symphysis of jaws tapering backward; few on head of vomer and palatines; and a fairly wide band of teeth on tongue (Ginsburg 1952a). First dorsal fin spines completely covered with skin and molded into body profile by 10–20 cm. Filaments of soft dorsal fin become progressively shorter with age although adults up to 59 cm have been found with fin rays twice as long as the body (Böhlke and Chaplin 1968). Transformation consists of losing filamentous fin rays, a decrease in body depth relative to length, a very considerable decrease in relative size of pelvic fins, and assumption of a more falcate shape by pectoral fins.

Meristics. Dorsal fin rays VII (covered by skin by 17 cm FL), I, 18 or 19; anal fin rays II (detached, covered by skin), I, 15–17; lateral line scales 120–140, 12–30 in posterior part modified as scutes; gill rakers 4–6 + 13–17; vertebrae 10 + 14 = 24 (Berry and Smith-Vaniz 1978).

Color. Upper surface bluish, sides silvery, with traces of darker bars and blotches that tend to disappear with age; the prolonged parts of the dorsal and anal fins are bluish black; pelvic fins mostly black; the fins otherwise more or less yellowish.

Size. Possibly reach 130–150 cm TL; common to 90 cm FL; a 109-cm FL specimen weighed 16.5 kg (Berry and Smith-Vaniz 1978). The all-tackle game fish record is a 22.90-kg fish taken at Daytona Beach, Fla., in April 1990 (IGFA 2001).

Distinctions. Combination of a head strongly convex in dorsal profile and the first few rays of the soft dorsal and anal fins being extremely long and threadlike distinguishes African pompano at a glance from other jacks of the northeastern coast.

Habits. The young are usually pelagic and drifting, adults generally near the bottom to at least 60 m (Berry and Smith-Vaniz 1978).

Breeding Habits. Little is known about reproduction and early life history. The eggs and larvae have not been described. The smallest juveniles collected were 13–15 mm and were taken offshore from July to September (Berry in McClane 1974).

General Range. Worldwide in tropical and subtropical waters. Known on both coasts of tropical America; in the western Atlantic from Massachusetts, Bermuda, and the Gulf of Mexico to Santos, Brazil (Berry and Smith-Vaniz 1978). Adults have not been reported north of southern Florida.

Occurrence in the Gulf of Maine. The only records of this tropical fish from the Gulf are of one about 85 mm long taken in a trap at Sagamore on the southern shore of Cape Cod Bay and another taken in a trap at North Truro, Mass. These may have come through the Cape Cod Canal.

Importance. Adults are reported to be a fine food and game fish, usually caught by trolling (Böhlke and Chaplin 1968).

BLUE RUNNER / *Caranx crysos* (Mitchill 1815) / Yellow Jack, Hardtail / Bigelow and Schroeder 1953:376–377

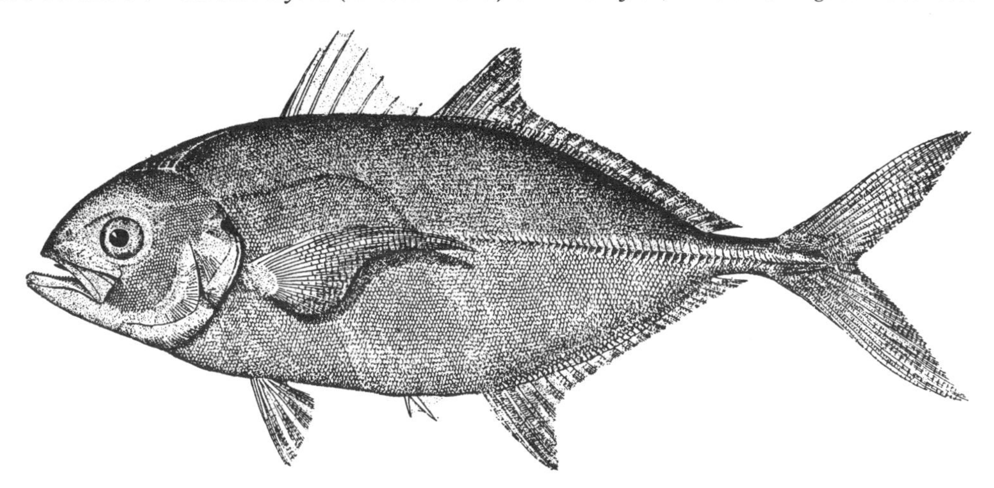

Figure 221. Blue runner *Caranx crysos.* Woods Hole, Mass. Drawn by H. L. Todd.

Description. Body elongate and compressed, depth contained about 3.2–3.5 times in FL (Fig. 221). Head short with a slightly pointed snout and a strong convex dorsal profile. Mouth oblique and terminal, maxilla extending to about mideye; supramaxilla present. Teeth comparatively large; upper jaw with an inner series of smaller teeth, a single series in lower jaw, canines absent. Teeth present on vomer, palatines, and tongue (Hildebrand and Schroeder 1928). Dorsal and anal fin [illegible] lobes moderately produced, first dorsal lobe shorter than second. Pectoral fins falcate, longer than head, extending back barely to anal origin; pelvic fins do not quite reach anus. Caudal fin broadly forked; caudal peduncle with paired keels. Scales small and cycloid; chest scaly. Anterior curve of lateral line moderately high; most of scales in posterior part scutelike, transversely expanded with a long keel ending in a [illegible]

Meristics. Dorsal fin rays VIII, I, 22–25; anal fin rays II (short detached spines in front of fin), 19–21; pectoral rays 19 or 20; lateral line scales 86–98, the last 46–56 scutes; vertebrae 10 + 15 = 25 (Berry and Smith-Vaniz 1978).

Color. Greenish bronze above and golden or silvery below. The fins may show dusky cloudings, tips of caudal fin lobe blackish, pelvic fins mostly white with a tinge of yellow, pectorals plain or slightly yellowish without a black spot (Hildebrand and Schroeder 1928), and there usually is a dark spot on the gill cover near the margin. Young are more or less distinctly cross-barred on the sides, but these bars disappear with growth.

Size. Maximum size 68 cm TL, 62 cm FL; common to 35 cm FL (Berry and Smith-Vaniz 1978), but northern examples are seldom more than 30.5 cm FL. The all-tackle game fish record was a 5.05-kg fish taken at Dauphin Island, Ala., in June 1997 (IGFA 2001).

Distinctions. Blue runner resemble crevalle jack, rough scad, and bigeye scad in relative size and arrangement of the fins, deeply forked tail, slender caudal peduncle, and presence of a row of bony scutes along at least the rear part of the lateral line. The scaly chest, lack of canine teeth in the lower jaw, and lack of a black spot on the pectoral fin separate them from crevalle jack; the fact that the bony plates increase in size rearward along the lateral line marks them off from rough scad; and the strongly arched lateral line separates them from bigeye scad.

Habits. Adults are primarily warm-water schooling fish that occur in shelf and inshore waters in summer and fall in New York and southern New England and move south or into warmer waters of the Gulf Stream in winter and spring. They have been taken in salinities of 26.0–36.2 ppt (McKenney et al. 1958; Franks 1970) and temperatures of 20°–30.8°C (McKenney et al. 1958).

Blue runner produce sound by stridulation of pharyngeal teeth amplified by the large, thin-walled swim bladder (Fish and Mowbray 1970). These sounds may be grating, loud grunts or hornlike. Sound production is associated with alarm or social behavior.

Food. Blue runner are reported to eat fishes and shrimps (Fish and Mowbray 1970).

Parasites. An isopod, *Cymothoa oestrum,* was found in the mouth of a blue runner in Bermuda (Williams et al. 1994) and a digenetic trematode, *Alcicornis carangis,* was reported from the intestine of one from Puerto Rico (Dyer et al. 1998).

Breeding Habits. Based on the occurrence of small young (18–21 mm), spawning occurs in offshore, subtropical waters, but the actual location is unknown (Berry 1959b; Able and Fahay 1998). Some information concerning reproduction is available from the eastern Gulf of Mexico (Goodwin and Fincane 1985). Monthly mean gonadosomatic indices indicate that peak spawning occurs in June–August. Fecundity varies from 41,000 eggs in a 243-mm FL fish to 1,546,000 eggs in a 385-mm FL fish. Females outnumber males with sex ratios of 1.15–1.91 females to 1 male.

Early Life History. Eggs and yolk-sac larvae have not been described. The smallest larva collected was 2.6 mm TL. The mouth is large and nearly vertical. Preopercular spines present in two rows with the largest spine at the angle; reach maximum relative size at 4–5 mm SL; an occipital crest with serrate edge forms at 3.8 mm NL and disappears at about 8 mm SL; tiny posttemporal spines are present from 4.2 mm NL until about 8.5 mm SL; and a few spines occur along the upper jaw in larvae 4–5 mm (Berry 1959b). Dorsal and anal spines and principal caudal rays are complete at about 5.4 mm SL; pelvic buds form at 5.4 mm SL; dorsal, anal, and pectoral counts are complete at 7.5–8 mm SL. Pigmentation consists of a row of spots along bases of second dorsal and anal fins and along the lateral midline between them. Spots appear over the gut at 4 mm NL and increase with growth; spots on the head increase with development; the caudal peduncle remains unpigmented throughout the larval stage.

Age and Growth. In the northern Gulf of Mexico, the largest blue runner was 460 mm FL and the oldest fish was 11 years (Goodwin and Johnson 1986). The von Bertalanffy equation for combined sexes was $FL_t = 1 - e^{-0.35(t + 1.07)}$, where FL = fork length (mm) and t = age (years). Weight-length relationship for combined sexes was $W = 0.0000251355FL^{2.94593}$, where W = body weight (g).

General Range. In the western Atlantic from Halifax Harbour, N.S., throughout the Gulf of Mexico and Caribbean Sea to São Paulo, Brazil (Berry and Smith-Vaniz 1978). The northernmost record is from the Gulf of St. Lawrence off Caribbean Head, St. Georges Bay, N.S. (MacKay and Thomas 1969). In the eastern Atlantic, known from Senegal to Angola, Ascension and St. Helena islands, and the Mediterranean Sea (W. Smith-Vaniz, pers. comm.), north to the south coasts of Great Britain (Swaby et al. 1996). There is a closely allied species in the eastern Pacific.

Occurrence in the Gulf of Maine. Reported at Chatham on Cape Cod in 1933, at Provincetown, in Boston Harbor, off Gloucester, and in Ipswich Bay. Eleven specimens were taken in a fish trap at Barnstable on the shore of Cape Cod Bay on 6 September 1950, and several in a gill net near the Pilgrim Nuclear Power Plant at Plymouth, Mass. (Lawton et al. 1984). They are more likely than crevalle to round Cape Cod. Young fish are often found around Woods Hole from July to November. They are also encountered more frequently than most other jacks during late summer and fall along the coast of Nova Scotia.

Importance. Blue runner are an important commercial and sport fish in northwest Florida waters (Fable et al. 1981), but they are not present in sufficient numbers to be of any significance in the Gulf of Maine.

CREVALLE JACK / *Caranx hippos* (Linnaeus 1766) / Bigelow and Schroeder 1953:375–376

Figure 222. Crevalle jack *Caranx hippos*. Woods Hole, Mass. Drawn by H. L. Todd.

Description. Body flattened and oblong, only about two and a half times as long as deep, but with a slender caudal peduncle and a blunt head with a very steep anterior profile making a strongly convex curve (Ginsburg 1952a) (Fig. 222). Mouth oblique and terminal with a broad villiform band of teeth in upper jaw and an outer series of large, wide-set conical teeth. Upper jaw extends to below posterior margin of eye or beyond. Teeth in lower jaw in one row, with a distinct canine on each side of symphysis. Villiform teeth on vomer, palatines, and tongue. Dorsal fins well separated; soft dorsal and anal fins with high lobes. Pectoral fins falcate and long (longer than head). Caudal fin broadly forked, a pair of lateral keels on caudal peduncle. Anterior curve in lateral line moderately high, most scales in posterior straight part scutelike, transversely expanded with a long keel ending in a sharp backward-directed point. Small scales absent from most of opercle and head except for cheek and upper part of opercle, absent along a narrow strip on midback tapering to dorsal fin origin, a large area in front of pelvic base except for a small patch directly in front of pelvic base and below pectoral base.

Meristics. Dorsal fin rays VIII (posterior I–IV separate and covered by skin in specimens larger than 45 cm), I, 19–21; anal fin rays II (detached spines in front of fin), I, 16 or 17; posterior lateral line scutes 23–35; gill rakers 6–9 + 16–19; vertebrae 10 + 14 = 24 (Berry 1959b; Berry and Smith-Vaniz 1978).

Color. Greenish, bluish green, or greenish bronze above with large black blotch on the gill cover, a fainter dark spot on the lower rays of the pectorals (in adults), and a black blotch in their axils. Fins more or less yellowish. Very young fish have five or six dark crossbars.

Size. Maximum recorded length 101 cm TL, common to 60 cm FL (Berry and Smith-Vaniz 1978). The all-tackle game fish record is a 26.50-kg fish taken at Barra do Kwanza, Angola, in December 2000 (Anon. 2001).

Distinctions. The rounded black spot on the pectoral fins and the small patch of scales on the otherwise naked breast distinguish crevalle jack from all other Gulf of Maine jacks. Presence of a well-developed first dorsal fin combined with an anal fin nearly as long as the second dorsal but without detached finlets separates crevalle jack from all other jacks known from the Gulf of Maine, except bigeye scad, blue runner, and rough scad. The arched lateral line and presence of (usually) two pairs of small but plainly visible canine teeth in the lower jaw distinguish them from bigeye scad, and the partially naked breast and canine teeth distinguish them from blue runner and rough scad. The dorsal profile of the head of crevalle jack is also characteristic, and the long scimitar-shaped pectoral fins are convenient field marks to separate them from pilotfish, rudderfish, and mackerel scad, in which the pectorals are short and blunter.

Habits. Adults occur inshore, even in brackish waters (Berry 1959b). They are frequently found upstream in coastal rivers,

waters offshore. Crevalle jack are a schooling fish, but large individuals may become solitary (Berry in McClane 1974). They are common off Maryland from July to September (Schwartz 1964b). Crevalle jack have been taken from freshwater (Herald and Strickland 1949) to salinities of 43.8 ppt, but are most common in salinities higher than 30 ppt (Gunter 1945). They tolerate temperatures of 18°–33.6°C (Roessler 1970).

Food. In the Gulf of Mexico, the favored food is small fishes such as clupeids and carangids, but penaeoid shrimps, portunid crabs, and other invertebrates are also eaten (Saloman and Naughton 1984). Clupeoids (herrings and anchovies) are the major food in the Gulf of Guinea (Kwei 1978).

Predators. The only report of predation on crevalle jack in the NEFSC survey of food habits of 180 species of fishes (Rountree 1999; Bowman et al. 2000) was by one weakfish.

Breeding Habits. Spawning probably occurs offshore from March to September (Berry 1959b), primarily south of the Florida Straits (Fahay 1975).

Early Life History. There is no description of crevalle jack eggs although Able and Fahay (1998) include a literature description purported to be this species from the Indian Ocean, where it does not occur (see General Range). Larvae of *C. hippos* and *C. latus* are not presently separable. Dorsal and anal spines form by 5.4 mm, pelvic buds form at 5.4 mm, and rays are complete at 6.9 mm; principal caudal rays are complete at 6.9 mm and dorsal, anal, and pectoral rays at 8.3 mm; preopercle and posttemporal spines are present at 5.4 mm and lost at 6.2 mm. Pigmentation consists of spots on the membrane of the anal spine and first four dorsal spines; a midline row of spots is present at 5.4–6.3 mm but absent by 6.9 mm; spots become scattered on the body and head (Fahay 1983).

Pelagic juveniles have been collected from surface waters of the South Atlantic Bight from May through August (Fahay 1975). They have been taken in salinities of 35.2–36.7 ppt and temperature ranges of 20.4°–29.4°C (Berry 1959b). Juveniles probably migrate inshore at an early stage. They are found from July to November in the Hudson River, Haverstraw Bay, and Jamaica Bay, N.Y. (McBride 1995). Young-of-the-year are found under a wide range of estuarine environmental conditions: bottom temperatures 9°–30°C, salinities 1.3–32.0 ppt, and dissolved oxygen 2.0–13.6 ppm (McBride 1995). However, they leave the Mid-Atlantic Bight estuaries in the fall apparently because they cannot tolerate those winter temperatures and some of them may migrate south along the continental shelf to return to the spawning population (McBride and McKown 2000). Mortalities have been reported at 7.4°–9.0°C in October in Massachusetts (Hoff 1971).

General Range. Western Atlantic from Nova Scotia (Vladykov 1935a; Scott and Scott 1988) throughout the Gulf of Mexico to Uruguay (Berry and Smith-Vaniz 1978), with several records from the Bahamas. Also occur in the eastern Atlantic from Morocco to Angola and in the Mediterranean Sea (Smith-Vaniz et al. 1990). A closely related species occurs in the eastern Pacific.

Occurrence in the Gulf of Maine. There are only two records of this southern fish from the Gulf, one picked up on Lynn Beach on the shore of Massachusetts Bay during the summer of 1847 and a second taken at Provincetown in 1933. They are regular summer visitors at Woods Hole, although they are not common there.

Importance. A famous game fish, but not present in sufficient numbers to be of any importance in the Gulf of Maine.

MACKEREL SCAD / *Decapterus macarellus* (Cuvier 1833) / Bigelow and Schroeder 1953:374–375

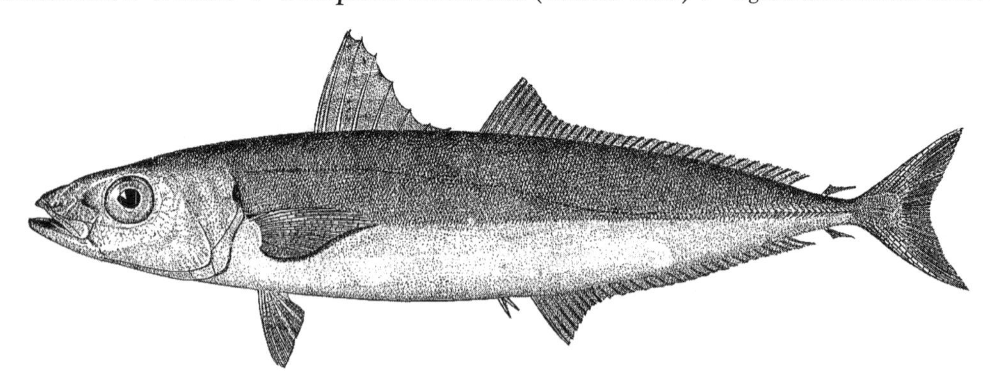

Figure 223. Mackerel scad *Decapterus macarellus*. Woods Hole, Mass. Drawn by H. L. Todd.

Description. Body elongate and fusiform, one-fifth as deep as long (Fig. 223). Mouth small, maxilla not reaching as far as eye, premaxilla protractile. Small teeth present on jaws, tongue, and palatines, and medially on shaft of vomer. Triangular first dorsal fin originates over middle of pectoral fins. Second dorsal fin separated from first by a very short space, fin extending back nearly to caudal base. Anal fin similar to second dorsal in shape but shorter, originating roughly under seventh or eighth ray of second and preceded by two short spines. Pelvic fins shorter than pectoral fins and situated under them. Caudal fin less deeply forked than in most jacks. In place of fleshy keels on caudal peduncle, rear half of lateral line armed with a series of 31 keeled scutes, largest on peduncle, and all much larger than ordinary scales.

Meristics. Dorsal fin rays VIII, I, 31–37 (including the terminal finlet); anal fin rays II, I, 27–31 (including the terminal finlet); gill rakers 9–13 + 32–39; scales in curved part of lateral line 68–79, scutes in straight part 23–32, total scales and scutes 119–133; vertebrae 10 + 14 = 24 (Berry and Smith-Vaniz 1978).

Color. Slate blue, blue-green, or leaden above, silvery to white below, with a small black spot near the upper margin of the gill cover, and the axil of the pectoral black.

Size. Reported to reach about 35 cm SL; common to 20 cm SL (Berry and Smith-Vaniz 1978).

Distinctions. Mackerel scad are easily recognized among jacks found in the Gulf of Maine by presence of a small detached finlet between the second dorsal and the base of the caudal fin and a similar finlet behind the anal fin. The great length of the second dorsal fin and the fact that there is only one dorsal finlet and one anal finlet separate mackerel scad from mackerels.

Habits. Adults are primarily pelagic, offshore, and associated with oceanic islands (Berry 1968); occasionally seen in schools over outer reefs.

General Range. Circumtropical, in the western Atlantic, Bermuda to northern Brazil, rarely straying northward to the Gulf of Maine and Nova Scotia (Berry and Smith-Vaniz 1978). In the eastern Atlantic, known from the Cape Verde, St. Helena, and Ascension islands and the Gulf of Guinea (Berry 1968).

Occurrence in the Gulf of Maine. A specimen caught with smelt in Casco Bay, and another taken in a trap at Richmond Island, off Cape Elizabeth, are the only Gulf of Maine records, although they have been taken at Canso and at Port Mouton Bay, N.S. They are common in autumn about Woods Hole so it would not be surprising to find them north of Cape Cod in summer.

PILOTFISH / *Naucrates ductor* (Linnaeus 1758) / Rudderfish, Shark Pilot / Bigelow and Schroeder 1953:372–373

Figure 224. Pilotfish *Naucrates ductor.* New Bedford, Mass. Drawn by H. L. Todd.

Description. One of the more slender jacks (body about one-quarter as deep as long), round-sided, about two-thirds as thick as deep, and somewhat mackerel-like in appearance though with a blunter, more rounded snout and smaller mouth (Fig. 224). Edge of gill cover rounded in adult but with a spine in juveniles. Caudal peduncle with a conspicuous keel on each side. First dorsal fin reduced to four or five short inconspicuous spines, connected by a membrane in young fish, but membrane lost with growth. Second dorsal weakly concave in outline, originating midway between tip of snout and spines. Pelvic fins, situated far forward under pectoral fins, as large as the latter. Caudal fin large and deeply forked.

Meristics. Dorsal fin rays IV or V, I, 25–29; anal fin rays II (in front of fin), I, 15–17; gill rakers 6 or 7 + 15–20 = 21–27; vertebrae 10 + 15 = 25 (Berry and Smith-Vaniz 1978).

Color. Bluish, cross-barred with five to seven dark bands, two or three of which run up on the dorsal fin and down on the

Size. Maximum length 70 cm TL, 63 cm FL; common to 35 cm FL; weight 0.5 kg at 33 cm FL (Berry and Smith-Vaniz 1978).

Distinctions. The long second dorsal fin separates them from mackerels. They resemble rudderfish in this but the first dorsal fin has only four or five spines instead of seven or eight.

Habits. Pilotfish are frequently sighted in the presence of sharks, tunas, and billfishes in tropical seas, either picking up a living from their protectors' scraps or feeding on parasites with which the latter are infested. They often follow sailing vessels, hence the common name pilotfish.

Food. Pilotfish associated with sharks and tunas off southern Brazil fed on small fishes and zooplankton such as gastropod veligers, pteropods, heteropods, cephalopods, amphipods, copepods, and euphausiids (Vaske 1995). Stomachs of small pilotfish (mostly 20–27 cm FL) caught as by-catch of the seasonal dolphin fisheries carried out in the western Mediterranean Sea showed the main prey to be pelagic planktonic invertebrates such as hyperiid amphipods, gastropods, and decapod larvae (Reñones et al. 1998).

Predators. Found in the stomach of one spiny dogfish (Rountree 1999).

Early Life History. Eggs are pelagic, spherical, and transparent, with a single oil globule and a narrow perivitelline space (Sanzo 1931b). Egg diameter is 1.32 mm and oil globule 0.28 mm.

Larvae are deep-bodied with a large, deep head with numerous spines evident by 5.3 mm. There are two heavy post-temporal spines and pronounced preopercular spines, especially the marginal row. There is no occipital crest, but a large orbital crest with three spines is present. Notochord flexion occurs at 6.6 mm SL, and dorsal and anal fins are complete by about 6.9 mm SL. The body is heavily pigmented except for the peduncle; pelvic fins and dorsal spines are pigmented at about 11.7 mm SL. There are pigmented areas on the dorsal and anal fin by about 20.8 mm SL (Aboussouan 1975; Fahay 1983).

General Range. A circumtropical fish of the high seas, rarely straying as far north as outer Nova Scotia (Sable and Sambro banks) south to Bermuda and 35°30′ S off Argentina (Berry and Smith-Vaniz 1978). The northernmost record is of nine pilotfish about 30–40 cm in length that accompanied a leatherback turtle entangled in a fisherman's nets about 8 km from Fox Harbour, Nfld. (47°19′ N, 53°55′ W) on 17 September 1987 (Goff et al. 1994).

Occurrence in the Gulf of Maine. Early records of this species from within the Gulf are of one taken in a mackerel net in Provincetown Harbor, the fish probably having followed a whaling ship that had arrived a few days earlier; one caught near Seguin Island; two off Portland; one from a mackerel net at Provincetown; four from the Georges Bank area, one from the northern edge of Georges Bank, two from the South Channel southeast of Cape Cod; and one picked up in a trawl on the northern slope of Georges Bank (42°10′ N, 66°32′ W). There are more recent reports from Richmond Island and Southport, Maine (Scattergood 1959).

LEATHERJACK / *Oligoplites saurus saurus* (Bloch and Schneider 1801) / Leatherjacket /

Bigelow and Schroeder 1953:380–381

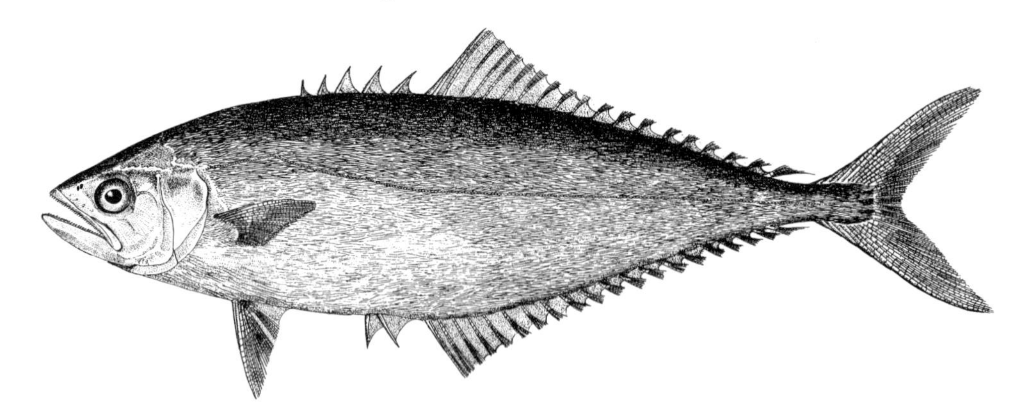

Figure 225. Leatherjack *Oligoplites saurus saurus.* Martha's Vineyard, Mass. Drawn by H. L. Todd.

Description. Body elongate, spindle-shaped, three and a half times as long as deep, strongly compressed laterally, and thin, only about one-third as thick as deep (Fig. 225). Mouth large, upper jaw reaching back about as far as rear edge of eye; lower jaw slightly projecting; supramaxilla absent, premaxilla of adults not protrusible, connected anteriorly to snout at midline by a wide, fleshy bridge. Jaw teeth moderately small; premaxillary teeth in two distinct rows. Additional row of minute teeth sometimes occurs anteriorly. Specialized teeth on lower jaw of juveniles (Smith-Vaniz and Staiger 1973). Teeth on vomer in a wide band tapering posteriorly, tongue nearly covered with granular teeth. Snout moderately pointed. Caudal peduncle very slender. First dorsal fin reduced to about five separate spines, each with a small fin membrane. Rear part of soft dorsal fin posterior to seventh ray; anal fin posterior to fifth ray, broken into a series of 12 low, nearly separate finlets, the ray in each subdivided at tip like hairs of a little brush. Anal fin preceded by two stout conspicuous spines. Venom glands apparently associated with first seven dorsal and first two anal spines (Smith-Vaniz and Staiger 1973). Lateral line nearly straight. Scales very small, long, narrow and embedded in skin, which is corrugated with a great number of short, fine, longitudinal ridges, giving it a leathery appearance, hence its common name. Leatherjack replaces leatherjacket to avoid confusion with filefishes, which are called leatherjackets in Australia.

Meristics. Dorsal fin rays V (rarely IV or VI), I, 19–21; anal fin rays II, I, 19–22; gill rakers 5–8 + 13–16 = 19–23; vertebrae 10 + 16 = 26 (Berry and Smith-Vaniz 1978).

Color. Bluish or greenish silver above, silvery below, with yellow fins, sometimes with seven or eight irregular broken silvery bars along middle of sides (Berry and Smith-Vaniz 1978).

Size. Maximum size 29.7 cm FL, 0.287 kg; common to 27 cm FL (Berry and Smith-Vaniz 1978).

Distinctions. Readily distinguishable from other jacks by the long, narrow, needle- or spindle-shaped embedded scales, lack of lateral line scutes, nonprotractile upper jaw, and semidetached finlet-like posterior dorsal and anal soft rays. Leatherjack vaguely resemble mackerel except for the five detached spines preceding the second dorsal fin and lack of wavy dark pigment bands characteristic of mackerel.

Habits. Leatherjack adults are found along sandy beaches, inlets, and bays, where they seem to prefer tidal currents. They are schooling fish occurring more often in turbid than clear water (Randall 1968; Berry in McClane 1974). They are taken in shallow waters from spring or summer through late fall (Reid 1954; Springer and Woodburn 1960). They tolerate low-salinity waters, 0–45.2 ppt, and temperatures from 16.1° to 34.5°C (Roessler 1970). Escape thumps and weak knocks were produced by the swim bladder in several fish collected and held for sonar tracing (Fish and Mowbray 1970).

Food. The main category of food throughout the year in stomachs of 440 leatherjack (69–285 mm TL) from Caribbean Colombia (Duque-Nivia et al. 1996) was fishes, particularly anchovies. Other prey included penaeid shrimps, fish scales, and insects. Leatherjack are known to practice lepidophagy, scale-eating, with their specialized teeth, particularly at sizes

Juveniles ate plankton (calanoid copepods, parasitic and nonparasitic crustaceans, chaetognaths, fish larvae) and fish scales. More of the schooling fish ate scales and ectoparasites. Scale-feeding was believed to be preadaptive for cleaning behavior (Carr and Adams 1972).

Parasites. A survey of metazoan parasites of three species of *Oligoplites* from Rio de Janeiro, Brazil (Takemoto et al. 1996), revealed 13 species of parasites from 37 specimens of *O. saurus:* five species of digenean trematodes from the digestive tract, three monogenean trematodes from gill filaments, two cestodes, one nematode, and two copepods.

Breeding Habits. On the Caribbean coast of Colombia (Duque-Nivia et al. 1995), spawning apparently occurs in the sea, with a peak from February through April. Sexual maturity in females was first reached at 198 mm TL. Ripe females, 167–265 mm TL, produced 21,923–144,641 eggs, mean 65,119, or 898 eggs per gram of total weight.

Early Life History. Eggs are pelagic, 0.87–0.88 mm diameter, smooth membrane, perivitelline space narrow, yolk bright yellow, unsegmented, one oil globule (0.33–0.34 mm) ventrally located on the egg. Late embryo with stellate melanophores along back and upper sides of body, a large melanophore present at posteroventral midline (Aprieto 1974); pigment on yolk and oil globule. Larvae hatch at 1.87–1.97 mm; the body is slender, and the gut straight; dorsal and anal finfold completely surround fish except for the mouth. Melanophores are present along sides and back, and a characteristic large one occurs over the vent. In older larvae, the snout is convex and the head deep; flexion occurs at 4–6 mm and metamorphosis at 7–10 mm SL. There is no occipital crest; serrate orbital crest present from 4 mm to metamorphosis. Preopercular angle spines with one to three secondary serrations on dorsal surface. Pigmentation consists of uniform dense spots with a prominent midline row; unpigmented U-shaped area on peduncle from 7.2 to 20 mm SL; spots on membrane of dorsal and anal spines after metamorphosis (Aprieto 1974).

Juveniles appear on the west coast of Florida during June and July (Ginsburg 1952a). Juveniles between 26 and 40 mm observed near Crystal River, Fla., showed interspecific schooling, with rough and tidewater silversides *Membras martinicus* and *Menidia peninsulae;* pinfish, *Lagodon rhomboides;* and menhaden, *Brevoortia* sp. Cleaning behavior on redfin needlefish, *Strongylura notata,* was noted by Carr and Adams (1972), and swimming in a leaf-mimic position or close to algae or debris with the head down has been observed as well.

General Range. Common on both coasts of tropical America, in the western Atlantic from the Gulf of Maine to Montevideo (Berry in McClane 1974) and throughout the Gulf of Mexico and most of the Caribbean. Replaced by another subspecies, *O. saurus inornatus,* in the eastern Pacific (Smith-Vaniz and

Occurrence in the Gulf of Maine. Reach the southwestern part of the Gulf of Maine as a stray; the only record of this southern fish within the Gulf is of one taken in a trap off the outer beach at Chatham.

BIGEYE SCAD / *Selar crumenophthalmus* (Bloch 1793) / Goggle-Eyed Scad /

Bigelow and Schroeder 1953:377–378 (as *Trachurops crumenopthalmus*)

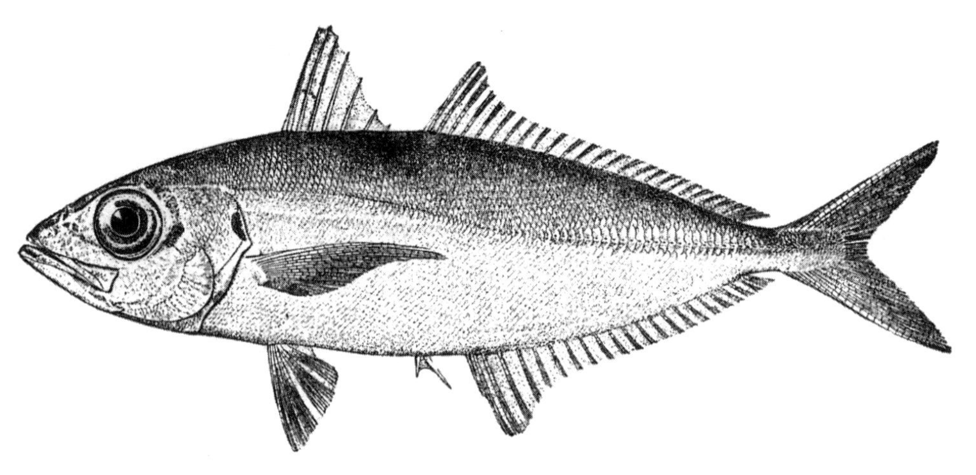

Figure 226. Bigeye scad *Selar crumenopthalmus.* Woods Hole, Mass. Drawn by H. L. Todd.

Description. Body moderately slender, spindle-shaped, depth contained 3.7–4.1 times in FL (Fig. 226). Snout moderate to obtuse, eye very large with a well-developed adipose eyelid covering most of it. Mouth medium-sized, well-inclined with lower jaw projecting; maxilla of moderate width, ending under anterior margin of pupil; supramaxilla present. Teeth small and recurved, in a single row in lower jaw, two rows in upper jaw, in a narrow band on palatines and head of vomer, and in a wide-to-medium band on tongue. First dorsal fin rather high, spines connected by membranes, anterior spine flexible. Pelvic fins originate a little behind pectorals. Pectoral fins shorter than head, reaching nearly to vent. Caudal fin forked; no keels on caudal peduncle. Scales small; entire chest and parts of head scaly. Lateral line slightly curved in anterior part, which lacks scutes. Scutes in straight part of lateral line about half an eye's diameter in height. Accessory lateral line ending under origin or middle of first dorsal fin. Pectoral girdle with two papillae on vertical limb and a furrow or groove below lower (viewed by raising and bending the gill cover forward).

Meristics. Dorsal fin rays VIII, I, 24–27; anal fin rays II (stout detached spines), I, 21–23; gill rakers 9–12 + 27–31 = 37–42; curved portion of lateral line with 48–58 scales, straight part with total 30–43 scutes; vertebrae 10 + 14 = 24 (Berry and Smith-Vaniz 1978).

Color. Bluish or gray-green above, silvery below, the two regions separated by a brassy yellow band (Randall 1968); fins, snout, and tip of the lower jaw have dusky markings. In Hawaii, bigeye scad more than 220 mm SL were sexually dichromatic during the spawning season (Clarke and Privitera 1995). The soft portion of the anal fin was black in males, almost pure white in females.

Size. Unsubstantiated record of 60 cm SL; documented record of 27 cm SL; common to about 24 cm FL and weights of about 0.23 kg (Berry and Smith-Vaniz 1978).

Distinctions. Bigeye scad resemble mackerel scad in general appearance, but have larger eyes and lack detached finlets behind dorsal and anal fins. The high first dorsal fin separates them from pilotfish, whereas the slightly instead of strongly arched forward half of the lateral line distinguishes them from crevalle jack, blue runner, and rough scad.

Habits. Bigeye scad occur in small aggregations or large schools, often near shore. They may occur over shallow reefs, but are usually found where the water is turbid (Randall 1968). They have been taken in waters of 15.1°–26.5°C and salinities of 27–37.4 ppt (Franks et al. 1972). They are common in fall in New York, Massachusetts, and as far north as Nova Scotia, but leave the area after September (Nichols and Breder 1927; Scott and Scott 1988).

Bigeye scad produce sounds—a sustained irregular series of toothy grating sounds and a few knocks—by means of tooth stridulation and swim bladder vibration (Fish and Mowbray 1970).

Food. Feed primarily on planktonic or benthic invertebrates, including shrimps, crabs, and foraminifera, as well as on fishes (Berry and Smith-Vaniz 1978).

Breeding Habits. In Hawaii, bigeye scad mature at about 200 mm SL (Clarke and Privitera 1995). Spawning occurs from April through September or October. Frequency of post-ovulatory follicles indicates that females spawn every 3 days. Batch fecundity for 23 females (199–256 mm SL) ranged from 48,000 to 262,000 eggs (mean 92,000). Males and females in breeding condition were taken in June at Tortugas, Fla. (Longley and Hildebrand 1941).

Early Life History. Eggs are pelagic, average diameter 0.77 mm, with a segmented yolk that is not evident in preserved eggs. The segments are larger toward the center of the yolk. There is a single yellow oil globule 0.24 mm in diameter located at the anterior end of the yolk sac. Black and brown pigment spots are present on the embryo, oil globule, and yolk, especially on the dorsal side. Larvae hatch at 1.3 mm and the yolk is completely absorbed after 64 h. Three patches of chromatophores appear along the dorsal portion of the myotomes at about 1.9 mm larval length (Delsman 1926).

General range. Cosmopolitan in warm seas. In the western Atlantic, straying north to St. Georges Bay, Gulf of St. Lawrence (MacKay and Thomas 1969), and south to Rio de Janeiro, including Bermuda, the West Indies, and the Gulf of Mexico (Berry and Smith-Vaniz 1978).

Occurrence in the Gulf of Maine. The only positive records of this species in the Gulf are of one taken in a trap at Provincetown; a second 13 km off Chatham; and a third from Sandwich. They are taken in summer and fall as far northward and eastward as Woods Hole, on Georges Bank, and in the Canadian Atlantic (Scott and Scott 1988).

ATLANTIC MOONFISH / *Selene setapinnis* (Mitchill 1815) / Bigelow and Schroeder 1953:378–379 (as *Vomer setapinnis*)

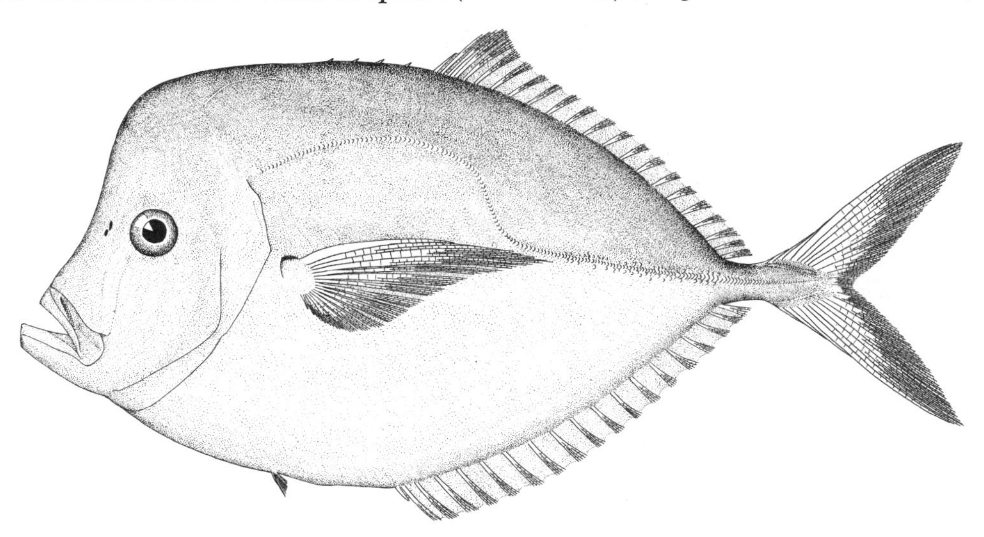

Figure 227. Atlantic moonfish *Selene setapinnis*. Fulton Fish Market, New York, 237 mm. Drawn by H. L. Todd.

Description. Body very deep and strongly compressed, depth contained 1.8–2.3 times in FL (Fig. 227). Outline of head square, slightly concave; dorsal profile from soft dorsal origin to above eye nearly horizontal. Teeth very small, one row in lower jaw, a narrow band in upper, an extensive patch on vomer and a broad band tapering anteriorly on tongue (Ginsburg 1952a). Lower profile to anal fin origin moderately rounded. Body inclines to a slender caudal peduncle with a deeply forked caudal fin. First dorsal fin of adults reduced to eight very short, inconspicuous, detached spines, the first two prolonged and filamentous in young. Second dorsal and anal fins about equal in length, both low and tapering very slightly from front to rear. Dorsal fin lobe better developed than that of anal. Fold at dorsal and anal fin bases slightly developed for looked except in young, in which pelvic rays and dorsal spines more or less filamentous. Pectoral fins long, scythe-shaped. Lateral line curved for half its length. Accessory lateral line ends under beginning of first dorsal fin. Scales along lateral line not large enough to be conspicuous; scutes confined to caudal peduncle, not evident in smaller specimens; keel only moderately developed.

Meristics. Dorsal fin rays VIII, I, 21–24; anal fin rays II, I, 16–19; lateral line scales 62–81, only 7–17 poorly differentiated scutes on caudal peduncle; gill rakers 7–10 + 27–35 = 34–44; vertebrae 10 + 14 = 24 (Berry and Smith-Vaniz 1978).

Color. Bluish green above, sides bright silvery; smaller speci-

pale gray, sometimes light yellow at its base; caudal greenish yellow; pectorals light yellow or dusky greenish.

Size. Maximum length 33.2 cm FL; common to 24 cm FL (Berry and Smith-Vaniz 1978).

Distinctions. The very deep, thin, sharp-edged body of Atlantic moonfish tapering to a slender caudal peduncle and the concave upper anterior profile of the head are enough to separate them at a glance from pilotfish, mackerel scad, crevalle jack, blue runner, rough scad, or bigeye scad; the very low dorsal and anal fins distinguish them from lookdown, which are of somewhat the same shape. Atlantic moonfish juveniles closely resemble those of lookdown, but can be distinguished by shorter first dorsal filaments (which do not exceed the length of the body) and the prominent dark body spot. Minute pelvic fins and soft dorsal fin and anal fin that are nearly even in height from end to end separate them from African pompano.

Habits. Adults are usually found in water of salinities from 17.4 to 37.9 ppt and temperatures from 13.3° to 30°C (Franks et al. 1972; Gunter 1945). Moonfish are reported to be nocturnal feeders and are common around docks and piers at night in the southern part of their range (Ursin 1977).

They produce almost continuous piglike grunting noises when kept in tanks and are easily stimulated by handling (Fish and Mowbray 1970). The sonic mechanisms are the swim bladder and pharyngeal teeth.

Predators. Skipjack and yellowfin tuna prey upon moonfish off the coast of West Africa (Dragovich and Potthoff 1972).

Early Life History. Little is known of the breeding habits of Atlantic moonfish. A female with nearly ripe gonads was taken in July in the western Gulf of Mexico, indicating that they reach maturity by at least 16.5 cm (Hildebrand 1954).

Larvae and juveniles have been described from 2.8 to 60 mm TL. Notochord flexion occurs from 4.3 to 6.3 mm NL. A supraoccipital crest is present from 2.9 to 5.25 mm NL. There are two series of spines on the preopercle, the longest at the posterior angle of the preopercle. Beginnings of the fins can be seen at 2.90 mm NL. Dorsal fin spines form by 3.40 mm NL. The first four spines increase in length throughout larval growth and the longest second spine can reach 75% of body length during late postflexion (Katsuragawa 1997).

Early juveniles (35 mm) have the first four dorsal spines produced as filaments, the second longest reaching more or less beyond the beginning of the soft dorsal, sometimes to its middle. These filaments are lost early, and in medium-sized specimens (50–60 mm), four anterior spines are short, slender, and flexible, the first about half as long as the second; the second to fourth graduated. At 35 mm, two anterior anal spines are present as fixed points, the third spine broad and short with a procumbent spur. In very small fish the second to fourth rays of the second dorsal fin are more or less prolonged. Pelvic fins are very elongate in 30-mm specimens but gradually shorten (Ginsburg 1952a).

Pigmentation gradually increases during flexion and postflexion (Katsuragawa 1997). A patch of melanophores forms on the dorsal margin of the trunk by 4.25 mm NL, spreads, and becomes more dense, forming an oval spot in juveniles. The transversely elongate black spot becomes faint between 60–80 mm and then disappears. A transversely oblique narrow dusty band occurs over the eye but disappears before the lateral line spot.

General Range. Warm seas off the east coast of America from Mar del Plata, Argentina, through the West Indies to Bermuda and Cape Cod, straying to Nova Scotia (Berry and Smith-Vaniz 1978). The northeast limit is off Sable Bank, Scotian Shelf (Scott and Scott 1988). Moonfish are common from Chesapeake Bay southward. Replaced by closely related species in the eastern Atlantic and eastern Pacific.

Occurrence in the Gulf of Maine. This waif from warmer waters has been recorded from the South Channel off Cape Cod; off Cape Cod (96.5 km south by east from Highland Light); Gloucester (several specimens); Magnolia, Danvers, Salem, and South Boston around Massachusetts Bay; Saco Beach; and Casco Bay. MCZ specimens from Quincy and Dorchester were examined by Collette and Hartel (1988). Impinged on traveling screens at Salem Power Plant but not found during field sampling (Anderson et al. 1975). They have even been reported once or twice as far east as Liverpool and Halifax. Thus, they appear to reach the Gulf more often than any of their relatives do. They appear more often (if irregularly) at Woods Hole, where young fish are sometimes common in August and September.

LOOKDOWN / *Selene vomer* (Linnaeus 1758) / Bigelow and Schroeder 1953:379–380

Description. Very peculiar and characteristic form with a deep, rhomboid, but very thin flat body (Fig. 228). Trunk only about one and a quarter times as long as deep, abruptly truncate in front, slightly concave upper anterior profile, tapering rearward to a slender caudal peduncle. Mouth set very low and eye very high, producing characteristic expression on its face. Teeth minute, conical, and recurved in jaws, granular elsewhere. A narrow band of teeth in upper jaw; a narrow band in lower jaw tapering posteriorly to an irregular row; an arrow-shaped tooth patch on vomer; a band on tongue

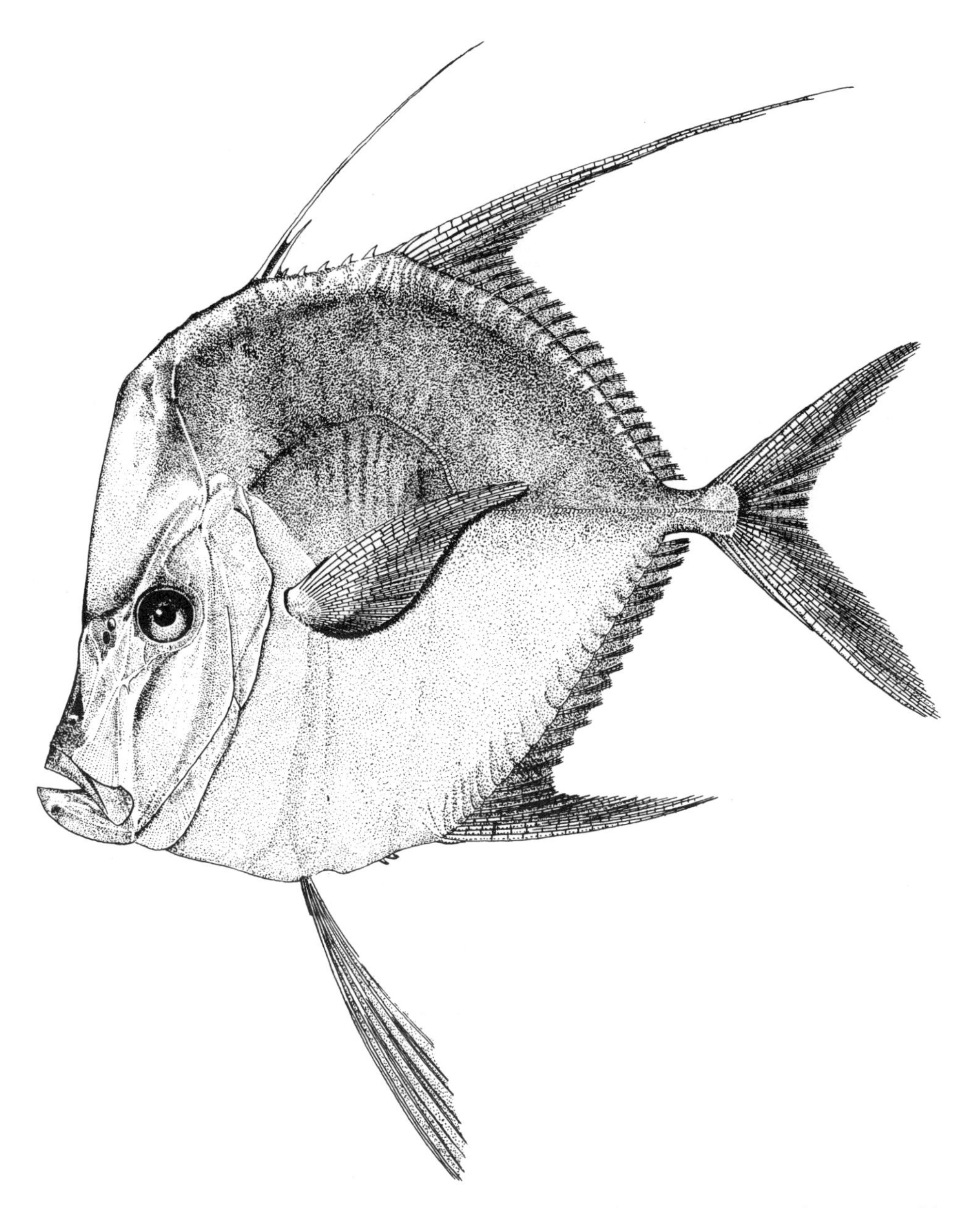

Figure 228. Lookdown *Selene vomer,* Staten Island, New York, 123 mm. Drawn by H. L. Todd.

broader posteriorly; no teeth on palatines (Ginsburg 1952a). In adults, first dorsal fin reduced to seven or eight short inconspicuous spines, only the first three connected by a membrane. Pelvic fins very small. Some spines of first dorsal very long in fry up to 101–127 mm in length, pelvic fins much longer than in adults, and anal fin preceded by two short detached spines that disappear with growth. Caudal fin deeply forked, as in other jacks. Pectoral fins pointed and falciform, reaching back behind middle of second dorsal fin. Scales tiny, covering almost entire body except for predorsal area and head. Boundary of scales an irregular line running approximately from origin of the first dorsal fin to upper angle of gill opening. Lateral line curved moderately anteriorly. Scutes obsolescent in straight part; some posterior scales with bulges suggesting a keel but not actually keeled or spinous, no larger than adjacent scales. Accessory lateral line ends under spinous dorsal fin origin (Ginsburg 1952a).

Meristics. Dorsal fin rays VIII, I, 20–23; anal fin rays II, I, 17–20; scales in curved lateral line 72–84, in straight part 73–75 plus 5–10 on caudal fin; gill rakers 6–9 + 23–27 = 31–35; vertebrae 10 + 14 = 24 (Ginsburg 1952a; Aprieto 1974; Berry and Smith-Vaniz 1978).

Color. Small specimens, and northern strays usually are small, silvery above, as well as below, with the ground tint of the back leaden; sides barred with several faint crossbands, variously described as dark or golden. These bands fade with growth. Adults are silvery or golden; the back above the lateral line with a metallic bluish tinge; first prolonged dorsal and anal fin rays often blackish (Ginsburg 1952a).

Size. Reaches a length of 48.3 cm TL (Pounds 1962); common to 24 cm FL. The all-tackle game fish record is a 2.10-kg fish from Rio de Janeiro caught in November 1993 (IGFA 2001).

Distinctions. Lookdown are unlikely to be confused with any fish other than Atlantic moonfish and African pompano. The very high second dorsal and anal fins of lookdown and their peculiar falcate outline with the second ray much the longest and the next four or five rays successively shorter make it easy to distinguish them from moonfish, and smaller fins and steep forehead distinguish them from African pompano.

Habits. Adults are common southward on sandy shores and are occasionally seen around harbor piers. They are seen in Ocean City, Md., in late July and August and in the Chesapeake Bay in September–October (Hildebrand and Schroeder 1928). Lookdown have been collected in salinities from 8.6 to 45.2 ppt and temperatures from 16° to 31.3°C (Roessler 1970). Several stunned or dead individuals were found in Tampa Bay at 10.4°C (Solomon and Rinkey 1963). A group of juveniles was captured over a sand-silt bottom in Wickford Cove, Narragansett Bay, R.I., in association with common summer resident fishes at a temperature and salinity of 22.6°C and 30 ppt. Lookdown juveniles occasionally occur in Rhode Island waters in summer, but this was the first time they were found well up in the bay itself (Oviatt and Gray 1968).

Lookdown produce loud grunts with the swim bladder and pharyngeal teeth when handled (Fish and Mowbray 1970).

Predators. Found in the stomach of one spiny dogfish (Rountree 1999).

Early Life History. Little is known of the reproductive habits of lookdown, and eggs and yolk-sac larvae have not been described. Larvae have been taken in the western Atlantic in all months except June, October, and December; in the Gulf of Mexico they are abundant mainly in northeast waters in August (Aprieto 1974).

Larvae are deep-bodied with a very deep head and a steep, slightly concave profile in early larvae. Flexion occurs at 4–5.5 mm and metamorphosis at greater than 12 mm. Occipital and orbital crests are well developed in early larvae and decline by about 5 mm. Preopercular spines are in two rows with marginal spines (four to seven) longest. These are lost at metamorphosis. Second and third dorsal spines develop into long filaments about twice the length of the body; pelvic fins are elongate, often extending to the anal fin. These shorten after the fish metamorphose (Aprieto 1974).

Pigmentation consists of spots on the lower side of the tail; other spots are scattered along jaw tips, head, pelvic fins, dorsal spines, and caudal base. Rows of spots occur along dorsal and anal fin bases and at the posterior midline. Pigment forms bars at metamorphosis.

They resemble juvenile African pompano, but the eye is farther from the upper edge of the mouth and the gill raker count is higher, 23–27 on lower limb vs. 14–17 in African pompano (Böhlke and Chaplin 1968).

General Range. Confined to the western Atlantic from Uruguay, around Bermuda, and the West Indies, north rarely to Cape Cod, straying into the Gulf of Maine and to Nova Scotia (Berry and Smith-Vaniz 1978); common from Chesapeake Bay southward. Replaced by two related species in the eastern Pacific.

Occurrence in the Gulf of Maine. There were only three records for lookdown in the Gulf up to 1933: two for Casco Bay and one for Boston Harbor (Dorchester). The Dorchester record is based on a specimen (now MCZ 41234) re-identified as an Atlantic moonfish by Collette and Hartel (1988). Many small lookdown were reported from traps at the mouth of Casco Bay during the autumn of 1933, one from Beverly on the north shore of Massachusetts Bay and one from North Truro on Cape Cod. Evidently this was an unusual incursion. Another individual was collected at the Pilgrim Nuclear Station in 1984 (Lawton et al. 1984).

BANDED RUDDERFISH / *Seriola zonata* (Mitchill 1815) / Amberjack / Bigelow and Schroeder 1953:373–374

Description. Rather slender (body depth contained 3.8–4.3 times in FL), laterally compressed, with a moderately pointed snout (Fig. 229). Mouth medium-sized and terminal, maxilla reaching roughly to posterior margin of eye. Teeth minute, subequal, in moderately broad bands in both jaws and palatines; teeth on vomer in an anchor-shaped pattern and on tongue in a medium-width band (Ginsburg 1952a). First dorsal fin reduced, low, anterior spines connected by a membrane, last two partly disconnected (Aprieto 1974). Second dorsal fin base about twice anal fin base; pelvic fins extend more than half the distance from base of first dorsal to soft anal fin origin. In young of 5–7.6 cm second dorsal originates a little in front of pectoral fins, but by 20–23 cm it originates slightly behind tips of pectoral fins, and still farther back in larger specimens. Young fish have two short spines in front of anal fin; spines also present in adults but may be covered with skin. Pelvic fins a little longer than pectorals and more pointed in large fish than in small. Caudal fin deeply forked; slender caudal peduncle with a longitudinal keel on each side.

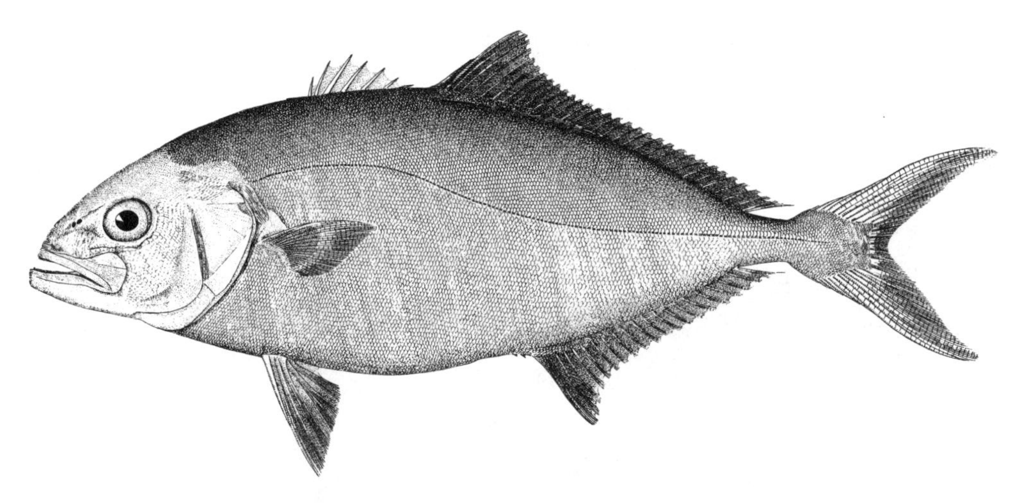

Figure 229. Banded rudderfish *Seriola zonata,* Woods Hole, Mass., 391 mm.

Meristics. Dorsal fin rays VIII (first spine covered with skin in large adults), I, 33–40; anal fin rays II (in front of fin), I, 19–21; lateral line scales 160–187; gill rakers 12–16 in large specimens (anterior gill rakers may become rudimentary with growth); vertebrae 11 + 13 = 24 (Ginsburg 1952a; Berry and Smith-Vaniz 1978).

Color. Bluish or silvery brown above, paler on sides, and white below. Young fish (2–30 cm FL), with a dark nuchal bar through the eye to origin of the first dorsal fin, six broad dark blue or brown vertical bars, the third, fourth, and fifth extending onto the soft membranes of the dorsal and anal fins. These bands fade with growth and disappear in large fish. The first dorsal is black; the anal white at the base; the pelvic fins black above, pale below; and the caudal dusky green with white tips.

Size. Maximum size at least 69 cm FL and 5.2 kg; common to 47 cm FL, unsubstantiated reports to 80 cm FL (Berry and Smith-Vaniz 1978).

Distinctions. Banded rudderfish are deeper-bodied, relatively, than pilotfish and the pelvic fins are relatively much longer than in the latter. There are 33–40 rays in the second dorsal fin vs. only 25–29 in pilotfish.

Habits. Banded rudderfish juveniles are common in summer and fall in Massachusetts and New York (Nichols and Breder 1927). They often follow sharks and other large fishes and may be found under jellyfish and drifting weed.

Breeding Habits. Reproduction occurs mainly in offshore waters south of Cape Hatteras. The season is continuous, or possibly in two parts, winter-spring and fall (Aprieto 1974).

Early Life History. There is no information on eggs or yolk-13–14 mm. They are fairly deep-bodied with a large head and concave snout that becomes straight. There is no occipital crest; a low orbital crest with a weak spine is present in early larvae, with smooth spines on preopercular margin and ridge until metamorphosis, after which two secondary spines form on the longest spine. Banded rudderfish have the highest second dorsal ray count (33–40) of any western Atlantic *Seriola* at sizes greater than 9 mm. There are five or six spots along the base of the dorsal fin against a background of smaller spots in early larvae and rows of spots along the dorsal and anal fin bases and at the posterior midline. Flexion occurs at 4.7–7.5 mm and metamorphosis at about 13 mm (Fahay 1983). Planktonic larvae are apparently carried along the Florida Current and Gulf Stream to reach northern limits as juveniles (Aprieto 1974). A few 35-mm specimens were taken in June and August in Indian River Inlet, Del. (de Sylva et al. 1962).

Juveniles of 15–17 mm show a distinct notch separating the first and second dorsal fins. Scales form at 20 mm along the posterior end of the lateral line. Marginal and lateral surface preopercular spines are present, which decrease in size and become overgrown by the preopercle with age.

At 17 mm, the body is banded with a distinct nuchal bar running from the eye straight to the dorsal fin origin and six black bands starting behind the head and ending on the caudal peduncle. The body is light brown or steely between the bars. Tips of the caudal lobes are a distinctive white (Aprieto 1974). This pattern persists to 220–260 mm (Ginsburg 1952a).

General Range. Atlantic coast of America from Halifax through the Gulf of Mexico to Santos, Brazil (Berry and Smith-Vaniz 1978).

Occurrence in the Gulf of Maine. Banded rudderfish are ordinarily rare visitors to the Gulf of Maine, and most of the ones seen there have been small individuals, made conspicuous by their cross-barred pattern. Two have been taken at

two in Portland Harbor; one on Nantucket Shoals; and several from Boothbay Harbor, the Sheepscot River, and at Gloucester (Bigelow and Schroeder). In the summer and fall of the years 1949–1951, large numbers of banded rudderfish were caught or observed in and around traps at Barnstable, Cape Cod Bay (Mather 1952). Several were taken by gill net near the Pilgrim Nuclear Power Station (Lawton et al. 1984). Small fish (3.8–17.8 cm), are regular summer visitors at Woods Hole.

ROUGH SCAD / *Trachurus lathami* Nichols 1920 / Saurel, Jurel /

Bigelow and Schroeder 1953:377 (as *Trachurus trachurus*)

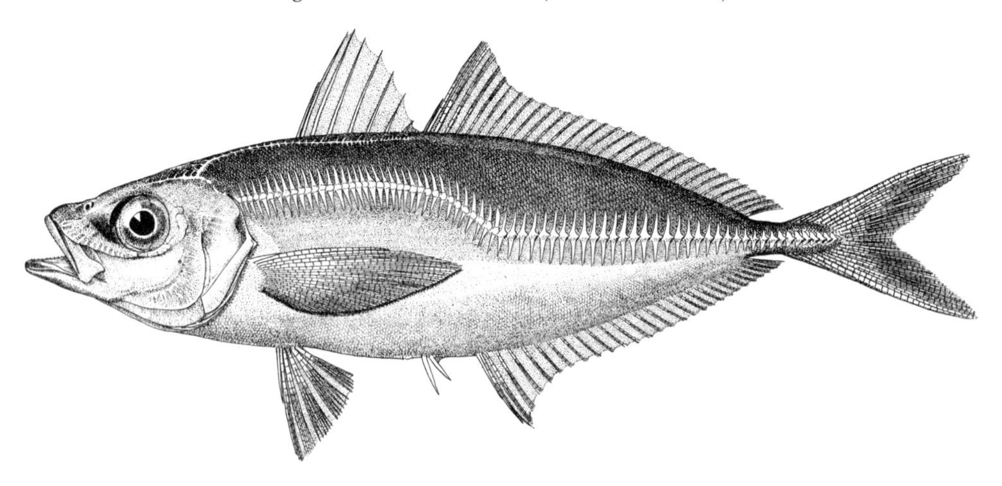

Figure 230. Rough scad *Trachurus lathami*, Newport, R.I., 151 mm. Drawn by H. L. Todd.

Description. Body slender, spindle-shaped with a very slender caudal peduncle (Fig. 230). Snout moderate, obtuse. Mouth medium-sized, well-inclined, and terminal or subterminal. Lower jaw subequal to upper or a little projecting; maxilla of moderate width, ending under anterior margin of eye; supramaxilla present (Berry and Cohen 1974). Teeth small, almost all in a single row in jaws, on head and shaft of vomer, palatines, and in a narrow band on tongue (Ginsburg 1952a). Eye with well-developed adipose eyelid. Spinous dorsal fin high, spines connected by membranes, anterior spines flexible. Dorsal and anal fins with moderate lobes and folds. Low, scaly sheath along bases of dorsal and anal soft rays. Two well-developed disconnected anal spines. Pectoral fins shorter than head length, reaching to soft dorsal fin origin in large specimens, shorter in small specimens. Pectoral girdle without papillae. Caudal fin deeply forked. Scales small, covering body except for a small area behind pectoral fins. Anterior part of lateral line curved moderately over pectoral fins, posterior part straight. Scales in anterior part transversely expanded but not keeled or spinous; scales in straight part scutelike, keeled and spinous. Accessory lateral line extending back to between eighth spine and fourth soft ray of second dorsal fin.

Meristics. Dorsal fin rays VIII, I, 28–34; anal fin rays II, I, 24–30; lateral line scales 68–77, 31–42 scales and scutes in curved lateral line, 33–39 scutes in straight lateral line; gill rakers 12–16 + 33–41 = 46–54 total; vertebrae 10 + 14 = 24 (Ginsburg 1952a; Berry and Cohen 1974 Berry and Smith-Vaniz 1978).

Color. Bluish green above, silvery or golden below, with a black spot on the edge of the gill cover above its rear angle on larger specimens.

Size. Maximum size 39.7 cm TL and 0.5 kg; common to 20 cm SL (Berry and Smith-Vaniz 1978).

Distinctions. Rough scad are distinguishable by having about 75 bony plates along the lateral line, compared with 30 or fewer in other Gulf of Maine jacks. They are somewhat deeper than mackerel scad but more slender than blue runner or crevalle jack.

Habits. Adults are found offshore, most commonly between 50–90 m (Klima 1971), usually on or near the bottom but probably also in the water column. They have been taken in salinities of 14–38.7 ppt (Mansueti 1960; Isaacson 1964; Franks et al. 1972) and temperatures of 14.4°–20°C (Franks et al. 1972).

Food. On the continental shelf off Rio Grande de Sul, Brazil (Saccardo and Katsuragawa 1995), prey consisted mainly of zooplankton, mostly *Euphausia similis,* which occurred in half the stomachs, and copepods *Calanoides carinatus* and *Calanus australis.*

Predators. The only records of predation on rough scad in the NEFSC survey of food habits were by smooth dogfish and bluefish (Rountree 1999).

Parasites. *Trachurus lathami* is the final host of a digenean trematode *Monascus filiformis* in the southwest Atlantic Ocean (Martorelli and Cremonte 1998); the first intermediate host is a bivalve, *Nucula obliqua;* the second intermediate hosts are chaetognaths and hydromedusae. A copepod, *Lernanthropus trachuri,* has been reported from the gills of Argentinean rough scad (Timi and Etchegoin 1996).

Breeding Habits. Rough scad are reported to spawn in the dry season (January to June) in the northwest waters of Trinidad (Manickchand-Dass et al. 1984) and in spring and summer, with a peak in spring (November) at water temperatures of 15°–17°C on the southeastern coast of Brazil (Saccardo and Katsuragawa 1995). In Brazil, first maturity occurs at about 115 mm TL at 2 years old, and all fish were mature by 155 mm TL and 4 years of age.

Early Life History. There is no information on eggs. Three larvae were taken in Indian River Inlet in June (de Sylva et al. 1962). Juveniles have been taken offshore, sometimes associated with jellyfish (Berry in McClane 1974). They have been collected in New Jersey at salinities of 29–30.5 ppt and temperatures of 14°–18.5°C (Milstein and Thomas 1976). Most larvae in Brazil were caught at 18°–25°C and 33.5–36.0 ppt (Saccardo and Katsuragawa 1995).

Juveniles have large mouths and small sharp teeth and are described as having numerous chromatophores, usually dispersed conspicuously between the posterior lateral line and anal fin base up to a size of at least 50 mm (Schekter 1972).

Age and Growth. Maximum age in Brazil was 8 years (Saccardo and Katsuragawa 1995). Length-weight relationships for fish 110–210 mm TL and 13–80 g from Brazil showed significant sexual dimorphism in winter-spring: $W = 0.000008431L^{3.0042}$ for females; $W = 0.00001136L^{2.9434}$ for males. The von Bertalanffy growth equation was $L_t = 258.97(1 - e^{-0.16\,(t + 1.85)})$.

General Range. Known from nearly all warm and temperate seas, sometimes common off the Florida Keys, in the Gulf of Mexico, and off southeastern Brazil (Saccardo and Katsuragawa 1995). Western Atlantic from St. Margaret Bay, N.S. (Scott and Scott 1988) to northern Argentina (Berry and Cohen 1974). Rare in the northern part of their range—New York, Rhode Island, and the Gulf of Maine.

Occurrence in the Gulf of Maine. A specimen of this rare fish was reported from Casco Bay, a second at Castine, and a third at Sandwich. MCZ specimens from the Annisquam River in Gloucester, Sandwich, and Provincetown were examined by Collette and Hartel (1988). Several specimens (identified as *Trachurus trachurus*) were reported from Georges Bank (Halliday and Scott 1969) and off the Scotian shelf in St. Margaret Bay (Scott and Scott 1988). Most of these were collected in late summer or autumn.

ACKNOWLEDGMENTS. A draft of the *Caranx hippos* account was reviewed by Richard S. McBride, and a draft of the entire carangid section was reviewed by William Smith-Vaniz.

POMFRETS. FAMILY BRAMIDAE

Bruce A. Thompson and Grace Klein-MacPhee

Pomfrets or sea breams, are circumglobal in temperate and tropical open ocean waters and are probably only encountered by longline fishermen working the outer edge of the Gulf of Maine. This distinctive family consists of seven genera with 19 or 20 species. Most species are laterally compressed and oval in outline with a single dorsal fin. The biology of many bramids is still poorly known and the remarkable changes in body morphology with growth plagues efforts to delineate the various forms. The only worldwide revision was by Mead (1972), but Thompson and Russell (1996) provided a recent review of Atlantic species.

Taractichthys longipinnis is the only bramid recorded from the Gulf of Maine. However, Gilhen and McAllister (1980) reported *Pterycombus brama* from off southern Nova Scotia (44°48′ N, 62°34′ W) so it might be found in the Gulf of Maine. *Brama brama* has been reported from the Grand Banks twice, in 1881 or earlier and in 1969 (Scott and Scott 1988), and Mead and Haedrich (1965) showed several records of *B. brama* in the Atlantic

KEY TO ADULT BRAMIDAE IN THE GULF OF MAINE REGION

1a. Dorsal and anal fins broadly expanded, no scales on dorsal or anal fins; modified scales form sheath along base of both dorsal and anal fins; dorsal fin insertion on or close to head (subfamily Pteraclinae) . . . . **Atlantic fanfish**

1b. Dorsal and anal fins not broadly expanded, scales present along most of the length of each rays in dorsal and anal fins; no sheath of modified scales along base of dorsal and anal fins; dorsal fin insertion distinctly posterior to head (subfamily Braminae) **2**

2a. Caudal fin with wide white posterior margin; dorsal and anal fin lobes greatly to moderately elongate (dorsal 33–50% [mean 39%] and anal 32–54% [mean 41%] of SL), with lobes becoming proportionally shorter with size; pelvic fins short, less than 9% of SL; precaudal grooves well developed; belly and breast distinctly rounded between pelvic fins . . . . . **Bigscale pomfret**

2b. Caudal fin black, no white posterior margin; dorsal and anal fin lobes not elongate: D 16–26% and A 14–15% of SL; pelvic fins longer, more than 10% of SL; precaudal grooves absent; belly and breast more flattened between pelvic fins . . . . **Atlantic pomfret**

BIGSCALE PROMFRET / *Taractichthys longipinnis* (Lowe 1843) /

Bigelow and Schroeder 1953:361–362 (as *Taractes princeps*)

Figure 231. Bigscale pomfret *Taractichthys longipinnis.*

Description. Body massive, compressed, oval, about one-half as deep as long to base of tail (Fig. 231). Dorsal and anal fins long, scythe-shaped with long rays in front, followed by lower rays, detached at their distal ends but covered by scaly skin at base, both fins proportionally shorter in larger specimens. Caudal peduncle relatively narrow; caudal fin deeply lunate. Well-developed precaudal grooves present. Long pectoral fins originate just in front of very small pelvic fins. Dorsal profile of head strongly arched and rounded, producing a blunt snout. Eyes large, oval, taller than wide. Lateral line present in all specimens examined by Thompson and Russell (1996), although Mead (1972) stated that the lateral line was absent. Large scales cover dorsal and anal fins, body scales varying greatly in size, largest along sides of body and smallest on back, breast, and fins. Scales vary greatly in shape, basically rhomboid, but with exposed margins concave, convex, notched, or straight. Scales in juveniles with a ridge and spine oriented horizontally across scale; these structures lost in larger individuals. Gill rakers lathlike (Mead 1972). Several bands of small, pointed, recurved teeth on jaws, some still exposed with mouth closed. Inner, anterior premaxillary and mandibular teeth longest. A small, narrow band of teeth on palatines (Mead 1972).

Meristics. Dorsal rays 33–38, the first 5 or 6 rays thickened, graduated, and unbranched, the rest branched; anal rays 27–30, similar to dorsal fin; pectoral rays 20–22; gill rakers on first arch 1–3 + 6–9, most often 2 + 7; vertebrae 19–21 + 25 or 26 = 44–47 (Mead 1972; Thompson and Russell 1996).

Color. In fresh specimens the body is suffused with dark crimson or violet, snout pale to flesh-colored (Fowler 1956). Life colors fade very rapidly with death. Overall, body and fins are black. Free ends of dorsal and anal fin rays are pale or silver. Some specimens are lightly speckled with silver on the body. Caudal fin is black with a wide white concave posterior margin. Fishermen report that the eye is red when the fish is taken alive.

Size. The largest *Taractichthys longipinnis* reported was 852 mm SL (Wheeler 1962; from Great Britain). Bigelow and Schroeder (1929) reported a 618-mm SL specimen from the Gulf of Maine and Thompson and Russell (1996) reported on 17 specimens, 367–740 mm SL, from the Gulf of Mexico.

Distinctions. The elevated, sickle-shaped dorsal and anal fins with a series of low rays distinguish this species from others in the Gulf of Maine. The black, compressed, oval body is distinctive. It is the only Gulf of Maine species with a wide white band across the posterior margin of the caudal fin; this band is absent from the other two bramids found in the region. Small pelvic fins and large precaudal grooves are also distinctive.

Habits. Adults are typically oceanic and occur down to considerable depths, but occasionally are seen at the surface and in shallow waters. The lower temperature limit is probably about 10°C. Mead (1972) suggested that they might school. They are commonly captured in the Gulf of Mexico as by-catch from pelagic longlines, mostly between November and March.

Food. No published information but fishes were found in several specimens from the Gulf of Mexico (B. A. Thompson, unpubl.).

Predators. Juvenile bigscale pomfret have been found in stomachs of tunas and lancetfishes, *Alepisaurus* by Mead (1972).

Reproduction. There is little published information but a summer spawning season is suggested based on examination of histological sections of ovaries (Thompson and Russell 1996). There is no published information on eggs. The smallest described larvae was 5.1 mm. The yolk sac is retained until about 7 mm, during which time fin ray formation is complete in the dorsal and anal fins, pectoral and pelvic fin rays form, and notochord flexion occurs. There is a diagnostic fringelike series of short uniform spines along anterior and ascending arms of the preopercle. By 6.5 mm a few melanophores are seen on the gular fold, and there is a sharp demarcation between the white caudal fin and peduncle and the dark anterior body. Body squamation is complete by 8–9 mm and a few large recurved canine teeth are present on the jaws. Major changes in body form occur between 20 and 200 mm.

General Range. This species is found on both sides of the Atlantic in tropical and temperate regions. In the western Atlantic, they range from Nova Scotia to the Gulf of Mexico (Thompson and Russell 1996) and in the eastern Atlantic from Great Britain and Madeira south to South Africa (Mead 1972). Early records of this species from the Pacific and Indian oceans are considered to be *Taractichthys steindachneri* (Mead 1972).

Occurrence in the Gulf of Maine. Bigelow and Schroeder (1929) reported (as *Taractes princeps*) a 618-mm SL (830 mm TL) specimen (MCZ 31598) taken 10 January 1928 by the fishing schooner *Wanderer* while "line trawling" on Browns Bank about 81 km south of Cape Sable, N.S. A 700-mm TL adult was caught by a Canadian swordfish longliner *Dorothy and Gail* on northeast Georges Bank (Scott and Scott 1988).

Importance. This species is a rare straggler in the Gulf of Maine and has no commercial or recreational importance although it is used for food in the eastern Atlantic.

PORGIES. FAMILY SPARIDAE

GRACE KLEIN-MACPHEE

Structure of the fins is essentially the same in the family Sparidae as in sea basses; both spiny and soft portions of the dorsal are well developed and pelvic fins are situated below the pectorals. The edge of the gill cover does not end with a sharp spine in porgies but is rounded or at most bluntly angular, and the maxilla is sheathed and hidden by the preorbital bone when the mouth is closed. Long, pointed pectoral fins are likewise characteristic of the family; spiny and soft portions of the dorsal fin are continuous, and the soft-rayed anal fin is about as long as the soft part of the dorsal. Of 29 genera and about 100 species in the family (Nelson 1994), scup and sheepshead are the two species that occur in the Gulf of Maine. Only scup are of importance in the Gulf and recent information on the species has been summarized by Steimle et al. (1999e).

KEY TO GULF OF MAINE PORGIES

1a. Outline of caudal fin deeply lunate, with sharp corners. . . . . **Scup**
1b. Outline of caudal fin only slightly concave, with rounded corners . . . . **Sheepshead**

SHEEPSHEAD / *Archosargus probatocephalus* (Walbaum 1792) / Bigelow and Schroeder 1953:416–417

Description. Body deep, compressed (Fig. 232). Eyes located high on sides of head. One long dorsal fin, anterior two-thirds tion of dorsal, under which it stands. Dorsal and anal fins depressible into deep grooves. Pectoral fins long, pointed. Pelvic

Figure 232. Sheepshead *Archosargus probatocephalus.* North Carolina, 192 mm. Drawn by H. L. Todd.

emarginate as in scup, with rounded corners instead of pointed ones. Scales large.

Meristics. Dorsal fin rays XI or XII, 11–13; anal fin rays III, 10 or 11; pectoral fin rays 15; lateral line scales about 62 (Scott and Scott 1988).

Color. The sides have seven broad, dark brown or black crossbars on a gray or greenish yellow ground, instead of being plain-colored like the sides of scup.

Size. Although Bigelow and Schroeder and others have reported lengths up to 91 cm, the all-tackle world record sheepshead, caught in New Orleans in April 1982, was only 596 mm SL (730 mm TL) and weighed 9.63 kg (Schwartz 1990b; IGFA 2001).

Distinctions. Sheepshead differ from scup by the fact that the caudal fin is not so deeply emarginate and has rounded corners instead of pointed ones; the dorsal spines are alternately stout and slender; the second anal spine is much stouter than that of scup; the dorsal profile of the head is steeper; the snout is blunter; and the teeth are much broader. Furthermore, the body of sheepshead is noticeably thicker, the back is rounded, and the sides show seven broad dark brown or black crossbars on a gray or greenish yellow ground, instead of being plain like scup.

Food. Sheepshead have a diverse diet in the Mississippi Sound, including more than 113 species of polychaetes, mollusks, crustaceans, and fishes (Overstreet and Heard 1982). They undergo an ontogenetic shift in diet, adding larger and more robust prey to the small invertebrate species. This increase in durophagus prey is associated with increased oral-jaw crushing force produced by the adductor mandibulae muscle complex (Hernandez and Motta 1997).

Age and Growth. Schwartz (1990b: Fig. 1) presented a figure for the length-weight relationship of 282 North Carolina sheepshead: $\log W = -4.5287 + 3.0446 \log SL$. Based on scales, Schwartz estimated the maximum age of North Carolina specimens at more than 8 years.

General Range. Sheepshead are found on the Atlantic and Gulf of Mexico coasts of the United States from Texas to Cape Cod and north to the Bay of Fundy as strays. They were formerly abundant as far north as New York and not uncommon around Woods Hole; they are still common to the south, although catches in North Carolina decreased drastically around 1918 and have only increased somewhat since 1981 (Schwartz 1990b). They have been decidedly rare east of New York for many years, although a number, about 150 mm in length, were taken off Onset at the head of Buzzards Bay in late August 1951.

Occurrence in the Gulf of Maine. A 63-mm SL specimen was collected in Shelburne Harbour, N.S. (43°41′ N, 65°20′ W), in July 1973 (Gilhen et al. 1976). The only previous record of sheepshead north or east of the elbow of Cape Cod is from Saint John Harbour, N.B. (Cox 1895), but positive identification cannot be verified (Scott and Scott 1988).

SCUP / *Stenotomus chrysops* (Linnaeus 1766) / Porgy / Bigelow and Schroeder 1953:411–416 (as *Stenotomus versicolor*)

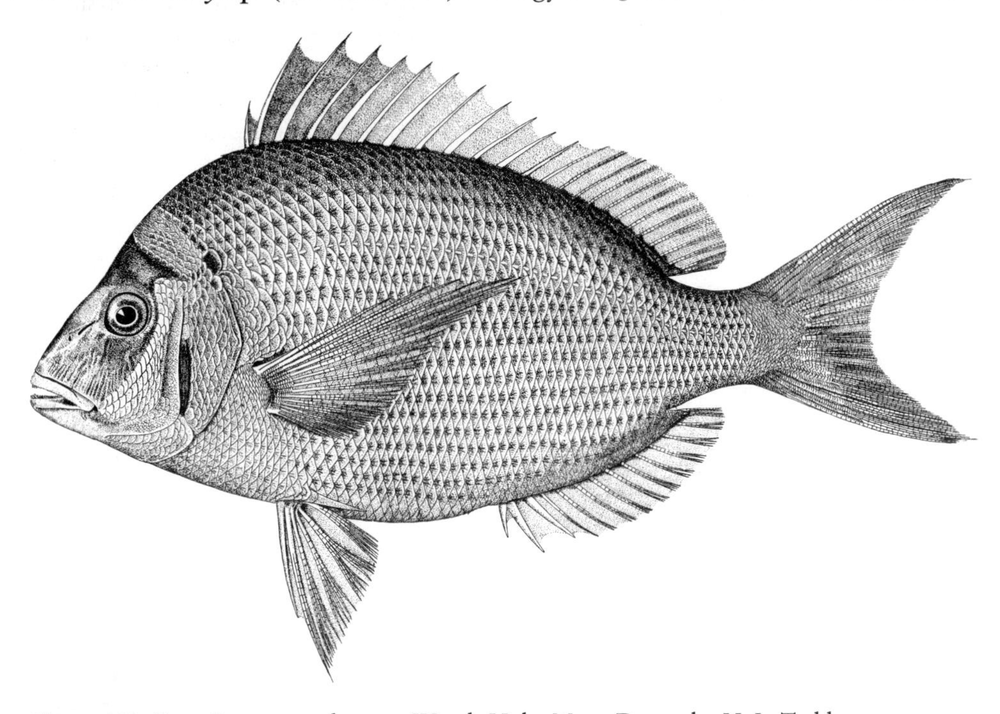

Figure 233. Scup *Stenotomus chrysops.* Woods Hole, Mass. Drawn by H. L. Todd.

Description. Body ovate-elliptical, about half as deep as long, laterally compressed (Fig. 233). Mouth small. Teeth small, narrow elliptical, almost conical; molars in two rows. Eyes situated high up on side of head. Margins of gill covers rounded. One long dorsal fin originating over pectoral fins, preceded by a forward-pointing spine; spiny and soft parts of fin continuous. Dorsal fin moderately high, first spine much shorter than others, rear corner rounded; fin fits in groove along midline of back. Anal fin under soft portion of dorsal fin nearly as long as soft part of dorsal, almost even in height from front to rear, but with first spine shorter than others. Anal fin depressible in conspicuous groove, like dorsal. Pectoral fins very long, reaching to below soft part of dorsal fin, sharply pointed, with slightly concave lower rear margins. Pelvic fins situated below pectorals, moderate-sized. Caudal fin deeply concave with sharp corners. Lateral line complete, slightly arched. Scales large, firmly attached.

Meristics. Dorsal fin rays XII, 12; anal fin rays III, 11 or 12; lateral line scales 46–55; gill rakers 12–20; vertebrae 10 precaudal + 14 caudal (Miller and Jorgenson 1973).

Color. Dull silvery and iridescent, somewhat darker above than flecked with light blue, with a light blue streak following the base of the dorsal fin. Head silvery, marked with irregular dusky blotches; belly white. Dorsal, caudal, and anal fins dusky, flecked with blue; pectoral fins of a brownish tinge; pelvics white and bluish, and very slightly dusky; iris silvery; pupil black. Live fish can display black, vertical blotched bands along midbody.

Distinctions. Scup are easily recognizable by the fact that the spiny portion of the dorsal fin is considerably longer and higher than the soft-rayed portion, which, with its deeply lunate caudal fin, separates them from all other Gulf of Maine fishes of similarly deep and compressed bodies. They differ from butterfish in that the dorsal profile of the rather short head is slightly concave instead of convex and the scales are rather large, thick, and firmly attached, rather than small, thin, and easily detached as in butterfish. Small teeth and lunate pointed tail separate scup from their closest local relative, sheepshead.

Size. Scup reach a maximum length of about 40 cm (Gabriel 1998). The all-tackle world record scup, 2.06 kg, was caught in Nantucket Sound in June 1992 (IGFA 2001).

Habits. Scup occur from 2 to 180 m and are rarely caught any

the coast so closely that a line drawn 8–10 km beyond the outermost headlands would probably enclose the great majority of the total population at that time of year. They occur inshore in spring and summer and move offshore in fall. Scup are inshore from early April at the mouth of Chesapeake Bay and from early May northward to southern Massachusetts. Most of them withdraw from the coast late in October, although a few fish linger through November and an occasional one into December even as far north as the vicinity of Woods Hole. Scup usually congregate in schools. The young come close to land in less than 1 m of water, but large fish are seldom caught in water shallower than 2–4 m (occasionally at the surface) or deeper than 27–36 m in summer. They prefer smooth to rocky bottom, which results in a very local distribution. They appear to avoid water temperatures below 7°C and are found in greatest abundance at temperatures of 13°–16°C (Fritz 1965). They are so sensitive to low temperatures that considerable numbers (both large ones and small) have perished during sudden cold spells in shallow water. The critical thermal maximum for juvenile scup is dependent on acclimation temperature and varies from 30.2° to 35.6°C at acclimation temperatures of 14.8°–22.2°C. During extreme thermal stress, scup make rapid darting movements to the surface, experience complete loss of equilibrium, stop moving, and die (Everich and Gonzalez 1977).

Scup are schooling fish and, based on the pound net catches, probably school by size (Morse 1978). They appear to school more closely at night. Catch records showed that most fish have been collected near midnight and the fewest around noon (Fritz 1965). Steady swimming in schools is powered by a combination of red and pink muscles (Coughlin and Rome 1996, 1999). Pink muscle is used to augment red muscle power production at higher swimming speeds, allowing a higher aerobically based steady swimming speed than is possible using red muscle alone.

Scup produce sound in response to electrical stimulation and when kept together in tanks, but not when solitary. The sounds, which have been described as guttural grunts, knocks, and stridulatory rasping, are probably produced by the simple thin-walled swim bladder and the upper and lower incisors (Fish and Mowbray 1970).

Food. Scup are bottom feeders, seldom rising far above the bottom, preying on cnidarians, squids, polychaetes, crustaceans, and fishes (Maurer and Bowman 1975; Bowman and Michaels 1984; Bowman et al. 2000). Smaller scup, up to 25 cm TL, eat a larger proportion of cnidarians, polychaetes, amphipods, and mysids, whereas larger scup (over 26 cm TL) consume more squids and fishes. Fish prey include *Ammodytes dubius* and butterfish (Bowman et al. 2000). Adult scup, as many other fishes, may cease feeding during spawning time, although there is a rod-and-reel fishery for spawning scup in May in Vineyard Sound (J. Galbraith, pers. comm., April 1994).

Juvenile scup in Narragansett Bay feed on polychaetes, mysids and other crustaceans, mollusks, and fish eggs and larvae (Michaelman 1988). Polychaetes and crustacea made up more than 50% of the diet by dry weight and averaged 88% of the identifiable seasonal diet. Crustaceans made up 41% and polychaetes 47% of the diet during the summer. Fish larvae accounted for 6–19% of the diet in August. Important species consumed were *Nephtys, Nereis, Pherusa affinis* (polychaetes); *Leptochirus* (amphipod); *Neomysis* (mysid shrimp), and *Ceri-anthiopsis* (coelenterate). Juvenile scup collected in Mid-Atlantic and southern New England coastal waters on groundfish surveys ate similar food (Bowman et al. 1987). They were divided into two size-groups, 6–10 cm and 11–15 cm. Both groups fed on polychaetes (31 and 32%, respectively). The scup measuring 6–10 cm ate amphipods (16% net weight); decapods (10%), mainly *Cancer irroratus;* mysids (9%), mainly *Neomysis americana,* and copepods. Larger scup ate mollusks (18%), mainly squids; amphipods (6%), and decapods (2%). Food habits of scup on the outer continental shelf (Sedberry 1983), were similar: amphipods and polychaetes were the two major components of the diet, with amphipods decreasing in importance with size and polychaetes increasing.

Juvenile scup in Narragansett Bay were daytime feeders. Feeding energetics of juvenile scup (average wet weight 67 g) fed *Crangon* shrimp showed that the daily ration was 3.99% dry weight. They ate an average of 5% of their wet weight per day. The exponential estimate of gastric evacuation rate (stomach contents in percent dry weight) was $0.34 \cdot h^{-1}$. The metabolic rate of scup was 0.23 ml $O_2 \cdot g^{-1}$ wet weight at 20°C for 1-year-old scup. Their metabolic expenditure was 1.86 dry weight·day^{-1}, and their estimated rate of growth was 0.84% dry weight·day^{-1} (Michaelman 1988).

Predators. Scup are eaten by elasmobranchs and bony fishes throughout their range, including 19 species, of which the three most frequent predators are spiny dogfish, smooth dogfish, and bluefish (Rountree 1999).

Parasites. The parasitic branchiuran *Argulus intectus* has been found on scup (Yamaguti 1963).

Breeding Habits. Median length at maturity is 15.5 cm and 15.6 cm for females and males, respectively (O'Brien et al. 1993). Male and female scup mature at age 2 (Finkelstein 1969b), corresponding to a length of about 16 cm (Finkelstein 1969a). Maturity is complete by age 3 at 21 cm TL (Gabriel 1998). Scup spawn once a year in coastal waters from May to August, with the peak during May and June (O'Brien et al. 1993). They spawn close to shore during daylight hours, usually in the morning (Ferraro 1980), and at temperatures of 10°–18°C on the bottom and 13°–23°C at the surface (Wheatland 1956). It appears that spawning does not occur over the continental shelf, as MARMAP surveys from 1977 to 1987 collected only 14 larvae from three stations near the mouth of Narragansett Bay (Able and Fahay 1998). Mean fecundity of scup 17.5–23.0 cm FL is about 7,000 eggs (Gray 1990).

Scup eggs are buoyant, transparent, spherical, and rather

small, 0.85–1.15 mm average diameter (Wheatland 1956). Pigmentation appears on the embryo at the 15- to 20-myomere stage as black and yellow spots scattered sparsely over the embryo and oil globule. As hatching time approaches, yellow chromatophores aggregate to form heavily pigmented areas (Hildebrand and Schroeder 1928). Eggs resemble those of silver hake, windowpane flounder, and fourspot flounder, but the oil globule is much smaller, and the pigmentation on the embryo and oil globule is sparse, whereas it is heavier and darker on hake (Colton and Marak 1969). Pigmentation is also heavier and darker on windowpane embryos. The fourspot egg is a bit larger with a mean size of 1.04 mm.

Incubation occupies only 70–75 h at 18°C, 44–54 h at 21°C (Griswold and McKenney 1984), and 40 h at 22°C (probably 2–3 days in the June temperatures of Massachusetts Bay), judging from the spawning season at Woods Hole. It is not likely that development can proceed normally in water colder than about 10°C. At hatching larvae are about 2 mm long, the eyes are not pigmented, and the mouth is not functional. The head projects slightly beyond the anterior end of the yolk sac, the oil globule is in the posterior end. The larvae have a small group of melanophores scattered over the dorsal and dorsolateral part of the body. There are small areas of yellow pigment on the dorsal and lateral part of the head, above the vent, and opposite the vent dorsally; a transverse band halfway from the vent to the posterior end of the body extends from the base of the pelvic finfold onto the dorsal finfold. There are black and yellow chromatophores on the oil globule (Kuntz and Radcliffe 1917).

The yolk is fully absorbed within 3 days at 18°C, when the larva is about 2.8 mm long, and there is then a characteristic row of black pigment spots along the ventral margin of the trunk. Fin rays first appear at 2.9 mm SL in the pectoral fin, at 4.3 mm in the caudal fin, at 5.5 mm in the dorsal and anal fins, and at 8.8 mm in the pelvic fin. The full complement of fin rays is present by 14 mm SL. The dorsal row of preopercular spines was first noted at 4.1 mm and increased gradually until they formed a serrate edge at 16.9 mm. Ossification begins in the skull at 6.1 mm and is completed by 18–19 mm, except for some sutures in the skull. Scales first appear between 9.9 and 10.8 mm, and larvae are completely scaly by 12.3–13 mm (Griswold and McKenney 1984). Larval pigmentation consists of a small group of melanophores over the dorsal area of the head, the dorsolateral aspects of the body along the myotomes, and on the oil globule. After yolk-sac absorption, the pigment pattern changes. There are a few areas of pigment on the head and lateral trunk, a conspicuous spot on the anterior edge of the vent, and a series of pigment spots along the ventral region of the tail. Pigmentation increases on the trunk and then turns into a barred pattern characteristic of the juveniles by 18.7–19 mm (Colton and Marak 1969; Griswold and McKenney 1984). At 25 mm, the pectoral fins assume their pointed outline and the caudal fin is slightly forked, but the pelvic fins are still very small and the body very slender.

Length-weight relationships of larval scup reared in the lab-

New England waters, juveniles of 5.1–7.6 cm, evidently the product of that season's spawning, have been taken in September; they are 6.4–8.3 cm long in October and may be as long as 10.1 cm at Woods Hole in November. Apparently young scup grow very little during the winter, for many of 10.1 cm are seen in the spring, probably the crop of the preceding season.

Age and Growth. Finkelstein (1969a) investigated growth of scup in Long Island waters using fish-length to scale-length relationships. He determined the von Bertalanffy growth functions by sex, where length is FL in mm: $L_t = 342.5(1 - e^{-0.2688(t_n + 0.40531)})$ for males; $L_t = 374.1(1 - e^{-0.2247(t_n + 0.47047)})$ for females.

Scup live for 13–20 years (Finkelstein 1969b; Hamer 1970; Gabriel 1998). Smith and Norcross (1968) had difficulty aging fish older than 2 years. Length-at-age data are given in Penttila et al. (1989) (Fig. 234).

General Range. Scup live in the western North Atlantic, from Sable Island Bank, N.S., to Cape Hatteras, N.C., but are infrequent north of Cape Cod (Scott and Scott 1988).

Occurrence in the Gulf of Maine. Although scup are among the most familiar shore fishes right up to the elbow of Cape Cod, few find their way past Monomoy Point into the colder waters of the Gulf of Maine. Massachusetts Bay records are from Rockport, Swampscott, and Cohasset Narrows (Collette and Hartel 1988). Northern records are from Eastport, Maine; St. Mary's Bay, St. Margaret Bay, and Sable Island Bank, N.S.; and the St. Croix River estuary and Murr Ledges, south of Grand Manan, Bay of Fundy, N.B. (Leim and Scott 1966; Scott and Scott 1988).

The first definite mention of scup caught north of Cape Cod is Storer's statement that one was taken at Nahant in 1835, and another in 1836, but that they had never been seen there before. Possibly these and one picked up dead at Cohasset in 1833 were survivors of a smack load that had been liberated in Boston Harbor a year or two earlier, and a similar plant was made in Plymouth Bay in 1834 or 1835. There is no reason to suppose that these planted fish established themselves. When the practice of setting mackerel nets outside Provincetown Harbor was first adopted (about 1842), a few scup were taken in them from year to year. Fish were caught in Cape Cod Bay yearly and between Boston and Cape Ann during the period from 1860 to 1867; a number were taken in a weir on Milk Island near Gloucester in 1878. There were a few scup in northern Massachusetts waters in most years (or terms of years) down to the first decade or so of the twentieth century, alternating with other years (or terms of years) when only an occasional fish was taken.

The cataclysmic shrinkage that took place in the stock of scup off southern Massachusetts between 1896 (prior to which the annual catch had usually run from 1 to 3 million lb) and

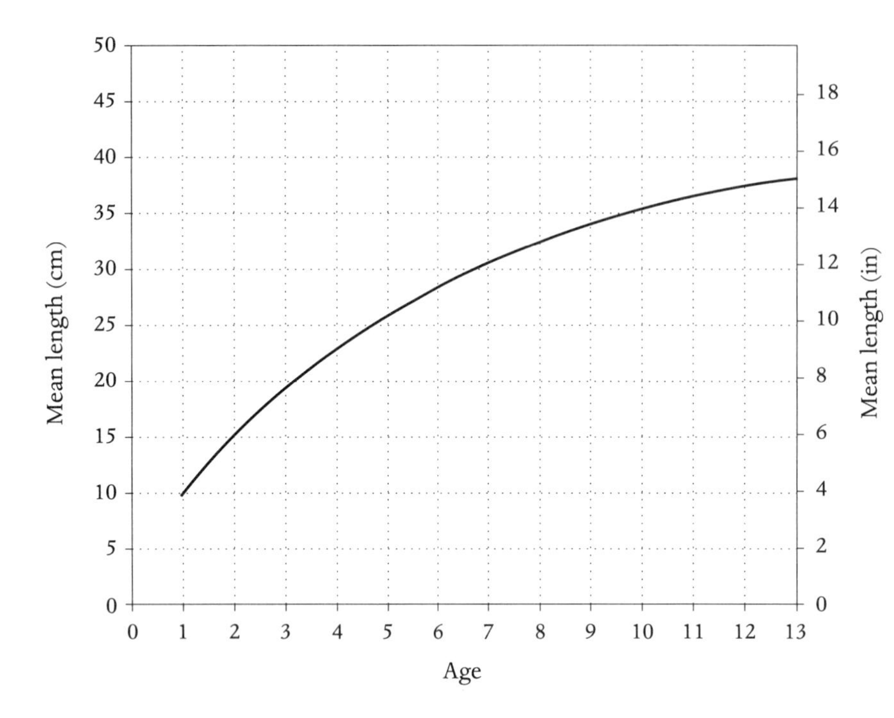

Figure 234. Length-age curve for scup *Stenotomus chrysops,* all areas combined. (From Penttila et al. 1989:25.)

have involved scup in Cape Cod Bay as well for none was reported there from 1907 through 1911 or from 1918 to 1920, but there was an unusually large run there in 1917. However, 1908, 1909, and 1919 were good scup years for the north shore of Massachusetts—good, that is, for those northerly waters, suggesting that when conditions are favorable, a small independent population may be present there. Perhaps the fact that larger catches than usual are not always registered in both these regions in the same year is also an indication.

No scup were reported from Essex County for 1919, 1928, or 1930, nor were enough taken in Cape Cod Bay in those years to cause any local comment. Lawton et al. (1984) reported 245 individuals taken by otter trawl and several in gill nets near the Pilgrim Nuclear Power Station at Plymouth in western Cape Cod Bay. Scup are taken sporadically in Cape Cod Bay off Wellfleet on the bay side (J. Galbraith, pers. comm.). It is not known whether they come through the Cape Cod Canal or around the Cape.

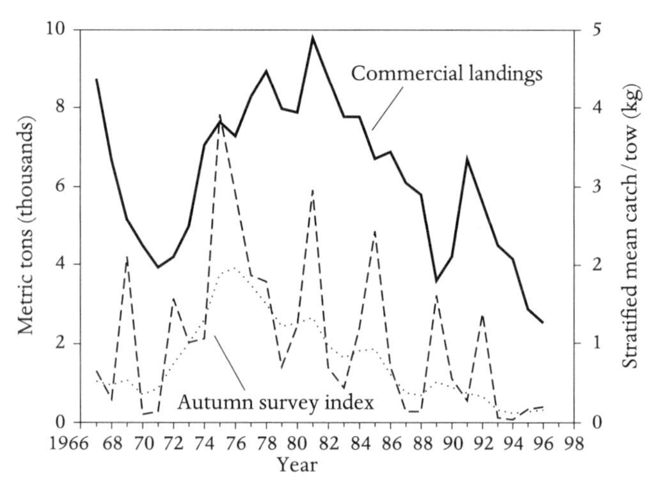

Figure 235. Commercial landings from Massachusetts to Virginia and NEFSC autumn survey index for scup *Stenotomus chrysops* from 1966 to 1996. (From Gabriel 1998.)

Seasonal Migrations and Movements. Scup are summer migrants along the New England coast (Pearcy and Richards 1962; Richards 1963b). They move into estuaries with bluefish, kingfish, mullet, weakfish, and butterfish, and are the most important migrant species in both Long Island Sound (Richards 1963b) and Narragansett Bay (Oviatt and Nixon 1973).

Migration patterns are seasonal and have both a north-south and an inshore-offshore component. Scup winter in offshore waters between New Jersey and North Carolina. In spring they migrate northward and inshore to New Jersey, New York, and southern New England. They remain there until fall, when a reverse migration occurs (Finkelstein 1971). Scup have been taken during winter in depths of 81–126 m off southern New England in numbers large enough to show that part of the northern contingent of the species simply moves offshore in autumn, to come inshore again in spring. Differences in locations of the largest catches in cool vs. warm winters make it likely that a preference for water at least as warm as about 21°C is the factor that determines how far seaward the scup move off any part of the coast in any particular winter.

Importance. Scup are never plentiful enough anywhere north of the elbow of Cape Cod to be of importance either commercially or to the angler, but they are an important food fish to the west and south, where they are plentiful. A summary of scup landings from Massachusetts to Virginia from 1879 to 1973 was given in Morse (1978), and commercial landings from 1966 to 1996 are shown in Fig. 235.

Scup, black sea bass, and summer flounder are harvested by several types of gear in the Mid-Atlantic Bight and southern New England (Shepherd and Terceiro 1994). For each of the three species, otter trawls produce more than 60% of total landings. The commercial trawl fishery is primarily in winter, generally targeting summer flounder and, less frequently, scup. Significant landings of scup are also made by pound nets, floating trap nets, and fish traps.

Historical catch statistics for the trap net fishery for Cape Cod and Massachusetts bays are summarized in Bigelow and Schroeder. A popular account of the Narragansett Bay trap fishery is given by Page (1978). Trap fishing still continues in the bay but the number of traps has decreased from a high of 283 in 1910 to only a few. Sissenwine and Saila (1974) investigated the possible effects of dredge spoil disposal on the scup fishery in Narragansett Bay but did not find anything that could not be attributed to natural population cycles of abundance and scarcity.

Annual commercial landings fluctuated between 18,000 and 27,000 mt (including the distant-water fleet) between 1953 and 1963, but declined to about 4,000 mt in during the early 1970s (Gabriel 1998). Beginning in the early 1970s, commercial catches increased steadily, reaching a peak of 9,800 mt in 1981 (Fig. 235). Thereafter, commercial landings generally continued to decline, except for 1990–1991, to a low of 2,179 mt in 1997.

Spawning stock biomass declined steadily from 1990 to a record low in 1995–1996 (Gabriel 1998). Recruitment at age 0 also declined to a record low level in 1996. Fishing mortality rates have been very high since 1990, far in excess of biological reference points. The stock of scup is overexploited and at a low biomass level. The age structure of the stock is highly truncated, which is likely a reflection of prolonged high fishing mortality.

By-catch and discard mortality is at times very high for scup (Kennelly 1999). Discard rates measured by observers in the trawl fishery in the offshore Mid-Atlantic and southern New England fishery showed variability in time and space, but one area off Long Island was consistently high, averaging 319 $lb \cdot h^{-1}$. The overall average throughout the region was 37.5 $lb \cdot h^{-1}$. This area appears to be frequented by migrating juveniles, which are particularly vulnerable.

A 16-year time series of research trawl catches, commercial landings, and effort data was used to evaluate two areas protected from mobile gear fishing off Cape Cod and assess effects of the spring otter trawl fishery for longfin squid on local abundance of finfish frequently taken as by-catch (Cadrin et al. 1995). Catch rates among a seasonal closure, a permanent closure, and adjacent waters open to mobile gear fishing were compared. Winter flounder and scup were more abundant in the two protected areas. Decreased local density of finfish in open areas was not related to inshore spring squid trawling effort or landings. Regional trawl effort on Georges Bank and in southern New England did have significantly negative effects on local finfish density.

The fishery is managed under Amendment 8 to what is now the Summer Flounder, Scup, and Black Sea Bass Fishery Management Plan. Management measures include moratorium permits, gear and minimum-size restrictions, commercial quotas, and recreational harvest limits (Gabriel 1998).

Recreational Fishery. Porgy, as they are commonly called along that part of the coast, are also a favorite with anglers, for they bite readily and are a good pan fish. Porgy bite very greedily throughout the summer on clams, bits of crab, and sea worms (*Nereis*), as do immature fish throughout their stay. Many are caught on hook and line for home consumption. The recreational fishery for scup peaks during spring and fall, when fish are found in estuaries and coastal waters (Shepherd and Terceiro 1994). Recreational catches have accounted for 20–50% of the total annual catch since 1990. The 1995 recreational catch of 600 mt and the 1996 catch of 1,000 mt were the lowest in the 1979–1996 time series (Gabriel 1998).

Utilization. Jhaveri et al. (1984) compared chemical composition and protein quality of several underutilized species. Protein quality of scup was superior to the other fishes compared and to squids, but cholesterol level was somewhat higher and mineral values were variable.

ACKNOWLEDGMENTS. Drafts of the sparid section were reviewed by John Galbraith and Mark Terceiro.

CROAKERS, DRUMS, AND WEAKFISHES. FAMILY SCIAENIDAE

GRACE KLEIN-MACPHEE

In croakers both the spiny and the soft portions of the dorsal fin are well developed (either separately or as one continuous fin). Croakers are readily distinguished from sea basses (Serranidae), porgies (Sparidae), and wrasses (Labridae) by the fact that their anal fin has only one or two spines instead of three and is much shorter than the soft portion of the dorsal; from redfishes (Scorpaenidae) and sculpins (Cottidae) by their rela- jacks (Carangidae) by their stout caudal peduncles and rounded or only slightly concave caudal fins. Most of them produce loud drumming sounds by rapid contractions of certain abdominal muscles against the gas-filled swim bladder (hence the common names "croaker" and "drum"), with kingfish as an exception. Croakers also have very large sagittal otoliths associated with enhanced hearing. The Sciaenidae

which 22 genera and 58 species are known from the western Atlantic Ocean (Chao 1978) and 4 species occasionally extend north into the Gulf of Maine.

KEY TO GULF OF MAINE SCIAENIDS

1a. No barbel on chin 2
1b. Chin bears one or more barbels 3
2a. Body only about one-quarter as deep as it is long (to base of caudal fin); anterior profile of head sloping only moderately; snout pointed; no dark spot behind upper corner of gill opening . . . . **Weakfish**
2b. Body about one-third as deep as it is long (to base of caudal fin); anterior profile of head sloping steeply; snout blunt; dark spot present close behind upper corner of gill opening . . . . **Spot**
3a. Several barbels on chin; snout ends about even with front of lower jaw; cheek smooth . . . . **Black drum**
3b. Only one barbel on chin; snout projects considerably beyond lower jaw; cheek with two short, toothlike serrations . . . . **Northern kingfish**

WEAKFISH / *Cynoscion regalis* (Bloch and Schneider 1801) / Squeteague, Sea Trout, Gray Trout /

Bigelow and Schroeder 1953:417–423

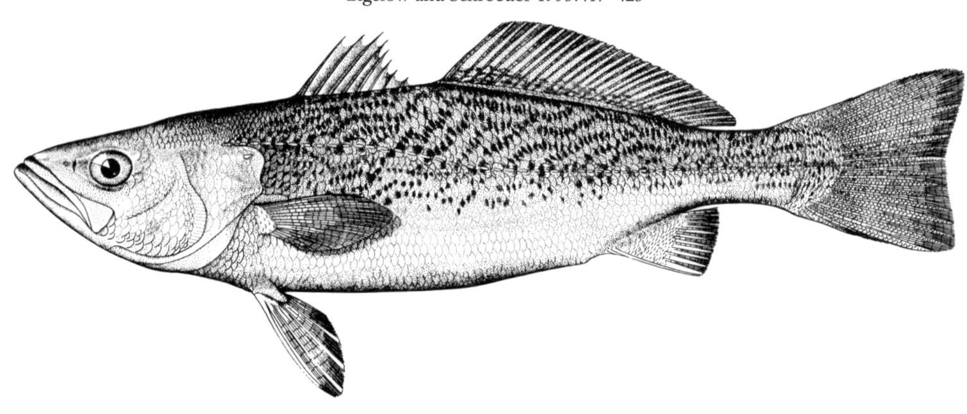

Figure 236. Weakfish *Cynoscion regalis.* North Carolina. Drawn by L. E. Cable.

Description. Body slim, shapely, about four times as long as deep (to caudal fin base), only slightly compressed with rather stout caudal peduncle (Fig. 236). Head about one-third as long as body, snout moderately pointed, mouth large, with projecting lower jaw. Upper jaw armed with two large canine teeth; narrow bands of teeth on sides of upper and lower jaws. First dorsal fin triangular, originating a little behind pectoral fins. Second dorsal originating close behind first, more than twice as long as first and roughly rectangular. Anal fin less than half as long as second dorsal. Pelvic fins below pectorals, which they resemble in size and pointed outline. Caudal fin moderately broad, slightly concave in outline. Body covered with moderate-sized scales. Lateral line distinct and slightly arched.

Meristics. Dorsal fin rays X, I, 24–29; anal fin rays II, 10–13; lateral line scales 76–86; 4 or 5 gill rakers on upper arch and 10–12 on lower (Tagatz 1967); 13 precaudal + 12 caudal = 25 vertebrae (Miller and Jorgenson 1973).

Color. Dark olive green above with back and sides variously burnished with purple, lavender, green, blue, gold, or copper and marked with a large number of small black, dark green, or bronze spots, which are vaguely outlined and more or less run together, especially on the back, forming irregular lines that run downward and forward. Spots most numerous above the lateral line; none on lower part of sides or belly. Lower surface, forward to tip of jaw, either chalky or silvery white. Dorsal fins dusky, usually more or less tinged with yellow; caudal olive or dusky with its lower edge yellowish at the base; pelvics and anal yellow; pectorals olive on outer side, but usually yellow on inner side.

Size. Fish heavier than 7.0 kg or longer than 100 cm TL are rare. Off southern Massachusetts, the largest fish run 2.7–4.5 kg, whereas most taken there weigh 0.5–3 kg and are 35–70 cm long. The all-tackle game fish record is shared by two 8.67-kg fish, one taken at Jones Beach Inlet, Long Island, N.Y., in October 1984 and the second from Delaware Bay in May 1989 (IGFA 2001).

Distinctions. The relative sizes and shapes of the fins of weakfish and their color are such ready field marks that they are among the most easily identified fishes. The slightly emarginate tail distinguishes them from mackerels and jacks, and this same character combined with a short anal fin and a first dorsal fin higher than the second gives them an appearance quite different from bluefish. The second dorsal being much longer than the first, the presence of only two anal spines, and the slender body together obviate all possibility of confusing them with striped bass or white perch. The shape of the head and of the dorsal and caudal fins and the absence of a chin barbel distinguish them at a glance from kingfish. The lack of chin barbels separates them from drum.

Habits. Adult weakfish are usually found in shallow waters along open sandy shores and in larger bays and estuaries, including salt-marsh creeks. They even run up into river mouths, but never into freshwater. Depth distribution is surf zone to 100 m. They have been taken in salinities from 6.6 to 32.3 ppt (Dahlberg 1972) and temperatures from 17° to 26.5°C (Merriner 1976). They cannot tolerate temperatures below 5°C, as many dead and numb ones were found at this temperature in North Carolina. Schwartz (1964a) subjected weakfish collected at 20.7°C to normal winter water temperatures and found swimming speed decreased as temperature approached 10°C, feeding ceased at 7.9°C, and all fish died at 3.3°C.

Weakfish move in schools, often small but sometimes consisting of many thousands. They have been described as swimming near the surface, this being the general rule near New York and along the southern New England coast, where great numbers are caught on hook and line within a few feet of the surface. They usually remain in the upper 10 m in summer.

Male weakfish produce sounds, especially during the breeding season. These are deep thumps like a drumbeat and bursts of high-pitched croaking. They also produce loud knocks and rapid bursts of croaking associated with alarm, defense, and offense. Weakfish have a large carrot-shaped swim bladder with three anterior diverticula that are vibrated by external paired sonic muscles. Sonic muscle mass increased threefold concomitant with increased plasma androgen levels at the peak of the spawning season (April and May) in Delaware Bay (Connaughton and Taylor 1994). Females lack these specialized muscles and do not make drumming sounds. Both sexes have well-developed pharyngeal patches of strong recurved canines close to the swim bladder that are associated with bursts of higher-pitched clucking (Fish and Mowbray 1970).

Food. Weakfish feed on a wide variety of prey, including crabs, amphipods, mysid and decapod shrimps, squids, shelled mollusks, and annelid worms, but chiefly on smaller fishes such as anchovies, menhaden, herring, silverside, jacks, spot, Atlantic croaker, scup, butterfish, sand lance, and flounders (Welsh and Breder 1924; Merriner 1975; Chao and Musick 1977; Wilk 1979; Bowman et al. 2000). The precise diet varies with locality, that is, with what is most readily available, although small clupeids and anchovies are probably the most important components. Weakfish smaller than 20 cm TL feed mostly on crustaceans (especially *Neomysis americana* and *Crangon septemspinosa*), 74–96% of the diet by weight for two smaller size-classes (Bowman et al. 2000). Larger weakfish prey mostly on fishes, 60–94% of the diet for six larger size-classes.

Merriner (1975) studied the diet of weakfish in North Carolina and reviewed much of the previous food habit data. He concluded that they feed on any locally abundant organism, and clupeid fishes throughout the species range. In North Carolina the dominant food items were penaeid and mysid shrimps, anchovies, and clupeid fishes. The importance of shrimps in the diet decreased from age 2 on, when older fish began to feed mainly on clupeids that were dominant in a given area. There was some gear bias in stomach content studies as weakfish caught in pound nets fed on fishes trapped in the net. For example, thread herring occurrence in the diet was high, but only in weakfish taken from pound nets. Weakfish are very visually oriented when feeding and have a well-developed chemosensory response mechanism (Wilk 1979). On open coasts they often feed on bottom right in the surf zone. They also feed on bottom in estuarine waters when preying on bottom-living animals, but in upper water layers when preying on small fishes. The prey-capture sequence includes visual fixation and orientation toward the prey; active pursuit; and, once within striking distance (20–50 cm), rapid beating of the caudal fin and a forward and upward lunge with jaws agape and opercles spread (Lascara 1981). In the Chesapeake Bay area, weakfish are important top carnivores in areas of eelgrass beds and appear to forage along the periphery of these beds during the low-light periods of dusk and dawn (Lascara 1981).

Predators. The main predators on young weakfish are older weakfish, bluefish (Maurer and Bowman 1975), and striped bass (Wilk 1979). Other fish predators include dusky shark, spiny dogfish, smooth dogfish, clearnose skate, angel shark, goosefish, and summer flounder (Rountree 1999). Cannibalism by adults on young weakfish was considerable in the Delaware River (Thomas 1971), probably in shallow estuarine areas where adults and juveniles co-occur.

Parasites. A parasitic isopod *Lironeca* sp. is a common ectoparasite of weakfish in Delaware (Thomas 1971). Linton (1905) described 14 parasites of weakfish from North Carolina, including the cestodes *Scolex polymorphous, Rhinobothrium* sp., *Rhynchobothrium speciosum, Otobothrium crenacolle, Tetrarhynchus bisulcatus,* and *Symbothrium* sp.; trematodes *Distomum vitellosum, Distomum polyorchis,* and *Microcotyle* sp.; acanthocephalan *Echinorhynchus pristis;* and nematodes, *Ascaris* sp.

Breeding Habits. Weakfish are multiple spawners with indeterminate fecundity and a spawning season from May to August in the Chesapeake Bay region (Lowerre-Barbieri 1994). Fecundity estimates ranged from 4,593 eggs for a fish of 203 mm TL (Shepherd and Grimes 1984) to 4,969,940 eggs for a fish of 569 mm SL (Merriner 1976). Length and weight are equally predictive of fecundity (Shepherd and Grimes 1984). The number of eggs produced is a function of size and geographic location of the stock. There appears to be latitudinal variation in age at maturity, growth, and fecundity, which suggests physiological responses to different environmental conditions or, perhaps, distinct stocks. Fish from North Carolina have greater fecundity and mature earlier than fish from the

whereas northern fish have the potential to reproduce up to age 10, thereby realizing the same reproductive potential over a lifetime (Shepherd and Grimes 1984).

Both male and female weakfish in North Carolina begin to mature at age 0. Some 52% of age-0 females and 62% of males were mature. By age 1 all females larger than 175 mm SL were mature, and all age-2 females were mature. Size of individual fish rather than age was the dominant determining factor for sexual maturity. In the vicinity of Morehead City, N.C., at least 50% of males about 130 mm SL and 50% of females 150 mm SL were mature. In Pamlico Sound, N.C., at least 50% of males 145 mm SL and 50% of females 190 mm SL were mature (Merriner 1976). In the New York Bight, the calculated length at which 50% of the collected individuals reached maturity was 256 mm TL for females and 230 mm TL for males. The smallest mature male and female both measured 200 mm TL; the maximum size for immature fish was 400 mm TL for females and 330 mm TL for males. This corresponded to 230 mm and 180 mm for females and males, respectively, from North Carolina (Shepherd and Grimes 1984). In New Jersey, males mature at 2–3 years and females at 3–4 years, with the majority on the spawning grounds at Cape May being 4–6 years old (Welsh and Breder 1924).

There is no specific information on courtship, although male weakfish possess specialized drumming muscles that are not present in females, which indicates that there is probably some type of auditory display as males do make drumming noises. Large schools of weakfish were observed assembling on the eastern side of Delaware Bay in depths of 5–9 m, and spawning took place over mud and sand bottom in the early evening and at night (Welsh and Breder 1924). A "milling" behavior was observed during spawning in Great South Bay, Long Island, N.Y., at which time weakfish broke the surface (Poole in Mercer 1983).

Weakfish eggs were collected from March to August between Cape Hatteras, N.C., and Narragansett Bay, R.I., predominantly in the inner one-half to one-third of shelf waters (Berrien and Sibunka 1999: Fig. 58). Spawning takes place chiefly at night in inlets, bays, or sounds in larger estuaries or close to mouths of estuaries (Ferraro 1980). Spawning begins as early as March or April off Cape Hatteras and spreads rapidly in nearshore waters extending to New Jersey in May, Long Island in June, and in some years to Narragansett Bay in July. In the New York Bight, the peak spawning period varied with the size of the fish. The largest individuals entered estuaries first, in May, and spawned by mid-May, whereas smaller ones arrived later and reached a spawning peak in June (Shepherd and Grimes 1984).

Batch spawning was found in weakfish in North Carolina. Peak activity occurred from late April through June, but ripe females were found from March to September, and there was evidence of two batches of oocytes in the ovaries (Merriner 1976). Multiple spawns were not evident in fish from the New York Bight (Shepherd and Grimes 1984).

Spawning occurs after spring migration and appears to be regulated by water temperature and photoperiod, as it has been induced in the laboratory by manipulating these parameters. Water temperature was lowered to 13°–14°C and day length shortened from 12 h light/12 h dark to 8 h light/16 h dark over a 3-week period. Fish were held under these conditions for 11 weeks; then the temperature was raised to 22°–23°C and the light exposure changed gradually to 14 h light/10 h dark gradually, at which point they began to spawn (Epifanio et al. 1988).

Early Life History. The eggs are buoyant, spherical, transparent, 0.68–1.18 mm in diameter (Welsh and Breder 1924; Merriman and Sclar 1952; Lippson and Moran 1974), with one to six (usually one) pale amber oil globules that coalesce into a single large one as development progresses. Average diameter of the single oil globule is 0.22 mm and it has pigment in the later stages (Joseph et al. 1964). There is a narrow perivitelline space. Incubation takes 36–40 h at a temperature of 18°–24°C. The eggs can tolerate 12°–31.5°C and 10–33 ppt salinity. Reduction of dissolved oxygen to 4.3 ppm reduces successful hatching, and 2.4 ppm prevents hatching. Changes in temperature and salinity of 6°C and 5–6 ppt may have deleterious effects on embryos (Harmic 1958). Eggs have been taken in tow nets at various localities in temperatures of 12°–24°C and salinities of 28–30.9 ppt.

Weakfish eggs resemble those of mackerel, but the latter have only one oil globule, a larger mean diameter (1.13–1.15 mm vs. 0.84–0.96 for weakfish), and usually occur in salinities over 30 ppt. Weakfish eggs resemble scup and butterfish eggs; both of the latter have only one oil globule but are otherwise very similar. All sciaenid eggs have similar general characteristics, but the only sciaenid likely to spawn north of Cape Cod is the northern kingfish, *Menticirrhus saxatilis,* and they have 1–18 oil globules and a slightly larger perivitelline space (Hardy 1978b).

Newly hatched larvae are 1.49–1.99 mm long (mean 1.8 mm TL) (Harmic 1958). The yolk sac is large and contains the oil globule at the posterior end, adjacent to the anus. The body is elongate and slender. Pigmentation at hatching shows yellow chromatophores grouped behind the eye, in a transverse band behind the otocyst, on the underside of the snout, and on the yolk sac. There are scattered black melanophores on the dorsal surface of the body and the oil globule. About 8 h after hatching, the yellow pigment is more aggregated around the eye and behind the otocyst and forms two bands on the body behind the anus. Some 24 h later there is another band of yellow chromatophores posteriorly (Welsh and Breder 1924). At about 3 mm, larvae resemble silver perch and Atlantic croaker but are distinguishable by their small teeth. The only sciaenid larva likely to co-occur with weakfish north of Cape Cod are northern kingfish, which are more heavily pigmented on the body and have an anal spine (Fahay 1983).

At 30 mm, young weakfish have attained most of the structural characters of adults, but until they are 15–20 cm long they are much deeper and more compressed, the head and eyes are relatively larger, and the caudal fin is obtusely pointed with center rays much the longest, instead of concave. Smaller fish (38–46 mm) are marked with four dark, saddle-shaped patches extending downward on the sides to a little below the lateral line; these are not lost until a length of about 114 mm is reached. As young fish grow, other bands of pigment are interpolated below the lateral line, adult coloration not being fully developed until they are 178–203 mm.

Larval and juvenile weakfish have been collected from near shore to 70 km offshore (Hildebrand and Cable 1934; Berrien et al. 1978), as well as in estuaries and tidal passes. Larvae become demersal at 1.5–8 mm TL and probably utilize subsurface currents to move to lower-salinity nursery grounds (Hildebrand and Cable 1934; Harmic 1958). Juveniles are euryhaline and have been collected in freshwater (Thomas 1971). They are more frequently found in deeper waters of rivers, bays, and sounds than close to shore or on shallow flats (Dahlberg 1972; Chao and Musick 1977). Sampling of North Carolina sounds by that state's Division of Marine Fisheries showed that juvenile weakfish occurred most often in shallow bays or navigation channels at moderate depths, in slightly higher salinities, and on sand or sand-grass bottoms (Mercer 1983).

Food of larval weakfish in Delaware Bay consists primarily of cyclopoid copepods and their eggs. Other organisms consumed were calanoid copepods, tintinnids, and polychaete larvae (Goshorn and Epifanio 1989). Juveniles 67–183 mm feed primarily on clupeids, *Anchoa mitchilli,* and shrimp, *Neomysis americanus.* Smaller weakfish feed mainly on crustaceans and larger ones on fishes (Chao and Musick 1977).

Age and Growth. Age composition and growth have been estimated from annual rings on scales, otoliths, and vertebrae, and from length frequencies. Estimates vary considerably from season to season, year-to-year, and area of collection (Wilk 1979). Scale analysis, the most frequently used aging technique, demonstrated that annulus formation occurred from April to June (Shepherd and Grimes 1983). Lowerre-Barbieri et al. (1994) found that transverse otolith sections are the best to age weakfish and that the scale method was imprecise and apparently inaccurate at more advanced ages. Maximum ages in Chesapeake and Delaware bays in the 1990s were 11 or 12 years but one fish collected in Delaware Bay in 1985 was 17 years old (Lowerre-Barbieri 1994).

There are differentials in growth and life span of weakfish from six geographical regions of the Mid-Atlantic Bight (Shepherd and Grimes 1983), with three distinct regions with respect to growth: I, Delaware Bay and Cape Cod; II, Chesapeake Bay; and III, Cape Hatteras. Length-at-age was greater for females in all regions, and the difference increased with age. These differences were not statistically significant until age 6 except for Region III where it was noted at age 2. Growth from Region III was highest in the first year, but slower thereafter, and these fish did not live as long. Length-weight coefficients approach the cubic power of fish length (2.67–2.98) for both males and

Weakfish young grow at so variable a rate during the first summer that they may be anywhere between 92.6 and 152 mm in the fall, when they are about 6 months old. The smallest fish seen in spring are 193–254 mm. Thereafter the rate of annual growth is slower. Variation in the length attained by the young during their first summer and autumn, consequent to the protracted spawning season, combined with the fact that scale studies of this species have proved puzzling, makes it difficult to group the older age-classes by size. Growth data for young-of-the-year and age-1 weakfish from different areas along the Atlantic coast were summarized by Chao and Musick (1977).

General Range. Eastern coast of the United States from Florida to Massachusetts Bay, straying northward to the Bay of Fundy and Nova Scotia. The center of abundance for weakfish is along the coast of the Mid-Atlantic states from the Virginia Capes to New York.

Occurrence in the Gulf of Maine. Weakfish occur regularly as far north and east as Cape Cod. However, abundance fluctuates widely on the southern New England coast, and it is only during periods of great abundance there that they appear in any numbers in Cape Cod and Massachusetts bays, which may be set as the northern limit of the range except for strays. In years in which they have passed Cape Cod in appreciable numbers, they have always been far more plentiful along the inner side of the Cape and in Cape Cod Bay at Provincetown, Truro, Brewster, and Sandwich than north of Boston, where they have been found at Nahant, Manchester, and Cape Ann. However, even in the Cape Cod Bay region, there has been only one period of abundance in the past century and a half and that lasted for about 9 years.

Migrations and Movements. In the southern part of their range, weakfish are considered to be resident; however, from the offshore waters of Virginia and the Carolinas, they undergo a spring migration northward to estuarine spawning areas, spending summer inshore and withdrawing again in autumn to overwintering grounds (Nesbit 1954). Weakfish younger than age 4 move out of inshore and estuarine areas and south to Florida along the coast in fall and north again in spring and summer. Larger fish, usually older than age 4, move south but offshore in fall, probably no farther than North Carolina, and return to inshore grounds in spring. The largest fish move the fastest and tend to gather in the northern part of the range (Wilk 1979).

Seasonal migrations occur in conjunction with movements of the 16°–24°C isotherms (Shepherd and Grimes 1983). For Chesapeake Bay, the fishing season usually lasts from the middle of April, commencing a week or two later up the bay, to the middle of November, with good catches occasionally made as late as December. On the southern New England coast, as typified by Woods Hole, weakfish are caught from May (some years as early as April, other years not until

expected north of the elbow of Cape Cod until June nor later than October for most disappear from the Mid-Atlantic coast before the end of October.

Capture of weakfish in some numbers between the offings of Chesapeake Bay and off Cape Hatteras by otter trawlers during winter months has dispelled some of the mystery regarding the wintering grounds of this species. The fact that several were caught in the 100-m zone off Rhode Island and Martha's Vineyard is evidence that some of those that summer to the north only move offshore to escape falling temperatures.

Importance. Weakfish were plentiful off southern New England toward the end of the eighteenth century and, from fishermen's reports, were well known in Massachusetts Bay at that time, but they vanished so completely sometime prior to 1800 that when a stray specimen was taken at Provincetown in June 1838, it was sent to Boston for identification. This disappearance evidently involved the whole northern part of the range, for they also vanished from the Nantucket–Martha's Vineyard region sometime between 1800 and 1837. By 1867 they had reappeared off southern Massachusetts, and the catch in the Cape Cod Bay–Massachusetts Bay region, localized chiefly on the outer side of Cape Cod and in Cape Cod Bay, was larger for the next few years. They appeared in such numbers in Cape Cod Bay in 1900 that the catch there for that year jumped to more than 49,000 kg, and a few were taken even as far north as Boston Harbor and Gloucester. This marked the start of a period of local abundance, which was entirely unexpected and which, with its equally sudden eclipse, is perhaps the most interesting event in the history of local fisheries. Unfortunately, reliable catch statistics are not available for the crucial years, but weakfish were so plentiful in Cape Cod Bay in 1901 as to glut the market; while in 1902 and 1903 pound nets in Cape Cod Bay were often filled with schools of large weakfish, averaging about 2.3 kg. This abundance continued through 1904, by which time it seems to have been accepted as the normal state of affairs and no longer worth comment. But it seems to have culminated in that summer or the next, for weakfish were reported as less plentiful in 1906. There is no reason to suppose that they have entered Cape Cod Bay in any numbers since that time.

It is doubtful whether any large numbers of weakfish have reached the northern side of Massachusetts Bay since 1909, when 90 kg were reported from a pound net off Gloucester. Large landings were reported from the northern part of the Massachusetts coast (Essex County) in the 1940s, but one cannot assume that any of them were caught north of Cape Cod, for all fish taken by vessels sailing out of Gloucester during those years were credited to that port, irrespective of where caught or landed.

There is no explanation for this unexpected invasion of weakfish north of Cape Cod about the turn of the twentieth century or for its equally sudden eclipse, the opportunity having passed long since for obtaining any information as to sizes and ages of the fish, their movements, and the physical state of the water at the time. It was not a local event, however, but part of a corresponding fluctuation in the population as a whole east and north of New York. Thus the catch for the southern coast of New England was more than eight times as great in 1904, but thereafter declined so markedly that in 1908 both the commercial fishermen and the anglers of Rhode Island and of southern Massachusetts complained of the scarcity. Weakfish nearly vanished from the southern shores of Massachusetts by 1920 and 1921. Partial recovery took place off the southern Massachusetts coast during the period 1931–1938, but did not bring weakfish back to Cape Cod Bay.

Currently, weakfish are of no importance in the Gulf of Maine, commercially or to the angler, although they were a very valuable addition to shore fisheries of Cape Cod Bay during their brief period of plenty there. However, they are one of the most important food fishes along more southern coasts, and a favorite game fish that has been the subject of many accounts from the angler's standpoint. A summary of their commercial and recreational importance (Wilk 1981) and management and protection of the fishery (Mercer 1983) are further summarized in the Atlantic States Marine Fisheries Commission management plan (ASMFC 1985).

Stocks. There is evidence that there may be three stocks based on body length–scale length relationships and differences in growth and longevity (Shepherd and Grimes 1983): Region I, Cape Cod to Ocean City, Md.; II, Ocean City to Virginia Beach, Va.; III, Virginia Beach to Cape Fear, N.C. Other authors have hypothesized the existence of two (Pearson 1941; Joseph 1972) or three (Seguin 1960) subpopulations. Most thought that the separation point was at Cape Hatteras. Results of electrophoretic studies of young-of-the-year and adults from Long Island, Delaware Bay, York River in Virginia, and Cape Hatteras (Crawford et al. 1989) and mitochondrial DNA from populations along the Atlantic coast of the United States (Graves et al. 1992a) indicate that weakfish comprise a single gene pool, so the fishery should be managed as a single, interdependent unit.

SPOT / *Leiostomus xanthurus* Lacepède 1802 / Bigelow and Schroeder 1953:423

Description. Body rather deep, compressed, with a strongly elevated back (Fig. 237). Head short and obtuse, snout blunt, mouth inferior reaching to about middle of eye. Teeth small, villiform, and set in bands in jaws. Underside of lower jaw with five pores at tip (Chao 1978). Dorsal fin continuous, notch between spinous and soft portions. Dorsal spines slen-

Figure 237. Spot *Leiostomus xanthurus.* Rhode Island. Drawn by H. L. Todd.

der, third and fourth longest. Pectoral fins long, pointed, reaching well beyond tip of pelvic fin. Scales small, ctenoid, extending on to base of fins (Hildebrand and Schroeder 1928; Chao 1978).

Meristics. Dorsal fin rays X, I, 30–34; anal fin rays, II, 12 or 13; lateral line scales 72–77; gill rakers 8–12 + 20–23; vertebrae 10 + 14 or 15 (Chao 1978).

Color. Bluish gray above with golden reflections, silvery below. Medium-sized fish marked on both sides with 12–15 yellowish crossbars dipping obliquely forward, but these fade with age. A conspicuous black spot close behind upper corner of each gill opening. Fins partly yellowish, partly dusky.

Size. Spot grow to a length of about 34 cm and have a maximum weight of less than 0.7 kg, but adults average only about 26 cm long, and few weigh more than 0.34 kg. The all-tackle game fish record is a 0.45-kg fish caught at Nags Head, N.C., in July 2000 (IGFA 2001).

Distinctions. Spot resemble weakfish in arrangement, general shape, and relative sizes of the fins, and in lacking chin barbels. They are much deeper fish (body about one-third as high as it is long), with a blunt instead of a pointed snout; they have no large canine teeth; the tail is more forked; and they are marked on either side with a conspicuous black spot close behind the upper corner of each gill opening. Spot most closely resemble black drum but lack chin barbels and have forked rather than squared caudal fins and the characteristic black spot.

Habits. Spot are found most often over mud and sand bottoms in inshore and estuarine waters and offshore to a depth

haline and can tolerate salinities of 0–34.2 ppt and temperatures of 6°–36.7°C. The lower lethal temperature is near 5°C (Thomas 1971).

Spot are schooling fish and often travel in very large numbers. Like most drums, they make sounds. Males produce croaks during feeding and under electrical stimulation. Females produce knocks when alarmed. Sounds are created by the swim bladder and associated drumming muscles in males but are weaker than those of most other sciaenids, probably owing to the thin walls of the swim bladder and poorly developed muscles (Fish and Mowbray 1970).

Food. Based on examination of 442 stomachs, spot feed on polychaetes (53.9% by weight), crustaceans (13.9%), bivalves (11.6%), and fishes (2.9%) (Bowman et al. 2000). The diet changes with size in young-of-the-year spot (O'Neill and Weinstein 1988). Small spot, less than 30 mm SL, consume more planktonic food items such as calanoid copepods, polydora, *Callinectes* larvae, and plant matter and are more selective; larger spot, greater than 30 mm SL, are more opportunistic and eat more benthic prey such as harpacticoids, maldanid and nereid polychaetes, nematodes, oligochaetes, clam siphons, and ostracods.

Predators. Spot are preyed on by several species of sharks, including sand tiger, sandbar shark, dusky shark, smooth dogfish, spiny dogfish, and angel shark, of which spiny dogfish is the most frequent predator, and bony fishes such as goosefish, bluefish, weakfish, and summer flounder (Lascara 1981; Medved et al. 1985; Rountree 1999; Bowman et al. 2000).

Reproduction. Spot become sexually mature by the end of their second year or the beginning of their third at about

1958). Ova of several sizes have been found in developing ovaries (Hildebrand and Cable 1930). Nothing is known of their spawning behavior, but males produce sounds and there is probably some kind of auditory courtship. Spot spawn in offshore waters during fall and early winter throughout their range (Hildebrand and Cable 1930; Nelson 1969). Principal spawning areas and times of spawning are Chesapeake Bay–Cape Hatteras from October to February (Colton et al. 1979).

Early Life History. Spot eggs are pelagic, spherical, and transparent with a diameter of 0.72–0.87 mm. The chorion is smooth, perivitelline space narrow, yolk homogeneous and unpigmented with one to twelve oil globules usually coalescing into one. The single oil globule diameter is 0.18–0.28 mm, located posteriorly in the yolk and has black pigment anteriorly.

Larvae hatch in 48 h at 20°C and are 0.6–1.7 mm SL. The eye is unpigmented and the body is slightly pigmented. Pigment spots are scattered on the anterior body; there are a few spots on the dorsal surface of the oil globule; and two to five dorsal spots on the tail, which disappear in older larvae (Powell and Gordy 1980). Characteristic pigment develops at about 6 mm SL. In ventral view, there is often a triangular three-spot pattern on the abdomen (Fahay 1983). Spot co-occur with several other sciaenids and can be distinguished from black drum by the absence of dark markings along the back, and from silver perch by the lack of a broad dark patch behind the head and over the abdomen (Lippson and Moran 1974).

Spot spawn in offshore waters and larvae are transported toward shore and into estuaries that serve as nursery areas (Fahay 1975; Chao and Musick 1977). They enter estuaries at an age of about 59 days. Migrating fish are segregated by size (or age); younger, smaller larvae migrate at the beginning and end of the migration period (Warlen and Chester 1986).

Larval spot eat tintinnids; pteropods; pelecypods; ostracods; and the egg, naupliar, copepodid, and adult stages of copepods (Govoni et al. 1983).

Age and Growth. Growth of fish sampled in New Jersey is 8–10 cm at the end of the first year, 17–22 cm at age 2, and 24–29 cm at age 3 (Welsh and Breder 1924). Few spot live longer than 3 years (Hildebrand and Cable 1930). Information on age and growth was summarized by Barger and Williams (1980).

General Range. Inshore waters from Mexico to southern New England, and recorded from Massachusetts Bay as a stray. Spot are plentiful in some years as far north as New York, and young ones are described as common in autumn around Woods Hole.

Occurrence in the Gulf of Maine. The normal range is bounded so sharply by Cape Cod that they have been reported from the Gulf of Maine only twice: a single specimen from Massachusetts Bay in November 1936 (Smith and Goffin 1937) and one from the Pilgrim Nuclear Power Station at Plymouth in western Cape Cod Bay (Lawton et al. 1984).

Migration and Seasonal Movements. Spot migrate into sounds, bays, and estuaries in spring and remain there during summer. In fall they migrate south and offshore toward winter spawning grounds. In late winter, most are found south of Cape Hatteras (Silverman 1982).

Importance. Spot are so rare in the Gulf of Maine that they are of no commercial or recreational importance; however, they are very important in the Chesapeake Bay area and off the Carolinas. They are often abundant as far north as the New York Bight (Silverman 1982) and, when abundant, are an important recreational species there (McHugh and Ginter 1978).

NORTHERN KINGFISH / *Menticirrhus saxatilis* (Bloch and Schneider 1801) /

King Whiting, Whiting / Bigelow and Schroeder 1953:423–425

Description. Body elongate, rounded, with an elevated back and rather flat ventral sides (Fig. 238). Head low with blunt nose and conical snout that projects beyond mouth. Mouth small and inferior with protrusible upper jaw. Teeth villiform, set in bands with outer ones in upper jaw slightly enlarged. Pharyngeal plate lacks molariform teeth. Chin with single short barbel, perforated by an apical pore with four lateral pores. Dorsal fins continuous with notch between spinous and soft portions. Dorsal spines slender and flexible, second long and slightly tapering, third even more elongate, especially in fish larger than 20 cm. Pectoral fins relatively long and pointed. Anal fin originates under middle of soft dorsal. Caudal fin uneven, lower portion slightly longer. Scales small, ctenoid. Lateral line complete, slightly arched.

Meristics. Dorsal fin rays IX or X, I, 22–27; anal fin rays I or II, 7–9; gill rakers tuberculate, 3–5 + 0–7; vertebrae 10 + 15 = 25 (Irwin 1970; Miller and Jorgenson 1973).

Color. Leaden or dusky gray above (sometimes so dark as to be almost black) with silvery and metallic reflections; milky or yellowish white below. Sides cross-marked irregularly with dark bars that run obliquely forward and downward behind the spiny dorsal fin, with the foremost one or two bars running in the opposite direction, so that they form a V-shaped blotch or two dark V's below the fin. Pale belly bounded above by a dark longitudinal streak on either side that extends onto the lower lobe of the caudal fin in large specimens. Fins are

Figure 238. Northern kingfish *Menticirrhus saxatilis.* Pensacola, Fla., 244 mm. Drawn by H. L. Todd.

dusky or blackish; the first dorsal fin, anal, pectorals, and pelvics are tipped with dirty white.

Size. Northern kingfish grow to a maximum length of 55 cm and a weight of about 1.5 kg (Schaefer 1965) but they generally run from 25.4 to 30.6 cm and weigh between 0.2 and 0.7 kg. The all-tackle game fish record is a 1.11-kg fish caught at Salvo, N.C., in September 2000 (IGFA 2001).

Distinctions. Northern kingfish resemble weakfish in the general arrangement and relative sizes of their fins, the second dorsal being much longer than the first or the anal; but the first dorsal spine is much higher and more pointed than that of weakfish, with the third spine not only prolonged but filamentous at the tip in adults. They also differ from weakfish in that they have a chin barbel and fleshy lips and lack canine teeth. The rather blunt snout overhanging the mouth gives northern kingfish a very characteristic countenance. The upper jaw projects beyond the lower, whereas the reverse is the case in weakfish. The body is about as slender, proportionally, as that of weakfish, but northern kingfish carry their weight farther forward (they are deepest below the first dorsal fin). The tail, too, is of very characteristic outline, with the lower half rounded but the upper half concave, suggesting the tail of black sea bass but without a filament.

Habits. Northern kingfish are confined to the immediate vicinity of the coast, frequenting enclosed as well as open waters, even entering river mouths; but are unknown on offshore banks. They are commonly taken in 7–45 m, their maximum depth being 126 m (Irwin 1970). They run in schools, keep close to the bottom, and prefer hard or sandy bottoms. They have been found at temperatures of 7.8°–35.8°C (Irwin 1970). Juveniles can tolerate salinities of 5–35.1 ppt (Hardy 1978b). Unlike most drums, they do not have a swim bladder and do not produce sounds.

Food. Based on examination of 78 stomachs, northern kingfish feed on crustaceans (84.8% by weight, mostly decapods, 82.7%), fishes (9.4%), and polychaetes (3.8%) (Bowman et al. 2000). The most important decapods include *Callianassa setimanus, Crangon septemspinosa, Ovalipes ocellatus,* and *Pinnixa chaetopterana.* Stomach contents of northern kingfish in the York River consisted mainly of crustaceans (*Neomysis, Crangon,* and *Palaemonetes,* amphipods, and copepods), polychaetes, and organic detritus (Chao and Musick 1977). The main percentage by volume of food of kingfish from Cape May, N.J., consisted of amphipods, polychaetes, detritus, decapod shrimps, and isopods (Welsh and Breder 1924). Known fish prey include margined snake eel, anchovies, and flounders (Bowman et al. 2000). Northern kingfish appear to be bottom feeders that locate food by olfaction and touch with their inferior mouth and pored barbel (Chao and Musick 1977).

Breeding Habits. Both males and females begin to mature at age 1 in New York waters (Schaefer 1965). Half of all age-1 and age-2 males were running ripe, and a large percentage of age-1 and age-2 females were ripening. No young-of-the-year showed signs of gonadal development. There is no information on fecundity.

Northern kingfish apparently spawn in outside waters in the southern part of their range from April to May or June (Hildebrand and Cable 1934), and may spawn in bays and sounds farther north from May to August or September (de Sylva et al. 1962). Eggs are found regularly as far north as Narragansett Bay (Bourne and Govoni 1988), but it is unlikely that young hatched in the Gulf of Maine from eggs spawned by the occasional visitor would survive to adulthood.

Early Life History. The eggs are pelagic, spherical, and almost colorless with a faint yellow tinge. The average diameter is 0.80–0.85 mm. Between 1 and 18 oil globules are present, coalescing to one by hatching. Single oil globules are 0.19–0.26 mm diameter (Welsh and Breder 1924). Eggs and larvae of various species of kingfish are difficult to distinguish, but any found in the Gulf of Maine would be northern kingfish. Eggs hatch in 45–50 h at 20°–21°C (Welsh and Breder 1924). Larvae are 2–2.5 mm TL at hatch. The oil globule is pigmented and located in the posterior part of the yolk sac. Pigment consists of three vertical bands of black and gold chromatophores, one above the anus and two posterior to it, dividing the caudal region into three nearly equal parts. By 5 mm TL, anal fin rays are differentiating and pigmentation consists of a row of melanophores along the ventral surface of the abdomen, a few at the nape and in rows midventrally, some on the upper lip, and a patch on the roof of the mouth that is visible externally (Chao 1978). By 10 mm TL all fins have a full complement of rays and juveniles resemble adults, readily recognizable as kingfish although they vary widely in color, ranging from the pattern of the adult to almost uniform blackish brown.

Juveniles probably live on or near the bottom like adults. Some are found in salinities of 5–35 ppt but they are common only above 16 ppt (de Sylva et al. 1962; Thomas 1971). The temperature range is 15.5°–30.7°C (Springer and Woodburn 1960). Food of juveniles consists mainly of crustaceans and polychaetes (Chao and Musick 1977).

Age and Growth. Age and growth of northern kingfish in New York waters was determined by identifying scale annuli (Schaefer 1965). Annulus formation took place between June and August. Growth was very rapid in the first summer of life; the modal length of age-0 fish reached 250 mm in October. Growth appeared to be slower in the Woods Hole region, reaching a maximum of 20 mm by July, 80 mm by August, and 150 mm by September (Welsh and Breder 1924). Growth rates of males were slower than those of females over their 4-year life span (Schaefer 1965). Length-weight relationships were calculated for northern kingfish from New York waters (Schaefer 1965).

General Range. Atlantic coast of the United States from Florida (Key West, Pensacola) northward regularly to Cape Cod; most numerous from Chesapeake Bay to New York; known as far north as Casco Bay, Maine, as a stray and as far south as Progreso, Yucatán (Irwin 1970).

Occurrence in the Gulf of Maine. This excellent food and game fish reaches the Gulf of Maine only as a stray from the south. The only positive records of northern kingfish within the limits of the Gulf are as follows, south to north: Monomoy and North Truro on Cape Cod in 1896 (collected by Dr. W. C. Kendall); one taken at Provincetown in July 1847, another there in November of that same year, and many small ones, apparently chilled by the cold, that appeared in that harbor in 1879; one taken at the entrance to Boston Harbor in a lobster pot sometime before 1833; one at Lynn in 1840; one 20 cm long off Marblehead on 15 October 1872; one 16.5 cm at Danvers on 28 October 1874; others at Nahant (one record), and in Casco Bay.

Importance. Northern kingfish are not common enough in the Gulf to interest either commercial fishermen or anglers. They are one of the better table fishes, and a favorite with surf anglers along the coasts of New York, New Jersey, and southward, as they bite readily and fight well.

BLACK DRUM / *Pogonias cromis* (Linnaeus 1766) / Bigelow and Schroeder 1953:425–426

Description. Body short and deep, less than three times as long as deep, with high-arched back and flattish belly (Fig. 239). Head moderately short, mouth horizontal and set very low. Chin with five pores and 12–13 pairs of barbels along inner edges of lower jaw, usually extending back to below middle of eye. Eye relatively small and set high. Jaw teeth small, pointed, set in broad bands; throat armed with large, flat, pavementlike pharyngeal teeth surrounded with strong conical teeth. Dorsal fin continuous with deep notch between spinous and soft portions. Arrangement and sizes of fins essentially as in weakfish, except second dorsal relatively shorter and anal spine much stouter. First dorsal fin rounded-triangular with stiff, slender spines, third spine longest; second dorsal oblong. Anal fin very short with the second spine long and stout, especially in young drum. Caudal fin square-tipped with moderately high peduncle. Pectoral fins sharp pointed, about as long as head. Scales firm and ctenoid.

Meristics. Dorsal fin rays X, I, 19–23; anal fin rays II, 5–7; lateral line scales 41–45; gill rakers 4–6 + 12–16; vertebrae 10 + 14 (Hildebrand and Schroeder 1928; Lippson and Moran 1974; Chao 1978).

Color. Variable, blackish to silvery with a brassy luster, grayish white below, turning to a dark gray after death. Young fish have four or five broad dark vertical bars that fade out with age. Fins dusky or black. This drum occurs in two color phases, grayish and reddish.

Size. Black drum grow to a huge size. The largest recorded weighed 66.3 kg (Hildebrand and Schroeder 1928); adults usu-

Figure 239. Black drum *Pogonias cromis.* Matanzas Inlet, Fla., 228 mm.

ally run from 9 to 18 kg. The all-tackle world game fish record is a 51.28-kg fish caught off Lewes, Del., in September 1975 (IGFA 2001).

Distinctions. Black drum resemble spot, but have chin barbels and a smaller eye. Arrangement and size of fins similar to that of weakfish, but the high arched back and flat ventral aspect distinguish black drum, as do the square caudal fin and numerous barbels.

Habits. Black drum are an inshore schooling fish and inhabit bays, estuaries, lagoons, and shoreline areas over sandy bottoms. They also occur over oyster, clam, and mussel beds, in high marsh areas, and around piers and breakwaters. They move inshore and northward along the Mid-Atlantic coast each spring and southward and offshore by late fall (Richards 1973). They can tolerate salinities of 10–85 ppt on the Gulf of Mexico coast but their usual range is 25–50 ppt (Simmons and Breuer 1962), and they are most abundant at 10–15 ppt (Gunter 1945). They have been taken at temperatures of 3°–35°C (Simmons and Breuer 1962).

Black drum are schooling fish. As one might expect from their name, they produce sounds, which can be quite loud and vary from loud croaking, high pitched growls, spontaneous knocks and sustained musical bursts of noise to short, rapid croaks and stridulations. Sounds are created by the large swim bladder, the associated drumming muscles, and the large, flat, pavementlike pharyngeal teeth, in response to fright and during the spawning season. Both males and females make sounds, but those of the female are usually softer. Maximum sound production occurs in the late afternoon and early evening. High-amplitude rumbling sounds in the vicinity of large schools can be heard or felt through a boat's hull (Fish and Mowbray 1970).

Food. Black drum are reported to feed on mollusks and crustaceans.

Breeding Habits. Black drum mature at about 32 cm at the end of their second year (Pearson 1929; Simmons and Breuer 1962). A 111-cm female has about 6 million eggs (Pearson 1929). Black drum spawn in or near the estuaries where larval and juvenile development occurs (Peters and McMichael 1990). They enter Chesapeake Bay in early spring and spawn throughout the day from late April through May (Joseph et al. 1964; Daniel 1995); they spawn in Delaware Bay the second to fourth weeks of May, with a possible second spawning in September (Thomas 1971). "Drumming" is associated with the breeding season and may be important in courtship.

Early Life History. The eggs are pelagic, 0.82–1.02 mm in diameter, with multiple oil globules (usually two to four) through the germ-ring stage in about half the eggs. If multiple, the diameter varies; if single it is 0.22–0.26 mm. By hatching there is only one oil globule located in the rear portion of the yolk sac. The oil globule is pigmented with discrete black dots, but the yolk is unpigmented. Embryonic pigmentation is not distinctive (Joseph et al. 1964).

Eggs hatch in 24 h at 20°C. Newly hatched larvae are 1.9–2.4 mm TL and pigmentation is still not distinctive. At 2.8 mm, pectoral buds are present and pigment has developed on the head and trunk. There is a group of ventral chromatophores present posterior to the anus; two large chromatophores on the midcaudal region, one dorsal, one ventral;

and a smaller chromatophore halfway between the midventral chromatophore and the terminal myomere (Joseph et al. 1964). By 6 mm, weak spines are present on the opercle and dorsal and ventral finfolds are separated from the tail. Pigmentation is very stellate and branching in the tail region, two to five spots on ventral midline and one or two spots along the second dorsal fin base. At 15 mm the general adult shape is attained, scales begin to form, barbels are present, and six black bars are pronounced, extending vertically from the back to slightly below the lateral line (Thomas 1971).

Among prey items in stomachs of larval black drum in Tampa Bay, copepods were dominant, making up 95% of the numbers and 58% of the volume (Peters and McMichael 1990). With increasing size, copepods became less important in the diet and were replaced by bivalves and gastropods, although polychaetes, shrimps, and crabs were also present.

Age and Growth. Thomas (1971) noted the growth of black drum in the Delaware River estuary during the first year as follows: early June, 10.1–11.7 mm; late June, 17.3–23.4 mm; July, 61.1–71.1 mm; and August, 112.6–127.3 mm. Richards (1973) gave mean length for fish in Chesapeake Bay as age 1, 223 mm; age 2, 406 mm; and age 3, 562 mm. The length-weight relationship in Chesapeake Bay is: $\log W = 3.0655 \log L - 4.903$, where W is weight in kg and L is cm FL (Richards 1973).

General Range. Atlantic and Gulf of Mexico coasts of America from Argentina to Nova Scotia (Scott and Scott 1988); common from New York southward, abundant from the Carolinas to the Rio Grande, straying north to Massachusetts Bay and Nova Scotia.

Occurrence in the Gulf of Maine. Black drum are rare visitors to the Gulf of Maine, where two or three individuals have been taken at Provincetown and one in the Mystic River, which empties into Boston Harbor. They have been reported from near the Pilgrim Nuclear Power Station (Lawton et al. 1984). The northernmost record is a specimen from the Bay of Fundy (Bleakney 1963a).

Importance. Black drum are too scarce to be of importance in the Gulf but are valuable sport fish south of New Jersey.

ACKNOWLEDGMENT. A draft of the weakfish account was reviewed by Stuart Wilk.

BUTTERFLYFISHES. FAMILY CHAETODONTIDAE

RICHARD S. MCBRIDE

Butterflyfishes are compressed, deep-bodied coral reef fishes that resemble angelfishes (Pomacanthidae), but angelfishes have a strong preopercular spine, which butterflyfishes lack, and do not exhibit the distinctive tholichthys prejuvenile stage characterized by large, smooth, bony plates that encase the head (Burgess 1978). These head plates develop during the planktonic period and diminish in size soon after settlement. The suborbital pair of plates is particularly modified and extends posteriorly in the shape of horns. Ten genera of butterflyfishes are recognized with 89 species in the genus *Chaetodon* (Burgess 1978; Allen 1979; Nelson 1994). Juveniles of one species of butterflyfish occasionally venture into the Gulf in late summer.

SPOTFIN BUTTERFLYFISH / *Chaetodon ocellatus* Bloch 1787

Description. Body oval, laterally compressed, body depth approximately 60–75% of SL (Fig. 240). Mouth small, terminal, with numerous rows of slender teeth (Motta 1989). Lachrymal bone smooth (Burgess 1978). Eye moderate-sized, approximately 9–13% of SL. Spinous and soft-rayed components of dorsal fin connected. Dorsal and anal fins extend back to caudal peduncle. Caudal fin moderate-sized, slightly rounded. Pectoral and pelvic fins extend back to about anus. Lateral line high on body, ending under soft dorsal fin. Large, ctenoid scales cover body in oblique pattern, smaller scales cover head, caudal peduncle, and parts of dorsal and anal fins.

Meristics. Dorsal fin rays XII or XIII (rarely XIV), 18–20 (rarely 21); anal fin rays III, 15–17; pectoral fin rays 14 or 15 (rarely 16); lateral line scales 33–40 (usually 35–39); lateral line pores 30–38; scales above lateral line 6–8 (usually 7); scales below lateral line 15–18 (usually 16 or 17); gill rakers 14–17, usually 16 (Burgess 1978).

Color. The side of the body is white, with a brilliant yellow fringe along the medial fins and covering the paired fins. A prominent bar extends from the origin of the dorsal, through the eye, to the ventral edge of the preopercle. It appears jet black and its margins may be yellow. An eyespot appears at the base of the soft dorsal fin, but, despite its common name and Bloch's original description, it is not ocellated (Burgess 1978). A second, smaller spot may be present in fish larger than 50 mm SL at the angle of the soft dorsal fin margin. Presence of these spots is variable, which may confound identification. Color may fade under certain conditions, particularly at night (Longley and Hildebrand 1941; Burgess 1978; Böhlke and Chaplin 1993).

Figure 240. Spotfin butterflyfish *Chaetodon ocellatus.* Beesleys Point, N.J., 38 mm. Drawn by W. S. Haines.

Size. Maximum size 150 mm SL (Burgess 1978).

Distinctions. No other butterflyfish has been reported from the Gulf of Maine, but foureye butterflyfish (*C. capistratus*) may occur there because they have been repeatedly collected as far north as Woods Hole (McBride and Able 1998). Spotfin butterflyfish develop an unocellated eyespot at the base of the soft dorsal and, eventually, a second smaller spot (also unocellated) along the margin of the soft dorsal. Foureye butterflyfish may have one or two eyespots as well, but both are ocellated. A permanent eyespot appears at the caudal peduncle of foureye butterflyfish and a more transitory eyespot appears only in young juveniles on the soft dorsal fin. Illustrations of young fish can be found in Bean (1888), Smith (1898a), Fowler (1945), Schaefer and Doheny (1973), Burgess (1978), Fritzsche (1978), Allen (1979), Wang and Kernehan (1979), Böhlke and Chaplin (1993), and Able and Fahay (1998).

Habits. Spotfin butterflyfish live in association with live-bottom and shelf-edge habitats offshore of the South Atlantic Bight (Struhsaker 1969a). Butterflyfishes are typically restricted to mortality of spotfin butterflyfish has occurred in natural habitats near Tampa Bay, Fla., during unusually cold winters (Gilmore et al. 1978). Spotfin butterfly fish are rare and highly seasonal in the Gulf of Maine (reported only for August), consisting of expatriates from southern spawning grounds (McBride and Able 1998). Juveniles have been collected in a variety of habitats in New England, including salt-marsh tidal creeks (Murphy 1991) and eelgrass (Sumner et al. 1913).

Butterflyfishes are generally territorial or restricted to a small home range (Reese 1975). Spotfin often occur in pairs or in small groups (Longley and Hildebrand 1941). Juvenile spotfin remove parasites from other fishes (Burgess 1978).

Food. The diet of spotfin butterflyfish in Bahamian coral habitats consisted primarily of anthozoan tissue, polychaetes, eggs, decapod crustaceans, and other miscellaneous invertebrates (Pitts 1991).

Predators. Larval butterflyfishes are relatively common in stomachs of large pelagic fishes (e.g., tunas and dolphinfishes) but there are very few reports of older fish as prey (Burgess

Parasites. Various parasites have been noted for species of *Chaetodon* but there are no specific reports for spotfin butterflyfish (Burgess 1978).

Breeding Habits. There is no specific reference for spotfin butterflyfish breeding habits (Burgess 1978). Spawning by three sympatric butterflyfishes has been observed near sunset during February–April, with some evidence of a lunar spawning cycle (Colin 1989).

Early Life History. The eggs are buoyant, transparent, and spherical, approximately 0.6–0.7 mm in diameter (Thresher 1984). Larvae may not have clearly defined pigment patterns, but fish as small as 12 mm SL develop an eyebar and a second bar along the posterior edge of the body (Able and Fahay 1998). The second bar, present only in fish smaller than 20–22 mm SL (Longley and Hildebrand 1941; R. McBride, unpubl. data), can appear more indefinite but generally spans the entire posterior edge of the fish (i.e., along the base of the soft dorsal fin, across the caudal peduncle, and along the base of the anal fin). As this posterior bar fades, a black spot remains at the base of the soft dorsal and is regarded as a false eyespot. Head plates, which are conspicuous during the planktonic phase, become reduced during settlement and are lost by 20–22 mm SL. Comparisons of dip net and bottom trap collections indicate that spotfin butterflyfish settle by 18 mm SL (McBride and Able 1998).

Age and Growth. The sizes encountered north of Cape Hatteras (i.e., <70 mm TL) appear to be exclusively young-of-the-year fish (McBride and Able 1998). Length-weight relationship for 48 live fish (21.0–68.2 mm TL) collected in unbaited traps in a southern New Jersey estuary (R. McBride, unpubl. data), where W = weight (g), and L = TL (mm): $\log_{10} W = -4.62 + 3.02(\log_{10} L)$.

General Range. Spotfin butterflyfish occur along the eastern coast of North America from Tortugas and the Gulf of Mexico to Nova Scotia (Longley and Hildebrand 1941; Scott and Scott 1988). It is a tropical species that is a perennial summer visitor to temperate waters, including southern New England (McBride and Able 1998).

Occurrence in the Gulf of Maine. Published records are limited to two fish: one (31 mm TL) collected in the Bay of Fundy during the summer of 1933 (Vladykov 1935a) and the other (19 mm TL) collected in southern Maine in August 1989 (Murphy 1991). Three more fish were collected that August, near the mouth of Portsmouth Harbor, N.H. (P. S. Levin, pers. comm.).

Labroid Fishes. Suborder Labroidei

Wrasses (or labrids) are perciform fishes thought to be closely related to parrotfishes, damselfishes, surfperches, and cichlids (Kaufman and Liem 1982; Stiassny and Jensen 1987). Collectively, members of the suborder Labroidei are referred to as pharyngognath teleosts, characterized by shared specializations of the pharyngeal jaw apparatus and associated musculature. Wrasses have strong pharyngeal jaws located on the roof of the mouth and floor of the throat (pharynx), armed with opposing patches of conical or knoblike teeth with which they grind the hard-shelled mollusks and crustaceans on which they feed.

WRASSES. FAMILY LABRIDAE

THOMAS A. MUNROE

Wrasses comprise a speciose family with about 68 genera and approximately 453 species (Parenti and Randall 2000). The Labridae is the second most diverse family of marine fishes (in terms of numbers of species) and the third largest perciform family. Most species are small (<6 cm), brightly colored, inhabitants of shallow water tropical reef areas. Some wrasses, however, attain relatively large sizes reaching 1–3 m in length.

Wrasses are generally associated with structurally complex environments, inhabiting man-made as well as naturally occurring structures, such as reefs, rocky outcroppings, shipwrecks, or heavily vegetated areas. Many species especially depend on structure during their nighttime sleep phase, lying quiescently in a crevice or hole in the reef or wreck or even burying themselves in the sand adjacent to the reef.

Wrasses are well known for their complex social biology and reproductive strategies (Warner and Robertson 1978). Many species are protogynous, changing sex from female to male. Males, females, and juveniles of protogynous species often display markedly different color phases, making identification extremely difficult. Other species may not undergo sex changes as part of their reproductive repertoire, although their reproductive behavior can still be quite complicated, with some species utilizing more than one reproductive strategy. In these species, territorial males combine aggressive pair-spawning with individual females, and several individuals of both sexes (including nonterritorial males) may participate in nonaggressive, simultaneous, group spawning.

Only two species of wrasses, cunner and tautog, are found

in the Gulf of Maine. They are distinguished from other Gulf of Maine fishes in having a single, long dorsal fin with its forward part spiny and its posterior part soft-rayed, with no evident demarcation between the two. Both cunner and tautog are resident species, but cunner are the more commonly occurring and abundant of the two and also range further north. The center of abundance for tautog is south of Cape Cod, but they are frequently taken in the Gulf of Maine, although not usually in substantial quantities.

KEY TO GULF OF MAINE WRASSES

1a. Gill covers scaly; snout somewhat pointed; dorsal profile of head rather flat; dorsal fin with 18 spines and 9 or10 soft rays; anal fin rays usually 9 or 10 (rarely 7 or 8) . . . . **Cunner**
1b. Gill covers largely naked; snout blunt; dorsal profile of head high-arched; dorsal fin with 16 or 17 strong spines and 10 or 11 soft rays; anal fin usually with 7 or 8 soft rays. . . . . **Tautog**

TAUTOG / *Tautoga onitis* (Linnaeus 1758) / Blackfish, White Chin, Tog / Bigelow and Schroeder 1953:478–484

Figure 241. Tautog *Tautoga onitis.* Woods Hole, Mass. Drawn by H. L. Todd.

Description. Body stout; deep, 2.55–3.00 times in SL; somewhat laterally compressed (Fig. 241). Caudal peduncle deep; head rather short, blunt, with steep upper profile and somewhat rounded snout. Head length 3.25–3.56 in SL, somewhat shorter than body depth. Mouth small, slightly subinferior with thick lips and protractile upper jaw; posterior end of maxilla slightly anterior to vertical through front margin of eye in juveniles, reaching vertical through anterior nostril in adults. Each jaw with two to three series of stout, conical teeth, which gradually decrease in size posteriorly; strong, rounded, crushing pharyngeal teeth at rear of mouth. Snout short, rather rounded, nearly equal to eye diameter. Eye moderate, 3.05–6.00 in HL. Body covered with small, thin, cycloid scales and a tough skin; membrane of dorsal and anal fins partly covered with small scales; top of head, preorbital, maxilla, lower jaw, interopercle, and posterior portions of preopercle naked; series of five to six small scales extend below and posterior to eye and onto anterior region of preopercle; ventral and posterior part of gill cover scaleless; scales on body between gill opening and bases of pectoral and pelvic fins reduced in size. Dorsal fin continuous; originating over upper

and continuing posteriorly to caudal peduncle. Anteriormost four or five dorsal fin rays somewhat graduated in size; others about equal in height; soft-rayed dorsal fin with rays somewhat higher than spines, about a third as long as spiny part, rounded in outline; second dorsal fin ray highest. Anal fin originating somewhat posterior to vertical through midpoint of dorsal fin; rounded in outline. Caudal fin broad, truncate, or slightly rounded. Pelvic fin moderate, at vertical through midpoint of pectoral fin. Pectoral fin relatively large, broad, somewhat rounded. Lateral line complete and continuous, arched anteriorly. Gill rakers short, blunt.

Meristics. Dorsal fin rays XVI or XVII, 10 or 11; anal fin rays III, 7 or 8; pectoral fin rays 16; lateral line scales 60–68; gill rakers 9 on lower limb of first arch; branchiostegals 5 or 6; vertebrae (16) 17 + 18 = 34 or 35.

Color. Tautog on different bottoms, like cunner, vary greatly in color and in their markings. Adults are often rather darkly colored, ranging from a generally mouse-colored background to one of chocolate gray, deep dusky, olive green, or dull

darker pigment. Lateral mottlings are more evident in young fish than in adults and are usually grouped as three pairs of more or less continuous bars or as a series of interconnecting dark blotches. Large fish are often mostly plain brownish or blackish. The belly is only slightly paler than the sides, but the chin is usually white on larger fish, a very conspicuous character. Sexually mature fish vary markedly in coloration. Females and some nondimorphic males tend to be dull mottled brown, usually with a series of lateral blotches. In contrast, dimorphic males are typically grayish or blackish, with prominent white markings on ventral and dorsal margins of pectoral and caudal fins and on the chin, and are often without darkly pigmented lateral markings. Adult males also have a single small, whitish spot on their midsides just posterior to the posterior margin of the pectoral fins (roughly at the height of the lateral line). Newly recruited, young-of-the-year fish, usually those smaller than 35 mm TL, collected in habitats containing large quantities of sea lettuce (*Ulva lactuca*) are often cryptically mint green in coloration.

Size. Tautog grow to about 90 cm and 10.2 kg; most weigh 0.56–1.8 kg (Bigelow and Schroeder; Fahay 1983; Scott and Scott 1988). The all-tackle game fish record (IGFA 2001) is an 11.33-kg fish taken off Ocean City, N.J., in January 1998. Most tautog caught by sport fishermen are 0.9–1.8 kg, and fish 5.4–6.4 kg are unusual.

Distinctions. The body plan of tautog suggests that of an overgrown cunner, but heavier and stouter, about three times as long as deep. The most obvious differences between tautog and cunner are that the tautog head is high-arched, steep, and rounded in dorsal profile vs. not high-arched, more pointed in profile in cunner; the snout is much more blunt, and the lips are much thicker. The cheek region anterior to the gill opening is naked in tautog (scaly in cunner) and velvety to the touch. Tautog are easily distinguished from cunner by counts of fin rays: dorsal spines XVI or XVII vs. XVIII in cunner; anal fin rays 7 or 8 vs. 8 or 9, but rarely 7 or 9. Also, the pelvic fins in tautog are located under the midpoint of the pectoral fins (vs. at a vertical through anterior base of pectorals in cunner). Tautog also grow much larger than cunner.

Habits. Tautog are a strictly coastal fish, especially in the northern part of their range. Northward from Cape Cod tautog rarely occur more than 5–6 km from land or deeper than 9–18 m and are unknown from offshore banks (Bigelow and Schroeder). South of New England, they occur inshore and range farther seaward into deeper waters. They are commonly caught in 18–24 m on Cholera Bank, 18–22 km offshore of Long Island, and on Seventeen Fathom Bank, 13 km off New Jersey. Off Virginia, they occur inshore seasonally and are present in 12–25 m year-round in reeflike areas (*Triangle, Tiger,* and *Fish Haven* wrecks; Chesapeake Light Tower), 33–60 km offshore (Musick 1972; Hostetter and Munroe 1993). At the other extreme, they follow the flood tide up above low-water level around ledges to prey on the abundant supply of blue mussels in the intertidal zone, dropping back into deeper water during ebb tide. Bigelow and Schroeder seined many small tautog close to shore in water less than a meter deep at Provincetown as well as at other southern locations in the Gulf. Newly settled juveniles recruit to shallow waters of the estuary, whereas with increasing size, young-of-the-year fish move from shallow to deeper areas in the estuary (Sogard et al. 1992; Dorf and Powell 1997). Smaller juveniles utilize sea lettuce (*Ulva*) and other macroalgal habitats, and may be dependent on them (Heck et al. 1989; Sogard and Able 1991; Dorf and Powell 1997), moving to eelgrass and rocky habitats as they grow (Olla et al. 1979; Sogard et al. 1992). The importance of this habitat for young stages of tautog is reflected in the fact that growth rates of young fish were higher in areas where habitat quality (presence of sea lettuce) was also high (Sogard 1992). Young-of-the-year and age-1+ juveniles also utilize empty oyster and clam shells (Bigelow and Schroeder).

Tautog do not school but many individuals often congregate in the same habitat. Tautog, like cunner, live near the bottom, strongly associated with cover. They are most numerous along steep, rocky shores; around breakwaters, off-lying ledges, and submerged wrecks; around piers, docks, and jetties; over boulder-strewn bottoms; and on mussel beds in shallow water (usually well under 30 m). However, in some places (e.g., the eastern side of Cape Cod Bay) considerable numbers are caught on smooth bottom, away from any structure or relief (Bigelow and Schroeder; Schwartz 1964b). Small juveniles, 19–155 mm long, are sometimes seined on sandy beaches, but young tautog usually inhabit vegetated areas such as *Zostera* beds or mats of macrophytic algae (Tracy 1910; Bigelow and Schroeder; Briggs and O'Connor 1971; Sogard 1989; Heck et al. 1989; Sogard and Able 1991; Able and Fahay 1998). In Narragansett Bay, R.I., juvenile abundances varied significantly with submerged vegetation cover density (Dorf and Powell 1997), with high and medium cover densities harboring significantly greater numbers of tautog than sites with low cover densities. Tautog usually enter poly- and mesohaline regions of estuaries, but as yet are unrecorded from freshwater.

Tautog produce deep thumps and barklike grunts in the field, probably in response to alarming stimuli (Fish 1954). They also produce escape sounds and thumps during feeding in the laboratory. The sonic mechanism is their large swim bladder vibrated by contraction of skeletal muscle, striking or rubbing of the ribs, and drum-beating motions of the opercles (Fish and Mowbray 1970).

Tautog exhibit a typical labrid diel behavioral pattern of high activity during daylight hours and nearly complete inactivity at night (Olla et al. 1974; Arendt et al. 2001a). From ultrasonic tracking studies, Olla et al. (1974) found that initiation and cessation of activity varied somewhat relative to morning and evening civil twilight. Activity began from 10 min before to 69 min after the start of morning twilight. Cessation of activity was more variable, ranging from 222 min before to 69 min after the end of evening twilight. In Chesapeake Bay, mean daily activity began at sunrise and ceased at sunset, with daily peaks in activity occurring during slack tide in the early morning or late afternoon (Arendt et al. 2001a). Tautog of all sizes require a suitable physical structure (home area or shelter) during their quiescent (sleep) nighttime phase (Olla et al. 1974), and up to the age of 3 years are totally dependent upon the home area at all activity levels.

Daily movements consist of fish leaving a home area and moving to a feeding area. They remain at the feeding area (usually a mussel bed) until the approach of twilight, whereupon they return to the home area, settle in one location, and remain there throughout the night in an inactive state (Olla et al. 1974). Tautog smaller than 25 cm do not venture further than 2–3 m from their home area to forage and, unlike the larger fish, feed within proximity of the shelter.

Adult tautog undertake seasonal inshore-offshore migrations, at least in northern regions of the range. Migration occurs when water temperatures approach 8°–12°C. Throughout the northern parts of their range tautog are seldom observed inshore before late April or after November (Bigelow and Schroeder; Cooper 1966; Briggs and O'Connor 1971). Tagging studies off Rhode Island (Cooper 1966) and New York (Olla et al. 1974; Briggs 1977b) indicated marked seasonal inshore-offshore migrations of adults depending on temperature. There was little if any north-south component to this migration. Juvenile tautog overwinter in inshore areas (Bigelow and Schroeder; Olla et al. 1974; Briggs 1977b) and are inaccessible to most conventional collecting gear during this time. In some regions, a portion of the adult population remains offshore in deep water throughout the year (Olla and Samet 1977; Hostetter and Munroe 1993). In Chesapeake Bay adult tautog occupy inshore habitats year-round (Arendt et al. 2001b). Some fish in this region evidently do not make seasonal inshore-offshore movements.

Tautog are not usually caught in Massachusetts Bay or at Duxbury in Cape Cod Bay before late April or early May (Bigelow and Schroeder). In 1950, which appears to have been an early season, they were reported as biting well in Cape Cod Bay by 25 May and at Duxbury by the last days of the month. Further up the Bay, however, at Cohasset and Swampscott, very few were caught before July. In most years the best catches are made in August, September, and into October, and Bigelow and Schroeder reported that tautog are not usually taken anywhere in the Gulf after early November at the latest.

Based upon field observations and tag returns, Cooper (1966) showed that adult tautog inhabited Narragansett Bay from early May until late October. Offshore migration of adults began in late October or early November. With onset of colder temperatures, adults migrated approximately 3 km offshore to areas of rugged bottom topography 24–60 m in depth.

Seasonal movements of adults and juveniles in an inshore area near Long Island, N.Y., were observed by ultrasonic tracking and diving studies (Olla et al. 1974) over a 2-year period (two autumns, one spring). In August (water temperature November (water temperature 10°C), many small tautog were observed, but none of the larger ones was seen in the study area. When water temperatures decreased below about 6°C (2.0°–5.5°C), small tautog overwintering inshore were lethargic or torpid and partially covered with silt. The following spring (May), with water temperatures about 10°C, active tautog of all sizes were again observed inshore.

Tautog of all sizes become lethargic and even torpid in low water temperatures. The lack of escapement behavior in their behavioral repertoire, other than to move offshore (adults only), makes this species susceptible to injury or death from cold-shock. Mass mortalities, presumably owing to cold-shock, were reported (1841, 1857, 1875, 1901, and probably others) for populations off Rhode Island and Massachusetts (Bigelow and Schroeder), and tautog mortalities were reported in a major anoxia-hypoxia event off New Jersey (Azarovitz et al. 1979).

In the northern parts of their range, tautog apparently do not feed during seasonal thermal minima. Young fish overwintering inshore remain in a nonfeeding, dormant state at their home site (Olla et al. 1974). Adults observed in waters 24–28 m deep off Rhode Island during winter (water temperature 7.5°C) were torpid, with some lying on their sides and others upright in bottom crevices (Cooper 1966). Intestinal tracts of three captured specimens were greatly reduced in size and devoid of food. Of 12 additional adults that Cooper obtained from commercial fishermen during December, only two had food in their digestive tracts (remains of *Cancer irroratus*). Further south, for example, off Virginia, where winter temperatures are not as cold, tautog at offshore sites actively feed throughout much of the winter (Hostetter and Munroe 1993). In inshore areas of Chesapeake Bay, adult tautog were seasonally not active for 1–7 days during annual minimum temperatures (5°–7°C) (Arendt et al. 2001a)

Sublethal thermal stress studies (Olla and Studholme 1975; Olla et al. 1975a, 1978) indicated that feeding behavior for both juvenile and adult tautog is depressed at elevated temperatures, and activity, aggression, and feeding are reduced, whereas association with shelter is increased. Upper lethal temperatures are about 31°–33°C (Pearce 1969; McCormack 1976).

Food. Tautog are opportunistic sight feeders (Olla et al. 1974, 1975a), feeding throughout the day on a variety of invertebrates, chiefly mollusks (both univalves and bivalves), especially mussels (*Mytilus edulis* in northern parts of the range, *Brachiodontes exustus* at Masonboro Inlet, N.C. [Lindquist et al. 1985]); barnacles that they pick off rocks and pilings; various other crustaceans including amphipods, isopods, and decapods, echinoderms; and occasionally small fishes. Feeding begins shortly after morning twilight and continues up to evening twilight. With the approach of evening they become inactive, and they cease feeding during night time.

Tautog jaws are equipped with large canine teeth on both

jaw teeth is to capture and manipulate prey, not mastication of prey items. Tautog possess a highly evolved pharyngeal jaw apparatus for crushing and grinding their hard-shelled prey (Liem and Sanderson 1986).

Food habit studies conducted in the New York area indicate that blue mussels (*Mytilus edulis*) were the principal prey item of tautog of all sizes (Olla et al. 1974, 1975a). Tautog (34–53 cm) held in an aquarium fed on clumps of mussels by grasping them from the substrate using their canine teeth, and if the shell clump was too large, the shells were separated using the canines (Olla et al. 1974). Initial ingestion of mussels did not involve any crushing by the canines. Instead, shells were crushed after ingestion by the pharyngeal teeth. The average size of ingested mussels was 11.9 mm (1–2 years old). All fish observed in the field and aquariums selected young, small-sized mussels, although the mouth could accommodate much larger ones. Preference for small mussels results from size limitations imposed by the pharyngeal jaw apparatus. Fish 34–53 cm could crush mussels in the pharyngeal apparatus that were only 0.47 times the maximum size of mussels that could be accommodated in the buccal region (Olla et al. 1974).

Other food items in the tautog diet include rock crabs, hermit crabs, mud crabs, sand dollars, gammarid amphipods (sand fleas), scallops, clams, shrimps, isopods, and lobsters (Hildebrand and Schroeder 1928; Steimle and Ogren 1982; Lindquist et al. 1985). Rock crabs constituted 78% of the diets of tautog at an artificial reef off New York (Steimle and Ogren 1982). Small lobsters are generally swallowed whole, but shells of larger ones are first cracked by using the pharyngeal crushing teeth. It is also likely that tautog living in shallow bays (e.g., Duxbury) prey to a considerable extent on sea worms (*Nereis*); certainly they take these freely as bait (Bigelow and Schroeder). Young-of-the-year fish feed on copepods (mostly harpacticoids), amphipods, isopods, and small decapod crustaceans (Dorf 1994). Beyond lengths of about 120 mm, tautog gradually change their diets to include greater numbers of small mussels.

Predators. No species is known to preferentially feed on tautog. Juveniles are preyed upon by piscivorous birds such as cormorants (Nichols and Breder 1927) and undoubtedly by piscivorous fishes. In a laboratory study of predator avoidance, Dixon (1994) showed that juvenile tautog avoided an active predator (juvenile bluefish), but did not react to the presence of inactive, camouflaged predators (toadfish) or to a moderately active predator (longhorn sculpin). Bigelow and Schroeder reported the following fish species as predators on juvenile or adult tautog: smooth dogfish, barn-door skate, red hake, sea raven, and goosefish. Schaefer (1960) reported tautog in the diet of silver hake collected off northern New Jersey.

Parasites. Linton (1901b) noted that tautog, unlike other benthic omnivorous fishes in the Woods Hole region, did not host large or diverse populations of helminth parasites. He suggested that the low helminth parasite loads he observed may have resulted from the "antihelminthic" diet of tautog, which includes the abrasive fragments from barnacles, decapod crustaceans, and blue mussels crushed and macerated during feeding. These sharp fragments may prevent establishment, or may actually destroy, the delicate larval stages of helminths ingested with the food supply. Despite this potentially difficult environment, five helminth species have been recovered from the alimentary tract of tautog: the cestode *Bothrimonus intermedius* (Linton 1941) and digenetic trematodes *Zoogonoides laevis, Homalometron pallidum, Opecoeloides vitellosus* from fishes at Woods Hole (see Yamaguti 1958), and the acanthocephalan *Echinorhynchus gadi* from tautog at Narragansett Bay (T. A. Munroe, unpubl. data).

Tautog also host a variety of external parasites, including a monogenetic trematode, *Microcotyle hiatulae,* which has been reported in heavy infections at Woods Hole (Linton 1940), Newport, R.I. (Goto 1900), and coastal Virginia (Thoney and Munroe 1987). Tautog apparently are also parasitized by the leech *Calliobdella vivida,* because blood samples from tautog in Chesapeake Bay revealed infestations with a hemoflagellate, *Trypanoplasma bullocki,* which is transmitted to fish via feeding of a marine leech (Burreson and Zwerner 1982). The most conspicuous and frequent external parasites of tautog are the encysted metacercariae of a digenetic trematode *Cryptocotyle lingua.* Linton (1901b, 1928, 1940) recorded heavy and pervasive infestations of these metacercarial cysts in tautog collected at Woods Hole, noting that it was unusual to find a fish not infected with this larval trematode. Cheung et al. (1979) reported tautog lethally infected with an intestinal coccidian parasite.

Reproductive Biology. Tautog are gonochoristic (Bigelow and Schroeder; Chenoweth 1963; Cooper 1967). The sex of adults can easily be recognized by the strong dimorphism in mandible structure and dichromatic coloration; males have more pronounced mandibles than females (Cooper 1967). Little is known concerning reproductive biology of populations in the Gulf of Maine and further north so information presented is based on studies conducted in more southern parts of the range.

Sexual maturation in tautog occurs at the beginning of the third year of life for males and at the beginning of the fourth year for females, corresponding to sizes of approximately 14–25 cm (Chenoweth 1963; Cooper 1967; Stolgitis 1970; Hostetter and Munroe 1993). Males off Long Island (Briggs 1977b) were ripe at 215 mm (Cooper's age 3) and the smallest ripe female was 230 mm (age 4). Briggs concluded that females may mature somewhat later in the more southerly waters of New York.

Olla and Samet (1977) reported capturing small (sizes unstated), reproductively mature individuals in which the distinctive dimorphism in mandible structure characteristic of older, larger fishes was not developed. It is not known whether these smaller fish participate in or contribute to the reproductive success of the population.

Tautog are indeterminate serial spawners with a protracted spawning season (Hostetter and Munroe 1993; White 1996), and estimating fecundity poses several problems. Fecundity data for this species contained in unpublished theses (Chenoweth 1963; Stolgitis 1970; White 1996) indicate that they are relatively prolific. However, as White (1996) has shown, previous fecundity estimates were for batch, not total, fecundity.

Batch fecundity estimates for tautog from these three studies provide the following range of estimates by age-group: females age 3 produced 3,400–24,000 eggs per batch; age 4–6, 46,000–54,000; age 7–9, 103,000–117,000; age 13–16, 209,000–457,000; and age 20, about 150,000–483,000. Maximum batch fecundity was observed in females ages 7–9 with reduced fecundity in fishes older than age 16 (Chenoweth 1963). Stolgitis found that batch fecundity for 49 females he studied was more closely related to length and weight than to age. Fecundity increased in proportion to the weight of the fish and to the cube of its length. White (1996) estimated a spawning frequency of about 1.2 days, resulting in 58 spawning days per female per season. Based on batch fecundity and spawning frequency estimates, potential annual fecundity for tautog ages 3–9 ranges from 160,000 to 10,500,000 eggs.

Spawning in a natural environment has never been described for tautog primarily because these fish are easily disturbed by diver intrusions (Olla and Samet 1977). Bridges and Fahay (1968) observed possible courtship behavior of two laboratory-held tautog. Olla and Samet (1977) believed that the tautog observed by Bridges and Fahay were exhibiting aggressive behavior and not courtship displays, either because the fish were not in complete reproductive synchrony or because the confines of the aquarium produced behavioral artifacts.

Olla and Samet (1977) successfully spawned tautog in the laboratory and provided the following observations concerning the behavior of the spawning fish. Two separate groups of tautog, each with two males and one female, were studied over an entire spawning season. The larger male was dominant over the other two fish and, once reaching reproductive readiness, was the primary spawning partner of the female. The subordinate male did spawn with the female, although only infrequently. Each day the female exhibited dynamic and transient shading changes that became maximally developed as the time of each spawning approached in the afternoon. Actual gamete release occurred after 6–8 h of courtship, as male and female moved upward in synchrony and spawned at or near the surface. Often, both fish broke the surface while simultaneously releasing gametes. Breaking of the water's surface at gamete release may be critical for maximizing fertilization.

During one set of observations, Olla and Samet (1977) observed 37 spawnings, and in a second set noted 23 spawnings. Laboratory spawning took place almost exclusively in pairs, but whether or not a true pair bonding between mates occurs in nature is still unclear. Other spawning behaviors, such as aggregate spawning, were infrequently observed in this species, but cannot be eliminated because spawning has not been ob-

In some areas, tautog were reported to begin spawning in the afternoon (Olla and Samet 1977; Perry 1994). Based on the ages of eggs collected in ichthyoplankton samples, Ferraro (1980) reported that tautog apparently spawn into the night. However, more accurate estimates of egg release based on hydrated oocyte methods (White 1996) indicated that age estimates based on ichthyoplankton samples are not reliable. Based on hydrated oocytes in the ovaries, White (1996) reported that tautog daily ovarian development followed a general pattern of hydration early in the day (0700–0930 EST), running ripe (spawning) midday (0930–1530 EST), and partially spent/redeveloping in late afternoon and early evening (1430–1830 EST). Differences in spawning windows among groups of tautog sampled on different days indicated that changes in environmental conditions such as water depth, ambient light, or tidal stage may also influence ovarian cycles.

Spawning Seasonality. Tautog eggs were collected by the MARMAP surveys from April to September in coastal and midshelf waters (Berrien and Sibunka 1999: Fig. 64). Spawning begins as early as April in the southern portion of the Mid-Atlantic and by May progresses northward into southern New England waters. Peak spawning occurred in inshore waters from Chesapeake Bay to Nantucket Shoals in June and July. Based on egg and larval occurrences in plankton collections, it is evident that tautog begin spawning at water temperatures of 9.0°–10.0°C and continue throughout the summer until early October.

Austin (1976) noted that developing embryos were absent and no larvae were taken from ichthyoplankton samples in waters warmer than 21°C. He suggested that effectual spawning of tautog may therefore be restricted to a narrower, early summer range of temperatures and that conceivably tautog do not spawn in any numbers when temperatures exceed 22°C. It is interesting to note that from a larval energetics study, Laurence (1973) also found that 22°C approaches the upper temperature limit for normal metabolism of embryonic and prolarval tautog. He demonstrated that tautog prolarvae may encounter a potential energy deficit at higher temperatures (19°–22°C) within their spawning range. Larvae produced at a lower spawning temperature (16°C) did not appear to be as susceptible to a potential energy deficit as those produced at the higher temperatures. Also, prolarvae hatched at 16°C were larger at completion of yolk absorption and had a longer period between hatching and the initiation of feeding. In the laboratory, when incubation temperatures were gradually raised from 20.0°C, anatomical deformities, including stunted embryos and/or abnormal body curvatures, as well as increased mortality, occurred between 24.2° and 26.3°C (Olla and Samet 1978).

Spawning Location. Tagging studies in Rhode Island (Cooper 1967) and New York (Briggs 1977b) indicate that tautog form discrete, localized groups that spawn in the same areas for several seasons. North of Cape Cod, tautog eggs

(Elliott and Jimenez 1981), but it is uncertain to what degree tautog populations in the Gulf of Maine are maintained by local spawning vs. immigration of fishes spawned further south (Bigelow and Schroeder). Bigelow and Schroeder did not collect eggs, larvae, or juvenile tautog north of Cape Cod Bay, and other authors have also noted the absence of eggs and larvae in areas north of Massachusetts Bay (Nova Scotia [Dannevig 1919] and Central Maine [Chenoweth 1973; Berrien and Sibunka 1999]). It seems likely that some limited spawning probably does occur in northern portions of range, at least during periods of favorably warm temperatures. Hauser (1973), for example, collected two larval tautog from the Sheepscott River estuary–Montsweag Bay, Maine, and Bleakney (1963b) collected a gravid female in Nova Scotia, although little is known of tautog spawning activities in Canadian waters (Scott and Scott 1988).

Spawning occurs in inshore and nearshore waters, at the mouths of estuaries, and offshore on wrecks and reefs on the continental shelf (Herman 1963; Cooper 1966; Eklund and Targett 1990; Sogard et al. 1992; Hostetter and Munroe 1993), at temperatures of 9°–26°C in Long Island Sound (in salinities of 26–29 ppt) and 13°–14°C (in polyhaline water) in Rhode Island (Fritzsche 1978).

Early Life History. Detailed descriptions of tautog eggs and larvae appear in Kuntz and Radcliffe (1917), Fritzsche (1978), and Fahay (1983). Laboratory culturing of tautog has been accomplished successfully (Perry 1994; Mercaldo-Allen et al. 1997; Perry et al. 1997a,b, 1998) and described in detail (Schoedinger and Epifanio 1997).

Tautog eggs are pelagic, buoyant, spherical, lack an oil globule, and have a diameter of 0.9–1.0 mm (Able and Fahay 1998). Identifying them in ichthyoplankton samples is complicated by the fact that eggs of cunner and tautog are similar in size and seasonal occurrence and both decrease in diameter during the course of the spawning season (Williams 1967). However, the decreases in size are parallel in the two species, with the smaller eggs being those of cunner. Eggs of the two species can also be distinguished by immunodiffusion (Orlowski et al. 1972).

Hatching occurs in 42–45 h at 20°–21°C (Kuntz and Radcliffe 1917), whereas at 16°C it takes 7 days (Laurence 1973), compared with 5 days at 19°C and 4 days at 22°C. Larvae hatch at 1.7–2.2 mm with unpigmented eyes and unformed mouth parts (Schoedinger and Epifanio 1997; Able and Fahay 1998). Yolk-sac absorption occurs at 3.2–3.5 mm (Fritzsche 1978) and the mouth is fully formed at 4 days (temperature ca. 20°–22°C). At 5 mm, the first traces of caudal fin rays are evident. At 10 mm, dorsal and anal fins are differentiated. By about 30 mm, the fins, body form, deep caudal peduncle, and blunt nose of adult tautog are evident. Prey density has a strong effect on growth of tautog larvae (Schoedinger and Epifanio 1997), with mean growth at high prey density (1,000 liter^{-1}) approximately four times that at low prey density (100 liter^{-1}). Sogard et al. (1992) estimated that tautog larvae are planktonic for about 3 weeks before settling.

Larvae and small juveniles of tautog and cunner resemble each other closely in general form. However, the arrangement of pigment on the body offers a ready means for identification at all but the very earliest stages, for black pigment cells remain more or less uniformly scattered over the whole trunk in tautog, whereas they soon cluster into two definite patches in cunner.

Age and Growth. No aging studies have been published for tautog populations from the Gulf of Maine. Cooper (1967) successfully used opercular bones for aging tautog in Narragansett Bay, and Hostetter and Munroe (1993) examined age-growth characteristics of tautog occurring off Virginia. Scales were not used to age tautog in these studies because annuli were difficult to discern and scales were often pitted by metacercariae of a digenetic trematode (*Cryptocotyle*) and regenerated. Otoliths can be used to age fish during their first few years of life, but beyond that there are difficulties with interpretation of annuli on the outer margin of these structures (Hostetter and Munroe 1993).

Estimates of daily growth using otolith increments were provided for juveniles by Sogard et al. (1992), who found that young tautog in New Jersey grow relatively rapidly, at an average rate of 0.5 mm·day^{-1}, during the summer. They reach a modal size of about 75 mm SL (40–100 mm SL) after their first summer and 155 mm (110–170 mm SL) by the end of their second summer. Only minor growth was evident during fall, winter, and spring. Based on formation of a settlement mark, it was estimated that young tautog spend about 3 weeks in the plankton before settling to the benthos. In one study, growth rates of juveniles were also affected by habitat quality (Sogard 1992), with highest rates occurring in areas with abundant quantities of sea lettuce, whereas Phelan et al. (2000) reported that growth was relatively independent of whether a habitat was vegetated or adjacent to vegetation.

Annulus formation for tautog in Rhode Island waters occurs during middle or late May (spawning season) and only one annulus is formed per year. Age estimates for tautog populations off Rhode Island and Virginia reveal that tautog are relatively long-lived, with estimates of longevity ranging up to 34 years (Cooper 1965b, 1967; Hostetter and Munroe 1993). Tautog grow slowly and attain relatively large sizes. Males grow faster and bigger than females ($L_\infty = 655$ mm for males; $L_\infty = 506$ mm for females). Hostetter and Munroe (1993) estimated parameters of the von Bertalanffy growth equation for tautog (both sexes) in Virginia waters as: $L_\infty = 742(1 - e^{-0.085(t - 1.316)})$, with L in mm and t in years.

Using eviscerated weights of tautog collected in June 1961, Cooper derived the following length-weight relationships for tautog collected in Rhode Island: $\log W = 4.35670 + 2.77660 \log L$ for males; $\log W = -4.80357 + 3.01607 \log L$ for females.

Back-calculated length increments agreed well with observed increments for tagged fish. The largest annual increments in growth occurred during years 1–4. Males (average length 73 mm) and females (66 mm) grew the most in length during their second year of life. Annual increments of 12–36 mm occurred in years 6–12; increments decreased to 23 mm (males) and 22 mm (females) in year 10; and after age 12 growth of both sexes continued, but at much slower rates (11→2 mm·year^{-1}).

Briggs and O'Connor (1971) measured seasonal size progression of young-of-the-year tautog in Long Island. Young-of-the-year (19–35 mm TL) first appeared in the study area in July. By November these fish had grown to 76–134 mm, and these lengths lie within back-calculated values for age-1 tautog from Rhode Island (Cooper 1967).

With derived length-weight relationships (see above), Cooper calculated annual increments of weight for males and females. He found that females were heavier than males at a given size. Body weight (eviscerated) increased approximately as the cube of the length for females and less rapidly for males. Annual increments increased to a maximum of 169 g for males age 9 and 208 g for females age 11, despite the fact that the greatest annual increments in length were attained during the second year. Males required 10 years and females 9 years to reach a body weight (excluding viscera) of 908 g (2 lb). From his length-weight formula, Cooper calculated that tautog reach 1,820 g (4 lb) in approximately 25 and 15 years for males and females, respectively.

A length-weight relationship based on uneviscerated weights was derived by Briggs (1969b) for tautog from Long Island: $\log W = -5.99220 + 2.9162 \log L$. This relationship represents data from 3,156 fish sampled from sport catches during May through November for the period 1964–1966. This equation differs from those provided by Cooper primarily because fish were unsexed, uneviscerated, and collected throughout the season. From his growth equation (assuming growth rates to be similar for tautog from Rhode Island and Long Island) a more reasonable estimate can be made for age at entry into the recreational fishery. From Briggs's weight data, a 908-g fish corresponds to a 7- to 8-year-old and a 1,820-g fish to a 9- to 11-year-old.

General Range. Tautog occur along the eastern coast of North America from Halifax, N.S. (Scott and Scott 1988), to northern South Carolina (Bearden 1961), but are most abundant between Cape Cod and Chesapeake Bay (Hildebrand and Schroeder 1928; Bigelow and Schroeder; Hostetter and Munroe 1993). Previous to 1957, tautog were rarely reported from Canadian waters (Bleakney 1963b; Leim and Scott 1966), and they are not usually taken in commercial numbers north of Massachusetts or south of Chesapeake Bay (Bigelow and Schroeder; Scott and Scott 1988; Hostetter and Munroe 1993).

Occurrence in the Gulf of Maine. Tautog are extremely local, perhaps more so than any other Gulf of Maine fish that are of interest either to the angler or to the commercial fisherman. Apart from Mitchill's statement that by 1814 the Boston market had a full supply (which may have come from south and not north of Cape Cod), the first positive record in Massachusetts Bay is of several that were caught along the Cohasset rocks in 1824, which local fishermen said was a species new to them (Bigelow and Schroeder). By 1839 tautog were caught in numbers in inner parts of Massachusetts Bay (e.g., Lynn, Nahant, Boston Harbor), were even more abundant around Manomet Headland in Plymouth, and supported a considerable hook-and-line fishery at Wellfleet. A few years later their presence was established for the coast of Maine, and in 1851 they were reported as common (according to Perley) in Saint John Harbor, N.B., though these Bay of Fundy fish were introduced (not native). In 1876, weirs north of Cape Cod took 1,034 kg of tautog, and in 1879 Goode and Bean described them as abundant in many localities around Cape Ann.

Presently, the regular range of tautog includes the whole coastline from Cape Cod around to Cape Ann, in suitable localities (Bigelow and Schroeder). They are less regular northward from Cape Ann, less abundant, and more local. There are some tautog grounds around the Isles of Shoals, off Cape Porpoise, and around Casco Bay, where Kendall (1931) reported them as having been locally numerous for some time. Tautog were also taken along the ledges near Boothbay Harbor and in Penobscot Bay. They are uncommon east of Penobscot Bay and so scarce in the Passamaquoddy region (they have long since vanished from Saint John Harbor) that only three specimens are known to have been taken there within recent years: one near the head of the Bay of Fundy on the Nova Scotian side (Scotts Bay, Kings County), one on the Nova Scotian shore of the open Gulf of Maine (Cranberry Head, Yarmouth County), and one on the outer coast of Nova Scotia near Halifax (Petpeswick Harbor, Halifax County), the most northerly record for tautog.

The more productive tautog grounds north of the elbow of Cape Cod are the Cape Cod Bay shore southward from Wellfleet; the Sandwich-Sagamore shore with the jetties at the mouth of the Cape Cod Canal; boulder habitat around Manomet headland and nearby; Gurnet Point at Duxbury; ledges off Scituate and Cohasset, and especially those off Swampscott; the Nahant, Marblehead, and Magnolia Rocks; and along the rocky shore from Gloucester Harbor around Cape Ann. The Cape Cod Bay grounds are exceptional for tautog are caught there on smooth bottom, not among ledges, which are their usual haunts. Bigelow and Schroeder also reported that good-sized tautog were taken inside of Nauset Inlet (where there are scattered boulders only), one in a lobster pot during the summer of 1949, and quite a number, large and small, within Duxbury Bay, especially around the pilings of Powder Point Bridge.

Bigelow and Schroeder questioned whether or not the stock north of Cape Cod is maintained wholly by local reproduction or is reinforced by recruitment from fish produced further south. Bleakney (1963b) suggested that because of their rarity in Nova Scotia and New Brunswick, tautog may exist only in relict, disjunct populations in more protected

supporting Bleakney's hypothesis of disjunct populations includes the sporadic nature of catches of this species in the northern regions of the species range, including the rare capture of a single gravid female and other unusual sightings in Mahone Bay, N.S. (Bleakney 1963b); sport fish catches in 1957 of over 2,000 tautog from Eel Brook Lake, N.S. (Leim and Day 1959); and collection of only two larval tautog from the Sheepscot River estuary–Montsweag Bay (Hauser 1973).

Tagging studies of tautog from Rhode Island and Long Island (Cooper 1966; Briggs 1977b) provide estimates of population sizes and indicated that tautog apparently form localized populations that mix very little with adjacent coastal populations. Analysis of mitochondrial and nuclear genes (Orbacz and Gaffney 2000), however, suggests that tautog from Rhode Island to Virginia represent a single genetic stock.

Importance and Utilization. Unlike most labrids, tautog are a highly valued recreational species and an excellent table fish. A small commercial fishery has existed for at least the past 100 years (Goode 1884), although, until only recently, they were never a primary target species of commercial fishermen. Historically, the tautog fishery resulted from by-catches of other commercial fisheries, with the greatest numbers of fish being landed in inshore and nearshore fisheries utilizing nonselective gear such as handlines, weirs, and fyke, pound, and floating trap nets. Although this species has traditionally supported a small commercial fishery, especially south of Cape Cod, harvesting was done primarily by rod-and-reel fishermen and spear fishermen (Cooper 1966; Briggs 1969a, 1975, 1977b). In recent years, however, a directed commercial fishery has developed, primarily in Massachusetts (south of Cape Cod), Rhode Island, Connecticut, and New York. Fish are now taken commercially by a number of gear types including gill nets, trawls, and fish pots.

Although tautog are harvested in commercial quantities from Massachusetts to Virginia, the greatest numbers are landed in the area extending from Massachusetts to New York. Rhode Island generally has the highest reported landings. Few if any tautog are taken commercially north of Cape Cod or south of Virginia, primarily because the species is never abundant enough to support directed fisheries in these areas. Within the Gulf of Maine, historical fluctuations have occurred in commercial catches for various areas. Bigelow and Schroeder noted in particular that the average yield per pound from net or trap was from 2 to 20 times as great from Cape Cod Bay as from the north shore of Massachusetts Bay. Other areas along the coast from Boston Harbor to Gloucester reported catches of tautog from time to time, and the chief center of abundance for tautog for Cape Cod Bay in some years has been along the Sagamore shore. In other years, unusual concentrations appeared at Sandwich and Brewster, whereas in some years some of the best tautog fishing was reported from the Wellfleet region.

As early as 1884, Goode reported that annual catches of tautog were difficult to estimate because, on the one hand, they were generally landed as by-catch of other, more valuable fisheries and, on the other, much of the catch was consumed locally. In 1883 about 90,909 kg of tautog, landed by 200 fishermen, was shipped to New York, but the total catch for 1883 was estimated to have been between 182,000 and 227,000 kg (Goode 1884). Annual fishery statistics from 1883 to 1919 are scattered and incomplete. Most tautog taken commercially have been landed in Massachusetts, Rhode Island, and Connecticut. Annual landings from Massachusetts to Virginia for 1919–1974 were highest during the 1920s and 1930s, peaking at about 408,219 kg in 1930. Landings declined to about 136,986 kg in 1935 and to 91,324 kg in 1950. Prior to 1950, annual landings reported from Massachusetts, Rhode Island, and Connecticut combined ranged from 41,552 kg (1919) to a peak of 252,968 kg in 1930. Since 1950, 60,731–95,434 kg of tautog have been landed yearly. According to the most recent Fishery Management Plan for Tautog, spawning stock biomass and landings appear to be declining. The annual recreational catch peaked in 1986 at 7.6 million kg and declined to 2.4 million kg in 1990 and 1993. Annual commercial landings peaked in 1987 at 0.5 million kg and began declining in 1991. The 1993 commercial harvest of 0.2 million kg was the lowest since 1983 (ASMFC 1996).

Reasons for declines in landings of tautog since 1930 are difficult to assess because there are no historical data on size or age composition of commercially caught fish. Tautog are a slow-growing species that forms localized populations (Cooper 1966, 1967; Briggs 1977b), which are vulnerable to commercial overexploitation. It is possible that overfishing occurred during the 1920s and 1930s and that tautog populations were not able to recover to former levels. It is also possible that the eelgrass decline during the early 1930s severely affected populations of young tautog.

Another explanation for earlier declines in landings is that fisheries that employ the most successful gears for taking tautog (handlines, fykes, weirs, trap nets) have declined since the 1920s and 1930s. For example, in 1905 about 265 fish traps were operated in Rhode Island waters, whereas in 1958 there were fewer than a dozen (Gordon 1960). About a dozen traps are still in operation in Rhode Island. There have been similar declines in fishing effort in the other New England states as well as in the Mid-Atlantic and Chesapeake states.

The effects of recent extensive harvesting on stocks of a long-lived, slow-growing species such as tautog can be particularly significant (Hostetter and Munroe 1993). In fact, once the recent directed fishery for tautog began, negative impacts on tautog populations were noted almost immediately. In response to the dramatic impacts on tautog stocks, a coastwise management plan was adopted by the Atlantic States Marine Fisheries Commission in an attempt to restore stock to prefishery levels by imposing size limits and catch restrictions for both commercial and recreational fishermen. Availability of tautog to the commercial fishery generally corresponds to periods of high activity during seasons of inshore-offshore migrations (in the north) of adults (including the spring spawning migration). Most of the commercial catch is landed from May to July and from late September to early November in New England and south to New Jersey. Further south, tautog are caught earlier in the spring and later into the fall and during the winter, depending upon water temperatures. Few tautog are taken offshore by commercial trawlers during the winter, primarily because these fish seek out areas of rugged bottom topography and are dormant within crevices within this structure. Little information is available on size-at-entry and size-age composition of commercially landed fish. Cooper (1966) reported that tautog entered the sport and commercial fisheries in Rhode Island at about 275 mm. Bigelow and Schroeder estimated that fish landed in the Gulf of Maine and Massachusetts Bay averaged about 0.9–1.8 kg.

Recently, interest has developed in investigating the suitability of tautog as a candidate for aquaculture (Perry 1994; Mercaldo-Allen et al. 1997; Perry et al. 1998), especially as a year-round source of live or fresh fish for ethnic markets. With enhanced nutrition and elevated water temperature, they might be cultured to market size, which is a 0.5–1.0 kg or "wok-sized" fish, within a 2-year period (Perry et al. 1998). Suggestions to release cultured tautog into the wild to enhance diminished natural populations (Perry et al. 1998) should be carefully considered before implementation.

South of the Gulf of Maine, tautog are among the most important game fishes sought by shore fishermen along rocky areas (Migdalski 1958; Wilcoxson 1975), and they are also a primary species at artificial reefs in the northeast (Zawacki 1969; Briggs 1975). Active sport fisheries for tautog exist along the coast south to Virginia. The sport fishing season tends to be bimodal (May–July and September–October) in New York and northward. Tautog are generally available to nearshore fishermen throughout the spring, summer, and fall seasons, but many of the larger fish apparently move to deeper water during periods of summer high temperatures and winter low temperatures in inshore waters (Briggs 1969a), and move inshore again with the onset of favorable temperatures. In 1990 and 1991, tautog were the most frequently caught fish in the recreational fishery in the North Atlantic, with an estimated catch of 1.1 million fish, of which some 289,000 were caught in Massachusetts (Voorhees et al. 1992). The principal mode of capture was by anglers on private and rental boats, followed by party / charter and shore fishing anglers.

Along the stretch from Manomet Headland, Plymouth, to Cape Ann, tautog are caught either from a boat at anchor over submerged ledges or bottoms covered with boulders or by casting with a long rod from dry ledges or from the rocky coastline. In either case, the fish are so local and irregular in distribution (depending on food supply and the contour of the rocks) and so stationary that it is worth fishing for them only in certain spots. Tautog are a challenge to sport fishermen. A few feet from their structure one way or the other may mean the difference between success and failure. In Cape Cod Bay, however, where the tautog occur on smooth bottom, they lie in little openings among eelgrass (whenever there is any), and, if a fiddler or hermit crab is lowered in a clear spot in front of them, they can be caught in very shallow water.

Fishing the Cohasset rocks, Bigelow and Schroeder found green crabs (*Carcinus maenas*) the most attractive bait, whole if small enough, cut if larger; rock crabs (*Cancer*) or hermit crabs second best; and large snails or cockles (*Polynices*) fairly good. They also noted that lobster would perhaps be best of all, were it not so expensive. Mussels are often successful bait for capturing tautog, but are difficult to keep on the hook. Small, whole clams, hooked through the siphon with the shell cracked so as to let the juices escape, are also a good bait. However, mussels and clams are next to worthless if shelled because they are stolen almost immediately by the swarms of cunner frequenting the same habitats as tautog. In Cape Cod Bay, where tautog are caught on smooth bottom, the baits most used are hermit and fiddler crabs. Bigelow and Schroeder also noted that tautog will occasionally strike sea worms (*Nereis*) as well.

Tautog are not the easiest fish to catch on hook and line. When a tautog bites, it passes the bait back to the pharyngeal teeth to crush the shells before swallowing them, and in doing so it gives several distinctive jerks or twitches. This is the time to hook it. Many fish are missed by being struck too soon by anglers not experienced in the ways of the tautog.

CUNNER / *Tautogolabrus adspersus* (Walbaum 1792) / Bergall, Chogset / Bigelow and Schroeder 1953:473–478

Description. Body moderately deep, 3.00–3.33 in SL; rather robust; moderately compressed (Fig. 242). Caudal peduncle very deep; head flat-topped, with nearly straight upper and lower profiles. Head length 3.25–3.50 in SL, nearly equal to body depth. Mouth small, terminal, with thick lips and protractile upper jaw; posterior end of maxilla slightly anterior to vertical through front margin of eye. Each jaw with several series of unequal, conical teeth; outermost row of canines strongest; several bands of smaller concave teeth situated behind canines in each jaw; teeth on sides of jaw enlarging ante- four canine teeth in front row, respectively; pharyngeal teeth strong. Snout moderate, rather pointed, longer than eye. Eye moderate, 4.5 in HL. Body and opercle covered with large, cycloid scales and a tough skin. Top of head, preorbital, maxilla, lower jaw, interopercle, and posterior margins of preopercle and opercle naked; anterior and median portions of preopercle with about five rows of small scales; anterior and median opercle with four or five rows of larger scales. Body scaly between gill opening and bases of pectoral and pelvic fins. Fins without scales. Dorsal fin continuous; originating over upper

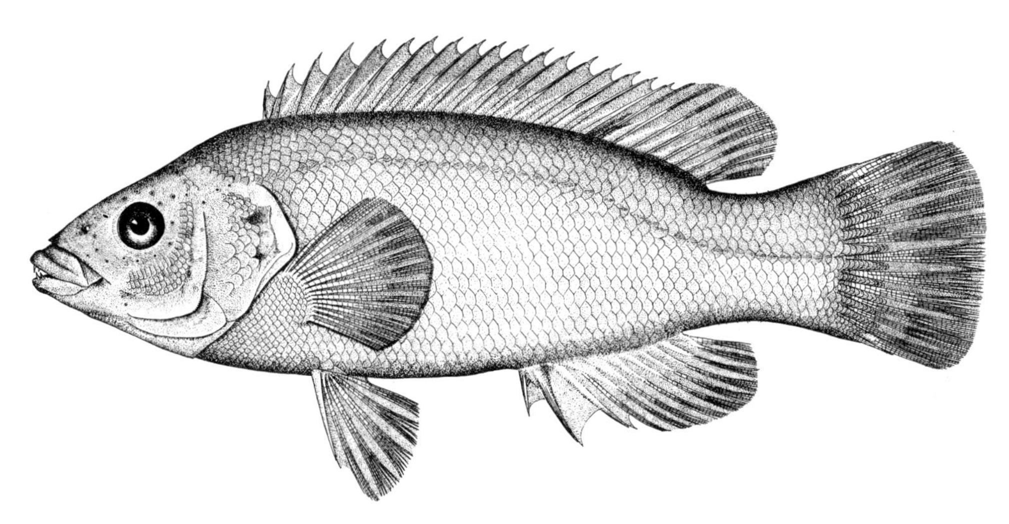

Figure 242. Cunner *Tautogolabrus adspersus*. Woods Hole, Mass., 182 mm. Drawn by H. L. Todd.

of pectoral fins, and continuing posteriorly to caudal peduncle. Anteriormost four or five dorsal fin spines graduated in size; others about equal in height. Soft-rayed dorsal fin only a little more than one-third as long as spiny part, rounded in outline; second dorsal fin ray highest. Anal fin originating under or slightly posterior to middle of dorsal fin, rounded in outline. Caudal fin slightly convex with rounded corners. Pelvic fin moderate, directly under or slightly posterior to vertical through origin of pectoral fin. Pectoral fin moderate in size, somewhat rounded. Lateral line complete, arched.

Meristics. Dorsal fin rays XVIII, 9 or 10; anal fin rays III, 9 or 10 (rarely 7 or 8); pectoral fin rays 14; lateral line scales 46; gill rakers very short, about 6 + 11; branchiostegals 5 or 6; vertebrae 17 + 19 = 36.

Color. Cunner are among the most variably colored fishes in the Gulf of Maine, often closely matching the background color of the substrate on which they live. As a rule, upper parts of the body range from lighter to darker shades of reddish brown with a bluish cast to blue with brownish tinge, variously mottled with blue, brown, and reddish pigment. Some fish are uniformly brown, whereas others caught over mud bottom are often very deep sepia. Occasionally, coloration is a dull olive green mingled with blue, brown, rust-colored, or slate gray. Reddish or rust tones prevail in those living among red seaweeds. Cunner caught in deep water are often almost as red as redfishes, whereas those taken over sandy bottoms are often very pale and more or less entirely speckled with blackish dots resulting from encysted trematode metacercariae. The sides of the fish usually have a more or less brassy luster; sometimes the head has a brassy luster as well. The ventral surface is invariably of a more or less vivid bluish cast or sometimes whitish, dusky, or slightly paler than the sides. The lips and lining of the mouth are sometimes bright yellow. Young fish (up to about 10 cm) often have dark bars and blotches and possess a conspicuous black spot at the anterior end of the soft-rayed dorsal fin.

Size. Cunner grow to about 43.2 cm long and can weigh up to 1.5 kg, but are not usually more than 30.5 cm (Bigelow and Schroeder; Scott and Scott 1988). The all-tackle game fish record is 0.99 kg for a fish caught at Two Lights State Park, Maine, in July 1999 (IGFA 2001).

Distinctions. Cunner co-occur with tautog throughout most of their range but are easily distinguished from the latter by differences in counts (dorsal: XVIII, 9 or 10 vs. XVI or XVII, 10 or 11 in tautog; anal: III, 8 or 9 vs. III, 7 or 8; gill rakers usually 6 + 11 vs. 3 + 6). Cunner have a rather elongate, pointed profile to the dorsal aspect of the head (vs. dorsal profile arched in tautog) and have a more streamlined body (less robust than tautog). Cunner also have the entire operculum covered with scales and the pelvic fins situated nearly directly under the pectoral fins, whereas in tautog the ventral half of the operculum is naked and the pelvic fins are located posterior to the pectoral fin base. Both species have thick, fleshy lips, although the lips of cunner are somewhat thinner.

Habits. Cunner occur primarily in coastal habitats, usually within 3 km of the shoreline. Some individuals occur on offshore grounds, such as Stellwagen Bank, Jeffreys and Cashes ledges, and even on Georges and Browns banks, where they are captured with otter trawls. However, large catches are seldom made far out at sea, neither along southern New England nor to the north (Bigelow and Schroeder). In northern portions of the range, cunner are most plentiful from just below the low tide mark to about 18–30 m. On offshore ledges and banks, they commonly range as deep as 46–65 m, whereas on Georges Bank, they have been taken as deep as 128 m (Bigelow and Schroeder). Most cunner caught in deep water and far offshore are large. The origins of these large cunner are not known, but it is speculated that they move offshore from inshore reef communities (Bigelow and Schroeder; Dew 1976; Scott and Scott 1988). Cunner are rapid colonizers of manmade reefs and wrecks (Brewer 1965; Unger 1966). South of New York, most cunner are found in deeper water, hence somewhat farther offshore, depending on topography of the coastline and of the bottom.

Cunner are not a schooling species although many individuals are often found congregating in the same habitat. They live near the bottom and are strongly associated with cover, where they are frequently observed in the daytime swimming among eelgrass (*Zostera*), seaweed beds, rocky outcroppings, pilings, wharves, or floats in harbors, and swarming around just about any other object offering shelter (Olla et al. 1979). Although they are generally abundant in areas they inhabit, their numbers drop off rapidly a short distance from cover. They also move into poly- and mesohaline regions of deeper tidal creeks, with smaller fish ranging farther upstream than larger ones. Apparently neither size-group enters water that is appreciably brackish (Bigelow and Schroeder; Leim and Scott 1966). Young cunner (9–30 mm) commonly occupy extremely shallow-water habitats with dense vegetation (algae or eelgrass) including tide pools, where they sometimes are a numerically dominant component of the fish community (Collette 1986; Levin 1991). Newly recruiting cunner are positively associated with microhabitats containing macrophytes and negatively associated with microhabitats of little structural complexity (Levin 1991).

Cunner are diurnally active fishes and the onset and cessation of their daily activities coincide with the rising and setting of the sun (Olla et al. 1975a; Bradbury et al. 1995, 1997). Activity begins in the morning approximately 16–55 min following morning civil twilight and ceases with the onset of a nocturnal quiescent phase initiated approximately 5–55 min prior to the end of evening civil twilight. During nighttime, cunner lie quiescently in, alongside, or among rocks, boulders, and other objects offering cover such as algal mats or kelp beds. This behavior presumably reduces energy expenditure and may provide protection from predators (Bradbury et al. 1995).

Daily activity patterns of cunner are dependent upon a variety of abiotic and biotic factors, including seasonal and daily water temperatures and sea conditions (Bradbury et al. 1997), reproductive activity (Pottle and Green 1979a,b), and foraging behavior (Whoriskey 1983). In early summer in northern regions, where water temperatures can suddenly drop below 5°C, cunner cease activity until temperatures increase again (Bradbury et al. 1997). At water temperatures above 5°C, no significant effects on the onset or cessation of activity or on the percentage of time cunner were inactive were observed. In exposed areas, sea state also affects activity patterns, with periods of inactivity tending to be longer on days with high surface waves (Bradbury et al. 1997). There is a marked seasonal decrease in duration of diurnal activity with decreasing photoperiod. Off Newfoundland, daily activity for cunner ranged from 16.5 $h \cdot day^{-1}$ in June and July (mean 12.5 $h \cdot day^{-1}$) down to 11.0 $h \cdot day^{-1}$ (mean 3.0 $h \cdot day^{-1}$) or less in October and November (Bradbury et al. 1997). Decreases in daily activity in fall are apparently cued by an endogenous mechanism sensitive to day length or some other environmental factor.

During the day, cunner also undergo periods of quiescence when they seek cover under rocks and boulders or hide in crevices (Pottle and Green 1979a,b; Whoriskey 1983; Bradbury et al. 1997). Off Newfoundland, diurnal activity was interrupted by periods of inactivity usually lasting 5–15 min. Some authors have assumed that temperate wrasses use cover to avoid predation (Olla et al. 1979; Whoriskey 1983), although as Bradbury et al. (1997) pointed out, the threat of predation on cunner has not been well documented, and their observations during 400 h of diving indicated that predation rates on cunner in Conception Bay, Nfld., were very low. Other possible hypotheses to explain shelter-seeking by cunner include avoidance of conspecifics and reduction in energy expenditure.

Adult cunner are closely associated with structure (Bigelow and Schroeder) and have a rather limited home range (Green 1975; Olla et al. 1975a; Bradbury et al. 1995), with daytime movements restricted to within several meters of the nighttime shelter (Olla et al. 1975a). Generally, central portions of the home range were occupied more intensively than other areas, with relatively little time spent on outside excursions (Bradbury et al. 1995). Adult female cunner tracked with transmitters generally used different night resting shelters on consecutive nights, but these shelters were always within the daily home range (Bradbury et al. 1995). Despite having a limited home range, cunner have the ability to return to home sites after displacements of up to 4 km (Green 1975). Home-range size for females (mean daily home ranges of 299.6–2,252.6 m^2) did not seem to be influenced by tidal stage, and percent cloud cover and seawater temperature, individually, accounted for less than 3.3% of the variance. Also, changing energy requirements associated with spawning and overwintering torpor appeared to play a more important role than reproductive behavior per se in determining home-range size.

Like many other fishes associated with cover, cunner spend much time resting quietly or swimming slowly around structure. When moving in and about their habitat, they utilize the labriform swimming style: body held rigid while they propel themselves through rowing motions of the pectoral fins. For fast starts or escape movements, however, a cunner can swim rapidly by stroking powerfully with its broad caudal fin.

Generally speaking, cunner are year-round residents wherever they are found and apparently do not undertake any extensive seasonal migrations. There are small-scale movements associated with seasonal temperature changes, but these are relatively small when compared with migrations of other Gulf of Maine fishes. During the coldest months, for example, cunner may descend into deeper water, especially along exposed coastal areas (Chao 1973; Ojeda and Dearborn 1990), and in midsummer they leave the shoalest parts of certain enclosed bays, thereby avoiding the high temperatures produced there as the sun strikes the flats at low tide. It has been observed that during periods of adverse sea conditions, some cunner take shelter under bottom relief structures, whereas others remain close to the bottom in deeper, offshore waters where the effects of water movement are reduced (Bradbury et al. 1995). Olla et al. (1979) found that a small portion of the resident population of cunner located off Fire Island, N.Y., dispersed from an overwintering habitat in late spring but returned there in late fall (not necessarily to the exact location from which they dispersed in the spring). They believed that this seasonal dispersion could also have been made in response to fluctuations in food availability.

As year-round residents in nearshore habitats, especially in protected coves and bays, cunner must tolerate a wide range of temperatures. They can withstand temperatures as low as 0°C in the laboratory (Green and Farwell 1971) and inhabit inshore environments in Newfoundland where the minimum temperature is below −1.5°C (J. Green, pers. comm.). Recently, Valerio et al. (1990) demonstrated the presence of antifreeze polypeptides in the skin, but not the blood plasma, of cunner in Newfoundland, thus indicating a metabolic adaptation to prevent freezing during winter. Based on geographic distribution of the species, an upper thermal limit is around 21°–23°C.

Although able to survive a wide range of temperatures, cunner do not maintain the same activity levels throughout the year (Haugaard and Irving 1943). When seasonal temperatures fall below about 7°–8°C on Fishers Island, Conn. (Dew 1976), and 5°C in Newfoundland (Green and Farwell 1971; Bradbury et al. 1997), they begin an overwintering period, which they pass in a state of torpor, hidden under loose rocks and boulders. In some areas, larger fish may move into deeper waters at this time (Ojeda and Dearborn 1990). Larger cunner become inactive 1–3 weeks earlier and remain inactive 1–3 weeks longer than the smallest ones (Dew 1976). In Conception Bay cunner are active for only 6 months (June–November), remaining in a torpid state for the rest of the year (Green and Farwell 1971). During this overwintering period, they cease feeding (Olla et al. 1975a; Dew 1976). In deeper waters, where winter temperatures are somewhat higher than those in shallower areas, they may remain active further into winter. Bigelow and Schroeder reported that some cunner were caught by fishermen in precisely the same locations in wintertime as in summer, and they noted that a few fish, taken by trawling, were landed in Boston even during the coldest months of the year.

It is well known that cunner are vulnerable to very low temperatures and that severe winters occasionally produce mass mortalities of this species (Bigelow and Schroeder). During the winter of 1834–1835, dead cunner were washed ashore in large numbers between Marblehead and Gloucester (Goode 1888), and again in 1900–1901 numerous dead cunner were reported washing ashore (Johansen 1925; Bigelow and Schroeder). A localized mass mortality of cunner also occurred in February 1973 (Green 1974) at Upper Gullies, Conception Bay. Green (1974) noted that sea ice placed in crevices containing torpid cunner killed the fish and hypothesized that the presence of ice crystals in the storm surge was responsible for the mass mortality by facilitating formation of ice crystals in the bodies of these torpid fish.

Food. Cunner are omnivorous, opportunistic predators that feed on a wide variety of organisms, but primarily on small mollusks and crustaceans (Bowman et al. 2000). They are sight feeders, feeding continuously throughout daylight hours regardless of tidal stage. With the approach of evening, they become inactive and at night they cease feeding and can be found lying in a quiescent state in, under, and around various objects (Dew 1976).

Cunner jaws are equipped with relatively large canine-like or incisor-like teeth on both premaxillae and dentaries. The protrusible premaxillae bear a row of conical teeth directed slightly anteriorly, as are the conical dentary teeth. The anteriormost rows of both premaxillary and dentary teeth are two to four times larger than the other teeth, and the number of teeth increases with the size of the fish (Chao 1973). The primary function of jaw teeth is for food capture and handling, not for mastication. Cunner, like other labrids, possess a highly evolved pharyngeal jaw apparatus, which is the primary structure for shearing and mastication (Liem and Sanderson 1986), with the upper and lower pharyngeal jaws making complex orbital movements that shear and crush food items. Shearing is accomplished by the forward moving upper jaw colliding with the dorsally held lower jaw, whereas the crushing action is effected by the extreme posterodorsal movements of the lower jaw against the retracted upper jaw.

In subtidal areas, cunner feed among seaweeds, stones, or dock piles, biting off barnacles and small blue mussels, fragments of which often fill their digestive tracts (Shumway and Stickney 1975). During flood tide stages, they may even move into the intertidal zone to forage, returning to subtidal habitats during ebb tide (Whoriskey 1983). They devour enormous numbers of amphipods, shrimps, young lobsters (Lavalli and Barshaw 1986), small crabs, and other small crustaceans of all kinds, as well as gastropod mollusks, smaller bivalves, hydroids, and annelid worms. They sometimes eat small sea urchins, bryozoans, and ascidians (Osman and Whitlatch 1995) and occasionally capture small fishes, including silversides, sticklebacks, pipefish, mummichog, and the young of other larger species, as well as fish eggs. Cunner are also busy scavengers in harbors, congregating and feeding on any animal refuse, as well as on amphipods and other crustaceans attracted by the same morsels. Apart from animal food, fragments of eelgrass and algae are often found in cunner diets, but Chao (1973) considered that plant material was probably ingested incidentally during feeding because when held in the laboratory, starved cunner do not actively feed on algae.

Diets of cunner vary considerably with fish size and availability of prey organisms. Juveniles less than 100 mm SL feed on small motile crustaceans, particularly amphipods, in the water column, whereas cunner larger than 100 mm feed primarily on sessile organisms (Chao 1973; Bowman et al. 2000). Within prey types, cunner are often selective for specific sizes. For example, Johnson and Mann (1986) reported that they fed preferentially on small-sized snails (*Lacuna vincta*) and Harris (1986) demonstrated similar size-selective predation on nudibranchs.

The significance of the ecological role of cunner as a mobile predator in the intertidal and nearshore subtidal zones has been debated. Edwards et al. (1982) regarded them as high-level mobile carnivores that played a significant role in structuring the intertidal community, whereas Menge (1982) maintained that effects of cunner were insignificant in the structuring of intertidal communities. Cunner have also been implicated as a primary source of predation mortality in subtidal communities (Sand 1982).

Seasonal shifts occur in diets of cunner from Long Island (Olla et al. 1975a). During May and June, they feed primarily on mussels, *Mytilus edulis*, whereas from July to September, dietary preference switched to an isopod, *Idotea baltica*. Whether cunner prefer *I. baltica* to *M. edulis* or switch their diet in response to competition from juvenile tautog is unknown. Sand (1982) found that their diet in Narragansett Bay varied seasonally at a site where prey densities varied, whereas at a site where densities of prey items were stable seasonally, their diet also remained stable.

Feeding activity and prey selection are also influenced by reproductive stage and courtship behaviors during the spawning season. In Conception Bay, Nfld., cunner foraging strategies were related to differences in reproductive behavior of the sexes (Green et al. 1984, 1985). During courting behavior and spawning territorial males fed more actively in the morning than in the afternoon. Conversely, females foraged continuously throughout the day. Diet selection was also different among territorial males, nonterritorial males, and females. Territorial males were termed time-minimizers because they fed on abundant prey items (gastropods and echinoderms) that had low handling times. Females, on the other hand, were termed energy-maximizers because they more often ingested less abundant or evenly dispersed food items (mussels), but those that had higher energy content. Nonterritorial males were intermediate between the other sexual stages with respect to the spectrum of food items they consumed.

Cunner cease feeding altogether when water temperatures drop to 5°–6°C in late fall and winter. Digestive tracts of torpid cunner collected at these times are usually empty (Olla et al. 1975a; Dew 1976). Feeding is resumed shortly after emergence from the torpid state when water temperatures rise again in the following spring (Green and Farwell 1971).

Of particular interest regarding the trophic biology of cunner is that wrasses lack a morphologically or physiologically distinct stomach (Chao 1973). The alimentary tract consists of the pharynx, followed by a short esophagus, which is joined to the intestine at the esophageal-intestinal valve. A pH of 7.0–8.5 exists throughout the digestive tract with trypsin and a trypsinlike enzyme probably responsible for proteolysis (Simpson and Haard 1985). No hydrochloric acid is produced so shells of mussels cannot be digested but must be excreted.

Predators. The NEFSC food habits survey (Rountree 1999) found cunner in the stomachs of four species of fishes, Atlantic cod, sea raven, smooth dogfish, and white hake, of which the first two were the most frequent predators. Other fish predators include skates (Cornish 1907), sculpins (Scott and Scott 1988), tomcod (Morrow 1951), and cod (Johansen 1925). They are also preyed upon by piscivorous birds such as double-breasted cormorant (Gross 1923; Scott and Scott 1988; Blackwell et al. 1995). Predation on larval stages has not been assessed, but Levin (1991) noted that sculpins (*Myoxocephalus aenaeus*) prey on cunner recruits in high numbers.

Parasites. Cunner are host to, and play an important role in, the life cycles of a variety of parasites, including protozoans, monogenetic and digenetic trematodes, nematodes, cestodes, and at least one species of acanthocephalan. Earlier works on helminth parasites of cunner were those of Linton (1901a,b, 1928, 1940, 1941). Serchuk and Frame (1973) and Margolis and Arthur (1979) listed parasites known to infect cunner: 12 species of trematodes, 6 cestodes, 7 nematodes, and 1 acanthocephalan (*Echinorhynchus gadi*). The ciliate *Uronemia marinum* has caused extensive damage to the mesentery and body muscle tissue in cunner and several other fishes in the New York Aquarium (Cheung et al. 1980), but infections by this ciliate have not been reported in wild cunner. The monogenetic trematode *Gyrodactylus adspersi* has been reported from the gills, fins, and body surface of cunner from Avalon Peninsula, Nfld. (Cone and Wiles 1983; Cone and Odense 1984). Sekhar and Threlfall (1970b) noted that cunner were not infected with leeches, despite their close association with the bottom and other fishes commonly infected with them. In an aquarium, cunner in direct contact with sculpins infected with leeches did not become infected, suggesting some physiological or morphological barrier that prevents leeches from being transferred to cunner.

Cunner serve as second intermediate hosts in life cycles of several species of cestodes and nematodes (Sekhar and Threlfall 1970b) whose primary hosts are piscivorous fishes and elasmobranchs. They are among the many shallow-water fishes serving as second intermediate hosts in the life cycle of the digenetic trematode *Cryptocotyle lingua* (Linton 1940; Sekhar and Threlfall 1970a). The immature encysted stage (metacercaria) of this digenetic trematode, whose definitive hosts are piscivorous birds, frequently appears as small black spots on fins, skin, cornea, and gills of cunner. Infections of *C. lingua* are widespread among cunner and at times can be quite heavy. For example, in the Weweantic River estuary, Mass., all cunner examined that were longer than 60 mm were infected with metacercariae (Serchuk and Cole 1974), and in Newfoundland the rate of infection in 642 cunner was nearly 90%, with infection intensities ranging from 134 to 2,541 (mean 945) metacercariae per fish (Sekhar and Threlfall 1970a).

Reproduction. Sexes are distinct, without marked dimorphism in morphology. In the Gulf of St. Lawrence, females were slightly larger than males of the same age (Johansen 1925), but just the opposite was true (males larger than females of the same age) in populations in Newfoundland, Massachusetts, and Fishers Island Sound, Conn. (Dew 1976; W. Chaisson et al. pers. comm.). While no distinct morphological differences exist between the sexes, there is marked sexual dichromatism associated with spawning (Johansen 1925; Pottle and Green 1979a,b). From 4 to 5 weeks prior to the spawning season, both territorial and nonterritorial males gradually develop a blue-phase coloration. Females and immature males are brownish. Coloration of females does not change during spawning activities.

Sexual maturation in cunner appears related to size of fish and latitude. Cunner in the Gulf of St. Lawrence mature at 70–90 mm TL (age 2+) and do not spawn until the following summer, at age 3 (Johansen 1925). Dew (1976) reported that in Fishers Island Sound nearly 50% of the 0-year-group fish sampled after September had developed sexually distinct gonads and that 1-year-old fish (50–60 mm) were sexually mature and capable of spawning.

Nitsche et al. (2001) estimated relationships between the age of a female, size, and weight for cunner occurring in Cape Cod Bay. Their models assumed that, due to a short spawning season, only mature or maturing oocytes present in ovaries prior to the spawning season were actually spawned during that spawning season. By counting only the number of oocytes in these developmental stages, these authors derived an estimate of annual fecundity. Of several models developed, the quadratic form of the length-specific fecundity model, $\log_{10} F = 24.954 \log_{10} L - 5{,}348 \log_{10} L^2 - 24.421$, $r^2 = 0.71$, was most appropriate for modeling cunner fecundity. Fecundity increased substantially with small increases in fish length, ranging from 1,192 eggs for a 76-mm fish to 84,403 eggs for a 171-mm fish. Whether this formula reliably represents annual fecundity or only estimates batch fecundity depends on whether or not the cunner is a serial spawner, which its spawning behavior strongly indicates, and whether or not smaller size-classes of ova mature during a spawning season. With the use of histological techniques, seasonal production of hydrated oocytes can be accurately determined, and this method will also determine whether or not the cunner is a determinate serial spawner.

Cunner utilize dual patterns of reproductive behavior. Small, nonterritorial males take part in nonaggressive group spawning (Wicklund 1970b; Pottle et al. 1981), whereas larger, territorial males engage in pair spawning (Pottle and Green 1979b; Pottle et al. 1981; Martel and Green 1987). Wicklund (1970) observed that only group spawning occurred at Shrewsbury Rocks, N.J., whereas Pottle and Green (1979a) found only pair spawning occurring in Conception Bay. At Bonne Bay, Nfld., both forms of reproductive behavior were observed (Pottle et al. 1981).

In Conception Bay male cunner defend permanent territories during their entire active period, from May to early or late November, and females only initiate spawning with territorial males (Martel and Green 1987). Possession of a territory (16–103 m^2) is essential for successful courtship. However, territory size is not the only factor contributing to spawning success. A female's choice among territorial males was also based on characteristics of the males, such as courtship and aggression frequencies and relative size (TL). In this area, therefore, male attributes accounted for more of the variance in spawning success than territory size (Martel and Green 1987).

In pair spawning, reproductive behavior is initiated by a territorial male's "courtship approach" to a female within his territory (Pottle and Green 1979b). In this display, the male swims toward and to one side of the female, with his bright blue median fins erect and his head turned toward her at a sharp angle to his body. His belly is inclined toward the female and locomotion is carangiform rather than the usual labriform, although rapid beating of the pectorals is a component of the display. In the vertical plane, the male's course describes a shallow arc above the substrate to a maximum height of about 1.5 m.

Receptive females rest on the bottom or move slowly over the substrate while the male makes one or more courtship approaches. If the female does not flee, the male begins to circle her at a height of less than 1 m, either displaying on each side as in a courtship approach, displaying briefly while circling, swimming in a complete circle without raising the median fins or turning the head, or holding the courtship approach posture during the entire circuit. After circling, the male slowly approaches the female from behind and passes over her several times (usually fewer than ten) at an angle of 20–60° to her longitudinal axis. As he passes her, he comes into contact with her dorsal surface while stiffly flexing his head and tail from side to side during some of the passes. He then places his belly against her dorsal surface, with his snout just behind her head, and begins to quiver from snout to tail. Within 5–15 s, the pair begins to move forward and the male's quivering is transformed into slow, high-amplitude, carangiform swimming, succeeded by rapid acceleration of both fish. Approximately 2 m from the starting point, both fish make a sharp upward turn, with the female less than a body length ahead of the male. The spawning pair rises at high speed at an angle of 60–80° to the substrate, with median fins erect, both gaping widely just before and during gamete release. Duration of the vertical run is under 1.5 s. The entire process from courtship to gamete release ranges from 3 to 10 min. Fertilization is external (Johansen 1925). Interruption and/or termination of the spawning sequence can be caused by the presence of other fishes.

Both males and females are capable of multiple spawning on the same and successive days (Pottle and Green 1979b). Courtship approaches may take place any time during the day (0600–2100). Spawning, however, was recorded only after midday (1300–2100), with peak occurrence between 1700 and 1900.

Group spawning may involve aggregations of 30–150 fish ranging in size from 80 to 180 mm (Wicklund 1970b; Pottle et al. 1981). During any single spawning event, however, the number of fish actually spawning is much smaller, usually

3–15 fish. Group spawning begins with courtship of a brown-phase female by a male with nuptial (blue-phase) coloration. Courtship is identical to that described for pair spawning. Spawnings follow a chase in which the courted female swims rapidly through the aggregation pursued by the courting male. Several fish of undetermined sex closely follow the courting male as he chases the female. Ultimately the courted female initiates a vertical spawning run (often the chase sequence does not culminate in this vertical run). At the apex of the spawning run, which always takes place above the aggregation, all fish involved in the chase release gametes nearly simultaneously. Within one aggregation of cunner studied off Newfoundland (Pottle et al. 1981), all group spawnings took place between 1500 and 1900 with as many as 26 spawnings observed over a 2-h period. Group spawning was also reported to take place in the afternoon (between 1200 and 1700) in a population off New Jersey (Wicklund 1970b).

Spawning Seasonality. Cunner eggs were collected by MARMAP surveys from May to November (Berrien and Sibunka 1999: Fig. 67). Spawning began inshore during May in Mid-Atlantic and southern New England waters and in some years in the western Gulf of Maine near Cape Cod. Peak intensity occurred during June and July and expanded in area from south of Chesapeake Bay to the northern Gulf of Maine and onto Georges Bank. In Canadian waters, spawning occurs from mid-June through August with peaks in mid-July (Reid 1929; Pottle and Green 1979a) and mid-August (Faber 1976a).

Recently spawned eggs, larvae, and juvenile cunner, generally are closely confined to the coastline. Bigelow and Schroeder noted that all of their catches of 100 or more specimens in the Gulf were made either in harbors or, at most, at sites not more than 3 km from shore. Some successful spawning may take place on Cashes and Jeffreys ledges. Bigelow and Schroeder found no evidence that large cunner that wandered offshore to Georges Bank actually spawned there, but cunner eggs and larvae are occasionally collected offshore both on Georges Bank and in the Gulf of Maine (Marak et al. 1962a,b; Able and Fahay 1998).

In inshore areas throughout much of their geographic range, cunner eggs are very abundant in plankton collections, especially during late spring and early summer. For example, in two estuarine systems on Prince Edward Island, cunner and yellowtail flounder eggs (not distinguished) were the most abundant (83.5% of total), and cunner larvae, collected from late June to mid-July, were the most abundant of those taken (Johnston and Morse 1988). Cunner eggs (43.7%) and larvae (25.3%) were the most abundant collected in the Merrimac River estuary, Mass. (Peterson 1975). Eggs were present in this system in May and June, whereas larvae were taken in July and August.

Fish (1925) cited cunner (and tautog) as being the most abundant larvae around Woods Hole. Herman (1963) and Bourne and Govoni (1988) reported that cunner and tautog Bay. Cunner eggs and larvae are also common components of the ichthyoplankton assemblage in Block Island Sound (Merriman and Sclar 1952; Marak et al. 1962a,b) and Long Island Sound (Wheatland 1956; Richards 1959). Pearcy and Richards (1962) did not distinguish between tautog and cunner eggs, but they reported that labrid eggs were the most abundant pelagic eggs in the Mystic River estuary of eastern Connecticut.

Cunner eggs were collected most often when water temperatures were 10°–26°C and salinities 26–29 ppt (Wheatland 1956; Williams 1967, 1968). Bigelow and Schroeder suggested that failure of cunner to spawn successfully within the Bay of Fundy indicated that the lower thermal limit for successful reproduction might be 13°–15°C, although in Newfoundland cunner appear to be able to spawn successfully at temperatures below 13°C (J. Green, pers. comm.).

Early Life History. Cunner eggs are nonadhesive, buoyant, transparent, spherical, and lack an oil droplet (Kuntz and Radcliffe 1917; Hildebrand and Schroeder 1928; Bigelow and Schroeder; Colton and Marak 1969; Fahay 1983; Able and Fahay 1998). Cunner eggs and larvae are strongly vertically stratified in Long Island Sound, with 96% floating in the upper 5 m of the water column (Williams 1968). Egg diameters, 0.75–1.03 mm (mean 0.85 mm), are correlated with temperature. Richards (1959) noted that average diameters of cunner eggs decreased progressively through the summer. Williams (1967) confirmed that cunner (and tautog) eggs were largest in May, when the water was coldest, and smallest in August, when water temperatures were at their highest levels.

Cunner eggs resemble those of tautog, although eggs of the two species can be separated on the basis of size (those of cunner are smaller). Unfortunately, overlap in size over a season makes distinctions difficult. Williams (1967) noted that although egg diameters of both species changed with respect to season, they changed in parallel, with those of cunner always being smaller. Williams (1967) found that the eggs of the two species could be separated using size frequency distributions, and Orlowski et al. (1972) described an immunodiffusion technique for separating them.

Cunner eggs can be easily confused with those of winter and yellowtail flounders (Fahay 1983). In southern areas of the range, spawning times do not overlap seasonally with those of winter flounder. Cunner eggs can also be distinguished from those of winter flounder by the fact that they are slightly larger and because winter flounder eggs are demersal, adhesive, and typically found in aggregates rather than singly and pelagically, as are those of cunner (Fritzsche 1978). Cunner eggs closely resemble those of yellowtail flounder but differ in pigmentation of the developing embryo (Markle and Frost 1985).

Incubation takes 2–6 days depending on temperature, but lasts about 40 h at 21°–22°C (Agassiz and Whitman 1885; Cunningham 1888; Johansen 1925). Incubation at 22°C was complete in 42–45 h (Kuntz and Radcliffe 1917). An incubation period of 4.5–5 days was noted by Agassiz (1882), but temper-

Embryology and larval development of cunner have been described by Agassiz and Whitman (1885), Kuntz and Radcliffe (1917), and Fahay (1983). Pigmentation is first observed in embryos with 10–15 myomeres; early chromatophores appear as minute black dots distributed over the dorsum; pigment spots increase in size, but not in number, up to hatching.

Newly hatched larvae are planktonic; 2.00–3.43 mm in length (Miller 1958), with unpigmented eyes and a nonfunctional mouth. The yolk sac is relatively large. The mouth and the caudal and pectoral fin rays appear the second day after hatching. Dorsal and ventral finfold depths are greater than the body depth posterior to the vent. Newly hatched larvae have small chromatophores limited almost entirely to dorsal and dorsolateral aspects of the head and body and extending to the tail. Shortly after hatching, some chromatophores aggregate into spots, one at the turn in the gut, a second ventrally halfway between the anus and the tip of the tail, and a third near the tip of the tail. Posterior caudal regions and finfolds are unpigmented.

By 3 mm, pigment cells gather into a pair of black spots (characteristic of the species), located dorsally and ventrally about halfway between the anus and the base of the caudal fin rays. Spots persist to about 10–20 mm. By 3 days posthatch (2.8–3.3 mm), the yolk sac is absorbed. The pectoral and caudal fins become distinct at sizes greater than 3 mm. At 4.2 mm, one or more small pigmented areas appear dorsally just posterior to the head, another on the dorsal aspect of the body opposite the one on the ventral aspect and halfway from the vent to the tip of the tail, and one or two very small areas at the base of the ventral finfold near the tip of the tail. The mouth is completely terminal. Pectoral fins continue to develop.

In 7–8 mm juveniles, dorsal, anal, and caudal fins are well differentiated; spinous and soft portions of the dorsal fin begin differentiation. A definite band of pigment overlies the gut, a single large melanophore is on the back of the head, a dorsoventral pair of large melanophores is halfway between the anus and the tip of the tail, and an aggregate of melanophores is on the ventral side of the tip of the tail. By 9–10 mm, pigmentation is more elaborate and extensive in the form of eight crossbands on the side of the body and tail, continuing out onto the dorsal and anal fins. A cross arrangement of dark bars is present on the side of the head, one of which is from the maxilla through the eye to the corner of the opercle and one from the eye to the isthmus. Some pigment is present on the jaws, branchiostegal membranes, and pelvic and pectoral fins. Metamorphosis is complete by about 10 mm. At 15 mm, young cunner are practically of adult form. Eight dark crossbands become somewhat dissolved laterally and begin to amalgamate, first ventrally. At about 25 mm, color patterns of juveniles are as variable as those of the adults.

Larval Ecology and Behavior. In St. Georges Bay, N.S., cunner larvae (3–11 mm TL) were present in relatively shallow waters (5 m) during the day and dispersed throughout the upper mixed layers at dusk, but with the center of abundance at 15 m by 0900 (Harding et al. 1986). Nighttime dispersion and descent may serve to remove them from surface areas, where predation by jellyfish may occur at night. Larval cunner are visual predators, with peak feeding occurring during the daytime (Harding et al. 1986).

At 8–14 mm, cunner assume a benthic life style characteristic of larger juveniles and adults (Collette 1986; Tupper and Boutilier 1995). Off York, Maine, newly recruiting cunner settle into areas containing macrophytes and tall filamentous (especially *Codium fragile*) and foliose algae and generally avoid areas without cover (Levin 1991). While the presence of recruits is associated with these microhabitat features, other factors, such as small-scale variation in larval supply (Levin 1996) or variability in food availability (Levin 1994a), may contribute to variation in density of settling individuals.

Levin (1993, 1994b) found no evidence for relationships between density-dependent mortality and settlement rates. There was a strong relationship initially between settlement and recruitment, but it weakened over time. So much so, that within 2 months after cessation of settlement, postsettlement loss was greater than 99%, and no correlation remained between recruitment and the initial pattern of settlement. Postsettlement loss was significantly greater on randomly dispersed habitats compared with that on clumped habitats, indicating that mortality of new recruits varied with structural aspects of the habitat. In St. Margaret Bay, N.S., natural populations of cunner are sufficiently large to result in density-dependent growth and postsettlement mortality (Tupper and Boutilier 1995). Here, settlement of cunner was unaffected by the density of prior resident adults, but at adult densities twice that of natural levels there was complete mortality of newly settled fish. In contrast, removal of adults resulted in enhanced growth and recruitment success over natural populations. Growth and mortality of newly settled fish were also suppressed by high recruit densities and enhanced by low densities of conspecific recruits. On all study reefs, newly settled cunner formed social dominance hierarchies, and both growth rate and postsettlement survival were positively correlated with fish size. Settlement in this region appeared unaffected by habitat type or adult density (Tupper and Boutilier 1997). However, postsettlement survival and adult density varied with habitat and were positively correlated with habitat complexity. Since initial patterns of settlement were dramatically altered within a very short time, these authors concluded that habitat-mediated postsettlement processes play an important role in population dynamics of cunner.

T. R. Gleason and C. Recksiek (pers. comm.) directly observed newly settled (8–9 mm) cunner around a rocky adult habitat from early July through August in Narragansett Bay. These fish were often observed swimming at the base of boulders on the interface between open bottom and the rocky area itself. Larger juveniles (10–20 mm) were also noted in these same areas, suggesting that some fish settled from the plankton as early as June in Narragansett Bay. Young-of-the-year cunner were generally not observed swimming among adults until they were somewhat larger than 20 mm (T. R. Gleason and C. Recksiek, pers. comm.). Small juvenile cunner taken at

Woods Hole fed chiefly on minute crustacea such as copepods, amphipods, and isopods (Bigelow and Schroeder). Tetracycline-labeled juvenile cunners (15–40 mm SL) released and recaptured on an artificial reef in Narragansett Bay demonstrated daily patterns (on average) of increment formation on lapillar otoliths under natural conditions (Gleason and Recksiek 1990).

Age and Growth. Scales, otoliths, and opercle bones have been used to age cunner (Johansen 1925; Serchuk and Cole 1974; Dew 1976). Some authors report difficulties in aging cunner using scales, especially in attempting to resolve the first annulus and in determining the ages of older individuals. Johansen (1925) used otoliths to corroborate age estimates made from scales for younger fish, but provided no details concerning suitability of otoliths for aging older fish. Dew (1976) noted that in older fish there is an increased frequency of regenerated scales, and nonregenerated scales are often pitted because of increased incidence of infection with trematode (*Cryptocotyle*) metacercariae with age. Dew attempted to age cunner using opercle bones, but found that the posterior margins of this bone were severely eroded in fish over 12 cm TL.

Johansen's study (1925) on cunner from the Gulf of St. Lawrence and the outer Nova Scotian coast was the first to examine age structure in a population of this species. The oldest and largest fish (10 years old and 37 cm TL) were from Nova Scotian waters and the Bay of Fundy. The largest cunner recorded was 44 cm TL, but this fish was not aged (Leim and Day 1959). Females, especially those greater than 15 cm TL, attained larger sizes than males of the same age. Most cunner reached maturity in their third summer (i.e., when 2 full years old) at 7.0–8.5 cm. Johansen noted that the largest cunner aged was a female (37 cm TL) and all cunner larger than 25 cm TL were females.

Bigelow and Schroeder estimated that in the Gulf of Maine adult cunner usually measure 15–25 cm in length and weigh less than 230 g. They considered a cunner of 30 cm to be very large; fish measuring to 38 cm and weighing about 1.15 kg are caught occasionally. Juveniles 2.5–3.1 cm have often been taken in August in the Gulf, and young fish up to 5.2 cm in September occur in southern New England waters (Bigelow and Schroeder). Based on these findings and results of Johansen's study, they provided estimates of the growth rate for Gulf of Maine cunner. Young-of-the-year fish (probably hatched somewhat later than those occurring in more southern areas) average about 5.0–6.4 cm by their first autumn, and 6.4–7.0 cm by the following June when they are age 1. Subsequent growth rates have not been studied for cunner in the Gulf of Maine. Based on Johansen's age determinations for cunner from the Gulf of St. Lawrence, probable lengths at age for Gulf of Maine cunner are: 7.6–10.5 cm at age 2; 10.5–12.6 cm at 2 or 3 years; 12.6–15.0 cm at age 3; 15.0–17.5 cm at 3–4 years; 17.5–20.0 cm at 4–5 years; 20.0–23.0 cm at 5–6 years; 23.0–25.0 cm at about 6 years; and 25.0–28.0 cm at 6–7 years. Although Johansen (1925) [illegible]

light of results presented for other cunner populations (Dew 1976; W. Chiasson et al., pers. comm.).

Serchuk and Cole (1974) and Dew (1976) used scales to age cunner collected in the Weweantic River estuary and Fishers Island Sound, respectively. None of the fish examined by these authors approached sizes or ages of those examined by Johansen. Serchuk and Cole estimated that annulus formation occurred before July (most probably during late May or June), whereas Dew estimated that in the cunner he studied annulus formation was completed during May. In the Weweantic estuary, the oldest fish ($n = 3$) were age 6 (sex not determined), and in Fishers Island Sound most fish were found to be 4 years or less in age. Only two fish (both age 5) were older than 4 years (a 215-mm female and a 225-mm male) in Dew's study. In contrast to Johansen's study (1925), Dew (1976) also found that male cunner in Fishers Island Sound grew faster than females. W. Chaisson et al. (pers. comm.) validated the use of otoliths for aging cunners and concluded that earlier studies probably underestimated their ages. In all four populations studied (three in Newfoundland, one in Massachusetts), males were significantly larger at age than females.

Serchuk and Cole (1974) provided length-weight relationships and estimated parameters of the von Bertalanffy growth equation for their sample (sexes combined) as: $\log W = -5.559675 + 3.3783109 \log L$ and $L_t = 284.77(1 - e^{-0.1979(t - 0.1044)})$, with W in g, L in mm, and t in years. Length-weight estimates determined separately for males and females (eviscerated weights) were not significantly different (Dew 1976). Length-weight relationship (sexes combined) for cunner occurring in Fishers Island Sound are: $\log W = -5.2512 + 3.2169 \log L$.

General Range. Cunner are widespread along the Atlantic coast of North America and offshore banks, from the eastern coast of northern Newfoundland, and western and southern parts of the Gulf of St. Lawrence, southward in abundance to New Jersey, and as far south as the mouth of Chesapeake Bay. In Canadian Atlantic areas, adults and juveniles have been reported from off Newfoundland, in the Gulf of St. Lawrence, around all the Maritime Provinces, and in the Bay of Fundy and Passamaquoddy Bay. Few cunner are found in the Bay of Fundy, and these are usually only larger members of the species (Huntsman 1922).

Occurrence in the Gulf of Maine. Cunner are among the most common and familiar members of the Gulf of Maine fish fauna that can be found all around the shoreline of the Gulf (Map 26). Bigelow and Schroeder reported that they were generally less numerous but, on average, larger in size in waters east of Casco Bay and that along the shore from Penobscot Bay toward the Bay of Fundy they become progressively less abundant. Cunner are occasionally taken along the coast eastward to the Grand Manan Channel. Sometimes large numbers are taken, but ordinarily they are so scarce around Grand Manan and within Passamaquoddy Bay that

down to the early 1920s. While cunner are reported from Black River east of Saint John this species seems to be unknown farther in along the New Brunswick shore of the Bay of Fundy or in Chignecto Bay and Minas Basin at the head. Annapolis Basin on the Nova Scotian side of the Bay harbors a few, while cunner of all sizes are so numerous in St. Mary's Bay that this may be an important center of reproduction and the source of the few large (i.e., old) ones that are caught farther up the Bay of Fundy. They are also reported along the western shore of Nova Scotia, at Pubnico, for example.

There are large cunner in small numbers on the offshore fishing grounds in the Gulf, Stellwagen at the mouth of Massachusetts Bay, Cashes Ledge, and Georges and Browns banks, in depths down to 90 m, but it is not likely that they ever descend into the deep basins of the Gulf.

Farther east and north, cunner are reported as numerous all along the outer coast of Nova Scotia, including the many bays and inlets, in the southern side of the Gulf of St. Lawrence from Cape Breton to the Gaspé Peninsula, and in bays all around the Newfoundland coast, including those on the northern peninsula, but they tend not to occur in sites exposed to high or persistent wave action (J. Green, pers. comm.). They have not been reported from either the St. Lawrence estuary or anywhere along the north shore of the Gulf of St. Lawrence (Bigelow and Schroeder).

Importance. Historical accounts indicated that cunner were considered a very tasty fish by residents from Eastport to the vicinity of Boston (Goode 1888) and fried cunner were much esteemed as table fare (Field 1907). Elsewhere, however, they were rarely eaten and usually regarded with disgust.

During the 1870s, the annual commercial catch of cunner by small boats fishing out of Boston was estimated at almost 136,364 kg, and the fact that 47,318 kg of cunner were reported for Maine in 1889; 67,409 kg in 1898; and 127,955 kg in 1905 shows that the annual harvest was still considerable at that time. By 1919, the reported catch had fallen to 13,952 kg for Maine and to only about 4,545 kg for the entire coast of Massachusetts, south as well as north of Cape Cod. Maine reported only 4,545 kg for 1928 and 789 kg for 1929, while only 14 and 20 kg were reported for Massachusetts for those 2 years. From 1928 to 1947, commercial catches of cunner were reported for Maine in only 3 years. Landings reported for Massachusetts during this period ranged from 1,409 to 8,500 kg (average 3,386 kg) for 1944–1947, indicating that there was a small demand for cunner until as recently as the late 1940s.

Although usually not regarded as a game fish, cunner afford amusement to thousands of vacationers near seaside resorts in the Gulf of Maine. In 1990 and 1991, about 171,000 cunner were caught in the recreational fishery in Massachusetts, primarily in the private and rental boat mode (Voorhees et al. 1992). Probably more cunner are caught on bits of clam than on other baits, but they will also take snails broken from their shells, bits of crab, lobster, or pieces of sea worms (*Nereis*) almost as freely. Small cunner are a great nuisance to fishermen in pursuit of larger fishes, often stealing the bait as fast as it is offered. Since cunner are year-round inshore residents living on or near the bottom in shallow waters, they have the potential of being an indicator species for both short- and long-term changes in nearshore environmental conditions (Fletcher et al. 1971; Horst 1977; Walton et al. 1978, 1983; Williams and Kiceniuk 1987; Mercer et al. 1997).

The potential usefulness of cunner as a candidate for removing sea lice (*Caligus elongatus*) from farmed Atlantic salmon was explored, but they performed poorly as cleaner fish (MacKinnon 1995).

ACKNOWLEDGMENTS. Drafts of the cunner account were reviewed by John Green and Fred Serchuck.

Suborder Zoarcoidei

Most zoarcoids are elongate and all have a single nostril on each side but there is no known diagnostic character or simple combination of characters that distinguishes this suborder from other blennylike perciforms (Nelson 1994). There are nine families with about 98 genera and 318 species. All species are marine with the largest concentration of species in the North Pacific Ocean. Representatives of five families occur in the Gulf of Maine: eelpouts (Zoarcidae—three species), pricklebacks (Stichaeidae—four), wrymouths (Cryptacanthodidae—one), gunnels (Pholidae—one), and wolffishes (Anarhichadidae—two).

EELPOUTS. FAMILY ZOARCIDAE

Grace Klein-MacPhee and Bruce B. Collette

Eelpouts and wolf eels are slender eel-like fishes with the anal fin continuous with the caudal. In most members of the family the dorsal fin also joins the caudal equally, making one continuous fin extending around the tip of the tail, but in the common Gulf of Maine species, ocean pout, the rear portion of the dorsal is so low that there seems to be a space between it and the caudal. Pelvic fins, when present, are small and jugular, located in front of the pectoral fins. Scales very small and embedded, or absent. Gill membranes are joined to the isthmus. Based on an osteological revision of the family (Anderson 1994), there are four subfamilies, 45 genera, and about 220 species, mostly from mud bottoms of the continental shelves and slopes of the cold waters of the North Atlantic and North Pacific oceans. Almost all eelpouts are oviparous, laying relatively few, large eggs (Anderson 1984b). The closest affinities of eelpouts among Gulf of Maine fishes are with pricklebacks (Stichaeidae), wrymouths (Cryptacanthodidae), and wolffishes (Anarhichadidae). They are easily separable from pricklebacks and wrymouths by the fact that at least the major part of the dorsal fin is soft-rayed, not spiny; and from wolffishes by their more slender form and smaller teeth.

Only three species are definitely known from the Gulf of Maine; one, ocean pout, *Zoarces americanus,* is very plentiful (a summary of recent information on this species was published by Steimle et al. 1999b), whereas the others, wolf eel, *Lycenchelys verrillii,* and Atlantic soft pout, *Melanostigma atlanticum,* are much less so. Four other species, *Lycodes esmarkii* Collett 1875, *Lycodes atlanticus* Jensen 1904, *Lycodonus mirabilis* Goode and Bean 1879, and *Lycenchelys paxillus* (Goode and Bean 1879), are well documented off New England north of Cape Cod, but outside the Gulf of Maine and generally in deeper water (K. Sulak, pers. comm., June 1994). Two additional boreal/Arctic species *Lycodes lavalaei* Vladykov and Tremblay 1936 and *L. vahlii* Reinhardt 1831, occur well north of New England, beginning off central to northern Nova Scotia and continuing north. An additional species included in Bigelow and Schroeder, *Lycodes reticulatus* Reinhardt 1835, Arctic Ocean pout, is exclusively Arctic, found only in the deep Saguenay Fjord at the head of the St. Lawrence estuary. This species has long been confused with the closely related *L. lavalaei.* The figure in Bigelow and Schroeder is actually *L. lavalaei,* and it appeared in subsequent books as well, such as Scott and Scott (1988). There are no confirmed records of *L. reticulatus* south of 44° N.

In addition to the three resident Gulf of Maine eelpouts, three more species are included in the key: *Lycodes esmarkii, Lycodes atlanticus,* and *Lycenchelys paxilla.* The various species of the genus *Lycodes* resemble one another so closely that their identification is very difficult. If one should be taken in the Gulf that does not key out, we suggest that it be sent either to the Systematics Laboratory or the Division of Fishes, National Museum of Natural History, Washington, D.C. 20560, or to the Department of Fishes, Museum of Comparative Zoology, Cambridge, Mass., to be identified.

KEY TO GULF OF MAINE AND NOVA SCOTIAN EELPOUTS AND WOLF EELS

1a. Pelvic fins absent; scales absent; body with loose gelatinous skin. . . . . ***Melanostigma atlanticum***
1b. Pelvic fins present but reduced; scales present; body with firm skin . . . **2**
2a. Dorsal fin separated from caudal fin by considerable gap; posterior part of dorsal fin composed of 16–24 short, stiff spines . . . . . ***Zoarces americanus***
2b. Dorsal, caudal, and anal fins form one continuous fin, without short, stiff spines posteriorly **3**
3a. Trunk extremely slender, at least 14–17 times as long as it is deep; dorsal fin originates over tips of pectoral fins (*Lycenchelys*) **4**
3b. Trunk stouter, 7–13 times as long as it is deep (*Lycodes*) **5**
4a. Dorsal fin rays 85–92; anal fin rays 86–88; lower surface of body with only a few scales . . . . ***Lycenchelys verrillii***
4b. Dorsal fin rays about 118; anal fin rays about 110; lower surface of body uniformly scaly, like upper surface . . . . ***Lycenchelys paxillus***
5a. Body without markings, usually uniform brown, but without stripes, bars, or reticulations; lateral line complete, ventral branch only . . . . ***Lycodes atlanticus***
5b. Body with 5–9 white or yellowish bands on back, extending onto dorsal fin, often with a light band at nape; lateral line inconspicuous but double, midlateral and ventral branches . . . . ***Lycodes esmarkii***

WOLF EELPOUT / *Lycenchelys verrillii* (Goode and Bean 1877) / Bigelow and Schroeder 1953:515–516

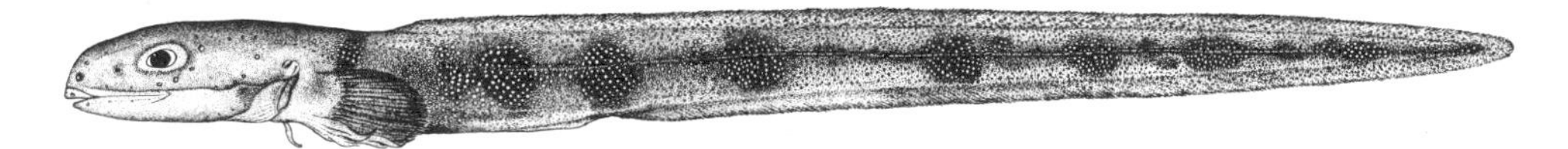

Figure 243. Wolf eelpout *Lycenchelys verrillii.* Gulf of Maine, *Delaware II,* sta. 99-02-069. Drawn by K. H. Moore.

Description. Body slender, 14–17 times as long as deep, tapering near tail (Fig. 243). Head seven in TL, except in old males transformed almost beyond specific recognition by an extraordinary development of entire head in front of eyes. Snout becomes shovel-shaped, its length equal to two-fifths of HL, normally one-quarter. Head depressed, profile in front of eye rather steep. Mouth subterminal, angle of lower jaw under or behind eye, depending on head length. Two rows of uniform teeth in lower jaw, one row in upper jaw, teeth on vomer and palatines. Six large pores on side of upper jaw, seven on lower jaw. Eye six in HL, more in old males. No separation between dorsal, caudal, and anal fins. Dorsal fin originates over or slightly behind tip of pectoral fin, of uniform height, rays extend to and continuous with rounded caudal fin. Anal fin rays slightly longer than dorsal fin rays. Pectoral fins rounded, moderately large, base low on side, a short distance behind gill openings. Pelvic fins small, located in front of pectoral fins, under gill openings. Lateral line single, median, straight. Body covered with widely separated, deeply embedded scales, more numerous on upper part of body and along dorsal fin base.

Meristics. Dorsal fin rays 86–92; anal fin rays 86–88; vertebrae 107–111 (Scott and Scott 1988; B. B. Collette, pers. obs., April 1992).

Color. Sides light brown above the lateral line, yellowish white below it, with a series of seven to ten irregular dark brown patches bisected by the lateral line. In addition, there are two dark patches on the ventral surface forward of the dorsal fin. The belly is blue, its lining jet black.

Size. Reach about 25.4 cm; usual length 10.2–15.2 cm.

Distinctions. Wolf eelpout differ from true eels in possessing pelvic fins and from blennies by having no separate caudal fin. They differ from ocean pout in having all soft dorsal rays with no apparent gap between dorsal and caudal fin. The dorsal fin originates farther back, over the tip of the pectoral instead of in front of the base of the latter as it does in ocean pout. They differ from members of the genus *Lycodes* in being extremely slender.

Habits. Wolf eelpout are a bottom fish, living on mud or sand and confined to considerable depths. Normally, 45–54 m is the upper limit; to the south, on the continental slope, they have been trawled down to 1,190 m in Norfolk Canyon (Wenner 1978). Eeelpouts, most likely wolf eelpout, were the most abundant fish observed on videotapes of the seafloor taken at 420–853 m off Cape Hatteras, N.C., by submersibles (Felley and Vecchione 1994). Individuals tended to be small (<15 cm) and lay in a sinusoidal position, usually near small anemones. Wolf eelpout may account for a large proportion of benthic fishes: 11.9% of benthic fishes taken at 393–1,095 m off southern New England (Haedrich et al. 1975), 36.7% in transects by the submersible *Alvin* in deep basins of the Gulf of Maine (Rowe et al. 1975), and one of two dominant species at 530–840 m on the middle continental slope off Virginia and Cape Hatteras (Moser et al. 1996).

Breeding Habits. Wolf eelpout are seasonal spawners showing maximum gonadal activity from June through September (Wenner 1978). Fecundity varied from 15 eggs in a 122-mm TL fish to 50 in a 133-mm fish, with the mean fecundity for 103 gravid fish being 29.3. The fecundity–total length relationship was not significant. The most sexually mature specimens had eggs 3.0–4.2 mm in diameter.

Food. The bulk of the diet, by percent frequency and percent volume, in 13 digestive tracts of wolf eelpout from Norfolk Canyon was composed of bivalves (mainly Nuculanidae) and gastropods (mainly Scaphandridae and Retusidae), with lesser amounts of scaphopods, amphipods, cumaceans, and polychaetes (Wenner 1978).

Predators. Recorded from stomachs of goosefish, white hake, and sea raven, of which white hake are the most frequent predator (Rountree 1999); also American plaice (Bowman and Michaels 1984).

Parasites. A chondrocanthid copepod, *Diocus lycenchelus,* was described from the nasal cavity of wolf eelpout from the continental slope off North Carolina (Hogans and Sulak 1992). Wolf eelpout are also parasitized by a pennellid copepod *Haemobaphaes cyclopterina* (K. Sulak, pers. comm., June 1994).

General Range. Known off the south coast of Newfoundland in Hermitage Bay, N.S., New England, and southward along the continental slope to Beaufort, N.C., in rather deep water.

Occurrence in the Gulf of Maine. Wolf eelpout were formerly regarded as very rare within the Gulf of Maine. A few specimens were trawled off the mouth of Passamaquoddy Bay, one was taken off Monhegan Island, and several were collected by the U.S. Fish Commission many years ago off Cape Ann, in the Western Basin, and off Cape Cod. Capture of 61 specimens of wolf eelpout in the trough west of Jeffreys Ledge by *Albatross II* and observations in transects by the *Alvin* in Wilkinson and Murray basins (Rowe et al. 1975) show that they are more plentiful in the deeper parts of the Gulf than previous records suggested. On a 1994 *Albatross IV* cruise in the Gulf of Maine, we collected specimens at nine deep stations using a flat Yankee trawl that stays close to the bottom but none was collected at the subsequent eleven stations using the same style net, but with rollers that prevent the net from reaching the bottom. Wolf eelpout are routinely captured in the Bay of Fundy in depths greater than about 50 m (K. Sulak, pers. comm., June 1994).

ATLANTIC SOFT POUT / *Melanostigma atlanticum* Koefoed 1952

Description. Body elongate, tapering from pectoral fins to tip of tail (Fig. 244). Head blunt, snout rounded. Eye large. Mouth small, terminal, oblique. Elongate teeth on vomer and palatines, in a single series on jaws. Pelvic fins absent. Skin loose and without scales. Lateral line absent.

Meristics. Dorsal fin rays 92–99; anal rays 77–84; pectoral rays 6–9; gill rakers 11–13; vertebrae about 83–93, of which 62–76 are caudal (McAllister and Rees 1964).

Color. In life, the head and anterior parts of body are a brilliant silvery blue, the remainder of the body is translucent

Figure 244. Atlantic soft pout *Melanostigma atlanticum*. Drawn by Todd and Stackhouse. (From Scott and Scott 1988:414.)

(McAllister and Rees 1964). Living individuals taken in bottom sediments with a box core were cream-colored with a slight blue iridescence (Silverberg et al. 1987). The anterior part of the head of specimens taken on *Delaware II* in February 1999 was black, the peritoneum iridescent blue shining through the translucent body, the posterior part of the body was pink, and there were black blotches on the tail (B. B. Collette, pers. obs., 2 Feb. 1999). The dorsal and ventral thirds of the body were completely translucent.

Size. The largest specimen reported by McAllister and Rees (1964) was a 138-mm individual from the southwest edge of the Grand Banks. A larger specimen, 150 mm SL and 156 mm TL, was collected by *Delaware II* at sta. 99-02-52, 37°05′ N, 74°32′ W between 500 and 600 m on 15 February 1999 (B. B. Collette, pers. obs.).

Distinctions. A delicate translucent appearance coupled with the absence of pelvic fins and scales distinguish the Atlantic soft pout from other eelpouts.

Habits. Atlantic soft pout are mesopelagic, living in midwater away from the ocean floor but descending to the seafloor for spawning (Markle and Wenner 1979; Silverberg et al. 1987). Daytime observations from the *Alvin* confirm that adults are mesopelagic in early June off Virginia (Wenner 1978). Many individuals were seen between 200–590 m at temperatures of 4°–12°C drifting passively in the water column, some head down, some head up, and others horizontal. Temperatures at presumed depth of capture in Canadian waters varied from 3.0° to 5.2°C, with salinities from about 33.4 to 34.7 ppt (McAllister and Rees 1964).

Food. Little was found in the stomachs of 100 bottom-trawled specimens from the continental slope off the eastern United States (Wenner 1978). Crustacean remains were found in 29, including ostracods, copepods, and euphausiids. Gulf of St. Lawrence specimens from about 360 m contained several species of copepods in their stomachs, *Metridia longa, Pseudocalanus minutus, Calanus finmarchius,* and *C. hyperboreas* (McAllister and Rees 1964).

Parasites. Of 100 specimens from off the eastern United States, 58 contained digenetic trematodes in their intestines; all that were identified were *Fellodistomum melanostigmum* (Markle and Wenner 1979).

Predators. Found in stomachs of little skate, pollock, and red and white hakes, most frequently white hake (Rountree 1999); also redfish, *Sebastes "marinus"* from the Gulf of St. Lawrence (Steele 1957).

Reproduction. A 98-mm specimen from off Trois Pistoles, Quebec, captured in July contained 30 large eggs, up to 2.8 mm in diameter (McAllister and Rees 1964). Eggs can be seen through the abdominal wall of a 117-mm specimen illustrated by Goode and Bean (1896: Fig. 284). There have been reports of similar-sized eggs from July through September on the continental slope off Virginia and of much smaller ones (<1 mm) in January and November (Markle and Wenner 1979). Fecundity varied from 26 in a 117-mm TL fish to 106 in a 132-mm fish (Wenner 1978). Male *M. atlanticum* have sexually dimorphic fanglike teeth on both jaws and vomer (McAllister and Rees 1964) and are significantly larger at sexual maturity than females: mean lengths 137 and 126 mm, respectively (Wenner 1978). This normally mesopelagic species has been collected with box cores 15–32 cm deep within sediments from 350 m in the Laurentian Trough of the maritime estuary of the St. Lawrence in July and August (Silverberg et al. 1987). Six individuals, one male and five females, were found in a burrow with a cluster of 50 large eggs, 3.9 mm in diameter.

General Range. North Atlantic. In the western North Atlantic from the Gulf of St. Lawrence southward along the south coast of Newfoundland, Grand Bank, along the edge of the continental shelf to off Cape Hatteras (McAllister and Rees 1964; Scott and Scott 1988). In the eastern Atlantic, from the Faeroe-Icelandic Ridge to Cap Blanc, Mauritania, including the Mediterranean Sea (Anderson 1994).

Occurrence in the Gulf of Maine. Not reported by Bigelow and Schroeder although they did include several additional species of eelpouts that they thought might occur in the Gulf. Known from 11 MCZ series taken off Jeffreys Ledge and Stellwagen Bank. On board *Albatross IV* in April 1994, we took them at three stations off Stellwagen with midwater trawls and with a bottom trawl at one of those stations as well. The specimens were sometimes wrapped around the twine of the large mesh part of the net rather than being in the cod end.

OCEAN POUT / *Zoarces americanus* (Bloch and Schneider 1801) /

Bigelow and Schroeder 1953:510–515 (as *Macrozoarces americanus*)

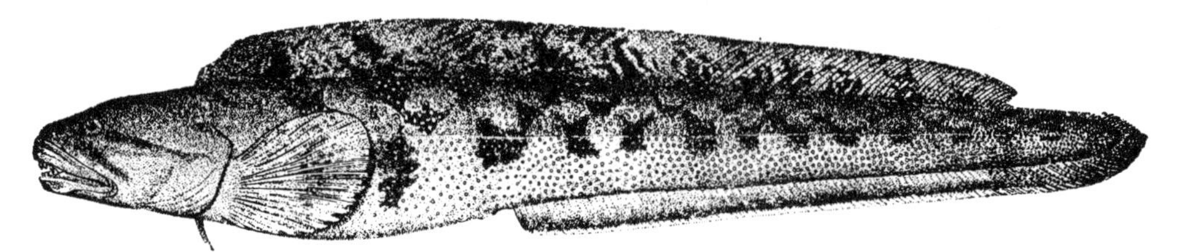

Figure 245. Ocean pout *Zoarces americanus.* Eastport, Maine. Drawn by H. L. Todd.

Description. Body about eight times as long as deep (10–11 times in young fish up to about 20 cm long), moderately compressed, tapering backward from abreast of pectoral fins to a pointed tail (Fig. 245). Body soft, scales very small, embedded in a mucus-covered skin like an eel. Upper jaw projects a little beyond lower; soft, fleshy upper lip somewhat farther still, enclosing tip of lower lip when the mouth is closed. Mouth wide, gaping back beyond small eyes, set low with thick and fleshy lips that give a distinctive profile. Both jaws armed anteriorly with two rows of strong, blunt conical teeth, largest in front. Dorsal fin runs from nape along length of trunk, consisting of soft rays followed by 16–24 short spines, hardly visible, creating apparent free gap between dorsal and caudal fins; then about 16–31 more soft rays. Anal fin originates a little in front of midlength of fish, continuous with caudal fin. Dorsal and anal fins of nearly even height from end to end except as just noted, but dorsal nearly twice as high as anal. Pectoral fins large and rounded. Pelvic fins very small, under throat, well in front of pectoral fins.

Meristics. Dorsal fin rays 92–103, followed by 16–24 short spines, then 16–31 more soft rays; anal fin rays 105–124; pectoral fin rays 18–21; gill rakers 4–6; vertebrae 129–146, 25–28 precaudal, 103–118 caudal (Miller and Jorgensen 1973; Anderson 1984a, 1994; Scott and Scott 1988).

Color. Although described as reddish brown mottled with olive or salmon-colored, most ocean pout are some shade of muddy yellow, paler or darker; some are tinged with brown, salmon, or orange; and a few have been pure olive green. Fishermen usually describe them as yellow, which is evidently the prevailing hue in offshore parts of the Gulf. Other ocean pout caught inshore along the coast of Maine, however, have shown yellow only on margins of the fins, particularly the lower edge of the pectorals, with the general ground tint of sides and back ranging from pale gray (sometimes with purplish tinge) to dull brown or to dark dusky olive; the belly ranges from dirty white, yellow, or pink to the same dark shade as the back. Whatever the ground tint, the sides are marked with small dark irregular crossbars, extending out on the dorsal fin, and there is a dark brown stripe running from the eye to the edge of the gill cover.

Size. Reach 118 cm and more than 6 kg (Anderson 1984a). The largest of 2,500 specimens examined by Olsen and Merriman (1946) was 97.8 cm and weighed 5.2 kg. Average is 40.6–71.2 cm and 0.45–1.8 kg.

Distinctions. The most useful field marks for identification of ocean pout among the several eellike fishes with which they might be confused are their fins. There is no notch between the anal and the caudal as there is in true eels. The mouth lacks the crushing teeth that are so characteristic of wolffishes. The predominantly soft dorsal fin rays distinguish small eelpout from blennies. The short row of dorsal fin spines near the caudal fin separates them from other eelpouts and cusk-eels.

Taxonomic Note. The generic name *Macrozoarces* was placed in the synonymy of *Zoarces* on the basis of a comprehensive osteological revision of the family (Anderson 1994).

Habits. Ocean pout are ground fish, as might be expected from the fact that they lack a swim bladder, as well as from their food. The habits of fish kept in aquariums, where they are described as remaining coiled up in the darkest parts, suggest that they spend most of their lives hiding among seaweeds and stones.

The vertical range of ocean pout extends at least as deep as 363 m. At the opposite extreme, young ones are sometimes found around rocks and in seaweed along the shore in the Bay of Fundy during ebb tide (Clemens and Clemens 1921). The average depth range is 15–80 m (Clark and Livingstone 1982). On the Scotian Shelf and in the Bay of Fundy the depth range is 27–363 m. Ocean pout also frequent different types of bottom in different localities. They prefer continental shelf areas with sand and gravel substrates (Orach-Meza 1975). Observations from submersibles show them to be positively associated with shells, and a young-of-the-year ocean pout was photographed sheltering under an overturned ocean quahog valve (Auster et al. 1991). Bigelow and Schroeder took ocean pout in the Gulf of Maine on sandy mud, sticky sand, broken bottom, and pebbles and gravel. In fact, the only type of bottom where they were not caught in the Gulf is the soft oozy mud with high organic content that floors certain of the deeper depressions, such as the trough west of Jeffreys Ledge. They are commonly caught on stony ground farther east along the coast of Maine, and Huntsman (1922) described them as taken on hard bottom in the Bay of Fundy. On the Scotian Shelf they prefer silty bottoms (Scott 1982a).

They are a cold-water fish preferring temperatures of 6°–9°C (Scott 1982b) but can tolerate a range of 0°–16°C. Ocean pout blood contains antifreeze proteins that protect them from freezing (Hew et al. 1984). These are of high molecular weight (6,000) and are similar to the antifreeze proteins produced by sea raven (Kao et al. 1986). Unlike many other temperate fishes that produce antifreeze proteins only during colder months, ocean pout produce them year round. Ocean pout from Newfoundland produce very high levels even when the water temperature is at its highest (14°C). New Brunswick pout also produce antifreeze peptides all year, but at reduced levels (Fletcher et al. 1985).

Salinity preferences are 32–34 ppt (Scott 1982b), but they are known to run into rivers for some distance, though always holding to the bottom, in the undercurrent of higher-salinity water.

Ocean pout are described as moving slowly backward and forward by undulations of the fanlike pectoral fins or of swimming more rapidly by undulating motions of the rear part of the trunk and tail, with the pectorals widely spread and held horizontally and the dorsal and anal fins close to the body. They are normally slow moving but can swim actively when disturbed (Beamish 1966b). They do not school but congregate during the spawning season (winter).

Food. Adult ocean pout feed on a wide variety of shelled mollusks, univalve and bivalve, large and small crustaceans, echinoderms, other invertebrates, and to a lesser extent on fishes and fish eggs. Diet varies somewhat in different geographic regions and by size. Amphipod crustaceans (mostly of the family Corophiidae) accounted for a large percentage of the diet of small (11–35 cm TL) ocean pout (Bowman and Michaels 1984). Also important in this size range were a crab, *Cancer irroratus;* polychaete worms; mollusks; and echinoderms, especially a sand dollar, *Echinarachnius parma.* Echinoderms contributed more than 40% by weight to the diet of ocean pout in all length categories from 31 to 85 cm TL (Bowman et al. 2000). Other echinoderms identified in stomach contents were ophiuroid brittle stars *Euryale* sp. and *Ophiopholis aculeata.* Green sea urchin (*Strongylocentrotus droebachiensis*) constituted 62% of the overall diet by weight of 151 ocean pout in eastern Newfoundland and the brittle star *Ophiopholus aculeata* constituted 7% of the diet (Keats et al. 1987b). Prey groups of lesser importance in the diet of ocean pout over 35 cm TL (Bowman et al. 2000) included polychaetes, amphipods, decapods, and mollusks (primarily *Placopecten*).

The most important food items in the diet of 46 ocean pout from the Gulf of Maine (Hacunda 1981) were echinoderms (20.6% by weight) and crustaceans (relative overall importance by weight, numbers, and occurrence). The most important

sis; a sand dollar *Echinarachnius parma;* and a brittle star, *Amphipholis squamata.* Principal crustacean prey were amphipods and cumaceans. Mollusks were also important by weight, particularly sea scallop, soft-shell clam, and cockle. Ocean pout stomachs also contained large quantities of bottom sediment and organic material. Sand dollars were the chief items in the stomachs of 850 ocean pout taken in the southwestern part of the Gulf and off southern New England, with crabs (*Cancer*) and isopod crustaceans (*Unicola*) second; while some had eaten bivalve mollusks (*Yoldia* and *Pecten*) in large amounts and eggs of longhorn sculpin, which are often laid among branches of finger sponge (*Chalina*) (Olsen and Merriman 1946).

On the outer continental shelf of the Mid-Atlantic Bight, ocean pout diet consists of amphipod and decapod crustaceans in all seasons, polychaetes in winter and spring, echinoids in winter, and cumaceans and pelecypods in small quantities in all seasons (Sedberry 1983). Bay of Fundy fish feed chiefly on two common mussels, *Mytilis* and *Modiolus,* whelks (*Buccinum*), periwinkles (*Littorina*), and scallops (*Pecten*) as well as various other bottom-living mollusks, sea urchins, brittle stars, and barnacles (Clemens and Clemens 1921). In the rocky subtidal zone off eastern Newfoundland they eat mainly sea urchins (62% of the diet by weight), brittle stars (*Ophiophalus;* 7% by weight), miscellaneous invertebrates, fishes such as capelin, and fish eggs (capelin and lumpfish).

The diet of juveniles 110–150 mm from the Gulf of Maine and western Nova Scotia is composed of amphipods (*Unicola* 40%, gammarids 31%, *Melita* 11%) and ophiuroids 2% (Bowman et al. 1987). Small ocean pout also eat polychaetes, scallops, hermit crabs, and sand dollars (Bowman et al. 2000). Stomachs of 0-group ocean pout from Newfoundland all contained harpacticoid copepods, which made up 82% of the diet (Keats and Steele 1993). Other items include, in order of importance, amphipods, mysids, isopods, ostracods, and polychaetes.

Numerous studies have been done on food resource utilization of ocean pout and other associated benthic fishes. Studies in Passamaquoddy Bay (Tyler 1972; MacDonald and Green 1986) and in Johns Bay, Maine (Hacunda 1981) indicated that there is considerable dietary overlap, but that ocean pout appear to utilize more benthic infauna, particularly mollusks and sedentary polychaetes from deeper strata by using the feeding adaptation of scooping and digging with their mouth and digesting shelled invertebrates more effectively in their intestine.

Ocean pout feed mainly during the day, although limited feeding also occurs at night (MacDonald and Waiwood 1987). They reduce or stop feeding during breeding season. Females stop feeding prior to spawning as gonads mature and do not resume until the eggs hatch. Males stop feeding prior to spawning, but resume before the females do (Keats et al. 1985).

Laboratory observations of feeding behavior showed that ocean pout rarely leave the bottom to feed; and when they do so, they maneuver very poorly. They showed no strong response to moving food. When foraging, they swam a few centimeters from the bottom, settled on the sediment with

coiled tail served as a base from which they extended their head to capture prey. Food was captured by a rapid scooping motion of the lower jaw together with opercular suction. Alternately, some pout folded the lower lip outward to turn over stones or scoop up mouthfuls of sediment, which were sorted for food. Sight appears to be relatively unimportant for food capture as they frequently do not see prey swimming near their head or miss them when they lunge. Ocean pout possess taste bud organs and pressure-sensitive tactile papillae on the ventral surface of the upper lip, and taste bud organs and mucous glands in the lower lip (MacDonald 1983).

Food passed through the digestive tract quickly (MacDonald et al. 1982), 15 h for mollusks, 8 h for crustaceans, and 7 h for polychaetes) compared to other groundfishes such as cod (30 h for mollusks, 17 h for crustaceans and polychaetes). The stomach acts as a holding organ and most digestion occurs in the intestine. The intestine has a large diameter and a thick wall that may have a higher absorptive area making up for the fact that ocean pout lack pyloric caeca. Daily ration of a bivalve, polychaete, and amphipod diet was 1.68% of the fish's body weight (MacDonald and Waiwood 1987).

Predators. Found in stomachs of 18 species of fishes (Rountree 1999), most frequently in spiny dogfish, skates (little, winter, and thorny), cod, hakes (red, white, and spotted), sea raven, and bluefish.

Parasites. Ocean pout are parasitized by protozoans *Pleistophora macrozoarcidis, Sarcocystis,* and *Chloromyxum,* which are present in the muscle and *Ceratomyxa acadiensis* in the gall bladder. Helminth parasites include a digenic trematode *Cryptocotyle lingua* in the skin; a cestode *Bothrimonus intermedius* in the intestine; trematodes *Porrocaecum decipiens* in the muscle, and *Contracaecum* sp. and an acanthocephalan *Echinorhynchus gadi* in the intestine. A leech *Platybdella buccalis* was found in the mouth (Nigrelli 1946). An amphipod *Lafystius morhuanus* was found on ocean pout (Bousfield 1987) but the location was not given.

The most serious parasite from a human standpoint is *Pleistophora,* as the lesions that it causes in fillets makes them unappealing for human consumption and has relegated ocean pout to a trash fishery. Sheehy et al. (1974) found that 29% of the specimens collected in Rhode Island Sound had parasitic lesions. Incidence of infestation was significantly correlated with age, length, and weight. There was no difference in infestation rates between the sexes. Historical data showed that fish from Block Island Sound had 64% incidence of infestation (Olsen and Merriman 1946) and fish from Cape May, N.J., to Cape Cod had infestation rates of 4–38% (Sandholzer et al. 1945). There is no evidence that humans have any ill effects from eating infested ocean pout (Sheehy et al. 1974).

Breeding Habits. Median length at maturity for female and male ocean pout from the Gulf of Maine was 26.2 and 30.3 cm TL, respectively (O'Brien et al. 1993). Ocean pout from southern New England had L_{50} values of 31.3 and 31.9 cm for females and males, respectively. Larger L_{50} values were calculated by Olsen and Merriman (1946) for ocean pout from southern New England in 1944. They found that ocean pout males matured between 2 and 4 years of age at lengths ranging from 25 to 39 cm, and females between 5 and 9 years of age. Differences in L_{50} values for females between the two studies may be due to difficulties in identifying mature females (O'Brien et al. 1993).

Although ocean pout are egg layers unlike their viviparous eastern Atlantic relative, recent observations in the field and in the laboratory indicate that they practice internal fertilization (Mercer et al. 1993; Yao and Crim 1995a,b). In the laboratory, fertilized females spawned spontaneously 6–17 h following copulation (Yao and Crim 1995b). Fertilized eggs adhere to each other forming an egg ball, 10–15 cm in diameter, containing 1,300–1,700 eggs (Yao and Crim 1995b). Large females lay more eggs than small ones: numbers of maturing eggs range from 1,306 in a fish 55 cm long to 4,161 in one of about 87.5 cm.

Adults congregate in rocky areas prior to spawning. They frequently occupy nesting holes under rocks or in crevices in pairs or solitary. Females lay egg masses encased in a gelatinous matrix and then guard them for 2.5–3 months until they hatch, during which time they do not feed. Males apparently do not guard nests, but move off and begin feeding after spawning (Keats et al. 1985).

Ocean pout breed between New Jersey and Newfoundland. Spawning occurs in September and October in southern New England (Olsen and Merriman 1946) and Passamaquoddy Bay (Clemens and Clemens 1921); and in August and September in Newfoundland (Keats et al. 1985). Production of the highest motility sperm is restricted to the middle of the breeding season in Newfoundland (Wang and Crim 1997). Copulation in the field was observed in Newfoundland in late August (Mercer et al. 1993). Egg masses were collected in Conception Bay, Nfld., in October and November (Methven and Brown 1991). Adults move into shallower water (50 m or less) at temperatures of about 10°C (Clark and Livingstone 1982).

Early Life History. The eggs are spherical and yellow, 6–9.2 mm in diameter, and are laid in masses held together by a gelatinous substance (Methven and Brown 1991; Yao and Crim 1995a). The eggs contain abundant tiny oil droplets (Yao and Crim 1995a). After 3–5 days of incubation in ambient seawater, fertilized eggs showed little change in oil droplet distribution, but the oil droplets in unfertilized eggs fused into one to four large oil globules (Yao and Crim 1995a) matching the description of one large (3.2 mm) light yellowish brown oil globule given by Methven and Brown (1991). The yolk is clear, homogeneous, unsegmented, and unpigmented. The egg mass is slightly eccentric to cucumber-shaped, with the eggs closely coherent but not attached to the substrate. The incubation period is 2.5–3.5 months (Methven and Brown 1991; Yao and Crim 1995a). Keats et al. (1985) observed egg masses in nesting holes first in late August and well-developed embryos into late November.

No real larval stage is present; ocean pout hatch as juveniles (Methven and Brown 1991; Yao and Crim 1995a). Newly hatched individuals are well developed, have adult-type pigmentation, the adult complement of fin rays, and a well-developed and functional jaw that allows them to feed within 2 days of hatching. They are about 30 mm at hatching with a 5- to 8-mm yolk sac that shrinks and is taken into the abdomen within 20 s after hatching according to White (1939) or within hours or days according to Methven and Brown (1991). They are so much larger than larvae of other fishes and so far advanced in development that they are easily identified as ocean pout. They probably remain on the bottom from the time they hatch. All catches of immature fish recorded so far have, indeed, been on the bottom.

Young, up to 7–10 cm, are checkered along the sides and irregularly blotched on the back with light and dark brown. They have a small but prominent black spot on the forward part of the dorsal fin until about 30 cm, but this spot fades out with growth. There is a dark streak extending across the cheek from the eye to the posteroventral edge of the operculum. The anal area is light with a slight yellow tinge; the belly, pectoral and pelvic fins, and ventral parts are light-colored.

Ocean pout reach a length of about 41–60 mm by April, 45–70 mm by May, and 54–75 mm by June, and they are about 127 mm at 1 year (Olsen and Merriman 1946).

In juveniles 74–295 mm, the gill rakers are short, the body is elongate and slender, the head short and less heavy than in adults, and the snout is blunter (Olsen and Merriman 1946). Pigmentation is similar to larvae but more mottled than checkered. Color is light and dark brown with a suggestion of yellow-green on the dorsal fin. The dark spot fades and disappears as the fish grows.

Age and Growth. Age and growth of fish from Massachusetts, Connecticut (Olsen and Merriman 1946), and Rhode Island (Sheehy et al. 1974) was determined by otolith analysis. Fish from Rhode Island appeared to have a slower growth rate, and those from the Bay of Fundy slower yet (Clemens and Clemens 1921). Fish have been aged up to 16 years, but few survive beyond this (Olsen and Merriman 1946).

General Range. Coast of North America from near Battle Harbor, Labrador (Backus 1957b), the Strait of Belle Isle, Gulf of St. Lawrence, and southeastern Newfoundland south to Chesapeake Bay. They are common from the southern side of the Gulf of St. Lawrence and northern Nova Scotia to New Jersey. There is a doubtful record for North Carolina (Smith 1907).

Occurrence in the Gulf of Maine. Ocean pout are familiar fish in the Gulf in moderate depths both near shore and on the offshore banks (Map 27). Adult ocean pout are among the

chusetts Bay in spring (Steimle et al. 1999b). They are abundant locally off western Nova Scotia, the Bay of Fundy, all along the coasts of Maine and Massachusetts, and on Georges Bank. Very small ones have been collected off Chatham, Cape Cod; on Stellwagen Bank at the mouth of Massachusetts Bay; near Mt. Desert Island, Maine; in the Bay of Fundy and Passamaquoddy Bay. The NEFSC bottom trawl survey collected juvenile ocean pout south and west of Cape Cod in winter and within the Gulf of Maine and on Georges Bank in other seasons (Steimle et al. 1999b).

In the Massachusetts Bay region, they are known as a comparatively deepwater fish living at depths greater than 18 m, although they have been recorded from Gloucester Harbor and the Mystic River. It was sixth in abundance in Quincy Bay (Jerome et al. 1966) and twenty-first in Salem Harbor (Anderson et al. 1975). Ocean pout on offshore grounds usually live deeper than 36 m. *Albatross II* collected a number in the basin of the Gulf down to 162 m, and a large number have been trawled on Georges Bank at depths of 36–100 m.

Migrations and Movements. There is no evidence that ocean pout carry out any extensive migrations. Tagging studies off Block Island Sound (Sheehy et al. 1977) found that 90% were recaptured within 4 km and none further than 12.8 km away. Seasonal changes in distribution occur that reflect localized movements associated with temperature changes and spawning behavior. In the southern New England–Georges Bank area, they move into cooler rocky areas in summer, returning in late autumn (Orach-Meza 1975), and move similarly out of inshore areas in the Gulf of Maine later in summer, returning the following spring. Adults congregate through summer, autumn, and early winter on rocky bottoms, where eggs are deposited and guarded, to disperse again in midwinter (after the eggs have hatched) over smoother grounds.

Importance. Although ocean pout have few bones and have sweet meat, there was no regular market prior to the early 1930s. Nearly all of those that were caught incidentally by larger vessels were thrown overboard. A small demand then developed for them resulting in landings for Massachusetts ranging between 21–52 mt from 1935 to 1942. A concerted attempt was made to market ocean pout as fillets during World War II, and it was so successful that landings peaked at 2,000 mt in 1943 (Wigley 1998a). However, infestation by a parasitic protozoan and incidence of lesions in the flesh and the subsequent public health embargo caused it to be relegated to a trash fishery (Olsen and Merriman 1946; Sheehy et al. 1977). During the 1950s and 1960s most of the U.S. landings were used for industrial purposes. From 1966 to 1969 distant-water fleets took significant numbers of ocean pout and total landings peaked at 27,000 mt in 1969 (Wigley 1998a) (Fig. 246). All landings since 1974 have been by the United States, with catches declining to an average of 600 mt annually in 1975–1983. Marketing efforts in the 1970s had considerable success and sales volumes in-

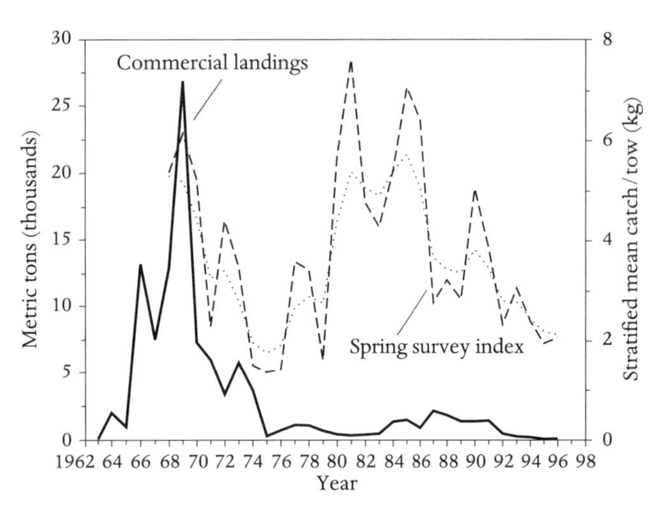

Figure 246. Commercial landings and index of abundance of ocean pout *Zoarces americanus* from the Gulf of Maine and the Mid-Atlantic. (From Wigley 1998a.)

to 1,300–1,500 mt in 1984 and 1985 owing to development of a small directed fishery in Cape Cod Bay supplying the fresh fillet market (Wigley 1998a). Landings dropped to 225 mt in 1993 and have continued to decline.

The principal fishing gear is the otter trawl, and the fishery occurs primarily between December and May (Wigley 1998a). The fishery in the Exclusive Economic Zone (EEZ) is managed under the New England Fishery Management Council's Multispecies Fishery Management Plan; the state of Massachusetts regulates the inshore fishery in Cape Cod Bay. Total landings in 1996 were only 51 mt, the lowest since 1963 (Wigley 1998a). Declines in commercial landings and the NEFSC spring bottom trawl survey biomass index to new lows show the population to be overexploited and at a low biomass level (Wigley 1998a).

Ocean pout are a very lean fish with an average of 1% lipid, 81% moisture, 17% protein, and 1% ash (Sheehy et al. 1977). Ocean pout fillets were rated just as acceptable as flounder fillets in terms of taste, texture, appearance, and odor and were acceptable after frozen storage (Jhaveri et al. 1985).

Studies on the feasibility of ocean pout culture were initiated at the Marine Finfish Research Hatchery in Newfoundland in 1989 (Brown et al. 1992). Survival of juveniles hatched from egg masses collected in the wild was comparatively high (75–80%) over the first year, so these initial results were encouraging. Subsequent success in artificial insemination of ocean pout (Yao and Crimm 1995a) led to the desirability of long-term storage of sperm and successful use of cryopreservation (Yao et al. 2000).

Stocks. Tagging studies (Sheehy et al. 1977) and meristic and morphometric data (Orach-Meza 1975) suggest the existence of distinct unit stocks with little apparent mixing. Vertebral count data, growth rates, and condition factor indices show there are at least two independent stocks; one centered in the Bay of Fundy–northern Gulf of Maine, east of Cape Elizabeth, the other from Cape Cod Bay, Georges Bank, and south to Delaware (Olsen and Merriman 1946; Orach-Meza 1975; Wigley 1998a). The southern stock is characterized by faster growth rates and has supported the commercial fishery.

ACKNOWLEDGMENTS. The food habits paragraphs were revised by Ray Bowman, a draft of the ocean pout account was reviewed by Susan Wigley, and drafts of the entire zoarcid section were reviewed by Eric Anderson and Ken Sulak.

PRICKLEBACKS. FAMILY STICHAEIDAE

Bruce B. Collette

Pricklebacks are characterized among Gulf of Maine fishes by the position of their pelvic fins, which are under or in front of the pectoral fins, combined with a single dorsal fin that is spiny throughout its length and extends the whole length of the trunk, and with a slender form, eel-like in some of them. Of the approximately 36 genera and about 65 species in the family (Nelson 1994), three genera and four species occur in the Gulf. Larvae of all four species of Gulf of Maine pricklebacks have been taken in tow nets in the Gulf of St. Lawrence from April through June (Faber 1976b). Prickleback larvae are compared and illustrated by Faber (1976b) and Fahay (1983). Pricklebacks most resemble rock gunnels (Pholidae), which have rudimentary pelvic fins and much longer spiny dorsal fins. Pricklebacks are somewhat similar to wolffishes (Anarhichadidae) and wrymouths (Cryptacanthodidae), but these families lack pelvic fins, which are present in all Gulf pricklebacks, although they may be very small. The canine tusks and molar teeth of wolffish have no counterpart among pricklebacks, and the peculiar face of wrymouth is equally distinctive. Eelpouts (Zoarcidae) are somewhat similar in appearance, but the greater part of their dorsal fin is soft-rayed, not spiny, and their anal fin is continuous with their caudal fin.

KEY TO GULF OF MAINE PRICKLEBACKS

1a. Body very slender, about 18–20 times as long as it is deep . . . . **Snakeblenny**
1b. Body only moderately slender, not more than 8–10 times as long as it is deep **2**
2a. A row of conspicuous roundish black or dusky spots present along the dorsal fin . . . . **Arctic shanny**
2b. Only one large and conspicuous dark spot on the dorsal fin, or none . . **3**

3a. Pectoral fins evenly rounded, middle rays longest; dorsal fin marked with one large, conspicuous, anterior dark blotch; two lateral lines; 43–44 dorsal fin spines . . . . **Radiated shanny**

3b. Pectoral fins with lower rays longer than upper rays and free at their tips; no conspicuous blotch in dorsal fin; lateral line single, 55–61 dorsal fin spines . . . . **Daubed shanny**

SNAKEBLENNY / *Lumpenus lumpretaeformis* (Walbaum 1792) / Serpent Blenny /

Bigelow and Schroeder 1953:494–497

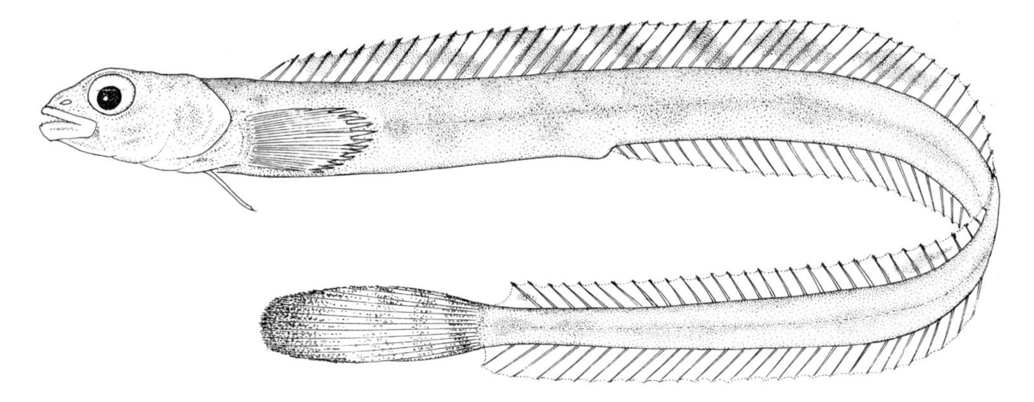

Figure 247. Snakeblenny *Lumpenus lumpretaeformis.* Gulf of Maine, *Albatross IV,* sta. 94-02-24, USNM 329958. Drawn by K. H. Moore.

Description. Slender, one-twentieth as deep as long, slightly compressed (Fig. 247). Head about one-eighth as long as body to base of caudal fin. Eyes large, mouth wide, snout blunt. Pelvic fins well developed with short spine and three longer rows, located slightly in advance of large pectoral fins. Caudal fin pointed, central rays longer than outer ones.

Color. Whitish or pale brown on the back and sides, with darker brown markings. On a 30.5-cm fish taken off the coast of Maine, the head was pale brown, sides of the body blotched with brown, dorsal fin marked obliquely with 18 pale bars, caudal marked transversely with eight bars, anal rays pale brown against a colorless membrane, pelvic fins white, and pectoral fins tinged with brown.

Meristics. Dorsal fin spines 68–85; anal fin rays I, 47–53; pelvic fin rays I, 3; pectoral fin rays 14–16; vertebrae 80–85 (Fahay 1983; Scott and Scott 1988).

Size. Maximum known size 48.3 cm.

Distinctions. Snakeblenny somewhat suggest sand lance in general form, but they are much more slender and eel-like. The rounded tail (that of sand lance is forked), large pectoral fins, spiny dorsal fin (sand lance have a soft dorsal only), and the fact that the lower jaw does not project beyond the upper, together with color, also serve to separate them from sand lance. The very slender body is the most obvious difference between this species and its allies, daubed shanny, Arctic shanny, and radiated shanny, which are rather robust. The oval pointed caudal fin is also diagnostic. The chief anatomical feature (apart from its slenderness) distinguishing them from rock gunnel is that the pelvic fins, each of one spine and three longer rays (one spine and one ray in rock gunnel), are well developed, one-third to one-half as long as the pectorals. The pectoral fins are much larger than those of rock gunnel, and the dorsal and anal fins are fully twice as high relative to body depth, while the anal fin originates farther forward; separation of dorsal and anal fins from the caudal is more evident; and the eyes are noticeably larger.

Taxonomic Remarks. Vladykov (1935b) noted that Newfoundland specimens have more dorsal fin spines (85) and anal fin rays (62) and a longer caudal fin (longer than the head) than others from the St. Lawrence estuary (75–79 dorsal spines, 52–56 anal rays), showing that snakeblenny tend to break up into local populations. He described the St. Lawrence population as a subspecies, *americanus,* and the Newfoundland population as another, *terraenovae,* both of which have more spines and rays than have been recorded for some eastern Atlantic specimens. Gulf of Maine specimens, with 77–83 dorsal spines and 56–59 anal rays, are intermediate between the Newfoundland and Gulf of St. Lawrence populations in this respect.

Habits. Snakeblenny are usually taken with trawls over soft sediments (Sim 1890; Gordon and Duncan 1979; Nash 1980; Pelster et al. 1988a), but may also occur on hard bottoms. They burrow in soft bottoms of sublittoral waters (Pelster et al. 1988a). Although they are not found in the littoral zone, they are fish of

according to Bigelow and Schroeder, and apparently most common from 2 m or so below tide markings down to 73–91 m. In aquariums, snakeblenny construct and maintain a Y-shaped burrow in soft sediments (Nash 1980; Atkinson et al. 1987). Partial pressure of oxygen in the burrows may be very low and carbon dioxide values high, a problem alleviated by active ventilation of the burrow (Atkinson et al. 1987) and by the ability to shift to a predominantly anaerobic metabolism (Pelster et al. 1988a,b).

Food. The main food items (as percent frequency) of snakeblenny on the west coast of Scotland are meiobenthic organisms such as polychaetes, harpacticoid copepods, nematodes, and ostracods (Gordon and Duncan 1979: Table 4). Other food items include foraminifera, bivalves, gastropods, calanoid copepods, isopods, euphausiids, gastrotrichs, and fishes.

Predators. Snakeblenny are eaten by larger fishes, such as cod and halibut in Massachusetts Bay (Goode and Bean 1879), winter skate, spiny dogfish, goosefish, spotted and white hake, cod, and sea raven (Rountree 1999). They are also among the fishes caught by marine birds, such as double-crested cormorant in the Gulf of St. Lawrence, to feed their young (Rail and Chapdelaine 1998).

Parasites. Three digenean trematodes, *Helicometra insolata, Stenakron vetustum,* and *Progonus muelleri,* have been reported from Canadian snakeblenny (Bray 1979). A copepod, *Haemobaphes cyclopterina,* was found on the gills (Gooding and Humes 1963).

Reproduction. Bigelow and Schroeder's tow-nettings during March, April, and May of 1920 yielded drifting snakeblenny larvae off Seguin Island, near Cape Elizabeth, over Platts Bank, near the Isles of Shoals, off Ipswich Bay, off Cape Ann, off Boston Harbor, and in the southwest basin of the Gulf off Cape Cod—evidence that they breed successfully throughout the southern part of the range. Yolk-sac larvae were found in the upper Damariscotta and Sheepscot river estuaries along the central coast of Maine, suggesting that spawning occurs there (Chenoweth 1973). Larvae were collected from January through April with a peak in March, and they were the fourth most abundant larvae in this area. Larvae have also been taken in Narragansett Bay, R.I., in April with shallow plankton tows at less than 10 m (G. Klein-MacPhee, pers. comm., Feb. 1997).

The eggs are demersal and adhesive and number 600–1,100 (Andriashev 1954). Apparently larvae are of considerable size at hatching, for the smallest Bigelow and Schroeder saw were about 11 mm long, though they still lacked any trace of dorsal or anal fin rays. Snakeblenny larvae have the most myomeres of the four species of Gulf of Maine pricklebacks (80–85 vs. 45–72), postanal myomeres (58–63 vs. 28–44) but fewer than rock gunnel (86–89 vs. 49–51) (Faber 1976b; Fahay 1983). Snakeblenny larvae resemble larvae of daubed shanny and differ from radiated and Arctic shannies in having lateral external pigment on the intestine (instead of dorsal internal) and no pigment on the head or dorsal edge. They have 58–63 postanal ventral spots compared to 38–44 in daubed shanny. Ratio of preanal length to total length (De Lafontaine 1986) averages 0.42 as in radiated shanny but less than in daubed shanny (0.49) and more than in Arctic shanny (0.37).

Age and Growth. Snakeblenny live for up to 9 years and grow rapidly in the first 2 years (Gordon and Duncan 1979). Mean total lengths by age-group for Loch Linnhe, Scotland, were: 1, 12.7 cm; 2, 13.3 cm; 3, 14.2 cm; 4, 15.4 cm; 5, 16.2 cm; 6, 16.6 cm; 7, 17.6 cm; 8, 18.5 cm; and 9, 20.3 cm (Gordon and Duncan 1979: Table 2).

General Range. Arctic and northern Atlantic Ocean; Davis Strait, Baffin Island, and Labrador south to southern New England on the western side of the Atlantic (Scott and Scott 1988); Greenland and Iceland south to Scotland, the Baltic, and the southern North Sea on the eastern side.

Occurrence in the Gulf of Maine. Snakeblenny probably occur in small numbers around the coastline of the Gulf at moderate depths. Goode and Bean (1879) described them long ago as common residents in the deeper parts of Massachusetts Bay. Huntsman reported them from St. Mary's Bay, N.S., in August and September; from Passamaquoddy Bay, N.B., from April to August; and in open waters of the Bay of Fundy from January on. They were recorded off Eastport in 1872; *Albatross II* trawled a specimen 4.8 km south of Great Duck Island near Mt. Desert, Maine, in 50–60 m, April 1927; two from 21 km east of Boone Island, in 160 m, August 1928; one off the Isles of Shoals at 132–143 m, August 1926; and one at 77 m on the eastern slope of Stellwagen Bank, about 27 miles off Cape Ann in July 1931. A total of 23 specimens was recorded at six BIOME stations in Massachusetts Bay at depths of 34–73 m (Collette and Hartel 1988).

DAUBED SHANNY / *Lumpenus maculatus* (Fries 1837) /

Bigelow and Schroeder 1953:497 (as *Leptoclinus maculatus*)

Description. Caudal fin rounded to subtruncate (Fig. 248). Lower five pectoral fin rays longest, separate at tips. Scales small, covering body. Lateral line complete but indistinct.

Color. Dirty yellow, paler below, back marked with indistinct yellowish brown blotches of various sizes. The dorsal fin is barred obliquely with about ten rows of brownish dots and

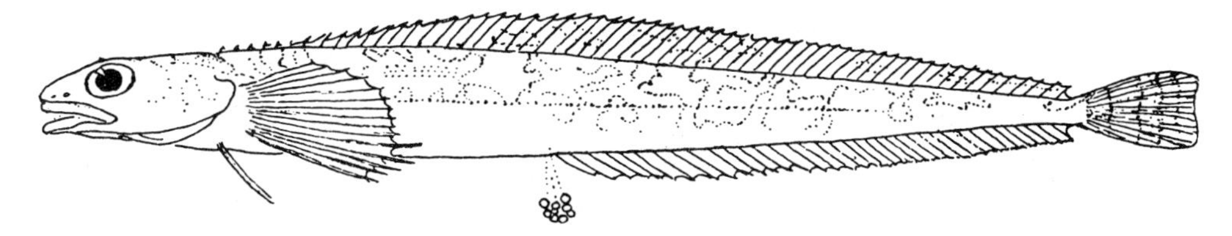

Figure 248. Daubed shanny *Lumpenus maculatus.* (From Bigelow and Schroeder: Fig. 261.)

the pectorals are cross-barred with about five rows. These fins show no distinct markings on preserved specimens; the caudal fin, however, shows one or two dark crossbars, even after preservation.

Meristics. Dorsal fin spines 58–61; anal fin rays I, 35–38; pelvic fin rays I, 3; pectoral fin rays 14 or 15; vertebrae 66–68 (Andriashev 1954; Scott and Scott 1988).

Size. Maximum length about 17.7 cm.

Distinctions. Daubed shanny resemble snakeblenny in general appearance and in location and shape of the dorsal and anal fins, but are not as slender (only 10–12 times as long as they are deep instead of about 20 times). The tail is rounded to subtruncate instead of narrowly oval or pointed as in snakeblenny. The lower five rays of the pectoral fin are separate at their tips, not united with the rest of the fin. Daubed shanny have fewer dorsal fin spines (58–61) and anal fin rays (35–38) than do snakeblenny.

Habits. A benthic species found over mud bottom in 38–55 m of water in Passamaquoddy Bay (Tyler 1971a). They live deeper, 75–475 m, usually 300–400 m off west Greenland (Jensen 1944a). In Scandinavian waters, daubed shanny spend most of the year in deep water, probably coming up to the shallows to spawn. In an aquarium they keep close to the bottom, with the body extended and the pectoral fins expanded, apparently supporting themselves on the free lower rays of those fins (Smitt 1892).

Food. Annelid worms and pelagic amphipods.

Predators. Eaten by bottom-feeding fishes such as cod and white hake in Passamaquoddy Bay (Tyler 1972), cod off east Greenland (Jensen 1944a), and thorny skate and sea raven in the Gulf of Maine (Bowman et al. 2000). Other fish predators include red, white, and silver hakes, longhorn sculpin, and halibut (Rountree 1999). After the decline of capelin in southern Newfoundland just prior to 1996, common and thick-billed murres on the Gannet Islands switched from feeding their chicks primarily on capelin to feeding mostly daubed shanny (Bryant et al. 1999). In 1996, when their diets were composed principally of daubed shanny, Bryant et al.(1999) estimated that these chicks consumed 2.8 million of these fish (about 290 tons) before departure.

Reproduction. Female daubed shanny contain about 970 eggs (Andriashev 1954). Daubed shanny larvae have fewer myomeres than snakeblenny but more than radiated and Arctic shannies (66–72 vs. 80–85 and 45–55 respectively), postanal myomeres (38–44 vs. 58–63 and 28–37) and fewer than rock gunnel (86–89 vs. 49–51) (Faber 1976b; Fahay 1983). Daubed shanny larvae resemble larvae of snakeblenny and differ from radiated and Arctic shannies in having lateral external pigment on the intestine (instead of dorsal internal) and no pigment on the head and dorsal edge. They have 38–44 postanal ventral spots compared to 58–63 in snakeblenny. Ratio of preanal length to total length (De Lafontaine 1986) averages 0.49, more than in snakeblenny or radiated and daubed shannies (0.37–0.42) but less than in rock gunnel (0.52).

General Range. An Arctic fish, known south to Norway and Sweden on the eastern side of the Atlantic and to Cape Cod on the western side.

Occurrence in the Gulf of Maine. Widespread in the Gulf of Maine, particularly in inshore waters in spring (Map 28). Distribution seems to be more offshore in fall, based on comparison of NEFSC spring and fall bottom trawl surveys. A total of 484 specimens was taken at 14 BIOME stations in Massachusetts Bay at depths of 55–88 m (Collette and Hartel 1988).

ARCTIC SHANNY / *Stichaeus punctatus* (Fabricius 1780) / Bigelow and Schroeder 1953:497–498

Description. Dorsal profile of head straight, from tip of snout to origin of dorsal fin (Fig. 249). Dorsal fin single, long, originating directly over edge of gill cover, of uniform height throughout its length, except first and last two or three spines shorter than others. Dorsal fin extends back nearly to caudal fin, but fins separated by a conspicuous notch. Anal fin about two-thirds as long as dorsal fin, of approximately same height, and with similar relation to caudal fin. Caudal fin

Figure 249. Arctic shanny *Stichaeus punctatus.* Halifax, N.S., 186 mm. Redrawn by E. N. Fischer.

gently rounded in outline. Pectoral fins broadly rounded, a little longer than depth of body. Pelvic fins somewhat less than half as long as pectoral fins. Lateral line single, conspicuous, ending at about midlength of body, under dorsal spines 19–25.

Color. A single row of five to nine round black spots with pale margins on the dorsal fin and irregular dark bars on the cheeks and chin are the most conspicuous markings. In a fresh Gulf of Maine specimen, the posterior four (of five) dorsal spots consisted of an anteroventral concentration of melanophores around an orange-red ocellus (Collette and MacPhee 1969). The ground color of the body was a mottled brownish shortly after preservation, with darker cloudings extending from close behind the head to the caudal fin base. The lower surface of the body (except the head) is plain whitish. The anal fin is dusky, edged with white; the pectorals and caudal are crossed by pale bars, and the pelvics are plain yellow. Color changes associated with agonistic behavior in juveniles are described by Farwell and Green (1973).

Meristics. Dorsal spines 48–50; anal fin rays I or II, 32–35; pectoral fin rays 15 or 16; pelvic fin rays I, 4; vertebrae 14–16 precaudal + 36–41 caudal = 51–56 total (Jensen 1944b; Makushok 1958; Collette and MacPhee 1969; Scott and Scott 1988).

Size. Reported to reach 220 mm (Cox 1921), but of more than 1,000 specimens measured by John Green (pers. comm., March 1997), none was larger than 180 mm.

Distinctions. Arctic shanny resemble rock gunnel in color pattern but are distinguished by having well-developed pelvic fins, considerably larger pectoral fins, and fewer dorsal fin spines (only 48–50). Presence of a series of large roundish spots on the dorsal fin separates them at a glance from radiated shanny, which are similar in form but have only a single large blotch in the dorsal. The lateral line is single, whereas it is double in radiated shanny. The spotted dorsal fin and evenly rounded pectoral fins distinguish Arctic from daubed shanny; and the much less slender body (only about one-seventh as deep as long) separate them from snakeblenny.

Habits. Arctic shanny are benthic, living in cold waters to depths of 55 m (Vladykov 1933) over pebble and cobble bottoms and boulder bedrock (Farwell et al. 1976). This was the most abundant benthic fish, 46.7% of specimens of seven species, in rotenone collections made at depths of 3–10 m near the Nuvuk Islands in northern Hudson Bay in 1982–1986 (J. M. Green, pers. comm., March 1997). They are territorial for at least several months following metamorphosis from a planktonic larva to a benthic juvenile, but adults are not territorial (Farwell and Green 1973; Brown and Green 1976). Analysis of length-frequency distributions of Nuvuk Islands collections indicated that recruitment was highly variable between years and that recruitment failure occurred prior to settlement of larvae from the plankton (J. M. Green, pers. comm., March 1997).

Food. In Newfoundland, juveniles eat copepods and amphipods, whereas adults feed on amphipods, polychaetes, isopods, mysids, and ostracods (Farwell et al. 1976). Epiphytic harpacticoid copepods are the most important food for small Arctic shanny (<55 mm) in Newfoundland, but these decrease in importance with increasing size, and amphipods and decapods become more important (Keats et al. 1993). They also consume significant quantities of mysids and polychaetes. The stomach of a 47.5-mm specimen collected with rotenone at Nahant, Mass., contained 22 amphipods (three *Jassa marmorata,* eighteen *Calliopius laeviusculus,* one orchestid), three ostracods, a juvenile prosobranch gastropod, six juvenile mytilid pelecypods, a cumacean (*Diastylis polita*), and a zoeal larval decapod (Collette and MacPhee 1969).

Predators. Often found in stomachs of cod, Greenland halibut, and other fishes off Greenland (Jensen 1944b). The first record from the Gulf of Maine was based on a specimen taken from a cod stomach (Schroeder 1931). Although no Arctic shanny were found in sculpin stomachs in Newfoundland waters (Farwell et al. 1976), cunner and longhorn and shorthorn sculpins were seen pursuing them (LeDrew and Green 1975). Arctic shanny are the principal food that black guillemot use to feed their chicks in the Hudson Bay region (Cairns 1987).

Reproduction. In Newfoundland, spawning apparently occurs during February and March, between spawning periods for rock gunnel and radiated shanny (Farwell et al. 1976). In early February, an 11-cm female extruded an egg mass with 1,624 eggs averaging 1.7 mm in diameter (Farwell et al. 1976). A 15.7-cm female from Greenland contained 2,475 eggs about 1 mm in diameter (Jensen 1944b). The eggs are demersal and adhesive and are deposited in an ovoid egg mass (Farwell et al. 1976).

Planktonic larvae 25–31 mm SL were taken at the surface on the Grand Banks in July and August (Farwell et al. 1976). Arctic shanny larvae have fewer myomeres than either species of *Lumpenus* but more than radiated shanny (51–55 vs. 66–85 and 45–49), postanal myomeres (33–37 vs. 38–63 and 28–37) and fewer than rock gunnel (86–89 and 49–51) (Faber 1976b; Fahay 1983). Arctic shanny larvae resemble larvae of radiated shanny and differ from snakeblenny and daubed shanny in having dorsal internal pigment on the intestine (instead of lateral external) and have pigment on the head and dorsal edge. Ratio of preanal length to total length (De Lafontaine 1986) averages 0.37, less than in any of the other three pricklebacks (0.42–0.49) and rock gunnel (0.52).

Age and Growth. Growth rates in Newfoundland were approximately 16 mm·year^{-1} for the first 4 years, leveling off with little growth after 7–8 years (Keats et al. 1993). Maximum age recorded was 16 years.

General Range. Arctic and circumpolar, known from Hudson Bay, western Greenland, Labrador, Newfoundland, and Nova Scotia south to Massachusetts Bay in the Atlantic and from the Bering, Okhotsk, and Japanese seas in the Pacific (Farwell et al. 1976; Scott and Scott 1988).

Occurrence in the Gulf of Maine. There are two records from the Gulf of Maine (Farwell et al. 1976). A 114-mm specimen (MCZ 34576) was taken from the stomach of a cod half a mile off Little Duck Island near Mt. Desert on 30 April 1930 (Schroeder 1931). The second specimen is a 47.5 mm juvenile (USNM 203405) collected with rotenone in a small rocky cove at Nahant on 22 August 1968 (Collette and MacPhee 1969).

RADIATED SHANNY / *Ulvaria subbifurcata* (Storer 1839) / Bigelow and Schroeder 1953:498–500

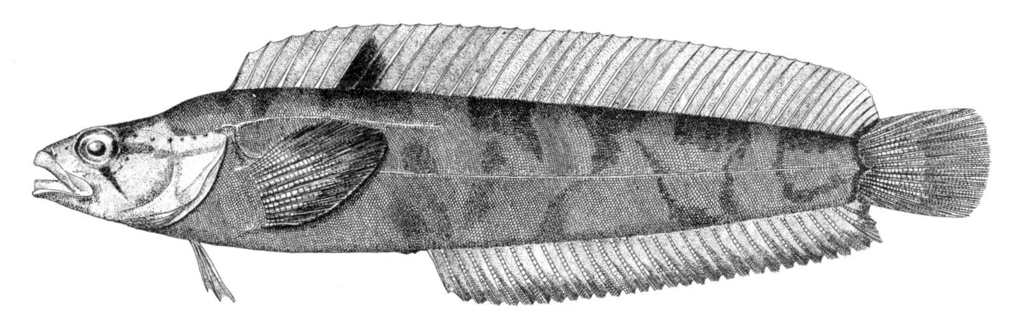

Figure 250. Radiated shanny *Ulvaria subbifurcata*. Halifax, N.S., 126 mm. Drawn by H. L. Todd.

Description. Body stout, moderately elongate, about one-sixth to one-seventh as deep as long (Fig. 250). Dorsal profile of head more convex than ventral profile. Gill openings wide, extending forward under throat, upper corner of gill cover terminating in a rounded flap. Dorsal fin high, practically continuous with caudal fin. Anal fin about half as long as dorsal fin and separated from caudal by a short but definite interspace, made obvious by abrupt rear angle of anal. Pectoral and pelvic fins relatively large, pectorals evenly rounded in outline, reaching back about as far as eighth dorsal fin spine. Pelvic fins situated well in advance of pectorals. Rear margin of caudal fin evenly rounded. Lateral line double, lower branch running length of body but upper branch (more obvious of the two) reaching only about as far back as tips of pectoral fins.

Color. The most distinctive feature is the presence of a large, oval dusky blotch on the dorsal fin extending from the fifth or sixth spine to the eighth or tenth spine. The back and upper parts of the sides are dull brown, obscurely barred or with pale and dark alternate blotches; sides of head marked with a dark bar running obliquely downward and backward from the

white). Color changes can occur fairly rapidly (J. Green, pers. comm., March 1997). The caudal fin is cross-barred with three or four series of dark dots, and the dorsal fin is marked with many tiny dark dots in addition to the dorsal blotch. Territorial males and males protecting egg masses are darker than subdominant males or females (Green et al. 1987).

Meristics. Dorsal spines 43–44; anal fin rays II, 30–31; pectoral fin rays 15; pelvic fin rays I, 3; vertebrae 45–49 (Fahay 1983).

Size. The largest recorded specimen is 16.5 mm long.

Distinctions. The most conspicuous feature of radiated shanny is that the forward part of the dorsal fin is marked with a large, elongate, dark blotch. They are much stouter of body than snakeblenny; pectoral fins and especially the pelvic fins are considerably larger, relatively, than those of rock gunnel, while the pelvics are situated farther in advance of the pectorals; the gill openings are much wider and extend forward under the throat (confined to the sides of the neck in rock gunnel). The evenly rounded outline of the pectoral fins dis-

are the longest and are separate at their tips. They are stouter than Arctic shanny, and the lateral line is double, with an upper and a lower branch, whereas the lateral line of Arctic shanny is single. The margin of the gill covers, with the upper corner terminating in a rounded flap concealing a sharp angle, is diagnostic, for they are rounded in other Gulf of Maine pricklebacks.

Habits. Much has been learned about the biology and behavior of radiated shanny since Bigelow and Schroeder, particularly through the field and laboratory studies of John Green and his colleagues at the Memorial University of Newfoundland. They are bottom fish like other pricklebacks, living among seaweed and stones from the low-tide mark down at least to 55 m, and very likely much deeper. Homing studies show that adults occupy and home to sites that are generally less than 3 m^2, which they occupy year-round (Green and Fisher 1977). Unlike Arctic shanny, radiated shanny are nocturnal (LeDrew and Green 1975; Green and Fisher 1977; Goff and Green 1978).

Food. Copepods are the main food item in radiated shanny up to 60 mm in Newfoundland (LeDrew and Green 1975). Amphipods (four species) are consumed by radiated shanny of all sizes. In larger fish, nereid and polynoid (scaleworms) polychaetes become of increasing importance in the diet. Other food items found in 304 specimens from Newfoundland included foraminifera, actinarian coelenterates, oligochaetes, isopods, decapods, ostracods, Acarina, Diptera, gastropods (six species), pelecypods (*Mytilus*), echinoderms, sand lance, and capelin eggs (LeDrew and Green 1975: Table 1). In Conception Bay, Nfld., 80% of prey items in larvae (3–14 mm SL) consisted of copepod nauplii, 150–650 µm in length (Dower et al. 1998). Copepodites and adults of smaller-bodied copepods also occurred regularly.

Predators. Bigelow and Schroeder found many radiated shanny in stomachs of cod caught on Nantucket Shoals, Cashes Ledge, and other offshore grounds. Juveniles in Newfoundland are frequently eaten by grubby, *Myoxocephalus aenaeus* (LeDrew and Green 1975). Other predators include little skate, *Leucoraja erinacea* (Bowman and Michaels 1984), smooth dogfish, red and white hakes, summer flounder (Rountree 1999), and a sheldrake (*Merganser*) shot near Robinson Hole in the Woods Hole region (Sumner et al. 1913).

Parasites. A protozoan, *Trichodina elizabethae,* was found on the gills of a radiated shanny off Newfoundland (Lom and Laird 1969).

Reproduction. Spawning occurs from early May to June in Newfoundland at temperatures of 1.5°–4°C, with females spawning once and males up to four times per season (LeDrew and Green 1975). The mean number of eggs found in six females was 1,512, and the number in eleven egg masses was 2,706 (LeDrew and Green 1975). Mean diameter of fertilized eggs was 1.55 mm. The granular yolk contained a single, large, highly refractile oil globule. No obvious courtship was recorded in laboratory studies (Green et al. 1987). Olfaction may be important in attracting females to the male and the spawning site as it has been shown to have a role in homing (Green and Fisher 1977; Goff and Green 1978). Males care for the eggs by fanning, nudging, and encircling them, as well as by chasing potential brood predators away (LeDrew and Green 1975; Green et al. 1987). In an aquarium, eggs spawned on 27 May hatched 35 days later as water temperatures rose from 4° to 9°C (LeDrew and Green 1975). By 4 h after hatching, the pelagic larvae had a median finfold but fin rays had not yet developed (LeDrew and Green 1975: Fig. 3). Pectoral fins were well developed at hatching with visible fin rays. The mouth was open and appeared functional. Five melanophores were located on the head dorsal to the eyes; four were present anterior to the median finfold; and four more were present along the body wall within the body cavity. A row of 20 very small pigment spots was present along the dorsal fin base and there were 33 similar spots along the anal fin base.

Radiated shanny larvae have fewer total myomeres (45–49) than any other Gulf species of prickleback (51–85) and fewer postanal myomeres (28–33 vs. 33–63) (Faber 1976b; Fahay 1983). Radiated shanny larvae resemble larvae of Arctic shanny and differ from snakeblenny and daubed shanny in having dorsal internal pigment on the intestine (instead of lateral external) and pigment on the head and dorsal edge. Ratio of preanal length to total length (De Lafontaine 1986) averages 0.42 as in snakeblenny, less than in daubed shanny (0.49), and greater than in Arctic shanny (0.37).

Pelagic larvae (mean length 12 mm) were observed in the water column in August and settlement took place shortly thereafter in Newfoundland waters (LeDrew and Green 1975). Bigelow and Schroeder collected pelagic larvae in tow nets near Seal Island, N.S., in the Grand Manan Channel, at the mouth of Casco Bay, near Cape Porpoise, off the Isles of Shoals, near Cape Ann, and in Massachusetts Bay. Radiated shanny were the dominant larvae in spring and summer in the Damariscotta and Sheepscot estuaries of Maine (Chenoweth 1973). Planktonic larvae peak in abundance in Beverly–Salem Harbor from the end of April into May (Elliott et al. 1979).

Age and Growth. In Newfoundland, males reach about 155 mm and 10 years of age, and females 105 mm and 7 years (LeDrew and Green 1975: Fig. 4). At 2 years they are about 55 mm, at 4 years 90 mm, and at 6 years 110 mm. At 150 mm, radiated shanny weigh about 30 g (LeDrew and Green 1975: Fig. 5). Males weigh slightly less than females of the same length. The length-weight relationship is $\log W = 3.22 \log L - 1.345$ for males; $\log W = 3.384 \log L - 1.476$ for females.

General Range. Known off the boreal coasts of eastern North America, from the Strait of Belle Isle (Jeffries 1932), off Newfoundland, the Gulf of St. Lawrence, off Nova Scotia at St.

Georges Bay and Canso (Cornish 1907; Kenchington 1980) to Nantucket Shoals and Vineyard Sound in southern Massachusetts (Sumner et al. 1913).

Occurrence in the Gulf of Maine. Radiated shanny, first described by Storer from a specimen found among the seaweed at Nahant on the north shore of Massachusetts Bay at an unusually low tide in 1838, was once supposed to be rare. They are, however, not at all unusual in Nahant, being taken in 12 of 22 rotenone collections made in two tide pools in Nahant over a 20-year period, usually only one or two specimens per collection, but one collection contained 12 specimens (Collette 1986). They are widespread in the Gulf of Maine, particularly in relatively shallow waters (Map 29).

GUNNELS. FAMILY PHOLIDAE

BRUCE B. COLLETTE

Gunnels are small littoral fishes most often found around rocks or in tide pools in the North Atlantic and North Pacific. They are elongate and strongly laterally compressed. They have long spiny dorsal fins, with 75–100 spines, about twice as long as the anal fin. Vertebrae are also numerous, 84–107. Pelvic fins are rudimentary or absent. Of four genera and about 14 species (Yatsu 1985), one species of *Pholis* extends into the Gulf of Maine.

ROCK GUNNEL / *Pholis gunnellus* (Linnaeus 1758) / Bigelow and Schroeder 1953:492–494

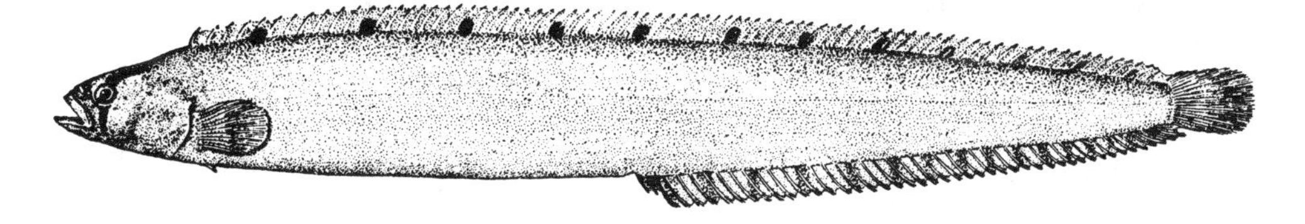

Figure 251. Rock gunnel *Pholis gunnellus*. Gloucester, Mass. Drawn by H. L. Todd.

Description. Trunk slender, flexible, only about one-tenth as deep as long, compressed, about half as thick as deep (Fig. 251). Head short, snout rounded. Upper jaw armed with several conical teeth, lower jaw with only a single row. Dorsal fin long, extending from nape of neck along length of trunk to base of caudal fin, of uniform height throughout. Dorsal separated from caudal fin by shallow notch. Anal fin originates opposite midlength of dorsal fin, to which it corresponds in height and outline; extends posteriorly similarly to meet caudal fin, separated by a distinct notch, but no free space between fins. Caudal fin small, rounded. Pelvic fins tiny, set near each other, in front of or under pectoral fins, each reduced to one very short spine and one rudimentary ray. Pectoral fins smaller than in Gulf of Maine pricklebacks, a little longer than one-half the head length. Trunk clothed with very small cycloid scales, hardly visible through thick layer of mucus; head naked.

Meristics. Dorsal fin spines 73–86, modes in Newfoundland collections 77–80; anal fin rays II (very short), 37–44, modes in Newfoundland 42–44; pectoral fin rays 10–12; pelvic fin rays I, 1; vertebrae 86–89, modes in Newfoundland 82–86 (Proudfoot 1975; Fahay 1983; Scott and Scott 1988).

Color. A row of 10–14 round, black-centered and pale-edged

and spreading out onto the dorsal fin, is the most characteristic feature of the color pattern. The ground tint of the upper part is yellowish, olive brown, reddish, or light red, matching the seaweed or the bottom, with pale, irregularly rounded cloudings on the sides, and with an oblique streak from the eye to the angle of the jaw. The belly varies from pale gray to yellowish white. Pectoral, caudal, and anal fins are yellowish. Bigelow and Schroeder reported a specimen (from Boothbay) that was brick red above and below, light and dark mottled, flecked with tiny black dots and with dark red rather than black spots on the dorsal fin. Change in color to bright orange was associated with the reproductive season of October to April in Newfoundland (Proudfoot 1975).

Size. Reach about 30.5 cm but few are more than 15–20 cm.

Distinctions. Pelvic fins of rock gunnel are much smaller and have only one spine and one ray whereas Gulf pricklebacks have longer pelvic fins with rays I, 3. Pectoral fins are also shorter than in Gulf pricklebacks, only a little longer than one-half the head length.

Habits. Rock gunnel are often found along the low-tide mark, left by the ebb in little pools of water under stones, or among

diurnal. They are not confined to very shoal water as once supposed, for they have also been collected in considerable numbers both within the Gulf and on the offshore banks down to 73 m; Schroeder (1933) reported one from 183 m on Georges Bank.

They are rather local throughout their range. In some places one is to be found under almost every stone; in others they are rare; their presence or absence along any particular stretch of shore probably depends on the character of the bottom, for this fish prefers pebbly, gravelly, or stony ground or shell beds, not mud or eelgrass. Rock gunnel were taken in all 22 collections made with rotenone in two tide pools in Nahant over a 20-year period, 2–232 individuals per sample, 34–170 mm SL (Collette 1986). Bigelow and Schroeder did not find them near the steep ledges so numerous along the rockbound coasts in the Gulf of Maine. In Scandinavian waters according to Smitt (1892), they often take refuge inside large empty mussel shells.

When disturbed they squirm like eels. Eel-like, they swim by sidewise undulations, and they are so active and so slippery (hence the name "butterfish") that it calls for quick work to catch one by hand, even in a very small pool.

Food. Amphipods, isopods, and polychaetes are major food items; other crustaceans, gastropods, bivalves, echinoids, insects, and algae have also been found in stomachs of rock gunnel on both sides of the Atlantic (Stroud 1939; Backus 1957b; Qasim 1957; Sawyer 1967; Proudfoot 1975; Cheetham and Fives 1990). Amphipods accounted for the greatest volume of any single food and showed the highest frequency of occurrence (76% and 89% in two samples) in Newfoundland waters (Proudfoot 1975). Similarly, about 64% of the food eaten by 30 specimens from New Hampshire was composed of several species of gammarid amphipods and two species of isopods (Sawyer 1967).

Predators. Rock gunnel have been found in stomachs of various larger fishes in moderate depths, especially cod and pollock, and are often prey of seabirds and inshore fishes (Scott and Scott 1988). They have been reported from the stomachs of 13 species of fishes, of which cod, longhorn sculpin, sea raven, and Atlantic halibut are the most frequent predators (Rountree 1999). Rock gunnel are part of a core of five benthic taxa consistently present in lists of the ten highest ranking prey taxa for nestling double-crested cormorants in Penobscot Bay, Maine (Blackwell et al. 1995) and are one of three major prey items of these birds in the Gulf of St. Lawrence (Rail and Chapdelaine 1998). They are also eaten by marine mammals such as harp seal (Beck et al. 1993).

Parasites. Nine species of helminths were recorded from rock gunnel from three sites in western Ireland (Cheetham and Fives 1990). The most commonly occurring helminths were the trematode *Lecithochirium furcolabiata* in the body cavity and embedded in body wall muscles and the nematode *Hysterothylacium aduncum* free in the body cavity or on the surface of the liver.

Reproduction. In Newfoundland, rock gunnel egg masses have been found from October to April (Proudfoot 1975) and their larvae are in the water column and on the bottom during February and March (Farwell et al. 1976). In Newfoundland, egg masses were found as shallow as 2.4 m (Proudfoot 1975). In the eastern Atlantic and the North Sea, they have been reported to spawn from between tide marks down to 22 m or more, from November to February or March (McIntosh and Masterman 1897; Ehrenbaum 1904). A female from Peconic Bay, N.Y., contained 686 eggs (Nichols and Breder 1927). The eggs, by European accounts, are about 2 mm in diameter, opaque, whitish, but iridescent on the surface, with a single globule of about 0.6 mm, and they are laid in holes or crannies. The eggs are adhesive, and both parents have been observed forming them into balls or clumps a few millimeters or so across by coiling around them. In European waters incubation occupies 8–10 weeks, during which period parent fish of both sexes have been seen lying close beside the egg masses (Gudger 1927; Wheeler 1969). Two guardians were usually found with each egg mass in Newfoundland waters, but sometimes there was only one (Proudfoot 1975).

Larvae are much larger at hatching (about 9 mm) and further advanced in development than most fishes that lay buoyant eggs. Older larvae of rock gunnel resemble corresponding stages of sand lance and snakeblenny in their extremely slender form. They are easily distinguishable from both these species by the presence of a row of small black pigment spots below the intestine, instead of above it, and from herring (the only other very slender larvae apt to be met in any numbers in the Gulf in the same season) by location of the vent about midway up the body and by the fact that their tails are rounded, not forked. The 12 black dorsal fin spots characteristic of adults are first noticeable against the transparent trunk in young fry of 25–30 mm.

Bigelow and Schroeder reported pelagic larvae (20–39 mm long) in their tow nets off Seal Island, on German Bank, near Mt. Desert Island, off Matinicus Island, and off Ipswich Bay, in April, May, June, and August, and Huntsman stated that they were caught in early summer in the Bay of Fundy—evidence that they breed all around the Gulf from early spring on into the summer. Yolk-sac larvae were found in the upper Damariscotta and Sheepscot estuaries along the central Maine coast, suggesting that they spawned there (Chenoweth 1973). Larvae were present from January to June with a peak in February and March and were the second most abundant species, representing 20–50% of the catch. The larvae are a common component of the ichthyoplankton, ranking seventeenth in abundance of the total larvae taken in Beverly–Salem Harbor (Elliott et al. 1979). The young fish sink to the bottom when 30–40 mm long, in late summer or early autumn in the Gulf of Maine.

Age and Growth. Otoliths were used to age rock gunnel in both the eastern Atlantic (Qasim 1957; Cheetham and Fives 1990) and western Atlantic (Sawyer 1967; Proudfoot 1975).

Growth rates of populations from southern Maine, New Hampshire, and Wales were comparable. Fish 3 and 4 years old attained lengths of about 130 mm and 150 mm, respectively, on both sides of the Atlantic. A 180-mm rock gunnel weighs about 15 g (Sawyer 1967). The length-weight relationship in New England is $\log W = -13.5 + 3.1 \log L$ for males; $\log W = -12.9 + 3.0 \log L$ for females. In Newfoundland, males tend to be larger than females after the first year of growth and reach a larger maximum size (Proudfoot 1975). Ages up to 14+ were recorded from Newfoundland but most individuals were 2+ (Proudfoot 1975).

General Range. Shoal waters on both sides of the North Atlantic, from Hudson Strait to off Delaware Bay on the American coast and south to France on the European coast, most numerous north of Cape Cod and north of the English Channel. Replaced in northern Hudson Bay by *Pholis fasciatus.*

Occurrence in the Gulf of Maine. Rock gunnel are to be found all around the shores of the Gulf from Nova Scotia to Cape Cod. They are definitely recorded at Yarmouth, N.S.; at various localities on both sides of the Bay of Fundy; at half a dozen points along the Maine coast; on Cashes Ledge; at Portsmouth, N.H.; at Hampton; in Ipswich Bay; along the northern shore of Massachusetts Bay; at Nahant; at Cohasset, on the southern shore; among the stones and boulders of the Gurnet, off Plymouth; and at Provincetown.

They also occur in considerable numbers on Nantucket Shoals, as well as on Georges and Browns banks, down to 73 m; one was even as deep as 183 m (Schroeder 1933). They are described as common inshore along outer Nova Scotia waters, eastward from the limits of the Gulf (Vladykov and McKenzie 1935); and as "rather common" on the southern side of the Gulf of St. Lawrence (Cox 1921).

Movements. Bigelow and Schroeder believed that rock gunnel are resident throughout the year wherever they are found, but Sawyer (1967) contended that there is an offshore migration in December off New Hampshire and southern Maine. In Newfoundland, they move out of the intertidal zone in the fall but can still be found in shallow (about 3 m) water throughout the winter (J. M. Green, pers. comm., March 1997). Short-term tag and recapture studies in Scotland showed that recaptured fish moved a mean of 2.1 m in 1.6 days (Koop and Gibson 1991).

WRYMOUTHS. FAMILY CRYPTACANTHODIDAE

BRUCE B. COLLETTE

Wrymouths are slender, eel-like fishes, close relatives of pricklebacks but much larger. Like pricklebacks, they have a long dorsal fin that is spiny throughout its length, but the demarcation between dorsal, caudal, and anal fins is so vague that the median fins are practically continuous. They have no pelvic fins, and their mouth is so strongly oblique that it is nearly vertical. Four monotypic genera are known (Nelson 1994) from inshore waters of the northwest Atlantic and northern Pacific oceans, one species of which occurs in the Gulf of Maine.

WRYMOUTH / *Cryptacanthodes maculatus* Storer 1839 / Bigelow and Schroeder 1953:500–502

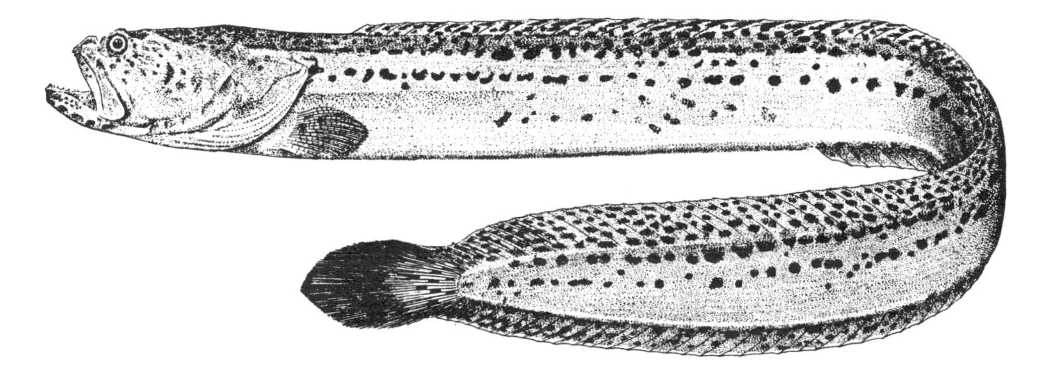

Figure 252. Wrymouth *Cryptacanthodes maculatus.* Drawn by H. L. Todd.

Description. Body eel-like, about 13 times as long as deep but much compressed (Fig. 252). Head flat-topped, eyes set high up heavy. Gill openings wide, running forward under throat. No definite demarcation between dorsal and caudal fins or be-

a continuous fin around the tail. Dorsal fin extends from close behind pectoral fins back to caudal fin, spiny for entire length, as in pricklebacks. Both dorsal and anal fins low (less than half as high as body depth in large specimens, relatively higher in small ones), and of uniform height throughout most of length, with anal about two-thirds as long as dorsal. Pectoral fins small, rounded. Pelvic fins absent. Caudal fin elongate oval.

Color. Varying shades of brown or reddish brown; the upper part of the sides is marked with two or three irregular rows of small darker brown spots that run from head to tail; the top of the head is thickly speckled; dorsal and anal fins are spotted with similar but smaller dots. The belly is grayish white. A few spotless specimens have been seen.

Meristics. Dorsal fin spines 73–77; anal fin rays 47–50; branchiostegals 6; vertebrae 85–86 (Scott and Scott 1988).

Size. Reach nearly 1 m.

Distinctions. Similar to pricklebacks and rock gunnel in having a long spiny dorsal fin but differing in complete absence of pelvic fins. The peculiar profile is a useful field mark, the head flat-topped, the eyes set high up in very prominent orbits, and the mouth strongly oblique with so heavy a lower jaw that it gives the face a bulldoglike expression when the mouth is closed.

Habits. Wrymouth are bottom fish living from the intertidal zone down to about 183 m in the Gulf and to 450–595 m off New Jersey. Willey and Huntsman (1921) found 40-cm wrymouth in burrows in the mud on the flats at the mouth of the Magaguadavic River, a tributary of Passamaquoddy Bay. These burrows "were found in very soft mud from the lower part of the Fucus zone downward; that is, as far up as 4 feet above low-water mark," and "each system of burrows, inhabited by only one fish, consisted of branching tunnels about 5 cm in diameter and from 3 to 8 cm below the surface" of the mud, originating from a more or less centrally placed mound, where the main entrance was, with other smaller openings along the tunnels and at their terminations. Willey and Huntsman also give interesting data on wrymouth respiration and on their responses to various stimuli.

It seems that the burrowing instinct is strong, for a fish kept in a tank constantly inhabited a piece of hard rubber tubing. They seem as likely to be inshore in shoal water in winter as in summer, for one was speared many years ago in Marblehead Harbor in December (Putnam 1874).

Food. Willey and Huntsman (1921) found amphipods (*Gammarus*), shrimps (*Crangon*) and fragments of winter flounder in several wrymouth that they opened, and one kept in captivity ate sand-hoppers, hermit crabs, small herring, and mollusks such as limpets, periwinkles, whelks, clams, and mussels. Apparently they locate food as much by sight as by smell.

Predators. Wrymouth have been found in the stomachs of spiny dogfish, thorny skate, goosefish, cod, haddock, white hake, sea raven, and Acadian redfish, of which sea raven is the most frequent predator (Rountree 1999). Thorny skate and Acadian redfish prey on wrymouth in the Gulf of Maine (Bowman et al. 2000). Wrymouth are part of a core of five benthic taxa consistently present in lists of the ten highest ranking prey taxa for nestling double-crested cormorants in Penobscot Bay (Blackwell et al. 1995).

Parasites. The myxosporidian *Ceratomyxa acadiensis* and four species of trematodes, *Derogenes varicus, Fellodistomum furcigerum, Stephanostomum baccatum,* and *Hemiurus* sp., have been reported from wrymouth (Margolis and Arthur 1979).

Reproduction. Ripe wrymouth have yet to be seen, but presence of larvae early in spring in Passamaquoddy Bay, as reported by Huntsman, with seasonal occurrence of fry noted below, indicates they are winter spawners in the Gulf of Maine. They may breed later in the Gulf of St. Lawrence, for Dannevig (1919) recorded a 38-mm young wrymouth taken as late as 10 June. The localities where young fish have been taken suggest that wrymouth spawn all around the coast of the Gulf of Maine and wherever they occur on offshore banks. The most northern locality record that Bigelow and Schroeder found for its drifting larvae was off the outer coast of Labrador, about 32 km north of Belle Isle; the most southern are for one trawled by *Albatross II* off northern New Jersey, 40°04′ N, 73°32′ W, August 1936, at 64 m; and of another dredged by the *Atlantis* 48 km farther south (39°31′ N, 72°16′ W) between 446 and 592 m, that same year. Bigelow and Schroeder collected late larvae and fry in tow nets (11 specimens 18–40 mm long) in Massachusetts Bay off Boston Harbor, over Jeffreys Bank, in the trough near the Isles of Shoals, in the western basin a few miles west of Cashes Ledge, off Penobscot Bay, near Mt. Desert Island, and in the deep basin off Machias, Maine, in May 1915 and in March and April 1920. Wrymouth larvae are found in the Damariscotta and Sheepscot estuaries along the central Maine coast in late winter and early spring (Chenoweth 1973).

Neither eggs nor early larval stages have been described. By the time the young have grown to a length of 21–22 mm, they show the long dorsal and anal fins and lack of pelvic fins characteristic of their parents, although they are much less slender, relatively, their caudal fins are larger and square instead of rounded, and their mouths are nearly horizontal. Pigmentation of the fry is likewise characteristic, the upper sides from the eye back to the caudal fin being thickly speckled with dark brown dots, which are sparser on the lower part of the sides.

General Range. Atlantic coast of North America, from southeastern Labrador, the coasts and banks of Newfoundland, and the Gulf of St. Lawrence south to Long Island Sound and off central New Jersey.

Occurrence in the Gulf of Maine. Published records are from the Bay of Fundy, Eastport, Casco Bay, Portland, the mouth of the Piscataqua River, Gloucester, Marblehead Harbor, Swampscott, Nahant, and Dorchester in Boston Harbor; and in the outer waters of Massachusetts Bay. There are specimens in the Museum of Comparative Zoology from Trenton, Maine, from outer Boston Harbor, and from near Provincetown. Two were taken in the central basin of the Gulf in July 1931 at a depth of 161–174 m; one was trawled by the *Atlantis* in the deep trough west of Jeffreys Ledge at 132–143 m and another in the southwestern basin of the Gulf off Cape Cod at about 183 m, in August 1936; *Albatross II* trawled one on the eastern slope of Nantucket Shoals (40 °05′ N, 69°22′ W) at 95 m in May 1950. One of the crew of the dragger *Eugene H* reported capture of four on the northeastern part of Georges Bank on 12 October 1951.

ACKNOWLEDGMENTS. Drafts of the stichaeid, pholid, and cryptacanthodid sections were reviewed by John Green and Thomas A. Munroe.

WOLFFISHES. FAMILY ANARHICHADIDAE

RODNEY A. ROUNTREE

Wolffishes are large, blennylike fishes of the suborder Zoarcoidei and are most similar to wrymouths (Cryptacanthidae) and gunnels (Pholidae). Lack of pelvic fins distinguishes wolffishes from all other blennylike fishes except wrymouth, but they can be readily separated from the latter by the large canine tusks, deeper, more robust body, and presence of a separate caudal fin. Wolffishes differ from other blennioid fishes, and from most other acanthopterygian fishes, in that the posterior dorsal spines are short and rigid, whereas all anterior spines have flexible tips (Gill 1911). However, the particularly wide separation of the sacculus and lagena from the utriculus of the inner ear is the most distinctive feature of the family (Berg 1940; Barsukov 1959).

Wolffishes are the largest of the blennylike fishes, reaching 1.8 m in the western Atlantic (Robins et al. 1986), and more than 2.5 m worldwide (Nelson 1994). There are five species in the family, two of which occur in the Gulf of Maine; one, Atlantic wolffish (*Anarhichas lupus*), commonly, and the other, spotted wolffish (*A. minor*), only as a stray from the north. Another cold-water species, northern wolffish (*A. denticulatus*), is included in the following key, as it has been recorded repeatedly from nearby Nova Scotian waters; there is one record from the "Gulf of Maine" (MCZ 99513), and it occurs south to Block Canyon. Owing to strong ontogenetic changes, there is currently no reliable key for all sizes of wolffishes. Barsukov (1959) provided separate keys for wolffish 2–6 cm, 7–20 cm, and larger than 20 cm. The following key has been compiled based on information from Barsukov 1959 and Templeman 1984b, 1986c.

Wolffishes are among the most interesting fishes in the Gulf of Maine region. Apart from their impressive physical appearance, aspects of wolffish reproductive biology, including large egg size, prolonged incubation period, egg brooding behavior, probable internal fertilization, and possible spawning migrations, together with the annual loss of their entire set of teeth and the apparently prolonged fasting period associated with spawning and tooth replacement, make wolffishes unique among Gulf of Maine fishes. In addition to arousing our interest and curiosity with their unique biology, wolffish are good eating. In fact, extensive research is under way in Europe and Canada to develop aquaculture of Atlantic and spotted wolffishes because of the relative ease of raising the young and the growing market demand for the species.

KEY TO GULF OF MAINE AND NOVA SCOTIAN WOLFFISHES LARGER THAN 20 CM TL

1a. Bands of molar teeth on vomer and palatine bones about equal length; body distinctly spotted, spots extending well onto sides of head . . . . **Spotted wolffish**

1b. Bands of molar teeth on vomer and palatine bones not of equal length; body usually plain-colored, barred, or indistinctly spotted; if spotted, spots not extending onto sides of head 2

2a. Central band of molar teeth much shorter than bands flanking it on the palatines; canine teeth small, not prominent; body usually plain or blotched; if spotted, spots not extending onto sides of head; head pointed, body deepest at the midpoint; dorsal fin rays 76–81; vertebrae 78–82 . . . **Northern wolffish**

2b. Central band of molar teeth much longer than those flanking it on the palatines; canine teeth very large and prominent, forming tusks; body distinctly barred; head blunt, body deepest at head; dorsal fin rays 69–79; vertebrae 72–78 . . . . **Atlantic wolffish**

ATLANTIC WOLFFISH / *Anarhichas lupus* Linnaeus 1758 / Catfish, Ocean Whitefish /

Bigelow and Schroeder 1953:503–507

Description. Body deepest close behind heavy, blunt head, tapering back to slender caudal peduncle (Fig. 253). Dentition formidable, a row of about six very large, stout, conical canine hind them. Roof of mouth armed with three series of crushing teeth. Central series of vomerine teeth a double row of about four pairs of large, rounded molars united (but not

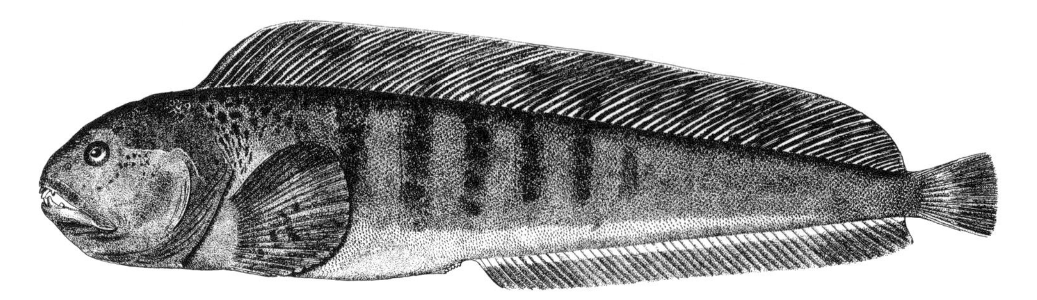

Figure 253. Atlantic wolffish *Anarhichas lupus.* Georges Bank. Drawn by H. L. Todd.

with two alternating rows of blunt conical teeth. Lower jaw with four to six large tusks in front, two longitudinal diverging rows of rounded molars behind. Dorsal fin extends from nape to caudal fin base, uniform in height except for rounded corners. Anal fin about half as high as dorsal, little more than half as long. Pectoral fins large, rounded. Caudal fin small, weak, slightly convex in outline. Head naked but poorly developed scales present on body. Swim bladder absent. Presence of lateral lines previously overlooked in North American literature (e.g., Bigelow and Schroeder; Scott and Scott 1988; Nelson 1994) but two branches of the lateral line present on body (Barsukov 1959). Both branches begin just behind head, at about level of pectoral fins. Dorsal branch extends to about midpoint of body, running approximately one eye diameter below dorsal fin. Lower branch runs laterally down middle of flank to at least caudal fin base. In living specimens, lateral lines clearly visible as two rows of faint black (posteriorly) and white (anteriorly) bicolor spots (R. Rountree, pers. obs.).

Meristics. Dorsal fin spines 69–79; anal fin rays 42–48; pectoral fin rays 18–22; branchiostegal rays 6 or 7; vertebrae 72–78 (Barsukov 1959; Scott and Scott 1988).

Color. Atlantic wolffish are dull-colored and vary widely in tint. Adults are usually marked with a variable number (8–13) of dark bars (formed by numerous irregular patches and spots) that extend onto the dorsal fin. However, early juveniles lack these bars and could be confused with spotted wolffish, which have similar bars until they reach a maximum length of 20 cm (Barsukov 1959). Throat and belly back to the vent are dirty white tinged with the general ground tint of the upper parts. Wolffish vary in color geographically from dull olive green, purplish, brownish, bluish gray, to slate (Bigelow and Schroeder). Bigelow and Schroeder suggested that the color varies with that of the fish's surroundings, purplish and brown tints ruling among red seaweeds and olive gray on clean bottom. Specimens living in aquariums tend to be somewhat lighter than described by Bigelow and Schroeder (R. Rountree, pers. obs.). Lateral line pores on the head, particularly those around the eye, stand out as bright white spots. The margins of the dark bars are often boldly marked off in white above the dorsal branch of the lateral line. These white margins sometimes expand dorsally to form strong white bars on the dorsal fin that alternate with the often fainter black bars. Usually the anterior white margin of the first dark bar forms a bold white patch over the first three spines of the dorsal fin. Another bold white patch occurs on the base of the pectoral fin. Presence of these white markings appears to be behaviorally controlled (R. Rountree, pers. obs.).

Size. Reach 1.5 m and 18 kg (Goode 1884; Idoine 1998b), but wolffish longer than 1.2 m are seldom seen, and the larger fish brought in run less than 1 m (Bigelow and Schroeder). European authors write of Atlantic wolffish of 1.8 m and even longer but lengths over 1.25 m have been disputed (Barsukov 1959). A fish 0.8 m long weighs about 4.5 kg and one of 0.9 m about 7.3 kg. The largest fish recorded from Icelandic waters was 1.16 m (Jonsson 1982). A general north to south increase in wolffish size has been reported for southwestern Greenland (Riget and Messtorff 1988) and in the northwest Atlantic (Templeman 1986b). The all-tackle game fish record is 23.58 kg for a wolffish caught on Georges Bank in June 1986 (IGFA 2001).

Distinctions. The great projecting tusks, blunt snout, massive head, and small eyes give Atlantic wolffish a singularly savage aspect. Wolffish resemble a huge blenny in general makeup, except that the dorsal fin spines are flexible at their tips instead of stiff; pelvic fins are absent; and the mouth is armed with a set of teeth more formidable than those of any other Gulf of Maine fish, except for spotted wolffish.

Habits. Wolffish are widely renowned among fishermen for their ferocious appearance, temperament, and biting ability. They have been known to bite through a broom stick, break steel knives (see comments in Gill 1911), and crush the sides of steel fish baskets (R. Rountree, pers. obs.). Their fierce appearance has inspired many legends over the centuries (Smitt 1892; Barsukov 1959). *Anarhichas* comes from a Greek word meaning "to climb up" because it was believed that they climb up onto rocks and cliffs. Some legends claim that wolffish gnawed scars onto anchor chains on which they had climbed (see, e.g., Smith 1833).

Atlantic wolffish are benthic fish distributed throughout the Gulf of Maine in depths of 40–240 m, but are concentrated between 80–120 m (Nelson and Ross 1992). Bigelow and Schroeder noted reports of the species from tide pools at Eastport,

Maine, but never heard of it in such situations or at low-water mark anywhere else in the Gulf. They have been reported from depths of 0–600 m off western Greenland (Riget and Messtorff 1988) and Newfoundland (Albikovskaya 1982; Templeman 1984a); 8–450 m off Iceland (Jonsson 1982); and 10–215 m off Norway and the Barents Sea (Falk-Petersen and Hansen 1991). They have been collected at depths of less than 2–10 m with seines, gill nets, and pound nets in the White Sea (Pavlov and Novikov 1986, 1993; Pavlov 1994). They were most abundant at depths of 40–180 m in Icelandic waters (Jonsson 1982), and 101–350 m off Newfoundland (Albikovskaya 1982).

Albikovskaya (1982) collected Atlantic wolffish at temperatures of −1.9° to 11.0°C, with the biggest catches between −0.4° and 4.0°C off Newfoundland. In Norwegian waters they are found at temperatures of −1.3° to 11°C (Falk-Petersen and Hansen 1991). Beese and Kandler (1969, cf. Jonsson 1982) suggested a temperature tolerance of −1.3° to 10.2°C for Icelandic and northern European waters. Bigelow and Schroeder reported a similar temperature range for the Gulf of Maine. However, data from laboratory predation experiments indicate survival at temperatures as high as 17°C, although feeding activity was negatively correlated with tem- [illegible] and appeared to cease by 17°C (Hagen and Mann [illegible] wolffish can survive temperatures as low as [illegible] atory, owing to a high concentration of [illegible] ood (King et al. 1989). In the White Sea, [illegible] e been collected in water temperatures as [illegible] Pavlov and Radzikhovskaya 1991).

[illegible] e usually solitary except during the mating season, when bonded pairs occur. However, some apparent "colonial settlements" have been reported (Pavlov and Novikov 1993), the occurrence of which may be related to limited shelter availability. In a scuba survey of shelter sites Pavlov and Novikov (1993) found that Atlantic wolffish prefer sites with complex bottom relief such as rocks and large stones and were only rarely observed in algae or over sand bottom. The entrance to a shelter was frequently marked by piles of crushed shells that had apparently been evacuated from the gut. They suggest that feeding takes place largely in areas away from the shelter sites. Adults did not occupy permanent shelters and did not display territorial behavior. In fact, they were sometimes observed sharing shelters with other conspecifics and with large cod (Pavlov and Novikov 1993). However, other researchers suggest that shelter sites may be limiting during the breeding season owing to competition with conspecifics (Johannessen et al. 1993) and ocean pout (Keats et al. 1985).

Although scuba observations suggest that wolffish may remain inactive at night and forage only during the day (Bernstein et al. 1981), other researchers consider wolffish to be essentially nocturnal (Gill 1911; Barsukov 1959; Jonsson 1982); Pavlov and Novikov (1986) found that Atlantic wolffish feed most actively at night in large ponds and fish pens. Wolffish may be able to vocalize by grinding their pharyngeal teeth [illegible]

Food. The food of Atlantic wolffish was first examined as early as 1772 from Icelandic waters (Ólafsson 1772, cf. Palsson 1983) and 1784 in European waters (Andre 1784). Gill (1911) reviewed other early accounts of wolffish food habits. The best food habits data available for Atlantic wolffish is from the Labrador-Newfoundland region of the northwest Atlantic (Albikovskaya 1983). In this region, all three wolffish species had very similar diets (70% similarity), with northern wolffish exhibiting the greatest differences. Atlantic and spotted wolffish fed primarily on benthic fauna, whereas northern wolffish fed mainly on bathypelagic forms. Diet composition exhibited strong regional variation. The most important food groups were crabs, starfish, brittle stars, sea urchins, and bivalve and gastropod mollusks. In the North Sea (Liao and Lucas 2000a), diet was dominated by decapods (particularly Paguridae), bivalves (particularly Pectinidae), and gastropods (particularly Buccinidae). Although Bigelow and Schroeder reported that the diet of Atlantic wolffish consisted wholly of hard-shelled mollusks, crustaceans, and echinoderms, fishes have also been reported (Smith 1892; Fulton 1903a; Sæmundsson 1949; Barsukov 1959; Albikovskaya 1983; Palsson 1983; Templeman 1985; Keats et al. 1986a,b; Liao and Lucas 2000a,b), and can be important locally (Albikovskaya 1983; Palsson 1983). Even large fishes such as Atlantic cod, cunner, redfishes, and skates have occasionally been found in Atlantic wolffish stomachs (Gill 1911; Jonsson 1982; Albikovskaya 1983; Templeman 1985; Keats et al. 1986b).

More recently, bivalves (including the economically important sea scallop, Icelandic scallop, and ocean quahog), gastropods, decapods, and echinoderms (particularly an ophiuroid *Ophiura sarsi*), were the most frequently observed food items in the Gulf of Maine–Georges Banks area (Nelson and Ross 1992; Bowman et al. 2000). Larger individuals (51–80 cm TL and larger) fed primarily on bivalves; medium-sized individuals (11–50 cm TL) consumed a much larger proportion of echinoderms; and the smallest individuals (1–10 cm TL) ate more amphipods (particularly *Parathemisto* spp.) and euphausiids (Bowman et al. 2000).

Cannibalism of eggs and larvae has been noted under both natural and laboratory conditions (Jonsson 1982; Keats et al. 1986b; Moksness et al. 1989; Moksness and Stefanussen 1990; Pavlov and Radzikhovskaya 1991; Johannessen et al. 1993). A high incidence of fishing boat discards has been reported in stomachs of wolffish (Barsukov 1959).

Feeding by Larval and Juvenile Stages. Food habits of the pelagic larval stage in northern European waters have been described by Baranenkova et al. (1960), Pavlov et al. (1987), Falk-Petersen et al. (1990), and Orlova et al. (1990), and initial feeding and rearing studies have been examined in Europe by Ringø et al. (1987) and Moksness et al. (1989). Off northern Norway, 250 stomachs of larvae 20–40 mm TL were found to contain 1- to 3-mm crustaceans and 6- to 10-mm fish larvae (Falk-Petersen et al. 1990). The diet was dominated by am- [illegible] crustacean larvae (16%), and fish larvae (8%). Similarly, pelagic fry (24–26 mm) fed on pelagic larvae of benthic invertebrates and on fish larvae and eggs, but benthic juveniles (7–19 cm) fed exclusively on benthic forms such as echinoderms and mollusks (Baranenkova et al. 1960).

Feeding Behavior. The behavior of Atlantic wolffish feeding on green sea urchins in Newfoundland has been described by Keats et al. (1986b): "When an urchin is 'sighted' the wolffish generally turns slightly on its side and grasps the urchin with its canine teeth, while its body is at a 45° angle laterally to the bottom. A side-to-side motion is sometimes used to remove the urchin from the substratum. . . . There is a violent dorsoventral jerking of the wolffish as the urchin test is being crushed." Similar observations have been made by Russian scientists from submersibles (Barsukov 1959).

Several studies have examined the highly specialized dental morphology of wolffishes and discussed implications for food habits (Andre 1784; Crisp 1853; Gill 1911; Luhmann 1954; Barsukov 1961; Verigina 1974; Le Cabellec et al. 1978; Albikovskaya 1983). The large hook-shaped teeth are used to tear food from the bottom, whereas the heavy conic and round teeth on the vomer and palate are used to crush hard skeletons and shells of prey. Although these studies imply that the food of wolffishes is completely crushed and macerated before it reaches the stomach (except Crisp 1853), intestines full of small whole crabs and sand dollars have been observed (R. Rountree, pers. obs.). Similar observations were made in early studies (Verrill 1871; Gill 1911).

Wolffish teeth are quickly worn down by the grinding action used for crushing hard-shelled prey and are replaced annually (Luhmann 1954; Barsukov 1959, 1961; Albikovskaya 1983). According to Barsukov (1961), peak shedding of teeth of Barents Sea Atlantic wolffish probably occurs from December to January, but can take place anytime between October and May. Detailed anatomical descriptions of tooth replacement are provided by Luhmann (1954) and Barsukov (1959). Jonsson (1982) provided data from a survey of the tooth replacement stage of Atlantic wolffish from Icelandic waters. Teeth are exchanged from September to December, rarely into January; hence exchange occurs during, or just after, the spawning season. Females appear to undergo tooth replacement shortly before males. New teeth are loose and surrounded by soft gum tissue, but by January or February they have become firm. New and developing teeth are bright red in all three North Atlantic wolffish species (Luhmann 1954; Barsukov 1959), but the color fades within a few weeks in Atlantic wolffish (R. Rountree, pers. obs.). It may take more than 2–3 months after tooth loss before new teeth are fully functional (Barsukov 1959). Nine adult and late juvenile Atlantic wolffish captured on Georges Bank in early December 1994 exhibited a wide range of tooth replacement stages from the presence of scattered old broken teeth, to the absence of all teeth, to the presence of new teeth in various stages of development, including scattered red teeth (R. Rountree, pers. obs.). It appears, therefore, that there is a great deal of individual variation in timing of tooth replacement in Gulf of Maine wolffish.

Wolffish appear to fast for up to several months during the tooth replacement, spawning, and brooding periods. The extent to which these periods overlap is not clear. Male wolffish appear to reduce feeding or fast during up to several months of egg brooding between July and November (Keats et al. 1985; Ringø and Lorentsen 1987). Wild-caught Atlantic wolffish held in captivity during breeding studies stopped feeding 1–2 weeks prior to spawning (Pavlov and Novikov 1986; Johannessen et al. 1993). Gut indices indicate that feeding activity is higher during spring and summer and drops in winter (Albikovskaya 1983; Templeman 1986b; Falk-Petersen and Hansen 1991; Liao and Lucas 2000b). However, in Norwegian waters feeding activity was not correlated with the reproductive cycle or the egg-guarding period (Falk-Petersen and Hansen 1991). Most fish collected in Icelandic waters during September and December had empty stomachs and were in poor condition, suggesting that Atlantic wolffish do not feed during the tooth replacement–spawning period (Jonsson 1982). Baranenkova et al. (1960) also reported that benthic juveniles (7–19 cm) do not feed during the period of tooth replacement. Barsukov (1959) and Luhmann (1954) independently hypothesized that feeding is restricted to soft-bodied animals during the tooth replacement period to prevent damage to the developing new teeth. They further suggest that the nearly simultaneous loss of all teeth is an adaptation to feeding on hard-shelled animals, since a developing individual tooth would not be able to withstand grinding pressures during feeding. The unusually low resting metabolism, among the lowest known, of Atlantic wolffish (Liao and Lucas 2000a) may be an adaptation to extended fasting periods required by tooth replacement and brooding.

A short, large-diameter esophagus empties directly into a small sacklike stomach (13% of body length), and then into an enlarged, but relatively short (52–86% of body length) intestine (Verigina 1974). After crushing and maceration in the oral cavity and pharynx, calcareous exterior material (shells and tests) are evacuated through the large anus (Crisp 1853; Bray 1987). Most digestion occurs in the enlarged intestines, which have an increased surface area resulting from heavy folds in the intestinal wall, rather than in the stomach (Verigina 1974; Bray 1987). Much of the material filling the digestive tract is not digestible, so a large amount of food must be processed to obtain an adequate diet (Orlova et al. 1989). Digestion rates for various prey under different temperatures have been examined in Europe (Orlova et al. 1989). Wolffish have been noted to have an unusually large gall bladder (Crisp 1853).

Predators. Atlantic wolfish have been reported from stomachs of Greenland shark (Bigelow and Schroeder 1948b; Barsukov 1959), Atlantic cod (Sæmundsson 1949; Barsukov 1959; Orlova et al. 1990), haddock (Orlova et al. 1990), and gray seal, *Halichoerus grypus* (Pierce et al. 1989). Eggs are frequently preyed on by spotted wolffish (Jonsson 1982). Atlantic wolffish were rarely found in an extensive demersal fish food habitats data set collected by the Northeast Fisheries Science Center between 1973 and 1990, occurring only in spiny dogfish, thorny skate, Atlantic cod, red hake, pollock, haddock, and sea raven stomachs, of which spiny dogfish, sea raven, and cod had the most frequent occurrences (Rountree 1999). Most prey were 3- to 10-cm juveniles. Sea raven were the only species to prey on wolffish larger than 25 cm, with prey ranging from 7 to 57 cm TL.

Parasites. Jonsson (1982) reviewed some early accounts. Most notably, a sporozoan parasite infecting wolffish muscle tissue causes a condition referred to as "hairy catfish" and can sometimes significantly adversely affect the marketability of catches in Iceland (Jonsson 1982). Other information on parasites is contained in Margolis and Arthur (1979), Khan et al. (1980), Zubchenko (1980), Appy and Dadswell (1981), Bray (1987), and Bray and Gibson (1991). Wolffish are infected mainly by parasites that have developmental connections with benthic organisms (Zubchenko 1980). Bray (1987) examined the relationship between digestive system physiology and function and parasite location. A parasitic fungoid microorganism *Mycelites ossifragus* has been found to burrow into wolffish teeth (Schmidt 1954; Kerebel et al. 1979) and may play a role in wolffish tooth destruction (Barsukov 1959).

Reproductive Habits. Wolffish ovaries are paired, elongate, and yellowish, located in the dorsal half of the abdominal cavity (Falk-Petersen and Hansen 1991). Males have a pair of elongate testes ventral to the kidneys. Sexually mature males can be distinguished from females by an enlarged urogenital papilla (Falk-Petersen and Hansen 1991), which probably functions in copulation (Johannessen et al. 1993; Pavlov 1994; Pavlov and Moksness 1994). Ripe females have a pronounced pot-bellied appearance up to a few weeks before spawning (Pavlov and Novikov 1986; Johannessen et al. 1993, Pavlov and Moksness 1994). The oral cavity and lower lips of both sexes may become pink, red, or orange 1–2 weeks prior to spawning (Pavlov and Novikov 1986). A few hours before spawning a transverse sickle-shaped genital pore opens into the oviduct.

In Norway gonadosomatic indices (GSI) of females are highest during summer and fall, whereas males show little seasonal variation (Falk-Petersen and Hansen 1991). However, testes contain active spermatozoa through much of the year (Falk-Petersen and Hansen 1991; Johannessen et al. 1993; Pavlov and Moksness 1994). In the western Atlantic, female GSI appear to be negatively correlated with feeding index, with reduced or eliminated feeding during the period of highest GSI (Keats et al. 1985). Male GSI and feeding index were negatively correlated with brooding activity, with lowest GSI and lowest feeding during brooding.

Most individuals are mature by age 6 and about 40 cm TL (Falk-Petersen and Hansen 1991). In the White Sea, males begin to mature at about 40 cm and age 6–7 (Pavlov and

ally owing to temperature effects (Templeman 1986b). In the northwest Atlantic females reach sexual maturity at 43 cm off Labrador and northeast Newfoundland and at 58 cm in St. Pierre Bank and southern Grand Bank (Templeman 1986b). Sizes at 50% maturity were 51.4 cm for northern areas and 68.2 cm for southern areas (Templeman 1986b). Mature females have been found as small as 25 cm in the relatively warm waters of Iceland (Jonsson 1982). Some females mature in under 2 years (0.5–1.0 kg) under culture (Moksness and Stefanussen 1990).

Females appear to be group-synchronous, with several egg generations present at any given time (Falk-Petersen and Hansen 1991). Some individuals may not spawn every year (Pavlov and Novikov 1986; Falk-Petersen and Hansen 1991; Pavlov and Moksness 1996). Egg masses vary in size according to the size of the female, but generally are about 10–14 cm in diameter (Jonsson 1982); hence, fecundity increases with fish size from about 5,000 eggs at 60 cm to 12,000 eggs at 80–90 cm (Falk-Petersen and Hansen 1991). Jonsson (1982) reported 338 eggs in a 25-cm (age-7), and 4,900–5,000 eggs in a 59-cm (age-9) Atlantic wolffish from Iceland. In the Canadian Atlantic, the relationship between fecundity and length has been reported as $F = 0.3090L^{2.4239}$ (Templeman 1986b). Individual fecundity ranged from 2,300 eggs in a 44-cm fish to 37,920 eggs in a 117-cm fish. Gusev and Shevelev (1997) recently described and compared fecundity among all three Barents Sea wolffishes. Absolute fecundity increases in all three species with increase in body length and weight. Atlantic wolfish have the lowest absolute fecundity. Dzerzhinskiy and Pavlov (1992), Pavlov and Novikov (1993), Pavlov and Moksness (1994, 1996) have described gonad maturation and reproduction in White Sea populations.

Spawning Season. Information on the spawning period of Atlantic wolffish is somewhat contradictory, partly because of geographic differences (McIntosh and Prince 1890; Barsukov 1959; Beese and Kandler 1969; Jonsson 1982; Templeman 1984a,b; Keats et al. 1985; Pavlov 1986; Falk-Petersen and Hansen 1991; Johannessen et al. 1993, Pavlov and Novikov 1993; Pavlov and Moksness 1994). The long incubation time (4–9 months, depending on temperature) and the tendency for mechanical stimuli to induce late-stage embryos to hatch prematurely (Pavlov and Novikov 1986) are probably major sources of this confusion. Findings that wolffish can spawn over much of the year in the laboratory (Pavlov and Moksness 1994) further complicates interpretation of field evidence. In other words, there is an inadequate distinction between the spawning and hatching periods in most studies.

In the northwest Atlantic eggs have been collected in February (McKenzie and Homans 1938) and March (Powles 1967), whereas early larvae have been collected from January to March (Bigelow and Schroeder). In addition, scuba observations of recently hatched larval wolffish have been reported during October and November from the coast of Nova Scotia

July through April (Barsukov 1959; Jonsson 1982; Falk-Petersen et al. 1990, and citations therein), whereas recently hatched larvae have been collected from January to June (Barsukov 1959; Baranenkova et al. 1960; Falk-Petersen et al. 1990; Pavlov and Novikov 1993; except Collett 1883, cf. Barsukov 1959). A single wolffish larvae was reported in November off Norway (Collett 1883).

Based on egg and larvae collections, a late fall and winter spawning period was suggested for the Gulf of Maine (Bigelow and Schroeder) and Icelandic waters (Sæmundsson 1949). Subsequent studies, however, have concluded this to be in error and that peak spawning occurs in September and October in these regions (Jonsson 1982 and citations below). However, the only indisputable evidence for a spawning period was provided by scuba observations off Nova Scotia (Keats et al 1985). Nested pairs of adults or brooding males, together with eggs, were observed from August to October in 5–15 m of water off Nova Scotia (Keats et al. 1985). Based on egg size, Templeman (1986b) concluded that spawning occurred mainly in autumn off Newfoundland; however, precise identification of the spawning period is uncertain owing to the small sample sizes during this period. A similar July–October spawning period was later reported for Norwegian waters based on back-calculated growth of spring-captured larvae (Falk-Petersen et al. 1990) and on female gonadosomatic indices (Falk-Petersen and Hansen 1991). Jonsson (1982) states that Atlantic wolffish spawn mainly in September and October, but as late as December in Icelandic waters (based on implied captures of ripe females). He further discussed apparent misidentification of the spawning period by earlier researchers based on egg and larval collections. Finally, White Sea populations have also been reported to spawn in late summer and autumn based on laboratory studies (Pavlov and Novikov 1986; Pavlov and Radzikhovskaya 1991), and later on laboratory spawning, larval collections, and scuba observations of pair bonding (Pavlov and Novikov 1993). In direct contradiction to these studies, some of these same authors later concluded from laboratory observations that spawning occurs from October to February (Johannessen et al. 1993; Pavlov and Moksness 1994).

Two different studies found that wolffish are capable of spawning over much of the year in the laboratory (Johannessen et al. 1993; Pavlov and Moksness 1994). In the first, wild brood stock from Norway were spawned under laboratory and seminatural conditions from October to February. In the second, eggs and larvae collected off the coast of Norway and reared to maturity spawned mainly from December through March at 5–8°C, although some spawning occurred from October through July (Pavlov and Moksness 1994). Many males were found to be ripe through most of this period. The authors suggest that the protracted spawning period resulted from the constant photoperiod in the laboratory, since spawning occurred in water temperature too high for the eggs to remain viable, and that timing of maturation and spawning is dependent more on photoperiod than temperature. These observations suggest the intriguing possibility that timing of egg maturation and spawning of Atlantic wolffish may be highly plastic and dependent on local conditions. Reports of apparent seasonal spawning migrations may result from inshore-offshore seasonal migrations related to foraging and tooth replacement episodes, in which some members of the population spawn inshore during the late summer and fall and offshore during late fall and winter.

Spawning Behavior. Male and female pairs form during the spring and summer before spawning (Keats et al. 1985; Pavlov and Novikov 1986, 1993). Data from scuba studies in 5–15 m of water off Newfoundland indicate that wolffish first move into the shallows in mid to late spring, form mated pairs during summer, and spawn from August to October (Keats et al. 1985). Egg masses were observed in nests under rocks from August to November; however, no mated pairs were observed after October. Pavlov and Novikov (1993) made similar observations of prespawning behavior in the White Sea, but noted that prespawning pairs left their shallow-water nesting sites and migrated to deeper water immediately prior to spawning in late July to September.

Johannessen et al. (1993) and Pavlov and Moksness (1994) examined female spawning behavior. The pot-belly appearance of ripe females, which is due to a gradual increase in egg size, begins developing 3–5 months prior to spawning (Johannessen et al. 1993), corresponding to the time of pair-bonding observed in the field (Keats et al. 1985). The pot-belly becomes pronounced 1–2 weeks prior to spawning (Pavlov and Novikov 1986; Johannessen et al. 1993; Pavlov and Moksness 1994). A 12- to 24-h side-laying phase occurs 30–50 h prior to spawning, in which the females rest motionless on the bottom. This is followed by 3- to 6-h "labor" phase, in which females undergo violent twisting and bending motions and convulsive shivering. By the end of the labor phase a 2–10 mm opening into the oviduct has appeared. Copulation probably occurs at this time (Johannessen et al. 1993). An 8- to 15-h resting phase occurs after the labor phase, in which eggs probably become fertilized (Pavlov and Moksness 1994). Extrusion of eggs occurred in 3–7 min at the end of the resting phase. The female then curls up around the eggs, which are embedded in mucus (Pavlov and Radzikhovskaya 1991; Johannessen et al. 1993). After 6–10 h the mucus has dissolved and the eggs are firmly attached to each other in a ball. The female continually turns the eggs, which are not attached to the substrate. The probable internal fertilization of wolffish has been discussed by several authors (Johannessen and Moksness 1990; Kvalsund 1990; Pavlov and Radzikhovskaya 1991; Pavlov et al. 1992; Pavlov 1994; Pavlov and Moksness 1994).

Little is known of male spawning behavior, as they show little interest in spawning in the laboratory (Johannessen et al. 1993; Pavlov and Moksness 1994). Although females initially guard the eggs after spawning in aquariums and under laboratory conditions, they lose interest after a few hours and may cannibalize them (Hognestad 1965; Ringø and Lorentsen 1987; Johannessen et al. 1993; Pavlov and Moksness 1994). The male is believed to begin brooding the eggs shortly after spawning. Based on sex determination of eight adults guarding eggs in the wild, Keats et al. (1985) suggested that in this species only males brood eggs. Since eggs can take up to 9 months to develop, it is not clear whether parental care always lasts through hatching, although brooding for up to several months has been observed (Keats et al. 1985; Ringø and Lorentsen 1987). Male parental care of eggs has also been suggested for White Sea populations based on sex ratio data from trawl catches (Barsukov 1953, 1959; Pavlov and Novikov 1986). Males appear to reduce or stop feeding during the brooding period (Keats et al. 1985). They become very aggressive during the spawning season, and paired adults must be isolated when held in captivity (Pavlov and Novikov 1986). It has been speculated that males periodically release milt, even prior to the spawning season, as a pheromone to mark territories and attract females to nesting sites (Johannessen et al. 1993).

Early Life History. Egg and larval development are described in detail for Atlantic wolffish from Scotland (McIntosh and Prince 1890) and the White Sea (Pavlov 1986; Pavlov et al. 1992). The effect of temperature on the development of the skeletal system has been described in detail (Pavlov 1997; Pavlov and Moksness 1997). Eggs of Atlantic wolffish are large (5.5–6.8 mm in diameter), yellowish, opaque, and with an oil globule of 1.75 mm (Bigelow and Schroeder; McIntosh and Prince 1890; McKenzie and Homans 1938; Powles 1967). Eggs are laid in large tight clusters in nests guarded by the parental male (Keats et al. 1985; Ringø and Lorensen 1987). They may be confused with the eggs of ocean pout (*Zoarces americanus*), which lay similar egg masses at the same time as wolffish (Keats et al. 1985). However, ocean pout eggs are 8–9 mm in diameter and have a 3.2 mm oil globule (Methven and Brown 1991), distinguishing them from the smaller wolffish eggs. Ringø et al. (1987) described the amino acid and lipid composition of eggs, and Loenning et al. (1988) compared information on Atlantic wolffish egg morphology and physiology with that of several other pelagic and demersal eggs from northern Norway.

Egg and larval development has been described from a series of studies in Europe (Pavlov 1986; Pavlov and Novikov 1986; Pavlov et al. 1987, Ringø et al. 1987; Pavlov and Moksness 1994). Incubation time is highly dependent on water temperature and varies between 3 and 9 months, with 5°–7°C optimal. Larvae hatch at 17–20 mm. Prolarvae (about 22–24 mm) retain remnants of the yolk sac and oil globule and have a large eye with a diameter that is one-half the head depth. Small teeth are present. The finfold is completely differentiated. Melanophores are uniformly distributed over the body, but transverse pigment bands begin to appear on the trunk in some individuals. Narrow longitudinal bands of pigment are located along the base of the dorsal and anal fins, and a third band may be present along the midline of the body. These three pigment bands were also described by McIntosh and lantic wolffishes by Baranenkova et al. (1960). A few melanophores are present along the bases of the dorsal and anal fins. Canals of the lateral line appear on the head. The prolarval stage lasts from a few hours to 6 days.

In the larvae (>28 mm) the yolk sac has been reabsorbed and the relative depth of the body has increased. In White Sea populations 9–11 transverse pigment bands (bars) are present at this time (Pavlov et al. 1987). However, in Atlantic populations bars do not begin to develop until 130–150 mm (McIntosh and Prince 1890; Barsukov 1959). The feeding larval stage lasts only about 10–15 days at 5°–7°C. In the laboratory, fingerlings begin settling to the bottom at 25–30 mm, where they appear to defend territories. At this stage fingerlings become deeper bodied, pigment spots appear at the base of the dorsal (9–10) and anal fins (4–5), and conical teeth appear on the premaxillary, palatine, vomer, and dentary bones. The total pelagic stage lasts about 20 days, but might last up to 2 months at colder temperatures (Pavlov 1986).

The descriptions of newly hatched Atlantic wolffish larvae first provided by McIntosh and Prince (1890) and then by Bigelow and Schroeder were probably of individuals that had hatched prematurely owing to disturbance in the trawl and laboratory (Ringø et al. 1987). This would account for the long larval developmental period they reported and bias their description of larval behavior. Bigelow and Schroeder described wolffish 31.8–44.5 mm long as silvery on the sides in life, but the metallic hue fades after preservation, leaving only the dark brown pigment granules with which the sides are thickly dotted.

Fry of all three species of northwest Atlantic *Anarhichas* were described by Baranenkova et al. (1960). Pelagic larvae and juveniles of *A. minor* can be distinguished by presence of six or seven dark transverse stripes (or bars), giving them a somewhat mottled appearance. With growth these bars break up into large spots. *Anarhicas lupus* and *A. denticulatus* are more darkly and evenly pigmented, with vertical bars forming in the largest (over 10 cm) *A. lupus*. Smaller sizes of these two species can be distinguished chiefly by the length of a rayless membrane connecting the anal and caudal fins. In *A. lupus* this membrane is longer than the last anal ray, whereas in *A. denticulatus* it is shorter. The arrangement of teeth, though different from that of the adults, might be a useful for distinguishing among species of pelagic larvae (Barsukov 1959; Baranenkova et al. 1960).

Larval and Juvenile Distribution. In the Gulf of Maine, Atlantic wolffish larvae have been reported from the channel between Browns Bank and Cape Sable, near Seal Island (Nova Scotia), on German Bank and off its slope, off Lurcher Shoal, off Machias (Maine), on Jeffreys Bank (off Penobscot Bay), and in Massachusetts Bay a few miles off Gloucester. Bigelow and Schroeder interpreted this distribution data as indicating that Atlantic wolffish breed in the Gulf wherever they are found. They made similar observations for more northern parts of

have been reported off northeastern Newfoundland, in the Strait of Belle Isle, and off Sandwich Bay on the Atlantic coast of Labrador by the Newfoundland Fisheries Research Commission. In the Gulf of Maine, small numbers of pelagic juveniles (30–90 mm TL) were collected from stomachs of groundfishes captured at depths of 54–268 m, primarily during April and May, and which were widely distributed throughout the region (NEFSC, unpubl. data). A notably large number (102) of pelagic juveniles (37–49 mm TL) were found in six Atlantic cod stomachs from a single trawl tow made in 100 m off the western coast of Nova Scotia in the vicinity of Cape St. Mary on 20 April 1984.

Orlova et al. (1990) determined that pelagic juveniles are rare at depths of 0–50 m in the water column throughout northern Europe. Based on data from infrequent catches from various gears and from occurrences in haddock stomachs, they concluded that there is little relationship between juvenile size and depth of capture. Recently hatched larvae (22–26 mm) were collected in near-bottom layers in areas 68–385 m deep. Larger individuals (19–55 mm) have been collected in the upper 100 m over depths of 80–3,100 m. Atlantic wolffish juveniles (27–70 mm) were also found in haddock stomachs from 100 m (Orlova et al. 1990). Falk-Petersen et al. (1990) collected larvae 13–42 mm TL in northern Norway during March, May, and July. They found no significant length relationship with depth, but most larvae were collected in the upper 25 m. An earlier study in the region reported that pelagic larvae occurred mainly in depths of 100–200 m, and juveniles at 68–385 m (Baranenkova et al. 1960). Little is known of the distribution or habitat use of juveniles after settlement (Keats et al. 1986a). Wolffish larger than 50 cm TL have not been collected in shallow water (<30 m) off Newfoundland, where spawning activity has been documented (Keats et al. 1985, 1986a). After a review of distribution records in the literature, Keats et al. (1986a) hypothesized that juveniles probably inhabit deeper offshore waters and move inshore only when they become sexually mature. Atlantic wolffish size-depth distribution data for the Gulf of Maine (Nelson and Ross 1992) support this idea. However, small numbers of juveniles 11–50 cm are consistently collected throughout the Gulf of Maine during spring and fall NEFSC groundfish surveys, with the greatest concentrations occurring in the vicinity of Jeffreys Ledge (NEFSC, unpubl. data). In northern European waters, benthic juveniles (7–19 cm) were most abundant at 100–200 m depths (Baranenkova et al. 1960). However, juvenile wolffish have been reported in shallow kelp beds off Greenland (Fabricius 1780); and Iceland (Sæmundsson 1949).

Larval and Juvenile Habits. Atlantic wolffish larvae and early juveniles are pelagic between 20 and 40 mm TL, with settlement beginning by 50 mm TL (Falk-Petersen and Hansen 1990, 1991). In the laboratory they appear to undergo a transitional period between 50–100 mm TL in which they spend part of their time on the bottom and part in the water column (Moksness et al. 1989). At this time they also become aggressive and show signs of territorial behavior. However, juveniles larger than 100 mm TL were more gregarious and spent most of their time on the bottom. Based on limited trawl data, Bigelow and Schroeder concluded that larvae probably remain near the bottom and do not disperse widely from the hatching locality. Changes in juvenile coloration (Baranenkova et al. 1960; Pavlov 1994) and teeth arrangement may be associated with the shift from a pelagic to a benthic habitat (Baranenkova et al. 1960).

Age and Growth. In the Gulf of Maine, Atlantic wolffish mean length at age has been reported as 4.7 cm TL at age 0 and 98 cm TL at age 22 (Nelson and Ross 1992). In Iceland, fish 10.5–98.5 cm ranged from age 0 to age 20 (Jonsson 1982). Similarly, fish from Norwegian waters of 26–112 cm TL were age 5 to age 23 (Falk-Petersen and Hansen 1991). Data provided by Jonsson (1982) indicate that they reach 10.5 cm during the first year (age 0), and 13.6 cm by age 1. A comparison with other age-growth studies (Jonsson 1982) indicates a size of 21.8–28.7 cm by age 4 (Barsukov 1959; Beese and Kandler 1969, cf. Jonsson 1982). In contrast, growth rates of cultured fish have been much higher (Moksness et al. 1989; Moksness and Stefanussen 1990). Atlantic wolffish fry cultured in Norwegian studies grew 2–3.6% body weight per day over the first 150 days posthatch and grew best below 10°C (Moksness et al. 1989). More recently, larvae reared from 2.6 g at start-feeding grew to 900 g within 34 months, for a growth rate of 0.58% body weight per day (Moksness and Stefanussen 1990). Wolffish raised on dry and moist pellets grew more slowly, 0.16–0.26 $g \cdot day^{-1}$ (Stefanussen et al. 1993).

Atlantic wolffish from the Gulf of Maine apparently grow faster than those from Iceland (Nelson and Ross 1992) but those from the North Sea grow fastest (Liao and Lucas 2000a). The von Bertalanffy parameters for the North Sea population are (Liao and Lucas 2000a) $L_\infty = 111.2$ cm, $t_0 = -0.43$, $K = 0.12$ for males; $L_\infty = 115.1$ cm, $t_0 = -0.39$, $K = 0.11$ for females. Males have been reported to grow faster and reach a larger size than females in Icelandic and Norwegian waters (Jonsson 1982; Falk-Petersen and Hansen 1991), although differences do not become obvious until after age 12 (Jonsson 1982). Nelson and Ross (1992) noted that studies in the western Atlantic have lacked sufficient data to detect growth differences between the sexes. They found that growth begins to slow at about age 5 to age 6, corresponding approximately to the size at maturity (47 cm based on Templeman 1986b), and hypothesized that the decrease in growth rate was due to a diversion of resources to gonadal development.

General Range. Both sides of the North Atlantic; north to Davis Strait in American waters; south regularly to Cape Cod; less often west along southern New England, and exceptionally to New Jersey; also Greenland, Iceland, and northern Europe south to northern France and Ireland (Briggs 1988).

Occurrence in the Gulf of Maine. Atlantic wolffish are widely scattered throughout the Gulf of Maine (Map 30).

West of the Scotian Shelf, abundance is highest in the southwestern portion of the Gulf of Maine from Jeffreys Ledge to the Great South Channel at depths of 80–120 m (Nelson and Ross 1992). High abundances are also found on the northeast peak of Georges Bank and on Browns Bank. Somewhat smaller concentrations appear off of southwestern Nova Scotia in the vicinity of Wedgeport, and throughout the central Gulf. Populations of Atlantic wolffish in the western Gulf of Maine are probably discrete from those on the Browns Bank and Scotian Shelf areas (Idoine 1998b).

Movements. Whether or not wolffish undergo seasonal movements between deep and shallow waters has been argued for some time (see reviews of early accounts in Gill 1911, Jonsson 1982, Templeman 1984a, and Keats et al. 1985). Bigelow and Schroeder considered it a relatively stationary species based on monthly commercial catches and a generalized account of its behavior. Nelson and Ross (1992) reported a weak seasonal shift in the depth distribution between shallow water (<120 m) in spring and deeper water in fall in the Gulf of Maine. They also reported a weak size stratification among depths during spring, but data from less than 40 m are sparse, and they reported no significant corresponding difference during fall. They interpreted these data as an indication of inshore movement by mature adults in spring. However, their own data show that large numbers of wolffish remain in deep water during both seasons. The spatial patterns they observed could alternatively be interpreted as resulting from decreased catchability in the fall owing to behavioral changes (discussed below), compounded by a nonseasonal depth stratification of size classes (i.e., juveniles inhabit deeper water). Keats et al. (1985) also suggested that mature wolffish move inshore during spring, spawn during summer and fall, and move back offshore during late fall after the eggs hatch, based on scuba observations conducted in 5–15 m off Nova Scotia. However, as Keats et al. (1985) themselves recognize, these movements must involve only part of the spawning population since egg masses have been collected in deeper waters (100–130 m) in the region (McKenzie and Homans 1938; Powles 1967). Further, an extensive trawl survey in nearby Newfoundland waters found that Atlantic wolffish were most abundant in depths greater than 100 m during April–July, the time of presumed inshore migrations. A tagging study off Newfoundland (Templeman 1984a) lends little support for a migration hypothesis. Atlantic wolffish were usually recaptured within 8 km of the release site after 5–7 years, although a few individuals migrated as far as 338–853 km (Templeman 1984a).

Inshore-offshore migrations have been suggested in other geographic areas. Strong evidence of short (<185 km) seasonal movements, based on long-term tag recapture and catch data, has been reported in Icelandic waters (Jonsson 1982). Some individuals, however, have been reported to travel distances ranging from about 200 to 670 km within a year of release and adults move inshore during January–March to feed after a prolonged fasting period associated with spawning and/or with annual tooth loss (see discussion below). In late summer they begin migrating to offshore spawning grounds in 140–200 m, where they remain between September and January (Jonsson 1982). White Sea populations are thought to make similar movements, with spawning occurring from late July to September (Pavlov and Novikov 1986, 1993). Barsukov (1959) suggests that the seasonal movement pattern may switch between northern and southern populations, with the northern populations moving inshore during summer and offshore during winter, whereas southern populations do the opposite.

Movements may be influenced by local seasonal temperature regimes. For example, Pavlov and Novikov (1986, 1993) suggest that spawning movements in the White Sea may result from temperature requirements for the developing eggs. They hypothesized that mature adults leave inshore feeding grounds, where water temperatures are too high for normal egg development, in late July and move to offshore spawning grounds, where water temperatures range from −1° to 5°C and are more suitable for egg development. Interpretation of distribution data may also be complicated if some members of the adult population skip spawning years (Pavlov and Novikov 1986, 1993).

More extensive tagging data on movements as well as better spatial and temporal data on tooth replacement and spawning are needed to determine to what extent wolffish populations undergo seasonal migrations in all geographic areas and why they do so. It is clear that any presumed inshore-offshore migrations cannot be attributed to spawning activity alone, since eggs are found in both inshore and offshore areas and large segments of the populations are found in deep water during all seasons. At the present time, the data are not sufficient to discount the possibility that spawning occurs in all depths of their distribution and that apparent seasonal movements result from behavioral changes affecting catchability.

Importance. The market demand for wolffish is comparatively recent. It is an excellent table fish, selling readily as "ocean catfish" or as "whitefish." The commercial catch of Atlantic wolffish in the Gulf of Maine–Georges Bank area results mainly as by-catch from demersal otter trawl fisheries (Idoine 1998b). The total nominal catch increased from about 200 mt in 1970 to about 1,200 mt in 1983 (Fig. 254) and then declined steadily to a low of 400 mt in 1996. The NEFSC spring bottom trawl survey biomass index has also shown a consistent downward trend; the 1997 index value of 0.13 kg per tow, is the lowest in the time series (Idoine 1998b). Population dynamics and stock assessment parameters are currently unknown. Although the species is currently not managed, it is clearly overexploited and depleted.

The Icelandic fishery for wolffish has historically been the largest Atlantic wolffish fishery (Jonsson 1982). Catches of wolffish in Iceland and Europe increased greatly after World

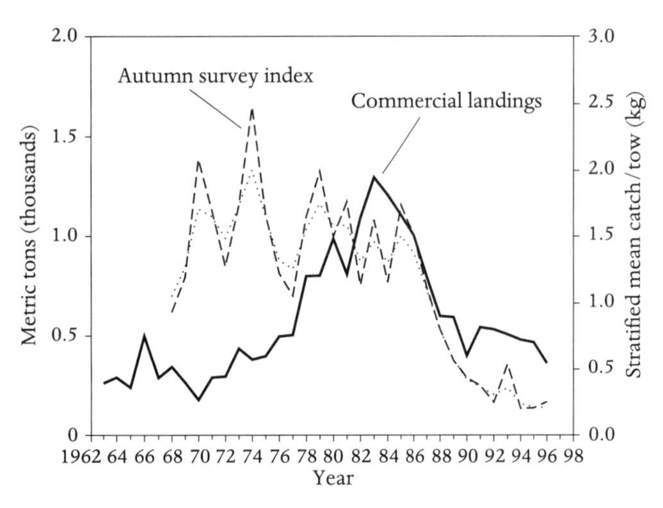

Figure 254. Commercial landings and NEFSC autumn survey index for Atlantic wolffish *Anarhichas lupus,* Gulf of Maine–Georges Bank, 1963–1996. (From Idoine 1998b.)

year^{-1} (Jonsson 1982). Catches in the northwest Atlantic have historically been small compared to Icelandic and European catches, averaging about 7,257 mt·year^{-1} between 1954 and 1973 (Jonsson 1982).

West Greenland waters have supported a local and international fishery for wolffish (including both spotted and Atlantic wolffish) since 1938 (Smidt 1981). The local longline fishery originally targeted spotted wolffish for production of skins. Frozen spotted wolffish fillets were targeted by inshore longliners beginning in 1951. The offshore trawl fishery began to increase substantially in the 1970s owing to declining Atlantic cod catches. Total nominal catches fluctuated between 4,000 and 6,000 tons during the 1950s and 1960s and appeared to be declining by the late 1970s (Smidt 1981).

It has been suggested that wolffish might be important predators on economically important bivalves (Sivertsen and Bjørge 1980, cf. Hawkins and Angus 1986; Nelson and Ross 1992; Stokesbury and Himmelman 1995). Several studies have used Atlantic wolffish as a model species in describing predation on sea urchins and attempt to examine its role in sea urchin population control (Breen and Mann 1976; Bernstein et al. 1981; Keats et al. 1986b; Hagen and Mann 1992). Sea urchin aggregation behavior is strongly influenced by wolffish and American plaice abundance (Bernstein et al. 1981).

Wolffish are the subject of extensive culture efforts in northern Europe (Pavlov and Novikov 1986; Ringø et al. 1987; Moksness et al. 1989; Johannessen et al. 1993; Pavlov and Moksness 1994; Tulloch et al. 1996).

SPOTTED WOLFFISH / *Anarhichas minor* Ólafsen 1772 / Spotted Catfish /

Bigelow and Schroeder 1953:507–508

Figure 255. Spotted wolffish *Anarhichas minor.* Off La Have Bank. Drawn by H. L. Todd.

Description. Body stout, moderately compressed (Fig. 255). Head heavy, blunt, profile rounded. Central (vomerine) band of teeth on roof of mouth about equal in length to bands on either side (palatine).

Meristics. Dorsal fin spines 74–80; anal fin rays 44–48; pectoral fin rays 20–23; branchiostegal rays 6 or 7; vertebrae 76–79 (Barsukov 1959; Scott and Scott 1988).

Size. Notwithstanding their Latin name, spotted wolffish are fully as large as Atlantic wolffish and are said to reach 1.83 m, but lengths above 1.35 m have been disputed (Barsukov 1959). One 94-cm-long individual weighed 5.9 kg eviscerated. Spotted wolffish lengths tend to increase along a north to south cline in western Greenland waters (Smidt 1981; Riget and Messtorff 1988). No length differences between males and females have been reported (Templeman 1986a).

Color. Body pale olive to chocolate brown, upper parts including dorsal and caudal fins thickly sprinkled with blackish brown spots of different sizes and irregular shapes (Scott and Scott 1988).

Distinctions. Spotted wolfish resemble Atlantic wolffish closely in general form and in arrangement of fins. The chief difference is that while the central (vomerine) band of teeth on the roof of the mouth is longer than the band on either side (palatine) in Atlantic wolffish, they are of about equal length in spotted wolffish. The red teeth described by Bigelow and Schroeder cannot be taken as a distinctive character since new teeth developing after the annual tooth loss are red in all wolffish species (Luhmann 1954; Barsukov 1959). Presence of thickly sprinkled irregularly shaped blackish brown spots on the dorsal and caudal fins, as well as on the head and flank, is the most useful field mark for spotted wolffish. A spotted form of northern wolffish (Templeman 1986c) can be distinguished from spotted wolffish by the lack of spots on the sides of the head and by differences in the number of vertebrae and dorsal fin rays. Young spotted wolffish smaller than 20 cm have 7–11 bars on their sides that break up into scattered spots with growth (Barsukov 1959), and hence could be confused with Atlantic wolffish. However, bars are lacking in Atlantic wolffish smaller than 7 cm and are absent or weak in individuals up to 20 cm. If present, bars on Atlantic wolffish are brownish and much longer than wide, whereas in spotted wolffish they are dark gray to black and nearly as wide as long.

Habits. Spotted wolffish were the least abundant species of wolffish collected off Newfoundland by Albikovskaya (1982) during April–July. They occurred at all depths from 1 to 600 m and at temperatures from −1.4° to 9.0°C, but were most common between 101 and 350 m and at temperatures under 5°C. In western Greenland waters, spotted wolffish were evenly distributed across three depth zones (0–200 m, 200–400 m, and 400–600 m) and between northern and southern areas (Riget and Messtorff 1988). Pelagic fry (24–26 mm) of spotted wolffish occur farther offshore and at greater depths (264–455 m) than Atlantic wolffish in the Barents Sea (Baranenkova et al. 1960). Juveniles (9–19 cm) were most frequently captured at depths of 200–250 m. Pelagic larvae (25–43 mm) were abundant in offshore waters in western (Smidt 1981) and eastern Greenland (Hansen 1968, cf. Scott and Scott 1988) during the summer. In the Barents and Norwegian seas, extensive collections have only rarely taken larvae or pelagic juveniles in the upper 100 m (Orlova et al. 1990, and citations therein). Orlova et al. (1990) collected 11 juvenile spotted wolffish (14–22 cm) in 220 m of water off Spitsbergen during August. Seasonal migrations have been suggested based on limited tag return data from the Barents Sea (Konstantinov 1961) and western Greenland (Riget and Messtorff 1988). Most individuals moved less than 100 km from the release site, although the longest migration was 777 km (Konstantinov 1961; Riget and Messtorff 1988). Little movement was noted for spotted wolffish tagged off Newfoundland, based on a limited sample size (Templeman 1984a). The pattern of longline catches of spotted wolffish from coastal, inshore, and fjord areas of Greenland suggests movements inshore and into the fjords during June and July (Riget and Messtorff 1988).

Food. Spotted wolffish from the Labrador-Newfoundland region have diets similar to Atlantic wolffish (Albikovskaya 1983). Both species feed heavily on echinoderms, but spotted wolffish eat fewer mollusks and more fishes, including *Sebastes* and *Gadus morhua*. Diet was highly variable among locations, man (1986a) reported a similar diet of 52% echinoderms, 23% fish, 15% decapod crustaceans, and 10% other invertebrates, based on more limited data. Fishes preyed on included *Amblyraja radiata, Gadus morhua, Melanogrammus aeglefinus,* and *Sebastes* sp. No significant differences were found among diets of pelagic fry of all three wolffish species, based on limited data (Baranenkova et al. 1960; Orlova et al. 1990). However benthic juveniles had diets similar to adults, with more fishes in spotted wolffish (Baranenkova et al. 1960; Orlova et al. 1990). Frequent collections of Atlantic wolffish eggs in stomachs of spotted wolffish captured on Atlantic wolffishes' Icelandic spawning grounds (Jonsson 1982) suggest strong interspecific interactions. Spotted wolffish have been noted to feed frequently on fish offal discarded from fishing boats (Sæmundsson 1949).

Tooth Replacement. Spotted and Atlantic wolffish have similar dentition; teeth are replaced annually in both species (Albikovskaya 1983), and replacement in the two is similar (Luhmann 1954). Tooth shedding takes place later in the Barents Sea (January–February) for spotted and northern wolffish (February–March), compared with Atlantic wolffish (October–May, peaking December and January) (Barsukov 1961). Juveniles do not feed during tooth replacement (Baranenkova et al. 1960).

Parasites. Eleven species of parasites were reported from spotted wolffish, with heavy infestations of several trematodes (Zubchenko 1980). Many parasites were common to both spotted and Atlantic wolffishes. Heavy infestations of *Acanthopsolus anarhichae* in both wolffishes may be an indication of migrations to shallow waters, since the intermediate gastropod host, *Buccinum undatum,* is only found in depths less than 150 m (Zubchenko 1980).

Predators. Reported from stomachs of Greenland shark, cod, and pollock (Barsukov 1959).

Reproduction. Females mature at 48–62 cm and males at 53–71 cm in western Greenland waters (Smidt 1981). In the Barents Sea, female spotted wolffish first mature at 53 cm and age 7, and 50% of the females are mature at 75 cm and age 9 (Shevelev 1988). Males first mature at 66 cm and age 9, with 50% maturity at 95 cm and age 12. All females larger than 100 cm and males larger than 110 cm are mature (Shevelev 1988). Templeman (1986a) examined trawl data from throughout the northwest Atlantic between 1946 and 1967 and concluded that spawning probably occurs during or soon after July–August based on egg size and ovarian weights. Unusually large catches (11–59,000 kg) of spotted wolffish in 146–192 m of water on the Grand Bank during August and September may be an indication of spawning aggregations (Templeman 1986a). First maturity occurred at 75–80 cm for females, but exhibited some geographic variation. Fecundity varies from 4,200 to 35,200 eggs, averaging 5,504 eggs at an average fish length of 66.2 cm, 9,415 at 75.5 cm, 14,219 at 85.4 cm, 18,867 at 97.1 cm, 24,829 at

Age and Growth. Spotted wolffish from the Barents Sea average 15.2 cm and 0.03 kg at age 0, 19.4 cm and 0.07 kg at age 1, 46.4 cm and 0.99 kg at age 5, and 118.5 cm at age 17 to age 21, based on analysis of hard parts (Shevelev 1988 and citations therein). Moksness and Stefanussen (1990) observed much higher growth rates for cultured spotted wolffish, with the fry growing faster (0.62% per day) than those of Atlantic wolffish (0.58% per day), increasing from 6.3 g on capture to 3,400 g after 34 months. They predicted that, if raised from start-feeding at an optimal temperature of 6°C, spotted wolffish should reach 6 kg in 2 years. However, an individual tagged in Norwegian waters increased in size from 78 cm to 120 cm in 1 year (Østvedt 1963, cf. Shevelev 1988), corresponding to an increase from 5 to 19 kg (0.21% body weight per day).

General Range. Chiefly north of the Arctic circle; north coast of Russia, White and Barents seas, and Iceland, south to middle Norway (vicinity of Bergen) on the European coast; Greenland; and southward occasionally to the Gulf of Maine on the American coast.

Occurrence in the Gulf of Maine. Goode and Bean's (1879) statement that "the Fish Commission has specimens from off the mouth of Gloucester Harbor and from Eastport, Maine," long remained the only notice of this northern fish for the Gulf of Maine, and fishermen have either never seen it there or have failed to distinguish it from Atlantic wolffish, which is unlikely, so striking is its color pattern. The late Walter Rich, of the U.S. Bureau of Fisheries, obtained a specimen that had been taken in 64 m off Cape Elizabeth (then in the collection of the Portland Society of Natural History); another, weighing 1.5 kg, was caught on a longline off the Portland lightship on 23 April 1927. Evidently spotted wolffish reach the Gulf of Maine only as accidental waifs from their Arctic home, to be watched for but hardly to be expected. They appear to occur regularly on Sable Island Bank and Banquereau Bank off outer Nova Scotia (Bean 1881; Vladykov 1935a; McKenzie and Homans 1938; McKenzie 1939).

Importance. Spotted wolffish are taken as by-catch from trawl and longline fisheries on both sides of the Atlantic, but are not usually reported separately from Atlantic wolffish (Smidt 1981; Jonsson 1982 [see Atlantic wolffish]). A local fishery for spotted wolffish skins off west Greenland began in 1938 and switched to mainly a filet fishery in 1951 (Smidt 1981). It has been found that spotted wolffish off the coast of Greenland in the vicinity of a zinc-lead mine have concentrations of lead in their organs (Bollingberg and Johansen 1979).

Suborder Trachinoidei

Thirteen families with a total of 51 genera are currently placed in this suborder (Nelson 1994), of which one family, the Ammodytidae, occurs in the Gulf of Maine. Systematic placement of this family has persistently plagued researchers studying these fishes. Many years of discussion concerning the phylogenetic placement of ammodytids resulted in proposed affiliations of the Ammodytidae with several different families and family groups (see Pietsch and Zabetian [1990] for historical background), but no completely acceptable conclusions were derived from these earlier studies. The current hypothesis of relationships places Ammodytidae as the sister group to Trachinidae and Uranoscopidae (Pietsch and Zabetian 1990), although Ida et al. (1994) noted inconsistencies in the number of shared characters between Trachinidae and Ammodytidae reported by Pietsch and Zabetian (1990).

SAND LANCES. FAMILY AMMODYTIDAE

MARTHA S. NIZINSKI

Sand lances are slender, schooling marine fishes found in both littoral and offshore environments in all major oceans and seas. The Ammodytidae currently includes seven genera and approximately 27 species (Ida et al. 1994; Collette and Randall 2000).

Confusion also surrounds species recognition within this group, and much of it can be attributed to morphological similarities and clinal variation, both with latitude and distance offshore, in meristic features among species. For example, 23 nominal species have been described in the genus *Ammodytes*. However, only six species are currently recognized (Reay 1970), two of which occur in the western North Atlantic.

Members of this family are characterized by a narrow, elongate body; small head, with the lower jaw protruding beyond the upper; jaws toothless, no teeth on roof of mouth; more abdominal than caudal vertebrae; a small, deeply forked caudal fin; pelvic fins minute or absent; low ridge of skin present on either side along abdomen; and dorsal and anal fins without spines (Pietsch and Zabetian 1990). Additionally, these fishes possess distinctive rows of oblique folds of skin, called plicae (Fig. 256), which occur on the lateral body surface and are lined on the underside by cycloid scales. Characteristically, plicae run downward and backward in a regular serial arrangement from the area above the pectoral fin base to the caudal peduncle.

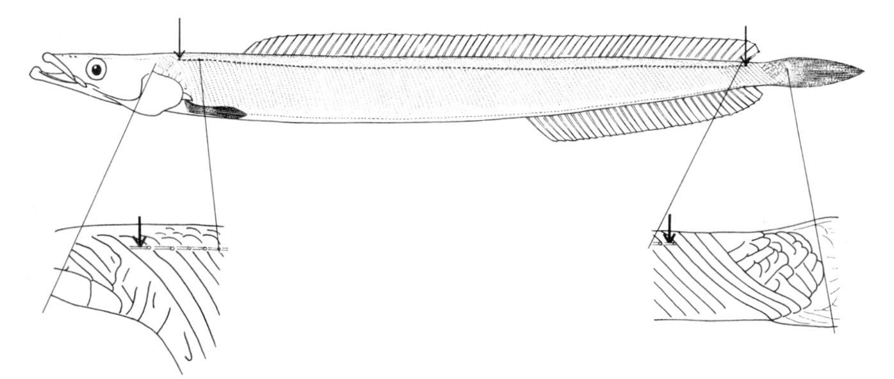

Figure 256. Outline drawing of offshore sand lance *Ammodytes dubius* to show serial arrangement of plicae and how they are counted (between arrows). (From Nizinski et al. 1990: Fig. 3.)

Bigelow and Schroeder questioned the validity of two western North Atlantic species but a later study (Nizinski et al. 1990) provides strong support for recognition of two morphologically similar species of *Ammodytes* in the Gulf of Maine, *A. americanus* and *A. dubius*. It is difficult to distinguish between these fishes and they are frequently mistaken for one another. Much uncertainty exists regarding the identity of specimens examined by various researchers. Therefore, citing specific differences in previously published literature concerning these species becomes a dubious task because the possibility exists that one species or a combination of both species has been included in the material of these earlier studies. Whenever possible, previously reported findings have been evaluated to determine which species was the one most likely collected. As far as can be determined, the habits of these fishes are quite similar, so some sections in the species accounts are written more generally.

The continued occurrence of both species inshore suggests that the western North Atlantic species are reproductively isolated. The mechanism, however, is not known. Okamoto et al. (1989) found that two sympatric populations of *Ammodytes* off Japan have different spawning peaks and the range of the spawning season for each population is thought not to overlap. Eastern North Atlantic *Ammodytes* (*A. marinus* and *A. tobianus*) may have staggered spawning seasons as well (Okamoto et al. 1989). Possibly, the reproductive isolation mechanism is similar for western North Atlantic species.

The readiest field marks distinguishing sand lances among Gulf of Maine fishes are their slender form and sharply pointed snouts, coupled with one long dorsal fin (separated from the caudal) and the absence of pelvic fins. The only fishes with which one would likely confuse *Ammodytes* are young eels, but dorsal, caudal, and anal fins in the latter are confluent, and the caudal fin is rounded, not forked as it is in *Ammodytes.*

Sand lances swim as eels do, by lateral undulations that run along the body from front to rear, which makes them easy to recognize in the water. They have the habit of burrowing, with great speed, several centimeters deep into the sand, opening the way with their sharply pointed snouts.

Sand lance are of commercial value in the Gulf of Maine only as bait, for which purpose 30,818 kg were landed from the traps in Massachusetts in 1919 and 9,091 kg in 1946.

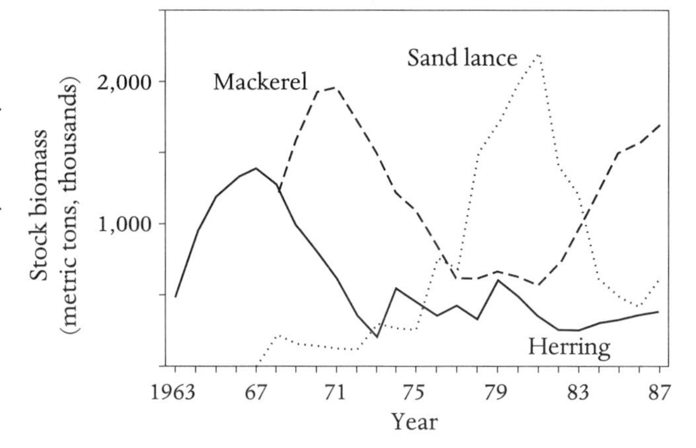

Figure 257. Abundance (in thousands of metric tons) of mackerel, herring, and sand lance populations from 1963 to 1987 in the western North Atlantic. (Population Dynamics Branch, Conservation and Utilization Division, NMFS/NEFC.)

fishes are often considered an underutilized resource. Interestingly, the former NMFS Gloucester Laboratory packed some experimentally like sardines and found them to be of good quality. In the North Sea, sand lance are part of an important fish meal industry (Macer 1966). Despite their low commercial value, they are of great ecological importance and play a significant role in the community structure and predator/prey relations in the Gulf of Maine.

Western North Atlantic populations of sand lance increased dramatically in the early 1980s (Sherman et al. 1981; Winters 1983). This population explosion was correlated with a decline in herring (*Clupea harengus*) and mackerel (*Scomber scombrus*) stocks along the eastern coast of the United States (Fig. 257). Sand lance seemed to have replaced these stocks. Concurrently, piscivorous fishes increased their consumption of sand lance. Peak abundance of sand lance in this region was reached in 1981 and numbers have since decreased (Nelson and Ross 1991). Again, this shift in sand lance abundance was correlated with mackerel numbers; mackerel populations have been steadily increasing since 1983.

Fogarty et al. (1991) further examined the large-scale shifts

northeastern United States. Results of their exploratory analysis support the hypothesis that predation by pelagic predators has played an important role in regulating sand lance populations in this area. Stocks of mackerel and herring overwinter in the southern New England–Mid-Atlantic region and can potentially affect sand lance populations in this area during the winter-spring period (Fogarty et al. 1991). Sand lance constituted up to 40% of stomach contents by weight of mackerel in southern New England and approximately 10% by weight of the herring diet (Fogarty et al. 1991). Although predation by mackerel, herring, and/or other predators (piscivorous fishes, marine mammals, seabirds) seems to play a major role in regulating sand lance abundance, physical transport mechanisms also may be an important factor in stock regulation (Fogarty et al. 1991).

KEY TO GULF OF MAINE SAND LANCES

1a. Lateral plicae 106–126, usually 112–124; vertebrae (including hypural plate) 63–71, usually 65–70; found in shallow, coastal waters and estuaries .... **Inshore sand lance**

1b. Lateral plicae 124–147, usually 128–138; vertebrae 69–77, usually 70–72; found in deeper offshore waters, but occasionally collected inshore .... **Offshore sand lance**

INSHORE SAND LANCE / *Ammodytes americanus* DeKay 1842 /

American Sand Lance, Sand Eel, Lance / Bigelow and Schroeder 1953:488–491

Description. Slender, elongate, small- to medium-sized fishes (49–168 mm SL). Body of nearly uniform width, about one-tenth as deep as long, from gill opening to anus, with a gradual taper in posterior quarter of body to caudal peduncle. Head long with sharply pointed snout. Gill openings wide. Mouth large with lower jaw projecting far beyond upper. Dorsal fin low and long, with anteriormost dorsal fin rays slightly in advance of vertical line through tips of pectorals and extending posteriorly nearly to caudal fin base. Anal fin, similar in outline to dorsal, extending posteriorly nearly to caudal fin base. Anteriormost anal fin rays slightly posterior to vertical through middle of dorsal fin. Pectorals set very low on body. Cycloid scales small, lying in cross series on sides of body between plicae. Lateral line straight and incomplete, terminating under posterior rays of dorsal fin.

Meristics. Dorsal fin rays 52–61, usually 55–59; anal fin rays 26–33, usually 27–31; pectoral fin rays 11–15, usually 13; lateral plicae 106–126, usually 112–124; vertebrae 63–71, usually 65–70; gill rakers 21–28, usually 24–26 (Nizinski et al. 1990).

Color. Sand lance are usually olive, brownish, or bluish green above, with silvery sides and a duller white belly. Some have a longitudinal stripe of steel blue iridescence along each side, but others lack this coloration. Color varies in shade among individuals taken over different substrates and their iridescent luster fades at death.

Size. In excess of 100 mm SL, to at least 168 mm SL (Nizinski et al. 1990).

Distinctions. Variation in six meristic characters, especially when considered in combination, provide a good basis for species identification. *Ammodytes americanus* tend to have lower counts of lateral plicae, vertebrae, dorsal fin rays, anal fin rays, pectoral fin rays, and gill rakers than *A. dubius*. The number of lateral plicae is the most useful single character for distinguishing *A. americanus* from its western North Atlantic congener. Also, most individuals can be identified to species using number of vertebrae plotted against number of plicae. This bivariate plot separates individuals of the two species with little overlap among them (Nizinski et al. 1990) (Fig. 258). However, for those individuals whose counts fall in the area of overlap the following equation is useful for placing them into species groups: $(8.09 \times \text{number of plicae}) + (2.44 \times \text{total number of dorsal rays}) + (2.31 \times \text{total number of vertebrae}) + (1.23 \times \text{total number of gill rakers}) + (1.03 \times \text{total number of anal rays}) + (0.45 \times \text{total number of pectoral rays})$. Individuals with scores lower than 1,378 are considered *A. americanus* (Nizinski et al. 1990).

Habits. *Ammodytes americanus* are found chiefly in shallow (usually 2 m or less) coastal waters and estuaries and are seldom seen along rocky shores. They are found in habitats with substrates conducive to burrowing (i.e., sand, sand with crushed shell, or fine gravel).

During high tide, sand lance may burrow above the low-water mark. Once the tide recedes, they may remain buried in these exposed tidal flats until the next tide. On one tidal flat, Bigelow and Schroeder saw them vanish with surprising rapidity when alarmed by clam diggers, and these authors could not improve on Goode's 1884 account of seeing "a great section of the beach" in Provincetown harbor become "alive with dancing forms of dozens of these agile fishes" when he stuck his clam-hoe into the sand. Rapid burial in shallow areas rather than inshore-offshore movements may explain their sudden appearances and disappearances. It is uncertain whether they exhibit this burying behavior only in shoal water, where they often come under direct observation. Sand lance have also been observed with just their heads protruding from the substrate (Reay 1970; Meyer et al. 1979; Scott and Scott 1988). When disturbed, they generally escape into the water column with a rapid sideways motion out of the sand (Reay 1970; Meyer et al. 1979). It has been suggested that they spend a large part of the time so buried, particularly at night and over winter (Reay 1970; Winslade 1974a,c). Mass mortal-

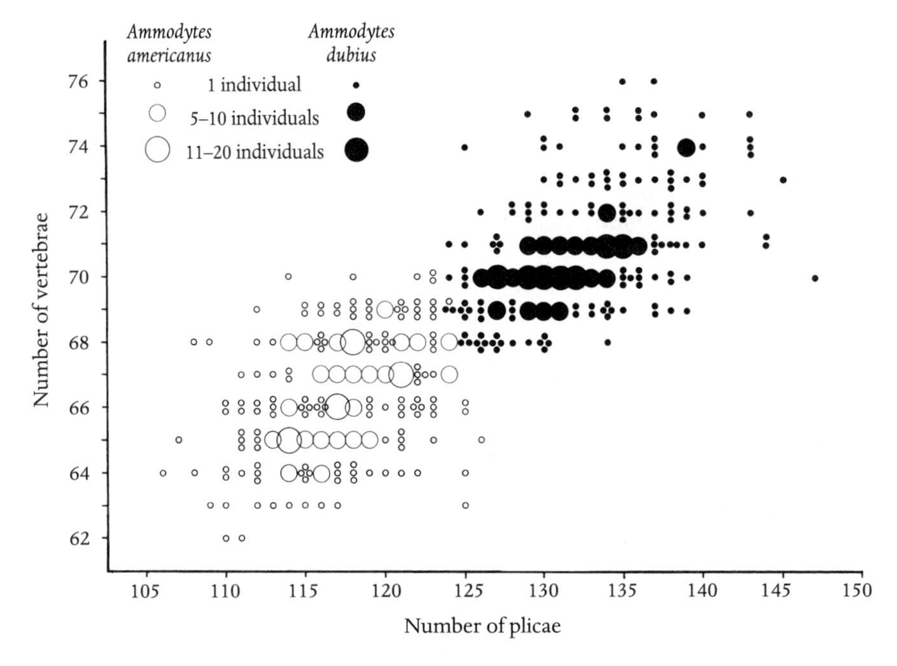

Figure 258. Scattergram of vertebral count plotted against number of lateral plicae for *Ammodytes americanus* and *A. dubius*. (From Nizinski et al. 1990: Fig. 6.)

ity, presumably caused by temperature stress, has been observed in intertidal areas (Scott and Scott 1988).

Sand lance congregate in dense schools, which have been observed to range from hundreds to several thousand individuals on the Provincetown slope (identification of fishes as *A. americanus* is questionable [Meyer et al. 1979]). Within a single school, the fish were of similar size (mean 15 cm), but slightly larger individuals were observed either at the head of the school or in the central core (Meyer et al. 1979). Schools of nonfeeding individuals held a constant shape, vertically compressed, tightly compacted, and bluntly linear in lateral appearance (1–5 m wide, 3–20 m long, 0.5–1.5 m high) (Meyer et al. 1979). Distance between individuals decreased during escape and increased during feeding.

Sand lance postlarvae have been observed schooling with herring (*Clupea harengus*) postlarvae (Richards 1976b). Unlike adult sand lance schools, which comprise individuals of similar sizes, postlarval sand lance and herring in mixed schools differed in both length and width of individuals. These two species apparently school together during their first year (Richards 1976b).

Migrations and daily movements of sand lance are not well understood. There is some evidence for an inshore (summer) and offshore (winter) movement, particularly in northern latitudes (Reay 1970). But since only a stray *A. americanus* has been found offshore (Nizinski et al. 1990), this migratory pattern may only pertain to *A. dubius*.

Larval *A. americanus* have been collected in waters with salinities ranging from 4.3 to less than 1.8 ppt in White Creek, Indian River, Del. (Chamberlin in Norcross et al. 1961). As *A. americanus* frequent estuaries, it is clear that they can tolerate low salinities. However, Chamberlain suggested that sand lance become less tolerant of low salinity waters with increas-

Food. Food habits of the two species of western North Atlantic sand lances have not been studied in detail. Further information is given in the account of *A. dubius* below.

Predators. Predation on western North Atlantic sand lances has only been studied in general terms, and published studies probably include information on a combination of both species. Sand lances are preyed upon by several different kinds of predators, including sea birds and fishes. Sundry predacious, commercial, and recreational fishes such as little skate, goosefish, cod, haddock, halibut, silver hake, red hake, Atlantic salmon, mackerel, striped bass, bluefish, pollock, American plaice, yellowtail flounder, and Gulf Stream flounder, (Reay 1970; Meyer et al. 1979; Sedberry 1983; Winters 1983; Bowman and Michaels 1984; Bowman et al. 1987; Safina and Burger 1989; Zamarro 1992; Armstrong et al. 1996) include sand lance in their diets. Bigelow and Schroeder reported that sand lance, fleeing from their pursuers, especially silver hake, which do not hesitate to follow them up onto the sand, often strand in such numbers as to cover the flats.

Sand lances are also preyed upon by various kinds of sea birds (Reay 1970; Powers and Backus 1987). Terns (*Sterna hirundo*) feeding near Fire Island Inlet, Long Island, feed almost exclusively on *A. americanus* and bay anchovies with *Ammodytes* being by far the most important prey item (Safina and Burger 1985). In the Gulf of St. Lawrence, sand lance are the most numerous fish caught by double-crested cormorant to feed their young (Rail and Chapdelaine 1998).

Breeding Habits. *Ammodytes* spawn from the Gulf of Maine to Cape Hatteras, N.C.; spawning is widespread over the continental shelf (Richards and Kendall 1973; Sherman et al. 1984), and occurs near the end of their second year in late fall

however, fish autopsied after spawning in the laboratory had very few eggs remaining in the ovaries (Smigielski et al. 1984). Although sand lance have never been observed spawning naturally, it is believed that they spawn in the Merrimack River estuary, Mass., in early December when water temperatures are 4.4°–6.3°C (Smigielski et al. 1984). Embryos develop and larvae survive and grow for extended periods of time at temperatures as low as 2°C (Buckley et al. 1984).

In Massachusetts, gonads of females were slightly further developed in December, whereas those of males remained virtually unchanged. Some males were running ripe in late November, but most had ripened by the middle of December, which coincided with the females' spawning (Smigielski et al. 1984). In the Gulf of St. Lawrence, however, 82% of gonads examined were at prespawning or spawning stages in September (Brêthes et al. 1992). GSI for both sexes was approximately 15% in September, with females having slightly lower values than males (Brêthes et al. 1992). Mean weight loss after spawning was calculated as more than 35% (Smigielski et al. 1984), but decrease in body weight of up to 45% during the spawning season has been reported (Westin et al. 1979).

Little information on fecundity is available. Richards (1982) estimated the number of eggs produced by an *A. americanus* female to be between 1,855–5,196. However, this estimate may be low since only a limited number of females from Long Island Sound (eight, ranging in size from 89 to 141 mm) were examined and most of the fish were age 0. Westin et al. (1979) determined the following fecundity estimates for sand lance from the Merrimack River based on length, weight, and age, of which length may be the most convenient estimator: $\log F = 3.857 \log L - 0.484$; $F = 1236.7W - 1698.7$; $F = 2940.4A - 783.6$.

Spawning peaks about December–January and ends February–March (Scott 1968; Richards 1982; Winters 1983), possibly later to the north. Bigelow and Schroeder collected larvae only a few days old (7–8 mm), with the yolk still showing, off Newburyport, Mass., in March, whereas the Canadian Fisheries Expedition obtained an abundance of slightly older stages (7–15 mm) off the southeast coast of Nova Scotia in May. Spawning may be temperature-mediated. Norcross et al. (1961) hypothesized 9°C as the maximum temperature for spawning.

Early Life History. Sand lance were formerly thought to spawn on sandy beaches above the low-water mark while burrowing in the sand, but their eggs have never been found in such a situation. Ammodytid eggs have, however, been collected in nearshore habitats. Williams et al. (1964) collected eggs (presumably those of *A. americanus*) over sandy or gravelly bottoms in shallow (<2 m) channels, where current speeds were low. They suggested that these eggs must have been spawned in quite shallow water, near where they were collected since only in such locations would wave turbulence be sufficient to keep demersal, slightly adhesive eggs in suspension and permit their collection by plankton nets.

Sand lance eggs are demersal and adhesive. They are irregularly shaped, not quite spherical, contain one oil globule (0.28–0.38 mm, mean 0.35 mm), and range in diameter from 0.94 to 1.03 mm (mean 1.00 mm) (Smigielski et al. 1984). Williams et al. (1964) reported slightly smaller measurements for both egg diameter (0.67–0.91 mm, mean 0.83 mm) and oil globule (0.17–0.33 mm, mean 0.22). The egg membrane is pitted, the perivitelline space narrow, the interior of the egg cloudy, and the yolk dull yellow to white (Williams et al. 1964).

Incubation and hatch duration are unusually long. Smigielski et al. (1984) collected adults from the Merrimack River estuary and successfully incubated both eggs and larvae in the laboratory at a range of temperatures (2°, 4°, 7°, 10°C). Start of hatching ranged from 61 days (2°C) to 25 days (10°C) after fertilization, with hatch duration from 74 days (2°C) to 30 days (10°C). Mean length of larvae at hatching was 5.71–6.34 mm SL.

Mortality is strongly influenced by plankton density but not temperature. Plankton densities as low as 200 liter^{-1} resulted in reduced mortality compared to unfed larvae (Buckley et al. 1984, 1987). Survival to metamorphosis is estimated to be 0.12% at low food levels (200 rotifers·liter^{-1}) and 11.74% at high food levels (1,000 rotifers·liter^{-1}) (Buckley et al. 1984).

Larvae. *Ammodytes americanus* larvae are fully developed at hatching and possess pigmented eyes, a complete gut, and a functional mouth (Smigielski et al. 1984). They are slender and range from 3 to 6 mm at hatching (Richards 1982). Yolk-sac absorption occurs between 2 and 7 days (6.3–6.8 mm SL), and oil globule absorption between 5 and 14 days (7.2–7.4 mm SL) (Smigielski et al. 1984). Although some individuals commenced feeding several hours after hatching, time to first feeding is usually 1 or 2 days posthatch (Smigielski et al. 1984). After hatching, laboratory-reared larvae demonstrated low mortality rates at low plankton densities in the absence of predation (Buckley et al. 1984). Young sand lance larvae demonstrated the ability to survive starvation for up to 2 weeks under laboratory conditions (Buckley et al. 1984). It appears, therefore, that *A. americanus* is well adapted for spawning during the winter minimum in zooplankton abundance (Buckley et al. 1984, 1987).

Dorsal and anal fin rays begin to develop when the larva is about 9–16 mm long, but it is not until the fish is upward of 25 mm that the tail begins to assume its forked outline. Transformation to juvenile morphology occurs at 29 mm (102–131 days after hatching, depending on temperature [Smigielski et al. 1984]). Larvae of different lengths are characterized by differences in numbers and distribution of melanophores. Richards (1982) described larval pigment development, but her descriptions may be based on both species of western North Atlantic sand lance. There are differences between western North Atlantic species in the rate of melanophore development but they are not obvious. Early larval stages are easily recognizable by their slender form in combination with the vent opening at one side, not at the margin of the larval finfold, similar to fishes in the cod tribe. The forked tail is a convenient field mark for distinguishing between sand lance and herring, the tail of the latter being deeply forked from a much earlier stage. Older larvae resemble the corresponding stages of rock gunnel in their slim form and in the location of the vent slightly behind the middle of the trunk (vs. farther back in the similarly slender larvae of the herring tribe), but may be recognized by the row of black pigment cells along the dorsal side of the intestine (vs. along the ventral side) and by their pointed snouts.

Larval fish reared at 7°C in tanks with a sandy bottom exhibited schooling behavior after attaining lengths of 25–30 mm (approximately 90 days after hatching) (Smigielski et al. 1984). These fish were also observed burrowing into the bottom 133 days after hatching (35–40 mm) (Smigielski et al. 1984). Fish remained buried for several days only to emerge in groups of two and three or several dozen to begin actively seeking food on the surface.

Sand lance larvae consume phytoplankton and various developmental stages of a variety of commonly occurring copepods (Covill 1959; Monteleone and Peterson 1986). Diet is influenced by food availability (diet shifts to copepods after the spring bloom) and larval swimming ability (small larvae consumed phytoplankton and larger larvae ate copepods of increasingly older developmental stages [Monteleone and Peterson 1986]). Covill (1959) examined gut contents of 200 larvae collected in Long Island Sound and concluded that *A. americanus* larvae take suitable food organisms more or less randomly and that several pelagic copepods are probably equally vulnerable to predation by the larvae. The absence of harpacticoids and other benthic organisms strengthens the argument that these larvae are pelagic feeders (Covill 1959). Feeding behavior of small larvae is passive, but becomes more aggressive as they mature (Monteleone and Peterson 1986). Biomass of food consumed increased with increased prey density and water temperature. Based on laboratory experiments, calculations of the predatory impact of sand lance larvae on copepod nauplii indicate that they are insignificant consumers, taking an estimated maximum of 13.4% of the copepod production per day, which is only about 0.27% of the standing stock (Monteleone and Peterson 1986). It is probable that food biomass on the order of 4–5% of body weight would have to be ingested daily to ensure normal growth and activity (Covill 1959).

Age and Growth. Age-length curves based on otolith analysis were given by Brêthes et al. (1992) for male and female *Ammodytes* from the Gulf of St. Lawrence. The most rapid growth occurred in the first 3 years; growth rates decreased between age 4 and 5. Seasonally, growth is much more rapid during the summer months (July–September) (Brêthes et al. 1992).

Estimated short-term growth rates for larvae collected on Georges Bank ranged from 4.9 to 23.9% per day; *A. americanus* reared in the laboratory attained growth rates of 0.73–1.53% increase in SL per day or between 2.73 and 11.31 mm per month (Smigielski et al. 1984). Growth appears to be influenced by temperature. Specific growth rates in dry weight per

Length-frequency modes identified by modal analysis corresponded closely to mean lengths of fish based on otolith annuli counts, supporting the hypothesis that the number of annuli corresponds to the age of the fish (Brêthes et al. 1992). Otolith analysis, however, may not be conclusive as both Scott (1973c) and Winters (1981) had problems with back-calculations of age at length. Deposition of opaque material in otoliths for males, females, and immature individuals occurs during the summer (to September). An opaque ring on the otolith is laid down annually, with increase in growth during summer accounting for 75–100% of the annual increase in size (Brêthes et al. 1992).

Based on samples collected in August and September, 50% of observed fish were mature at 95 mm. Of those individuals in which sex could be determined, 50% of females were mature at 90 mm and 83% of males were mature at 85 mm. The youngest mature fish observed was an age-0 female 69 mm in length (Brêthes et al. 1992). The maximum life span for *A. americanus* is estimated to be 12 years (Brêthes et al. 1992), which is much longer than the estimates for other species of *Ammodytes*. It is possible that low growth rates correspond to a longer life span (Brêthes et al. 1992).

Parameters of the von Bertalanffy growth equation calculated for sand lance collected in the Gulf of St. Lawrence (Brêthes et al. 1992) were estimated to be: $K = 0.24$ year^{-1}; $L_\infty = 188.2$ mm; and $t_0 = -1.14$ year. The length-weight relationship for both male and female prespawning *A. americanus* in the Merrimack River estuary was calculated as: $\log W = 3.772 \log L - 3.347$ (Westin et al. 1979).

Analysis of the length-weight relationship indicates isometric growth (Brêthes et al. 1992). Growth of *A. americanus* in the Gulf of St. Lawrence appears to be slower than that of *A. dubius* or even of conspecifics in more southern areas (Brêthes et al. 1992), but growth patterns seem to be affected by changes in growth rate with age, selective mortality, and growth compensation, as well as environmental factors (Winters 1981).

General Range. *Ammodytes americanus* is found along the Atlantic coast of North America from at least as far south as Chesapeake Bay (possibly southward to Cape Hatteras) to Newfoundland and northern Labrador (Nizinski et al. 1990).

Occurrence in the Gulf of Maine. Sand lance are common on sandy beaches throughout the Gulf. They swarm on the strands of Cape Cod Bay throughout the summer. In some years they continue to be plentiful there during winter and great numbers are sometimes cast on the beach in stormy weather. There is a general decrease in their numbers close inshore during the cold months, which suggests that a considerable proportion of the local population may move out into deeper water for winter or bury and overwinter in the sand to return in spring as do most of the sand lance in northern European seas (Reay 1970; Winslade 1974a,c). Since most of the available information on distribution is based on offshore

OFFSHORE SAND LANCE / *Ammodytes dubius* Reinhardt 1837 / Northern Sand Lance, Sand Eel

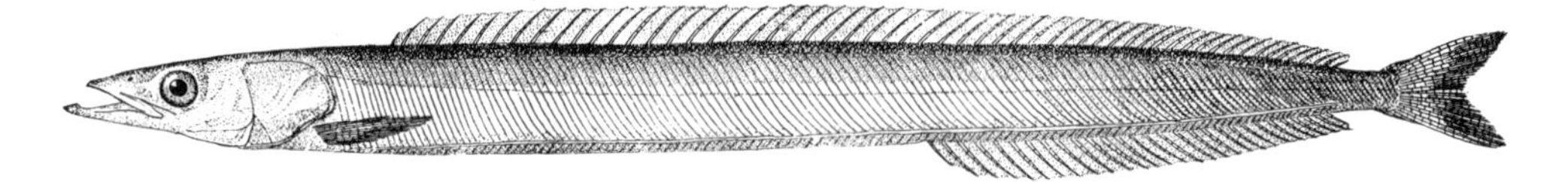

Figure 259. Offshore sand lance *Ammodytes dubius.* Nantucket. Drawn by H. L. Todd.

Description. Slender, elongate, small- to medium-sized fishes (77–253 mm SL) (Fig. 259). Body of nearly uniform width, about one-tenth as deep as long, from gill opening to anus, gradually tapering in posterior quarter to caudal peduncle. Head long with sharply pointed snout. Gill openings wide. Mouth large with lower jaw projecting far beyond upper. Dorsal fin low and long, with anteriormost rays slightly in advance of vertical through tips of pectorals and extending posteriorly nearly to caudal fin base. Anal fin similar in outline to dorsal fin. Anteriormost anal fin rays slightly posterior to a vertical through middle of dorsal fin; anal fin extends posteriorly nearly to caudal fin base. Pectorals set very low on body. Cycloid scales small, lying in cross series on sides of body between plicae. Lateral line straight and incomplete, terminating on the sides under posteriormost dorsal fin rays.

Meristics. Dorsal fin rays 56–67, usually 61–62; anal fin rays 28–35, usually 30–32; pectoral fin rays 12–16, usually 14; lateral plicae 124–147, usually 129–138; vertebrae 69–77, usually 70–72; gill rakers 23–31, usually 25–28 (Nizinski et al. 1990).

Color. Bluish to bluish green above, with silvery lower sides and a duller white belly. Some with a longitudinal stripe of steel blue iridescence along each side, but others lack this pattern.

Size. In excess of 300 mm SL, reaching at least 372 mm (Scott 1968).

Distinctions. Meristic features for *A. dubius* tend to be higher than those of *A. americanus.* The number of lateral plicae is the best single character for distinguishing the former species from the latter; however, the combination of the number of vertebrae plotted against the number of plicae separates individuals of the two species with little overlap (Nizinski et al. 1990) (Fig. 258).

Habits. *Ammodytes dubius* are typically found in more offshore locations including shoaler areas of offshore fishing banks; however, this species has also been captured inshore (Winters and Dalley 1988). Sand lance are found in habitats with substrates conducive to burrowing (i.e., sand, sand with crushed shell, or fine gravel) and are seldom seen off rocky coasts or over muddy bottoms in deep water.

Tolerable temperatures range from −2° to 11°C but 0° to 6°C is most common (Scott 1968). Richards et al. (1963) reported temperatures of 0° to 4°C at time of egg development in the Gulf of Maine and hypothesized a salinity range of 26–36 ppt. Few individuals, however, are found in salinities below 30 ppt (Norcross et al. 1961; Richards et al. 1963).

It has been suggested that sand lance spend a large part of the time buried, particularly at night and over winter (Reay 1970; Winslade 1974a,c). Thus their sudden appearances and disappearances may be attributed to this behavior rather than indicating migration. It is not known whether they exhibit this burying behavior only in shoal water, where they come under direct observation, or whether they also burrow in deeper waters. If the burrowing habit is for refuge, it is not always successful, for Bigelow and Schroeder reported an observation of porpoises rooting sand lance out of the sand. It has also been hypothesized that temperature, light intensity and photoperiod, and food availability all affect their swimming activity (Winslade 1974a,b,c) and would, therefore, have an effect on when and how long these fish are in the substrate. Sand lance have also been observed with just their heads protruding from the substrate (Reay 1970; Meyer et al. 1979; Scott and Scott 1988). When disturbed, they generally retreat back into the substrate (Reay 1970; Meyer et al. 1979).

Sand lance congregate in dense schools, ranging from hundreds to tens of thousands of individuals on Stellwagen Bank (Meyer et al. 1979). Offshore, deeper-water schools were observed to have more individuals (Reay 1970). Within a single school, the fish were of similar size (7.4–24.0 cm FL), but slightly larger individuals were observed either at the head of the school or in the central core (Meyer et al. 1979). Schools of nonfeeding individuals held a constant shape, vertically compressed, tightly compacted, and bluntly linear in lateral appearance (Meyer et al. 1979). Distance between individuals decreased during escape and increased during feeding.

Sand lance postlarvae have been observed schooling with Atlantic herring postlarvae (Richards 1976b). Unlike adult sand lance schools, which have individuals of similar sizes, postlarval sand lance and herring in mixed schools differed in both length and width of individuals. These two species apparently school together during their first year (Richards 1976b).

Migrations and daily movements of sand lance are not well understood. There is some evidence for an inshore (summer) and offshore (winter) movement, particularly in northern latitudes (Reay 1970). *Ammodytes dubius* have also been collected inshore particularly in Newfoundland waters (Winters and Dalley 1988). Nizinski et al. (1990) also found both species in coastal collections from Labrador and New England southward to New Jersey. It has been suggested that *A. dubius* move inshore to spawn (Dalley and Winters 1987; Winters and Dalley 1988). Juvenile *A. dubius* appear to undergo diel migration, being found on or near the bottom during daytime and moving up to surface waters at night, but avoidance of sampling gear by larger sand lance precludes a firm conclusion on diel movement (Potter and Lough 1987). Larvae and small juveniles (<10 mm SL), however, are found throughout the water column regardless of the time of day. At sizes greater than 10 mm SL, individuals begin to undertake diel migrations (Potter and Lough 1987).

Food. Sand lance generally feed on various types of small marine animals, especially copepods. Mysids, euphausiids, chaetognaths, salps, urochordates, animal eggs and larvae, dinoflagellates and diatoms (Meyer et al. 1979; Bowman et al. 2000), and fish fry commonly occur in the diet as well. Diet changes with ontogeny (Monteleone and Peterson 1986). Feeding behavior of adults, other than references to feeding schools (Meyer et al. 1979), has not been described. It has been hypothesized, however, that *Ammodytes* is a selective feeder for larger prey items and a filter feeder for smaller prey (Scott 1973a).

Most researchers agree that sand lance prey upon macrozooplankton, but where and when these fishes feed is a matter of controversy. It has been suggested that these visual feeders forage only during the day (Winslade 1974a); however, Scott (1973a) indicated that *Ammodytes* spp. off the Scotian shelf fed close to the surface at night and on the bottom during the day. Scott and Scott (1988) proposed that toward evening "*A. dubius*" leave the bottom to follow and feed upon the plankton as it rises through the water column and return to the bottom at dawn. However, their "*A. americanus*" was thought to feed on and off the bottom, both day and night, but perhaps more intensively at night. Meyer et al. (1979) observed feeding schools in midwater and near the surface but not on the bottom, presumably during the day.

Little is known about seasonal variation in feeding (Reay 1970). Seasonal changes in diet of *Ammodytes* in Long Island Sound were noted, with prey composition mostly affected by fluctuations in abundance and diversity of prey within the plankton (Richards 1963b). It has been suggested that some species (e.g., *A. marinus*) overwinter in the sand; therefore, feeding intensity increases in the fall so that individuals can reach a certain minimum fat content to ensure survival and gonadal development until feeding resumes in the spring (Winslade 1974c). Based on population energy budgets hypothesized for sand lance on Georges Bank (1977–1986), *A. dubius* consumed significant proportions of the total annual zooplankton production (Gilman 1994). Total production (growth + reproduction) for an individual adult is estimated at 10.53 $\text{kcal}\cdot\text{year}^{-1}$, and total consumption is estimated to be 52.62 $\text{kcal}\cdot\text{year}^{-1}$; therefore, ecological efficiency is estimated to be 20% (Gilman 1994).

Predators. Fin and humpback whales consume large quantities of sand lance (Overholtz and Nicolas 1979). Payne et al. (1986) hypothesized that the observed humpback whale distribution in the Gulf of Maine was due to distribution of sand lance, since sightings of humpback whale were significantly correlated with the number of sand lance captured. Bigelow and Schroeder reported that fin whale devoured sand lance greedily when they found them in abundance. For example, they reported that sand lance appeared in great numbers early in June 1880 and that their arrival was followed by a return of fin whales just a few days later. Atlantic harp seal also include *A. dubius* in both their inshore and offshore diets (Lawson and Stenson 1997), while porpoise deem them a staple article of food. Various predacious fishes include sand lance in their diets as well. Fish predators of offshore sand lance taken in the NEFSC trawl surveys include 22 species, seven of which fed on sand lance in the Gulf of Maine: spiny dogfish, winter skate, silver hake, haddock, red hake, white hake, and halibut (Bowman et al. 2000).

Parasites. No extensive investigation has been conducted on parasites of *Ammodytes;* however, Scott (1973b) examined 438 specimens and found six intestinal helminth parasites. These included three species of trematodes (*Derogenes varicus, Brachyphallus crenatus,* and *Lecithaster gibbosus*); two types of cestodes (*Bothriocephalus scorpii* and tetraphyllid larva); and a nematode, *Thynnascaris* (*Contracaecum*) *aduncum.* All parasites occurred at very low levels of intensity and low to moderate incidence.

Breeding Habits. A significant percentage of sand lance mature at the end of their second year of growth but a small proportion of individuals mature at the end of the first year (Winters 1989; Nelson and Ross 1991). On Georges Bank onset of gonadal maturation began in late July. By late October both Stage II (developing) and Stage III (ripe) individuals constituted 79% of sand lance collected from Georges Bank (Nelson and Ross 1991). Stages IV (spent) and V (recovering) individuals were identified in March–May in all regions studied, including the Gulf of Maine (Nelson and Ross 1991). Based on monthly changes in the GSIs of all mature sand lance over time, it has been suggested that males mature earlier than females (Nelson and Ross 1991). Mean weight loss after spawning has been estimated at 30% (Scott 1972; Smigielski et al. 1984).

Fecundity estimates of 1,169–22,904 eggs per female were reported for *A. dubius* 1–4+ years old and ranging in size from 137 to 213 mm TL (Nelson and Ross 1991). Nelson and Ross (1991) also demonstrated that fecundity was related best to total weight, total length, and age: $F = 709.632W_T - 683.651$; $\log F = 3.867 \log L_T - 4.728$; and $F = 5582.750A - 6759.560$.

Spawning has not been observed. However, based on maturity and fecundity observations, sand lance spawn once a year in mid-autumn to early winter (Nelson and Ross 1991). Occurrence of specimens with ripe gonads in October (Nelson and Ross 1991) and the presence of spent gonads in November suggest that spawning begins as early as October (Winters 1989). Spawning peaks about December, January and ends

February–March (Scott 1968; Richards 1982; Winters 1983). Larvae are abundant and widely distributed in the plankton over the Scotian Banks particularly in February–March and remain until July (Scott 1972). Mean GSI for females is always higher than that for males in spring and summer but the converse is true in autumn, suggesting the males mature earlier than females (Nelson and Ross 1991).

Larval sand lance are widespread, occurring over the Nova Scotian Banks, in the Gulf of St. Lawrence northward nearly to the Strait of Belle Isle, throughout the Grand Bank region, off the east coast of Newfoundland and the outer coast of Labrador, north to Sandwich Bay, and at least as far south as lower Chesapeake Bay. It has been suggested that sand lance spawn coastally in Newfoundland waters (Dalley and Winters 1987; Winters and Dalley 1988); but based on the distribution of 5- to 6-mm larvae Norcross et al. (1961) estimated that spawning took place in 9–12 m of water. Larvae have also been collected in abundance in the Georges Bank area at least 200 km offshore (NEFSC survey data). Sherman et al. (1984) regard sand lance as ubiquitous spawners that appear to maintain relatively high densities of eggs over a wide temporal and spatial range within the continental shelf ecosystem, thereby enabling them to respond rapidly to favorable environmental conditions. They also suggest that, owing to their relatively short life span and generation time, sand lance are capable of rapid population expansion under favorable conditions.

Early Life History. No studies similar to those conducted for *A. americanus* on early life–history parameters have been done for *A. dubius,* so information is limited and more speculative. Presumably early life–history stages and habits of the western North Atlantic species are similar. Since *A. dubius* is found most frequently in offshore environments (Nizinski et al. 1990), this sand lance may spawn in deeper water on sandy substrates, where eggs can stick fast to the grains of sand, as was reported for the European *A. tobianus* (Ehrenbaum 1904). Eggs are demersal, adhesive, and irregularly shaped, being generally subspherical and ranging from 0.87 to 1.23 mm in diameter with a single yellow oil globule (Scott 1972).

Larvae from northern regions or from offshore develop more slowly, are generally thinner, and have higher meristic values (e.g., numbers of vertebrae and dorsal and anal fin rays) than similar larvae developing in inshore regions (Richards 1965). Larvae are slender and range from 3 to 6 mm at hatching (Richards 1982); oil-globule absorption occurs between 5 and 7.5 mm. Dorsal and anal fin rays begin development when a larva is about 9–16 mm but it is not until the little fish is upward of 25 mm that the tail begins to assume its forked outline. Larvae of different lengths are characterized by differences in numbers and distribution of melanophores. Scott (1972) has described larval pigment development. The most obvious differences between western North Atlantic species are differences in the rate of melanophore development. Thus, identification of individuals at the species level is difficult. The *A. americanus* account above should be consulted for more information regarding early life history and larval development.

Age and Growth. Age-length curves based on otolith analysis were given by Winters (1989) for male and female *Ammodytes* from Newfoundland and by Nelson and Ross (1991) for the New England region. The most rapid growth occurred in the first 3 years, with growth rates decreasing between ages 4 and 5. Seasonally, growth is much more rapid during the summer months on Georges Bank (Nelson and Ross 1991). Consequently, deposition of opaque material in otoliths of *A. dubius* occurs from spring through early autumn but peaks during the summer (Nelson and Ross 1991). Otolith analysis, however, may not be conclusive since both Scott (1973c) and Winters (1981) had problems back-calculating age at length. Nelson and Ross (1991) suggested that vertebrae are more suitable for age determination. Mean length at maturity is estimated at 175–180 mm (Winters 1989). Onset of gonadal maturation began on Georges Bank in late July. By October both Stage II (developing) and III (ripe) individuals represented 79% of the Georges Bank samples; Stage IV (spent) and V (recovering) individuals were identified in March–May throughout the range (Nelson and Ross 1991). Males mature earlier each year than females (Nelson and Ross 1991).

Parameters of the von Bertalanffy growth equation calculated for sand lance collected in the Gulf of Maine (Nelson and Ross 1991) were estimated to be: $K = 0.31$ year^{-1}; $L_\infty = 239$ mm; and $t_0 = -0.65$ year.

The length-weight relationship exhibits seasonal and regional differences owing to abiotic (temperature) and biotic (food availability) factors (Nelson and Ross 1991). Body growth should thus be more rapid in the season and region with higher temperatures and temporally more abundant food resources. The length-weight relationship may also be affected by the reproductive maturity cycle (Nelson and Ross 1991). For example, increased gonadal weights during autumn result in length-weight curves similar to those calculated during the summer months (Nelson and Ross 1991).

Growth patterns in sand lance seem to be affected by apparent changes in growth rate with age and growth compensation (attributed to differing ages of maturity based on initial growth rate and discriminatory mortality of faster-growing individuals), as well as environmental factors (Winters 1981). *Ammodytes dubius* from Georges Bank appear to grow more slowly than sand lance living in more northern regions, thus implying a decrease in maximum size, age, length at maturity, and growth rate of *A. dubius* with decreasing latitude (Nelson and Ross 1991). Sand lance have been reported as attaining sizes as large as 256 mm on Georges Bank (Nizinski et al. 1990), 372 mm in Nova Scotian waters (Scott 1968), and 288 mm off Newfoundland (Winters 1983) and reaching 9–10 years of age (Scott 1973c; Winters 1983). Comparisons of growth curves among these three regions indicate that Georges Bank sand lance are younger and smaller in length than sand lance captured in Canadian waters (Nelson and Ross 1991).

General Range. *Ammodytes dubius* are found in deeper, more offshore waters ranging from Cape Hatteras to Greenland (Nizinski et al. 1990). Sand lance are locally plentiful, both inshore and offshore, southward through New Jersey to Cape May at the entrance to Delaware Bay.

Occurrence in the Gulf of Maine. Sand lance are abundant along the coast from Cape Cod to Cape Sable wherever there are sandy shores but are seldom seen off rocky portions of the coast. There is no reason to suppose that they inhabit the deep central waters of the Gulf of Maine regularly since most landings occur in water less than 100 m (Map 31), but they are plentiful on Nantucket Shoals and over the shallows of Georges and Browns banks.

Further north, fishermen are familiar with sand lance all along the outer coast of Nova Scotia and on the Scotian Banks; they are reported from Prince Edward Island and the Magdalens, the Strait of Belle Isle, and Sandwich Bay and Sloop Harbor in southeastern Labrador, and they can be expected all along the outer Labrador coast. They have also been found on the southern side of Hudson Strait and in southern and western parts of Hudson Bay.

ACKNOWLEDGMENTS. Drafts of the ammodytid section were reviewed by Thomas A. Munroe and Gary Nelson.

Mackerel-like Fishes. Suborder Scombroidei

The Scombroidei is a suborder of the Perciformes containing 7 families, 46 genera, and about 140 species (Collette et al. 1984; Johnson 1986; Nakamura and Parin 1993). All species are marine; most are pelagic, some epipelagic and some meso- to bathypelagic.

Scombroids are perciform fishes with epiotics separated by the supraoccipital, gill membranes free from the isthmus, premaxillae beaklike, upper jaw nonprotrusile (except in *Scombrolabrax*), predorsal bones lost (except for a small one in three genera of Gempylidae and three well-developed ones in *Gasterochisma*), second epibranchial extending over the top to the third infrapharyngobranchial (except in *Gasterochisma*), vertebrae 24 or more, interorbital commissure of the supraorbital canals widely incomplete or absent. The six families recognized (Collette et al. 1984b; Nakamura and Parin 1993) are: Scombrolabracidae (monotypic); Gempylidae (16 genera, 23 species); Trichiuridae (9 genera, about 32 species); Xiphiidae (monotypic); Istiophoridae (3 genera, 10–11 species); and Scombridae (15 genera, 53 species [Collette 1999]). Barracudas, Sphyraenidae, are included in the Scombroidei as the sister group of all other scombroids in an alternative phylogeny of the suborder (Johnson 1986).

Representatives of all the families except the Scombrolabracidae occur in the Gulf of Maine. There are reports of nine species of Scombridae, one each from the Sphyraenidae, Gempylidae, Trichiuridae, and Xiphiidae, and three Istiophoridae.

BARRACUDAS. FAMILY SPHYRAENIDAE

RODNEY A. ROUNTREE

Barracudas resemble freshwater pikes (Esocidae) in general appearance but differ in having two dorsal fins, the first with five spines, the second with one spine and eight or nine soft rays. Gill rakers are absent. There are about 20 tropical and temperate species of Sphyraenidae, all in *Sphyraena*, four or five of which occur in the western Atlantic (de Sylva 1963). Barracudas have traditionally been thought to be most closely related to the Mugilidae, Polynemidae, and Atherinidae. However, Johnson (1986) argued for their placement within the Scombroidei, with closest relations to scombrids and *Pomatomus*. Two species, *S. borealis* and *S. guachancho*, have been collected in or near the Gulf of Maine. An adult 541-mm guaguanche, *Sphyraena guachancho* Cuvier 1829, was reported from Woods Hole by Goode and Bean (1880a; USNM 21226) so it is included in the key.

KEY TO BARRACUDAS LARGER THAN 12 MM NEAR THE GULF OF MAINE

1a. Dorsal fin origin above pelvic fins; fleshy knob at tip of lower jaw; pectoral fin does not reach origin of pelvic fin; 115–130 lateral line scales; teeth conical, widely spaced .... **Northern sennet**

1b. Dorsal fin origin behind pelvic fins; no fleshy knob at tip of lower jaw; pectoral fin reaches origin of pelvic fins; 108–114 lateral line scales; teeth flattened, angled backward......................... **Guaguanche**

NORTHERN SENNET / *Sphyraena borealis* DeKay 1842 / Northern Barracuda /

Bigelow and Schroeder 1953:306–307

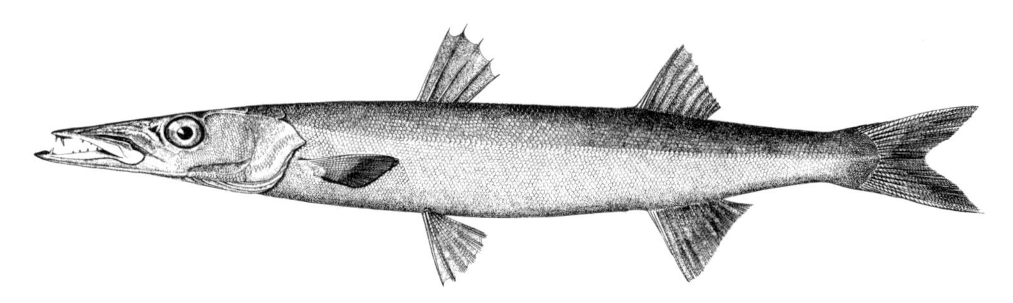

Figure 260. Northern sennet *Sphyraena borealis.* Woods Hole, Mass. Drawn by H. L. Todd.

Description. Body slender, elongate, and fusiform (Fig. 260). Head long, lower jaw projecting beyond upper, fleshy knob at tip of lower jaw. Both jaws studded with large pointed teeth of unequal sizes. First dorsal fin origin above pelvic fins. Anal fin roughly opposite second dorsal. Pelvic fins opposite first dorsal. Pectoral fins short, not reaching origin of pelvic fins. Caudal fin forked. Gill covers scaly; no spines on opercular bones. Gill membranes free from isthmus. Lateral line well developed. Scales cycloid and deciduous.

Meristics. Dorsal fin rays V or VI, I, 8 or 9; anal fin rays I or II, 8 or 9; pectoral fin rays 11 or 12; lateral line scales 115–130; branchiostegals 7; vertebrae 12 + 12 = 24 (Ditty et al. 1999).

Color. The adult is olivaceous above, silvery below, with dusky dorsal and anal fins. The young have dusky blotches along the back and along the lateral line.

Size. The smallest of the northwest Atlantic barracudas, reaching a maximum size of 46 cm but most average less than 30 cm.

Distinctions. Combination of slender shape, with a long head, projecting lower jaw, a first dorsal with five spines opposite the pelvic fins, a second dorsal opposite the anal, and a forked tail separates northern sennet from all other Gulf of Maine fishes, except for the very similar guaguanche (see key).

Habits. Little is known about the behavior or habitat preferences of adult northern sennet but they are generally assumed to be pelagic inhabitants of the continental shelf (de Sylva 1963, 1984).

Juveniles are frequently reported from estuaries on the east coast during the summer (Bean 1903; Smith 1907; Tracy 1910; Pearcy and Richards 1962; Dahlberg 1972; Derickson and Price 1973; Wang and Kernehan 1979). Juveniles (46–188 mm SL) have been commonly collected in subtidal and intertidal marsh creek habitats of southern New Jersey from June to September (Rountree and Able 1992a; Rountree et al. unpubl.), almost exclusively in day samples (Rountree and Able 1993), which suggests diurnal activity. Northern sennet collected in marsh creeks grew an estimated 1.2 mm SL per day and 0.3 g body weight per day during estuarine residence, based on length-frequency analysis (Rountree 1992). Juveniles travel in small schools of up to six individuals (R. Rountree, pers. obs.); however, solitary individuals often hunted among the *Spartina* grass along the marsh fringe at high tide (K. J. Smith, Rutgers University, pers. comm.). Individuals or small schools appear to move into marsh creek habitats during flood tide stages. Occasionally, they may even move up onto the high marsh during spring tides (R. Rountree, pers. obs.).

Food. Although little is known of feeding habits, northern sennet are known to feed on silversides, *Menidia* spp., other fishes, mysids, and gastropods (Smith 1907; Sumner et al. 1913; de Sylva 1963; R. Rountree, pers. obs.).

Reproduction. Spawning probably occurs offshore in the winter, since Houde (1972) collected eggs and larvae in the Florida Current during that season. However, no data on the temporal or spatial range of spawning are currently available. Development of wild-caught eggs and larvae was described by Houde (1972). Eggs hatch at 2.6 mm SL and larvae reach 13.5 mm SL by 21 days. The characteristic fleshy tip on the lower jaw develops after 5 mm, as do the teeth. Transfer to the juvenile stage occurs about 13.5 mm SL. Juvenile pigment pattern develops around 12.5–14.5 mm. Larvae began swimming two days after hatching and began to feed by day 3. They assume an S-flex position and dart out at prey. Captured prey were vigorously shaken before being turned and swallowed either head or tail first. Larvae larger than 9 mm appeared to prefer fish larvae as prey. Northern sennet larvae became aggressive toward each other after 11 days.

General Range. Northern sennet are distributed from Nova Scotia to Panama and questionably to South America (Martin and Drewry 1978), but their distribution south of North Carolina is unclear because of taxonomic confusion with *S. picudilla,* which is sometimes synonymized with *S. borealis.*

Occurrence in the Gulf of Maine. Bigelow and Schroeder reported a single 51-mm specimen from Nauset Beach, Cape Cod. They also reported that young fry "a few inches long" are sometimes collected in Vineyard Sound and in Buzzards Bay during the summer and fall. Young (5.1–15.2 cm) have been reported as very common in the summer as far north as Cape Cod (Bean 1903) and Woods Hole, Mass. (Tracy 1910). A single 62-mm specimen was taken near Schoodic Point, Maine (DeWitt et al. 1981) and another 57 mm-specimen from Halifax Harbor, N.S. (Scott and Scott 1988).

SNAKE MACKERELS. FAMILY GEMPYLIDAE

BRUCE B. COLLETTE

Snake mackerels are closely allied to true mackerels, the most obvious differences being that they lack the keels on the sides of the caudal peduncle so characteristic of mackerels. Body oblong or elongate, compressed; maxilla exposed; strong anterior canine teeth present; base of spinous dorsal fin longer than soft dorsal; three anal spines except *Rexea* and *Nealotus* with two spines and *Ruvettus* with no anal spines; pelvic fins I, 5, reduced to only a spine or absent; caudal fin present; vertebrae 32–58; anterior precaudal vertebrae without parapophyses, with sessile ribs; posterior precaudal vertebrae with ribs attached at the extremities of closed hemal arches. The family currently includes 23 species in 16 genera (Nakamura and Parin 1993). Only *Ruvettus,* oilfish, have been reported from the Gulf of Maine.

OILFISH / *Ruvettus pretiosus* Cocco 1830 / Escolar / Bigelow and Schroeder 1953:349–350

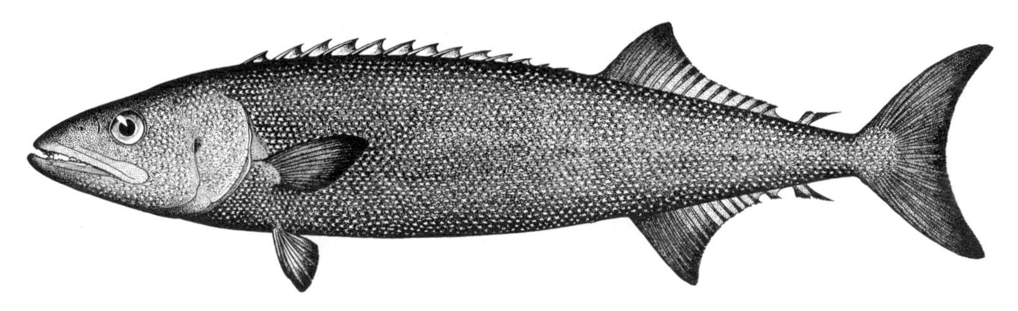

Figure 261. Oilfish *Ruvettus pretiosus.* Georges Bank. Drawn by J. C. van Hook. (From Bigelow and Schroeder: Fig. 185.)

Description. Body slender, fusiform, slightly compressed, depth 4.3–4.9 in SL (Fig. 261). First (spiny) dorsal fin much longer and much lower than second dorsal; both fins followed by two finlets. Caudal fin deeply forked, without caudal keels. Anal fin situated below second dorsal, which it parallels in its outline. Small cycloid scales on body interspersed with rows of sharp, spiny tubercles. Abdominal row of scales forming keel between pelvic fins and anus.

Meristics. Dorsal fin rays XIII–XV, 15–18, plus 2 finlets; anal fin rays 15–18 plus 2 finlets; pectoral fin rays 14 or 15; vertebrae 16 + 16 – 32 (Nakamura and Parin 1993).

Color. Purplish brown, darkest above, with blackish patches; tip of pectoral and pelvic fins black; inside of mouth dusky.

Size. Oilfish grow to 3 m TL and a weight of at least 63 kg. The all-tackle game fish record is a 63.50-kg fish caught at [illegible]

Distinctions. Oilfish are separable at a glance from all Gulf of Maine scombroids by having only two dorsal and two anal finlets and the skin set with bony plates armed with short spines instead of being velvety with small scales, as is the case with mackerels. No keels at the base of the caudal fin.

Biology. Oceanic, benthopelagic on continental slope and sea rises from 100 to 700 m (Nakamura and Parin 1993). Usually solitary or in pairs near sea bottom. Feed on fishes, squids, and crustaceans. Stomach contents of one fish (identified only as a gempylid) contained a well-digested unidentifiable fish (Maurer and Bowman 1975). Gempylidae are preyed on by swordfish (Bowman et al. 2000).

General Range. Widely distributed in tropical and temperate waters of the world (Nakamura and Parin 1993: Fig. 93). They are plentiful around Cuba though not reported from Puerto [illegible]

Ireland in the east and to the Grand Banks of Newfoundland in the west.

Occurrence in the Gulf of Maine. Two specimens, 1.2 m and 1.9 m long, were brought in to the U.S. Fish Commission from Georges Bank during autumn of 1891 (Goode and Bean 1896). They have not been seen in the Gulf of Maine since then. The nearest record to the south is an 882-mm SL specimen (MCZ 145256) taken on the southern New England slope near Block Canyon in 1995. A specimen was taken in 1952 off Sable Island Bank, 43°22′ N, 60°32′ W in 600 m (Leim and Scott 1966).

Importance. Of no importance in the Gulf of Maine region but appear as by-catch in the tuna longline fishery. Flesh very oily, with purgative properties. There are regular fisheries for oilfish off Cuba, about the Canaries, and in the Pacific.

CUTLASSFISHES. FAMILY TRICHIURIDAE

BRUCE B. COLLETTE

Cutlassfishes are characterized by a scaleless, band-shaped body tapering to a slim, pointed tail, with one dorsal fin extending the whole length of the body; the anal is also long but very low. The large mouth is armed with strong teeth of various sizes. Body elongate, strongly compressed; maxilla sheathed by preorbital; anterior canine teeth strong; only one nostril on each side of the snout; spinous dorsal not longer than soft dorsal (very slightly longer in occasional specimens of *Aphanopus*); two anal spines immediately posterior to the vent; pelvic fins reduced to I, 1 or absent; dorsal spines and interneurals correspond in number to vertebrae; dorsal soft rays correspond to or are slightly more numerous than vertebrae; vertebrae numerous, 98–192; ribs feeble, sessile. Most species are benthopelagic on shelves and slopes of tropical seas.

The family contains nine genera and at least 32 species (Nakamura and Parin 1993). There are early records of Atlantic cutlassfish from the Gulf. *Benthodesmus simonyi* and *B. tenuis* occur on the slope of Georges Bank and in southern New England waters and the former may venture into the Gulf (J. Moore, pers. comm., Dec. 1998). The recently described crested scabbardfish, *Lepidopus altifrons,* is also included in the key because there are records of it off Nova Scotia and on the edge of Georges Bank (Parin and Collette 1993).

KEY TO GULF OF MAINE CUTLASSFISHES

1a. Caudal and pelvic fins absent; dorsal fin rays 130–135; anal fin reduced to 100–105 minute spinules usually embedded in skin . . . . **Atlantic cutlassfish**

1b. Small forked caudal fin present; pelvic fins with one small, scalelike spine and 1 or 2 tiny soft rays, inserted behind end of pectoral fin base; dorsal fin rays 90–109; anal fin with two spines and 52–102 rays. **2**

2a. Head profile rising very gradually from tip of snout to dorsal fin origin without forming a sagittal crest; a notch between spinous and soft dorsal fins; second dorsal fin rays 104–109. . . . . **Simony's frostfish**

2b. Head profile with a prominent sagittal crest; no notch between spinous and soft dorsal fins; total dorsal fin elements 90–96 . . . . **Crested scabbardfish**

ATLANTIC CUTLASSFISH / *Trichiurus lepturus* Linnaeus 1758 /

Cutlassfish, Hairtail, Scabbardfish / Bigelow and Schroeder 1953:350–351

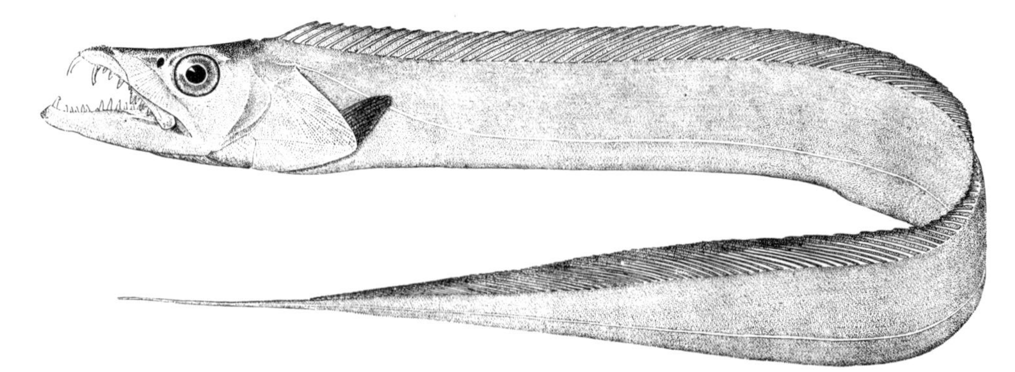

Figure 262. Atlantic cutlassfish *Trichiurus lepturus.* Florida. Drawn by H. L. Todd.

Description. Body bandlike, tapering to a pointed whiplike tail without a caudal fin (Fig. 262). Body depth about one-thirteenth to one-fifteenth in TL. Snout pointed, mouth gaping back to below eye and lower jaw projecting beyond upper. Jaws with long barbed fangs in front, four in upper, two in lower jaw, seven to ten smaller teeth behind fangs. Dorsal fin long, originating close behind eyes, about two-thirds as high at midlength as body depth, diminishing to nothing some distance in front of tip of tail. Anal fin composed of minute spinules, usually embedded in skin. Pectoral fins small, situated a little in advance of rear corners of gill covers. Skin naked.

Meristics. Dorsal fin rays III, 130–135; anal fin rays II, 100–105; pectoral fin rays I, 11–13; gill rakers 10–22; vertebrae (39 or 40) + (123–128) – 162–168 (Nakamura and Parin 1993).

Color. Plain silvery all over. Dorsal fin plain yellowish or dusky green in life, dark-edged or speckled along the margin with black; tips of jaws dusky.

Size. Maximum 120 cm total length, commonly 50–100 cm. The all-tackle game fish record is a 3.68-kg fish caught at Rio de Janeiro in September 1997 (IGFA 2001).

Distinctions. The only other cutlassfishes that might occur in the Gulf, *Lepidopus altifrons* and *Benthodesmus simonyi,* both have small caudal and pelvic fins.

Habits. Adults are benthopelagic, inhabiting the continental shelf to 350 m (Nakamura and Parin 1993).

Food. Young and immature fish feed mostly on euphausiids, small pelagic crustaceans, and small fishes (Nakamura and Parin 1993). Adults become more piscivorous and feed on a wide variety of fishes and occasionally on squids and crustaceans. Stomachs of 11 cutlassfish caught between Cape Hatteras and Cape Cod contained mostly unidentified fishes (69.2% by weight) and a sergestid shrimp, *Acetes* (20.4%), as well as mysids and a mud shrimp *Callianassa* (Bowman et al. 2000).

Breeding Habits. Spawning occurs offshore in the Gulf of Mexico at depths greater than 46 m (Dawson 1967). Adult females produce 33,000–85,000 eggs (Tsukahara 1961). The eggs are pelagic, 1.7–1.9 mm in diameter with one oil droplet 0.4 mm. Yolk-sac larvae hatch at 5.5–6.5 mm and reach 7 mm in 3 days. A summary of development is presented by Fritzsche (1978: 57–58).

Age and Growth. Growth, estimated from length-frequency analyses, approximates 250 mm for age-0 fish in the Gulf of Mexico (Dawson 1967). Early age-1 and age-2 fish attain lengths of 400 and 700 mm, respectively. Length-weight data on 47 fish, 217–607 mm, were fitted to the length-weight equation: $\log W = \log c + n \log L$, with W = weight in grams and L = total length in mm. Calculated values for $\log c$ and n were −7.38983 and 3.43407, respectively (Dawson 1967).

General Range. Worldwide in warm seas (Nakamura and Parin 1993: Fig. 200); abundant in the West Indies and Gulf of Mexico; not rare along the south Atlantic coast of the United States, occasionally straying as far north as Massachusetts Bay.

Occurrence in the Gulf of Maine. Only an occasional cutlassfish is seen north of Cape Cod. One was taken at Wellfleet in the summer of 1845 and one in Salem Harbor many years ago, and it was recorded from Lynn by Kendall (1908). The Massachusetts Bay and Provincetown records listed by Kendall (1908) are based on the Wellfleet specimen. Kendall (1908) also recorded it from Monhegan Island, Maine, quoting Storer as his authority, but Storer stated that only two had come to his notice: the Wellfleet specimen and one taken at the head of Buzzards Bay. Two specimens (CU 22582, 25542) were taken in Buzzards Bay in July 1951. The most recent record from the Gulf of Maine is a 28-cm TL specimen captured on NMFS survey cruise 8102, station 394, 43°54′ N, 67°43′ W at 208–236 m in May 1981, but no voucher was saved.

Importance. Although not utilized for food in the United States, cutlassfish are marketed in the Caribbean, the Orient, and elsewhere. They represent a significant portion of the industrial bottom-fish production in the Gulf of Mexico.

SWORDFISH. FAMILY XIPHIIDAE

BRUCE B. COLLETTE

The Xiphiidae is monotypic, containing only *Xiphias gladius.* The upper jaw and snout are greatly prolonged, forming a flat, sharp-edged sword. The first dorsal fin is very high, the second is very small, and both are soft-rayed; the tail is broad and lunate; two separate anal fins are present, the second very small, with a strong longitudinal keel on either side of the caudal peduncle. Pelvic fins and girdle are absent. Adults have neither teeth nor scales.

SWORDFISH / *Xiphias gladius* Linnaeus 1758 / Broadbill / Bigelow and Schroeder 1953:351–357

Description. Body stout, only slightly compressed, deepest just behind the gill openings, and tapering rearward to a slender keel on either side (Fig. 263). Upper jaw prolonged into long, flattened, sharp-edged and pointed "sword" occupying nearly

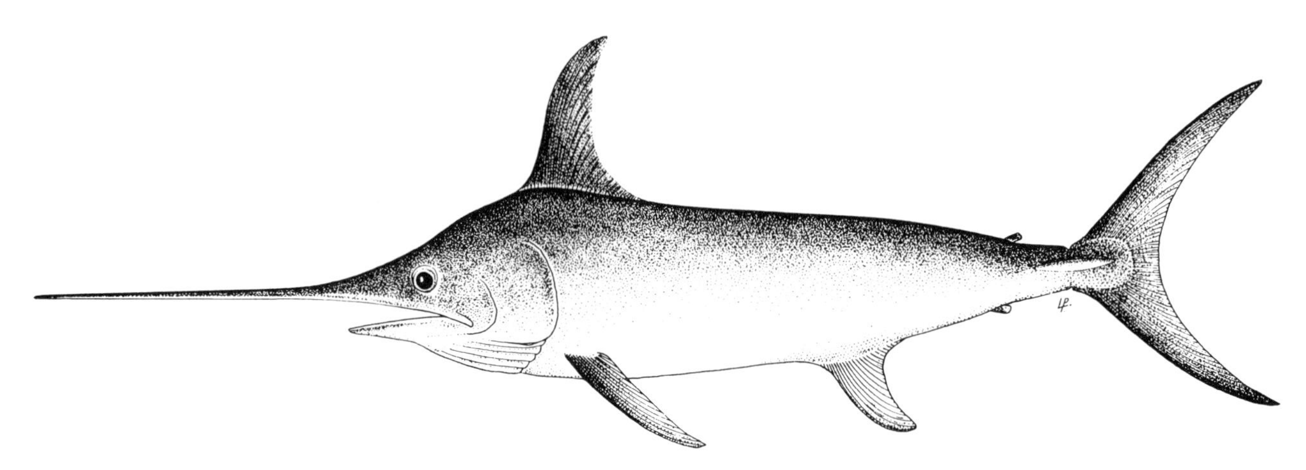

Figure 263. Swordfish *Xiphias gladius.* Drawn by P. Lastrico. (Reprinted from Nakamura 1985:48. © Food and Agriculture Organization of the United Nations.)

pointed and mouth so wide that it gapes far back behind very large eyes, which are set close to base of sword. Fine filelike teeth and scales with small spines present in juveniles but are lost with growth at about 1 m in length. First dorsal fin originates over upper angle of gill openings, much higher than long, with deeply concave rear margin. Second dorsal fin very small, set far back on caudal peduncle. Two anal fins, second as small as second dorsal, located below latter, first similar to first dorsal in outline but shorter and located well behind it, close to second anal. Pectoral fins narrow, very long, scythe-shaped, and set very low down on sides below first dorsal. Caudal fin short, as broad as one-half the length of the fish from tip of lower jaw to base of caudal fin, with deeply lunate margin and pointed tips.

Meristics. First dorsal fin rays 34–49, second dorsal fin rays 4–6; first anal fin rays 13 or 14, second anal fin rays 3 or 4; pectoral fin rays 16–18; vertebrae (15 or 16) + (10 or 11) = 26 (Nakamura 1985).

Color. Swordfish are dark above and whitish with silvery sheen below, but the upper surface varies from purplish to a dull leaden blue or even to black. The eye has been described as blue. Very young swordfish, like very young tuna, are transversely barred, but none small enough to show this pattern has ever been found within the limits of the Gulf of Maine. The colors fade soon after death.

Size. Swordfish reach a maximum length of about 445 cm TL and weight of about 540 kg (Nakamura 1985). Only females attain weights of 540 kg; the males reach about 120 kg. The all-tackle game fish record is a 536.15-kg fish caught off Iquique, Chile, in May 1953 (IGFA 2001). The average weight of fish seen in commercial landings from the northwest Atlantic is only 54 kg (Beckett 1974). The largest fish from the western Atlantic was a dressed fish that must have weighed about 550 kg alive, which was landed in Cape Breton (Beckett 1974). A fish of 2.2 m weighs about 55 kg, one of 3.1–3.4 m weighs about 114 kg, and one of 4–4.2 m weighs about 273–318 kg.

Distinctions. Billfishes (Istiophoridae) are the only other family represented in the Gulf of Maine fauna that resemble swordfish, but they have long pelvic fins, minute teeth, bills that are round in cross section, and either one long continuous dorsal fin or two, with the first several times as long, relatively, as in swordfish.

Habits. Swordfish are oceanic, not dependent in any way either on the coast (except as this offers a supply of food) or on the bottom; they are also warm-water fish, most plentiful in localities and at depths where temperatures are higher than about 13°C. Occasional captures of swordfish on halibut lines set near bottom as deep as 372 m, together with the fact that swordfish are by no means rare on the Newfoundland Banks, whence several fish were brought back by the American cod fleet in 1920, proves that temperatures as low as 10°–13°C do not preclude their presence, at least for brief stays. Swordfish descend occasionally into waters of 5°–10°C at depths of 650 m (Nakamura 1985). They have been acoustically tracked in the western Atlantic and eastern Pacific to depths of 617 m (Carey and Robison 1981; Carey 1990). A swordfish tagged and tracked on the Northeast Peak of Georges Bank swam along the bottom at 200 m associated with echo returns from schools of demersal fish (Carey 1990). The next day, this same swordfish followed the deep-scattering layer as it moved down from the surface just before dawn.

The brain and eyes of swordfish are warmer than the water in which they live (Carey 1982). The tissue that heats the brain is developed from the superior rectus eye muscle (Block 1991). The brain heater is rich in mitochondria and cytochrome *c* and is supplied with blood through a vascular heat exchanger. It protects the central nervous system from rapid cooling during daily vertical excursions of as much as 300 m that may take swordfish through a temperature range as broad as 19°C in less than 2 h. Brain heaters in swordfish permit extended periods of foraging beneath the thermocline, opening vast resources of the mesopelagic zone (Block 1991).

Although swordfish may gather in certain localities they do not school, but are always scattered, either singly or at most two fish swimming together (Tibbo et al. 1961). On calm days they often lie quietly on the surface, with both the high first dorsal fin and the tip of the caudal fin above water, so they are easily harpooned, as they often allow a vessel to approach until the pulpit projecting from the bow comes directly above them. When a swordfish is swimming at the surface, its first dorsal fin and the upper part of its tail fin both show above the water, whereas a marlin shows only its caudal. One can tell a surfacing swordfish from a shark by its sharp-pointed dorsal (that of a shark is more broadly triangular) and by the fact that its caudal fin seems to cut the water in a direct line, not wobbling from side to side as the tips of the tails of most sharks do (other than the mackerel shark tribe), if they show above the water at all.

When swordfish are at the surface, they jump a good deal, perhaps in vain attempts to shake off the remoras that so often cling to them. Bigelow and Schroeder saw one leap clear of the water four or five times in rapid succession close to the *Grampus,* off Shelburne, N.S., on 28 July 1914. Reports by fishermen, and Bigelow and Schroeder's experience, indicate that they surface only in daylight.

Many tales are current of swordfish attacking slow-moving vessels without any provocation and driving their swords through the planking, either in "fits of temporary insanity," as Goode (1883) expressed it or, more likely, while pursuing dolphins or other fishes. Most of the attacks of this sort reported from tropical seas seem actually to have been by billfishes but some in northern waters were almost certainly by swordfish (see review by Gudger 1940). A case in point is that of the schooner *Volunteer,* out of Gloucester, which received a strong blow near Block Island, 7 August 1887, apparently from a 136-kg swordfish that was seen swimming alongside, which, when it was harpooned and brought on board, proved to have lost its entire sword. Bigelow and Schroeder found no information on swordfish making unprovoked attacks on any of the fishing vessels that pursued them every summer or on any of the other craft, large or small, that cruise off the coasts of the Gulf of Maine. But fish that have been harpooned often turn on their pursuers, and it is a common event for one to pierce the thin bottom of a dory. Bigelow and Schroeder knew of several fishermen wounded in the leg in this way, but always after the fish had been struck with a harpoon. Under these circumstances swordfish have been known to drive their swords right through the planking of a fishing vessel. A swordfish even attacked the submersible *Alvin* at 654 m (Zarudski and Haedrich 1974), apparently without provocation.

Stories of swordfish attacking whales are time-honored traditions of the sea, mostly with no more foundation than the myth that they ally themselves with the harmless thresher

the nostril of a mackerel shark caught at Gloucester, probably picked up somewhere off southern New England. Swordfish are easily frightened, and they will not often allow a small boat to come within striking range, which made harpooning from dories difficult in the old days (Rich 1947). But they will allow themselves to be almost run down by a larger vessel without paying the least attention to its approach until aroused by its shadow or by the swirl of water under its bow. Storer wrote that swordfish sometimes sound with such speed and force as to drive the sword into the bottom, which fishermen say is by no means uncommon. Bigelow and Schroeder saw this off Halifax in August 1914, when a fish more than 3.1 m long, which they had harpooned from the *Grampus,* plunged with such force that it buried itself in the mud beyond its eyes in 104 m of water. When finally hauled alongside, it brought up enough mud plastered to its head to yield a good sample of the bottom.

Food. Small larvae (7.8 and 9.0 mm) feed on zooplankton whereas larger larvae feed mainly on fish larvae (Arata 1954). Juvenile swordfish feed on squids, fishes, and pelagic crustaceans.

Adult swordfish are opportunistic feeders, known to forage from the surface to the bottom over a wide depth range. Over deep water, they feed primarily on pelagic fishes and squids, while in relatively shallow waters they take chiefly neritic pelagic fishes such as mackerels, herrings, and sauries (see summary of food items from different regions in Palko et al. 1981: Table 1) (Bowman et al. 2000). During their stay in American waters, they feed on mackerel, menhaden, bluefish, silver hake, butterfish, herring, argentines, rattails (*Macrourus bairdii*), and indeed on any smaller fishes, buckets of which have been taken from swordfish stomachs. Squids are often found in their stomachs and may be their chief diet at times. The jaws of one of the giant squids (perhaps *Architeuthis*) taken from the stomach of a swordfish harpooned on the northern edge of Georges Bank was an especially interesting find. Swordfish have been described as rising through schools of mackerel, menhaden, and other fishes, striking right and left with their swords, then turning to gobble the dead or mangled fish, and Bigelow and Schroeder saw them so employed on more than one occasion. Slashes found on the bodies of prey in swordfish stomachs support the idea that they use their swords in prey capture (Nakamura 1985; Stillwell and Kohler 1985).

In the northwestern Atlantic, swordfish feed mostly on squids (67.4% by weight, mostly *Illex illecebrosus*) (Bowman et al. 2000) and fishes (32.5%) as shown by Scott and Tibbo (1968, 1974), Stillwell and Kohler (1985), and Bowman et al. (2000). Fish prey include menhaden, Atlantic herring, barracudinas (Paralepididae), silver hake, bluefish, Atlantic mackerel, butterfish, and Arcadian redfish (Bowman et al. 2000).

Swordfish taken on the offshore banks frequently contain deep-sea fishes in their stomachs, sometimes swallowed so recently that they are still in good condition when the swordfish is opened. Rich (1947) reported the following genera from

Chauliodus, Chiasmodon, Lampadaena, Macrostoma, Myctophum, Notoscopelus, and *Stomias.*

Predators. Longfin and shortfin mako sharks are known predators of swordfish (Bowman et al. 2000). However, full-grown swordfish are so active, so powerful, and so well armed that they have few enemies. Sperm and killer whales and larger sharks alone could menace them. Bigelow and Schroeder found no evidence that swordfish ever fall prey to the first two, but they stated that Captain Atwood found a good-sized swordfish in the stomach of a mako shark. A swordfisherman described seeing two large sharks bite or tear off the tail of a 159-kg swordfish, which he afterward harpooned. A 55-kg swordfish, nearly intact with sword still attached, was found in the stomach of a 332-kg mako taken near Bimini, Bahamas, while another mako of about 364 kg, harpooned off Montauk, Long Island, was seen attacking a swordfish and was found to have about 68 kg of the flesh of the latter in its stomach when it was landed. Rich (1947) noted other similar cases. Young swordfish are a common food source for other fishes, including larger swordfish, marlins, tunas, and dolphins.

Parasites. Swordfish are infested with many parasites besides being accompanied by remoras, several of which are often found clinging to one fish. No less than 12 species of worms and 6 of copepods have been reported from fish taken off Woods Hole. A list of 25 species of parasites was compiled by Palko et al. (1981: Table 1).

Breeding Habits. In the western Atlantic, spawning apparently occurs throughout the year in the Caribbean, the Gulf of Mexico, and waters off Florida, with the peak of spawning from April through September (Nakamura 1985). Spawning also occurs in the Mediterranean Sea and the Sea of Marmara. Swordfish spawn in the upper water layers at depths of 0–75 m, at temperatures around 23°C, and salinities of 33.8–37.4 ppt. Females produce 2–5 million eggs.

Early Life History. The eggs are buoyant and pelagic, 1.60–1.87 mm in diameter with a single oil globule 0.50–0.52 mm. Swordfish larvae occur in all tropical seas closely associated with surface temperatures of 24°–29°C (Tåning 1955). In the western Atlantic, the greatest densities of swordfish larvae occur from the Straits of Florida to Cape Hatteras and in the Virgin Islands–Leeward Islands area (Markle 1974). The fry differ in appearance from their parents, having only one long dorsal fin and one long anal fin, a rounded tail, both jaws prolonged and toothed, and skin covered with rough spiny plates and scales. A swordfish larva is easily recognized by its long snout, heavily pigmented elongate body, and prominent supraorbital crest. Larvae develop four rows of spiny scales on each side with smaller "scatter" scales between the rows (Potthoff and Kelley 1982). Spiny scales begin to appear on the ventral surface at 5.3–6.1 mm orbit to notochord length. By 187 mm orbit to hypural length, scales cover the body and fin rays (Potthoff and Kelley 1982: Fig. 30) and remnants of scales are present at least to 668 mm. Young swordfish differ from billfish larvae (Istiophoridae) in lacking the strong pterotic and preopercular spines so prominent in young billfishes. Figures of juveniles are included in Fritzsche (1978:199–202), Palko et al. (1981: Fig. 5), Potthoff and Kelley (1982: Fig. 30), and Fahay (1983:325).

Age and Growth. Reports of age estimates and growth rates of swordfish are many and appear to vary according to the method of aging used, geographical location, and size and range of samples obtained. Beckett (1974), using weight frequencies from the Canadian swordfish fishery, together with examination of vertebral rings and tagging data, suggested rapid growth rate of female swordfish with weights of 4, 15, 40, 70, and 110 kg for ages 1–5 years. There were three subsequent studies of swordfish aging. A total of 439 swordfish from the Florida Straits, collected between 1978 and 1980, were aged from bands present on thin sections of the second element of the anal fin (Berkeley and Houde 1983). Bands assumed to be annual events were counted under a dissecting scope and used to back-calculate lengths at estimated age, which were used to fit the von Bertalanffy growth model. Differences in size and growth rates of males and females are shown by the resulting parameter estimates: $L_\infty = 217.4$ cm LJFL (lower jaw fork length), $k = 0.19$, and $t_0 = -2.04$ years for males; $L_\infty = 340.0$ cm LJFL, $k = 0.09$, and $t_0 = -2.59$ years for females.

Otoliths (sagitta, lapillus, and asteriscus) from 121 Atlantic swordfish were collected from commercial vessels along the Atlantic coast between Cape Hatteras and Florida during 1981 and used to estimate age (Wilson and Dean 1983). Age and growth were estimated based on external features of otoliths from 303 swordfish captured by longline from Cape Hatteras to the Grand Banks of Newfoundland during the summer and fall of 1980 (Radtke and Hurley 1983). Scanning electron microscope examination of whole and sectioned sagittae otoliths revealed well-defined external ridges and finely spaced internal increments. These finely spaced internal increments were similar to those found in other fish species and their total number in two swordfish support annual formation of the external ridges. LJFL-at-age estimates were analyzed using the von Bertalanffy growth equation and produced estimates of k, t_0, and L_∞ of 0.07, −3.94 years, and 277 cm LJFL for males and 0.12, −1.68 years, and 267 cm LJFL for females, respectively. Swordfish grow very rapidly during the first year of life and then growth slows down considerably so the standard von Bertalanffy growth functions fitted to these data do not express growth correctly (Ehrhardt 1992).

Swordfish live for at least 9 years. Males apparently mature at a smaller size than females, about 21 kg for males and 74 kg for females (Palko et al. 1981). It is not clear whether the larger size attained by females relative to males is due to more rapid growth or to a considerably longer life span (Beckett 1974).

General Range. Cosmopolitan in tropical, temperate, and sometimes cold waters of all oceans including the Mediterranean and Black seas from about 45° N to 45° S (Nakamura 1985:49). In the Atlantic Ocean, north to northern Norway, southern and western parts of the Gulf of St. Lawrence, south coast of Newfoundland and Grand Banks, south to Argentina. Localities where swordfish have been caught or sighted in the northwest Atlantic were mapped by Tibbo et al. (1961: Fig. 6).

Occurrence in the Gulf of Maine. Swordfish seem to have attracted little attention in the Gulf in colonial days, and although they long supported a lucrative fishery off New England, little more of their life there is known today than was known in 1883 when Goode published his *Materials for a History of the Sword-fish*. The outer half of the continental shelf off Block Island and southern Massachusetts, offshore parts of the Nantucket Shoals region, Georges Bank, the deep channel between Georges and Browns banks, Browns Bank and La Have, and banks off the outer coast of Cape Breton were its chief centers of abundance off the coasts of the Gulf of Maine. In the early days of the fishery, some 10,000–20,000 were harpooned every summer off the New England coast, with as many more off eastern Nova Scotia. A few have been harpooned off the Gulf of St. Lawrence shore of Cape Breton. The only other definite report of swordfish in the Gulf of St. Lawrence is from Bonne Bay, on the west coast of Newfoundland.

Migrations. Swordfish, like tunas, are summer fish on the North American coast, and their presence in blue water between the outer edge of the continent and the inner edge of the Gulf Stream proper, off southern New England and the Gulf of Maine, added to the fact that few are seen along the coast south of New York, makes it likely that they come directly in from offshore. They appear almost simultaneously off New York, off Block Island, on Nantucket Shoals, and on Georges Bank, sometime between 25 May and 20 June, but seldom on the Scotian Banks until somewhat later, or in the inner parts of the Gulf of Maine before July. They are most numerous in July and August, and they vanish at the approach of cold weather. So far as Bigelow and Schroeder were aware none was reported east of Cape Cod after the first half of November (in 1875 one was taken on Georges Bank in November in a snowstorm [Rich 1947]), and most of them are gone by the last week in October, although some fish have been taken off New York and New Jersey in December and even in January. Nearly all the swordfish that visit the Gulf are subadults or adults weighing 23–27 kg.

Importance. Appreciation of swordfish as a food fish is rather recent. Down to the middle of the nineteenth century they were unsalable in Boston and brought a very low price in New York, but in recent years demand greatly exceeds available supply. In 1919, the price to the fishermen averaged about $0.24 per pound, in 1946 about $0.60. Today it is a highly de- [illegible]

pound ex-vessel. In the 10-year period 1967–1977, some 25 countries reported total swordfish landings from the Atlantic of 10,000–15,600 mt·year^{-1} (Palko et al. 1981: Table 4).

Practically all swordfish brought to market in Bigelow and Schroeder's time were harpooned; they never heard of one caught in net or seine. Swordfish were also taken from time to time on handlines and on longlines baited for cod or halibut with mackerel or other fish or on tuna longlines. Rich (1947) gave an interesting account of the methods of the New England swordfishery. Now, most swordfish are caught on longlines or in pelagic gill nets.

Bigelow and Schroeder reviewed annual landings for the earlier days of the fishery: The catch reported from Portland, Gloucester, and Boston within the period 1904–1929 ranged from 401,365 to 2,087,725 kg, averaging about 909,090 kg, or between 4,000 and 18,000 fish per year. Landings in New England ports ran from 779,545 to 2,304,545 kg during the decade 1930–1939 for southern New England and the Gulf of Maine. Interruption of swordfishing by German submarines and diversion of manpower was reflected in much lower landings during the first 2 years of World War II. Swordfishing picked up again after the war, to landings of between 900,000 and 1,400,000 kg for southern New England and the Gulf of Maine, including western Browns Bank. Now, about 95% of swordfish caught by U.S. fishermen in the Atlantic are on longlines.

Status of Stocks. In the western North Atlantic, catch-per-unit effort (CPUE) in the Canadian longline fishery declined from 2.88 fish per 100 hooks in 1963 to 0.92 fish per 100 hooks in 1965 (Palko et al. 1981). This decline in CPUE was accompanied by a decrease in average size from 120 kg round weight to less than 60 kg. Part of this decline in average size may have been due to an expansion of the fishery to more southern grounds. In 1975, however, following a 4-year period in which there was little or no fishing because of restrictions on the sale of swordfish owing to mercury contamination, CPUE had risen to 2.31 fish per 100 hooks. CPUE in the Japanese longline fishery for the total Atlantic increased steadily from 1956 to 1968 and then stabilized through 1975 (Palko et al. 1981). East coast landings of swordfish have declined (e.g., 10.23 million lb in 1988 to 6.3 million lb in 1995) and average size also declined from 266 lb in 1963 to 90 lb in 1995. The fishable biomass declined by 68% from 1960 to 1995, and the population is currently 58% of that needed to produce maximum sustained yield (Draft Amendment 1 to the Fishery Management Plan for Atlantic Swordfish).

Management. Restrictions on the sale of swordfish containing levels of mercury greater than 0.5 ppm in Canada and the United States in the early 1970s caused collapse of the Canadian fishery and severely restricted landings in the United States. The mercury guidelines were raised to 1.0 ppm in 1979, and by 1980 catch and effort reached a new high in the north- [illegible]

Recreational Fishery. Until 1976, trolling, after observing swordfish basking at the surface, was the technique used by sports fishermen (Palko et al. 1981). Then, it was discovered that swordfish could be caught by drifting baited lines at night. This substantially increased fishing success and is now a standard method of recreational fishing along the Atlantic and Gulf coasts of the United States.

BILLFISHES. FAMILY ISTIOPHORIDAE

BRUCE B. COLLETTE

Billfishes, like swordfish, have bills formed by prolongation of the snout and upper jaw. But their bills are rounded in cross section, not flattened, and narrower than that of swordfish. Their bodies, too, are closely clothed with narrow lanceolate scales, generally pointing rearward and embedded in the skin, either wholly or with their sharp tips projecting slightly; the first dorsal fin is also much longer, occupying the greater part of the back behind the nape, and can be depressed into a groove along the back. They have two small longitudinal keels on either side of the caudal peduncle instead of the single broad one found in the swordfish. Pectoral fins are placed low on the body and pelvic fins consisting of one spine and two long rays are present. Three genera: *Tetrapturus,* the spearfishes (six species); *Makaira,* the marlins (three species); and *Istiophorus,* the sailfish (one or two species, depending on whether or not one considers the Atlantic sailfish, *Istiophorus albicans* Latreille, a separate species from the Indo-Pacific *Istiophorus platypterus* [Shaw and Nodder 1792], as Nakamura [1985] does).

Two species of marlins, the blue and the white, are known off the middle and north Atlantic coasts of the United States. The sailfish, so common in warmer parts of the Atlantic, is included in the following key because it has been taken at Woods Hole on several occasions, though not yet recorded from the Gulf of Maine.

KEY TO GULF OF MAINE BILLFISHES

1a. First dorsal fin sail-like, much higher than body depth; pelvic fin rays very long, reaching nearly to anal fin origin . . . . **Sailfish**
1b. First dorsal fin not higher than body depth; pelvic fins short, not reaching near anal fin **2**
2a. Apex of first dorsal fin and tips of pectorals pointed . . . . . . **Blue marlin**
2b. Apex of first dorsal and tips of pectorals rounded . . . . **White marlin**

BLUE MARLIN / *Makaira nigricans* Lacepède 1802 / Bigelow and Schroeder 1953:358–360 (as *Makaira ampla*)

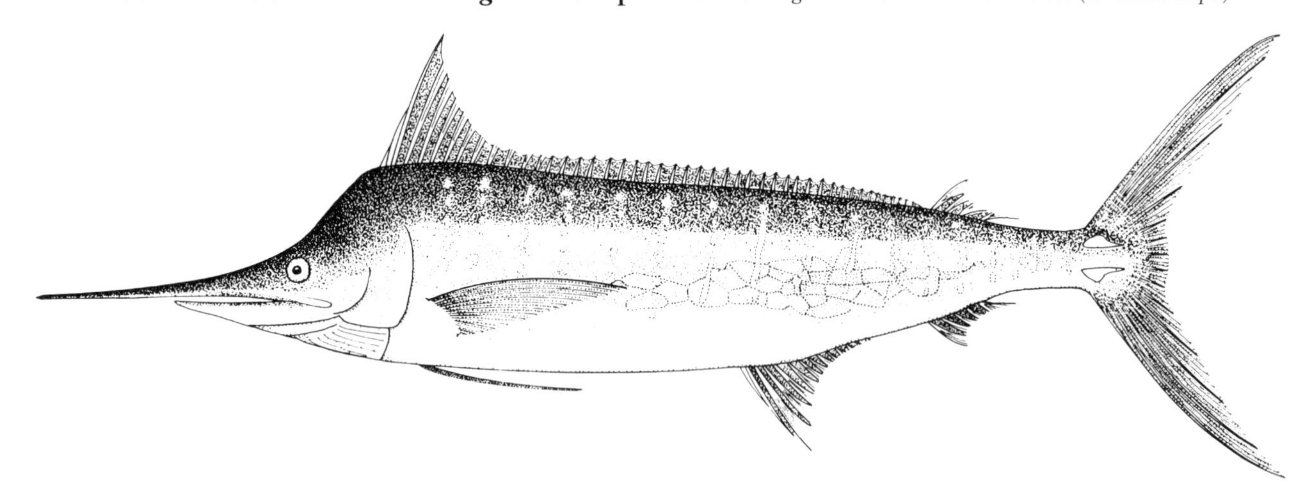

Figure 264. Blue marlin *Makaira nigricans.* Drawn by P. Lastrico. (Reprinted from Nakamura 1985:33. © Food and Agriculture Organization of the United Nations.)

Description. Body deepest at level of pectoral fins, about six and a quarter times as long, not counting caudal fin, as deep; tapering evenly to caudal peduncle (Fig. 264). Upper jaw in front of eye (including bill) about twice as long as length of head behind eye. First dorsal fin slightly separated from second dorsal. First anal fin triangular, situated below rear part of first dorsal. First dorsal and first anal fins with falcate anterior lobes. Short second anal similar to second dorsal fin in position and number of fin rays. Pelvic fins below pectoral fins. Caudal fin resembles that of swordfish in lunate outline. Lateral line visible only after skin is removed and dried, and then a complex reticulate pattern appears.

Meristics. First dorsal fin rays 39–43, second dorsal fin rays 6 or 7; anal fin rays 13–16, second anal fin rays 6 or 7; pectoral fin rays 18–23; pelvic fin rays I, 2; no gill rakers; vertebrae 11 + 13 – 24 (Nakamura 1985).

Color. Dark dull blue on the back and sides down about to the level of the eyes, washed with coppery reflections, also on the bill, with a rather abrupt transition to much paler gray-blue lower down on the sides and on the lower surface, the belly being as dark as the lower part of the sides; the sides cross-marked with some 13–15 indistinct violet-blue stripes, about 2.5–3.8 cm wide on a fish 2.5 m long, showing pale against the dark blue of the upper parts of the body, but dark against the paler blue of the lower part of the sides. First and second dorsal fins, pectoral and pelvic fins, and first anal fin dark, rather vivid blue. Caudal fin of about the same color as upper part of trunk; second anal fin of same pale gray-blue as the belly.

Size. Maximum size exceeds 375 cm body length and 636 kg in weight. The all-tackle Atlantic game fish record is 636.0 kg for a blue marlin taken off Vitória, Brazil, in February 1992 (IGFA 2001). Many weighing more than 227 kg are caught off the north coast of Cuba and on the Bahamas side of the Straits of Florida every year, and one taken on the southern part of Browns Bank weighed 261 kg dressed, when landed, or about 318 kg alive. Fish caught by commercial longliners in the Atlantic range from about 230–345 cm TL (Nakamura 1985).

Distinctions. Blue marlin differ from white marlin in the shape of the apex of the first dorsal fin and of the tips of the pectoral fins, both of which are pointed, as well as in the much darker color of the sides and belly, and in the fact that they grow much larger. Blue marlin differ from black marlin, *M. indica,* by having a flexible pectoral joint so that the fin folds back against the body, instead of being rigid and fixed in position.

Habits. This is an epipelagic oceanic species usually found in waters with surface temperatures of 22°–31°C. Little is known about spawning grounds and seasons, but larvae and juveniles have been taken in warm waters well to the south of the Gulf. A synopsis of biological data was presented by Rivas (1975) and developmental information was summarized by Fritzsche (1978:184–186).

Food. Blue marlin feed mostly in near-surface waters on fishes such as tunas and dolphins and on squids.

General Range. Distributed mainly in tropical and temperate waters of the Atlantic, from about 40° to 45°C in the North Atlantic to about 40°C in the South Atlantic (Nakamura 1985:34).

Occurrence in the Gulf of Maine. This southern, warm-water fish was reported from the South Channel, between Georges Bank and Nantucket Shoals, between 1877 and 1880, by the fishing schooner *Phoenix.* No other marlins that Bigelow and Schroeder could be sure were blues were reported within the limits of the Gulf of Maine until 5 September 1930, when a small one, 2.1 m long, was harpooned on the southern part of Browns Bank. A very large one was caught in that same vicinity by the *Col. Lindbergh* the following July and brought into the Boston Fish Pier. A marlin about 1.5 m long was taken on Georges Bank by the schooner *Ethel Merriam,* on 5 August 1925, but this may have been a white (Scott and Scott 1988). Blue marlin are occasionally sighted off Martha's Vineyard. Fishermen also report them as occurring now and then along the southern edge of Georges Bank (any very large marlin is a blue).

Importance. Blue marlin are game fish par excellence, and much sought after off Cuba and on the Bahamas side of the Straits of Florida. They also support a considerable commercial fishery off the north coast of Cuba and are sought by Japanese longliners.

WHITE MARLIN / *Tetrapturus albidus* Poey 1860 / Bigelow and Schroeder 1953:360 (as *Makaira albida*)

Description and Distinctions. White marlin differ from their larger relative, blue marlin in the rounded first dorsal and pectoral fins, the pale color of the lower part of the sides, the white belly (Fig. 265), and the smaller size.

Meristics. First dorsal fin rays 38–43, second dorsal fin rays 5–7; anal fin rays 13–16, second anal fin rays 5 or 6; pectoral fin rays 17–21; no gill rakers; vertebrae 12 + 12 – 24 (Nakamura 1985).

Size. Reach a maximum size of over 280 cm TL, and over 82 kg in weight, but few grow larger than 57 kg. White marlin caught by commercial longliners range from 130 to 210 cm body length, mostly around 165 cm. The all-tackle game fish record is 82.50 kg for a fish caught off Vitória in December 1979 (IGFA 2001).

Biology. White marlin feed on surface fishes such as herrings, dolphins, jacks, and mackerels and on squids. Spawning is concentrated along the coasts of Cuba, the Greater Antilles, and southern Brazil in deep blue, oceanic waters with high surface temperatures (20°–29°C) and high surface salinities (>35 ppt). A synopsis of biological data was presented by Mather et al.

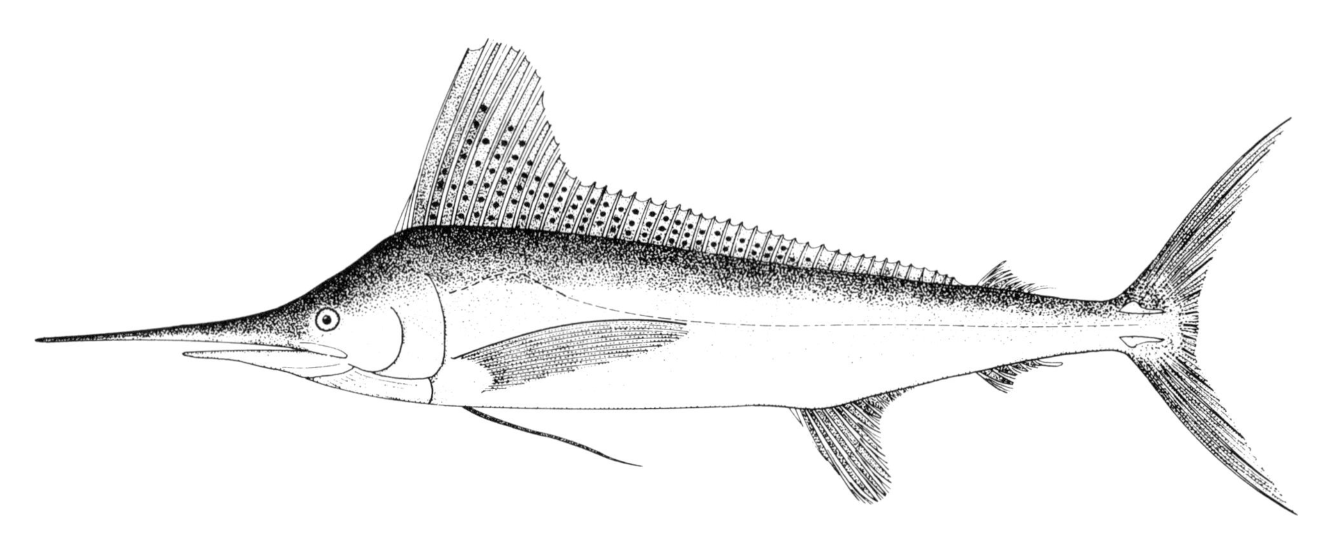

Figure 265. White marlin *Tetrapturus albidus.* Drawn by P. Lastrico. (Reprinted from Nakamura 1985:36. © Food and Agriculture Organization of the United Nations.)

(1975) and developmental information was summarized by Fritzsche (1978:187–189).

General Range. Based on Japanese longliners' catches, the range extends over most of the Atlantic from about 45° N to 45° S in the western Atlantic (Nakamura 1985:36). Common in Cuban and Bahamian waters and off southern Florida; north regularly in summer to the offing of Delaware Bay in abundance and to southern New England waters in lesser numbers.

Occurrence in the Gulf of Maine. So many white marlin come northward as far as New York that about 500 were taken off Montauk, Long Island, on rod and reel during the 11 years from 1925 to 1936, and more than 150 in 1935 alone. A few are caught off southern Massachusetts islands in most summers but their usual turning point is west of Nantucket. Farrington (in Vesey-Fitzgerald and LaMonte 1949) wrote of "great quantities" of them as seen on Georges Bank; but Bigelow and Schroeder could not find any marlin caught there that had been identified positively as white, although one about 1.5 m long taken on 5 August 1925 may have been one. The meager record of occurrences in the Gulf of Maine suggests that they may stray more often to outer Nova Scotian waters, for a 1.5-m fish weighing 9.5 kg, caught on Sable Island Bank, 18 August 1931, was probably a white marlin, while Farrington reported one harpooned off Glace Bay, N.S., in 1945, and others were sighted off Halifax that same year. A 180.7-cm FL individual was caught at 40°49′ N, 63°41′ W, southeast of Georges Bank in July 1966 and another, 164.5 cm TL, was caught at 40°38′ N, 66°18′ W in July 1964 (Scott and Scott 1988).

MACKERELS. FAMILY SCOMBRIDAE

BRUCE B. COLLETTE

The mackerels are a homogeneous group, all of them having a spiny dorsal as well as a soft dorsal fin, 5–12 small finlets behind the latter and behind the anal, a very slender caudal peduncle with two small keels at the base, a deeply forked or lunate caudal fin, a very shapely form tapering both to snout and tail, and velvety skin that is naked or covered with very small scales. The hypural plate is mostly covered by caudal fin rays; caudal fin rays are supported by three to five centra. Pectoral fins are placed high on the body, with 19–36 rays; pelvic fins I, 5; vertebrae 31–64 (Collette et al. 1984b). All are predacious, swift swimmers, powerfully muscled, and are fish of the open sea that are more or less migratory.

The Scombridae contains 15 genera and 53 species in two subfamilies (Collette 1999), the monotypic Gasterochismatinae and the Scombrinae. Scombrinae is composed of two groups of tribes. Primitive mackerels (Scombrini) and Spanish mackerels (Scomberomorini) have a distinct notch in the hypural plate, lack any bony support for the fleshy keels on the caudal peduncle, and do not have preural centra two and three greatly shortened. The more advanced bonitos (Sardini) and tunas (Thunnini) form a monophyletic group showing loss of the notch between fused lower and fused upper hypural bones, bony support for the medial caudal peduncle keel, anterior corselet of enlarged scales, and preural centra two and three greatly shortened. Scomberomorini, like the two more advanced tribes, have a median fleshy keel on the caudal peduncle between the pair of small keels but there is no bony support for it as there is in bonitos and tunas. Sardini differ from Thunnini (*Auxis, Euthynnus, Katsuwonus,* and *Thunnus*) in lacking any trace of the subcutaneous vascular system that enables species of Thunnini to be warmer than the water around them.

Summaries of biological information are included with accounts of species of scombrids in Collette and Nauen (1983). Development for all scombrid species found in the Gulf is summarized by Fritzsche (1978) and Fahay (1983). Identification of early stages is presented by Richards (1989). Parasitic copepods are listed by Cressey and Cressey (1980) and Cressey et al. (1983). Recent information on the biology of Atlantic mackerel and of chub mackerels (*S. colias* and *S. japonicus*) was summarized by Studholme et al. (1999) and Castro-Hernández and Santana Ortega (2000), respectively. Information from these sources is included in the species accounts.

In the following key all species actually recorded from within the limits of the Gulf of Maine so far are included, but it would not be astonishing if others were to stray in from the open Atlantic. For example, the albacore (*Thunnus alalunga* Bonnaterre 1788) has been taken at Woods Hole, on Banquereau Bank, off eastern Nova Scotia (Goode and Bean 1879:15), and at Corsair Canyon on Georges Bank, so it may show up in the Gulf of Maine. It is easily recognizable among North Atlantic scombrids by its very long pectoral fins, which reach back past its second dorsal fin. Both albacore and bigeye tuna, *Thunnus obesus,* are long-lined from canyons on the south side of Georges Bank (B. Chase, Mass. Div. Mar. Fish., pers. comm.).

KEY TO GULF OF MAINE SCOMBRIDAE

1a. Only two small keels at base of caudal fin on each side; 5 dorsal and 5 anal finlets; adipose eyelids cover front and rear of eye 2
1b. A large median keel on each side of caudal peduncle in addition to the pair of small keels; 7–10 dorsal and 7–10 anal finlets; adipose eyelids absent. 3
2a. Sides below midline silvery, not spotted; distance between end of first dorsal fin groove and origin of second dorsal fin greater than length of groove, about one and a half times as long. . . . . **Atlantic mackerel**
2b. Sides below midline mottled with dusky blotches; distance between end of first dorsal fin groove and origin of second dorsal fin equal to or less than length of groove . . . . **Atlantic chub mackerel**
3a. Teeth in jaws strong, compressed, almost triangular; no prominent anterior corselet 4
3b. Teeth in jaws slender, conical, not compressed; anterior corselet of enlarged scales extends posteriorly to origin of dorsal, pectoral, and pelvic fins 6
4a. Lateral line with an abrupt downward curve under second dorsal fin; total gill rakers on first arch usually 7–10; no bronze spots on side in life . . . . **King mackerel**
4b. Lateral line descending gradually; gill rakers on first arch 11–18; bronze spots present on sides in life 5
5a. Sides with a few rows of oval bronze or yellowish spots plus one or two longitudinal dark streaks; gill rakers on first arch usually 15–18; scales present on pectoral fin. . . . . **Cero**
5b. Sides with bronze spots but without longitudinal dark streaks; gill rakers on first arch usually 11–14; no scales on pectoral fin . . **Spanish mackerel**
6a. Upper part of body with 5–10 narrow dark longitudinal stripes; first dorsal fin with 20–22 spines; upper surface of tongue without a pair of cartilaginous longitudinal ridges. . . . . **Atlantic bonito**
6b. Upper part of body without longitudinal stripes; first dorsal fin with 9–16 spines; upper surface of tongue with a pair of cartilaginous longitudinal ridges 7
7a. First and second dorsal fins widely separated; process between bases of pelvic fins large and undivided, at least as long as longest pelvic fin ray; first dorsal spines 10–12. . . . . **Bullet tuna**
7b. First and second dorsal fins barely separated; interpelvic process short and divided at tip; first dorsal spines 12–16 8
8a. Lower part of sides, below lateral line marked with 4–6 dark longitudinal bands, but no definite dark markings on back; total gill rakers on first arch 53–63 . . . . **Skipjack tuna**
8b. No dark markings on lower sides below lateral line, but the back may have dark markings; total gill rakers on first arch 19–45 9
9a. Body naked behind corselet of enlarged and thickened scales; several black spots usually present between bases of pectoral and pelvic fins; back dark blue-green with a complex pattern of wavy lines. . . . . **Little tunny**
9b. Body covered with small scales behind corselet; no black spots on body; back dark blue without any lines . . . . **Atlantic bluefin tuna**

BULLET TUNA / *Auxis rochei rochei* (Risso 1810) / Frigate Mackerel

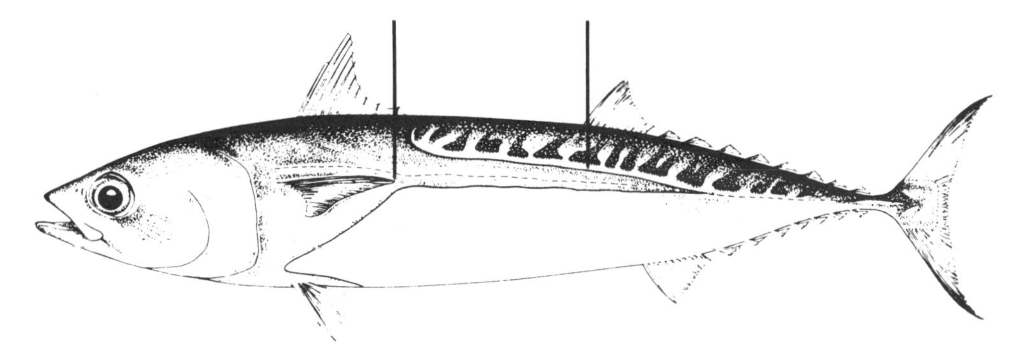

Figure 266. Bullet tuna *Auxis rochei rochei* showing anterior extent of dorsal scaleless area in relation to distal end of pectoral fin (left line) and width of corselet under origin of second [illegible]

Description. Body robust, elongate, and rounded (Fig. 266). Teeth small and conical, in a single series. Two dorsal fins separated by a long interspace, at least equal to length of first dorsal fin base. Large, undivided fleshy interpelvic process, longer than pelvic fins, between bases of pelvic fins. Strong central keel on each side of caudal peduncle. Body naked except for anterior corselet, which extends posteriorly along lateral line past origin of second dorsal fin.

Meristics. Dorsal fin rays X–XII, 10–12 + 8 finlets; anal fin rays 11–14 + 7 finlets; pectoral fin rays 23–25; gill rakers on first arch 39–46 in the western North Atlantic; corselet scales under origin of second dorsal fin 5–24, usually 6–19; vertebrae 20 + 19 = 39 (Collette and Aadland 1996).

Color. Back bluish, turning deep purple or almost black on the head. A pattern of 15 or more fairly broad, nearly vertical bars present in the dorsal naked area under the second dorsal fin. Black spots occasionally found on sides between pectoral and pelvic fins, as in little tunnies, *Euthynnus*.

Size. Maximum size 50 cm FL; common to 35 cm.

Distinctions. *Auxis* have a single, very long interpelvic process, equal to or longer than the pelvic fins themselves; the interpelvic process is shorter than the pelvic fins in all other Scombridae.

Taxonomic Note. Two subspecies are recognized, *Auxis r. rochei* in the Atlantic and Indo–west Pacific and *A. r. eudorax* in the eastern Pacific (Collette and Aadland 1996).

Biology. Bullet tuna feed on a large variety of fishes, particularly anchovies and other clupeoids (Etchevers 1976), crustaceans, and squids. Bullet tuna constitute a significant part of the food of adult tunas and billfishes (Uchida 1981). They are parasitized by a variety of worms and copepods (Uchida 1981: Table 19). Size at first spawning is about 350 mm FL for females and 365 mm FL for males (Rodríguez-Roda 1966). Spawning occurs in warm waters. The eggs are pelagic, 0.81–1.10 mm in diameter (Fahay 1983:312–313). A synopsis of biological data was presented by Uchida (1981).

General Range. An epipelagic neritic and oceanic species cosmopolitan in warm waters (Collette and Aadland 1996). In the western Atlantic, found from Cape Cod south through the Gulf of Mexico and Caribbean Sea south to Mar del Plata, Argentina.

Occurrence in the Gulf of Maine. Cape Cod had been considered the northern limit of the range so Bigelow and Schroeder did not include this species in the fauna of the Gulf. The only record for the Gulf of Maine is a 33-cm FL specimen (MCZ 39682) from a fish trap at Barnstable, Cape Cod Bay, on 1 August 1954 (Mather and Gibbs 1957). They also reported specimens from 40°1158′ N, 64°1106′ W and 41°1106′ N, 64°1112′ W, off Nova Scotia (ROM 22066), and near No Mans Land off Martha's Vineyard.

LITTLE TUNNY / *Euthynnus alletteratus* (Rafinesque 1810) / Bigelow and Schroeder 1953:336–337

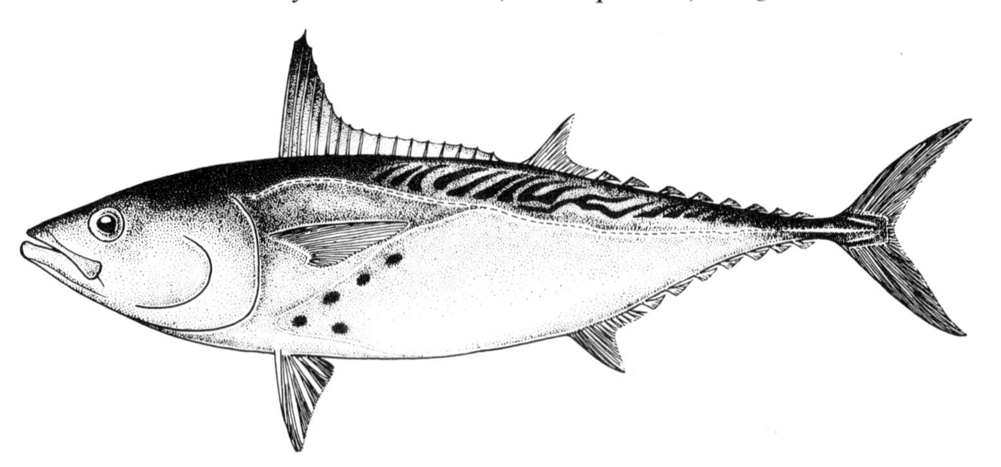

Figure 267. Little tuna *Euthynnus alletteratus.* Drawn by P. Lastrico. (Reprinted from Collette and Nauen 1983:34. © Food and Agriculture Organization of the United Nations.)

Description and Distinctions. Little tunny resemble skipjack very closely in body form, in size and arrangement of the fins, and in the fact that the body is naked except for scales on the forward and upper part of the trunk, the corselet, and along the lateral line (Fig. 267). They are distinguishable from skipjack by the color pattern, for it is above the lateral line that there are dark markings on the sides, not below, and by fewer gill rakers, 37–45 vs. 51–63.

Meristics. Dorsal fin rays XIII–XVII, 11–13 + 8–9 finlets; anal fin rays 11–15 + 7–8 finlets; pectoral fin rays 25–29; gill rakers on first arch 37–45; vertebrae 20 + 19 = 39 (Collette and Nauen 1983).

Color. Steel blue above and glistening white lower down on the sides and on the belly. Sides without markings below the lateral line, except for a few large, dark spots between the pectoral and pelvic fins, but marked above the lateral line with dark, wavy bands in various patterns.

Size. About the same as skipjack, that is, 100 cm FL. The all-tackle game fish record is a 15.95-kg fish caught at Cap de Garde, Algeria, in December 1988 (IGFA 2001).

Biology. Little tunny are opportunistic predators that feed on fishes, particularly clupeids (e.g., round herring), anchovies, and jacks; squids; and crustaceans such as amphipods, shrimps, and stomatopods (Manooch et al. 1985; Bowman et al. 2000). They are preyed upon by tunas, billfishes, and sharks such as shortfin mako. Ten species of parasitic copepods are known from little tunny (Cressey et al. 1983) and additional parasites were listed by Yoshida (1979). Off Florida, maturity is reached at about 35 cm. A 75-cm FL female from Senegal contained about 1.75 million eggs. In the Florida Straits, spawning extends from April to November. Eggs are pelagic, spherical, and transparent, 0.84–1.08 mm in diameter, with a single oil droplet 0.28 mm in diameter. Development was summarized by Fritzsche (1978) and Fahay (1983). A synopsis of biological data on the three species of *Euthynnus* was presented by Yoshida (1979).

General Range. An inshore neritic species of warm waters of the Atlantic Ocean, including the Gulf of Mexico and the Caribbean, Mediterranean, and Black seas (Collette and Nauen 1983:35).

Occurrence in the Gulf of Maine. Little tunny are caught from time to time near Woods Hole in July or August. The only records known to Bigelow and Schroeder within the Gulf were of 200–300 taken in a trap at Barnstable in the autumn of 1948 and of 28 taken in another trap in Cape Cod Bay, near Sandwich, on 11 September 1949 (Schuck 1951a). NMFS observers recorded a little tunny caught south of Cape Ann (42°1128′ N, 70°1146′ W). Local anglers report occasional catches of little tunny in the Cape Cod Canal (J. Moore, pers. comm., Dec. 1998).

SKIPJACK TUNA / *Katsuwonus pelamis* (Linnaeus 1758) / Bigelow and Schroeder 1953:335–336 (as *Euthynnus pelamis*)

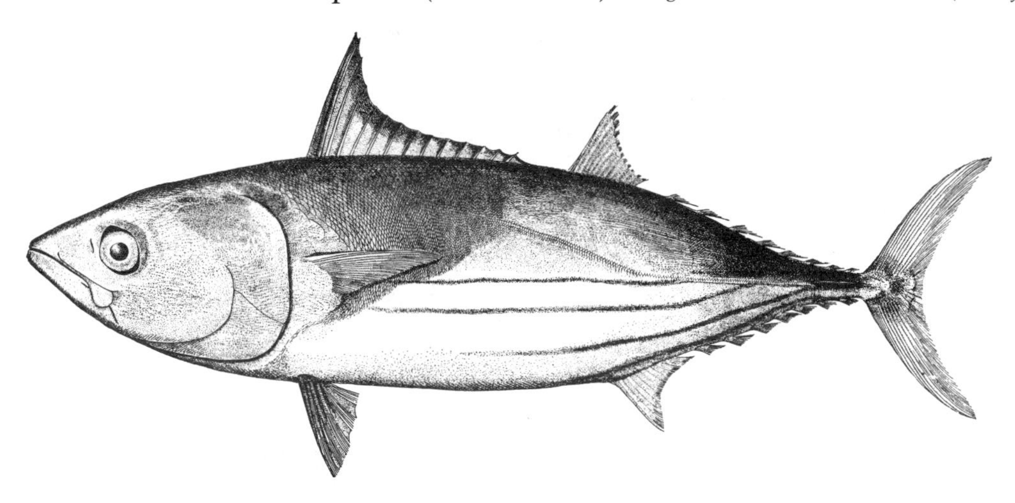

Figure 268. Skipjack tuna *Katsuwonus pelamis.* Drawn by P. Lastrico. (Reprinted from Collette and Nauen 1983:42. © Food and Agriculture Organization of the United Nations.)

Description. Body fusiform, stout, about one-fourth as deep as long, tapering to a pointed snout and to an extremely slender caudal peduncle (Fig. 268). First dorsal fin long, its upper edge abruptly concave behind second spine, last nine or ten spines much shorter. Second dorsal fin triangular, originating close to posterior end of first dorsal. Anal fin as large as second dorsal, about the same shape. Pectoral fins very short, broad, and lunate in outline. Caudal peduncle with a prominent median longitu-

Meristics. Dorsal fin rays XIV–XVI, 14–16 + 7 or 8 finlets; anal fin rays 14–16 + 6–8 finlets; pectoral fin rays 26–28; gill rakers on first arch 51–63; vertebrae 20 + 21 = 41 (Collette and Nauen 1983).

Color. Deep steel blue above, with the lower part of sides, throat, and belly shining silvery white. Each side is marked behind the corselet with four to six longitudinal dark stripes, the

line, the lower three or four fading out as they near the caudal peduncle.

Size. Skipjack grow to a length of about 108 cm FL and a weight of 32.5–34.5 kg; common to 80 cm and 8–10 kg. The all-tackle angling record is a 20.54-kg fish caught on Flathead Bank, Baja California, in November 1996 (IGFA 2001).

Distinctions. The very deeply concave contour of the first dorsal fin separates skipjack at a glance from Atlantic bonito or young tuna and from Spanish and king mackerels. The four to six conspicuous wavy longitudinal stripes below the lateral line are the readiest field mark by which to distinguish skipjack from little tunny. Skipjack have 53–63 gillrakers on the first arch vs. only 37–45 for little tunny.

Biology. Skipjack are opportunistic visual predators that feed on a wide variety of pelagic prey including fishes, crustaceans, and squids, representing 11 orders of invertebrates and 80 families of fishes (Matsumoto et al. 1984: Table 12). Skipjack predators include tunas, billfishes, and seabirds (Matsumoto et al. 1984: Table 9). More than 40 species of parasitic worms and 13 parasitic copepods have been reported from skipjack (Matsumoto et al. 1984: Table 10). Minimum size at maturity is 40 cm for females (Matsumoto et al. 1984). Fecundity ranges from 141,000 to 1,331,000 eggs for fishes 46.5–80.9 cm long from the Atlantic Ocean. Eggs are buoyant, spherical, and smooth, 0.80–1.17 mm in diameter, with a single yellow oil droplet 0.22–0.27 mm in diameter. A developmental series of eggs, prelarvae, and larvae was presented by Fritzsche (1978: 88–98) and by Fahay (1983:317).

General Range. Skipjack are an epipelagic oceanic species whose adults may be found between the 15°C isotherms of all the great oceans, the Atlantic, Pacific, and Indian (Matsumoto et al. 1984: Fig. 10; Collette and Nauen 1983:43).

Occurrence in the Gulf of Maine. Specimens obtained at Provincetown in 1880 by J. Henry Blake and in 1954 (Mather and Gibbs 1957) are the only records for this oceanic fish in the Gulf, but it sometimes appears in numbers about Woods Hole, where 2,000–3,000 were taken in 1878, but it did not show there again until October 1905. A 54-cm TL skipjack was taken in a mackerel net in Eastern Passage off Devils Island, Halifax, N.S., in October 1972 (Scott and Scott 1988). NMFS groundfish surveys document capture of two skipjack south of Nantucket Shoals (40°1115′ N, 69°1133′ W) on 26 October 1977. NMFS observers report numerous captures of skipjack along the southern edge of Georges Bank.

ATLANTIC BONITO / *Sarda sarda* (Bloch 1793) / Bigelow and Schroeder 1953:337–338

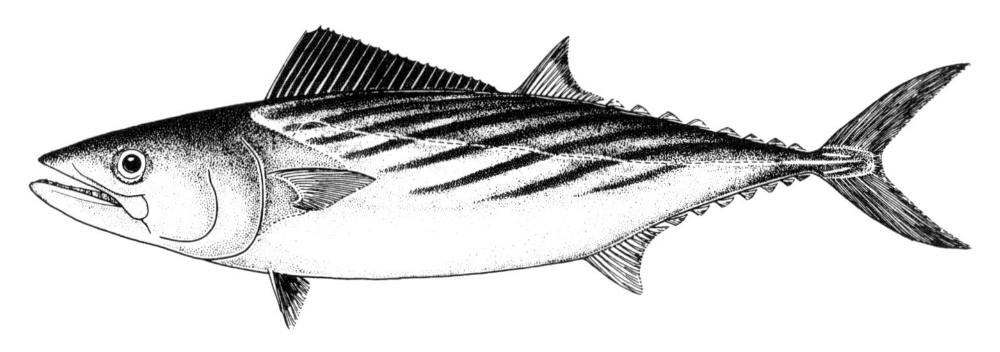

Figure 269. Atlantic bonito *Sarda sarda*. Drawn by P. Lastrico. (Reprinted from Collette and Nauen 1983:53. © Food and Agriculture Organization of the United Nations.)

Description. Body tuna-shaped, thick, and stout, one-fourth as deep as long, tapering to a pointed snout and very slender caudal peduncle (Fig. 269). Dorsal fins close together, practically confluent. First dorsal fin long, triangular, tapering regularly backward, with a nearly straight upper edge. Margins of second dorsal and anal fins deeply concave. Caudal fin lunate, much broader than long. Caudal peduncle with a median longitudinal keel. Body scaly all over. Lateral line not deeply bowed below second dorsal, only wavy.

Meristics. Dorsal fin rays XX–XXIII, 13–18, usually 15 or 16 + 8 or 9 finlets; anal fin rays 14–17 + 6 or 7 finlets; pectoral fin rays 23–26; gill rakers on first arch 16–20; vertebrae (26–28) + (23–26) = (50–53), usually 51. Tables of meristic and morphometric data were presented by Collette and Chao (1975).

Color. Steely blue above with a silvery hue on the lower part of its sides and abdomen. Upper part of sides marked with 7–20 narrow dark bluish bands running obliquely downward and forward across the lateral line. While young, back transversely barred with 10–12 dark blue stripes, but these crossbars usually disappear before maturity.

Distinctions. The shape of the fins distinguish Atlantic bonito at a glance from small bluefin tuna, the only regular member of the Gulf of Maine fish fauna with which they are apt to be confused: the first dorsal is relatively much longer than that of tuna (about a third as long as the body, not counting the caudal, with 20–23 spines) and the second dorsal is considerably longer than high, whereas in bluefin tuna the second dorsal is at least as high as it is long. The mouth is relatively larger than that of bluefin, gaping back as far as the posterior margin of the eye, and the 12–26 jaw teeth are larger, with the two to four in the front of the lower jaw noticeably bigger than the others. The shape of the first dorsal fin, which has a nearly straight upper margin, distinguishes bonito from skipjack and little tunny, in which this fin is deeply concave in outline.

Size. Atlantic bonito grow to a length of about 91.4 cm FL and a weight of 5.4 kg; commonly to 50 cm and 2 kg. The all-tackle angling record is an 8.30-kg fish caught in the Azores in July 1953 (IGFA 2001).

Food. Atlantic bonito are a strong, swift, predacious inhabitant of the open sea and, like all its tribe they travel in schools. When these fish visit northern waters they prey upon alewife, menhaden, mackerel, and other small fishes such as sand lance and silversides, as well as on squids and shrimps. They are very likely to be noticed, for they jump a great deal when in pursuit of their prey.

Biology. Four species of parasitic copepods are known from Atlantic bonito and additional parasites were listed by Yoshida (1980: Table 13). Minimum length at first maturity is about 39.5 cm FL for males and 40.5 cm for females. Fecundity estimates range from 450,000 to 3 million eggs. Further south, bonito spawn in June, but they are not likely to spawn in the Gulf of Maine. The eggs are pelagic, spherical, and transparent, 1.15–1.57 mm in diameter, with one to nine oil droplets. Young 13–15 cm have long been reported as common off Orient Point, Long Island, N.Y., early in September (Nichols and Breder 1927). Development is summarized by Fritzsche (1978: 100–105) and Fahay (1983:314–315). Biological data on all four species of *Sarda* were presented by Yoshida (1980).

General Range. An epipelagic, neritic, schooling species found in the warmer parts of the Atlantic and the Black and Mediterranean seas; north to outer Nova Scotia on the American coast and to Scandinavia on the European coast (Collette and Chao 1975: Fig. 75).

Occurrence in the Gulf of Maine. Cape Ann is the northern limit for the usual occurrence of bonito within the Gulf. They have been taken occasionally in Casco Bay; one was recorded from the mouth of the Kennebec River in September 1930 and two more in July 1932. Bigelow and Schroeder found no definite records of this species east of this on the coast of Maine or in the Bay of Fundy, although there are several records from the outer coast of Nova Scotia (e.g., off Shag Harbour [Mather and Gibbs 1957]). Bonito have been known to reach Cape Ann in larger numbers in the past, as happened in 1876, when 73 were taken on one August day in a weir near Gloucester. They may be more plentiful every year out at sea in the southern part of the Gulf than these meager returns would suggest, for fishermen often mention schools of them. Capt. Solomon Jacobs reported them as very plentiful in August 1896, for instance, in deep water to the north of Georges Bank. Bigelow and Schroeder reported schools of large scombroids, probably bonito, splashing and jumping off Cape Cod more than once in August.

Apparently bonito visit New England shores only in summer and fall. Thus the earliest catch made by a set of pound nets at Provincetown over a period of about 10 years was in July 1915 and the latest on 4 October 1919.

Bonito are more regular in their occurrence west and south of the Cape, being common in some years at Woods Hole and especially off Martha's Vineyard, whence about 26 mt were marketed in 1945. Party-boat captains have described Buzzards Bay and waters around the Vineyard and Nantucket as full of them in some summers.

Importance. Bonito are a good food fish. In 1919, pound nets, traps, and other gear accounted for landings of approximately 15 mt in Cape Cod Bay, but only 41 kg about Cape Ann, whereas the entire catch landed in the fishing ports of Maine during that year was only half a dozen fish (20 kg). Bonito are also game fish, readily biting a bait trolled from a moving boat once one has the lure that it will strike on the particular occasion. A good many are caught in this way off southern New England, and Bigelow and Schroeder asserted that a bonito is one of the strongest fish that swims, weight for weight. Bonito are picked up now and then in Cape Cod Bay by anglers trolling for other fish; Bigelow and Schroeder heard of two taken in this way off Wellfleet on 29 August 1950. Anglers report bonito from the Cape Cod Canal up to the power plant in Sandwich so the Canal may be a point of entry into the Gulf (J. Moore, pers. comm., Dec. 1998).

ATLANTIC CHUB MACKEREL / *Scomber colias* Gmelin 1789 / Hardhead, Bullseye /

Bigelow and Schroeder 1953:333–335 (as *Pneumatophorus colias*)

Description and Distinctions. Chub mackerel resemble Atlantic mackerel so closely that only the differences need to be which Atlantic mackerel lack. It is not necessary to open the fish to identify it for there is a characteristic color difference

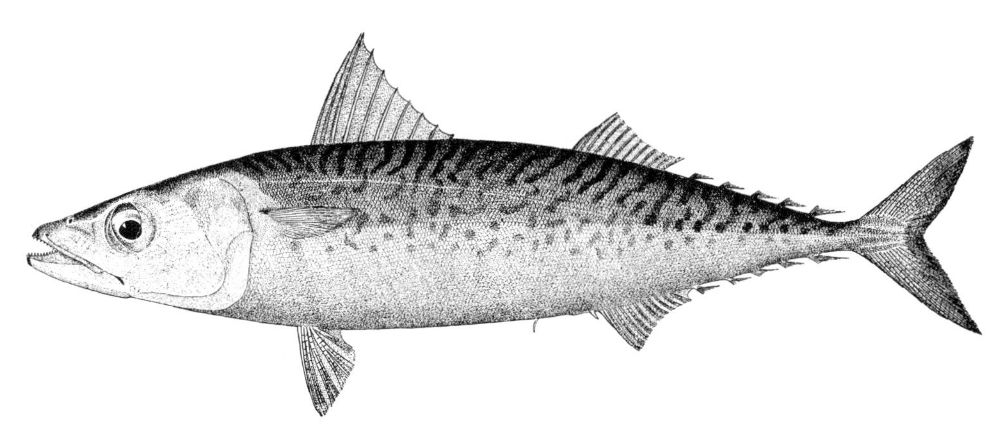

Figure 270. Atlantic chub mackerel *Scomber colias*. Provincetown, Mass. Drawn by H. L. Todd.

(otherwise colored somewhat like Atlantic mackerel) are mottled with small dusky blotches, and chub have a larger eye than Atlantic mackerel. Less obvious differences are that the dorsal fins are closer together in chub and that there are only nine or ten spines in the first dorsal fin instead of eleven or more, the usual count in Atlantic mackerel.

Meristics. Dorsal fin rays IX or X, 9–15 + 5 finlets; conspicuous but small anal spine anterior to anal fin, anal fin rays I, 9–11 + 5 finlets; pectoral fin rays 17–21; 12–15 interneural bones under first dorsal fin; gill rakers (10–12) + (25–29); vertebrae 14 + 17 = 31 (Collette and Nauen 1983).

Size. Reach 50 cm FL and common to 30 cm, but chub mackerel are smaller fish than Atlantic mackerel in the Gulf of Maine.

Taxonomic Note. Bigelow and Schroeder recognized Atlantic chub mackerel as a species, *S. colias,* separate from Indo-Pacific *S. japonicus*. Matsui (1967) placed *S. colias* in synonymy and subsequent researchers (e.g., Collette and Nauen 1983) have followed him. Recent molecular data (Scoles et al. 1998) show that chub mackerel from the Atlantic differ strongly from those in the Indo-Pacific leading to reappraisal of the status of Atlantic chub mackerel and a return to its recognition as a distinct species (Collette 1999).

Biology. Feeding habits of chub mackerel are similar to those of Atlantic mackerel, for Kendall found fish on Georges Bank in August 1896 full of the same species of pelagic crustacea and sagitta that Atlantic mackerel had taken at the same time and place, whereas specimens taken at Woods Hole dieted chiefly on copepods and to a lesser extent on amphipods, salps, appendicularians, and young herring. They follow thrown bait as readily and bite quite as greedily as mackerel do. The few examined by Bowman et al. (2000) contained copepods, larvaceans, and fishes. Studies of their breeding habits were summarized by Fritzsche (1978:108–114). At 4 to 11 mm, larval *Scomber colias* are deeper bodied, and at 3 to 15 mm have greater preanal lengths than *S. scombrus* of comparable sizes (Berrien 1978). *Scomber colias* larvae are less heavily pigmented and acquire pigmentation later than *S. scombrus.*

General Range. Atlantic chub mackerel are a coastal pelagic species inhabiting warm and temperate transition waters of the Atlantic Ocean and adjacent seas (Collette and Nauen 1983:57). In the western Atlantic, outer Nova Scotia (Vladykov 1935a) and the Gulf of St. Lawrence south to Florida, the Bahamas, the Gulf of Mexico, and Venezuela.

Occurrence in the Gulf of Maine. The early history of chub mackerel in Gulf waters was summarized first by Goode (1884) and then by Bigelow and Schroeder. They were tremendously abundant toward the end of the eighteenth century and early years of the nineteenth, down to 1820–1830. Capt. E. E. Merchant, an experienced and observant fisherman, described them as so plentiful off Provincetown from 1812 to 1820 that three men and a boy could catch 3,000 in a day on hook and line. They practically disappeared from the U.S. coast some time between 1840 and 1850. So completely did they vanish that the Smithsonian Institution tried in vain for 10 years prior to 1879 to obtain a single specimen. Chub mackerel were taken singly and in schools by the mackerel fleet on Georges Bank during August 1896 by Dr. W. C. Kendall, while many were caught on hook and line from the *Grampus* in Block Island Sound during the first week of that September. Kendall found them at Monomoy, the southerly elbow of Cape Cod in 1898, and they were sufficiently reestablished by then for Smith (1898a) to describe them as uncommon to abundant at Woods Hole. They then dropped out of the published record again until 1900, when they were found in the Casco Bay region. In 1906 and 1908, many were taken in traps near Woods Hole. The mackerel fleet found great schools of chub mackerel on Georges Bank in 1909, when vessels brought in 50,000–100,000 of them during the first week of July.

Other definite Gulf of Maine records are mostly about Casco Bay and one from Johns Bay, Maine. Bigelow and Schroeder found no records farther east along the coast of Maine; chub mackerel are not known in the Bay of Fundy, nor do they seem to reach the west Nova Scotian coast. Individuals are caught in the summer and fall of almost every year in mackerel traps in St. Margaret Bay, N.S., mixed with the regular catch of Atlantic mackerel (Scott and Scott 1988). In good "hardhead" years, they can be expected all along Georges Bank and on Browns Bank as well.

Importance. Chub mackerel are as choice a table fish as Atlantic mackerel. Owing to their low abundance compared to *S. scombrus,* they are not very important in the Gulf. Some 6 mt were taken in traps at North Truro between 11 August and 5 October 1952.

ATLANTIC MACKEREL / *Scomber scombrus* Linnaeus 1758 / Bigelow and Schroeder 1953:317–333

Figure 271. Atlantic mackerel *Scomber scombrus.* Washington, D.C., market, 469 mm.

Description. Body fusiform, tapering rearward to a very slim caudal peduncle and forward to a pointed snout (Fig. 271). Body 4.5–5.5 times as long as deep, oval in section, thick, and firm-muscled. Head long, one-quarter length to caudal. Mouth large, gaping back to middle of eye. Jaws of equal length, armed with small, sharp, slender teeth. Eye large, hollows in front of and behind it filled with so-called "adipose eyelid," two transparent, gelatinous masses, an anterior and a posterior, which cover eye except for a perpendicular slit over pupil. Two medium-sized dorsal fins: first originating over middle of pectoral fin when latter laid back, triangular, of rather weak spines that can be laid down in deep groove along midline of back; second dorsal separated from first by interspace longer than length of latter, smaller, followed by five small finlets. Anal fin similar to second dorsal in shape and size, originates slightly behind it, followed by five small finlets that correspond to dorsal finlets in size and shape. Caudal fin broad, short, deeply forked. Caudal peduncle bears two small longitudinal keels on either side but lacks the median lateral keel characteristic of Spanish mackerels and tunas. Pelvic fins small, originating below origin of first dorsal fin. Scales so small that skin feels velvety; hardly visible on belly without magnification. Scales around pectoral fins and on shoulders somewhat larger, forming poorly developed corselet.

Meristics. Dorsal fin rays XI–XIII, 9–15 (usually 12) + 5 finlets; anal fin preceded by small but distinct anal spine, anal fin rays I, 11 or 12 + 5 finlets; pectoral fin rays 20–22; 21–28 interneural bones under first dorsal fin; gill rakers 30–36 on lower limb of first gill arch; vertebrae 13 + 18 = 31 (Collette and Nauen 1983).

Color. Upper surface dark steely to greenish blue, often al- [illegible]

27–30) dark transverse bands that run down in an irregular wavy course nearly to the midlevel of the body, below which there is a narrow dark streak running along each side from pectoral to tail. Pectoral fins black or dusky at the base, the dorsal and caudal fins gray or dusky. Jaws and gill covers silvery. Lower parts of the sides white with silvery, coppery, or brassy reflections and iridescence; the belly is silvery white. The iridescent colors fade so rapidly after death that a dead fish gives little idea of the brilliance of a live one.

Size. Reach 56 cm FL; common to 30 cm. Most adult fish are 35–46 cm long. The all-tackle angling record is a 1.20-kg fish caught in Kraakvaag Fjord, Norway, in June 1992 (IGFA 2001). Fish about 35 cm weigh about 0.5 kg in the spring and about 0.6 kg in the fall when they are fat; 46-cm fish weigh about 1 kg; a 56-cm mackerel would likely weigh 2 kg.

Distinctions. Atlantic mackerel differ from Atlantic chub mackerel in lacking spots below the midline, in having the space between the end of the first dorsal fin groove and the origin of the second dorsal fin clearly longer than the length of the groove, in lacking a swim bladder, and in having 13 instead of 14 precaudal vertebrae.

Habits. Atlantic mackerel are fish of the open sea; although numbers of them, small ones especially, often enter estuaries and harbors in search of food, they never run up into freshwater. Neither are they directly dependent either on the coastline or on the bottom in any way at any stage in their lives. They are often encountered far out over outer parts of the continental shelf. They are most numerous within the inner [illegible] not to extend oceanward beyond the upper part of the continental slope.

The depth range of Atlantic mackerel is from the surface down to perhaps 183 m at one season or another. From spring through summer and well into the autumn, mackerel are in upper water layers mostly shoaler than 46–55 m; schools of all sizes come to the surface more or less regularly. They frequently disappear from the surface, often for considerable periods. Larger fish tend to swim deeper than smaller ones, on the whole, especially in mid and late summer (Sette 1950). Their vertical movements during the warmer parts of the year, when they are feeding actively, are probably governed by the level at which food is most abundant, which for the most part is shoaler than about 91 m, at least in the western Atlantic.

The highest temperature at which Atlantic mackerel commonly occur is about 20°C. At the opposite extreme they are sometimes found in abundance in water of 8°C; commercial catches are sometimes made in water as cold as 7°C, but few mackerel have been taken in temperatures lower than that in American waters. However, large catches of mackerel are made by trawlers in the North Sea in winter in water as cold as 6°–7°C.

The Gulf of St. Lawrence (where ice sometimes forms), outer Nova Scotian waters, and the upper 36 m or so within the Gulf of Maine, which chill to 2°–4°C or colder, are all too cold by late winter for Atlantic mackerel. In most years this is also true of the inner part of the continental shelf as a whole, and south as far as northern Virginia, for the water usually cools there to 3°–4°C at the time of the winter minimum. Atlantic mackerel need only move out to the so-called warm zone at the outer edge of the shelf to find a more suitable environment, for the bottom water there is warmer than 7°–8°C year-round as far north and east as the central part of Georges Bank and about 5°C along outer Nova Scotia.

Mackerel are swift-moving, swimming with very short sidewise movements of the rear part of the body and the powerful caudal fin. When caught they beat a rapid tattoo with their tails on the bottom of the boat until exhausted. They require so much oxygen for their vital processes that when the water is warm (hence its oxygen content low), they must swim constantly to bring a sufficient flow of water to their gill filaments.

Mackerel, like herring, gather in dense schools of many thousands. Members of any given school are usually all about the same size, that is, the same age. This tendency to separate according to size is probably due to the fact that larger fish swim faster than smaller ones (Sette 1950). Mackerel school by themselves, as a rule, but they are sometimes found mingled with herring, alewife, or shad (Kendall 1910:287). Schools of mackerel are often seen at the surface. In the daytime they can be recognized by the appearance of the ripple they make, for this is less compact than that made either by herring or by menhaden. Mackerel do not ordinarily "fin" or raise their noses above the surface, as is the common habit of menhaden. An observer at masthead height might see a school of mackerel as deep as 14–18 m by day, if the water is calm and the sun is behind him. On dark nights the schools are likely to be betrayed by the "firing" of the water, caused by the luminescence of the tiny organisms that they disturb in their progress. Sette (1950:267) reported one case of a school recognized by its firing as deep as 46 m, but the water is seldom (if ever) clear enough in the Gulf of Maine for a submerged light more than 27 m down to be visible from above. The trail of bluish light left behind by individual fish as they dart to one side or the other, while one rows or sails through a school on a moonless, overcast night when the water is firing, is a beautiful spectacle in Gulf coastal waters. The speed at which a school travels when it is not disturbed probably depends on the size of the fish of which it is composed. Mackerel less than 1 year old swim at about 11 $km \cdot h^{-1}$ (3 $m \cdot s^{-1}$) while circling inside a live car; yearlings at a rate of about 21 $km \cdot h^{-1}$ (6 $m \cdot s^{-1}$), or nearly twice as fast (Sette 1950). They have a preferred swimming speed of between 0.9 and 3.5 body lengths per second (Wardle and He 1988).

Food. Atlantic mackerel are opportunistic feeders that swallow prey whole. Their food consists primarily of zooplankters captured by active pursuit of individual animals or by passive filtering (Pepin et al. 1988). Practically all floating animals that are neither too large nor too small regularly serve to nourish mackerel, and a dietary listing for any given locality would include all the local pelagic Crustacea and their larvae. The diet of Atlantic mackerel changes markedly during ontogeny (Peterson and Ausubel 1984; Fortier and Villeneuve 1996). First-feeding larvae (3.5 mm in length) are phytophagous. The diet of larvae 4.5 mm is composed of nauplii of *Acartia hudsonica, Temora longicornis,* and *Pseudocalanus* sp. Larvae larger than 5 mm eat copepodites of *A. hudsonica* and *T. longicornis* and smaller proportions of phytoplankton and copepod nauplii, and when larger than 6.5 mm add conspecifics and other fish larvae to their diet. Larval fish prey (Fortier and Villeneuve 1996) include conspecifics (66%), yellowtail flounder (18%), silver hake (12%), and redfish (4%) larvae. Stomach content weight averages 1.8% of an individual's body weight. In order to satisfy its daily energy requirement, a mackerel larva consumes 25–75% of its body weight per day. The diet of young mackerel is much the same in the English Channel (Lebour 1920).

Young fish depend more and more upon larger prey as they grow. Juveniles eat mostly small crustaceans such as copepods, amphipods, mysid shrimps, and decapod larvae. They also take large quantities of small pelagic mollusks (*Limacina* and *Clione*) when available. A series of small fish examined by Vinal Edwards contained copepods, shrimps, crustacean and molluscan larvae, annelid worms, appendicularians, squid, fish eggs, and fish fry such as herring, silversides, and sand lance.

Adults continue to feed on the same food as juveniles, but their diet also includes a wider assortment of organisms and larger prey items. For example, euphausiid, pandalid, and crangonid shrimps are common prey; chaetognaths, larvaceans, pelagic polychaetes, and the larvae of many different marine species have also been identified in mackerel stomachs.

Larger prey such as squids (*Loligo*) and fishes (silver hake, sand lance, herring, hakes, and sculpins) are not uncommon, especially in large mackerel (Bowman et al. 1984). Analysis of stomach contents of 3,617 mackerel gathered during 1963–1983 in waters off the northeastern United States showed that the types of prey eaten by Atlantic mackerel vary enormously depending on the year, the season, and the area (Bowman et al. 1984). Copepods, amphipods, and euphausiids can be considered staples in their diet but other prey types are taken whenever they are available.

Bigelow and Schroeder examined many Gulf of Maine mackerel packed full of *Calanus,* the "red feed" or "cayenne" of fishermen, as well as with other copepods. Mackerel also feed greedily, as do herring, on euphausiid shrimps, especially in the northeastern part of the Gulf, where these crustaceans come to the surface in abundance. Various other planktonic animals also enter regularly into their diet. Kendall wrote in his field notes that some of the fish caught on the northern part of Georges Bank in August 1896 were packed with crab larvae and others were full of sagitta, amphipods (*Euthemisto*), small copepods (*Temora*), or red feed (*Calanus*), so that even fish from the same school had selected members of the drifting community in varying proportions. Similarly, 1,000 mackerel caught near Woods Hole from June to August contained pelagic amphipods (*Euthemisto*), copepods, squid, and sand lance; others taken off No Mans Land have been found full of shelled pteropods (*Limacina*).

Mackerel have often been seen to bite the centers out of large medusae, but they probably do this for the amphipods (*Hyperia*) that live commensal within the cavities of the jellyfish, not for the jellyfish themselves (Nilsson 1914). Under laboratory conditions, mackerel feed on *Aglantha digitale,* a small transparent medusa common in temperate and boreal waters (Runge et al. 1987).

Most authors describe mackerel as feeding by two methods: either by filtering smaller pelagic organisms from the water using their gill rakers or by selecting individual animals by sight. Mackerel have long rakers with spines on the foremost gill arch only, and these are not fine enough to retain the smallest organisms (see Bigelow 1926: Fig. 42C,D for photographs of the gill rakers). Much discussion has centered on the relative serviceability of these two methods of feeding. It is not yet known how small an object a fish is able to select; they do take fish individually, as well as such large Crustacea as euphausiid shrimps and amphipods, just as herring do, and also larger copepods, to judge from the fact that mackerel stomachs are often full of *Calanus* or of one or two other sorts in localities where indiscriminate feeding would yield them a variety. Whether they select smaller copepods and crustacean larvae is not so clear.

In laboratory experiments, adult mackerel acclimated at 17.5°C cleared their stomachs of fish (silverside, *Menidia menidia*) in approximately 38 h independent of ration size or body weight (Lambert 1985). Assimilation efficiency was high

Predators. Mackerel fall prey to all the larger predacious sea animals. Whales, porpoises, sharks, tunas, bonito, bluefish, and striped bass take a heavy toll. Atlantic cod often eat small mackerel; squid destroy great numbers of young fish less than 10 or 13 cm long, and seabirds of various kinds follow and prey upon the schools when these are at the surface. Other predators include bigeye thresher, thresher, shortfin mako, tiger, blue, spiny dogfish, and dusky sharks, thorny and winter skates, silver, red, and white hakes, pollock, goosefish, weakfish, and king mackerel (Maurer and Bowman 1975; Bowman and Michaels 1984; Rountree 1999; Bowman et al. 2000). Summaries of NEFSC food habits data indicate that spiny dogfish, Atlantic cod, and silver hake are the most important fish predators on Atlantic mackerel (Langton and Bowman 1980; Bowman and Michaels 1984; Bowman et al. 1984; Overholtz et al. 1991a; Rountree 1999). Cannibalism by larval Atlantic mackerel 3–14 mm long is significant but does not appear to contribute to density-dependent regulation of the species (Fortier and Villeneuve 1996).

Parasites. One species of parasitic copepod, *Caligus pelamydis,* is found on Atlantic mackerel (Cressey et al. 1983). A considerable list of parasitic worms, both round and trematode, are known to infest the digestive tract of mackerel.

Breeding Habits. Median lengths at maturity of female and male Atlantic mackerel from the northeast coast of the United States were 25.7 and 26.0 cm FL, respectively (O'Brien et al. 1993). Median age at maturity was 1.9 years for both sexes. Mackerel sampled in Newfoundland waters had higher median lengths at first maturity: 34 cm for females and 35 cm for males (Moores et al. 1975).

Spawning of Atlantic mackerel in the northwest Atlantic occurs from April to August and progresses from south to north as surface waters warm and the fish migrate (Sette 1943). There are two spawning contingents. The southern contingent spawns from April to July in the Mid-Atlantic Bight and the Gulf of Maine and the northern contingent spawns in the southern Gulf of St. Lawrence in June and July (Berrien 1982; Morse et al. 1987; O'Brien et al. 1993).

Most spawn in the shoreward half of continental shelf waters, although some spawning extends to the shelf edge and beyond (Berrien 1982: Map 12; Morse et al. 1987: Fig. 16; Berrien and Sibunka 1999: Fig. 79). Available data point to the oceanic bight between Chesapeake Bay and southern New England as the most productive area, the Gulf of St. Lawrence as considerably less so, and the Gulf of Maine and coast of outer Nova Scotia as ranking third (Sette 1950:158–164; Morse et al. 1987). Maximum concentrations of Atlantic mackerel eggs in the Gulf of Maine occur in May and June (Berrien and Sibunka 1999: Fig. 79).

Mackerel do not begin spawning until the water has warmed to about 8°C, with the chief production of eggs taking place at temperatures of 9°–14°C, so the spawning season

Thus the chief production takes place as early as mid-April off Chesapeake Bay; during May off New Jersey; in June off southern Massachusetts and in the region of Massachusetts Bay; through June off outer Nova Scotia; and from late June through early July on the southern side of the Gulf of St. Lawrence, where eggs have been taken from early June to mid-August (Sette 1943:158–163).

It seems from the relative numbers of eggs that Cape Cod Bay is the only subdivision of the Gulf that has rivaled the more southern spawning grounds in egg production during the particular years that intensive studies were done. Subsequent information and especially the results of early tow-nettings on the southern grounds (Sette 1943) have shown that the Gulf of Maine as a whole is much less productive than the more southern spawning grounds, not more so as Bigelow and Welsh (1925:206) believed. Mackerel also spawn to some extent northward, as far as Casco Bay, but Bigelow and Schroeder believed very few do so farther east than along the coast of Maine. It is also unlikely that mackerel breed successfully in the northern side of the Bay of Fundy, as neither eggs nor larvae have been taken there; however, some production may take place on the Nova Scotian side, as Huntsman reported eggs at the mouth of the Annapolis River. While a moderate amount of spawning takes place along the outer coast of Nova Scotia (Sparks 1929), it seems that the eggs do not hatch at the low temperatures prevailing there, for no larvae have been found. But the southern side of the Gulf of St. Lawrence, where the surface waters warm to a high temperature in summer, is an extremely productive spawning ground.

Estimates of fecundity range from 285,000 to 1.98 million eggs for southern contingent mackerel between 307 and 438 mm FL (Morse 1980). Analyses of egg diameter frequencies indicate that mackerel spawn between five and seven batches of eggs per year.

Early Life History. The eggs are 1.09–1.39 mm in diameter, have one oil globule 0.19–0.53 mm in diameter, and generally float in the surface water layer above the thermocline or in the upper 10–15 m (Berrien 1975, 1982; Markle and Frost 1985). Incubation depends primarily on temperature; it takes 7.5 days at 11°C, 5.5 days at 13°C, and 4 days at 16°C (Worley 1933).

Newly hatched larvae are 3.1–3.3 mm long and have a large yolk sac. Numerous black pigment cells are scattered over the head, trunk, and oil globule. The yolk is absorbed and the mouth formed, teeth are visible, and the first traces of the caudal fin rays form by the time the larva is about 6 mm long. Rays of the second dorsal, anal, and pelvic fins appear at about 9 mm TL and those of the first dorsal when the larva is about 14–15 mm. Dorsal and anal finlets are distinguishable in fry of 22 mm, and the caudal fin has begun to assume its lunate shape, but the head and eyes are still relatively much larger than in the adult, the snout blunter, and the teeth longer. Studies on development were summarized by Fritzsche (1978:116–120). At 50 mm, little mackerel resemble their parents so closely that their identity is evident. Late-stage eggs and early yolk-sac larvae have fewer melanophores on the yolk surface than *S. colias,* and larvae during and at the end of yolk absorption have several dorsal trunk melanophores, whereas few *S. scombrus* at this stage have any such pigmentation (Berrien 1975, 1978).

Age and Growth. Aging methods for Atlantic mackerel were first described for mackerel from the English Channel and Celtic Sea (Steven 1952). Growth patterns on mackerel otoliths are not as complex as those of some other fish species but older mackerel (>10 years) can be difficult to age because annuli are extremely thin and closely spaced near the edge of the otolith (Dery 1988b). Calculated growth curves and von Bertalanffy growth parameters from several previous studies were compared by Anderson and Paciorkowski (1980). A length-at-age graph is presented in Fig. 272.

Atlantic mackerel are about 3 mm long at hatching, grow to about 50 mm in 2 months, and reach approximately 20 cm in December, near the end of their first year of growth (Berrien 1982). Gulf mackerel run about 25–28 cm long in spring and early summer of their second year of growth (known then as tinkers). The brood of 1923 averaged almost 36 cm in their third autumn, about 38 cm in their fourth year, about 39 cm in their fifth, about 40 cm in their sixth, 41 cm in their seventh, and about 42 cm in their eighth year. Fish 10–13 years old reach a length of 39–40 cm. Thus American mackerel, like European, grow very slowly after their third summer, although they are long-lived. The two sexes grow about equally fast.

MacKay (1967) studied growth in several year-classes of Atlantic mackerel in Canada and theorized that growth is population density–dependent; that is, abundant year-classes grow more slowly than less abundant year-classes. Moores et al. (1975) did not find this relationship for the same year-classes of Newfoundland fish. Overholtz (1989) found the 1982 cohort to be one of the slowest growing on record and it is also one of the largest recruiting year-classes that has been observed. Early growth may be related to year-class size, whereas stock size may be more influential after juveniles join adults further offshore.

General Range. North Atlantic Ocean, including the Baltic Sea; eastern Atlantic including the Mediterranean and Black seas; and western Atlantic from Black Island, Labrador (Parsons 1970) to Cape Lookout, N.C. (Collette and Nauen 1983:59).

Occurrence in the Gulf of Maine. Once mackerel enter the Gulf, schools are to be expected anywhere around its coastal belt and on Nantucket Shoals, the western part of Georges Bank, and Browns Bank (Map 32). Their presence in the Gulf is closely related to seasonal movements for this is a migratory fish, appearing at the surface and near the coasts in spring and vanishing late in the autumn. There has been much discussion as to whether the main bodies of mackerel merely sink when they leave the coast in autumn and move directly out to the nearest deep water or whether they combine their offshore and onshore journeys with extensive north and south migra-

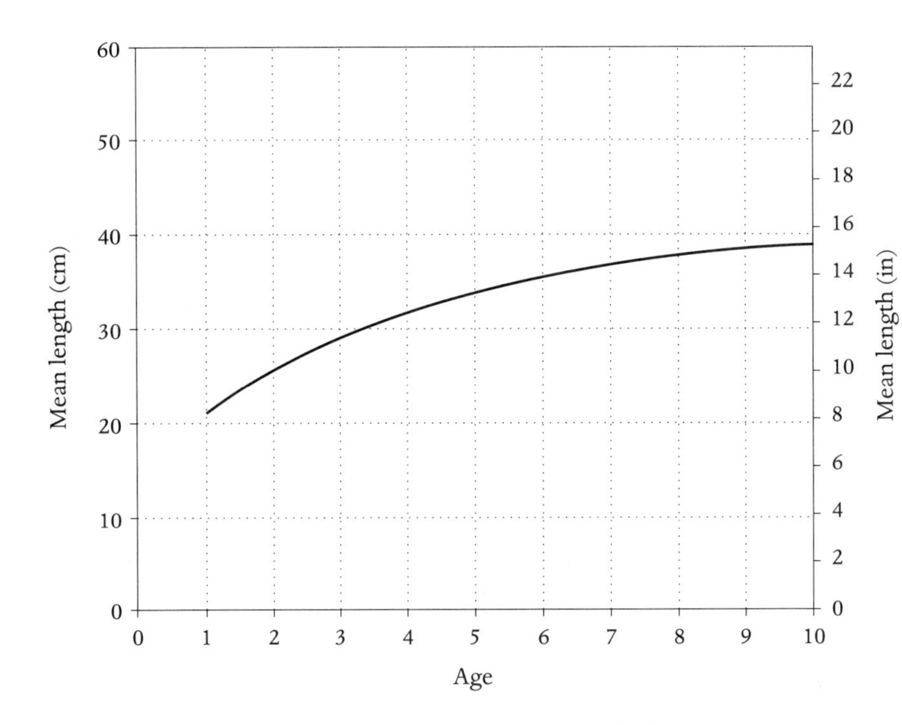

Figure 272. Length-at-age curve for Atlantic mackerel *Scomber scombrus,* all areas combined. (From Penttila et al. 1989:11.)

tions. The literature dealing with this subject is extensive (see Goode et al. 1884; Tracy 1907; Sette 1950:268–313; Berrien 1982). Most mackerel withdraw from the coast by the end of December, not only from the Gulf of St. Lawrence, but also from the entire inshore belt, not to be seen there again until the following spring or early summer.

Migrations. Based on separation of spawning areas, tagging studies, and the size and age composition of commercial catches, Sette (1950) described northern and southern population contingents with different spring and autumn migration patterns and summer distributions. Apparent mixing of the two contingents during migrations led him to suggest that they may not be genetically distinct. Subsequent biochemical and meristic studies did not detect significant genetic differences between the two groups (MacKay 1967; MacKay and Garside 1969). However, there are significant genetic differences between Atlantic mackerel from the eastern and the western Atlantic (Scoles et al. 1998).

Atlantic mackerel apparently overwinter in moderately deep water, 70–200 m, along the continental shelf from Sable Island Bank, off Nova Scotia, to the Chesapeake Bay region (Sette 1950; Leim and Scott 1966; MacKay 1967). In spring there is a general inshore, then northeastward migration, and in autumn the pattern is reversed (Berrien 1982: Map 44).

The southern contingent begins the spring spawning migration by moving inshore between Delaware Bay and Cape Hatteras, usually between mid-March and mid-April. They then move northeast along the coast, continually joined by schools from the northern contingent also moving inshore. The overwintering area and timing of spring migration varies from year to year, probably owing to differences in the water temperature

and Long Island in late April and May. By early June, schools are off southern Massachusetts and by later in June and July have moved to the western side of the Gulf of Maine, where they remain for the summer. The southern contingent leaves the Gulf of Maine in October and returns to deep water to overwinter, apparently near the shelf edge, where the bottom temperature is above 7°C, probably between Long Island and Chesapeake Bay (Sette 1950; Berrien 1982).

The northern contingent usually begins to move inshore off southern New England in late May, mixing temporarily with part of the southern contingent. Northern fish then migrate eastward along the coast of Nova Scotia, are joined by other mackerel schools from offshore, and move into the Gulf of St. Lawrence, where they spawn in June and July and then spend the rest of the summer there. Small fish (>30 cm) tend to lag behind large ones during the spring migration and spawn later (MacKay 1967; Stobo and Hunt 1974; Moores et al. 1975). Some northern fish separate from the main body and remain along the coasts of Maine, Nova Scotia, and Cape Breton Island for the summer. The main body of northern fish leaves the Gulf of St. Lawrence in September and October via Cabot Strait, moves west along Nova Scotia, and passes through the Gulf of Maine from October to December, where northern and southern contingents again mix temporarily. Northern fish leave the Gulf of Maine near Cape Cod in December and are assumed to overwinter in deep water over the outer shelf between Sable Island Bank and Long Island; further mixing probably occurs then (Sette 1950; Parsons and Moores 1974; Moores et al. 1975).

Tagging experiments on Atlantic mackerel have demonstrated extensive movements (Sette 1950; Moores et al. 1975; Berrien 1982). Tagged fish have moved from Newfoundland

Mackerel tagged in the Gulf of St. Lawrence in October were recaptured on Georges Bank and off New Jersey and Delaware in winter. Mackerel tagged off Nova Scotia were recaptured off New England, off Sable Island and Delaware in winter, and off New Jersey. One Atlantic mackerel tagged near Woods Hole in June was recaptured in Nova Scotian waters. Recapture south of Long Island of an Atlantic mackerel tagged in Newfoundland waters represents the farthest documented migrant in the northwest Atlantic, at least 2,260 km between 30 August and 28 December 1972 (Parsons and Moores 1974). These studies indicate probable mixing of fish from different contingents during the winter.

Importance. Mackerel are delicious fish, but they do not keep as well as fishes that have less oil in their tissues. When mackerel were plentiful they were among the most valuable fishes of the Gulf commercially, surpassed in dollar value only by haddock, cod, and redfish for the years 1943–1947.

It has been common knowledge since early colonial days that mackerel fluctuate widely in abundance in the Gulf from year to year, perhaps more widely than any other important food fishes, with periods of great abundance alternating with terms of scarcity or almost total absence. Bigelow and Schroeder reviewed fluctuations in catch of Atlantic mackerel in the Gulf of Maine. From 1925 to 1946, catch ranged between a low of about 9,090,900 (1937) and a high of 26,818,000 kg (1932). Average annual Gulf of Maine catch for the period 1933–1946 was about 16,818,000 kg.

Formerly, most mackerel were caught with hook and line, ground bait being thrown out to lure the fish close enough to the vessel (Goode and Collins 1887:275–294). This way of fishing was given up about 1870 and replaced by the purse seine. Practically the entire catch of mackerel in the northwest Atlantic for the past 100 years has been made with purse seines, pound nets, weirs and floating traps, and gill nets. In 1943, for example, when the total Gulf of Maine catch was about 24,000 mt, about 80% was taken in purse seines; 12–13% in pound nets, weirs, and floating traps combined; and 3–4% (450–900 mt) in gill nets (anchored or drifting), but only 770 kg on handlines.

In the mid-1960s, an extensive offshore otter trawl fishery began in the northwest Atlantic by distant-water fleets, chiefly from Europe. They began fishing on mackerel, partly because of its high abundance and partly because of declining herring stocks. The estimated total mackerel biomass (age 1 and older) in the northwest Atlantic increased from about 600,000 mt in the early 1960s to a peak of 2.4 million mt in 1969, and then declined rapidly to less then 500,000 mt in 1978 (Fig. 273). The big rise and fall in abundance was due chiefly to the recruitment of four consecutive strong year-classes in 1966–1969, particularly the extraordinary one of 1967, followed by a series of relatively small year-classes from 1970 to 1977, generally less than half the strength of those of 1966, 1968, and 1969 and less than one-quarter the size of the one from 1967 (Anderson and Paciorkowski 1980). Total commercial landings from this stock increased from only 7,300 mt in 1960 to a peak of 419,000 mt in 1973, and then declined rapidly to 77,000 mt in 1977, the first year of management under extended jurisdiction in both U.S. and Canadian waters inside the 200-mile zone and then declined further to 27,400 mt in 1995 (Overholtz 1998b). The bulk of this harvest came from the area between Georges Bank and Cape Hatteras (e.g., 90% during the peak 1970–1974 period). For the same period, about 60% came from the Mid-Atlantic Bight alone. In 1978 the commercial catch from the Bight dropped to about 1,000 mt, most of which was taken by U.S. fishermen. Coincident with the extension of U.S. management responsibility to 200 miles offshore, the Mid-Atlantic Fishery Management Council established controls on the fishery in the early 1980s under authority of the Magnuson Fishery Conservation and Management Act (Overholtz et al. 1991a). The fishery management plan for the United States developed by the Management Council in 1984 relies on annual catch quotas developed from an assessment of the stock based on a virtual population analysis and an estimate of incoming recruitment at age 1 (Overholtz 1993).

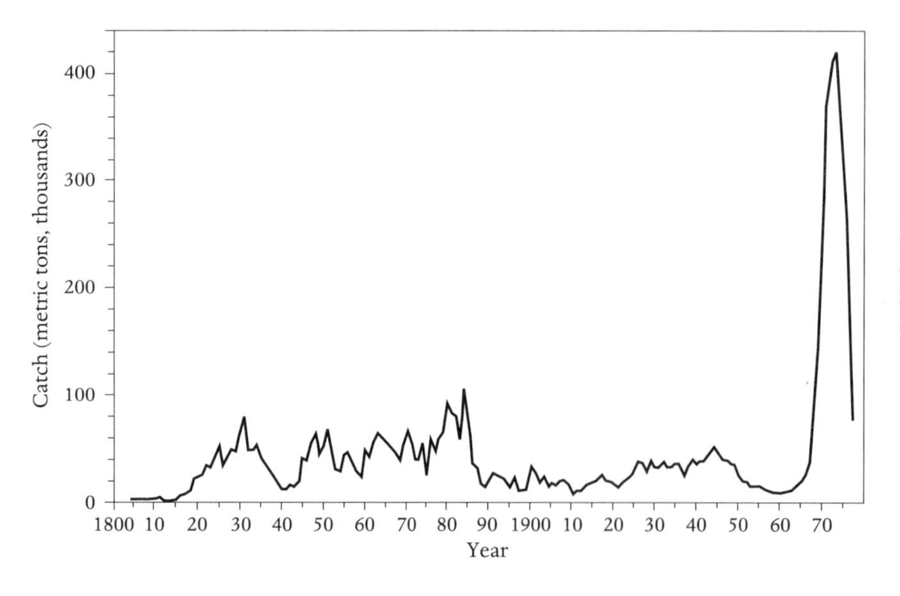

Figure 273. Commercial catch of Atlantic mackerel *Scomber scombrus* in the northwest Atlantic, 1804–1977. (From Anderson and Paciorkowski 1980: Fig. 3.)

Fluctuations in mackerel year-classes are believed to be due to variations in larval survival (Sette 1943). Factors influencing mortality of larvae may include water temperature, predation, zooplankton abundance, wind-driven surface currents, epizootics, and abundance of mackerel larvae relative to their prey (Sette 1943; Taylor et al. 1957; MacKay 1967; Berrien 1982). Average recruitment levels may be reduced when the spawning stock drops below some critical level. There was concern that heavy fishing coupled with poor recruitment in the 1970s would drive the spawning stock down below such a level, and catch restrictions have been imposed since 1976 to promote rebuilding of the stock (Anderson and Paciorkowski 1980).

Recreational Fishery. Many anglers troll or baitfish for mackerel all along the coast from Cape Cod to Penobscot Bay; as far as Mt. Desert if mackerel are on the coast that far east. In good years Bigelow and Schroeder reported that it was not unusual for three or four anglers fishing from a party boat to bring in 100 or 200 fish. In summers when young tinkers are plentiful inshore many of them are caught from the wharves in various harbors. If one chooses to troll, an ordinary no. 3 pickerel spinner serves well, especially if tipped with a small piece of pork rind or with mackerel skin, a small metal jig similarly adorned, or any small bright spoon. Mackerel will also take a bright artificial fly and bite greedily on almost any bait, such as a piece of clam, a piece of mackerel belly, or a sea worm (*Nereis*), especially if attracted by ground bait.

The U.S. recreational catch of Atlantic mackerel is substantially greater than the U.S. commercial catch, averaging 13,600 mt·year^{-1} (1960–1977) compared to only 2,200 mt·year^{-1} for the commercial fishery (Anderson and Paciorkowski 1980). The estimated sport catch has also followed major trends in mackerel abundance, rising from 5,000 mt in 1960 to 13,000–16,000 mt in 1969–1972, then declining to the lowest estimated catch of 522 mt in 1977 (Overholtz et al. 1991b). The recreational catch stabilized to between 3,000 and 5,000 mt·year^{-1} in the next decade. U.S. recreational catches are made mainly from April to October, whereas catches in Canadian waters off Nova Scotia and Newfoundland have typically been from May to November.

KING MACKEREL / *Scomberomorus cavalla* (Cuvier 1829) / Bigelow and Schroeder 1953:349

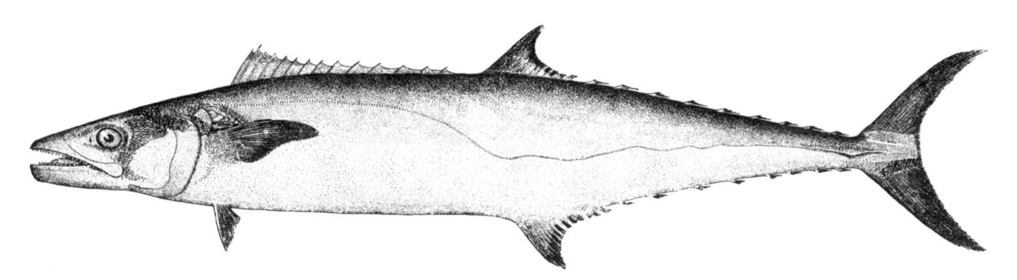

Figure 274. King mackerel *Scomberomorus cavalla.* Woods Hole, Mass. Drawn by H. L. Todd.

Description. The lateral line dips downward abruptly in the king mackerel under the forepart of the second dorsal fin, but only curves down gradually in the cero and Spanish mackerel (Fig. 274).

Meristics. Dorsal fin rays XII–XVII (usually XV), 15–18 + 7–10 finlets; anal fin rays 16–20 + 7–10 finlets; pectoral fin rays 21–23; gill rakers on first arch (1 or 2) + (6–11) = 7–13; vertebrae (16 or 17) + (24–26) = (41–43). Tables of meristic and morphometric data were presented by Collette and Russo (1985).

Color. Iron gray above; silvery lower down on the sides and on the belly; sides marked with darker gray or yellowish spots, which tend to disappear in large fish. First dorsal fin clear anteriorly, lacking the black area found in cero and Spanish mackerel.

Size. Reach 172.5 cm FL and 41 kg; common to 70 cm. The all-tackle game fish record is a 42.18-kg fish caught off San Juan, Puerto Rico, in April 1999 (IGFA 2001).

Distinctions. King mackerel differ from the other two species of *Scomberomorus* that might occur in the Gulf by having a prominent dip in the lateral line under the second dorsal fin, fewer gill rakers (usually 10 or fewer compared to 11 or more), fewer vertebrae (41–43 vs. 47 or more), and fewer dorsal spines (usually 15 or fewer vs. 16 or more).

Biology. King mackerel feed primarily on fishes and, in smaller quantities, on penaeoid shrimps and squids (DeVane 1978). Herringlike fishes such as round herring and menhaden are particularly important prey. They also eat mackerel (Bowman et al. 2000). Four species of parasitic copepods are known from king mackerel (Cressey et al. 1983) and additional parasites are listed by Berrien and Finan (1977a). In Florida, females usually mature in their fourth summer at a mean length of 83.7 cm FL and males in their third summer at 73 cm. Fecundity of females 68–123 cm FL ranges from 345,000 to 2.28 million eggs. Spawning takes place in late

General Range. Warm parts of the western Atlantic; south to Brazil (Collette and Russo 1985: Fig. 51), north regularly to North Carolina (June to November), occasionally to southern Massachusetts, and as strays to the southern Gulf of Maine.

Occurrence in the Gulf of Maine. The only Gulf of Maine record appears to be a 560-mm FL specimen (MCZ 37041) taken in a trap at North Truro, Cape Cod, in August 1949. At least four specimens have been recorded from Canadian Atlantic waters off Nova Scotia (Scott and Scott 1988). There are several summer records from the southern side of Cape Cod, from traps in Buzzards Bay and at Quissett in 1953 and 1957 (Mather 1954; Mather and Gibbs 1957).

SPANISH MACKEREL / *Scomberomorus maculatus* (Mitchill 1815) / Bigelow and Schroeder 1953:347–348

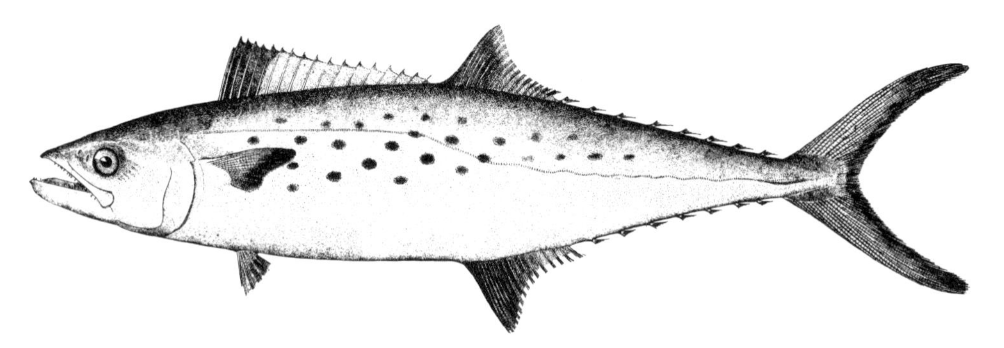

Figure 275. Spanish mackerel *Scomberomorus maculatus*. New York market. Drawn by H. L. Todd.

Description. Body shaped more like a slender mackerel than a stout bonito, four and a half to five times as long as deep (Fig. 275). First dorsal fin triangular. Second dorsal concave, originating a short distance in front of anal fin. Anal fin similar to second dorsal in form and size. Pectoral fins naked. Caudal fin deeply lunate, outer rays decidedly longer than those of mackerel. Caudal peduncle with median keel. Lateral line wavy.

Meristics. Dorsal fin rays XVII–XIX, 17–20 + 7–9 finlets; anal fin rays 17–20 + 7–10 finlets; pectoral fin rays 20–23; gill rakers on first arch (1–4) + (8–13) = 12–14; teeth 7–32 in each jaw; vertebrae (21 or 22) + (30 or 31) = 51–53. Tables of meristic and morphometric characters were presented by Collette and Russo (1985).

Color. Spanish mackerel are dark bluish or blue green above, pale below, like all scombrids, and silvery; sides marked with about three rows of round to elliptical dull orange or yellowish, spots, both above and below the lateral line. The membrane of the anterior third of the first dorsal fin is black, whereas the rear part is white, a useful field mark. Second dorsal and pectoral fins are pale yellowish with dusky edges; anal and pelvic fins are white.

Size. Reach about 77 cm FL and 4.8 kg. The all-tackle game fish record is a 5.89-kg fish caught in Ocracoke Inlet, N.C., in November 1987 (IGFA 2001).

Distinctions. There is no danger of confusing Spanish mackerel with either of the true mackerels: first because the two dorsal fins (like those of the bonitos) are close together and second because of their color pattern. The high second dorsal, slender form, and spotted sides separate them from bonitos, whereas the color, slender form, long first dorsal fin, and outline of the second dorsal distinguish them from small tuna. The most distinctive anatomical characters of Spanish, king, and cero among local scombroids are the large compressed jaw teeth. Of the three species of *Scomberomorus* that might occur in the Gulf, *S. maculatus* have an intermediate number of gill rakers, 10–16, usually 12–14. King mackerel have fewer, 6–11, usually 10 or fewer, and cero have more, 14–17. King mackerel also have fewer spines in the first dorsal fin, 14–16 vs. 17–19, and an abrupt downward dip in the lateral line under the second dorsal fin, which is absent in Spanish mackerel and cero. Cero have one or two longitudinal stripes in addition to the spots of Spanish mackerel; king mackerel lack spots and stripes.

Food. Food consists mainly of small fishes with lesser quantities of pandalid and penaeoid shrimps and squids. Clupeoids, such as round herring, menhaden, and alewives, along with anchovies, are particularly important forage off North Carolina and Florida (Bigelow and Schroeder; Bowman et al. 2000).

Biology. Four species of parasitic copepods are known from Spanish mackerel (Cressey et al. 1983) and a number of additional parasites were listed by Berrien and Finan (1977b). In Florida, females attain sexual maturity at 25–37 cm FL and males at 28–34 cm (Klima 1959). Spawning takes place in New York–New Jersey in late August to early September, earlier to the south (Earll 1883). Fecundity ranges from 300,000 to

1,500,000 eggs in specimens weighing 0.5–2.7 kg (Earll 1883). Summaries of biological information have been presented by Berrien and Finan (1977b) and Collette and Russo (1985). Development is summarized by Fritzsche (1978) and Fahay (1983:320–321). There is also a useful annotated bibliography by Manooch et al. (1978).

General Range. An epipelagic, neritic species found along Atlantic and Gulf of Mexico coasts of North America, north commonly as far as Chesapeake Bay. Spanish mackerel are summer visitors along the Atlantic coast of the United States north to New York; less regularly along the southern coasts of New England, although a few are taken during most summers at Woods Hole. The NMFS groundfish survey caught two 16-cm individuals in Buzzards Bay in September 1994. Replaced from Belize south to Brazil by *S. brasiliensis* (Collette and Russo 1985).

Occurrence in the Gulf of Maine. Spanish mackerel are only strays in the colder waters of the Gulf of Maine, where occasional fish are taken in Cape Cod Bay every year or two. In 1896, the local catch was 37 fish (Provincetown and Truro traps), and there are early records from Monhegan Island, Maine (Goode and Bean 1879), and Lynn, Mass. (Storer 1843). The northern record is off Sauls Island, Halifax, N.S. (Gilhen and McAllister 1989).

CERO / *Scomberomorus regalis* (Bloch 1793) / Bigelow and Schroeder 1953:348

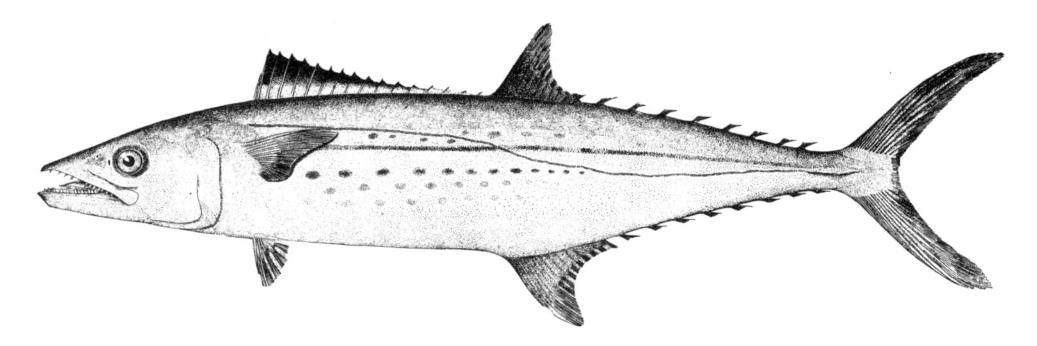

Figure 276. Cero *Scomberomorus regalis.* Key West, Fla. Drawn by H. L. Todd.

Description and Distinctions. Cero resemble Spanish mackerel closely in general appearance, but are marked by a narrow yellow-orange stripe running from close behind each pectoral fin to the base of the caudal, crossing the lateral line as the latter curves downward below the second dorsal fin (Fig. 276). Cero also have scales on their pectoral fins.

Meristics. Dorsal fin rays XVI–XVIII, 16–19 + 7–9 finlets; anal fin rays 15–20 + 7–10 finlets; pectoral fin rays 20–24; gill rakers on first arch (2–4) + (10–14) = 12–18; vertebrae (19 or 20) + 28 = (47 or 48). Tables of meristic and morphometric data were presented by Collette and Russo (1985).

Size. Reach 83.5 cm FL. The all-tackle game fish record is a 7.76-kg fish caught in Islamadora, Fla., in April 1986 (IGFA 2001).

Biology. In the West Indies, 96% of the food of cero consists of small schooling fishes, particularly herringlike fishes (Randall 1967). Seven species of parasitic copepods are known from cero (Cressey et al. 1983). Males mature between 32.5 and 34.9 cm FL and females at about 38 cm. Fecundity varies from 160,000 to 2.23 million eggs in females 38–80 cm FL.

General Range. Atlantic coast of North America, Cape Cod to Brazil, particularly abundant in coral reef habitats in the West Indies and around southern Florida (Collette and Russo 1985).

Occurrence in the Gulf of Maine. Cero were recorded by Dr. W.C. Kendall at Monomoy at the southern elbow of Cape Cod but have not been recorded elsewhere in the Gulf.

ATLANTIC BLUEFIN TUNA / *Thunnus thynnus* (Linnaeus 1758) / Horse Mackerel /

Bigelow and Schroeder 1953:338–347

Description. Body robust, one-fourth to one-sixth as deep as long, tapering to a pointed snout and to a very slender caudal peduncle that bears a strong median longitudinal keel on either

backward from first spine, depressible into a groove, last spine very short. Second dorsal fin almost confluent with first, a little lower than latter in young fish and a little higher in older

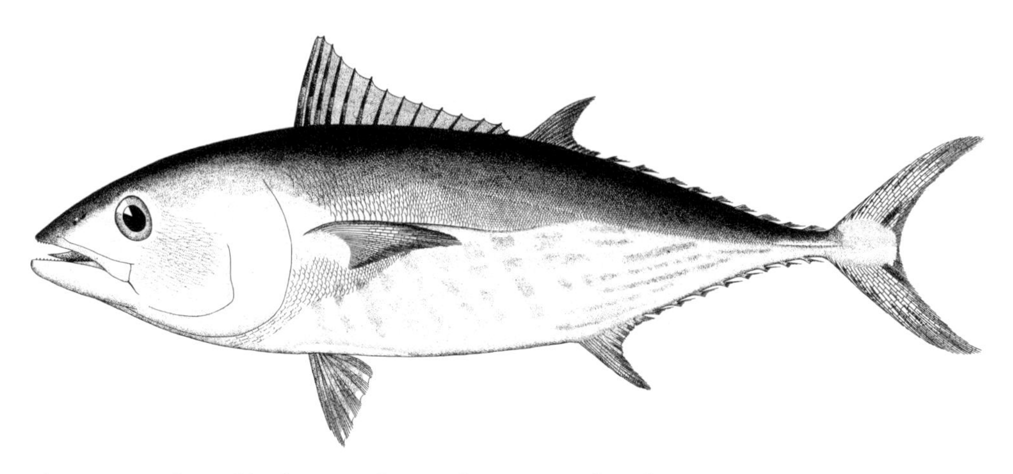

Figure 277. Atlantic bluefin tuna *Thunnus thynnus.* Woods Hole, Mass., 667 mm.

end of second dorsal, similar in outline and size to second dorsal. Caudal fin much broader than long, margin evenly lunate, both lobes sharply pointed. Pectoral fins very short for a tuna, less than 80% of HL, never reaching interspace between dorsal fins. Minute scales cover body behind anterior corselet of enlarged scales. Swim bladder large. Liver with three subequal lobes, ventral surface striated.

Meristics. Dorsal fin rays XII–XIV, 13–15 + 8–10 finlets; anal fin rays 13–16 + 7–9 finlets; pectoral fin rays 30–36; gill rakers on first arch 34–43; vertebrae 18 + 21 = 39. Meristic and morphometric characters were summarized by Gibbs and Collette (1967).

Color. Back dark lustrous steel blue or nearly black, with gray or green reflections; cheeks silvery; sides and belly silvery gray, often with large silvery spots and bands, and iridescent with pink. First dorsal fin dusky to blackish; second dorsal dusky to reddish brown; dorsal finlets yellow with dark edgings. Anal fin silvery gray; anal finlets the same, or yellow; caudal fin dusky but more or less silvery; pelvic and pectoral fins blackish above and silvery gray below. The median caudal keel is black in adults.

Size. Reach over 300 cm FL; common to 200 cm. This is the largest Gulf of Maine fish, except for some sharks; a length of 4 m or more and a weight of 726 kg are rumored. The all-tackle angling record is a 679.0-kg fish of 304 cm FL taken off Aulds Cove, N.S., in October 1979 (IGFA 2001). The heaviest Rhode Island fish on record, taken about 1913, weighed 556 kg; four or five fish were brought into Boston that weighed approximately 545 kg each, and one in 1924 was said to have reached 591 kg; and Sella (1931) mentioned a "fairly well-authenticated instance" of one caught 60–70 years ago off Narragansett Pier, R.I., that weighed in the neighborhood of 682 kg, which was divided among the various hotels and fed 1,000 people.

Distinctions. The two dorsal fins of bluefin tuna are practically continuous, a character (with the numerous finlets) sufficient to separate young bluefin from either of the true mackerels. Small bluefin are readily separable from skipjack and little tunny because the entire trunk of the tuna, including the belly, is scaly and the upper outline of the first dorsal fin only very slightly concave; they can be distinguished from bonito by a second dorsal that is considerably higher than it is long, the shape of the anal with only a weakly concave margin, the small size of the jaw teeth, and the midline of the roof of the mouth armed with fine teeth. The plain coloration of tuna, without dark markings, is yet another convenient field mark for separating small ones from any scombrid reported from the Gulf. Bluefin have more gill rakers on the first arch (34–43) than either of the other two tunas of the genus *Thunnus* that might possibly wander into the Gulf, the albacore, *T. alalunga,* and the bigeye tuna, *T. obesus,* which would have, at most, 33 gill rakers.

Taxonomic Note. Bluefin tuna in the North Atlantic and North Pacific oceans have been considered as subspecies *Thunnus thynnus thynnus* in the North Atlantic and *T. t. orientalis* in the North Pacific (Gibbs and Collette 1967; Collette and Nauen 1983). Molecular data coupled with morphological differences have led to a reappraisal of the status of the two populations and a decision to recognize both as separate species (Collette 1999).

Habits. Bluefin are epipelagic, usually oceanic but depending on the season may approach close to shore. They are found in moderately warm seas but are more tolerant of cold water than most of their relatives. Offshore in the northwest Atlantic, large bluefin are taken at surface temperatures of 6.4°–28.8°C, 6.5°–26.9°C at the estimated fishing depth on longline sets (Squire 1963). Of 854 bluefin caught on longlines, 50% were taken at 12.8°–15.5°C and 87.5% at 12.8°–23.9°C (Squire 1963: Table 2).

Smaller-sized tuna seem rather closely restricted to regions where the surface layer is warmer than 15°–17°C, and whereas large ones are regular summer visitors to the eastern side of the Gulf, where the water warms only to about 10°–12°C, this is probably about the lower limit to the thermal range they favor. Tuna that visit the west coast of Newfoundland find summer temperatures as high as 15°–16°C along the southern coast of Newfoundland, and 13°–14°C in Trinity and Conception bays on the southeastern part of the Newfoundland coast. Few tuna, for example, whether large or small, are seen in the Passamaquoddy region during most summers: although the multitudes of small herring would seem to offer ideal feeding conditions, the temperature there, even in August, when it is at its highest, rises only to about 11°–12°C. Seasonal chilling is generally accepted as the factor that drive tuna from northern coasts in autumn.

Tuna tolerate a wide range of salinity; they run well up into bays in pursuit of herring, (e.g., Bras D'or "lake," Cape Breton, on the outer Nova Scotian coast; Bonne Bay on the west coast of Newfoundland; and Trinity and Conception bays) and even into harbors.

Tuna that are seen or caught in the Gulf of Maine are all near the surface or at least where the water is not more than 65–74 m deep. Eleven adult bluefin were tagged on Stellwagen Bank or in Cape Cod Bay and tracked with ultrasonic telemetry (Lutcavage et al. 2000). All the fish remained within the Gulf. Mean swimming depth was 14 ± 4.7 m and maximum depth for individuals ranged from 22 to 215 m but 90% of their time was spent in the uppermost 30 m. Mean speed over the bottom was 5.9 $km \cdot h^{-1}$ but for brief periods surpassed 20–31 $km \cdot h^{-1}$.

Tunas have evolved elaborate retia mirabile in the circulatory system that act as countercurrent heat exchangers. These heat exchangers form thermal barriers that prevent metabolic heat loss, thus enabling these fish to maintain a high internal temperature, as high as 28.8°C for a bluefin taken from 7.3°C water. Bluefin can thermoregulate and can maintain a relatively constant muscle temperature over a wide range of ambient water temperatures (Carey and Teal 1969; Block 1991; Stevens et al. 2000).

Small, medium, and fairly large-sized fish, up to 159–227 kg or so, commonly travel in small schools of half a dozen to 30 or 40 fish, but sometimes in much larger schools, each school usually composed of fish of about the same size. Bigelow and Schroeder never heard of large and small tuna schooling together, and it also seems that very large fish are usually solitary (Crane 1936).

When tuna are at the surface, as they often are, they are proverbial for their habit of jumping, either singly or in schools; they may do this when swimming about, or harrying smaller fishes, or, less often, when traveling in a definite direction, in which case all the fish that are jumping are doing so in the same direction. Frank Mather, for instance, reported seeing a school of 90-kg fish jumping in unison 60–90 cm clear of the water. When large tuna jump, they sometimes fall flat, making a great splash. When schools at the surface are not jumping, they often splash a good deal, which makes them conspicuous. Bigelow and Schroeder once sighted a large school so employed, 5 km off the Cohasset shore. Even if they case, if the sea is smooth the wakes that large ones leave behind them betray their presence.

On calm days, they sometimes cut the surface with the sickle-shaped second dorsal fin and the tip of the caudal fin, and they have been photographed that way. Bigelow and Schroeder and experienced tuna fishermen noted that it is rare for tuna to fin in the Gulf of Maine. They often break the surface when striking a bait and then may even leap clear. For some reason they do not ordinarily jump after they are hooked, but first make one or more swift shallow runs and then tend to bore deep unless they are in very shallow water.

Food. Bluefin eat fishes, squids, crustaceans, salps, and other invertebrates. Fishes and squids are the principal food for two size groups of bluefin, 52–102 cm and 160–267 cm, occurring from the Carolinas to New York and from the Bahamas (Dragovich 1970). By volume, fish families that made up the major portion of the diet were: Scombridae (34.6%), Bramidae (6.2%), Myctophidae (5.5%), and Gempylidae (2.3%). In the Gulf of Maine bluefin devour great numbers of herring, large and small, as well as mackerel, which often fill their stomachs. Tuna pursue silver hake, and at Portland, Maine, these were found in the stomachs of 26 out of 30 individuals (Crane 1936). Crane also found squids in two at Portland and quantities of euphausiid shrimps (*Meganyctiphanes*) in two others. A tuna has been known to swallow a whole dogfish as large as 3.6 kg. South of Cape Cod they prey intensively on menhaden. Tuna sometimes strand in pursuit of their prey.

Acoustic telemetry has shown that the stomach temperature of large bluefin tuna held in an impoundment changes markedly during feeding (Carey et al. 1984). The stomach cools rapidly on ingesting cold food. It then warms to a maximum of 10°–15°C above water temperature over a period of 12–20 h. Temperature decreases slowly over the next 20–30 h to a final state in which it remains 3°–6°C above the temperature of the water. Conductive heat losses are reduced by an overlying gas bladder and by the thick fatty muscle of the body wall. Convective heat losses are prevented by heat exchangers in the circulatory system. The temperature rise can be accounted for by heat released in the hydrolytic processes of digestion and by an increase in metabolic rate. The advantage of a warm stomach is that protein is digested in about one-third the time, so tuna can process about three times as much food per day as would otherwise be possible under ambient conditions (Carey et al. 1984; Stevens and McLeese 1984).

Predators. Tuna have no serious enemies in the Gulf of Maine, but killer whales take a toll of them in Newfoundland waters, where, "one or more times annually, usually in September, orcas will ravage the tuna schools in the bays they frequent most." Other known predators are white shark, longfin and shortfin mako, and bluefish (Bowman et al. 2000).

Parasites. Seven species of parasitic copepods are known

of parasites are listed by Tiews (1963: Table 11). Parasite infestation increases with host age (Walters 1980).

Breeding Habits. In the western Atlantic, spawning takes place principally in the Gulf of Mexico from 15 April to 15 June, with some larval occurrence in the Straits of Florida (Richards 1977; McGowen and Richards 1986, and included references). Juvenile bluefin were found in tern stomachs in the vicinity of the Dry Tortugas (Potthoff and Richards 1970). Major spawning grounds in the eastern Atlantic are in the central Mediterranean Sea from late April through about mid-July (Tiews 1963).

For bluefin 205–269 cm FL and 156–324 kg round weight, estimates for the average number of ovarian eggs measuring 0.33 mm in diameter and larger and 0.47 mm and larger were, respectively, 60.3 million and 34.2 million (Baglin 1982).

Early Life History. Fertilized eggs are buoyant, spherical, transparent, and small for so large a fish (1.05–1.12 mm in diameter) with one oil globule of about 0.27 mm.

Larval *Thunnus* are identified to species based solely on the distribution of melanophores and erythrophores found on the tip of the lower jaw and on the trunk. Bluefin larvae have melanophores on both ventral and dorsal margins of the body (Matsumoto et al. 1972; Richards and Potthoff 1975). Development has been summarized by Fritzsche (1978:161–166), with figures of a developmental series from a 3.0-mm yolk-sac larva to 16.8-mm juvenile.

Age and Growth. Growth of bluefin tuna based on specimens captured mostly in the Cape Cod and Long Island areas was studied by counting annuli on scales and vertebrae and by analyzing length-frequency data (Mather and Schuck 1960). Similar results obtained by these two methods support their validity for ages 0–4. Older ages were determined by counting annuli only, but tag returns and weight-frequency data afford some corroboration for ages 5–7. Growth is believed to be extremely rapid during the first summer and about 20 mm per month in the first winter. During the next 3.5 years, they grow at a rate of about 32 mm per month in summer and about 0.8 mm per month in winter, or about 178 mm per year. Growth rate appears to decline gradually to about 100 mm per year in this region. Only slight differences were found between sizes and growth rates of fish of the same age taken in different years.

Recent aging studies on bluefin have utilized otoliths and vertebrae (Hurley and Iles 1983; Lee et al. 1983; Prince et al. 1985). Bluefin are relatively long-lived with a maximum longevity of more than 30 years (Hurley and Iles 1983). Based on otolith sections of 1,416 northwest Atlantic bluefin, individual length-at-age data were fitted to the von Bertalanffy growth curve and produced estimates for L_{∞}, k, and t of 278 cm FL, 0.17, and 0.25 year for males and 266 cm FL, 0.17, and 0.11 year for females, respectively (Hurley and Iles 1983).

General Range. Atlantic bluefin inhabit warmer parts of the Atlantic (and the Mediterranean) north regularly to Hamilton Inlet, Labrador, and the west, south, and southeast coasts of Newfoundland in the western Atlantic; to Iceland and northern Norway (Lofoten Islands) on the European side.

Occurrence in the Gulf of Maine. Bluefin tuna are yearly visitors to the Gulf. Tuna can be found all around the shores of the Gulf from Cape Cod to eastern Maine, in the Bay of Fundy, and along the west coast of Nova Scotia. Fishermen often report them on Nantucket Shoals and Georges and Browns banks. Most of the 126 schools of giant bluefin photographed by airplane spotter pilots were in four areas traditionally fished for them: Great South Channel, Wilkinson Basin, Platts' Bank, and Jeffreys Ledge (Lutcavage and Kraus 1995; Lutcavage et al. 1997).

In ordinary years the first bluefin are likely to be seen as early in the season between Cape Ann and the Maine state line as they are off Cape Cod. In 1950, for example, the earliest report of them was off Hampton, N.H., 26 May and the next off Plum Island, Mass., on 9 June; it was not until about 16 June that word came of one hooked in Cape Cod Bay. Tuna are to be expected throughout the western side of the Gulf generally by the middle or end of June, which is about as early as they ordinarily appear in any numbers off southern New England, and on the Nova Scotian side of the Gulf by 1 July if not earlier. The peak season is usually from about the middle or end of July to the middle of September off Massachusetts, July and August off Casco Bay, and through August and September along western Nova Scotia.

The vicinity of Provincetown, within Cape Cod Bay, has long been known as a center of abundance for tuna. Other well-known centers are from Cape Ann north to Boone Island and from the Ipswich Bay–Plum Island shore out to Jeffreys Ledge, some 50 km offshore; off the mouth of Casco Bay and for some distance thence eastward; and the vicinity of Wedgeport, on the west coast of Nova Scotia, where international tuna tournaments are held. Fewer are seen along the eastern coast of Maine.

So many tuna used to come so close inshore in Cape Cod Bay that nearly all of the commercial catch made there was taken in traps; large schools were even been sighted within Provincetown Harbor (e.g., on 11 October 1950). Tuna taken north of Cape Ann are farther out; all of them, however, are caught within 50 km or so of land. A great concentration of tuna was encountered by *Albatross III* on the southwestern part of Georges Bank, on 18 September 1950, when 25 were hooked and landed, all very small, about 5 kg apiece, but it is unusual to see large numbers on the offshore banks.

Tuna may occur in good numbers along the outer Nova Scotian coast, off Shelburne, the vicinity of Liverpool at the mouth of the Mersey River, the mouth of the La Have River, Mahone Bay, and St. Margaret Bay being centers of abundance as reviewed by Bigelow and Schroeder.

Migrations. Tuna are definitely migratory; those that visit the Gulf of Maine coasts work northward in spring and drop out of sight late in autumn. They usually arrive in Bonne Bay, on the Gulf of St. Lawrence coast of Newfoundland, in late June or early July, and a week or two later in Trinity and Conception bays, on the southeastern part of the Newfoundland coast. Bigelow and Schroeder reviewed when and where bluefin were caught in the Gulf of Maine up through the early 1950s.

Giant bluefin (over 123 kg) annually pass northward through the Straits of Florida in May and June during or just after spawning (Schuck 1982). They follow the Gulf Stream northward and usually appear in coastal waters off New Jersey and Long Island and in Cape Cod Bay about June or July and off the Maine coast, Nova Scotia, Prince Edward Island, and Newfoundland shortly thereafter. Medium-size bluefin (32–123 kg) normally appear in the Bight area in June and then move inshore. Juveniles or school tuna (3–32 kg) typically appear in early July first off Virginia, Delaware, and Maryland, next off New Jersey, then off Long Island, and finally south of Cape Cod. In some years, late season concentrations of bluefin of various sizes have occurred off the New York Bight.

Tag returns show that there is some interchange between populations of the eastern and western Atlantic. Two small individuals tagged off Martha's Vineyard were recaptured in the Bay of Biscay 2 and 5 years later (Mather 1960), two giant bluefin tagged off Cat Cay, Bahamas, were recaptured off Bergen, Norway, more than 4,000 miles away after 118 and 119 days at large (Mather 1962), and 14 small bluefin tagged in the New York Bight in the summer of 1965 were recaptured in the Bay of Biscay in 1966 (Mather et al. 1967).

Importance. Tuna were once regarded as a nuisance on the Atlantic coast, for bands of them made trouble for fishermen by following herring or mackerel into the traps and pounds, to tear their way out again through the net unless harpooned. Many years ago, when fish oil was more valuable than now, a few were sometimes harpooned for oil, which was rendered out of the heads and bellies, but there was no market for their meat. However, they have been highly valued as a food fish for many years, not only in the Mediterranean but also on the west coast of the United States. More recently a local demand developed in the Gulf of Maine, supplied chiefly by local fisheries off Casco Bay, in the Cape Ann–Boone Island region, and in the Cape Cod Bay region.

Bigelow and Schroeder reviewed annual landings on the Maine and Massachusetts coasts as they rose from about 42,700 kg in 1919, to around 114,000 kg in the early 1930s, and up to nearly 910,000 kg for 1945–1948, representing around 3,000–6,000 fish, if they averaged 136–182 kg in weight. The average value to the fisherman in 1946 was about $0.70 to $0.90 per pound. In the 1950s, bluefin was selling for about the same price paid for the Atlantic mackerel that were used as bait for exploratory longline fishing, about $0.50 a pound.

Bluefin are caught with different types of gear, such as purse seines, longlines, trolling lines, trap nets, and others.

trap fisheries (Maggio 2000). The commercial catch off the coasts of Maine was made mostly by harpoon; that off northern Massachusetts by hook and line and harpoon; and that off the Cape Cod Bay region mostly in traps. In 1945 about 60% of the catch reported for Maine was by harpoon, with almost all the remainder on handlines; in 1946 about 95% was harpooned. About 56% of the Massachusetts catch was taken in traps in 1945 and about 90% in 1946.

After World War II, commercial and sport fishing efforts for bluefin tuna increased on both sides of the Atlantic; new markets developed, more efficient fishing methods were introduced, smaller sizes were more actively sought, and the total fishing pressure on all sizes increased dramatically (Schuck 1982). Increased effort led to peak landings in 1955 from all fisheries combined of about 32,000 mt, declining thereafter to about 10,000 mt by 1970 and 7,000 mt by 1973. Declines first became obvious in European waters; by 1973 landings of some fisheries had dropped by 99% and several had gone out of business owing to lack of fish.

In the northwest Atlantic, U.S. demand for bluefin tuna was low prior to the mid-1950s. Giants were taken in modest numbers from New York Bight to Newfoundland by harpoon, handline, rod and reel, and shore traps. Medium-sized tuna were taken along with giants (except at Bimini) and with school fish in the Bight. A few young-of-the-year were also caught in the New York Bight area by sport fishermen. Catches of all sizes in the western Atlantic were modest, and the stock appeared stable. However, three developments rapidly changed the situation (Schuck 1982):

1. Long-lining for bluefin was introduced into the Atlantic by the Japanese in 1956, and they (with other nations) began heavy fishing in midocean throughout the North and South Atlantic.
2. U.S. purse-seining for bluefin tuna was introduced into New York Bight in 1962, a method to which bluefin proved extremely vulnerable. School tuna (ages 1–4) were the main targets.
3. The Japanese learned that giant tuna taken in late summer off New England and Canada had the ideal fat content for their raw fish delicacy *sashimi* and could be shipped by air to Tokyo. Their offering to buy giants at up to $1.45 a pound (instead of the previous $0.20 to $0.50 a pound) in autumn 1972 greatly increased U.S. fishing pressure on giants, particularly by rod-and-reel fishermen. Individual giant bluefin in prime condition have sold for as much as $68,000, about $45 a pound, but a new record price of $173,600 (20 million yen) was reached for a 444-lb bluefin sold in Tokyo's Tsukiji Central Fish Market in January 2001 (CNN 2001)

Management. As a result of the three changes in the fishery noted above, fishing mortality rates increased substantially in the northwest Atlantic, particularly those of school tuna

maturity, and with a declining spawning stock the status of the stock was considered serious (Schuck 1982). Since 1974, measures have been taken by the International Commission for the Conservation of Atlantic Tunas (ICCAT) to prevent any further increase in fishing mortality. In late 1982, ICCAT set the 1983 catch limit for the western Atlantic to 2,660 mt divided among the contracting parties (Canada, Japan, and the United States). Concern about the continued low level of abundance of small bluefin resulted in an ICCAT decision to limit the catch of fish smaller than 120 cm to 15% by weight of the total catch in the western Atlantic. In these waters, fisheries are also controlled through the number of licenses, limitation of fishing season, minimum size and maximum-catch-per-boat-and-day-regulations. Sport-fishing boats are also obliged to report a descriptive log of their operations on a weekly basis and use prescribed gear.

A bluefin tuna ranching operation was established in St. Margaret Bay, a deep fjordlike bay just south of Halifax, N.S. (Carey et al. 1984). Numbers of bluefin tuna come there for a brief period in early summer and are taken in mackerel traps (weirs). These tuna have just migrated from their spawning area in the Gulf of Mexico and are emaciated and of little commercial value when they arrive. By late summer the fish are fat and in prime condition for the Japanese market, but few are caught in St. Margaret Bay at that time. In the late 1970s a ranching operation developed in which tuna trapped early in the season were transferred to holding pounds and fed until the fall when they were fat enough to be sold on the *sashimi* market in Japan. The pounds were about 50 m across, 15 m deep, and made of heavy netting suspended on floats. During 1977 and 1978, each pound contained about 30 tuna, with a maximum of 98 in one larger pound.

ACKNOWLEDGMENTS. The food habits paragraphs were reviewed by Ray Bowman; a draft of the Atlantic mackerel account was reviewed by William J. Overholz; and drafts of the entire scombroid section were reviewed by Jon A. Moore, Thomas A. Munroe, and William J. Richards.

Suborder Stromateoidei

This suborder contains six families, 16 genera, and about 65 species of marine fishes (Nelson 1994). It is defined by the presence of pharyngeal pouches, toothed saccular outgrowths behind the last gill arch. Five species in three families, Centrolophidae, Ariommatidae, and Stromateidae, have been reported from the Gulf of Maine. In recent years, specimens of *Cubiceps,* a genus of a fourth stromateoid family, the Nomeidae, have been found off the northeast coast of the United States. The species of this genus most likely to stray into Gulf waters is *C. capensis* (see Agafona and Kukuyev 1990). One specimen was taken southeast of Browns Bank (Scott and Scott 1988) and another (MCZ 131847) south of the Grand Banks. Specimens of *Tetragonurus atlanticus,* belonging to a fifth stromateoid family, the Tetragonuridae, have been taken on the southern flank of Georges Bank in 70 m (MCZ 135324) and so might also occur in the Gulf.

KEY TO GULF OF MAINE FAMILIES OF STROMATEOIDEI

1a. Two dorsal fins, distinctly, though scarcely separated, the first with 10–20 spines; pelvic fins always present 2
1b. Dorsal fin continuous, or if two dorsal fins, scarcely separated, the first with fewer than 10 spines; pelvic fins present or absent 3
2a. Vomer and palatines with small, often almost indistinguishable, teeth; caudal peduncle compressed, least depth greater than 5% SL, without lateral keels; usually more than 15 rays in both dorsal and anal fins. . . . . **Nomeidae**
2b. Vomer and palatines toothless; caudal peduncle square in cross section, least depth less than 5% SL, with two low lateral keels on each side near caudal fin base; 14 or 15 rays (rarely 13 or 16) in dorsal and anal fins. . . . . **Ariommatidae**
3a. First dorsal fin with 10–20 short spines, the longest only half the length of the longest fin ray in the second dorsal fin; anal fin rays 10–14; scales keeled, heavy, very adherent; modified scales form two well-developed lateral keels on each side of caudal peduncle . . . . **Tetragonuridae**
3b. First dorsal fin with about 10 long, slender spines, often folded into a groove, the longest spine nearly as long as, or longer than, the longest fin ray in the second dorsal fin; anal fin rays 14–30; scales cycloid, thin, deciduous; no fleshy lateral keels on caudal peduncle 4
4a. Pelvic fins always present; anal fin rays 15–30; median fins never falcate, their bases rarely the same length; jaw teeth all conical and simple . . . . **Centrolophidae**
4b. Pelvic fins never present; anal fin rays 30–50; median fins often falcate, their bases about equal in same length; jaw teeth laterally compressed, either simple or with 3–5 cusps . . . . **Stromateidae**

MEDUSAFISHES. FAMILY CENTROLOPHIDAE

GRACE KLEIN-MACPHEE

Medusafishes have moderately stout bodies, short blunt snouts with convex profiles, and a moderately deep caudal peduncle without longitudinal keels. The single dorsal fin extends from over the pectoral fins to the caudal peduncle. The spiny dorsal is either reduced to a few flexible spines covered by skin so that it is hard to find them or represented by several detached spines so short that they might be overlooked. The caudal fin is slightly emarginate; the anal fin is similar to the dorsal in shape but much shorter; the pelvic fins are below the pectoral fins and are smaller than the latter; pelvic fins are retained in adults. The mouth is small, with small teeth in the jaws. The family includes seven genera with about 27 species (Nelson 1994), of which only two species are known off the Atlantic coast of the United States, both reported from the Gulf of Maine.

KEY TO GULF OF MAINE CENTROLOPHIDAE

1a. Single dorsal fin preceded by 6–8 short detached spines; sides of head scaly . . . . **Barrelfish**
1b. Dorsal fin not preceded by any detached spines; sides of head naked . . . . **Black ruff**

BLACK RUFF / *Centrolophus niger* (Gmelin 1789) / Blackfish / Bigelow and Schroeder 1953:370–371

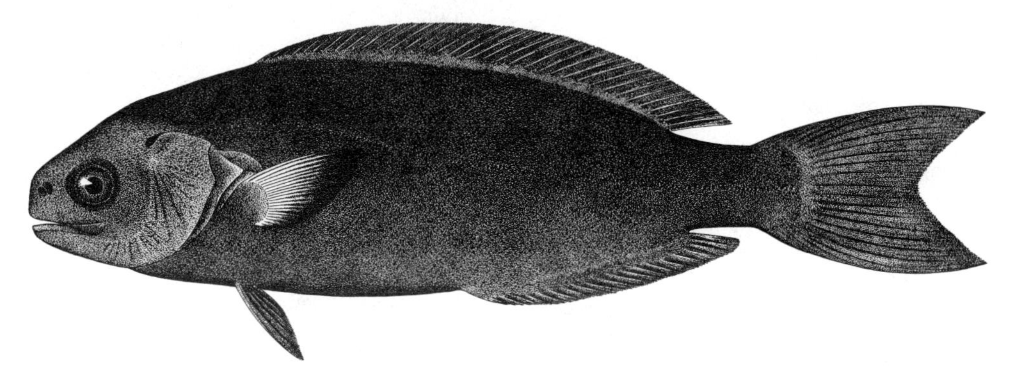

Figure 278. Black ruff *Centrolophus niger.* Dennis, Mass. Drawn by S. F. Denton.

Description. Body elongate, head small with prominent pores, eyes medium-sized with no adipose tissue, snout rounded with nostrils near tip, angle of gape extending below eye, and premaxilla not protractile (Fig. 278). Jaw with small, pointed, uniserial teeth; vomer, palatines, and basibranchials toothless. Opercle and preopercle with finely denticulate margins; opercle with two weak, flat spines. Gill rakers heavy and toothed on inner edge. Dorsal fin originates a little behind insertion of pectoral fins, continuous, with weak spines graduating to rays. Anal fin of same basic shape, originating a little behind middle of body. Pectoral fin rounded in young, pointed in adult; relative length decreasing slightly with growth. Pelvic fins inserting under posterior portion of pectoral base and folding into shallow groove. Caudal peduncle broad, thick, long. Caudal fin broad, moderately long. Scales very small, cycloid, deciduous, covering fleshy bases of median fins, absent on head. Lateral line slightly arched anteriorly, extending onto caudal peduncle. Skin fairly thick (Haedrich 1967).

Meristics. Dorsal fin with 4 or 5 weak spines, 37–43 total spines + rays; anal fin rays III, 20–27; pectoral fin rays 19–23; gill rakers 5 or 6 + 1 = 12–15, usually 19 total; lateral line scales 160–205; vertebrae 10 + 15 = 25 (Haedrich 1967; Borodulina 1989; Robins and Ray 1986).

Color. Live fish were described as dark bluish gray, sometimes almost black, with a pale peritoneum (Robins and Ray 1986). After a few weeks in alcohol, they are dark leaden brown on sooty; the fins are darker than the back, and the belly only a little paler. Other specimens have been described as brownish pink or brown all over, darkest above, some with irregular and obscure markings, either yellow or dark blue (Bigelow and Schroeder). Small specimens have been described as light cinnamon in color, with small white spots (Borodulina 1989) and vertical bands that disappear by the time they reach 10 cm (Haedrich 1967). Fraser-Brunner (1935) noted sexual dimorphism in coloration, females being lighter than males.

Size. Attain 110 cm.

Distinctions. Black ruff can be distinguished from other stromateoids by the slender body, prominent head pores, and the continuous dorsal fin with weak spines and numerous scales (160–230) in the lateral line (Haedrich 1967). They resemble pilot fish in general body form more than barrelfish, which is a closer relative, but their bodies (about two and a half times as high as it is thick) are more compressed than those of pilot fish; the caudal peduncle is much deeper and has no lateral keels; the dorsal fin is considerably longer; and there are no detached spines in front of the dorsal fin. Distinctions from barrelfish are described under that species.

Habits. Black ruff are oceanic; young fish are found near the surface, whereas larger individuals become mesopelagic. Maximum depths for Australian and New Zealand fish were 400–600 m (McDowall 1982). Capture of small groups (6–10)

taken under jellyfish (Collett 1896) and swimming with *Mola* (MacKay 1972).

Food. Food items include euphausiids, hyperiid amphipods (especially *Euthemisto* sp.), squids (Templeman and Haedrich 1966), and jellyfish (Lo Bianco 1909).

Predators. Found in stomachs of hake (Blacker 1962).

Breeding Habits. Little is known of breeding habits or spawning times, but a spent female was collected in December in the western North Atlantic (Haedrich 1967), and eggs and small fish occur in the Mediterranean from October into winter (Padoa 1956).

Early Life History. Eggs are pelagic and spherical, 1.2 mm diameter with a pigmented oil globule. Hatching occurs at about 4.4 mm, eye pigment and mouth are undeveloped. Flexion occurs at 5.7 mm SL; pelvic buds form at 6.7 mm; adult body form is attained by 17.2 mm TL, but pectoral fins are not complete until about 20 mm (Sanzo 1932). Eggs and larvae do not occur in the Gulf of Maine.

Growth. This is reported to be a fast-growing fish but there are few details. A Mediterranean specimen grew from 2 to 17 cm in 5 months (Padoa 1956).

General Range. Oceanic and widespread in low latitudes in the western North Atlantic from the Grand Banks, Nova Scotia, Massachusetts, Georges Bank (Bigelow and Schroeder; Templeman and Haedrich 1966) south to New Jersey (Wilk and Schmidt 1971). Present in the eastern North Atlantic and the Mediterranean Sea. Black ruff have also been reported from the Pacific Ocean off Australia and New Zealand (Haedrich 1967) and in the South Pacific, including the coast of South America (Borodulina 1989).

Occurrence in the Gulf of Maine. Rare in the Gulf of Maine. One specimen, about 77.5 cm, was taken in a trap at North Truro, on Cape Cod Bay on 6 September 1890; one 53 cm was brought in from the northern edge of Georges Bank in September 1936; one from east of Georges Bank in 1963 (Templeman and Haedrich 1966); one about 33 cm FL was taken in a trap at North Truro on 23 June 1951; and another about 23 cm was taken in 1888 at Dennis, Mass., but it is not known whether this last record should be credited to the Gulf or to the southern coast of Massachusetts, since that township fronts on both Cape Cod Bay and Nantucket Sound. The most recent record is a 66.5-cm TL specimen caught in a gill net in about 90 m off Cape Elizabeth, Maine, on 20 April 1980 (R. E. Goode, S. Maine Vocational Tech. Inst., pers. comm., 14 June 1989).

Importance. This is a rare stray in the Gulf of Maine with no commercial or recreational value. Black ruff were offered for sale in a market in England, but no one bought them (Blacker 1962).

BARRELFISH / *Hyperoglyphe perciformis* (Mitchill 1818) / Logfish, Rudderfish, Black Pilot /

Bigelow and Schroeder 1953:369–370 (as *Palinurichthys perciformis*)

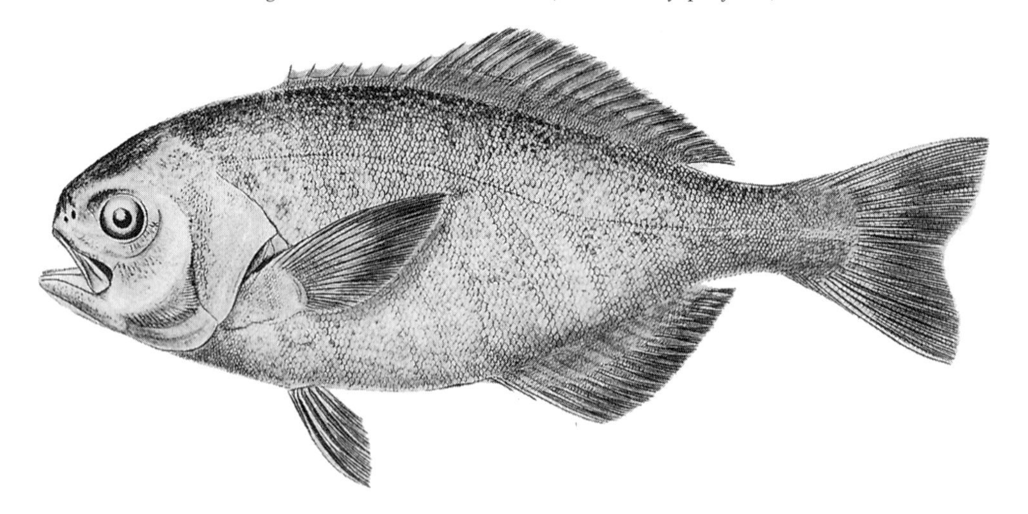

Figure 279. Barrelfish *Hyperoglyphe perciformis.*

Description. Body stout, oblong, and moderately compressed, with a bluntly rounded snout, convex forehead, and small mouth with small jaw teeth (Fig. 279). Dorsal fin consists of a series of short detached spines with very small triangular fin membranes closely followed by a long soft-rayed dorsal fin. Both fins moderately high, tapering slightly from front to rear. Dorsal fin origin just behind level of gill opening, ending at beginning of caudal peduncle. Anal fin preceded by three short spines, embedded in skin and hardly visible. Pectoral and pelvic fins large with rounded tips. Body covered with moderate-sized, somewhat deciduous, cycloid scales; top of head naked. Bases of median fins and sides of head scaly. Lateral line arched anteriorly, straightening out over middle of anal fin and extending onto caudal peduncle. Caudal fin slightly forked; caudal peduncle moderately stout, without keels.

Meristics. Dorsal fin rays VI–VIII, the first 2 almost embedded in the skin of larger specimens, 19–23; anal fin rays III, often embedded in the skin, 15–17; pectoral fin rays 18–22; lateral line scales 89–95; gill rakers 5–7 + 1 + 15–17; vertebrae 10 + 14 or 15 = 24–25 (Haedrich 1967; Haedrich and Horn 1969; Merriner et al. 1970).

Color. Barrelfish vary from blackish to green in life and either dark below and above or paling to bluish white on the belly, the latter variously mottled with darker dots and bars. They are said to change color according to their surroundings.

Size. Reach 91 cm and 12.3 kg (Robins and Ray 1986).

Distinctions. Barrelfish remotely resemble tautog in general appearance, but the rudimentary spiny dorsal fin and forked caudal fin are distinctive. Presence of dorsal fin spines and the scaly sides of the head distinguish them from black ruff.

Habits. Barrelfish owe their common name to the habit of congregating under floating planks or any drifting wreckage or inside of barrels and boxes. Off southern New England they are often found under gulfweed or any other raft of drifting seaweed or eelgrass (*Zostera*); they sometimes gather about slow-moving vessels (Bigelow and Schroeder). Adults are thought to be deep-living over the continental slope and in submarine canyons, with juveniles drifting in surface water sometimes quite close to shore. Depth range is from the surface to 244 m (Schwartz 1963).

Food. Barrelfish feed on small crustaceans, barnacles, hydroids, young squids, small mollusks, and salps, which they find near or attached to their floating homes; ctenophores; and small fishes. The diet of individuals taken at Woods Hole includes herring, menhaden, silverside, scup, mackerel, and sand lance. Sometimes stomachs contain seaweed, but this probably is eaten for the attached animals.

Reproduction. Nothing is known of the breeding habits, eggs, or larvae. Juveniles occur under flotsam, but not under jellyfish, in surface waters near the edge of the continental shelf (Haedrich 1967), and are common inshore around Massachusetts in the summer (Smith 1921).

General Range. Barrelfish occur off the Atlantic coast of North America, from Key West, Fla. (Schwartz 1963) to outer Nova Scotia and the southwestern edge of the Grand Banks (Scott and Scott 1988); most abundant south of Cape Cod.

Occurrence in the Gulf of Maine. Barrelfish are so rare in the Gulf of Maine that the only published records are one from Boston Harbor, one from Salem, one from Annisquam, one vaguely described as brought in from the fishing banks off the coast of Maine, one from the northern edge of Georges Bank (Bigelow and Schroeder), and two from the Cape Cod Canal in 1967 (Collette and Hartel 1988).

RAGFISHES. FAMILY ARIOMMATIDAE

GRACE KLEIN-MACPHEE

The family Ariommatidae includes one genus with about six species (Nelson 1994). Marine deepwater fishes, they retain their pelvic fins into adulthood. Two distinct dorsal fins are present, the first with 10–12 slender spines. The caudal peduncle has two low, fleshy lateral keels on each side.

SILVER-RAG / *Ariomma bondi* Fowler 1930 / Haedrich 1967:93–94

Description. Body elongate, caudal peduncle short with poorly defined low, fleshy keels on each side at base of caudal fin (Fig. 280). Eye large, bony supraorbital ridge pronounced; adipose tissue around eye well developed, extending over lacrimals and around nostrils. Mouth relatively small; premaxilla not protractile. Jaw teeth minute, covered basally with a membrane, uniserial; vomer, palatines, and basibranchials toothless. Gill rakers slender, one-half length of filaments with brittle spines folding into a deep groove; second dorsal with rays about half as long as first dorsal spine. Anal fin originates behind middle of body and behind origin of second dorsal fin, two or three spines preceding rays. Pectoral fin rounded in young fish, becoming pointed with growth. Pelvic fins insert under end or behind base of dorsal fin, attached to abdomen with a membrane and folding into a pronounced groove that reaches the anus. Caudal fin stiff, deeply forked

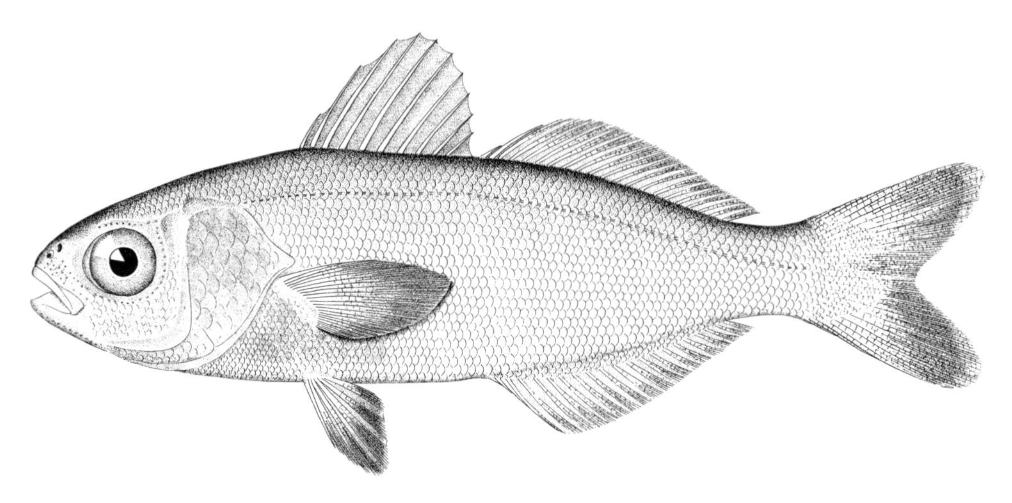

Figure 280. Silver-rag *Ariomma bondi*. Drawn by H. L. Todd.

median fins, extending over nape to level of anterior border of eye. Lateral line high on body, not extending onto caudal peduncle, a branch extending forward over eye in a bony tract; scales with branched tubes.

Meristics. Dorsal fin rays XI or XII, 14–17; anal fin rays II or III, 12–16; lateral line scales 30–45; gill rakers 8 + 15 (Haedrich 1967; Robins and Ray 1986).

Color. Bluish above and silver on the sides and below (Robins and Ray 1986).

Size. Reach 30 cm (Robins and Ray 1986).

Distinctions. Silver-rag superficially resemble jacks and herrings, but the lack of scutes, adipose tissue around the eye, and a rounded snout distinguish them from the former, and the presence of two dorsal fins and caudal keels separate them from the latter. They can be distinguished from butterfish by their slender shape, two dorsal fins, and the presence of pelvic fins and lateral keels.

Taxonomic Note. The junior synonym *Cubiceps nigriargenteus* Ginsburg 1954 was based on specimens from the Sandwich fish trap in Cape Cod Bay.

Habits. The genus *Ariomma* contains bottom or near-bottom fishes most often taken in deep water, although juveniles are taken in surface collections. All large specimens have been taken in bottom trawls, usually at depths greater than 100 m (Poll 1959; Lowe 1962). Their depth range has been reported as 60–180 m (Robins and Ray 1986). Little else is known but they are said to occur in large numbers in some regions (Robins and Ray 1986).

Predators. Found in the stomach of one weakfish (Rountree 1999).

Early Life History. The eggs of *Ariomma* have not been described, and the larvae of only one species, *A. regulus,* has been described (McKenney 1961). The larvae collected probably represented both deep-bodied and elongate forms of *Ariomma,* but there is no definitive description of silver-rag (McKenney 1961).

General Range. Western North Atlantic from Nova Scotia (Scott and Scott 1988) south to the Gulf of Mexico and Caribbean Sea to Uruguay. They also occur in the eastern Atlantic (Robins and Ray 1986).

Occurrence in the Gulf of Maine. There are two records from Nova Scotia: a 15-cm specimen from a trap net at 8 m in Georges Bay, Caribbean Head, in September 1967 (MacKay and Thomas 1969) and an 11.5-cm SL specimen from a mackerel trap in St. Margaret Bay in September 1981 (Scott and Scott 1988); it was also reported from Sandwich by Ginsburg (1954b).

BUTTERFISHES. FAMILY STROMATEIDAE

Grace Klein-MacPhee

In adult butterfishes the body is usually very deep and compressed and the pelvic fins are lost. The dorsal fin is continuous. The snout is blunt and the mouth is small. There are three genera of butterfishes and about 13 species (Nelson 1994), two of which occur in the Gulf of Maine: butterfish are common summer visitors and harvestfish are rare strays from the south. Recent information on butterfish has been summarized by Cross et al. (1999).

KEY TO GULF OF MAINE STROMATEIDAE

1a. Anterior part of anal fin about two to three times as high as posterior portion of fin; margins of anal and dorsal fins only slightly concave in outline . . . . **Butterfish**

1b. Anterior part of anal fin at least seven times as high as posterior portion of fin; margins of anal and dorsal fins very deeply concave in outline . . . . **Harvestfish**

HARVESTFISH / *Peprilus paru* (Linnaeus 1758) / Bigelow and Schroeder 1953:368 (as *Peprilus alepidotus*)

Figure 281. Harvestfish *Peprilus paru*. New York. Drawn by H. L. Todd.

Description. Body almost as deep as long, ovate in outline (Fig. 281). Snout rounded, mouth very small, and head very short. Teeth tiny, uniserial, laterally compressed with three subequal cusps (Horn 1970). Outlines of dorsal and anal fins high in front forming a falcate lobe, continuing nearly straight rearward. Scales small, thin, cycloid, and deciduous.

Meristics. Dorsal fin rays II–IV, 38–49; anal fin rays II or III, 35–45; pectoral fin rays 18–24; gill rakers 14–16; vertebrae 29–31.

Color. Harvestfish are greenish silvery above, silvery sometimes tinged with yellow on the sides and belly, and the fins of some specimens are slightly dusky or yellowish.

Size. Reach 28 cm (Robins and Ray 1986)

Distinctions. Harvestfish are similar to butterfish in general appearance but differ in the shape of the dorsal and anal fins,

these fins being much higher and falcate. The pores along the back that are so conspicuous in butterfish are absent in harvestfish and the color is greenish rather than blue.

Predators. Found in the stomach of a spiny dogfish (Rountree 1999).

Reproduction. Harvestfish spawn offshore in spring and early summer. Eggs are pelagic and about 1 mm in diameter. Larvae are 1.8 mm TL at hatching. Neither eggs nor larvae have been reported from the Gulf of Maine. Larval stages were described and compared with butterfish by Ditty and Truesdale (1983).

General Range. Cape Elizabeth, Maine, south to Uruguay, including the West Indies and Gulf of Mexico.

Cape Cod coast. There are four Gulf of Maine records: one taken at Monomoy Point by Dr. W. C. Kendall in 1896; five or six caught in floating traps at Richmond Island, off Cape Elizabeth, in July 1929; one from the Damariscotta River, Maine in August 1933; and one taken at Race Point at the tip of Cape Cod in October 1949.

BUTTERFISH / *Peprilus triacanthus* (Peck 1804) / Bigelow and Schroeder 1953:363–367 (as *Poronotus triacanthus*)

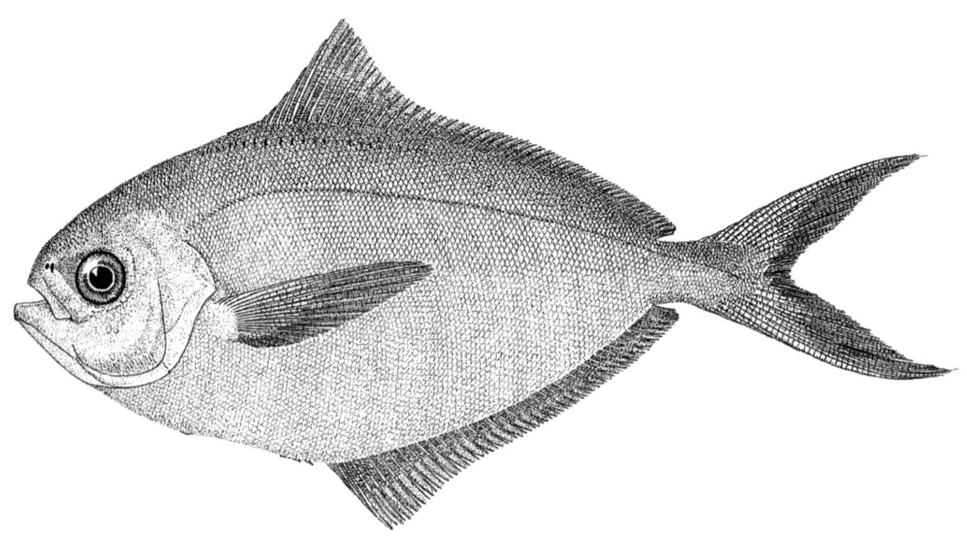

Figure 282. Butterfish *Peprilus triacanthus.* Beesleys Point, N.J., 187 mm. Drawn by H. L. Todd.

Description. Body very thin and deep, only about twice as long as deep to base of caudal fin (Fig. 282). Head short; snout blunt. Mouth small, terminal, with weak teeth. Dorsal fin single, long, soft-rayed, originating close behind axils of pectoral fins, tapering at first abruptly and then gradually backward. Anal fin almost as long as dorsal, narrowing evenly from front to rear. A very short forward-pointing spine close in front of dorsal fin and three very short spines almost wholly embedded in skin in front of anal fin, first of which points forward. Both dorsal and anal fins extend posteriorly almost to caudal base. Pectoral fins long and pointed. Caudal fin deeply forked; caudal peduncle short, slender, without keels. Scales very small, cycloid, and easily detached. A row of conspicuous pores below anterior half of dorsal fin. Lateral line high on sides, slightly arched.

Meristics. Dorsal fin rays II–IV, 40–48; anal fin rays III, 37–44; pectoral fin rays 17–22; gill rakers 22–25; lateral line scales 96–105; vertebrae 30–33 (Horn 1970).

Color. Leaden bluish above and pale on the sides, with numerous irregular dark spots that fade after death. Belly silvery.

Size. Reaches about 30.5 cm (Hildebrand and Schroeder 1928); average length is 15–23 cm.

Distinctions. Absence of pelvic fins in adults separates them from jacks; lack of obvious dorsal spines distinguishes them from scup and John Dory. They are easily distinguishable from harvestfish by the much lower dorsal and anal fins and the presence of pores along the back.

Habits. Butterfish show a decided preference for sandy rather than rocky or muddy bottoms. General experience is that butterfish keep near the surface over depths of 22–55 m during their stay near the coast. They often come close inshore, where schools are frequently seen on shallow flats, sheltered bays, and estuaries. They appear to spend winter and early spring near bottom in depths down to 183 m. They are eurythermal and euryhaline, tolerating temperatures of 4.4°–21.6°C (Fritz 1965; Schaefer 1967; Horn 1970) and salinities from 5 ppt to full-strength seawater (Musick 1972). They travel in small bands or loose schools. Draggers reported catching several times as many by night as by day, whereas NEFSC survey results indicated that day catches were higher (Waring 1975).

Food. Identifiable stomach contents of 852 butterfish (Bowman et al. 2000) showed a predominance of urochordates (29% by weight: Larvacea 14%, Ascidacea 9.2%, and Thaliacea 4.8%) and thecosome mollusks (18.6%, mostly *Clione*). Minor food items included cnidarians; ctenophores; chaetognaths; polychaetes (mainly Tomopteridae); and crustaceans such as amphipods, copepods, mysids, and euphausiids (Fritz 1965; Horn 1970; Bowman and Michaels 1984; Bowman et al. 1987, 2000).

A ctenophore, *Mnemiopsis leidyi,* has been shown to be an important component of the diet of butterfish juveniles in Narragansett Bay, R.I., and feeding behavior of butterfish on ctenophores was observed in the laboratory (Oviatt and Kremer 1977). They fed both day and night and their usual mode was to bite at the ctenophores until they were consumed, rather than ingesting them whole. Night feeding may be facilitated by florescence of the ctenophores. Consumption of ctenophores by butterfish was 8.2 mg carbon·g dry $wt^{-1} \cdot h^{-1}$ (Oviatt and Kremer 1977). Basal respiration rate of juvenile butterfish has been calculated as 0.2 mg O_2·g wet $wt^{-1} \cdot h^{-1}$.

Predators. Butterfish are preyed upon by 30 species of fishes and squids (Rountree 1999). They form an important part of the diet of a number of commercially important fishes, including haddock, silver hake, swordfish, bluefish, weakfish, and summer flounder (Horn 1970; Rountree 1999); goosefish and hammerhead shark (Maurer and Bowman 1975; Armstrong et al. 1996); as well as longfin squid (Tibbetts 1977; Rountree 1999).

Parasites. Butterfish are infected by a monogenetic trematode, *Microstyle poronoti* and a digenetic trematode, *Lepidapedon elongatum* (Murawski et al. 1978).

Breeding Habits. Butterfish begin to mature in their second summer (age 1) at about 180 mm SL (Horn 1970) and are fully recruited by their third summer (age 2) (DuPaul and McEachran 1973). Median length at maturity for female and male butterfish was 120 and 114 mm, respectively, and median age at maturity was 0.9 year for both sexes (O'Brien et al. 1993). They are reported to be broadcast spawners with no special courtship behavior (Horn 1970), although no direct observations of spawning have been made. There is no information on fecundity. They spawn once a year.

Spawning occurs primarily in the early evening (Ferraro 1980) in late spring and summer, with most activity in June and July (O'Brien et al. 1993; Berrien and Sibunka 1999). Butterfish eggs were collected by MARMAP surveys from April to September (Berrien and Sibunka 1999: Fig. 19). As the season advances and water temperatures rise, spawning proceeds shoreward in a south to north progression extending onto Georges Bank and into the Gulf of Maine (Murawski et al. 1978; Berrien and Sibunka 1999); it probably does not take place below 15°C (Colton 1972). Spawning in the Gulf of Maine begins in June, soon after the butterfish arrive; the height of the reproductive season is in July, and maximum egg concentrations occur in August (Berrien and Sibunka 1999). On the Scotian Shelf, spawning occurs from July to October (Markle and Frost 1985).

Early Life History. Butterfish eggs are buoyant, transparent, spherical, 0.68–0.84 mm in diameter, usually with a single oil globule of about 0.17–0.21 mm (Markle and Frost 1985; Able globules that coalesce as development advances. The yolk is homogeneous and unpigmented, and the perivitelline space is narrow. Butterfish eggs resemble those of red hake and fourbeard rockling, but have unpigmented yolks, whereas the two other species show dark pigment on the yolk. Scup eggs are very similar to those of butterfish but are larger, 0.85–1.15 mm (Elliott and Jimenez 1981). The incubation period is about 48 h at 18°C and 72 h at an average temperature of 14.6°C (Colton and Honey 1963).

MARMAP surveys made from 1977 to 1984 from Cape Hatteras, N.C., to Cape Sable, N.S., showed that butterfish larvae ranked 10th of the 19 numerically dominant species collected in the Gulf of Maine and 18th out of 20 collected on Georges Bank. They were most abundant from July through October (Morse et al. 1987).

Larvae are about 1.68–1.75 mm TL at hatching and are characterized shortly after by their short, deep form, their 30 myomeres, and the row of black spots along the ventral edge in the postanal region. The oil globule is at the posterior end of the yolk. The vent, located some distance from the posterior end of the yolk sac a little more than halfway down the body, opens at the finfold margin (Colton and Honey 1963). Butterfish larvae resemble those of scup, but the latter have more closely spaced melanophores in the ventral row of pigment (Colton and Marak 1969). Dorsal, anal, and caudal fin rays are visible in larvae of 6 mm, when the body has already begun to assume the deep, thin form characteristic of adults. At a length of 15 mm, the caudal fin is deeply forked, the dorsal and anal fins are formed, and the little fish resembles an adult sufficiently well for ready identification. A detailed description of larval development is given by Ditty and Truesdale (1983).

During the first summer young butterfish often live in the shelter of large jellyfishes and Goode graphically described larvae of 50–65 mm as swimming among the tentacles of the red jellyfish (*Cyanea*), sometimes ten to fifteen little fish under one jellyfish, where they find protection from predators but to which they sometimes fall prey. This association, however, is not essential to their welfare, for larvae are often seen living independently at the surface, particularly in sheltered bays west and south of Cape Cod. They end their association with jellyfish by about 60 mm and assume the adult schooling behavior.

Age and Growth. The maximum age reported for butterfish is 6 years (Draganik and Zukowski 1966); the typical life span is probably 2–3 years. Fish sampled from waters of the Gulf of Maine to Cape Fear show two types of otolith growth patterns (Dery 1989c), although they are not always clearly differentiated. The offshore pattern is characteristic of butterfish sampled in waters deeper than 27 m. These are predominant in survey and commercial catches and exhibit more clearly defined annular zones. Checks may be prominent but can be distinguished from annular zones and do not normally complicate age determination. The inshore type of otolith growth pattern is char-

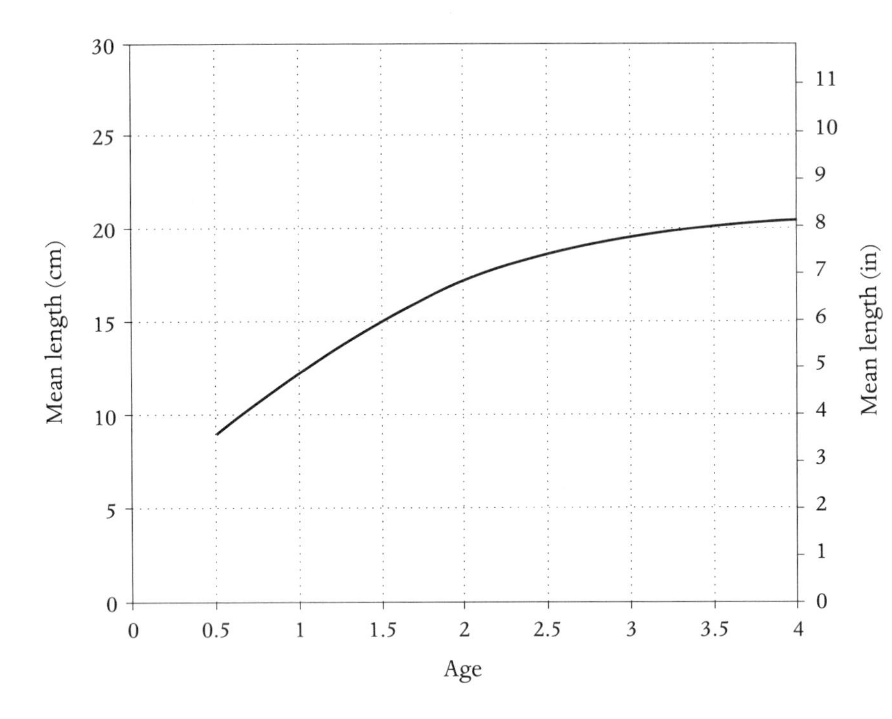

Figure 283. Length-age relationship of butterfish *Peprilus triacanthus,* Georges Bank–Mid-Atlantic stock. (From Penttila et al. 1989:15.)

especially from the New York Bight to Cape Fear. These are often difficult to age. The pattern has numerous checks and diffuse annuli, and the otoliths are often poorly calcified. The fish most difficult to age are those that occur in shallow waters off Maryland, south to Cape Fear. Age-length curves have been published for fish from southern New England (Waring 1975); the York River, Va. (DuPaul and McEachran 1973); and ICNAF Subarea 5 and Statistical Area 6 off the northwest Atlantic (Kawahara 1978). Otoliths examined by Waring (1975) and Kawahara (1978) exhibited the offshore pattern and are most typical of fish found in the Gulf of Maine.

Butterfish grow to about half their adult size in their first year. There is a difference in growth between those spawned early in the year and those that hatch later. Modal sizes were June, 0–4 mm, July–August 0–4 and 5–9 mm, September 15–19 and 35–39 mm, and October 70–74 mm (Perlmutter 1939). Larval growth rates of 0.23 mm·day^{-1} for fish up to 30 mm have been reported, based on analyses of otolith increments (Rotunno 1992). Estimates for length-at-age of butterfish are given in Fig. 283 (Penttila et al. 1989).

General Range. Atlantic coast of North America from off South Carolina and coastal North Carolina waters to the outer coast of Nova Scotia and Cape Breton; northward as a stray to the Gulf of St. Lawrence and to the south and east coasts of Newfoundland; southward to Florida in deep water.

Occurrence in the Gulf of Maine. Butterfish are regular summer visitors to the Gulf of Maine, locally abundant along the shores of Massachusetts off Barnstable, Plymouth, and Essex counties; less common along the coast of Maine (Map 33). They are also common in some years along the Nova Scotian coast of the Gulf but they appear only irregularly and in small numbers on the New Brunswick shore of the Bay of Fundy, although they have been taken frequently in Passamaquoddy Bay.

Butterfish also appear in the Nantucket Shoals region and on Georges Bank in summer, often in large numbers. They occur off the coast of Nova Scotia in St. Margaret Bay and Halifax Harbor in summer and autumn, and are said to be common eastward as far as Canso. This appears to be the normal limit to their range, for only strays have been taken in the Gulf of St. Lawrence and on the Newfoundland coast.

Migrations. Butterfish north of Cape Hatteras display definite migratory patterns in response to water temperature (Colton 1972). Summer movements are both inshore and northward (Horn 1970; Fritz 1965). South of Cape Hatteras, there is no inshore-offshore migration (Caldwell 1961; Horn 1970).

Butterfish are warm-season fish in the Gulf of Maine. They may appear off Rhode Island by the last half of April and at Woods Hole by the middle of May, although they are not plentiful in the Woods Hole region until June. It is likely that these early comers move in across the shelf from offshore, rather than following along the coast. They occur on Georges Bank in early June, but it is not until the end of that month or early in July that they are plentiful anywhere north of the elbow of Cape Cod. From that time on they are found in the inner parts of the Gulf and on Georges Bank throughout the late summer and autumn. They are exceedingly irregular and unpredictable in their appearances and disappearances. Most leave the Gulf by November, but they all vanish from the coast by the end of December at the latest, not only from the Gulf but also from the more southerly part of their range. They overwinter along the 183-m contour off the continental shelf, but the offshore movement is not so extensive south of Delaware Bay and there is apparently some movement to the south in shallower waters (Waring and Murawski 1982).

Importance. Bigelow and Schroeder described butterfish as one of the best table fish: fat, oily, and delicious. They were often used to enrich land in planting during the first half of the nineteenth century, and appreciation of the fact that they were too good for this use was a later development.

Butterfish have been landed by domestic fishermen since the 1800s. From 1920 to 1962, the annual domestic harvest averaged 3,500 mt (Overholtz 1998a). An historical summary of catches in Massachusetts is given in Bigelow and Schroeder and for New England from 1880 to 1965 in Murawski et al. (1978). In the 1960s distant-water fleets began to exploit butterfish and landings peaked in 1973 at 19,500 mt (Murawski and Waring 1979; Fig. 284). Following the phase-out of foreign fishing, annual landings declined and since 1985 have averaged less than 5,000 mt·year^{-1} (Fig. 284). The commercial catch is made mostly (98%) by otter trawl. Butterfish and longfin squid (*Loligo pealei*) exhibit high degrees of overlap in temperature and depth preferences but seasonal and areal management measures could result in significant reductions of the by-catch of either species if this becomes desirable (Lange and Waring 1992). Other gears used are pound nets, gill nets, floating traps, and purse seines.

Butterfish are managed by the Mid-Atlantic Fishery Management Council under provisions of the Atlantic Mackerel, Squid, and Butterfish Fishery Management Plan. Demand for butterfish for export to the Japanese market has decreased and overall it appears that the butterfish stock is underexploited and at a medium abundance level (Overholtz 1998a).

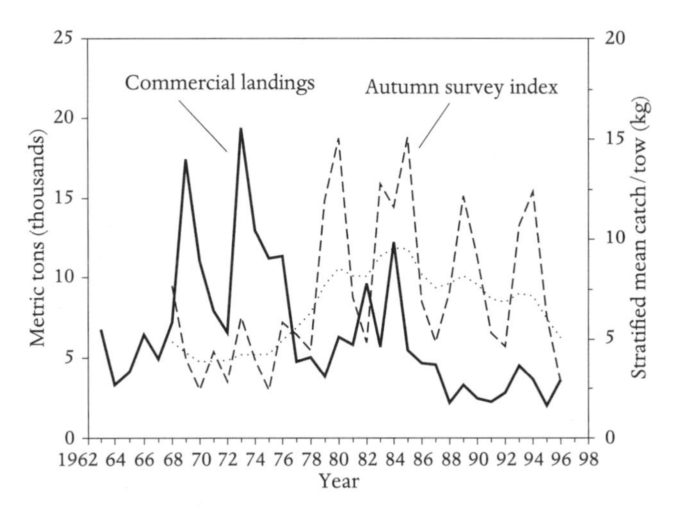

Figure 284. Commercial landings and autumn survey index for butterfish *Peprilus triacanthus* from the Gulf of Maine–Mid-Atlantic. (From Overholtz 1998a.)

Stocks. The population of butterfish from Cape Hatteras to the Gulf of Maine is considered to be a unit stock for management purposes (Overholtz 1998a). Meristic and morphometric studies indicate two separate stocks south of Cape Hatteras, which appear to be depth-isolated (Caldwell 1961; Horn 1970). The inshore population occurs to depths of 20 m from Cape Hatteras south to Florida. The deeper water form is shallower-bodied with 18 or 19 caudal vertebrae and spots. The inshore form is deeper-bodied with 17 or 18 caudal vertebrae and no spots. There appears to be a certain amount of genetic mixing but not enough to group them as a single stock.

ACKNOWLEDGMENTS. Drafts of the butterfish account were reviewed by Jon Brodziak and Gordon Waring, and drafts of the entire stromateoid section were reviewed by Richard Haedrich and Karsten E. Hartel.

FLATFISHES. ORDER PLEURONECTIFORMES

Flatfishes are a very homogeneous group, so different from all other fishes that they are unlikely to be mistaken for any other kind. What strikes one first is their flatness; less obvious is the fact that they do not lie on their bellies but on one side, right or left. Their skulls twist in the course of development so that the eye that was originally on the underside migrates around the head, until both eyes finally come to lie close together on the side that is uppermost as the fish lies on the bottom. The mouth more nearly retains its original position, so that it is often described as opening sidewise. Larval flounders swim on edge like other fishes; migration of the eye takes place shortly before the fry become demersal.

All flatfishes have a single long fin on each edge, one the dorsal and the other the anal; they also have well-developed pelvic fins (at least on the eyed side) that are on either the right- the fish lies on the bottom and the other on the lower side. The pelvic fins are in front of the pectorals or in the line with them. The abdominal cavity is very short and in some species is armed with a stout anal spine.

Currently, more than 570 species are recognized in approximately 123 genera and 11 families (Nelson 1994). Almost all species are marine. The 14 Gulf of Maine flatfishes look much alike and are often confused with one another even though they are presently placed in five families.

KEY TO GULF OF MAINE FLATFISHES

1a. Eyes on the left side and guts at left-hand edge, as the fish lies on bottom 2
1b. Eyes on the right side and guts at right-hand edge 6

3a. Upper jaw moderate to long, maxilla extending posteriorly to middle of eye; osseous protuberance on snout . . . . **Gulf Stream flounder**
3b. Upper jaw very short, maxilla extending just past anterior margin of eye; no osseous protuberance on snout . . . . **Smallmouth flounder**
4a. Pelvic fins not alike; those on eyed side continuous with the anal fin; those on blind side separate from it . . . . **Windowpane**
4b. Both pelvic fins alike 5
5a. Eyed side marked with four large oblong black eye-spots; fewer than 82 dorsal fin rays . . . . **Fourspot flounder**
5b. Eyed side marked with many small spots; more than 84 dorsal fin rays. . . . . **Summer flounder**
6a. Pectoral fin well developed on eyed side 7
6b. Pectoral fins absent (Achiridae). . . . . **Hogchoker**
7a. Mouth large, gaping back as far as lower eye; jaws and teeth nearly equally developed on both sides 8
7b. Mouth small, not gaping back as far as lower eye; jaws nearly straight on eyed side, curved on blind side 10
8a. Margin of tail fin rounded . . . . **American plaice**
8b. Margin of tail fin slightly concave, with angular corners 9
9a. Lateral line arched close behind gill opening . . . . **Halibut**
9b. Lateral line nearly straight . . . . **Greenland halibut**
10a. Blind side of head with large open mucous pits; 100 or more dorsal fin rays. . . . . **Witch**
10b. Blind side of head lacks open mucous pits; fewer than 90 dorsal fin rays 11
11a. Lateral line arched behind gill opening. . . . . **Yellowtail**
11b. Lateral line nearly straight. 12
12a. Interorbital space rough with scales. . . . . **Winter flounder**
12b. Interorbital space smooth and naked. . . . . **Smooth flounder**

LEFTEYE FLOUNDERS. FAMILY BOTHIDAE

Grace Klein-MacPhee

The family Bothidae is left-eyed. The pelvic fin base on the eyed side is longer than on the blind side, and the pectoral and pelvic fin rays are unbranched. The eggs have a single oil globule (Nelson 1994). The family comprises 20 genera with at least 115 species found in the oceans of the world (Nelson 1994), of which one species strays into the Gulf of Maine.

EYED FLOUNDER / *Bothus ocellatus* (Agassiz 1831) / Gutherz 1967:41

Description. Body ovate, deep anteriorly (Fig. 285). Head with slightly concave profile in front of widely separated eyes in males, but not in females (Jutare 1962). A fleshy ridge on posterodorsal area of upper eye and posteroventral area of lower eye. Interorbital width greater than eye diameter in adult males, less than eye diameter in adult females. Mouth small, oblique. Teeth conical and fine; two series in upper and lower jaws (Jutare 1962). Dorsal fin originates opposite anterior nostril; anal fin below insertion of pectoral fins. Left pelvic fin begins under middle of lower eye (Jutare 1962; Gutherz 1967). Scales moderate-sized, a few ctenoid on ocular side, all cycloid on blind side (Jutare 1962). Lateral line distinctly arched anteriorly. Body proportions: body depth 58.8–70.1% SL; head length 24.5–30% SL; eye 22.2–30% HL; upper jaw length 24–27% HL (Cervigón 1966; Gutherz 1967; Randall 1968).

Meristics. Dorsal fin rays 76–91; anal fin rays 58–68; pectoral fin rays on ocular side 8–10, on blind side 6 or 7; pelvic fin rays on left 6, on right 5; lateral line scales 70–78; vertebrae 35–37; gill rakers 0–6 + 7–10 (Jutare 1962; Gutherz 1967; Martin and Drewry 1978).

Color. Color pattern highly variable, but the ocular side is pale brown or gray with spots and mottling. Two dark spots on the anterior portion of the caudal fin, one below and one above the midline; these are sometimes absent or indistinct (Gutherz 1967). In males the preopercular region is sometimes streaked with blue lines with clusters of yellow between (Longley and Hildebrand 1941). Eyed flounder can change their markings to match their background with surprising fidelity (Ramachandran et al. 1996). By adjusting the contrast of different sets of "splotches" of different grain size (or spatial frequency) on the skin, a fish can blend into a wide range of background textures in just 2–8 s.

Size. Reach 160 mm (Böhlke and Chaplin 1968).

Distinctions. Large eyes, widely spaced in males; small mouth; and ovate outline distinguish this flatfish.

Habits. Adults inhabit clear shallow water over sand (Jutare 1962), along sandy shores (Nichols and Breder 1927), or in muddy or sandy bays (Cervigón 1966). They are most frequently found at temperatures of 20°–32°C and salinities of 24.1–37.6 ppt (Christensen 1965).

Breeding Habits. Spawning probably occurs inshore with eggs carried offshore by currents (Jutare 1962). In the South Atlantic Bight, larvae are present at all seasons, indicating year-round spawning (Fahay 1975), but peaks occur in July and December (Jutare 1962).

Mating groups of eyed flounder were observed by divers in the Caribbean (Konstantinou and Shen 1995). Each group consisted of a male and one to six females, each female having a distinct subunit within the male's territory. Courtship activity began about 1 h before sunset and continued until sunset, when the male retired for the night. Both males and females retired into the sand at locations outside their daytime terri-

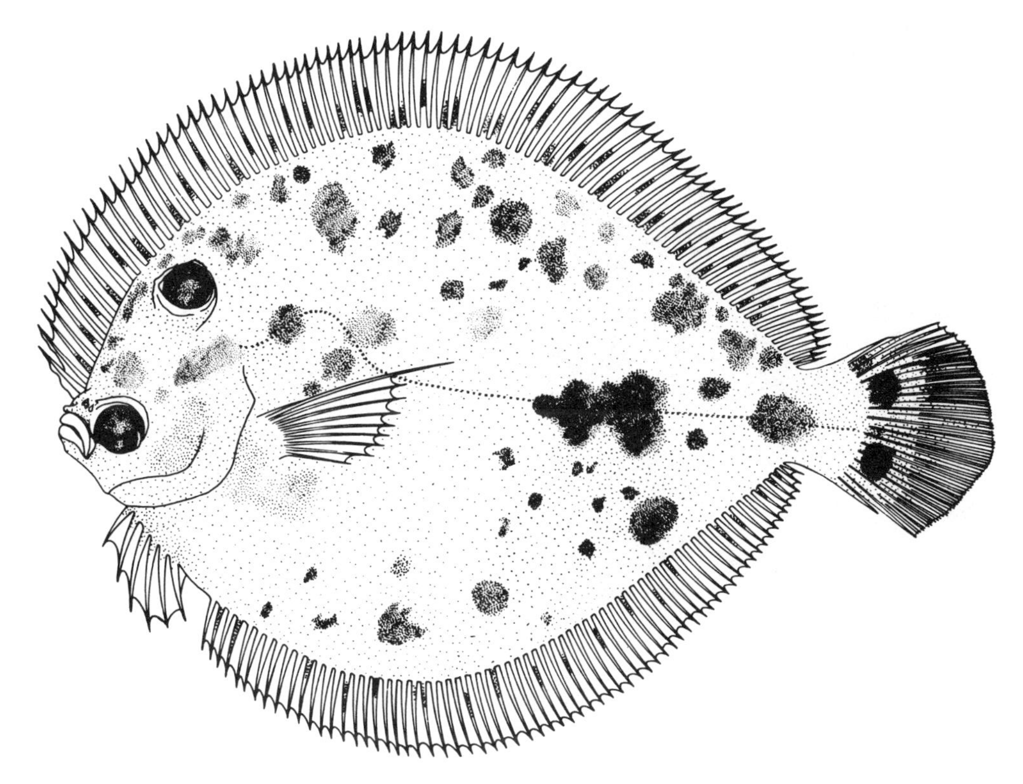

Figure 285. Eyed flounder *Bothus ocellatus.* Male, 122 mm SL. Drawn by G. Reinhardt. (From Gutherz 1967: Fig. 43.)

tory. Spawning began with a male moving under a female who was resting on the sandy bottom. The pair then made a slow upward rise approximately 15–75 cm above the substrate, which culminated in the release of a cloud of gametes.

Early Life History. There is no information on eggs or yolk-sac larvae. The smallest larvae described were 3 mm, somewhat stout-bodied, and had an elongate first dorsal fin ray (Jutare 1962), which sometimes appeared split (Yevseyenko 1976), and a few pigment spots (Kyle 1913). By 5.4 mm, there is a prominent swim bladder, the intestine is looped, the anus opens about midbody, dorsal and anal fin rays are almost complete, and flexion of the urostyle occurs (Yevseyenko 1976). By about 13 mm the tentacle-like first dorsal ray, pigment spots, and swim bladder disappear (Jutare 1962), and the larvae are very deep-bodied and strongly compressed laterally (Yevseyenko 1976). There is no information on the duration of metamorphosis, but symmetrical larvae of 40–42 mm TL have been collected (Colton 1961; Evseenko 1976), indicating that the larval stage is very long.

General Range. Atlantic coast of the United States from Long Island to Florida, Bermuda, the Bahamas, West Indies, Caribbean Sea, and South America to Rio de Janeiro (Kyle 1913; Beebe and Tee-Van 1933; Gutherz 1967; Randall 1968). Larvae are more widely distributed, from the Grand Banks and Georges Bank (Yevseyenko 1976) to just west of the Azores (Jutare 1962).

Occurrence in the Gulf of Maine. Larvae are carried by the Gulf Stream currents and occur in the upper 100 m from April to September (Yevseyenko 1976). A collection of 129 larval eyed flounder (4.5–42 mm TL) was made on Georges Bank in 1956 from September to November in the upper 10 m. Water temperature was above 14.6°C and contained Sargassum weed, indicating Gulf Stream origin. None of the larvae showed eye migration (Colton 1961).

FAMILY SCOPHTHALMIDAE

GRACE KLEIN-MACPHEE

The family Scophthalmidae is left-eyed. The pelvic fin bases are elongate, extending anteriorly to the urohyal. The mouth is and Mediterranean and Black seas (Chapleau 1993; Nelson 1994); one species, windowpane, occurs in the Gulf of Maine.

WINDOWPANE / *Scopthalmus aquosus* (Mitchill 1815) / Sand Dab, Spotted Flounder, Brill /

Bigelow and Schroeder 1953:290–294 (*Lophopsetta maculata*)

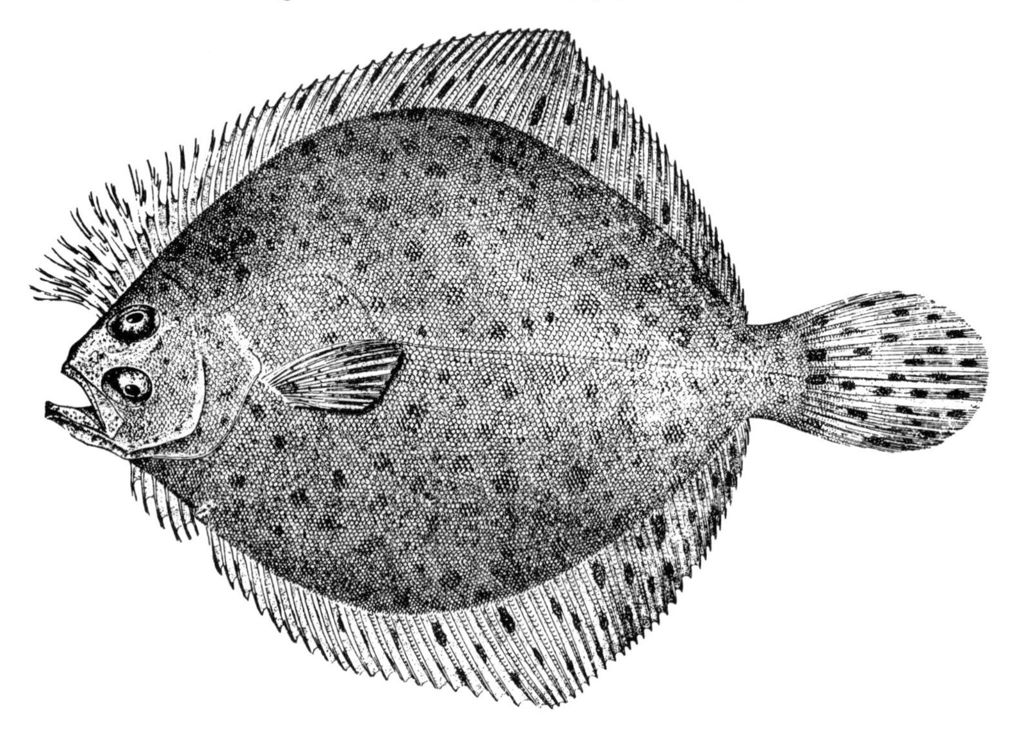

Figure 286. Windowpane *Scopthalmus aquosus.* Drawn by H. L. Todd.

Description. Body nearly round in outline, thin, almost translucent (Fig. 286). Lower jaw projects, with bony knob on chin. Teeth in both jaws small, in a single series laterally and a band anteriorly, although the gape of the mouth is wide. First 10–12 rays of dorsal fin free from fin membrane along outer half of their lengths, branched toward their tips, forming a conspicuous fringe that is without parallel among Gulf of Maine flatfishes. Dorsal fin tapers toward tail; anal fin tapers toward head and tail. Both fins thick and fleshy at base. No free anal spine. Pectoral fin on eyed side longer and more pointed than pectoral on blind side. Pelvic fins as wide at base as at tip, each simulating a detached segment of anal fin. Left and right pelvic fins differ: pelvic fin on eyed side longer, practically a continuation of anal fin, whereas pelvic fin on blind side situated a short distance up right-hand side of throat. Caudal fin rounded. Scales cycloid, smooth to touch. Lateral line arched above pectoral fin.

Meristics. Dorsal fin rays 63–71; anal fin rays 46–55; lateral line scales 85–102; vertebrae 33–36 (Hildebrand and Schroeder 1928; Moore 1947; Gutherz 1967).

Color. Windowpane vary less in color than most shoal-water flatfishes. The general ground tint of its eyed side is a pale and rather translucent greenish olive or slightly red or light slate brown; they are mottled darker and paler and usually dotted with many small irregularly shaped brown spots. Some fish are also marked on the body and on the bases of the dorsal, anal, and caudal fins with white spots that vary in number and size. Others lack these spots. The dorsal, anal, and caudal fins are of the general body tint, whereas the pectoral fin on the eyed side is dark cross-barred or speckled. The blind side is white, but individuals have been seen on which it was irregularly dark-blotched (Moore 1947), and partial ambicoloration has been reported (Burgess and Schwartz 1975).

Size. Reach 51 cm TL (*Albatross IV* cruise 64-1, sta. 118) and a weight of over 1 kg. The largest seen by Bigelow and Schroeder was 46 cm, but adult fish average about 25–30.5 cm, although seven fish with lengths of 45–49 cm were caught on *Albatross IV* cruise 80-3, sta. 610.

Distinctions. Windowpane can be distinguished from all other left-eyed Gulf of Maine flatfishes by the free first dorsal fin rays and by pelvic fins that are almost continuous with the anal fin.

Habits. Windowpane are shoal-water fish. In the Gulf of Maine, the depth range extends from the high-tide mark down to 200 m, with the greatest abundance at depths less than 55 m. They occur regularly down to 50 m off Connecticut (Moore 1947). They are caught on sandy bottoms off southern New England and southward, but their comparative abundance in Casco Bay and Minas Channel shows that they also frequent softer and muddier grounds in the Gulf of Maine. A summary of habitat characteristics was presented by Chang et al. (1999c).

Adult windowpane can tolerate a wide temperature range, 0°–26.8°C (Moore 1947), occurring as they do over many degrees of latitude and in shallow waters, where they are exposed to the extremes of winter chilling and summer warming. Temperature appears to govern their northerly range and local abundance, for they usually occur where surface waters warm to 13°C or higher in summer. They appear to be euryhaline, having been collected at salinities of 5.5–36 ppt (Tagatz 1967).

Windowpane are sensitive to hypoxic conditions. In Long Island Sound, abundance was low at dissolved oxygen concentrations less than 3 mg·liter^{-1}, presumably because the fish avoided the area (Howell and Simpson 1994). Windowpane collected in Hempstead Harbor, N.Y., which experiences chronic summer hypoxia, had higher blood hematocrit levels and slightly higher hemoglobin levels than those sampled from nearby sites that did not experience hypoxia, thus indicating some degree of adaptation to this condition (Dawson 1990).

Food. Based on examination of 1,092 stomachs (Bowman et al. 2000), the three major components of the windowpane diet are mysids (41.7% by weight), especially *Neomysis americana,* fishes (31.4%), and decapods (14.3%), especially *Crangon septemspinosa.* Other prey items include chaetognaths, squids, mollusks, ascidians, polychaetes, cumaceans, isopods, amphipods, euphausiids, and salps (Hacunda 1981; Bowman et al. 2000). Windowpane over 20 cm feed on these items but also prey on juvenile fishes such as anchovies, margined snake eel, silver hake, tomcod, cusk, killifishes, silversides, pipefish, blackbelly rosefish, longhorn sculpin, striped bass, sand lance, flatfishes, and various fish larvae (Langton and Bowman 1981; Bowman et al. 2000).

A study by Moore (1974) on the life history of windowpane in Long Island Sound found that larval windowpane fed on copepods such as *Temora* and *Centropages.* Mysids were by far the most important food, and fishes were not identified as major prey items. In Johns Bay, Maine (Hacunda 1981), the diet consisted mainly of crustaceans, primarily *Mysis mixta* (79.3% by weight, 99% by numbers), but also included finfishes such as herring (20.3% by weight).

Windowpane forage on actively swimming prey. A comparison of the diets of winter flounder, yellowtail flounder, and windowpane in Johns Bay showed that windowpane consumed the largest prey (13–17 mm) and fed exclusively on nektonic organisms (Hacunda 1981).

Predators. Windowpane are consumed by 16 species of sharks, skates, and bony fishes, of which only spiny dogfish and Atlantic cod are frequent predators (Rountree 1999). They were found in more than one stomach of five additional species: thorny and little skates, cobia, bluefish, and windowpane.

Parasites. Windowpane flounder are host to numerous parasites, including two protozoan species; three acanthocepha- 1905; Margolis and Arthur 1979; Marcogliese and McClelland 1992). From Virginia to Massachusetts they were found to have a condition known as epitheliocystis (Lewis et al. 1992), which occurs as gill lesions whose causative agent has not yet been identified; 28% of the examined fish were infected.

Breeding Habits. Windowpane mature between ages 3 and 4 (Moore 1947) but males may mature at age 2 (Grosslein and Azarovitz 1982). There is little difference between the sexes for median length at maturity from Georges Bank (males 22.2 cm, females 22.5 cm) and southern New England (males 21.5 cm, females 21.2 cm); median age at maturity is slightly above age 3 (O'Brien et al. 1993).

Windowpane eggs were collected by the MARMAP surveys from February to November (Berrien and Sibunka 1999: Fig. 94). Spawning begins in February or March in inner-shelf waters, usually between Cape Hatteras and New Jersey, expands in April into deeper waters and onto Georges Bank in some years. Throughout the peak spawning months of May through October, eggs are found over a broad area of the Mid-Atlantic, southern New England, Georges Bank, and the Gulf of Maine.

Windowpane spawn at temperatures between 6°–17°C, and most larvae occur at bottom temperatures between 8.5°–13.5°C (Smith et al. 1975). The optimal temperature range was 16°–19°C in the Mid-Atlantic Bight and 13°–16°C on Georges Bank (Morse and Able 1995). There appears to be a split spawning season off Virginia and North Carolina when temperatures exceeded 15° C, and there is a strong correlation between water temperature and spawning. Fish stop spawning at temperatures over 15°C and under 5°C. In the New York Bight, windowpane typically have a split spawning season but have continuous spawning only during colder summers (Wilk et al. 1990). Eggs have been found in salinities of 18.2–30.0 ppt (Wheatland 1956).

Early Life History. Windowpane eggs are transparent, 0.9–1.4 mm in diameter (Able and Fahay 1998), with a single colorless or pale lemon oil globule 0.15–0.28 mm in diameter (Wheatland 1956). The oil globule may be pigmented (Elliott and Jimenez 1981). The eggs resemble those of hake, fourspot, and scup. However, they usually occur closer to shore than hake; the oil globule is larger than that of fourspot, although sizes may overlap. The diameter of the egg is larger in windowpane than in scup but the sizes overlap (windowpane 0.9–1.38 mm, mean 1.07; scup 0.85–1.15 mm, mean 0.94). Pigment is heavier on the embryo and oil globule in windowpane than in scup. In later-stage embryos, windowpane have denser pigment, which is concentrated on the dorsal finfold over the pectoral region, whereas it is present midway down the tail in fourspot (Elliott and Jimenez 1981).

Incubation takes 8 days at 10°–13°C (Bigelow and Schroeder). Hatching length is 1.8–2.3 mm (Wheatland 1956; Colton and Marak 1969). Yolk is absorbed at 5.5 mm, at which

from behind the head to about midlength (Colton and Marak 1969). There are three or more pigment bars present (Moore 1947). Eye migration begins at 6.5 mm TL (Colton and Marak 1969) and proceeds very rapidly, being completed at 10–13 mm TL (Williams 1902; Martin and Drewry 1978). Laboratory-reared windowpane showed evidence of eye migration 17 days after hatching at water temperatures of about 17°C. Eye migration was complete 7–9 days later (24–26 days after hatching) when the fish were a mean size of 12 mm TL (G. Klein-MacPhee, unpubl. data). Young windowpane tend to move offshore to deeper waters (Grosslein and Azarovitz 1982).

Laboratory experiments on sediment preference of transitional (8–18 mm SL) and larger juvenile windowpane (32–89 mm SL) showed that windowpane of all sizes prefer sand over mud, but transitional windowpane were observed on sand less frequently, buried less often, and exhibited larval pigmentation more often than juveniles (Neuman et al. 1998). Transitional fish had a higher probability of moving from the preferred sediment (sand) during hours of darkness, and both stages had a higher probability of moving onto mud in the absence of food. Juveniles were also more active when there was no food, but to a lesser extent than transitional fish. All sediment choices were within the range observed in the field.

Age and Growth. Windowpane grow faster in offshore waters than in the waters off Block Island (Moore 1947; Grosslein and Azarovitz 1982). Females grow larger and faster than males after age 3. Growth after settlement is markedly different between spring- and fall-spawned cohorts in New Jersey waters (Morse and Able 1995). Spring-spawned fish reached 11–19 cm by September, whereas fall-spawned fish did not grow in winter and reached 4–8 cm by March; thus spawning time influences growth rates and the age and size composition of young-of-the-year.

General Range. Windowpane inhabit coastal waters of eastern North America, from the Gulf of St. Lawrence to Florida (Gutherz 1967; Wenner and Sedberry 1989) and are most abundant on Georges Bank and in the New York Bight.

Occurrence in the Gulf of Maine. This flounder is not abundant in the Gulf of Maine, except in the inner waters of Massachusetts Bay (Map 34). In the BIOME surveys conducted in Massachusetts Bay, windowpane ranked 32nd in abundance among 43 species (Lux and Kelly 1978).

Movements. Windowpane are year-round residents on Georges Bank and in inshore waters south of Cape Cod to Cape Hatteras. NEFSC bottom trawl surveys indicate that windowpane are also year-round residents in the Gulf of Maine. The young, which settle in shallow water inshore, tend to move into deeper, offshore waters as they grow. Tagging studies off southern New England have shown that individuals may travel along the coast for considerable distances or across open

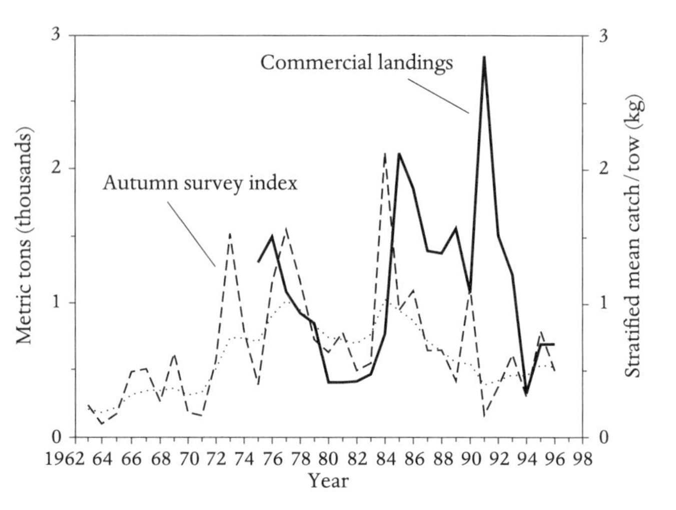

Figure 287. Commercial landings and autumn survey index for windowpane *Scopthalmus aquosus,* Gulf of Maine–Georges Bank, 1963–1996. (From Hendrickson 1998.)

water. Some of them have moved as far as 129 km in 3 months (Moore 1947). The movements of a few adults may play an important part in the intermingling of local populations. Trawl survey data indicate that windowpane on Georges Bank aggregate in shallow water during summer and early fall and move offshore for the winter and early spring (Grosslein and Azarovitz 1982).

Importance. Windowpane flounder are not commercially important in the Gulf of Maine owing to low abundance and their thin, small bodies. A market developed for them in New York during World War II but demand fell as the war drew to a close. Foodfish landings of windowpane began again, particularly in New Bedford, Mass., probably owing to fishing restrictions imposed on the declining yellowtail fishery and the increased abundance of windowpane (Lange and Lux 1978). Windowpane were also landed as one of the "industrial" species that were made into fish meal (Grosslein and Azarovitz 1982).

Increased windowpane landings during the mid-1980s probably reflected an expansion of the fisheries offshore and increased targeting of windowpane as an alternative to other depleted flatfish stocks. Landings from the Gulf of Maine–Georges Bank region peaked in 1991, then showed a 40% decrease in 1992 (Fig. 287). In 1994, total landings reached their lowest level (500 mt) since 1975 and have remained at less than 1,000 mt (Hendrickson 1998). Preliminary indices of commercial catch per unit effort (CPUE) for this region show a declining trend since 1975 and indicate that the stock continues to be overexploited (Hendrickson 1998). The principal commercial gear utilized in this fishery is the otter trawl. There is no recreational fishery.

Windowpane from Long Island Sound are used as an indicator species for environmental contamination (Greig et al. 1983).

SAND FLOUNDERS. FAMILY PARALICHTHYIDAE

GRACE KLEIN-MACPHEE

The family Paralichthyidae is left-eyed. The pelvic fin bases are short and nearly symmetrical, and the pectoral fin rays are branched. The family comprises about 16 genera and at least 85, mostly marine, species (Nelson 1994). Three subgroups are recognized within the family but there is no evidence that the family is monophyletic (Chapleau 1993; Nelson 1994). Four species in three genera have been found in the Gulf of Maine, one of which, summer flounder, is important enough to warrant preparation of an Essential Fish Habitat Source Document (Packer et al. 1999).

GULF STREAM FLOUNDER / *Citharichthys arctifrons* Goode 1880 / Bigelow and Schroeder 1953:294–296

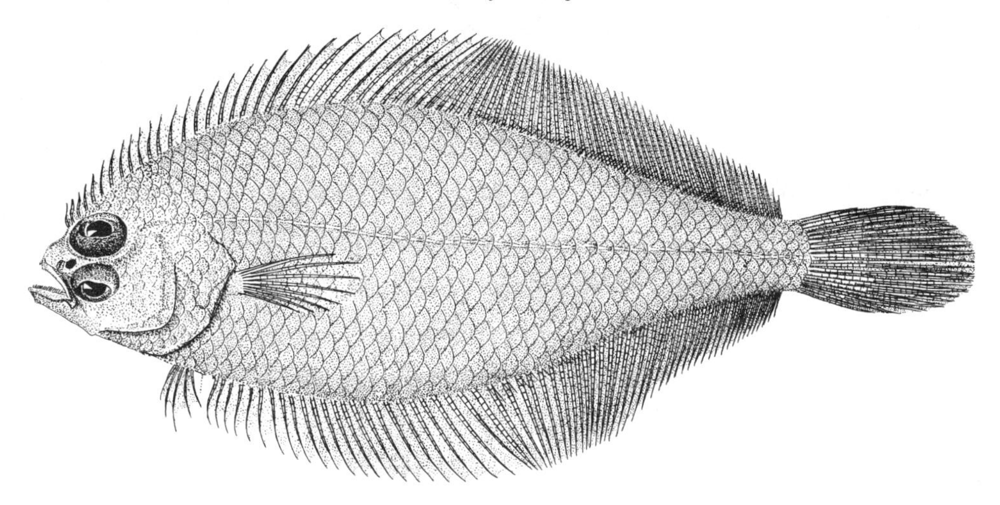

Figure 288. Gulf Stream flounder *Citharichthys arctifrons.* Off Rhode Island, 115 mm. Drawn by H. L. Todd.

Description. Body ovate, very thin (Fig. 288). Mouth wide, gaping back as far as anterior edge of lower eye. Dorsal and anal fins of moderate breadth; dorsal fin originates over anterior margin of eye; anal fin originates a little in advance of pectoral fins. Both pectoral fins well developed but pectoral on eyed side considerably larger than its mate on blind side. Pelvic fin on eyed side at midline of body; blind side pelvic fin a short distance above midline. Pelvic fins alike in females; pelvic fin on blind side much longer in males. Caudal fin rounded. Lateral line nearly straight. Scales large, cycloid or feebly ctenoid. Body proportions: body depth 34–43% SL; head length 23–38% SL; pectoral fin on ocular side 16–19% SL; orbit length 26–29% HL; upper jaw 27–32% HL (Gutherz 1967).

Meristics. Dorsal fin rays 75–83; anal fin rays 58–67; pectoral fin rays on ocular side 9–11, blind side 7; 37–43 scale rows along lateral line; gill rakers short, 5 + 6–8 (Goode and Bean 1896, Gutherz 1967; Richardson and Joseph 1973).

Color. Light brown above, with scales usually outlined with a darker brown; brownish white below.

Distinctions. Gulf Steam flounder resemble summer, fourspot, and windowpane flounders in having eyes on the left side of the head. They differ from all of these by their nearly straight lateral line, the great disparity in size between the pectoral fins, and the very large scales. They are also much smaller at maturity.

Habits. Little is known of their habits. They are found chiefly in water deeper than 37 m in the Chesapeake Bight (Richardson and Joseph 1973). Bigelow and Schroeder gave their depth distribution as 21–360 m, but they have been recorded down to 682 m over a temperature range of 8°–11.5°C (Edwards et al. 1962).

Food. Examination of 224 stomachs (Bowman et al. 2000) showed polychaetes (47.2% by weight) and crustaceans (41.3%) as the two major categories of food. Some of the polychaetes identified as prey include *Nicomache, Lumbrineris, Ophelina,* Nephtyidae, and Eunicidae. Predominant amphipod genera include *Ampelisca, Byblis, Erichthonius, Unciola,* and *Leptocheirus.* Other food items include copepods, ostracods, euphausiids, mysids, decapod shrimps, stomatopods, cumaceans,

(Langton and Bowman 1981; Sedberry 1983; Bowman et al. 2000).

The most important prey taxa during all seasons on the outer continental shelf in the Mid-Atlantic Bight were amphipods and polychaetes (Sedberry 1983). Other prey groups vary in importance as food according to the season. Larvaceans (primitive chordates) and cumaceans are important food in spring; fishes are very important prey in summer, and copepods and ostracods are of minor importance as prey in the fall. There is a gradual change in diet with an increase in predator size. Copepods are taken as food mostly by smaller fish, whereas polychaetes and cumaceans are eaten by larger ones. Small fishes are consumed by intermediate-sized individuals. Sedberry (1983) gave a detailed species list of food items and various measures of their importance in the Gulf Stream flounder diet.

Predators. Gulf Stream flounder were found in the stomachs of 14 species of fishes by the NEFSC food habits survey (Rountree 1999), seven of which had the most occurrences: spiny dogfish, spotted hake, windowpane, red hake, silver hake, little skate, and Atlantic cod, in the order of the number of occurrences.

Breeding Habits. Maturation occurs at 50 mm SL (Richardson and Joseph 1973). Spawning takes place from July to October, but larvae have been collected in May and December so spawning may occur sporadically throughout the year (Richardson and Joseph 1973). Eggs are difficult to distinguish from those of smallmouth flounder, so egg distributions of the two species had to be combined (Berrien and Sibunka 1999). Larvae were collected in all months except January and February (Smith et al. 1975), which also suggests year-round spawning.

Early Life History. Richardson and Joseph (1973) first described development of Gulf Stream flounder. The eggs are almost transparent, spherical to slightly ovoid with a smooth surface and no apparent oil globule. They range in size from 0.7 to 0.82 mm in diameter, with an average of 0.74 mm. Larvae are estimated to be 2 mm at hatch. Yolk is absorbed by 2.5 mm. Pigmentation consists of melanophores scattered along the ventral body margin from the angle of the jaw to the cleithrum, becoming less conspicuous with age. There is a heavy concentration of pigment over the swim bladder, which persists until metamorphosis. Smaller larvae (2–3 mm) may have a postanal band of pigment, which becomes a dorsal and anal bar of pigment. A notch appears in the dorsal fin at 4 mm, and at 5 mm three elongate rays, characteristic of many bothids, appear and persist until metamorphosis. These become pigmented at the tips. Eye migration begins at about 11 mm and metamorphosis is completed between 13–15 mm. At this time the swim bladder and the elongate dorsal rays disappear. Gulf Stream flounder larvae have been collected on Georges Bank from May to November, peaking in July and August. They ranked 19th out of 20 dominant species caught on Georges Bank and 17th out of 19 in the Gulf of Maine (Morse et al. 1987).

Larvae of this flounder resemble smallmouth flounder (*Etropus microstomus*). They can be distinguished at less than 4 mm TL by the presence of preopercular spines in *E. microstomus;* larvae larger than 4 mm by pigment patterns (*C. arctifrons* has less) and the presence of three elongate dorsal fin rays in *C. arctifrons.*

General Range. Eastern coast of America, along the outer part of the continental shelf from the southwestern part of Georges Bank (Bigelow and Schroeder) to the Gulf coast of Florida and Yucatán, Mexico (Gutherz 1967). Their main area of abundance is Cape Cod to Cape Hatteras.

Occurrence in the Gulf of Maine. This little flatfish has never been reported from inner parts of the Gulf, nor is it to be expected there to judge from its general distribution. *Albatross I* took one in a tow net over the southwestern part of Georges Bank at about the 150-m contour line. Other collections in this general area show that its regular range extends eastward far enough to include not only the slope of Nantucket Shoals, but the southwestern sector of Georges Bank arc as well, at the appropriate depth. It is plentiful on the outer part of the shelf off southern New England (Map 35), for *Albatross III* trawled a considerable number of them there, including one catch of 100 specimens off Montauk Point, Long Island, and another just as large off Rhode Island. NEFSC groundfish surveys collected individuals on the northern slope of Georges Bank and in Cape Cod Bay (Rodney Rountree, NEFSC, pers. comm.).

SMALLMOUTH FLOUNDER / *Etropus microstomus* (Gill 1864) / Gutherz 1967:28

Description. Body ovate; head small, upper profile a bit concave; snout blunt (Fig. 289). Mouth small with a very oblique gape (Goode and Bean 1896); teeth close-set, larger in front (Fowler 1906). Dorsal fin originates above front rim of orbit (Fowler 1906); anal fin of same general shape and height. Pectoral fin on ocular side much larger than that on blind side (Goode and Bean 1896). Caudal fin rounded. Scales large and deciduous, ctenoid on eyed side near lateral line, cycloid on blind side with secondary squamation present (Goode and Bean 1896; Fowler 1906; Gutherz 1967). Lateral line nearly straight (Martin and Drewry 1978). Body proportions: body depth 43–51% SL; head length 21–27% SL; eye diameter 22–30% HL; upper jaw length 24–28% HL (Gutherz 1967).

Meristics. Dorsal fin rays 67–83; anal fin rays 50–63; pectoral fin rays on ocular side 9–12, on the blind side 8 or 9; caudal fin

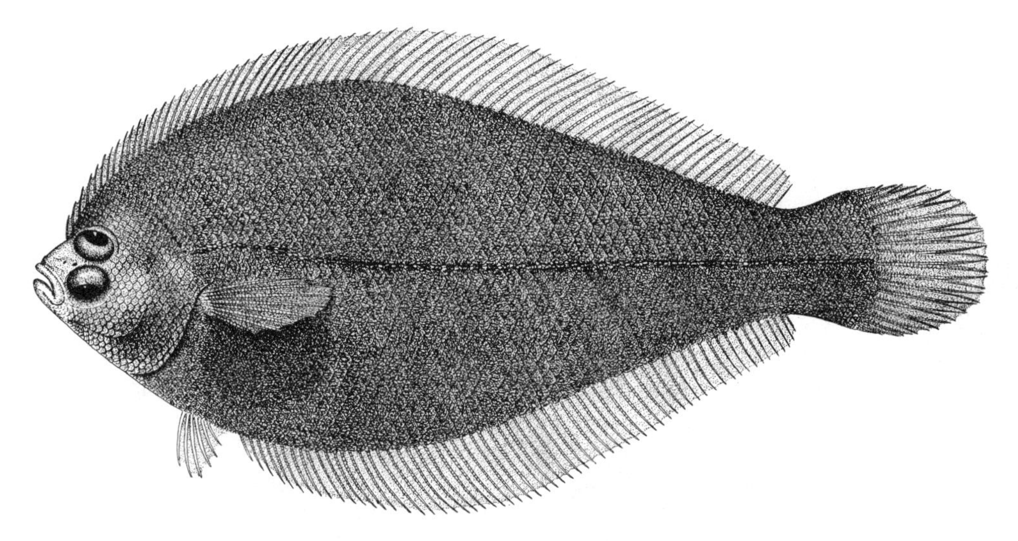

Figure 289. Smallmouth flounder *Etopus microstomus.* Ocean City, N.J., 112 mm. Drawn by W. S. Haines.

rays 17 or 18; lateral line scales 37–45; gill rakers 3–6 + 4–7; teeth 6 on eyed side, 16 on blind; vertebrae 10 + 24–26 (Fowler 1906; Norman 1934; Gutherz 1967; Miller and Jorgenson 1973; Richardson and Joseph 1973; Leslie and Stewart 1986).

Color. The eyed side is a uniform brown or reddish brown sometimes spotted or blotched with two or three dark-pigmented blotches along basal portions of the anal and dorsal fins. The blind side is white (Hildebrand and Schroeder 1928).

Size. Maximum 152 mm (Nichols and Breder 1927).

Distinctions. Smallmouth flounder can be distinguished from other Gulf of Maine paralichthyids such as fourspot, summer flounder, and windowpane by the small mouth, which does not extend much below the anterior margin of the lower orbit, and the straight to slightly curved lateral line. They most closely resemble Gulf Stream flounder but have no horizontal protuberance on the snout and fewer teeth (6/16 for smallmouth flounder vs. 40/25 for Gulf Stream flounder).

Habits. Adults are found from near shore to about 35 m (Richardson and Joseph 1973) and have been taken as deep as 91.4 m off Cape Hatteras (Gutherz 1967). Smallmouth flounder tolerate temperatures from 11° (Miller and Jorgenson 1973) to 25°C (Leslie and Stewart 1986), and salinities of 7–32.5 ppt (Musick 1972). In the lower Chesapeake Bay, they frequent channels and mud bottoms (Reiss and McConaugha 1999).

Predators. The NEFSC food habits survey found smallmouth flounder in the stomachs of seven species of fishes, of which spiny dogfish and winter flounder were the most important (Rountree 1999).

takes place near shore from Cape Cod to North Carolina. Gravid females were collected in July and August from the Chesapeake Bight, and the spawning peak appeared to be July through October; most larvae were collected within the 37-m isobath but some were found as far out as 138 m, and they appeared to concentrate at the surface from 0 to 12 m (Richardson and Joseph 1973). Eggs were found in Block Island samples from June to September. Water temperatures ranged from 11.9° to 22.3°C. Larvae were taken from July to October (Scherer and Bourne 1980). Smallmouth flounder are characteristic of the ichthyoplankton assemblages of the inner continental shelf waters in the Chesapeake Bay area, where they are considered shelf-spawned taxa (Reiss and McConaugha 1999).

Early Life History. The eggs are small, 0.56–0.74 mm in diameter (mean 0.64), with a single small oil globule, 0.051–0.165 mm (mean 0.12), and spherical, with a transparent, unsculptured chorion. Occasionally two oil globules or multiple small ones were found but this was believed to be an artifact of collection or preservation. Faint pigment spots occur on the dorsal surface of middle-stage embryos, but there are none on the yolk or oil globule. By the late stage, melanophores enlarge; migrate onto the finfold; and coalesce into four distinct spots, two in the dorsal and two in the ventral finfolds, with another group aggregating near the tip of the notochord. The oil globule is located near the posterior part of the yolk sac and now has one or two pigment spots.

Smallmouth flounder eggs resemble those of fourbeard rockling, *Enchelyopus cimbrius;* hakes, *Urophycis* sp.; butterfish, *Peprilus triacanthus;* and possibly Gulf stream flounder, *Citharichthys arctifrons.* Most early- and middle-stage smallmouth eggs can be distinguished by their smaller egg and oil globule diameters. Only 2% of oil globule diameters exceeded 0.13 mm, the smallest reported for the other species. Once

which are always scattered on the yolk and oil globules of hakes and on oil globules and occasionally on the yolk in fourbeard. Butterfish have two distinct rows of pigment on the embryo from the eyes to the tail but none on the finfolds (Scherer and Bourne 1980).

Larvae hatch at 1.4 mm NL (Scherer and Bourne 1980). They are identical to late-stage embryos described above. Between 2.1 and 2.3 mm NL, the yolk sac is absorbed, eye pigmentation develops, and small teeth and preopercular spines appear. Pigment is present on the posteroventral margin of the lower jaw and on the ventral body margin from the angle of the lower jaw to the cleithrum, from the cleithrum to the hindgut, and from two-thirds of the distance from the anus to the notochord tip. By about 3 mm, this band becomes two horizontal lines of pigment on the dorsal and ventral margin, and a third appears along the notochord. These persist until metamorphosis. By 4 mm, characteristic internal pigment spots begin to appear along the dorsal margin of the notochord from the abdominal region to the tail. Pigmentation appears on the swim bladder at 2.5 mm and persists until the swim bladder disappears after metamorphosis, then spreads onto the hindgut. The urostyle begins to turn upward by 5–6 mm; metamorphosis occurs at 10–12 mm (Richardson and Joseph 1973). Larvae most closely resemble Gulf Stream flounder, and differences are discussed under that species.

General Range. Smallmouth flounder adults occur primarily from Cape Cod to Cape Hatteras (Parr 1931; Richardson and Joseph 1973). They have been found as far south as South Carolina and Florida but reports of this fish from west of the Mississippi delta, the Caribbean, and South America are believed to be misidentifications (Leslie and Stewart 1986).

Occurrence in the Gulf of Maine. The larvae are fairly common off southern New England and Cape Cod ranking fifteenth in abundance (Morse et al. 1987). The species was first reported from Cape Cod Bay as larvae (Scherer 1984), and then two adult specimens 25 and 130 mm SL were collected by Steven Correia off Wellfleet in Cape Cod Bay. In addition, the Massachusetts Department of Marine Fisheries trawl surveys collected 61 specimens off Nantucket and Martha's Vineyard and in Buzzards Bay (Collette and Hartel 1988).

Importance. Smallmouth flounder are too small to be commercially important even where they are abundant (Richardson and Joseph 1973) and they are rare in the Gulf of Maine.

SUMMER FLOUNDER / *Paralichthys dentatus* (Linnaeus 1766) / Fluke / Bigelow and Schroeder 1953:267–270

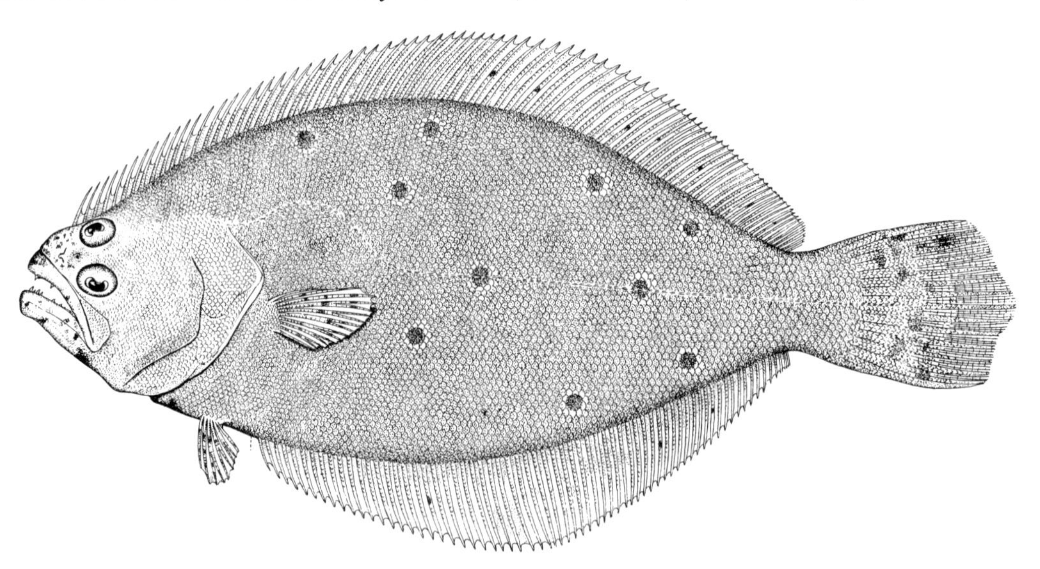

Figure 290. Summer flounder *Paralichthys dentatus*. St. Wales, Miss., 263 mm.

Description. Body narrow; mouth large, with strong canine teeth, 16–28 on eyed side (Woolcott et al. 1968) (Fig. 290). Dorsal fin originates opposite forward margin of upper eye. Pelvic fins alike, separated from long anal fin by a considerable space. Caudal fin with rounded margin. Lateral line arched over pectoral fin. Scales small, cycloid, secondary squamation present (Norman 1934). Body proportions: body depth 47% SL; head length 24–34% SL; maxilla length 12–16% SL (Ginsburg 1952b; Gutherz 1967).

Meristics. Dorsal fin rays 80–98; anal fin rays 60–74; pectoral fin rays 18 or 19; lateral line scales 91–108; gill rakers on lower part of anterior arch 14–17; vertebrae 10 or 11 + 30 or 31 = 41 or 42 (Norman 1934; Gutherz 1967; Miller and Jorgenson 1973; Able and Fahay 1998).

Color. Summer flounder are among the most variable flatfishes in color of all the local species and they adapt their pattern most closely to the ground on which they lie. Like most flatfishes they are white on the blind side, with shades of brown, gray, or drab on the eyed side. They can assume a wide range of tints, from nearly white on white sand through various hues of gray, blue, green, orange, pink, and brown to almost black (Mast 1916). Upper surface variegated with pale and dark, with the pattern fine or coarse according to the bottom. They may or may not be marked with small eyespots, a darker tint than the general ground color. Mast's experiments showed that they are slower in adapting their coloration to actual colors of the bottom than to the general pattern, and that they respond more rapidly to yellows and browns than to reds, greens, or blues. He also observed that the skin simulates the pattern of the background and does not reproduce it.

Incidences of ambicoloration are much rarer than in winter flounder; however, several specimens have been described that were pigmented on both sides, a condition that is often accompanied by incomplete eye migration and a hooked dorsal fin (Gudger 1935a; Dawson 1962; Powell and Schwartz 1972).

Size. Richards (1980) estimated maximum size to be 61 cm in length, 2.6 kg weight for males and 94 cm, 13.4 kg for females. Average size is 40–56 cm and 1–2.3 kg. The all-tackle world record summer flounder weighed 10.17 kg and was caught at Montauk Point, Long Island, in September 1975 (IGFA 2001).

Distinctions. The only Gulf of Maine flatfish with which summer flounder share eyes on the left side, large mouth, and symmetrical pelvic fins are fourspot flounder, but the latter has a distinctive color pattern (four large spots) and fewer dorsal fin rays (72–81) and gill rakers (7–10).

Habits. Summer flounder spend most of their lives on or close to the bottom , as other flatfishes do. During their stay in shoal water they prefer sandy bottom or mud, where they are often seen; it takes them only an instant to bury themselves to the eyes in the sand, although they do not do this as often or as deeply as winter flounder do (Keefe and Able 1993). Summer flounder often lurk in sand patches near eelgrass beds or among dock pilings. In Mid-Atlantic estuaries, they have been found in salt marshes and seagrass beds that have muddy or silty substrates (Dahlberg 1972; Orth and Heck 1980; Rountree and Able 1992a).

Summer flounder are sensitive to oxygen concentrations of less than 3 ppm and may move to avoid these hypoxic areas (Murawski and Festa 1974) or may be killed by such events (Freeman and Turner 1977a; Swanson and Sindermann 1979). Some summer flounder come close inshore during the warm half of the year, where they are caught regularly both along open coasts and in bays and harbors, the smaller sizes often from docks and bridges. Some even run up into freshwater rivers. The great majority of the population, especially of larger ones, lie farther offshore even at that season, in depths of 70–155 m (Grosslein and Azarovitz 1982) and deeper, at least in the northern part of the range.

Food. The main [illegible]

over all size-classes of 655 summer flounder, mostly from southern New England south (Bowman et al. 2000: Table B-53a). Cephalopods become important at 31 cm TL and constitute 34.2% of the diet across all size-classes. Crustaceans are important for small size-classes, 43.3% for summer flounder less than 21 cm, decreasing to 22.4% by 36 cm, and then becoming insignificant. Mysids, particularly *Neomysis americana,* and decapods, particularly *Cancer irroratus,* are the most important crustaceans in the diet of smaller summer flounder. Fish prey include a wide assortment of bottom and surface species, with sand lance being most significant starting with summer flounder 26 cm and longer and constituting 22.3% by weight of the total diet. Other fish prey important for particular size-classes of summer flounder include anchovies, round herring, silver hake, and flatfishes.

Summer flounder are active in pursuit of prey, often following schools of small fishes up to the surface and jumping clear of the water in their dashes, actions very different from those of the sluggish dab and winter flounder. However, summer flounder also feed on the bottom. They were reported as being active during daytime (Olla et al. 1972) but other studies in the laboratory (Klein-MacPhee 1978) and in the field using telemetry (Szedlmayer and Able 1993) indicated that juveniles, at least, are active at night. Their feeding behavior has been described in detail by Olla et al. (1972).

The most important food items for young-of-the-year summer flounder (100–200 mm) in Pamlico Sound, N.C., were mysids (*Neomysis americanus*), fishes (especially anchovies and sciaenids), amphipods, and crabs (Powell and Schwartz 1979). In New Jersey, young-of-the-year also eat silverside, mummichog, and shrimps such as *Palamonetes vulgaris* and *Crangon septemspinosus* (Rountree and Able 1992b). Food consumption generally increases with temperature and falls in winter (Powell and Schwartz 1979). Summer flounder move into marsh creeks to feed, and these tidal-mediated movements typically occur up the creeks at night on flood tide and down the creeks following ebb tide (Rountree and Able 1992b). Feeding rates of juvenile summer flounder maintained in the laboratory at various temperature-salinity regimes ranged from $1\% \cdot day^{-1}$ at 2°C to $24\% \cdot day^{-1}$ at 18°C. Salinity had no apparent effect. Assimilation efficiency averaged 60.3% over a range of temperatures (2°–18°C) and salinities (10–30 ppt). Mean specific growth rates were not significantly different between 2° and 10°C ($0.14\% \cdot day^{-1}$) and growth rates ranged from 2.4 to $3.9\% \cdot day^{-1}$ at 14°–18°C (Malloy and Targett 1991).

Feeding habits of pelagic larval summer flounder were examined in relation to larval stage (Grover 1998). Incidence of feeding and gut-fullness data indicated that larvae begin feeding near sunrise and continue throughout daylight hours. Incidence of feeding reached its lowest point, 8.3% at 0400–0559 hours, then dramatically increased to 54.6% at 0600–0759. Maximum gut-fullness was seen between 1200 and 1559. Immature copepodites were the primary prey for oceanic larval stages, tintinnids and copepod nauplii made major contribu- [illegible]

of premetamorphic and metamorphic larvae. At 1800–1859 and 2000–2159 hours, the incidence of feeding in estuarine larvae was significantly lower than in oceanic larvae. The estuarine diet was dominated by a calanoid copepod, *Temora longicornis*. Incidence of feeding was observed to decline as metamorphosis progressed.

Predators. Summer flounder have been found in the stomachs of spiny dogfish, blue shark, little skate, Atlantic cod, silver hake, goosefish, northern sea robin, spot, bluefish, and winter flounder (Bowman and Michaels 1984; Kohler 1988; Rountree 1999; Bowman et al. 2000), of which spiny dogfish are the most significant predator.

Parasites. The following parasites have been recorded from summer flounder: a protozoan, *Trypanoplasma bullocki* (Sypek and Burreson 1983); an acanthocephalan, *Echinorhynchus sagittifer;* a nematode, immature *Ascaris;* a cestode, *Rhyncobothrium sp.;* two trematodes, *Distomum dentatus, D. monticelii;* and a copepod, *Lernaeonema* (Linton 1905). A systematic bacterial infection by *Vibrio sp.* caused an ulcerative disease in summer flounder collected in Connecticut (Robohm and Brown 1978).

Although summer flounder are tolerant of a wide range in temperature, temperature has been indirectly responsible for unusual mortality of summer flounder juveniles in the York River estuary, Chesapeake Bay, Va., and Pamlico Sound, N.C. (Goldstein 1985). A combination of low temperatures and infection by the blood parasite *Trypanoplasma bullocki* is fatal. The parasite is transferred to summer flounder by an estuarine leech, *Calliobdella vivida*. Under normal temperature conditions the juveniles are able to combat the infection and survive but at temperatures below 5°C they succumb (Sypek and Burreson 1983; Burreson and Zwerner 1984).

Breeding Habits. Median length at maturity for female and male summer flounder is 28.0 cm and 24.9 cm, respectively (O'Brien et al. 1993) corresponding to ages 2.5 for females and age 2 for males (Penttila et al 1989). Based on size at the end of the first year of life, many fish may reach maturity by age 1 (Almeida et al. 1992; Szedlmayer et al. 1992). Spawning occurs during autumn migration to offshore wintering grounds on or near the bottom, where temperatures range from 12° to 19°C (Morse 1981).

Fecundity estimates range from 463,000 to 4,188,000 eggs per female for fish between 36.6 and 68 cm total length. Larger fish mature first during the spawning period and produce more eggs (Morse 1981). Females are serial spawners, continuously producing egg batches, which are shed over a period of several months (September to February or March).

Summer flounder have two distinct spawnings each year: intense spawning in autumn and winter over much of the Mid-Atlantic and southern New England regions and a lesser spawning during spring in the southern part of the Mid-Atlantic region (Berrien and Sibunka 1999: Fig. 88). Major spawning begins in September in inshore waters of the Mid-Atlantic and southern New England. Peak spawning occurs in October when egg distribution broadens to include much of Georges Bank. Spawning in deeper waters continues into November, when eggs occur across the entire breadth of the shelf.

Early Life History. Embryonic development was first described by Smith and Fahay (1970) from artificially reared eggs, which were compared to eggs collected at sea. The eggs are pelagic and contain a single oil globule. The egg diameter ranges from 0.91 to 1.1 mm, mean 1.02 mm, and that of the oil globule from 0.18 to 0.31 mm, mean 0.25 mm (Smith and Fahay 1970). Fertilized eggs are spherical with a rigid transparent shell. The perivitelline space occupies 6% of the egg radius. Eggs hatch between 48 and 71 h postfertilization, depending on incubation temperatures (Johns and Howell 1980).

Newly hatched larvae are 2.41–2.82 mm long. The body is thin and long except at the yolk-sac region. The head is flexed downward. Black pigment spots are present on the anterior half of the anal finfold and anterior two-thirds of the dorsal finfold. The posterior quarter of the larva is clear. Pigmented areas on the head include the snout, midbrain, anterior to the yolk, and ventral to the eye. There are a few pigment spots on the upper part of the yolk but the oil globule is clear.

Yolk is absorbed at 3.16 mm NL, at which time the mouth has developed, the eyes become pigmented, and the anus is formed at the margin of the finfold. Dorsal fin rays begin to form at 8.64 mm NL and the first few are elongate. The eye begins to migrate at 9.5 mm SL and completes migration by about 14–21.4 mm SL. Total time required for metamorphosis of laboratory-reared fish from the start of eye migration to the time the eye reaches its final position is 20–32 days, mean 24.5 days. Details of metamorphosis were described by Keefe and Able (1993).

Duration of metamorphosis in laboratory-reared summer flounder was dramatically affected by temperature, averaging 46.5 days at a mean temperature of 14.5°C and 92.2 days at a mean of 6.6°C (Keefe and Able 1993). Mortality rates during metamorphosis ranged from 17 to 83% and were significantly greater at cold temperatures (4°C).

Eggs and larvae of summer flounder are found only at sea, whereas young-of-the-year juveniles are found in or near mouths of estuaries (Smith 1973). Eggs were collected north of Chesapeake Bay from September to December and south of the Bay from November to February at temperatures ranging from 9.1° to 22.9°C. The most productive spawning grounds were off New York and New Jersey in 1965–1966, but during 1980–1986 the center of reproduction was from Massachusetts to New York (Able et al. 1990).

Larvae were collected at the same time as eggs from 22 to 83 km offshore. Concentrations occurred off Martha's Vineyard, Long Island, northern New Jersey, and in Delaware Bay at temperatures of 0°–22°C. Larvae have been collected as early as September off Cape Cod and as late as May in North Carolina (Smith 1973; Smith et al. 1975; Bolz et al. 1981; Able et al. 1990; Able and Fahay 1998). Most postlarvae were collected at night (Smith 1973). It is believed that postlarvae begin to migrate southward and into estuaries during morphological transformation and fin development and that complete metamorphosis and settlement take place in the estuaries (Keefe and Able 1993). Metamorphic summer flounder enter Great Bay–Little Egg Harbor estuary, N.J., prior to completion of metamorphosis and permanent settlement (Keefe and Able 1994). Laboratory experiments indicate a preference by both juvenile and metamorphic summer flounder for sand over mud substrate. Metamorphic individuals do not appear to be capable of burying themselves in the substrate when they enter the estuary. Complete burial was only observed in late metamorphic and juvenile stages. A diel pattern of burying behavior was observed that was dependent upon several environmental variables, including substrate, water temperature, tide, and presence and type of predator. In general, postlarvae begin entering estuaries from New Jersey to North Carolina from October to April, beginning earlier in the northern part of the range, October–April from Long Island Sound to Chesapeake Bay (Olney 1983; Able et al. 1990; Szedlmayer et al. 1992; Norcross and Wyanski 1994), and later in the southern part (December–April in North and South Carolina [Weinstein 1979; Bozeman and Dean 1980; McGovern 1986; Hettler and Chester 1990]). Larval densities outside of Beaufort Inlet, N.C., were correlated with the north component of the wind, nearshore water temperature, and distance to the midshelf front (Hettler and Hare 1998). Differences in larval density across the inlet were significantly correlated with the east component of the wind.

Larvae are eurythermal and euryhaline and have been collected at temperatures ranging from −2.0° (Szedlmayer et al. 1992) to 23.4°C (McGovern and Wenner 1990) and in salinities from 0 (McGovern and Wenner 1990) to 35 ppt (Williams and Deubler 1968). Laboratory-reared summer flounder showed high mortality rates at temperatures less than 2°C and estuarine larvae may suffer increased mortality owing to both severity and duration of cold water temperatures (Malloy and Targett 1991; Szedlmayer et al. 1992). Postlarval *Paralichthys* maintain a preferred position in the Cape Fear River estuary, N.C., by migrating to the surface at night and using tidal drift to enter salt marshes (Weinstein et al. 1980). Young-of-the-year are well adapted to estuaries because they can withstand a wide range of temperatures and salinities (Deubler and White 1962; Szedlmayer et al. 1992), but generally prefer salinities higher than 12 ppt (Powell and Schwartz 1977). During spring and summer in New Jersey marsh creeks (Szedlmayer and Able 1993), they usually remain within narrow limits of temperature (22.3°–24.9°C), salinity (27–31 ppt), and dissolved oxygen (5.9–6.8 ppt).

They remain in estuaries until their second year of life in the southern part of their range but to the north move just outside them in winter. A large portion of juveniles tagged in estuaries return to the same system the next summer (Hamer and Lux 1962; Poole 1962; Jesien et al. 1992). At maturity, they join the adult population in their migration (Grosslein and Azarovitz 1982).

Age and Growth. Summer flounder grow rapidly in their first year. Growth rates of 1.9 mm·day^{-1} have been measured in New Jersey estuaries and young-of-the-year ranged from 200 to 326 mm TL by September (Szedlmayer et al. 1992). Length-at-age was calculated by Poole (1961), Eldridge (1962), and Smith and Daiber (1977) using otoliths, but this proved problematic because of difficulties in interpreting the first opaque zone (Dery 1988d). Poole believed that the first ring was laid down at age 1 when fish were 250–320 mm, whereas Smith and Daiber believed that it was laid down at age 2 and that the fish reached sizes of 170–180 mm at age 1. Poole's interpretation was supported by Szedlmayer et al. (1992). Currently, scales are used for aging and the first distinct annulus measured from the focus is used to represent first-year growth (Almeida et al. 1992). Females grow faster than males and are larger than males after age 2; males attain a maximum age and length of about 7 years and 60 cm, as compared with 12 years and 82 cm for females (Dery 1988d) (Fig. 291).

General Range. Continental waters of eastern North America from Nova Scotia (Vladykov and McKenzie 1935) to South Carolina, possibly to Florida (Wilk et al. 1980; Gilmore et al. 1981) chiefly south of Cape Cod. Two tag recoveries reported from the Gulf of Mexico (Briggs 1958) seem very unlikely (Wilk et al. 1980).

Occurrence in the Gulf of Maine. Summer flounder are plentiful offshore in the southern Gulf of Maine, eastward to Nantucket Shoals, and to the western part of the South Channel. NMFS surveys and commercial catch records show that they are present on the westernmost edges of Georges Bank and in shallow waters on Georges Bank (Able and Kaiser 1994). They are rare north of Cape Cod, but occasionally occur north as far as Brown's Bank, and to outer Nova Scotia waters on La Have Bank and Passamaquoddy Bay, N.B. (Scott and Scott 1988).

Movements. In inshore waters of the New York Bight, summer flounder appear in April and continue to move inshore during May and June to reach peak numbers in July or August (Grosslein and Azarovitz 1982). They appear in early May near Woods Hole (Bigelow and Schroeder). Those that come close inshore from Chesapeake Bay northward move offshore again during autumn, presumably to escape winter chilling, although some overwinter in Delaware Bay (Smith and Daiber 1977). They begin offshore migrations in September and are usually gone from the northern part of their range by October or November. Tagging on both inshore and offshore grounds (Hamer and Lux 1962), showed that fish tagged in New Jersey inshore waters moved northward along the coast and eastward along Long Island toward Martha's Vineyard. They scattered over the Mid-Atlantic Bight during their offshore migration. Summer flounder tagged on offshore grounds east of Hudson Canyon were recaptured between lower New York Bay and Cape Cod and seldom strayed south

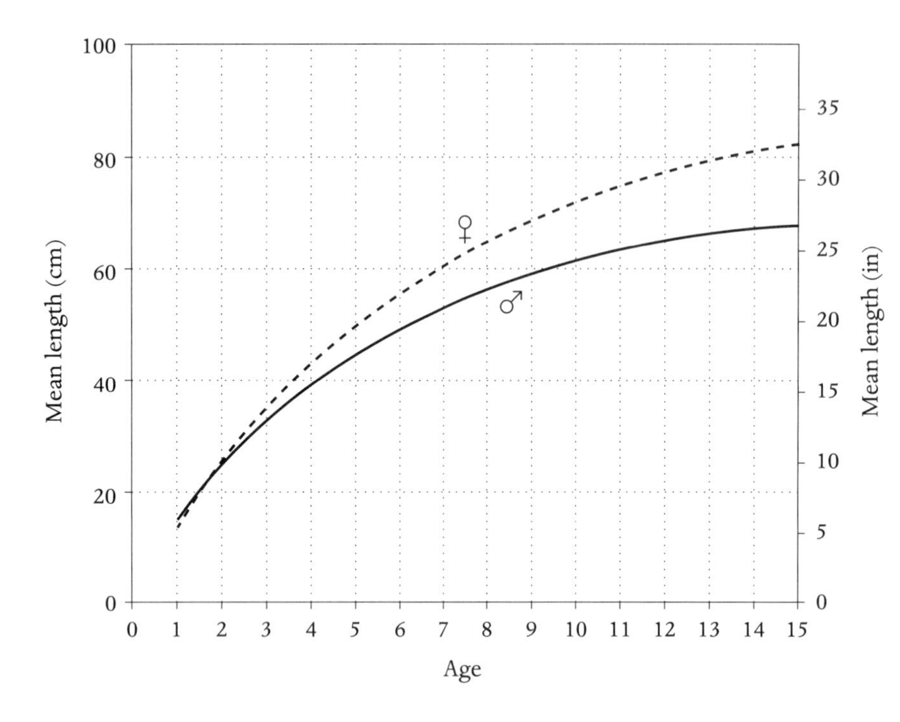

Figure 291. Age-length curves for male and female summer flounder *Paralichthys dentatus,* all areas combined. (From Penttila et al. 1989:31.)

Hudson Canyon. This is probably the source of fish summering in Rhode Island and Massachusetts waters.

Importance. This is the most commercially important flatfish to the west and south of Rhode Island and the most sought after by sportsmen there. They occur in commercial quantities sporadically on Georges Bank. They were landed in the 1940s and 1950s but were essentially absent from 1963 to 1973. Numbers then increased but only to low levels compared to other areas (Henderson 1979). Georges Bank–Mid-Atlantic commercial landings peaked at about 15,000 mt from 1976 to 1988 and then declined to about 5,000 mt (Fig. 292). The otter trawl is the most common commercial gear used (accounting for about 90% of the commercial landings), but handlines, haul seines, various traps and nets, and spears are also used (Byrne and Azarovitz 1982). Summer flounder are an important by-catch and discard in the small-mesh fishery for squid in Nantucket and Vineyard sounds (Glass et al. 1999).

The recreational fishery probably accounts for more than half the total catch and considerably exceeds it from New York to Maine (McHugh and Ginter 1978). Most fishermen are anglers fishing from shore, piers, docks, and small boats, but in the southern part of the range they use spears as well (Byrne and Azarovitz 1982; Manooch 1984). This is the gamest of the local flatfishes, biting freely on almost any bait, even taking artificial lures at times, and large ones put up a strong resistance when hooked.

Commercial aquaculture of summer flounder began in 1996 (Bengston 1999). Both the research leading to commercialization and the production itself have been heavily oriented toward the hatchery phase. Producers are experimenting with both recirculation systems and net pens to identify the equipment that optimizes grow-out production. Current operations are located in New England and New York.

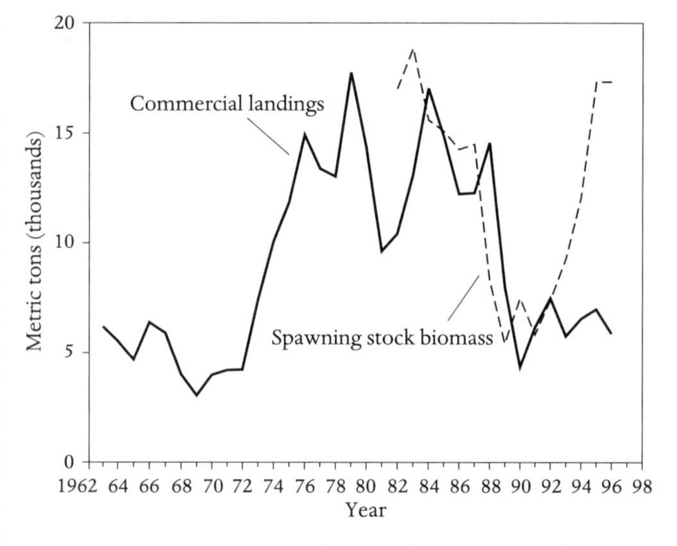

Figure 292. Commercial landings and spawning stock biomass for summer flounder *Paralichthys dentatus,* Mid-Atlantic–Georges Bank, 1963–1996. (From Terceiro 1998.)

Stocks. There have been three interpretations of stock structure of summer flounder. (1) Two distinct populations, one in the Mid-Atlantic Bight between Cape Cod and Cape Hatteras and the other between Cape Hatteras and Florida, based on meristic and morphological differences (Smith and Daiber 1977; Wilk et al. 1980; Fogarty 1981). This is the concept currently accepted for stock assessment and management purposes (Terceiro 1998b). (2) Two Mid-Atlantic Bight stocks, one of which appears to make a consistent offshore migration in summer and another that appears to spend the summer in estuaries and inner-shelf areas from Virginia to Maryland, but overwinters near Cape Hatteras. This trans-Hatteras stock is supported by electrophoretic (Van Housen 1984), meristic and morphometric

(Delaney 1986) analyses and by tagging studies (Able and Kaiser 1994). (3) Three stocks, Mid-Atlantic Bight, South Atlantic Bight, and trans-Hatteras. Support for distinction of local populations north and south of Cape Hatteras comes from tagging studies (Mercer et al. 1987; Monaghan 1992).

A recent mitochondrial DNA study (Jones and Quattro 1999) of summer flounder from Massachusetts to South Carolina evaluated the effect of Cape Hatteras on gene flow and found no significant population subdivision centered around Cape Hatteras.

FOURSPOT FLOUNDER / *Paralichthys oblongus* (Mitchill 1815) / Bigelow and Schroeder 1953:270–271

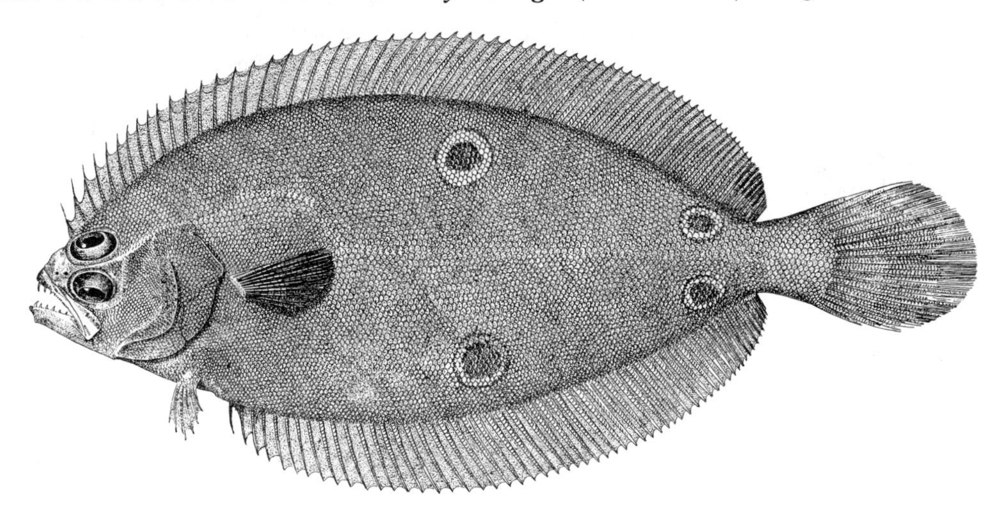

Figure 293. Fourspot flounder *Paralichthys oblongus.* Woods Hole, Mass., 321 mm. Drawn by H. L. Todd.

Description. Mouth large, extending back below posterior margin of lower eye (Fig. 293). Upper jaw with four or five moderate-sized canines along with numerous small teeth. Lower jaw with seven or eight teeth on each side. Dorsal fin originates just posterior to upper eye, nearly equal in height throughout its length. Anal fin a bit lower in anterior part. Pelvic fins alike on both sides of body, separated from anal fin by a gap. Pectoral fins equal in size. Gill rakers moderately long and thick. Lateral line arched over pectoral fin. Scales cycloid, accessory scales present (Gutherz 1967). Body depth 38–44% SL; head length 25–29% SL; eye diameter 25–30% HL; upper jaw length 44–50% HL (Gutherz 1967).

Meristics. Dorsal fin rays 71–86; anal fin rays 58–72 (rarely 76); pectoral fin rays 10–12; lateral line scales about 95; gill rakers on lower part of anterior arch 7–10; vertebrae 41 or 42 (Norman 1934; Gutherz 1967; Miller and Jorgenson 1973).

Color. Fourspot flounder are mottled gray or brown and the back is marked with four large, oblong and very conspicuous eye spots edged with a paler color. Two spots are located at midbody and the other two are at the anterior part of the caudal peduncle. Bigelow and Schroeder reported two fish on which the blind side, posterior to the gill openings, was as dark as the eyed side, and marked similarly with four eye spots; also others that were more or less dark below.

Distinctions. Fourspot flounder resemble summer flounder except for the four large spots.

Habits. Although rather common about Woods Hole in May and June, and still more numerous along the coast of New York, little is known of their habits. Fourspot do not come into as shoal water as summer flounder often do. In the northern Mid-Atlantic, they occur over most of the shelf and inshore waters, but in the southern Mid-Atlantic they are found mostly offshore (Gutherz 1967). Trawl surveys indicate that north of Delaware Bay there is a seasonal movement into deeper waters in winter, particularly on Georges Bank, and a return to shallower water in summer (Grosslein and Azarovitz 1982). They are resident year-round in the New York Bight (Wilk et al. 1975). Seasonal movements are thought to be related to changes in bottom temperature. They occur at depths of 27–400 m, the greater depths are off Florida (Ginsburg 1952b; Wilk et al. 1990). The average range of water temperatures is 8.9°–13.9°C (Martin and Drewry 1978), but they have been encountered at temperatures of 6°–22°C (Smith et al. 1975).

Food. The stomachs of 178 fourspot flounder contained three major categories of food (Bowman et al. 2000): cephalopods (54.3% by weight), crustaceans (29.8%), and bony fishes (11.8%), but their proportions changed with size. Crustaceans and fishes were major components up to 35 cm TL,

Leptocheirus), mysids (mostly *Neomysis*), and decapod shrimps (especially *Crangon* and *Dichelopandalus*). Older fish (>20 cm in length) continue to prey on decapod shrimps, but their diet also includes large quantities of crabs (mostly *Cancer*), squids (mainly *Loligo* and *Rossia*), and small fishes such as juvenile silver hake, Atlantic cod, sculpins, butterfish, and Gulf Stream and yellowtail flounders (Langton and Bowman 1981; Sedberry 1983; Bowman and Michaels 1984; Bowman et al. 2000). Fourspot also take polychaete worms, particularly *Asabellides oculata*, sipunculids, and small lobsters as food (Bowman et al. 2000).

Fourspot are active during daylight hours (Bowman 1988). Their stomachs contained only small quantities of well-digested food at dawn and large amounts of fresh food at dusk. During darkness, when they cease feeding, their behavior is such that they are more readily caught by bottom trawls. In July 1980 in waters south of Nantucket the estimated food intake of adult fourspot was 1.3% of their body weight per day (Bowman 1988).

The diet of the few fourspot examined from the Gulf of Maine (Langton and Bowman 1981) gives some indication that decapod shrimps and small fishes are probably important food in the Gulf.

Predators. Spiny dogfish are the most important predator of fourspot flounder; other predators are goosefish, silver hake, and windowpane and summer flounders (Bowman et al. 1984; Rountree 1999; Bowman et al. 2000).

Breeding Habits. In the New York Bight, ripe females occur from April to September and the percent of ripe females peaks in June and July (Wilk et al. 1990). There is no information on age at maturity or fecundity but ripe females there range in size from 153 to 419 mm (Wilk et al. 1980). Eggs were collected by the MARMAP surveys from April to November and were abundant during most of the 8 months over the entire breadth of the shelf (Berrien and Sibunka 1999: Fig. 91). Spawning began offshore in April or May, then spread onto the shelf. Spawning generally began in the Mid-Atlantic and by May was widespread throughout the Mid-Atlantic, southern New England, and southern Georges Bank. Eggs were found in low concentrations in the Gulf of Maine from June to September. Distribution of eggs indicates spawning takes place at temperatures between 6° and 9°C in waters of 35–80 m (Smith et al. 1975).

Early Life History. The eggs are buoyant, 0.9–1.12 mm in diameter, with a single oil globule of 0.16–0.19 mm. They occur with hake and scup but may be differentiated from the former by the unpigmented oil globule and from the latter by the late-stage embryo, which has black pigment scattered over the embryo and the yolk mass in the fourspot and black pigment only on the head and oil globule in the scup (Colton and Marak 1969). These melanophores appear on the embryo about 36 h after fertilization, and yellow spots (xanthophores) at 42 h. Yellow spots are more numerous than black ones on the larva, but there are fewer on the oil globule. The incubation period is 8 days at 10°–13°C (Ginsburg 1952b); and 54 h at 21.1°C (Miller and Marak 1962). Larvae are long and thin at hatching (2.7–3.2 mm). The anus opens at the margin of the finfold about midway down the body. Black spots are evenly scattered over head, trunk, and yolk sac. Yellow spots on the dorsal finfold form brown-yellow blotches. Halfway between the vent and the tail there are black spots that form a slight vertical band. Metamorphosis begins at 10–11 mm SL, and at 12 mm (about 3 months after hatching), they may begin to go to the bottom, although larvae larger than 8 mm are rarely caught (Smith et al. 1975). A detailed study of larval development was presented by Leonard (1971).

There are two centers of abundance of larvae from Cape Hatteras to Block Island (Smith et al. 1975), midway out on the shelf off New Jersey and on the outer shelf off southern New England. Eggs and larvae have also been collected in July and August in Salem Harbor, Mass. (Elliott and Jimenez 1981).

General Range. Found from the eastern part of Georges Bank to Tortugas, Fla. (Gutherz 1967).

Occurrence in the Gulf of Maine. Fourspot flounder are plentiful along the continental shelf as far eastward as the general vicinity of Nantucket and to the southeastern part of Georges Bank (Map 36). They have also been collected in Cape Cod Bay and occasionally along the coast of Maine. They ranked 24th in abundance of 43 species in Massachusetts Bay during the BIOME Survey (Collette and Hartel 1988). Other records for fourspot in the coastal waters of the Gulf of Maine are from Monomoy at the southern angle of Cape Cod; the vicinity of Provincetown, Gloucester and Salem; Sheepscot Bay (Scattergood and Coggins 1958); Passamaquoddy Bay; and the Bay of Fundy (Leim and Day 1959).

Importance. There has been no directed fishery for fourspot either commercially or recreationally, but they are often combined with other landings of miscellaneous flatfishes as "unclassified flounders" (Ralph 1982).

RIGHTEYE FLOUNDERS. FAMILY PLEURONECTIDAE

Grace Klein-MacPhee

Pleuronectidae is a large family, 21 genera and about 60 species (Nelson 1994; Cooper and Chapleau 1998), with eyes almost always on the right side and eggs without oil globules. Gulf of Maine pleuronectids are also characterized by the presence of a neural arch on the first precaudal vertebra, a well-developed lateral line on both sides of the body, and olfactory lamellae that are parallel without a central rachis (Chapleau 1993). This group has undergone a number of taxonomic changes. In a detailed phenetic study, Sakamoto (1984) lumped many genera (including *Limanda, Pseudopleuronectes,* and *Liopsetta*) into *Pleuronectes,* a conclusion that is not accepted by all flatfish researchers (Wheeler 1992; Rass 1996). More recently, the monophyly, intrarelationships, and taxonomy were redefined based on a cladistic analysis of the group (Cooper and Chapleau 1998), which resulted in inclusion of *Liopsetta* in *Pleuronectes* but continued recognition of *Limanda* and *Pseudopleuronectes* as valid genera. Seven pleuronectid species belonging to the subfamilies Hippoglossinae (*Hippoglossus hippoglossus, Reinhardtius hippoglossoides*), Hippoglossoidinae (*Hippoglossoides platessoides*), and Pleuronectinae (*Glyptocephalus cynoglossus, Limanda ferruginea, Pleuronectes putnami,* and *Pseudopleuronectes americanus*) are found in the Gulf of Maine (Cooper and Chapleau 1998). There are summaries of recent information on western North Atlantic populations of the five most economically important species in the area: witch flounder (Cargnelli et al. 1999d), American plaice (Johnson et al. 1999a), halibut (Cargnelli et al. 1999a), yellowtail flounder (Johnson et al. 1999b), and winter flounder (Pereira et al. 1999).

WITCH FLOUNDER / *Glyptocephalus cynoglossus* (Linnaeus 1758) / Gray Sole /

Bigelow and Schroeder 1953:285–290

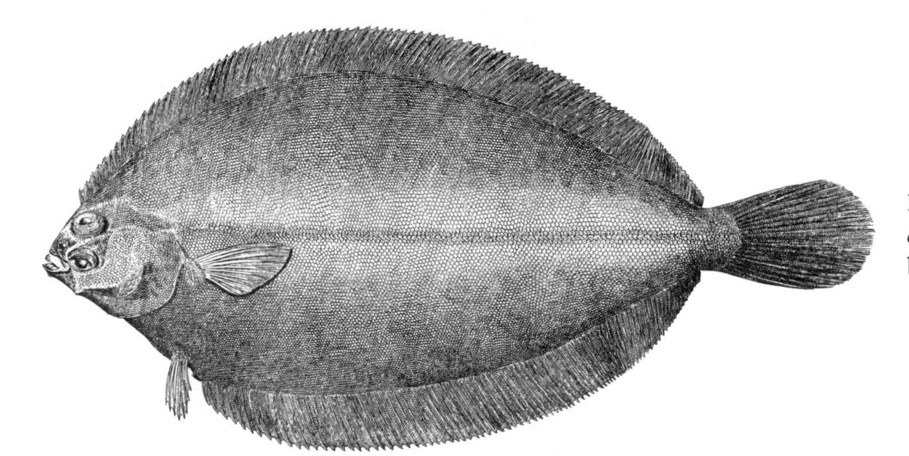

Figure 294. Witch flounder *Glyptocephalus cynoglossus.* Maine. Drawn by H. L. Todd.

Description. Body two and half to three times as long as it is broad, elliptical in outline, very thin (Fig. 294). Head only about one-fifth TL; dorsal profile convex. Mouth very small. Teeth small, incisor-like, and in a single series. About 12 mucous pits or depressions on blind side of head, and less obvious ones on eyed side. Dorsal and anal fins of uniform width throughout most of their length, but narrow gradually toward head and tail. Anal fin preceded by short, sharp spine pointing forward, a prolongation of the postabdominal bone. Pectoral and pelvic fins alike on both sides. Caudal fin small, rounded. Lateral line usually straight, but somewhat arched above pectoral fin in some individuals. Entire body and head (except for tip of snout and lower jaw) covered with cycloid scales.

Meristics. Dorsal fin rays 97–117; anal fin rays 87–101; pectoral fin rays 9–13 on eyed side; gill rakers 6–9 on lower arch; lateral line scales 110–140; branchiostegal rays 7; vertebrae 57 or 58 (Goode and Bean 1896; Norman 1934; Miller and Jorgenson 1973; Scott and Scott 1988).

Color. Witch flounder are less variable in color than most other flatfishes. They are brown or russet gray on the eyed

tinged with violet and either plain or spotted. The pectoral fin membrane on the eyed side is dusky or even black with a narrow light distal border. The blind side is white and more or less dotted with minute dark points. Most of the witch 205–455 mm TL collected by *Albatross IV* in April 1994 were pigmented on the blind side as well as the eyed side (pers. obs.).

Size. Maximum length 78 cm (Powles 1965b). Bigelow and Schroeder reported a maximum size of 63.5 cm and an average size of 30.5–51 cm for the Gulf of Maine. Maximum size 65 cm in Nova Scotia (Powles and Kennedy 1967).

Distinctions. Witch resemble winter and yellowtail flounders. They can be distinguished by their smooth scales and their greater number of fin rays, narrower body, smaller head, and prominent mucous pits on the blind side of the head.

Habits. Witch generally occur at depths of 90–330 m in U.S. waters (Smith et al. 1975) and 46–274 m off Canada (Scott and Scott 1988). Maximum depth reported by Bigelow and Schroeder was 1,569 m off southern Nova Scotia. They are frequently caught on fine muddy sand, clay, or mud. They

they are common there on the smooth ground between rocky patches. In Newfoundland they occur on muddy bottom (Pinhorn 1976).

Witch occur in the Gulf of Maine at temperatures of 1.7°–14.5°C, according to precise locality and depth. In the Gulf of St. Lawrence, they occur at −1° to 5.5°C. They are apparently never found in any numbers in waters warmer than 10°C.

Food. Polychaete worms account for 85.7% by weight of the witch flounder diet (Bowman et al. 2000). Families making up most of the diet, in terms of percentage weight, are the Lumbrineridae, Goniadidae, Onuphidae, Nephtyidae, and Capitellidae. Witch diet also includes echinoderms (especially holothurians), amphipods and isopods, squids, and small-shelled mollusks (Langton and Bowman 1981; Bowman and Michaels 1984; Bowman et al. 2000). Several studies provide comprehensive listings of food organisms for witch caught off the northeastern United States (Langton and Bowman 1981; Bowman and Michaels 1984; Bowman et al. 2000). Off Scotland, polychaetes were also the major prey group (Rae 1969) but crustaceans, amphipods, and cumaceans were the most important food of fish up to 20 cm.

Predators. Fish predators include thorny and smooth skates, spiny dogfish, goosefish, white hake, and halibut (Bowman et al. 2000); harp seal also prey on witch (Mansfield 1967).

Parasites. Witch flounder are infected by 3 species of protozoans; 13 trematodes; 5 nematodes; 2 acanthocephalans, *Corynosoma* sp. and *Echinorhynchus gadi;* and a copepod, *Acanthochondria cornuta* (Bray 1979; Margolis and Arthur 1979; Margolis and Kabata 1988; Lile et al. 1994).

Breeding Habits. Median lengths at maturity for female and male witch were 30.4 and 25.3 cm, respectively, corresponding to ages of 4.4 and 3.6 years (O'Brien et al. 1993). Witch in Newfoundland waters reach sexual maturity at 4–5 years of age for males at 25–30 cm and 8–10 years for females at 40–50 cm (Bowering 1976). Egg production ranges from 350,000 eggs for a 45-cm female to a maximum of 3 million eggs for a 65-cm female (Bowering 1978). Fecundity was also correlated with age and weight, with size being more important than age. Significant variations in fecundity occurred between different areas sampled. Eastern Atlantic fish produce more eggs per length-group than western Atlantic fish, but overall egg production was the same because west Atlantic fish grow to a larger size.

Witch eggs were collected by MARMAP surveys from February to October with a peak in June (Berrien and Sibunka 1999: Fig. 97). Eggs were found first in February or March in the vicinity of Chesapeake Bay, then further north in southern New England in April, and then on Georges Bank and in the Gulf of Maine in May. Eggs were found in the Gulf of Maine from May to October.

Larvae were present from April to October from Cape Cod to North Carolina (Smith et al. 1975). They were first collected in April from North Carolina to Long Island at depths of 10–90 m. Bottom water temperatures were then 5°C and the center of distribution was off eastern Long Island and southern New England (Smith et al. 1975). Bigelow and Schroeder reported that eggs are shed at water temperatures of 4°–9°C.

Early Life History. Witch eggs are buoyant, spherical, and transparent, with a narrow perivitelline space (the space is broad in eggs of American plaice, which overlap witch eggs in diameter), without an oil globule, and measure 1.07–1.45 mm in diameter. The chorion is thick with many folds (Evseenko and Nevinsky 1975). Newly spawned witch eggs may be confused with those of cod and haddock, for they overlap in size and season. Identification is possible after a few days' incubation, for black pigment is seen in gadoid eggs soon after the embryo is visible, but does not appear in witch embryos until after hatching.

Incubation occupies 7–8 days at temperatures varying from 7.8° to 9.4°C, and newly hatched larvae are about 4.9 mm long with a larger yolk sac than those of other Gulf flatfishes. The range in hatching length is 3.52–5.59 mm (Colton and Marak 1969). The yellow and black pigment aggregates into five transverse bands on the body, yolk, and finfolds within a few days after hatching, when a larva is 5–6 mm long. One of these bands is in the region of the pectoral fin, one at the vent, and three on the trunk posterior to the vent. The yolk is absorbed about 10 days after hatching; caudal rays begin to appear at a length of 15 mm; rays of the vertical fins are well advanced at 21 mm and reach their final number at about 30 mm. Witch can be distinguished from other pleuronectid larvae at this stage by their high dorsal (100–115) and anal (87–100) fin ray counts. The eyes are still symmetrical up to this stage, but the left eye has moved to the dorsal surface of the head in larvae of about 40 mm. Migration of the eye is complete at a length of 40–50 mm, at which time the young fish take to the bottom. The pelagic larval stage is lengthy compared with other flounders, lasting 4–6 months (Bigelow and Schroeder) to 1 year (Evseenko and Nevinsky 1975).

Juvenile witch nursery areas have not been definitely determined for most localities. Powles and Kohler (1970) and Markle (1975) collected juveniles from Nova Scotia and Virginia, respectively, and discovered what appears to be a complex life-history pattern. Powles and Kohler (1970) found 10–30 cm juveniles (about 2–5 years of age) in deep benthic habitats (144–450 m). No adults were captured here, but commercial catches are made chiefly at 108–144 m. They also reported that G. Kelly and R. Marak had collected larval witch up to 6.8 cm long above the thermocline in 30–40 m in the Gulf of Maine. These fish were very close to metamorphosis.

Markle (1975) collected young witch 11–25 cm long on the continental slope off Virginia at depths between 256–1,080 m with bottom temperatures of 4.1°–11.3°C and salinities of 34.9–35.7 ppt, indicating that this area is an important nursery for 1- to 4-year-old witch. Markle (1975) postulated that the young originate from a spawning stock north of Cape Cod and that the larvae and juveniles travel southward along the deeper isotherms being carried by the water currents flowing west from Georges Bank. Egg and larval collections (Smith et al. 1975) show that significant numbers of eggs and larvae are produced south of Cape Cod, and these probably settle on the Virginia slope. Both of these studies indicate that juveniles remain separated from adults and occupy deeper areas until the onset of sexual maturity.

Age and Growth. Relative to other flatfishes in the region, witch are slow-growing, late-maturing, and long-lived (Burnett 1988). Maximum observed length and age for the Gulf of Maine–Georges Bank region are 72 cm TL and 30 years. Powles and Kennedy (1967) used whole otoliths for aging their Scotian Shelf samples, validating their interpretation of annuli by using modal analysis of back-calculated lengths. They found that females grow faster and larger than males in the Gulf of Maine–Georges Bank region (Fig. 295) and Nova Scotia, and also determined that the maximum size and age of witch in Canadian waters is 65 cm and 20 years. Growth is reported to be slower at greater depths (Molander 1925), and females are heavier than males for a given size. Lux (1969) provided length-weight relationships for witch flounder caught south of Nantucket Shoals. Maximums for age and growth seem to be lower in the eastern Atlantic: the largest specimen recorded there was 48 cm and was estimated to be 14 years old (Bowers 1960). Significant changes in age composition, growth, and maturation have been noted for witch off southern Newfoundland after 20 years of commercial exploitation. In general the trend has shifted from older fish and a broader age range to younger fish with a narrower age range (Bowering 1989). The age structure of the Gulf of Maine–Georges Bank stock has shown similar age truncation (Wigley et al. 1998).

General Range. Witch occur in moderately deep water on both sides of the North Atlantic. In American waters free-drifting witch larvae are reported from as far north as the Strait of Belle Isle, around the coast of Newfoundland, and over the Grand Banks region in general. Adults are known from the Hamilton Bank, northern half of Newfoundland, the Grand Banks (Bowering 1987), Gulf of St. Lawrence south to northern Virginia in moderate depths, and to Cape Hatteras in deep water (Goode and Bean 1896). In Europe, they are caught from Murmansk to the west coast of France.

Occurrence in the Gulf of Maine. Distribution of witch flounder is governed by water depth. They can be expected anywhere the water is deeper than 27–36 m, if the bottom is suitable (Map 37). They are common on Browns Bank, Georges Bank, and Nantucket Shoals. Areas of maximum density according to trawl records are the South Channel grounds off eastern Massachusetts and western Maine. They were the third most abundant fish in the BIOME surveys in Massachusetts Bay (Lux and Kelly 1978).

Movements. Witch are more stationary in the Gulf of Maine than many other flounders: they are caught year-round, with no evidence of movement inshore or offshore with the change of seasons. In Swedish waters, however, they are said to work up into shoaler water in autumn and deeper again in late winter and spring (Molander 1925).

Importance. Witch were of no commercial importance in the Gulf three-quarters of a century ago; few fishermen distinguished them from other flounders and no records were kept of the catch. They are an excellent table fish and now are in such demand that they bring a higher price than either

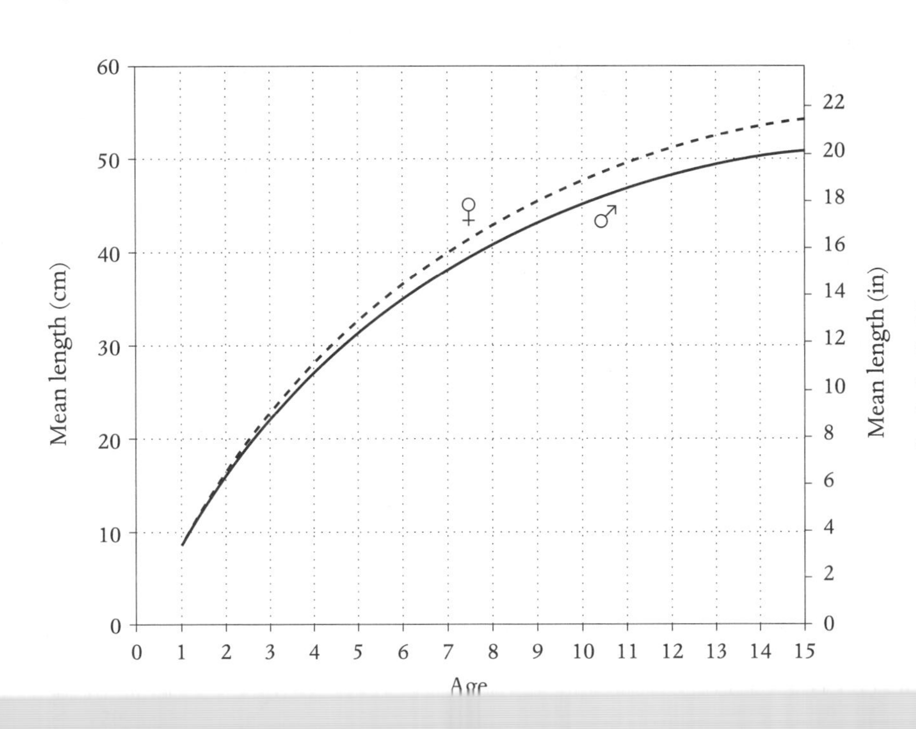

Figure 295. Age-length curve for male and female witch flounder *Glyptocephalus cynoglossus*, Georges Bank–Gulf of Maine region. (From Penttila et al. 1989:37.)

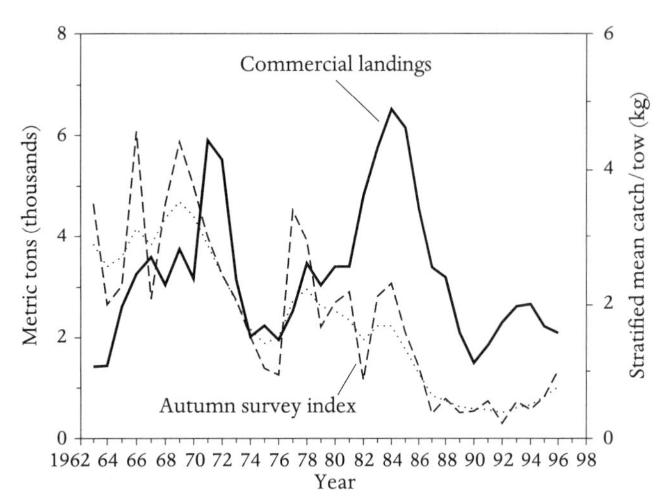

Figure 296. Commercial landings and NMFS autumn survey index for witch flounder *Glyptocephalus cynoglossus* from Georges Bank–Gulf of Maine, 1963–1996. (From Wigley 1998b.)

yellowtail or American plaice. Figure 296 shows the catch of witch flounder for the Gulf of Maine and Georges Bank. The history of the fishery in Newfoundland was summarized by Bowering (1976). The fishery started as a by-catch of the fisheries in the 1940s and, owing to the decline in catches of cod and Greenland halibut, has become an important groundfish resource. The otter trawl is the only gear now in use in the United States that has been adapted to catch witch flounder on a commercial scale.

In the Gulf of Maine–Georges Bank fishery, landings of witch were primarily a by-catch until the 1980s, when it became a directed fishery. Landings peaked in 1983 and then declined to record lows in 1990. Spawning stock biomass declined from 26,000 mt in 1982 to about 6,300 mt in 1990. There was a strong year-class in 1990 that may increase the spawning stock biomass in the short term but it is expected to decline again unless fishing mortality is reduced. The stock is at a low biomass and is overexploited (Wigley 1998b).

Discards occur in the Gulf of Maine shrimp fishery and the large mesh otter trawl fishery. In the shrimp fishery, about 76% witch are discarded per tow per fish caught on a weight basis. The discards are mostly age 3 (Howell and Langan 1992). Mortality is caused by capture, on-deck sorting, and bird predation when the discards are thrown overboard (Ross and Hokenson 1997). In the otter trawl fishery, fish age 4 and older are affected (Wigley 1998b).

Witch flounder is being investigated as a potential candidate for aquaculture (Rabe et al. 1999). Recreational catches are insignificant (Wigley 1998b).

AMERICAN PLAICE / *Hippoglossoides platessoides* (Fabricius 1780) /

Canadian Plaice, Dab, American Dab / Bigelow and Schroeder 1953:259–267

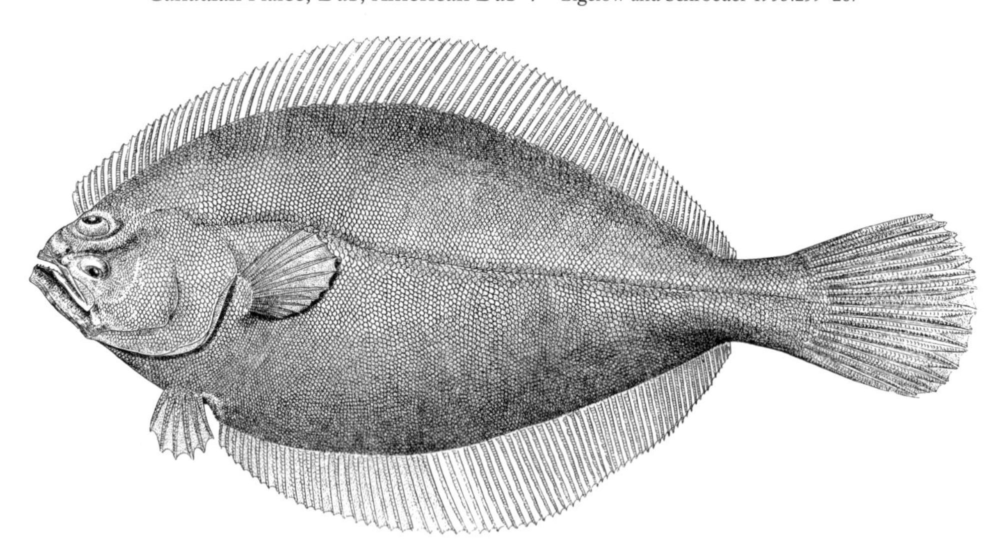

Figure 297. American plaice *Hippoglossoides platessoides*. La Have Bank. Drawn by H. L. Todd.

Description. Body comparatively broad, about two and a half times as long to base of caudal fin as it is broad (Fig. 297). Snout pointed; mouth gaping back to middle of lower eye. One irregular row of sharp conical teeth in each jaw. Dorsal fin originates in front of middle of upper eye. Anal fin arises slightly in advance of pectoral fin base. Both long fins taper toward head and tail, a short, sharp spine (prolongation of postabdominal bone) pointing forward in front of anal fin. Pectoral fin on eyed side longer and more rounded than corresponding fin on blind side, usually with one or two more rays. Pelvic fins alike in size, shape, and location. Margin of caudal fin convex, either rounded or with middle rays longest, forming a blunt angle. Lateral line nearly straight. Free edges of scales on eyed side of body and head ctenoid, giving a characteristic rough feeling when handled. Scales on blind side cycloid except on rear part of body and along fin bases.

Meristics. Dorsal fin rays 76–101, usually 78–98; anal fin rays 60–79; pectoral fin rays 9–12; lateral line scales 85–97; gill rakers on lower part of anterior arch 9–12; branchiostegals 8; vertebrae 42–48 (Pitt 1963; Scott and Scott 1988).

Color. American plaice are more uniform in color than most of the smaller flatfishes in the Gulf of Maine, ranging from red to gray-brown (darker or paler) above and pure or bluish white below. The tips of the rays of the dorsal and anal fins are white. On one specimen the right edge of the eyed side was white from the gill opening to the last ray of the pelvic fin (Bigelow and Schroeder). Two cases of ambicolored fish have been reported from the Gulf of St. Lawrence (Brunel 1971).

Size. Adults measured by Welsh off Cape Ann, Mass., ranged from 29.4 to 61 cm, and few of those caught in the Gulf of Maine are any longer. Nova Scotian fish ranged from 29.4 to 61 cm in length (Huntsman 1918a), whereas those caught in colder waters off Newfoundland averaged 45.7 cm. The largest fish recorded from American waters, taken near Sable Island in May 1939, was 82.5 cm long and weighed 6.4 kg (McKenzie 1940). The next largest was taken in the Gulf of Maine (43°30′ N, 69°16′ W) at 90–100 m and was 81 cm long (D. Flescher, pers. comm., NMFS Woods Hole lab trawl survey records April 1969).

Distinctions. The most obvious characters of American plaice are that they are right-eyed and large-mouthed, with a rounded tail and a nearly straight lateral line. They most closely resemble Atlantic halibut and Greenland halibut. They differ from both in having a rounded caudal fin; from the former in having a straight lateral line; and from the latter in having a shorter, more rounded head, eyes closer together, and smaller teeth.

Taxonomic Remarks. American plaice tend to differentiate into local races in different seas. Fin rays are more numerous on the average in fish from high latitudes than in those from low latitudes, and the body is relatively wider in fish caught off Greenland and North America than in those from Scandinavia and the North Sea (Bigelow and Schroeder). They have been regarded as two distinct subspecies that intergrade around Iceland (Norman 1934; Wheeler 1969; Scott and Scott 1988): *H. platessoides platessoides* in North America and *H. p. limandoides* (Bloch 1787) in Europe.

Habits. American plaice are bottom fish like other flatfishes, but they rise some distance off bottom on occasion and move about to a considerable extent. Bigelow and Schroeder caught one in a tow net at least 9–10 m above the bottom off Ipswich Bay, where the water was 90 m deep, and many are caught in gill nets. In one part of their range or another they are found from the tide line down to 713 m (Scott and Scott 1988). The maximum depth recorded was 1,400 m off the Grand Banks (Iglesias and Paz 1996). They are often found in very shoal water in colder seas. They are often seen under wharves around Newfoundland, for example, and some are seined right on the beach on the west Greenland coast. Adults have never been caught in less than 18 m of water in the Gulf of Maine, and their preferred depth off the coast of Maine is 100–119 m (Sherman et al. 1993). The preference of plaice for moderately deep water in the southern part of their range bars them from most of the Gulf of Maine harbors and river mouths, but they enter the deeper estuaries and passages between the islands in the northeastern part of the Gulf and those near Mt. Desert Island, Passamaquoddy Bay, N.B., and St. Mary's Bay, N.S. Depth alone appears to be a better determinant of habitat for plaice than temperature alone, but both factors are important (Swain and Morin 1997). Females tend to occupy warmer waters than males, a preference that is density-independent. Males occupy colder water at higher levels of abundance and both sexes appear to occupy warmer waters when Atlantic cod, a competitor, are in higher abundance (Swain 1997). The lower limit to their occurrence in commercial concentrations, which applies to the whole North American coastline including the Scotian Shelf, is 182 m (O'Brien 1998b). In the Grand Banks region, they are most common between 100–200 m, but in spring they are found mostly in waters less than 100 m (Morgan and Brodie 1991).

Like some other flatfishes, plaice avoid rocky or hard bottoms, preferring a fine, sticky but gritty mixture of sand and mud. They are also caught in numbers on the soft oozy mud of the deeper basins in the western side of the Gulf.

Based on their temperature range, this is an arctic-boreal species. Most development occurs in waters of 1.7°–7.7°C, but they are able to live in the lowest polar temperatures (–1.5°C); the upper temperature limit is about 10°–13°C.

They live in a wide range of salinities in different waters, from 30 ppt or lower in the Baltic Sea to 34 ppt in the open Atlantic. They are never found in water that could be described as brackish along the coasts of New England, but they have been found in water 20–22 ppt in Hamilton Inlet, Labrador (Backus 1957b).

Commercial catch data indicate that American plaice move off the bottom at night and thus are less subject to capture in otter trawls then than they are during the day. There was less diurnal variability in catches from depths less than 100 m than from water 155 m and deeper (Pitt 1967). Echograms also indicated that plaice move off the bottom at night (Beamish 1966a).

Food. In the northwest Atlantic, the diet of adult plaice (Bowman et al. 2000) consists chiefly of echinoderms (49.8% by weight), mostly brittle stars [illegible]

Ophiura and bivalves (17.7%). They also eat polychaetes, various shrimps and other crustaceans, and sand dollars, in fact, they prey on practically any bottom-living animals that are small enough for them to devour. Occasionally they catch small fishes like sand lance (Pitt 1967). Their diet depends, for the most part, on where in the northwest Atlantic they are located. For example, in southern New England they eat large quantities of amphipods, shrimp (*Crangon*), polychaetes, and bivalves. On Georges Bank they consume mostly echinoderms (92.3% by weight), brittle stars (75.4%), and sand dollars (16.9%). The brittle star *Ophiura sarsi* is a very important component in plaice diet at selected sites in the Gulf of Maine, occurring in 65% of fish stomachs and accounting for 82% of the stomach contents by wet weight. Plaice selected brittle stars of 4–10 mm disk size probably because of the greater abundance and accessibility of this size in the area studied (Packer et al. 1994). Plaice from the Gulf of Maine also consumed a large proportion (24.6%) of bivalves (Bowman et al. 2000). Those from western Nova Scotia eat large quantities of sand dollars, brittle stars, and hermit crabs (Langton and Bowman 1981; Bowman and Michaels 1984). In the Magdalen Shallows, N.B., echinoderms and bivalves were the most important food items. Amphipods and polychaetes were eaten but were volumetrically less significant (Powles 1965a). Plaice are daytime feeders whose maximum food consumption occurs around 1500–2000 hr. Their estimated daily ration in Passamaquoddy Bay is 1.28% on a per-gram-body-weight basis (MacDonald and Waiwood 1987).

In the planktonic stage, small plaice larvae feed on diatoms but consume copepods when they are larger and more active (Huntsman 1918a). Juveniles, after first settling to the bottom, eat small shrimps and other crustaceans. Their diet overlaps that of juvenile cod in the Magdalen Shallows; at this stage there is competition between species, but none between younger and older plaice because they have different diets (Powles 1965a). Juveniles also feed extensively on polychaete worms in the families Ampharetidae, Nephtyidae, and Sabellidae. They switch to feeding primarily on echinoderms by 41–45 cm (Bowman et al. 2000).

Feeding intensity varies seasonally. It almost ceases in January and February, is moderate in November and April, increases rapidly in May, and is intense through September (Powles 1965a). High rates of feeding in summer produce energy stores that are used for metabolism and gonad maturation in winter and early spring when feeding ceases (MacKinnon 1972).

Predators. All large predacious fishes that feed near the bottom probably prey upon American plaice. They were eaten by eight species of fishes in the NEFSC food habits study: goosefish, Atlantic cod, white hake, spiny dogfish, sea raven, thorny skate, Acadian redfish, and halibut, in order of number of occurrences (Rountree et al. 1999). In the Magdalen Shallows (Powles 1965a), cod fed primarily on plaice 35 cm or less in length; cod larger than 100 cm contained more plaice in their stomachs than smaller cod. In more northern seas, Greenland shark prey regularly upon them (Bigelow and Schroeder). Gray seal and harp seal are also predators (Bowen et al. 1993; Lawson et al. 1998). Plaice occurred in 2.2% of gray seal stomachs collected from the Scotian Shelf. The size range eaten averaged 14.7 cm and 30 g wet weight. In the Gulf of St. Lawrence, plaice occurred in 10.9% of gray seal stomachs in 1986 and 1987 (Benoit and Bowen 1990a).

Parasites. Plaice are hosts to: 2 protozoans, *Trypanosoma murmanensis* and *Haemohormidium terraenovae* (Khan et al. 1991); 2 cestodes, *Bothrimonus sturionis* and *Scolex pleuronectis;* 17 trematodes (northeastern Atlantic); 3 acanthocephalans, *Corynosoma* sp., *Echinorhynchus gadi, E. laurentianus;* 7 nematodes; and 2 copepods, *Acanthochondria cornuta* and *Lernocera branchialis* (Margolis and Kabata 1988; Køie 1993; Lile et al. 1994).

Breeding Habits. Median length at maturity of American plaice sampled from the Gulf of Maine region during spring in 1986–1990 was 26.8 cm for females and 21.1 for males. Median age at maturity for females and males was 3.6 and 3 years, respectively (O'Brien et al. 1993). These values are similar to those reported by Sullivan (1981) for females but not for males: samples taken in the Gulf of Maine in 1980 had a median length of 27.4 cm and age of 3.4 years for females and 26.5 cm and 3.4 years for males. A decline in median length and age at maturity of both female and male American plaice in the Scotian Shelf area was reported from 31.0 cm and 6 years for females during 1970–1974 to 30.8 cm and 4.7 years during 1975–1779. Values for males declined from 24.8 cm and 4.5 years during 1970–1974 to 21.9 cm and 3.5 years during 1975–1979 (Beacham 1983b).

An average female plaice of 50 cm from the Grand Banks and Newfoundland produces 500,000 eggs and the maximum for a 68-cm fish was 2,200,000 eggs (Pitt 1964). Fecundity was related primarily to length and weight and secondarily to age. Bigelow and Schroeder gave fecundity as only 30,000–60,000 eggs, but Pitt (1964) believed that this referred to European stocks, which comprise smaller fish.

Eggs were collected year-round by the MARMAP surveys but there was a strong seasonal cycle with fewest eggs occurring in autumn and early winter and peak densities during spring and early summer (Berrien and Sibunka 1999: Fig. 100). Eggs were most abundant in shallower portions of the Gulf of Maine and Georges Bank and were present in at least low concentrations in the Gulf of Maine in all months.

American plaice reproductive performance under laboratory conditions showed that they are batch spawners over a period of 26 days, with periodic peaks of egg production in increasing intervals at about 8, 20, and 34 days. Mean relative fecundity was 1.5×10^5 eggs per kg body weight. There was no relationship between actual fecundity and either length or weight. Mean percentages of viable and fertile eggs were 44 and 13%, respectively, with a significant decline in fertile eggs with time but no significant decline in viable eggs. Hatch success rate varied during the spawning period, but the mean success rate was 23%. The number of larvae hatched showed a positive correlation with the percentage of fertile eggs (Nagler et al. 1999). Histological observations of ovarian development are detailed in Maddock and Burton (1998) and indicate that plaice may have the ability to push oocytes through vitellogenesis from a previtellogenic condition during the spawning season.

The temperatures and salinities at which eggs are produced can be stated rather definitely for the Gulf of Maine because plaice lie close to the bottom, if not actually on it. The earliest spawning takes place at about 2.7°C, and no eggs have been found at a bottom temperature higher than about 4.4°C. At locations where eggs have been taken in large numbers, salinities range between 31.8 and 32.8 ppt at the bottom. Plaice spawn in more saline waters on the other side of the Atlantic.

Early Life History. American plaice eggs are buoyant and have no oil globule, but the perivitelline space around the yolk is so broad that they are not likely to be confused with those of any other Gulf of Maine fish. This space is created by water entering between the egg proper and its covering membrane after the eggs are shed and it about doubles the total diameter of the egg. Eggs taken in the Gulf of Maine averaged about 2.5 mm in diameter, but they have been reported as 1.38–3.2 mm in other seas, depending on size of the perivitelline space. Colton and Marak (1969) found them to be 1.5–2.8 mm, average 2.28 mm.

Incubation takes 11–14 days at a temperature of 3.9°C. The eggs change specific gravity as development proceeds, for Huntsman found that in the Gulf of St. Lawrence, newly spawned eggs floated at the surface but eggs nearly ready to hatch drifted suspended at a depth of about 18 m.

During development, minute black and yellow pigment cells are scattered over the embryo, not aggregated into any diagnostic clusters. Pigment gathers in five definite groups very soon after hatching, one on the gastric region, one about the vent, and three behind the vent, a pattern similar to that of larval witch flounder. Larvae are 4–6 mm long at hatching. The yolk is absorbed about 5 days after hatching, when the larva has grown to 6.2–7.5 mm in length. Caudal rays appear shortly after this; dorsal and anal rays appear at about 11–12 mm; and the three vertical fins are differentiated at about 15–18 mm. By this stage the body has begun to assume the deep but very thin form characteristic of all young flounders, and the jaws have developed sufficiently to show that the little fish belongs to one of the large-mouthed species. The left eye may begin its migration when a larva is about 20 mm long. Welsh (in Bigelow and Schroeder) found it visible above the outline of the snout in Gulf of Maine specimens of 24 mm and almost at the dorsal edge at 34 mm. Metamorphosis is usually complete by 30–40 mm (Colton and Marak 1969).

The only other Gulf of Maine species for which larval plaice might be mistaken are witch flounder and halibut. Witch are longer at corresponding stages of development, but

shorter, and the outlines of the throat and abdomen are sufficiently different to distinguish them from halibut.

Young plaice drift freely up to the time of metamorphosis, close to the surface at first but sinking deeper as they grow, until they seek the bottom. The period occupied by larval growth and metamorphosis varies with temperature, probably covering 3–4 months in the Gulf of Maine, where Bigelow and Schroeder took pelagic larvae as early in the season as 26 May and as late as 2 August.

Immature American plaice are strongly concentrated along and inside the 100-m isobath in the western Gulf of Maine during spring and summer. They are also found scattered in deeper waters of the western and central Gulf of Maine and on the northern edge of Georges Bank (Wigley and Gabriel 1991).

Howell and Caldwell (1984) studied the bioenergetics of plaice embryos and prolarvae at four temperatures 2°, 6°, 10°, and 14°C. Temperatures of 14°C and above were lethal to the eggs. The rate of growth was fastest at 10°C and slowest at 2°C. Embryos and larvae survived at 10°C but were smaller then those incubated at 2° and 6°C, and yolk utilization efficiency was lower.

Age and Growth. Age is calculated from otoliths, which often become difficult to interpret when the fish become older. Otoliths can often exhibit complex zone formation, which requires cross-verification of age using both the thin section and/or the whole otolith or sectioned otolith half. Young fish can be aged by examining the whole otolith in alcohol if the hyaline zones are well defined. If not, preparation of a thin-sectioned otolith, preferably the left, is necessary (Dery 1988e). Age and growth of American plaice from the Magdalen Shallows were studied by Powles (1965a) using otolith ring counts. He also compared data of fish from Cape Cod (Lux 1969), Labrador (Yanulov 1962b), and Iceland (Sæmundsson 1925). His investigation showed that growth rates from different localities in the Magdalen Shallows were similar, with a tendency for slower growth in deeper waters. Males and females grew at similar rates up to the fourth year, after which females grew faster. Growth rates differed among geographic areas, being more rapid to the south. American plaice on Georges Bank grow faster than plaice in the Gulf of Maine (Esteves and Burnett 1993). An age-length graph for American plaice from the Gulf of Maine is presented in Fig. 298.

In the Gulf of St. Lawrence American plaice live from 24 to 30 years. The oldest found in the Gulf of Maine NEFSC survey cruises was 24 years old (Penttila et al. 1989). In the Gulf of St. Lawrence, most fish age 3 and younger were males, but females outnumbered males among older fish. All fish 14 years and older were females (Huntsman 1918a). No explanation has yet been devised for the apparently higher mortality rate for males.

General Range. American plaice are common on both sides

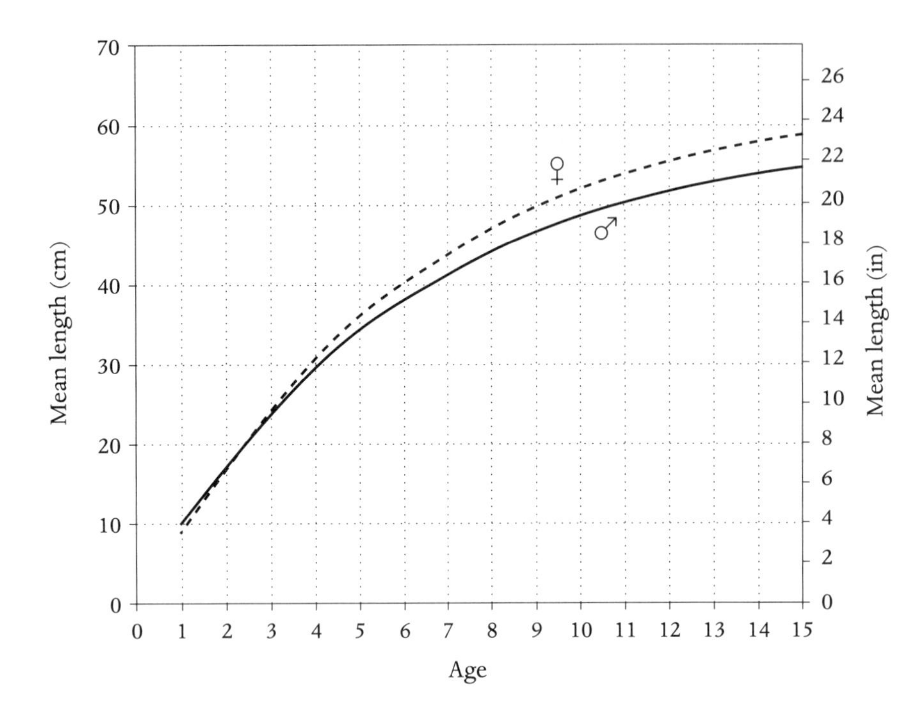

Figure 298. Age-length curves for male and female American plaice *Hippoglossoides platessoides*, Gulf of Maine region. (From Penttila et al. 1989:5.)

as far north as Upernavik near the Arctic Circle, south from Frobisher Bay, Baffin Island, and western Hudson Bay (Hunter et al. 1984), along the Labrador coast, Newfoundland Banks, and into the Gulf of Maine. A few are caught in the Woods Hole region, off Martha's Vineyard, and off Narragansett Bay, R.I. The most southerly record is one caught off Montauk Point, Long Island, in 200 m of water (Bigelow and Schroeder) although larvae are found as far south as Delaware (Johnson et al. 1999a).

In European waters they range from Iceland and Spitsbergen southward to the North Sea, where they are an important commercial fish, and to the west Baltic. The English Channel is the southern boundary of their regular occurrence.

Occurrence in the Gulf of Maine. Plaice are probably the most abundant of all Gulf of Maine flatfishes at depths greater than 54–90 m (Map 38), except for witch flounder. They are caught all around the inner parts of the Gulf wherever the water is more than 27 m deep and the bottom is smooth. They are also caught on Stellwagen Bank, Cashes Ledge, and right up to the head of the Bay of Fundy. They are also widespread on Georges Bank and were the most abundant fish captured during resource surveys in Massachusetts Bay during the 1970s (Lux and Kelly 1978). Plaice are still widespread but are now less abundant.

Migrations. Tagging studies of stocks in the Gulf of St. Lawrence, Nova Scotia, Labrador, and the Grand Banks show that plaice are relatively sedentary and form discrete populations with no significant migration between coastal and offshore areas (Powles 1965a; Pitt 1969). Short seasonal migrations occur with fish moving into shallow water from June to September and offshore from October to May. This appears to be a response to water temperature changes (Powles 1965a). Populations tend to return to the same areas each summer, with little mixing. The results of Pitt (1969) for the Grand Banks, were confirmed by Zubchenko (1985), who studied parasites of plaice. He found that the Grand Banks stocks were geographically isolated with a specific parasite fauna and different degrees of infestation. Populations were separated from each other by deep channels, but were not genetically isolated because eggs and larvae are transported from north to south by the Labrador Current. Analysis of genetic population structure of plaice from the Gulf of St. Lawrence showed them to be a single, randomly mating population (Stott et al. 1992).

Importance. The Gulf of Maine produced 72% of American plaice landings since 1976 (Sullivan 1981). The principal commercial fishing gear used is the otter trawl. Landings increased from an average of 2,300 mt in the 1970s to an average of 12,700 mt in 1979–1984. Annual landings then declined and in the 1990s ranged between 4,000 and 7,000 mt (Fig. 299). Annual CPUE indices declined from a high in 1977 to a low in 1988 but have remained relatively stable at low values since then. Fishing mortality is estimated to be above the overfishing level and the stock is overexploited (O'Brien 1998b). Recreational catches are insignificant.

Discard estimates of American plaice in the Gulf of Maine shrimp fishery averaged 81% (Howell and Langan 1992). Fish age 2 and 3 were most commonly affected and fewer than 10% of those more than 30 cm long survived after 30 min of deck exposure. This indicates that the shrimp fishery causes substantial reduction in abundance and yield of finfish, including

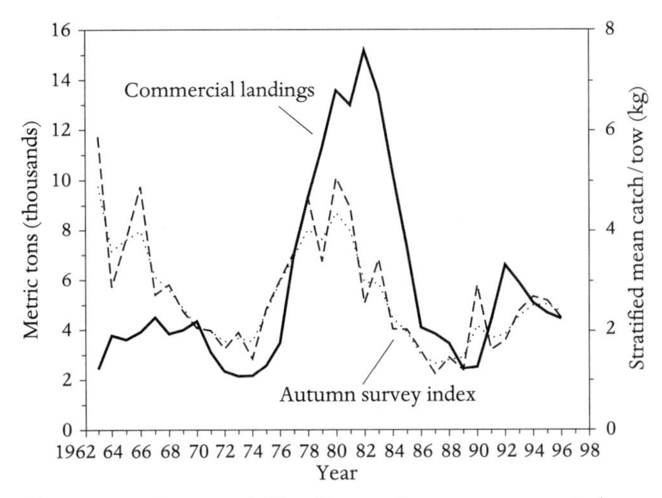

Figure 299. Commercial landings and autumn survey index for American plaice *Hippoglossoides platessoides,* Gulf of Maine–Georges Bank, 1963–1996. (From O'Brien 1998b.)

plaice, owing to the large numbers of fish discarded, the high mortality of the discards, and the large numbers of vessels engaged in shrimping. American plaice age 3 and 4 have the highest discard rates in the large mesh otter trawl fishery (O'Brien 1998b).

Stocks. Research survey distributions indicated continuous occurrence of American plaice over the Gulf of Maine and Georges Bank so they have been considered a single stock. However, comparison of growth rates suggests there may be differences between the population from the Gulf of Maine, which is slower growing, and the population from Georges Bank (Esteves and Burnett 1993). There appears to be a single, randomly mating population in the Gulf of St. Lawrence based on genetic structure (Stott et al. 1992). On the Scotian Shelf there appear to be two stocks based on distributions of sexually mature females and egg distributions (Neilson et al. 1988).

ATLANTIC HALIBUT / *Hippoglossus hippoglossus* (Linnaeus 1758) / Bigelow and Schroeder 1953:249–258

Figure 300. Atlantic halibut *Hippoglossus hippoglossus.* Eastport, Maine. Drawn by H. L. Todd.

Description. Dorsal fin origin above upper eye, extending length of fish, broadening slightly for first third of length and then narrowing again toward caudal peduncle (Fig. 300). Anal fin similar to dorsal in shape but shorter, originating close behind pectoral fins. Anal fin preceded by a sharp spinelike extension of postabdominal bone, which projects in young fish but is hidden by skin in older ones. Pectoral fins of different shapes: pectoral on eyed side obliquely pointed, fin on blind side rounded. Pelvic fins alike; small, located in front of pectoral fins, separated from anal by a considerable space. Head and body scaly.

Meristics. Dorsal fin rays 92–107; anal fin rays 69–84; lateral line scales about 160; gill rakers on lower limb of first arch 7 or 8; vertebrae 50 or 51 (Norman 1934; Martin and Drewry 1978; Scott and Scott 1988).

Color. Halibut are chocolate to olive or slaty brown on the eyed side. Small fish are paler and more or less mottled, whereas larger ones are more uniform and darker, sometimes almost black. The blind side is usually pure white in smaller fish, but larger ones are often blotched or clouded with gray. Occasionally a halibut is taken with the blind side marked with patches of the same color as the eyed side. Bigelow and Schroeder saw one fish in which the rear third of the lower surface was a dark brown. Gudger (1935a) and Gudger and Firth (1937) reported two partially ambicolored reversed halibut and one almost totally ambicolored with structural anomalies.

Size. Among Gulf of Maine fishes only swordfish, tuna, and some of the larger sharks grow bigger than halibut. The record fish, taken in 1917 by Capt. A. S. Ree about 83 km northeast of Cape Ann, weighed 280 kg eviscerated with the head still attached when brought in and must have been 318 kg while alive. The next largest was one about 205 kg caught on a handline in the deep water between Browns and Georges Bank in 1908 by W. F. Clapp. A 186-kg halibut brought in to the Boston fish pier in 1941 was spoken of as the largest that had been landed there in a score of years. According to Bigelow and Schroeder, halibut heavier than 136 kg are rarities anywhere in the North Atlantic. Full-grown females average 45.5–68 kg. Males run smaller, and most of the large fish landed in New England ports weigh 22.7–91 kg. The all-tackle world record halibut weighed 161.20 kg and was caught in Valevag, Norway, in 1997 (IGFA 2001).

Distinctions. Apart from the very large size, halibut can be distinguished from most other right-eyed flatfishes by the large mouth, gaping back as far as the anterior half of the lower eye; the arched lateral line; and the distinctly concave caudal fin. They most closely resemble Greenland halibut but differ in the arched lateral line, larger mouth, and more widely spaced eyes.

Habits. Halibut, like all flatfishes, are normally groundfish once the young settle to the bottom, but they come to the surface on occasion. They are very powerful fish when hooked. When caught in shallow water they are very active, usually starting off at great speed in their attempts to escape when they are hauled up from the bottom. They are usually found on sand, gravel, or clay, not on soft mud or on rock bottom; 720–900 m may be set as the lower boundary at which they are found in any numbers; they have been reported from 25 to 1,000 m (Scott and Scott 1988; Miller et al. 1991), but their absolute depth limit is not known. Halibut are boreal fish. Large catches are made only at times and places where the water is at least as warm as 2.2°–3.3°C. In the Grand Banks region, for instance, they are mostly caught either far enough down the slope to be below the Labrador Current or at times and places where the current does not touch the bottom if the fish are on the bank. The lower limit to their temperature range is not clear-cut. Bigelow and Schroeder recorded one halibut trawled in bottom water as cold as 0.6°C. At the opposite extreme, only a few are taken in the parts of the North Sea where the bottom water is warmer than 8°C and none where it is warmer than 15°C. Most halibut taken from the Gulf of St. Lawrence and Nova Scotia were trawled at temperatures of 3°–9°C (McCracken 1958) and from U.S. waters at 4°–6°C in spring and 9°–13°C in autumn (Cargnelli et al. 1999a). Halibut move into deeper water in winter and shallower water in summer. They have also been described as shifting ground in the same way in the coastal belt of the Gulf of Maine from season to season.

Food. Halibut are voracious, preying chiefly on other fishes, including skates, herring, capelin, cod, cusk, haddock, grenadiers, silver hake, redfishes, sculpins, sand lance, wolffish, mackerel, and flounders. Half a bushel of flatfishes have been taken from the stomach of one individual. They also eat crabs, lobsters, clams, and mussels, and even seabirds. Fishermen have reported finding halibut with the discarded heads and backbones of cod thrown overboard, a variety of indigestible objects such as pieces of wood or iron, and even fragments of drift ice in their stomachs.

Food of 125 halibut had three main components: bony fishes (65.7% by weight), cephalopods (18.5%), and crustaceans (15.4%); crustaceans are important up to 60 cm TL and fishes and cephalopods are more important in larger halibut (Bowman et al. 2000).

The diet of halibut in any particular locality depends on what other groundfishes are abundant. In the Gulf of Maine, squid (*Illex*), crabs (*Cancer*), alewife, silver hake, sand lance, and ocean pout were the most important food in 1977–1980 (Bowman et al. 2000). On Georges Bank, they feed chiefly on longhorn sculpin and ocean pout. Jensen (1925) found that halibut caught off west Greenland had fed chiefly on large shrimps (*Pandalus borealis*).

Halibut, like other flounders, must be nearly invisible as they lie on the bottom, capturing by a sudden rush any fish that passes within reach. On one occasion a halibut of about 32 kg was seen at the surface trying to kill a small cod with its tail. Fishermen said that in the days when they were still plentiful on the shoaler banks, the appearance of a school of halibut drove away the cod and haddock.

Halibut eat invertebrates, mostly annelid worms and crustaceans, almost exclusively up to a length of about 30 cm (Scott and Scott 1988). They feed on both invertebrates and fishes from 30 to 80 cm long, and then almost exclusively on fishes. Bowman et al. (2000) also found that halibut less than 60 cm FL eat large quantities of shrimps (e.g., *Dichelopandalus, Pandalus,* and *Crangon*), hermit crabs (*Pagurus*), and rock crabs (*Cancer*). Fish prey identified included many of the species noted above. However, squid, hitherto not noted as prey, are also taken, especially large ones over 70 cm.

Halibut captured in the field and maintained in the laboratory fed in the water column and would not touch food that fell to the bottom of the tank. There appeared to be a size hierarchy in feeding as larger halibut took most of the food during a feeding session. Meal size averaged 12% of body weight, and it took about 4 days to completely process a meal. The first feces were produced 24–33 h after a chromic oxide–labeled meal and most material was voided 49 h after ingestion, but total gut transit time took up to 120 h (Davenport et al. 1990).

Predators. Halibut, in their turn, fall prey to seals and especially to Greenland shark, for which they are a staple food, as well as to spiny dogfish and goosefish (Bowman et al. 2000).

Parasites. Halibut are host to: three monogenean species; ten digeneans; an isopod, *Aega psora;* four copepods; and an amphipod, *Opisa eschrichti* in the northwestern Atlantic (Margolis and Kabata 1988). In the northeastern Atlantic, they are also host to two more digeneans; five nematodes; and an acanthocephalan, *Echinorhynchus gadi* (Bray 1979; Køie 1993; Lile 1998).

Breeding Habits. Average age at maturity for halibut was thought to be about 10 years. Off Nova Scotia females were thought to mature at ages 10–12 and males at 8–11 (McCracken 1958), but these conclusions were based on only a few specimens. Subsequent studies (1972–1984) showed that males matured as early as 4 years and females at 6 and that the 50% levels of maturity were 8 and 12 years, respectively (Bowering 1986).

Large halibut are very prolific and may produce up to 7 million eggs (Haug and Gulliksen 1988). An Atlantic female of about 90 kg contained about 2,182,772 eggs (Bigelow and Schroeder). Atlantic halibut are annual group synchronous spawners (Neilson et al. 1993) and females are batch spawners, ovulating several batches of eggs in a season (Methven et al. 1992). In Canadian waters Atlantic halibut spawn from late winter to early spring with the chief production of eggs in November and December (Neilson et al. 1993). In the eastern Atlantic, they spawn from March to May with the chief production of eggs in April. A few females may ripen as early as the end of January and some not until June. Off west Greenland they spawn late in spring. Off the American coast the spawning season continues until September at various localities from Georges Bank to the Grand Banks. Individual halibut may spawn over a considerable period of time. Presumably halibut spawn on the bottom like other flatfishes, but the depths at which spawning occurs are not known for certain. East of Cape Cod fishermen report ripe males and females on the slopes of all the offshore banks, suggesting that American fish may spawn at least as shoal as Pacific fish do, that is, at depths of 270–405 m (Tåning 1936), and perhaps even shoaler. In Norway, spawning halibut were captured in a fjord north of Bergen at depths of 600–700 m (Blaxter et al. 1983).

Early Life History. Although halibut eggs are buoyant they do not float on the surface, but drift suspended in the water at depths greater than 54–90 m. They are 2.96–3.8 mm in diameter and do not have an oil globule (Tåning 1936; Lønning et al. 1982). The only other equally large buoyant fish eggs that are likely to be found in the Gulf of Maine are those of argentine, but these have a large oil globule, so there is no danger of mistaking them. The buoyant eggs of Greenland halibut are even larger and readily distinguished from those of Atlantic halibut. At least under laboratory conditions (Lønning et al. 1982; Blaxter et al. 1983), both the incubation period and development of halibut are very unusual. Ultrastructure studies of the chorion of the large egg and embryonic development suggest that the eggs are pelagic; however, their neutral buoyancy is only achieved at salinities of 36 ppt or greater, leading to the conclusion that they are bathypelagic, as indicated earlier (Tåning 1936; Rollefsen 1934). The high density of the egg may be an adaptation to reduce mortality (Blaxter et al. 1983), since the eggs are less subject to predation if they are closer to, but not actually on, the bottom, as opposed to being in the pelagic zone.

The eggs divide slowly, there is only a small perivitelline space, the yolk is very large, and the larva is thin and unpigmented. Incubation times are 20 days at 4.7°C, 18 days at 5°C and 13 days at 7°C (Blaxter et al. 1983). They hatch at a very immature stage. There are no functional eyes or mouth and little pigment, heartbeats are difficult to observe, and the yolk sac is very large. During late organogenesis the larval tail is bent, and this curvature is present in newly hatched larvae for about 7 days. Size at hatching is 6–7 mm (Lønning et al. 1982; Rollefsen 1934). Prominent neuromasts were present, and their ultrastructure was described by Blaxter et al. (1983). Yolk absorption took 50 days at 5.3°C. The mouth first opened 3 weeks posthatch and became functional at 25 days. The rectum developed after 4 weeks, and 10 days later the rest of the gut had developed and had started to coil. Anatomical evidence indicated that halibut can feed 28–35 days after hatching at 5°C, while they still have a large yolk sac (Blaxter et al. 1983).

The smallest naturally hatched Atlantic halibut yet seen was 13.5 mm long and had vertical fin rays. At about 22 mm, dorsal and anal fins develop, and the pelvic fins are visible. By this size the eye has moved upward until its margin is just visible above the contour of the head. Fish of this size also show the large-mouth characteristic of the species. Up to this stage there is little pigment. About a quarter of the eye appears above the profile when the little halibut is about 27 mm long, but even at 34 mm the eye has not completed its migration. Larvae up to 25 mm in length are recognizable as halibut by their curiously upturned snout. Older ones are separable from American plaice by the outlines of the head and the abdomen.

How long young halibut live in the plankton is not known. Young, so small (47–64 mm long) that they were evidently spawned the preceding spring or summer, have been trawled off Iceland during the last week of July. The smallest bottom stages have been taken only in water shoaler than 50 m, evidence that larvae of Atlantic halibut tend to rise toward the surface and drift inshore during their pelagic stage as do Pacific halibut.

Age and Growth. Female halibut from the Gulf of St. Lawrence grow faster than males after 5 years and reach a larger maximum size (McCracken 1958). Growth rates are similar in female halibut from southwestern Nova Scotia, but males grow more slowly than males in the Gulf of St. Lawrence.

General Range. Boreal and subarctic Atlantic, in continental waters. Stragglers have been found off New Jersey, New York, and Block Island. The most southerly record on the western side of the Atlantic is from Reedville, Va. (Walford 1946). They were once fairly common from Nantucket Shoals to Labrador. The northernmost record for North America is Cutthroat Harbour, Labrador (Backus 1951). Halibut are not known off the Arctic coasts of Asia or North America, but the Greenland side of the Davis Strait supports a halibut fishery as far north as Disco Bay. In the eastern Atlantic, halibut have been reported from the Bay of Biscay north to Iceland, the Norwegian coast and Spitsbergen, as well as the Barents Sea (Haug 1990).

Occurrence in the Gulf of Maine. The history of halibut in the Gulf of Maine must be written largely in the past tense, for they have been depleted by overfishing. In colonial days, they were familiar fish and very abundant on the coast of northern New England. They seem to have maintained their numbers down to the first quarter of the nineteenth century, when contemporary writers described them as numerous around the whole coastline of the Gulf of Maine. They were discovered in abundance on Nantucket Shoals, Georges Bank, Browns Bank, and on the Seal Island ground as soon as fishing was regularly undertaken offshore. In the 1980s, a 77-kg halibut was caught of Scituate and an 83-kg fish was caught on Cashes Ledge (Chamberlain 1980), but there is no present-day spawning population within the Gulf of Maine. The fact that the inshore grounds were fished out so soon with little apparent tendency to recover when the fishery slackened and that depletion by overfishing has not been accompanied by a corresponding decrease in average size of the fish that are caught suggest that the halibut population of the inner parts of the Gulf always depended more on immigration from east and north of Cape Sable for its maintenance than it did on local production.

Movements. In the coastal belt of the Gulf of Maine halibut move into deeper water in winter and shallower water in summer. They have been described as shifting ground in the same way in the Gulf of St. Lawrence and Nova Scotian waters (McCracken 1958). Seasonal movement of halibut onto the Greenland Banks as early in summer as temperature allows seems to be a search of food, for a much richer supply of small fish is available on these shoaler bottoms than deeper down the Davis Strait slope. Food supply is probably equally important in influencing their seasonal movements in the Gulf of Maine (Bigelow and Schroeder).

Tagging studies in the Gulf of St. Lawrence and off western Nova Scotia show little movement of fish; most were caught in the same area in which they were tagged (McCracken 1958). Recaptures, size and age composition, and differences in growth rates suggest that these two regions support separate stocks. Several long-distance recaptures were made. One halibut tagged off the Gulf of St. Lawrence was recaptured 2,575 km away in Iceland; one was recaptured 241 km away off Quebec, and one from Nova Scotia was captured 805 km away on the Grand Banks.

Juveniles begin emigration from nursery areas when they are 3–4 years old and undergo their most intensive migrations (Stobo et al. 1988). Most remain within the same general region, but some make very long migrations—Labrador to the western coast of Greenland, Gulf of St. Lawrence to Iceland, Scotian Shelf to the Grand Banks, and western Greenland to the Grand Banks (summarized in Cargnelli et al. 1999a).

Importance. Halibut, because of their present-day scarcity, are of little commercial importance in the Gulf of Maine. Bigelow and Schroeder gave an historical account of the fishery and its unfortunate decline. Since 1953 U.S. landings have been below 100 mt and the stock is overfished. Almost all halibut caught in NEFSC survey trawls are juveniles (Cargnelli et al. 1999a). They are still an important commercial species in Canada (Scott and Scott 1998), but have declined there as well (Cargnelli et al. 1999a).

Halibut are being studied for their aquaculture potential in Europe and North America (Haug 1990; Waiwood et al. 1997). Their biology has been summarized (Haug 1990; Trumble et al. 1993).

Stocks. Genetic data suggest there may be more than one panmictic population from Greenland eastward (Foss et al. 1998). To the west two stocks, Gulf of St. Lawrence and Scotian Shelf, are recognized (Trumble et al. 1993). Microsatellite markers have recently been developed to further understanding of the genetic structure of Atlantic halibut in the western Atlantic (McGowan and Reith 1999).

YELLOWTAIL FLOUNDER / *Limanda ferruginea* (Storer 1839) / Yellowtail, Rusty Flounder /

Bigelow and Schroeder 1953:271–275

Description. Body oval, comparatively wide, nearly half as broad as it is long (Fig. 301). Dorsal outline of head more deeply concave than any other Gulf of Maine flounder. Eyes set so close together that their rounded orbits almost touch each other. Each side of jaw has a single series of teeth in about equal numbers. Teeth small, conical, close set. Dorsal fin originates over upper eye, middle rays longest. Anal fin similar in outline to dorsal, but much shorter, preceded by a short, sharp spine (postabdominal bone) pointing forward. Pelvic fins alike, each separated from anal fin by a considerable space. Pectoral fin on blind side slightly shorter than mate on eyed side. Gill rakers of moderate length. Lateral line distinctly arched over pectoral fin. Scales ctenoid on eyed side, cycloid on blind side.

Meristics. Dorsal fin rays 73–91; anal fin rays 51–68; pectoral fin rays 10; lateral line scales 88–100; gill rakers 10–12 on lower arch; vertebrae 40–45 (Norman 1934; Scott and Scott 1988).

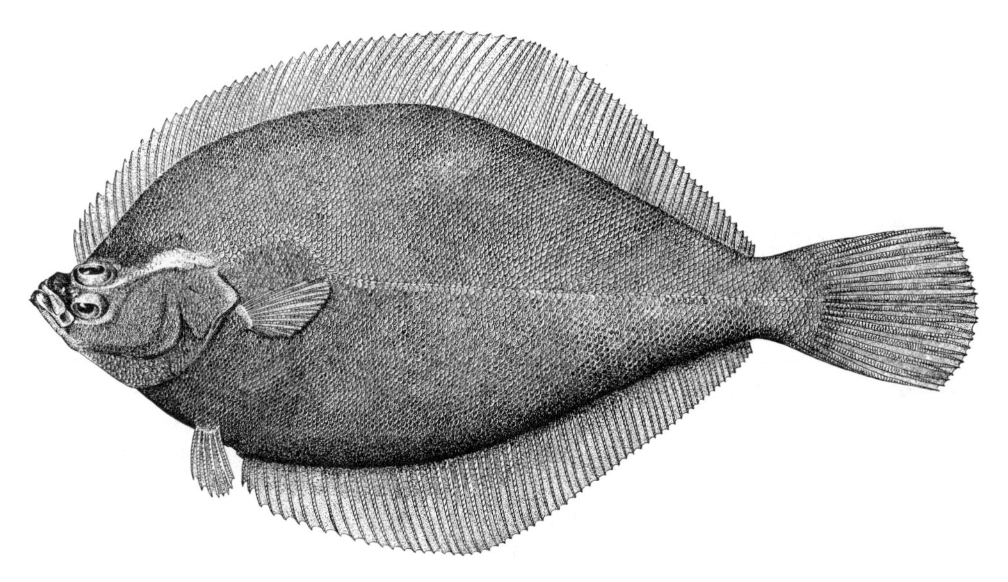

Figure 301. Yellowtail flounder *Limanda ferruginea*. Drawn by H. L. Todd.

Color. Yellowtail are more constant in color than most other Gulf of Maine flatfishes. The eyed side, including the fins, is brown or slaty olive, tinged with red and marked with large irregular rusty red spots. The caudal fin and margins of the dorsal and anal fins are yellow; the yellow tail in particular is a diagnostic character. The blind side is white, except for the caudal peduncle and margins of the dorsal and anal fins, which are yellow.

Size. This is a medium-sized flatfish. Males average 40 cm (30–47.6 cm) and females 46 cm (39.4–55.3 cm). The largest specimens taken were a 55.4-cm female caught off Cape Cod (Penttila et al. 1989) and a 62.7-cm fish caught off Newfoundland (Scott and Scott 1988).

Distinctions. Small mouth and thick fleshy lips separate yellowtail from the large-mouthed flounders (halibut and plaice). They are easily distinguished from winter flounder by their more pointed snout, thin body, arched lateral line, and more numerous fin rays; from smooth flounder by the last two characters, as well as by the concave dorsal profile of the head and being scaly between the eyes; and from witch by the arched lateral line, fewer fin rays, concave dorsal profile of the head, and lack of mucous pits on the blind side of the head.

Habits. Yellowtail keep to deeper water than either winter or smooth flounder. They are generally an offshore species, found from 10 to 100 m. From the Mid-Atlantic Bight to Georges Bank they are found most often at depths of 37–73 m (Overholtz and Cadrin 1998), but are common in shallow waters (9–64 m) off Cape Cod (Lux 1964). Almost any sandy bottom or mixture of sand and mud suits them, but they shun rocks, stony ground, and very soft mud.

Yellowtail tolerate a wide [illegible]

18°C), but water temperature and year-class strength are correlated—cooler temperatures promoting larger year-classes (Sissenwine 1974). They can also tolerate a considerable range of combinations (Laurence and Howell 1981), but survival is maximized at intermediate temperatures and high salinities (8°–14°C, 32–38 ppt). On the Scotian Shelf the preferred temperature and salinity ranges are 2°–6°C and 32–38 ppt (Scott 1982b).

Food. Adult yellowtail feed chiefly on amphipods (41.5% by weight), especially *Ericthonius rubricornis,* and polychaetes (38.5%) (Bowman et al. 2000). They also eat cnidarians, small crabs and shrimps, mysids, cumaceans, isopods, univalve and bivalve mollusks, echinoderms, and sometimes even small fishes such as sculpins, sand lance, and cusk (Langton and Bowman 1981; Langton 1983; Bowman et al. 1987; Collie 1987; Bowman et al. 2000).

Yellowtail feed primarily during daylight hours. Stomachs contain the least amount of food at dawn and the most at dusk, an indication that feeding begins near sunrise and ends, for the most part, around sunset (Bowman et al. 2000). Stomachs of spawning fish contain on average only small quantities of food; stomachs of fish with developing gonads are the fullest (Langton 1983). Yellowtail caught on Georges Bank had more food in their stomachs than those caught in the Mid-Atlantic or southern New England (Efanov and Vinogradov 1973; Langton 1983). In the Gulf of Maine, the diet of this flounder is generally the same as in other areas except that it includes larger quantities of a sea cucumber, *Stereoderma unisemita* (Langton and Bowman 1981). Comprehensive lists of prey and other information on yellowtail feeding habits and behavior can be found in reports by Langton and Bowman (1981), Langton (1983), Bowman and Michaels (1984), Collie (1987), Martel and [illegible]

Predators. Yellowtail flounder were found in stomachs of 11 species of fishes in the NEFSC food habits study, of which only spiny dogfish and Atlantic cod are significant predators (Rountree 1999). Other fish predators include blue shark, skates (little, winter, and smooth), goosefish, hakes (spotted, white, and silver), longhorn sculpin, bluefish, Atlantic halibut, and fourspot flounder (Lux and Mahoney 1972; Maurer and Bowman 1975; Bowman and Michaels 1984; Rountree 1999; Bowman et al. 2000). On Sable Island Bank, Atlantic mackerel 6–14 mm in length prey on larval yellowtail flounder, which account for 18% of their diet (Fortier and Villeneuve 1996). Predation on fish larvae declined with increasing density of copepod nauplii in the environment. Yellowtail are also preyed upon by gray seal on the Scotian Shelf (Bowen et al. 1993).

Parasites. Yellowtail are parasitized by 3 species of protozoans, *Haemohormidium terraenovae, Glugea stephani,* and *Trypanosoma murmanesis;* 15 trematodes; 7 nematodes; 5 cestodes; 2 acanthocephalans; and a copepod, *Acanthochondria cornuta* (Bray 1979; Margolis and Arthur 1979).

Yellowtail from Delaware to Georges Bank have been found with epitheliocystis, an infection that is visible as cystlike lesions on the gills and fins. About 9% of the fish collected were infected. Some yellowtail caught on Sable Island Bank off Nova Scotia were infected with a fungus, *Ichthyophonus,* which was evident as white cysts on the liver and other internal organs (Powles et al. 1968).

Breeding Habits. Yellowtail are batch spawners (Howell 1983). Median age at maturity for females is 1.6 years in southern New England (Royce et al. 1959), and 1.8 years on Georges Bank, 2.6 years off Cape Cod (O'Brien et al. 1993), but 5.0 years on the Grand Banks (Pitt 1970). Females produce 350,000–4,570,000 eggs, depending on body length and, to a lesser extent, age (Pitt 1971). The southern New England stock has a higher fecundity at a given length and age (Howell and Kesler 1977) than the Grand Banks stock described by Pitt (1971), correlated with higher water temperatures (4.9° to 12.3°C vs. −1° to 6.5°C) in southern New England.

Yellowtail eggs were collected by the MARMAP surveys from February to September with peak abundances from April to June (Berrien and Sibunka 1999: Fig. 103). Spawning began in February or March, occurring first in the northern half of the Mid-Atlantic and then extending rapidly into southern New England and Georges Bank. In April and May spawning increased in intensity in these areas and began in the Gulf of Maine. Eggs were found in the Gulf of Maine from April to September. Spawning occurs at water temperatures ranging from 4.5° to 8.1°C and over depths of 45–75 m (Smith et al. 1975).

Laboratory studies of estimates of the ovulatory periodicity of yellowtail flounder indicate that a 1-day interval may characterize regular ovulation patterns. Females produced a mean number of 14–22 batches for about a month, and batch fecundities usually remained within a range of 10,000–60,000 eggs. Maternal variation in egg production and egg quality was large and independent of size differences among females. Some females had disrupted ovulation patterns that affected the realization of potential fecundity contained within the prespawning ovary. High interbatch variation in egg quality was not related to progressive decreases in egg diameter and dry weight over time. Batches with high survival rates appeared at random during a female's period of ovulation (Manning and Crim 1998). Temperature interacts with maternal contributions to egg size to affect development time and size of yellowtail flounder larvae at hatch. Maternal effects contributed significantly to differences in egg sizes but development time was most affected by temperature. Average length at hatch varied significantly among females and with temperature, as did the variance in hatching length within a population. Overall, the nonadditive interaction between maternal contributions and the environment suggests that female effects must be considered over the entire range of environmental conditions experienced by their progeny. The results also suggest that it is inappropriate to quantify female effects among eggs and extrapolate these difference to larvae (Benoit and Pepin 1999).

Early Life History. Yellowtail eggs are buoyant, spherical, very transparent, have a narrow perivitelline space, and do not have an oil globule. Eggs range from 0.68 to 1.01 mm in diameter, averaging about 0.9 mm (Howell 1980; Colton and Marak 1969). The surface of the egg is covered with very minute striations, and the germinal disc is a very pale buff color in life. Embryonic pigment gathers in three groups shortly before hatching (which takes place in 5 days at a temperature of 10°–11°C); one group in the region of the head, one near the vent, and a third halfway between the vent and the tip of the tail (Bigelow and Schroeder).

Laurence and Howell (1981) gave more detailed descriptions of the early life–history stages. They did not see striations on the egg surface. Newly hatched larvae are 2.1–2.5 mm and display the three pigment patches described above, although head pigment may be absent. Larvae have unpigmented eyes, a large yolk sac, no functional mouth, and fin buds only in the pectoral region. Pigmentation increases during yolk-sac absorption. A group of five or six melanophores is seen near the anus, which is just posterior to the yolk sac; a second group of black pigment spots is seen midway between the anus and the tip of the tail; and a row of fine pigment spots is seen in the ventral finfold below the notochord. These are useful in distinguishing yellowtail from other flatfish larvae. When larvae reach 4–14 mm, four vertical bars are present at the pectoral fin and evenly spaced caudally (Martin and Drewry 1978). In the New England area metamorphosis occurs at about 14 mm.

Yellowtail and winter flounder larvae co-occur in March–May in New England, but the former can be distinguished at the yolk-sac stage by their lighter pigmentation and more extensive dorsal pigmentation. Ventral pigmentation extends to the tail tip in yellowtail but not in winter flounder. The midtail band extends onto the finfolds in winter flounder but not in yellowtail, and there is little or no lateral line pigmentation. Winter flounder metamorphosis occurs at 9 mm and that of yellowtail at 14 mm (Elliott and Jimenez 1981). Yellowtail larvae co-occur with American plaice on the Scotian Shelf. Yellowtail have 41–44 vertebrae, 38–41 myomeres, and 51–67 anal fin rays, and metamorphose at about 16 mm SL (Van Guelpen 1980); American plaice have 45–48 vertebrae, 44–47 myomeres, and 62–76 anal fin rays, and metamorphose at greater than 25 mm SL.

Yellowtail eggs and larvae are pelagic. Larvae are near the surface at night and move down to 20 m with ascent and descent occurring at sunset and sunrise, respectively (Smith et al. 1978). Larger larvae make longer migrations. This habit may assist in dispersal of larvae because circulation at the surface transports the larvae further. Distributional maps of yellowtail eggs and larvae are given in Colton and St. Onge (1974) and Smith et al. (1975). After young yellowtail become benthic they are not readily caught by trawlers, so nursery grounds for this species are not well defined. Recent studies on the continental shelf of the New York Bight found that age-0 yellowtail flounder (minimum size 5.7 mm SL, mean 17.4 mm SL, maximum 34.9 mm SL) were the most abundant fish collected. Collections included larval, age-0, and adult fish. Settlement took place in the summer primarily in the midshelf area (41–70 m), and fish tended to remain in this vicinity, which was both a settlement and a nursery area. Mortality was high after settlement. Bottom temperature and depth were the most important parameters for determining habitat associations. Higher abundances occurred above 3°C and the preferred temperature appeared to be between 4° and 8°C. The midshelf area is a "cold pool" and catches decreased in the fall when the bottom temperature rose owing to cold pool turnover. Densities in the midshelf area averaged 49.81 fish per 1,000 m^2. Given the large areal extent of settlement, this represents a very large number of fish (Steves et al. 2000).

Immature yellowtail flounder (<20 cm TL) are found well inshore of 100 m. In the Gulf of Maine, they are concentrated between Massachusetts Bay and Cape Cod Bay in spring and autumn. They are found along the southern edge of Georges Bank in spring, but the concentration shifts northward to near the Great South Channel and primarily eastward to slightly deeper waters along the southeastern part of Georges Bank. In southern New England, immature fish are found midshelf at depths around 45 m in spring and 60 m in autumn. Mean depth distribution for immature yellowtail in all regions is 47.3 m in spring (range 9–179 m) and 63 m in fall (14–287 m) (Wigley and Gabriel 1991).

Age and Growth. Historically, scales have been used for age determinations of yellowtail flounder (Royce et al. 1959; Lux and Nichy 1969; Penttila 1988). Their growth off southern New England is rapid in the first 3 years, then slows down (Lux and Nichy 1969). In the third year, females begin to grow faster than males; they also live longer. Cape Cod yellowtail reach greater maximum lengths than those from Georges Bank and southern New England (Fig. 302). The maximum age recorded for yellowtail is 17 years, although individuals older than age 7 are not common (Penttila et al. 1989). Canadian stocks grow more slowly but reach a greater maximum length; both sexes have the same growth rate for 7 years, after which females grow faster (Scott and Scott 1988).

General Range. North American continental waters from the Labrador side of the Strait of Belle Isle (Backus 1957b), northern Newfoundland and the Newfoundland Banks, the north shore of the Gulf of St. Lawrence, southward to the lower part of Chesapeake Bay (Bigelow and Schroeder 1939).

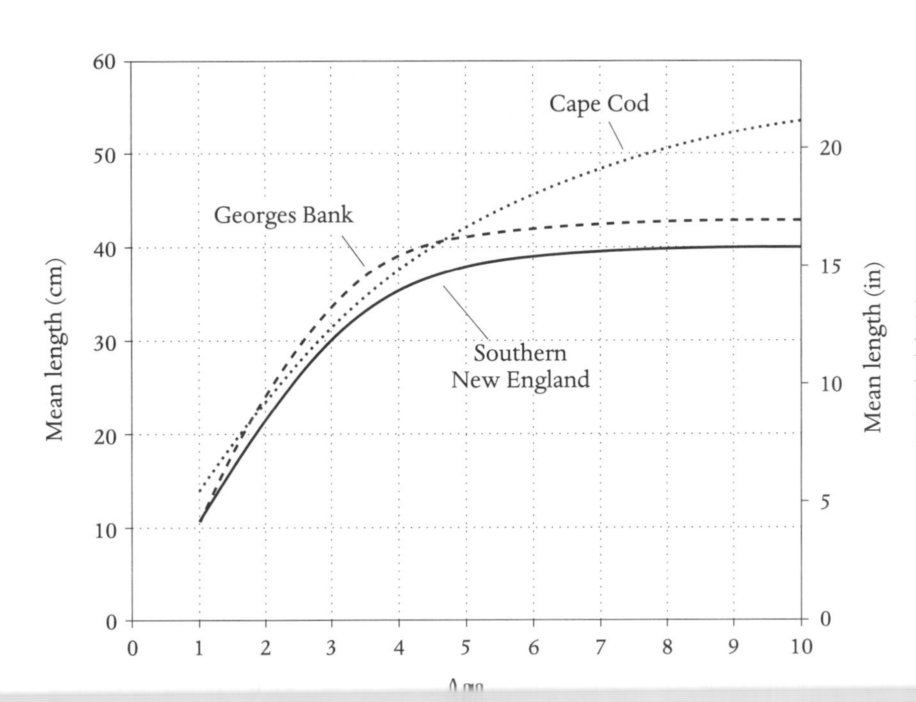

Figure 302. Age-length curves for three populations of yellowtail flounder *Limanda ferruginea:* Cape Cod, Georges Bank, and southern New England. (From Penttila et al. 1989:39.)

Occurrence in the Gulf of Maine. Yellowtail occur throughout the Gulf of Maine from 10 to 100 m. They are abundant in the region of Nantucket Shoals in the South Channel, on the western part of Georges Bank, along the eastern side of Cape Cod Bay and Stellwagen Bank, and in deeper parts of Massachusetts Bay. They are also numerous between the Isle of Shoals and Great Boars Head, Maine, but are less common to the north and northeast. They are less common in the Bay of Fundy and Nova Scotia.

Migrations. Tagging experiments (Royce et al. 1959; Lux 1963) show that yellowtail may migrate for considerable distances. Fish tagged off Block Island moved eastward in spring and summer to the vicinity of southern Nantucket Shoals, some as far as Georges Bank. Those on the Bank moved westward in the winter and eastward in summer. Some moved as far south as the southern New England grounds. Cape Cod fish dispersed northward but none were captured on the other two grounds. Lux (1963) also noted that fish on the Cape Cod grounds are often infested with a trematode parasite that is not found in fish from the other two grounds, indicating that this group does not mingle with the others. Lux found no difference in meristic characters among the three groups and some mixing did occur seasonally when some fish from southern New England migrated to the Cape Cod grounds. Scott (1954a) compared yellowtail groups from three Atlantic fishing areas and did find meristic differences in dorsal and anal fin ray counts between Nova Scotian yellowtail and those from southern New England.

Importance. Yellowtail are among the most valuable flatfishes in the Gulf of Maine. They compare favorably in quality with summer and winter flounder, but because the body is thinner, they bring a lower price to fishermen. The commercial catch is made by otter trawls, and most of the fish are caught from Georges Bank to the New York Bight. The fishery began in the 1930s, and catches increased rapidly until 1934 before declining. Catches have fluctuated considerably since then, rising in the late 1950s to early 1970s, but since then recruitment has declined, especially in southern New England and Georges Bank. There was a brief increase in the early 1980s owing to a strong year-class, but catches declined thereafter.

Catch quotas were imposed on the southern New England and Mid-Atlantic stocks in 1971 and 1975, respectively, but recruitment continued to be low. There are three yellowtail fishing grounds in the region: the southern New England region, extending from eastern Long Island to south of Nantucket; Georges Bank; and the Cape Cod grounds, which extend north along the Massachusetts coast. A 1997 assessment indicated that the stock was at very low levels on Georges Bank and in southern New England. On Georges Bank, spawning stock biomass suffered a severe decline from 1973 to 1988, fluctuated from 1989 to 1994, and began to increase in 1997. During this time fishing mortality decreased from 1.7 (76% exploitation rate) to 0.2 (16% exploitation rate in 1996–1997).

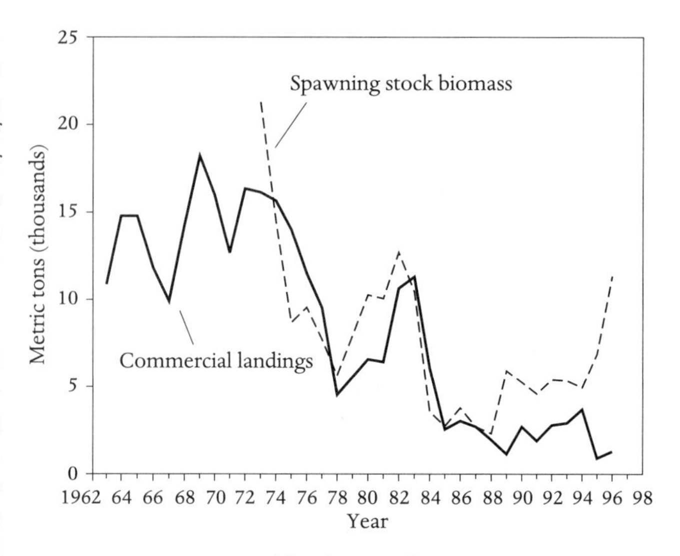

Figure 303. Commercial landings and spawning stock biomass for yellowtail flounder *Limanda ferruginea* from Georges Bank, 1963–1996. (From Overholtz and Cadrin 1998.)

The stock is considered fully exploited and rebuilding. In southern New England, landings declined abruptly after 1969 and have continued to decline except for some increased landings in 1981–1983 and in 1990 owing to two strong year-classes. Fishing mortality averaged 1.6 (74% exploitation rate) from 1980 to the early 1990s and has since declined to 0.12 (10% exploitation rate). In 1997 the stock was assessed at low levels with a slight increase and is considered to be fully exploited. The Cape Cod stock has shown increased landings since 1993, probably indicative of the high degree of decimation of the other stocks. Stock biomass was reduced by high catches in the 1970s and 1980s. Trends in the Mid-Atlantic stock are similar to those in the southern New England stock (Overholtz and Cadrin 1998). Commercial landings and spawning stock biomass are shown in Fig. 303 for Georges Bank (Overholtz and Cadrin 1998).

Because of the depth at which most of the resource occurs combined with their small mouth in relation to the size of hooks used in the coastal recreational fishery, there is very little recreational fishery for yellowtail. Work on developing the species for commercial aquaculture production in Newfoundland has been going on since 1994. Principal areas of work have focused on developing broodstock, larviculture, and juvenile grow-out protocols (Brown and Crim 1998).

Stocks. Based on tagging data, larval distributions, geographical patterns of landings, and bottom trawl survey data, there are four relatively discrete stocks in U.S. waters: southern New England, Georges Bank, Cape Cod, and Mid-Atlantic. Intermingling among these groups has not been quantified but is probably very limited (Overholtz and Cadrin 1998). Based on the distribution of sexually mature females and eggs, there appear to be two stocks on the Scotian Shelf (Neilson et al. 1988). There is also an isolated stock on the Grand Banks.

SMOOTH FLOUNDER / *Pleuronectes putnami* (Gill 1864) / Smoothback Flounder, Eelback /

Bigelow and Schroeder 1953:283–285 (as *Liopsetta putnami*)

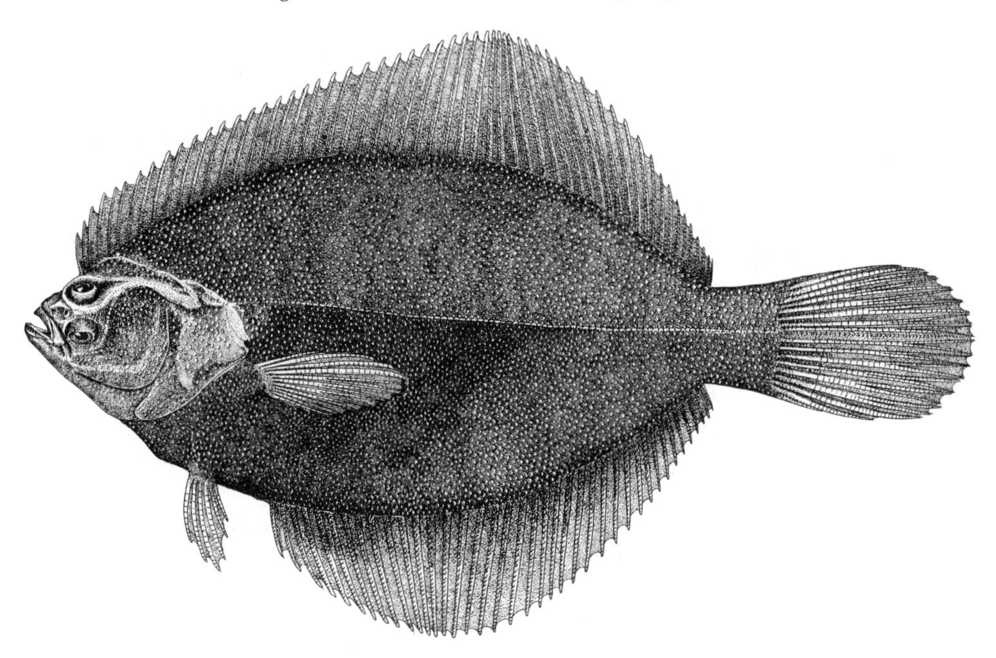

Figure 304. Smooth flounder *Pleuronectes putnami*. Salem, Mass. Drawn by H. L. Todd.

Description. Mouth small, jaws with two sets of teeth on blind side, one on eyed side (Fig. 304). Eyes separated by scaleless ridge. Dorsal and anal fins highest at midpoint of body, tapering toward head and tail. Pectoral fins unequal in size, shorter on blind side. Pectoral fins on eyed side longer (about four-fifths as long as head) and more pointed in males than in females. Caudal fin long and rounded. Lateral line nearly straight. Smooth flounder are peculiar among local flatfishes in their sexual dimorphism: scales are ctenoid on both sides of males, cycloid on both sides of females. Fifth pharyngeal teeth molariform, borne on triangular plate.

Meristics. Dorsal fin rays 48–59; anal fin rays 35–41; pectoral fin rays 10 or 11; branchiostegals 7; vertebrae 34–40 (Norman 1934; Backus 1957b; Laroche 1981; Scott and Scott 1988).

Color. Smooth flounder vary from gray to muddy or slaty brown to almost black above, either uniform or variously mottled with a darker shade of the same tint; dorsal, anal, and caudal fins are of the same general ground color. These fins were mottled darker or paler in specimens Bigelow and Schroeder examined, but Storer described them as black-spotted. The blind side is white.

Size. The smallest flatfish common in the Gulf of Maine, smooth flounder rarely exceed 20 cm in the Great Bay estuary, N.H. (Armstrong and Starr 1994). The largest female captured in the Miramichi estuary was 33 cm and the largest male was

Distinctions. Smooth flounder can be separated from yellowtail by the fact that the lateral line is straight, not arched; the dorsal profile of the head is straight, not concave; and there are fewer fin rays. They have fewer than half as many dorsal and anal rays as witch flounder; their long fins are highest midway along the body and taper toward the head and tail, whereas they are nearly uniform in height in witch; and they lack the mucous pits characteristic of the blind side of the head of the latter. They resemble winter flounder closely in general outline and in body thickness but are distinguished from the latter by the fact that the head is scaleless between the eyes, they have longer pectoral fins and fewer anal fin rays (35–40 vs. 45–58), and males have smooth scales on both sides of the body.

Habits. Smooth flounder are confined to the close vicinity of the coast throughout their geographic range, occurring chiefly in estuaries or river mouths and in sheltered bays and harbors, mostly on soft mud bottom. In the sea, they are found from the tide line down to a maximum depth of perhaps 27 m, with 3.6–9 m as the zone of greatest abundance in the Gulf of Maine. Their shoal water habitat exposes them to temperatures close to the freezing point of saltwater in winter and as high as 15.5°C or more in summer. They can tolerate temperatures up to 32°C (Huntsman and Sparks 1927), and larvae can tolerate a change in temperature ranging from 20° to 30°C above their ambient temperature for 5 min to 1 h (Burke et al. 1981). Smooth flounder were present in shallower parts of Monts-

which is cooler and more saline; hence, they may prefer temperatures above 14°C and salinities below 28.5 ppt (Fried 1973). They are most abundant in the oligomesohaline areas of the Great Bay estuary (Armstrong and Starr 1994).

Food. Smooth flounder eat amphipods, mollusks, small crabs, shrimps, and marine worms (Scott and Scott 1988). In New Hampshire, annelids were the most important food category volumetrically for winter flounder throughout the year, whereas mollusks were most important for smooth flounder, both volumetrically and numerically, all through the year except winter when annelids were most important (Laszlo 1972). In the Miramichi estuary, the principal prey of all sizes of smooth flounder was small bivalves, mostly *Macoma balthica* (Hanson and Courtenay 1997). Small numbers of sand shrimp (*Crangon septemspinosa*) and crabs (*Rhithopanopeus harrisi*) were eaten by larger individuals. Feeding activity there was highest during June and July, declined during October and November, and all stomachs were empty during January and February (Hanson and Courtenay 1997). Juvenile smooth flounder 6.6–74.1 mm caudal length in Minas Basin, Bay of Fundy, fed on harpaticoid copepods, mysid shrimps, calanoid copepods, salt-marsh insects, and trochophore larvae (Imrie and Daborn 1981).

Predators. In the Gaspereau River estuary, N.S., smooth flounder accounted for 98% of the identified prey of blue heron (Quinney and Smith 1980). Occasionally, great black-backed gull stole fish from the herons. Potential fish predators in the Miramichi estuary, such as shorthorn sculpin and Greenland cod, were not found to have eaten smooth flounder (Hanson and Courtenay 1997).

Parasites. Many different parasites have been recorded from smooth flounder, including five species of protozoans, two monogeneans, ten digeneans, two cestodes, four nematodes, two acanthocephalans, two crustaceans, and an unidentified leech (Margolis and Arthur 1979; Burn 1980). Nine digenean trematodes were found in the Great Bay estuary (Burn 1980).

Breeding Habits. Winter is the breeding season, as nearly ripe females have been taken in Salem Harbor in December and spent fish at Bucksport, Maine, in the first week of March. February is probably the peak month of spawning in Nova Scotia. The spawning season in New Hampshire extends from December to March. Spawning occurs in early winter in the Miramichi estuary, when water temperatures are typically near or below 0°C (Hanson and Courtenay 1997). Males appear to ripen before females. Males with ripening gonads have occasionally been collected in October in Great Bay, N.H. (Laszlo 1972). Fecundity of females 87–172 mm from Great Bay ranged from 4,600 to 52,000 eggs (Armstrong and Starr 1994). Males mature at a smaller size than females. In the Miramichi estuary, 50% of males were mature at 96 mm compared with 135 mm for females (Hanson and Courtenay 1997). In New Hampshire, females mature at age 0 or 1 at a mean length of 97.7 mm TL; males begin to mature at age 0 at about 73 mm TL and are probably all mature at the end of age 0. The larger size at maturity in the Miramichi estuary compared with that in New Hampshire is consistent with the need to balance reproduction against growth and overwinter survival (Armstrong and Starr 1994; Hanson and Courtenay 1997). Sex ratio was about 1:1 (Armstrong and Starr 1994) in Great Bay.

Early Life History. Eggs are demersal and nonadhesive (Laroche 1981). They are colorless and 1.1–1.4 mm in diameter, average 1.2 mm. The yolk diameter is 0.9–1.2 mm, average 1.1 mm. There is no oil globule. Hatching time is 25 days at 5°C. Newly hatched larvae are 3.1–3.6 mm TL. They resemble winter flounder larvae but can be recognized by their pigmented eyes at hatching, the lack of internal melanophores over the notochord, the lack of melanophores on the fins, and a higher ratio of snout to anus per standard length.

Larvae were present in the Damariscotta estuary from March to May with peak abundance in April (Townsend 1984). They were most abundant in the Miramichi estuary at salinities of 6–10 ppt in May (Locke and Courtney 1995) and in the Gulf of St. Lawrence at distances less than 0.8 km from shore in May and June (Powles et al. 1984).

Age and Growth. In Great Bay males were longer than females at age 1, but thereafter females grew faster and were heavier than males of the same length (Laszlo 1972).

General Range. Smooth flounder are arctic-boreal. They have been recorded from as far north as Ungava Bay, Labrador (Backus 1957b), although this has been disputed (Scott and Scott 1988), and are described as the most plentiful flatfish along the coasts of the Strait of Belle Isle in all seasons (Jeffries 1932). They are the second most plentiful flatfish after winter flounder on the Cape Breton shore of the Gulf of St. Lawrence. The southernmost records are from Providence, R.I. (Bigelow and Schroeder) and Connecticut (P. Howell, pers. comm.). The range may be continuous in the north with that of its polar relative, *P. glacialis* (Pallas 1776) of the Arctic coasts of North America and Siberia. There is a question about validity of the two species.

Occurrence in the Gulf of Maine. Smooth flounder are found in estuaries, river mouths, and harbors all along the shores of the Gulf from the Bay of Fundy to the northern side of Massachusetts Bay. They are very abundant in the Sheepscot River estuary, less so in Great Bay, and rare off Salem and Boston. They are not found along the shores of Cape Cod or in the Woods Hole region.

Importance. Smooth flounder are neither large enough nor plentiful in the open Gulf waters to be of commercial importance. There was a small fishery for them in Canada as a food for foxes (Leim and Scott 1966).

WINTER FLOUNDER / *Pseudopleuronectes americanus* (Walbaum 1792) /

Blackback, Georges Bank Flounder, Lemon Sole, Rough Flounder / Bigelow and Schroeder 1953:276–283

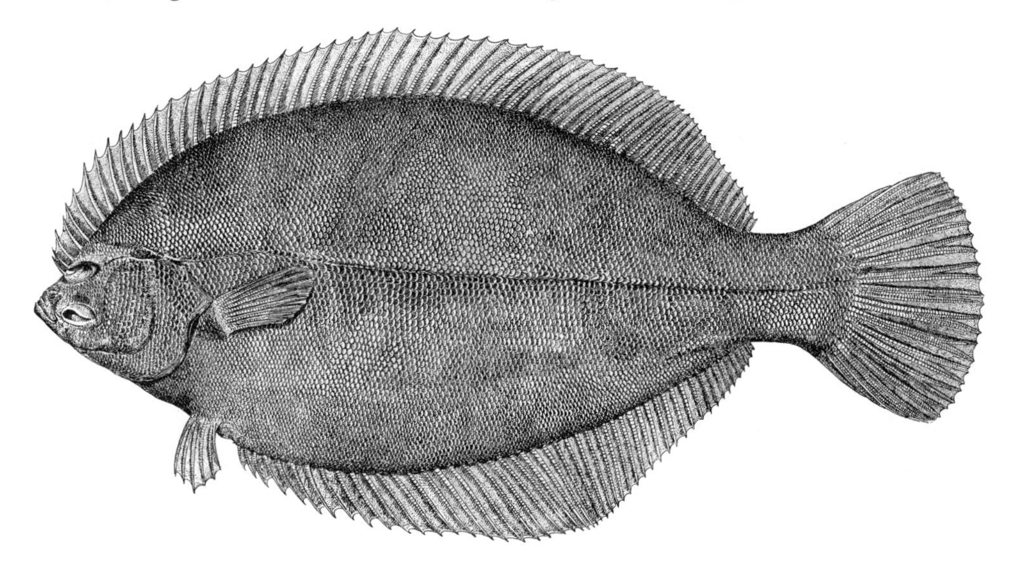

Figure 305. Winter flounder *Pseudopleuronectes americanus.* Drawn by H. L. Todd.

Description. Body oval, about two and a quarter times as long to base of caudal fin as it is wide, thick-bodied, and with a proportionately broader caudal peduncle and tail than any of the other local small flatfishes (Fig. 305). Mouth small, not gaping back to eye, lips thick and fleshy. Blind side of each jaw armed with one series of close-set incisor-like teeth. Eyed side with only a few teeth, or even toothless. Dorsal fin originates opposite forward edge of upper eye, of nearly equal height throughout its length. Anal fin highest about midway, preceded by a short, sharp spine (postabdominal bone). Pelvic fins alike on both sides of body, separated from long anal fin by a considerable gap. Gill rakers short and conical. Lateral line nearly straight, with a slight bow above pectoral fin. Scales rough on eyed side, including interorbital space; smooth on blind side in juveniles and mature females, rough in mature males when rubbed from caudal peduncle toward head. An occasional large adult female may also be rough ventrally.

Meristics. Dorsal fin rays 59–76; anal fin rays 44–58; pectoral fin rays 10 or 11; branchiostegals 7; lateral line scales 88–100; gill rakers 7 or 8 on lower anterior arch; vertebrae 36 (Norman 1934; Scott and Scott 1988).

Color. Winter flounder, like other flatfishes, vary in hue according to the bottom on which they lie, but as a rule they are the darkest of the Gulf of Maine flatfishes. Large ones are usually some shade of muddy or slightly reddish brown, olive green, or dark slate above, sometimes almost black. They vary from plain or more or less mottled to definitely marked with smaller and larger spots of a darker shade of the general ground tone. There usually is a wide variation in this respect among any lot of flounders. Fish caught on Georges Bank

The blind side is white, more or less translucent toward the edge, where there is often a faintly bluish tinge; the lower side of the caudal peduncle is yellowish in some specimens and pure white in others. Dorsal and anal fins are usually tinged with pink, red, or yellow on the eyed side; the pelvics and pectorals of the eyed side are the general ground tone, but their mates on the blind side are pure white. Small fish are usually paler and more blotched or mottled than large ones.

Various color abnormalities have been recorded, for example, fish that are partially white on the eyed as well as on the blind side (Lux 1973), and it is not uncommon to see specimens with dark blotches on the blind side. In fact, one-third of the fish caught near Providence during the winter of 1897–1998 were these "black-bellies," as fishermen called them (Sherwood and Edwards 1901).

Winter flounder change color to some extent to match their surroundings, usually being very dark on mud and pale on bright sand bottoms, but field experience suggests they have less control than summer flounder over shade and pattern. Pattern-related responsiveness of melanophores is neurally controlled and xanthophores are regulated by the pituitary. An overview of color-related changes is given in Burton (1998).

Size. The largest winter flounder on record from inshore was 57.8 cm long (Scattergood 1952). Bigelow and Schroeder handled one Georges Bank fish of 63.5 cm, weighing 3.6 kg. The NEFSC groundfish bottom trawl survey collected two 64-cm females, one on 22 October 1986 at 42°67′ N, 67°32′ W and a second on 15 April 1997 at 41°01′ N, 68°25′ W (R. Brown, pers. comm., 25 Sept. 1998). Port samplings of commercial catches by NMFS since 1964 have recorded at least six winter flounder of 67 cm TL. Fish longer than 46 cm or heavier than

0.9 kg. They grow larger on Georges Bank, where many fish of 2–3 kg are taken.

Distinctions. Winter flounder are easily separable from yellowtail by the fact that the lateral line is nearly straight, the dorsal profile of the head is less concave, the snout is blunter, the eyes farther apart, there are fewer fin rays, and the fins are less tapered in outline. The most obvious difference between winter flounder and smooth flounder is that winter flounder have more anal fin rays. They differ from witch flounder in that they have only two-thirds as many dorsal rays, they lack the mucous pits that are conspicuous on the left (blind) side of the head of witch, and the tail is proportionately much larger.

Taxonomic Remarks. Winter flounder show some tendency to break up into local races based on the number of fin rays and the size to which the fish grow. The most interesting of these races, from the fisheries standpoint, is the population on Georges Bank, for flounder there tend to grow larger than they do inshore. This fact was first brought to the attention of scientists in 1912, when some of the large flounder from Georges Bank were received by the Bureau of Fisheries and made the basis of a new species, *Pseudopleuronectes dignabilis* Kendall (1912). Kendall felt that the greater number of dorsal and anal fin rays, color, larger size, and different spawning season merited species status. Bigelow and Schroeder concluded that it was just a larger more rusty brown local race of winter flounder. Lux et al. (1970) counted dorsal and anal fin rays on winter flounder from inshore waters off Massachusetts north and south of Cape Cod and from Georges Bank. They then compared historic water temperature records during winter flounder spawning seasons and found them to be higher on Georges Bank. Since environmental conditions influence phenotypic expressions of traits, including fin ray numbers, growth rates, and spawning time, differences in water temperatures might account for the differences in the Georges Bank stock.

Habits. Winter flounder range up into brackish water of river mouths and estuaries and have even been caught in the Susquehanna River, tributary to Chesapeake Bay, in essentially freshwater. They are plentiful at 18–37 m in Cape Cod Bay, Stellwagen Bank, and around Boon Island. Few, if any are caught deeper than this in the inner parts of the Gulf, but in the Bay of Fundy they are taken on soft bottom down to 55–92 m in winter. On Georges Bank they are taken between 46 and 82 m; 143 m is the deepest definite record (McCracken 1954). There appears to be an ontogenetic shift in depth distribution, with larger and older fishes inhabiting deeper water.

Typical inshore habitat consists of muddy sand, especially where this is broken by patches of eelgrass. Winter flounder are also distributed over cleaner sand, on clay, and even on pebbly and gravelly ground. Populations on the offshore banks (Georges Bank and Nantucket Shoals) are on hard bottom. When fish are on muddy bottom, they usually lie buried, all but the eyes, working themselves down into the mud soon after settling on the bottom. Flounder that live on the flats usually lie motionless over the low tide and actively search for food during flood tide. Tyler (1972) observed that they entered the intertidal zone in the Passamaquoddy Bay region with the rising tide, occupied the area for up to 8 h, then moved back to the sublittoral zone 2.5–0.5 h before the next low tide. Small (4–15 cm) and large (25–49 cm) flounder moved with the same synchrony; intermediate flounder did not move into the intertidal area. These may be responses to light, since intermediate-sized flounder in laboratory and field situations preferred lower light intensities than small flounder or mature ones (McCracken 1963).

Local physical conditions appear to determine inshore distribution patterns, whereas offshore movements appear to be associated with extreme summer and winter conditions. In general, in summer months adult winter flounder stay in the shallow shore zone when water temperature is not excessive and food availability is adequate. If these conditions are not met, they may move into deeper channels or offshore, or may take evasive action (see the discussion under temperature tolerance). In the fall, as gonads ripen, adult winter flounder remain in or move into shallow water to spawn—in southern locations in winter and northern locations in spring. During winter in the southern parts of their range they remain or move into shallow water to spawn, whereas in northern regions they remain inshore in protected areas and move offshore in exposed areas to avoid turbulence and drifting pack ice (Van Guelpen and Davis 1979). Winter flounder in deeper waters such as Georges Bank remain at these depths year-round.

The normal distribution of winter flounder covers a wide range of temperatures in one season or another: a minimum close to the freezing point of saltwater around Newfoundland, in Nova Scotian waters, and in the Gulf of Maine in late winter; a maximum of about 18°–19°C in shallow water in the southwestern part of the Gulf in summer; and 20°–21°C in the southern part of the range.

Lethal temperatures appear to be −1.4°C (Duman and De Vries 1974b) and 19.3°C. Winter flounder often recover from cold shock, but never from heat shock (Hoff and Westman 1966). Their blood contains an AFP (antifreeze protein) that helps protect them against freezing (Duman and De Vries 1974b, 1976). This protein is produced seasonally, and the cycles differ in populations from different geographical areas (Fletcher and Smith 1980). Both shorthorn sculpin and winter flounder secrete AFP into the blood serum, which serves to depress the freezing temperature of their extracellular fluid and both produce skin-type AFPs that act as the first line of defense against freezing (Low et al. 1998). Winter flounder bury in sediment when water temperatures are below 0°C and ice crystals are present in the water but may succumb to anchor ice in winter if overtaken in very shoal water in a severe freeze.

Divers observing flounder in Great South Bay, Long Island, where bottom temperatures ranged from 17° to 24°C reported that flounder were active up to 22°C and became inactive at 23°C. Temperatures measured 50–60 mm below the sand were 2°–3°C below ambient (Olla et al. 1969), so the flounder could avoid the heat by burying themselves. If they are trapped in shallow enclosed bays they can perish by the thousands during spells of very hot summer weather, as happened in Moriches Bay, Long Island, in 1917, when temperatures rose to about 30°C (Nichols 1918).

Winter flounder are sensitive to dissolved oxygen (DO) concentrations lower than 3 mg·liter^{-1}. They were present in significantly lower numbers in Long Island Sound, where the DO concentration was 2–2.2 mg·liter^{-1}, and showed reduced lengths at DO concentrations higher than 2 mg·liter^{-1}. There was a 4-cm difference in mean lengths between fish taken between concentrations less than 2 mg·liter^{-1} and greater than 5 mg·liter^{-1} (Howell and Simpson 1994). Although this might indicate a sorting by size-class differential to DO, winter flounder growth rates in the laboratory were twice as high in fish held at high DO (6.7 mg·liter^{-1}) than in those held at low DO (2.2 mg·liter^{-1}) and reduced at fluctuating levels of DO compared to high levels (Bejda et al. 1992).

It has long been believed that winter flounder are among the most stationary fishes, apart from seasonal movements of the sorts just mentioned and a general tendency for the young that are produced in bays and estuaries to work offshore as they grow older (Perlmutter 1947). This essentially stationary nature has been demonstrated by extensive marking experiments in Long Island Sound, along southern New England, and on the coast of Maine: about 94% of the recaptures were made in the general areas where the fish had been tagged (Lobell 1939; Perlmutter 1947). In a 10-year tagging study at 21 locations off Massachusetts (Howe and Coates 1975), winter flounder showed the following movements: north of Cape Cod they were localized and confined to inshore waters and south of Cape Cod they were seasonally dispersed in a southeast direction with little mixing between Georges Bank and inshore areas. These movements appear to be related to water temperature. In Passamaquoddy Bay, mature winter flounder leave the shore zone only in areas where temperatures rise above 15°C (McCracken 1963). This movement is restricted to depths at which the temperature does not go below 12°C. They return to the shore zone in fall after the temperature falls below 15°C. In spring, both immature and mature fish are found along the shore, with spawning fish concentrated in shallow water where the temperature has warmed to 3°–4°C. In the southern Gulf of St. Lawrence, winter flounder move into the Miramichi estuary during October and November and remain there until spring (Hanson and Courtenay 1996). Thus the population consists of many independent localized stocks inhabiting bays and estuaries along the coast, with fish tending to scatter from population centers, although a few may move considerable distances.

Food. Adult winter flounder are limited by their small gape to a diet of small invertebrates and, rarely, fishes such as sand lance. Principal stomach contents of 1,746 winter flounder

anthozoans (33.4%), and amphipods (6.7%). Other food items included shrimps, small crabs, and other crustaceans; ascidians; holothurians; squids; and bivalve and univalve mollusks. They often break off clam siphons that protrude from the sand. Feeding habits, with a comprehensive list of food organisms found in their stomachs, were summarized by Klein-MacPhee (1978). More recent studies of offshore fish indicate they eat more hydrozoans and anthozoans than inshore fish (Langton and Bowman 1981; Bowman 1988; Bowman et al. 2000). Frank and Leggett (1984) described the importance of capelin eggs in the diet of winter flounder from Conception Bay, Nfld.

Winter flounder are sight feeders and are diurnally active in both inshore and offshore waters (Pearcy 1962; Bowman 1988). At night they lie flat, heads resting on the bottom and eye turrets retracted, becoming active at sunrise. While feeding, a winter flounder lies with its head raised off the bottom and 12–17 dorsal fin rays braced vertically into the substrate. The left pelvic fin and several anal fin rays are used to support the head. Eye turrets are extended and the eyes move independently of one another. After sighting prey, the fish remains stationary, pointed toward the target, and then lunges forward and downward to seize the prey. Debris is expelled from the right gill covering. The fish then resumes the feeding position. If no food is sighted it swims to another location less than a meter away (Olla et al. 1969).

Daily ration estimates in the field vary from 1.27% body weight per day for a spring population from a Rhode Island salt pond (Worobec 1984), to 1.3–1.5% for a late-fall population from Georges Bank (Huebner and Langton 1982), to 1.77% for a spring population from Passamaquoddy Bay (MacDonald and Waiwood 1987).

Diatoms are the first food larvae take after yolk-sac absorption (Sullivan 1915; Pearcy 1962; and Klein-MacPhee, unpubl. data). Later, they feed on rotifers, tintinnids, and invertebrate eggs and finally on bivalve and polychaete larvae, copepod nauplii, and copepodites. Newly metamorphosed juveniles eat small isopods, amphipods, polychaetes, other crustaceans, annelids, and mollusks (Linton 1921; Pearcy 1962). With a progressive increase in size, young winter flounder tend to prefer larger prey organisms (Pearcy 1962; Richards 1963b; Mulkana 1966).

Predators. Winter flounder were found in stomachs of 11 species of fishes in the NEFSC food habits survey, of which four species had multiple occurrences: Atlantic cod, spiny dogfish, goosefish, and winter skate, in order of number of occurrences (Rountree 1999). Other known fish predators include little skate, smooth dogfish, hakes (spotted, white, and silver), sea raven, striped sea robin, striped bass, bluefish, and wrymouth (Willey and Huntsman 1921; Dickie and McCracken 1955; Derickson and Price 1973; Bowman and Michaels 1984; Manderson et al. 1999; Rountree 1999; Bowman et al. 2000). They are also eaten by harbor, harp, and gray seals (Fisher and Mackenzie 1955; Selzer et al. 1986; Bowen

flounder are eaten by blue heron and cormorant (Tyler 1971b), sand lance and moon jellies (Grove 1982), and toadfish and summer flounder (Pearcy 1962).

Parasites. Winter flounder are host to a wide variety of parasites (summarized by Klein-MacPhee [1978] and Margolis and Arthur [1979]): 7 species of protozoans, 2 myxosporidians, 18 trematodes, 5 cestodes, 13 nematodes, 6 acanthocephalans, 4 branchiurans, 4 copepods, and an isopod.

The microsporidian *Glugea stephani* is a common parasite that primarily infects the intestinal wall and pyloric caeca (Stunkard and Lux 1965). The infection is temperature-dependent (above 15°C) and probably enters the host by way of a crustacean vector (Takvorian and Cali 1981, 1984). It can be fatal, especially in young-of-the-year; it also causes a decline in immunoglobulin levels and suppresses the host's immune response to other antigens (Laudan et al. 1987).

Breeding Habits. Winter flounder breed in the winter and early spring, spawning from January to May (inclusive) in New England. Spawning activity peaks during February and March south of Cape Cod and in the Massachusetts Bay region, but later along the coast of Maine. Near Boothbay, spawning commences in early March and continues until mid-May, with peak egg deposition occurring in early to mid-April (Bigelow and Schroeder). Spawning occurs earlier in the southern part of the range: November to April in the Indian River Bay, Del. (Fairbanks et al. 1971). On Georges Bank, the timing of spawning is unclear but it has been reported to take place in April and May (Bigelow and Schroeder).

Spawning occurs in inshore waters at close to minimal seasonal water temperatures: 0° to 1.7°C in the Woods Hole region, 0° to 2.8°C near Gloucester, and −0.5° to 1.7°C near Boothbay. Most egg deposition occurs before the water has warmed above 3.3°C, with about 4.4°–5.6°C as the maximum for any extensive spawning in the inner parts of the Gulf of Maine. Spawning on Georges Bank occurs at higher water temperatures, ranging from about 3.3°C to perhaps 5.5°C.

Winter flounder spawn on sandy bottom and algal mats (Anonymous 1972; Grove 1982) often in water as shoal as 1.8–5.4 m, but as deep as 45–72 m off Georges Bank. Most eggs are deposited at salinities of 31–32.3 ppt in the inner parts of the Gulf to somewhere between 32.7–33 ppt on Nantucket Shoals and Georges Bank. Estuary spawning occurs in more brackish water, in salinities as low as 11.4 ppt near Woods Hole.

Winter flounder in the New York region mature at age 2 for males and age 3 for females, when they are 20–25 cm TL (Perlmutter 1947). There is a clinal gradient in maturity with later maturation occurring in more northerly latitudes, at age 3 for females and males south of Cape Cod, age 3.3 for males and 3.5 for females north of Cape Cod, and 6 for males and 7 for females in Newfoundland (Kennedy and Steele 1971). Maturity appears to be a function of size rather than age. The Georges Bank stock has a mean age of 1.9 years for both sexes (O'Brien et al. 1993). Maturation of winter flounder on the Scotian Shelf and in the southern Gulf of St. Lawrence was found to be highly variable from year to year (Beacham 1982) and large numbers of nonreproductive individuals were found in any given year (Burton and Idler 1984). This appears to result from restricted feeding prior to and immediately subsequent to the current spawning period, which indicates a nutritionally sensitive critical period (Burton 1994). There is a 2- to 3-year cycle in oocyte maturation (Burton and Idler 1984). Individual females produce an average of 500,000 eggs annually, but 3,329,000 were taken from a 5-year-old (40-cm) fish (Topp 1968).

Sex ratios tend to favor males, especially older ones, but vary with the population examined. Ratios were 7:3 female to male in Green Hill Pond, R.I. (Saila 1961), 3:2 in Narragansett Bay (Saila 1962), 3:1 on fishing grounds south of Rhode Island and Massachusetts (Lux 1969), 2.3:1 in Massachusetts (Howe and Coates 1975), and 1:1 in Long Pond, Nfld.(Kennedy and Steele 1971). Explanations for this vary from catch selectivity for larger fish, which favors larger, faster-growing females (Saila 1962) to higher mortality rates for males (Witherell and Burnett 1993).

Winter flounder migrate into shallow water or estuaries and coastal ponds to spawn, and tagging studies show that most return repeatedly to the same spawning grounds (Lobell 1939; Saila 1961; Grove 1982). Winter flounder are batch spawners. Continuous observations in a large research aquarium monitored their behavior during and after the spawning season. Fish spawned over a 60-day period, with an average of 40 spawns per female and 147 spawns per male. Males initiated all observed spawning events, which occurred throughout the night, but primarily between sunset and midnight. Spawning by one pair frequently elicited sudden convergence and spawning by secondary males; consequently, strictly paired spawning was uncommon. Male and female activity patterns were almost entirely nocturnal during the reproductive season but became increasingly diurnal during the postspawning season. Field and laboratory results indicate that male spawning strategy is adapted to maximize the number of eggs fertilized. There is probably high genetic diversity in the offspring from any one female, owing to frequent spawning and to multiple males participating in individual spawning events (Stoner et al. 1999).

The pattern of sperm release and changes in sperm quality were investigated throughout the normal spawning season of male winter flounder caught in Conception Bay. Whereas a lengthy period of spermiation, lasting about 6 months (December–July), can be detected in some males, the major period for sperm release was from May to July in the summer spawning season. By late July, sperm production fell rapidly along with a notable deterioration in sperm motility (Shangguan and Crim 1999).

Early Life History. This species is peculiar among Gulf of Maine flatfishes in that their eggs are not buoyant but sink to the bottom, where they stick together in clusters, usually so closely matted that individual eggs are forced into irregular outlines. They are 0.74–0.85 mm in diameter, and newly shed eggs have no oil globule. Eggs vary in color, with most pale to bright yellow, but some are pink to salmon-colored (pers. obs.). A description of developmental events is given in Martin and Drewry (1978).

Incubation occupies 15–18 days at a temperature of 2.8°–3.3°C, which is about what they encounter in nature. Young larvae, which are about 3–3.5 mm long at hatching, are marked by a broad vertical band of pigment cells that subdivides the postanal part of the body, a characteristic feature; the end of the gut is also heavily pigmented. In water of about 3.8°C larvae grow to 5 mm in length, and the yolk is absorbed in 12–14 days. Vertical fin rays begin to appear 5–6 weeks after hatching, at a length of about 7 mm, and the left eye has moved upward by then until about half of it is visible above the dorsal outline of the head, while the whole left eye shows from the right side and the fins are fully formed in larvae of 8 mm. Metamorphosis continues rapidly. The left eye moves from this position to the right side of the head, pigment fades from the blind side, the eyed side becomes uniformly pigmented, and the little fish now lies and swims with the blind side down, its metamorphosis complete when it is only 8–9 mm long.

The youngest larval stages are identifiable as winter flounder by the pigment bar just mentioned. They are distinguished from smooth flounder by the lack of pigment in the eyes of yolk-sac larvae and the presence of internal pigment spots over the notochord in larvae larger than 3.2 mm, the presence of dark pigment cells on the fins of yolk sac preflexion and flexion larvae, and a lower snout to anus length–standard length ratio (Laroche 1981). After the fin rays appear, the winter flounder's small mouth separates them from any of the large-mouthed flounders; their short, deep body, combined with small number of fin rays, separates them from witch; and the number of fin rays marks them off from yellowtail. Winter flounder also complete metamorphosis at a smaller size than either of these other small-mouthed flatfishes. A large series of larvae from newly hatched through metamorphosis is illustrated in Laroche (1981).

Rate of larval development is governed by temperature, occupying 2.5–3.5 months, and larvae that hatch later may exhibit compensatory growth. Average age at metamorphosis is 59.5 days but may vary by as much as 25 days. Larvae that metamorphose late do so at a larger size. Length at metamorphosis is significantly less variable than age at metamorphosis and both length and age at metamorphosis are positively correlated (Chambers and Leggett 1987).

Larvae have been taken in bays, estuaries, and salt ponds from Delaware to Newfoundland (Frank and Leggett 1983). The larvae are less at the mercy of the tide and current than other Gulf of Maine flatfishes, for they have been described as alternately swimming upward then sinking (Pearcy 1962), instead of remaining constantly adrift near the surface, as the larvae of most of the flatfishes do at a corresponding stage in their development. Larval, metamorphosing, and newly metamorphosed swimming behavior observed in the laboratory showed constant upright swimming after yolk-sac absorption

of fins for propulsion at metamorphosis. After metamorphosis there was a decrease in overall activity level (Jearld et al. 1993).

Young-of-the-year winter flounder exhibit little movement from the areas where they settle, although this may be site-specific. Winter flounder collected during the first 2 months of settlement in Waquoit Bay, Mass., and marked with acrylic paint were mostly recaptured (98%) within 100 m of their release site (Saucerman and Deegan 1991), indicating limited movement after settlement during their first summer. Age-0+ individuals in Point Judith Pond, R.I., were also recaptured within the area they were marked during the first summer postsettlement (O'Connor 1997). On the other hand, age-0+ juveniles in Great Salt Pond, Block Island, R.I., appeared to move out of certain locations in the late summer, perhaps in response to high temperatures (Neuman 1993). Age-0+ winter flounder in Great South Bay appeared to be relatively stationary until they reached a size of 30–50 mm, at which time they began to move into the upper estuary. They exhibited little segregation by depth, but few were found on the intertidal mud flats (Armstrong 1997). In a New Jersey estuary, age-0+ juveniles were more abundant in unvegetated areas than in vegetated locations (eelgrass beds and macroalgae accumulations) and showed better growth when caged in unvegetated habitat (Sogard 1992). In contrast, in the Damariscotta River estuary, winter flounder abundance was greater in eelgrass beds than in unvegetated areas and they dominated fish collections at night when the invertebrate epifaunal abundance was also higher (Mattila et al. 1999). In Connecticut, abundance of young-of year winter flounder in habitat types within nursery areas was correlated with sediment type in shallow embayments. Highest densities occurred in mud-shell litter habitat, and sites with mud combinations had higher densities than sandy sites. This may reflect an abundance of prey in these areas (Howell et al. 1999).

A large series from Casco Bay, measured by Welsh, and others seen near Boothbay Harbor and at Mt. Desert indicate that off southern New England juveniles of the previous winter grow to an average length of 4–8 cm by August, are 5–10 cm long by the end of September, and 10–15 cm long in January and February, when they are almost 1 year old; this is probably true north of Cape Cod as well. They may grow somewhat faster in more southern (warmer) waters, as in Chesapeake Bay, where young-of-the-year are 11–18 cm long in January and February. Daily growth rates of young-of-the-year were 0.18 mm·day^{-1} in a Rhode Island salt pond (O'Conner 1997) and 0.09–0.11 mm·day^{-1} in a New Jersey estuary (Sogard 1992).

Age and Growth. Winter flounder are relatively long-lived, reaching a maximum age of about 15 years and a length of 58 cm (Fields 1988). Scales and otoliths have been used for determining age and growth (Fields 1988). Age verification of winter flounder in Narragansett Bay using sectioned otoliths showed that age determination was probably accurate for ages

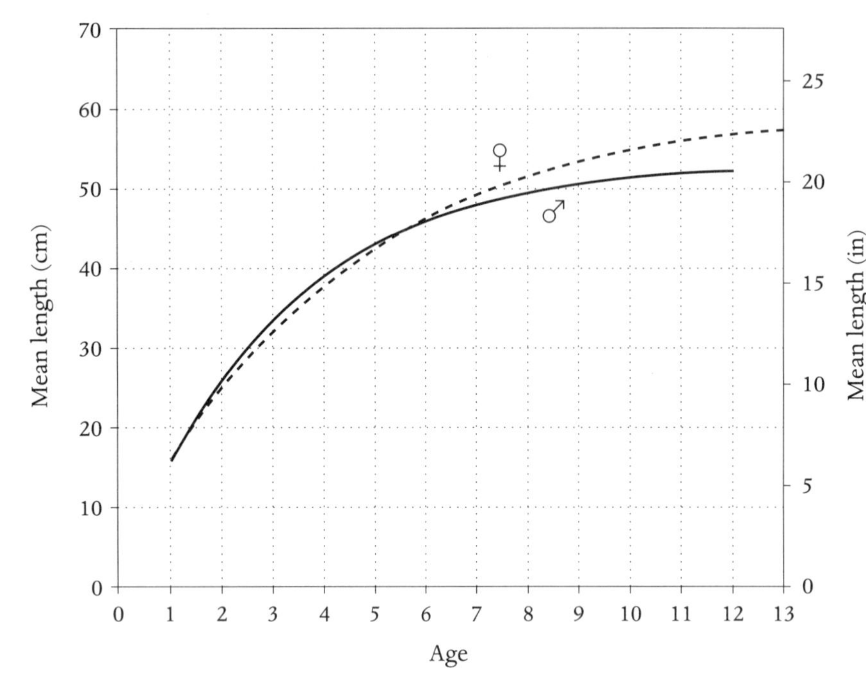

Figure 306. Age-length curves for male and female winter flounder *Pseudopleuronectes americanus,* Georges Bank region. (From Penttila et al. 1989:35.)

otoliths, and scales, so any of these techniques can be used (Haas and Recksiek 1995). Growth rate up to age 5 is the same for both sexes, after which females grow faster and live longer than males (Fig. 306). Witherell and Burnett (1993) determined that total length for female flounder in Massachusetts calculated from scale samples was 19.9 mm at age 2, 26.8 mm at age 3, 32 mm at age 4, 36.1 mm at age 5, 39.1 mm at age 6, 41.4 mm at age 7, and 43.2 mm at age 8. Their findings are consistent with those of Lobell (1939) and Perlmutter (1947). Winter flounder on Georges Bank grow faster than fish from inshore areas (Lux 1973).

General Range. Atlantic coast of North America from Windy Tickle, Labrador, 55°45′ N (Backus 1957b) south to North Carolina and Georgia. They are most abundant from the Gulf of St. Lawrence to Chesapeake Bay.

Occurrence in the Gulf of Maine. Winter flounder are the most common shoal-water flounder and perhaps the most familiar of all the groundfishes of the Gulf of Maine. Most of the winter flounder population of the inner parts of the Gulf live shoaler than 55 m (Map 39); the zone occupied around the coast north of the elbow of Cape Cod is about 13–16 km wide, measured from the outer headlands or islands, except for Stellwagen Bank, which lies several kilometers farther out, and off Cape Sable, where their outer-depth limit is 24 km offshore. Their range extends out along the offshore rim of the Gulf, in somewhat deeper water, to include the Nantucket Shoals region as a whole and the shoaler parts of Georges Bank.

Importance. Winter flounder are the thickest and meatiest of all the Gulf flatfishes. Most of today's commercial catch is made by otter trawlers, with a small part taken with hook and line or various types of nets. The commercial fishing grounds include the inner parts of the Gulf, from the tip of Cape Cod around to Cape Sable, Nantucket Shoals, and Georges and Browns banks. In Canadian waters, they are caught off the Scotian Shelf and at the mouth of the Bay of Fundy. The fishing season generally begins in the spring after adults spawn and ends in late fall. The peak is in spring and early summer (Klein-MacPhee 1978).

Recreational fishing for winter flounder occurs in harbors, estuaries, and other sheltered situations all around the shores of the Gulf, from bridges, piers, and small boats. Most anglers bottom fish with small hooks baited with sea worms, clams, snails, squid, and other bait. When they are not actively feeding, flounders can sometimes be attracted by stirring the bottom.

Commercial landings in the Gulf of Maine increased from 1,000 mt in 1961 to nearly 3,000 mt in 1982 and then declined. The recreational catch almost equals the commercial landings. In 1979 the combined catch totaled 7,100 mt. This dropped to 3,100 mt in 1983 owing to a 70% reduction in recreational catches and a 25% reduction in commercial landings. Since 1989 landings in both fisheries have trended downward. Bottom trawl survey indices decreased from 1983 to 1988 and reached a record low in 1994 (Fig. 307). Low landings and CPUE and survey indices show that winter flounder abundance has been substantially reduced, and the stock is considered overexploited and at a low biomass level (Brown and Gabriel 1998).

Commercial landings from Georges Bank increased from 1,900 mt in 1976 to 3,800 mt during 1980–1984. They began to decline in 1985–1988, averaging 2,400 mt, and from 1991 to 1992 averaged 1,700 mt. In 1993 landings were near the lowest on record (1,700 mt) and CPUE indices were also low. The autumn survey stock biomass has trended downward since 1977. The stock has declined to record lows and is overexploited. No recreational catches have been reported from the Georges

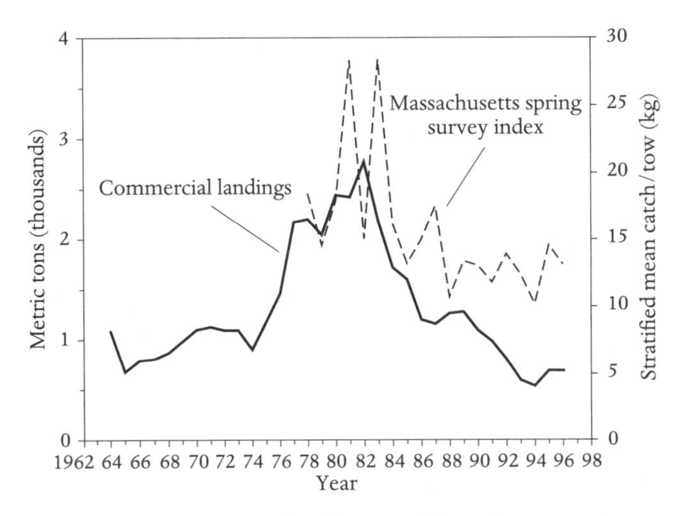

Figure 307. Commercial landings and Massachusetts spring survey index for winter flounder *Pseudopleuronectes americanus*, Gulf of Maine, 1963–1996. (From Brown and Gabriel 1998.)

Bank stock (Brown and Gabriel 1998). The southern New England–Mid-Atlantic stock is also at low biomass levels and remains overexploited (Brown and Gabriel 1998).

By-catch of winter flounder occurs in the Gulf of Maine shrimp fishery, and there is an 11% discard (by number) in the size range of 17–28 cm TL (Howell and Langan 1992). Mortality of winter flounder discarded is estimated at 10%, which is lower than that for other fish species (Ross and Hokenson 1997). In the spring inshore fishery for squid in Nantucket and Vineyard sounds, high catches and discards of undersized flounder occur. By-catch rates vary spatially and temporally but over 30% by weight of the total catch is discarded at sea, 3% of which is winter flounder (Glass et al. 1999).

Winter flounder have been identified as a candidate for aquaculture in Atlantic Canada. Their resistance to low temperatures provides an opportunity for aquaculture development in some of the colder waters along the Atlantic Canadian coast previously deemed unsuitable for fish culture. The focus is on development of techniques and technology for larval rearing and juvenile grow-out. Recent advances have been summarized by Litvak (1999).

Stocks. Winter flounder appear to form relatively discrete local groups (see Habits). For assessment purposes these are divided into three groups: Gulf of Maine, southern New England–Mid-Atlantic, and Georges Bank but additional studies of stock structure are needed (Brown and Gabriel 1998).

GREENLAND HALIBUT / *Reinhardtius hippoglossoides* (Walbaum 1792) /

Greenland Turbot, Newfoundland Turbot / Bigelow and Schroeder 1953:258–259

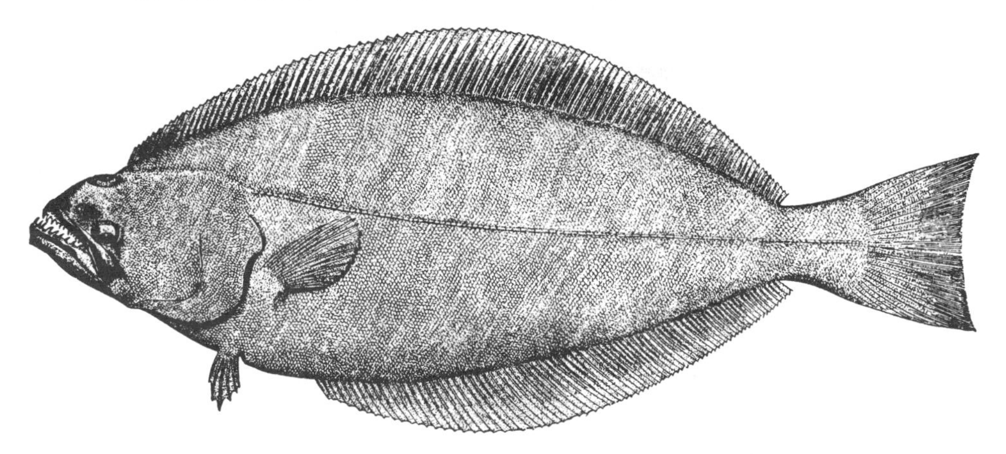

Figure 308. Greenland halibut *Reinhardtius hippoglossoides.* Drawn by H. L. Todd.

Description. Mouth very large, its angle below posterior edge of eye (Fig. 308). Large conical teeth in both jaws, two rows at the front of upper jaw, one elsewhere. Central rays of dorsal fin longest. No preanal bony spine. Pectoral fins rounded; pelvic fins symmetrical; caudal fin slightly concave. Lateral line almost straight.

Meristics. Dorsal fin rays 92–102; anal fin rays 71–76; pectoral rakers on lower part of anterior arch 10–12; vertebrae 59–64 (Norman 1934; Templeman 1970; Scott and Scott 1988).

Color. Greenland halibut are yellowish or grayish brown and may be almost black on the eyed side. They are pale-colored on the blind side but not white except when young. Pelagic larvae are pigmented on both sides, although darker on the right side (Jensen 1925). After metamorphosis, pigment dis-

then the blind side becomes soot-colored and eventually quite dark (Boyar 1964). Although adults are pigmented on both sides, the melanophores on the blind side are different from those on the eyed side, being much larger and morphologically different (Burton 1988). This is not true for ambicolored winter flounder, which have the same kind of melanophores on eyed and blind sides.

Size. Greenland halibut are among the largest of North Atlantic flatfishes, next to Atlantic halibut, reaching a weight of 25 kg and a length of 120 cm (Scott and Scott 1988). Fish caught on the Grand Banks are considerably smaller, weighing 2.3–4.5 kg.

Distinctions. The only fish with which Greenland halibut could be confused are Atlantic halibut from which they can be distinguished by their straight lateral line, larger mouth, and smaller eyes, which are farther apart.

Habits. Greenland halibut are deepwater fish known from depths of 63–1,216 m (Boyar 1964; Khan et al. 1991). They appear to be active swimmers that make extensive vertical migrations and spend less time on the bottom than other flatfishes (Bowering and Brodie 1991). They are found at water temperatures of −1.0° to 7°C and are most abundant between 0° and 7°C (Bowering and Chumakov 1989). Peak abundance in the northwest Atlantic was in 400–700 m at water temperatures of 2°–6°C, compared to 0°–4°C in the northeast Atlantic (Bowering and Nedreaas 2000).

Food. Fishes are the predominant food of Greenland halibut. Other prey such as pandalid shrimps, octopus, and squids are also consumed. Stomachs of more than 76,000 specimens caught between Davis Strait and eastern Newfoundland contained roundnose grenadier, beaked redfish, Atlantic cod, Greenland halibut, capelin, sand lance, squid, and various crustaceans (Chumakov and Podrazhanskaya 1986). In a more recent study (Orr and Bowering 1997), the most important food items in Davis Strait were *Pandalus borealis,* cephalopods, mysids, amphipods, redfish (*Sebastes mentella*), Greenland halibut, and Arctic cod. On the slope of the northeast Newfoundland continental shelf at depths of 1,000–1,250 m, a squid, *Gonatus* sp., predominated in the diet (Dawe et al. 1998). The change from predation primarily upon pelagic prey may be related to annual variability in abundance and distribution of both the predator and capelin, its principal pelagic fish prey species. A linear size relationship in observed predation may reflect a common size-related depth distribution pattern between predator and prey or prey selectivity. At sizes over 60 cm, Greenland halibut switched from *Gonatus* sp. to groundfish as their principal prey. Average daily consumption by 5-year-old fish is about 1.2% of body weight per day, but females eat more than males. Minimum consumption by Greenland halibut was 750,000 tons of food annually, which provides some indication of their voracious appetites (Chumakov and Podrazhanskaya 1986).

Predators. Greenland shark is the most important predator of adult Greenland halibut, but they are also eaten by marine mammals such as white whale, narwhale, hooded seal (Mansfield 1967; Bowering 1983), and harp seal (Lawson and Stenson 1997). The young are preyed on by cod, salmon, and other Greenland halibut (Dunbar and Hildebrand 1952; Bowering 1983b; Orr and Bowering 1997).

Parasites. There are over 76 parasite taxa reported from Greenland halibut throughout its range, but only 2, the myxosporan *Myxoproteus reinhardtii* and the copepod *Hatschekia reinhardtii* are specific (Arthur and Albert 1994). The parasite fauna from fish collected in the western Atlantic include 6 protozoan species; a monogenean trematode, *Entobdella hippoglossi,* 18 digenean trematodes, seven nematodes, 6 cestodes, 5 acanthocephalans, 4 copepods, 2 isopods, and a leech, *Notostomum laeve.*

Breeding Habits. Greenland halibut females have been reported to produce 30,000–300,000 eggs, the number being positively correlated with length (Lear 1970; Bowering 1980). Recent studies have shown that gametogenesis is not synchronous, suggesting that the spawning season is not well defined. There are indications that the spawning pattern is not determinate, so fecundity estimates may be low. Greenland halibut may be capable of fast-tracking oocytes to maturity thus increasing yearly reproductive output (Rideout et al. 1999). Little is known about reproductive behavior but they are thought to spawn in the deep waters of Davis Strait south of 66°45′ N in winter or early spring at depths of 650–1,000 m (Jensen 1935; Templeman 1970; Bowering 1983).

Data on Greenland halibut stocks throughout the Canadian northwest Atlantic show large spatial and temporal variability in age at maturity (Bowering 1983; Morgan and Bowering 1997): age at 50% maturity was 9.5–15 and 8.2–11.6 years for females and males, respectively. Overall length at maturity for females from a variety of sources was 71.5 cm, but this was not weighted and covered only part of the area. Differences in age at first maturity in different areas make management difficult.

Early Life History. There are few data for fish in the western North Atlantic. In the eastern Bering Sea (Bulatov 1963), the egg membrane is reddish brown and the yolk is colorless and large, a narrow perivitelline space occupying only 1.3–2.9% of its diameter. The embryo is not pigmented. Eggs range in size from 3.71 to 4.10 mm, averaging 3.84 mm. Eggs were collected between February and March and April to June. Larvae were first collected in May when they were 16–17 mm long and the yolk sac was almost completely absorbed. In June they were 25–28 mm long. By the end of June they averaged 33 mm.

In the North Atlantic, eggs and larvae drift northward with the West Greenland current toward Greenland and Baffin Island. Those captured by the Labrador Current drift south to the continental slopes of Labrador and Newfoundland as far as Georges Bank (Templeman 1973b). Larvae of 27 mm and longer move vertically near the surface layer until they reach about 70 mm, then live at depths of about 250 m and drift with the current. Later, they descend to greater depths but do not become as closely associated with the bottom as most flatfishes (Bowering 1983).

Age and Growth. Males and females from seven different regions off eastern Canada grow at about the same rate for the first 5–7 years until they reach a length of 45 cm; then females grow faster (Bowering 1983; Bowering and Nedreaas 2001). All fish over 90 cm were females. Maximum known age of females was 20 years (105 cm) in 1988, compared with 12 years (85 cm) for males (Bowering and Nedreaas 2001), but older fish are difficult to age (Lear and Pitt 1975; Bowering 1983).

General Range. Arctic parts of the Atlantic Ocean south to Iceland, Greenland, the Grand Banks, off Newfoundland, the Gulf of St. Lawrence, and the British Isles in deep water (Norman 1934). The southernmost record in the Atlantic seems to be the Gulf of Maine (Boyar 1964). They are also found in the Pacific Ocean, the Bering Sea, and the Sea of Okhotsk (D'Yakov 1991).

Occurrence in the Gulf of Maine. Bigelow and Schroeder considered Greenland halibut to be a fish of the Arctic and sub-Arctic Atlantic and noted that they had no real place in the Gulf of Maine fish fauna because they had only been captured in the gully between La Have and Georges banks. Four specimens were trawled from near Boothbay Harbor, Maine (Boyar 1964); two from depths of 90–100 m near Sequin Island and Pemaquid, Maine, and the other two at depths of 63–72 m near Pemaquid. These were rare strays into the Gulf of Maine and appeared to have moved in during a winter of unusually cold water temperatures.

Importance. There is no commercial fishery in the Gulf of Maine but there is an important fishery in Canada (Scott and Scott 1988). Recently, the exploitation rate of Greenland halibut has increased beyond sustainable levels (Bowering and Brodie 1995).

Stocks. Helminth parasites have been used as biological tags in population studies of Greenland halibut (Boje et al. 1997). Based on spatial infestation patterns of three digeneans and three nematodes, populations off Labrador, Davis Strait, and the fjords of Umanak, west Greenland show strong similarities, whereas populations in southwest Greenland fjords and in Denmark Strait appear to be isolated. However, comparisons of mitochondrial DNA among fish throughout the northwestern Atlantic (Flemish Pass, Grand Banks, Davis Strait, and northwest Greenland) with samples from the Gulf of St. Lawrence, Iceland, and Norway suggest that there is sufficient mixing of Greenland halibut to preclude development of genetically independent stocks (Vis et al. 1997).

AMERICAN SOLES. FAMILY ACHIRIDAE

GRACE KLEIN-MACPHEE

Soles have small eyes on the right side of the body. The dorsal and anal fins are free of the caudal fin and the right pelvic fin is joined to the anal fin. Most soles are marine, but many tropical species enter freshwater or even live there permanently. Nine genera with about 28 species are known (Nelson 1994), only one of which has been found in the Gulf of Maine.

HOGCHOKER / *Trinectes maculatus* Bloch and Schneider 1801 /

Bigelow and Schroeder 1953:296–297 (as *Achirus fasciatus*)

Description. Body oval, without definite caudal peduncle (Fig. 309). Eyes small, set flat instead of in prominent orbits. Gape of mouth shorter and much more crooked on blind side than on eyed side. Dorsal fin originates at tip of snout. Dorsal and anal fins highest posteriorly. Preanal spine absent. Pectoral fins absent on both sides. Pelvic fin on eyed side continuous with anal fin. Lateral line straight. Skin very slimy with mucus. Scales very rough on both sides, those on upper part of head and chin on eyed side and on entire head on blind side larger than body scales. Scales of unusual structure, their ctenii with large hinged joints (De Lamater and Courtenay 1973). Body proportions: body depth 54.3–57.2% SL; head length 21.3–28.8% SL; and orbit length 2.1–2.7% SL (Cervigón

Meristics. Dorsal fin rays 50–56; anal fin rays 36–42; lateral line scales 66–75; vertebrae 28–29 (Tracy 1908; Hildebrand and Schroeder 1928; Miller and Jorgenson 1973).

Color. Dusky or slaty olive to dark brown on the eyed side, barred transversely with a varying number (usually seven or eight) of indistinct darker stripes, with a dark longitudinal stripe along the lateral line, and sometimes with pale mottling. The dorsal, caudal, and anal fins are of the general body tint. The blind side is dirty white, usually marked with dark round spots that vary in size and number, but some individuals lack these spots.

Size. Hogchoker probably do not exceed 20 cm TL; com-

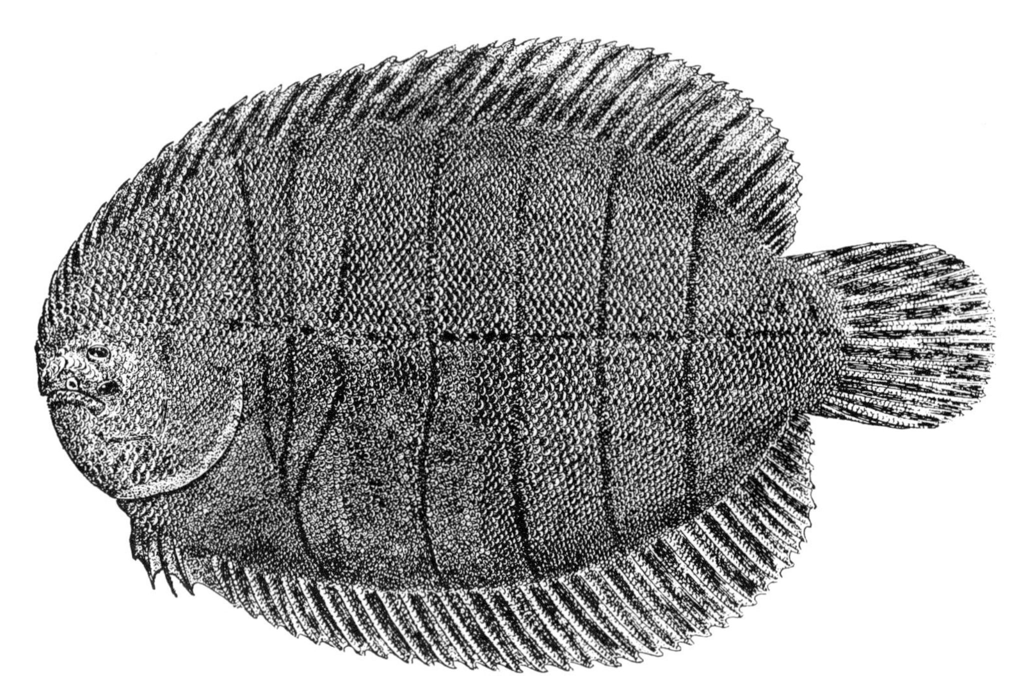

Figure 309. Hogchoker *Trinectes maculatus.* Woods Hole, Mass. Drawn by H. L. Todd.

Distinctions. Hogchoker cannot be mistaken for any other Gulf of Maine right-eyed flatfish because they have no pectoral fins, their eyes are tiny, and their mouth is asymmetrical.

Habits. Hogchoker are most common in bays and estuaries where the water is more or less brackish, and they sometimes run up into freshwater. They are common over mud, sand, or silt (Martin and Drewry 1978). They can tolerate salinities from 0 to 50 ppt (Raney and Massmann 1953) and temperatures from 1.1° to 35.1°C (Dovel et al. 1969; de Sylva et al. 1962).

Hogchoker are tolerant of low oxygen levels; they can survive oxygen concentrations of 1 mg·liter^{-1} for at least 10 days, but die at concentrations of 0.4 mg·liter^{-1} (Pihl et al. 1991). They moved out of hypoxic areas in the York River, Va., and Chesapeake Bay when oxygen levels fell to 2 mg·liter^{-1} and returned when levels increased.

Hogchoker in the Patuxent River estuary show an interesting movement pattern (Dovel et al. 1969). They spawn in the lower river, and larvae move upstream to a low-salinity nursery area, where they remain during winter. In spring they move toward the spawning area. This cyclical movement continues until at least their fourth season, ranging further downstream toward higher-salinity water.

Food. In the York River, Va., hogchoker eat amphipods (particularly *Leptochirus plumulosus*), clam siphons, nereid worms, copepods, and fishes (S. M. Smith et al. 1984; Hines et al. 1990). There is a shift in diet with increasing size of the fish: gammarid amphipods are more important for fish smaller than 61 mm SL, while nereid worms and clam siphons are more important for larger fish. Clam siphons form a significant part of the diet of hogchokers in the Cape Fear River estuary, N.C. (Schwartz 1997a). There is usually a large proportion of unidentifiable matter in their stomachs because these fish tend to macerate their food.

Hogchoker may depend upon chemical stimuli for detecting food, and the barbels around the mouth are chemosensory organs (O'Connor 1972). In the laboratory under a 15:9 light-dark period, they were active only in the dark period (O'Conner 1972), with peaks associated with slack tide in the natural habitat. Under continuous dim light, activity peaks coincided with slack tide and they were active in the diurnal as well as the nocturnal phase of the cycle. O'Connor believed that this activity pattern might have a selective advantage because food organisms might be more available at slack tide, and after dark the hogchoker would be safe from visually oriented predators.

Feeding rate was higher at higher temperatures but food conversion efficiency was influenced by a temperature-salinity interaction (Peters and Boyd 1972). Maximum growth occurred at 25°C and 30 ppt; food conversion efficiency was highest at 15°C and 30 ppt.

Predators. The only predator found by the NEFSC food habits survey was a single record for smooth dogfish (Rountree 1999).

Parasites. Hogchoker are host to a variety of parasites, including a nematode, *Ichthyonema* sp.; cestodes; unidentified cysts; larvae, *Dibothrium* larvae, *Scolex polymorphus, Synbothrium filicolle,* and *Tetrarhynchus bisulcatus;* and trematodes, *Distomum appendiculatum* and *D. corpulentum* (Linton 1905).

Breeding Habits. Hogchoker mature at ages 2–4 (Dovel et al. 1969). The smallest gravid female recorded was 4.78 cm SL, and fecundity ranged from 11,075 eggs for an individual 8.7 cm SL to 23,075 for one 10.8 cm SL (Castagna 1955). Bigelow and Schroeder reported 54,000 eggs from a 16.5-cm female.

The spawning season is April to October but eggs have been reported as early as January, and in the Gulf of Mexico spawning appears to go on year-round (Martin and Drewry 1978). Spawning takes place in estuaries, but eggs may be carried as far as 11 km offshore. Spawning occurs in the evening between 1800 and 2000 hours (Martin and Drewry 1978). It begins when water temperatures reach 20°C and peaks at 25°C. Eggs have been found at salinities of 0–24 ppt, but most occur at 10–16 ppt (Dovel et al. 1969). Larvae 3.6–4.1 mm were collected in August in the Weweantic River estuary, Mass. (Lebida 1969).

Early Life History. Hogchoker eggs are pelagic at higher salinities and near the bottom at lower ones. They range in size from 0.67 to 0.84 mm or 1.05 to 1.22 mm, being smaller at higher salinities. There are 15–50 oil droplets in the yolk, which is yellow to greenish in color. The perivitelline space is not noticeable in marine eggs but is definitely present in estuarine eggs. The eggs are distinguished by greenish granules and many chromatophores on both embryo and yolk (Martin and Drewry 1978). Incubation takes 26–36 h at 23–24°C (Hildebrand and Cable 1938). Larvae are 1.7–1.9 mm at hatching. They can be distinguished by the prominent hump on the head; multiple oil droplets in the yolk; the anus separated from the yolk mass and later the stomach, and the greenish specks on the body, yolk sac, and finfold. These specks become concentrated on the finfolds to form four or five blotches that expand to form bands as the larva grows. Pectoral fins appear after yolk-sac absorption (48 h after hatching) and remain until metamorphosis, when they are lost. Eye migration begins at 5 mm, 34 days after hatching.

Age and Growth. Patuxent River, Md., specimens ranging from 50 to 148 mm SL weighed 6.0–123 grams and were 1–7 years old (Mansueti and Pauly 1956). In the Hudson River, growth is rapid during the first 2 years of life and females grow larger and live longer than males (Koski 1978).

General Range. Hogchoker are found off the Atlantic and Gulf coasts of North America, from Maine (Peters and Boyd 1972) to the Gulf of Mexico and Venezuela (Martin and Drewry 1978). They are abundant in Chesapeake Bay and south, and moderately common as far north as southern New England, but are rare north of Cape Cod.

Occurrence in the Gulf of Maine. There are old records from Provincetown, Boston Harbor (several deposited in the MCZ), the mouth of the Charles River, and Nahant. They are not known north or east of Cape Ann, nor on the offshore banks. Juveniles were taken in a stream tributary to Cape Cod Bay (Collette and Hartel 1988), and adults were recorded near the Pilgrim Nuclear Station in western Cape Cod Bay (Lawton et al. 1984).

Importance. Hogchoker are said to be delicious eating, but they are so small that they are of no commercial value, even in Chesapeake Bay, where they are plentiful. The rumored origin of the name "hogchoker" is that hogs that "feed on fish discarded on the beaches have difficulty in swallowing the sole because of its extremely hard, rough scales" (Hildebrand and Schroeder 1928).

ACKNOWLEDGMENTS. Drafts of flatfish species accounts were reviewed as follows: winter and witch flounders by Jay Burnett, yellowtail by Steven X. Cadrin, hogchoker by Karsten E. Hartel, summer flounder by Anne Richards and Mark Terceiro, and windowpane by Laura Thorpe. A draft of the entire flatfish section was reviewed by Andrew Cooper.

ORDER TETRAODONTIFORMES

Tetraodontiformes is probably the most derived order of fishes. Bones such as parietals, nasals, and infraorbitals have been lost. Gill openings are restricted. Scales are usually modified as spines, shields, or plates. The order contains ten families with approximately 100 genera and 500 species, which are mostly tropical and marine (Matsuura and Tyler 1994; Nelson 1994). Five families have representatives that venture north into the Gulf of Maine.

KEY TO GULF OF MAINE TETRAODONTIFORM FAMILIES

1a. First dorsal fin of 1 large and 2 or no small spines; skin hard and scaly or rough like sandpaper 2
1b. First dorsal fin not consisting of 1 large and 2 or no small spines; skin leathery or naked with small prickles or large spines 3
2a. First dorsal fin with 1 large and 2 small spines; body covered with hard, interlocking platelike scales; teeth well developed and crowded but not fused . . . . . **Balistidae**
2a. First dorsal fin with 1 large spine only; body covered with small denticles, like sandpaper; teeth well developed and fused. . . . . . **Monacanthidae**
3a. Body terminating abruptly behind erect dorsal and anal fins; without rayed caudal fin; skin leathery . . . . . **Molidae**
3b. A rayed caudal fin present; skin either prickly or covered with stout spines 4
4a. Skin covered with very small prickles . . . . . **Tetraodontidae**
4b. Skin with stout spines . . . . . **Diodontidae**

TRIGGERFISHES. FAMILY BALISTIDAE

GRACE KLEIN-MACPHEE

Triggerfishes have three dorsal spines: the third may be minute and the second acts as a trigger by fitting into the first to lock the large first dorsal spine into place—hence the common name. The scales are platelike, in regular series. The upper jaw has four teeth in the outer series and three in the inner one on each premaxilla and four outer teeth and no inner teeth on each dentary. They lack pelvic fins and have 18 vertebrae. The family comprises 11 genera with about 40 species, all marine fishes (Nelson 1994). One species has been reported from the Gulf of Maine.

GRAY TRIGGERFISH / *Balistes capriscus* Gmelin 1789 / Bigelow and Schroeder 1953:520–521 (as *Balistes carolinensis*)

Figure 310. Gray triggerfish *Balistes capriscus*. New York. Drawn by H. L. Todd.

Description. Body deep, laterally compressed, with slender caudal peduncle (Fig. 310). Mouth small, terminal. Dorsal and ventral profiles of snout nearly straight. Eyes situated high on head, preceded by narrow nasal groove. Teeth large, protruding incisors. Gill slits very short, wholly above insertions of pectoral fins. Armor of thick scales covering entire head and body. Spiny dorsal fin triangular, with three spines: first spine so stout that it is more like a horn, situated close behind eyes; second spine acts as a trigger to lock first spine erect. Soft dorsal fin separated from first by a considerable interspace, rhomboid in outline with third or fourth ray longest, tapering back to caudal peduncle base. Anal fin corresponds to soft dorsal in outline and in location. Caudal fin moderate-sized; rear margin moderately concave, with sharp, pointed, somewhat prolonged corners. Pectoral fins short, rounded, situated below gill slits. Pelvic fins reduced to one short, median, stout, blunt spine protruding from abdomen, encased in spiny scales and connected to body by a sort of dewlap strengthened by elongate scales. Four incisor-like teeth exposed on each side of each jaw, three others internal on each side of upper jaw. Scales thick, covered with asperities; edges not free. Two large, modified scales above pectoral fin base and several smaller modified scales above these. Lateral line poorly developed, irregular, undulating with branches over nape and over and under orbit, which unite behind the orbit to form a single line that proceeds posteriorly and slightly dorsally to dorsalmost point just below third dorsal spine. From there lateral line bends abruptly downward to its ventralmost position three-quarters of distance to anal fin base; line then turns upward to middle of side and runs posteriorly along middle to caudal base (Moore 1967). Body proportions: Head length 33.3–37.4% SL, decreasing with length; depth 45.4% SL; eye 20.0–22.7% HL; 7.0–7.9% SL (Cervigón 1966; Leim and Scott 1966; Randall 1968).

Meristics. Dorsal fin rays III, 26–29; anal fin rays 23–26; pectoral fin rays 13–15; pelvic fin, 1 modified spine, encased by scales and protruding as the pelvic process; lateral scales 48–53; gill rakers 31–35; vertebrae 18 (Longley and Hildebrand 1941; Moore 1967; Randall 1968; Miller and Jorgenson 1973).

Color. Colors vary widely. In general the body is gray with four dark patches at the second dorsal fin base (Moore 1967). Other examples have been described as olive gray, marked with violet dots and dark crossbars, the vertical fins tinted with yellow, blue, and olive, and the pectoral fins greenish.

Size. Average size of large adults is 35–40 cm FL (Johnson and Saloman 1984) and about 0.5 kg, but one was recorded as 2.3 kg (Scott and Scott 1988). The all-tackle game fish record is a 6.15-kg fish caught at Murrell's Inlet, S.C., in May 1989 (IGFA 2001).

Distinctions. Gray triggerfish are distinguished from other Gulf of Maine fishes by the large, thick dorsal spine, the small terminal mouth with thick lips and large incisor-like outer teeth, the location of the eyes high on the head, and the compressed body. They differ from the closely related filefishes by having three dorsal spines rather than one or two and by the large platelike scales on the body.

Habits. Adults inhabit coral or rocky reefs down to 55 m (Franks et al. 1972) or seaweed-covered areas near shore (Lythgoe and Lythgoe 1992). Gray triggerfish have been reported from salinities of 16.6–40.1 ppt and temperatures of 14°–33.2°C (Roessler 1970; Franks et al. 1972).

Triggerfish are not strong swimmers. They propel themselves by undulating movements of dorsal and anal fins, known as balistiform locomotion, and only use the caudal fin for short bursts (Lythgoe and Lythgoe 1992). They blow water into sand bottoms to uncover the invertebrates on which they feed, as well as to create a nest (Garnaud 1960). Gray triggerfish produce small grunts and hissing noises, as well as scratchy clicks and rasping sounds by tooth stridulation, resonated by the adjacent swim bladder (Fish and Mowbray 1970). They can give a nasty nip to fishermen who do not handle them carefully (Straughan 1958).

Food. Gray triggerfish feed diurnally on algae, bryozoans, worms, amphipods, barnacles, mollusks, echinoderms, fishes, and unidentified eggs (Aiken 1975; Randall 1968; Vose and Nelson 1994). While foraging over a sandy bottom they assume a relatively vertical position and direct a jet of water at the bottom, displacing sand and revealing sand dollars; they then pick up a sand dollar and drop it, exposing its oral surface, crush the center, and bite chunks out of the damaged area (Frazer et al. 1991). They prey on three species of sand dollars, *Mellita tenuis, Leodia sexiesperforata,* and *Encope michelini.*

Predators. Juvenile triggerfishes fall prey to tunas but they are often not identifiable to species in stomach contents. A gray triggerfish was found in the stomach of a blue marlin captured off Port Antonio, Jamaica (de Sylva in Aiken 1975)

Parasites. Three trematodes, *Apocreadium coili, A. balistes,* and (Sogandares-Bernal 1959), and a copepod, *Taeniacanthus balistae,* on the gill filaments of a West African individual (Ho and Rokicki 1987).

Breeding Habits. Spawning occurred in October–December, peaking in November and December off Ghana. First-time spawners were 13.3–15.7 cm FL. Fecundity was correlated with body length. Linear regression analysis gave $\log F = 1.176 + 1.642 \log L$ (Ofori-Danson 1990). Spawning in the Mediterranean occurs during summer when temperatures reach 21°C. Adults excavate a hollow nest in the seabed, deposit an egg mass in it, and guard and aerate the mass (Lythgoe and Lythgoe 1992).

Early Life History. The eggs are demersal and hatch in 2–5 days; the larvae first float to the surface and then settle to the bottom, remaining quiescent there during the day but becoming more active at night (Garnaud 1960; Lythgoe and Lythgoe 1992). Development of fins and their supports was studied in field-collected larvae ranging from 3.2 to 23.5 mm NL. The first dorsal spine appears at 3.2 mm and the second at 3.3 mm. Barbs appear on the first dorsal spine and the second spine's locking mechanism is apparent by 3.9 mm. Notochord flexion occurs by 4.5 mm. A tuft of setae that appears to be a characteristic of balistoid larvae is present on the cheek at 3.2 mm, but it is transient and disappears after larvae reach 4.7 mm. Pectoral fins are present at 3.2 mm and by 5.8 mm all components of the pectoral girdle and suspensorium except the posterior radial are ossified. The pelvis and rudimentary fin ray elements appear at about 4.3–4.5 mm NL and are similar to those of the adult by 6.7 mm SL. Fusion of the four rudimentary fin ray elements (two from each side) is complete by 23.5 mm SL. Notochord flexion is evident from drawings of a 4.5 mm NL larva and is complete by 5.5 mm. The caudal complex shows almost the same form as the adult by 23.5 mm SL (Matsuura and Katsuragawa 1981). Larvae resemble those of the filefish *Stephanolepis hispidus,* with which they co-occur in Brazil. At about 3 mm they can be distinguished from similar-sized larvae of this filefish by the presence of two dorsal spines, a patch of small spines on the cheek, and the lack of small spines on the forehead region (Matsuura and Katsuragawa 1985).

Juveniles resemble adults in general shape. Snout length is 19–21% SL at 10–20 mm, increasing with size; eye diameter is 15–18% SL at 10–20 mm SL, decreasing with size; and they have scales with single spines that are lost at 24–40 mm SL (Moore 1967). A specimen 50 mm long taken on Georges Bank was yellowish, with many small blue-violet spots on the sides and dusky-blotched along the back, with one broad, irregular dusky band extending from the base of the dorsal fin almost to the anal fin. The caudal fin was pale yellow. Others have been described as having translucent second dorsal and anal fins, with saddle markings interspersed with light spots (Moore 1967).

Juveniles are pelagic, accompanying flotsam (Longley and

the surface of the water column (Fahay 1975). They have also been collected among mangrove prop roots, floating *Sargassum*, and *Thalassia testudinales* near Port Royal, Jamaica (Aiken 1975).

Age and Growth. Gray triggerfish from the Gulf of Mexico were aged using sections of the first dorsal spine (Johnson and Saloman 1984). The maximum estimated ages were 12 and 13 years for females and males, respectively. Mean annual mortality rates ranged from 0.32 to 0.53. The von Bertalanffy growth equations, using weighted means, were $l_t = 491.9(1 - e^{-0.382(t - 0.227)})$ for males; $l_t = 437.5(1 - e^{-0.383(t - 0.150)})$ for females, where l = FL in millimeters and t = age in years. The weight-length relationships are $W = 6.7150 \times 10^{-6}L^{3.187}$ for males; $W = 1.3939 \times 10^{-5}L^{3.065}$ for females, where W = weight in grams and L = FL in millimeters. Two gray triggerfish tagged in Florida increased 8.7–17.4% SL over the 15 months they were at liberty; three other individuals grew 3.6–12.2 mm·month^{-1} (Beaumariage 1964, 1969).

General Range. Both sides of the tropical Atlantic. In the west from Nova Scotia to Argentina, including the Gulf of Mexico, the Bahamas, and Bermuda (Böhlke and Chaplin 1968). The northernmost record is of a fish taken from a fishing net in the vicinity of Port Saunders, Daniels Harbor, Nfld., 50°39′ N, 57°18′ W (Scott and Scott 1988). In the east, straying north to England and Ireland on the European coast; also in the Mediterranean and along the coast of Africa south to Angola (Tortonese 1986).

Occurrence in the Gulf of Maine. Several gray triggerfish have been reported from the Gulf of Maine, probably having drifted in over the offshore rim of the Gulf: One specimen taken in the "Squam River" at Annisquam, near Gloucester years ago; two in the Annisquam River in the early 1980s (Collette and Hartel 1988); two small fish of 5 and 7.6 cm picked up on the northeast part of Georges Bank among *Sargassum* weed from *Albatross II* in mid-September 1927; a large one about 38 cm long gaffed at the surface 22.5 km southeast of Highland Light, Cape Cod, on 19 July 1929; one, now in the Museum of Comparative Zoology, picked up at Plymouth on 5 September 1932; one reported from Casco Bay, Small Point, Maine; and one near Boothbay Harbor, Linekin Bay in 1949 (Bigelow and Schroeder).

Importance. Gray triggerfish are utilized by commercial and recreational fisheries in the Gulf of Mexico (Johnson and Saloman 1984) and the Florida Keys (Smith 1907); they are also sold occasionally in fish markets as far north as Rhode Island (pers. obs.), but they are rare strays in the Gulf of Maine and have no commercial or recreational importance there.

FILEFISHES. FAMILY MONACANTHIDAE

GRACE KLEIN-MACPHEE

Filefishes usually have two dorsal spines, the second much smaller than the first and sometimes absent. The body is prickly or furry to the touch owing to shagreenlike spinules on the scales. The upper jaw usually has three teeth in the outer series and two in the inner one on each premaxilla, and the lower jaw has three teeth (rarely two) in a single series on each dentary. There are 19–31 vertebrae. The family comprises about 31 genera and 95 species, all marine (Nelson 1994). Three genera and four species have been reported from the Gulf of Maine.

KEY TO GULF OF MAINE FILEFISHES

1a. Prominent external pelvic spine at anterior margin of ventral dewlap; gill slits nearly vertical **2**
1b. No external pelvic spine at anterior margin of ventral dewlap; gill slits oblique **3**
2a. Dorsal profile of head anterior to eyes straight or only slightly concave; no spines on sides of caudal peduncle . . . . . . **Planehead filefish**
2b. Dorsal profile of head anterior to eyes conspicuously concave; about 6 anteriorly directed spines on sides of caudal peduncle . . . . **Fringed filefish**
3a. Dorsal fin rays 32–39; anal fin rays 35–41 . . . . . . **Orange filefish**
3b. Dorsal fin rays 43–49; anal fin rays 46–52 . . . . . . **Scrawled filefish**

ORANGE FILEFISH / *Aluterus schoepfi* (Walbaum 1792) / Filefish, Turbot, Hogfish, Unicornfish /

Bigelow and Schroeder 1953:524–525 (as *Alutera schoepfii*)

Description. Body somewhat oblong, very compressed (Fig. 311). Snout bluntly rounded; nearly straight to noticeably concave profile in small to medium-sized individuals; concave over snout and convex over eyes in individuals over 400 mm. Mouth small, almost terminal, lower jaw projecting. Upper and lower jaw teeth incisor-like, usually notched. Body covered with minute, rough scales of uniform size. Lateral line indistinct. First dorsal spine inserted over eye. Soft dorsal fin separated from spinous dorsal by a gap about equal to length of its base. Soft dorsal originates behind middle of trunk, rounded in outline. Anal fin corresponds to soft dorsal in size, shape, and position. Pectoral fins short, rounded, sit-

Figure 311. Orange filefish *Aluterus schoepfi*. Woods Hole, Mass., 228 mm.

uated opposite lower half of oblique gill slits. Caudal fin relatively narrower than in other filefishes and triggerfishes; longest rays more than three times as long as body in small specimens, but only one-fourth to one-fifth as long as body in fish half-grown and larger. Body proportions: body depth 17.3–47.4% SL; head length 23.3–34.2% SL; snout length 12.0–28.6% SL; eye diameter 4.8–8.8% SL (Hildebrand and Schroeder 1928; Berry and Vogele 1961; Randall 1968).

Meristics. First dorsal fin rays II, 32–39; anal fin rays 35–41; pectoral fin rays 11–14; pelvic fins absent; gill rakers 21–27; vertebrae 7 + 16 = 23 (Berry and Vogele 1961; Leim and Scott 1966; Randall 1968; Miller and Jorgenson 1973).

Color. Varying from uniform olive gray to rich orange yellow or milky white above, mottled with darker hues of the same tints; bluish white beneath (Bigelow and Schroeder); head and body usually with numerous small orange spots in life (Berry and Vogele 1961). The caudal fin is usually yellow in adults but sometimes it is dusky, edged with white (Bigelow and Schroeder). Lips may be blackish (Randall 1968). Juveniles often exhibit an obliquely dark banded pattern (Townsend 1917).

Size. Maximum length about 61 cm.

Distinctions. The orange filefish resembles its relatives of the genera *Monacanthus* and *Stephanolepis* in most respects, but while they are all equally compressed, the former are relatively shallower, being not more than half as deep as long. The pelvic bone is as prolonged as it is in other filefishes, but there is no median pelvic spine projecting externally, nor is there a ventral dewlap, which is the readiest field mark by which to distinguish *Aluterus* from *Monacanthus*. The eyes are also set lower on the sides of the head; the gill slits are relatively longer and more oblique; the dorsal spine is relatively shorter and more slender; and the lower jaw projects considerably beyond the upper.

Habits. The orange filefish is a relatively shallow-water species found from the surface (Springer and Woodburn 1960) and shallow flats (Longley and Hildebrand 1941) to 88 m (Böhlke and Chaplin 1968). Adults are found over sand and mud, around seagrasses (Randall 1968; Nichols and Breder 1927), around pilings and jetties (Schwartz 1964b), and on coral or reefs (Böhlke and Chaplin 1968). They have been found in salinities of 16.4–42.9 ppt and at temperatures of 12.8°–32.5°C (Schwartz 1964a; Roessler 1970).

Orange filefish may remain in one spot for varying periods of time, sometimes tilted with the head downward. They may also drift along slowly, propelled only by the dorsal and anal fins (Townsend 1912; Böhlke and Chaplin 1968). Captive, acclimated filefish in tanks with other species produced toothy scratching sounds spontaneously and when mildly annoyed. They also make scraping, rasping, and faint scratching noises. During competitive feeding in a test tank, they produced scratching, wheezing, and rasping sounds, sometimes in two-tone pairs. The sonic mechanism is stridulation by specially modified incisors and the pharyngeal teeth (Fish and Mowbray 1970).

Food. Orange filefish feed mainly on vegetation, but also eat bryozoans, hermit crabs, gastropods, and amphipods. The diet of specimens 34.5–35 cm SL from the West Indies consisted by volume of 67% seagrasses, 31.8% algae, 0.6% hermit crabs, and 0.6% gastropods (Randall 1967).

Parasites. A copepod, *Anchistratos occidentalis,* parasitizes orange filefish in west Africa (Ho and Rokicki 1987). Heavy infestations of a protozoan, *Trichodina spheroidesi,* apparently killed orange filefish in Puerto Rico (Bunkley-Williams and Williams 1994).

Reproduction. Little is known about the reproductive process in orange filefish but partial gonad development was noted in late July in Narragansett Bay, R.I. (Fish and Mowbray 1970). In the Caribbean, ripe fish are found throughout the year (Munro et al. 197[illegible])

Early Life History. Filefish eggs in general are demersal and adhesive with several oil droplets. Larval stages of *Aluterus* sp. were described by Aboussouan and Leis (1984). Juveniles are more elongate than adults. Depth changes from 20.8% TL at 38 mm to 38.5% TL at 100 mm (Longley and Hildebrand 1941). They have pronounced barbs on the dorsal spine and color begins to develop by 32.5 mm (Berry and Vogele 1961). Juveniles are pelagic and often found in *Sargassum* mats (Böhlke and Chaplin 1968).

General Range. Atlantic and Gulf of Mexico coasts of the United States, Bermuda, Bahamas, and Brazil (Böhlke and Chaplin 1968), not uncommon in summer as far north as Cape Cod; reported to Portland, Maine, and Halifax, N.S. A poorly preserved specimen 10.1 cm long was caught at Herring Cove, N.S., in August 1938; a second fish 47.2 cm long was taken at Peggy's Cove, N.S., in August 1955 (Leim and Scott 1966).

Occurrence in the Gulf of Maine. Although orange filefish have been described as "rather common" at Woods Hole during summers, only three specimens have been reported within the Gulf of Maine: two from Salem, Mass., and one from Portland, Maine, all of them many years ago. They were also found near the Pilgrim Nuclear Power Station in Plymouth (Lawton et al. 1984).

SCRAWLED FILEFISH / *Aluterus scriptus* (Osbeck 1765) / Unicornfish /

Bigelow and Schroeder 1953:525 (as *Alutera scripta*)

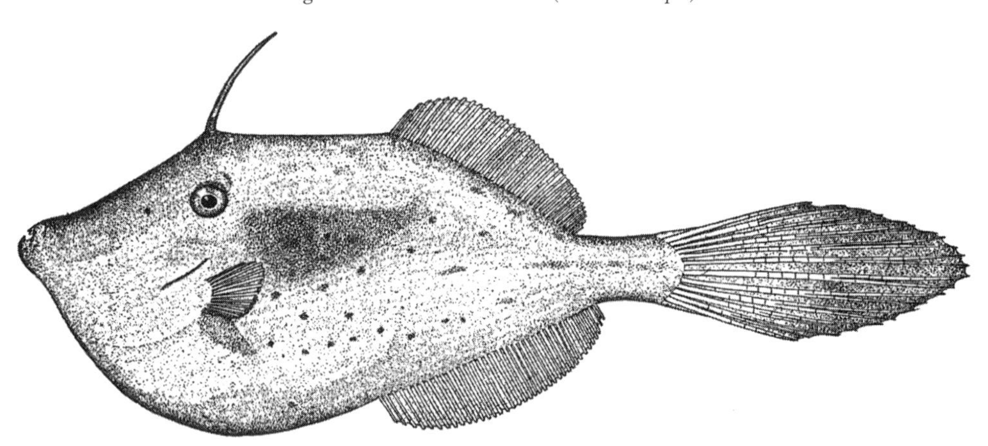

Figure 312. Scrawled filefish *Aluterus scriptus.*

Description. Scrawled filefish resemble orange filefish except that the body is more slender (two to three times as long as deep) and the dorsal, anal, and caudal fins are longer and have a distinctly concave profile (Fig. 312). The color pattern is also different.

Meristics. First dorsal fin rays II, 43–49; anal fin rays 46–52; vertebrae 21 (Böhlke and Chaplin 1968; Miller and Jorgenson 1973).

Color. The body is olive to brown; lines, dashes, and spots of intermediate intensity are bright blue; round spots are deep maroon to black. Most of the caudal fin is dusky to black. Another color phase has been described that shows rows of polygons, some of them with dark centers; the horizontal components of the polygons are interrupted by blue lines; the transverse divisions are broken light lines. Specimens from North Carolina were described as having dark green rather than blue markings. Scrawled filefish can change color to match their surroundings and also change color at night (Clark 1950) and during spawning (Patton 1965). The young possess a greater ability to change color than adults.

Size. Said to reach 91 cm (Böhlke and Chaplin 1968). The all-tackle game fish record is a 2.15-kg fish caught in Pompano Beach, Fla., in January 1998 (IGFA 2001).

Distinctions. Scrawled filefish differ from orange filefish by having more than 42 dorsal rays and more than 45 anal fin rays. The color pattern lacks orange spots and has blue lines and deep maroon to black spots.

Habits. Scrawled filefish are found in grass beds and over reefs (Randall 1967) and in lagoons and outer reef slopes to 20 m (Randall et al. 1990). The young mimic blades and fronds in seagrass beds, often resting with the head downward. They also have been seen drifting along at oblique angles, nose downward, propelled only by their dorsal and anal fins (Böhlke and Chaplin 1968). They produce knocks and scrapes singly or in combination by stridulation of incisors and pharyngeal teeth amplified by the swim bladder (Fish and Mowbray 1970).

Food. Food of scrawled filefish in the West Indies consisted, by volume, of hydrozoans (39.4%), algae (34.2%), gorgonians (12.6%), seagrasses (9.0%), zoantharians (2.4%), and tunicates (1.1%). Unusual items in the diet were large quantities of "stinging coral," *Millepora alcicornis,* and gorgonians (Randall 1967).

Breeding Habits. Spawning has been observed, with a long string of eggs being laid while the nearby male was in a much different color pattern than the female (Patton 1965).

General Range. Scrawled filefish are found worldwide in temperate and tropical waters (Robins and Ray 1986) except for the eastern Pacific (Briggs 1960a). In the western Atlantic from Sable Island in Canadian waters to Bermuda, the Bahamas, Gulf of Mexico, and West Indies south to Brazil.

Occurrence in the Gulf of Maine. Two scrawled filefish, 12.7 cm and 14 cm long, caught on the western edge of Georges Bank on 15 September 1930 by the schooner *Old Glory* are the only reports from the Gulf of Maine. A third, 12.7 cm long, was taken by *Atlantis,* south of Sable Island in Canadian waters (40°55′ N, 59°55′ W) on 18 August 1941 (Bigelow and Schroeder).

FRINGED FILEFISH / *Monacanthus ciliatus* (Mitchill 1818) / Bigelow and Schroeder 1953:523–524

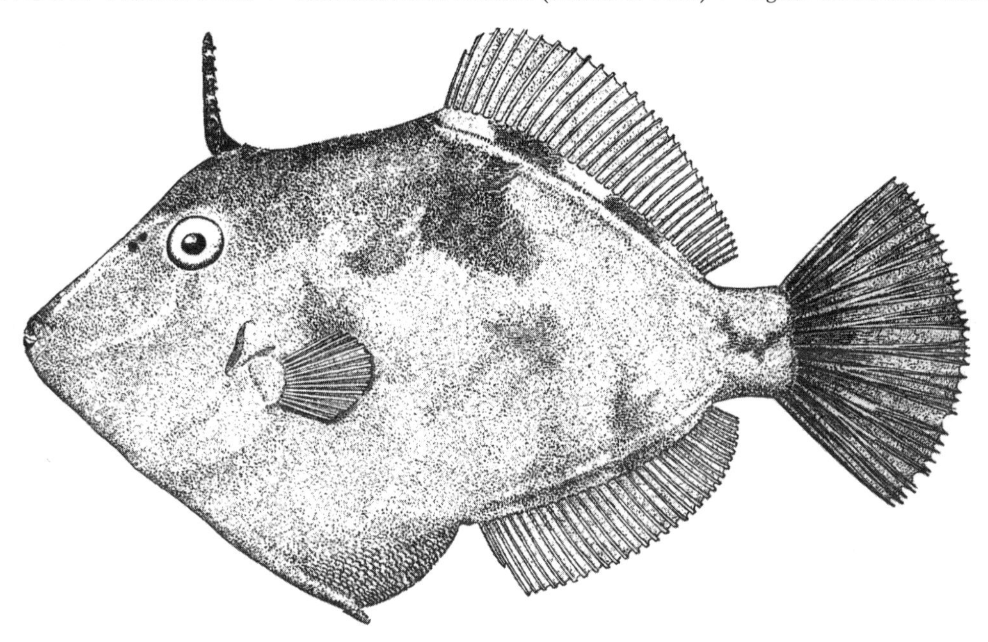

Figure 313. Fringed filefish *Monacanthus ciliatus.* Bahama Islands, 85 mm. Drawn by H. B. Bigelow. (From Bigelow and Schroeder: Fig. 277.)

Description. Body strongly compressed (Fig. 313). Notably concave profile over snout; eyes high on head. Mouth slightly superior; lower jaw slightly longer than upper. Gill openings small, oblique, slits shorter than eye. Jaw teeth broad incisors with sharp cutting edge. Body covered with short spines; adults with recurved spines and bristles at base of caudal fin. Median pelvic spine prominent. Ventral dewlap variously developed, usually extending beyond pelvic spine even in young. First dorsal fin consists of a large spine placed over posterior part of eye, bearing two series of barbs posteriorly, its length about equal to that of snout, and a much smaller second spine without barbs. Second dorsal and anal fins similar. Caudal fin no longer than snout, convex at all ages. Pectoral fins short

54% SL, head length 29–38.7% SL, eye to dorsal spine distance 6.7–10.1% SL (Berry and Vogele 1961).

Meristics. First dorsal fin rays II, 29–32; anal fin rays 28–36; pectoral fin rays 9–13; pelvic fin rays I, modified, encased by scales, and protruding as a large and movable strut; gill rakers 15–23; vertebrae 6 + 13 = 19 (Berry and Vogele 1961; Randall 1968; Miller and Jorgenson 1973).

Color. Color is variable, changing considerably to match the surroundings (Böhlke and Chaplin 1968). It varies from olive gray or grass green to yellowish brown, with darker blotches or crossbands. Dorsal and anal fins are pink, usually with three

and the caudal fin is green, mottled dark and pale (Bigelow and Schroeder). The margin of the ventral flap is bright golden yellow in males and a dull and pale greenish yellow in females. Males also have a fine dark line near the margin of the ventral flap, a marking lacking in females and immature specimens (Böhlke and Chaplin 1968). Fringed filefish from the Bahamas have been described as having a striped color phase with the main color stripe extending from the snout tip to the midbase of the caudal fin, with one or two additional dark streaks above this stripe and two or three diagonal ones below it (Böhlke and Chaplin 1968).

Size. Ranges from 10.2 to 20.3 cm (Bigelow and Schroeder).

Distinctions. The fringed filefish resembles *S. hispidus* except that the first dorsal ray is prolonged, the dorsal profile of the head in front of the eyes is conspicuously concave, the ventral dewlap extends somewhat further behind the tip of the pelvic spine, the caudal peduncle in the adult is armed with two or three pairs of strong forward-curving hooks on either side (Bigelow and Schroeder), and the body is shallower and less robust (Meek and Hildebrand 1928).

Habits. The fringed filefish is a warm-water species that occurs around reefs, seagrass meadows, coral rubble, and sand bottoms and lives in close association with benthic plants such as turtle grass (Randall 1968). These fish have been observed camouflaged, head down, among clumps of algae (Randall and Randall 1960). Adults, especially males, head-stand in the water and flare their ventral dewlaps in establishing dominance, slapping their tails and changing colors at the same time (Clark 1950).

Food. Adult fringed filefish are omnivorous, most commonly feeding on algae, organic detritus, and small planktonic crustaceans (Randall 1968). A study of juvenile feeding habits in Apalachee Bay, Fla., showed that five groups of food organisms accounted for more than 80% of their diet: amphipods, seagrass, copepods, polychaetes, and bivalves (Clements and Livingston 1984). Food of fish ranging in size from 4.7 to 9.7 cm SL from West Indian reefs consisted, by volume, of algae and organic detritus 21.2%, seagrass (*Thalassia testudinum*) 15.4%, copepods 14.6%, shrimp and shrimp larvae 13.1%, amphipods 5.4%, tanaids 4.6%, polychaetes 4.2%, stomatopod larvae 3.9%, isopods 3.1%, pelecypods 2.3%, unidentified animal materials 1.9%, gastropods 1.5%, and hydroids 0.37% (Randall 1967).

Laboratory experiments in prey selectivity showed that a strikingly pigmented amphipod species, *Lembos unicornis,* which made up a large proportion of the fringed filefish diet in Apalachee Bay, was disproportionately consumed by these fish when offered with two other less distinctively pigmented gammarid amphipods (Clements and Livingston 1984).

General Range. Fringed filefish occur in the warmer parts of the Atlantic, from Brazil to Cape Cod on the American coast, and have been reported occasionally in Canadian waters in late summer. A straggler was even reported from Argentia, on the southern coast of Newfoundland (Scott and Scott 1988).

Occurrence in the Gulf of Maine. A 17.8-cm specimen was taken in a trap at Provincetown, on 9 November 1929. Earlier reports from Massachusetts Bay referred to in Storer's (1863) description and illustration were actually based on a specimen of *S. hispidus.* Several specimens were collected east of Georges Bank in 1964 by dip net from the *A. T. Cameron* (Scott and Scott 1988).

PLANEHEAD FILEFISH / *Stephanolepis hispidus* (Linnaeus 1766) /

Bigelow and Schroeder 1953:522–523 (as *Monacanthus hispidus*)

Description. Body rhomboidal, laterally compressed with triangular head, large dorsal spine located over rear edge of eye, and short caudal peduncle (Fig. 314). Rear edge of dorsal spine armed with double series of barbs. Soft dorsal fin originates behind middle of body, rounded in outline, narrowing from front to rear with prolonged first soft dorsal ray bearing filamentous tip in adult males (young 2.5–5 cm long lack this filamentous ray). Anal fin below soft dorsal, of same shape except that no rays prolonged. Caudal fin rounded. Caudal peduncle of mature males with bristly recurved spines. Pectoral fins short, rounded, situated lower than gill openings, which are nearly vertical slits. Body depth 43.3–65.8% SL; head length 29.5–41.4% SL; snout length 14.4–27.5% SL; eye diameter 6.9–17.1% SL; eye to dorsal spine distance 7.3–17.1% SL (Berry and Vogele 1961).

Meristics. First dorsal fin rays II, 29–35; anal fin rays 28–35; pectoral fin rays 12–14; pelvic fin process, 1 modified median spine, encased by scales and protruding as a large and movable strut; gill rakers 30–41; vertebrae 7 + 12 = 19 (Longley and Hildebrand 1941; Berry and Vogele 1961; Randall 1968; Miller and Jorgenson 1973).

Color. Color is variable depending on habitat: green over turtle grass, olive or gray over sand, or brown in *Sargassum* weed. The back and sides of young fish are mottled with irregular darker blotches, usually two dark areas along bases of the anal and dorsal fins, and a dark blotch on the upper side below the front part of the dorsal fin (Robins and Ray 1986), but adults are plain. The dorsal spine and the caudal fin are green. The soft dorsal fin and the anal fin are pale and translucent.

Size. The maximum length is about 25.4 cm.

Distinctions. Planehead filefish differ from fringed filefish in that the first dorsal ray is not prolonged in males, the profile is

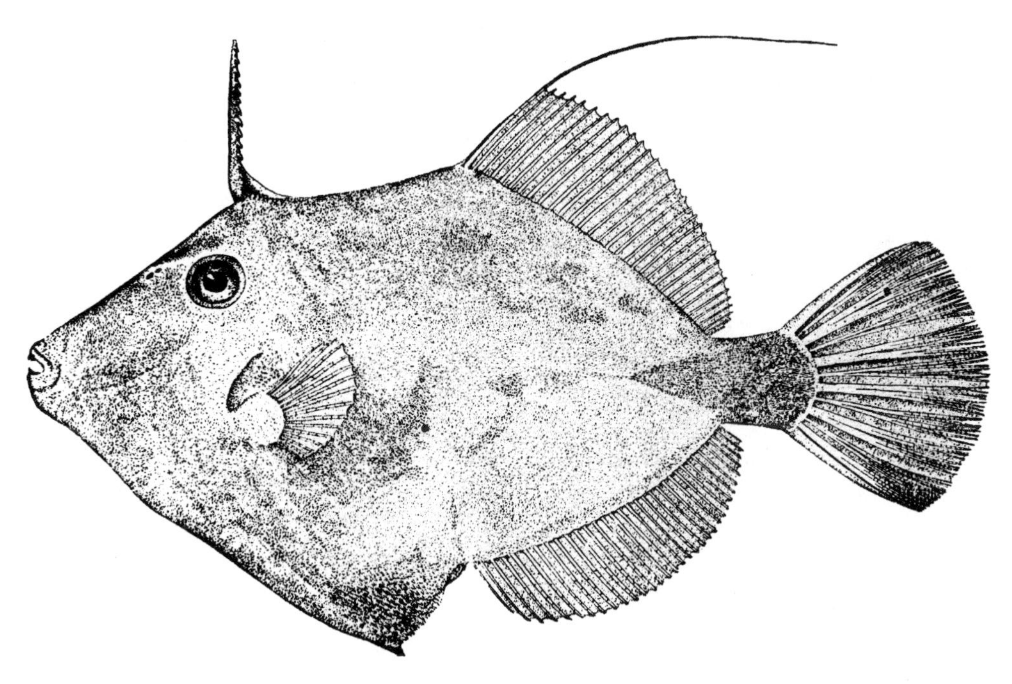

Figure 314. Planehead filefish *Stephanolepis hispidus.*

triangular, the snout is not elongate, and the ventral dewlap does not extend back as far beyond the pelvic spine. There is no deep groove behind the dorsal spine (Berry and Vogele 1961).

Habits. Their habits of planehead filefish are similar to those of triggerfish, but they are not such strong swimmers and are more closely associated with the bottom (Randall 1967). They are benthic as adults, found on reefs and around seagrass meadows (Clements and Livingston 1984), but are also found, especially as juveniles, among floating vegetation and debris. They are mostly shallow-water fishes, living to depths of 44 m (Hildebrand 1954) and are most abundant in clear water and around vegetation (Tabb and Manning 1961). They have been collected in salinities of 11–42.9 ppt and temperatures of 10°–38.8°C (Tagatz 1967; Roessler 1970). They are sometimes trawled in great quantities, up to 8,000–9,000 lb in 90-min tows off Florida and the Carolinas (Anonymous 1964a,b).

Food. Planehead filefish juveniles collected in Apalachee Bay fed on a wide variety of organisms, but the most important groups were amphipods, particularly *Erichthonius;* seagrass, *Thalassia testudinum;* copepods; polychaetes, particularly Serpulidae; and mollusks, especially *Brachiodontes exustus.* There was an ontogenetic shift in diet from gammarid amphipods in the stomachs of individuals 21–40 mm long to a bivalve *B. exustus* and seagrasses with their associated epifauna in fishes 41–60 mm (Clements and Livingston 1984).

Predators. Known predators include bluefish (Buckel et al. 1999b) and Atlantic sharpnose shark, *Rhizoprionodon terraenovae*

Breeding Habits. Little is known of their breeding habits. Females with eggs have been collected at 81 mm (Berry and Vogele 1961). Small juveniles are present year-round along the southern U.S. coast (Fahay 1975), indicating a prolonged spawning period (Martin and Drewry 1978).

Early Life History. There is little information on planehead filefish eggs. Unfertilized eggs have been described as 0.7 mm, adhesive, pale green with a group of small oil droplets on the side of the yolk (Ryder 1887b). In general, filefish eggs are attached to vegetation, and there is no record of parental care (Aboussouan and Leis 1984).

Descriptions of yolk-sac and early-stage larvae 1.7–2.8 mm may have been misidentified (Martin and Drewry 1978). In general, newly hatched larvae are cylindrical and somewhat compressed (Aboussouan 1966; Aboussouan and Leis 1984). The body becomes deeper at 3 mm, the mouth is small and terminal, the dorsal spine is prominent with a few barbs, traces of dorsal and anal rays are present, and scales are represented by small prickles. There is dark pigment on the head extending to the back, on the sides above and behind the abdomen with the ventral periphery slightly pigmented. By 5 mm the dorsal and anal rays are well-formed and the skin becomes covered with prickles. Pigmentation consists of dark dots scattered over the body (Hildebrand and Cable 1930).

Juveniles 6.5 mm SL have the dorsal spine barbs very well developed and the lateral color patterns forming (Berry and Vogele 1961). By 8 mm the lateral color pattern is more complete and the slope of the forehead is less steep; anal, caudal, and pectoral fins are well developed and shaped as in the adult; the pelvic spine is largely free and prickly. By 15 mm the body

blunter but elongates as the fish grows; the dorsal spine barbs are proportionately smaller, disappearing by 40–50 mm; the pelvic spine is attached to the body by a membranous flap that increases in size until it reaches the tip of the spine by 75 mm. Pigment is brown spots with dark centers everywhere on the body and head, with some dark marbling evident, and the fins are almost colorless (Hildebrand and Cable 1930).

Juveniles are very abundant in pelagic *Sargassum* communities, accounting for 84.5% of the ichthyofauna in the eastern Gulf of Mexico (Bortone et al. 1977) and 61 and 69% in Miami and Cape Lookout, respectively (Dooley 1972).

General Range. *Stephanolepsis hispidus* is a tropical species, common along the south Atlantic coast of the United States, Bermuda, the Gulf of Mexico, and the West Indies, as far south as Brazil (Cervigón 1966). They have been taken from time to time as far north as Woods Hole and have been recorded from St. Margaret Bay and from off Halifax on the outer coast of Nova Scotia. They also occur around the Canaries and Madeira in the eastern Atlantic and appear to be present in East Indian waters as well.

Occurrence in the Gulf of Maine. Occasional specimens of planehead filefish were recorded from Hingham, Lynn, Nahant, and Boston Harbor in Massachusetts Bay and from Cape Cod, all many years ago. Later records from the Gulf are of 181 juveniles, 2.5–5 cm long, picked up by *Albatross II* on the northeastern part of Georges Bank among floating *Sargassum* in September 1927; a larger one taken off Seguin Island, Maine, on 12 September 1927; one 15 cm at Provincetown on 6 November 1929; one picked up among floating rockweed (*Fucus* or *Ascophyllum*) and *Sargassum* on the western part of Georges Bank on 15 September 1930; one taken off Portland Lightship on 17 July 1931; one taken in a trap at Provincetown on 6 October 1950; two, about 15 cm, taken off Wood End, Provincetown, on 30 October 1951 (Bigelow and Schroeder); and 20 near the Pilgrim Nuclear Power Station at Plymouth in western Cape Cod (Lawton et al. 1984). An occasional filefish straying from the south may be expected anywhere on Georges Bank or on the western side of the Gulf, but there is no evidence that they ever enter its eastern side or reach the Bay of Fundy.

PUFFERS. FAMILY TETRAODONTIDAE

GRACE KLEIN-MACPHEE

Puffers can inflate their bodies by inhaling water or, if out of water, air. Their scales are usually modified into short prickles. They have a total of four fused tooth plates in both jaws (teeth fused on each side of each jaw but separated by a median suture), forming a parrotlike beak. The "flesh" (especially the viscera) of some puffers, especially in the tropical Indo–West Pacific, contains the alkaloid poison tetraodotoxin, which can be fatal to humans. The family comprises 19 genera with about 121 species (Nelson 1994), mostly marine, but with some freshwater species in South America and Asia. One species strays into the Gulf of Maine.

NORTHERN PUFFER / *Sphoeroides maculatus* (Bloch and Schneider 1801)

Puffer, Swellfish, Swell Toad, Toad, Baloonfish, Blowfish / Bigelow and Schroeder 1953:526–527

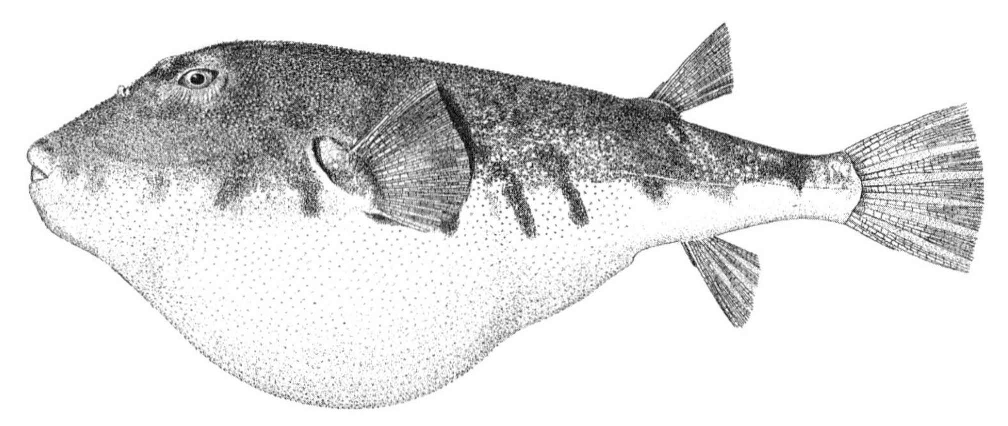

Figure 315. Northern puffer *Sphoeroides maculatus*. Noank, Conn. Drawn by W. S. Haines.

Description. When not inflated, body moderately slender (about three times as long as deep), about as thick as it is deep, tapering from gill opening to a rather slender caudal peduncle posteriorly, to a rounded snout anteriorly (Fig. 315). Mouth small, located at tip of snout. Teeth incorporated into bones of upper and lower jaws, forming nibbling or cutting edge, each divided in the middle by suture, appearing as two large incisors above and two below. Gill openings very small, set obliquely backward and downward. Eyes set very high, horizontal, oval in outline. Skin naked, but sides of head and body, dorsum from snout to dorsal fin, and belly posterior to vent rough with small, stiff, close-set prickles. Prickles on back blunt, nearly vertical; those on sides and belly rather sharp, erect or pointing backward depending, respectively, on whether the fish is inflated or not. Lateral line inconspicuous. Spiny dorsal fin absent. Soft dorsal fin very short, rhomboid in outline, about twice as high as long, set far back close to caudal peduncle. Anal fin similar to dorsal in shape and size and arises close behind it. Pectoral fins fan-shaped, situated close behind gill openings. Pelvic fins absent. Caudal fin of moderate size, weakly rounded, with angular corners. Body proportions: head 33.3–37% SL; snout 47.6–58.8% HL; eye 12.5–25% HL (Shipp 1974).

Meristics. Dorsal and anal fin rays 6–8; pectoral fin rays 15 or 16, rarely 17; vertebrae 8 + 11 = 17 (Shipp and Yerger 1969; Miller and Jorgenson 1973; Shipp 1974).

Color. Dark olive green above, sometimes ashy or dusky, the sides greenish yellow to orange, cross-barred with six to eight indefinite dark bands or blotches (Bigelow and Schroeder). Tiny jet black spots about 1 mm in diameter are scattered over pigmented areas, especially evident on the cheeks; these may be absent in fish shorter than 100 mm. The belly is white. The base and distal half of the caudal fin may be dusky with a lighter central region, but often the entire caudal fin appears uniformly dusky; other fins are devoid of pigment (Shipp and Yerger 1969; Shipp 1974).

Size. Reported to reach 36.6 cm, but few are more than 25.4 cm.

Distinctions. This fish can be separated from other Gulf of Maine fish by its inflatable body, fused cutting plate teeth, and prickly skin. The only other fish that can inflate its body is the porcupine fish, but the latter has large spines with three-pronged roots rather than short, slender prickles and is oval, not fusiform, in outline.

Habits. Northern puffer are inshore fish, often coming in to the tide line. They enter slightly brackish water in various estuaries, and can tolerate salinities of 6.7–34 ppt (Tagatz and Dudley 1961; Shipp 1974). They are found on sandy shores (Nichols and Breder 1927); over shell, gravel, silt, and mud (de Sylva et al. 1962); around piers (Schwartz 1964b); in estuaries (Shipp 1974) and bays, and along the Atlantic coast (Shipp and Yerger 1969); they are seldom are caught at depths more than 20 m or more than a few kilometers from land (Bigelow and Schroeder). Northern puffer are considered demersal, but are known to move actively throughout the water column (Nichols and Breder 1927). They can tolerate temperatures of 10.0°–34.1°C in the field (Tagatz 1967; Richards and Castagna 1970). Laboratory experiments show they have an upper temperature tolerance limit of 32.5°C and a lower one of 8°C for a period of 72 h (Hoff and Westman 1966). A puffer kill off the New Jersey coast was attributed to an influx of cold waters that caused bottom temperatures to drop to 4°C coupled with their nocturnal resting behavior (Wicklund 1970a). Northern puffer move through the water column by day, descending to the bottom and becoming quiescent at night. If taken off the bottom and released, they soon return to the bottom again (Wicklund 1970a).

The most striking behavioral characteristic is that if annoyed or frightened, puffers are capable of inflating their bellies to balloonlike proportions and of deflating again at will. They do this by swallowing water rapidly and forcing the fluid into a sac that is a ventral diverticulum of the stomach (Breder and Clark 1947). If a fish is at the surface when disturbed, it may ingest air and float upside down on the water for a few moments or sometimes longer until it can expel the air and return to its normal swimming position. Defensive swelling usually occurs with the accompaniment of sounds that are described as long bursts of creaking erk-erks, produced by a grinding of the upper and lower jaw plates and amplified by the swim bladder. Similar sounds are made during feeding (Fish 1954). Electric stimulation produced sounds consisting of low dull thumps and did not generally induce inflation.

Northern puffer usually swim slowly by moving their dorsal, anal, and pectoral fins rather than laterally wagging the caudal fin (Townsend 1916). They are known to partially bury themselves in soft bottoms by a shoveling motion of their specially developed postclavicular apparatus (Parr 1927). Young fish are found in large loose schools, but adults are more solitary (Fish 1954).

Food. Puffer are opportunistic feeders, eating small crustaceans of all sorts, especially crabs, shrimps, isopods, and amphipods, as well as small mollusks, worms, barnacles, and sea urchins; other invertebrates such as bryozoans, sponges, sea anemones, and sea squirts; and algae (Linton 1905; Townsend 1916; Hildebrand and Schroeder 1928; Isaacson 1963; Randall 1967). They are daytime feeders and have been known to unite in a group to attack a blue crab (Nichols and Breder 1927). Young of 7–10 mm examined by Linton at Woods Hole had eaten copepods as well as crustacean and molluscan larvae.

Predators. Juveniles are frequently found in stomachs of large pelagic fishes such as tunas, bluefish (Buckel et al. 1999b), and Atlantic sharpnose shark (Bowman et al. 2000). Fish hawk,

Pandion halietus, may prey on puffer, based on an observation of an unsuccessful attempt by one of these birds to capture a puffer, which subsequently inflated itself out of the bird's talons (Nichols and Breder 1927). Northern puffer have also been observed to successfully avoid predation by scup or porgy through inflation (Townsend 1916).

Parasites. A cestode, *Tetrarhynchus bisulcatus;* two trematodes, *Distomum vibex* and *Gasterostomum gracilescens;* two nematodes, *Ascaris* sp. and *A. habena;* and two copepods *Tucca corpulentus* and *Pseudochondracanthus diceraus* were found to be parasitic on puffer from Woods Hole (Linton 1905; Wilson 1932; Ho 1968).

Reproduction. Puffer have dissimilar-sized ovaries, the left being considerably larger than the right (Welsh and Breder 1922). Fecundity was estimated at 176,000 eggs from a 26.5-cm fish (Hildebrand and Schroeder 1928), but a subsequent study by Merriner and Laroche (1977) showed this figure to be low by about 273,000 eggs. Total fecundity increases both as a linear function of fish weight ($F = -735.28 + 754.21 \times$ body weight in grams) and as a function of total length ($F = 5957.7e^{0.015}$ TL). Average relative fecundity was 5,204 eggs per gram ovarian net weight and 751 eggs per gram body weight (Merriner and Laroche 1977).

Male and female puffer are sexually mature by age 1 and individuals as small as 88 mm are capable of spawning (Merriner and Laroche 1977). Puffer spawn in shoal water close to shore, from mid-May in Chesapeake Bay and from early June off southern Massachusetts through the summer. Mating has not been observed in the field, but captive puffer were observed to deposit eggs in the sand, partially burying them in a circular pattern. There is no parental care (Breder and Clark 1947).

The transparent eggs (about 0.85–0.91 mm in diameter, with many small, pale yellow oil globules in a foamy cluster averaging 0.34 mm diameter) sink and stick fast to each other or to whatever they chance to touch. Surfaces of the eggs are faintly reticulated (Welsh and Breder 1922). Incubation takes 3–5 days at a temperature of about 20°C. Larvae are about 2.4 mm long at hatching and are brilliantly pigmented with red, orange, yellow, and black. Numerous small tubules are present over most of the body. At 48 h the yolk material is reduced, mouth and vent are open, and green pigmentation appears. In 72 h the mouth functions; when they are 7.4 mm long fin rays are complete, the young fish show most of the diagnostic characters of adults, and they can inflate themselves even more, until the bulging skin entirely hides the dorsal and anal fins. They are, however, thicker-bodied, have large eyes, and lack adult color patterns. Larvae in general are characterized by a chunky body with a low number of myomeres (19), no pelvic fins, and heavy pigmentation except on the last third of the body. Their caudal fin ray count is unique ($0 + 5 + 6 + 0$) and it is the last fin to form (Welsh and Breder 1922). Juveniles lack the dark black "pepper spots," which appear at 40–100 mm. The larger dark lateral marks on juveniles become oblong in shape when they become adults. Prickles develop in individuals as small as 10 mm (Shipp and Yerger 1969).

Age and Growth. Growth marks on the vertebrae are used for northern puffer age determination. For all age-groups, females are larger than males. Most growth takes place from June to October of the first growing season, and it is during this period that the females grow faster, becoming larger than the males; after this there is no significant difference in growth rates between the sexes (Laroche and Davis 1973). Linear growth rate for young-of-the-year was approximately 1.11 mm·day^{-1} over a 45-day period for fish of 16 mm TL (Marcellus 1972).

Length-weight relationships for puffer in prespawning condition (April–July) are: $\log W = -4.414 + 2.901 \log L$ for females; $\log W = -4.643 + 2.993 \log L$ for males. Postspawning (late July to November): $\log W = -4.133 + 2.787 \log L$ for females; $\log W = -3.985 + 2.726 \log L$ for males.

Males weigh less than females at a given length in fish from New Jersey (Welsh and Breder 1922). Fish collected from Virginia waters showed the greatest variation in weight associated with pre- and postspawning. Puffer recovering from spawning from July to November were heavier than fish of comparable lengths prior to spawning (Laroche and Davis 1973). A comparison of length-weight data on fish collected from both Virginia and New Jersey showed that Virginia puffer are heavier for a corresponding length, although feeding habits are similar (Isaacson 1963).

General Range. Atlantic coast of the United States from Florida to Cape Cod in abundance, to Casco Bay in small numbers, and to the Bay of Fundy as strays. Their northern limit is Newfoundland (Leim and Day 1959) and their southern limit is Flagler County, Fla., although offshore populations may extend south to latitude 27°30′ N (Shipp 1974).

Occurrence in the Gulf of Maine. Anglers find puffer plentiful along the southern shores of Massachusetts, but the elbow of Cape Cod marks the eastern and northern limit to their presence in any numbers. They have been reported at Monomoy, Truro, and Provincetown. Cape Cod Bay may support a small resident population. Storer described them as common at Nahant, a few miles northeast of Boston, but this seems to have been an error, for Wheatland (1852), writing at about the same time, not only spoke of them as seldom seen in Massachusetts Bay but considered a single specimen taken in Salem Harbor in the summer of 1848 as worthy of a note. This remained the only positive record for a puffer for Essex County until 24 August 1920, when one was caught at Gloucester. One was dropped by a seagull at Nahant in 1980 (Collette and Hartel 1988). There are records of puffers north of Cape Ann—two taken in a trap in Casco Bay in 1896 and one taken near Long Island, off Portland Harbor, Maine, on 24 July 1933. More recently, David Norman (pers. comm., 7 Oct. 1997) and his dive buddy saw four northern puffer and photographed one at Folley Cove in the Cape Ann region on 28 Sept. 1997. There may be small local populations in Casco Bay and in the vicinity of Boothbay Harbor, Maine, for pound net fishermen were acquainted with them there in the 1950s, and Scattergood (in Bigelow and Schroeder) received three specimens from Pemaquid Point, where fishermen reported them as most common in June. A skeleton, apparently of a puffer, was found on the shore of Minas Basin at the head of the Bay of Fundy on the Nova Scotian side.

Seasonal Migrations and Movements. Throughout the northern part of the range, northern puffer belong in the rather large and varied category of "summer" fishes, taken from April to November in Chesapeake Bay and from late May or early June to October or early November along southern New England. It is probable that when puffer disappear from their usual summer haunts with the onset of cold weather, they merely descend into somewhat deeper water nearby, to spend the winter on bottom in a more or less quiescent state.

Importance. Northern puffer became an important food fish during World War II, with the largest catches from the Chesapeake area. Catches peaked in 1963–1967 and have declined substantially since 1973. Puffer are fished primarily with pound nets, otter trawls, haul seines, and pot traps. They are an important export species in New York (McHugh 1969). They are marketed as "sea squab," the protein quality of the meat being about equal to beef (Darling and Nilson 1946); some are used for fishmeal (Lyles 1966).

Northern puffer are taken in the sport fishery in the Middle, South, and North Atlantic but are not abundant enough to be important as a sport or commercial species in the Gulf of Maine.

They were tested as a potential aquaculture species at the New York Ocean Science Laboratory with success at inducing spawning and growing puffers from egg to adult (Sibunka and Pacheco 1981).

Toxicity. The family Tetraodontidae includes many toxic species. The poison is tetraodotoxin, which acts as a neurotoxin causing, in the worst cases, suffocation and cardiac paralysis. A species that is toxic in one area may not be in another. The toxicity is largely influenced by the puffers' reproductive cycle. They are most poisonous prior to and during the spawning season. The skin, liver, ovaries, and intestines are usually the most toxic parts of the body although the musculature may at times be poisonous as well. The toxin is not destroyed by cooking, the degree of toxicity cannot be determined from the appearance or size of the fish, and there is no known antidote for the poison (Halstead 1978). Although the toxicity of Pacific puffers has been extensively tested, Atlantic species have received comparatively little attention. Extracts of tissues from northern puffer from Florida were toxic when injected into mice, rats, chicks, and frogs. Extracts from the skin were highly toxic to white mice, being fatal in 93% of the animals tested. Before dying all the test animals displayed symptoms of central nervous system involvement, particularly the respiratory center. Other tissue samples (liver, testes, ovaries, and muscle) showed a lower degree of toxicity. There was a death reported in Homestead, Fla., of a woman who had eaten a puffer, possibly a northern puffer (Halstead 1967). Puffers from Chesapeake Bay, adjacent waters, and off Morehead City, N.C., were found to be nontoxic (Robinson and Schwartz 1968).

PORCUPINEFISHES. FAMILY DIODONTIDAE

GRACE KLEIN-MACPHEE

Porcupinefishes can inflate their bodies like puffers but are they covered with well-developed sharp spines. In many species of porcupinefishes, the spines are erect only when the body is inflated, whereas in burrfishes, the spines, which are much shorter and stouter, are permanently erect. Teeth are fused with the jaw bones, which themselves are fused across the midline, with a single large tooth plate in both the upper and lower jaw. Pelvic fins are absent. The family comprises six genera and 19 species of marine fishes (Nelson 1994). Striped burrfish, *Chilomycterus schoepfi,* have been recorded from the Gulf of Maine.

STRIPED BURRFISH / *Chilomycterus schoepfi* (Walbaum 1792) /

Burrfish, Porcupinefish, Rabbitfish, Oysterfish / Bigelow and Schroeder 1953:527–528

Description. Body broad, oval, and slightly depressed (Fig. 316). Caudal peduncle small. Head short, broad, with a short, broad snout and large terminal mouth. Teeth incorporated into a continuous plate with a continuous edge and no median suture. Lateral line not apparent. Scales modified into stout, cover most of head and body. Dorsal fin rounded, inserted far back on body, just in front of caudal peduncle. Caudal fin long, narrow, and rounded. Anal fin under dorsal. Pectoral fins broad, truncate with rounded corners. Body proportions: head length 36.4–47.6% TL; eye 22.2–26.3% HL (Fowler 1906;

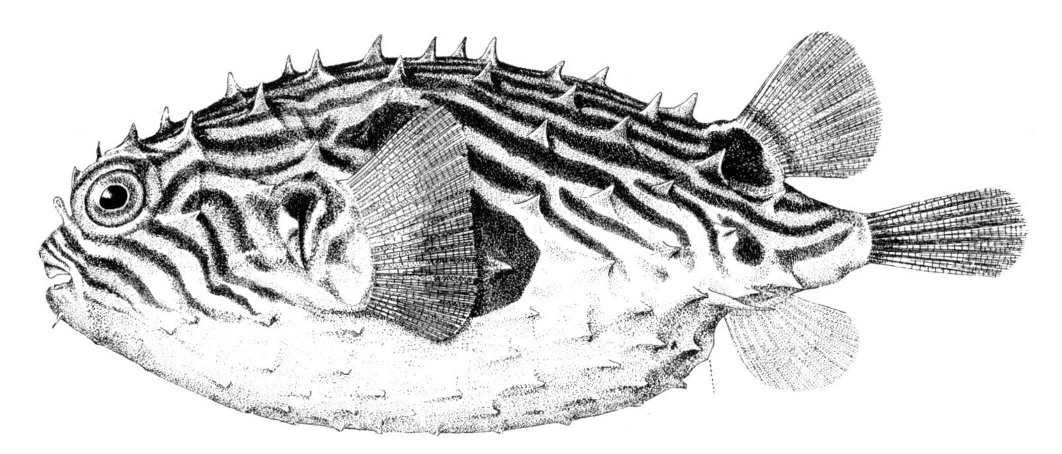

Figure 316. Striped burrfish *Chilomycterus schoepfi*. Noank, Conn. Drawn by H. L. Todd.

Meristics. Dorsal fin rays 10–12; anal fin rays 9–12; vertebrae 18–22 (Moffett 1957; Leim and Scott 1966; Miller and Jorgenson 1973).

Color. The background color varies from green to olive or brown above with a pale belly usually tinted with yellow or orange. The back and sides are irregularly striped with olive brown, dusky or black lines running downward and backward, roughly parallel to one another. There is a dark blotch on each side at the base of the dorsal fin, a smaller one between the latter and the anal fin, one above the base of each pectoral fin, and a fourth close behind the latter.

Size. Maximum length reported is 25.4 cm. The all-tackle game fish record is a 0.63-kg fish caught in Delaware Bay, N.J., in August 1989 (IGFA 2001).

Distinctions. Burrfish resemble puffer in the positions of the dorsal and anal fins, but the skin is armed with short, stout, three-rooted fixed spines instead of prickles; they have one broad incisor instead of two in each jaw; the pectoral fins are much larger and their upper edge is level with the upper corner of the gill openings, not below it; nostril openings are prolonged in a single tubular tentacle; and they are oval in outline, not fusiform.

Habits. Adults live most often on grass beds, but they are also found in bays and harbors, on shallow reefs, and along sandy beaches (Böhlke and Chaplin 1968). They usually occur in less than 18 m water but have been found to 91 m (Franks et al. 1972). They can tolerate salinities of 6.9–47 ppt and temperatures of 12.4°–38°C (Roessler 1970), but are able to survive in water even colder than 5.8°C (Moore 1976).

Food. They are known to eat hermit crabs, *Pagurus longicarpus* (Kuhlmann 1992) and snails, *Littorina irrorata* (Hamilton 1976).

Parasites. Commonly parasitized by a copepod, *Tucca impressus*, from Woods Hole south to Jamaica and the Gulf of Mexico (Ho 1968). A dinoflagellate (*Oodinium ocellatum*) sometimes attacks the skin, causing oozing of a reddish fluid (Nigrelli 1935).

Reproduction. A 190-mm female was reported with ripe eggs (Nichols and Breder 1927). Spawning is believed to be offshore (Springer and Woodburn 1960) in early spring (Miller 1965) or July (Nichols and Breder 1927), but individuals with nearly ripe gonads have been taken in October (Hildebrand and Schroeder 1928).

Early Life History. Unfertilized eggs are demersal, nonadhesive, and transparent, with an average diameter of 1.8 mm (Nichols and Breder 1927; Breder and Clark 1947), but when spawned they are probably pelagic (Leis 1978).

Little is known about this species except that they probably have a vesicular dermal sac that can be more or less inflated. Some Atlantic species of *Chilomycterus* have a presettlement stage known as "lyosphaera," which lacks dermal spines but has fleshy protuberances in the location the spines will occupy and other enlarged protuberances unassociated with spines (Evermann and Kendall 1898; Heck and Weinstein 1978; Humann 1994:331). In general, the larvae are deep and wide in head and trunk, the tail is small and compressed, the head is large and rounded, the gut coiled and massive, and the eye particularly large (Leis 1984).

General Range. Along the east coast of the United States from Florida to New York, occasionally to Cape Cod and Massachusetts Bay. The northernmost record is off Sambra near Halifax in 1896 (Piers 1899). They are most abundant from the Carolinas southward, and are also found in the Bahamas to the Gulf of Mexico and southeastern Brazil (Böhlke and Chaplin 1968).

Occurrence in the Gulf of Maine. The only published Gulf records of burrfish are of an individual taken in Massachusetts Bay (Kendall 1908) as noted by Bigelow and Schroeder and a 159-mm specimen caught in a West Point, Maine fish trap in August 1949 (Scattergood et al. 1951).

OCEAN SUNFISHES. FAMILY MOLIDAE

GRACE KLEIN-MACPHEE

The teeth of ocean sunfish are incorporated with the jaw bone as one large plate in both the upper and the lower jaw. The caudal fin is absent but rays of the dorsal and anal fins may extend around the posterior part of the body to form a pseudocaudal fin or clavus (Gregory and Raven 1934). There is neither a lateral line nor a swim bladder. They have 16–18 vertebrae. The major locomotory thrust is provided by the dorsal and anal fins. The family comprises three monotypic marine genera (Nelson 1994), of which two are reported from the Gulf of Maine.

KEY TO GULF OF MAINE OCEAN SUNFISHES

1a. No caudal fin evident . . . . **Ocean sunfish**
1b. Pseudocaudal fin present . . . . **Sharptail sunfish**

SHARPTAIL SUNFISH / *Masturus lanceolatus* (Liénard 1840) / Bigelow and Schroeder 1953:531–532

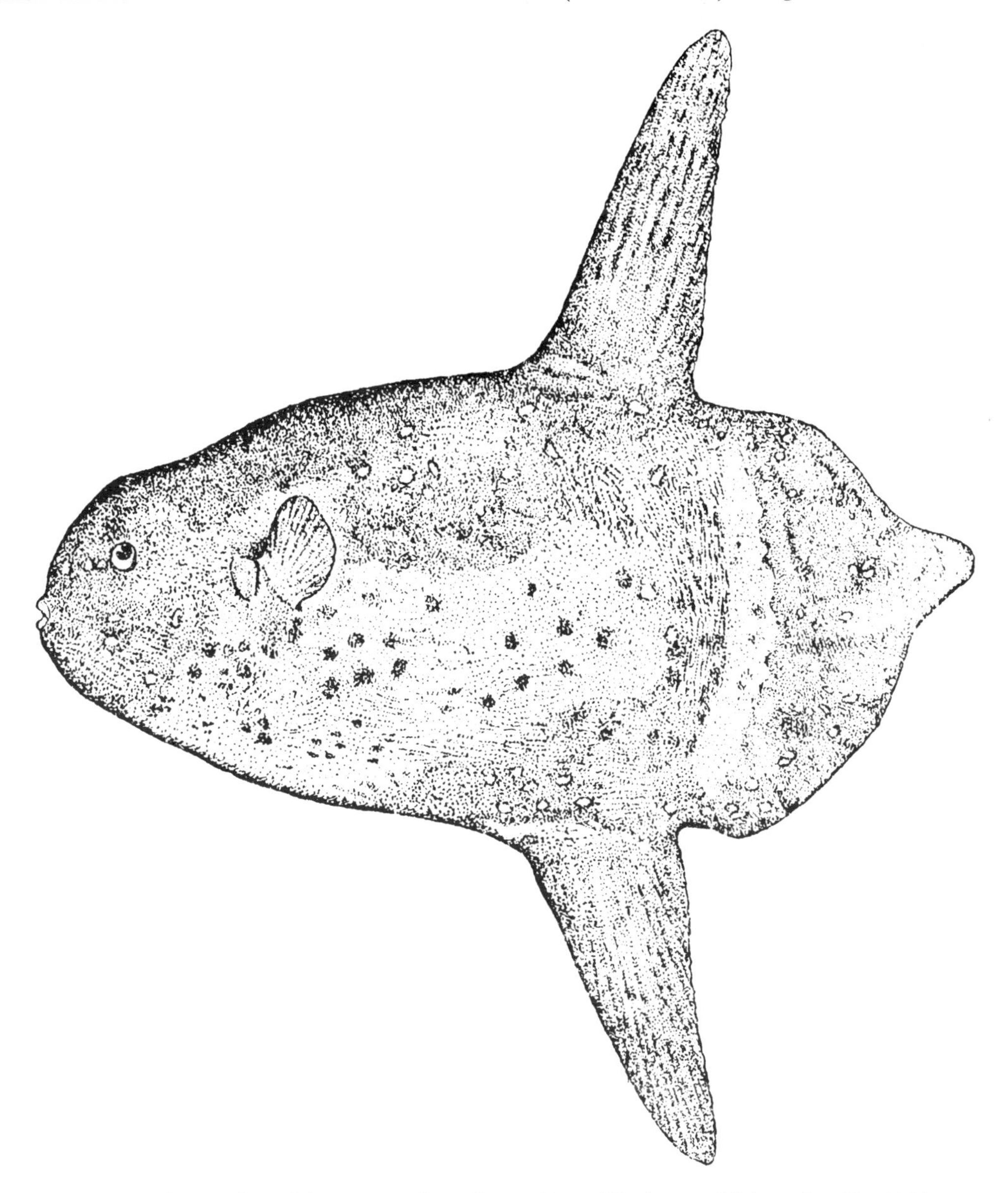

Description. Body compressed, elongate, and ovoid (Fig. 317). Head wider than body without ridges above or below eyes. Snout rounded, interorbital region broadly convex. Eye oblong with a nictitating membrane. Nostrils inconspicuous. Mouth small, terminal. Tongue broad, hard, not free. Gill openings reduced, semicircular, and directed backward (Smedley 1932); gill rakers concealed in thick skin. Pectoral fins short and rounded. Pelvic fins and pelvic girdle absent. Dorsal and anal fins similar and opposed, both united with clavus by broad membranes. Dorsal and ventral portions of clavus meet in a median fleshy flap that contains rays unsupported by radial elements (Collett 1896; Gudger 1935b). Claval rays not branched distally in adults; all median fin rays obscured by thick integument. Body covered with thick collagenous hide bearing fine integumentary scutes with a narrow oblong base and several conical, multipointed cusps (Howell-Rivero 1936). Skin not covered with mucus (Brimley 1939). Skeleton poorly ossified. Ovary single (Glover 1966). Proportions are as follows: distal tip of dorsal fin to distal tip of anal fin about 140–154% PCL (preclaval length), greatest depth of body 54.6% PCL, pectoral fin length 12.1–14.2% PCL, snout 80% HL, eye 4.3–5.8% PCL, greatest body width 20.2–29% PCL, horizontal length of clavus 232.1% TL (Collett 1896; Fowler 1928; Smedley 1932; Brimley 1939; Funderburg and Eaton 1952).

Meristics. Dorsal fin rays 18–20; pectoral fin rays 7–10; clavus 4–8 rays in median lobe, 6 or 7 above lobe (derived from dorsal fin), and 7–11 below the lobe (derived from anal fin), total dorsal + anal + caudal 55–64; branchiostegals 6; vertebrae 8 + 8 = 16 (Fowler 1928; Gudger 1935b; Raven 1939; Fraser-Brunner 1951; Funderburg and Eaton 1952; Glover 1966).

Color. Entire trunk is more or less silvery; upper parts of the sides grayish brown to black or blue-black, lower parts paler; sides either plain or variously marked with ill-defined dark spots; dorsal and anal fins with slaty dark spots; caudal fin sometimes with pale blotches. Metallic reflections on the sides and blue of claval spots are intense in life; lips and interior of mouth dark rose with a trace of purple in life (Howell-Rivero 1936).

Size. Appear to grow as large as the more common *Mola,* perhaps even larger. The largest reported specimens have been from the Atlantic, 259.1 cm TL from Cuba (Palmer 1936) and 254 cm TL from Miami (Hubbs and Giovanoli 1931).

Distinctions. Sharptail sunfish can be distinguished from other Gulf of Maine species by their huge size, truncate body, and high dorsal and anal fins. They differ from common sunfish chiefly in the fact that the rear margin of the body is edged by a short but evident pseudocaudal fin of 18–20 soft rays, which extends from close behind the dorsal fin to close behind the anal fin, with a lobelike, blunt-tipped projection a little above the midlevel of the body. Scales are much finer and less evident to the touch than those of *Mola.*

Habits. Adults and juveniles are oceanic in nature but come close inshore on occasion, even into estuarine situations; considered epipelagic, but occasionally found at great depths. A sharptail sunfish was encountered by observers in the *Johnson Sea Link* at 670 m, and it may be more than a transient member of the midwater community (Harbison 1987). Sharptail sunfish often have remoras attached to them (Gudger 1935b), even inside the gill cavity (Palmer 1936).

Food. Little is known regarding feeding habits although they are probably similar to those of ocean sunfish. Benthic sponges and annelids have been found in the stomachs of young fish (Yabe 1953).

General Range. The sharptail sunfish, like *Mola mola,* appears to be cosmopolitan in tropical–warm temperate latitudes. In the western Atlantic reported from near Havana, Cuba (seven specimens), east coast of Florida (nine specimens), and Nova Scotia to North Carolina. Young have been taken off the Azores, in the Sargasso Sea, west of the Canaries, in the Caribbean, and in the Gulf of Mexico.

Occurrence in the Gulf of Maine. The only record for sharptail sunfish in the Gulf is for four young, about 5 cm long, that were taken many years ago in Massachusetts Bay. These were originally reported by Putnam (1871) as young ocean sunfish, but have been shown by subsequent workers to be sharptail sunfish (Bigelow and Schroeder). The nearest locality record for an adult of the species to date is from Pamlico Sound, N.C.

OCEAN SUNFISH / *Mola mola* (Linnaeus 1758) / Bigelow and Schroeder 1953:529–531

Description. Body oblong, suggesting head and fore trunk of some enormous fish cut off short, truncate immediately back of dorsal and anal fins; caudal peduncle absent (Fig. 318). Tapering in front of fins toward snout so that forward half of trunk is oval in profile. Body less than twice as long as deep, strongly compressed. Snout overhangs upper jaw as a kind of rough, mobile wart or pad, nearly vertical in females but projecting in males (Fraser-Brunner 1951). Mouth very small, at tip of snout. Teeth completely united in each jaw. Teeth present on fourth pharyngeal bone. Eye small, oblong with its major axis longitudinal, very movable in its orbit (Storer 1863; Whitley 1931), with well-developed nictitating membrane (Gregory and Raven 1934) in line with mouth. Gill openings remarkably short. Broad supraoccipital ridge extends from

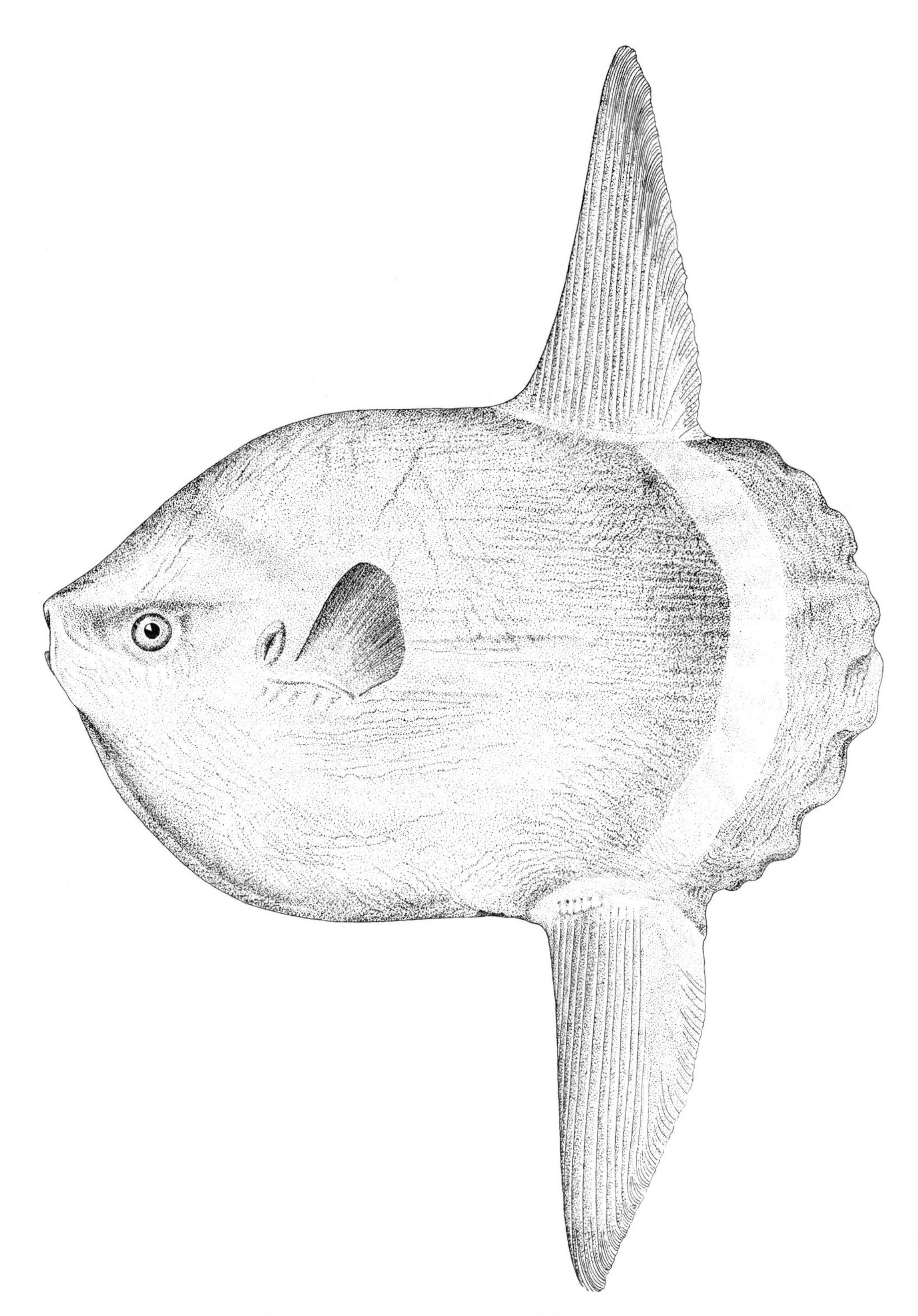

Figure 318. Ocean sunfish *Mola mola.* Drawn by H. L. Todd.

snout back over eyes; a less pronounced ridge on cheek extending back below pectoral fin (Whitley 1931). No spines in dorsal or anal fins. Dorsal fin located over anal fin close behind midlength of fish. Both fins very much higher than long, triangular with sharply rounded tips, seventh ray longest. Clavus extends around whole posterior margin of body. Clavus confluent with dorsal and anal fins in young and hardly separated from them in adult, so short its rays are hidden by thick opaque skin, making it look more like a fold of skin than a typical fin. General outline rounded, paralleling rear outline of body, but margin scalloped, with a rounded bony prominence or knob in line with each caudal ray and with a notch between every two of these prominences. Pectoral fins small, rounded, situated about halfway up body close behind tiny gill slits.

Skin unusually thick, very tough, elastic in texture, criss-crossed with low ridges. Fins and trunk covered with small bony tubercles or scutes, polygonal in shape, each with a rugose central spine projecting straight outward (Whitley 1931), similar to shark skin. Skeleton poorly ossified (Gregory and Raven 1934). Typical teleost otoliths absent; maculae containing instead aggregations of small, white, rounded otoconia (Thompson 1888). Ovary single (Cleland 1862). Body proportions: distal tip of dorsal fin to distal tip of anal fin 98.5–137.0% TL, greater in females, decreasing with increasing TL in both sexes; body depth (dorsal fin origin to anal fin origin) 50.0–65.0% TL; pectoral fin length 9.2–15.0% TL; head 25.0–30.0% TL; eye 2.8–4.2% TL; horizontal length of clavus 8.9–22.0% TL (Heilner 1920; Whitley 1931; Roon and ter Pelkwijk 1939; Roon 1942; Holt 1965; Anderson and Cupka 1973).

Meristics. Dorsal fin rays 15–20, modally 18 or 19; anal fin rays 14–18, modally 16–18; pectoral fin rays 11–13; clavus (pseudocaudal) 12–15, males with 8–11 distal claval notches; total dorsal, anal, and clavus rays 40–52, mean 47.4; branchiostegals 6; vertebrae 17 (Steenstrup and Lütken 1898; Schmidt 1921; Gregory and Raven 1934; Fraser-Brunner 1951; Anderson and Cupka 1973; Tyler 1980).

Color. Dark gray above, the back with a brownish to blackish cast, sides paler with silvery reflections, belly dusky to dirty white; small round or elliptical bright spots on the sides of males (Lidth de Jeude 1892). Some descriptions mention a broad black bar along bases of the dorsal and anal fins; median fins colored like adjacent part of body (Storer 1863); lips and area around nostrils black; inside of mouth pink; teeth dirty white (Whitley 1931); pupil of eye black, iris brown or smoky, with an inner silver ring around pupil (Storer 1863).

Size. Ocean sunfish reach a great size. The record capture is of one 367.7 cm long off Avalon, Cal. (Heilner 1920); Jordan and Evermann (1898) recorded another California specimen 251 cm and about 816 kg; and a 244-cm specimen was taken off Cape Lookout, N.C., in 1904, but large ones such as these are exceptional, the usual size being 91.4–152.4 cm and 79.5–227 kg.

Distinctions. Ocean sunfish are easily recognized by their large size, compressed truncate form, high matching dorsal and anal fins, rough leathery skin, porelike gill opening, and absence of pelvic fins. They differ from sharptail sunfish in that the latter have a projecting lobe in the middle of the clavus.

Habits. Ocean sunfish are wanderers in the high seas, drifting with ocean currents, but capable of strong swimming. Although seen only infrequently as single individuals or in pairs at or near the surface, up to a dozen were encountered in a day off Nahant (Binney 1842). They have been observed by scuba divers at about 40 m on the bottom in the Mediterranean Sea, motionless with the head directed upward (Roghi 1961).

The dorsal and anal fins cannot be laid back as they can in most bony fishes, and the sunfish propels itself along by waving them from side to side. When unlucky vagrants are sighted in cool northern waters they have usually been chilled into partial insensibility. They float awash on the surface, feebly fanning with one or the other fin, the personification of helplessness. Usually they pay no attention to the approach of a boat, but may come to life with surprising suddenness and sound swiftly, sculling with strong fin strokes, when pursued. When one is struck it struggles and thrashes vigorously while the tackle is being slung to hoist it aboard, suggesting they are far more active in their native haunts than their feeble movements in fatally cold surroundings might suggest (Bigelow and Schroeder). In support of this, the gill area of *Mola mola* falls within the range of the intermediate activity group of fishes (Adenay and Hughes 1977).

Food. Ocean sunfish have an unusual diet, for as a rule the stomach contains jellyfish, Portuguese man-o-war, ctenophores, or salps, or a slimy liquid that probably represents partially digested remains of these. This was true of a sunfish brought in to Woods Hole, but various crustacean, molluscan, hydroid, and serpent-star remains; bits of algae and eelgrass (*Zostera*); fishes, including flounder and a gadid, *Molva macrophthalmus;* and leptocephalus larvae have been found in sunfish stomachs in European waters, proving that at times they feed either on the bottom in shoal water or among patches of floating weeds (Norman and Fraser 1949).

Predators. Nothing is known of predators on adult ocean sunfish, but their large size and tough skin are probably serious deterrents. Prejuveniles have been reported from albacore and dolphin, *Coryphaena,* stomachs (Kuronuma 1940).

Parasites. Ocean sunfish are host to a great variety of parasites, external and internal, with copepods and trematodes clinging to its skin and infesting its gills, its muscles harboring round worms, and various round and flat worms inhabiting its intestines. A list of known parasites includes: cestodes, *Floriceps saccatus* and *Molicola horridus* (Andersen 1987); monogenean *Capsala martinieri;* digenean *Cirkennedya poilockensis* (Gibson and Bray 1979); copepods *Cecrops latereilli, Lepeophtheirus nortmanni, L.* sp., *Orthagoriscicola muricatus, Pennella filosa,* and *Philorthagoriscus serratus* (Margolis and Kabata 1984, 1988). Ocean sunfish frequent "cleaning stations," where a variety of fishes pick at surface parasites (Gottschall 1967).

Breeding Habits. A female 150 cm long was estimated to contain about 300 million eggs (Schmidt 1921), making ocean sunfish one of the most fecund fishes. There is no direct information on spawning location or seasonality. Prejuveniles have been taken in Japan in March, August, and September (Sokolovskaya and Sokolovskaya 1975) and larvae off Brazil in January and February (Schmidt 1921).

Early Life History. Fertilized eggs have not been described, but they are probably pelagic and large, with multiple oil droplets like other molid eggs (Leis 1984). The smallest larva described was 5 mm TL and was round with stout spines, a huge eye, and relatively normal tail. The tail atrophies and a true caudal fin never forms. The clavus is evident at 11 mm. Young differ from their parents in being armed with eight short, stout spines on either side and with a single median row of four spines along the back and seven along the ventral margin of the body. As the massive spines decrease in size, smaller spines form, as well as ossifications within the skin, which eventually make up the carapace-like skin covering. The genus *Mola* passes through a fairly long ontogenetic stage between larvae and juveniles, which is characterized by retention of reduced massive spines; a deep, compressed body with a ventral keel; and a shape quite different from the adult (Leis 1984). A more detailed description of juvenile stages is given in Martin and Drewry (1978).

General Range. Ocean sunfish are oceanic and cosmopolitan in tropical and temperate seas. In the western Atlantic, known from the Newfoundland banks, Gulf of St. Lawrence, and the outer coast of Nova Scotia south to the Gulf of Mexico and Argentina (Briggs 1958).

Occurrence in the Gulf of Maine. Ocean sunfish are only stray visitors to the Gulf, which they enter now and then from warmer waters outside the continental slope. There are published records from Saint John Harbor, N.B., near Birch Harbor; near Seguin Island; off Small Point; and off Cape Elizabeth, Maine, where they have been reported repeatedly; off Cape Ann; and from various localities in Massachusetts Bay. They have even been seen in Boston Harbor: one 124 cm long, weighing 234.5 kg was caught on 14 August 1922; and one 1.5 m long was killed in a narrow creek in Quincy, Mass. (Bigelow and Schroeder).

Importance. Ocean sunfish stray into the Gulf of Maine and are of no commercial or recreational significance. Although the family has many toxic members, this species showed no traces of tetraodotoxin or its derivatives in its flesh or organs (Saito et al. 1991).

ACKNOWLEDGMENT. Drafts of the tetraodontiform section were reviewed by James Tyler.

DISTRIBUTION MAPS

Distribution maps for 38 species are included in this appendix. Map 1 shows the coverage of the fall bottom trawl stations occupied by NEFSC, 1968–1996. Some maps are based on the spring surveys and some on the fall surveys; for some rarer species, data for the spring and fall surveys have been combined. The maps are grouped together in this appendix to facilitate comparison of different distribution patterns, for example, shallow-water versus deepwater species. Distribution plots of 99 species of groundfishes from Cape Chidley, Labrador, to Cape Hatteras, N.C., based on NMFS and Department of Fisheries and Oceans data, are available in the Groundfish Atlas at http://www-orca.nos.noaa.gov/projects/ecnasap/ecnasap.html.

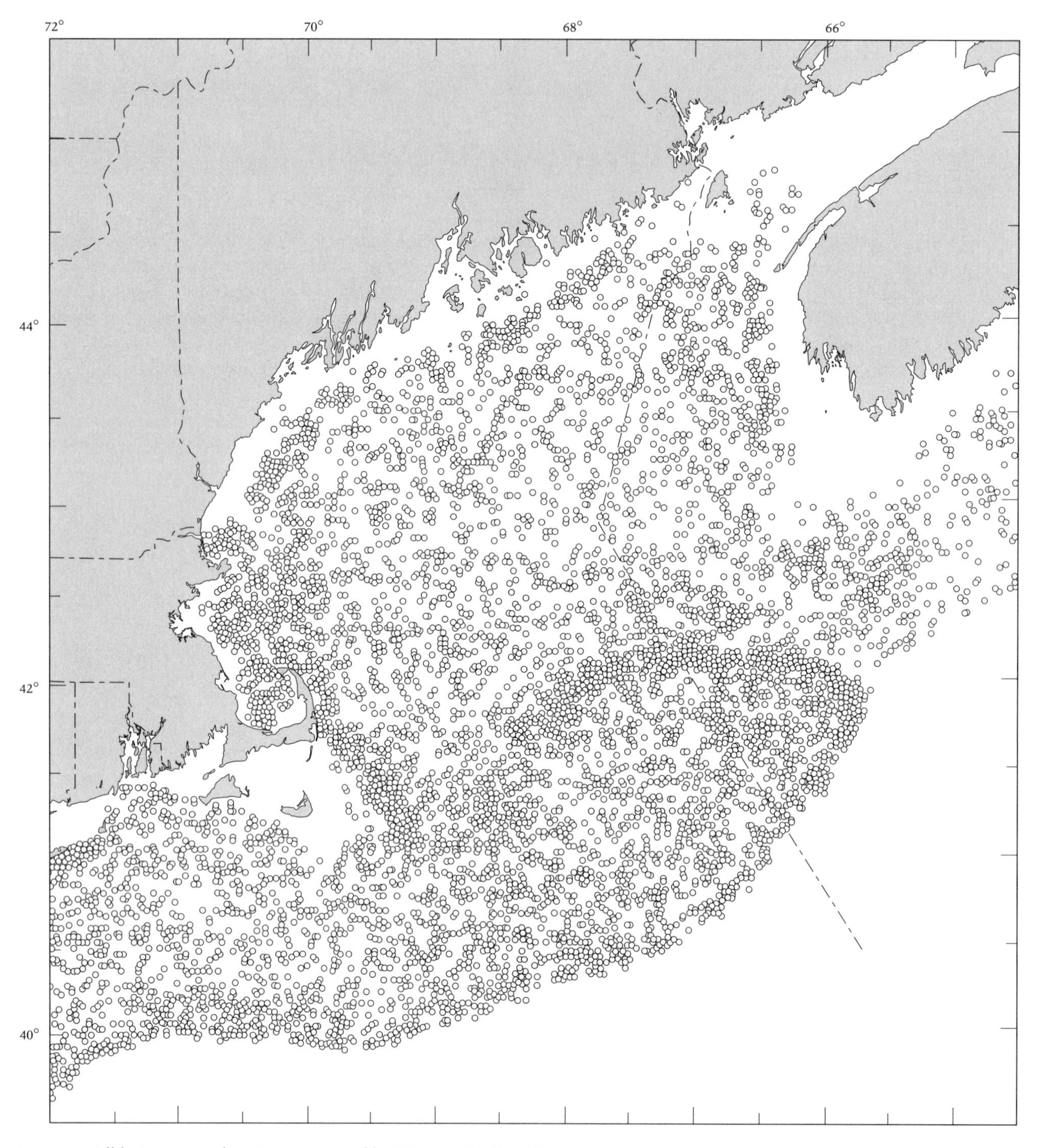

Map 1. Fall bottom trawl stations occupied by NEFSC, 1968–1996.

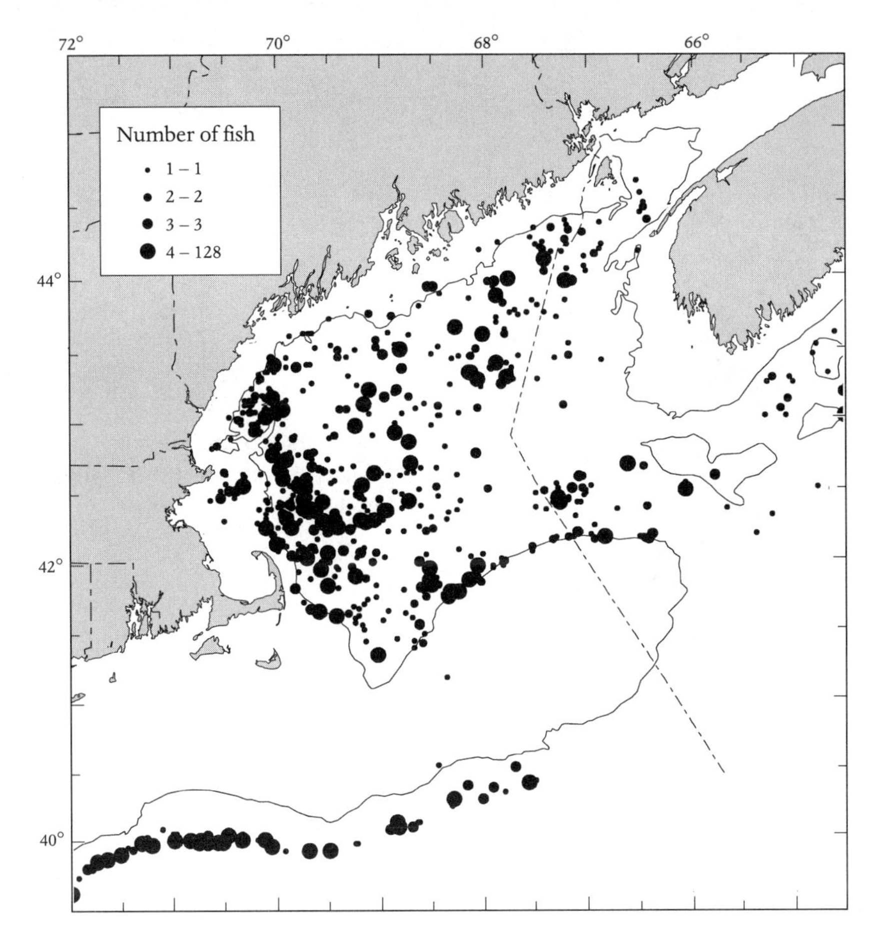

Map 2. Distribution of Atlantic hagfish *Myxine glutinosa,* based on NEFSC fall and spring bottom trawl surveys, 1968–1996.

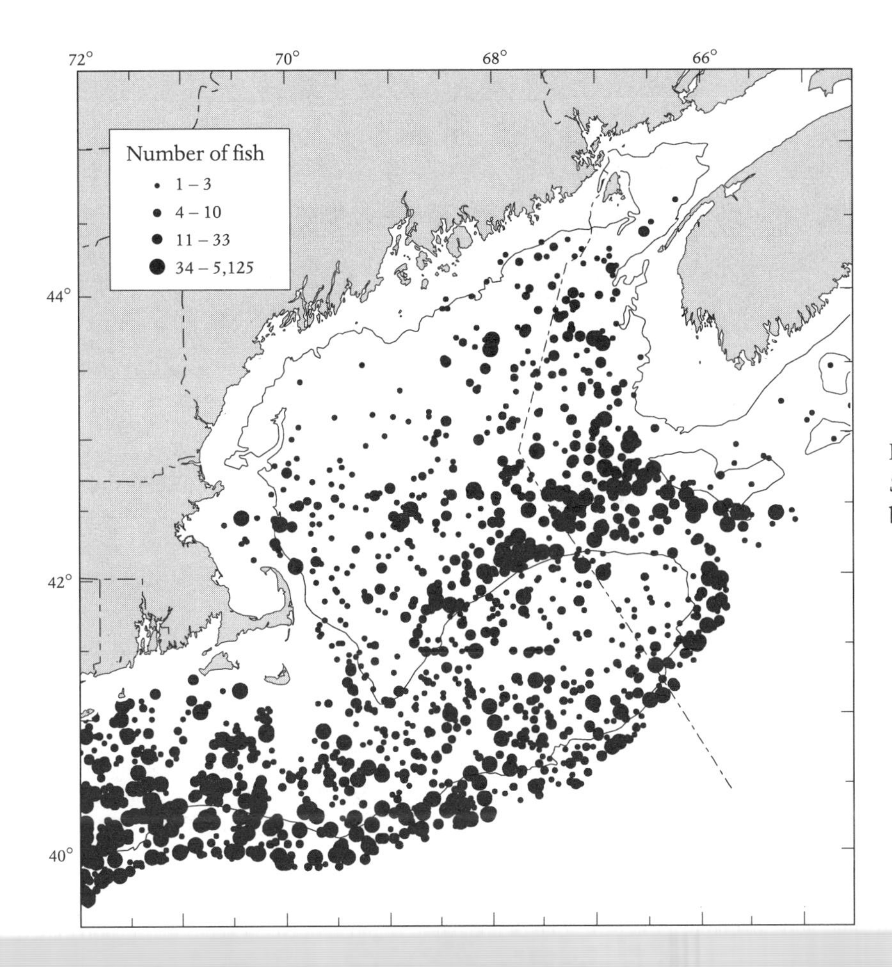

Map 3. Distribution of spiny dogfish *Squalus acanthias,* based on NEFSC fall bottom trawl surveys, 1968–1996.

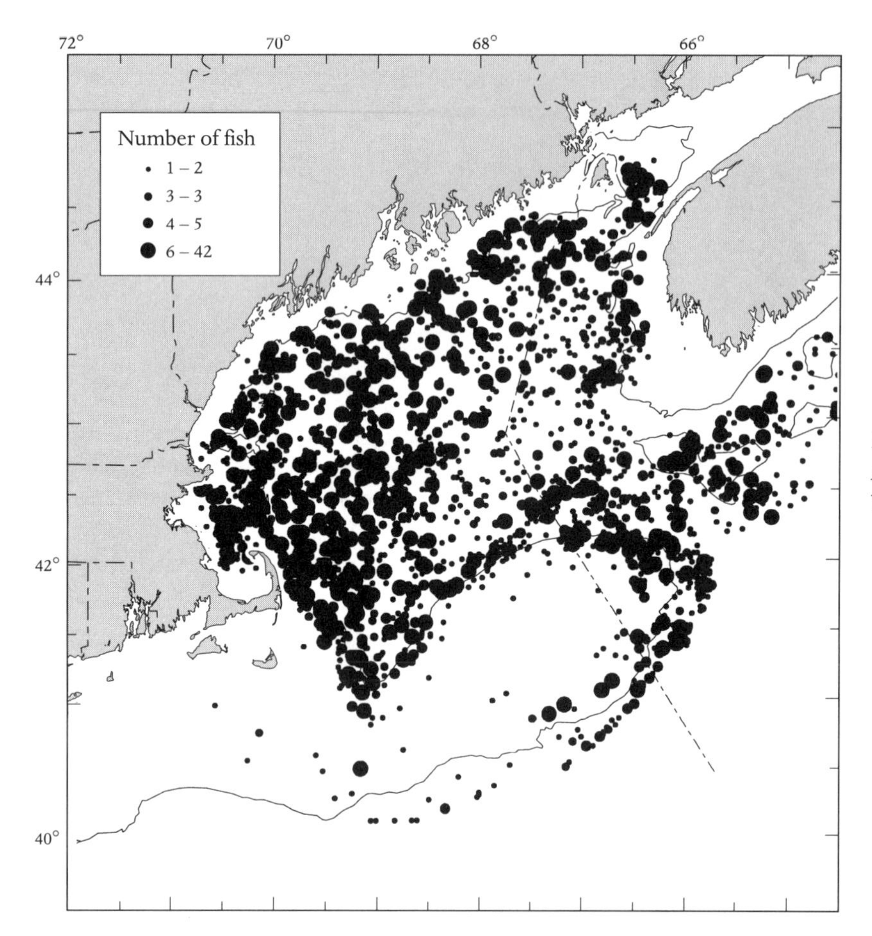

Map 4. Distribution of thorny skate *Amblyraja radiata,* based on NEFSC spring bottom trawl surveys, 1968–1996.

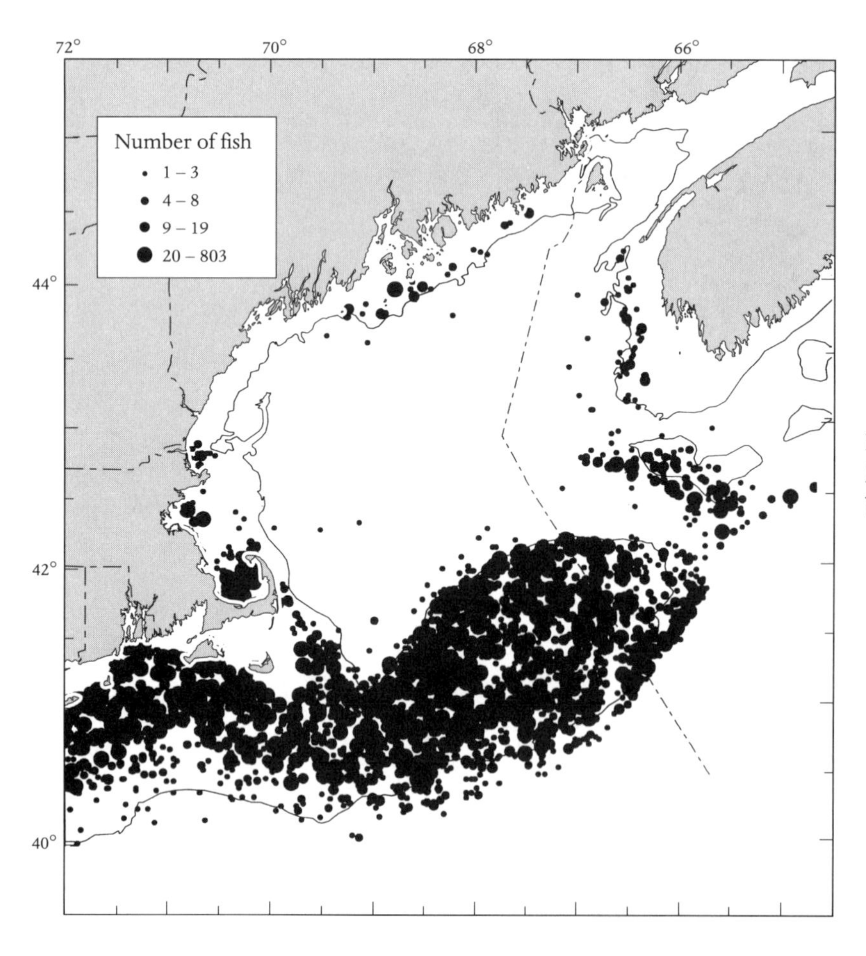

Map 5. Distribution of little skate *Leucoraja erinacea,* based on NEFSC fall bottom trawl surveys, 1968–1996.

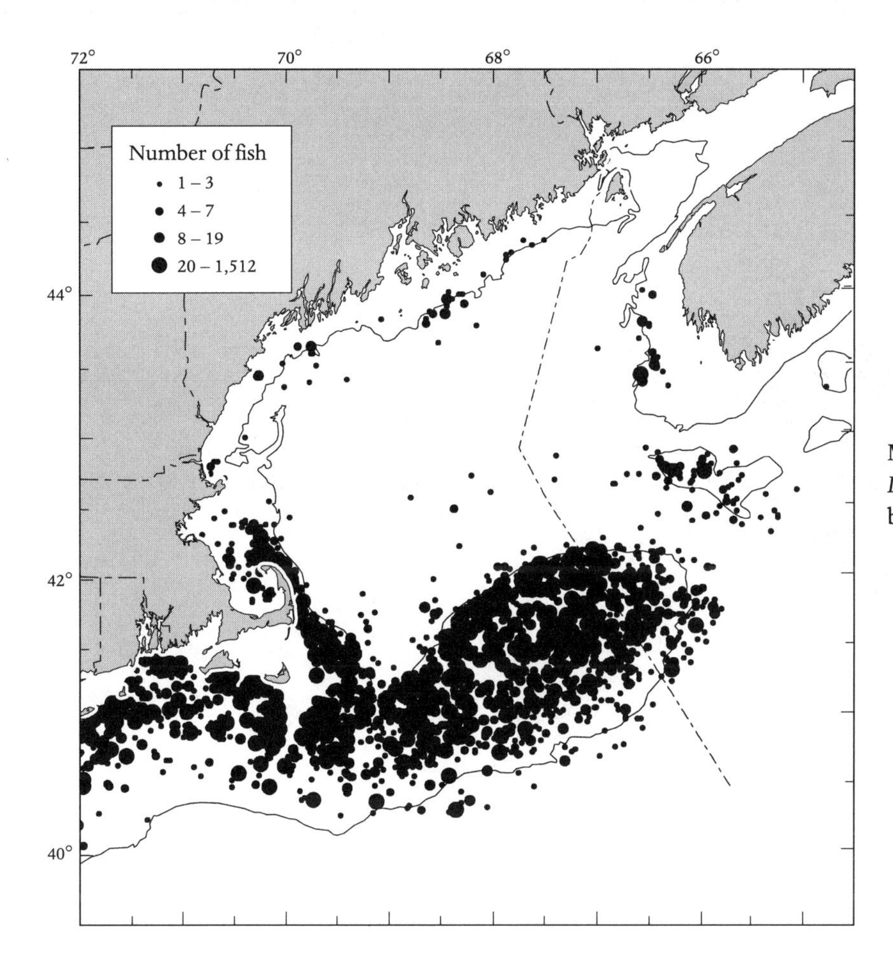

Map 6. Distribution of winter skate *Leucoraja ocellata,* based on NEFSC spring bottom trawl surveys, 1968–1996.

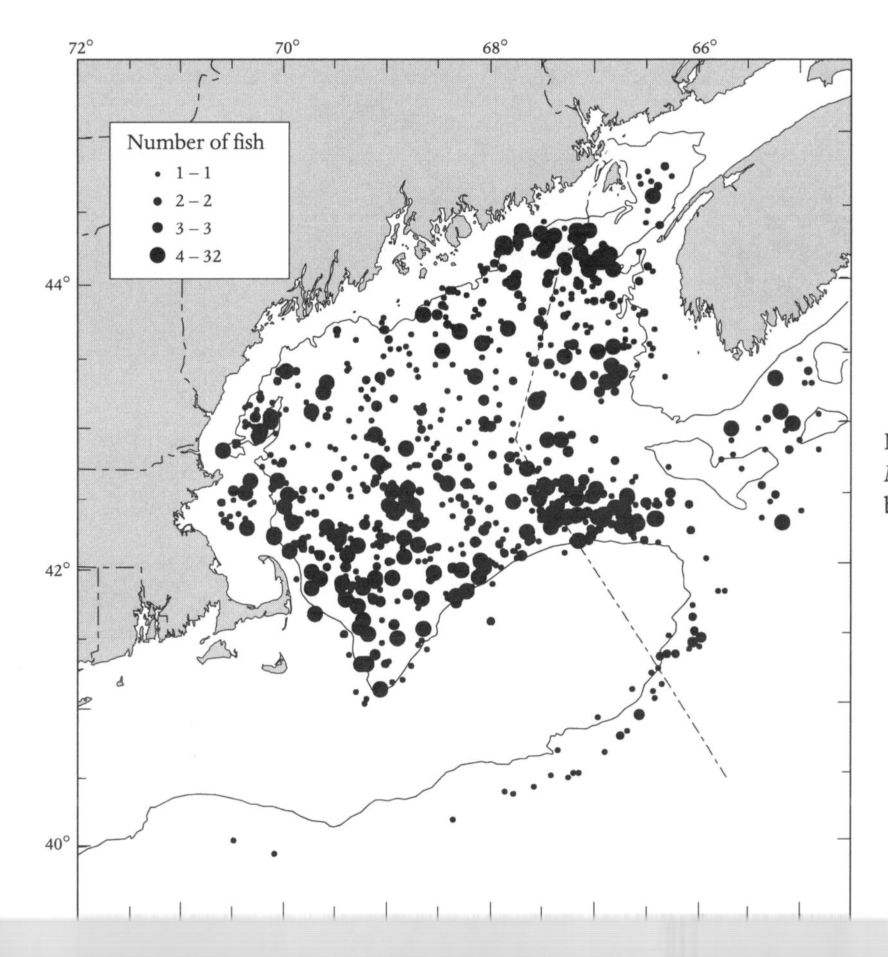

Map 7. Distribution of smooth skate *Malacoraja senta,* based on NEFSC spring bottom trawl surveys, 1968–1996.

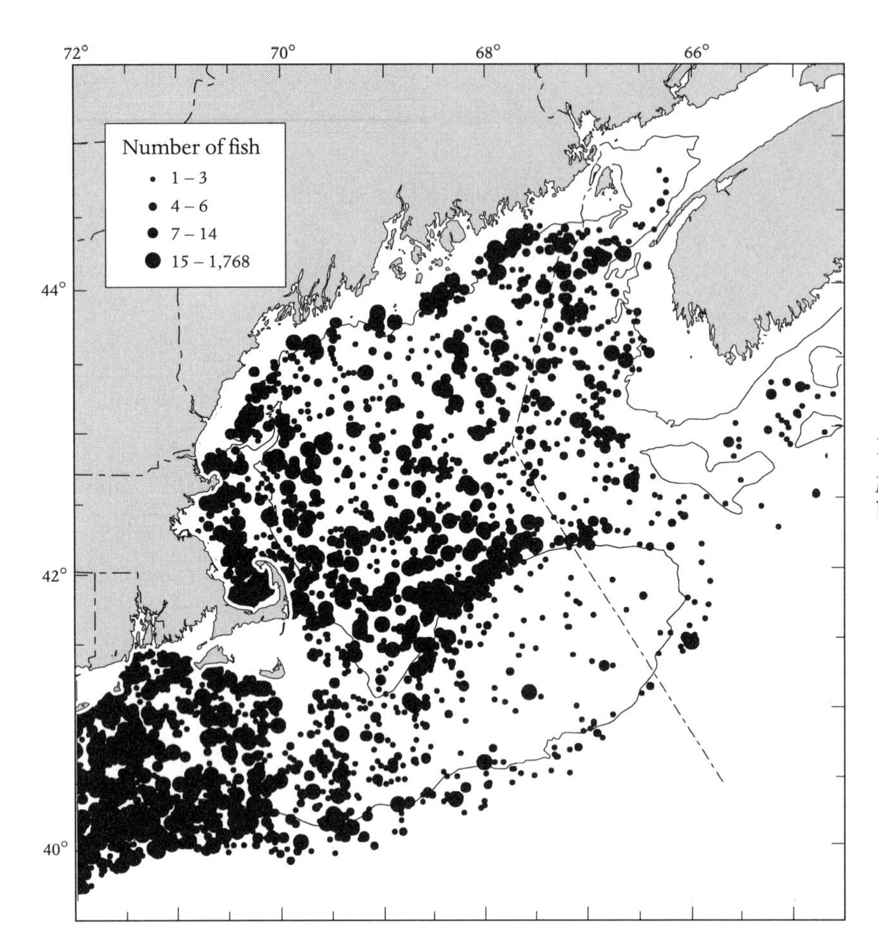

Map 8. Distribution of alewife *Alosa pseudoharengus,* based on NEFSC spring bottom trawl surveys, 1968–1996.

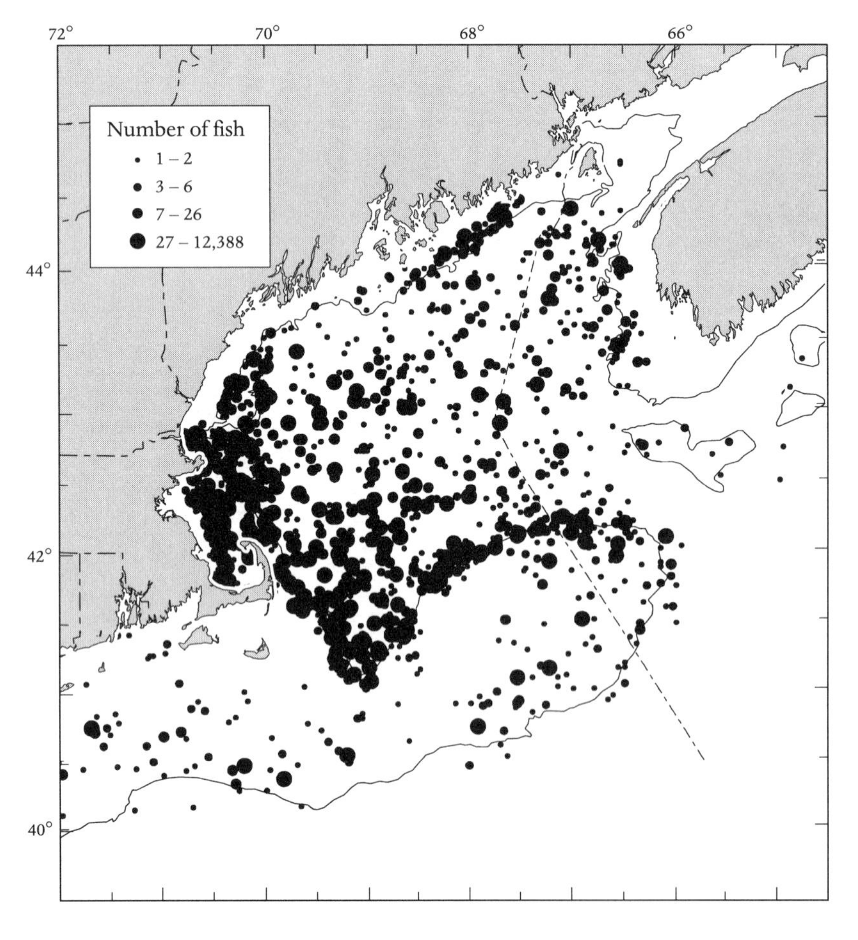

Map 9. Distribution of Atlantic herring *Clupea harengus,* based on NEFSC fall bottom trawl surveys, 1968–1996.

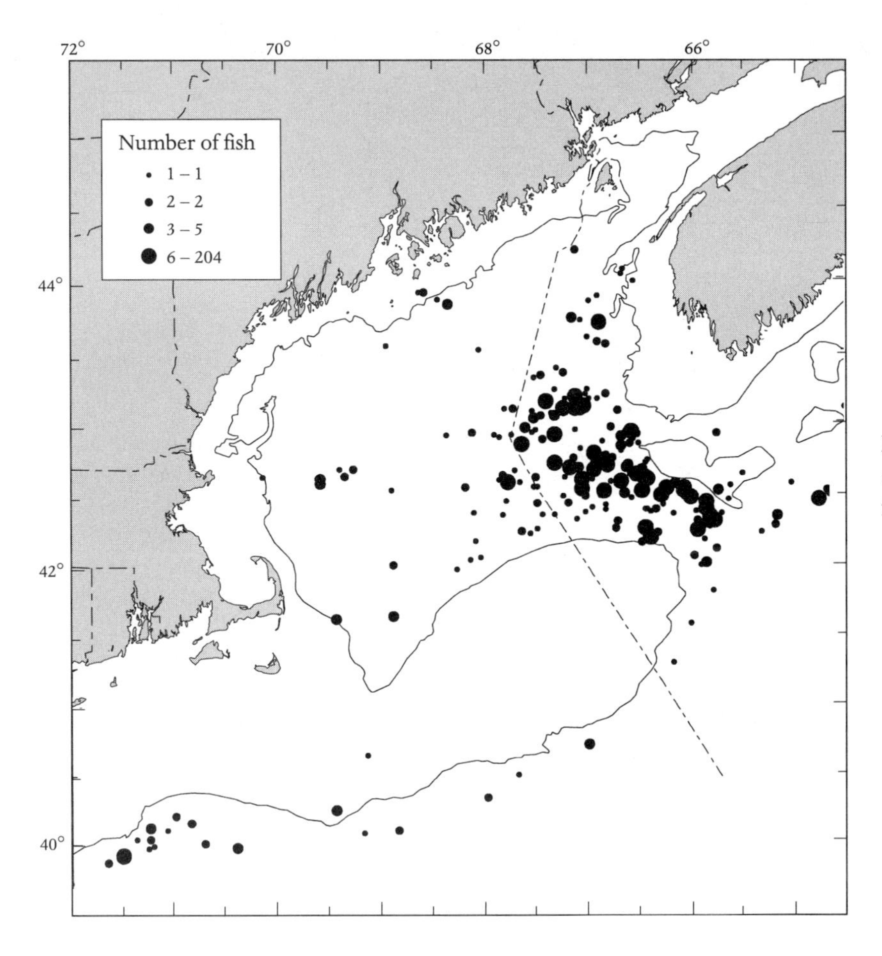

Map 10. Distribution of Atlantic argentine *Argentina silus,* based on NEFSC fall bottom trawl surveys, 1968–1996.

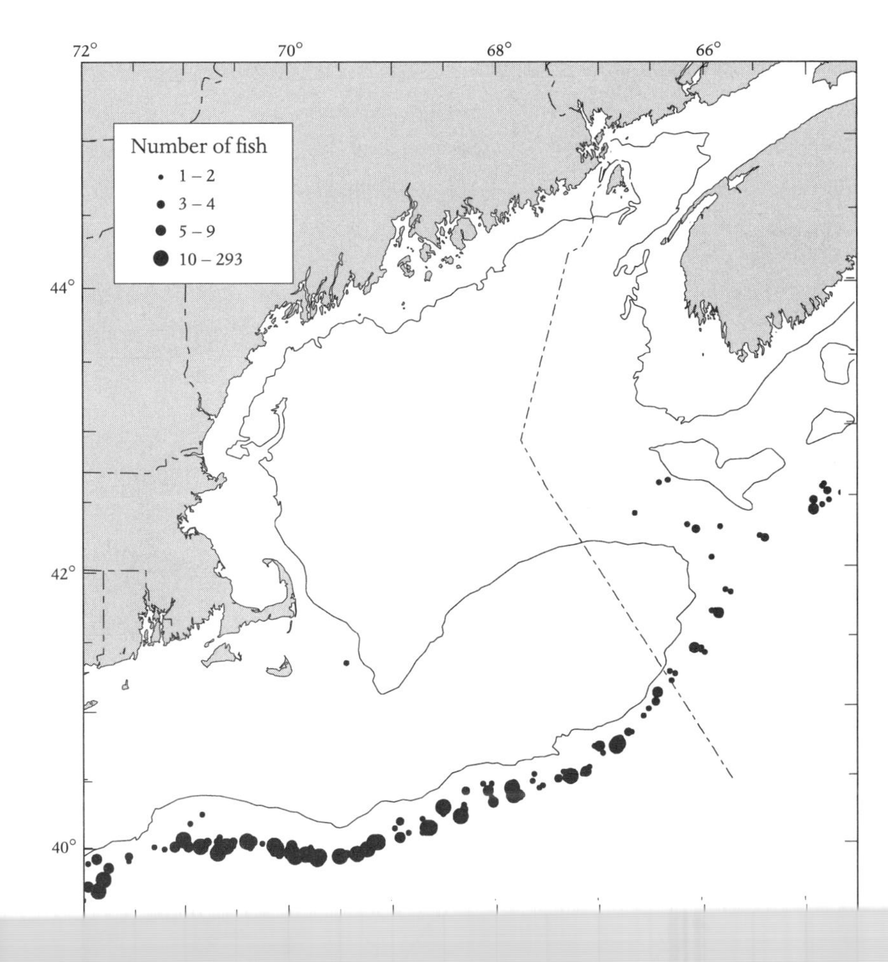

Map 11. Distribution of shortnose greeneye *Chlorophthalmus agassizi,* based on NMFS spring and fall bottom trawl surveys, 1968–1996.

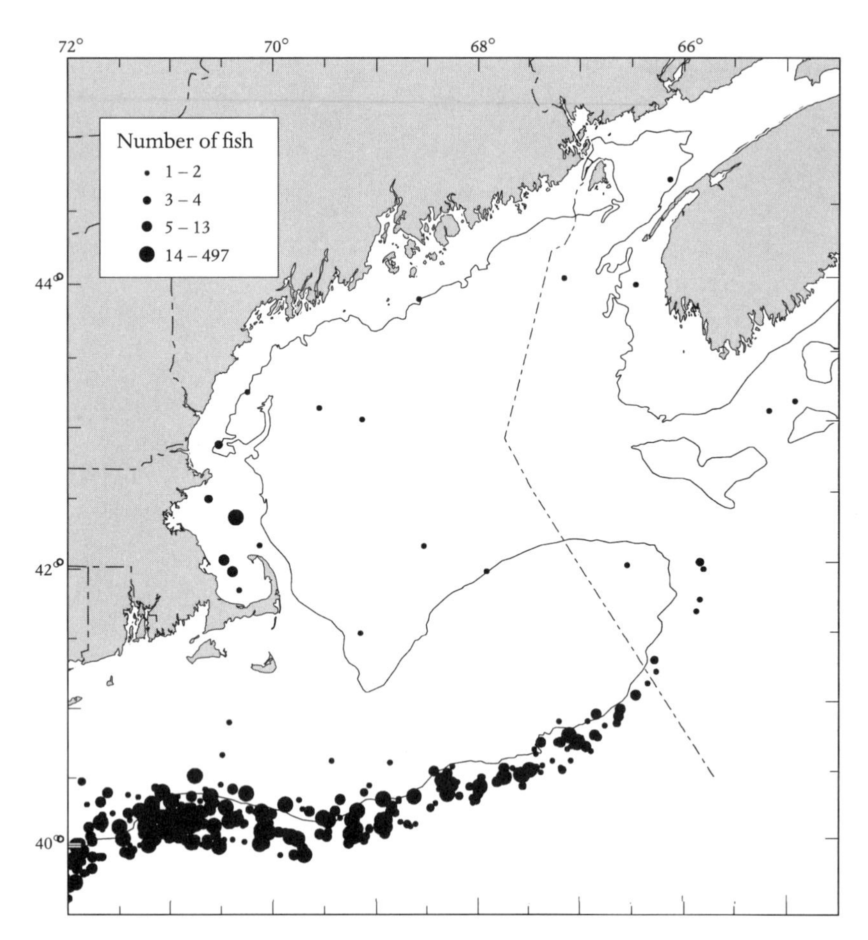

Map 12. Distribution of fawn cusk-eel *Lepophidium profundorum,* based on NMFS spring bottom trawl surveys, 1968–1996.

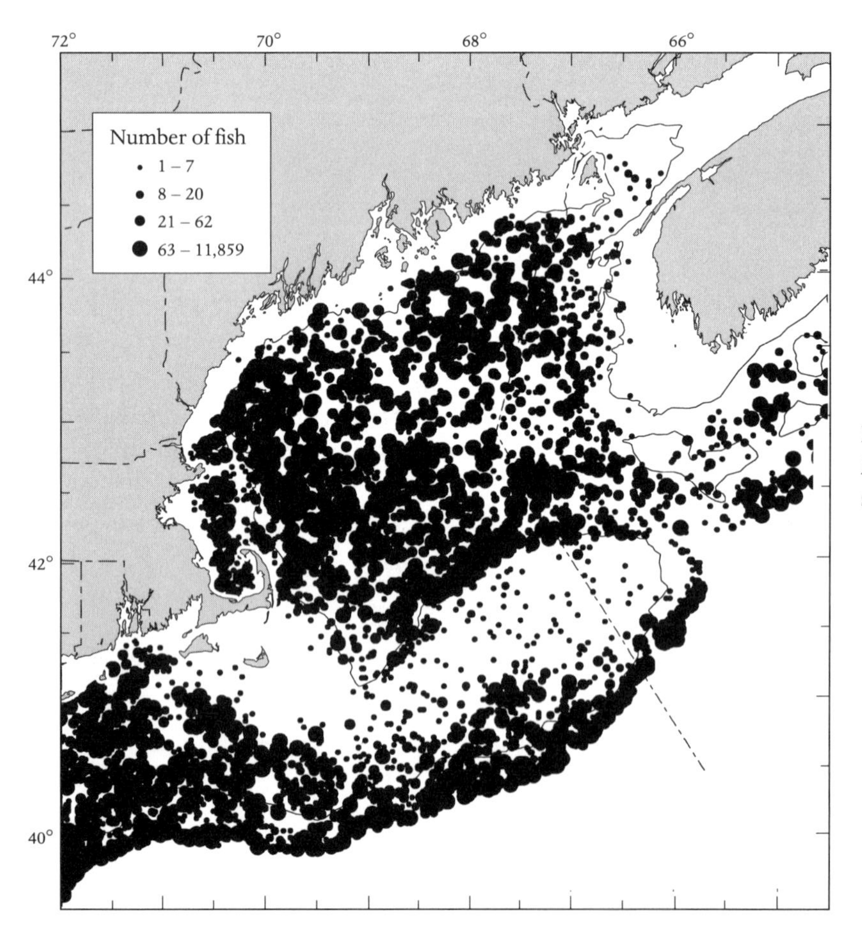

Map 13. Distribution of silver hake *Merluccius bilinearis,* based on NEFSC spring bottom trawl surveys, 1968–1996.

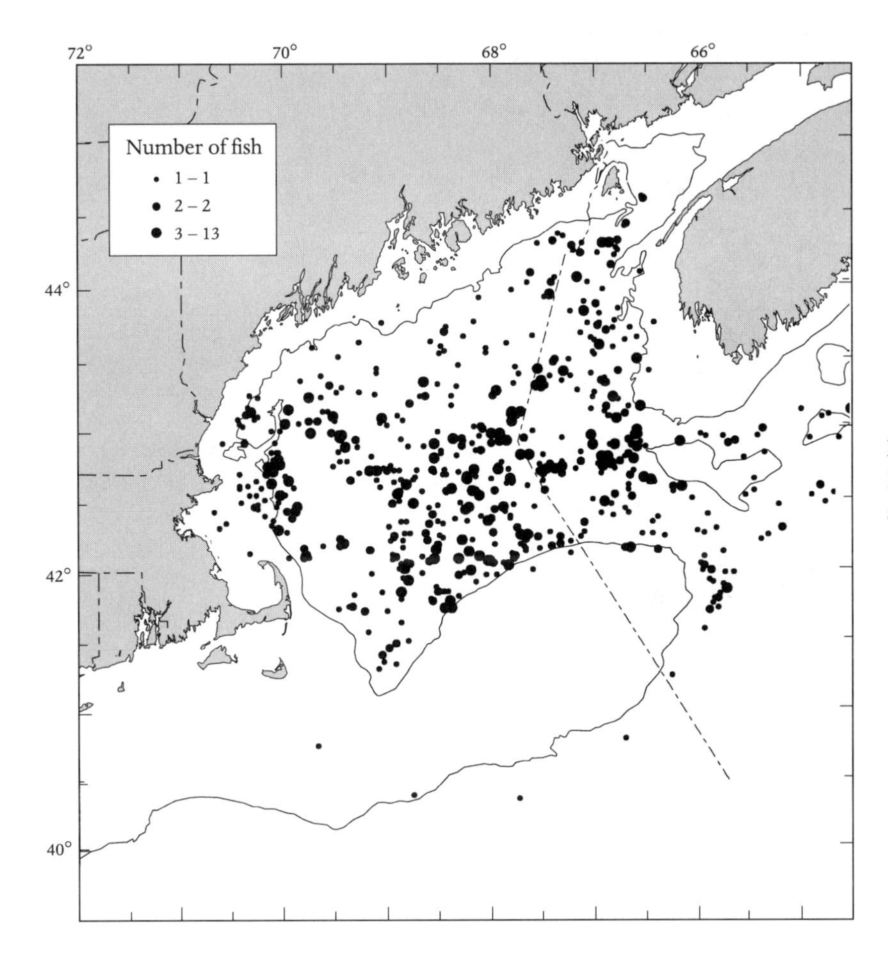

Map 14. Distribution of cusk *Brosme brosme,* based on NEFSC fall bottom trawl surveys, 1968–1996.

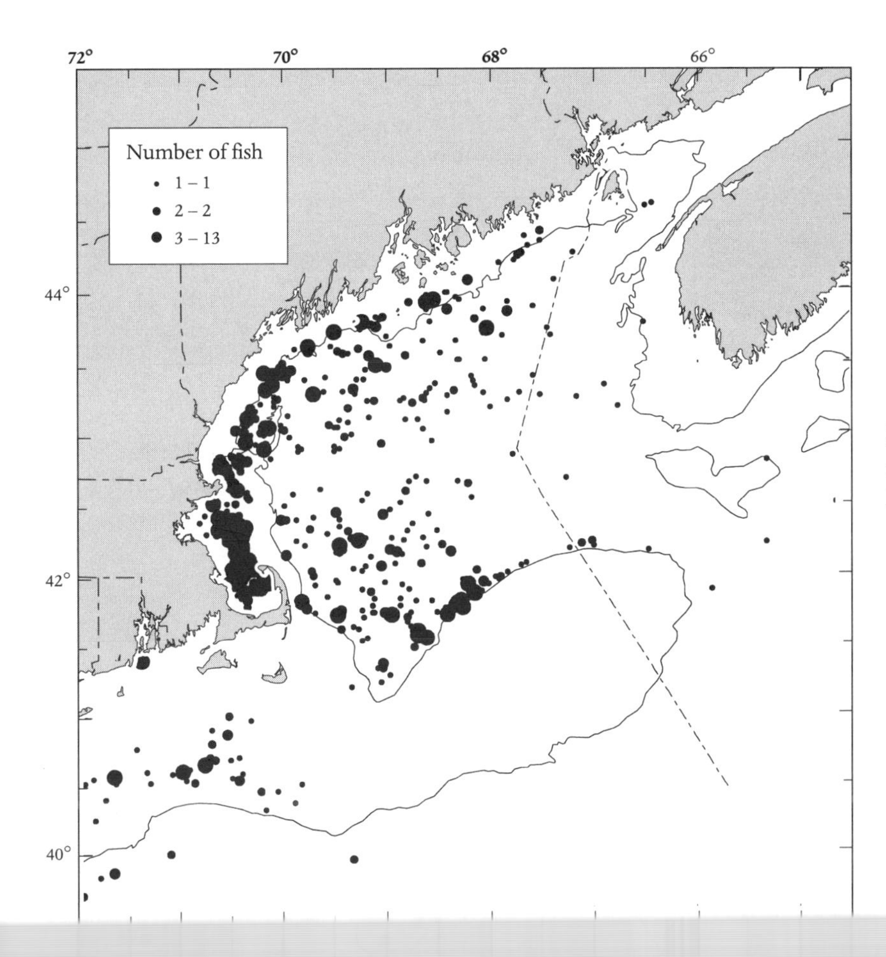

Map 15. Distribution of fourbeard rockling *Enchelyopus cimbrius,* based on NEFSC fall bottom trawl surveys, 1968–1996.

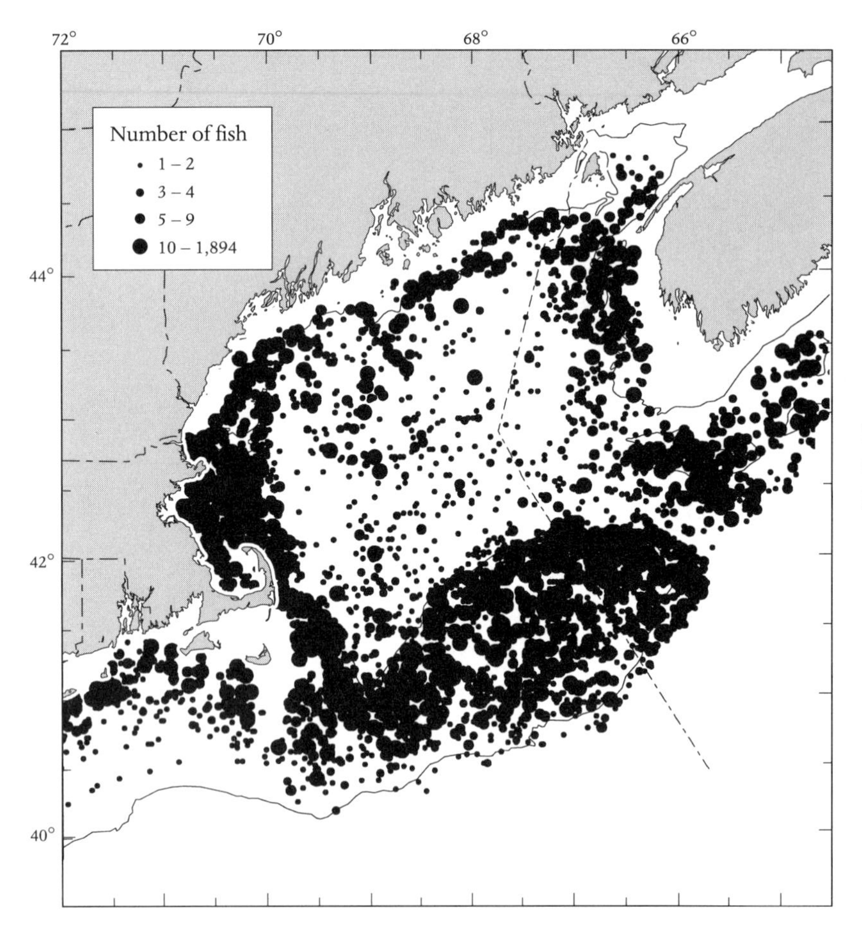

Map 16. Distribution of Atlantic cod *Gadus morhua,* based on NEFSC spring bottom trawl surveys, 1968–1996.

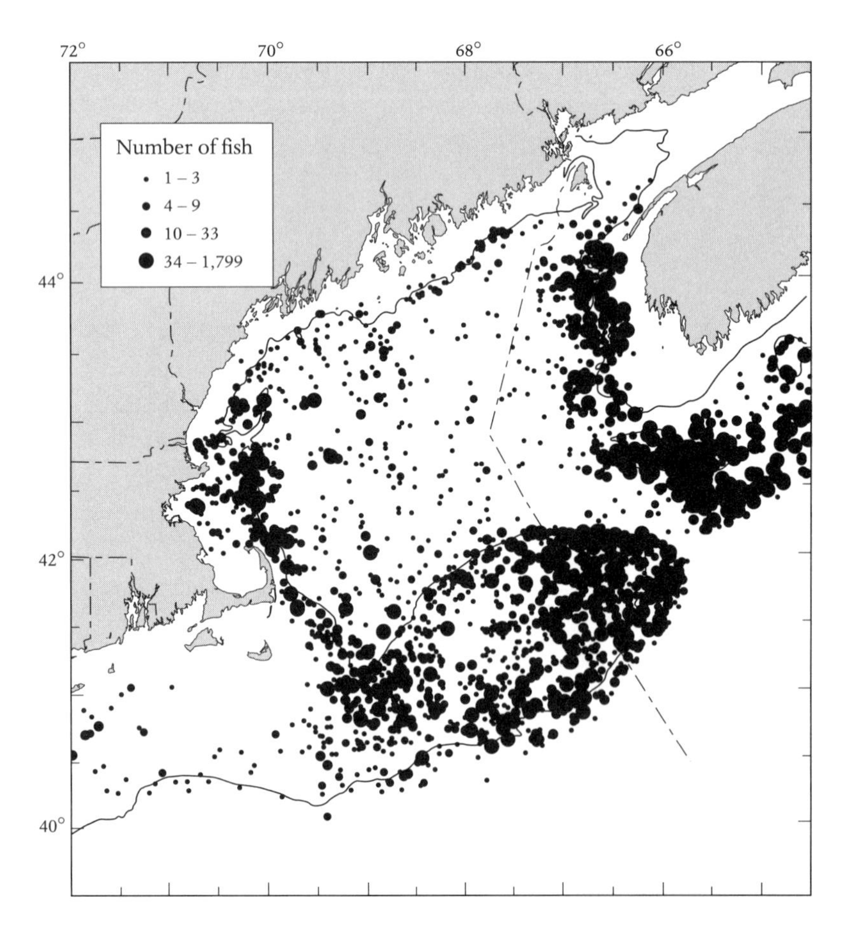

Map 17. Distribution of haddock *Melanogrammus aeglefinus,* based on NEFSC spring bottom trawl surveys, 1968–1996.

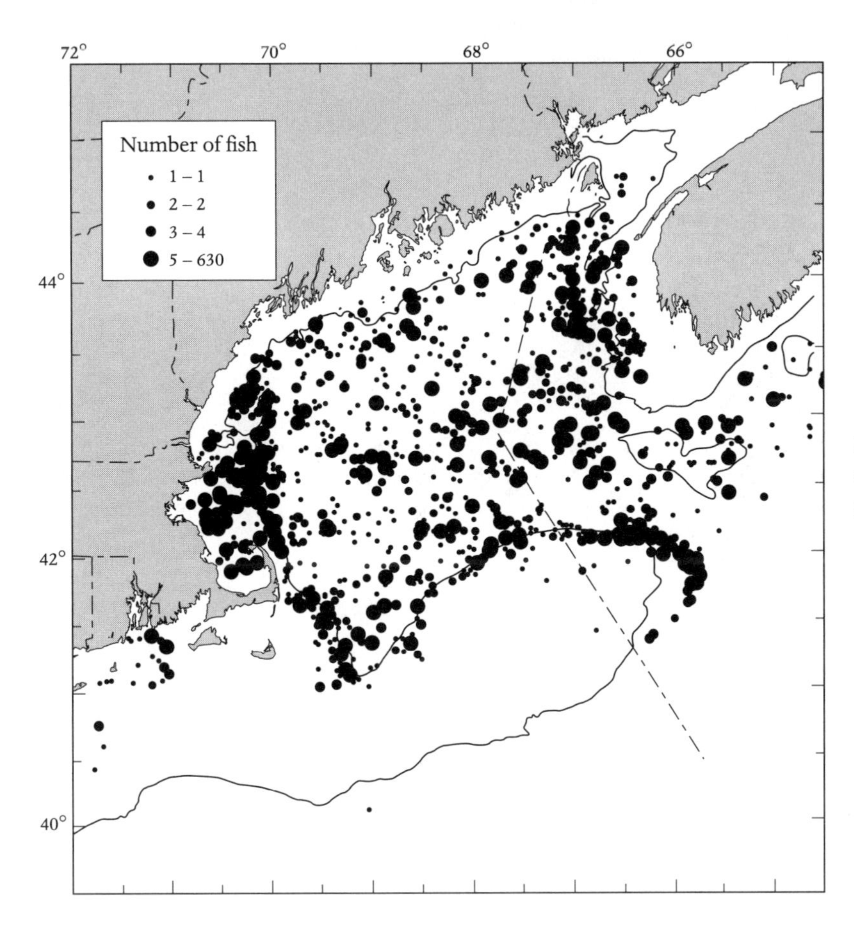

Map 18. Distribution of pollock *Pollachius virens,* based on NEFSC fall bottom trawl surveys, 1968–1996.

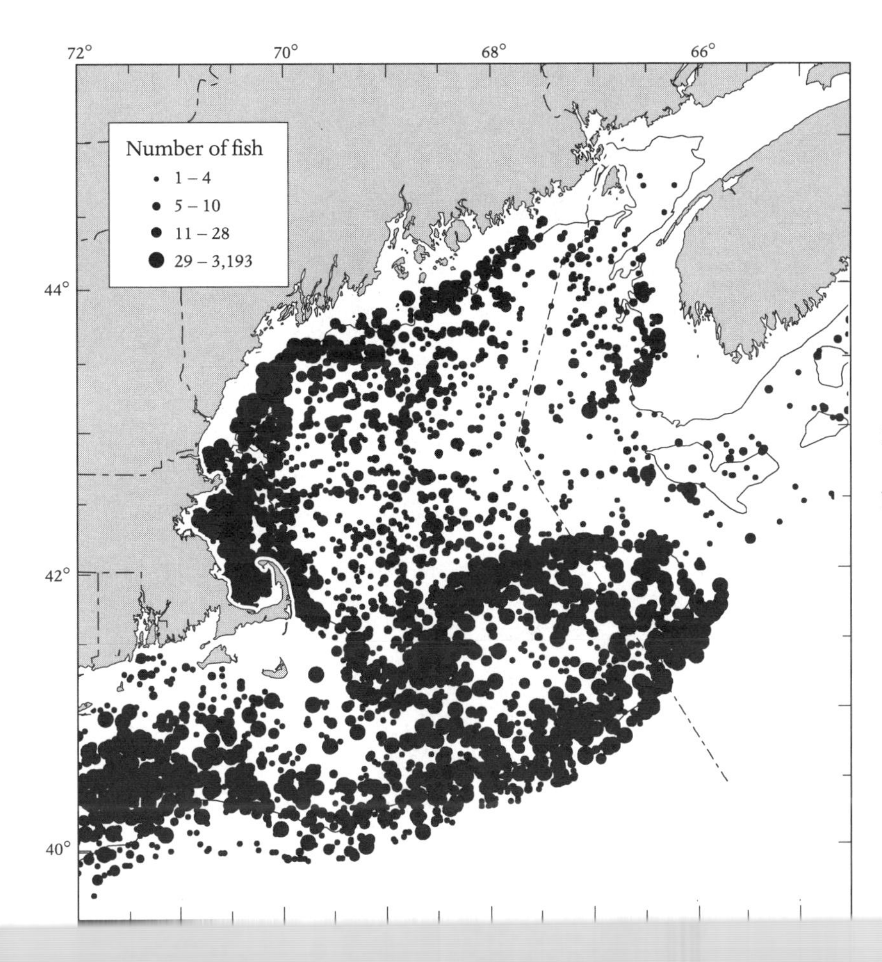

Map 19. Distribution of red hake *Urophycis chuss,* based on NEFSC fall bottom trawl surveys, 1968–1996.

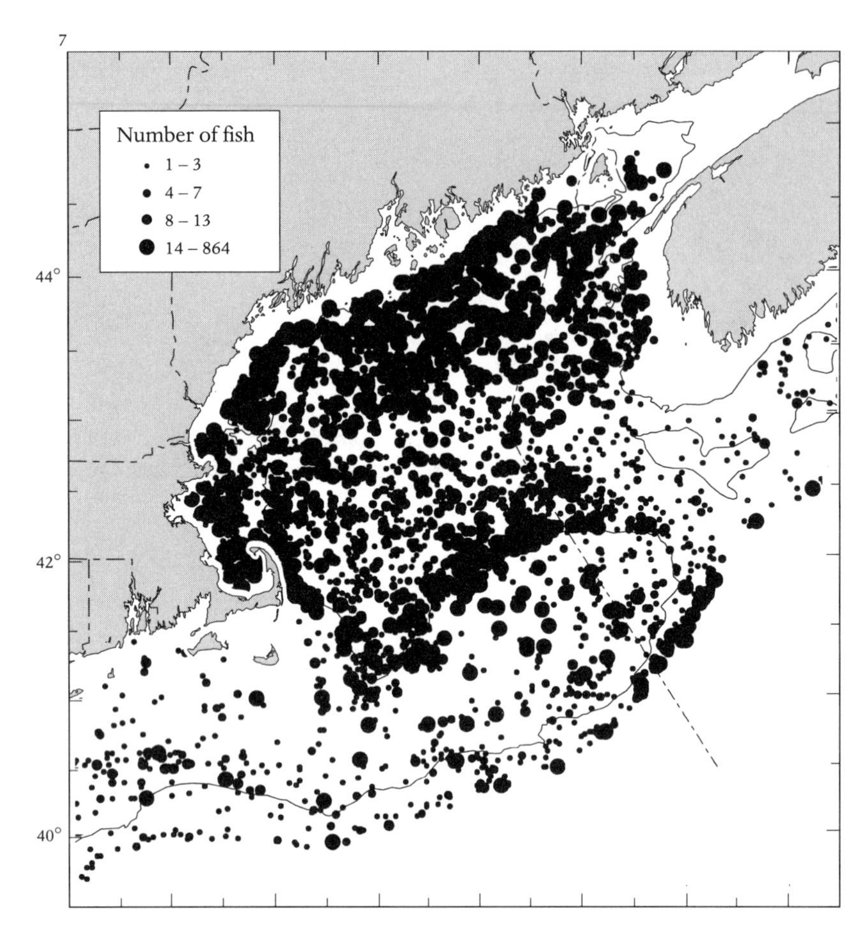

Map 20. Distribution of white hake *Urophycis tenuis,* based on NEFSC fall bottom trawl surveys, 1968–1996.

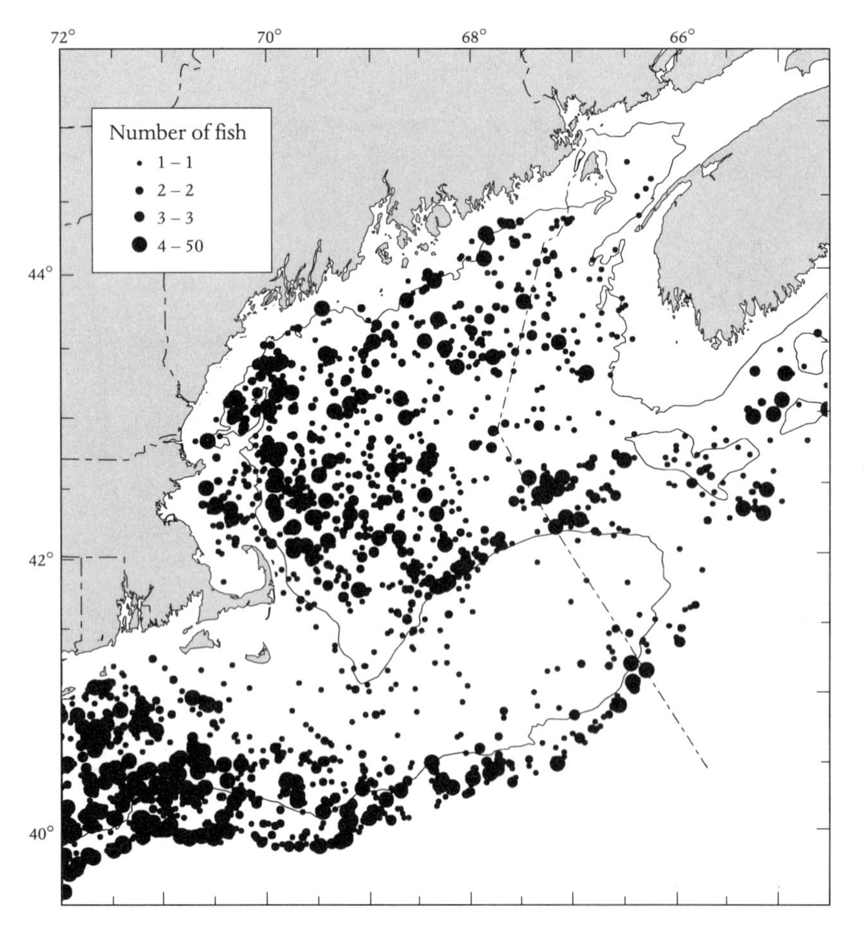

Map 21. Distribution of goosefish *Lophius americanus,* based on NEFSC spring bottom trawl surveys, 1968–1996.

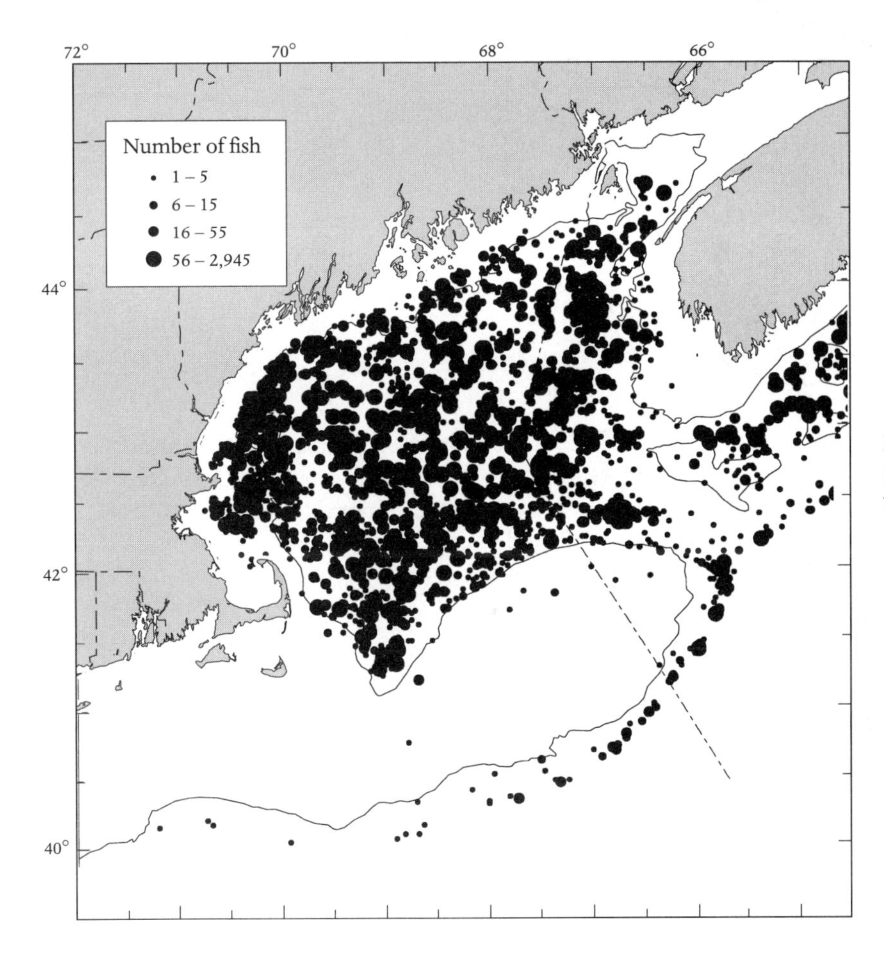

Map 22. Distribution of Acadian redfish *Sebastes fasciatus,* based on NEFSC fall bottom trawl surveys, 1968–1996.

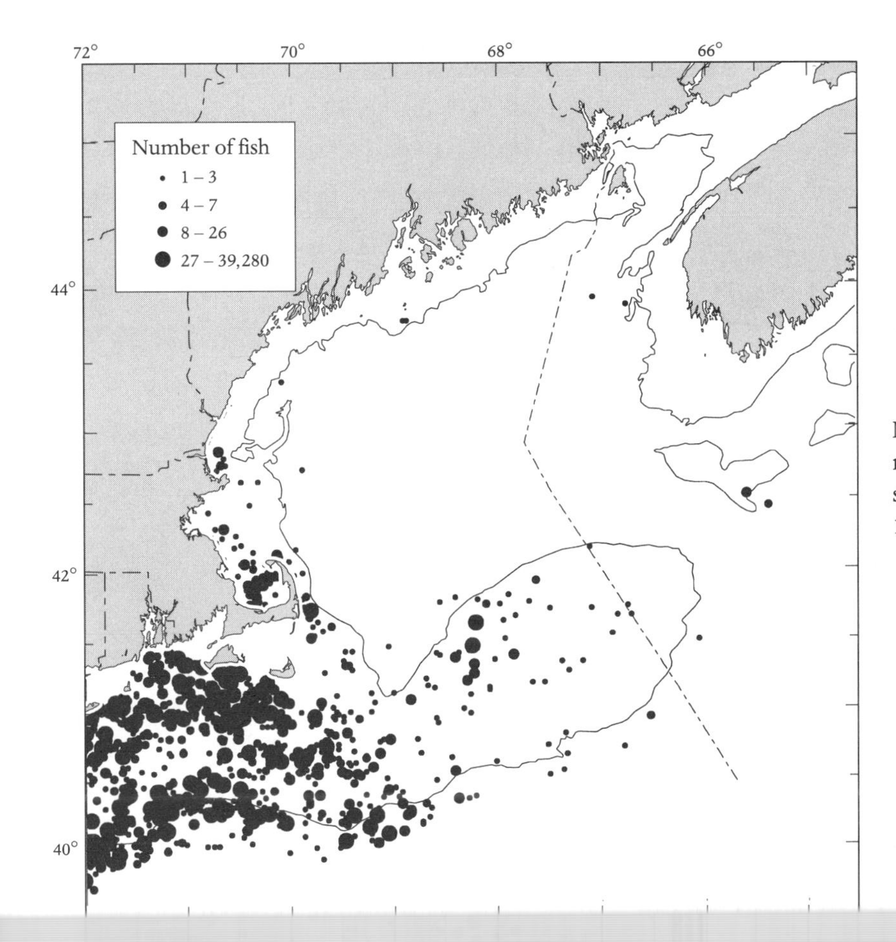

Map 23. Distribution of northern sea robin *Prionotus carolinus,* based on NEFSC spring and fall bottom trawl surveys, 1968–1996.

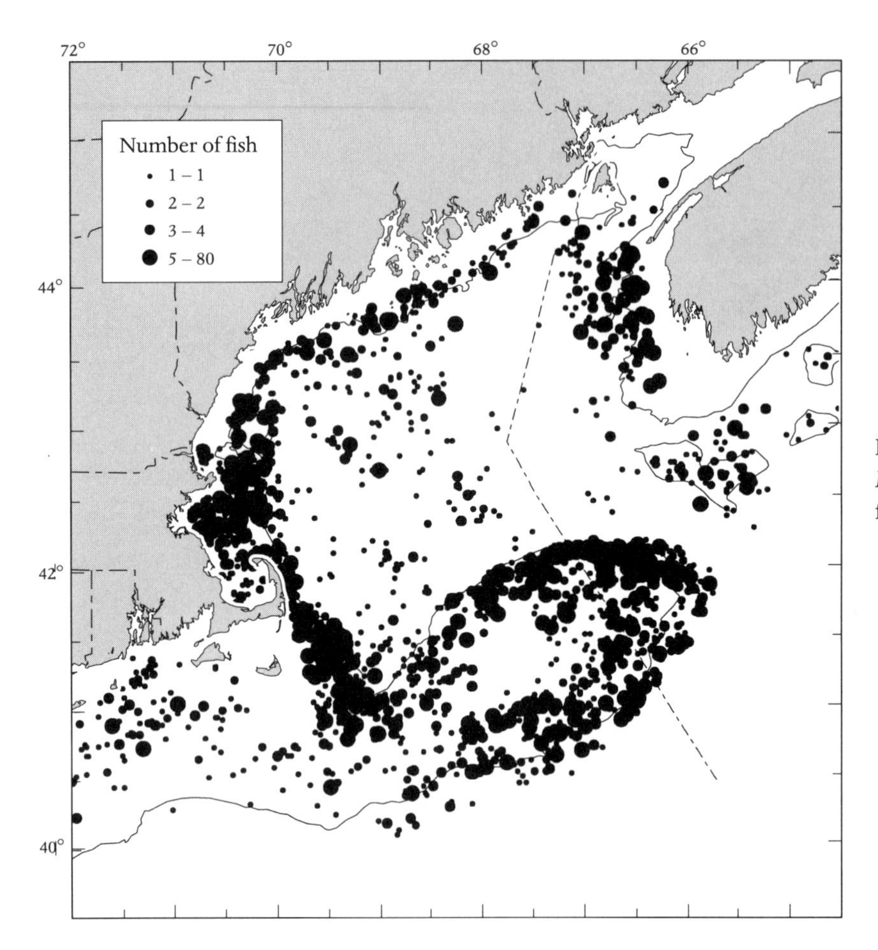

Map 24. Distribution of sea raven *Hemitripterus americanus,* based on NEFSC fall bottom trawl surveys, 1968–1996.

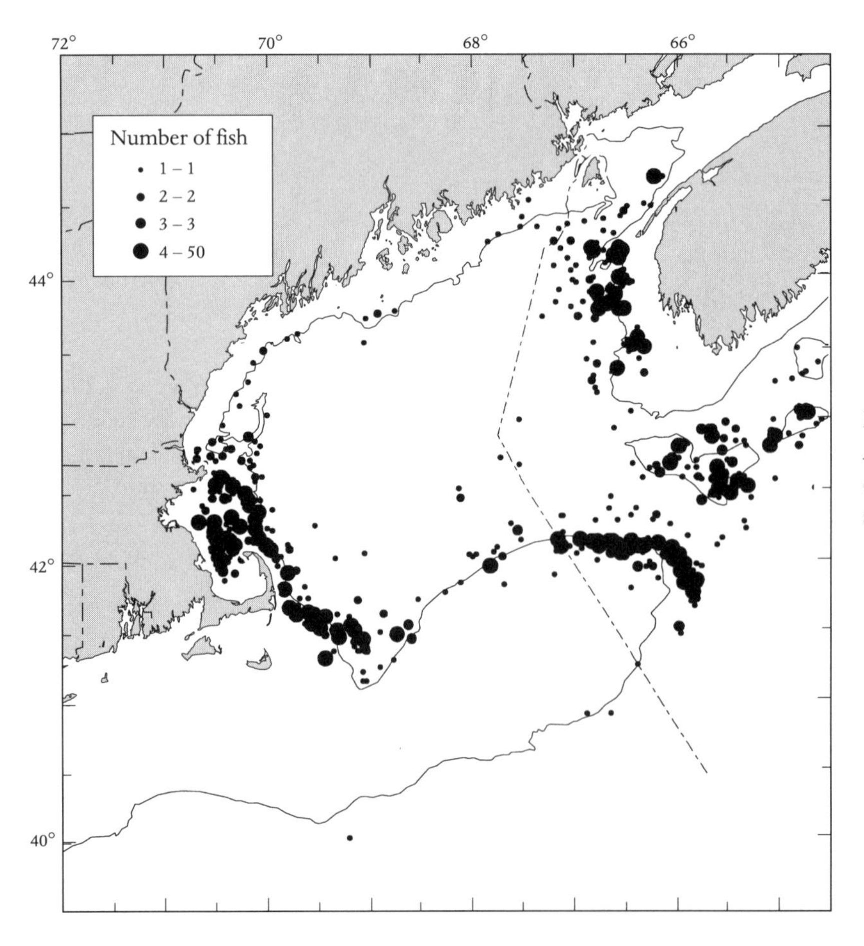

Map 25. Distribution of alligatorfish *Aspidophoroides monopterygius,* based on NEFSC spring and fall bottom trawl surveys, 1968–1996.

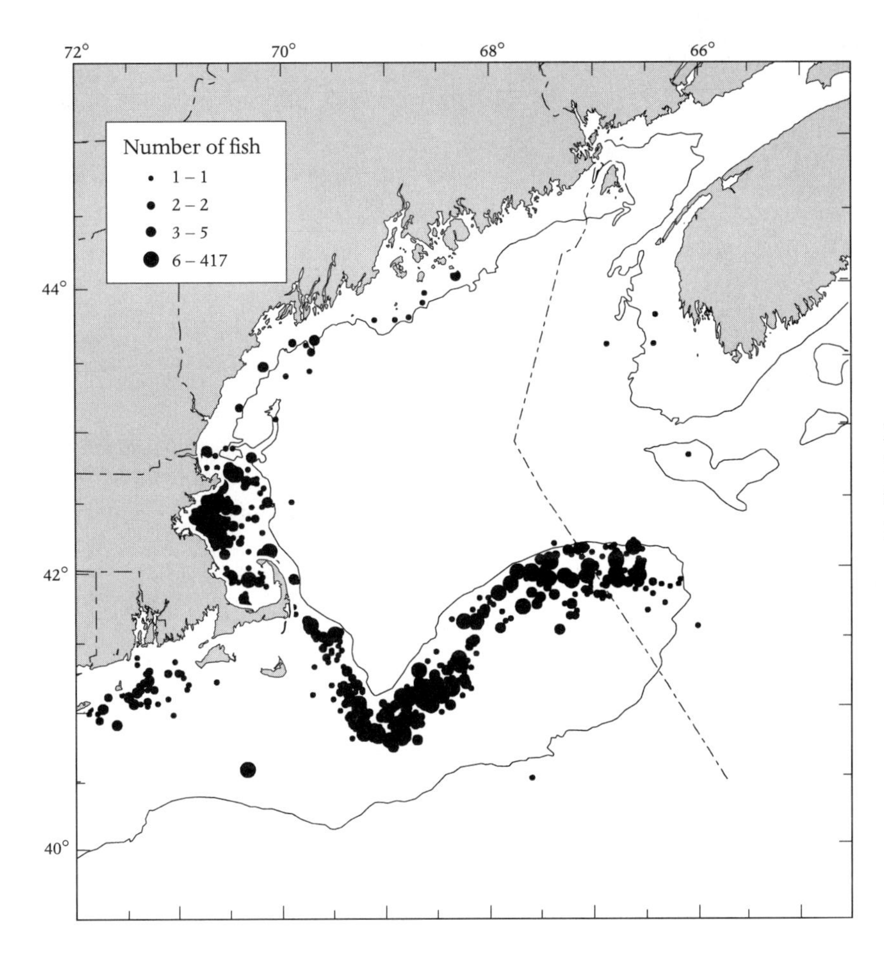

Map 26. Distribution of cunner *Tautogolabrus adspersus,* based on NEFSC fall and spring bottom trawl surveys, 1968–1996.

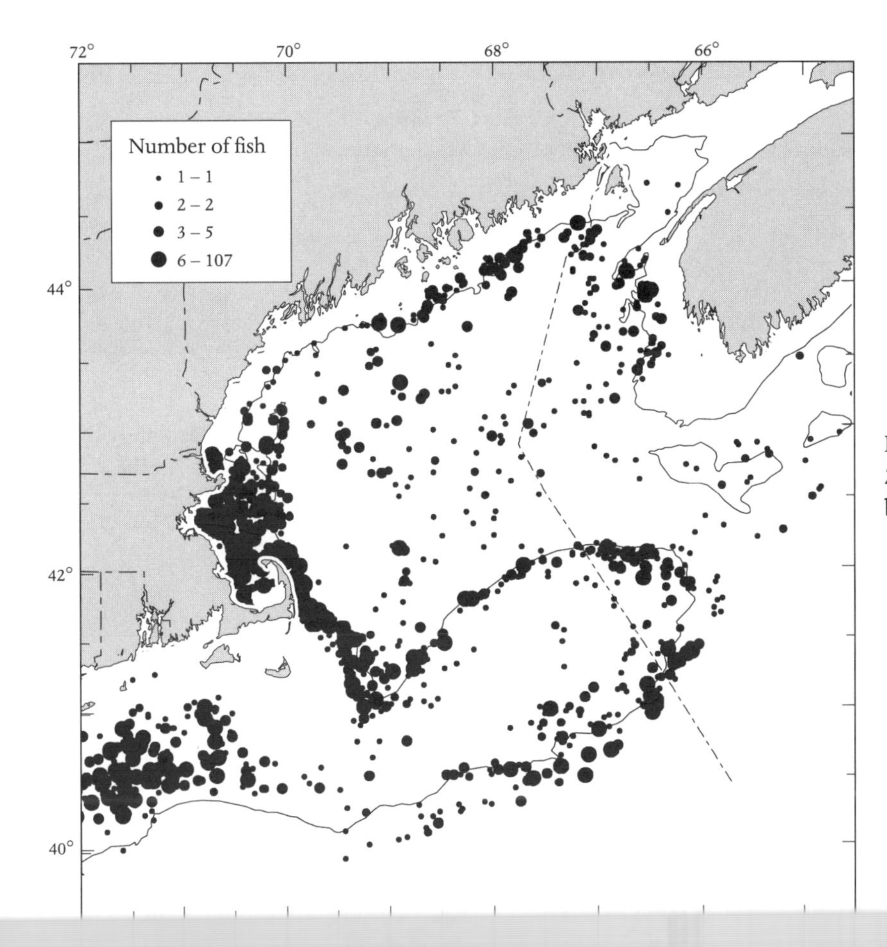

Map 27. Distribution of ocean pout *Zoarces americanus,* based on NEFSC fall bottom trawl surveys, 1968–1996.

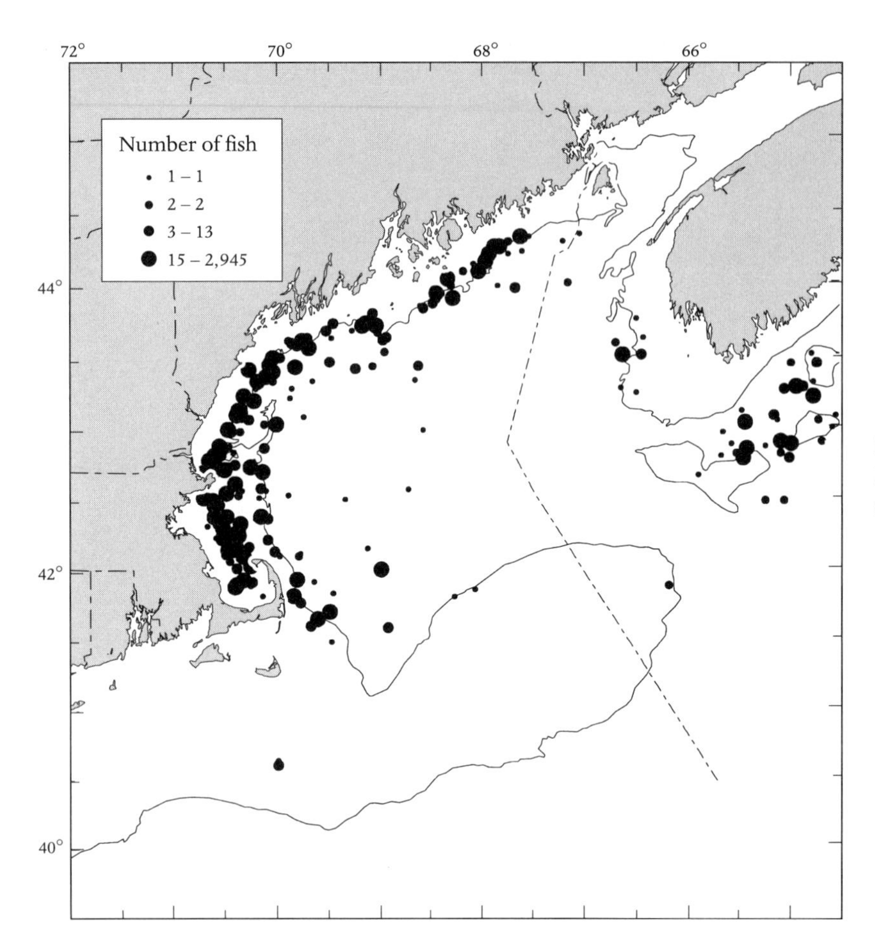

Map 28. Distribution of daubed shanny *Lumpenus maculatus,* based on NEFSC spring bottom trawl surveys, 1968–1996.

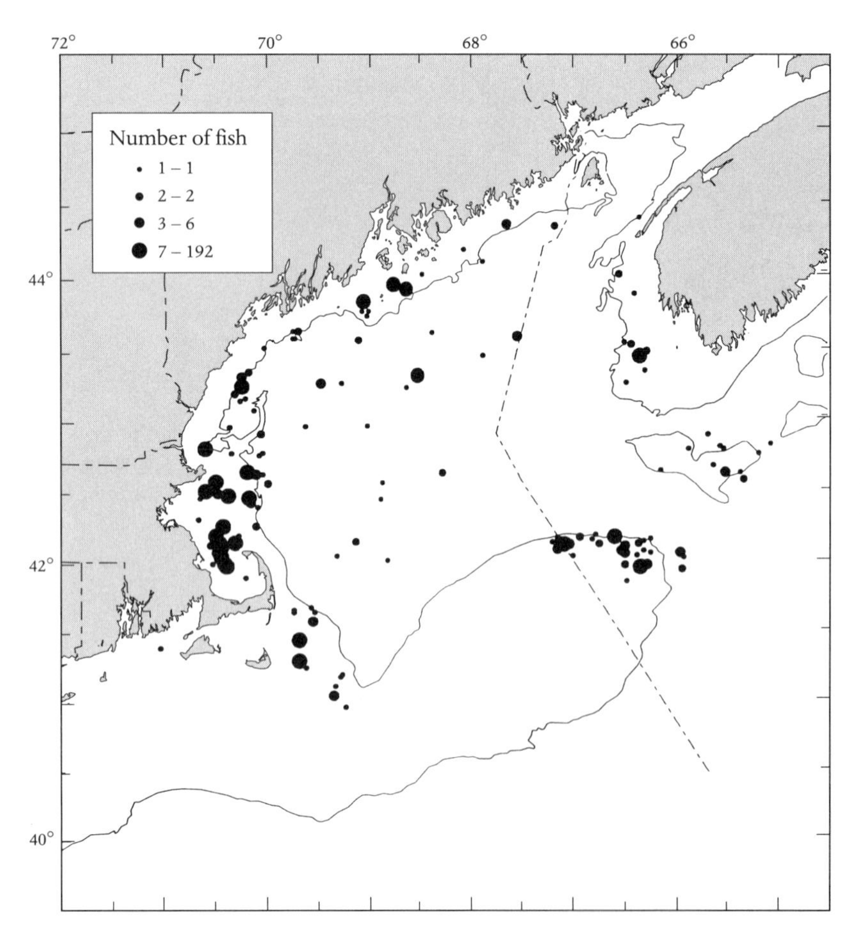

Map 29. Distribution of radiated shanny *Ulvaria subbifurcata,* based on NEFSC spring and fall bottom trawl surveys, 1968–1996.

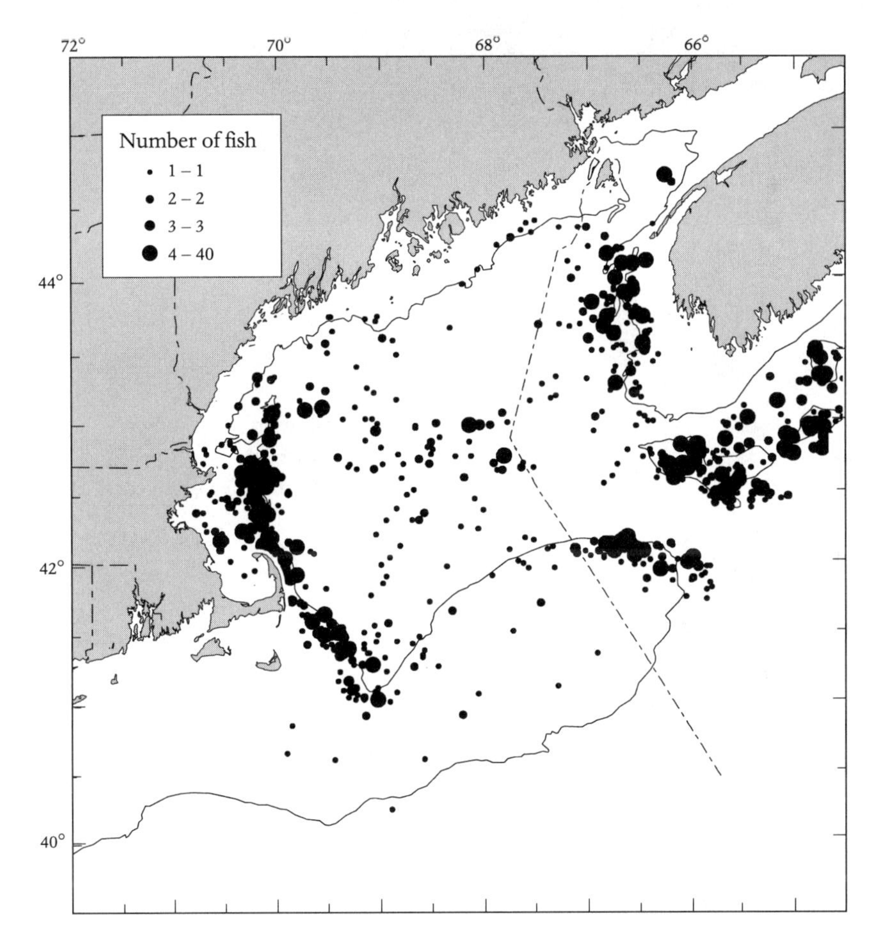

Map 30. Distribution of Atlantic wolffish *Anarhichas lupus,* based on NEFSC spring bottom trawl surveys, 1968–1996.

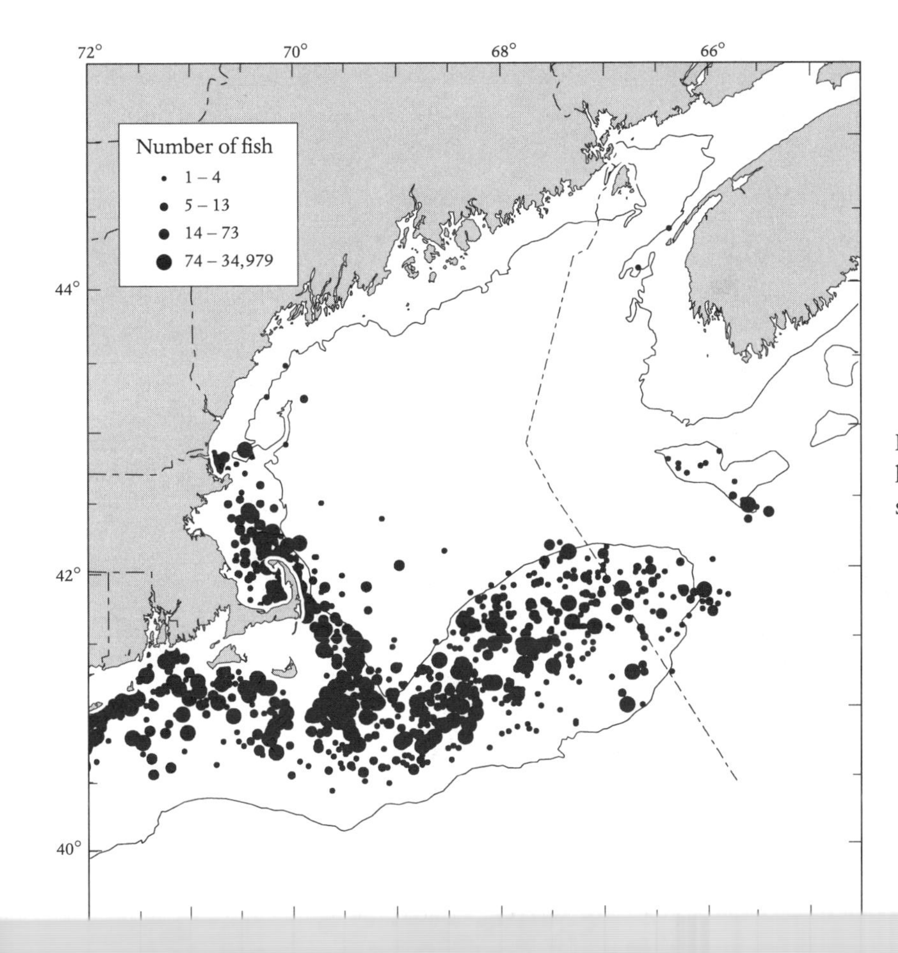

Map 31. Distribution of offshore sand lance *Ammodytes dubius,* based on NEFSC spring bottom trawl surveys, 1968–1996.

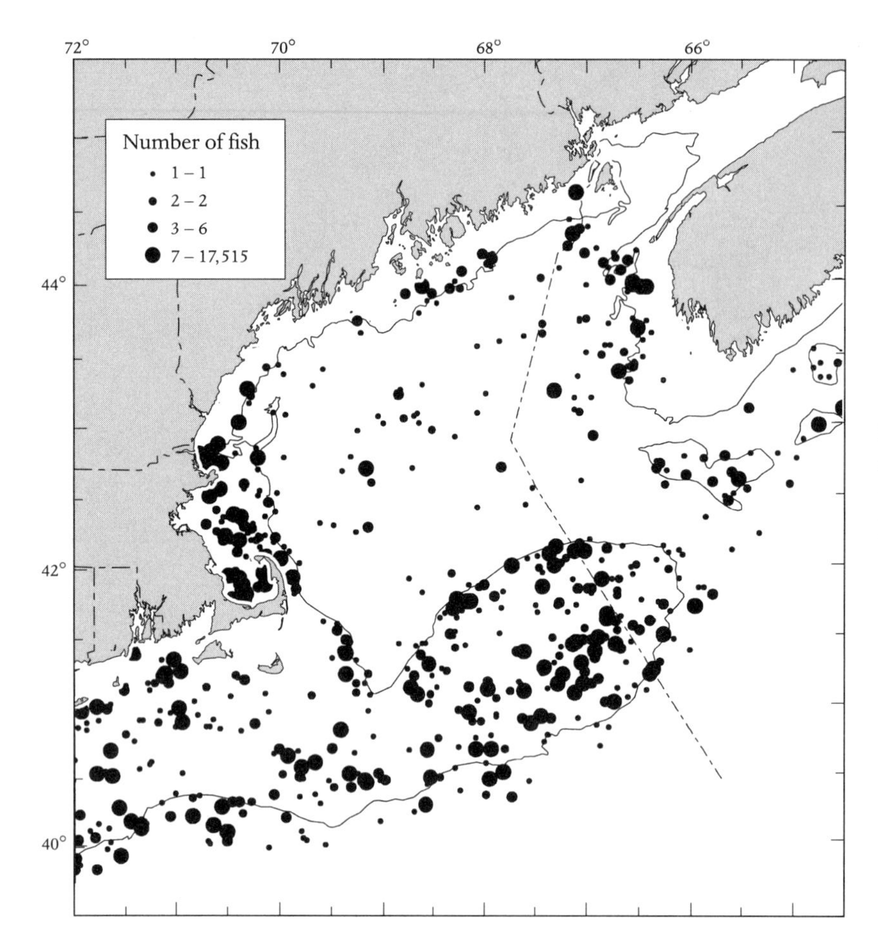

Map 32. Distribution of Atlantic mackerel *Scomber scombrus,* based on NEFSC fall bottom trawl surveys, 1968–1996.

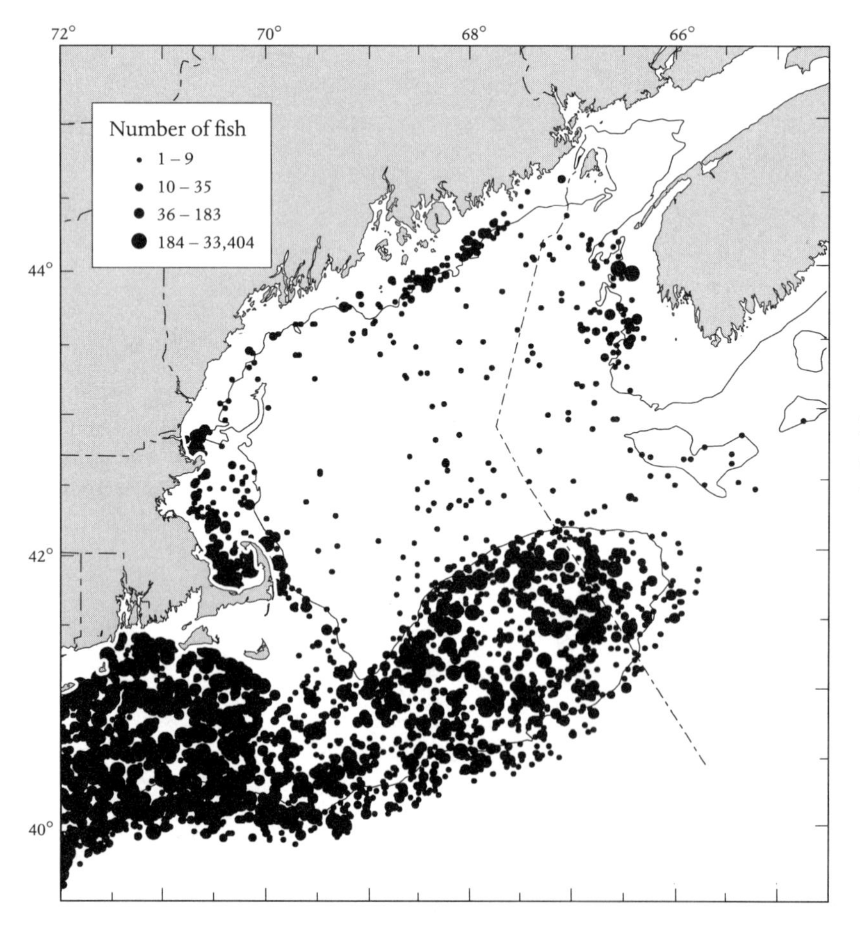

Map 33. Distribution of butterfish *Peprilus triacanthus,* based on NEFSC fall bottom trawl surveys, 1968–1996.

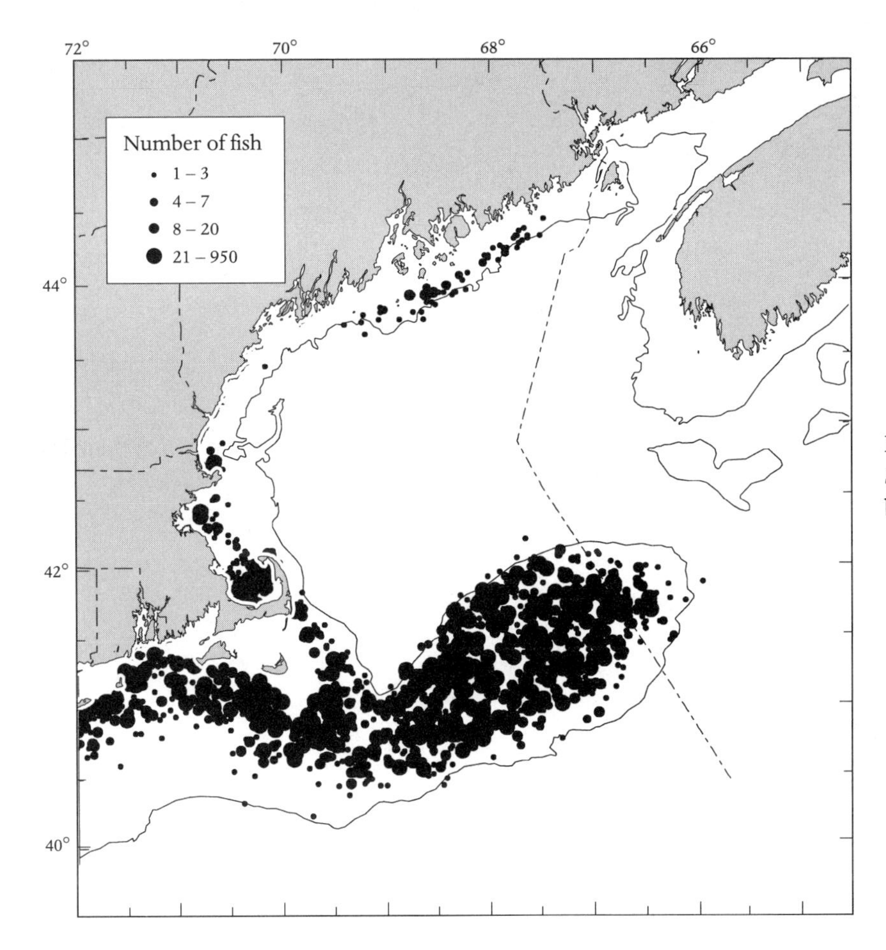

Map 34. Distribution of windowpane *Scopthalmus aquosus,* based on NEFSC fall bottom trawl surveys, 1968–1996.

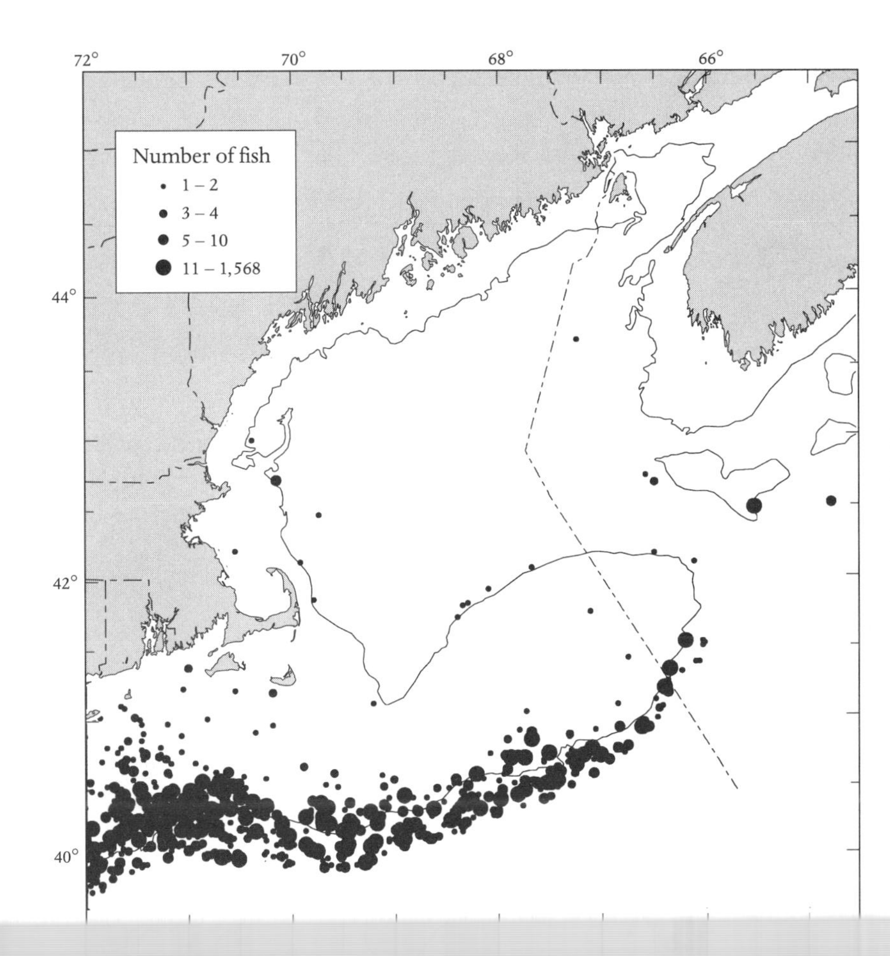

Map 35. Distribution of Gulf Stream flounder *Citharichthys arctifrons,* based on NEFSC spring bottom trawl surveys, 1968–1996.

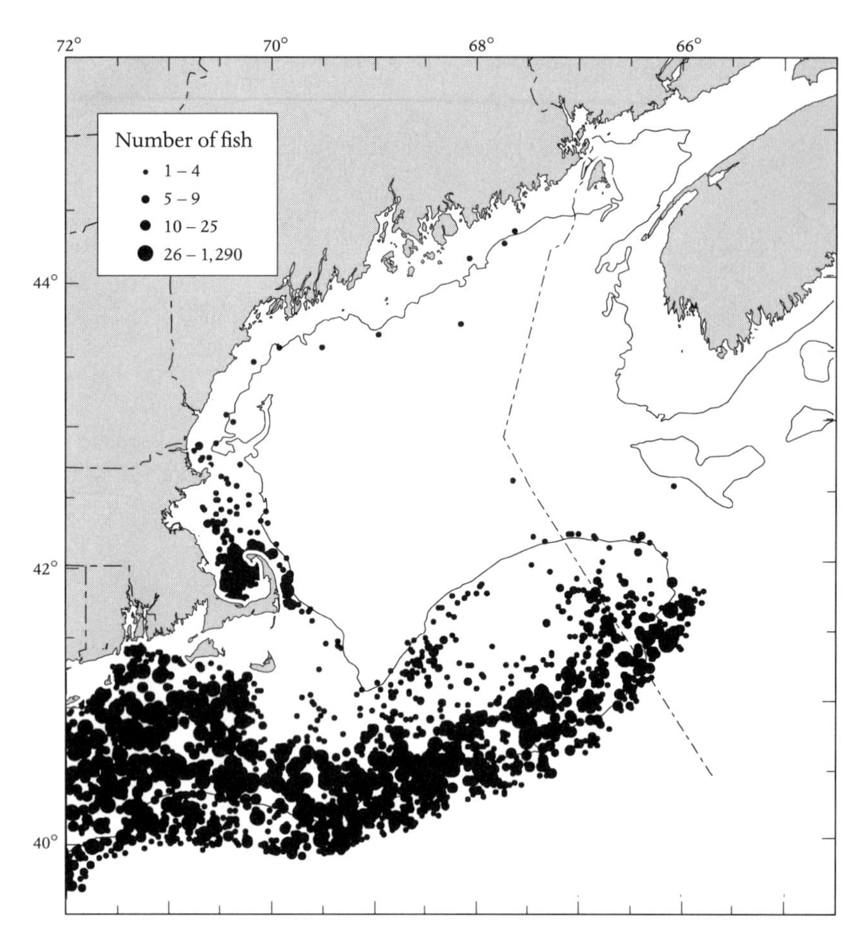

Map 36. Distribution of fourspot flounder *Paralichthys oblongus,* based on NEFSC spring bottom trawl surveys, 1968–1996.

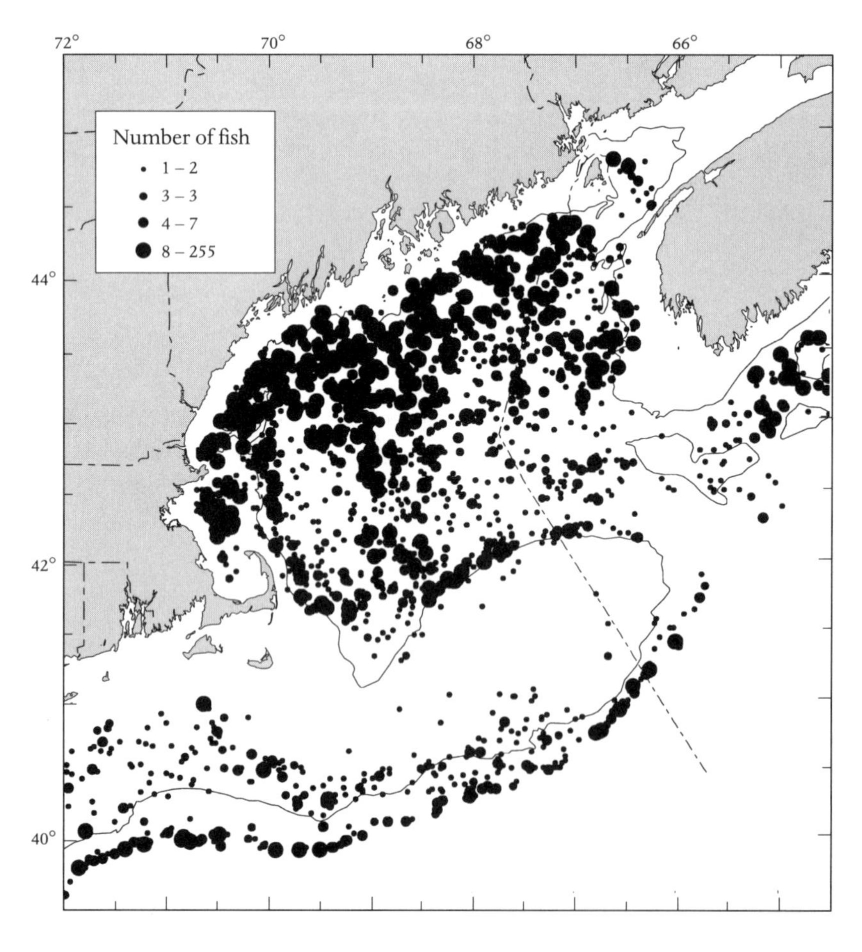

Map 37. Distribution of witch flounder *Glyptocephalus cynoglossus,* based on NEFSC fall bottom trawl surveys, 1968–1996.

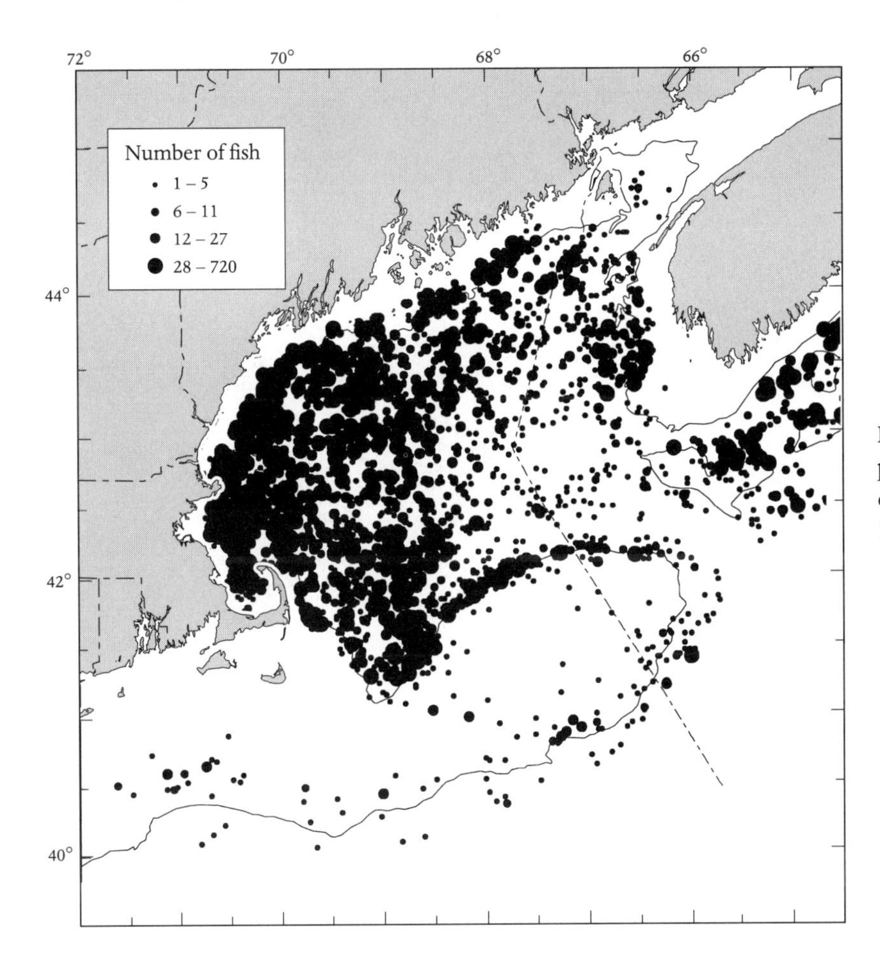

Map 38. Distribution of American plaice *Hippoglossoides platessoides,* based on NEFSC spring bottom trawl surveys, 1968–1996.

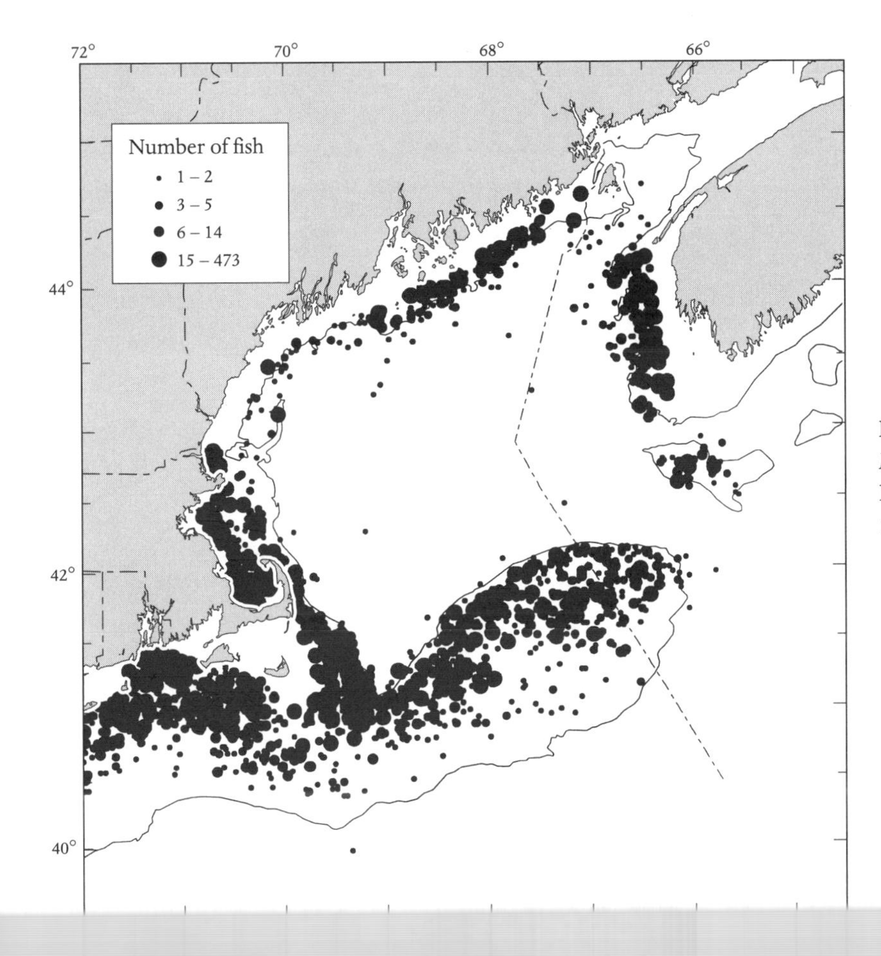

Map 39. Distribution of winter flounder *Pseudopleuronectes americanus,* based on NEFSC fall bottom trawl surveys, 1968–1996.

LITERATURE CITED

Aasen, O. 1963. Length and growth of the porbeagle (*Lamna nasus* Bonnaterre) in the northwest Atlantic. *Fisk. Havunder.* 13:20–37.

Aasen, O. 1966a. Brugde, *Cetorhinus maximus* (Gunnerus) 1765. *Saertrykk Fiskets Gang* 49:909–919.

Aasen, O. 1966b. Blahaien, *Prionace glauca* (Linnaeus) 1758. *Fisken. Havet.* 1:1–16.

Able, K. W. 1973. A new cyclopterid fish, *Liparis inquilinus,* associated with the sea scallop, *Placopecten magellanicus,* in the western North Atlantic, with notes on the *Liparis liparis* complex. *Copeia* 1973:787–794.

Able, K. W. 1976a. Cleaning behavior in the cyprinodontid fishes: *Fundulus majalis, Cyprinodon variegatus* and *Lucania parva. Chesapeake Sci.* 17:35–39.

Able, K. W. 1976b. A new cyclopterid fish *Liparis coheni* from the western North Atlantic with notes on life history. *Copeia* 1976:515–521.

Able, K. W. 1984a. Cyprinodontiformes: development. In: H. G. Moser et al., eds. *Ontogeny and Systematics of Fishes.* Am. Soc. Ich. Herp. Spec. Publ. No. 1:362–368.

Able, K. W. 1984b. Variation in spawning site selection of the mummichog, *Fundulus heteroclitus. Copeia* 1984:522–525.

Able, K. W. 1990a. A revision of Arctic snailfishes of the genus *Liparis* (Scorpaeniformes: Cyclopteridae). *Copeia* 1990:476–492.

Able, K. W. 1990b. Life history patterns of salt marsh killifishes in New Jersey. *Bull. N.J. Acad. Sci.* 35:23–30.

Able, K. W., and M. P. Fahay. 1998. *The First Year in the Life of Estuarine Fishes in the Middle Atlantic Bight.* Rutgers University Press, New Brunswick, 342 pp.

Able, K. W., and J. D. Felley. 1986. Geographical variation in *Fundulus heteroclitus:* tests for concordance between egg and adult morphologies. *Am. Zool.* 26:145–157.

Able, K. W., and D. Flescher. 1991. Distribution and habitat of chain dogfish, *Scyliorhinus retifer,* in the Mid-Atlantic Bight. *Copeia* 1991:231–234.

Able, K. W., and S. M. Hagan. 2000. Effects of common reed (*Phragmites australis*) invasion on marsh surface macrofauna: response of fishes and decapod crustaceans. *Estuaries* 23:633–646.

Able, K. W., and L. S. Hales, Jr. 1997. Movements of juvenile black sea bass, *Centropristis striata* (Linnaeus), in a southern New Jersey estuary. *J. Exper. Mar. Biol. Ecol.* 213:153–167.

Able, K. W., and D. Hata. 1984. Reproductive behavior in the *Fundulus heteroclitus–F. grandis* complex. *Copeia* 1984:820–825.

Able, K. W., and W. Irion. 1985. Distribution and reproductive seasonality of snailfishes and lumpfishes in the St. Lawrence River estuary and the Gulf of St. Lawrence. *Can. J. Zool.* 63:1622–1628.

Able, K. W., and S. C. Kaiser. 1994. Synthesis of summer flounder habitat parameters. NOAA Coastal Ocean Office Prog. Decision Analysis Ser. No. 1, 68 pp.

Able, K. W., and D. E. McAllister. 1980. Revision of the snailfish genus *Liparis* from Arctic Canada. *Can. Bull. Fish. Aquatic Sci.* 208:1–52.

Able, K. W., and J. A. Musick. 1976. Life history, ecology, and behavior of *Liparis inquilinus* (Pisces: Cyclopteridae) associated with the sea scallop, *Plagopecten magellanicus. Fish. Bull., U.S.* 74:409–421.

Able, K. W., and R. E. Palmer. 1988. Salinity effects on fertilization success and larval mortality of *Fundulus heteroclitus. Copeia* 1988:345–350.

Able, K. W., C. B. Grimes, R. A. Cooper, and J. R. Uzmann. 1982. Burrow construction and behavior of tilefish, *Lopholatilus chamaeleonticeps,* in Hudson Submarine Canyon. *Environ. Biol. Fishes* 7:199–205.

Able, K. W., D. F. Markle, and M. P. Fahay. 1984. Cyclopteridae: development. In: H. G. Moser et al., eds. *Ontogeny and Systematics of Fishes.* Am. Soc. Ich. Herp. Spec. Publ. No. 1:428–437.

Able, K. W., M. P. Fahay, and D. F. Markle. 1986. Development of larval snailfishes (Pisces: Cyclopteridae: Liparidinae) from the western North Atlantic. *Can. J. Zool.* 64:2294–2316.

Able, K. W., D. C. Twichell, C. B. Grimes, and R. S. Jones. 1987. Sidescan sonar as a tool in fisheries exploration. *Fish. Bull., U.S.* 85:725–736.

Able, K. W., R. E. Matheson, W. W. Morse, M. P. Fahay, and G. Shepherd. 1990. Patterns of summer flounder *Paralichthys dentatus* early life history in the Mid-Atlantic Bight and New Jersey estuaries. *Fish. Bull., U.S.* 88:1–12.

Able, K. W., C. B. Grimes, R. S. Jones, and D. C. Twitchell. 1993. Temporal and spatial variation in habitat characteristics of tilefish, *Lopholatilus chamaeleonticeps* off the east coast of Florida. *Bull. Mar. Sci.* 53:1013–1026.

Able, K. W., M. P. Fahay, and G. R. Shepherd. 1995. Early life history of black sea bass, *Centropristis striata,* in the Mid-Atlantic Bight and a New Jersey estuary. *Fish. Bull., U.S.* 93:429–445.

Able, K. W., D. A. Witting, R. S. McBride, R. A. Rountree, and K. J. Smith. 1996. Fishes of polyhaline estuarine shores in Great Bay–Little Egg Harbor, New Jersey: a case study of seasonal and habitat influences. In: K. F. Nordstrom and C. T. Roman, eds. *Estuarine Shores: Environments and Human Alterations.* John Wiley & Sons, Chichester, England, pp. 335–353.

Aboussouan, A. 1966. Oeufs et larves de téléostéens de l'ouest africain. III. Larves de *Monacanthus hispidus* (L.) et de *Balistes forcipatus* Gm. *Bull. Inst. Fran. Afr. Noire, Ser. A* 28(1):276–282.

Aboussouan, A. 1975. Oeufs et larves de téléostéens de l'ouest africain. XIII. Contribution à l'identification des larves de Carangidae. *Bull. Inst. Fondam. Afr. Noire (A: Sci. Nat.)* 37:899–938.

Aboussouan, A., and J. M. Leis. 1984. Balistoidei: development. In: H. G. Moser et al., eds. *Ontogeny and Systematics of Fishes.* Am. Soc. Ich. Herp. Spec. Publ. No. 1:450–459.

Adam, H. 1960. Different types of body movement in the hagfish, *Myxine glutinosa* L. *Nature* 188:595–596.

Adam, H., and R. Strahan. 1963. Notes on the habitat, aquarium maintenance, and experimental use of hagfishes. In: A. Brodal and R. Fänge, eds. *The Biology of Myxine.* Universitetsforlaget, Oslo, pp. 33–41.

Adams, J. A. 1960. A contribution to the biology and postlarval development of the sargassum fish, *Histrio histrio* (Linnaeus), with a discussion of the *Sargassum* complex. *Bull. Mar. Sci. Gulf Carib.* 10:55–82.

Adeney, R. J., and G. M. Hughes. 1977. Some observations on the gills of the oceanic sunfish, *Mola mola. J. Mar. Biol. Assoc. U.K.* 57:825–837.

Agafona, T. B., and E. I. Kukuyev. 1990. New data on the distribution of *Cubiceps gracilis. Voprosy Ikhtiol.* 30:1028–1031.

Agassiz, A. 1882. On the young stages of some osseous fishes. III. *Proc. Am. Acad. Arts Sci.* 17:271–303.

Agassiz, A., and C. O. Whitman. 1885. The development of osseous fishes. I. The pelagic stages of young fishes. Mus. Comp. Zool. Mem. No. 14(1):1–56.

Ahlstrom, E. H. 1974. The diverse patterns of metamorphosis in gonostomatid fishes—an aid to classification. In: J. H. S. Blaxter, ed. *The Early Life History of Fish.* Springer-Verlag, Berlin, pp. 659–674.

Ahlstrom, E. H., W. J. Richards, and S. H. Weitzman. 1984. Families Gonostomatidae, Sternoptychidae, and associated stomiiform groups: development and relationships. In: H. G. Moser et al., eds. *Ontogeny and Systematics of Fishes.* Am. Soc. Ich. Herp. Spec. Publ. No. 1:184–198.

Ahrenholz, D. W. 1991. Population biology and life history of the North American menhadens, *Brevoortia* spp. *Mar. Fish. Rev.* 53(4):3–19.

Ahrenholz, D. W., J. F. Guthrie, and R. M. Clayton. 1987a. Observations of ulcerative mycosis infections on Atlantic menhaden *Brevoortia tyrannus.* NOAA Tech. Memo. NMFS-SEFC-196, 11 pp.

Ahrenholz, D. W., W. R. Nelson, and S. P. Epperly. 1987b. Population and fishery characteristics of Atlantic menhaden, *Brevoortia tyrannus. Fish. Bull., U.S.* 85:569–600.

Ahrenholz, D. W., J. F. Guthrie, and C. W. Krouse. 1989. Results of abundance surveys of juvenile Atlantic and Gulf menhaden *Brevoortia tyrannus* and *B. patronus.* NOAA Tech. Memo. NMFS No. 84, 14 pp.

Ahrenholz, D. W., G. R. Fitzhugh, J. A. Rice, S. W. Nixon, and W. C. Pritchard. 1995. Confidence of otolith ageing through the juvenile stage for Atlantic menhaden, *Brevoortia tyrannus. Fish. Bull., U.S.* 93:209–216.

Aiken, K. A. 1975. The biology, ecology and bionomics of the triggerfishes, Balistidae. In: J. L. Munro, ed. *Caribbean Coral Reef Fishery Resources.* ICLARM Stud. Rev. No 7. Manila, pp. 191–205.

Albikovskaya, L. K. 1982. Distribution and abundance of Atlantic wolffish, spotted wolffish and northern wolffish in the Newfoundland area. Northw. Atl. Fish. Org., Sci. Coun. Stud. No. 3:29–32.

Albikovskaya, L. K. 1983. Feeding characteristics of wolffishes in the Labrador-Newfoundland region. Northw. Atl. Fish. Org., Sci. Coun. Stud. No. 6:35–38.

Albrecht, A. B. 1964. Some observations on factors associated with survival of striped bass eggs and larvae. *Calif. Fish Game* 50:100–113.

Aldridge, R. J., and P. C. J. Donoghue. 1998. Conodonts: a sister group to hagfishes? In: J. M. Joergensen et al., eds. *The Biology of Hagfishes.* Chapman and Hall, London, pp. 15–31.

Alexander, L. C. 1971. Feeding chronology and food habits of the tomcod *Microgadus tomcod* (Walbaum) and winter flounder *Pseudopleuronectes americanus* Walbaum in Montsweag Bay (Sheepscot River), Maine. M.S. thesis, University of Maine, Orono, 36 pp.

Allen, D. M., W. S. Johnson, and V. Ogburn-Matthews. 1995. Trophic relationships and seasonal utilization of salt-marsh creeks by zooplanktivorous fishes. *Environ. Biol. Fishes* 42:37–50.

Allen, G. M. 1916. The whalebone whales of New England. Boston Soc. Nat. Hist. Mem. No. 8: 107–322.

Allen, G. R. 1979. *Butterfly and Angelfishes of the World,* Vol. 2. John Wiley & Sons, New York, pp.145–352.

Almeida, F. P. 1987. Stock definition of silver hake in the New England–Middle Atlantic area. *No. Am. J. Fish. Manage.* 7:169–186.

Almeida, F. P., R. E. Castaneda, R. Jesien, R. E. Greenfield, and J. M. Burnett. 1992. Proceedings of the NEFC/ASMFC summer flounder, *Paralichthys dentatus,* aging workshop, 11–13 June 1990, Northeast Fisheries Center, Woods Hole, Mass. NOAA Tech. Memo. NMFS-F/NEC-89, 7 pp.

Almeida, F. P., D-L. Hartley, and J. Burnett. 1995. Length-weight relationships and sexual maturity of goosefish off the northeast coast of the United States. *No. Am. J. Fish. Manage.* 15:14–25.

Alperin, I. M. 1987. Management of migratory Atlantic coast striped bass: an historical perspective, 1938–1986. *Fisheries* 12(6):2–3.

Altukhov, Yu. P., and E. A. Salmenkova. 1981. Applications of the stock concept to fish populations in the USSR. *Can. J. Fish. Aquat. Sci.* 38:1591–1600.

Andersen, K. 1987. S. E. M. observations on plerocercus larvae of *Floriceps saccatus* Cuvier, 1817 and *Molicola horridus* (Goodsir, 1814) (Cestoda; Trypanorhyncha) from sunfish (*Mola mola*). Fauna Norvegica, Ser. A, No. 8:25–28.

Anderson, C. O., Jr., D. J. Brown, B. A. Ketschke, E. M. Elliott, and P. L. Rule. 1975. The effects of the addition of a fourth generating unit at the Salem Harbor Electric Generating Station on the marine ecosystem of Salem Harbor. Mass. Div. Mar. Fish., 47 pp.

Anderson, D. M., and D. Wall. 1978. Potential importance of benthic cysts of *Gonyaulax tamarensis* and *G. excavata* in initiating toxic blooms. *J. Phycol.* 14:224–234.

Anderson, E. D. 1982. Silver hake, *Merluccius bilinearis.* In: M. D. Grosslein and T. R. Azarovitz, eds. *Fish Distribution.* MESA New York Bight Atlas Monogr. No.15:72–74.

Anderson, E. D. 1990. Fisheries models as applied to elasmobranchs. NOAA Tech. Rept. NMFS No. 90:473–484.

Anderson, E. D. 1995. Atlantic mackerel. In: Status of the fishery resources off the northeastern United States for 1994. NOAA Tech. Memo. NMFS-NE-108:100–101.

Anderson, E. D., and A. L. Paciorkowski. 1980. A review of the northwest Atlantic mackerel fishery. *Rapp. Proc-Verb. Réun. Cons. Int. Explor. Mer* 177:175–211.

Anderson, E. D., F. E. Lux, and F. P. Almeida. 1980. The silver hake stocks and fishery of northeastern United States. *Mar. Fish. Rev.* 42(1):12–20.

Anderson, H. G. 1970. Annotated list of parasites of the bluefish. Bur. Sport Fish Wildl. Tech. Rept. No. 54, 15 pp.

Anderson, J. M., F. G. Whoriskey, and A. Goode. 2000. Atlantic salmon on the brink. *Endangered Species Update* 17(1):15–21.

Anderson, M. E. 1984a. On the anatomy and phylogeny of the Zoarcidae (Teleostei: Perciformes). Ph.D. dissertation, College of William and Mary, Williamsburg, 254 pp.

Anderson, M. E. 1984b. Zoarcidae: development and relationships. In: H. G. Moser et al., eds. *Ontogeny and Systematics of Fishes.* Am. Soc. Ich. Herp. Spec. Publ. No. 1:578–582.

Anderson, M. E. 1994. Systematics and osteology of the Zoarcidae (Teleostei: Perciformes). *J. L. B. Smith Inst. Ichthyol., Ichthyol. Bull.* 60, 120 pp.

Anderson, W. D., Jr., and C. C. Baldwin. 2000. A new species of *Anthias* (Teleostei: Serranidae: Anthiinae) from the Galápagos Islands, with keys to *Anthias* and eastern Pacific Anthiinae. *Proc. Biol. Soc. Washington* 113:369–385.

Anderson, W. D., Jr., and D. M. Cupka. 1973. Records of the ocean sunfish, *Mola mola,* from the beaches of South Carolina and adjacent waters. *Chesapeake Sci.* 14:295–298.

Anderson, W. D., Jr., and E. J. Gutherz. 1964. New Atlantic coast ranges for fishes. *Quart. J. Florida Acad. Sci.* 27:299–306.

Anderson, W. W. 1957. Early development, spawning, growth, and occurrence of the silver mullet (*Mugil curema*) along the south Atlantic coast of the United States. *Fish. Bull., U.S.* 57:397–414.

Anderson, W. W. 1958. Larval development, growth, and spawning of the striped mullet *Mugil cephalus* along the south Atlantic coast of the United States. *Fish. Bull., U.S.* 58:501–519.

Andersson, L., N. Ryman, R. Rosenberg, and G. Stahl. 1981. Genetic variability in Atlantic herring *Clupea harengus:* description of protein loci and population data. *Hereditas* 95:69–78.

Andre, W. 1784. A description of the teeth of *Anarchichas lupus* Linnaei and those of *Choetodon nigricans* of the same author, to which is added an attempt to prove that the teeth of cartilaginous fishes are perpetually renewed. Phil. Trans. Roy. Soc. London No. 15:539–543.

Andriashev, A. P. 1954. Fishes of the northern seas of the U.S.S.R. *Zool. Inst. Akad. Nauk SSSR, No. 53:1-566* (translated by Israel Program for Scientific Translation, Jerusalem, No. OTJ63–11160, 1964).

Andriashev, A. P. 1986. *Review of the Snailfish Genus* Paraliparis *(Scorpaeniformes: Liparididae) of the Southern Ocean.* Koeltz Scientific

Anonymous. 1964a. Trawling survey off Florida east coast. *Comm. Fish. Rev.* 26(6):27–29.

Anonymous. 1964b. Exploratory trawling off North and South Carolina continued. *Comm. Fish. Rev.* 26(7):31–32.

Anonymous. 1972. Algae form a nursery for winter flounder. *Maritimes* 16:12–14.

Anonymous. 2001. New world records. *Int. Angler* 63(5):2.

Anthony, V. C. 1971. The density dependence of growth of the Atlantic herring in Maine. *Rapp. Proc.-Verb. Réun. Cons. Int. Explor. Mer* 160:196–205.

Anthony, V. C. 1972. Population dynamics of the Atlantic herring in the Gulf of Maine. Ph.D. dissertation, University of Washington, Seattle, 266 pp.

Anthony, V. C. 1977. June 1977 assessments of herring from the Gulf of Maine and Georges Bank areas. NMFS, NEFSC, Woods Hole Lab., Lab. Ref. No. 77-16, 16 pp.

Anthony, V. C. 1981a. The use of meristic counts in indicating herring stocks in the Gulf of Maine and adjacent waters. Northw. Atl. Fish. Org. SCR Doc. 81/IX/127, Ser. No. N433, 37 pp.

Anthony, V. C. 1981b. Status of herring tagging research in the Gulf of Maine. Northw. Atl. Fish. Org. SCR Doc. 81/IX/133, Ser. No. N439, 3 pp.

Anthony, V. C., and H. C. Boyar. 1968. Comparison of meristic characters of adult Atlantic herring from the Gulf of Maine and adjacent waters. *Int. Comm. Northw. Atl. Fish. Res. Bull.* 5:91–98.

Anthony, V. C., and M. J. Fogarty. 1985. Environmental effects on recruitment, growth and vulnerability of Atlantic herring (*Clupea harengus harengus*) in the Gulf of Maine region. *Can. J. Fish. Aquat. Sci.* 42(Suppl. 1):158–173.

Anthony, V. C., and G. T. Waring. 1978. The assessment and management of the Georges Bank herring fishery. *Rapp. Proc.-Verb. Réun. Cons. Int. Explor. Mer* 177:72–111.

Anthony, V. C., and G. T. Waring. 1980. Estimates of herring spawning stock biomass and egg production for the Georges Bank–Gulf of Maine region. NAFO SCR Doc. No. 135, Ser. No. N209, 38 pp.

Aoki, M., and S. Ueyanagi. 1989. Larval morphology and distribution of the dolphin-fishes *Coryphaena hippurus* and *C. equiselis* (Coryphaenidae). *Tokai Univ. Dept. Oceanogr. Bull.* 28:157–174.

Appenzeller, A. R., and W. C. Leggett. 1995. An evaluation of light-mediated vertical migration of fish based on hydroacoustic analysis of the diel vertical movements of rainbow smelt (*Osmerus mordax*). *Can. J. Fish. Aquat. Sci.* 52:504–511.

Applegate, V. C. 1950. Natural history of the sea lamprey, *Petromyzon marinus,* in Michigan. U.S. Fish. Wildl. Serv. Spec. Sci. Rept. Fish. No. 55, 237 pp.

Appleton, T. E., and M. D. B. Burt. 1991. Biochemical characterization of third-stage larval sealworm, *Pseudoterranova decipiens* (Nematoda, Anisakidae) in Canadian Atlantic waters using isoelectric focusing of soluble proteins. *Can. J. Fish. Aquat. Sci.* 48:1800–1803.

Appy, R. G. 1979. Parasites of cod, *Gadus morhua* L. in the northwestern Atlantic Ocean. Ph.D. dissertation, University of New Brunswick, New Brunswick.

Appy, R. G., and M. J. Dadswell. 1978. Parasites of *Acipenser brevi-*

Acipenseridae) in the Saint John River Estuary, N.B., with a description of *Caballeronema pseudoargumentosus* sp. n. (Nematoda: Spirurida). *Can. J. Zool.* 56:1382–1391.

Appy, R. G., and M. J. Dadswell. 1981. Marine and estuarine piscicolid leeches (*Hirudinea*) of the Bay of Fundy and adjacent waters with a key to species. *Can. J. Zool.* 59:183–192.

Aprieto, V. L. 1974. Early development of five carangid fishes of the Gulf of Mexico and the south Atlantic coast of the United States. *Fish. Bull., U.S.* 72:415–443.

Arata, G. F., Jr. 1954. A contribution to the life history of the swordfish, *Xiphias gladius* Linnaeus, from the south Atlantic coast of the United States and the Gulf of Mexico. *Bull. Mar. Sci. Gulf Carib.* 4:183–243.

Arendt, M. D., J. A. Lucy, and D. A. Evans. 2001a. Diel and seasonal activity patterns of adult tautog, *Tautoga onitus,* in lower Chesapeake Bay, inferred from ultrasonic telemetry. *Environ. Biol. Fishes* 62:379–391.

Arendt, M. D., J. A. Lucy, and T. A. Munroe. 2001b. Seasonal occurrence and site-utilization pattterns of adult tautog, *Tautoga onitus* (Labridae), at manmade and natural structures in lower Chesapeake Bay. *Fish. Bull., U.S.* 99:519–527.

Armstrong, M. J., and R. M. Prosch. 1991. Abundance and distribution of the mesopelagic fish *Maurolicus muelleri* in the southern Benguela system. *S. Afr. J. Mar. Sci.* 10:13–28.

Armstrong, M. P. 1997. Seasonal and ontogenetic changes in distribution and abundance of smooth flounder, *Pleuronectes putnami,* and winter flounder, *Pleuronectes americanus,* along estuarine depth and salinity gradients. *Fish. Bull., U.S.* 95:414–430.

Armstrong, M. P., and B. A. Starr. 1994. Reproductive biology of the smooth flounder in Great Bay Estuary, New Hampshire. *Trans. Am. Fish. Soc.* 123:112–114.

Armstrong, M. P., J. A. Musick, and J. A. Colvocoresses. 1992. Age, growth, and reproduction of the goosefish *Lophius americanus* (Pisces: Lophiiformes). *Fish. Bull., U.S.* 90:217–230.

Armstrong, M. P., J. A. Musick, and J. A. Colvocoresses. 1996. Food and ontogenetic shifts in feeding of the goosefish, *Lophius americanus. J. Northw. Atl. Fish. Sci.* 18:99–103.

Armstrong, P. B., and J. S. Child. 1965. Stages in the normal development of *Fundulus heteroclitus. Biol. Bull., Woods Hole* 128:143–168.

Arnold, E. L., and J. R. Thompson. 1958. Offshore spawning of the striped mullet, *Mugil cephalus,* in the Gulf of Mexico. *Copeia* 1958:130–132.

Arruda, L. M. 1988. Maturation cycle in the female gonad of the snipefish, *Macrorhamphosus gracilis* (Lowe, 1839) (Gasterosteiformes, Macrorhamphosidae), off the western coast of Portugal. *Inv. Pesq., Barcelona* 52:355–374.

Arthur, J. R., and E. Albert. 1994. A survey of the parasites of Greenland halibut (*Reinhardtius hippoglossoides*) caught off Atlantic Canada, with notes on their zoogeography in this fish. *Can. J. Zool.* 72:765–778.

Arthur, J. R., and H. P. Arai. 1984. Annotated checklist and bibliography of parasites of herring (*Clupea harengus* L.). Can. Fish. Aquat. Sci. Spec. Publ. No. 70, 26 pp.

Arthur, J. R., E. Albert, and F. Boily. 1995. Parasites of capelin (*Mallotus villosus*) in the St. Lawrence Estuary and Gulf. *Can. J. Fish. Aquat. Sci.* 55:631–638.

Artyukhin, E., and P. Vecsei. 1999. On the status of Atlantic sturgeon: conspecificity of European *Acipenser sturio* and North American *Acipenser oxyrinchus. J. Appl. Ichthyol.* 15(4/5):35–37.

ASMFC (Atlantic States Marine Fisheries Commission). 1985. Fishery management plan for weakfish. ASMFC Fish. Manage. Rept. No. 7, 129 pp.

ASMFC (Atlantic States Marine Fisheries Commission). 1996. Fishery management plan for tautog. ASMFC Fish. Manage. Rept. No. 25, 56 pp.

ASMFC (Atlantic States Marine Fisheries Commission). 1998. Atlantic sturgeon stock assessment peer review report, 125 pp.

Assis, C. A. 1993. On the systematics of *Macrorhamphosus scolopax* (Linnaeus, 1758) and *Macrorhamphosus gracilis* (Lowe, 1839). II. Multivariate morphometric analysis. *Arq. Mus. Bocage, n. sér.* 2:383–402.

Atkins, C. G. 1887. The river fisheries of Maine. In: G. B. Goode, ed. *The Fisheries and Fishery Industries of the United States,* Vol. 1, Sect. 5. Government Printing Office, Washington, D.C., pp. 673–728.

Atkinson, D. B. 1984. Distribution and abundance of beaked redfish in the Gulf of St. Lawrence, 1976–81. *J. Northw. Atl. Fish. Sci.* 5:189–197.

Atkinson, D. B. 1987. The redfish resources off Canada's east coast. In: *Melteff Bed. Proc. Int. Rockfish Symposium, Oct. 1986, Anchorage.* Univ. Alaska, Sea Grant Rept. No. 87, 2 pp.

Atkinson, R. J. A., B. Pelster, C. R. Bridges, A. C. Taylor, and S. Morris. 1987. Behavioural and physiological adaptations to a burrowing lifestyle in the snake blenny, *Lumpenus lampraetaeformis,* and the red band-fish, *Cepola rubescens. J. Fish Biol.* 31:639–659.

Atz, J. W. 1986. *Fundulus heteroclitus* in the laboratory: a history. *Am. Zool.* 26:111–120.

Audet, C., G. J. FitzGerald, and H. Guderley. 1985. Salinity preferences of four sympatric species of sticklebacks (Pisces: Gasterosteidae) during their reproductive season. *Copeia* 1985:209–213.

Auditore, P. J., R. G. Lough, and E. A. Broughton. 1994. A review of the comparative development of Atlantic cod (*Gadus morhua* L.) and haddock (*Melanogrammus aeglefinus* L.) based on an illustrated series of larvae and juveniles from Georges Bank. NAFO Sci. Coun. Stud. No. 20:7–18.

Auer, N. A., ed. 1982. Identification of larval fishes of the Great Lakes basin with emphasis on the Lake Michigan drainages. Great Lakes Fish. Comm. Spec. Publ. Nos. 82–83, 744 pp.

Auster, P. J., R. J. Malatesta, S. C. LaRosa, R. A. Cooper, and L. L. Stewart. 1991. Microhabitat utilization by the megafaunal assemblage at a low relief outer continental shelf site—Middle Atlantic Bight, USA. *J. Northw. Fish. Sci.* 11:59–69.

Auster, P. J., R. J. Malatesta, and C. L. S. Donaldson. 1997. Distributional responses to small-scale variability by early juvenile silver hake, *Merluccius bilinearis. Environ. Biol. Fishes* 50:195–200.

Austin, H. M. 1976. Distribution and abundance of ichthyoplankton in the New York Bight during the fall in 1971. *N.Y. Fish Game J.* 23:58–72.

Austin, H. M., J. Dickinson, and C. R. Hickey. 1973. An ecological study of the ichthyofauna at the Northport Power Station, Long

Island, New York. Reports of the Long Island Lighting Co., Hicksville, 248 pp.

Austin, H. M., A. D. Sosnow, and C. R. Hickey Jr. 1975. The effects of temperature on the development and survival of the eggs and larvae of the Atlantic silversides, *Menidia menidia*. *Trans. Am. Fish. Soc.* 104:762–765.

Avise, J. C., C. A. Reeb, and N. C. Saunders. 1987. Geographic population structure and species differences in mitochondrial DNA of mouthbrooding marine catfishes (Ariidae) and demersal spawning toadfishes (Batrachoididae). *Evolution* 41:991–1002.

Avise, J. C., B. W. Bowen, and T. Lamb. 1989. DNA fingerprints from hypervariable mitochondrial genotypes. *Mol. Biol. Evol.* 6:258–269.

Ayvazian, S. G., and W. H. Krueger. 1992. Lateral plate ontogeny in the North American ninespine stickleback, *Pungitius occidentalis*. *Copeia* 1992:209–214.

Ayvazian, S. G., L. A. Deegan, and J. T. Finn. 1992. Comparison of habitat use by estuarine fish assemblages in the Acadian and Virginian zoogeographic provinces. *Estuaries* 15:368–383.

Azarovitz, T. R. 1988. A brief historical review of the Woods Hole Laboratory trawl survey time series. Can. Fish. Aquat. Sci. Spec. Publ. No. 58:62–67.

Backus, R. H. 1951. New and rare records of fishes from Labrador. *Copeia* 1951:288–294.

Backus, R. H. 1957a. Notes on western North Atlantic sharks. *Copeia* 1957:246–248.

Backus, R. H. 1957b. The fishes of Labrador. *Bull. Am. Mus. Nat. Hist.* 113:279–337.

Backus, R. H., J. E. Craddock, R. L. Haedrich, D. L. Shores, J. M. Teal, A. S. Wing, G. W. Mead, and W. D. Clarke. 1968. *Ceratoscopelus maderensis:* peculiar sound-scattering layer identified with this myctophid fish. *Science* 160:991–993.

Backus, R. H., J. E. Craddock, R. L. Haedrich, and B. H. Robison. 1977. Atlantic mesopelagic zoogeography. Fishes of the western North Atlantic. Sears Found. Mar. Res. Mem. No. 1(7):266–287.

Badcock, J. 1984a. Gonostomatidae. In: P. J. P. Whitehead et al., eds. *Fishes of the North-eastern Atlantic and the Mediterranean,* Vol. 1. UNESCO, Paris, pp. 284–301.

Badcock, J. 1984b. Stenoptychidae. In: P. J. P. Whitehead et al., eds. *Fishes of the North-eastern Atlantic and the Mediterranean,* Vol 1. UNESCO, Paris, pp. 302–317.

Badcock, J. 1984c. Photichthyidae. In: P. J. P. Whitehead et al., eds. *Fishes of the North-eastern Atlantic and the Mediterranean,* Vol. 1. UNESCO, Paris, pp. 318–324.

Badcock, J., and N. R. Merrett. 1976. Midwater fishes in the eastern North Atlantic. I. Vertical distribution and associated biology in 30° N, 23° W, with developmental notes on certain myctophids. *Prog. Oceanog.* 7:3–58.

Baglin, R. E., Jr. 1982. Reproductive biology of western Atlantic bluefin tuna. *Fish. Bull., U.S.* 80:121–134.

Bailey, K. M. 1984. Comparison of laboratory rates of predation on five species of marine fish larvae by three planktonic invertebrates: effects of larval size on vulnerability. *Mar. Biol.*

Bailey, M. M. 1964. Age, growth, maturity, and sex composition of the American smelt, *Osmerus mordax* (Mitchill), of western Lake Superior. *Trans. Am. Fish. Soc.* 93:382–395.

Bailey, R. F. J., K. W. Able, and W. C. Leggett. 1977. Seasonal and vertical distribution and growth of juvenile and adult capelin (*Mallotus villosus*) in the St. Lawrence estuary and western Gulf of St. Lawrence. *J. Fish. Res. Bd. Can.* 34:2030–2040.

Bailey, R. S., and Batty, R. S. 1984. Laboratory study of predation by *Aurelia aurita* on larvae of cod, flounder, plaice and herring: development and vulnerability to capture. *Mar. Biol.* 83:287–291.

Baird, D., and D. E. Ulanowicz. 1989. The seasonal dynamics of the Chesapeake Bay ecosystem. *Ecol. Monogr.* 59:329–364.

Baird, F. T., Jr. 1967. The smelt, *Osmerus mordax*. Fish Educ. Ser., Unit No. 5, *Bull. Maine Dept. Sea Shore Fish.,* 7 pp.

Baird, R. C. 1971. The systematics, distribution, and zoogeography of the marine hatchetfishes (family Sternoptychidae). *Bull. Mus. Comp. Zool.* 142:1–128.

Baird, S. F. 1871. Spawning of the goose fish (*Lophius americanus*). *Am. Nat.* 5:785–786.

Baird, S. F. 1889. The sea fisheries of eastern North America. U.S. Fish Comm. Fish. Rept. 1886, Part 14, App. A, pp. 3–224.

Bakke, T. A., and P. D. Harris. 1998. Diseases and parasites in wild Atlantic salmon (*Salmo salar*) populations. *Can. J. Fish. Aquat. Sci.* 55(Suppl. 1):247–266.

Baldwin, C. C., and G. D. Johnson. 1996. Interrelationships of Aulopiformes. In: M. L. J. Stiassny, L. R. Parenti, and G. D. Johnson, eds. *Interrelationships of Fishes*. Academic Press, San Diego, pp. 355–404.

Ball, A. O., G. R. Sedberry, M. S. Zatcoff, R. W. Chapman, and J. L. Carlin. 2000. Population structure of the wreckfish *Polyprion americanus* determined with microsatellite genetic markers. *Mar. Biol.* 137:1077–1090.

Bane, G. W., Jr. 1965. The opah (*Lampris regius*), from Puerto Rico. *Carib. J. Sci.* 5:63–66.

Banford, H. M., and B. B. Collette. 1993. *Hyporhamphus meeki,* a new species of halfbeak (Teleostei: Hemiramphidae) from the Atlantic and Gulf coasts of the United States. *Proc. Biol. Soc. Washington* 106:369–384.

Baragi, V., and R. T. Lovell. 1986. Digestive enzyme activities in striped bass from first feeding through larva development. *Trans. Am. Fish. Soc.* 115:478–484.

Baranenkova, A. S., V. V. Barsukov, I.Ya. Ponomarenko, T. K. Sysoeva, and N. S. Khokhlina. 1960. Morphological peculiarities, distribution, and feeding of young Baerents Sea wolffishes (*Anarhichas lupus* L., *A. minor* Olafsen, *A. latifrons* Steenstrup et Hallgrimsson). *Zool. Zh.* 39:1186–1200 (in Russian).

Barans, C. A. 1972. Spotted hake, *Urophycis regius,* of the York River and lower Chesapeake Bay. *Chesapeake Sci.* 13:59–62.

Barans, C. A., and A. C. Barans. 1972. Eggs and early larval stages of the spotted hake, *Urophycis regius. Copeia* 1972:188–190.

Barans, C. A., and V. G. Burrell, Jr. 1976. Preliminary findings of trawling on the continental shelf off the southeastern United States during four seasons (1973–1975). S.C. Mar. Resource Cen-

Barans, C. A., and B. W. Stender. 1993. Trends in tilefish distribution and relative abundance off South Carolina and Georgia. *Trans. Am. Fish. Soc.* 122:165–178.

Barber, S. B., and W. H. Mowbray. 1956. Mechanisms of sound production in the sculpin. *Science* 124:219–220.

Barbour, T. 1942. More concerning ceratioid fishes. *Proc. New England Zool. Club* 21:77–86.

Bardach, J. E., and J. Case. 1965. Sensory capabilities of the modified fins of squirrel hake (*Urophycis chuss*) and searobins (*Prionotus carolinus* and *P. evolans*). *Copeia* 1965:194–206.

Bardack, D. 1998. Relationships of living and fossil hagfishes. In: J. M. Joergensen et al., eds. *The Biology of Hagfishes*. Chapman and Hall, London, pp. 3–14.

Barger, L. E., and M. L. Williams. 1980. A summarization and discussion of age and growth of spot, *Leiostomus xanthurus* Lacépède, sand seatrout, *Cynoscion arenarius* Ginsburg, and silver seatrout, *Cynoscion nothus* (Holbrook), based on a literature review. NOAA Tech. Memo. NMFS-SEFC-14, 15 pp.

Barry, T., and B. Kynard. 1986. Attraction of adult American shad to fish lifts at Holyoke Dam, Connecticut River. *No. Am. J. Fish. Manage.* 6:233–241.

Barsukov, V. V. 1953. On the biology of reproduction of the White Sea wolffish (*Anarhichas lupus* L.). *Zool. Zh.* 32:1211–1216 [transl. from Russian, Fish. Res. Bd. Can. Transl. Ser. No. 62, 1956].

Barsukov, V. V. 1959. The wolffish (Anarhichadidae). *Trudy Zool. Inst. Acad. Sci. USSR: Fishes.* 5(5):173 pp. [transl. from Russian, U. S. Dept. Commerce, Nat. Tech. Info. Svc. No. TT67–59074, 292 pp.].

Barsukov, V. V. 1961. On the time of teeth changing in Atlantic wolffish (Anarhichadidae). *Zool. Zh.* 40:462–465.

Barsukov, V. V. 1979. Subspecies of the Atlantic blackbelly rosefish, *Helicolenus dactylopterus* (De la Roche, 1809). *Voprosy Ikhtiol.* 19:579–595 [in Russian, translation in *J. Ichthyol.* 19:1–17].

Barsukov, V. V., N. I. Litvenenko, and V. P. Serebryakov. 1984. Manual for the identification of redfish species of the North Atlantic and adjacent areas. USSR Min. Fish., Atl. Sci.-Res. Inst., Fish. Oceanogr., Kaliningrad, USSR, 28 pp. [transl. from Russian, Can. Transl. Fish. Aquat. Sci. 5168].

Barsukov, V. V., I. A. Oganin, and A. I. Pavlov. 1991. Morphological and ecological differences between *Sebastes fasciatus* and *S. mentella* on the Newfoundland Shelf and Flemish Cap. *J. Ichthyol.* 31:1–17.

Bason, W. H. 1971. Ecology and early life history of striped bass, *Morone saxatilis,* in the Delaware Estuary. *Ichthy. Assoc. Bull.* 4:122 pp.

Bass, A. H. 1990. Sounds from the intertidal zone: vocalizing fish. *BioScience* 40:249–258.

Bass, A. H., M. Weiser, and R. G. Baker. 1987. Functional organization of the sonic motor system in sea robins. *Biol. Bull., Woods Hole* 173:435–437.

Bath, D. W., and J. M. O'Connor. 1982. The biology of the white perch, *Morone americana,* in the Hudson River Estuary. *Fish. Bull., U.S.* 80:599–610.

Bath, D. W., and J. M. O'Connor. 1985. Food preference of white perch in the Hudson River Estuary. *N.Y. Fish Game J.* 32:63–70.

Bath, D. W., J. M. O'Connor, J. B. Alber, and L. G. Arvidson. 1981. Development and identification of larval Atlantic sturgeon (*Acipenser oxyrhynchus*) and shortnose sturgeon (*A. brevirostrum*) from the Hudson River Estuary, New York. *Copeia* 1981:711–717.

Battle, H. I. 1929. Effects of extreme temperatures and salinities on the development of *Enchelyopus cimbrius. Contrib. Canad. Biol.* 5(6):109–192.

Battle, H. I. 1934. Laboratory feeding of the herring. Biol. Bd. Can. Ann. Rept. for 1933, pp. 14–15.

Battle, H. I. 1951. Contributions to a study of the life history of the hake-spawning with notes on age determinations. Fish. Res. Bd. Can., St. Andrews, N.B. Manuscript Rept. Ser. No. 434.

Battle, H. I., A. C. Huntsman, A. M. Jeffers, G. W. Jeffers, W. H. Johnson, and N. H. McNairn. 1936. Fatness, digestion and food of Passamaquoddy young herring. *J. Biol. Bd. Can.* 2:401–429.

Batty, R. S. 1994. The effect of temperature on the vertical distribution of larval herring (*Clupea harengus* L.). *J. Exp. Mar. Biol. Ecol.* 177:269–276.

Batty, R. S., J. H. S. Blaxter, and D. A. Libby. 1986. Herring *Clupea harengus* filter-feeding in the dark. *Mar. Biol.* 91:371–375.

Batty, R. S., J. H. S. Blaxter, and J. M. Richard. 1990. Light intensity and the feeding behaviour of herring, *Clupea harengus. Mar. Biol.* 107:383–388.

Bauchot, M.-L. 1959. Etude des larves leptocéphales du groupe *Leptocephalus lanceolatus* Strömman et identification à la famille de Serrivomeridae. Dana Rept. No. 48, 148 pp.

Baum, E. T. 1997. Maine Atlantic salmon: a national treasure. Atlantic Salmon Unlimited, Hermon, Maine, 240 pp.

Baumiller, T. K., M. LaBarbera, and J. D. Woodley. 1991. Ecology and functional morphology of the isocrinid *Cenocrinus asterius* (Linneaus) (Echinodermata: Crinoidea): in situ and laboratory experiments and observations. *Bull. Mar. Sci.* 48:731–748.

Bayless, J. D. 1972. Artificial propagation and hybridization of striped bass, *Morone saxatilis* (Walbaum). S.C. Wildl. Res. Dept. Rept., 135 pp.

Bayliff, W. H. 1950. The life history of the Atlantic silverside, *Menidia menidia* (L.). Chesapeake Biol. Lab. Publ. No. 90, 27 pp.

Beacham, T. D. 1982. Biology and exploitation of winter flounder *Pseudopleuronectes americanus* in the Canadian Maritimes area of the northwest Atlantic Ocean. Can. Fish. Aquatic Sci. Tech. Rept. No. 1113, 33 pp.

Beacham, T. D. 1983a. Variability in size and age at sexual maturity of argentine, *Argentina silus,* on the Scotian Shelf in the northwest Atlantic Ocean. *Environ. Biol. Fishes* 8:67–72.

Beacham, T. D. 1983b. Variability in size and age at sexual maturity of witch flounder, *Glyptocephalus cynoglossus,* in the Canadian Maritimes region of the northwest Atlantic Ocean. *Can. Field-Nat.* 97:409–422.

Beacham, T. D. 1983c. Variability in median size and age at sexual maturity of Atlantic cod, *Gadus morhua,* on the Scotian Shelf in the Northwest Atlantic Ocean. *Fish. Bull., U.S.* 81:303–321.

Beacham, T. D., and S. J. Nepszy. 1980. Some aspects of the biology of white hake, *Urophycis tenuis,* in the southern Gulf of St. Lawrence. *J. Northw. Atl. Fish. Sci.* 1:49–54.

Beamish, F. W. H. 1966a. Vertical migration by demersal fish in the northwest Atlantic. *J. Fish. Res. Bd. Can.* 23:109–139.

Beamish, F. W. H. 1966b. Swimming endurance of some northwest Atlantic fishes. *J. Fish. Res. Bd. Can.* 23:341–347.

Beamish, F. W. H. 1980. Biology of the North American anadromous sea lamprey, *Petromyzon marinus. Can. J. Fish. Aquat. Sci.* 37:1924–1943.

Beamish, F. W. H., and T. E. Medland. 1988. Age determination for lampreys. *Trans. Am. Fish. Soc.* 117:63–71.

Bean, T. H. 1880. Check-list of duplicates of North American fishes distributed by the Smithsonian Institution in behalf of the United States National Museum, 1877–1880. *Proc. U.S. Nat. Mus.* 3:75–116.

Bean, T. H. 1888. Report on the fishes observed in Great Egg Harbor Bay, New Jersey, during the summer of 1887. *Bull. U.S. Fish Comm.* 7:129–150.

Bean, T. H. 1903. Catalogue of the fishes of New York. *New York State Mus. Bull.* 60, 784 pp.

Bearden, C. M. 1961. List of marine fishes recorded from South Carolina. Bears Bluff Lab Wadmalaw Island, S.C., 12 pp.

Beardsley, G. L., Jr. 1967. Age, growth, and reproduction of the dolphin, *Coryphaena hippurus,* in the Straits of Florida. *Copeia* 1967:441–451.

Beaumariage, D. S. 1964. Returns from the 1963 Schlitz tagging program. Fla. State Bd. Cons. Mar. Lab., Tech. Ser. No. 43, 34 pp.

Beaumariage, D. S. 1969. Returns from the 1965 Schlitz tagging program, including a cumulative analysis of previous results. Fla. Dept. Nat. Res., Mar. Res. Lab. Tech. Ser. No. 59, 38 pp.

Beaven, M., and J. A. Mihursky. 1980. Food and feeding habits of larval striped bass: an analysis of larval striped bass stomachs from the 1976 Potomac Estuary collection. Final Rept. Maryland Power Plant Siting Program. University of Maryland, Chesapeake Biol. Lab. UMCEES 79-45-CBC, Solomons, Md.

Beck, G. G., M. O. Hammil, and T. G. Smith. 1993. Seasonal variation in the diet of harp seals (*Phoca groenlandica*) from the Gulf of St. Lawrence and western Hudson Strait. *Can. J. Fish. Aquat. Sci.* 50:1363–1371.

Beckett, J. S. 1974. Biology of swordfish, *Xiphias gladius* L., in the northwest Atlantic Ocean. In: R. S. Shomura and F. Williams, eds. *Proc. Int. Billfish Symp., Hawaii, 9–12 Aug. 1972.* NOAA Tech. Rept. NMFS SSRF-675, 2:103–106.

Beebe, W. 1935. Deep-sea fishes of the Bermuda Oceanographic Expeditions: Family Derichthyidae. *Zoologica, N.Y.* 20:1–23.

Beebe, W., and J. Crane. 1936. Deep-sea fishes of the Bermuda Oceanographic Expeditions: Family Serrivomeridae. I. Genus *Serrivomer. Zoologica, N.Y.* 20:53–102.

Beebe, W., and J. Tee-Van. 1933. *Field Book of the Shore Fishes of Bermuda.* G. P. Putnam's Sons, New York, 337 pp.

Beese, G. Von, and R. Kandler. 1969. Beiträge zur Biologie der drei nordatlantischen Katfischarten *Anarhichas lupus* L., *A. minor* Olafsen und *A. denticulatus. Berichte der Deutschen Wissenschaftlichen Kommission für Meeresforschung* 20:21–59 [not seen].

Begg, G. A. 1998. A review of stock identification of haddock, *Melanogrammus aeglifinus,* in the northwest Atlantic Ocean. *Mar. Fish. Rev.* 60(4):1–15.

Bejda, A. J., A. L. Studholme, and B. L. Olla. 1987. Behavioral responses of red hake, *Urophycis chuss,* to decreasing concentrations of dissolved oxygen. *Environ. Biol. Fishes* 19:261–268.

Bejda, A. J., B. A. Phelan, and A. L. Studholme. 1992. The effect of dissolved oxygen on the growth of young-of-the-year winter flounder, *Pseudopleuronectes americanus. Environ. Biol. Fishes* 34:321–327.

Bekker, V. E. 1983. Mytophid fishes of the world ocean. Inst. Okean. Shirshova, Akad. Nauk SSSR, 247 pp. [in Russian].

Belding, D. L. 1920. The preservation of the alewife. *Trans. Am. Fish. Soc.* 49:92–104.

Belding, D. L. 1921. A report upon the alewife fisheries of Massachusetts. Div. Fish. Game, Dept. Conserv., Mass. Contrib. No. 11, 135 pp.

Bell, M. A. 1976. Evolution of phenotypic diversity in *Gasterosteus aculeatus* superspecies on the Pacific coast of North America. *Syst. Zool.* 25:211–227.

Bell, M. A. 1981. Lateral plate polymorphism and ontogeny of the complete plate morph of threespine sticklebacks (*Gasterosteus aculeatus*). *Evolution* 35:67–74.

Bell, M. A. 1984. Evolutionary phenetics and genetics: the threespine stickleback, *Gasterosteus aculeatus,* and related species. In: B. J. Turner, ed. *Evolutionary Genetics of Fishes.* Plenum Press, New York, pp. 431–528.

Bell, M. A., and J. V. Baumgartner. 1984. An unusual population of *Gasterosteus aculeatus* from Boston, Massachusetts. *Copeia* 1984:258–262.

Bell, M. A., and S. A. Foster, eds. 1994. *The Evolutionary Biology of the Threespine Stickleback.* Oxford University Press, New York, 571 pp.

Belveze, H., and J. Bravo de Laguna. 1980. Les ressources halieutiques de l'Atlantique centre-est. Deuxième partie. Les ressources de la côte ouest-africaine entre 24° N et le détroit de Gibralter. F.A.O. Fish. Tech. Pap. No. 186, 64 pp.

Benfey, T. J., and D. A. Methven. 1986. Pilot-scale rearing of larval and juvenile lumpfish (*Cyclopterus lumpus* L.), with some notes on early development. *Aquaculture* 56:301–306.

Bengston, D. A. 1984. Resource partitioning by *Menidia menidia* and *Menidia beryllina* (Osteichthyes: Atherinidae). *Mar. Ecol. Progr. Ser.* 18:21–30.

Bengston, D. A. 1985. Laboratory experiments on mechanisms of competition and resource partitioning between *Menidia menidia* (L.) and *Menidia beryllina* (Cope) (Osteichthyes: Atherinidae). *J. Exper. Mar. Biol. Ecol.* 92:1–18.

Bengston, D. A. 1999. Aquaculture of summer flounder (*Paralichthys dentatus*): status of knowledge, current research and future research priorities. *Aquaculture* 176:39–49.

Bennett, M. V. L. 1971. Electric organs. In: W. S. Hoar and D. J. Randall, eds. *Fish Physiology,* Vol. 5. Academic Press, New York, pp. 347–491.

Benoit, D., and W. D. Bowen. 1990a. Summer diet of grey seals (*Halichoerus grypus*) at Anticosti Island, Gulf of St. Lawrence, Canada. *Bull. Can. Fish. Aquat. Sci.* 222:227–242.

Benoit, D., and W. D. Bowen. 1990b. Seasonal and geographic variation in the diet of grey seals (*Halichoerus grypus*) in eastern

Benoit, H. P., and P. Pepin. 1999. Interaction of rearing temperature and maternal influence on egg development rates and larval size at hatch in yellowtail flounder (*Pleuronectes ferrugineus*). *Can. J. Fish. Aquat. Sci.* 56:785–794.

Bentzen, P., W. C. Leggett, and G. G. Brown. 1988. Length and restriction site heteroplasmy in the mitochondrial DNA of American shad (*Alosa sapidissima*). *Genetics* 118:509–518.

Bentzen, P., G. G. Brown, and W. C. Leggett. 1989. Mitochondrial DNA polymorphism, population structure, and life history variation in American shad (*Alosa sapidissima*). *Can. J. Fish. Aquat. Sci.* 46:1446–1454.

Benz, G. W., and S. A. M. Adamson. 1990. Disease caused by *Nemesis robusta* (van Beneden, 1851) (Eudactylinidae: Siphonostomatoida: Copepoda) infections on gill filaments of thresher sharks (*Alopias vulpinus* (Bonnaterre, 1758)), with notes on parasite ecology and life history. *Can. J. Zool.* 68:1180–1186.

Benz, G. W., Z. Lucas, and L. F. Lowry. 1998. New host and ocean records for the copepod *Ommatokoita elongata* (Siphonostomatoida: Lernaeopodidae), a parasite of the eyes of sleeper sharks. *J. Parasitol.* 84:1271–1274.

Berg, L. S. 1940. System of living and fossil fishes. *Trudy Zool. Inst. Akad. Nauk SSSR, V* 2:87–517 [in Russian and English].

Bergeron, N. E., A. G. Roy, D. Chaumont, Y. Mailhot, and E. Guay. 1998. Winter geomorphological processes in the Sainte-Anne River (Quebec) and their impact on the migratory behaviour of Atlantic tomcod (*Microgadus tomcod*). *Regul. Rivers: Res. Manage.* 14:95–105.

Bergert, B. A., and P. C. Wainwright. 1997. Morphology and kinematics of prey capture in the syngnathid fishes *Hippocampus erectus* and *Syngnathus floridae*. *Mar. Biol.* 127:563–570.

Berggren, T. J., and J. T. Lieberman. 1978. Relative contribution of Hudson, Chesapeake and Roanoke striped bass, *Morone saxatilis*, stocks to the Atlantic coast fishery. *Fish. Bull., U.S.* 76:335–345.

Bergin, J. D. 1984. Massachusetts coastal trout management. *Wild Trout* III:137–142.

Bergstad, O. A., and J. D. M. Gordon. 1994. Deep-water ichthyoplankton of the Skagerrak with special reference to *Coryphaenoides rupestris* Gunnerus, 1765 (Pisces, Macrouridae) and *Argentina silus* (Ascanius, 1775) (Pisces, Argentinidae). *Sarsia* 79:33–43.

Berkeley, S. A., and E. D. Houde. 1983. Age determination of broadbill swordfish, *Xiphias gladius*, from the Straits of Florida, using anal fin spine sections. NOAA Tech. Rept. NMFS No. 8:137–143.

Berland, B. 1961. Copepod *Ommatokoita elongata* (Grant) in the eyes of the Greenland shark: a possible cause of mutual dependence. *Nature* 191:829–830.

Bernstein, B. B., B. E. Williams, and K. H. Mann. 1981. The role of behavioral responses to predators in modifying urchins' (*Strongylocentrotus droebachiensis*) destructive grazing and seasonal foraging patterns. *Mar. Biol.* 63:39–49.

Berrien, P. L. 1975. A description of Atlantic mackerel, *Scomber scombrus*, eggs and early larvae. *Fish. Bull., U.S.* 73:186–192.

Berrien, P. L. 1978. Eggs and larvae of *Scomber scombrus* and *Scomber japonicus* in continental shelf waters between Massachusetts and Florida. *Fish. Bull., U.S.* 76:95–115.

Berrien, P. L. 1982. Atlantic mackerel, *Scomber scombrus*. In: M. D. Grosslein and T. R. Azarovitz, eds., *Fish Distribution*. MESA New York Bight Atlas Monogr. No. 15:99–102.

Berrien, P. L., and D. Finan. 1977a. Biological and fisheries data on king mackerel, *Scomberomorus cavalla* Cuvier. Sandy Hook Lab. Tech. Ser. Rept. No. 8, 40 pp.

Berrien, P. L., and D. Finan. 1977b. Biological and fisheries data on Spanish mackerel, *Scomberomorus maculatus* (Mitchill). Sandy Hook Lab. Tech. Ser. Rept. No. 9, 52 pp.

Berrien, P. L., and J. Sibunka. 1999. Distribution patterns of fish eggs in the U.S. northeast continental shelf ecosystem, 1977–1987. NOAA Tech. Rept. NMFS No. 145, 310 pp.

Berrien, P. L., M. P. Fahay, A. W. Kendall Jr., and W. G. Smith. 1978. Ichthyoplankton from the RV *Dolphin* survey of continental shelf waters between Martha's Vineyard, Massachusetts and Cape Lookout, North Carolina, 1965–66. NMFS NEFC Sandy Hook Lab. Tech. Ser. Rept. No. 15, 152 pp.

Berrien, P. L., N. A. Naplin, and M. R. Pennington. 1981. Atlantic mackerel, *Scomber scombrus*, egg production and spawning population estimates for 1977 in the Gulf of Maine, Georges Bank, and Middle Atlantic Bight. *Rapp. Proc.-Verb. Réun. Cons. Int. Explor. Mer* 178:279–288.

Berrill, N. J. 1929. The validity of *Lophius americanus* Val. as a species distinct from *L. piscatorius* Linn., with notes on the rate of development. *Contrib. Canad. Biol. Fish., n. ser.* 4:145–151.

Berry, F. H. 1959a. Boarfishes of the genus *Antigonia* of the western Atlantic. *Bull. Fla. State Mus., Biol. Sci.* 4:205–250.

Berry, F. H. 1959b. Young jack crevalles (*Caranx* species) off the southeastern Atlantic coast of the United States. *Fish. Bull., U.S.* 59:417–535.

Berry, F. H. 1968. A new species of carangid fish (*Decapterus tabl*) from the western Atlantic. *Contr. Mar. Sci.* 13:145–167.

Berry, F. H. 1978. Zeidae. In: W. Fischer, ed. *FAO Species Identification Sheets for Fishery Purposes: Western Central Atlantic (Fishing Area 31)*, Vol. 5. FAO, Rome, 3 pp.

Berry, F. H., and L. Cohen. 1974 (1972). Synopsis of the species of *Trachurus* (Pisces: Carangidae). Quart. *J. Fla. Acad. Sci* 35:177–211.

Berry, F. H., and L. R. Rivas. 1962. Data on six species of needlefishes (Belonidae) from the western Atlantic. *Copeia* 1962:152–160.

Berry, F. H., and W. F. Smith-Vaniz. 1978. Carangidae. In: W. Fischer, ed. *FAO Species Identification Sheets for Fishery Purposes: Western Central Atlantic (Fishing Area 31)*, Vol. 1. FAO, Rome, 48 pp.

Berry, F. H., and L. E. Vogele. 1961. Filefishes (Monacanthidae) of the western North Atlantic. *Fish. Bull., U.S.* 61:60–109.

Bertelsen, E. 1943. Notes on the deep-sea angler-fish *Ceratias holbölli* Kr. based on specimens in the Zoological Museum of Copenhagen. *Vidensk. Medd. Dansk Naturh. Foren.* 107:185–206.

Bertelsen, E. 1951. The ceratioid fishes: ontogeny, taxonomy, distribution and biology. Dana Rept. No. 39, 276 pp.

Bertelsen, E. 1984. Ceratioidei: development and relationships. In: H. G. Moser et al., eds. *Ontogeny and Systematics of Fishes*. Am. Soc. Ichth. Herp. Spec. Publ. No. 1:325–334.

Bertelsen, E., and T. W. Pietsch. 1998. Anglerfishes. In: J. R. Paxton and W. N. Eschmeyer, eds. *Encyclopedia of Fishes*, 2nd Ed. Weldon Owen Pty Ltd, New South Wales, pp. 137–141.

Beverley-Burton, M., and J. H. C. Pippy. 1977. Morphometric variations among larval *Anisakis simplex* (Nematoda: Ascaridoidea) from fishes of the North Atlantic and their use as biological indicators of host stocks. *Environ. Biol. Fishes* 2:309–314.

Beverton, R. J. H. 1963. Maturation, growth and mortality of clupeid and engraulid stocks in relation to fishing. *Rapp. Proc.-Verb. Réun. Cons. Int. Explor. Mer* 154:44–67.

Bibko, P. N., L. Wirtenon, and P. E. Kuser. 1974. Preliminary studies on the effects of air bubbles and intense illumination of the swimming behavior of the striped bass (*Morone saxatilis*) and the gizzard shad (*Dorosoma cepedianum*). In: L. D. Jensen, ed. *Entrainment and Intake Screening: Proc. 2nd Workshop Entrainment Intake Screening.* The Johns Hopkins University, Baltimore, pp. 293–304.

Biette, R. M., D. P. Dodge, R. L. Hassinger, and T. M. Stauffer. 1981. Life history and timing of migrations and spawning behavior of rainbow trout (*Salmo gairdneri*) populations of the Great Lakes. *Can. J. Fish. Aquatic Sci.* 38:1759–1771.

Bigelow, H. B. 1926. Plankton of the offshore waters of the Gulf of Maine. *Bull. U.S. Bur. Fish.* 40(2):1–509.

Bigelow, H. B. 1927. Physical oceanography of the Gulf of Maine. *Bull. U.S. Bur. Fish.* 40(2):511–1027.

Bigelow, H. B., and T. Barbour. 1944a. A new giant ceratioid fish. *Proc. New England Zool. Club* 23: 9–15.

Bigelow, H. B., and T. Barbour. 1944b. *Reganula gigantea* to replace *Reganichthys giganteus. Copeia* 1944:123.

Bigelow, H. B., and W. C. Schroeder. 1929. A rare bramid fish (*Taractes princeps* Johnson) in the northwestern Atlantic. *Bull. Mus. Comp. Zool.* 69:41–50.

Bigelow, H. B., and W. C. Schroeder. 1939. Notes on the fauna above mud bottoms in deep water in the Gulf of Maine. *Biol. Bull., Woods Hole* 76:305–324.

Bigelow, H. B., and W. C. Schroeder. 1940. Notes on New England fishes. *Copeia* 1940:139.

Bigelow, H. B., and W. C. Schroeder. 1948a. Cyclostomes. Fishes of the western North Atlantic. Sears Found. Mar. Res. Mem. No. 1(1):29–58.

Bigelow, H. B., and W.C. Schroeder. 1948b. Sharks. Fishes of the western North Atlantic. Sears Found. Mar. Res. Mem. No. 1(1):59–546.

Bigelow, H. B., and W. C. Schroeder. 1953a. Sawfishes, guitarfishes, skates and rays. Fishes of the western North Atlantic. Sears Found. Mar. Res. Mem. No. 1(2):1–514.

Bigelow, H. B., and W. C. Schroeder. 1953b. Chimaeroids: fishes of the western North Atlantic. Sears Found. Mar. Res. Mem. No. 1(2):515–562.

Bigelow, H. B., and W. C. Schroeder. 1953c. Fishes of the Gulf of Maine. *Fish. Bull., U.S.* 53:1–577.

Bigelow, H. B., and W. C. Schroeder. 1954. Deep water elasmobranchs and chimaeroids from the northwestern Atlantic slope. *Bull. Mus. Comp. Zool.* 112:35–87.

Bigelow, H. B., and W. C. Schroeder. 1955. Occurrence off the middle and north Atlantic United States of the offshore hake *Merluccius albidus* (Mitchill) 1818, and the blue whiting *Gadus (Micromesistius) poutassou* (Risso) 1826. *Bull. Mus. Comp. Zool.*

Bigelow, H. B., and W. C. Schroeder. 1962. New and little known batoid fishes from the western Atlantic. *Bull. Mus. Comp. Zool.* 128:159–244.

Bigelow, H. B., and W. C. Schroeder. 1963. Family Osmeridae. Fishes of the western North Atlantic. Sears Found. Mar. Res. Mem. No. 1(3):553–594.

Bigelow, H. B., and W. C. Schroeder. 1965. A further account of batoid fishes from the western Atlantic. *Bull. Mus. Comp. Zool.* 132:443–477.

Bigelow, H. B., and W. C. Schroeder. 1968. Additional notes on batoid fishes from the western Atlantic. *Breviora* 281, 23 pp.

Bigelow, H. B., and W. W. Welsh. 1925. Fishes of the Gulf of Maine. *Bull. U.S. Bur. Fish.* 40(1):1–567.

Billerbeck, J. M., G. Orti, and D. O. Conover. 1997. Latitudinal variation in vertebral number has a genetic basis in the Atlantic silverside, *Menidia menidia. Can. J. Fish. Aquat. Sci.* 54:1796–1801.

Binney, A. 1842. Observations made during two successive summers at Nahant, on the habits of the *Orthagoriscus mola,* or short sunfish. *Proc. Boston Soc. Nat. Hist.* 1:93.

Bird, J. L., D. T. Eppler, and D. M. Checkley, Jr. 1986. Comparisons of herring otoliths using Fourier series shape analysis. *Can. J. Fish. Aquat. Sci.* 43:1228–1234.

Birstein, V. J., J. R. Waldman, and W. E. Bemis, eds. 1997. Sturgeon biodiversity and conservation. *Environ. Biol. Fishes* 48, 444 pp.

Bjelland, O., O. A. Bergstad, J. E. Skjæraasen, and K. Meland. 2000. Trophic ecology of deep-water fishes associated with the continental slope of the eastern Norwegian Sea. *Sarsia* 85:101–117.

Björke, H. J., J. Gsjösaeter, and R. Saetre. 1972. Undersøkelsei pa"loddos gytefelti 1972 (Investigations at the spawning grounds of capelin in 1972). *Fiskets Gang* 58:710–716.

Blacker, R. W. 1962. Rare fishes from the Atlantic Slope fishing grounds. *Ann. Mag. Nat. Hist., Ser. 13,* 5:261–271.

Blacker, R. W. 1971. Synopsis of biological data on haddock *Melanogrammus aeglefinus* (Linnaeus) 1758. FAO Fish. Synop. No. 84, 51 pp.

Blacker, R. W. 1983. Pelagic records of the lumpsucker, *Cyclopterus lumpus* L. *J. Fish Biol.* 23:405–417.

Blackwell, B. F., W. B. Krohn, and R. B. Allen. 1995. Foods of nestling double-crested cormorants in Penobscot Bay, Maine, USA: temporal and spatial comparisons. *Colonial Waterbirds* 18(2):199–208.

Blackwood, G. 1983. Lumpfish roe fishery development in Newfoundland, 1982–83. Newfoundland Dept. Fish., Ind. Support Serv., Develop. Rept. No. 31, 29 pp.

Blaxter, J. H. S. 1964. Spectral sensitivity of the herring *Clupea harengus* L. *J. Exp. Biol.* 41:155–162.

Blaxter, J. H. S. 1966. The effect of light intensity on the feeding ecology of herring. In: R. Bainbridge, G. C. Evans, and O. Rackham, eds. *Light as an Ecological Factor.* Symp. Brit. Ecol. Soc., 30 March–1 April 1965, Cambridge. John Wiley & Sons, New York, pp. 393–409.

Blaxter, J. H. S. 1968. Visual thresholds and spectral sensitivity of

Blaxter, J. H. S. 1985. The herring: a successful species? *Can. J. Fish. Aquat. Sci.* 42(Suppl. 1):21–30.

Blaxter, J. H. S. 1990. The herring. *Biologist (London)* 37:27–31.

Blaxter, J. H. S., and F. G. T. Holliday. 1963. The behavior and physiology of herring and other clupeids. *Adv. Mar. Biol.* 1:261–393.

Blaxter, J. H. S., and J. R. Hunter. 1982. The biology of the clupeoid fishes. *Adv. Mar. Biol.* 20:1–223.

Blaxter, J. H. S., and B. B. Parrish. 1965. The importance of light in shoaling, avoidance of nets and vertical migration by herring *Clupea harengus. J. Conseil, Conseil Int. Explor. Mer* 30:40–57.

Blaxter, J. H. S., and M. Staines. 1971. Food searching potential in marine fish larvae. In: D. J. Crisp, ed. *Fourth European Marine Biology Symposium.* Cambridge University Press, Cambridge, pp. 467–485.

Blaxter, J. H. S., D. Danielssen, E. Moksness, and V. Øiestad. 1983. Description of the early development of the halibut *Hippoglossus hippoglossus* and attempts to rear the larvae past first feeding. *Mar. Biol.* 73:99–107.

Bleakney, J. S. 1963a. First record of the fish *Pogonias cromis* from Canadian waters. *Copeia* 1963:173.

Bleakney, J. S. 1963b. Notes on the distribution and reproduction of the fish *Tautoga onitis* in Nova Scotia. *Can. Field-Natur.* 77:64–65.

Bleakney, J. S., and D. E. McAllister. 1973. Fishes stranded during extreme low tides in Minas Basin, Nova Scotia. *Can. Field-Natur.* 87:371–376.

Bley, P. W., and J. R. Moring. 1988. Freshwater and ocean survival of Atlantic salmon and steelhead: a synopsis. U.S. Fish Wild. Serv., Biol. Rept. No. 88(9):1–20.

Block, B. A. 1991. Endothermy in fish: thermogenesis, ecology and evolution. In: P. W. Hochachka and T. Mommsen, eds. *Biochemistry and Molecular Biology of Fishes,* Vol. 1:269–311. Elsevier Science Publishers, Amsterdam.

Blouw, D. M., and D. W. Hagen. 1981. Ecology of the fourspine stickleback, *Apeltes quadracus,* with respect to a polymorphism for dorsal spine number. *Can. J. Zool.* 59:1677–1692.

Blouw, D. M., and D. W. Hagen. 1984a. The adaptive significance of dorsal spine variation in the fourspine stickleback, *Apeltes quadracus.* IV. Phenotypic covariation within closely related species. *Heredity* 53:383–396.

Blouw, D. M., and D. W. Hagen. 1984b. The adaptive significance of dorsal spine variation in the fourspine stickleback, *Apeltes quadracus.* III. Correlated traits and experimental evidence on predation. *Heredity* 53:371–382.

Blouw, D. M., and D. W. Hagen. 1990. Breeding ecology and evidence of reproductive isolation of a widespread stickleback fish (Gasterosteidae) in Nova Scotia, Canada. *Biol. J. Linn. Soc.* 39:195–217.

Boehlert, G. W. 1984. Scanning electron microscopy. In: H. G. Moser et al., eds. *Ontogeny and Systematics of Fishes.* Am. Soc. Ich. Herp. Spec. Publ. No. 1:43–48.

Boëtius, J., and E. F. Harding. 1985. A re-examination of Johannes Schmidt's Atlantic eel investigations. *Dana* 4:129–162.

Böhlke, J. E., and C. C. G. Chaplin. 1968. *Fishes of the Bahamas and Adjacent Tropical Waters.* Livingston Publ. Co., Wynnewood, Pa., 771 pp.

Böhlke, J. E., and C. C. G. Chaplin. 1993. *Fishes of the Bahamas and Adjacent Tropical Waters,* 2nd Ed. University of Texas Press, Austin, 771 pp.

Boileau, M. G. 1985. The expansion of the white perch, *Morone americana,* in the lower Great Lakes. *Fisheries* 10(1):6–10.

Boje, J., F. Riget, and M. Koie. 1997. Helminth parasites as biological tags in population studies of Greenland halibut (*Reinhardtius hippoglossoides* (Walbaum)), in the north-west Atlantic. *ICES J. Mar. Sci.* 54:886–895.

Bolduc, F., and G. J. FitzGerald. 1989. The role of selected environmental factors and sex ratio upon egg production in three-spine sticklebacks, *Gasterosteus aculeatus. Can. J. Zool.* 67:2013–2020.

Bolles, K. L., and G. A. Begg. 2000. Distinction between silver hake (*Merluccius bilinearis*) stocks in U.S. waters of the northwest Atlantic based on whole otolith morphometrics. *Fish. Bull., U.S.* 98:451–462.

Bollingberg, H. J., and P. Johansen. 1979. Lead in spotted wolffish, *Anarhichas minor,* near a zinc-lead mine in Greenland. *J. Fish. Res. Bd. Can.* 36:1023–1028.

Bolz, G. R., and R. G. Lough. 1983. Growth of larval Atlantic cod, *Gadus morhua,* and haddock, *Melanogrammus aeglefinus,* on Georges Bank, spring 1981. *Fish. Bull., U.S.* 81:827–836.

Bolz, G. R., R. G. Lough, and D. C. Potter. 1981. Autumn and winter abundance and distribution of ichthyoplankton on Georges Bank and Nantucket Shoals, 1974–76 with special emphasis on dominant species. *Rapp. P-V Réun. Cons. Perm. Int. Explor. Mer* 178:168–170.

Bonn, E. W., W. M. Bailey, J. D. Bayless, K. E. Erickson, and R. E. Stevens, eds. 1976. *Guidelines for Striped Bass* Culture. Striped Bass Comm., S. Div. Am. Fish. Soc., 103 pp.

Booker, R. G., D. J. Baumgartner, J. A. Hutcheson, R. H. Roy, and T. L. Wellborin Jr. 1969. Striped bass *Morone saxatilis* (Walbaum) 1968 report on the development of essential requirements for production. U.S. Fish. Wildl. Ser. Publ., 112 pp.

Booth, R. A. 1967. A description of the larval stages of the tomcod, *Microgadus tomcod,* with comments on its spawning ecology. Ph.D. thesis, University of Connecticut, Storrs, 53 pp.

Boreman, J. 1983. Status of bluefish along the Atlantic Coast 1982. NMFS Woods Hole Lab. Ref. Doc. No. 83-28, 35 pp.

Boreman, J. 1997. Sensitivity of North American sturgeon and paddlefish to fishing mortality. *Environ. Biol. Fishes* 48: 399–405.

Boreman, J., and R. R. Lewis. 1987. Atlantic coastal migration of striped bass. In: M. J. Dadswell et al., eds. *Common Strategies of Anadromous and Catadromous Fishes. Am. Fish. Soc. Symp.* 1:331–339.

Boreman, J., B. S. Nakashima, J. A. Wilson, and R. L. Kendall, eds. 1997. *Northwest Atlantic Groundfish: Perspectives on a Fishery Collapse.* Am. Fish. Soc., 242 pp.

Borges, L. 2000. Age and growth of the snipefish, *Macrorhamphosus* spp., in the Portuguese continental waters. *J. Mar. Biol. Assoc. U.K.* 80:147–153.

Borges, L. 2001. A new maximum length for the snipefish, *Macroramphosus scolopax. Cybium* 25:191–192.

Borodin, N. A. 1924. Age of shad estimated from examination of scales. *Science* 60:477.

Borodin, N. 1925. Biological observations on the Atlantic sturgeon *Acipenser sturio. Trans. Am. Fish. Soc.* 55:184–190.

Borodulina, O. D. 1964. Some data on the biology of the argentine *Argentina silus* (Ascanius). *Voprosy Ikhtiol.* 4:68–81 [transl. from Russian, Fish. Lab., Lowestoft, Suffolk, n.s. 59 (1965) 4:68–81].

Borodulina, O. D. 1968. *Argentina silus* in the Atlantic Ocean. Section 3: Greater silver smelt (or argentine) *Argentina silus* (Ascanius, 1775). *Rapp. P.-Verb. Réun. Cons. Int. Explor. Mer* 158:54–57.

Borodulina, O. D. 1989. On the distribution of *Centrolophus niger* and *Schedophilus huttoni* (Centrolophidae) in the southern part of the Pacific Ocean. *Voprosy Ikhtiol.* 29:656–658 [in Russian, translation in *J. Ichthyol.* 29(6): 157–160].

Bortone, S. A., P. A. Hastings, and S. B. Collard. 1977. The pelagic-*Sargassum* ichthyofauna of the eastern Gulf of Mexico. *Northeast Gulf Sci.* 1:60–67.

Borucinska, J. D., G. W. Benz, and H. E. Whiteley. 1998. Ocular lesions associated with attachment of the parasitic copepod *Ommatokoita elongata* (Grant) to corneas of Greenland sharks, *Somniosus microcephalus* (Bloch & Schneider). *J. Fish Diseases* 21:415–422.

Botta, N. 1990. Prickly and slimy: urchins and eels bolster west coast fisheries. *Pacific Fishing,* Nov. 1990:93–97.

Boughton, D. A., B. B. Collette, and A. R. McCune. 1991. Heterochrony in jaw morphology of needlefishes (Teleostei: Belonidae). *Syst. Zool.* 40:329–354.

Bourgeois, C. E., and I-H. Ni. 1984. Metazoan parasites of northwest Atlantic redfishes (*Sebastes* spp.). *Can. J. Zool.* 62:1879–1885.

Bourne, D. W., and J. J. Govoni. 1988. Distribution of fish eggs and larvae and patterns of water circulation in Narragansett Bay, 1972–1973. In M. P. Weinstein, ed. *Larval Fish and Shellfish Transport through Inlets. Am. Fish. Soc. Symp.* 3:132–148.

Bourne, N. 1964. Scallops and the offshore fishery of the maritimes. *Bull. Fish. Res. Bd. Can.* 145, 60 pp.

Bousfield, E. L. 1987. Amphipod parasites of fishes of Canada. *Bull. Can. Fish. Aquatic Sci.* 217, 37 pp.

Bowen, B. W., and J. C. Avise. 1990. Genetic structure of Atlantic and Gulf of Mexico populations of sea bass, menhaden, and sturgeon: influences of zoogeographic factors and life history patterns. *Mar. Biol.* 107:371–381.

Bowen, W. D., and G. D. Harrison. 1996. Comparison of harbour seal diets in two inshore habitats of Atlantic Canada. *Can. J. Zool.* 74:125–135.

Bowen, W. D., J. W. Lawson, and B. Beck. 1993. Seasonal and geographic variation in the species composition and size of prey consumed by grey seals (*Halichoerus grypus*) on the Scotian Shelf. *Can. J. Fish. Aquat. Sci.* 50:1768–1778.

Bowering, W. R. 1976. Distribution, age and growth, and sexual maturity of witch flounder (*Glyptocephalus cynoglossus*) in Newfoundland waters. *J. Fish. Res. Bd. Can.* 33:1574–1584.

Bowering, W. R. 1978. Fecundity of witch flounder (*Glyptocephalus cynoglossus*) from St. Pierre Bank and the Grand Bank of Newfoundland. *J. Fish. Res. Bd. Can.* 35:1199–1206.

Bowering, W. R. 1980. Fecundity of Greenland halibut, *Reinhardtius hippoglossoides* (Walbaum), from southern Laborador and southeastern Gulf of St Lawrence. *J. Northw. Atl. Fish. Sci.* [illegible]

Bowering, W. R. 1983. Turbot (Greenland halibut). Dept. Fish. Oceans, Underwater World, 5 pp.

Bowering, W. R. 1986. The distribution, age, growth and sexual maturity of Atlantic halibut (*Hippoglossus hippoglossus*) in the Newfoundland and Laborador area of the northwest Atlantic. Can. Fish. Aquatic Sci. Tech. Rept. No. 1432, 34 pp.

Bowering, W. R. 1987. Distribution of witch flounder *Glyptocephalus cynoglossus,* in the southern Laborador and eastern Newfoundland area and changes in certain biological parameters after 20 years of exploitation. *Fish. Bull., U.S.* 85:611–629.

Bowering, W. R. 1989. Witch flounder distribution off southern Newfoundland, and changes in age, growth, and sexual maturity patterns with commercial exploitation. *Trans. Am. Fish. Soc.* 118:659–669.

Bowering, W. R., and W. B. Brodie. 1991. Distribution of commercial flatfishes in the Newfoundland–Laborador region of the Canadian northwest Atlantic and changes in certain biological parameters since exploitation. *Neth. J. Sea Res.* 27:407–422.

Bowering, W. R., and W. B. Brodie. 1995. Greenland halibut (*Reinhardtius hippoglossoides*): a review of the dynamics of its distribution and fisheries off eastern Canada and Greenland. In: A. G. Hopper, ed. *Deep Water Fisheries of the North Atlantic Slope.* NATO ASI Ser. No. 20, pp. 113–160. Kluwer Academic Publishers.

Bowering, W. R., and G. R. Lilly. 1992. Greenland halibut (*Reinhardtius hippoglossoides*) off southern Labrador and northeastern Newfoundland (northwest Atlantic) feed primarily on capelin (*Mallotus villosus*). *Neth. J. Sea Res.* 29:211–222.

Bowering, W. R., and K. H. Nedreaas. 2000. A comparison of Greenland halibut (*Reinhardtius hippoglossoides* (Walbaum)) fisheries and distribution in the Northwest and Northeast Atlantic. *Sarsia* 85:61–76.

Bowering, W. R., and K. H. Nedreaas. 2001. Age validation and growth of Greenland halibut (*Reinhardtius hippoglossoides* (Walbaum)): a comparison of populations in the Northwest and Northeast Atlantic. *Sarsia* 86:53–68.

Bowers, A. B. 1960. Growth of the witch (*Glyptocephalus cynoglossus* (L.) in the Irish Sea. *J. Conseil, Conseil Perm. Int. Explor. Mer* 25(2):168–176.

Bowers, A. B., and A. R. Brand. 1973. Stock size and recruitment in Manx herring. *Rapp. Proc.-Verb. Réun, Cons. Int. Explor. Mer* 164:37–41.

Bowman, A. 1919. The eggs and larvae of the angler (*Lophius piscatorius*) in Scottish waters. Fish. Bd. Scotland Rept. *Sci. Invest.* 1919:(2):21–23.

Bowman, R. E. 1980. Food of northwest Atlantic juvenile haddock. M.A. thesis, Bridgewater State College, Bridgewater, Mass., 95 pp.

Bowman, R. E. 1981. Food of 10 species of northwest Atlantic juvenile groundfish. *Fish. Bull., U.S.* 79:200–206.

Bowman, R. E. 1984. Food of silver hake, *Merluccius bilinearis. Fish. Bull., U.S.* 82:21–35.

Bowman, R. E., and E. W. Bowman. 1980. Diurnal variation in the feeding intensity and catchability of silver hake (*Merluccius bilinearis*). *Can. J. Fish. Aquat. Sci.* [illegible]

Bowman, R. E., and W. L. Michaels. 1984. Food of seventeen species of northwest Atlantic fish. NOAA Tech. Memo. NMFS-F/NEC-28, 183 pp.

Bowman, R. E., R. Eppi, and M. D. Grosslein. 1984. Diet and consumption of spiny dogfish in the northwest Atlantic. ICES Demersal Fish. Comm., ICES C.M. 1984/G:27, 8 pp.

Bowman, R. E., T. R. Azarovitz, E. S. Howard, and B. P. Hayden. 1987. Food and distribution of juveniles of seventeen northwest Atlantic fish species, 1973–1976. NOAA Tech. Memo. NMFS-F/NEC-45, 57 pp.

Bowman, R. E., C. E. Stillwell, W. L. Michaels, and M. D. Grosslein. 2000. Food of Northwest Atlantic fishes and two common species of squid. NOAA Tech. Memo. NMFS-F/NE-155, 138 pp.

Boyar, H. C. 1964. Occurrence of the Greenland halibut, *Reinhardtius hippoglossoides* (Walbaum), in shallow waters in the Gulf of Maine. *Copeia* 1964:232–233.

Boyar, H. C. 1968. Age, length and gonadal stages of herring from Georges Bank and the Gulf of Maine. *Int. Comm. Northw. Atl. Fish. Res. Bull.* 5:49–61.

Boyar, H. C., R. A. Cooper, and R. A. Clifford. 1973a. A study of the spawning and early life history of herring (*Clupea harengus harengus* L.) on Jeffrey's Ledge in 1972. Int. Comm. Northw. Atl. Fish. Res. Doc. 73/96, Ser. No. 3054, 27 pp.

Boyar, H. C., R. R. Marak, F. E. Perkins, and R. A. Clifford. 1973b. Seasonal distribution and growth of larval herring (*Clupea harengus* L.) in the Georges Bank-Gulf of Maine area from 1962 to 1970. *J. Cons. Int. Explor. Mer* 35:36–51.

Boynton, W. R., T. T. Polgar, and H. H. Zion. 1981. Importance of juvenile striped bass food habits in the Potomac estuary. *Trans. Am. Fish. Soc.* 110:56–63.

Bozeman, E. L., Jr., and J. M. Dean. 1980. The abundance of estuarine larval and juvenile fish in a South Carolina intertidal creek. *Estuaries* 3:89–97.

Bozhkov, A. T. 1975. An instance of an attack by a swordfish (*Xiphius gladius*) on a mako shark (*Isurus glaucus*). *Voprosy Ikhtiol.* 15:934–935 [in Russian, translation in *J. Ichthyol.* 15:842–843].

Bradbury, C., J. M. Green, and M. Bruce-Lockhart. 1995. Home ranges of female cunner, *Tautogolabrus adspersus* (Labridae), as determined by ultrasonic telemetry. *Can. J. Zool.* 73:1268–1279.

Bradbury, C., J. M. Green, and M. Bruce-Lockhart. 1997. Daily and seasonal activity patterns of female cunner, *Tautogolabrus adspersus* (Labridae), in Newfoundland. *Fish. Bull., U.S.* 95:646–652.

Bradford, A. D., J. G. Miller, and K. Buss. 1968. Bioassays of egg and larval stages of American shad, *Alsoa sapidissima*. In: *Suitability of the Susquehanna River for Restoration of Shad.* U.S. Dept. Inter., Washington, D.C., 60 pp.

Bradford, R. G., G. Chaput, T. Hurlbut, and R. Morin. 1997. Bycatch of striped bass, white hake, winter flounder, and Atlantic tomcod in the autumn "open water" smelt fishery of the Miramichi River Estuary. Can. Fish. Aquat. Sci. Tech. Rept. No. 2195, 37 pp.

Branstetter, S. 1981. Biological notes on the sharks of the north central Gulf of Mexico. *Contrib. Mar. Sci.* 24:13–34.

Branstetter, S. 1986. Biological parameters of the sharks of the northwestern Gulf of Mexico in relation to their potential as a commercial fishery resource. Ph.D. dissertation, Texas A & M University, College Station, Tex., 138 pp.

Branstetter, S. 1987. Age and growth validation of newborn sharks held in laboratory aquaria, with comments on the life history of the Atlantic sharpnose shark, *Rhizoprionodon terraenovae. Copeia* 1987:291–300.

Branstetter, S. 1990. Early life-history implications of selected carcharhinoid and lamnoid sharks of the northwest Atlantic. NOAA Tech. Rept. NMFS No. 90:17–28.

Branstetter, S., and G. Burgess. 1997. Commercial shark fishery observer program: 1996. Final report to NOAA/NMFS (MARFIN Award NA57FF0286).

Branstetter, S., and J. D. McEachran. 1983. A first record of the bigeye thresher, *Alopias superciliosus,* the blue shark, *Prionace glauca,* and the pelagic stingray, *Dasyatis violacea,* from the Gulf of Mexico. *Northeast Gulf Sci.* 6:59–61.

Branstetter, S., and J. A. Musick. 1994. Age and growth estimates for the sand tiger in the northwestern Atlantic Ocean. *Trans. Am. Fish. Soc.* 123:242–254.

Branstetter, S., J. A. Musick, and J. A. Colvocoresses. 1987. A comparison of the age and growth estimates of the tiger shark, *Galeocerdo cuvieri,* from off Virginia and from the northwestern Gulf of Mexico. *Fish. Bull., U.S.* 85:269–279.

Braune, B. M. 1987. Mercury accumulation in relation to size and age of Atlantic herring (*Clupea harengus harengus*) from the southwestern Bay of Fundy, Canada. *Arch. Environ. Contam. Toxicol.* 16:311–320.

Braune, B. M., and D. E. Gaskin. 1982. Feeding ecology of nonbreeding populations of larids off Deer Island, New Brunswick. *Auk* 99:67–76.

Brawn, V. M. 1960a. Temperature tolerance of unacclimated herring (*Clupea harengus* L.). *J. Fish. Res. Bd. Can.* 17:721–723.

Brawn, V. M. 1960b. Underwater television observations of the swimming speed and behaviour of captive herring. *J. Fish. Res. Bd. Can.* 17:689–698.

Brawn, V. M. 1960c. Temperature tolerance of unacclimated herring *Clupea harengus* L. *J. Fish. Res. Bd. Can.* 17:721–723.

Brawn, V. M. 1960d. Survival of herring (*Clupea harengus* L.) in water of low salinity. *J. Fish. Res. Bd. Can.* 17:725–726.

Brawn, V. M. 1961. Reproductive behavior of the cod (*Gadus callarias* L.). *Behavior* 18:177–198.

Brawn, V. M. 1969. Feeding behavior of cod (*Gadus morhua*). *J. Fish. Res. Bd. Can.* 26:583–596.

Bray, R. A. 1979. Digenea in marine fishes from the eastern seaboard of Canada. *J. Nat. Hist.* 13:399–431.

Bray, R. A. 1987. A study of the helminth parasites of *Anarhichas lupus* (Perciformes: Anarhichadidae) in the North Atlantic. *J. Fish Biol.* 31:237–264.

Bray, R. A., and D. I. Gibson. 1991. The Acanthocolpidae (Digenea) of fishes from the north-east Atlantic: the status of *Neophasis* Stafford, 1904 (Digenea) and a study of North Atlantic forms. *Syst. Parasitol.* 19:95–117.

Bray, R. A., and K. MacKenzie. 1990. *Aponurus laguncula* Looss, 1907 (Digenea: Lecithasteridae): a report from herring, *Clupea harengus*

L., in the eastern English Channel and a review of its biology. *Syst. Parasitol.* 17:115–124.

Breder, C. M., Jr. 1932. On the habits and development of certain Atlantic Synentognathi. Carnegie Inst. Wash. Publ. No. 435, Tortugas Lab. Pap. No. 28, 35 pp.

Breder, C. M., Jr. 1933. Young tarpon on Andros Island. *Bull. N.Y. Zool. Soc.* 36:65–67.

Breder, C. M., Jr. 1942. Social and respiratory behavior of large tarpon. *Zoologica, N.Y.* 27:1–4.

Breder, C. M., Jr. 1967. On the survival value of fish schools. *Zoologica, N.Y.* 50:97–114

Breder, C. M., Jr., and E. Clark. 1947. A contribution to the visceral anatomy, development, and relationships of the Plectognathi. *Bull. Am. Mus. Nat. Hist.* 88:287–319.

Breder, C. M., Jr., and R. F. Nigrelli. 1936. The winter movements of the landlocked alewife, *Pomolobus pseudoharengus* (Wilson). *Zoologica, N.Y.* 21:165–175.

Breder, C. M., Jr., and D. E. Rosen. 1966. *Modes of Reproduction in Fishes.* Natural History Press, Garden City, N.Y., 941 pp.

Breen, P. A., and K. H. Mann. 1976. Destructive grazing of kelp by sea urchins in eastern Canada. *J. Fish. Res. Bd. Can.* 33:1278–1283.

Brêthes, J. C. 1979. Contribution à l'étude des populations de *Macrorhamphosus scolopax* (L., 1758) et *Macrorhamphosus gracilis* (Lowe, 1839) des côtes Atlantiques Marocaines. *Bull. Inst Pêches Maritimes* 26:1–62

Brêthes, J.-C. F., R. Saint-Pierre, and G. Desrosiers. 1992. Growth and sexual maturation of the American sand lance (*Ammodytes americanus* Dekay) off the north shore of the Gulf of St. Lawrence. *J. Northw. Atl. Fish. Sci.* 12:41–48.

Brewer, J. 1965. Marine life on the artificial reef off Fire Island, New York. *Underwater Natur.* 3:11–14.

Bridges, D. W., and M. P. Fahay. 1968. Sexual dichromatism in the tautog, *Tautoga onitis* (Linnaeus), with an observation of possible courtship behavior. *Trans. Am. Fish. Soc.* 97:208–209.

Bridges, D. W., D. R. Pedro, and M. Laird. 1975. The occurrence of *Haemogregarina* sp. in the goosefish (*Lophius americanus* Valenciennes) from the Gulf of Maine. *Trans. Am. Fish. Soc.* 104: 284–285.

Bridges, W. L., and R. D. Anderson. 1984. A brief survey of Pilgrim Nuclear Power Plant effects upon the marine aquatic environment. In: J. D. Davis and D. Merriman, eds. *Observations on the Ecology and Biology of Western Cape Cod Bay, Massachusetts.* Springer-Verlag, Berlin, pp. 263–271.

Briggs, J. C. 1958. A list of Florida fishes and their distribution. *Bull. Fla. State Mus. Biol. Ser.* 2:223–318.

Briggs, J. C. 1960a. Fishes of worldwide (circumtropical) distribution. *Copeia* 1960:171–180.

Briggs, J. C. 1960b. The nomenclature of *Centropristis,* Cuvier, 1829. *Copeia* 1960:358.

Briggs, P. T. 1969a. The sport fisheries for tautog in the inshore waters of eastern Long Island. *N.Y. Fish Game J.* 16:238–254.

Briggs, P. T. 1969b. A length-weight relationship for tautog from the inshore waters of eastern Long Island. *N.Y. Fish Game J.* 16:[illegible]

Briggs, P. T. 1975. An evaluation of artificial reefs in New York's marine waters. *N.Y. Fish Game J.* 22:51–56.

Briggs, P. T. 1977a. An unusually large striped searobin from Long Island Sound, New York. *N.Y. Fish Game J.* 24:96–97.

Briggs, P. T. 1977b. Status of tautog populations at artificial reefs in New York waters and effect of fishing. *N.Y. Fish Game J.* 24:154–167.

Briggs, P. T., and J. S. O'Connor. 1971. Comparison of shore-zone fishes over naturally vegetated and sand-filled bottoms in Great South Bay. *N.Y. Fish Game J.* 18:15–41.

Briggs, R. P. 1988. The occurrence of two rare fish species off the northern Irish coast. *Irish Nat. J.* 22:480–482.

Brimley, H. H. 1939. The ocean sun-fishes on the North Carolina coast: the pointed-tailed *Masturus lanceolatus* and the round-tailed *Mola mola. J. Elisha Mitchill Sci. Soc.* 15: 295–303.

Brodal, A., and R. Fänge, eds. 1960. *The Biology of Myxine.* Universitetsforlaget, Oslo, 588 pp.

Brooks, J. L., and S. I. Dodson. 1965. Predation, body size and composition of the plankton. *Science* 150:28–35.

Browder, J. A., J. C. Davis, and E. Sullivan. 1981. Paying-passenger recreational fisheries of the Florida Gulf coast and Keys. *Mar. Fish. Rev.* 43(8):12–26.

Brown, B. E., and G. P. Patil. 1986. Risk analysis in the Georges Bank haddock fishery—a pragmatic example of dealing with uncertainty. *No. Am. J. Fish. Manage.* 6:183–191.

Brown, J. A., and J. M. Green. 1976. Territoriality, habitat selection, and prior residency in underyearling *Stichaeus punctatus* (Pisces: Stichaeidae). *Can. J. Zool.* 54:1904–1907.

Brown, J. A. 1986. The development of feeding behaviour in the lumpfish, *Cyclopterus lumpus. J. Fish Biol.* 29(Suppl.A):171–178.

Brown, J. A., and L. W. Crimm. 1998. Progress on the development of yellowtail flounder (*Pleuronectes ferrugineus*) for aquaculture. *Bull. Aquacult. Assoc. Can.* 98–1:16–17.

Brown, J. A., D. C. Somerton, P. J. Hambrook, and D. A. Methven. 1989. Why Atlantic lumpfish and ocean pout are potential candidates for cold-water marine finfish aquaculture. Aquacult. Assoc. Canada, St. Johns, Nfld., pp. 50–52.

Brown, J. A., D. C. Somerton, D. A. Methven, and J. R. Watkins. 1992. Recent advances in lumpfish *Cyclopterus lumpus* and ocean pout *Macrozoarces americanus* larviculture. *J. World Aquacult. Soc.* 23:271–276.

Brown, R. W. 1998. Haddock. In: S. H. Clark, ed. *Status of Fishery Resources off the Northeastern United States for 1998.* NOAA Tech. Memo. NMFS-NE-115:53–56.

Brown, R. W., and W. L. Gabriel. 1998. Winter flounder. In: S. H. Clark, ed. *Status of Fishery Resources off the Northeastern United States for 1998.* NOAA Tech. Memo. NMFS-NE-115:81–84.

Brown, S. K., R. Mahon, K. C. T. Zwanenburg, K. R. Buja, L. W. Claflin, R. N. O'Boyle, B. Atkinson, M. Sinclair, G. Howell, and M. E. Monaco. 1996. East coast of North America groundfish: Initial explorations of biogeography and species assemblages. NOAA, Silver Spring, Md., 111 pp.

Brown, W. W., and C. Cheng. 1946. Investigations into the food of the cod (*Gadus callarias* L.) off Bear Island, and of the cod and haddock (*G. aeglefinus* L.) off Iceland and the Murman coast. *Hull* [illegible]

Bruce, N. L. 1986. Revision of the isopod crustacean genus *Mothocya* Costa, in Hope, 1851 (Cymothoidae: Flabellifera), parasitic on marine fishes. *J. Nat. Hist.* 20:1089–1192.

Brunel, P. 1965. Food as a factor or indicator of vertical migrations of cod in the western Gulf of St. Lawrence. Int. Comm. Northw. Atlan. Fish. Spec. Publ. No. 6:439–448.

Brunel, P. 1971. Deux cas d'ambicoloration partielle chez la plie, *Hippoglossoides platessoides* (Pisces, Heterosomata), dans le golfe du Saint-Laurent. *Naturaliste Canad.* 98:937–939.

Bruun, A. F. 1937. Contributions to the life histories of the deep sea eels: Synaphobranchidae. Dana Rept. No. 9, 31 pp.

Bryant, R., I. L. Jones, and J. M. Hipfner. 1999. Response to the change in prey availablity by Common Murres and Thick-billed Murres at the Gannet Islands, Labrador. *Can. J. Zool.* 77:1278–1287.

Buchanan, C. C., R. B. Stone, and F. W. Steimle. 1988. Marine recreational boat fishery of the New York Bight Apex in 1971. *Mar. Fish. Rev.* 50(2):6–19.

Buckel, J. A., N. D. Steinberg, and D. O. Conover. 1995. Effects of temperature, salinity, and fish size on growth and consumption of juvenile bluefish. *J. Fish Biol.* 47:696–706.

Buckel, J. A., D. O. Conover, N. D. Steinberg, and K. A. McKown. 1999a. Impact of age-0 bluefish (*Pomatomus saltatrix*) predation on age-0 fishes in the Hudson River estuary: evidence for density-dependent loss of juvenile striped bass (*Morone saxatilis*). *Can. J. Fish. Aquat. Sci.* 56:275–287.

Buckel, J. A., M. J. Fogarty, and D. O. Conover. 1999b. Foraging habits of bluefish, *Pomatomus saltatrix,* on the U.S. east coast continental shelf. *Fish. Bull., U.S.* 97:758–775.

Buckel, J. A., M. J. Fogarty, and D. O. Conover. 1999c. Mutual prey of fish and humans: a comparison of biomass consumed by bluefish, *Pomatomus saltatrix,* with that harvested by fisheries. *Fish. Bull., U.S.* 97:776–785.

Buckley, J. 1989. Species profiles: life histories and environmental requirements of coastal fishes and invertebrates. (North Atlantic)—Rainbow smelt. U.S. Fish Wildl. Serv. Biol. Rept., U.S. Army Corps Eng. TR EL-82-4, 11 pp.

Buckley, J., and B. Kynard. 1985a. Yearly movements of shortnose sturgeons in the Connecticut River. *Trans. Am. Fish. Soc.* 114:813–820.

Buckley, J., and B. E. Kynard. 1985b. Vertical distribution of outmigrating juvenile American shad and blueback herring. Final Report to Northeast Utilities Service Company, Hartford.

Buckley, L. J., and G. R. Lough. 1987. Recent growth, biochemical composition, and prey field of larval haddock (*Melanogrammus aeglifinus*) and Atlantic cod (*Gadus morhua*) on Georges Bank. *Can. J. Fish. Aquatic Sci.* 44:14–25.

Buckley, L. J., S. I. Turner, T. A. Halavik, A. S. Smigielski, S. M. Drew, and G. C. Laurence. 1984. Effects of temperature and food availability on growth, survival, and RNA-DNA ratio of larval sand lance (*Ammodytes americanus*). *Mar. Ecol. Prog. Ser.* 15:91–97.

Buckley, L. J., T. A. Halavik, A. S. Smigielski, and G. C. Laurence. 1987. Growth and survival of the larvae of three species of temperate marine fishes reared at discrete prey densities. In R. D. Hoyt, ed. *10th Annual Larval Fish Conference. Am. Fish. Soc. Symp.* 2:82–92.

Bulatov, O. A. 1983. Distribution of eggs and larvae of Greenland halibut, *Reinhardtius hippoglossoides* (Pleuronectidae), in the eastern Bering Sea. *Voprosy Ikhtiol.* 23:162–164 [in Russian, translation in *J. Ichthyol.* 23(1):157–159].

Bullis, H. R., Jr. and J. R. Thompson. 1965. Collections by the exploratory fishing vessels *Oregon, Silver Bay, Combat,* and *Pelican* made during 1956 to 1960 in the southwestern North Atlantic. U. S. Fish Wildl. Serv. Spec. Sci. Rept. Fish. No. 510, 130 pp.

Bullis, H. R., Jr., J. S. Carpenter, and C. M. Roithmayr. 1971. Untapped West-Central Atlantic fisheries. In: S. Shapiro, ed.*Our Changing Fisheries.* Govt. Print. Off., Washington, D.C., pp. 374–391.

Bumpus, H. C. 1899. The reappearance of the tilefish. *Bull. U.S. Fish. Comm.* 18:321–333.

Bunkley-Williams, L., and E. H. Williams. 1994. Diseases caused by *Tichodina spheroidesi* and *Cryptocaujon irritans* (Ciliophora) on wild coral reef fishes. *J. Aquat. Animal Health* 6:360–361.

Burbidge, R.G. 1969. Age, growth, length-weight relationship, sex ratio, and food habits of American smelt, *Osmerus mordax* (Mitchill), from Gull Lake, Michigan. *Trans. Am. Fish. Soc.* 98:631–640.

Burbidge, R. G. 1974. Distribution, growth, selective feeding, and energy transformation of young-of-the-year blueback herring, *Alosa aestivalis* (Mitchill), in the James River, Virginia. *Trans. Am. Fish. Soc.* 103:297–311.

Burd, A. C. 1984. Density-dependent growth in North Sea herring. Int. Comm. Explor. Sea CM 1984/H:4.

Burdick, G. E. 1954. An analysis of the factors, including pollution, having possible influence on the abundance of shad in the Hudson River. *N.Y. Fish Game J.* 1:188–205.

Burgess, G. H. 1976. Aquarium feeding behavior of the cornetfish, *Fistularia tabacaria* and the southern stargazer *Astroscopus y-graecum. Fla. Sci.* 39:5–7.

Burgess, G. H. 1980. *Alosa mediocris,* Hickory shad. p. 64. In: D. S. Lee, C. R. Gilbert, C. H. Hocutt, R. E. Jenkins, D. E. McAllister, and J. R. Stauffer, Jr., eds. *Atlas of North American Freshwater Fishes.* N.C. St. Mus. Nat. Hist., Publ. No. 1980-12, N.C. Biol. Surv., 854 pp.

Burgess, G. H., and F. J. Schwartz. 1975. Anomalies encountered in freshwater and marine fishes from the eastern United States. *Assoc. Southeast. Biol. Bull.* 22:1–44.

Burgess, W. E. 1978. *Butterflyfishes of the World: Monograph of the Family Chaetodontidae.* T. F. H. Publications, Inc., Neptune City, N.J., 832 pp.

Burke, V. 1930. Revision of the fishes of the family Liparidae. *Bull. U.S. Nat. Mus.* 150, 204 pp.

Burkholder, J. M., E. J. Noga, C. H. Hobbs, and H. B. Glasgow, Jr. 1992. New "phantom" dinoglagellate is the causative agent of major estuarine fish kills. *Nature* 358:407–410.

Burn, P. R. 1980. The parasites of smooth flounder, *Liopsetta putnami* (Gill), from the Great Bay Estuary, New Hampshire. *J. Parasitol.* 66:532–541.

Burnett, J. M. 1988. Witch flounder, *Glyptocephalus cynoglossus.* In: J. Penttila and L. M. Dery, eds. *Age Determination Methods for Northwest AtlanticSspecies.* NOAA Tech. Rept. NMFS No. 72: 109–110.

Burnett-Herkes, J. 1974. Parasites of the gills and buccal cavity of the dolphin, *Coryphaena hippurus,* from the Straits of Florida. *Trans. Am. Fish. Soc.* 103:101–106.

Burr, B. M., and R. L. Mayden. 1980. Dispersal of rainbow smelt, *Osmerus mordax,* into the upper Mississippi River (Pisces: Osmeridae). *Am. Midl. Nat.* 104:198–201.

Burr, B. M., and F. J. Schwartz. 1986. Occurrence, growth, and food habits of the spotted hake, *Urophycis regia,* in the Cape Fear estuary and adjacent Atlantic Ocean, North Carolina. *Northeast Gulf Sci.* 8:115–127.

Burreson, E. M., and D. E. Zwerner. 1980. Host range, life cycle, and pathology of *Trypanoplasma bullocki* in lower Chesapeake Bay fishes. *J. Protozool.* 27:23A–23B (abstract).

Burreson, E. M., and D. Zwerner. 1982. The role of host biology, vector biology, and temperature in the distribution of *Trypanoplasma bullocki* infections in the lower Chesapeake Bay. *J. Parasitol.* 68:306–313.

Burreson, E. M., and D. E. Zwerner. 1984. Juvenile summer flounder, *Paralichthys dentatus,* mortalities in the western Atlantic Ocean caused by the hemoflagellate *Trypanoplasma bullocki:* evidence from field and experimental studies. *Helgolander Meeresunter.* 37:343–352.

Burton, D. 1988. Melanophore comparisons in different forms of ambicoloration in the flatfish *Pseudopleuronectes americanus* and *Reinhardtius hippoglossoides. J. Zool.* 214:353–360.

Burton, D. 1998. The chromatic biology of flatfish (Pleuronectidae). *Italian J. Zool.* 65(Suppl.):399–403.

Burton, D. T., L. B. Richardson, and C. J. Moore. 1980. Effect of oxygen reduction rate and low dissolved oxygen concentration on two estuarine fishes. *Trans. Am. Fish. Soc.* 109:552–557.

Burton, M. P. 1994. A critical period for nutritional control of early gametogenesis in female winter flounder, *Pseudopleuronectes americanus* (Pisces: Teleostei). *J. Zool.* 233:405–415.

Burton, M. P., and D. R. Idler. 1984. The reproductive cycle in winter flounder, *Pseudopleuronectes americanus* (Walbaum). *Can. J. Zool.* 62:2563–2567.

Burton, M. P., and S. R. Flynn. 1998. Differential postspawning mortality among male and female capelin (*Mallotus villosus* Müller) in captivity. *Can. J. Zool.* 76:588–592.

Busakhin, S. V. 1982. Systematics and distribution of the family Berycidae (Osteichthyes) in the world ocean. *Voprosy Ikhtiol.* 22:904–921 [in Russian, translation in *J. Ichthyol.* 22:1–21].

Butler, J. A. 1961. Development of a thread herring fishery in the Gulf of Mexico. *Comm. Fish. Rev.* 23:12–17.

Butner, A., and B. H. Brattstrom. 1960. Local movement in *Menidia* and *Fundulus. Copeia* 1960:139–141.

Byrne, C. J. 1982. Anchovy. In: M. D. Grosslein and T. R. Azarowitz, eds. *Fish Distribution.* MESA New York Bight Atlas Monogr. No. 15, New York Sea Grant Institute, Albany, pp. 65–67.

Byrne, C. J., and T. R. Azarovitz. 1982. Summer flounder *Paralichthys dentatus.* In: Grosslein, M. D., and T. R. Azarovitz, eds. *Fish Distribution.* MESA New York Bight Atlas Monogr. No. 15:109–111.

Caddy, J. F., and T. D. Iles. 1973. Underwater observations on herring spawning grounds on Georges Bank. *Int. Comm. Northw. Atl. Fish. Res. Bull.* 10:131–139.

Cadrin, S. X., A. B. Howe, S. J. Correia, and T. P. Currier. 1995. Evaluating the effects of two coastal mobile gear fishing closures on finfish abundance off Cape Cod. *No. Am. J. Fish. Manage.* 15:300–315.

Cailliet, G. M., and D. W. Bedford. 1983. The biology of three pelagic sharks from California waters, and their emerging fisheries: a review. Calif. Coop. Fish. Invest. Rept. No. 24:57–69.

Cailliet, G. M., L. J. Natanson, B. A. Weldon, and D. A. Ebert. 1985. Preliminary studies on the age and growth of the white shark, *Carcharodon carcharias,* using vertebral bands. S. Cal. Acad. Sci. Mem. No. 9:49–60.

Cailliet, G. M., L. K. Martin, J. T. Harvey, D. Kusher, and B. A. Weldon. 1983. Preliminary studies on the age and growth of blue, *Prionace glauca,* common thresher, *Alopias vulpinus,* and shortfin mako, *Isurus oxyrinchus,* sharks from California waters. NOAA Tech. Rept. NMFS No. 8:179–188.

Cailliet, G. M., K. G. Yudin, S. Tanaka, and T. Taniuchi. 1990. Growth characteristics of two populations of *Mustelus manazo* from Japan based upon cross-readings of vertebral bands. NOAA Tech. Rept. NMFS No. 90:167–176.

Cairns, D. K. 1987. The ecology and energetics of chick provisioning by black guillemots. *Condor* 80:627–635.

Caldwell, D. K. 1961. Populations of the butterfish, *Poronotus triacanthus* (Peck), with systematic comments. *Bull. S. Cal. Acad. Sci.* 60:19–31.

Caldwell, D. K. 1962a. Development and distribution of the short bigeye *Pseudopriacanthus altus* (Gill), in the western North Atlantic. *U.S. Fish Wildl. Serv. Fish. Bull.* 62:103–150.

Caldwell, D. K. 1962b. Western Atlantic fishes of the family Priacanthidae. *Copeia* 1962:417–424.

Caldwell, D. K., and W. W. Anderson. 1959. Offshore occurrence of larval silver mullet, *Mugil curema,* in the western Gulf of Mexico. *Copeia* 1959:252–253.

Campbell, R .A., and R. A. Bray. 1993. *Lepidapedon* spp. (Digenea: Lepocreadiidae) from deep-sea gadiform fishes of the NW Atlantic Ocean, including four new species. *Syst. Parasitol.* 24:99–110.

Campbell, R. A., R. L. Haedrich, and T. A. Munroe. 1980. Parasitism and ecological relationships among deep-sea benthic fishes. *Mar. Biol.* 57:301–313.

Campeau, S., H. Guderley, and G. FitzGerald. 1984. Salinity tolerances and preferences of fry of two species of sympatric sticklebacks: possible mechanisms of habitat segregation. *Can. J. Zool.* 62:1048–1051.

Carey, F. G. 1982. A brain heater in the swordfish. *Science* 216:1327–1329.

Carey, F. G. 1990. Further acoustic telemetry observations of swordfish. *Mar. Recreation. Fish.* 13:103–122.

Carey, F. G., and B. H. Robison. 1981. Daily patterns in the activities of swordfish, *Xiphias gladius,* observed by acoustic telemetry. *Fish. Bull., U.S.* 79:277–292.

Carey, F. G., and J. V. Scharold. 1990. Movements of blue sharks (*Prionace glauca*) in depth and course. *Mar. Biol.* 106:329–342.

Carey, F. G., and J. M. Teal. 1969. Regulation of body temperature by the bluefin tuna. *Comp. Biochem. Physiol.* 28:205–213.

Carey, F. G., J. W. Kanwisher, O. Brazier, G. Gabrielsen, J. G. Casey, and H. L. Pratt, Jr. 1982. Temperature and activities of a white shark, *Carcharodon carcharias. Copeia* 1982:254–260.

Carey, F. G., J. W. Kanwisher, and E. D. Stevens. 1984. Bluefin tuna warm their viscera during digestion. *J. Exper. Biol.* 190:1–20.

Carey, F. G., J. G. Casey, H. L. Pratt, D. Urquhart, and J. E. Mc-Cosker. 1985. Temperature, heat production, and heat exchange in lamnid sharks. S. Cal. Acad. Sci. Mem. No. 9:92–108.

Cargnelli, L. M., S. J. Griesbach, and W. W. Morse. 1999a. Essential Fish Habitat Source Document: Atlantic halibut, *Hippoglossus hippoglossus,* life history and habitat characteristics. NOAA Tech. Memo. NMFS-NE-125, 17 pp.

Cargnelli, L. M., S. J. Griesbach, P. L. Berrien, W. W. Morse, and D. L. Johnson. 1999b. Essential Fish Habitat Source Document: Haddock, *Melanogrammus aeglefinus,* life history and habitat characteristics. NOAA Tech. Memo. NMFS-NE-128, 31 pp.

Cargnelli, L. M., S. J. Griesbach, D. B. Packer, P. L. Berrien, D. L. Johnson, and W. W. Morse. 1999c. Essential Fish Habitat Source Document: Pollock, *Pollachius virens,* life history and habitat characteristics. NOAA Tech. Memo. NMFS-NE-131, 30 pp.

Cargnelli, L. M., S. J. Griesbach, D. B. Packer, P. L. Berrien, W. W. Morse, and D. L. Johnson. 1999d. Essential Fish Habitat Source Document: Witch flounder, *Glyptocephalus cynoglossus,* life history and habitat characteristics. NOAA Tech. Memo. NMFS-NE-139, 29 pp.

Carlander, K. D. 1969. *Handbook of Freshwater Fishery Biology,* Vol. I. *Life History Data on Freshwater Fishes of the United States and Canada, Exclusive of the Perciformes.* Iowa State University Press, Ames, 752 pp.

Carlson, D. M., and K. C. Simpson. 1987. Gut contents of juvenile shortnose sturgeon in the upper Hudson Estuary. *Copeia* 1987:796–802.

Carlson, F. T. 1968. Report on the biological findings. In: *Suitability of the Susquehanna River for Restoration of Shad.* U.S. Dept. Inter., Washington, D.C., pp. 4–41.

Caron, F., and S. Tremblay. 1999. Structure and management of an exploited population of Atlantic sturgeon (*Acipenser oxyrinchus*) in the St. Lawrence Estuary, Québec, Canada. *J. Appl. Ichthyol.* 15:153–156.

Carr, W. E. S., and C. A. Adams. 1972. Food habits of juvenile marine fishes: evidence of the cleaning habit in the leatherjacket, *Oligoplites saurus,* and the spottail pinfish, *Diplodus holbrooki. Fish. Bull., U.S.* 70:1111–1120.

Carr, W. E. S., and C. A. Adams. 1973. Food habits of juvenile marine fishes occupying seagrass beds in the estuarine zone near Crystal River, Florida. *Trans. Am. Fish. Soc.* 102:511–540.

Carscadden, J. E., and W. C. Leggett. 1975a. Life history variations in populations of American shad, *Alosa sapidissima* (Wilson), spawning in tributaries of the St. John River, New Brunswick. *J. Fish Biol.* 7:595–609.

Carscadden, J. E., and W. C. Leggett. 1975b. Meristic differences in spawning populations of American shad, *Alosa sapidissima:* evidence for homing to tributaries in the St. John River, New Brunswick. *J. Fish. Res. Bd. Can.* 32:653–660.

Carson, R. L. 1943. Food from the sea: fish and shellfish from New England. *U.S. Fish Wildl. Ser. Cons. Bull.* 33, 74 pp.

Caruso, J. H. 1977. The systematics of the fish family Lophiidae. Ph.D. dissertation, Tulane University, New Orleans, 220 pp.

Caruso, J. H. 1981. The systematics and distribution of the lophiid anglerfishes: I. A revision of the genus *Lophiodes* with the description of two new species. *Copeia* 1981:522–549.

Caruso, J. H. 1983. The systematics and distribution of the lophiid anglerfishes: II. Revisions of the genera *Lophiomus* and *Lophius. Copeia* 1983:11–30.

Caruso, J. H. 1985. The systematics and distribution of the lophiid anglerfishes: III. Intergeneric relationships. *Copeia* 1985:870–875.

Caruso, J. H. 1989. Systematics and distribution of the Atlantic chaunacid anglerfishes (Pisces: Lophiiformes). *Copeia* 1989:153–165.

Casey, J. G. 1964. Angler's guide to sharks off the northeastern United States, Maine to Chesapeake Bay. U.S.Fish Wildl. Serv. Circ. No. 179, 32 pp.

Casey, J. G., and N. E. Kohler. 1990. Long distance movements of Atlantic sharks from the NMFS cooperative shark tagging program. In: S. Gruber, ed. *Discovering Sharks.* Spec. Publ. 14:87–91 [vols. 19(4) and 20(1)], American Littoral Society, Highlands, N.J.

Casey, J. G., and H. L. Pratt. 1985. Distribution of the white shark, *Carcharodon carcharias,* in the western North Atlantic. S. Cal. Acad. Sci. Mem. No. 9:2–14.

Casey, J. G., H. L. Pratt, Jr., and C. E. Stillwell. 1985. Age and growth of the sandbar shark (*Carcharhinus plumbeus*) from the western North Atlantic. *Can. J. Fish. Aquat. Sci.* 42:963–975.

Casey, J. M., and R. A. Myers. 1998. Near extinction of a large, widely distributed fish. *Science* 281:690–692.

Castagna, M. 1955. A study of the hogchoker, *Trinectes maculatus* (Bloch and Schneider), in the Wakulla River, Florida. M.A. thesis, Florida State University, Tallahassee, 39 pp.

Castle, P. H. J. 1970. Distribution, larval growth, and metamorphosis of the eel *Derichthys serpentinus* Gill, 1884 (Pisces: Derichthyidae). *Copeia* 1970:444–452.

Castro, J. I. 1983. *The Sharks of North American Waters.* Texas A&M University Press, College Station, 180 pp.

Castro, J. I., P. M. Bubucis, and N. A. Overstrom. 1988. The reproductive biology of the chain dogfish, *Scyliorhinus retifer. Copeia* 1988:740–746.

Castro, L. R., and R. K. Cowen. 1991. Environmental factors affecting the early life history of bay anchovy *Anchoa mitchilli* in Great South Bay, New York. *Mar. Ecol. Prog. Ser.* 76:235–247.

Castro-Hernández, J. J., and A. T. Santana Ortega. 2000. Synopsis of biological data on the chub mackerel (*Scomber japonicus* Houttuyn, 1782). FAO Fish. Synop. No. 157, 77 pp.

Cating, J. P. 1953. Determining the age of Atlantic shad from their scales. *U.S. Fish Wildl. Serv. Fish. Bull.* 54:187–199.

Cau, A., and A. M. Deiana. 1982. Contributo alla conoscenza della biologia di *Hoplostethus mediterraneus* (Valenciennes 1829) (Osteitti, Bericiformi). *Boll. Soc. Sarda Sci. Natur.* 21:185–192.

Cervigón, M. F. 1966. Los peces marinos de Venezuela. Estac. Invest. Mar. Margarita, Fund. La Salle Cienc. Nat. Monogr. No. 11, 951 pp.

Chacon Chaverri, D., and W. O. McLarney. 1992. Desarrollo temparano del sábalo, *Megalops atlanticus* (Pisces: Megalopidae). *Rev. Biol. Trop.* 40:171–178.

Chadwick, E. M. P., and R. R. Claytor. 1989. Run timing of pelagic fishes in Gulf of St. Lawrence: area and species effects. *J. Fish Biol.* 35(Suppl. A):215–223.

Chadwick, E. M. P., D. K. Cairns, H. M. C. Dupuis, K. V. Ewart, M. H. Kao, and G. L. Fletcher. 1990. Plasma antifreeze levels reflect the migratory behaviour of Atlantic herring (*Clupea harengus harengus*) in the southern Gulf of St. Lawrence. *Can. J. Fish. Aquat. Sci.* 47:1534–1536.

Chadwick, H. C. 1929. Feeding habits of the angler-fish, *Lophius piscatorius*. *Nature* 124:337.

Chakrabartty, A., C. L. Hew, M. Shears, and G. Fletcher. 1988. Primary structures of the alanine-rich antifreeze polypeptides from grubby sculpin, *Myoxocephalus aenaeus*. *Can. J. Zool.* 66:403–408.

Chamberlain, T. 1980. It's a year for halibut stories. *Boston Sunday Globe,* September, p. 9.

Chambers, J. R., J. A. Musick, and J. Davis. 1976. Methods of distinguishing larval alewife from larval blueback herring. *Chesapeake Sci.* 17:93–100.

Chambers, R. C., and W. C. Leggett. 1987. Size and age at metamorphosis in marine fishes: an analysis of laboratory-reared winter flounder (*Pseudopleuronectes americanus*) with a review of variation in other species. *Can. J. Fish. Aquat. Sci.* 44:1936–1947.

Chang, S. 1990. Seasonal distribution patterns of commercial landings of 45 species off the northeastern United States during 1977–88. NOAA Tech. Memo. NMFS-F/NEC-78, 130 pp.

Chang, S., P. L. Berrien, D. L. Johnson, and C. A. Zetlin. 1999a. Essential Fish Habitat Source Document: offshore hake, *Merluccius albidus,* life history and habitat characteristics. NOAA Tech. Memo. NMFS-NE-130, 24 pp.

Chang, S., W. W. Morse, and P. L. Berrien. 1999b. Essential Fish Habitat Source Document: white hake, *Urophycis tenuis,* life history and habitat characteristics. NOAA Tech. Memo. NMFS-NE-136, 23 pp.

Chang, S., P. L. Berrien, D. L. Johnson, and W. W. Morse. 1999c. Essential Fish Habitat Source Document: windowpane, *Scophthalmus aquosus,* life history and habitat characteristics. NOAA Tech. Memo. NMFS-NE-137, 32 pp.

Chao, L. N. 1973. Digestive system and feeding habits of the cunner, *Tautogolabrus adspersus,* a stomachless fish. *Fish. Bull., U.S.*

Chao, L. N. 1978. A basis for classifying western Atlantic Sciaenidae (Teleostei: Perciformes). NOAA Tech. Rept. NMFS Circ. No. 415, 65 pp.

Chao, L. N., and J. A. Musick. 1977. Life history, feeding habits, and functional morphology of juvenile sciaenid fishes in the York River estuary, Virginia. *Fish. Bull., U.S.* 75:657–702.

Chapleau, F. 1993. Pleuronectiform relationships: a cladistic reassessment. *Bull. Mar. Sci.* 52:516–540.

Chapman, R. W. 1993. Genetic investigation of the resurgence of American shad in the Susquehanna River. Maryland Dept. Nat. Resour., Ches. Bay Res. Monitor. Div., Power Plant Topical Res. Progr., Final Rept. No. CBRM-TR-94-2, 22 pp.

Chapman, R. W., J. C. Patton, and B. Eleby. 1994. Comparisons of mitochondrial DNA variation in four alosid species as revealed by the total genome, the NADH dehydrogenase I and cytochrome b regions. In: A. R. Beaumont, ed. *Genetics and Evolution of Aquatic Organisms.* Chapman and Hall, New York, pp. 249–263.

Chave, E. H., and Mundy, B. C. 1994. Deep-sea benthic fish of the Hawaiian Archipelago, Cross Seamount, and Johnston Atoll. *Pacific Sci.* 48:367–409.

Checkley, D. M., Jr. 1982. Selective feeding by Atlantic herring *Clupea harengus* larvae on zooplankton in natural assemblages. *Mar. Ecol. Prog. Ser.* 9:245–253.

Checkley, D. M., Jr., S. Raman, G. L. Maillet, and K. M. Mason. 1988. Winter storm effects on the spawning and larval drift of a pelagic fish. *Nature* 335:346–348.

Cheek, R. P. 1968. The American shad. U.S. Fish Wildl. Serv. Leaflet No. 614, 13 pp.

Cheetham, C., and J. M. Fives. 1990. The biology and parasites of the butterfish *Pholis gunnellus* (Linnaeus, 1758) in the Galway Bay area. *Proc. Roy. Irish Acad.* 90B:127–149.

Chen, M. Y. 1970. Reproduction of American smelt *Osmerus mordax* (Mitchill). Ph.D. dissertation, University of Waterloo, Canada, 254 pp.

Chen, W., J. J. Govoni, and S. M. Warlen. 1992. Comparison of feeding and growth of larval round herring (*Etrumeus teres*) and Gulf menhaden (*Brevoortia patronus*). *Fish. Bull., U.S.* 90:183–189.

Chenoweth, J. F., S. E. McGladdery, C. J. Sindermann, T. K. Sawyer, and J. W. Bier. 1986. An investigation into the usefulness of parasites as tags for herring *Clupea harengus* stocks in the western North Atlantic, with emphasis on use of the larval nematode *Anisakis simplex*. *J. Northw. Atl. Fish. Sci.* 7:25–33.

Chenoweth, S. B. 1963. Spawning and fecundity of tautog, *Tautoga onitis* (L.). M.S. thesis., University of Rhode Island, Kingston, 60 pp.

Chenoweth, S. B. 1973. Fish larvae of the estuaries and coast of central Maine. *Fish. Bull., U.S.* 71:105–113.

Chernoff, B., J. V. Conner, and C. F. Bryan. 1981. Systematics of the *Menidia beryllina* complex (Pisces: Atherinidae) from the Gulf of Mexico and its tributaries. *Copeia* 1981:319–336.

Chesmore, A. P., S. A. Testaverde, and F. P. Richards. 1971. A study of the marine resources of Dorchester Bay. Mass. Div. Mar. Fish. Monogr. Ser. No. 10, 41 pp.

Chesmore, A. P., D. J. Brown, and R. D. Anderson. 1972. A study of the marine resources of Lynn-Saugus Harbor. Mass. Div. Mar.

Chesmore, A. P., D. J. Brown, and R. D. Anderson. 1973. A study of the marine resources of Essex Bay. Mass. Div. Mar. Fish. Monogr. Ser. No. 13, 38 pp.

Cheung, P. J., R. F. Nigrelli, and G. D. Ruggieri. 1979. Coccidian parasite of blackfish, *Tautoga onitis* (L.): life cycle and histopathology. *Am. Zool.* 19:963.

Cheung, P. J., R. F. Nigrelli, and G. D. Ruggieri. 1980. Studies on the morphology of *Uronemia marinum* Dujardin (Ciliata: Uronematidae) with a description of the histopathology of the infection in marine fishes. *J. Fish Dis.* 3:295–303.

Chiarella, L. A., and D. O. Conover. 1990. Spawning season and first-year growth of adult bluefish from the New York Bight. *Trans. Am. Fish. Soc.* 119:455–462.

Chidester, F. E. 1916. A biological study of the more important of the fish enemies of the salt marsh mosquitoes. *New Jersey Agr. Exp. Stn. Bull.* 300:1–16.

Chidester, F. E. 1920. The behavior of *Fundulus heteroclitus* on the salt marshes of New Jersey. *Am. Nat.* 57:244–245.

Chittenden, M. E., Jr. 1969. Life history and ecology of the American shad, *Alosa sapidissima,* in the Delaware River. Ph.D. dissertation, Rutgers University, New Brunswick, 471 pp.

Chittenden, M. E., Jr. 1971a. Transporting and handling young American shad. *N.Y. Fish Game J.* 18:123–128.

Chittenden, M. E., Jr. 1971b. Effects of handling and salinity on oxygen requirements of the striped bass, *Morone saxatilis. J. Fish. Res. Bd. Can.* 28:1823–1830.

Chittenden, M. E., Jr. 1972a. Responses of young American shad to low temperatures. *Trans. Am. Fish. Soc.* 101:680–685.

Chittenden, M. E., Jr. 1972b. Salinity tolerance of young blueback herring, *Alosa aestivalis. Trans. Am. Fish. Soc.* 101:123–125.

Chittenden, M. E., Jr. 1973. Effects of handling on oxygen requirements of American shad *Alosa sapidissima. J. Fish. Res. Bd. Can.* 30:105–110.

Christensen, R. F. 1965. An ichthyological survey of Jupiter Inlet and Loxahatchee River, Florida. M.A. thesis, Florida State University, Tallahassee, 318 pp.

Christensen, V. 1983. Predation by sand-eel on herring larvae. *Int. Counc. Explor. Sea* 27:1–10.

Christmas, J. Y., and R. S. Waller. 1973. Estuarine vertebrates, Mississippi. In: J. Y. Christmas, ed. *Cooperative Gulf of MexicoEestuarine Inventory and Study, Mississippi.* Gulf Coast Res. Lab., pp. 323–406.

Chumakov, A. K., and S. G. Podrazhanskya. 1986. Feeding of Greenland halibut (*Reinhardtius hippoglossoides*) in the Northwest Atlantic. NAFO Sci. Coun. Stud. No. 10:47–52.

Cianci, J. M. 1969. Larval development of the alewife, *Alosa pseudoharengus* Wilson, and the glut herring, *Alosa aestivalis* Mitchill. M.S. thesis, University of Connecticut, Storrs, 62 pp.

Cislo, P. R., and J. N. Caira. 1993. The parasite assemblage in the spiral intestine of the shark *Mustelus canis. J. Parasitol.* 79:886–899.

Clark, E. 1950. Notes on the behavior and morphology of some West Indian plectognath fishes. *Zoologica, N.Y.* 35:159–168.

Clark, E. 1963. The maintenance of sharks in captivity, with a report on their instrumental conditioning. In: P. W. Gilbert, ed. *Sharks and Survival.* Heath & Co., Boston, pp. 115–149.

Clark, E., and K. von Schmidt. 1965. Sharks of the central Gulf coast of Florida. *Bull. Mar. Sci.* 15:13–83.

Clark, J. R. 1959. Sexual maturity of haddock. *Trans. Am. Fish. Soc.* 88:212–213.

Clark, S. H., ed. 1998. Status of fishery resources off the northeastern United States for 1998. NOAA Tech. Memo. NMFS-NE-115, 149 pp.

Clark, S. H., and R. Livingstone, Jr. 1982. Ocean pout *Macrozoarces americanus.* In: M. D. Grosslein and T. R. Azarovitz, eds. *Fish Distribution.* MESA New York Bight Atlas Mongr. No. 15: 76–79.

Clark, S. H., W. J. Overholtz, and R. C. Hennemuth. 1982. Review and assessment of the Georges Bank and Gulf of Maine haddock fishery. *J. North Atl. Fish. Sci.* 3:1–27.

Clarke, M. R., D. C. Clarke, H. R. Martins, and H. M. da Silva. 1996. The diet of the blue shark (*Prionace glauca* L.) in Azorean waters. *Arquipélago, Bull. Univ. Azores* 14A:41–56.

Clarke, T. A. 1984. Diet and morphological variation in snipefishes, presently recognized as *Macrorhamphosus scolopax,* from southeast Australia: evidence for two sexually dimorphic species. *Copeia* 1984:595–608.

Clarke, T. A., and L. A. Privitera. 1995. Reproductive biology of two Hawaiian pelagic carangid fishes, the bigeye scad, *Selar crumenopthalmus,* and the round scad, *Decapterus macarellus. Bull. Mar. Sci.* 56:33–47.

Clayton, G. R. 1976. Reproduction, first year growth, and distribution of anadromous rainbow smelt, *Osmerus mordax,* in the Parker River and Plum Island Sound estuary, Massachusetts. M.S. thesis, University of Massachusetts, Amherst, 105 pp.

Clayton, G. R., C. F. Cole, S. Murawski, and J. Parrish. 1976. Common marine fishes of coastal Massachusetts. Mass. Coop. Fish. Res. Unit Contrib. No. 54, 231 pp.

Cleland, J. 1862. On the anatomy of the short sun-fish (*Orthagorescus mola*). *Nat. Hist. Rev.* 1862:170–185.

Clemens, W. A., and L. S. Clemens. 1921. Contribution to the biology of the muttonfish, *Zoarces anguillaris. Contrib. Can. Biol.* 1918–1920:69–83.

Clements, W. H., and R. J. Livingston. 1984. Prey selectivity of the fringed filefish *Monacanthus ciliatus* (Pisces: Monacanthidae): role of prey accessability. *Mar. Ecol. Prog. Ser.* 16:291–295.

Clemmer, G. H., and F. J. Schwartz. 1964. Age, growth, and weight relationships of the striped killifish, *Fundulus majalis,* near Solomons, Maryland. *Trans. Am. Fish. Soc.* 93:197–198.

Cleveland, A. 1994. Nest site habitat preference and competition in *Gasterosteus aculeatus* and *G. wheatlandi. Copeia* 1994:698–704.

Clymer, J. P. 1978. The distributions, trophic activities and competitive interactions of three salt marsh killifishes (Pisces: Cyprinodontidae). Ph.D. dissertation, Lehigh University, Bethlehem, Pa.

CNN.com. 2001. Giant tuna sells for record $173,600. cnn.com/2001/FOOD/news/0l/05/pricy.tuna.ap/index.html.

Coad, B. W. 1983. Plate morphs in freshwater samples of *Gasterosteus aculeatus* from Arctic and Atlantic Canada: complementary comments on a recent contribution. *Can. J. Zool.* 61:1174–1177.

Coad, B. W. 1986. The shortbarbel dragonfish, *Stomias brevibarbatus,* new to the fish fauna of the Atlantic coast of Canada. *Can. Field-Nat.* 100:394–395.

Coad, B. W., and G. Power. 1973a. Observations on the ecology and phenotypic variation of the threespine stickleback, *Gasterosteus aculeatus* L., 1758, and the blackspotted stickleback, *G. wheatlandi* Putnam, 1867 (Osteichthyes: Gasterosteidae) in Amory Cove, Québec. *Can. Field-Nat.* 87:113–121.

Coad, B. W., and G. Power. 1973b. Observations on the ecology and meristic variation of the ninespine stickleback, *Pungitius pungitius* (L., 1758) of the Matamek River system, Quebec. *Am. Midl. Nat.* 90:498–503.

Coad, B. W., and G. Power. 1973c. Life history notes and meristic variation in the freshwater fourspine stickleback, *Apeltes quadracus* (Mitchill), near Sept-Iles, Quebec. *Nat. Can.* 100:247–251.

Cohen, D. M. 1958. A revision of the fishes of the subfamily Argentininae. *Bull. Florida State Mus. Biol. Sci.* 3:93–172.

Cohen, D. M. 1964. Suborder Argentinoidae. Fishes of the western North Atlantic. Sears Found. Mar. Res. Mem. No. 1(4):1–70.

Cohen, D. M., ed. 1989. Papers on the systematics of gadiform fishes. Los Angeles Co. Nat. Hist. Mus. Sci. Ser. No. 32, 262 pp.

Cohen, D. M., and J. L. Russo. 1979. Variation in the fourbeard rockling, *Enchelyopus cimbrius,* a North Atlantic gadid fish, with comments on the genera of rocklings. *Fish. Bull., U.S.* 77:91–104.

Cohen, D. M., T. Inada, T. Iwamoto, and N. Scialabba. 1990. *FAO Species Catalogue,* Vol. 10. *Gadiform Fishes of the World* (Order Gadiformes). FAO Fish. Synop. No. 125, 442 pp.

Cohen, R. E., and R. G. Lough. 1983. Prey field of larval herring *Clupea harengus* on a continental shelf spawning area. *Mar. Ecol. Prog. Ser.* 10:211–222.

Cole, F. J. 1913. A monograph on the general morphology of the myxinoid fishes, based on a study of *Myxine.* Part V. The anatomy of the gut and its appendages. *Trans. Royal Soc. Edinburgh* 49:293–344.

Colin, P. L. 1974. Observation and collection of deep-reef fishes off the coasts of Jamaica and British Honduras (Belize). *Mar. Biol.* 24:29–38.

Colin, P. L. 1989. Aspects of the spawning of western Atlantic butterfly fishes (Pisces: Chaetodontidae). *Env. Biol. Fishes* 25:131–141.

Collette, B. B. 1968. *Strongylura timucu* (Walbaum): a valid species of western Atlantic needlefish. *Copeia* 1968:189–192.

Collette, B. B. 1978a. Belonidae. In: W. Fischer, ed. *FAO Species Identification Sheets for Fishery Purposes: Western Central Atlantic (Fishing Area 31),* Vol. 1. FAO, Rome, 14 pp.

Collette, B. B. 1978b. Hemiramphidae. In: W. Fischer, ed. *FAO Species Identification Sheets for Fishery Purposes: Western Central Atlantic (Fishing Area 31),*.Vol. 2. FAO, Rome, 12 pp.

Collette, B. B. 1986. Resilience of the fish assemblage in New England tidepools. *Fish. Bull., U.S.* 84:200–204.

Collette, B. B. 1990. Problems with gray literature in fishery science. In: J. Hunter, ed. *Writing for Fishery Journals.* Am. Fish. Soc.,

Collette, B. B. 1999. Mackerels, molecules, and morphology. *Proc. 5th Indo-Pacif. Fish Conf., Nouméa, Soc. Fr. Ictyol:* 149–164.

Collette, B. B. 2001. *Opsanus dichrostomus,* a new toadfish (Teleostei: Batrachoididae) from the western Caribbean Sea and southern Gulf of Mexico. Mus. Zool., Univ. Mich. Occ. Pap. No. 731, 16 pp.

Collette, B. B. in press. Family Echeneidae In: K. E. Carpenter, ed. *FAO Identification Guide to Living Marine Resources of the West Central Atlantic.* FAO, Rome.

Collette, B. B., and C. R. Aadland. 1996. Revision of the frigate tunas (Scombridae, *Auxis*), with descriptions of two new subspecies from the eastern Pacific. *Fish. Bull., U.S.* 94:423–441.

Collette, B. B., and L. N. Chao. 1975. Systematics and morphology of the bonitos (*Sarda*) and their relatives (Scombridae, Sardini). *Fish. Bull., U.S.* 73:516–625.

Collette, B. B., and K. E. Hartel. 1988. An annotated list of the fishes of Massachusetts Bay. NOAA Tech. Memo. NMFS-F/NEC-51, 70 pp.

Collette, B. B., and J. A. MacPhee, Jr. 1969. First Massachusetts Bay record of the Arctic shanny *Stichaeus punctatus. J. Fish. Res. Bd. Can.* 26:1375–1377.

Collette, B. B., and C. E. Nauen. 1983. *FAO Species Catalogue,* Vol. 2. *Scombrids of the World.* FAO Fish. Synop. No. 125, 137 pp.

Collette, B. B., and J. E. Randall. 2000. *Ammodytoides leptus,* a new species of sand lance (Teleostei: Ammodytidae) from Pitcairn Island. *Proc. Biol. Soc. Washington* 113:397–400.

Collette, B. B., and J. L. Russo. 1985. Morphology, systematics, and biology of the Spanish mackerels (*Scomberomorus,* Scombridae). *Fish. Bull., U.S.* 82:545–692.

Collette, B. B., R. H. Gibbs, Jr., and G. E. Clipper. 1969. Vertebral numbers and identification of the two species of dolphin (*Coryphaena*). *Copeia* 1969:630–631.

Collette, B. B., G. E. McGowen, N. V. Parin, and S. Mito. 1984a. Beloniformes: development and relationships. In: H. G. Moser, et al., eds. *Ontogeny and Systematics of Fishes.* Am. Soc. Ich. Herp. Spec. Publ. No. 1:334–354.

Collette, B. B., T. Potthoff, W. J. Richards, S. Ueyanagi, J. L. Russo, and Y. Nishikawa. 1984b. Scombroidei: development and relationships. In: H. G. Moser et al., eds. *Ontogeny and Systematics of Fishes.* Am. Soc. Ich. Herp. Spec. Publ. No. 1:591–620.

Collett, R. 1883. Meddelelser om Norges fiske i Aarene 1879–83. *Nyt Mag. Naturv. Kab. Christiania:* 29:47–123.

Collett, R. 1896. *Poissons provenants des campagnes du yacht "l'Hirondelle" (1885–88).* Monaco, 198 pp.

Collie, J. 1987. Food selection by yellowtail flounder (*Limanda ferruginea*) on Georges Bank. *Can. J. Fish. Aquat. Sci.* 44:357–367.

Colligan, M. A., J. F. Kocik, D. C. Kimball, G. Marancik, J. F. McKeon, and P. R. Nickerson. 1999. Review of the status of anadromous Atlantic salmon (*Salmo salar*) under the U.S. Endangered Species Act. Joint NMFS/USFWS Publ., Gloucester, Mass., 232 pp.

Collins, G. B. 1952. Factors influencing the orientation of migrating anadromous fishes. *U.S. Fish Wildl. Serv. Fish. Bull.* 52:373–396.

Collins, J. W. 1884. History of the tilefish. *Bull. U.S. Fish. Comm.* 10:237–295.

Collins, M. A. J. 1976. The lumpfish (*Cyclopterus lumpus* L.) in New-

Collins, M. R., and B. W. Stender. 1987. Larval king mackerel (*Scomberomorus cavalla*), Spanish mackerel (*S. maculatus*), and bluefish (*Pomatomus saltatrix*) off the southeast coast of the United States, 1973–1980. *Bull. Mar. Sci.* 41:822–834.

Collins, M. R., and B. W. Stender. 1989. Larval striped mullet (*Mugil cephalus*) and white mullet (*Mugil curema*) off the southeastern United States. *Bull. Mar. Sci.* 45:580–589.

Colton, J. B., Jr. 1955. Spring and summer distribution of haddock on Georges Bank. U. S. Fish Wildl. Serv. Spec. Sci. Rept., Fish. No. 156, 65 pp.

Colton, J. B., Jr. 1961. The distribution of eyed flounder and lanternfish larvae in the Georges Bank area. *Copeia* 1961: 274–279.

Colton, J. B., Jr. 1965. The distribution and behavior of pelagic and early demersal stages of haddock in relation to sampling techniques. ICNAF Spec. Publ. No. 6:317–333.

Colton, J. B., Jr. 1972. Temperature trends and the distribution of groundfish in continental shelf waters, Nova Scotia to Long Island. *Fish. Bull., U. S.* 70:637–657.

Colton, J. B., Jr., and K. A. Honey. 1963. The eggs and larval stages of the butterfish, *Poronotus triacanthus*. *Copeia* 1963: 447–450.

Colton, J. B., Jr., and R. R. Marak. 1969. Guide for identifying the common planktonic fish eggs and larvae of continental shelf waters, Cape Sable to Block Island. Bur. Comm. Fish., Woods Hole Lab. Ref. Doc. No. 69-9, 72 pp.

Colton, J. B., Jr., and J. M. St. Onge. 1974. Distribution of fish eggs and larvae in continental shelf waters, Nova Scotia to Long Island. Am. Geogr. Soc., Serial Atlas Mar. Environ. Folio No. 23.

Colton, J. B., Jr., and R. F. Temple. 1961. The enigma of Georges Bank spawning. *Limnol. Oceanogr.* 6:280–291.

Colton, J. B., Jr., W. G. Smith, A. W. Kendall, Jr., P. L. Berrien, and M. P. Fahay. 1979. Principal spawning areas and times of marine fishes, Cape Sable to Cape Hatteras. *Fish. Bull., U.S.* 76:911–915.

Colvocoresses, J. A., and J. A. Musick. 1984. Species associations and community composition of Middle Atlantic Bight continental shelf demersal fishes. *Fish. Bull., U.S.* 82:295–313.

Comeau, N. A. 1909. Life and sport on the north shore of the lower St. Lawrence and Gulf, containing chapters on salmon fishing, trapping, the folk-lore of the Montagnais Indians and tales of adventure on the fringe of the Labrador peninsula. Daily Telegraph Printing House, Quebec. 404 pp.

Compagno, L. J. V. 1984. *FAO Species Catalogue,* Vol. 4. *Sharks of the World: An Annotated and Illustrated Catalogue of the Shark Species Known to Date.* FAO Fish. Synop. No. 125, 655 pp.

Comyns, B. H., and G. C. Grant. 1993. Identification and distribution of *Urophycis* and *Phycis* (Pisces, Gadidae) larvae and pelagic juveniles in the U.S. Middle Atlantic Bight. *Fish. Bull., U.S.* 91:210–223.

Cone, D. K., and P. H. Odense. 1984. Pathology of five species of *Gyrodactylus* Nordmann, 1832 (Monogenea). *Can. J. Zool.* 62:1084–1088.

Cone, D. K., and M. Wiles. 1983. Comparative morphology of *Gyrodactylus groenlandicus* Levinson, 1881, *G. nainum* Hanek and Threlfall 1970, *G. pleuronecti* Cone, 1981 and *G. adsperila* sp. n. (Monogenea) from northwest Atlantic fishes. *Can. J. Zool.* 61:417–422.

Connaughton, M. A., and M. H. Taylor. 1994. Seasonal cycles in the sonic muscles of the weakfish, *Cynoscion regalis. Fish. Bull., U.S.* 92:697–703.

Conniff, R. 1992. They come, they die, they stink to high heaven. *Yankee Mag.* 56:82–87, 114–116.

Connolly, C. J. 1921. On the development of the angler (*Lophius piscatorius* L.). *Contrib. Can. Biol.* 7:115–124.

Conover, D. O., and B. E. Kynard. 1984. Field and laboratory observations of spawning periodicity and behavior of a northern population of the Atlantic silverside, *Menidia menidia* (Pisces: Atherinidae). *Env. Biol. Fishes* 11:161–171.

Conover, D. O., and S. A. Murawski. 1982. Offshore winter migration of the Atlantic silverside, *Menidia menidia. Fish. Bull., U.S.* 80:145–150.

Conover, D. O., and T. M. C. Present. 1990. Countergradient variation in growth rates: compensation for length of the growing season among Atlantic silversides from different latitudes. *Oecologia* 83:316–324.

Conover, D. O., and M. R. Ross. 1982. Patterns in seasonal abundance, growth and biomass of the Atlantic silverside, *Menidia menidia,* in a New England estuary. *Estuaries* 5:275–286.

Conover, D. O., J. J. Brown, and A. Ehtisham. 1997. Countergradient variation in growth of young striped bass (*Morone saxatilis*) from different latitudes. *Can. J. Fish. Aquat. Sci.* 54:2401–2409.

Conover, J. T., R. L. Fritz, and M. Vieira. 1961. A morphometric study of silver hake. U.S. Fish Wildl. Serv. Spec. Sci. Rept., Fish. No. 368, 13 pp.

Consi, T. R., P. A. Seifert, M. S. Triantafyllou, and E. R. Edelman. 2001. The dorsal fin engine of the seahorse (*Hippocampus* sp.). *J. Morph.* 248:80–97.

Cooper, G. P., and J. L. Fuller. 1945. A biological survey of Moosehead Lake and Haymock Lake, Maine. Maine Dept. Fish. Wildl. Fish Survey Rept. No. 6, 108 pp.

Cooper, J. A., and F. Chapleau. 1998. Monophyly and intrarelationships of the family Pleuronectidae (Pleuronectiformes) with a revised classification. *Fish. Bull., U.S.* 96:686–726.

Cooper, J. C. 1984. *Proceedings of a Workshop on Critical Data Needs for Shad Research on Atlantic Coast of North America.* Hudson River Foundation, NewYork.

Cooper, J. E. 1978. Identification of eggs, larvae, and juveniles of the rainbow smelt, *Osmerus mordax,* with comparisons to larval alewife *Alosa pseudoharengus,* and gizzard shad, *Dorosoma cepedianum. Trans. Am. Fish. Soc.* 107:56–62.

Cooper, R. A. 1961. Early life history and spawning migration of the alewife *Alosa pseudoharengus.* M.S. thesis, University of Rhode Island, Kingston, 58 pp.

Cooper, R. A. 1965a. An unusually large menhaden, *Brevoortia tyrannus* (Latrobe), from Rhode Island. *Trans. Am. Fish. Soc.* 94:412

Cooper, R. A. 1965b. Life history of the tautog, *Tautoga onitis* (Linnaeus), from Rhode Island. Ph.D. dissertation, University of Rhode Island, Kingston, 153 pp.

Cooper, R. A. 1966. Migration and population estimation of the tautog, *Tautoga onitis* (Linnaeus), from Rhode Island. *Trans. Am. Fish. Soc.* 95:239–247.

Cooper, R. A. 1967. Age and growth of the tautog, *Tautoga onitis* (Linnaeus), from Rhode Island. *Trans. Am. Fish. Soc.* 96:134–142.

Cooper, R. A., J. R. Uzmann, R. A. Clifford, and K. J. Recci. 1975. Direct observations of herring *Clupea harengus harengus* L. egg beds on Jeffreys Ledge, Gulf of Maine, in 1974. Int. Comm. Northw. Atl. Fish. Res. Doc., No. 93, Ser. No. 3573, 6 pp.

Cormier, S. M. 1986. Fine structure of hepatocytes and hepatocellular carcinoma of the Atlantic tomcod, *Microgadus tomcod* (Walbaum). *J. Fish Dis.* 9:179–194.

Cornish, G. A. 1907. Notes on fishes of Canso. *Contrib. Can. Biol., Ann. Rept. Dept. Mar. Fish.* 39:81–90.

Côté, G., P. Lamoureux, J. Boulva, and G. Lacroix. 1980. Séparation des populations de hareng de l'Atlantique *(Clupea harengus harengus)* de l'estuaire du Saint-Laurent et de la péninsule gaspésienne. *Can. J. Fish. Aquat. Sci.* 37:66–71.

Coughlin, D. J., and L. C. Rome. 1996. The roles of pink and red muscle in powering steady swimming in scup, *Stenotomus chrysops. Am. Zool.* 36:666–677.

Coughlin, D. J., and L. C. Rome. 1999. Muscle activity in steady swimming scup, *Stenotomus chrysops,* varies with fiber type and body position. *Biol. Bull., Woods Hole* 196:145–152.

Courtenay, S. C. 1985. Simultaneous multinesting by the fourspine stickleback, *Apeltes quadracus. Can. Field-Nat.* 99:360–363.

Courtenay, W. R., Jr. 1971. Sexual dimorphism of the sound producing mechanisms of the striped cusk–eel, *Rissola marginata* (Pisces: Ophidiidae). *Copeia* 1971:259–268

Courtois, R., and J. J. Dodson. 1986. Régime alimentaire et principaux facteurs influençant l'alimentation des larves de capelan *Mallotus villosus,* d'éperlan *Osmerus mordax* et de hareng *Clupea harengus harengus* dans un estuaire partiellement mélangé. *Can. J. Fish. Aquat. Sci.* 43:968–979.

Coutant, C. C. 1985. Striped bass, temperature, and dissolved oxygen: a speculative hypothesis for environmental risk. *Trans. Am. Fish. Soc.* 114:31–61.

Coutant, C. C. 1986. Thermal niches of striped bass. *Sci. Am.* 255: 98–105.

Covill, R. W. 1959. Food and feeding habits of larvae and postlarvae of *Ammodytes americanus,* 1952–1955. *Bull. Bingham Oceanog. Coll.* 17(1):125–146.

Cowan, G. I. M. 1971. Comparative morphology of the cottid genus *Myoxocephalus* based on meristic, morphometric, and other anatomical characters. *Can. J. Zool.* 49:1479–1496.

Cowan, J. H., Jr., and E. D. Houde. 1990. Growth and survival of bay anchovy *Anchoa mitchilli* larvae in mesocosm enclosures. *Mar. Ecol. Prog. Ser.* 68:47–57.

Cowen, R. K., L. A. Chiarella, C. J. Gomez and M. A. Bell. 1991. Offshore distribution, size, age, and lateral plate variation of late larval/early juvenile sticklebacks (*Gasterosteus*) off the Atlantic coast of New Jersey and New York. *Can. J. Fish. Aquat. Sci.* 48:1679–1684.

Cox, P. 1895. Catalogue of the marine and fresh-water fishes of New Brunswick. *Bull. Nat. Hist. Soc. New Brunswick* 13:62–75.

Cox, P. 1920. Histories of new food fishes. II. The lumpfish. *Bull.*

Cox, P. 1921. List of the fishes collected in 1917 off Cape Breton coast and the Magdalen Islands. *Contrib. Can. Biol. 1918–1920,* 11:109–114.

Cox, P., and M. Anderson. 1922. A study of the lumpfish (*Cyclopterus lumpus* L.). *Contrib. Can. Biol.* 1:1–20.

Crabtree, R. E. 1995. Relationship between lunar phase and spawning activity of tarpon, *Megalops atlanticus,* with notes on the distribution of larvae. *Bull. Mar. Sci.* 56:895–899.

Crabtree, R. E., E. C. Cyr, R. E. Bishop, L. M. Falkenstein, and J. M. Dean. 1992. Age and growth of tarpon, *Megalops atlanticus,* larvae in the eastern Gulf of Mexico, with notes on relative abundance and probable spawning areas. *Environ. Biol. Fishes* 35:361–370.

Crabtree, R. E., E. C. Cyr, and J. M. Dean. 1995. Age and growth of tarpon, *Megalops atlanticus,* from south Florida waters. *Fish. Bull., U.S.* 93:619–628.

Crabtree, R. E., E. C. Cyr, D. C. Chaverri, W. O. McLarney, and J. M. Dean. 1997. Reproduction of tarpon, *Megalops atlanticus,* from Florida and Costa Rican waters and notes on their age and growth. *Bull. Mar. Sci.* 61:271–285.

Craddock, J. E., and K. E. Hartel. in press. Family Myctophidae. In: K. E. Carpenter, ed. *FAO Identification Guide to Living Marine Resources of the West Central Atlantic.* FAO, Rome.

Craddock, J. E., R. H. Backus, and M. A. Daher. 1992. Vertical distribution and species composition of midwater fishes in warm-core Gulf Stream meander/ring 82-H. *Deep-Sea Res.* 39(Suppl. 1): S203–S218.

Craig, D., and G. J. FitzGerald. 1982. Reproductive tactics of four sympatric sticklebacks (Gasterosteidae). *Environ. Biol. Fishes* 7:369–375.

Crane, J. 1936. Notes on the biology and ecology of the giant tuna, *Thunnus thynnus* Linnaeus, observed at Portland, Maine. Zoologica, N.Y. 21:207–212.

Crawford, M. K., C. B. Grimes, and N. E. Buroker. 1989. Stock identification of weakfish, *Cynoscion regalis,* in the Middle Atlantic Region. *Fish. Bull., U.S.* 87:205–211.

Crawford, P., and G. R. Daborn. 1986. Seasonal variations in body size and fecundity in a copepod of turbid estuaries. *Estuaries* 9:133–141.

Crawford, R. H., R. R. Cusack, and T. R. Parlee. 1986. Lipid content and energy expenditure in the spawning migration of alewife (*Alosa pseudoharengus*) and blueback herring (*Alosa aestivalis*). *Can. J. Zool.* 64:1902–1907.

Crawford, R. J. M. 1981. Distribution, availability and movements of round herring, *Etrumeus teres,* off South Africa, 1964–1976. *Fish. Bull. So. Afr.* 14:141–181.

Creaser, E. P., and D. A. Libby. 1986. Tagging of Age 1 herring *Clupea harengus* L. and their movement along the Maine and New Brunswick coasts. *J. Northw. Atl. Fish. Sci.* 7:43–46.

Creaser, E. P., and D. A. Libby. 1988. Seasonal movements of juvenile and adult herring (*Clupea harengus* L.) tagged along the Maine and New Hampshire coasts in 1976–82. *J. Northw. Atl. Fish. Sci.* 8:33–42.

Creaser, E. P., and H. C. Perkins. 1994. The distribution, food, and age of juvenile bluefish, *Pomatomus saltatrix,* in Maine. *Fish. Bull.,*

Creaser, E. P., D. A. Libby, and G. D. Speirs. 1984. Seasonal movements of juvenile and adult herring, *Clupea harengus* L., tagged along the Maine coast. *J. Northw. Atl. Fish. Sci.* 5:71–78.

Crecco, V. A. 1982. Overview of alewife and blueback herring runs in Connecticut. Report to Atlantic States Marine Fisheries Commission, Alosid Scientific and Statistical Committee, Washington, D.C.

Crecco, V. A., and M. M. Blake. 1983. Feeding ecology of coexisting larvae of American shad and blueback herring in the Connecticut River. *Trans. Am. Fish. Soc.* 112:498–507.

Crecco, V. A., and T. F. Savoy. 1984. Effects of fluctuations in hydrographic conditions on year-class strength of American shad (*Alosa sapidissima*) in the Connecticut River. *Can. J. Fish. Aquat. Sci.* 41:1216–1223.

Crecco, V. A., and T. F. Savoy. 1985. Effects of biotic and abiotic factors on growth and relative survival of young American shad, *Alosa sapidissima,* in the Connecticut River. *Can. J. Fish. Aquat. Sci.* 42:1640–1648.

Crecco, V. A., and T. F. Savoy. 1987. Review of recruitment mechanisms of American shad: the critical period and match-mismatch hypotheses reexamined. In: M. J. Dadswell et al., eds., *Common Strategies of Anadromous and Catadromous Fishes. Am. Fish. Soc. Symp.* 1:455–468.

Crecco, V. A., T. F. Savoy, and L. Gunn. 1983. Daily mortality rates of larval and juvenile American shad (*Alosa sapidissima*) in the Connecticut River with changes in year-class strength. *Can. J. Fish. Aquat. Sci.* 40:1719–1728.

Creed, R. P., Jr. 1985. Feeding, diet, and repeat spawning of blueback herring, *Alosa aestivalis,* from the Chowan River, North Carolina. *Fish. Bull., U.S.* 83:711–716.

Cressey, R. F. 1971. Two new argulids (Crustacea: Branchirua) from the eastern United States. *Proc. Biol. Soc. Washington* 84:253–258.

Cressey, R. F., and B. B. Collette. 1970. Copepods and needlefishes: a study in host-parasite relationships. *Fish. Bull., U.S.* 68:347–432.

Cressey, R. F., and H. B. Cressey. 1980. Parasitic copepods of mackerel- and tuna-like fishes (Scombridae) of the world. *Smithsonian Contrib. Zool.* 311, 186 pp.

Cressey, R. F., and E. A. Lachner. 1970. The parasitic copepod diet and life history of diskfishes (Echeneidae). *Copeia* 1970:310–318.

Cressey, R. F., B. B. Collette, and J. L. Russo. 1983. Copepods and scombrid fishes: a study in host-parasite relationships. *Fish. Bull., U.S.* 81:227–265.

Crestin, D. S. 1973. Some aspects of the biology of adults and early life stages of the rainbow smelt, *Osmerus mordax* (Mitchill), from the Weweantic River Estuary, Wareham-Marion, Massachusetts, 1968. M.S. thesis, University of Massachusetts, Amherst, 108 pp.

Crisp, E. 1853. On some points relating to the structure and mechanism of the wolffish (*Anarhichas lupus*). *Ann. Mag. Nat. Hist., Ser. 2,* 11:463–466.

Cross, J. N., C. A. Zetlin, P. L. Berrien, D. L. Johnson, and C. McBride. 1999. Essential Fish Habitat Source Document: Butterfish, *Peprilus triacanthus,* life history and habitat characteristics. NOAA Tech. Memo. NMFS-NE-145, 42 pp.

Crystall, B. 2000. Monstrous mucus. *New Sci.* 165:38–41.

Cunningham, J. T. 1891. On the reproduction and development of the conger. *J. Mar. Biol. Assoc. U. K., n. ser.* 2:16–42.

Cunningham, J. T. 1888. The eggs and larvae of teleosteans. *Trans. R. Acad. Edinb.* 33:97–136.

Curley, J. R., R. P. Lawton, M. J. Hickey, and J. D. Fiske. 1970. A study of the marine resources of the Waquoit Bay-Eel Pond Estuary. Mass. Div. Mar. Fish. Monogr. Ser. No. 9, 40 pp.

Curley, J. R., R. P. Lawton, D. K. Whittaker, and M. J. Hickey. 1972. A study of the marine resources of Wellfleet Harbor. Mass. Div. Mar. Fish. Monogr. Ser. No. 12, 37 pp.

Curran, H. W., and D. T. Ries. 1937. Fisheries investigations in the lower Hudson River. In: *A Biological Survey of the Lower Hudson Watershed.* New York State Cons. Dept. Ann. Rept.No. 26(Suppl.):124–145.

Curran, S., and J. N. Caira. 1995. Attachment site specificity and the tapeworm assemblage in the spiral intestine of the blue shark (*Prionace glauca*). *J. Parasitol.* 81:149–157.

Cushing, D. H. 1967. The grouping of herring populations. *J. Mar. Biol. Assoc. U.K.* 47:193–208.

Cushing, D. H. 1969. The regularity of the spawning season in some fishes. *J. Cons. Cons. Int. Explor. Mer* 33:81–92.

Cushing, D. H. 1973. Food and the stabilization mechanism in fishes. Mar. Biol. Assoc. India. (Special Publication dedicated to Dr. N. K. Panekkar, May 1973), pp. 29–39.

Cushing, D. H. 1975. Marine ecology and fisheries. Cambridge University Press, London, 278 pp.

Cushing, D. H. 1985. Reporter's summary. *Can. J. Fish. Aquat. Sci.* 42(Suppl. 1):275–278.

Cyr, E. C. 1991. Aspects of the life history of the tarpon, *Megalops atlanticus,* from south Florida. Ph.D. dissertation, University of South Carolina, Columbia, 138 pp.

D'Yakov Yu. P. 1991. Population structure of Pacific black halibut, *Reinhardtius hippoglossoides. Voprosy Ikhtiol.* 31:404–414.

Daborn, C. R., and R. S. Gregory. 1983. Occurrence, distribution and feeding habits of juvenile lumpfish, *Cyclopterus lumpus* L. in the Bay of Fundy. *Can. J. Zool.* 61:797–801.

Dadswell, M. J. 1975. Further new localities for certain coldwater fishes in eastern Ontario and western Quebec. *Can. Field-Natur.* 89:447–450.

Dadswell, M. J. 1979. Biology and population characteristics of the shortnose sturgeon, *Acipenser brevirostrum,* LeSueur 1818 (Osteichthyes: Acipenseridae), in the Saint John River Estuary, New Brunswick, Canada. *Can. J. Zool.* 57:2186–2210.

Dadswell, M. J. 1984. Status of the shortnosed sturgeon, *Acipenser brevirostrum,* in Canada. *Can. Field-Natur.* 98:75–79.

Dadswell, M. J. 1985. Status of the blueback herring, *Alosa aestivalis,* in Canada. *Can. Field-Natur.* 99:409–412.

Dadswell, M. J., G. D. Melvin, and P. J. Williams. 1983. Effect of turbidity on the temporal and spatial utilization of the inner Bay of Fundy by American shad *Alosa sapidissima* (Pisces: Clupeidae) and its relationship to local fisheries. *Can. J. Fish. Aquat. Sci.* 40(Suppl. 1):322–330.

Dadswell, M. J., B. D. Taubert, T. S. Squiers, D. Marchette and J. Buckley. 1984. Synopsis of biological data on shortnose sturgeon,

Acipenser brevirostrum LeSueur 1818. NOAA Tech. Rept. NMFS 14, 45 pp.

Dadswell, M. J., G. D. Melvin, P. J. Williams, and D.E. Themelis. 1987. Influences of origin, life history, and chance on the Atlantic coast migration of American shad. In: M. J. Dadswell et al., eds., *Common Strategies of Anadromous and Catadromous Fishes. Am. Fish. Soc. Symp.* 1:313–330.

Dahl, E., D. S. Danielssen, E. Moksness, and P. Solemdal, eds. 1984. The propagation of cod *Gadus morhua* L. *Flødevigen Rapp.* 1:123–143.

Dahlberg, M. D. 1970. Atlantic and Gulf of Mexico menhadens, genus *Brevoortia* (Pisces: Clupeidae). *Bull. Fla. St. Mus. Biol. Sci.* 15:91–162.

Dahlberg, M. D. 1972. An ecological study of Georgia coastal fishes. *Fish. Bull., U.S.* 70:323–353.

Daiber, F. C. 1960. A technique for age determination in the skate, *Raja eglanteria. Copeia* 1960:258–260.

Daiber, F. C. 1982. *Animals of the Tidal Marsh.* Van Nostrand Reinhold Co., New York, 422 pp.

Dalley, E. L., and G. H. Winters. 1987. Aspects of the early life history of sand launce (*Ammodytes* sp.) in Fortune Bay, Newfoundland. *Fish. Bull., U.S.* 85:631–641.

Dalton, P. D. 1987. Ecology of bay anchovy (*Anchoa mitchilli*) eggs and larvae in the mid-Chesapeake Bay. M.S. thesis, University of Maryland, College Park, 104 pp.

Damas, D. 1909. Contribution à la biologie des Gadides. *Rapp. Proc.-Verb. Réun. Cons. Int. Explor. Mer* 10B(3):1–277.

Daniel, L. B. III. 1995. Spawning and ecology of early life stages of black drum, *Pogonias cromis,* in lower Chesapeake Bay. Ph. D. dissertation, College of William and Mary, Gloucester Point, 167 pp.

Dannevig, A. 1919. Biology of Atlantic waters of Canada: Canadian fish eggs and larvae. In: J. Hjort, ed. *Canadian Fisheries Expedition 1914–1915: Investigations in the Gulf of St. Lawrence and Atlantic Waters of Canada.* Ottawa Dept. Naval. Serv. Mem. No. 2, 49 pp.

Darling, D. B., and H. W. Nilson. 1946. Nutritive value of the protein of swellfish. *U.S. Fish Wildl. Serv. Comm. Fish. Rev.* 8(8):10–11.

Darnell, R. M. 1958. Food habits of fishes and larger invertebrates of Lake Pontchartrain, Louisiana, an estuarine community. Inst Mar. Sci. Univ. Tex. Publ. No. 5:353–416.

Davenport, D., and M. Warmuth. 1965. Notes on the relationship between the freshwater mussel, *Anodonta implicata* Say and the alewife *Pomolobus pseudoharengus* (Wilson). *Limnol. Oceanogr.* 10(Suppl.):R74–R78.

Davenport, J. 1983. Oxygen and the developing eggs and larvae of the lumpfish, *Cyclopterus lumpus. J. Mar. Biol. Assoc. U.K.* 63:633–640.

Davenport, J. 1985. Synopsis of biological data on the lumpsucker, *Cyclopterus lumpus* (Linnaeus, 1758). FAO Fish Synop. No. 147, 31 pp.

Davenport, J., and C. Bradshaw. 1995. Observations on skin colour changes in juvenile lumpsucker. *J. Fish Biol.* 47:143–154.

Davenport, J., and E. Kjorsvik. 1986. Buoyancy in the lumpsucker, *Cyclopterus lumpus. J. Mar. Biol. Assoc. U.K.* 66:159–174.

Davenport, J., and V. Thorsteinsson. 1989. Observations on the colours

Davenport, J., S. Lonning, and E. Kjorsvik. 1983. Ammonia output by eggs and larvae of the lumpsucker, *Cyclopterus lumpus* L., the cod, *Gadus morhua* and the plaice, *Pleuronectes platessa. J. Mar. Biol. Assoc. U.K.* 63:713–723.

Davenport, J., E. Kjorsvik, and T. Haug. 1990. Appetite, gut transit, oxygen uptake and nitrogen excretion in captive Atlantic halibut, *Hippoglossus hippoglossus* L., and lemon sole, *Microstomus kitt* (Walbaum). *Aquaculture* 90:267–277.

Davies, W. D. 1970. The effects of temperature, pH and total dissolved solids on the survival of immature striped bass, *Morone saxatilis* (Walbaum). Ph.D. thesis, North Carolina State University, Raleigh, 100 pp.

Davis, J., and E. B. Joseph. 1964. Southern record of *Sebastes marinus,* ocean perch. *Chesapeake Sci.* 5:212.

Davis, J. R., J. P. Miller, and W. L. Wilson. 1970. Completion Report-embodying annual progress report of the anadromous fish project: biology and utilization of anadromous alosids. Virginia Inst. Mar. Sci., Proj. No. VA AFC-1, 1969–70, 185 pp.

Davis, P. S. 1990. The first occurrence of torsk, *Brosme brosme* (Ascanius, 1772) for the coast of Northumberland, and additional records of the swordfish, *Xiphias gladius* Linnaeus, 1758. *Trans. Nat. Hist. Soc. Northumbria* 55:149.

Davis, R. H., Jr. 1966. Population studies of striped bass, *Roccus saxatilis* (Walbaum), in Maine based on age distributions and growth rate. M.S. thesis, University of Maine, Orono, 50 pp.

Dawe, E. G., E. L Dalley, and W. W. Lidster. 1997. Fish prey spectrum of short-finned squid (*Illex illecebrosus*) at Newfoundland. *Can. J. Fish. Aquat. Sci.* 54(Suppl. 1):200–208.

Dawe, E. G., W. R. Bowering, and J. B. Joy. 1998. Predominance of squid (*Gonatus* spp.) in the diet of Greenland halibut (*Reinhardtius hippoglossoides*) on the deep slope of the northeast Newfoundland continental shelf. *Fish. Res. (Amsterdam)* 36:267–273.

Dawson, C. E. 1958. A study of the biology and life history of the spot, *Leiostomus xanthurus* Lacépède, with special reference to South Carolina. Contrib. Bears Bluff Lab. No. 28, 48 pp.

Dawson, C. E. 1962. Notes on anomalous American Heterosomata with descriptions of five new records. *Copeia* 1962:138–146.

Dawson, C. E. 1967. Contributions to the biology of the cutlassfish (*Trichiurus lepturus*) in the northern Gulf of Mexico. *Trans. Am. Fish. Soc.* 96:117–121.

Dawson, C. E. 1982. The pipefishes (sufamilies Doryrhamphinae and Syngnathinae). Fishes of the western North Atlantic. Sears Found. Mar. Res. Mem. No. 1(8):4–172.

Dawson, C. E., and R. A. Fritzsche. 1975. Odontoid processes in pipefish jaws. *Nature* 257:390.

Dawson, J. A. 1960. The oral cavity, the 'jaws' and the horny teeth of *Myxine glutinosa.* In: A. Brodal and R. Fänge, eds. *The Biology of Myxine.* Universitetsforlaget, Oslo, pp. 231–255.

Dawson, M. A. 1990. Blood chemistry of the windowpane flounder *Scopthalmus aquosus* in Long Island Sound: geographical, seasonal, and experimental variation. *Fish. Bull., U.S.* 88:429–437.

Day, F. 1880–1884. *The Fishes of Great Britain and Ireland.* Williams and Norgate, London, Edinburgh, 2 vols., 336 pp., 388 pp.

Day, L. R. 1957a. Populations of herring in the northern Gulf of St.

Day, L. R. 1957b. Populations of herring in the southern Gulf of St. Lawrence. *Bull. Fish. Res. Bd. Can.* 111:121–137.

De Lafontaine, Y. 1986. Useful morphometric index for the identification of northern blennioid larval fishes. *Nat. Can.* 113:219–222.

De Lamater, E. D., and W. R. Courtenay, Jr. 1973. Studies on scale structure of flatfishes. I. The genus *Trinectes,* with notes on related forms. Proc. 27th Ann. Conf. Southeast. Assoc. Game Fish Comm., pp. 591–608.

de Menezes, M. F. 1968. Sôbre a alimentação do camurupium, *Tarpon atlanticus* (Valenciennes), no Estado do Ceará. Arq. Est. Biol. Mar. Univ. Fed. Ceará 8:145–149.

de Menezes, M. F., and M. P. Paiva. 1966. Notes on the biology of tarpon, *Tarpon atlanticus* (Cuvier & Valenciennes), from coastal waters of Ceará State, Brazil. *Arq. Est. Biol. Mar. Univ. Fed. Ceará* 6:83–98.

de Oliveira, R. F., V. C. Almada, and M. de Fátima Gil. 1993. The reproductive behavior of the longspine snipefish, *Macrorhamphosus scolopax* (Syngnathiformes, Macrorhamphosidae). *Environ. Biol. Fishes* 36:337–343.

De Silva, C. 1974. Development of the respiratory system in herring and plaice larvae. In: J. H. S. Blaxter, ed. *The Early Life History of Fish.* Springer-Verlag, Berlin, pp. 465–485.

de Sylva, D. P. 1963. Systematics and life history of the great barracuda, *Sphyraena barracuda* (Walbaum). Stud. Trop. Oceanogr., Miami, Fla.

de Sylva, D. P. 1984. Sphyraenoidei: development and relationships. In: H. G. Moser et al., eds. *Ontogeny and Systematics of Fishes.* Am. Soc. Ich. Herp. Spec. Publ. No. 1:534–540.

de Sylva, D. P., F. A. Kalber, Jr., and C. N. Shuster, Jr. 1962. Fishes and ecological conditions in the shore zone of the Delaware River estuary, with notes on other species collected in deeper water. Univ. Del. Mar. Lab. Info. Ser. Publ. No. 51, 164 pp.

De Vries, M. C., R. B. Forward, Jr., and W. F. Hettler. 1995a. Behavioral response of larval Atlantic menhaden *Brevoortia tyrannus* (Latrobe) and spot *Leiostomus xanthurus* (Lacépède) to rates of salinity change. *J. Exp. Mar. Biol. Ecol.* 185:93–108.

De Vries, M. C., R. B. Forward, Jr., and W. F. Hettler. 1995b. Behavioural response of larval Atlantic menhaden to different rates of temperature change. *J. Fish Biol.* 47:1081–1095.

Dean, B. 1893. Recent experiments in sturgeon hatching on the Delaware. *Trans. N.Y. Acad. Sci.* 13:69–75.

Dean, B. 1895. The early development of gar-pike and sturgeon. *J. Morphol.* 11:1–62.

Dean, B. 1900. The egg of the hagfish, *Myxine glutinosa. N.Y. Acad. Sci.* 2:33–45.

DeArmon, I. A., Jr. 1948. Sampling technique on the ovary of the striped bass (*Roccus saxatilis* Walbaum). M.S. thesis, Virginia Polytechnic Institute, Blacksburg.

Debelius, H. 1997. *Mediterranean and Atlantic Fish Guide.* IKAN Book Service, Ridgecrest, Cal., 309 pp.

Deblois, E. M., and W. C. Leggett. 1993. Impact of amphipod predation on the benthic eggs of marine fish-an analysis of *Calliopius laeviusculus* bioenergetics demands and predation on the eggs of a beach spawning osmeriid (*Mallotus villosus*). *Mar. Ecol. Prog. Ser.* 93:205–216.

Deegan, L. A., B. J. Peterson, and R. Portier. 1990. Stable isotopes and cellulase activity as evidence for detritus as a food source for juvenile gulf menhaden. *Estuaries* 13:14–19.

De Kay, J. E. 1842. Zoology of New York. IV. Fishes. Nat. Hist. New York Geol. Survey, 415 pp.

Delaney, G. R. 1986. Morphometric and meristic stock identification of summer flounder (*Paralichthys dentatus*). M.A. thesis, College of William and Mary, Williamsburg, 47 pp.

Delbeek, J. C., and D. D. Williams. 1987a. Food resource partitioning between sympatric populations of brackishwater sticklebacks. *J. Anim. Ecol.* 56:949–967.

Delbeek, J. C., and D. D. Williams. 1987b. Morphological differences among females of four species of stickleback (Gasterosteidae) from New Brunswick and their possible ecological significance. *Can. J. Zool.* 65:289–295.

Delbeek, J. C., and D. D. Williams. 1988. Feeding selectivity of four species of sympatric stickleback in brackish-water habitats in eastern Canada. *J. Fish Biol.* 32:41–62.

Delsman, H. C. 1926. Fish eggs and larvae from the Java Sea. V. *Caranx kurra, macrosoma,* and *crumenophthalmus. Treubia* 8:199–211.

Dempsey, C. H. 1978. Chemical stimuli as a factor in feeding and intraspecific behaviour of herring larvae. *J. Mar. Biol. Assoc. U.K.* 58:739–747.

Dempsey, C. H., and R. N. Bamber. 1983. Spawning of herring (*Clupea harengus* L.) in the Blackwater Estuary, spring 1979. *J. Cons. Int. Explor. Mer* 41:85–92.

Dempson, J. B., and T. R. Porter. 1993. Occurrence of sea lamprey, *Petromyzon marinus,* in a Newfoundland River, with additional records from the northwest Atlantic. *Can. J. Fish. Aquat. Sci.* 50:1265–1269.

Dempson, J. B., L. J. Ledrew, and G. Furey. 1983. Occurrence of American shad (*Alosa sapidissima*)in northern Labrador waters. *Naturaliste Can. (Rev. Écol. Syst.).* 110:217–221.

Denoncourt, R. F., J. C. Fisher, and K. M. Rapp. 1978. A freshwater population of the mummichog, *Fundulus heteroclitus* from the Susquehanna River drainage in Pennsylvania. *Estuaries* 1:269–272.

Deree, H. L. 1999. Age and growth, dietary habits, and parasitism of the fourbeard rockling, *Enchelyopus cimbrius,* from the Gulf of Maine. *Fish. Bull., U.S.* 97:39–52.

Derickson, W. K., and K. S. Price, Jr. 1973. The fishes of the shore zone of Rehoboth and Indian River bays, Delaware. *Trans. Am. Fish. Soc.* 102:552–562.

Dery, L. M. 1988a. Atlantic herring *Clupea harengus.* In: J. Penttila and L. M. Dery, eds. *Age Determination Methods for Northwest Atlantic Species.* NOAA Tech. Rept. NMFS No. 72:17–22.

Dery, L. M. 1988b. Atlantic mackerel, *Scomber scombrus.* In: J. Penttila and L. M. Dery, eds. *Age Determination Methods for Northwest Atlantic Species.* NOAA Tech. Rept. NMFS No. 72:77–84.

Dery, L. M. 1988c. Butterfish, *Peprilus triacanthus.* In: J. Pentilla and L. M. Dery, eds. *Age Determination Methods for Northwest Atlantic Species.* NOAA Tech. Rept. NMFS No. 72:85–92.

Dery, L. M. 1988d. Summer flounder, *Paralichthys dentatus.* In: J. Penttila and L. M. Dery, eds. *Age Determination Methods for Northwest Atlantic Species.* NOAA Tech. Rept. NMFS No. 72:97–102.

Dery, L. M. 1988e. American plaice, *Hippoglossoides platessoides*. In: J. Penttila and L. M. Dery, eds. *Age Determination Methods for Northwest Atlantic Species*. NOAA Tech. Rept. NMFS No. 72: 111–118.

Dery, L. M., and J. Chenoweth. 1979. Recent problems in ageing sea herring from the Gulf of Maine. Natl. Mar. Fish. Serv., Woods Hole Lab., Ref. Doc. No. 79-38, 7 pp.

Des Clers, S., and K. Andersen. 1995. Sealworm (*Pseudoterranova decipiens*) transmission to fish trawled from Hvaler, Oslofjord, Norway. *J. Fish Biol.* 46:8–17.

DesFosse, J. C., and J. G. Loesch. 1985. Multivariate analysis of meristic characters of juvenile *Alosa*. Anadromous Fishes, Compl. Rept. Proj. No. AFC 15-1, Virginia Institute of Marine Sciences, 36 pp.

DesFosse, J. C., N. M. Burkhead, and R. E. Jenkins. 1994. Herrings, Family Clupeidae. In: R. E. Jenkins and N. M. Burkhead, eds. *Freshwater Fishes of Virginia*. Am. Fish. Soc., Bethesda, pp. 209–228.

Detwyler, R. 1963. Some aspects of the biology of the seasnail, *Liparis atlanticus* (Jordan and Evermann). Ph.D. dissertation, University of New Hampshire, 103 pp.

Detwyler, R., and E. D. Houde. 1970. Food selection by laboratory-reared larvae of the scaled sardine *Harengula pensacolae* (Pisces, Clupeidae) and the bay anchovy *Anchoa mitchilli* (Pisces, Engraulidae). *Mar. Biol.* 7:214–222.

Deubler, E. E., Jr., and J. C. White, Jr. 1962. Influence of salinity on growth of postlarvae of the summer flounder, *Paralichthys dentatus*. *Copeia* 1962:468–469.

Deuel, D. G. 1964a. Evidence of spawning of tagged bluefish. *Underwater Nat.* 2:24.

Deuel, D. G. 1964b. A note on a bluefish eating a sea lamprey. *Underwater Nat.* 2:32.

Deuel, D. G., J. R. Clark, and A. J. Mansueti. 1966. Description of embryonic and early larval stages of bluefish, *Pomatomus saltatrix*. *Trans. Am. Fish. Soc.* 95:264–271.

DeVane, J. C., Jr. 1978. Food of king mackerel, *Scomberomorus cavalla,* in Onslow Bay, North Carolina. *Trans. Am. Fish. Soc.* 107: 583–586.

Dew, C. B. 1976. A contribution to the life history of the cunner, *Tautogolabrus adspersus,* in Fishers Island Sound, Connecticut. *Chesapeake Sci.* 17:101–113.

Dew, C. B. 1981. Biological characteristics of commercially caught Hudson River striped bass 1973–75. In: J. W. Racklin and G. Tauber, eds. *Papers of the Fifth Symposium on Hudson River Ecology 1980.,* Hudson River Environ. Soc., New York, pp. 82–102.

Dew, C. B. 1988. Biological characteristics of commercially caught Hudson River striped bass, 1973–1975. *No. Am. J. Fish. Manage.* 8:75–83.

Dew, C. B., and J. H. Hecht. 1994a. Hatching, estuarine transport, and distribution of larval and early juvenile Atlantic tomcod, *Microgadus tomcod,* in the Hudson River. *Estuaries* 17:472–488.

Dew, C. B., and J. H. Hecht. 1994b. Recruitment, growth, mortality, and biomass production of larval and early juvenile Atlantic tomcod in the Hudson River estuary. *Trans. Am. Fish. Soc.* 123:

Dewar, A. B., L. Lipton, and G. Mack. 1971. Processing lumpfish caviar: a progress report of work in 1970. Applied Research Development Lab., Prog. Rept. Tech. Rept. No. 7, 21 pp.

DeWitt, H. H., P. A. Grecay, J. S. Hacunda, B. P. Lindsay, R. F. Shaw, and D. W. Townsend. 1981. An addition to the fish fauna of the Gulf of Maine with records of rare species. *Proc. Biol. Soc. Washington* 94:669–674.

Dey, W. P. 1981. Mortality and growth of young-of-the-year striped bass in the Hudson River Estuary. *Trans. Am. Fish. Soc.* 110:151–157.

Dey, W. P., T. H. Peck, C. E. Smith, S. Cormier, and P. Baumann. 1984. Status of hepatomas in Hudson River tomcod. Conference on the fate and transport of toxicants on the Hudson River, Troy, N.Y.

Dey, W. P., T. H. Peck, C. E. Smith, and G.-L. Kreamer. 1993. Epizoology of hepatic neoplasia in Atlantic tomcod (*Microgadus tomcod)* from the Hudson River Estuary. *Can. J. Fish. Aquat. Sci.* 50:1897–1907.

Dick, M. M., and W. C. Schroeder. 1968. Henry Bryant Bigelow, 1879–1967. *Copeia* 1968:657–659.

Dickie, L. M., and F. D. McCracken. 1955. Isopleth diagrams to predict equilibrium yields of a small flounder fishery. *J. Fish. Res. Bd. Can.* 12:187–209.

Didier, D. A. 1995. Phylogenetic systematics of extant chimaeroid fishes (Holocephali, Chimaeroidei). Am. Mus. Nat. Hist. Novitates No. 3119, 86 pp.

Dietrich, C. S. 1979. Fecundity of the Atlantic menhaden, *Brevoortia tyrannus. Fish. Bull., U.S.* 77:308–311.

Dillon, W. A. 1966. Provisional list of parasites occurring on *Fundulus* spp. *Virginia J. Sci.* 17:21–31.

Di Michele, L., and D. A. Powers. 1982. LDH-B genotype-specific hatching times of *Fundulus heteroclitus. J. Exp. Zool.* 214:181–187.

Di Michele, L., and M. H. Taylor. 1980. The environmental control of hatching in *Fundulus heteroclitus. J. Exp. Zool.* 214:181–187.

Di Michele, L., and M. E. Westerman. 1997. Geographic variation in development rate between populations of the teleost *Fundulus heteroclitus. Mar. Biol.* 128:1–7.

Di Michele, L., D. A. Powers, and M. H. Taylor. 1986. Introduction to the symposium: the biology of *Fundulus heteroclitus. Am. Zool.* 26:109.

Din, Z. B., and G. Gunter. 1986. The food and feeding habits of the common bay anchovy, *Anchoa mitchilli* (Valenciennes). *Pertanika* 9:99–108.

Ditty, J. G., and F. M. Truesdale. 1983. Comparative larval development of *Peprilus burti, P. triacanthus* and *P. paru* (Pisces: Stromateidae) from the western North Atlantic. *Copeia* 1983:397–406.

Ditty, J. G., R. F. Shaw, C. B. Grimes, and J. S. Cope. 1994. Larval development, distribution, and abundance of common dolphin, *Coryphaena hippurus,* and pompano dolphin, *C. equiselis* (family: Coryphaenidae), in the northern Gulf of Mexico. *Fish. Bull., U.S.* 92:275–291.

Ditty, J. G., R. F. Shaw, and T. Farooqi. 1999. Preliminary guide to the identification of the early life stages of sphyraenid fishes of the western central Atlantic. NOAA Tech. Memo. NMFS-SEFSC-

Ditty, J. G., T. Farooqi, and R. F. Shaw. 2000. Preliminary guide to identification of the early life stages of mullet (Pisces: Mugilidae) from the western central Atlantic. NOAA Tech. Memo. NMFS-SEFSC-424, 10 pp.

Dixon, M. S. 1994. Habitat selection in juvenile tautog, *Tautoga onitis,* and juvenile cunner, *Tautogolabrus adspersus.* M.S. thesis, University of Connecticut, Storrs.

Dobbs, D. 2000. The great Gulf: fishermen, scientists, and the struggle to revive the world's greatest fishery. Island Books, Washington, D.C, 206 pp.

Dodrill, J. W. 1977. A hook and line survey of the sharks found within five hundred meters of shore along Melbourne Beach, Brevard County, Florida. M.S. thesis, Florida Inst. Technol., Melbourne, 304 pp.

Dodson, J. J., and W. C. Leggett. 1973. The behavior of American shad *Alosa sapidissima* homing to the Connecticut River from Long Island Sound. *J. Fish. Res. Bd. Can.* 30:1847–1860.

Dodson, J. J., and W. C. Leggett. 1974. Role of olfaction and vision in the behavior of American shad *Alosa sapidissima* homing to the Connecticut River from Long Island Sound. *J. Fish. Res. Bd. Can.* 31:1607–1619.

Domermuth, R. B., and R. J. Reed. 1980. Food of juvenile American shad, *Alosa sapidissima,* juvenile blueback herring, *Alosa aestivalis,* and pumpkinseed, *Lepomis gibbosus,* in the Connecticut River below Holyoke Dam, Massachusetts. *Estuaries* 3:65–68.

Dominy, C. L. 1973. Effects of entrance pool weir elevation and fish density on passage of alewives in a pool and weir fishery. *Trans. Am. Fish. Soc.* 102:398–404.

Dooley, J. K. 1972. Fishes associated with the pelagic *Sargassum* complex, with a discussion of the *Sargassum* community. *Contr. Mar. Sci., Univ. Texas* 16:1–32.

Dooley, J. K. 1978. Systematics and biology of the tilefishes (Perciformes: Branchiostegidae and Malacanthidae), with descriptions of two new species. NOAA Tech. Rept. NMFS Circ. No. 411, 78 pp.

Dorf, B. A. 1994. Ecology of juvenile tautog (*Tautoga onitis*) in Narragansett Bay, Rhode Island. Ph.D. dissertation, University of Rhode Island, Kingston, 228 pp.

Dorf, B. A. and J. C. Powell. 1997. Distribution, abundance, and habitat characteristics of juvenile tautog (*Tautoga onitis,* Family Labridae) in Narragansett Bay, Rhode Island, 1988–1992. *Estuaries* 20:589–600.

Dorfman, D. 1970. Responses of some anadromous fishes to varied oxygen concentrations and increased temperatures. Ph.D. dissertation, Rutgers University, New Brunswick, 76 pp.

Dornheim, H. 1975. Mean weight at age of the 1970 year-class herring in ICNAF Div. 5Z. Int. Comm. Northw. Atl. Fish. Res. Doc. No. 75/114, Ser. No. 3607, 3 pp.

Doroshev, S. I. 1970. Biological features of the eggs, larvae and young of the striped bass [*Roccus saxatilis* (Walbaum)] in connection with the problem of its acclimatization in the USSR. *Voprosy Ikhtiol.* 10:341–359 [in Russian, translation in *J. Ichthyol.* 10:235–248].

Dorr, J. A. I., D. J. Jude, F. J. Tesar, and N. J. Thurber. 1976. Identification of larval fishes taken from the inshore waters of southeastern Lake Michigan near D.C. Cook Nuclear plant 1973–1975. In: J. Boreman, ed. *Great Lakes Fish Egg and Larvae Identifications: Proceedings of a Workshop.* U.S. Fish Wildl. Serv., Nat. Power Plant Team, Ann Arbor, Mich.

Doss, M. A., and M. M. Farr. 1969. Index-Catalogue of medical and veterinary zoology. Subjects: Trematoda and trematode diseases. Parts 11 and 12:. Hosts. U.S. Dept. Agr., 694 pp.

Doubleday, W. G. 1985. Managing herring fisheries under uncertainty. *Can. J. Fish. Aquat. Sci.* 42(Suppl. 1):245–257.

Dovel, W. 1960. Larval development of the oyster toadfish, *Opsanus tau. Chesapeake Sci.* 1:187–195.

Dovel, W. L. 1968. Predation by striped bass as a possible influence on population size of the Atlantic croaker. *Trans. Am. Fish. Soc.* 97:313–319.

Dovel, W. L. 1971. Fish eggs and larvae of the upper Chesapeake Bay. Univ. Md. Nat. Resour. Inst., Spec. Rept. No. 4, 71 pp.

Dovel, W. L. 1978. Sturgeons of the Hudson River, New York. Final and Performance Report for New York Department of Environmental Conservation, 181 pp.

Dovel, W. L., and T. J. Berggren. 1983. Atlantic sturgeon of the Hudson River estuary, New York. *N.Y. Fish Game J.* 30:140–172.

Dovel, W. L., T. A. Mihursky, and A. J. McErlean. 1969. Life history aspects of the hogchoker, *Trinectes maculatus,* in the Patuxent River estuary, Maryland. *Chesapeake Sci.* 10:104–119.

Dower, J. F., P. Pepin, and W. C. Leggett. 1998. Enhanced gut fullness and an apparent shift in size selectivity by radiated shanny (*Ulvaria subbifurcata*) larvae in response to increased turbulence. *Can. J. Fish. Aquatic Sci.* 55:128–142.

Doyle, M. J. 1977. A morphological staging system for the larval development of the herring, *Clupea harengus* L. *J. Mar. Biol. Assoc. U.K.* 57:859–867.

Draganik, B., and C. Zukowski. 1966. The rate of growth of butterfish [*Poronotus triacanthus* (Peck)] and ocean pout [*Macrozoarces americanus* (Bloch and Schneider)] from the region of Georges Bank. Int. Comm. Northw. Atl. Fish. Res. Doc. No. 66-42.

Dragovich, A. 1970. The food of bluefin tuna (*Thunnus thynnus*) in the western North Atlantic Ocean. *Trans. Am. Fish. Soc.* 99:726–731.

Dragovich, A., and T. Potthoff. 1972. Comparative study of food of skipjack and yellowfin tunas off the coast of West Africa. *Fish. Bull., U.S.* 70:1087–1110.

Drapeau, G. 1973. Sedimentology of herring spawning grounds on Georges Bank. *Int. Comm. Northw. Atl. Fish. Res. Bull.* 10:151–162.

Drewry, G. E. 1981. Externally visible features useful for separating young of striped bass (*Morone saxatilis*) from those of white perch (*Morone americanus*) in the Chesapeake Bay Region. *Proc. Verb. Réun. Cons. Int. Explor. Mer.* 178:590–592.

Dreyer, G. D., ed. 2000. Proceedings of a symposium: the history, status, and future of the New England offshore fishery. *Northeast. Nat.* 7:315–436.

Driedzic, W. R., J. L. West, D. H. Sephton, and J. A. Raymond. 1998. Enzyme activity levels associated with the production of glycerol as an antifreeze in liver of rainbow smelt (*Osmerus mordax*). *Fish Physiol. Biochem.* 18:125–134.

Dryfoos, R. L., R. P. Cheek, and R. L. Kroger. 1973. Preliminary analysis of Atlantic menhaden, *Brevoortia tyrannus*, migration, population structure, survival and exploitation rates and availability as indicated by tag returns. *Fish. Bull., U.S.* 71:719–734.

Duarte, R., M. Azevedo, and P. Pereda. 1997. Study of the growth of southern black and white monkfish stocks. *ICES J. Mar. Sci.* 54:866–874.

Dubois, A. B., G. A. Cavogna, and R. S. Fox. 1976. Locomotion of bluefish. *J. Exp. Zool.* 195:223–236.

Dudnik, Yu. I., V. K. Zilanov, V. D. Kudrin, V. A. Nesvetov, and A. S. Nesterov. 1981. Distribution and biology of Atlantic saury, *Scomberesox saurus* (Walbaum), in the northwest Atlantic. NAFO Sci. Coun. Stud. No. 1:23–29.

Duman, J. G., and A. L. DeVries. 1974a. The effects of temperature and photoperiod on antifreeze production in cold water fishes. *J. Exp. Zool.* 190:89–97.

Duman, J. G., and A. L. DeVries. 1974b. Freezing resistance in winter flounder, *Pseudopleuronectes americanus*. *Nature* 247:237–238.

Duman, J. G., and A. L. DeVries. 1976. Isolation, characterization and physical properties of protein antifreeze from the winter flounder, *Pseudopleuronectes americanus*. *Comp. Biochem. Physiol.* 54B:375–380.

Dunbar, M. J. and H. H. Hildebrand. 1952. Contribution to the study of the fishes of Ungava Bay. *J. Fish. Res. Bd. Can.* 9:83–128.

Dunn, J. R., and A. C. Matarese. 1984. Gadidae: development and relationships. In: H. G. Moser et al., eds. *Ontogeny and Systematics of Fishes.* Am. Soc. Ich. Herp. Spec. Publ. No. 1:283–299.

DuPaul, W. D., and J. D. McEachran. 1973. Age and growth of the butterfish, *Peprilus triacanthus*, in the lower York River. *Chesapeake Sci.* 14:205–207.

Duque-Nivia, G., A. Acero P., and A. Santos–Martinez. 1995. Aspectos reproductivos de *Oligoplites saurus* y *O. palometa* (Pisces: Carangidae) en la Cíenaga Grande de Santa Marta, Caribe Colombiano. *Carib. J. Sci.* 31:317–326.

Duque-Nivia, G., A. Acero P., A. Santos–Martinez, and E. R. Rubio. 1996. Food habits of the species of the genus *Oligoplites* (Carangidae) from the Cíenaga Grande de Santa Marta-Colombian Caribbean. *Cybium* 20:251–260.

Durbin, A. G., and E. G. Durbin. 1975. Grazing rates of the Atlantic menhaden, *Brevoortia tyrannus*, as a function of particle size and concentration. *Mar. Biol.* 33:265–277.

Durbin, A. G., E. G. Durbin, P. G. Verity, and T. J. Smayda. 1981. Voluntary swimming speeds and respiration rates of a filter-feeding planktivore, the Atlantic menhaden, *Brevoortia tyrannus* (Pisces: Clupeidae). *Fish. Bull., U.S.* 78:877–886.

Durbin, A. G., E. G. Durbin, T. J. Smayda, and P. G. Verity. 1983. Age, size, growth, and chemical composition of Atlantic menhaden, *Brevoortia tyrannus*, from Narragansett Bay, Rhode Island. *Fish. Bull., U.S.* 81:133–141.

Durbin, E. G., and A. G. Durbin. 1981. Assimilation efficiency and nitrogen excretion of a filter-feeding planktivore, the Atlantic menhaden, *Brevoortia tyrannus* (Pisces: Clupeidae). *Fish. Bull., U.S.* 79:601–616.

Durbin, E. G., A. G. Durbin, R. W. Langton, and R. E. Bowman. Atlantic cod, *Gadus morhua*, and estimation of their daily rations. *Fish. Bull., U.S.* 81:437–454.

Dutcher, B. W., and F. J. Schwartz. 1962. A preferential parasitic copepod–oyster toadfish association. *Chesapeake Sci.* 3:213–215.

Dutil, J. D., and J.-M. Coutu. 1988. Early marine life of Atlantic salmon, *Salmo salar*, post-smolts in the northern Gulf of St. Lawrence. *Fish. Bull., U.S.* 86:197–212.

Dutil, J. D., and G. Power. 1980. Coastal populations of brook trout, *Salvelinus fontinalis*, in Lac Guillaume-Delisle (Richmond Gulf) Québec. *Can. J. Zool.* 58:1828–1835.

Dutil, J. D., M. Fortin, and Y. Vigneault. 1982. L'importance des zones littorales pour les ressources halieutiques. Can. Fish. Aquatic Sci. Manuscript Rept. 1653F.

Dyer, B. S., and B. Chernoff. 1996. Phylogenetic relationships among atheriniform fishes (Teleostei: Atherinomorpha). *Zool. J. Linnean Soc.* 117:1–69.

Dyer, W. G., E. H. Williams, Jr., and L. Bunkley-Williams. 1998. Some digenetic trematodes of marine fishes from Puerto Rico. *Carib. J. Sci.* 34:141–146.

Dykstra, M. J., J. F. Levine, E. J. Nega, J. H. Hawkins, P. Gerdes, W. J. Haigis, Jr., H. J. Grier, and D. Te Strake. 1989. Ulcerative mycosis: a serious menhaden disease of the southeastern coastal fisheries of the United States. *J. Fish. Dis.* 12:175–178.

Dymond, J. R. 1963. Family Salmonidae. Fishes of the western North Atlantic. Sears Found Mar. Res. Mem. No. 1(3):457–546.

Dzerzhinskiy, K. F., and D. A. Pavlov. 1992. Gametogenesis in the White Sea wolffish, *Anarchichas lupus marisalbi*. *Voprosy Ikhtiol.* 32(2):107–118 [in Russian, translation in *J. Ichthyol.* 32(6):87–99].

Eales, J. G. 1968. The eel fisheries of eastern Canada. *Bull. Fish. Res. Bd. Can.* 166, 79 pp.

Earll, R. E. 1883. The Spanish mackerel, *Cybium maculatum* (Mitch.) Ag.; its natural history and artificial propagation, with an account of the origin and development of the fishery. U.S. Comm. Fish Fisher. Rept., Part. 8, App. E, 1880, pp. 395–426.

Earll, R. E. 1887. The coast of Maine and its fisheries in 1880. In: G. B. Goode, ed. *The Fisheries and Fishery Industries of the United States.* Sect. 2, Part. 1. Govt. Print. Off., Washington, D.C.

Edgar, R. K., and J. G. Hoff. 1976. Grazing of freshwater and estuarine, benthic diatoms by adult Atlantic menhaden, *Brevoortia tyrannus*. *Fish. Bull., U.S.* 74:689–693.

Edsall, T. A. 1964. Feeding by three species of fishes on the eggs of spawning alewives. *Copeia* 1964:226–227.

Edsall, T. A. 1970. The effect of temperature on the rate of development and survival of alewife eggs and larvae. *Trans. Am. Fish. Soc.* 102:376–380.

Edwards, D. C., D. O. Conover, and F. Sutter, III. 1982. Mobile predators and the structure of marine intertidal communities. *Ecology* 63:1175–1180.

Edwards, R. L. 1965. Relation of temperature to fish abundance and distribution in the southern New England area. ICNAF Spec. Publ. No. 6:95–110.

Edwards, R. L., and R. E. Bowman. 1979. Food consumed by continental shelf fishes. In: H. Clepper, ed. *Predator-Prey Systems in Fish Communities and Their Role in Fisheries Management.* Sport Fishing

Edwards, R. L., and K. O. Emery. 1968. The view from a storied sub: the "Alvin" off Norfolk, Va. *Comm. Fish. Rev.* 30(8–9): 48–55.

Edwards, R. L., R. Livingstone, Jr., and P. E. Hamer. 1962. Winter water temperatures and an annotated list of fishes: Nantucket Shoals to Cape Hatteras, *Albatross III* Cruise no. 126. U.S. Fish Wildl. Serv. Spec. Sci. Rept. Fish. No. 397, 31 pp.

Efanov, V. N., and V. I. Vinogradov. 1973. Feeding patterns of yellowtail of two New England stocks. Int. Comm. Northw. Atl. Fish. *Redbook* III:75–77.

Ehrenbaum, E. 1904. Eier und Larven von Fischen der deutschen Bucht. 3. Fische mit festsitzenden Eiern. *Wissenschaft. Meeresunters., Helgoland.* n. ser. 6:127–200.

Ehrenbaum, E. 1909. Eier und Larven von Fischen. 2. Teil. Nordisches Plankton. 10:217–413.

Ehrhardt, N. M. 1992. Age and growth of swordfish, *Xiphias gladius,* in the northwestern Atlantic. *Bull. Mar. Sci.* 50:292–301.

Ehrich, S. 1976. Zur Taxonomie, Ökologie und Wachstum von *Macroramphosus scolopax* (Linnaeus, 1758) (Pisces, Syngnathiformes) aus dem subtropischen Nordostatlantik. *Ber. Deut. Wissen. Komm. Meeresforschung* 24:251–266.

Ehrich, S. 1986. Macroramphosidae. In: P. J. P. Whitehead et al. *Fishes of the North-eastern Atlantic and the Mediterranean,* Vol. 2. UNESCO, Paris, 627 pp.

Ehrich, S., and H.-Ch. John. 1973. Zur Biologie und Ökologie der Schnepfenfische (Gattung *Macrorhamphosus*) vor Nordwestafrika und Überlegungen zum Altersaufbau der adulten Bestände der Grossen Meteorbank. *Meteor Forsch.-Ergebn. Berlin* 14:87–98.

Ehrich, S., H.-Ch. John, and P. Westhaus-Ekau. 1987. Southward extension of the reproductive range of *Macroramphosus scolopax* in the upwelling area off north-west Africa. In: A. I. L. Payne, J. A. Gulland, and K. H. Brink, eds. *The Benguela and Comparable Ecosystems. S. Afr. J. Mar. Sci.* 5:95–105.

Ehrlich, K. F., J. H. S. Blaxter, and R. Pemberton. 1976. Morphological and histological changes during growth and starvation of herring and plaice larvae. *Mar. Biol.* 35:105–118.

Eigenmann, C. H. 1902. The egg and development of the conger eel. *Bull. U. S. Fish Comm.* 21:37–44.

Eigenmann, C. H., and C. H. Kennedy. 1902. The leptocephalus of the American eel and other American leptocephali. *Bull. U.S. Fish Comm.* 21:81–92.

Einarsson, H. 1951. Racial analysis of Icelandic herrings by means of otoliths. *Rapp. Proc.-Verb. Réun. Cons. Int. Explor. Mer* 128(App.): 55–74.

Eisler, R. 1970. Acute toxicities of organochloride and organophosphorous insecticides to estuarine fish. U.S. Bur. Sport Fish. Wildl. Tech. Pap. No. 46, 12 pp.

Eisler, R. 1986. Use of *Fundulus heteroclitus* in pollution studies. *Am. Zool.* 26:283–288.

Eisler, R., and D. G. Deuel. 1965. Acute toxicity of soaps to estuarine fishes. *Progr. Fish-Cult.* 27:45–48.

Eklund, A.-M., and T. E. Targett. 1990. Reproductive seasonality of fishes inhabiting hard bottom areas in the Middle Atlantic Bight. *Copeia* 1990:1180–1184.

Eldred, B., and W. G. Lyons. 1966. Larval ladyfish, *Elops saurus* Linnaeus 1766, (Elopidae) in Florida and adjacent waters. Dept. Nat. Res. Lab. Leafl. Ser. No. 4(2), 6 pp.

Eldridge, M. B., J. A. Whipple, D. Eng, M. J. Bowers, and B. M. Jarvis. 1981. Effects of food and feeding factors on laboratory-reared striped bass larvae. *Trans. Am. Fish. Soc.* 110:111–120.

Eldridge, P. J. 1962. Observations on the winter trawl fishery for summer flounder, *Paralichthys dentatus.* M.A. thesis, College of William and Mary, Williamsburg, Va., 58 pp.

Eldridge, P. J., and G. M. Meaburn. 1992. Potential impact of PCB's on bluefish, *Pomatomus saltatrix,* management. *Mar. Fish. Rev.* 54(4):19–24.

Ellertsen, B., P. Solemdal, T. Strømme, S. Tilseth, T. Westgard, E. Moksness, and V. Olestad. 1980. Some biological aspects of cod larvae (*Gadus morhua* L.). *Fiskerdir. Skr. Ser. Havunders.* 17:29–47.

Elliott, E. M., and D. Jimenez. 1981. Laboratory manual for the identification of ichthyoplankton from the Beverly-Salem Harbor area. Mass. Div. Mar. Fish., 230 pp.

Elliott, E. M., D. Jimenez, C. O. J. Anderson, Jr., and D. J. Brown. 1979. Ichthyoplankton abundance and distribution in Beverly-Salem Harbor, March 1975 through February 1977. Mass. Div. Mar. Fish., 174 pp.

Ellis, J. R., M. G. Pawson, and S. E. Shackley. 1996. The comparative feeding ecology of six species of sharks and four species of ray (Elasmobranchii) in the north-east Atlantic. *J. Mar. Biol. Assoc. U.K.* 76:89–106.

Ellis, M. M., B. A. Westfall, D. K. Meyer, and W. S. Platner. 1947. Water quality studies of the Delaware River with reference to shad migration. U.S. Fish Wildl. Serv., Spec. Sci. Rept. No. 38, 19 pp.

Emery, A. R., and F. D. McCracken. 1966. Biology of the Atlantic argentine (*Argentina silus* Ascanius) on the Scotian Shelf. *J. Fish. Res. Bd. Can.* 23:1145–1160.

Engel, D. W., W. F. Hettler, L. Coston-Clements, and D. E. Hoss. 1987. The effect of abrupt salinity changes on the osmoregulatory abilities of the Atlantic menhaden, *Brevoortia tyrannus. Comp. Biochem. Physiol.* 86A:723–727.

Engen, F., and I. Folstad. 1999. Cod courtship song: a song at the expense of dance? *Can. J. Zool.* 77:542–550.

Ennis, G. P. 1968. Occurrences of the staghorn sculpin (*Gymnocanthus tricuspis*) in Newfoundland waters. *J. Fish. Res. Bd. Can.* 25:2729–2731.

Ennis, G. P. 1969. Occurrences of the little sculpin, *Myoxocephalus aeneus,* in Newfoundland waters. *J. Fish. Res. Bd. Can.* 26:1689–1694.

Ennis, G. P. 1970. Age, growth and sexual maturity of the shorthorn sculpin, *Myoxocephalus scorpius,* in Newfoundland waters. *J. Fish. Res. Bd. Can.* 27:2155–2158.

Epifanio, C. E., D. Goshorn, and T. E. Targett. 1988. Induction of spawning in the weakfish, *Cynscion regalis. Fish. Bull., U.S.* 86:168–171.

Epperly, S. P. 1989. A meristic, morphometric and biochemical investigation of Atlantic menhaden, *Brevoortia tyrannus* (Latrobe). *J. Fish Biol.* 35:139–152.

Erdman, D. S. 1960. Larvae of tarpon, *Megalops atlantica,* from the Anasco River, Puerto Rico. *Copeia* 1960:146.

Erickson, D. L., and G. D. Grossman. 1986. Reproductive demography of tilefish from the South Atlantic Bight with a test for the presence of protogynous hermaphroditism. *Trans. Am. Fish. Soc.* 115:279–285.

Eschmeyer, W. N. 1969. A systematic review of the scorpionfishes of the Atlantic Ocean (Pisces: Scorpaenidae). S. Cal. Acad. Sci. Occas. Pap. No. 79, 130 pp.

Eschmeyer, W. N., ed. 1998. *Catalog of Fishes.* California Academy of Science, San Francisco, 3 vol.

Essig, R. J., and C. F. Cole. 1986. Methods of estimating larval fish mortality from daily increments in otoliths. *Trans. Am. Fish. Soc.* 115:34–40.

Esteves, C., and J. Burnett. 1993. A comparison of growth rates for American plaice, *Hippoglossoides platessoides,* in the Gulf of Maine-Georges Bank region derived from two different data sources. NEFSC Ref. Doc. No. 93-09, 8 pp.

Esteves, E., J. Aníbal, H. Krug, and H. M. da Silva. 1997. Aspects of age and growth of bluemouth, *Helicolenus dactylopterus dactylopterus* (Delaroche, 1809) from the Azores. *Arquipélago* 15A:83–95.

Etchevers, S. L. 1976. Incidencia de clupeoideos en la alimanentacion de las cabañas: *Euthynnus alletteratus* (Rafinesque) y *Auxis thazard* (Lacepède) en la costa noreste de Margarita. Lagena 37–38:9–11.

Evans, R. R. 1973. The swimbladder and associated structures in western Atlantic sea robins (Triglidae). *Copeia* 1973:315–321.

Everich, D., and J. G. Gonzalez. 1977. Critical thermal maxima of two species of estuarine fish. *Mar. Biol.* 41:140–145.

Evermann, B. W. 1905. Report of inquiry respecting food-fishes and the fishing grounds. U.S. Bur. Fish. Rept. for 1904, pp. 81–120.

Evermann, B. W., and W. C. Kendall. 1898. Descriptions of new or little-known genera and species of fishes from the United States. *Bull. U. S. Fish Comm.* 17:125–133.

Evseenko, S. A. 1976. Larvae of *Bothus ocellatus* (Agassiz) from the northwestern Atlantic. *Voprosy Ikhtiol.* 16:661–669 [in Russian, translation in *J. Ichthyol.* 16:598–605].

Evseenko, S. A., and M. M. Nevinsky. 1975. Spawning and development of witch flounder, *Glyptocephalus cynoglossus* L.,in the northwest Atlantic. *Int. Comm. Northw. Atl. Fish. Res., Bull.* 11:111–123.

Ewart, K. V., and G. L. Fletcher. 1990. Isolation and characterization of antifreeze proteins from smelt (*Osmerus mordax*) and Atlantic herring (*Clupea harengus harengus*). *Can. J. Zool.* 68:1652–1658.

Faber, D. J. 1976a. Hyponeustonic fish larvae in Northumberland Strait during summer 1962. *J. Fish. Res. Bd. Can.* 33:1167–1174.

Faber, D. J. 1976b. Identification of four northern blennioid fish larvae in the Canadian Atlantic Ocean (Stichaeidae, Lumpenidae). *J. Fish. Res. Bd. Can.* 33:1798–1802.

Fable, W. A. Jr., H. A. Brusher, L. Trent, and J. Finnegan, Jr. 1981. Possible temperature effects on charter boat catches of king mackerel and other coastal pelagic species in northwest Florida. *Mar. Fish. Rev.* 43(8):21–26.

Fabricius, D. 1780. Fauna Groenlandica. Hafniae et Lipsiae. 452 pp.

Fabrizio, M. C. 1987. Contribution of Chesapeake Bay and Hudson River stocks of striped bass to Rhode Island coastal waters as estimated by isoelectric focusing of eye lens proteins. *Trans. Am. Fish. Soc.* 116:588–593.

Facey, D. E., and M. J. Van den Avyle. 1987. Life histories and environmental requirements of coastal fishes and invertebrates (North Atlantic): American eel. Fish Wildl. Serv. Biol. Rept. No. 82, 28 pp.

Fahay, M. P. 1974. Occurrence of silver hake, *Merluccius bilinearis,* eggs and larvae along the middle Atlantic continental shelf during 1966. *Fish. Bull., U.S.* 72:813–834.

Fahay, M. P. 1975. An annotated list of larval and juvenile fishes captured with surface-towed meter net in the South Atlantic Bight during four RV *Dolphin* cruises between May 1967 and February 1968. NOAA Tech. Rept. NMFS SSRF-685, 39 pp.

Fahay, M. P. 1978. Biological and fisheries data on American eel, *Anguilla rostrata* (LeSueur). NMFS Tech. Ser. Rept. No. 17, 82 pp.

Fahay, M. P. 1983. Guide to the early stages of marine fishes occurring in the western North Atlantic Ocean, Cape Hatteras to the southern Scotian Shelf. *J. Northw. Atl. Fish. Sci.* 4, 423 pp.

Fahay, M. P. 1992. Development and distribution of cusk eel eggs and larvae in the Middle Atlantic Bight with a description of *Ophidion robinsi* n. sp. (Teleostei: Ophidiidae). *Copeia* 1992: 799–819.

Fahay, M. P., and K. W. Able. 1989. White hake, *Urophycis tenuis,* in the Gulf of Maine: spawning seasonality, habitat use, and growth in young of the year and relationships to the Scotian Shelf populations. *Can. J. Zool.* 67:1715–1724.

Fahay, M. P., and P. Berrien. 1981. Preliminary description of larval tilefish, *Lopholatilus chamaeleonticeps. Rapp. Proc.-Verb. Réun. Cons. Int. Explor. Mer* 178:600–602.

Fahay, M. P., and D. F. Markle. 1984. Gadiformes: development and relationships. In: H. G. Moser et al., eds. *Ontogeny and Systematics of Fishes.* Am. Soc. Ich. Herp. Spec. Publ. No. 1:265–283.

Fahay, M. P., and C. L. Obenchain. 1978. Leptocephali of the ophichthid genera *Ahlia, Myrophis, Ophichthus, Pisodonophis, Callechelys, Letharchus* and *Apterichtus* on the Atlantic continental shelf of the United States. *Bull. Mar. Sci.* 28:442–486.

Fahay, M. P., P. L. Berrien, D. L. Johnson, and W. W. Morse. 1999a. Essential Fish Habitat Document: Atlantic cod, *Gadus morhua,* life history and habitat characteristics. NOAA Tech. Memo. NMFS-NE-124, 41 pp.

Fahay, M. P., P. L. Berrien, D. L. Johnson, and W. W. Morse. 1999b. Essential Fish Habitat Document: Bluefish, *Pomatomus saltatrix,* life history and habitat characteristics. NOAA Tech. Memo. NMFS-NE-144, 68 pp.

Fahy, W. E. 1978. The influence of crowding upon the total number of vertebrae developing in *Fundulus majalis* (Walbaum). *J. Cons. Int. Explor. Mer* 38:252–256.

Fahy, W. E. 1979. The influence of temperature change on number of anal fin rays developing in *Fundulus majalis* (Walbaum). *J. Cons. Int. Explor. Mer* 38:280–285.

Fahy, W. E. 1980. The influence of temperature change on number of dorsal fin rays developing in *Fundulus majalis* (Walbaum). *J. Cons. Int. Explor. Mer* 39:104–109.

Fahy, W. E. 1982. The influence of temperature change on number of pectoral fin rays developing in *Fundulus majalis* (Walbaum). *J. Cons. Int. Explor. Mer* 40:21–26.

Fairbanks, R. B., and R. P. Lawton. 1977. Occurrence of large striped mullet, *Mugil cephalus,* in Cape Cod Bay, Massachusetts,

Fairbanks, R. B., W. S. Collins, and W. T. Sides. 1971. An assessment of the effects of electrical power generation on marine resources in the Cape Cod Canal. Mass. Div. Mar. Fish., 48 pp.

Falk-Petersen, I. R., and T. K. Hansen. 1991. Reproductive biology of wolffish *Anarhichas lupus* from north-Norwegian waters. Int. Coun. Explor. Mer, Demersal Fish. Comm. ICES CM 1991/G:14:17.

Falk-Petersen, I. R., T. Haug, and E. Moksness. 1990. Observations on the occurrence, size, and feeding of pelagic larvae of the common wolffish (*Anarhichas lupus*) in western Finnmark, northern Norway. *J. Conseil* 46:148–154.

Fänge, R., and J. B. Wittenberg. 1958. The swimbladder of the toadfish (*Opsanus tau* L.). *Biol. Bull., Woods Hole* 115:172–179.

Farley, T. C. 1966. Striped bass, *Roccus saxatilis,* spawning in the Sacramento - San Josquin River systems during 1963 and 1964. *Calif. Dept. Fish Game, Fish Bull.* 136:28–43.

Farwell, M. K., and J. M. Green. 1973. Agonistic behaviour of juvenile *Stichaeus punctatus* (Pisces: Stichaeidae). *Can. J. Zool.* 51: 449–456.

Farwell, M. K., J. M. Green, and V. A. Pepper. 1976. Distribution and known life history of *Stichaeus punctatus* in the northwest Atlantic. *Copeia* 1976:598–602.

Fearing, D. B. 1903. Some early notes on striped bass. *Trans. Am. Fish. Soc.* 32:90–98.

Felley, J. D., and M. Vecchione. 1994. Assessing habitat use by nekton on the continental slope using archived videotapes from submersibles. *Fish. Bull., U.S.* 93:262–273.

Fernholm, B. 1969. A third embryo of *Myxine:* considerations on hypophysial ontogeny and phylogeny. *Acta Zool.* 50:169–177.

Fernholm, B. 1975. Ovulation and eggs of the hagfish, *Eptatretus burgeri. Acta Zool.* 56:199–204.

Fernholm, B. 1998. Hagfish systematics. In: *The Biology of Hagfishes.* J. M. Joergensen et al., eds. Chapman and Hall, London, pp. 33–44.

Ferraro, S. P. 1980. Daily time of spawning of 12 fishes in the Peconic Bays, New York. *Fish. Bull., U.S.* 78:455–464.

Ferraro, S. P. 1981a. Embryonic development of Atlantic menhaden, *Brevoortia tyrannus,* and a fish embryo age estimation method. *Fish. Bull., U.S.* 77:943–949.

Ferraro, S. P. 1981b. Eggs and larvae of Atlantic menhaden *Brevoortia tyrannus* in the Peconic Bays, New York in 1972–74. *Rapp. Proc.-Verb. Réun. Cons. Int. Explor. Mer* 178:181–182.

Field, I. A. 1907. Unutilized fishes and their relation to the fishing industries. U.S. Comm. Fish. Rept. for 1906, 50 pp.

Field, J. D. 1997. Atlantic striped bass management: where did we go right? *Fisheries* 22(7):6–8.

Fields, B. 1988. Winter flounder, *Pseudopleuronectes americanus.* In: J. Penttila and L. M. Dery, eds. *Age Determination Methods for Northwest Atlantic Species.* NOAA Tech. Rept. NMFS No. 72:103–104.

Fine, M. L. 1975. Sexual dimorphism of the growth rate of the swimbladder of the toadfish *Opsanus tau. Copeia* 1975:483–490.

Fine, M. L. 1978. Seasonal and geographical variation of the mating call of the oyster toadfish *Opsanus tau* L. *Oecologia* 36:45–57.

Fine, M. L., N. M. Burns, and T. M. Harris. 1990. Ontogeny and sexual dimorphism of sonic muscle in the oyster toadfish. *Can. J. Zool.* 68:1374–1381.

Fine, M. L., J. W. McKnight, Jr., and C. R. Blem. 1995. Effect of size and sex on buoyancy in the oyster toadfish. *Mar. Biol.* 123:401–409.

Finger, T. E. 1982. Somatopy in the representation of the pectoral free finrays in the spinal cord of the searobin *Prionotus carolinus. Biol. Bull., Woods Hole* 163:154–161.

Finger, T. E., and K. Kakil. 1985. Organization of motoneuronal pools in the rostral spinal cord of the searobin, *Prionotus carolinus. J. Comp. Neurology* 239:384–390.

Fink, W. L. 1985. Phylogenetic interrelationships of the stomiid fishes (Teleostei: Stomiiformes). Misc. Publ. Mus. Zool. Univ. Mich. No. 171, 127 pp.

Finkelstein, S. L. 1969a. Age and growth of scup in the waters of eastern Long Island. *N. Y. Fish Game J.* 16:84–110.

Finkelstein, S. L. 1969b. Age and maturity of scup from New York waters. *N.Y. Fish Game J.* 16:224–237.

Finkelstein, S. L. 1971. Migration, rate of exploitation and mortality of scup from the inshore waters of eastern Long Island. *N.Y. Fish Game J.* 18:97–111.

Finucane, J. H., and R. N. Vaught. 1986. Species profile of Atlantic thread herring *Opisthonema oglinum* (LeSueur 1818). NOAA Tech. Memo. NMFS-SEFC-182, 30 pp.

Firth, F. E. 1931. Some marine fishes collected recently in New England waters. *Bull. Boston Soc. Nat. Hist.* 61:8–14.

Firth, F. E. 1933. *Anthias nicholsi,* a new fish taken off Virginia in the deep-water trawl fishery. *Copeia* 1933:158–160.

Fischer, W., ed. 1978. *FAO Species Identification Sheets for Fishery Purposes: Western Central Atlantic (Fishing Area 31),* Vols. 1–7. FAO, Rome.

Fischler, K. J. 1959. Contributions of Hudson and Connecticut rivers to New York-New Jersey shad catch of 1956. *Fish. Bull., U.S.* 60:161–174.

Fish, C. J. 1925. Seasonal distribution of the plankton of the Woods Hole region. *Bull. U.S. Bur. Fish.* 41:91–179.

Fish, M. P. 1954. The character and significance of sound production among fishes of the western North Atlantic. *Bull. Bingham Oceanogr. Coll.* 14(4):1–109.

Fish, M. P., and W. H. Mowbray. 1970. *Sounds of Western North Atlantic Fishes.* Johns Hopkins Press, Baltimore, 207 pp.

Fisher, H. D., and B. A. MacKenzie. 1955. Food habits of seals in the Maritimes. Prog. Rept. Fish. Res. Bd. Can. No. 61:5–9.

Fiske, J. D., C. E. Watson, and P. G. Coates. 1966. A study of the marine resources of the North River. Mass. Div. Mar. Fish. Monogr. Ser. No. 3, 52 pp.

Fiske, J. D., C. E. Watson, and P. G. Coates. 1967. A study of the marine resources of Pleasant Bay. Mass. Div. Mar. Fish. Monogr. Ser. No. 5, 56 pp.

Fiske, J. D., J. R. Curley, and R. P. Lawton. 1968. A study of the marine resources of the Westport River. Mass. Div. Mar. Fish. Monogr. Ser. No. 7, 52 pp.

Fitz, E. S., Jr., and F. C. Daiber. 1963. An introduction to the biology of *Raja eglanteria* Bosc 1802 and *Raja erinacea* Mitchill 1825 as they occur in Delaware Bay. *Bull. Bingham Oceanogr. Coll.* 18(3):69–97.

FitzGerald, G. J. 1983. The reproductive ecology and behavior of three sympatric sticklebacks (Gasterosteidae) in a saltmarsh. *Biol. Behav.* 8:67–79.

FitzGerald, G. J., and J.-D. Dutil. 1981. Evidence for differential predation on an estuarine stickleback community. *Can. J. Zool.* 59:2394–2395.

FitzGerald, G. J., and F. G. Whoriskey. 1985. The effects of interspecific interactions upon male reproductive success in two sympatric sticklebacks, *Gasterosteus aculeatus* and *G. wheatlandi. Behaviour* 93:112–126.

Fives, J. M., S. M. Warlen, and D. E. Hoss. 1986. Aging and growth of larval bay anchovy, *Anchoa mitchilli,* from the Newport River estuary, North Carolina. *Estuaries* 9:362–367.

Fletcher, G. L., and J. C. Smith. 1980. Evidence for permanent population differences in the annual cycle of plasma "antifreeze" levels of winter flounder. *Can. J. Zool.* 58:507–512

Fletcher, G. L., R. J. Hoyle, and D. A. Horne. 1971. The relative toxicities of yellow phosphorous production wastes to seawater-maintained fish. Fish. Res. Bd. Can. Tech. Rept. No. 255, 18 pp.

Fletcher, G. L., R. F. Addison, D. Slaughter, and C. L. Hew. 1982. Antifreeze proteins in the Arctic shorthorn sculpin (*Myoxocephalus scorpius*). *Arctic* 35:302–306.

Fletcher, G. L., M. H. Kao, and K. Haya. 1984. Seasonal and phenotypic variations in plasma protein antifree levels in a population of marine fish, sea raven (*Hemitripterus americanus*). *Can. J. Fish. Aquat. Sci.* 41:819–824.

Fletcher, G. L., C. L. Hew, X. Li, K. Haya, and M. H. Kao. 1985. Year-round presence of high levels of plasma antifreeze peptides in temperate fish, ocean pout (*Macrozoarces americanus*). *Can. Zool.* 63:488–493.

Fletcher, G. L., M. J. King, and M. H. Kao. 1987. Low temperature regulation of antifreeze glycopeptide levels in Atlantic cod (*Gadus morhua*). *Can. J. Zool.* 65:227–233.

Fogarty, M. J. 1981. Review and assessment of the summer flounder (*Paralichthys dentatus*) fishery in the northwest Atlantic. NMFS Woods Hole Lab. Ref. Doc. No. 81-25, 54 pp.

Fogarty, M. J., E. B. Cohen, W. L. Michaels, and W. W. Morse. 1991. Predation and the regulation of sand lance populations: an exploratory analysis. *ICES Mar. Sci. Symp.* 193:120–124.

Folkvord, A., G. Blom, O. Dragesund, A. Johannessen, O. Nakken, and G. Naevdal. 1994. A conceptual framework for enhancing and studying recruitment of marine fish stocks. *Aquaculture Fish. Manage.* 25(Suppl. 1):245–258.

Fontaine, P.-M., M. O. Hammill, C. Barrette, and M. C. Kingsley. 1994. Summer diet of the harbour porpoise (*Phocoena phocoena*) in the estuary and northern Gulf of St. Lawrence. *Can. J. Fish. Aquat. Sci.* 51:172–178.

Fore, P. L. 1971. The distribution of eggs and larvae of the round herring, *Etrumeus teres,* in the northern Gulf of Mexico. *Assoc. Southeast Biol. Bull.* 18: 34.

Forrester, C. R., A. E. Peden, and R. M. Wilson. 1972. First records of the striped bass, *Morone saxatilis,* in British Columbia waters. *J. Fish. Res. Bd. Can.* 29:337–339.

Forster, G. R. 1965. *Raja richardsoni* from the continental slope off south-west England. *J. Mar. Biol. Assoc. U. K.* 45:773–777.

Fortier, L., and C. Quiñonez-Velazquez. 1998. Dependence of survival ... *Melanogrammus aeglefinus:* a field study based on individual hatch-dates. *Mar. Ecol. Prog. Ser.* 174:1–12.

Fortier, L., and A. Villeneuve. 1996. Cannibalism and predation on fish larvae by larvae of Atlantic mackerel, *Scomber scombrus:* trophodynamics and potential impact on recruitment. *Fish. Bull., U.S.* 94:268–281.

Forward, R. B., Jr., L. M. McKelvey, W. F. Hettler, and D. E. Hoss. 1993. Swimbladder inflation of the Atlantic menhaden *Brevoortia tyrannus. Fish. Bull., U.S.* 91:254–259.

Forward, R. B., W. E. Hettler, and D. E. Hoss. 1994. Swimbladder deflation in the Atlantic menhaden, *Brevoortia tyrannus. Fish. Bull., U.S.* 92:641–646.

Foss, A., A. K. Imsland, and G. Nævdal. 1998. Population genetic studies of the Atlantic halibut in the North Atlantic Ocean. *J. Fish Biol.* 53:901–905.

Foss, G. 1963. Some observations on the ecology of *Myxine glutinosa* L. *Sarsia* 7:17–22.

Foss, G. 1968. Behaviour of *Myxine glutinosa* L. in natural habitat: investigation of the mud biotype by a suction technique. *Sarsia* 31:1–13.

Foster, J. R. 1977. The role of breeding behavior and habitat preferences on the reproductive isolation of three allopatric populations of ninespine stickleback, *Pungitius pungitius. Can. J. Zool.* 55:1601–1611.

Foster, N. E. 1967. Comparative studies on the biology of killifishes (Pisces, Cyprinodontidae). Ph.D. dissertation, Cornell University, Ithaca, 369 pp.

Fourmanoir, P. 1976. Formes post-larvaires et juvéniles de poissons côtiers pris au chalut pélagique dans le sud-ouest Pacifique. Cahiers Pacif. No. 19:47–88.

Fournie, J. W., S. S. Foss, L. A. Courtney, and A. H. Undeen. 1990. Testing of insect microsporidians (Microspora: Nosematidae) in nontarget aquatic species. *Diseases Aquatic Org.* 8:137–144.

Fowler, H. W. 1906. The fishes of New Jersey. New Jersey State Museum, Ann. Rept., pp. 35–466.

Fowler, H. W. 1928. Fishes of Oceania. Bernice P. Bishop Mus. Mem. No. 10, 540 pp.

Fowler, H. W. 1937. Notes on fishes from the Gulf Stream and the New Jersey coast. *Proc. Acad. Nat. Sci. Philadelphia* 89:297–308

Fowler, H. W. 1945. A study of the fishes of the southern Piedmont and coastal plain. Acad. Nat. Sci. Philadelphia Monogr. No. 7, 408 pp.

Fowler, H. W. 1946. Description of a new deep-water angler (Chaunacidae) from off New Jersey. *Not. Nat.* Vol.180, 4 pp.

Fowler, H. W. 1956. A rare pomfret, *Taractes princeps* (Johnson), on the New Jersey coast. New Jersey State Mus., 4 pp.

Francis, M. P. 1981a. Meristic and morphometric variation in the lancet fish, *Alepisaurus,* with notes on the distribution of *A. ferox* and *A. brevirostris. New Zealand J. Zool.* 8(3):403–408.

Francis, M. P. 1981b. Von Bertalanffy growth rates in species of *Mustelus* (Elasmobranchii: Triakidae). *Copeia* 1981:189–192.

Frank, K. T., and W. C. Leggett. 1981a. Prediction of egg development and mortality rates in capelin (*Mallotus villosus*) from meteorological, hydrographic, and biological factors. *Can. J. Fish.*

Frank, K. T., and W. C. Leggett. 1981b. Wind regulation of emergence times and early larval survival in capelin (*Mallotus villosus*). *Can. J. Fish. Aquat. Sci.* 38:215–223.

Frank, K. T., and W. C. Leggett. 1983. Multispecies larval fish associations: accident or adaptation? *Can. J. Fish. Aquat. Sci.* 40:754–762.

Frank, K. T., and W. C. Leggett. 1984. Selective exploitation of capelin (*Mallotus villosus*) eggs by winter flounder (*Pseudopleuronectes americanus*): capelin egg mortality rates, and contribution of egg energy to the annual growth of flounder. *Can. J. Fish. Aquat. Sci.* 41:1294–1302.

Franks, J. S. 1970. An investigation of the fish population within the inland waters of Horn Island, Mississippi, a barrier island in the northern Gulf of Mexico. Gulf. Res. Rept. No. 3:3–104.

Franks, J. S., J. Y. Christmas, W. L. Siler, R. Combs, R. Waller, and C. Burns. 1972. A study of nektonic and benthic faunas of the shallow Gulf of Mexico off the state of Mississippi as related to some physical, chemical and geographical factors. Gulf Res. Rept. No. 4:1–148.

Franzin, W. G., B. A. Barton, R. A. Remnant, D. B. Wain, and S. J. Pagel. 1994. Range extension, present and potential distribution, and possible effects of rainbow smelt in Hudson Bay drainage waters of northwestern Ontario, Manitoba, and Minnesota. *No. Am. J. Fish. Manage.* 14:65–76.

Fraser, P. J. 1987. Atlantic salmon, *Salmo salar* L., feed in Scottish coastal waters. *Aquacult. Fish. Manage.* 18: 243–247.

Fraser-Brunner, A. 1935. New or rare fishes from the Atlantic Irish slope. *Proc. Roy. Irish Acad.* 42B:319–326.

Fraser-Brunner, A. 1951. The ocean sunfishes (family Molidae). *Bull. British Mus. (Nat. Hist.), Zool.* 1:87–121.

Frazer, T. K., W. J. Lindberg, and G. R. Stanton. 1991. Predation on sand dollars by gray triggerfish, *Balistes capriscus,* in the northeastern Gulf of Mexico. *Bull. Mar. Sci.* 48:159–164.

Freadman, M. A. 1979. Swimming energetics of striped bass (*Morone saxatilis*) and bluefish (*Pomatomus saltatrix*): gill ventilation and swimming metabolism. *J. Exp. Biol.* 83:217–230.

Freadman, M. A. 1981. Swimming energetics of the striped bass (*Morone saxatilis*) and bluefish (*Pomatomus saltatrix*): hydrodynamic correlates of locomotion and gill ventilation. *J. Exp. Biol.* 90:253–265.

Frèchet, A., J. J. Dodson, and H. Powles. 1983. Use of variation in biological characters for the classification of anadromous rainbow smelt (*Osmerus mordax*) groups. *Can. J. Fish. Aquat. Sci.* 40:718–727.

Freeman, B. L., and S. C. Turner. 1977a. The effects of anoxic water on the summer flounder (*Paralichthys dentatus*), a bottom dwelling fish. NOAA Tech. Ser. No. 3:451–462.

Freeman, B. L., and S. C. Turner. 1977b. Biological and fisheries data on tilefish, *Lopholatilus chamaeleonticeps* Goode and Bean. Nat. Mar. Fish. Serv., Sandy Hook Lab. Tech. Ser. Rept. No. 5, 41 pp.

Freeman, B. L., and S. C. Turner. 1982. Tilefish *Lopholatilus chameleonticeps*. In: M. D. Grosslein and T. R. Azarovitz, eds. *Fish Distribution*. MESA New York Bight Atlas Monogr. No. 15:83–86. New York Sea Grant Inst., Albany.

Fried, S. M. 1973. Distribution of demersal fishes in Montsweag Bay-Back River and lower Sheepscot estuaries, Wiscasset, Maine. M.S. thesis, University of Maine, Orano, 47 pp.

Fried, S. M., and J. D. McCleave. 1973. Occurrence of the shortnose sturgeon (*Acipenser brevirostrum*), an endangered species, in Montsweag Bay, Maine. *J. Fish. Res. Bd. Can.* 30:563–564.

Fried, S. M., J. D. McCleave, and G. W. Labar. 1978. Seaward migration of hatchery-reared Atlantic salmon, *Salmo salar,* smolts in the Penobscot River estuary, Maine: riverine movements. *J. Fish. Res. Bd. Can.* 35:76–87.

Friedland, K. D. 1985. Functional morphology of the branchial basket structures associated with feeding in the Atlantic menhaden, *Brevoortia tyrannus* (Pisces: Clupeidae). *Copeia* 1985:1018–1027.

Friedland, K. D. 1998a. Atlantic and shortnose sturgeons. In: S. H. Clark, ed. *Status of Fishery Resources off the Northeastern United States for 1998*. NOAA Tech. Memo. NMFS-NE-115:142–143.

Friedland, K. D. 1998b. Atlantic herring. In: S.H. Clark, ed. *Status of Fishery Resources off the Northeastern United States for 1998*. NOAA Tech. Memo. NMFS-NE-115:104–105.

Friedland, K. D., L. W. Haas, and J. V. Merriner. 1984. Filtering rates of the juvenile Atlantic menhaden *Brevoortia tyrannus* (Pisces: Clupeidae), with consideration of the effects of detritus and swimming speed. *Mar. Biol.* 84:109–117.

Friedland, K. D., G. C. Garman, A. J. Bejda, A. L. Studholme, and B. Olla. 1988. Interannual variation in diet and condition in juvenile bluefish during estuarine residency. *Trans. Am. Fish. Soc.* 117: 474–479.

Friedland, K. D., D. W. Ahrenholz, and J. F. Guthrie. 1989. Influence of plankton on distribution patterns of the filter-feeder *Brevoortia tyrannus* (Pisces: Clupeidae). *Mar. Ecol. Prog. Ser.* 54:1–11.

Friedland, K. D., R. E. Haas, and T. F. Sheehan. 1996. Post-smolt growth, maturation, and survival in two stocks of Atlantic salmon. *Fish. Bull., U.S.* 94:654–663.

Friedland, K. D., D. G. Reddin, N. Shimizu, R. E. Haas, and A. F. Youngson. 1998. Strontium:calcium ratios in Atlantic salmon (*Salmo salar*) otoliths and observations on growth and maturation. *Can. J. Fish. Aquat. Sci.* 55:1158–1168.

Frisbie, C. M. 1967. Age and growth of the striped bass *Roccus saxatilis* (Walbaum), in Massachusetts coastal waters. M.S. thesis, University of Massachusetts, Amherst, 58 pp.

Fritz, E. S., and E. T. Garside. 1975. Comparison of age composition, growth, and fecundity between two populations each of *Fundulus heteroclitus* and *F. diaphanus* (Pisces: Cyprinodontidae). *Can. J. Zool.* 53:361–369.

Fritz, E. S., W. H. Meredith, and V. A. Lotrich. 1975. Fall and winter movements and activity level of the mummichog, *Fundulus heteroclitus,* in a tidal creek. *Chesapeake Sci.* 16:211–215.

Fritz, R. L. 1962. Silver hake. U.S. Fish Wildl. Serv. Fish. Fishery Leaflet No. 538, 7 pp.

Fritz, R. L. 1965. Autumn distribution of groundfish species in the Gulf of Maine and adjacent waters, 1955–1961. Am. Geol. Soc., Serial Atlas Mar. Environ. Folio No. 10, 48 pp.

Fritzsche, R. A. 1976. A review of the cornetfishes, genus *Fistularia* (Fistulariidae), with a discussion of intergeneric relationships and zoogeography. *Bull. Mar. Sci.* 26:196–204.

Fritzsche, R. A. 1978. *Development of Fishes of the Mid-Atlantic Bight,* Vol. 5. *Chaetodontidae through Ophidiidae.* U.S. Fish Wildl. Serv., Biol. Serv. Prog. FWS/OBS–78/12, 340 pp.

Fritzsche, R. A., and G. D. Johnson. 1980. Early osteological development of white perch and striped bass with emphasis on identification of their larvae. *Trans. Am. Fish. Soc.* 109:387–406.

Fuiman, L. A. 1976. Notes on the early development of the sea raven, *Hemitripterus americanus. Fish. Bull., U.S.* 74:467–470.

Fuiman, L. A., and J. C. Gamble. 1988. Predation by Atlantic herring, sprat, and sandeels on herring larvae in large enclosures. *Mar. Ecol. Prog. Ser.* 44:1–6.

Fulton, T. W. 1898. The ovaries and ovarian eggs of the angler or frog-fish (*Lophius piscatorius*), and of the John Dory (*Zeus faber*). Fish. Bd. Scotland Ann. Rept.No. 16:125–134.

Fulton, T. W. 1903a. Investigations on the abundance, distribution, and migrations of food fishes. Fish. Bd. Scotland Ann. Rept. No. 21:15–108.

Fulton, T. W. 1903b. The distribution, growth and food of the angler (*Lophius piscatorius*). Fish. Bd. Scotland Ann. Rept. No. 21:186–217.

Fulton, T. W. 1906. On the spawning of the lumpsucker (*Cyclopterus lumpus*) and the paternal guardianship of the eggs. Fish. Bd. Scotland Ann. Rept. 1906, Part III:169–178.

Funderburg, J. B., and T. H. Eaton. 1952. A new record of the pointed-tailed ocean sunfish, *Masturus lanceolatus,* from North Carolina. *Copeia* 1952:200.

Gabaudan, J. 1984. Posthatching morphogenesis of the digestive system of striped bass. Ph.D. dissertation, Auburn University, Auburn, 167 pp.

Gabriel, W. L. 1998. Scup. In: S. H. Clark, ed. *Status of Fishery Resources off the Northeastern United States for 1998.* NOAA Tech. Memo. NMFS-NE-115:90–91.

Gaevskaya, A. V., and B. A. Umnova. 1977. On the parasite fauna of the principal commercial fishes of the Northwest Atlantic. *Biol. Morya (Vladivostok)* 4:40–48 [in Russian].

Gallego, A., and M. R. Heath. 1994. The development of schooling behaviour in Atlantic herring, *Clupea harengus. J. Fish Biol.* 45:569–588.

Galligan, J. P. 1962. Depth distribution of lake trout and associated species in Cayuga Lake, New York. *N.Y. Fish Game J.* 9:44–68.

Gannon, D. P., J. E. Craddock, and A. J. Read. 1998. Autumn food habits of harbor porpoises, *Phocoena phocoena,* in the Gulf of Maine. *Fish. Bull., U.S.* 96:428–437.

Garnaud, J. 1960. La ponte, l'éclosion, la larve du baliste *Balistes capriscus* Linné 1758. *Bull. Inst. Océanogr., Monaco* 1169:1–6.

Garrick, J. A. F., and L. P. Schultz. 1963. A guide to the kinds of potentially dangerous sharks. In: P. W. Gilbert, ed. *Sharks and Survival.* D.C. Heath & Co., Boston, pp. 3–60.

Gaskin, D. E., and G. J. D. Smith. 1979. Observations on marine mammals, birds and environmental conditions in the Head Harbour region of the Bay of Fundy. In: D. J. Scarratt, ed. *Evaluation of Recent Data Relative to Potential Oil Spills in the Passamaquoddy Area.* Can. Fish. Mar. Serv. Tech. Rept. No. 901:69–86.

Gaston, A. J., D. K. Carins, R. D. Elliot, and D. G. Noble. 1985. A natural history of Digges Sound. Can. Wild. Serv. Rept. Ser. No. 46, 63 pp.

Gehringer, J. W. 1959. Early development and metamorphosis of the ten-pounder, *Elops saurus* Linnaeus. *Fish. Bull., U.S.* 59: 619–647.

Geiger, S. P. 1993. Respiratory physiology of juvenile tarpon, *Megalops atlanticus.* M.S thesis, University of South Florida, Tampa, 96 pp.

Geiger, S. P., J. J. Torres, and R. E. Crabtree. 2000. Air-breathing and gill ventilation frequencies in juvenile tarpon, *Megalops atlanticus:* responses to changes in dissolved oxygen, temperature, hydrogen sulfide, and pH. *Environ. Biology Fishes* 59:181–190.

Geistdoerfer, P. 1976. Alimentation de deux Macrouridae de l'Atlantique nord: *Macrourus berglax* et *Coryphaenoides rupestris* (téléostéens Gadiformes). *Rev. Trav. Inst. Pêch. Marit.* 40:579–580.

Gelsleichter, J., J. A. Musick, and S. Nichols. 1999. Food habits of the smooth dogfish, *Mustelus canis,* dusky shark, *Carcharhinus obscurus,* Atlantic sharpnose shark, *Rhizoprionodon terraenovae,* and the sand tiger, *Carcharias taurus,* from the northwest Atlantic ocean. *Environ. Biol. Fishes* 54:205–217.

Geoghegan, P., J. N. Strube, and R. A. Sher. 1998. The first occurrence of the striped cusk eel, *Ophidion marginatum* (DeKay), in the Gulf of Maine. *Northeast Nat.* 5:363–366.

Gibbs, R. H., Jr. 1960. *Alepisaurus brevirostris,* a new species of lancetfish from the western North Atlantic. *Mus. Comp. Zool., Breviora* 123, 14 p.

Gibbs, R. H., Jr. 1969. Taxonomy, sexual dimorphism, vertical distribution, and evolutionary zoogeography of the bathypelagic fish genus *Stomias* (Stomiatidae). Smithsonian Contrib. Zool. No. 31, 25 pp.

Gibbs, R. H., Jr. 1978. Exocoetidae. In: W. Fischer, ed. *FAO Species Identification Sheets for Fishery Purposes: Western Central Atlantic (Fishing Area 31),*Vol. 2. FAO, Rome, 32 pp.

Gibbs, R. H., Jr. 1984. Stomiidae. In: P. J. Whitehead et al., eds. *Fishes of North-eastern Atlantic and Mediterranean.* UNESCO, Paris, Vol. 1:338–340.

Gibbs, R. H., Jr., and B. B. Collette. 1959. On the identification, distribution, and biology of the dolphins, *Coryphaena hippurus* and *C. equiselis. Bull. Mar. Sci. Gulf Carib.* 9:117–152.

Gibbs, R. H., Jr., and B. B. Collette. 1967. Comparative anatomy and systematics of the tunas, genus *Thunnus. U. S. Fish Wildl. Serv. Fish. Bull.* 66:65–130.

Gibbs, R. H., Jr., and J. C. Staiger. 1970. Eastern tropical Atlantic flyingfishes of the genus *Cypselurus* (Exocoetidae). *Stud. Trop. Oceanogr.* 4:432–466.

Gibbs, R. H., Jr., and N. J. Wilimovsky. 1966. Family Alepisauridae. Fishes of the western North Atlantic. Sears Found. Mar. Res. Mem. No. 1(5):482–497.

Gibson, D. I., and R. A. Bray. 1979. *Cirkennedya porlockensis,* a new genus and species of digenean from the sunfish *Mola mola* L. *J. Helminth.* 53:245–250.

Gibson, R. N. 1988. Development, morphometry and particle retention capability of the gill rakers in the herring, *Clupea harengus* L. *J. Fish Biol.* 32:949–962.

Gibson, R. N., and I. A. Ezzi. 1985. Effect of particle concentration on filter-and particulate-feeding in the herring *Clupea harengus.* [illegible]

Gibson, R. N., and I. A. Ezzi. 1990. Relative importance of prey size and concentration in determining the feeding behaviour of the herring *Clupea harengus*. *Mar. Biol.* 107:357–362.

Gifford, V. M., and M. F. Crawford. 1988. Redfish, *Sebastes fasciatus*. In: J. Pentilla and L. M. Dery, eds. *Age Determination Methods for Northwest Atlantic Species*. NOAA Tech. Rept. NMFS No. 72:93–94.

Gilbert, C. H. 1913. Descriptions of two new fishes of the genus *Triglops* from the Atlantic coast of North America. *Proc. U.S. Nat. Mus.* 44:465–468.

Gilbert, C. R., ed. 1992. *Rare and Endangered Biota of Florida*, Vol. 2. *Fishes*. Florida University Press, Gainesville, 247 pp.

Gilhen, J. 1969. Striped mullet *Mugil cephalus* records from Halifax County, Nova Scotia, in 1966 and 1968. *Can. Field-Nat.* 83:161–162.

Gilhen, J. 1972. The white mullet, *Mugil curema*, added to and the striped mullet, *M. cephalus*, deleted from the Canadian Atlantic fish fauna. *Can. Field-Nat.* 86:74–77.

Gilhen, J. 1986. First three records of the wreckfish, *Polyprion americanus*, for Nova Scotia. *Can. Field-Nat.* 100:381–382.

Gilhen, J., and D. E. McAllister. 1980. First Nova Scotian and second Canadian record of Atlantic fanfish, *Pterycombus brama* Fries, 1837 (Bramidae). *Proc. Nova Scotia Inst. Sci.* 30:191–193.

Gilhen, J., and D. E. McAllister. 1981. First Canadian record of yellowfin bass, *Anthias nicholsi* (Firth) taken off Nova Scotia. *Proc. Nova Scotia Inst. Sci.* 31:251–254.

Gilhen, J., and D. E. McAllister. 1989. The Atlantic Spanish mackerel, *Scomberomorus maculatus*, new to Nova Scotia and Canada. *Can. Field-Nat.* 103:287–289.

Gilhen, J., C. G. Gruchy, and D. E. McAllister. 1976. The sheepshead, *Archosargus probatocephalus*, and the feather blenny, *Hypsoblennius hentzi:* two new additions to the Canadian ichthyofauna. *Can. Field-Nat.* 90:42–46.

Gill, T. N. 1905. The life history of the angler. *Smithsonian Misc. Coll.* 47:500–516.

Gill, T. N. 1907. Life histories of toadfishes (batrachoidids), compared with those of weevers (trachinids) and stargazers (uranoscopids). *Smithsonian Misc. Coll.* 48:388–427.

Gill, T. N. 1909. Angler fishes: their kinds and ways. Smithsonian Inst. Ann. Rept. for 1908, pp. 565–615.

Gill, T. N. 1911. Notes on the structure and habits of the wolffishes. *Proc. U.S. Nat. Mus.* 39:157–187.

Gilman, S. L. 1994. An energy budget for northern sand lance, *Ammodytes dubius*, on Georges Bank, 1977–1986. *Fish. Bull., U.S.* 92:647–654.

Gilmore, R. G. 1977. Fishes of the Indian River Lagoon and adjacent waters, Florida. *Bull. Fla. State Mus.* 22:101–147.

Gilmore, R. G. 1990. The reproductive biology of lamnoid sharks. In: S. Gruber, ed. *Discovering Sharks*. Am. Littoral Soc. Spec. Publ. No. 14, Highlands, N.J., pp. 64–67.

Gilmore, R. G., L. H. Bullock, and F. H. Berry. 1978. Hypothermal mortality in marine fishes of south-central Florida, January, 1977. *Northeast Gulf Sci.* 2:77–97.

Gilmore, R. G., Jr., C. J. Donohoe, D. W. Cooke, and D. J. Herrema. 1981. Fishes of the Indian River Lagoon and adjacent waters. Harbor Branch Found. Inc. Tech. Rept. No. 41, 64 pp.

Gilmore, R. G., J. W. Dodrill, and P. A. Linley. 1983. Reproduction and embryonic development of the sand tiger shark, *Odontaspis taurus* (Rafinesque). *Fish. Bull., U.S.* 81:201–225.

Ginsburg, I. 1937. Review of the seahorses (*Hippocampus*) found on the coasts of the American continents and of Europe. *Proc. U.S. Nat. Mus.* 83:497–594.

Ginsburg, I. 1950. Review of the western Atlantic Triglidae (fishes). *Texas J. Sci.* 4:489–527.

Ginsburg, I. 1952a. Fishes of the family Carangidae of the northern Gulf of Mexico and three related species. *Publ. Inst. Mar. Sci. Univ. Tex.* 2:43–117.

Ginsburg, I. 1952b. Flounders of the genus *Paralichthys* and related genera in American waters. *U.S. Fish Wildl. Serv. Fish. Bull.* 52:267–351.

Ginsburg, I. 1954a. Whitings on the coasts of the American continent. *Fish. Bull., U.S.* 56:187–208.

Ginsburg, I. 1954b. Four new fishes and one little-known species from the east coast of the United States including the Gulf of Mexico. *J. Washington Acad. Sci.* 44:256–264.

Gjösæter, J. 1973a. Age, growth, and mortality of the myctophid fish, *Benthosema glaciale* (Reinhardt), from western Norway. *Sarsia* 52:1–14.

Gjösæter, J. 1973b. The food of the myctophid fish, *Benthosema glaciale* (Reinhardt), from western Norway. *Sarsia* 52:53–58.

Glass, C. W., B. Sarno, O. M. Henry, G. D. Morris, and H. A. Carr. 1999. Bycatch reduction in Massachusetts inshore squid (*Loligo pealeii*) trawl fisheries. *Mar. Tech. Soc. J.* 33:35–42.

Gleason, T. R., and D. A. Bengtson. 1996. Growth, survival and size-selected predation mortality of larval and juvenile inland silversides, *Menidia beryllina* (Pisces; Atherinidae). *J. Exper. Mar. Biol. Ecol.* 199:165–177.

Gleason, T. R., and C. Recksiek. 1990. Preliminary field verification of daily growth increments in the lapillar otoliths of juvenile cunners. In N. C. Parker, ed. *International Symposium and Educational Workshop on Fishmarking Techniques. Am. Fish. Soc. Symp.* 7:562–565.

Glebe, B. D., and W.C. Leggett. 1981a. Latitudinal differences in energy allocation and use during the freshwater migrations of American shad *Alosa sapidissima* and their life history consequences. *Can. J. Fish. Aquat. Sci.* 38:806–820.

Glebe, B. D., and W. C. Leggett. 1981b. Temporal, intra-population differences in energy allocation and use by American shad *Alosa sapidissima* during the spawning migration. *Can. J. Fish. Aquat. Sci.* 38:795–805.

Glover, C. J. M. 1966. Three fishes previously unrecorded from South Australia. *Rec. South Australia Mus.* 15:353–355.

Goff, G. P., and J. M. Green. 1978. Field studies of the sensory basis of homing and orientation to the home site in *Ulvaria subbifurcata* (Pisces: Stichaeidae). *Can. J. Zool.* 56:2220–2224.

Goff, G. P., J. Lien, G. B. Stenson, and J. Fretey. 1994. The migration of a tagged leatherback turtle, *Dermochelys coriacea*, from French Guiana, South America, to Newfoundland, Canada, in 128 days. *Can. Field-Nat.* 108:72–73.

Goldman, K. J. 1997. Regulation of body temperature in the white shark, *Carcharodon carcharias*. *J. Comp. Physiol. B* 167:423–429.

Goldstein, R. J. 1985. Winter kill of summer flounder. *Sea Frontiers* 31:104–111.

Gon, O. 1990a. Gonostomatidae. In: O. Gon and P. C. Heemstra, eds. *Fishes of the Southern Ocean.* J. L. B. Smith Inst. Ichthyol., Grahamstown, South Africa, pp. 116–122.

Gon, O. 1990b. Stomiidae. In: O. Gon and P. C. Heemstra, eds. *Fishes of the Southern Ocean.*, J. L. B. Smith Inst. Ichthyol., Grahamstown, South Africa, pp. 127–133.

Gonzalez, G. and L. Alvarez-Lajonchere. 1978. Alimentacion natural de *Mugil liza, M. curema, M. trichodon* and *M. hospes,* (Pisces, Mugilidae) en las Lagunas Costeras de Tunas de Zaza, Cuba. *Investig. Marin. Ser.* No. 8(41):1–40.

Gonzalez-Villasenor, L. I., and D. A. Powers. 1989. Mitochondrial-DNA restriction-site polymorphisms in the teleost *Fundulus heteroclitus* support secondary intergradation. *Evolution* 44:27–37.

Goodbred, C. O., and J. E. Graves. 1996. Genetic relationships among geographically isolated populations of bluefish (*Pomatomus saltatrix*). *Mar. Freshwater Res.* 47:347–355.

Goode, G. B. 1883. Materials for a history of the sword-fish. U.S. Comm. Fish. Rept. for 1880, pp. 287–394.

Goode, G. B. 1884. The food fishes of the United States. In: *The Fisheries and Fishery Industries of the United States.* U.S. Comm. Fish Fish. Rept., Sect. 1, Part. 3, pp. 163–682.

Goode, G. B. 1888. *American Fishes: A Popular Treatise upon the Game and Food Fishes of North America with Special Reference to Habits and Methods of Capture.* Standard Book Co., New York, 496 pp.

Goode, G. B., and T. H. Bean. 1879. A catalogue of the fishes of Essex county, Massachusetts, including the fauna of Massachusetts bay and the contiguous deep waters. *Bull. Essex Inst.* 11:1–38.

Goode, G. B. and T. H. Bean. 1880a. Catalogue of a collection of fishes sent from Pensacola, Florida, and vicinity, by Mr. Silas Stearns, with descriptions of six new species. *Proc. U.S. Nat. Mus.* 2:121–156.

Goode, G. B. and T. H. Bean. 1880b. Description of a new genus and species of fish, *Lopholatilus chamaeleonticeps,* from the south coast of New England. *Proc. U.S. Nat. Mus.* 2:205–209.

Goode, G. B., and T. H. Bean. 1883. Reports on the results of dredging under the supervision of Alexander Agassiz, on the east coast of the United States, during the summer of 1880, by the U.S. coast Survey Steamer "Blake," Commander J. R. Bartlett, U.S.N., Commanding. XIX. Report on the fishes. *Bull. Mus. Comp. Zool.* 10(5):183–226.

Goode, G. B., and T. H. Bean. 1896. Oceanic ichthyology. Deep-sea and pelagic fishes of the world. *Spec. Bull. U.S. Nat. Mus.* 2, 553 pp.

Goode, G. B., and J. W. Collins. 1887. Part III. The mackerel fishery of the United States. In: G. B. Goode, ed. *Fisheries and Fishery Industries of the United States,* Sect. 5, Vol.1., pp. 247–313.

Goode, G. B., J. W. Collins, R. E. Earll, and A. H. Clark. 1884. Materials for a history of the mackerel fishery. Comm. Fish Fisher. Ann. Rept. for 1881, App. B:91–531.

Gooding, R. U., and A. G. Humes. 1963. External anatomy of the female *Haemobaphes cyclopterina,* a copepod parasite of marine fishes. *J. Parasitol.* 49:663–677.

Goodwin, J. M., IV, and J. H. Finucane. 1985. Reproductive biology of blue runner (*Caranx crysos*) from the eastern Gulf of Mexico. *Northeast Gulf Sci.* 7:139–146.

Goodwin, J. M., and A. G. Johnson. 1986. Age, growth, and mortality of blue runner, *Caranx crysos* from the northern Gulf of Mexico. *Northeast Gulf Sci.* 8:107–114.

Gopalakrishnan, V. 1971. Taxonomy and biology of tropical finfish for coastal aquaculture in the Indo–Pacific region. In: T. V. R. Pillay, ed. *Coastal Aquaculture in the Indo-Pacific Region.* Fishing News Limited, London, pp. 120–149.

Gorbman, A., H. Kobayashi, Y. Honma, and M. Matsuyama. 1990. The hagfishery of Japan. *Fisheries* 15(4):12–18.

Gordon, B. L. 1956. The paternal pipefish. *Nature Magazine* 49:243–244, 276.

Gordon, B. L. 1960. *The Marine Fishes of Rhode Island.* The Book and Tackle Shop, Watch Hill, 136 pp.

Gordon, D. J., D. F. Markle, and J. E. Olney. 1984. Ophidiiformes: development and relationships. In: H. G. Moser et al., eds. *Ontogeny and Systematics of Fishes.* Am. Soc. Ich. Herp. Spec. Publ. No. 1:308–319.

Gordon, J. D. M., and J. A. R. Duncan. 1979. Some notes on the biology of the snake blenny, *Lumpenus lampretaeformis* on the west coast of Scotland. *J. Mar. Biol. Assoc. U.K.* 59:413–419.

Gordon, J. D. M., and J. A. R. Duncan. 1987. Aspects of the biology of *Hoplostethus atlanticus* and *H. mediterraneus* (Pisces: Berycomorphi) from the slopes of the Rockall Trough and the Porcupine Sea Bight (north-eastern Atlantic). *J. Mar. Biol. Assoc. U.K.* 67: 119–133.

Gordon, M. 1938. Animals of the Sargasso merry-go-round. *Nat. Hist., N.Y.* 42:12–20, 78.

Gordon, M. 1955. *Histrio:* The fish on the Sargasso Sea merry-go-round. *Aquarium* 24:386–393.

Gordon, M. S. 1950. Wintering mackerel off New York. *Marine Life* 1:39–40.

Gordon, M. S., and R. H. Backus. 1957. New records of Laborador fishes with special reference to those of Hebron Fjord. *Copeia* 1957:17–20.

Gordon, M. S., B. H. Andrews, and P. H. Scholander. 1962. Freezing resistance in some northern fishes. *Biol. Bull., Woods Hole* 122:56–62.

Gordon, W. G. 1961. Food of the American smelt in Saginaw Bay, Lake Huron. *Trans. Am. Fish. Soc.* 90:439–443.

Gorham, S. W., and D. E. McAllister. 1974. The shortnose sturgeon, *Acipenser brevirostrum,* in the Saint John River, New Brunswick, Canada, a rare and possibly endangered species. *Syllogeus* 5:5–18.

Goshorn, D. M., and C. E. Epifanio. 1991. Diet of larval weakfish and prey abundance in Delaware Bay. *Trans. Am. Fish. Soc.* 120: 684–692.

Gosselin, S., L. Fortier, and J. Gagne. 1989. Vulnerability of marine fish larvae to the toxic dinoflagellate *Protogonyaulax tamarensis.* *Mar. Ecol. Prog. Ser.* 57:1–10.

Goto, S. 1900. Notes on some exotic species of ectoparasitic trematodes. *J. Coll. Sci. Imper. Univ. Tokyo* 12:263–295.

Gotschall, D. W. 1967. Cleaning symbioses in Monterey Bay, Cali-

Goulet, D., and J. M. Green. 1988. Reproductive success of the male lumpfish (*Cyclopterus lumpus* L.) (Pisces: Cyclopteridae): evidence against female mate choice. *Can. J. Zool.* 66:2513–2519.

Goulet, D., J. M. Green, and T. H. Shears. 1986. Courtship, spawning, and parental care behavior of the lumpfish, *Cyclopterus lumpus* L., in Newfoundland. *Can. J. Zool.* 64:1320–1325.

Govender, A., N. Ksitnasamy, and R. P. van der Elst. 1991. Growth of spotted ragged-tooth sharks *Carcharias taurus* (Rafinesque) in captivity. *S. Afr. J. Mar. Sci.* 11:15–19.

Govoni, J. J. 1993. Flux of larval fishes across frontal boundaries: examples from the Mississippi River plume front and the western Gulf Stream in winter. *Bull. Mar. Sci.* 53:538–566.

Govoni, J. J., and L. J. Pietrafesa. 1994. Eulerian views of layered water currents, vertical distribution of some larval fishes, and inferred advective transport over the continental shelf off North Carolina, USA, in winter. *Fish. Oceanogr.* 3:120–132.

Govoni, J. J., D. E. Hoss, and A. J. Chester. 1983. Comparative feeding of three species of larval fishes in the northern Gulf of Mexico: *Brevoortia patronus, Leiostomus xanthurus,* and *Micropogonias undulatus. Mar. Ecol. Prog. Ser.* 13:189–199.

Grabe, S. A. 1978. Food and feeding habits of juvenile, Atlantic tomcod, *Microgadus tomcod,* from Haverstraw Bay, Hudson River. *Fish. Bull., U.S.* 76:89–94.

Grabe, S. A. 1980. Food of age 1 and 2 Atlantic tomcod *Microgadus tomcod,* from Haverstraw Bay, Hudson River, New York. *Fish. Bull., U.S.* 77:1003–1006.

Graham, J. J. 1956. Observations on the alewife, *Pomolobus pseudoharengus* (Wilson), in fresh water. Univ. Toronto Stud., Biol. Ser., Ont. Fish. Res. Lab. Publ. No. 74:1–43.

Graham, J. J. 1982. Production of larval herring, *Clupea harengus,* along the Maine coast, 1964–78. *J. Northw. Atl. Fish. Sci.* 3:63–85.

Graham, J. J., and H. C. Boyar. 1965. Ecology of herring larvae in the coastal water of Maine. Int. Comm. Northw. Atl. Fish. Spec. Publ. No. 6:625–634.

Graham, J. J., and K. M. Sherman. 1984. Estimates of spawning period and spawning intensity of Maine herring. Maine Dept. Mar. Resour., Res. Ref. Doc. No. 84/15, 4 pp.

Graham, J. J., S. B. Chenoweth, and C. W. Davis. 1972. Abundance, distribution, movements and lengths of larval herring along the western coast of the Gulf of Maine. *Fish. Bull., U.S.* 70: 307–321.

Graham, M. 1968. Henry Bryant Bigelow (3 October 1879–11 December 1967). *Deep-Sea Res.* 15:125–132.

Grande, L. 1985. Recent and fossil clupeomorph fishes with materials for revision of the subgroups of clupeoids. *Bull. Am. Mus. Nat. Hist.* 181:231–372.

Grande, L., and G. J. Nelson. 1985. Interrelationships of fossil and Recent anchovies (Teleostei: Engrauloidea) and description of a new species from the Miocene of Cyprus. Am. Mus. Novitates No. 2826, 16 pp.

Grandperrin, R., and M. Legand. 1970. Contribution à la connaissance des *Alepisaurus* (Pisces) dans le Pacifique equatorial et sud-tropical. *Cah. Off. Rec. Sci. Tech. Outre-Mer (Oceanogr.)* 8:11–34.

Grant, G. C. 1962. Predation by bluefish on young Atlantic menhaden in Indian River, Delaware. *Chesapeake Sci.* 3:45–47.

Grant, G. C. 1974. The age composition of striped bass catches in Virginia Rivers, 1967–1971, and a description of the fishery. *Fish. Bull., U.S.* 72:193–199.

Grant, S. M., and J. A. Brown. 1998. Diel foraging cycles and interactions among juvenile Atlantic cod (*Gadus morhua*) at a nearshore site in Newfoundland. *Can. J. Fish. Aquat. Sci.* 55:1307–1316.

Grant, W. S. 1984. Biochemical population genetics of Atlantic herring, *Clupea harengus. Copeia* 1984:357–364.

Grant, W. S. 1986. Biochemical genetic divergence between Atlantic, *Clupea harengus,* and Pacific, *C. pallasi,* herring. *Copeia* 1986:714–719.

Grant, W. S, and F. Utter. 1984. Biochemical population genetics of Pacific herring *Clupea pallasi. Can. J. Fish. Aquat. Sci.* 41: 856–864.

Graves, J. E., J. R. McDowell, and M. L. Jones. 1992a. A genetic analysis of weakfish *Cynoscion regalis* stock structure along the mid-Atlantic coast. *Fish. Bull., U.S.* 90:469–475.

Graves, J. E., J. R. McDowell, A. M. Beardsley, and D. R. Scoles. 1992b. Stock structure of the bluefish *Pomatomus saltatrix* along the mid-Atlantic coast. *Fish. Bull., U.S.* 90:703–710.

Gray, C. L. 1990. Scup (*Stenotomus chrysops*): species profile. Rhode Island Dept. Environ. Manage., Div. Fish. Wildl., Mar. Fish. Sect., 38 pp.

Gray, G-A., and H. E. Winn. 1961. Reproductive ecology and sound production of the toadfish, *Opsanus tau. Ecology* 42:274–282.

Gray, R. W., and C. W. Andrews. 1970. Sex ratio of the American eel (*Anguilla rostrata* (LeSueur)) in Newfoundland waters. *Can. J. Zool.* 48:483–487.

Greeley, J. R. 1937. Fishes of the area with annotated list. In: A biological survey of the lower Hudson watershed. New York State Cons. Dept. Ann. Rept. Suppl. No. 26:45–103.

Green, J. M. 1974. A localized mass winter kill of cunners in Newfoundland. *Can. Field-Nat.* 88:96–97.

Green, J. M. 1975. Restricted movements and homing of the cunner, *Tautogolabrus adspersus* (Walbaum) (Pisces: Labridae). *Can. J. Zool.* 53:1427–1431.

Green, J. M., and M. Farwell. 1971. Winter habits of the cunner, *Tautogolabrus adspersus* (Walbaum, 1792), in Newfoundland. *Can. J. Zool.* 49:1497–1499.

Green, J. M., and R. Fisher. 1977. A field study of homing and orientation to the home site in *Ulvaria subbifurcata* (Pisces: Stichaeidae). *Can. J. Zool.* 55:1551–1556.

Green, J. M., G. Martel, and D. W. Martin. 1984. Comparison of the feeding activity and diets of male and female cunners *Tautogolabrus adspersus* (Pisces: Labridae). *Mar. Biol.* 84:7–11.

Green, J. M., G. Martel, and E. A. Kingsland. 1985. Foraging time allocation in a territorial fish: influence of reproductive activities. *Mar. Ecol. Prog. Ser.* 24:23–26.

Green, J. M., A-L. Mathisen, and J. A. Brown. 1987. Laboratory observations on the reproductive and agonistic behaviour of *Ulvaria subbifurcata* (Pisces: Stichaeidae). *Natur. Can.* 114:195–202.

Greenwood, P. H., D. E. Rosen, S. H. Weitzman, and G. S. Myers. 1966. Phyletic studies of teleostean fishes, with a provisional classification of living forms. *Bull. Am. Mus. Nat. Hist.* 131: 341–455.

Gregory, R. S., and G. R. Daborn. 1982. Notes on adult lumpfish *Cyclopterus lumpus* L. from the Bay of Fundy. *Proc. Nova Scotia Inst. Sci.* 32:321–326.

Gregory, R. S., G. S. Brown, and G. R. Daborn. 1983. Food habits of young anadromous alewives, *Alosa pseudoharengus*, in Lake Ainslie, Nova Scotia. *Can. Field-Nat.* 97:423–426.

Gregory, W. K., and H. C. Raven. 1934. Notes on the anatomy and relationships of the ocean sunfish (*Mola mola*). *Copeia* 1934: 145–151.

Greig, R. A., S. Schurman, J. Pereira, and P. Naples. 1983. Metals and PCB concentrations in windowpane flounder from Long Island Sound. *Bull. Environ. Contam. Toxicol.* 31:257–262.

Grey, M. 1956. The distribution of fishes found below a depth of 2000 meters. *Fieldiana, Zool.* 36:73–337.

Grey, M. 1964. Family Gonostomatidae.Fishes of the western North Atlantic. Sears Found. Mar. Res. Mem. No. 1(4):78–240.

Griffith, R. W. 1974. Environment and salinity tolerance in the genus *Fundulus*. *Copeia* 1974:319–331.

Grimes, C. B., and S. C. Turner. 1999. The complex life history of tilefish *Lopholatilus chamaelonticeps* and vulnerability to exploitation. In: J. A. Musick, ed. *Life in the Slow Lane: Ecology and Conservation of Long-Lived Animals. Am. Fish. Soc. Symp.* 23:17–26.

Grimes, C. B., K. W. Able, and S. C. Turner. 1980a. A preliminary analysis of the tilefish, *Lopholatilus chamaeleonticeps*, fishery in the Mid-Atlantic Bight. *Mar. Fish. Rev.* 41(11):13–18.

Grimes, C. B., K. W. Able, S. C. Turner, and S. J. Katz. 1980b. Tilefish: its continental shelf habitat. *Underwater Nat.* 12(4):34–38.

Grimes, C. B., K. W. Able, and S. C. Turner. 1982. Direct observation from a submersible vessel of commercial longlines for tilefish. *Trans. Am. Fish. Soc.* 111:94–98.

Grimes, C. B., S. C. Turner, and K. W. Able. 1983. A technique for tagging deepwater fish. *Fish. Bull., U.S.* 81:663–666.

Grimes, C. B., K. W. Able, and R. S. Jones. 1986. Tilefish (*Lopholatilus chamaeleonticeps*) habitat, behavior and community structure in Mid-Atlantic and southern New England waters. *Environ. Biol. Fishes* 15:273–292.

Grimes, C. B., C. F. Idelberger, K. W. Able, and S. C. Turner. 1988. The reproductive biology of tilefish, *Lopholatilus chamaeleonticeps* Goode and Bean, from the United States Mid-Atlantic Bight, and the effects of fishing on the breeding system. *Fish. Bull., U.S.* 86:745–762.

Griswold, B. L., and L. L. Smith, Jr. 1972. Early survival and growth of the ninespine stickleback, *Pungitius pungitius*. *Trans. Am. Fish. Soc.* 101:350–352.

Griswold, B. L., and L. L. Smith, Jr. 1973. The life history and trophic relationship of the ninespine stickleback, *Pungitius pungitius*, in the Apostle Islands area of Lake Superior. *Fish. Bull., U.S.* 71:1039–1060.

Griswold, C. A., and T. W. McKenney. 1984. Larval development of the scup, *Stenotomus chrysops* (Pisces: Sparidae). *Fish. Bull., U.S.* 82:77–84.

Gross, A. O. 1923. The black-crowned night heron (*Nycticorax nycticorax nacuicus*) off Sandy Hook. *Auk* 40:1–30.

Grosslein, M. D. 1969. Groundfish survey program of BCF Woods

Grosslein, M. D., and T. R. Azarovitz. 1982. Fish distribution. MESA New York Bight Atlas Monogr. No. 15, 182 pp.

Grosslein, M. D., R. W. Langton, and M. P. Sissenwine. 1980. Recent fluctuations in pelagic fish stocks of the northwest Atlantic, Georges Bank region, in relation to species interactions. *Rapp. Proc.-Verb. Réun. Cons. Int. Explor. Mer* 177:374–404.

Grossman, G. D., M. J. Harris, and J. E. Hightower. 1985. The relationship between tilefish (*Lopholatilus chamaeleonticeps*) abundance and sediment composition off Georgia. *Fish. Bull., U.S.* 83: 443–447.

Grout, D. E., and B. W. Smith. 1994. Anadromous fish investigations and marine recreational fishery evaluation: New Hampshire anadromous and inland fisheries operational management investigation, July 1, 1988 to December 31, 1988. State of New Hampshire, Fish and Game Dept., Durham, Final Rept., Proj. No. F-50-R-8, 85 pp.

Grove, C. A. 1982. Population biology of the winter flounder, *Pseudopleuronectes americanus*, in a New England estuary. M.S. thesis, University of Rhode Island, Kingston, 95 pp.

Grover, J. J. 1998. Feeding habits of pelagic summer flounder, *Paralichthys dentatus*, larvae in oceanic and estuarine habitats. *Fish. Bull., U.S.* 96:248–257.

Gruchy, C. G., and B. Parker. 1980. *Acipenser oxyrhynchus* Mitchill, Atlantic sturgeon. In: D. S. Lee et al., eds. *Atlas of North American Freshwater Fishes.* North Carolina State Mus. Nat. Hist., p. 41.

Gubanov, E. P. 1972. On the biology of the common thresher shark [*Alopias vulpinus* (Bonnaterre)] from the northwest Indian Ocean. *Voprosy Ihktiol.* 12:646–656 [in Russian, translation in *J. Ichthyol.* 12:591–600].

Gubanov, E. P. 1978. The reproduction of some species of pelagic sharks from the equatorial zone of the Indian Ocean. *Voprosy Ikhtiol.* 18:879–891 [in Russian, translation in *J. Ichthyol.* 18:781–792].

Gudger, E. W. 1905. A note on the eggs and egg laying of *Pterophryne histrio*, the gulfweed fish. *Science*, n.s. 22:841–843.

Gudger, E. W. 1906. The breeding habits and segmentation of the egg of the pipefish, *Siphostoma floridae*. *Proc. U.S. Nat. Mus.* 29: 447–500.

Gudger, E. W. 1910. Habits and life history of the toadfish (*Opsanus tau*). *Bull. U. S. Bur. Fish.* 28:1071–1107.

Gudger, E. W. 1927. The nest and nesting habits of the butterfish or gunnel, *Pholis gunnellus*. *Am. Mus. Nat. Hist.* 27:65–72.

Gudger, E. W. 1930. The opah or moonfish, *Lampris luna*, on the eastern coast of North America. *Am. Nat.* 64:168–178.

Gudger, E. W. 1935a. Two partially ambicolorate flatfishes (Heterostoma). I. A summer flounder, *Paralichthys dentatus*. II.A rusty dab, *Limanda ferruginea*. Am. Mus. Novitates No. 768, 8 pp.

Gudger, E. W. 1935b. A photograph and description of *Masturus lanceolatus* taken at Tahiti, May, 1930. The sixteenth adult specimen on record. Am. Mus. Novitates No. 778, 7 pp.

Gudger, E. W. 1940. The alleged pugnacity of the swordfish and the spear-fishes as shown by their attacks on vessels. Roy. Asiatic Soc. Bengal Mem. No. 12:215–315.

Gudger, E. W. 1945. The angler-fishes, *Lophius piscatorius* et ameri-

Gudger, E. W., and F. E. Firth. 1935. An almost totally ambicolorate halibut, *Hippoglossus hippoglossus,* with partially rotated eye and hooked dorsal fin—the only recorded specimen. Am. Mus. Novitates No. 811, 7 pp.

Gudger, E. W., and F. E. Firth. 1937. Two reversed partially ambicolorate halibuts: *Hippoglossus hippoglossus.* Am. Mus. Novitates No. 925, 10 pp.

Gulland, J. A., and G. R. Williamson. 1962. Transatlantic journey of tagged cod. *Nature* 195:921.

Gunning, G. E., and C. R. Shoop, 1962. Restricted movements of the American eel, *Anguilla rostrata* (Lesueur), in freshwater streams, with comments on growth rate. *Tulane Stud. Zool.* 9:265–272.

Gunter, G. 1945. Studies on marine fishes of Texas. *Publ. Inst. Mar. Sci. Univ. Tex.* 1:1–100.

Günther, A. C. L. G. 1861. *Catalogue of the Acanthopterygian Fishes in the Collection of the British Museum,* Vol. 3. British Museum, London, 586 pp.

Gusev, E. V., and M. S. Shevelev. 1997. New data on the individual fecundity of the wolffishes of the genus *Anarchichas* in the Barents Sea. *Voprosy Ikhtiol.* 37(3):377–384 [in Russian, translation in *J. Ichthyol.* 37(5):381–388].

Gustafson, G. 1935. On the biology of *Myxine glutinosa* L. *Arkiv Zool.* 28(2):1–8.

Gutherz, E. J. 1967. Field guide to the flatfishes of the family Bothidae in the western North Atlantic. U.S. Fish Wildl. Serv. Circ. No. 263, 47 pp.

Guthrie, J. F., and F. J. Schwartz. 1990. Occurrence of the barnacle *Balanus venustus* on an Atlantic menhaden, *Brevoortia tyrannus,* off South Carolina. *J. Elisha Mitchill Sci. Soc.* 106:21–23.Haas, R. E., and C. W. Recksiek. 1995. Age verification of winter flounder in Narragansett Bay. *Trans. Am. Fish. Soc.* 124:103–111.

Hacunda, J. S. 1981. Trophic relationships among demersal fishes in a coastal area of the Gulf of Maine. *Fish. Bull., U.S.* 79:775–788.

Haedrich, R. L. 1964. Food habits and young stages of North Atlantic *Alepisaurus* (Pisces, Iniomi). *Breviora* 201, 15 pp.

Haedrich, R. L. 1967. The stromateoid fishes: systematics and a classification. *Bull. Mus. Comp. Zool.* 135: 31–139.

Haedrich, R. L., and S. O. Haedrich. 1974. A seasonal survey of the fishes in the Mystic River, a polluted estuary in downtown Boston, Massachusetts. *Est. Coast. Mar. Sci.* 2:59–73.

Haedrich, R. L., and M. H. Horn. 1969. A key to the stromateoid fishes. Woods Hole Oceanogr. Inst. Tech. Ref. No. 69-70, 46 pp.

Haedrich, R. L., and J. G. Nielsen. 1966. Fishes eaten by *Alepisaurus* (Pisces, Iniomi) in the southeastern Pacific Ocean. *Deep-Sea Res.* 13:909–919.

Haedrich, R. L., G. T. Rowe, and P. T. Polloni. 1975. Zonation and faunal composition of epibenthic populations on the continental slope south of New England. *J. Mar. Res.* 33:191–212.

Haedrich, R. L., G. T. Rowe, and P. T. Polloni. 1980. The megabenthic fauna in the deep sea south of New England, USA. *Mar. Biol.* 57:165–179.

Haegele, C. W., and J. F. Schweigert. 1985. Distribution and characteristics of herring spawning grounds and description of spawning behavior. *Can. J. Fish. Aquat. Sci.* 42(Suppl. 1):39–55.

Hagen, D. W., and L. G. Gilbertson. 1972. Geographic variation and environmental selection in *Gasterosteus aculeatus* L. in the Pacific Northwest, America. *Evolution* 26:32–51.

Hagen, D. W., and G. E. E. Moodie. 1982. Polymorphism for plate morphs in *Gasterosteus aculeatus* on the east coast of Canada and an hypothesis for their global distribution. *Can. J. Zool.* 60:1032–1042.

Hagen, N. T., and K. H. Mann. 1992. Functional response of the predators American lobster *Homarus americanus* (Milne-Edwards) and Atlantic wolffish *Anarhichas lupus* (L.) to increasing numbers of the green sea urchin *Strongylocentrotus droebachiensis* (Müller). *J. Exp. Mar. Biol. Ecol.* 159:89–112.

Haglund, T. R., D. G. Buth, and D. M. Blouw. 1990. Allozyme variation and the recognition of the "white stickleback". *Biochem. Syst. Ecol.* 18:559–563.

Haglund, T. R., D. G. Buth, and R. Lawson. 1993. Allozyme variation and phylogenetic relationships of Asian, North American, and European populations of the ninespine stickleback, *Pungitius pungitius.* In: R. L. Mayden, ed. *Systematics, Historical Ecology, and North American Freshwater Fishes.* Stanford University Press, Stanford, pp. 438–452.

Hain, J. H. W. 1975. The behaviour of migratory eels, *Anguilla rostrata,* in response to current, salinity and lunar period. *Helgoländer Wissen. Meeresunt.* 27: 211–233.

Hain, J. H. W., G. R. Carter, S. D. Kraus, C. A. Mayo, and H. E. Winn. 1982. Feeding behavior of the humpback whale, *Megaptera novaeangliae,* in the western North Atlantic. *Fish. Bull., U.S.* 80: 259–268.

Haley, A. J. 1952. Preliminary observations on a severe epidemic of microsporidiosis in the smelt, *Osmerus mordax* (Mitchell). *J. Parasitol.* 38:183.

Haley, A. J. 1954. Microsporidian parasite, *Glugea hertwigi,* in the American smelt from the Great Bay Region, New Hampshire. *Trans. Am. Fish. Soc.* 83:84–90.

Halkett, A. 1913. *Check List of the Fishes of the Dominion of Canada and Newfoundland.* King's Printer, Ottawa, 138 pp.

Halkett, A. 1922. The fishes of the Bay of Fundy. *Contrib. Can. Biol. 1921,* 3:49–72.

Hall, J. W., T. I. J. Smith, and S. D. Lamprecht. 1991. Movements and habitats of shortnose sturgeon, *Acipenser brevirostrum* in the Savannah River. *Copeia* 1991:695–702.

Hall, L. W., Jr. 1987. Acidification effect on larval striped bass *Morone saxatilis* in Chesapeake Bay tributaries: a review. *Water Air Soil Pollut.* 35:87–96.

Hall, L. W., Jr. 1988. Studies of striped bass in three Chesapeake Bay spawning habitats. *Mar. Pollut. Bull.* 19:478–486.

Hall, L. W., Jr., A. E. Pinkney, L. O. Horseman, and S. E. Finger. 1985. Mortality of striped bass larvae in relation to contaminants and water quality in a Chesapeake Bay tributary. *Trans. Am. Fish. Soc.* 114:861–868.

Hall, S. J. 1987. Maximum daily ration and the pattern of food consumption in haddock, *Melanogrammus aeglefinus* (L.), and dab, *Limanda limanda* (L.). *J. Fish Biol.* 31:479–491.

Hall-Arber, M. 1996. Workshop probes hagfish processing potential in US: fishery discards worry fishermen. *Comm. Fish. News,* Feb. 29–Mar. 3 issue, pp. 18B–19B.

Hallacher, L. E. 1974. The comparative morphology of extrinsic gasbladder musculature in the scorpionfish genus *Sebastes* (Pisces: Scorpaenidae). *Proc. S. Cal. Acad. Sci., Ser. 4,* 40:59–86.

Halliday, R. G. 1970. Growth and vertical distribution of the glacier lanternfish, *Benthosema glaciale,* in the northwestern Atlantic. *J. Fish. Res. Bd. Can.* 27:105–116.

Halliday, R. G. 1991. Marine distribution of the sea lamprey (*Petromyzon marinus*) in the northwest Atlantic. *Can. J. Fish. Aquat. Sci.* 48:832–842.

Halliday, R. G., and W. B. Scott. 1969. Records of mesopelagic and other fishes from the Canadian Atlantic with notes on their distribution. *J. Fish. Res. Bd. Can.* 26:2691–2702.

Halliday, R. G., J. McGlade, R. Mohn, R. N. O'Boyle, and M. Sinclair. 1986. Resource and fishery distributions in the Gulf of Maine area in relation to the Subarea 4/5 Boundary. NAFO Sci. Coun. Stud. No. 10:67–92.

Halpin, P. M. 1997. Habitat use patterns of the mummichog, *Fundulus heteroclitus,* in New England. I. Intramarsh variation. *Estuaries* 20:618–625.

Hals, G. D. 1981. Mechanisms of coexistence in two species of *Fundulus. Biol. Bull., Woods Hole* 161:326.

Halstead, B. W. 1978. *Poisonous and Venomous Marine Animals of the World,* Rev. Ed. Darwin Press, Princeton, 1043 pp.

Hamer, P. E. 1966. The occurrence of striped cusk-eel, *Rissola marginata,* in the body cavity of the striped bass, *Roccus saxatilis. Chesapeake Sci.* 7:214–215.

Hamer, P. E. 1970. Studies of the scup, *Stenotomus chrysops,* in the Middle Atlantic Bight. N.J. Div. Fish Game Shellfish Misc. Rept. No. 5M, 14 pp.

Hamer, P. E., and F. E. Lux. 1962. Marking experiments on fluke (*Paralichthys dentatus*) in 1961, North Atlantic Section. Atlantic States Mar. Fish. Comm., Minutes 21st Meeting, Appendix MA–6.

Hamilton, P. V. 1976. Predation on *Littorina irrorata* (Mollusca: Gastropoda) by *Callinectes sapidus* (Crustacea: Portunidae). *Bull. Mar. Sci.* 26:403–409.

Hamlett, W. C., and F.J. Schwartz. 1979. TEM and SEM of the axillary mucous gland of the oyster toadfish, *Opsanus tau. Am. Zool.* 19:931 (abstract).

Hansen, P. M. 1943. Capelin (*Mallotus villosus*). *J. Cons. Perm. Int. Explor. Mer Ann. Biol.* 1(1939–1941):121–124.

Hansen, P. M. 1968. Report on wolffish larvae in West Greenland waters. ICNAF Spec. Publ. No. 7:183–185.

Hanson, J. M., and S. C. Courtenay. 1996. Seasonal use of estuaries by winter flounder in the southern Gulf of St. Lawrence. *Trans. Amer. Fish. Soc.* 125:705–718.

Hanson, J. M., and S. C. Courtenay. 1997. Seasonal distribution, maturity, condition, and feeding of smooth flounder (*Pleuronectes putnami*) in the Miramichi estuary, southern Gulf of St. Lawrence. *Can. J. Zool.* 75:1226–1240.

Harbison, G. R. 1987. Encounters with a swordfish (*Xiphias gladius*) and sharptail mola (*Mola lanceolatus*) at depths greater than 600 meters. *Copeia* 1987:511–513.

Hardcastle, A. B. 1944. *Eimeria brevoortiana,* a new sporozoan parasite from menhaden *Brevoortia tyrannus,* with observations on its life history. *J. Parasitol.* 30:60–68.

Harden-Jones, F. R. 1968. *Fish Migration.* Edward Arnold Publishers, Ltd., London, 325 pp.

Harding, G. C., W. P. Vass, B. T. Hargrave, and S. J. Pearre. 1986. Diel vertical movements and feeding activity of zooplankton in St. Georges Bay, N.S., using net tows and a newly developed passive trap. *Can. J. Fish. Aquat. Sci.* 43:952–967.

Hardisty, M. W., and I. C. Potter. 1972. The general biology of adult lampreys. In: M. W. Hardisty and I. C. Potter, eds. *The Biology of Lampreys,* Vol. 1. Academic Press, New York, pp. 127–206.

Hardisty, M. W., and I. C. Potter, eds. 1972–1982. *The Biology of Lampreys.* Academic Press, New York, 4 vols.

Hardy, J. D., Jr., ed. 1978a. *Development of Fishes of the Mid-Atlantic Bight: An Atlas of the Egg, Larval and Juvenile Stages. Vol. II. Anguillidae through Syngnathidae.* U.S. Fish Wildl. Serv. Biol. Serv. Prog. FWS/OBS-78/12, 458 pp.

Hardy, J. D., Jr., ed. 1978b. *Development of Fishes of the Mid-Atlantic Bight: An Atlas of the Egg, Larval and Juvenile Stages. Vol. III. Aphredoderidae through Rachycentridae.* U.S. Fish Wildl. Serv. Biol. Serv. Prog. FWS/OBS-78/12, 394 pp.

Hardy, J. D., and R. K. Johnson. 1974. Descriptions of halfbeak larvae and juveniles from Chesapeake Bay (Pisces: Hemiramphidae). *Chesapeake Sci.* 15:241–246.

Hare, G. M., and H. P. Murphy. 1974. First record of the American shad *Alosa sapidissima* from Labrador waters. *J. Fish. Res. Bd. Can.* 31:1536–1537.

Hare, J. A., and R. K. Cowen. 1993. Ecological and evolutionary implications of the larval transport and reproductive strategy of bluefish *Pomatomus saltatrix. Mar. Ecol. Prog. Ser.* 98:1–16.

Hare, J. A., and R. K. Cowen. 1994. Ontogeny and otolith microstructure of bluefish *Pomatomus saltatrix* (Pisces: Pomatomidae). *Mar. Biol.* 118:541–550.

Hare, J. A., and R. K. Cowen. 1996. Transport mechanisms of larval and pelagic juvenile bluefish (*Pomatomus saltatrix*) from South Atlantic Bight spawning grounds to Middle Atlantic Bight nursery habitats. *Limnol. Oceanogr.* 41:1264–1280.

Hargis, W. J., Jr. 1985. Quantitative effects of marine diseases on fish and shellfish populations. *Trans. No. Am. Wildlfe. Nat. Resour. Conf.* 50:608–640.

Harkonen, T. 1986. Guide to the otoliths of the bony fishes of the Northeast Atlantic. Danbiu Aps. Biolog. Consults. Henningsens Allé 58, DK-2900, Hellerup, Denmark, 256 pp.

Harman, R. I. 1988. Squamous cell carcinoma in rainbow smelt *Osmerus mordax. Dis. Aquat. Org.* 5:71–73.

Harmic, J. L. 1958. Some aspects of the development and ecology of the pelagic phase of the gray squeteague, *Cynscion regalis* (Bloch and Schneider), in the Delaware Estuary. Ph.D. dissertation, University of Delaware, Newark, 84 pp.

Haro, A., W. Richkus, K. Whalen, A. Hoar, W.-Dieter Busch, S. Lary, T. Brush, and D. Dixon. 2000. Population decline of the American eel: implications for research and management. *Fisheries* 25(9):7–16.

Harold, A. S. 1994. A taxonomic revision of the sternoptychid genus *Polyipnus* (Teleostei: Stomiiformes) with an analysis of phyloge- [illegible]

Harold, A. S., and S. H. Weitzman. 1996. Interrelationships of stomiiform fishes. In: M. L. J. Stiassny, L. R. Parenti, and G. D. Johnson, eds. *Interrelationships of Fishes.* Academic Press, San Diego, pp. 333–353.

Harrington, R. W. 1958. Morphometry and ecology of small tarpon, *Megalops atlantica* Valenciennes from transitional stage through onset of scale formation. *Copeia* 1958:1–10.

Harrington, R. W. 1966. Changes through one year in the growth rates of tarpon, *Megalops atlanticus* Valenciennes, reared from mid-metamorphosis. *Bull. Mar. Sci.* 16:863–883.

Harrington, R. W., and E. S. Harrington. 1960. Food of larval and young tarpon, *Megalops atlantica. Copeia* 1960:311–319.

Harris, L. G. 1986. Size selective predation in a sea anemone, nudibranch, and fish food chain. *Veliger* 29:38–47.

Harris, M. J., and G. D. Grossman. 1985. Growth, mortality, and age composition of a lightly exploited tilefish substock off Georgia. *Trans. Am. Fish. Soc.* 114:837–846.

Harrison, I. J., and G. J. Howes. 1991. The pharyngobranchial organ of mugilid fishes; its structure, variability, ontogeny, possible function and taxonomic utility. *Bull. Brit. Mus. Nat. Hist. (Zool.)* 57(2):111–132.

Hart, J. L. 1973. Pacific fishes of Canada. *Bull. Fish. Res. Bd. Can.* 180, 740 pp.

Hartel, K. E., D. B. Halliwell, and A. E. Launer. 2002. *Inland Fishes of Massachusetts.* Massachusetts Audubon Society, Lincoln, Mass.

Harvey, R., and R. S. Batty. 1998. Cutaneous taste buds in cod. *J. Fish Biol.* 53:138–149.

Hassler, W. W., and W. T. Hogarth. 1977. The growth and culture of dolphin, *Coryphaena hippurus,* in North Carolina. *Aquaculture* 12:115–122.

Hastings, J. W., and J. G. Morin. 1991. Bioluminescence. In: C. L. Prosser, ed. *Neural and Integrative Animal Physiology.* Wiley-Liss, New York, pp. 131–170.

Hastings, R. W., and R. E. Good. 1977. Population analysis of the fishes of a freshwater tidal tributary of the lower Delaware River. *Bull. New Jersey Acad. Sci.* 22:13–20.

Hastings, R. W., J. C. O'Herron II, K. Schick, and M. A. Lazzari. 1987. Occurrence and distribution of shortnose sturgeon, *Acipenser brevirostrum,* in the upper tidal Delaware River. *Estuaries* 10:337–341.

Haug, T. 1990. Biology of the Atlantic halibut, *Hippoglossus hippoglossus* (L., 1758). *Adv. Mar. Biol.* 26:1–70.

Haug, T., and B. Gulliksen. 1988. Fecundity and oocyte sizes in ovaries of female Atlantic halibut, *Hippoglossus hippoglossus* (L.). *Sarsia* 73:259–261.

Haugaard, N., and L. Irving. 1943. The influence of temperature upon the oxygen consumption of the cunner (*Tautogolabrus adspersus* Walbaum) in summer and winter. *J. Cell. Comp. Physiol.* 21:19–26.

Hauser, W. J. 1973. Larval fish ecology of the Sheepscot River–Montsweag Bay Estuary, Maine. Ph.D. thesis, University of Maine, Orono, 79 pp.

Havey, K. A. 1950. The freshwater fisheries of Long Pond and Echo Lake, Mount Desert Island, Maine. M.S. thesis, University of Maine, Orono, 83 pp.

Havey, K. A. 1961. Restoration of anadromous alewives at Long Pond, Maine. *Trans. Am. Fish. Soc.* 90:281–286.

Havey, K. A. 1973. Production of juvenile alewives, *Alosa pseudoharengus,* at Love Lake, Washington County, Maine. *Trans. Am. Fish. Soc.* 102:434–437.

Hawke, J. P. 1976. A survey of the diseases of striped bass, *Morone saxatilis,* and pompano, *Trachinotus carolinus,* cultured in earthen ponds. *Proc. 7th Ann. World Mariculture Soc.,* pp. 495–509.

Hawkins, A. D., and M. C. P. Amorim. 2000. Spawning sounds of the male haddock, *Melanogrammus aeglefinus. Environ. Biol. Fishes* 59:29–41.

Hawkins, A. D., and C. J. Chapman. 1966. Underwater sounds of the haddock, *Melanogrammus aeglefinus. J. Mar. Biol. Assoc. U.K.* 46:241–247.

Hawkins, A. D., K. J. Chapman, and D. J. Symonds. 1967. Spawning of haddock in captivity. *Nature* 215:923–925.

Hawkins, C. M., and R. B. Angus. 1986. Preliminary observations of predation on ocean quahaugs, *Arctica islandica,* by Atlantic wolffish, *Anarhichas lupus. Nautilus* 100:126–129.

Hay, D. E. 1985. Reproductive biology of Pacific herring *Clupea harengus pallasi. Can. J. Fish. Aquat. Sci.* 42(Suppl. 1):111–126.

Hayden, T. 1998. Saving the seahorses. *Newsweek,* Dec. 21, 1998, p. 76.

Hazin, F. H. V., K. Kihara, K. Otsuka, C. E. Boeckman, and E. C. Leal. 1994. Reproduction of the blue shark *Prionace glauca* in the south-western equatorial Atlantic Ocean. *Fish. Sci.* 60: 487–491.

Heard, W. R. 1991. Life history of pink salmon (*Oncorhynchus gorbuscha*). In: C. Groot and L. Margolis, eds. *Pacific Salmon Life Histories.* University of British Columbia Press, Vancouver, pp. 119–230.

Heath, M., and N. Nicoll. 1991. Infection of larval herring by helminth parasites in the North Sea and the effect on feeding incidence. *Cont. Shelf Res.* 11:1477–1489.

Heck, K. L., Jr., and M. P. Weinstein. 1978. Mimetic relationships between tropical burrfishes and opisthobranchs. *Biotropica* 10:78–79.

Heck, K. L., Jr., K. W. Able, M. P. Fahay, and C. T. Roman. 1989. Fishes and decapod crustaceans of Cape Cod eelgrass meadows: species composition, seasonal abundance patterns and comparison with unvegetated substrates. *Estuaries* 12:59–65.

Heemstra, P. C. 1980. A revision of the zeid fishes (Zeiformes: Zeidae) of South Africa. *J. L. B. Smith Inst. Ichthyol. Bull.* 41, 18 pp.

Heemstra, P. C. 1986a. Family No. 126: Berycidae. In: M. M. Smith and P. C. Heemstra, eds. *Smith's Sea Fishes.* MacMillan South Africa, Johannesburg, pp. 409–410.

Heemstra, P. C. 1986b. Family No. 165: Polyprionidae. In: M. M. Smith and P. C. Heemstra, eds. *Smith's Sea Fishes.* MacMillan South Africa, Johannesburg, p. 509.

Heemstra, P. C. 1997. A review of the smooth-hound sharks (genus *Mustelus,* family Triakidae) of the western Atlantic Ocean, with descriptions of two new species and a new subspecies. *Bull. Mar. Sci.* 60:894–928.

Heilner, V. C. 1920. Notes on the taking of an ocean sunfish (*Mola mola*) off Santa Catalina Island, California September 3, 1919. *Bull. New York Zool. Soc.* 23(b):126–127.

Helser, T. E., and F. P. Almeida. 1997. Density-dependent growth and sexual maturity of silver hake in the north-west Atlantic. *J. Fish Biol.* 51:607–623.

Hemmer, M. J., D. P. Middaugh, and J. C. Moore. 1990. Effects of temperature and salinity on *Menidia beryllina* embryos exposed to terbufos. *Diseases Aquatic Org.* 8:127–136.

Hempel, G. 1971. Egg production and egg mortality in herring. *Rapp. Proc.-Verb. Réun. Cons. Int. Explor. Mer* 160:8–11.

Hempel, G., and J. H. S. Blaxter. 1967. Egg weight in Atlantic herring (*Clupea harengus* L.). *J. Cons. Perm. Int. Explor. Mer* 31:170–195.

Henderson, A. C., K. Flannery, and J. Dunne. 2001. Observations on the biology and ecology of the blue shark in the north-east Atlantic. *J. Fish Biol.* 58:1347–1358.

Henderson, E. M. 1979. Summer flounder (*Paralichthys dentatus*) in the northwest Atlantic. NMFS Woods Hole Lab. Ref. Doc. No. 79-31, 13 pp.

Hendricks, J. D. 1972. Two new host species for the parasitic fungus *Ichthyophonus hoferi* in the Northwest Atlantic. *J. Fish. Res. Bd. Can.* 29:1776–1777.

Hendrickson, L. 1998. Windowpane. In: S. H. Clark, ed. *Status of Fishery Resources off the Northeastern United States for 1998*. NOAA Tech. Memo. NMFS-NE-115:85–87.

Hennemuth, R. C., M. D. Grosslein, and F. D. McCracken. 1964. Abundance, age composition of landings, and total mortality of haddock caught off southern Nova Scotia, 1956–1961. *Res. Bull. Int. Comm. Northw. Atl. Fish.* 1:43–73.

Henry, K. A. 1971. Atlantic menhaden *(Brevoortia tyrannus)*, resource and fishery: analysis of decline. NOAA Tech. Rept. NMFS-SSRF-642, 32 pp.

Henry, R. 1997. Overview of recent developments in haddock culture in New Brunswick. Conference on Coldwater Aquaculture to the Year 2000. Bull. Aquacult. Assoc. Can. Spec. Publ. No. 2:37–40.

Herald, E. S. 1943. Studies on the classification and interrelationships of the American pipefishes. Ph.D. dissertation, Stanford University, Stanford, 339 pp.

Herald, E. S. 1951. Stable requirements for raising sea horses. *Aquar. J.* 22:234–242.

Herald, E. S., and R. R. Strickland. 1949 (1948). Annotated list of the fishes of Homosassa Springs, Florida. *Quart. J. Fla. Acad. Sci.* 11:99–101.

Hergenrader, G. L. 1980. Current distribution and potential for dispersal of white perch (*Morone americana*) in Nebraska and adjacent waters. *Amer. Midl. Nat.* 103:404–406.

Herman, R. L., C. N. Burke, and S. Perry. 1997. Epidermal tumors of rainbow smelt with associated virus. *J. Wild. Dis.* 33:925–929.

Herman, S. S. 1963. Planktonic fish eggs and larvae of Narragansett Bay. *Limnol. Oceanogr.* 8:103–109.

Hermes, R. 1985. Distribution of neustonic larvae of hakes *Urophycis* spp. and fourbeard rockling *Enchelyopus cimbrius* in the Georges Bank area. *Trans. Am. Fish. Soc.* 114:604–608.

Hernandez, L. P., and P. J. Motta. 1997. Trophic consequences of differential performance: ontogeny of oral jaw-crushing performance in the sheepshead, *Archosargus probatocephalus* (Teleostei, Sparidae). J. Zool. [illegible]

Herrick, C. J. 1904. The organ and sense of taste in fishes. *Bull. U.S. Fish Comm.* 22:237–272.

Hettler, W. F., Jr. 1968. Artificial fertilization among yellowfin and Gulf menhaden *Brevoortia* and their hybrid. *Trans. Am. Fish. Soc.* 97:119–123.

Hettler, W. F., Jr. 1976. Influence of temperature and salinity on routine metabolic rate of young Atlantic menhaden. *J. Fish Biol.* 8:55–65.

Hettler, W. F., Jr. 1981. Spawning and rearing Atlantic menhaden. *Prog. Fish-Cult.* 43:80–84.

Hettler, W. F., Jr., and A. J. Chester. 1990. Temporal distribution of ichthyoplankton near Beaufort Inlet, North Carolina. *Mar. Ecol. Prog. Ser.* 68:157–168.

Hettler, W. F., Jr., and J. A. Hare. 1998. Abundance and size of larval fishes outside the entrance to Beaufort Inlet, North Carolina. *Estuaries* 21:476–499.

Hettler, W. F., Jr., and A. B. Powell. 1981. Egg and larval production at the NMFS Beaufort Laboratory, Beaufort, N.C., U.S.A. *Rapp. Proc.-Verb. Réun. Cons. Int. Explor. Mer* 178:501–503.

Hew, C. L., D. Slaughter, G. L. Fletcher, and S. B. Jashi. 1981. Antifreeze glycoproteins in the plasma of Newfoundland Atlantic cod (*Gadus morhua*). *Can. J. Zool.* 59:2186–2192.

Hew, C. L., D. Slaughter, S. B. Ioski, G. L. Fletcher, and V. S. Ananthanarayanan. 1984. Antifreeze polypeptides from the Newfoundland Ocean pout, *Macrozoarces americanus:* presence of multiple and compositionally diverse components. *J. Comp. Physiol.* 155B:81–88.

Heyerdahl, E. G., and R. Livingstone, Jr. 1982. Atlantic cod, *Gadus morhua*. In: M. D. Grosslein and T. R. Azarovitz. *Fish Distribution*. MESA New York Bight Atlas Monogr. No. 15:70–72.

Higham, J. R., and W. R. Nicholson. 1964. Sexual maturation and spawning of Atlantic menhaden. *Fish. Bull., U.S.* 63:255–271.

Hightower, J. E., and G. D. Grossman. 1988. Status of the tilefish, *Lopholatilus chameleonticeps,* fishery off South Carolina and Georgia and recommendations for management. *Fish. Bull., U.S.* 87:177–188.

Hildebrand, H. H. 1954. A study of the fauna of the brown shrimp (*Penaeus aztecus* Ives) grounds in the Western Gulf of Mexico. *Publ. Inst. Mar. Sci. Univ. Tex.* 3:233–366.

Hildebrand, S. F. 1934. The capture of a young tarpon, *Tarpon atlanticus,* at Beaufort, North Carolina. *Copeia* 1934:45–46.

Hildebrand, S. F. 1943. Notes on the affinity, anatomy, and development of *Elops saurus* Linnaeus. *J. Washington Acad. Sci.* 33:90–94.

Hildebrand, S. F. 1948. A review of the American menhaden, genus *Brevoortia,* with a description of a new species. *Smithsonian Misc. Coll.* 107:1–39.

Hildebrand, S. F. 1963a. Family Elopidae. Fishes of the western North Atlantic. Sears Found. Mar. Res. Mem. No. 1(3):111–131.

Hildebrand, S. F. 1963b. Family Engraulidae. Fishes of the western North Atlantic. Sears Found. Mar. Res. Mem. No. 1(3):152–249.

Hildebrand, S. F. 1963c. Family Clupeidae: Fishes of the western North Atlantic. Sears Found. Mar. Res. Mem. No. 1(3): 257–454.

Hildebrand, S. F., and L. E. Cable. 1930. Development and life history of fourteen teleostean fishes at Beaufort, N. C. *Bull. U.S. Bur.*

Hildebrand, S. F., and L. E. Cable. 1934. Reproduction and development of whitings or kingfishes, drums, spot, croaker, and weakfishes or sea trouts, family Sciaenidae, of the Atlantic coast of the United States. *Bull. U.S. Bur. Fish.* 48:41–117.

Hildebrand, S. F., and L. E. Cable. 1938. Further notes on the development and life history of some teleosts at Beaufort, N.C. *Bull. U.S. Bur. Fish.* 48:505–642.

Hildebrand, S. F., and W. C. Schroeder. 1928. Fishes of Chesapeake Bay. *Bull. U.S. Bur. Fish.* 43(1):1–366.

Hill, D. R. 1959. Some uses of statistical analysis in classifying races of American shad *Alosa sapidissima*. *Fish. Bull., U.S.* 59: 269–286.

Hilton, E. J., and W. E. Bemis. 1999. Skeletal variation in shortnose sturgeon (*Acipenser brevirostrum*) from the Connecticut River: implications for comparative osteological studies of fossil and living fishes. In: G. Arratia and H.-P. Schultze, eds. *Mesozoic Fishes 2:– Systematics and Fossil Record*. Verlag F. Pfeil, Munich, pp. 69–94.

Hines, A. H., A. M. Haddon, and L. A. Wiechert. 1990. Guild structure and foraging impact of blue crabs and epibenthic fish in a subestuary of Chesapeake Bay. *Mar. Ecol. Prog. Ser.* 67:105–126.

Hislop, J. R. G., and W. S. MacDonald. 1989. Damage to fish by seabirds in the Moray Firth. *Scott. Birds* 15:151–155.

Hislop, J. R. G., and R. G. J. Shelton. 1993. Marine predators and prey of Atlantic salmon. In: D. Mills, ed. *Salmon in the Sea and New Enhancement Strategies.*, Fishing News Books, Oxford, pp. 104–118.

Hislop, J. R. G., and A. F. Youngson. 1984. A note on the stomach contents of salmon caught by longline north of the Faroe Islands in March, 1983. ICES C. M. 1984/M:17.

Hjørt, J. 1914. Fluctuations in the great fisheries of northern Europe viewed in the light of biological research. *Rapp. Proc.-Verb. Réun. Cons. Int. Explor. Mer* 20:1–228.

Hjørt, J. 1938. Studies of growth in the north-eastern area. *Rapp. Proc.-Verb. Réun. Cons. Int. Explor. Mer* 108:1–8.

Ho, J.-S. 1968. Cyclopoid copepods of the genus *Tucca* (Tuccidae), parasitic on diodontid and tetradontid fishes. *Fish. Bull., U.S.* 66:285–298.

Ho, J.-S., and J. Rokicki. 1987. Poecilostomatoid copepods parasitic on fishes off the west coast of Africa. *J. Nat. Hist.* 21:1025–1034.

Hoberman, J. M., and A. C. Jensen. 1962. The growth rate of New England pollock. *Trans. Am. Fish. Soc.* 91:227–228.

Hodder, V. M. 1963. Fecundity of Grand Bank haddock. *J. Fish. Res. Bd. Can.* 30:1465–1487.

Hodder, V. M. 1965. The possible effects of temperature on the fecundity of Grand Bank haddock. Int. Comm. Northw. Atl. Fish. Spec. Publ. No. 6:515–522.

Hodder, V. M. 1972. The fecundity of herring in some parts of the Newfoundland area. *Int. Comm. Northw. Atl. Fish. Res. Bull.* 9:99–107.

Hodder, V. M., and L. S. Parsons. 1971. Some biological features of southwest Newfoundland and northern Scotian Shelf herring stocks. *Int. Comm. Northw. Atl. Fish. Res. Bull.* 8:72–73.

Hoese, H. D., and R. H. Moore. 1977. *Fishes of the Gulf of Mexico, Texas, Louisiana, and Adjacent Waters*. Texas A&M University Press, College Station, 327 pp.

Hoff, J. G. 1970. Artificial spawning of black sea bass, *Centropristes striatus melanus* Ginsburg, aided by chorionic gonadotrophic hormones. Fla. Dept. Nat. Res. Mar. Res. Lab. Spec. Sci. Rept. No. 25, 17 pp.

Hoff, J. G. 1971. Mass mortality of the crevalle jack, *Caranx hippos* (Linnaeus) on the Atlantic coast of Massachusetts. *Chesapeake Sci.* 12:49.

Hoff, J. G. 1972. Movements of adult tidewater silverside, *Menidia beryllina* (Cope), tagged in New England waters. *Am. Midland Nat.* 88:499–502.

Hoff, J. G. 1979. Annotated bibliography and subject index on the shortnose sturgeon, *Acipenser brevirostrum*. NOAA Tech. Rept. NMFS SSRF No. 731, 16 pp.

Hoff, J. G. 1980. Review of the present status of the stocks of the Atlantic sturgeon *Acipenser oxyrhynchus* Mitchill. Report to NMFS Northeast Reg., Gloucester, Mass.

Hoff, J. G., and R. M. Ibara. 1977. Factors affecting the seasonal abundance, composition and diversity of fishes in a southeastern New England estuary. *Est. Coast. Mar. Sci.* 5:665–678.

Hoff, J. G., D. McGill, and P. Barrows. 1968. Some aspects of the hydrography of the Slocum River. *Proc. 23rd Indust. Waste Conf., Purdue University,* pp. 87–98.

Hoff, T. B., R. J. Klauda, and J. R. Young. 1988. Contribution to the biology of shortnose sturgeon in the Hudson River estuary. In: C. L. Smith, ed. *Fisheries Research in the Hudson River: Hudson River Environmental Society,* State University of New York Press, Albany, pp. 171–189.

Hoffmann, G. L. 1967. *Parasites of North American Freshwater Fishes*. University of California Press, Los Angeles, 486 pp.

Hogans, W. E., and K. J. Sulak. 1992. *Diocus lycenchelus* new species (Copepoda: Chondracanthidae) parasitic on the eelpout *Lycenchelys verrillii* (Zoarcidae) from the Hatteras slope of the northwest Atlantic Ocean. *Bull. Mar. Sci.* 51:301–308.

Hogans, W. E., M. J. Dadswell, L. S. Uhazy, and R. G. Appy. 1993. Parasites of American shad, *Alosa sapidissima* (Osteichthyes: Clupeidae), from rivers of the North American Atlantic coast and the Bay of Fundy, Canada. *Can. J. Zool.* 71:941–9.

Hognestad, P. 1965. Iakttagelse av gyting hos gråsteinbit (*Anarhichas lupus*). *Fauna* 18:31–32.

Holden, M.J. 1974. Problems in the rational exploitation of elasmobranch populations and some suggested solutions. In: F. R. Hardin-Jones, ed. *Sea Fisheries Research*. John Wiley & Sons, New York, pp. 117–137.

Holland, B. F., Jr., and G. F. Yelverton. 1973. Distribution and biological studies of anadromous fishes offshore North Carolina. N.C. Dept. Nat. Econ. Res., Spec. Sci. Rept. No. 24, 132 pp.

Holliday, F. G. T., and J. H. S. Blaxter. 1960. The effects of salinity on the developing eggs and larvae of the herring. *J. Mar. Biol. Assoc. U.K.* 39:591–603.

Hollis, E. H. 1948. The homing tendency of the American shad. *Science* 108:332–333.

Hollis, E. H. 1952. Variations in the feeding habits of the striped bass, *Roccus saxatilis* (Walbaum), in Chesapeake Bay. *Bull. Bingham Oceanogr. Coll.* 14(1):111–131.

Holmgren, N. 1946. On two embryos of *Myxine glutinosa*. *Acta Zool.* 27:1–90.

Holsapple, J. G., and L. E. Foster. 1975. Reproduction of white perch in the lower Hudson River. *N.Y. Fish Game J.* 22:122–127.

Holt, D. E. 1965. Un nuevo pez para les costas de Venezuela: *Mola mola* L. (Pisces: Tetraodontiformes). *Lagena* No. 5:2–7.

Homans, R. E. S., and A. W. H. Needler. 1944. Food of the haddock. *Proc. Nova Scotia Inst. Sci.* 21:15–49.

Hood, P., K. W. Able, and C. B. Grimes. 1988. Biology of the conger eel (*Conger oceanicus*), in the Mid-Atlantic Bight. I. Distribution, age, growth, and reproduction. *Mar. Biol.* 98:587–596.

Horn, M. H. 1970. Systematics and biology of the stromateoid fishes of the genus *Peprilus*. *Bull. Mus. Comp. Zool.* 140:165–261.

Horn, P. L., and B. R. Massey. 1989. Biology and abundance of alfonsino and bluenose off the lower east coast North Island, New Zealand. N.Z. Fisheries Tech. Rept. No. 15:1–32.

Horst, T. J. 1977. Use of the Leslie matrix for assessing environmental impact with an example for a fish population. *Trans. Am. Fish. Soc.* 106:253–257.

Horton, D. B. 1965. A study of population behavior, homing orientation, and movement in the common mummichog, *Fundulusheteroclitus*. Ph.D. dissertation, University of Rhode Island, 148 pp.

Hoss, D. E, and J. H. S. Blaxter. 1982. Development and function of the swim bladder-inner ear-lateral line system in the Atlantic menhaden, *Brevoortia tyrannus* (Latrobe). *J. Fish Biol.* 20:131–142.

Hoss, D. E, W. F. Hettler, Jr., and L. C. Coston. 1974. Effects of thermal shock on larval estuarine fish-ecological implications with respect to entrainment in power plant cooling systems. In: J. H. S. Blaxter, ed. *The Early Life History of Fish*. Springer-Verlag, New York, pp. 357–371.

Hoss, D. E, D. M. Checkley, Jr., and L. R. Settle. 1989. Diurnal buoyancy changes in larval Atlantic menhaden *Brevoortia tyrannus*. *Rapp. Proc.-Verb. Réun. Cons. Int. Explor. Mer* 191:105–111.

Hostetter, E. B., and T. A. Munroe. 1993. Age, growth, and reproduction of tautog *Tautoga onitis* (Labridae: Perciformes) from coastal waters of Virginia. *Fish. Bull., U.S.* 91:45–64.

Houde, E. D. 1972. Development and early life history of the northern sennet, *Sphyraena borealis* DeKay (Pisces: Sphyraenidae) reared in the laboratory. *Fish. Bull., U.S.* 70:185–195.

Houde, E. D. 1976. Abundance and potential for fisheries development of some sardine-like fishes in the eastern Gulf of Mexico. *Proc. Gulf Carib. Fish. Inst.* 28:73–82.

Houde, E. D. 1977. Abundance and potential yield of the round herring, *Etrumeus teres,* and aspects of its early life history in the eastern Gulf of Mexico. *Fish. Bull., U.S.* 75:61–89.

Houde, E. D. 1978. Critical food concentrations for larvae of three species of subtropical marine fishes. *Bull. Mar. Sci.* 28:395–411.

Houde, E. D., and P. L. Fore. 1973. Guide to identity of eggs and larvae of some Gulf of Mexico clupeid fishes. Fla. Dept. Nat. Resour., Mar. Res. Lab., Leafl. Ser. IV, Part 1, No. 23, 14 pp.

Houde, E. D., and J. A. Lovdal. 1984. Seasonality of occurrence, foods and food preferences of ichthyoplankton in Biscayne Bay, Florida. *Est. Coast. Shelf Sci.* 18:403–419.

Houde, E. D., and C. E. Zastrow. 1991. Bay anchovy, *Anchoa mitchilli*. In: S .L. Funderburk, S. L. Mihursky, J. A. Jordan, and S. D. Riley, eds. *Habitat Requirements for Chesapeake Bay Living Resources,* 2nd Ed. Living Resources Subcommittee, Chesapeake Bay Program, Annapolis, pp. 8.1–8.14.

Hourston, A. S. 1958. Population studies on juvenile herring in Barkley Sound, British Columbia. *J. Fish. Res. Bd. Can.* 15:909–960.

Houseman, L. 0., and R. Kernehan. 1976. An indexed bibliography of the striped bass, *Morone saxatilis* 1670–1976. Ichthyological Assoc. Bull. No. 13. Delmarva Ecological Study, Middletown, Del., 118 pp.

Houston, K. A., and R. L. Haedrich. 1986. Food habits and intestinal parasites of deep demersal fishes from the upper continental slope east of Newfoundland, northwest Atlantic Ocean. *Mar. Biol.* 92:563–574.

Howe, A. B. 1971. Biological investigations of Atlantic tomcod *Microgadus tomcod* (Walbaum), in the Weweantic River Estuary, Massachusetts. M.S. thesis, University of Massachusetts, Amherst, 82 pp.

Howe, A. B., and P. G. Coates. 1975. Winter flounder movements, growth, and mortality off Massachusetts. *Trans. Am. Fish. Soc.* 104:13–29.

Howell, P., and D. Simpson. 1994. Abundance of marine resources in relation to dissolved oxygen in Long Island Sound. *Estuaries* 17:394–402.

Howell, P., D. R. Molnar, and R. B. Harris. 1999. Juvenile winter flounder distribution by habitat type. *Estuaries* 22:1090–1095.

Howell, W. H. 1980. Temperature effects on growth and yolk utilization in yellowtail flounder, *Limanda ferruginea,* yolk-sac larvae. *Fish. Bull., U.S.* 78:731–739.

Howell, W. H. 1983. Seasonal changes in the ovaries of adult yellowtail flounder, *Limanda ferruginea*. *Fish. Bull., U.S.* 81: 341–355.

Howell, W. H., and M. A. Caldwell. 1984. Influence of temperature on energy utilization and growth of embryonic and prolarval American plaice, *Hippoglossoides platessoides* (Fabricius). *J. Exp. Mar. Biol. Ecol.* 79:173–189.

Howell, W. H., and D. H. Kesler. 1977. Fecundity of the southern New England stock of yellowtail flounder, *Limanda ferruginea*. *Fish. Bull., U.S.* 75:877–880.

Howell, W. H., and W. H. Krueger. 1987. Family Sternoptychidae, marine hatchetfishes and related species. In: R. H. Gibbs, Jr., and W. H. Krueger, eds. *Biology of Midwater Fishes of the Bermuda Ocean Acre. Smithsonian Contrib. Zool.* 452:32–50.

Howell, W. H., and R. Langan. 1992. Discarding of commercial groundfish species in the Gulf of Maine shrimp fishery. *No. Am. J. Fish. Manage.* 12:568–580.

Howell, M. H., J. P. Mowrer, R. J. Hochberg, A. A. Jarzynski, and D. R. Weinrich. 1990. *Investigation of Anadromous Alosids in Chesapeake Bay.* Maryland Dept. Natural Resources, Annapolis.

Howell-Rivero, L. 1936. Six records of the pointed-tail ocean sunfish near Havana, Cuba. *Am. Nat.* 70:92–95.

Howey, R. G. 1985. Intensive culture of juvenile American shad. *Prog. Fish-Cult.* 47:203–212.

Hubbs, C. L. 1929. The Atlantic American species of the fish genus *Gasterosteus*. Mus. Zool. Univ. Mich. Occ. Pap. No. 200, 9 pp.

Hubbs, C. L. 1933. Observations on the flight of fishes, with a sta-

evolution of the flight of fishes. Mich. Acad. Sci., Arts, Lett. Pap. No. 17:575–611.

Hubbs, C. L. 1955. Review: Fishes of the Gulf of Maine. *Quart. Rev. Biol.* 30:182.

Hubbs, C. L., and L. Giovanoli. 1931. Records of the rare sunfish *Masturus lanceolatus* for Japan and Florida. *Copeia* 1931:135–137.

Hubbs, C. L., and R. L. Wisner. 1980. Revision of the sauries (Pisces, Scomberesocidae) with descriptions of two new genera and one new species. *Fish. Bull., U.S.* 77:521–566.

Huber, M. E. 1978. Adult spawning success and emigration of juvenile alewives *Alosa pseudoharengus* from the Parker River, Massachusetts. M.S. thesis, University of Massachusetts, Amherst, 67 pp.

Hudson, I. L., and J. D. Hardy Jr. 1975. Eggs and larvae of the Atlantic seahorse, *Hippocampus hudsonius.* Univ. Md., CEES Ref. Doc. No. 75-12CBL, 4 pp.

Huebner, J. D., and R. W. Langton. 1982. Rate of gastric evacuation for winter flounder, *Pseudopleuronectes americanus. Can. J. Fish. Aquat. Sci.* 39:356–360.

Huff, J. A. 1975. Life history of the Gulf of Mexico sturgeon, *Acipenser oxyrhynchus desotoi,* in the Suwannee River, Florida. Fla. Mar. Res. Publ. No. 16, 32 pp.

Hulbert, P. J. 1974. Factors affecting spawning site selection and hatching success in anadromous rainbow smelt, *Osmerus mordax.* M.S. thesis, University of Maine, Orono, 43 pp.

Hulley, P. A. 1981. Results of the research cruises of FRV "Walther Herwig" to South America. LVIII. Family Myctophidae (Osteichthyes, Myctophiformes). *Arch. Fisch. Wiss.* 31:1–300.

Hulley, P. A. 1984. Myctophidae. In: P. J. Whitehead et al., eds. *Fishes of the North-eastern Atlantic and the Mediterranean,* Vol. 1. UNESCO, Paris, pp. 429–483.

Humann, P. 1994. Reef fish identification, Florida, Caribbean, Bahamas. New World Publications, Inc., Jacksonville, 396 pp.

Hunninen, A. V., and R. M. Cable. 1940. Studies on the life history of *Anisoporus manteri* sp. nov. (Trematoda: Allocreadiidae). *Biol. Bull., Woods Hole* 79:373–374.

Hunt, J. J. 1980. Guidelines for age determination of silver hake, *Merluccius bilinearis,* using otoliths. *J. Northw. Atl. Fish. Sci.* 1:65–80.

Hunt, J. J., L. S. Parsons, J. E. Watson, and G. H. Winters. 1973. Report of the herring ageing workshop. Int. Comm. Northw. Atl. Fish. Res. Doc. No. 73/2, Ser. No. 2901, 27 pp.

Hunt, J. J., W. T. Stobo, and F. Almeida. 1999. Movement of Atlantic cod, *Gadus morhua,* tagged in the Gulf of Maine area. *Fish. Bull., U.S.* 97:842–860.

Hunter, J. G., S. T. Leach, D. E. McAllister, and M. B. Steigerwald. 1984. A distributional atlas of records of the marine fishes of Arctic Canada in the National Museums of Canada and Arctic Biological Station. *Syllogeus* 52:1–35.

Huntsman, A. G. 1918a. Histories of new food fishes. I. The Canadian plaice. *Bull. Biol. Bd. Can.* 1:1–22.

Huntsman, A. G. 1918b. The effects of the tide on the distribution of the fishes of the Canadian Atlantic coast. *Trans. Roy. Soc. Can., Ser. 3,* 12:61–67.

Huntsman, A. G. 1919. Growth of the young herring (so-called sardines) of the Bay of Fundy. Can. Fish. Exped. 1914–15, Dept. Naval Service, Canada, pp. 165–171.

Huntsman, A. G. 1922. The fishes of the Bay of Fundy. *Contrib. Can. Biol.* 1921, 3:49–72.

Huntsman, A. G. 1931. The maritime salmon of Canada. *Bull. Biol. Bd. Can.* 21, 99 pp.

Huntsman, A. G. 1948. Atlantic salmon investigations. Fish. Res. Bd Can. Ann. Rept. for 1947, pp. 37–38.

Huntsman A. G., and M. I. Sparks. 1924. Limiting factors for marine animals. III. Relative resistance to high temperatures. Contrib. Can. Biol. n.s. 2:97–114.

Hurley, P. C. F., and T. D. Iles. 1983. Age and growth estimation of Atlantic bluefin tuna, *Thunnus thynnus,* using otoliths. In: E. D. Prince and L. M. Pulos, eds. *Proc. Int. Workshop Age Determination of Oceanic Pelagic Fishes: Tunas, Billfishes, and Sharks.* NOAA Tech. Rept. NMFS No. 8:71–76.

Huse, G. 1998. Sex-specific life history strategies in capelin (*Mallotus villosus*)? *Can. J. Fish. Aquat. Sci.* 55:631–638.

Hutchins, J. A., T. D. Bishop, and C. R. McGregor-Shaw. 1999. Spawning behaviour of Atlantic cod, *Gadus morhua:* evidence of mate competition and mate choice in a broadcast spawner. *Can. J. Fish. Aquat. Sci.* 56:97–104.

Hutchinson, G. E. 1953. The concept of pattern in ecology. *Proc. Acad. Nat. Sci. Philadelphia* 105:1–12.

Hvidsten, N. A., and R. A. Lund. 1988. Predation on hatchery-reared and wild smolts of Atlantic salmon, *Salmo salar* L., in the estuary of River Orkla, Norway. *J. Fish Biol.* 3:121–126.

Hvidsten, N. A., and P. I. Møkkelgjerd. 1987. Predation on salmon smolts, *Salmo salar* L., in the estuary of the River Surna, Norway. *J. Fish Biol.* 30:273–280.

ICES (International Council for the Exploration of the Sea). 1995. Report of the study group on the biology and assessment of deep-sea fisheries resources. ICES C.M.1995/Assess:4:91

Ida, H., P. Sirimontaporn, and S. Monkolprasit. 1994. Comparative morphology of the fishes of the family Ammodytidae, with a description of two new genera and two new species. *Zool. Stud.* 33:251–277.

Idoine, J. S. 1998a. Goosefish. In: S. H. Clark, ed. *Status of Fishery Resources of the Northeastern United States for 1998.* NOAA Tech. Memo. NMFS-NE-115:88–89.

Idoine, J. S. 1998b. Atlantic Wolffish. In: S. H. Clark, ed. *Status of Fishery Resources off the Northeastern United States for 1998.* NOAA Tech. Memo. NMFS-NE-115:100–101.

IGFA (International Game Fish Association). 2001. *World Record Game Fishes, 2001Edition.* IGFA, Danio Beach, Fla. 344 pp.

Iglesias, S., and J. Paz. 1996. Occurrence of American plaice (*Hippoglossoides platessoides*) at non-habitual depths in the northwest Atlantic, 1990–93. NAFO Sci. Coun. Stud. No. 24:91–95.

Iles, T. D. 1964. The duration of maturation stages in herring. *J. Cons. Perm. Int. Explor. Mer* 29:166–183.

Iles, T. D. 1967. Growth studies on North Sea herring. I. The second year's growth (1-group) of East Anglian herring, 1939–63. *J. Cons. Perm. Int. Explor. Mer* 31:56–76.

Iles, T. D., and M. Sinclair. 1982. Atlantic herring: stock discreteness and abundance. *Science* 215:627–633.

Iles, T. D., M. J. Power, and R. L. Stephenson. 1985. Evaluation of the use of larval survey data to tune herring stock assessments in the Bay of Fundy/Gulf of Maine. Northw. Atl. Fish. Org., SCR Doc. No. 107, 16 pp.

Imamura, H. 2000. An alternative hypothesis on the phylogenetic position of the family Dactylopteridae (Pisces: Teleostei), with a proposed new classification. *Ichthyol. Res.* 47:203–222.

Imrie, D. M. G., and G. R. Daborn. 1981. Food of some immature fish of Minas Basin, Bay of Fundy. *Proc. Nova Scotia Inst. Sci.* 31:149–153.

Irwin, R. 1970. Geographical variation, systematics, and general biology of shore fishes of the genus *Menticirrhus,* family Sciaenidae. Ph.D. thesis, Tulane University, 295 pp.

Isaacs, J. D. and R. A. Schwartzlose. 1975. Active animals of the deep-sea floor. *Sci. Am.* 233 (Oct.):84–91.

Isaacson, P. A. 1963. Length-weight relationship and stomach contents of the swellfish (*Spheroides maculatus*) in the York River, Virginia. *Comm. Fish. Rev.* 25(9):5–7.

Isaacson, P. A. 1964. A second estuarine record for the rough scad (*Trachurus lathami*). *N.Y. Fish Game J.* 11:67.

Iwanowicz, H. R., R. D. Anderson, and B. A. Ketschke. 1973. A study of the marine resources of the Hingham Bay. Mass. Div. Mar. Fish. Monogr. Ser. No. 14, 40 pp.

Iwanowicz, H. R., R. D. Anderson, and B. A. Ketschke. 1974. A study of the marine resources of the Plymouth, Kingston and Duxbury Bay. Mass. Div. Mar. Fish. Monogr. Ser. No. 17, 37 pp.Jackson, C. F. 1953. Northward occurrence of southern fish (*Fundulus, Mugil, Pomatomus*) in coastal waters of New Hampshire. *Copeia* 1953:192.

Jackson, H. W., and R. E. Tiller. 1952. Preliminary observation on spawning potential in the striped bass *Roccus saxatilis* (Walbaum). Md. Dept. Res. Ed. No. 93, 16 pp.

Jacquaz, B., K. W. Able, and W. C. Leggett. 1977. Seasonal distribution, abundance and growth of larval capelin (*Mallotus villosus*) in the Gulf of St. Lawrence. *J. Fish. Res. Bd. Can.* 34:2018–2029.

James, A. G. 1988. Are clupeid microphagists herbivorous or omnivorous? A review of the diets of some commercially important clupeids. *So. Afr. J. Mar. Sci.* 7:161–177.

Jangaard, P. M. 1970. The role played by the Fisheries Research Board of Canada in the 'red' herring phosphorous pollution crisis in Placentia Bay, Newfoundland. Fish. Res. Bd. Can., A.R.O. Circ. No. 1, 20 pp.

Jangaard, P. M. 1974. The capelin (*Mallotus villosus*): biology, distribution, exploitation, utilization, and composition. *Bull. Fish. Res. Bd. Can.* 186, 70 pp.

Janssen, J. 1976. Feeding modes and prey size selection in the alewife *Alosa pseudoharengus. J. Fish. Res. Bd. Can.* 33:1972–1975.

Janssen, J. 1978a. Feeding-behavior repertoire of the alewife, *Alosa pseudoharengus,* and ciscoes, *Coregonus hoyi* and *C. artedii. J. Fish. Res. Bd. Can.* 35:249–253.

Janssen, J. 1978b. Will alewives *Alosa pseudoharengus* feed in the dark? *Environ. Biol. Fishes* 3:239–240.

Janssen, J. 1982. Comparison of searching behavior for zooplankton in an obligate planktivore, blueback herring *Alosa aestivalis* and a facultative planktivore, bluegill *Lepomis macrochirus. Can. J. Fish. Aquat. Sci.* 39:1640–1654.

Janssen, J., and S. B. Brandt. 1980. Feeding ecology and vertical migration of adult alewives *Alosa pseudoharengus* in Lake Michigan. *Can. J. Fish. Aquat. Sci.* 37:177–184.

Janssen, J., W. R. Jones, A. Whang, and P. E. Oshel. 1995. Use of the lateral line in particulate feeding in the dark by juvenile alewife *Alosa pseudoharengus. Can. J. Fish. Aquat. Sci.* 52:358–363.

Jean, Y. 1956. A study of spring- and fall-spawned herring *Clupea harengus* L. from the estuary and the Gulf of St. Lawrence. *Nature (Can.)* 94:7–27.

Jean, Y. 1965. Seasonal distribution of monkfish along the Canadian Atlantic mainland. *J. Fish. Res. Bd. Can.* 22:621–624.

Jearld, A. J., S. L. Sass, and M. F. Davis. 1993. Early growth, behavior, and otolith development of the winter flounder, *Pleuronectes americanus. Fish. Bull., U.S.* 91:65–75.

Jeffries, G. W. 1932. Fishes observed in the Strait of Belle Isle. *Contrib. Can. Biol., Fishes of Nova Scotia* 7:203–211.

Jeffries, H. P. 1972. Fatty acid ecology of a tidal marsh. *Limnol. Ocean.* 17:433–440.

Jeffries, H. P. 1975. Diets of juvenile Atlantic menhaden *Brevoortia tyrannus* in three estuarine habitats as determined from fatty acid composition of gut contents. *J. Fish. Res. Bd. Can.* 32: 587–592.

Jenkins, J. T. 1925. *The Fishes of the British Isles.* Frederick Warne & Co., Ltd., London, 408 pp.

Jensen, A. C. 1970. Validation of age determined from otoliths of Gulf of Maine cod. *Trans. Am. Fish. Soc.* 197:359–362.

Jensen, A. C. 1972. *The Cod.* Thomas Y. Crowell Co., New York, 182 pp.

Jensen, A. C., and J. P. Wise. 1962. Determining age of young haddock from their scales. *Fish. Bull., U.S.* 61:439–450.

Jensen, A. S. 1925. On the fishery of the Greenlanders. Medd. Komm. Danmarks Fisk. *Havunders. Ser.: Fisk.* 7:17–29.

Jensen, A. S. 1935. The Greenland halibut (*Reinhardtius hippoglossoides*): its development and migrations. *K. Dan. Vidensk. Seldk. Skr.* 6:1–32.

Jensen, A. S. 1942. Contributions to the ichthyofauna of Greenland. II. The Greenland species of the genera *Leptagonus* and *Aspidophoroides. Spolia Zool. Mus. Haun.* 2:29–36.

Jensen, A. S. 1944a. Contributions to the ichthyofauna of Greenland. V. The Greenland species of the genus *Lumpenus. Spolia Zool. Mus. Haun.* 4:31–35.

Jensen, A. S. 1944b. Contributions to the ichthyofauna of Greenland. VI. The Greenland species of the genera *Stichaeus* and *Eumesogrammus. Spolia Zool. Mus. Haun.* 4:36–39.

Jensen, A. S. 1944c. Contributions to the ichthyofauna of Greenland. VII. The Greenland species of the genera *Cyclopterus, Eumicotremus,* and *Cyclopterus. Spolia Zool. Mus. Haun.* 4:40–60.

Jensen, A. S. 1948. Contributions to the ichthyofauna of Greenland. VIII–XXIIII. *Spolia Zool. Mus. Haun.* 9:1–182.

Jensen, A. S. 1952. On the Greenland species of the genera *Artediellus, Cottunculus, Gymnocanthus (Teleostei, Scleroparii, Cottidae). Medd. Grønland.* 142:1–21.

Jensen, A. S., and H. Volsøe. 1949. A revision of the genus *Icelus* (Cottidae) with remarks on the structure of its urogenital papilla.

Jensen, C. F., L. J. Natanson, H. L.Pratt, N. E. Kohler, and S. E. Campana. in press. The reproductive biology of the porbeagle shark, *Lamna nasus,* in the western North Atlantic. *Fish. Bull., U.S.*

Jensen, T., K. Andersen, and S. Des Clers. 1994. Sealworm (*Pseudoterranova decipiens*) infections in demersal fish from two areas in Norway. *Can. J. Zool.* 72:598–608.

Jerome, W. C., Jr., A. P. Chesmore, C. O. Anderson, Jr., and F. Grice. 1965. A study of the marine resources of the Merrimack River Estuary. Mass. Div. Mar. Fish. Monogr. Ser. No. 1, 90 pp.

Jerome, W. C., Jr., A. P. Chesmore, and C. O. Anderson, Jr. 1966. A study of the marine resources of Quincy Bay. Mass. Div. Mar. Fish. Monogr. No. 2, 62 pp.

Jerome, W. C., Jr., A. P. Chesmore, and C. O. Anderson, Jr. 1967. A study of the marine resources of Beverley-Salem Harbor. Mass. Div. Mar. Fish. Monogr. Ser. No. 4, 74 pp.

Jerome, W. C., Jr., A. P. Chesmore, and C. O. Anderson, Jr. 1968. A study of the marine resources of the Parker River-Plum Island Sound Estuary. Mass. Div. Mar. Fish. Monogr. Ser. No. 6, 79 pp.

Jerome, W. C., Jr., A. P. Chesmore, and C. O. Anderson, Jr. 1969. A study of the marine resources of the Anisquam River-Gloucester Harbor Coastal System. Mass. Div. Mar. Fish. Monogr. Ser. No. 8, 62 pp.

Jesien, R. V., C. H. Hocutt, and S. K. Gaichas. 1992. Tagging studies and stock characterizations of summer flounder, *Paralichthys dentatus,* in Maryland coastal waters near Ocean City, MD. Progress Rept. to Md. Dept. Nat. Res., Tidewater Admin. Contract F25292-008, Horn Point Environ. Lab., Md.

Jespersen, A. 1975. Fine structure of spermiogenesis in eastern Pacific species of hagfish (Myxinidae). *Acta Zool.* 56:189–198.

Jessop, B. M. 1990a. Diel variation in density, length composition, and feeding activity of juvenile alewife, *Alosa psedoharengus* Wilson, and blueback herring, *A. aestivalis* Mitchill, at near-surface depths in a hydroelectric dam impoundment. *J. Fish Biol.* 37:813–822.

Jessop, B. M. 1990b. Passage and harvest of river herring at the Mactaquac Dam, Saint John River: an attempt at active fishery management. *No. Am. J. Fish. Manage.* 10:33–38.

Jessop, B. M. 1990c. Stock-recruitment relationships of alewives and blueback herring returning to the Mactaquac Dam, Saint John River, New Brunswick. *No. Am. J. Fish. Manage.* 10:19–32.

Jessop, B. M. 1993. Fecundity of anadromous alewives and blueback herring in New Brunswick and Nova Scotia. *Trans. Am. Fish. Soc.* 122:85–98.

Jessop, B. M. 1994. Relations between stock and environmental variables, and an index of abundance, for juvenile alewives and blueback herring. *No. Am. J. Fish. Manage.* 14:564–579.

Jessop, B. M., and W. E. Anderson. 1989. Effect of heterogeneity in the spatial and temporal pattern of juvenile alewife *Alosa pseudoharengus* and blueback herring *A. aestivalis* density on estimation of an index of abundance. *Can. J. Fish. Aquat. Sci.* 46: 1564–1574.

Jessop, B. M., A. H. Vromans, and W. E. Anderson. 1982. Life-history data on alewife and blueback herring, Mactaquac Dam, 1975–81. Can. Fish. Aquat. Sci. Data Rept. No. 367.

Jessop, B. M., W. E. Anderson, and A. H. Vromans. 1983. Life-history data on alewife and blueback herring of the Saint John River, New Brunswick, 1981. Can. Fish. Aquat. Sci. Data Rept. No. 426.

Jhaveri, S. N., P. A. Karakoltsidis, J. Montecalvo Jr., and S. M. Constantinides. 1984. Chemical composition and protein quality of some southern New England marine species. *J. Food Sci.* 49: 110–113

Jhaveri, S. N., P. A. Karakoltsidis, S. Y. K. Shenouda, and S. M. Constantinides. 1985. Ocean pout (*Macrozoarces americanus*): nutrient analysis and utilization. *J. Food Sci.* 50:719–722.

Jilek, R., B. Cassell, D. Peace, Y. Garza, L. Riley, and T. Siewart. 1979. Spawning population dynamics of smelt, *Osmerus mordax. J. Fish Biol.* 15:31–35.

Jimenez, D., J. E. Pelczarski, and H. R. Iwanowicz. 1982. Incidence of piscine erythrocytic necrosis (PEN) and *Glugea hertwigi* (Weisseberg) in rainbow smelt (*Osmerus mordax* Mitchill) in selected Massachusetts streams. *Estuaries* 5:145–149.

Johannessen, T., and E. Moksness. 1990. Gyteadferd hos gråsteinbit.–Norsk Fiskeoppdrett 15(10A):8–9 [not seen].

Johannessen, T., J. Gjosaeter and E. Moksness. 1993. Reproduction, spawning behaviour and captive breeding of the common wolffish *Anarhichas lupus* L. *Aquaculture* 115:41–51.

Johansen, F. 1925. Natural history of the cunner. *Contrib. Can. Biol.* 2:423–468.

Johansen, T., K. Nedreaas, and G. Nævdal. 1993. Electrophoretic discrimination of blue mouth, *Helicolenus dactylopterus* (De La Roche, 1809), from *Sebastes* spp. in the northeast Atlantic. *Sarsia* 78:25–29.

Johns, D. M., and W. H. Howell. 1980. Yolk utilization in summer flounder (*Paralichthys dentatus*) embryos and larvae reared at two temperatures. *Mar. Ecol. Prog. Ser.* 2:1–8.

Johnson, A. G., and C. H. Saloman. 1984. Age, growth, and mortality of gray triggerfish, *Balistes capriscus,* from the northeastern Gulf of Mexico. *Fish. Bull., U.S.* 82:485–492.

Johnson, C. R., and K. H. Mann. 1986. The importance of plant defense abilities to the structure of subtidal seaweed communities: the kelp *Laminaria longicruris* de la Pylaie survives grazing by snail *Lacuna vincta* (Montagu) at high population densities. *J. Exp. Mar. Biol. Ecol.* 97:231–267.

Johnson, D. L., P. L. Berrien, W. W. Morse, and J. J. Vitaliano. 1999a. Essential Fish Habitat Document: American plaice, *Hippoglossoides platessoides,* life history and habitat characteristics. NOAA Tech. Memo. NMFS-NE-123, 31 pp.

Johnson, D. L., W. W. Morse, P. L. Berrien, and J. J. Vitaliano. 1999b. Essential Fish Habitat Document: Yellowtail flounder, *Limanda ferruginea,* life history and habitat characteristics. NOAA Tech. Memo. NMFS-NE-140, 29 pp.

Johnson, G. D. 1984. Percoidei: development and relationships. In: H. G. Moser et al., eds. *Ontogeny and Systematics of Fishes.* Am. Soc. Ich. Herp. Spec. Publ. No. 1:464–498.

Johnson, G. D. 1986. Scombroid phylogeny: an alternative hypothesis. *Bull. Mar. Sci.* 39:1–41.

Johnson, G. D. 1992. Monophyly of the euteleostean clades—Neoteleostei, Eurypterygii, and Ctenosquamata. *Copeia* 1992:8–25.

Johnson, G. D., and C. Patterson. 1993. Percomorph phylogeny: a survey of acanthomorphs and a new proposal. *Bull. Mar. Sci.* 52:554–626.

Johnson, H. D. 1987. Potential fish egg predation by *Mnemiopsis leidyi,* determined by hydrography at the Chesapeake Bay mouth. M.S. thesis, Virginia Institute of Marine Science, College of William and Mary, Gloucester Point, 52 pp.

Johnson, J. H. 1983. Summer diet of juvenile fish in the St. Lawrence River. *N.Y. Fish Game J.* 30:91–99.

Johnson, J. H., and N. H. Ringler. 1980. Diets of juvenile coho salmon (*Oncorhynchus kisutch*) and steelhead trout (*Salmo gairdneri*) relative to prey availability. *Can. J. Zool.* 58:553–558.

Johnson, J. H., D. S. Dropkin, B. E. Warkentine, J. W. Rachlin, and W. D. Andrews. 1997. Food habits of Atlantic sturgeon off the central New Jersey coast. *Trans. Am. Fish. Soc.* 126:166–170.

Johnson, J. R., and J. G. Loesch. 1983. Morphology and development of hatchery-cultured American shad, *Alosa sapidissima* (Wilson). *Fish. Bull., U.S.* 81:323–338.

Johnson, J. R., and J. G. Loesch. 1986. A morphometrical comparison between cultured and wild juvenile American shad. *Prog. Fish-Cult.* 48:168–170.

Johnson, M. S. 1975. Biochemical systematics of the atherinid genus *Menidia. Copeia* 1975:662–691.

Johnston, C. E., and J. C. Cheverie. 1988. Observations on the diel and seasonal drift of eggs and larvae of anadromous rainbow smelt, *Osmerus mordax,* and blueback herring, *Alosa aestivalis,* in a coastal stream on Prince Edward Island. *Can. Field-Nat.* 102:508–514.

Johnston, C. E., and M. Morse. 1988. Summer ichthyoplankton communities of two estuarine systems of Prince Edward Island. *Can. J. Zool.* 66:737–745.

Johnston, I. A. 1983. pCa-tension characteristics of skinned muscle cells isolated from fish atria (*Myoxocephalus scorpius:* Teleostei). *J. Physiol.* 340:640.

Jones, C., and E. B. Brothers. 1987. Validation of the otolith increment aging technique for striped bass, *Morone saxatilis,* larvae reared under suboptimal feeding conditions. *Fish. Bull., U.S.* 85:171–178.

Jones, D. H., and A. W. G. John. 1978. A three-spined stickleback, *Gasterosteus aculeatus* L. from the North Atlantic. *J. Fish Biol.* 13: 231–236.

Jones, J. M. 1879. List of fishes of Nova Scotia. *Proc. Nova Scotia Inst. Sci.* 5:87–97.

Jones, P. W., J. S. Wilson, R. P. Morgan, II, H. R. Lunsford, Jr., and J. Lawson. 1977. Potomac River fisheries study; striped bass spawning stock assessment: interpretive report 1974–1976. Univ. Md. CEES Ref. Doc. No. 77-56-CBL. Chesapeake Biology Lab., Solomons, Md.

Jones, P. W., F. D. Martin, and J. D. Hardy, Jr. 1978. Development of Fishes of the Mid-Atlantic Bight: An Atlas of Egg, Larval and Juvenile Stages, Vol. 1. Acipenseridae through Ictaluridae. U.S. Fish Wildl. Serv., Biol. Serv. Prog. FWS/OBS-78/12, 366 pp.

Jones, R. S., E. J. Gutherz, W. R. Nelson, and G. C. Matlock. 1989. Burrow utilization by yellowedge grouper, *Epinephalus flavolimbatus,* in the northwestern Gulf of Mexico. *Environ. Biol. Fishes*

Jones, W. J., and J. M. Quattro. 1999. Genetic structure of summer flounder (*Paralichthys dentatus*) populations north and south of Cape Hatteras. *Mar. Biol.* 133:129–135.

Jonsson, B. 1997. A review of ecological and behavioural interactions between cultured and wild Alantic salmon. *ICES J. Mar. Sci.* 54:1031–1039.

Jonsson, G. 1982. Contribution to the biology of catfish (*Anarhichas lupus*) at Iceland. *Rit Fiskideild* 6:3–26.

Jordan, C. M. and E. T. Garside. 1972. Upper lethal temperatures of threespine stickleback, *Gasterosteus aculeatus* (L.), in relation to thermal and osmotic acclimation, ambient salinity, and size. *Can. J. Zool.* 50:1404–1411.

Jordan, D. S., and B. W. Evermann. 1896. The fishes of North and Middle America. Part I. *Bull. U.S. Nat. Mus.* 47:1–1240.

Jordan, D. S., and B. W. Evermann. 1898. The fishes of North and Middle America. Part II. *Bull. U.S. Nat. Mus.* 47:1241–2183.

Jørgensen, J. M., J. P. Lomholt, R. E. Weber, and H. Malte. 1998. *The Biology of Hagfishes.* Chapman and Hall, London, 578 pp.

Joseph, E. B. 1972. The status of the sciaenid stocks of the middle Atlantic Coast. *Chesapeake Sci.* 13:87–100.

Joseph, E. B., and J. Davis. 1965. A preliminary assessment of the river herring stocks of lower Chesapeake Bay: a progress report to the herring industry. Virginia Institute of Marine Science, Spec. Sci. Rept. No. 51, Gloucester Point, Va., 23 pp.

Joseph, E. B., and U. P. Sakesena. 1966. Determination of salinity tolerances in mummichog (*Fundulus heteroclitus*) larvae obtained from hormone-induced spawning. *Chesapeake Sci.* 7:193–197.

Joseph, E. B., W. H. Massmann, and J. J. Norcross. 1964. The pelagic eggs and early larval stages of the black drum from Chesapeake Bay. *Copeia* 1964:425–434.

Juanes, F., R. E. Marks, K. A. McKown, and D. O. Conover. 1993. Predation by age-0 bluefish on age-0 anadromous fishes in the Hudson River Estuary. *Trans. Am. Fish. Soc.* 122:348–356.

Juanes, F., J. A. Buckel, and D. O. Conover. 1994. Accelerating the onset of piscivory: intersection of predator and prey phenologies. *J. Fish Biol.* 45:41–54.

Judy, M. H. 1961. Validity of age determinations from scales of marked American shad. *Fish. Bull., U.S. Fish. Wildl. Serv.* 61:161–170.

Judy, M. H., and R. M. Lewis. 1983. Distribution of eggs and larvae of Atlantic menhaden, *Brevoortia tyrannus,* along the Atlantic Coast of the United States. NOAA Tech. Rept. NMFS-SSRF No. 774, 23 pp.

June, F. C. 1958. Variation in meristic characters of young Atlantic menhaden, *Brevoortia tyrannus. Rapp. Proc.-Verb. Réun. Cons. Int. Explor. Mer* 143:26–35.

June, F. C. 1965. Comparison of vertebral counts of Atlantic menhaden. U.S. Fish Wildl. Serv. SSRF–513, 12 pp.

June, F. C., and F.T. Carlson. 1971. Food of young Atlantic menhaden, *Brevoortia tyrannus,* in relation to metamorphosis. *Fish. Bull., U.S.* 68:493–512.

June, F. C., and W. R. Nicholson. 1964. Age and size composition of the menhaden catch along the Atlantic coast of the United States, 1958, with a brief review of the commercial fishery. U.S. Fish

June, F. C., and J. W. Reintjes. 1960. Age and size composition of the menhaden catch along the Atlantic coast of the United States, 1956, with a brief review of the commercial fishery. U.S. Fish Wildl. Serv. SSRF-336, 38 pp.

June, F. C. and J. W. Reintjes. 1976. The menhaden fishery. In: M. E. Stansby, ed. *Industrial Fishery Technology.* R. E. Krieger Publ. Co., New York, pp. 136–149.

June, F. C., and C. M. Roithmayr. 1960. Determining age of Atlantic menhaden from their scales. *Fish. Bull., U.S.* 60:323–342.

Jungerson, H. F. E. 1910. Ichthyotomical contributions. II. The structure of the Aulostomidae, Syngnathidae and Solenostomidae. *K. Danske Vidensk. Skrift. Natur., Ser. No. 8,* 7:269–374.

Jury, S. H., J. D. Field, S. L. Stone, D. M. Nelson, and M. E. Monaco. 1994. Distribution and abundance of fishes and invertebrates in North Atlantic estuaries. N0AA/NOS Strategic Assessments Div., ELMR Rept. No. 13, 221 pp.

Jutare, T. V. 1962. Studies on the biology of *Bothus ocellatus,* with a description of a related new species. M.S. thesis, University of Miami, 91 pp.

Jutila, E., and J. Toivonen. 1985. Food composition of salmon post-smolts (*Salmo salar* L.) in the northern part of the Gulf of Bothnia. ICES C.M. 1985/M: 21.

Kabata, Z. 1958. *Lernaeocera obtusa* n. sp. its biology and its effects on the haddock. *Mar. Res.* 3:1–26.

Kahnle, A., K. Hattala, and K. McKown. 1998. Hudson River. In: A. W. Kahnle, K. A. Hattala, K. A. McKown, C. A. Shivey, M. R. Collins, T. S. Quires, and T. Savoy. *Stock Status of Atlantic Sturgeon of Atlantic Coast Estuaries.* Atlantic States Mar. Fish. Comm. Rept., Washington, D.C, 141 pp.

Kanayama, T. 1991. Taxonomy and phylogeny of the family Agonidae (Pisces: Scorpaeniformes). Fac. Fish. Hokkaido Univ. Mem. No. 38, 199 pp.

Kane, J. 1984. The feeding habits of co-occuring cod and haddock larvae from Georges Bank. *Mar. Ecol. Prog. Ser.* 16:9–20.

Kao, M. H., G. L. Fletcher, N. C. Wang, and C. L. Hew. 1986. The relationship between molecular weight and antifreeze polypeptide activity in marine fish. *Can. J. Zool.* 64:578–582.

Karlsbakk, E. 1995. The occurrence of metacercariae of *Bucephaloides gracilescens* (Digenea: Gasterostomata) in an intermediate host, the four-bearded rockling, *Enchelyopus cimbrius* (Gadidae). *J. Fish Biol.* 46:18–27.

Karmovskaya, E. S. 1985. Mesopelagic eels of family Derichthyidae (Anguilliformes). *Voprosy Ikhtiol.* 25:883–898 [in Russian, translation in *J. Ichthy.* 25(6):119–134]

Karnella, C. 1973. The systematic status of *Merluccius* in the tropical western Atlantic Ocean including the Gulf of Mexico. *Fish. Bull., U.S.* 71:83–91.

Karnella, C. 1987. Family Myctophidae, lanternfishes. In: R. H. Gibbs, Jr. and W. H. Krueger, eds. *Biology of Midwater Fishes of the Bermuda Ocean Acre. Smithsonian Contrib. Zool.* No. 452:51–168.

Karnella, C., and R. H. Gibbs, Jr. 1977. The lanternfish *Lobianchia dofleini:* an example of the importance of life-history information in prediction of oceanic sound scattering. In: N. R. Anderson and B. J. Zahuranec, eds. *Oceanic Sound Scattering Predictions.* Plenum Press, New York, pp. 361–379.

Karrer, C., and Heemstra, P. C. 1986. Family No. 140: Grammicolepididae. In: M. M.Smith and P. C. Heemstra, eds. *Smith's Sea Fishes.* MacMillan South Africa, Johannesburg, pp. 440–441.

Kato, S., S. Springer, and M. H. Wagner. 1967. Field guide to eastern Pacific and Hawaiian sharks. U.S. Fish Wildl. Serv. Circ. No. 271, 47 pp.

Katsuragawa, M. 1997. Larval development of the Atlantic moonfish *Selene setapinnis* (Osteichthyes, Carangidae) from southeastern Brazil. *Bull. Mar. Sci.* 61:779–789.

Katz, S. J., C. B. Grimes, and K. W. Able. 1983. Delineation of tilefish, *Lopholatilus chamaeleonticeps,* stocks along the United States east coast and in the Gulf of Mexico. *Fish. Bull., U.S.* 81:41–50.

Kaufman, L. S., and K. F. Liem. 1982. Fishes of the suborder Labroidei (Pisces: Perciformes): phylogeny, ecology, and evolutionary significance. *Breviora* 472, 19 pp.

Kawahara, S. 1978. Age and growth of butterfish, *Poronotus triacanthus* (Peck), in ICNAF Subarea 5 and Statistical Area 6. Int. Comm. Northw. Atl. Fish., Select Pap. No. 3:73–78.

Keats, D. W., and D. H. Steele. 1990. The fourbeard rockling, *Enchelyopus cimbrius* (L.), in eastern Newfoundland. *J. Fish Biol.* 37:803–811.

Keats, D. W., and D. H. Steele. 1993. Food of 0-group ocean pout [*Macrozoarces americanus* (Schneider)], in eastern Newfoundland: the importance of harpacticoid copepods. *J. Fish Biol.* 42:145–148.

Keats, D. W., G. R. South, and D. H. Steele. 1985. Reproduction and egg guarding by Atlantic wolffish (*Anarhichas lupus:* Anarhichidae) and ocean pout (*Macrozoarces americanus:* Zoarcidae) in Newfoundland waters. *Can. J. Zool.* 63:2565–2568.

Keats, D. W., G. R. South, and D. H. Steele. 1986a. Where do juvenile Atlantic wolffish, *Anarhichas lupus,* live? *Can. Field-Nat.* 100:556–558.

Keats, D. W., D. H. Steele, and G. R. South. 1986b. Atlantic wolffish (*Anarhichas lupus* L.; Pisces: Anarhichidae) predation on green sea urchins (*Strongylocentrotus droebachiensis* (O. F. Mull.); Echinodermata: Echinoidea) in eastern Newfoundland. *Can. J. Zool.* 64:1920–1925.

Keats, D. W., D. H. Steele, and G. R. South. 1987a. The rôle of fleshy macroalgae in the ecology of juvenile cod (*Gadus morhua* L.) in inshore waters off eastern Newfoundland. *Can. J. Zool.* 65:49–53.

Keats, D. W., D. H. Steele, and G. R. South. 1987b. Ocean pout (*Macrozoarces americanus* (Bloch and Schneider) (Pisces: Zoarcidae)) predation on green sea urchins (*Strongylocentrotus droebachiensis* (O.F. Mull.) (Echinodermata, Echinoidea)) in eastern Newfoundland. *Can. J. Zool.* 65:1515–1521.

Keats, D. W., D. H. Steele, J. M. Green, and G. M. Martel. 1993. Diet and population size structure of the Arctic shanny, *Stichaeus punctatus* (Pisces: Stichaeidae), at sites in eastern Newfoundland and the eastern Arctic. *Environ. Biol. Fishes* 37:173–180.

Kedney, G. I., V. Boulé, and G. J. FitzGerald. 1987. The reproductive ecology of threespine sticklebacks breeding in fresh and brackish water. In: M. J. Dadswell et al., eds., *Common Strategies of Anadromous and Catadromous Fishes. Am. Fish. Soc. Symp.* 1:151–161.

Keefe, M., and K. W. Able. 1993. Patterns of metamorphosis in summer flounder, *Paralichthys dentatus. J. Fish Biol.* 42:713–728.

Keefe, M., and K. W. Able. 1994. Contributions of abiotic and biotic factors to settlement in summer flounder, *Paralichthys dentatus*. *Copeia* 1994:458–465.

Keefe, M. L., and H. E. Winn. 1991. Chemosensory attraction to home stream water and conspecifics by native brook trout, *Salvelinus fontinalis*, from two southern New England streams. *Can. J. Fish. Aquat. Sci.* 48:938–944.

Keirans, W. J., S. S. Herman, and R. G. Malsberger. 1986. Differentiation of *Prionotus carolinus* and *Prionotus evolans* eggs in Hereford Inlet estuary, southern New Jersey, using immunodiffusion. *Fish. Bull., U.S.* 84:63–68.

Keivany, Y., and J. S. Nelson. 2000. Taxonomic review of the genus *Pungitius*, ninespine sticklebacks (Gasterosteidae). *Cybium* 24:107–122.

Keller, A. A., P. H. Doering, S. P. Kelly, and B. K. Sullivan. 1990. Growth of juvenile Atlantic menhaden, *Brevoortia tyrannus* (Pisces: Clupeidae) in MERL mesocosms: effects of eutrophication. *Limnol. Oceanogr.* 35:109–122.

Kelley, J. R., Jr. 1969. Investigations on the propagation of striped bass, *Roccus saxatilis* (Walbaum). Ph.D. dissertation, Auburn University, 113 pp.

Kellogg, R. L. 1982. Temperature requirements for the survival and early development of the anadromous alewife. *Prog. Fish-Cult.* 44:63–73.

Kellogg, R. L., and J. J. Gift. 1983. Relationship between optimum temperatures for growth and preferred temperatures for the young of four fish species. *Trans. Am. Fish. Soc.* 112:424–430.

Kelly, C. J., P. L. Connolly, and J .J. Bracken. 1999. Age estimation, growth, maturity, and distribution of the bluemouth rockfish *Helicolenus d. dactylopterus* (Delaroche 1809) from the Rockall Trough. *ICES J. Mar. Sci.* 56:61–74.

Kelly, G. F., P. M. Earl, J. D. Kaylor, F. E. Lux, H. R. McAvoy, and E. D. McRae. 1972. Redfish. NOAA NMFS Ext. Fish. Facts Publ. No. 1, 18 pp.

Kelly, K. H., and D. K. Stevenson. 1985. Fecundity of Atlantic herring (*Clupea harengus*) from three spawning areas in the western Gulf of Maine, 1969 and 1982. *J. Northw. Atl. Fish. Sci.* 6:149–155.

Kenchington, T .J. 1980. The fishes of St. Georges Bay, Nova Scotia. Can. Fish. Aquatic Sci. Tech. Rept. No. 955, 154 pp.

Kenchington, T. J. 1986. Morphological comparisons of two northwestern Atlantic redfishes, *Sebastes fasciatus* and *S. mentella* and techniques for their identification. *Can. J. Fish. Aquat. Sci.* 43: 781–787.

Kendall, A. W., Jr. 1972. Description of black sea bass, *Centropristis striata* (Linnaeus), larvae and their occurrences north of Cape Lookout, North Carolina, in 1966. *Fish. Bull., U. S.* 70:1243–1260.

Kendall, A. W., Jr. 1977. Biological and fisheries data on black sea bass, *Centropristis striata* (Linnaeus). Sandy Hook Lab. Tech. Ser. Rept. No. 7, 29 pp.

Kendall, A. W., Jr., and N. A. Naplin. 1981. Diel-depth distribution of summer ichthyoplankton in the Middle Atlantic Bight. *Fish. Bull., U.S.* 79:705–726.

Kendall, A. W., Jr., and J. W. Reintjes. 1975. Geographic and hydrographic distribution of Atlantic menhaden eggs and larvae along the middle Atlantic coast from R/V *Dolphin* cruises, 1965–66. *Fish. Bull., U.S.* 73:317–335.

Kendall, A. W., Jr., and L. A. Walford. 1979. Sources and distribution of bluefish, *Pomatomus saltatrix*, larvae and juveniles off the east coast of the United States. *Fish. Bull., U.S.* 77:213–227.

Kendall, W. C. 1896. Description of a new stickleback, *Gasterosteus gladiunculus*, from the coast of Maine. *Proc. U.S. Nat. Mus.* 18:623–624.

Kendall, W. C. 1902. Notes on the silversides of the genus *Menidia* of the east coast of the United States, with descriptions of two new sub-species. U.S. Comm. Fish. Rept. 1901, pp. 241–267.

Kendall, W. C. 1908. Fauna of New England. 8. List of the Pisces. Bost. Soc. Nat. Hist. Occ. Pap. No. 7:1–152.

Kendall, W. C. 1909. The fishes of Labrador. *Proc. Portland Soc. Nat. Hist.* 2:207–243.

Kendall, W. C. 1910. Effects of the menhaden and mackerel fisheries upon the fish supply. *Bull. U.S. Bur. Fish.* 28:279–293.

Kendall, W. C. 1912. Notes on a new species of flatfish from off the coast of New England. *Bull. U.S. Bur. Fish.* 30:391–394.

Kendall, W. C. 1914. An annotated catalogue of the fishes of Maine. *Proc. Portland Soc. Nat. Hist.* 3:1–198.

Kendall, W. C. 1931. Remarks on additions to the marine fauna of the coast of Maine. *Bull. Bost. Soc. Nat. Hist.* 9(1):1–166.

Kendall, W. C. 1935. The fishes of New England. The salmon family. Part 2. The salmons. Boston Soc. Nat. Hist. Mem. 9(1):1–166.

Kennedy, V. S., and D. H. Steele. 1971. The winter flounder (*Pseudopleuronectes americanus*) in Long Pond, Conception Bay, Newfoundland. *J. Fish. Res. Bd. Can.* 28:1153–1165.

Kennelly, S. J. 1999. Areas, depths, and times of high discard rates of scup, *Stenotomus chrysops*, during demersal fish trawling off the northeastern United States. *Fish. Bull., U.S.* 97:185–192.

Kerebel, B., M. T. Le Cabellec, and L. M. Kerebel. 1979. Structure and ultrastructure of intra-vitam parasitic destruction of the external dental tissue in the fish, *Anarhichas lupus* L. *Archiv. Oral Biol.* 24:147–153.

Kernehan, R. J., M. R. Headrick, and R. E. Smith. 1981. Early life history of striped bass in the Chesapeake and Delaware Canal and vicinity. *Trans. Am. Fish. Soc.* 110:137–150.

Kerr, J. E. 1953. Studies of fish preservation at the Contra Costa Steam Plant of the Pacific Gas and Electric Company. *Calif. Dept. Fish Game Fish Bull.* 92, 66 pp.

Keysler, H. D. 1968. Investigations on the stocks of *Argentina silus* in the waters off Norway, Iceland, and Newfoundland. *Rapp. Proc.-Verb. Réun. Cons. Int. Explor. Mer* 158:58–64.

Khan, N. Y. 1971. Comparative morphology and ecology of the pelagic larvae of nine Cottidae (Pisces) on the Northwest Atlantic and St. Lawrence drainage. Ph.D. thesis, University of Ottawa, 234 pp.

Khan, R. A., and E. M. Lee. 1989. Influence of *Lernaeocera branchialis* (Crustacea: Copepoda) on growth rate of Atlantic cod, *Gadus morhua*. *J. Parasitol.* 75:449–454.

Khan, R. A., and M. W. Newman. 1982. Blood parasites from fish of the Gulf of Maine to Cape Hatteras, northwest Atlantic Ocean, with notes on the distribution of fish hematozoa. *Can. J. Zool.*

Khan, R. A., M. Barrett, and J. Murphy. 1980. Blood parasites of fish from the northwestern Atlantic Ocean. *Can. J. Zool.* 58: 770–781.

Khan, R. A., E. M. Lee, and W. S. Whitty. 1991. Blood protozoans of fish from Davis Strait in the northwestern Atlantic Ocean. *Can. J. Zool.* 69:410–413.

Kieffer, M. C., and B. Kynard. 1993. Annual movements of shortnose and Atlantic sturgeons in the Merrimack River, Massachusetts. *Trans. Am. Fish. Soc.* 122:1088–1103.

Kieffer, M. C., and B. Kynard. 1996. Spawning of the shortnose sturgeon in the Merrimack River, Massachusetts. *Trans. Am. Fish. Soc.* 125:179–186.

King, D. P. F. 1985. Morphological and meristic differences among spawning aggregations of north-east Atlantic herring, *Clupea harengus* L. *J. Fish Biol.* 26:591–607.

King, M. J., M. H. Kao, J. A. Brown, and G. L. Fletcher. 1989. Lethal freezing temperatures of fish: limitations to seapen culture in Atlantic Canada. *Proc. Ann. Aquacult. Assoc. Can.* 89–3:47–49.

Kiorboe, T., and P. Munk. 1986. Feeding and growth of larval herring, *Clupea harengus,* in relation to density of copepod nauplii. *Environ. Biol. Fishes* 17:133–139.

Kissil, G. W. 1974. Spawning of the anadromous alewife, *Alosa pseudoharengus,* in Bride Lake, Connecticut. *Trans. Am. Fish. Soc.* 103:312–317.

Kjelson, M. A., G. N. Hohnson, R. L. Garner, and J. P. Johnson. 1976. The horizontal-vertical distribution and sample variability of ichthyoplankton populations within the nearshore and offshore ecosystems of Onslow Bay. In: Atlantic Estuarine Fisheries Center Ann. Rept. to ERDA (Energy, Research and Development Administration), NMFS, Beaufort, N.C., pp. 287–341.

Kjesbu, O. S. 1989. The spawning activity of cod, *Gadus morhua* L. *J. Fish Biol.* 34:195–206.

Kjorsvik, E., T. van der Meeren, H. Kryvi, J. Arnfinnson, and P. G. Kvenseth. 1991. Early development of the digestive tract of cod larvae *Gadus morhua* L. during start-feeding and starvation. *J. Fish Biol.* 38:1–15.

Klauda, R. J., T. H. Peck, and G. K. Rice. 1981. Accumulation of polychlorinated biphenyls in Atlantic tomcod (*Microgadus tomcod*) collected from the Hudson River Estuary, New York. *Bull. Environ. Contam. Toxicol.* 27:829–835.

Klauda, R. J., S. A. Fischer, L. W. Hall, Jr., and J. A. Sullivan. 1991. Alewife and blueback herring: *Alosa pseudoharengus* and *Alosa aestivalis.* In: S. L. Funderburk, J. A. Mihursky, S. J. Jordan, and D. Riley, eds. *Habitat Requirements for Chesapeake Bay Living Resources,* 2nd Ed. Chesapeake Bay Program, Living Resources Subcommittee, Annapolis, pp. 10.1–10.29.

Klawe, W. L. 1957. Common mummichog and newt in a lake on Digby Neck, Nova Scotia. *Can. Field-Nat.* 71:154–155.

Klein-MacPhee, G. 1978. Synopsis of biological data for the winter flounder, *Pseudopleuronectes americanus* (Walbaum). NOAA Tech. Rept. Circ. No. 414, 43 pp.

Klein-MacPhee, G. 1979. Growth, activity, and metabolism studies of summer flounder *Paralichthys dentatus* (L.) under laboratory conditions. Ph.D. dissertation, University of Rhode Island, Kingston, 99 pp.

Klepaker, T. 1996. Lateral plate polymorphism in marine and estuarine populations of the threespine stickleback (*Gasterosteus aculeatus*) along the coast of Norway. *Copeia* 1996:832–838.

Klima, E. F. 1959. Aspects of the biology and fishery for Spanish mackerel, *Scomberomorus maculatus* (Mitchill), of southern Florida. Fla. State Board Cons., Tech. Ser. No. 27, 39 pp.

Klima, E. F. 1971. Distribution of some coastal pelagic fishes in the western Atlantic. *Comm. Fish. Rev.* 33(6):21–34.

Klyashtorin, L. B., and A. A. Yarzhombek. 1975. Some aspects of the physiology of the striped bass, *Morone saxatilis. Voprosy Ikhtiol.* 15:1101–1106 [in Russian, translation in *J. Ichthyol.* 15:985–989].

Kneib, R. T. 1982. The effects of predation by wadingbirds (*Arderdae*) and blue crabs (*Callinectes sapidus*) on the population size structure of the common mummichog *Fundulus heteroclitus. Est. Coast. Shelf Sci.* 14:159–165.

Kneib, R. T. 1987. Predation risk and use of intertidal habitats by young fishes and shrimp. *Ecology* 68:379–386.

Kneib, R. T. 1997. The role of tidal marshes in the ecology of estuarine nekton. *Oceanogr. Mar. Biol. Ann. Rev.* 35:163–220.

Kneib, R. T., and A. E. Stiven. 1978. Growth, reproduction, and feeding of *Fundulus heteroclitus* (L.) on a North Carolina salt marsh. *J. Exp. Mar. Biol. Ecol.* 31:121–140.

Knutsen, G. M., and S. Tilseth. 1985. Growth, development, and feeding success of Atlantic cod larvae *Gadus morhua* related to egg size. *Trans. Am. Fish. Soc.* 114:507–511.

Knutsen, H., E. Moksness, and N. B. Vogt. 1985. Distinguishing between one-day-old cod (*Gadus morhua*) and haddock (*Melanogrammus aeglefinus*) eggs by gas chromatography and SIMCA pattern recognition. *Can. J. Fish. Aquatic Sci.* 42:1823–1826.

Kocik, J. F. 1998a. River herring. In: S. H. Clark, ed. *Status of Fishery Resources off the Northeastern United States for 1998.* NOAA Tech. Memo. NMFS-NE-115:134–135.

Kocik, J. F. 1998b. American shad. In: S. H. Clark, ed. *Status of Fishery Resources off the Northeastern United States for 1998.* NOAA Tech. Memo. NMFS-NE-115:136–137.

Kocik, J. F., and M. L. Jones. 1997. Pacific salmonines in the Great Lakes Basin. In: W. W. Taylor and C. P. Ferreri, eds. *Great Lakes Fishery Policy and Management: A Binational Perspective,* pp. 91–124.

Kocik, J. F., W. W. Taylor, and W. C. Wagner. 1991. Abundance, size, and recruitment of pink salmon (*Oncorhynchus gorbuscha*) in selected Michigan tributaries of the upper Great Lakes, 1984–1988. *J. Great Lakes Res.* 17:203–213.

Koeller, P. A., P. C. F. Hurley, P. Perley, and J. D. Neilson. 1986. Juvenile fish surveys on the Scotian Shelf: implications for year-class size assessments. *J. Conseil* 43:59–76.

Koeller, P. A., L. Coates-Markle, and J. D. Nielson. 1989. Feeding ecology of juvenile (age-0) silver hake (*Merluccius bilinearis*) on the Scotian shelf. *Can. J. Fish. Aquat. Sci.* 46:1762–1768.

Kohler, A. C. 1964. Variations in the growth of Atlantic cod (*Gadus morhua* L.). *J. Fish. Res. Bd. Can.* 21:57–100.

Kohler, A. C. 1971. Tagging of white hake, *Urophycis tenuis* Mitchill, in the southern Gulf of St. Lawrence. *Int. Comm. Northw. Atl. Fish. Res. Bull.* 8:21–25.

Kohler, A. C., and J. R. Clark. 1958. Haddock scale-otolith comparisons. *J. Fish. Res. Bd. Can.* 15:1239–1246.

Kohler, A. C., and D. N. Fitzgerald. 1969. Comparisons of food of cod and haddock in the Gulf of St. Lawrence and on the Nova Scotia Banks. *J. Fish. Res. Bd. Can.* 26:1273–1287.

Kohler, N. 1988. Aspects of feeding ecology of the blue shark *Prionace glauca* in the western North Atlantic. Ph.D. dissertation, University of Rhode Island, Kingston, 163 pp.

Kohler, N. E., J. G. Casey, and P. A. Turner. 1996. Length-length and length-weight relationships for 13 shark species from the western North Atlantic. NOAA Tech. Memo. NMFS-NE-110, 22 pp.

Køie, M. 1993. Nematode parasites in teleosts from 0–1540 m depth off the Faroe Islands (the North Atlantic). *Ophelia* 38:217–243.

Konstantinou, H., and D. C. Shen. 1995. The social and reproductive behavior of the eyed flounder, *Bothus ocellatus,* with notes on the spawning of *Bothus lunatus* and *Bothus ellipticus. Eviron. Biol. Fishes* 44:311–324.

Konstantinov, K. G. 1961. Tagging of bottom fishes in the Barents Sea. *Voprosy Ikhtiol.* 1(19):275–280 [in Russian].

Koop, J. H., and R. N. Gibson. 1991. Distribution and movements of intertidal butterfish *Pholis gunnellus. J. Mar. Biol. Assoc. U.K.* 71:127–136.

Kornfield, I., and S. M. Bogdanowicz. 1987. Differentiation of mitochondrial DNA in Atlantic herring, *Clupea harengus. Fish. Bull., U.S.* 85:561–568.

Kornfield, I., P. S. Gagnon, and B. D. Sidell. 1981. Inheritance of allozymes in Atlantic herring *Clupea harengus harengus. Can. J. Genet. Cytol.* 32:715–720.

Kornfield, I., B. D. Sidell, and P. S. Gagnon. 1982. Stock definition in Atlantic herring *Clupea harengus harengus:* genetic evidence for discrete fall and spring spawning populations. *Can. J. Fish. Aquat. Sci.* 39:1610–1621.

Koski, R. T. 1978. Age, growth, and maturity of the hogchoker, *Trinectes maculatus,* in the Hudson River, New York. *Trans. Am. Fish. Soc.* 107:449–453.

Kosmath, I., R. A. Patzner, and H. Adam. 1981. The cloaca of *Myxine glutinosa* (Cyclostomata): a scanning electron microscopical and histochemical investigation. *Z. Mikrosk.-anat. Forsch., Leipzig* 95:936–942.

Koster, J. 1977. Creature feature. *Oceans* 10:56–59.

Kotlyar, A. N. 1980. Systematics and distribution of trachichthyid fishes (Trachichthyidae, Beryciformes) from the Indian Ocean. *Trudy Inst. Okean.* 110:177–224 [in Russian].

Kotlyar, A. N. 1984. Systematics and the distribution of fishes of the family Polymixiidae (Polymixioidei, Beryciformes). *Voprosy Ikhiol.* 24:691–708 [in Russian, translation in *J. Ichthyol.* 24(6):1–20].

Kotlyar, A. N. 1986. Systematics and distribution of species of the genus *Hoplostethus* Cuvier (Beryciformes, Trachichthyidae). *Trudy Inst. Okean.* 121:97–140 [in Russian].

Kotlyar, A. N. 1987. Systematics and distribution of fishes of the family Diretmidae (Beryciformes). *Voprosy Ikhtiol.* 27:883–897 [in Russian, translation in *J. Ichthyol.* 28(2):1–15].

Kotlyar, A. N. 1990. *Diretmichthys*—a new genus of diretmid fishes (Diretmidae, Beryciformes). *Voprosy Ikhtiol.* 30:144–151 [in Russian, translation in *J. Ichthyol.* 30(2):153–162].

Kotlyar, A. N. 1992. A new species of the genus *Polymixia* from the of the genus (Polymixiidae, Beryciformes). *Voprosy Ikhtiol.* 32:11–26 [in Russian, translation in *J. Ichthyol.* 33(3):30–49].

Kotlyar, A. N. 1996. *Beryciform Fishes of the World Ocean.* VNIRO, Moscow, 368 pp.

Kotlyar, A. N. 1998. Species composition and distribution of holocentrids in the oceans of the world (Holocentridae, Beryciformes). *Voprosy Ikhtiol.* 38(2):199–217 [in Russian, translation in *J. Ichthyol.* 38(2):170–189].

Krauthamer, J., and W. A. Richkus. 1987. Source document for characterization of the biology and fisheries for Maryland stocks of alewife and river herring. Prepared for Maryland Department of Natural Resources, Annapolis.

Krefft, G. 1961. A contribution to the reproductive biology of *Helicolenus dactylopterus* (De la Roche, 1809) with remarks on the evolution of the Sebastinae. *Rapp. Proc.-Verb. Réun. Cons. Int. Explor. Mer* 150:243–244.

Krochmal, S. B. 1949. Ecology of the smelt, *Osmerus mordax* (Mitchill) in Great Bay, New Hampshire. M.S. thesis, University of New Hampshire, Durham, 78 pp.

Kroger, R. L., and J. F. Guthrie. 1972. Incidence of a parasitic isopod, *Olencira paraegastator,* in juvenile Atlantic menhaden. *Copeia* 1972:370–374.

Kroger, R. L., R. L. Dryfoos, and G. R. Huntsman. 1971. Movement of juvenile Atlantic menhaden tagged in New England waters. *Chesapeake Sci.* 12:114–115.

Krueger, W. H. 1994. Lateral plate ontogeny and evolution of the low phenotype in the blackspotted stickleback, *Gasterosteus wheatlandi. Copeia* 1994:508–511.

Kruger, R. L., and R. W. Brocksen. 1978. Respiratory metabolism of striped bass, *Morone saxatilis* (Walbaum), in relation to temperature. *J. Exp. Mar. Biol. Ecol.* 31:55–66.

Krzynowek, J., and J. Murphy. 1987. Proximate composition, energy, fatty acid, sodium, and cholesterol content of finfish, shellfish, and their products. NOAA Tech. Rept. NMFS No. 55, 53 pp.

Krzynowek, J., J. Murphy, R. S. Maney, and L. J. Panunzio. 1989. Proximate composition and fatty acid and cholesterol content of 22 species of northwest Atlantic finfish. NOAA Tech. Rept. NMFS No. 74, 35 pp.

Kubota, T., and T. Uyeno. 1970. Food habits of lancetfish *Alepisaurus ferox* (Order Myctophiformes) in Suruga Bay, Japan. *Japan. J. Ichthyol.* 17:22–28.

Kuenstner, S. E. 1996. *Harvesting the Value-Added Potential of Atlantic Hagfish.* New England Fish. Develop. Assoc., Boston, 46 pp.

Kuhlmann, M. L. 1992. Behavioral avoidance of predation in an intertidal hermit crab. *J. Exper. Mar. Biol. Ecol.* 157:143–158.

Kuntz, A. 1914. The embryology and larval development of *Bairdiella chrysura* and *Anchovia mitchilli. Bull. U.S. Bur. Fish.* 33:1–19.

Kuntz, A., and L. Radcliffe. 1917. Notes on the embryology and larval development of twelve teleostean fishes. *Bull. U.S. Bur. Fish.* 35:87–134.

Kurlansky, M. 1997. *Cod: A Biography of the Fish that Changed the World.* Walker and Co., New York, 294 pp.

Kuronuma, K. 1940. A young of ocean sunfish, *Mola mola,* taken

lanceolatus as the second record from Japanese waters. *Bull. Biogeogr. Soc. Japan* 10:25–28.

Kvalsund, R. 1990. Reproduksjon hos steinbit. -Norsk Fsikeoppdrett 15:4–5 [not seen].

Kyle, H. M. 1913. Flat-fishes (Heterosomata). *Dana Oceanogr. Exp. 1908–1910,* 2(A.1):96–114.

Kynard, B., M. Horgan, M. Kieffer, and D. Seibel. 2000. Habitats used by shortnose sturgeon in two Massachusetts rivers, with notes on estuarine Atlantic sturgeon: a hierarchical approach. *Trans. Am. Fish. Soc.* 129:487–503.

Kwei, E. A. 1978. Food and spawning activity of *Caranx hippos* (L.) off the coast of Ghana. *J. Nat. Hist.* 12:195–215.

La Bar, G. W., J. D. McCleave, and S. M. Fried. 1978. Seaward migration of hatchery-reared Atlantic salmon (*Salmo salar*) smolts in the Penobscot River estuary, Maine: open-water movements. *J. Cons. Int. Mer* 38:257–269.

LaChance, S., P. Magnan, and G. J. FitzGerald. 1987. Temperature preferences of three sympatric sticklebacks (Gasterosteidae). *Can. J. Zool.* 65:1573–1576.

Lachner, E. A. 1955. Populations of the berycoid fish family Polymixiidae. *Proc. U.S. Nat. Mus.* 105:189–206.

Lachner, E. A. 1986. Echeneididae. In: P. J. P. Whitehead et al., eds. *Fishes of the North-eastern Atlantic and the Mediterranean,* Vol. III, UNESCO, Paris, pp. 1329–1333.

Laird, M., and W. L. Bullock. 1969. Marine fish haematozoa from New Brunswick and New England. *J. Fish. Res. Bd. Can.* 26:1075–1102.

Lambert, T. C. 1984. Larval cohort succession in herring *Clupea harengus* and capelin *Mallotus villosus. Can. J. Fish. Aquat. Sci.* 41:1552–1564.

Lambert, T. C. 1985. Gastric emptying time and assimilation efficiency in Atlantic mackerel (*Scomber scombrus*). *Can. J. Zool.* 63:817–820.

Lambert, T. C., and S. N. Messieh. 1989. Spawning dynamics of Gulf of St. Lawrence herring *Clupea harengus. Can. J. Fish. Aquat. Sci.* 46:2085–2094.

Lambert, T. C., and D. M. Ware. 1984. Reproductive strategies of demersal and pelagic spawning fish. *Can. J. Fish. Aquat. Sci.* 41:1565–1569.

Lambert, T. C., D. M. Ware, and J. K. McRuer. 1982. Spawning and early life history of herring and capelin in St. Georges Bay, Nova Scotia. Can. Fish. Aquat. Sci. Tech. Rept. No. 1128, 82 pp.

Lametschwandtner, A., T. Weiger, U. Lametschwandtner, V. Georgieva-Hanso, R.A. Patzner, and H. Adam. 1989. The vascularization of the skin of the Atlantic hagfish, *Myxine glutinosa* L. as revealed by scanning electron microscopy of vascular corrosion casts. *Scan. Micros.* 3:305–314.

Landry, T., A. D. Boghen, and G. M. Hare. 1992. Les parasites de l'alose d'été *Alosa aestivalis* et du gaspereau *Alosa pseudoharengus* de la rivière Miramichi, Nouveau-Brunswick. *Can. J. Zool.* 70:1622–1624.

Lange, A. M., and F. E. Lux. 1978. Review of the other flounder stocks (winter flounder, American plaice, witch flounder and windowpane flounder) off the northeast United States. NMFS NEFC Woods Hole Lab. Ref. Doc. No. 78-44, 52 pp.

Lange, A. M., and G. T. Waring. 1992. Fishery interactions between long-finned squid (*Loligo pealei*) and butterfish (*Peprilus triacanthus*) off the northeast USA. *J. Northw. Atl. Fish. Sci.* 12:49–62.

Langton, R. W. 1982. Diet overlap between Atlantic cod, *Gadus morhua,* silver hake, *Merluccius bilinearis,* and fifteen other northwest Atlantic finfish. *Fish. Bull., U.S.* 80:745–759.

Langton, R. W. 1983. Food habits of yellowtail flounder, *Limanda ferruginea* (Storer), from off the northeastern United States. *Fish. Bull., U.S.* 81:15–22.

Langton, R. W., and R. E. Bowman. 1980. Food of fifteen northwest Atlantic gadiform fishes. NOAA Tech. Rept. NMFS SSRF No. 740, 23 pp.

Langton, R. W., and R. E. Bowman. 1981. Food of eight northwest Atlantic pleuronectiform fishes. NOAA Tech. Rept. NMFS SSRF No. 749, 16 pp.

Langton, R. W., and L. Watling. 1990. The fish benthos connection: a definition of prey groups in the Gulf of Maine. In: M. Barnes and R. N. Gibson, eds. *Trophic Relationships in the Marine Environment.* Proc. European Mar. Biol. Symp., Aberdeen University Press, Aberdeen, pp. 424–438.

Lanier, T. C. 1984. Suitability of red hake, *Urophycis chuss* and silver hake, *Merluccius bilinearis,* for processing into surimi. *Mar. Fish. Rev.* 46(2):43–48.

Lapointe, D. F. 1957. Age and growth of the American shad from three Atlantic coast rivers. *Trans. Am. Fish. Soc.* 87:139–150.

Laprise, R., and J. J. Dodson. 1989a. Ontogenetic changes in the longitudinal distribution of two species of larval fish in a turbid well-mixed estuary. *J. Fish Biol.* 35:39–47.

Laprise, R., and J. J. Dodson. 1989b. Ontogeny and importance of tidal vertical migrations in the retention of larval smelt *Osmerus mordax* in a well-mixed estuary. *Mar. Ecol. Progr. Ser.* 55:101–111.

Laroche, J. L. 1982. Trophic patterns among larvae of five species sculpin (family: Cottidae) in a Maine estuary. *Fish. Bull., U.S.* 80:827–840.

Laroche, J. L., and J. Davis. 1973. Age, growth, and reproduction of the northern puffer, *Sphoeroides maculatus. Fish. Bull., U.S.* 71:955–963

Laroche, W. A. 1981. Development of larval smooth flounder, *Liopsetta putnami,* with a redescription of development of winter flounder, *Pseudopleuronectes americanus* (family Pleuronectidae). *Fish. Bull., U.S.* 78:897–909.

Lascara, J. 1981. Fish predator-prey interactions in areas of eelgrass (*Zostera marina*). M.S. thesis, College of William and Mary, Williamsburg, Va., 81 pp.

Lassiter, R.R. 1962. Life history aspects of the bluefish, *Pomatomus saltatrix* (Linnaeus), from the coast of North Carolina. M.S. thesis, North Carolina State College, Raleigh,103 pp.

Last, J. M. 1989. The food of herring, *Clupea harengus,* in the North Sea, 1983–1986. *J. Fish Biol.* 34:489–501.

Laszlo, P. S. 1972. Age-growth, food, and reproduction of the smooth flounder, *Liopsetta putnami* (Gill), in Great Bay, New Hampshire. Ph.D. dissertation, University of New Hampshire, Durham, 75 pp.

Latham, R. 1917. Migration notes of fishes, 1916, from Orient, Long Island. *Copeia* No.41:17–23.

Laudan, R., J. S. Stolen, and A. Cali. 1987. The immunomodulating effect of the microsporidian *Glugea stephani* on the humoral response and immunoglobulin levels in winter flounder, *Pseudopleuronectes americanus. J. Fish Biol.* 31(Suppl. A):155–160.

Laurence, G. C. 1973. Influence of temperature on energy utilization of embryonic and prolarval tautog, *Tautoga onitis. J. Fish. Res. Bd. Can.* 30:435–442.

Laurence, G. C. 1974. Growth and survival of haddock (*Melanogrammus aeglefinus*) larvae in relation to planktonic prey concentration. *J. Fish. Res. Bd. Can.* 31:1415–1419.

Laurence, G. C. 1978. Comparative growth, respiration and delayed feeding abilities of larval cod (*Gadus morhua*) and haddock (*Melanogrammus aeglefinus*) as influenced by temperature during laboratory studies. *Mar. Biol.* 50:1–7.

Laurence, G. C. 1979. Larval length-weight relationships for seven species of northwest Atlantic fishes reared in the laboratory. *Fish. Bull., U.S.* 76:890–895.

Laurence, G. C., and W. H. Howell. 1981. Embryology and influence of temperature and salinity on early development and survival of yellowtail flounder *Limanda ferruginea. Mar. Ecol. Prog. Ser.* 6:11–18.

Laurence, G. C., and C. A. Rogers. 1976. Effects of temperature and salinity on comparative embryo development and mortality of Atlantic cod (*Gadus morhua* L.) and haddock (*Melanogrammus aeglefinus* L.). *J. Conseil* 36:220–228.

Laurence, G. C., A. S. Smigielski, T. A. Halovek, and B. R. Burns. 1981. Implication of direct competition between larval cod (*Gadus morhua*) and haddock (*Melanogrammus aeglefinus*) in laboratory growth and survival studies at different food densities. *Rapp. Proc.-Verb. Réun. Cons. Int. Explor. Mer* 178:304–311.

Lavalli, K. L., and D. E. Barshaw. 1986. Burrows protect postlarval lobsters *Homarus americanus* from predation by the nonburrowing cunner *Tautogolabrus adsperus,* but not from the burrowing mud crab *Neopanope taxana. Mar. Ecol. Prog. Ser.* 32:13–16.

Lavenda, N. 1949. Sexual differences and normal protogynous hermaphroditism in the Atlantic sea bass, *Centropristes striatus. Copeia* 1949:185–194.

Lawler, E. F. 1976. The biology of the sandbar shark, *Carcharhinus plumbeus* (Nardo, 1827), in the lower Chesapeake Bay and adjacent waters. M.A. thesis, Virginia Inst. Mar. Sci., College of William and Mary, Williamsburg, Va., 49 pp.

Lawson, J. W., and G. B. Stenson. 1997. Diet of northwest Atlantic harp seals (*Phoca groenlandica*) in offshore areas. *Can. J. Zool.* 75:2095–2106.

Lawson, J. W., G. B. Stenson, and D. G. McKinnon. 1994. Diet of harp seals *Phoca groenlandica* in divisions 2J and 3KL during 1991–93. NAFO Sci. Counc. Stud. No. 21:143–154.

Lawson, J. W., J. T. Anderson, E. L. Dalley, and G. B. Stenson. 1998. Selective foraging by harp seals, *Phoca groenlandica,* in nearshore and offshore waters of Newfoundland, 1993 and 1994. *Mar. Ecol. Prog. Ser.* 163:1–10.

Lawton, R., R. D. Anderson, P. Brady, C. Sheehan, W. Sides, E. Kauloheras, M. Borgatti, and V. Malkoski. 1984. Fishes of the western inshore Cape Cod Bay: studies in the vicinity of the *servations on the Ecology and Biology of Western Cape Cod Bay, Massachusetts: Lecture Notes on Coastal and Estuarine Studies.* Springer-Verlag, New York, pp. 191–230.

Lawton, R., P. Brady, C. Sheehan, S. Correia, and M. Borgatti. 1990. Final report on spawning sea-run rainbow smelt (*Osmerus mordax*) in the Jones River and impact assessment of Pilgrim Station on the population, 1979–81. Pilgrim Nuclear Power Sta. Mar. Environ. Monitoring Prog. Rept. Ser. No. 4, 72 pp.

Layman, C. A., D. E. Smith, and J. D. Herod. 2000. Seasonally varying importance of abiotic and biotic factors in marsh-pond fish communities. *Mar. Ecol. Progr. Ser.* 207:155–169.

Layzer, J. B. 1974. Spawning sites and behavior of American shad, *Alosa sapidissima* (Wilson), in the Connecticut River between Holyoke and Turners Falls, Massachusetts, 1972. M.S. thesis, University of Massachusetts, Amherst, 46 pp.

Lazzari, M. A., and K. W. Able. 1990. Northern pipefish, *Syngnathus fuscus,* occurrences over the Mid-Atlantic Bight continental shelf: evidence of seasonal migration. *Environ. Biol. Fishes* 27:177–185.

Lazzari, M. A., J. C. O'Herron, II, and R. W. Hastings. 1986. Occurrence of juvenile Atlantic sturgeon, *Acipenser oxyrhynchus,* in the upper tidal Delaware River, USA. *Estuaries* 9:356–361.

Lazzari, M. A., K. W. Able, and M. P. Fahay. 1989. Life history and food habits of the grubby, *Myoxocephalus aeneus* (Cottidae) in a Cape Cod estuary. *Copeia* 1989:7–12.

Lazzari, M. A., S. Sherman, C. S. Brown, J. King, B. J. Joule, S. B. Chenoweth, and R. W. Langton. 1999. Seasonal and annual variations in abundance and species composition of two nearshore fish communities in Maine. *Estuaries* 22:636–647.

Lea, E. 1919. Age and growth of herring in Canadian waters. Can. Fish. Exped. 1914–1915, Dept. Naval Service [Canada], pp. 75–164.

Leach, G. C. 1925. Artificial propagation of shad. U.S. Comm. Fish. Doc. No. 981, App. 8:459–486.

Leak, J. C., and E. D. Houde. 1987. Cohort growth and survival of bay anchovy *Anchoa mitchilli* larvae in Biscayne Bay, Florida. *Mar. Ecol. Prog. Ser.* 37:109–122.

Leaning, J. 1998. Stranded fish die along bay beaches. *Cape Cod Times* (Nov. 4), A–3.

Lear, W. H. 1970. Fecundity of Greenland halibut (*Reinhardtius hippoglossoides*) in the Newfoundland-Laborador area. *J. Fish. Res. Bd. Can.* 27:1880–1882.

Lear, W. H. 1972. Food and feeding of Atlantic salmon in coastal areas and over oceanic depth. *ICNAF Res. Bull.* 9:27–39.

Lear, W. H., and J. M. Green. 1984. Migration of the "northern" Atlantic cod and the mechanisms involved. In: J. D. McCleave, ed. *Mechanisms of Migration in Fishes.* Plenum, New York, pp. 309–315.

Lear, W. H., and T. K. Pitt. 1975. Otolith age validation of Greenland halibut (*Reinhardtius hippoglossoides*). *J. Fish. Res. Bd. Can.* 32:289–292.

Leard, R. et al. 1995. The striped mullet fishery of the Gulf of Mexico, United States: a regional management plan. Gulf States Mar. Fish. Comm. No. 33.

Lebida, R. C. 1969. The seasonal abundance and distribution of

Massachusetts, 1966. M.S. thesis, University of Massachusetts, Amherst, 59 pp.

Lebour, M. V. 1920. The food of young fish. No. III (1919). *J. Mar. Biol. Assoc. U. K.* 12:261–324.

Lebour, M. V. 1925. Young anglers in captivity and some of their enemies: a study in a plunger jar. *J. Mar. Biol. Assoc. U.K.* 13: 721–734.

LeBrasseur, R. J. 1966. Stomach contents of salmon and steelhead trout in the northeastern Pacific Ocean. *J. Fish. Res. Bd. Can.* 23: 85–100.

Le Cabellec, M.-T., G. Daculsi, and P. Geistdoerfer. 1978. Rapports de la morphologie et de l'histologie dentaires d'*Anarchichas lupus* L. (poisson téléostéen perciforme) avec son mode d'alimentation: apport de la microradiographie et du marquage par la tétracycline. *Can. J. Zool.* 56:1103–1109.

LeDrew, B. R., and J. M. Green. 1975. Biology of the radiated shanny *Ulvaria subbifurcata* Storer in Newfoundland (Pisces: Stichaeidae). *J. Fish Biol.* 7:485–495.

Lee, D. W., E. D. Prince, and M. E. Crow. 1983. Interpretation of growth bands on vertebrae and otoliths of Atlantic bluefin tuna, *Thunnus thynnus.* In: E. D. Prince and L. M. Pulos, eds. *Proc. Int. Workshop Age Determin. Oceanic Pelagic Fishes: Tunas, Billfishes, and Sharks.* NOAA Tech. Rept. NMFS No. 8:61–69.

Legaré, J. E. H., and D. C. MacClellan. 1960. A qualitative and quantitative study of the plankton of the Quoddy region in 1957 and 1958 with special reference to the food of the herring. *J. Fish. Res. Bd. Can.* 17:409–448.

Leggett, W. C. 1969. Studies on the reproductive biology of the American shad *Alosa sapidissima* (Wilson). A comparison of populations from four rivers of the Atlantic seaboard. Ph.D. dissertation, McGill University, Montreal, 125 pp.

Leggett, W. C. 1973. The migrations of the shad. *Sci. Am.* 228: 92–98.

Leggett, W. C. 1976. The American shad *Alosa sapidissima,* with special reference to its migration and population dynamics in the Connecticut River. In: D. Merriman and L. M. Thorpe, eds. *The Connecticut River Ecological Study: The Impact of a Nuclear Power Plant.* Am. Fish. Soc. Monogr. No. 1:169–225.

Leggett, W. C. 1977a. Ocean migration rates of American shad *Alosa sapidissima. J. Fish. Res. Bd. Can.* 34:1422–1426.

Leggett, W. C. 1977b. The ecology of fish migrations. *Ann. Rev. Ecol. Syst.* 8:285–308.

Leggett, W. C., and J. E. Carscadden. 1978. Latitudinal variation in reproductive characteristics of American shad *Alosa sapidissima:* evidence for population specific life history strategies in fish. *J. Fish. Res. Bd. Can.* 35:1469–1478.

Leggett, W. C., and R. R. Whitney. 1972. Water temperature and the migrations of American shad. *Fish. Bull., U.S.* 70:659–670.

Lehmann, D., H. Hettwer, and H. Taraschewski. 2000. RAPD-PCR investigations of systematic relationships among four species of eels (Teleostei: Anguillidae), particularly *Anguilla anguilla* and *A. rostrata. Mar. Biol.* 137:195–204.

Lehodey, P., and R. Grandperrin. 1996. Age and growth of the alfonsino *Beryx splendens* over the seamounts off New Caledonia. *Mar. Biol.* 125:249–258.

Leidy, J. 1868. Remarks on shad brought to our markets during the late autumnal months, which were caught in salt water, and their food. *Proc. Acad. Nat. Sci. Philadelphia* 20:228.

Leim, A. H. 1924. The life history of the shad, *Alosa sapidissima* (Wilson), with special reference to factors limiting its abundance. *Contrib. Can. Biol. Fish.* 2:163–284.

Leim, A. H. 1937. Progress reports of Atlantic Biological Station, St. Andrews, N.B., Atlantic Fisheries Experimental Station, Halifax, Gaspe Fisheries Experimental Station, Grand River, Que. Fish. Res. Bd. Can. Progr. Rept. No. 21, 5 pp.

Leim, A. H. 1957. Fatness of herring in Canadian Atlantic waters. Fish. Res. Bd. Can., Atl. Prog. Rept. No. 111:177–184.

Leim, A. H., and L. R. Day. 1959. Records of uncommon and unusual fishes from eastern Canadian waters, 1950–1958. *J. Fish. Res. Bd. Can.* 16:503–514

Leim, A. H., and W. B. Scott. 1966. Fishes of the Atlantic coast of Canada. *Bull. Fish. Res. Bd. Can.* 155, 485 pp.

Leis, J. M. 1978. Systematics and zoogeography of the porcupinefishes (*Diodon,* Diodontidae, Tetraodontiformes), with comments on egg and larval development. *Fish. Bull., U.S.* 76:535–567.

Leis, J. M. 1984. Tetraodontoidei: development. In: H. G. Moser et al., eds. *Ontogeny and Systematics of Fishes.* Am. Soc. Ich. Herp. Spec. Publ. No. 1:447–450.

Leis, J. M., and D. S. Rennis. 1983. *The Larvae of Indo-Pacific Coral Reef Fishes.* University of Hawaii Press, Honolulu, 269 pp.

Leonard, S. B. 1971. Larvae of the fourspot flounder, *Hippoglossina oblonga* (Pisces: Bothidae), from the Chesapeake Bight, western North Atlantic. *Copeia* 1971:676–681.

Leslie, A. J., and D. J. Stewart. 1986. Systematics and distributional ecology of *Etropus* (Pisces, Bothidae) on the Atlantic coast of the United States with description of a new species. *Copeia* 1986:140–156.

Lesser, M. P., F. H. Martini, and J. B. Heiser. 1996. Ecology of the hagfish, *Myxine glutinosa* L., in the Gulf of Maine. I. Metabolic rates and energetics. *J. Exper. Mar. Biol. Ecol.* 208:215–225.

LeSueur, C. A. 1818. Description of several species of chondropterygious fishes, of North America, with their varieties. *Trans. Am. Philos. Soc.* 1:383–394.

Lett, P. F., and A. C. Kohler. 1976. Recruitment: a problem of multispecies interaction and environmental pertubations with special reference to Gulf of St. Lawrence Atlantic herring *Clupea harengus harengus. J. Fish. Res. Bd. Can.* 33:1353–1371.

Levesque, R. C., and R. J. Reed. 1972. Food availability and consumption by young Connecticut River shad *Alosa sapidissima. J. Fish. Res. Bd. Can.* 29:1495–1499.

Levin, P. S. 1991. Effects of microhabitat on recruitment variation in a Gulf of Maine reef fish. *Mar. Ecol. Prog. Ser.* 75:183–189.

Levin, P. S. 1993. Habitat structure, conspecific presence and spatial variation in the recruitment of a temperate reef fish. *Oecologia* 94:176–185.

Levin, P. S. 1994a. Fine-scale temporal variation in recruitment of a temperate demersal fish: the importance of settlement versus post-settlement loss. *Oecologia* 97:124–133.

Levin, P. S. 1994b. Small-scale recruitment variation in a temperate fish: the roles of macrophytes and food supply. *Environ. Biol. Fishes* 40:271–281.

Levin, P. S. 1996. Recruitment in a temperate demersal fish: does larval supply matter? *Limnol. Oceanogr.* 41:672–679.

Levine, J. F., J. H. Hawkins, M. J. Dykstra, E. J. Noga, D. W. Moye, and R. S. Cone. 1990. Epidemiology of ulcerative mycosis in Atlantic menhaden. *J. Aquat. Anim. Health* 2:162–171.

Levy, A., K. W. Able, C. B. Grimes, and P. Hood. 1988. Biology of the conger eel *Conger oceanicus* in the Mid-Atlantic Bight. II. Foods and feeding ecology. *Mar. Biol.* 98:597–600.

Lewis, E. J., S. M. McLaughlin, J. E. Bodammer, and T. K. Sawyer. 1992. Epitheliocystis in ten new host species of marine fish. *J. Fish Dis.* 15:267–271.

Lewis, R. M. 1965. The effect of minimum temperature on the survival of larval Atlantic menhaden, *Brevoortia tyrannus. Trans. Am. Fish. Soc.* 94:409–412.

Lewis, R. M., and R. R. Bonner, Jr. 1966. Fecundity of the striped bass, *Roccus saxatilis* (Walbaum). *Trans. Am. Fish. Soc.* 95:328–331.

Lewis, R. M., and W. F. Hettler, Jr. 1968. Effect of temperature and salinity on survival of young Atlantic menhaden, *Brevoortia tyrannus. Trans. Am. Fish. Soc.* 97:344–349.

Lewis, R. M., D. W. Ahrenholz, and S. P. Epperly. 1987. Fecundity of Atlantic menhaden, *Brevoortia tyrannus. Estuaries* 10:347–350.

Lewis, V. P., and D. S. Peters. 1984. Menhaden:a single step from vascular plant to fishery harvest. *J. Exp. Mar. Biol. Ecol.* 84:95–100.

Liao, Y.-Y., and M. C. Lucas. 2000. Growth, diet and metabolism of common wolf-fish in the North Sea, a fast-growing population. *J. Fish Biol.* 56:810–825.

Libby, D. A. 1981. Difference in sex ratios of the anadromous alewife, *Alosa pseudoharengus,* between the top and bottom of a fishway at Damariscotta Lake, Maine. *Fish. Bull., U.S.* 79:207–211.

Licciardello, J. J., and E. M. Ravesi. 1988. Frozen storage characteristics of cownose ray (*Rhinoptera bonasus*). *J. Food. Qual.* 11:71–76.

Lidster, W. W., G. R. Lilly, and E. G. Dawe. 1994. Otoliths of Arctic cod *Boreogadus saida,* small Atlantic cod *Gadus morhua,* and three other fish species from Newfoundland waters: description and relationship of body length to otolith length. *J. Northw. Atl. Fish. Sci.* 16:33–40.

Lidth de Jeude, T. W. van. 1892. On *Orthagoriscus nasus,* Ranzani. *Notes Leyden Mus.* 14:127–128.

Lie, U. 1961. On the growth and food of O-group coalfish, *Pollachius virens* (L.), in Norwegian waters. *Sarsia* 3:1–36.

Liem, K. F., and S. L. Sanderson. 1986. The pharyngeal jaw apparatus of labrid fishes: a functional morphological perspective. *J. Morph.* 187:143–158.

Lile, N. K. 1998. Alimentary tract helminths of four pleuronectid flatfish in relation to host phylogeny and ecology. *J. Fish Biol.* 53:945–953.

Lile, N. K., O. Halvorsen, and W. Hemmingsen. 1994. Zoogeographical classification of the macroparasite faunas of four flatfish species from the northeastern Atlantic. *Polar Biol.* 14:137–141.

Lindenberg, J. G. 1976. Seasonal depth distribution of landlocked alewives, *Alosa pseudoharengus* (Wilson), in a shallow, eutrophic lake. *Trans. Am. Fish. Soc.* 105:395–399.

Lindquist, D. G., M. V. Ogburn, W. B. Stanley, H. L. Troutman, and S. M. Pereira. 1985. Fish utilization patterns on temperate rubble-

Linkowski, T. B., R. L. Radtke, and P. H. Lenz. 1993. Otolith microstructure, age and growth of two species of *Ceratoscopelus* (Osteichthyes: Myctophidae) from the eastern North Atlantic. *J. Exp. Mar. Biol. Ecol.* 167:237–260.

Linton, E. 1901a. Fish parasites collected at Woods Hole in 1898. *Bull. U. S. Fish. Comm.* (1889) 19:267–304.

Linton, E. 1901b. Parasites of fishes of the Woods Hole region. *Bull. U. S. Fish. Comm.* (1899. 19:405–492.

Linton, E. 1905. Parasites of fishes of Beaufort, North Carolina. *Bull. U.S. Bur. Fish.* 24:321–428.

Linton, E. 1921. Food of young winter flounders. U.S. Fish Comm. Rept., App. IV, 14 pp.

Linton, E. 1928. Notes on trematode parasites of birds. *Proc. U.S. Nat. Mus.* 73:1–36.

Linton, E. 1940. Trematodes from fishes mainly from the Woods Hole region, Massachusetts. *Proc. U.S. Nat. Mus.* 88:1–172.

Linton, E. 1941. Cestode parasites of teleost fishes of the Woods Hole region, Massachusetts. *Proc. U.S. Nat. Mus.* 90:417–421.

Linton, J. R., and B. L. Soloff. 1964. The physiology of the brood pouch of the male seahorse *Hippocampus erectus. Bull. Mar. Sci. Gulf Carib.* 14:45–61.

Lippson, A. J., and R. L. Moran. 1974. Manual for the identification of early developmental stages of fishes of the Potomac River Estuary. Md. Dept. Nat. Res. Power Plant Siting Prog. PPSP MP-13, 282 pp.

Liquori, V. M., and G. D. Insler. 1985. Gill parasites of the white perch: phenologies in the lower Hudson River. *N.Y. Fish Game J.* 32:71–76

Little, M. J. 1995. A report on the historic spawning grounds of the striped bass, *Morone saxatilis. Maine Nat.* 3:107–113.

Litvak, M. K. 1999. The development of winter flounder (*Pleuronectes americanus*) for aquaculture in Atlantic Canada: current status and future prospects. *Aquaculture* 176:55–64.

Litvinenko, N. I. 1979. *Sebastes fasciatus kellyi* (Scorpaenidae) from coastal waters off Eastport, Maine, USA. *Voprosy Ikhtiol.* 19:387–401 [in Russian, translation in *J. Ichthyol.* 19:1–14].

Litvinenko, N. I. 1980. The structure, function and origin of the drumming muscles in the North Atlantic ocean perches of the genus *Sebastes* (Scorpaenidae). *Voprosy Ikhtiol.* 20:866–876 [in Russian, translation in *J. Ichthyol.* 20:89–98].

Livingstone, D. A. 1953. The fresh water fishes of Nova Scotia. *Proc. Nova Scotia Inst. Sci.* 23:1–90.

Livingstone, R., Jr., and P. Hamer. 1978. Age and length at first maturity of herring in the Georges Bank and Gulf of Maine stocks: an update. NEFSC Woods Hole Lab., Unpubl. Rept., 8 pp.

Livingstone, R. Jr., and L. Dery. 1976. An observation on the age and length at maturity of cod in the Georges and Browns bank stocks. Int. Comm. Northw, Atl., Fish. Res. Doc. No. 76/41/42 Ser. No. 3826, 2 pp.

Lo Bianco, S. 1909. *Notizie biologiche riguardanti specialmente il period di maturita sessuale degli animali del Golfo di Napoli, Vol. 19,* pp. 515–761.

Lobell, M. J. 1939. A biological survey of the salt waters of Long Island, 1938. Report on certain fishes: Winter flounder (*Pseudopleuronectes americanus*). New York Conserv. Dept., 28th Ann.

Locke, A., and S. C. Courtney. 1995. Zooplankton and ichthyoplankton of the Miramichi Estuary, Gulf of St. Lawrence. *J. Plankton Res.* 17:333–349.

Lockwood, S. 1867. The seahorse and its young. *Am. Nat.* 1:225–234.

Loenning, S., E. Kjoersvik, and I. B. Falk-Peterson. 1988. A comparative study of pelagic and demersal eggs from common marine fishes in northern Norway. *Sarsia* 73:49–60.

Loesch, J. G. 1968. A contribution to the life history of *Alosa aestivalis* (Mitchill). M.S. thesis, University of Connecticut, Storrs, 31 pp.

Loesch, J. G. 1969. A study of the blueback herring, *Alosa aestivalis* (Mitchill), in Connecticut waters. Ph.D. dissertation, University of Connecticut, Storrs, 78 pp.

Loesch, J. G. 1981. Weight relation between paired ovaries of blueback herring. *Prog. Fish-Cult.* 43:77–79.

Loesch, J. G. 1987. Overview of life history aspects of anadromous alewife and blueback herring in freshwater habitats. In: M. J. Dadswell et al., eds., *Common Strategies of Anadromous and Catadromous Fishes. Am. Fish. Soc. Symp.* 1:89–103.

Loesch, J. G., and W. H. Kriete, Jr. 1980. Anadromous fisheries research program, Virginia. Ann. Rept. 1980. NMFS Proj. No. AFC10–1, Virginia Inst. Mar. Sci., Gloucester Point, 96 pp.

Loesch, J. G., and W. A. Lund, Jr. 1977. A contribution to the life history of the blueback herring, *Alosa aestivalis. Trans. Am. Fish. Soc.* 106:583–589.

Loesch, J. G., W. H. Kriete, Jr., J. C. Travelstead, E. J. Foell, and M. A. Hennigar. 1979. Biology and management of Mid-Atlantic anadromous fishes under extended jurisdiction. II. Virginia. Virginia Inst. Mar. Sci. Spec. Rept., 204 pp.

Loesch, J. G., W. H. Kriete, Jr., and E. J. Foell. 1982. Effects of light intensity on the catchability of juvenile anadromous *Alosa* species. *Trans. Am. Fish. Soc.* 111:41–44.

Lokkeborg, S., and A. Ferno. 1999. Diel activity pattern and food search behaviour in cod, *Gadus morhua. Environ. Biol. Fishes* 54:345–353.

Lokkeborg, S., A. Bjordal, and A. Ferno. 1989. Response of cod (*Gadus morhua*) and haddock (*Melanogrammus aeglefinus*) to baited hooks in the natural environment. *Can. J. Fish. Aquat. Sci.* 46: 1478–1483.

Lom, J. 1970. Protozoa causing diseases in marine fishes. In: S. F. Snieszko, ed. *A Symposium on Diseases of Fishes and Shellfishes.* Am. Fish. Soc. Spec. Publ. No. 5:101–123.

Lom, J., and M. Laird. 1969. Parasitic protozoans from marine and euryhaline fish of Newfoundland and New Brunswick. I. Peritrichous ciliates. *Can. J. Zool.* 47:1367–1380.

Lom, J., E. R. Noble, and M. Laird. 1975. Myxosporidia from the deep-sea fish *Macrourus berglax,* off Newfoundland and Iceland. *Folia Parasitol.* 22:105–109.

Lomond, T. M., D. C. Schneider, and D. A. Methven. 1998. Transition from pelagic to benthic prey for age group 0-1 Atlantic cod, *Gadus morhua. Fish. Bull., U.S.* 96:908–911.

Longley, W. H., and S. F. Hildebrand. 1941. Systematic catalogue of the fishes of Tortugas, Florida with observations on color, habits and local distribution. Tortugas Lab. Pap. No. 34, 331 pp.

Lønning, S., E. Kjørsvik, T. Haug, and B. Gulliksen. 1982. The early development of the halibut, *Hippoglossus hippoglossus* (L.), compared with other marine teleosts. *Sarsia* 67:85–91.

Lønning, S., E. Kjørsvik, and I.-B. Folk-Peterson. 1988. A comparative study of pelagic and demersal eggs from common marine fishes in northern Norway. *Sarsia* 73:49–60.

Lotrich, V. A. 1975. Summer home range and movements of *Fundulus heteroclitus* (Pisces: Cyprinodontidae) in a tidal creek. *Ecology* 56:191–198.

Lough, R. G. 1976. Mortality and growth of Georges Bank-Nantucket Shoals herring larvae during three winters. ICES CM Doc. No. L:37, 25 pp.

Lough, R. G., and G. R. Bolz. 1989. The movement of cod and haddock larvae onto the shoals of Georges Bank. *J. Fish Biol.* 35(Suppl. A):71–79.

Lough, R. G., and D.C. Potter. 1993. Vertical distribution patterns and diel migrations of larval and juvenile haddock *Melanogrammus aeglefinus* and Atlantic cod *Gadus morhua* on Georges Bank: 1982 vs. 1985. *ICES Mar. Sci. Symp.* 198:356–378.

Lough, R. G., G. R. Bolz, M. R. Pennington, and M. D. Grosslein. 1980. Larval abundance and mortality of Atlantic herring (*Clupea harengus* L.) spawned in the Georges Bank and Nantucket Shoals areas, 1971–78 seasons, in relation to spawning stock size. *J. Northw. Atl. Fish. Sci.* 6:21–35.

Lough, R. G., P. C. Valentine, D. C. Potter, P. J. Auditore, G. R. Bolz, J. D. Nielson, and R. I. Perry. 1989. Ecology and distribution of juvenile cod, and haddock in relation to sediment type and bottom currents on eastern Georges Bank. *Mar. Ecol. Prog. Ser.* 56:1–12.

Lourie, S. A., A. C. J. Vincent, and H. J. Hall. 1999. *Seahorses: An Identification Guide to the World's Species and their Conservation.* Project Seahorse, London, 214 pp.

Low, W.-K., M. Miao, K. Vanya Ewart, D. S. C. Yang, G. L. Fletcher, and C. L. Hew. 1998. Skin-type antifreeze protein from the shorthorn sculpin, *Myoxocephalus scorpius. J. Biol. Chem.* 273:23098–23103.

Lowe, C. G., and K. J. Goldman. 2001. Thermal and bioenergetics of elasmobranchs: bridging the gap. *Environ. Biol. Fishes* 60:251–266.

Lowe, R. H. 1962. The fishes of British Guiana continental shelf, Atlantic coast of South America, with notes on their natural history. *J. Linnean Soc. London, Zool.* 44: 669–700.

Lowerre-Barbieri, S. K. 1994. Life history and fisheries ecology of weakfish, *Cynoscion regalis,* in the Chesapeake Bay region. Ph.D. dissertation, College of William and Mary, Gloucester Point, Va., 224 pp.

Lowerre-Barbieri, S. K., M. E. Chittenden Jr., and C. M. Jones. 1994. A comparison of a validated otolith method to age weakfish, *Cynoscion regalis,* with the traditional scale method. *Fish. Bull., U.S.* 92:555–568.

Lowry, L. F., K. J. Frost, and J. J. Burns. 1978. Food of ringed seals and bowhead whales near Point Barrow, Alaska. *Can. Field-Nat.* 92:67–70.

Lubieniecki, B. 1973. Note on the occurrence of larval *Anisakis* in adult herring and mackerel from Long Island to Chesapeake Bay. *Int. Comm. Northw. Atl. Fish. Res. Bull.* 10:79–82.

Luczkovich, J. J., and B. L. Olla. 1983. Feeding behavior, prey consumption, and growth of juvenile red hake. *Trans. Am. Fish. Soc.* 112:629–637.

Luczkovich, J. J., G. M. Watters, and B. L. Olla. 1991. Seasonal variation in usage of a common shelter resource by juvenile inquiline snailfish (*Liparis inquilinus*) and red hake (*Urophycis chuss*). *Copeia* 1991:1004–1109.

Lühmann, M. 1954. Die histogenetischen Grundlagen des periodischen Zahnweschsels der Katfische und Wasserkatzen (Fam. Anarrhichidae, Teleostei). Zeits. Zellforschung, 40:470–509.

Lukmanov, E. G., T. B. Nikiforova, and V. P. Ponomarenko. 1985. Distribution of menek, *Brosme brosme* (Gadidae), in waters of the Norwegian and Barents Seas. *Voprosy Ikhtiol.* 25:427–432 [in Russian, translation in *J. Ichthyol.* 25:133–139].

Lund, W. A., Jr. 1961. A racial investigation of the bluefish, *Pomatomus saltatrix* (Linnaeus) of the Atlantic coast of North America. *Boll. Inst. Oceanogr. Cumaná* 1:73–129.

Lund, W. A. Jr., and G. C. Maltezos. 1970. Movements and migrations of the bluefish, *Pomatomus saltatrix,* tagged in waters of New York and southern New England. *Trans. Am. Fish. Soc.* 99: 719–725.

Lund, W. A., Jr., and B. C. Marcy Jr. 1975. Early development of the grubby, *Myoxocephalus aenaeus* (Mitchill). *Biol. Bull., Woods Hole* 149:373–383.

Luo, J. 1993. Tidal transport of the bay anchovy, *Anchoa mitchilli,* in darkness. *J. Fish Biol.* 42:531–539.

Luo, J., and J. A. Musick. 1991. Reproductive biology of the bay anchovy in Chesapeake Bay. *Trans. Am. Fish. Soc.* 120:701–710.

Lutcavage, M. E., and S. Kraus. 1995. The feasibility of direct photographic assessment of giant bluefin tuna, *Thunnus thynnus,* in New England waters. *Fish. Bull., U.S.* 93:495–503.

Lutcavage, M. E., J. L. Goldstein, and S. Kraus. 1997. Distribution, relative abundance, and behavior of giant bluefin tuna in New England waters, 1995. ICCAT Collect. Vol. Sci. Pap. 46(2):332–347.

Lutcavage, M. E., R. W. Brill, G. B. Skomal, B. C. Chase, J. L. Goldstein, and J. Tutein. 2000. Tracking adult North Atlantic bluefin tuna (*Thunnus thynnus*) in the northwestern Atlantic using ultrasonic telemetry. *Mar. Biol.* 137:347–358.

Lux, F. E. 1963. Identification of New England yellowtail flounder groups. *Fish. Bull., U.S.* 63:1–10.

Lux, F. E. 1964. Landings, fishing effort, and apparent abundance in the yellowtail flounder fishery. *ICNAF Res. Bull.* 1:5–21.

Lux, F. E. 1969. Length-weight relationships of six New England flatfishes. *Trans. Am. Fish. Soc.* 98:617–621.

Lux, F. E. 1973. Age and growth of the winter flounder, *Pseudopleuronectes americanus,* on Georges Bank. *Fish. Bull., U.S.* 71: 505–512.

Lux, F. E., and G. F. Kelly. 1978. Fisheries resources of the Cape Cod and Massachusetts Bay region. Environmental Assessment Rept. 5-78. NMFS Woods Hole Lab. Ref. No. 78-15, 23 pp.

Lux, F. E. and J. V. Mahoney. 1972. Predation by bluefish on flatfishes. *Mar. Fish. Rev.* 34(7–8):30–35.

Lux, F. E., and F. E. Nichy. 1969. Growth of yellowtail flounder, *Limanda ferruginea* (Storer), on three New England fishing

Lux, F. E., and F. E. Nichy. 1971. Number and lengths, by season, of fishes caught with an otter trawl near Woods Hole, Massachusetts, September 1961 to December 1962. NOAA–NMFS Spec. Sci. Rept. No. 622, 15 pp.

Lux, F. E., A. E. Peterson, Jr., and R. F. Hutton. 1970. Geographical variation in fin ray number in winter flounder, *Pseudopleuronectes americanus* (Walbaum), off Massachusetts. *Trans. Am. Fish. Soc.* 99:483–488.

Lydersen, C., I. Gjertz, and J. M. Weslowski. 1989. Stomach contents of autumn feeding marine vertebrates from Hornsund Svalbord. *Polar Research* 25:107–114.

Lyles, C. H. 1966. Fishery statistics of the United States 1964. *Statistical Digest* 58, 541 pp.

Lyman, H. 1987. *Bluefishing.* Nick Lyons Books, New York, 154 pp.

Lyman, H., and F. Woolner. 1954. *The Complete Book of Striped Bass Fishing.* A.S. Barnes and Co., New York.

Lythgoe, J., and G. Lythgoe 1992. *Fishes of the Sea.* M.I.T. Press, Cambridge, 256 pp.

MacCoy, C. V. 1931. Fishes: museum notes. *Bull. Bost. Soc. Nat. Hist.* 58:16–18.

MacCrimmon, H. R. 1971. World distribution of rainbow trout (*Salmo gairdneri*). *J. Fish. Res. Bd. Can.* 28:663–704.

MacDonald, J. S. 1983. Laboratory observations of feeding behaviour of the ocean pout (*Macrozoarces americanus*) and winter flounder (*Pseudopleuronectes americanus*) with reference to niche overlap of natural populations. *Can. J. Zool.* 61:539–546.

MacDonald, J. S., and R. H. Green. 1986. Food resource utilization by five species of benthic feeding fish in Passamaquoddy Bay, New Brunswick. *Can. J. Fish. Aquat. Sci.* 43:1534–1546.

MacDonald, J. S., and K. G. Waiwood. 1987. Feeding chronology and daily ration calculations for winter flounder (*Pseudopleuronectes americanus*), American plaice (*Hippoglossoides platessoides*), and ocean pout (*Macrozoarces americanus*) in Passamaquoddy Bay, New Brunswick. *Can. Zool.* 65:499–503.

MacDonald, J. S., K. G. Waiwood, and R. H. Green. 1982. Rates of digestion of different prey in Atlantic cod (*Gadus morhua*), ocean pout (*Macrozoarces americanus*) and American plaice (*Hippoglossoides platessoides*). *Can. J. Fish. Aquat. Sci.* 39:651–659.

Macer, C. T. 1966. Sand eels (Ammodytidae) in the south-western North Sea; their biology and fishery. *Fish. Invest., Ser. II,* 24(6):55.

MacFarland, W. E. 1931. A study of the Bay of Fundy herring. Biol. Bd. Can. Ann. Rept. for 1930, pp. 23–24.

MacFarlane, R. D., J. J. McLaughlin, and G. L. Bullock. 1986. Quantitative and qualitative studies of gut flora in striped bass from estuarine and coastal marine environments. *J. Wildl. Dis.* 22: 344–348.

MacKay, K. T. 1967. An ecological study of mackerel, *Scomber scombrus* (Linnaeus), in the coastal waters of Canada. Can. Fish. Res. Board Tech. Rept. No. 31, 127 pp.

MacKay, K. T. 1972. Further records of the stromateoid fish *Centrolophus niger* from the northwestern Atlantic, with comments on body proportions and behavior. *Copeia* 1972:185–187.

MacKay, K. T., and E. T. Garside. 1969. Meristic analyses of Atlantic mackerel, *Scomber scombrus,* from the North American coastal

MacKay, K. T., and G. Thomas. 1969. First records of *Ariomma bondi, Caranx crysos,* and *Selar crumenophthalmus* (Pisces) in the Gulf of St. Lawrence. *J. Fish. Res. Bd. Can.* 26:2769–2771.

MacKenzie, K. 1987. Relationships between the herring, *Clupea harengus* L., and its parasites. *Adv. Mar. Biol.* 24:263–319.

MacKinnon, B. M. 1995. The poor potential of cunner, *Tautogolabrus adspersus,* to act as cleaner fish in removing sea lice (*Caligus elongatus*) from farmed salmon in eastern Canada. *Can. J. Fish. Aquat. Sci.* 52(Suppl. 1):175–177.

MacKinnon, J. C. 1972. Summer storage of energy and its use for winter metabolism and gonad maturation in American plaice (*Hippoglossoides platessoides*). *J. Fish. Res. Bd. Can.* 29:1749–1759.

MacLellan, P., G. E. Newsome, and P. A. Dill. 1981. Discrimination by external features between alewife *Alosa pseudoharengus* and blueback herring *A. aestivalis. Can. J. Fish. Aquat. Sci.* 38:544–546.

MacNeill, D. B., and S. B. Brandt. 1990. Ontogenetic shifts in gill-raker morphology and predicted prey capture efficiency of the alewife, *Alosa pseudoharengus. Copeia* 1990:164–171.

Macpherson, E. 1979. Estudio sobre el régimen alimentario de algunos peces en el Mediterráneo occidental. *Misc. Zool. Barcelona* 5:93–107.

Macpherson, E. 1985. Daily ration and feeding periodicity of some fishes off the coast of Namibia. *Mar. Ecol. Prog. Ser.* 26:253–260.

Maddock, D. M., and M. P. P. Burton. 1998. Gross and histological observations of ovarian development and related condition changes in American plaice. *J. Fish Biol.* 53:928–944.

Maggio, T. 2000. *Mattanza, Love and Death in the Sea of Sicily.* Perseus Publishing, Cambridge, Mass., 263 pp.

Magnin, E. 1964. Croissance en longeur de trois esturgeons d'Amérique du Nord: *Acipenser oxyrhynchus* Mitchill, *Acipenser fulvescens* Rafinesque et *Acipenser brevirostris* LeSueur. Verhand. Int. Verein. Theoret. Angew. Limnol. 15:968–974.

Magnin, E., and G. Beaulieu. 1963. Étude morphométrique comparée de l'*Acipenser oxyrhynchus* Mitchill du Saint-Laurent et de l'*Acipenser sturio* Linné de la Gironde. *Natur. Can.* 90:5–38.

Magnuson, J. J., and J. G. Heitz. 1971. Gill raker apparatus and food selectivity among mackerels, tunas and dolphins. *Fish. Bull., U.S.* 69:361–370.

Magnússon, J. V. 1996. Greater silver smelt, *Argentina silas,* in Icelandic waters. *J. Fish Biol.* 49(Suppl. A):259–275.

Mahon, R., and J. D. Neilson. 1987. Diet changes in Scotian Shelf haddock during the pelagic and demersal phases of the first year of life. *Mar. Ecol. Prog. Ser.* 37:123–130.

Mahoney, J. B., F. H. Midlige, and D. G. Deuel. 1973. A fin rot disease of marine and euryhaline fishes in the New York Bight. *Trans. Am. Fish. Soc.* 102:596–605.

Maina, J. N., C. M. Wood, C. Hogstrand, T. E. Hopkins, Y.-H. Luo, P. D. L. Gibbs, and P. J. Walsh. 1998. Structure and function of the axillary organ of the gulf toadfish, *Opsanus beta* (Goode and Bean). *Comp. Biochem. Physiol.* 119A:17–26.

Makushok, V. M. 1958. The morphology and classification of the northern blennioid fishes (Stichaeoidae, Blennioidei, Pisces). Trudy Zool. Inst. Akad. Nauk SSSR 25:3–129 [in Russian, Bur. Comm. Fish. Syst. Lab. Translation No. 1].

Malloy, K. D., and T. E. Targett. 1991. Feeding, growth and survival of juvenile summer flounder *Paralichthys dentatus:* experimental analysis of the effects of temperature and salinity. *Mar. Ecol. Prog. Ser.* 72:213–223.

Manderson, J. P., B. A. Phelan, A. J. Bejda, L. L. Stehlik, and A. W. Stoner. 1999. Predation by striped searobin (*Prionotus evolans,* Triglidae) on young-of-the-year winter flounder (*Pseudopleuronectes americanus,* Walbaum): examining prey size selection and prey choice using field observations and laboratory experiments. *J. Exper. Mar. Biol. Ecol.* 242:211–231.

Manickchand-Dass, S. M. Julien, and G. del Stern. 1984. Seasonality and breeding activity of a multispecies fish stock in Trinidad with a brief report on trawl net mesh trials. *Proc. Assoc. Isl. Mar. Lab Carib.* 18:9.

Mann, D. A., J. Bowers-Altman, and R. A. Rountree. 1997. Sounds produced by the striped cusk-eel *Ophidion marginatum* (Ophidiidae) during courtship and spawning. *Copeia* 1997:610–612.

Manning, A. J., and L. W. Crim. 1998. Maternal and interannual comparison of the ovulatory periodicity, egg production and egg quality of the batch-spawning yellowtail flounder. *J. Fish Biol.* 53:954–972.

Manooch, C. S., III. 1984. *Fisherman's Guide: Fishes of the Southeastern United States.* North Carolina Mus. Nat. Hist., Raleigh, 326 pp.

Manooch, C. S., III, E. Nakamura, and A. B. Hall. 1978. Annotated bibliography of four Atlantic scombrids: *Scomberomorus brasiliensis, S. cavalla, S. maculatus,* and *S. regalis.* NOAA Tech. Rept. NMFS Circ. No. 418, 166 pp.

Manooch, C. S., III, L. E. Abbas, and J. L. Ross. 1981. A biological and economic analysis of the North Carolina charter boat fishery. *Mar. Fish. Rev.* 43(8):1–11.

Manooch, C. S. III, D. L. Mason, and R. S. Nelson. 1984. Food and gastrointestinal parasites of dolphin *Coryphaena hippurus* collected along the southeastern and Gulf coasts of the United States. *Bull. Japan. Soc. Sci. Fish.* 50:1511–1525.

Manooch, C. S., III, D. L. Mason, and R. S. Nelson. 1985. Foods of little tunny *Euthynus alletteratus* collected along the southeastern and Gulf coasts of the United States. *Bull. Japan. Soc. Sci. Fish.* 51:1207–1218.

Mansfield, A. W. 1967. Seals of arctic and eastern Canada. *Bull. Fish. Res. Bd. Can.* 137, 35 pp.

Mansueti, A. J., and J. D. Hardy, Jr. 1967. *Development of Fishes of the Chesapeake Bay Region: An Atlas of Egg, Larval, and Juvenile Stages.* Part 1. Nat. Res. Inst., Univ. Md., 202 pp.

Mansueti, R. J. 1956. Alewife herring eggs and larvae reared successfully in lab. *Md. Tidewater News* 13:2–3.

Mansueti, R. J. 1958a. Eggs, larvae, and young of the striped bass, *Roccus saxatilis* (Walbaum). Chesapeake Biol. Lab. Contrib. No. 112, 35 pp.

Mansueti, R. J. 1958b. The hickory shad unmasked. *Nat. Mag.* 51: 351–354, 386.

Mansueti, R. J. 1960. The occurrence of the rough scad, *Trachurus lathami,* in Chesapeake Bay, Maryland. *Chesapeake Sci.* 1:117–118.

Mansueti, R. J. 1961a. Movements, reproduction, and mortality of the white perch, *Roccus americanus,* in the Patuxent Estuary, Maryland. *Chesapeake Sci.* 2:142–205.

Mansueti, R. J. 1961b. Age, growth, and movements of the striped bass, *Roccus saxatilis,* taken in size selective fishing gear in Maryland. *Chesapeake Sci.* 2:9–36.

Mansueti, R. J. 1962a. Eggs, larvae, and young of the hickory shad, *Alosa mediocris,* with comments on its ecology in the estuary. *Chesapeake Sci.* 3:173–205.

Mansueti, R. J. 1962b. Distribution of small, newly metamorphosed sea lampreys, *Petromyzon marinus,* and their parasitism on menhaden, *Brevoortia tyrannus,* in mid-Chesapeake Bay during winter months. *Chesapeake Sci.* 3:137–139.

Mansueti, R. J. 1964. Eggs, larvae, and young of the white perch *Roccus americanus,* with comments on its ecology in the estuary. *Chesapeake Sci.* 5:3–45.

Mansueti, R. J., and E. H. Hollis. 1963. Striped bass in Maryland tidewater. University of Maryland Nat. Res. Inst. Educ. Ser. No. 61, 28 pp.

Mansueti, R. J., and H. Kolb. 1953. A historical review of the shad fisheries of North America. Maryland Dept. Res. Educ., Ches. Lab. Publ. No. 97, 293 pp.

Mansueti, R. J., and A. J. Mansueti. 1955. White perch eggs and larvae studied in lab. *Md Tidewater News* 12:1–3.

Mansueti, R. J., and R. Pauly. 1956. Age and growth of the northern hogchoker, *Trinectes maculatus maculatus,* in the Patuxent River, Maryland. *Copeia* 1956:60–62.

Manzer, J. I. 1968. Food of Pacific salmon and steelhead trout in the northeast Pacific Ocean. *J. Fish. Res. Bd. Can.* 25:1085–89.

Marak, R. R. 1960. Food habits of larval cod, haddock, and coalfish in Gulf of Maine and Georges Bank area. *J. Cons. Int. Explor. Mer* 25:147–157.

Marak, R. R. 1967. Eggs and early larval stages of the offshore hake, *Merluccius albidus. Trans. Am. Fish. Soc.* 96:227–228.

Marak, R. R. 1973. Food and feeding of larval redfish in the Gulf of Maine. In: J. H. S. Blaxter, ed. *The Early Life History of Fish.* Springer-Verlag, Berlin, pp. 267–275.

Marak, R. R., and J. B. Colton, Jr. 1961. Distribution of fish eggs and larvae, temperature, and salinity in the Georges Bank-Gulf of Maine area, 1953. U.S. Fish. Wildl. Serv. SSR-Fish. No. 398, 61 pp.

Marak, R. R., and R. J. Livingstone, Jr. 1970. Spawning dates of Georges Bank haddock. *Int. Comm. Northw. Atl. Fish Res. Bull.* 7:56–58.

Marak, R. R., J. B. Colton, Jr., and D. B. Foster. 1962a. Distribution of fish eggs and larvae, temperature, and salinity in the Georges Bank-Gulf of Maine area, 1955. U.S. Fish. Wildl. Serv. SSR-Fish. No. 411, 66 pp.

Marak, R. R., J. B. Colton, Jr., D. B. Foster, and D. Miller. 1962b. Distribution of fish eggs and larvae, temperature, and salinity in the Georges Bank-Gulf of Maine area, 1956. U. S. Fish. Wildl. Serv. SSR-Fish. No. 412, 95 pp.

Marcellus, K. L. 1972. Fishes of Barnegat Bay, New Jersey, with particular reference to seasonal influences and the possible effects of thermal discharges. Ph.D. dissertation, Rutgers University, New Brunswick, 190 pp.

Marcogliese, D. J. 1992. First report of the threespine stickleback, *Gasterosteus aculeatus,* from Sable Island. *Can. Field-Nat.*

Marcogliese, D. J. 1995. Comparison of parasites of mummichogs and sticklebacks from brackish and freshwater ponds on Sable Island, Nova Scotia. *Amer. Midl. Nat.* 133:333–343.

Marcogliese, D. J., and G. McClelland. 1992. *Corynosoma wegeneri* (Acanthocephala: Polymorphida) and *Pseudoterranova decipiens* (Nematoda: Ascaridoidea) larvae in Scotian Shelf groundfish. *Can. J. Fish. Aquat. Sci.* 49:2062–2069.

Marcy, B. C., Jr. 1969. Age determination from scales of *Alosa pseudoharengus* (Wilson) and *Alosa aestivalis* (Mitchill) in Connecticut waters. *Trans. Am. Fish. Soc.* 98:622–630.

Marcy, B. C., Jr. 1971. Survival of young fish in the discharge canal of a nuclear power plant. *J. Fish. Res. Bd. Can.* 28:1057–1060.

Marcy, B. C., Jr. 1972. Spawning of the American shad, *Alosa sapidissima,* in the lower Connecticut River. *Chesapeake Sci.* 13:116–119.

Marcy, B. C., Jr. 1973. Vulnerability and survival of young Connecticut River fish entrained at a nuclear power plant. *J. Fish. Res. Bd. Can.* 30:1195–1203.

Marcy, B. C., Jr., and F. P. Richards. 1974. Age and growth of the white perch, *Morone americana,* in the lower Connecticut River. *Trans. Am. Fish. Soc.* 103:117–120.

Marcy, B. C., Jr. 1976a. Fishes of the lower Connecticut River and the effects of the Connecticut Yankee Plant. In: D. Merriman and L. M. Thorpe, eds. *The Connecticut River Ecological Study: The Impact of a Nuclear Power Plant.* Am. Fish. Soc. Monogr. No. 1: 61–114.

Marcy, B. C., Jr. 1976b. Planktonic fish eggs and larvae of the lower Connecticut River and effects of the Connecticut Yankee plant including entrainment. In: D. Merriman and L. M. Thorpe, eds. *The Connecticut River Ecological Study: The Impact of a Nuclear Power Plant.* Am. Fish. Soc. Monogr. No. 1:115–139.

Marcy, B. C., Jr. 1976c. Early life history studies of American shad in the lower Connecticut River and the effects of the Connecticut Yankee plant. In: D. Merriman and L. M. Thorpe eds. *The Connecticut River Ecological Study: The Impact of a Nuclear Power Plant.* Am. Fish. Soc. Monogr. No. 1:141–168.

Margolis, L., and J. R. Arthur. 1979. Synopsis of the parasites of fishes of Canada. *Bull. Fish. Res. Bd. Can.* 199, 269 pp.

Margolis, L., and Z. Kabata, eds. 1984. Guide to the parasites of fishes of Canada. I. Monogenea and Turbellaria. Can. Fish. Aquat. Sci. Spec. Publ. No. 74, 209 pp.

Margolis, L., and Z. Kabata, eds. 1988. Guide to the parasites of fishes of Canada. II. Crustacea. Can. Fish. Aquat. Sci. Spec. Publ. No. 101, 184 pp.

Markle, D. F. 1975. Young witch flounder, *Glyptocephalus cynoglossus,* on the slope off Virginia. *J. Fish. Res. Bd. Can.* 32:1447–1450.

Markle, D. F. 1982. Identification of larval and juvenile Canadian Atlantic gadoids with comments on the systematics of gadid subfamilies. *Can. J. Zool.* 60:3420–3438.

Markle, D. F., and L.-A. Frost. 1985. Comparative morphology, seasonality, and a key to planktonic fish eggs from the Nova Scotian shelf. *Can. J. Zool.* 63:246–257.

Markle, D. F., and G. C. Grant. 1970. The summer food habits of young-of-the-year striped bass in three Virginia rivers. *Chesapeake*

Markle, D. F., and J. A. Musick. 1974. Benthic-slope fishes found at 900 m depth along a transect in the western North Atlantic Ocean. *Mar. Biol.* 26:225–233.

Markle, D. F., and C. A. Wenner. 1979. Evidence of demersal spawning in the mesopelagic zoarcid fish *Melanostigma atlanticum* with comments on demersal spawning in the alepocephalid fish *Xenodermichthys copei. Copeia* 1979:363–366.

Markle, D. F., W. B. Scott, and A. C. Kohler. 1980. New and rare records of Canadian fishes and the influence of hydrography on resident and nonresident Scotian Shelf ichthyofauna. *Can. J. Fish. Aquat. Sci.* 37:49–65.

Markle, D. F., D. A. Methven, and L. J. Coates-Markle. 1982. Aspects of spatial and temporal concurrence in the life history stages of the sibling hakes, *Urophycis chuss* (Walbaum 1792) and *Urophycis tenuis* (Mitchill 1815) (Pisces: Gadidae). *Can. J. Zool.* 60:2057–2078.

Markle, G. E. 1974. Distribution of larval swordfish in the northwest Atlantic Ocean. In: R. S. Shomura and F. Williams, eds. *Proc. Int. Billfish Symp. Kailua–Kona, Hawaii 9–12 Aug. 1972.* NOAA Tech. Rept. NMFS SSRF No. 675(2):252–260.

Marks, R. E., and D. O. Conover. 1993. Ontogenetic shift in the diet of young-of-year bluefish, *Pomatomus saltatrix,* during the oceanic phase of the early life history. *Fish. Bull., U.S.* 91:97–106.

Marsh, R., B. Petrie, C. R. Weidman, R. R. Dickson, J. W. Loder, C. G. Hannah, K. Frank, and K. Drinkwater. 1999. The 1882 tilefish kill: a cold event in shelf waters off the north-eastern United States? *Fish. Oceanogr.* 8:39–49.

Marshall, M. D. 1976. Anadromous fisheries research program; Tar River, Pamlico River and northern Pamlico Sound. Completion Rept. Proj. AFCS-10, 15 May 1974–30 June 1976. 90 pp. North Carolina Div. Mar. Fish., Morehead City,.

Marshall, N. 1946. Observations on the comparative ecology and life history of two sea robins, *Prionotus carolinus* and *Prionotus evolans strigatus. Copeia* 1946:118–144.

Marshall, N. B. 1965. Systematic and biological studies of the macrourid fishes (Anacanthini-Teleostei). *Deep-Sea Res.* 12:299–322.

Marshall, N. B., and T. Iwamoto. 1973. Family Macrouridae. Fishes of the western North Atlantic. Sears Found. Mar. Res. Mem. No. 1(6):496–665.

Marshall, S. M., A. G. Nicholls, and A.P. Orr. 1937. On the growth and feeding of the larval and post-larval stages of the Clyde herring. *J. Mar. Biol. Assoc. U.K.* 22:245–268.

Marteinsdottir, G., and K. W. Able. 1988. Geographic variation in egg size among populations of the mummichog, *Fundulus heteroclitus* (Pisces: Fundulidae). *Copeia* 1988:471–478.

Marteinsdottir, G., and K. W. Able. 1992. Influence of egg size on embryos and larvae of *Fundulus heteroclitus* (L.). *J. Fish Biol.* 41: 883–896.

Martel, G., and J. M. Green. 1987. Differential spawning success among territorial cunners, *Tautogolabrus adspersus,* Labridae. *Copeia* 1987:643–648.

Martell, D. J., and G. McClelland. 1994. Diets of sympatric flatfishes, *Hippoglossoides platessoides, Pleuronectes ferrugineus, Pleuronectes americanus,* from Sable Island Bank, Canada. *J. Fish Biol.* 44:821–848.

Martin, F. D., and G. E. Drewry. 1978. Development of Fishes of the Mid-Atlantic Bight: an Atlas of Egg, Larval, and Juvenile Stages, Vol. 6. Stromateidae through Ogcocephalidae. U.S. Fish Wildl. Serv., Biol. Ser. Prog. FWS/OBS-78/12, 416 pp.

Martin, F. D., D. A. Wright, J. C. Means, and E. M. Setzler-Hamilton. 1985. Importance of food supply to nutritional state of larval striped bass in the Potomac River Estuary. *Trans. Am. Fish. Soc.* 114:137–145.

Martini, F. H. 1998a. The ecology of hagfishes. In: J. M. Jørgensen et al., eds. *The Biology of Hagfishes.* Chapman and Hall, London, pp. 57–77.

Martini, F. H. 1998b. Secrets of the slime hag. *Sci. Am.* 279(4):70–75.

Martini, F. H. 2000. The evidence for and potential significance of cutaneous respiration in hagfishes. *FASEB J.* 14(4):A436.

Martini, F. H., and J. B. Heiser. 1989. Field observations on the Atlantic hagfish, *Myxine glutinosa,* in the Gulf of Maine. *Am. Zool.* 29(4):38a (abstract).

Martini, F. H., J. B. Heiser, and M. P. Lesser. 1997a. A population profile for hagfish, *Myxine glutinosa* (L.), in the Gulf of Maine. Part 1. Morphometrics and reproductive state. *Fish. Bull., U.S.* 95:311–320.

Martini, F. H., M. P. Lesser, and J. B. Heiser. 1997b. Ecology of the hagfish, *Myxine glutinosa* L., in the Gulf of Maine: II. Potential impacts on benthic communities and commercial fisheries. *J. Exper. Mar. Biol. Ecol.* 214:97–106.

Martini, F. H., M. P. Lesser, and J. B. Heiser. 1998. A population profile for hagfish, *Myxine glutinosa,* in the Gulf of Maine. Part 2. Morphological variation in populations of *Myxine* in the North Atlantic Ocean. *Fish. Bull., U.S.* 96:516–524.

Martorelli, S. R., and F. Cremonte. 1998. A proposed three-host life history of *Monascus filiformis* (Rudolphi, 1819) (Digenea: Fellodistomidae) in the southwest Atlantic Ocean. *Can. J. Zool.* 76:1198–1203.

Marwitz, S. R. 1986. Young tarpon in a roadside ditch near Matagorda Bay in Calhoun County, TX. Texas Parks Wildl. Dept., Coastal Fish. Branch, Man. Data Ser. No. 100, 8 pp.

Mason, H. W. 1882. Report of operations on the Navesink River, New Jersey, in 1879, in collecting living striped bass for transportation to California. U.S. Fish. Comm. Rept. No. 7:663–666.

Massicotte, B., and J. J. Dodson. 1991. Endogenous activity rhythms in tomcod (*Microgadus tomcod*) post-yolk-sac larvae. *Can. J. Zool.* 69:1010–1016.

Massmann, W. H. 1952. Characteristics of spawning areas of shad *Alosa sapidissima,* Wilson in some Virginia streams. *Trans. Am. Fish. Soc.* 81:78–93.

Massmann, W. H. 1954. Marine fishes in fresh and brackish waters of Virginia. *Ecology* 35:75–78.

Massmann, W. H. 1963. Summer food of juvenile American shad in Virginia waters. *Chesapeake Sci.* 4:167–171.

Massmann, W. H., and A. L. Pacheco. 1957. Shad catches and water temperatures in Virginia. *J. Wildl. Manage.* 21:351–352.

Massmann, W. H., J. J. Norcross, and E. B. Joseph. 1962. Atlantic menhaden larvae in Virginia coastal waters. *Chesapeake Sci.* 3:42–45.

Mast, S. O. 1915. The behavior of *Fundulus,* with especial reference to overland escape from tidepools and locomotion on land. *J. Anim. Behav.* 5:341–350.

Mast, S. O. 1916. Changes in shade, color, and pattern in fishes, and their bearing on the problems of adaptation and behavior with especial reference to the flounders *Paralichthys* and *Ancylopsetta*. *Bull. U.S. Bur. Fish.* 34:177–238.

Mather, F. J., III. 1952. Three species of fishes, genus *Seriola*, in the waters of Cape Cod and vicinity. *Copeia* 1952:209–210.

Mather, F. J., III. 1954. Northerly occurrences of warmwater fishes in western Atlantic. *Copeia* 1954:292–293.

Mather, F. J., III. 1960. Recaptures of tuna, marlin and sailfish tagged in the western North Atlantic. *Copeia* 1960:149–151.

Mather, F. J., III. 1962. Transatlantic migration of two large bluefin tuna. *J. Conseil* 27:325–327.

Mather, F.J., III., and R. H. Gibbs, Jr. 1957. Distributional records of fishes from waters off New England and the middle Atlantic states. *Copeia* 1957:242–244.

Mather, F. J., III., and H. A. Schuck. 1960. Growth of bluefin tuna of the western North Atlantic. *U. S. Fish Wildl. Serv. Fish. Bull.* 61:39–52.

Mather, F. J., III., M. R. Bartlett, and J. S. Beckett. 1967. Trans-Atlantic migrations of young bluefin tuna. *J. Fish. Res. Bd. Can.* 24:1991–1997.

Mather, F. J., III., H. L. Clark, and J. M. Mason Jr. 1975. Synopsis of the biology of the white marlin, *Tetrapturus albidus* Poey (1861). In: *Proc. Int. Billfish Symp., Kailua-Kona, Hawaii, 9–12 August 1972.* NOAA Tech. Rept. NMFS SSRF No. 675(3):55–94.

Matlock, G. C. 1992. Life history aspects of seahorses, *Hippocampus*, in Texas. *Tex. J. Sci.* 44:213–222.

Matsui, T. 1967. Review of the mackerel genera *Rastrelliger* and *Scomber* with description of a new species of *Rastrelliger. Copeia* 1967:71–83.

Matsumoto, W. M., E. H. Ahlstrom, S. Jones, W. L. Klawe, W. J. Richards, and S. Ueyanagi. 1972. On the clarification of larval tuna identification particularly in the genus *Thunnus. Fish. Bull., U. S.* 70:1–12.

Matsumoto, W. M., R. A. Skillman, and A. E. Dizon. 1984. Synopsis of biological data on skipjack tuna, *Katsuwonus pelamis*. NOAA Tech. Rept. NMFS Circ. No. 451, 92 pp.

Matsuura, Y., and M. Katsuragawa. 1981. Larvae and juveniles of grey triggerfish, *Balistes capriscus* from southern Brazil. *Japan. J. Ichthyol.* 28:267–275.

Matsuura, Y., and M. Katsuragawa. 1985. Osteological development of fins and their supports of larval grey triggerfish, *Balistes capriscus*. *Japan. J. Ichthyol.* 31:411–421.

Matsuura, Y., and J. C. Tyler. 1994. Triggerfishes and their allies. In: J. R. Paxton and W. N. Eschmeyer, eds. *Encyclopedia of Fishes.* University of New South Wales Press, Sydney, pp. 229–233.

Matthews, F. D., D. M. Damkaer, L. W. Knapp, and B. B. Collette. 1977. Food of western North Atlantic tunas (*Thunnus*) and lancetfishes (*Alepisaurus*). NOAA Tech. Rept. NMFS SSRF No. 706, 19 pp.

Matthews, L. H. 1950. Reproduction in the basking shark, *Cetorhinus maximus* (Gunner.). *Phil. Trans. Roy. Soc. London* (B) 234:247–316.

Mattila, J., G. Chaplin, M. R. Eilers, K. L. Heck, Jr., J. P. O'Neal, and J. F. Valentine. 1999. Spatial and diurnal distribution of invertebrated sediments in Damariscotta River, Maine (USA). *J. Sea Res.* 41:321–332.

Mattson, S. 1981. The food of *Galeus melanostomus, Gadiculus argenteus thori, Trisopterus esmarkii, Rhinonemus cimbrius* and *Glyptocephalus cynoglossus* (Pisces) caught during the day with shrimp trawl in a West-Norwegian fjord. *Sarsia* 66:109–127.

Mattson, S. 1992. Food and feeding habits of fish species over a soft sublittoral bottom in the northeast Atlantic. III. Haddock (*Melanogrammus aeglefinus* (L.)) (Gadidae). *Sarsia* 77:33–45.

Maul, G. E. 1946. Monografia dos peixes du Museu Municipal do Funchal. *Bol. Mus. Munic. Funchal* 24:1–53.

Maul, G. E. 1986a. Berycidae. In: P. J. P. Whitehead et al., eds. *Fishes of the North-eastern Atlantic and the Mediterranean,* Vol. 2. UNESCO, Paris, pp. 740–742.

Maul, G. E. 1986b. Trachichthyidae. In: P. J. P. Whitehead et al., eds. *Fishes of the North-eastern Atlantic and the Mediterranean,* Vol. 2. UNESCO, Paris, pp. 749–752.

Maurer, R. 1976. A preliminary analysis of interspecific trophic relationships between the sea herring (*Clupea harengus* L.) and the Atlantic mackerel (*Scomber scombrus* L.). Int. Comm. Northw. Atl. Fish. Res. Doc. No. 76/121, Ser. No. 3967, 22 pp.

Maurer, R. O., and R. E. Bowman. 1975. Food chain investigations: food habits of marine fishes of the northwest Atlantic: data report. NMFS Woods Hole Lab. Ref. Doc. No. 75-3, 90 pp.

Maurice, K. R., R. W. Blye, P. L. Harmon, and D. Lake. 1987. Increased spawning by American shad coincident with improved dissolved oxygen in the tidal Delaware River. In: M. J. Dadswell et al., eds., *Common Strategies of Anadromous and Catadromous Fishes. Am. Fish. Soc. Symp.* 1:79–88.

Maxfield, G. H. 1953. The food habits of hatchery-produced pond-cultured shad *Alosa sapidissima* reared to a length of two inches. Chesapeake Biol. Lab. Publ. No. 98, Solomons, Md. 38 pp.

May, A. W. 1967. Fecundity of Atlantic cod. *J. Fish. Res. Bd. Can.* 24:1531–1551.

Mayo, R. K. 1974. Population structure, movement, and fecundity of the anadromous alewife, *Alosa pseudoharengus* (Wilson), in the Parker River, Massachusetts. M.S. thesis, University of Massachusetts, Amherst, 118 pp.

Mayo, R. K. 1998a. Silver hake. In: S. H. Clark, ed. *Status of Fishery Resources off the Northeastern United States for 1998.* NOAA Tech. Memo. NMFS-NE-115:60–63.

Mayo, R. K. 1998b. Pollock. In: S. H. Clark, ed. *Status of Fishery Resources off the Northeastern United States for 1998.* NOAA Tech. Memo. NMFS-NE-115:67–69.

Mayo, R. K. 1998c. Redfish. In: S. H. Clark, ed. *Status of Fishery Resources off the Northeastern United States for 1998.* NOAA Tech. Memo NMFS-NE-115:57–59.

Mayo, R. K., and L. O'Brien. 1998. Atlantic cod. In: S. H. Clark, ed. *Status of Fishery Resources off the Northeastern United States for 1998.* NOAA Tech. Memo. NMFS-NE-115:49–52.

Mayo, R. K., V. M. Gifford, and A. Jearld, Jr. 1981. Age validation of redfish, *Sebastes marinus* (L.), from the Gulf of Maine–Georges

Mayo, R. K., U. B. Dozier, and S. H. Clark. 1983. An assessment of the redfish, *Sebastes fasciatus*, stock in the Gulf of Maine–Georges Bank region. Woods Hole Lab. Ref. Doc. No. 83-22, 39 pp.

Mayo, R. K., J. M. McGlade, and S. H. Clark. 1989. Patterns of exploitation and biological status of pollock (*Pollachius virens* L.) in the Scotian Shelf, Georges Bank, and Gulf of Maine area. *J. Northw. Atl. Fish. Sci.* 9:13–36.

Mayo, R. K., J.Burnett, T. D. Smith, and C. A. Muchant. 1990. Growth-maturation interaction of Acadian redfish (*Sebastes fasciatus* Storer) in the Gulf of Maine–Georges Bank region of the northwest *Atlantic. J. Cons.* 46:287–305.

McAllister, D. E. 1963. A revision of the smelt family, Osmeridae. *Bull. Nat. Mus. Can.* 191, 53 pp.

McAllister, D. E., and E. I. S. Rees. 1964. A revision of the eelpout genus *Melanostigma* with a new genus and with comments on *Maynea. Bull. Nat. Mus. Can.* 199:85–110.

McBride, R. S. 1994. Comparative ecology and life history of two temperate, northwestern Atlantic searobins, *Prionotus carolinus* and *P. evolans* (Pisces: Triglidae). Ph.D. dissertation, Rutgers University, New Brunswick, 176 pp.

McBride, R. S. 1995. Perennial occurrence and fast growth rates by crevalle jacks (Carangidae: *Caranx hippos*) in the Hudson River estuary. In: E. A. Blair and J. R. Waldman, eds. *Final Reports of the Tibor T. Polgar Fellowship Program, 1994.* Hudson River Foundation, Section VI, pp. 1–34.

McBride, R. S., and K. W. Able. 1994. Reproductive seasonality, distribution, and abundance of *Prionotus carolinus* and *P. evolans* (Pisces: Triglidae) in the New York Bight. *Est. Coast. Shelf Sci.* 38:173–188.

McBride, R. S., and K. W. Able. 1998. Ecology and fate of butterflyfishes, *Chaetodon* spp., in the temperate, western North Atlantic. *Bull. Mar. Sci.* 63:401–416.

McBride, R. S., and D. O. Conover. 1991. Recruitment of young-of-the-year bluefish *Pomatomus saltatrix* to the New York Bight: variation in abundance and growth of spring- and summer-spawned cohorts. *Mar. Ecol. Prog. Ser.* 78:205–216.

McBride, R. S., and K. A. McKown. 2000. Consequences of dispersal of subtropically spawned crevalle jacks, *Caranx hippos,* to temperate estuaries. *Fish. Bull., U.S.* 98:528–538.

McBride, R. S., J. B. O'Gorman, and K. W. Able. 1998. Interspecific comparisons of searobin (*Prionotus* spp.) movements, size structure, and abundance in the temperate western North Atlantic. *Fish. Bull., U.S.* 96:303–314.

McCarthy, K., C. Cross, R. Cooper, R. Langton, K. Pecci, and J. Uzmann. 1979. Biology and geology of Jeffreys Ledge and adjacent basins: an unpolluted inshore fishing area, Gulf of Maine, NW Atlantic. Int. Comm. Explor. Sea CM1979/E:44, 12 pp.

McClane, A. J., ed. 1974. *McClane's New Standard Fishing Encyclopedia.* Holt, Rinehart and Winston, New York. 1156 pp.

McCleave, J. D., S. M. Fried, and A. K. Towt. 1977. Daily movements of shortnose sturgeon, *Acipenser brevirostrum,* in a Maine estuary. *Copeia* 1977:149–157.

McCormack, W. H. 1976. Laboratory behavior of young tautog (*Tautoga onitis*) at acclimation temperature and under a temperature increase. M.S. thesis, Long Island University, Brookville, 72 pp.

McCormick, S. D., R. J. Naiman, and E. T Montgomery. 1985. Physiological smolt characteristics of anadromous and non-anadromous brook trout (*Salvelinus fontinalis*) and Atlantic salmon (*Salmo salar*). *Can. J. Fish. Aquat. Sci.* 42:529–538.

McCosker, J. E., E. B. Böhlke, and J. E. Böhlke. 1989. Family Ophichthidae. Snake eels and worm eels. Fishes of the western North Atlantic. Sears Found. Mar. Res. Mem. No. 1(9):254–412.

McCracken, F. D. 1954. Seasonal movements of the winter flounder, *P. americanus* (Walbaum), on the Atlantic coast. Fish. Res. Bd. Can. MS Rept. Biol. Sta. No. 582, 167 pp.

McCracken, F. D. 1958. On the biology and fishery of the Canadian Atlantic halibut, *Hippoglossus hippoglossus* L. *J. Fish. Res. Bd. Can.* 15:1269–1311.

McCracken, F. D. 1963. Seasonal movements of the winter flounder, *Pseudopleuronectes americanus* (Walbaum), on the Atlantic coast. *J. Fish. Res. Bd. Can.* 20:551–586.

McCracken, F. D. 1965. Distribution of haddock off the eastern Canadian mainland in relation to season, depth bottom and temperature. ICNAF Spec. Publ. No. 6:113–129.

McCullough, R. D., and J. G. Stanley. 1981. Feeding niche dimensions in larval rainbow smelt (*Osmerus mordax*). *Rapp. Proc.-Verb. Réun. Cons. Int. Explor. Mer* 178:352–354.

McDermott, J. J. 1965. Food habits of the toadfish, *Opsanus tau* (L.), in New Jersey waters. *Proc. Penn. Acad. Sci.* 38:64–71.

McDonald, M. 1884. The shad and alewifes. In: G. B. Goode et al. *The Fisheries and Fishery Industries of the United States, Sect. I. Natural History of Useful Aquatic Animals.* U.S. 47th Cong. 1st Sess. Senate Misc. Doc.No. 124, Washington, D.C., pp. 594–607.

McDowall, R. M. 1982. The centrolophid fishes of New Zealand (Pisces, Stromateoidei). *J. Roy. Soc. New Zealand* 12: 103–142.

McDowall, R. M. 1988. *Diadromy in Fishes: Migrations Between Freshwater and Marine Environments.* Timber Press, Portland, Ore, 308 pp.

McEachran, J. D. 1970. Egg capsules and reproductive biology of the skate *Raja garmani* (Pisces: Rajidae). *Copeia* 1970:197–199.

McEachran, J. D. 1977. Variation in *Raja garmani* and the status of *Raja lentiginosa* (Pisces:Rajidae). *Bull. Mar. Sci.* 27:423–439.

McEachran, J. D., and C. Capape. 1984. Dasyatidae. In: P. J. P. Whitehead et al., eds. *Fishes of the North-eastern Atlantic and Mediterranean,* Vol. 1. UNESCO, Paris, pp. 197–202.

McEachran, J. D., and J. Davis. 1970. Age and growth of the striped searobin. *Trans. Am. Fish. Soc.* 99:343–352

McEachran, J. D., and M. R. de Carvalho. in press. Rajidae. In: K. E. Carpenter and P. Oliver, eds. *FAO Species Identification Sheets for Fishery Purposes: Western Central Atlantic.* FAO, Rome.

McEachran, J. D., and K. A. Dunn. 1998. Phylogenetic analysis of skates, a morphologically conservative clade of elasmobranchs (Chondrichthyes: Rajidae). *Copeia* 1998:271–290.

McEachran, J. D., and J. D. Fechhelm. 1998. *Fishes of the Gulf of Mexico: Myxiniformes to Gasterosteiformes,* Vol. 1. University of Texas Press, Austin, 1112 pp.

McEachran, J. D., and H. Konstantinou. 1996. Survey of the variation in alar and malar thorns in skates: phylogenetic implications (Chrondrichthyes: Rajoidei). *J. Morph.* 228:165–178.

McEachran, J. D., and C. O. Martin. 1977. Possible occurrence of character displacement in the sympatric skates *Raja erinacea* and *R. ocellata* (Pisces: Rajidae). *Env. Biol. Fishes* 2:121–130.

McEachran, J. D., and J. A. Musick. 1973. Characters for distinguishing between immature specimens of the sibling species, *Raja erinacea* and *R. ocellata* (Pisces: Rajidae). *Copeia* 1973:238–250.

McEachran, J. D., and J. A. Musick. 1975. Distribution and relative abundance of seven species of skates (Pisces: Rajidae) which occur between Nova Scotia and Cape Hatteras. *Fish. Bull., U.S.* 73:110–136.

McEachran, J. D., D. F. Boesch, and J. A. Musick. 1976. Food division within two sympatric species-pairs of skates (Pisces: Rajidae). *Mar. Biol.* 35:301–317.

McGill, E. M., Jr. 1967. Pond water for rearing striped bass fry, *Roccus saxatilis* (Walbaum) in aquaria. *Proc. 20th Ann. Conf. Southeast Assoc. Game Fish Comm.* 1966:331–340.

McGladdery, S. E. 1986. *Anisakis simplex* (Nematoda: Anisakidae) infection of the musculature and body cavity of Atlantic herring *Clupea harengus harengus*. *Can. J. Fish. Aquat. Sci.* 43:1312–1317.

McGladdery, S. E. 1987. Potential of *Eimeria sardinae* (Apicomplexa: Eimeridae) oocysts for distinguishing between spawning groups and between first- and repeat-spawning Atlantic herring *Clupea harengus harengus*. *Can. J. Fish. Aquat. Sci.* 44:1379–1385.

McGladdery, S. E., and M. D. B. Burt. 1985. Potential of parasites for use as biological indicators of migration, feeding, and spawning behavior of northwestern Atlantic herring *Clupea harengus*. *Can. J. Fish. Aquat. Sci.* 42:1957–1968.

McGlade, J. M., M. C. Annand, and T. J. Kenchington. 1983. Electrophoretic identification of *Sebastes* and *Helicolenus* in the northwestern Atlantic. *Can. J. Fish. Aquat. Sci.* 40:1861–1876.

McGovern, J. C. 1986. Seasonal recruitment of larval and juvenile fishes into impounded and non-impounded marshes. M.S. thesis, College of Charleston, Charleston, S.C., 123 pp.

McGovern, J. C., and J. E. Olney. 1988. Potential predation by fish and invertebrates on early life history stages of striped bass in the Pamunkey River, Virginia. *Trans. Am. Fish. Soc.* 117:152–161.

McGovern, J. C., and C. A. Wenner. 1990. Seasonal recruitment of larval and juvenile fishes into impounded and non-impounded marshes. *Wetlands* 10:203–222.

McGowan, C., and M. E. Reith. 1999. Polymorphic microsatellite markers of Atlantic halibut, *Hippoglossus hippoglossus*. *Molec. Ecol.* 8:1761–1763.

McGowan, M. F., and W. J. Richards. 1986. Distribution and abundance of bluefin (*Thunnus thynnus*) larvae in the Gulf of Mexico in 1982 and 1983 with estimates of the biomass and population size of the spawning stock for 1977, 1978 and 1981–1983. Int. Comm. Cons. Atlantic Tunas. Collect. Vol. Sci. Pap. 24:182–195

McHugh, J. L. 1967. Estuarine nekton. In: G. H. Lauff, ed. *Estuaries*. Am. Assoc. Adv. Sci., Washington, D C, pp. 581–620.

McHugh, J. L. 1969. Fisheries of Chesapeake Bay. Proc. Governors Conf., Chesapeake Bay. Wye Inst., Centreville, Md., 2:135–160.

McHugh, J. L., and J. J. C. Ginter. 1978. Fisheries. MESA New York Bight Atlas Monogr. No. 16. New York Sea Grant Inst., Albany, 120 pp.

McHugh, J. L., and A. D. Williams. 1976. *Historical Statistics of the Fisheries of the New York Bight Area*. New York Sea Grant Institute, Albany, 73 pp.

McInerny, J. E. 1969. Reproductive behaviour of the blackspotted stickleback, *Gasterosteus wheatlandi*. *J. Fish. Res. Bd. Can.* 26: 2061–2075.

McIntosh, W. C., and A. T. Masterman. 1897. *The Life Histories of British Marine Food Fishes*. C. J. Clay and Sons, London, 516 pp.

McIntosh, W. C., and E. E. Prince. 1890. On the development and life-histories of the teleostean food and other fishes. *Trans. Roy. Soc. Edinburgh* 35, Part III, No. 19: 665–946.

McKenney, T. W. 1961. Larval and adult stages of the stromateoid fish *Psenes regulus*, with comments on its classification. *Bull. Mar. Sci. Gulf Carib.* 11:210–236.

McKenney, T. W., E. C. Alexander, and G. L. Voss. 1958. Early development and larval distribution of the carangid fish *Caranx crysos* (Mitchill). *Bull. Mar. Sci. Gulf Carib.* 8:167–200.

McKenzie, J. A. 1973. Comparative electrophoresis of tissues from blueback herring and gaspereau. *Comp. Biochem. Physiol., B. Comp. Biochem.* 44:65–68.

McKenzie, J. A., and M. H. A. Keenleyside. 1970. Reproductive behavior of ninespine sticklebacks [*Pungitius pungitius* (L.)] in South Bay, Manitoulin Island, Ontario. *Can. J. Zool.* 48:55–61.

McKenzie, R. A. 1939. Some marine and salp records. *Proc. Nova Scotia Inst. Sci.* 20:13–20.

McKenzie, R. A. 1940. Some marine records from Nova Scotian fishing waters. *Proc. Nova Scotia Inst. Sci.* 20:42–46.

McKenzie, R. A. 1959. Marine and freshwater fishes of the Miramichi River and estuary, New Brunswick. *J. Fish. Res. Bd. Can.* 16:807–833.

McKenzie, R. A. 1964a. Observations on herring spawning off southwest Nova Scotia. *J. Fish. Res. Bd. Can.* 21:203–205.

McKenzie, R. A. 1964b. Smelt life history and fishery in the Miramichi River, New Brunswick. *Bull. Fish. Res. Bd. Can.* 144, 77 pp.

McKenzie, R. A. 1967. Fawn cusk-eel, *Lepophidium cervinum*, in the northwest Atlantic. *J. Fish. Res. Bd. Can.* 24:213–214.

McKenzie, R. A., and R. E. S. Homans. 1938. Rare and interesting fishes and salps in the Bay of Fundy and off Nova Scotia. *Proc. Nova Scotia Inst. Sci.* 19:277–281.

McKenzie, R. A., and S. N. Tibbo. 1963. An occurrence of opah, *Lampris regius* (Bonnaterre), in the northwest Atlantic. *J. Fish. Res. Bd. Can.* 20:1097–1099.

McMahon, J. W. 1963. Monogenetic trematodes from some Chesapeake Bay fishes. I. The superfamilies Capsaloidea Price, 1936 and Diclidophoroidea Price, 1936. *Chesapeake Sci.* 4:151–160.

McMillan, D. G., and W. W. Morse. 1999. Essential Fish Habitat Document: spiny dogfish, *Squalus acanthias*, life history and habitat characteristics. NOAA Tech. Memo. NMFS-NE-150, 19 pp.

McPhail, J. D. 1963. Geographic variation in North American ninespine sticklebacks, *Pungitius pungitius*. *J. Fish. Res. Bd. Can.* 20:27–44.

McQuinn, I. H. 1996. Year-class twinning in sympatric seasonal spawning populations of Atlantic herring, *Clupea harengus*. *Fish. Bull., U.S.* 95:126–136.

McQuinn, I. H. 1997. Metapopulations and the Atlantic herring. *Rev. Fish Biol. Fish.* 7:297–329.

McQuinn, I. H., G. J. Fitzgerald, and H. Powles. 1983. Environmental effects on embryos and larvae of the Isle Verte stock of Atlantic herring *Clupea harengus harengus. Naturaliste Can. (Rev. Écol. Syst.)* 110:343–355.

Mead, G.W. 1966. Family Chlorophthalmidae. Fishes of the western North Atlantic. Sears Found. Mar. Res. Mem. No. 1(5):162–189.

Mead, G. W. 1972. Bramidae. Dana Rept. No. 81, 166 pp.

Mead, G. W., and R. L. Haedrich. 1965. The distribution of the oceanic fish *Brama brama. Bull. Mus. Comp. Zool.* 134:29–68.

Meador, M. R. 1982. Occurrence and distribution of larval fish in the Santee River system. M.S. thesis, Clemson University, Clemson, S.C.

Medcof, J. C. 1957. Nuptial or pre-nuptial behavior of the shad, *Alosa sapidissima* (Wilson). *Copeia* 1957:252–253.

Medved, R. J., and J. A. Marshall. 1981. Feeding behavior and biology of young sandbar sharks, *Carcharhinus plumbeus* (Pisces, Carcharhinidae) in Chincoteague Bay, Virginia. *Fish. Bull., U.S.* 79:441–447.

Medved, R. J., C. E. Stillwell, and J. G. Casey. 1985. Stomach contents of young sandbar sharks, *Charcharhinus plumbeus,* in Chincoteague Bay, Virginia. *Fish. Bull., U.S.* 83:395–402.

Meehan, W. E. 1910. Experiments in sturgeon culture. *Trans. Am. Fish. Soc.* 39:85–91.

Meehean, O. L. 1940. A review of the parasitic crustacea of the genus *Argulus* in the collections of the United States National Museum. *Proc. U.S. Nat. Mus.* 88:459–522.

Meek, S. E. 1883. A note on the Atlantic species of the genus *Anguilla. Bull. U.S. Fish Comm.* 3:430.

Meek, S. E., and S. F. Hildebrand. 1928. Fishes of Panama. Field Mus. Nat. Hist,. Vol. 15, Zool. Ser., Part III, 1045 pp.

Mehrle, P. M., D. Beedler, S. Finger, and L. Ludke. 1984. Impact of contaminants on striped bass. U.S. Fish Wildl. Serv. Rept., Columbia Nat. Fish. Res. Lab., Colombia, Md., 28 pp.

Meinz, M. 1978. Improved method for collecting and transporting young American shad. *Prog. Fish-Cult.* 42:150–151.

Meister, A. L. 1984. The marine migrations of tagged Atlantic salmon (*Salmo salar* L.) of USA origin. ICES CM 1984/M:27.

Meister, A. L., and L. N. Flagg. 1997. Recent developments in the American eel fisheries of eastern North America. *Focus* 22(1): 25–26.

Melvin, G. D., M. J. Dadswell, and J. D. Martin. 1985. Impact of lowhead hydroelectric tidal power development on fisheries. I. A preparation study of the spawning population of American shad *Alosa sapidissima* (Pisces: Clupeidae) in the Annapolis River, Nova Scotia, Canada. Can. Fish. Aquat. Sci. Tech. Rept. No. 1340, 33 pp.

Melvin, G. D., M. J. Dadswell, and J. D. Martin. 1986. Fidelity of American shad, *Alosa sapidissima* (Clupeidae), to its river of previous spawning. *Can. J. Fish. Aquat. Sci.* 43:640–646.

Melvin, G. D., M. J. Dadswell, and J. A. McKenzie. 1992. The usefulness of meristic and morphometric characters in discriminating populations of American shad, *Alosa sapidissima,* (Osteichthys: Clupeidae) inhabiting a marine environment. *Can. J. Fish. Aquat. Sci.* 49:266–280.

Menge, B. A. 1982. Reply to comment by Edwards, Conover, and Sutter. *Ecology* 63:1180–1184.

Menni, R. C., and H. L. López. 1979. Biological data and otolith (sagitta) morphology of *Polyprion americanus* and *Schedophilus grisolineatus* (Osteichtyes, Serranidae and Centrolophidae). *Stud. Neotrop. Fauna Environ.* 14:17–32.

Menni, R. C., G. H. Burgess, and M. L. Garcia. 1993. Occurrence of *Centroscyllium fabricii* (Reinhardt, 1825) (Elasmobranchii, Squalidae) in the Beagle Channel, southern South America. *Bull. Mar. Sci.* 52:824–832.

Mercado, J. E., and A. Ciardelli. 1972. Contribucíon a la morfología y organogenésis de los leptocéfalos del sábalo *Megalops atlanticus* (Pisces: Megalopidae). *Bull. Mar. Sci.* 22:153–184.

Mercaldo-Allen, R., D. M. Perry, C. Kuropat, and J. Hughes. 1997. Tautog culture: preliminary studies. *J. Shellfish Res.* 16(1):292.

Mercer, I. R. G., D. E. Barker, and R. A. Khan. 1997. Stress-related changes in cunner, *Tautogolabrus adspersus,* living near a paper mill. *Bull. Environ. Contam. Toxicol.* 58:442–447.

Mercer, L. P. 1973. The comparative ecology of two species of pipefish (Syngnathidae) in the York River, Virginia. M.A. thesis, College of William and Mary, Gloucester Point, Va., 37 pp.

Mercer, L. P. 1978. The reproductive biology and population dynamics of the black sea bass, *Centropristis striata.* Ph.D. dissertation, College of William and Mary, Williamsburg, Va., 196 pp.

Mercer, L. P. 1983. A biological and fisheries profile of weakfish, *Cynoscion regalis.* North Carolina Dept. Nat. Res. Comm. Dev., Div. Mar. Fish., Spec. Sci. Rept. No. 39, 107 pp.

Mercer, L. P. 1989. Species profiles: life histories and environmental requirements of coastal fishes and invertebrates (South Atlantic)—black sea bass. U.S. Fish Wildl. Ser. Biol. U.S. Corps Engineers TR EL-82-4, 16 pp.

Mercer, L. P., J. P. J. Monaghan, and J. L. Ross. 1987. Marine Fisheries Research. North Carolina Div. Mar. Fish., Morehead City. Ann. Prog. Rept. Proj. F-29-1.

Mercer, S., G. E. Brown, S. Clearwater, and Z. Yao. 1993. Observations of the copulatory behaviour of the ocean pout, *Macrozoarces americanus. Can. Field-Nat.* 107:243–244.

Meredith, W. H. 1975. Production dynamics of a tidal creek population of *Fundulus heteroclitus* (Linnaeus). M.S. thesis, University of Delaware, Newark, 98 pp.

Meredith, W. H., and V. A. Lotrich. 1979. Production dynamics of a tidal creek population of *Fundulus heteroclitus* (Linnaeus). *Est. Coast. Mar. Sci.* 8:99–118.

Merrett, N. R. 1986. Macrouridae of the eastern North Atlantic. Fiches d'Identification du Plancton. ICES, Nos. 173/174/175, 14 pp.

Merrett, N. R. 1989. The elusive macrourid alevin and its seeming lack of potential in contributing to intrafamilial systematics. In: D. M. Cohen, ed. *Papers on the Systematics of Gadiform Fishes.* Los Angeles Co. Nat. Hist. Mus. Sci. Ser. No. 32:175–185.

Merrett, N. R., and P. A. Domanski. 1985. Observations on the biology of deep-sea bottom-living fishes collected off northwest Africa. II. The Moroccan slope (27–34° N), with special reference to *Synaphobranchus kaupii. Biol. Oceanogr.* 3:349–399.

Merrett, N. R., and N. B. Marshall. 1981. Observations on the ecology of deep-sea bottom-living fishes collected off northwest Africa (08° N, 27° W). *Prog. Oceanogr.* 9:185–244.

Merriman, D. 1941. Studies on the striped bass (*Roccus saxatilis*) of the Atlantic coast. *U.S. Fish Wildl. Serv. Fish. Bull.* 50:1–77.

Merriman, D. 1973. William Charles Schroeder. Fishes of the western North Atlantic. Sears Found. Mar. Res. Mem. No. 1(6): xii–xiii.

Merriman, D., and R. C. Sclar. 1952. Hydrographic and biological studies of Block Island Sound: the pelagic eggs and larvae of Block Island Sound. *Bull. Bingham Oceanogr. Coll.* 13(3):165–219.

Merriner, J. V. 1975. Food habits of the weakfish, *Cynoscion regalis,* in North Carolina waters. *Chesapeake Sci.* 16:74–76.

Merriner, J. V. 1976. Aspects of the reproductive biology of the weakfish, *Cynoscion regalis* (Sciaenidae), in North Carolina. *Fish. Bull., U.S.* 74:18–26.

Merriner, J. V., and J. L. Laroche. 1977. Fecundity of the northern puffer, *Sphoeroides maculatus* from Chesapeake Bay. *Chesapeake Sci.* 18:81–84.

Merriner, J. V., W. A. Foster, and F. J. Schwartz. 1970. The barrelfish, *Hyperoglyphe perciformis* (Pisces, Stromateidae), in Pamlico Sound, N.C., and adjacent Atlantic Ocean. *J. Elisha Mitchell Sci. Soc.* 8: 28–30.

Messieh, S. N. 1972. Use of otoliths in identifying herring stocks in the southern Gulf of St. Lawrence and adjacent waters. *J. Fish. Res. Bd. Can.* 29:1113–1118.

Messieh, S. N. 1975. Delineating spring and autumn herring populations in the southern Gulf of St. Lawrence by discriminant function analysis. *J. Fish. Res. Bd. Can.* 32:471–477.

Messieh, S. N. 1976. Fecundity studies on Atlantic herring from the southern Gulf of St. Lawrence and along the Nova Scotia coast. *Trans. Am. Fish. Soc.* 105:384–394.

Messieh, S. N. 1977. Population structure and biology of alewives *Alosa pseudoharengus* and blueback herring *A. aestivalis* in the Saint John River, New Brunswick. *Environ. Biol. Fishes* 2:195–210.

Messieh, S. N. 1980. A bibliography of herring *Clupea harengus* in the northwest Atlantic. Fish. Mar. Serv. Tech. Rept. No. 919, 25 pp.

Messieh, S. N. 1988. Spawning of Atlantic herring in the Gulf of St. Lawrence. In R. D. Hoyt, ed. *11th Annual Larval Fish Conference. Am. Fish. Soc. Symp.* 5:31–48.

Messieh, S. N., and S. N. Tibbo. 1970. A critique on the use of otoliths for ageing Gulf of St. Lawrence herring *Clupea harengus* L. *J. Cons. Int. Explor. Mer* 33:181–191.

Messieh, S. N., and S. N. Tibbo. 1971. Discreteness of herring populations in spring and autumn fisheries on the southern Gulf of St. Lawrence. *J. Fish. Res. Bd. Can.* 28:1009–1014.

Messieh, S. N., R. Pottle, P. MacPherson, and T. Hurlbut. 1985. Spawning and exploitation of Atlantic herring (*Clupea harengus*) at Escuminac in the southwestern Gulf of St. Lawrence, spring 1983. *J. Northw. Atl. Fish. Sci.* 6:125–133.

Messieh, S. N., D. S. Moore, and P. Rubec. 1987. Estimation of age and growth of larval Atlantic herring as inferred from examination of daily growth increments of otoliths. In: R. C. Summerfelt and G. E. Hall, eds. *Age and Growth of Fish.* Iowa State University Press, Ames, pp. 433–442.

Methven, D. A. 1985. Identification and development of larval and juvenile *Urophycis chuss, U. tenuis* and *Phycis chesteri* (Pisces: Gadidae) from the northwest Atlantic. *J. Northw. Atl. Fish. Sci.* 6:9–20.

Methven, D. A., and J. A. Brown. 1991. Time of hatching affects development, size, yolk volume, and mortality of newly hatched *Macrozoarces americanus* (Pisces: Zoarcidae). *Can. J. Zool.* 69: 2161–2167.

Methven, D. A., and D. S. McKelvie. 1986. Distribution of *Phycis chesteri* (Pisces: Gadidae) on the Grand Bank and Labrador Shelf. *Copeia* 1986:886–891.

Methven, D. A., and J. F. Piatt. 1991. Seasonal abundance and vertical distribution of capelin (*Mallotus villosus*) in relation to water temperature at a coastal site off eastern Newfoundland. *ICES J. Mar. Sci.* 48:187–193.

Methven, D. A., L. W. Crim, B. Norberg, J. A. Brown, G. P. Goff, and I. Huse. 1992. Seasonal reproduction and plasma levels of sex steroids and vitellogenin in Atlantic halibut (*Hippoglossus hippoglossus*). *Can. J. Fish. Aquat. Sci.* 49:754–759.

Meyer, T. L., R. A. Cooper, and R. W. Langton. 1979. Relative abundance, behavior, and food habits of the American sand lance, *Ammodytes americanus,* from the Gulf of Maine. *Fish. Bull., U.S.* 77:243–253.

Michelman, M. S. 1988. The biology of juvenile scup (*Stenotomus chrysops* [L.]) in Narragansett Bay, RI: food habits, metabolic rate and growth rate. M.S. thesis, University of Rhode Island, Kingston, 106 pp.

Middaugh, D. P. 1981. Reproductive ecology and spawning periodicity of the Atlantic silverside, *Menidia menidia* (Pisces: Atherinidae). *Copeia* 1981:766–776.

Middaugh, D. P., and P. W. Lempesis. 1976. Laboratory spawning and rearing of a marine fish, the silverside *Menidia menidia menidia. Mar. Biol.* 35:295–300.

Middaugh, D. P., and T. Takita. 1983. Tidal and diurnal spawning cues in the Atlantic silverside, *Menidia menidia. Environ. Biol. Fishes* 8:97–104.

Middaugh, D. P., G. I. Scott, and J. M. Dean. 1981. Reproductive behavior of the Atlantic silverside, *Menidia menidia* (Pisces: Atherinidae). *Environ. Biol. Fishes* 6:269–276.

Middaugh, D. P., M. J. Hemmer, and Y. Lamadrid-Rose. 1986. Laboratory spawning cues in *Menidia beryllina* and *M. peninsulae* (Pisces, Atherinidae) with notes on survival and growth of larvae at different salinities. *Environ. Biol. Fishes* 15:107–117.

Migdalski, E. C. 1958. *Angler's Guide to the Salt Water Game Fishes, Atlantic and Pacific.* Ronald Press Co., New York, 506 pp.

Miller, D. 1958. A key to some of the more common larval fishes of the Gulf of Maine. Woods Hole Lab. MS Rept. No. 58-1, 56 pp.

Miller, D., and R. R. Marak. 1959. The early larval stages of the red hake, *Urophycis chuss. Copeia* 1959:248–250.

Miller, D., and R. R. Marak. 1962. Early larval stages of the fourspot flounder, *Paralichthys oblongus. Copeia* 1962:454–455.

Miller, D. S., and J. E. Carscadden. 1989. Biomass estimates from two hydroacoustic surveys for capelin (*Mallotus villosus*) in NAFO division 3L + 3N and observations of the Soviet fishery for capelin in Division 3N0. NAFO, SCR Doc No. 89/52.

Miller, G. L., and S. C. Jorgenson. 1973. Meristic characters of some marine fishes of the western Atlantic Ocean. *Fish. Bull., U.S.* 71:301–312.

Miller, J. M. 1965. A trawl survey of the shallow Gulf fishes near Port Aransas, Texas. *Publ. Inst. Mar. Sci., Univ. Tex.* 10:80–107.

Miller, J. M., J. S. Burke, and G. R. Fitzhugh. 1991. Early life history patterns of Atlantic North American flatfish: likely (and unlikely) factors controlling recruitment. *Neth. J. Sea Res.* 27:261–275.

Miller, J. P., F. R. Griffiths, and P. S. Thurston-Rodgers. 1982. The American shad *Alosa sapidissima* in the Delaware River Basin. Del. Basin Fish. Wildlfe. Coop., Trenton, N.J.

Miller, R. J. 1959. A review of the seabasses of the genus *Centropristes* (Serranidae). *Tulane Stud. Zool.* 7:33–68.

Milliman, J. D., and F. T. Manheim. 1968. Observations in the deep-scattering layers off Cape Hatteras, U.S.A. *Deep-Sea Res.* 15:505–507.

Mills, D. 1989. *Ecology and Management of Atlantic Salmon.* Chapman and Hall, London, 351 pp.

Milstein, C. B. 1981. Abundance and distribution of juvenile *Alosa* species off southern New Jersey. *Trans. Am. Fish. Soc.* 110:306–309.

Milstein, C. B., and D. L. Thomas. 1976. Fishes new or uncommon to the New Jersey coast. *Chesapeake Sci.* 17:198–204.

Minchin, D. 1988. A record of the deep-sea anglerfish, *Cryptosaras couesi* Gill, from the north–eastern Atlantic. *J. Fish Biol.* 32:313.

Mironova, N. V. 1957. Biology and industry of pollock. *Trudy Murmansk. Biol. Sta.* 3:114–129 [in Russian].

Mitchill, P. H., and Staff. 1925. A report of investigations concerning shad in the rivers of Connecticut. I. Conn. St. Bd. Fish Game, Hartford, pp.7–44.

Mito, S. 1960. Egg development and hatched larvae of the common dolphin-fish, *Coryphaena hippurus,* Linné. *Bull. Japan. Soc. Sci. Fish.* 26:223–226 [in Japanese with English abstract].

Mito, S. 1961. Pelagic eggs from Japanese waters—I. Clupeina, Chanina, Stomiatina, Myctophida, Anguillida, Belonida and Syngnathida. *Sci. Bull. Fac. Agric. Kyushu Univ.* 18:285–310.

Miya, M. and M. Nishida. 1996. Molecular phylogenetic perspective on the evolution of the deep-sea genus *Cyclothone* (Stomiiformes: Gonostomatidae). *Ichthyol. Res.* 43:375–398.

Miya, M. and M. Nishida. 2000. Molecular systematics of the deep-sea fish genus *Gonostoma* (Stomiiformes: Gonostomatidae): two paraphyletic clades and resurrection of *Sigmops. Copeia* 2000: 378–389.

Mochek, A. D. 1973. Spawning behavior of the lumpsucker [*Cyclopterus lumpus* (L.)]. *Voprosy Ikhtiol.* 13:733–739 [in Russian, translation in *J. Ichthyol.* 13(4):615–619].

Moffitt, C. M., B. Kynard, and S. G. Rideout. 1982. Fish passage facilities and anadromous fish restoration in the Connecticut River basin. *Fisheries* 7(6):2–11.

Moksness, E., and D. Stefanussen. 1990. Growth rates in cultured common wolffish (*Anarhichas lupus*) and spotted wolffish (*A. minor*). Int. Counc. Explor. Sea, Mariculture Comm. ICES-Council-Meeting-1990. CM 1990/F:2 REF. G, 9 pp.

Moksness, E., J. Gjøsæter, A. Reinert, and I. S. Fjallstein. 1989. Start-feeding and on-growing of wolffish (*Anarhichas lupus*) in the laboratory. *Aquaculture* 77:221–228.

Molander, A. R. 1925. Observations on the witch (*Pleuronectes cynoglossus* L.) and its growth. Cons. Int. Explor. Mer Publ. Circ. No. 85:1–15.

Möller, H. 1984. Reduction of a larval herring population by jellyfish predator. *Science* 224:621–622.

Mollet, H. F., G. Cliff, H. L. Pratt, Jr., and J. D. Stevens. 2000. Reproductive biology of the female shortfin mako, *Isurus oxyrinchus,* Rafinesque, 1810, with comments on the embryonic development of lamnoids. *Fish. Bull., U.S.* 98:299–318.

Mollomo, P. 1998. The white shark in Maine and Canadian Waters. *Northeast. Nat.* 5:207–214.

Molloy, J. 1984. Density-dependent growth in Celtic Sea herring. Int. Comm. Explor. Sea CM 1984/H:30.

Monaghan, J. P. J. 1992. Migration and population dynamics of summer flounder (*Paralichthys dentatus*) in North Carolina. North Carolina Div. Mar. Fish., Morehead City, Mar. Fish. Res., Completion Rept. Proj. F-29.

Monteleone, D. M. 1992. Seasonality and abundance of ichthyoplankton in Great South Bay, New York. *Estuaries* 15:230–238.

Monteleone, D. M., and L. E. Duguay. 1988. Laboratory studies of predation by the ctenophore *Mneniopsis leidyi* on the early stages in the life history of the bay anchovy, *Anchoa mitchilli. J. Plank. Res.* 10:359–372.

Monteleone, D. M., and W.T. Peterson. 1986. Feeding ecology of American sand lance *Ammodytes americanus* larvae from Long Island Sound. *Mar. Ecol. Prog. Ser.* 30:133–143.

Montevecchi, W. A., D. K. Cairns, and V. L. Birt. 1988. Migration of postsmolt Atlantic salmon, *Salmo salar,* off northeastern Newfoundland, as inferred by tag recoveries in a seabird colony. *Can. J. Fish. Aquat. Sci.* 45:568–571.

Moore, C. J., and D. T. Burton. 1975. Movements of striped bass, *Morone saxatilis,* tagged in Maryland waters of Chesapeake Bay. Trans. *Am. Fish. Soc.* 104:703–709.

Moore, D. 1967. Triggerfishes (Balistidae) of the western Atlantic. *Bull. Mar. Sci.* 17:689–722.

Moore, E. 1947. Studies on the marine resources of southern New England. VI. The sand flounder *Lophopsetta aquosa* (Mitchill): A general study of the species with special emphasis on age determination by means of scales and otoliths. *Bull. Bingham Oceanogr. Coll.* 11(3):1–79.

Moore, H. F. 1898. Observations on the herring and herring fisheries of the northeast coast, with special reference to the vicinity of Passamaquoddy Bay. U.S. Comm. Fish Fisheries Rept. for 1896, pp. 387–442.

Moore, I. A., and J. W. Moore. 1974. Food of shorthorn sculpin, *Myoxocephalus scorpius,* in the Cumberland Sound area of Baffin Island. *J. Fish. Res. Bd. Can.* 31:355–359.

Moore, R. H. 1974. General ecology, distribution and relative abundance of *Mugil cephalus* and *Mugil curema* on the south Texas coast. *Contr. Mar. Sci.* 18:241–255.

Moore, R. H. 1976. Observations on fishes killed by cold at Port Aransas, Texas 11–12 January 1973. *Southwest Nat.* 20:461–466.

Moores, J. A., and G. H. Winters. 1982. Growth patterns in a Newfoundland Atlantic herring *Clupea harengus harengus* stock. *Can. J. Fish. Aquat. Sci.* 39:454–461.

Moores, J. A., and G. H. Winters. 1984. Migration patterns of Newfoundland west coast herring, *Clupea harengus,* as shown by tagging studies. *J. Northw. Atl. Fish. Sci.* 5:17–22.

Moores, J. A., G. H. Winters, and L. S. Parsons. 1975. Migrations and biological characters of Atlantic mackerel (*Scomber scombrus*) occurring in Newfoundland waters. *J. Fish. Res. Bd. Can.* 32: 1347–1357.

Morais, R. 1981. Sobre a pescaria e biologia do apara-lapis ou trombeteiro. Bol. Inst. Nac. Pescas, Lisbon 6:5–35.

Moran, J. D. W., J. R. Arthur, and M. D. E. Burt. 1996. Parasites of sharp-beaked redfishes (*Sebastes fasciatus* and *Sebastes mentella*) collected from the Gulf of St. Lawrence, Canada. *Can. J. Fish. Aquat. Sci.* 53:1821–1826.

Morgan, A. R., and A. R. Gerlach. 1950. Striped bass studies on Coos Bay, Oregon in 1949 and 1950. Oregon Fish Comm., Contrib. No. 14, 31 pp.

Morgan, M. J., and W. R. Bowering. 1997. Temporal and geographic variation in maturity at length and age of Greenland halibut (*Reinhardtius hippoglossoides*) from the Canadian northwest Atlantic with implications for fisheries management. *ICES J. Mar. Sci.* 54:875–885.

Morgan, M. J., and W. B. Brodie. 1991. Seasonal distribution of American plaice on the northern Grand Banks. *Mar. Ecol. Prog. Ser.* 75:101–107.

Morgan, M. J., J. T. Anderson, and J. A. Brown. 1995. Early development of shoaling behaviour in larval capelin (*Mallotus villosus*). *Mar. Behav. Physiol.* 24:197–206.

Morgan, R. P., II, and V. J. Raisin, Jr. 1973. Effects of salinity and temperature on the development of egg and larvae of striped bass and white perch. App. X. Hydrographic and ecological effects of enlargement of the Chesapeake and Delaware Canal. Contract No. DACV–61–71–C–0062, Army Corps Eng., Philadelphia Dist. NRI Ref. No. 73.

Morgan, R. P., II, and V. J. Raisin, Jr. 1982. Influence of temperature and salinity on development of white perch eggs. *Trans. Am. Fish. Soc.* 111:396–398.

Moring, J. R. 1989. Food habits and algal associations of juvenile lumpfish, *Cyclopterus lumpus* L., in intertidal waters. *Fish. Bull., U.S.* 87:233–237.

Moring, J. R. 1990. Seasonal absence of fishes in tidepools of a boreal environment (Maine, USA). *Hydrobiologia* 194:163–168.

Moring, J. R., and S. W. Moring. 1991. Short-term movements of larval and juvenile lumpfish, *Cyclopterus lumpus* L., in tidepools. *J. Fish Biol.* 38:845–850.

Mork, J., P. Solemdal, and G. Sundnes. 1983. Identification of marine fish eggs: a biochemical genetics approach. *Can. J. Fish. Aquat. Sci.* 40:361–369.

Mork, J. P., N. Ryman, G. Stahl, F. Utter, and G. Sundnes. 1985. Genetic variation in Atlantic cod (*Gadus morhua*) throughout its range. *Can. J. Fish. Aquat. Sci.* 42:1580–1587.

Morrill, A. D. 1895. The pectoral appendages of *Prionotus* and their innervation. *J. Morph.* 11:177–191.

Morris, R. 1972. Osmoregulation. In: M. W. Hardisty and I. C. Potter, eds. *The Biology of Lampreys,* Vol. 2. Academic Press, New York, pp. 193–239.

Morrison, C. M., and W. E. Hawkins. 1984. Coccidians in the liver and testis of the herring *Clupea harengus* L. *Can. J. Zool.* 62: 480–493.

Morrison, C. M., and C. A. MacDonald. 1995. Epidermal tumors on rainbow smelt and Atlantic salmon from Nova Scotia. *J. Aquat. Animal Health* 7:241–250.

Morrow, J. E., Jr. 1951. Studies on the marine resources of southern New England. VIII. The biology of the longhorn sculpin, *Myxocephalus octodecimspinosus* Mitchill, with a discussion of the southern New England "trash fishery". *Bull. Bingham Oceanogr. Collect.* 8(2):1–89.

Morrow, J. E., Jr. 1964a. Family Chauliodontidae. Fishes of the western North Atlantic. Sears Found. Mar. Res. Mem. No. 1(4):274–289.

Morrow, J. E., Jr. 1964b. Family Stomiatidae. Fishes of the western North Atlantic. Sears Found. Mar. Res. Mem. No. 1(4):290–310.

Morrow, J. E., Jr., and R. H. Gibbs, Jr. 1964. Family Melanostomiatidae. Fishes of the western North Atlantic. Sears Found. Mar. Res. Mem. No. 1(4):351–511.

Morse, W. W. 1978. Biological and fisheries data on scup, *Stenotomus chrysops* (Linnaeus). Sandy Hook Lab. Tech. Ser. Rept. No. 12, 41 pp.

Morse, W. W. 1979. An analysis of maturity observations of 12 groundfish species collected from Cape Hatteras, North Carolina to Nova Scotia in 1977. NMFS/NEFC, Sandy Hook Lab. Rept. SHL 79-32, 20 pp.

Morse, W. W. 1980. Spawning and fecundity of Atlantic mackerel, *Scomber scombrus,* in the Middle Atlantic Bight. *Fish. Bull., U.S.* 78:103–108.

Morse, W. W. 1981. Reproduction of the summer flounder, *Paralichthys dentatus* (L.). *J. Fish Biol.* 19:189–203.

Morse, W. W. 1989. Catchability, growth, and mortality of larval fishes. *Fish. Bull., U.S.* 87:417–446.

Morse, W. W., and K. W. Able. 1995. Distribution and life history of windowpane, *Scophthalmus aquosus,* off the northeastern United States. *Fish. Bull., U.S.* 93:675–693.

Morse, W. W., M. P. Fahay, and W. G. Smith. 1987. MARMAP surveys of the continental shelf from Cape Hatteras, North Carolina, to Cape Sable, Nova Scotia (1977–1984). Atlas No. 2. Annual distribution patterns of fish larvae. NOAA Tech. Memo. NMFS-F/NEC-47, 215 pp.

Morse, W. W., D. L. Johnson, P. L. Berrien, and S. J. Wilk. 1999. Essential Fish Habitat Source Document: silver hake, *Merluccius bilinearis,* life history and habitat characteristics. NOAA Tech. Memo. NMFS-NE-135, 42 pp.

Morsell, J. W., and C. R. Norden. 1968. Food habits of the alewife, *Alosa pseudoharengus* (Wilson), in Lake Michigan. *Proc. 11th Conf. Great Lakes Res.,* pp. 96–102.

Morton, T. 1989. Species profiles: life histories and environmental requirements of coastal fishes and invertebrates (Mid-Atlantic)—bay anchovy. U.S. Fish Wildl. Serv., Biol. Rept. No. 82(11.97), 13 pp.

Moser, H. G., E. H. Ahlstrom, and E. M. Sandknop. 1977. Guide to the identification of scorpionfish larvae (family Scorpaenidae) in the eastern Pacific with comparative notes on species of *Sebastes* and *Helicolenus* from other oceans. NOAA Tech. Rept. NMFS Circ. No. 402, 71 pp.

Moser, H. G., E. H. Ahlstrom, and J. R. Paxton. 1984. Myctophidae: development. In: H. G. Moser et al., eds. *Ontogeny and Systematics of Fishes.* Am. Soc. Ich. Herp. Spec. Publ. No. 1:218–244.

Moser, M., J. Whipple, J. Sakanari, and C. Reilly. 1983. Protandrous hermaphroditism in striped bass from Coos Bay, Oregon. *Trans. Am. Fish. Soc.* 112:567–569.

Moser, M. L., S. W. Ross, and K. J. Sulak. 1996. Metabolic responses to hypoxia of *Lycenchelys verrillii* (wolf eelpout) and *Glyptocephalus cynoglossus* (witch flounder): sedentary bottom fishes of the Hatteras/Virginia Middle Slope. *Mar. Ecol. Prog. Ser.* 144:57–61.

Mosher, C. 1954. Observations on the behavior and the early larval development of the Sargassum fish *Histrio histrio* (Linnaeus). *Zoologica, N.Y.* 39:141–152.

Moss, S. A. 1970. The response of young American shad to rapid temperature changes. *Trans. Am. Fish. Soc.* 99:381–384.

Moss, S. A. 1972. Tooth replacement and body growth rates in the smooth dogfish, *Mustelus canis* (Mitchill). *Copeia* 1972:808–811.

Moss, S. A., W. C. Leggett, and W. A. Boyd. 1976. Recurrent mass mortalities of the blueback herring, *Alosa aestivalis,* in the lower Connecticut River. In: D. Merriman and L. M. Thorpe, eds. *The Connecticut River Ecological Study: The Impact of a Nuclear Power Plant.* Am. Fish. Soc. Monogr. No. 1:227–234.

Motta, P. J. 1989. Dentition patterns among Pacific and western Atlantic butterflyfishes (Perciformes, Chaetodontidae): relationship to feeding ecology and evolutionary history. *Environ. Biol. Fishes* 25:159–170.

Mouland, K. and M. N. Voigt, eds. 1990. Use of semi-cryoprotested mince surimi and lumpfish in formulating seafood nuggets. *Advances Fish. Tech. Biotech. Increased Profit.* 1:121–142.

Moulton, J. M. 1958. A summer silence of sea robins, *Prionotus* spp. *Copeia* 1958:234–235.

Mukhacheva, V. A. 1974. Cyclothones (genus *Cyclothone,* family Gonostomatidae) of the world ocean and their distribution. *Trudy Inst. Okeanogr.* 96:205–249 [in Russian].

Mulkana, M. S. 1966. The growth and feeding habits of juvenile fishes in two Rhode Island estuaries. Gulf Res. Rept. No. 2:97–168.

Mullan, J. W. 1958. The sea-run or "salter" brook trout (*Salvelinus fontinalis*) fishery of the coastal streams of Cape Cod, Massachusetts. *Mass. Div. Fish. Game Bull.* 17:1–25.

Mundy, B. C. 1990. Development of larvae and juveniles of the alfonsins, *Beryx splendens* and *B. decadactylus* (Berycidae, Beryciformes). *Bull. Mar. Sci.* 46:257–273.

Munk, O., and E. Bertelsen. 1980. On the esca light organ and its associated light-guiding structures in the deep-sea anglerfish *Chaenophryne draco* (Pisces, Ceratioidei). *Vidensk. Meddr. Dansk Naturh. Foren.* 142:103–129.

Munk, O., and E. Bertelsen. 1983. Histology of the attachment between the parasitic male and the female in the deep-sea anglerfish *Haplophryne mollis* (Brauer, 1902) (Pisces, Ceratioidei). *Vidensk. Meddr Dansk Naturh. Foren.* 144:49–74.

Munk, P., and T. Kiørboe. 1985. Feeding behaviour and swimming activity of larval herring (*Clupea harengus*) in relation to density of copepod nauplii. *Mar. Ecol. Prog. Ser.* 24:15–21.

Muñoz, M., M. Casadevall, S. Bonet, and I. Quaqio-Grassiotto. 2000. Sperm storage structures in the ovary of *Helicolenus dactylopterus* (Teleostei: Scorpaenidae): an ultrastructural study. *Environ. Biol. Fishes* 58:53–59.

Munro, J. L., V. C. Gaut, R. Thompson, and P. H. Reeson. 1973. The spawning seasons of Caribbean reef fish. *J. Fish Biol.* 5:69–84.

Munroe, T. A. 2000. An overview of the biology, ecology, and fisheries of the clupeoid fishes occurring in the Gulf of Maine. NMFS NEFSC Ref. Doc. No. 00-02, 226 pp.

Munson, D. A. 1970. A study of the parasites of *Liparis atlanticus* Jordan and Evermann with emphasis on the histochemical morphology of *Spathebothrium simplex* Linton 1922 (Cestoda). Ph.D. dissertation, University of New Hampshire, Dartmouth.

Murawski, S. A. 1976. Population dynamics and movement patterns of anadromous rainbow smelt, *Osmerus mordax* (Mitchill), in the Parker River estuary. M.S. thesis, University of Massachusetts, Amherst, 125 pp.

Murawski, S. A. 1993. Climate change and marine fish distributions: forecasting from historical analogy. *Trans. Am. Fish. Soc.* 122: 647–658.

Murawski, S. A., and C. F. Cole. 1978. Population dynamics of anadromous rainbow smelt, *Osmerus mordax,* in a Massachusetts River system. *Trans. Am. Fish. Soc.* 107:535–542.

Murawski, S. A., and A. L. Pacheco. 1977. Biological and fisheries data on Atlantic sturgeon, *Acipenser oxyrhynchus* (Mitchill). Sandy Hook Lab. Tech. Ser. Rept. No. 10, 69 pp.

Murawski, S. A., and G. T. Waring. 1979. A population assessment of butterfish, *Peprilus triacanthus,* in the northwestern Atlantic Ocean. *Trans. Am. Fish. Soc.* 108:427–439.

Murawski, S. A., D. G. Frank, and S. Chang. 1978. Biological and fisheries data on butterfish, *Peprilus triacanthus* (Peck). Sandy Hook Lab. Tech. Ser. Rept. No. 6, 39 pp.

Murawski, S. A., G. R. Clayton, R. J. Reed, and C. F. Cole. 1980. Movements of spawning rainbow smelt, *Osmerus mordax,* in a Massachusetts estuary. *Estuaries* 3:308–314.

Murawski, W. S. 1969. The distribution of striped bass, *Roccus saxatilis,* eggs and larvae on the lower Delaware River. N.J. Dept. Conserv. Econ. Dev. Div. Fish Game, Nacote Creek Res. Stn. Misc. Rept. No. 1M, 39 pp.

Murawski, W. S., and P. J. Festa. 1974. A study of young and larval summer flounder in New Jersey estuarine waters. New Jersey Div. Fish, Game, Shellfish, Macote Creek Res. Sta., Misc. Rept. No. 11M.

Murawski, W. W. 1970. Study of the ichthyoplankton associated with two of New Jersey's coastal inlets. N.J. Div. Fish Game Shellfish, 34 pp.

Murphy, S. C. 1991. The ecology of estuarine fishes in southern Maine high salt marshes; access corridors and movement patterns. M.S. thesis, University of Massachusetts, Amherst, 87 pp.

Murray, J., and J. Hjort. 1912. *The Depths of the Ocean: A General Account of the Modern Science of Oceanography Based Largely on the Scientific Results of the Norwegian Steamer "Michael Sars" in the North Atlantic.* MacMillan, London, 821 pp.

Musick, J. A. 1966. The distribution of *Helicolenus dactylopterus* in the Gulf of Maine. *Copeia* 1966:877.

Musick, J. A. 1967. Designation of the hakes, *Urophycis chuss* and *Urophycis tenuis,* in ICNAF statistics. Int. Comm. Northw. Atl. Fish. Res. Doc. No. 67/76.

Musick, J. A. 1969. The comparative biology of two American Atlantic hakes, *Urophycis chuss* and *U. tenuis* (Pisces: Gadidae). Ph.D. dissertation, Harvard University, Cambridge, 150 pp.

Musick, J. A. 1972. Fishes of Chesapeake Bay and the adjacent Coastal Plain. In: M. L. Wass et al., eds. *A Checklist of the Biota of Llower Chesapeake Bay.* Virginia Inst. Mar. Sci., Spec. Sci. Rept. No. 65:175–206.

Musick, J. A. 1973a. A meristic and morphometric comparison of the hakes, *Urophycis chuss* and *U. tenuis* (Pisces, Gadidae). *Fish. Bull., U.S.* 71:479–488.

Musick, J. A. 1973b. Mesopelagic fishes from the Gulf of Maine and the adjacent continental slope. *J. Fish. Res. Bd. Can.* 30: 134–137.

Musick, J. A. 1974. Seasonal distribution of sibling hakes, *Urophycis chuss* and *U. tenuis* (Pisces, Gadidae). *Fish. Bull., U.S.* 72:481–495.

Musick, J. A., and K. W. Able. 1969. Occurrence and spawning of the sculpin *Triglops murrayi* (Pisces, Cottidae) in the Gulf of Maine. *J. Fish. Res. Bd. Can.* 26:473–475.

Musick, J. A., and L. P. Mercer. 1977. Seasonal distribution of black sea bass, *Centropristis striata* in the Mid-Atlantic Bight with comment on the ecology and fisheries of the species. *Trans. Am. Fish. Soc.* 106:12–25.

Musick, J. A., C. A. Wenner, and G. R. Sedberry. 1975. Archibenthic and abyssal benthic fishes. In: *May 1974 Baseline Investigation of Deepwater Dumpsite 106.* NOAA Dumpsite Evaluation Rept. 75-1, 388 pp.

Musick, J. A., R. E. Jenkins, and N. M. Burkhead. 1993a. Sturgeons: family Acipenseridae. In: *Freshwater Fishes of Virginia.* Am. Fish. Soc., Bethesda, pp. 183–190.

Musick, J. A., S. Branstetter, and J. A. Colvocoresses. 1993b. Trends in shark abundance from 1974 to 1991 for the Chesapeake Bight region of the U.S. mid-Atlantic coast. NOAA Tech. Rept. NMFS No. 115:1–18.

Musick, J. A., M. M. Harbin, S. A. Berkeley, G. H. Burgess, A. M. Eklund, L. Findley, R. G. Gilmore, J. T. Golden, D. S. Ha, G. R. Huntsman, J. C. McGovern, S. J. Parker, S. G. Poss, E. Sala, T. W. Schmidt, G. R. Sedberry, H. Weeks, and S. G. Wright. 2000. Marine, estuarine, and diadromous fish stocks at risk of extinction in North America (exclusive of salmonids). *Fisheries* 25(11): 6–30.

Musick, J. C., Jr. 1974. Observations on the spot (*Leiostomus xanthurus*) in Georgia estuaries and close inshore ocean waters. Georgia Dept. Nat. Res., Coast. Fish. Off. Contrib. Ser. No. 28, 29 pp.

Myrberg, A. A., Jr., and S. H. Gruber. 1974. The behavior of the bonnethead shark, *Sphyrna tiburo. Copeia* 1974:358–374.

Nafpaktitis, B. G., R. H. Backus, J. E. Craddock, R. L. Haedrich, B. H. Robison, and C. Karnella. 1977. Family Myctophidae. Fishes of the western North Atlantic. Sears Found. Mar. Res. Mem. No. 1(7):13–265.

Nagler, J. J., B. A. Adams, and D. G. Cyr. 1999. Egg production, fertility, and hatch success of American plaice held in captivity. *Trans. Am. Fish. Soc.* 128:727–736.

Naiman, R. J., S. D. McCormick, W. L. Montgomery, and R. Morin. 1987. Anadromous brook charr, *Salvelinus fontinalis:* opportunities and constraints for population enhancement. *Mar. Fish. Rev.* 49(4): 1–13.

Nakamura, I. 1985. *FAO Species Catalogue,* Vol. 5. *Billfishes of the World: An Annotated and Illustrated Catalogue of Marlins, Sailfishes, Spearfishes and Swordfishes Known to Date.* FAO Fish. Synop. No. 125, 65 pp.

Nakamura, I. 1986. *Important Fishes Trawled off Patagonia.* Japan Marine Fishery Resources Research Center, Tokyo, 369 pp.

Nakamura, I., and N. V. Parin. 1993. *FAO Species Catalogue,* Vol. 15. *Snake Mackerels and Cutlassfishes of the World (Families Gempylidae and Trichiuridae).* FAO Fish. Synop. No. 125, 136 pp.

Nakano, H., M. Okazaki, and H. Okamoto. 1997. Analysis of catch depth by species for tuna longline fishery based on catch by branch lines. *Bull. Nat. Res. Inst. Far Seas Fish.* 34:43–62.

Nakano, S., and M. Kaeiryama. 1995. Summer microhabitat use and diet of four sympatric stream-dwelling salmonids in a Kamchatkan stream. *Fish. Sci.* 61:926–30.

Naplin, N. A., and C. L. Obenchain. 1980. A description of eggs and larvae of the snake eel, *Pisodonophis cruentifer* (Ophichthidae). *Bull. Mar. Sci.* 30:413–423.

Nash, R. D. M. 1980. Laboratory observations on the burrowing of the snake blenny, *Lumpenus lampraetaeformis* (Walbaum), in soft sediment. *J. Fish Biol.* 16:639–648.

Natanson, L. J. 1990. Relationship of vertebral band deposition to age and growth in the dusky shark, *Carcharhinus obscurus,* and the little skate, *Raja erinacea.* Ph.D. dissertation, University of Rhode Island, Kingston, 153 pp.

Natanson, L. J., J. J. Mello, and S. E. Campana. 2002. Validated age and growth of the porbeagle shark, *Lamna nasus,* in the western North Atlantic. *Fish. Bull., U.S.* 100.

Naughton, S. P., and C. H. Saloman. 1984. Food of bluefish (*Pomatomus saltatrix*) from the U.S. south Atlantic and Gulf of Mexico. NOAA Tech. Memo. NMFS-SEFC-150, 37 pp.

Neave, F. 1965. Transplants of pink salmon. Fish. Res. Bd. Can. Manuscript Rept. No. 830, 23 pp.

Needler, A. W. H. 1930. The migrations of haddock and the interrelationships of haddock populations in North American waters. *Contrib. Can. Biol. Fish., N.S.* 6(10):241–314.

Needler, A. W. H. 1940. A preliminary list of the fishes of Malpeque Bay. *Proc. Nova Scotia Inst. Sci.* 20:33–41.

Neilson, J. D., R. I. Perry, J. S. Scott, and P. Valerio. 1987. Interactions of caligid ectoparasites and juvenile gadids on Georges Bank. *Mar. Ecol. Prog. Ser.* 39:221–232.

Neilson, J. D., E. M. DeBlois, and P. C. F. Hurley. 1988. Stock structure of Scotian Shelf flatfish as inferred from ichthyoplankton survey data and the geographic distribution of mature females. *Can. J. Fish. Aquat. Sci.* 45:1674–1685.

Neilson, J. D., P. Perley, and H. Sampson. 1993. Reproductive biology of Atlantic halibut (*Hippoglossus hippoglossus*) in Canadian waters. *Can. J. Fish. Aquat. Sci.* 50:551–563.

Nelson, G. A., and M. R. Ross. 1991. Biology and population changes of northern sand lance (*Ammodytes dubius*) from the Gulf of Maine to the Middle Atlantic Bight. *J. Northw. Atl. Fish. Sci.* 11:11–27.

Nelson, G. A., and M. R. Ross. 1992. Distribution, growth and food habits of the Atlantic wolffish (*Anarhichas lupus*) from the Gulf of Maine–Georges Bank region. *J. Northw. Atl. Fish. Sci.* 13:53–61.

Nelson, G. A., and M. R. Ross. 1995. Gastric evacuation in little skate. *J. Fish Biol.* 46:977–986.

Nelson, G. J. 1984. Notes on the rostral organ of anchovies. *Japan J. Ichthyol.* 31:86–87.

Nelson, J. S. 1968a. Salinity tolerance of brook sticklebacks, *Culaea inconstans,* freshwater ninespine sticklebacks, *Pungitius pungitius,* and freshwater fourspine sticklebacks, *Apeltes quadracus. Can. J. Zool.* 46:663–667.

Nelson, J. S. 1968b. Ecology of the southernmost sympatric population of the brook stickleback, *Culaea inconstans,* and the ninespine stickleback, *Pungitius pungitius,* in Crooked Lake, Indiana. *Proc. Ind. Acad. Sci.* 77:185–192.

Nelson, J. S. 1968c. Deep-water ninespine sticklebacks, *Pungitius pungitius,* in the Mississippi drainage, Crooked Lake, Indiana. *Copeia* 1968:326–334.

Nelson, J. S. 1971. Absence of the pelvic complex in ninespine sticklebacks, *Pungitius pungitius,* collected in Ireland and Wood Buffalo National Park region, Canada, with notes on meristic variation. *Copeia* 1971:707–717.

Nelson, J. S. 1994. *Fishes of the World,* 3rd Ed. John Wiley & Sons, New York, 600 pp.

Nelson, W. R. 1969. Studies on the croaker, *Micropogon undulatus* Linnaeus and the spot, *Leiostomus xanthurus* Lacepede, in Mobile Bay, Alabama. *J. Mar. Sci. Alabama* 1:4–92.

Nelson, W. R., and J. S. Carpenter. 1968. Bottom longline explorations in the Gulf of Mexico: a report on "Oregon II's" first cruise. *Comm. Fish. Rev.* 30(10):57–62.

Nelson, W. R., M. C. Ingham, and W. E. Schaaf. 1977. Larval transport and year-class strength of Atlantic menhaden, *Brevoortia tyrannus. Fish. Bull., U.S.* 75:23–41.

Nesbit, R. A. 1954. Weakfish migration in relation to its conservation. U.S. Fish Wildl. Serv., Spec. Sci. Rept. Fish. No. 115, 81 pp.

Nesterov, A. A., and T. A. Shiganova. 1976. The eggs and larvae of the Atlantic saury, *Scomberesox saurus* of the North Atlantic. *Voprosy Ikhtiol.* 16:315–322 [in Russian, translation in *J. Ichthyol.* 16:277–283].

Netboy, A. 1974. *The Salmon: Their Fight for Survival.* Houghton Mifflin Co., Boston, 613 pp.

Netzel, J., and E. Stanek. 1966. Some biological characteristics of blueback, *Alosa aestivalis* (Mitchill), and alewife, *Alosa pseudoharengus* (Wilson), from Georges Bank, July and October, 1964. *Int. Comm. Northw. Atl. Fish. Res. Bull.* 3:106–110.

Neuman, M. J. 1993. Distribution, abundance and diversity of shoreline fishes in the Great Salt Pond, Block Island, Rhode Island. M.S. thesis, University of Rhode Island, Kingston, 88 pp.

Neuman, M. J., K. W. Able, R. Berghahn, A. D. Rijnsdorp, and H. W. Van Der Veer. 1998. Experimental evidence of sediment preference by early life history stages of windowpane (*Scopthalmus aquosus*). *Neth. J. Sea Res.* 40:33–41.

Neumann, D. A., J. M. O'Connor, and J. A. Sherk, Jr. 1981. Oxygen consumption of white perch (*Morone americana*), striped bass (*M. saxatilis*) and spot (*Leiostomus xanthurus*). *Comp. Biochem. Physiol.* 69A:467–478.

Neves, R. J. 1981. Offshore distribution of alewife, *Alosa pseudoharengus,* and blueback herring, *Alosa aestivalis,* along the Atlantic coast. *Fish. Bull., U.S.* 79:473–485.

Neves, R. J. , and L. Depres. 1979. The oceanic migration of American shad, *Alosa sapidissima,* along the Atlantic coast. *Fish. Bull., U.S.* 77:199–212.

Newberger, T. A. 1989. Relative abundance, age, growth, and mortality of bay anchovy in the Mid-Chesapeake Bay. M.S. thesis, University of Maryland, College Park, 119 pp.

Newberger, T. A., and E. D. Houde. 1995. Population biology of bay anchovy *Anchoa mitchilli* in the mid Chesapeake Bay. *Mar. Ecol. Prog. Ser.* 116:25–37.

Newman, H. N. 1907. Spawning behavior and sexual dimorphism in *Fundulus heteroclitus* and allied fish. *Biol. Bull., Woods Hole* 12:314–346.

Newman, H. N. 1908. The process of heredity as exhibited by the development of *Fundulus* hybrids. *J. Exp. Zool.* 5:503–561.

Newman, H. N. 1909a. Contact organs in the killifishes of Woods Hole. *Biol. Bull., Woods Hole* 17:170–180.

Newman, H. N. 1909b. The question of viviparity in *Fundulus majalis. Science* 30:769–771.

Newman, H. N. 1914. Modes of inheritance in teleost hybrids. *J. Exp. Zool.* 16:447–499.

Newth, D. R, and D. M. Ross. 1955. On the reaction to light of *Myxine glutinosa* L. *J. Exp. Biol.* 32:4–21.

Ni, I-H. 1981a. Separation of sharp-beaked redfish, *Sebastes fasciatus* and *S. mentella,* from northeastern Grand Bank by morphology of extrinsic gas bladder musculature. *J. Northw. Atl. Fish. Sci.* 2:7–12.

Ni, I-H. 1981b. Numerical classification of sharp-beaked redfishes, *Sebastes mentella* and *S. fasciatus,* from northeastern Grand Bank. *Can. J. Fish. Aquat. Sci.* 38:873–879.

Ni, I-H. 1982. Meristic variation in beaked redfishes, *Sebastes mentella* and *S. fasciatus,* in the northwest Atlantic. *Can. J. Fish. Aquat. Sci.* 39:1664–1685.

Ni, I-H. 1984. Meristic variation in golden redfish, *Sebastes marinus,* compared to beaked redfishes of the northwest Atlantic. *J. Northw. Atl. Fish. Sci.* 5:65–70.

Ni, I-H., and E. J. Sandeman. 1984. Size at maturity for northwest Atlantic redfishes (*Sebastes*). *Can. J. Fish. Aquat. Sci.* 41: 1753–1762.

Ni, I-H., and E. J. Templeman. 1985. Reproductive cycles of redfishes (*Sebastes*) in southern Newfoundland waters. *J. Northw. Atl. Fish. Sci.* 6:57–63.

Nichols, J. T. 1918. An abnormal winter flounder and others. *Copeia* No. 55:37–39.

Nichols, J. T., and C. M. Breder, Jr. 1927. The marine fishes of New York and southern New England. *Zoologica, N.Y.* 9:1–192.

Nichols, J. T., and F. E. Firth. 1939. Rare fishes off the Atlantic coast including a new grammicolepid. *Proc. Biol. Soc. Washington* 52:85–88.

Nichols, P. R. 1966. Comparative study of juvenile American shad populations by fin ray and scute counts. U.S. Fish Wildl. Serv. SSRF-525, 10 pp.

Nichols, P. R., and R. V. Miller. 1967. Seasonal movements of striped bass, *Roccus saxatilis* (Walbaum) tagged and released in the Potomac River, Maryland, 1959–1961. *Chesapeake Sci.* 8:102–124.

Nicholson, W. R. 1971. Coastal movements of Atlantic menhaden as inferred from changes in age and length distributions. *Trans. Am. Fish. Soc.* 100:708–716.

Nicholson, W. R. 1972. Population structure and movement of Atlantic menhaden, *Brevoortia tyrannus,* as inferred from back-calculated length frequencies. *Chesapeake Sci.* 13:161–174.

Nicholson, W. R. 1978. Movements and population structures of Atlantic menhaden indicated by tag returns. *Estuaries* 1: 141–150.

Nichy, F. 1969. Growth patterns on otoliths from young silver hake, *Merluccius bilinearis* (Mitch.). *ICNAF Res. Bull.* 6:107–117.

Nielsen, J. G. 1963. Marine fishes new or rare to the Danish fauna (from the period 1937–1961). *Vidensk. Medd. Dansk Naturh. Foren.* 125:147–165.

Nielsen, J. G., and D. G. Smith. 1978. The eel family Nemichthyidae (Pisces, Anguilliformes). Dana Rept. No. 88, 71 pp.

Nielsen, J. G., D. M. Cohen, D. F. Markle, and C. R. Robins. 1999. *FAO Species Catalogue,* Vol. 18. *Ophidiiform Fishes of the World (Order Ophidiiformes).* FAO Fish. Synop. No. 125, 178 pp.

Nigrelli, R. F. 1935. The morphology, cytology, and life history of *Oodinium ocellatum* Brown, a dinoflagellate parasite on marine fishes. *Zoologica, N.Y.* 21:129–164.

Nigrelli, R. F. 1946. Studies on the marine resources of southern New England. V. Parasites and diseases of the ocean pout, *Macrozoarces americanus. Bull. Bingham Oceanogr. Coll.* 9(5):187–221.

Nigrelli, R. F., and F. E. Firth. 1939. On *Sphyrion lumpi* (Krøyer), a copepod parasite on the redfish, *Sebastes marinus* (Linnaeus), with special reference to the host-parasite relationships. *Zoologica, N.Y.* 24:1–10.

Nigrelli, R. F., and H. Vogel. 1963. Spontaneous tuberculosis in fishes and in other coldblooded vertebrates with special reference to *Mycobacterium fortuitum* Cruz from fish and human lesions. *Zoologica, N.Y.* 48:131–144.

Nigrelli, R. F., K. S. Pokorny, and G. D. Ruggieri. 1975. Studies on parasitic kinetoplastids. II. Occurrence of a biflagellate kinetoplastid in the blood of *Opsanus tau* (toadfish), transmitted by the leech (*Piscicola funduli*). *J. Protozool.* 22:43A (abstract).

Nilsson, D. 1914. A contribution to the biology of the mackerel; investigations in Swedish waters, Publ. Circ., Copenhagen, No. 69, 68 pp.

Nitschke, P., M. Mather, and F. Juanes. 2001. A comparison of length-, weight-, and age-specific fecundity relationships for cunner in Cape Cod Bay. *No. Am. J. Fish. Manage.* 21:86–95.

Nixon, S. W., and C. A. Oviatt. 1973. Ecology of a New England salt marsh. Ecol. Monogr. 43:463–498.

Nizinski, M. S., B. B. Collette, and B. B. Washington. 1990. Separation of two species of sand lances, *Ammodytes americanus* and *A. dubius,* in the western North Atlantic. *Fish. Bull., U.S.* 88: 241–255.

NMFS (National Marine Fisheries Service). 1998a. Recovery plan for the shortnose sturgeon (*Acipenser brevirostrum*). Office of Protected Resources, NMFS, Silver Spring, Md., 97 pp.

NMFS (National Marine Fisheries Service). 1998b. Status review of Atlantic sturgeon (*Acipenser oxyrinchus oxyrinchus*). Office of Protected Resources, NMFS, Silver Spring, Md., 125 pp.

Noga, E. J., and M. J. Dykstra. 1986. Oomycete fungi associated with ulcerative mycosis in menhaden, *Brevoortia tyrannus* (Latrobe). *J. Fish Dis.* 9:47–53.

Nolan, K., J. Grossfield, and I. Wirgin. 1991. Discrimination among Atlantic coast populations of American shad (*Alosa sapidissima*) using mitochondrial DNA. *Can. J. Fish. Aquat. Sci.* 48:1724–1734.

Norcross, B. L., and D. M. Wyanski. 1994. Interannual variation in the recruitment pattern and abundance of age-0 summer flounder, *Paralichthys dentatus,* in Virginia estuaries. *Fish. Bull., U.S.* 92:591–598.

Norcross, J. J., W. H. Massmann, and E. B. Joseph. 1961. Investigations of inner continental shelf waters off lower Chesapeake Bay. Part II. Sand lance larvae, *Ammodytes americanus. Chesapeake Sci.* 2:49–59.

Norcross, J. J., S. L. Richardson, W. H. Massmann, and E. B. Joseph. 1974. Development of young bluefish (*Pomatomus saltatrix*) and distribution of eggs and young in Virginia coastal waters. *Trans. Am. Fish. Soc.* 103:477–497.

Norden, C. R. 1967. Age, growth and fecundity of the alewife, *Alosa pseudoharengus* (Wilson), in Lake Michigan. *Trans. Am. Fish. Soc.* 96:387–393.

Norden, C. R. 1968. Morphology and food habits of larval alewife, *Alosa pseudoharengus* in Lake Michigan. *Proc. Conf. Great Lakes Res.* 10:70–78.

Nordøy, E. S., and A. S. Blix. 1992. Diet of minke whales in the northeastern Atlantic. Int. Whaling Comm. Rept. No. 42, pp. 393–398.

Norman, J. R. 1934. *A Systematic Monograph of the Flatfishes (Heterosomata),* Vol. 1. *Psettodidae, Bothidae, Pleuronectidae.* British Museum of Natural History, London, 459 pp.

Norman, J. R., and F. C. Fraser. 1949. *Field Book of Giant Fishes.* G. P. Putnam's Sons, New York, 376 pp.

Norris, D. E., and R. M. Overstreet. 1975. *Thynnascaris reliquens* sp. n. and *T. habena* (Linton, 1900) (Nematoda: Ascaridoidea) from fishes in the northern Gulf of Mexico and eastern U. S. seaboard. *J. Parasitol.* 61:330–336.

Norvillo, G. V., and N. J. Zhuravleva. 1989. Early post embryonic development of the arctic staghorn sculpin *Gymnacanthus tricuspis. Voprosy Ikhtiol.* 29:331–333 [in Russian, translation in *J. Ichthyol.* 29(5):148–150.

Nyman, R. M., and D. O. Conover. 1988. The relation between spawning season and the recruitment of young-of-the-year bluefish, *Pomatomus saltatrix,* to New York. *Fish. Bull., U.S.* 86:237–250.

O'Boyle, R. N., M. Sinclaire, R. J. Conover, K. H. Mann, and A. C. Kohler. 1984. Temporal and spatial distribution of ichthyoplankton communities of the Scotian Shelf and relation of biological, hydrological and physiographic features. *ICES Rapp. Proc.-Verb.* 183:27–40.

O'Brien, L. 1998a. Cusk. In: S. H. Clark, ed. *Status of Fishery Resources off the Northeastern United States for 1994.* NOAA Tech. Memo. NMFS-NE-115:98–99.

O'Brien, L. 1998b. American plaice. In: S. H. Clark, ed. *Status of Fishery Resources off the Northeastern United States for 1998.* NOAA Tech. Memo. NMFS-NE-115:77–78.

O'Brien, L. 1998c. Factors influencing rates of maturation in the Georges Bank and the Gulf of Maine Atlantic cod stocks. NAFO Sci. Counc. Res. Doc. No. 98/104, 34 pp.

O'Brien, L., J. Burnett, and R. K. Mayo. 1993. Maturation of nineteen species of finfish off the northeast coast of the United States, 1985–1990. NOAA Tech. Rept. NMFS No. 113, 66 pp.

O'Connor, J. M. 1972. Tidal activity rhythm in the hogchoker *Trinectes maculatus* (Bloch and Schneider). *J. Exp. Mar. Biol. Ecol.* 9:173–177.

O'Connor, M. 1997. Assessment of abundance and distribution of young-of-the-year winter flounder, *Pleuronectes americanus,* in Point Judith Pond. M.S. thesis, University of Rhode Island, Kingston, 41 pp.

O'Grady, S., J. D. Schrag, J. A. Raymond, and A. L. DeVries. 1982. Comparison of antifreeze glycopetides from Arctic and Antarctic fishes. *J. Exper. Zool.* 224:177–185.

O'Herron, J. C., II, K. W. Able, and R. W. Hastings. 1993. Movements of shortnose sturgeon *Acipenser brevirostrum* in the Delaware River. *Estuaries* 16:235–240.

O'Neil, S. P., and M. P. Weinstein. 1988. Feeding habitats of spot, *Leiostomus xanthurus,* in polyhaline versus meso-oligohaline tidal creeks and shoals. *Fish. Bull., U.S.* 85:785–796.

O'Neill, J. T. 1980. Aspects of the life histories of anadromous alewife and the blueback herring, Margaree River and Lake Aninsle, Nova Scotia, 1978–1979. M.S. thesis, Acadia University, Wolfville, N.S.

Odell, T. T. 1934. The life history and ecological relationships of the alewife *Pomolobus pseudoharengus* (Wilson) in Seneca Lake, New York. *Trans. Am. Fish. Soc.* 64:118–24.

Odense, P. H. 1980. Herring isoenzymes and population. *Anim. Blood Groups Biochem. Genet.* 11(Suppl.): 66.

Odense, P. H., and M. C. Annand. 1980. Herring (*Clupea harengus*) isoenzyme studies. Int. Counc. Explor. Sea CM 1980/H:25.

Odense, P. H., and V. H. Logan. 1976. Prevalence and morphology of *Eimeria gadi* (Feibiger, 1913) in the haddock. *J. Protozool.* 23:564–571.

Odense, P. H., T. C. Leung, and C. Amand. 1973. Isozyme systems of some Atlantic herring populations. ICES Doc. 1973/H:21, 13 pp.

Odum, W. E. 1970. Utilization of the direct grazing and plant detritus food chains by the striped mullet, *Mugil cephalus.* In: J. H. Steele, ed. *Marine Food Chains.* University of California Press, Berkeley, p. 222–240.

Oelschläger, H. 1974. Das Jugendstadium von *Lampris guttatus* (Brünnich, 1788) (Osteichthyes, Allotriognathi), ein Beitrag zur Kenntnis seiner Entwicklung. *Arch. Fisch. Wiss.* 25:3–19.

Ofori-Danson, P.K. 1990. Reproductive ecology of the triggerfish, *Balistes capriscus* from the Ghanaian coastal waters. *Trop. Ecol.* 31:1–11.

Ogden, J. C. 1970. Relative abundance, food habits, and age of the American eel, *Anguilla rostrata* (LeSueur) in certain New Jersey streams. *Trans. Am. Fish. Soc.* 99:54–59.

Ogren, L., J. Chess, and J. Lindenberg. 1968. More notes on the behavior of young squirrel hake, *Urophycis chuss. Underwater Nat.* 5:38–39.

Ojeda, F. P., and J. H. Dearborn. 1990. Diversity, abundance, and spatial distribution of fishes and crustaceans in the rocky subtidal zone of the Gulf of Maine. *Fish. Bull., U.S.* 88:403–410.

Ojeda, F. P., and J. H. Dearborn. 1991. Feeding ecology of benthic mobile predators: experimental analysis of their influence in rocky subtidal communities of the Gulf of Maine. *J. Exp. Mar. Biol. Ecol.* 149:13–44.

Okamoto, H., H. Sato, and K. Shimazaki. 1989. Comparison of reproductive cycle between two genetically distinctive groups of sand lance (genus *Ammodytes*) from northern Hokkaido. *Nippon Suisan Gakkaishi* 55:1935–1940.

Ólafsson, E. 1772. Ferðabok Eggerts Ólafssonar og Bjarna Pálssonar um ferðir peirra á Íslandi árin 1752–1757 (The travels of Eggert Ólafsson and Bjarni Pálsson in Iceland during 1752–1757. In Icelandic). Ísfoldarprentsmiðja h.f. Reykjavík 1943. [not seen]

Oldham, W. S. 1972. Biology of Scotian shelf cusk, *Brosme brosme. Int. Comm. North Atlantic Fish., Res. Bull.* 9:85–98.

Olla, B. L., and C. Samet. 1977. Courtship and spawning behavior of the tautog, *Tautoga onitis* (Pisces: Labridae), under laboratory conditions. *Fish. Bull., U.S.* 75:585–599.

Olla, B. L., and C. Samet. 1978. Effects of elevated temperature on early embryonic development of the tautog, *Tautoga onitis. Trans. Am. Fish. Soc.* 107:820–824.

Olla, B. L., and A. L. Studholme. 1971. The effect of temperature on the activity of bluefish, *Pomatomus saltatrix,* L. *Biol. Bull., Woods Hole* 141:337–349.

Olla, B. L., and A. L. Studholme. 1972. Daily and seasonal rhythms of activity in the bluefish, *Pomatomus saltatrix,* L. In: H. E. Winn and B. L. Olla, eds. *Behavior of Marine Animals: Recent Advances,* Vol. II. Plenum Press, New York, pp. 303–326.

Olla, B. L., and A. L. Studholme. 1975. The effect of temperature on the behavior of young tautog, *Tautoga onitis* (L.). *Proc. 9th Europ. Mar. Biol. Symp.* 1975:75–93.

Olla, B. L., R. Wicklund, and S. Wilk. 1969. Behavior of winter flounder in a natural habitat. *Trans. Am. Fish. Soc.* 98:717–720.

Olla, B. L., H. L. Katz, and A. L. Studholme. 1970. Prey capture and feeding motivation in the bluefish, *Pomatomus saltatrix. Copeia* 1970:360–362.

Olla, B. L., C. E. Samet, and A. L. Studholme. 1972. Activity and feeding behavior of the summer flounder (*Paralichthys dentatus*) under controlled laboratory conditions. *Fish. Bull., U.S.* 70: 1127–1136.

Olla, B. L., A. J. Bejda, and A. D. Martin. 1974. Daily activity, movements, feeding, and seasonal occurrence in the tautog, *Tautoga onitis. Fish. Bull., U.S.* 72:27–35.

Olla, B. L., A. J. Bejda, and A. D. Martin. 1975a. Activity, movements, and feeding behavior of the cunner, *Tautogolabrus adspersus,* and comparison of food habits with young tautog, *Tautoga onitis,* off Long Island, New York. *Fish. Bull., U.S.* 73:895–900.

Olla, B. L., A. L. Studholme, A. J. Bejda, C. Samet, and A. D. Martin. 1975b. The effect of temperature on the behavior of marine fishes: a comparison among Atlantic mackerel, *Scomber scombrus,* bluefish, *Pomatomus saltatrix,* and tautog, *Tautoga onitis.* In: *Combined Effects of Radioactive, Chemical and Thermal Releases to the Environment, Vienna: IAEA 1975.* Internat. Atom. Ener. Agen. SM–197/4, pp. 299–308.

Olla, B. L., A. L. Studholme, A. J. Bejda, C. Samet, and A. D. Martin. 1978. Effect of temperature on activity and social behavior of the adult tautog *Tautoga onitis* under laboratory conditions. *Mar. Biol.* 45:369–378.

Olla, B. L., A. J. Bejda, and A. D. Martin. 1979. Seasonal dispersal and habitat selection of cunner, *Tautogolabrus adspersus,* and young tautog, *Tautoga onitis,* in Fire Island Inlet, Long Island, New York. *Fish. Bull., U.S.* 77:255–261.

Olla, B. L., A. L. Studholme, and A. L. Bejda. 1985. Behavior of juvenile bluefish *Pomatomus saltatrix* in vertical thermal gradients: influence of season, temperature acclimation and food. *Mar. Ecol. Prog. Ser.* 23:165–177.

Olney, J. E. 1983. Eggs and early larvae of the bay anchovy, *Anchoa mitchilli,* and the weakfish, *Cynoscion regalis,* in lower Chesapeake Bay with notes on associated ichthyoplankton. *Estuaries* 6:20–35.

Olney, J. E. 1984. Lampriformes: development and relationships. In: H. G. Moser et al., eds. *Ontogeny and Systematics of Fishes.* Am. Soc. Ich. Herp. Spec. Publ. No. 1:368–379.

Olney, J. E., and G. W. Boehlert. 1988. Nearshore ichthyoplankton associated with seagrass beds in the lower Chesapeake Bay. *Mar. Ecol. Prog. Ser.* 45:33–43.

Olney, J. E., G. C. Grant, F. E. Schultz, C. L. Cooper, and J. Hageman. 1983. Pterygiophore–interdigitation patterns in larvae of four *Morone* species. *Trans. Am. Fish. Soc.* 112:525–531.

Olney, J. E., G. D. Johnson, and C. C. Baldwin. 1993. Phylogeny of lampridiform fishes. *Bull. Mar. Sci.* 52:137–169.

Olsen, Y. H., and D. Merriman. 1946. Studies on the marine resources of southern New England. IV. The biology and economic importance of the ocean pout, *Macrozoarces americanus* (Bloch and Schneider). *Bull. Bingham Oceanogr. Coll.* 9(4):1–184.

Orach-Meza, F. L. 1975. Distribution and abundance of ocean pout, *Macrozoarces americanus* (Bloch and Schneider) 1801 in the western North Atlantic Ocean. M.S. thesis, University of Rhode Island, Kingston, 143 pp.

Orbacz, E. A., and P. M. Gaffney. 2000. Genetic structure of tautog (*Tautoga onitis*) populations assayed by RFLP and DGGE analysis of mitochondrial and nuclear genes. *Fish. Bull., U.S.* 98: 336–344.

Oren, O. H., ed. 1981. *Aquaculture of Grey Mullets.* Cambridge University Press, Cambridge, 507 pp.

Orlova, E. L., L. I. Karamushko, E. G. Berestovskii, and E. A. Kireeva. 1989. Studies on feeding in the Atlantic wolffish, *Anarhichas lupus,* and the spotted wolffish, *A. minor,* under experimental conditions. *Voprosy Ihktiol.* 29:792–801 [in Russian, translation in *J. Ichthyol.* 29:91–101].

Orlova, E. L., E. G. Berestovskii, O. V. Karamushko, and G. V. Norvillo. 1990. On feeding and distribution of young wolffishes, *Anarhichas lupus* and *A. minor,* in the Barents and Norwegian seas. *Voprosy Ihktiol.* 30:867–870 [in Russian, translation in *J. Ichthyol.* 30:126–131].

Orlowski, S. J., S. S. Herman, R. G. Malsberger, and H. N. Pritchard. 1972. Distinguishing cunner and tautog eggs by immunodiffusion. *J. Fish. Res. Bd. Can.* 29:111–112.

Orr, D. C., and W. R. Bowering. 1997. A multivariate analysis of food and feeding trends among Greenland halibut (*Reinhardtius hippoglossoides*) sampled in Davis Strait, during 1986. *ICES J. Mar. Sci.* 54:819–829.

Orth, R. J., and K. L. Heck, Jr. 1980. Structural components of eelgrass (*Zostera marina*) meadows in the lower Chesapeake Bay—fishes. *Estuaries* 3:278–288.

Osman, R. W., and R. B. Whitlach. 1995. Ecological factors controlling the successful invasion of three species of ascidians into marine subtidal habitats of New England. In: N. Balcom, ed. *Proceedings of the Northeast Conference on Non-Indigenous Aquatic Nuisance Species.* Cromwell, Groton, Conn., pp. 49–60.

Østvedt, Ó. J. 1963. On the life history of the spotted catfish (*Anarhichas minor* Olafsen). *Fiskeridir. Skr. (Havunders)* 13:54–72.

Otwell, W. S., and J. V. Merriner. 1975. Survival and growth of juvenile striped bass, *Morone saxatilis,* in a factorial experiment with temperature, salinity, and age. *Trans. Am. Fish. Soc.* 104:560–566.

Overholtz, W. J. 1989. Density-dependent growth in the northwest Atlantic stock of Atlantic mackerel (*Scomber scombrus*). *J. Northw. Atl. Fish. Sci.* 9:115–121.

Overholtz, W. J. 1993. Harvesting strategies and fishing mortality reference point comparisons for the northwest Atlantic stock of Atlantic mackerel (*Scomber scombrus*). *Can. J. Fish. Aquat. Sci.* 50:1749–1756.

Overholtz, W. J. 1998a. Butterfish. In: S. H. Clark, ed. *Status of Fishery Resources off the Northeastern United States for 1998.* NOAA Tech. Memo. NMFS-NE-115:108–109.

Overholtz, W. J. 1998b. Atlantic mackerel. In: S. H. Clark, ed. *Status of Fishery Resources off the Northeastern United States for 1998.* NOAA Tech. Memo. NMFS-NE-115:106–107.

Overholtz, W. J., and S. X. Cadrin. 1998. Yellowtail flounder. In: S. H. Clarke, ed. *Status of Fishery Resources off the Northeastern United States for 1998.* NOAA Tech. Memo. NMFS-NE-115:70–74.

Overholtz, W. J., and J. R. Nicolas. 1979. Apparent feeding by the fin whale, *Balaenoptera physalus,* and humpback whale, *Megaptera novaengliae,* on the American sand lance, *Ammodytes americanus,* in the northwest Atlantic. *Fish. Bull., U.S.* 77:285–287.

Overholtz, W. J., S. A. Murawski, and W. L. Michaels. 1991a. Impact of compensatory responses on assessment advice for the northwest Atlantic mackerel stock. *Fish. Bull., U.S.* 89:117–128.

Overholtz, W. J., R. S. Armstrong, D. G. Mountain, and M. Terceiro. 1991b. Factors influencing spring distribution, availability, and recreational catch of Atlantic mackerel (*Scomber scombrus*) in the middle Atlantic and southern New England regions. NOAA Tech. Memo. NMFS-F/NEC-85, 13 pp.

Overstreet, R. M., and R. W. Heard. 1982. Food contents of six commercial fishes from Mississippi Sound. Gulf Res. Rept. No. 7: 137–149.

Oviatt, C. A. 1977. Menhaden, sport fish, and fishermen. Grad. School Oceanogr., Univ. Rhode Island, NOAA Sea Grant Mar. Tech. Rept. No. 60, 24 pp.

Oviatt, C. A., and G. W. Gray, Jr. 1968. Juvenile lookdowns, *Selene vomer,* in Wickford Cove, Narragansett Bay, Rhode Island. *Trans. Am. Fish. Soc.* 97:64.

Oviatt, C. A., and P. M. Kremer. 1977. Predation on the ctenophore, *Mnemiopsis leidyi,* by butterfish, *Peprilus triacanthus,* in Narragansett Bay, Rhode Island. *Chesapeake Sci.* 18:236–240.

Oviatt, C. A., and S. W. Nixon. 1973. The demersal fish of Narragansett Bay: an analysis of community structure, distribution and abundance. *Est. Coast. Mar. Sci.* 1:361–378.

Oviatt, C. A., A. L. Gall, and S. W. Nixon. 1971. Environmental effects of Atlantic menhaden on surrounding waters. *Chesapeake Sci.* 13:321–323.

Oxenford, H. A. 1999. Biology of the dolphinfish (*Coryphaena hippurus*) in the western central Atlantic: a review. *Sci. Mar.* 63:277–301.

Oxenford, H. A., and W. Hunte. 1983. Age and growth of dolphin, *Coryphaena hippurus,* as determined by growth rings in otoliths. *Fish. Bull., U.S.* 81:241–244.

Pacheco, A. L., and G. C. Grant. 1965. Studies of the early life history of Atlantic menhaden in estuarine nurseries. I. Seasonal occurrence of juvenile menhaden and other small fishes in a tributary creek of Indian River, Delaware, 1957–58. U.S. Fish Wildl. Ser. Spec. Sci. Rept. No. 504, 32 pp.

Packard, W. H. 1905. On resistance to lack of oxygen and on a method of increasing this resistance. *Am. J. Physiol.* 15:30–41.

Packer, D. B., L. Watling, and R. W. Langton. 1994. The population structure of the brittle star *Ophiura sarsi* Lütken in the Gulf of Maine and its trophic relationships to American plaice (*Hippoglossoides platessoides* Fabricus). *J. Exp. Mar. Biol. Ecol.* 179:207–222.

Packer, D. B., S. J. Griesbach, P. L. Berrien, C. A. Zetlin, D. L. Johnson, and W. W. Morse. 1999. Essential Fish Habitat Source Document: Summer flounder, *Paralichthys dentatus,* life history and habitat characteristics. NOAA Tech. Memo. NMFS-NE-151, 88 pp.

Padgett, H. R. 1967. Very young bluefish found in Florida. *Underwater Nat.* 4:42–43.

Padoa, E. 1956. Centrolophidae, Nomeidae. In: Fauna e Flora Golfo di Napoli. Monogr. No. 38, pp. 538–545.

Page, C. 1978. Trap fishing. *Oceans* 11:44–50.

Page, F. H., F. T. Frank, and K. R. Thompson. 1989. Stage dependent vertical distribution of haddock (*Melanogrammus aeglefinus*) eggs in a stratified water column: observations and model. *Can. J. Fish. Aquat. Sci.* 46(Suppl. 1):55–67.

Palko, B. J., G. L. Beardsley, and W. J. Richards. 1981. Synopsis of the biology of the swordfish, *Xiphias gladius* Linnaeus. NOAA Tech. Rept. NMFS Circ. No. 441, 21 pp.

Palko, B. J., G. L. Beardsley, and W. J. Richards. 1982. Synopsis of the biological data on dolphin-fishes, *Coryphaena hippurus* Linnaeus and *Coryphaena equiselis* Linnaeus. NOAA Tech. Rept. NMFS Circ. No. 443, 28 pp.

Palmer, G., and H. A. Oelschläger. 1976. Use of the name *Lampris guttatus* (Brünnich, 1788) in preference to *Lampris regius* (Bonnaterre, 1788) for the opah. *Copeia* 1976:366–367.

Palmer, R. E., and K. W. Able. 1987. Effect of acclimation salinity on fertilization success in the mummichog, *Fundulus heteroclitus. Physiol. Zool.* 60:614–621.

Palmer, R. H. 1936. Ocean sunfish in Habana waters. *Science* 83:597.

Palsson, J., and M. Beverly-Burton. 1983. *Laminiscus* n.g. (Monogenea: Gyrodactylidae) from capelin, *Mallotus villosus* (Müller), (Pisces: Osmeridae) in the northwest Atlantic with redescriptions of *L. gussevi* n. comb., *Gyrodactyloides petruschewskii,* and *G. andriaschewi. Can. J. Zool.* 61:298–306.

Pálsson, Ó. K. 1983. The feeding habits of demersal fish species in Icelandic waters. *Rit. Fiskideild* 7(1):1–60.

Pankratov, A. M., and I. K. Sigajev. 1973. Studies on Georges Bank herring spawning in 1970. *Int. Comm. Northw. Atl. Fish. Res. Bull.* 10:125–129.

Pannella, G. 1971. Fish otoliths: daily growth layers and periodical patterns. *Science* 173:1124–1127.

Paperna, I., and R. M. Overstreet. 1981. Parasites and diseases of mullets (Mugilidae). In: O. H. Oren, ed. *Aquaculture of Grey Mullets.* Cambridge University Press, Cambridge, pp. 411–493.

Paperna, I., and D. E. Zwerner. 1976. Parasites and diseases of striped bass, *Morone saxatilis* (Walbaum), from the lower Chesapeake Bay. *J. Fish Biol.* 9:267–287.

Paradis, M., R. G. Ackman, J. Hingley, and C. A. Eaton. 1975. Utilization of wastes from lumpfish, *Cyclopterus lumpus,* roe harvesting operations: an examination of the lipid and glue potential, and comparison of meal with that from Nova Scotia–caught menhaden. *J. Fish. Res. Bd. Can.* 32:1643–1648.

Parenti, L. R. 1981. A phylogenetic and biogeographic analysis of cyprinodontiform fishes (Teleostei, Atherinomorpha). *Bull. Am. Mus. Nat. Hist.* 168:335–557.

Parenti, L. R. 1993. Relationships of atherinomorph fishes (Teleostei). *Bull. Mar. Sci.* 52:170–196.

Parenti, P., and J. E. Randall. 2000. An annotated checklist of the species of the labroid fish families Labridae and Scaridae. *J. L. B. Smith Inst. Ichthyol. Bull.* 68, 97 pp.

Parin, N. V., and O. D. Borodulina. 1990. Survey of the genus *Polymetme* (Photichthyidae) with a description of two new species. *Voprosy Ikhtiol.* 30:733–743 [in Russian, translation in *J. Ichthyol.* 30(6):108–121].

Parin, N. V., and B. B. Collette. 1993. Results of the research cruises of FRV *Walther Herwig* to South America. LXIX. *Lepidopus altifrons,* a new species of cutlassfish (Pisces, Scombroidei, Trichiuridae) from the western Atlantic Ocean. *Arch. Fisch. Wiss.* 41:187–195.

Parin, N. V., and S. G. Kobyliansky. 1993. Review of the genus *Maurolicus* (Sternoptychidae, Stomiiformes) with reestablishing validity of five species considered junior synonyms of *M. muelleri* and descriptions of nine new species. *Trudy P.P. Shirshov Inst. Okean., Akad. Nauk SSSR* 128:69–107 [in Russian].

Parin, N. V., and S. G. Kobyliansky. 1996. Diagnoses and distribution of fifteen species recognized in the genus *Maurolicus* Cocco (Sternoptychidae, Stomiiformes) with a key to their identification. *Cybium* 20:185–195.

Parin, N. V., and Ye. I. Kukuev. 1983. Reestablishment of the validity of *Lampris immaculata* Gilchrist and the geographical distribution of opahs (Lampridae). *Voprosy Ikhtiol.* 23:3–14 [in Russian, translation in *J. Ichthyol.* 23(1):1–12].

Parin, N. V., K. N. Nesis, and M. Ye. Vinogradov. 1969. Data on the feeding of *Alepisaurus* in the Indian Ocean. *Voprosy Ikhtiol.* 9(3):526–538 [in Russian, translation in *Prob. Ichthyol.* 9:418–427].

Parker, G. H. 1925. Melanism and color changes in killifishes. *Copeia* 1925:81–83.

Parker, G. H. 1945. Melanophore activators in the common American eel, *Anguilla rostrata* Le Sueur. *J. Exper. Zool.* 98: 211–234.

Parker, H. W., and F. C. Stott. 1965. Age, size and vertebral calcification in the basking shark, *Cetorhinus maximus*(Gunnerus). *Zool. Meded.* 40:305–319.

Parker, R. O, Jr., and S. W. Ross. 1986. Observing reef fishes from submersibles off North Carolina. *Northeast Gulf Sci.* 8:31–49.

Parr, A. E. 1927a. On the functions and morphology of the postclavicular apparatus in *Sphaeroides* and *Chilomycterus. Zoologica, N.Y.* 9:245–269.

Parr, A. E. 1927b. A contribution to the theoretical analysis of the schooling behavior fishes. Bingham Oceanogr. Coll. Occas. Pap. No. 1:1–32.

Parr, A. E. 1931. A practical revision of the western Atlantic species of the genus *Citharichthys* (including *Etropus*). *Bull. Bingham Oceanogr. Coll.* 4(1):1–24.

Parrish, B. B, and A. Saville. 1965. The biology of northeast Atlantic herring populations. *Oceanogr. Mar. Biol. Ann. Rev.* 3:323–373.

Parsons, G. R. 1983. The reproductive biology of the Atlantic sharpnose shark, *Rhizoprionodon terraenovae* (Richardson). *Fish. Bull., U.S.* 81:61–73.

Parsons, G. R. 1985. Growth and age estimation of the Atlantic sharpnose shark, *Rhizoprionodon terraenovae:* a comparison of techniques. *Copeia* 1985:80–85.

Parsons, G. R. 1987. Life history and bioenergetics of the bonnethead shark, *Sphyrna tiburo* (Linnaeus): a comparison of two populations. Ph.D. dissertation, University of South Florida, Tampa, 170 pp.

Parsons, L. S. 1970. Northern range extension of the Atlantic mackerel, *Scomber scombrus,* to Black Island, Labrador. *J. Fish. Res. Bd. Can.* 27:610–613.

Parsons, L. S. 1972. Use of meristic characters and a discriminant function for classifying spring and autumn-spawning Atlantic herring. *Int. Comm. Northw. Atl. Fish. Res. Bull.* 9:5–9.

Parsons, L. S. 1973. Meristic characteristics of Atlantic herring, *Clupea harengus harengus* L., stocks in Newfoundland and adjacent waters. *Int. Comm. Northw. Atl. Fish. Res. Bull.* 10:37–52.

Parsons, L. S., and V. M. Hodder. 1971. Variation in incidence of larval nematodes in herring from Canadian Atlantic waters. *Int. Comm. Northw. Atl. Fish. Res. Bull.* 8:5–14.

Parsons, L. S., and V. M. Hodder. 1974. Some biological characteristics of the Fortune Bay, Newfoundland, herring stock, 1966–71. *Int. Comm. Northw. Atl. Fish. Res. Bull.* 10:15–22.

Parsons, L. S., and V. M. Hodder. 1981. Meristic differences between spring- and autumn-spawning Atlantic herring (*Clupea harengus harengus*) from southwestern Newfoundland. *J. Fish. Res. Bd. Can.* 28:553–558.

Parsons, L. S., and J. A. Moores. 1974. Long-distance migration of an Atlantic mackerel (*Scomber scombrus*). *J. Fish. Res. Bd. Can.* 31: 1521–1522.

Parsons, L. S., and G. H. Winters. 1972. ICNAF herring otolith exchange 1971–72. Int. Comm. Northw. Atl. Fish. Res. Doc. No. 72/92, Ser. No. 2816, 23 pp.

Pascoe, P. L. 1987. Monogenean parasites of deep-sea fishes from the Rockall Trough (N.E. Atlantic) including new species. *J. Mar. Biol. Assoc. U.K.* 67:603–622.

Pate, P. P. 1972. Life history aspects of the hickory shad, *Alosa mediocris* (Mitchill), in the Neuse River, North Carolina. M.S. thesis, North Carolina State University, Raleigh.

Patterson, C., and D. E. Rosen. 1977. A review of ichthyodectiform and other Mesozoic teleost fishes, and the theory and practice of classifying fossils. *Bull. Am. Mus. Nat. Hist.* 158:81–172.

Patton, D. K. 1965. Apparent mating color phase in the male scrawled filefish. *Underwater Nat.* 3(3):22–23.

Patzner, R. A., and H. Adam. 1981. Changes in the weight of the liver and the relationship to reproduction in the hagfish *Myxine glutinosa* (Cyclostomata). *J. Mar. Biol. Assoc. U.K.* 61:461–464.

Pauly, D. 1978. A critique of some literature data on the growth, reproduction and mortality of the lamnid shark, *Cetorhinus maximus* (Gunnerus). Report to Pelagic Fish Comm., Int. Council Explor. Seas. CM 1978/H:17, 10 pp.

Pavlov, D. A. 1986. Developing the biotechnology of culturing White Sea wolffish, *Anarhichas lupus marisalbi* Barsukov. II. Ecomorphological peculiarities of early ontogeny. *Voprosy Ikhtiol.* 26:835–849 [in Russian, translation in *J. Ichthyol.* 26:156–169].

Pavlov, D. A. 1994. Fertilization in the wolffish, *Anarhichas lupus:* external or internal? *Voprosy Ikhtiol.* 33:664–670 [in Russian, translation in *J. Ichthyol.* 34(1):140–151].

Pavlov, D. A., and E. Moksness. 1994. Production and quality of eggs obtained from wolffish (*Anarhichas lupus* L.) reared in captivity. *Aquaculture* 122:295–312.

Pavlov, D. A., and E. Moksness. 1996. Repeat sexual maturation of wolffish (*Anarhichas lupus* L.) broodstock. *Aquaculture* 139:249–263.

Pavlov, D. A., and G. G. Novikov. 1986. On the development of biotechnology for rearing of White sea wolffish, *Anarhichas lupus marisalbi.* I. Experience on obtaining mature sex products, incubation of eggs and rearing of the young fish. *Voprosy Ikhtiol.* 26:476–487 [in Russian, translation in *J. Ichthyol.* 26:95–106].

Pavlov, D. A., and G. G. Novikov. 1993. Life history peculiarities of common wolffish (*Anarhichas lupus*) in the White Sea. *ICES J. Mar. Sci.* 50:271–277.

Pavlov, D. A., and Ye. K. Radzikhovskaya. 1991. Reproduction biology of White Sea wolffish, *Anarhichas marisalbi,* based on experimental studies. *Voprosy Ikhtiol.* 31:433–441 [in Russian, translation in *J. Ichthyol.* 31(7):52–62].

Pavlov, D. A., Yu. B. Burykin, and L. A. Konoplya. 1987. Pelagic young of the White Sea wolffish, *Anarhichas lupus marisalbi* Barsukov. *Voprosy Ikhtiol.* 27:163–166 [in Russian, translation in *J. Ichthyol.* 27:175–179].

Pavlov, D. A., K. F. Dzerzhinskiy, and E. K. Radzikhovskaya. 1992. Assessing the quality of roe from White Sea wolffish (*Anarhichas lupus marisalbi*), obtained under experimental conditions. *Voprosy Ikhtiol.* 31:743–755 [in Russian, translation in *J. Ichthyol.* 32(1):88–104].

Paxton, J. R., and J. E. Hanley. 1989. Trachichthyidae (255). In: J. R. Paxton, D. F. Hoese, G. R. Allen, and J. E. Hanley, eds.. *Zoological Catalogue of Australia,* Vol. 7. *Pisces: Petromyzontidae to Carangidae.* Australian Gov. Publ. Serv., Canberra, pp. 365–367.

Payne, P. M., J. R. Nicolas, L. O'Brien, and K. D. Powers. 1986. The distribution of the humpback whale, *Megaptera novaeangliae,* on Georges Bank and in the Gulf of Maine in relation to densities of the sand eel, *Ammodytes americanus. Fish. Bull., U.S.* 84:271–277.

Payne, R. H., and I-H. Ni. 1982. Biochemical population genetics of redfishes (*Sebastes*) off Newfoundland. *J. Northw. Atl. Fish. Sci.* 3:169–172.

Pearce, J. B. 1969. Thermal addition and the benthos, Cape Cod Canal. *Chesapeake Sci.* 10:227–233.

Pearcy, W. C. 1962. Ecology of an estuarine population of winter flounder, *Pseudopleuronectes americanus* (Walbaum). *Bull. Bingham Oceanogr. Coll.* 18(1):1–78.

Pearcy, W. G., and S. W. Richards. 1962. Distribution and ecology of fishes of the Mystic River estuary, Connecticut. *Ecology* 43:248–259.

Pearse, A. S. 1949. Observations on flatworms and nemerteans collected at Beaufort, N.C. *Proc. U.S. Nat. Mus.* 100:25–38.

Pearson, J. C. 1929. Natural history and conservation of the redfish and other commercial sciaenids on the Texas coast. *U.S. Bur. Fish. Bull.* 44:129–214.

Pearson, J. C. 1938. The life history of the striped bass, or rockfish, *Roccus saxatilis* (Walbaum). *Fish. Bull., U.S.* 49:825–851.

Pearson, J. C. 1941. The young of some marine fishes taken in lower Chesapeake Bay, Virginia, with special reference to the gray sea trout, *Cynoscion regalis* (Bloch). *Fish. Bull., U.S.* 50:77–102.

Pearson, W. H., S. E. Miller, and B. L. Olla. 1980. Chemoreception in the food-searching and feeding behavior of the red hake, *Urophycis chuss* (Walbaum). *J. Exper. Mar. Biol. Ecol.* 48:139–150.

Peck, J. I. 1894. On the food of the menhaden. *Bull. U.S. Fish Comm.* 13:113–126.

Pedersen, S. A. 1995. Feeding habits of starry ray (*Raja radiata*) in west Greenland waters. *ICES J. Mar. Sci.* 52:43–53.

Pelster, B., and W. E. Bemis. 1992. Structure and function of the external gill filaments of embryonic skates (*Raja erinacea*). *Resp. Physiol.* 89:1–13.

Pelster, B., C. R. Bridges, and M. K. Grieshaber. 1988a. Respiratory adaptations of the burrowing marine teleost *Lumpenus lampraetiformis* (Walbaum). II. Metabolic adaptations. *J. Exp. Mar. Biol. Ecol.* 124:43–55.

Pelster, B., C. R. Bridges, A. C. Taylor, S. Morris, and R. J. A. Atkinson. 1988b. Respiratory adaptations of the burrowing marine teleost *Lumpenus lampraetiformis* (Walbaum) I. O_2 and CO_2 transport, acid-base balance: a comparison with *Cepola rubescens* L. *J. Exp. Mar. Biol. Ecol.* 124:31–42.

Penney, R. W. 1985. Comparative morphology of pre-extrusion larvae of the North Atlantic sharp-beaked redfishes, *Sebastes mentella* and *Sebastes fasciatus* (Pisces: Scorpaenidae). *Can. J. Zool.* 63:1181–1188.

Penney, R. W., and G. T. Evans. 1985. Growth histories of larval redfish (*Sebastes* spp.) on an offshore Atlantic Fishing bank determined by otolith increment analysis. *Can. J. Fish. Aquat. Sci.* 42: 1452–1464.

Penttila, J. A. 1988. Yellowtail flounder, *Limanda ferruginea.* In: J. A. Penttila and L. M. Dery, eds. 1988. *Age Determination Methods for Northwest Atlantic Species.* NOAA Tech. Rept. NMFS No. 72:119–124.

Penttila, J. A., and L. M. Dery, eds. 1988. *Age Determination Methods for Northwest Atlantic Species.* NOAA Tech. Rept. NMFS No. 72, 135 pp.

Penttila, J. A., and V. M. Gifford. 1976. Growth and mortality rates for cod from Georges Bank and Gulf of Maine areas. *ICNAF Res. Bull.* 12:29–36.

Penttila, J. A., G. A. Nelson, and J. M. Burnett, III. 1989. Guidelines for estimating lengths at age for 18 northwest Atlantic finfish and shellfish species. NOAA Tech. Memo. NMFS-F/NEC-66, 39 pp.

Pepin, P., and S. M. Carr. 1993. Morphological, meristic, and genetic analysis of stock structure in juvenile Atlantic cod (*Gadus morhua*) from the Newfoundland Shelf. *Can. J. Fish. Aquat. Sci.* 50:1924–1933.

Pepin, P., J. A. Koslow, and S. Pearre, Jr. 1988. Laboratory study of foraging by Atlantic mackerel, *Scomber scombrus,* on natural zooplankton assemblages. *Can. J. Fish. Aquat. Sci.* 45:879–887.

Pereira, J. J., R. Goldberg, J. J. Ziskowski, P. L. Berrien, W. W. Morse, and D. L. Johnson. 1999. Essential Fish Habitat Source Document: Winter flounder, *Pseudopleuronectes americanus,* life history and habitat characteristics. NOAA Tech. Memo. NMFS-NE 138, 39 pp.

Perez-Garcia, M. A., and A. L. Ibanez-Aguirre. 1992. Morfometria de los peces *Mugil cephalus* y *M. curema* (Mugiliformes: Mugilidae) en Veracruz, Mexico. *Rev. Biol. Trop.* 40:325–333.

Perkins, F. E., and V. C. Anthony. 1969. A note on the fecundity of herring (*Clupea harengus* L.) from Georges Bank, the Gulf of Maine and Nova Scotia. *Int. Comm. Northw. Atl. Fish. Redbook 1969,* Part III:33–38.

Perkins, P. J. 2001. Drumming and chattering sounds recorded underwater in Rhode Island. *Northeast Nat.* 8:359–370.

Perlmutter, A. 1939. A biological survey of the salt waters of Long Island, 1938: an ecological survey of young fish and eggs identified from tow-net collections. New York Cons. Dept. Ann. Rept. No. 28, Part 2, pp. 11–71.

Perlmutter, A. 1947. The blackback flounder and its fishery in New England and New York. *Bull. Bingham Oceanogr. Coll.* 11(2): 1–92.

Perlmutter, A. 1953. Population studies of rosefish. *Trans. N.Y. Acad. Sci.,* Ser. No. 2, 5:189–191.

Perlmutter, A. 1963. Observations on fishes of the genus *Gasterosteus* in the waters of Long Island, New York. *Copeia* 1963:168–173.

Perlmutter, A., and G. M. Clarke. 1949. Age and growth of immature rosefish (*Sebastes marinus*) in the Gulf of Maine and off western Nova Scotia. *Fish. Bull., U.S.* 51:207–228.

Perry, D. M. 1994. Artificial spawning of tautog under laboratory conditions. *Prog. Fish-Cult.* 56:33–36.

Perry, D. M., R. Mercaldo-Allen, C. A. Kuropat, and J. B. Hughes. 1997a. Laboratory culture of tautog: a pilot study. *J. Shellfish Res.* 15(2):459.

Perry, D. M., R. Mercaldo-Allen, C. A. Kuropat, and J. B. Hughes. 1997b. Green-water culture of tautog. *J. Shellfish Res.* 16(1):293.

Perry, D. M., R. Mercaldo-Allen, C. A. Kuropat, and J. B. Hughes. 1998. Laboratory culture of tautog. *Progr. Fish-Cult.* 60:50–54.

Peters, D. S., and M. T. Boyd. 1972. The effect of temperature, salinity, and availability of food on the feeding and growth of the hogchoker *Trinectes maculatus* (Bloch and Schneider). *J. Exp. Mar. Biol. Ecol.* 9:201–207.

Peters, D. S., and M. A. Kjelson. 1975. Composition and utilization of food by various postlarval and juvenile fishes of North Carolina estuaries. *Est. Res.* 1:447–472.

Peters, D. S., and W. E. Schaaf. 1981. Food requirements and sources for juvenile Atlantic menhaden. *Trans. Am. Fish. Soc.* 110:317–324.

Peters, K. M., and R. H. McMichael, Jr. 1990. Early life history of the black drum *Pogonias cromis* (Pisces: Sciaenidae) in Tampa Bay, Florida. *Northeast Gulf Sci.* 11:39–58.

Peterson, R. H., P. H. Johansen, and J. L. Metcalfe. 1980. Observations on early life stages of Atlantic tomcod, *Microgadus tomcod. Fish. Bull., U.S.* 78:147–158.

Peterson, S. J. 1975. The seasonal abundance and distribution of fish eggs, larvae, and juveniles in the Merrimack River Estuary, Massachusetts, 1974–75. M.S. thesis, University of Massachusetts, Amherst, 170 pp.

Peterson, W. T., and S. J. Ausubel. 1984. Diets and selective feeding by larvae of Atlantic mackerel *Scomber scombrus* on zooplankton. *Mar. Ecol. Prog. Ser.* 17:65–75.

Petrillo, A. P. 1987. Trophic structure of a tidal marsh finfish community (Branford, Connecticut). M.S. thesis, Southern Connecticut State University, New Haven, 101 pp.

Phelan, B. A., R. Goldberg, A. J. Bejda, J. Pereira, S. Hagan, P. Clark, A. L. Studholme, A. Calabrese, and K. W. Able. 2000. Estuarine and habitat-related differences in growth rates of young-of-the-year winter flounder (*Pseudopleuronectes americanus*) and tautog (*Tautoga onitis*) in three northeastern U.S. estuaries. *J. Exper. Mar. Biol. Ecol.* 247:1–28.

Phillips, R. R., and S. B. Swears. 1981. Diel activity cycles of two Chesapeake Bay fishes, the striped blenny (*Chasmodes bosquianus*) and the oyster toadfish (*Opsanus tau*). *Estuaries* 4:357–362.

Piatt, J. F. 1987. Behavioural ecology of common murre and Atlantic puffin predation on capelin: implications for population biology. Ph.D. thesis, Memorial University, St. John's, Newfoundland., 311 pp.

Piavis, G. W. 1972. Embryology. In: M. W. Hardisty and I. C. Potter, eds. *The Biology of Lampreys,* Vol. 1. Academic Press, New York, pp. 361–400.

Picard, P., Jr., J. J. Dodson, and G. J. FitzGerald. 1990. Habitat segregation among the age groups of *Gasterosteus aculeatus* in the middle St. Lawrence estuary, Canada. *Can. J. Zool.* 68:1202–1208.

Pierce, G. J., J. S. W. Diack, and P. R. Boyle. 1989. Digestive tract contents of seals in the Moray Firth area of Scotland. *J. Fish Biol.* 35(Suppl. A):341–343.

Piers, H. 1899. Observations on a fish (*Chilomycterus schoepfi*) new to the fauna of Nova Scotia. *Proc. Nova Scotia Inst. Sci.* 10:110–111.

Pietsch, T. W. 1974. Osteology and relationships of ceratioid anglerfishes of the family Oneirodidae, with a review of the genus *Oneirodes* Lütken. *Nat. Hist. Mus. Los Angeles Co., Sci. Bull.* 18, 113 pp.

Pietsch, T. W. 1976. Dimorphism, parasitism and sex: reproductive strategies among deep-sea ceratioid anglerfishes. *Copeia* 1976: 781–793.

Pietsch, T. W. 1986. Systematics and distribution of bathypelagic anglerfishes of the family Ceratiidae (order Lophiiformes). *Copeia* 1986:479–493.

Pietsch, T. W. 1993. Systematics and distribution of cottid fishes of the genus *Triglops* Reinhardt (Teleostei: Scorpaeniformes). *Zool. J. Linnean Soc.* 109:335–393.

Pietsch, T. W., and D. B. Grobecker. 1987. *Frogfishes of the World: Systematics, Zoogeography, and Behavioral Ecology.* Stanford University Press, Stanford, 420 pp.

Pietsch, T. W., and C. P. Zabetian. 1990. Osteology and interrelationships of the sand lances (Teleostei: Ammodytidae). *Copeia* 1990:78–100.

Pietsch, T. W., D. B. Groebecker, and B. Stockley. 1992. The sargassum frogfish, *Histrio histrio* (Linnaeus)(Lophiiformes: Antennariidae), on the Pacific Plate. *Copeia* 1992:247–248.

Pihl, L., S. P. Baden, and R. J. Diaz. 1991. Effects of periodic hypoxia on distribution of demersal fish and crustaceans. *Mar. Biol.* 108: 349–360.

Pikanowski, R. A., W. W. Morse, P. L. Berrien, D. L. Johnson, and D. G. McMillan. 1999. Essential Fish Habitat Document: Redfish, *Sebastes* spp., life history and habitat characteristics. NOAA Tech. Memo. NMFS-NE-132, 19 pp.

Pillay, S. R. 1962. A revision of Indian Mugilidae. Part II. *J. Bombay Nat. Hist. Soc.* 59:547–576.

Pinhorn, A. T. 1976. Living marine resources of Newfoundland-Labrador: status and potential. *Fish. Res. Bd. Can., Bull.* 194, 64 pp.

Pippy, J. H. C., and P. Van Banning. 1975. Identification of *Anisakis* larvae as *Anisakis simplex* (Rudolphi, 1809, det. Krabb 1878) (Nematoda: Ascaridata). *J. Fish. Res. Bd. Can.* 32:29–32.

Pitcher, T. J., and J. K. Parrish. 1993. Functions of shoaling behaviour in teleosts. In: T. J. Pitcher, ed. *Behaviour of Teleost Fishes,* 2nd Ed. Chapman and Hall, London, pp. 363–439.

Pitcher, T. J., A. E. Magurran, and J. I. Eduards. 1985. Schooling mackerel and herring choose neighbours of similar size. *Mar. Biol.* 86:319–322.

Pitt, T. K. 1958. Distribution, spawning and racial studies of the capelin, *Mallotus villosus* (Müller), in the offshore Newfoundland area. *J. Fish. Res. Bd. Can.* 15:275–293.

Pitt, T. K. 1963. Vertebral numbers of American plaice, *Hippoglossoides platessoides* (Fabricius) in the northwest Atlantic. *J. Fish. Res. Bd. Can.* 20:1159–1181.

Pitt, T. K. 1964. Fecundity of the American plaice, *Hippoglossoides platessoides* (Fabr.) from Grand Bank and Newfoundland areas. *J. Fish. Res. Bd. Can.* 21:597–612.

Pitt, T. K. 1967. Diurnal variation in the catches of American plaice, *Hippoglossoides platessoides* (Fabr.) from Grand Bank. *ICNAF Res. Bull.* 4:53–58.

Pitt, T. K. 1969. Migrations of American plaice on the Grand Bank and in St. Mary's Bay, 1954, 1959, and 1961. *J. Fish. Res. Bd. Can.* 26:1301–1319.

Pitt, T. K. 1970. Distribution, abundance and spawning of yellowtail flounder, *Limanda ferruginea,* in the Newfoundland area of the northwest Atlantic. *J. Fish. Res. Bd. Can.* 27:2261–2271.

Pitt, T. K. 1971. Fecundity of the yellowtail flounder (*Limanda ferruginea*) from the Grand Bank, Newfoundland. *J. Fish. Res. Bd. Can.* 28:456–457.

Pitts, P. A. 1991. Comparative use of food and space by three Bahamian butterflyfishes. *Bull. Mar. Sci.* 48:749–756.

Pohle, G., T. J. Kenchington, and R. G. Halliday. 1992. Potentially exploitable deepwater resources off Atlantic Canada. Can. Fish. Aquat. Sci. Tech. Rept. No. 1843, 79 pp.

Poll, M. 1959. Poissons V. Téléostéens acanthoptérygiens (deuxième partie). Expédition Océanographique Belge dans les eaux côtières Africaines de l'Atlantique Sud (1948–1949). Inst. Roy. Sci. Nat. Belg. Mem. No. 4(3B):1–417.

Ponomarenko, V. P. 1984. Discovery of *Rhinonemus cimbius* (L.) in the Barents Sea and *Gaidropsarus argentatus* (Reinhart) (Gadidae) in the Greenland Sea. *Voprosy Ikhtiol.* 24:344 [in Russian, translation in *J. Ichthyol.* 24:131].

Poole, J. C. 1961. Age and growth of the fluke in Great South Bay and their significance to the sport fishery. *N.Y. Fish Game J.* 8:1–18.

Poole, J. C. 1962. The fluke population in Great South Bay in relation to the sport fishery. *N.Y. Fish Game J.* 9:93–116.

Popova, O. A. 1962. Some data on the feeding of cod in the Newfoundland area of the northwest Atlantic. In: Yu. Yu. Marti, ed. *Soviet Fisheries Investigations in the Northwest Atlantic.* [In Russian, translated by Israel Prog. Sci. Trans.], pp. 228–248.

Post, A. 1986a. Family No. 130: Diretmidae. In: M. M. Smith and P. C. Heemstra, eds. *Smith's Sea Fishes.* MacMillan South Africa, Johannesburg, pp. 414–415.

Post, A. 1986b. Diretmidae. In: P. J. P. Whitehead et al., eds. *Fishes of the North-eastern Atlantic and the Mediterranean, Vol. 2.* UNESCO, Paris, pp. 743–746.

Post, A. 1987. Results of the research cruises of FRV "Walther Herwig" to South America. LXVII. Revision of the subfamily Paralepidinae (Pisces, Aulopiformes, Alepisauroidei, Paralepididae). I. Taxonomy, morphology and geographical distribution. *Arch. Fisch. Wiss.* 38(1/2):75–131.

Post, A., and J.-C. Quéro. 1981. Révision des Diretmidae (Pisces, Trachichthyoidei) de l'Atlantique avec description d'un nouveau genre et d'une nouvelle espèce. *Cybium* 5:33–60.

Potoschi, A. 1996. Observations about some biological aspects of *Scomberesox saurus* (Walbaum, 1792) in the area of the Straits of Messina (Italy). *Ophelia* 22:139–146.

Potter, D. C., and R. G. Lough. 1987. Vertical distribution and sampling variability of larval and juvenile sand lance (*Ammodytes* sp.) on Nantucket Shoals and Georges Bank. *J. Northw. Atl. Fish. Sci.* 7:107–116.

Potthoff, T. 1980. Development and structure of fins and fin supports in dolphin fishes *Coryphaena hippurus* and *Coryphaena equiselis* (Coryphaenidae). *Fish. Bull., U.S.* 78:277–312.

Potthoff, T., and S. Kelley. 1982. Development of the vertebral column, fins and fin supports, branchiostegal rays, and squamation in the swordfish, *Xiphias gladius. Fish. Bull., U.S.* 80:161–186.

Potthoff, T., and W. J. Richards. 1970. Juvenile bluefin tuna, *Thunnus thynnus* (Linnaeus), and other scombrids taken by terns in the Dry Tortugas, Florida. *Bull. Mar. Sci.* 20:389–413.

Pottle, R. A. 1979. A field study of territorial and reproductive behaviour of the cunner, Tautogolabrus adspersus, and in Conception Bay, Newfoundland. M.S. thesis, Memorial University, St. John's, Newfoundland, 104 pp.

Pottle, R. A., and J. M. Green. 1979a. Field observations on the reproductive behavior of the cunner, *Tautogolabrus adspersus* (Walbaum), in Newfoundland. *Can. J. Zool.* 57:247–256.

Pottle, R. A., and J. M. Green. 1979b. Territorial behavior of the north temperate labrid, *Tautogolabrus adspersus. Can. J. Zool.* 57: 2337–2347.

Pottle, R. A., P. A. Macpherson, S. N. Messieh, and D. S. Moore. 1980. A SCUBA survey of herring (*Clupea harengus* L.) spawning bed in Miramichi Bay, N.B. Can. Fish. Aquat. Sci. Tech. Rept. No. 984, 14 pp.

Pottle, R. A., J. M. Green, and G. Martel. 1981. Dualistic spawning behavior of the cunner, *Tautogolabrus adspersus* (Pisces: Labridae), in Bonne Bay, Newfoundland. *Can. J. Zool.* 59:1582–1585.

Poulin, R., and G. J. FitzGerald. 1987. The potential of parasitism in the structuring of a salt marsh stickleback community. *Can. J. Zool.* 65:2793–2798.

Poulin, R., and G. J. FitzGerald. 1989. Early life histories of three sympatric sticklebacks in a salt-marsh. *J. Fish Biol.* 34:207–221.

Pounds, S. 1962. An unusually large lookdown, *Selene vomer* from Texas. *Copeia* 1962:444–445.

Powell, A. B. 1993. A comparison of early-life-history traits in Atlantic menhaden *Brevoortia tyrannus* and gulf menhaden *B. partonus. Fish. Bull., U.S.* 91:119–128.

Powell, A. B. 2000. Preliminary guide to the identification of the early life history stages of priacanthid fishes of the western central Atlantic. NOAA Tech. Memo. NMFS-SEFSC-439, 4 pp.

Powell, A. B., and H. R. Gordy. 1980. Egg and larval development of the spot, *Leiostomus xanthurus* (Sciaenidae). *Fish. Bull., U.S.* 78:701–714.

Powell, A. B., and G. Phonlor. 1986. Early life history of Atlantic menhaden, *Brevoortia tyrannus,* and gulf menhaden, *B. patronus. Fish. Bull., U.S.* 84:991–995.

Powell, A. B., and F. J. Schwartz. 1972. Anomalies of the genus *Paralichthys* (Pisces, Bothidae), including an unusual double-tailed southern flounder *Paralichthys lethostigma. J. Elisha Mitchell Sci. Soc.* 88:155–161.

Powell, A. B., and F. J. Schwartz. 1977. Distribution of paralichthid flounders (Bothide: *Paralichthys*) in North Carolina estuaries. *Chesapeake Sci.* 18:334–339.

Powell, A. B., and F. J. Schwartz. 1979. Food of *Paralichthys dentatus* and *P. lethostigma* (Pisces: Bothidae) in North Carolina estuaries. *Estuaries* 2:276–279.

Powers, D. A., M. Smith, I. Gonzalez-Villasenor, L. DiMichele, D. Crawford, G. Bernardi, and T. Lauerman. 1993. A multidisciplinary approach to the selection/neutralist controversy using the model teleost, *Fundulus heteroclitus.* In: D. Futuyma and J. Antonovics, eds. *Oxford Surveys in Evolutionary Biology,* Vol. 9. Oxford University Press, New York, pp. 43–108.

Powers, K. D., and E. H. Backus. 1987. Energy transfer to seabirds. In: R. H. Backus, ed. *Georges Bank.* M.I.T. Press, Cambridge, pp. 372–374.

Powles, H., F. Auger, and G. J. FitzGerald. 1984. Nearshore ichthyoplankton of a north temperate estuary. *Can. J. Fish. Aquat. Sci.* 41:1653–1663.

Powles, P. M. 1958. Studies of reproduction and feeding of Atlantic cod (*Gadus callarias* L.) in the southwestern Gulf of St. Lawrence. *J. Fish. Res. Bd. Can.* 15:1383–1402.

Powles, P. M. 1965a. Life history and ecology of American plaice (*Hippoglossoides platessoides* F.) in the Magdalen Shallows. *J. Fish. Res. Bd. Can.* 22:565–598.

Powles, P. M. 1965b. New size record for greysole (*Glyptocephalus cynoglossus*). *J. Fish. Res. Bd. Can.* 22:1565–1566.

Powles, P. M. 1967. Atlantic wolffish (*Anarhichas lupus* L.) eggs off southern Nova Scotia. *J. Fish. Res. Bd. Can.* 24:207–208.

Powles, P. M., and V. S. Kennedy. 1967. Age determination of Nova Scotian greysole, *Glyptocephalus cynoglossus* L., from otoliths. *Int. Comm. Northw. Atl. Fish. Bull.* 4:91–100.

Powles, P. M., and A. C. Kohler. 1970. Depth distributions of various stages of witch flounder (*Glyptocephalus cynoglossus*) off Nova Scotia and in the Gulf of St. Lawrence. *J. Fish. Res. Bd. Can.* 27: 2053–2062.

Powles, P. M., D. G. Garnett, G. D. Ruggieri, and R. F. Nigrelli. 1968. *Ichthyophonus* infection in yellowtail flounder (*Limanda ferruginea*) off Nova Scotia. *J. Fish. Res. Bd. Can.* 25:597–598.

Pratt, H. L., Jr. 1979. Reproduction in the blue shark, *Prionace glauca*. *Fish. Bull., U.S.* 77:445–470.

Pratt, H. L., and J. G. Casey. 1983. Age and growth of the shortfin mako, *Isurus oxyrinchus*, using four methods. *Can. J. Fish. Aquat. Sci.* 40:1944–1957.

Pratt, H. L., and J. G. Casey. 1990. Shark reproductive strategies as a limiting factor in directed fisheries, with a review of Holden's method of estimating growth parameters. NOAA Tech. Rept. NMFS No. 90:97–109.

Pratt, H. L., J. G. Casey, and R. B. Conklin. 1982. Observations on large white shark, *Carcharodon carcharias*, off Long Island, New York. *Fish. Bull., U.S.* 80:153–156.

Price, W. S. 1978. Otolith comparison of *Alosa pseudoharengus* (Wilson) and *Alosa aestivalis* (Mitchill). *Can. J. Zool.* 56:1216–1218.

Prince, E. D., D. W. Lee, and J. C. Javech. 1985. Internal zonations in sections of vertebrae from Atlantic bluefin tuna, *Thunnus thynnus*, and their potential use in age determination. *Can. J. Fish. Aquat. Sci.* 42:938–946.

Procter, W., H. C. Tracy, E. Helwig, C. H. Blake, J. E. Morrison, and S. Cohen. 1928. A contribution to the life-history of the angler (*Lophius piscatorius*). *Biol. Surv. Mount Desert Region* 2:3–13.

Proudfoot, L. A. 1975. The biology of the rock gunnel *Pholis gunnellus* (Linnaeus). B.S. Hon. thesis, Memorial University, St. John's, Newfoundland, 40 pp.

Purcell, M. K., I. Kornfield, M. Fogarty, and A. Parker. 1996. Interdecadal heterogeneity in mitochondrial DNA of Atlantic haddock (*Melanogrammus aeglefinus*) from Georges Bank. *Mol. Mar. Ecol. Biotech.* 5:185–192.

Putnam, F. W. 1871. On the young of *Orthagoriscus mola*. *Proc. Am. Assoc. Adv. Sci.* 19:255–260.

Putnam, F. W. 1874. Rare fishes taken in Marblehead, Salem, and Beverly Harbor. *Bull. Essex Inst.* 6:11–13.Qasim, S. Z. 1957. The biology of *Centronotus gunnellus* (L.) (Teleostei). *J. Animal Ecol.* 26:389–401.

Quéro, J.-C. 1979. Remarques sur le *Grammicolepis brachiusculus* (Pisces, Zeiformes, Grammicolepididae) observé au port de La Rochelle. *Ann. Soc. Sci. Nat. Charente-Maritime* 6:573–576.

Quéro, J.C. 1982. Trachichthyidae. In: C. Maurin and J.-C. Quéro, eds. *Poissons des côtes nord-ouest africaines (campagnes de la 'Thalassa' 1962, 1968, 1971 et 1973)*, pp. 29–43. *Rev. Trav. Inst. Pêches Marit.* 45:5–71.

Quéro, J.-C. 1986a. Zeidae. In: P. J. P. Whitehead et al., eds. *Fishes of the North-eastern Atlantic and the Mediterranean*, Vol. 2. UNESCO, Paris, pp. 769–772.

Quéro, J.-C. 1986b. Grammicolepididae. In: P. J. P. Whitehead et al., eds. *Fishes of the North-eastern Atlantic and the Mediterranean*, Vol. 2. UNESCO, Paris, pp. 773–774.

Quéro, J.-C. 1986c. Caproidae. In: P. J. P. Whitehead et al., eds. *Fishes of the North-eastern Atlantic and the Mediterranean*, Vol. 2. UNESCO, Paris, pp. 777–779.

Quinn, T. P., and J. T. Light. 1989. Occurrences of threespine sticklebacks (*Gasterosteus aculeatus*) in the open North Pacific Ocean: migration or drift? *Can. J. Zool.* 67:2850–2852.

Quinney, T. E., and P. C. Smith. 1980. Comparative foraging behaviour and efficiency of adult and juvenile great blue herons. *Can. J. Zool.* 58:1168–1173.

Quiñonez-Velázquez, C. 1999. Age validation and growth of larval and juvenile haddock, *Melanogrammus aeglefinus*, and pollock, *Pollachius virens*, on the Scotian Shelf. *Fish. Bull., U.S.* 97:306–319.

Rabe, J. H., J. A. Brown, D. A. Bidwell, and W. H. Howell. 1999. Preliminary observations on the larviculture of witch flounder (*Glyptocephalus cynoglossus*). *Bull. Aquacult. Assoc. Can.* 98–2:19–20.

Radcliffe, L. 1916. An extension of the recorded range of three species of fishes in New England waters. *Copeia* No. 26:2–3.

Radcliffe, L. 1922. Fisheries of the New England states in 1919. U. S. Comm. Fish. Rept. for 1921, pp. 120–187.

Radtke, R. L., and J. M. Dean. 1982. Increment formation in the otoliths of embryos, larvae, and juveniles of the mummichog, *Fundulus heteroclitus*. *Fish. Bull., U.S.* 80:201–215.

Radtke, R. L., and P. C. F. Hurley. 1983. Age estimation and growth of broadbill swordfish, *Xiphias gladius*, from the northwest Atlantic based on external features of otoliths. NOAA Tech. Rept. NMFS No. 8:145–150.

Radtke, R. L., M. L. Fine, and J. Bell. 1985. Somatic and otolith growth in the oyster toadfish (*Opsanus tau* L.). *J. Exp. Mar. Biol. Ecol.* 90:259–275.

Rae, B. B. 1969. *The Food of the Witch*. Mar. Res. Dept., Agr. Fish., Scotland, 23 pp.

Rail, J. F., and G. Chapdelaine. 1998. Food of double-crested cormorants, *Phalacrocorax auritus*, in the Gulf and Estuary of the St. Lawrence River, Quebec, Canada. *Can. J. Zool.* 76:635–643.

Ralph, D. 1982. Fourspot flounder *Paralichthys oblongus*. In: Grosslein, M. D., and T. R. Azarovitz, eds. *Fish Distribution*. MESA New York Bight Atlas Monogr. No. 15:113–116.

Ramachandran, V. S., C. W. Tyler, R. L. Gregory, D. Rogers-Ramachandran, S. Duensing, C. Pillsbury, and C. Ramachandran.

1996. Rapid adaptive camouflage in tropical flounders. *Nature* 379:815–818.

Randall, J. E. 1967. Food habits of reef fishes of the West Indies. Univ. Miami Trop. Oceanogr. Stud. No. 5:655–847.

Randall, J. E. 1968. *Caribbean Reef Fishes.* THF Publications, Neptune City, N.J., 318 pp.

Randall, J. E. 1987. Refutation of lengths of 11.3, 9.0, and 6.4 m attributed to the white shark, *Carcharodon carcharias. Calif. Fish Game* 73:163–168.

Randall, J. E. 1992. Review of the biology of the tiger shark (*Galeocerdo cuvier*). *Austr. J. Mar. Freshw. Res.* 43:21–31.

Randall, J. E., and H. A. Randall. 1960. Examples of mimicry and protective resemblance in tropical marine fishes. *Bull. Mar. Sci. Gulf Carib.* 10:444–480.

Raney, E. C. 1952. The life history of the striped bass, *Roccus saxatilis* (Walbaum). *Bull. Bingham Oceanogr. Coll.* 14(1):5–97.

Raney, E. C. 1954. The striped bass in New York waters. *N.Y. State Conservationist* 8:14–17.

Raney, E. C. 1957. Subpopulations of the striped bass *Roccus saxatilis* (Walbaum) in tributaries of Chesapeake Bay. U.S. Fish Wildl. Serv. Spec. Sci. Rept. Fish. No. 208:85–107.

Raney, E. C., and W. H. Massmann. 1953. The fishes of the tidewater section of the Pamunkey River, Virginia. *J. Washington Acad. Sci.* 43:424–432.

Rangeley, R. W., and D. L. Kramer. 1995. Use of rocky intertidal habitats by juvenile pollock *Pollachius virens. Mar. Ecol. Prog. Ser.* 126:9–17.

Rangeley, R. W., and D. L. Kramer. 1998. Density-dependent antipredator tactics and habitat selection in juvenile pollock. *Ecology* 79:943–952.

Rankine, P. W., and J. A. Morrison. 1989. Predation on herring larvae and eggs by sand-eels *Ammodytes marinus* (Rait) and *Hyperoplus lanceolatus* (Lesauvage). *J. Mar. Biol. Assoc. U.K.* 69:493–498.

Rasquin, P. 1958. Ovarian morphology and early embryology of the pediculate fishes *Antennarius* and *Histrio. Bull. Am. Mus. Nat. Hist.* 114:331–371.

Rass, T. S. 1996. On taxonomy of Pleuronectini (Pleuronectidae). *Voprosy. Ikhtiol.* 36(4):569–771 [in Russian, translation in *J. Ichthyol.* 36(7):546–548].

Raven, H. C. 1939. On the anatomy and evolution of the locomotor apparatus of the nipple-tailed ocean sunfish (*Masturus lanceolatus*). *Bull. Amer. Mus. Nat. Hist.* 76:143–150.

Ray, C. 1961. Spawning behavior and egg raft morphology of the ocellated fringed frogfish, *Antennarius numifer* (Cuvier). *Copeia* 1961:230–231.

Read, K. R. H. 1975. Black sea bass in Hodgkins Cove: a rarity in the Gulf of Maine. *Aquasphere* 9:18–22.

Reay, P. J. 1970. Synopsis of the biological data on North Atlantic sand eels of the genus *Ammodytes* (*A. tobianus, A. dubius, A. americanus,* and *A. marinus*). FAO Fish. Synop. No. 82, 28 pp.

Recksiek, C. W., and J. D. McCleave. 1973. Distribution of pelagic fishes in the Sheepscot River-Back River estuary, Wiscasset, Maine. *Trans. Am. Fish. Soc.* 102:541–551.

Reddin, D. G. 1985. Atlantic salmon (*Salmo salar*) on and east of the Grand Bank. *J. Northw. Atl. Fish. Sci.* 6: 157– 164.

Reddin, D. G., and K. D. Friedland. 1993. Marine environmental factors influencing the movement and survival of Atlantic salmon. In: D. E. Mills, ed. *Salmon in the Sea and New Enhancement Strategies.* Fishing News Books, Oxford, pp. 79–103.

Reddin, D. G., and P. B. Short. 1991. Postsmolt Atlantic salmon (*Salmo salar*) in the Labrador Sea. *Can. J. Fish. Aquatic Sci.* 48:2–6.

Reese, E. S. 1975. A comparative field study of the social behavior and related ecology of reef fishes of the family Chaetodontidae. *Z. Tierpsychol.* 37:37–61.

Regan, C. T. 1925. Dwarfed males parasitic on the females in oceanic angler-fishes (Pediculati, Ceratioidea). *Proc. Roy. Soc. B* 97:386–400.

Regehr, H. M., and W. A. Montevecchi. 1997. Interactive effects of food shortage and predation on breeding failure of black-legged kittiwakes: indirect effects of fisheries activities and implications for indicator species. *Mar. Ecol. Prog. Ser.* 155:249–260.

Reid, G. K., Jr. 1954. An ecological study of the Gulf of Mexico fishes, in the vicinity of Cedar Key, Florida. *Bull. Mar. Sci. Gulf Carib.* 4:1–94.

Reid, M. E. 1929. The distribution and development of the cunner (*Tautogolabrus adspersus* Walbaum) along the eastern coast of Canada. *Contrib. Can. Biol.* 4:431–441.

Reid, R. N., F. P. Almeida, and C. A. Zetlin. 1999a. Essential Fish Habitat Source Document: fishery-independent surveys, data sources, and methods. NOAA Tech. Memo. NMFS-NE-122, 39 pp.

Reid, R. N., L. M. Cargnelli, S. J. Griesbach, D. B. Packer, D. L. Johnson, C. A. Zetlin, W. W. Morse, and P. L. Berrien. 1999b. Essential Fish Habitat Source Document: Atlantic herring, *Clupea harengus,* life history and habitat characteristics. NOAA Tech. Memo. NMFS-NE-126, 48 pp.

Reimchen, T. E. 1992. Extended longevity in a large-bodied stickleback, *Gasterosteus,* population. *Can. Field-Nat.* 106: 122–125.

Reintjes, J. W. 1960. Continuous distribution of menhaden along the Atlantic and Gulf coasts of the United States. *Proc. Gulf Carib. Fish. Inst.* 12:31–35.

Reintjes, J. W. 1969. Synopsis of biological data on Atlantic menhaden, *Brevoortia tyrannus.* FAO Species Synopsis No. 42, U.S. Fish Wildl. Serv. Circ. No. 320, 30 pp.

Reintjes, J. W. 1979. Coastal herrings and associated species: a profile of species or groups of species, their biology, ecology, current exploitation with economic and social information. Prepared for Gulf of Mexico Fish. Managem. Counc., by Southeast Fish. Cent., NMFS/NOAA, Beaufort, N.C., 170 pp.

Reintjes, J. W. 1980. Marine herring and sardine resources of the northern Gulf of Mexico. In: M. Flandorfer and L. Skuplen, eds. *Workshop for Potential Fishery Resources of the Northern Gulf of Mexico, New Orleans, 1980.* Miss.-Alabama Sea Grant Consort. Publ. MASGP-80-012, 14 pp.

Reintjes, J. W. 1982. Atlantic menhaden, *Brevoortia tyrannus.* In: M. D. Grosslein and T. R. Azarovitz, eds. *Fish Distribution.* Mesa New York Bight Atlas Monogr. No. 15:61–63.

Reintjes, J. W., and A. L. Pacheco. 1966. The relation of menhaden to estuaries. Am. Fish. Soc. Spec. Publ. No. 3:50–58.

Reintjes, J. W., R. B. Chapoton, W. R. Nicholson, and W. E. Schaaf. 1979. Atlantic menhaden: a most abundant fish. Marine Resources of the Atlantic Coast, Leaflet No. 2. Atlantic States Marine Fisheries Comm., Washington, D.C.

Reis, R. R., and J. M. Dean. 1981. Temporal variation in the utilization of an intertidal creek by the bay anchovy (*Anchoa mitchilli*). *Estuaries* 4:16–23.

Reish, R. L., R. B. Deriso, D. Ruppert, and R. J. Carroll. 1985. An investigation of the population dynamics of Atlantic menhaden *Brevoortia tyrannus*. *Can. J. Fish. Aquat. Sci.* 42(Suppl. 1):147–157.

Reisman, H. M., M. H. Kao, and G. L. Fletcher. 1984. Antifreeze glycoprotein in a "southern" population of Atlantic tomcod, *Microgadus tomcod*. *Comp. Biochem. Physiol.* 78A:445–447.

Reisman, H. M., G. L. Fletcher, M. H. Kao, and M. A. Shears. 1987. Antifreeze proteins in the grubby sculpin, *Myoxocephalus aeneus* and the tomcod, *Microgadus tomcod:* comparisons of seasonal cycles. *Environ. Biol. Fishes* 18:295–301.

Relyea, K. 1983. A systematic study of two species complexes of the genus *Fundulus* (Pisces: Cyprinodontidae). *Bull. Florida State Mus., Biol. Sci.* 29:1–64.

Remnant, R. A., P. G. Graveline, and R. L. Bretecher. 1997. Range extension of the rainbow smelt, *Osmerus mordax,* in the Hudson Bay drainage of Manitoba. *Can. Field-Nat.* 111:660–662.

Render, J. H., and C. A. Wilson. 1992. Reproductive biology of sheepshead in the northern Gulf of Mexico. *Trans. Am. Fish. Soc.* 121:757–764.

Render, J. H., B. A. Thompson, and R. L. Allen. 1995. Reproductive development of striped mullet in Louisiana estuarine waters with notes on the applicability of reproductive assessment methods for isochronal species. *Trans. Am. Fish. Soc.* 124:26–36.

Reno, P. W., M. Philippon-Fried, B. L. Nicholson, and S. W. Shelburne. 1978. Ultrastructural studies of piscine erythrocytic necrosis of herring (*Clupea harengus*). *J. Fish. Res. Bd. Can.* 35:148–154.

Reñones, O., E. Massutí, S. Deudero, and B. Morales-Nín. 1998. Biological characterization of pilotfish (*Naucrates ductor*) from the FADS fishery of the island of Mallorca (western Mediterranean). *Bull. Mar. Sci.* 63:249–256.

Rich, W. H. 1947. The swordfish and swordfishery of New England. *Proc. Portland Soc. Nat. Hist.* 4:1–102.

Richards, C. E. 1970. Analog simulation in fish population studies. Virginia Inst. Mar. Sci. Contrib. 345, 4 pp.

Richards, C. E. 1973. Age, growth and distribution of the black drum (*Pogonias cromis*) in Virginia. *Trans. Am. Fish. Soc.* 102:584–590.

Richards, C. E., and M. Castagna. 1970. Marine fishes of Virginia's eastern shore (inlet and marsh, seaside waters). *Chesapeake Sci.* 11:235–248.

Richards, R. A., and D. G. Deuel. 1987. Atlantic striped bass: stock status and the recreational fishery. *Mar. Fish. Rev.* 49(2):58–66.

Richards, S. W. 1959. IV. Pelagic fish eggs and larvae of Long Island Sound. In: Oceanography of Long Island Sound. *Bull. Bingham Oceanogr. Coll.* 17(1):95–124.

Richards, S. W. 1963a. The demersal fish population of Long Island Sound. I. Species composition and relative abundance in two localities, 1956–1957. *Bull. Bingham Oceanogr. Coll.* 18(2):5–31.

Richards, S. W. 1963b. The demersal fish population of Long Island Sound. II. Food of the juveniles from a sand-shell locality (Station I). *Bull. Bingham Oceanog. Coll.* 18(2):32–72.

Richards, S. W. 1963c. The demersal fish population of Long Island Sound. III. Food of the juveniles from a mud locality (Station 3A). *Bull. Bingham Oceanogr. Coll.* 18(2):73–101.

Richards, S. W. 1965. Description of the postlarvae of the sand lance (*Ammodytes*) from the east coast of the North America. *J. Fish. Res. Bd. Can.* 22:1313–1317.

Richards, S. W. 1976a. Age, growth, and food of bluefish (*Pomatomus saltatrix*) from east-central Long Island Sound from July through November 1975. *Trans. Am. Fish. Soc.* 105:523–525.

Richards, S. W. 1976b. Mixed species schooling of postlarvae of *Ammodytes hexapterus* and *Clupea harengus harengus*. *J. Fish. Res. Bd. Can.* 33:843–844.

Richards, S. W. 1982. Aspects of the biology of *Ammodytes americanus* from the St. Lawrence river to Chesapeake Bay, 1972–75, including a comparison of the Long Island Sound postlarvae with *Ammodytes dubius*. *J. Northw. Atl. Fish. Sci.* 3:93–104.

Richards, S. W., and A. W. Kendall, Jr. 1973. Distribution of sand lance, *Ammodytes* sp., larvae on the continental shelf from Cape Cod to Cape Hatteras from R.V. Dolphin surveys in 1966. *Fish. Bull., U.S.* 71:371–386.

Richards, S. W., and A. M. McBean. 1966. Comparison of postlarvae and juveniles of *Fundulus heteroclitus* and *Fundulus majalis* (Pisces: Cyprinodontidae). *Trans. Am. Fish. Soc.* 95:218–226.

Richards, S. W., D. Merriman, and L. H. Calhoun. 1963a. Studies on the marine resources of southern New England. IX. The biology of the little skate, *Raja erinacea* Mitchill. *Bull. Bingham Oceanogr. Coll.* 18(3):1–67.

Richards, S. W., A. Perlmutter, and D. C. McAneny. 1963b. A taxonomic study of the genus *Ammodytes* from the east coast of North America (Teleostei: *Ammodytes*). *Copeia* 1963:358– 377.

Richards, S. W., J. M. Mann, and J. A. Walker. 1979. Comparison of spawning seasons, age, growth rates, and food of two sympatric species of searobins, *Prionotus carolinus* and *Prionotus evolans,* from Long Island Sound. *Estuaries* 2:255–268.

Richards, W. J. 1977. A further note on the spawning of bluefin tuna. Int. Comm. Cons. Atlantic Tunas Collect. Vol. Sci. Pap. 6:335–336.

Richards, W. J. 1989. Preliminary guide to the identification of the early life history stages of scombroid fishes of the western central Atlantic. NOAA Tech. Memo. NMFS-SEFC-240, 101 pp.

Richards, W. J., and T. Potthoff. 1975. Analysis of the taxonomic characters of young scombrid fishes, genus *Thunnus*. In: J. H. S. Blaxter, ed. *The Early Life History of Fish*. Springer-Verlag, Berlin, pp. 632–648.

Richards, W. J., R. V. Miller, and E. D. Houde. 1974. Egg and larval development of the Atlantic thread herring, *Opisthonema oglinum*. *Fish. Bull., U.S.* 72:1123–1136.

Richards, W. L. 1968. Ecology and growth of juvenile tarpon, *Megalops atlanticus,* in a Georgia salt marsh. *Bull. Mar. Sci.* 18:220–239.

Richardson, S. L. 1974. Eggs and larvae of the ophichthid eel, *Pisodonophis cruentifer,* from the Chesapeake Bight, western North Atlantic. *Chesapeake Sci.* 15:151–154.

Richardson, S. L., and E. B. Joseph. 1973. Larvae and young of western North Atlantic bothid flatfishes *Etropus microstomus* and *Citharichthys arctifrons* in the Chesapeake Bight. *Fish. Bull., U.S.* 71:735–767.

Richkus, W. A. 1974. Factors influencing the seasonal and daily patterns of alewife (*Alosa pseudoharengus*) migration in a Rhode Island river. *J. Fish. Res. Bd. Can.* 31:1485–1497.

Richkus, W. A., and G. DiNardo. 1984. Current status and biological characteristics of the anadromous alosid stocks of the eastern United States: American shad, hickory shad, alewife, and blueback herring. Interstate Fisheries Management Program, Atlantic States Marine Fisheries Commission, Washington, D.C.

Richkus, W. A., and K. Whalen. 2000. Evidence for a decline in the abundance of the American eel, *Anguilla rostrata* (LeSueur), in North America since the early 1980s. *Dana* 12:83–97.

Richmond, A. M., and B. Kynard. 1995. Ontogenetic behavior of shortnose sturgeon, *Acipenser brevirostrum*. *Copeia* 1995:172–182.

Ricker, W. E. 1954. Pacific salmon for Atlantic waters? *Can. Fish Cult.* 16:6–14.

Ricker, W. E. 1972. Heredity and environmental factors affecting certain salmonid populations. In: R. C. Simon, ed. *The Stock Concept in Pacific Salmon.*, H. R. MacMillan Lectures in Fisheries. Inst. Fish., Univ. British Columbia, Vancouver, pp. 27–160.

Rideout, R. M., D. M. Maddock, and M. P. M. Burton. 1999. Preliminary observations on the larviculture of witch flounder (*Glyptocephalus cynoglossus*). *Bull. Aquacult. Assoc. Can.* 98–2:19–20.

Rideout, S. G. 1974. Population estimate, movement, and biological characteristics of anadromous alewives *Alosa pseudoharengus* (Wilson), utilizing the Parker River, Massachusetts, in 1971–1972. M.S. thesis, University of Massachusetts, Amherst, 183 pp.

Ridgway, G. J. 1975. A conceptual model of stocks of herring (*Clupea harengus*) in the Gulf of Maine. Int. Comm. Northw. Atl. Fish. Res. Doc. 75/100, Ser. No. 3586, 17 pp.

Ridgway, G. S., S. W. Sherburne, and R. D. Lewis. 1970. Polymorphism in the esterases of Atlantic herring. *Trans. Am. Fish. Soc.* 99:147–151.

Riget, F., and J. Messtorff. 1988. Distribution, abundance and migration of Atlantic wolffish (*Anarhichas lupus*) and spotted wolffish (*Anarhichas minor*) in west Greenland waters. NAFO-SCI. Counc. Stud. No. 12:13–20.

Rilling, G. C. 1996. Temporal and spatial variability in distribution dynamics of early life stages of bay anchovy (*Anchoa mitchilli*) in Chesapeake Bay. M.S. thesis, University of Maryland, College Park, 132 pp.

Ringø, E., and H. Lorentsen. 1987. Brood protection of wolf fish (*Anarhichas lupus* L.) eggs. *Aquaculture* 65:239–241.

Ringø, E., R. E. Olsen, and B. Bøe. 1987. Initial feeding of wolf fish (*Anarhichas lupus* L.) fry. *Aquaculture* 62:33–34.

Ritter, J. A. 1989. Marine migration and natural mortality of North American Atlantic salmon (*Salmo salar* L.). Can. Man. Fish. Aquat. Sci. Rept. 2041, 146 pp.

Ritzi, C. F. 1959. Eastern brook trout populations of two Maine coastal streams. M.A. thesis, University of Maine, Orono, 72 pp.

Rivas, L. R. 1975. Synopsis of biological data on blue marlin, *Makaira nigricans* Lacépède, 1802. In: R. S. Shomura and F. Williams, eds. *Proceedings of the International Billfish Symposium Kailura-Kona, Hawaii, 9–12 August 1972.* III. Species synopses. NOAA Tech. Rept. NMFS SSRF No. 675:1–16.

Rivière, D., D. Roby, A. C. Horth, M. Arnac, and M. F. Khalil. 1985. Structure génétique de quatre populations de hareng de l'estuaire du Saint-Laurent et de la Baie des Chaleurs. *Nat. Can. Rev. Écol. Syst.* 112:105–112.

Robbins, T. W. 1969. A systematic study of the silversides *Membras Bonaparte* and *Menidia* (Linnaeus) (Atherinidae, Teleostei). Ph.D. dissertation, Cornell University, Ithaca, 282 pp.

Roberts, C. D. 1977. The wreckfish *Polyprion americanus* (Schneider, 1801) in Irish waters: an underwater sighting and review of the Irish records. *Irish Nat. J.* 19:108–112.

Roberts, C. D. 1989. Reproductive mode in the percomorph fish genus *Polyprion* Oken. *J. Fish Biol.* 34:1–9.

Roberts, C. D. 1993. Comparative morphology of spined scales and their phylogenetic significance in the Teleostei. *Bull. Mar. Sci.* 52: 60–113.

Roberts, S. C. 1978. Biological and fisheries data on northern searobin, *Prionotus carolinus* (Linnaeus). NMFS Sandy Hook Lab. Tech. Ser. Rept. No. 13, 53 pp.

Roberts-Goodwin, S. C. 1981. Biological and fisheries data on striped searobin, *Prionotus evolans* (Linnaeus). NMFS Sandy Hook Lab. Tech. Ser. Rept. No. 25, 50 pp.

Robinette, H. R. 1983. Species profiles: life histories and environmental requirements of coastal fishes and invertebrates (Gulf of Mexico)-bay anchovy and striped anchovy. U.S. Fish Wildl. Serv., Div. Biol. Serv. Rept., FWS/OBS-82/11.12, 15 pp.

Robins, C. H., and C. R. Robins. 1989. Family Synaphobranchidae. *Fishes of the western North Atlantic.* Sears Found. Mar. Res. Mem. No. 1(9):207–253.

Robins, C. R. 1986. The status of the ophidiid fishes *Ophidium brevibarbe* Cuvier, *Ophidium graellsi* Poey, and *Leptophidium profundorum* Gill. *Proc. Biol. Soc. Washington* 99:384–387.

Robins, C. R., and G. C. Ray. 1986. *A Field Guide to Atlantic Coast Fishes of North America.* Houghton Mifflin Co., Boston, 354 pp.

Robins, C. R., R. M. Bailey, C. E. Bond, J. R. Brooker, E. A. Lachner, R. N. Lea, and W. B. Scott. 1986. Names of the Atlantic redfishes, genus *Sebastes*. *Fisheries* 11(1):28–29.

Robins, C. R., R. M. Bailey, C. E. Bond, J. R. Brooker, E. A. Lachner, R. N. Lea, and W. B. Scott. 1991. Common and scientific names of fishes from the United States and Canada. Am. Fish. Soc. Spec. Publ. No. 20, 183 pp.

Robinson, P. F., and F. J. Schwartz. 1965. A revised bibliography of papers dealing with the oyster toadfish, *Opsanus tau*. Chesapeake Biol. Lab., Solomons, Md. Contrib. No. 284, 18 pp.

Robinson, P. F., and F. J. Schwartz. 1968. Toxicity of the northern puffer, *Sphaeroides maculatus*, in the Chesapeake Bay and its environs. *Chesapeake Sci.* 9:136–137.

Robohm, R. A., and C. Brown. 1978. A new bacterium (presumptive *Vibrio* species) causing ulcers in flatfish. *Mar. Fish. Rev.* 40(10):5–7.

Rodríguez-Roda, J. 1966. Estudio de la bacoreta, *Euthynnus alletteratus* (Raf.), bonito, *Sarda sarda* (Bloch) y melva, *Auxis thazard* (Lac.), capturados por las almadrabas españolas. Inves. Pesq., Barcelona 30:247–292.

Roe, H. S. J. 1969. The food and feeding habits of the sperm whale (*Physeter catodon* L.) taken off the west coast of Iceland. *J. Conseil Perm. Int. Explor. Mer* 33:93–102.

Roe, H. S. J., and J. Badcock. 1984. The diel migrations and distributions within a mesopelagic community in the north east Atlantic. 5. Vertical migrations and feeding of fish. *Prog. Oceanogr.* 13:389–424.

Roelke, D. L., and S. M. Sogard. 1993. Gender-based differences in habitat selection and activity level in the northern pipefish (*Syngnathus fuscus*). *Copeia* 1993:528–532.

Roessler, M. A. 1970. Checklist of fishes in Buttonwood Canal, Everglades National Park, Florida, and observations on the seasonal occurrence and life histories of selected species. *Bull. Mar. Sci.* 20:860–893.

Rofen, R. R. 1966. Family Paralepididae. Fishes of the western North Atlantic. Sears Found. Mar. Res. Mem. No. 1(5):205–461.

Rogers, B. A., and D. T. Westin. 1980. A culture methodology for striped bass, *Morone saxatilis.* EPA Ecological Research Series 660/3-80-000. Narragansett ERC–EPA, 286 pp.

Rogers, B. A., and D. T. Westin. 1981. Laboratory studies on effects of temperature and delayed initial feeding on development of striped bass larvae. *Trans. Am. Fish. Soc.* 110:100–110.

Rogers, B. A., D. T. Westin., and S. B. Saila. 1977. Life stage duration in Hudson River striped bass. Univ. Rhode Island Mar. Tech. Rept. No. 31, 111 pp.

Roghi, G. 1961. Osservazioni sul pesce luna, *Mola mola* (L.). Riv. Sci. Nat., Soc. Ital. Sci. Nat., Milano 52(3):109–112.

Rohr, B. A., and E. J. Gutherz. 1977. Biology of offshore hake, *Merluccius albidus,* in the Gulf of Mexico. *Fish. Bull., U.S.* 75:147–158.

Rojo Lucio, A. 1955. Datos sobre la edad del bacalao (*G. callarias* L.), eglefino (*Melanogrammus aeglefinus* L.), colin (*Pollachius virens* L.) y locha (*Urophycis tenuis* Mitch.) con indicacion de las technicas usadas. *Bol. Inst. Españ. Oceanogr.* 73, 16 pp.

Rollefsen, G. 1934. The eggs and larvae of the halibut (*Hippoglossus vulgaris*). *Det Kongelige Norske Videnskabers Selskab Forhandlinger* 7:20–23.

Roon, J. M. van. 1942. Some additional notes on external features and on the jaw muscles of *Orthagoriscus mola* (L). *Zool. Meded. Rijksmus. Nat. Hist. Leiden* 23:313–317.

Roon, J. M. van., and J. J. ter Pelkwijk. 1939. Mechanisms of the jaw and body muscles of *Orthagoriscus mola* L. *Zool. Meded Rijksmus. Nat. Hist. Leiden* 22:65–75.

Rose, C. D., and W. W. Hassler. 1968. Age and growth of the dolphin, *Coryphaena hippurus* (Linnaeus), in North Carolina waters. *Trans. Am. Fish. Soc.* 97:271–276.

Rose, C. D., and W. W. Hassler. 1974. Food habits and sex ratios of dolphin *Coryphaena hippurus* captured in the western North Atlantic Ocean off Hatteras, North Carolina. *Trans. Am. Fish. Soc.* 103:94–100.

Rose, G. A., and W. C. Leggett. 1988. Atmosphere-ocean coupling and Atlantic cod migrations: effects of wind-forced variations in sea temperatures and currents on nearshore distributions and catch rates of *Gadus morhua. Can. J. Fish. Aquat. Sci.* 45:1234–1243.

Rosen, D. E. 1973. Interrelationships of higher euteleosteans. In: *Interrelationships of Fishes.* P. H. Greenwood, R. S. Miles, and C. Patterson, eds. Academic Press, London, pp. 397–513.

Rosenblatt, R. H., and G. D. Johnson. 1976. Anatomical considerations of pectoral swimming in the opah, *Lampris guttatus. Copeia* 1976:367–370.

Ross, M. R. 1991. Recreational fisheries of coastal New England. University of Massachusetts Press, Amherst, 279 pp.

Ross, M. R., and F. P. Almeida. 1986. Density-dependent growth of silver hakes. *Trans. Am. Fish. Soc.* 115:548–554.

Ross, M. R., and S. R. Hokenson. 1997. Short-term mortality of discarded finfish bycatch in the Gulf of Maine fishery for northern shrimp *Pandalus borealis. No. Am. J. Fish. Manage.* 17:902–909.

Ross, S. W., G. W. Link, Jr., and K. A. MacPherson. 1981. New records of marine fishes from the Carolinas, with notes on additional species. *Brimleyana* 6:61–72.

Ross, S. W., K. J. Sulak, and T. A. Munroe. 2001 Association of *Syscenus infelix* (Crustacea: Isopoda: Aegidae) with benthopelagic rattail fishes, *Nezumia* spp. (Macrouridae), along the western North Atlantic continental slope. *Mar. Biol.* 138:595–601.

Rotunno, T. K. 1992. Species identification and temporal spawning patterns of butterfish, *Peprilus* spp., in the South and Mid-Atlantic Bights. M.S. thesis, State University of New York, Stony Brook, 77 pp.

Rounsefell, G. A., and L. D. Stringer. 1943. Restoration and management of the New England alewife fisheries with special reference to Maine. *Trans. Am. Fish. Soc.* 73:394–424.

Rountree, R. A. 1992. Fish and macroinvertebrate community structure and habitat use patterns in salt marsh creeks of southern New Jersey, with a discussion on marsh carbon export. Ph.D. dissertation, Rutgers University, New Brunswick, 303 pp.

Rountree, R. A. 1999 Nov. Diets of NW Atlantic fishes and squid. <http://www.fishecology.org> Accessed 17 Aug. 2000.

Rountree, R. A., and K. W. Able. 1992a. Fauna of polyhaline subtidal marsh creeks in southern New Jersey: composition, abundance and biomass. *Estuaries* 15:171–185.

Rountree, R. A., and K. W. Able. 1992b. Foraging habits, growth, and temporal patterns of salt-marsh creek habitat use by young-of-year summer flounder in New Jersey. *Trans. Am. Fish. Soc.* 121: 765–776.

Rountree, R. A., and K. W. Able. 1993. Diel variation in decapod crustacean and fish assemblages in New Jersey polyhaline marsh creeks. *Est. Coast. Shelf Sci.* 37:181–201.

Rountree, R. A., and K. W. Able. 1996. Seasonal abundance, growth, and foraging habits of juvenile smooth dogfish, *Mustelus canis,* in a New Jersey estuary. *Fish. Bull., U.S.* 94:522–534.

Rountree, R. A., and J. Bowers-Altman. 2002. Soniferous behavior of the striped cusk-eel, *Ophidion marginatum. Bioacoustics* 12.

Rowe, G. T., P. T. Polloni, and R. L. Haedrich. 1975. Quantitative biological assessment of the benthic fauna in deep basins of the Gulf of Maine. *J. Fish. Res. Bd. Can.* 32:1805–1812.

Rowland, W. J. 1974. Reproductive behavior of the fourspine stickleback, *Apeltes quadracus. Copeia* 1974:183–194.

Rowland, W. J. 1983. Interspecific aggression and dominance in *Gasterosteus. Environ. Biol. Fishes* 8:269–277.

Rowland, W. J., C. L. Baube, and T. T. Horan. 1991. Signalling of sexual receptivity by pigmentation pattern in female sticklebacks. *Anim. Behav.* 42:243–249.

Royce, W. F., R. J. Buller, and E. D. Premetz. 1959. Decline of the yellowtail flounder (*Limanda ferruginea*) off New England. *Fish. Bull., U.S.* 59:169–267.

Rubec, L. A. 1991. Redescription of *Diclidophoroides macallumi* (Monogenea: Diclidophoridae) from the gills of longfin hake, *Physis chesteri,* from the Gulf of St.Lawrence. *Can. J. Zool.* 69: 146–150.

Rubec, L. A., and W. E. Hogans. 1987. Redescription of *Clavellisa cordata* Wilson, 1915 (Copepoda: Lernaeopodidae) from anadromous clupeids in eastern Canada. *Can. J. Zool.* 65:1559–1563.

Rubec, P. J., J. M. McGlade, B. L. Trottier, and A. Ferron. 1991. Evaluation of methods for separation of Gulf of St. Lawrence beaked redfishes, *Sebastes fasciatus* and *S. mentella:* malate dehydrogenase mobility patterns compared with extrinsic gasbladder muscle passages and anal fin ray counts. *Can. J. Fish. Aquat. Sci.* 48: 640–660.

Rulifson, R. A., and S. A. McKenna. 1987. Food of striped bass in the upper Bay of Fundy, Canada. *Trans. Am. Fish. Soc.* 116:119–122.

Runge, J. A., P. Pepin, and W. Silvert. 1987. Feeding behavior of the Atlantic mackerel *Scomber scombrus* on the hydromedusa *Aglantha digitale. Mar. Biol.* 94:329–333.

Rupp, R. S. 1959. Variation in the life history of the American smelt in inland waters of Maine. *Trans. Am. Fish. Soc.* 88:241–252.

Rupp, R. S. 1965. Shore-spawning and survival of eggs of the American smelt. *Trans. Am. Fish. Soc.* 94:160–168.

Rupp, R. S. 1968. Life history and ecology of the smelt (*Osmerus mordax*) in the inland waters of Maine. Maine Dept. Inland Fish Game, 36 pp.

Russell, F. S. 1976. *The Eggs and Planktonic Stages of British Marine Fishes.* Academic Press, London., 524 pp.

Russell, M., M. Grace, and E. J. Gutherz. 1992. Field guide to the searobins (*Prionotus* and *Bellator*) in the western North Atlantic. NOAA Tech. Rept. NMFS No. 107, 26 pp.

Ryall, P. J. C., and B. T. Hargrave. 1984. Attraction of the Atlantic wreckfish (*Polyprion americanus*) to an unbaited camera on the mid-Atlantic Ridge. *Deep Sea Res.* 31:79–83.

Ryder, J. A. 1882a. A contribution to the development and morphology of the lophobranchiates; (*Hippocampus antiquorum,* the seahorse). *Bull. U.S. Fish Comm.* 1:191–199.

Ryder, J. A. 1882b. Development of the silver gar (*Belone longirostris*) with observations on the genesis of the blood in embryo fishes, and a comparison of fish ova with those of other vertebrates. *Bull. U. S. Fish Comm.* 1:283–302.

Ryder, J. A. 1887a. Preliminary notice of the development of the toadfish, *Batrachus tau. Bull. U. S. Fish Comm.* 6:4–8.

Ryder, J. A. 1887b. On the development of osseous fishes, including marine and freshwater forms. U.S. Fish Comm. Rept. No. 13:488–605.

Ryder, J. A. 1888. On the development of the common sturgeon (*Acipenser sturio*). *Am. Nat.* 22:659–660.

Ryder, J. A. 1890. The sturgeons and sturgeon industries of the eastern coast of the United States, with an account of experiments bearing upon sturgeon culture. *Bull. U.S. Fish Comm.* 8:231–328.

Ryer, C. H. 1988. Pipefish foraging: effects of fish size, prey size and altered habitat complexity. *Mar. Ecol. Prog. Ser.* 48:37–45.

Ryer, C. H., and G. W. Boehlert. 1983. Feeding chronology, daily ration, and the effects of temperature upon gastric evacuation in the pipefish, *Syngnathus fuscus. Environ. Biol. Fishes* 9:301–306.

Ryer, C. H., and R. J. Orth. 1987. Feeding ecology of the northern pipefish, *Syngnathus fuscus,* in a seagrass community of the lower Chesapeake Bay. *Estuaries* 10:330–336.

Ryman, N., U. Lagercrantz, L. Andersson, R. Chakraborty, and R. Rosenberg. 1984. Lack of correspondence between genetic and morphologic variability patterns in Atlantic herring (*Clupea harengus*). *Heredity* 53:687–704.

Ryther, J. H. 1997. *Anadromous Brook Trout: Biology, Status, and Enhancement.* Trout Unlimited, Arlington, Va., 34 pp.

Saccardo, S. A., and M. Katsuragawa. 1995. Biology of the rough scad *Trachurus lathami,* on the southeastern coast of Brazil. *Sci. Mar.* 59:265–277.

Sæmundsson, B. 1922. Zoologiske meddelelser fra Island. XIV. 11. Fiske, ny for Island, og supplerende Oplysninger om andre, tidligere kendte. *Vidensk. Medd. Dansk Naturh. Foren.* 74:159–201.

Sæmundsson, B. 1925. Fiskirannsoknir 1923–1924. Andvari Reykjavik 33–71.

Sæmundsson, B. 1949. Marine Pisces. In: A. Frioriksson et al., eds. *The zoology of Iceland,* Vol. 4(72):1–150. Ejnar Munksgaard, Copenhagen.

Safford, S. E., and H. Booke. 1992. Lack of biochemical genetic and morphometric evidence for discrete stocks of northwest Atlantic herring *Clupea harengus harengus. Fish. Bull., U.S.* 90:203–210.

Safina, C., and J. Burger. 1985. Common tern foraging: seasonal trends in prey fish densities and competition with bluefish. *Ecology* 66:1457–1463.

Safina, C., and J. Burger. 1989. Population interactions among free-living bluefish and prey fish in an ocean environment. *Oecologia* 79:91–95.

Saila, S. B. 1961. The contribution of estuaries to the offshore winter flounder fishery in Rhode Island. *Proc. Gulf Carib. Fish. Inst.* 14:95–109.

Saila, S. B. 1962. Proposed hurricane barriers related to winter flounder movements in Narragansett Bay. *Trans. Am. Fish. Soc.* 91:189–195.

Saila, S. B., and R. G. Lough. 1981. Mortality and growth estimation from size data-an application to some Atlantic herring larvae. *Rapp. Proc.-Verb. Réun. Cons. Int. Explor. Mer* 178:7–14.

Saito, T., Y. Noguchi, Y. Shida, T. Abe, and K. Hashimoto. 1991. Screening of tetrodotoxin and its derivatives in puffer-related species. *Nippon Suisan Gakkaishi* 57:1573–1577.

Sakamoto, K. 1984. Interrelationships of the family Pleuronectidae (Pisces: Pleuronectiformes). Fac. Fish. Hokkaido Univ. Mem. No. 31:95–215.

Sakanari, J. A., C. A. Reilly, and M. Moser. 1983. Tubercular lesions in Pacific coast populations of striped bass. *Trans. Am. Fish. Soc.* 112:565–566.

Saksena, V. P., and E. D. Houde. 1972. Effect of food level on the growth and survival of laboratory-reared larvae of bay anchovy (*Anchoa mitchilli* Valenciennes) and scaled sardine (*Harengula pensacolae* Goode and Bean). *J. Exp. Mar. Biol. Ecol.* 8:249–258.

Saldanha, L. 1980. Régime alimentaire de *Synaphobranchus kaupii* Johnson, 1862 (Pisces Synaphobranchidae) au large des côtes européennes. *Cybium* 8: 91–98.

Saldanha, L., and P. J. Whitehead. 1990. Megalopidae. In: J. C. Quéro et. al., eds. *Checklist of the Fishes of the Eastern Tropical Atlantic. Clofeta.* 2:120–121. JNICT, Lisbon.

Salinas, I. M., and I. A. McLaren. 1983. Seasonal variation in weight-specific growth rates, feeding rates, and growth efficiency in *Microgadus tomcod. Can. J. Fish. Aquat. Sci.* 40:2197–2200.

Saloman, C. H., and S. P. Naughton. 1984. Food of crevalle jack (*Caranx hippos*) from Florida, Louisiana, and Texas. NOAA Tech. Memo. NMFS-SEFC-134, 34 pp.

Samaritan, J. M., and R. E. Schmidt. 1982. Aspects of the life history of a freshwater population of the mummichog, *Fundulus heteroclitus* (Pisces: Cyprinodontidae), in the Bronx River, New York, U.S.A. *Hydrobiologia* 94:149–154.

Sameoto, D. D. 1988. Feeding of lantern fish *Benthosema glaciale* off the Nova Scotia shelf. *Mar. Ecol. Prog. Ser.* 44:113–129.

Sampson, R. 1981. Connecticut marine recreational fisheries survey 1979–1980. Conn. Dept. Environ. Protec. Mar. Fish., 49 pp.

Sánchez, R. P., and E. M. Acha. 1988. Development and occurrence of embryos, larvae and juveniles of *Sebastes oculatus* with reference to two southwest Atlantic scorpaenids: *Helicolenus dactylopterus lahillei* and *Pontinus rathbuni. Meeresforschung* 32: 107–133.

Sand, R. L. 1982. Aspects of the feeding ecology of the cunner, *Tautogolabrus adspersus,* in Narragansett Bay. M.S. thesis, University of Rhode Island, Kingston, 94 pp.

Sandeman, E. J. 1969. Age determination and growth rate of redfish, *Sebastes* sp., from selected areas around Newfoundland. *ICNAF Res. Bull.* 6: 79–105.

Sandercock, F. K. 1991. Life history of coho salmon (*Oncorhynchus kisutch*). In: C. Groot and L. Margolis, eds. *Pacific Salmon Life Histories.* University of British Columbia Press, Vancouver, pp. 395–445.

Sandholzer, L. A., T. Norstrand, and L. Young. 1945. Studies of an ichthyosporidian-like parasite of ocean pout, (*Zoarces anguillaris*). U.S. Fish Wildl. Serv. Spec. Sci. Rept. No. 31, 12 pp.

Sandström, O. 1980. Selective feeding by Baltic herring. *Hydrobiologia* 69:199–207.

Sandy, J. M., and J. H. S. Blaxter. 1980. A study of retinal development in larval herring and sole. *J. Mar. Biol. Assoc. U.K.* 60:59–71.

Sanzo, L. 1915. Contributo alla conoscenza dello sviluppo negli Scopelini Müller (*Saurus griseus* Lowe, *Chlorophthalmus agassizi* Bp., *Aulopus filamentosus* Cuv.). R. Com. Talassogr. Ital. Mem. No. 49, 21 pp.

Sanzo, L. 1931a. Sottordine: Salmonoidei. In: *Uova, Larve e Stadi Giovanili di Teleostei.* Fauna Flora Golfo Napoli Monogr. No. 38:21–92.

Sanzo, L. 1931b. Uovo, stadi embrionali e postembrionali di *Naucrates ductor* L. Com. Talassogr. Ital. Mem. No. 185, 14 pp.

Sanzo, L. 1932. Uovo, stadi larvali e giovanili di *Centrolophus pompilius.* C.V. Com. Talissogr. Ital. Mem. No. 196, 16 pp.

Sanzo, L. 1933. Uovo, stadi larvali e giovanili di *Dactylopterus volitans* L. R. Com. Talissogr. Ital. Mem. No. 207, 26 pp.

Sanzo, L. 1939. Rarisimi stadi larvali de Teleostei. *Arch. Zool. Ital.* 26(121–150):1–26.

Sanzo, L. 1940. Sviluppo embrionale e larva appena sciasa de *Scomberesox saurus* (Flem.). Com. Talassogr. Ital. Mem. No. 276, 6 pp.

Sargent, R. C., M. A. Bell, W. H. Krueger, and J. V. Baumgartner. 1984. A lateral plate cline, sexual dimorphism, and phenotypic variation in the blackspotted stickleback, *Gasterosteus wheatlandi. Can. J. Zool.* 62:368–376.

Saucerman, S. E., and L. A. Deegan. 1991. Lateral and cross-channel movement of young-of-the-year winter flounder (*Pseudopleuronectes americanus*) in Waquoit Bay, Massachusetts. *Estuaries* 14:440–446.

Saunders, R. L. 1991. Salmonid mariculture in Atlantic Canada and Maine, USA. Special Session on Salmonid Aquaculture, World Aquaculture Society. Can. Fish. Aquat. Sci. Tech. Rept. No. 1831:21–36.

Sauskan, V. I. 1964. Results of Soviet observations on the distribution of silver hake *Merluccius bilinearis* Mitchill on Georges Bank and off Nova Scotia in 1962–63. ICNAF Res. Doc. No. 64/61.

Sauskan, V. I., and G. N. Semenov. 1969. Saury: the North Atlantic saury. *Ann. Biol.* 25:250–252.

Sauskan, V. I., and V. P. Serebryakov. 1968. Reproduction and development of the silver hake (*Merluccius bilinearis* Mitchill). *Voprosy Ikhtiol.* 8:500–524 [in Russian, translation in *Prob. Ichthyol.* 8(3):398–414].

Saville, A. 1978. The growth of herring in the northwestern North Sea. *Rapp. Proc.-Verb. Réun. Cons. Int. Explor. Mer* 172: 164–171.

Saville, A., and S. H. Jackson. 1974. Recent changes in growth of Clyde spring-spawning herring. Int. Comm. Explor. Sea CM 1974/H:61.

Savvatimsky, P. I. 1971. Determination of the age of grenadiers (order Macrouriformes). *Voprosy Ikhtiol.* 11:495–501 [in Russian, translation in *J. Ichthyol.* 11(2):397–403].

Savvatimsky, P. I. 1985. Biological characteristics of the roughhead grenadier *Macrourus berglax* Lacépède, near the Lofoten Islands. *Voprosy Ikhtiol.* 25:610–616 [in Russian, translation in *J. Ichthyol.* 25(5):23–29].

Savvatimsky, P. I. 1989a. Investigations of roughhead grenadier (*Macrourus berglax* L.) in the northwest Atlantic, 1967–1983. NAFO Sci. Coun. Stud. No. 13:59–75.

Savvatimsky, P. I. 1989b. Distribution and biology of common grenadier (*Nezumia bairdi*) from trawl surveys in the northwest Atlantic, 1969–83. NAFO Sci. Coun. Stud. No. 13:53–58.

Sawyer, P. J. 1967. Intertidal life-history of the rock gunnel, *Pholis gunnellus,* in the western Atlantic. *Copeia* 1967:55–61.

Sayers, R. E., J. R. Moring, P. R. Johnson, and S. A. Roy. 1989. Importance of rainbow smelt in the winter diet of landlocked Atlantic salmon in four Maine lakes. *No. Am. J. Fish. Manage.* 9:298–302.

Scattergood, L. W. 1952. The maturity of Maine herring (*Clupea harengus*). *Maine Dept. Sea and Shore Fish., Res. Bull.* 7, 11 pp.

Scattergood, L. W. 1953. Notes on Gulf of Maine fishes in 1952. *Copeia* 1953:194–195.

Scattergood, L. W. 1958. Western North Atlantic records of *Beryx splendens* Lowe and B. *decadactylus* Cuvier and Valenciennes. *Copeia* 1958:231.

Scattergood, L. W. 1959. New records of Gulf of Maine fishes. *Maine Field Nat.* 15:107–109.

Scattergood, L. W. 1962a. White sharks, *Carcharodon carcharias,* in Maine, 1959–1960. *Copeia* 1962:446–447.

Scattergood, L.W. 1962b. First record of mako, *Isurus oxyrinchus,* in Maine waters. *Copeia* 1962:462.

Scattergood, L. W., and G. W. Coffin. 1957. Records of some Gulf of Maine fishes. *Copeia* 1957:155–156.

Scattergood, L. W., and P. L. Coggins. 1958. Unusual records of Gulf of Maine fishes. *Maine Field Nat.* 14:40–43.

Scattergood, L. W., P. S. Trefethen, and G. W. Coffin. 1951. Notes on Gulf of Maine fishes in 1949. *Copeia* 1951:297–298.

Schaefer, R. H. 1960. Growth and feeding habits of the whiting or silver hake in the New York Bight. *N.Y. Fish Game J.* 7:85–98.

Schaefer, R. H. 1965. Age and growth of the northern kingfish in New York waters. *N.Y. Fish. Game J.* 12:199–216.

Schaefer, R. H. 1967. Species composition, size and seasonal abundance of fish in the surf waters of Long Island. *N.Y. Fish Game J.* 14:1–46.

Schaefer, R. H. 1970. Feeding habits of striped bass from the surf waters of Long Island. *N.Y. Fish Game J.* 17:1–17.

Schaefer, R. H., and T. Doheny. 1973. First record of the four-eye butterflyfish from New York waters. *N.Y. Fish Game J.* 20: 74–75.

Schaeffer, J. S., and F. J. Margraf. 1986a. Food of white perch (*Morone americana*) and potential for competition with yellow perch (*Perca flavescens*) in Lake Erie. *Ohio J. Sci.* 86:26–28.

Schaeffer, J. S., and F. J. Margraf. 1986b. Population characteristics of the invading white perch (*Morone americana*) in western Lake Erie. *J. Great Lakes Res.* 12:127–131.

Schaeffer, J. S., and F. J. Margraf. 1987. Predation on fish eggs by white perch, *Morone americana,* in western Lake Erie. *Environ. Biol. Fishes* 18:77–80.

Schaner, E., and K. Sherman. 1960. Observations on the fecundity of the tomcod, *Microgadus tomcod* (Walbaum). *Copeia* 1960: 347–348.

Scharf, F. S., J. A. Buckel, F. Juanes, and D. O. Conover. 1997. Estimating piscine prey size from partial remains: testing for shifts in foraging mode by juvenile bluefish. *Environ. Biol. Fishes* 49:377–388.

Schekter, R. C. 1972. Food habits of some larval and juvenile fishes from the Florida Current near Miami, Florida. U.S. Environ. Prot. Agency Tech. Rept., 85 pp.

Schenk, R. 1981. Population identification of silver hake (*Merluccius bilinearis*) using isoelectric focusing. NMFS Woods Hole Lab. Ref. Doc. No. 81-44, 31 pp.

Scherer, M. D. 1972. The biology of the blueback herring, *Alosa aestivalis* (Mitchill), in the Connecticut River above the Holyoke Dam, Holyoke, Massachusetts. M.S. thesis, University of Massachusetts, Amherst, 90 pp.

Scherer, M. D. 1984. The ichthyoplankton of Cape Cod Bay. In: J. D. Davis and D. Merriman, eds. *Observations on the Ecology and Biology of Western Cape Cod Bay, Massachusetts.* Springer-Verlag, Berlin, pp. 151–190.

Scherer, M. D., and D. W. Bourne. 1980. Eggs and early larvae of the smallmouth flounder, *Etropus microstomus. Fish. Bull., U.S.* 77: 708–712.

Schmelz, G. W. 1970. Some effects of temperature and salinity on the life processes of the striped killifish, *Fundulus majalis* (Walbaum). Ph.D. dissertation, University of Delaware, Newark, 104 pp.

Schmid, T. H., F. L. Murru, and F. McDonald. 1991. Feeding habits and growth rate of bull (*Carcharhinus leucas* {Valenciennes}), sandbar (*Carcharhinus plumbeus* {Nardo}), sandtiger (*Eugomphodus taurus* {Rafinesque}), and nurse (*Ginglymostoma cirratum* {Bonnaterre}) sharks maintained in captivity. *J. Aquaricult. Aquat. Sci.* 4:100–105.

Schmidt, J. 1905. On the larval and post-larval stages of the torsk (*Brosmius brosme* (Ascan.). *Medd. Komm. Havunders. Ser. Fisk.* 1(8):1–10.

Schmidt, J. 1921. New studies of sun-fishes made during the "Dana" expeditions 1920. *Nature* 107:76–79.

Schmidt, J. 1922. The breeding places of the eel. *Philos. Trans. Royal Soc. London, B* 211: 179–208.

Schmidt, J. 1930. *Nessorhamphus,* a new cosmopolitan genus of oceanic eels. *Vidensk. Dansk Naturhist. For.* 90:371–376.

Schmidt, J. 1931. Eels and conger eels of the North Atlantic. *Nature* 128: 602–604.

Schmidt, W. J. 1954. Über Bau und Entwicklung der Zähne des Knochenfisches *Anarrchichas lupus* L. und ihren Befall mit *"Mycelites ossifragus." Zeit. Zellforsch. Microsk. Anat.* 40(1):25–48.

Schneppenheim, R., and H. Theede. 1982. Freezing point depressing peptide and glycoproteins from Arctic-boreal and Antarctic fishes. *Polar Biol.* 1:115–123.

Schoedinger, S. E., and C. E. Epifanio. 1997. Growth, development and survival of larval *Tautoga onitis* (Linnaeus) in large laboratory containers. *J. Exp. Mar. Biol. Ecol.* 210:143–155.

Schreiner, K.E. 1955. Studies on the gonad of *Myxine glutinosa* L. Universitet Bergen. Naturvit. Rekke 8, 40 pp.

Schroeder, W. C. 1930. A record of *Polyprion americanus* (Bloch and Schneider) from the northwestern Atlantic. *Copeia* 1930:46–48.

Schroeder, W. C. 1931. Notes on certain fishes collected off the New England coast from 1924 to 1930. *Bull. Boston Soc. Nat. Hist.* 58:3–8.

Schroeder, W. C. 1933. Unique records of the brier skate and rock eel from New England. *Bull. Boston Soc. Nat. Hist.* 66:5–6.

Schroeder, W. C. 1937. Records of *Pseudopriacanthus altus* (Gill) and *Fundulus majalis* (Walbaum) from the Gulf of Maine. *Copeia* 1937:238.

Schroeder, W. C. 1942. Results of haddock tagging in the Gulf of Maine from 1923–1932. *J. Mar. Res.* 5:1–19.

Schroeder, W. C. 1947. Notes on the diet of the goosefish, *Lophius americanus. Copeia* 1947:201.

Schroeder, W. C. 1955. Report on the results of exploratory otter-trawling along the continental shelf and slope between Nova Scotia and Virginia during the summers of 1952 and 1953. *Deep-Sea Res.* 3(Suppl.):358–372

Schuck, H. A. 1951a. Northern record for the little tuna, *Euthynnus alletteratus. Copeia* 1951:98.

Schuck, H. A. 1951b. New Gulf of Maine record for occurrence of dolphin, *Coryphaena hippurus,* and data on small specimens. *Copeia* 1951:171.

Schuck, H. A. 1982. Bluefin tuna, *Thunnus thynnus.* In: M. D. Grosslein and T. R. Azarovitz, eds. *Fish Distribution.* MESA New York Bight Atlas Monogr. No. 15:102–105.

Schuijf, A., and M. E. Siemelink. 1974. The ability of cod (*Gadus morhua*) to orient towards a sound source. *Experientia* 30:773–774.

Schultz, E. T., and D. O. Conover. 1997. Latitudinal differences in somatic energy storage: adaptive responses to seasonality in an estuarine fish (Atherinidae: *Menidia menidia*). *Oecologia* 109:516–529.

Schultz, L. P. 1931. Hermaphroditism in the striped bass. *Copeia* 1931:64.

Schwartz, A. 1971. Swimbladder development and function in the haddock, *Melanogrammus aeglefinus* L. *Biol. Bull., Woods Hole* 141:176–188.

Schwartz, F. J. 1961. Fishes of Chincoteague and Sinepuxent bays. *Am. Midl. Natur.* 65:384–408.

Schwartz, F. J. 1963. The barrelfish from Chesapeake Bay and the Middle Atlantic Bight, with comments on its zoogeography. *Chesapeake Sci.* 4:147–149.

Schwartz, F. J. 1964a. Effects of winter water conditions on fifteen species of captive marine fishes. *Am. Midl. Nat.* 71:434–444.

Schwartz, F. J. 1964b. Fishes of the Isle of Wight and Assawoman bays near Ocean City, Maryland. *Chesapeake Sci.* 5:172–193.

Schwartz, F. J. 1974. Movements of the oyster toadfish (Pisces: Batrachoididae) about Solomons, Maryland. *Chesapeake Sci.* 15:155–159.

Schwartz, F. J. 1983. Shark ageing methods and age estimation of scalloped hammerhead, *Sphyrna lewini,* and dusky, *Carcharhinus obscurus,* sharks based on vertebral ring counts. NOAA Tech. Rept. NMFS No. 8:167–174.

Schwartz, F. J. 1990a. Mass migratory congregations and movements of several species of cownose rays, genus *Rhinoptera:* a world-wide review. *J. Elisha Mitchell Sci. Soc.* 106:10–13.

Schwartz, F. J. 1990b. Length-weight, age and growth, and landings observations for sheepshead *Archosargus probatocephalus* from North Carolina. *Fish. Bull., U.S.* 88:829–832.

Schwartz, F. J. 1996. Biology of the clearnose skate, *Raja eglanteria,* from North Carolina. *Fla. Sci.* 59:82–95.

Schwartz, F. J. 1997a. Clam siphon tip nipping by fishes in the estuarine Cape Fear River, North Carolina. *Brimleyana* 24:33–45.

Schwartz, F. J. 1997b. Status of the Atlantic sturgeon, *Acipenser oxyrinchus* (Pisces, Acipenseridae) in North Carolina. *J. Elisha Mitchell Sci. Soc.* 113:46–52.

Schwartz, F. J. 1997c. Recent capture of an adult bigeye soldierfish, *Ostichthys trachypoma* (Holocentridae) from North Carolina. *J. Elisha Mitchell Sci. Soc.* 113:183–185.

Schwartz, F. J. 1997d. Biology of the striped cusk-eel, *Ophidion marginatum,* from North Carolina. *Bull. Mar. Sci.* 61:327–342.

Schwartz, F. J., and B. B. Bright. 1982. A bibliography of papers dealing with the oyster toadfish, *Opsanus tau,* 1965 through April 1982. Inst. Mar. Sci., Univ. N.C., Morehead City, Spec. Publ., 33 pp.

Schwartz, F. J., and B. W. Dutcher. 1963. Age, growth, and food of the oyster toadfish near Solomons, Maryland. *Trans. Am. Fish. Soc.* 92:170–173.

Schwartz, F. J., and C. Jensen. 1996. Cornetfishes (*Fistularia,* Fistulariidae) as food of the dusky shark, *Carcharhinus obscurus* (Carcharhinidae), from North Carolina, and problems identifying fistularids from only heads. *J. Elisha Mitchell Sci. Soc.* 112:40–44.

Schwartz, F. J., and D. G. Lindquist. 2000. Flying gurnards, *Dactylopterus volitans* (Pisces: Dactylopteridae) from North Carolina and adjacent states. *J. Elisha Mitchell Sci. Soc.* 116:146–152.

Schwartz, F. J., and J. Purifoy. 1997. Coloration and benthic occurrence of a very small short bigeye, *Pristigenys alta* (Priacanthidae, Pisces), in North Carolina. *J. Elisha Mitchell Sci. Soc.* 113:1–3.

Schwartz, F. J., and P. F. Robinson. 1963. Survival of exposed oyster toadfish and biological clocks. *Prog. Fish-Cult.* 25:151–154.

Scofield, E. C. 1928. Striped bass studies. *Calif. Fish Game* 14:29–37.

Scofield, E. C. 1931. The striped bass of California (*Roccus lineatus*). *Calif. Div. Fish Game Fish Bull.* 29, 82 pp.

Scoles, D. R., B. B. Collette, and J. E. Graves. 1998. Global phylogeography of mackerels of the genus *Scomber. Fish. Bull., U.S.* 96:823–842.

Scott, D. M. 1954a. A comparative study of the yellowtail flounder from three Atlantic fishing areas. *J. Fish. Res. Bd. Can.* 11:171–197.

Scott, D. M. 1954b. Experimental infection of Atlantic cod with a larval marine nematode from smelt. *J. Fish. Res. Bd. Can.* 11:894–900.

Scott, J. S. 1968. Morphometrics, distribution, growth, and maturity of offshore sand launce (*Ammodytes dubius*) on the Nova Scotia banks. *J. Fish. Res. Bd. Can.* 25:1775–1785.

Scott, J. S. 1969a. *Lampritrema nipponicum* (Trematoda) from west Atlantic argentines. *Can. J. Zool.* 47:139–140.

Scott, J. S. 1969b. Morphology and morphometric variation in *Lecithophyllum botryophorum* (Trematoda: Hemiuridae) in *Argentina silus. Can. J. Zool.* 47:213–216.

Scott, J. S. 1969c. Trematode populations in the Atlantic argentine, *Argentina silus,* and their use as biological indicators. *J. Fish. Res. Bd. Can.* 26:879–891.

Scott, J. S. 1972. Eggs and larvae of northern sand lance (*Ammodytes dubius*) from the Scotian Shelf. *J. Fish. Res. Bd. Can.* 29:1667–1671.

Scott, J. S. 1973a. Food and inferred feeding behavior of northern sand lance (*Ammodytes dubius*). *J. Fish. Res. Bd. Can.* 30:451–454.

Scott, J. S. 1973b. Intestinal helminth parasites of northern sand lance (*Ammodytes dubius*). *J. Fish. Res. Bd. Can.* 30:291–292.

Scott, J. S. 1973c. Otolith structure and growth in northern sand lance, *Ammodytes dubius,* from the Scotian Shelf. *Int. Comm. Northw. Atl. Fish. Res. Bull.* 10:107–115.

Scott, J. S. 1975. Meristics of herring (*Clupea harengus*) from the Canadian maritimes area. Dept. Environ., Fish. Mar. Serv. Res. Dev. Dir. Tech. Rept. No. 599, 24 pp.

Scott, J. S. 1980. Occurrence of pollock, *Pollachius virens,* and sand lance, *Ammodytes* sp., larvae in the Bay of Fundy. *J. Northw. Atl. Fish. Sci.* 1:45–48.

Scott, J. S. 1982a. Selection of bottom type by groundfishes of the Scotian Shelf. *Can. J. Fish. Aquat. Sci.* 39:943–947.

Scott, J. S. 1982b. Depth, temperature and salinity preferences of common fishes of the Scotian Shelf. *J. Northw. Atl. Fish. Sci.* 3:29–39.

Scott, J. S. 1985. Occurrence of alimentary tract helminth parasites of pollock (*Pollachius virens* L.) on the Scotian Shelf. *Can. J. Zool.* 63:1695–1698.

Scott, J. S. 1987. Helminth parasites of the alimentary tract of the hakes (*Merluccius, Urophycis, Phycis:* Teleostei) of the Scotian Shelf. *Can. J. Zool.* 65:304–311.

Scott, J. S. 1988. Helminth parasites of redfish (*Sebastes fasciatus*) from the Scotian Shelf, Bay of Fundy, and eastern Gulf of Maine. *Can. J. Zool.* 66:617–621.

Scott, W. B., and W. J. Christie. 1963. The invasion of the lower Great Lakes by the white perch, *Roccus americanus* (Gmelin). *J. Fish. Res. Bd. Can.* 20:1189–1195.

Scott, W. B., and E. J. Crossman. 1964. *Fishes Occurring in the Fresh Waters of Insular Newfoundland.* Dept. of Fisheries, Ottawa, 124 pp.

Scott, W. B., and E. J. Crossman. 1973. Freshwater fishes of Canada. *Bull. Fish. Res. Bd. Can.* 184, 966 pp.

Scott, W. B., and M. G. Scott. 1988. Atlantic fishes of Canada. *Can. Bull. Fish. Aquat. Sci.* 219, 731 pp.

Scott, W. B., and S. N. Tibbo. 1968. Food and feeding habits of swordfish, *Xiphias gladius,* in the western North Atlantic. *J. Fish. Res. Bd. Can.* 25:903–919.

Scott, W. B., and S. N. Tibbo. 1974. Food and feeding habits of swordfish, *Xiphias gladius* Linnaeus, in the northwest Atlantic Ocean. In: R. S. Shomura and F. Williams, eds. *Proc. Int. Billfish Symp.* NOAA Tech. Rept. NMFS SSRF No. 675(2):138–141.

Scotton, L. N., R. E. Smith, N. S. Smith, K. S. Price, and D. P. de Sylva. 1973. Pictorial guide to fish larvae of Delaware Bay. Univ. Del., Del. Bay Rept. Ser. No. 7, 205 pp.

Scruggs, G. D., Jr., and J. C. Fuller, Jr. 1955. Indications of a freshwater population of striped bass, *Roccus saxatilis* (Walbaum) in Santee-Cooper reservoir. *Proc. Southeastern Assoc. Game Fish Comm. Nov.* 12:64–69.

Secor, D. H., E. J. Niklitschek, J. T. Stevenson, T. E. Gunderson, S. P. Minkkinen, B. Richardson, B. Florence, M. Mangold, J. Skjeveland, and A. Henderson-Arzapalo. 2000. Dispersal and growth of yearling Atlantic sturgeon, *Acipenser oxyrinchus,* released into Chesapeake Bay. *Fish. Bull., U.S.* 98:800–810.

Sedberry, G. R. 1983. Food habits and trophic relationships of a community of fishes on the outer continental shelf. NOAA Tech. Rept. NMFS SSRF No. 773, 56 pp.

Sedberry, G. R. 1988. Food and feeding of black sea bass, *Centropristis striata,* in live bottom habitats in the South Atlantic Bight. *J. Elisha Mitchell Sci. Soc.* 104:35–50.

Sedberry, G. R., and J. A. Musick. 1978. Feeding strategies of some demersal fishes of the continental slope and rise off the mid-Atlantic coast of the USA. *Mar. Biol.* 44:357–375.

Sedberry, G. R., J. L. Carlin, R. W. Chapman, and B. Eleby. 1996. Population structure in the pan–oceanic wreckfish, *Polyprion americanus* (Teleostei: Polyprionidae), as indicated by mtDNA variation. *J. Fish Biol.* 49(Supl. A):318–329.

Sedberry, G. R., C. A. P Andrade, J. L. Carlin, R. W. Chapman, B. E. Luckhurst, C. S. Manooch, III, G. Menezes, B. Thomsen, and G. F. Ulrich. 1999. Wreckfish *Polyprion americanus* in the North Atlantic: fisheries, biology, and management of a widely distributed and long-lived fish. In: J. A. Musick, ed., Life in the slow lane: ecology and conservation of long-lived animals. *Am. Fish. Soc. Symp.* 23:27–50.

Seelbach, P. W. 1993. Population biology of steelhead in a stable-flow, low-gradient tributary of Lake Michigan. *Trans. Am. Fish. Soc.* 122:179–198.

Seguin, R. T. 1960. Variation in the Middle Atlantic coast population of the grey squetague, *Cynscion regalis* (Bloch and Schneider), 1801. Ph.D. dissertation, University of Delaware, Newark, 70 pp.

Sekavec, G. B. 1974. Summer foods, length-weight relationship, and condition factor of juvenile ladyfish, *Elops saurus* Linnaeus, from Louisiana coastal streams. *Trans. Am. Fish. Soc.* 103:472–476.

Sekhar, S. C., and W. Threlfall. 1970a. Infection of the cunner, *Tautogolabrus adspersus* (Walbaum), with metacercariae of *Cryptocotyle lingua* (Creplin, 1825). *J. Helminthol.* 44:189–198.

Sekhar, S. C., and W. Threlfall. 1970b. Helminth parasites of cunner, *Tautogolabrus adspersus* (Walbaum) in Newfoundland. *J. Helminthol.* 44:169–188.

Sella, M. 1929. Estese migrazioni dell'anguilla in acque sotterranee. R. Com. Talissogr. Ital. Mem. No. 158, 17 pp.

Sella, M. 1931. The tuna (*Thunnus thynnus* L.) of the western Atlantic: an appeal to fishermen for the collection of hooks found in tuna fish. *Int. Rev. Gesam. Hydrobiol.* 25:46–67.

Selman, K., R. A. Wallace, and D. Player. 1991. Ovary of the seahorse, *Hippocampus erectus. J. Morphol.* 209:285–304.

Serchuk, F. M., and C. F. Cole. 1974. Age and growth of the cunner, *Tautogolabrus adspersus* (Walbaum) (Pisces: Labridae), in the Weweantic River estuary, Massachusetts. *Chesapeake Sci.* 15:205–213.

Serchuk, F. M., and D. W. Frame. 1973. An annotated bibliography of the cunner, *Tautogolabrus adspersus* (Walbaum). NOAA Tech. Rept. NMFS-SSR No. 668, 43 pp.

Serchuk, F. M., and S. E. Wigley. 1986. Assessment and status of the Georges Bank and Gulf of Maine Atlantic cod stocks—1986. NMFS Woods Hole Lab. Ref. Doc. No. 86-12, 84 pp.

Serchuk, F. M., and S. E. Wigley. 1992. Assessment and management of the Georges Bank cod fishery: an historical review and evaluation. *J. Northw. Atl. Fish. Sci.* 13:25–52.

Serchuk, F. M., and P. W. Wood. 1979. Review and status of the southern New England-Middle Atlantic Atlantic cod, *Gadus morhua,* populations. Woods Hole Lab. Ref. No. 79-37, 77 pp.

Serchuk, F. M., M. D. Grosslein, R. G. Lough, D. G. Mountain, and L. O'Brien. 1994. Fishery and environmental factors affecting trends and fluctuations in the Georges Bank and Gulf of Maine Atlantic cod stocks: an overview. *ICES Mar. Sci. Symp.* 198:77–109.

Serebryakov, V. P. 1978. Development of the spotted hake, *Urophycis regius,* from the northwestern Atlantic. *Voprosy Ikhtiol.* 18:892–899 [in Russian, translation in *J. Ichthyol.* 18(5):793–799].

Sergeant, D. E. 1963. Minke whales, *Balaenoptera acutorostrata* Lacépède, of the western North Atlantic. *J. Fish. Res. Bd. Can.* 20:1489–1504.

Sergeant, D. E. 1973. Feeding, growth, and productivity of northwest Atlantic harp seals (*Pagophilus groenlandicus*). *J. Fish. Res. Bd. Can.* 30:17–29.

Sette, O. E. 1943. Biology of the Atlantic mackerel (*Scomber scombrus*) of North America. Part I: Early life history including growth, drift, and mortality of the egg and larval populations. *U.S. Fish Wildl. Serv. Fish. Bull.* 50:149–237.

Sette, O. E. 1950. Biology of the Atlantic mackerel (*Scomber scombrus*) of North America. Part II: Migrations and habits. *U.S. Fish Wildl. Serv. Fish. Bull.* 51:251–358.

Setzler, E. M., W. R. Boynton, K. V. Wood, H. H. Zion, L. Lubbers, N. K. Mountford, P. Frere, L. Tucker, and J. A. Mihursky. 1980. Synopsis of biological data on striped bass, *Morone saxatilis* (Walbaum). NOAA Tech. Rept. NMFS Circ. No. 433, 69 pp.

Setzler, E. M., J. H. Mihursky, K. V. Wood, W. R. Boynton, T. T. Polgar, and G. E. Drewry. 1981. Major features of ichthyoplankton populations in the upper Potomac estuary: 1974–76. *Rapp. Proc.-Verb. Réun. Cons. Int. Explor. Mer* 178:204–206.

Setzler-Hamilton, E. M., W. R. Boynton, J. A. Mihursky, T. T. Polgar, and K. V. Wood. 1981. Spatial and temporal distribution of striped bass eggs, larvae, and juveniles in the Potomac Estuary. *Trans. Am. Fish. Soc.* 110:121–136.

Setzler-Hamilton, E. M., J. A. Whipple, and R. B. MacFarlane. 1988. Striped bass populations in Chesapeake and San Francisco bays: two environmentally impacted estuaries. *Mar. Pollut. Bull.* 19: 466–477.

Shafer, T. H., D. E. Padgett, and D. A. Celio. 1990. Evidence for enhanced salinity tolerance of a suspected fungal pathogen of Atlantic menhaden, *Brevoortia tyrannus* Latrobe. *J. Fish. Dis.* 13: 335–344.

Shangguan, B., and L. W. Crim. 1999. Seasonal variations in sperm production and sperm quality in male winter flounder, *Pleuronectes americanus:* the effects of hypophysectomy, pituitary replacment therapy, and GnRH-A treatment. *Mar. Biol.* 134:19–27.

Shannon, E. H., and W. B. Smith. 1967. Preliminary observations of the effect of temperatures on striped bass eggs and sac fry. *Proc. 21st. Ann. Conf. Southeastern Assoc. Game Fish Comm.*, pp. 257–260.

Sharrock, E. S. 1934. The private life of a sea-horse. *Home Aquar. Bull., East Orange, N.J.* 4:11–12.

Shaw, E. 1970. Schooling in fishes: critique and review. In: L. R. Aronson, D. L. Lehrman, and J. S. Rosenblatt, eds. *Development and Evolution of Behavior.* W. H. Freeman, New York, pp. 452–480.

Shaw, E., and Drullinger, D. L. 1990. Early-life-history profiles, seasonal abundance, and distribution of four species of clupeid larvae from the northern Gulf of Mexico, 1982 and 1983. NOAA Tech. Rept. NMFS No. 88, 60 pp.

Shcherbachev, Yu. N. 1973. The biology and distribution of the dolphins (Pisces, Coryphaenidae). *Voposy Ikhtiol.* 13:219–230 [in Russian, translation in *J. Ichthyol.* 13:182–191].

Sheehy, D. J., M. P. Sissenwine, and S. B. Saila. 1974. Ocean pout parasites. *Mar. Fish. Rev.* 36(5):29–33.

Sheehy, D. J., S. Y. K. Shenouda, A. J. Alton, S. B. Saila, and S. M. Constantinides. 1977. The ocean pout: an example of underutilized fisheries resource development. *Mar. Fish. Rev.* 39(6):5–15.

Shelford, V. E., and E. B. Powers. 1915. An experimental study of the movements of herring and other marine fishes. *Biol. Bull., Woods Hole* 28:315–334.

Shelton, R. G. J. 1978. On the feeding of the hagfish *Myxine glutinosa* in the North Sea. *J. Mar. Biol. Assoc. U.K.* 58:81–86.

Shepherd, G. R. 1991. Meristic and morphometric variation in black sea bass north of Cape Hatteras, North Carolina. *No. Am. J. Fish. Manage.* 11:139–148.

Shepherd, G. R. 1998a. Striped bass. In: S. H. Clark, ed. *Status of Fishery Resources off the Northeastern United States for 1998.* NOAA Tech. Memo. NMFS-NE-115:138–139.

Shepherd, G. R. 1998b. Black sea bass. In: S. H. Clark, ed. *Status of Fishery Resources off the Northeastern United States for 1998.* NOAA Tech. Memo. NMFS-NE-115:92–93.

Shepherd, G. R. 1998c. Tilefish. In: S. H. Clark, ed. *Status of Fishery Resources off the Northeastern United States for 1998.* NOAA Tech. Memo. NMFS-NE-115:102–103.

Shepherd, G. R., and C. B. Grimes. 1983. Geographic and historic variations in growth of weakfish, *Cynoscion regalis,* in the Middle Atlantic Bight. *Fish. Bull., U.S.* 81:803–813.

Shepherd, G. R., and C. B. Grimes. 1984. Reproduction of weakfish, *Cynoscion regalis,* in the New York Bight and evidence for geographically specific life history characteristics. *Fish. Bull., U.S.* 82:501–511.

Shepherd, G. R., and M. Terceiro. 1994. The summer flounder, scup, and black sea bass fishery of the Middle Atlantic Bight and southern New England waters. NOAA Tech. Memo. NMFS-122, 13 pp.

Sherburne, S. W. 1977. Occurrence of piscine erythrocytic necrosis (PEN) in the blood of anadromous alewife, *Alosa pseudoharengus,* from Maine coastal streams. *J. Fish. Res. Bd. Can.* 34:281–286.

Sherburne, S. W., and L. L. Bean. 1979. Incidence and distribution of piscine erythrocytic necrosis and the microsporidian, *Glugea hertwigi,* in rainbow smelt, *Osmerus mordax,* from Massachusetts to the Canadian Maritimes. *Fish. Bull., U.S.* 77:503–509.

Sheri, A. N., and G. Power. 1968. Reproduction of white perch, *Roccus americanus,* in the Bay of Quinte, Lake Ontario. *J. Fish. Res. Bd. Can.* 25:2225–2231.

Sherman, K. 1980. MARMAP, a fisheries ecosystem study in the northwest Atlantic: fluctuations in ichthyoplankton-zooplankton components and their potential impact on the system. In: F. P. Diemar, F. J. Vernberg, and D. Z. Mirkes, eds. *Advanced Concepts in Ocean Measurements for Marine Biology.* Belle W. Baruch Inst. Mar. Biol. Coast. Res., Univ. S.C. Press, Columbia, pp. 9–37.

Sherman, K., and K. A. Honey. 1971. Seasonal variations in the food of larval herring in coastal waters of central Maine. *Rapp. Proc.-Verb. Réun. Cons. Int. Explor. Mer* 160:121–124.

Sherman, K., and H. C. Perkins. 1971. Seasonal variation in the food of juvenile herring in coastal waters of Maine. *Trans. Am. Fish. Soc.* 100:121–124.

Sherman, K., and J. P. Wise. 1961. Incidence of the cod parasite *Lernaeocera branchialis* L. in the New England area, and its possible use as an indicator of cod populations. *Limnol. Oceanogr.* 6:61–67.

Sherman, K., C. Jones, L. Sullivan, W. Smith, P. Berrien, and L. Ejsymont. 1981. Congruent shifts in sand eel abundance in western and eastern North Atlantic ecosystems. *Nature* 291:486–489.

Sherman, K., W. Smith, W. Morse, M. Berman, J. Green, and L. Ejsymont. 1984. Spawning strategies of fishes in relation to circulation, phytoplankton production, and pulses in zooplankton off the northeastern United States. *Mar. Ecol. Prog. Ser.* 18:1–19.

Sherman, S., R. Langton, D. Schick, M. Brown, J. Burnett, and F. Almeida. 1993. Distribution and abundance of groundfish along the coast of Maine, U.S.A. *J. Fish Biol.* 43(Suppl. A):334 (abstract).

Sherwood, G. H., and V. N. Edwards. 1901. Notes on the migration, spawning, abundance, etc., of certain fishes in 1900. Biological Notes No. 2. *Bull. U.S. Fish Comm.* 21:27–31.

Shevelev, M. S. 1988. Ontogenic stages in spotted wolffish (*Anarhichas minor* Olaf) from the Barents Sea. Int. Counc. Explor. Sea, Demersal Fish Comm. ICESCM1988/G:31.

Shipp, R. L. 1974. Pufferfishes (Tetraodontidae) of the Atlantic Ocean. Gulf Coast Res. Lab. Mus. Publ. 4, 162 pp.

Shipp, R. L., and R. W. Yerger. 1969. Status, characters and distribution of the northern and southern puffers of the genus *Sphaeroides*. *Copeia* 1969:425–433.

Shoubridge, E. A. 1978. Genetic and reproductive variation in American shad. M.S. thesis, McGill University, Montreal, 73 pp.

Shoubridge, E. A., and W. C. Leggett. 1978. Occurrence and adaptive signficance of distinct reproductive strategies in local populations of American shad. In: *Genetic and Reproductive Variation of American Shad*. Proj. AFC-10 (Connecticut). Natl. Mar. Fish. Serv., U.S. Final Rept., pp. 25–73.

Showell, M. A., and C. G. Cooper. 1997. Development of the Canadian silver hake fishery, 1987–96. NAFO Sci. Counc. Res. Doc., 10 pp.

Shumway, S. E., and R. R. Stickney. 1975. Notes on the biology and food habits of the cunner. *N.Y. Fish Game J.* 22:71–79.

Sibunka, J. D., and A. L. Pacheco. 1981. Biological and fisheries data on northern puffer *Sphoeroides maculatus* (Bloch and Schneider). NOAA/NMFS, Sandy Hook Tech Rept. No. 26, 56 pp.

Sidwell, V. D. 1981. Chemical and nutritional composition of finfishes, whales, crustaceans, mollusks and their products. NOAA Tech. Memo. NMFS F/SEC-11, 432 pp.

Silver, W. L., and T. E. Finger. 1984. Electrophysiological examination of a non-olfactory, non-gustatory chemosense in the sea robin, *Prionotus carolinus*. *J. Comp. Physiol.* 154A:167–174.

Silverberg, N., H. M. Edenborn, G. Oullet, and P. Béland. 1987. Direct evidence of a mesopelagic fish, *Melanostigma atlanticum* (Zoarcidae) spawning within bottom sediments. *Environ. Biol. Fishes* 20:195–202.

Silverman, M. J. 1975. Scale development in the bluefish *Pomatomus saltatrix*. *Trans. Am. Fish. Soc.* 104:773–774.

Silverman, M. J. 1982. Spot, *Leiostomus xanthurus*. In: M. D. Grosslein and T. R. Azarovitz, eds. *Fish Distribution*. MESA New York Bight Atlas Monogr. No. 15:93–95.

Sim, G. 1887. Occurrence of *Lupenus lampetriformis* on the north coast of Scotland; with notes on its habits, foods and the ground it frequents. *J. Linnaean Soc., Zool.* 20:38–48.

Simmons, E. G., and J. P. Breuer. 1962. A study of redfish, *Sciaenops ocellata* Linnaeus and black drum, *Pogonias cromis* Linnaeus. *Publ. Inst. Mar. Sci. Univ. Tex.* 8:184–211.

Simpson, B. K., and N. F. Haard 1985. Characterization of the trypsin fraction from cunner (*Tautogolabrus adspersus*). *Comp. Biochem. Physiol.* 80B:475–480.

Simpson, D. G. 1954. Two small tarpon from Texas. *Copeia* 1954: 71–72.

Sinclair, M. 1988. *Marine Populations: An Essay on Population and Regulation and Speciation*. University of Washington Press, Seattle, 252 pp.

Sinclair, M., and T. D. Iles. 1985. Atlantic herring (*Clupea harengus*) distributions in the Gulf of Maine-Scotian shelf area in relation to oceanographic features. *Can. J. Fish. Aquat. Sci.* 42: 880–887.

Sinclair, M., and M. J. Tremblay. 1984. Timing of spawning of Atlantic herring (*Clupea harengus harengus*) populations and the match-mismatch theory. *Can. J. Fish. Aquat. Sci.* 41:1055–1065.

Sinclair, M., A. Sinclair, and T. D. Iles. 1982. Growth and maturation of southwest Nova Scotia Atlantic herring (*Clupea harengus harengus*). *Can. J. Fish. Aquat. Sci.* 39:288–295.

Sindermann, C. J. 1957. Diseases of fishes of the western North Atlantic. V. Parasites as indicators of herring movements. *Maine Dept. Sea Shore Fish., Res. Bull.* 27:1–30.

Sindermann, C. J. 1958. An epizootic in Gulf of St. Lawrence fishes. *Trans. No. Am. Wildl. Conf.* 23:349–360.

Sindermann, C. J. 1961. Parasites as tags for marine fish. *J. Wildl. Manage.* 25:41–47.

Sindermann, C. J. 1963. Diseases in marine populations. *Trans. No. Am. Wildl. Conf.* 28:221–245.

Sindermann, C. J. 1965. Effects of environment on several diseases of herring from the western North Atlantic. Int. Comm. Northw. Atl. Fish. Spec. Publ. No. 6:603–610.

Sindermann, C. J. 1970. *Principal Diseases of Marine Fish and Shellfish*. Academic Press, New York, 369 pp.

Sindermann, C. J. 1979. Status of Northwest Atlantic herring stocks of concern to the United States. NEFSC, Sandy Hook Lab., Tech. Ser. Rept. No. 23, 449 pp.

Sismour, E. N., and R. S. Birdsong. 1986. Biochemical and genetic analysis of American shad migrating into the Chesapeake Bay. In: M. J. Dadswell et al., eds., *Common Strategies of Anadromous and Catadromous Fishes. Am. Fish. Soc. Symp.* 1:555 (abstract).

Sissenwine, M. P. 1974. Variability in recruitment and equilibrium catch of the southern New England yellowtail flounder fishery. *J. Cons. Intl. Explor. Mer* 36:15–26.

Sissenwine, M. P., and S. B. Saila. 1974. Rhode Island dredge spoil disposal and trends in the floating trap industry. *Trans. Am. Fish. Soc.* 103:498–506.

Sivertsen, K., and A. Bjørge. 1980. Reduksjon av tareskogen Pa Helgelandskysten (Reduction of algal vegetation in Helgeland Coastal Waters). *Fisken Hav.* 4:1–9 [not seen].

Skoglund, C. R. 1961. Functional analysis of swim-bladder muscles engaged in sound production of the toadfish. *J. Biophys. Biochem. Cytol.* 10:187–200.

Skomal, G. B. 1990. Age and growth of the blue shark, *Prionace glauca,* in the north Atlantic. M.S. thesis, University of Rhode Island, Kingston, 103 pp.

Sleggs, G. S. 1933. Observations upon the economic biology of the capelin (*Mallotus villosus* O.F. Muller). Nfld. Fish. Res. Comm. Rept. No. 1:1–66.

Smedley, N. 1932. An ocean sunfish, *Mola lanceolata* (Liénard), in Malaysian waters. *Bull. Raffles Mus.* 7:17–21.

Smidt, E. 1981. The wolffish fishery of West Greenland. Northw. Atl. Fish. Org. Sci. Council Stud. No. 1:35–39.

Smigielski, A. S., T. A. Halavik, L. J. Buckley, S. M. Drew, and G. C. Laurence. 1984. Spawning, embryo development and growth of the American sand lance *Ammodytes americanus* in the laboratory. *Mar. Ecol. Prog. Ser.* 14:287–292.

Sminkey, T. R., and J. A. Musick. 1995. Age and growth of the sandbar shark, *Carcharhinus plumbeus,* before and after population depletion. *Copeia* 1995:871–883.

Sminkey, T. R., and C. R. Tabit. 1992. Reproductive biology of the chain dogfish, *Scyliorhinus retifer,* from the Mid-Atlantic Bight. *Copeia* 1992:251–253.

Smith, B. A. 1971. The fishes of four low-salinity tidal tributaries of the Delaware River Estuary. M.S. thesis. Cornell University, Ithaca, 304 pp.

Smith, B. R. 1972. Sea lampreys in the Great Lakes of North America. In: M. W. Hardisty and I. C. Potter, eds. *The Biology of Lampreys,* Vol. 1. Academic Press, New York, pp. 207–247.

Smith, B. R., and J. J. Tibbles. 1980. Sea lamprey (*Petromyzon marinus*) in lakes Huron, Michigan, and Superior: history of invasion and control 1936–78. *Can. J. Fish. Aquat. Sci.* 37:1780–1801.

Smith, C. L. 1985. *The Inland Fishes of New York State.* New York State Dept. Environ. Cons., 522 pp.

Smith, D. G. 1980. Early larvae of the tarpon, *Megalops atlantica* Valenciennes (Pisces: Elopidae), with notes on spawning in the Gulf of Mexico and the Yucatán Channel. *Bull. Mar. Sci.* 30:136–141.

Smith, D. G. 1989a. Family Anguillidae: freshwater eels. Fishes of the western North Atlantic. Sears Found. Mar. Res. Mem. No. 1(9):25–47.

Smith, D. G. 1989b. Family Congridae: conger eels. Fishes of the western North Atlantic. Sears Found. Mar. Res. Mem. No. 1(9):460–567.

Smith, D. G. 1989c. Order Elopiformes. Fishes of the western North Atlantic. Sears Found. Mar. Res. Mem. 1(9):961–972.

Smith, D. G., and J. G. Nielsen. 1989. Family Nemichthyidae: snipe eels. Fishes of the western North Atlantic. Sears Found. Mar. Res. Mem. No. 1(9):441–459.

Smith, D. G., and B. D. Taubert. 1980. New records of leeches (Annelida: Hirudinea) from the short shortnose sturgeon *Acipenser brevirostrum*) in the Connecticut River. *Proc. Helminthol. Soc. Washington* 47:147–148.

Smith, G. R., and R. F. Stearley. 1989. The classification and scientific names of rainbow and cutthroat trouts. *Fisheries* 14(1):4–10.

Smith, H. M. 1895. Notes on the capture of Atlantic salmon at sea and in the coast waters of the eastern states. *Bull. U.S. Fish Comm.* 14:95–100.

Smith, H. M. 1898a. The fishes found in the vicinity of Woods Hole. *Bull. U.S. Fish Comm.* 17:85–111.

Smith, H. M. 1898b. Fishes new to the fauna of southern New England recently collected at Woods Hole. *Science, N.S.* 8:543–544.

Smith, H. M. 1898c. Notes on the extent and condition of the alewife fisheries of the United States in 1896. In: *U.S. House of Representatives Document 221, Part 24: Report of the Commissioner for the Year Ending June 30, 1898.* Washington, D.C.

Smith, H. M. 1907. The fishes of North Carolina. *N.C. Geol.Econ. Surv.* 2:1–453.

Smith, H. M. 1921. Rudderfishes at Woods' Hole in 1920. *Copeia* No. 91:9–10.

Smith, H. M., and R. A. Goffin. 1937. A fish new to Massachusetts Bay. *Copeia* 1937:236.

Smith, J. V. C. 1833. *Natural History of the Fishes of Massachusetts, Embracing a Practical Essay on Angling.* Allen and Ticknoe, Boston. Reprinted 1970, New York Freshet Press, Inc., Rockville Centre, 399 pp.

Smith, J. W. 1991. The Atlantic and Gulf menhaden purse seine fisheries: origins, harvesting technologies, biostatistical monitoring, recent trends in fisheries statistics, and forecasting. *Mar. Fish. Rev.* 53(4):28–41.

Smith, J. W. 1994. Biology and fishery for Atlantic thread herring, *Opsthonema oglinum,* along the North Carolina coast. *Mar. Fish. Rev.* 56(4):1–7.

Smith, J. W., and R. Wootten. 1978. *Anisakis* and anisakiasis. *Adv. Parasitol.* 16:93–163.

Smith, J. W., W. R. Nicholson, D. S. Vaughan, D. L. Dudley, and E. A. Hall. 1987. Atlantic menhaden, *Brevoortia tyrannus,* purse seine fishery, 1972–84, with a brief discussion of age and size composition of the landings. NOAA Tech. Rept. NMFS No. 59, 23 pp.

Smith, K. J. 1995. Processes regulating habitat use by salt marsh nekton in a southern New Jersey estuary. Ph.D. dissertation, Rutgers University, New Brunswick, N.J., 166 pp.

Smith, K. J., and K. W. Able. 1994. Salt-marsh tide pools as winter refuges for the mummichog, *Fundulus heteroclitus,* in New Jersey. *Estuaries* 17:226–234.

Smith, K. J., G. Taghon, and K. W. Able. 2000. Trophic linkages in marshes: ontogenetic changes in diet for young-of-the-year mummichog, *Fundulus heteroclitus.* In: M. P. Weinstein and D. A. Kreeger, eds. *Concepts and Controversies in Tidal Marsh Ecology.* Kluwer Academic Publishers, Dordrecht.

Smith, M. W., and J. W. Saunders. 1958. Movements of brook trout, *Salvelinus fontinalis* (Mitchill), between and within fresh and salt water. *J. Fish. Res. Bd. Can.* 15:1403–1449.

Smith, P. E. 1985. Year-class strength and survival of 0-group clupeoids. *Can. J. Fish. Aquat. Sci.* 42(Suppl. 1):69–82.

Smith, P. J., and A. Jamieson. 1986. Stock discreteness in herrings: a conceptual revolution. *Fish. Res.* 4:223–234.

Smith, R. E., and R. J. Kernehan. 1981. Predation by the free–living copepod, *Cyclops bicuspidatus thomasi,* on larvae of the striped bass and white perch. *Estuaries* 4:81–83.

Smith, R. W., and F. C. Daiber. 1977. Biology of the summer flounder, *Paralichthys dentatus,* in Delaware Bay. *Fish. Bull., U.S.* 75:823–830.

Smith, S. M., J. G. Hoff, S. P. O'Neil, and M. P. Weinstein. 1984. Community and trophic organization of nekton utilizing shallow marsh habitats, York River, Virginia. *Fish. Bull., U.S.* 82:455–467.

Smith, T. I. J. 1985. The fishery, biology, and management of Atlantic sturgeon, *Acipenser oxyrhynchus,* in North America. *Environ. Biol. Fishes* 14:61–72.

Smith, T. I. J., and J. P. Clugston. 1997. Status and management of Atlantic sturgeon, *Acipenser oxyrinchus,* in North America. *Environ. Biol. Fishes* 48:335–346.

Smith, T. I. J., D. E. Marchette, and G. F. Ulrich. 1984. The Atlantic sturgeon fishery in South Carolina. *No. Am. J. Fish. Manage.* 4:164–176.

Smith, T. I. J., E. K. Dingley, R. D. Lindsey, S. B. Van Sant, R. A. Smiley, and A. D. Stokes. 1985. Spawning and culture of shortnose sturgeon, *Acipenser brevirostrum. J. World Maric. Soc.* 16:104–113.

Smith, W., P. Berrien, and T. Potthoff. 1994. Spawning patterns of bluefish, *Pomatomus saltatrix,* in the northeast continental shelf ecosystem. *Bull. Mar. Sci.* 54:8–16.

Smith, W. G. 1973. The distribution of summer flounder, *Paralichthys dentatus,* eggs and larvae on the continental shelf between Cape Cod and Cape Lookout, 1965–66. *Fish. Bull., U.S.* 71:527–548.

Smith, W. G., and M. P. Fahay. 1970. Description of eggs and larvae of the summer flounder, *Paralichthys dentatus.* U.S. Fish Wildl. Serv. Res. Rept. No. 75, 21 pp.

Smith, W. G., and W. W. Morse. 1985. Retention of larval haddock *Melanogrammus aeglefinus* in the Georges Bank region, a gyre-influenced spawning area. *Mar. Ecol. Prog. Ser.* 24:15–21.

Smith, W. G., and J. J. Norcross. 1968. The status of the scup (*Stenotomus chrysops*) in winter trawl fishery. *Chesapeake Sci.* 9: 207–216.

Smith, W. G., J. D. Sibunka, and A. Wells. 1975. Seasonal distributions of larval flatfishes (Pleuronectiformes) on the continental shelf between Cape Cod, Massachusetts, and Cape Lookout, North Carolina, 1965–66. NOAA Tech. Rept. NMFS SSRF No. 691, 68 pp.

Smith, W. G., J. D. Sibunka, and A. Wells. 1978. Diel movements of larval yellowtail flounder, *Limanda ferruginea,* determined from discrete depth sampling. *Fish. Bull., U.S.* 76:167–178.

Smith, W. R. 1892. On the food of fishes. Fish. Board Scotland Ann. Rept. No. 10:211–231.

Smith-Vaniz, W. F. 1978. Dactylopteridae. In: W. Fischer, ed. *FAO Species Identification Sheets for Fishery Purposes: Western Central Atlantic (Fishing Area 31),* Vol. 2. FAO. Rome, 4 pp.

Smith-Vaniz, W.F. 1984. Carangidae: relationships. In: H. G. Moser et al., eds. *Ontogeny and Systematics of Fishes.* Am. Soc. Ich. Herp. Spec. Publ. No. 1:522–530.

Smith-Vaniz, W. F., and J. C. Staiger. 1973. Comparative revision of *Scomberoides, Oligoplites, Parona* and *Hypacanthus* with comments on the phylogenetic position of *Campogramma* (Pisces: Carangidae). *Proc. Calif. Acad. Sci., Ser. 4,* 39:185–256.

Smith-Vaniz, W. F., J.-C. Quéro, and M. Desoutter. 1990. Carangidae. In J.-C. Quéro et al., eds. *Check-List of the Fishes of the Eastern Tropical Atlantic. Clofeta* 2:729–755. JNICT, Lisbon.

Smith-Vaniz, W. F., B. B. Collette, and B. E. Luckhurst. 1999. Fishes of Bermuda: history, zoogeography, annotated checklist, and identification keys. Am. Soc. Ich. Herp. Spec. Publ. No. 4, 424 pp.

Smitt, F. A., ed. 1892. Scandinavian fishes. In: B. Fries, C. U. Ekström, and C. Sundevall, eds. *A History of Scandinavian Fishes,* 2nd Ed. Revised and completed by F.A. Smitt. Part I. P.A. Norstedt & Soner, Stockholm, pp. 231–238.

Snieszko, S. F. 1978. Mycobacteriosus (tuberculosis) of fishes. U. S. Fish Wildl. Serv. Fish Disease Leaflet No. 55.

Snieszko, S. F., G. L. Bullock, E. Hollis, and J. G. Boone. 1964. *Pastuerella* sp. from an epizootic of white perch (*Roccus americanus*) in Chesapeake Bay tidewater areas. *J. Bacteriol.* 88:1814–1815.

Snyder, D. E. 1988. Description and identification of shortnose and Atlantic sturgeon larvae. In R. D. Hoyt, ed. *11th Annual Larval Fish Conference. Am. Fish. Soc. Symp.* 5:7–30.

Sogandares-Bernal, F. 1959. Digenetic trematodes of marine fishes of the Gulf of Panama and Bimini, British West Indies. *Tulane Stud. Zool.* 7:69–117.

Sogard, S. M. 1989. Colonization of artificial seagrass by fishes and decapod crustaceans: importance of proximity to natural eelgrass. *J. Exp. Mar. Biol. Ecol.* 133:15–37.

Sogard, S. M. 1992. Variability in growth rates of juvenile fishes in different estuarine habitats. *Mar. Ecol. Prog. Ser.* 85:35–53.

Sogard, S. M., and K. Able. 1991. A comparison of eelgrass, sea lettuce macroalgae, and marsh creeks as habitats for epibenthic fishes and decapods. *Est. Coast. Shelf Sci.* 33:501–519.

Sogard, S. M., G. V. N. Powell, and J. G. Holmquist. 1989. Utilization by fishes of shallow, seagrass-covered banks in Florida Bay: a. Diel and tidal patterns. *Environ. Biol. Fishes* 24:81–92.

Sogard, S. M., K. W. Able, and M. P. Fahay. 1992. Early life history of the tautog *Tautoga onitis* in the Mid-Atlantic Bight. *Fish. Bull., U.S.* 90:529–539.

Sokolovskaya, T. S., and A. S. Sokolovskiy. 1975. New data on expansion of the area of reproduction of ocean sunfishes (Pisces, Molidae) in the northwestern part of the Pacific Ocean. *Voprosy Ikhtiol.* 15:750–752 [in Russian, translation in *J. Ichthyol.* 15(4):675–678].

Solomon, C. H., and G. R. Rinckey. 1963. Large lookdowns from Tampa Bay, Florida. *J. Fla. Acad. Sci.* 26:192–193.

Sosebee, K. A. 1998a. Spiny dogfish. In: S. H. Clark, ed. *Status of Fishery Resources off the Northeastern United States for 1998.* NOAA Tech. Memo. NMFS-NE-115:112–113.

Sosebee, K. A. 1998b. Skates. In: S. H. Clark, ed. *Status of Fishery Resources off the Northeastern United States for 1998.* NOAA Tech. Memo. NMFS-NE-115:114–115.

Sosebee, K. A. 1998c. Red hake. In: S. H. Clark, ed. *Status of Fishery Resources off the Northeastern United States for 1998.* NOAA Tech. Memo. NMFS-NE-115:64–66.

Sosebee, K. A. 1998d. White hake. In: S. H. Clark, ed. *Status of Fishery Resources off the Northeastern United States for 1998.* NOAA Tech. Memo. NMFS-NE-115:96–97.

Sparks, M. I. 1929. The spawning and development of mackerel on the outer coast of Nova Scotia. *Contrib. Can. Biol. Fish., n.s.* 4:443–452.

Specker, J. L., D. L. Berlinsky, H. D. Bibb, and J. F. O'Brien. 1987. Oocyte development in striped bass: factors influencing estimates of age at maturity. In: M. J. Dadswell et al., eds., *Common Strategies of Anadromous and Catadromous Fishes. Am. Fish. Soc. Symp.* 1:162–174.

Speirs, H. J. 1977. Status of some finfish stocks in the Chesapeake Bay. *Water, Air, Soil, Pollut.* 35:49–62.

Spraker, H., and H. M. Austin. 1997. Diel feeding periodicity of Atlantic silverside, *Menidia menidia,* in the York River, Chesapeake Bay, Virgina. *J. Elisha Mitchell Sci. Soc.* 113:171–182.

Springer, S. 1951. The effect of fluctuations on the availability of sharks on a shark fishery. *Proc. Gulf Carib. Fish. Inst. 4th Ann. Sess.,* pp. 140–145.

Springer, S. 1960. Natural history of the sandbar shark, *Eulamia milberti. Fish. Bull., U.S.* 61:1–38.

Springer, S. 1979. A revision of the catsharks, family Scyliorhinidae. NOAA Tech. Rept. NMFS Circ. No. 422, 152 pp.

Springer, S., and G. H. Burgess. 1985. Two new dwarf dogsharks (*Etmopterus,* Squalidae) found off the Caribbean coast of Colombia. *Copeia* 1985:584–591.

Springer, V. G. 1964. A revision of the carcharhinid shark genera *Scoliodon, Loxodon,* and *Rhizoprionodon. Proc. U.S. Nat. Mus.* 115:559–632.

Springer, V. G., and J. A. F. Garrick. 1964. A survey of vertebral numbers in sharks. *Proc. U.S. Nat. Mus.* 116:73–96.

Springer, V. G., and K. D. Woodburn. 1960. An ecological study of the fishes of the Tampa Bay area. Fla. Board Conserv. Mar. Res. Lab. Prof. Pap. Ser. No. 1, 104 pp.

Squiers, T. S., and T. Savoy. 1998. New England. In: A. W. Kahnle, K. A. Hattala, K. A. McKown, C. A. Shivey, M. R. Collins, T. S. Squiers, and T. Savoy, eds. *Stock Status of Atlantic Sturgeon of Atlantic Coast Estuaries.* Atlantic States Mar. Fish. Comm. Rept., Washington, D.C., 141 pp.

Squiers, T. S., and M. Smith. 1979. Distribution and abundance of shortnose sturgeon in the Kennebec River Estuary. Final Rept. AFS No. 19 to Maine Dept. Mar. Res., Augusta.

Squiers, T. S., Jr., and J. Stahlnecker, III. 1994. State of Maine. Shad and river herring 1994 activities: summary sheet; 4 pp.

Squiers, T. S., L. Flagg, M. Smith, K. Sherman, and D. Ricker. 1981. American shad enhancement and status of sturgeon stocks in selected marine waters. Maine Dept. Mar. Res. Rept. to NMFS, Gloucester, Mass.

Squire, J. L., Jr. 1963. Thermal relationships of tuna in the oceanic northwest Atlantic. FAO Fish. Rept. No. 6(3):1639–1657.

St. Pierre, R., ed. 1977. *Proceedings of a Workshop on American Shad, Dec. 14–16, 1976.* U.S. Fish Wildl. Serv. and Natl. Mar. Fish. Serv., Amherst, Mass., 350 pp.

Starnes, W. C. 1988. Revision, phylogeny and biogeographic comments on the circumtropical marine percoid fish family Priacanthidae. *Bull. Mar. Sci.* 43:117–203.

State of Maine. 1982. Anadromous fisheries river management plan. In: *Statewide River Fisheries Management Plan.* Prepared by Department of Marine Resources for The Governor's Cabinet Committee on Hydropower Policy, Sect. 1, pp.1–29.

Steele, D. H. 1957. The redfish (*Sebastes marinus* L.) in the western Gulf of St. Lawrence. *J. Fish. Res. Bd. Can.* 14:899–924.

Steele, D. H. 1963. Pollock (*Pollachius virens* (L.)) in the Bay of Fundy. *J. Fish. Res. Bd. Can.* 20:1267–1314.

Steele, D. H. 1967. The occurrence of the pearlsides, *Maurolicus mulleri* (Gmelin), in the northwestern Atlantic. *Can. Field-Nat.* 81:184–186.

Steenstrup, J., and C. Lütken. 1898. Spolia Atlantica. Bidrag til Kundskab om klump-eller maane-fiskene (Molidae). Oversigt Dansk. Vid. Selsk. Kjobenhavn 6, Ser. 9, No. 1, 102 pp.

Stefanescu, C., and J. E. Cartes. 1992. Benthopelagic habits of adult specimens of *Lampanyctus crocodilus* (Risso, 1810) (Osteichthyes, Myctophidae) in the western Mediterranean deep slope. *Sci. Mar.* 56:69–74.

Stefanescu, C., D. Lloris, and J. Rucabado. 1994. Revalidation of *Lampanyctus gemmifer* (Goode & Bean, 1879), a junior synonym of *Lampanyctus crocodilus* (Risso, 1810) in the Atlantic Ocean (Myctophidae). *Cybium* 18:315–323.

Stefanussen, D., Ø. Lie., E. Moksness, and K. I. Ugland. 1993. Growth of juvenile common wolffish (*Anarhichas lupus*) fed practical fish feeds. *Aquaculture* 114:103–111.

Stehmann, M. 1978. *Raja "bathyphila,"* eine Doppelart des Subgenus *Rajella:* Weiderbeschreibung von *R. bathyphila* Holt & Byrne, 1908 und *Raja bigelowi* spec. nov. (Pisces, Rajiformes, Rajidae). *Arch. Fisch. Wiss.* 29:23–58.

Stehmann, M., and D. L. Bürkel. 1984a. Torpedinidae. In: P .J. P. Whitehead et al., eds. *Fishes of the North–eastern Atlantic and the Mediterranean,* Vol. 1. UNESCO, Paris, pp. 159–162.

Stehmann, M., and D. L. Bürkel. 1984b. Rajidae. In: P. J. P. Whitehead et al., eds. *Fishes of the North–eastern Atlantic and the Mediterranean,* Vol. 1. UNESCO, Paris, pp. 163–196.

Stehmann, M., and D. L. Bürkel. 1984c. Chimaeridae, Rhinochimaeridae. In: P. J. P. Whitehead et al., eds. *Fishes of the North–eastern Atlantic and the Mediterranean,* Vol. 1. UNESCO, Paris, pp. 212–218.

Stehmann, M., and J. D. McEachran. 1978. Rajidae. In: W. Fischer, ed. *FAO Species Identification Sheets for Fishery Purposes: Western Central Atlantic (Fishing Area 31),* Vol. V. FAO, Rome, 8 pp.

Stehmann, M., and N. V. Parin. 1994. Deepest capture of a thorny skate, *Raja radiata,* from the northeastern region of the Norwegian Sea. *Voprosy Ikhtiol.* 34:280–283 [in Russian, translation in *J. Ichthyol.* 34(6):143–148].

Steimle, F. W., and L. Ogren. 1982. Food of fish collected on artificial reefs in the New York Bight and off Charleston, South Carolina. *Mar. Fish. Rev.* 44(6–7):49–52.

Steimle, F. W., W. W. Morse, and D. L. Johnson. 1999a. Essential Fish Habitat Source Document: Goosefish, *Lophius americanus,* life history and habitat characteristics. NOAA Tech. Memo. NMFS-NE-127, 31 pp.

Steimle, F. W., W. W. Morse, P. L. Berrien, D. L. Johnson, and C. A. Zetlin. 1999b. Essential Fish Habitat Source Document: Ocean pout, *Macrozoarces americanus,* life history and habitat characteristics. NOAA Tech. Memo. NMFS–NE-129, 26 pp.

Steimle, F. W., W. W. Morse, P. L. Berrien, and D. L. Johnson. 1999c. Essential Fish Habitat Source Document: Red hake, *Urophycis chuss,* life history and habitat characteristics. NOAA Tech. Memo. NMFS–NE-133, 34 pp.

Steimle, F. W., C. A. Zetlin, P. L. Berrien, and S. Chang. 1999d. Essential Fish Habitat Source Document: Black sea bass, *Centropristis striata,* life history and habitat characteristics. NOAA Tech. Memo. NMFS–NE-143, 42 pp.

Steimle, F. W., C. A. Zetlin, P. L. Berrien, D. L. Johnson, and S. Chang. 1999e. Essential Fish Habitat Source Document: Scup, *Stenotomus chrysops,* life history and habitat characteristics. NOAA Tech. Memo. NMFS–NE-149, 39 pp.

Steimle, F. W., C. A. Zetlin, P. L. Berrien, D. L. Johnson, and S. Chang. 1999f. Essential Fish Habitat Source Document: Tilefish, *Lopholatilus chamaeleonticeps,* life history and habitat characteristics. NOAA Tech. Memo. NMFS–NE-152, 30 pp.

Steiner, W. W., and B. L. Olla. 1985. Behavioral responses of prejuvenile red hake, *Urophycis chuss,* to experimental thermoclines. *Environ. Biol. Fishes* 14:167–173.

Steiner, W. W., J. J. Luczkovich, and B. L. Olla. 1982. Activity, shelter usage, growth and recruitment of juvenile red hake *Urophycis chuss. Mar. Ecol. Prog. Ser.* 7:125–135.

Stephens, E. B., M. W. Newman, A. L. Zachary, and F. M. Hetrick. 1980. A viral aetiology for the annual spring epizootics of Atlantic menhaden *Brevoortia tyrannus* (Latrobe) in Chesapeake Bay. *J. Fish Dis.* 3:387–398.

Stephenson, R. L., and I. Kornfield. 1990. Reappearance of spawning Atlantic herring (*Clupea harengus harengus*) on Georges Bank: population resurgence not recolonization. *Can. J. Fish. Aquat. Sci.* 47:1060–1064.

Stephenson, R. L., and M. J. Power, 1989. Observations on herring larvae retained in the Bay of Fundy: variability in vertical movement and position of the patch edge. *Rapp. Proc.-Verb. Réun. Cons. Int. Explor. Mer* 191:177–183.

Stephenson, R. L., M. J. Power, and T. D. Iles. 1987. Assessment of the 1986 4WX herring fishery. Can. Atl. Fish. Sci. Advis. Comm. Res. Doc. No. 87/75, 39 pp.

Steven, D. M. 1955. Experiments on the light sense of the hag, *Myxine glutinosa* L. *J. Exp. Biol.* 32:22–38.

Steven, G. A. 1952. Contributions to the biology of the mackerel, *Scomber scombrus* L. III. Age and growth. *J. Mar. Biol. Assoc. U.K.* 30:549–568.

Stevens, E. D., and J. M. McLeese. 1984. Why bluefin tuna have warm tummies: temperature effect on trypsin and chymotrypsin. *Am. J. Physiol.* 246:R487–R494.

Stevens, E. D., J. W. Kanwisher, and F. G. Carey. 2000. Muscle temperature in free-swimming giant Atlantic bluefin tuna (*Thunnus thynnus* L.). *J. Thermal Biol.* 25:419–423.

Stevens, J. D. 1975. Vertebral rings as a means of age determination in the blue shark (*Prionace glauca* L.). *J. Mar. Biol. Assoc. U.K.* 55:657–665.

Stevens, J. D. 1983. Observations on reproduction in the shortfin mako *Isurus oxyrinchus. Copeia* 1983:126–130.

Stevens, R. E. 1966. Hormone-induced spawning of striped bass for reservoir stocking. *Progr. Fish Cult.* 28:19–27

Stevenson, C. H. 1899. The shad fisheries of the Atlantic coast of the United States. U.S. Comm. Fish Fish., Rept. Comm., Part 24, pp. 101–269.

Stevenson, D. K. 1984. Locations of Atlantic herring, *Clupea harengus* L., egg beds in eastern Maine. Maine Dept. Mar. Res. Lab. Ref. Doc. 84/2, 3 pp.

Stevenson, D. K. 1989. Spawning locations and times for Atlantic herring on the Maine coast. Maine Dept. Mar. Res. Lab. Ref. Doc. 89/5, 16 pp.

Stevenson, D. K., K. M. Sherman, and J. J. Graham. 1989. Abundance and population dynamics of the 1986 year class of herring along the Maine coast. *Rapp. Proc.-Verb. Réun. Cons. Int. Explor. Mer* 191:345–350.

Stevenson, J. T., and D. H. Secor. 1999. Age determination and growth of Hudson River Atlantic sturgeon, *Acipenser oxyrinchus. Fish. Bull., U.S.* 98:153–166.

Stevenson, R. A., Jr. 1958. The biology of the anchovies *Anchoa mitchilli* Cuvier and Valenciennes 1848 and *Anchoa hepsetus hepsetus* Linnaeus 1758 in Delaware Bay. M.S. thesis, University of Delaware, Newark.

Stevenson, S. C., and J. W. Baird. 1988. The fishery for lumpfish (*Cyclopterus lumpus*) in Newfoundland waters. Can. Fish. Aquat. Sci. Tech. Rept. No. 1595, 18 pp.

Steves, B. P., R. K. Cowen, and M. H. Malchoff. 2000. Settlement and nursery habitats for demersal fishes on the continental shelf of the New York Bight. *Fish. Bull., U.S.* 98:167–188.

Stiasny, G. 1911. Ueber einige postlarvale Entwicklungsstadien von *Lophius piscatorius* L. *Arbeit. Zool. Inst. Vienna* 19:54–74.

Stiasny, G. 1913. Ueber einige vorgeschrittene Entwicklungsstadien von *Lophius piscatorius* L. *Arbeit. Zool. Inst. Vienna* 20:1–6.

Stiassny, M. L. J. 1993. What are grey mullets? *Bull. Mar. Sci.* 52:197–219.

Stiassny, M. L. J., and J. S. Jensen. 1987. Labroid interrelationships revisited: morphological complexity, key innovations, and study of comparative diversity. *Bull. Mus. Comp. Zool.* 151:269–319.

Stickney, A. P. 1969. Orientation of juvenile Atlantic herring (*Clupea harengus* L.) to temperature and salinity. In: *Proceedings of the FAO Conference on Fish Behavior in Relation to Fishing Techniques and Tactics.* FAO Fish. Rept. No. 62:323–342.

Stickney, A. P. 1972. The locomotor activity of juvenile herring (*Clupea harengus harengus* L.) in response to changes in illumination. *Ecology* 53:438–445.

Stier, K., and B. Kynard. 1986. Movement of sea-run sea lampreys, *Petromyzon marinus,* during the spawning migration in the Connecticut River. *Fish. Bull., U.S.* 84:749–753.

Stillwell, C. E. 1990. The ravenous mako. In: S. H. Gruber, ed. *Discovering Sharks,* Vols. 19(4) and 20(1). Am. Littoral Soc., Highlands, N.J. Spec. Publ. No. 14, pp. 77–78.

Stillwell, C. E., and N. E. Kohler. 1982. Food, feeding habits, and estimates of daily ration of the shortfin mako (*Isurus oxyrinchus*) in the northwest Atlantic. *Can. J. Fish. Aquat. Sci.* 39:407–414.

Stillwell, C. E., and N. E. Kohler. 1985. Food and feeding ecology of the swordfish *Xiphias gladius* in the western North Atlantic Ocean with estimates of daily ration. *Mar. Ecol. Prog. Ser.* 22:239–247.

Stillwell, C. E., and N. E. Kohler. 1993. Food habits of the sandbar shark *Carcharhinus plumbeus* off the U.S northeast coast, with estimates of daily ration. *Fish. Bull., U.S.* 91:138–150.

Stobo, W. T. 1976. Movements of herring tagged in the Bay of Fundy-update. ICNAF Res. Doc. No. 76/6/48, Ser. No. 3834, 16 pp.

Stobo, W. T. 1983. Report of the *ad hoc* working group on herring tagging. Northw. Atl. Fish. Org. SCS Doc. No. 83/VI/18, 41 pp.

Stobo, W. T., and J. J. Hunt. 1974. Mackerel biology and history of the fishery in Subarea 4. Int. Comm. Northw. Atl. Fish. Res. Doc. No. 74/9, Ser. No. 3155.

Stobo, W. T., J. D. Neilson, and P. G. Simpson. 1988. Movements of Atlantic halibut (*Hippoglossus hippoglossus*) in the Canadian North Atlantic. *Can. J. Fish. Aquat. Sci.* 45:484–491.

Stokesbury, K. D. E., and J. H. Himmelman. 1995. Biological and physical variables associated with aggregations of the giant scallop *Placopecten magellanicus. Can. J. Fish. Aquat. Sci.* 52:743–753.

Stolgitis, J. A. 1970. Some aspects of the biology of the tautog, *Tautoga onitis* (Linnaeus), from the Weweantic River Estuary, Massachusetts, 1966. M.S. thesis, University of Massachusetts, Amherst, 48 pp.

Stone, H. H. 1986. Composition, morphometric characteristics and feeding ecology of alewives (*Alosa pseudoharengus*) and blueback herring (*Alosa aestivalis*) (Pisces: Clupeidae) in Minas Basin. M.S. thesis, Acadia University, Wolfville, Nova Scotia, 191 pp.

Stone, H. H., and G. R. Daborn. 1987. Diet of alewives, *Alosa pseudoharengus* and blueback herring, *A. aestivalis* (Pisces: Clupeidae) in Minas Basin, Nova Scotia, a turbid, macrotidal estuary. *Environ. Biol. Fishes* 19:55–67.

Stone, H. H., and B. M. Jessop. 1992. Seasonal distribution of river herring *Alosa pseudoharengus* and *A. aestivalis* off the Atlantic coast of Nova Scotia. *Fish. Bull., U.S.* 90:376–389.

Stone, H. H., and B. M. Jessop. 1993. Feeding habits of anadromous alewives, *Alosa pseudoharengus,* off the Atlantic coast of Nova Scotia. *Fish. Bull., U.S.* 92:157–170.

Stoner, A. W., A. J. Bejda, J. P. Manderson, B. A. Phelan, L. L. Stehlik, and J. P. Pessutti. 1999. Behavior of winter flounder, *Pseudopleuronectes americanus,* during the reproductive season: laboratory and field observations on spawning, feeding, and locomotion. *Fish. Bull., U.S.* 97:999–1016.

Storer, D. H. 1843. First record of *Cybium maculatus* at Lynn. *Proc. Boston Soc. Nat. Hist.* 1:40.

Storer, D. H. 1846. A synopsis of the fishes of North America. Am. Acad. Arts Sci. Mem. No. 2:253–550.

Storer, D. H. 1863. A history of the fishes of Massachusetts. Am. Acad. Arts Sci. Mem. No. 15: 389–434.

Stott, W., M. M. Ferguson, and R. F. Tallman. 1992. Genetic population structure of American plaice (*Hippoglossoides platessoides*) from the Gulf of St. Lawrence, Canada. *Can. J. Fish. Aquat. Sci.* 49:2538–2545.

Strahan, R. 1963. The behaviour of myxinoids. *Acta Zool.* 44:1–30.

Straughan, R. P. L. 1954. The Sargassum fish, *Histrio pictus. Aquarium* 23:277–279.

Straughan, R. P. L. 1958. The triggerfish: piranha of the sea. *Aquarium* 27(8):236–239.

Stroud, R. H. 1939. Survey of the food of four fishes of the Bay of Fundy. Bowdoin Sci. Sta., Bowdoin College, 4th Ann. Rept. No. 6:19–24.

Struhsaker, P. 1969a. Demersal fish resources: composition, distribution and commercial potential of the continental shelf stocks off southeastern United States. *U.S.Fish. Wildl. Serv. Ind. Res.* 4:261–300.

Struhsaker, P. 1969b. Observations on the biology and distribution of the thorny stingray, *Dasyatis centroura* (Pisces: Dasyatidae). *Bull. Mar. Sci.* 19:456–481.

Studholme, A. L., D. B. Packer, P. L. Berrien, D. L. Johnson, C. A. Zetlin, and W. W. Morse. 1999. Essential Fish Habitat Source Document: Atlantic mackerel, *Scomber scombrus,* life history and habitat characteristics. NOAA Tech. Memo. NMFS-NE 141, 35 pp.

Stunkard, H. W. 1976. The life cycles, intermediate hosts, and larval stages of *Rhipidocotyle transversale* Chandler, 1935 and *Rhipidocotyle lintoni* Hopkins, 1954: life cycles and systematics of bucephalid trematodes. *Biol. Bull., Woods Hole* 150:294–317.

Stunkard, H. W., and F. E. Lux. 1965. A microsporidian infection of the digestive tract of the winter flounder, *Pseudopleuronectes americanus. Biol. Bull., Woods Hole* 129:371–387.

Sulak, K. J., and Y. N. Shcherbachev. 1997. Zoogeography and systematics of six deep-living genera of synaphobranchid eels, with a key to taxa and description of two new species of *Ilyophis. Bull. Mar. Sci.* 60:1158–1194.

Sullivan, L. F. 1982. American plaice, *Hippoglossoides platessoides,* in the Gulf of Maine: I. The fishery. II. Age and growth. III. Spawning and larval distribution. M.S. thesis, University of Rhode Island, Kingston, 95 pp.

Sullivan, W. E. 1915. A description of the young stages of the winter flounder (*Pseudopleuronectes americanus* Walbaum). *Trans. Am. Fish. Soc.* 44:125–136.

Sullivan, W. L., Jr. 1982. Ocean science in relation to living resources. *Fisheries* 7(4):18–19.

Sumner, F. B, R. C. Osburn, and L. J. Cole. 1913. A biological survey of the waters of Woods Hole and vicinity. Part II., Sect. III. A catalogue of the marine fauna. *Bull. U.S. Bur. Fish.* 31:545–794.

Sund, O. 1943. Et bugdebarsel. *Naturen* 67:285–286.

Sutherland, D. F. 1963. Variation in vertebral numbers of juvenile Atlantic menhaden. U.S. Fish Wildl. Serv. No. SSRF-435, 21 pp.

Sutter, F. C. 1980. Reproductive biology of anadromous rainbow smelt, *Osmerus mordax,* in the Ipswich Bay area, Massachusetts. M.S. thesis, University of Massachusetts, Amherst, 49 pp.

Sutton, T. T., and T. L. Hopkins. 1996. Species composition, abundance, and vertical distribution of the stomiid (Pisces: Stomiiformes) fish assemblage of the Gulf of Mexico. *Bull. Mar. Sci.* 59:530–542.

Svetovidov, A. N. 1952. *Fauna of the U.S.S.R. Fishes: Clupeidae,* Vol. 2, No. 1. Zool. Inst. Acad. Sci. USSR, 323 pp. [transl. from Russian, Israel Program Sci. Transl., 1963].

Svetovidov, A. N. 1962. *Fauna of the USSR: Fishes,* Vol. 9, No. 4. Gadiformes. Zool. Inst., Akad. Nauk SSSR, n.s. No. 34, 221 pp. [in Russian, Off. Tech. Serv. Translation 63–11071].

Svetovidov, A. N. 1986. Gadidae. In: P. J. P.Whitehead et al., eds. *Fishes of the North-eastern Atlantic and the Mediterranean,* Vol. 2. UNESCO, Paris, pp. 680–710.

Swaby, S. E., G. W. Potts, and J. Lees. 1996. The first records of the blue runner *Caranx crysos* (Pisces: Carangidae) in British waters. *J. Mar. Biol. Assoc. U.K.* 76:543–544.

Swain, D. P. 1997. Sex-specific temperature distribution of American plaice (*Hippoglossoides platessoides*) and its relation to age and abundance. *Can. J. Fish. Aquat. Sci.* 54:1077–1087.

Swain, D. P., and R. Morin. 1997. Effects of age, sex and abundance on the bathymetric pattern of American plaice in the southern Gulf of St. Lawrence. *J. Fish Biol.* 50:181–200.

Swanson, R. L., and C. J. Sindermann. 1979. Oxygen depletion and associated benthic mortalities in the New York Bight, 1976. NOAA Prof. Pap. No. 11, 345 pp.

Swartz, R. C., and W. A. Van Engel. 1968. Length, weight, and girth relations in the toadfish, *Opsanus tau. Chesapeake Sci.* 9:249–253.

Sykes, J. E., and B. A. Lehman. 1957. Past and present Delaware River shad fishery and considerations for its future. U.S. Fish Wildl. Serv. Res. Rept. No. 46:1–25.

Symons, P. E. K., and J. D. Martin. 1978. Discovery of juvenile Pacific salmon (coho) in a small coastal stream of New Brunswick. *Fish. Bull., U.S.* 76:487–489.

Sypek, J. P., and E. M. Burreson. 1983. Influence of temperature on the immune response of juvenile summer flounder *Paralichthys dentatus* and its role in the elimination of *Trypanosoma bullocki* infections. *Dev. Comp. Immunol.* 7:277–286.

Szedlmayer, S. T., and K. W. Able. 1993. Ultrasonic telemetry of age-0 summer flounder, *Paralichthys dentatus,* movements in a southern New Jersey estuary. *Copeia* 1993:728–736.

Szedlmayer, S. T., and K. W. Able. 1996. Patterns of seasonal availablity and habitat use by fishes and decapod crustaceans in a southern New Jersey estuary. *Estuaries* 19:697–709.

Szedlmayer, S. T., K. W. Able, and R. A. Rountree. 1992. Growth and temperature-induced mortality of young-of-the-year summer flounder (*Paralichthys dentatus*) in southern New Jersey. *Copeia* 1992:120–128.

Tabb, D. C., and R. B. Manning. 1961. A checklist of the flora and fauna of northern Florida Bay and adjacent brackish waters of the Florida mainland collected during the period July, 1957 through September, 1960. *Bull. Mar. Sci. Gulf Carib.* 11:526–649.

Tagatz, M. E. 1961. Tolerance of striped bass and American shad to changes of temperature and salinity. U.S. Fish Wildlfe. Serv. Spec. Sci. Rept. Fish. No. 388, 8 pp.

Tagatz, M. E. 1967. Fishes of the St. Johns River, Florida. *Quart. J. Fla. Acad. Sci.* 30:25–50.

Tagatz, M. E., and D. L. Dudley. 1961. Seasonal occurrence of marine fishes in four shore habitats near Beaufort, N.C., 1957–60. U.S. Fish Wildl. Serv. Spec. Sci. Rept. Fish. No. 390, 19 pp.

Takao, Y. 1990. Survey of Anisakidae larvae from marine fish caught in the sea near Kyushu Island, Japan. In: H. Ishikura and K. Kikuchi, eds. *Intestinal Anisakiasis in Japan: Infected Fish, Sero–Immunological Diagnosis, and Prevention.* Springer-Verlag, Tokyo, pp. 61–72.

Takemoto, R. M., J. F. R. Amato, and J. L. Luque. 1996. Comparative analysis of the metazoan parasite communities of leatherjackets, *Oligoplites palometa, O. saurus,* and *O. saliens* (Osteichthyes: Carangidae), from Sepetiba Bay, Rio de Janeiro, Brazil. *Rev. Brasil. Biol.* 56:639–650.

Takvorian, P. M., and A. Cali. 1981. The occurrence of *Glugea stephani* (Hagenmuller, 1899) in American winter flounder, *Pseudopleuronectes americanus* (Walbaum) from the New York–New Jersey lower bay complex. *J. Fish. Biol.* 18:491–501.

Takvorian, P. M., and A. Cali. 1984. Seasonal prevalence of the microsporidian, *Glugea stephani* (Hagenmuller), in winter flounder, *Pseudopleuronectes americanus* (Walbaum), from the New York–New Jersey lower bay complex. *J. Fish Biol.* 24:655–663.

Takvorian, P. M., and A. Cali. 1986. The ultrastructure of spores (Protozoa: Microsporida) from *Lophius americanus,* the angler fish. *J. Protozool.* 33:570–575.

Talbot, C. W., and K. W. Able. 1984. Composition and distribution of larval fishes in New Jersey high marshes. *Estuaries* 7: 434–443.

Talbot, G. B., and J. E. Sykes. 1958. Atlantic coast migrations of American shad. *Fish. Bull., U.S.* 58:473–490.

Tåning, Å. V. 1918. Mediterranean Scopelidae (*Saurus aulopus, Chlorophthalmus,* and *Myctophum*). Dana Oceanogr. Exped. Mediterr. Rept. No. 2, Biology, A.7, 154 pp.

Tåning, Å. V. 1923. *Lophius.* Danish Oceanogr. Exped., 1908–10, Rept. No. 7, Vol. 2 (Biol.), A 10, 30 pp.

Tåning, Å. V. 1936. On the eggs and young stages of halibut. *Medd. Komm. Danmarks Fisk. Havunders. Ser.: Fisk.* 10:1–23.

Tåning, Å. V. 1955. On the breeding areas of the swordfish (*Xiphias*). *Deep Sea Res.,* 3(Suppl.):438–450.

Taub, S. H. 1969. Fecundity of the white perch. *Prog. Fish-Cult.* 31:166–168.

Taubert, B. D. 1980. Reproduction of shortnose sturgeon (*Acipenser brevirostrum*) in Holyoke Pool, Connecticut River, Massachusetts. *Copeia* 1980:114–117.

Taubert, B. D., and M. J. Dadswell. 1980. Description of some larval shortnose sturgeon (*Acipenser brevirostrum*) from the Holyoke Pool, Connecticut River, Massachusetts, U.S.A., and the Saint John River, New Brunswick, Canada. *Can. J. Zool.* 58:1125–1128.

Taylor, C. C., H. B. Bigelow, and H. W. Graham. 1957. Climatic trends and the distribution of marine animals in New England. *U.S. Fish. Wildl. Serv. Fish Bull.* 57:291–345.

Taylor, C. E. 1951. A survey of former shad streams in Maine. U.S. Fish Wildl. Serv., Spec. Sci. Rept. Fish. No. 66, 29 pp.

Taylor, E. B., ed. 2000. Proceedings of the Third International Conference on Stickleback Behaviour and Evolution. *Behaviour* 137(7–8):827–1140.

Taylor, E. B., and J. D. McPhail. 1986. Prolonged and burst swimming in anadromous and freshwater threespine stickleback, *Gasterosteus aculeatus. Can. J. Zool.* 64: 416–420.

Taylor, M. H. 1986. Environmental and endocrine influences on reproduction of *Fundulus heteroclitus. Am. Zool.* 26:159–171.

Taylor, M. H., L. Di Michele, and G. J. Leach. 1977. Egg stranding in the life cycle of the mummichog, *Fundulus heteroclitus. Copeia* 1977:397–399.

Taylor, M. H., G. J. Leach, L. Di Michele, W. M. Levitan, and W. F. Jacob. 1979. Lunar spawning cycle in the mummichog, *Fundulus heteroclitus* (Pisces: Cyprinodontidae). *Copeia* 1979:291–297.

Tchernavin, V. V. 1953. *The Feeding Mechanisms of a Deep Sea Fish,* Chauliodus sloani *Schneider.* British Museum (Natural History), London, 101 pp.

Teague, G. W. 1951. The sea-robins of America: a revision of the triglid fishes of the genus *Prionotus.* Comun. Zool. Mus. Hist. Nat., Montevideo 3(61), 58 pp.

Teague, G. W. 1961. The armored sea-robins of America: a revision of the American species of the family Peristidiidae. An. Mus. Hist. Nat., Montevideo Ser. No. 2, 7(2), 27 pp.

Templeman, W. 1948. The life history of the capelin (*Mallotus villousus* O.F. Müller) in Newfoundland waters. *Bull. Nfld. Govt. Lab. (Res.)* 17, 151 pp.

Templeman, W. 1965a. Rare skates of the Newfoundland and neighbouring areas. *J. Fish. Res. Bd. Can.* 22:259–279.

Templeman, W. 1965b. Some resemblances and differences between *Raja erinacea* and *Raja ocellata,* including a method of separating mature and large-immature individuals of these two species. *J. Fish. Res. Bd. Can.* 22:899–912.

Templeman, W. 1965c. Mass mortalities of marine fishes in the Newfoundland area presumably due to low temperature. Int. Comm. Northw. Atl. Fish., Spec. Publ. No. 6:137–147.

Templeman, W. 1965d. Some instances of cod and haddock behavior and concentration in the Newfoundland and Labrador areas in relation to food. Int. Comm. Northw. Atl. Fish., Spec. Publ. No. 6, pp. 449–461.

Templeman, W. 1968. Review of some aspects of capelin biology in the Canadian area of the northwest Atlantic. *Rapp. Proc.-Verb. Réun. Cons. Int. Explor. Mer* 158:41–53.

Templeman, W. 1970. Vertebral and other meristic characteristics of Greenland halibut, *Reinhardtius hippoglossoides,* from the northwest Atlantic. *J. Fish. Res. Bd. Can.* 27:1549–1562.

Templeman, W. 1973a. First records, description, distribution, and notes on the biology of *Bathyraja richardsoni* (Garrick) from the northwest Atlantic. *J. Fish. Res. Bd. Can.* 30:1831–1840.

Templeman, W. 1973b. Distribution and abundance of the Greenland halibut, *Reinhardtius hippoglossoides* (Walbaum) in the northwest Atlantic. *ICNAF Res. Bull.* 10:83–98.

Templeman, W. 1984a. Migrations of wolffishes, *Anarhichas* sp., from tagging in the Newfoundland area. *J. Northw. Atl. Fish. Sci.* 5:93–97.

Templeman, W. 1984b. Vertebral and dorsal fin-ray numbers in Atlantic wolffish (*Anarhichas lupus*) of the Northwest Atlantic. *J. Northw. Atl. Fish. Sci.* 5:207–212.

Templeman, W. 1984c. Migrations of thorny skate, *Raja radiata,* tagged in the Newfoundland area. *J. Northw. Atl. Fish. Sci.* 5: 55–63.

Templeman, W. 1985. Stomach contents of Atlantic wolffish (*Anarhichas lupus*) from the northwest Atlantic. *Northw. Atl. Fish. Org., Sci. Council Stud.* No. 8:49–51.

Templeman, W. 1986a. Contribution to the biology of the spotted wolffish (*Anarchichas minor*) in the northwest Atlantic. *J. Northw. Atl. Fish. Sci.* 7:47–55.

Templeman, W. 1986b. Some biological aspects of Atlantic wolffish (*Anarhichas lupus*) in the Northwest Atlantic. *J. Northw. Atl. Fish. Sci.* 7:57–65.

Templeman, W. 1986c. Spotted forms of the northern wolffish (*Anarhichas denticulatus*). *J. Northw. Atl. Fish. Sci.* 7:77–80.

Templeman, W., and A. M. Fleming. 1962. Cod tagging in the Newfoundland area during 1947 and 1948. *J. Fish. Res. Bd. Can.* 19: 445–487.

Templeman, W., and R. L. Haedrich. 1966. Distributions and comparisons of *Centrolophus niger* (Gmelin) and *Centrolophus britannicus* Günther (Centrolophidae) from the North Atlantic. *J. Fish. Res. Bd. Can.* 23:1161–1185.

Templeman, W., and H. J. Squires. 1960. Incidence and distribution of infestation by *Sphyrion lumpi* (Krøyer) on the redfish, *Sebastes fasciatus* (L.), of the western North Atlantic. *J. Fish. Res. Bd. Can.* 17:9–31.

Templeman, W., H. J. Squires, and A. M. Fleming. 1957. Nematodes in the fillets of cod and other fishes in Newfoundland and neighbouring waters. *J. Fish. Res. Bd. Can.* 14:831–897.

Teo, S. L.-H., and K. W. Able. unpublished data. Habitat use and movement of the mummichog (*Fundulus heteroclitus*) in a restored salt marsh.

Terceiro, M. E. 1998a. Bluefish. In: S. H. Clark, ed. *Status of Fishery Resources off the Northeastern United States for 1998.* NOAA Tech. Memo. NMFS–NE–115:110–111.

Terceiro, M. E. 1998b. Summer flounder. In: S. H. Clark, ed. *Status of Fishery Resources off the Northeastern United States for 1998.* NOAA Tech. Memo. NMFS-NE-115:75–76.

Texas Instruments Inc. 1973. Hudson River ecology study in the area of Indian Point. First Annual Report to Consolidated Edison Co. of N.Y., Inc., 348 pp.

Thoits, C. F., III. 1958. A compendium of the life history and ecology of the white perch, *Morone americana* (Gmelin). *Mass. Div. Fish Game, Fish. Bull.* 24, 16 pp.

Thomas, D. L. 1971. The early life history and ecology of six species of drum (Sciaenidae) in the lower Delaware River, a brackish tidal estuary. III. In: *An Ecological Study of the Delaware River in the Vicinity of Artificial Island.* Ichthyol. Assoc. Del. Prog. Rept. No. 3, 247 pp.

Thompson, B. A., and S. J. Russell. 1996. Pomfrets (family Bramidae) of the Gulf of Mexico and nearby waters. Espec. Inst. Esp. Oceanogr. Publ. No. 21:185–198.

Thompson, D. W. 1888. On the auditory labyrynth of *Orthagoriscus mola* L. *Anat. Anzeig.* 3:93–96.

Thompson, H. 1929. General features in the biology of the haddock (*Gadus aeglefinus* L.) in Icelandic waters in the period 1903–1926. *Rapp. Proc.–Verb. Réun. Cons. Int. Explor. Mer* 57:1–73.

Thomson, J. M. 1963. Synopsis of biological data on the grey mullet, *Mugil cephalus* Linnaeus 1758. Aust. C.S.I.R.O. Div. Fish. Oceanogr. Fish. Synop. No. 1, 68 pp.

Thomson, J. M. 1966. The grey mullets. *Oceanogr. Mar. Biol. Ann. Rev.* 4:301–335.

Thomson, J. M. 1981. The taxonomy of grey mullets. In: O. H. Oren, ed. *Aquaculture of Grey Mullets.* Cambridge University Press, Cambridge, pp. 1–15.

Thoney, D. A., and T. A. Munroe. 1987. *Microcotyle hiatulae* Goto, 1900 (Monogenea), a senior synonym of *M. furcata* Linton, 1940, with a redescription and comments on postlarval development. *Proc. Helminthol. Soc. Washington* 54:91–95.

Thorpe, J. E., L. G. Ross, G. Struthers, and W. Watts. 1981. Tracking Atlantic salmon smolts, *Salmo salar* L., through Loch Voil, Scotland. *J. Fish Biol.* 19:519–537.

Thorsteinsson, V. 1981. The aging validation of the lumpsucker (*Cyclopterus lumpus*) and the age composition of the lumpsucker in Icelandic lumpsucker fisheries. Int. Counc. Explor. Sea CM 1981/G:58.

Threlfall, W. 1982. In vitro culture of *Anisakis* spp. larvae from fish and squid in Newfoundland. *Proc. Helminthol. Soc. Washington* 49:65–70.

Thresher, R. E. 1984. *Reproduction in Reef Fishes.* T. F. H. Publ. Inc., Ltd., Neptune City, N.J., 399 pp.

Thunberg, B. E. 1971. Olfaction in parent stream selection by the alewife (*Alosa pseudoharengus*). *Anim. Beh.* 19:217–225.

Thurston-Rogers, P. A., and C. F. Baren. 1978. Present configuration of the Delaware River pollution block and its relationship to shad migrations. Delaware River Basin Anadromous Fishery Project, U.S. Fish Wildlfe. Serv., Spec. Rept., Rosemont, N.J.

Tibbetts, A. M. 1977. Squid fisheries (*Loligo pealei* and *Ilex illecebrosus*) off the northeastern coast of the United States of America, 1963–74. Int. Comm. Northw. Atl. Fish., Selected Pap. No. 2:85–109.

Tibbo, S. N. 1956. Populations of herring (*Clupea harengus* L.) in Newfoundland waters. *J. Fish. Res. Bd. Can.* 13:449–466.

Tibbo, S. N. 1962. New records for occurrence of the white-tip shark, *Pterolamiops longimanus* (Poey), and the dolphin, *Coryphaena hippurus* L., in the northwest Atlantic. *J. Fish. Res. Bd. Can.* 19:517–518.

Tibbo, S. N. 1964. Effect of light on movements of herring in the Bay of Fundy. Int. Comm. Northw. Atl. Fish. Spec. Pub. No. 6:579–582.

Tibbo, S. N., and T. R. Graham. 1963. Biological changes in herring stocks following an epizootic. *J. Fish. Res. Bd. Can.* 20:435–439.

Tibbo, S. N., and R. D. Humphreys. 1966. An occurrence of capelin (*Mallotus villosus*) in the Bay of Fundy. *J. Fish. Res. Bd. Can.* 23:463–467.

Tibbo, S. N., T. R. Graham, L. W. Scattergood, and R. F. Temple. 1958. On the occurrence and distribution of larval herring (*Clupea harengus* L.) in the Bay of Fundy and Gulf of Maine. *J. Fish. Res. Bd. Can.* 15:1451–1469.

Tibbo, S. N., L. R. Day, and W. F. Doucet. 1961. The swordfish (*Xiphius gladius* L.) its life-history and economic importance in the northwest Atlantic. *Fish. Res. Bd. Can. Bull.* 130, 47 pp.

Tibbo, S. N., D. J. Scarratt, and P. W. G. McMullon. 1963. An Investigation of herring (*Clupea harengus* L.) spawning using free-diving techniques. *J. Fish. Res. Bd. Can.* 20:1067–1079.

Tiews, K. 1963. Synopsis of biological data on bluefin tuna *Thunnus thynnus* (Linnaeus) 1758 (Atlantic and Mediterranean). FAO Fish. Rept. No. 6(2):422–481.

Tighe, K. A. 1975. The systematics, vertical distribution, and life histories of serrivomerid eels in the Bermuda Ocean Acre. M.S. thesis, University of Rhode Island, Kingston, 62 pp.

Tighe, K. A. 1989. Family Serrivomeridae. Fishes of the western North Atlantic. Sears Found. Mar. Res. Mem. No. 1(9):613–627.

Tilseth, S. 1990. New marine fish species for cold-water farming. *Aquaculture* 85:235–245.

Tilseth, S., G. Blom, and K. Naas. 1992. Recent progress in research and development of marine cold water species for aquaculture production in Norway. *J. World Aquacult. Soc.* 23:277–285.

Timi, J. T., and J. A. Etchegoin. 1996. A new species of *Lernanthropus* (Copepoda: Lernanthropidae) parasite of *Cynoscion striatus* (Pisces: Sciaenidae) from Argentinean waters, and new records of *Lernanthropus trachuri*. *Folia Parasitol.* 43:71–74.

Topp, R. W. 1968. An estimate of fecundity of the winter flounder, *Pseudopleuronectes americanus*. *J. Fish. Res. Bd. Can.* 25:1299–1302.

Tortonese, E. 1986. Balistidae. In: P. J. P. Whitehead et al., eds. *Fishes of the North–eastern Atlantic and the Mediterranean,* Vol. 3. UNESCO, Paris, pp. 1335–1337.

Towne, S. A. 1940. State of Maine Striped Bass Survey. Maine Develop. Comm. Dept. Sea Shore Fish, 30 pp.

Townsend, C. H. 1912. The orange filefish. *Bull. New York Zool. Soc.* 16(54):933–934.

Townsend, C. H. 1916. The puffers: its defense by inflation. *Bull. New York Zool. Soc.* 19:1331–1333.

Townsend, C. H. 1917. Young orange filefish. *Bull. New York Zool. Soc.* 20(6):1554–1555.

Townsend, D. W. 1984. Comparison of inshore zooplankton and ichthyoplankton populations of the Gulf of Maine. *Mar. Ecol. Prog. Ser.* 15:79–90.

Townsend, D. W. 1992. Ecology of larval herring in relation to the oceanography of the Gulf of Maine. *J. Plank. Res.* 14:467–493.

Townsend, D. W., and J. J. Graham. 1981. Growth and age structure of larval Atlantic herring, *Clupea harengus harengus,* in the Sheepscot River Estuary, Maine, as determined by daily growth increments in otoliths. *Fish. Bull., U. S.* 79:123–130.

Townsend, D. W., J. J. Graham, and D. K. Stevenson. 1986. Dynamics of larval herring (*Clupea harengus* L.) production in tidally mixed waters of the eastern coastal Gulf of Maine. In: M. J. Bowman, C. M. Yentsch, and W. T. Peterson, eds. *Tidal Mixing and Plankton Dynamics.* Springer-Verlag, Berlin, pp. 253–277.

Townsend, D. W., J. P. Christensen, D. K. Stevenson, J. J. Graham, and S. B. Chenoweth. 1987. The importance of a plume of tidally-mixed water to the biological oceanography of the Gulf of Maine. *J. Mar. Res.* 45:699–728.

Townsend, D. W., L. M. Mayer, Q. Dortch, and R. W. Spinrad. 1992. Vertical structure and biological activity in the bottom nepheloid layer of the Gulf of Maine. *Cont. Shelf Res.* 12:367–387.

Tracy, H. C. 1906. A list of the fishes of Rhode Island. Rhode Island Comm. Inland Fish Ann. Rept. No. 36:38–99.

Tracy, H. C. 1907. The fishes of Rhode Island. III. The fishes of the mackerel family. Rhode Island Comm. Inland Fish. Ann. Rept. No. 37:33–64.

Tracy, H. C. 1908. The fishes of Rhode Island. V. The flat-fishes. Rhode Island Comm. Inland Fish. Ann. Rept. No. 38, pp. 47–84.

Tracy, H. C. 1910. Annotated list of fishes known to inhabit the waters of Rhode Island. Rhode Island Comm. Inland Fish. Ann. Rept. No. 40:35–176.

Trent, W. L., and W. W. Hassler. 1966. Feeding behavior of adult striped bass, *Roccus saxatilis,* in relation to stages of sexual maturity. *Chesapeake Sci.* 7:189–192.

Tresselt, E. F. 1952. Spawning grounds of the striped bass or rock, *Roccus saxatilis* (Walbaum), in Virginia. *Bull. Bingham Oceanogr. Coll.* 14(1):98–110.

Trippel, E. A. 1998. Egg size and viability and seasonal offspring production of young Atlantic cod. *Trans. Am. Fish. Soc.* 127:339–359.

Trottier, B. L., P. J. Rubec, and A. C. Ricard. 1988. Biochemical separation of Atlantic Canadian redfish: *Sebastes mentella* and *Sebastes norvegicus. Can. J. Zool.* 67:1332–1335.

Trumble, R. J., J. D. Neilson, W. R. Bowering, and D. A. McCaughran. 1993. Atlantic halibut (*Hippoglossus hippoglossus*) and Pacific halibut (*H. stenolepis*) and their North American fisheries. *Bull. Can. Fish. Aquat. Sci.* 227, 84 pp.

Truveller, C. A. 1971. A study of blood groups in herring (*Clupea harengus* L.) from the North Sea in connection with the problem of race differentiation. *Rapp. Proc.Verb. Réun. Cons. Int. Explor. Mer* 161:33–39.

Tsukahara, H. 1961. Biology of the cutlassfish, *Trichiurus lepturus* Linnaeus. 1. Early life history. Rec. Oceanogr. Works, Japan, Special No. 5:117–121.

Tsuneki, K., M. Ouji, and H. Saito. 1983. Seasonal migration and gonadal changes in the hagfish, *Eptatretus burgeri. Japan. J. Ichthyol.* 29:429–440.

Tucker, D. W. 1959. A new solution to the Atlantic eel problem. *Nature* 183:495–501.

Tucker, J. W., Jr. 1984. Hormone-induced ovulation of black sea bass and rearing of larvae. *Progr. Fish. Cult.* 46:201–203.

Tucker, J. W., Jr. 1988. Energy utilization in bay anchovy, *Anchoa mitchilli,* and black sea bass, *Centropristis striata striata,* eggs and larvae. *Fish. Bull., U.S.* 78:279–293.

Tucker, J. W., Jr., and R. G. Hodson. 1976. Early and mid-metamorphic larvae of the tarpon, *Megalops atlanticus,* from the Cape Fear River estuary, North Carolina 1973–1974. *Chesapeake Sci.* 17:123–125.

Tulloch, S., S. Goddard, and J. Watkins. 1996. A preliminary investigation of food and feeding requirements of striped wolffish, *Anarhichas lupus. Bull. Aquacult. Assoc. Can.* 96–3:21–23.

Tully, O., and P. Ó'Céidigh. 1989. The ichthyoneuston of Galway Bay (Ireland) I. The seasonal, diel and spatial distribution of larval, post-larval and juvenile fish. *Mar. Biol.* 100:27–41.

Tupper, M., and K. W. Able. 2000. Movements and food habits of striped bass (*Morone saxatilis*) in Delaware Bay (USA) salt marshes: comparison of a restored and a reference marsh. *Mar. Biol.* 137:1049–2004.

Tupper, M., and R. G. Boutilier. 1995. Effects of conspecific density on settlement, growth and post-settlement survival of a temperate reef fish. *J. Exp. Mar. Biol. Ecol.* 191:209–222.

Tupper, M., and R. G. Boutilier. 1997. Effects of habitat on settlement, growth, predation risk and survival of a temperate reef fish. *Mar. Ecol. Progr. Ser.* 151:225–236.

Turner, J. L., and T. C. Farley. 1971. Effects of temperature, salinity, and dissolved oxygen on the survival of striped bass eggs and larvae. *Calif. Fish Game* 57:268–273.

Turner, J. T., P. A. Tester, and W. F. Hettler. 1985. Zooplankton feeding ecology: a laboratory study on predation on fish eggs and larvae by the copepods *Anomalocera ornata* and *Centropages typicus. Mar. Biol.* 90:1–8.

Turner, S. C. 1986. Population dynamics of and impact of fishing on, tilefish, *Lopholatilus chamaeleonticeps,* in the Middle Atlantic-southern New England region during the 1970's and early 1980's. Ph.D. dissertation, Rutgers University, New Brunswick, 289 pp.

Turner, S. C., C. B. Grimes, and K. W. Able. 1983. Growth, mortality, and age/size structure of the fisheries for tilefish, *Lopholatilus chamaeleonticeps,* in the Middle Atlantic–southern New England region. *Fish. Bull., U.S.* 81:751–763.

Twitchell, D. C., C. B. Grimes, R. S. Jones, and K. W. Able. 1985. The role of erosion by fish in shaping topography around Hudson Submarine Canyon. *J. Sediment. Petrol.* 55:712–719.

Tyler, A. V. 1971a. Periodic and resident components in communities of Atlantic fishes. *J. Fish. Res. Bd. Can.* 28:935–946.

Tyler, A. V. 1971b. Surges of winter flounder, *Pseudopleuronectes americanus,* into the intertidal zone. *J. Fish. Res. Bd. Can.* 28:1727–1732.

Tyler, A. V. 1972. Food resource division among northern, marine, demersal fishes. *J. Fish. Res. Bd. Can.* 29:997–1003.

Tyler, J. C. 1980. Osteology, phylogeny and higher classification of the fishes of the order Plectognathi (Tetraodontiformes). NOAA Tech. Rept. NMFS Circ. No. 434, 422 pp.

Tyus, H. M. 1974. Movements and spawning of anadromous alewives, *Alosa pseudoharengus* (Wilson) at Lake Mattamuskeet, North Carolina. *Trans. Am. Fish. Soc.* 103:392–395.

Uchida, R. N. 1981. Synopsis of biological data on frigate tuna, *Auxis thazard,* and bullet tuna, *A. rochei.* NOAA Tech. Rept. NMFS Circ. No. 436, 63 pp.

Uchiyama, J. H., R. K. Burch, and S. A. Kraul, Jr. 1986. Growth of dolphins, *Coryphaena hippurus* and *C. equiselis,* in Hawaiian waters as determined by daily increments on otoliths. *Fish. Bull., U.S.* 84:186–191.

Unger, I. 1966. Artificial reefs: a review. Am. Littoral Soc. Spec. Publ. No. 4, 74 pp.

Ursin, M. J. 1977. *A Guide to Fishes of the Temperate Atlantic Coast.* E. P. Dutton, New York, 262 pp.

Valentine, P. C., J. R. Uzmann, and R. A. Cooper. 1980. Geology and biology of Oceanographer Submarine Canyon. *Mar. Geol.* 38:283–312.

Valerio, P. F., M. H. Kao, and G. L. Fletcher. 1990. Thermal hysteresis activity in the skin of the cunner, *Tautogolabrus adspersus. Can. J. Zool.* 68:1065–1067.

Valiela, I., J. E. Wright, J. M. Teal, and S. B. Volkmann. 1977. Growth, production and energy transformations in the salt-marsh killifish *Fundulus heteroclitus. Mar. Biol.* 40:135–144.

Valtonen, E. T., H.-P. Fagerholm, and E. Helle. 1988. *Contracaecum osculatum* (Nematoda: Anisakidae) in fish and seals in Bothnian Bay (Northeastern Baltic Sea). *Intl. J. Parasitol.* 18:365–370.

Van Den Avyle, M. J. 1983. Species profiles: life histories and environmental requirements (South Atlantic):–Atlantic sturgeon. U.S. Fish Wildl. Serv. Biol. Serv. Prog. No. FWS-OBS-82/11.

van der Meeren, T. 1991. Algae as first food for cod larvae, *Gadus morhua* L.: filter feeding or ingestion by accident. *J. Fish Biol.* 39:225–237.

Van Eenennaam, J. P., S. I. Doroshov, G. P. Moberg, J. G. Watson, D. S. Moore, and J. Linaras. 1996. Reproductive conditions of the Atlantic sturgeon (*Acipenser oxyrinchus*) in the Hudson River. *Estuaries* 19:769–777.

Van Guelpen, L. 1980. Taxonomic variability and criteria for distinguishing metamorphosing larval and juvenile stages of *Limanda ferruginea* and *Hippoglossoides platessoides* (Pisces: Pleuronectidae) from the Scotian shelf. *Can. J. Zool.* 58:202–206.

Van Guelpen, L. 1986. Hookear sculpins (genus *Artediellus*) of the North American Atlantic: taxonomy, morphological variability, distribution, and aspects of life history. *Can. J. Zool.* 64:677–690.

Van Guelpen, L. 1993. Substantial northward range extension for *Gephyroberyx darwini* (Berycoidei, Trachichthyidae) in the western North Atlantic, possibly explained by habitat preference. *J. Fish Biol.* 42:807–810.

Van Guelpen, L., and C. C. Davis. 1979. Seasonal movements of the winter flounder, *Pseudopleuronectes americanus,* in two contrasting inshore locations in Newfoundland. *Trans. Am. Fish. Soc.* 108:26–37.

Van Housen, G. 1984. Electrophoretic stock identification of summer flounder, *Paralichthys dentatus.* M.A. thesis, College of William and Mary, Williamsburg, Va., 66 pp.

Van Thiel, P. H., F. C. Kuipers, and R. T. Roskam. 1960. A nematode parasite of herring, causing acute abdominal syndromes in man. *Trop. Geogr. Med.* 2:97–113.

Van Vliet, W. H. 1970. Shore and freshwater fish collections from Newfoundland. *Natl. Mus. Can. Publ. Zool.* 3, 30 pp.

Vari, R. P. 1982. The seahorses (subfamily Hippocampinae). Fishes of the western North Atlantic. Sears Found. Mar. Res. Mem. No. 1(8):173–189.

Vaske, T., Jr. 1995. Alimentaçao de rêmora *Remora osteochir* (Cuvier, 1829), e peixe–piloto *Naucrates ductor* (Linnaeus, 1758) no sul do Brasil. *Rev. Brasil. Biol.* 55:315–321.

Vaske, T., Jr. and G. Rincón-Filho. 1998. Conteúdo estomacal dos tubarões azul (*Prinace glauca*) e anequim (*Isurus oxyrinchus*) em águas oceânicas no sul do Brasil. *Rev. Brasil. Biol.* 58:445–452.

Vasquez Rojas, A. V. 1989. Energetics, trophic relationships and chemical composition of bay anchovy, *Anchoa mitchilli,* in the Chesapeake Bay. M.S. thesis, University of Maryland, College Park, 166 pp.

Vaughan, D. S. 1990. Assessment of the status of the Atlantic menhaden stock with reference to internal waters processing. NOAA Tech. Memo. NMFS-SEFC-262, 20 pp.

Vecsei, P., and D. Peterson. 2000a. Threatened fishes of the world: *Acipenser oxyrinchus* Mitchill, 1815 (Acipenseridae). *Environ. Biol. Fishes* 59:98.

Vecsei, P., and D. Peterson. 2000b. Threatened fishes of the world: *Acipenser brevirostrum* Lesueur, 1818 (Acipenseridae). *Environ. Biol. Fishes* 59:270.

Verigina, I. A. 1974. The structure of the alimentary canal in some of the northern Blennioidei. I. The alimentary canal of the Atlantic wolffish (*Anarhichas lupus*). *Voprosy Ikhtiol.* 14:1098–1110 [in Russian, translation in *J. Ichthyol.* 14:954–959].

Vernberg, J. 1977. A short analysis of stock enhancement possibilities for certain commercially important marine species. In: *Working Papers Establishing a 200-Mile Fisheries Zone.* Office of Technology Assessment, Washington, D.C., Working Pap. No. 4.

Vernick, S. H., and G. B. Chapman. 1968. Ultrastructure of axillary glands of the toadfish *Opsanus tau. Chesapeake Sci.* 9:182–197.

Verreault, G., and R. Courtois. 1989. Seasonal changes in the stomach contents of anadromous brook trout (*Salvelinus fontinalis*) in the Matapedia and Ristigouche rivers (Quebec). *Nat. Can.* 116:251–260.

Verrill, A. E. 1871. On the food and habits of some of our marine fishes. *Am. Nat.* 5:397–400.

Verrill, A. E. 1882. Notice of the remarkable marine fauna occupying the outer banks off southern coast of New England, No. 7, and some of the additions to the fauna of Vineyard Sound. *Am. J. Sci.* 3:360–371.

Vesey-Fitzgerald, B., and F. LaMonte. 1949. *Game Fish of the World.* Harper & Brothers, New York, 446 pp.

Vesin, J.-P., W. C. Leggett, and K. W. Able. 1981. Feeding ecology of capelin (*Mallotus villosus*) in the estuary and western Gulf of St. Lawrence and its multispecies implications. *Can. J. Fish. Aquat. Sci.* 38:257–267.

Vincent, A. C. V., and J. R. S. Clifton-Hadley. 1989. Parasitic infection of the seahorse (*Hippocampus erectus*) - a case report. *J. Wildl. Dis.* 25:404–406.

Vinnichenko, V. I. 1996. Russian investigations and deep water fishery on the Corner Rising. Northw. Atl. Fish. Org. SCR Doc. No. 96/38, 16 pp.

Vinogradov, V. I. 1984. Food of silver hake, red hake and other fishes on Georges Bank and adjacent waters, 1968–74. NAFO Sci. Counc. Stud. No. 7:87–94.

Vis, M. L., S. M. Carr, W. R. Bowering, and W. S. Davidson. 1997. Greenland halibut (*Reinhardtius hippoglossoides*) in the North Atlantic are genetically homogeneous. *Can. J. Fish. Aquat. Sci.* 54:1813–1821.

Vladykov, V. D. 1933. Biological and oceanographic conditions in Hudson Bay. 9. Fishes from the Hudson Bay region (except the Coregonidae). *Contrib. Can. Biol. Fish., n. s.,* 8:13–61.

Vladykov, V. D. 1935a. Some unreported and rare fishes for the coast of Nova Scotia. *Proc. Nova Scotian Inst. Sci.* 19:1–8.

Vladykov, V. D. 1935b. Two new subspecies of *Lumpenus lampetraeformis* (Walbaum) from North America. Newfoundland Fish. Res. Lab. Rept. No. 2:75–78.

Vladykov, V. D. 1935c. Haddock races along the North America coast. Atlantic Biol. Sta. Progress Rept. No. 14:3–7.

Vladykov, V. D. 1936a. Capsules d'oeufs de raies de l'Atlantique canadien appartenant au genre *Raja. Natur. Can.* 63:211–2310.

Vladykov, V. D. 1936b. Occurrence of three species of anadromous fishes on the Nova Scotian banks during 1935 and 1936. *Copeia* 1936:168.

Vladykov, V. D. 1955a. *Cods: Fishes of Quebec.* Quebec Dept. Fish. Album No. 4, 12 pp.

Vladykov, V. D. 1955b. A comparison of Atlantic sea sturgeon with a new subspecies from the Gulf of Mexico (*Acipenser oxyrhynchus de sotoi*). *J. Fish. Res. Bd. Can.* 12:754–761.

Vladykov, V. D. 1955c. *Eels: Fishes of Canada*. Quebec Dept. Fish. Album No. 6, 12 pp.

Vladykov, V. D., and J. R. Greeley. 1963. Order Acipenseroidei. Fishes of the western North Atlantic. Sears Found. Mar. Res. Mem. No. 1(3):24–60.

Vladykov, V. D., and E. Kott. 1980. First record of the sea lamprey, *Petromyzon marinus* L., in the Gulf of Mexico. *Northeast Gulf Sci.* 4:49–50.

Vladykov, V. D., and R. A. McKenzie. 1935. The marine fishes of Nova Scotia. *Proc. Nova Scotian Inst. Sci.* 19:17–113.

Vladykov, V. D., and D. H. Wallace. 1952. Studies of the striped bass, *Roccus saxatilis* (Walbaum), with special reference to the Chesapeake Bay region during 1936–1938. *Bull. Bingham Oceanogr. Coll.* 14(1):132–177.

von Dorrien, C. F. 1996. Reproduction and larval ecology of the Arctic fish species *Artediellus atlanticus* (Cottidae). *Polar Biol.* 6:401–407.

Voorhees, D. A., J. F. Witzig, M. F. Osborn, M. C. Holliday, and R. J. Essig. 1992. Marine recreational fishery statistics survey, Atlantic and Gulf coasts, 1990–91. NOAA/NMFS, Marine Recreational Fishery Statistics Survey, Atlantic and Gulf Coasts. Current Fisheries Statistics No. 9204, 275 pp.

Vose, F. E., and W. G. Nelson. 1994. Grey triggerfish (*Balistes capriscus* Gmelin) feeding from artificial and natural substrates in shallow Atlantic waters of Florida. *Bull. Mar. Sci.* 55:1316–1323.

Vouglitois, J. J., K. W. Able, R. J. Kurtz, and K. A. Tighe. 1987. Life history and population dynamics of the bay anchovy in New Jersey. *Trans. Am. Fish. Soc.* 116:141–153.

Wade, R. A. 1962. The biology of the tarpon, *Megalops atlanticus,* and the ox-eye, *Megalops cyprinoides,* with emphasis on larval development. *Bull. Mar. Sci.* 12:545–622.

Wade, R. A. 1969. Ecology of juvenile tarpon and effects of dieldrin on two associated species. Bur. Sport Fish. Wildl. Tech. Pap. No. 41, 85 pp.

Waiwood, K. G., S. J. Smith, and M. R. Petersen. 1991. Feeding of Atlantic cod (*Gadus morhua*) at low temperatures. *Can. J. Fish. Aquat. Sci.* 48:824–831.

Waiwood, K. G., J. Reid, and K. Howes. 1997. Canadian perspective on halibut culture. Proc. Huntsman Mar. Sci. Cent. Symp. Coldwater Aquaculture to 2000. *Bull Aquacult. Assoc. Spec. Publ.* 2:43–45.

Walburg, C. H. 1957. Observations on food and growth of juvenile American shad, *Alosa sapidissima*. *Trans. Am. Fish. Soc.* 86:302–306.

Walburg, C. H. 1960. Abundance and life history of American shad, St. Johns River, Florida. *Fish. Bull., U.S.* 60:487–501.

Walburg, C. H., and P. R. Nichols. 1967. Biology and management of American shad and status of the fisheries along the coast of the United States, 1960. U.S. Fish Wildl. Serv. SSRR-550, 105 pp.

Waldman, J. R., and I. I. Wirgin. 1998. Status and restoration options for Atlantic sturgeon in North America. *Conserv. Biol.* 12:631–638.

Waldman, J. R., J. Grossfield, and I. Wirgin. 1988. Review of stock discrimination techniques for striped bass. *No. Am. J. Fish. Manage.* 8:410–425.

Waldman, J. R., J. T. Hart, and I. I. Wirgin. 1996a. Stock composition of the New York Bight Atlantic sturgeon fishery based on analysis of mitochondrial DNA. *Trans. Am. Fish. Soc.* 125:364–371.

Waldman, J. R., K. Nolan, J. Hart, and I. I. Wirgin. 1996b. Genetic differentiation of three key anadromous fish populations of the Hudson River. *Estuaries* 19:759–768.

Waldron, M. E., R. M. Prosch, and M. J. Armstrong. 1991. Growth of juvenile round herring *Etrumeus whiteheadi* in the Benguela system. *S. Afr. J. Mar. Sci.* 10:83–89.

Walford, L. A. 1946. New southern record for Atlantic halibut. *Copeia* 1946:100–101.

Walker, P. A., G. Howlett, and R. Millner. 1997. Distribution, movement and stock structure of three ray species in the North Sea and eastern English Channel. *ICES J. Mar. Sci.* 54:797–808.

Wallace, L. B. 1893. The structure and development of the axillary gland of *Batrachus*. *J. Morph.* 8:563–568.

Wallace, R. A., and K. Selman. 1981. The reproductive activity of *Fundulus heteroclitus* females from Woods Hole, Massachusetts, as compared with more southern locations. *Copeia* 1981:212–215.

Wallace, S. D., and D. M. Lavigne. 1992. A review of stomach contents of harp seals (*Phoca groenlandica*) from the Northwest Atlantic. Intl. Mar. Mam. Assoc. 92-03.

Walsh, F. G., and G. J. FitzGerald. 1984. Resource utilization and coexistence of three species of sticklebacks (Gasterosteidae) in tidal salt-marsh pools. *J. Fish Biol.* 25:405–420.

Walsh, W. A., and W. A. Lund, Jr. 1983. Early development of the longhorn sculpin, *Myoxocephalus octodecimspinosus*. *Fish. Bull., U.S.* 81:781–788.

Walters, V. 1980. Parasitic Copepoda and Monogenea as biological tags for certain populations of Atlantic bluefin tuna. *Int. Comm. Cons. Atlantic Tunas, Coll. Vol. Sci. Pap.* 9(2):491–498.

Walton, C. J. 1983. Growth parameters for typical anadromous and dwarf stocks of alewives, *Alosa pseudoharengus* (Pisces, Clupeidae). *Environ. Biol. Fishes* 9:277–287.

Walton, C. J. 1987. Parent-progeny relationship for an established population of anadromous alewives in a Maine lake. In: M. J. Dadswell et al., eds., *Common Strategies of Anadromous and Catadromous Fishes. Am. Fish. Soc. Symp.* 1:451–454.

Walton, C. J., and M. E. Smith. 1974. Population biology and management of the alewife (*Alosa pseudoharengus*) in Maine. Maine Dept. Mar. Resour., Compl. Rept., Proj. No. AFC-18-2.

Walton, D. G., W. R. Penrose, and J. M. Green. 1978. The petroleum-inducible mixed-function oxidase of cunner (*Tautogolabrus adspersus* Walbaum 1792): some characteristics relevant to hydrocarbon monitoring. *J. Fish. Res. Bd. Can.* 35:1547–1552.

Walton, D. G., L. L. Fancey, J. M. Green, J. W. Kiceniuk, and W. R. Penrose. 1983. Seasonal changes in aryl hydrocarbon hydroxylase activity of a marine fish *Tautogolabrus adspersus* (Walbaum) with and without petroleum exposure. *Comp. Biochem. Physiol. C,* 76:247–253.

Walvig, F. 1967. Experimental marking of hagfish (*Myxine glutinosa* L.). *Norwegian J. Zool.* 15:35–39.

Wang, J. C. S., and R. J. Kernehan. 1979. *Fishes of Delaware Estuaries: A Guide to the Early Life Histories.* Ecological Analysts, Inc., Towson, Md, 410 pp.

Wang, S. B., and E. D. Houde. 1994. Energy storage and dynamics in bay anchovy *Anchoa mitchilli*. *Mar. Biol.* 121:219–227.

Wang, Z., and L. W. Crim. 1997. Seasonal changes in the biochemistry of seminal plasma and sperm motility in the ocean pout, *Macrozoarces americanus*. *Fish Physiol. Biochem.* 16:77–83.

Wardle, C. S., and P. He. 1988. Burst swimming speeds of mackerel, *Scomber scombrus* L. *J. Fish Biol.* 32:471–478.

Warfel, H. E., and D. Merriman. 1944a. Studies on the marine resources of southern New England. I. An analysis of the fish population of the shore zone. *Bull. Bingham. Oceanogr. Coll.* 9(2):1–91.

Warfel, H. E., and D. Merriman. 1944b. The spawning habits, eggs and larvae of the sea raven, *Hemitripterus americanus,* in southern New England. *Copeia* 1944:197–205.

Warfel, H. E., and Y. H. Olson. 1947. Vertebral counts and the problem of races in the Atlantic shad. *Copeia* 1947:177–183.

Warfel, H. E., T. P. Frost, and W. H. Jones. 1943. The smelt, *Osmerus mordax,* in Great Bay, New Hampshire. *Trans. Am. Fish. Soc.* 72:257–262.

Waring, G. 1975. A preliminary analysis of the status of butterfish in ICNAF subarea 5 and statistical area 6. Int. Comm. Northw. Atl. Fish., Selected Pap. No. 2, 32 pp.

Waring, G. 1981. Results of the international herring tagging program conducted by USA in the Gulf of Maine, Georges Bank and contiguous waters from 1976–78. NAFO SCR Doc. No. 122, Ser. No. N428, 24 pp.

Waring, G., and S. Murawski. 1982. Butterfish, *Peprilus triacanthus.* In: M. D. Grosslein and T. R. Azarovitz, eds. *Fish distribution.* MESA New York Bight Atlas Mongr. No. 15:105–107.

Waring, G. T. 1984. Age, growth and mortality of the little skate off the northeast coast of the United States. *Trans. Am. Fish. Soc.* 113:314–321.

Warlen, S. M., and A. J. Chester. 1986. Age, growth, and distribution of larval spot, *Leiostomus xanthurus,* off North Carolina. *Fish. Bull., U.S.* 83:587–599.

Warme, J. E., R. A. Slater, and R. A. Cooper. 1977. Bioerosion in submarine canyons. In: D. J. Stanley and G. Kelling, eds. *Sedimentation in Submarine Canyons, Fans, and Trenches.* Dowden, Hutchinson, and Ross, Inc., Stroudsburg, Pa., pp. 65–70.

Warner, J., and B. Kynard. 1986. Scavenger feeding by subadult striped bass, *Morone saxatilis,* below a low-head hydroelectric dam. *Fish. Bull., U.S.* 84:220–222.

Warner, R. R., and D. R. Robertson. 1978. Sexual patterns in the labroid fishes of the western Caribbean. I. The wrasses (Labridae). *Smithsonian Contrib. Zool.* 254, 27 pp.

Warner, R. W., and S. C. Katkansky. 1970. An inflammatory lesion in an American shad, *Alosa sapidissima. J. Fish. Res. Bd. Can.* 27:191–193.

Waters, S. C. 1994. Menhaden. *Virg. Mar. Res. Bull. (Virg. Sea Grant Progr.).* 26:1–22.

Watkins, W. A., and W. E. Schevill. 1979. Aerial observation of feeding behavior in four baleen whales: *Eubalena glacialis, Balaenoptera borealis, Megaptera novaeangliae,* and *Balenoptera physalus. J. Mammal.* 60:155–163.

Watson, J. E. 1964. Determining the age of young herring from their otoliths. *Trans. Am. Fish. Soc.* 93:11–20.

Watson, J. F. 1968. The early life history of the American shad, *Alosa sapidissima* (Wilson), in the Connecticut River above Holyoke, Massachusetts. M.S. thesis, University of Massachusetts, Amherst, 55 pp.

Weaver, J. E. 1975. Food selectivity, feeding chronology, and energy transformation of juvenile alewife (*Alosa pseudoharengus*) in the James River near Hopewell, Virginia. Ph.D. dissertation, University of Virginia, Charlottesville.

Weinrich, D. R., N. H. Butowski, E. W. Franklin, and J. P. Mowrer. 1987. *Investigation of Anadromous Alosids.* Maryland Dept. of Natural Resources, Annapolis.

Weinstein, M. P. 1979. Shallow marsh habitats as primary nurseries for fishes and shellfish, Cape Fear River, North Carolina. *Fish. Bull., U.S.* 77:339–357.

Weinstein, M. P., S. L. Weiss, R. G. Hodson, and L. R. Gerry. 1980. Retention of three taxa of postlarval fishes in an intensely flushed tidal estuary, Cape Fear River, North Carolina. *Fish. Bull., U.S.* 78:419–436.

Weis, J. S., and P. Weis. 1976. Optical malformations induced by insecticides in embryos of the Atlantic silversides, *Menidia menidia. Fish. Bull., U.S.* 74:208–211.

Weisberg, S. B. 1986. Competition and coexistence among four estuarine species of *Fundulus. Am. Zool.* 26:249–257.

Weisberg, S. B., and W. H. Burton. 1993. Spring distribution and abundance of ichthyoplankton in the tidal Delaware River. *Fish. Bull., U.S.* 91:788–797.

Weisberg, S. B., and V. A. Lotrich. 1982. The importance of an infrequently flooded intertidal marsh surface as an energy source for the mummichog *Fundulus heteroclitus:* an experimental approach. *Mar. Biol.* 66:307–310.

Weisberg, S. B., and V. A. Lotrich. 1986. Food limitation of a Delaware salt marsh population of the mummichog, *Fundulus heteroclitus* (L.). *Oecologia* 68:168–173.

Weiss, G., Hubold, G., and Bainy, A. C. D. 1987. Larval development of the zeiform fishes *Antigonia capros* Lowe, 1843 and *Zenopsis conchifer* (Lowe, 1852) from the southwest Atlantic. *Cybium* 11:79–91.

Weitzman, S. H. 1974. Osteology and evolutionary relationships of the Sternoptychidae with a new classification of stomiatoid families. *Bull. Am. Mus. Nat. Hist.* 153:329–478.

Weitzman, S. H. 1986. Order Stomiiformes: introduction. In: M. M. Smith and P. C. Heemstra, eds. *Smith's Sea Fishes.* Macmillan South Africa, Johannesburg, pp. 227–229.

Welsh, W. W., and C. M. Breder, Jr. 1922. A contribution to the life history of the puffer, *Spheroides maculatus* (Schneider). *Zoologica, N.Y.* 2:261–276.

Welsh, W. W., and C. M. Breder, Jr. 1924. Contributions to the life histories of Sciaenidae of the eastern United States coast. *Bull. U.S. Bur. Fish.* 39:141–201.

Wenner, C. A. 1976. Aspects of the biology and morphology of the snake eel, *Pisodonophis cruentifer* (Pisces, Ophichthidae). *J. Fish. Res. Bd. Can.* 33:656–665.

Wenner, C. A. 1978. Making a living on the continental slope and in the deep-sea: life history of some dominant fishes of the Norfolk Canyon area. Ph.D. dissertation, College of William and Mary, Williamsburg, Va., 294 pp.

Wenner, C. A. 1983a. Biology of the longfin hake, *Phycis chesteri,* in the western North Atlantic. *Biol. Oceanogr.* 3:41–75.

Wenner, C. A. 1983b. Species associations and day-night variability of trawl-caught fishes from the inshore sponge-coral habitat, South Atlantic Bight. *Fish. Bull., U.S.* 81:537–552.

Wenner, C. A., and J. A. Musick. 1974. Fecundity and gonad observations of the American eel, *Anguilla rostrata,* migrating from Chesapeake Bay, Virginia. *J. Fish. Res. Bd. Can.* 31:1387–1391.

Wenner, C. A., and J. A. Musick. 1975. Food habits and seasonal abundance of the American eel, *Anguilla rostrata,* from the lower Chesapeake Bay. *Chesapeake Sci.* 16:62–66.

Wenner, C. A., and G. R. Sedberry. 1989. Species composition, distribution, and relative abundance of fishes in the coastal habitat off the southeastern United States. NOAA Tech. Rept. NMFS No. 79, 47 pp.

Wenner, C. A., C. A. Barans, B. W. Stender, and F. H. Berry. 1979. Results of MARMAP otter trawl investigations in the South Atlantic Bight. III. Summer 1974. South Carolina Res. Cent. Tech. Rept. No. 41, 62 pp.

Wenner, C. A., W. A. Roumillat, and C. W. Waltz. 1986. Contributions to the life history of the black sea bass, *Centropristis striata,* off the southeastern United States. *Fish. Bull., U.S.* 84:723–741.

Werme, C. 1981. Resource partitioning in a salt marsh fish community. Ph.D. dissertation, Boston University, 126 pp.

Werner, F. E., F. H. Page, D. R. Lynch, J. W. Loder, R. G. Lough, R. I. Perry, D. A.Greenberg, and M. Sinclair. 1993. Influences of mean advection and simple life behavior on the distribution of cod and haddock early life stages on Georges Bank. *Fish. Oceanogr.* 2:43–64.

Wertz, M. 1977. Observazioni sull' alimentazione di *Helicolenus dactylopterus* (Delaroche 1809) (Osteichthys, Scorpaenidae) dei fondi batiali strascicabili del Mer Ligure. Atti 9 Congr. F. Cinelli, E. Fresi, and L. Mazzella, eds. *Soc. Ital. Biol. Mar.* 1: 463–469.

Westin, D. T. 1978. Serum and blood from adult striped bass, *Morone saxatilis. Estuaries* 1:126–128.

Westin, D. T., and B. A. Rogers. 1978. Synopsis of biological data on the striped bass *Morone saxatilis* (Walbaum 1792). Univ. Rhode Island Grad. School Oceanogr. Mar. Tech. Rept. No. 67, 154 pp.

Westin, D. T., K. J. Abernethy, L. E. Meller, and B. A. Rogers. 1979. Some aspects of biology of the American sand lance, *Ammodytes americanus. Trans. Am. Fish. Soc.* 108:328–331.

Westman, J. R., and R. F. Nigrelli. 1955. Preliminary studies of menhaden and their mass mortalities in Long Island and New Jersey waters. *N.Y. Fish Game J.* 2:142–153.

Wheatland, R. H. 1852. Notice of several fishes of rare occurrence. *J. Essex Co. Nat. Hist. Soc.* 1:122–125.

Wheatland, S. B. 1956. Pelagic fish eggs and larvae. In: *Oceanography of Long Island Sound, 1952–1954. Bull. Bingham Oceanogr. Coll.* 15(7):234–314.

Wheeler, A. 1962. A rare British, *Taractes (Taractichthys) longipinnis* (Lowe). *Ann. Mag. Nat. Hist., Ser. No. 13,* 5:257–260.

Wheeler, A. 1969. *The Fishes of the British Isles and North-West Europe.* Michigan State University Press, East Lansing, 613 pp.

Wheeler, A. 1973. Macroramphosidae. In: J. C. Hureau and Th. Monod, eds. *Checklist of the Fishes of the North-eastern Atlantic and of the Mediterranean,* Vol. 1. UNESCO, Paris, p. 273.

Wheeler, A. 1977. Macroramphosidae. In: W. Fischer, ed. *FAO Species Identification Sheets for Fishery Purposes: Western Central Atlantic,* Vol 3. FAO, Rome, 3 pp.

Wheeler, A. 1992. A list of the common and scientific names of fishes of the British Isles. *J. Fish Biol.* 41(Suppl. A), 37 pp.

Wheeler, A., S. J. de Groot, and H. Nijssen. 1974. The occurrence of a second species of *Lophius* in northern European waters. *J. Mar. Biol. Assoc. U.K.* 54:619–623.

Wheeler, J. P., and G. H. Winters. 1984a. Migrations and stock relationships of east and southeast Newfoundland herring (*Clupea harengus*) as shown by tagging studies. *J. Northw. Atl. Fish. Sci.* 5:121–129.

Wheeler, J. P., and G. H. Winters. 1984b. Homing of Atlantic herring (*Clupea harengus harengus*) in Newfoundland waters as indicated by tagging data. *Can. J. Fish. Aquat. Sci.* 41:108–117.

White, A. W. 1977. Dinoflagellate toxins as probable cause of an Atlantic herring (*Clupea harengus harengus*) kill, and pteropods as apparent vector. *J. Fish. Res. Bd. Can.* 34:2421–2424.

White, A. W. 1980. Recurrence of kills of Atlantic herring (*Clupea harengus harengus*) caused by dinoflagellate toxins transferred through herbivorous zooplankton. *Can. J. Fish. Aquat. Sci.* 37:2262–2265.

White, A. W. 1981. Sensitivity of marine fishes to toxins from the red-tide dinoflagellate *Gonyaulax excavata* and implications for fish kills. *Mar. Biol.* 65:255–260.

White, D. B., D. M. Wyanski, and G. R. Sedberry. 1998. Age, growth, and reproductive biology of the blackbelly rosefish from the Carolinas, U.S.A. *J. Fish Biol.* 53:1274–1291.

White, D. S., C. D'Avanzo, I. Valiela, C. Lasta, and M. Pascual. 1986. The relationship of diet to growth and ammonium excretion in salt marsh fish. *Environ. Biol. Fishes* 16:105–111.

White, G. G. 1996. Reproductive biology of tautog, *Tautoga onitis,* in the lower Chesapeake Bay and coastal waters of Virginia. M.S. thesis, College of William and Mary, Williamsburg, Va., 100 pp.

White, H. C. 1939. The nesting and embryo of *Zoarces anguillaris. J. Fish. Res. Bd. Can.* 4:337–338.

White, H. C. 1942. Sea life of the brook trout (*Salvelinus fontinalis*). *J. Fish. Res. Bd. Can.* 5:471–473.

White, H. C. 1953. The eastern belted kingfisher in the Maritime Provinces. *Bull. Fish. Res. Bd. Can.* 97, 44 pp.

White, H. C. 1957. Food and natural history of mergansers of salmon waters in the Maritime Provinces of Canada. *Bull. Fish. Res. Bd. Can.* 116, 61 pp.

Whitehead, P. J. P. 1962. The species of *Elops* (Pisces: Elopidae). *Ann. Mag. Nat. Hist.* 13:321–329.

Whitehead, P. J. P. 1963. A revision of the recent round herrings (Pisces: Dussumieriidae). *Bull. Br. Mus. Nat. Hist. (Zool.).* 10:305–380.

Whitehead, P. J. P. 1985a. *FAO Species Catalogue,* Vol. 7. Clupeoid Fishes of the World (Suborder Clupeoidei). Part I. Chirocentridae, Clupeidae and Pristigasteridae. FAO Fish Synop. No. 125, 7(1):1–303.

Whitehead, P. J. P. 1985b. King herring: his place amongst the clupeoids. *Can. J. Fish. Aquat. Sci.* 42(Suppl. 1):3–20.

Whitehead, P. J. P., G. J. Nelson, and T. Wongratana. 1988. *FAO Species Catalogue,* Vol. 7. Clupeoid Fishes of the World (Suborder

Clupeoidei). Part II. Engraulididae. FAO Fish Synop. No. 125, 7(2):305–579.

Whitley, G. P. 1931. Studies in ichthyology No. 4. *Rec. Austral. Mus.* 18:96–133.

Whitworth, W. R., and D. H. Bennett. 1970. A limnological study of the lower Farmington River with special reference to the ability of the river to support American shad. Inst. Water Res., Univ. Conn., Rept. No. 9, 57 pp.

Whoriskey, F. G. 1983. Intertidal feeding and refuging by cunners, *Tautogolabrus adspersus* (Labridae). *Fish. Bull., U.S.* 81:426–428.

Whoriskey, F. G., and G .J. FitzGerald. 1985a. The effects of bird predation on an estuarine stickleback (Pisces: Gasterosteidae) community. *Can. J. Zool.* 63:301–307.

Whoriskey, F. G., and G. J. FitzGerald. 1985b. Sex, cannibalism and sticklebacks. *Behav. Ecol. Sociobiol.* 18:15–18.

Whoriskey, F. G., and R. J. Wootton. 1987. The swimming endurance of threespine sticklebacks, *Gasterosteus aculeatus* L., from the Afon Rheidol, Wales. *J. Fish Biol.* 30:335–339.

Whoriskey, F. G., R. J. Naiman, and W. L. Mongomery. 1981. Experimental sea ranching of brook trout *Salvelinus fontinalis* Mitchill. *J. Fish Biol.* 19:637–651.

Whoriskey, F. G., G. J. FitzGerald, and S. G. Reebs. 1986. The breeding-season population structure of three sympatric, territorial sticklebacks (Pisces: Gasterosteidae). *J. Fish Biol.* 29:635–648.

Wicklund, R. I. 1970a. A puffer kill related to nocturnal behavior and adverse environmental changes. *Underwater Nat.* 6:28–29.

Wicklund, R. I. 1970b. Observations on the spawning of the cunner in waters of northern New Jersey. *Chesapeake Sci.* 11:137.

Wicklund, R. I., S. J. Wilk, and L. Ogren. 1968. Observations on wintering locations of the northern pipefish and spotted seahorse. *Underwater Nat.* 5:26–28.

Wiggins, T. A., T. R. Bender, Jr., V. A. Mudrak, and J. A. Coll. 1985. The development, feeding, growth, and survival of cultured American shad larvae through the transition from endogenous to exogenous nutrition. *Prog. Fish-Cult.* 47:87–93.

Wiggins, T. A., T. R. Bender, Jr., V. A. Mudrak, J. A. Coll, and J. A. Whittington. 1986. Effect of initial feeding rates of *Artemia* nauplii and dry-diet supplements on the growth and survival of American shad larvae. *Prog. Fish–Cult.* 48:290–293.

Wigley, R. L. 1956. Food habits of Georges Bank haddock. U.S. Fish Wildl. Serv. Spec. Sci. Rept. Fish. No. 165, 26 pp.

Wigley, S. E. 1998a. Ocean pout. In: S. H. Clark, ed. *Status of Fishery Resources off the Northeastern United States for 1998*. NOAA Tech. Memo. NMFS-NE-115:94–95.

Wigley, S. E. 1998b. Witch flounder. In: S. H. Clark, ed. *Status of Fishery Resources off the Northeastern United States for 1998*. NOAA Tech. Memo. NMFS-NE-115:79–80.

Wigley, S. E., and W. L. Gabriel. 1991. Distribution of sexually immature components of 10 northwest Atlantic groundfish species based on Northeast Fisheries Center bottom trawl surveys, 1968–86. NOAA Tech. Memo. NMFS-F/NEC-80, 17 pp.

Wigley, S. E., J. M. Burnett, and P. J. Rago. 1998. An evaluation of maturity estimates derived from two diferent sampling schemes: are the observed changes fact or artifact? NAFO Sci. Counc. Res. Doc. 98/100.

Wilcoxson, K. H. 1975. *Angler's Guide to Salt Water Fishing in the Northeast*. Book Products Services, Inc., Danvers, Mass., 234 pp.

Wilder, D. G. 1952. A comparative study of anadromous and freshwater populations of brook trout (*Salvelinus fontinalis* (Mitchill)). *J. Fish. Res. Bd. Can.* 9:169–203.

Wilk, S. J. 1977. Biological and fisheries data on bluefish, *Pomatomus saltatrix* (Linnaeus). NMFS Sandy Hook Lab., Tech. Ser. Rept. No. 11, 56 pp.

Wilk, S. J. 1979. Biological and fisheries data on weakfish *Cynoscion regalis* (Bloch and Schneider). NMFS Sandy Hook Lab., Tech. Ser. Rept. No. 21, 49 pp.

Wilk, S. J. 1981. The fisheries for Atlantic croaker, spot, and weakfish. In: H. Clepper, ed., *Proceedings of the Sixth Annual Marine Recreational Fisheries Symposium*. Sport Fishing Institute, Washington, D.C., pp. 59–68.

Wilk, S. J. 1982. Bluefish, *Pomatomus saltatrix*. In: M. D. Grosslein and T. R. Azarovitz, eds. *Fish Distribution*. MESA New York Bight Atlas Mongr. No. 15:86–89.

Wilk, S. J., and R. E. Schmidt. 1971. A note on the capture of the black ruff, *Centrolophus niger* (Gmelin), in New Jersey waters. *Chesapeake Sci.* 12:185.

Wilk, S. J., W. W. Morse, D. E. Ralph, and E. J. Steady. 1975. Life history aspects of New York Bight fishes. NMFS Sandy Hook Lab. Ref. SHL 75-1.

Wilk, S. J., W. W. Morse, and D. E. Ralph. 1978. Length-weight relationships of fishes collected in the New York Bight. *Bull. New Jersey Acad. Sci.* 23:58–64.

Wilk, S. J., W. G. Smith, D. E. Ralph, and J. Sibunka. 1980. Population structure of summer flounder between New York and Florida based on linear discriminant analysis. *Trans. Am. Fish. Soc.* 109:265–271.

Wilk, S. J., W. W. Morse, and L. L. Stehlik. 1990. Annual cycles of gonad-somatic indices as indicators of spawning activity for selected species of finfish collected from the New York Bight. *Fish. Bull., U S.* 88:775–786.

Wilkens, E. P. H., and R. M. Lewis. 1971. Abundance and distribution of young Atlantic menhaden, *Brevoortia tyrannus,* in the White Oak River Estuary, North Carolina. *Fish. Bull., U.S.* 69: 783–789.

Willey, A. 1923. Notes on the distribution of free-living Copepoda in Canadian waters. *Contrib. Can. Biol. (New Ser.)* 1:305–334.

Willey, A., and A. G. Huntsman. 1921. Faunal notes from the Atlantic Biological Station (1920). *Can. Field-Nat.* 35:1–7.

Williams, A. B., and E. E. Deubler, Jr. 1968. A ten-year study of meroplankton in North Carolina estuaries: assessment of environmental factors and sampling success among bothid flounders and penaeid shrimps. *Chesapeake Sci.* 9:27–41.

Williams, D. D., and J. C. Delbeek. 1989. Biology of the threespine stickleback, *Gasterosteus aculeatus,* and the blackspotted stickleback, *G. wheatlandi,* during their marine pelagic phase in the Bay of Fundy, Canada. *Environ. Biol. Fishes* 24:33–41.

Williams, E. H., Jr., L. Bunkley-Williams, and T. G. Rand. 1994. Some copepod and isopod parasites of Bermuda marine fishes. *J. Aquat. Animal Health* 6:279–280.

Williams, G. C. 1960. Dispersal of young marine fishes near Woods Hole, Massachusetts. Mich. State Univ. Biol. Ser. Publ. No. 1:329–367.

Williams, G. C. 1967. Identification and seasonal size changes of eggs of the labrid fishes, *Tautogolabrus adspersus* and *Tautoga onitis,* of Long Island Sound. *Copeia* 1967:452–453.

Williams, G. C. 1968. Bathymetric distribution of planktonic fish eggs in Long Island Sound. *Limnol. Oceanogr.* 13:382–385.

Williams, G. C., S. W. Richards, and E. G. Farnworth. 1964. Eggs of *Ammodytes hexapterus* from Long Island, New York. *Copeia* 1964:242–243.

Williams, P. J. 1985. Use of otoliths for stock differentiation of American shad (*Alosa sapidissima,* Wilson). M.S. thesis, Acadia University, Wolfville, Canada.

Williams, R. G., and G. Daborn. 1984. Spawning of American shad in the Annapolis River, Nova Scotia, Canada. *Proc. Nova Scotian Inst. Sci.* 34:9–14.

Williams, S. R. 1902. Changes accompanying the migration of the eye and observations on the tractus opticus and tectum opticum in *Pseudopleuronectus americanus. Bull. Mus. Comp. Zool.* 40:1–57.

Williams, U. P., and J. W. Kiceniuk. 1987. Feeding reduction and recovery in cunner, *Tautogolabrus adspersus,* following exposure to crude oil. *Bull. Environ. Contam. Toxicol.* 38(6):1044–1048.

Williamson, E. G., and L. Di Michele. 1997. An ecological simulation reveals balancing selection acting on development rate in the teleost *Fundulus heteroclitus. Mar. Biol.* 128:9–15.

Wilson, C. A., and J. M. Dean. 1983. The potential use of sagittae for estimating age of Atlantic swordfish, *Xiphias gladius.* NOAA Tech. Rept. NMFS No. 8:151–156.

Wilson, C. A., J. M. Dean, and R. Radtke. 1982. Age, growth rate and feeding habits of the oyster toadfish, *Opsanus tau* (Linnaeus) in South Carolina. *J. Exp. Mar. Biol. Ecol.* 62:251–259.

Wilson, C. B. 1915. North American parasitic Copepoda belonging to the Lernaeopodidae, with a revision of the entire family. *Proc. U.S. Nat. Mus.* 47:656–729.

Wilson, C. B. 1932. The copepods of the Woods Hole region, Massachusetts. *Bull. U. S. Nat. Mus.* 158, 635 pp.

Wilson, D. P. 1937. The habits of the angler-fish, *Lophius piscatorius* L., in the Plymouth Aquarium. *J. Mar. Biol. Assoc. U.K.* 21:477–496.

Wilson, H. V. 1891. The embryology of the sea bass (*Serranus atrarius*). *Bull. U.S. Fish Comm.* 9:209–277.

Wilson, P. C., and J. S. Beckett. 1970. Atlantic Ocean distribution of the pelagic stingray, *Dasyatis violacea. Copeia* 1970:696–707.

Winn, H. E. 1972. Acoustic discrimination by the toadfish with comments on signal systems. In: H. E. Winn and B. Olla, eds. *Behavior of Marine Animals: Current Perspectives in Research,* Vol. 2. Plenum Press, New York, pp. 361–385.

Winslade, P. 1974a. Behavioural studies on the lesser sandeel *Ammodytes marinus* (Raitt). I. The effect of food availability on activity and the role of olfaction in food detection. *J. Fish Biol.* 6:565–576.

Winslade, P. 1974b. Behavioural studies on the lesser sandeel *Ammodytes marinus* (Raitt). II. The effect of light intensity on activity. *J. Fish Biol.* 6:577–586.

Winslade, P. 1974c. Behavioural studies on the lesser sandeel *Ammodytes marinus* (Raitt). III. The effect of temperature on activity and the environmental control of the annual cycle of activity. *J. Fish Biol.* 6:587–599.

Winters, G. H. 1969. Capelin (*Mallotus villosus*). In: F. E. Firth, ed. *Encyclopedia of Marine Resources.* Van Nostrand, New York, pp. 94–101.

Winters, G. H. 1970. Record size and age of Atlantic capelin, *Mallotus villosus. J. Fish. Res. Bd. Can.* 27:393–394.

Winters, G. H. 1975. Review of capelin ecology and estimation of surplus yield from predator dynamics. ICNAF Res. Doc. No. 75/2, Ser. No. 3430, 25 pp.

Winters, G. H. 1976. Recruitment mechanisms of southern Gulf of St. Lawrence Atlantic herring (*Clupea harengus harengus*). *J. Fish. Res. Bd. Can.* 33:1751–1763.

Winters, G. H. 1981. Growth patterns in sand lance, *Ammodytes dubius,* from the Grand Bank. *Can. J. Fish. Aquat. Sci.* 38:841–846.

Winters, G. H. 1983. Analysis of the biological and demographic parameters of northern sand lance, *Ammodytes dubius,* from the Newfoundland Grand Bank. *Can. J. Fish. Aquat. Sci.* 40:409–419.

Winters, G. H. 1989. Life history parameters of sand lances (*Ammodytes* spp.) from the coastal waters of eastern Newfoundland. *J. Northw. Atl. Fish. Sci.* 9:5–11.

Winters, G. H., and J. E. Carscadden. 1978. Review of capelin ecology and estimation of surplus yield from predator dynamics. ICNAF Res. Doc. No. 13:21–30.

Winters, G. H., and E. L. Dalley. 1988. Meristic composition of sand lance (*Ammodytes* spp.) in Newfoundland waters with a review of species designations in the Northwest Atlantic. *Can. J. Fish. Aquat. Sci.* 45:516–529.

Winters, G. H., J. A. Moores, and R. Chaulk. 1973. Northern range extension and probable spawning of gaspereau (*Alosa pseudoharengus*) in the Newfoundland area. *J. Fish. Res. Bd. Can.* 30:860–861.

Wirgin, I. I., J. R. Waldman, L. Maceda, J. Stabile, and V. J. Vecchio. 1997. Mixed-stock analysis of Atlantic coast striped bass (*Morone saxatilis*) using nuclear DNA and mitochondrial DNA markers. *Can. J. Fish. Aquat. Sci.* 54:2814–2826.

Wirgin, I. I., J. R. Waldman, J. Rosko, R. Gross, M. R. Collins, S. G. Rogers, and J. Stabile. 2000. Genetic structure of Atlantic sturgeon populations based on mitochondrial DNA control region sequences. *Trans. Am. Fish. Soc.* 129:476–486.

Wise, J. P. 1962. Cod groups in the New England area. *Fish. Bull., U.S.* 63:189–203.

Wisner, R. L., and C. B. McMillan. 1995. Review of the new world hagfishes of the genus *Myxine* (Agnatha, Myxinidae) with descriptions of nine new species. *Fish. Bull., U.S.* 93:530–550.

Witherell, D. B., and J. Burnett. 1993. Growth and maturation of winter flounder, *Pluronectes americanus,* in Massachusetts. *Fish. Bull., U.S.* 91:816–820.

Wolke, R. E., D. S. Wyand, and L. H. Khairallah. 1970. A light and electron microscopic study of epitheliocystis disease in the gills of Connecticut striped bass (*Morone saxatilis*) and white perch (*Morone americanus*). *J. Comp. Pathol.* 80:559–563.

Wong, R. S. P. 1968. Age and growth of the northern searobin, *Prionotus carolinus* (Linnaeus). M.A. thesis, Virginia Inst. Mar. Sci., Gloucester Point, Va., 48 pp.

Wood, R. J., and D. F. S. Raitt. 1968. Some observations on the biology of the greater silver smelt, particularly in the North-eastern Atlantic Ocean. *Rapp. Proc.-Verb. Réun. Cons. Int. Explor. Mer* 158:64–73.

Wood, P. W., Jr. 1982. Goosefish, *Lophius americanus*. In: M. D. Grosslein and T. R. Azarovitz. MESA New York Bight Atlas Monogr. No. 15:67–70.

Woodhead, P. M. J. 1965. Effects of light upon behavior and distribution of demersal fishes of the North Atlantic. Int. Comm. Northw. Atl. Fish., Spec. Publ. pp. 267–287.

Woods, L. P., and P. M. Sonoda. 1973. Order Berycomorphi (Beryciformes). Fishes of the western North Atlantic. Sears Found. Mar. Res. Mem. No. 1 (6):263–396.

Woolcott, W. S., C. Beirne, and W. M. Hall, Jr. 1968. Descriptive and comparative osteology of the young of three species of flounders, genus *Paralichthys*. *Chesapeake Sci.* 9:109–120.

Wootton, R. J. 1976. *The Biology of the Sticklebacks*. Academic Press, New York, 387 pp.

Wootton, R. J. 1984. *A Functional Biology of Sticklebacks*. University of California Press, Berkeley and Los Angeles, 265 pp.

Worgan, J. P., and G. J. FitzGerald. 1981a. Habitat segregation in a salt marsh among adult sticklebacks (Gasterosteidae). *Environ. Biol. Fishes* 6:105–109.

Worgan, J. P., and G. J. FitzGerald. 1981b. Diel activity and diet of three sympatric sticklebacks in tidal salt marsh pools. *Can. J. Zool.* 59:2375–2379.

Worley, L. G. 1933. Development of the egg of mackerel at different constant temperatures. *J. Gen. Physiol.* 16:841–857.

Worobec, M. N. 1984. Field estimates of the daily ration of winter flounder, *Pseudopleuronectes americanus* (Walbaum) in a southern New England salt pond. *J. Exp. Mar. Biol. Ecol.* 77:183–196.

Worth, S. G. 1882. The artificial propagation of the striped bass (*Roccus lineatus*) on Albemarle Sound. *Bull. U. S. Fish Comm.* 1:174–177.

Worthington, J. 1905. Contribution to our knowledge of the myxinoids. *Am. Nat.* 39:625–663.

Wourms, J. P. 1991. Reproduction and development of *Sebastes* in the context of the evolution of piscine viviparity. *Environ. Biol. Fishes* 30:111–126.

Wright, G. M., F. W. Keeley, J. H. Youson, and D. L. Babineau. 1984. Cartilage in the Atlantic hagfish, *Myxine glutinosa*. *Am. J. Anat.* 169:407–424.

Wroblewski, J. S., W. L. Bailey, and J. Russel. 1999. Grow-out cod farming in southern Labrador. *Bull. Aquacult. Assoc. Can.* 98-2:47–49.

Yabe, H. 1953. Juvenile of the pointed-tailed ocean sunfish *Masturus lanceolatus*. *Contrib. Nankai Reg. Fish. Res. Lab.* 4:40–42 [in Japanese].

Yabe, M. 1985. Comparative osteology and myology of the superfamily Cottoidea (Pisces: Scorpaeniformes), and its phylogenetic classification. Fac. Fish. Hokkaido Univ. Mem. No. 32:1–130.

Yamaguti, S. 1958. *Systema Helminthum*, Vol. I. *The Digenetic Trematodes of Vertebrates*.I. Interscience Publ., New York, 979 pp.

Yamaguti, S. 1963. *Parasitic Copepoda and Brachiura of Fishes*. Interscience Publ., New York, 1104 pp.

Yano, K. 1995. Reproductive biology of the black dogfish, *Centroscyllium fabricii*, collected from waters off western Greenland. *J. Mar. Biol. Assoc. U.K.* 75:285–310.

Yano, K., and S. Tanaka. 1984. Some biological aspects of the deep-sea squaloid shark *Centroscymmus* from Suruga Bay, Japan. *Bull. Jap. Soc. Sci. Fish.* 50:249–256.

Yano, K., and S. Tanaka. 1988. Size at maturity, reproductive cycle, fecundity, and depth segregation of the deep-sea squaloid sharks *Centroscymmus owstoni* and *C. coelolepis* in Suruga Bay, Japan. *Nippon Suisan Gakkaishi* 54:167–174.

Yanulov, K. P. 1962a. On the reproduction of the rough headed grenadier (*Macrourus berglax* Lacépède). *Zool. Zhur.* 41(8):1259–1262 [in Russian].

Yanulov, K. P. 1962b. Age and growth of the American plaice in the northwest Atlantic. In: Y. Y. Marti, ed. *Soviet Fisheries Investigations in the Northwest Atlantic*. Israel Prog. Sci. Transl. OTS 63-11103, Jerusalem, 1963, pp. 355–360.

Yao, Z., and L. W. Crim. 1995a. Spawning of ocean pout (*Macrozoarces americanus* L.): evidence in favour of internal fertilization of eggs. *Aquaculture* 130:361–372.

Yao, Z., and L. W. Crim. 1995b. Copulation, spawning and parental care in captive ocean pout. *J. Fish Biol.* 47:171–173.

Yao, Z., L. W. Crim, G. F. Richardson, and C. J. Emerson. 2000. Motility, fertility and ultrastructual changes of ocean pout (*Macrozoarces americanus* L.) sperm after cryopreservation. *Aquaculture* 181:361–375.

Yarrow, H. C. 1877. Notes on the natural history of Fort Macon, N.C., and vicinity (No.3). *Proc. Acad. Nat. Sci. Philadelphia* 29:203–218.

Yatsu, A. 1985. Phylogeny of the family Pholididae (Blennioidei) with a redescription of *Pholis* Scopoli. *Japan. J. Ichthyol.* 32:273–282.

Yoshida, H. O. 1979. Synopsis of biological data on tunas of the genus *Euthynnus*. NOAA Tech. Rept. NMFS Circ. No. 429, 57 pp.

Yoshida, H. O. 1980. Synopsis of biological data on bonitos of the genus *Sarda*. NOAA Tech. Rept. NMFS Circ. No. 432, 50 pp.

Yoshin, T. P., and E. R. Noble. 1973. Myxosporidia in macrourid fishes of the North Atlantic. *Can. J. Zool.* 51:745–752.

Young, B. H., I. H. Morrow, and S. R. Wanner. 1982. First record of the bluespotted cornetfish in the Hudson River. *N.Y. Fish Game J.* 29:106.

Young, J. R., T. B. Hoff, W. P. Dey, and J. G. Hoff. 1988. Management recommendations for a Hudson River Atlantic sturgeon fishery based on an age-structured population model. In: C. L. Smith, ed. *Fisheries Research in the Hudson River*. Hudson River Environ. Soc., State University New York Press, Albany. pp. 353–365.

Young, J. S., and C. I. Gibson. 1973. Effect of thermal effluent on migrating menhaden. *Mar. Poll. Bull.* 4:94–96.

Young, K. M. 1950. Observations on the distribution and growth of the genus *Fundulus* in the upper Chesapeake Bay. M.S. thesis, University of Maryland, College Park, 49 pp.

Youngson, A.F., and E. Verspoor. 1998. Interactions between wild and introduced Atlantic salmon (*Salmo salar*). *Can. J. Fish. Aquat. Sci.* 55(Suppl. 1):153–160.

Youson, J. H., and I. C. Potter. 1979. A description of the stages in the metamorphosis of the anadromous sea lamprey, *Petromyzon marinus* L. *Can. J. Zool.* 57:1808–1817.

Yudin, K. G., and G. M. Cailliet. 1990. Age and growth of the gray smoothhound, *Mustelus californicus,* and the brown smoothhound, *M. henlei,* sharks from central California. *Copeia* 1990: 191–204.

Yuschak, P. 1985. Fecundity, eggs, larvae and osteological development of the striped searobin, (*Prionotus evolans*) (Pisces, Triglidae). *J. Northw. Atl. Fish. Sci.* 6:65–85.

Yuschak, P., and W. A. Lund, Jr. 1984. Eggs, larvae and osteological development of the northern searobin, *Prionotus carolinus* (Pisces, Triglidae). *J. Northw. Atl. Fish. Sci.* 5:1–15.

Zale, A. V., and S. G. Merrifield. 1989. Species profiles: life histories and environmental requirements of coastal fishes and invertebrates (South Florida): ladyfish and tarpon. U.S. Fish Wildl. Serv. Biol. Rept. No. 82(11.104). U.S. Army Corps Eng. TR ELO-82-4, 17 pp.

Zalewski, B. R., and J. B. Weir. 1981. Range extensions for 15 teleost fishes in the Hudson Bay lowlands, Ontario. *Can. Field-Natur.* 95:212–214.

Zamarro, J. 1992. Feeding behaviour of the American plaice (*Hippoglossoides platessoides*) on the southern Grand Bank of Newfoundland. *Neth. J. Sea Res.* 29:229–238.

Zarudski, E. F. K., and R. L. Haedrich. 1974. Swordfish (*Xiphias gladius*) attacks submarine (ALVIN). *Oceanology* 3:111–116.

Zastrow, C. E., E. D. Houde, and L. G. Morin. 1991. Spawning, fecundity, hatch-date frequency and young-of-the-year growth of bay anchovy *Anchoa mitchilli* in mid-Chesapeake Bay. *Mar. Ecol. Prog. Ser.* 73:161–171.

Zawacki, C. S. 1969. Long Island's artificial fishing reefs. *N.Y. State Conservationist.* 24:18–21.

Zehren, S. J. 1987. Osteology and evolutionary relationships of the boarfish genus *Antigonia* (Teleostei: Caproidae). *Copeia* 1987: 564–592.

Zhitenev, A. N. 1970. Ecological and morphological features of the reproduction of the lumpsucker *Cyclopterus lumpus* (L.). *Voprosy Ikhtiol.* 10:94–102 [in Russian, translation in *J. Ichthyol.* 10(1):77–88].

Zhudova, A. M. 1971. Material on the study of the eggs and larvae of some species of fishes from the Gulf of Guinea and the adjacent waters of the open ocean. *Trudy AtlantNIRO* 22:135–163, 1969 [in Russian, English translation issued by Inter-Amer. Trop. Tuna Comm., 1971].

Zilanov, V. K. 1970. Investigations on the biology of *Scomberesox saurus* in the North Atlantic in 1969. *Ann. Biol.* 26:263–265.

Zilanov, V. K. 1977. Peculiarities of the behavior of some fish species of the North Atlantic in the zone of artificial lights. In: *Fish Behavior in Relation to Fishing Gear.* Moscow, pp. 40–46 [in Russian].

Zilanov, V. K., and S. I. Bogdanov. 1969. Results of research on *Scomberesox saurus* in the northeastern Atlantic in 1968. *Ann. Biol.* 25:252–255.

Zinkevich, V. N. 1967. Observations on the distribution of herring, *Clupea harengus* L., on Georges Bank and in adjacent waters in 1962–65. *Int. Comm. Northw. Atl. Fish. Res. Bull.* 4:101–115.

Zolotnitskij, A. P., and V. F. Romanenko. 1981. Gametogenesis in artificially reared American striped bass. In N. E. Sal'nikov, ed. *Ecologo-Physiological Foundations of Aquaculture in the Black Sea.* Kerch', pp. 68–79 [in Russian].

Zubchenko, A. V. 1980. Parasitic fauna of Anarhichadidae and Pleuronectidae families in the northwest Atlantic. ICNAF 1980 Sci. Pap. No. 6, pp. 41–46.

Zubchenko, A. V. 1981. Parasitic fauna of some Macrouridae in the northwest Atlantic. *J. Northw. Atl. Fish. Sci.* 2:67–72.

Zubchenko, A. V. 1985. Parasitic fauna of American plaice (*Hippoglossoides platessoides*) from the northwest Atlantic. *J. Northw. Atl. Fish. Sci.* 6:165–171.

Zukowski, C. 1972. Growth and mortality of Atlantic argentine, *Argentina silus* Ascanius, on the Nova Scotia banks. *ICNAF Res. Bull.* 9:109–115.

Zurbrigg, R. E., and W. B. Scott. 1972. Evidence for expatriate populations of the lanternfish *Myctophum punctatum* in the northwest Atlantic. *J. Fish. Res. Bd. Can.* 29:1679–1683.

INDEX OF SCIENTIFIC NAMES

INDEX OF COMMON NAMES

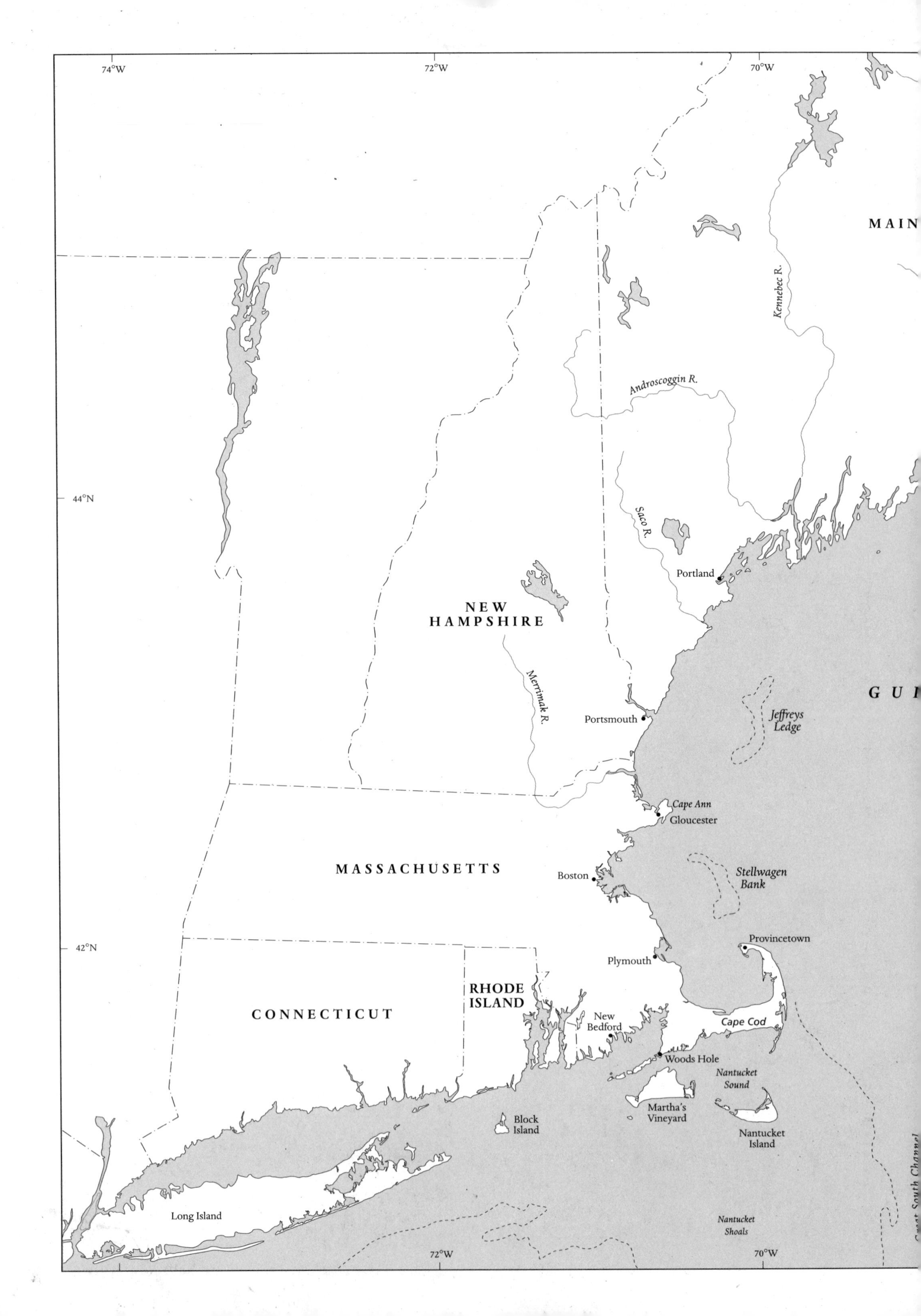

74°W
72°W
70°W
MAIN
Kennebec R.
Androscoggin R.
Saco R.
Portland
44°N
NEW HAMPSHIRE
Merrimak R.
Portsmouth
GU
Jeffreys Ledge
Cape Ann
Gloucester
MASSACHUSETTS
Boston
Stellwagen Bank
42°N
Provincetown
Plymouth
RHODE ISLAND
CONNECTICUT
New Bedford
Cape Cod
Woods Hole
Nantucket Sound
Martha's Vineyard
Block Island
Nantucket Island
Long Island
Nantucket Shoals
72°W
70°W